Historical Unrest at
Large Calderas of the World

Ruins of the Roman marketplace Serapeo near the center of the Phlegraean Fields Caldera, Italy, where marine borings record two millenia of ground subsidence and uplift. Photograph taken in October 1983, during the peak of the most recent episode of caldera unrest.

"In the year 1749, the upper portions of three marble columns that had been nearly concealed by underwood, were discovered, in the neighbourhood of the town of Pozzuoli. In the following year excavations were made, and ultimately it was found that they formed part of a large temple [now known to be a marketplace] which was supposed to have been dedicated to the god Serapis***.

"The most remarkable circumstance which first attracts the attention of the observer is the state of the remaining three large columns. Throughout a part of their height, commencing at nearly 11 feet above the floor of the temple, and continuing about 8 feet, they are perforated in all directions by a species of boring marine animal, the Modiola lithophaga of Lamarck—which still exists in the adjacent parts of the Mediterranean***.

"It appears***that the joint action of certain existing and admitted causes must necessarily produce on the earth's surface a continual but usually slow change in the relative levels of the land and the water. Large tracts of its surface must be slowly susiding through ages, whilst other portions must be rising irregularly at various rates: some, though perhaps few, may remain stationary."

C. Babbage, in a paper read before the Geological Society of London in 1834 and published in 1847.

Rabaul Caldera

Predicted fate of a quarantine station that was proposed to be built on Vulcan Island, west side of Rabaul Caldera, Papua New Guinea. The island was known to rise, tilt, and subside over time and was devastated by pyroclastic flows and tephra fall during the eruption of Vulcan in 1937. Sketch by Bobbie Myers.

"Between four and five years ago, when the erection of a Quarantine Station was being considered, some Heads of Departments, including myself, inspected Vulcan Island with the object of seeing whether it might not be made into an animal or plant Quarantine Station, but turned it down as there was evidence that it was subject to frequent rising and subsidence.

"On my last visit a casarina tree which had been well above high water mark was particularly noticeable as being about twelve inches below water level, and the shore line seemed to have tilted. For these reasons I stated that I did not consider the island suitable for a Quarantine Station on account of the danger of its possible sudden subsidence, a statement trewated with some levity by my companions, but having had personal experience with volcanic eruptions in other parts of the world I know that it was no joke. My final remark was 'it came up in a night and might go down in a night and I would suggest that life buoys be hung round the verandah rails of the buildings where they would at least be ornamental, but would not be required if the island went up instead of down'."

George H. Murray (1939), in a discussion of the 1937 eruption of Vulcan, Rabaul Caldera, Papua New Guinea.

Historical Unrest at Large Calderas of the World

By CHRISTOPHER G. NEWHALL and DANIEL DZURISIN

Volume 1

U.S. GEOLOGICAL SURVEY BULLETIN 1855

DEPARTMENT OF THE INTERIOR
DONALD PAUL HODEL, Secretary

U.S. GEOLOGICAL SURVEY
Dallas L. Peck, Director

Library of Congress Catalog Number 88-600541

UNITED STATES GOVERNMENT PRINTING OFFICE, WASHINGTON : 1988

For sale by the
Books and Open-File Reports Section
U.S. Geological Survey
Federal Center, Box 25425
Denver, CO 80225

CONTENTS

Volume 1.

Volume 2.

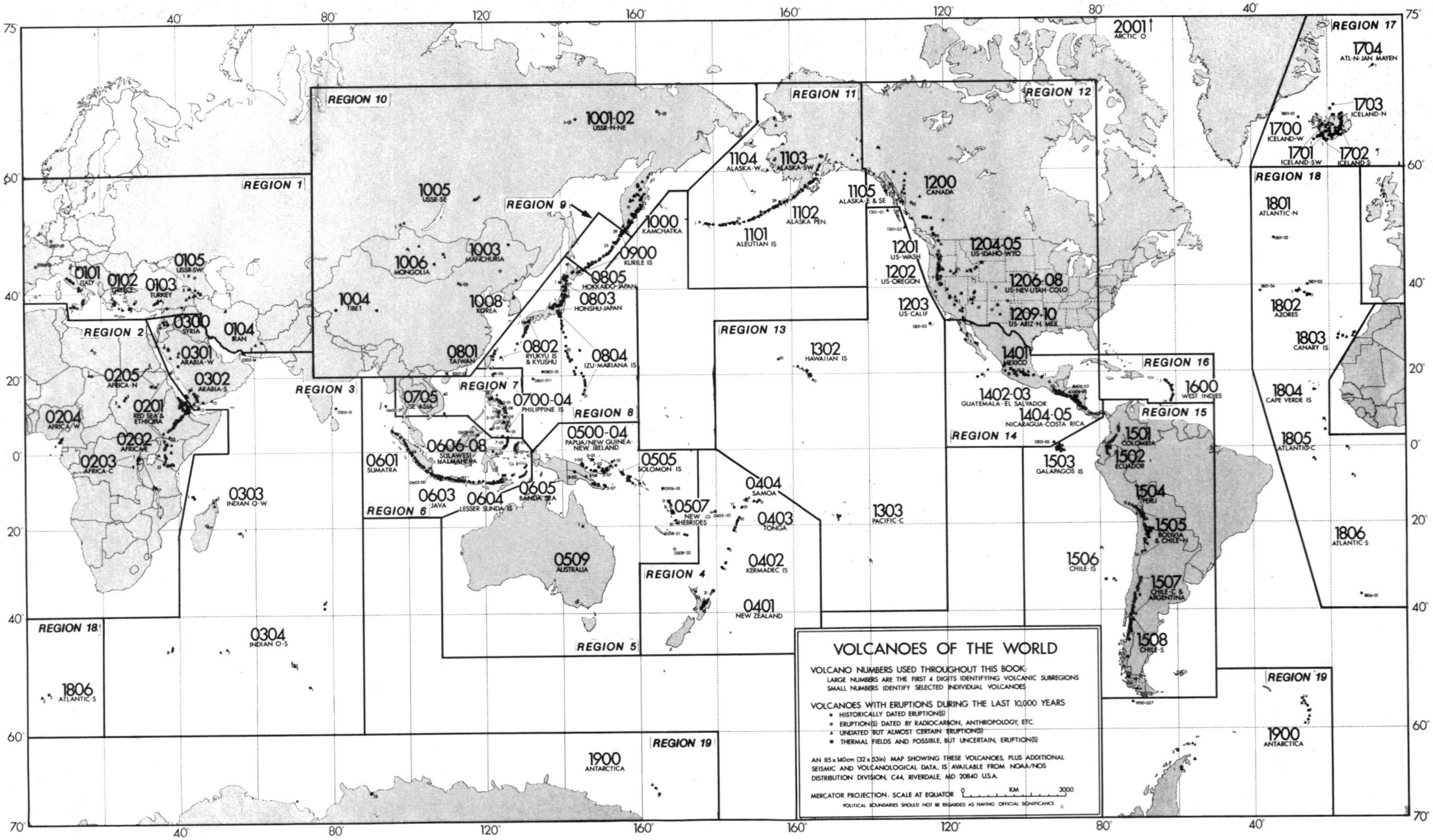

Figure 1. Outlines of the 19 regions of the world addressed in text (from Simkin and others, 1981). First two digits denote region number (for example, Region 5, Papua New Guinea, Australia, the Solomon Islands, and Vanuatu [New Hebrides]; Region 11, Alaska, including the Aleutian Islands; and so on).

PART I: APPROACH, DATA SUMMARY, AND CONCLUSIONS

By Christopher G. Newhall *and* Daniel Dzurisin

Ang hindi marunong lumingon sa pinanggalingan, ay hindi makakarating sa paroro'onan.

Tagalog proverb, meaning
"He who does not know how to look back cannot reach where he is going."

ABSTRACT

Historical seismicity, ground deformation, thermal activity, and eruptions at large Quaternary calderas of the world are common, spatially and temporally complex, and usually caused by combinations of tectonic, magmatic, and hydrologic processes. Study of historical unrest helps to define what is ordinary and what is extraordinary, and to document processes of unrest that act over periods of hours to millennia.

At least 1,299 episodes of unrest have occurred at 138 calderas ≥5 km in diameter during historical time. The most commonly reported unrest includes local earthquakes or earthquake swarms; volcanic tremor; uplift; subsidence; tilt; ground fissuring; changes in the temperature of soil, water, or gas; changes in fumarolic activity; and eruptions. Of 94 well-reported episodes of unrest within the past 30 years, 45 culminated in eruptions. However, at silicic calderas that had not erupted for at least 100 years, only 10 of 60 episodes of unrest led to eruptions.

Most unrest involves interaction between the local tectonic stress field, subcaldera magma, and ground water. In most unrest, the subcaldera environment gains or loses energy; often, magma and volatiles also are added or lost. A few processes of unrest occur within a closed subcaldera environment, with neither import nor export of energy or mass. We do not attempt to prove that specific processes have acted in specific episodes of unrest. Instead, we demonstrate the general plausibility of many processes and argue that any given episode of unrest likely involves a combination of these processes.

The subcaldera environment is dynamic and in delicate balance between quiet and unrest. The stimuli for unrest need not be large; most often, they are everyday geologic processes such as regional tectonic strain or small magmatic intrusions.

Many, perhaps most, episodes of unrest end without an eruption, and eruptions that do occur are generally small. Nonetheless, caldera unrest is worrisome because the largest eruptions from calderas devastate large areas and, in populous countries, could kill several million people. Caldera systems are complex, and forecasting the outcome of unrest is more difficult than at smaller volcanic centers. Residents of restless calderas will continue to be faced with difficult uncertainties and must learn to cope with false alarms.

Publication authorized by the Director, U.S. Geological Survey, April 12, 1988.

INTRODUCTION

Earthquakes, ground deformation, fumarolic and hydrologic changes, and threats of eruptions are a part of life for millions of people living in or near calderas. Some aspects of caldera unrest are truly remarkable and unique to the caldera setting. Where but at calderas does the ground rise and fall tens or even hundreds of meters within decades, or several meters within hours or days? Where else are millennia-old archaeological ruins now covered by the sea, or coral reefs of equally young age now hundreds of meters above sea level? Where else do earthquake swarms occur repeatedly in response to stimuli as slight as earth tides or distant earthquakes? And where but at calderas can tens of cubic kilometers of magma accumulate in just a few millenia, or hundreds to thousands of cubic kilometers of magma erupt within a few days or weeks, devastating an entire region?

In a typical year some form of unrest (earthquakes, anomalous ground deformation, changes in fumarolic activity, other noneruptive unrest, or eruptions) occurs at about 18 large calderas worldwide, and eruptions occur within or near at least 5 of them. During historical time, at least 1,299 episodes of unrest have been reported at 138 restless calderas worldwide, and many other episodes were unreported.

Recently, relatively intense unrest has occurred at Long Valley Caldera, California (1978–present), the Phlegraean Fields Caldera, Italy (1969–72, 1983–present), and Rabaul Caldera, Papua New Guinea (1971–present). Each of these three calderas has had a catastrophic eruption in the geologically recent past and now hosts a large human population. Local residents have been threatened by both volcanic and earthquake hazards and economic and social disruption related to the geologic hazards. Various interpretations of each episode have been advanced, but reliable forecasts of the outcomes of the unrest have not been possible.

This report reviews geophysical and geochemical unrest at large calderas of the world. More fundamentally, it compares unrest in and above large magma reservoirs, as manifested by calderas. We examine the nature, frequency, duration, outcomes, and possible causes of caldera unrest, in much the same way as epidemiologists examine the general occurrence and possible causes and controls of a puzzling disease. This "epidemiological" approach cannot conclusively identify the cause of a specific episode of caldera unrest, but it can provide information about what is common and uncommon, and ideas about processes that might not otherwise have been considered. This report provides a context in which future episodes of unrest can be interpreted.

We consider a large variety of calderas, from those on basaltic shield volcanoes to those erupting rhyolite magma and forming great volcano-tectonic depressions. Within this large set, we distinguish between gross differences in magma composition and in the nature of unrest, and between differences in caldera age, diameter, and hydrothermal activity. Our goal in considering a large and diverse set of calderas is to be able to assess how differences between calderas affect unrest. For example, Kilauea and Mauna Loa by themselves might not tell us much about processes at silicic calderas, but we will learn more about the effects of magma composition, density, and viscosity on the nature of unrest by considering both mafic and silicic magma bodies.

Our literature search has reached a variety of disciplines, including geology, geochemistry, geothermics, limnology, neotectonics, seismology, geodesy, and other subfields of geophysics—virtually all disciplines pertinent to large volcanic systems. Phenomena reported within one discipline often are related to those reported within another, but too often the connection is not recognized. Too many changes at calderas have been considered either unrelated to the caldera (for example, "tectonic" earthquakes at or near a caldera) or purely volcanic in origin, when in fact most changes reflect a combination of tectonic, magmatic, and hydrologic processes. Because we cast a wide net, we note some changes that others might consider unrelated to calderas. We assume that changes that are temporally and spatially close to each other are related, even though we might not understand that relation.

Most of what we report is taken directly from the published literature. However, in a few cases, we reinterpreted features or events on the basis of our experience with the large number of calderas covered in this report and our integration of data from several disciplines. We offer some new inferences regarding processes responsible for unrest in the "Comments" sections of Parts 3 and 4. When published interpretations differed, we usually cited each interpretation without prejudice; in a few cases we stated our preference. We hope that this report provides a useful tool in coping with future unrest at calderas. Most forecasts of geologic events have been based on the geologic record and on the monitoring of contemporary processes. Less reliance has been put on the historical past, perhaps because historical information about past eruptions and noneruptive unrest is sparse, scattered through a diverse literature, and difficult to relate between calderas. However, the historical record is the sole source for information about eruptions and other unrest that occurred before recent advances in monitoring but that were not preserved in the geologic record. Historical reports contain the only available information about precursors to infrequent, relatively large eruptions, or about the incidence of increased activity that does not culminate in eruptions. Future crises will not likely have exact analogues in this compilation, but partial analogues and lessons about processes of unrest should help to narrow the list of likely outcomes.

One of the most important tasks facing volcanologists today is recognition of the precursors to what are potentially the most devastating of all eruptions—explosive eruption of tens, hundreds, or even thousands of cubic kilometers of magma. Such eruptions have occurred at most of the volcanic centers in this compilation, as evidenced by the calderas that formed or were enlarged during or after these eruptions. Similar huge eruptions could recur at most of these same calderas. Most of the unrest described here may be minor compared with that which will precede the next colossal, caldera-forming eruption. Or is it? Simultaneous eruptions from multiple ring-fracture vents or remarkable caldera-wide uplift and seismicity around an entire ring-fracture system, as have already occurred at some calderas, seem likely precursors to future caldera-forming eruptions. This compilation offers a chance to examine what little is known about precursors to historical caldera-forming eruptions and about changes in similar settings that did not culminate in caldera-forming eruptions.

ACKNOWLEDGMENTS

We are indebted to all of the meticulous observers from Roman time to the present who, often under hazardous or trying conditions, produced written records of caldera unrest. Good observations will outlive most interpretations!

We have received helpful information from many current observers, some of whom have contributed data as members of the IAVCEI Working Group on Mitigation of Volcanic Disasters.

Tito Aguila (Philippines)
Shigeo Aramaki (Japan)
Peter Baker (Great Britain)
Franco Barberi (Italy)
Jorge Barquero (Costa Rica)
Russell Blong (Australia)
Sutikno Bronto (Indonesia)
Michele Caputo (Italy)
Juan Carracedo (Canary Islands)
Tom Casadevall (U.S.)
Jean-Louis Cheminee (France)
Craig Chesner (U.S.)
Robert Decker (U.S.)
Patrice de St. Ours (P.N.G.)
Ray Dibble (New Zealand)
Wendell Duffield (U.S.)
John Dvorak (U.S.)
A.C. Effendi (Indonesia)
Pall Einarsson (Iceland)
Edward Erlich (U.S.)
Sergei Fedotov (U.S.S.R.)
Giday W-Gabriel (Ethiopia)
Cynthia Gardner (U.S.)
Harry Glicken (U.S.)
Oscar Gonzalez-Ferran (Chile)
John Guest (Great Britain)
Minard Hall (Ecuador)
Dave Harlow (U.S.)
Grant Heiken (U.S.)
Wes Hildreth (U.S.)
Dave Hill (U.S.)
Bruce Houghton (New Zealand)
Kazuhiro Ishihara (Japan)
Wally Johnson (Australia)
Kosuke Kamo (Japan)
Yoshio Katsui (Japan)
Jorg Keller (Fed. Rep. Germany)
Juergen Kienle (U.S.)
Christopher Kilburn (U.K.)
Robert Koyanagi (U.S.)
Teiji Kumagai (Japan)
Philip Kyle (U.S.)
Ken Lajoie (U.S.)
Gudrun Larsen (Iceland)
John Latter (New Zealand)
Ted Lloyd (New Zealand)
Jack Lockwood (U.S.)
Peter Lowenstein (P.N.G.)
Jim Luhr (U.S.)
Eduardo Malavassi (Costa Rica)
Marino Martini (Italy)
Lindsay McClelland (U.S.)
Chris McKee (P.N.G.)
Steve McNutt (U.S.)
Tom Miller (U.S.)
Dick Moore (U.S.)
Hugo Moreno Roa (Chile)
Jim Mori (P.N.G./U.S.)
Ichio Moriya (Japan)
Ian Nairn (New Zealand)
Kazuaki Nakamura (Japan)
Yasunori Nishida (Japan)
Jacques Nougier (France)
Hiromu Okada (Japan)
Koji Ono (Japan)
Giovanni Orsi (Italy)
Peter Otway (New Zealand)
Norio Oyagi (Japan)
Liek Pardyanto (Indonesia)
Ray Punongbayan (Philippines)
Emmanuel Ramos (Philippines)
Jim Riehle (U.S.)
John Roobol (Wales)
Bill Rose (U.S.)
Mauro Rosi (Italy)
Rodrigo Saenz (Costa Rica)
Moto Sato (U.S.)
Yosihiro Sawada (Japan)
Roberto Scandone (Italy)
O. Seliangin (U.S.S.R.)
John Shepherd (Trinidad)
Daisuke Shimozuru (Japan)
Lee Siebert (U.S.)
Haraldur Sigurdsson (U.S.)
Gudmundur Sigvaldason (Iceland)
Tom Simkin (U.S.)
Genrich Steinberg (U.S.S.R)
Yasuaki Sudo (Japan)
Adjat Sudradjat (Indonesia)
David Sussman (U.S.)
Ben Talai (Papua New Guinea)
Haroun Tazieff (France)
P. Tokarev (U.S.S.R.)
John Tomblin (Trinidad-U.N.)
R. van der Laat (Costa Rica)
Geoff Wadge (U.K.)
George P.L. Walker (U.S.)
Stanley Williams (U.S.)
Chuck Wood (U.S.)
Rick Wunderman (U.S.)
Izumi Yokoyama (Japan)
Betsy Yount (U.S.)

Willie Lee kindly searched his bibliographies and discovered many important references. U.S. Geological Survey librarians Trudy Sinnott, Jackie Freeberg, and Nancy Blair located many references that we could not find in libraries of Oregon and Washington.

Jesse Johnston and Ed Brown assisted with computer programming and applications, and Elva Klassen, Joy Seeley, Matt Mosner, Garren Mayer, and Rachel Carlin translated numerous papers. Chris Janda and Bill Dize prepared photographic prints, and Mark Uhrich produced computer-generated maps. Ken Yamashita, Victoria Avery, Rick Wessels, Gary Stoopes, and Dave Schneider helped with drafting and other work on figures.

Nine individuals deserve special recognition. Lauren Mullineaux volunteered much hard work as we were defining the scope of this project. Hiromu Okada familiarized us with the extensive Japanese literature on restless calderas and contributed much to our discussions of how to treat volcano monitoring data. Roy Torley, linguist *par excellence,* helped us to understand an enormous volume of Dutch, French, German, Russian, and other foreign literature. Robin Sutton and Alan Vaughn led the formidable task of figure preparation with much appreciated optimism. L.J. Patrick Muffler provided encouragement at critical moments. Tom Simkin, Lee Siebert, Lindsay McClelland, and colleagues at the Smithsonian Institution provided many leads to unrest through the Scientific Event Alert Network Bulletin and through their Volcano Reference File and its derivative publication, "Volcanoes of the World." In addition, we have benefited greatly from their careful thought about how to present incomplete data on volcanoes.

In a large compilation, errors and omissions are inevitable. We tried to minimize them by seeking reviews from numerous colleagues around the world who are familiar with the volcanic systems in question. We thank the technical reviewers and editors for their considerable help. However, we also invite readers to call our attention to remaining errors and to new work that bears on processes of unrest at large calderas.

Finally, we thank all of the publishers who granted permission to reproduce copyrighted illustrations; if requested, specific acknowledgments are given in the figure captions.

Special Acknowledgment.—Our late friend and esteemed colleague, Prof. Kazuaki Nakamura of the Earthquake Research Institute, Tokyo University, in his own quiet and unobtrusive way, made a profound contribution to this study. Nakamura-san, or Kazu as we knew him, had a special knack for questioning conventional wisdom and suggesting alternate and novel explanations of geological phenomena. Where many of us might try to limit the number of variables in a problem, Kazu tried to relate all of those variables.

Several of Nakamura-san's ideas have strongly influenced our own thinking about processes of caldera unrest. Foremost among these is his proposal that a magmatic system can act as a volumetric strain gauge and an indicator of regional crustal strain (as at Izu-Oshima Volcano). We believe many large magma bodies beneath calderas act as volumetric strain gauges. Another of his ideas concerned the importance of ground water in stimulating seismicity and causing uplift, as at Matsushiro. We agree, and think that the role of ground water is underestimated in too many instances of unrest. Other influential work included a classic volcano-stratigraphic study of Izu-Oshima Volcano that recognized multiple episodes of magmatic and phreatomagmatic explosions, an explanation of how cinder cone alignments can indicate regional tectonic stress directions, proposals concerning unconventional mechanisms for generating a linear distribution of earthquake epicenters, and an interpretation that pinching of the tip of the Philippine Sea plate between the Sagami and Suruga Troughs causes uplift of a central ridge and unrest on the Izu Peninsula.

If Nakamura-san were here he would tell us that we overestimate his contributions. If anything, we might be underestimating them.

DEFINITIONS

We consider only *historical unrest* that occurred at *large, Quaternary calderas*, as defined below.

Historical—Events that occurred during the period of recorded history. In some areas, such as Japan and the Mediterranean, history has been recorded for more than a millennium; elsewhere, as in the Aleutian Islands, history has been recorded continuously for less than one century.

Unrest—Changes from the normal state at a caldera, including its associated magmatic and hydrothermal systems. Most changes are geophysical or geochemical events, such as earthquake swarms or an increase in the sulfate content of ground water. Often these events are precursory to specific eruptions, but sometimes they cease without eruption or cannot be related to any particular eruption. Indeed, we use the term "unrest" as a deliberate alternative to "precursor" because not all changes at a caldera can be related to the next eruption at that caldera. All eruptions are included as episodes of unrest, even if no other information on unrest associated with the eruption is available. We distinguish three types of unrest:

Eruption only—An eruption, without *reported* precursory seismicity, uplift, or other anomalous geophysical or geochemical activity.

Broadly defined unrest—A significant change in seismicity *or* ground deformation, fumarolic activity, hydrologic activity, and so on, sometimes culminating in an eruption.

Narrowly defined unrest—Increased seismicity *and* uplift, sometimes culminating in an eruption. This category is intended to highlight the type of unrest that has occurred recently at the Phlegraean Fields, Rabaul, Aira (Japan), Long Valley, and Yellowstone (U.S.) Calderas.

In many cases these distinctions are dictated by the quality of available reports. Much of the unrest in the first two categories might properly belong in the third if historical observations and reports were more complete.

Other necessary definitions are the following:

Episode of unrest—A period of unrest of unspecified duration, preceded and followed by relative quiescence of equal or longer duration. Typically, unrest lasts several weeks to years and quiescence lasts several years to centuries.

Caldera-year of unrest (c-y)—A calendar year at one caldera during which unrest is known to have occurred. The duration is not considered; one day counts the same as 365 days. The purpose of this definition is to normalize observational data to a common, reproducible basis that does not depend on whether the observer chose to report unrest as many discrete events or as a smaller number of episodes of closely related events. For example, we can refer to 10 caldera-years of increased seismicity at Long Valley Caldera, from 1978 to 1987, without needing to distinguish between seismic swarms, smaller seismic "flurries," and background activity. The same principle applies to ground deformation: We can refer to approximately 8 caldera-years of uplift at Long Valley, from 1980 to 1987, without having to guess whether the uplift occurred continuously or in several discrete episodes.

Large caldera—A closed volcanic depression with a diameter of at least 5 km, many times the diameter of vent(s) therein. The term "caldera" is further restricted to depressions formed largely by collapse into a magma reservoir. This restricted, genetic definition is adopted so that the diameter of a caldera can be used as a rough guide to the size of the underlying magma body. Calderas formed as a result of large volcanic avalanches or erosion are excluded. A few calderas smaller than 5 km in diameter (for example, Izu-Oshima) are included to illustrate processes that probably act, but have not been documented, at larger calderas.

Quaternary caldera—A caldera listed in the "Catalogue of Active Volcanoes of the World including Solfatara Fields" (IAVCEI, 1951–present), in the "Data Sheets of the Post-Miocene Volcanoes of the World" (IAVCEI, 1973–present), in "Volcanoes of the World" (Simkin and others, 1981) and the unpublished Volcano Reference File on which that book is based, and in various primary sources, unless there is positive evidence that the caldera formed more than 2.5 m.y. ago. All historically restless calderas are probably Quaternary; the distinction of Quaternary becomes important for identifying quiet calderas of comparable age.

SOURCES AND QUALITY OF DATA

General Remarks

Data in this compilation come from many different sources: Roman historians and philosophers, monks, priests, and other clerics, explorers and colonizers, traders in seal furs and spices, sea and airline captains, traveling naturalists, long-time residents of communities near volcanoes, and scientists from many disciplines. Many observations were made by accident—being in view of an eruption, feeling an earthquake, noticing that barnacles were cleansed from one's ship by hot acidic waters, and so on. Beginning around the start of this century and increasingly in the past few decades, instruments have been used to systematically monitor a wide range of geophysical, geochemical, and geologic changes at many volcanoes, and remote satellite- and land-based cameras now maintain a constant watch around the globe. Gaps in early records resulted from infrequent observations and reliance on visual sightings that are easily foiled by clouds. Gaps in recent records have the same causes, plus instrument failures and funding problems.

Sometimes even pertinent observations have not been reported. Observers or intermediate links in a chain of reporting might judge an observation to be insignificant, especially in the face of difficulties in communication or of other more compelling events, such as an ensuing volcanic eruption or war. Simkin and others (1981) and Simkin and Fiske (1983) show a clear relation between the completeness of reporting and various human events, such as establishment of transoceanic telegraph links, world wars, severe economic depression, or even fresh memories of another, major eruption like that of Krakatau in 1883. Reporting of volcanic events since 1960 has been greatly improved by the efforts of the Volcanological Society of Japan (Bulletin of Volcanic Eruptions), joined in 1968 by the Smithsonian Institution's Center for Short-Lived Phenomena (SICSLP) and in 1975 by SICSLP's successor, the Scientific Event Alert Network (SEAN).

Clearly, incomplete recording and reporting of events dictate great caution in interpreting what is known about caldera unrest. In general, the following can be stated:

(1) Estimates of the frequency of unrest are underestimates.

(2) Estimates of the relative frequency of events, or correlations of events, are strongly biased by differences in the ease with which the events are detected, the human and scientific interest they stir, and hence the likelihood that they will be recorded. Easily detected, interesting events such as eruptions and earthquakes are reported frequently, whereas subtle, less spectacular events such as slight changes in level of the water table are only rarely reported.

(3) Reports of precursors to eruptions or strong earthquakes are sometimes exaggerated by observers who, after the major event, attach significance to everyday events that occurred shortly before the event. Studies of possibly anomalous animal behavior before earthquakes are particularly noted for this bias (Lott, 1976).

(4) Events that leave a trace for later discovery, such as eruptions that leave deposits, are better reported than those that leave subtle or no traces, such as swarms of small earthquakes and indeed most aspects of caldera unrest. This and the second and third factors suggest that subtle unrest is seriously underreported relative to eruptions, and that a great deal of unrest that ends without an eruption is never reported.

(5) Apparent increases in the frequency of events through time, both at single calderas and in the global sample, are mostly the result of changes in recording and reporting. The march of telecommunications, not Armageddon, is at hand!

How can these difficulties be circumvented? One approach is to consider only data from calderas that are intensively monitored, and only for those periods in which systematic observations are reasonably assured. Another approach is to consider only that unrest that is sufficiently strong, interesting, or alarming to be completely recorded. Both approaches have merit, depending on the purpose of the investigation. In general, we consider only data from recent decades and well-studied calderas to estimate the frequencies at which changes occur, but we consider data from throughout historical time for cumulative totals or averages that require a large sample size. The few "statistics" we cite are based on an incomplete data set, and we urge appropriate caution in their interpretation.

Data About Calderas

Caldera name and number

In general, caldera names used are listed in the "Catalogue of Active Volcanoes of the World" (CAVW) (IAVCEI, 1951–present), "Data Sheets of the Post-Miocene Volcanoes of the World" (IAVCEI, 1973–present), or "Volcanoes of the World" (Simkin and others, 1981). Calderas not listed in these publications are referred to by their most commonly used name. Some calderas have the same name as that of their most active vent. Where a group of closely related calderas lie within a large volcano-tectonic depression (for example, Karymsky, Akademii Nauk, and several other calderas within the Zhupanovsky volcano-tectonic depression), all are described together under the name of the largest feature.

Calderas are listed geographically in Parts 2 and 3 according to the number assigned to their active vent(s) in the CAVW or as modified by Simkin and others (1981). The order of presentation, by region, is (1) Europe, (2) Africa, (3) the Indian Ocean, (4) New Zealand, Tonga and the Kermadec Islands, (5) Melanesia (Papua New Guinea and Vanuatu), (6) Indonesia, (7) the Philippines, (8) Japan and the Mariana Islands, (9) the Kurile Islands, (10) Kamchatka and mainland Asia, (11) Alaska, (12) western North America, (13) Hawaii,

(14) Central America, (15) South America, (16) the Caribbean Sea, (17) Iceland, (18) Atlantic islands (Azores, Canary, and Cape Verde Islands), and (19) Antarctica (figs. 1, 1.1, 2–10, 10.8, 11–19). Figures pertinent to each caldera follow the tabular and descriptive material for that caldera. The figures are numbered sequentially by region, caldera number within that region, and figure number for that caldera. Thus, figure 8.10.19 refers to the 19th figure for the 10th caldera in region 8. Figures 20.1.1 through 20.6.10 in Part 4 refer to selected noncaldera settings from throughout the world.

Latitude and longitude

We have listed the latitude and longitude of the center of each caldera or the coordinates of an active intracaldera or caldera-rim vent. Thus the coordinates in this report are adequate for locating a caldera on a regional map but not for relating the coordinates of a well-located local earthquake or bench mark to the center of the caldera.

Caldera diameter

The diameters of most calderas can be easily measured, but the diameters of others are poorly known or complicated. In Parts 3 and 5, we list a single diameter for calderas that are nearly circular and the length and width for calderas that are elongate or irregularly shaped. For statistical purposes, we use the average of length and width. We measure rim-to-rim distance rather than the diameter of the ring fault, because often the position of ring faults is not known or else the caldera formed by chaotic collapse without discrete ring faults. As a result of mass wasting, the topographic diameter of a caldera is usually greater than its structural diameter. For a few calderas, neither a topographic nor structural boundary is obvious, so we have used the distribution of vents and thermal areas to estimate caldera diameter. Our assumption in such cases is that postcaldera vents and major thermal features lie near or within the caldera rim.

Local tectonic setting

The regional tectonic setting is listed for those calderas where local tectonic structure or the pattern of unrest at a caldera is consistent with the regional tectonic setting. The local, or both local and regional, tectonic settings are indicated for those calderas where local tectonic features differ from those of the regional pattern (for example, an extensional graben along a regionally convergent plate margin, or local shear in a region of tectonic extension).

Precaldera edifice

On the premise that processes beneath calderas might vary according to the structure of the precaldera edifice(s), we have classed those edifices as strato (stratovolcano), shield (shield volcano), LL-strat (low, lava-dominated stratovolcano), LP-strat (low, pyroclastic-dominated stratovolcano), dome (one or more volcanic domes), and none (no known preeruption edifice). Information about some edifices has been inferred from topographic or geologic maps if not given explicitly in published literature.

Composition of products

Details of the petrography and petrochemistry of products from caldera-forming and recent eruptions are beyond the scope of this report. In general, we note only the silica content of those products, a crude indicator of magma viscosity, density, and explosive potential. The alkali, volatile, and crystal contents of magma also exert important controls on these parameters, as do confining pressure and involvement of ground water.

Our discussion distinguishes only between silicic (SiO_2 >57 percent, water-free) and mafic ($SiO_2 \leq 57$ percent) compositions, because silica content is only a crude indicator of magma properties, because a range of compositions can be erupted during a single eruption, and because no chemical analyses are available for the products of many eruptions in this compilation. Specific rock names, including b=basalt, a=andesite, d=dacite, and r=rhyolite, are given in the data tables if they were originally reported without chemical analyses.

Compositions of caldera-forming and postcaldera products are indicated by C and R, respectively. We show the range of compositions of compositionally zoned or heterogeneous magmas and also note any dominant composition.

Age of latest collapse(s)

Collapse itself does not produce datable products. Ages reported here are radiometric, fission-track, or other kinds of absolute ages of eruption products (or charcoal within deposits) from caldera-forming eruptions. The products of caldera-forming eruptions are usually so much more voluminous than those of other eruptions that the correlation of eruption and collapse is clear. At a few calderas, uncertainty about which deposits correlate with collapse dictates corresponding uncertainty about the age of collapse. Some ages of collapse are inferred from geomorphic or stratigraphic relations with dated glacial deposits; they are listed as preglacial or postglacial. Other ages of collapse are constrained by stratigraphic relations with precollapse or postcollapse volcanic deposits of known age; in such cases, the possible range in age is listed. The few ages of collapse known from historical records are listed exactly.

Data About Unrest

Dates of unrest

Estimates of starting and ending dates vary widely from one report to another, even for modern episodes of unrest. No standard convention is used worldwide. We try to define the beginning and end of each episode of unrest as the dates encompassing all changes that are genetically related to a single disturbance. For an unusually clear-cut magmatic intrusion and eruption, unrest might begin with increased seismicity, thermal changes, or ground deformation and end in an eruption or series of eruptions. Another common scenario is an intrusion that causes anomalous changes that peak and gradually decline to background levels without an eruption.

In general, we have considered one episode of unrest to have ended and another to have begun if the period of quiet between those episodes is longer than one year. Conversely, unrest or eruptions within a single year are treated as single, composite events unless, as at Kilauea, monitoring suggests a logical, more detailed definition of episodes. We selected a one-year period as a compromise between the value of resolving details of individual phases of unrest and the value of showing closely related changes together. One year can be an insignificant pause in unrest that is affecting a large silicic caldera, but the same year is long in relation to unrest of most stratovolcanoes associated with calderas. Coincidentally, in the lay public's perception of related events, unrest following a year of quiet will probably be considered a new event.

Many episodes of unrest, especially from past centuries, are defined solely on the basis of eruptions. For them we followed the same one-year-of-repose criterion for distinguishing episodes, unless some distinctive pattern of eruptions suggested a more logical division. We also used caldera-years of unrest in parts of our analysis owing to the ambiguities and subjectivity inherent in defining the beginning and end of unrest.

Characteristics of unrest

The characteristics of unrest are summarized in coded form in the tabular section for each caldera, followed by details (if known) in the narrative text. In the tabular section, columns headed by "**ESTU STHF MGTF H**" contain information about the occurrence and duration of various types of unrest. The columns refer to:

E, earthquakes (of probable local origin);
S, earthquake swarms;
T, volcanic tremor;
U, uplift;
S, subsidence;
T, tilt;
H, horizontal strain;
F, surface fractures, faults, or fissures;
M, magnetic changes;
G, changes in the gravitational field;
T, changes in the temperature of ground or surface water, soil, or fumaroles;
F, changes in fumarolic intensity, area, or composition;
H, changes in hydrologic parameters, including discharge or composition of ground water or springs, or water level (stage) in wells or lakes.

To find data about a particular type of unrest and to avoid confusion from duplicate letters, look for the appropriate letter *within* the appropriate set of four letters. For example, to find data about subsidence, look for entries in the column labeled S *within* the set labeled **STHF**.

Each type of unrest is coded according to its maximum *preeruption* duration: A (lasted only minutes before an eruption), B (hours), C (days), D (weeks), E (months), F (years), or G (decades). Unrest that occurred at an unknown time relative to an eruption is coded as "x"; that which occurred during an eruption as "Y"; and that which occurred after an eruption as "Z." For those episodes of unrest that did not include an eruption, the code letters refer to the total duration of that aspect of the unrest. Where unrest occurred before, during, and after an eruption, preeruption duration is indicated; where unrest occurred during and after an eruption, a "Y" is shown to indicate that change during the eruption. Aspects of unrest that were explicitly reported as not having occurred during a given episode are coded "N"; those for which we found no data are coded "-." Note that the latter designation does not indicate that a certain aspect of unrest did not occur, but rather that we found no mention of its occurrence or nonoccurrence during a given episode.

The columns headed by "**Te**" contain information about temporal relations between unrest and regional earthquakes tens to hundreds of kilometers from the caldera, if the time between the earthquakes and unrest was short relative to the normal recurrence frequency of both phenomena. "UQ" indicates unrest followed by an earthquake; "QU" indicates an earthquake followed by caldera unrest; "UU," unrest, a large earthquake, and then more unrest; and "QQ," an earthquake, unrest, and then another large earthquake. No causal relations are implied, although some of the events are almost surely related.

Seismicity

Until this century, almost all reports of seismic activity dealt only with felt earthquakes, which generally indicated a local magnitude of at least 2 and usually 3 or more. In remote, sparsely populated areas, detection thresholds were much higher, and even today the detection threshold of the Worldwide Network of Standard Seismic Stations (WWNSS; Oliver and Murphy, 1971) is about magnitude 5 in remote areas.

Instrumental records of seismicity near calderas come from regional seismic networks, most of which are designed to record large, regional earthquakes, and from local networks or single seismometers operated for volcano-monitoring purposes. Some local networks cover an entire caldera; an equal or greater number are limited to, or focused on, an active postcaldera vent.

The spatial resolution of seismic networks in volcanic areas varies greatly. Most epicenters associated with caldera collapse at Fernandina (Galápagos Islands) in 1968 fall far outside the caldera rim because epicentral determinations are based on the World-wide Network of Standard Seismic Stations, which affords wide coverage but limited spatial resolution (Filson and others, 1973). Even local networks sometimes yield poor epicentral locations unless the network is carefully designed and the local seismic velocity structure is well known. Only the best local networks give reliable depth determinations.

Earthquake counts are notoriously ambiguous because the number of earthquakes detected depends strongly on the location, detection threshold, and frequency response of seismometers in the network. Thus, earthquake counts from two episodes of unrest at the same caldera can be compared meaningfully only if records from a single, unmodified station are used; comparisons of earthquake counts from one caldera to another are rarely meaningful. Earthquake counts also are suspect when earthquakes become so closely spaced in time that individual events no longer can be distinguished. Counts of earthquakes above a specified magnitude threshold, or records of seismic energy release, are better for most comparisons.

Magnitude, a measure of the strength of an earthquake and the strain energy released by it, is a single value that does not vary with the observer's location relative to an earthquake. Intensity, a measure of the local effects of an earthquake, generally decreases with distance from an earthquake and varies with local ground conditions. Most recent reports indicate Richter magnitude *M* of earthquakes, and a few cite the Russian energy class *K* (K=5.2 + 1.6M). Many older reports give only intensity, for which the most common scale is the 12-point Modified Mercalli Intensity (MMI) scale. Other intensity scales in use are the Rossi-Forel (10 point), JMA (Japan Meteorological Agency) (8 point), and MSK (Medvedev, Sponheuer, and Karnik) (12 point) scales. The approximate relations among various intensity scales are shown below.

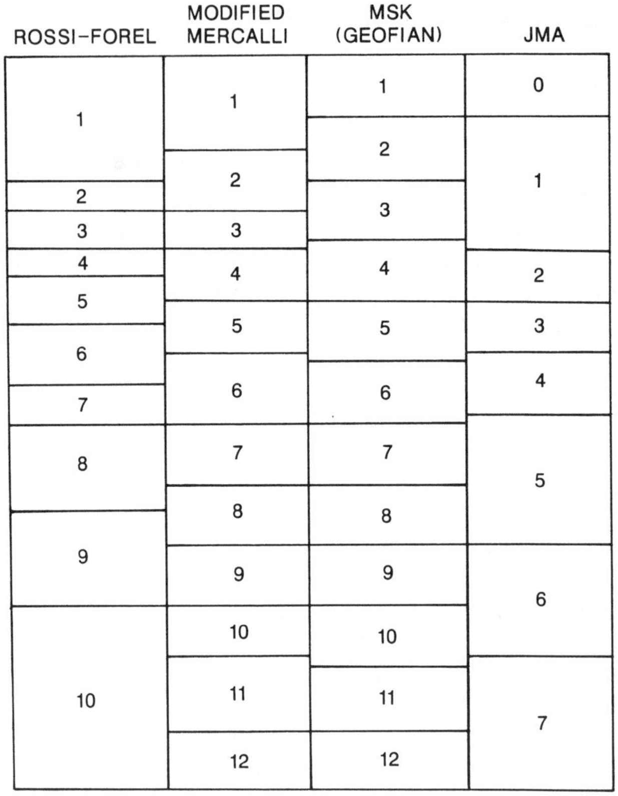

The terminology of seismic events differs from one country to another and from one era to the next. Except in direct quotations, we have translated "tremors" as "earthquakes" to refer to discrete events. For continuous tremor, we usually have avoided the term "harmonic" unless the signal was clearly harmonic in nature; in most cases, "volcanic tremor" is a more suitable, less specific term that includes continuous signals caused by overlap of frequent discrete earthquakes (Finch, 1949; S. McNutt, oral commun., 1987). The most widely used categories of volcanic earthquakes are the A-type, B-type, and C-type categories of Minakami (1960) and Minakami and others (1951, 1969). A-type earthquakes occur between 1 and 10 (sometimes 20) km depth; B-type earthquakes occur <1 km beneath a volcano; and C-type earthquakes, also shallow and similar to the B-type, occur repeatedly during viscous dome emplacement. Other descriptive terms, such as "low-" or "high-frequency earthquakes," "surface events," "volcano-tectonic," and "tectonic earthquakes" have been included as originally reported.

Many events in or near calderas that have been termed volcanic earthquakes differ from tectonic earthquakes only by their occurrence in swarms. Conversely, many events in or near calderas that have been termed tectonic would not have occurred were it not for volcanic processes. We suspect that many earthquakes in or near calderas are hybrid volcano-tectonic events, but we have not attempted to change terminology from the original reports. Minakami (1960), Shimozuru (1971), Latter (1979), and Malone (1983) discussed these and other complexities of terminology for volcano-related seismic events.

In descriptions of seismic unrest, we included several items of information whenever they were reported. These items are, for each episode, total seismic energy release (ergs), maximum rate of seismic energy release (ergs/day), magnitude of the strongest earthquake ("M_{max}"), maximum hypocentral depth (km), and maximum reported intensity (usually "MMI_{max}").

Ground deformation

Occasionally, ground deformation at calderas is so dramatic that it can be noticed and measured by laymen, such as fishermen who note that docks are rising out of the sea, farmers who realize that fields are bulging, or highway crews who discover offsets in roadways. Normally, however, ground deformation is subtle and can be detected only by precise surveying (Decker and Kinoshita, 1971). Vertical displacement is often measured in millimeters or centimeters, and horizontal strain rarely exceeds 100 parts per million (100 ppm = 10 cm/km). Ground tilts are commonly a few microradians to a few hundred microradians, and the noise level in tilt measurements can easily be several tens of microradians. We have included information about maximum measured uplift, subsidence, tilt, strain, and the maximum rate of change of these parameters whenever such data are available.

Magnetic and gravity changes

Magnetic changes are reported as changes in total field strength, strength of one component of the magnetic field, declination or inclination of magnetic north, or the difference in any of these parameters between a reference station and a field station. Most of the magnetic changes are presumably of piezomagnetic or thermomagnetic origin; we list a specific mechanism only on the basis of original reports.

Gravity changes are reported as changes in the strength of the gravitational field at a particular point or, less commonly, in the contours of a gravity anomaly map. Such changes are the result of uplift, subsidence, or changes in the density of nearby rock (for example, intrusion of magma into country rock of different density).

Thermal changes

Thermal changes are easily detected and can be described simply as increases or decreases from the norm. Most ambiguities in thermal measurements are caused by infrequent measurements (some changes are not detected until months or even years after they occur) and by inconsistencies in measurement technique, such as the depth to which a probe is inserted in a fumarole. Temperatures of hot springs, crater lakes, soil, and low-temperature fumaroles are easily and reproducibly measured. Published temperatures of fumaroles above a few hundred degrees Celsius are in some instances minimum temperatures, limited by equipment, protective clothing, or changes in the orifice of a fumarole. The interpretation of thermal changes is often complicated by uncertainties in the role of ground water in buffering temperatures measured at the surface.

Temperatures within long-standing hydrothermal systems at calderas and other geothermal areas commonly reach 200 400 °C. Temperatures above 400 °C usually indicate recent magma intrusion. Temperatures near 100 °C (less at higher elevations) may be buffered by ground water and therefore might not indicate a change in heat input.

Fumarolic and hot-spring changes

Commonly, changes are noted in fluid composition or flux of fumaroles or hot springs, level and composition of ground water, or water level in crater lakes. The most commonly measured chemical species at volcanoes are CO_2, CO, CH_4, SO_2 and H_2S (or S or SO_4), HCl (or Cl), HF (or F), Rn, and Hg. The sampling and analysis of volcanic waters are essentially the same as for other naturally occurring waters. In samples of fumarolic gases, atmospheric contamination can introduce large differences between individual samples, but corrections can be applied after analysis to eliminate most contamination effects. Perhaps the greatest shortcoming of most chemical studies at calderas is that sampling is done monthly, annually, or even less frequently—not often enough to detect rapid changes during caldera unrest. Recent technological advances, such as the correlation spectrometer (COSPEC) for measuring SO_2 flux and a fuel-cell sensor for monitoring the reducing capacity of volcanic gases (mostly H_2), make frequent measurement of chemical parameters more practical.

In most reports, information about the flux or intensity of fumarolic activity is based on visual inspection, supplemented in recent years by COSPEC measurements of SO_2. Visual observations are nonquantitative, reports vary among observers, and appearances depend on atmospheric conditions. Measurements of water level in wells and flow rates of hot springs are straightforward, though for maximum precision the effects of earth tides must be considered.

Eruptions

This compilation emphasizes noneruptive unrest and gives few details of eruptions beyond the dates of their occurrence. Readers who seek details of specific eruptions are referred to papers cited under each caldera, to the CAVW (IAVCEI, 1951-present), and to Simkin and others (1981).

For this compilation, we have adopted a few simple terms for eruptions and associated phenomena. Three types of explosions are distinguished: (1) phreatic explosions, driven by the flashing of hot ground water to water vapor and ejecting fragments of country rock (pex); (2) "normal" explosions of unspecified type, including magmatic explosions (ex); and (3) demonstrably hybrid phreatomagmatic explosions (p-ex). The first two categories are from the CAVW; we have added the third to note explosions driven by explosive magma-ground water interaction. Unfortunately, few reports distinguish between phreatic and magmatic activity, and many reportedly "normal" explosions have phreatic as well as magmatic components. Other phenomena noted are pyroclastic flows (pf), lava flows (lf), lava lakes (ll), domes of viscous lava (dm), cryptodomes (cdm), landslides (landsl), submarine eruptions (sub), subglacial eruptions (subgl); cinder cones (cc), mudflows (mf), and debris avalanches (da).

We use a scale of explosive activity that follows the Volcanic Explosivity Index of Newhall and Self (1982), an index defined in several ways to make use of whatever information is available. Defined by the volume of pyroclastic ejecta, the VEI scale is as follows:

VEI 8	1000–10000 km^3 (bulk volume)
VEI 7	100–1000 km^3 (bulk volume)
VEI 6	10–100 km^3 (bulk volume)
VEI 5	1–10 km^3 (bulk volume)
VEI 4	0.1–1.0 km^3 (bulk volume)
VEI 3	0.01–0.1 km^3 (bulk volume)
VEI 2	0.001–0.01 km^3 (bulk volume)
VEI 1	0.0001–0.001 km^3 (bulk volume)
VEI 0	nonexplosive—no pyroclastic debris.

We have adopted a modified system in this report in order to accommodate generalized reports of individual explosions ("large,""small," and similar descriptors), to make the scale of explosions more readily understandable to nonvolcanologists, and to simultaneously specify the type of explosion where known. VEI magnitudes 1–8 are reduced to three, and magnitude is indicated by upper- and lower-case letters (for example, EX=major explosion (VEI 4 and above); Ex=moderate-size explosion (VEI 3); ex=small to moderate-size explosion (VEI 1 or 2)). A similar convention is used for phreatic explosions (PEX, Pex, and pex), phreatomagmatic explosions (P-EX, P-Ex, and p-ex), and lava flows (LF, Lf, and lf). Lava flows were assigned magnitudes based on a relative scale for each caldera (large, average, or small for that caldera); lava-flow magnitudes at different calderas are not necessarily comparable.

In summary, the code

ESTU	STHF	MGTF	H	Te	Erup type
FEBF	YD-Y	- - xx	Z	QU	pex,Ex,dm

contains the following information about a hypothetical episode of unrest: An eruption (see description below) was preceded by several years of local earthquakes (F), months of earthquake swarms (E), hours of volcanic tremor (B), and years of uplift (F). Subsidence occurred during the eruption (Y), but horizontal strain was not reported (-); ground tilt preceded the eruption by several weeks (D), and surface fractures, faults, or fissures formed during the eruption (Y). No magnetic or gravity changes were reported (--), but thermal and fumarolic changes occurred at some unspecified time during the unrest (xx). Hydrologic changes followed the eruption after an unspecified interval (Z). Unrest was preceded by a large tectonic earthquake (QU) and culminated with a small phreatic explosion (pex), a moderate explosion of unspecified type (probably magmatic) (Ex), and extrusion of a small dome (dm). Additional information might be included elsewhere in the tabular material for the caldera (for example, composition of erupted products, if known) or in the text under the heading "Historical Activity."

DESCRIPTION OF CALDERA UNREST

Our summary description of caldera unrest is organized around the questions, What, When, Where, and in What Associations? Section headings are as follows: Noneruptive Unrest; Eruptive Unrest; Prevalence and Duration of Caldera Unrest; Distribution of Unrest Within Individual Calderas; Correlations Between Characteristics of Calderas and Their Unrest; and Temporal Relations With Regional Earthquakes and Earth Tides. Details of unrest at specific calderas are given in Part 3.

Noneruptive Unrest

Unrest is often a composite of many different types of changes—seismic, geodetic, thermal, fumarolic, hydrologic, and others—and of course eruptions. First, we consider characteristics of each type of noneruptive unrest observed at calderas.

1. What types of seismicity are observed at calderas?

Most seismic events within or near calderas are discrete, relatively small earthquakes, with magnitudes typically less than 3 and rarely greater than 6. Some result from brittle failure of country rock in response to magma intrusion; other events reflect release of regional tectonic stress, shear of viscous magma along the walls of a conduit, explosions in magma, or collapse as magma is erupted or drains laterally into a rift zone.

Calderas develop in regions that are already seismically active (for example, above subduction zones, midocean ridges, and hotspots). Focal-mechanism studies indicate that most earthquakes in caldera settings, including those associated with magma intrusion, are similar in character to tectonic earthquakes—that is, they reflect sudden frictional sliding in response to shear stress across a fracture surface. Some volcano-tectonic earthquakes occur outside caldera boundaries but in swarms or with signatures or focal mechanisms resembling those of volcanic origin (for example, swarms at Ontake, Japan, in 1976–84 and earthquakes near the south margin of Long Valley Caldera, California, in 1980–83). A broad group of earthquakes at calderas probably results from both tectonic and volcanic processes. At least at some calderas, failure is facilitated by high pore pressures within a hydrothermal system (for example, at Hakone, Japan).

Volcanic tremor is another common feature of seismicity at restless calderas. Some harmonic tremor occurs when magma or gases are flowing through constricted space; other tremor may result from bubble implosions and other mechanisms. Much tremor consists of earthquakes that are so closely spaced on seismograms as to be indistinguishable from one another. One notable pattern is bursts of tremor lasting a few minutes to a few tens of minutes, alternating with comparable or longer periods of quiet. The result on a seismogram is a rhythmically banded pattern. McKee and others (1981) at Karkar, Kieffer (1984) at Yellowstone, and McNutt (1987) at Pavlof noted a correlation of banded tremor with hydrothermal and geyser activity.

In some unrest, the number of earthquakes per unit time increases steadily and then decreases either suddenly or gradually until quiet returns. In many other episodes, earthquakes occur in short, hour- to week-long bursts (seismic swarms), separated by longer periods (weeks, months) of relative quiet. Eruptions can occur at any stage in the waxing or waning of seismic energy release. One common pattern is for an eruption to begin at the peak of rapidly increasing seismic energy release; another less common but notable pattern is for an eruption to begin hours to days after a sudden decrease of previously increasing seismic energy release. However, patterns are too varied and too poorly understood to conclude that any specific pattern will lead to an eruption. The intensity of precursory seismic activity is not always proportional to the size of associated eruptions.

A few remarkably strong earthquakes have occurred near calderas in close temporal association with caldera unrest. These include magnitude 7+ earthquakes near Kilauea Caldera in 1868 and 1975; a "strong" tectonic earthquake near Rabaul Caldera in 1940 and a magnitude 7 earthquake there in 1941; a magnitude 7 earthquake near the rim of Aira Caldera during the 1914 eruption of Sakurajima; and a magnitude 7.3 earthquake about 25 km northwest of Yellowstone Caldera in 1959. Other notable examples include a magnitude 6.1 earthquake along the northern rim of Yellowstone Caldera in 1975 and four magnitude 6+ earthquakes within and just south of Long Valley Caldera in 1980. Each of these large earthquakes was at least partly tectonic in origin, and each bears a different relationship to eruptions or other aspects of caldera unrest. At Kilauea, large earthquakes reflect lateral spreading and gravitational collapse of part of the volcano's south flank, caused by repeated intrusions of basaltic magma into shallow rift zones. At Rabaul, the cause of the 1940 and 1941 earthquakes may have been regional tectonic stress, but the earthquakes also occurred during the 1937–41 episode of unrest at Rabaul Caldera (see Part 3). The earthquake at Aira occurred 8 hours after a major eruption of Sakurajima began but before the largest volume of lava was extruded. This eruption may have caused a relaxation of regional stress, which in turn triggered the earthquake. The primary cause of the 1959 and 1975 earthquakes at Yellowstone was probably regional tectonic stress, but both earthquakes triggered dramatic hydrothermal changes within Yellowstone Caldera and may have temporarily influenced the rate of crustal uplift within the caldera.

2. What types of ground deformation are observed?

All types and scales of ground deformation occur at calderas. The most commonly reported changes are uplift, horizontal strain, and resulting tilt around a postcaldera vent as magma intrudes beneath that vent. Such inflation usually is measurable on and a short distance away from the postcaldera vent structure. During or following an eruption, corresponding deflation typically occurs.

Calderas in the summit regions of large shield volcanoes have a distinctive deformation pattern. The entire summit areas of Kilauea and Mauna Loa inflate until lava is intruded or erupted in the summit caldera or along a rift zone. Summit deformation is readily tracked by tiltmeters on the caldera rim, by measurements of selected EDM (electronic distance meter) lines spanning the caldera, and by leveling surveys before and after episodes of inflation or deflation.

Occasionally, localized but dramatic ground deformation occurs directly above a volcanic conduit. Two examples illustrate the range of such activity. Ascent of viscous magma at Usu Volcano, on the south margin of Toya Caldera, Japan, pushes the ground surface as much as 300 m skyward, forming cryptodomes (hidden domes) (Minakami and others, 1951; Yokoyama and others, 1981). Ascent of fluid basaltic magma in the central conduit of Oshima Volcano (Izu Islands, Japan) raises the crusted crater floor as much as 400 m over the course of several decades, but upon release of regional tectonic stress by great earthquakes, the magma in the conduit and the crater floor suddenly drop by a few hundred meters (Yamashina and Nakamura, 1978). Other episodes of dramatic local deformation and fissure formation have occurred above shallow dike intrusions (for example, in rift zones radiating from Kilauea and Mokuaweoweo (Mauna Loa) Calderas, and from Krafla Caldera).

One of the most intriguing types of ground deformation is subtle, caldera-wide uplift and subsidence (and associated horizontal strain and tilt). Such uplift and subsidence are noted only where precise leveling surveys, gravity surveys, a tide gauge, or a changing shoreline permits their recognition. Broad, subtle, years-to-decades-long uplift has occurred within at least 10 of 14 large calderas where adequate surveys have been made. Calderas known to show such uplift (alternating in most cases with subsidence) are the Phlegraean Fields, Rabaul, Aira, Iwo-Jima, Long Valley, Yellowstone, Kilauea, Mauna Loa, Askja, and Krafla. The most rapid sustained uplift is at Iwo-Jima, where the caldera floor has risen 15 to 20 cm/yr for at least the past 200 years. Most caldera-wide uplift and subsidence have been modeled as inflation or deflation of "point sources" at depths of 5 to 15 km (for example, Yellowstone and Long Valley); in a few instances such as Rabaul and the Phlegraean Fields, modeling suggests sources as shallow as 2–3 km. In at least two calderas, rapid uplift and subsidence have been superimposed on slower, longer term uplift and subsidence: 4 to 6 m of uplift occurred one to two days before historical eruptions of the Phlegraean Fields and Rabaul Calderas.

Other calderas that have been carefully surveyed show irregular changes. Leveling surveys around the margin of Toya Caldera several years before and after eruptions show only minor caldera-wide deformation, but broad uplift may have occurred immediately before an eruption in 1910. Surveys spanning only a few years at Taupo and Kakuto show subsidence or irregular deformation. The important point is that even where dramatic or sustained deformation is not occurring, at least some caldera-wide deformation usually *is* occurring. Subtle, caldera-wide deformation is more the rule than an exception.

Only a few calderas are known to have formed in historical time, and even fewer collapses have been observed. The best-documented collapse occurred at Fernandina Caldera (Galápagos Islands) in 1968 (Simkin and Howard, 1970; Filson and others, 1973). At that time, a hinged, trapdoor-like piece of a preexisting caldera floor collapsed by as much as 350 m in about 9 days.

3. What other types of geophysical changes are observed?

The strength and orientation of the Earth's magnetic field are known to vary slightly through time near some volcanoes, in response to thermal demagnetization of country rock or tectonic strain (Yokoyama, 1971, 1974). Both increases and decreases in field strength have been reported, over periods of a few hours to a few months. However, the number of reports is too small to generalize about these changes.

Gravity measurements at calderas show secular and spatial variations in response to vertical displacements or changes in subsurface density (Yokoyama, 1971, 1974). Most changes relate directly to uplift or subsidence and closely follow the free-air gradient (0.3086 mGal/m). In some episodes, however, the gravity gradient differs from the free-air gradient owing to changes in the subsurface distribution of mass, usually caused by magmatic intrusion or changes in ground water level. Reported examples include Kilauea (Dzurisin and others, 1980; Johnson, 1987), Rabaul (McKee and others, 1987), the Phlegraean Fields (Corrado and others, 1977, 1983; Berrino and others, 1984), Krafla (Johnsen and others, 1980), Amatitlán (Pacaya Volcano) (Eggers, 1983, 1987), Poas (Rymer and Brown, 1987), and Long Valley (Jachens and Roberts, 1984).

For the case of magmatic intrusion, the density contrast between country rock and intruding magma can be calculated from the observed gravity gradient. However, uncertainty arises because the roof rocks of large magma reservoirs (that is, the floors of calderas) are structurally complex and magma is a compressible fluid (Johnson, 1987). The first complication means that newly intruded magma might be accommodated within newly created extension space at shallow depth without causing proportionate uplift. The compressibility of magma adds a similar complication, because the relationship between intruded mass and surface uplift depends on density changes (expansion or compression) within the intruding magma.

Geoelectrical changes occur at some calderas as a result of ground water heating, magma movement, or tectonic stress changes. However, the number of reported instances is small, so we have not included them in Part 3.

4. What changes occur in fumaroles or hot springs?

Several calderas in this compilation have major thermal areas on their floors or along their rims (for example, Yellowstone and Uzon-Geyzernaya). Fumaroles and hot springs in these areas typically change in discharge, temperature, or composition during episodes of caldera unrest. The most common changes are increases in total discharge, in the discharge and proportion of acid gases (SO_2, H_2S, HCl, and HF), and in CO_2 emission. Several workers also have reported increases in H_2 and Rn gases derived from country rock, hydrothermal systems, or magmatic sources. Hot springs have also changed their discharge during caldera unrest in relation to changes in the water table or opening of cracks. For example, hot springs have flowed profusely just before eruptions of Sakurajima (Aira Caldera) and Monte Nuovo (Phlegraean Fields Caldera), and geyser activity at Yellowstone increased dramatically after the 1959 M=7.3 Hebgen Lake earthquake.

Temperatures of hot springs and crater lakes are never above and are usually well below the boiling temperature of water. Temperature changes have been reported for several caldera lakes, such as Batur's, but these changes have for the most part been localized near hot springs or fumaroles. Temperatures of fumaroles range from <100 °C to >950 °C, and temperature changes of several hundred degrees are not uncommon during unrest. Examples include fumaroles at Tavurvur (Rabaul) in 1941 (300 °C increase), Poas (870 °C increase), and Karkar in 1978–79 (>900 °C increase). Remarkably, temperatures have risen above 500 °C at a number of calderas, but only a few eruptions have ensued.

Crater lakes also exhibit changes in color (usually reflecting algal blooms or changes in the amount of finely disseminated sulfur in the lake water) and in density stratification. Surface and bottom temperatures at tropical caldera lakes and at lakes in temperate zones near times of seasonal lake overturn differ only slightly. Unusually high winds, cool surface temperatures, or heating through the caldera floor can destabilize them and cause overturn (Newhall and others, 1987). During overturn, oxygen-poor, H_2S-rich bottom water rises to the surface, displacing oxygenated surface water and often leaving a telltale sign of dead fish (for example, several instances at Lake Ranau). Overturn can also release dangerous amounts of CO_2 gas, as at Lake Nyos (Kling and others, 1987) and Lake Monoun (Sigurdsson and others, 1987).

Eruptive Unrest

1. What proportion of unrest includes an eruption?

The percentage of episodes of specified unrest that culminated in an eruption, together with sample size *N*, is given in tables 1a–c. For example, the first column and second row of table 1a show that at mafic calderas, 83 percent of 107 episodes of broadly defined unrest culminated in eruptions.

Table 1a. Proportion of unrest that included eruptions
[Sample size *N* in parentheses; values given in percent]

Type of unrest	Magma of caldera-forming eruption			
	Mafic	Silicic	Unknown	Total
Eruption only	100(193)	100(468)	100(75)	100(736)
Broadly defined unrest.	83(107)	52(327)	71(52)	61(486)
Narrowly defined unrest.	77(30)	69(42)	100(5)	74(77)
Total	92(330)	80(837)	89(132)	84(1299)

The apparent implication is that a large proportion (84 percent) of all caldera unrest includes eruptions. However, this conclusion is surely biased by underreporting of unrest that does not include an eruption. In fact, noneruptive unrest is relatively common at most well-monitored calderas (table 1b)

Unrest at mafic calderas includes an eruption more often than deos unrest at silicic calderas. In the subset of rows 2 and 3 (table 1a), eruptions occurred in 82 percent of 137 episodes of unrest at mafic calderas versus 54 percent of 369 cases at silicic calderas. This conclusion is not seriously affected by underreporting; indeed, underreporting is less of a problem at well-monitored mafic calderas, such as Kilauea, Mauna Loa, and Piton de la Fournaise, than at many remote, poorly monitored silicic calderas.

Another interesting conclusion from table 1a is that approximately 1,090 eruptions have occurred at calderas during historical time, or 18 percent of the roughly 6,000 eruptions known from all types of volcanoes worldwide throughout historical time (Simkin and others, 1981).

2. What proportion of recent unrest included eruptions?

To answer this question, we considered only unrest during the past 30 years at well-studied calderas: the Phlegraean Fields, Taupo, Okataina/Haroharo, Karkar, Rabaul, Krakatau, Sunda, Tengger, Taal, Aira, Kakuto, Aso, Izu-Oshima, Toya, Shikotsu, Kutcharo, Akan, Long Valley, Yellowstone, Kilauea, Mauna Loa, Krafla, and Askja. The criteria for selection were that (a) all significant unrest *with or without* eruption is likely to have been reported, and (b) no incompletely reported, "eruption only" unrest was reported within this period. The results are shown in table 1b.

Table 1b. Proportion of unrest at well-studied calderas that included eruptions
[Sample size *N* in parentheses; values given in percent]

Type of unrest	Magma of caldera-forming eruption			
	Mafic	Silicic	Unknown	Total
Broadly defined unrest.	53(19)	34(35)	100(4)	45(58)
Narrowly defined unrest.	57(21)	43(14)	100(1)	53(36)
Total	54(39)	38(50)	100(5)	48(94)

Eruptions occurred during 54 percent (21 of 39 episodes) of unrest at well-monitored mafic calderas, and during 38 percent (19 of 50 episodes) of unrest at well-monitored silicic calderas.

Unrest at *long-quiet* silicic calderas usually does not lead to eruptions. Six of the above-mentioned, well-studied silicic calderas have not erupted within the past century (Phlegraean Fields, Taupo, Okataina/Haroharo, Kutcharo, Long Valley, and Yellowstone), yet during this period they collectively have had 28 episodes of unrest, including 17 within the past 30 years of relatively better reporting. For all silicic calderas with unrest in the past 100 years, at least 60 episodes of unrest occurred after repose of more than 100 years, and 50 of these did not lead to an eruption. Episodes that led to eruption were at Suoh in 1933, Geger Halang in 1937, Kikai in 1934, Ontake in 1979, Onikobe/Kurikoma in 1944, Akan in 1927, Nemo in 1906, Rasshua in 1957, Semisopochnoi in 1987, and Las Cañadas in 1909. Episodes that did not lead to eruption were at the Phlegraean Fields (6), Ischia (1), Taupo (6), Toba (4), Geger Halang (1), Ata (3), Hakone (10), Kutcharo (9), Opala (1), Long Valley (4), Yellowstone (1), Apoyo (1), Miravalles (1), Qualibou (1), and Agua de Pau (1). Some of this unrest was minor, but we believe that the data nonetheless indicate that unrest at long-quiet silicic calderas usually does not lead to an eruption.

3. What proportion of each unrest type is followed by an eruption?

At mafic calderas, 84 percent of 90 cases of local seismicity, 81 percent of 37 cases of uplift, and 73 percent of 44 fumarolic changes were associated with eruptions. However, as in table 1a, unrest that does not lead to eruptions is underreported. Given this underreporting, the result from table 1c that much unrest at silicic calderas is *not* associated with eruptions is especially interesting. At silicic calderas, only tremor, uplift, and ground fracturing seem to be strong indicators of an impending eruption, and even these, individually, are imperfect indicators.

Table 1c. Proportions of various aspects of unrest associated with eruptions

[Sample size *N* in parentheses; values given in percent]

Aspect of Unrest	Magma of caldera-forming eruption		
	Mafic	Silicic	Unknown
Local seismicity	84(90)	55(264)	86(29)
Earthquake swarms	84(44)	39(113)	88(8)
Tremor	93(44)	88(16)	100(2)
Uplift	81 (37)	75(60)	93(15)
Subsidence	79(52)	53(55)	—(1)
Tilt	85 (40)	42(19)	75(4)
Horizontal strain	86(22)	53(15)	(0)
Ground fracturing	90(42)	77(47)	83(6)
Magnetic changes	86(7)	71(7)	—(0)
Gravity changes	83(6)	17(6)	—(1)
Thermal changes	67(15)	52(75)	69(13)
Fumarole, gas changes	73(44)	58(156)	69(36)
Spring, lake changes	67(12)	55(67)	54(13)

4. What types of eruptions have occurred recently at large calderas?

Lava-flow and lava-lake activity are predictably common during eruptions at mafic calderas (table 2a); explosions, lava flows, and dome growth are common during eruptions at silicic calderas.

Table 2a. Types of historical explosive eruptions

[Data for past 100 years only]

Type of eruption[a]	Magma of caldera-forming eruption			
	Mafic	Silicic	Unknown	Total
"Normal" explosions (includes all explosions not reported as phreatic).	87	225	50	362
Phreatic explosions	8	72	22	102
Phreatomagmatic explosions	3	2	1	6
Lava flows	124	73	11	208
Lava lakes	22	0	3	25
Domes (including) cryptodomes).	0	25	1	26
Submarine, subglacial	10	13	3	26
Unknown	12	13	2	27

[a]Note: Two or more of these types may occur during a single eruption.

Most explosive eruptions at calderas are relatively small (table 2b), and the relative proportions of VEI's in the caldera sample for the past 100 years (column labeled "Total") are similar to those for the worldwide historical sample (in parentheses). Large calderas hold a greater potential for catastrophic eruptions than do small centers, but most eruptions at calderas are no larger than eruptions at other volcanic centers.

Table 2b. Explosive magnitude of eruptions, past 100 years

[Number of eruptions of each volcanic explosivity (VEI) during past 100 years; values in parentheses are numbers of eruptions of each VEI in historical time]

Size of explosions	Magma of caldera-forming eruption			
	Mafic	Silicic	Unknown	Total
Maximum VEI[a]:				
4+(EX, PEX, P-EX)	1	10	1	12(183)[b]
3 (Ex, Pex, P-Ex)	16	43	6	66(720)
1,2 (ex, pex, p-ex)	81	230	59	370(3,469)
0 (non-explosive)	73	11	0	84(443)
Unknown	12	13	2	27

[a]Volcanic Explosivity Index (Newhall and Self, 1982); values correspond roughly to the bulk volume of tephra erupted. VEI 0, less than 10^4 m^3; VEI 1, 10^4-10^6 m^3; VEI 2, 10^6-10^7 m^3; VEI 3, 10^7-10^8 m^3; VEI 4, 10^8-10^9 m^3.

[b]Numbers in parentheses are totals for all historical eruptions, including those associated with calderas (from Simkin and others, 1981; T. Simkin, 1988, written commun.). They are probably biased toward larger eruptions by poor early reporting of small eruptions.

The magma of postcaldera eruptions need not resemble that of the caldera-forming eruption. In fact, most postcaldera eruptions at silicic calderas are of relatively mafic magma (table 2c).

Table 2c. Composition of postcaldera and caldera-forming products

Minimum SiO_2 of most recently erupted magma	Magma of caldera-forming eruption			
	Mafic	Silicic	Unknown	Total
Mafic (< 57 percent)	297	377	75	749
Silicic (≥57 percent)	7	255	32	294
Unknown	1	33	16	50
Total	305	665	123	1,093

Data of table 2c are biased toward lower silica contents because we used the minimum of any range in reported silica values. However, use of the maximum for any range in reported silica values leads to the same conclusion: Mafic magmas prevail in postcaldera eruptions, even at silicic calderas.

Prevalence and Duration of Caldera Unrest

1. What proportion of the world's large Quaternary calderas are reported to have been restless in historical time? In the past 100 years?

More than half of our sample of 225 large Quaternary calderas have experienced unrest within historical time (tables 3a, 3b). Of these 225 calderas, 138 had reported unrest within historical time (61 percent), and at least 120 of these have been restless within the past century (53 percent of total).

Table 3a. Numbers of quiescent and restless calderas in historical time

Number of large calderas with[a]	Magma of caldera-forming eruption			
	Mafic	Silicic	Unknown	Total
No historical unrest	14	60	13	87
Eruption only	1	22	2	25
Broadly defined unrest (with or without eruption).	13	52	15	80
Narrowly defined unrest (with or without eruption).	8	22	3	33
Total	36	156	33	225

Table 3b. Numbers of quiescent and restless calderas in past 100 years

Number of large calderas with[a]	Magma of caldera-forming eruption			
	Mafic	Silicic	Unknown	Total
No unrest in past 100 yr	16	75	14	105
Eruption only	1	13	2	16
Broadly defined unrest (with or without eruption).	10	49	17	76
Narrowly defined unrest (with or without eruption).	8	18	2	28
Total	35	155	35	225

[a]Restless calderas were counted as having narrowly defined unrest, broadly defined unrest, or eruption only, in the first applicable category in that order of consideration.

The proportion of restless calderas in the set of all large Quaternary calderas (known and unknown) might be higher or lower—raised because quiet calderas are less likely to be discovered, or lowered because calderas with frequent eruptions fill themselves with postcaldera eruptive products and are therefore unrecognized.

2. How many calderas are restless in a typical year?

The average number of restless calderas per year in the past 100 years is the number of caldera-years of unrest in that period divided by 100. In a typical year, some form of unrest occurs at approximately 19 calderas (1,906 c-y/100 yr), including narrowly defined unrest at approximately 4 calderas (436 c-y/100 yr) (table 4). Eruptions occur at more than 5 calderas (558 c-y of "eruption only" unrest/100 yr) and perhaps at as many as 14 calderas (1,404 c-y of eruptive unrest/100 yr).

Table 4. Number of caldera-years of unrest in the past 100 years
[Values in parentheses are caldera-years of unrest that included an eruption]

Type of unrest	Magma of caldera-forming eruption			
	Mafic	Silicic	Unknown	Total
Eruption only	147 (147)	368 (368)	43 (43)	558 (558)
Broadly defined unrest	195 (160)	619 (353)	98 (66)	912 (579)
Narrowly defined unrest	81 (72)	344 (184)	11 (11)	436 (267)
Total	423 (379)	1331 (905)	152 (120)	1906 (1404)

3. How often, and for how long, does each type of unrest occur as a potential eruption precursor?

The average duration and corresponding sample size for each main type of unrest are shown in table 5. The sample sizes give a rough idea of the frequency with which each type of unrest is *reported*. However, reporting of most geophysical and geochemical activity is so incomplete, even within the past 100 years, that the actual number of episodes of unrest must be several times greater than shown. Apparent differences in the relative frequency of each type of unrest mean little.

Table 5. Average duration of various types of unrest during the past 100 years
[Sample size in parentheses]

Type of unrest	Magma of caldera-forming eruption		
	Mafic	Silicic	Unknown
Local earthquakes	Weeks (60)	Weeks (108)	Days-weeks (11)
Earthquake swarms	Days-weeks (37)	Weeks-months (62)	Days (6)
Tremor	Hours-days (23)	Days-weeks (7)	— (2)
Uplift	Months-years (22)	Months-years (31)	Years-decades (13)
Subsidence	Weeks (18)	Months-years (27)	— (0)
Tilt	Weeks-months (36)	Months (14)	— (0)
Horizontal strain	Months (7)	Months-years (5)	— (0)
Ground fracturing	Days-weeks (8)	Weeks (20)	— (1)
Magnetic changes	— (3)	— (2)	— (0)
Gravity changes	— (2)	Years (5)	— (0)
Thermal changes	Months (4)	Months (40)	Months (6)
Fumarole, gas changes	Months (13)	Weeks-months (57)	Months (13)
Spring, lake changes	Months (4)	Weeks-months (36)	— (3)

We excluded from table 5 unrest that occurred during or after eruptions, so what is shown may be considered precursory or potentially precursory to eruptions. In order to supplement quantitative information about the duration of unrest with semiquantitative information (for example, earthquakes lasting "a few days"), we assigned numerical values to each of the duration codes A–G. A (minutes) became 1, B (hours) became 2, C (days) became 3, and so on. We then calculated a weighted average duration for each type of unrest. For example, 3 episodes of local earthquakes at mafic calderas lasted minutes (3×1=3), 8 episodes lasted hours (8×2=16), 10 episodes lasted days (10×3=30), 15 episodes lasted weeks (15×4=60), 20 episodes lasted months (20×5=100), and 4 lasted years (4×6=24). The aggregate count for 60 episodes of seismicity at mafic calderas was 233, or 3.9 per episode, which corresponds most closely to an average duration of weeks (assigned value 4).

Most types of unrest last weeks or longer before they culminate in an eruption or otherwise stop. The fact that caldera unrest typically lasts for months, years, or even decades makes socially useful, short-range forecasts difficult. The most consistently short indicator of an impending eruption is volcanic tremor, especially at mafic calderas.

Unrest lasts marginally longer at silicic calderas than at mafic calderas. Factors may include a relatively lesser fluidity of silicic magma, which slows any response to disturbance, and the density shadow effect of silicic magma on mafic magma (Smith, 1979; Hildreth, 1981).

4. What are the average and maximum durations of whole episodes of unrest?

Some unrest at large calderas, such as the Phlegraean Fields and Iwo-Jima, lasts for centuries, millennia, or longer. Uplift and subsidence at Aira Caldera have been occurring at least since measurements began in the 1890's. Other episodes of broad, caldera-wide unrest lasted several years or several decades (for example, several episodes of unrest at Rabaul Caldera within the past 150 years).

A list of (*a*) sample size and average duration (c-y) of all reported episodes of unrest in each compositional category (including episodes that lasted less than one year, which are counted as 1 caldera-year), (*b*) sample size and average duration of those episodes that lasted more than one year, and (*c*) maximum duration of unrest in each category is given in table 6. The first average overestimates the true duration of unrest, because episodes that lasted less than one year are counted as a full caldera-year. The second average is more nearly correct for those episodes that lasted more than one year.

Table 6. Average and maximum durations of unrest

[Values given in caldera-years (c-y)]

Duration of unrest (c-y)	Magma of caldera-forming eruption		
	Mafic	Silicic	Unknown
Average (all)	1.9	2.1	1.7
Sample size	315	785	124
Average (> 1 year)	4.7	5.0	3.4
Sample size	75	218	32
Maximum (yr)	42	100+	19
Caldera name	Masaya	Yellowstone, Phlegraean Fields, Iwo-Jima	Nevados de Chillan

No significant difference is apparent between the average durations of unrest at mafic and silicic calderas, although the longest episodes at silicic calderas are probably longer than those at mafic calderas. The calculated durations are affected by several biases and complications. These include a bias toward longer average durations introduced by counting each episode as at least 1 caldera-year in duration; a bias toward shorter average durations caused by underreporting of noneruptive unrest; and an ambiguity because some observers report many short episodes of unrest while others report fewer, longer episodes. Improved reporting would show a longer duration of unrest at most calderas, but in a few cases (for example, Kilauea), improved reporting suggests a short duration of individual episodes of unrest within a background of persistent activity.

5. Is unrest typically continuous or intermittent?

During an episode of unrest, activity may wax or wane several times before culminating in an eruption or a shallow intrusion or before returning to quiescence. Seismicity often occurs in repeated swarms rather than in a continuous pattern, and uplift often alternates with subsidence (for example, subsidence at Yellowstone during 1985–87 following decades of net uplift, or uplift at the Phlegraean Fields Caldera alternating with decades to centuries of subsidence). Ground deformation can be episodic too, as suggested by the existence of multiple terraces at Toba and Iwo-Jima. The intermittent nature of most unrest at calderas poses a particularly difficult challenge for eruption forecasting, because "successful" forecasts must distinguish between short lulls and long periods of quiescence in order to be socially useful.

Distribution of Unrest Within Individual Calderas

1. Does unrest usually affect an entire caldera?

Most *documented* unrest is limited to parts of calderas, most often beneath an active, postcaldera vent. Much of the documented seismicity in this compilation is beneath such active vents. Likewise, most of the reported geodetic changes are on the slopes or in the crater regions of postcaldera vents, and most fumarolic changes are at active postcaldera vents.

However, seismic studies that deliberately look at a whole caldera commonly find seismic activity around caldera margins and elsewhere within the caldera. Similarly, geodetic networks that extend all the way to stable bench marks outside a caldera usually record some deformation throughout the caldera. Ten of fourteen large calderas for which precise geodetic measurements were available showed significant caldera-wide uplift and subsidence over the course of decades, and an additional three showed significant but irregular deformation, such as block faulting as part of regional extension at Taupo.

2. Are caldera seismicity, ground deformation, thermal changes, and eruptions typically concentric?

Seismicity, uplift, increased thermal activity, and eruptions at a given caldera do not always share a common center. Furthermore, centers of each may shift from one episode of unrest to the next, or even within a single episode. Contemporary uplift in Rabaul Caldera bears only a crude spatial relation to eruption sites, which are structurally determined by the caldera's ring fractures (McKee and others, 1984, 1985). Seismic swarms preceding the 1979 eruption of On-take Volcano were 7–15 km southeast of On-take, so far away that they were not considered at the time to be related (see details in section on On-take, Part 2). Uplift at Long Valley Caldera has centered on the resurgent dome, but most seismicity has occurred within the south moat of the caldera or even farther south, in the Sierran tectonic block. Seismicity and ground deformation at Katmai were not monitored in 1912, but the lateral migration of magma suggested by Hildreth (1987) would surely have created a complex pattern of changes in those parameters.

3. How prevalent is unrest near but outside calderas?

Magma reservoirs beneath some summit calderas feed magma outward into rift zones; examples are at Hawaiian, Ethiopian, and Icelandic volcanoes, and at Izu-Oshima, Kakuto, Ambrym, and others. Unrest affects areas outside such calderas whenever magma is intruded into the rift zones. Caldera-wide uplift at Kilauea, Mauna Loa, and Krafla is usually followed by eruptions many kilometers downrift.

A related process also occurs, in which regional extension allows (or initiates?) dike intrusion into areas previously lacking such intrusions. Examples include the emplacement of a silicic dike and domes in and north of Long Valley Caldera (Miller, 1985; Eichelberger and others, 1985; Sieh and Bursick, 1986) and the Cordón Caulle lava flows in and near the Cordillera Nevada caldera in Chile (Katsui and Katz, 1967; Moreno Roa, 1977).

Even where no rift zone or new dike is present, earthquakes, ground deformation, and thermal, fumarolic, or hydrologic changes occur outside some calderas. This observation is biased by our convention of including all changes that occur in *or near* calderas, but the level of tectonic unrest in areas north and northwest of Yellowstone, south of Long Valley, north of Taupo, west and northwest of Aso, southwest of On-take, west and northwest of Izu-Oshima, south of Kutcharo, southeast of Toba, and both southwest and northeast of Taal is higher than in surrounding regions. We suspect that some of this unrest near calderas occurs because regional stresses are concentrated at weaknesses in the crust, and calderas act like weak holes in crustal plates (see "Processes of Caldera Unrest" below; also Lekhnitskii, 1968).

4. What is the vertical distribution of caldera unrest?

Strictly speaking, unrest reaches from a caldera down to the zone of magma generation beneath that caldera. In a few instances, earthquakes as deep as 60 km have preceded, accompanied, and followed shallow-seated unrest, as at Kilauea in 1959 (Eaton and others, 1987). More commonly, earthquakes around calderas are <15 km deep, and deformation can be modeled with point sources at depths of <15 km and usually <10 km. Seismic imaging and other geophysical studies of the crust beneath several large calderas indicate that the tops of large magma reservoirs lie 5–15 km below the surface (for example, Aira, Aso, Long Valley, Yellowstone, and Krafla) (Eaton and others, 1975; Ono and others, 1978; Einarsson, 1978; Hirata and Uchiyama, 1981; Sanders, 1984; Sudo, 1987; Rundle and Hill, 1988).

5. Does unrest ever affect most or all of a caldera ring-fault system?

Earthquakes at Rabaul from 1983 to 1985 were concentrated along steeply outward-dipping ring faults of an inner, relatively young caldera (Mori and McKee, 1987). Simultaneous eruptions from multiple vents along ring faults are known from Deception Island (1967, 1969) and Niuafo'ou (1853, 1867), as are simultaneous eruptions on opposite sides of Rabaul in 1878 and 1937. At Tondano Caldera, eruptions occurred at two or more ring vents in 1952, 1958–59, 1971, 1973, 1985, and 1987. Similarly, eruptions occurred at four different vents of Bayonnaise during 1915. In the Galápagos Islands magma rising along circumferential fissures around caldera rims fed simultaneously active vents at Sierra Negra in 1979 and at Fernandina in 1981 (Simkin, 1984).

Correlations Between Characteristics of Calderas and Their Unrest

1. Does unrest differ from silicic to mafic centers?

Unrest is equally common at silicic and mafic calderas (tables 3a, 6). However, unrest at silicic centers leads to eruptions much less consistently than does unrest at mafic centers (tables 1a–c), and eruptions at silicic centers are more explosive than those at mafic centers (table 2b). In addition, precursors to eruptions last marginally longer at silicic centers than at mafic centers (table 5).

2. Do systematic differences exist between unrest at calderas of various diameters?

Judging from the number of caldera-years of unrest per caldera during historical time and during the past 100 years, calderas with diameters greater than 15 km are slightly more restless than smaller calderas (table 7). No consistent trends are apparent in the average repose between eruptions or the average duration of unrest as a function of caldera diameter. We expected that average repose periods and durations of unrest would be longer at large calderas than at small calderas, but available information does not support this idea.

Table 7. Relation between caldera diameter and unrest

[Sample size in parentheses]

	Caldera diameter					
	5–9	10–14	15–19	20–24	25+	Unknown
Historically quiet calderas.	37	18	5	5	4	18
Historically restless calderas.	63	33	17	9	7	0
Caldera-years of unrest per caldera in past 100 yr.	10	13	19	16	20	—
Average repose (yr) between postcaldera eruptions.[a]	28 (381)	27 (227)	26 (160)	26 (106)	8 (49)	56 (16)
Average duration of seismic unrest.	Weeks (48)	Weeks (40)	Wks-mo (21)	Days (18)	Months (10)	— (6)

[a]Does not include reposes that are longer than the span of historical time. Low value for calderas with diameter 25 km is further influenced by frequent eruptions at Tondano Caldera and the Zhupanovsky complex.

3. Does unrest differ between older and younger calderas?

All calderas eventually become quiet over geologic time. Indeed, 73 percent (41/56) of the calderas in table 8 are less than 50,000 yr old, implying that most older Quaternary calderas are quiet, unrecognized, or both. However, if judged by the number of caldera-years of unrest per caldera, younger restless calderas are neither more nor less active than older restless calderas. The average repose between eruptions possibly increases with increasing caldera age, but the data are not compelling. Actual reposes might be longer or shorter than shown here, because this table does not include potentially long reposes preceding the first historically recorded eruption nor potentially shorter reposes that could be revealed by better monitoring and reporting.

Table 8. Caldera age[a] and unrest

[Sample size in parentheses]

	Age of most recent collapse (ka)			
	0-9	10-49	50-99	100+
Caldera-years of unrest per caldera in past 100 yr.	21 (25)	14 (16)	3 (5)	19 (10)
Average repose (yr) between postcaldera eruptions.[a]	27 (234)	27 (125)	160 (15)	32 (101)
Average duration of seismic unrest.	Weeks (63)	Months (31)	Weeks (7)	Wks-mo (22)

[a]Caldera age is taken as the age of the most recent major collapse and is given in thousands of years before present (ka).

Many calderas in our compilation first formed hundreds of thousands of years ago, have had multiple episodes of collapse since then, and still show evidence (from historical unrest or geophysical studies) that a large volume of magma remains beneath them. Some of the oldest calderas in this compilation are among the most restless (for example, Yellowstone and Long Valley). Even if we consider only the most recent collapse at each caldera, 10 of 138 restless calderas are at least 100,000 yr old, and at least 3 are more than 500,000 yr old (Long Valley, Yellowstone, and those of the Daisetsu-Tokachi area). If earlier calderas at the same or a nearby site are also included, the oldest still-restless calderas are at least 11–12 million years old (Atitlán). As above, aging does not necessarily result in decreasing activity; other factors, perhaps the contemporary rate of supply of basaltic magma from depth and the rate of regional tectonic strain, counterbalance or override the effects of age.

4. Do earthquakes or uplift show any correlation with the extent of surficial hydrothermal activity at calderas?

We distinguished three qualitative degrees of surficial hydrothermal activity: minor, moderate, and extensive. Calderas with few or no hot springs or fumaroles have "minor" hydrothermal activity, calderas with thermal areas in two or three parts of the caldera have "moderate" hydrothermal activity, and calderas with numerous vigorous thermal areas have "extensive" hydrothermal activity. Examples are Kilauea and Tengger (minor), Aso and Okmok (moderate), and Yellowstone and Haroharo (extensive).

Calderas with extensive hydrothermal activity have no greater number of caldera-years of unrest (per caldera) than do calderas with minor hydrothermal activity (table 9). However, calderas with extensive hydrothermal activity may have more frequent seismic unrest than calderas with little or no hydrothermal activity. A positive correlation of hydrothermal activity and seismicity could result from high pore pressures and hydrofracting; from easy ascent of hydrothermal fluids along active faults (Mase and Smith, 1987); from magma intrusions; or from the fact that people like to live near thermal springs and report changes in their natural environment, including earthquakes. Historical uplift seems equally prevalent in each of the three classes of hydrothermal activity. The average durations of seismic unrest and uplift at calderas seem to increase with increasing hydrothermal activity, but the reporting bias noted earlier might apply here too.

Table 9. Seismicity, uplift, and hydrothermal activity
[Sample size in parentheses; percentage in brackets]

	Surface hydrothermal manifestations		
	Minor	Moderate	Extensive
Number of caldera-years of unrest per caldera in past 100 years.	24 (19)	12 (58)	15 (29)
Episodes of seismic unrest and percentage of restless calderas in each group with historical seismicity.	78 [60]	151 [69]	117 [90]
Average duration of seismic unrest.	Weeks (68)	Weeks (63)	Wks-mo (56)
Episodes of uplift and percentage of restless calderas in each group with historical uplift.	32 [35]	42 [31]	23 [29]
Average duration of uplift.	Months-yrs (22)	Years (23)	Years (19)

Temporal Relations with Regional Earthquakes and Earth Tides

Caldera unrest occurred at least 79 times in close temporal association with regional earthquakes or, in a few instances, with swarms of regional earthquakes (table 10). By close temporal association we mean within a timespan that is short in relation to the usual recurrence intervals of both the regional earthquakes and the unrest, usually within a few months or less.

Table 10. Temporal relations between unrest and regional earthquakes

Temporal relation	Number of cases
Regional earthquake, then unrest	50
Unrest, then regional earthquake	17
Unrest, regional earthquake, then more unrest	12
Total	79

Fifty regional earthquakes (most $M \geq 6$) were followed within hours to months by unrest at nearby calderas. Notable examples include Askja, 1875; Krakatau, 1880(?); Kakuto, 1913, 1961, 1966–67; Rabaul, 1910, 1916, 1919, 1971; Suoh, 1933; Akan, 1933, 1982; Zhupanovsky-Karymsky, 1940, 1952, 1960–67; Zhupanovsky-Maly Semiachik, 1952; Tao-Rusyr, 1952; Yellowstone, 1959, 1975; Cordillera Nevada, 1960; Batur, 1963; and Daisetsu-Tokachi, 1962, 1968. Twenty-seven of these episodes culminated in eruptions, and three others are continuing without eruptions as yet (Rabaul, Wrangell, and Yellowstone). Some of the eruptions followed a strong earthquake by only a few days (for example, the Suoh eruption in 1933, the Tao-Rusyr, Karymsky, and Maly Semiachik eruptions in 1952, and the Cordón Caulle - Cordillera Nevada eruption of 1960). Other, less certain examples could be added, such as the renewed unrest at the Phlegraean Fields in 1982 less than two years after the 1980 Naples (Calabria) earthquake, and unrest at Rabaul beginning about a month after two large, distant earthquakes in 1971.

Even earth tides and ocean loading can disturb caldera systems. Eruptions at Kilauea and several Japanese calderas show positive correlations with fortnightly earth tides (Mauk and Johnston, 1973; Dzurisin, 1980; Shimozuru, 1987), as did volcanic earthquakes at Ilopango during 1879–80 (Golombek and Carr, 1978). Ocean unloading directly atop a caldera with ring fractures can trigger earthquakes, as at Rabaul in 1983–84 (Lowenstein, 1986), or eruptions, as perhaps at Kikai in 1934 (Tanakadate, 1936) or at Pantelleria over periods of millennia (Wallmann and others, 1987). In these cases, the compressive stress acting normal to ring fractures might decrease in response to unloading. Conversely, seasonal ocean loading outside and adjacent to a caldera and magma reservoir, as at Pavlof on the Alaska Peninsula, can apparently increase the frequency of eruptions (McNutt and Beavan, 1987). At Pavlof, ocean loading may squeeze magma up the conduit.

At least 27 regional earthquakes occurred within 100 km of a restless caldera during or shortly after caldera unrest. Notable examples include Rabaul, 1938–43; Aira, 1913–15; Izu-Oshima, early 1920's, 1950–51, 1986–87; Akan, 1973, 1982; Kutcharo, 1959; Kilauea and Mauna Loa, 1868, 1975; Poas, 1910; Nevados de Chillan, 1906; and Krafla, 1975–76. Unrest at Izu-Oshima may foreshadow large earthquakes in the Sagami Bay-Tokyo area, as discussed in the sections on Izu-Oshima in Part 3 and the Izu Peninsula in Part 4.

We conclude that magma bodies beneath young calderas often react to changes in regional tectonic strain, and that unrest at calderas is sometimes a general, long-range precursor to regional earthquakes. To our knowledge, only a few workers (Yamashina and Nakamura, 1978) have attempted to test this premise. We doubt that caldera unrest is by itself a reliable indicator of an impending regional earthquake, but suggest that if a strong earthquake is imminent, careful monitoring at nearby calderas would likely detect changes that might aid future earthquake prediction.

PROCESSES OF CALDERA UNREST

Caldera unrest reflects tectonic, magmatic, and hydrologic processes. For the purposes of this discussion, we define tectonic processes as those that occur in country rock and dominantly involve changes in mechanical energy with little or no movement of mass into or out of the subcaldera environment. Magmatic processes are those that occur within a magma reservoir, and in which thermal energy, magma, and magmatic volatiles can (though need not) move into or out of the subcaldera environment. Hydrologic processes are those involving movement of subcaldera ground water or in which the physical or chemical state of subcaldera ground water is changed. Probably no episode of unrest is purely tectonic, purely magmatic, or purely hydrologic, because tectonic and magmatic changes invariably influence a ground water system and vice versa, and magma (if present) invariably interacts with the local tectonic stress field.

One possible classification of subcaldera processes is as follows:

1. Predominantly tectonic processes:
- (a) strain that is uninfluenced by the presence of a caldera or magma body (that is, strain that occurs only coincidentally at a caldera);
- (b) strain that is focused at a caldera because the caldera and the underlying magma body act as a weak hole in a plate;
- (c) strain that causes constant-volume deformation of a magma body; and
- (d) decrease of lithostatic confining pressure.

2. Predominantly magmatic processes:
- (a) intrusions into the base of a subcaldera magma reservoir;
- (b) intrusions outward or upward from a subcaldera magma reservoir;
- (c) in situ expansion or contraction of magma, from exsolution of volatiles (vesiculation), heating, or cooling; and
- (d) partial melting (assimilation) of country rock.

3. Predominantly hydrologic processes:
- (a) convective heating of country rock;
- (b) changes in pore-water pressures; and
- (c) changes in the discharge/recharge balance of a ground water system.

Below, we discuss observational evidence and other arguments that each of these processes is not only plausible but also likely to be involved in caldera unrest.

(1) Predominantly tectonic processes:

(a) Unrest caused by regional strain, uninfluenced by the presence of a caldera

Two observations from this compilation support the idea that some unrest at calderas is of primarily regional tectonic origin, uninfluenced by the presence of a caldera or large magma reservoir.

(i) Seismicity and ground deformation similar to that which occurs at calderas also occurs in noncaldera and even nonvolcanic settings (see Part 4 of this compilation). For example, most Japanese workers interpret seismic swarms and uplift at Matsushiro, Japan, in 1897, 1941–44, and 1965–68 as responses to regional tectonic strain and ground water movement without any clear involvement of the small, old volcanic systems of the area (Ichikawa, 1969; Kasahara, 1970; Nakamura, 1971; Ohtake, 1974, 1976) (see section on Matsushiro in Part 4 for alternative interpretations). Similar unrest of probable tectonic or tectono-magmatic origin has occurred on the Izu Peninsula, Japan, in 1930 and 1974–present (Tsumura, 1977; Tsumura and others, 1978; Thatcher and Savage, 1982; Fujita and Tada, 1983; Mogi, 1985), and on the Kii Peninsula, Japan, in the 1920's and in 1952–53 (Kanamori, 1972; Wakita and others, 1987).

(ii) Many episodes of unrest do not culminate in eruptions, and some have patterns of seismicity and deformation that parallel regional patterns and crosscut a caldera without obvious influence from the caldera. For example, seismic swarms near On-take Caldera, Japan (Fujii and Aoki, 1985), and Apoyo Caldera, Nicaragua (Shepherd, 1986), are thought to reflect regional tectonic processes and to be largely uninfluenced by the nearby caldera.

(b) Unrest caused by regional strain focused by the presence of a caldera

As is well known to structural engineers, stress is concentrated around a hole or other flaw in a plate subjected to external forces (Lekhnitskii, 1968, p. 157–216). At the Timber Mountain caldera in Nevada, bending of regional Basin and Range faults near the northern and southern margins of the caldera can be explained in this way (Christiansen and others, 1965; Cummings, 1968). If, for example, extensional stress is applied in an east-west direction in the vicinity of a caldera, the extension will be amplified at the northern and southern margins of the caldera, and lesser north-south compression will result at the western and eastern margins of the caldera.

Focusing of regional strain as a cause of caldera unrest is suggested by at least three observations:

(i) Some episodes of caldera unrest show no obvious relation to magmatic or volcanic processes (for example, earthquake swarms and extension near Taupo).

(ii) The intersections of calderas and regional faults tend to be areas of high seismicity (see list under "Distribution of Unrest Within Individual Calderas" (question 3, above)).

(iii) Some periods of caldera unrest are also periods of increased regional unrest, such as near Long Valley Caldera from 1978 to the present.

Regional tectonic strain also contributes to the ultimate in caldera unrest, caldera formation itself, by focusing magma supply from depth and by creating a structural environment in which collapse can readily occur. Volcano-tectonic depressions such as Toba, Taupo, and Ilopango are clear examples of this influence, and many calderas intersect regional faults or fault zones.

(c) Strain that causes constant-volume deformation of a magma body

Horizontal compression of a magma body ("squeezing") causes uplift of its roof and/or lateral intrusion of magma, and extension causes subsidence of the roof. If a conduit of relatively small diameter rises from a reservoir of much larger diameter, small horizontal strain at depth may cause large vertical movements in the conduit—just as slight squeezing of a carton of milk or juice causes that fluid to rise rapidly through a straw. A large block of crust that has collapsed into a magma reservoir to form a caldera may act as a piston with a diameter roughly equal to that of the magma reservoir. Horizontal strain in this case may cause an approximately equal vertical strain. Observations supporting constant-volume deformation include the following:

(i) Nakamura (1971, 1975) and Yamashina and Nakamura (1978) inferred that a narrow conduit and crater floor at Mihara Volcano, within the Izu-Oshima Caldera, act as a volumetric strain gauge for regional compression in the Sagami Bay area just south of Tokyo. More than 400 m of piston-like rise of the crater floor from 1933 to 1950 corresponded to approximately one microstrain of regional compression, and comparable subsidence of the floor followed stress release during the 1953 Boso-Oki earthquake. Similarly, McNutt and Beavan (1987) inferred that Pavlof Volcano behaves as a volumetric strain gauge and erupts more frequently during seasonal ocean loading.

(ii) Some episodes of uplift are followed by comparable amounts of subsidence, without eruption or other explanation for the reversal. For example, uplift at Yellowstone Caldera recently changed to subsidence, a change that we tentatively attribute to recent dominance of regional extension over continued but slow supply of basaltic magma into the system.

(d) Decrease of lithostatic confining pressure

Magma is generally buoyant and will tend to rise through most crustal rock unless it is confined by overlying rock (lithostatic pressure). Massive landslides of several cubic kilometers or more, melting of ice cover, or major changes in ocean loading may trigger unrest. Observations consistent with unrest caused by decreased lithostatic pressure include the following:

(i) Pantelleria was unusually active during low Pleistocene stands of eustatic sea level (Wallmann and others, 1987).

(ii) A passing storm surge in 1934 might have triggered an eruption of Kikai Volcano (Tanakadate, 1936).

(iii) Even weak semidiurnal ocean tides bear an inverse relation to shallow earthquakes at Rabaul (Lowenstein, 1986).

(2) Predominantly magmatic processes:

(a) Magmatic intrusions into a subcaldera reservoir

Observations and calculations consistent with magmatic intrusion into subcaldera magma reservoirs include the following:

(i) Models of caldera-wide deformation often suggest inflation of a large, 5 to 15-km-deep subcaldera reservoir (see sections on Aira, Long Valley, Yellowstone, and Krafla in Part 3).

(ii) Deep, lower crust or mantle earthquakes accompanied inflation of the shallow reservoir of Kilauea from 1955 through 1959 (Eaton and others, 1987), suggesting gradual resupply of magma over that period.

(iii) Eruption of relatively undifferentiated basaltic magma from many calderas in this compilation implies historical supply of that magma from depth (see sections on Kilauea and Krafla in Part 3).

(iv) Mixed mafic and silicic pumice, disequilibrium mineral assemblages, and other evidence of magma mixing suggest intrusions of mafic magma into large silicic reservoirs (see sections on Krakatau, Batur, Long Valley/Inyo Craters, Atitlán, Askja, Torfajökull, Agua de Pau, among others, in Part 3).

(v) Periodic resupply of basaltic magma is essential to maintain subcaldeara reservoirs in a molten state for more than 10^4–10^6 yr, as some are known to have existed, and to replenish magma that has been erupted. Calculations by Smith and Shaw (1975) and Lachenbruch and others (1976) show that silicic magma reservoirs beneath calderas solidify in 10^4–10^6 yr unless they have a continuing source of heat, probably in the form of basaltic intrusions from depth.

(b) Intrusion of magma upward or laterally from a subcaldera reservoir

Observations suggesting intrusion of magma upward or laterally from a subcaldera reservoir include the following:

(i) Many cases of unrest culminate in magmatic eruptions.

(ii) Some earthquake swarms have migrated upward or outward from a several-kilometers-deep magma reservoir and have then been followed by magmatic eruptions (see sections on Aira, Toya, Kilauea, and Krafla in Part 3).

(iii) Some deformation can be modeled as shallow, relatively small intrusions of magma upward from a presumed larger reservoir (see sections on Rabaul, Kilauea, and Krafla in Part 3).

(iv) Some unrest occurs along linear chains of vents or fissures, clearly situated above a dike (see sections on Haroharo, Cordillera Nevada, and all calderas with associated rift zones in Part 3).

(v) Dike swarms can be seen in the walls of many calderas (for example, Hakone).

(c) In situ expansion or contraction of magma

Magma expands if it is heated or exsolves a separate fluid phase (gas or liquid). Coefficients of volumetric thermal expansion are on the order of 10^{-5} to 10^{-3}/°C for silicate melt(s), and on the order of 20–40×10^{-6}/°C for typical minerals of crystallizing magma (Skinner, 1966). Thus, a 10 °C heating of magma could cause 0.1 percent expansion and 1 m of uplift above a 1-km-thick lens of magma. Larger reservoirs can produce the same amount of uplift with an even smaller percent expansion. Hot basaltic magma that underplates or intrudes a relatively cool magma reservoir is the most likely source of new heat, though frictional heating may be important along fault zones. Cooling, by itself, will result in comparable contraction and subsidence.

Thermal expansion of magma occurs only, and automatically, as a result of magmatic intrusion. The corresponding parameters of unrest are therefore the same as those of magmatic intrusions. Whenever a relatively cool magma is intruded by a hotter magma, the cooler magma must undergo some amount of thermal expansion.

Exsolution begins whenever magma becomes saturated with respect to a volatile component, principally H_2O or CO_2. The exsolved phase will exist as a gas at depths of a few kilometers and less, and as a liquid at greater depths (Greenland, 1987a,b). Primary causes of exsolution are decreasing lithostatic pressure, upward migration of volatiles through a magma column, absorption of meteoric ground water into magma (Hildreth and others, 1984), and second boiling, a process in which cooling and crystallization concentrate volatiles in a diminishing volume of residual melt until saturation is reached (Bowen, 1928; Burnham, 1979). Vesiculation by any means increases the volume of magma and decreases its density; exsolution of a liquid phase has a similar but less pronounced effect.

The following observations suggest exsolution and vesiculation as processes of caldera unrest:

(i) The occurrence of fumarolic emissions at calderas indicates that at least the tops of many subcaldera magma reservoirs are saturated with H_2O, CO_2, and other species.

(ii) Comparison of fumarolic gas compositions to calculated equilibrium compositions in magma suggests pre-eruption saturation of at least CO_2 and sulfur gases (Harris and Anderson, 1984; Gerlach and Graeber, 1985; Greenland, 1987a, b).

(iii) Volatile contents of glass inclusions in phenocrysts (Anderson, 1975) indicate that magma at Kilauea and probably other calderas in this compilation is volatile saturated before eruption.

(iv) Separation of a CO_2-rich fluid phase can begin at pressures as high as 10 kbar (40 km) (Gerlach and Graeber, 1985), and this fluid phase can be degassed from depths of 2–6 km in mafic magma (Greenland, 1987a,b). Separation of a fluid phase results in a small volume increase; transformation of that fluid into a gas results in a substantial volume increase. At high lithostatic pressures, a separated fluid phase can hydrofracture country rock as it seeks to rise.

(v) Transient fluctuation of gravity without concomitant vertical displacements at Poás Caldera (Rymer and Brown, 1987) and at Pacaya Volcano on the rim of Amatitlán Caldera (Eggers, 1987) has been interpreted by those authors as vesiculation or flashing of ground water to steam.

(d) Partial melting (assimilation) of country rock

Partially melted xenoliths and elevated $^{87}Sr/^{86}Sr$ ratios suggest that many subcaldera magmas assimilate country rock. The phase transformation from solid to melt is accompanied by a volume increase of 5–10 percent (Skinner, 1966). Although most geologists think of assimilation as a process that occurs over many thousands of years, recent experimental work suggests that by a process of heat exchange between convecting layers of mafic and silicic melt, a cubic kilometer of anatectic melt can form within a few months to a few decades (R.S.J. Sparks, oral commun., 1987).

(3) Predominantly hydrologic processes:

(a) Convective heating of country rock

Heat is transferred from magma through a hydrothermal convection system to country rock. Heated country rock expands linearly on the order of 8–10×10^{-6}/°C for typical country rock between 20 °C and 100 °C (Skinner, 1966). Thus, a 10 °C heating of country rock could cause 0.01 percent linear expansion or 0.1 m of uplift above a 1-km-thick body of heated country rock. Babbage (1847) postulated heating of country rock as the cause of uplift in the Phlegraean Fields Caldera; we doubt that such heating is the sole cause of uplift, but think that heating may well contribute to uplift in this and other calderas.

(b) Changes in pore-water pressure

Any mechanism that increases pore-water pressure will cause subtle uplift above a hydrothermal system (Oliveri del Castillo and Quagliariello, 1969; Grindley, 1976), lubricate faults, and permit regional strain release in small increments, as in earthquake swarms rather than main shock-aftershock sequences. Heating of ground water will cause pore pressures to increase if permeabilities are relatively low and prevent free escape of heated water. Grindley (1976) suggested that "* * * even a 10 °C heating of a 3 km column of [water-saturated] rock with 20 percent porosity would lead to a gradual rise in ground level of 2 metres, assuming one third of the stresses were relieved upwards." Self-sealing (precipitation of silica and other solutes in the pores and cracks of a hydrothermal system) or tectonic closure of cracks will also reduce permeability and increase pore-water pressures.

Pore-water pressures decrease by cooling, breaking of self-sealed rock or opening of cracks, or by excessive fluid withdrawal by man. Subsidence and microearthquakes occur during significant decreases in pore-water pressure.

(c) Changes in the discharge/recharge balance of a ground water system

Changes in the balance of discharge and recharge of ground water may be reflected by changes in the temperature or discharge of thermal springs and fumaroles. Factors influencing discharge and recharge include changes in fracture and matrix permeability, changes in precipitation, and man's withdrawal of geothermal fluids.

The following observations suggest that hydrologic processes contribute to unrest:

(i) Many eruptions at calderas begin as phreatic eruptions, and some never progress beyond phreatic activity.

(ii) Shallow seismicity and ground deformation are common in areas of known hydrothermal systems. Seismic swarms have been reported in many geothermal areas and interpreted as focused release of regional strain in small increments, lubricated by high pore pressures (Minakami, 1960; Ward and Bjornsson, 1971; Hiraga, 1972; Hill and others, 1975). Some hypocentral volumes correspond to high-temperature parts of hydrothermal systems, as at Hakone Caldera (Hiraga, 1972; Oki and others, 1978). Geothermal fields subside as a result of excessive fluid withdrawal and at least partially recover upon reinjection. For example, fluid withdrawal caused subsidence at Kakuto Caldera and lowered ground water temperatures at Hakone Caldera (Oki and others, 1978). Remarkable subsidence owing to geothermal fluid withdrawal is occurring at Wairakei, New Zealand, near Taupo Caldera; cumulative subsidence at Wairakei exceeded 8 m by 1982 (Allis and Barker, 1982). Matsushiro is a classic example (in a cool, noncaldera environment) of interaction between ground water pore pressures and regional tectonic strain; during 1965–68, pore pressures apparently increased in response to tectonic strain, lubricated a fault, and were the major cause of uplift (Ichikawa, 1969; Kasahara, 1970; Nakamura, 1971; Ohtake, 1974, 1976) (see also the section on Matsushiro in Part 4).

(iii) Many episodes of seismic or deformation unrest include an increase in the temperature of fumaroles and thermal waters, or development of artesian hot springs. For example, hot ground water was expelled shortly before eruptions at the Phlegraean Fields in 1538, Aira and Sakurajima in 1914, and Tokachi in 1962. At Yellowstone, the M=7.3 1959 Hebgen Lake earthquake severely disturbed the hydrothermal system, cracked self-sealed caps, and brought deep, hot water into shallow zones where it heated shallow country rock and increased pore pressures and hot spring discharge (Marler, 1964). These hydrologic changes were initially in response to tectonic or magmatic processes, but they also contributed to continued unrest at those calderas.

(iv) Some uplift reverses to subsidence without any obvious tectonic or magmatic explanation (for example, at the Phlegraean Fields in 1972 and 1985).

UNREST AND THE STRUCTURE OF CALDERAS AND LARGE MAGMA RESERVOIRS

Patterns of unrest at calderas give information about the depth, size, and shape of large magma bodies, attitudes of ring faults, and the process of resurgent doming. Most of the caldera earthquakes noted in this compilation occurred above 15 km depth, and most uplift that has been modeled implies point sources at 10 km depth or less. Magma reservoirs presumably lie beneath or are surrounded by zones of seismicity, at or below the levels indicated by the point-source modeling.

Details of unrest suggest complex structures in magma reservoirs. Not all unrest at calderas is concentric. Seismicity may occur in one part of a caldera, uplift in another, and eruptions in yet another. Some of this complexity reflects different processes in different parts of a caldera—for example, strike-slip faulting along the southern moat of Long Valley and simultaneous inflation of a magma reservoir beneath the resurgent dome of that caldera. Other complexity is predicted by theoretical models—for example, parallel ridges of uplift and a central zone of subsidence that are expected on either side of a rising dike (Dieterich and Decker, 1975; Pollard and others, 1983). We suspect that much of the remaining complexity reflects the complex geometry and plumbing of large magma reservoirs. Geologic mapping of the dissected roots of ancient volcanoes has shown that magma reservoirs beneath large calderas commonly had irregular tops or cupolas (Lipman, 1984). Seismic studies also suggest complexity in the structure of large magma reservoirs (Hill, 1976; Ryall and Ryall, 1981; Sanders, 1984). We envision that large magma reservoirs consist of more molten and less molten parts with some interconnection of the more fluid parts. A disturbance in any one part of a magma reservoir by shaking, squeezing, or injection of melt may eventually be distributed in complex ways throughout the entire reservoir. Judging from the long duration of unrest at silicic calderas, it may take years to decades for a disturbance to propagate through a viscous silicic magma body and its overlying hydrothermal system.

One long-standing debate in volcanology is whether caldera ring faults dip inward or outward and, if inward, how collapse can occur. Earthquakes along the southern boundary of Long Valley in 1983 define a vertical plane (Savage and Cockerham, 1984), and earthquakes at Rabaul define a ring fault that dips about 80° outward (Mori and McKee, 1987). Collapse at Fernandina in 1968 occurred along a steeply inward-dipping ring fault (Simkin and Howard, 1970). Ring faults in older calderas of the Western United States dip vertically or steeply inward (Lipman, 1984). Simultaneous eruptions from multiple vents around the rims of calderas, as at Niuafo'ou, Deception Island, and Rabaul, also imply the existence of ring dikes but do not indicate their dip. No historical collapse at a small Krakatau- or Crater Lake-type caldera has been sufficiently well documented to show the existence or dip of ring faults.

Some unrest noted in this compilation is related to structural doming at young calderas. Historical uplift of several-hundred-thousand-year-old resurgent domes at Long Valley and Yellowstone has been too rapid to have been continuous for the entire lifetimes of those domes. Episodic uplift (and subsidence) are likely, and we doubt that those domes are still growing. In contrast, progressively uplifted coral reefs and terraces, geodetically measured uplift, and accompanying seismicity at Iwo-Jima suggest that a resurgent dome in that caldera is still growing. Resurgent blocks at 75,000-yr-old Toba Caldera also might be rising, but present data are inadequate to judge the rate of any current uplift.

FORECASTING ACTIVITY AT RESTLESS CALDERAS

Unrest at calderas and in large magma reservoirs can lead to truly devastating eruptions and therefore requires particularly careful monitoring. However, calderas are complex systems, so critical forecasts are among the most difficult to make. Lessons from this compilation that bear on forecasting are listed below.

1. Can one tell whether a particular episode of unrest will lead to an eruption?

Slightly more than half of the episodes of unrest at mafic calderas are followed by an eruption within one year or less (table 1b); fewer than half of the episodes of unrest at silicic calderas are followed by an eruption within one year or less. Within the past 100 years at silicic calderas that had not erupted for at least 100 years, only 10 of 60 episodes of unrest led to eruptions.

Characteristics of a specific episode of unrest rarely signal a unique outcome. Dramatically accelerated uplift of a large area of caldera floor, to a rate of several meters per day, is the only change in this compilation that has always been followed by an eruption. Between 4 and 5 m of uplift occurred in the second and third days before the 1538 eruption of Monte Nuovo in the Phlegraean Fields; 6 m of uplift occurred in 2 days before the 1878 eruption of Rabaul; and 2 m of uplift occurred in 8 hours before the 1937 eruption of Rabaul. In each of this admittedly small number of instances, an eruption began within 3 days of rapidly accelerating uplift.

No other set of changes is known to *always* or *never* lead to an eruption. A few changes and combinations of changes *usually* lead to eruptions, generally within hours to days after the changes are noted. Among the more diagnostic signs of an impending eruption are the following:

- volcanic tremor;
- an exponentially escalating rate of seismic energy release or a sudden, pronounced drop in seismic energy release;
- a sudden increase in the rate of uplift (to tens of centimeters per day or more) or sudden subsidence after an extended period of uplift;
- opening of, and intense fumarolic emissions from, new fissures; or
- a sudden, sharp increase in hot-spring discharge.

A pronounced increase in the discharge of sulfur gases is a good indication that magma is near the surface and could easily erupt; however, an absence of change in gas discharge or composition at the surface does not imply that magma is still deep and unlikely to erupt, because buffering by ground water and other processes can mask such changes. Minor seismic swarms, changes in the apparent volume of steaming, thermal changes, and several other changes are poor indicators of an eruption and occur many times without subsequent eruptions. At silicic calderas, only tremor, uplift, and ground fracturing seem to be strong indicators of an impending eruption, and even these, individually, are imperfect indicators.

2. What is the minimum level of precursory activity that can be expected before an eruption?

Probably none of the eruptions in this compilation occurred entirely without precursors, but the intensity and character of precursors varied greatly. Many eruptions have occurred without *reported* precursors and a few reportedly occurred without precursors, but such instances are probably attributable to inadequate monitoring or failure to recognize the significance of changes that did occur. No eruption at a well-monitored caldera has occurred without precursors.

To design a monitoring and forecasting program where little risk can be accepted, the question of minimum expectable change might be reworded as, "What are the smallest changes that led to *any* eruption at a caldera? Are those changes at a microscale (strain of 1 part in 10^6, and earthquakes no stronger than M=0), or are the changes larger?" Precursors to individual phases of an ongoing eruption can be very small—for example, tenths of a microradian of tilt before recurring explosions of Sakurajima (Kamo and Ishihara, 1986). Precursors to other eruptions have also been subtle or brief. Some phreatic eruptions were preceded only by changes that are common during noneruptive periods and therefore do not seem exceptional or worrisome. Precursors to some other eruptions were recognized only with hindsight, such as seemingly unrelated changes in hot springs at one site and mild earthquake swarms at another, both located many kilometers from the eventual eruption site (as occurred at Tarawera and the Haroharo Caldera in 1886 and at On-take in 1976–84). Except for the above-mentioned special cases, we know of no eruptions that occurred after a repose of months or longer without precursors substantially above the microlevel. Historically, failures in forecasting eruptions have been failures to monitor even large changes, or else failures in the interpretation rather than the measurement of small changes.

However, two cautions are required. First, most eruptions occur where there is no microlevel monitoring. Although we do not know of any eruptions that demonstrably occurred after repose of months or longer with only microlevel changes, the possibility remains that some such eruptions have occurred. The data are simply too sparse to ignore that possibility. Second, precursors to large regional earthquakes in non-caldera settings are thought to be at the microlevel, and we noted earlier that some large regional earthquakes have apparently triggered dike intrusion and eruptions within hours. We suspect that calderas will show above-microlevel changes before a regional earthquake but caution that the available data barely show, much less guarantee, that possibility. Furthermore, it may be many years before accumulated experience in earthquake prediction near calderas permits confident interpretation of such changes.

Where higher levels of risk are acceptable or the consequences of false alarms are unacceptable, the question of minimum expectable unrest can be reworded, "What are the thresholds in rates of seismic energy release, rates of uplift, and similar measures, above which an eruption is likely?" Such thresholds for the entire set of calderas are high and dominated by unusual cases. A few large volcano-tectonic earthquakes and earthquake swarms have not been followed by eruptions, so the critical threshold in seismic energy release would be very high. Similarly, a few examples of rapid uplift (such as the cryptodome intrusion at Usu in Toya Caldera, or the rise of the crater floor at Mihara-yama in Izu-Oshima Caldera) have not been followed by eruptions, so a generalized threshold would be meaningless. One could exclude extreme cases and then search for threshold values among the remaining examples; alternatively, and we believe more defensibly, one can define critical thresholds for each caldera individually.

The minimum level of precursory unrest is by no means the norm. In fact, most magmatic eruptions from new or long-inactive vents at calderas were preceded at least by pronounced seismicity and ground deformation, and usually by thermal and hydrologic changes too.

3. What is the minimum expectable duration of precursors to an eruption within a caldera?

Precursory unrest begins a few hours to a few decades before eruptions at calderas (for example, 32 hours before the Usu eruption of 1977 and about 38 years before the Phlegraean Fields eruption of 1538). The minimum duration of precursory unrest is probably longer at a volcano that has been in repose for centuries or millennia than at a volcano that erupts every few years. However, after a repose of 102 years, Masaya Caldera erupted in 1772 within minutes of the first reported precursory earthquakes. In 1704–5 after a repose of 212 years, new vents at Las Cañadas Caldera erupted after only a few hours to a few days of seismicity. After a repose of 365 years, the 1727 eruption of Oraefajökull was preceded by only 1 day of precursory seismicity (as felt 9 km from the summit). Other caldera volcanoes that have erupted on short notice after long (but imprecisely known) repose periods include Suoh (1933, 14 days after a regional earthquake and 13 days after increased steaming), Kikai (1934, about 5 days after the first reported earthquakes), and On-take (1979, after 6 hours of immediate seismic precursor and 3 years of related, regional seismicity). Periods of a few days are inadequate for civil defense preparations in most communities and are especially short where residents are not accustomed to volcanic threats.

4. Can one tell whether an impending eruption will be of caldera-forming magnitude?

Recognition of precursors to a caldera-forming eruption remains as difficult as ever, if not more so. This compilation includes events that we might once have considered good indicators of an impending caldera-forming eruption, such as simultaneous eruptions from multiple ring-fracture vents, pronounced caldera-wide uplift, and seismicity along entire ring-fracture systems, but none of these events led to a caldera-forming eruption. Similarly, we find events from the small sample of historical caldera-forming eruptions that alerted observers that the caldera was restless but were not sufficiently distinctive or intense to indicate that a caldera-forming eruption was imminent. For example, increased steaming and phreatic eruptions at Tambora from 1812 to 1815 clearly showed that the volcano was restless but did not suggest the imminent caldera-forming eruption.

5. Can one tell when a particular episode of unrest is finished?

This is one of the most important and difficult questions posed in this study. Residents of a caldera may be well aware of when unrest begins, even without advice from scientists. A greater challenge is to know when unrest is finished, because in the meantime social and economic disruption is almost inevitable.

Much of the unrest described in this compilation, especially at silicic calderas, is intermittent—a little activity, then a pause, then more activity, and so on. Intermittent seismic swarms are a good example; intermittent ground deformation is also known (such as the uplift at Iwo-Jima), and fumarolic activity can probably be intermittent also. The obvious question during intermittent unrest is, "Which event is the last of a series?"

In principle, unrest can be deemed finished when the causes of that unrest have been removed or greatly reduced. For example, unrest resulting from magmatic intrusion is finished when all of the eruptible magma has been erupted and resulting changes in other parameters have diminished to preeruption levels. In practice, very few instances of unrest end as cleanly as this.

The next simplest indicator of the end of unrest is the cessation of changes. However, an apparent end of unrest is not always a good sign, especially if it occurs suddenly after a long buildup. Some intrusions pause after the magma has reached a shallow depth. It may then be simply a matter of time before volatile accumulation in the magma, a strong earthquake, or other factors allow the magma to make a rapid final ascent to the surface with few additional precursors.

Another good indicator of the end of unrest is a gradual reversal of those changes that can be reversed—elastic ground deformation, magnetic and gravity changes, thermal changes, and fumarolic and hydrologic changes. Local subsidence may occur around a vent immediately before an eruption, but otherwise a reversal from uplift to subsidence is generally a good indicator of a lessening of unrest, especially if the original changes are fully recovered. Reversals of other changes are equally diagnostic. One particularly interesting reversal was noted at the Phlegraean Fields from 1982–85, in which H_2, CH_4, and S gases began to decrease before seismic energy release slowed and before uplift reversed to subsidence (Gruppo Nazionale per la Vulcanologia, 1986). We do not know if similar gas changes always occur, but volcanologists would be wise to look for them in the absence of other definitive indicators of the end of unrest. One exception to reversals as an indicator of lessening unrest is if two processes of unrest with opposite symptoms alternate in dominance—for example, alternating uplift (from magma intrusion?) and subsidence (from tectonic extension?) at Yellowstone.

6. Are false alarms at calderas likely and, if so, more or less likely than at smaller volcanic centers?

The task of forecasting eruptions at calderas is more difficult than at smaller volcanic centers because the complexity of magmatic and hydrothermal systems beneath calderas increases the likelihood that influxes of magma into those reservoirs might not be erupted immediately. Precursors at small volcanic centers usually intensify as the eruption nears and culminate within a period rarely longer than a few months. Unrest at large calderas, in contrast, often proceeds intermittently and may continue for years or decades, so it is very difficult to know whether any particular increase in activity will soon lead to an eruption. Earthquakes and other changes at calderas are large, owing in part to the large size of many calderas. Thus, residents and scientists are faced with a difficult problem: Unrest is abundantly clear but hard to interpret and more likely than not to subside rather than lead directly to an eruption. On the other hand, the potential always exists for a truly devastating eruption to occur with little additional warning. Worrisome changes that do not lead to an eruption are likely, and the public must be willing to accept false alarms if they wish to receive timely warning of an eruption.

PRINCIPAL CONCLUSIONS

1. The most common processes causing seismic swarms, uplift, and other elements of caldera unrest are regional tectonic strain, magmatic intrusion and processes within magma reservoirs, and hydrologic processes, as listed in the section on "Processes of Caldera Unrest."
2. Unrest resembling that at calderas occurs in noncaldera and even nonvolcanic settings, calling special attention to

the role of regional tectonic and hydrothermal processes as agents of unrest.

3. Most unrest is a hybrid of several processes acting simultaneously or in sequence, and in many instances interacting with each other. Because of feedback mechanisms between processes, it is rarely possible or meaningful to classify specific unrest as purely tectonic, magmatic, or hydrothermal.
4. Calderas are dynamic and delicately balanced systems, and they can be disturbed by even small stimuli. Most unrest at silicic calderas occurs in response to routine, relatively minor tectonic strain or small-volume basaltic underplating of silicic reservoirs.
5. Different processes with different implications for hazards produce similar symptoms of unrest. Even a well-understood hybrid of processes can lead to many different outcomes.
6. The most common types of unrest (for example, seismic swarms and thermal, fumarolic, and hydrologic changes) are not reliable indicators of impending eruptions. Changes that are more diagnostic of impending eruptions are harmonic tremor; an exponentially escalating rate of seismic energy release or a sudden, pronounced drop in seismic energy release; a sudden, pronounced increase in the rate of uplift or sudden deflation after an extended period of inflation; opening of, and intense fumarolic emissions from, new fissures; or a sudden, sharp increase in hot-spring discharge.
7. Although large calderas hold a greater potential for catastrophic eruptions than small centers, most eruptions at calderas are relatively small.
8. Caldera unrest is likely to persist for months to years, sometimes even for decades to centuries, and to be intermittent rather than to show a systematic increase until culmination.
9. Much unrest does not culminate in an eruption, but eruptions are likely enough that many persons will choose to take precautionary measures. The choice of such measures is complicated by the long periods over which unrest can persist and the wide range of scales of potential eruptions, from minor to colossal eruptions that change the course of human history.
10. The outcome of a specific episode of unrest cannot and should not be forecast solely on the basis of patterns of unrest at other calderas, given the uncertainties in generalizing from one caldera to another or from this entire compilation to a single restless caldera. Rather, we suggest that this compilation can (*a*) show what is ordinary and extraordinary (qualitatively and in a few instances semiquantitatively), (*b*) suggest multiple interpretations of unrest, and (*c*) suggest new investigations to constrain interpretations of unrest.
11. The delicate equilibrium of calderas may cause them to become restless if regional stress increases, as in advance of major regional earthquakes. Calderas should be monitored for unrest that might forewarn of strong regional earthquakes.

There is no simple solution to the interpretation and forecasting of complex events at calderas. However, the rich record of historical caldera unrest is important for basic research and crisis response alike. Processes that last millennia and longer are recorded in the geological record, and day-to-day or year-to-year processes are recorded by modern monitoring. Many of the processes in large magma reservoirs and hydrothermal systems, as seen at large calderas, operate over periods of months to millennia. The historical records in this study are a bridge between classical geology and modern volcano-tectonic monitoring. We recommend the fascinating walk across this bridge, before and during future crises at calderas.

REFERENCES CITED IN PART 1

(Additional references are given under each caldera in Part 3)

Allis, R.G., and Barker, P., 1982, Update on subsidence at Wairakei: Proc. Pacific Geotherm. Conf. 1982, pt. 2, p. 365-370.

Anderson, A.T., 1975, Some basaltic and andesitic gases: Rev. Geophys. Space Phys., v. 13, p. 37-55.

Babbage, C., 1847, Observations on the Temple of Serapis, at Pozzuoli, near Naples, with remarks on certain causes which may produce geological cycles of great extent: Quart. Jour. Geol. Soc. London, v. 3, p. 186-217.

Barberi, F., Corrado, G., Innocenti, F., and Luongo, G., 1984, Phlegraean Fields 1982-1984: Brief chronicle of a volcano emergency in a densely populated area: Bull. Volcanol., v. 47-2, p. 175-185.

Berrino, G., Corrado, G., Luongo, G., and Toro, B., 1984, Ground deformation and gravity changes accompanying the 1982 Pozzuoli uplift: Bull. Volcanol., v. 47-2, p. 187-200.

Bowen, N.L., 1928, The evolution of the igneous rocks: Princeton, N.J., Princeton Univ. Press, 332 p.

Burnham, C.W., 1979, The importance of volatile constituents, in Yoder, H.S., Jr., ed., The evolution of the igneous rocks, fiftieth anniversary perspectives: Princeton, N.J., Princeton Univ. Press, p. 439-482.

Christiansen, R.L., Lipman, P.W., Orkild, P.P., and Byers, F.M., Jr., 1965, Structures of the Timber Mountain caldera, southern Nevada, and its relation to basin-range structure: U.S. Geol. Surv. Prof. Paper 535-B, p. B43-48.

Corrado, G., Grimaldi, M., LoBascio, A., and Luongo, G., 1983, Ground deformation in the Phlegraean Fields volcanic area, Italy: Proc. 7th Ann. Mtg., European Geophys. Soc., Budapest.

Corrado, G., Guerra, I., LoBascio, A., Luongo, G., and Rampoldi, R., 1977, Inflation and microearthquake activity of Phlegraean Fields, Italy: Bull. Volcanol., v. 40, p. 169-188.

Cummings, D., 1968, Mechanical analysis of the effect of the Timber Mountain Caldera on Basin and Range faults: Jour. Geophys. Res., v. 73, p. 2787-2794.

Decker, R.W., and Kinoshita, W.T., 1971, Geodetic measurements, *in* The surveillance and prediction of volcanic activity, Paris, UNESCO, p. 47-74.

Dieterich, J.H., and Decker, R.W., 1975, Finite element modeling of surface deformation associated with volcanism: Jour. Geophys. Res., v. 80, p. 4094-4102.

Dzurisin, D., 1980, Influence of fortnightly earth tides at Kilauea Volcano, Hawaii: Geophys. Res. Lett., v. 7, p. 925-928.

Dzurisin, D., Anderson, L.A., Eaton, G.P., Koyanagi, R.Y., Lipman, P.W., Lockwood, J.P., Okamura, R.T., Puniwai, G.S., Sako, M.K., and Yamashita, K.M., 1980, Geophysical observations of Kilauea Volcano, Hawaii, 2. Constraints on the magma supply during November 1975 - September 1977: Jour. Volcanol. Geotherm. Res., v. 7, p. 241-269.

Eaton, G.P., Christiansen, R.L., Iyer, H.M., Pitt, A.M., Mabey, D.R., Blank, H.R., Jr., Zietz, I., and Gettings, M.E., 1975, Magma beneath Yellowstone National Park: Science, v. 188, p. 787-796.

Eaton, J.P., Richter, D.H., and Krivoy, H.L., 1987, Cycling of magma between the summit reservoir and Kilauea Iki lava lake during the 1959 eruption of Kilauea Volcano, *in* Decker, R.W., Wright, T.L., and Stauffer, P.H., eds., Volcanism in Hawaii: U.S. Geol. Surv. Prof. Paper 1350, v. 2, p. 1307-1335.

Eggers, A.A., 1983, Temporal gravity and elevation changes at Pacaya Volcano, Guatemala: Jour. Volcanol. Geotherm. Res., v. 19, p. 223-237.

Eggers, A.A., 1987, Residual gravity changes and eruption magnitude: Jour. Volcanol. Geotherm. Res., v. 33, p. 201-216.

Eichelberger, J.C., Lysne, P.C., Miller, C.D., and Younker, L.W., 1987, Research drilling at Inyo Domes, California: 1984 results: Eos, Trans. Am. Geophys. Union, v. 66, p. 186-187.

Einarsson, P., 1978, S-wave shadows in the Krafla caldera in NE-Iceland: evidence for a magma chamber in the crust: Bull. Volcanol., v. 41, p. 187-195.

Filson, J., Simkin, T., and Leu, L., 1973, Seismicity of a caldera collapse: Galápagos Islands 1968: Jour. Geophys. Res., v. 78, p. 8591-8622.

Finch, R.H., 1949, Volcanic tremor (part I): Bull. Seismol. Soc. Am., v. 39, p. 73-78.

Fujii, I., and Aoki, H., 1985, Seismicity in the southern part of Hida Mountains: *in* Okada, Hm., ed., Comparative study of physical background of volcanic activity and its relation to prediction of eruption disasters: Special Rep. on a Grant-in-Aid Natural Disaster Science, for the Ministry of Education, publ. by Hokkaido Univ., Sapporo, in cooperation with Tohoku Univ., Univ. of Tokyo Earthq. Res. Inst., Nagoya Univ., Kyoto Univ., and Kyoto Univ. Disaster Prev. Res. Inst., p. 88-91.

Fujita, N., and Tada, T., 1983, Anomalous crustal uplift in the northeastern part of the Izu Peninsula, central Honshu, Japan: Tectonophysics, v. 97, p. 147-152.

Gerlach, T.M., and Graeber, E.J., 1985, Volatile budget of Kilauea Volcano: Nature, v. 313, p. 273-277.

Golombek, M.P., and Carr, M.J., 1978, Tidal triggering of seismic and volcanic phenomena during the 1879-1880 eruption of Islas Quemadas Volcano in El Salvador, Central America: Jour. Volcanol. Geotherm. Res., v. 3, p. 299-307.

Greenland, L.P., 1987a, Hawaiian eruptive gases, *in* Decker, R.W., Wright, T.L., and Stauffer, P.H., eds., Volcanism in Hawaii: U.S. Geol. Surv. Prof. Paper 1350, v. 1, p. 759-770.

———1987b, Composition of gases from the 1984 eruption of Mauna Loa Volcano, *in* Decker, R.W., Wright, T.L., and Stauffer, P.H., eds., Volcanism in Hawaii: U.S. Geol. Surv. Prof. Paper 1350, v. 1, p. 781-790.

Grindley, G. W., 1976, Relation of volcanism to earth movements, Bay of Naples, Italy: in Gonzales-Ferran, O., ed., Proc. Symp. on Andean and Antarctic volcanology problems: Rome, IAVCEI, p. 598-612.

Gruppo Nazionale per la Vulcanologia C.N.R. Italia, 1986, Phlegraean Fields: The 1982-1984 Bradyseismic crisis (posters prepared for the 1986 IAVCEI Int. Volcanol. Congr., New Zealand).

Harris, D.M., and Anderson, A.T., Jr., 1984, Volatiles H_2O, CO2, and Cl in a subduction related basalt: Contrib. Mineral. Petrol., v. 87, p. 120-128.

Hildreth, W., 1981, Gradients in silicic magma chambers: Implications for lithospheric magmatism: Jour. Geophys. Res., v. 86, p. 10,153-10, 192.

—1987, New perspectives on the eruption of 1912 in the Valley of Ten Thousand Smokes, Katmai National Park, Alaska: Bull. Volcanol., v. 49, p. 680-693.

Hildreth, W., Christiansen, R.L., and O'Neil, J.R., 1984, Catastrophic modification of rhyolitic magma at times of caldera subsidence, Yellowstone Plateau Volcanic Field: Jour. Geophys. Res., v. 89, p. 8339-8370.

Hill, D.P., 1976, Structure of Long Valley caldera, California, from a seismic refraction experiment: Jour. Geophys. Res., v. 81, no. 5, p. 745-753.

Hill, D.P., Mowinckel, P., and Peake, L.G., 1975, Earthquakes, active faults, and geothermal areas in the Imperial Valley, California: Science, v. 188, p. 1306-1308.

Hiraga, S., 1972, Earthquake swarms of geothermal fields in Japan: Jour. Japan Geotherm. Res. Assoc., v. 9, no. 1, ser. 32, p. 30-39.

Hirata, T., and Uchiyama, T., 1981, Damping area in the Aira Caldera of South Kyushu: Zisin, v. 34, no. 3, p. 435-437.

IAVCEI (International Association of Volcanology and Chemistry of the Earth's Interior, various authors), 1951-present, Catalogue of active volcanoes of the world including solfatara fields: Naples and Rome, IAVCEI, 22 vol. to date.

IAVCEI (International Association of Volcanology and Chemistry of the Earth's Interior, Working Group on the World Volcanological Map), 1973-present, Data sheets of the post-Miocene volcanoes of the world, with index maps: Rome, IAVCEI, unpaginated.

Ichikawa, M., 1969, Matsushiro earthquake swarm: Geophys. Mag., v. 34, p. 307-331.

Jachens, R.C., and Roberts, C.W., 1984, Gravity investigations at Long Valley Caldera, California: U.S. Geol. Surv. Open-File Rep. 84-939, p. 248-279.

Johnsen, G.V., Bjornson, A., and Sigurdsson, S., 1980, Gravity and elevation changes caused by magma movement beneath Krafla Caldera, northeast Iceland: Jour. Geophys., v. 47, p. 132-140.

Johnson, D., 1987, Elastic and inelastic magma storage at Kilauea Volcano, *in* Decker, R.W., Wright, T.L., and Stauffer, P.H., eds., Volcanism in Hawaii: U.S. Geol. Surv. Prof. Paper 1350, v. 2, p. 1297-1306.

Kamo, K., and Ishihara, K., 1986, Precursor of summit eruption observed by water-tube tiltmeters and extensometers: Ann. Rep. Disaster Prev. Res. Inst., Kyoto Univ., no. 29B-1, p. 1-12.

Kanamori, H., 1972, Relation between tectonic stress, great earthquakes and earthquake swarms: Tectonophysics, v. 14, p. 1-12.

Kasahara, K., 1970, The source region of the Matsushiro earthquake

swarm: Bull. Earthq. Res. Inst., v. 48, p. 581-602.
Katsui, Y., and Katz, H.R., 1967, Lateral fissure eruptions in the southern Andes of Chile: Jour. Fac. Sci., Hokkaido Univ., ser. 4, v. 13, no. 4, p. 433-448.
Kieffer, S.W., 1984, Seismicity at Old Faithful Geyser: An isolated source of geothermal noise and possible analogue of volcanic seismicity: Jour. Volcanol. Geotherm. Res., v. 22, nos. 1-2, p. 59-95.
Kling, G.W., Clark, M.A., Compton, H.R., Devine, J.D., Evans, W.C., Humphrey, A.M., Koenigsberg, E.J., Lockwood, J.P., Tuttle, M.L., and Wagner, G.N., 1987, The 1986 Lake Nyos gas disaster in Cameroon, West Africa: Science, v. 236, p. 169-175.
Lachenbruch, A.H., Sass, J.H., Munroe, R.J., and Moses, T.H., Jr., 1976, Geothermal setting and simple heat conduction models for the Long Valley Caldera: Jour. Geophys. Res., v. 81, p. 769-784.
Latter, J.H., 1979, Volcanological observations at Tongariro National Park, part 2. Types and classification of volcanic earthquakes, 1976-1978: New Zealand Dept. Sci. Indust. Res., Geophys. Div. Rep. 150, 60 p.
Lekhnitskii, S.G., 1968, Anisotropic plates (2nd ed.), tr. by S.W. Tsai and T. Cheron: New York, Gordon and Breach Sci. Publ., 534 p.
Lipman, P.W., 1984, The roots of ash flow calderas in western North America: windows into the tops of granitic batholiths: Jour. Geophys. Res., v. 89, no. B10, p. 8801-8841.
Lott, D.F., 1976, A behaviorist's perspective on reports that animals respond to earthquake precursors, *in* Evernden, J., ed., Abnormal animal behavior prior to earthquakes (I): U.S. Geol. Surv., unpubl. rep., p. 63-84.
Lowenstein, P., 1986, Tidal triggering of volcanic earthquakes during the Rabaul seismic crisis of 1983-84 [abs.]: Abstracts, Int. Volcanol. Congr., New Zealand, February 1986, p. 255.
Malone, S.D., 1983, Volcanic earthquakes: Examples from Mount St. Helens, *in* Kanamori, H., and Boschi, E., eds., Earthquakes: observations, theory, and interpretation: Bologna, Soc. Ital. di Fisica, p. 436-455.
Marler, G.D., 1964, Effects of the Hebgen Lake earthquake of August 17, 1959, on the hot springs of the Firehole Geyser Basins, Yellowstone National Park: U.S. Geol. Surv. Prof. Paper 435-Q, p. 185-197.
Mase, C.W., and Smith, L., 1987, The role of pore fluids in tectonic processes: Rev. Geophys., v. 25, no. 6, p. 1348-1358.
Mauk, F.J., and Johnston, M.J.S., 1973, On the triggering of volcanic eruptions by earth tides: Jour. Geophys. Res., v. 78, p. 3356-3362.
McKee, C.O., Johnson, R.W., Lowenstein, P.L., Riley, S.J., Blong, R.J., de Saint Ours, P., and Talai, B., 1985, Rabaul Caldera, Papua New Guinea: Volcanic hazards, surveillance, and eruption contingency planning: Jour. Volcanol. Geotherm. Res., v. 23, p. 195-237.
McKee, C.O., Lowenstein, P.L., de Saint Ours, P., and Talai, B., 1984, Seismic and deformation crises at Rabaul Caldera -- prelude to an eruption?: Bull. Volcanol., v. 47, no. 2, p. 397-410.
McKee, C.O., Mori, J., and Talai, B., 1987, Temporal microgravity changes accompanying structural deformation at Rabaul Caldera: Bull. Volcanol., in press.
McKee, C.O., Wallace, D.A., Almond, R.A., and Talai, B., 1981, Fatal hydro-eruption of Karkar volcano in 1979: development of a maar-like crater, *in* Johnson, R.W., ed., Cooke-Ravian volume of volcanological papers: Geol. Surv. Papua New Guinea Memoir 10, p. 63-84.
McNutt, S.R., 1987, Volcanic tremor at Pavlof Volcano, Alaska, 1973-1986 [abs.]: Abstracts, Hawaii Symposium on How Volcanoes Work, Hilo, January 1987, p. 172.
McNutt, S.R., and Beavan, R.J., 1987, Eruptions of Pavlof Volcano and their possible modulation by ocean load and tectonic stresses: Jour. Geophys. Res., v. 92, no. B11, p. 11,509-11,523.
Miller, C.D., 1985, Holocene eruptions at the Inyo volcanic chain, California: implications for possible eruptions in Long Valley caldera: Geology, v. 13, p. 14-17.
Minakami, T., 1960, Fundamental research for predicting volcanic eruptions, Pt. I: Bull. Earthq. Res. Inst., v. 38, p. 497-544.
Minakami, T., Hiraga, S., Miyazaki, T., and Uchibori, S., 1969, Fundamental research for predicting volcanic eruptions, Pt. II: Bull. Earthq. Res. Inst., v. 47, p. 893-949.
Minakami, T., Ishikawa, T., and Yagi, K., 1951, The 1944 eruption of Volcano Usu, Hokkaido, Japan: Bull. Volcanol., v. 11, p. 45-157.
Mogi, K., 1985, Earthquake prediction: Tokyo, Academic Press, 355 p.
Moreno Roa, H., 1977, Geologia del area volcanica Puyehue-Carran: Santiago, Univ. Chile, Memoria de Titulo, 181 p.
Mori, J., and McKee, C.O., 1987, Outward-dipping ring-fault structure at Rabaul Caldera as shown by earthquake locations: Science, v. 235, p. 193-195.
Murray, G.H., 1939, In attendance at the birth of a volcano: Pacific Islands Monthly, v. 9, no. 10, p. 54-55.
Nakamura, K., 1971, Volcano as a possible indicator of crustal strain: Bull. Volcanol. Soc. Japan, v. 16, p. 63-71.
———1975, Volcano structure and possible mechanical correlation between volcanic eruptions and earthquakes: Bull. Volcanol. Soc. Japan, v. 20, p. 229-240.
Newhall, C.G., and Self, S., 1982, The Volcanic Explosivity Index (VEI) An estimate of explosive magnitude for historical volcanism: Jour. Geophys. Res., v. 87, no. C2, p. 1231-1238.
Newhall, C.G., Paull, C.K., Bradbury, J.P., Higuera-Gundy, A., Poppe, L.J., Self, S., Bonar Sharpless, N., and Ziagos, J., 1987, Recent geologic history of Lake Atitlán, a caldera lake in western Guatemala: Jour. Volcanol. Geotherm. Res., v. 33, p. 81-107.
Ohtake, M., 1974, Seismic activity induced by water injection at Matsushiro, Japan: Jour. Phys. Earth, v. 22, p. 163-176.
Ohtake, M., 1976, The Matsushiro earthquakes after 10 years: Kagaku (Science), v. 46, p. 306-313 (in Japanese).
Oki, Y., Aramaki, S., and Nakamura, K., 1978, Volcanoes of Hakone, Izu and Oshima: Hakone, Hakone Town Office, 88 p.
Oliver, J., and Murphy, L., 1971, WWNSS: Seismology's global network of observing stations: Science, v. 174, p. 254-261.
Oliveri del Castillo, A., and Quagliariello, M.T., 1969, Sulla genesi del bradisismo flegreo: Atti Assoc. Geofis. Ital., XVIII Conv., Naples, p. 557-594.
Ono, K., Ito, K., Hasegawa, I., Ichikawa, K., Iizuka, S., Kakuta, T., and Suzuki, H., 1978, Explosion seismic studies in south Kyushu, especially around Sakurajima Volcano: Jour. Phys. Earth, v. 26 suppl., p. S309-S319.
Pollard, D.D., Delaney, P.T., Duffield, W.A., Endo, E.T., and Okamura, A.T., 1983, Surface deformation in volcanic rift zones: Tectonophysics, v. 94, p. 541-584.
Rundle, J.B., and Hill, D.P., 1988, The geophysics of a restless

caldera—Long Valley, California: Ann. Rev. Earth Planet. Sci., v. 16, p. 251-271.

Ryall, A., and Ryall, F., 1981, Attenuation of P and S waves in a magma chamber in the Long Valley caldera, California: Geophys. Res. Lett., v. 8, no. 6, p. 557-560.

Rymer, H., and Brown, G.C., 1987, Causes of microgravity change at Poas Volcano, Costa Rica: an active but non-erupting system: Bull. Volcanol., v. 49, p. 389-398.

Sanders, C.O., 1984, Location and configuration of magma bodies beneath Long Valley, California, determined from anomalous earthquake signals: Jour. Geophys. Res., v. 89, no. B10, p. 8287-8302.

Savage, J.C., and Cockerham, R.S., 1984, Earthquake swarm in Long Valley, California, January 1983: Evidence for dike inflation: Jour. Geophys. Res., v. 89, no. B10, p. 8315-8324.

Shepherd, J.B., 1986, Report on a mission to Nicaragua, Feb 9 to Feb 22, 1986: unpubl. rep. to United Nations Development Programme (UNDP) and United Nations Educational, Scientific, and Cultural Organization (UNESCO), 13 p.

Shimozuru, D., 1971, A seismological approach to the prediction of volcanic eruptions, *in* The surveillance and prediction of volcanic activity: Paris, UNESCO, p. 19-45.

———1987, Tidal effects on Hawaiian volcanism, *in* Decker, R.W., Wright, T.L., and Stauffer, P.H., eds., Volcanism in Hawaii: U.S. Geol. Surv. Prof. Paper 1350, v. 2, p. 1337-1343.

Sieh, K., and Bursick, M., 1986, Most recent eruptions of the Mono Craters, eastern central California: Jour. Geophys. Res., v. 91, no. B12, p. 12,539-12,571.

Sigurdsson, H., Devine, J.D., Tchoua, F.M., Presser, T.S., Pringle, M.K.W., and Evans, W.C., 1987, Origin of the lethal gas cloud from Lake Monoun, Cameroon: Jour. Volcanol. Geotherm. Res., v. 31, p. 1-16.

Simkin, T., 1984, Geology of Galápagos Islands, *in* R. Perry, ed., Galápagos (Key Environments): Oxford, Pergamon Press, p. 15-41.

Simkin, T., and Fiske, R.S., 1983, Krakatau 1883: The volcanic eruption and its effects: Washington, D.C., Smithsonian Institution Press, 464 p.

Simkin, T., and Howard, K.A., 1970, Caldera collapse in the Galápagos Islands, 1968: Science, v. 169, p. 429-437.

Simkin, T., Siebert, L., McClelland, L., Bridge, D., Newhall, C., and Latter, J.H., 1981, Volcanoes of the world: Stroudsburg, Pa., Hutchinson Ross Publ. Co., 232 p.

Skinner, B.J., 1966, Thermal expansion, *in* Clark, S.P., Jr., ed., Handbook of physical constants: Geol. Soc. Am. Memoir 97, p. 78-96.

Smith, R.L., 1979, Ash-flow magmatism, *in* Chapin, C.E., and Elston, W.E., eds., Ash-flow tuffs: Geol. Soc. Am. Spec. Paper 180, p. 5-27.

Smith, R.L., and Shaw, H.R., 1975, Igneous-related geothermal systems, *in* White, D.E., and Williams, D.L., eds., Assessment of geothermal resources of the United States: U.S. Geol. Surv. Circular 726, p. 58-83.

Sudo, Y., 1987, Attenuation of seismic waves in the Aso Caldera, Kyushu, Japan; evidence for an attenuating body in the crust [abs.]: Abstracts, Hawaii Symposium on How Volcanoes Work, Hilo, January 1987, p. 243.

Tanakadate, H., 1936, Volcanic activity in Japan and vicinity during the period between July 1934 and October 1935: Japan Jour. Astron. Geophys., v. 13, p. 121-139.

Thatcher, W., and Savage, J.C., 1982, Triggering of large earthquakes by magma chamber inflation, Izu Peninsula, Japan: Geology, v. 10, p. 637-640.

Tsumura, K., 1977, Anomalous crustal activity in the Izu Peninsula, Central Honshu: Jour. Phys. Earth, v. 25 supp., p. S51-S68.

Tsumura, K., Karakama, I., Ogino, I., and Takahashi, M., 1978, Seismic activities before and after the Izu-Oshima-Kinkai earthquake of 1978: Bull. Earthq. Res. Inst., v. 53, p. 675-706.

Wakita, H., Sano, Y., and Mizoue, M., 1987, High $^3He/^4He$ ratios and seismic swarms observed in a nonvolcanic, forearc region: Jour. Geophys. Res., v. 92, no. B12, p. 12,539-12,546.

Wallmann, P.C., Mahood, G.A., and Pollard, D.D., 1987, Mechanical models for correlation of ring-fracture eruptions at Pantelleria, Strait of Sicily, with glacial sea level drawdown [abs.]: Abstracts, Hawaii Symposium on How Volcanoes Work, Hilo, January 1987, p. 263.

Ward, P.L., and Bjornsson, S., 1971, Microearthquakes, swarms, and the geothermal areas of Iceland: Jour. Geophys. Res., v. 76, p. 3953-3982.

Yamashina, K., and Nakamura, K., 1978, Correlations between tectonic earthquakes and volcanic activity of Izu-Oshima Volcano, Japan: Jour. Volcanol. Geotherm. Res., v. 4, p. 233-250.

Yokoyama, I., 1971, Gravimetric, magnetic, and electrical methods, *in* The surveillance and prediction of volcanic activity: Paris, UNESCO, p. 75-101.

Yokoyama, I., 1974, Geomagnetic and gravity anomalies in volcanic areas, *in* Civetta, L., Gasparini, P., Luongo, G., and Rapolla, A., eds., Physical volcanology: Amsterdam, Elsevier, p. 167-194.

Yokoyama, I., Yamashita, H., Watanabe, H., and Okada, H., 1981, Geophysical characteristics of dacite volcanism—the 1977-78 eruption of Usu Volcano: Jour. Volcanol. Geotherm. Res., v. 9, p. 335-358.

PART 2: Quick Reference to Parts 3 and 4

Readers may use this section as a supplement to the table of contents to compare summary tabular data for two or more calderas, or to scan for specific patterns of unrest. Narrative details of unrest are given in Parts 3 and 4.

PART 2: QUICK REFERENCE TO PARTS 3 AND 4
HISTORICAL UNREST AT LARGE CALDERAS AND SELECTED NONCALDERA SETTINGS

See inside back cover for explanation and abbreviations

CAVW number (active vent)	Latitude Longitude (degrees)	Diameter (km)	Local tectonic setting	Precaldera edifice	SiO_2 content (wt pct)	Age of collapse (yr B.P.)	Date of unrest	Type and duration of unrest ESTU STHF MGTF H Te	Eruption type
VULSINI (LATERA/CONCA DI BOLSENA)									
01-01-003 (Vulsini)	42.58N 11.92E	Lat: 11 x 8 CdiB: 16			Leucite phonolite, trachyte	Lat: 100,000 CdiB: 300,000	B.C. 104?	---- ---- ---- - --	ex?
ALBAN HILLS/ARTEMISIO									
01-01-004	41.75N 12.75E	11	Exten?	Strato	R = 46 C = ?	37,000 ca. 60,000?	first millennium B.C.	---- ---- ---- - --	unknown, ex?
ROCCAMONFINA									
01-01-004A	41.30N 13.93E	5 x 7	Exten?	Strato	R = 49-63 C = silicic	Main: ca.385,000 gli Stagli: ca.300,000	B.C. 276 or B.C. 269	---- ---- ---- - --	ex (uncertain)
PHLEGRAEAN FIELDS									
01-01-01	40.83N 14.14E	13	Exten?	None?	R = 50-61 C = 62	35,000 (Campanian) 4,300 (Agnano)	A.D. 1198	---x ---- ---- - --	pex?
							1488?-1538	GG-G B--- ---- B --	Ex, pex
							1564-70	F--- G--- ---- - --	none
							1582	x--- G--- ---- - --	none
							1594	x--? G--- ---- - --	none
							1621?	x--- G--- ---- - --	none
							1832	x--- G--- ---- - --	none
							pre-1901	x--- G--- ---- - --	none
							1908	x--- G--- ---- - --	none
							1913	x--- G--- ---- - --	none
							1917	x--- G--- ---- - --	none
							1930	x--- G--- ---- D QU	none
							1969-74	FF-F FFF- -F?- ? --	none
							1982-85	FF-F -FFF -F-E F QU?	none

PART 2: QUICK REFERENCE TO PARTS 3 AND 4 (continued)
HISTORICAL UNREST AT LARGE CALDERAS AND SELECTED NONCALDERA SETTINGS
See inside back cover for explanation and abbreviations

CAVW number (active vent)	Latitude Longitude (degrees)	Diameter (km)	Local tectonic setting	Precaldera edifice	SiO_2 content (wt pct)	Age of collapse (yr B.P.)	Date of unrest	Type and duration of unrest ESTU STHF MGTF H Te	Eruption type
ISCHIA									
01-01-03	40.73N	ca. 8 km	Exten	Strato?	R = 57	ca. 53,000	11th century B.C.	---- ---- ---- - --	ex?
	13.90E				C = trachyte		B.C. 470	---- ---- ---- - --	ex?
						between	B.C. 400	x--- ---- ---- - --	ex?
						and	B.C. 352		
							A.D. 1228	x--- ---- ---- - =?	none
							1302	EE-- ---- ---- - --	ex,lf
							1557	x--- ---- ---- - =?	none
							1762	x--- ---- ---- - =?	none
							1767	x--- ---- ---- - =?	none
							1796	x--- ---- ---- - =?	none
							1827-28	xx-- ---- ---- - =?	none
							1841	x--- ---- ---- - =?	none
							1863	x--- ---- ---- - =?	none
							1867	x--- ---- ---- - =?	none
							1880-83	xx-- ---- ---- - =?	none
							1961	x--- FF-- ---- - =?	none
CINQUE DENTI (MONASTERO) and LA VECCHIA									
01-01-071	36.78N	6 (CD)	Exten	Strato	R = 45-46	45,000+	1890-91	F--F ---x ---F - --	ex
(Pantelleria, Foerstner vent)	12.00E	ca. 6 (LV)			C = 65-70	114,000			(submarine)
SANTORINI (THIRA)									
01-02-04	36.40N	10	Exten	Graben,	R = 64-68	100,000;	B.C. 1390	F--- ---- ---- - QU	pex,EX,PEX
	25.40E			shields	C = 68	and	B.C. 197	---- ---- ---- - --	EX
						ca.	A.D. 19	---- ---- ---- - --	ex,dm
						B.C. 1390	46	---- ---- ---- - --	ex,dm
							726	---- ---- ---- - --	ex,dm
						between	1570-73?	---- ---- ---- - --	ex,dm
							1649-50	E--- ---- ---- - --	ex,dm
							1707-11	C--D ---- --DD - --	ex,cdm
						between	1810-30	---x ---- --FF - --	cdm?
							1866-70	C--x CC-C --CC C --	dm,ex
							1925-26	D--- G--- ---- - --	ex,dm
							1928	C--- ---F ---C - --	ex,dm
							1939-41	---- ---- --D- - --	ex,dm
							1950	---- ---- ---- - --	ex,dm

PART 2: QUICK REFERENCE TO PARTS 3 AND 4 (continued)
HISTORICAL UNREST AT LARGE CALDERAS AND SELECTED NONCALDERA SETTINGS
See inside back cover for explanation and abbreviations

CAVW number (active vent)	Latitude Longitude (degrees)	Diameter (km)	Local tectonic setting	Precaldera edifice	SiO_2 content (wt pct)	Age of collapse (yr B.P.)	Date of unrest	Type and duration of unrest ESTU STHF MGTF H Te	Eruption type
NISYROS									
01-02-05	36.58N 27.18E	4	Compr, intersection of regional faults	Strato	R = d,r C = ?	?	1422?	---- ---- ---? - --	ex,lf?
							1830	---- ---- ---x x --	none
							1871	x--- ---- ---- - --	ex
							1873	x--- ---- ---- y --	ex,pex
							1888	---- ---- ---y - --	pex
FANTALE									
02-01-19	8.98N 39.93E	3 x 4	Exten	?	R = b-r C = 70?	?	13th cent.	---- ---- ---- - --	"major"
							ca.1820	---- ---- ---- - --	lf
							1981	EE-- ---E ---- - =?	none
K'ONE (GARIBOLDI)									
02-01-20	8.80N 39.69E	ca. 5	Exten	?	R = b C = silicic	?	1820?	---- ---- ---- - --	ex,lf
AWASA/CORBETTI									
02-01-29 (Chabbi)	7.18N 38.48E (Chabbi)	12 x 15 (Corbetti) 30 x 40 (Awasa)	Exten		R = 73-76 C = 73	<1,000,000?	1957-64+	F--- ---- ---F - --	none
							1983-84	xx-- ?--- ---- - =?	none
THE BARRIER									
02-02-03 (Teleki, Likaiyu)	2.33N 36.62E	5 x 6 (outer) 3 (inner)	Exten	Shield	R = 50-51 C = silicic		between 1868-73?	---- ---- ---- - --	ex,lf
							1888	---- ---- ---- - --	lf
							1895	---- ---- ---- - --	ex,lf
							1897?	---- ---- ---- - --	ex?
							1906	---- ---- ---- - --	ex
							1917(-18?)	---- ---- ---- - --	ex
							1921-22	---- ---- ---- - --	ex,lf

PART 2: QUICK REFERENCE TO PARTS 3 AND 4 (continued)
HISTORICAL UNREST AT LARGE CALDERAS AND SELECTED NONCALDERA SETTINGS
See inside back cover for explanation and abbreviations

CAVW number (active vent)	Latitude Longitude (degrees)	Diameter (km)	Local tectonic setting	Precaldera edifice	SiO_2 content (wt pct)	Age of collapse (yr B.P.)	Date of unrest	Type and duration of unrest ESTU STHF MGTF H Te	Eruption type
DERIBA									
02-05-03 (Jebel Marra)	12.95N 24.27E	5	?	Strato	R = 54? C = inter-mediate	ca. 4,000?	1958-66	---- ---- ---x x --	none
PITON DE LA FOURNAISE									
03-03-02	21.23S 55.71E	8 x 15	Hotspot?	Shield	R = 43-50 C = mafic	ca. 2,000	pre-1884	---- ---- ---- - --	lf,ex
							1884	---- ---- ---- - --	lf
							1889	---- ---- ---- - --	ex,lf
							1890-91	---- ---- ---- - --	ex,lf
							1897-1905	---- ---- ---- - --	ex,lf
							1907-10	---- ---- ---- - --	ex,lf
							1913	---- ---- ---- - --	ex,lf
							1915	---- ---- ---- - --	ex,lf
							1917	---- ---- ---- - --	lf
							1920-21	---- ---- ---- - --	ex,lf
							1924-27	---- ---- ---- - --	ex,lf
							1929-31	---- ---- ---- - --	ex,lf
							1933-34	---- ---- ---- - --	ex,lf
							1937-39	---- ---- ---- - --	ex,lf
							1942-46	---- ---- ---- - --	ex,lf
							1948-50	---- ---- ---- - --	ex,lf
							1952-53	---- ---- ---- - --	ex,lf
							1955-61	---- ---- ---- - --	ex,lf
							1963-66	---- ---- ---- - --	ex,lf
							1972-73	---- ---- ---- - --	ex,lf
							1975-77	---- ---- ---- - --	ex,lf
							1979-81	CB-- ---- ---- - --	lf
							1983-84	DBYD ZZZY ---- - --	ex,lf
							1985-86	DBYD -EEY C--- - --	lf
							1987	DBYD -EE- C--- - --	lf
HAROHARO/OKATAINA									
04-01-05,06 (Tarawera)	38.09S 176.51E(H) 38.23S 176.51E(O)	13 x 26	Exten	?	R = 51-75 C = 74-77	300,000-150,000; ca. 50,000	1886	BB-- B--D --E- E --	PEX,ex
							1896	---- ---- --x- x --	none
							1900's	xx-- ---- --EE E --	none
							1982-87+	Cx-- --x- ---F E --	none

PART 2: QUICK REFERENCE TO PARTS 3 AND 4 (continued)
HISTORICAL UNREST AT LARGE CALDERAS AND SELECTED NONCALDERA SETTINGS
See inside back cover for explanation and abbreviations

CAVW number (active vent)	Latitude Longitude (degrees)	Diameter (km)	Local tectonic setting	Precaldera edifice	SiO_2 content (wt pct)	Age of collapse (yr B.P.)	Date of unrest	Type and duration of unrest ESTU STHF MGTF H Te	Eruption type
TAUPO									
04-01-07	38.78S 176.12E	35? (poorly defined)	Exten	None?	R = 71-72 C = 71-72	330,000- 230,000; 20,000; ca. 1,800	1895	x--- ---x ---- - =?	none
							1922	EE-- E--E ---- - =?	none
							1950's	x--- E--- ---- - =?	none
							1964-65	EE-x ---- ---- - =?	none
							1974, 1981	x--- x--x ---- x =?	hydrothermal explosion
							1981-85+	FF-F FF-x ---- - =?	none
ROTORUA									
04-01-16 (Rotorua/ Whakarewarewa)	38.08S 176.25E	15	Exten	None?	R = ? C = 72	140,000	1886	---- ---- ---- ? --	(Tarawera)
							1932	---- ---- --xx x --	hydrothermal explosion
							1924-63	---- -x-- ---- - --	none
							1962	C--- ---- ---- - --	none
TOFUA									
04-03-06	19.80S 175.07W	5	Compr	Strato	R = 54-55 C = 55-58	?	1774	---- ---- ---- - --	ex
							1792	---- ---- ---- - --	lf
							1854	---- ---- ---- - --	ex
							1885	---- ---- ---- - --	ex
							1906	---- ---- ---- - --	ex,lf
							1937	---- ---- ---x - --	none
							1939	---- ---- ---x - --	none
							1955	---- ---- ---x - --	none
							1958-59	---- ---- ---x - --	ex
							1979	---- ---- ---x - --	none?
NIUAFO'OU									
04-03-11	15.60S 175.63W	5-6	Compr	Shield	R = 49-51 C = b?	?	1814	---- ---- ---- - --	pex,ex,lf
							1840	---- ---- ---- - --	unknown
							1853	---- ---- ---- - --	lf,Ex?
							1867	---- ---- ---- - --	lf
							1886-87	E--D ---- ---- - QU?	pex,Ex
							1912	---- ---Y ---- - --	ex,lf
							1929	x--- ---Y ---- - --	ex,lf
							1935-36	---- ---- ---E - --	ex,lf
							1943	---- ---- ---- - --	ex,lf
							1946-47	BB-Y ---Y ---- - --	ex,lf
							1959(?)	---- ---- ---- - --	unknown
							1985	BB-- ---B ---- - --	ex

PART 2: QUICK REFERENCE TO PARTS 3 AND 4 (continued)
HISTORICAL UNREST AT LARGE CALDERAS AND SELECTED NONCALDERA SETTINGS
See inside back cover for explanation and abbreviations

CAVW number (active vent)	Latitude Longitude (degrees)	Diameter (km)	Local tectonic setting	Precaldera edifice	SiO_2 content (wt pct)	Age of collapse (yr B.P.)	Date of unrest	Type and duration of unrest ESTU STHF MGTF H Te	Eruption type
ST. ANDREW STRAIT									
05-00-01	2.45S	16	Hotspot?	?	R = 69-71	?	1883	---- ---- ---- - --	ex?
(Tuluman)	147.32E				C = silicic		1953-57	---Y Y?-Y ---- - --	Ex,pex,lf
KARKAR									
05-01-03	4.65S	5.5, 3.2	Compr	Strato	R = 53-55	9,100,	1643	---- ---- ---- - --	Ex?
	145.96E				C = ?	1,400	1885	---- ---- ---- - --	ex
							1895	BB-- ---- ---- - --	ex,lf
							1962?	---- ---- ---- - --	ex
						between	1965/70	x--- ---- ---x - UQ?	none
							1974-75	---- ---- --FF - QU	ex,Lf
							1976-80	FFEY -Y-- -YEE E --	pex,ex
LONG ISLAND/LAKE WISDOM									
05-01-05	5.36S	10 x 12.5	NNE-SSW	LP-strat	R = 49-57	16,000	ca. 1700	x--- ---- ---- - --	EX
(Motmot Island)	147.12E		compr		C = 51-57	4,000	1933	---- ---- ---- x --	unknown
						300	1938	---- ---- ---- - --	unknown
							1943	---- ---- ---- - --	ex
							1953-55	---- ---- ---- - --	Ex
							1961?	---- ---- ---- - --	ex
							1968	---- ---- ---- - --	ex
							1973-76	---- E--E --EE E --	pex,ex,lf
							1979	---- ---- --EE E --	none
DAKATAUA									
05-02-04	5.06S	10 x 13	Compr	Strato	R = 59?	1,150?	ca. 1890	---- ---- ---- - --	ex,lf
(Benda, Makalia)	150.11E				C = 67?				
WITORI									
05-02-08	5.58S	7.5 x 5.5	NNW-SSE	LP-strat	R = 66	2,600	1900?	---- ---- ---- - --	uncertain
(Pago)	150.52E		compr		C = 68		1911-33	N--- ---- ---- - --	ex,lf

PART 2: QUICK REFERENCE TO PARTS 3 AND 4 (continued)
HISTORICAL UNREST AT LARGE CALDERAS AND SELECTED NONCALDERA SETTINGS
See inside back cover for explanation and abbreviations

CAVW number (active vent)	Latitude Longitude (degrees)	Diameter (km)	Local tectonic setting	Precaldera edifice	SiO_2 content (wt pct)	Age of collapse (yr B.P.)	Date of unrest	Type and duration of unrest ESTU STHF MGTF H Te	Eruption type
LOLOBAU ISLAND									
05-02-13 (Lolobau)	4.93S 151.17E	5.5	Compr	Strato	R = d C = silicic	?	1905	---- ---- ---- - --	lf,ex
RABAUL									
05-02-14	4.27S	14 x 9	Shear in nearby straits	LP-strat (3 cones)	R = 49-66	3,500	1767?	---- ---- ---- - --	ex
(Tavurvur,	152.20E				C = 64-72	1,400	1791	---- ---- ---- - --	ex
Sulphur Creek,							1850	x--x ---- ---- - --	pex
Rabalanakaia,							1878	B--F B--- --BB B --	Ex,pex
Vulcan)							1910	x--- ---- x--x - =?	none
							1916	x--- x--- ---- - =?	none
							1919-23	C--F ---- ---x - QU	none
							1937	C-CF B--B --BB - --	EX,pex
							1938-43	x--F FD-- --EF - UQ	pex,ex
							1971-present	FF-G EFFE -F-x - QU	none yet
UNNAMED, near Esa'Ala, D'Entrecasteaux Islands									
05-03-06,07	9.62S	7 x 11	?	Strato	R = 69-72	?	1953-57	FF-- ---- ---- - --	none
(Lamonai-Oiau,	150.88E				C = 70?		1969	xx-- ---- ---- - --	none
Dobu)							1981-82	---- ---- --F- F --	none
							1984	---- ---- --E- E --	none
GAUA ISLAND									
05-07-02	14.25S	8	Compr	Strato	R = mafic	?	1963-69	---- ---- ---- - --	ex,Ex
(Mt. Gharat)	167.50E				C = ?		1971	---- ---- ---- - --	ex
							1973-74	E-C- ---- ---x - QU?	pex
							1976-77	---- ---- ---x x --	ex
							1980-82	---- ---- ---x - --	pex, ex
AOBA									
05-07-03	15.33S	6	Compr, with transverse structure	Shield	R = 45-50	few thousand?	ca. 1670	---- ---- ---- - --	lf
	167.83E				C = mafic		ca. 1870	---- ---- ---- - --	ex
							1971	---- ---- ---E E --	none
							1976	---- ---- ---x x --	none

PART 2: QUICK REFERENCE TO PARTS 3 AND 4 (continued)
HISTORICAL UNREST AT LARGE CALDERAS AND SELECTED NONCALDERA SETTINGS

See inside back cover for explanation and abbreviations

CAVW number (active vent)	Latitude Longitude (degrees)	Diameter (km)	Local tectonic setting	Precaldera edifice	SiO_2 content (wt pct)	Age of collapse (yr B.P.)	Date of unrest	Type and duration of unrest ESTU STHF MGTF H Te	Eruption type
AMBRYM									
05-07-04	16.25S 168.08E	12	Compr, intersection of fault zone and arc	LL-strat (3 cones)	R = 50 C = mafic	ca. 2,000	1774?	---- ---- ---- - --	unknown
							1863-64	---- ---- ---- - --	ex
							1871	---- ---- ---- - --	ex
							1888	---- ---- ---- - --	lf
							1894-95	Y--- ---- ---- - --	Ex,lf
							1908	---- ---- ---- - --	unknown
							1912-15	x--- ---Y ---- - --	ex,Ex,lf
							1929	---- ---- ---- - --	ex,lf
							1935	x--- ---- ---- - --	ex,ll
							1937	x--- ---- ---- - --	ex,lf
							1942	---- ---- ---- - --	ex,lf
							1950-54	E--- ---- ---- - --	Ex,ex, ll,lf
							1959-81	---- ---- ---Y - --	Ex,ex, ll,lf
							1985-86	---- ---- ---- - --	ex
							1988	---- ---- ---- - --	ex
TOBA									
06-01-09 (Pusukbukit)	2.6 N 98.8 E	35 x 100	Shear, with graben; may also lie astride segment boundary	None	R = hb-a C = 68-77	75,000 <700,000 840,000	1892	x--- ---- ---- - --	none
							1916	x?-? ---- ---- - Q?	none
							1920-22	xx-? x--- ---x - QU	none
							1987	x--? ---- ---- - --	none
MANINJAU									
06-01-14 (Marapi= M) 06-01-15 (Tandikat= T) 06-01-151 (Maninjau= Mn)	0.33S 100.18E	11 x 22	Exten and right-lateral shear	3 strato?	R = 55 C = 74 with late-stage 55	280,000 or 70-80,000	1770 (M)	---- ---- ---- - --	ex
							1807 (M)	---- ---- ---- - --	ex
							1822 (M)	---- ---- ---- - --	ex
							1833-34(M)	---- ---- ---- - --	ex
							1845 (M)	---- ---- ---- - --	ex
							1854-56(M)	xx-- ---- ---- - --	ex
							1861 (M)	---- ---- ---- - --	ex
							1863 (M)	---- ---- ---- - --	ex
							1871 (M)	x--- ---- ---- - --	ex
							1876-77(M)	---- ---- ---- - --	ex

PART 2: QUICK REFERENCE TO PARTS 3 AND 4 (continued)
HISTORICAL UNREST AT LARGE CALDERAS AND SELECTED NONCALDERA SETTINGS
See inside back cover for explanation and abbreviations

CAVW number (active vent)	Latitude Longitude (degrees)	Diameter (km)	Local tectonic setting	Precaldera edifice	SiO_2 content (wt pct)	Age of collapse (yr B.P.)	Date of unrest	Type and duration of unrest ESTU STHF MGTF H Te	Eruption type
MANINJAU (continued)									
							1883 (M)	---- ---- ---- - --	ex
							1885-89(M,T)	x-x- ---- x--x - --	Ex,pex
							1904-19(M,T)	---- ---- ---- - --	ex
							1924 (T)	---- ---- ---x - --	none?
							1927 (M)	---- ---- ---- - --	ex
							1930 (M)	---- ---- ---- - --	ex
							1932? (M)	x--- ---- ---- - --	ex
							1939 (Mn)	---- ---- ---? - QU?	none
							1949-52(M)	---- ---- ---- - --	ex
							1954-58(M)	---- ---- ---- - --	ex
							1966-71(M)	---- ---- ---- - --	ex,pex
							1973 (M)	---- ---- ---- - --	ex
							1975-80(M)	---- ---- ---- - --	ex,pex
							1982-84(M)	---- ---- ---- - --	ex
							1987-88(M)	---- ---- ---- - --	ex
RANAU									
06-01-25F (Seminung)	4.54S 103.53E	16 x 20	Shear, with exten	None	C = silicic		1887-88	---- ---- --?? ? --	none
							1903	---- ---- --?? ? --	none
SUOH									
06-01-27 (Pematang Bata)	5.25S 104.27E	8 x 13	Exten, with lateral shear	None?	R = ? C = ?		1933	D--- ---x ---D ? QU	Pex
KRAKATAU									
06-02-00 (Anak Krakatau)	6.10S 105.42E	8	Compr, boundary between two subducted slabs?	Strato	R = 51-70 C = 65-69	A.D. 416 A.D. 1883	3rd century	---- ---- ---- - --	Ex?
							416	---- ---- ---- - --	EX
							9th-16th centuries	---- ---- ---- - --	ex,Ex?
							1680-81	x--- ---- ---- - --	Ex,lf
							1880-84	D--- ---- ---- - QU?	ex,pex,EX
							1913	?--- ---- --?- - --	none?
							1927-81	xx-- ---- ---E - --	ex,pex,lf
							1988	?--- ---- ---- - --	pex,ex

PART 2: QUICK REFERENCE TO PARTS 3 AND 4 (continued)
HISTORICAL UNREST AT LARGE CALDERAS AND SELECTED NONCALDERA SETTINGS

See inside back cover for explanation and abbreviations

CAVW number (active vent)	Latitude Longitude (degrees)	Diameter (km)	Local tectonic setting	Precaldera edifice	SiO_2 content (wt pct)	Age of collapse (yr B.P.)	Date of unrest	Type and duration of unrest ESTU STHF MGTF H Te	Eruption type
SUNDA									
06-03-09 (Tangkuban-parahu)	6.77S 107.60E	6-8	Compr	Strato	R = 54-58 C = silicic	>40,000	1826	---- ---- ---- - --	unknown
							1829	?--- ---- ---- - --	ex
							1842?	---- ---- ---- - --	unknown
							1846	---- ---- ---- - --	Ex,surge?
							1862	---- ---- --xx - --	none
							1887-88	---- ---- --xx - --	none
							1896	---- ---- ---Y Y --	pex
							1908	?--- ---- ---- - --	none
							1910	?--- ---- ---- - --	ex
							1913	---- ---- --?C - --	none
							1926	---- ---- ---- - --	pex
							1929	---- ---- --EE - --	pex
							1952	D--- ---- --DD - --	pex
							1957	---- ---- --x- - --	none
							1961-62	---- ---- --E- - --	none
							1965	---- ---- --x- - --	none
							1967	---- ---- --xx - --	none
							1969	---- ---- --xx - --	pex
							1971	---- ---- --xx x --	none
							1983-present	EE-E -E-- --?? ? --	none yet
GEGER HALANG									
06-03-17 (Cereme)	6.89S 108.40E	5.5 x 4	Compr	Strato	R = 54-58 C = silicic		1698	---- ---- ---- - --	Ex
							1772	---- ---- ---- - --	ex
							1775	---- ---- ---- - --	ex
							1805	---- ---- ---- - --	ex
							1917	---- ---- ---x - --	none
							1924	---- ---- ---x - --	none
							1937-38	Y--- ---x --FF - --	pex,ex
							1941	---- ---- ---E - --	none
							1951	---- ---- ---- - --	ex
PRAHU (DIENG)									
06-03-20 (Dieng)	7.2S 109.92E	6.5	Compr	Strato	R = 58-63 C = 61?	>16,000	late 1300's	---- ---- ---- - --	Ex
							1786	EE-- ---Y ---Y - --	ex,landsl
							1825-26	---- ---- ---- - --	Ex
							1847	---- ---- ---- - --	ex?
							1884	---- ---- --xx - --	none
							1895	---- ---x ---- - --	none

PART 2: QUICK REFERENCE TO PARTS 3 AND 4 (continued)
HISTORICAL UNREST AT LARGE CALDERAS AND SELECTED NONCALDERA SETTINGS
See inside back cover for explanation and abbreviations

CAVW number (active vent)	Latitude Longitude (degrees)	Diameter (km)	Local tectonic setting	Precaldera edifice	SiO_2 content (wt pct)	Age of collapse (yr B.P.)	Date of unrest	Type and duration of unrest ESTU	STHF	MGTF	H	Te	Eruption type
PRAHU (DIENG) (continued)													
							1924	x---	----	--xx	-	--	none
							1928	DD--	---D	---Y	-	--	pex
							1939	BB--	---B	----	-	--	pex
							1943-44	NN--	---E	----	-	--	pex
							1952-56	NN--	----	--??	?	--	pex
							1964	NN--	----	---x	-	--	pex
							1965	DD--	----	---?	-	--	none
							1979	xx--	---?	--xY	-	--	pex
							1981	DD--	----	--NN	-	--	none
							1984	DD--	----	--N-	-	--	none
							1986	D---	----	--N-	?	QU	none
TENGGER/NGADISARI													
06-03-31	7.94S	16	Compr	Strato?/	R = 54-56		1804	----	----	----	-	--	ex
(Bromo)	112.95E			LP-strat?	C =		1815	----	----	----	-	--	ex
					intermediate?		1820	----	----	----	-	--	ex
							1822-23	??--	----	----	-	--	ex
							1825	----	----	----	-	--	ex
							1829-30	----	----	----	-	--	ex
							1835	----	----	----	-	--	ex
							1837-38	----	----	----	x	--	none
							1842-43	----	----	----	-	--	ex
							1844	----	----	----	-	--	ex
							1856-60	x---	----	----	-	--	ex
							1865-68	?---	----	----	-	--	ex
							1877	----	----	----	-	--	ex
							1885-88	----	----	---Y	-	--	ex
							1890	----	----	----	-	--	ex
							1893	----	----	----	-	--	ex
							1896	----	----	----	-	--	ex
							1898	x---	----	----	-	--	none
							1906-10	----	----	----	-	--	ex
							1915-16	----	----	----	-	--	ex
							1921-22	----	----	----	-	--	ex
							1928-30	----	----	----	-	--	ex
							1935	----	----	----	-	--	ex
							1939-40	----	----	----	-	--	ex
							1948	----	----	----	-	--	Ex
							1950	----	----	----	-	--	ex
							1955-56	----	----	----	-	--	ex
							1972	x---	----	----	-	--	ex
							1980	----	----	---C	-	--	ex
							1983-84	----	----	---x	-	--	ex

PART 2: QUICK REFERENCE TO PARTS 3 AND 4 (continued)
HISTORICAL UNREST AT LARGE CALDERAS AND SELECTED NONCALDERA SETTINGS
See inside back cover for explanation and abbreviations

CAVW number (active vent)	Latitude Longitude (degrees)	Diameter (km)	Local tectonic setting	Precaldera edifice	SiO_2 content (wt pct)	Age of collapse (yr B.P.)	Date of unrest	Type and duration of unrest ESTU	STHF	MGTF	H	Te	Eruption type
IJEN													
06-03-34	8.12S	20	Compr	Strato	R = 51-59		1586 (RA)	----	----	----	-	--	Ex
(Raung= RA)	114.04E			complex	C = silicic		1593 (RA)	----	----	----	-	--	Ex
				(3 or more			1597 (RA)	----	----	----	-	--	Ex
06-03-35	8.06S			cones)			1638 (RA)	Y---	----	----	-	--	Ex
(Merapi,	114.24E						1730 (RA)	----	----	----	-	--	Ex
Kawah							between 1787-99 (RA)	----	----	----	-	--	ex
Ijen= IJ)							1796 (IJ)	----	----	----	-	--	pex
							between 1800-08 (RA)	----	----	----	-	--	ex
							1812-15? (RA)	----	----	----	-	--	Ex
							1817 (IJ)	----	----	----	-	--	Ex or Pex
							1817 (RA)	----	----	----	-	--	Ex
							1838 (RA)	----	----	----	-	--	ex
							1859-60 (RA)	----	----	----	-	--	ex
							1864 (RA)	----	----	----	-	--	ex
							1881? (RA)	----	----	----	-	--	ex
							1885 (RA)	----	----	----	-	--	ex
							1890 (RA)	----	----	----	-	--	ex
							1896-97 (RA)	----	----	----	-	--	ex
							1902-04 (RA)	Y---	----	---Y	-	--	ex,lf
							1913-17 (RA)	----	----	----	-	--	ex
							1917 (IJ)	----	----	--DD	-	QU	pex
							1921 (RA)	----	----	----	-	--	ex,lf
							1921 (IJ)	----	----	--DD	D	--	none
							1924 (RA)	----	----	----	-	--	ex,lf
							1927-29 (RA)	----	----	----	-	--	ex
							1931 (RA)	A---	----	---x	-	--	none?
							1933 (RA)	----	----	----	-	--	ex
							1936 (IJ)	----	----	----	-	--	pex
							1936-45 (RA)	----	----	----	-	--	ex,lf
							1938-39 (IJ)	----	----	--xx	-	--	none
							1952 (IJ)	----	----	----	-	--	ex
							1953 (RA)	----	----	----	-	--	ex,lf
							1955-56 (RA)	----	----	----	-	--	ex
							1958 (IJ)	x---	----	--xx	x	--	none
							1971 (RA)	----	----	----	-	--	pex
							1973-77 (RA)	----	----	----	-	--	ex,pex
							1982 (RA)	B---	----	----	-	--	ex
							1985 (RA)	x---	----	----	-	--	ex

PART 2: QUICK REFERENCE TO PARTS 3 AND 4 (continued)
HISTORICAL UNREST AT LARGE CALDERAS AND SELECTED NONCALDERA SETTINGS
See inside back cover for explanation and abbreviations

CAVW number (active vent)	Latitude Longitude (degrees)	Diameter (km)	Local tectonic setting	Precaldera edifice	SiO_2 content (wt pct)	Age of collapse (yr B.P.)	Date of unrest	Type and duration of unrest ESTU STHF MGTF H Te	Eruption type
BATUR									
06-04-01	8.24S	10 x 14	Compr	LL-strat	R = 50-55	23,700	1804	---- ---- ---- - --	ex
(Batur)	115.38E	(outer)			C = 62-66		1821	---- ---- ---- - --	ex
		7.5					1849	---- ---- ---- - --	ex,lf
		(inner)					1854	---- ---- ---- - --	ex
							1888	---- ---- ---- - --	ex,lf
							1897	---- ---- ---- - --	ex
							1902	?--- ---- ---- - --	none?
							1904-05	---- ---- ---- - --	ex,lf
							1917	xx-- ---- ---- - --	none
							1921-26	x--- ---Y --Y- Y QU?	ex,lf
							1959	---- ---- --?? ? --	none
							1963-64	B--- ---- ---- - QU	ex,lf
							1965-66	---- ---- ---- - --	ex
							1968	---- ---- ---- - --	Ex,lf
							1969	---- ---- --?? ? --	none
							1970-74	---- ---- ---- - --	ex,lf
SEGARA ANAK									
06-04-03	8.42S	5	Compr	Strato	R = 52-53		1847	---- ---- ---- - --	ex
(Rinjani,	116.47E				C = 65(?)		1884	Y--- ---- ---- - --	ex
G. Baru)							1900-01	---- ---- ---- - --	ex,lf
							1903	x--- ---- ---- - Q-	none
							1906	---- ---- ---- - --	ex
							1915	---- ---- ---- - --	ex
							1941?	---- ---- ---- - --	unknown
							1944	---- ---- ---- - --	ex,dm
							1949-50	---- ---- ---- - --	dm
							1965-66	---- ---- ---- - --	dm,ex
TAMBORA									
06-04-04	8.25S	6	Back-arc	Double	R = 48-57	A.D. 1815	1812-15	C--- ---- ---F - --	ex,EX,pex
(Tambora)	118.00E		exten?	strato	C = 56-57		1819	---- ---- ---- - --	ex
				on		between	1847-1913	x--- ---- ---- - --	ex,lf
				shield					

PART 2: QUICK REFERENCE TO PARTS 3 AND 4 (continued)
HISTORICAL UNREST AT LARGE CALDERAS AND SELECTED NONCALDERA SETTINGS
See inside back cover for explanation and abbreviations

CAVW number (active vent)	Latitude Longitude (degrees)	Diameter (km)	Local tectonic setting	Precaldera edifice	SiO_2 content (wt pct)	Age of collapse (yr B.P.)	Date of unrest	Type and duration of unrest ESTU STHF MGTF H Te	Eruption type
LOBOBUTA/BADJAWA									
06-04-09 (Inielika)	8.73S 120.98E	7.5	Compr			Pleisto-cene(?)	1905	N--- ---- ---- - --	pex
SUKARIA									
06-04-12 (Sukaria) 06-04-13 (Ndete Napu) 06-04-14 (Keli Mutu)	8.47S 121.46E	6 x 9 or 12 x 18?	Compr				1860's	---- ---- ---- - --	pex
							1938	---- ---- ---E - --	pex
							1968	---- ---- ---- - --	pex?
							1986	C--- ---- --?x - --	none
BANDA API									
06-05-09 (Banda Api)	4.52S 129.87E	7 (outer) 3 (inner)	Compr	Submarine strato	R = 52-66 C = silicic?		1586	---- ---- ---- - --	ex
							1598-1602	---- ---- ---- - --	Ex
							1609	---- ---- ---- - --	ex
							1615	---- ---- ---- - --	Ex
							1632	x--- ---- ---- - --	ex
							1635?	---- ---- ---x - --	ex? or none?
							1683	x--- ---x ---- - --	Ex
							1690-96	---- ---- ---- - --	Ex
							1712	---- ---- ---- - --	ex
							1722(-23?)	---- ---- ---- - --	ex
							1749	---- ---- ---- - --	ex
							1762	---- ---- ---- - --	ex
							1765-66	---- ---- ---- - --	ex
							1773	---- ---- ---- - --	ex
							1775	---- ---- ---- - --	ex
							1778	---- ---- ---- - --	ex
							1816	x--- ---- ---- - --	ex
							1820	---- ---- ---- - --	Ex,lf
							1824	---- ---- ---- - --	ex
							1835	D--- ---- ---D - UQ?	none?
							1852-53	x--- ---- ---x - --	none
							1855-56	x--- ---- ---x - --	none
							1859-60	E--- ---- ---x - QU?	none?
							1877	E?-- ---- ---- - --	none
							1887	x--- ---- ---x - QU	none
							1890	E--- ---- ---E - --	ex
							1901-02	x--- ---- ---- - --	ex,lf
							1908	---- ---- ---x - --	none or ex?
							1987-88	C--- ---- ---x - QU	ex,lf

PART 2: QUICK REFERENCE TO PARTS 3 AND 4 (continued)
HISTORICAL UNREST AT LARGE CALDERAS AND SELECTED NONCALDERA SETTINGS
See inside back cover for explanation and abbreviations

CAVW number (active vent)	Latitude Longitude (degrees)	Diameter (km)	Local tectonic setting	Precaldera edifice	SiO_2 content (wt pct)	Age of collapse (yr B.P.)	Date of unrest	Type and duration of unrest ESTU STHF MGTF H Te	Eruption type
TONDANO							late 1300's(LO)	---- ---- ---- - --	Ex
06-06-03	1.25N	20 x 30?	Compr		R = b-a,		late 1700's(LO)	---- ---- ---- - --	Ex
(Soputan= SO)	124.85E				ol-d,		1785 (SO)	---- ---- ---- - --	ex
					r		before 1788 (MA)	---- ---- ---- - --	ex
06-06-04							1789 (MA)	---- ---- ---- - --	ex,pex?
(Sempu= SE)							1819? (SE)	---- ---- ---- - --	ex?
					(LO) 50-67		1819 (SO)	---- ---- ---- - --	ex
06-06-10							1829 (LO)	---- ---- ---- - --	ex
(Lokon-Empung= LO)					C = silicic		1833? (SO)	---- ---- ---- - --	ex
							1845 (SO)	---- ---- ---- - --	ex
06-06-11							1846 (MA)	---- ---- ---- - --	pex
(Mahawu= MA)							1890 (SO)	---- ---- ---- - --	ex
							1893 (LO)	x--- ---- ---- - --	ex
							1894 (LO)	x--- ---- ---x - --	none
							1898? (SO)	---- ---x ---x - --	none
							1901 (SO)	xx-- ---x ---x x --	pex
							1904 (MA)	---- ---- ---- - --	ex
							1906-13 (SO)	DD-- ---D ---F - --	ex,lf
							1915 (SO)	---- ---- ---- - --	lf,dm
							1917 (SO)	---- ---- ---- - --	ex
							1923-24 (SO)	---- ---- ---- - --	ex,lf
							1930 (LO)	---- ---- ---- - --	ex
							1942 (LO)	---- ---- ---- - --	ex
							1947 (SO)	---- ---- ---- - --	ex
							1949 (LO)	---- ---- ---- - --	pex
							1951-53 (LO)	---- ---- --EE - --	ex
							1952 (MA)	---- ---- ---- - --	?
							1953 (SO)	---- ---- ---- - --	?
							1958-59 (MA)	---- ---- ---- - --	ex,pex
							1958-59 (LO)	xx-- ---- ---x - --	ex
							1961-62 (LO)	---- ---- ---x - --	ex,pex
							1963-66 (LO)	---- ---- ---- - --	ex
							1966-68 (SO)	YY-- ---- ---B - --	Ex,lf,pf,dm
							1969-71 (LO)	YY-- ---- ---x - --	ex,pex,pf,dm
							1970 (SO)	?--- ---- ---x - --	none
							1971 (SO)	---- ---- ---- - --	pex
							1973 (SO)	---- ---- ---- - --	ex
							1973-74 (LO)	---- ---- ---- - --	ex
							1975-79 (LO)	DD-- ---- ---- - --	ex,pf,dm
							1978 (MA)	---- ---- --x- - --	none
							1982 (SO)	A--- ---- ---- - --	Ex
							1984 (SO)	N--- ---- ---- - --	Ex
							1985 (SO)	EE-Y ---- ---- - --	Ex
							1982-87 (LO)	x--- ---- ---F - --	pex, ex
							1987 (MA)	E--- ---- --EE E --	none

PART 2: QUICK REFERENCE TO PARTS 3 AND 4 (continued)
HISTORICAL UNREST AT LARGE CALDERAS AND SELECTED NONCALDERA SETTINGS
See inside back cover for explanation and abbreviations

CAVW number (active vent)	Latitude Longitude (degrees)	Diameter (km)	Local tectonic setting	Precaldera edifice	SiO_2 content (wt pct)	Age of collapse (yr B.P.)	Date of unrest	Type and duration of unrest ESTU STHF MGTF H Te	Eruption type
BULUSAN (IROSIN)									
07-03-01 (Bulusan)	12.77N 124.05E	11	Compr	Strato	R = 59? or basaltic andesite? (1918)	>36,000	1852	---- ---- ---- - --	ex
							1886	---- ---- ---- - --	ex
							1889	---- ---- ---- - --	unknown
							1892	---- ---- ---- - --	ex
							1894	---- ---- ---- - --	ex
					C = 69-71		1916	x--- ---- ---x - --	pex,ex
							1918-22	x--- ---- --xx - --	pex,ex,lf
							1928	---- ---- ---- - --	ex
							1933	---- ---- ---- - --	ex
							1966	---- ---- --x? ? --	none
							1978-83	Cx-- ---- --DD C --	pex,ex,Ex
							1986-88+	FF-- ---- ---F - --	pex
TAAL									
07-03-07	14.00N 121.00E	15 x 20	NE-SW fault normal to volcanic chain	Over-lapping strato-volcanoes	R = 50-59		1572	---- ---- ---- - --	Ex
					C = 58-64		1591	---- ---- ---- - --	ex
							1605-11	---- ---- ---- - --	ex
							1634-35	---- ---- ---- - --	ex
							1641	---- ---- ---- - --	Ex
							1645	---- ---- ---- - --	ex
							1707	---- ---- ---- - --	ex
							1709	---- ---- ---- - --	ex
							1715-16	x--- ---- ---? - --	P-Ex
							1729	---- ---- ---- - --	ex
							1731	x--- ---- ---- - --	p-ex
							1749	xx-? Y-?Y --B- Y --	Ex
							1754	Y--? ---x ---Y - --	P-EX
							1790	---- ---- ---- - --	ex
							1808	---- ---- ---- - --	ex
							1825	---- ---- ---- - --	ex
							1842	x--- ---- ---- - --	ex
							1873-74	---- ---- ---- - QU?	ex
							1878	D--- ---- ---- - --	ex
							1880	BB-- ---- ---- - --	none
							1881	xx-- ---- ---- - --	none
							1883	A--- ---- ---- - --	none
							1887	xx-- ---- ---- - --	none
							1903-04	---- ---- ---E - --	ex,pex
							1911	CC-x x-?x ---C - --	ex,P-EX
							1953	---- ---- --EE E --	none
							1965-70	A--C -?-- x-EE E --	P-EX,lf
							1971-77	N--- ---C --FF - --	ex,pex
							1986-87	DD-- x-?D ---- - --	none

PART 2: QUICK REFERENCE TO PARTS 3 AND 4 (continued)
HISTORICAL UNREST AT LARGE CALDERAS AND SELECTED NONCALDERA SETTINGS
See inside back cover for explanation and abbreviations

CAVW number (active vent)	Latitude Longitude (degrees)	Diameter (km)	Local tectonic setting	Precaldera edifice	SiO_2 content (wt pct)	Age of collapse (yr B.P.)	Date of unrest	Type and duration of unrest ESTU STHF MGTF H Te	Eruption type
KIKAI									
08-02-06	30.78N	19	Exten?	LP-strat?	R = 65-72	6,300	1934-36	CC-- Z--C --ZZ - --	ex,lf
(Tokara-Iwo Jima	130.28E		(graben)		C = 71-72		1988	---- ---- ---Z - --	ex
=Satsuma-Iwo-Jima;									
=Shin-Iwo-Jima)									
ATA									
08-02-07	31.20N	15 x 25	Exten?	None?	R = 51-57	>40,000	A.D. 91	?--- ---- ---- - --	ex?
(Kaimon-dake;	130.55E				C = 68-73	(Ata);	860	---- ---- ---- - --	ex
Ibusuki						ca. 5,000	866	---- ---- ---- - --	ex
Volcanic Field)						(Ikeda)	874	---- ---- ---- - --	ex
08-02-071							882	---- ---- ---- - --	ex
(Ikeda Caldera)							885	---- ---- ---- - --	ex
							1615?	---- ---- ---- - --	ex?
							1967-68	CC-- ---- ---? ? --	none
							1970	EE-- ---- ---- - --	none
							1981?	BB-- ---- ---- - --	none
AIRA									
08-02-08	31.58N	20	Exten?	None?	R = 57-67	22,000	708	---- ---- ---- - --	Ex
(Sakurajima)	130.67E		(NNE-SSW		C = 74-75		712	---- ---- ---- - --	Ex?
			graben or				716-18	---- ---- ---- - --	Ex
			"volcano-				764	---- ---- ---- - --	Ex
			tectonic				766	x--- ---- ---- - --	Ex
			depression")				778	---- ---- ---- - --	unknown
							1468	---- ---- ---- - --	Ex
							1471	---- ---- ---- - --	Ex,lf
							1473	---- ---- ---- - --	ex
							1475-76	C--x ---- ---- - --	Ex,lf,cdm
							1478	---- ---- ---- - --	ex
							1642	---- ---- ---- - --	ex
							1670?	---- ---- ---- - --	unknown
							1678	---- ---- ---- - --	ex
							1706	---- ---- ---- - --	ex
							1742	---- ---- ---- - --	ex
							1749	---- ---- ---- - --	ex
							1756	---- ---- ---- - --	ex
							1766	---- ---- ---- - --	ex
							1779-81	C--C Z--- ---C C --	Ex,lf,cdm
							1785	---- ---- ---- - --	ex

PART 2: QUICK REFERENCE TO PARTS 3 AND 4 (continued)
HISTORICAL UNREST AT LARGE CALDERAS AND SELECTED NONCALDERA SETTINGS
See inside back cover for explanation and abbreviations

CAVW number (active vent)	Latitude Longitude (degrees)	Diameter (km)	Local tectonic setting	Precaldera edifice	SiO_2 content (wt pct)	Age of collapse (yr B.P.)	Date of unrest	Type and duration of unrest ESTU STHF MGTF H Te	Eruption type
AIRA (continued)									
							1790-92	---- ---- ---- - --	ex
							1794	---- ---- ---- - --	ex
							1797	---- ---- ---- - --	ex
							1799	---- ---- ---- - --	ex
							1860	---- ---- ---- - --	ex?
							1878-79	---- ---- --x- ? --	none
							1899	---- ---- ---- - --	ex
							1913-15	E--G Z--- --EB E UQ	EX,lf
							1931	---- x--- ---- - --	none
							1935	C--G ---- ---- - --	ex
							1938-42	---G Z--- ---- - --	pex,ex
							1946	---G Z--- ---- - --	ex,lf
							1948	---G ---- ---- - --	unknown
							1950	---G ---- ---- - --	pex,ex
							1954	---G ---- ---- - --	pex
							1955-88+	CCBG ZDx- ---D x --	pex,ex,lf
KAKUTO									
08-02-09	32.03N	10 x 15	Exten?	Strato?	R = 50-67	110,000	742	---- ---- ---- - --	ex
(Kirishima)	130.82E	(Kakuto)			C = 73-75	65,000	788	---- ---- ---- - --	Ex
		12 x 16				12,000	837	---- ---- ---- - --	ex
		(Kobayashi)					843	---- ---- ---- - --	ex
		12 x 30					857-58	---- ---- ---- - --	ex
		(combined)					945	---- ---- ---- - --	ex
							1112	---- ---- ---- - --	ex
							1167	---- ---- ---- - --	ex
							1184	---- ---- ---- - --	ex
							1235	---- ---- ---- - --	ex
							1524	---- ---- ---- - --	ex
							1554	---- ---- ---- - --	ex
							1566	---- ---- ---- - --	ex
							1574	---- ---- ---- - --	ex
							1576-78	---- ---- ---- - --	ex
							1585	---- ---- ---- - --	ex
							1587-88	---- ---- ---- - --	ex
							1598-1600	---- ---- ---- - --	ex
							1617	---- ---- ---- - --	ex
							1628	---- ---- ---- - --	ex
							1637-38	---- ---- ---- - --	ex
							1659-64	---- ---- ---- - --	ex
							1677-78	---- ---- ---- - --	ex
							1706	---- ---- ---- - --	ex

PART 2: QUICK REFERENCE TO PARTS 3 AND 4 (continued)
HISTORICAL UNREST AT LARGE CALDERAS AND SELECTED NONCALDERA SETTINGS
See inside back cover for explanation and abbreviations

CAVW number (active vent)	Latitude Longitude (degrees)	Diameter (km)	Local tectonic setting	Precaldera edifice	SiO_2 content (wt pct)	Age of collapse (yr B.P.)	Date of unrest	Type and duration of unrest ESTU STHF MGTF H Te	Eruption type
KAKUTO (continued)									
							1716-17	---- ---- ---- - --	Ex,ex
							1719	---- ---- ---- - --	unknown
							1768	---- ---- ---- - --	lf,dm
							1771	---- ---- ---- - --	ex?
							1822	---- ---- ---- - --	ex
							1832	---- ---- ---- - --	ex
							1880	---- ---- ---- - --	ex
							1887-91	---- ---- ---- - --	ex
							1894-1900	---- ---- ---- - --	ex
							1903	---- ---- ---- - --	ex
							1913-14	EE-- ---- ---- - QU	ex
							1915	EE-- ---- ---- - --	none
							1923	---- ---- ---- - --	ex
							1934	---- ---- ---x x --	none
							1946	---- ---- ---- - --	ex
							1959	ZZ-- ---- ---- - --	pex
							1961	EE-- ---- ---- - QU	none
							1966-69	FFC- x-?- --F- F QU	none
							1971	NZ-- ---- --F- F --	pex,ex
							1974-76	EE-- xx-- ---- - --	none
							1977-79	FF-- xx-- --xx x --	pex
							1980-83	E--- ---- --FF - --	pex
ASO									
08-02-11	32.88N	21	Right-	Strato	R = 52-54	300,000	pre-1884	---- ---- ---- - --	ex,Ex
(Naka-dake)	131.10E		lateral	(several	C = 70-50	150,000	1884	---- ---- ---- - --	ex
			NE-SW	cones)	(Aso-4	100,000	1894	---- ---- ---- - --	ex
			fault with		glass	29,000(?)	1897	---- ---- ---- - --	ex
			associated		dominantly		1906	---- ---- ---- - --	Ex
			N-S exten		69-70 pct		1916	---- ---- ---- - --	ex
					SiO_2)		1920	---- ---- ---- - --	ex
							1925-85+	EEEE FC-- --EE E --	ex,Ex,pex
HAKONE									
08-03-02	35.22N	10	Compr	Strato	R = 54-59	180,000	1786	xx-- ---- ---- - --	none
(Kamiyama)	139.02E				C = 58-63	66,000	1917	xx-- ---- ---- - --	none
						49,000	1920	xx-- ---- ---- - --	none
							1923	---- ---- ---x - --	none

PART 2: QUICK REFERENCE TO PARTS 3 AND 4 (continued)
HISTORICAL UNREST AT LARGE CALDERAS AND SELECTED NONCALDERA SETTINGS
See inside back cover for explanation and abbreviations

CAVW number (active vent)	Latitude Longitude (degrees)	Diameter (km)	Local tectonic setting	Precaldera edifice	SiO_2 content (wt pct)	Age of collapse (yr B.P.)	Date of unrest	Type and duration of unrest ESTU STHF MGTF H Te	Eruption type
HAKONE (continued)									
							1933-35	EE-- ---- --EE - --	none
							1940-46	FF-- ---- ---F F --	none
							1952-53	EE-- ---- --?? ? --	none
							1959-60	EE-- ---- ---- - --	none
							1966-67	EE-- ---- --FF F --	none
							1972-73	EE-- ---- --FF F --	none
							1981-83+	FF-- ---- ---- - --	none
ON-TAKE									
08-03-04 (On-take)	35.90N 137.48E	4 x 5	Compr	Strato	R = 51-69 C = 69?	between 70,000 and 90,000	1976-84	FF-F ---- ---F F UQ	pex
ONIKOBE and NARUGO									
08-03-20 (Narugo) 08-03-20A (Onikobe) 08-03-21 (Kurikoma)	38.73N 140.73E 38.80N 140.67E 38.95N 140.78E	10 (Onikobe) 5? (Narugo)			C = 56-65 (Onikobe) R = 69-71 (Narugo) R = 55-58 (Kurikoma)	2.3×10^6	1716-36	---- ---- ---- - --	pex
							1744	---- ---- ---- - --	pex
							1783	---- ---- ---- - --	pex
							1944	---- ---- ---- - --	pex
							1946	---- ---- ---- - --	pex
							1950	---- ---- ---- - --	pex
							1957	x--- ---- ---- - --	none
TOWADA									
08-03-271	40.50N 140.92E	11	Compr	Several small strato-volcanoes	C = 64-69 R = 68	11-13,000; 5,400; A.D. 915?	A.D. 915?	---- ---- ---- - --	EX,dm
IZU-OSHIMA									
08-04-01 (Izu-Oshima)	34.73N 139.38E	4	Compr	LL-strat	R = 52 C = 52-54	550±250	684	---- ---- ---- - --	lf?
							1112	---- ---- ---- - --	Ex,lf
							1338	---- ---- ---- - --	EX,lf
							1415	---- ---- ---- - --	Ex
							1416	---- ---- ---- - --	Ex
							1421	---- ---- ---- - --	Ex,lf,pex
							1442	---- ---- ---- - --	unknown

PART 2: QUICK REFERENCE TO PARTS 3 AND 4 (continued)
HISTORICAL UNREST AT LARGE CALDERAS AND SELECTED NONCALDERA SETTINGS
See inside back cover for explanation and abbreviations

CAVW number (active vent)	Latitude Longitude (degrees)	Diameter (km)	Local tectonic setting	Precaldera edifice	SiO_2 content (wt pct)	Age of collapse (yr B.P.)	Date of unrest	Type and duration of unrest ESTU STHF MGTF H Te	Eruption type
IZU-OSHIMA (continued)									
							1552	Z--- ---- ---- - --	Ex,lf
							1600/01	---- ---- ---- - --	ex
							1612/13	---- ---- ---- - --	ex
							1636-38	---- ---- ---- - --	ex
							1684-90	Y--- ---- ---- - --	Ex,lf
							1695	---- ---- ---- - --	unknown
							1777-92	---- ---- ---- - --	EX,lf
							1803	---- ---- ---- - --	ex
							1822-24	---- ---- ---- - --	ex
							1837	---- ---- ---- - --	ex
							1846	---- ---- ---- - --	ex
							1870	---- ---- ---- - --	ex
							1876-77	---? ---- ---- - --	ex,lf
							1910	---- ?--- ---- - --	ex
							1912-14	---F ---- ---- - --	ex,Lf
							1915	---F F--- ---- - --	ex
							1919-20	---F F--- ---- - UQ?	lf
							1922-23	---F ---- ---- - UQ	ex,pex,lf
							1928	---- F--- ---- - --	ex
							1933-35	---F ---- ---- - --	lf
							1938-40	EE-G ---- ---- - --	ex,lf
							1950-51	??YG ---- Z--- - UQ	ex,lf,ll
							1953-54	---- ---- E--- - --	ex,lf
							1955-71	BB-F F--- ---- - UU	ex,lf
							1974	---F ---- ---- - UQ	ex
							1977-78	EE-- ---- FF-- - --	none
							1983-88+	FFEF FEEC ---N - U?	Ex,lf
BAYONNAISE									
08-04-07 (Myojin-sho)	31.95N 140.01E (Myojin-sho)	10	Compr	Submarine strato?	R = 47-69 C = silicic		1896	---- ---- ---- - --	sub
							1906	---- ---- ---- - --	sub,ex
							1915	---- ---- ---- - --	sub
							1934	---- ---- ---- - --	sub
							1946	---- ---- ---- - --	sub,dm
							1952-54	---- ---- ---- - --	sub,ex,dm
							1955	---- ---- ---- - --	sub
							1957-59	---- ---- ---- - --	sub
							1960	---- ---- ---- - --	sub
							1970	---- ---- ---- - --	sub,ex
							1979?	---- ---- ---- - --	sub
							1980?	---- ---- ---- - --	sub

PART 2: QUICK REFERENCE TO PARTS 3 AND 4 (continued)
HISTORICAL UNREST AT LARGE CALDERAS AND SELECTED NONCALDERA SETTINGS

See inside back cover for explanation and abbreviations

CAVW number (active vent)	Latitude Longitude (degrees)	Diameter (km)	Local tectonic setting	Precaldera edifice	SiO_2 content (wt pct)	Age of collapse (yr B.P.)	Date of unrest	Type and duration of unrest ESTU STHF MGTF H Te	Eruption type
IWO-JIMA									
08-04-12	24.75N 141.33E	9	Compr	Strato	R = 58-61 C = ?	≥2,600	1889	---G ---- ---- - --	pex
							1922	---G ---- ---- - --	pex
							1935	---G ---- ---- - --	pex
							1944	---F ---- ---- - --	pex
							1955-57	---G ---G --F- F --	pex
							1967-69	---G ---- ---- - --	pex
							1975-76	---F ---- ---- - --	pex
							1978	---F ---- ---- - --	pex
							1980	xx-F -x-- ---- - --	pex
							1982	CC-F -x-- ---- - --	pex,sub
NORTH PAGAN									
08-04-17 (Mt. Pagan)	18.13N 145.80E	5 x 6		Strato	R = 48-52 C = mafic?		1669?	---- ---- ---- - --	Ex,1f
							1820's	---- ---- ---- - --	ex
							1872-73	---- ---- ---- - --	Ex,1f
							1909	---- ---- ---- - --	ex
							1917	x--- ---- ---- - --	ex
							1923-25	x--- ---- ---- - --	ex,1f
							1981-85	E--Y -Y-C ---C - --	Ex,ex,1f
							1987-88?	---- ---- ---- - --	ex?
TOYA									
08-05-03 (Usu)	42.53N 140.83E	10	Compr	?	R = 68-72 C = 73 precaldera = 49-53	110,000	1663	C--- ---- ---- - --	EX,PEX,dm
							1769	x--- ---- ---- - --	EX,dm
							1822	C--- ---- ---- - --	EX,cdm
							1853	D--- ---- ---- - --	EX,dm
							1910	C--C ---Y ---Z Z QU?	pex,cdm
							1932	---- x--- ---- - --	none
							1943-45	EE-E ZE-E Z--- E --	pex,cdm,dm
							1977-82	CCYA F--A --FA F --	Ex,ex,cdm
SHIKOTSU									
08-05-04 (Tarumai)	42.68N 141.38E	13 x 15	NW-SE line in caldera controls 3 vents	?	R = 55-62 C = 52-74	30,000-32,000	1667	---- ---- ---- - --	EX
							1739	x--- ---- ---- - --	EX
							1804-17	---- ---- ---- - --	Ex
							1867	x--- ---- ---- - --	ex,dm
							1874	---- ---- ---- - UQ	Ex

PART 2: QUICK REFERENCE TO PARTS 3 AND 4 (continued)
HISTORICAL UNREST AT LARGE CALDERAS AND SELECTED NONCALDERA SETTINGS
See inside back cover for explanation and abbreviations

CAVW number (active vent)	Latitude Longitude (degrees)	Diameter (km)	Local tectonic setting	Precaldera edifice	SiO_2 content (wt pct)	Age of collapse (yr B.P.)	Date of unrest	Type and duration of unrest ESTU STHF MGTF H Te	Eruption type
SHIKOTSU (continued)									
							1883-87	---- ---- ---- - --	ex,pex
							1894	---- ---- ---- - --	ex
							1909	x--- ---- ---- - QU?	Ex,dm
							1917-36	x--- --DD ---D - --	ex,pex
							1944-55	x--- ---- ---- - --	ex,pex?
							1967-81	FF-- --N- --?E - --	ex,pex
							1988	DD-- ---- ---? - --	none yet
DAISETSU-TOKACHI GRABEN									
08-05-05	43.42N	20 x 30?	Exten	Pyroclastic	R = 47-61		1670	---- ---- ---- - --	lf
(Tokachi-dake)	142.68E	(Daisetsu)		plateau	C = 72		1854-57+	---- ---- ---? - --	ex
08-05-06	(Tokachi)						1887	---- ---- ---- - --	ex
(Daisetsu)	43.68N						1889	---- ---- ---- - --	ex
	142.88E						1923-28	YY-- ---- --YY Y QU	Ex,ls,mf
	(Daisetsu)						1933	---- ---- ---x - --	none
							1936-37	---- ---- ---x - --	none
							1946-47	---- ---- ---x - --	none
							1952-62	EE-- ---F --EF B QU	pex, Ex
							1968-69	EE-- ---- ---- - QU	none
							1984-88	DDE- ---- --E- D --	pex
AKAN									
08-05-07	43.38N	24 x 13	Exten?		R = 52-63	>31,500	ca. 1800	---- ---- ---- - --	ex?
(Me-akan)	144.02E				C = 67		1927	E--- ---- ---- - --	pex?
							1933	x--- ---- ---- - QU	none
							1937	---- ---- ---C - --	none
							1951-66	ECC- ---- --EE E UU	pex
							1973	CC-- ---- ---- - UU	none
							1982	DD-- ---- ---- - QU	none
							1987-88+	DDC- ---- --FF - --	pex
KUTCHARO (including MASHU)									
08-05-08	43.55N	20x26	NE-SW	?	R = 58-73	Approxi-	1938	xx-- ---x ---x x =?	none
(Atosanupuri)	144.43E	(Kutcharo)	alignment		C = 71-74	mately	1959	xx-- ---- ---- - =?	none
		6	of vents			30,000	1961	xx-- ---- ---- - --	none
		(Mashu)				(Kutcharo)	1963	xx-- ---- ---- - --	none
						7,000	1965	xx-- ---- ---- - --	none
						(Mashu)	1967	xx-- ---- ---- - --	none
							1968	xx-- ---- ---- - --	none
							1971	xx-- ---- ---- - --	none
							1975	xx-- ---- ---- - --	none

PART 2: QUICK REFERENCE TO PARTS 3 AND 4 (continued)
HISTORICAL UNREST AT LARGE CALDERAS AND SELECTED NONCALDERA SETTINGS
See inside back cover for explanation and abbreviations

CAVW number (active vent)	Latitude Longitude (degrees)	Diameter (km)	Local tectonic setting	Precaldera edifice	SiO_2 content (wt pct)	Age of collapse (yr B.P.)	Date of unrest	Type and duration of unrest ESTU STHF MGTF H Te	Eruption type
GOLOVNIN									
09-00-01 (Golovnin)	43.82N 145.53E	7 (inner, 4)	Compr	Strato	R = 58-65 C = 66-72		1848?	---- ---- ---- - --	ex
MENDELEEV									
09-00-02 (Mendeleev)	43.90N 145.70E	6 (inner, 3)	"on 2 volcano-tectonic lines"	Strato (2 or more)	R = 53-66 C = 67?	pre-Würm inter-stadial	1880	---- ---- ---- - --	pex
							1900?	---- ---- ---- - --	pex
							1988	xx-- ---- ---x - --	none
MEDVEZHII									
09-00-10 (Kudriavy)	45.38N 148.80E	8	Compr	Strato	R = 49-55 C = 63-67	pre-glacial	1778 or 1779	---- ---- ---- - --	ex
							1883	---- ---- ---- - --	ex,lf
							1946	---- ---- ---- - --	pex
							1958	---- ---- ---? - --	unknown
CHIRPOI									
09-00-15 (Snow= S)	46.52N 150.88E	8-9	Southern and western parts cut by faults; compr	Ol-px basalt cone, submarine	R = 59-60 C = 68?		ca. 1712 (C)	---- ---- ---- - --	EX?
09-00-16 (Cherny= C)							between 1770 and 1810 (S)	---- ---- ---- - QU?	ex,lf
							1811 (S)	---- ---- ---- - --	ex
							1854 (?)	---- ---- ---- - --	ex
							1857 (C)	---- ---- ---- - --	Ex
							1879 (S)	---- ---- ---- - --	ex,lf
							1960 (S)	---- ---- ---- - --	ex
ZAVARITSKY									
09-00-18 (Zavaritsky)	46.92N 151.95E	10 (outer) 8 (middle) 3 (inner) (concentric)	Fault, west side of outer caldera; compr	Strato (3 generations)	R = 54-59 C = 64-72	Second caldera, inter-glacial	1916/1930	---- ---- ---- - --	ex,dm
							1957	B--- ---- Z--E E --	Ex,dm,lf

PART 2: QUICK REFERENCE TO PARTS 3 AND 4 (continued)
HISTORICAL UNREST AT LARGE CALDERAS AND SELECTED NONCALDERA SETTINGS
See inside back cover for explanation and abbreviations

CAVW number (active vent)	Latitude Longitude (degrees)	Diameter (km)	Local tectonic setting	Precaldera edifice	SiO_2 content (wt pct)	Age of collapse (yr B.P.)	Date of unrest	Type and duration of unrest ESTU STHF MGTF H Te	Eruption type
KETOI									
09-00-20 (Pallas Peak)	47.35N 152.48E	5 (outer) 1.5 (inner)	Compr	Shield	R= 58 C= silicic	pre-glacial?	1843-46	---- ---- ---- - QU?	ex,lf
							1924	---- ---- ---- - --	ex
							1960	---- ---- ---- - --	ex
RASSHUA									
09-00-22 (Rasshua)	47.77N 153.02E	6	Compr	Strato	R = 56± C = ?	pre-latest glaciation	1810?	---- ---- ---- - --	unknown
							1846	---- ---- ---- - --	Ex
							1946	---- ---- --?C - --	none?
							1957	---- ---- --?x - --	pex?
TAO-RUSYR									
09-00-31 (Krenitzyn Peak)	49.36N 154.71E	7.5	Inter-section of two volcano-tectonic lines	LL-strat	R = 49-64 C = 59-67	7,040	1846	---- ---- ---? - --	none
							1879	---- ---- ---? - --	none
							1952	?--- ---- C--- - QU	Ex,dm
NEMO PEAK									
09-00-32 (Nemo Peak)	49.57N 154.81E	11 each (older two) 5 (younger)	Two inter-secting volcano-tectonic lines; compr	LL-strat	R = 54-59 C = 64	inter-glacial (younger); pre-glacial (older)	early 1700's	---- ---- ---- - --	ex
							1906	---- ---- ---- - --	ex,dm?
							1938	---- ---- ---- - --	ex
KURILE LAKE/PAUZHETKA									
10-00-03 (Iliinsky)	51.45N 157.08E	20x25 (P) 10 (KL)	Inter-section of N-S graben and E-W strike-slip fault	None?	R = low 50's -63 C = d (KL); silicic (P)	8,000-8,340 (KL) 10,000-220,000 (P)	1901	---- ---- ---- - --	pex, Pex
							1973-77+	FF-- ---- ---- - --	none

PART 2: QUICK REFERENCE TO PARTS 3 AND 4 (continued)
HISTORICAL UNREST AT LARGE CALDERAS AND SELECTED NONCALDERA SETTINGS
See inside back cover for explanation and abbreviations

CAVW number (active vent)	Latitude Longitude (degrees)	Diameter (km)	Local tectonic setting	Precaldera edifice	SiO_2 content (wt pct)	Age of collapse (yr B.P.)	Date of unrest	Type and duration of unrest ESTU STHF MGTF H Te	Eruption type
KSUDACH									
10-00-05 (Stübel Cone)	51.83N 157.52E	7 x 13 (outer); 7 x 6, 4 x 5 4 x 4 4 x 4 (nested)	Inter-section of NE-trending graben and strike-slip fault	Shield	R = 60-69 C = d	>40,000 (outer); 30,000-40,000; 15,000; 8,800-3,600; 2,500-1,700 (nested)	1907	---- ---- ---- - --	EX
GORELY KHREBET									
10-00-07 (Gorely Khrebet)	52.45N 158.12E	10 x 13	WNW fracture normal to volcanic chain; compr	LL-strat	R = 51-71 (mostly b and a) C = 60-66	Upper Pleis-tocene	1828	---- ---- ---- - --	ex
							1832	---- ---- ---- - --	ex
							1855	---- ---- ---- - --	ex
							1869?	---- ---- ---- - --	unknown
							1929-31	---- ---- ---- - --	Ex
							1947	---- ---- ---x - --	ex?
							1960-61	---- ---- --EE - --	ex
							1977-81	---- ---- --FF F --	Pex,ex
							1984-85	---- ---- ---- - --	pex
OPALA									
10-00-08 (Opala)	52.43N 157.45E	10 x 12	E-W faults at north end of N-S graben	Shield? LL-strat?	R = 52-66 C = silicic?	22,000	1776	---- ---- ---- - --	ex
							1827	---- ---- ---? - --	none
							1854	---- ---- ---? - --	none
							1894	---- ---- ---? - --	none

PART 2: QUICK REFERENCE TO PARTS 3 AND 4 (continued)
HISTORICAL UNREST AT LARGE CALDERAS AND SELECTED NONCALDERA SETTINGS
See inside back cover for explanation and abbreviations

CAVW number (active vent)	Latitude Longitude (degrees)	Diameter (km)	Local tectonic setting	Precaldera edifice	SiO_2 content (wt pct)	Age of collapse (yr B.P.)	Date of unrest	Type and duration of unrest ESTU STHF MGTF H Te	Eruption type
ZHUPANOVSKY (Z) COMPLEX Including KARYMSKY (K), POLOVINKA (P), AKADEMII NAUK (AN), ODNOBOKY (O), STENA-SOBOLINY (SS), And MALY SEMIACHIK (MS) Calderas									
10-00-13	54.07N	Z = 40	Zone of	Shield?	K: R = a-d	8,400-	1771 (K)	---- ---- ---- - --	ex
(Karymsky)	159.60E		E-W right-		C = d	7,400	1804 (MS)	---- x--- ---- - --	Ex
	(K)	K = 5	lateral			(K)	1830 (K)	---- ---- ---- - --	ex
			strike-				1851-52 (MS)	---- ---- ---- - --	ex
		P = 15	slip		P: C = a-d	mid-	1852 (K)	---- ---- ---- - --	ex
10-00-14			faulting;			Pleistocene	1854 (K,MS?)	---- ---- ---- - --	ex
(Maly	54.12N	O = 4x5	compr			(P)	1908 (K)	---- ---- ---- - --	ex,LF
Semiachik)	159.88E						1911-12 (K)	---- ---- ---- - --	ex
	(MS)	AN = 3x5			AN: C =	8,100	1915 (K)	---- ---- ---- - --	ex
					silicic	(AN)	1921 (K)	---- ---- ---- - --	ex
		SS = 15x21					1923 (K)	---- ---- ---- - --	ex
					MS: R = mafic	10,000-	1925 (K)	---- ---- ---- - --	ex
		MS = 6x7			C = 61-64	12,000	1929 (K)	---- ---- ---- - --	ex
						(MS)	1932-35 (K)	---- ---- ---- - --	ex,lf
							1938 (K)	---- ---- ---- - --	pex
					SS: C = a-d	mid-	1940 (K)	---- ---- ---- - QU	ex
						Pleistocene	1943-47 (K)	---- ---- ---- - --	ex,lf
						(SS)	1945-46 (MS)	---- ---- ---- - QU	ex
							1952 (K)	---- ---- ---- - QU	ex
							1952 (MS)	---- ---- ---- - QU	ex
							1955-57 (K)	?--- ---- ---- - --	ex
							1960-67 (K)	E--- ---- ---- - QU	Ex,ex,lf
							1970-82 (K)	YY-Y YY-- ---x E --	Ex,ex,lf,dm
							1985 (K)	---- ---- ---- - --	lf
UZON-GEYZERNAYA									
10-00-161	54.62N	7 x 18	Intersect.	LL-strat?	R = 54-56+,	mid-	19th cent.	---- ---- ---x - --	none
(Unana)	159.70E	Uzon	of E-W	(Uzon)	some 65-71	Pleist-	20th cent.	xx-- ---- ---x - --	none
		and	strike-			tocene			
10-00-17	54.53N	Geyzer-	slip fault	Silicic	C = 68				
(Uzon)	160.08E	naya	zone and	edifice	(Uzon)				
		combined	NNE-SSW	(Geyzernaya)					
			graben						
KRASHENINNIKOV									
10-00-19	54.62N	9 x 11		Shield	C = d	39,600	1963?	---- ---- ---x - --	none
(Krasheninnikov)	160.60E				R = d,a,b				

PART 2: QUICK REFERENCE TO PARTS 3 AND 4 (continued)
HISTORICAL UNREST AT LARGE CALDERAS AND SELECTED NONCALDERA SETTINGS
See inside back cover for explanation and abbreviations

CAVW number (active vent)	Latitude Longitude (degrees)	Diameter (km)	Local tectonic setting	Precaldera edifice	SiO_2 content (wt pct)	Age of collapse (yr B.P.)	Date of unrest	Type and duration of unrest ESTU STHF MGTF H Te	Eruption type
TIEN-CHI									
10-03-02 (Changpaishan including = Baegdusan = Pai-tou-shan = P'aektu-san = Hakuto-san)	41.98N 128.08E	5	Exten?		R = 49-? C = 71	800-900	1597? 1702?	---- ---- ---- - -- ---- ---- ---- - --	unknown unknown
LITTLE SITKIN, CALDERA ONE AND CALDERA TWO									
11-01-05 (Little Sitkin)	51.95N 178.53E	5 (outer) 3 (inner)	Compr	Strato	R = 55-69 C = 58 (One) = 60-65 (Two)		1776 1828-30	---- ---- ---- - -- ---- ---- ---- - --	unknown ex
SEMISOPOCHNOI									
11-01-06 (Cerberus)	51.93N 179.60E	8	Compr	Strato	R = 56-63 C = 61-66	post- glacial	1772? 1790? 1792? 1830? 1873 1987	---- ---- ---- - -- ---- ---- ---- - -- ---- ---- ---- - -- ---- ---- ---- - -- ---- ---- ---- - -- ---- ---- ---- - --	"smoke" "smoke" "smoke" "smoke" unknown ex
TANAGA									
11-01-08 (Tanaga)	51.88N 178.13W	11	Compr	Strato	R = 52+		1763-70 1791? 1829 1914	---- ---- ---- - -- ---- ---- ---- - -- ---- ---- ---- - -- ---- ---- ---- - --	"active" "smoke" ex?, "smoke" lf
KANATON									
11-01-11 (Kanaga)	51.93N 177.15W	5.5	Compr	Strato	R = 55-60 C = 64		1763? 1768 1783-87 1790-91	---- ---- ---- - -- ---- ---- ---- - -- ---- ---- ---- - -- ---- ---- ---x - --	"active" "active" "active" "smoke"

PART 2: QUICK REFERENCE TO PARTS 3 AND 4 (continued)
HISTORICAL UNREST AT LARGE CALDERAS AND SELECTED NONCALDERA SETTINGS
See inside back cover for explanation and abbreviations

CAVW number (active vent)	Latitude Longitude (degrees)	Diameter (km)	Local tectonic setting	Precaldera edifice	SiO_2 content (wt pct)	Age of collapse (yr B.P.)	Date of unrest	Type and duration of unrest ESTU STHF MGTF H Te	Eruption type
KANATON (continued)									
							1827-29?	---- ---- ---- - --	"smoke"
							1904	---- ---- ---- - --	lf
							1906	x--- ---- ---- - --	lf
							1933	---- ---- ---- - --	unknown
							1942	---- ---- ---- - --	unknown
							1951	---- ---- ---x - --	ex?
OKMOK									
11-01-29 (Okmok)	53.42N 168.13W	10 (both calderas)	Compr	LL-strat	R = 51-54 C = 54-66	8,250, <2,400	1805	---- ---- ---- - --	unknown
							1817-20	---- ---- ---- - --	Ex
							1824-30	---- ---- ---- - --	unknown
							1878	Y--- ---- ---- - --	unknown, + tsunami?
							1899	---- ---- ---- - --	Ex
							1931	x--- ---- ---x - --	"smoke,"lf?
							1936?	---- ---- ---- - --	"smoke"
							1938?	---- ---- ---- - --	unknown
							1943	---- ---- ---- - --	ex
							1944-45	C--- ---- ---- - --	ex,lf
							1958	---- ---- ---- - --	lf
							1960-61	---- ---- ---- - --	Ex
							1983	---- ---- ---- - --	ex
							1986-88	---- ---- ---x - QU?	ex
FISHER									
11-01-35 (Fisher)	54.67N 164.35W	11 x 18	Compr	Strato	R = mafic? C = 67-68, minor 52-53	9,100	1795?	---- ---- ---- - --	ex
							1826-27	---- ---- ---- - --	Ex?
EMMONS LAKE									
11-02-03 (Pavlof)	55.35N 162.02W	19 x 10	Compr	Strato?	R = mafic C = 72	>10,000; possibly between 87,000 and 105,000	1762-86 PS?	---- ---- ---- - --	Ex,ex
							1790	---- ---- ---- - --	unknown
11-02-04 (Pavlof Sister=PS)	55.45N 161.87W						1817	---- ---- ---- - --	ex
							ca.1825?	---- ---- ---- - --	unknown
							1838	---- ---- ---- - --	"smoke"
							1844 or 1845	---- ---- ---- - ??	ex
							1846	---- ---- ---- - --	ex,lf

PART 2: QUICK REFERENCE TO PARTS 3 AND 4 (continued)
HISTORICAL UNREST AT LARGE CALDERAS AND SELECTED NONCALDERA SETTINGS
See inside back cover for explanation and abbreviations

CAVW number (active vent)	Latitude Longitude (degrees)	Diameter (km)	Local tectonic setting	Precaldera edifice	SiO_2 content (wt pct)	Age of collapse (yr B.P.)	Date of unrest	Type and duration of unrest ESTU	STHF	MGTF	H	Te	Eruption type
EMMONS LAKE (continued)													
							1852?	----	----	----	-	--	"smoke"
							1866?	----	----	----	-	UQ?	ex
							1880?	----	----	----	-	--	unknown
							1886	----	----	----	-	--	unknown
							1892	----	----	----	-	--	ex
							1894	----	----	----	-	--	ex
							1901	----	----	----	-	--	ex
							1906-11	YY--	---x	----	-	--	"smoke", Ex
							1914	----	----	----	-	--	ex
							1917	x---	----	----	-	--	ex
							1922-24	----	----	----	-	--	Ex, ex
							1929-31	----	----	----	-	--	ex
							1936-48	----	----	----	-	--	ex, Ex, lf?
							1950-53	----	----	----	-	--	ex
							1958	----	----	----	-	--	ex, lf
							1960-63	----	----	----	-	--	ex, lf
							1966	----	----	----	-	--	ex, lf
							1973-77+	C-Y-	----	----	-	UQ	Ex, ex, lf
							1980-83	D-Y-	----	----	-	UQ	Ex, ex, lf
							1986-88+	C-Y-	----	---x	-	--	Ex, ex
VENIAMINOF													
11-02-07 (Veniaminof)	56.17N 159.38W	10	Compr	Strato	R = 53-54 C = 65	3,700	1830-39	----	----	----	-	--	Ex?
							1852?	----	----	---?	-	--	"smoke"
							1874?	----	----	---?	-	--	"smoke"
							1892	----	----	----	-	--	Ex
							1930?	----	----	---?	-	--	"smoke"
							1939	----	----	----	-	--	ex
							1944	----	----	----	-	--	unknown
							1956	----	----	----	-	--	ex
							1983-84	Y-Y-	---Y	---Y	-	--	ex, lf
							1987	----	----	----	-	--	ex
ANIAKCHAK													
11-02-09 (Aniakchak)	56.88N 158.15W	9.5	Compr	Strato	R = 64-66 C = 68, late andesite	3,400	1931	----	----	----	-	--	Ex, dm

PART 2: QUICK REFERENCE TO PARTS 3 AND 4 (continued)
HISTORICAL UNREST AT LARGE CALDERAS AND SELECTED NONCALDERA SETTINGS
See inside back cover for explanation and abbreviations

CAVW number (active vent)	Latitude Longitude (degrees)	Diameter (km)	Local tectonic setting	Precaldera edifice	SiO_2 content (wt pct)	Age of collapse (yr B.P.)	Date of unrest	Type and duration of unrest ESTU STHF MGTF H Te	Eruption type
UGASHIK									
11-02-13A (Peulik=P)	57.75N 156.35W	5 x 6	Compr	Strato	R = b	40,000- 171,000	1814 (P) 1852? (P) 1977 (U)	---- ---- ---- - -- ---- ---- ---? - -- xx-- ---- ---- x --	Ex unknown Ex
11-02-13B (Ukinrek Maars =U)	57.83N 158.51W								
KATMAI/NOVARUPTA									
11-02-16 (Novarupta)	58.27N 155.16W	2 x 3	Compr	None?	R = C= r-a	A.D. 1912	1912	C--- Z--Z ---Z - --	EX,dm
11-02-19 (Katmai)	55.27N 154.98W	3 x 4		Strato	R = none C = d	A.D. 1912	1912 1929-31?	C--- Z--- ---Z - -- ---- ---- ---x - --	Pex none?
11-02-17 (Griggs)	58.35N 155.12W				R = ?		1965?	---- ---- ---x - --	none
11-02-18 (Trident)	58.23N 155.12W				R = a		1912-13? 1949-68 ca. 1974-75	---- ---- ---- - -- xx-- ---- ---F - -- ---- ---- ---- - --	ex? ex,Ex,lf Ex,lf
11-02-14 (Martin)	58.15N 155.38W				R = 63-64		1951? 1953?	---- ---- ---- - -- ---- ---- ---- - --	ex? ex?
11-02-15 (Mageik)	58.20N 155.25W						1912 1927? 1929? 1936? 1946?	---- ---- ---- - -- ---- ---- ---- - -- ---- ---- ---- - -- ---- ---- ---- - -- ---- ---- ---? - --	landsl ex? ex? Ex? none?
WRANGELL									
11-05-02 Mt. Wrangell	62.00N 144.02W	4 x 6 (modern) 15 (ancestral)		Shield	R = 58-62 C = a?	late Quaternary	1819? 1884? 1899 1902 1907 1965-present	---- ---- ---- - -- ---- ---- ---- - -- ---- ---- ---x - QU? ---- ---- ---- - -- ---- ---- --xx x -- xx-- ---- --xx - QU?	ex? unknown none ex none none yet

PART 2: QUICK REFERENCE TO PARTS 3 AND 4 (continued)
HISTORICAL UNREST AT LARGE CALDERAS AND SELECTED NONCALDERA SETTINGS

See inside back cover for explanation and abbreviations

CAVW number (active vent)	Latitude Longitude (degrees)	Diameter (km)	Local tectonic setting	Precaldera edifice	SiO_2 content (wt pct)	Age of collapse (yr B.P.)	Date of unrest	Type and duration of unrest ESTU STHF MGTF H Te	Eruption type
LONG VALLEY									
12-03-08, (Mono Craters)	37.70N 118.87W	17 x 32	Exten	Domes	R = 73-74 C = 75-77	700,000	1889	x--- ---- ---- - =?	none
12-03-09 (Inyo Craters)							1927	x--- ---- ---- - =?	none
							1941-42	x--- ---- ---- - =?	none
							1978-present	FFxF -FFC FFFF F UU	none yet
YELLOWSTONE									
12-05-01	44.58N 110.53W	85 x 45	E-W exten with complica-tions	None?	R = 70-77 C = 76-77	630,000	late 19th century and continuing	GG-G FG?x -GFx F QQ	none
KILAUEA									
13-02-01 (Kilauea)	19.42N 155.29W	8 (outer) 4 (inner)	Hotspot	Shield	R = 47-51 C = 47-51	ca. 1,500 200-250	1750(?)	---- ---- ---- - --	lf
							1790(?)	---- x--- ---- - --	lf, P-Ex
							1823	---- x--- ---- - --	lf
							1832	---- x--- ---- - --	lf
							1840	---- x--- ---- - --	lf
							1868	---- x--- ---- - UU	lf
							1877	---- ---- ---- - --	lf
							1884	---- ---- ---- - --	sub lf?
							1885	---- ---- ---- - --	lf, ll
							1886	---- x--- ---- - --	sub lf?
							1891	---- x--- ---- - --	none?
							Mar 1894	---- ---- ---- - --	lf
							Jul 1894	---- x--- ---- - --	lf
							1916	---- x--- ---- - --	none?
							1918	---- ---- ---- - --	lf
							Feb 1919	---- ---- ---- - --	lf, ll
							Dec 1919	---- x--- ---- - --	lf, ll
							1921	---- ---- ---- - --	lf
							1922	---- x--- ---- - --	lf
							1923	---- x--- ---- - --	lf
							May 1924	EED- ED-D ---- - --	pex
							Jul 1924	---- ---- ---- - --	lf
							1927	x-B- ---- ---- - --	lf, ll
							Feb 1929	B-x- -E-- ---- - --	lf, ll

PART 2: QUICK REFERENCE TO PARTS 3 AND 4 (continued)
HISTORICAL UNREST AT LARGE CALDERAS AND SELECTED NONCALDERA SETTINGS
See inside back cover for explanation and abbreviations

CAVW number (active vent)	Latitude Longitude (degrees)	Diameter (km)	Local tectonic setting	Precaldera edifice	SiO_2 content (wt pct)	Age of collapse (yr B.P.)	Date of unrest	Type and duration of unrest ESTU STHF MGTF H Te	Eruption type
KILAUEA (continued)									
							Jul 1929	x-B- -x-- ---- - --	lf
							1930	CBA- -C-- ---- - --	lf
							1931	D-Y- -E-- ---- - --	lf
							1934	B-B- -B-x ---- - --	lf
							Nov 1944	D--D xD-- ---- - --	none
							1950	x--E xE-x ---- - --	none
							1952	D--- -C-- ---- - --	lf,ll
							1954	EBB- -E-- ---- - --	lf
							1955	ECx- -C-x ---- - --	lf,cc
							1959	ECB- -E-- ---D - --	lf,cc
							1960	DCB- BD-B ---- - --	lf,cc
							Feb 1961	EBB- -E-- ---- - --	lf
							Mar 1961	D-Y- -D-- ---- - --	lf
							Jul 1961	D-Y- -E-- ---- - --	lf,ll
							Sep 1961	DBB- BE-Y ---- - --	lf
							1962	CBB- YE-- ---- - --	lf,ll
							May 1963	CxC- xD-x ---- - --	none
							Jul 1963	CCxx xC-x ---- - --	none
							Aug 1963	DDD- -E-x ---- - UU	lf,ll
							Oct 1963	B-B- BB-- ---- - --	lf
							Mar 1965	D-B- YEY- ---- - --	lf,ll
							Aug 1965	CC-- xD-- ---- - --	none
							Dec 1965	BBB- YE-Y ---- - --	lf
							1967-68	A-BF YFEY ---- - --	lf,ll
							Aug 1968	DCYx YDxx ---- - --	lf
							Oct 1968	BYBY YEx- ---x - --	lf
							1969-71	CYYE YEEY --YY - --	lf,ll
							1972-74	EYDY YEYY Y-YY - UU	lf,ll
							1975	DYY- YYxx -x-- - UU	lf
							Jun 1976	BB-- BBx- ---- - --	none
							Jul 1976	BB-- BB-- ---- - --	none
							Aug 1976	NN-x DDx- ---- - --	none
							Feb 1977	CBB- BBxx xx-- - --	none
							Sep 1977	BBB- BBxx -x-- - --	lf,cc
							May 1979	BB-- BB-- ---- - --	none
							Aug 1979	BBN- BB-- ---- - --	none
							Nov 1979	CCYY YDxx ---x - --	lf
							Mar 1980	CCY- YY-Y ---Y - --	lf (minor)
							Aug 1980	BB-x BB-- ---x - --	none
							Oct 1980	CBx- BB-- ---x - --	none
							Nov 1980	BB-- BB-- ---x - --	none
							Jan 1981	EE-- DDx- ---- - --	none
							Jun 1981	BBBx -B-- ---- - --	none
							Aug 1981	CCCx xCxx --xx - --	none

HISTORICAL UNREST AT LARGE CALDERAS AND SELECTED NONCALDERA SETTINGS

See inside back cover for explanation and abbreviations

CAVW number (active vent)	Latitude Longitude (degrees)	Diameter (km)	Local tectonic setting	Precaldera edifice	SiO_2 content (wt pct)	Age of collapse (yr B.P.)	Date of unrest	Type and duration of unrest ESTU STHF MGTF H Te	Eruption type
KILAUEA (continued)									
							Apr 1982	BBBY -BxY ---x - --	lf
							Jun 1982	CCC- CCxx ---C - --	none
							Sep 1982	BBBB -BxY ---Y - --	lf
							Dec 1982	BBN- BB-- --x- - --	none
							Jan 1983-88+	BBYY YYYY -xxx x --	lf,cc,ll
MAUNA LOA/MOKUAWEOWEO									
13-C2-02 (Mauna Loa)	19.48N 155.61W	6 x 8 (outer) 3 x 5 (inner)	Hotspot	Shield	R = 48-53 C = 48-53	< 590	1832	---- ---- ---- - --	lf
							1843	---- ---- ---- - --	Lf
							1849	---- ---- ---- - --	lf
							1851	---- ---- ---- - --	lf
							1852	---- ---- ---- - --	lf
							1855	---- ---- ---- - --	lf
							1859	---- ---- ---- - --	LF
							1865	---- ---- ---- - --	lf
							1868	---- ---- ---- - UU	lf
							1870	---- ---- ---- - --	lf
							1871	---- ---- ---- - --	lf
							1872	---- ---- ---- - --	lf
							1873	---- ---- ---- - --	lf
							1875	---- ---- ---- - --	lf
							1876	---- ---- ---- - --	lf
							1877	---- ---- ---- - --	lf
							1880	---- ---- ---- - --	Lf
							1887	---- ---- ---- - --	Lf
							1892	---- ---- ---- - --	lf
							1896	---- ---- ---- - --	lf
							1899	---- ---- ---- - --	lf
							1903	---- ---- ---- - --	lf
							1907	---- ---- ---- - --	lf
							1914	---- ---- ---- - --	lf
							1916	---- ---- ---- - --	lf
							1919	---- ---- ---- - --	Lf
							1926	E-B- -E-- ---- - --	lf
							1933	--Y- ---- ---Y - --	lf
							1935	B-Y- ---Y ---Y - --	lf
							1940	E-Y- -E-- ---Y - --	lf
							1942	E-Y- ---Y ---Y - --	lf
							1943	D-Y- ---- ---D - --	lf
							1949	Y-Y- -E-Y ---E - --	lf,cc
							1950	D-Y- -E-Y ---E - --	LF
							1975	FEYE YEEY ---Y - --	lf
							1984	FFYF YFFY -xEE - UU	Lf

PART 2: QUICK REFERENCE TO PARTS 3 AND 4 (continued)
HISTORICAL UNREST AT LARGE CALDERAS AND SELECTED NONCALDERA SETTINGS
See inside back cover for explanation and abbreviations

CAVW number (active vent)	Latitude Longitude (degrees)	Diameter (km)	Local tectonic setting	Precaldera edifice	SiO_2 content (wt pct)	Age of collapse (yr B.P.)	Date of unrest	Type and duration of unrest ESTU STHF MGTF H Te	Eruption type
ATITLAN									
14-02-06	14.59N	17 x 20	NW-SE	Strato	R = 51-56	84,000	1469	---- ---- ---- - --	ex
(V. Atitlán)	91.18W		right-	(several)	C = 72-76		1717-21	---- ---- ---- - --	"active"
			lateral				1826-28	---? ---- ---- - --	Ex,lf
			shear with				1833?	---? ---- ---- - --	ex
			N-S tension				1837	---? ---- ---- - --	ex
			cracks				1843?	---? ---- ---- - --	ex
							1852?-53	---? ---- ---- - --	Ex,ex
							1856?	---? ---- ---- - --	ex
AMATITLAN									
14-02-11	14.38N	14 x 16	NNE-SSW	?	R = 50-51	240,000	1565	x--- ---- ---- - --	Ex,lf?
(Pacaya)	90.60W		graben,		C = 68-75		1651	x--- ---- ---- - --	Ex
			transverse				1664-74	---- ---- ---- - --	Ex,ex
			break in				1690-99?	---- ---- ---x - --	ex
			volcanic				1775	N--- ---- ---- - --	ex,lf
			chain; inter-				1846	---- ---- ---- - --	ex
			section with				1891-1902	---- ---- ---x - --	none
			WNW-ESE trend				1961-88+	FFBx ---- -x-- - --	Ex,ex,lf
COATEPEQUE									
14-03-02,	13.87N	7 x 10	Compr,	Strato?	R = 51-62	10,000	1520 (SA)	---- ---- ---- - --	Ex
(Santa Ana=SA)	89.55W		astride		C = 70	to	1524 (SA)	---- ---- ---- - --	Ex
14-03-03			segment			40,000	1570? (SA)	---- ---- ---- - --	Ex
(Izalco=I)			boundary				1576 (SA)	---- ---- ---x - --	Ex
14-03-04							1874 (SA)	---- ---- ---x - --	Ex?
(San Marcelino=SM)							1878 (SA)	---- ---- ---- - --	ex
						Santa	1879 (SA)	---- ---- ---- - --	ex
						Ana	1880 (SA)	---- ---- ---- - --	Ex
							1882? (SA)	---- ---- ---- - --	ex
							1884 (SA)	---- ---- ---- - --	ex
							1904 (SA)	D--- ---- ---- - --	ex
							1920 (SA)	---- ---- ---- - --	ex
							1770-72 (I)	x--x ---- ---x - --	ex,lf
							1783 (I)	x--- ---- ---- - --	lf
							1793+ (I)	---- ---- ---- - --	ex,lf
							1798 (I)	---- ---- ---- - --	ex

PART 2: QUICK REFERENCE TO PARTS 3 AND 4 (continued)
HISTORICAL UNREST AT LARGE CALDERAS AND SELECTED NONCALDERA SETTINGS

See inside back cover for explanation and abbreviations

CAVW number (active vent)	Latitude Longitude (degrees)	Diameter (km)	Local tectonic setting	Precaldera edifice	SiO_2 content (wt pct)	Age of collapse (yr B.P.)	Date of unrest	Type and duration of unrest ESTU	STHF	MGTF	H	Te	Eruption type
COATEPEQUE (continued)													
							1802-03(I)	----	----	----	-	--	ex,lf
							1805-07(I)	----	----	----	-	--	ex
							1825? (I)	----	----	----	-	--	ex
							1836 (I)	----	----	----	-	--	ex
							1838 (I)	----	----	----	-	--	ex
							1840 (I)	----	----	----	-	--	ex
							1842 (I)	----	----	----	-	--	ex
							1844 (I)	----	----	----	-	--	ex,lf
						Izalco	1850 (I)	----	----	----	-	--	ex
							1854 (I)	----	----	----	-	--	ex,lf
							1856 (I)	----	----	----	-	--	ex,lf
							1858-60(I)	----	----	----	-	--	ex,lf
							1863-70(I)	----	----	---x	-	--	lf,ex
							1872-73(I)	----	----	----	-	--	ex
							1879-85(I)	----	----	----	-	--	ex,lf
							1887-1901(I)	----	----	----	-	--	ex,lf
							1902 (I)	----	----	----	-	QU?	ex,lf
							1903-05(I)	D---	----	----	-	--	ex
							1912-16(I)	x---	----	----	-	--	ex,lf
							1920-21(I)	N--A	A--x	---x	-	--	ex,lf
							1925-28?(I)	----	----	----	-	--	ex,lf
							1930-31(I)	----	----	----	-	--	lf,ex
							1933-34(I)	----	----	----	-	--	ex,lf
							1937-57(I)	----	----	----	-	--	ex,Ex,lf
							1966 (I)	----	----	-NZZ	-	--	lf
						San Marcelino	1600's(SM)	----	----	----	-	--	lf
							1722 (SM)	----	----	----	-	--	lf
ILOPANGO													
14-03-06 (Islas Quemadas)	13.67N 89.05W	8 x 11	Graben	?	R = 67 C = 63-68	A.D. 260	1765	xx--	----	----	-	--	none
							1769-70	xx--	----	----	-	--	none
							1873	---?	----	---x	-	QU	none
							1879-80	D--?	?---	----	-	--	dm,ex
							1884	----	----	---?	-	--	unknown

PART 2: QUICK REFERENCE TO PARTS 3 AND 4 (continued)
HISTORICAL UNREST AT LARGE CALDERAS AND SELECTED NONCALDERA SETTINGS
See inside back cover for explanation and abbreviations

CAVW number (active vent)	Latitude Longitude (degrees)	Diameter (km)	Local tectonic setting	Precaldera edifice	SiO_2 content (wt pct)	Age of collapse (yr B.P.)	Date of unrest	Type and duration of unrest ESTU STHF MGTF H Te	Eruption type
MASAYA									
14-04-10 (Masaya)	11.98N 86.15W	11 x 6	NW-SE-trending fissure	Shield	R = 51 C = 51-53	ca. 6,000; ca. 2,200	1524	---- ---- ---- - --	unknown
							1529-70	---- ---- ---- - --	ll
							1670	---- ---- ---- - --	Ex,lf
							1772-75	AA-- ---- ---- - --	lf,ex
							1852-59	EE-- ---- ---Y - --	ex,lf
							1902-06	D--- ---- ---Y - --	ex
							1913?	x--- ---- ---- - --	unknown
							1919-24	---- ---- ---Z - --	ex
							1946	---- ---- ---Z - --	lf
							1947-59	---- ---- ---F - --	none
							1965	---- ---- ---- - --	ex,lf
							1970-72+	---- ---- ---- - --	ex,ll
							1974-88+	x--- x--- ---Y - --	ll,ex
APOYO									
14-04-101 (Apoyo)	11.92N 86.03W	6.5	Exten	Shield	R = a C = 66	23,000 21,000	1974	DD-- ---- ---- - QU?	none
							1977	DD-- ---- ---- - --	none
							1978	DD-- ---- ---- - --	none
							1979	DD-- ---- ---- - --	none
							1981	DD-- ---- ---- - --	none
							1986	DD-- ---- ---- - QU?	none
RINCON DE LA VIEJA									
14-05-02 (Rincón de la Vieja)	10.83N 85.33W	15-20 (outer) 5 (inner)	Compr	Strato	R = 57-59 C = ?		1860-63	---- ---- ---- - --	ex
							1922	---- ---- ---C - --	ex
							1963	---- ---- ---x - --	none
							1965-70	xx-- ---- ---E - --	pex, Ex
							1983-87+	---- ---- ---- - --	p-ex
MIRAVALLES									
14-05-03	10.75N 85.15W	15 x 20	Compr; local extension, graben	Strato	R = 50-56 C = r	500,000 ±150,000	1946	xx-- ---- --?x - --	"steam eruption"

PART 2: QUICK REFERENCE TO PARTS 3 AND 4 (continued)
HISTORICAL UNREST AT LARGE CALDERAS AND SELECTED NONCALDERA SETTINGS
See inside back cover for explanation and abbreviations

CAVW number (active vent)	Latitude Longitude (degrees)	Diameter (km)	Local tectonic setting	Precaldera edifice	SiO_2 content (wt pct)	Age of collapse (yr B.P.)	Date of unrest	Type and duration of unrest ESTU STHF MGTF H Te	Eruption type
POAS									
14-05-04	10.20N	7 x 10	Compr	Strato	R = 50-63	<40,000;	1826-28	x--- ---- ---? - --	Pex,Ex?
(Poás)	84.22W	(outer)			C = 52-54	>7,540	1834	---- ---- ---- - --	Ex
		3					1838	---- ---- ---- - --	ex
		(inner)					1880	---- ---- ---- - --	Pex,Ex?
							1884	---- ---- ---- - --	pex
							1888-89	x--- ---- --xx x UQ/QU?	pex,geyser
							1899	---- ---- ---- x --	geyser
							1902-05	---- ---- ---- - --	geyser
							ca. 1907-08	---- ---- ---- - --	pex
							1910	x--- ---- ---- - UQ	Pex, Ex
							1912	?--- ---- ---- - QU	landsl
							1914-15	---- ---- ---- - --	pex,geyser
							1925	---- ---- ---- - --	pex
							1929?	---- ---- ---- - --	geyser?
							1946	---- ---- ---- - --	pex
							1952-57	B--- ---- ---- - --	geyser,pex,Ex
							1961	---- ---- ---- - --	pex
							1963-64	---- ---- ---- - --	pex, ex
							1967	---- ---- ---- - --	pex
							1969	---- ---- ---- - --	pex
							1972-73	---- ---- ---- - --	ex
							1974-75	---- ---- ---- - --	pex
							1976-80	---- ---- ---- - --	pex, ex
							1980-86	---- ---- --FF - --	none
							1987-88	x--- ---E --EE - --	pex
FERNANDINA									
15-03-01	0.37S	4 x 6.5	Intersection	Shield	R = 49	Renewed	1813-14	---- ---- ---- - --	lf,ex
(Fernandina)	91.55W		of lineaments		C = mafic	collapse	pre-1817	---- ---- ---- - --	ex,lf
			over "hotspot"			in A.D.	1819	---- ---- ---- - --	ex,lf
						1968	1825	---- ---- ---- - --	Ex,lf
							1846	---- ---- ---- - --	lf
							1888	---- ---- ---- - --	unknown
							1926-27	---- ---- ---- - --	lf
							1937	---- ---- ---- - --	lf
							1958	---- ---- ---- - --	lf,ll
							1961	---- ---- ---- - --	ex,lf
							1968	C--- Z--Z ---Z Z QU	ex,lf,EX
							1972	---- ---- ---- - --	ex,lf
							1973	---- ---- ---- - --	lf
							1974-76	E--F ---- ---- - --	none?
							1977-78	B--? ---- ---- - QU	ex,lf
							1981	---- ---- ---- - --	lf
							1984	---- ---- ---Y - --	lf,ex

PART 2: QUICK REFERENCE TO PARTS 3 AND 4 (continued)
HISTORICAL UNREST AT LARGE CALDERAS AND SELECTED NONCALDERA SETTINGS
See inside back cover for explanation and abbreviations

CAVW number (active vent)	Latitude Longitude (degrees)	Diameter (km)	Local tectonic setting	Precaldera edifice	SiO_2 content (wt pct)	Age of collapse (yr B.P.)	Date of unrest	Type and duration of unrest ESTU STHF MGTF H Te	Eruption type
V. WOLF									
15-03-02 (V. Wolf)	0.02N 91.35W	7 x 5.5	Intersection of lineaments over "hotspot"	Shield	R = mafic C = mafic		1797	---- ---- ---- - --	unknown
							1800	---- ---- ---- - --	ex
							1925-26	---- ---- ---- - --	lf
							1933	---- ---- ---- - --	lf
							1935	---- ---- ---- - --	unknown
							1938	---- ---- ---- - --	unknown
							1948	---- ---- ---- - --	ex,lf
							1963	---- ---- ---- - --	lf
							1973?	EC-- ---- ---- - QU	unknown
							1982	---- ---- ---Z - --	ex,lf
V. DARWIN									
15-03-03 (V. Darwin)	0.18S 91.28W	5	Intersection of lineaments over "hotspot"	Shield	R = 48 C = mafic		1813?	---- ---- ---- - --	ex
V. ALCEDO									
15-03-04 (V. Alcedo)	0.43S 91.12W	7	Intersection of lineaments over "hotspot"	Shield	R = mafic C = mafic		between 1945 and 1961	---- ---- ---- - --	lf
							1954	---x ---- ---- - --	none?
SIERRA NEGRA									
15-03-05 (Sierra Negra)	0.83S 91.17W	10.5 x 7	Intersection of lineaments over "hotspot"	Shield	R = 47-49 C = mafic		1813	---- ---- ---- - --	ex
							1817	---- ---- ---- - --	unknown
							1844	---- ---- ---- - --	lf
							1860	---- ---- ---- - --	ex
							1911 or 1912	---- ---- ---- - --	unknown
							1948	---- ---- ---- - --	ex,lf
							1953-54	---? ---- ---- - --	Ex,lf
							1957	---- ---- ---- - --	unknown
							1963	---- ---- ---- - --	ex,lf
							1979-80	A--- ---Y ---- - --	Ex,lf

PART 2: QUICK REFERENCE TO PARTS 3 AND 4 (continued)
HISTORICAL UNREST AT LARGE CALDERAS AND SELECTED NONCALDERA SETTINGS
See inside back cover for explanation and abbreviations

CAVW number (active vent)	Latitude Longitude (degrees)	Diameter (km)	Local tectonic setting	Precaldera edifice	SiO_2 content (wt pct)	Age of collapse (yr B.P.)	Date of unrest	Type and duration of unrest ESTU STHF MGTF H Te	Eruption type
CERRO AZUL									
15-03-06 (Cerro Azul)	00.90S 91.42W	4 x 5	Hotspot	Shield	R = mafic		1932	---- ---- ---- - --	unknown
					C = mafic		1940	---- ---- ---- - --	lf
							1943	---- ---- ---- - --	Ex, lf
							1948	---- ---- ---- - --	lf
							1949?	---- ---- ---- - --	lf
							1951	---- ---- ---- - --	unknown
							1959	---- ---- ---- - --	lf
							1979	---- ---- ---- - --	ex,lf
DIAMANTE									
15-07-021 (Maipo)	34.2S 69.8W	15 x 20	Compr		R = 60	450,000	1822?	---- ---- ---? - --	unknown
					C = 74-76	±60,000	1826?	---- ---- ---? - --	ex?
							1905?	---- ---- ---? - --	unknown
							1912?	---- ---- ---? - --	unknown
NEVADOS DE CHILLAN									
15-07-07 (Nevados de Chillán; V. Nuevo=N)	36.87S 71.38W	7 x 4.5	Compr		R = 60-69		1600's	---- ---- ---- - --	Ex
					C = ?		ca. 1750	---- ---- ---- - --	Ex
							1835	---- ---? ---? ? QU	none
							1861-62	---- ---- ---x - QU	lf,Ex
							1864-65	---- ---- ---- - --	Ex
							1872	---- ---- ---x - --	none
							1877	---- ---- ---x - --	none
							1883	---- ---- ---- - --	ex
							1891	---- ---- ---- - --	ex
							1898	---- ---- ---- - --	ex
							1906 (N)	---- ---- ---x - UQ	ex
							1923 (N)	---- ---- ---x - --	ex
							1929 (N)	---- ---- ---x - --	ex,pex?
							1934 (N)	---- ---- ---x - --	ex
							1945-47(N)	---- ---- ---x - --	ex?
							1965-87+(N)	---- ---- ---F - --	pex,ex,lf,dm

PART 2: QUICK REFERENCE TO PARTS 3 AND 4 (continued)
HISTORICAL UNREST AT LARGE CALDERAS AND SELECTED NONCALDERA SETTINGS
See inside back cover for explanation and abbreviations

CAVW number (active vent)	Latitude Longitude (degrees)	Diameter (km)	Local tectonic setting	Precaldera edifice	SiO_2 content (wt pct)	Age of collapse (yr B.P.)	Date of unrest	Type and duration of unrest ESTU STHF MGTF H Te	Eruption type
COPAHUE									
15-07-09 (Copahue)	37.85S 71.17W	ca. 10	Compr	Strato	silicic	Pleistocene	18th or 19th cent.	---- ---- ---- - --	unknown
							1867?	---- ---- ---- - --	unknown
							1937	---- ---- ---- - --	unknown
							1940-45	---- ---- --xx x --	none
VILLARRICA									
15-07-12 (Villarrica)	39.42S 71.95W	9 x 6 (outer) 2.2 (inner)	Compr	Strato	R = mafic C = 68-71		1558	---- ---- ---- - --	ex
							1575	---- ---- ---- - --	Ex?
							1640	---- ---- ---- - --	ex,lf
							1647	---- ---- ---- - --	lf,Ex
							1657	---- ---- ---- - --	unknown
							1730	---- ---- ---- - --	unknown
							1737	---- ---- ---- - --	unknown
							1777	---- ---- ---- - --	ex
							1787	---- ---- ---- - --	lf
							1790	x--- ---- ---- - --	ex
							1792	x--- ---- ---- - --	ex
							1796	x--- ---- ---- - --	ex
							1799	x--- ---- ---- - --	ex
							1801	x--- ---- ---- - --	ex
							1806	---- ---- ---- - --	ex
							1822*	---- ---- ---- - --	ex
							1832	---- ---- ---- - --	unknown
							1837	---- ---- ---- - --	unknown
							1852*	---- ---- ---- - --	ex
							1860*	---- ---- ---- - --	ex
							1867	---- ---- ---- - --	unknown
							1869*	---- ---- ---- - --	ex
							1874-76*	---- ---- ---- - --	ex
							1879	---- ---- ---- - --	unknown
							1883	---- ---- ---- - --	ex
							1893*	---- ---- ---- - --	ex
							1897-98	---- ---- ---- - --	ex
							1906-09	x--- ---- ---- - --	ex,lf
							1913	---- ---- ---- - --	Ex
							1915-18*	---- ---- ---- - --	ex
							1919	---- ---- ---- - --	unknown
							1920	B--- ---- ---- - --	ex
							1929*	---- ---- ---- - --	ex

PART 2: QUICK REFERENCE TO PARTS 3 AND 4 (continued)
HISTORICAL UNREST AT LARGE CALDERAS AND SELECTED NONCALDERA SETTINGS
See inside back cover for explanation and abbreviations

CAVW number (active vent)	Latitude Longitude (degrees)	Diameter (km)	Local tectonic setting	Precaldera edifice	SiO_2 content (wt pct)	Age of collapse (yr B.P.)	Date of unrest	Type and duration of unrest ESTU STHF MGTF H Te	Eruption type
VILLARRICA (continued)									
							1933*	---- ---- ---- - --	ex
							1935*	---- ---- ---- - --	ex
							1938*	---- ---- ---- - --	ex
							1948-49	---- ---- ---- - --	Ex,lf
							1959-61	---- ---- ---- - --	ex
							1963-64	---- ---- ---- - --	Ex,lf
							1971-72	---- ---- ---- - --	Ex,lf
							1979-80	---- ---- ---E - --	pex,ex
							1983-85	DD-Y ---- ---- - --	lf,ex
							* not shown in Riffo and others (1987)		
CORDILLERA NEVADA									
15-07-141 (Cordón Caulle)	40.46S 72.25W	9	Compr		R = 64-71 C = 62-74	Probably 100,000-200,000	1893?	---- ---- ---- - --	lf?
							1921-22	x--- ---- ---- - --	EX,lf
							1929	---- ---- ---- - --	ex
							1934	x--- ---- ---- - --	ex
							1960	C--- ---- ---- - QU	Ex,lf
MT. HUDSON									
15-08-057 (Mt. Hudson)	45.91S 72.92W	9	Several intersecting fault trends; youngest, E-W	Strato	R = 59-61 C = ?	Preglacial	1970-71	x--- ---- --EE E --	Ex
							1973	---- ---- --D- - --	none?, or subgl
UNNAMED, near Roseau, Dominica									
16-00-10 (Valley of Desolation; Micotrin)	15.30N 61.30W	10	Compr	Strato?	R = 59-63 C = silicic	30,000	1673	?--- ---- ---- - --	none
							1765	xx-- ---- ---x - --	none
							1843-49	xx-- ---- ---- - --	none
							1879-80	?--- ---- ---x - --	pex
							1967	CC-- ---- ---- - --	none
							1971	x--- ---- ---N N --	none
							1974	xx-- ---- ---N N --	none
							1976	xx-- ---- ---N N --	none

PART 2: QUICK REFERENCE TO PARTS 3 AND 4 (continued)
HISTORICAL UNREST AT LARGE CALDERAS AND SELECTED NONCALDERA SETTINGS
See inside back cover for explanation and abbreviations

CAVW number (active vent)	Latitude Longitude (degrees)	Diameter (km)	Local tectonic setting	Precaldera edifice	SiO_2 content (wt pct)	Age of collapse (yr B.P.)	Date of unrest	Type and duration of unrest ESTU STHF MGTF H Te	Eruption type
QUALIBOU									
16-00-14 (Qualibou)	13.83N 61.05W	5 x 7	N60°E-trending faults; compr	Strato-cones	R = d C = silicic	ca. 39,000	1766	---- ---- ---- - --	pex
							1906-08	xx-- ---- ---- - --	none
KATLA									
17-02-04 (Katla= K) including 17-02-14 (Eldgja= E)	63.63N 19.03W	11 x 14	Intraplate	LL-strat (subglacial)	R = 47-49		A.D. 934 (E)	---- ---x ---- - --	Ex,LF
							ca. 950?	---- ---- ---- - --	lf,subgl
							ca. 1179	---- ---- ---- - --	ex,subgl
							1245	---- ---- ---- - --	ex,subgl
							1262	---- ---- ---- - --	Ex,subgl
							1311	---- ---- ---- - --	Ex,subgl
							between 1354/1360	---- ---- ---- - --	Ex,subgl
							1416	---- ---- ---- - --	ex,subgl
							ca. 1490	---- ---- ---- - --	Ex,subgl
							1580	---- ---- ---- - --	ex,subgl
							1612	---- ---- ---- - --	ex,subgl
							1625	x--- ---- ---- - --	Ex,subgl
							1660-61	B--- ---- ---- - --	ex,subgl
							1721	x--- ---- ---- - --	EX,subgl
							1755-56	B--- ---- ---Y - --	EX,subgl
							1823	B--- ---- ---- - --	ex,subgl
							1860	B--- ---- ---- - --	ex,subgl
							1918	B--- ---- ---- - --	EX,subgl
							1951-88+	G--- ?--- ---- - --	subgl?
ORÆFAJÖKULL									
17-02-21 (Oræfajökull)	64.00N 16.65W	5 x 4	Intraplate	Strato	R = 70(1362)	<700,000	1362	?--- ---- ---- - --	EX,subgl
					R = 56(1727)		1727-28	C--- ---- ---- - --	Ex,subgl

PART 2: QUICK REFERENCE TO PARTS 3 AND 4 (continued)
HISTORICAL UNREST AT LARGE CALDERAS AND SELECTED NONCALDERA SETTINGS
See inside back cover for explanation and abbreviations

CAVW number (active vent)	Latitude Longitude (degrees)	Diameter (km)	Local tectonic setting	Precaldera edifice	SiO_2 content (wt pct)	Age of collapse (yr B.P.)	Date of unrest	Type and duration of unrest ESTU STHF MGTF H Te	Eruption type
BARDARBUNGA									
no CAVW number	64.62N	7 x 11	Exten	LL-strat	R = b		1477-ca.1480	---- --xx ---- - =?	EX
(Bárdarbunga)	17.50W			(subglacial)	C = b		1717	---- ---- ---- - --	EX
							1974-87	xx-- ---- ---- - --	none
17-02-18 (Trollagigar =Trollahraun)	64.43N 18.13W						1862-64	---- --xx ---- - --	lf,ex
within 17-02-11 (Vatnaöldur)	63.95N 19.28W						ca. 900	---- --xx ---- - --	EX,lf
17-02-11 (Veidivötn)	63.95N 19.28W						ca. 1480	---- --xx ---- - --	EX,lf
TORFAJÖKULL									
no CAVW number	63.93N	13 x 18	Intraplate		R = 70-72		ca. 900 A.D.	---- --xx ---- - --	lf,Ex
(Torfajökull)	19.15W				C = silicic		ca. 1480 A.D.	---- --xx ---- - --	lf,Ex
							1986-88+	CC-- ---- ---- - --	none
GRIMSVÖTN									
17-02-22	64.42N	6-8	Exten	LL-strat	R = mafic		1332	---- ---- ---- - --	ex,subgl
(Grimsvötn= G)	17.33W			(subglacial)			1341	---- ---- ---- - --	ex,subgl
17-02-15							1598	---- ---- ---- - --	Ex,subgl
(Lakagigar= L)							1603?	---- ---- ---- - --	ex,subgl
17-02-16							1619	---- ---- ---- - --	ex,subgl
(Sidhjökull= S)							1629	---- ---- ---- - --	ex,subgl
17-02-17							1638	---- ---- ---- - --	ex,subgl
(Hagongur= H)							1659	---- ---- ---- - --	ex,subgl
17-02-19							1681?	---- ---- ---- - --	subgl?
(Thordarhyrna= T)							1684-85	---- ---- ---- - --	ex,subgl
							1706	---- ---- ---- - --	ex,subgl
							1716	---- ---- ---- - --	ex,subgl
							1725-26	---- ---- ---- - --	subgl
							1753 (S)	---- ---- ---- - --	ex
							1766	---- ---- ---- - --	ex,subgl
							1774	---- ---- ---- - --	ex,subgl
							1783-84(L)	DD-- --xx ---- - --	EX,LF
							1783-85	---- ---- ---- - --	ex,subgl
							1797	---- ---- ---- - --	ex,subgl

PART 2: QUICK REFERENCE TO PARTS 3 AND 4 (continued)
HISTORICAL UNREST AT LARGE CALDERAS AND SELECTED NONCALDERA SETTINGS
See inside back cover for explanation and abbreviations

CAVW number (active vent)	Latitude Longitude (degrees)	Diameter (km)	Local tectonic setting	Precaldera edifice	SiO_2 content (wt pct)	Age of collapse (yr B.P.)	Date of unrest	Type and duration of unrest ESTU STHF MGTF H Te	Eruption type
GRIMSVOTN (continued)									
							1816	---- ---- ---- - --	ex,subgl
							1823?(G,T)	---- ---- ---- - --	ex,subgl
							1838	---- ---- ---- - --	ex,subgl
							1854?	---- ---- ---- - --	ex,subgl
							1861?	---- ---- ---- - --	ex,subgl
							1867	---- ---- ---- - --	ex,subgl
							1873	Y--- ---- ---- - --	Ex,subgl
							1883	---- ---- ---- - --	ex,subgl
							1887? (T)	---- ---- ---- - --	ex,subgl
							1891-92?	---- ---- ---- - --	ex,subgl
							1902-4(G,T)	x--- ---- ---- - --	ex,subgl
							1910?	---- ---- ---- - --	ex,subgl
							1922	---- ---- ---- - --	ex,subgl
							1933-34	Y--- ---- ---- - --	ex,subgl
							1938	---- ---- ---- - --	subgl
							1939	---- ---- ---- - --	subgl
							1941	---- ---- ---- - --	subgl
							1945	---- ---- --x- - --	pex,subgl
							1948	---- ---- --x- - --	none?
							1954	---- ---- ---- - --	pex,subgl
							1982-83	EBA- ---- ---- - --	subgl,ex
KVERKFJOLL									
17-02-23	64.65N	7.5 x 5.5	Exten	LL-strat,	R = mafic		1655	---- ---- ---- - --	subgl
(Kverkfjoll)	16.70W	8 x 5		partly			1684	---- ---- ---- - --	subgl,ex
				subglacial			1711	---- ---- ---- - --	subgl
							1716?-17	---- ---- ---- - --	subgl,ex?
							1726	---- ---- ---- - --	subgl
							1726	---- ---- ---- - --	subgl
							1729	---- ---- ---- - --	subgl
							1874-75?	---- ---- ---- - --	subgl,ex
ASKJA, OSKJUVATN									
17-03-03	65.03N	9 (Askja)	Exten	LL-strat	R = 50-51	A.D. 1875	1874-75	EE-- E--E --EE E QU	EX,lf
(Askja)	16.75W	4 (Oskj)			C = 51-75		1919?	---- ---- ---- - --	ex
							1921-26	---- ---- ---- - --	lf,ex
17-03-09							1929	---- ---- ---- - --	ex,lf
(Sveinagja)							1961	D--Z Z--D --DD D --	lf,ex
							late 1960's-present	---F FF-- ---- - --	none

PART 2: QUICK REFERENCE TO PARTS 3 AND 4 (continued)
HISTORICAL UNREST AT LARGE CALDERAS AND SELECTED NONCALDERA SETTINGS
See inside back cover for explanation and abbreviations

CAVW number (active vent)	Latitude Longitude (degrees)	Diameter (km)	Local tectonic setting	Precaldera edifice	SiO_2 content (wt pct)	Age of collapse (yr B.P.)	Date of unrest	Type and duration of unrest ESTU STHF MGTF H Te	Eruption type
KRAFLA									
17-03-11 (Krafla)	65.73N 16.68W	10, 8	Exten	Shield	R = b + minor r,d,a	<70,000	1724-29 1746 1975-88+	BB-Y Y--Y --YY Y UQ Y--Y ---- ---- Y -- EEBE BEEE -Exx Y UQ	P-ex,ex,lf lf? lf,pex
SETE CIDADES									
18-02-08	37.87N 25.78W	5	Spreading center exten	LP-LL strat	R = 44-48 C = 65	ca. ca. 22,000	1444 1638 1682 1713 1810-11 1861 1880	---- ---- ---- - -- EE-- ---- ---- - -- xx-- ---- ---- - -- xx-- ---- ---- - -- EE-E ---E ---- - -- ---- ---- ---- - -- ---- ---- ---- - --	P-ex ex,sub ex,sub sub? Ex,sub sub sub
AGUA DE PAU									
18-02-09	37.77N 25.47W	4 x 7 (outer) 2.5 x 3 (inner)	Spreading center exten	LL-LP strat	R= 45-65 C= ca. 65	15,200 and between 30,000 and 45,000	1563-64 1652 1980's	CC-- ---x ---- - -- DD-- ---B ---Y - -- xx-- ---- ---- - --	EX,pex,lf ex,lf none
FURNAS									
18-02-10	37.77N 25.32W	6	Spreading center; exten	LP-strat	R = ca. 65 C = 65	ca. 12,000	1630	B--- ---B ---B B --	EX,dm
LAS CAÑADAS									
18-03-03 (Pico de Teide, Chinyero, and others)	28.30N 16.63W	10 x 17	Hotspot?	LL-strat	R = 44-52 C = 60-61	ca. 150,000	1341? 1393/99? 1444? 1492 1704-06 1798 1909	---- ---- ---- - -- ---- ---- ---- - -- ---- ---- ---? - -- ---- ---- ---- - -- CC-- ---- ---- - -- ---- ---- ---- - -- FF-- ---- ---- - --	unknown unknown none? lf ex,lf ex,lf ex,lf

PART 2: QUICK REFERENCE TO PARTS 3 AND 4 (continued)
HISTORICAL UNREST AT LARGE CALDERAS AND SELECTED NONCALDERA SETTINGS
See inside back cover for explanation and abbreviations

CAVW number (active vent)	Latitude Longitude (degrees)	Diameter (km)	Local tectonic setting	Precaldera edifice	SiO_2 content (wt pct)	Age of collapse (yr B.P.)	Date of unrest	Type and duration of unrest ESTU STHF MGTF H Te	Eruption type
CHA									
18-04-01 (Fogo)	14.95N 24.35W	8	Intraplate	Shield	R = 40-50 C = mafic		<1500-1761	---- ---- ---- - --	ex,lf
							1769	---- ---- ---- - --	ex
							1785	---- ---- ---- - --	pex,lf
							1799	---- ---- ---- - --	ex,lf
							1817	---- ---- ---- - --	unknown
							1847	x--- ---- ---- - --	ex,lf
							1852	---- ---- ---- - --	ex,lf
							1857	---- ---- ---- - --	ex,lf
							1951	x--- ---- ---- - --	ex,lf
EREBUS									
19-00-02 (Erebus)	77.58S 167.17E	6 (older) 5 (younger)	Intraplate	Strato	R = 56 (anorthoclase phonolite)		1841	---- ---- ---- - --	ex,lf
							1900?	---- ---- ---- - --	ex
							1908	---- ---- ---E - --	ll
							1911-12	---- ---- ---- - --	ex
							1915	---- ---- ---- - --	ex
							1947	---- ---- ---- - --	ex,lf
							1955	---- ---- ---- - --	ex
							1963	---- ---- ---- - --	ll
							1972-87+	YYY- Y--- Y--- - --	ex,lf,ll
DECEPTION ISLAND									
19-00-03 (Deception Island)	62.93S 60.57W	7	Exten	Strato	R = 50-70 C = 51-54		1820/29	---- ---- ---- - --	ex
							1829	x--- ---- ---x x --	none?
							1839-42	---- ---- ---F F --	ex,lf
							ca. 1912	---- ---- ---- - --	ex,Ex
							ca. 1917	---- ---- ---- - --	ex,Ex
							1923-27	x--- x--- --xx - --	none?
							1930	x--- x--- ---- - --	none
							before 1956	---- ---- ---- - --	ex
							1967	FF-x ---x ---- - --	pex,Ex
							1969	x--? ---Y ---E - --	Ex,ex,pex
							1970	x--x Z--x ---- - --	Ex,ex,pex
							1972	---- ---- ---- - --	sub
							1987	x--- ---- ---- - --	Ex?

PART 2: QUICK REFERENCE TO PARTS 3 AND 4 (continued)
HISTORICAL UNREST AT LARGE CALDERAS AND SELECTED NONCALDERA SETTINGS

See inside back cover for explanation and abbreviations

Latitude Longitude (degrees)	Tectonic setting	Age of latest volcanism	Youngest volcanic rock types	Restless years	Type and duration of unrest ESTU STHF MGTF H Te	Associated eruption?
SELECTED NONCALDERA SETTINGS						
CHIJIWA						
32.7N 130.1E	East-west graben	A.D. 1791-92 1835?	a, d	860	---- ---- ---- - --	Ex?
				1657	---- ---- ---- - --	lf,ex?
				1663-64	---- ---- ---- - --	ex
				1791-92	EE-- ---Y ---Y Y --	ex,lf,LANDSL
				1835?	---- ---- ---- - --	ex?
				1922	CC-- ---- ---- - --	none
				1929-40	FF-- ---- ---- - --	none
				1951-66	FF-- ---- ---- - --	none
				1968-74	FF-- G-G- --F- F --	none
				1977-82	EE-- ---- ---- - --	none
				1984	EE-- ---- ---- - --	none
IZU PENINSULA						
34-35N 139E	Compr	Late Pleistocene	b, a, d	1905	xx-- ---- ---- - =?	none
				1930	EE-G ---x ---- - =?	none
				1967-80+	FF-G --Gx x--- F =?	none
KII PENINSULA						
34N 135-136E	Compr	Cretaceous?	?	1899	x--- ---- ---- - --	none
				1920-30+	xx-x xx-- ---- - --	none
				1944	x--- ---- ---- - --	none
				1946	x--x ---- ---- - --	none
				1948	x--- ---- ---- - --	none
				1952-53	xx-x xx-- ---- - --	none
MATSUSHIRO						
36.55N 138.20E	Compr	350,000 yr B.P.	d	887	x--- ---- ---- - --	none
				1718	x--- ---- ---- - --	none
				1847	x--- ---- ---- - --	none
				1853	x--- ---- ---- - --	none
				1858	x--- ---- ---- - --	none
				1897	EE-- ---- ---- - --	none
				1912	x--- ---- ---- - --	none
				1941-44	FF-- ---- ---- - --	none
				1952	x--- ---- ---- - --	none
				1958	x--- ---- ---- - --	none
				1965-68	FF-E EFFx FFF? F =?	none

PART 2: QUICK REFERENCE TO PARTS 3 AND 4 (continued)
HISTORICAL UNREST AT LARGE CALDERAS AND SELECTED NONCALDERA SETTINGS
See inside back cover for explanation and abbreviations

Latitude Longitude (degrees)	Tectonic setting	Age of latest volcanism	Youngest volcanic rock types	Restless years	Type and duration of unrest ESTU STHF MGTF H Te	Associated eruption?
			SELECTED NONCALDERA SETTINGS			
SOCORRO						
34.00N 107.00W	Extension, with left-lateral shear on ENE-trending fault(?)	Pliocene	b	1849-50	xx-- ---- ---- - --	none
				1904	xx-- ---- ---- - --	none
				1906-07	xx-- ---- ---- - --	none
				1960's-87+	GG-G ---- --?- ? --	none
MONTSERRAT (Soufriere Hills) CAVW # 16-00-05						
16.72N 62.18W	Compr	300 yr B.P.	58-63 pct SiO_2	1897-1900	FF-- ---- ---x x --	none
				1933-1937	FF-- ---- ---C x UQ	none
				1966-1967	FF-- -F-- ---F F --	none

PART 3: Historical Unrest at Large Quaternary Calderas: Details of Unrest

This section contains most of the data upon which Part 1 is based. Calderas are arranged geographically, according to their CAVW number; the Table of Contents, Part 2 (Quick Reference), and footers will help you find specific calderas. A tabular summary of unrest is given for each caldera, followed by a narrative of Tectonic Setting, Geologic History, Historical Unrest, Comments, and References. Figures are selected to show the location of the caldera, and representative or remarkable seismicity, ground deformation, and thermal, gas, or hydrologic changes, generally in that order. For a few calderas, figures are grouped separately for each major episode of unrest.

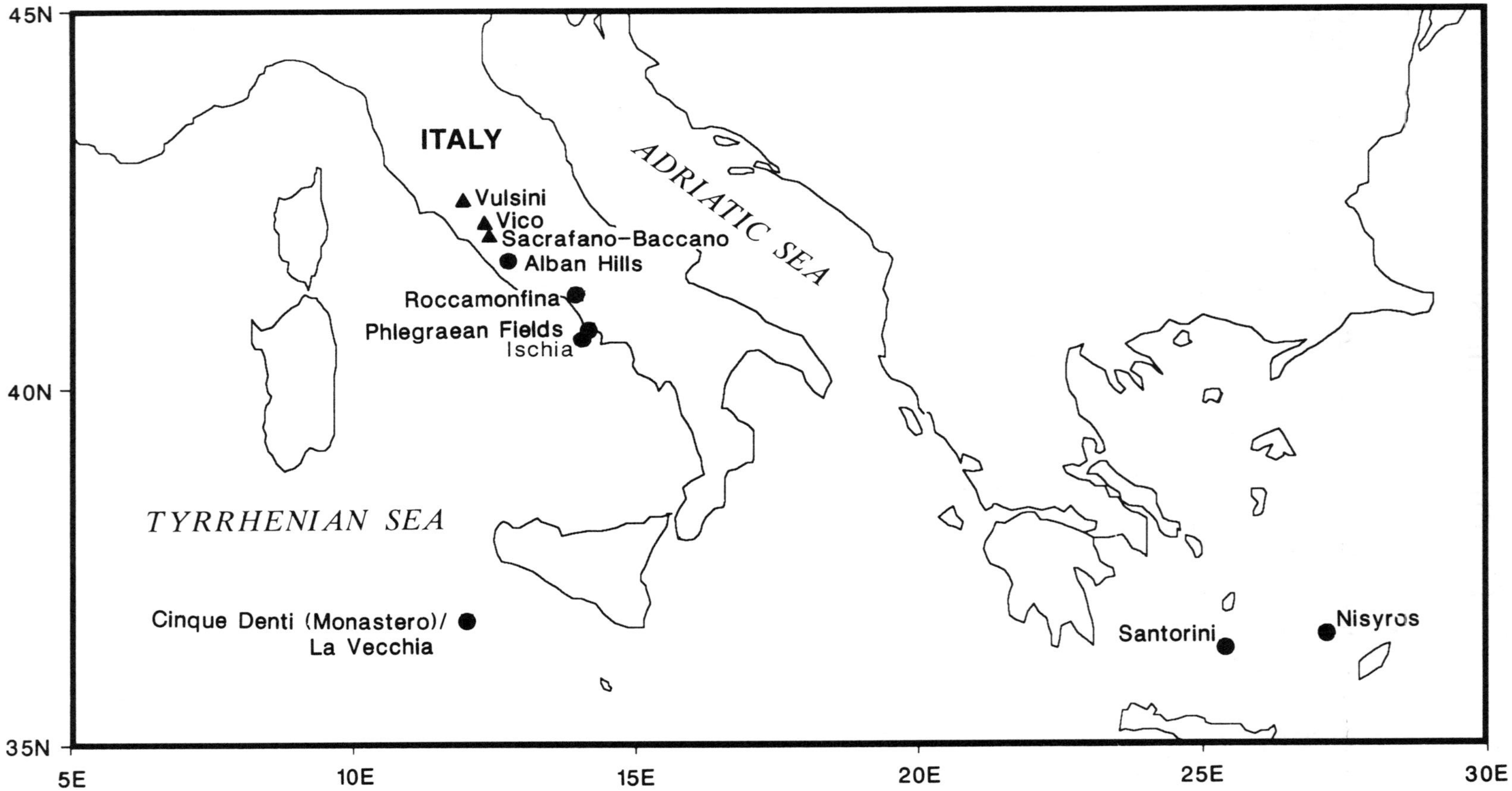

Figure 1.1. Locations of large, Quaternary restless calderas (solid circles) and nonrestless calderas (triangles) in Region 1.

VULSINI (LATERA/CONCA DI BOLSENA)

CAVW number (active vent)	Latitude Longitude (degrees)	Diameter (km)	Local tectonic setting	Precaldera edifice	SiO_2 content (wt pct)	Age of collapse (yr B.P.)	Date of unrest	Type and duration of unrest ESTU STHF MGTF H Te	Eruption type
01-01-003 (Vulsini)	42.58N 11.92E	Lat: 11 x 8 CdiB: 16			Leucite phonolite, trachyte	Lat: 100,000 CdiB: 300,000	B.C. 104?	---- ---- ---- - --	ex?

TECTONIC SETTING

Vulsini Volcano is in the Roman volcanic province, an alkaline province associated with extension along the Tyrrhenian margin of the Appenines chain.

GEOLOGIC HISTORY

Nappi (1969) and Sparks (1975) recognized three broad phases of activity: (1) lava extrusion and pyroclastic activity to build the main part of the Vulsini complex, (2) major ignimbrite eruptions and formation of the Latera collapse caldera, and (3) growth of postcaldera parasitic vents. Varekamp (1980) added details to this sequence and concluded that Lago di Bolsena is a volcano-tectonic depression that formed mainly as a NNW-SSE graben (figs. 1.1.1, 1.1.2).

HISTORICAL ACTIVITY

104 B.C.: "A flame shot up near Volsinii." (Stothers and Rampino, 1983, citing Obsequens 43). Also, legend has it that in about 500 B.C., "lightning" destroyed the town of Volsinii (Pliny the Elder 2.139-140; Tertullian, To the Nations 1.9, Apology 40, On the Pallium 2; all cited in Stothers and Rampino, 1983, p. 6368).

REFERENCES

Barton, M., Varekamp, J.C., and van Bergen, M.J., 1982, Complex zoning of clinopyroxene in the lavas of Vulsini, Latium, Italy: Evidence for magma mixing: Jour. Volcanol. Geotherm. Res., v. 14, nos. 3-4, p. 361-388.

Nappi, G., 1969, Stratigrafia e petrografia dei Vulsini sud-occidentali (Caldera di Latera): Boll. Soc. Geol. Ital., v. 88, p. 171-181.

VULSINI (LATERA/CONCA DI BOLSENA) (continued)

REFERENCES (continued)

Sparks, R.S.J., 1975, Stratigraphy and geology of the ignimbrites of Vulsini Volcano, central Italy: Geol. Rundschau, v. 64, p. 497-522.

Stothers, R.B., and Rampino. M.R., 1983, Volcanic eruptions in the Mediterranean before A.D. 630, from written and archaeological sources: Jour. Geophys. Res., v. 88, p. 6357-6371.

Varekamp, J.C., 1980, The geology of the Vulsinian area, Lazio, Italy: Bull. Volcanol., v. 43, p. 487-504.

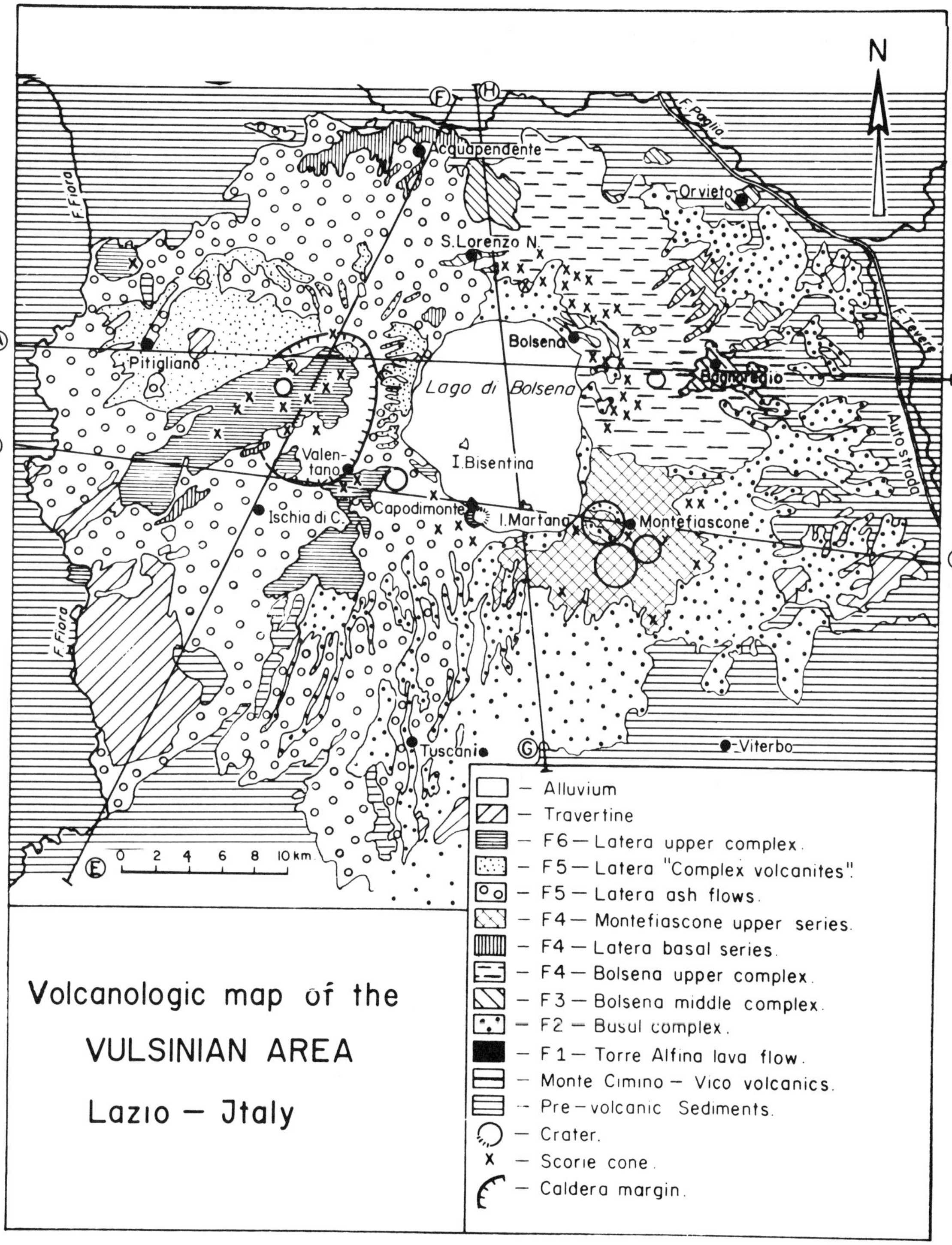

Figure 1.1.1. Volcanologic map of Vulsinian area, from Varekamp (1980). The products of six volcanic phases are indicated; letters for cross sections (A-B, C-D, and so on) refer to figure 1.1.2.

SECTIONS OF THE VULSINIAN AREA

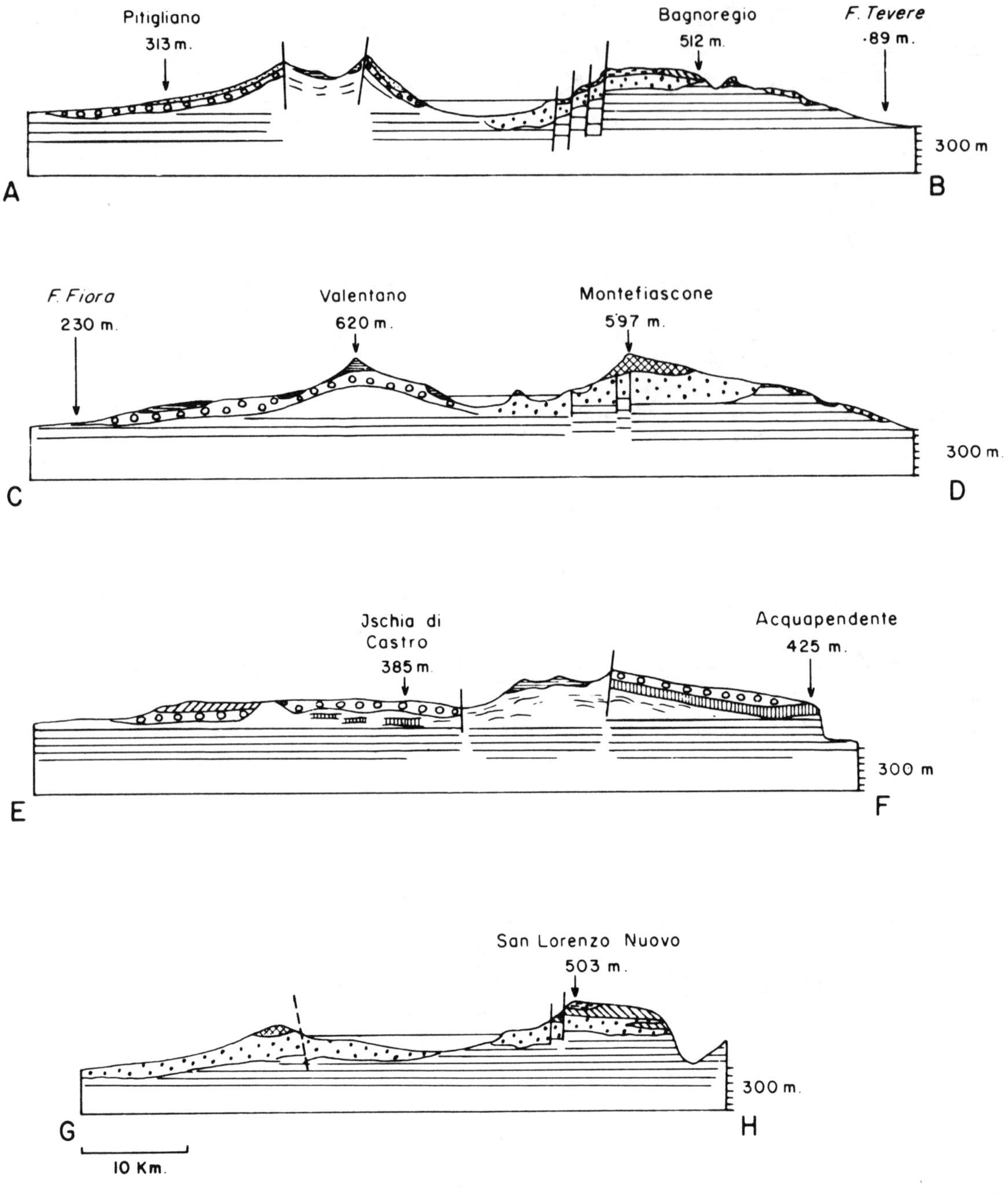

Figure 1.1.2. Schematic cross sections of Vulsinian area, from Varekamp (1980). Locations of sections are shown in figure 1.1.1. Vertical exaggeration is about 12.5X.

ALBAN HILLS/ARTEMISIO

CAVW number (active vent)	Latitude Longitude (degrees)	Diameter (km)	Local tectonic setting	Precaldera edifice	SiO_2 content (wt pct)	Age of collapse (yr B.P.)	Date of unrest	Type and duration of unrest ESTU STHF MGTF H Te	Eruption type
01-C1-004	41.75N 12.75E	11	Exten?	Strato	R = 46 C = ?	37,000 ca. 60,000? B.C.	first millennium	---- ---- ---- - --	unknown, ex?

TECTONIC SETTING

The Alban Hills lie about 50 km southeast of Rome, at the convergence of several structural elements including the Pontina Plain, the Sabina Mountains, and the Lepini Mountains (De Rita and Funiciello, 1982; fig. 1.2.1).

GEOLOGIC HISTORY

The oldest recognized volcanic deposits are granular pyroclastic materials of Pleistocene age interpreted as products of numerous explosive eruptions from eccentric vents (fig. 1.2.2). Earlier effusive activity is indicated by fragments of older lava flows in these pyroclastic materials. The explosive eruptions were followed by predominantly effusive eruptions that produced extensive lava flows. Explosive eruptions from a central vent then produced a large stratovolcano with a summit caldera about 10 km in diameter. Following a hiatus of unknown duration, explosive and effusive eruptions resumed from vents inside the caldera, producing a new central cone about 5 km in diameter. Explosive eruptions from eccentric vents occurred throughout the history of the central volcano. The most recent of these (less than 50,000 yr B.P.) may represent the waning phase of a volcanic cycle that began in the Pleistocene. Currently, the only surficial manifestations of volcanic activity are widespread cold mineral springs (De Rita and Funiciello, 1982).

HISTORICAL ACTIVITY

Nissen (1883, cited in Stothers and Rampino, 1983) described an early Alban necropolis covered by a mudflow of possible volcanic origin. T. Livius (cited in Krafft, 1974, and Stothers and Rampino, 1983) lists eruptions in 640 (or 540?) B.C. and 114 B.C. In 540 B.C., ash fell in Rome and there was a "rain of stones." In 114 B.C., Monte Albano was reported to be "on fire." Stothers and Rampino (1983, p. 6364) invoke nonvolcanic explanations for these reports, interpreting hail as the "stones" and a forest fire as the "fire" on Monte Albano. Thus, the Alban Hills might have erupted in the first millennium B.C. but did not certainly do so.

ALBAN HILLS/ARTEMISIO (continued)

REFERENCES

De Rita, D., and Funiciello, R., 1982, Guidebook for the field excursion to Alban Hills: CNR-NSF Workshop on explosive volcanism, 18 p.
Krafft, M., 1974, Guide des volcans d'Europe: Neuchatel, Delachaux and Niestle, 412 p.
Nissen, H., 1883, Italische landeskunde, v. 1: Berlin, Weidmann.
Rittmann, A., 1962, Volcanoes and their activity: New York, Interscience Publishers, 305 p.
Stothers, R.B., and Rampino. M.R., 1983, Volcanic eruptions in the Mediterranean before A.D. 630, from written and archaeological sources: Jour. Geophys. Res., v. 88, p. 6357-6371.

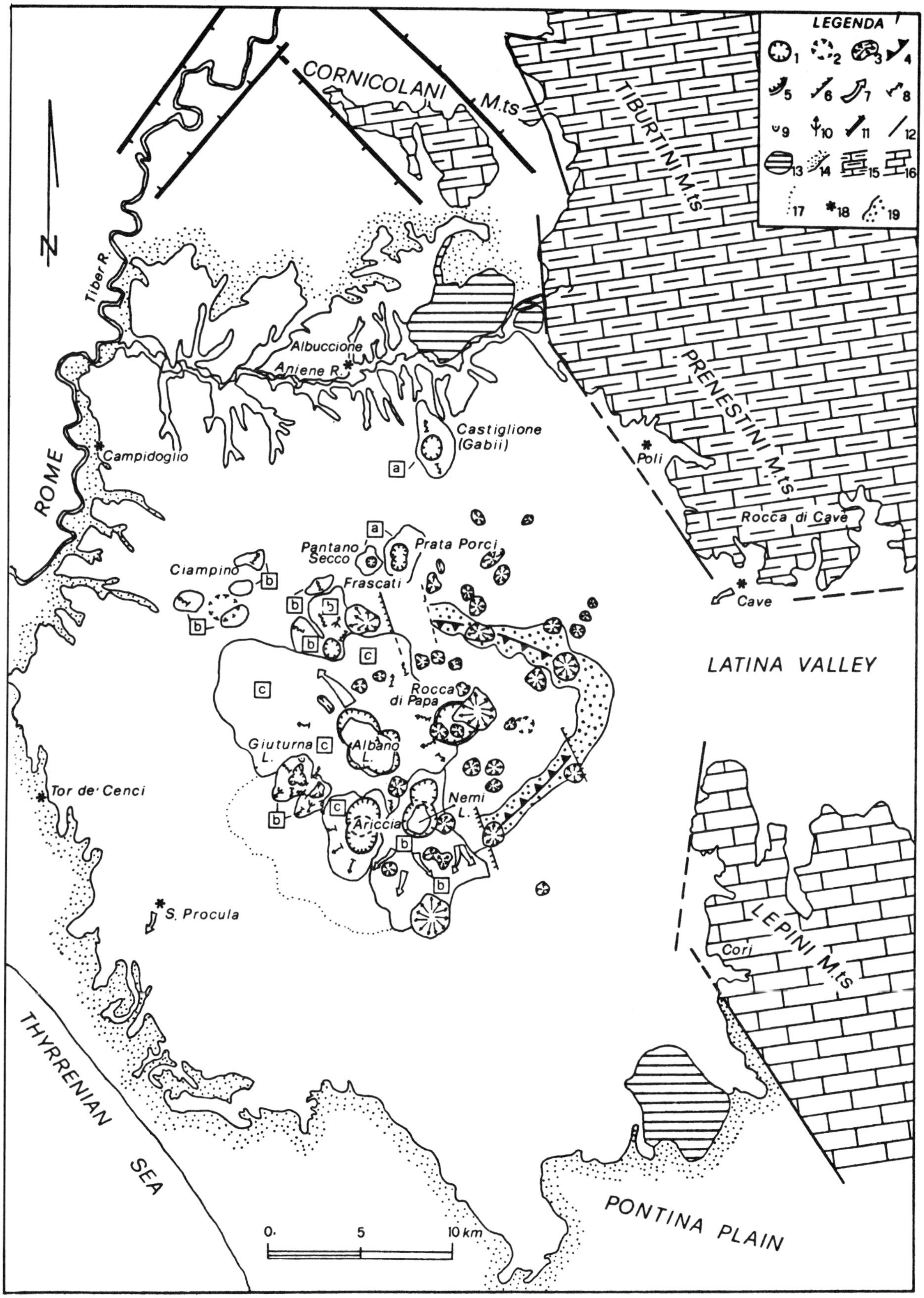

Figure 1.2.1. See next page for caption.

Figure 1.2.1. Structural geologic map of Alban Hills area, from De Rita and Funiciello (1982): (1) phreatomagmatic explosion craters, (2) crater locations inferred from distribution of their products, (3) cones of welded scoriae and leucitic lavas, (4) caldera rims, (5) central crater rims, (6) distensive faults in central volcanic system, (7-10) depositional structures in phreatomagmatic products, (7) prominent lahars and their directions, (8) sandwaves and their directions, (9) most significant impact sags, (10) flow channels, (11) recent distensive faults at margins of structural high of Monti Cornicolani and of small transverse structural low of Tiber River, (12) marginal distensive faults at boundaries between Apennine ridge and Campagna Romana, (13) outcrops of travertine of hydrothermal origin from Bagni di Tivoli and Cisterna, probably connected with late stages of magmatic-tectonic evolution of volcanic group, (14) boundary between Plio-Pleistocene sedimentary rocks and products from initial activity of volcanic system, (15) mainly pelagic carbonate units from Mt. Tiburtini and Mt. Prenestini and associated terrigenous flysch, (16) carbonate platform units of Mt. Lepini, (17) boundaries (not perfectly defined) of initial explosive products, (18) outcrops of phreatomagmatic eruption products, mainly from initial explosions, whose craters have not been located, (19) boundary of welded scoriae from Tuscolano-Artemisio activity, (a) most ancient explosions, (b) intermediate-age explosions, (c) most recent explosions.

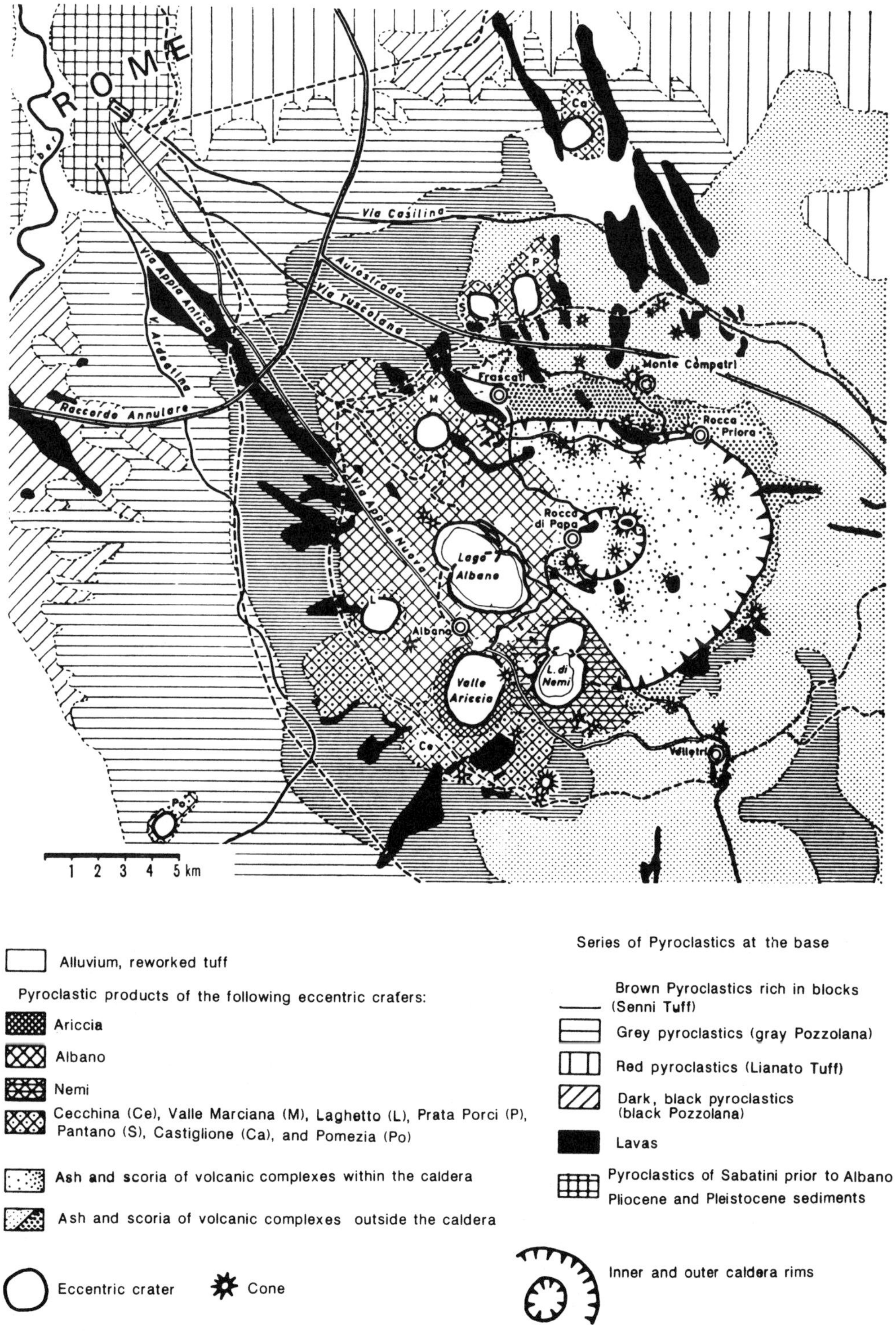

Figure 1.2.2. Geologic map of Alban Hills area, reproduced in Krafft (1974).

ROCCAMONFINA

CAVW number (active vent)	Latitude Longitude (degrees)	Diameter (km)	Local tectonic setting	Precaldera edifice	SiO_2 content (wt pct)	Age of collapse (yr B.P.)	Date of unrest	Type and duration of unrest ESTU STHF MGTF H Te	Eruption type
01-01-004A	41.30N 13.93E	5 x 7	Exten?	Strato	R = 49-63 C = silicic	Main: ca.385,000 gli Stagli: ca.300,000	B.C. 276 or B.C. 269	---- ---- ---- - --	ex (uncertain)

TECTONIC SETTING

Roccamonfina is a Quaternary volcano about 50 km north-northwest of Somma-Vesuvius along the west coast of Italy (fig. 1.3.1). It occupies a rift formed by the intersection of two fracture systems related to post-Miocene extensional tectonism: the Appenine (NW-SE) and anti-Appenine (NE-SW) systems.

GEOLOGIC HISTORY

Roccamonfina is a partially dissected stratovolcano with an average diameter of 18 km, resting on Pleistocene lacustrine sediments. The summit contains a central ("main") caldera 5.5 km by 6.5 km in diameter; a smaller ("gli Stagli") caldera lies near the northern rim of the main cone (fig. 1.3.1). Many parasitic cones are distributed on the flanks of the volcano, along the main caldera rim, and within the caldera (Giannetti, 1982a,b; Luhr and Giannetti, 1987). The history of Roccamonfina was divided into two main episodes by Giannetti (1982a,b). During the first episode from 1.54 Ma to 0.38 Ma, lavas and pyroclastics ranging in composition from leucite-tephrite to leucite-phonolite were erupted to form a large stratovolcano. This episode culminated 385,000 yr ago with a large explosive eruption of the compositionally zoned Brown Leucitic Tuff (BLT). Luhr and Giannetti (1987) suggest that the main caldera formed before eruption of the BLT, though the evidence presented is ambiguous. As much as 100-120 km^3 of magma may have been erupted during the first episode. In the second episode, basalts to trachytes were erupted to form prominent lava domes, lava flows, and ejecta cones. The smaller, "gli Stagli" caldera might have formed shortly after eruption of the compositionally zoned White Trachytic Tuff (WTT) about 300,000 yr ago (Giannetti and Luhr, 1983). Magmas of the first episode were leucite bearing, and magmas of the second episode were leucite free.

Alternate interpretations of the geology of Roccamonfina, not adopted here, include those of Tedesco (1965), Ghiara and others, (1973), and Chiesa and others (1985).

ROCCAMONFINA (continued)

HISTORICAL ACTIVITY

B.C. 269 or 276: "A flame shot up and burned for three days near Cales," in 269 B.C., possibly an eruption of Roccamonfina (Stothers and Rampino, 1983). Panichi (cited in Tedesco, 1965, p. 137) read the same original account by the Roman historian Orosius and concluded that this event occurred in B.C. 276. R. Scandone and C. Principe (written commun. to Smithsonian Institution, 1986) expressed skepticism that Roccamonfina erupted at either time, because the youngest products of Roccamonfina have been dated at 0.30 Ma. J. Guest (written commun., 1987) also doubts that an eruption has occurred in historical time, but accepts the possibility that some eruptions might have occurred more recently than 0.30 Ma.

COMMENTS

From geological evidence Roccamonfina is not a restless caldera. However, the record of many small eruptions is lost in the geologic record, and we include Roccamonfina here as an example of how historical records might reveal activity that otherwise would be missed. Not all old historical reports are accurate, but many contain some geological truth!

REFERENCES

Chiesa, S., Cornette, Y., Gillot, P.Y., and Vezzoli, L., 1985, New interpretation of Roccamonfina volcanic history [abs.]: Abstracts and Time Schedule, IAVCEI Scientific Assembly, Potassic Volcanism -- Mt. Etna Volcano, Catania, September 1985.

Ghiara, M.R., Lirer, L., and Stanzione, L., 1973, Contributo alla conoscenza vulcanologica e petrografica del vulcano di Roccamonfina: Period. Mineral., v. 42, p. 267-293.

Giannetti, B., 1979, The geology of Roccamonfina caldera (Campanian Province, Italy): Giornale Geol. (ser. 2), v. 43, p. 187-206.

----- 1982a, Roccamonfina Volcano, Italy: Volcano News, no. 12, p. 1-3.

----- 1982b, Cumulate inclusions from K-rich magmas, Roccamonfina Volcano, Italy: Earth Planet. Sci. Lett., v. 57, no. 2, p. 313-333.

Giannetti, B., and Luhr, J.F., 1933, The White Trachytic Tuff of Roccamonfina Volcano (Roman Region, Italy): Contrib. Mineral. Petrol., v. 84, p. 235-252.

Luhr, J.F., and Giannetti, B., 1987, The Brown Leucitic Tuff of Roccamonfina Volcano (Roman Region, Italy): Contrib. Mineral. Petrol., v. 95, p. 420-436.

Stothers, R.B., and Rampino, M.R., 1983, Volcanic eruptions in the Mediterranean before A.D. 630, from written and archaeological sources: Jour. Geophys. Res., v. 88, p. 6357-6371.

Tedesco, C., 1965, Main lines of the history of the Roccamonfina Volcano: Bull. Volcanol., v. 28, p. 119-137.

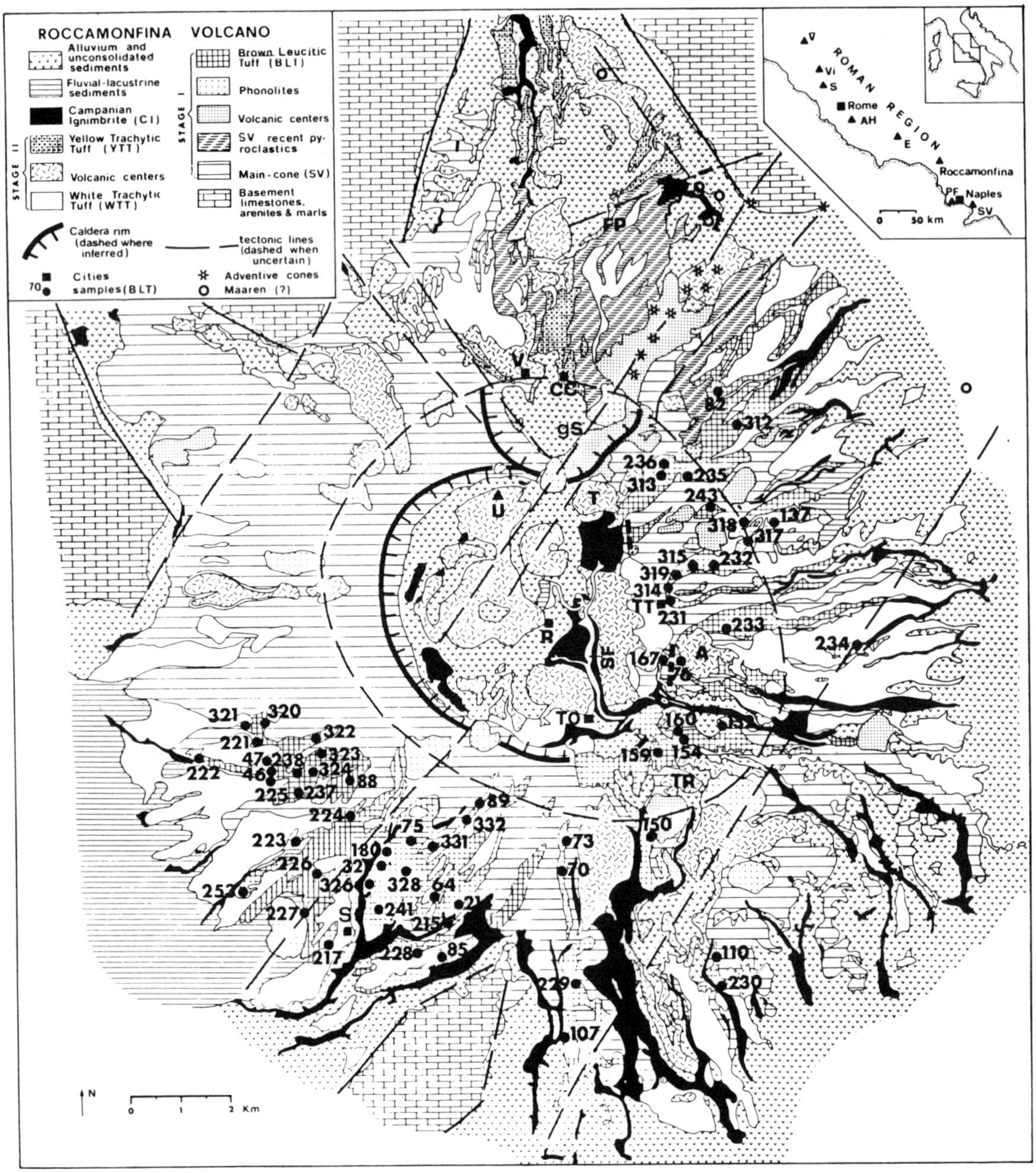

Figure 1.3.1. Geologic map of Roccamonfina Volcano, from Luhr and Giannetti (1987). TT, Tuoro di Tavola; V, Vezzara; CC, Conca Campania; R, Roccamonfina; TO, Torano; S, Sessa Aurunca; gS, "gli Stagli" caldera; TR, Mt. Triuci ejecta cone; A, Mt. Atano ejecta cone; T, Tuororame ejecta cone; FP, Fosso Pubblico; SF, Savone delle Ferriere ravine; U, Unocal Geothermal Division drillhole Gallo 85-1.

PART 3: HISTORICAL UNREST AT LARGE QUATERNARY CALDERAS (continued)

See inside back cover for explanation and abbreviations

PHLEGRAEAN FIELDS

CAVW number (active vent)	Latitude Longitude (degrees)	Diameter (km)	Local tectonic setting	Precaldera edifice	SiO_2 content (wt pct)	Age of collapse (yr B.P.)	Date of unrest	Type and duration of unrest ESTU	STHF	MGTF	H	Te	Eruption type
01-01-01	40.83N	13	Exten?	None?	R = 50-61	35,000	A.D. 1198	---x	----	----	-	--	pex?
	14.14E				C = 62	(Campanian)	1488?-1538	GG-G	B---	----	B	--	Ex, pex
						4,300	1564-70	F---	G---	----	-	--	none
						(Agnano)	1582	x---	G---	----	-	--	none
							1594	x--?	G---	----	-	--	none
							1621?	x---	G---	----	-	--	none
							1832	x---	G---	----	-	--	none
							pre-1901	x---	G---	----	-	--	none
							1908	x---	G---	----	-	--	none
							1913	x---	G---	----	-	--	none
							1917	x---	G---	----	-	--	none
							1930	x---	G---	----	D	QU	none
							1969-74	FF-F	FFF-	-F?-	?	--	none
							1982-85	FF-F	-FFF	-F-E	F	QU?	none

TECTONIC SETTING

The Phlegraean Fields is a volcanic area in southern Italy near the city of Naples; the town of Pozzuoli lies near the center of the Phlegraean Fields Caldera (fig. 1.4.1). Volcanic activity began in the late Pleistocene and has been localized at the edge of the Campanian Plain, which also includes Somma-Vesuvius and Roccamonfina. These volcanoes are located on the Tyrrhenian margin of the Apennine chain in a zone intensely affected by extensional tectonism during the Quaternary (Armienti and others, 1983).

GEOLOGIC HISTORY

The volcanic field consists of multiple phreatic explosion craters and cinder cones within a 13-km-diameter collapse caldera (fig. 1.4.2). The caldera formed about 35,000 yr B.P. during or soon after the eruption of 80 km^3 of Campanian tuff (fig. 1.4.3) (Rosi and others, 1983). Much of the postcaldera volcanism occurred between 10,000 yr B.P. and 8,000 yr B.P., and between 4,700 yr B.P. and 3,000 yr B.P. (Rosi and others, 1983); different periods of post-caldera volcanism are inferred by Di Girolamo and others (1984). Postcaldera vents have migrated toward the center of the caldera and erupted volumes have decreased, suggesting progressive

PHLEGRAEAN FIELDS (continued)

GEOLOGIC HISTORY (continued)

crystallization of a shallow magma reservoir (Armienti and others, 1983; Rosi and others, 1983). High temperatures (400 °C at 3-km depth) and contact metamorphism of rocks penetrated by geothermal wells suggest that the top of the magma reservoir is now 4-5 km deep. Early postcaldera volcanism was submarine; most subsequent volcanism has been subaerial. All periods of postcaldera volcanism have included frequent phreatic and phreatomagmatic explosions. Recent marine geophysical studies (Finetti and Morelli, 1974) improved definitions of several submarine vent areas, including a 2.5 km wide volcanic cone beneath Nisida Bank.

Different workers draw slightly different outlines of the Phlegraean Fields Caldera. In particular, Rosi and Sbrana (1987) draw the eastern and southern boundary of the caldera as in figure 1.4.1, whereas Lirer and others (1987) consider the boundary of a smaller caldera to pass from Pozzuoli through Nisida and thence across to Miseno (fig. 1.4.16).

Grindley (1976) inferred bradyseisms associated with at least three major eruptions in the past 40,000 years (fig. 1.4.4); a fourth was associated with the minor eruption that formed Monte Nuovo cinder cone in 1538 (figs. 1.4.5, 1.4.6). Cinque and others (1985) infer three periods of marine deposition alternating with subaerial deposits in the Starza terrace. The youngest marine sediments in the Starza terrace (interval F, inferred to be 6,750 yr old) had been uplifted about 40 m as of 1985 (Cinque and others, 1985); most of that uplift occurred between 5,000 yr B.P. and 2,000 yr B.P., before lesser uplift and subsidence that is recorded in historical records of the latest 2,000 years (see below).

HISTORICAL ACTIVITY

2nd century B.C. to A.D. 1198: The caldera has a long history of uplift and subsidence as recorded in the geologic record described above. The caldera is also a classic locality of so-called bradyseisms (Greek for slow earthquakes, referring to decades to centuries of uplift and subsidence), as recorded in historical documents and by once submerged, then elevated, and now once again submerged Roman structures (Palatino, 1826; Babbage, 1847; Parascandola, 1947). The famous Serapeo marketplace in Pozzuoli has two generations of floors, the second built when the first sank below sea level. Borings by <u>Lithodomus lithophagus</u> in the market's columns indicate as much as 11 m of subsidence by about A.D. 1000 (at an average rate of 1-2 cm/yr from the 1st century B.C) (Parascandola, 1947). Then, the same area rose about 2 m from A.D. 1000 to A.D. 1198, when an eruption occurred at Solfatara Crater (Parascandola, 1947; Caputo, 1979). A moderately strong earthquake occurred in 1198 in the general vicinity of Pozzuoli (Postpischl, 1985) in conjunction with (just before?) the eruption.

Slow uplift and subsidence are also known from other points in the Bay of Pozzuoli. Parascandola (1947, p. 54) cites accounts of a pier at Nisida, built in the second century B.C., which was 6 meters lower in the early 18th century than when built. Roman ruins are also found below sea level along the northern and western shores of the Bay of Pozzuoli, near Monte Nuovo (Baia) and Miseno respectively (Palatino, 1826; Parascandola, 1947).

PHLEGRAEAN FIELDS (continued)

HISTORICAL ACTIVITY (continued)

A.D. 1198-1538: The uplift that began in about A.D. 1000 continued unabated after the 1198 eruption. In 1456, a strong tectonic earthquake beneath the Campanian Apennines "ruined" Pozzuoli, and strong local earthquakes occurred near Pozzuoli in 1488 (Parascandola, 1947, citing various sources). Uplift accelerated in about 1500 (fig. 1.4.6), and by 1503 enough uplift had occurred that newly emerged shoreline was deeded to the local university (details in Parascandola, 1947). Net uplift from 1000 to 1538 was about 12 m, at an average rate of 2+ cm/yr. More than half of the uplift occurred after 1500 at an average rate of 18 cm/yr, and the uplift rate was considerably higher than that during the last few years before the eruption. Frequent earthquakes were felt in 1534 and in 1536-38, apparently from sources along the eastern and southern rim of the caldera (Caputo, 1979). Some of the earthquakes in 1534 occurred on the Friday and Saturday before Easter, one during a gospel reading about the earthquake that occurred when Christ rose again. "This greatly increased the devotion of the people" (Parascandola, 1946, citing an unpublished work by Cola Aniello Pacca). Fumarolic activity increased during 1537-38 in the Sudatoio area, where Monte Nuovo was soon to form (Gruppo Nazionale per la Vulcanologia, 1986). Earthquake intensities reached a climax during 26-27 September 1538: "On the day and in the night before the appearance of this eruption, above twenty shocks great and small were felt at (Pozzuoli and Naples)" (delli Falconi, 1538). The most rapid uplift also occurred during 26-27 September, when the ground level at Pozzuoli rose at least 4-5 m in 48 hours, causing the shoreline to recede 400 m (del Nero, ca. 1538). On 28 September, hot and cold water poured from fissures in the uplifted block, and on the morning of 29 September the uplifted ground subsided by as much as 4 m, especially along the plain between Lake Averno and Monte Barbaro (del Nero, ca. 1538; da Toledo, 1539; J. Dvorak, written commun., 1987). According to del Nero, land near what was soon to become the eruptive vent stopped subsiding about noon on 29 September and began to rise once again, so that "by eight o'clock it was as high as Monte Ruosi--i.e., it was as high as that hill where the little tower stands." The eruption site (Monte Nuovo) was about 3 km from the dramatic uplift of Pozzuoli; reports do not indicate whether a similar or greater amount of precursory uplift occurred near the eruption site, although reports of subsidence near the eruption site suggest prior uplift. A few hours after sunset, "fire issued forth with such force, noise, and light that I, who was standing in my garden, was struck with terror" (del Nero, ca. 1538). "It was on a Sunday, about an hour in the night; and, as I have been informed, they began to see on that spot, between the hot baths or sweating rooms, and Trepergule, flames of fire, which first made their appearance at the baths, then extended towards Trepergule, and fixing in the little valley that lies between the Monte Barbaro and the hillock called del Pericolo, ... in a short time the fire increased to such a degree that it burst open the earth in this place, and threw up so great a quantity of ashes and pumice stones mixed with water, as covered the whole country...the poor inhabitants of Pozzuolo, struck with so horrible a sight, quitted their habitations...carrying fish which they had found, and were to be met with in plenty upon the shore, the sea having been at that time considerably dried up....I saw likewise two springs in those lately discovered ruins, one before the house that was the Queen's, of hot and salt water, the other of fresh and cold water, on the shore, about 250 paces nearer to the eruption: some say, that, still nearer to the spot where the eruption happened, a stream of fresh water issued forth like a little river" (delli Falconi, 1538). The above-mentioned "fire" was apparently burning gas from the Sudatoio fumaroles; combustion moved along a fracture toward the village of Tre Pergole (Trepergule), where the vent opened explosively and juvenile pumice was erupted (Gruppo Nazionale per la Vulcanologia, 1986). The eruption continued five days, stopped for two days, and then culminated in an explosion and pyroclastic surges on 6 October that killed 24 people who had climbed the new cone. Trees as far as 5 km from the vent were blown down (Gruppo Nazionale per la Vulcanologia, 1986). The volume of erupted material was 0.03 km^3.

PHLEGRAEAN FIELDS (continued)

HISTORICAL ACTIVITY (continued)

1550-1969: A series of local earthquakes occurred beneath the Pozzuoli area in 1564, 1566, 1568-69, 1570, 1582, and 1594 (Baratta, 1901). The next earthquake to be felt widely in the Pozzuoli area, in 1832, was apparently centered near Monte Nuovo; additional, weaker earthquakes occurred in 1887-1892. Earthquakes in 1908, 1913, and 1930 were located along the west-southwest margin of the caldera. An earthquake in the Irpinia region (ca. 50 km east of the Phlegraean Fields) on 23 July 1930 was followed by a week of unusually large explosions at a mud volcano at Solfatara, Pozzuoli (Signore, 1929). From 1538 to 1969, the Pozzuoli area subsided at an average rate of 1.5 cm/yr (fig. 1.4.7), except for a brief period of uplift in 1951-52.

1969-74: Uplift of the caldera floor resumed in mid-1969 (Berrino and others, 1984; fig. 1.4.8). A dense seismic network installed in February 1970 recorded A-type earthquakes until autumn 1972, when uplift and seismicity waned (figs. 1.4.9-1.4.11). Most earthquakes in the 1969-72 episode were less than $\underline{M}$=1 ($\underline{M}_{max}$=3.0) and less than 4 km deep; the total seismic energy release in this episode was about 10^{17} ergs. Two bench marks near the center of the caldera (11 and 15 of Corrado and others, 1977) rose approximately 170 cm between 1969 and late 1972 at a maximum rate of 5 mm/day, then subsided about 20-30 cm by late 1974. From then until mid-1982, the bench marks rose and fell 10-15 cm with an annual period (Smithsonian Institution, 1984, from G. Luongo, R. Scandone, and F. Barberi).

1982-85: A magnitude 6.9 earthquake occurred on 23 November 1980, 100 km southeast of the Phlegraean Fields. Conceivably, this earthquake might have triggered or otherwise foreshadowed renewed unrest during 1982-85. Uplift resumed in July 1982 (figs. 1.4.8, 1.4.12, 1.4.13), and seismicity began to increase in March 1983 (Barberi and others, 1984; Berrino and others, 1984). A tide gauge at Pozzuoli Harbor near the deformation center indicated about 140 cm of uplift from July 1982 through January 1985, at an average rate of 1.5 mm/day. The uplift rate reached a maximum of 5 mm/day during periods of high seismicity in September-October 1983 and March-April 1984. Cumulative uplift at Pozzuoli was 185 cm from June 1982 to December 1984, and about 330 cm from 1969 to 1985. Uplift was detected to a radius of 7 km from Pozzuoli (figs. 1.4.14-1.4.16). Horizontal extension along a 4-km line that crosses the area of maximum uplift reached 40 cm by January 1984 (fig. 1.4.17). The patterns of vertical and horizontal deformation are best explained by a source at a constant depth of about 3 km. The volume of surface deformation from January 1982 to March 1984 was about 0.087 km^3 (Berrino and others, 1984), and that from 1969 to 1985 is estimated to be about 0.15 km^3. Shallow earthquakes up to magnitude 4 occurred during swarms in October 1983 and March 1984 (figs. 1.4.18-1.4.21). Total seismic energy release during 1982-84 (10^{21} ergs) was much greater than during 1969-72 (10^{17} ergs). Earthquake epicenters during 1982-85 were concentrated near Solfatara Crater, in Pozzuoli, along the north shoreline of the Bay of Pozzuoli, and beneath the Starza terrace just north of Pozzuoli. Focal mechanisms of 10 earthquakes near Solfatara Crater indicated normal faulting, with T axes that may be arranged tangentially to Solfatara (Smithsonian Institution, 1984, citing G. Luongo, R. Scandone, F. Barberi, and M. Carapezza). Typically during 1982-85, the rate of uplift increased for several weeks before each seismic swarm. However, decreases in uplift rate during constant seismic energy release in July-September 1984 indicated that uplift and seismicity are not directly linked. Activity began to wane in late 1984. The uplift rate steadily decreased starting in October 1984 and became null in December 1984. Slow deflation at an average rate of 0.4 mm/day that began in January 1985 totalled 5.1 cm by the end of April 1985. Seismicity decreased to very low levels in December 1984. Several months before the decrease in uplift and seismicity, the gas composition of a fumarole at Solfatara nearly

PHLEGRAEAN FIELDS (continued)

HISTORICAL ACTIVITY (continued)

returned to precrisis values. The episode of unrest that began in 1982 was considered exhausted, at least temporarily, at the end of 1984 (Barberi and others, 1984).

Other reported manifestations of unrest since 1969 include increased steaming, changes in vent geometry, temperature changes, and changes in the composition of fumarolic gases. At Solfatara Crater, the temperature of Bocca Grande fumarole increased from 150 °C to 159 °C from 1969 to 1979, continuing a trend that began at least as early as the early 1960's. During the 1982-84 episode of unrest, the temperature of Bocca Grande remained constant at 156-158 °C. The temperature of nearby Piccola Solfatara remained steady at 147 °C until 1973 or 1974, then increased 4 °C by 1979 (Palumbo and Battista, 1979). However, temperatures of both fumaroles have fluctuated historically by as much as 100 °C, and there is disagreement in the literature over the significance of recent changes. A new fumarolic vent opened in Solfatara Crater on 16-17 November 1984 with a violent emission of steam (Smithsonian Institution, 1984, citing G. Luongo, R. Scandone, and F. Barberi). The following geochemical changes were measured in fumaroles at Solfatara during 1982-84 (fig. 1.4.22): (1) increased steam/gas ratio; (2) increased S/C ratio, due mostly to increased H_2S; and (3) increased CH_4 emission (Barberi and others, 1984; Carapezza and others, 1984; Cioni and others, 1984; Martini and others, 1984; Martini, 1986). Interestingly, H_2, CH_4, and S began to decrease at the beginning of 1984, about 6 months before inflation and seismicity slowed notably (Gruppo Nazionale per la Vulcanologia, 1986) (fig. 1.4.22). Increased hydrogen gas emission preceded earthquakes in September-October 1983 by a few days, as they have during recent unrest at Long Valley Caldera (M. Sato, written commun., 1983). A small, seasonally adjusted increase in radon emission occurred in wells near Solfatara and Monte Nuovo from December 1983 through April 1984, accompanied by a small increase in temperature (Smithsonian Institution, 1985, citing G. Luongo, R. Scandone, R. Pece, and F. Barberi).

COMMENTS

The Phlegraean Fields' long history of dramatic ground movements is unsurpassed anywhere in the world. Since Roman times, the elevation of the caldera floor has varied by more than 12 m; in the 48 hours before the most recent eruption in 1538, the floor rose by at least 4-5 m. Several mechanisms have been invoked to explain historical unrest at the Phlegraean Fields. Oliveri del Castillo and Quagliariello (1969) and Grindley (1976) concluded that the 1503-11 and 1969-72 uplift episodes are best explained by increased pore pressure owing to heating of ground water. Grindley (1976) suggested that "even a 10°C heating of a 3 km column of rock with 20% porosity would lead to a gradual rise in ground level of 2 metres, assuming one third of the stresses were relieved upwards." Corrado and others (1977) inferred that the uplift was caused by intrusion of a relatively small-diameter, steeply dipping cylinder or narrow dike, with its top at about 2.5-km depth. More recently, Barberi and others (1984) concluded that the 1982-84 unrest was caused by "a pressure build up, and consequent energy release in the form of uplift, earthquakes and thermal flux, in a shallow magma chamber of small dimensions." In this model, heating of ground water may have contributed to the unrest by increasing pore pressures and inducing additional seismicity (G. Luongo and F. Barberi, unpublished briefing notes; Smithsonian Institution, 1985, citing G. Luongo, R. Scandone, R. Pece and F. Barberi). Many eruptions in the Phlegraean Fields have been phreatomagmatic, and the reversible character of uplift is consistent with an important role for ground water in recent unrest.

PHLEGRAEAN FIELDS (continued)

COMMENTS (continued)

At the Phlegraean Fields, the earliest indications of increasing unrest have been geochemical, followed by uplift, and finally by seismicity. Similarly, geochemical changes foreshadowed the decrease in unrest in late 1984. This sequence has important implications for monitoring of unrest in similar settings or with similar mechanisms.

REFERENCES

Armienti, P., Barberi, F., Bizouard, H., Clocchiatti, R., Innocenti, F., Metrich, N., Rosi, M., and Sbrana, A., 1983, The Phlegraean Fields: Magma evolution within a shallow chamber: Jour. Volcanol. Geotherm. Res., v. 17, p. 289-311.

Babbage, C., 1847, Observations on the Temple of Serapis, at Pozzuoli, near Naples, with remarks on certain causes which may produce geological cycles of great extent: Quart. Jour. Geol. Soc. London, v. 3, no. 1, p. 186-217.

Baratta, M., 1901, I Terremoti d'Italia: Torino, Fratelli Bocca Editori. (English translation by J. Dvorak)

Barberi, F., Corrado, G., Innocenti, F., and Luongo, G., 1984, Phlegraean Fields 1982-1984: Brief chronicle of a volcano emergency in a densely populated area: Bull. Volcanol., v. 47-2, p. 175-185.

Barberi, F., Innocenti, F., Lirer, L., Munno, R., Pescatore, T., and Santacroce, R., 1978, The Campanian Ignimbrite: a major prehistoric eruption in the Neapolitan area (Italy): Bull. Volcanol., v. 41, p. 10-31.

Berrino, G., Corrado, G., Luongo, G., and Toro, B., 1984, Ground deformation and gravity changes accompanying the 1982 Pozzuoli uplift: Bull. Volcanol., v. 47-2, p. 187-200.

Bullard, F.M., 1976, Volcanoes of the Earth (rev. ed.): Austin, Univ. Texas Press, 579 p.

Caputo, M., 1979, Two thousand years of geodetic and geophysical observations in the Phlegrean Fields near Naples: Geophys. Jour. Royal Astron. Soc., v. 56, p. 319-328.

Carapezza, M., Guerrieri, S., Nuccio, P.M., and Valenza, M., 1984, CO_2 and H_2S concentrations in the atmosphere at the Solfatara of Pozzuoli: Bull. Volcanol., v. 47-2, p. 287-293.

Casertano, L., Oliveri del Castillo, A., and Quagliariello, M.T., 1976, Hydrodynamics and geodynamics in the Phlegraean Fields area of Italy: Nature, v. 264, p. 161-164.

Cinque, A., Rolandi, G., and Zamparelli, V., 1985, L'estensione dei depositi marini olocenici nei Campi Flegrei in relazione alla vulcano-tettonica: Boll. Soc. Geol. Ital., v. 104, p. 327-348.

Cioni, R., Corazza, E., and Marini, L., 1984, The gas/steam ratio as indicator of heat transfer at the Solfatara fumaroles, Phlegraean Fields (Italy): Bull. Volcanol., v. 47-2, p. 295-302.

Corrado, G., Grimaldi, M., LoBascio, A., and Luongo, G., 1983a, Ground deformation in the Phlegraean Fields volcanic area, Italy: Proc. 7th Ann. Mtg., European Geophys. Soc., Budapest.

Corrado, G., Guerra, I., LoBascio, A., Luongo, G. and Rampoldi, R., 1977, Inflation and microearthquake activity of Phlegraean Fields, Italy: Bull. Volcanol., v. 40, p. 169-188.

Corrado, G., LoBascio, A., and Luongo, G., 1983b, A sudden uplift in the Phlegraean Fields (southern Italy): The short term volcanic risk in a densely inhabited area [abs.]: Abstracts Volume, IUGG Gen. Assem., Hamburg, 1983, v. 1, p. 360.

PHLEGRAEAN FIELDS (continued)

REFERENCES (continued)

Corrado, G., and Luongo, G., 1981, Ground deformation measurements in active volcanic areas by tide gauges: Bull. Volcanol., v. 44, no. 3, p. 505-511.

da Toledo, Pietro Giacomo di, 1539, Ragionamento del Terremoto, del Nuovo Monte, del Aprimento di Terra in Pozzuolo nell'Anno 1538, e della significatione d'essi: English translation in Hamilton, W., Sir, 1774, Observations on Mount Vesuvius, Mount Etna, and other volcanoes: London, T. Cadell, p. 137-141.

delli Falconi, Marco Antonio, 1538, Dell Incendio di Pozzuoli: English translation in Hamilton, W., Sir, 1774, Observations on Mount Vesuvius, Mount Etna, and other volcanoes: London, T. Cadell, p. 127-136.

del Nero, Francesco, ca. 1538, Lettera di Niccolo del Benino sul terremoto di Pozzuolo dal quale ebbe origine la Montagna Nuova nel 1538: Manuscript translated and published in English in Horner, L., 1846, Quart. Jour. Geol. Soc. London, v. 3, no. 2, p. 19 ff.

Di Girolamo, P., Ghiara, M.R., Lirer, L., Munno, R., Rolandi, G., and Stanzione, D., 1983, Vulcanologia e petrologia dei Campi Flegrei: Boll. Soc. Geol. Ital., v. 103, p. 349-413.

Di Vito, M., Lirer, L., Mastrolorenzo, G., and Rolandi, G., 1987, The 1538 Monte Nuovo eruption (Campi Flegrei, Italy): Bull. Volcanol., v. 49, p. 608-615.

Finetti, I. and Morelli, C., 1974, Esplorazione sismica a riflessione dei Golfi di Napoli e Pozzuoli: Boll. Geofis. Teor. Appl., v. 16, p. 175-222.

Grindley, G.W., 1976, Relation of volcanism to earth movements, Bay of Naples, Italy, in Gonzales-Ferran, O., ed., Proc. Symp. on Andean and Antarctic volcanology problems: Rome, IAVCEI, p. 598-612.

Gruppo Nazionale per la Vulcanologia C.N.R. Italia, 1986, Phlegraean Fields: The 1982-1984 bradyseismic crisis (posters prepared for the 1986 IAVCEI Int. Volcanol. Congr., New Zealand).

Imbo, G., 1965, Catalogue of active volcanoes of the world including solfatara fields, Pt. XVIII, Italy: Rome, IAVCEI, 72 p.

LaTorre, P., and Nannini, R., 1980, Geothermal well location in southern Italy: The contribution of geophysical methods: Boll. Geofis. Teor. Appl., v. 22, no. 87, p. 201-209.

Lirer, L., Luongo, G., and Scandone, R., 1987, On the volcanological evolution of Campi Flegrei: Eos, Trans. Am. Geophys. Union, v. 68, no. 16, p. 226-234.

Lyell, C., 1875, Principles of geology (12th ed.), 2 v.: London, John Murray.

Martini, M., 1986, Thermal activity and ground deformation at Phlegrean Fields, Italy: Precursors of eruptions or fluctuations of quiescent volcanism? A contribution of geochemical studies: Jour. Geophys. Res., v. 91, no. B12, p. 12,255-12,260.

Martini, M., Cellini Legittimo, P., Piccardi, G., and Giannini, L., 1984, Composition of hydrothermal fluids during the bradyseismic crisis which commenced at Phlegraean Fields in 1982: Bull. Volcanol., v. 47-2, p. 267-273.

Nunziata, C., and Rapolla, A., 1981, Interpretation of gravity and magnetic data in the Phlegraean Fields geothermal area, Naples, Italy: Jour. Volcanol. Geotherm. Res., v. 9, p. 209-225.

Oliveri del Castillo, A., and Quagliariello, M.T., 1969, Sulla genesi del bradisismo flegreo: Atti Assoc. Geofis. Ital., XVIII Conv., Naples, p. 557-594.

Palatino, L., 1826, Storia di Pozzuoli e Contorno: Napoli, published by the author (English translation by M. Capuano and J. Dvorak).

Palumbo, A., and Battista, R., 1979, Variazione geotermiche alla Solfatara di Pozzuoli: Boll. Soc. Nat. Napoli, v. 88, p. 319-330.

PHLEGRAEAN FIELDS (continued)

REFERENCES (continued)

Parascandola, A., 1946, Il Monte Nuovo ed il Lago Lucrino: Boll. Soc. Nat. Napoli, v. 55, p. 151-312 (English translation by G. Mastrolorenzo, M. Capuano, and J. Dvorak).

Parascandola, A., 1947, I fenomeni bradisismici del Serapeo di Pozzuoli: Napoli, Tip. Genovese, 156 p. (English translation by M. Capuano and J. Dvorak).

Porzio, Simone, 1551, De conflagratione agri puteolani Epistola Simoni Portii Neapolitan, published in Italian in I Tre Rarissimi Opuscoli di Simone Porzio, di Girolano Borgia, e di Marcantonio delli Falconi, scritti in occasione della Celebre Eruzione avvenuta in Pozzuoli nell'anno 1538, colle Memorie Storiche de' suddetti autore, Racotte da Lorenzo Giustiniani, Bibliotecario della Real Biblioteca borbonica, e Regio Revisore: Napoli, 1817. (English translation courtesy of J. Dvorak, 5 p.)

Postpischl, D., ed., 1985, Catalogo dei Terremoti Italiani dall'anno 1000 al 1980: Quaderni de "La Ricerca Scientifica" no. 114, v. 2B, Consiglio Nazionale delle Ricerche.

Rosi, M., and Santacroce, R., 1984, Volcanic hazard assessment in the Phlegraean Fields: a contribution based on stratigraphic and historical data: Bull. Volcanol., v. 47-2, p. 359-370.

Rosi, M., Sbrana, A., and Principe, C., 1983, The Phlegraean Fields: Structural evolution, volcanic history, and eruptive mechanisms: Jour. Volcanol. Geotherm. Res., v. 17, p. 273-288.

Rosi, M., and Sbrana, A., eds., 1987, Phlegrean Fields: Rome, Consiglio Nazionale delle Richerche (CNR), Monograph 9, 175 p. (CNR Quaderni de "La Ricerca Scientifica" no. 114).

Sicardi, L., 1979, Alcune osservazioni sulle tre fasi storiche del bradisismo puteolano rivelate dal Serapeo di Pozzuoli (Napoli): Boll. Soc. Nat. Napoli, v. 88, p. 27-47.

Signore, F., 1929, Sur la variation d'activite du volcan de Boue ("fangaia") de la Solfatare de Pouzzoles (Naples) par suite du grand tremblement de terre de l'Irpinia. 23 Juillet 1930: Bull. Volcanol., ser. I, nos. 19-22, p. 49-52.

Smithsonian Institution, Scientific Event Alert Network (SEAN), 1983-85, Campi Flegrei: SEAN Bull., v. 7-11 (with contributions from G. Luongo, R. Scandone, R. Pece, and F. Barberi).

Yokoyama, I., 1971, Pozzuoli event in 1970: Nature, v. 229, p. 532-533.

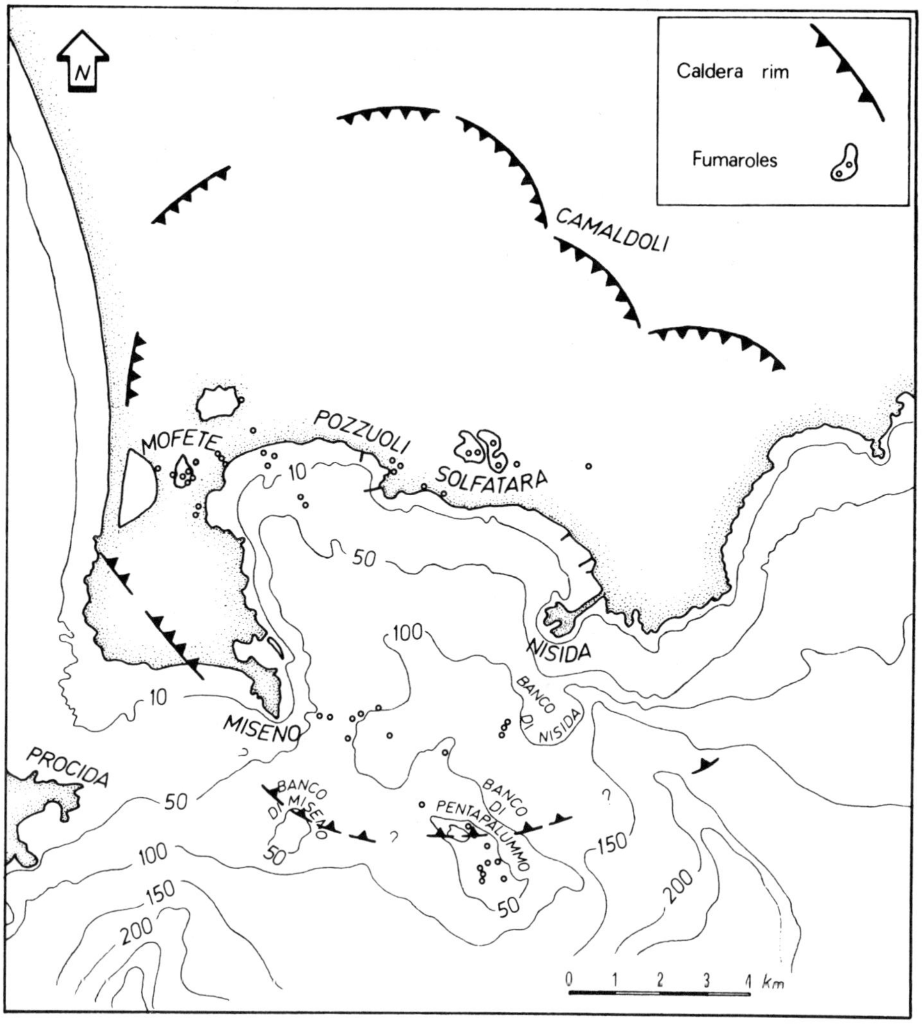

Figure 1.4.1. Outline of Phlegraean Fields Caldera and bathymetry in Pozzuoli Bay, from Rosi and Sbrana (1987). Lirer and others (1987) show an alternate outline with southeast rim of caldera passing through Nisida, thus making caldera smaller and more nearly circular.

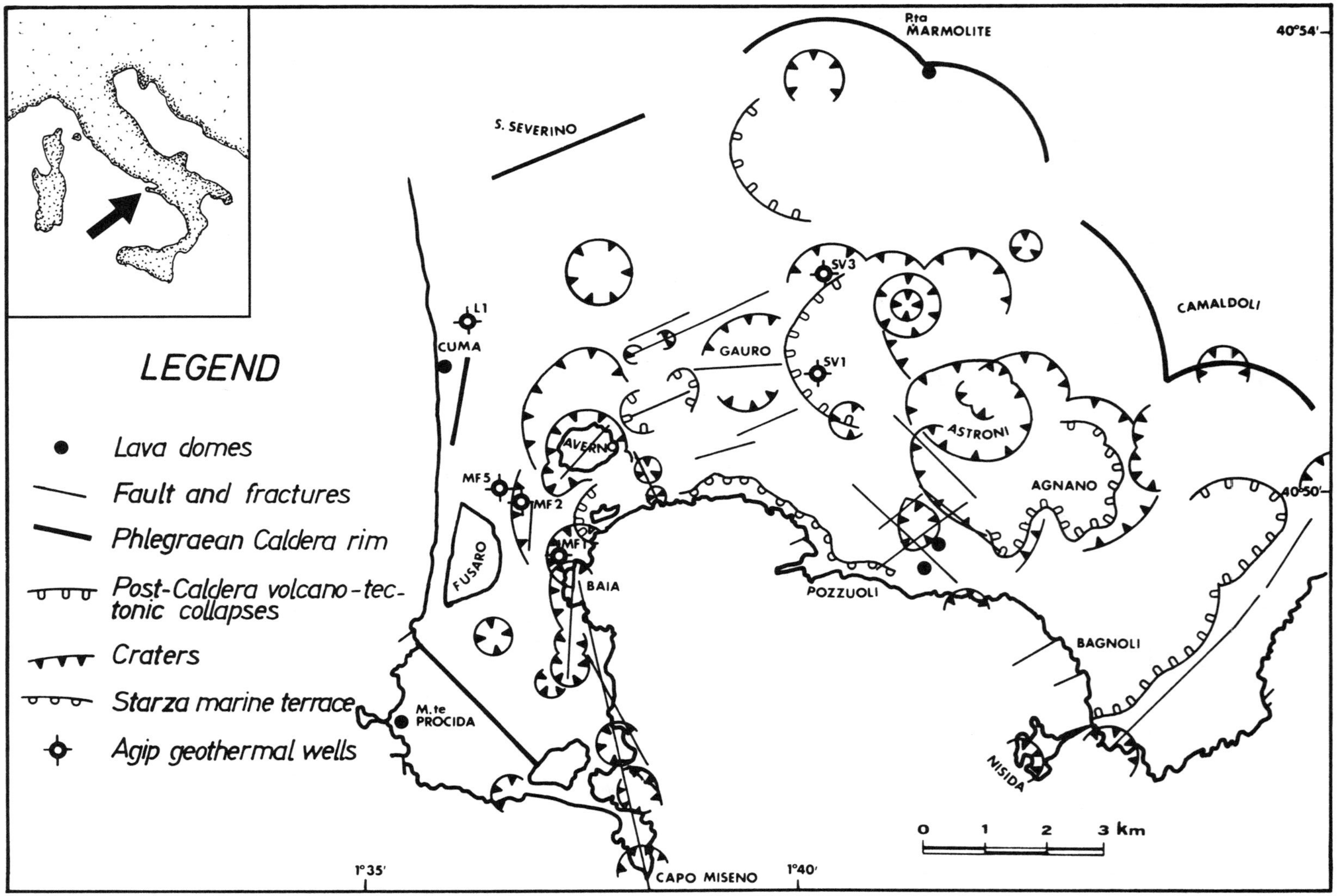

Figure 1.4.2. Volcano-tectonic sketch map of Phlegraean Fields showing location of postcaldera vents and deep geothermal wells, from Rosi and Sbrana (1987).

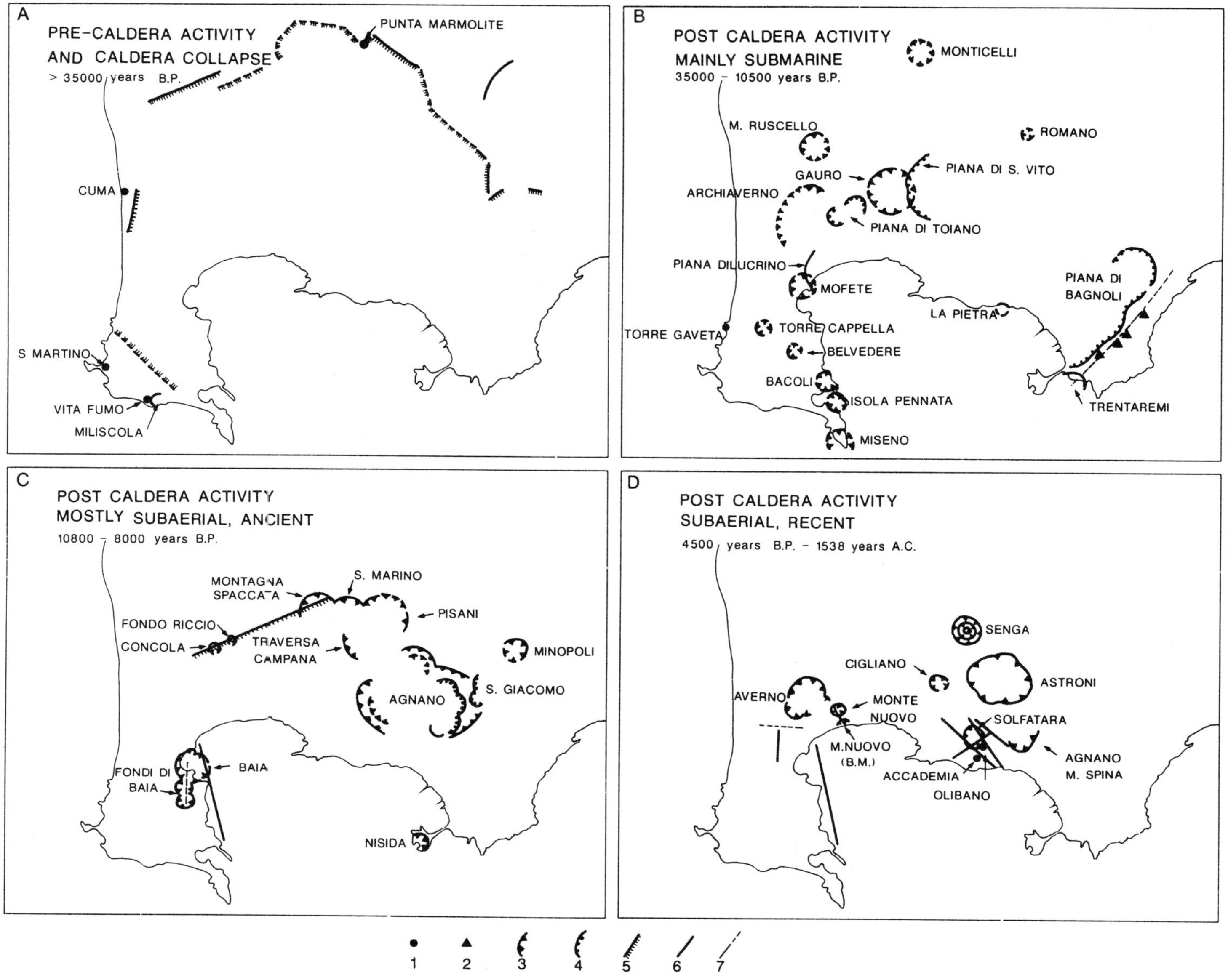

Figure 1.4.3. Main phases of Phlegraean Fields evolution, redrawn from Rosi and others (1983). Explanation: 1, lava dome; 2, inferred eruptive centers; 3, crater rim; 4, volcano-tectonic collapse; 5, caldera rim; 6, fractures; 7, faults.

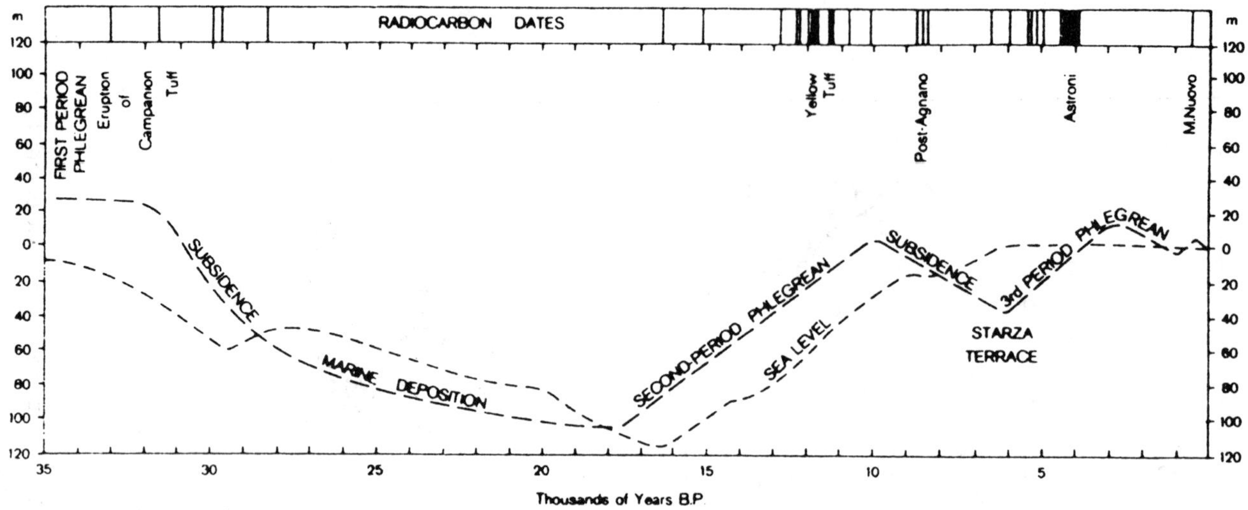

Figure 1.4.4. Late Quaternary earth movements in Bay of Naples region and their relation to major periods of volcanism in Phlegraean Fields, from Grindley (1976).

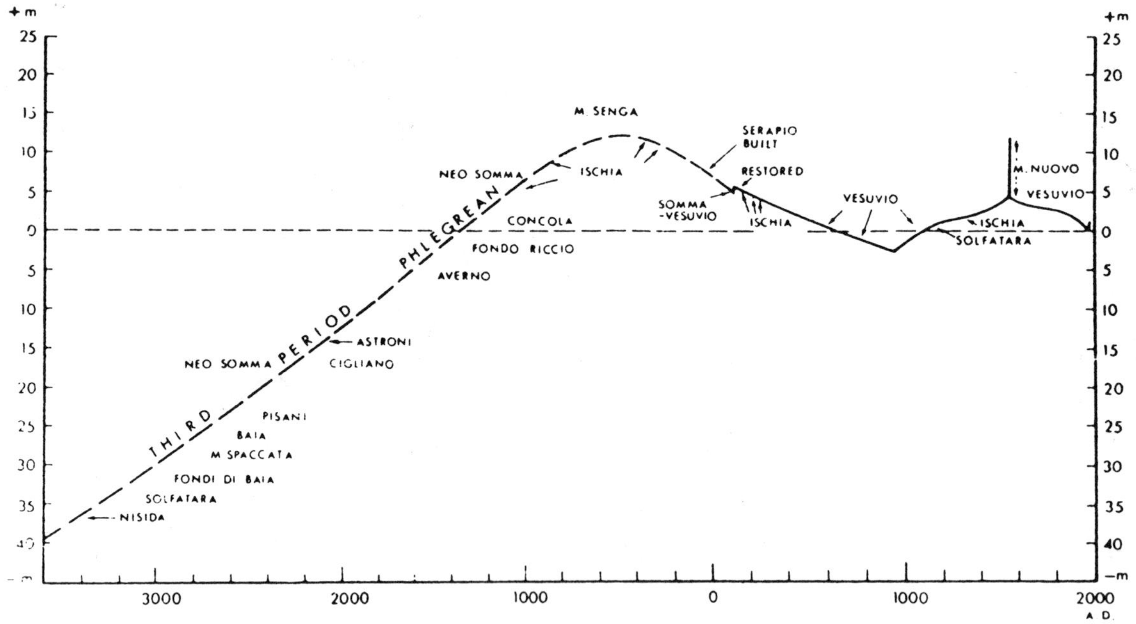

Figure 1.4.5. Uplift and subsidence at Pozzuoli during the last 5,000 years and relation to volcanism at Phlegraean Fields, Somma-Vesuvius, and Ischia, from Grindley (1976).

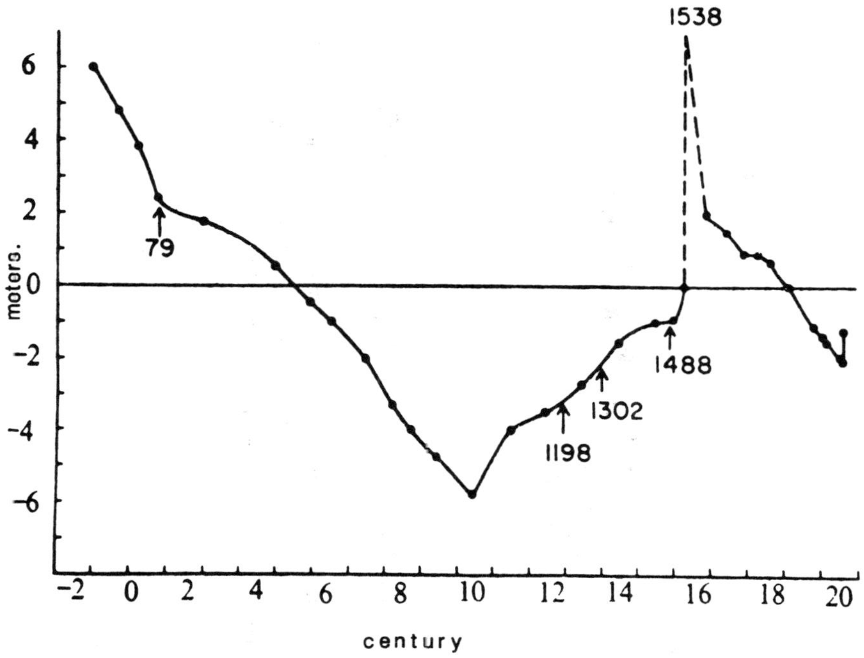

Figure 1.4.6. Changes in height of the floor of Roman marketplace Serapeo and relation to eruptions at Vesuvius (A.D. 79), Phlegraean Fields (1198 and 1538), and Ischia (1302), from A. Parascandola, cited by Yokoyama (1971). Also noted is occurrence of numerous earthquakes felt at Pozzuoli in 1488, followed by rapid uplift that culminated in the 1538 eruption of Monte Nuovo at Phlegraean Fields. Reprinted by permission from Nature; copyright 1971 Macmillan Magazines Ltd.

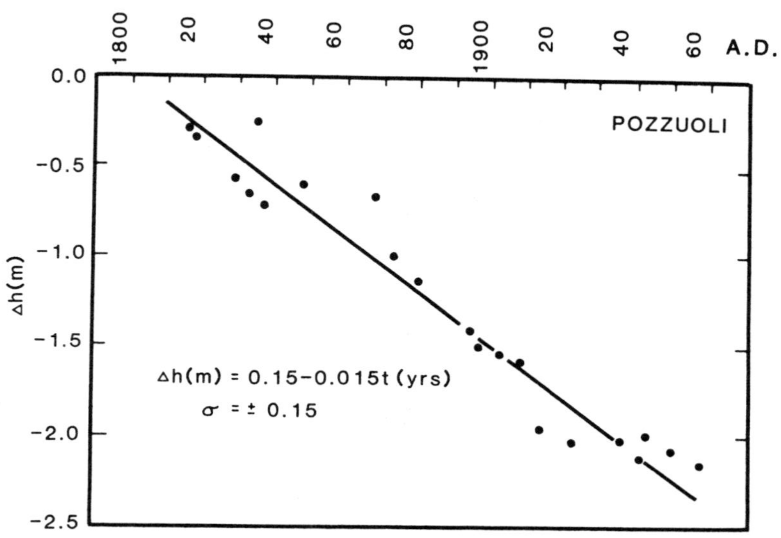

Figure 1.4.7. Subsidence at Pozzuoli, 1819 to 1969, from sea-level measurements at Serapeo (redrawn from Sicardi, 1979). Best-fit line yields an average subsidence rate of 0.015 m/yr.

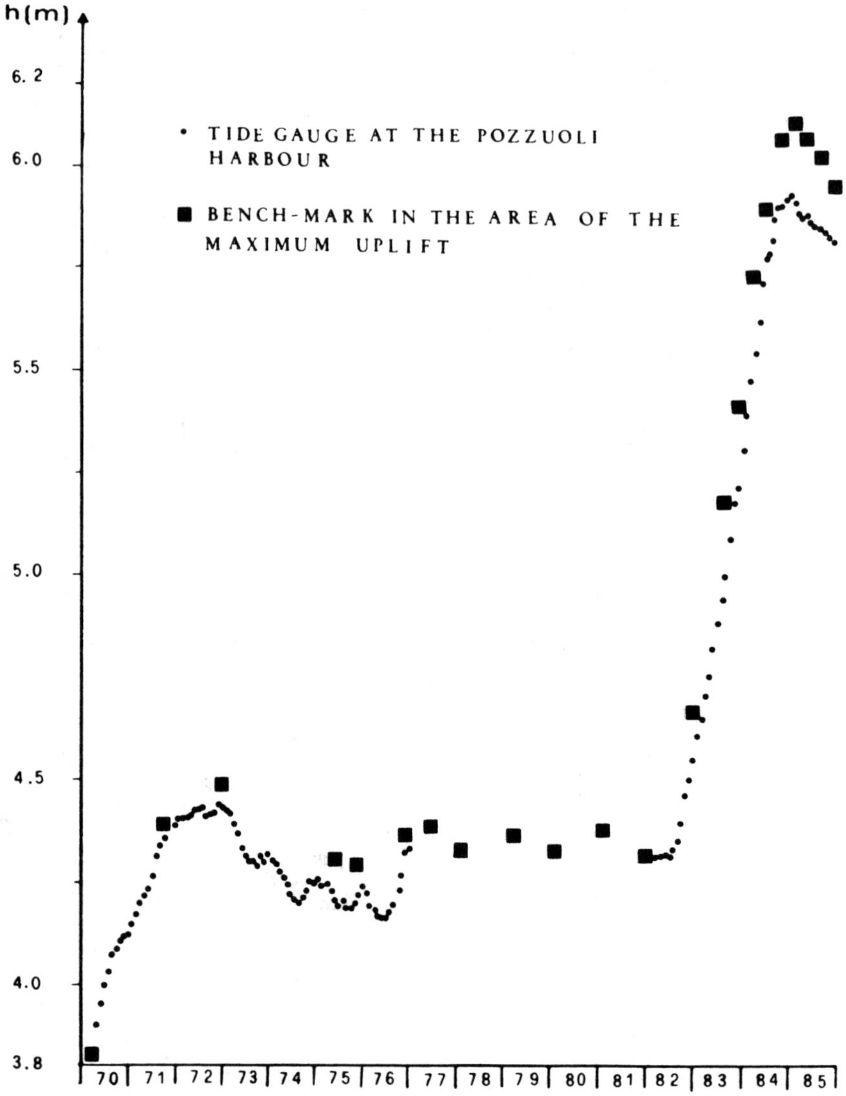

Figure 1.4.8. Vertical ground movements near center of Phlegraean Fields Caldera during 1970-85, from Gruppo Nazionale per la Vulcanologia (1986). About 1 m of additional uplift occurred from mid-1969 to early 1970 (Berrino and others, 1984).

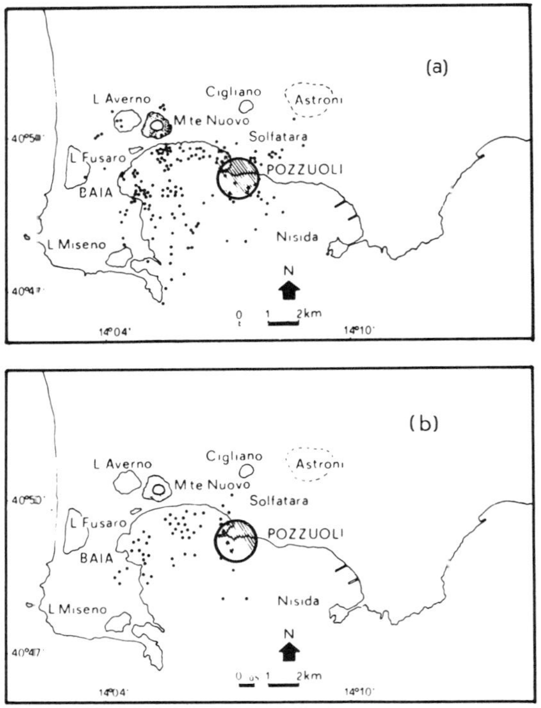

Figure 1.4.9. Epicenters of selected earthquakes at Phlegraean Fields: (a) 1970 to October 1972; (b) October 1972 to 1974, from Corrado and others (1977). Circles enclose epicentral area of B-type earthquakes.

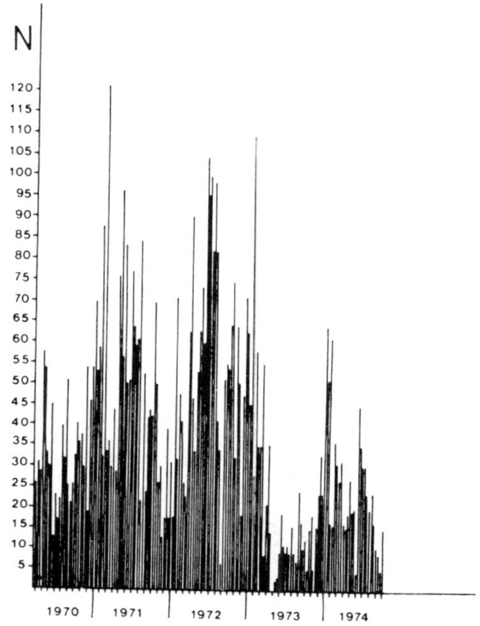

Figure 1.4.10. Number of earthquakes recorded at Pozzuoli in 10-day periods from April 1970 to September 1974, from Corrado and others (1977).

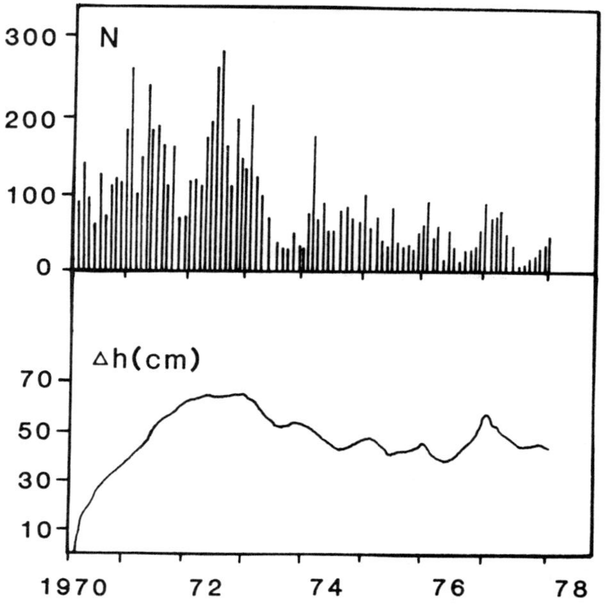

Figure 1.4.11. Number of earthquakes (top) and vertical ground deformation (bottom) in Pozzuoli area from 1970 to 1977, from Corrado and others (1983a).

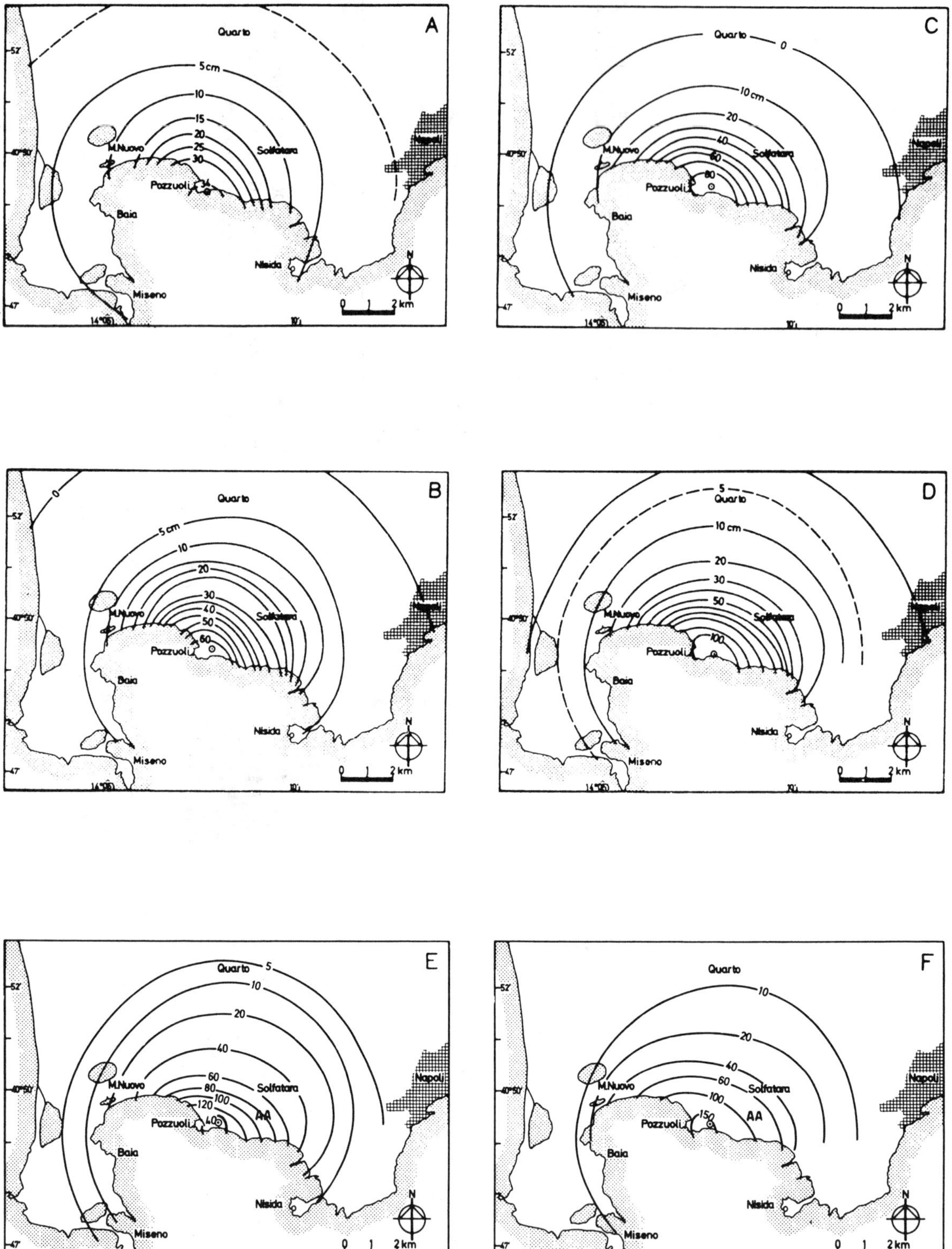

Figure 1.4.12. Areal distribution of vertical ground deformation at Phlegraean Fields during several intervals from January 1982 to June 1984, from Berrino and others (1984). Reference survey in January 1982; A, January 1983; B, June 1983; C, September 1983; D, December 1983; E, March 1984; F, June 1984.

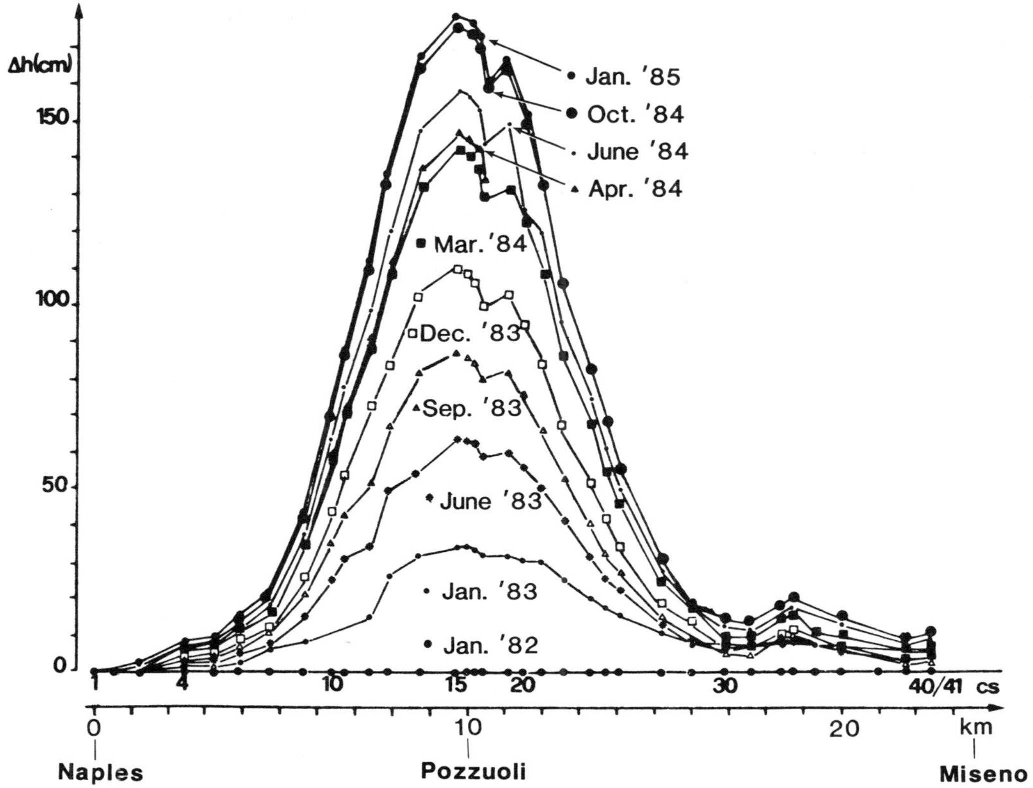

Figure 1.4.13. Uplift profiles along coastline from Naples to Pozzuoli to Miseno from January 1982 to January 1985 (Smithsonian Institution, February 1985, from G. Luongo, R. Scandone, R. Pece, and F. Barberi). Slow subsidence that began in January 1985 totaled 5.1 cm by the end of April 1985. Upper horizontal axis shows locations and numbers of bench marks along leveling route.

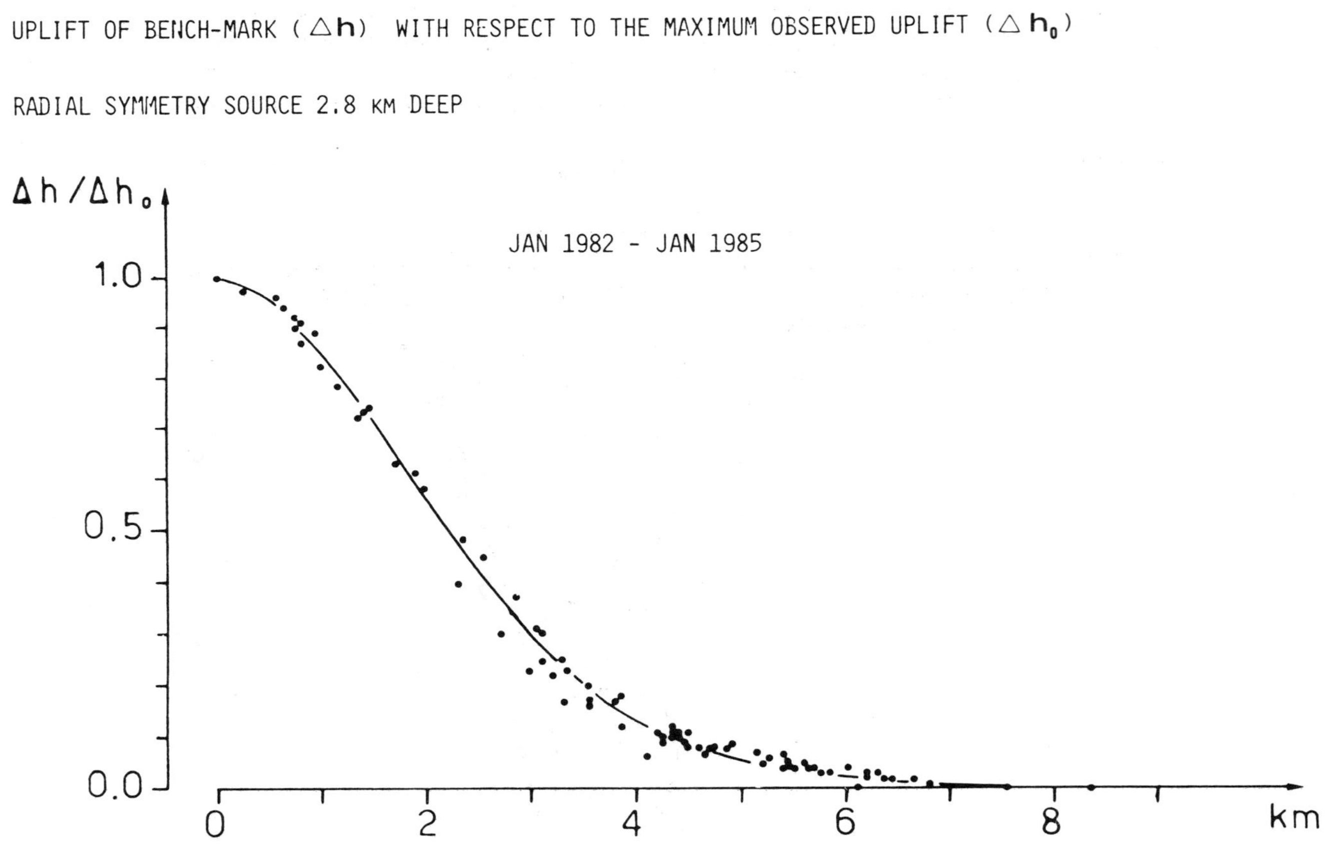

Figure 1.4.14. Uplift versus distance from point of maximum uplift for the period January 1982 to January 1985, from Gruppo Nazionale per la Vulcanologia (1986).

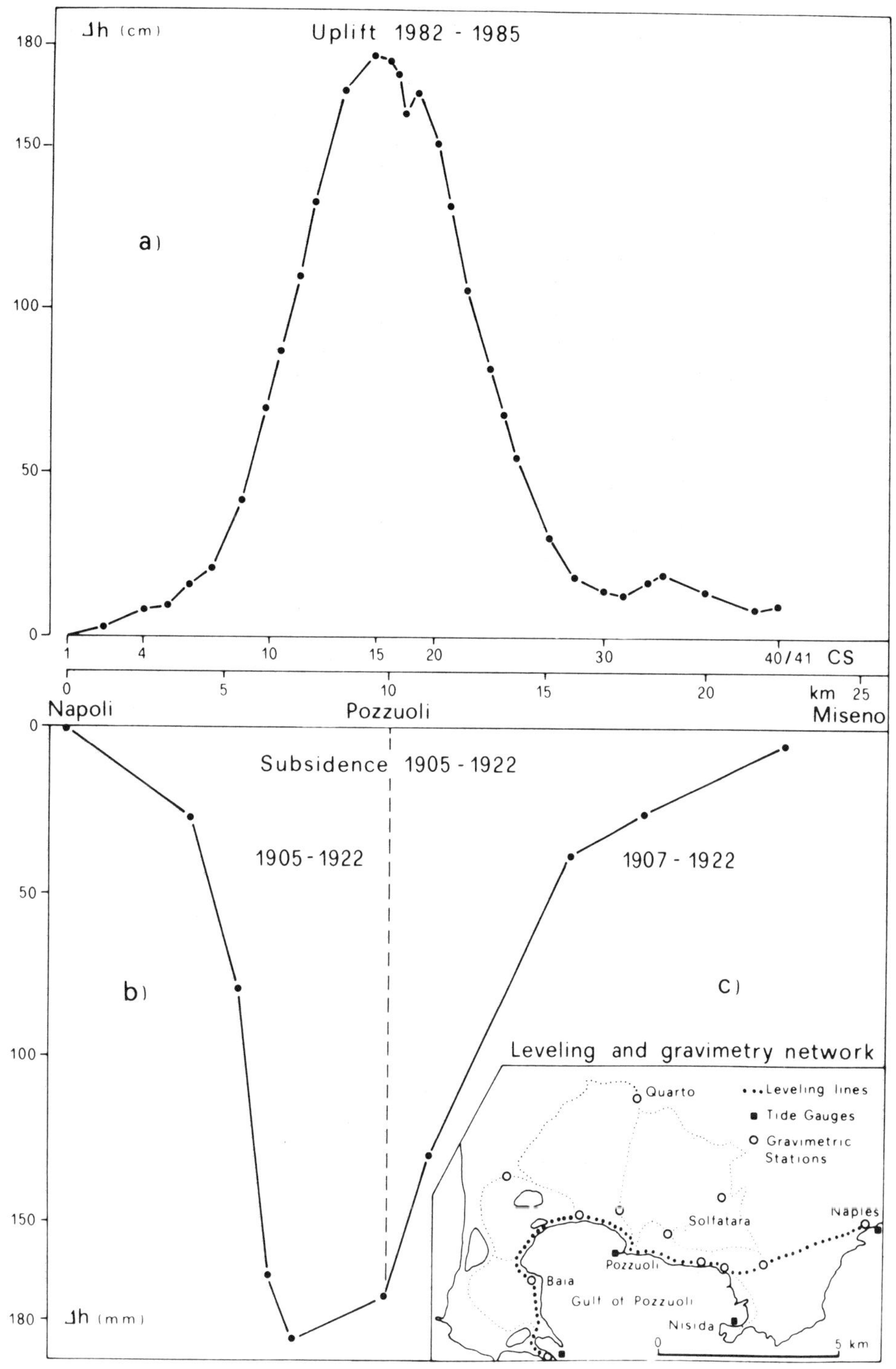

Figure 1.4.15. Vertical displacements at Phlegraean Fields for two intervals, from Lirer and others (1987). (a), Uplift during the period January 1982 to January 1985, measured along the Naples-Pozzuoli-Baia-Miseno leveling line. (b), Subsidence measured during 1905-22 along the Naples-Pozzuoli line and during 1907-22 along the Pozzuoli-Baia-Miseno line. Upper scale in centimeters; lower scale in millimeters. Inset (c), Locations of leveling lines, tide gauges, and gravity stations. Copyright by the American Geophysical Union.

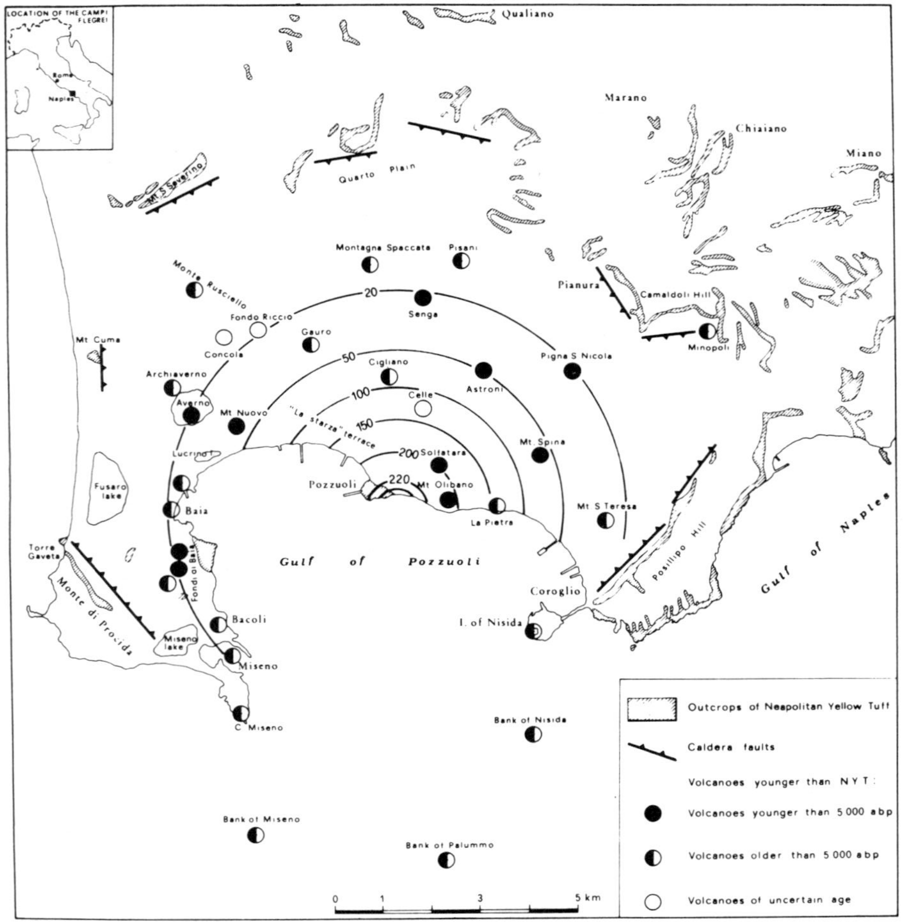

Figure 1.4.16. Contours of cumulative uplift, in centimeters, for the period 1970-85, from Lirer and others (1987). Maximum uplift near town of Pozzuoli is 240 cm. Copyright by the American Geophysical Union.

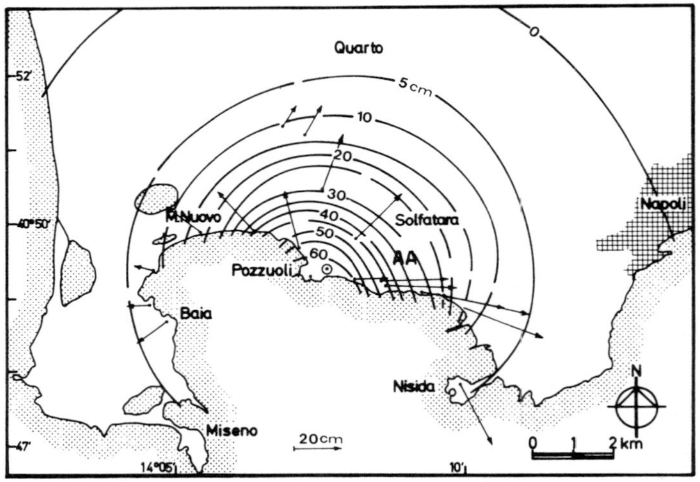

Figure 1.4.17. Horizontal displacements (arrows) and uplift contours (in centimeters) during 1980-83, from Berrino and others (1984). Contours are for uplift through September 1983.

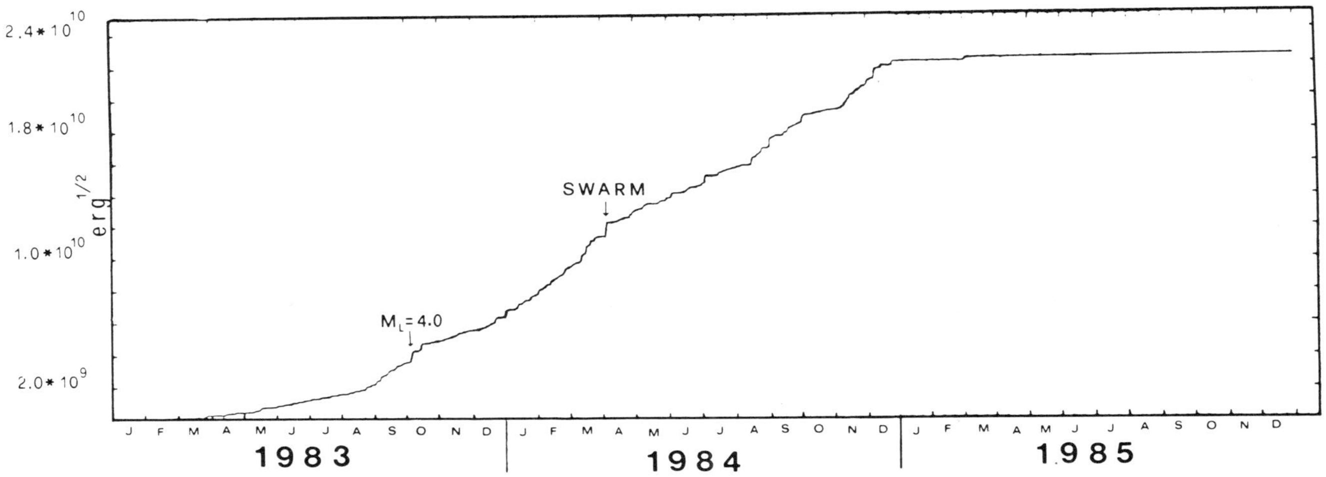

Figure 1.4.18. Strain energy release from earthquakes at Phlegraean Fields during 1983-85, from Gruppo Nazionale per la Vulcanologia (1986).

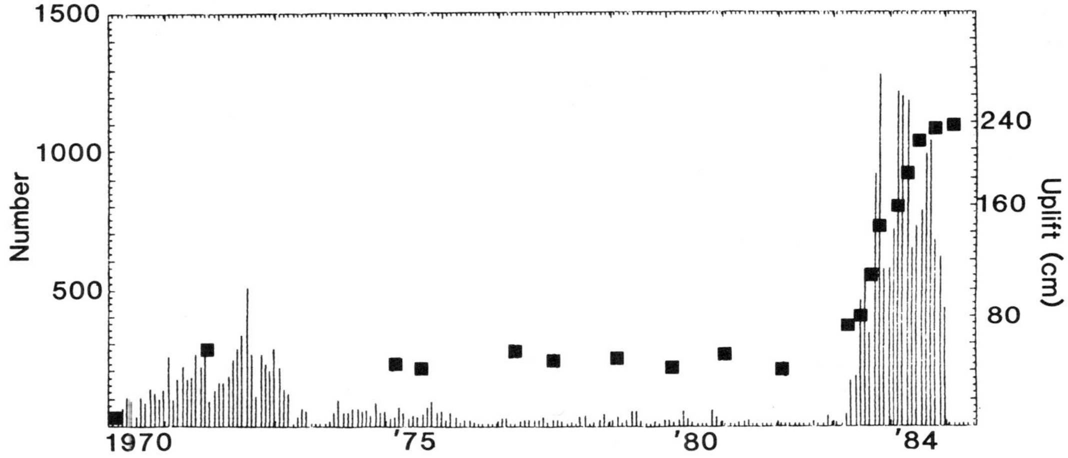

Figure 1.4.19. Uplift and seismicity at Phlegraean Fields from 1970 to 1984 (Smithsonian Institution, 1985, from G. Luongo, R. Scandone, R. Pece, and F. Barberi). Squares show relative height of bench mark 15 (Corrado and others, 1977); vertical lines show monthly number of recorded earthquakes.

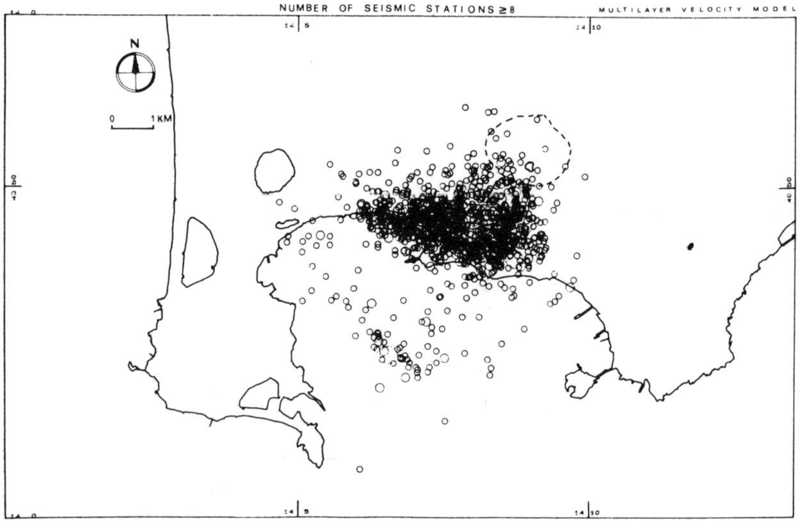

Figure 1.4.20. Distribution of epicenters at Phlegraean Fields for the period January 1983 to December 1984, from Gruppo Nazionale per la Vulcanologia (1986).

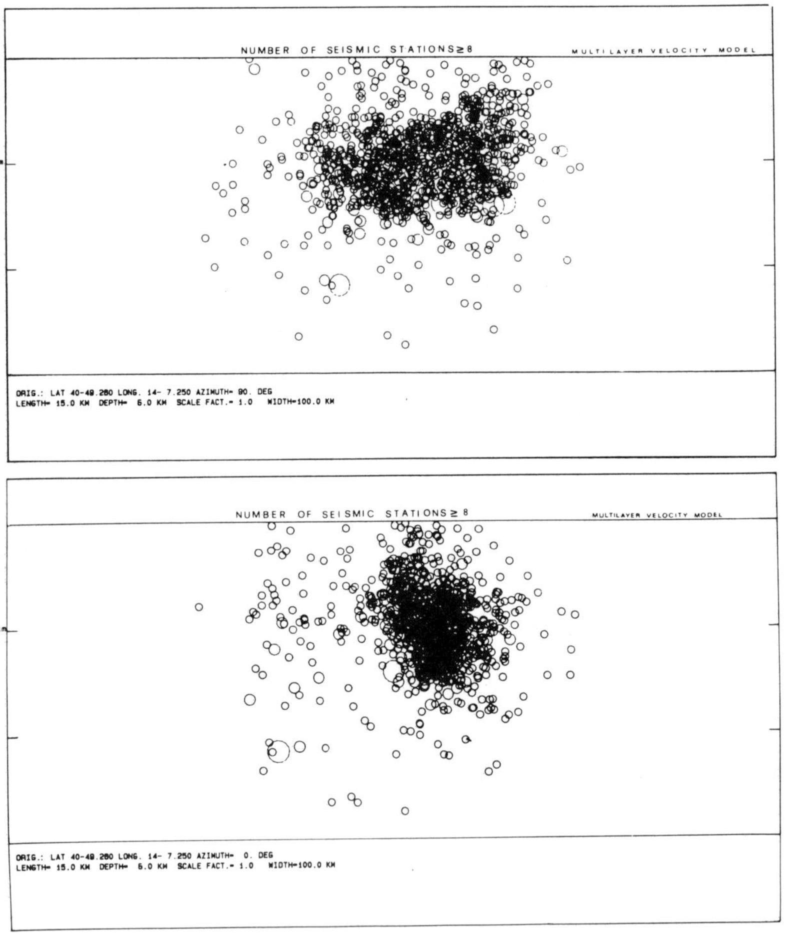

Figure 1.4.21. Epicenters from previous figure projected onto cross sections oriented east-west (top) and north-south (bottom), from Gruppo Nazionale per la Vulcanologia (1986).

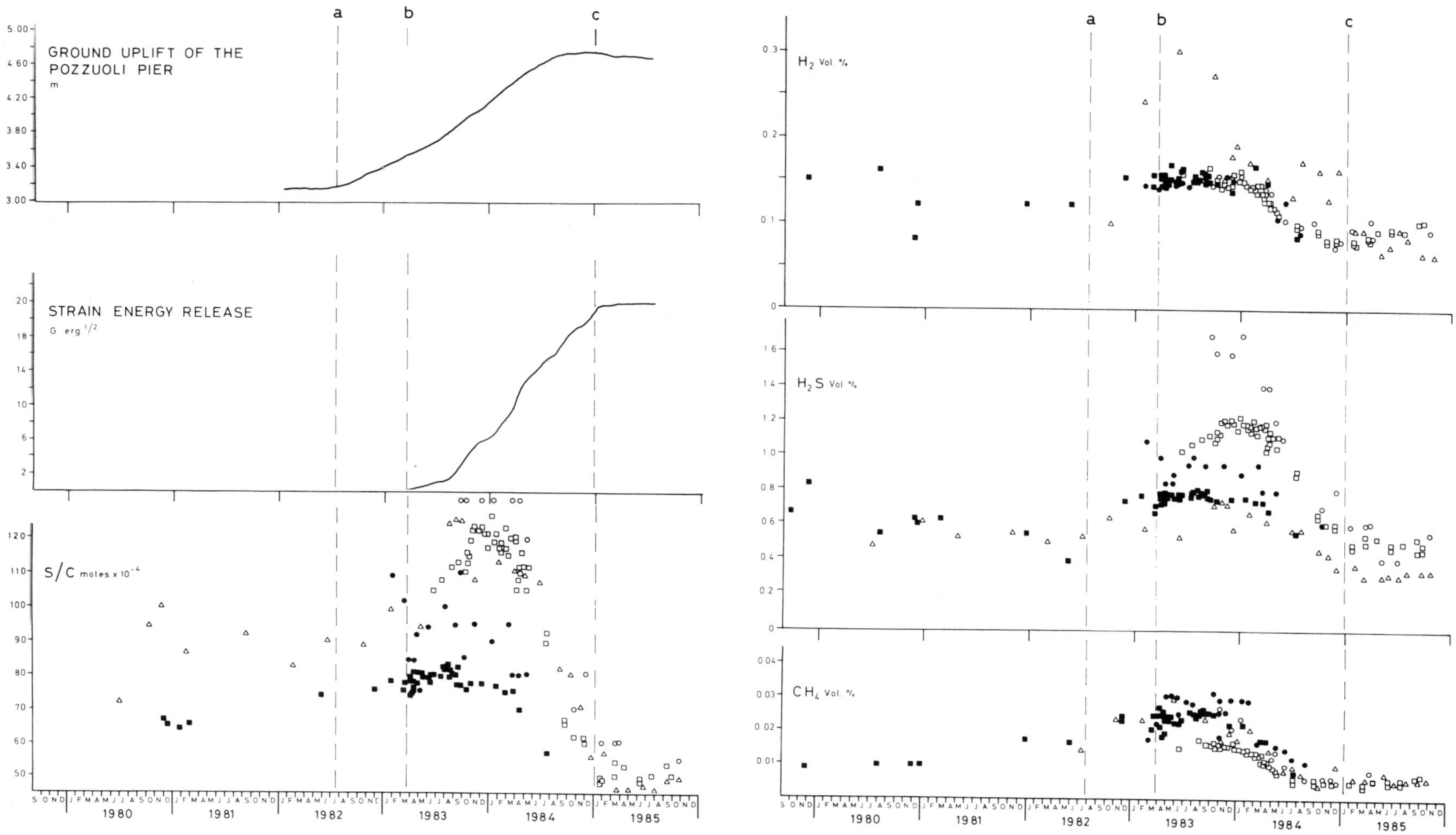

Figure 1.4.22. Temporal correlation between uplift, seismic energy release, and selected chemical parameters of fumarolic gases at Solfatara during 1979-85, from Gruppo Nazionale per la Vulcanologia (1986). Chemical changes were observed at Bocca Grande (open symbols) and Fumarola Circolare (solid symbols); dashed lines mark beginning of uplift (a), local seismicity (b), and cessation of both uplift and seismicity (c). Chemical data were produced by the Dipartimento di Scienze della terra, U. of Florence (triangles), jointly by Istituto di Geochimica dei Fluidi, C.N.R., Palermo and the Istituto di Geochimica Applicata, U. of Palermo (circles), and by the Istituto di Geocronologia e Geochimica Isotopica, C.N.R., Pisa (squares). Geophysical data were provided by the Osservatorio Vesuviano, Naples. There are systematic differences among data owing to different sampling and analytical techniques, but trends in data are similar. Significant changes in composition of fumarolic gases preceded both onset and cessation of uplift and seismicity.

PART 3: HISTORICAL UNREST AT LARGE QUATERNARY CALDERAS (continued)
See inside back cover for explanation and abbreviations

ISCHIA

CAVW number (active vent)	Latitude Longitude (degrees)	Diameter (km)	Local tectonic setting	Precaldera edifice	SiO_2 content (wt pct)	Age of collapse (yr B.P.)	Date of unrest	Type and duration of unrest ESTU STHF MGTF H Te	Eruption type
01-01-03	40.73N 13.90E	ca. 8 km	Exten	Strato?	R = 57 C = trachyte	ca. 53,000	11th century B.C.	---- ---- ---- - --	ex?
							B.C. 470	---- ---- ---- - --	ex?
							between B.C. 400 and B.C. 352	x--- ---- ---- - --	ex?
							A.D. 1228	x--- ---- ---- - =?	none
							1302	EE-- ---- ---- - --	ex, lf
							1557	x--- ---- ---- - =?	none
							1762	x--- ---- ---- - =?	none
							1767	x--- ---- ---- - =?	none
							1796	x--- ---- ---- - =?	none
							1827-28	xx-- ---- ---- - =?	none
							1841	x--- ---- ---- - =?	none
							1863	x--- ---- ---- - =?	none
							1867	x--- ---- ---- - =?	none
							1880-83	xx-- ---- ---- - =?	none
							1961	x--- FF-- ---- - =?	none

TECTONIC SETTING

Ischia Island lies on the northern edge of an ENE-WSW trending graben filled by the Gulf of Naples. The central, highest part of the island, Monte Epomeo, is a horst that has been interpreted by most as a dominantly tectonic feature.

GEOLOGIC HISTORY

Early eruptions of Ischia formed a volcanic complex that is now exposed in cliffs of the southern and especially southeastern shore of Ischia Island (fig. 1.5.1). Chiesa and others (1985) interpret a semicircular depression extending counterclockwise from Castello di Ischia to Punta Imperatore and perhaps to Monte di Vico as the relic moat of a caldera of unknown age. Large explosive eruptions about 53,000 yr ago (Gillot and others, 1982; Chiesa and others, 1985) produced the Tufo Verde (Green Tuff), a unit that is more than 1000 m thick and was originally deposited near or below sea level. Within the past 33,000 yr (Gillot and others, 1982) and probably within the past 20,000 yr (Chiesa and others), Monte Epomeo, a horst consisting largely of the Tufo Verde, has formed by resurgent upfaulting within the old caldera. At least 780 m of uplift in 20,000 yr implies an average uplift rate of at least 3.9 cm/yr.

ISCHIA (continued)

HISTORICAL ACTIVITY

Early 11th century B.C.: An Eretrian colony departed after their city was destroyed, apparently by an eruption of the Montagnone/Porto d'Ischia area (Fuchs, 1872; Johnston-Lavis, 1885). Rittmann (1930, p. 138) gives the date of this eruption as "before 500 B.C.", and the place as "Montagnone-Maschiatta?"

ca. B.C. 470: An eruption of Monte Marecocco and Zale (?) drove away the Syracusan colony (Johnston-Lavis, 1885).

B.C. 400-350: Sometime between the years B.C. 400 and B.C. 350, Monte Rotaro (?) erupted and drove away the Ischians. The eruption was preceded by earthquakes (Johnston-Lavis, 1885). Stothers and Rampino (1983) cite an eruption of the main peak, Epomeo, about B.C. 350, accompanied by earthquakes and a tsunami.

B.C. 91: Possible eruption of Rotaro I cone (Krafft, 1974); original sources cited in Stothers and Rampino (1983) are ambiguous about whether this eruption is from Ischia, Roccamonfina, or Vulsini.

A.D. 79-81, 180-222, and 284-305: Doubtful eruptions (Johnston-Lavis, 1885; Stothers and Rampino, 1983) or possible eruptions (Buchner, 1986). Krafft (1974) lists an age of A.D. 180 for the dome of Maschiatta-Montagnone, and Capaldi and others (1976-77) apparently adopt the same A.D. 180 date for the Montagnone-Maschiatta eruption. Neither Krafft (1974) nor Capaldi and others (1976-77) cite the original source of their age inference, and because Pliny the Elder described formation of Montagnone and Maschiatta (in Natural History, 2, 203, A.D. 77) (see Buchner, 1986), we conclude that the earlier, 11th century date is more likely.

A.D. 1228: In July a very large rock slide occurred on one flank of Epomeo, possibly caused by a shallow earthquake beneath Casamicciola (Baratta, 1901). About 700 persons were killed (Johnston-Lavis, 1885).

1302: In January or February, a moderate explosive eruption from the Cremate vent and extrusion of the lava flow of Arso. An earthquake may have preceded this eruption (Bonito, 1901), and one or more landslides occurred from the flank of Epomeo as a result of earthquakes in this year.

1557: An earthquake collapsed the church in the village of Campangnano (D'Ascia, 1868). (Johnston-Lavis, 1883, gives the year as 1559).

1762: On 23 July a strong earthquake occurred beneath Ischia. A church in Casamicciola (Campagnano) collapsed but no persons died. Perrey (1848a, p. 49) wrote that 62 events occurred during the night of 28-29 July.

1767: An earthquake in 1767 destroyed the church at Rotaro, Ischia (D'Ascia, 1868).

ISCHIA (continued)

HISTORICAL ACTIVITY (continued)

1796: On 18 March an earthquake shook the island of Ischia, with the greatest damage in Casamicciola (several houses collapsed; seven persons died in the ruins).

1827-28: On 11 April 1827, an earthquake was felt throughout Ischia, most strongly in Ponza. Following an aftershock the next day, inhabitants slept outside for 15 days. On 2 February 1828, a much stronger earthquake occurred beneath Fango or Casamicciola; many large boulders fell from Monte Epomeo. The earthquake of 2 February was barely felt in Ischia town and not felt in Procida, an island several km east of Ischia.

1841: Earthquakes on 6 March 1841 in the general vicinity of Ischia (Postpischl, 1985).

1863: Earthquakes.

1867: Casamicciola, earthquakes in August and November.

1880-1883: Earthquakes occurred on 25 July 1880, 4 March 1881, and 28 July 1883 (beneath Casamicciola, followed by aftershocks). Johnston-Lavis (1885) discounts most reports of higher-than-normal temperatures of wells immediately before the earthquakes of March 1881 and July 1883.

1961: 17 April, earthquake of $\underline{M}$=4.6. See comment below on subsidence and tilt.

As the preceding chronology indicates, most unrest of Ischia has been eruptive and seismic. However, some ground deformation accompanies the persistent seismicity. Ruins of a Roman lead and tin foundry are present near Castèllo d'Ischia (northeast Ischia) at a depth of 5-7 meters below sea level; Buchner (1986) inferred that the submersion occurred suddenly between the first century B.C. and the first century A.D. Along the southern coast, an elevated beach lies at a variable height of 15 to 30 m above current sea level, and thermal baths at Cava Scura have been uplifted 6-7 m since their construction sometime between 1757 and 1835 (average uplift rate about 4 cm/yr) (Buchner, 1986). Between 1890 and 1912, Ischia (exact locations unspecified) subsided about 5 mm/yr (Grablovitz, 1911, 1920-21). In the past half-century, at least the northern part of the island has been sinking at a rate of nearly 3.3 mm/yr (Oliveri del Castillo, 1962; Imbo, 1965). A more recent leveling survey by Maino and Tribalto (1971) showed that the southern part of the island subsided about 30 cm relative to the northern shore between 1913 and 1967. Subsidence of the southern coast in recent years is opposite to net uplift over the past several centuries. Major uplift of Monte Epomeo (possibly decoupled from the southern shore) would seem from seismic and rockfall activity to be continuing.

ISCHIA (continued)

COMMENTS

The idea that Ischia has a caldera is relatively new (Chiesa and others, 1985) and not universally accepted. Capaldi and others (1977) make no mention of a caldera. We have no first-hand knowledge with which to judge which interpretation of the geology of Ischia is correct. However, we note that geologically recent and probably continuing uplift of Monte Epomeo, including marine deposits on its highest portions, is similar in style and order of magnitude to uplift of the resurging dome of Iwo-Jima (see Iwo-Jima, region 8). We cannot rule out a strictly tectonic origin of Ischia uplift and seismicity, but consider that it might well reflect resurgent doming of Monte Epomeo within a 53,000 yr old (?) caldera that formed upon eruption of the Tufo Verde.

Unlike uplift, subsidence at Ischia is hard to explain by volcanic causes in the absence of an eruption. Elsewhere in the Gulf of Naples including Capri (an older island south of Ischia), slow "bradyseismic" subsidence similar to that on Ischia cannot be attributed to any volcanic cause. On Capri, baths of the Palazzo a Mare of Tiberius (built between A.D. 27 and A.D. 37) are now at least 2 m below sea level (Lyell, 1875; Gunther, 1903). A grotto/cave has steps that lead from 1 m above sea level to about (reportedly) 6 m below sea level (Gunther, 1903). Comparable subsidence is also recorded by ruins along the northern and southern shores of the Sorrentina Peninsula, which stretches between Capri and the area of Pompeii.

REFERENCES

Baratta, M., 1901, I terremoti d'Italia: Torino, Fratelli Bocca Editori. (English translation by J. Dvorak)

Buchner, G., 1986, Tremblements de terre, eruptions volcaniques et vie des hommes dans la Campanie antique: Bibliotheque de l'Institut Francais de Naples, ser. 2, v. 7, p. 145-169. (English translation by G. Mastrolorenzo and J. Dvorak)

Capaldi, G., Civetta, L., and Gasparini, P., 1976-77, Volcanic history of the island of Ischia (South Italy): Bull. Volcanol., v. 40, p. 11-22.

Chiesa, S., Gillot, P.Y., Pasquare, G., and Vezzoli, L., 1985, Collapse and resurgent calderic movements in a volcano-tectonic area: A new interpretation of the geological history of Ischia Island [abs.]: Abstracts and Time Schedule, IAVCEI Scientific Assembly, Potassic Volcanism -- Mt. Etna Volcano, Catania, September 1985.

D'Ascia, G., 1868, Storia dell'isola d'Ischia: Napoli, p. 527 p.

Friedlander, I., 1938, Sui bradisisma dell'Isola di Ischia e sulla "Grotta del Mago": Boll. Roy. Soc. Geogr. Ital., v. 75, p. 44-54.

Fuchs, C.W.C., 1872, L'Isola d'Ischia: monografia geologica: Firenze, G. Barbera (reprinted from v. 2, Memorie del Regio Comitato Geologico), 58 p.

Gillot, P.Y., Chiesa, S., Pasquare, G., Vezzoli, L, 1982, 33,000-yr K-Ar dating of the volcano-tectonic horst of the isle of Ischia, gulf of Naples: Nature, v. 229, p. 242-245.

Grablovitz, G., 1911, Il mareografo d'Ischia in relazione ai bradisismi: Boll. Soc. Sism. Ital, v. 15, p. 144-153.

----- 1920-21, Sul sospettato bradisismo Appenninico 1900-1908: Boll. Sism. Soc. Ital., v. 23, p. 147-191.

Gunther, R.T., 1903, Earth movements in the Bay of Naples: Geographical Jour., v. 22, p. 1-49.

Imbo, G., 1965, Catalogue of active volcanoes of the world including solfatara fields, Pt. XVIII, Italy: Rome, IAVCEI, 72 p.

ISCHIA (continued)

REFERENCES (continued)

Johnston-Lavis, H.J., 1885, Monograph of the earthquakes of Ischia: London, Dulau & Co., 107 p.

Lyell, C., 1875, Principles of geology (12th ed.), 2 v.: London, John Murray.

Maino, A. and Tribalto, G., 1971, Rilevamento gravimetrico di dettaglio dell'isola di Ischia (Napoli): Boll. Serv. Geol. Ital., v. 92, p. 109-123.

Oliveri del Castillo, A., 1962, Il bradisisma dell'Isola d'Ischia: Assoc. Geof. Ital., Atti XI Convegno, Rome, p. 319-324.

Parascandola, A., 1937, Il periodo sismico 1927-1936 nell'Isola d'Ischia: Boll. Soc. Nat. Naples, v. 49, p. 147-159.

Perrey, A., 1848, Memoire sur les tremblements de terre de la Peninsule Italique: Mem. Cour. et mem. Sav. etr. Acad. Roy. Belgique. v. 22, 1846-47, Bruxelles.

Platania, G., 1937, La grotta del Sole nell'Isola d'Ischia e il bradisismo: Boll. Soc. Nat. Napoli, v. 49, p. 115-121.

Postpischl, D., ed., 1985, Catalogo dei Terremoti Italiani dall'anno 1000 al 1980: Quaderni de "La Ricerca Scientifica" no. 114, v. 2B, Consiglio Nazionale delle Ricerche (CNR).

Rittmann, A., 1930, Geologie der Insel Ischia: Zeit. Vulkanol., Supplement, v. 6, 265 p.

Stothers, R.B. and Rampino, M.R., 1983, Volcanic eruptions in the Mediterranean before A.D. 630, from written and archaeological sources: Jour. Geophys. Res., v. 88, p. 6357-6371.

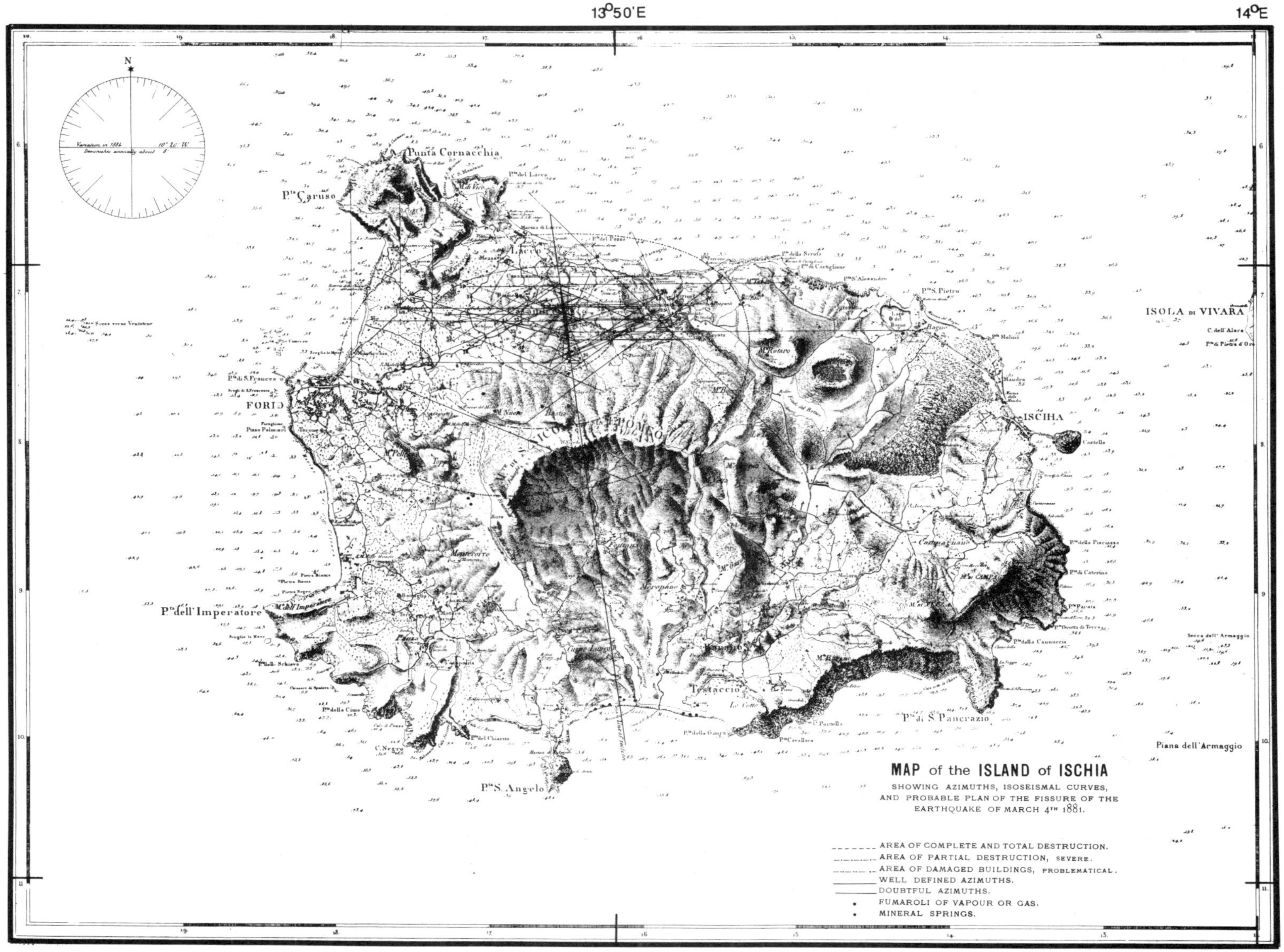

Figure 1.5.1. Map of the Island of Ischia from Johnston-Lavis (1885). Central, highest part of the island, Monte Epomeo, is a horst interpreted by most as a dominantly tectonic feature. However, semicircular depression south of Monte Epomeo may be relic moat of 53,000 yr B.P.(?) caldera and uplift of Monte Epomeo may reflect resurgent doming (see text).

CINQUE DENTI (MONASTERO) and LA VECCHIA

CAVW number (active vent)	Latitude Longitude (degrees)	Diameter (km)	Local tectonic setting	Precaldera edifice	SiO_2 content (wt pct)	Age of collapse (yr B.P.)	Date of unrest	Type and duration of unrest ESTU STHF MGTF H Te	Eruption type
01-01-071 (Pantelleria, Foerstner vent)	36.78N 12.00E	6 (CD) ca. 6 (LV)	Exten	Strato	R = 45-46 C = 65-70	45,000+ 114,000	1890-91	F--F ---x ---F - --	ex (submarine)

TECTONIC SETTING

The island of Pantelleria, best known in geology for its peralkaline rocks, sits astride a drowned continental rift in the Strait of Sicily, between Sicily and Tunisia (Mahood and Hildreth, 1983, 1986; fig. 1.6.1).

GEOLOGIC HISTORY

The island is a mosaic of coalescing pantelleritic centers, dominated by two calderas, Cinque Denti (or Monastero) and La Vecchia. The La Vecchia Caldera formed about 114,000 yr B.P. and was filled by three widespread welded units that were erupted 106,000, 94,000 and 79,000 yr B.P. (Mahood and Hildreth, 1986). Reactivation of the ring-fracture system about 67,000 yr B.P. ago is indicated by two large pantellerite centers and a chain of small shields along the ring fault. Eruption of the Green Tuff ca. 45,000-50,000 yr B.P. resulted in collapse of a second, nested caldera; this second caldera is outlined and named as the Cinque Denti Caldera (fig. 1.6.1, from Mahood and Hildreth, 1986) or the Monastero Caldera (fig. 1.6.2, from Civetta and others, 1988). Later, trachyte lavas built a broad cone that nearly filled this younger caldera (Civetta and others, 1988). Beginning about 18,000 yr B.P., inflation of the magma chamber opened old ring fractures and new radial fractures that served as vents for differentiated pantellerite pumice cones and shields. Continued uplift raised the northwest half of the intracaldera trachyte cone 275 m, creating the island's present summit by trapdoor uplift (Mahood and Hildreth, 1983, 1986).

HISTORICAL ACTIVITY

1890-91: An explosive submarine eruption of basalt occurred on the northwest slope of Pantelleria in October 1891. Earthquakes and uplift "of the whole island" preceded and accompanied the eruption (Butler, 1891, 1892b; Ricco, 1891, 1892; Washington, 1913-14). Fumarolic activity, uplift of much of the northeast coast, and earthquakes strong enough to crack cisterns on the island began about 24 May 1890. As much at 55 cm of uplift occurred along the northeast coast between Punta della Puzzolana and Punta dello Spadillo during 1890, and at least another 25 cm occurred during 1891 (fig. 1.6.1). At Punta Karuscia, just northwest of the aforementioned

CINQUE DENTI (MONASTERO) and LA VECCHIA (continued)

HISTORICAL ACTIVITY (continued)

section of coast, a 200-m-long fracture opened parallel to the elevated shoreline (Ricco, 1891). A sharp earthquake on the night of 14-15 October 1891 was accompanied by additional uplift. Additional strong earthquakes occurred on the night of 16-17 October. The eruption began on the morning of 17 October, accompanied by a rapid decrease in seismicity. Another submarine eruption may have occurred south of Pantelleria in December 1891 (Foerstner, 1891; disputed by Butler, 1892b).

COMMENTS

The 1891 activity at Pantelleria is a good example of caldera unrest apparently caused by basalt rising beneath a silicic magma reservoir, causing broad uplift and seismicity before erupting outside the "density shadow" of the silicic reservoir. Pantelleria is also a good example of asymmetric resurgent doming, but whether such doming has occurred in historical time is unknown.

REFERENCES

Butler, G.W., 1891, The October eruption north-west of Pantelleria: Nature, v. 45, no. 1155, p. 154.
----- 1892a, On the matter thrown up during the submarine eruption north-west of Pantelleria, October 1891: Nature, v. 45, no. 1159, p. 251-252.
----- 1892b, Abstract of Mr. A. Ricco's account of the submarine eruption north-west of Pantelleria, October 1891: Nature, v. 45, no. 1173, p. 584-585.
Civetta, L., Cornette, Y., Crisci, G., Gillot, P.Y., Orsi, G., and Requejo, C.S., 1984, Geology, geochronology and chemical evolution of the island of Pantelleria: Geol. Mag., v. 121, p. 541-562.
Civetta, L., Cornette, Y., Gillot, P.Y., and Orsi, G., 1988, The eruptive history of Pantelleria (Sicily Channel) in the last 50 ka: Bull. Volcanol., v. 50, p. 47-57.
Cornette, Y., Crisci, G.M., Gillot, P.Y., and Orsi, G., 1983, Recent volcanic history of Pantelleria: a new interpretation: Jour. Volcanol. Geotherm. Res., v. 17, p. 361-373.
Foerstner, H., 1891, Das Gestein der 1891 bei Pantelleria entstandenen Vulcaninsel und seine Beziehungen zu den jüngsten Eruptivgesteinen der Nachbarschaft: Tschermaks Mineral. Petrog. Mitt., v. 12, p. 510-521.
Mahood, G., 1984, Pyroclastic rocks and calderas associated with strongly peralkaline volcanism: Jour. Geophys. Res., v. 89, no. B10, p. 8540-8552.
Mahood, G., and Hildreth, W., 1983, Nested calderas and trapdoor uplift at Pantelleria, Strait of Sicily: Geology, v. 11, p. 722-726.
----- 1986, Geology of the peralkaline volcano at Pantelleria, Strait of Sicily: Bull. Volcanol., v. 48, p. 143-172.
Orsi, G., and Sheridan, M.F., 1984, The Green Tuff of Pantelleria: Rheoignimbrite or rheomorphic fall?: Bull. Volcanol., v. 47, no. 3, p. 611-626.

CINQUE DENTI (MONASTERO) and LA VECCHIA (continued)

REFERENCES (continued)

Ricco, A., 1891, Tremblement de terre, soulèvement et éruption sous-marine à Pantellaria: Comptes Rendus Acad. Sci. Paris, v. 113, p. 753-755.

----- 1892, Terremoti, sollevamenti e eruzione sottomarina a Pantelleria nella seconda metà dell'ottobre del 1891: Boll. Soc. Geograf. Ital., v. 29, 23 p. and in Annali Uff. Centr. Meteorol. Geodinam., v. 11, p. 5-23, parts 2 and 3 of 3.

Villari, L., 1974, The island of Pantelleria: Bull. Volcanol., v. 38, no. 3, p. 680-724 (plus references at end of volume).

Washington,, H.S., 1909, The submarine eruptions of 1831 and 1891 near Pantelleria: Am. Jour. Sci., v. 27, p. 131-150.

----- 1913-1914, The volcanoes and rocks of Pantelleria: Jour. Geol., I, II, and III, v. 21, p. 653-670, 683-713; v. 22, p. 16-27.

Wright, J.V., 1980, Stratigraphy and geology of the welded air-fall tuffs of Pantelleria, Italy: Geol. Rundschau, v. 69, no. 1, p. 263-291.

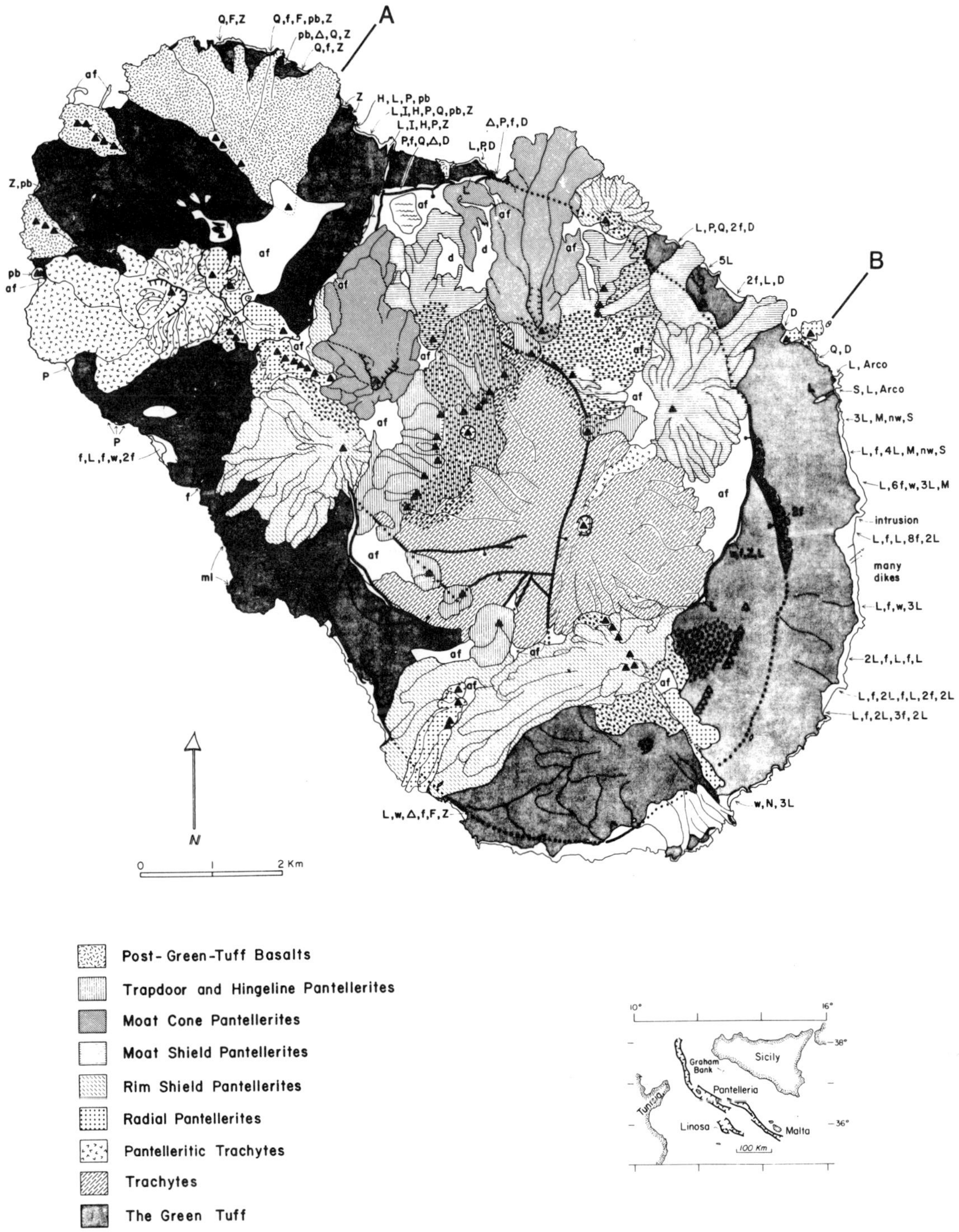

Figure 1.6.1. Generalized geologic map of Pantelleria, from Mahood and Hildreth (1986). Inset, location in Strait of Sicily rift system. Heavy lines outline two overlapping calderas and central trapdoor uplift. Solid triangles are eruptive vents; open triangles indicate vents mantled by the Green Tuff. Stippled overprint represents thick, young pumice-fall deposits. Unpatterned areas include alluvial and artificial fill (af), pumiceous debris flows (d), and pre-Green Tuff eruptive units (in ascending order at each section) designated L=pantelleritic lava; pb=basalt; f=pumice-fall deposit; nw,w=local or uncorrelated nonwelded or welded tuff; ml=Monastero lava shield; zl=Zichidi lava cone; zt=Zichidi tuff; open triangle=lithic-rich pumiceous breccia. Stratigraphically correlated tuffs are designated I, H, M, S, P, Q, F, D, Z, and N. Section of coast between A and B was uplifted during 1890-91 prior to the most recent eruption (from Butler, 1892).

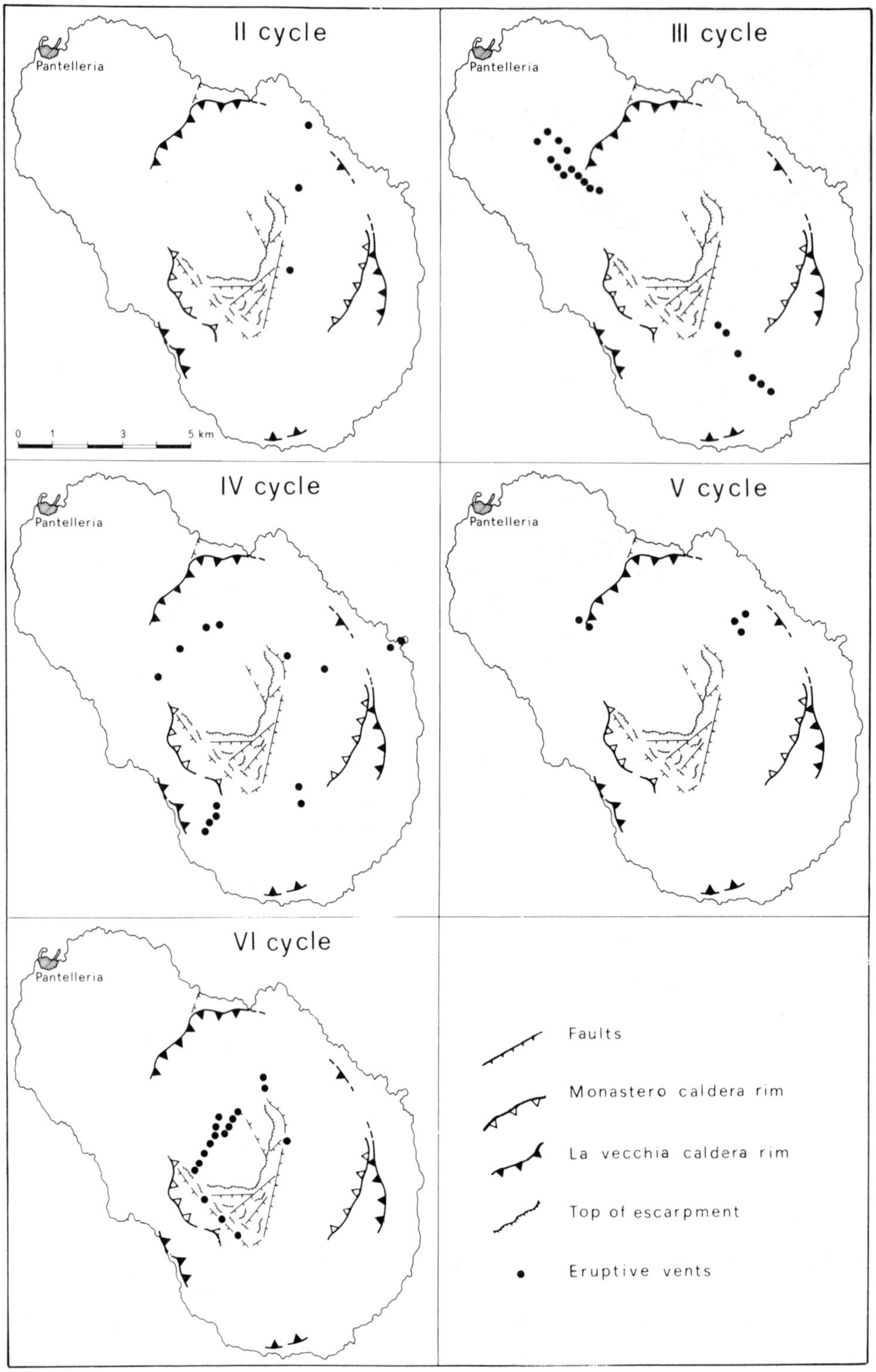

Figure 1.6.2. Alternative interpretation of caldera structures on Pantelleria, from Civetta and others (1988); compare with figure 1.6.1.

SANTORINI (THIRA)

CAVW number (active vent)	Latitude Longitude (degrees)	Diameter (km)	Local tectonic setting	Precaldera edifice	SiO_2 content (wt pct)	Age of collapse (yr B.P.)	Date of unrest	Type and duration of unrest ESTU STHF MGTF H Te	Eruption type
01-02-04	36.40N	10	Exten	Graben,	R = 64-68	100,000;	B.C. 1390	F--- ---- ---- - QU	pex,EX,PEX
	25.40E			shields	C = 68	and	B.C. 197	---- ---- ---- - --	EX
						ca.	A.D. 19	---- ---- ---- - --	ex,dm
						B.C. 1390	46	---- ---- ---- - --	ex,dm
							726	---- ---- ---- - --	ex,dm
						between	1570-73?	---- ---- ---- - --	ex,dm
							1649-50	E--- ---- ---- - --	ex,dm
							1707-11	C--D ---- --DD - --	ex,cdm
						between	1810-30	---x ---- --FF - --	cdm?
							1866-70	C--x CC-C --CC C --	dm,ex
							1925-26	D--- G--- ---- - --	ex,dm
							1928	C--- ---F ---C - --	ex,dm
							1939-41	---- ---- --D- - --	ex,dm
							1950	---- ---- ---- - --	ex,dm

TECTONIC SETTING

Santorini is located on the island of Thira in the Aegean Sea, about 110 km north of Crete. Thira is part of the Cyclades volcanic chain that also includes the young volcanoes Milos and Nisyros (fig. 1.7.1). Tectonic analyses cited by Heiken and McCoy (1984) indicate NW-SE extension in the area of Thira during Recent and Quaternary time, consistent with northeast-trending normal faults and vents within the volcanic field.

GEOLOGIC HISTORY

Santorini consists of a 100,000 yr B.P. caldera developed on a group of low volcanoes, and an overlapping ca. 3,380 yr B.P. (perhaps B.C. 1390) caldera of roughly the same size offset to the north (Heiken and McCoy, 1984). The younger caldera formed during or shortly after the famous Minoan eruption of 13-19 km^3 (Heiken and McCoy, 1984) or 27 km^3 (Druitt, 1984) of rhyodacite magma. Buildings were apparently destroyed by earthquakes several years or decades before the cataclysmic eruption, and were being rebuilt when they were abandoned (Vitaliano, 1973; Heiken and McCoy, 1984). The islands were abandoned shortly before the cataclysmic eruption, possibly because of small phreatic or phreatomagmatic eruptions that occurred just before the cataclysmic phase of the

SANTORINI (THIRA) (continued)

GEOLOGIC HISTORY (continued)

eruption (Heiken and McCoy, 1984). The Kameni Islands, a postcaldera dome complex in the center of the caldera, were erupted during the last 2,000 years along northeast-trending fissures in the area of overlap between the two calderas.

HISTORICAL ACTIVITY

B.C. 1390: see above.

B.C. 197 - A.D. 1650: According to Fouqué (1879), Hiera Island grew in B.C. 197, Thia dome (=Palaea Kameni?) grew in A.D. 19, further growth of Palaea Kameni occurred in A.D. 46 and A.D. 726, and Micra Kameni dome grew in 1570-73 (fig. 1.7.2). A cliff on Palaea Kameni failed in 1457, forming a reef of rubble and causing erroneous reports of an eruption in that year. Renewed earthquake activity began in 1649, causing inhabitants of Santorini to evacuate. Shocks became stronger in March 1650, and very strong shocks were felt on 14 and 27 September 1650. Seawater turned green at a location about 6 km northeast of Santorin, and six days later, on 26 or 27 September 1650, an eruption began that eventually formed Colombo Cone (now Colombo Banks, eroded by the sea). Following the eruption, seawater encroached onto the island of Thira at Nio and Sikino, suggesting subsidence (Fouqué, 1879).

1707-11: Earthquakes were felt again on 18 and 21 May 1707. A 6-8 m high shoal ("Nea Kameni") appeared off Micra Kameni on 23 May 1707. Bourguignon (1708) wrote: " ...Mariners, in hopes of booty, put out immediately to see what it was. Soon after we were amazed to hear by 'em, that it was a Shoal which began to spring up from the Bottom of the Sea; and was not as yet very plainly to be discerned. Next day (the 24th) several Persons, both Ecclesiasticks and Seculars, went out of Curiosity to satisfy themselves, not easily believing what the Mariners related: But they were no sooner got to the place, than their Eyes fully convinced them, that it was a Matter of Fact. Some of 'em were so bold as to get a-shoar on this new Shoal, which was still moving, and sensibly increased under their feet. They brought us back several Curiosities, and among others a kind of Oysters, very large, and of an Exquisite Taste, which they found sticking to the Rock, and raised out of the Water...". Water temperatures, water turbulence, and sulfur smell increased on 13-14 July 1708, and Micra Kameni and Thira subsided 2 m during(?) an explosive phase on 16 July 1708. The new shoal and dome continued to rise amid boiling water and thick smoke. By 20 November 1708, its diameter reached 5 km and its height 10 m; by 1711 the top of the new island was 70-80 m above sea level. As measured from the sea floor, Nea Kameni rose nearly 300 m. During the time that Nea Kameni was growing, Micra Kameni subsided by as much as 2 m (Fouqué, 1879, p. 140)

1810-30: von Humboldt (1863, p. 242, citing Virlet and Fiedler) reported that a new "island" that was 15 fathoms below sea level in 1810 had risen to within 3 or 4 fathoms of sea level by 1830. "It rises steeply like a giant cone, from the bottom of the sea, and the continuous activity of the submarine crater is obvious for the circumstance that sulphurous acid vapors are mixed with sea water, in the eastern bay of [Nea Kameni]...[and] coppered ships lie at anchor in the bay in order to get their bottoms cleaned and polished by this natural process."

SANTORINI (THIRA) (continued)

HISTORICAL ACTIVITY (continued)

1866-70: During late January 1866, fissures formed in the ground near the passage between Nea and Micra Kameni. On 30-31 January, the cracks widened further, earthquakes were felt, water temperature increased, gas rose through the bay south of Nea Kameni, and the southern end of Nea Kameni subsided by at least 60 cm in 4 hours. Houses around the cove of Vulcano (southeast side of Nea Kameni) "began to totter on the brink of ruin; they were losing their plumb and tilting toward the shoreline" (Fouqué, 1879, p. 38-39). Large rocks fell from the crater rim. The eruption began on or before 1 February, first with submarine lava effusion and later with small explosions; on 1 February at 5 o'clock in the morning, "flames appeared on the west side of the cove of Vulcano and ... formed, on the sea floor, a conical mound 4-5 m high..." (Fouqué, 1879). Continuous tremors were felt throughout the eruption. During February 1866 subsidence continued along the southwestern shore (2 m), the northern shore (1 m), and the eastern side of Nea Kameni (1.5 m). The lava dome continued to rise through March 1866 and an area between Georgios and Aphroessa rose 1.5-2 m between 10 and 24 March. The area between Aphroessa and Palaea Kameni grew to 108 m above sea level from a starting elevation 296 m below sea level. Between March and May 1866, a pier on the east coast of Nea Kameni near Micra Kameni sank 30 cm; it sank an additional 1.3 m between 15 May 1866 and 18 March 1867. By 1870, three new domes (Georgios, Aphroessa, and Reka) had grown on the southern edge of Nea Kameni, more than doubling the size of Nea Kameni and nearly joining it with Micra Kameni. Subsidence continued in the channel between Micra and Nea Kameni from 1866 to 1875. Concurrent subsidence of Nea Kameni was about 5 m (Fouqué, 1879).

1925-26: Earthquakes of local origin were felt in Phira on 28 July and 7, 8, and 9 August 1925. During the same period, water temperature rose around Micra Kameni. On 10 August, the temperature and turbulence in the bay were so great that fishermen and sailors left the aforementioned channel (Reck, 1936, p. 6). There is no record of subsidence of Nea or Micra Kameni as occurred in January 1866, but the islands were uninhabited in 1925, so the absence of a report means little. An eruption began explosively on 11 August 1925 and eventually formed the Dafni and Fouqué domes and flows.

1928-50: A north-south fissure began to form on the Kokkima-Nera lava flow, 300 m south-southeast of Dafni, sometime between observations on 8 May and 20 September 1926. Fissures appeared in the walls of Dafni crater from 1927 through 1933, the floor of the crater sank on one side, and sulfur-dioxide emissions were noted. The smell of hydrogen sulfide was noted in Phira in mid- to late January 1928. Fumarolic activity on the Kokkima-Nera flow increased on 20 January, and small earthquakes were noted on 22 January. Explosions and the growth of Naftilos dome began on 23 January 1928. In retrospect, the fissure on the lava flow was a precursor to the growth of Naftilos dome, though it was not recognized as such at the time (Reck, 1936). Renewed growth of Nea Kameni occurred in 1939, preceded by increasing temperatures in seawater from about 23 °C to 52 °C during 26 August to 22 September 1939 and by explosions that began on 23 September 1939. Liatsikas dome formed in 1950. A magnitude 7.8 earthquake that occurred about 165 km from Santorini on 9 July 1956 had no apparent effect on the volcanic system.

COMMENTS

Most documents regarding Santorini are unclear as to whether uplift is localized, as part of dome growth, or broader, involving the entire caldera. Seismic reflection (sparker) profiles indicate dome growth rather than structural resurgence (Heiken and McCoy,

SANTORINI (THIRA) (continued)

COMMENTS (continued)

1984; G. Heiken, written commun., 1987). Documents are also unclear as to whether submergence of land is due to subsidence or wave erosion. The several meters of subsidence observed after Santorini eruptions is comparable to that observed after eruptions at other calderas in this compilation, and can be explained as recovery of elastic preeruption uplift.

REFERENCES

Bourguignon, 1708, A relation of the New Island thrown up near the Island of Santerini: Phil. Trans. Roy. Soc. London, v. 26, p. 200-208.
Budetta, G., Condarelli, D., Fytikas, M., Kolios, N., Pascale, G., Rapolla, A., and Pinna, E., 1984, Geophysical prospecting on the Santorini Islands: Bull. Volcanol., v. 47, no. 3, p. 447-466.
Druitt, T.H., 1984, Silicic and intermediate magmatism on Santorini [abs.]: Eos, Trans. Am. Geophys. Union, v. 65, p. 1134.
Druitt, T.H., and Sparks, R.S.J., 1982, A proximal ignimbrite breccia facies on Santorini, Greece: Jour. Volcanol. Geotherm. Res., v. 13, p. 147-171.
Fouqué, F., 1879, Santorin et ses eruptions: Paris, G. Masson, 440 p.
Georgalas, G.C., 1962, Catalogue of active volcanoes of the world including solfatara fields, Pt. XII, Greece: Rome, IAVCEI, 40 p.
Georgalas, G.C., and Papastamatiou, J., 1951, Ueber den Ausbrüch der Santorin vulkanes von 1939-1941, Der Ktenas-Ausbrüch: Bull. Volcanol., ser. II, v. 11, p. 3-37.
Heiken, G., and McCoy, F., Jr., 1984, Caldera development during the Minoan eruption, Thira, Cyclades, Greece: Jour. Geophys. Res., v. 89, no. B10, p. 8441-8462.
Humboldt, A. von, 1863, Cosmos (English translation by E.C. Otte), v. 1: New York, Harper and Brothers, p. 241-242.
Pichler, H., and Kussmaul, S., 1980, Comments on the geologic map of the Santorin Islands, in Doumas, C., ed., Thera and the Aegean world: London, Thera and the Aegean World Publ., p. 413-426.
Reck, H., 1934, Nachrichtendienst über vulkanische ereignisse 1933/34, Santorin: Zeit. Vulkanol., v. 16, p. 61-62.
Reck, H., ed., 1936, Santorin, der Werdegang eines Inselvulkans und sein Ausbrüch 1925-1928: Berlin, D. Reimer, Andrews und Steiner, 3 v., 540 p.
Sparks, R.S.J., Sigurdsson, H., and Watkins, N.D., 1976, The Thera eruption and late Minoan-IB destruction on Crete: Nature, v. 271, p. 91.
Vitaliano, D.B., 1973, Legends of the Earth: New Jersey, Citadel Press, 305 p.
Vitaliano, D.B., and Vitaliano, C.J., 1971, Plinian eruptions, earthquakes, and Santorin: a review (with discussion): Acta, 1st Int. Sci. Congr. on the volcano of Thera, p. 88-108.
Washington, H.S., 1926, The Santorin eruption of 1925: Geol. Soc. Am. Bull., v. 37, p. 349-384.

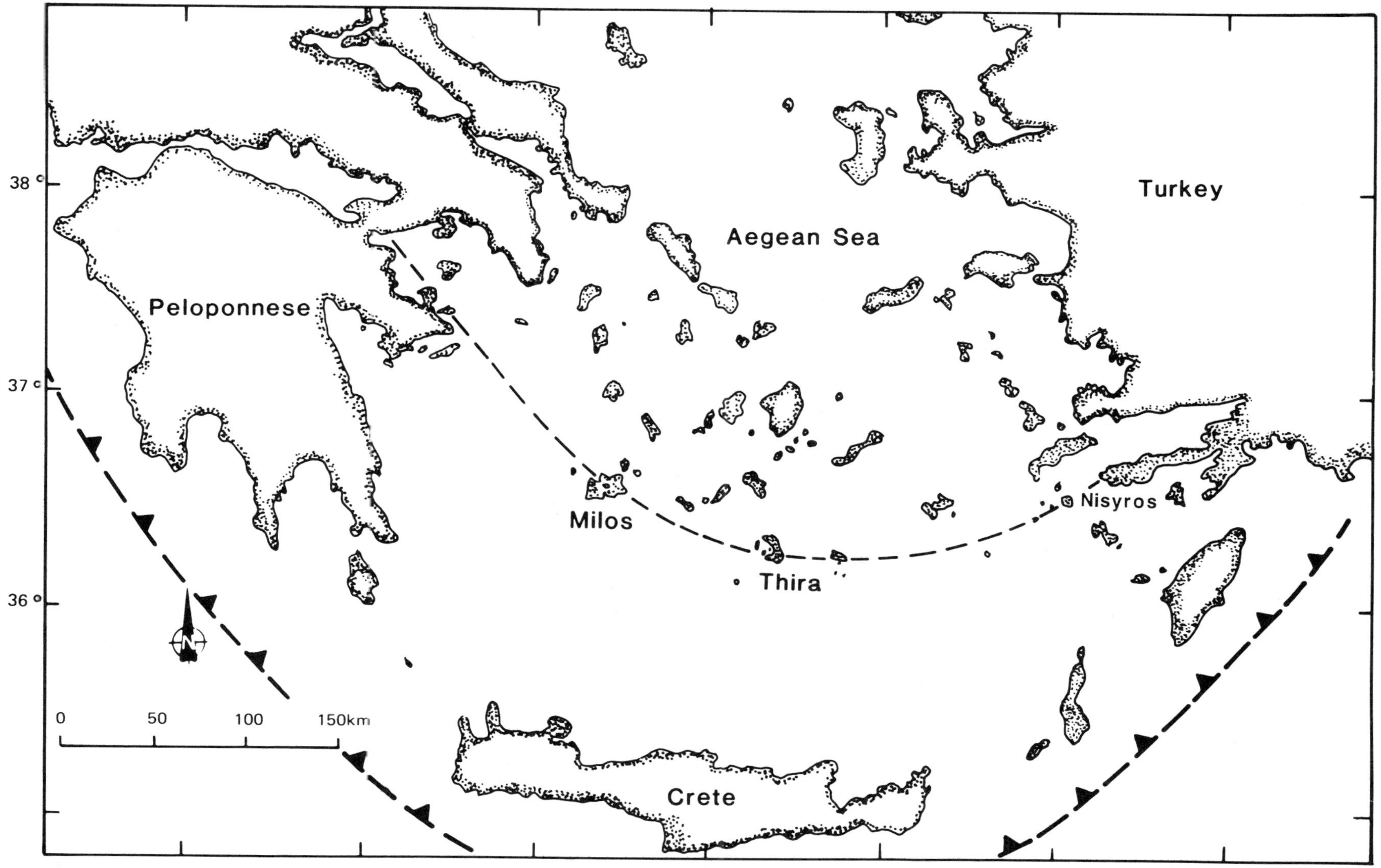

Figure 1.7.1. Map from Heiken and McCoy (1984) showing Thira and the Cyclades volcanic chain (upper dashed line), which includes young volcanoes Milos and Nisyros. Lower dashed line with triangles marks approximate boundary of Aegean plate. Copyright by the American Geophysical Union.

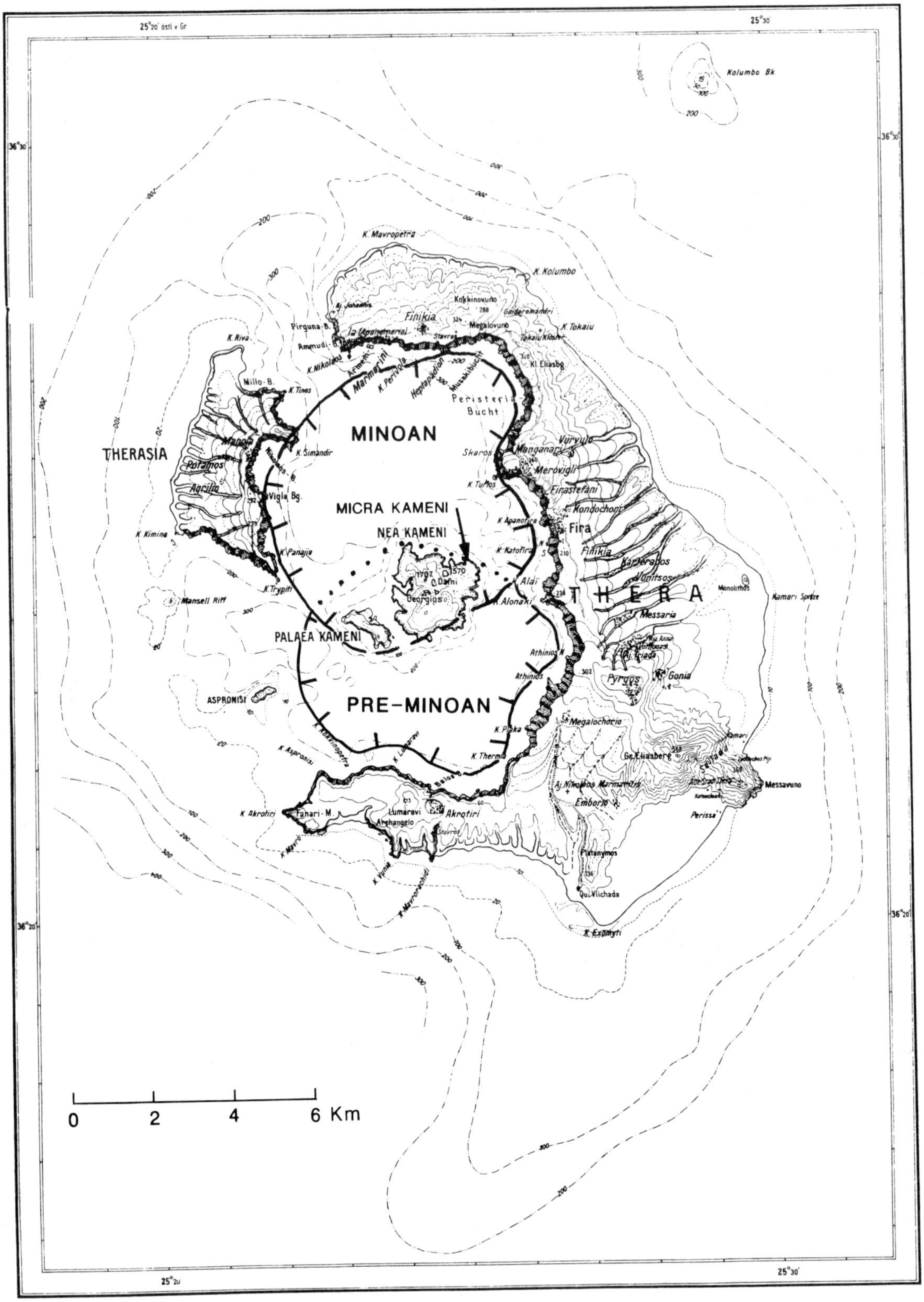

Figure 1.7.2. Topographic map of Santorini by H. Reck and M. Neumann van Padang (Reck, 1936), with locations used in text and outlines of two calderas (Minoan and pre-Minoan), from Heiken and McCoy (1984).

NISYROS

CAVW number (active vent)	Latitude Longitude (degrees)	Diameter (km)	Local tectonic setting	Precaldera edifice	SiO_2 content (wt pct)	Age of collapse (yr B.P.)	Date of unrest	Type and duration of unrest ESTU STHF MGTF H Te	Eruption type
01-C2-05	36.58N	4	Compr,	Strato	R = d,r	?	1422?	---- ---- ---? - --	ex,lf?
	27.18E		intersection		C = ?		1830	---- ---- ---x x --	none
			of regional				1871	x--- ---- ---- - --	ex
			faults				1873	x--- ---- ---- y --	ex,pex
							1888	---- ---- ---y - --	pex

TECTONIC SETTING

Davis (1968) asserts that Nisyros and Yali Island are located near the intersection of two regional fault lines: The first runs from Episcopi to Patmos, and the second through Soussaki, the Gulf of Saronic, Milos, Santorini, and Dodecanese.

GEOLOGIC HISTORY

Nisyros Caldera lies at the collapsed top of an andesitic stratovolcano (fig. 1.8.1). The caldera-forming eruption began with pyroclastic surges of silicic pumice and evolved to pyroclastic flows and tephra fall of mixed mafic and silicic pumice. Large rhyolitic and rhyodacitic postcaldera domes contain mafic inclusions, inferred to have been clots of mafic magma in silicic magma (Limburg and Varekamp, 1984).

Yali Island, 8 km north-northwest of the center of Nisyros Caldera, has undergone 160 m of uplift within the late Quaternary, possibly within the Holocene. That uplift apparently occurred as an obsidian dome lifted "roof rock" before and during dome extrusion. The uplift is inferred to have been like that which occurred at Santorini in 1707 and at Usu in 1943-44 (Keller, 1971).

HISTORICAL ACTIVITY

1422: Bondelmonte, cited in Georgalas (1962), wrote that "sulfureous ignis die ac nocte eructat in altum" (sulfur burned day and night, rising high) and compared the activity to that of Stromboli.

1830-88: All historical eruptions have been from vents within the caldera. Strong vapor emanations and hot-water ejections, accompanied by roarings and detonations, were reported in 1830. Local earthquakes, detonations, and "flames" occurred in 1871, when ash and lapilli destroyed fruit gardens on the southeast caldera floor. In 1873, a 6-7 m crater formed after an earthquake of

NISYROS (continued)

HISTORICAL ACTIVITY (continued)

unknown size and location. Ash and lapilli were ejected, and hot saline water overflowed to form a caldera lake. Eruptions of blackish mud mixed with hot water continued for three days. On 11 and 26 September 1873, explosions of saline water and mud accompanied numerous ejections of lapilli. By the end of September 1888, a strong eruption formed a cylindrical crater at least 25 m in diameter. Mud and lapilli were ejected, accompanied by a strong steam column. In July 1956, the only activity was solfataric, in the central and southern parts of the caldera (Georgalas, 1962).

REFERENCES

Davis, E.N., 1968, Zur Geologie und Petrologie der Inseln Nisyros und Jali (Dodekanes): Geol. Rundschau, v. 57, p. 811-821.

DiPaola, G.M., 1974, Volcanology and petrology of Nisyros Island (Dodecanese, Greece): Bull. Volcanol., v. 38, p. 944-987.

Georgalas, G.C., 1962, Catalogue of active volcanoes of the world including solfatara fields, Pt. XII, Greece: Rome, IAVCEI, 40 p.

Keller, J., 1971, The major volcanic events in recent Eastern Mediterranean volcanism and their bearing on the problem of Santorini ash layers: Acta, 1st Int. Sci. Congr. on the volcano of Thera, p. 152-169.

Limburg, E.L., and Varekamp, J.C., 1984, Volcanology of Nisyros, Greece [abs.]: Eos, Trans. Am. Geophys. Union, v. 65, no. 45, p. 1146.

St. Seymour, K., and Vlassopoulos, D., 1987, Explosive volcanic activity associated with dome growth, Nisyros, Aegean Volcanic Arc, Greece [abs.]: Abstracts, XIX IUGG Gen. Assembly, Vancouver, B.C., v. 2, p. 425.

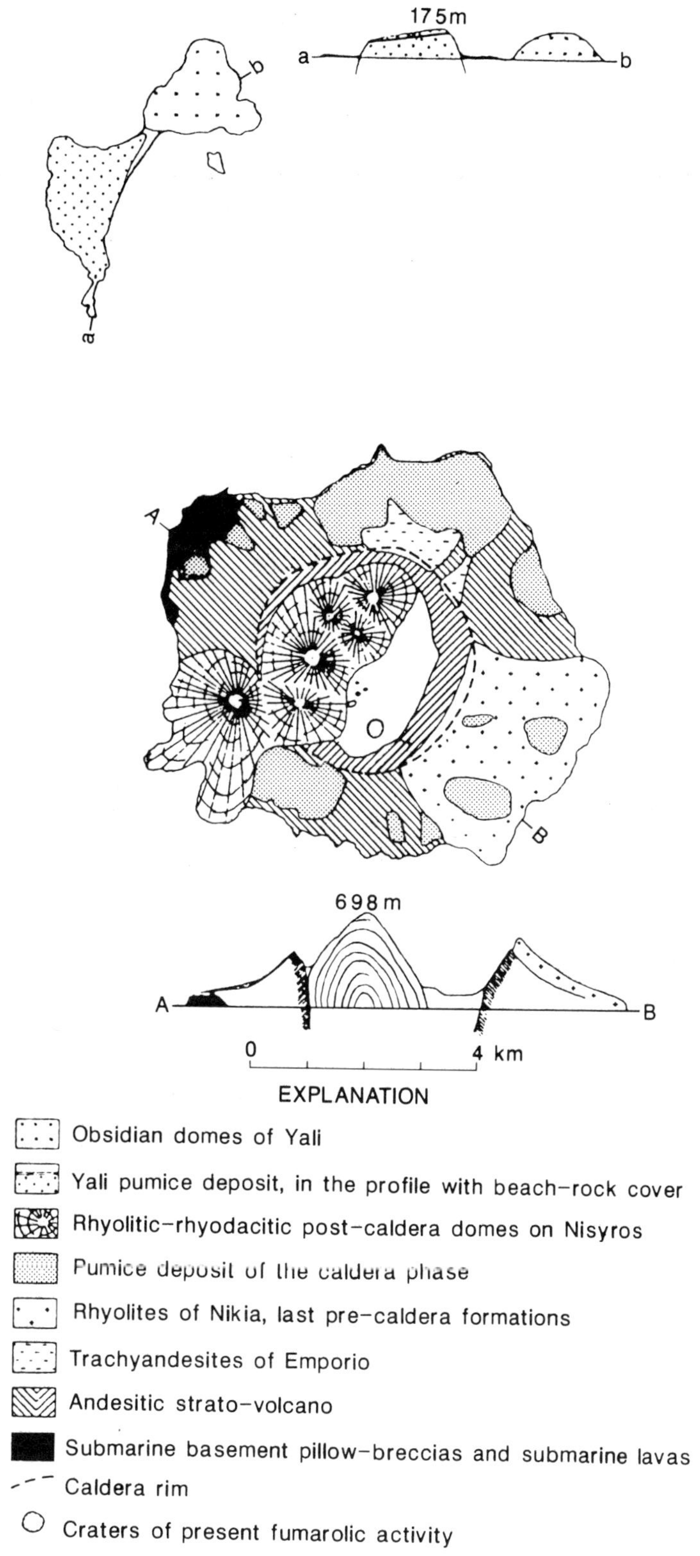

Figure 1.8.1. Geological interpretation of Nisyros-Yali group, from Keller (1971).

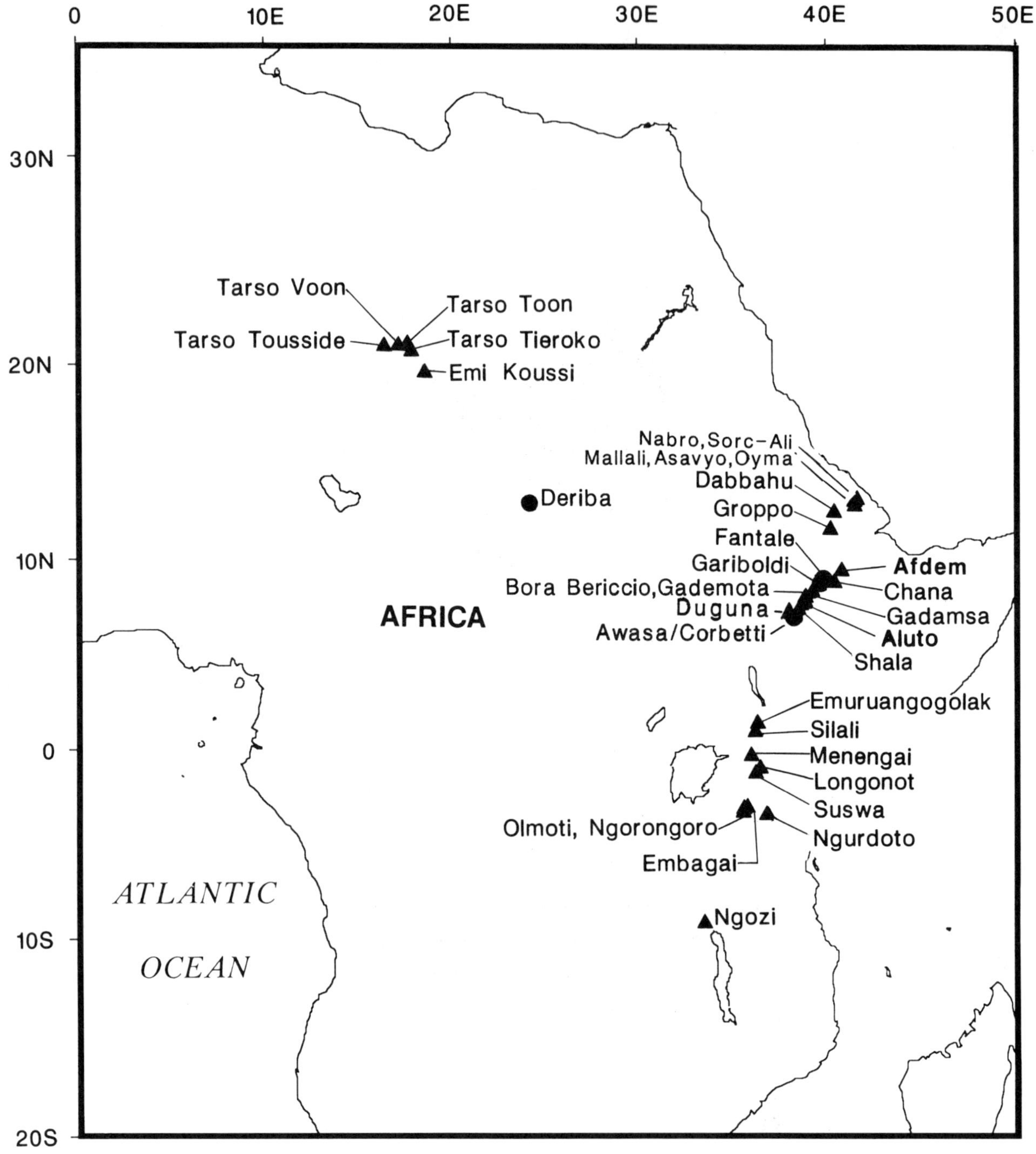

Figure 2. Locations of large, Quaternary restless calderas (solid circles) and nonrestless calderas (triangles) in Region 2.

FANTALE

CAVW number (active vent)	Latitude Longitude (degrees)	Diameter (km)	Local tectonic setting	Precaldera edifice	SiO_2 content (wt pct)	Age of collapse (yr B.P.)	Date of unrest	Type and duration of unrest ESTU STHF MGTF H Te	Eruption type
02-01-19	8.98N 39.93E	3 x 4	Exten	?	R = b-r C = 70?	?	13th cent.	---- ---- ---- - --	"major"
							ca.1820	---- ---- ---- - --	lf
							1981	EE-- ---E ---- - =?	none

TECTONIC SETTING

Fantale is in the northern end of the NNE-SSW-trending Main Ethiopian Rift, near the junction of the Rift Valley and the Afar Depression (Gibson, 1969). Structures of the Main Ethiopian Rift abut against the Fantale complex.

GEOLOGIC HISTORY

The main cone of Fantale consists of rhyolitic obsidian flows with rare intercalated pumice tuffs. Its caldera formed upon eruption of pantelleritic ash flows whose deposits are remarkably welded. Postcaldera rocks are trachyte and pantellerite lavas (Gibson, 1967, 1970, 1975) (fig. 2.1.1).

HISTORICAL ACTIVITY

13th century: Azais and Chambord (1931), cited by C. Wood (written commun., 1987), record an account by a local priest that "in the 13th century there was a church and an Abyssinian town where at present is the Lake Metahara. They were destroyed by a volcanic eruption."

1820: Several fluid olivine basalt flows were extruded in the Main Ethiopian Rift about 1820 (Gouin, 1979). One of these flows issued from a 4-km-long fissure related to the Fantale complex, where the fissure cuts a tuff ring on the flank of Fantale (C. Wood, written commun., 1979). Gibson (1975) considered these lavas to be more closely related to regional basaltic volcanism than to the Fantale system proper.

Tradition holds that young basalt flows destroyed a village and its monastery, and parents in the vicinity still warn misbehaving children that they will "melt like the Fantale" (Gouin, 1979, p. 93). C. Wood (written commun., 1987) believes that the young flows could be from either a 13th century or an 1820 eruption. Subsequent extension created fissures in those lava flows (Gibson, 1969).

FANTALE (continued)

HISTORICAL ACTIVITY (continued)

1930, 1960: Small fumaroles were noted at about 20 places in the caldera in 1930 (Reck, 1932), and intense fumarolic activity was noted in January 1960 (Mohr, 1962, p. 428). Observations are so infrequent in this area that neither the 1930 nor the 1960 observation suggests increased activity.

1981: An earthquake swarm with $\underline{M}_{max}$=4 occurred roughly 20 km north of Fantale in January to March 1981 (Asfaw, 1982). Fissures as much as 2 cm wide developed in the epicentral area (fig. 2.1.2); those that developed in sediment enlarged over the succeeding rainy season by erosion (Asfaw, 1982).

COMMENTS

Fantale Caldera is less than 5 km in diameter, but we have included it nonetheless because it lies astride an active rift zone and has experienced a historical injection of basaltic magma that did not trigger a silicic eruption. Mohr and Wood (1976) and C. Wood (written commun., 1987) indicate that coexistence of basaltic and silicic magma in the Main Ethiopian Rift is a normal occurrence, and that basaltic magma rises directly to the surface along fissures, independently of central volcanoes in the vicinity. Interesting comparisons may be made with rifting and magma supply at Icelandic calderas, described later in this compilation.

REFERENCES

Asfaw, M., 1982, Development of earthquake-induced fissures in the Main Ethiopian Rift: Nature, v. 297, p. 393-395.

Azais, R.P., and Chambord, P., 1931, Cinq anneés de recherches archaeologiques en Ethiopie: Paris, p. 142.

Gibson, I.L., 1967, Preliminary account of the volcanic geology of Fantale, Shoa, Ethiopia: Bull. Geophys. Obs. Addis Ababa, v. 10, p. 59-68.

----- 1969, The structure and volcanic geology of an axial portion of the main Ethiopian rift: Tectonophysics, v. 8, p. 561-565.

----- 1970, A pantelleritic welded ash-flow tuff from the Ethiopian Rift Valley: Contrib. Mineral. Petrol., v. 28, p. 89-111.

----- 1975, A review of the geology, petrology and geochemistry of the Volcano Fantale: Bull. Volcanol., v. 38, no. 3, p. 791-802.

Gouin, P., 1979, Earthquake history of Ethiopia and the Horn of Africa: Ottawa, Int. Development Res. Centre, Monograph 118e, 259 p.

Mohr, P.A., 1962, Surface cauldron subsidence with associated faulting and fissure basalt eruptions at Gariboldi Pass, Shoa, Ethiopia: Bull. Volcanol., v. 24, p. 421-428.

Mohr, P.A., and Wood, C.A., 1976, Volcano spacings and lithospheric attenuation in the eastern rift of Africa: Earth Planet. Sci. Lett., v. 33, p. 126-144.

Reck, H., 1932, Bilder und Bemerkungen zum jungen abessinischen Grabenvulkanismus: Zeit. Vulkanol., v. 14, p. 187-190.

Richards, J.J., and Neumann van Padang, M., 1957, Catalogue of active volcanoes of the world including solfatara fields, Pt. IV, Africa and the Red Sea: Rome, IAVCEI, 118 p.

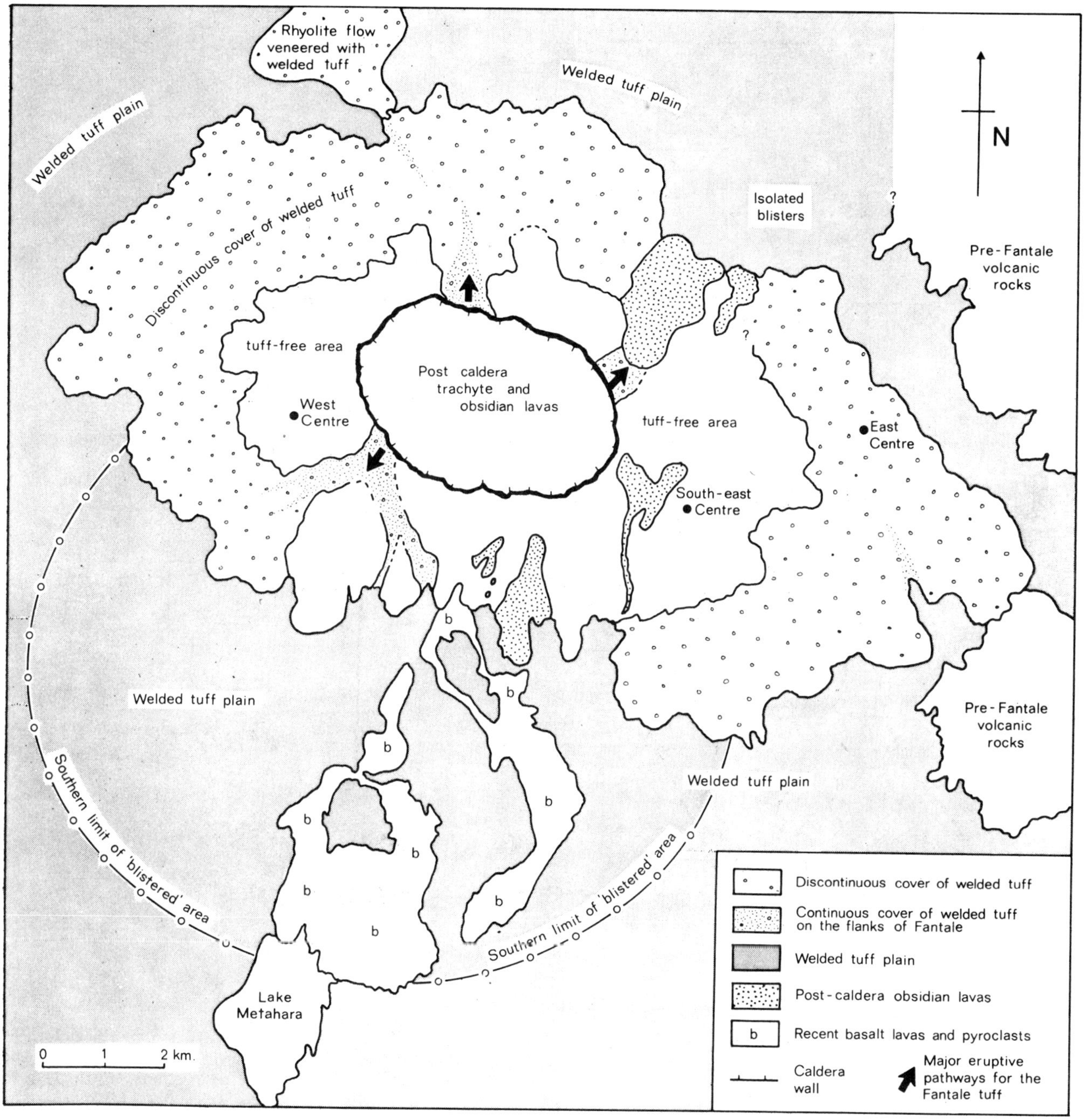

Figure 2.1.1. Geologic map of Fantale caldera, from Gibson (1970). "Blistered" area refers to zone in which postdepositional vesiculation of welded tuff causes "blisters" to form, in some cases up to 100 m in diameter (Gibson, 1970).

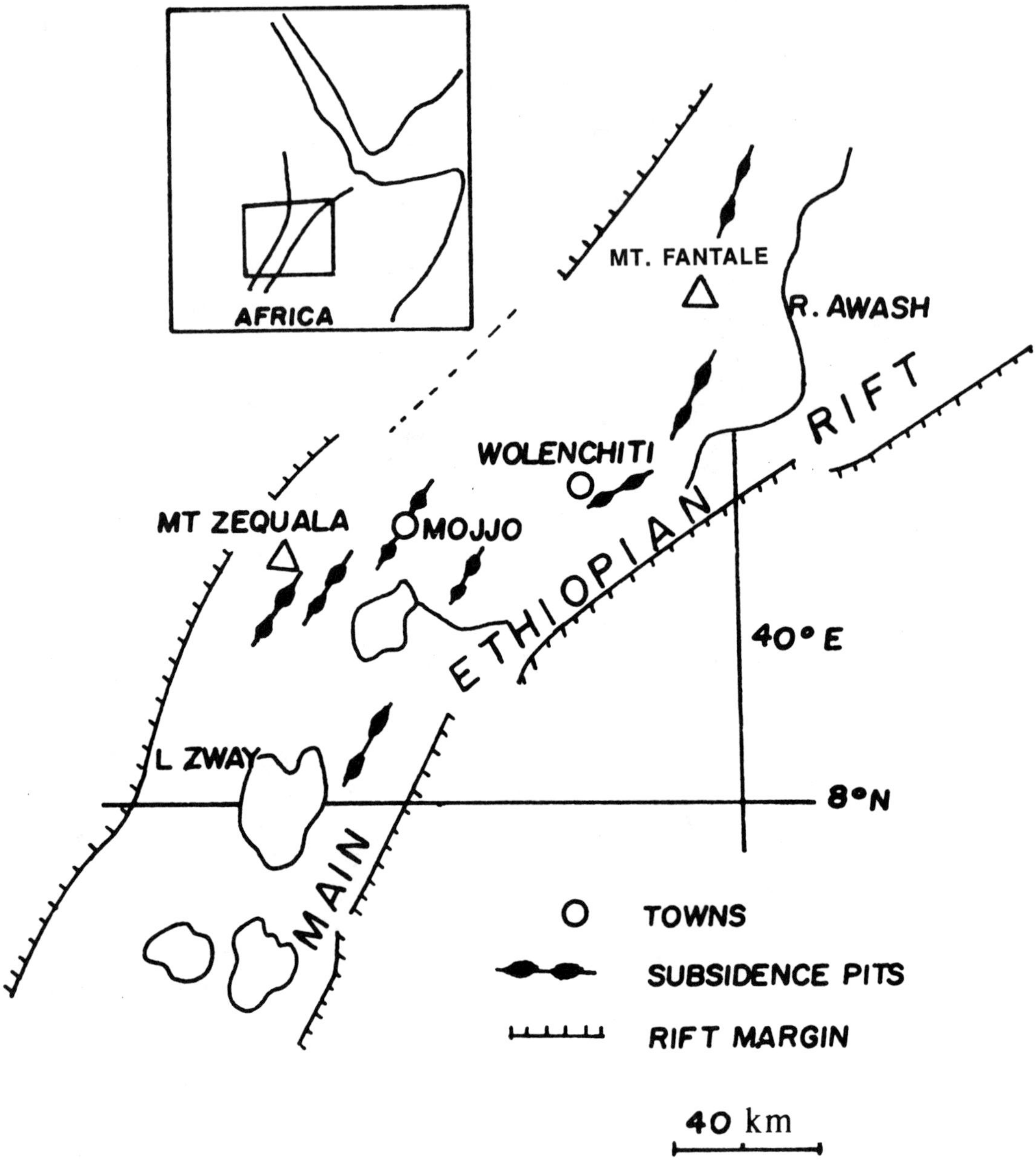

Figure 2.1.2. Sites of subsidence pits associated with fissures in Main Ethiopian Rift, including those associated with rifting event of January-March 1981 north of Fantale, from Asfaw (1982).

K'ONE (GARIBOLDI)

CAVW number (active vent)	Latitude Longitude (degrees)	Diameter (km)	Local tectonic setting	Precaldera edifice	SiO_2 content (wt pct)	Age of collapse (yr B.P.)	Date of unrest	Type and duration of unrest ESTU STHF MGTF H Te	Eruption type
02-01-20	8.80N 39.69E	ca. 5	Exten	?	R = b C = silicic	?	1820?	---- ---- ---- - --	ex,lf

TECTONIC SETTING

The K'one volcanic complex (called Gariboldi in some older reports) covers an area of 250 km^2 in the northern sector of the Main Ethiopian Rift, south-southwest of Fantale (fig. 2.2.1). K'One, Fantale, and Boseti Guda Volcanoes form part of a chain along the center of the rift associated with the Wonji Fault Belt, a zone of intense Quaternary faulting in which NNE-SSW-oriented rift-axis segments are arranged in a series of en echelon zones oblique to the rift margin (Cole, 1969).

GEOLOGIC HISTORY

The present form of the K'one complex is largely a result of silicic cone-building episodes, between which ignimbrites were erupted and areas collapsed to form calderas, and to a lesser extent of recent basalt eruptions. Cole (1969) identified eight silicic calderas with associated ignimbrite sheets, and a series of recent basaltic cinder cones and lava flows erupted along the hinge line between the two youngest calderas (fig. 2.2.2).

HISTORICAL ACTIVITY

1820?: The most recent basalt flows are dated from tradition at between 1810 and 1830, consistent with their lack of weathering or thick vegetation cover (Cole, 1969). Hot springs and fumaroles still exist in the area, which also has experienced historical seismicity (Gouin, 1960).

REFERENCES

Cole, J.W., 1969, Gariboldi volcanic complex, Ethiopia: Bull. Volcanol., v. 33, p. 566-578.
Gouin, P., 1960, Seismological notes: Bull. Geophys. Obs. Addis Ababa, v. 2, no. 1, p. 22-36.
Meyer, W., and Stets, J., 1976, Vulkanismus und Tektonik im "Wonji Fault Belt" (Main Ethiopian Rift) nach Luftbildauswertungen: Z. dt. geol. Ges., v. 127, p. 227-245.
Mohr, P.A., 1962, Surface cauldron subsidence with associated faulting and fissure basalt eruptions at Gariboldi Pass, Shoa, Ethiopia: Bull. Volcanol., v. 24, p. 421-428.

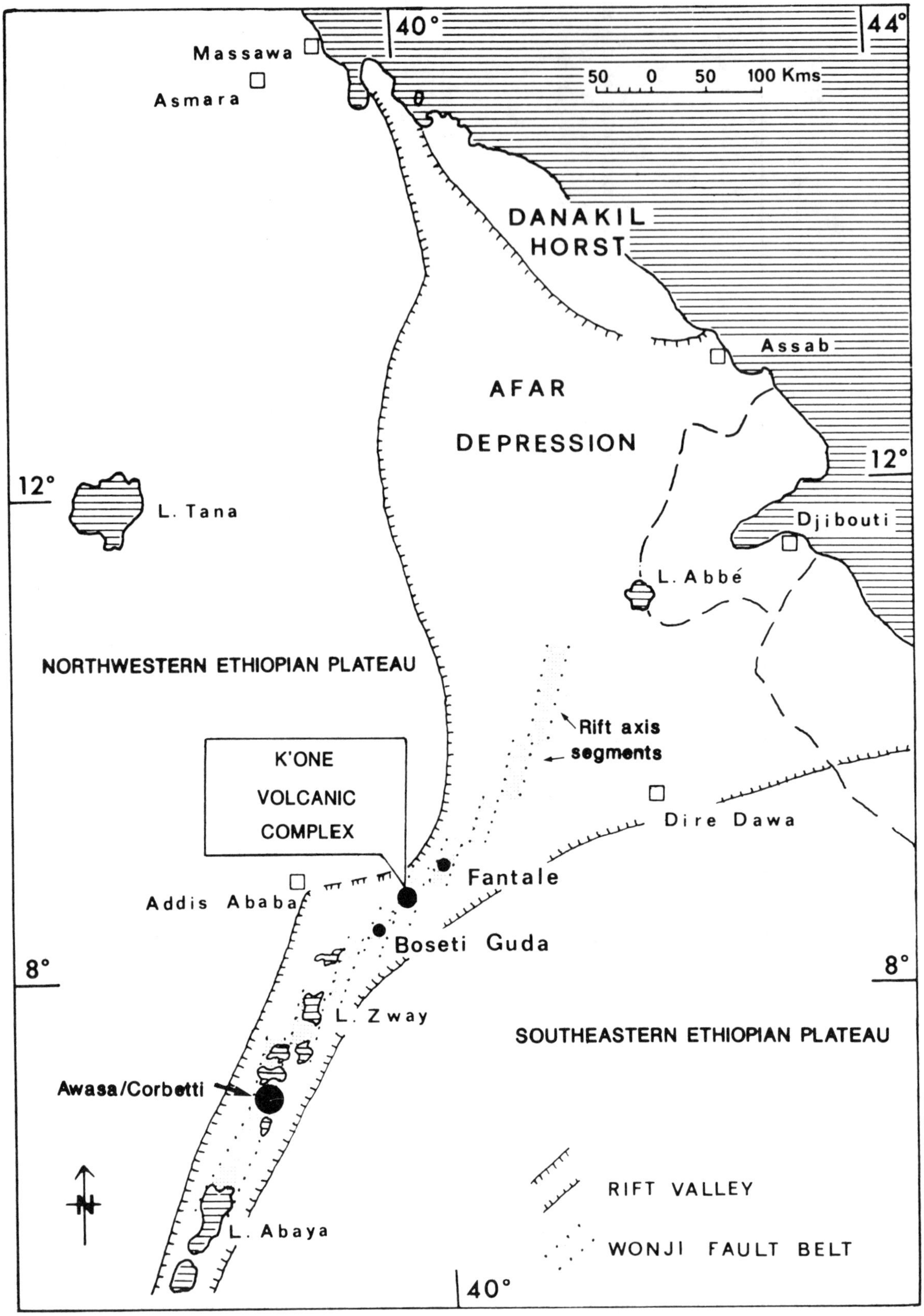

Figure 2.2.1. Position of K'one (Gariboldi) volcanic complex in relation to Main Ethiopian Rift and Wonji Fault Belt, from Cole (1969), modified slightly by Giday W-Gabriel (written commun., 1987).

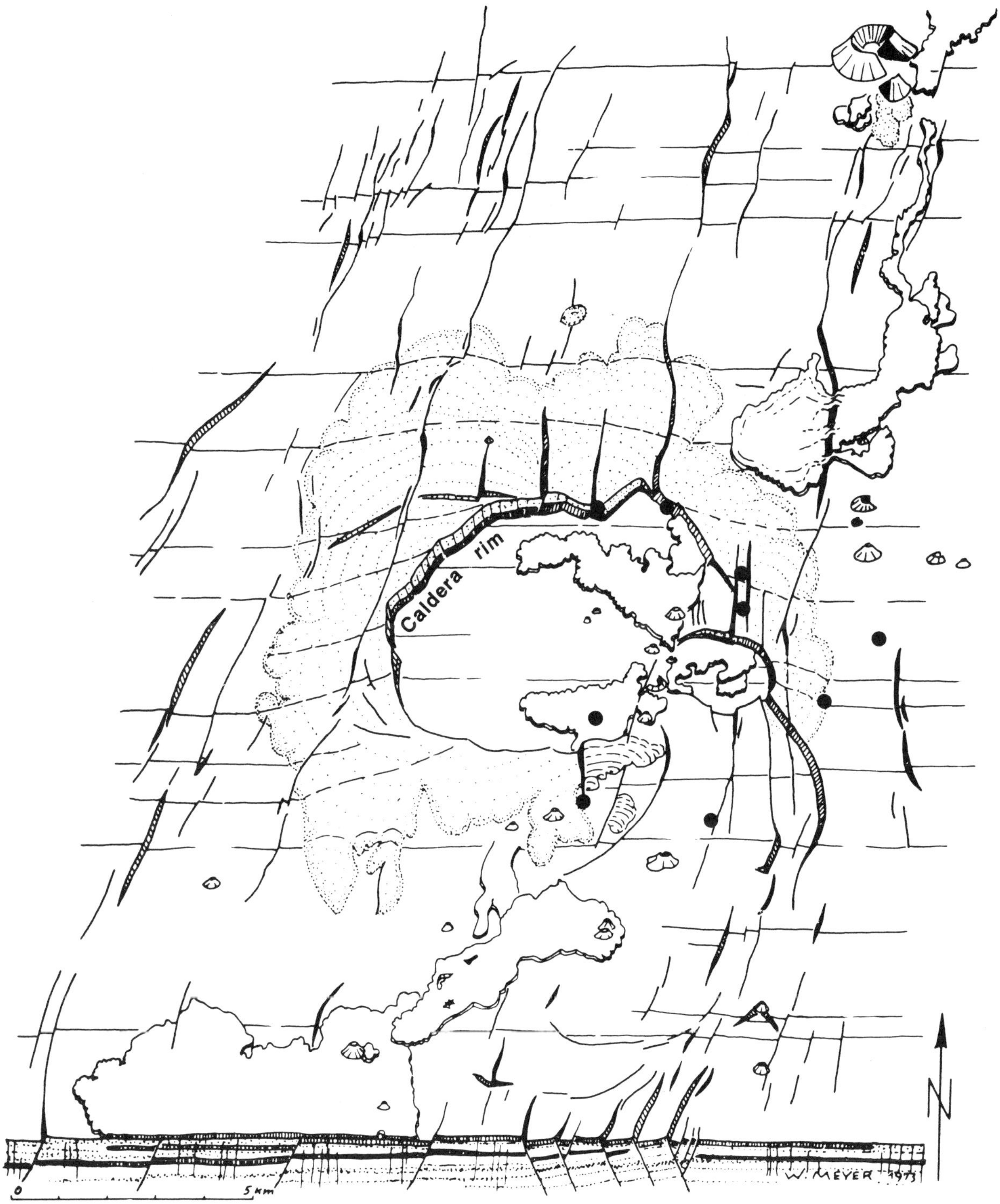

Figure 2.2.2. Structures of K'one (Gariboldi) volcanic complex, from Meyer and Stets (1976), with approximate locations of silicic vents (solid circles) after Cole (1969).

AWASA/CORBETTI

CAVW number (active vent)	Latitude Longitude (degrees)	Diameter (km)	Local tectonic setting	Precaldera edifice	SiO_2 content (wt pct)	Age of collapse (yr B.P.)	Date of unrest	Type and duration of unrest ESTU STHF MGTF H Te	Eruption type
02-01-29 (Chabbi)	7.18N 38.48E (Chabbi)	12 x 15 (Corbetti) 30 x 40 (Awasa)	Exten		R = 73-76 C = 73	<1,000,000?	1957-64+	F--- ---- ---F - --	none
							1983-84	xx-- ?--- ---- - =?	none

TECTONIC SETTING

The Awasa volcanic complex is located in the Main Ethiopian Rift Valley between Lake Shalla and Lake Abaya at the intersection of the rift axis with a major NW-trending lineament of stratovolcanoes and three small calderas on the western plateau (Gabriel, 1986) (see fig. 2.2.1 in section on K'one). The area has experienced geologically recent extension (Di Paola, 1972) and is 180 km south of another segment of the rift where 3-5 mm/yr of crustal extension was measured from 1969 to 1974 (Mohr and others, 1978).

GEOLOGIC HISTORY

The Awasa Caldera, approximately 30 km x 40 km, is a "Valles-type" caldera; the younger, 15-km-diameter Corbetti Caldera overlaps the northwest margin of the Awasa Caldera (figs. 2.3.1, 2.3.2). Collapse of Corbetti Caldera was associated with eruption of one or more ignimbrites and a large volume of unwelded pumice, probably from fissure vents buried within the caldera. Renewed activity built Urji Volcano inside the caldera, Chabbi Volcano on the east caldera rim, and a small unnamed pyroclastic cone on the west rim. Each of these vents is currently in the fumarolic stage (Di Paola, 1972). Mohr (1966) attributed E-W-trending fissures up to 15 m deep and 8 m wide on Chabbi Volcano to (prehistoric?) swelling of the volcanic edifice.

HISTORICAL ACTIVITY

1957-64+: Legends of volcanic eruptions as recent as 1930 exist, but these are discounted by Gouin (1979). Fumarolic activity was continuous from the earliest systematic observations (1957) through at least 1964 (Volcanological Society of Japan, 1964). Strong rhythmic steam emission from Chabbi was first observed in September 1959. A magnitude 6.3 earthquake occurred beneath or slightly southwest of Chabbi (in the Corbetti Caldera) on 14 July 1960 (Sykes and Landisman, 1964; Mohr, 1966; Gouin, 1979). No surface fracturing was observed, and fumarolic activity of Chabbi was not affected (Gouin, 1979).

AWASA/CORBETTI (continued)

HISTORICAL ACTIVITY (continued)

1983-84: In 1982 or 1983, levels of Lake Awasa increased and several houses along the shore were submerged. We do not know if the apparent change in lake level was uniform around the entire lake. Earthquakes occurred in 1984; buildings cracked, and students were evacuated from a Forestry school (G. Gabriel, written commun., 1987, citing information from M. Asfaw).

REFERENCES

Di Paola, G.M., 1972, Geology of the Corbetti Caldera area (main Ethiopian Rift Valley): Bull. Volcanol., v. 35, p. 497-506.

Gabriel, Giday W-, 1986, The Awasa Caldera in the Main Ethiopian Rift (MER) [abs.]: Abstracts, Int. Volcanol. Congr., New Zealand, p. 351.

Gouin, P., 1979, Earthquake history of Ethiopia and the Horn of Africa: Ottawa, Int. Development Res. Centre, Monograph 118e, 259 p.

Macdonald, R., and Gibson, I.L., 1969, Pantelleritic obsidians from the volcano Chabbi (Ethiopia): Contrib. Mineral. Petrol., v. 24, no. 3, p. 239-244.

Mohr, P.A., 1966, Chabbi Volcano (Ethiopia): Bull. Volcanol., v. 29, p. 797-815.

Mohr, P., Girnius, A., and Rolff, J., 1978, Present-day strain rates at the northern end of the Ethiopian Rift Valley: Tectonophysics, v. 44, p. 141-160.

Sykes, L.R., and Landisman, M., 1964, The seismicity of East Africa, the Gulf of Aden and the Arabian and Red Seas: Bull. Seismol. Soc. Am., v. 54, p. 1927-1940.

Volcanological Society of Japan, 1964, Chabbi Volcano: Bull. Volcanic Eruptions, no. 4, p. 10 (report from P.A. Mohr).

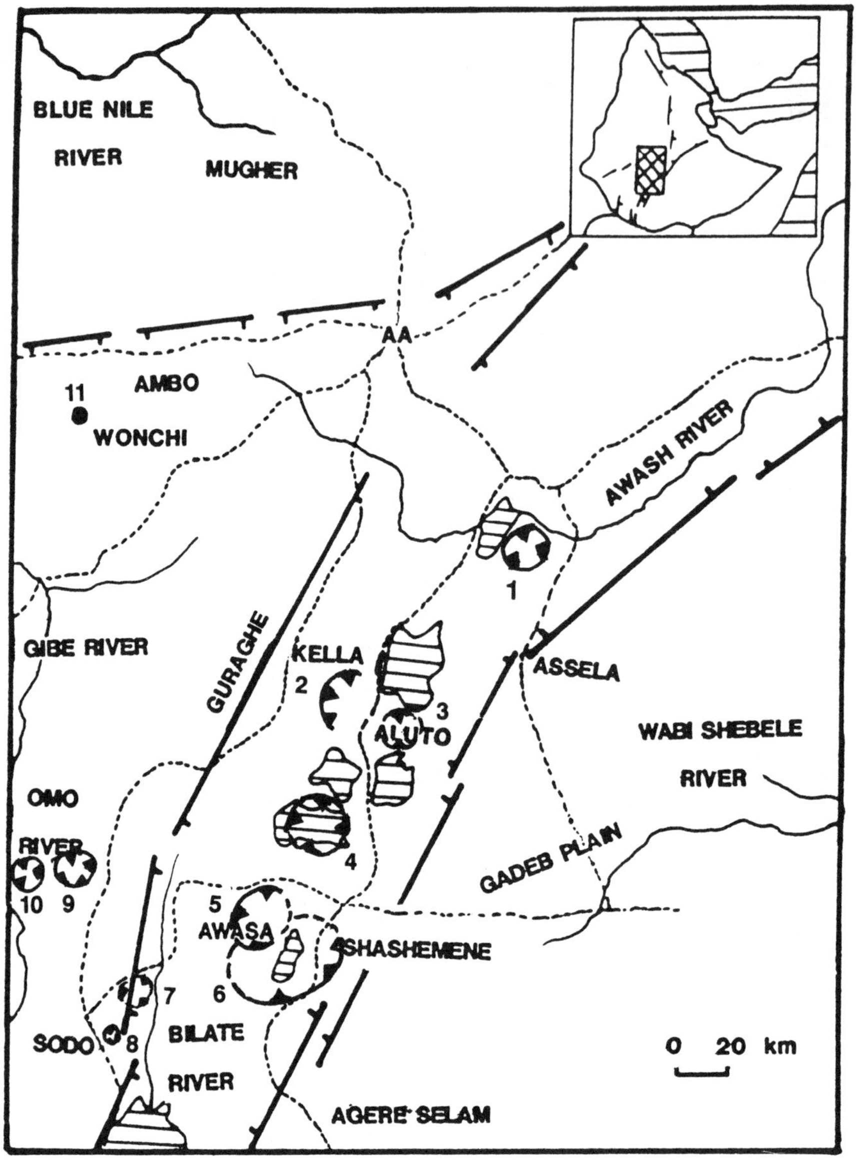

Figure 2.3.1. Pliocene-Pleistocene calderas in central sector of Main Ethiopian Rift, after Giday W-Gabriel (written commun., 1987). 1, Gadamsa; 2, Gademota; 3, Aluto; 4, Shalla; 5, Corbetti; 6, Awasa; 7, Duguna; 8, Hobicha; 9 and 10, Wagebeta Caldera Complex; 11, Wonchi.

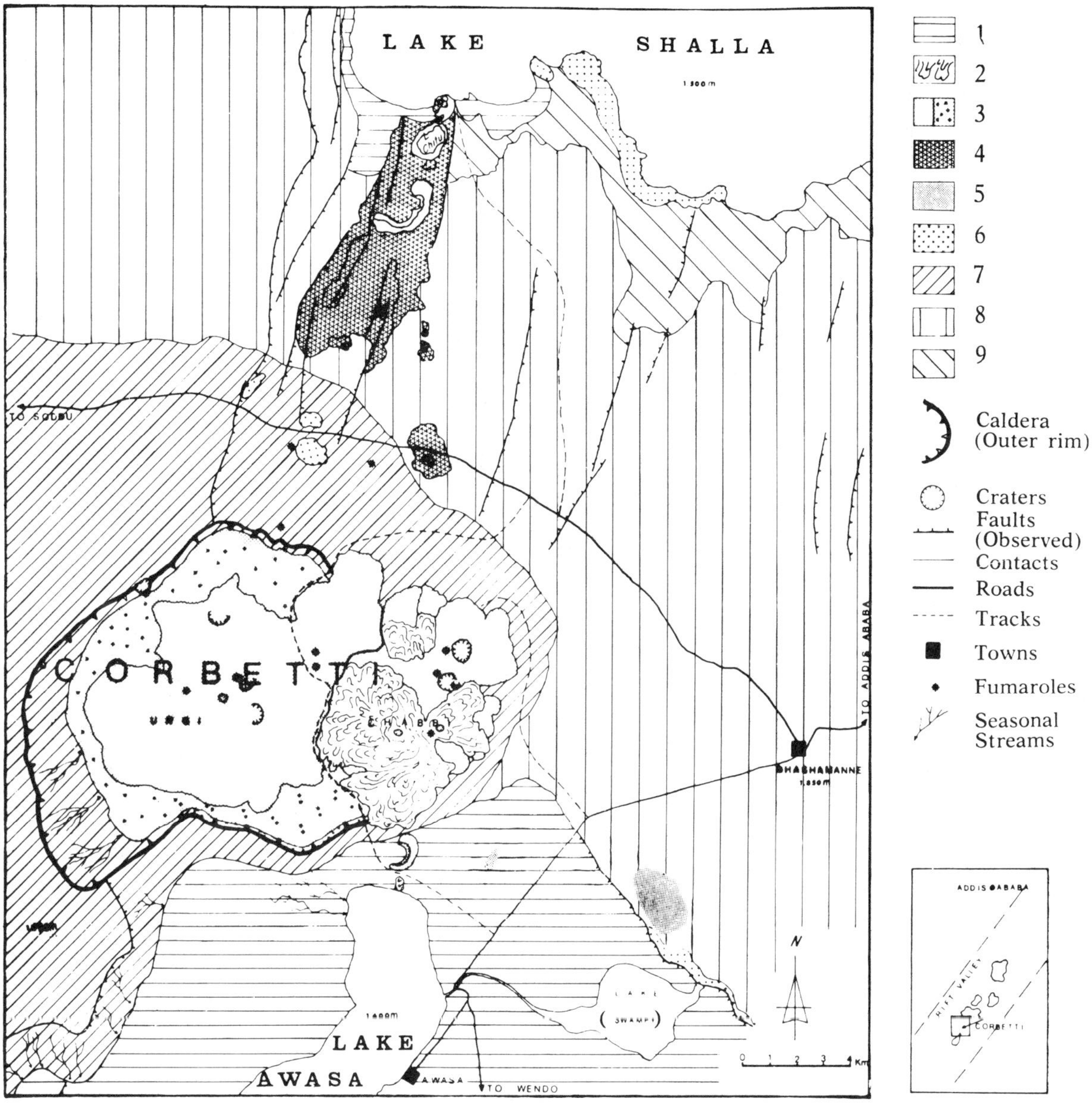

Figure 2.3.2. Geological sketch map of Awasa/Corbetti Caldera area, Ethiopia, from Di Paola (1972). Explanation: 1, lacustrine deposits; 2, obsidian lava flows; 3, recent pumice flows and pumice falls (dotted areas: detritus); 4, basaltic lava flows and scoria cones; 5, basaltic hyaloclastites; 6, peralkaline rhyolites (lava domes and flows); 7, old layered pumice; 8, ignimbrites; 9, unwelded ignimbritelike formation.

PART 3: HISTORICAL UNREST AT LARGE QUATERNARY CALDERAS (continued)
See inside back cover for explanation and abbreviations

THE BARRIER

CAVW number (active vent)	Latitude Longitude (degrees)	Diameter (km)	Local tectonic setting	Precaldera edifice	SiO_2 content (wt pct)	Age of collapse (yr B.P.)	Date of unrest	Type and duration of unrest ESTU STHF MGTF H Te	Eruption type
02-02-03 (Teleki, Likaiyu)	2.33N 36.62E	5 x 6 (outer) 3 (inner)	Exten	Shield	R = 50-51 C = silicic		between 1868-73?	---- ---- ---- - --	ex,lf
							1888	---- ---- ---- - --	lf
							1895	---- ---- ---- - --	ex,lf
							1897?	---- ---- ---- - --	ex?
							1906	---- ---- ---- - --	ex
							1917(-18?)	---- ---- ---- - --	ex
							1921-22	---- ---- ---- - --	ex,lf

TECTONIC SETTING/GEOLOGIC HISTORY

The Barrier lies in the Kenya Rift, at the southern end of Lake Turkana (fig. 2.4.1). An early lava shield was capped by trachytic tuff cones and agglomerate sheets. Then voluminous basaltic lava flows were extruded on the southeast flanks of the volcano, and the older caldera formed after eruption of trachyte on the northwest flanks. Eruption of voluminous postcaldera phonolitic lavas led to collapse of an inner caldera. By 9,600 yr B.P., The Barrier Volcano separated Lake Turkana and the nearby Suguta Valley, but it is not known whether one or both of the calderas had already formed by that time (Dodson, 1963; Williams and others, 1984)

HISTORICAL ACTIVITY

Eruptions were observed as early as 1888 (or perhaps 1868; see Sapper, 1917, p. 79); the latest known eruption was in 1921-22. No details are known.

REFERENCES

Dodson, R.G., 1963, Geology of the South Horr area: Geol. Surv. Kenya, Nairobi, no. 60.

Sapper, K., 1917, Katalog der geschichtlichen Vulkanausbrüche: Strasburg, Karl J. Trübner, 358 p.

Smith, W. Campbell, 1938, Petrographic description of volcanic rocks from Turkana, Kenya Colony, with notes on their field occurrence from the manuscript of Mr. A.M. Champion: Quart. Jour. Geol. Soc. London, v. 94, p. 507-550.

Williams, L.A.J., Macdonald, R., and Chapman, G.R., 1984, Late Quaternary caldera volcanoes of the Kenya Rift Valley: Jour. Geophys. Res., v. 89, no. B10, p. 8553-8570.

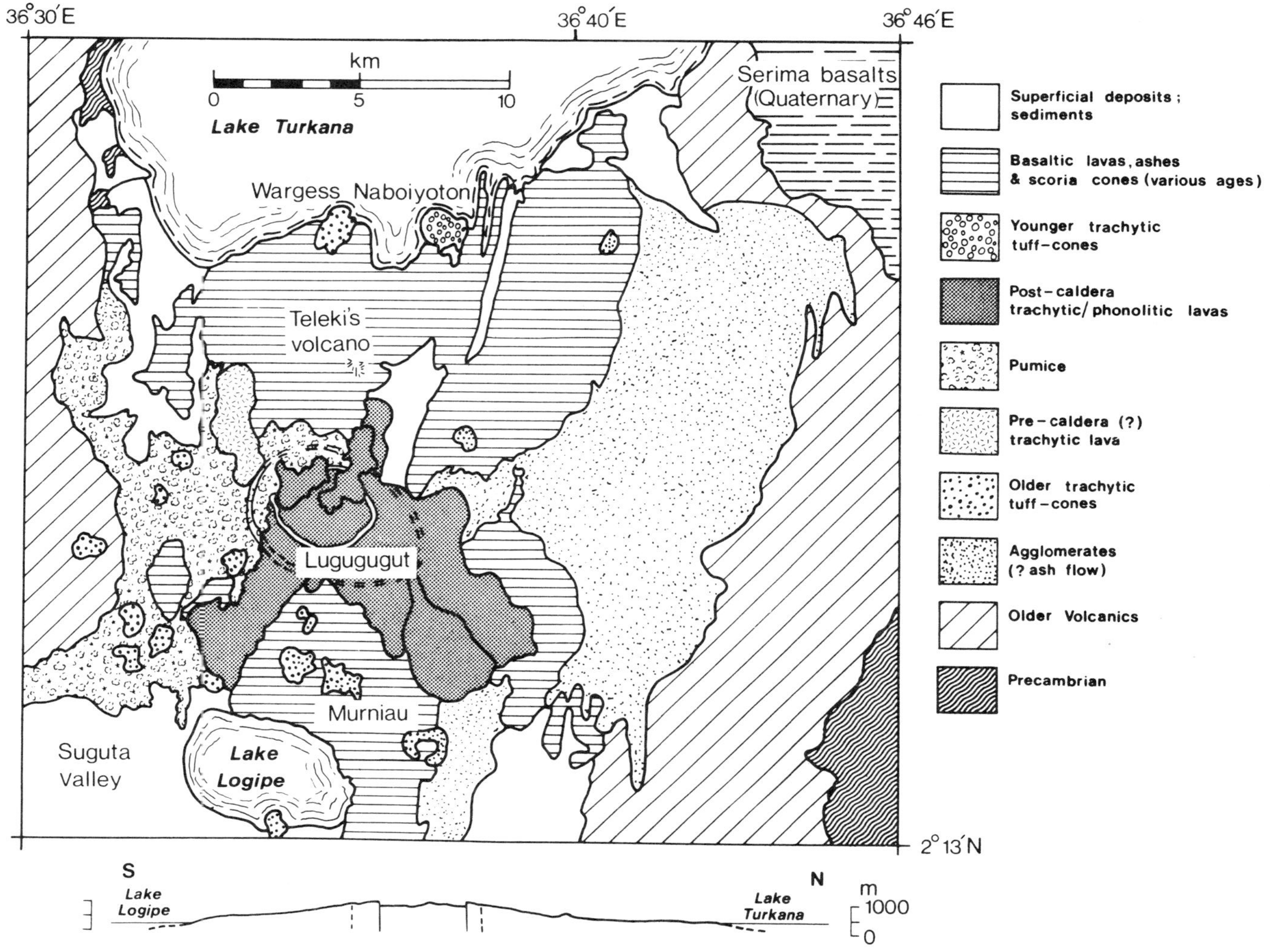

Figure 2.4.1. Simplified geologic map of The Barrier volcano and its summit caldera, from Williams and others (1984), modified from Dodson (1963). Faults are omitted; trends are N-NNE. Copyright by the American Geophysical Union.

DERIBA

CAVW number (active vent)	Latitude Longitude (degrees)	Diameter (km)	Local tectonic setting	Precaldera edifice	SiO_2 content (wt pct)	Age of collapse (yr B.P.)	Date of unrest	Type and duration of unrest ESTU STHF MGTF H Te	Eruption type
02-05-03 (Jebel Marra)	12.95N 24.27E	5	?	Strato	R = 54? C = intermediate	ca. 4,000?	1958-66	---- ---- ---x x --	none

TECTONIC SETTING

Deriba is a relatively small caldera at the summit of Jebel Marra Volcano in western Sudan (fig. 2.5.1). The caldera is surrounded by tuff deposits and trachytic lava, and lies within a much larger field of alkali basalt vents. The area lies at the intersection of two regional volcano-tectonic zones (Vail, 1972; fig. 2.5.2).

GEOLOGIC HISTORY

Vail (1973) outlined the following tentative sequence of events at Jebel Marra. Initial volcanic activity consisted of widespread outpourings of olivine basalt, perhaps alternating with small-scale pyroclastic activity. This initial activity built an edifice that was deeply eroded before additional basalt eruptions took place. Erupted products then became more acidic; most of the later lavas were trachytes. A third phase has been the emplacement of numerous secondary cones and plugs, many with central craters. Deriba Caldera probably formed during this third phase, followed by growth of a central cone and, in that, a deep crater lake (see also Burton and Wickens, 1966; Hammerton, 1966). Francis and others (1974) speculate that the caldera may have formed within historical time.

HISTORICAL ACTIVITY

mid-1964: Hot springs were observed in the crater lake in 1957 and February 1964 (Burton and Wickens, 1966). In mid-1966, a different observer reported fumaroles at four sites on the northwest flank of the central, intracaldera cone (Hammerton, 1966). Burton and Wickens (1966) suggest that these fumaroles might have developed after the time of their survey in February 1964.

DERIBA (continued)

REFERENCES

Burton, A.N., and Wickens, G.E., 1966, Jebel Marra Volcano, Sudan: Nature, v. 210, p. 1146-1147.

Francis, P.W., Thorpe, R.S., and Ahmed, F., 1973, Setting and significance of Tertiary-Recent volcanism in the Darfur province of western Sudan: Nature (Physical Sciences), v. 243, p. 30-31.

Francis, P.W., Thorpe, R.S., Smith, K., and Ahmed, F., 1974, Volcanism of the Jebel Marra area, W. Sudan [abs.]: Jour. Geol. Soc. London, v. 130, p. 180.

Hammerton, D., 1966, Hot springs and fumaroles at Jebel Marra, Sudan: Nature, v. 209, p. 1290.

Katsui, Y., 1971, ed., List of the world active volcanoes: Tokyo, Volcanol. Soc. Japan/IAVCEI/IUGG, 160 p.

Vail, J.R., 1973, Jebel Marra, a dormant volcano in Darfur Province, Western Sudan: Bull. Volcanol., v. 36, p. 251-265.

Volcanological Society of Japan, 1967, Jebel Marra: Bull. Volcanic Eruptions, no. 7, p. 4.

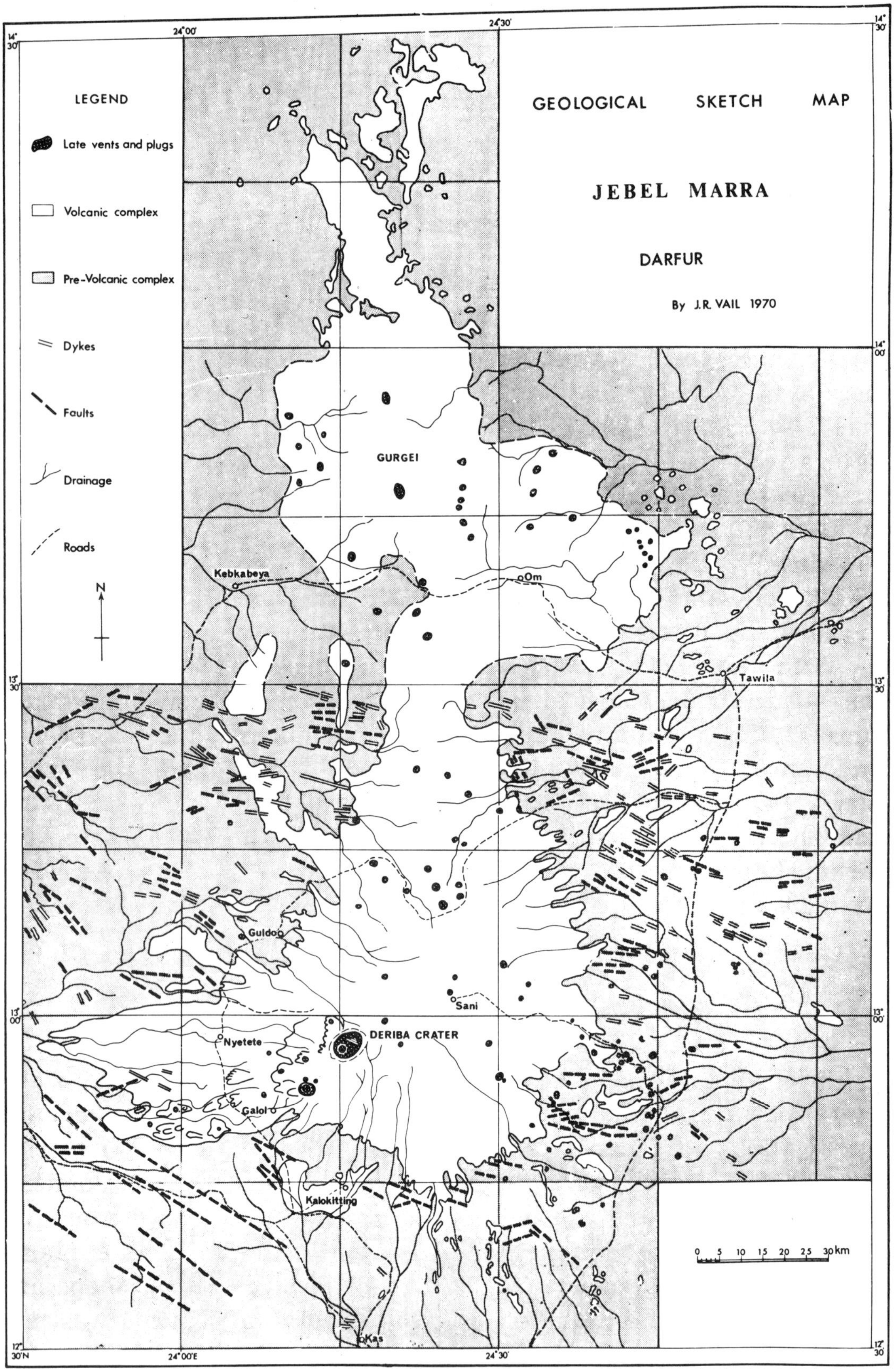

Figure 2.5.1. Geological sketch map of Jebel Marra Volcano and Deriba Caldera, Dafur Province, Sudan (from Vail, 1973).

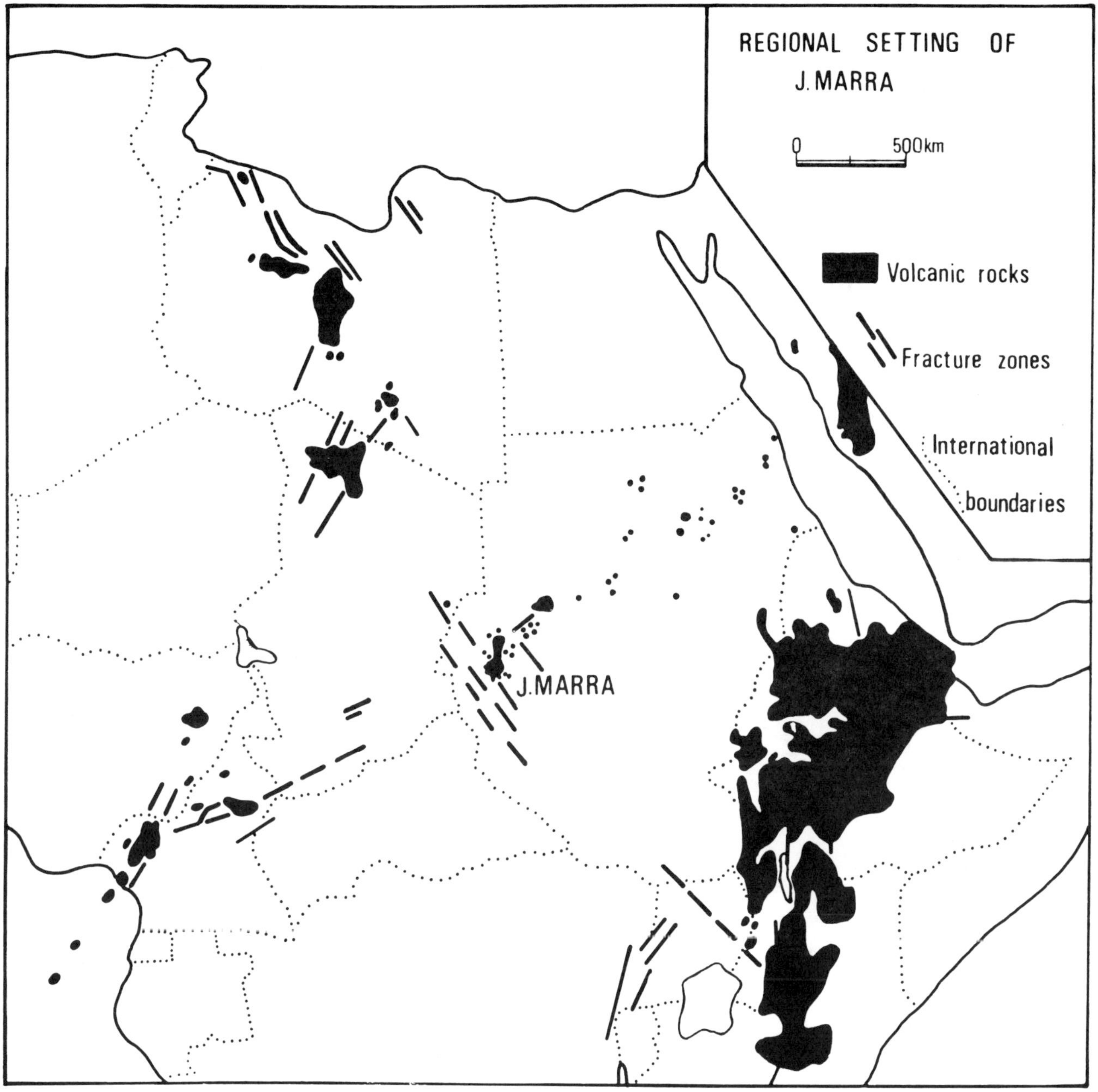

Figure 2.5.2. Regional setting of Jebel Marra Volcano and Deriba Caldera in northeast Africa, from Vail (1973). Note that area lies at intersection of two regional volcano-tectonic zones.

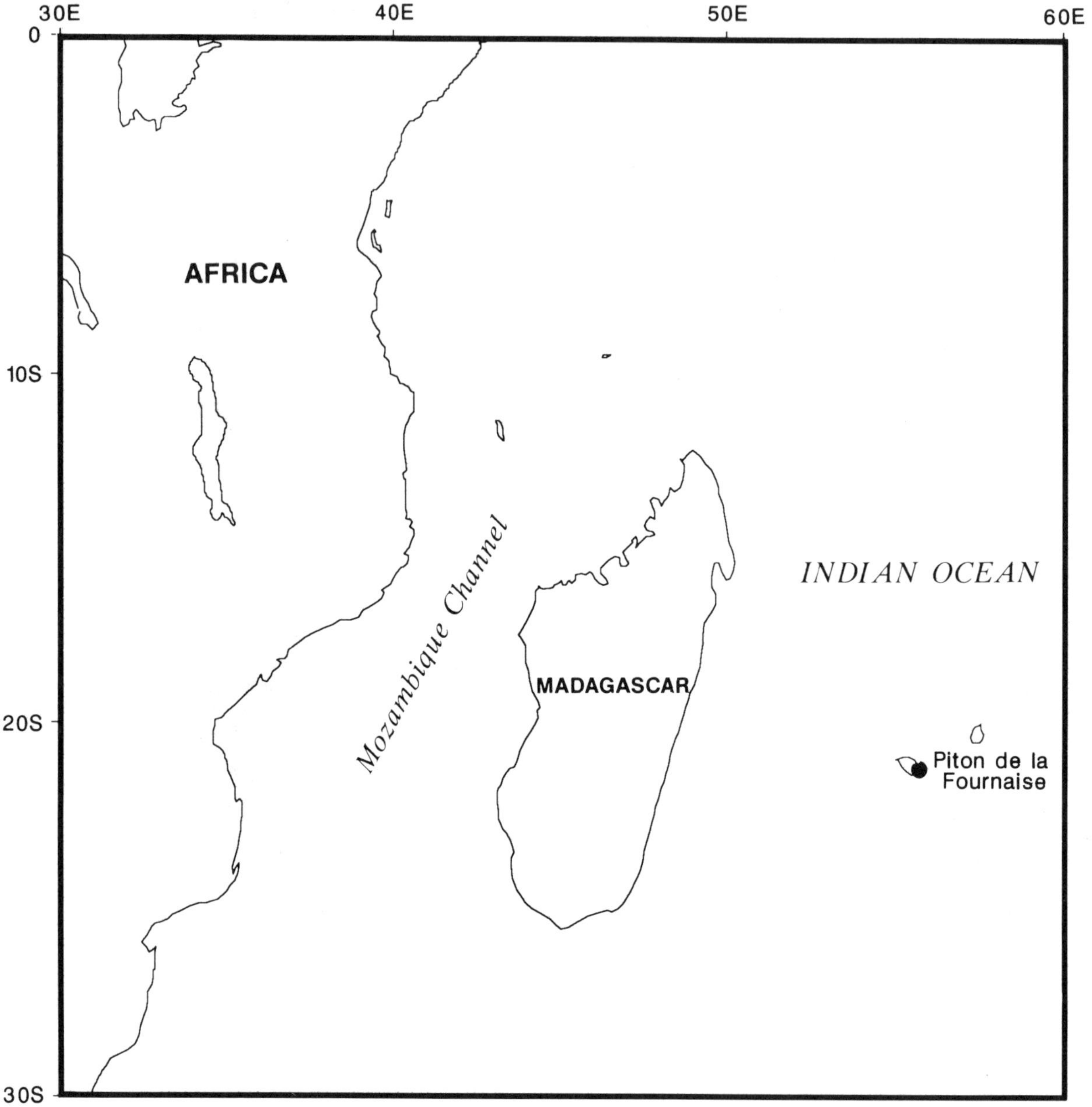

Figure 3. Location of Piton de la Fournaise (solid circle), a restless caldera on Réunion Island, Indian Ocean (Region 3).

PITON DE LA FOURNAISE

CAVW number (active vent)	Latitude Longitude (degrees)	Diameter (km)	Local tectonic setting	Precaldera edifice	SiO_2 content (wt pct)	Age of collapse (yr B.P.)	Date of unrest	Type and duration of unrest ESTU STHF MGTF H Te	Eruption type
03-03-02	21.23S	8 x 15	Hotspot?	Shield	R = 43-50	ca. 2,000	pre-1884	---- ---- ---- - --	lf,ex
	55.71E				C = mafic		1884	---- ---- ---- - --	lf
							1889	---- ---- ---- - --	ex,lf
							1890-91	---- ---- ---- - --	ex,lf
							1897-1905	---- ---- ---- - --	ex,lf
							1907-10	---- ---- ---- - --	ex,lf
							1913	---- ---- ---- - --	ex,lf
							1915	---- ---- ---- - --	ex,lf
							1917	---- ---- ---- - --	lf
							1920-21	---- ---- ---- - --	ex,lf
							1924-27	---- ---- ---- - --	ex,lf
							1929-31	---- ---- ---- - --	ex,lf
							1933-34	---- ---- ---- - --	ex,lf
							1937-39	---- ---- ---- - --	ex,lf
							1942-46	---- ---- ---- - --	ex,lf
							1948-50	---- ---- ---- - --	ex,lf
							1952-53	---- ---- ---- - --	ex,lf
							1955-61	---- ---- ---- - --	ex,lf
							1963-66	---- ---- ---- - --	ex,lf
							1972-73	---- ---- ---- - --	ex,lf
							1975-77	---- ---- ---- - --	ex,lf
							1979-81	CB-- ---- ---- - --	lf
							1983-84	DBYD ZZZY ---- - --	ex,lf
							1985-86	DBYD -EEY C--- - --	lf
							1987	DBYD -EE- C--- - --	lf

TECTONIC SETTING

Piton de la Fournaise is an active basaltic shield volcano on the island of La Réunion in the Indian Ocean (fig. 3.1.1). It is similar in many respects to Kilauea Volcano, Hawaii, including its intraplate setting over a possible "hotspot" or melting anomaly in the upper mantle.

PITON DE LA FOURNAISE (continued)

GEOLOGIC HISTORY

Piton de la Fournaise has grown by repeated extrusion of lava from the summit area and from fissures on all flanks of the volcano. The principal structural features of Fournaise are (1) a summit caldera and adjoining downfaulted trough, (2) a family of broadly curving faults that are nested around the caldera, and (3) three rift zones (fig. 3.1.2) (Duffield and others, 1982). A few dikes are visible in the walls of the summit caldera, trending perpendicular to the walls (Upton and Wadsworth, 1970). Mouginis-Mark (1980) and Chevallier and Bachelery (1981) identify at least three periods of caldera collapse. Duffield and others (1982) recognize a similar sequence, with the added interpretation that each of the calderas has been breached by large landslides (figs. 3.1.2, 3.1.3) like those that occur on the south flank of Kilauea Volcano in Hawaii.

HISTORICAL ACTIVITY

1708: An eruption along the northeast rift zone.

1800: The latest eruption along the southwest rift zone.

1800-1980: Earthquakes and other precursors to eruptions of Piton de la Fournaise are rarely noted by local residents, and local archives dating from 1800 contain no accounts of earthquakes that were felt outside the immediate vicinity of the volcano (Bachelery and others, 1982). However, small but measurable changes do precede most eruptions. Eruptions of June-November 1972 were preceded by formation of fissures within the caldera, distributed radially from Bory Crater (fig. 3.1.3). The eruption of July 1972 was also preceded by several days of frequent earthquakes (Montaggioni and others, 1972). Surface fissures are known to have formed before eruptions in 1975 and 1976 (Krafft and Gerente, 1977b), and undoubtedly formed before many previous eruptions as well.

1977: An eruption in April 1977 on the northeast flank of Piton de la Fournaise (figs. 3.1.2, 3.1.3), the first northeast-rift eruption since 1708, produced lava showing fractionation of olivine only. In contrast, products of an eruption in October-November 1977 showed plagioclase and clinopyroxene fractionation. The April eruption was therefore judged to be a "lateral" eruption (that is, one fed directly from a deep reservoir), and the October-November eruption was judged to be a "terminal" eruption (that is, one fed from a shallow central reservoir) even though it too occurred on the east flank of the volcano (Kieffer and others, 1977; Kieffer and Vincent, 1978).

1981-84: A volcano observatory was established at Piton de la Fournaise in 1979 to monitor seismicity, ground deformation, and other eruption precursors. An eruption that began on 3 February 1981 was preceded by (1) an earthquake 10-15 km beneath the caldera on 2 January, (2) 20 days of quiescence, and then (3) 13 days of shallow seismicity near the eruptive centers (fig. 3.1.4). Earthquakes were most frequent immediately before the eruption began. A geodetic network established only days before the 2 January earthquake showed little or no inflation as of 30 January and only local fissuring when lines were remeasured after the start of the eruption on 12 February (Bachelery and others, 1982). Stations just west of the caldera rim showed as much as 57 microradians of

PITON DE LA FOURNAISE (continued)

HISTORICAL ACTIVITY (continued)

inflationary tilt from 11 February to 5 May 1981, a period in which 10 x 10^6 m^3 of lava was erupted; tilting continued, though at a slower rate, during the noneruptive period that followed (Institut de Physique du Globe, 1984). Following the February-May 1981 eruption, background seismicity was 0 to 5 events/month until a 12-hour seismic swarm on 15 July 1981. Earthquake depths were shallow but not precisely determined. The seismic crisis prompted a resurvey of the deformation network, but no changes were detected. Seismicity returned to background but then increased again on 2 November. Earthquakes after 20 November clustered at 1.5 to 5 km beneath the summit. Ten earthquakes per day were recorded on 1 and 2 December; all were less than magnitude 1, so the total energy release was small. Another swarm started at 0642 hr on 4 December, with 300 events by 0859 hr, when tremor began. An eruption was seen a few minutes later on the south-southwest flank of the Central Cone. A resurvey of the deformation network immediately after the start of the eruption showed that there had been inflation with a shallow focus. A continuously recording borehole strainmeter indicated that most of the deformation had occurred within a period of 30 to 45 minutes, 2 hours before the start of the seismic swarm and intrusion (Smithsonian Institution, 1983). A second phase of this eruption, on 18 January, was preceded by inflation of the summit area in early January. A swarm of about 50 small earthquakes occurred between 0313 and 0454 hr on 18 January 1984, followed immediately by harmonic tremor and an eruption. Significant inflation occurred during, rather than before, the eruption of 18 January, as was also the case during the first phase in December. The 1983-84 eruption ended on 18 February following several days when tremor amplitude was irregular with periodic bursts.

1985-86: Following the 1983-1984 eruption and continuing until May 1985, seismicity was low (0-5 events/month) and tilt indicated slow deflation (1-7 microradians/month) of the 1983-84 vent area. Small, shallow earthquakes began on 15 May 1985, peaked on 8 June, then decreased until a 40-minute seismic swarm (intrusion) began at 1520 on 14 June. A 24-hour eruption began at 1604 hr. From mid-May to mid-June, summit tilt stations showed 20-50 microradians of inflationary tilt. Stronger seismic swarms began on 9 July beneath the summit, continued an hour later 3-6 km west-northwest of the summit, and then began again at the summit almost a month later. Deformation during this period was interpreted as inflation along a N15°E fissure. An eruption began on 5 August and continued until mid-October, when deflation began.

Renewed eruptions began on 29 December and continued through 7 February 1986. Tilt measurements showed summit deflation that began when the eruption began. Seismic records of the station at Bory Crater were saturated with tremor throughout the eruption, and a swarm of earthquakes occurred 3 km beneath the summit on 11-12 February, just after extrusion stopped. The total volume of lava extruded from June 1985 through February 1986 was 33 x 10^6 m^3. On 17 March 1986, additional earthquakes occurred 3 km beneath the summit. Additional fracturing and seismicity culminated with a small eruption in the southeast part of the caldera on 19 March, followed by eruptions on the upper southeast flank on 20 March and the lower southeast flank on 23 March. This was the first flank eruption since 1977. Tilt, elevation changes, and seismicity accompanied the intrusion southeastward from the summit. On 28 March, lava again rose beneath and then was erupted within the summit crater, accompanied by earthquakes, tremor, and localized inflation. Small, reversible and irreversible changes in total magnetic field strength have been noted at stations within and outside the caldera, relative to a station in the western part of the caldera. These changes, as much as -12 nT (gammas) inside the caldera and +4 nT outside the caldera, are attributed mainly to piezomagnetic effects (Zlotnicki and Le Mouel, 1987). A brief, minor eruption occurred on 13 July from the bottom of the March 1986 pit crater. Additional eruptions occurred through December 1986.

PITON DE LA FOURNAISE (continued)

HISTORICAL ACTIVITY (continued)

1987: The caldera floor inflated from mid-April through about 23 April, accompanied by low but constant seismicity and a decrease in magnetic field strength within the caldera. An intrusion occurred on 2 June, marked by an earthquake swarm but no eruption. After another 8 days of seismicity and inflation, lava broke out on 10 June. Another cycle of inflation and lava extrusion occurred in July. On 6-8 November, another eruption was preceded by 2 days of seismicity and a seismic swarm during 15 minutes before fissures opened north of Crater Dolomieu (J.-L. Cheminee, written commun., 1987)

COMMENTS

Piton de la Fournaise receives episodic supply of basaltic magma from depth, stores that magma in a shallow, central reservoir, and erupts it from the summit crater and flank rifts. Seismicity is limited to intermediate-depth earthquakes (1-5 km deep) in periods no longer and often less than 2-4 weeks before the start of eruptive sequences, to intense seismicity (50-100 events/hr) 1.5-2.5 hr just before the initial and subsequent episodes of each eruptive sequence, and to tremor during eruptions. Generalizations about deformation are difficult at this early stage of monitoring. However, it appears that the summit inflates for several weeks to months before the onset of each eruptive episode but does not begin to reinflate immediately after each eruption. The general style of activity is much like that in Hawaii, except that magma supply at Piton de la Fournaise is thought to be more episodic (Lenat and Bachelery, 1987).

REFERENCES

Aubert de la Rue, E., 1965, Le Piton de la Fournaise, Volcan actif de l'Ile de la Réunion: Bull. Volcanol., v. 28, p. 311-320.

Bachelery, P., Blum, P.A., Cheminee, J.-L., Chevallier, L., Gaulon, R., Girardin, N., Jaupart, C., Lalanne, F., Le Mouel, J.L., Ruegg, J.C., and Vincent, P., 1982, Eruption at Le Piton de la Fournaise volcano on 3 February 1981: Nature, v. 297, p. 395-397.

Benard, R., and Krafft, M., 1977, La Fournaise, volcan actif de l'Ile de la Réunion: Strasbourg, Istra, 121 p.

Billard, G., 1974, Carte géologique de l'Ile de la Réunion: Paris, Bur. Recherches Geol. Minieres.

Blum, P., Gaulon, R., Lalanne, F., Ruegg, J., 1981, Sur l'évidence de précurseurs de l'éruption du Volcan Piton de la Fournaise à la Réunion (Fevrier 1981): Comptes Rendus Acad Sci. Paris, v. 292, ser. ii, p. 1449-1455.

Chevallier, L., and Bachelery, P., 1981, Evolution structurale du volcan actif du Piton de la Fournaise, Ile de la Réunion--Océan indien occidental: Bull. Volcanol., v. 44, p. 723-741.

Chevallier, L., Lalanne, F.-X., Bachelery, P., and Vincent, P.M., 1981, L'éruption du mois de février 1981 au Piton de la Fournaise (Ile de la Réunion, Océan Indien). Phénomènologie et remarques structurales: Comptes Rendus Acad Sci. Paris, Ser. II, v. 3, p. 187-190.

Chevallier, L., and Vatin-Perignon, N., 1982, Volcano-structural evolution of Piton des Neiges, Réunion Island, Indian Ocean: Bull. Volcanol., v. 45, no. 4., p. 285-298.

PITON DE LA FOURNAISE (continued)

REFERENCES (continued)

Delisle, F., 1883, Les secousses de tremblement de terre a la Réunion et a maurice comme consequence de l'éruption volcanique du detroit de la sonde: Bull. Soc. Geogr., 1883, p. 524-526.

Ducrot, M., 1960, Le volcan actif de l'ile de la Réunion: Surveillance et etude de l'activité possibilites d'utilisation de l'energie emise: Bull. Volcanol., v. 23, p. 109-120.

Duffield, W.A., Stieltjes, L., and Varet, J., 1982, Huge landslide blocks in the growth of Piton de la Fournaise, La Réunion, and Kilauea Volcano, Hawaii: Jour. Volcanol. Geotherm. Res., v. 12, p. 147-160.

Institut de Physique du Globe, 1984, Observations volcanologiques, La Soufriere (Guadeloupe), La Montagne Pelee (Martinique), Le Piton de la Fournaise (Réunion), Année 1982: Paris, Institut de Physique du Globe, p. 44-69.

Kieffer, G., Tricot, B., and Vincent, P.M., 1977, Une éruption inhabituelle (avril 1977) du Piton de la Fournaise (Ile de la Réunion): ses enseignements volcanologiques et structuraux: Comptes Rendus Acad. Sci., Ser. D, v. 285, p. 957-960.

Kieffer, G., and Vincent, P.M., 1978, L'éruption d'octobre-novembre 1977 du Piton de la Fournaise (Ile de la Réunion): une éruption terminale sans cratere terminal?: Comptes Rendus Acad. Sci., Ser. D, v. 286, p. 1767-1770.

Kornprobst, J., Boivin, P., and Bachelery, P., 1979, L'alimentation des éruptions recentes du Piton de la Fournaise (Ile de la Réunion, Océan Indien): degré d'évolution et niveau de ségrégation des laves émises: Comptes Rendus Acad. Sci., Ser. D., v. 288, p. 1691-1694.

Krafft, M., and Gerente, A., 1977a, L'activité du Piton de la Fournaise entre Octobre 1972 et Mai 1973 (ile de la Réunion, océan Indien): Comptes Rendus Acad. Sci. Paris, Ser. D., v. 284, p. 607-610.

----- 1977b, L'activité du Piton de la Fournaise entre Novembre 1975 et Avril 1976 (ile de la Réunion, océan Indien): Comptes Rendus Acad. Sci. Paris, Ser. D., v. 284, p. 2091-2094.

Lacroix, A., 1936, Le volcan actif de l'Ile de la Réunion et ses products: Paris, Gauthier-Villars, 297 p.

----- 1938, Le volcan actif de la Réunion (Supplement) et celui de la Grande Comore: Paris, Gauthier-Villars, 57 p.

----- 1939, Les trasformations recentes du sommet du volcan actif (Piton de la Fournaise) de l'ile de la Réunion: Bull. Volcanol., v. 5, p. 3-18.

Lenat, J.-F., and Aubert, M., 1982, Structure of Piton de la Fournaise volcano (la Réunion Island, Indian Ocean) from magnetic investigations: An illustration of the analysis of magnetic data in a volcanic area: Jour. Volcanol. Geotherm. Res., v. 12, no. 3/4, p. 361-392.

Lenat, J.-F., and Bachelery, P., 1985, Ground deformation associated with the 1983-84 and June 1985 eruptions of Piton de la Fournaise, Isle de la Réunion: Eos, Trans. Am. Geophys. Union, v. 66, no. 46, p. 852.

----- 1987, Dynamics of magma transfers at Piton de la Fournaise: A model based on 5 years of monitoring and on the historic activity [abs.]: Abstracts, Hawaii Symposium on How Volcanoes Work, Hilo, January 1987, p. 153.

Ludden, J.N., 1977, Eruptive patterns for the volcano Piton de la Fournaise, Réunion Island: Jour. Volcanol. Geotherm. Res., v. 2, p. 385-396.

----- 1978, Magmatic evolution of the basaltic shield volcanoes of Réunion Island: Jour. Volcanol. Geotherm. Res., v. 4, p. 171-198.

Montaggioni, L., Nativel, P., and Billard, G., 1972, L'activité actuelle du Piton de la Fournaise (Ile de la Réunion, Océan Indien): Comptes Rendus Acad. Sci., Ser. D, v. 275, p. 2615-2618.

PITON DE LA FOURNAISE (continued)

REFERENCES (continued)

Mouginis-Mark, P., 1980, Piton de la Fournaise, Réunion: Volcano News, no. 2, p. 1-2.

Neumann van Padang, M., 1963, Catalogue of the active volcanoes of the world including solfatara fields, Pt. XVI, Arabia and the Indian Ocean: Rome, IAVCEI, p. 39-52.

Smithsonian Institution, Scientific Event Alert Network (SEAN), 1976-87, Piton de la Fournaise: SEAN Bull., v. 1, no. 2; v. 1, no. 14; v. 2, nos. 4-5; v. 2, no. 8; v. 2, nos. 10-12; v. 4, no. 5; v. 4, no. 7; v. 6, nos. 1-5; v. 8, nos. 11-12; v. 9, nos. 1-2; v. 10, nos. 5-10; v. 12, nos. 5,8 (communications from J.-F. Lenat and others).

Upton, B.G.J., and Wadsworth, W.J., 1965, Geology of Réunion Island, Indian Ocean: Nature, v. 207, p. 151-154.

----- 1970, Intra-volcanic intrusions of Réunion, in Newall, G., and Rast, N., eds., Mechanism of igneous intrusion: Liverpool, Gallery Press, p. 141-156.

Volcanological Society of Japan, 1961-1979, Piton de la Fournaise: Bull. Volcanic Eruptions, no. 12, p. 12-13; no. 15, p. 13; no. 16, p. 13-15; no. 17, p. 14-16; no. 19, p. 12.

Zlotnicki, J., and Le Mouel, J.L., 1987, Volcanomagnetic variations on La Fournaise Volcano (Réunion Island) [abs.]: Abstracts, Hawaii Symposium on How Volcanoes Work, Hilo, January 1987, p. 280.

Figure 3.1.1. Physiographic map of Piton de la Fournaise Volcano, from Lacroix (1936).

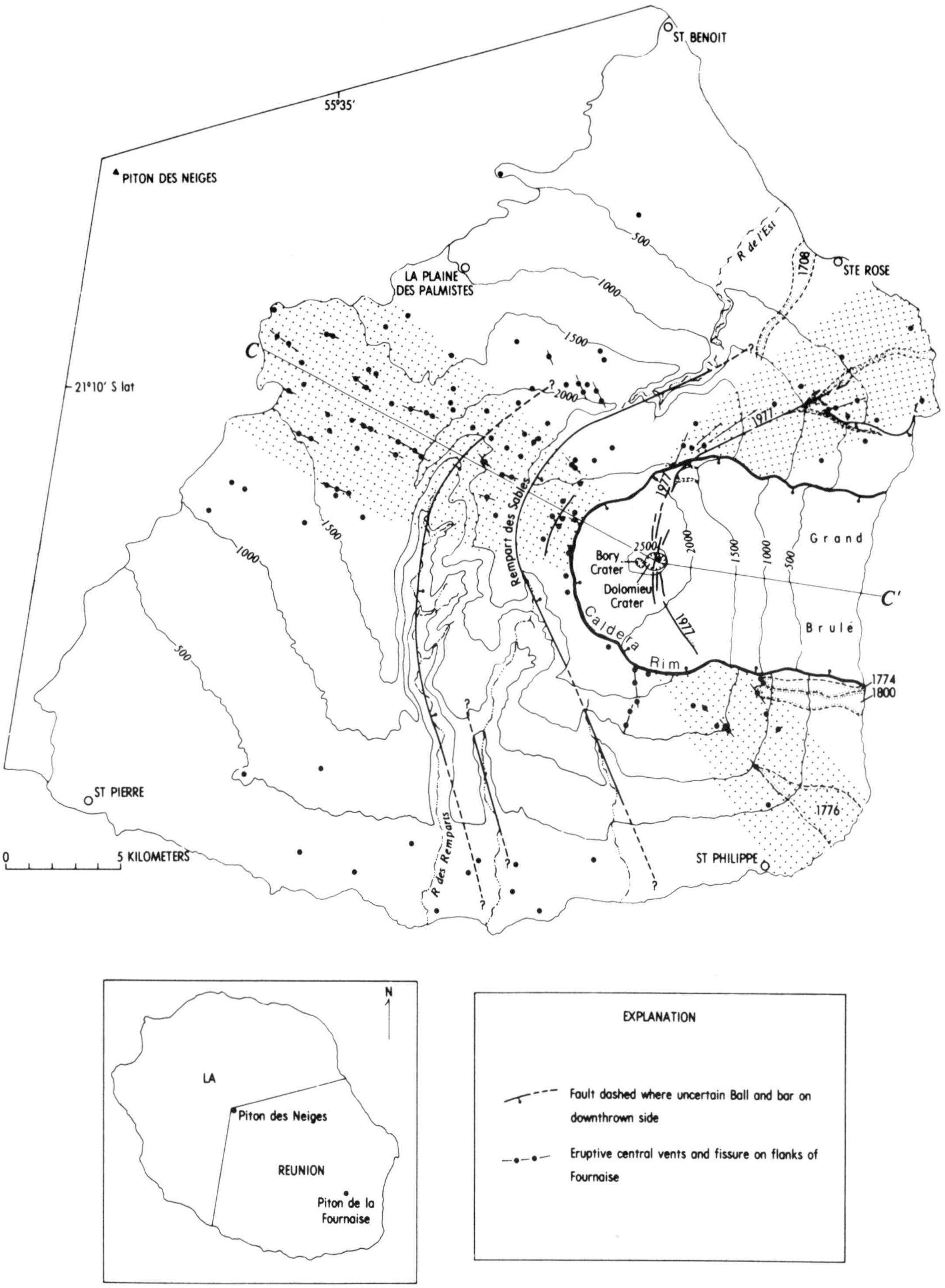

Figure 3.1.2. Piton de la Fournaise, from Duffield and others (1982). Contours in meters. Lines through flank vents represent obvious fissures on aerial photographs. Rift zones shaded. Only faults that arc from northeast to southeast rift zones shown in caldera. Large arcuate faults interpreted as headwalls of large prehistoric landslides.

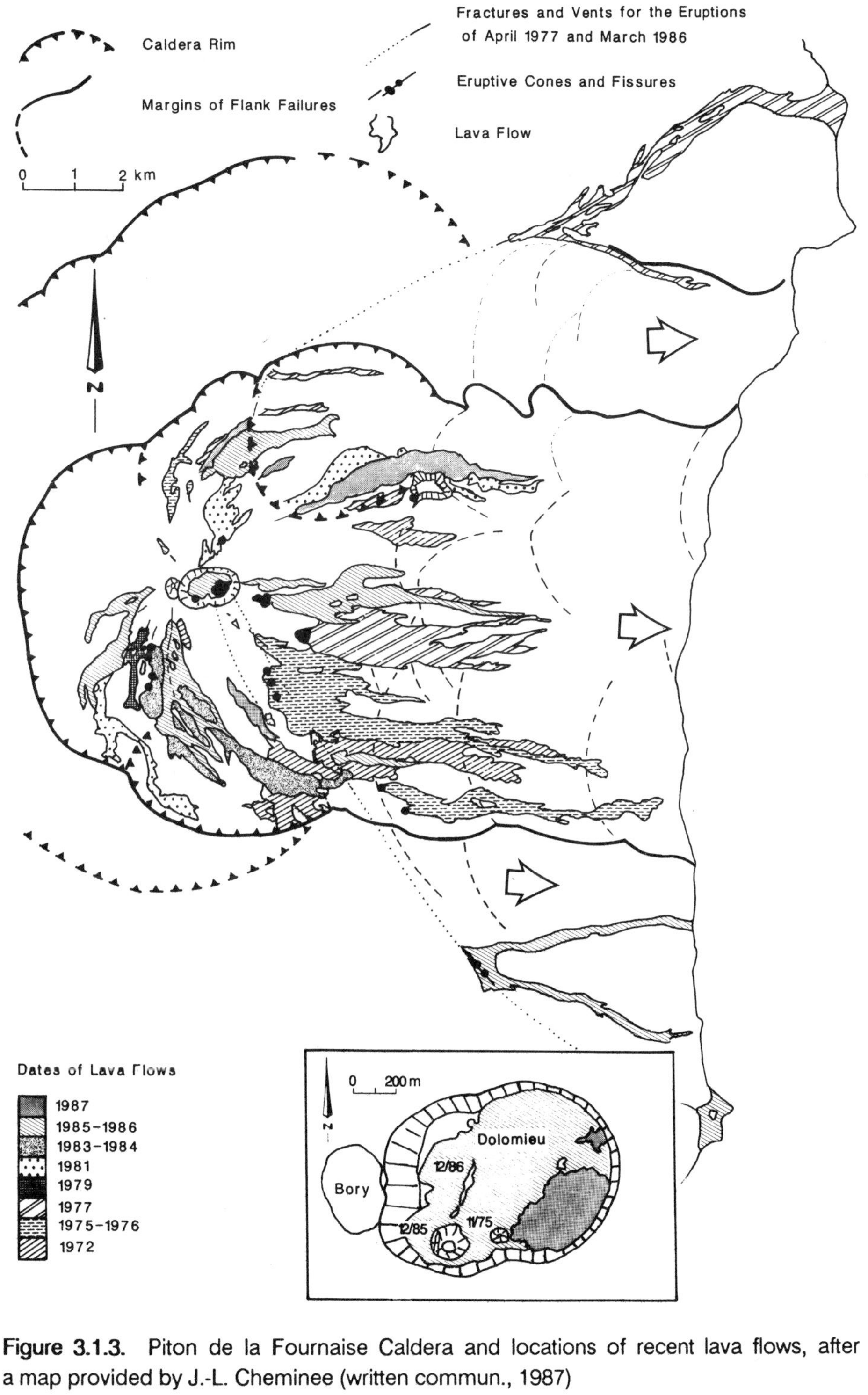

Figure 3.1.3. Piton de la Fournaise Caldera and locations of recent lava flows, after a map provided by J.-L. Cheminee (written commun., 1987)

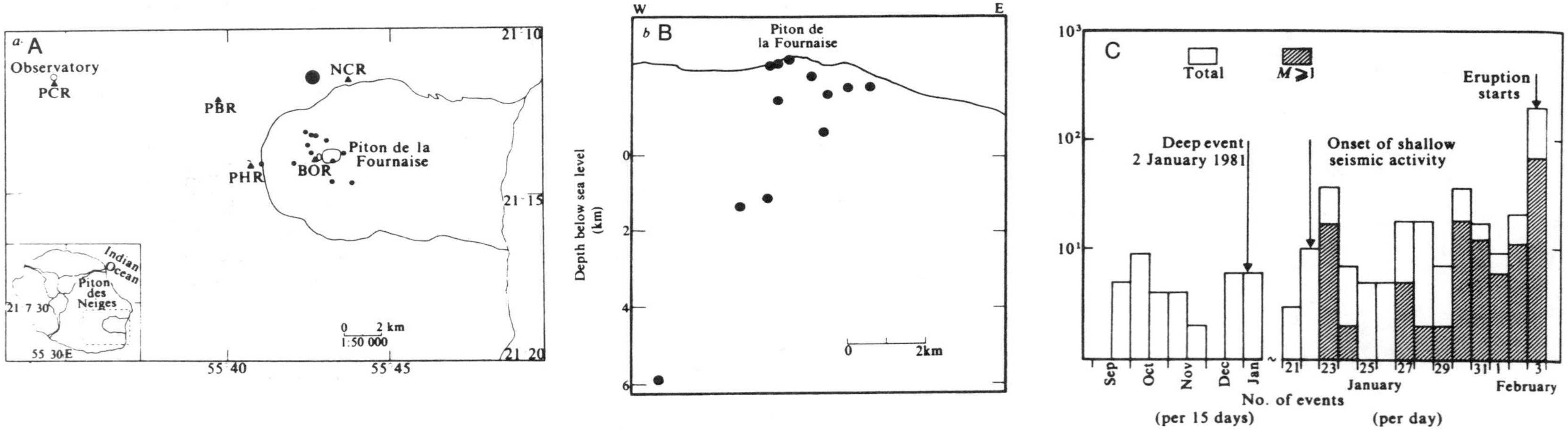

Figure 3.1.4. Seismicity associated with February 1981 eruption at Piton de la Fournaise, from Bachelery and others (1982). A, Epicenters of selected earthquakes during January-February 1981, including epicenter of deep event (10-15 km) of 2 January (large dot). B, Hypocenters of events shown in A (event at about 6 km depth occurred a few hours before start of eruption on 3 February). C, Number of earthquakes as a function of time (note scale change on 21 January). Reprinted by permission from Nature; copyright 1982 Macmillan Magazines Ltd.

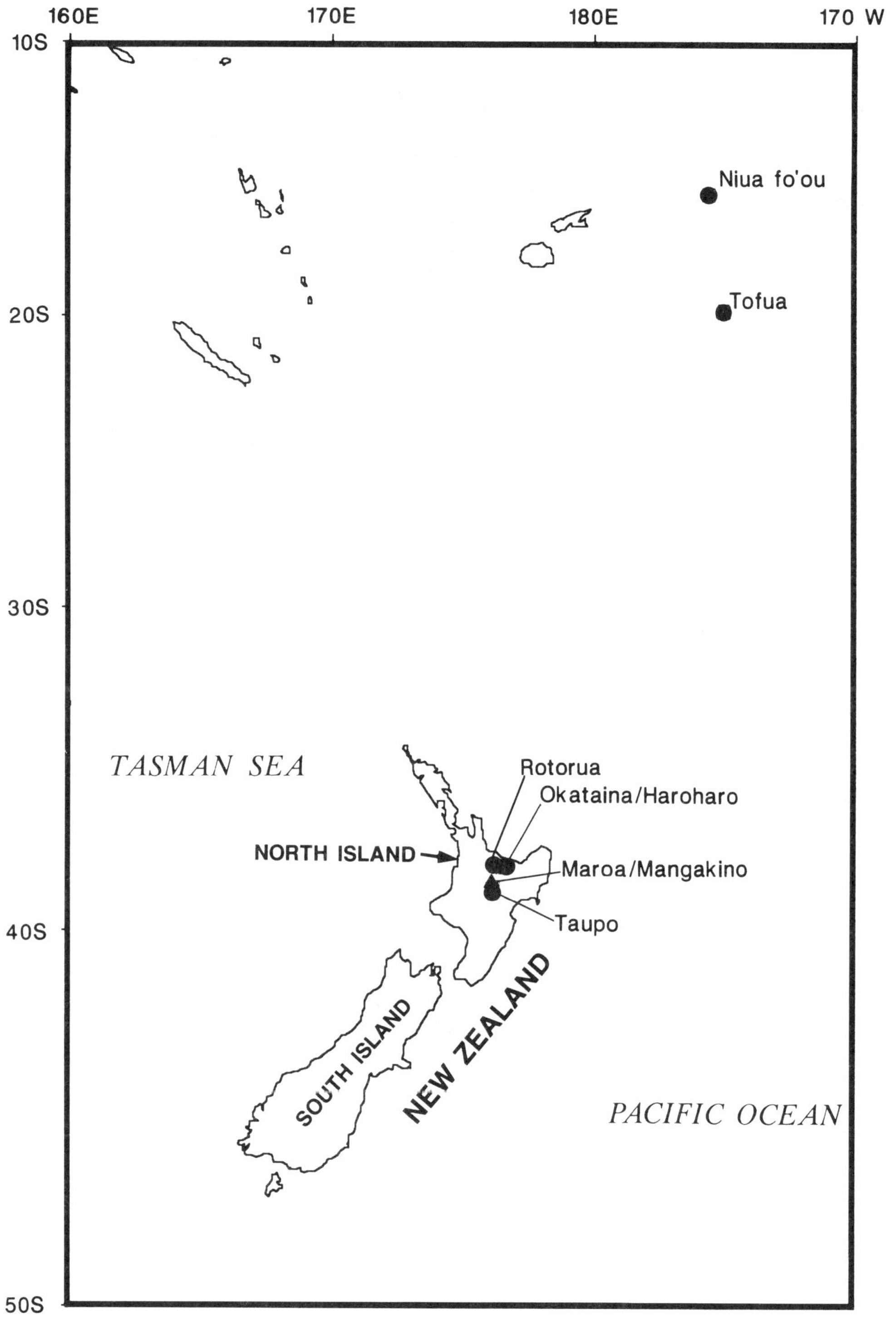

Figure 4. Locations of large, Quaternary restless calderas (solid circles) and nonrestless caldera (triangle) in Region 4.

See inside back cover for explanation and abbreviations

HAROHARO/OKATAINA

CAVW number (active vent)	Latitude Longitude (degrees)	Diameter (km)	Local tectonic setting	Precaldera edifice	SiO_2 content (wt pct)	Age of collapse (yr B.P.)	Date of unrest	Type and duration of unrest ESTU STHF MGTF H Te	Eruption type
04-01-05,06 (Tarawera)	38.09S 176.51E(H) 38.23S 176.51E(O)	13 x 26	Exten	?	R = 51-75 C = 74-77	300,000- 150,000; ca. 50,000	1886	BB-- B--D --E- E --	PEX,ex
							1896	---- ---- --x- x --	none
							1900's	xx-- ---- --EE E --	none
							1982-87+	Cx-- --x- ---F E --	none

TECTONIC SETTING

Haroharo Caldera and Okataina Volcanic Center lie in the northern half of the Taupo Volcanic Zone (TVZ), a region of intense volcanic and geothermal activity (figs. 4.1.1, 4.1.2). The TVZ is about 250 km long and up to 50 km wide; more than 16,000 km^3 of lavas and pyroclastics have been erupted from the TVZ since volcanism commenced in the late Pliocene. Rhyolites were erupted from the central part of the TVZ, a deep volcano-tectonic depression in which eruptions were associated with ring structures, cauldron subsidence, and calderas. A complex system of normal faults runs through the TVZ, and andesite volcanoes are found near its ends (Healy, 1962, 1964; Rogan, 1982). Dextral and normal movement has been inferred along the N57°E-trending Tarawera Rift, along which vents of the Tarawera complex are localized (Nairn and Cole, 1981; Scott, 1986).

GEOLOGIC HISTORY

Haroharo Caldera and Okataina Volcanic Center have been the sites of repeated large silicic eruptions during the past 300,000 years or more (Cole, 1970, 1979; Wilson and others, 1984). At least 20 additional explosive eruptions have occurred within the past 50,000 years (Froggatt, 1982), those in the past 20,000 years along fissures (Wilson and others, 1984). Rogan (1982) infers a large body of silicic magma beneath the Okataina Volcanic Center, based on comparison of gravity and magnetic anomalies. Recent eruptions have been from Tarawera, located on the southern margin of the caldera, and from thermal areas within and adjacent to the caldera. A major prehistoric eruption of Tarawera, the Kaharoa eruption of ca. A.D. 1250, produced thick pyroclastic deposits and four rhyolite domes at the summit of Tarawera that are now exposed in cross section along the 1886 rift (Cole, 1970).

HISTORICAL ACTIVITY

1886: The 1886 eruption of Tarawera was a particularly destructive eruption for its size, owing to phreatic explosions that occurred when basaltic melt rose into ground that was saturated with hot water (Nairn, 1979). Eruptions began in the summit region

HAROHARO/OKATAINA (continued)

HISTORICAL ACTIVITY (continued)

but spread to numerous vents along the new 17-km-long Tarawera Rift. A total of 0.7 km^3 of basaltic magma was erupted; the combined volume of juvenile plus older debris was considerably higher.

Early precursors were either unremarkable or seemingly unrelated to Tarawera, so they were not recognized as such. During April 1881 and October 1883, periods when little rain fell, Lake Rotokakahi (west of Lake Tarawera and about 10 km northwest of what was to become the Tarawera Rift) rose more than 1 m, causing flooding into Lake Tarawera. The water of Lake Rotokakahi also changed from olive green to a turbid, muddy appearance (Smith, 1886) and may have risen substantially in temperature (Conly, 1985, p. 32). Still farther to the northwest, the supply of hot water to baths of Rotorua may have failed for a period of several weeks sometime in the first half of 1886 (Conly, 1985), but we have not found this report corroborated by other sources. The level of Lake Rotoehu, 22 km north of Tarawera, dropped about 1 m in March or April 1886, but similar changes have occurred at other times (Thomas, 1888).

Hydrothermal activity in nearby Lake Rotomahana was greater in 1885 than in 1884, and in November 1885 one of the geysers at the lakeshore rose to a height of 150 ft, the highest ever recorded (Cole, 1970, citing Thomas, 1888). Hot springs of the Edgecumbe - Te Teko region, just east of the Okataina Complex, were more active and hotter than usual "for some time before the eruption" (Cole, 1970, citing Smith, 1886). Both Maoris and Europeans reported a warrior-filled spectral (ghost) canoe on Lake Tarawera on 31 May. Waves 9 to 12 in. high on Lake Tarawera were noted on 1 June 1886, and several more were noted over the next few days. The net lake level may have risen a few centimeters relative to the dock at a tourist resort on Lake Tarawera. A series of increasingly strong earthquakes, felt up to 250 km away, began at 0030 hr on 10 June 1886, and the eruption began one hour later.

1900's: Small earthquake swarms and changes in the temperature and discharge of fumaroles and hot springs are common in this area. The strongest geyser in the region, Waimangu ("black water"), began activity in late 1900. Unusually large hydrothermal explosions occurred at Waimangu on 12 April 1915 and 1 April 1917, and at least the latter occurred "without warning" (Lloyd and Keam, 1966). In the 1917 explosion, a steam blast laden with mud and rocks was directed toward the southwest; it partly destroyed the Government Tourist Department's Accommodation House 0.8 km away. A small earthquake swarm occurred in 1960 near Waimangu, a nearby hydrothermal area ($\underline{M}_{max}$=ca. 4). Five earthquakes occurred (or were felt?) in the area in 1962. Ground temperatures, fumarolic activity, and hot-spring activity increased in January 1973, about 40 days before, and at the site of, a small hydrothermal explosion at Frying Pan Lake, Waimangu, on 22 February 1973. Perhaps coincidentally, a magnitude 5.7 earthquake occurred 160 km south-southeast of Waimangu 34 minutes before the explosion.

1982-87+: Seismicity increased at Okataina in April 1982 and culminated with a $\underline{M}_L$ 5.2 earthquake on 14 December 1983, near the Waiotapu thermal area (Smithsonian Institution, 1982; Scott, 1986). As of January 1984, seismicity was still above background level. During an earthquake swarm from 23 to 29 September 1982, A-type earthquakes ($\underline{M}_{max}$=4.1) occurred at depths less than 5-10 km. Another $\underline{M}_L$ 4.1 quake occurred on 30 April 1983 along the NE-SW-trending Paeroa Fault. On 16 November 1987, a short but intense swarm of earthquakes occurred in the Lake Rotomahana area of the rift, with $\underline{M}_{max}$=3.8 (Smithsonian Institution, 1987).

HAROHARO/OKATAINA (continued)

HISTORICAL ACTIVITY (continued)

Monitoring of volcanic deformation has been upgraded since 1982, and early results suggest as much as 6-7 microstrain/yr of shear strain along the 1886 Tarawera Rift, where it passes beneath Lake Rotomahana. Tilt leveling over the same period, on the edifice of Tarawera and near Lake Rotomahana, has detected no significant changes. Minor tilt across Lake Tarawera in 1981 may have foreshadowed the earthquake swarms of 1981-82, but it was so slight as to be of questionable significance (Scott, 1986). No significant coseismic strain occurred in association with the November 1987 seismic swarm.

The level of Lake Rotomahana, situated in the southern caldera moat and formed in 1888, rose in the late 1960's and until 1973; since then, it has dropped about 2 m. The level of nearby Lake Tarawera, however, has not varied appreciably because it is has both a much larger catchment area and a large, permanent overflow. Activity at a well-known fumarolic area, the Steaming Cliffs, has started to change to geyser activity within the past few years (D. Sheppard, oral commun., 1983). These various hydrologic changes are attributed to climatic variation and have no obvious volcanic or tectonic implication (I.A. Nairn, written commun., 1983).

COMMENTS

Tarawera Volcano lies on the south margin of the Haroharo Caldera, but little is known of the geometry of magma supply to Tarawera and other parts of the caldera. Hydrologic changes over a broad region before the 1886 eruption suggest that activity at Tarawera is linked to activity of the caldera as a whole, but it is not known whether an intrusion beneath Tarawera triggered the broader unrest, or, alternatively, whether the eruption of Tarawera and the broader unrest were driven by a common, third factor (for example, regional tectonic strain).

REFERENCES

Cole, J.W., 1970, Structure and eruptive history of the Tarawera Volcanic Complex: New Zealand Jour. Geol. Geophys., v. 13, no. 4, p. 879-902.

----- 1979, Structure, petrology and genesis of Cenozoic volcanism, Taupo Volcanic Zone, New Zealand -- a review: New Zealand Jour. Geol. Geophys., v. 22, p. 631-657.

----- 1986, Distribution and tectonic setting of Late Cenozoic volcanism in New Zealand, <u>in</u> Smith, I.E.M., ed., Late Cenozoic volcanism in New Zealand: Roy. Soc. New Zealand Bull. 23, p. 7-20.

Cole, J.W., and Nairn, I.A., 1975, Catalogue of active volcanoes of the world including solfatara fields, Pt. XXII, New Zealand: Rome, IAVCEI, 156 p.

Conly, G., 1985, Tarawera: The destruction of the pink and white terraces: Wellington, Grantham House, 160 p.

HAROHARO/OKATAINA (continued)

REFERENCES (continued)

Froggatt, P.C., 1982, Review of methods of estimating rhyolitic tephra volumes; applications to the Taupo Volcanic Zone, New Zealand: Jour. Volcanol. Geotherm. Res., v. 14, p. 301-318.

Grayland, E., and Grayland, V., 1971, Tarawera: Auckland, Hodder and Stoughton, 158 p.

Healy, J., 1962, Structure and volcanism in the Taupo Volcanic Zone, New Zealand, in Crust of the Pacific Basin, Am. Geophys. Union, Geophys. Monogr., v. 6, p. 151-157.

----- 1964, Volcanic mechanisms in the Taupo Volcanic Zone: New Zealand Jour. Geol. Geophys., v. 7, no. 1, p. 6-22.

Hutton, F.W., 1887, The eruption of Mount Tarawera: Quart. Jour. Geol. Soc. London, v. 43, p. 178-189.

Lloyd, E.F., and Keam, R.F., 1966, Waimangu: geology, in Thompson, B.N., Kermode, L.O., and Ewart, A., eds., New Zealand volcanology, Central Volcanic Region: New Zealand Dept. Sci. Industrial Res. Info. Ser. 50, 211 p.

Nairn, I.A., 1979, Rotomahana-Waimangu eruption, 1886: base surge and basalt magma: New Zealand Jour. Geol. Geophys., v. 22, p. 363-378.

Nairn, I.A., and Cole, J.W., 1981, Basalt dikes in the 1886 Tarawera Rift: New Zealand Jour. Geol. Geophys., v. 24, p. 585-592.

Rogan, M., 1982, A geophysical study of the Taupo Volcanic Zone, New Zealand: Jour. Geophys. Res., v. 87, p. 4073-4088.

Scott, B.J., 1986, Monitoring at Okataina Volcanic Centre, in Gregory, J.G., and Watters, W.A., eds., Volcanic hazards assessment in New Zealand: New Zealand Geol. Surv. Rec. 10, p. 49-54.

Self, S., 1983, Large-scale phreatomagmatic silicic volcanism: a case study from New Zealand: Jour. Volcanol. Geotherm. Res., v. 17, p. 433-469.

Smith, S.P., 1886, The eruption of Tarawera: A report to the Surveyor-General: Wellington, Government Printer, 84 p.

Smithsonian Institution, Scientific Event Alert Network (SEAN), 1982, Haroharo Volcano: SEAN Bull., v. 7, no. 11, p. 8-10 (with reports from J.H. Latter, I.A. Nairn, B.J. Scott, and P.M. Otway).

----- 1987, Tarawera Rift: SEAN Bull., v. 12, no. 11, p. 5 (with report from B.J. Scott and S. Sherburn).

Thomas, A.P.W., 1888, Report on the eruption of Tarawera and Rotomahana, N.Z.: Wellington, Government Printer, 74 p.

Thompson, B.N., Kermode, L.O., and Ewart, A., eds., 1966, New Zealand volcanology, Central Volcanic Region: New Zealand Dept. Sci. Industrial Res. Info. Ser. 50, 211 p.

Volcanological Society of Japan, 1975, Echo Crater, Waimangu: Bull. Volcanic Eruptions, no. 13, p. 12.

Wilson, C.J.N., Rogan, A.M., Smith, I.E.M., Northey, D.J., Nairn, I.A., and Houghton, B.F., 1984, Caldera volcanoes of the Taupo Volcanic Zone, New Zealand: Jour. Geophys. Res., v. 89, p. 8463-8484.

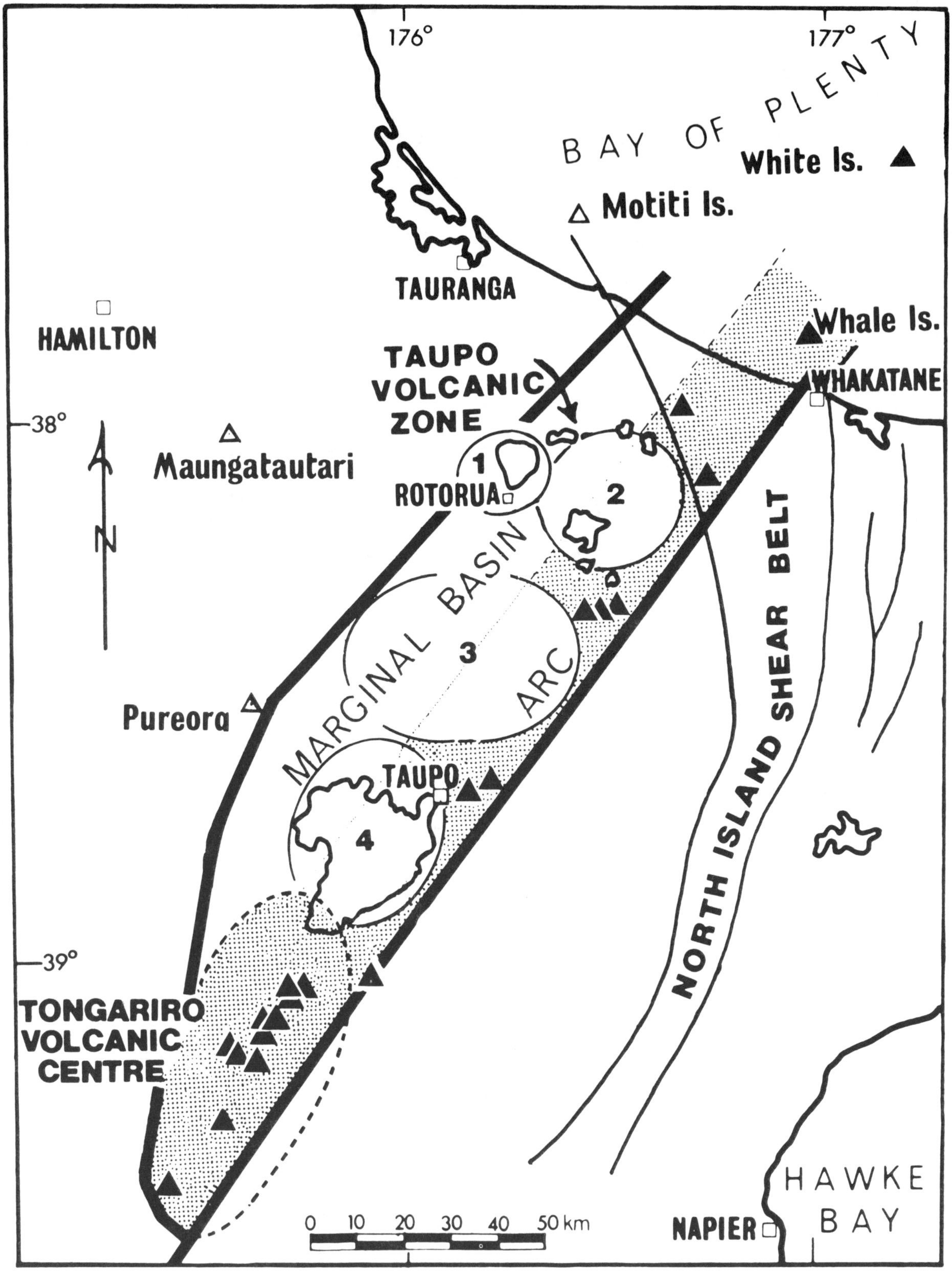

Figure 4.1.1. Sketch map of Taupo Volcanic Zone, from Cole (1986), showing locations of Rotorua Caldera (1), Haroharo Caldera and Okataina Volcanic Center (2), Maroa Caldera (3), and Taupo Caldera (4). Triangles show locations of basalt/medium-K andesite/dacite vents of volcanic arc.

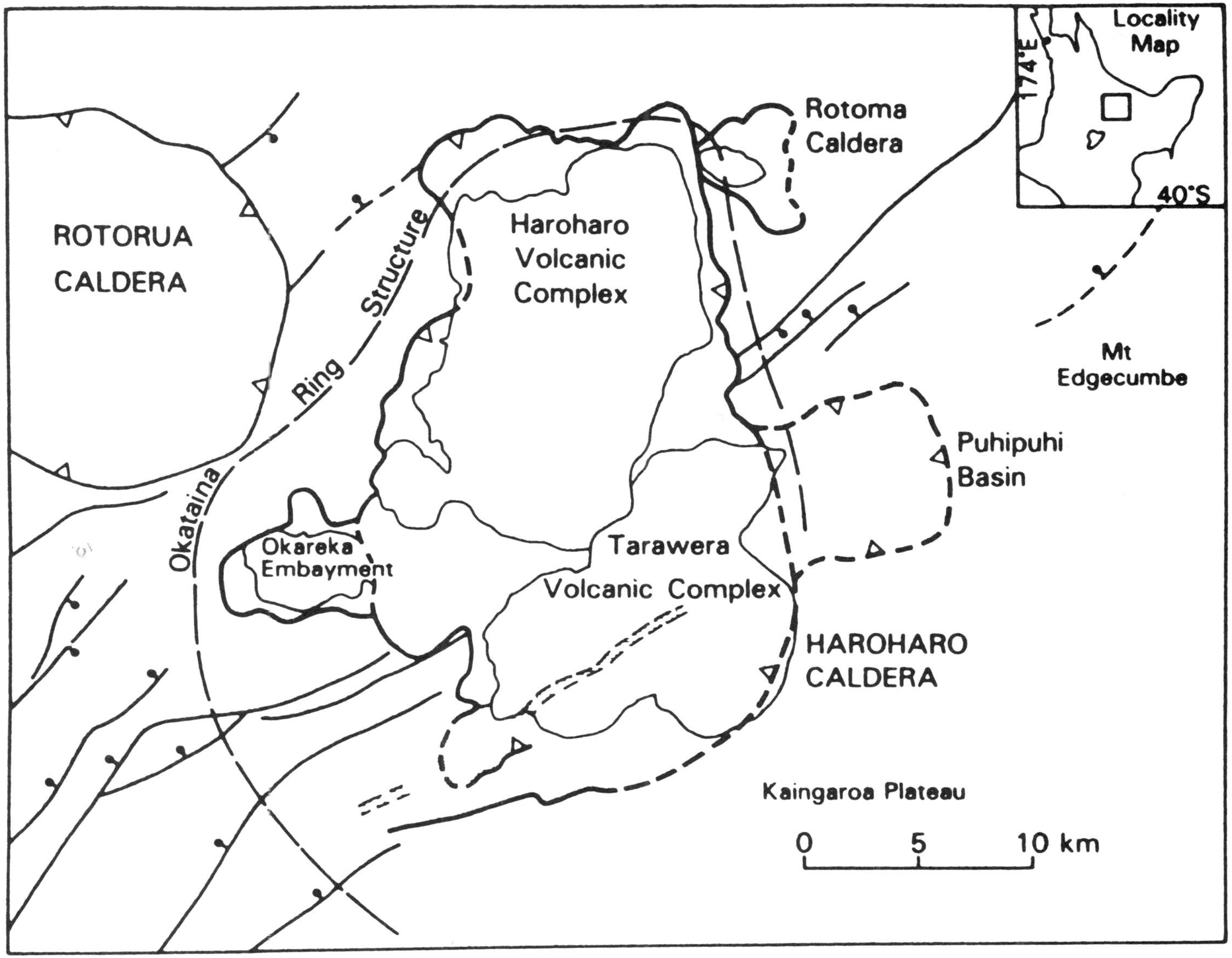

Figure 4.1.2. Sketch map of Haroharo Caldera and Okataina Volcanic Center, from Scott (1986).

PART 3: HISTORICAL UNREST AT LARGE QUATERNARY CALDERAS (continued)
See inside back cover for explanation and abbreviations

TAUPO

CAVW number (active vent)	Latitude Longitude (degrees)	Diameter (km)	Local tectonic setting	Precaldera edifice	SiO_2 content (wt pct)	Age of collapse (yr B.P.)	Date of unrest	Type and duration of unrest ESTU STHF MGTF H Te	Eruption type
04-01-07	38.78S 176.12E	35? (poorly defined)	Exten	None?	R = 71-72	330,000-	1895	x--- ---x ---- - =?	none
					C = 71-72	230,000;	1922	EE-- E--E ---- - =?	none
						20,000;	1950's	x--- E--- ---- - =?	none
						ca. 1,800	1964-65	EE-x ---- ---- - =?	none
							1974, 1981	x--- x--x ---- x =?	hydrothermal
							1981-85+	FF-F FF-x ---- - =?	none

TECTONIC SETTING

Taupo Caldera is a poorly defined caldera within the central graben of the southern Taupo Volcanic Zone (see fig. 4.1.1 in section on Haroharo/Okataina). Ring fractures are partly hidden by Lake Taupo (Northey, 1982; Wilson and others, 1984). Gravity data suggest that the caldera is deepest near the northernmost point of Lake Taupo, but this is poorly constrained because the lake makes gravity measurements difficult (Rogan, 1982). Rhyolite domes occur within and around the caldera, but magnetic data have been interpreted to show that there is no significant high-level silicic magma body beneath the caldera (Rogan, 1982). Wilson and others (1984) postulate that two deep-seated fractures control vent locations in the Taupo area: One trends north-northeast along the western shore and western bay of Lake Taupo, and the second trends northeast beneath the eastern side of Lake Taupo, beneath Horomatangi Reefs, to Tauhara, a late Pleistocene dacitic volcano.

Sissons (1979, cited in Bevin and others, 1984, and Otway and others, 1984) has inferred extension across the Taupo Volcanic Zone during this century at a rate of 7 mm/yr. Although extension is the dominant strain in the Taupo Volcanic Zone, a dextral strike-slip component is evident in focal mechanisms of deeper earthquakes (Smith and Webb, 1986).

GEOLOGIC HISTORY

Lake Taupo occupies the center of a ca. 40-km-diameter basin that has been the site of many rhyolite eruptions during the past 10,000 yr (Wilson and others, 1984; Wilson and Walker, 1985; Houghton and Wilson, 1986; Nairn and Wood, 1986) (fig. 4.2.1). At least three major ignimbrites were produced between 330,000 yr B.P. and 230,000 yr B.P. No major pyroclastic eruptions are recognized between 230,000 yr B.P. and 50,000 yr B.P., but at least five rhyolite eruptions occurred between 50,000 yr B.P. and 20,000 yr B.P.. A 10,000-yr hiatus followed the last of those eruptions, before the present sequence of eruptive activity began about 10,000 yr ago. A series of ten plinian or subplinian pyroclastic eruptions climaxed with the exceptionally violent Taupo Pumice eruptions about 1,800 yr ago (90+ km^3 bulk volume) (Healy and others, 1964; Wilson and others, 1980; Froggatt, 1981b, c; Wilson and Walker, 1985;

TAUPO (continued)

GEOLOGIC HISTORY (continued)

Wilson, 1985). Those eruptions are thought to have been from vents near the Horomatangi Reefs (Walker, 1980); the deepest point in the lake (164 m) is just south of the reefs (fig. 4.2.2).

Southwest of Lake Taupo, 5- to 20-cm-wide fissures opened during the middle stages of the 1,800 yr B.P. eruption and were infilled by ignimbrite of the late, most voluminous part of the eruption (Wilson and Walker, 1985). Wilson and Walker suggest that these fissures opened by severe shaking and localized downslope movement rather than by regional extension.

Tauhara Volcano is a complex of five dacite domes of late Pleistocene age (Lewis, 1968). Tauhara lies 2-3 km outside the northeast rim of Taupo Caldera, along a postulated deep-seated northeast-trending fracture that bounds the east side of the caldera (Wilson and others, 1984). The exact age of Tauhara is not known; its minimum age is 9,000 yr (Lewis, 1968). The location of the Tauhara geothermal field is shown in figures 4.2.3 and 4.2.6.

HISTORICAL ACTIVITY

1895: A strong earthquake on 18 August 1895 (estimated magnitude 6 to 7.5) caused landslides and fissuring in the Wairakei area and may have been associated with faulting east of Lake Taupo. Numerous aftershocks were felt (Eiby, 1968; Grindley and Hull, 1986).

1922: From April to December 1922, numerous felt earthquakes, including one estimated to be of $\underline{M}_L$ 5.5-6 (Smith and Webb, 1986) (fig. 4.2.3), were accompanied by up to 3 m of surface displacement along faults and up to 2.3-m net subsidence along the shores of Kaiapo, Whakaipo and Whangamata Bays (north shore, Lake Taupo). Several years later, Grange (1932) reported subsidence of 3.7 m at the tip of the peninsula between Whakaipo and Whangamata Bays, where 2.3 m of subsidence had earlier been measured. Displacements were along NNE-SSW-trending regional graben faults and were traceable at least 10 km north from the edge of the lake. Ground cracking, fissures, and dozens of water fountains were seen along the Kaiapo Fault. Eiby (1966) considered this to be a nonvolcanic sequence of foreshocks, main shock (9 June 1922), and aftershocks, rather than a true swarm like that which occurred in 1964-65. However, Grindley and Hull (1986) note that at least two other earthquakes as large as that of 9 June occurred on 14 July and 17 July. Grange (1932) related the displacements to the entire swarm; Eiby associated most of the faulting with the earthquake of 9 June 1922. Smith and Webb (1986) consider the 1922 events to be a nonvolcanic swarm during which displacements occurred in many small increments. The Puketarata Fault, active in 1922, is host to a line of shallow explosion craters and rhyolite domes where it passes north from Taupo through the Maroa Volcanic Center (Grindley and Hull, 1986).

1950's: A magnitude 5.3 earthquake occurred on 14 July 1953, centered at 38.8°S, 175.7°E, just west of Lake Taupo. No depth was reported; aftershocks followed (Eiby, 1968). Other earthquakes occurred west and southwest of Lake Taupo in 1956 and 1957 ($\underline{M}$=6.2, 5.3, 5.5, and 6.5). These were generally deeper than 150 km and not associated with swarms.

TAUPO (continued)

HISTORICAL ACTIVITY (continued)

1964-65: A swarm of 1,126 earthquakes larger than magnitude 2.5 began in December 1964 beneath the northwestern part of Lake Taupo. The swarm later migrated southeastward across the lake, where it continued until January 1965 (Gibowicz, 1973; figs. 4.2.4, 4.2.5). Focal depths were about 10 km or less. Maximum magnitude was 4.75, and the maximum rate was 35-40 shocks per 3 hour period. The $\underline{b}$-value was 1.2 during the first part of the swarm, but it decreased systematically to 0.6 near the end of the swarm (that is, the average magnitude increased with time). Seismicity increased sharply about 7 days after the swarm began, peaked within two days thereafter, and decreased more or less regularly from then through January 1965. As much as 7.9 cm of uplift occurred along the eastern shore of Lake Taupo between Waitahanui and Motuoapa between leveling surveys in 1956 and 1977. Because similar uplift occurred in conjunction with a seismic swarm in 1983, and because there is no evidence for continuous uplift along the eastern side of the lake, the 1956-77 uplift probably occurred in association with the 1964-65 earthquake swarms (Bevin and others, 1984; Otway and others, 1984; P. Otway, written commun., 1984; Grindley and Hull, 1986). No surface faulting was noted in association with the 1964-65 earthquake swarms.

1974, 1981: Hydrothermal explosions occurred in the Tauhara geothermal field, northeast of Lake Taupo, on 28 March 1974 and 20 June 1981. Aerial photographs show that an area of warm ground surrounding the explosion sites was larger in 1975 than it was in 1963; no further enlargement occurred between 1975 and 1981. Scott and Cody (1982) attributed the explosions to several factors: (1) a long-term increase in surface heat flow as geothermal steam production at Wairakei lowers the pressure in the deep Tauhara aquifer; (2) self-sealing of the near-surface deposits (Taupo Pumice) in the area of the explosions; (3) below-average rainfall and thus increased shallow ground temperatures. An earthquake was felt in the area on 12 June 1981, 8 days before the 1981 explosion, but it is not known whether that earthquake acted as a trigger for events that culminated with the explosion. No tectonic significance was supposed at the time.

1981-85+: Lake-level readings from three sites at the northern end of Lake Taupo showed no change from their start in April 1979 through an October 1981 survey, but showed >1 cm of subsidence in a trough between Kinloch and Horomatangi from October 1981 to October 1982 (Otway, 1986a). No unusual seismicity was noted at the time, but earthquake swarms occurred in February-March 1983 just northwest of Lake Taupo. Following the February-March swarm, lake-level measurements showed clear uplift of the north shore of Lake Taupo: from March to June this area rose as much as 5.7 cm relative to the southern end of the lake (Otway, 1986a). A new, stronger swarm began on 16 June ($\underline{M}_{max}$= 4.3) (fig. 4.2.6), and within days thereafter rapid subsidence was recorded northwest of the Kaiapo Fault (fig. 4.2.7), while a station southeast of the fault, Rangatira, continued to rise steadily (figs. 4.2.6-4.2.8). A week later, surface displacement occurred on the fault (about 5-cm upthrow to the southeast and 3-cm extension), in the same sense as the coseismic deformation already noted. Minor subsidence (2 cm) preceded a seismic swarm at the southern end of the lake in February-April 1984 and was followed by rebound of the same magnitude over the next several months. Uplift of 1-1.5 cm along the northeast shore of Lake Taupo occurred in late 1984, followed by a similar amount of subsidence that may have started when several small earthquakes were felt in the area during February-March 1985 (Otway, 1986a). Thus, with some variation, it appears that areas of impending swarms may rise slowly over periods of months and then subside after the earthquake swarms.

TAUPO (continued)

HISTORICAL ACTIVITY (continued)

On 1 September 1986, large amounts of fresh-looking pumice were noted in the Horomatangi Reefs area, and large gas bubbles were seen nearby. A routine lake-level survey shortly before these events showed no ground deformation in the Horomatangi area, but because small earthquakes had been recorded in the general area between 15 and 25 August, and because Horomatangi Reefs is the site of the most recent major eruption of Taupo, the matter received prompt attention. Upon close inspection, the pumice was found to harbor terrestrial mosses and mites and to be like that of the 3400 yr B.P. Waimihia Formation. The pumice probably collapsed into a stream that flows into the lake and was concentrated by currents at the Horomatangi Reefs, while the gas bubbles were probably like others that appear frequently at Taupo, unrelated to the pumice (Otway, 1986c; P. Otway, written commun., 1987).

COMMENTS

Seismic swarms and deformation at Taupo are presumably the manifestation of a slow process of graben formation, but volcanic and tectonic activity in the Taupo depression are so intimately linked that it would be difficult to discount some involvement of the Taupo magmatic system. Within the past century, earthquakes up to M=6 in the Taupo area have not triggered eruptions. Thus, Taupo presents a particularly difficult challenge for eruption forecasting, because precursors to the next eruption might be difficult to distinguish from noneruptive unrest that is common in the Taupo area.

REFERENCES

Bevin, A.J., Otway, P.M., and Wood, P.R., 1984, Geodetic monitoring of crustal deformation in New Zealand, in Walcott, R.I., compiler, An introduction to the recent crustal movements of New Zealand: Roy. Soc. New Zealand Misc. Ser., 7, p. 13-60.

Cole, J.W., 1979, Structure, petrology and genesis of Cenozoic volcanism, Taupo Volcanic Zone, New Zealand -- a review: New Zealand Jour. Geol. Geophys., v. 22, p. 631-657.

Cole, J.W., and Nairn, I.A., 1975, Catalogue of active volcanoes of the world including solfatara fields, Pt. XXII, New Zealand: Rome, IAVCEI, 156 p.

Eiby, G.A., 1966, Earthquake swarms and volcanism in New Zealand: Bull. Volcanol., v. 29, p. 61-73.

----- 1968, An annotated list of New Zealand earthquakes, 1460-1965: New Zealand Jour. Geol. Geophys., v. 11, p. 630-647.

Friedlander, B., 1899, Some notes on the volcanoes of the Taupo District: Trans. New Zealand Inst., v. 31, p. 498-510.

Froggatt, P.C., 1981a, Karapiti Tephra Formation: a 10,000 years B.P. rhyolitic tephra from Taupo: New Zealand Jour. Geol. Geophys., v. 24, p. 95-98.

----- 1981b, Stratigraphy and nature of Taupo Pumice Formation: New Zealand Jour. Geol. Geophys., v. 24, p. 231-248.

----- 1981c, Did Taupo's eruption enhance European sunsets?: Nature, v. 293, p. 491 (reply from Wilson and others, p. 491-492).

Gibowicz, S.J., 1973, Variation of frequency-magnitude relationship during Taupo earthquake swarm of 1964-65: New Zealand Jour. Geol. Geophys., v. 16, no. 1, p. 18-52.

Grange, L.I., 1932, Rents and faults formed during earthquake of 1922 in Taupo District: New Zealand Jour. Sci. Technol., v. 14, no. 3, p. 139-141.

TAUPO (continued)

REFERENCES (continued)

Grindley, G.W., 1961, Taupo, Sheet N94, in Geological map of New Zealand, 1:63,360: Wellington, Dept. Sci. Industrial Res.

Grindley, G.W., and Hull, A.G., 1986, Historical Taupo earthquakes and earth deformation: Roy. Soc. New Zealand Bull. 24, p. 173-186.

Healy, J., 1964, Volcanic mechanisms in the Taupo Volcanic Zone, New Zealand: New Zealand Jour. Geol. Geophys., v. 7, p. 6-23.

Healy, J., Vucetich, C.G., and Pullar, W.A., 1964, Stratigraphy and chronology of Late Quaternary volcanic ash in Taupo, Rotorua, and Gisborne districts: New Zealand Geol. Surv. Bull. 73, 88 p.

Houghton, B.F., and Wilson, C.J.N., 1986, Tour Guide A1: Explosive rhyolitic volcanism: The case studies of Mayor Island and Taupo volcanoes, in Houghton, B.F., and Weaver, S.D., eds., North Island volcanism: Int. Volcanol. Congr., New Zealand, 1986, p. 33-100.

Hull, A.G., and Grindley, G.W., 1984, Active faulting near Taupo: Eos, Trans. Am. Geophys. Union, v. 65, no. 7, p. 51-52.

Irwin, J., 1972, Provisional bathymetry, 1:50,000: New Zealand Oceanogr. Inst. Chart, Lake Ser.

Lewis, J.F., 1968, Tauhara Volcano, Taupo Zone. Pt. 1--Geology and structure: New Zealand Jour. Geol. Geophys., v. 11, p. 212-224.

Nairn, I. A., and Wood, C. P., 1986, Tour Guide C6: Active volcanoes of Taupo Volcanic Zone: Int. Volcanol. Congr., New Zealand, 1986, 87 p.

Northey, D.J., 1982, Seismic studies of the structure beneath Lake Taupo: Wellington, Victoria Univ., Ph.D. diss.

Otway, P.M., 1979-80, Tilt-level monitoring in the Taupo Volcanic Zone: New Zealand Volcanol. Rec., no. 8, p. 27-28; no. 9, p. 19-20.

----- 1980, Taupo lake level survey: New Zealand Volcanol. Rec., no. 9, p. 21-22.

----- 1982, Taupo Lake level survey: New Zealand Volcanol. Rec., no. 11, p. 15.

----- 1986a, Monitoring of vertical deformation in Taupo Volcanic Centre, in Gregory, J.G., and Watters, W.A., eds., Volcanic hazards assessment in New Zealand: New Zealand Geol. Surv. Rec. 10, p. 41-48.

----- 1986b, Vertical deformation associated with the Taupo earthquake swarm, June 1983: Roy. Soc. New Zealand Bull. 24, p. 187-200.

----- 1986c, Occurrence of pumice rafts and gas bubbles at Horomatangi Reefs, Lake Taupo, 26 August - 1 September 1986: Immediate Rep., New Zealand Geol. Surv., 2 September 1986, 14 p.

Otway, P.M., Grindley, G.W., and Hull, A.G., 1984, Earthquakes, active fault displacement and associated vertical displacement near Lake Taupo, Taupo Volcanic Zone: New Zealand Geol. Surv. Rep. 110, 73 p.

Rogan, M., 1982, A geophysical study of the Taupo Volcanic Zone, New Zealand: Jour. Geophys. Res., v. 87, p. 4073-4088.

Scott, B.J., and Cody, A.D., 1982, The 20 June 1981 hydrothermal explosion at Tauhara geothermal field, Taupo: New Zealand Geol. Surv. Rep. 103, 33 p.

Sissons, B.A., 1979, The horizontal kinematics of the North Island of New Zealand: Wellington, Victoria Univ., Ph.D. diss., 318 p.

Smith, E.G.C., and Webb, T.H., 1986, The seismicity and related deformation of the Central Volcanic Region, North Island, New Zealand, in Smith, I.E.M., ed., Late Cenozoic volcanism in New Zealand: Roy. Soc. New Zealand Bull. 23, p. 112-133.

Timperley, M.H., 1983, Geological history, in Forsyth, D.J., and Howard-Williams, C., eds., Lake Taupo: Ecology of a New Zealand lake: New Zealand Dept. Sci. Industrial Res. Info. Ser. 158, p. 5-15.

Walker, G.P.L., 1980, The Taupo Pumice: product of the most powerful known (Ultraplinian) eruption?: Jour. Volcanol. Geotherm. Res., v. 8, p. 69-94.

Ward, P.R., 1922, A note on the significance of the recent subsidence on the shore of Lake Taupo: New Zealand Jour. Sci. Technol., v. 5, p. 280-281.

TAUPO, Region 4, CAVW number 04-01-07

TAUPO (continued)

REFERENCES (continued)

Wilson, C.J.N., 1985, The Taupo eruption, New Zealand: II. The Taupo ignimbrite: Phil. Trans. Roy. Soc. London, Ser. A, v. 314, p. 229-310.

Wilson, C.J.N., Ambraseys, N.N., Bradley, J., and Walker, G.P.L., 1980, A new date for the Taupo eruption, New Zealand: Nature, v. 288, p. 252-253.

Wilson, C.J.N., Rogan, A.M., Smith, I.E.M., Northey, D.J., Nairn, I.A., and Houghton, B.F., 1984, Caldera volcanoes of the Taupo Volcanic Zone, New Zealand: Jour. Geophys. Res., v. 89, p. 8463-8484.

Wilson, C.J.N., and Walker, G.P.L., 1985, The Taupo eruption, New Zealand: I. General aspects: Phil. Trans. Roy. Soc. London, Ser. A, v. 314, p. 199-228.

GEOLOGY of the TAUPO VOLCANIC ZONE

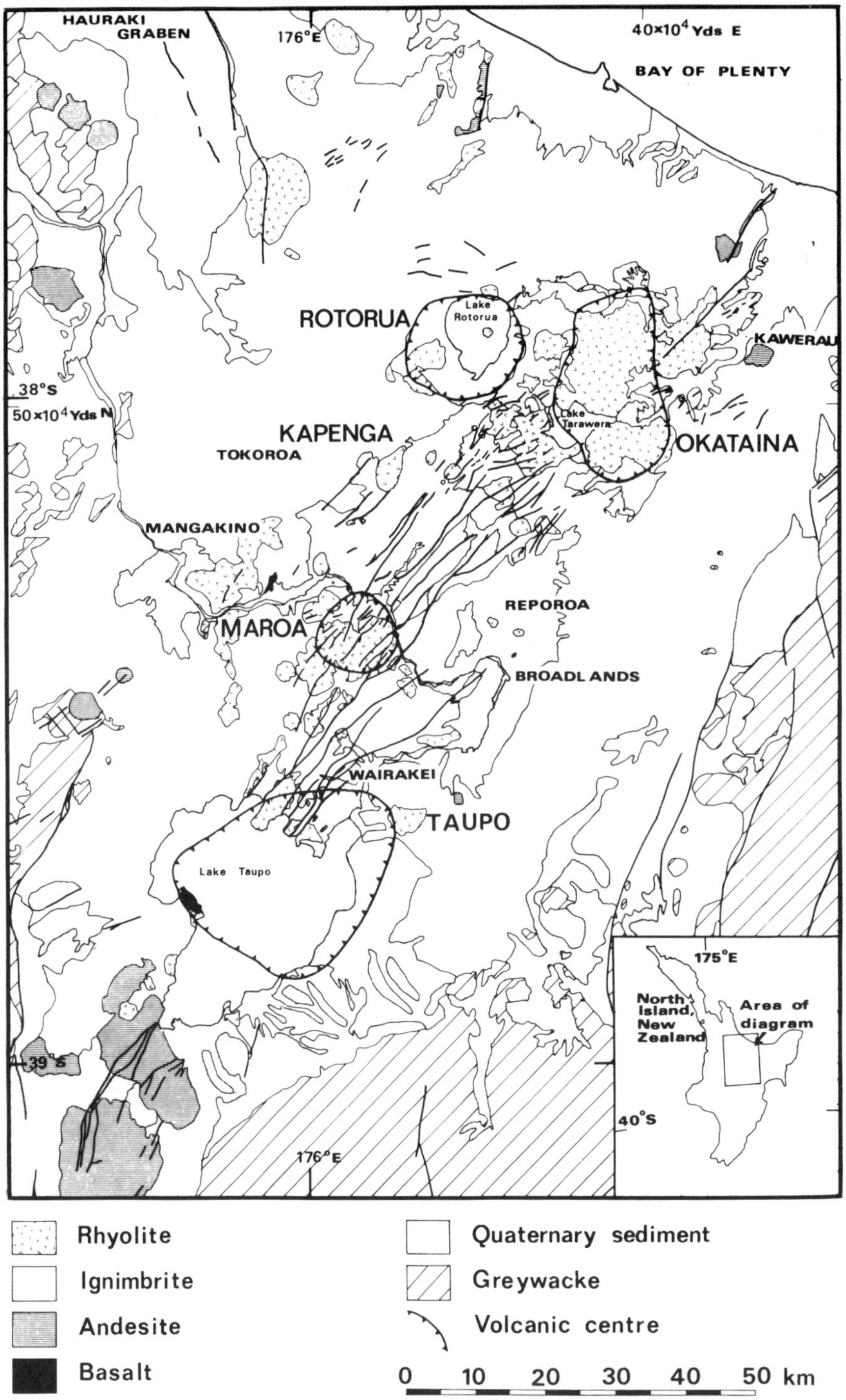

Figure 4.2.1. Simplified geology of Taupo Volcanic Zone and environs, from Rogan (1982). Location of region is shown in bottom right on a map of North Island. Grid values refer to New Zealand National Yard grid reference system. Copyright by the American Geophysical Union.

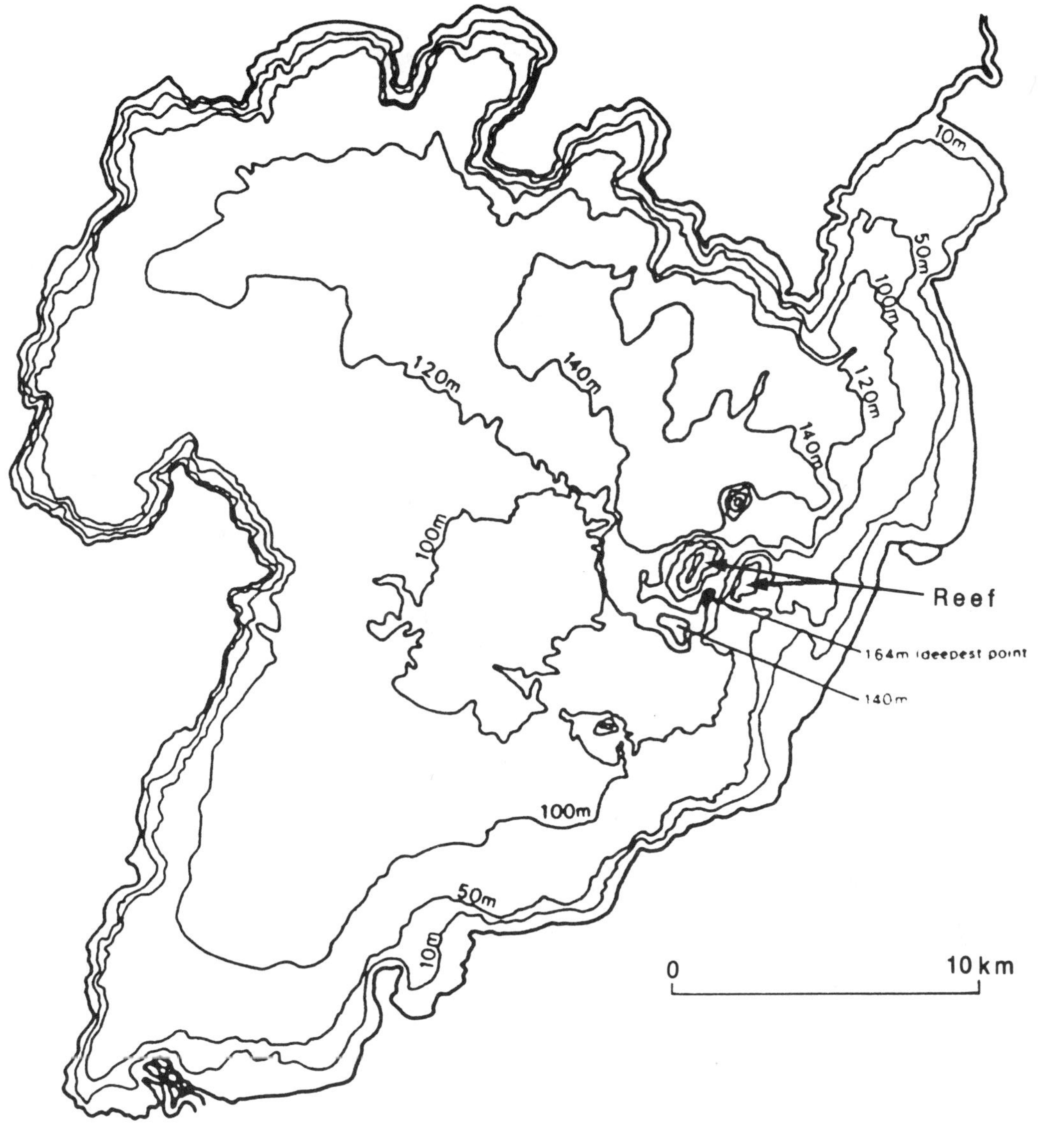

Figure 4.2.2. Bathymetry of Lake Taupo showing depths in meters, from Timperley (1983), after Irwin (1972). Vents for the 1,800 yr B.P. Taupo Pumice eruptions are thought to be near the Horomatangi Reefs ("Reef" in figure), in deepest part of lake.

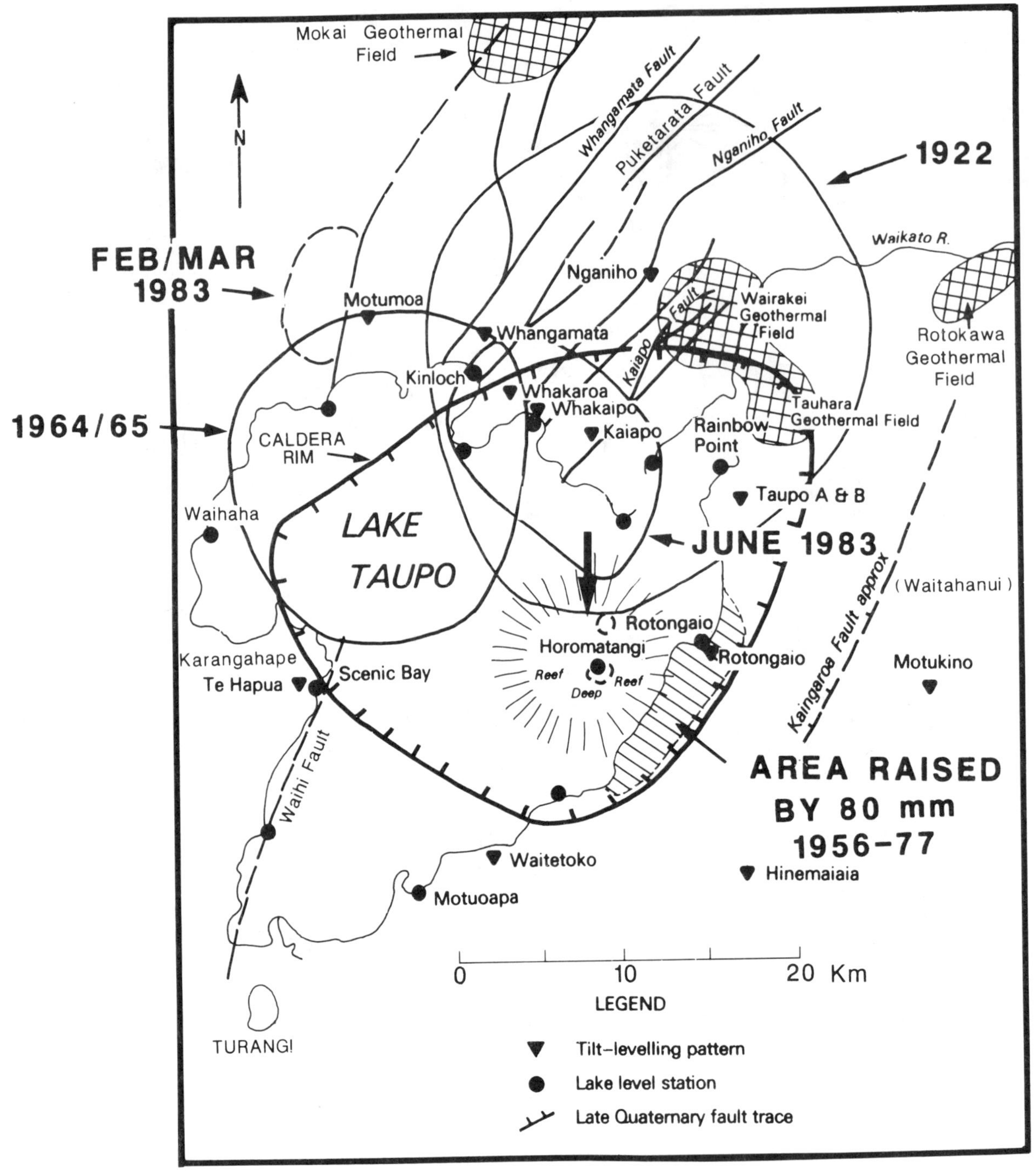

Figure 4.2.3. Epicentral areas of recent earthquake swarms near Taupo Caldera. Caldera rim after Rogan (1982); epicentral areas (for example, 1922, Feb/Mar 1983) and fault traces after Otway and others (1984) and Grindley and Hull (1986). Arrow near center of figure, probable vent area of Taupo Pumice (1,800 yr B.P.).

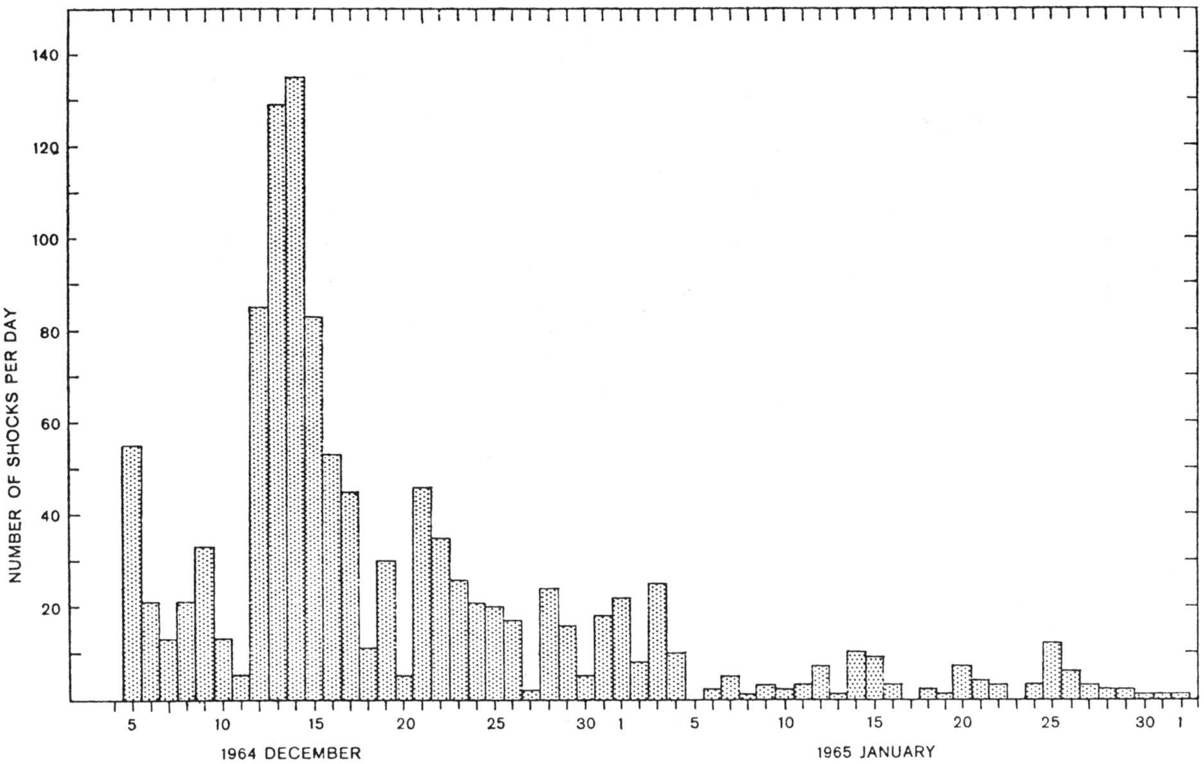

Figure 4.2.4. Daily number of earthquakes during December 1964 swarm at Lake Taupo, from Gibowicz (1973).

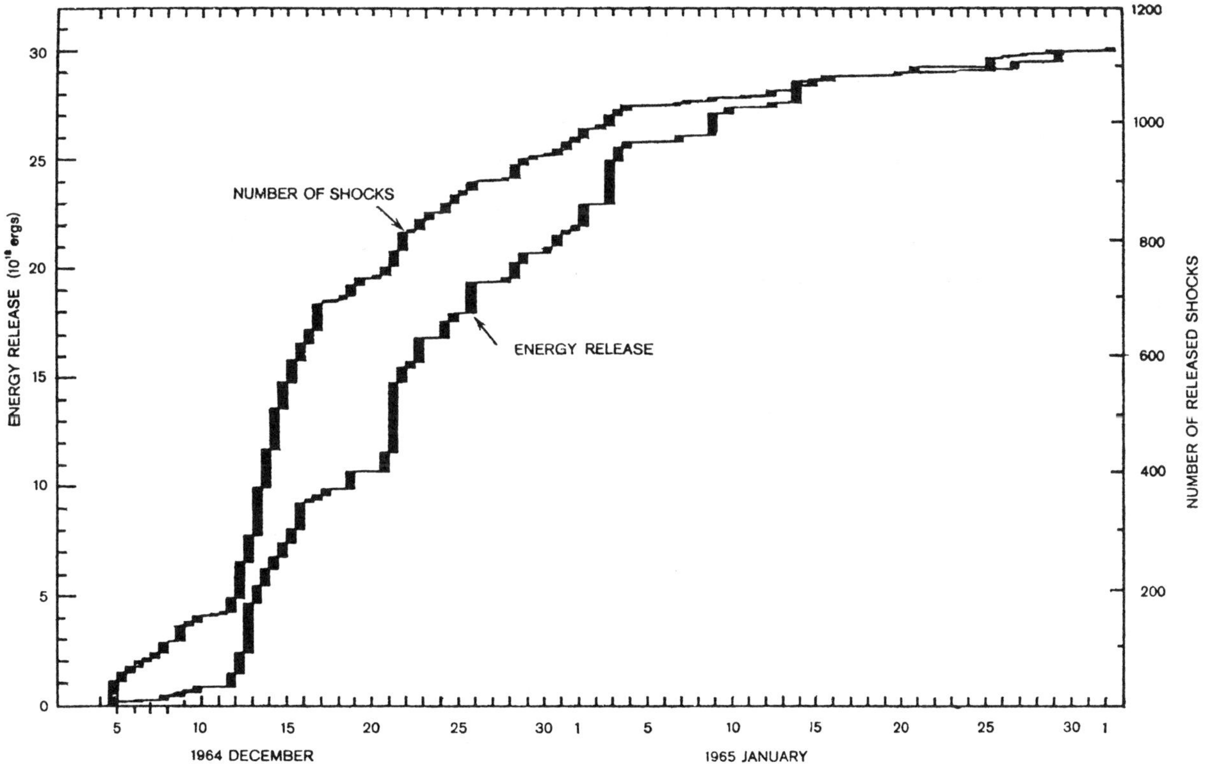

Figure 4.2.5. Cumulative energy release and number of earthquakes, at 12-hour intervals, during December 1964 swarm at Lake Taupo (from Gibowicz, 1973). Vertical scales chosen to bring both curves to same end points.

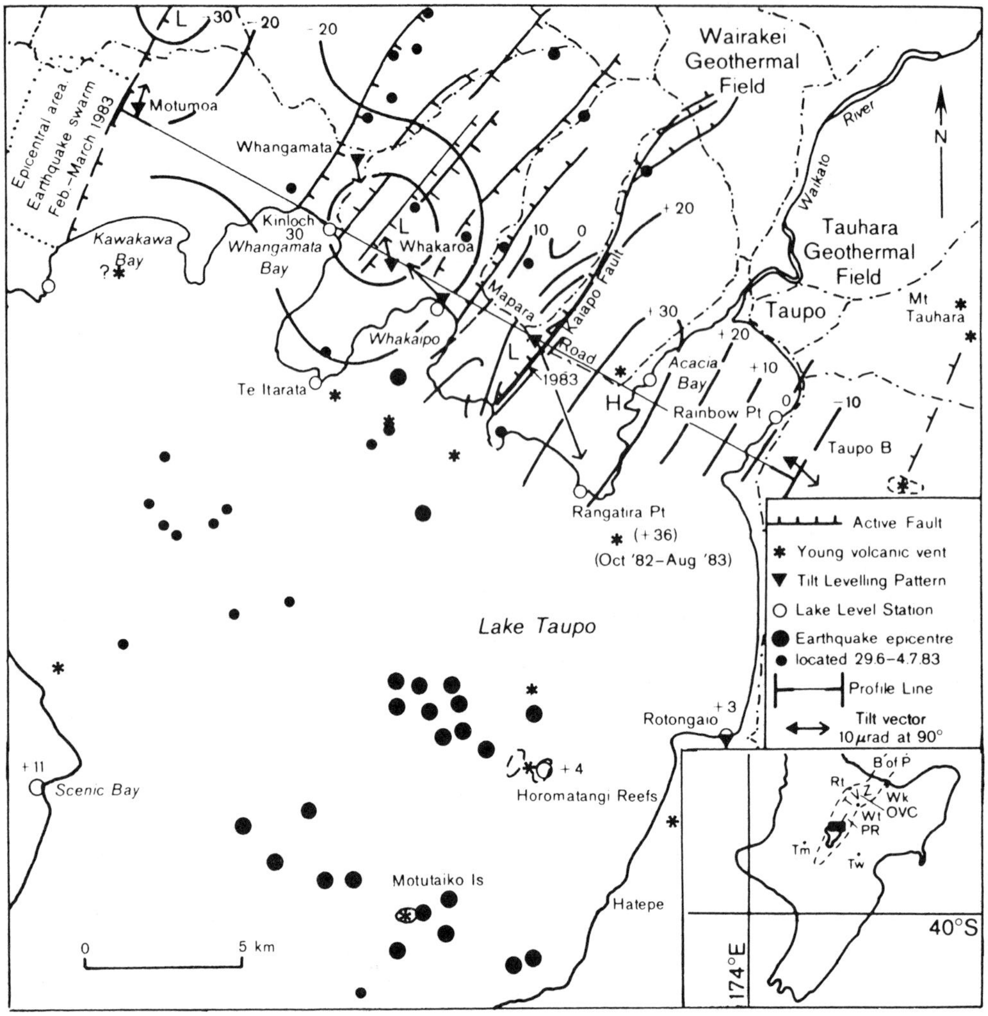

Figure 4.2.6. Northern Lake Taupo region showing epicenters of early-July 1983 earthquakes and height changes inferred from lake-level and tilt-leveling surveys, September 1979 - September 1983 (Otway, 1986b).

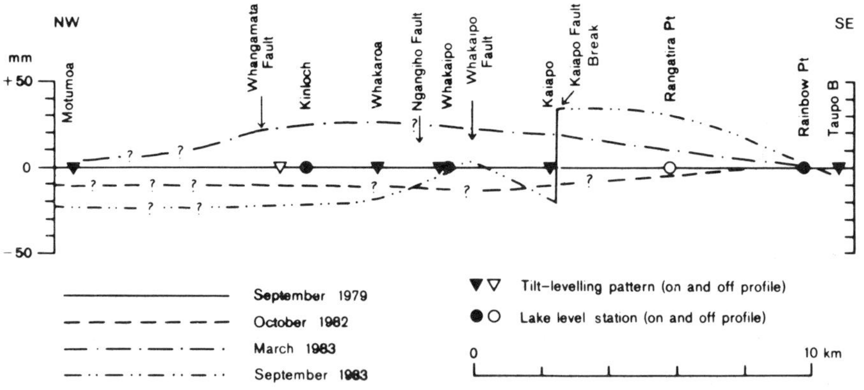

Figure 4.2.7. Profile between Motumoa and Rainbow Point showing level changes across Taupo Fault Belt during course of the 1983 earthquake swarms, from Otway (1986b). Based on lake-level and tilt-leveling surveys.

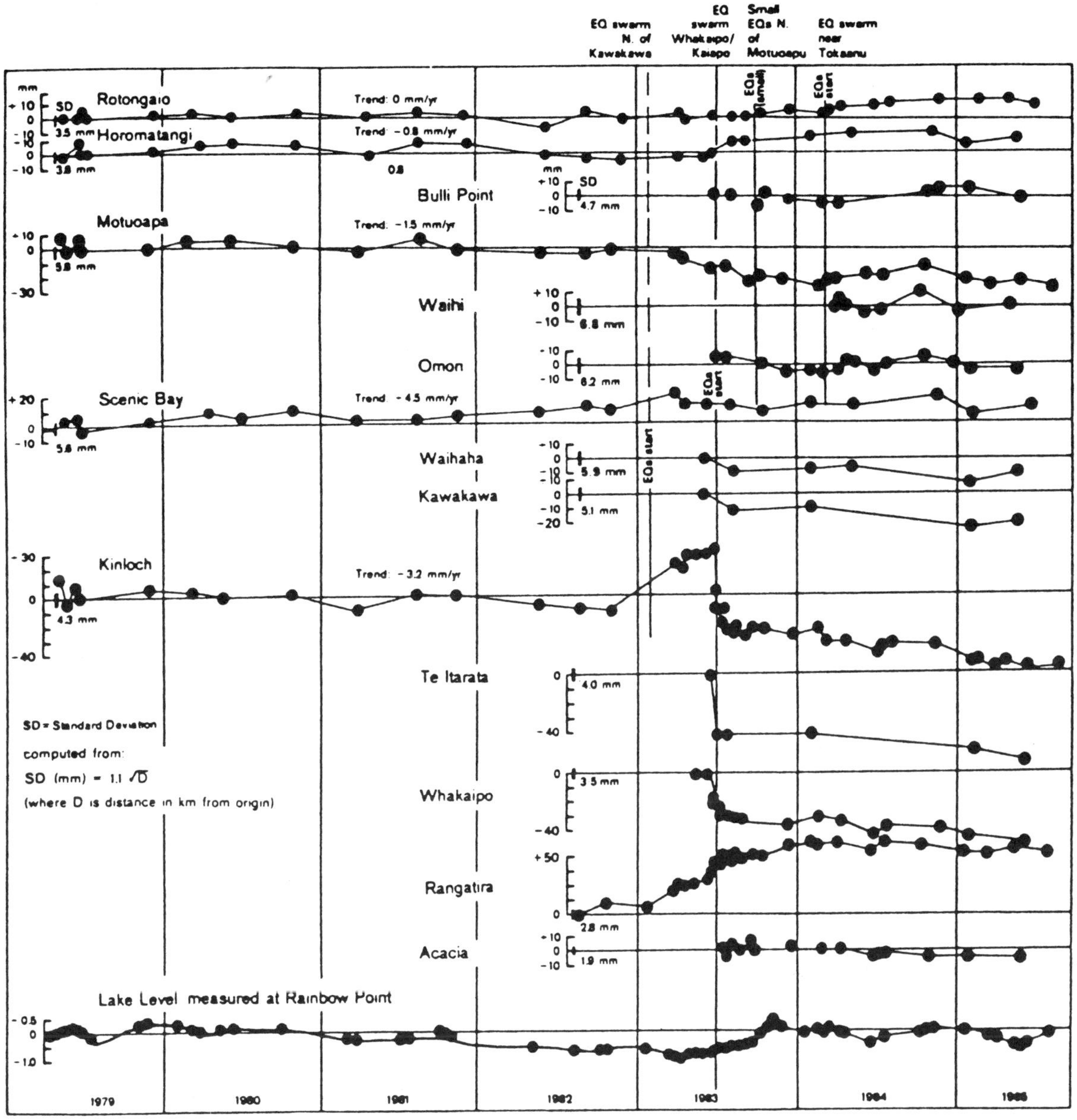

Figure 4.2.8. Apparent changes in bench-mark heights relative to Rainbow Point during 1979-85, from Otway (1986a).

ROTORUA

CAVW number (active vent)	Latitude Longitude (degrees)	Diameter (km)	Local tectonic setting	Precaldera edifice	SiO_2 content (wt pct)	Age of collapse (yr B.P.)	Date of unrest	Type and duration of unrest ESTU STHF MGTF H Te	Eruption type
04-01-16 (Rotorua/) Whakarewarewa)	38.08S 176.25E	15	Exten	None?	R = ? C = 72	140,000	1886	---- ---- ---- ? --	(Tarawera)
							1932	---- ---- --xx x --	hydrothermal explosion
							1924-63	---- -x-- ---- - --	none
							1962	C--- ---- ---- - --	none

TECTONIC SETTING

Rotorua Caldera lies near the western margin of the Taupo Volcanic Zone, west of Okataina Volcanic Center (fig. 4.1.1, in section on Haroharo/Okataina, and fig. 4.2.1, in section on Taupo).

GEOLOGIC HISTORY

Rotorua Caldera (fig. 4.3.1) formed as a result of eruption of the 200 km^3 (dense-rock equivalent) Mamaku ignimbrite. Subsequent lava domes in the caldera, from 1 to 10 km^3 in volume, have not been dated. No eruptions (other than hydrothermal explosions) are known within the past 10,000 years, but the Rotorua-Whakarewarewa area is known for numerous hot springs, geysers, and other geothermal features that support a significant tourist industry.

HISTORICAL ACTIVITY

1886: The supply of hot water to baths of Rotorua may have failed for a period of several weeks sometime in the first half of 1886 (Conly, 1985). Local changes in hot-water circulation are not uncommon, so we cannot be sure that the failure, even if it did occur, reflected any unrest on the scale of Rotorua Caldera. However, such a change is interesting in that it occurred about the time of hydrologic changes in nearby lakes, just before the eruption of Tarawera Volcano (see section on Haroharo/Okataina Calderas).

1932: A relatively large hydrothermal explosion occurred in the southern cove of Lake Rotorua on 18 January 1932. It was reportedly the largest mud-geyser eruption since the explosion of Waimangu (see section on Haroharo Caldera) in 1917 (Jaggar, 1932). Many other steam explosions have occurred, both before and since 1932, and geysers are routinely active. Changes in hydrothermal activity within the caldera are regarded as superficial in origin, related to climatic factors and, recently, to artificial withdrawal of hot waters (I.A. Nairn, written commun., 1987).

ROTORUA (continued)

HISTORICAL ACTIVITY (continued)

1924-63: Leveling of five bench marks around Lake Rotorua in 1924, 1953, and 1963 showed 0.91 m to 1.06 m of difference relative to a new (Moturiki) datum (figs. 4.3.2, 4.3.3). Most of this change reflects the new datum, but an apparent 30-80 mm uplift of bench marks just south of Lake Rotorua relative to bench marks just east of the lake suggested to Healy (1975) that there may have been slight (northeastward?) tilting between surveys. A 30-cm rise in the level of Lake Rotorua during 1940-70 and a concurrent rise in levels of nearby lakes were the results of abnormally high rainfall, not volcanic activity (Healy, 1975). Eiby (1966) reported that five earthquakes were felt at Rotorua on 24 January 1962; many other earthquakes are frequently felt, but we lack a systematic record and precise locations.

COMMENTS

Rotorua shows no evidence of magmatic or tectonic changes in historical time (Ian Nairn, written commun., 1987), but we include it here on the basis of its fluctuating hydrothermal activity, including a sizable hydrothermal explosion in 1932.

REFERENCES

Cole, J.W., and Nairn, I.A., 1975, Catalogue of active volcanoes of the world including solfatara fields, Pt. XXII, New Zealand: Rome, IAVCEI, 156 p.

Conly, G., 1985, Tarawera: The destruction of the pink and white terraces: Wellington, Grantham House, 160 p.

Eiby, G.A., 1966, Earthquake swarms and volcanism in New Zealand: Bull. Volcanol., v. 29, p. 61-73.

Grange, L.I., 1937, The geology of the Rotorua-Taupo subdivision: New Zealand Geol. Surv. Bull. 37, 138 p.

Healy, J., 1975, The gross effect of rainfall on lake levels in the Rotorua district: Jour. Roy. Soc. New Zealand, v. 5, no. 1, p. 77-100.

Healy, J., Vucetich, C.G., and Pullar, W.A., 1964, Stratigraphy and chronology of Late Quaternary volcanic ash in Taupo, Rotorua, and Gisborne districts: New Zealand Geol. Surv. Bull. 73, 88 p.

Houghton, B.F., 1982, Geyserland: A guide to the volcanoes and geothermal area Rotorua: Geol. Soc. New Zealand Guidebook 4, 48 p.

Jaggar, T.A., 1932, Eruption in Lake Rotorua, New Zealand: The Volcano Lett., no. 386, p. 1.

Nairn, I.A., 1980, Source, age, and eruptive mechanisms of Rotorua Ash: New Zealand Jour. Geol. Geophys., v. 23, p. 193-208.

Thompson, B.N., Kermode, L.O., and Ewart, A., eds., 1966, New Zealand volcanology, Central Volcanic Region: New Zealand Dept. Sci. Industrial Res. Info. Ser. 50, 211 p.

Wilson, C.J.N., Rogan, A.M., Smith, I.E.M., Northey, D.J., Nairn, I.A., and Houghton, B.F., 1984, Caldera volcanoes of the Taupo Volcanic Zone, New Zealand: Jour. Geophys. Res., v. 89, p. 8463-8484.

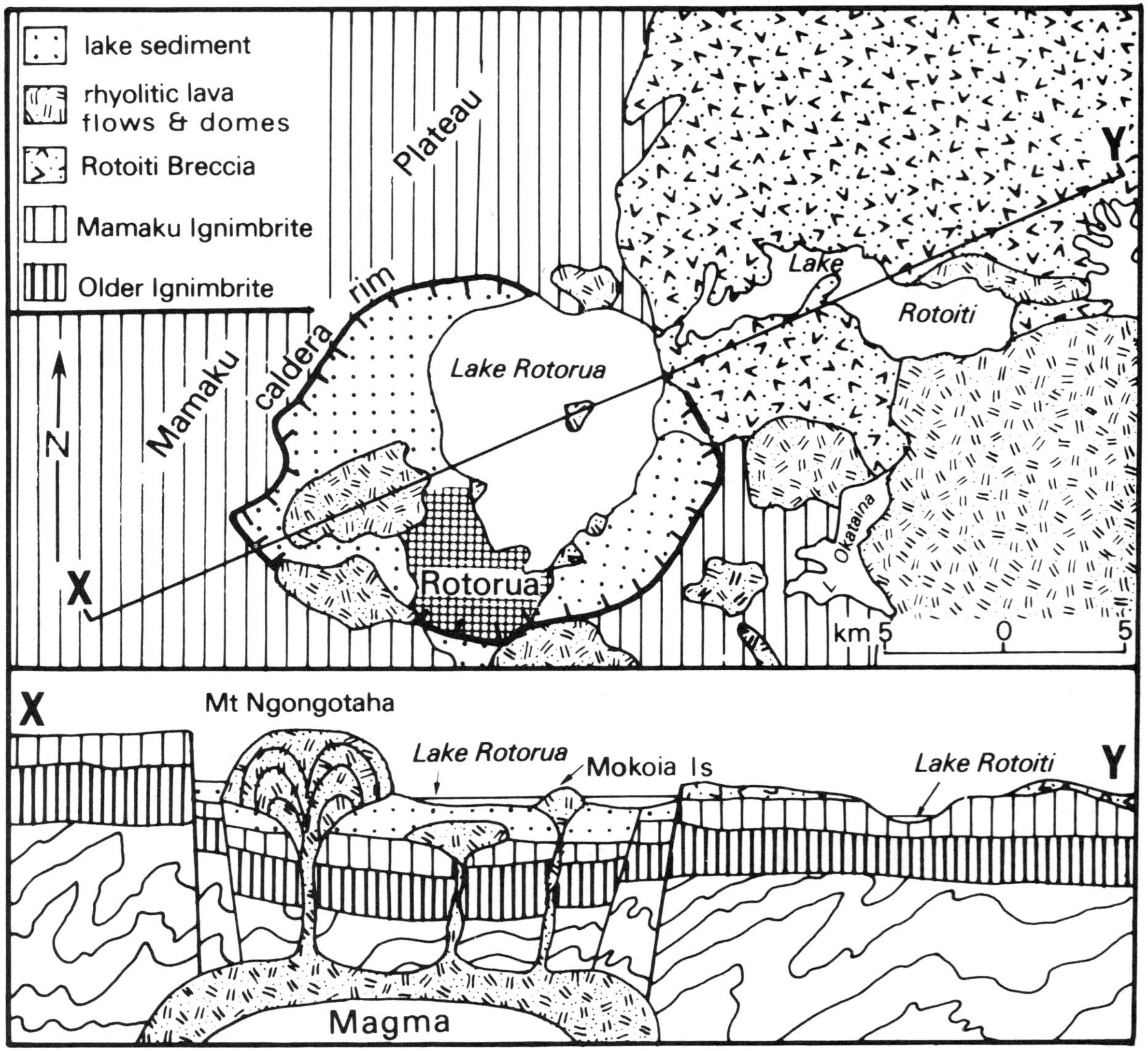

Figure 4.3.1. Geologic map of Rotorua Caldera and a simplified cross section along XY, from Houghton (1982). Hachured heavy line is caldera rim.

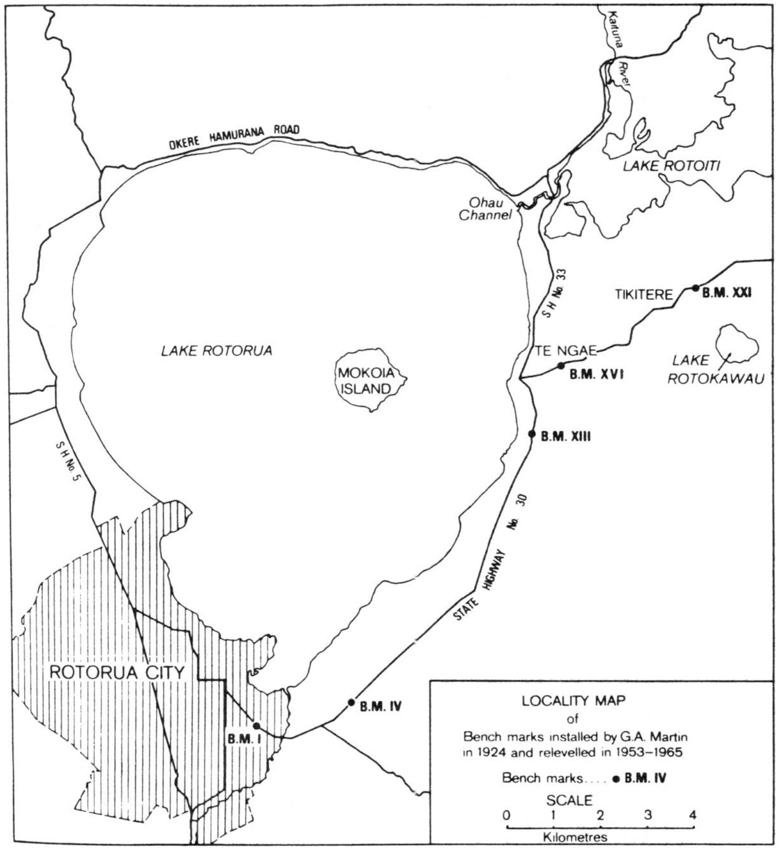

Figure 4.3.2. Locations of bench marks installed east of Lake Rotorua in 1924 and releveled during 1953-63, from Healy (1975).

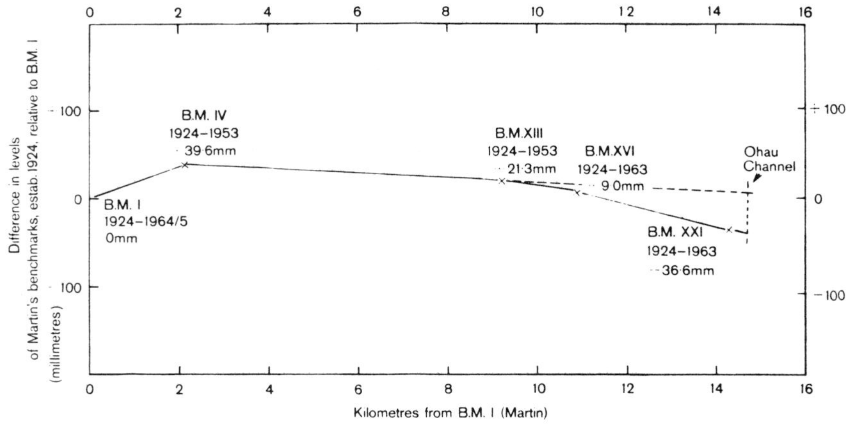

Figure 4.3.3. Apparent differences in level of bench marks near Lake Rotorua from 1924 to 1963. Figure from Healy (1975), who inferred that bench marks east of the lake may have risen 30-80 mm relative to BM 1 south of the lake between leveling surveys.

PART 3: HISTORICAL UNREST AT LARGE QUATERNARY CALDERAS (continued)
See inside back cover for explanation and abbreviations

TOFUA

CAVW number (active vent)	Latitude Longitude (degrees)	Diameter (km)	Local tectonic setting	Precaldera edifice	SiO_2 content (wt pct)	Age of collapse (yr B.P.)	Date of unrest	Type and duration of unrest ESTU STHF MGTF H Te	Eruption type
04-03-06	19.80S 175.07W	5	Compr	Strato	R = 54-55 C = 55-58	?	1774	---- ---- ---- - --	ex
							1792	---- ---- ---- - --	lf
							1854	---- ---- ---- - --	ex
							1885	---- ---- ---- - --	ex
							1906	---- ---- ---- - --	ex,lf
							1937	---- ---- ---x - --	none
							1939	---- ---- ---x - --	none
							1955	---- ---- ---x - --	none
							1958-59	---- ---- ---x - --	ex
							1979	---- ---- ---x - --	none?

TECTONIC SETTING

The Tonga Archipelago stretches 250 km along the western margin of the Tonga Trench in the southern Pacific Ocean. Along the volcanic chain there are several active subaerial and submarine volcanoes, Tofua being one of the more prominent. More than 35 eruptions in the Tongan Islands were recorded in the 200 years ending in 1970 (Ewart and others, 1977).

GEOLOGIC HISTORY

Tofua Caldera is the summit caldera of a steep-sided composite cone that forms Tofua Island (fig. 4.4.1). Pre-caldera activity is recorded by a sequence of pyroclastic deposits and lavas constituting the older cone, followed on the northern part of the island by froth lavas (Bauer, 1970) or welded and unwelded ignimbrite (Baker and others, 1971). Following caldera collapse, lavas were erupted from the northern part of the island and the caldera-rim fissure zone, scoria and lavas from the caldera-wall fissure zones, pyroclastics and lavas from intracaldera cones, and recent pyroclastic fall deposits on the outer cone. Eruptive products are mainly basaltic andesites and andesites, plus occasional dacite flows within the older cone (Ewart and others, 1977). A postcaldera cone with fumarolic activity (Lofia) is situated in the northern part of the caldera; a lake of unknown depth occupies most of the remainder. Bauer (1970) states that "the caldera is continuing to collapse along relatively minor faults on the northern, eastern, and southeastern sides. The fault displacements have given this part of the rim a hummocky topography, with small grabens and steps with escarpments averaging from 50 to 100 feet high." Such movements could be related to the original caldera collapse, or perhaps more likely to gravitational slumping of the caldera walls.

TOFUA (continued)

HISTORICAL ACTIVITY

Most historical eruptions have been small explosions from Lofia cone along the northern caldera rim. The eruption of 1958-59 caused most or all islanders to evacuate for a year or more (Baker and others, 1971).

REFERENCES

Baker, P.E., Harris, P.G., and Reay, A., 1971, The geology of Tofua Island, Tonga, in Cook Bicentenary Expedition in the South-West Pacific: Roy. Soc. New Zealand Bull., v. 8, p. 67-79.

Bauer, G.R., 1970, The geology of Tofua Island, Tonga: Pacific Sci., v. 24, p. 333-350.

Ewart, A., Brothers, R.N., and Mateen, A., 1977, An outline of the geology and geochemistry, and the possible petrogenetic evolution of the volcanic rocks of the Tonga-Kermadec-New Zealand island arc: Jour. Volcanol. Geotherm. Res., v. 2, p. 205-250.

Richard, J.J., 1962, Catalogue of active volcanoes of the world including solfatara fields, Pt. XIII, Kermadec, Tonga, and Samoa: Rome, IAVCEI, 38 p.

Smithsonian Institution, Scientific Event Alert Network (SEAN), 1979, Tofua: SEAN Bull., v. 4, no. 6 (with report from R.N. Richmond).

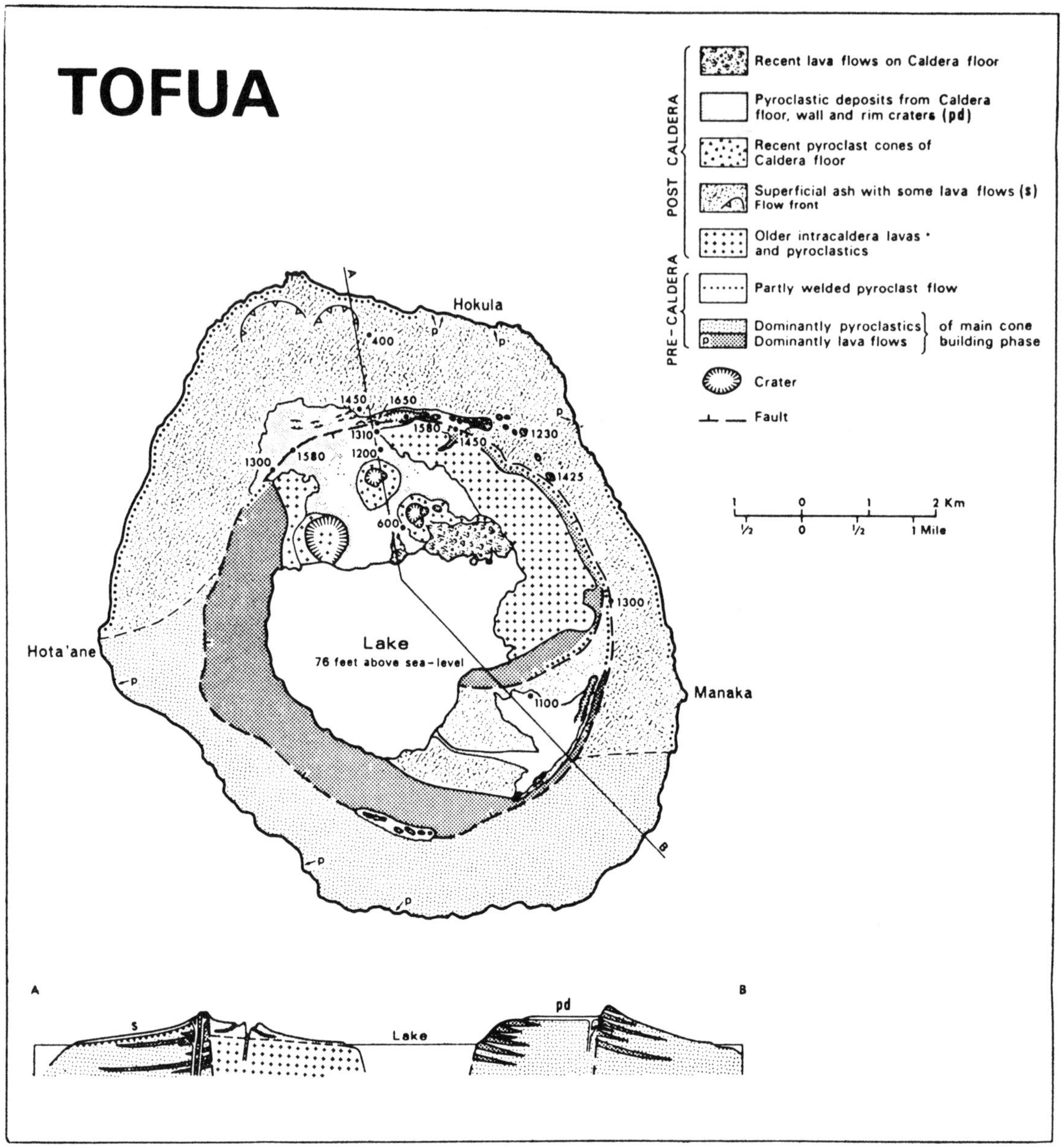

Figure 4.4.1. Geologic map of Tofua, from Baker and others (1971).

PART 3: HISTORICAL UNREST AT LARGE QUATERNARY CALDERAS (continued)
See inside back cover for explanation and abbreviations

NIUAFO'OU

CAVW number (active vent)	Latitude Longitude (degrees)	Diameter (km)	Local tectonic setting	Precaldera edifice	SiO_2 content (wt pct)	Age of collapse (yr B.P.)	Date of unrest	Type and duration of unrest ESTU STHF MGTF H Te	Eruption type
04-03-11	15.60S 175.63W	5-6	Compr	Shield	R = 49-51 C = b?	?	1814	---- ---- ---- - --	pex,ex,lf
							1840	---- ---- ---- - --	unknown
							1853	---- ---- ---- - --	lf,Ex?
							1867	---- ---- ---- - --	lf
							1886-87	E--D ---- ---- - QU?	pex,Ex
							1912	---- ---Y ---- - --	ex,lf
							1929	x--- ---Y ---- - --	ex,lf
							1935-36	---- ---- ---E - --	ex,lf
							1943	---- ---- ---- - --	ex,lf
							1946-47	BB-Y ---Y ---- - --	ex,lf
							1959(?)	---- ---- ---- - --	unknown
							1985	BB-- ---B ---- - --	ex

TECTONIC SETTING

Niuafo'ou Island ("Tin Can Island") lies at the northern end of the Tonga Archipelago. It is geographically within the Tongan group but consists of more mafic lavas (dominantly olivine tholeiites) than most Tongan volcanoes (Ewart and others, 1977).

GEOLOGIC HISTORY

Niuafo'ou Island is a broad lava shield capped by a composite cone (Taylor, 1986). The cone has been modified by development of a summit caldera, now filled by a large lake (fig. 4.5.1). Several cones have erupted within the caldera, and lava flows have emanated from fissures along the western outer slopes of the caldera, which are roughly concentric with the caldera rim. The latter erupted on five occasions from 1857 to 1946 (Ewart and others, 1977).

HISTORICAL ACTIVITY

pre-1886: The eruption of 1814 was from the caldera; eruptions in 1853 and 1867 were from concentric fissures on the southwest and south-southwest flanks, respectively (Richard, 1962).

NIUAFO'OU (continued)

HISTORICAL ACTIVITY (continued)

1886-87: Earthquakes on 8 and 11 June 1886 and on 12 August 1886 were followed by an eruption that began on 31 August 1886 (Sapper, 1917). "During the earthquake of August 12, the captain of a ship at anchor found that, whereas he had paid out twenty fathoms of chain over-night, he had only eight fathoms under him in the morning" (from a letter by C. Trotter, quoted by Bonney, 1886). Bonney (1886) also cites reports that in addition to the earthquakes noted above, the eruption was preceded by 24 hours of frequent smaller earthquakes. Phillips (1899) describes what was presumably the last few hours of seismic precursor: "About 7 P.M. the earthquake began by gently swaying the island, and continued until 12 P.M., to the fright of all the inhabitants--some seven hundred people--who aimlessly wandered about, carrying the old and the feeble to the highest land for safety. At 12 P.M. occurred a tremendous report like the discharge of a 60-ton gun, and a great rocket ascended from the lake to a distance of some 800 to 1,000 yards. The shaking of the island then ceased. From the spot where the rocket ascended an active volcano formed, and continued for seventeen days, throwing up sand, stones, and water." The eruption was accompanied by an unspecified amount of subsidence of the border benches of the caldera (Jaggar, 1930b).

1912, 1929: Coral was erupted together with lava in both 1912 and 1929, even though Jaggar (1931) found no modern coral. Thus, the southwestern side of the island may have subsided significantly during recent time.

1929-47: Only a few tremors were reported to have occurred before the eruption of 1929 (Jaggar, 1930b), but severe earth tremors were felt in Angaha about 1 hour and 15 minutes before a fissure eruption began near that town on 9 September 1946 (Malekamu, 1946). The British Consul, C.W.T. Johnson, reported on or about 27 September (after the eruption had begun) that "there are several craters at the site of the eruption on the north of the island which are still smoking, there are a number of cracks in the ground, some several inches wide, through which smoke still pours and one particularly wide crack at the top of the ridge overlooking the lake--the water in the lake is appreciably higher, and the reef on the north of the island, adjoining the landing place, has been lifted about 30-40 feet above the sea over a considerable area. There are still numerous earthquake shocks daily" (cited in Rogers, 1981).

Numerous vents have been active within historical time. Eruptions in 1853 and 1867 were on the south-southwest rim of the caldera, in 1886 just inside the north-northeast rim of the caldera, and in 1912 and 1929 through north-trending tension fractures (=ring fault?) on the western side of the island. In October 1912, lava flowed from "fully thirty points of eruption between the shore and the lake" (Thomson, 1926). Eruptions in 1935 and 1936 were on the south side of the island, in 1943 on the southwestern slope, and in 1946 on the northeastern side of the island, destroying the town of Angaha. The entire island was evacuated for more than 10 years following this eruption, at substantial social cost (Rogers, 1981).

1985: Moderately strong earthquakes (MMI_{max}=VII) were felt from 2050 hr on 21 March until 1132 hr on 22 March. None were strong enough to be recorded on the nearest seismometer, 500 km away. A diffuse, thunderous rumbling noise accompanied the earthquakes. A 250-m-long radial(?) crack formed across the northeast shore of the caldera lake, and although nothing was erupted subaerially, fresh black pumice was found floating on the caldera lake. No unusual temperatures or bubbles were noted (Smithsonian Institution, 1985; Taylor, 1986).

NIUAFO'OU, Region 4, CAVW number 04-03-11

NIUAFO'OU (continued)

COMMENTS

Historical eruptions at Niuafo'ou have occurred from multiple vents, possibly along the caldera ring-fracture system. In this regard, activity has been similar to that at Deception Island. At these two mafic centers, frequent historical leaks along ring fractures have yet to be followed by a catastrophic repeat of caldera formation.

REFERENCES

Bonney, T.G., 1886, Volcanic eruption in Niua-fu, Friendly Islands: Nature, v. 35, p. 127-128.
Ewart, A., 1976, Petrological study of younger Tongan andesites and dacites, and olivine tholeiites of Niua foou Island, S.W. Pacific: Contrib. Mineral. Petrol., v. 58, p. 1-22.
Ewart, A., Brothers, R.N., and Mateen, A., 1977, An outline of the geology and geochemistry, and the possible petrogenetic evolution of the volcanic rocks of the Tonga-Kermadec-New Zealand island arc: Jour. Volcanol. Geotherm. Res., v. 2, p. 205-250.
Jaggar, T.A., 1930a, Eruption on Niuafou, Tonga Islands: The Volcano Lett., no. 265, p. 3-4.
----- 1930b, The island volcano Niuafoou: The Volcano Lett., no. 312, p. 1-3.
----- 1931, Geology and geography of Niuafoou Volcano: The Volcano Lett., no. 318, p. 1-3.
----- 1945, Volcanoes declare war, logistics and strategy of Pacific volcano science: Honolulu, Paradise of the Pacific, Ltd., 166 p.
Macdonald, G.A., 1948, Notes on Niuafo'ou: Am. Jour. Sci., v. 246, p. 65-77.
Malekamu, S., 1946, Tongan wireless operator's story of the eruption (of Niuafo'ou): Pacific Islands Monthly, v. 17, no. 4, p. 28, 53.
Phillips, C., 1899, The volcanoes of the Pacific: Trans. New Zealand Inst., v. 31, p. 510-551.
Richard, J.J., 1962, Catalogue of active volcanoes of the world including solfatara fields, Pt. XIII, Kermadec, Tonga, and Samoa: Rome, IAVCEI, 38 p.
Rogers, G., 1981, The evacuation of Niuafo'ou, an outlier in the Kingdom of Tonga: Jour. Pacific Hist., v. 16, no. 4, p. 149-163.
Sapper, K., 1917, Katalog der geschichtlichen Vulkanausbrüche: Strasburg, K. Trübner, 358 p.
Smithsonian Institution, Scientific Event Alert Network (SEAN), 1985, Niuafo'ou Volcano: SEAN Bull., v. 10, no. 9, p. 6-7 (report from J.H. Latter, Mr. Fifita, R.J. Blong, and P. Taylor).
Taylor, P.W., 1986, Niuafo'ou Island, Tonga: Subaerial volcanism in an active back-arc basin [abs.]: Abstracts, Int. Volcanol. Congr., New Zealand, p. 123.
Thomson, J.A., 1926, Volcanoes of the New Zealand-Tonga volcanic zone: New Zealand Jour. Sci. Technol., v. 9, p. 354-371.

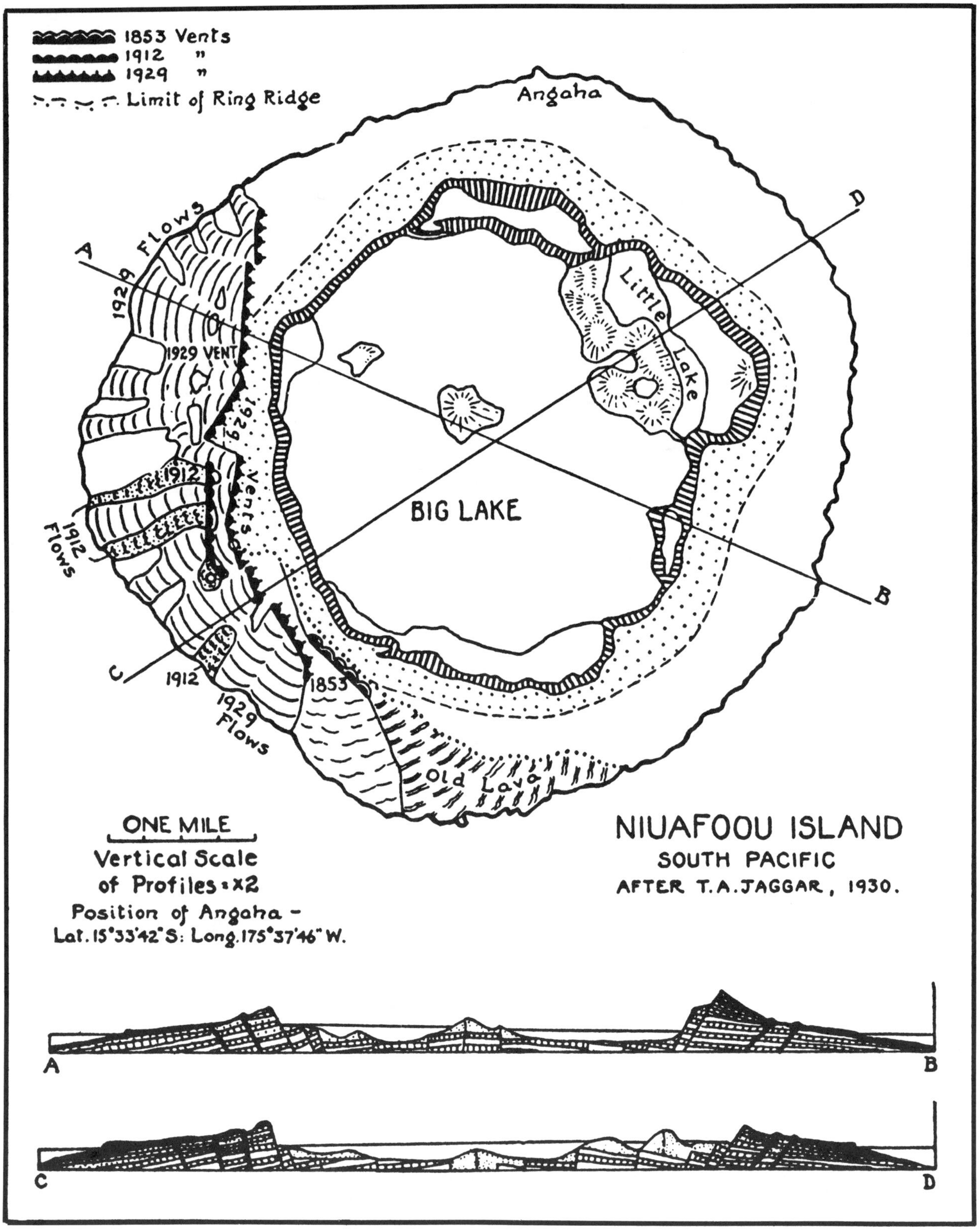

Figure 4.5.1. Map and cross sections of Niuafo'ou Island, from Jaggar (1945).

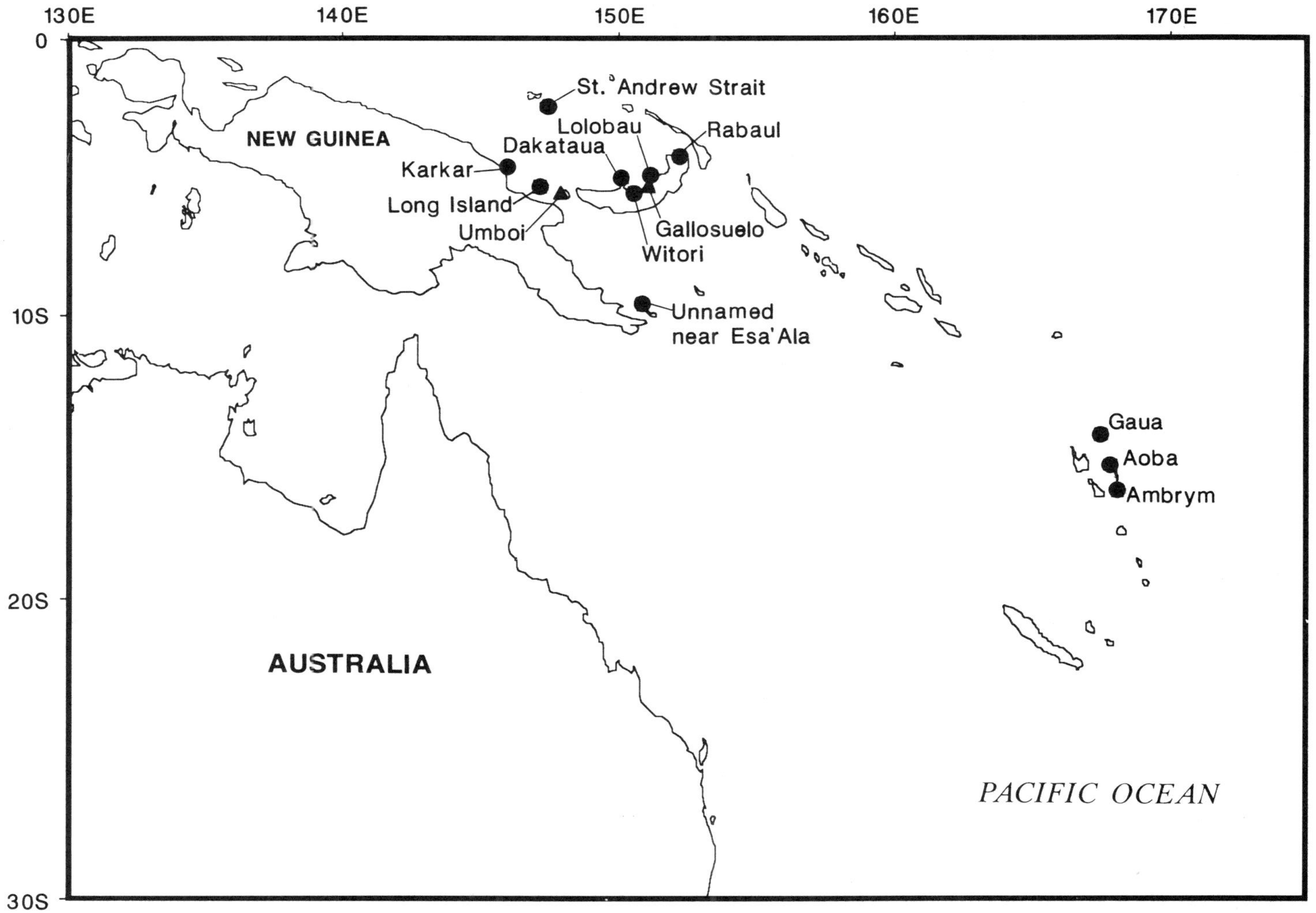

Figure 5. Locations of large, Quaternary restless calderas (solid circles) and nonrestless calderas (triangles) in Region 5.

ST. ANDREW STRAIT

CAVW number (active vent)	Latitude Longitude (degrees)	Diameter (km)	Local tectonic setting	Precaldera edifice	SiO_2 content (wt pct)	Age of collapse (yr B.P.)	Date of unrest	Type and duration of unrest ESTU STHF MGTF H Te	Eruption type
05-00-01 (Tuluman)	2.45S 147.32E	16	Hotspot?	?	R = 69-71 C = silicic	?	1883 1953-57	---- ---- ---- - -- ---Y Y?-Y ---- - --	ex? Ex,pex,lf

TECTONIC SETTING

Tuluman and Lou Islands lie on the northwestern side of St. Andrew Strait in the northern Bismarck Sea. Tuluman Island represents the youngest part of an extensive complex of rhyolitic volcanism (Smith and Johnson, 1981), the older part of which is developed on Lou Island. Tuluman may lie on what is either the rim of a submarine caldera or an incipient caldera ring fault, more likely the latter (fig. 5.1.1; Johnson and Smith, 1974; Reynolds and others, 1980).

GEOLOGIC HISTORY

Tuluman Island and Cone-3 Island ("Tuluman Islands") formed during the 1953-57 eruption, during which eight main centers were active; little is known about their earlier submarine history. Lou Island to the north is a chain of overlapping Quaternary cones consisting of rhyolitic lava flows and volcaniclastic deposits similar to those of the Tuluman Islands. Twelve main eruptive centers on Lou Island were identified by Johnson and Davies (1972), who suggested that volcanic activity may have progressed southwestward along the Lou Island chain, and that the 1953-57 Tuluman eruption represented the youngest stage in that progression (Reynolds and others, 1980).

HISTORICAL ACTIVITY

1883: An eruption of unspecified character, presumably explosive, was reported by Miklouho-Maclay (1885).

1953-57: No earthquakes were felt on nearby Lou Island before or during the 1953-57 eruption — remarkable in view of the composition of the magma (rhyolite) and the relatively long repose period (at least 70 years) preceding the eruption. A shock recorder on Baluan Island recorded no earthquakes from 23 October to 6 November 1954, and a recorder at the southern tip of Lou Island (1.5 km north of the vent) recorded only a few earthquakes from 29 November to 3 December 1956, all simultaneous with explosions. A tiltmeter installed on Baluan Island in March 1955 recorded approximately 70 arcseconds (350 microradians) of apparent northward tilt from 18 March to 21 April (fig. 5.1.2). This northward tilt correlated with the vigor of eruptions at Tuluman and was

ST. ANDREW STRAIT (continued)

HISTORICAL ACTIVITY (continued)

consistent with deflation in the St. Andrew Strait area (Reynolds and Best, 1957); however, the reliability of the tiltmeter has been subsequently questioned (J. Mori, written commun., 1988). Near the end of the 1953-57 eruption, points on the southern and western sides of the new island (those sides that face away from the inferred caldera) experienced local subsidence of up to several tens of meters, accompanied by formation of arcuate subsidence fractures. Initial, localized subsidence was followed by uplift of a somewhat broader area: 1 m of uplift along the eastern side of the island in January 1957 and 4-5 m of uplift along the southern and western sides of the island in February 1957. Owing to the uplift, precautions were taken against the possibility of a strong eruption, but none occurred (Reynolds, 1958). Since 1957, wave erosion has pushed shorelines, especially on the southern and western sides of the island, inland by more than 150 m.

REFERENCES

Johnson, R.W., and Davies, R.A., 1972, Volcanic geology of the St. Andrew Strait islands, Bismarck Sea, Papua New Guinea: Geol. Surv. Papua New Guinea Note on Invest. 72-002 (unpublished).

Johnson, R.W., Everingham, I.B., and Cooke, R.J.S., 1981, Submarine volcanic eruptions in Papua New Guinea: 1878 activity of Vulcan (Rabaul) and other examples, in Johnson, R.W., ed., Cooke-Ravian volume of volcanological papers: Geol. Surv. Papua New Guinea Memoir 10, p. 167-179.

Johnson, R.W., and Smith, I.E., 1974, Volcanoes and rocks of St. Andrew Strait, Papua New Guinea: Jour. Geol. Soc. Australia, v. 21, p. 333-351.

Miklouho-Maclay, N. de, 1885, On volcanic activity on the islands near the north-east coast of New Guinea and evidence of rising of the Maclay-coast in New Guinea: Proc. Linn. Soc. N.S.W., v. 9, no. 4, p. 963-967.

Pain, C.F., 1981, Stratigraphy and chronology of volcanic ash beds on Lou Island, in Johnson, R.W., ed., Cooke-Ravian volume of volcanological papers: Geol. Surv. Papua New Guinea Memoir 10, p. 221-225.

Reynolds, M.A., 1958, Activity of Tuluman Volcano, St. Andrew Strait, Admiralty Islands, September 1955-March 1957: Bur. Min. Res. Geol. Geophys., Australia, Rec. 1958/14, 34 p.

Reynolds, M.A., and Best, J.G., 1957, The Tuluman volcano, St. Andrew Strait, Admiralty Islands: Bur. Min. Res. Geol. Geophys., Australia, Rep. 33, 38 p.

----- 1976, Summary of the 1953-57 eruption of Tuluman volcano, Papua New Guinea, in Johnson, R.W., ed., Volcanism in Australasia: Amsterdam, Elsevier, p. 275-285.

Reynolds, M.A., Best, J.G., and Johnson, R.W., 1980, 1953-57 eruption Tuluman volcano: rhyolitic volcanic activity in the northern Bismarck Sea: Geol. Surv. Papua New Guinea Memoir 7, 44 p.

Smith, I.E.M., and Johnson, R.W., 1981, Contrasting rhyolite suites in the late Cenozoic of Papua New Guinea: Jour. Geophys. Res., v. 86, p. 10257-10272.

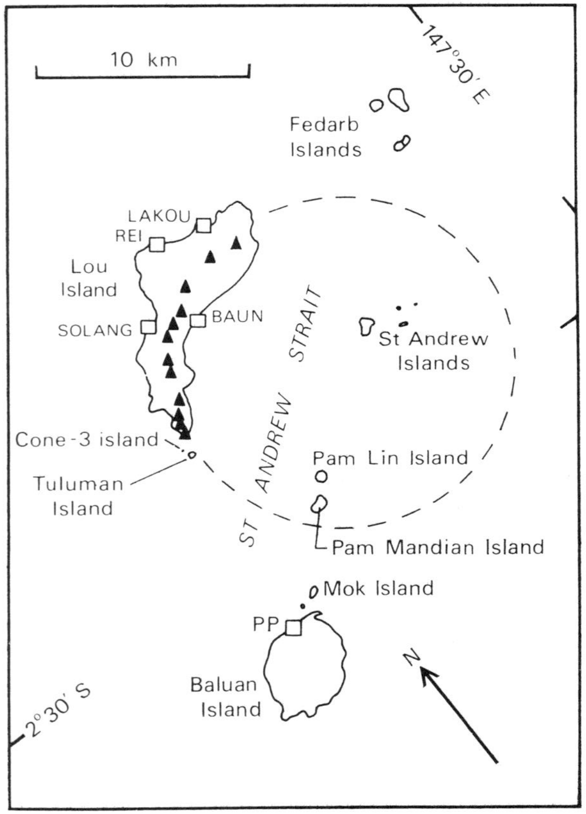

Figure 5.1.1. Map of St. Andrew Strait area, showing locations of Lou Island and Tuluman Island along arcuate volcanic chain that may be part of a caldera rim or nascent ring-fracture system (dashed line). Squares represent settlements; solid triangles represent Lou Island eruptive centers (map from Reynolds and others, 1980).

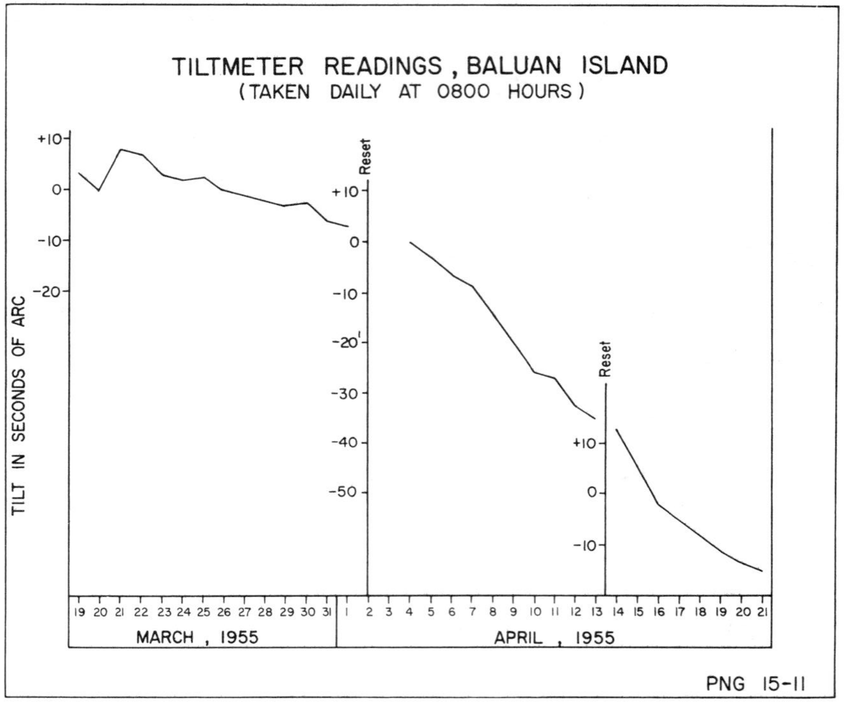

Figure 5.1.2. Record of ground tilt at Baluan Island during March-April 1955, from Reynolds and Best (1957). Those authors discuss possible sources of error but conclude that tilting by at least 70 arcseconds was related to ongoing eruption at Tuluman Island.

KARKAR

CAVW number (active vent)	Latitude Longitude (degrees)	Diameter (km)	Local tectonic setting	Precaldera edifice	SiO_2 content (wt pct)	Age of collapse (yr B.P.)	Date of unrest	Type and duration of unrest ESTU STHF MGTF H Te	Eruption type
05-01-03	4.65S 145.96E	5.5, 3.2	Compr	Strato	R = 53-55 C = ?	9100, 1400	1643	---- ---- ---- - --	Ex?
							1885	---- ---- ---- - --	ex
							1895	BB-- ---- ---- - --	ex,lf
							1962?	---- ---- ---- - --	ex
						between	1965/70	x--- ---- ---x - UQ?	none
							1974-75	---- ---- --FF - QU	ex,Lf
							1976-80	FFEY -Y-- -YEE E --	pex,ex

TECTONIC SETTING

Karkar is an active andesitic stratovolcano about 16 km off the northern coast of mainland Papua New Guinea and is a member of the western Bismarck volcanic arc (Pain and McKee, 1981).

GEOLOGIC HISTORY

Early eruptive events produced the basic shape of the volcano, but caldera formation and other explosive activity in the late Quaternary have modified the shape of the island. Karkar has a double caldera, one nested within the other, at the top of a truncated cone (fig. 5.2.1). Parasitic vents occur on the northern and southern flanks of the volcano. Five pyroclastic units are found over most of Karkar. Two ignimbrite units, the younger about 9,000 yr old, were probably associated with the collapse that formed the outer caldera. Airfall tephras represent a period when the caldera was being filled by a volcano consisting of tephras and lavas. Subsequent violent eruptions between 1,500 and 800 yr B.P. produced two more ignimbrite units that were almost certainly associated with the collapse(s) that formed the inner caldera. Tephra has been erupted more recently over the whole island, and tephra and lava are gradually filling the inner caldera (Pain and McKee, 1981). Most postcaldera eruptions have been from Bagiai, a small cone near the center of the inner caldera.

HISTORICAL ACTIVITY

pre-1895: A "great flame came steadily from the top of the mountain" in 1643, and "smoke and ash" were reported in 1885 (McKee and others, 1976).

KARKAR (continued)

HISTORICAL ACTIVITY (continued)

1895: Many earthquakes occurred on 17 June 1895; an eruption began on the same day and continued intermittently until August. Much of the lava on the pre-1974 floor of the inner caldera may have been from this 1895 eruption (McKee and others, 1976).

1930: A magnitude 6+ earthquake occurred beneath the northern tip of Karkar Island (location approximate) in 1930.

1970±: Between 1965 and 1970 the strongest fumarolic activity of Karkar shifted from the upper to the lower rim of the summit crater of Bagiai (Johnson and others, 1972). On 31 October 1970, a magnitude 7.0 earthquake occurred approximately 45 km southwest of Karkar (near Madang), at a depth of 41 km (Everingham, 1973). Aftershocks were located along a NE-SW trend, including some beneath Karkar Island. Fumarolic temperatures in June 1971 and April 1972 might have been a few degrees higher than in 1963, but uncertainties in measurement cast doubt on any temperature change (McKee and others, 1976).

1974-75: A slight increase in fumarolic activity was noticed 2-3 years before the 1974-75 eruption, during which about 70 million m^3 of andesitic lava were extruded (McKee and Wallace, 1981).

1976-80: B-type earthquakes, MMI_{max}=3 at 1.5 km from the vent, began to increase in mid-1976, increased further in November 1977, and reached a countable maximum in July 1978 (fig. 5.2.2). Overlapping B-type earthquakes became subcontinuous tremor in July 1978, and tremor amplitude increased to a maximum in late October 1978. A distinctive pattern of rhythmic, "banded tremor" (2-4 Hz) began in July 1978 but was particularly evident in October 1978 (photograph of trace in fig. 3 of McKee and others, 1981). This pattern of seismicity is similar to that of regular eruptions of some geysers (Kieffer, 1984). A decline in seismicity from October 1978 through the remainder of 1978 was interpreted to indicate that magma had stabilized at a shallow level. Phreatic eruptions began in January 1979, when tremor amplitude was relatively low. The largest of these eruptions, fatal to volcanologists Rob Cooke and Elias Ravian, occurred in March 1979 and was preceded by 1-2 days of slightly increased tremor. Beginning in February 1979, 15 to 40 b-type earthquakes/day were recorded (McKee and others, 1981).

Gravity and dry tilt measurements at Karkar were started in April 1979. About 40 microradians of cumulative tilt occurred near the caldera rim in 1979. Between April 1979 and November 1979, gravity measurements showed a possible 40-50 cm uplift at the rim and on slopes 4-5 km outside the caldera rim, and irregular deformation within the inner caldera. Beginning in about mid-1978, 6 months before the eruptions of 1979-80, ground temperatures increased from 100 °C to as much as 1000 °C (incandescent) over a 2-km^2 area. Fumarolic activity increased over the same period, on the central cone and at fumaroles on the floor of the inner caldera (McKee and others, 1981).

1984: A magnitude 6.5 earthquake of tectonic origin occurred on 27 March 1984, about 24 km north-northwest of Karkar at a depth of about 23 km. Surface fissures and subsidence occurred on Karkar as a direct result of earthquake shaking.

KARKAR (continued)

COMMENTS

Periodic or rhythmic tremor (so-called "banded tremor" because of its banded appearance on seismograms) occurred several months before the tragic eruption of 1979. Because of its similarity to tremor in geyser systems, and a correlation of later tremor and earthquakes with rainfall, the "banded tremor" is thought to represent subterranean geyserlike processes. If so, such tremor might forewarn of phreatic eruptions.

REFERENCES

Everingham, I.B., 1973, Seismological report on the Madang earthquake of 31 October 1970 and aftershocks: Bur. Min. Res. Geol. Geophys., Australia, Rec. 1973/66.

Fisher, N.H., 1957, Catalogue of active volcanoes of the world including solfatara fields, Pt. V, Melanesia: Rome, IAVCEI, 105 p.

Hantke, G., 1951, Uebersicht über die vulkanische Tätigkeit 1941-1947: Bull. Volcanol., ser. II, v. 11, p. 161-208.

Johnson, R.W., Taylor, G.A.M., and Davies, R.A., 1972, Geology and petrology of Quaternary volcanic islands off the north coast of New Guinea: Bur. Min. Res. Geol. Geophys., Australia, Rec. 1972/21.

Kieffer, S.W., 1984, Seismicity at Old Faithful Geyser: an isolated source of geothermal noise and possible analogue of volcanic seismicity: Jour. Volcanol. Geotherm. Res., v. 22, nos. 1-2, p. 59-95.

McKee, C.O., Cooke, R.J.S., and Wallace, D.A., 1976, 1974-75 eruptions of Karkar Volcano, Papua New Guinea, in Johnson, R.W., ed., Volcanism in Australasia: Amsterdam, Elsevier, p. 173-190.

McKee, C.O., and Wallace, D.A., 1981, Lava fields in the inner caldera of Karkar volcano, in Johnson, R.W., ed., Cooke-Ravian volume of volcanological papers: Geol. Surv. Papua New Guinea Memoir 10, p. 49-62.

McKee, C.O., Wallace, D.A., Almond, R.A., and Talai, B., 1981, Fatal hydro-eruption of Karkar volcano in 1979: development of a maar-like crater, in Johnson, R.W. ed., Cooke-Ravian volume of volcanological papers: Geol. Surv. Papua New Guinea Memoir 10, p. 63-84.

Mori, J., McKee, C., Talai, B., and Itikarai, I., 1986, Precursors to volcanic eruptions in Papua New Guinea: Geol. Surv. Papua New Guinea Rep. 86/27, 39 p.

Pain, C.F., and McKee, C.O., 1981, Late Quaternary eruptive history of Karkar Island, in Johnson, R.W., ed., Cooke-Ravian volume of volcanological papers: Geol. Surv. Papua New Guinea Memoir 10, p. 39-47.

Smithsonian Institution, Scientific Event Alert Network (SEAN), 1978-84, Karkar: SEAN Bull., v. 3, nos. 9-12; v. 4, nos. 1-12; v. 5, nos. 1,4-9; v. 6, nos. 1,3,7; v. 9, no. 3.

Taylor, G.A., 1956, Review of volcanic activity in the territory of Papua-New Guinea, the Solomon and New Hebrides Islands, 1951-1953: Bull. Volcanol., v. 18, p. 25-37.

Volcanological Society of Japan, 1976-82, Karkar: Bull. Volcanic Eruptions, no. 14, p. 20-21; no. 15, p. 18-19; no. 18, p. 17-19; no. 19, p. 20-24; no. 20, p. 22-23.

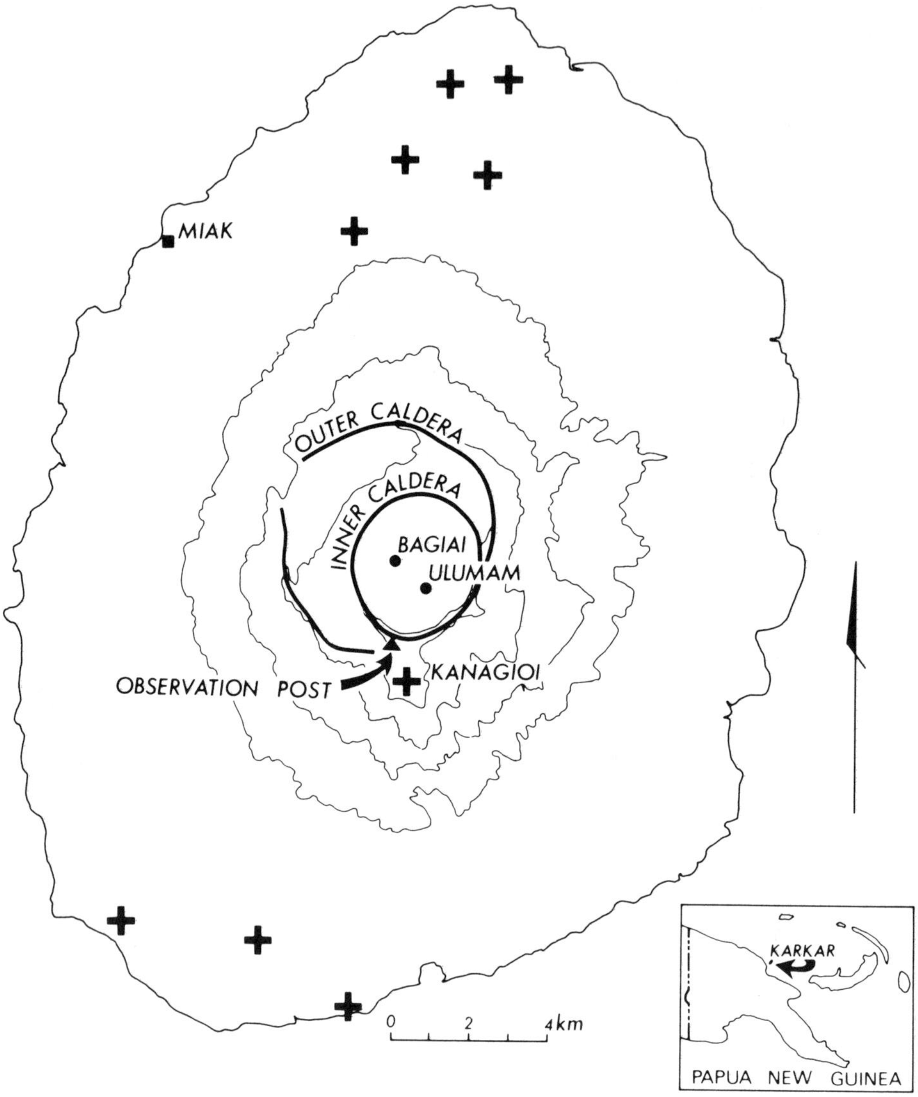

Figure 5.2.1. Karkar Volcano, from McKee and others (1976), showing form lines at approximately 300-m intervals. Crosses show parasitic cones; dots show recently active volcanic centers.

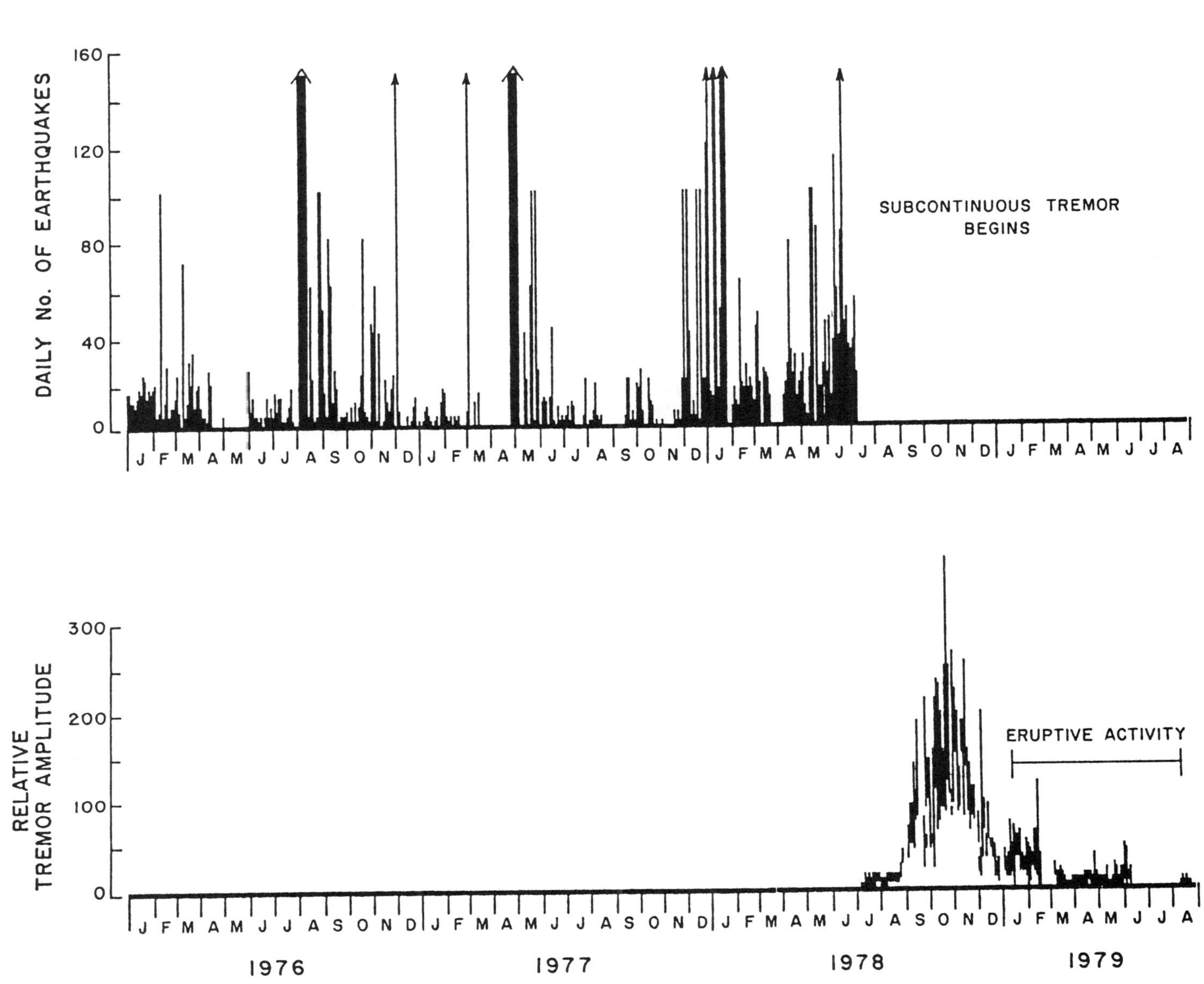

Figure 5.2.2. Daily number of earthquakes and tremor amplitude recorded at Karkar Volcano during 1976-79, from Mori and others (1986).

LONG ISLAND/LAKE WISDOM

CAVW number (active vent)	Latitude Longitude (degrees)	Diameter (km)	Local tectonic setting	Precaldera edifice	SiO_2 content (wt pct)	Age of collapse (yr B.P.)	Date of unrest	Type and duration of unrest ESTU STHF MGTF H Te	Eruption type
05-01-05	5.36S	10 x 12.5	NNE-SSW	LP-strat	R = 49-57	16,000	ca. 1700	x--- ---- ---- - --	EX
(Motmot Island)	147.12E		compr		C = 51-57	4,000	1933	---- ---- ---- x --	unknown
						300	1938	---- ---- ---- - --	unknown
							1943	---- ---- ---- - --	ex
							1953-55	---- ---- ---- - --	Ex
							1961?	---- ---- ---- - --	ex
							1968	---- ---- ---- - --	ex
							1973-76	---- E--E --EE E --	pex,ex,lf
							1979	---- ---- --EE E --	none

TECTONIC SETTING

Long Island is a Quaternary volcanic complex about 150 km east of Madang on the north coast of mainland Papua New Guinea. It is one of a series of volcanic islands that form the western half of the Bismarck volcanic arc. The western part of the chain is associated with a belt of seismicity that defines the boundary between the Indo-Australian plate in the south and the "South Bismarck" plate in the north (Ball and Johnson, 1976).

GEOLOGIC HISTORY

Long Island consists of two stratovolcanoes (Mt. Reaumur and Cerisy Peak) on opposite ends of a caldera complex containing a freshwater lake, Lake Wisdom (fig. 5.3.1). There have been at least three episodes of caldera formation--16,000 yr, 4,000 yr, and about 300 yr B.P. (Ball and Johnson, 1976; Pain and others, 1981). Marine fossils in deposits of the second of these eruptions indicate that part of the island was submerged at the time of that eruption; completely charred tree trunks in the basal section of the uppermost deposit suggest that uplift had preceded the third eruption by a period of many years (Best, 1956). The outer flanks of the island are covered by a thick pyroclastic mantle that is thought to have been deposited during the third caldera-forming eruption, about 300 yr B.P. (Polach, 1981; Blong, 1982). Historical volcanic activity built Motmot Island in the southern part of Lake Wisdom.

LONG ISLAND/LAKE WISDOM (continued)

HISTORICAL ACTIVITY

ca. 1700: The most recent caldera-forming eruption at Long Island occurred only about 300 yrs ago. Legends indicate that a strong earthquake occurred before tephra fell in the New Guinea highlands, but it is not clear whether the earthquake occurred before, during, or just after the eruption (Blong, 1982).

1933-79: A spring 6.5 km south of Malala, probably originating in Lake Wisdom, first began to flow in the 1930's (Ball and Glucksman, 1978). The birth of this spring might have resulted from ground deformation associated with the renewal of volcanic activity. The 1933, 1938, 1943, and 1953 eruptions were from one or several vents beneath the caldera lake, eventually building a small island (Motmot). Precursors to the 1973-74 eruption included, in late 1972, development of fissures and a small pit on Motmot, a change in fumarolic gas compositions, and an increase in the temperature of a small pond on Motmot and of Lake Wisdom at points along the shoreline of Motmot. Seismic recording during a brief lull in eruptive activity in April 1973 showed nothing that could be considered a precursor to the next phase, which began several days later (Cooke and Johnson, 1978). A patch of discolored water was noted 1.5 km west-southwest of Motmot, where water is 200-300 m deep, during inspections on 18 December 1974 and 6 February 1975 but not on 5 September 1975. A different patch of discolored water and a shift and intensification of fumarolic activity on Motmot were noted in mid-January 1979.

COMMENTS

Acid gases from the 1973 eruption at Long Island lowered the pH of Lake Wisdom from 8.2-8.5 to 8.0-8.1 (Ball and Glucksman, 1978, 1980). Although this is a small change, biological activity is very sensitive to changes in pH (and a related parameter, oxygen content of lake waters), and therefore the fossil record of biological activity in a caldera lake can be a useful record of volcanic unrest (see also the section on Atitlan Caldera).

The 1973-74 eruption of Motmot, the 1974-75 eruption of Karkar, and eruptions at several other nearby volcanoes followed a regional increase in seismicity beginning in 1970 (Cooke and others, 1976) (see also the section on Rabaul Caldera).

REFERENCES

Ball, E.E., and Glucksman, J., 1978, Limnological studies of Lake Wisdom, a large New Guinea lake with a simple fauna: Freshwater Biol., v. 8, p. 455-468.

----- 1980, A limnological study of Lake Dakataua, a large caldera lake on West New Britain, Papua New Guinea, with comparisons to Lake Wisdom, a younger nearby caldera lake: Freshwater Biol., v. 10, p. 73-84.

Ball, E.E., and Johnson, R.W., 1976, Volcanic history of Long Island, Papua New Guinea, in Johnson, R.W. ed., Volcanism in Australasia: Amsterdam, Elsevier, p. 133-147.

LONG ISLAND/LAKE WISDOM (continued)

REFERENCES (continued)

Best, J.G., 1956, Investigations of recent volcanic activity in the Territory of New Guinea: Proc. Eighth Pacific Sci. Congr., v. 2, p. 180-203.

Blong, R.J., 1982, The time of darkness: Seattle, Univ. Washington Press, 275 p.

Cooke, R.J.S., and Johnson, R.W., 1978, Volcanoes and volcanology in Papua New Guinea: Geol. Surv. Papua New Guinea Rep. 78/2, 37 p.

Cooke, R.J.S., McKee, C.O., Dent, V.F., and Wallace, D.A., 1976, Striking sequence of volcanic eruptions in the Bismarck volcanic arc, Papua New Guinea in 1972-1975, in Johnson, R.W., ed., Volcanism in Australasia: Amsterdam, Elsevier, p. 149-172.

Lowenstein, P.L., 1982, Problems of volcanic hazards in Papua New Guinea: Geol. Surv. Papua New Guinea Rep. 82/7, 62 p.

Mori, J., McKee, C., Talai, B., and Itikarai, I., 1986, Precursors to volcanic eruptions in Papua New Guinea: Geol. Surv. Papua New Guinea Rep. 86/27, 39 p.

Pain, C.F., Blong, R.J., and McKee, C.O., 1981, Pyroclastic deposits and eruptive sequences of Long Island, Pt. I, Lithology, stratigraphy, and volcanology, in Johnson, R.W., ed., Cooke-Ravian volume of volcanological papers: Geol. Surv. Papua New Guinea Memoir 10, p. 101-113.

Polach, H.A., 1981, Pyroclastic deposits and eruptive sequences of Long Island. Pt. 2, Radiocarbon dating of Long Island and Tibito tephras, in Johnson, R.W., ed., Cooke-Ravian volume of volcanological papers: Geol. Surv. Papua New Guinea Memoir 10, p. 101-113.

Volcanological Society of Japan, 1975-77, Long Island: Bull. Volcanic Eruptions, no. 13, p. 20-21; no. 14, p. 21; no. 15, p. 19-20; no. 19, p. 84.

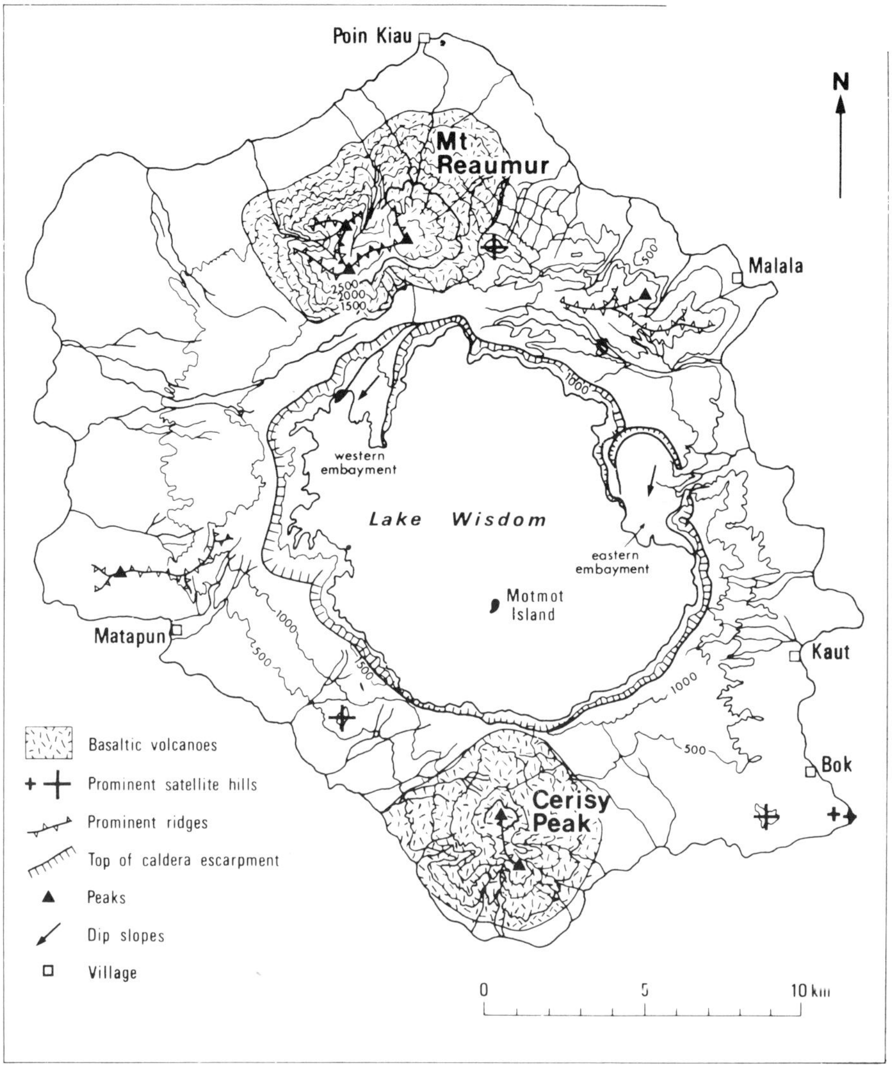

Figure 5.3.1. Map showing Long Island in 1972, from Ball and Johnson (1976). Contours in feet.

DAKATAUA

CAVW number (active vent)	Latitude Longitude (degrees)	Diameter (km)	Local tectonic setting	Precaldera edifice	SiO_2 content (wt pct)	Age of collapse (yr B.P.)	Date of unrest	Type and duration of unrest ESTU STHF MGTF H Te	Eruption type
05-02-04 (Benda, Makalia)	5.06S 150.11E	10 x 13	Compr	Strato	R = 59? C = 67?	1,150?	ca. 1890	---- ---- ---- - --	ex,lf

TECTONIC SETTING

Dakataua Caldera is located at the northern end of the Talasea Peninsula, which projects northward about 50 km into the Bismarck Sea from the north coast of New Britain Island, Papua New Guinea (fig. 5.5.1 in section on Witori Caldera). The volcanoes of New Britain and the island chain to the west, including Karkar and Long Island, comprise the Bismarck volcanic arc (Lowder and Carmichael, 1970).

GEOLOGIC HISTORY

Dakataua Caldera is roughly elliptical, 10 km x 13 km in diameter, with a volume of collapse of about 75 km^3. Unpublished preliminary work by C. Pain and R. Blong suggests that the (latest) collapse occurred about 1,150 yr B.P. The caldera has been partly filled by eruptions from an outer ring fracture and a central composite andesitic volcano, Mt. Makalia, and it contains a freshwater lake at least 120 m deep (fig. 5.4.1). Inner ring fractures are indicated by fault-bounded internal remnant blocks and by arcuate coalescing explosion craters (Lowder and Carmichael, 1970).

HISTORICAL ACTIVITY

ca. 1890: Makalia last erupted in the late nineteenth century, producing an andesitic lava flow from a summit crater. Warm springs issue from two sites at the base of Mt. Makalia (fig. 5.4.1; Branch, 1967), and a small amount of fumarolic gas, mostly steam, issues from one of three cinder cones formed in the summit crater during the last eruption (Lowder and Carmichael, 1970).

Lowder and Carmichael (1970) noted marine shells near the town of Talasea as much as 10 m above present sea level. High-water marks around the shoreline of Lake Dakataua indicate lake-level fluctuation of at least 1 m--easily explained by seasonal variation in rainfall. Lake Dakataua is effectively two lakes, separated by shallow water at Lalala (north-central part of the lake). Local people told Ball and Glucksman (1980) that "between Doko and Lalala the water has been lower and they have seen an island which they call 'Merara'." This is corroborated by aerial photographs taken in 1943 that show Lalala connected to the peninsula to the south,

DAKATAUA (continued)

HISTORICAL ACTIVITY (continued)

an area covered by water in 1974 (Ball and Glucksman, 1980). Apparent changes in water level can be caused by changes in water volume, erosion (as believed by Ball and Glucksman), or deformation of the caldera floor like that noted at a number of other calderas in this compilation.

COMMENTS

A careful study of strandlines and shorelines around the north, east, and west sides of Dakataua Caldera is needed to assess whether any volcanogenic deformation has occurred there, superimposed on what is known to be a high rate of regional tectonic uplift.

REFERENCES

Ball, E.E., and Glucksman, J., 1980, A limnological study of Lake Dakataua, a large caldera lake on West New Britain, Papua New Guinea, with comparisons to Lake Wisdom, a younger nearby caldera lake: Freshwater Biol., v. 10, p. 73-84.

Branch, C.D., 1965, Volcanic activity at Lake Dakataua Caldera, New Britain: Australia Bur. Min. Res. Rec. 65/67, 8 p. (not seen).

----- 1967, Short papers from the Volcanological Observatory, Rabaul, New Britain: Bur. Min. Res., Australia, Rep. 107, p. 21-25.

Fisher, N.H., 1957, Catalogue of active volcanoes of the world including solfatara fields, Pt. V, Melanesia: Rome, IAVCEI, 105 p.

Lowder, G.G., and Carmichael, I.S.E., 1970, The volcanoes and caldera of Talasea, New Britain: geology and petrology: Bull. Geol. Soc. Am., v. 81, p. 17-38.

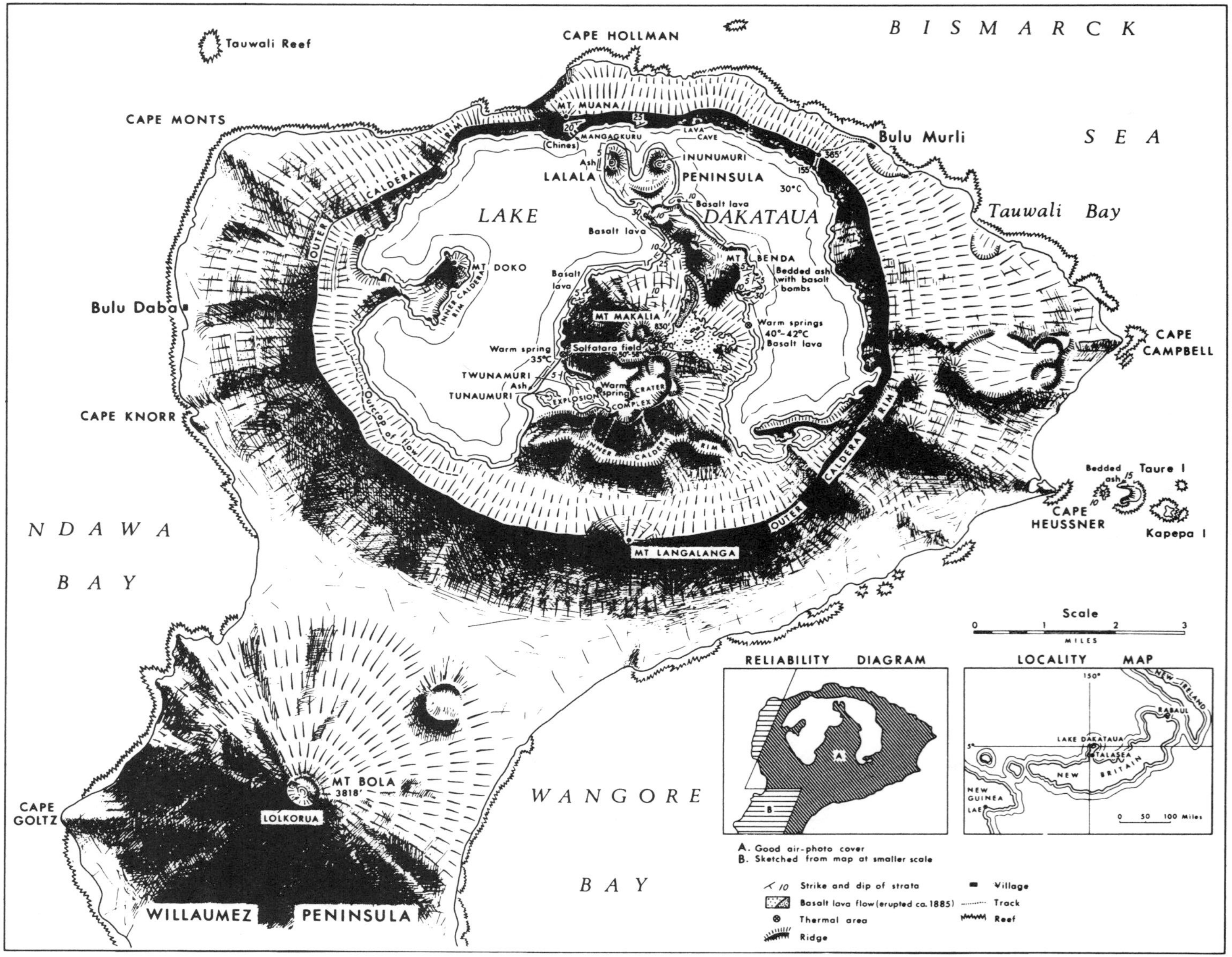

Figure 5.4.1. Sketch map of Dakataua Caldera, from Branch (1967), showing caldera rim, Lake Dakataua, Mt. Makalia, and locations of thermal springs.

WITORI

CAVW number (active vent)	Latitude Longitude (degrees)	Diameter (km)	Local tectonic setting	Precaldera edifice	SiO_2 content (wt pct)	Age of collapse (yr B.P.)	Date of unrest	Type and duration of unrest ESTU STHF MGTF H Te	Eruption type
05-02-08	5.58S	7.5 x 5.5	NNW-SSE	LP-strat	R = 66	2,600	1900?	---- ---- ---- - --	uncertain
(Pago)	150.52E		compr		C = 68		1911-33	N--- ---- ---- - --	ex,lf

TECTONIC SETTING

Witori Volcano is one of 11 Quaternary volcanoes in the Cape Hoskins area of New Britain Island, which forms part of the Bismarck volcanic arc (fig. 5.5.1; Blake, 1976).

GEOLOGIC HISTORY

Witori is a broad, low-angle cone about 700 m high with a well-preserved central caldera that contains an active cone, Pago (fig. 5.5.1). Pago consists of dacitic lava flows, domes, and pumiceous tephra. The flanks of Witori are made up almost entirely of pyroclastic deposits formed predominantly of dacitic pumice, but some andesitic lava flows are exposed in the walls of the caldera and in valleys on the upper southwest flank. The present caldera, which probably formed during a catastrophic eruption about 2,600 years ago, also contains a remnant of a dacitic dome, several pumice mounds, and narrow pumice ridges extending inward from the caldera wall (Blake, 1976).

HISTORICAL ACTIVITY

1900: Uncertain eruption; accounts of an eruption "near the turn of the century" may refer to the eruption of 1911-33.

1911-33: The only information we have about precursors to the 1911-33 eruption are of a negative sort -- that "the eruption was not accompanied by strong earth tremors" (Blake and Bleeker, 1970). The eruption produced a dacitic lava flow that originated at Pago and ponded on the caldera floor (fig. 5.5.2).

WITORI (continued)

REFERENCES

Blake, D.H., 1976, Pumiceous pyroclastic deposits of Witori Volcano, New Britain, Papua New Guinea, in Johnson, R.W., ed., Volcanism in Australasia: Amsterdam, Elsevier, p. 191-200.

Blake, D.H., and Bleeker, P., 1970, Volcanoes of the Cape Hoskins area, New Britain, Territory of Papua New Guinea: Bull. Volcanol., v. 34, p. 385-405.

Blake, D.H., and Ewart, A., 1974, Petrography and geochemistry of the Cape Hoskins Volcanoes, New Britain, Papua New Guinea: Jour. Geol. Soc. Australia, v. 21, no. 3, p. 319-331.

Cooke, R.J.S., 1981, Eruptions at Pago volcano, 1911-1933, in Johnson, R.W., ed., Cooke-Ravian volume of volcanological papers: Geol. Surv. Papua New Guinea Memoir 10, p. 135-146.

Fisher, N.H., 1957, Catalogue of active volcanoes of the world including solfatara fields, Pt. V, Melanesia: Rome, IAVCEI, 105 p.

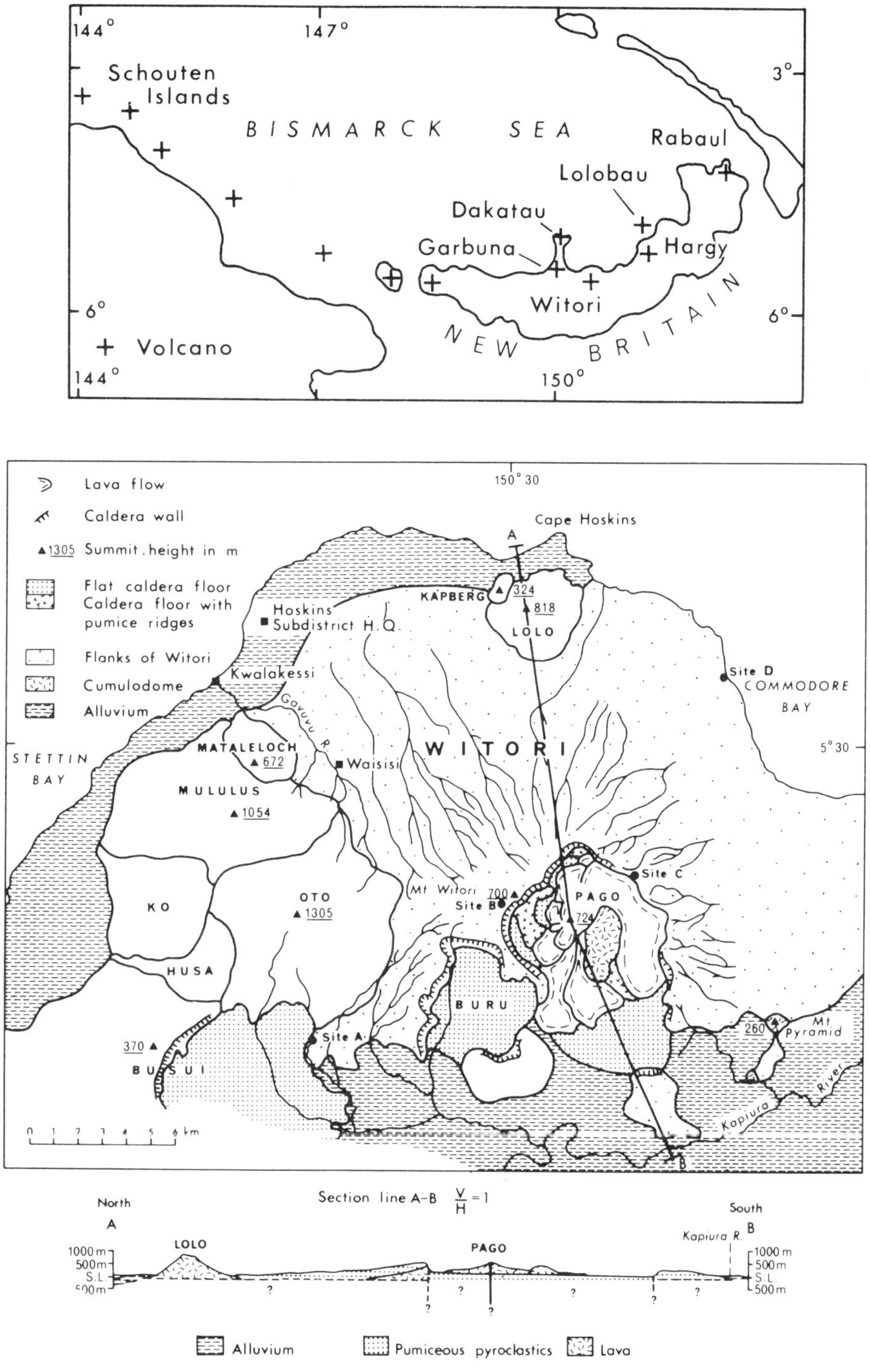

Figure 5.5.1. Locality map (top) and geologic map and cross section (bottom) of Witori Caldera, New Britain, from Blake (1976).

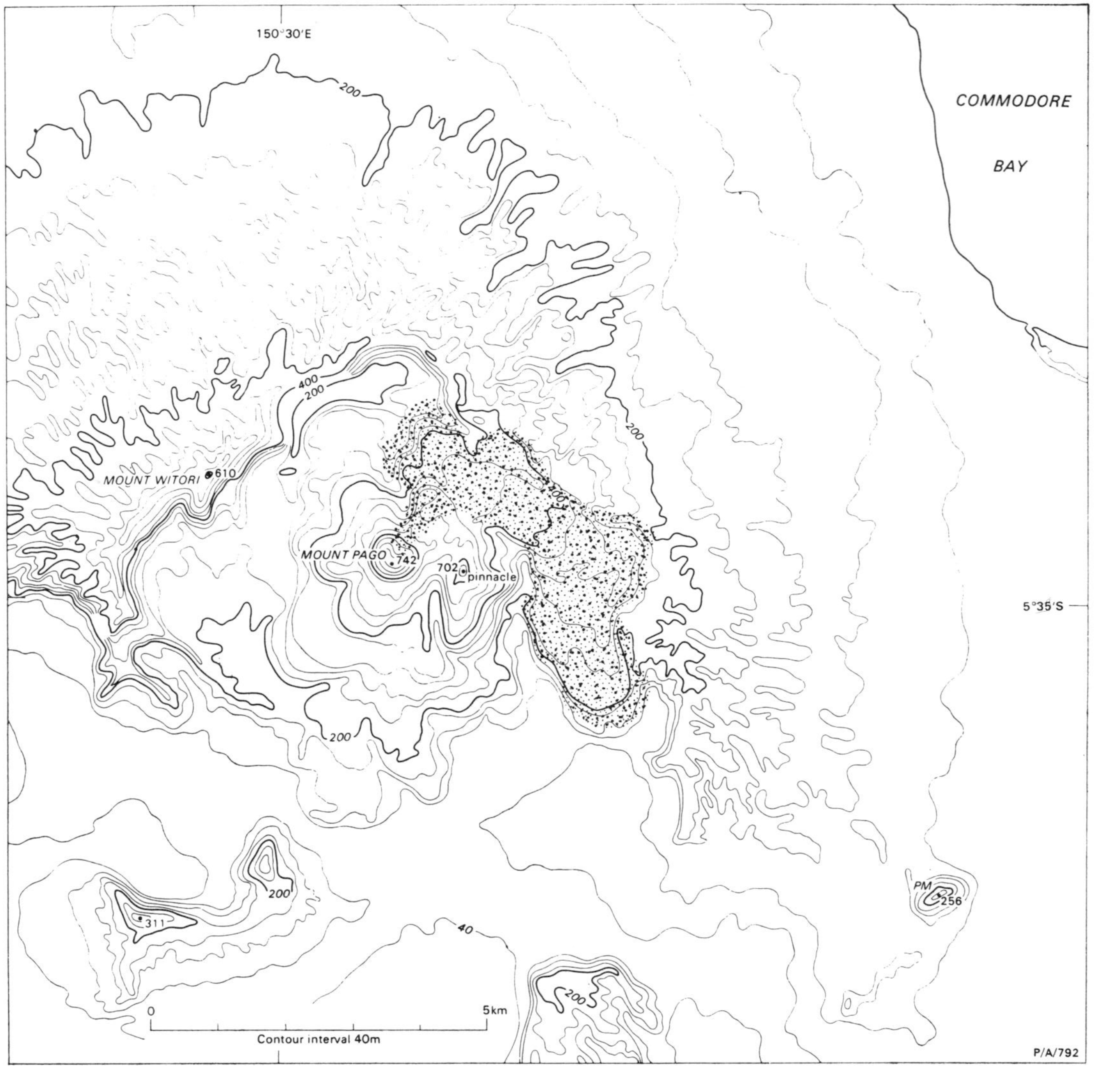

Figure 5.5.2. Topographic map of Witori Caldera, from Cooke (1981), showing location of dacitic lava flow (stippling) extruded from central vent, Pago, during 1911-18.

LOLOBAU ISLAND

CAVW number (active vent)	Latitude Longitude (degrees)	Diameter (km)	Local tectonic setting	Precaldera edifice	SiO_2 content (wt pct)	Age of collapse (yr B.P.)	Date of unrest	Type and duration of unrest ESTU STHF MGTF H Te	Eruption type
05-C2-13 (Lolobau)	4.93S 151.17E	5.5	Compr	Strato	R = d C = silicic	?	1905	---- ---- ---- - --	lf,ex

TECTONIC SETTING

Lolobau is a volcanic island off the north coast of New Britain Island, part of the Bismarck volcanic arc. Lolobau, Likuruanga Volcano, and several associated satellite cones lie on an east-west line canted about 60° to the chain of volcanoes on the New Britain mainland (Johnson, 1970).

GEOLOGIC HISTORY

The caldera of Lolobau Island formed in the summit of a dominantly andesitic stratovolcano (fig. 5.6.1), and at least four vents have been active in postcaldera time. Sulfur deposits occur in Lolobau Crater at the summit of postcaldera Lolobau Mountain (Johnson, 1970).

HISTORICAL ACTIVITY

1905: An eruption in 1905 produced ash and a dacitic lava flow. Solfatara activity in Lolobau Crater decreased notably between 1937 and 1950 (Fisher, 1957) and stopped entirely by 1969 (Johnson, 1970).

REFERENCES

Cooke, R.J.S., and Johnson, R.W., 1978, Volcanoes and volcanology in Papua New Guinea: Geol. Surv. Papua New Guinea Rep. 78/2, 37 p.
Fisher, N.H., 1957, Catalogue of active volcanoes of the world including solfatara fields, Pt. V, Melanesia: Rome, IAVCEI, 105 p.
Johnson, R.W., 1970, Likuruanga Volcano, Lolobau Island, and associated volcanic centres, New Britain: Geology and petrology: Bur. Min. Res. Geol. Geophys., Australia, Rec. 1970/42, 21 p.

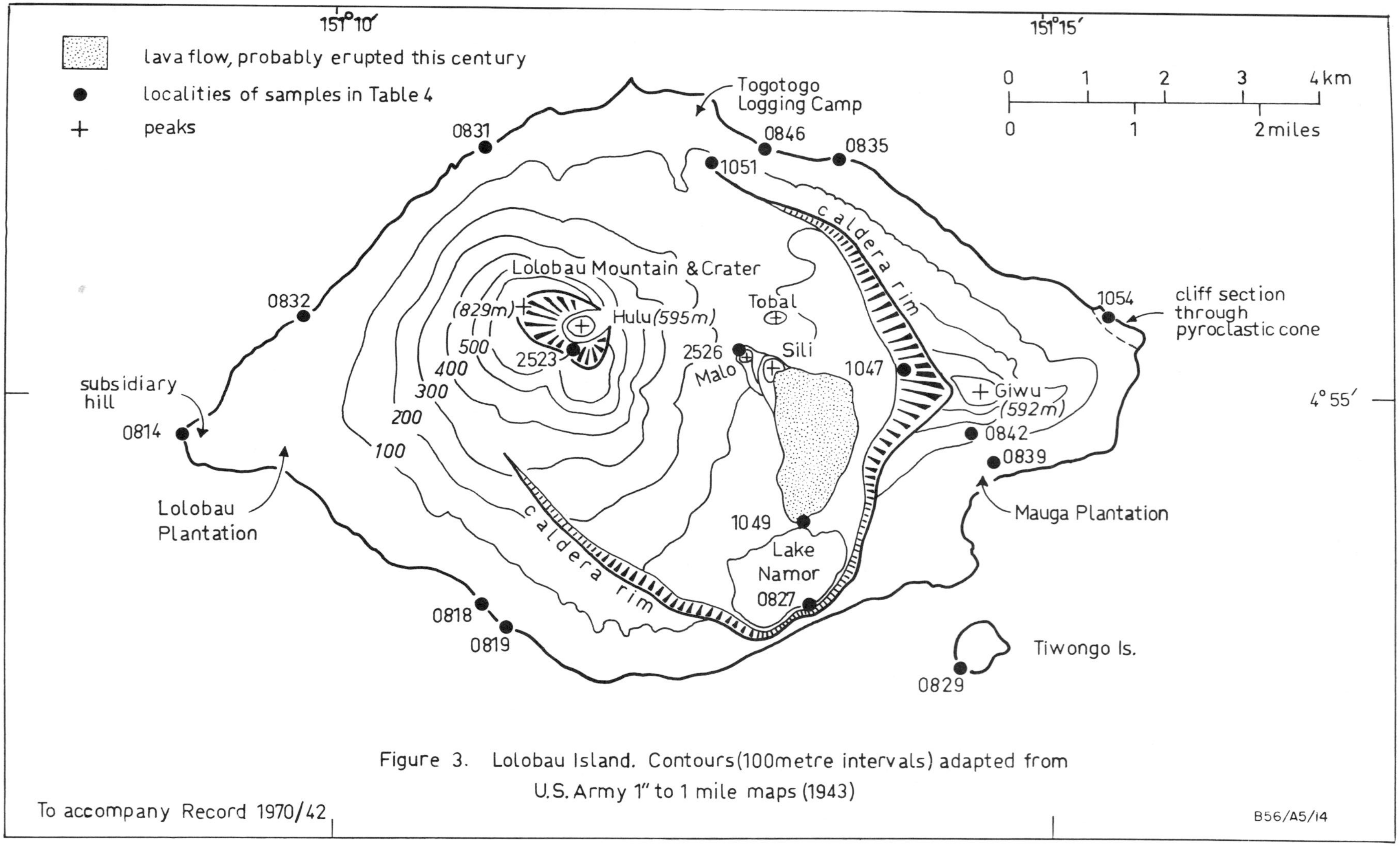

Figure 5.6.1. Sketch map of Lolobau Island, from Johnson (1970), showing caldera rim and 1905 lava flow (stippled).

RABAUL

CAVW number (active vent)	Latitude Longitude (degrees)	Diameter (km)	Local tectonic setting	Precaldera edifice	SiO_2 content (wt pct)	Age of collapse (yr B.P.)	Date of unrest	Type and duration of unrest ESTU STHF MGTF H Te	Eruption type
05-02-14	4.27S	14 x 9	Shear in	LP-strat	R = 49-66	3,500	1767?	---- ---- ---- - --	ex
(Tavurvur,	152.20E		nearby	(3 cones)	C = 64-72	1,400	1791	---- ---- ---- - --	ex
Sulphur Creek,			straits				1850	x--x ---- ---- - --	pex
Rabalanakaia,							1878	B--F B--- --BB B --	Ex,pex
Vulcan)							1910	x--- ---- x--x - =?	none
							1916	x--- x--- ---- - =?	none
							1919-23	C--F ---- ---x - QU	none
							1937	C-CF B--B --BB - --	EX,pex
							1938-43	x--F FD-- --EF - UQ	pex,ex
							1971-present	FF-G EFFE -F-x - QU	none yet

TECTONIC SETTING

The Bismarck volcanic arc begins west of Manam off the north coast of New Guinea and terminates at Rabaul Caldera. Rabaul lies close to a postulated junction between three crustal plates and is also near a transform fault separating the Pacific and Bismarck Sea plates (fig. 5.7.1, inset). Shallow earthquakes of large magnitude are concentrated along St. George's Channel, and earthquakes of intermediate foci occur north, west and south of Rabaul (Heming, 1974). A large number of earthquakes occur at a trench and along an inclined seismic zone south, southwest, and southeast of Rabaul (J. Mori, written commun., 1988).

GEOLOGIC HISTORY

Rabaul Caldera (figs. 5.7.1, 5.7.2) has been the source of many voluminous explosive eruptions in the past few hundred thousand years. Reliable dates are available for two such eruptions within the past 3,500 years. A caldera-modifying eruption about 3,500 yr B.P. produced primarily rhyolite but may have been triggered by an injection of basalt (Walker and others, 1981). A large eruption of dacitic magma roughly 1,400 yr B.P. may have resulted in further caldera collapse, perhaps of an inner caldera inferred from bathymetry and from recent seismicity. Postcaldera volcanics at Tavurvur, Sulphur Creek, and Rabalanakaia range from basalt to dacite; those at Vulcan and Davapia Rocks are dacite. The vent for the 3,500 yr B.P. eruption may have been in the north or northeast part of the present caldera (J. Mori, written commun., 1988); the vent for the 1,400 yr B.P. eruption may have been near the center of the caldera (Walker and others, 1981). Both the 3,500 and 1,400 yr B.P. eruptions "began with small scale

RABAUL (continued)

GEOLOGIC HISTORY (continued)

explosions... produced small-volume pyroclastic deposits... (and then) activity built up rather quickly to a climax" (Walker and others, 1981). "Rather quickly" in this context means only that there is no geologic evidence for a time gap between the small-scale and larger scale deposits, although Walker speculates that the gaps may have been as short as a few hours. Consideration of repose periods between large explosive eruptions at Rabaul suggests that another might occur in the geologically near future (McKee and others, 1985).

HISTORICAL ACTIVITY

1767?, 1791: The eruption of 1767 was probably at Tavurvur, though possibly also at Rabalanakaia; the eruption of 1791 was at Tavurvur.

1850: The eruption of 1850, at Sulphur Creek, was preceded by an unspecified period of strong earthquakes and several meters of uplift between Sulphur Creek and Matupit Island (McKee, 1981; McKee and others, 1988, citing Brown, 1878).

1878: Uplift over an unknown period before 1878 had exposed reefs in the Vulcan area (McKee and others, 1988, citing Brown, 1878). Strong earthquakes began on about 3 February 1878, causing several tsunamis on the morning of 4 February. Visible uplift also began on 3 February, raising shoals near Vulcan by "several feet" and raising the coastline immediately southeast of Tavurvur by as much as 6 m (McKee and others, 1988, citing Brown, 1878). Where the coast was raised 6 m, the coastline was shifted about 140 m seaward. Simultaneously, Davapia Rocks (within the caldera, 3-5 km north of Vulcan) subsided "several feet," partly submerging houses. Within hours(?) after the tsunamis, a cloud of steam and a "line of fire" (andesitic pumice) rose through the harbor, marking the beginning of what was to become Vulcan Island. Several hours after the Vulcan Island vent (distinct from Vulcan; see fig. 5.7.2) began to erupt, a phreatomagmatic eruption began at Tavurvur, on the opposite side of the caldera (composition of magma = high-silica andesite; J. Mori, written commun., 1988). Uplift continued to occur during the eruption, forming a new strip of land 6 m high and 45 m from the old shore near Escape Bay. Following the eruption, subsidence until 1905 caused Vulcan Island to decrease in size and height.

1910, 1916, 1919-23: A severe earthquake was felt at Rabaul on 24 February 1910, resulting in sulfur odor and compass deflection "of four points northeast to north" (Fisher, 1939, p. 17, citing a German Admiralty document). Another strong earthquake near Rabaul Harbour on 1 January 1916 ($\underline{M}$=7.9, shallow, 4.0°S, 154.0°E; Ripper, 1979) caused the causeway connecting Matupit Island and the Gazelle Peninsula to subside 2 m. Other large tectonic earthquakes occurred nearby in 1919 ($\underline{M}$=8.1, shallow, 5.0°S, 154.0°E), 1923 ($\underline{M}$=7.2, shallow, 4.5°S, 151°E; $\underline{M}$=7.2, shallow, 5.0°S, 152.0°E; Ripper, 1979) and 1933 (Fisher, 1944). A magnitude 7.7 earthquake occurred on 2 February 1920 in St. George's Channel, just northeast of Rabaul (Circum-Pacific Map Project, 1981). On 7 May 1919, the submerged Matupit causeway was elevated by almost 1 m, accompanied by great fumarolic activity at Matupit and Tavurvur. The causeway continued to rise: "...By 1924 it was once again possible to drive a motor car across to Matupit Island and this condition persisted

RABAUL (continued)

HISTORICAL ACTIVITY (continued)

until the 1937 eruption." (Fisher, 1939, p. 18). Uplift of the bay floor also occurred: Fr. Mayrhofer, quoted in Arculus and Johnson (1981), wrote that "In recent years, measurements have shown that the seabed between Vulcan Island and Kaia [on the west foot of Tavurvur] rose perceptibly every year and I often thought [that] if this goes on, one fine day Rabaul, the proud capital and port, will be cut off from the sea." General uplift was accompanied by at least local subsidence; Murray (1937) indicated that "between four and five years ago... a casuarina tree which had been well above high water mark was particularly noticeable as being about 12 inches below water level, and the shoreline seemed to have tilted."

1937: "Tremors" began on 26 or 27 May 1937, and a strong earthquake of local origin (MMI=7 or Rosi-Forel intensity 8) occurred at 1320 hr on 28 May. The intensity of this earthquake was strongest in Karavia, on the west shore of Blanche Bay and immediately west of Vulcan Island; one German missionary's home collapsed, injuring the occupant. Landslides blocked parts of the road from Rabaul to Kokopo, a village school collapsed in Raluana, and cracks opened in the southwest part of Vulcan Island. An account by one missionary (see Arculus and Johnson, 1981, p. 22) says that the seabed was raised 2 m during the earthquake, but this was probably an effect of a small tsunami. All other reports say that rapid uplift began on the following day. After the strong earthquake, no additional earthquakes were felt at Rabaul until the early morning of 29 May, but more or less continuous tremor was felt at Vulcan. Earthquakes became stronger at 5:50 a.m. on 29 May, and from then until the eruption began, they were felt throughout the caldera. During lulls between discrete shocks, a "gentle vibration was maintained"--perhaps volcanic tremor. "For much of the morning it was impossible to write during their maximum periods of vibration" (Fisher, 1939, p. 19). Tsunamis up to several meters high followed the larger earthquakes. Beginning about 8 a.m. on 29 May, Vulcan Island was uplifted by about 2 m and a reef at the southern end of Matupit Island (that had "never" before been exposed) was uplifted by at least 2 m. A reef on the west side of Vulcan Island began to rise at 1:10 p.m., disappeared at 1:15 p.m., and reappeared at 1:20 p.m. "During the afternoon many of the natives from the adjacent village engaged in looking for stranded fish on these exposed reefs, and about twenty of them are reported to have lost their lives when the eruption occurred" (Fisher, 1939, p. 19). At 12:45 p.m. on 29 May, narrow, northeast-southwest trending cracks up to 15 m long developed on Matupit Island. Also noted in the hours before the eruption were deep rumbling noises and a marked increase in gas emissions, especially gases bubbling through the sea water. The eruption of Vulcan began at 4:15 p.m. on the 29th and produced 3 x 10^8 m^3 of dacitic tephra. Pyroclastic flows caused approximately 500 deaths. The eruption from Tavurvur was smaller and exclusively phreatic, but it produced mud rains that caused considerable damage. Two new vents opened along the east caldera boundary, just north and south of Tavurvur; these emitted only a small volume of debris but a large volume of CO_2, killing animals in the vicinity for several months thereafter. Earthquakes ceased when the eruption began, but uplift continued to expose new land during the eruption. Posteruption subsidence occurred along the shore between Sulphur Creek and Matupit Island.

1938-43: SO_2 and HCl emissions at Tavurvur began to increase in late 1938 and increased further in 1940. Small steam explosions occurred in a mud pool at the bottom of Tavurvur crater on 2, 12, and 16 March 1940, and a strong tectonic earthquake occurred near Rabaul on 12 September 1940. In December 1940, temperatures of Tavurvur fumaroles (fig. 5.7.3) rose from 100 °C at a rate of 10 °C/day for two weeks, then rose an additional 22-28 °C during the morning before a large earthquake (M 7.0) on 14 January 1941 (13 January G.m.t.), centered approximately 30 km west of the caldera (4.5°S, 152.5°E; Ripper, 1979). Tilting at a rate of 1 sec (about

RABAUL (continued)

HISTORICAL ACTIVITY (continued)

5 microradians) per day, upward to the west-southwest, was observed at Rabaul Volcano Observatory (RVO) for one month before the earthquake (Fisher, 1944); the tilt direction was toward the inferred epicenter of the earthquake, not toward the center of the caldera. Fisher (1976) states that seismicity before the 1941 eruption was "tectonic" with certain volcanic connections. Temperatures of Tavurvur fumaroles continued to rise after the 14 January earthquake, reaching 400 °C by 3 June 1941. During the time of this remarkable temperature increase, amounts of acid gases (HCl and SO_2) also increased (Fisher, 1976). No change was noted in fumaroles along the Matupit Harbour (Greet Harbour) waterfront, 2 km away. A dominantly magmatic (dacitic) eruption began at Tavurvur on 6 June 1941 with small steam explosions. Ground deformation before the eruption was conspicuously minor. Tide gauges indicated 7.5 cm of subsidence of Tavurvur relative to Rabaul town during the 2 years preceding the eruption; there is no report of short-term uplift before the eruption. The tiltmeter at RVO and another located 1 km north of Tavurvur were sensitive to 1 arcsecond (5 microradians) and showed no tilt before the eruption. No immediate seismic precursor to this eruption was noted at RVO, using a seismometer with only 150x magnification. Minimal ground deformation and seismicity and the occurrence of a major tectonic(?) earthquake 6 months before the eruption suggest the possibility that the earthquake "short-circuited" the normal precursory buildup by easing the passage of magma to Tavurvur. Earthquakes were frequent between June 1941 and March 1942, then absent until another swarm in October-December 1943.

1967: On 14 August 1967, earthquakes of magnitude 6.2, 5.7, and 6.4 occurred beneath St. George's Channel, 35 km east of Rabaul. They shook Rabaul town and especially Kokopo town on the coast southeast of Rabaul. The quakes, with a maximum MMI of 7-8, were of tectonic origin and caused no immediate change in the Rabaul volcanoes (Heming, 1969; Denham, 1971). Unlike the 1941 earthquake, there was no noticeable effect on tiltmeters before or after these earthquakes.

1971-present: Seismic swarms and broad uplift of the caldera floor resumed in late-1971. Two $\underline{M}_L$=8.0 earthquakes occurred in the Solomon Sea in July 1971, 130 and 230 km from Rabaul. Seismic activity at Rabaul began to increase in November 1971 (fig. 5.7.4). Prior to November 1971, approximately 60 earthquakes/month were recorded at Rabaul (Everingham, 1975). That number increased to 220 earthquakes/month from November 1971 to August 1980, then to 320 earthquakes/month from August 1980 to August 1983. A dramatic increase in seismic activity in late 1983 (fig. 5.7.5, and described below) may have been linked to a major tectonic earthquake ($\underline{M}$=7.6, 200 km east of Rabaul) in March 1983. Activity generally declined from November 1984 to December 1985 (pre-1971 levels), increased slightly from early to mid-1986, and returned to precrisis levels through late 1986 and early 1987. Epicentral distributions of earthquakes from September 1983 through July 1985 define a striking pattern along a steeply outward-dipping ring-fracture system of the inner, 1,400 yr B.P. caldera (McKee and others, 1984; Mori and McKee, 1987) (figs. 5.7.6, 5.7.7). The largest local earthquakes from November 1971 to May 1984 occurred during swarms in September-October 1980 ($\underline{M}_L$ 5.2), January-March 1982 ($\underline{M}_L$ 5.1), and March 1984 ($\underline{M}_L$ 5.1). The total seismic energy release for the crisis period (September 1983 - July 1985) was 2.5 x 10^{19} ergs (Mori and others, 1986a).

Leveling from September 1973 to July 1983 showed about 1.0 m of uplift at the southern tip of Matupit Island (figs. 5.7.8, 5.7.9), 1 km northwest of an uplift center inferred from patterns of ground tilt to be beneath the mouth of Greet Harbour (fig.

RABAUL (continued)

HISTORICAL ACTIVITY (continued)

5.7.10). The average uplift rate from August 1983 through July 1985 was about 7 cm/month, compared to 0.8 cm/month for the preceding decade. Uplift extended out to the caldera boundaries but was most pronounced within 2-3 km of the inferred deformation center.

The maximum tilt rate measured before September 1983 was 4-8 microradians/month near the southeast coast of Matupit Island. A sharp increase in tilt rates on 19 September 1983 was preceded by one month of slightly higher rates. A considerable amount of the cumulative tilt occurred during rapid deformation events that accompanied seismic swarms--for example, about 50 microradians at stations about 1 km from the deformation source during a seismic crisis on 3 March 1984. In addition, progressive tilting at 25-30 microradians/month occurred around Greet Harbour. New tilt stations installed in October 1983 suggest a second deformation source immediately east of the Vulcan headland (fig. 5.7.10). Horizontal deformation measurements using a laser EDM (electronic distance meter) began in December 1983. Fairly steady, radially symmetric expansion of about 25 microstrain (parts per million, ppm) per month was observed at the mouth of Greet Harbour from 1 December 1983 through 1 March 1984 (fig. 5.7.11). Rapid changes of 30-40 microstrain accompanied seismic swarms on 3 March and 22 April 1984 (McKee and others, 1984).

Tide-gauge data at the Rabaul main wharf from 1969 through 1986 and tide staff data from Matupit Island from mid-1981 to mid-1983 showed that (1) Rabaul town proper has been stable with respect to sea level, and (2) the leveling results from Matupit are corroborated by the sea-level measurements. Tide-staff data from near Vulcan for the same period of mid-1981 to mid-1983 showed relatively little uplift (Mori and others, 1987).

Modeling of leveling data indicates that the focal depth of the Matupit source has been steady at about 1.8 km (McKee and others, 1988). This suggests that the magma body responsible for the deformation inflated but did not rise significantly (McKee and others, 1985, 1988). The estimated volume of deformation from 1971 to 1984 was about 50 million m^3 (0.05 km^3) for the Matupit (Greet Harbour) source (McKee and others, 1984). The relationship between gravity and uplift was -216 microgal/m, in contrast to the normal free-air gradient of -309 microgal/m. Thus an increase in subsurface mass has been inferred, presumably a magmatic intrusion (fig. 5.7.12) (McKee and others, 1988).

Marine geophysical studies (Greene and others, 1986) suggest a N-S fault zone at the crest of a N-S ridge of uplift (fig. 5.7.13). North-south elongation of uplift is consistent with the N-S elongation of the ring of recent seismicity (figs. 5.7.6, 5.7.7) and the inferred boundary of the 1,400 yr B.P. collapse but not wholly consistent with geodetically inferred centers of deformation in Greet Harbour and Karavia Bay (figs. 5.7.8-5.7.11). A faulted anticline at the southern shore of Matupit Island is probably related to the offshore structures. The folded and faulted rocks on Matupit Island are overlain by sea-rafted pumice deposits containing shells and coral, perhaps suggesting resurgence. Structures defined by the marine geophysical surveys probably predate the current unrest and may provide a new structural context for interpretation of recent changes.

CO_2 emissions occurred in an old crater on the south flank of Tavurvur in June and July 1981, killing a number of birds. Minor new fumarolic activity 1983 killed vegetation in a small area 2 km northwest of Tavurvur (Mori and others, 1986).

RABAUL (continued)

HISTORICAL ACTIVITY (continued)

Owing to the unusual intensity of the unrest at Rabaul and the high potential hazard there, monthly summaries of activity from August 1983 through February 1987 are given below (Smithsonian Institution, 1982-87, based on reports from C. McKee, P. Lowenstein, P. de Saint Ours, B. Talai, and J. Mori; update by J. Mori, written communication, 1988):

August 1983: Sharply increased activity began at Rabaul, possibly linked to a magnitude 7.6 earthquake that occurred 200 km to the east in March 1983.

September 1983: More than 2,135 earthquakes within the caldera, 621 of which occurred in a swarm on 19 September. The previous monthly maximum was 1,170 events in January 1982. Maximum $\underline{M}_L$ = 4.2; seismic energy release for September= 3.8 x 10^{17} ergs.

October 1983: 5,198 earthquakes recorded, including swarms on 15-16 October and 28-29 October with 1,173 and 851 earthquakes, respectively. 1.2 x 10^{18} ergs. Slight ground cracking observed on the west flank of Tavurvur after the swarms. Approximately 15-50 microradians of coseismic tilt during swarms in September and October 1983, superimposed on a background rate of about 20 microradians/month. Stage 2 eruption alert (eruption possible within months) declared on 29 October.

November 1983: 5,751 earthquakes, 3.2 x 10^{16} ergs.

December 1983: 7,117 earthquakes, 8.6 x 10^{16} ergs. Earthquakes more numerous than in previous months, but energy release less than in September and October 1983.

January 1984: 8,372 earthquakes, including 942 on 15 January, one $\underline{M}_L$ 4.6, and several others strongly felt around the caldera. Energy release 1.7 x 10^{18} ergs. Steady inflation of Karavia Bay and Greet Harbour magma reservoirs.

February 1984: 8,339 earthquakes, including several relatively strong events (maximum $\underline{M}_L$ 4.3). 1.7 x 10^{18} ergs. Increased rate of tilt around Vulcan (maximum 40 microradians, 4 times that of previous month). Tilt rates simultaneously decreased around Greet Harbour. Tilt pattern more complex than in 1983 and before. EDM measurements across Greet Harbour showed 60 ppm extension from December 1983 to February 1984.

March 1984: 8,729 earthquakes, including swarms of 932 and 726 events (maximum $\underline{M}_L$ 5.1). 1.9 x 10^{19} ergs. Increase in both the number and size of earthquakes. Increased incidence of slightly deeper (2-4 km) and more energetic earthquakes in the Vulcan area. During the larger seismic crisis on 3 March, inflationary tilt of up to 50 microradians took place, and approximately 40 ppm of extension was measured across Greet Harbour. Extension rate then returned to its previous level of 25 ppm/month. No significant deformation observed during the small seismic crisis on 25 March, but waterspouts up to 3 m high were observed near the northeast shore of Vulcan.

RABAUL (continued)

HISTORICAL ACTIVITY (continued)

April 1984: 13,749 earthquakes, including 1,011 and 1,717 in swarms on 21 and 22 April, respectively (maximum $\underline{M}_L$ 4.8, total energy release 2.7 x 10^{18} ergs). Up to 80 microradians of tilt in Greet Harbour area, including 30-50 microradians of coseismic tilt on 22 April. Likewise, 40-50 ppm of extension, including 20-30 ppm during the 21-22 April swarms. Rate of uplift increased sharply (7.6 cm at the southern tip of Matupit Island) during a four-week period from late March to late April.

May 1984: 8,938 earthquakes (decline), with one swarm near Vulcan (maximum $\underline{M}_L$ 3.6, total energy release 1.4 x 10^{17} ergs). Until a seismic and deformation crisis occurred on 29 May, the caldera showed mild inflation (10-20 microradians of inflationary tilt southeast of Vulcan and continued uplift of the southern tip of Matupit Island (4.1 cm in the first three weeks of May) and deflation of up to 40 microradians with slight subsidence (few millimeters) near Rapindik at the head of Greet Harbour. During the earthquake swarm of 29 May, unusual deflationary coseismic tilt of up to 30 microradians occurred around Vulcan, and apparently aseismic inflation of the same magnitude occurred at the head of Greet Harbour. Simultaneous changes on both sides of the caldera were reminiscent of simultaneous activity at Vulcan and Tavurvur in 1878 and 1937, though different in detail.

June 1984: 5,304 earthquakes, with a total seismic energy release of 1.9 x 10^{15} ergs (further decline). Tilt around Greet Harbour slowed (maximum 20 microradians); E-W extension across Greet Harbour continued at 25 ppm/month but north-south extension slowed. Uplift of southern tip of Matupit Island was 4.3 cm during 23 May - 6 June but only 1.0 cm during 6-26 June. No significant coseismic deformation. Slight deflationary tilt in Vulcan/Karavia area.

July 1984: 4,404 earthquakes, 1.3 x 10^{17} ergs. Extension across Greet Harbour continued at 20-25 ppm/month; tilt changes generally smaller than in previous months, mostly less than 8 microradians except at the northwest shore of Greet Harbour, where there was an inflationary tilt of 25 microradians. From late June through late July, 3.8 cm of uplift at the southern tip of Matupit Island.

August 1984: 5,285 earthquakes, 2.4 x 10^{16} ergs. Extension across Greet Harbour 10-20 ppm; relatively small tilt changes. Maximum measured uplift at Matupit Island: 3.5 cm.

September 1984: 4,048 earthquakes, 8 x 10^{14} ergs. Continued slow inflation beneath Greet Harbour. Uplift at southern tip of Matupit Island: 1.4 cm.

October 1984: 6,749 earthquakes, 5.5 x 10^{18} ergs. Brief increase in the monthly number of earthquakes, with 90 percent of energy release in swarm on 18 October (maximum $\underline{M}_L$ 4.9). Very localized uplift (10 cm), tilt (90 microradians), and northeast displacement of Sulphur Point (55 mm). Other deformation parameters: continuing, slow inflation.

RABAUL (continued)

HISTORICAL ACTIVITY (continued)

November 1984: 3,985 earthquakes, 6.0 x 10^{15} ergs, maximum $\underline{M}_L$ 2.9. Continuing slow inflation of both shallow magma reservoirs (Greet Harbour and immediately east of Vulcan). In Greet Harbour and at Matupit Island, 30 microradians of tilt, 20 ppm extension, and maximum uplift of 19 mm. Total volume increase: 1 x 10^6 m^3. Level of alert reduced to Stage 1 (eruption not anticipated before several months to a few years).

December 1984: 2,887 earthquakes, 1.3 x 10^{15} ergs, maximum $\underline{M}_L$ 2.5. Only 20 microradians of tilt at Greet Harbour.

January 1985: 1,297 earthquakes, 4.4 x 10^{15} ergs, maximum $\underline{M}_L$ 2.3.

February 1985: 1,672 earthquakes, 3.4 x 10^{15} ergs; maximum $\underline{M}_L$ 3.0. For both January and February, maximum of 15 microradians/month of tilt, 10 ppm of extension. Only 1.5 cm of uplift at the southern tip of Matupit Island between early December and late February, only slightly greater than the precrisis rate from 1971-83.

March 1985: 2,042 earthquakes, 6.9 x 10^{16} ergs, maximum $\underline{M}_L$ 3.6. Slight increase in seismicity with epicenters concentrated near Davapia Rocks and Rabalanakaia. Matupit Island - Greet Harbour area continued to be focus of ground deformation; maximum tilt: 30 microradians at Sulphur Point; leveling showed that the southern tip of Matupit Island rose 4.5 cm from late January to mid-March. Maximum horizontal extension: 10 ppm.

April 1985: 1,041 earthquakes, $<10^{14}$ ergs, maximum $\underline{M}_L$ 2.1. Tilt barely above noise level; maximum horizontal extension approximately 10 ppm.

May 1985: 723 earthquakes, 5.8 x 10^{16} ergs, >100 in response to $\underline{M}_L$ 7.1 earthquake in West New Britain on 10 May. Largest Rabaul earthquake: $\underline{M}_L$ 3.7 about one hour after a large regional earthquake. Little measurable ground deformation.

June 1985: 639 earthquakes, no strong concentration. Total seismic energy release $<10^{14}$ ergs, most associated with 10 May swarm. Continued slow (10 ppm/month) extension across Greet Harbour.

July 1985: 595 earthquakes, maximum $\underline{M}_L$ less than 2, seismic energy release $<10^{14}$ ergs. Local subsidence in response to a $\underline{M}_L$ 7.1 earthquake that occurred 80 km from Rabaul, but no apparent effect on the volcanic system.

August 1985 - March 1986: Seismicity at precrisis levels (between 48 and 317 earthquakes/month, average 166). Only a few earthquakes were large enough to be located, and none were of $\underline{M} \geq 2$ (C.O. McKee, in Smithsonian Institution, 1987). Seismic energy release in all months $<10^{14}$ ergs. Deformation was also at precrisis levels. Leveling in December 1985 showed the first consistent subsidence since the leveling program began in 1973 (1.4 cm subsidence from early October to early December). Minor inflation noted in February and March in Greet Harbour and Vulcan areas.

RABAUL (continued)

HISTORICAL ACTIVITY (continued)

April-July 1986: Renewed inflation and seismicity. In April, 1,768 earthquakes (highest monthly total since March 1985), maximum M_L 2.7. No swarms, but consistently higher daily earthquake counts. Counts in May, June, and July were 1,532, 1,570, and 1,312 per month; corresponding monthly energy release= 1.3 x 10^{15} ergs (April), 7.9 x 10^{14} ergs (May), 2.4 x 10^{14} ergs (June), and 5.4 x 10^{14} ergs (July). Most earthquakes in April were located near center of caldera and at less than 1 km depth, unlike most previous events located along ring-fracture zone at 1-3 km depth; however, events in May returned to the ring-shaped pattern. Increases in tilt rate (10-12 microradians/month), extension (10-15 ppm across Greet Harbour), and uplift (1.6 cm from 11 April to 7 May, and 0.9 cm from 7-23 May). Deformation returned to low rates in June and to precrisis rates in July. Deformation during this period can be modeled as a 1.5 x 10^6 m^3 inflation at 1.9 km depth about 1 km southeast of Matupit Island, with maximum uplift of 6.7 cm (C.O. McKee, in Smithsonian Institution, 1987).

August 1986 - February 1988 (time of writing): Low seismicity (<300 earthquakes/month and <10^{14} ergs/month except November 1986, which had 4.7 x 10^{15} ergs of seismic energy release). Low rates of ground deformation. Comparison of October 1986 and February 1987 leveling data suggested as much as 1 cm of subsidence of Matupit Island. Uplift of about 10 mm/month in June and July reverted to a comparable rate of subsidence in August 1987. As of the time of this writing (February 1988) seismicity and rates of deformation were very low (<100 earthquakes/month and no significant changes in tilt or horizontal distance measurements).

COMMENTS

The remarkable episode of unrest at Rabaul Caldera that began in 1971 is perhaps the most threatening example in this compilation. From 1971 until mid-1984 the intensity of seismic swarms generally increased with time, the maximum earthquake magnitudes stayed roughly constant or increased, and the rate of seismic energy release increased. Volcanologists at the Rabaul Volcano Observatory were justifiably concerned about the possibility of an eruption in the near future; the fact that none has yet occurred is a testimony to the complexity of large magma reservoirs and evidence that some reservoirs can accommodate a substantial intrusion of magma without erupting.

Another fact that has been graphically illustrated at Rabaul since 1971 is that some large magma reservoirs are subject to relatively minor external disturbances--in this case, shaking by regional tectonic earthquakes or regional tectonic strain changes. The susceptibility of the Rabaul volcanic system to minor strains is underscored by a strong negative correlation between diurnal ocean tides and volcanic earthquakes at Rabaul (Lowenstein, 1986). Low tides decrease load and seemingly trigger volcanic earthquakes. The difference in pressure from high to low tide is about 0.1 bars, or 0.025 bars of shear stress on the steeply dipping fault outlined by Mori and others (1986a, 1988), so in some cases very small stress changes can induce failure.

Caldera-forming eruptions only 3,500 and 1,400 years ago, simultaneous eruptions in 1878 and 1938-43 at Vulcan and Tavurvur (on opposite sides of the caldera ring fracture), and the annular pattern of seismicity since 1971 all raise concern over the possibility of another catastrophic eruption in the foreseeable future. Large silicic systems like Rabaul pose a difficult challenge -- frequent

RABAUL (continued)

COMMENTS (continued)

unrest at such systems is likely to cause many false alarms, but the sizes and consequences of possible eruptions are so great that each episode of unrest must be taken seriously, often at considerable social cost.

Lastly, volcanologists and public officials in Rabaul have prepared a clear and well-considered plan that defines four levels of volcanic unrest and corresponding alerts to be given to the general public. Stage 1 is a low level of unrest in which an eruption is possible within years to months. Because no immediate eruption is expected, no public announcements or public response is made. Stage 2 is significantly elevated activity, with tens to hundreds of earthquakes per day and measurable ground deformation; an eruption is judged to be likely within months to weeks. Quiet, low-key precautionary measures are taken. Stage 3 is an alarmingly high level of unrest, with several hundreds to thousands of earthquakes per day and marked ground deformation. In Stage 3, an eruption is expected within weeks to days, public announcements are given, and people are readied for evacuation. Stage 4 is critical, with thousands of earthquakes per day and rapid swelling of the ground, indicating that an eruption might begin within days to hours. At Stage 4, an evacuation is ordered. Stage 2 was invoked during a crisis period from October 1983 to November 1984; fortunately, unrest declined and as of early 1988, Stages 3 and 4 have not been required.

REFERENCES

Arculus, A., translator, and Johnson, R.W., compiler, 1981, 1937 Rabaul eruptions, Papua New Guinea: Translations of contemporary accounts by German missionaries: Bur. Min. Res. Geol. Geophys., Australia, Rep. 229, BMR Microform MF 158, 78 p.

Brown, G., 1878, Journal of the Rev. G. Brown, 1860-1902: Sydney, Mitchell Library, 2 v.

Circum-Pacific Map Project, 1981, Plate tectonic map of the circum-Pacific region, southwest quadrant: Tulsa, Circum-Pacific Counc. Energy Min. Res./Am. Assoc. Pet. Geol., scale 1:10,000,000.

Denham, D., 1971, Seismicity of the New Guinea/Solomon Islands region, 1967: Bur. Min. Res. Geol. Geophys., Australia, Rec. 1971/96, 14 p.

Everingham, I.B., 1975, Faulting associated with the major North Solomon Sea earthquakes of 14 and 26 July 1971: Jour. Geol. Soc. Australia, v. 22, pt. 1, p. 61-69.

Fisher, N.H., 1939, Geology and vulcanology of Blanche Bay, and the surrounding area, New Britain: Terr. New Guinea Geol. Bull. 1, 68 p.

----- 1944, The Gazelle Peninsula, New Britain, earthquake of January 14, 1941: Bull. Seismol. Soc. Am., v. 34, no. 1, p. 1-12.

----- 1957, Catalogue of active volcanoes of the world including solfatara fields, Pt. V, Melanesia: Rome, IAVCEI, 105 p.

----- 1959, Report of the sub-committee on vulcanology 1954-56, Australian National Committee on Geodesy and Geophysics: Bull. Volcanol., v. 21, p. 153-161.

----- 1976, 1941-42 eruption of Tavurvur Volcano, Rabaul, Papua New Guinea, in Johnson, R.W., ed., Volcanism in Australasia: Amsterdam, Elsevier, p. 201-210.

Greene, H.G., Tiffin, D.L., and McKee, C.O., 1986, Structural deformation and sedimentation in an active caldera, Rabaul, Papua New Guinea: Jour. Volcanol. Geotherm. Res., v. 30, nos. 3/4, p. 327-356.

RABAUL (continued)

REFERENCES (continued)

Heming, R.F., 1969, The Kokopo (New Britain) earthquakes of the 14 August 1967: Bur. Min. Res. Geol. Geophys., Australia, Rec. 1969/24, 11 p.

----- 1974, Geology and petrology of Rabaul Caldera, Papua New Guinea: Geol. Soc. Am. Bull., v. 85, p. 1253-1264.

Heming, R.F., and Carmichael, I.S.E., 1973, High-temperature pumice flows from the Rabaul Caldera, Papua New Guinea: Contrib. Mineral. Petrol., v. 38, p. 1-20.

Johnson, R.W., Everingham, I.B., and Cooke, R.J.S., 1981, Submarine volcanic eruptions in Papua New Guinea: 1878 activity of Vulcan (Rabaul) and other examples, in Johnson, R.W., ed., Cooke-Ravian volume of volcanological papers: Geol. Surv. Papua New Guinea Memoir 10, p. 167-179.

Johnson, R.W., and Threlfall, N.A., 1985, Volcano town: The 1937-43 Rabaul eruptions: Bathurst, N.S.W., Robert Brown and Assoc., 151 p.

Kizawa, T., 1951, Volcanic tremor and tilting of ground: Quart. Jour. Seismol., v. 15, p. 18-34.

Lowenstein, P.L., 1982, Problems of volcanic hazards in Papua New Guinea: Geol. Surv. Papua New Guinea Rep. 82/7, 62 p.

----- 1986, Tidal triggering of volcanic earthquakes during the Rabaul seismic crisis of 1983-84 [abs.]: Abstracts, Int. Volcanol. Congr., New Zealand, p. 255.

McKee, C.O., 1981, Recent eruptive history of the Rabaul volcanoes, present volcanic conditions, and potential hazards from future eruptions: Geol. Surv. Papua New Guinea Rep. 81/5, 15 p.

McKee, C.O., Johnson, R.W., Lowenstein, P.L., Riley, S.J., Blong, R.J., de Saint Ours, P., and Talai, B., 1985, Rabaul Caldera, Papua New Guinea: Volcanic hazards, surveillance, and eruption contingency planning: Jour. Volcanol. Geotherm. Res., v. 23, p. 195-237.

McKee, C.O., Lowenstein, P.L., de Saint Ours, P., and Talai, B., 1984 (published in 1986), Seismic and deformation crises at Rabaul Caldera--prelude to an eruption?: Bull. Volcanol., v. 47, no. 2, p. 397-410.

McKee, C.O., Mori, J., and Talai, B., 1988, Temporal microgravity changes accompanying structural deformation at Rabaul Caldera: Bull. Volcanol. [in press].

Mori, J., and McKee, C.O., 1987, Outward-dipping ring-fault structure at Rabaul Caldera as shown by earthquake locations: Science, v. 235, p. 193-195.

Mori, J., McKee, C.O., Itikarai, I., Lowenstein, P.L., de Saint Ours, P., and Talai, B., 1986a, Account and interpretation of the seismicity during the Rabaul seismo-deformational crisis September 1983 to July 1985: Geol. Surv. Papua New Guinea Rep. 86/26, 54 p. + appendices.

----- 1987, Sea level measurements for inferring ground deformation in Rabaul Caldera: Geo-Marine Lett., v. 6, no. 4, p. 241-246 (Also, Geol. Surv. Papua New Guinea Rep. 86/23, 14 p.).

----- 1988, Earthquakes of the Rabaul seismo-deformational crisis, September 1983 - July 1985: Seismicity on a caldera ring fault: Bull. Volcanol. [in press].

Mori, J., McKee, C., Talai, B., and Itikarai, I., 1986b, Precursors to volcanic eruptions in Papua New Guinea: Geol. Surv. Papua New Guinea Rep. 86/27, 39 p.

Murray, G.H., 1939, In attendance at the birth of a volcano: Pacific Islands Monthly, v. 9, no. 10, p. 54-55.

RABAUL (continued)

REFERENCES (continued)

Palfreyman, W.D., Johnson, R.W., Cooke, R.J.S., and Bultitude, R.J., 1986, Volcanic activity in Papua New Guinea before 1944: an annotated bibliography of reported observations: Bur. Min. Res. Geol. Geophys., Australia, Rep. 254, 6 p. + BMR Microform MF207.

Perry, W.J., and Crick, I.H., 1976, Aerial thermal infrared survey, Rabaul area, Papua New Guinea, 1973, in Johnson, R.W., ed., Volcanism in Australasia: Amsterdam, Elsevier, p. 211-219.

Ripper, I.D., 1979, Large earthquakes and seismic zones of the New Guinea region: Geol. Surv. Papua New Guinea Rep. 79/20, 18 p. + figures.

Smithsonian Institution, Scientific Event Alert Network (SEAN), 1982-87, Rabaul: SEAN Bull., v. 7, nos. 8-9; v. 8, nos. 9-11; v. 9, nos. 1-12; v. 10, nos. 3-4, 6-11; v. 11, nos. 1-12; v. 12, nos. 1-12; v. 13, nos. 1, 2 (communications from P.L. Lowenstein, C.O. McKee, P. de Saint Ours, B. Talai, and others).

Walker, G.P.L., Heming, R.F., Sprod, T.J., and Walker, H.R., 1981, Latest major eruptions of Rabaul volcano, in Johnson, R.W., ed., Cooke-Ravian volume of volcanological papers: Geol. Surv. Papua New Guinea Memoir 10, p. 181-193.

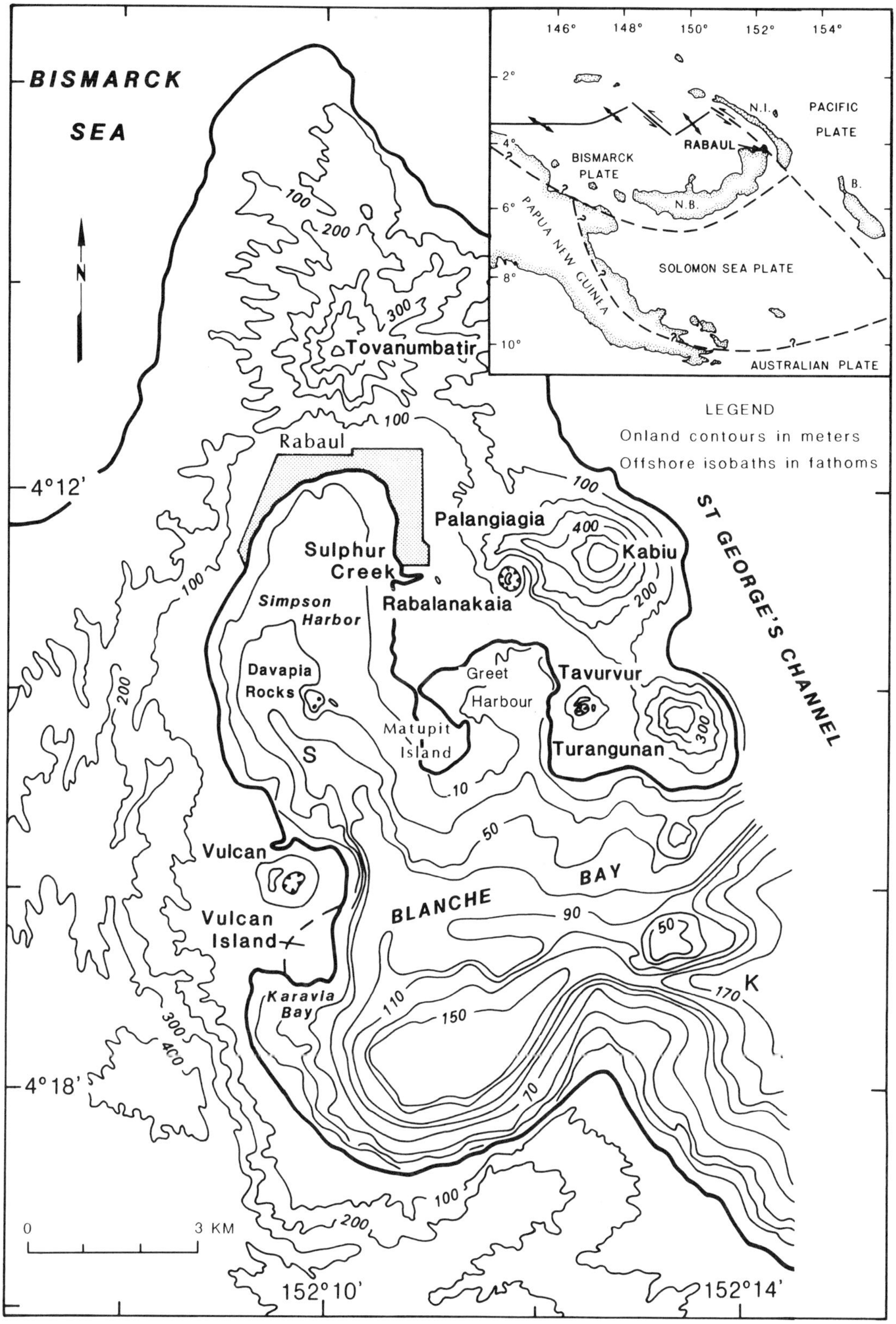

Figure 5.7.1. Topographic and bathymetric map of Rabaul Caldera, from Greene and others (1986). See next figure for location of caldera rim. Inset: tectonic-structure map of Papua New Guinea region. N.B., New Britain; N.I., New Ireland; B., Bougainville. Modified from Heming (1974).

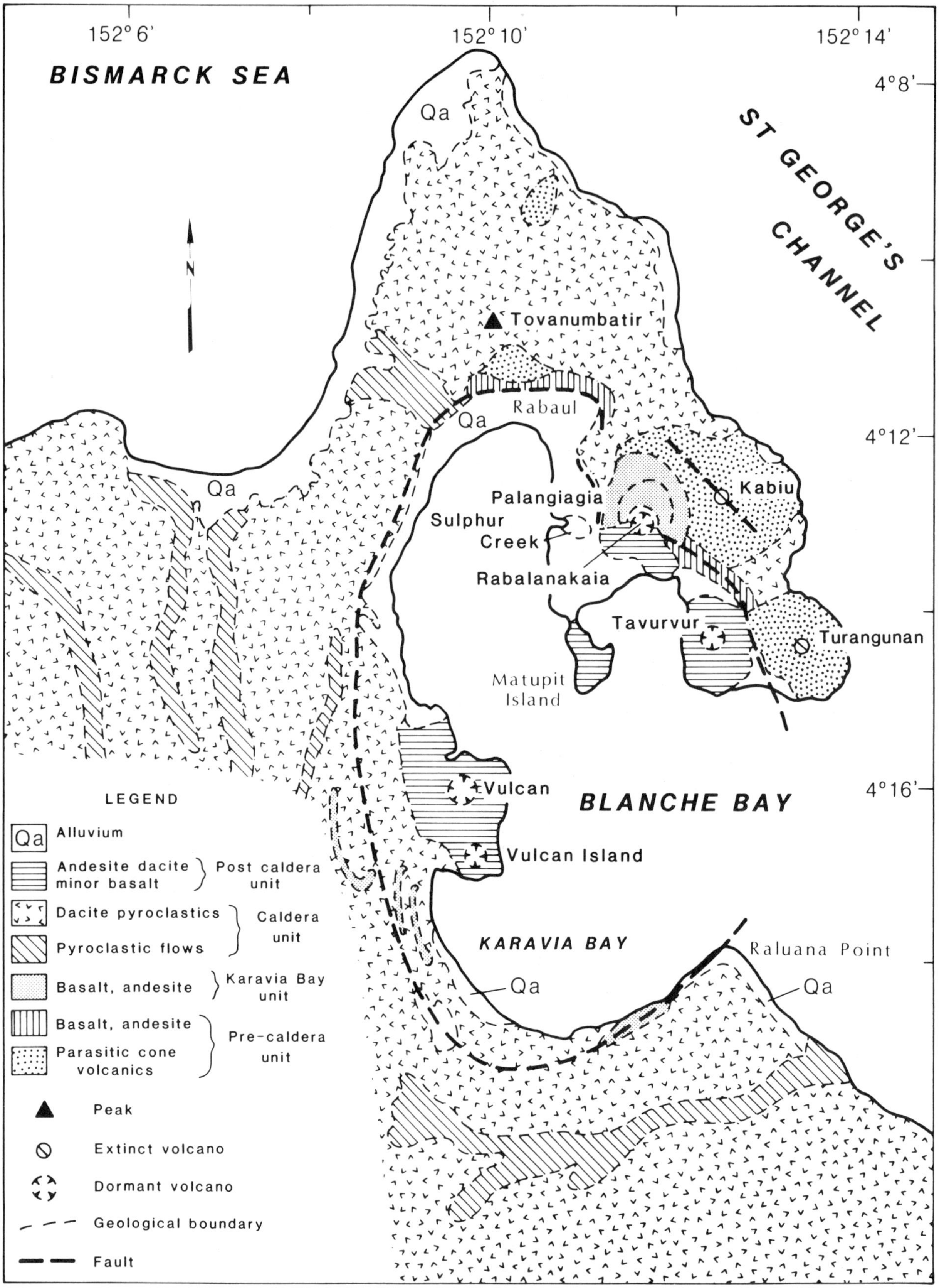

Figure 5.7.2. Geologic map of Rabaul Caldera, from Greene and others (1986), modified from Heming (1974).

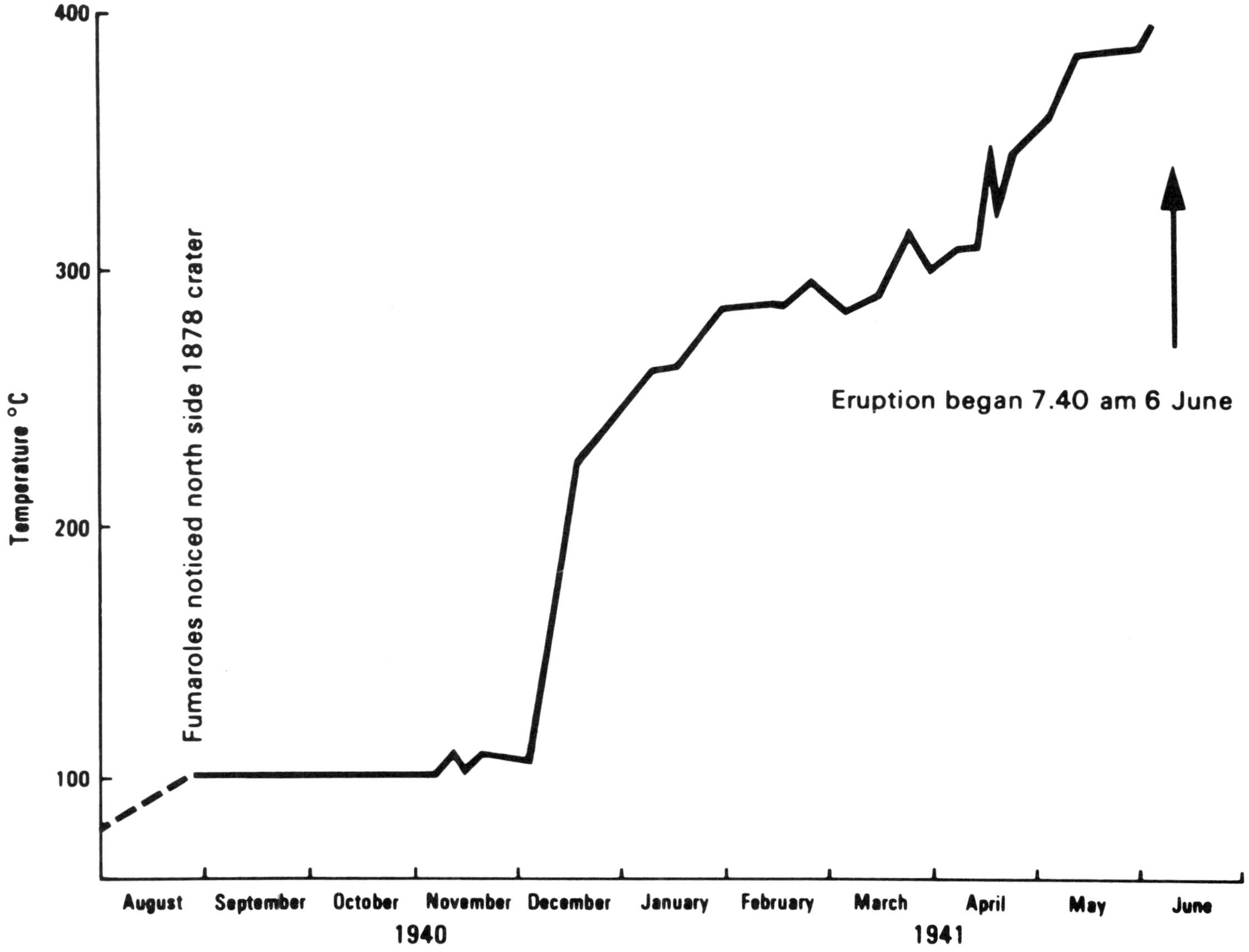

Figure 5.7.3. Plot showing temperature rise in fumaroles at Tavurvur before June 1941 eruption, from Fisher (1976).

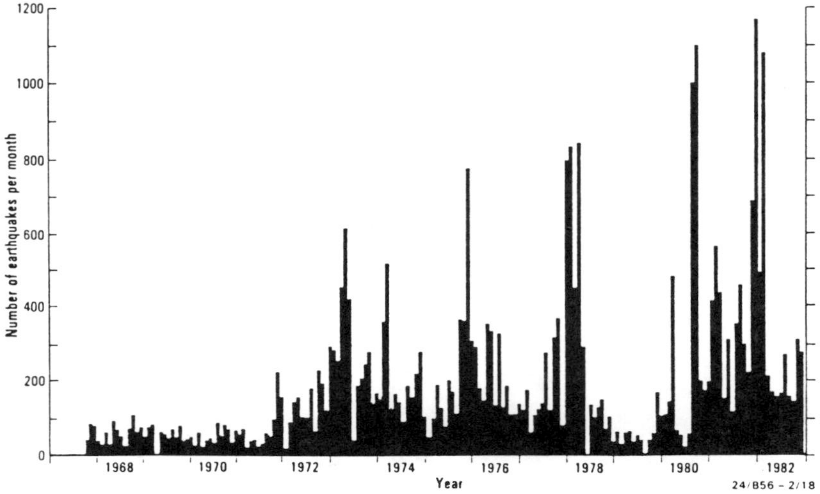

Figure 5.7.4. Monthly number of earthquakes at Rabaul Caldera during the period 1968-82, from McKee and others (1984). Note increase starting in August 1971, following two $\underline{M}_L$ 8.0 earthquakes in the Solomon Sea in July 1971.

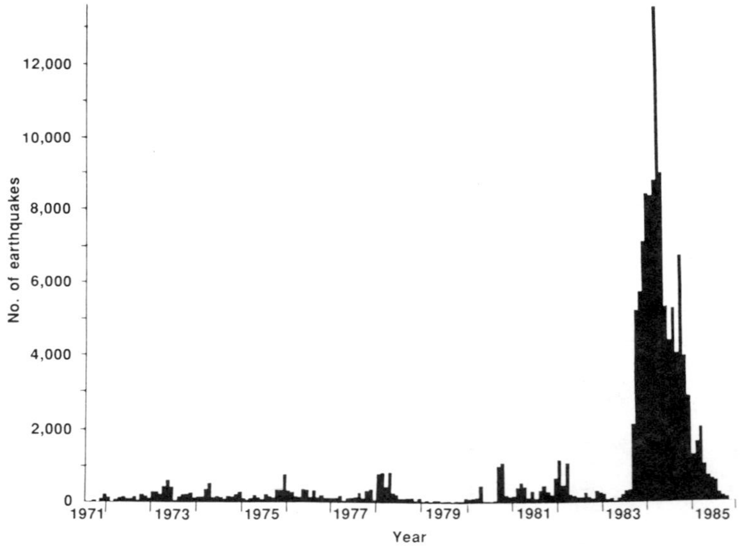

Figure 5.7.5. Monthly number of earthquakes (1971-85) recorded at two or more stations of Rabaul harbor seismic network, from Mori and McKee (1987). Detection limit for Rabaul network is about M_L -1. Note change in scale relative to figure 5.7.4, and dramatic increase in seismicity starting in late 1983. Copyright by the American Association for the Advancement of Science.

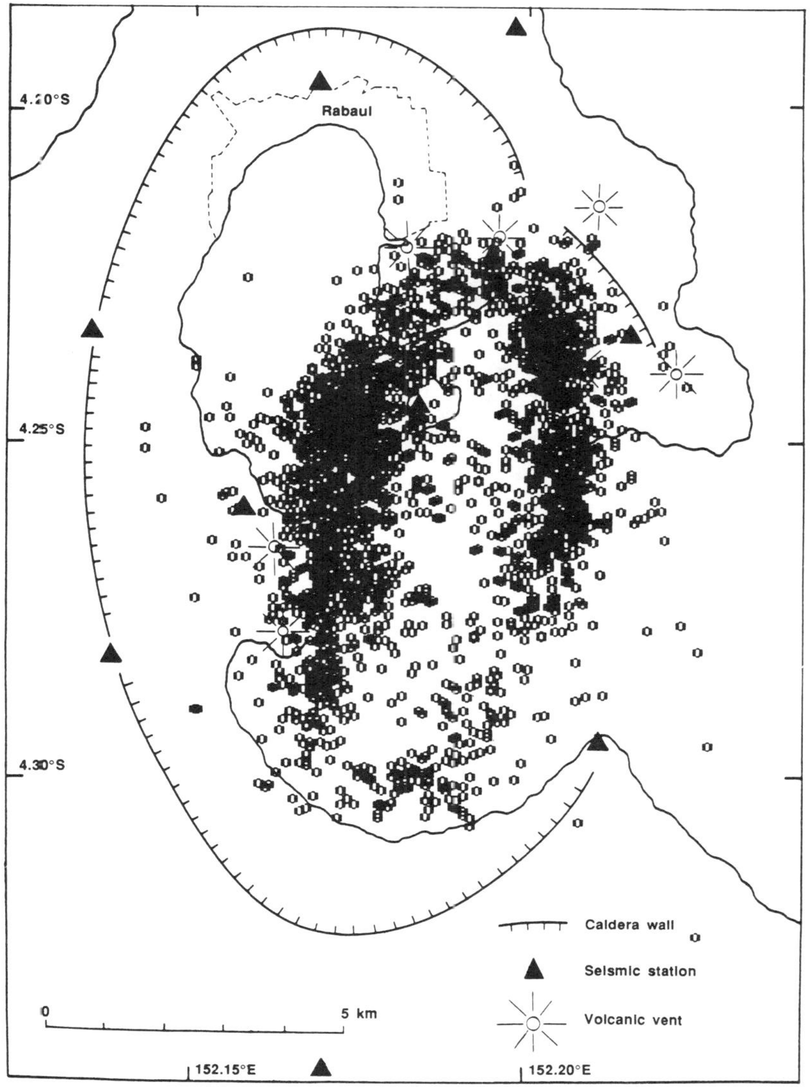

Figure 5.7.6. Epicentral locations of more than 2,500 earthquakes from September 1983 through July 1985, from Mori and McKee (1987). Triangles show positions of 10 seismic stations of Rabaul harbor network. See figure 5.7.7 for cross sections. Copyright by the American Association for the Advancement of Science.

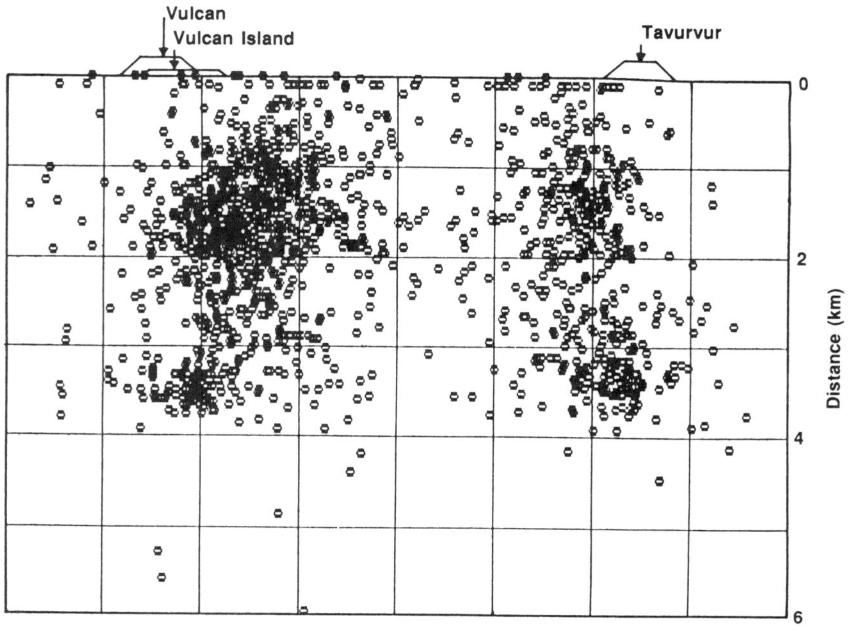

Figure 5.7.7. East-west cross section through central portion of Rabaul ring structure, showing locations of earthquakes from figure 5.7.6 along outward-dipping faults (Mori and McKee, 1987). Also shown are positions of three postcaldera vents within region of cross section. There is no vertical exaggeration. Copyright by the American Association for the Advancement of Science.

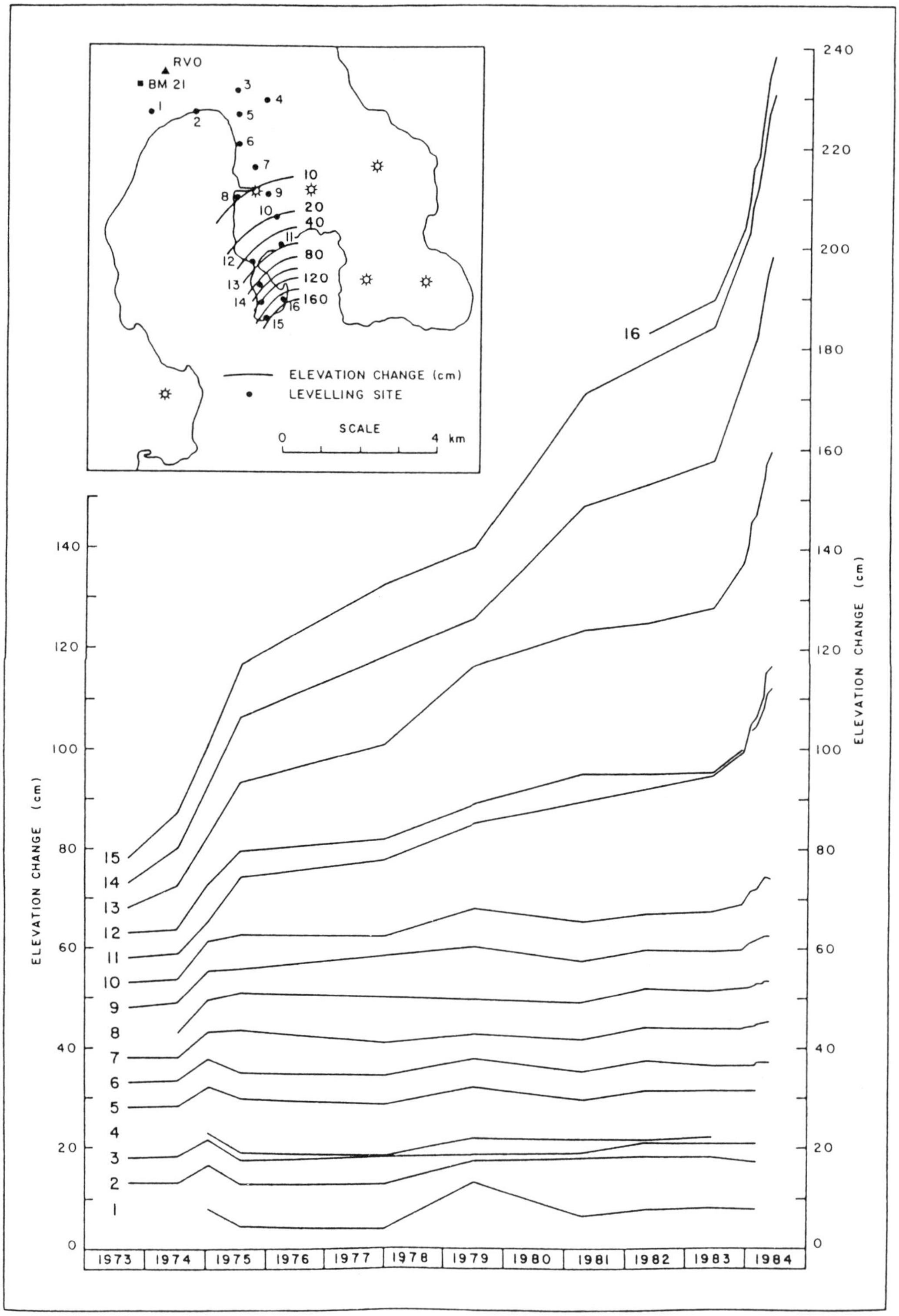

Figure 5.7.8. Elevation changes at Rabaul Caldera for the period September 1973 to May 1984, from McKee and others (1984). Partial contours show elevation changes, in centimeters, relative to BM 21 near the Rabaul Volcano Observatory (RVO) on northern caldera rim (inset).

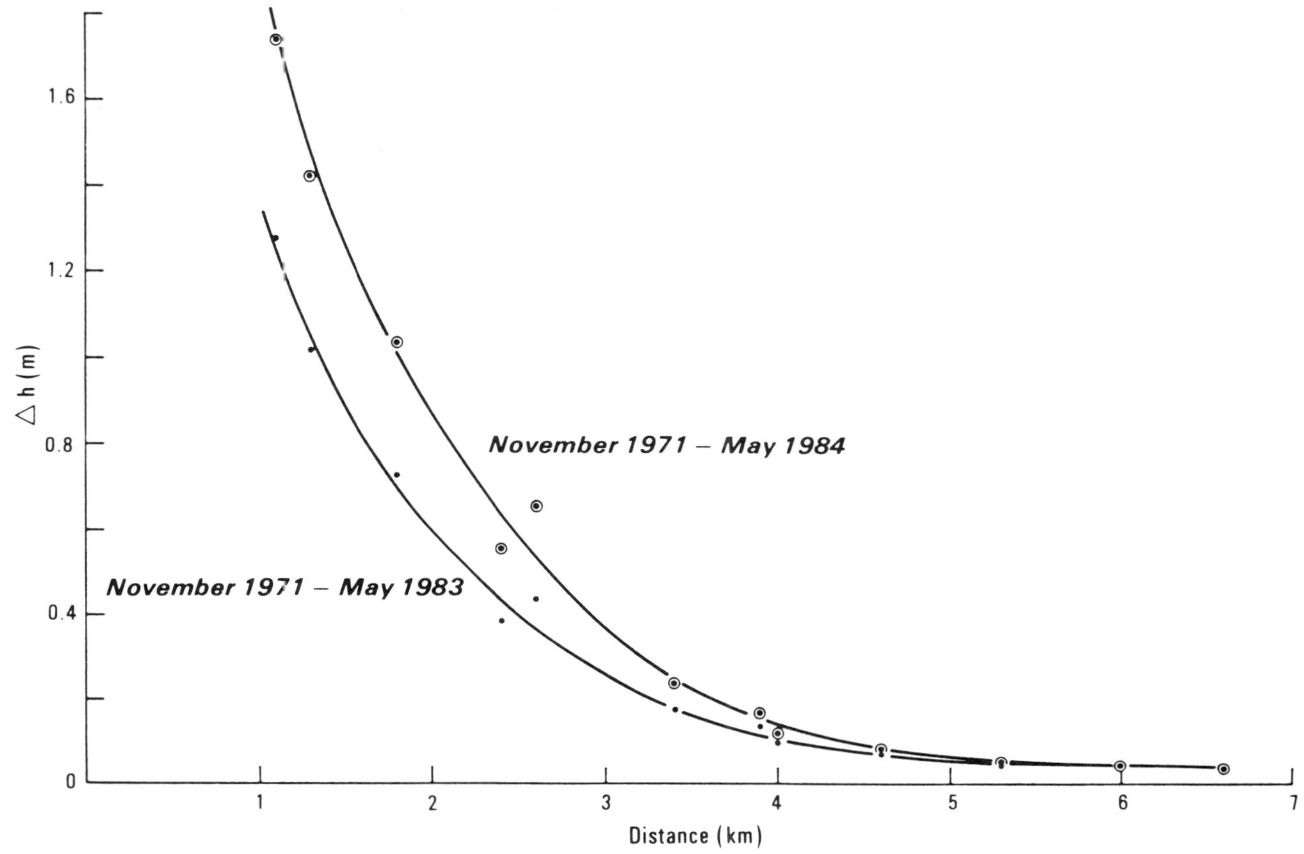

Figure 5.7.9. Uplift versus distance from an inflation center near Matupit Island for two periods shown on figure, from McKee and others (1984).

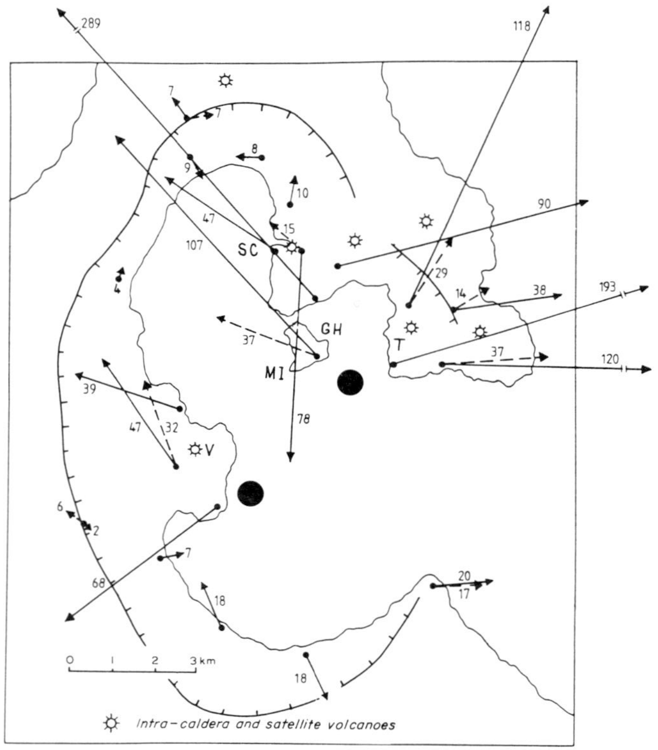

Figure 5.7.10. Tilt vectors (in microradians) at Rabaul Caldera for the periods 1 January - 30 September 1983 (broken arrows) and 1 October 1983 - 31 May 1984 (solid arrows), from McKee and others (1984). Inferred inflation centers shown as large filled circles. Anomalous southward vector (78 microradians) at head of Sulphur Creek inlet was associated with local subsidence initiated by a major seismic swarm on 22 April 1984. SC, Sulphur Creek; GH, Greet Harbor; T, Tavurvur; MI, Matupit Island; V, Vulcan.

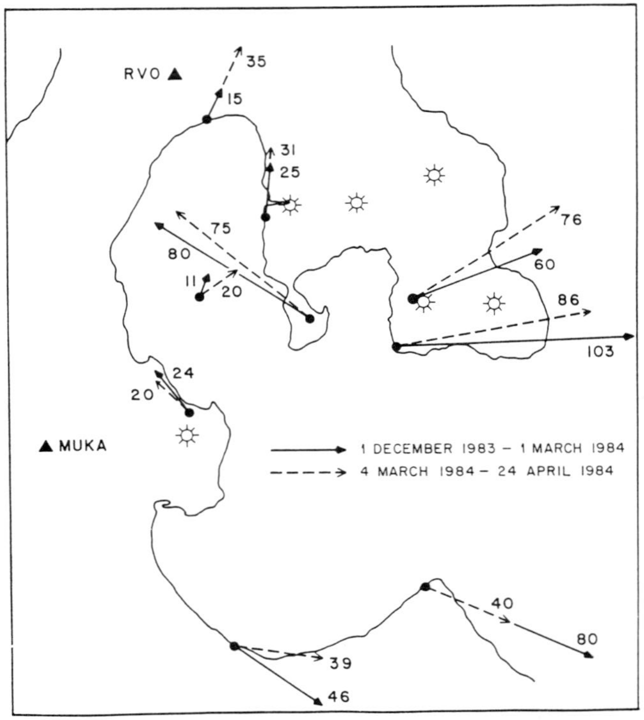

Figure 5.7.11. Horizontal distance changes (in millimeters) at Rabaul Caldera for the periods (1) 1 December 1983 to 1 March 1984 and (2) 4 March to 24 April 1984, from McKee and others (1984). Base stations RVO and MUKA held fixed.

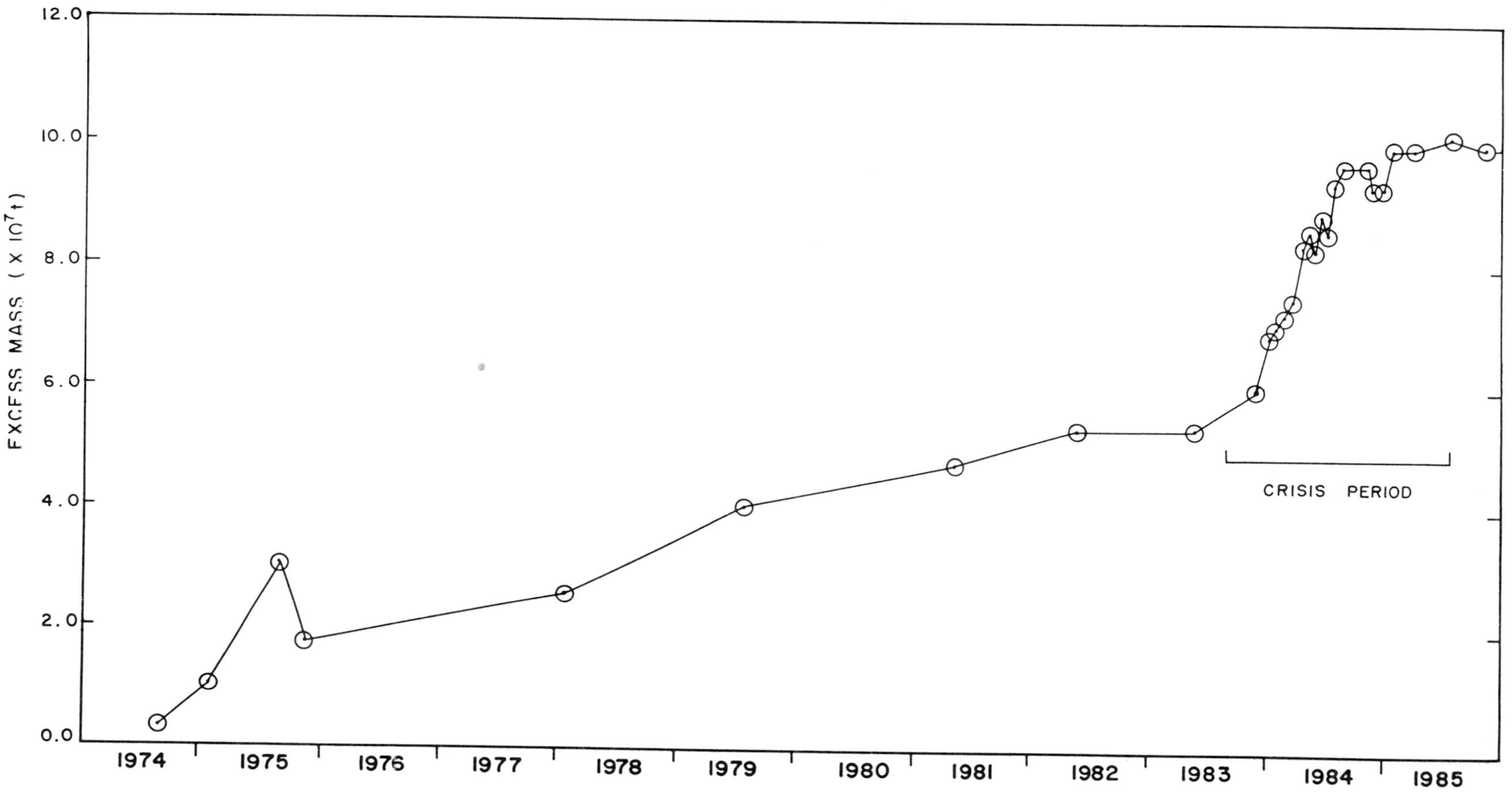

Figure 5.7.12. Excess mass at deformation center as a function of time (1974-85), inferred from precise gravity measurements (McKee and others, 1988).

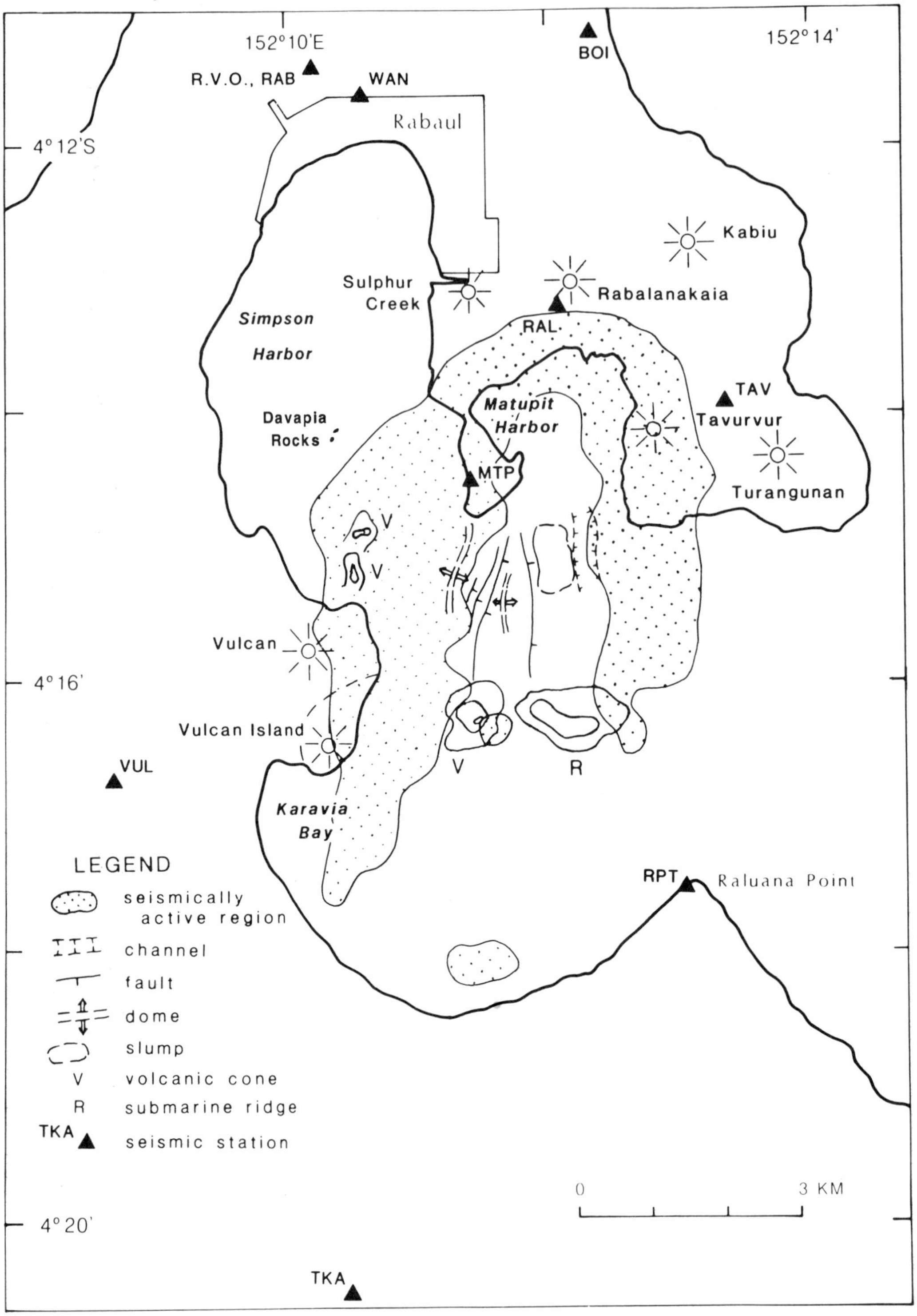

Figure 5.7.13. Sketch map of central part of Rabaul Caldera showing offshore geologic structure and outline of 1971-86 seismic zone (Greene and others, 1986, based on 1982 data from USGS S.P. Lee). Greatest amount of sediment disruption and sea floor disturbance is occurring along a north-south-trending ridge marked by recent faults and slumps, mainly within elliptical pattern defined by seismicity.

UNNAMED, near Esa'Ala, D'Entrecasteaux Islands

CAVW number (active vent)	Latitude Longitude (degrees)	Diameter (km)	Local tectonic setting	Precaldera edifice	SiO_2 content (wt pct)	Age of collapse (yr B.P.)	Date of unrest	Type and duration of unrest ESTU STHF MGTF H Te	Eruption type
05-03-06,07 (Lamonai-Oiau, Dobu)	9.62S 150.88E	7 x 11	?	Strato	R = 69-72 C = 70?	?	1953-57	FF-- ---- ---- - --	none
							1969	xx-- ---- ---- - --	none
							1981-82	---- ---- --F- F --	none
							1984	---- ---- --E- E --	none

TECTONIC SETTING

The D'Entrecasteaux Islands lie off the north shore of the eastern tip of mainland Papua New Guinea (fig. 5.8.1). Three main volcanic centers in the area are Mount Lamonai, Mount Oiau, and Dobu Island (fig. 5.8.2; Smith, 1973).

GEOLOGIC HISTORY

Davies and Ives (1965) reported outcrops of welded dacitic ashflow tuffs west of Mount Lamonai, from Gomwa Bay northward to Sebutia Bay (fig. 5.8.2), and the area is known for peralkaline rhyolite (Smith, 1973, 1976, 1981; Smith and Johnson, 1981). Lowenstein (1982b) suggested that a submarine caldera may lie between Dobu Island, Neumara Island, and the Mount Oiau area of Fergusson Island. No isotopic dates are available for the area, but Smith (1973) noted that the presence of young volcanic landforms and abundant solfataric activity indicate widespread volcanic activity in Recent time. The morphology of lava flows on Dobu Island and in the crater at Mount Oiau is so youthful that they were probably erupted within the past 100-200 years (Smith, 1973).

HISTORICAL ACTIVITY

1953-57: Several earthquake swarms were felt in the Esa'Ala area between 1953 and 1956 (Taylor, 1955; Reynolds, 1957). Additional swarms occurred in the Dobu Island area (probably the source area for those felt earlier in Esa'Ala) during July-September 1955 and in early 1957 (Fisher, 1959). In the intervening period, the level of seismicity was "higher than normal" (Fisher, 1959).

1966, 1969: An observatory was established at Esa'Ala in 1966, and observations by local residents are continuing. Seismic swarms in 1969 triggered a four-day, unofficial evacuation (Lowenstein, 1982a, p. 53, 57).

UNNAMED, near Esa'Ala, D'Entrecasteaux Islands (continued)

HISTORICAL ACTIVITY (continued)

1981-82: Between 1981 and June 1982, the temperature increased at a hot spring on Neumara Island, where warm water had been emerging for "several years." In June 1982, water reached the boiling point and was forcibly ejected to a height of about 30 cm. Thereafter, the temperature and intensity of hot-water emissions declined; as of late September 1982, the temperature was relatively steady at 55 °C (Lowenstein, 1982b).

1984: Residents of Dobu Island became concerned in January-February 1984 when hot water rising through a coral reef on the northeast shore of the island darkened the sea and killed many fish. Johnson and Kaita (1984) concluded that the phenomena were caused by a short-lived discharge of hot-spring water from a volcanic center that had been known to be thermally active for many years. They also noted that a minor, apparently new hot spring with a temperature of 47-48 °C had appeared just south of the affected reef (fig. 5.8.3). The hot springs on Neumara Island investigated earlier by Lowenstein (1982b) remained unchanged in 1984 relative to conditions in 1982.

REFERENCES

Baker, G., and Coulson, A., 1948, Metamorphic and volcanic rocks from D'Entrecasteaux Islands: Trans. Am. Geophys. Union, v. 29, p. 656-663

Davies, H.L., and Ives, D.J., 1965, The geology of Fergusson and Goodenough Islands, Papua: Bur. Min. Res., Australia, Rep. 82.

Eiby, G.A., 1966, Earthquake swarms and volcanism in New Zealand: Bull. Volcanol., v. 29, p. 61-73.

Fisher, N.H., 1959, Report of the sub-committee on vulcanology 1954-56, Australian National Committee on Geodesy and Geophysics: Bull. Volcanol., v. 21, p. 153-161.

Johnson, R.W., and Kaita, V., 1984, Volcanological report for the Dobu/Southeast Fergusson Island area, 24 February 1984, in Johnson, R.W., ed., Volcanological inspections in Papua New Guinea, February 1984: Geol. Surv. Papua New Guinea Rep. 84/4, p. 32-41.

Lowenstein, P.L., 1982a, Problems of volcanic hazards in Papua New Guinea: Geol. Surv. Papua New Guinea Rep. 82/7, 62 p.

----- 1982b, Report on the occurrence of a new thermal spring at Neumara Island in the D'Entrecasteaux Group, Milne Bay Province: Geol. Surv. Papua New Guinea Tech. Note 17-82, 5 p.

Reynolds, M.A., 1957, Volcano seismic phenomena in eastern Papua since 1939: Bur. Min. Res., Australia, Rec. 1957/14.

Smith, I.E.M., 1973, Late Cainozoic volcanism in the southeast Papuan Islands: Bur. Min. Res. Geol. Geophys., Australia, Rec. 973/67, 23 p. + 15 figures.

----- 1976, Peralkaline rhyolites from the D'Entrecasteaux Islands, Papua New Guinea, in Johnson, R.W., ed., Volcanism in Australasia: Amsterdam, Elsevier, p. 275-285.

----- 1981, Young volcanism in eastern Papua, in Johnson, R.W., ed., Cooke-Ravian volume of volcanological papers: Geol. Surv. Papua New Guinea Memoir 10, p. 257-265.

Smith, I.E.M., and Johnson, R.W., 1981, Contrasting rhyolite suites in the late Cenozoic of Papua New Guinea: Jour. Geophys. Res., v. 86, p. 10,257-10,272.

Taylor, G.A., 1955, Notes on volcanic activity and thermal areas in the D'Entrecasteaux Islands: Bur. Min. Res., Australia, Rec. 1955/75.

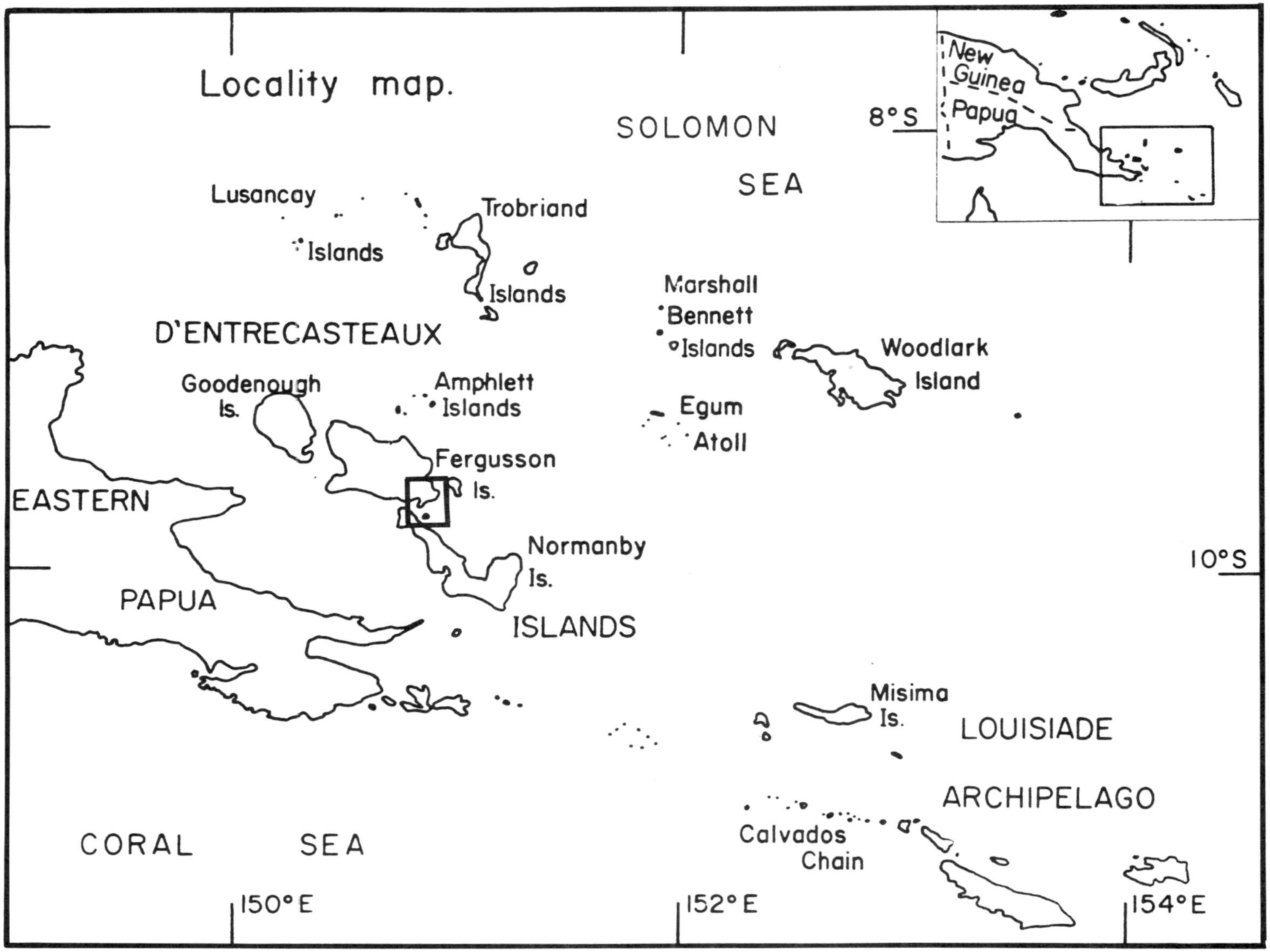

Figure 5.8.1. Locality map from Smith (1973), showing D'Entrecasteaux Islands off eastern tip of mainland Papua New Guinea (inset). Box that includes parts of Fergusson and Nomanby Islands marks approximate boundaries of figure 5.8.2.

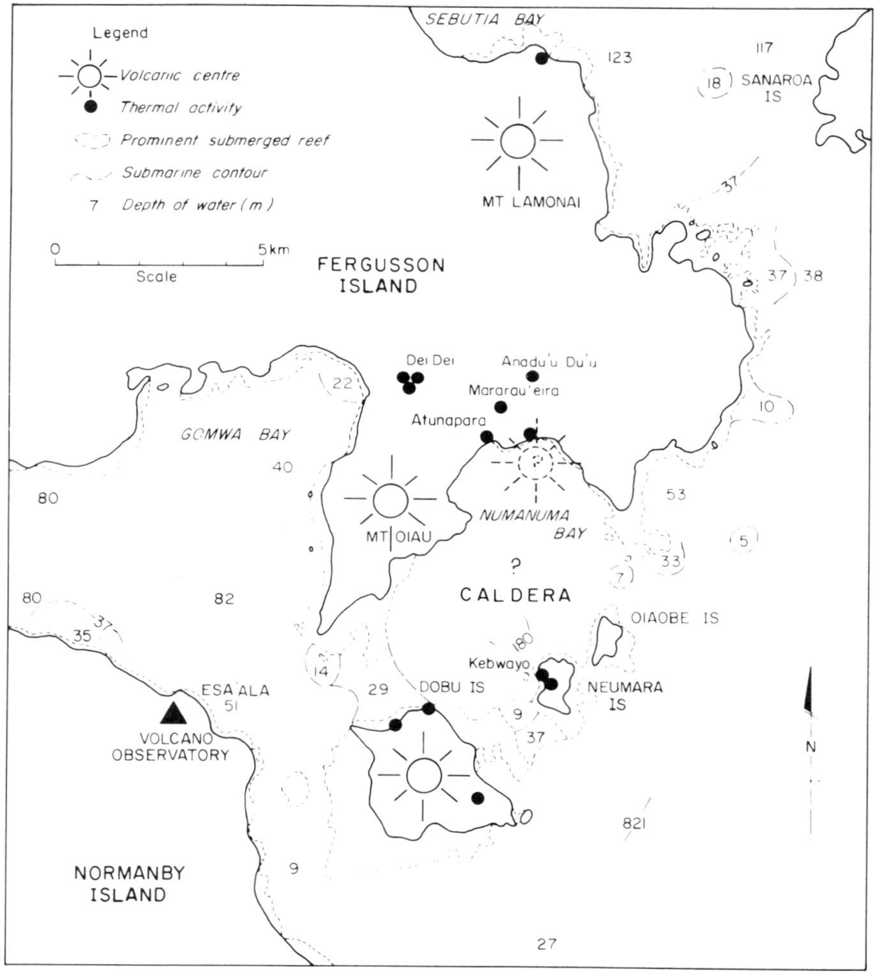

Figure 5.8.2. Sketch map of Mount Lamonai-Mount Oiau-Dobu Island area, from Lowenstein (1982b), showing location of volcanic centers, thermal activity, and a possible partly submerged caldera.

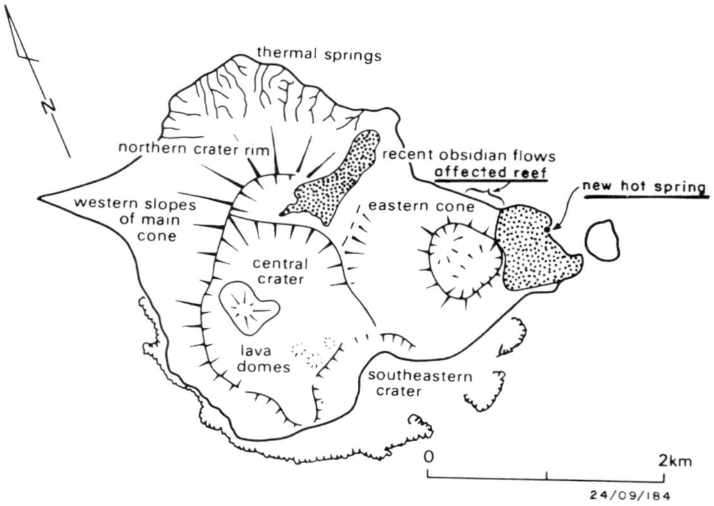

Figure 5.8.3. Sketch map of Dobu Island, showing coral reef affected by hot-water discharge in 1984 and location of an apparently new hot spring, from Johnson and Kaita (1984).

GAUA ISLAND

CAVW number (active vent)	Latitude Longitude (degrees)	Diameter (km)	Local tectonic setting	Precaldera edifice	SiO_2 content (wt pct)	Age of collapse (yr B.P.)	Date of unrest	Type and duration of unrest ESTU STHF MGTF H Te	Eruption type
05-07-02 (Mt. Gharat)	14.25S 167.50E	8	Compr	Strato	R = mafic C = ?	?	1963-69	---- ---- ---- - --	ex,Ex
							1971	---- ---- ---- - --	ex
							1973-74	E-C- ---- ---x - QU?	pex
							1976-77	---- ---- ---x x --	ex
							1980-82	---- ---- ---x - --	pex, ex

TECTONIC SETTING

Gaua Island lies in the northern part of the New Hebrides Archipelago, which Coleman (1970) described as a fractured island arc where patterns of straight-line fractures and block movements are the dominant structural features. Other aspects of the regional tectonics are discussed by Karig and Mammerickx (1972), Pascal and others (1978), Pascal (1979), and Ibrahim and others (1980).

GEOLOGIC HISTORY

Gaua is a small island stratovolcano with a summit caldera that is largely filled by Mt. Gharat, a postcaldera cone, and Steaming Hill Lake (fig. 5.9.1). It was thought to be in a dormant "solfataric stage" until 1963, when frequent eruptions began.

HISTORICAL ACTIVITY

1963-82+: Mt. Gharat erupted frequently from 1963 to 1977 (Volcanological Survey of Japan, 1964-84). On the basis of a magnitude 6.5 earthquake 200 km beneath Gaua on 1 August 1973, Dr. C. Blot predicted that Mt. Gharat would erupt at the end of December 1973 (Mallick and MacFarlane, 1975). Small phreatic explosions occurred on 18 October and on 12 and 17 November but not in December. A magnitude 7.5 earthquake occurred approximately 75 km south of Gaua on 28 December 1973, followed by numerous aftershocks. Another magnitude 7.5 earthquake occurred on 10 January 1974 in the immediate vicinity of Gaua, at a depth of only 10-15 km. This was followed by aftershocks during 10-14 January, by volcanic microtremor during 15-20 or 21 January, and by a small phreatic eruption on 21 January. Discoloration of Steaming Hill Lake at the base of Mt. Gharat was greater than normal in April 1977.

GAUA ISLAND (continued)

COMMENTS

Mallick and MacFarlane (1975) concluded that magma rose close to the surface under Mt. Gharat in late 1973, as predicted by Dr. C. Blot. At least two qualifications seem warranted. First, Mt. Gharat was frequently active from 1967 to 1973, so the timing of the October 1973, November 1973, and January 1974 phreatic eruptions relative to nearby large earthquakes may have been coincidental. Second, the earthquakes may have been of tectonic origin (the area is intensely active), and the eruptions thus may have been a relatively minor volcanic response to tectonic shaking or stress changes. Similar volcanic responses to tectonic processes are suspected at Akan, Rabaul, Tokachi, and several other calderas in this compilation.

REFERENCES

Coleman, P.J., 1970, Geology of the Solomon and New Hebrides Islands, as part of the Melanesian Re-entrant, Southwest Pacific: Pacific Sci., v. 24, no. 3, p. 289-314.

Fisher, N.H., 1957, Catalogue of active volcanoes of the world including solfatara fields, Pt. V, Melanesia: Rome, IAVCEI, 105 p.

Ibrahim, A.K., Pontoise, B., Latham, G., Larue, M., Chen, T., Isacks, B., Recy, J., and Louat, R., 1980, Structure of the New Hebrides arc-trench system: Jour. Geophys. Res., v. 85, p. 253-266.

Karig, D.E., and Mammerickx, J., 1972, Tectonic framework of the New Hebrides island arc: Jour. Marine Geol., v. 12, p. 187-205.

Mallick, D.I.J., and Ash, R.P., 1970, Gaua: New Hebrides Geol. Surv. Ann. Rep. (1968), p. 27-29.

Mallick, D.I.J., and MacFarlane, A., 1975, Eruption predictions by Dr. C. Blot and volcano-seismic activity of Gaua, late 1973 - early 1974: New Hebrides Geol. Survey Ann. Rep. (1973), p. 14-17.

Pascal, G., 1979, Seismotectonics of the Papua New Guinea-Solomon Islands region: Tectonophysics, v. 57, p. 7-34.

Pascal, G., Isacks, B.L., Barazangi, M., and Dubois, J., 1978, Precise relocations of earthquakes and seismotectonics of the New Hebrides Island arc: Jour. Geophys. Res., v. 83, p. 4957-4974.

Volcanological Society of Japan, 1964-84, Gaua: Bull. Volcanic Eruptions, nos. 4, 10, 12, 13, 14, 16, 17, 20, 21, 22 (information from A. MacFarlane, Geol. Survey Dept.).

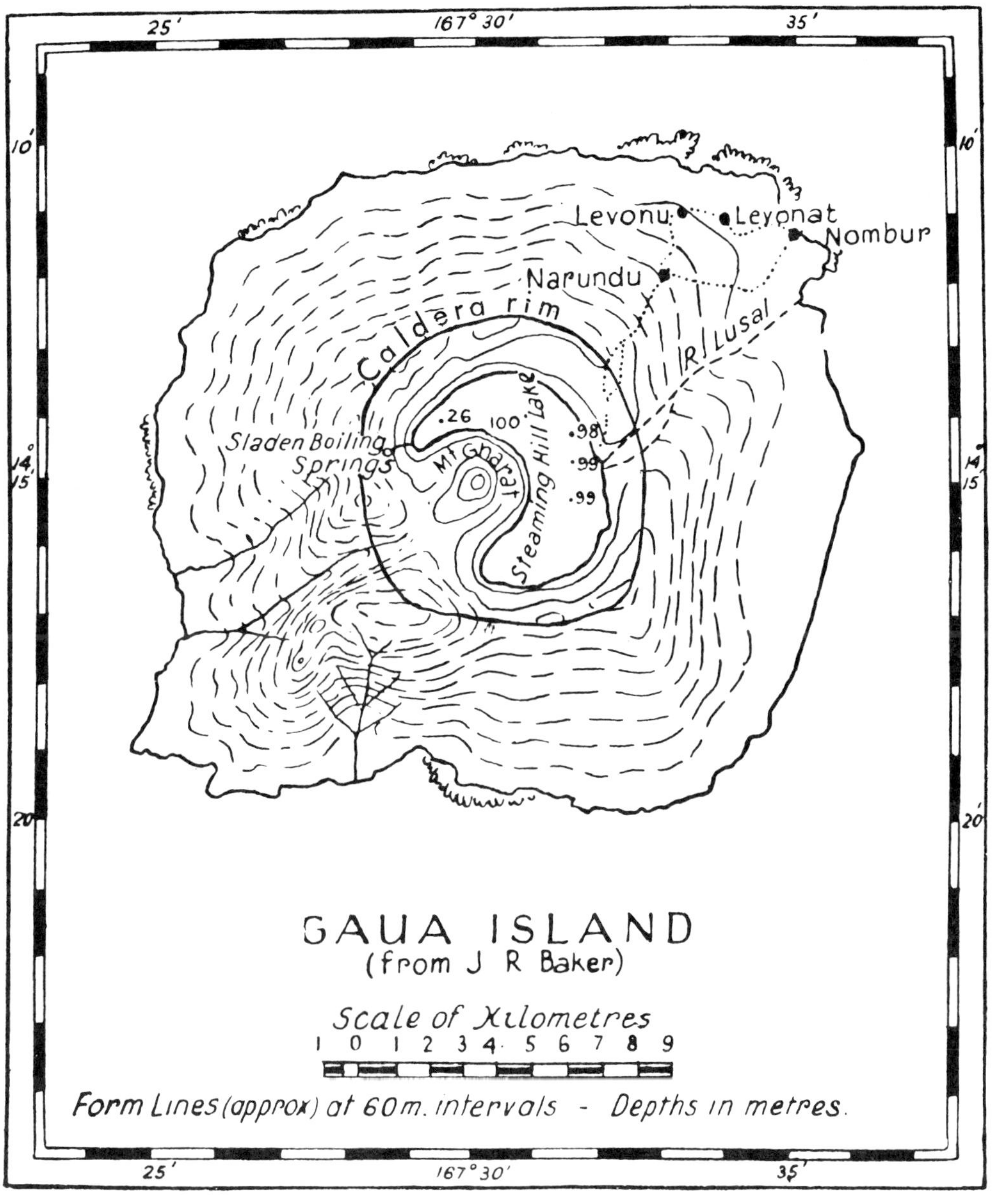

Figure 5.9.1. Sketch map of Gaua Island, from Fisher (1957), showing summit caldera mostly filled by Mt. Gharat and Steaming Hill Lake. Approximate location of caldera rim inferred from topography.

PART 3: HISTORICAL UNREST AT LARGE QUATERNARY CALDERAS (continued)
See inside back cover for explanation and abbreviations

AOBA

CAVW number (active vent)	Latitude Longitude (degrees)	Diameter (km)	Local tectonic setting	Precaldera edifice	SiO_2 content (wt pct)	Age of collapse (yr B.P.)	Date of unrest	Type and duration of unrest ESTU STHF MGTF H Te	Eruption type
05-07-03	15.33S 167.83E	6	Compr, with transverse structure	Shield	R = 45-50 C = mafic	few thousand?	ca. 1670	---- ---- ---- - --	lf
							ca. 1870	---- ---- ---- - --	ex
							1971	---- ---- ---E E --	none
							1976	---- ---- ---x x --	none

TECTONIC SETTING

Aoba is a basaltic shield volcano in the northern part of the New Hebrides Archipelago. The New Hebrides islands form a segment of the Circum-Pacific mobile belt, characterized by high seismicity and intense volcanic activity (Warden, 1970). The island owes its elongate shape (fig. 5.10.1) to multiple vents along a NE-SW-trending fissure system.

GEOLOGIC HISTORY

Growing from a depth of 2,400 m on the sea floor, Aoba probably emerged from the sea in the late Pliocene or early Pleistocene. Extrusion of relatively fluid lavas is recorded in growth of a shield. A later, more explosive episode is recorded by airfall pyroclastics and agglomerates. Activity continued with the effusion of picrite basalt, accompanied by ash eruptions. The most recent major explosive episode produced agglomerate, tuff, and occasional flows. Two nested summit calderas are apparently related to deep-seated, relatively quiet foundering during massive outpourings of lava. A broad pyroclastic cone occupies the center of the inner caldera (Warden, 1970).

HISTORICAL ACTIVITY

ca. 1670, ca. 1870: An effusive eruption from vents on the west flank about 300 yr B.P. was preceded (359 ± 54 ^{14}C yr) by explosive and effusive eruptions within the caldera; some of the early events may have accompanied collapse of the inner caldera. About 100 years ago, an explosive summit eruption deposited ash widely, and lahars annihilated villages on the southeast flank (Warden, 1970).

AOBA (continued)

HISTORICAL ACTIVITY (continued)

1971-76: Steam jetting in the summit caldera increased during February-April, 1971, then decreased to normal background by May 1971. Also by May, the water level of the upper of two crater lakes had fallen, and that of the lower lake had risen. Green discoloration of the crater lake, possibly due to increased fumarolic activity, was noted in June 1976.

REFERENCES

Fisher, N.H., 1957, Catalogue of active volcanoes of the world including solfatara fields, Pt. V, Melanesia: Rome, IAVCEI, 105 p.
Volcanological Society of Japan, 1974, 1978, Aoba: Bull. Volcanic Eruptions, no. 12, p. 43; no. 16, p. 69.
Warden, A.J., 1970, Evolution of the Aoba caldera volcano, New Hebrides: Bull. Volcanol., v. 34, p. 107-140.

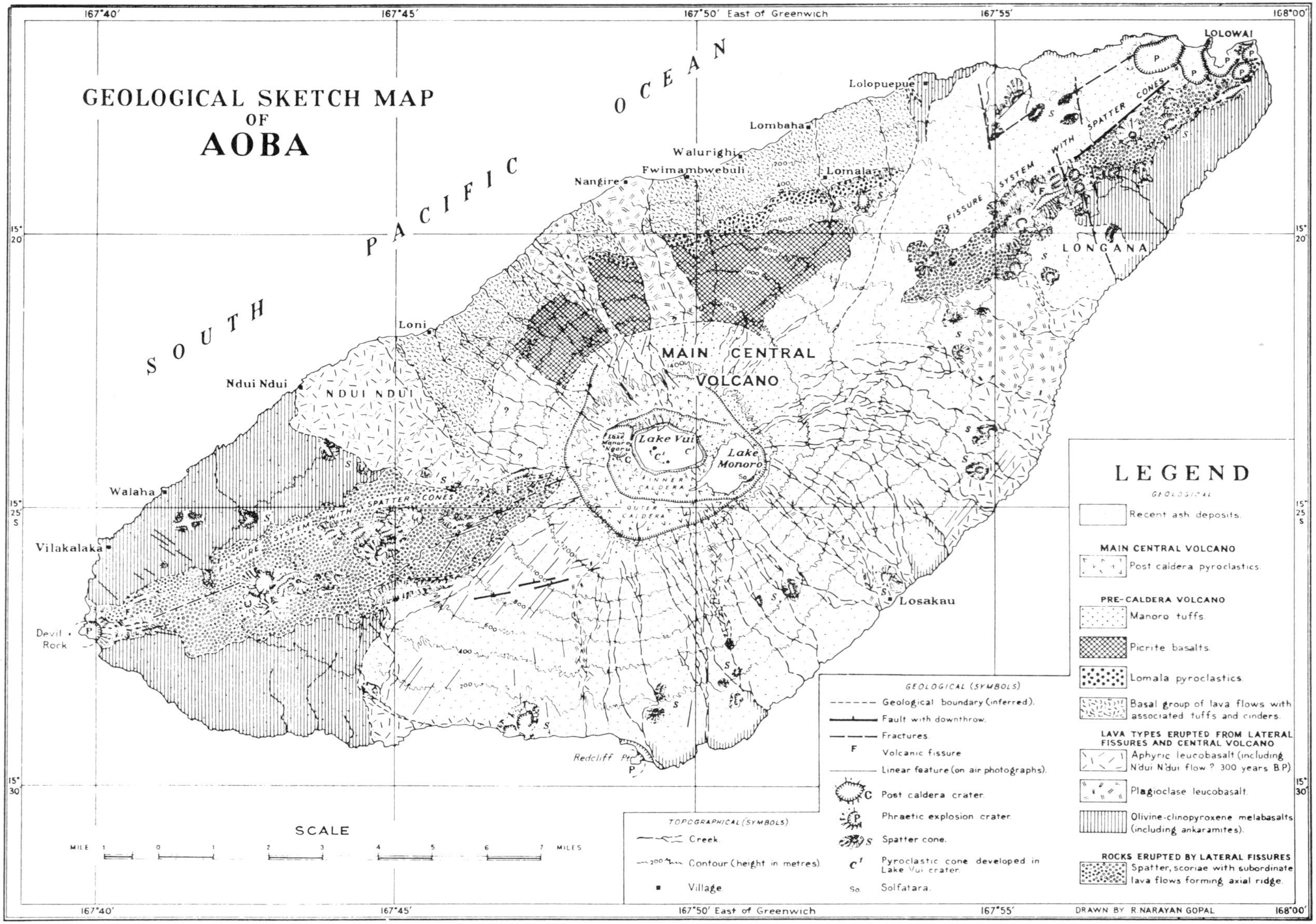

Figure 5.10.1. Geological sketch map of Aoba Volcano, from Warden (1970). Note nested calderas in summit area.

AMBRYM

CAVW number (active vent)	Latitude Longitude (degrees)	Diameter (km)	Local tectonic setting	Precaldera edifice	SiO_2 content (wt pct)	Age of collapse (yr B.P.)	Date of unrest	Type and duration of unrest ESTU STHF MGTF H Te	Eruption type
05-07-04	16.25S 168.08E	12	Compr, intersection of fault zone and arc	LL-strat (3 cones)	R = 50 C = mafic	ca. 2,000	1774?	---- ---- ---- - --	unknown
							1863-64	---- ---- ---- - --	ex
							1871	---- ---- ---- - --	ex
							1888	---- ---- ---- - --	lf
							1894-95	Y--- ---- ---- - --	Ex,lf
							1908	---- ---- ---- - --	unknown
							1912-15	x--- ---Y ---- - --	ex,Ex,lf
							1929	---- ---- ---- - --	ex,lf
							1935	x--- ---- ---- - --	ex,ll
							1937	x--- ---- ---- - --	ex,lf
							1942	---- ---- ---- - --	ex,lf
							1950-54	E--- ---- ---- - --	Ex,ex, ll,lf
							1959-81	---- ---- ---Y - --	Ex,ex, ll,lf
							1985-86	---- ---- ---- - --	ex
							1988	---- ---- ---- - --	ex

TECTONIC SETTING

Ambrym Island is one of a line of young volcanic islands stretching south from the Santa Cruz Islands through the New Hebrides (Varuatu) Archipelago (fig. 5.11.1). McCall and others (1970) suggest that volcanism at Ambrym occurs as a result of tension at the intersection of the E-W-trending D'Entrecasteaux Fracture Zone with the axis of the New Hebrides island arc.

GEOLOGIC HISTORY

Ambrym Caldera occurs in the central part of a broad composite cone that forms a N-S line with three smaller volcanoes (Tuvio, Rahoum, Tower Peak; fig. 5.11.2). After the caldera formed by relatively quiet subsidence less than 2,000 years ago, eruptions within it and along E-W-trending fissure zones built a broad shield. At about the time of caldera formation, the original N-S line of volcanoes was succeeded by the present E-W vent pattern. Two intracaldera cones (Mt. Benbow and Mt. Marum) remain active almost continuously, and there have been several historical flank eruptions (McCall and others, 1970).

AMBRYM (continued)

HISTORICAL ACTIVITY

pre-1894: Little is known of pre-1894 eruptions. Two columns of "smoke" were noted in 1774, and lava flowed from a vent along the southeast rift in 1888 (Fisher, 1957).

1894, 1912-15: The 1894 eruption was accompanied by numerous felt earthquakes (Purey-Cust, 1896; Aubert de la Rue, 1937, 1960). Frater (undated) writes that "a few years ago [1912-15] the regular, clockwork explosions (of Yasour, 500 km south of Ambrym) suddenly ceased....But no sooner had Tanna [Yasour] closed down than Mount Benbow ... burst out with increased activity. The eruptions of Mount Benbow, which had hitherto been spasmodic and irregular, now took place at regular intervals of three or four minutes, glowing at night....For a few weeks the Tanna volcano remained silent, and then, with a convulsive roar, the imprisoned giant broke its bonds and awoke to life. The clockwork regularity of Mount Benbow ceased, and the eruptions again became uncertain and spasmodic....As in the days of Captain Cook, the eruptions [of Benbow] were still taking place. There was the pillar of smoke by day and the pillar of fire by night....In December 1913, the age-long sleep of [extinct craters along Ambrym's east-west fissure] was broken....The outburst was heralded by a series of earthquake shocks which increased in frequency and severity until the solid earth reeled and tottered. The hospital buildings rocked like a ship at sea; the natives, in their picturesque mode of speech, saying that Ambrim danced. Then, from the newly-formed vents, was seen to rise, dark as the blackest London fog, a dense cloud which shot up like a pillar and spread out in all directions like a gigantic mushroom....Over an area of ten miles the earth seemed to have opened up, and out of this huge fissure tongues of living flame were shooting up into the sky. In one place, which seemed to be the centre of the disturbance, six volcanoes had burst out within a short distance of each other....Beginning at the extinct craters in the centre of the island, the line maintained a westerly direction, and every few miles a new volcano burst out...[and within 24 hours opened vents beneath the missionary hospital]." After the eruption, a 12-fathom-deep lagoon covered the former site of the hospital.

1937: Strong earthquakes occurred before the eruption of 28-30 March 1937 (Aubert de la Rue, 1937, 1960).

1950-54: The 1950 eruption was preceded by months of "tremors"; Taylor (1952, 1956) refers to records of an observer on the neighboring island of Paama, which suggest that the unusually prolonged and powerful eruption of 1951-54 was accompanied by abnormal seismic activity.

1959-80: No precursors were noted before a moderately large explosive event in 1972 that occurred during almost continuous eruptions from 1964 to 1980. Strong SO_2 emission during a small eruption of 10-11 February 1979 resulted in acid-rain fallout, damage to crops, and concern regarding human health.

1986: A small ash cloud was seen on 8 March, and more violent activity was reported to have occurred "three months earlier" (Smithsonian Institution, 1986).

1988: The crew of a commercial airliner reported an eruption from Benbow crater at 0357 hr on 13 February, with ash clouds below 12-km altitude. Activity had ceased by 19 February. Geologists in Vanuatu reported that activity from Benbow was "slightly stronger than usual" on 12-14 February (Smithsonian Institution, 1988).

AMBRYM, Region 5, CAVW number 05-07-04

AMBRYM (continued)

COMMENTS

The change from a N-S vent alignment before caldera collapse to an E-W alignment afterward may reflect development of Hawaiian-type rift zones under the influence of the D'Entrecasteaux Fracture Zone and a shallow magma reservoir.

REFERENCES

Aubert de la Rue, E., 1937, Le volcanisme aux Nouvelles Hebrides (Melanesie): Bull. Volcanol., ser. II, v. 2, p. 78-142.

----- 1960, Les manifestations actuelles du volcanisme aux Nouvelles Hebrides (Melanesie): Bull. Volcanol., v. 23, p. 197-206.

Fisher, N.H., 1957, Catalogue of active volcanoes of the world including solfatara fields, Pt. V, Melanesia: Rome, IAVCEI, 105 p.

Frater, M., undated, Midst volcanic fires: London, James Clarke and Co., 288 p.

McCall, G.J.H., LeMaitre, R.W., Malahoff, A., Robinson, G.P., and Stephenson, P.J., 1970, The geology and geophysics of the Ambrym Caldera, New Hebrides: Bull. Volcanol., v. 34, p. 681-696.

Purey-Cust, H.E., 1896, The eruption of Ambrym Island, New Hebrides (South West Pacific): Geogr. Jour., v. 8, p. 585-602.

Smithsonian Institution, Scientific Event Alert Network (SEAN), 1979, 1986, 1988, Ambrym: SEAN Bull., v. 4, nos. 3-4; v. 11, no. 3; v. 13, no. 2 (contributions from A. MacFarlane, R.E. Stoiber, J.H. Latter, and C. Clark).

Taylor, G.A., 1952, Report on volcanic activity on Ambrym Island: Bur. Min. Res., Australia, Rec. 1952/4.

----- 1956, Review of volcanic activity in the territory of Papua-New Guinea, the Solomon and New Hebrides Islands, 1951-1953: Bull. Volcanol., v. 18, p. 25-37.

Volcanological Society of Japan, 1974, 1981-82, Ambrym: Bull. Volcanic Eruptions, no. 12, p. 15-17; no. 19, p. 27; no. 20, p. 32 (contributions from A. MacFarlane and others).

Warden, A.J., 1970, Evolution of the Aoba caldera volcano, New Hebrides: Bull. Volcanol., v. 34, p. 107-140.

Williams, C.E.J., and Warden, A.J., 1964, Progress report of the Geological Survey for the period 1959-1962: British Service, New Hebrides Anglo-French Condominium, 75 p.

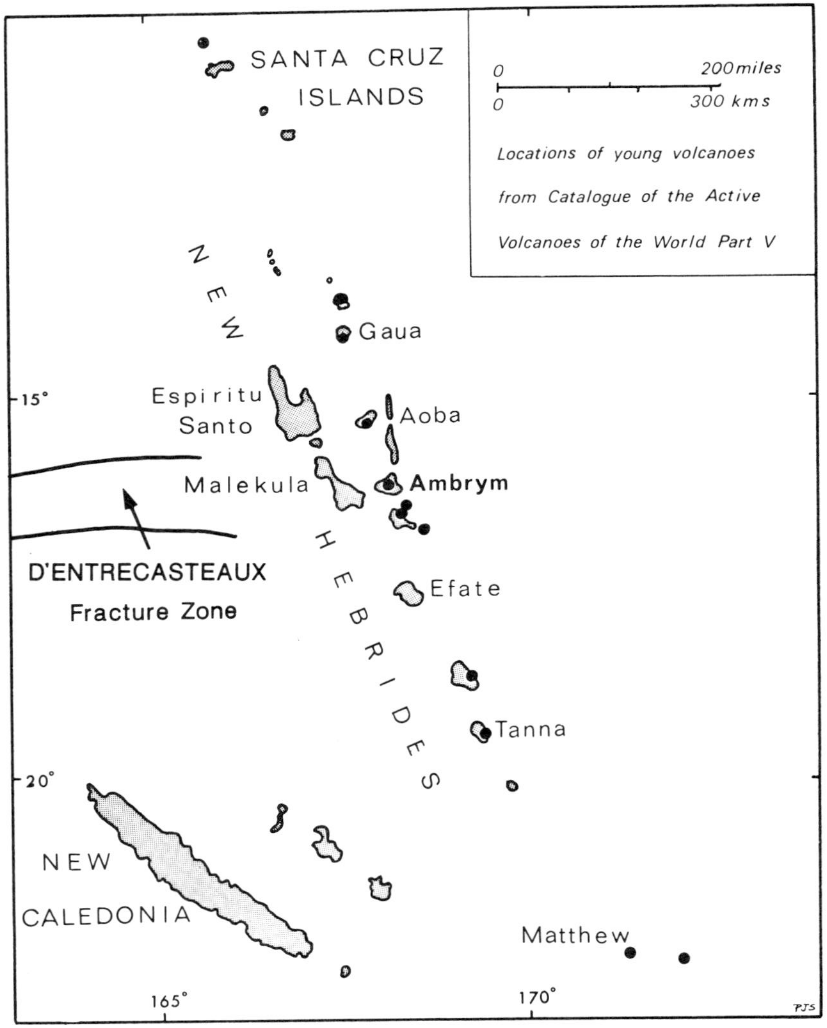

Figure 5.11.1. The Santa Cruz - New Hebrides volcanic chain in central New Hebrides Archipelago, from McCall and others (1970). Solid circles indicate active volcanoes. Note that Ambrym lies at intersection of volcanic arc with D'Entrecasteaux Fracture Zone.

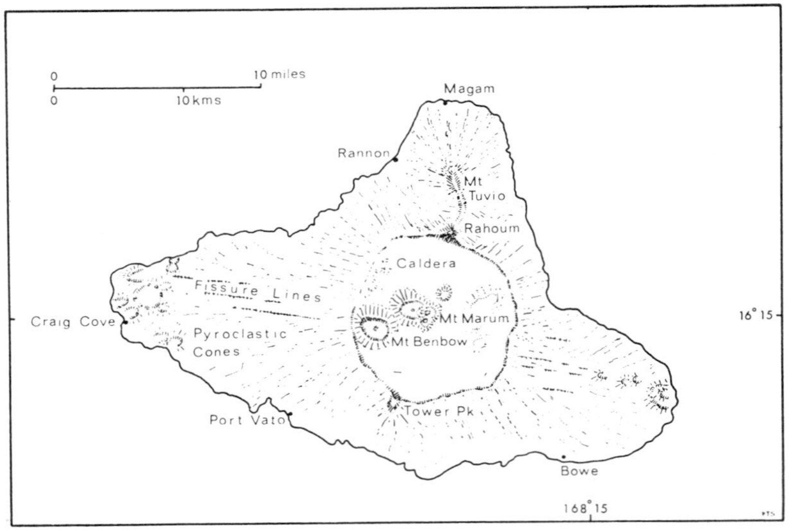

Figure 5.11.2. Sketch map of Ambrym Island, from McCall and others (1970).

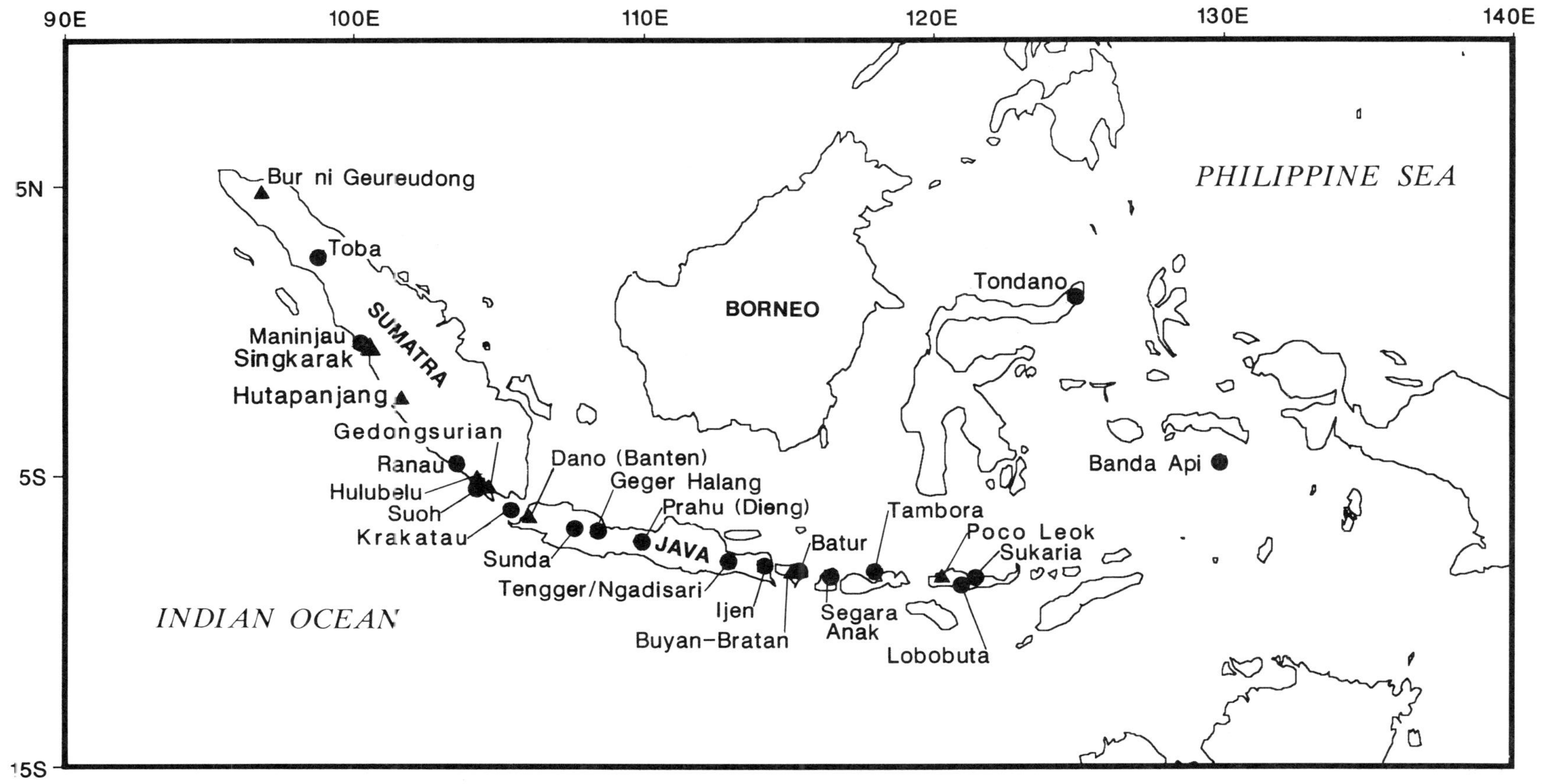

Figure 6. Locations of large, Quaternary restless calderas (solid circles) and nonrestless calderas (triangles) in Region 6.

PART 3: HISTORICAL UNREST AT LARGE QUATERNARY CALDERAS (continued)
See inside back cover for explanation and abbreviations

TOBA

CAVW number (active vent)	Latitude Longitude (degrees)	Diameter (km)	Local tectonic setting	Precaldera edifice	SiO_2 content (wt pct)	Age of collapse (yr B.P.)	Date of unrest	Type and duration of unrest ESTU STHF MGTF H Te	Eruption type
06-01-09 (Pusukbukit)	2.6 N 98.8 E	35 x 100	Shear, with graben; may also lie astride segment boundary	None	R = hb-a C = 68-77	75,000 <700,000 840,000	1892	x--- ---- ---- - --	none
							1916	x?-? ---- ---- - Q?	none
							1920-22	xx-? x--- ---x - QU	none
							1987	x--? ---- ---- - --	none

TECTONIC SETTING

Toba is one of the classic volcano-tectonic depressions of the world, where caldera collapse was strongly influenced by preexisting regional faults (figs. 6.1.1, 6.1.2). Collapse of the Toba depression occurred along the Ulu Air or great Sumatran (=Semangko) Fault Zone (van Bemmelen, 1929, 1939, 1949; Clarke and others, 1982; Aldiss and others, 1983; Aldiss and Ghazali, 1984). Westerveld (1947) and Verstappen (1961) challenged van Bemmelen's conclusion that collapse occurred because of eruptions, arguing instead that the depression is largely tectonic and that eruptions occurred along the tectonic faults. However, most workers now agree that eruption of the immense volumes of Toba tuffs (see below) must have been followed by major collapse, even if a tectonic depression existed before eruption of those tuffs. Some of the evidence for a graben that predates the latest Toba tuff is equally consistent with a caldera or volcano-tectonic depression associated with earlier Toba tuff eruptions.

A chain of andesitic volcanoes changes its relation to the Sumatran Fault Zone at Toba Caldera (fig. 6.1.3, inset, SFZ). South of Toba, the active volcanoes lie within or very near the fault zone, whereas from Toba northward they are displaced to the east of the fault zone and spread over a broader area. Page and others (1979) note that this change occurs across a N-S-trending zone of high seismicity, which itself is aligned along the northward projection of the Investigator Ridge transform fault. They interpret this zone of seismicity to lie above the boundary between two segments of subducted lithosphere and suggest (following Stoiber and Carr, 1973) that Toba owes its existence in part to extra magma generation along this segment boundary.

GEOLOGIC HISTORY

The Barisan Range of western Sumatra was uparched during Pliocene and early Pleistocene time, possibly by the rise of plutons beneath a chain of andesitic volcanoes (van Bemmelen, 1949). Andesitic volcanism continued along and near the fault zone, in some places to historical time. Locally, rhyolitic and dacitic pumice was erupted from silicic volcanic centers, as around Lakes Ranau, Maninjau, and Toba. Several tuffs are known at Toba (fig. 6.1.2), and Knight and others (1983) and Chesner (1984, 1985) suggest that the caldera formed in stages. Early Toba eruptions occurred about 840,000 yr B.P. and an indeterminate but lesser number of years ago; the most recent ignimbrite eruption produced the Young Toba Tuff (YTT) ca. 75,000 yr ago along a ring fracture around most or

TOBA (continued)

GEOLOGIC HISTORY (continued)

all of the present lake (Ninkovich and others, 1978; Rose and Chesner, 1987; C. Chesner, written commun., 1988). A still younger eruption (30,000 yr B.P.) is suggested by radiocarbon ages of a silicic ash on the Malay Peninsula (Stauffer and others, 1980, 1987), but deposits of that age are not known in the immediate Toba area. Aldiss and Ghazali (1984) suggested that the 30,000 yr B.P. tuff might have been erupted from Sibuatan, immediately northwest of Lake Toba; another possibility is that the Malaysian ash is in fact YTT ash (Rose and Chesner, 1987) and that one or the other of the inferred ages is incorrect. Rose and Chesner (1987) infer that at least 2800 km^3 of magma were erupted in the YTT eruption, by far the largest known eruption in the Quaternary.

Samosir Island and the Uluan Peninsula (just north of Samosir; figs. 6.1.2, 6.1.3) are parts of one or two resurgent domes. Knight and others (1983, 1986) interpret Samosir and Uluan as two generations of resurgent domes, both of which have been reactivated since the YTT eruption. C. Chesner and W.I. Rose (written commun., 1988) interpret Samosir and Uluan to be a single resurgent dome, cut by a medial graben. Verstappen (1961) and Clarke and others (1982) relate faults of the Latung Strait to the strike-slip Semangko Fault; alternatively, the Latung Strait graben is primarily a collapsed, axial graben above a Samosir-Uluan resurgent dome (C. Chesner, written commun., 1988). Based on the maximum elevation (above lake level) of lacustrine sediments on Samosir Island, net uplift of Samosir has been at least 450 m (Verstappen, 1961); an additional 400+ meters of uplift is suggested by the height of Samosir Island above deep basins to its northwest and southeast (fig. 6.1.4).

Klein (1917) and Verstappen (1961, 1973) recognized two major terraces along the southeastern side of the caldera (terraces I and II; figs. 6.1.5, 6.1.6), and at least five poorly developed younger terraces in the same area. Terraces I and II were uplifted and tilted outward during resurgence (fig. 6.1.7); the younger, poorly developed terraces are not known to be uparched. Terrace I can now be traced from 1,070 m to 1,420 m above sea level, and terrace II (younger) from 1,050 to 1,200 m above sea level. Lake level is presently about 905 m above sea level, but may have been 1,050 m above sea level before downcutting of the Asahan gorge. Postcaldera vertical movements clearly have been substantial, and may be continuing (Aldiss and Ghazali, 1984; see also, Historical Activity, 1987, below).

Pusukbukit is a small stratovolcano that has grown along the west margin of the caldera, connecting Samosir Island and the west shore of the lake. Shallow rhyolitic(?) intrusions beneath the west side of Samosir Island have deformed uplifted lake sediments (T. Casadevall, D. Sussman, and C. Chesner, written commun., 1988), and similar, local updoming is inferred in basement rock immediately west of Samosir Island (C. Chesner, written commun., 1988) (fig. 6.1.3). Boiling-temperature hot springs occur in several locations on the western ring fracture and along the west edge of Samosir Island.

HISTORICAL ACTIVITY

No eruption has occurred in the caldera in historical time. However, the southern rim of the caldera and nearby areas are seismically very active and have apparently been the site of strong earthquakes in 1892, 1916, 1920-22, and 1987 (Figee and Onnen,

TOBA (continued)

HISTORICAL ACTIVITY (continued)

1893; Koninklijk Magnetisch en Meteorologisch Observatorium te Batavia, 1922; Soetardjo and others, 1985; Smithsonian Institution, 1987).

1892: A great earthquake on 17 May 1892, centered approximately 125 km south of the caldera, destroyed many buildings in South Tapanuli and was accompanied by as much as 1.3 m of right-lateral displacement along the Semangko Fault (Muller, 1895; Reid, 1913). The area of Lake Toba was shaken but not otherwise disturbed. Soetardjo and others (1985) list an earthquake that was felt with MMI=VI on the same date, near Prapat on the shore of Lake Toba.

ca. 1914: Around 1914 a slump occurred near the upper end of the Asahan Valley, causing Lake Toba to rise about 1.5 m (Verstappen, 1973, citing an unpublished report by R.W. van Bemmelen). The obstruction was removed by dynamite to stop flooding of lakeside rice paddies. No tectonic or volcanologic significance is inferred.

1916: On 22 February 1916, offices and homes in Tarutung (40 km southeast of Toba Caldera) were destroyed by an earthquake.

1920-22: An earthquake in November 1920 was followed by major swarms beginning on 24 January and 1 April 1922. Damage was greatest along the southwest shore of Lake Toba, where ground subsided into the lake at Muara, Bakara, and near Nainggolan. The epicenter of the strongest shock was in the Toba basin. Fumaroles west and south of Lake Toba showed increased activity immediately following large earthquakes.

1987: On 25 April 1987, a M=6.5 earthquake destroyed numerous houses and public buildings in Tarutung. The earthquake was centered at the south end of Lake Toba, along the Semangko Fault near the town of Muara (Smithsonian Institution, 1987).

During at least several years leading up to the time of this writing (February 1988), lake level as measured near the Asahan Gorge (southeast Toba) hydroelectric dam intake dropped by as much as 2 m, leading to concern about the supply of water for power generation (Casadevall, 1987). A likely explanation for this drop in lake level is a change in the hydrologic balance between rainfall (low in recent years) and outflow (increased since 1983 for hydroelectric power generation). However, the drop in lake level might alternatively reflect uplift of the outlet site relative to other parts of the lake because the decrease in lake level has presently been confirmed at only one station, in an area of terraces that indicate geologically recent uplift (figs. 6.1.5-6.1.7).

TOBA (continued)

COMMENTS

Many historical earthquakes in the Toba and adjacent Tapanuli areas have been associated with shear along the Semangko Fault Zone. The earthquakes listed above hint at a crude northwesterly migration along the Semangko Fault Zone; if this migration is real, more activity might occur in the Toba region in the next few decades. Earthquakes near Lake Toba may release regional stress, preferentially at the structurally weak intersection of regional faults and the caldera-bounding fractures.

If resurgent uplift is still continuing, as hinted at by the terraces and the lake-level change noted above, additional microseismicity might be recorded in the area of resurgence. Seismic monitoring and repeated measurements of lake level at points all around Lake Toba would be of great interest and are currently planned. Information about possible uplift would be of considerable practical as well as scientific value.

REFERENCES

Aldiss, D.T., and Ghazali, S.A., 1984, The regional geology and evolution of the Toba volcano-tectonic depression, Indonesia: Jour. Geol. Soc. London, v. 141, p. 487-500.

Aldiss, D.T., Whandoyo, R., Sjaefudien, A.G., and Kusyono, 1983, The geology of the Sidikalang quadrangle, Sumatra: Bandung, Geol. Res. Devel. Centre, 41 p. plus geologic map, 1 sheet, scale 1:250,000.

Bemmelen, R.W. van, 1929, The origin of Lake Toba (N. Sumatra): Proc. 4th Pacific Sci. Congr., Bandung, v. IIA, p. 115-125.

----- 1939, The volcano-tectonic origin of Lake Toba (North Sumatra): De Ing. in Ned. Ind., v. 6, no. 9, p. 126-140.

----- 1949, The geology of Indonesia, v. 1A: The Hague, Government Printing Office, 732 p.

Braak, C., 1920, De waterstand van het Toba Meer: Nat. Tijdschr. v. Ned. Ind., pt. 74, p. 28-36; pt. 80, p. 77-80.

Casadevall, T.J., 1987, Recent lowering of water level of Lake Toba, northern Sumatra, Indonesia: unpubl. rep., U.S. Geol. Surv./Volcanol. Surv. Indonesia, 9 p.

Chesner, C.A., 1984, Ages of ignimbrites from Toba caldera, Sumatra: Lessons in dating Quaternary volcanic rocks [abs.]: Eos, Trans. Am. Geophys. Union, v. 65, no. 45, p. 1134.

----- 1985, Geochemistry of the Toba ignimbrites: Implications for silicic magma bodies, outflow patterns, and caldera collapse [abs.]: Eos, Trans. Am. Geophys. Union, v. 66, no. 46, p. 1141.

Chesner, C.A., and Rose, W.I., Jr., 1983, Geochemistry, mineralogy, and physical constraints of the Toba ignimbrites [abs.]: Eos, Trans. Am. Geophys. Union, v. 64, no. 45, p. 873.

----- 1987, Pre-eruptive conditions of magma bodies giving rise to the Toba Tuffs, Sumatra [abs.]: Abstracts, Hawaii Symposium on How Volcanoes Work, Hilo, January 1987, p. 36.

Clarke, M.C.G., Ghazali, S.A., Harahap, H., Kusyono, and Stephenson, B., 1982, The geology of the Pematangsiantar quadrangle: Bandung, Geol. Res. Devel. Centre, 26 p. plus geologic map, 1 sheet, scale 1:250,000.

Durham, J.W., 1940, Oeloe Aer Fault Zone, Sumatra: Bull. Am. Assoc. Pet. Geol., v. 24, p. 359-362.

TOBA (continued)

REFERENCES (continued)

Figee, S., and Onnen, H., 1893, Vulkanische verschijnselen en aardbevingen in den O.I. Archipel waargenomen gedurende het jaar 1892: Nat. Tijdschr. v. Ned. Ind., pt. 53, p. 132-161.

----- 1895, Vulkanische verschijnselen en aardbevingen in den O.I. Archipel waargenomen gedurende het jaar 1893: Nat. Tijdschr. v. Ned. Ind., pt. 54, p. 205-235.

Kemmerling, G.L.L., 1921, Vulkanen en vulkanische verschijnselen in de Residentien Sumatra's Westkust en Tapanoeli: Vulkanol. Meded., v. 1, Weltevreden.

Klein, W.C., 1917, De Oostoever van het Tobameer in Noord Sumatra. Jaarb. Mijnw. in Ned. Ind., v. 46, Verh. I, p. 136-187.

Knight, M.D., Walker, G.P.L., Caress, M., and Hehanussa, P.E., 1983, The most recent volcanic history of the Toba depression, Sumatra [abs.]: Eos, Trans. Am. Geophys. Union, v. 64, no. 45, p. 873.

Knight, M.D., Walker, G.P.L., Ellwood, B.B., and Diehl, J.F., 1986, Stratigraphy, paleomagnetism, and magnetic fabric of the Toba Tuffs: Constraints on their sources and eruptive styles: Jour. Geophys. Res., v. 91, no. B10, p. 10,355-10,382.

Koninklijk Magnetisch en Meteorologisch Observatorium te Batavia, 1922, Vulkanische verschijnselen en aardbevingen in den Oost-Indischen Archipel waargenomen gedurende het jaar 1921: Nat. Tijdschr. v. Ned. Ind., pt. 82, p. 222-269.

Marel, H.W. van der, 1948, Het Tobameer: Geol. Mijnb., v. 10, no. 4, p. 80-89.

Muller, J., 1895, Nota betreffende de verplaatsing van eenige triangulatie pilaren in de residentie Tapanuli tgv. de aardbeving van 17 Mei 1892: Nat. Tijdschr. v. Ned. Ind., pt. 54, p. 299-307.

Ninkovich, D., Sparks, R.S.J., and Ledbetter, M.T., 1978, The exceptional magnitude and intensity of the Toba eruption, Sumatra: An example of the use of deep-sea tephra layers as a geological tool: Bull. Volcanol., v. 41, p. 286-298.

Nishimura, S., Abe, E., Nishida, J., Yokoyama, T., Dharma, A., Hehanussa, P., and Hehuwat, F., 1984, A gravity and volcano-stratigraphic interpretation of the Lake Toba Region, North Sumatra, Indonesia: Tectonophysics, v. 109, nos. 3-4, p. 253-272.

Page, B.G.N., Bennett, B.N., Cameron, N.R., Bridge, D.M., Jeffery, D.H., Keats, W., and Thaib, J., 1979, A review of the main structural and magmatic features of northern Sumatra: Jour. Geol. Soc. London, v. 136, p. 569-579.

Reid, H.F., 1913, Sudden earth movements in Sumatra in 1892: Bull. Seismol. Soc. Am., v. 3, p. 72-79.

Rose, W.I., and Chesner, C.A., 1987, Dispersal of ash in the great Toba eruption, 75 ka: Geology, v. 15, p. 913-917.

Ruttner, F., 1931, Hydrographische und hydrochemische Beobachtungen auf Java, Sumatra und Bali: Arch. Hydrobiol., Supp., v. 8, p. 197-454.

----- 1935, Kielselgur und andere lakustrische Sedimente im Tobagebeit: Ein Beitrag zur Geschichte des Tobasees in Nordsumatra: Arch. Hydrobiol., Supp., v. 13, Tropische Binnengewasser, V., p. 399-461.

Smith, R.L., and Bailey, R.A., 1968, Resurgent cauldrons, in Coats, R.R., Hay, R.L., and Anderson, C.A., eds., Studies in volcanology: Geol. Soc. Am. Memoir 116, p. 613-662.

Smithsonian Institution, Scientific Event Alert Network (SEAN), 1987, Earthquakes of April 1987: SEAN Bull., v. 12, no. 4, p. 17 (information from Volcanological Survey of Indonesia, T. Casadevall, and USGS National Earthquake Information Center).

Soetardjo, R., Untung, M., Arnold, E.P., Soetadi, R., Isail, S., and Kertapati, E., 1985, Indonesia: SEASEE Series on Seismology, v. 5, 199 p.

TOBA (continued)

REFERENCES (continued)

Stauffer, P.H., Nishimura, S., and Batchelor, B.C., 1980, Volcanic ash in Malaya from a catastrophic eruption of Toba, Sumatra, 30,000 years ago, in Nishimura, S., ed., Physical geology of Indonesian island arcs: Kyoto Univ., p. 156-164.

----- 1987, Pleistocene volcanic ash on the Malay Peninsula: Evidence for magnitude and age of a catastrophic eruption of Toba, Sumatra [abs.]: Abstracts, Hawaii Symposium on How Volcanoes Work, Hilo, January 1987, p. 241.

Stehn, Ch.E., 1939, De nieuwe dieptekaart van het Toba meer (Noord Sumatra) van Drost & Beckering: De Ing. in Ned. Ind., v. 6, no. 9, sect. IV, p. 121-126.

Stoiber, R.E., and Carr, M.J., 1973, Quaternary volcanic and tectonic segmentation of Central America: Bull. Volcanol., v. 37, no. 3, p. 304-325.

Verstappen, H.Th., 1961, Some "volcano-tectonic" depressions of Sumatra, their origin and mode of development: Kon. Ned. Akad. Wetensch., Amsterdam, Proc. B, v. 64, p. 428-443.

----- 1973, A geomorphological reconnaissance of Sumatra and adjacent islands (Indonesia): Verh. Roy. Dutch Geogr. Soc., Gröningen, Wolters-Noordhoff Publ., 182 p.

Westerveld, J., 1947, On the origin of the acid volcanic rocks around Lake Toba, North Sumatra: Verh. Kon. Ned. Akad. v. Wetensch., Amsterdam, Afd. Natuurk., 2nd sect., pt. 43, no. 1, p. 1-51.

Zen, M.T., 1982, Toba resurgent cauldron: volcano-tectonic evolution and magmatic cycles: Bul. Dep. Teknik Geologi, Institut Teknologi Bandung (ITB), v. 7, p. 27-43.

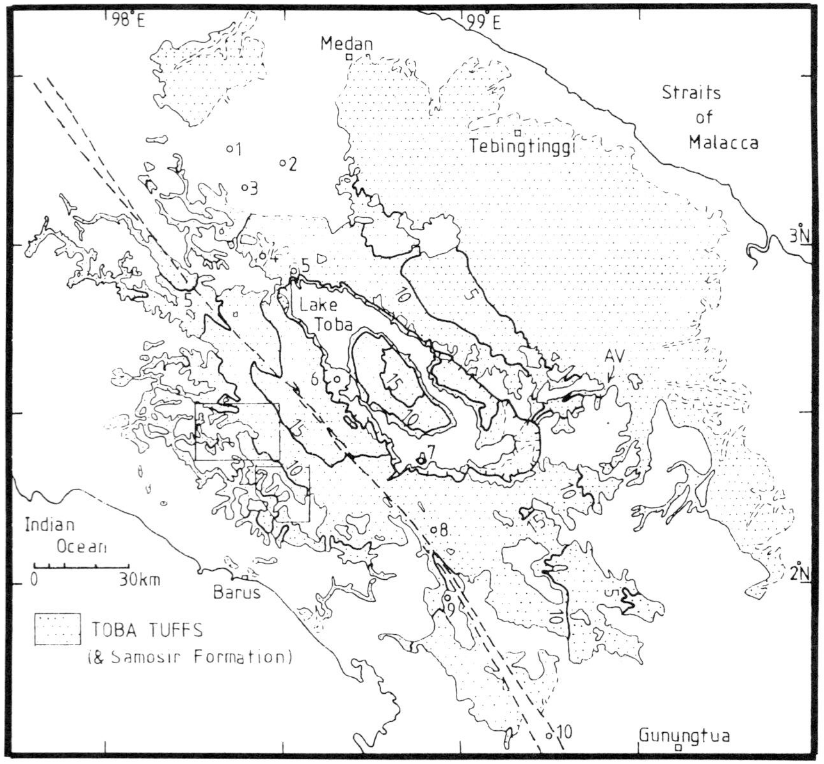

Figure 6.1.1. Distribution of Toba tuffs and relation of Toba Caldera to regional faults, from Aldiss and Ghazali (1984). Contact of Toba tuffs with older formations is shown with a solid line, otherwise with a broken line. Thick solid lines are topographic contours on surface of Toba tuffs, in hundreds of meters. Thick broken lines mark Sumatran (Semangko) fault zone. Topographic margin of Toba depression is shown by a thick line with ticks inside, equivalent to 1,500-m and 1,000-m contours. Small circles mark Quaternary volcanic centers: 1, Bekulap; 2, Sibayak; 3, Sinabun; 4, Sibuatan; 5, Tandukbenua; 6, Pusukbukit; 7, Pardepur; 8, Imun; 9, Martimban; 10, Sipirok. AV, Asahan Valley. Reproduced by permission of the Geological Society, London.

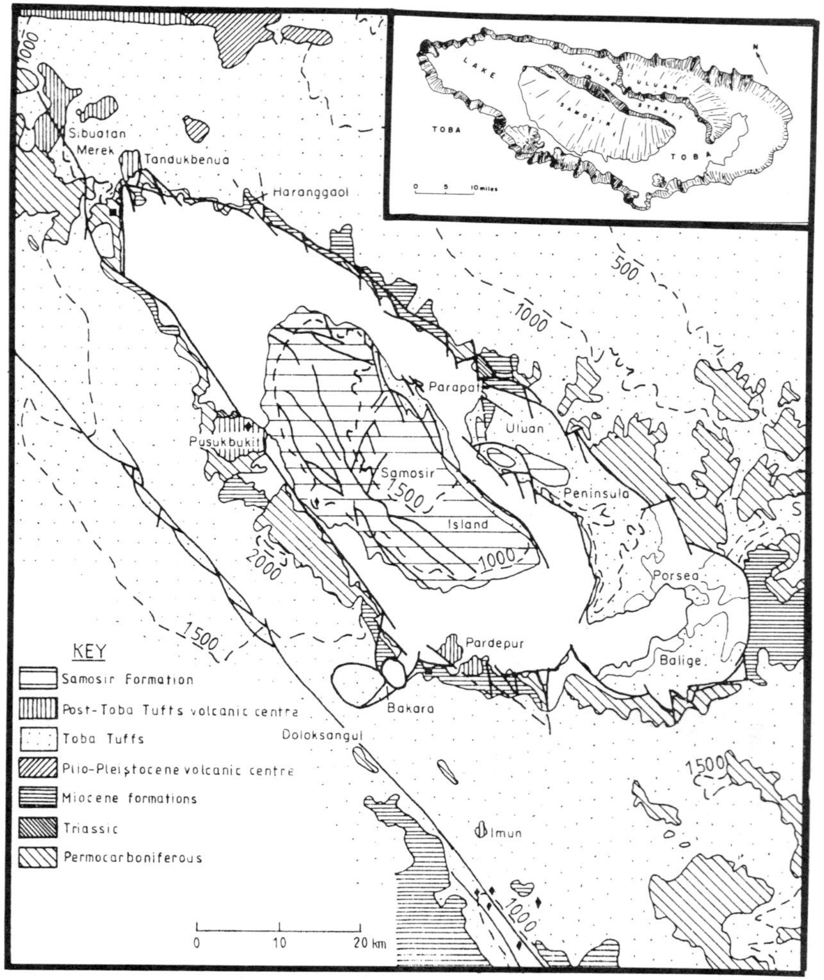

Figure 6.1.2. Geologic map of Toba volcano-tectonic depression, from Aldiss and Ghazali (1984). Thick lines are faults; broken lines are contours, in meters. Diamonds indicate clusters of hot springs. Small squares mark the intrusive phases of Toba ignimbrite. S, Siguragura waterfall (Asahan Valley). Inset from Smith and Bailey (1968). Reproduced by permission of the Geological Society, London.

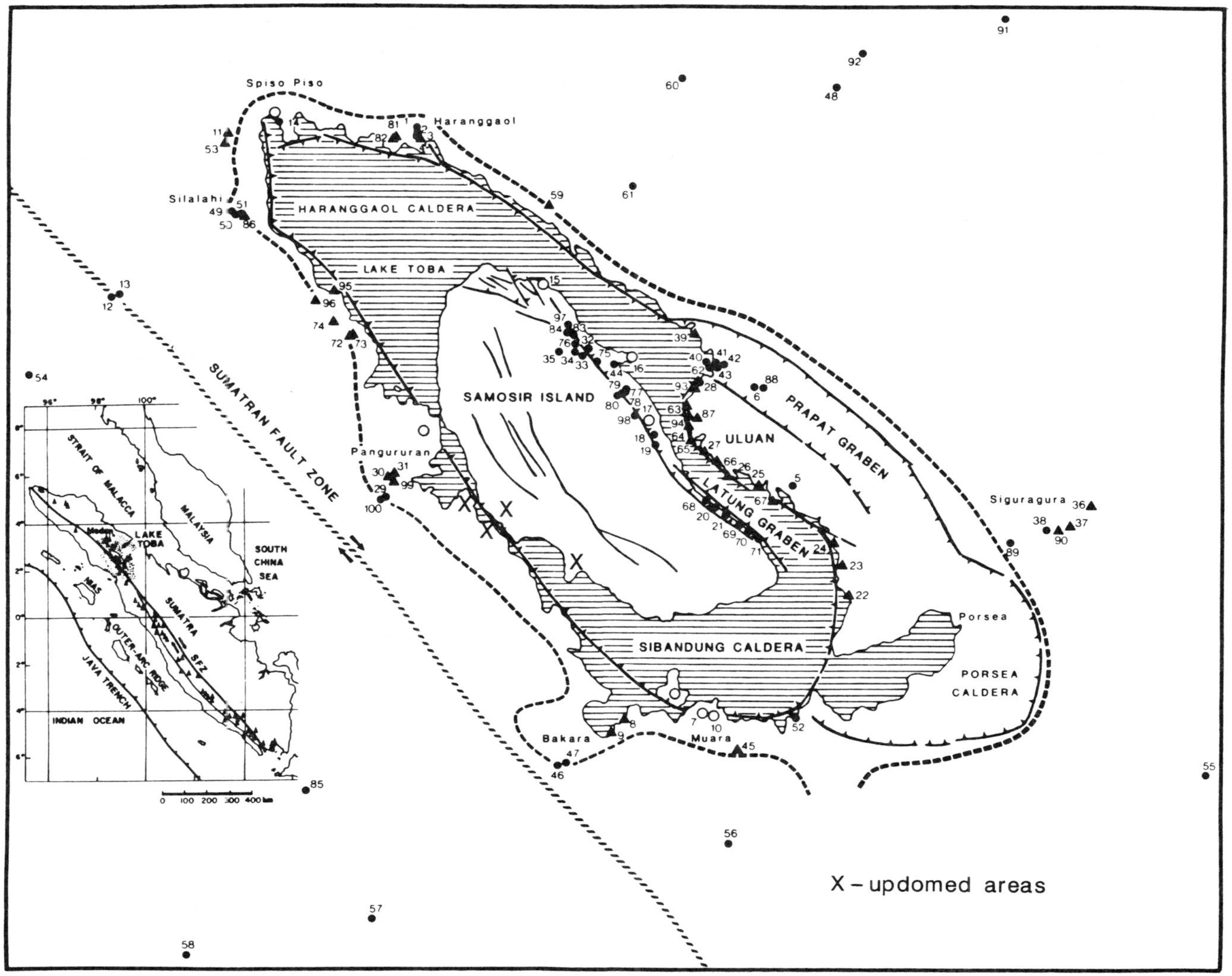

Figure 6.1.3. Toba depression showing multiple calderas as interpreted by Knight and others (1986). Solid circles represent sites of welded, normally magnetized ignimbrite; solid triangles represent reversely magnetized ignimbrite; and open circles represent andesitic to rhyolitic post-Toba-tuff lava domes. Locations of updomed areas marked by X's (Craig Chesner, written commun., 1988). Location map of Sumatra (inset) shows extent of Toba ignimbrite outflow sheets (stippled), Quaternary stratovolcanoes along Sunda arc (solid triangles), and Sumatran Fault Zone, in relation to Outer-Arc Ridge (accretionary prism), and Java Trench (Sunda Trench) in which the Indian Ocean lithospheric plate is subducting obliquely to northeast at 67 mm/yr under Eurasian plate. Copyright by the American Geophysical Union.

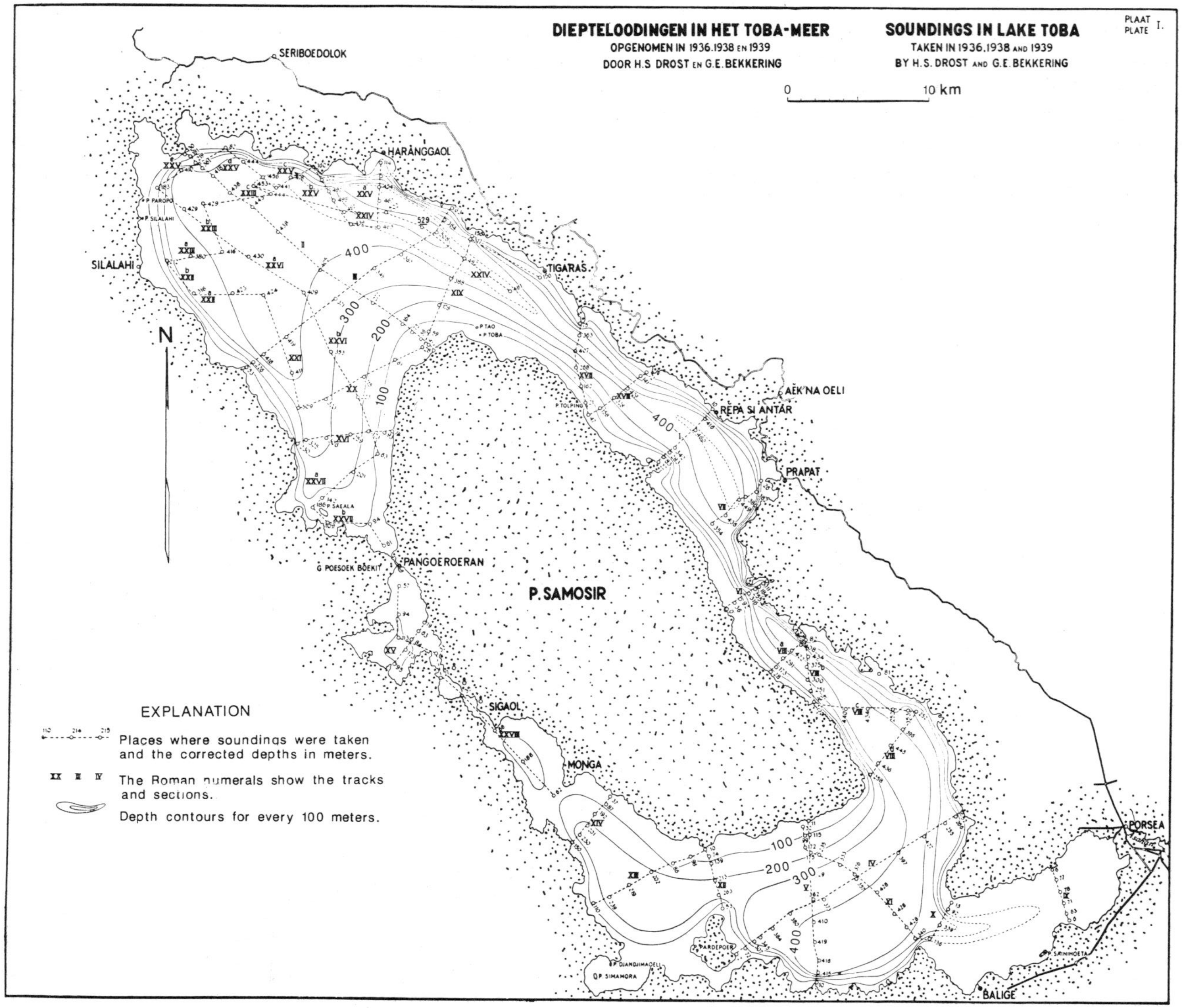

Figure 6.1.4. Bathymetry of Lake Toba, from Stehn (1939). Net uplift of Samosir Island has been at least 450 meters, based on maximum elevation (above lake level) of lacustrine sediments on the island (Verstappen, 1961). An additional 400+ meters of uplift is suggested by height of Samosir above deep basins to its northwest and southeast. Stippling marks edge of landmass.

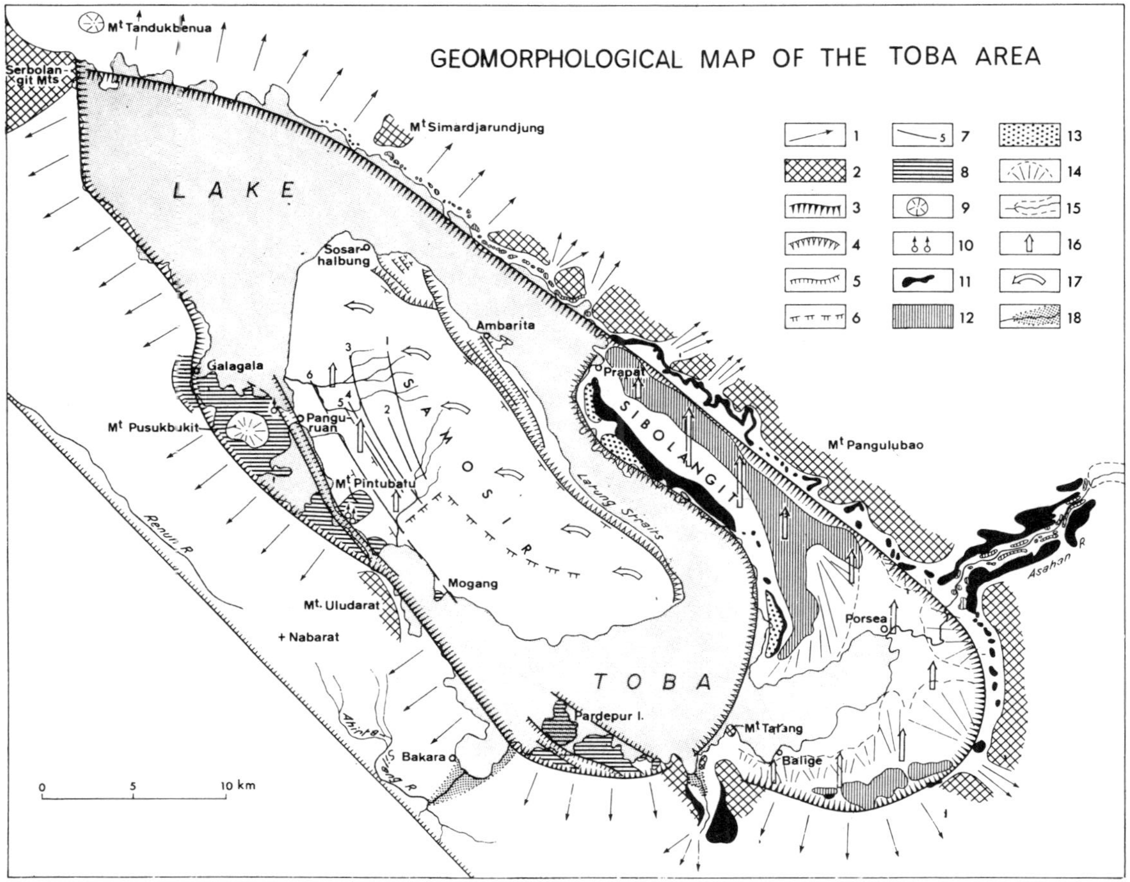

Figure 6.1.5. Geomorphological map of Toba graben, from Verstappen (1973). Explanation: 1, ignimbrites of most recent Toba eruption (arrows indicate places of overflow); 2, pre-Toba rocks emerging from tuff cover; 3, major fault scarps; 4, medium-size fault scarps; 5, minor fault scarps; 6, probable fault scarps; 7, minor faults or grabens; 8, slivers and other subsided blocks; 9, post-Toba volcanic cones; 10, solfataras; 11, Toba terrace I; 12, Toba terrace II; 13, areas emerging from first Toba terrace (cn Sibolangit Peninsula); 14, multicycle fluvio-volcanic fans; 15, Asahan amphitheatre; 16, post-Toba uplift/warping; 17, post-Toba tilting; 18, alluvial plains.

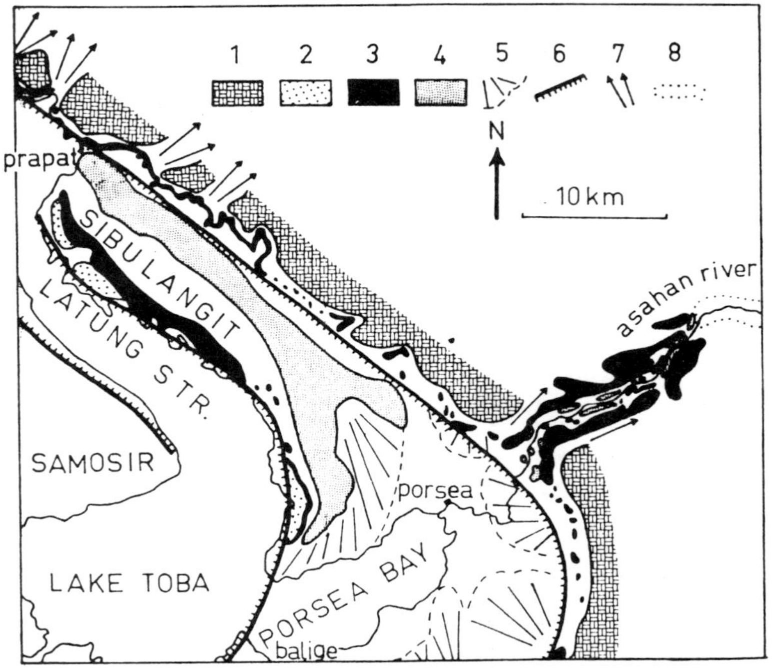

Figure 6.1.6. Geomorphic map of southeast part of Toba depression, showing location of warped lake terraces (from Verstappen, 1961). Explanation: 1, pre-Toba rocks emerging from tuff cover; 2, areas emerging from first Toba terrace; 3, first Toba terrace; 4, second Toba terrace; 5, multicycle alluvial fan; 6, fault scarp; 7, overflow of Toba tuffs; 8, escarpment of Asahan amphitheatre.

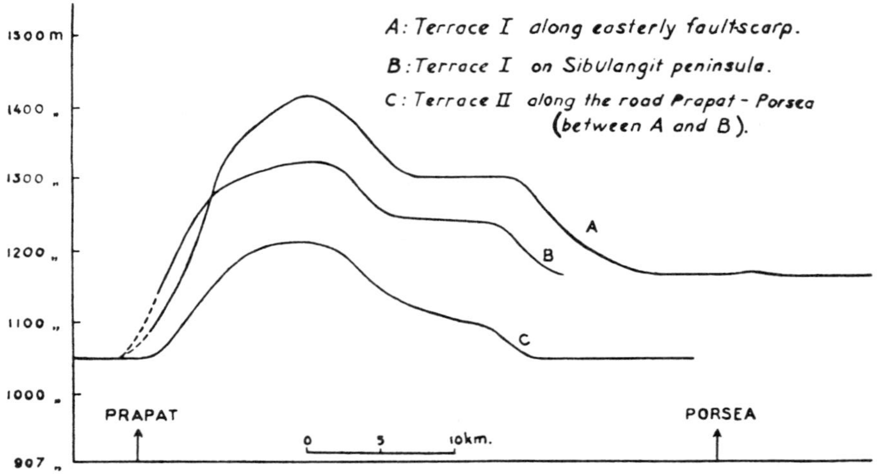

Figure 6.1.7. Warped lake terraces along eastern shore of Lake Toba between Prapat and Porsea, from Verstappen (1961).

MANINJAU

CAVW number (active vent)	Latitude Longitude (degrees)	Diameter (km)	Local tectonic setting	Precaldera edifice	SiO_2 content (wt pct)	Age of collapse (yr B.P.)	Date of unrest	Type and duration of unrest ESTU STHF MGTF H Te	Eruption type
06-01-14	0.33S	11 x 22	Exten and	3 strato?	R = 55	280,000	1770 (M)	---- ---- ---- - --	ex
(Marapi= M)	100.18E		right-		C = 74	or	1807 (M)	---- ---- ---- - --	ex
06-01-15			lateral		with	70-80,000	1822 (M)	---- ---- ---- - --	ex
(Tandikat= T)			shear		late-		1833-34(M)	---- ---- ---- - --	ex
06-01-151					stage		1845 (M)	---- ---- ---- - --	ex
(Maninjau= Mn)					55		1854-56(M)	xx-- ---- ---- - --	ex
							1861 (M)	---- ---- ---- - --	ex
							1863 (M)	---- ---- ---- - --	ex
							1871 (M)	x--- ---- ---- - --	ex
							1876-77(M)	---- ---- ---- - --	ex
							1883 (M)	---- ---- ---- - --	ex
							1885-89(M,T)	x-x- ---- x--x - --	Ex, pex
							1904-19(M,T)	---- ---- ---- - --	ex
							1924 (T)	---- ---- ---x - --	none?
							1927 (M)	---- ---- ---- - --	ex
							1930 (M)	---- ---- ---- - --	ex
							1932? (M)	x--- ---- ---- - --	ex
							1939 (Mn)	---- ---- ---? - QU?	none
							1949-52(M)	---- ---- ---- - --	ex
							1954-58(M)	---- ---- ---- - --	ex
							1966-71(M)	---- ---- ---- - --	ex, pex
							1973 (M)	---- ---- ---- - --	ex
							1975-80(M)	---- ---- ---- - --	ex, pex
							1982-84(M)	---- ---- ---- - --	ex
							1987-88(M)	---- ---- ---- - --	ex

TECTONIC SETTING

Maninjau lies along the Sunda island arc, about 250 km northeast of the Java Trench and 350 km southeast of Toba Caldera (fig. 6.2.1). The Semangko ("Great Sumatran" or "Central Barisan") Fault Zone passes about 10 km northeast of Maninjau (Katili, 1969, 1970; Posavec and others, 1973; Verstappen, 1973; Leo and others, 1980), between Tandikat and Marapi (Sumatra) Volcanoes.

MANINJAU (continued)

TECTONIC SETTING (continued)

Katili (1969, 1970) inferred approximately 6 cm/yr of right-lateral displacement along a section of the Semangko Fault in northern Sumatra, based on 20 km of displacement of a Toba tuff then believed to be 300,000 yr old. More recent age determinations (see Toba) might raise or lower this rate, but certainly displacements of several centimeters per year are reasonable in this setting. Regional tectonic calculations suggest right-lateral motion of 4 cm/yr (Circum-Pacific Map Project, Southwest Quadrant, 1981). At 4 cm/yr, a 280,000-yr-old Maninjau Caldera would have moved 11 km northwestward relative to Marapi Volcano, reducing though not eliminating the current separation of the two. A similar argument can be made for Tandikat, though tectonic separation of Tandikat and Maninjau has probably been less because of Tandikat's position within, rather than across, the shear zone. If we consider the young, active Marapi fixed in relation to the source of subduction-zone magma, Maninjau may have moved as much as 11 km northwestward from the focus of magma supply to this volcanic center.

GEOLOGIC HISTORY

Maninjau Caldera is the largest silicic center in the Padang volcanic area of south-central Sumatra (fig. 6.2.2, 6.2.3). Posavec and others (1973) inferred that the elongate caldera contains three explosion centers that decrease in age southward (fig. 6.2.4). Zen (1972), Verstappen (1973), Purbo-Hadiwidjoyo and others (1979), and Leo and others (1980) identified a sequence of precaldera andesites, rhyolitic tuff, and andesitic tuff related to Maninjau Caldera. Tuffs associated with caldera collapse are 0.28 ± 0.12 m.y. old (K-Ar)(Leo and others, 1980) or 0.07-0.08 ± 0.02 m.y. old (fission track dating)(Nishimura, 1980). Purbo-Hadiwidjoyo and others estimated the volume of the Maninjau tuffs to be 220 to 250 km^3, approximately equal to the volume of the caldera; Leo and others (1980) estimated the volume of the same tuff to be 100 km^3 (including 3-6 km^3 of late-stage andesitic ash-flow tuff), comparable to their estimate of 122 km^3 for the volume of the caldera.

Tandikat and Marapi, andesitic stratovolcanoes situated 20 and 30 km, respectively, southeast of Maninjau Caldera, are possibly related to Maninjau (see "Tectonic Setting," above).

HISTORICAL ACTIVITY

Marapi erupts every few years, most recently (as of the time of this writing) in January 1988; Tandikat is known to have erupted in 1889 and 1914. No sympathetic unrest has been reported in Maninjau Caldera, nor did an eruption occur shortly before or after the probable overturn of Lake Maninjau in 1939.

1854-55: A severe earthquake occurred on 29 August 1854, and explosions and ash columns were reported from Marapi in August 1854 and October 1855. Many "shuddering" earthquakes were felt in October 1854 (Anonymous, 1854, 1856)

1871: An earthquake occurred on 24 April 1871, and Marapi erupted on (and before?) that date (Bergsma, 1871).

MANINJAU (continued)

HISTORICAL ACTIVITY (continued)

1886-89: An eruption of Marapi was accompanied by horizontal shaking of the earth, underground rumbling, and an unspecified magnetic disturbance (Figee and Onnen, 1886). Continuing eruptions in 1888 were accompanied by "light vertical tremor" (Figee and Onnen, 1890). Tandikat began to steam profusely on 19 February 1889, a new fissure across the crater opened, and small ash eruptions began on 27 March or earlier (Martin, 1889; Figee and Onnen, 1891).

1939: In December 1939, observers noted a sulfurous smell and black mud at the surface of the caldera lake. The cause probably was not volcanic: "these phenomena might be the result of slight tremors which disturbed the equilibrium in the deposits of the bottom of the lake..." (Netherlands Indies Volcanological Survey, 1941).

1949-present: Eruptions of Marapi from 1949 to the present have been relatively small explosive eruptions, accompanied by increased steaming and rumbling sounds during eruptions (Suryo, 1961a,b, 1985).

COMMENTS

Maninjau is one of several calderas in this compilation that are not usually considered restless. We have included Maninjau for two reasons: (1) It is related to Tandikat and Marapi volcanoes but has progressively been separated from those volcanoes by movement along the Semangko Fault Zone; and (2) overturn of tropical caldera lakes is a potentially sensitive indicator of fumarolic heating of bottom water (see Newhall and others, 1987). Added heat input in Maninjau's setting need not imply magmatic intrusion; strain associated with the Semangko Fault could bring an increased flux of warm ground water into the lake. We have no direct evidence for influxes of thermal waters into Lake Maninjau, but we suspect that process when otherwise stable caldera lakes overturn.

REFERENCES

Anonymous, 1854, Berigten van verschillenden aard: Aardbevingen in den Indischen Archipel: Nat. Tijdschr. v. Ned. Ind., pt. 7, p. 331-332.

----- 1856, Berigten van verschillenden aard: Aardbevingen en berguitbarstingen in den Indischen Archipel: Nat. Tijdschr. v. Ned. Ind., pt. 10, p. 450-455.

Bergsma, P.A., 1871, Aardbevingen in den Indischen Archipel, gedurende het jaar 1871: Nat. Tijdschr. v. Ned. Ind., pt. 32, p. 437-449.

Circum-Pacific Map Project, 1981, Plate tectonic map of the circum-Pacific region, southwest quadrant: Tulsa, Circum-Pacific Counc. Energy Min. Res./Am. Assoc. Pet. Geol., scale 1:10,000,000.

de Waard, D., and Klompe, Th.H.F., 1952, The recent activity of G. Marapi in Central Sumatra: Nat. Tijdschr. v. Ned. Ind., pt. 108, nos. 1-2, p. 131-140.

MANINJAU (continued)

REFERENCES (continued)

Figee, S., and Onnen, H., 1886, Vulkanische verschijnselen en aardbevingen in den O.I. Archipel waargenomen gedurende de maanden Januari - Juni van het jaar 1886: Nat. Tijdschr. v. Ned. Ind., pt. 46, no. 1, p. 269-287.

----- 1890, Vulkanische verschijnselen en aardbevingen in den O.I. Archipel waargenomen gedurende het jaar 1888: Nat. Tijdschr. v. Ned. Ind., pt. 49, p. 111-159.

----- 1891, Vulkanische verschijnselen en aardbevingen in den O.I. Archipel waargenomen gedurende het jaar 1889: Nat. Tijdschr. v. Ned. Ind., pt. 50, p. 166-197.

Junghuhn, F., 1853/54, Java, deszelfs gedaante, bekleeding en inwendige structuur (in 2 volumes, 4 parts) [Java: its shape, covering, and internal structure], 1st ed. 1852-54; 2nd ed. 1853-54: Amsterdam, Uitgave C.W. Mieling.'s Gravenhage, p. 1238-1243.

Kastowo, and Leo, G.W., 1973, Peta geologi lembar Padang, Sumatra (Geologic map of the Padang quadrangle, Sumatra): Geol. Surv. Indonesia, 1:250,000, 1 sheet.

Katili, J.A., 1969, New results of radiometric age dating of some Indonesian Quaternary deposits: Bull. Natl. Inst. Geol. Min., Bandung, v. 2, no. 2, p. 29-31.

----- 1970, Large transcurrent faults in Southeast Asia with special reference to Indonesia: Geol. Rundschau, v. 59, p. 581-600.

Kemmerling, G.L.L., 1921, Vulkanen en vulkanische verschijnselen in de Residentien Sumatra's Westkust en Tapanoeli: Vulkanol. Meded., v. 1, Weltevreden.

Kusumadinata, K., ed., 1979, Data Dasar Gunungapi Indonesia: Bandung, Volcanological Survey of Indonesia, 820 p.

Leo, G.W., Hedge, C.E., and Marvin, R.F., 1980, Geochemistry, strontium isotope data, and potassium-argon ages of the andesite-rhyolite association in the Padang area, West Sumatra: Jour. Volcanol. Geotherm. Res., v. 7, p. 139-156.

Martin, K., 1889, Eruptie van den G. Tandikat op Sumatra: Tijdschr. Kon. Ned. Aardr. Gen., Leiden, p. 459-460.

Netherlands Indies Volcanological Survey, 1941, Lake Manindjau: Bull. Neth. Indies Volcanol. Surv., no. 90, p. 68.

Neumann van Padang, M., 1951, Catalogue of the active volcanoes of the world including solfatara fields, Pt. I, Indonesia: Rome, IAVCEI, 271 p.

Newhall, C.G., Paull, C.K., Bradbury, J.P., Higuera-Gundy, A., Poppe, L.J., Self, S., Sharpless, N.B., and Ziagos, J., 1987, Recent geologic history of Lake Atitlán, a caldera lake in western Guatemala: Jour. Volcanol. Geotherm. Res., v. 33, p. 81-107.

Nishimura, S., 1980, Re-examination of the fission-track ages of volcanic ashes and ignimbrites in Sumatra, <u>in</u> Nishimura, S., ed., Physical geology of Indonesian island arcs: Kyoto Univ., p. 148-153.

Posavec, M., Taylor, D., van Leeuwen, Th., and Spector, A., 1973, Tectonic controls of volcanism and complex movements along the Sumatran Fault system: Geol. Soc. Malaysia Bull., v. 6, p. 43-60.

Purbo-Hadiwidjoyo, M.M., Sjachrudin, M.L., and Suparka, S., 1979, The volcano-tectonic history of the Maninjau Caldera, western Sumatra, Indonesia: Geol. Mijnb., v. 58, no. 2, p. 193-200.

Ruttner, F., 1931, Hydrographische und hydrochemische Beobachtungen auf Java, Sumatra und Bali: Arch. Hydrobiol., Supp., v. 8, p. 197-454.

MANINJAU (continued)

REFERENCES (continued)

Suryo, I., 1961a, Merapi (Sumatra): Bull. Volcanol. Surv. Indonesia, no. 100, p. 20-24.
----- 1961b, Merapi (Sumatra): Bull. Volcanol. Surv. Indonesia, no. 101, p. 7.
----- 1985, Merapi: Bull. Volcanol. Surv. Indonesia, no. 106, p. 5-9.
Verstappen, H.Th., 1973, A geomorphological reconnaissance of Sumatra and adjacent islands (Indonesia): Verh. Roy. Dutch Geogr. Soc.: Gröningen, Wolters-Noordhoff Publ., 182 p.
Zen, M.T., 1972, The origin of several pyroclastic plateaux in the Padang Highlands (Central Sumatra): Inst. Teknol. Bandung, Proc., v. 6, p. 81-88.

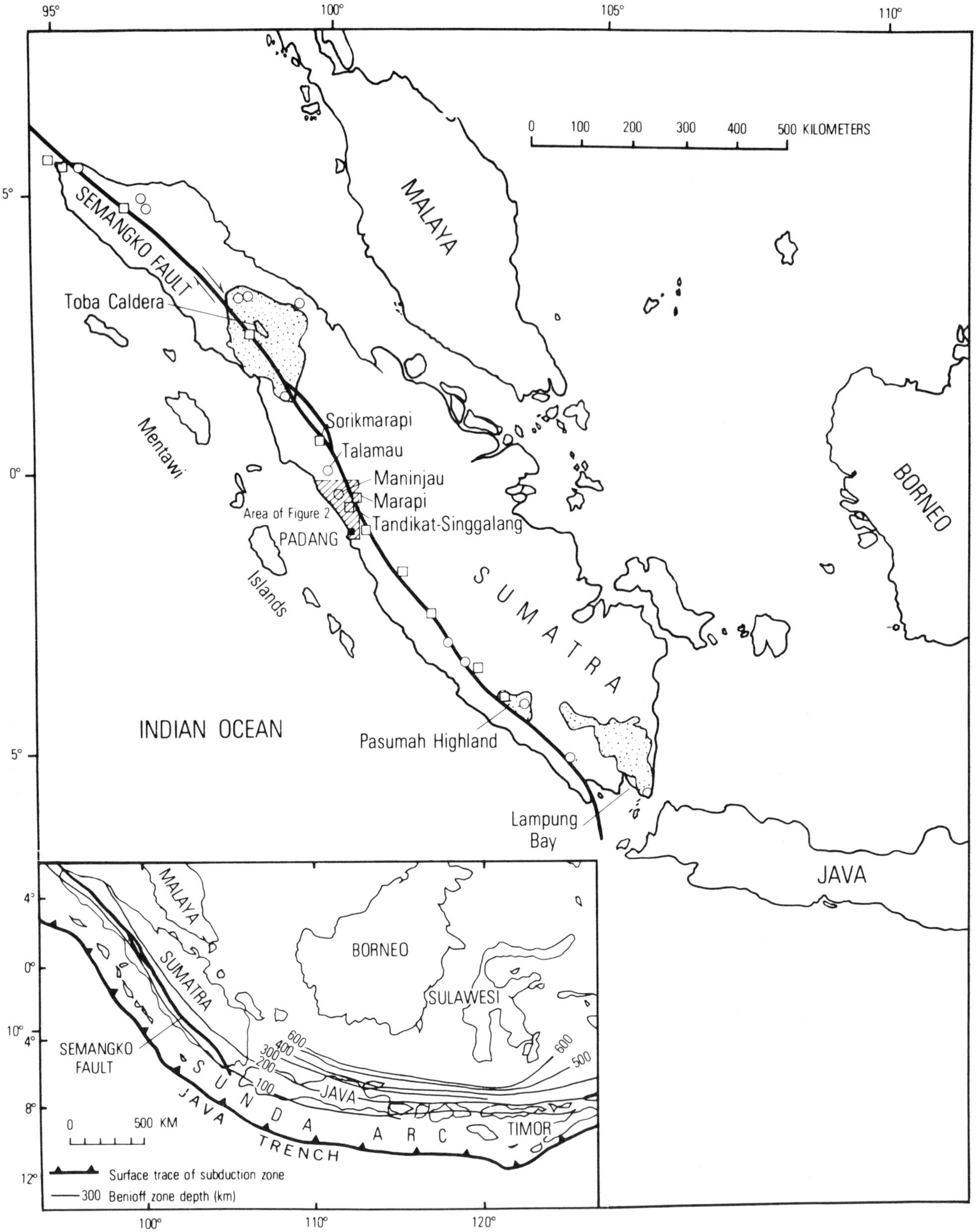

Figure 6.2.1. Location map of western Indonesia showing regional tectonics and volcanism, from Leo and others (1980). Squares, volcanoes that have erupted in historical time; circles, volcanoes in fumarolic states.

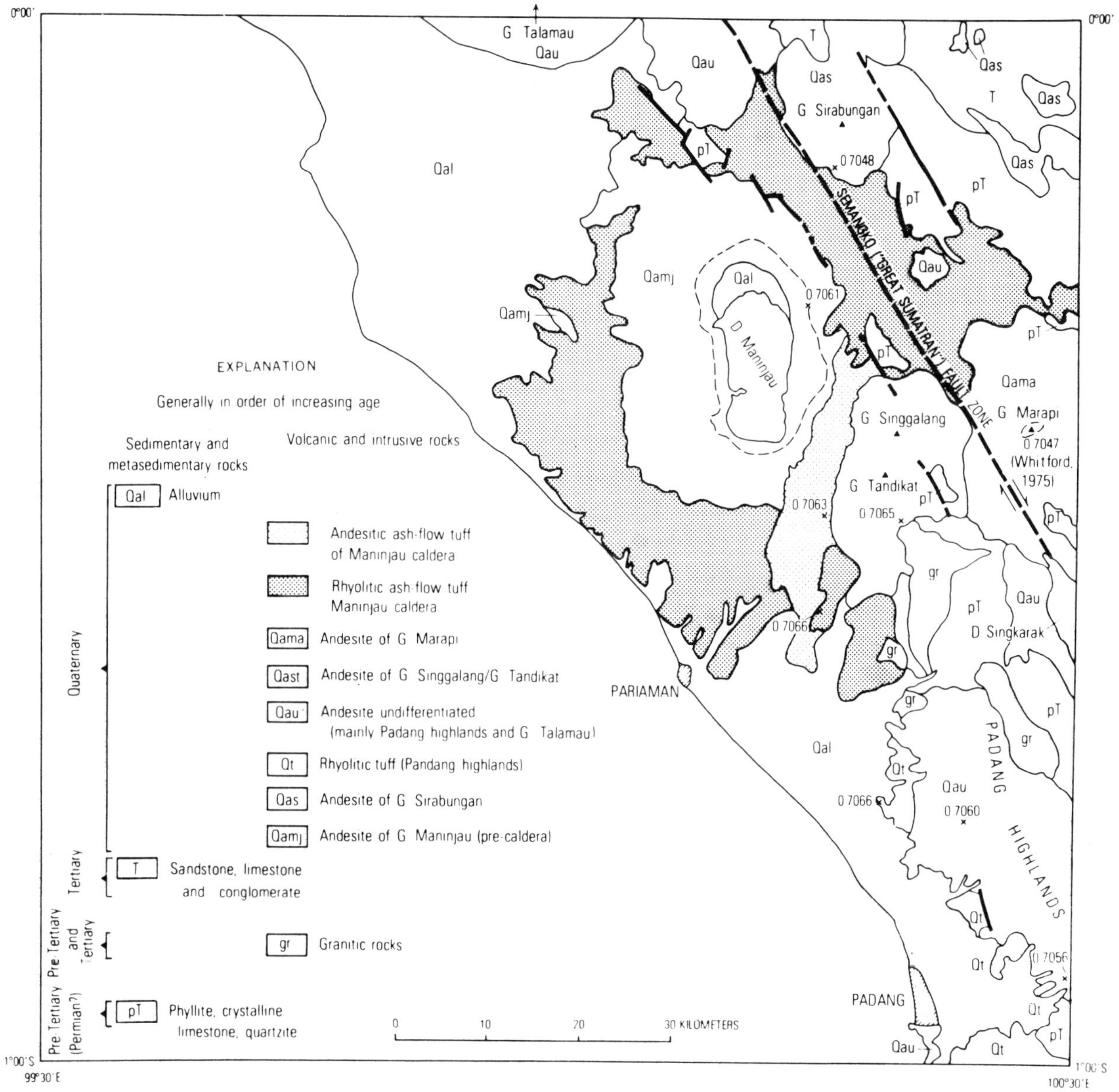

Figure 6.2.2. Geologic sketch map of Maninjau area, south-central Sumatra, from Leo and others (1980).

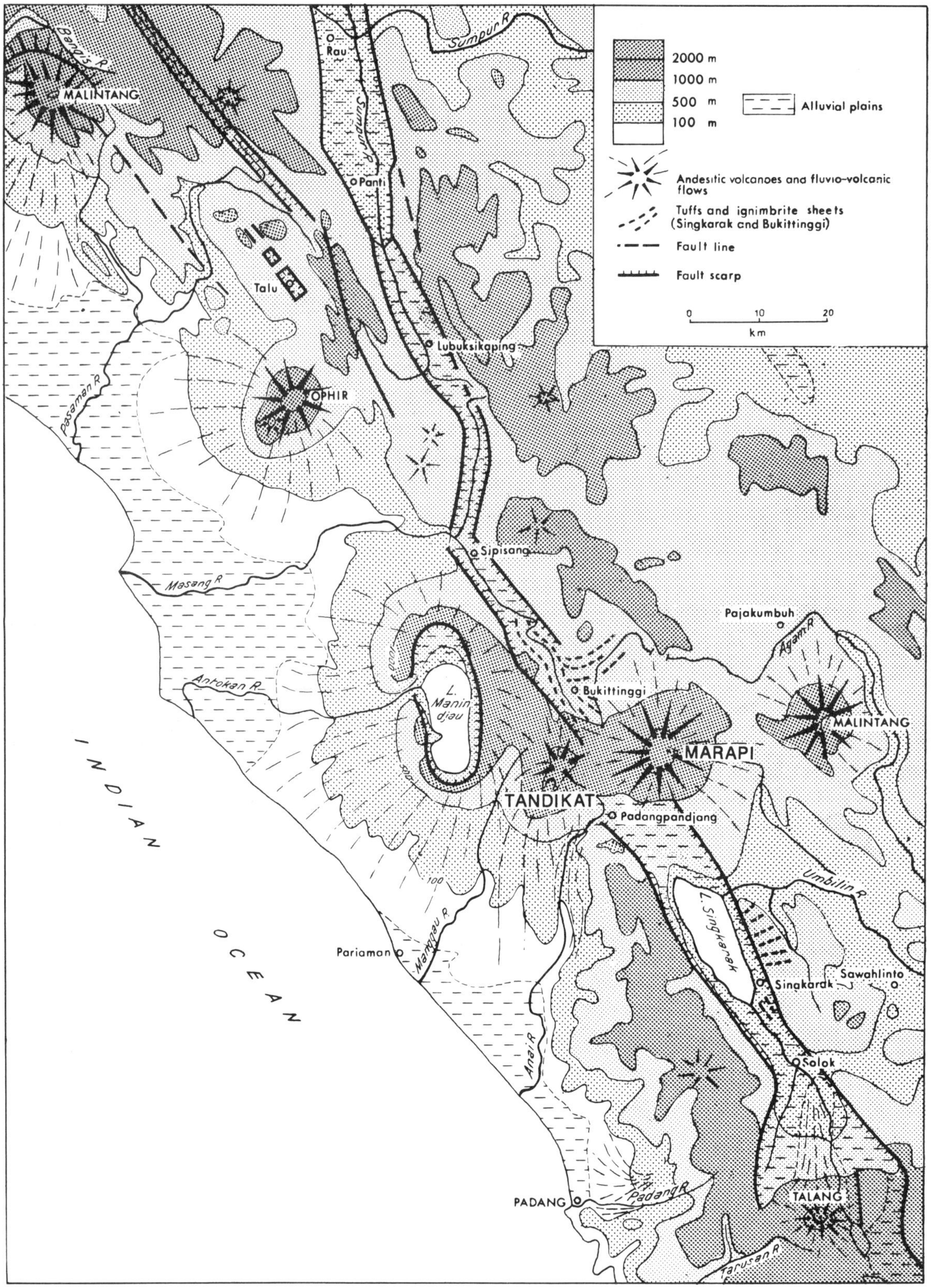

Figure 6.2.3. Sketch map of southwestern Sumatra, from Verstappen (1973), showing locations of Maninjau Caldera and nearby stratovolcanoes.

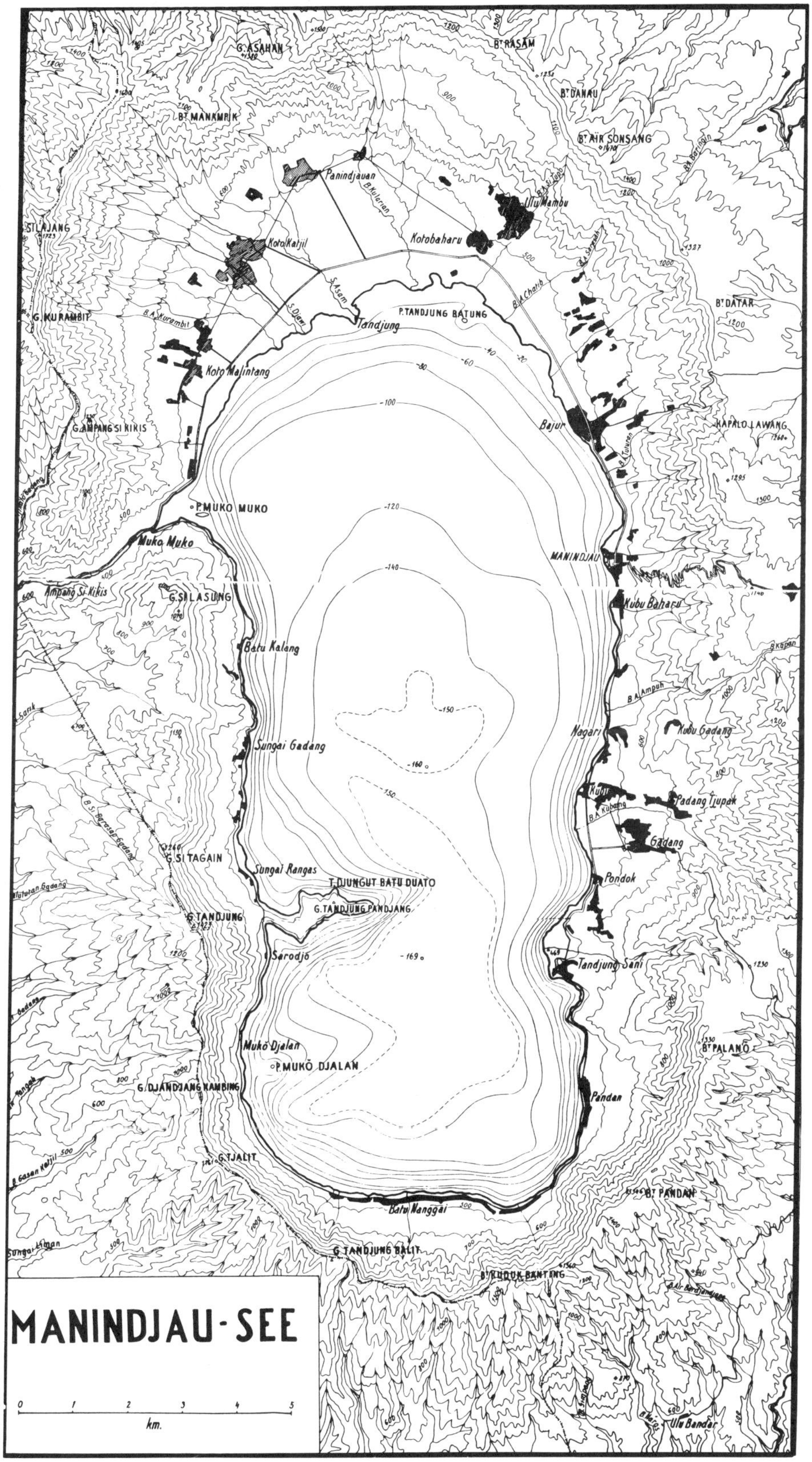

Figure 6.2.1. Topography and bathymetry of Maninjau Caldera, from Ruttner (1931).

RANAU

CAVW number (active vent)	Latitude Longitude (degrees)	Diameter (km)	Local tectonic setting	Precaldera edifice	SiO_2 content (wt pct)	Age of collapse (yr B.P.)	Date of unrest	Type and duration of unrest ESTU STHF MGTF H Te	Eruption type
06-01-25F (Seminung)	4.54S 103.53E	16 x 20	Shear, with exten	None	C = silicic		1887-88	---- ---- --?? ? --	none
							1903	---- ---- --?? ? --	none

TECTONIC SETTING

Ranau is a large caldera in south Sumatra along the great Semangko Fault Zone (figs. 6.3.1, 6.3.2). Ranau is 20 km northwest of the Liwa Plain, the epicenter of a M 7.5 tectonic earthquake in 1933 that triggered activity in nearby Suoh Caldera (see next entry in Part 2).

GEOLOGIC HISTORY

Little is known of the geology of this caldera. On the basis of morphology, the caldera is a simple collapse caldera whose southwest and northeast sides are controlled by NW-SE-trending regional faults (Verstappen, 1973) (figs. 6.3.1, 6.3.3). G. Seminung (fig. 6.3.3) is a postcaldera central cone.

HISTORICAL ACTIVITY

1887-88: On 7-8 October 1887 and again on 2-4 January and 17-20 January 1888, thousands of dead and dying fish were found on the surface and beaches of Lake Ranau. The lake water had various colors, including brown, yellow, and light and dark blue; it tasted and smelled sulfurous. On 2 January, most of the fish were near Pematang Ribu and Warku; on 3 January, near Banding Agung, and on 4 January, near Suka Negarie. This led residents to infer that the "poisoning" occurred near Warku and Pematang Ribu. "It was supposed that a volcanic eruption in the lake was the cause of that." (Figee and Onnen, 1889, 1890).

1895: A strong earthquake on 20 October 1895 caused seiches in Lake Ranau (Figee and Onnen, 1897); its epicenter is unknown.

1903: Another fish kill occurred in Lake Ranau on 9 December 1903; the water became partly gray, partly green, and thousands of fish were seen floating about as if intoxicated. Sulfur smell was noted as much as 100 m from the shore, but the water showed no increase in temperature (Koninklijk Magnetisch en Meteorologisch Observatorium te Batavia, 1905).

RANAU (continued)

COMMENTS

One mechanism for fish kill, known to have occurred at Ranau Lamongan in east Java, is overturn of a lake that normally has a layer of warmer, oxygen-rich water at the surface and a zone of H_2S-rich, oxygen-poor water below. If a stratified lake such as this is heated sufficiently from below (or cooled anomalously from above), the density stratification can be disturbed, the lake can overturn, and fish will be killed in the H_2S-rich, oxygen-poor water that has risen from depth (see Newhall and others, 1987).

REFERENCES

Figee, S., and Onnen, H., 1889, Vulkanische verschijnselen en aardbevingen in den O.I. Archipel waargenomen gedurende de maanden Juli - December van het jaar 1887: Nat. Tijdschr. v. Ned. Ind., pt. 48, p. 199-209.

----- 1890, Vulkanische verschijnselen en aardbevingen in den O.I. Archipel waargenomen gedurende het jaar 1888: Nat. Tijdschr. v. Ned. Ind., pt. 49, p. 111-159.

----- 1897, Vulkanische verschijnselen en aardbevingen in den O.I. Archipel waargenomen gedurende het jaar 1895: Nat. Tijdschr. v. Ned. Ind., pt. 56, p. 81-125.

Hustedt, F., 1937, Systematische und okologische Untersuchungen über die Diatomeen-Flora von Java, Bali und Sumatra nach dem Material der Deutschen Limnologischen Sunda-Expedition: Arch. Hydrobiol., Supp., v. 15, II. Allgemeiner Teil, p. 638-790.

Koninklijk Magnetisch en Meteorologisch Observatorium te Batavia, 1905, Vulkanische verschijnselen en aardbevingen in den Oost-Indischen Archipel waargenomen gedurende het jaar 1903: Nat. Tijdschr. v. Ned. Ind., pt. 64, p. 90-133.

Newhall, C.G., Paull, C.K., Bradbury, J.P., Higuera-Gundy, A., Poppe, L.J., Self, S., Sharpless, N.B., and Ziagos, J., 1987, Recent geologic history of Lake Atitlán, a caldera lake in western Guatemala: Jour. Volcanol. Geotherm. Res., v. 33, p. 81-107.

Philippi, H., 1916, Morphologische en geologische aanteekeningen bij de Kaart van Zuid-Sumatra: 1. Het Ranaumeer: Jaarverslag van den Topograph. Dienst en Ned. Ind., 1921, p. 182-207.

Stehn, Ch.E., 1934, Die semivulkanischen explosionen des Pematang Bata in der Soeoh-Senke (Sud-Sumatra) im Jahre, 1933: Nat. Tijdschr. v. Ned. Ind., pt. 94, p. 46-69.

Verstappen, H.Th., 1973, A geomorphological reconnaissance of Sumatra and adjacent islands (Indonesia): Verh. Roy. Dutch Geogr. Soc.: Gröningen, Wolters-Noordhoff Publ., 182 p.

Williams, H., 1941, Calderas and their origin: Univ. Calif. Publ. Geol. Sci., v. 25, p. 239-346.

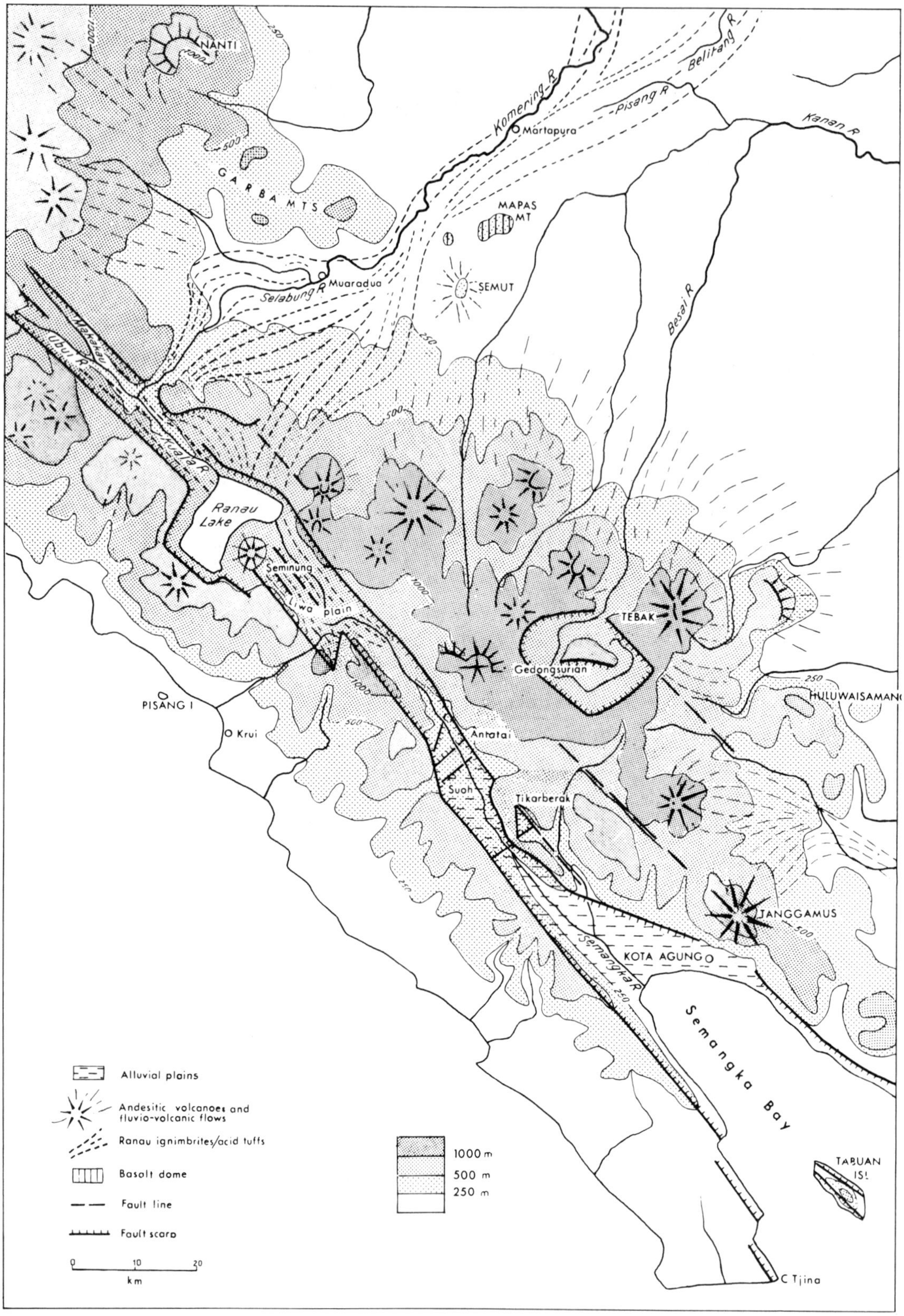

Figure 6.3.1. Sketch map of southwestern Sumatra, from Verstappen (1973), showing locations of Ranau Caldera ("Ranau Lake") and nearby stratovolcanoes.

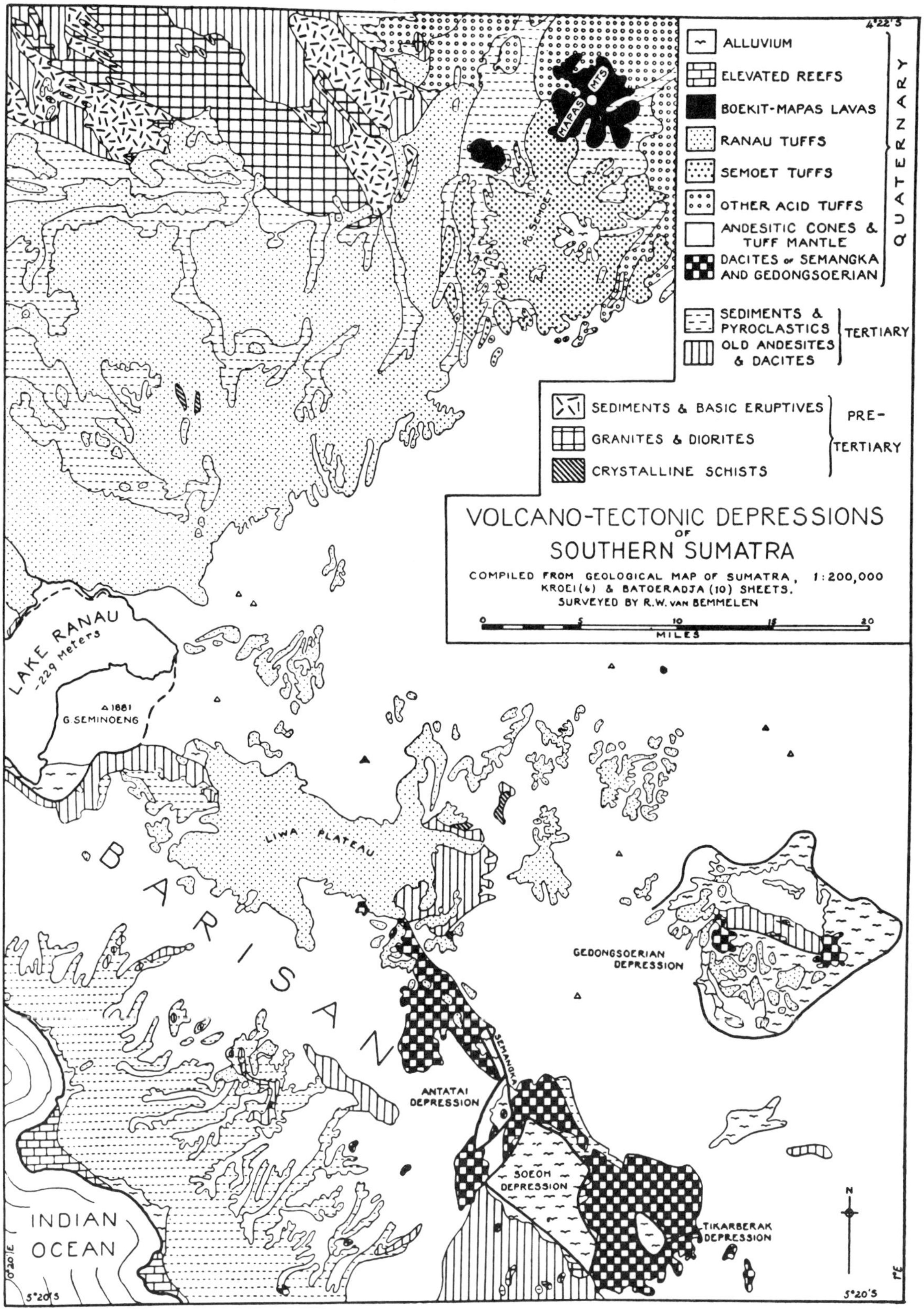

Figure 6.3.2. Geologic map of southern Sumatra, from Williams (1941), showing locations of Ranau ("Lake Ranau") and Suoh ("Soeoh") volcano-tectonic depressions.

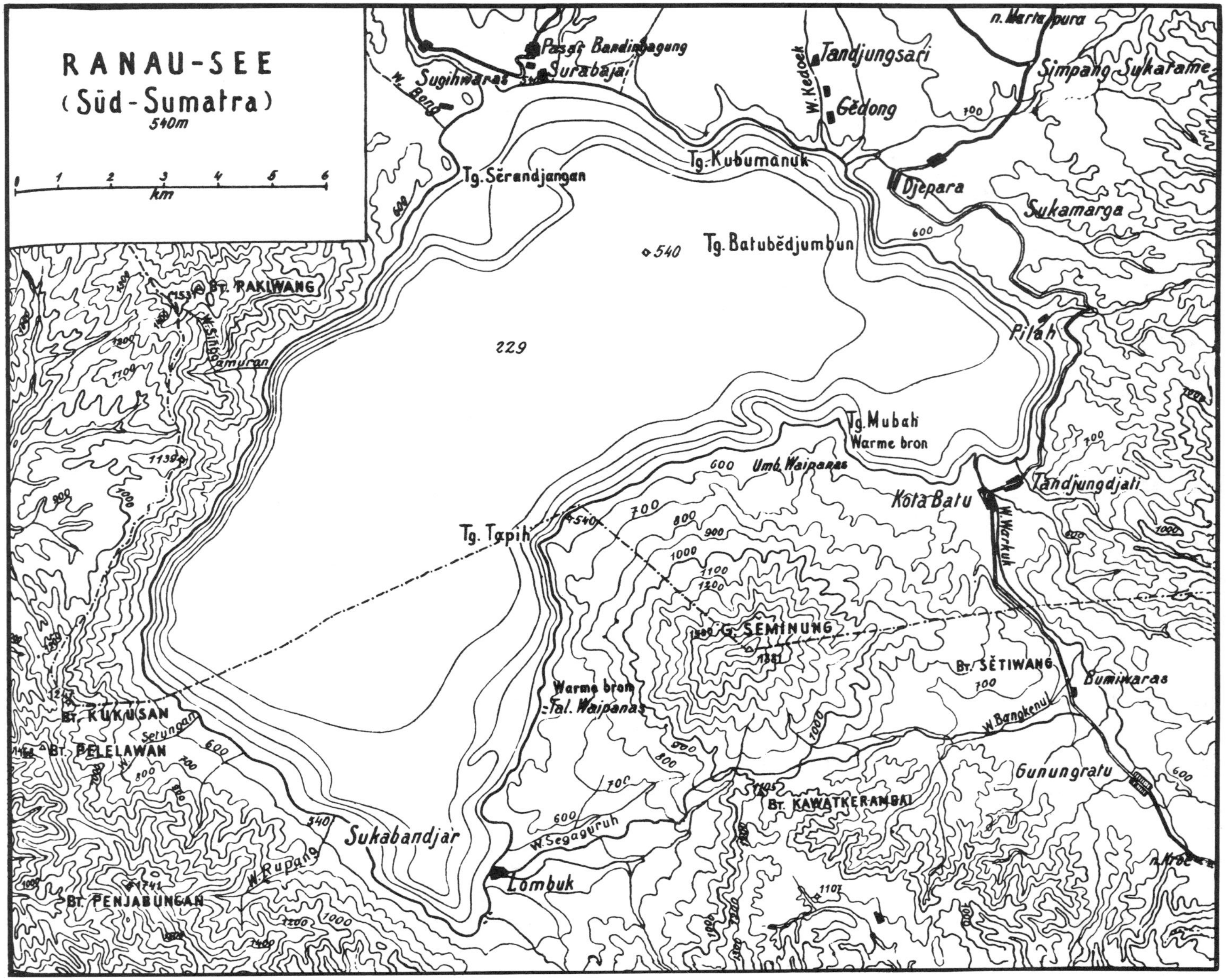

Figure 6.3.3. Topographic map of Lake Ranau area, from Hustedt (1937).

SUOH

CAVW number (active vent)	Latitude Longitude (degrees)	Diameter (km)	Local tectonic setting	Precaldera edifice	SiO_2 content (wt pct)	Age of collapse (yr B.P.)	Date of unrest	Type and duration of unrest ESTU STHF MGTF H Te	Eruption type
06-01-27 (Pematang Bata)	5.25S 104.27E	8 x 13	Exten, with lateral shear	None?	R = ? C = ?		1933	D--- ---x ---D ? QU	Pex

TECTONIC SETTING

The Suoh "caldera" is a depression along (and bounded by faults of) the Semangko (Sumatra) Fault Zone, a major strike-slip zone parallel to the volcanic chain of Sumatra (fig. 6.4.1, and fig. 6.3.2 in section on Ranau Caldera).

GEOLOGIC HISTORY

Little is known of the origin and subsequent history of the Suoh depression. Van Bemmelen (1933, 1949) considered it to be a volcano-tectonic depression of the same type as Toba, but Verstappen (1961) and Westerveld (1962) argued that the Suoh depression is fundamentally of tectonic origin, and that eruptions have occurred along preexisting faults. None of the large-volume tuffs in southern Sumatra are known with certainty to have been erupted from the Suoh depression, but silicic domes occur along its margins and numerous hot springs occur along faults within the depression.

HISTORICAL ACTIVITY

1933: A $\underline{M}$ 7.5 earthquake (the "Liwa earthquake") on 25 June 1933 occurred at shallow depth along the Semangko Fault Zone, at approximately 5.09°S, 104.7°E (Newcomb and McCann, 1987). Steaming in the Suoh depression was seen to have increased 13 hours after the earthquake, and 14 days later, on 10 July, a relatively large phreatic explosion occurred in the Suoh depression. Two explosion craters formed, and a pyroclastic surge(?) devastated areas within 10 km of the vents (figs. 6.4.2, 6.4.3). The sound of the explosion was heard as far as 660 km away. Additional small phreatic explosions continued during observations from mid-July to early August 1933 (Stehn, 1934).

Seismic swarms near Sungei Langka (South Sumatra) in June and August 1931 might have been precursors to the great Liwa earthquake. Newcomb and McCann (1987) report a $\underline{M}$ 7.2 earthquake on 10 February 1931 at 5.45°S, 102.88°E, and a $\underline{M}$ 7.5 earthquake on

SUOH (continued)

HISTORICAL ACTIVITY (continued)

25 September 1931 at 5.10°S, 102.61°E, both in the zone of shallow thrust earthquakes just offshore from South Sumatra. Together, these and the Liwa earthquake represent a major strain release in South Sumatra over the period 1931-33.

A M 6.6 shallow earthquake occurred on 27 December 1985 at 5.84°S, 104.14°E (Smithsonian Institution, 1985); we have not seen any reports of increased thermal activity around that time in the Suoh depression. Other large earthquakes have occurred within a 100-km radius of Pematang Bata about once every 10-20 years since 1903 (Newcomb and McCann, 1987).

COMMENTS

The 1933 activity is one of the best-known examples of a phreatic explosion occurring as a direct consequence of a large tectonic earthquake. Even though some doubt exists about whether the Suoh depression is a caldera, many calderas in this compilation contain hydrothermal systems like the one that exploded at Suoh in 1933.

REFERENCES

Bemmelen, R.W. van, 1933, Geologische kaart van Sumatra, sheet 6, 1:200,000. Dienst van den Mijnb., Bandung.
----- 1949, The geology of Indonesia, v. 1A: The Hague, Government Printing Office, 732 p.
Netherlands Indies Volcanological Survey, 1933, Steam explosion in the Soeoh Plain (S. Sumatra): Pematang Bata: Bull. Neth. Ind. Volcanol. Surv., no. 64, p. 39-41.
Netherlands Indies Volcanological Survey, 1935, Pematang Bata (Soeoh): Bull. Neth. Ind. Volcanol. Surv., no. 73, p. 152.
Newcomb, K.R., and McCann, W.R., 1987, Seismic history and seismotectonics of the Sunda arc: Jour. Geophys. Res., v. 92, no. B1, p. 421-439.
Smithsonian Institution, Scientific Event Alert Network (SEAN), 1985, Earthquakes in December 1985: SEAN Bull., v. 10, no. 12, p. 19.
Stehn, Ch.E., 1934, Die semivulkanischen explosionen des Pematang Bata in der Soeoh-Senke (Sud-Sumatra) im Jahre, 1933: Nat. Tijdschr. v. Ned. Ind., pt. 94, p. 46-69.
Verstappen, H.Th., 1961, Some "volcano-tectonic" depressions of Sumatra, their origin and mode of development: Kon. Ned. Akad. Wetensch., Amsterdam, Proc. B, v. 64, p. 428-443.
Westerveld, J., 1952, Quaternary volcanism on Sumatra: Bull. Geol. Soc. Am., v. 63, p. 561-594.

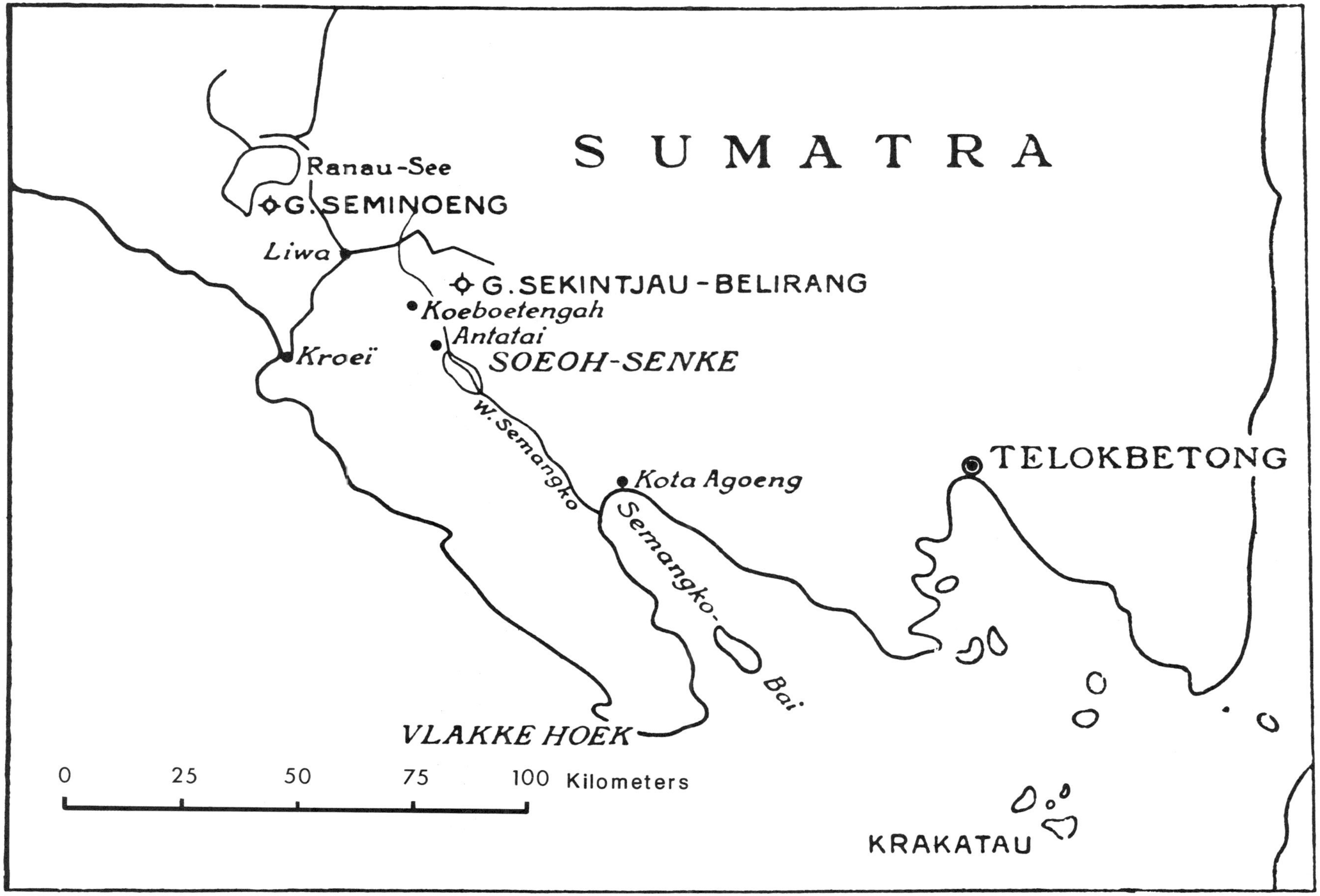

Figure 6.4.1. Southern Sumatra, from Stehn (1934), showing locations of Suoh ("Soeoh-Senke") and Ranau ("Ranau-See") volcano-tectonic depressions.

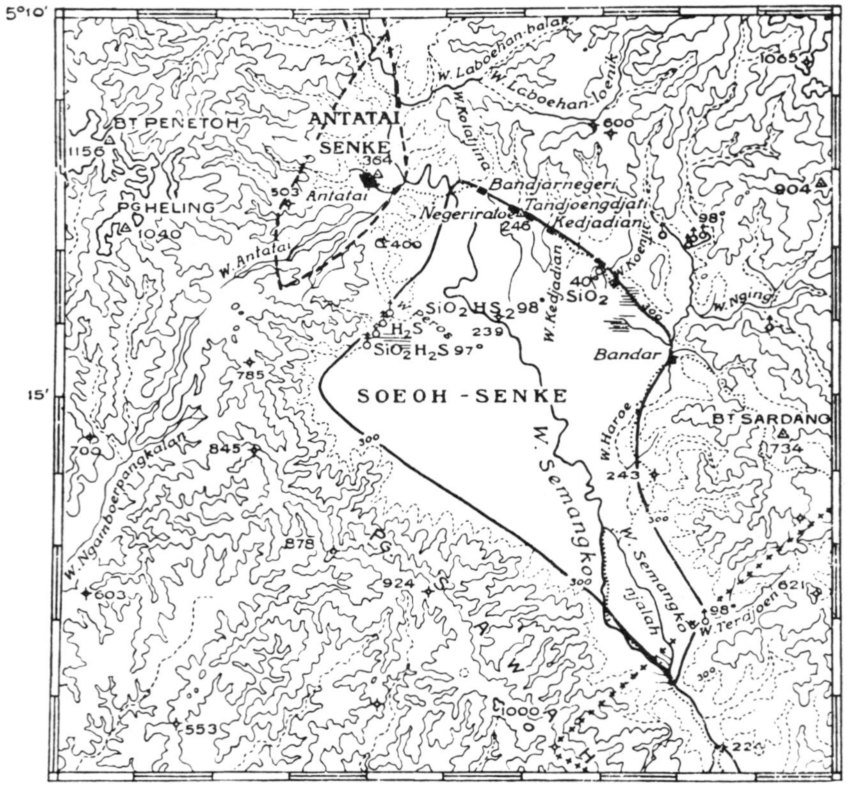

Figure 6.4.2. Configuration of Suoh depression before a phreatic explosion on 10 July 1933 (from Stehn, 1934).

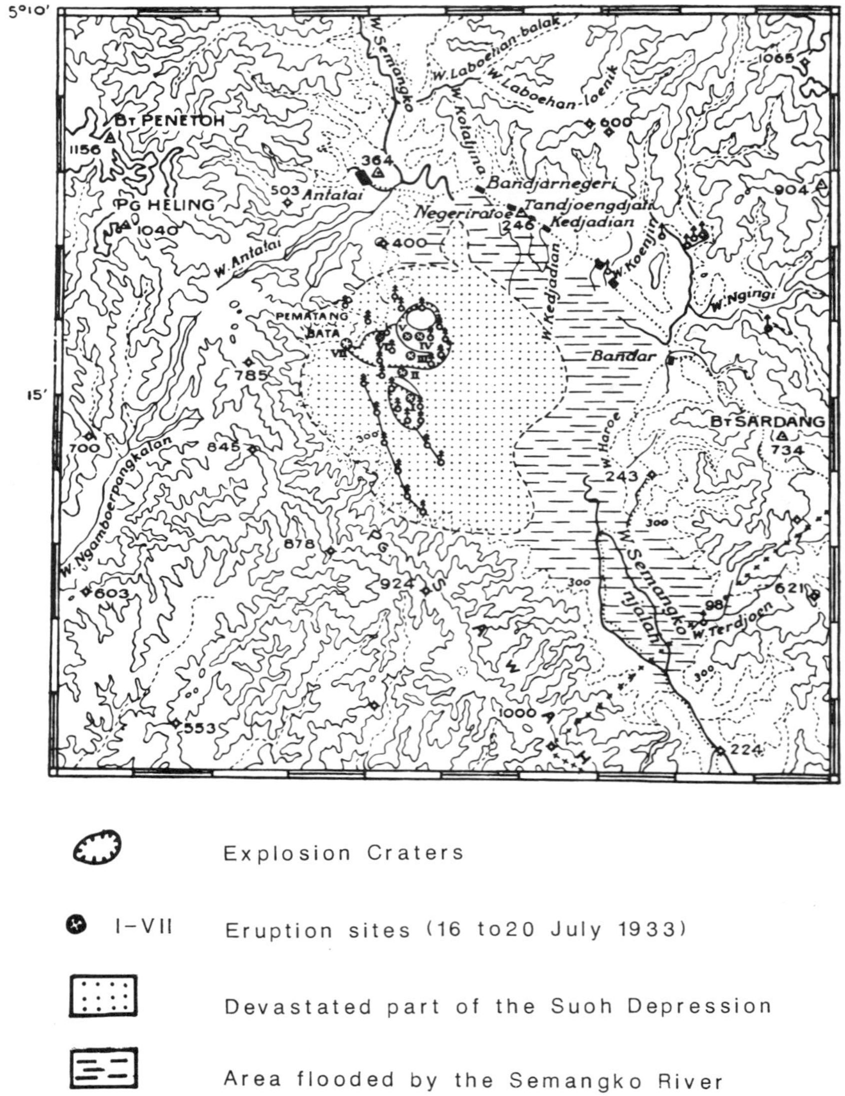

Figure 6.4.3. Configuration of Suoh depression after a phreatic explosion on 10 July 1933 (from Stehn, 1934).

KRAKATAU

CAVW number (active vent)	Latitude Longitude (degrees)	Diameter (km)	Local tectonic setting	Precaldera edifice	SiO_2 content (wt pct)	Age of collapse (yr B.P.)	Date of unrest	Type and duration of unrest ESTU STHF MGTF H Te	Eruption type
06-C2-00	6.10S	8	Compr,	Strato	R = 51-70	A.D. 416	3rd century	---- ---- ---- - --	Ex?
(Anak Krakatau)	105.42E		boundary		C = 65-69	A.D. 1883	416	---- ---- ---- - --	EX
			between				9th-16th	---- ---- ---- - --	ex,Ex?
			two				centuries		
			subducted				1680-81	x--- ---- ---- - --	Ex,lf
			slabs?				1880-84	D--- ---- ---- - QU?	ex,pex,EX
							1913	?--- ---- --?- - --	none?
							1927-81	xx-- ---- ---E - --	ex,pex,lf
							1988	?--- ---- ---- - --	pex,ex

TECTONIC SETTING

Krakatau lies in the Sunda Strait between Java and Sumatra, apparently at the intersection of the WNW-ESE-trending Sumatran Fault Zone (right-lateral strike-slip) and a N-S- or NNE-SSW-trending transverse fault zone that may lie above a break in the subducted plate (Ranneft, 1976; Zen, 1985).

GEOLOGIC HISTORY

The Krakatau edifice grew as one or more stratovolcanoes of dominantly hypersthene-augite andesite composition. The geology of Krakatau has been described by Effendi and others (1985, 1986), who identified five main evolutionary periods (fig. 6.5.1). Period 1 was an early growth phase that included accumulation of lavas and pyroclastics. Period 2 was marked by caldera formation, accompanied by pyroclastic flows and partly welded tuffs (ignimbrites). The cones of Rakata, Danan, and Perbuwatan grew during the third period and were largely destroyed during the fourth period, which included the paroxysmal eruption of August 1883 (figs. 6.5.2, 6.5.3). Another growth phase (period 5) began by December 1927, when Anak Krakatau was first noticed in submarine eruption.

Although much of the stratigraphic record of Krakatau is obscured by water of the Sunda Straits, accessible outcrops suggest that basaltic and andesitic magmas initially fed several overlapping stratovolcanoes. Following long repose periods, large explosive eruptions of dacitic magma occurred, followed by caldera collapse and, after a period of several decades, renewed mafic stratovolcano growth.

KRAKATAU (continued)

HISTORICAL ACTIVITY

3rd century: There are Chinese records of a violent outburst of Krakatau (de Neve, 1985, p. 38).

416: "The world was shaken with violent thunderings that were accompanied by heavy rains of stone... the God of Fire, Kapi [=Krakatau] burst into pieces and sank into the depths of the earth. Then, the sea rose and Java and Sumatra were divided into two parts..." (de Neve, 1985, quoting the Javanese chronicle, Pararaton). Judd (1889), de Neve (1985), and Newhall and others (1985) inferred that this eruption might have been the penultimate caldera-forming eruption of Krakatau.

9th-16th centuries: de Neve (1985) cites eruptions in the 9th, 10th (twice?), 11th, 12th, 14th, 16th, and 17th centuries.

1680-81: Strong earthquakes were felt in the vicinity of the Sunda Straits in May 1680, and a ship's captain reported in November 1681 that Krakatau had been erupting for about a year. Records do not indicate whether the earthquakes occurred before or after the start of the eruption, nor whether they were felt locally or over a broader area of Java and Sumatra. A lava flow on Perbuwatan, inferred by Verbeek (1886) to be from the 1680-81 eruption, had 69 percent SiO_2 (Neumann van Padang, 1951) and was apparently a leak of silicic magma that foreshadowed the large eruption of 1883.

1880-84: Several years of regional seismicity culminated in the famous caldera-forming eruption of August 1883 (Verbeek, 1886; Judd, 1888). Earthquakes were felt in west Java and eastern Sumatra 5-6 years before the 1883 eruption, and in northern Australia in the 3 years before that eruption. On 1 September 1880, a strong earthquake damaged the First Point lighthouse on the west end of Java (70 km south-southwest of Krakatau). More earthquakes were felt at First Point on 9-10 May 1883 and at Katimbang (40 km north-northeast) during 15-20 May 1883. None of the above-mentioned earthquakes is known to have been of volcanic origin and centered beneath Krakatau; more likely, they reflect a buildup and release of tectonic stress in the Sunda Strait area.

A mild ash and steam eruption began from Perbuwatan on 20 May 1883, and similar eruptions continued for 3 months. At times several vents were in eruption simultaneously. On 11 August, vigorous ash-laden gas columns rose from the main crater of Perbuwatan, the former vent of Danan, and from the foot of Danan; in addition, "no fewer than eleven other foci of eruption could be observed on the visible portions of the island, from which smaller steam-columns issued and ejections of ash took place" (Judd, 1888, p. 13). At least some of the early ash was basaltic, suggesting that intrusion of basaltic melt into the silicic reservoir may have played a role in initiating these eruptions (Francis and Self, 1983). The small eruptions intensified on 24 August 1883 and culminated in the famous, climactic explosion on 27 August 1883 (Verbeek, 1886; Judd, 1888; Simkin and Fiske, 1983).

1913: Earthquakes and rumbling, not demonstrably from Krakatau, were noted in 1898 (Koninklijk Magnetisch en Meteorologisch Observatorium te Batavia, 1899). On 13 August 1913, a strong tectonic earthquake occurred in the Bantam-Krakatau region, damaging the First Point lighthouse. Fishermen who were on the islands of Krakatau reported that part of the caldera wall collapsed. They also reported that at 1000 hrs a "column of fire" rose out of the crater, whereupon they heard a cannonlike sound and fled (Visser, 1920). There are no independent reports of eruptions of Krakatau in 1913, and we suspect that only a landslide occurred.

KRAKATAU (continued)

HISTORICAL ACTIVITY (continued)

1927-81: Eruptions from December 1927 to 1981, mostly of basaltic andesite, built the Anak Krakatau ("child of Krakatau") cone in the center of the 1883 caldera. Gas emissions increased 6 months before the 1927 eruption began, and small seismic swarms have been noted before recent eruptions of Anak Krakatau (for example, one or probably more earthquakes occurred in the 10-day period before an eruption of 9 September 1980; Siswowidjojo, 1985). Some periods, such as those from 1958 to 1963 and from 1972 to 1981, had frequent eruptions; other periods, such as that from 1963 to 1972, had relatively little activity.

1988: An increase in the volume of white fume from the summit crater of Anak Krakatau was observed beginning on 11 February 1988. Similar activity, stronger than it had been for several years, continued intermittently until 28 February, when plume color darkened and emissions became more frequent. Plumes apparently rose a few hundred meters above the crater. On 1 March 1988, glow was visible from the Java coast, suggesting the presence of new lava. Earthquakes were felt on 5-6 March in Anyer, 55 km east of the volcano (Smithsonian Institution, 1988).

COMMENTS

The August 1883 eruption of Krakatau is often cited as a classic example of caldera formation by collapse following eruption of large volumes of pumice (Williams, 1941; Williams and McBirney, 1979; Self and Rampino, 1981, 1982; Francis and Self, 1983). However, other workers have suggested alternate mechanisms for formation of the Krakatau caldera. Yokoyama (1981, 1982) concluded that the caldera formed by explosive destruction and reaming of the preeruption edifice, and Camus and Vincent (1983) and Francis (1985) favored an origin by large-scale collapse of the northern part of Krakatau Island (similar to the volcanic landslide at Mount St. Helens on 18 May 1980). Regrettably, much of the evidence is submarine and inaccessible, but we are impressed by the similarity of Krakatau and other, better-exposed calderas (for example, Crater Lake) that are thought to have formed by simple collapse following voluminous pumice eruptions. The volume of magma erupted in the plinian eruption (9 km^3) is adequate to explain the caldera without invoking a landslide origin.

Krakatau is the only volcano in this compilation to have experienced two episodes of caldera formation in historical time--first in A.D. 416, and again in 1883. Precursors to the first eruption are unknown; the eruption of 1883 was preceded by an intracaldera leak of silicic magma in 1680, by strong earthquakes of uncertain origin, and by three months of small ash eruptions, sometimes from several linearly aligned intracaldera vents simultaneously. Intrusion of basaltic magma may have contributed to triggering the 1883 eruption.

KRAKATAU (continued)

REFERENCES

Bullard, F. M., 1976, Volcanoes of the earth, rev. ed.: Austin and London, Univ. Texas Press, 579 p.

Camus, G., and Vincent, P.M., 1983, Discussion of a new hypothesis for the Krakatau volcanic eruption in 1883: Jour. Volcanol. Geotherm. Res., v. 19, nos. 1-2, p. 167-174.

de Neve, G.A., 1985, Earlier eruptive activities of Krakatau in historic time and during the Quaternary, in Sastrapradja, D., and others, eds., Proc., Symp. on 100 Years Development of Krakatau and its Surroundings: Jakarta, Lembaga Ilmu Pengetahuan Indonesia, v. 1, p. 35-46.

Effendi, A.C., Bronto, S., and Sukhyar, R., 1986, Geologic map of Krakatau Volcano Complex, Sunda Strait, Lampung Province: Bandung, Volcanological Survey of Indonesia, 1 sheet, scale 1:25,000.

Effendi, A.C., Kartakusumah, R.S., and Bronto, S., 1985, Geology of the Krakatau Complex, in Sastrapradja, D., and others, eds., Proc., Symp. on 100 Years Development of Krakatau and its Surroundings: Jakarta, Lembaga Ilmu Pengetahuan Indonesia, v. 1, p. 247-252.

Escher, B.G., 1919, De Krakatau groep als vulkaan: Handelingen, Nederlandsch-Indisch Natuurwetenschappelijk Congres, Weltevreden, p. 28-35.

Francis, P.W., 1985, The origin of the 1883 Krakatau tsunamis: Jour. Volcanol. Geotherm. Res., v. 25, no. 3/4, p. 349-363.

Francis, P., and Self, S., 1983, The eruption of Krakatau: Sci. Am., v. 249, p. 172-187.

Judd, J.W., 1888, On the volcanic phenomena of the eruptions, and on the nature and distribution of the ejected materials, in Symons, G.J., ed., The eruption of Krakatoa and subsequent phenomena: Report of the Krakatoa Committee of the Royal Society, London: London, Trübner and Co., p. 1-56

----- 1889, The earlier eruptions of Krakatoa: Nature, v. 40, p. 365-366.

Kartakusumah, R.S., and Effendi, A.C., 1985, Volcano stratigraphy of the Krakatau complex preceding Anak Krakatau, in Sastrapradja, D., and others, eds., Proc., Symp. on 100 Years Development of Krakatau and its Surroundings: Jakarta, Lembaga Ilmu Pengetahuan Indonesia, v. 1, p. 210-216.

Koninklijk Magnetisch en Meteorologisch Observatorium te Batavia, 1899, Vulkanische verschijnselen en aardbevingen in den Oost-Indischen Archipel waargenomen gedurende het jaar 1898: Nat. Tijdschr. v. Ned. Ind., pt. 59, p. 79-139.

----- 1914, Vulkanische verschijnselen en aardbevingen in den Oost-Indischen Archipel waargenomen gedurende het jaar 1913: Nat. Tijdschr. v. Ned. Ind., pt. 74, no. 1, p. 67-115.

Kusumadinata, K., ed., 1979, Data Dasar Gunungapi Indonesia: Bandung, Volcanological Survey of Indonesia, 820 p.

Neumann van Padang, M., 1951, Catalogue of the active volcanoes of the world including solfatara fields, Pt. I, Indonesia: Rome, IAVCEI, 271 p.

Newhall, C.G., Decker, R.W., Sumartadipura, A.S., Tilling, R.I., and Peterson, D.W., 1985, A comparison of eruptions and magma reservoirs of Krakatau, Mount St. Helens and Galunggung, in Sastrapradja, D., and others, eds., Proc., Symp. on 100 Years Development of Krakatau and its Surroundings: Jakarta, Lembaga Ilmu Pengetahuan Indonesia, v. 1, p. 11-19.

Ranneft, T.S.M., 1976, Understanding segmentation of island arcs aids exploration: Oil and Gas Jour., Nov. 15, 1976, p. 114-118.

KRAKATAU (continued)

REFERENCES (continued)

Self, S., and Rampino, M.R., 1981, The 1883 eruption of Krakatau: Nature, v. 294, p. 699-704.

----- 1982, Comments on "A geophysical interpretation of the 1883 Krakatau eruption", by I. Yokoyama, Jour. Volcanol. Geotherm. Res., v. 13, p. 379-383.

Simkin, T., and Fiske, R.S., eds., 1983, Krakatau; The volcanic eruption and its effects, a centennial retrospective: Washington, D.C., Smithsonian Institution Press, 464 p.

Siswowidjoyo, S., 1985, The renewed activity of Krakatau Volcano after its catastrophic eruption in 1883, in Sastrapradja, D., and others, eds., Proc., Symp. on 100 Years Development of Krakatau and its Surroundings: Jakarta, Lembaga Ilmu Pengetahuan Indonesia, v. 1, p. 192-198.

Smithsonian Institution, Scientific Event Alert Network (SEAN), 1988, Anak Krakatau: SEAN Bull., v. 13, no. 2 (report from Volcanological Survey of Indonesia).

Stehn, Ch.E., 1929, The geology and volcanism of the Krakatau group: Proc. 4th Pacific Sci. Congr., Bandung, p. 1-55.

Verbeek, R.D.M., 1886, Krakatau: Batavia, Imprimerie de l'Etat, 495 p.

Visser, S.W., 1920, Aardbevingen in en om West-Java: Nat. Tijdschr. v. Ned. Ind., pt. 79, p. 181-201.

Williams, H., 1941, Calderas and their origin: Univ. Calif. Publ. Geol. Sci., no. 25, p. 239-346.

Williams, H., and McBirney, A.R., 1979, Volcanology: San Francisco, Freeman Cooper, 397 p.

Yokoyama, I., 1981, A geophysical interpretation of the 1883 Krakatau eruption: Jour. Volcanol. Geotherm. Res., v. 9, p. 359-378.

----- 1982, Author's reply to the comment by S. Self and M. R. Rampino: Jour. Volcanol. Geotherm. Res., v. 13, p. 384-386.

Zen, M.T., 1985, Krakatau and the tectonic importance of the Sunda Strait, in Sastrapradja, D., and others, eds., Proc., Symp. on 100 Years Development of Krakatau and its Surroundings: Jakarta, Lembaga Ilmu Pengetahuan Indonesia, v. 1, p. 100-106. Also in Bul. Jurusan Geol., Inst. Teknol. Bandung, v. 12, 1983, p. 9-22.

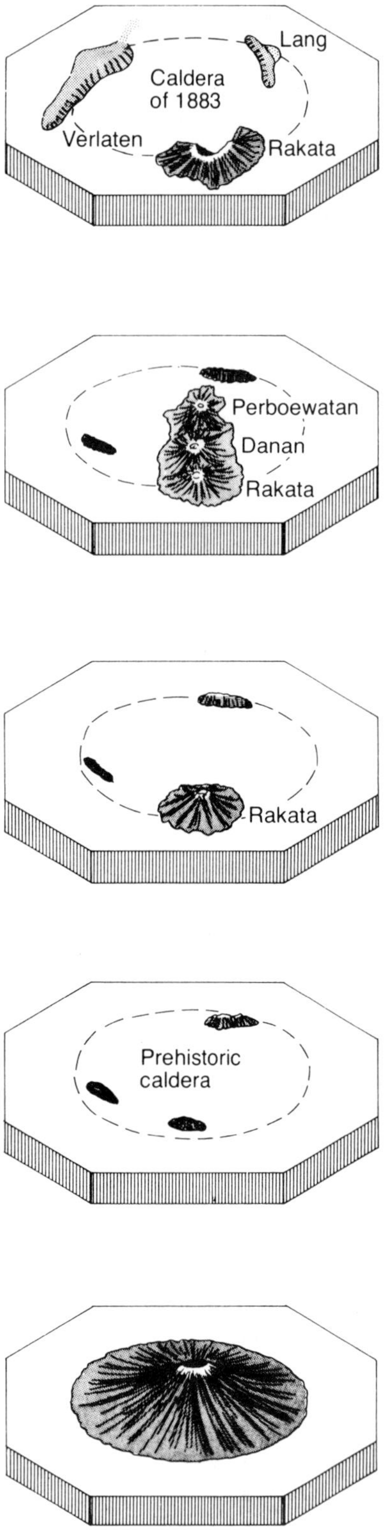

5. After the 1883 eruptions. Later eruptions have since built up the island of Anak Krakatoa within the caldera.

4. Krakatoa before 1883, after two later andesitic cones had coalesced with Rakata.

3. Growth of Rakata, a basaltic cone.

2. After explosive evisceration, probably accompanied by collapse of the superstructure, a great caldera was formed, rimmed by three small islands.

1. Original andesitic cone of Krakatoa.

Figure 6.5.1. Schematic history of Krakatau, from Bullard (1976), modified from Escher (1919).

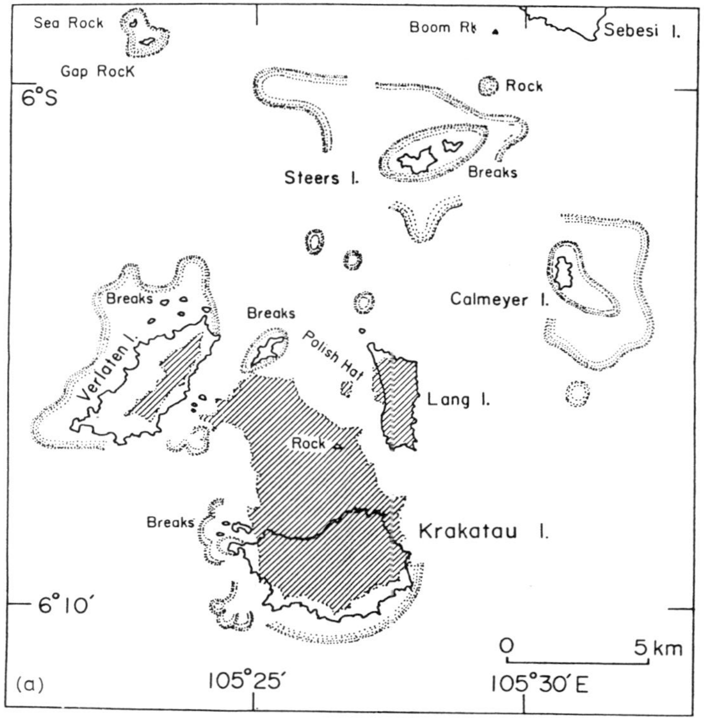

Figure 6.5.2. Krakatau and surrounding islands immediately following 1883 eruption, after Judd (1888), reproduced in Yokoyama (1981). Shaded areas indicate preeruption form of Krakatau Islands.

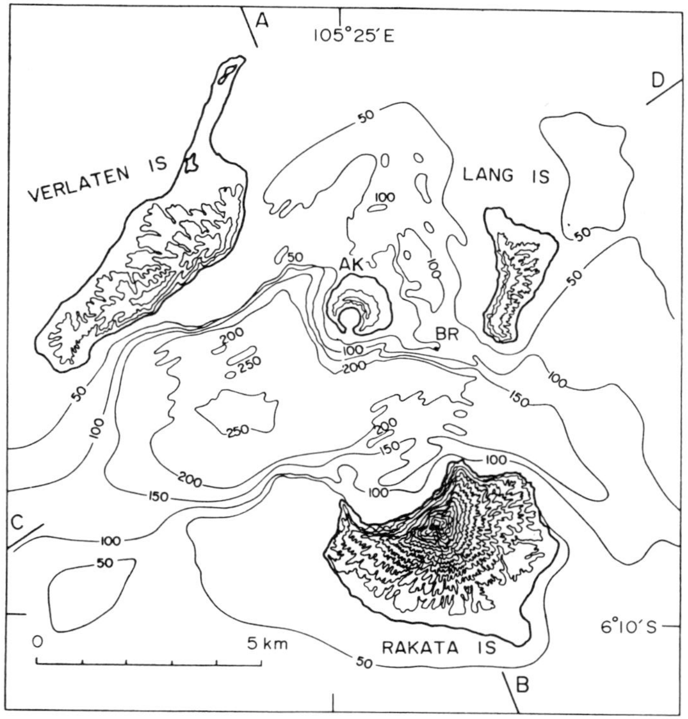

Figure 6.5.3. Geological Survey of Indonesia bathymetric chart of Krakatau Islands in 1940, reproduced by Yokoyama (1981). Isobaths in meters; AK, Anak Krakatau.

PART 3: HISTORICAL UNREST AT LARGE QUATERNARY CALDERAS (continued)
See inside back cover for explanation and abbreviations

SUNDA

CAVW number (active vent)	Latitude Longitude (degrees)	Diameter (km)	Local tectonic setting	Precaldera edifice	SiO_2 content (wt pct)	Age of collapse (yr B.P.)	Date of unrest	Type and duration of unrest ESTU STHF MGTF H Te	Eruption type
06-03-09 (Tangkuban-parahu)	6.77S 107.60E	6-8	Compr	Strato	R = 54-58 C = silicic	>40,000	1826	---- ---- ---- - --	unknown
							1829	?--- ---- ---- - --	ex
							1842?	---- ---- ---- - --	unknown
							1846	---- ---- ---- - --	Ex, surge?
							1862	---- ---- --xx - --	none
							1887-88	---- ---- --xx - --	none
							1896	---- ---- ---Y Y --	pex
							1908	?--- ---- ---- - --	none
							1910	?--- ---- ---- - --	ex
							1913	---- ---- --?C - --	none
							1926	---- ---- ---- - --	pex
							1929	---- ---- --EE - --	pex
							1952	D--- ---- --DD - --	pex
							1957	---- ---- --x- - --	none
							1961-62	---- ---- --E- - --	none
							1965	---- ---- --x- - --	none
							1967	---- ---- --xx - --	none
							1969	---- ---- --xx - --	pex
							1971	---- ---- --xx x --	none
							1983-present	EE-E -E-- --?? ? --	none yet

TECTONIC SETTING

The Sunda Caldera, in west Java, lies in the northern part of a wide belt (or double chain?) of volcanoes, where the Indo-Australian lithosphere is subducting northward beneath Java and Sumatra (Hamilton, 1979). The Lembang Fault, a major E-W structure down-dropped to the north, passes about 4 km south of the buried southern rim of the Sunda Caldera (Tjia, 1968).

GEOLOGIC HISTORY

Van Bemmelen (1934, 1971) concluded that a gravitational, landslide failure of Sunda Volcano triggered a voluminous pumice eruption; we have found no evidence for gravitational failure and believe instead that Sunda Caldera formed by collapse after

SUNDA (continued)

GEOLOGIC HISTORY (continued)

eruption of voluminous pumiceous ash flows. The age of caldera formation is known only approximately. Van Bemmelen (1949, 1971) suggested an age of about 2,000 years, based on the existence of an oral tradition about damming of a nearby river and on the age of artifacts on a terrace around the lake that formed in the Bandung basin behind this river blockage. However, radiocarbon dating of ash-flow and lake deposits has shown an age slightly older than the range of the radiocarbon measuring technique, >40,000 yr B.P. (C. Newhall, R. Hadisantono, and M. Rubin, unpubl. data).

Tangkubanparahu is a basaltic shieldlike stratovolcano in the center or on the eastern rim of the Sunda Caldera (van Bemmelen, 1934; Hadisantono and others, 1988) (fig. 6.6.1). The summit of Tangkubanparahu contains three large craters (from east to west, Kawah Ratu, Kawah Upas, and Kawah Baru) and several smaller craters (fig. 6.6.2).

HISTORICAL ACTIVITY

Historical eruptions have been small, mostly phreatic explosions from one of several craters in the summit area. Slight increases in ground cracking, gas emissions, and temperature are common, in part related to seasonal variations in rainfall. CO_2 and H_2S collect in the craters and pose a hazard to small animals and humans.

1829: Tangkubanparahu "began to burn" on the night of 1 May 1829, "after a few days of violent noises." (Weitzel, 1859).

1846: The 1846 eruption was more violent than most from Tangkubanparahu, and sent a pyroclastic surge(?) down the east slope of the volcano. Trees were snapped off a few feet above the ground, and the entire forest was killed (Bleeker, 1850). Stehn (1929) writes that the ash and mud were ejected obliquely in an easterly direction and flowed as a "mudstream" down toward the valley, destroying shrubs and trees.

1896: On 22 May 1896, activity resumed in Kawah Upas. The next day, brooks and rivulets on the north slope of the volcano were dirty with mud and had an irritating smell. Small explosions during 22-29 May were (each?) preceded by steady rise of steam from the crater (Figee, 1897). A new crater, Kawah Baru, was formed.

1908, 1910, 1913: In 1908, rumbling sounds were heard from the direction of Bukit Tunggul on 7, 17, 21, and 28 May and 10 June. No rain was falling, so thunder was thought not to be the cause, although neither was any change evident in the craters of Tangkubanparahu (Koninklijk Magnetisch en Meteorologisch Observatorium te Batavia, 1910). On 7 April 1910, activity increased suddenly: there were reports of a heavy crash, underground rumbling, and an ash column from Kawah Ratu. A rather severe earthquake was felt at Lembang and Bandung on 22 April (2155 hr), and recorded at the Observatory at Weltevreden. Another light earthquake was felt on the morning of 30 April; further ash eruptions (including a juvenile, magmatic component) occurred in early May (Stehn, 1929). In 1913, increased solfatara activity and rumbling were reported, with especially strong steaming from the 1910 vent (since clogged) and the 1896 vent (Kawah Baru, on the walls of Kawah Upas) (Koninklijk Magnetisch en Meteorologisch Observatorium te Batavia, 1914).

SUNDA (continued)

HISTORICAL ACTIVITY (continued)

1926: Phreatic explosions formed Kawah Ecoma, a 50-m-diameter crater within Kawah Ratu (Stehn, 1929).

1967, 1969, 1971: Solfataras in Kawah Ratu grew wider and more active in April 1967; in May a new solfatara ejected mud, and in July several vents ejected a minor amount of ash. In September, the temperature of some of these solfataras increased from 250 to 400 °C, but no eruption occurred (Suryo, 1985). From July through October 1969, phreatic eruptions from Kawah Ratu (the same crater that had shown unrest in 1967) produced a thin ash layer on all sides of the volcano (Smithsonian Institution, 1983; Suryo, 1985). Increased thermal activity and small-volume ejection of mud occurred in 1971 (Smithsonian Institution, 1983).

1983-present: Seismicity began to increase in June 1983 (3 earthquakes/day), then increased sharply in August 1983 (as many as 10 events/day in early August, 14 events/day in the last week of August, 60-70 events/day in early September, and 100-1100 events/day in mid-September). In mid-October 1983, the most restless month to this time of writing (March, 1988), the daily earthquake counts were 100 to 200x those of May-July 1983, with an average of 72 B-type earthquakes/day and a maximum of 127 earthquakes/day (fig. 6.6.3). Nearly all earthquakes were shallow, centered near the Baru ("new") fumarole field; a few A-type and tectonic earthquakes were also counted (A.C. Effendi, oral commun., 1983; Smithsonian Institution, 1983; Volcanological Survey of Indonesia, 1984). Fumarole temperatures rose only slightly, from 94 °C to 98 °C by the end of September 1983, probably buffered to boiling point by ground water. Fumarole temperatures in Kawah Baru (fig. 6.6.4) rose from 90 °C to 140 °C in the two weeks prior to 9 December 1985 and were as high as 173 °C by 3 January 1986; they fluctuated between 150 °C and 180 °C from January until at least September 1986 (the date of our most recent report) (Smithsonian Institution, 1985, 1986; Volcanological Survey of Indonesia, written commun., 1986). Minor tremor was noted in March 1986 (Smithsonian Institution, 1986). Electronic distance and spirit-level tilt data suggest inflation of Tangkubanparahu since the first measurements in 1981 until at least May 1986; during that period, approximately 8×10^6 m^3 of magma was added to a reservoir at a depth of about 1.5 km beneath Kawah Upas (Dvorak and others, 1987; J. Dvorak, written commun., 1988; Matahelumual and others, 1987).

REFERENCES

Bemmelen, R.W. van, 1934, Geologische kaart van Java, 1:100,000, Toelichting bij Blad 36 (Bandoeng): Dienst van den Mijnb. in Ned. Ind., Bandung, 95 p. + map.

----- 1949, The geology of Indonesia, v. 1A: The Hague, Government Printing Office, 732 p.

----- 1971, Four volcanic outbursts that influenced human history: Toba, Sunda, Merapi, and Thera, in Kaloyeropoyloy, A., ed., Acta, 1st Int. Sci. Congr. on the Volcano of Thera, Athens, Archaeol. Serv. Greece, p. 5-50.

Bleeker, P., 1850, Berigten van verschillenden aard: Vulkanische verschijnselen: Nat. Tijdschr. v. Ned. Ind., pt. 1, p. 153-154.

Dvorak, J., Said, H., Hadisantono, R.D., Raharja, N., Mulyadi, D., Reksowirogo, D., and Restikadjaja, K., 1987, Geodetic measurements at Indonesian volcanoes: U.S. Geol. Surv. Open-File Rep. 87-130, 21 p.

SUNDA (continued)

REFERENCES (continued)

Figee, S., 1897, Vulkanische verschijnselen en aardbevingen in den O. I. Archipel waargenomen gedurende het jaar 1896: Nat. Tijdschr. v. Ned. Ind., pt. 57, no. 1, p. 377-431.

Hadikusumo, D., 1961, Tangkuban Prahu: Bull. Volcanol. Surv. Indonesia, no. 100, p. 31-32.

Hadisantono, R.D., Juwarna, H., and Soetoyo, 1988, Geologic map of Tangkubanparahu Volcano, West Java: Bandung, Volcanological Survey of Indonesia, 1 sheet, scale 1:50,000 [in press].

Hamilton, W., 1979, Tectonics of the Indonesian region: U.S. Geol. Surv. Prof. Paper 1078, 345 p.

Koninklijk Magnetisch en Meteorologisch Observatorium te Batavia, 1910, Vulkanische verschijnselen en aardbevingen in den Oost-Indischen Archipel waargenomen gedurende het jaar 1908: Nat. Tijdschr. v. Ned. Ind., pt. 69, p. 107-153.

----- 1914, Vulkanische verschijnselen en aardbevingen in den Oost-Indischen Archipel waargenomen gedurende het jaar 1913: Nat. Tijdschr. v. Ned. Ind., pt. 74, no. 1, p. 67-115.

Kusumadinata, K., ed., 1979, Data dasar gunungapi Indonesia: Bandung, Volcanological Survey of Indonesia, 820 p.

Matahelumual, J., Mulyady, D., Rahardja, N., Reksowirogo, L., Restikadjaja, K., Said, H., Dvorak, J., and Okamura, A., 1987, Analysis of tilt measurements at Tangkubanparahu Volcano, West Java, Indonesia [abs.]: Abstracts, Hawaii Symposium on How Volcanoes Work, Hilo, January 1987, p. 167.

Neumann van Padang, M., 1951, Catalogue of the active volcanoes of the world including solfatara fields, Pt. I, Indonesia: Rome, IAVCEI, 271 p.

----- 1971, Two catastrophic eruptions in Indonesia, comparable with the Plinian outburst of the Volcano of Thera (Santorini) in Minoan time, in Kaloyeropoyloy, A., ed., Acta, 1st Int. Sci. Congr. on the Volcano of Thera, Athens, Archaeol. Serv. Greece, p. 51-63.

Siswowidjoyo, S., 1975, Tangkubanparahu Volcano: mimeographed guidebook, 6 p. + figures.

Smithsonian Institution, Scientific Event Alert Network (SEAN), 1983, 1985, 1986, Tangkubanparahu: SEAN Bull., v. 8, 10, 11 (reports from A. Sudradjat, L. Pardyanto, and T. Casadevall).

Stehn, Ch.E., 1929, Tangkoeban Prahoe: Excursion guide, 4th Pacific Sci. Congr., Bandung, 1929. (excursion B3, 22 p.)

Suryo, I., 1981, Tangkubanparahu: Bull. Volcanol. Surv. Indonesia, no. 104, p. 8.

----- 1985, Tangkubanparahu: Bull. Volcanol. Surv. Indonesia, no. 106, p. 15-19.

Tjia, H.D., 1968, The Lembang Fault, West Java: Geol. Mijnb., v. 47, p. 136-140.

Volcanological Survey of Indonesia, 1984, Tangkubanparahu excursion: A short guide: Bandung, Volcanological Survey of Indonesia, 15 p.

Weitzel, A.W.P., 1859, Berigten over aardbevingen en beruitbarstingen, vermeld in de Bataviasche Koloniale Courant, de Java-Governement Gazette, de Bataviasche Courant en de Javasche Courant, van 1810 tot 1830: Nat. Tijdschr. v. Ned. Ind., pt. 18, p. 152-244.

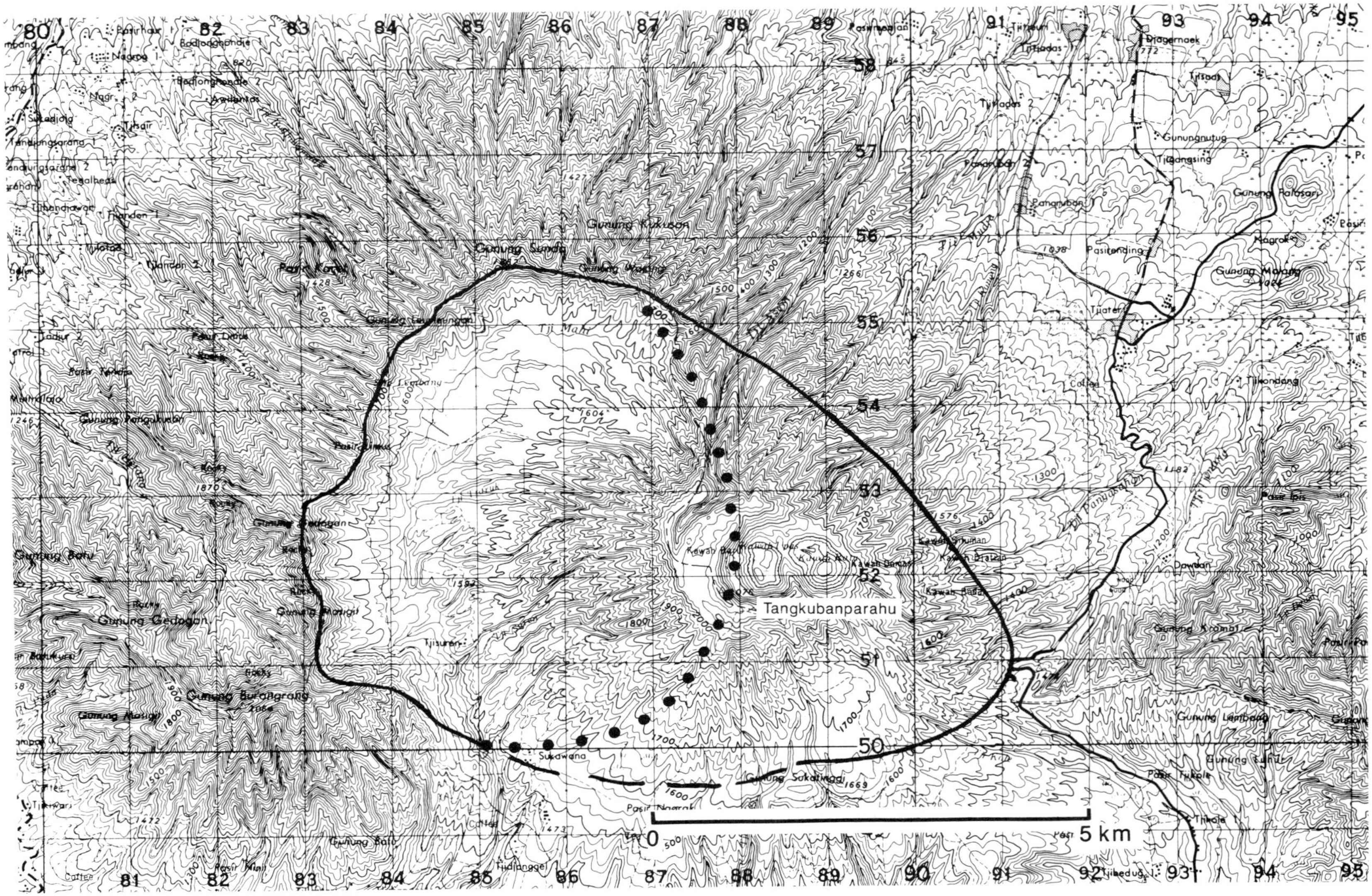

Figure 6.6.1. Topographic map of an area about 15 km north of the city of Bandung, central Java, showing two possible outlines of Sunda Caldera atop an ancient stratovolcano. Also shown, postcaldera cone of Tangkubanparahu. Larger-diameter outline (solid and dashed) is from van Bemmelen (1934); smaller-diameter outline (dotted) shows another possible configuration with Tangkubanparahu along east rim of caldera.

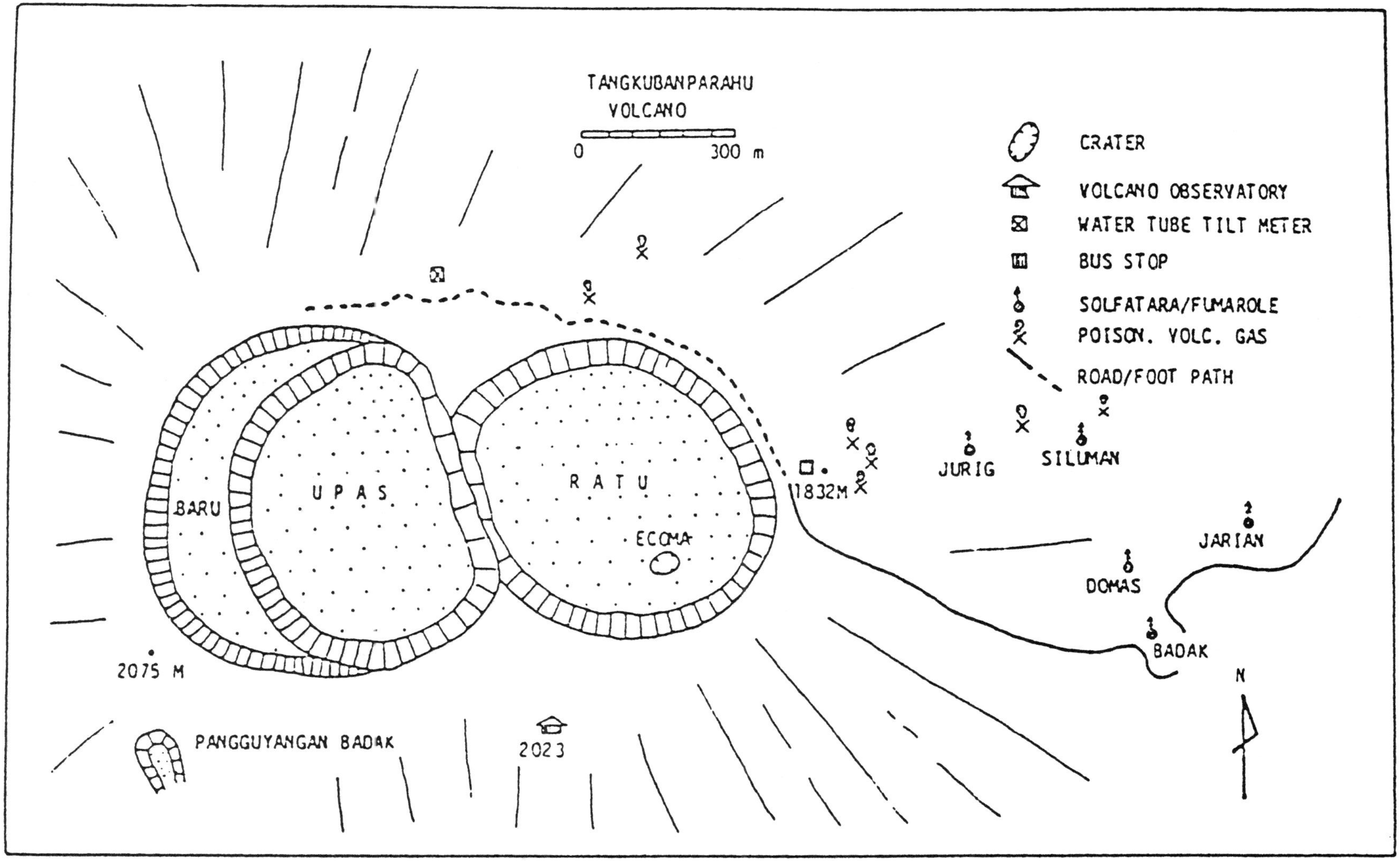

Figure 6.6.2. Sketch map of summit area of Tangkubanparahu Volcano, showing configuration of summit craters, monitoring sites, and fumarolic areas, from Volcanological Survey of Indonesia (1984).

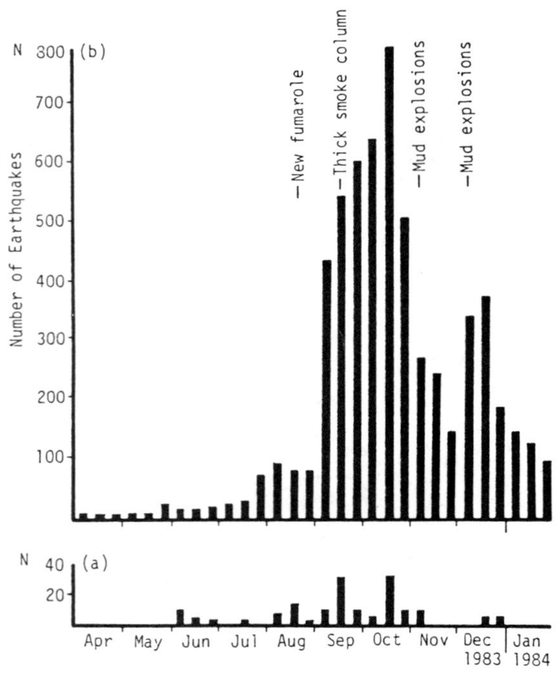

Figure 6.6.3. Number of volcanic earthquakes (a, deep; b, shallow) at Tangkubanparahu from April 1983 to January 1984, after Volcanological Survey of Indonesia (1984).

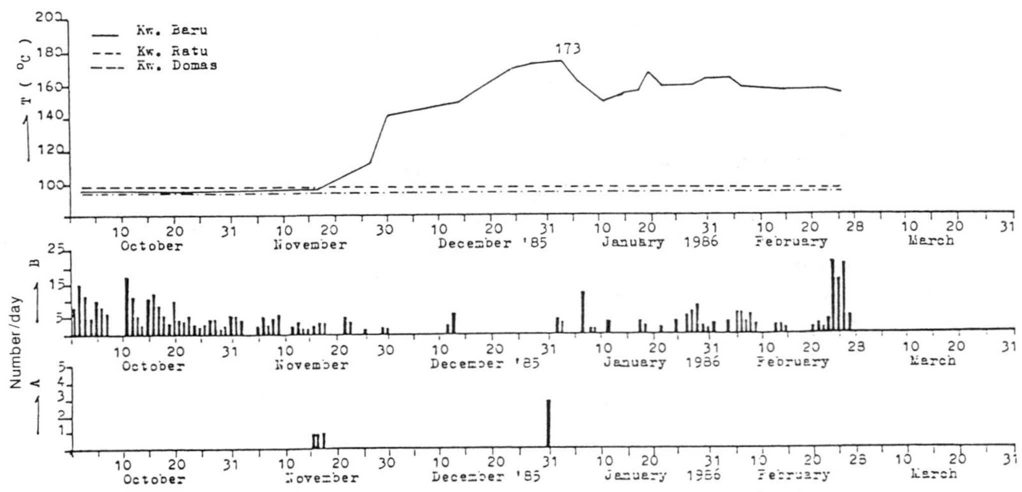

Figure 6.6.4. Relation between solfatara temperatures and volcanic earthquakes at Tangkubanparahu from October 1985 to March 1986, from Volcanological Survey of Indonesia (1984). Top, solfatara temperatures; middle, daily frequency of shallow volcanic earthquakes; bottom, daily frequency of deep volcanic earthquakes.

GEGER HALANG

CAVW number (active vent)	Latitude Longitude (degrees)	Diameter (km)	Local tectonic setting	Precaldera edifice	SiO_2 content (wt pct)	Age of collapse (yr B.P.)	Date of unrest	Type and duration of unrest ESTU STHF MGTF H Te	Eruption type
06-03-17	6.89S	5.5 x 4	Compr	Strato	R = 54-58		1698	---- ---- ---- - --	Ex
(Cereme)	108.40E				C = silicic		1772	---- ---- ---- - --	ex
							1775	---- ---- ---- - --	ex
							1805	---- ---- ---- - --	ex
							1917	---- ---- ---x - --	none
							1924	---- ---- ---x - --	none
							1937-38	Y--- ---x --FF - --	pex,ex
							1941	---- ---- ---E - --	none
							1951	---- ---- ---- - --	ex

TECTONIC SETTING

Cereme lies behind the volcanic front of West Java, in an area of several left-lateral NW-SE faults (Situmorang and others, 1988).

GEOLOGIC HISTORY

Cereme is a stratovolcano on the northern rim of the Geger Halang Caldera in central Java (Situmorang and others, 1988) (fig. 6.7.1). Presumably the caldera formed as a result of voluminous eruptions of silicic magma. However, the northwest rim of the caldera is now buried by Cereme, and a large landslide deposit occurs farther to the northwest, raising the alternate possibility that the caldera is of landslide origin. We suspect an origin by collapse after a voluminous eruption but cannot rule out a landslide origin.

HISTORICAL ACTIVITY

1698: A major explosive eruption damaged land and killed many persons (Brascamp, 1919).

1876: Earthquakes and associated subsidence occurred in Kuningan, Cirebon (approximately 10 km southeast of the Geger Halang Caldera). There was no obvious relation to the volcanic system.

1917, 1924: Sulfur-gas emission increased in 1917, and the strength of fumarolic activity increased in September 1924 (Kusumadinata, 1979).

GEGER HALANG (continued)

HISTORICAL ACTIVITY (continued)

1937-38: Eruptions resumed in 1937 without felt precursory earthquakes (Neumann van Padang, 1937). After the eruptions began, frequent earthquakes that rattled doors in Cirebon were followed sometimes by increases in steaming of Cereme.

1951: "In the evening of March 1, 1951, at 9 P.M., one single detonation coming from the direction of this volcano was heard by the inhabitants.... Next morning thick smoke was observed above the volcano which lasted until 9:30 A.M." (Hadikusumo, 1961).

1973: Epicenters of tectonic earthquakes in April 1973 were (possibly) about 15 km southwest of G. Cereme; there was no apparent relation to volcanic activity (Kusumadinata, 1979).

REFERENCES

Bemmelen, R.W. van, 1943, Tjereme: Bull. East Ind. Volcanol. Surv. for the year 1941, no. 95-98, p. 62-63.
Brascamp, E.H.B., 1919, Uitbarsting van den Tjeremai in 1698?: Tijdschr. Kon. Ned. Aardr. Genoot., ser. 2, v. 36, p. 487.
Hadikusumo, D., 1961, Tjerme: Bull. Volcanol. Surv. Indonesia, no. 100, p. 31-32.
Koninklijk Magnetisch en Meteorologisch Observatorium te Batavia, 1939, Aardbevingen in den Oost Indischen Archipel waargenomen gedurende het jaar 1937: Nat. Tijdschr. v. Ned. Ind., pt. 99, no. 3, p. 101-131.
Kusumadinata, K., ed., 1979, Data Dasar Gunungapi Indonesia: Bandung, Volcanological Survey of Indonesia, 820 p.
Neumann van Padang, M., 1937, De uitbarsting van den Tjerimai in 1937 (verschijnselen van Juni t/m September): De Ing. in Ned. Ind., v. 4, no. 12, p. 211-227.
----- Catalogue of the active volcanoes of the world including solfatara fields, Pt. I, Indonesia: Rome, IAVCEI, 271 p.
Situmorang, T., Hadisantono, R.D., and Rachmat, H., 1988, Geologic map of Cereme Volcano, Kuningan, West Java: Bandung, Volcanological Survey of Indonesia, 1 sheet, scale 1:50,000 [in press]. (Preliminary summary in: Laporan Tahunan Direktorat Vulkanologi (Annual Report, Volcanological Survey of Indonesia), 1983-84, p. 34-42.)

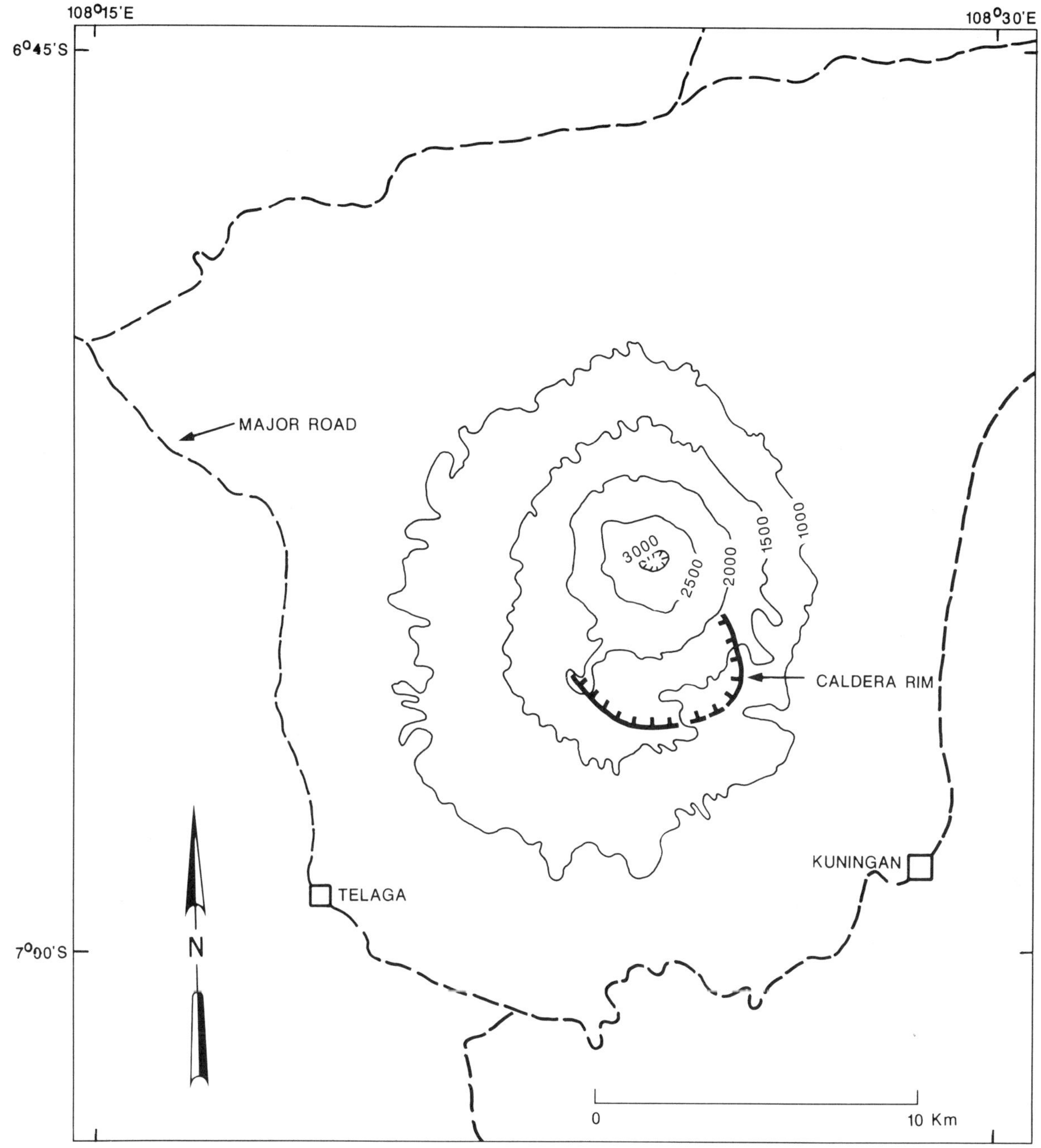

Figure 6.7.1. Sketch map of Geger Halang Caldera and Cereme Volcano. Location of caldera rim from Situmorang and others (1988). Contour interval is 500 meters.

PRAHU (DIENG)

CAVW number (active vent)	Latitude Longitude (degrees)	Diameter (km)	Local tectonic setting	Precaldera edifice	SiO_2 content (wt pct)	Age of collapse (yr B.P.)	Date of unrest	Type and duration of unrest ESTU STHF MGTF H Te	Eruption type
06-03-20 (Dieng)	7.2S 109.92E	6.5	Compr	Strato	R = 58-63 C = 61?	>16,000	late 1300's	---- ---- ---- - --	Ex
							1786	EE-- ---Y ---Y - --	ex,landsl
							1825-26	---- ---- ---- - --	Ex
							1847	---- ---- ---- - --	ex?
							1884	---- ---- --xx - --	none
							1895	---- ---x ---- - --	none
							1924	x--- ---- --xx - --	none
							1928	DD-- ---D ---Y - --	pex
							1939	BB-- ---B ---- - --	pex
							1943-44	NN-- ---E ---- - --	pex
							1952-56	NN-- ---- --?? ? --	pex
							1964	NN-- ---- ---x - --	pex
							1965	DD-- ---- ---? - --	none
							1979	xx-- ---? --xY - --	pex
							1981	DD-- ---- --NN - --	none
							1984	DD-- ---- --N- - --	none
							1986	D--- ---- --N- ? QU	none

TECTONIC SETTING

The Dieng Plateau contains two intersecting sets of faults--an E-W or ESE-WNW set that cuts straight through the plateau, and a N-S or NNE-SSW set that is best expressed (and offset?) in the Butak Petarangan/Dringo area in the northwest part of the plateau and near Pakuwojo in the southeast part of the plateau (Condon and others, 1975; Delarue, 1980; Sukhyar and others, 1986).

GEOLOGIC HISTORY

The Dieng Plateau (fig. 6.8.1) is an area of thermal activity and frequent phreatic eruptions, surrounded by the relics of two or more stratovolcanoes (Sukhyar and others, 1986). Junghuhn (1853/54), Verbeek and Fennema (1893), and several subsequent workers considered it to be a caldera, but Umbgrove (1929) doubted this because he considered the composite nature of the rim (pieces of several stratovolcanoes) to be inconsistent with the idea of a caldera. Delarue (1980) shows what he interpreted to be products of a caldera-forming eruption (SiO_2=61 percent), but no one has identified a large volume of deposits from a single eruption. According

PRAHU (DIENG) (continued)

GEOLOGIC HISTORY (continued)

to Delarue, eruptions shortly after caldera formation were of basaltic andesite (as were most precaldera eruptions), but the most recent magmatic product, a lava flow in the southeast part of the Plateau near Pakuwojo, is biotite andesite or dacite (SiO_2=63 percent).

Roughly 1,000-yr-old Hindu temples in the Dieng plateau are overlain by a small thickness of phreatic products (age=755 ± 100 ^{14}C yr; N.G. Banks and M. Rubin, written commun., 1984), but not by any clearly juvenile products. Delarue (1980) reported a radiocarbon date of 2,000±100 yr for the young lava flow near Pakuwojo; C. Dan Miller and Meyer Rubin (unpubl. data) report ages of 2,450±80, 8,540±150, and 16,770±250 for three postcaldera deposits.

A hydrothermal system is being explored for geothermal power in the east-central part of the plateau.

HISTORICAL ACTIVITY

1786: Earthquakes occurred for four months in 1786, followed by eruption not only through "existing volcanic apertures, but through the neighboring territory to the extent of 2 miles in diameter. Various vents were formed which emitted a sulphureous vapour, separate tracks sunk away and were swallowed up by the earth..." (Horsfield, 1816, p. 280-281). Three craters near Gunung Jimat were aligned NNW-SSE (165°).

1825-26: Between 1400 and 1500 hr on 9 October 1826, strong "cannonade" sounds were heard as far away as Yogyakarta. For two hours on 11 October, strong explosions and earthquakes were felt as far away as Pekalongan (Siswowidjoyo, 1980). The source may have been Pakuwojo, but Delarue (1980) found no trace of 1826 eruption products there.

1847: A small eruption occurred from an unknown vent on 4 October 1847, without any report of earthquakes.

1884, 1895: After unusually low solfataric and fumarolic activity during the eruptive activity of nearby Sundoro Volcano in August 1883, solfataric and fumarolic activity at Dieng increased again in late 1884, including mud eruptions at Kawah Sikidang and Kawah Banteng (De Commissie tot het Organiseeren en Verzamelen van Aardbevings- waarnemingen, 1885). Fissuring at Dieng was reported in 1395 (Figee and Onnen, 1897).

1924, 1928: On 2 December 1924, a strong earthquake occurred near Wonosobo; Taverne (1925) says that earthquake originated along N-S faults in the Dieng region. A seismic swarm began on 3 May 1928 at 1100 hr; another earthquake occurred at 1430 hr, and two strong earthquakes occurred on 4 May at 2330 hr, accompanied by opening of NNE-SSW fissures near Timbang. The intensity of earthquakes increased until 13 May at 0940 hr, when a series of shocks occurred that were strong enough to damage houses in Batur village and to be registered on seismographs at Merapi (Central Java), Papandayan (West Java), and Jakarta (West Java). An eruption began shortly thereafter from three new craters along a north-northwest-trending fissure. Earthquakes continued until 18 June 1928; strong CO_2 emission continued at least until 1937.

PRAHU (DIENG) (continued)

HISTORICAL ACTIVITY (continued)

1939: Another seismic swarm occurred in October 1939; the largest shocks at 1530 hr on 13 October were followed by small phreatic eruptions from at least 15 craters along a NNE-SSW-trending fissure that overlaps and is essentially an extension of the 1928 fissure. CO_2 emission was moderate, less than in 1928. An average of several earthquakes per day continued to be felt in nearby villages until December 1939 (Siswowidjoyo, 1980).

1943-44: In 1943, many small N-S fissures appeared northwest of the Sileri crater. The fissures were 5-6 m long, 1-1.5 m wide, and deeper than could be determined. On 3 November 1944, two phreatic explosions formed small craters, and on 4 December, a strong phreatic explosion killed as many as 114 persons. No earthquakes were felt before or during this eruption (Siswowidjoyo, 1980; Delarue, 1980, p. 159).

1952-56: Increased CO_2 emission began near Timbang in August 1952, accompanied by several small felt earthquakes. Increased gas emission and light ashfall occurred near Keputjukan on 21 March 1953. There was a small eruption on 6 December 1954, and many local earthquakes occurred in November 1955. Increased steaming and gas (CO_2?) emission from Sikendang crater occurred in March 1956, and from Sileri crater in June 1956 (Hadikusumo, 1961).

1964: A fumarole field at Bitingan enlarged during December 1964, and a small phreatic eruption occurred from Sileri crater on 13 December 1964. Earthquakes were not felt during this episode (Miller and others, 1981; Suryo, 1985).

1965: In May 1965, seven weak earthquakes were centered near Timbang and felt in nearby Batur. There was an apparent increase in emission of CO_2 and H_2S, which produced a ground-hugging gas layer reaching 1 m above ground level in August 1965.

1979: A seismograph at Karangtengah, 4 km east of Sinila crater, recorded six weak earthquakes on 16 February and one on 19 February 1979 (Siswowidjoyo, 1980). On 20 February, earthquakes at 0200, 0220, and 0330 hr were felt over a radius of 10 km, and four other earthquakes were felt over a smaller area. An eruption began at 0500 hr from Telago Sinila (Sinila lake) in the area of N-S faulting along the side of the Butak Petarangan complex. At 0615 hr, another small eruption occurred from the Sigludug zone, 250 m west of Sinila. Copious CO_2 was emitted from a N-S fissure, together with minor amounts of H_2S and CH_4. CO_2 displaced oxygen in the air near the ground, killing 150 persons (fig. 6.8.2; Le Guern and others, 1982). Many more earthquakes (134) occurred on 24 March, just after the eruption and strong gas emission (Siswowidjoyo, 1980) (fig. 6.8.3). Another small earthquake swarm occurred 1 month after the eruption, but there was no further increase in volcanic activity.

1981, 1984: A small earthquake swarm occurred in 1981, centered near the Sinila (1979) crater. Another swarm began on 10 October 1984 and peaked on 20 October 1984 (fig. 6.8.4). The 1984 swarm was centered beneath the Sileri crater (last active in 1943-44), and damaged a few buildings in the Dieng Plateau. Gas chemistry and temperatures (buffered at the boiling temperature of water) remained unchanged during both episodes, but from September 1984 to January 1985, mud pots in Sileri crater were more active than usual, spouting mud 2-3 m above mud-pond level (Volcanological Survey of Indonesia, 1986).

PRAHU (DIENG), Region 6, CAVW number 06-03-20

PRAHU (DIENG) (continued)

HISTORICAL ACTIVITY (continued)

1986: A widely felt earthquake occurred near Dieng on 13 April 1986, followed by 10 minutes of tremor and two weeks of small earthquakes. No changes in thermal activity were noted (Smithsonian Institution, 1986). In July 1986, minor mud spouting followed a moderate tectonic earthquake that was reportedly far outside the caldera.

Delarue (1980) distinguished four types of phreatic and phreatomagmatic eruptions at Dieng:

(1) phreatic eruptions as a result of self sealing (for example, Kawah Sileri, 1943-44, Candradimuka, 1945, and Sikidang);

(2) phreatic eruptions as a result of tectonic fissuring that allows hot ground water to flash to steam (for example, Timbang, 1928, 1939);

(3) phreatomagmatic eruptions as a result of fissuring during dike intrusion, which allows hot ground water to flash to steam (for example, at Kendil); and

(4) rootless phreatic explosions as a result of lava entering a body of surface water (for example, at Pakuwojo).

Eruptions in the central and northeastern parts of the Dieng Plateau are of type 1, and those in the western part are mainly of type 2. The best examples of types 3 and 4 are in the southeastern part of the plateau. Eruptions of type 1 are not preceded by notable seismicity, whereas those of types 2 and 3 are preceded by notable seismicity and surface faulting.

COMMENTS

We suspect that Dieng is a caldera, but we do not know of any deposit that can be correlated with caldera formation. We include Dieng in this compilation for three reasons. First, it is a caldera in the descriptive sense--a large depression mostly surrounded by relics of volcanic edifices. Second, activity at Dieng illustrates relations between rifting and phreatic eruptions. Third, Dieng illustrates the hazard posed by accumulations of volcanic gases in low-lying areas.

CO_2 may be concentrated in pockets or an aureole around the hydrothermal system, as in several other geothermal areas (Muffler, 1968; Hedenquist, 1986). If so, the most hazardous concentrations of CO_2 might develop around the margins of high-temperature geothermal areas. Such concentrations of CO_2 can be released by tectonic fracturing without accompanying volcanic activity.

PRAHU (DIENG) (continued)

REFERENCES

Condon, W.H., Pardyanto, L., and Ketner, K.B., 1975, Geologic map of the Banjarnegara and Pekalongan Quadrangles, Java, 1:100,000, 1 sheet: Bandung, Geological Survey of Indonesia.

De Commissie tot het Organiseeren en Verzamelen van Aardbevings- waarnemingen, 1885, Uitbarstingen van vulkanen en aardbevingen in den O.I. Archipel waargenomen gedurende het jaar 1884: Nat. Tijdschr. v. Ned. Ind., pt. 45, no. 1, p. 458-471.

Delarue, C., 1980, Exploration geothermique dans la region de Dieng (Java): Univ. Paris XI, unpubl. Ph.D. thesis, 247 p.

Figee, S., and Onnen, H., 1897, Vulkanische verschijnselen en aardbevingen in den O.I. Archipel waargenomen gedurende het jaar 1895: Nat. Tijdschr. v. Ned. Ind., pt. 56, p. 81-125.

Gunawan, R., 1968, Geological investigations in the Dieng area, Central Java: Inst. Teknol. Bandung, unpubl. M.Sc. thesis, 123 p.

Hadikusumo, D., 1961, Report on the volcanological research and volcanic activity in Indonesia for the period 1950-1957: Bull. Volcanol. Surv. Indonesia, no. 100, p. 34.

Hedenquist, J.W., 1986, Geothermal systems in the Taupo Volcanic Zone: Their characteristics and relation to volcanism and mineralization, in Smith, I.E.M., ed., Late Cenozoic volcanism in New Zealand: Roy. Soc. New Zealand Bull. 23, p. 134-168.

Horsfield, Th., 1816, Over aardbevingen en een eruptie in het Dieng-gebergte, in 1786, in Essay on the geography, mineralogy, and botany of the western portion of the territory of the natives princes of Java: Verh., Batavia. Gen. VIII, 1816, 1st Dr. 140., 2nd Dr. 281.

Junghuhn, F., 1853/54, Java, deszelfs gedaante, bekleeding en inwendige structuur (in 2 volumes, 4 parts): Amsterdam, Uitgave C.W. Mieling's Gravenhage.

Le Guern, F., Tazieff, H., and Faivre-Pierret, R., 1982, An example of health hazard: People killed by gas during a phreatic eruption: Dieng Plateau (Java, Indonesia), February 20th 1979: Bull. Volcanol., v. 45, no. 2, p. 153-156.

Miller, C.D., Sukhyar, R., Santoso, and Hamidi, S., 1981, Eruptive history of the Dieng Mountains region, Central Java, and potential hazards from future eruptions: U.S. Geol. Surv. Open-File Rep. 83-68, 26 p.

Muffler, L.J.P., 1968, Origin of CO_2 in the Salton Sea geothermal system, southeastern California, U.S.A.: Proc., 23rd Int. Geol. Congr., Prague, v. 17, p. 185-191.

----- 1970, Geothermal potential of the Dieng Mountains, Central Java, Indonesia: U.S. Geol. Surv. Project rep. (IR) IND-3, 22 p.

----- 1971, Evaluation of initial investigations, Dieng geothermal area, Central Java, Indonesia: U.S. Geol. Surv. Project rep. (IR) IND-10, 20 p.

Siswowidjoyo, S., 1980, Kegiatan vulkanik di daerah Dieng, Jawa Tengah: unpublished ms., presented at 9th meeting of the Indonesian Association of Geologists (IAGI), Yogyakarta, December 1980, 14 p.

Smithsonian Institution, Scientific Event Alert Network (SEAN), 1986, Dieng Volcano: SEAN Bull., v. 11, no. 4, p. 22 (contribution from Olas, Suratman, Suparto, Kaswanda, and Adjat Sudradjat, Volcanological Survey of Indonesia).

Stehn, Ch.E., 1928, Eruption near the village of Batoer: Bull. Neth. East Ind. Volcanol. Surv., no. 9, p. 59-60.

----- 1939, Steam explosions at Timbang District, Batur, Dieng: Bull. Neth. East Ind. Volcanol. Surv., no. 90, p. 69-71.

Sukhyar, R., and others, 1982, Laporan Kegiatan II, Pemetaan Geologi G. Dieng: Rep. 41/DV/82, Direktorat Vulkanologi, Bandung.

Sukhyar, R., Sumartadipura, N.S., and Effendi, W., 1986, Geologic map of Dieng Volcano complex, Central Java: Bandung, Volcanological Survey of Indonesia, 1 sheet, scale 1:50,000.

Suryo, I., 1985, Dieng: Bull. Volcanol. Surv. Indonesia, no. 106, p. 25-26.

PRAHU (DIENG) (continued)

REFERENCES (continued)

Taverne, N.J.M., 1925, Vulkanologische Berichten, No. 42-52.: Nat. Tijdschr. v. Ned. Ind., pt. 85, no. 2, p. 102-205.

Umbgrove, J.H.F., 1929, Het ontstaan van het Diëngplateau: Leidsche Geol. Meded., v. 3, no. 3, p. 131-150.

Verbeek, R.D.M., and Fennema, R., 1893, Geologische beschrijving van Java en Madoera, 2 vol. + folio with maps: Amsterdam, J.G. Stemler.

Volcanological Society of Japan, 1981, Dieng: Bull. Volcanic Eruptions, no. 19, p. 33-36 (contribution from A. Sudradjat and S. Siswowidjoyo).

Volcanological Survey of Indonesia, 1986, Laporan Tahunan Direktorat Vulkanologi 1984-85: Bull. Volcanol. Surv. Indonesia, no. 113, 259 p.

Zen, M.T., 1971, Geothermal system of the Dieng-Batur volcanic complex: Proc. Inst. Teknol. Bandung, v. 6, no. 1, p. 23-38.

Zen, M.T., and Alzwar, M., 1980, Hydrothermal eruption at Sinila crater, Dieng Volcanic complex (central Java): Bul. Dept. Teknik Geol., Inst. Teknol. Bandung, v. 2, p. 1-20.

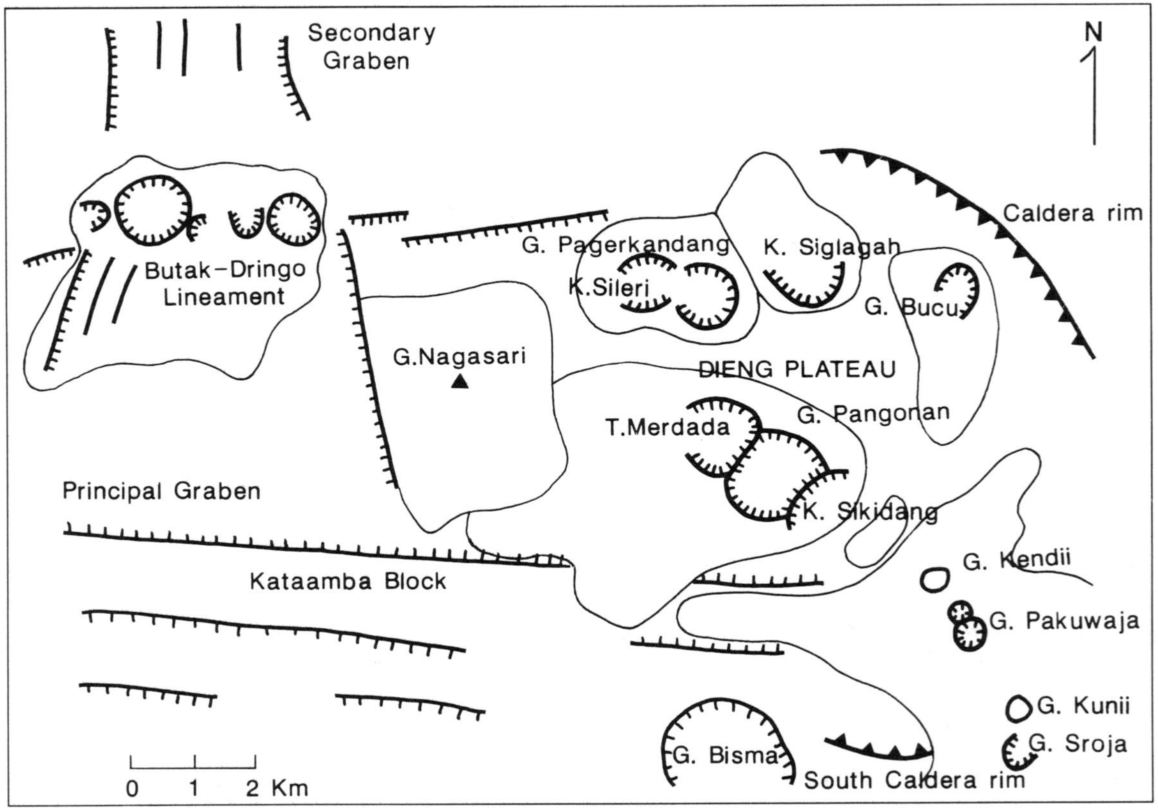

Figure 6.8.1. Sketch map of Dieng Plateau and Caldera, modified from Delarue (1980).

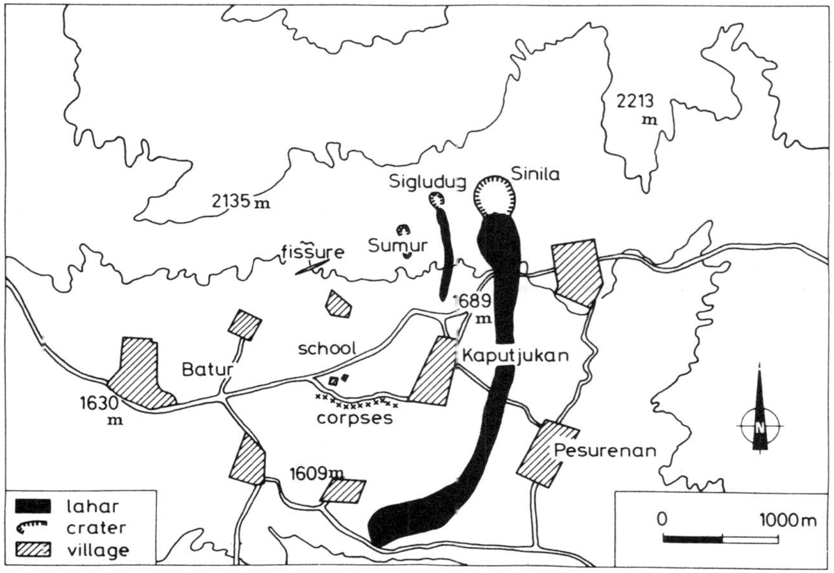

Figure 6.8.2. Area in which 150 people were killed by volcanic gases on 20 February 1979, from Le Guern and others (1982).

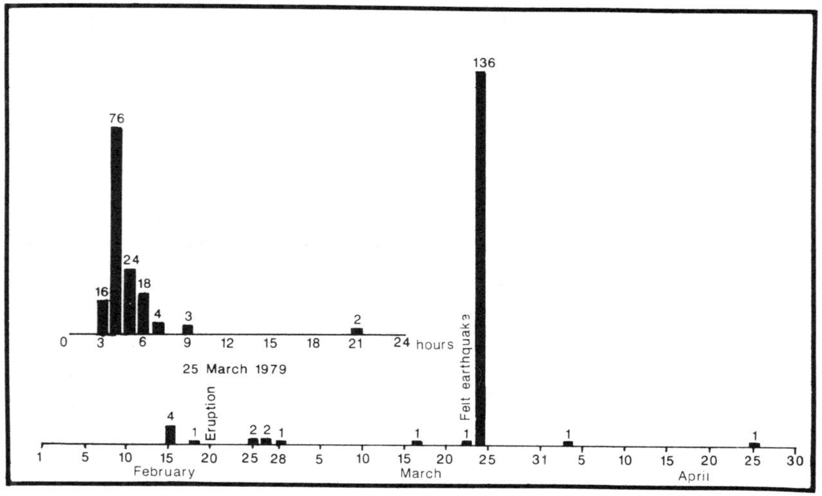

Figure 6.8.3. Daily and hourly numbers of earthquakes before, during, and after phreatic and gas eruption of 20 February 1979, from Siswowidjoyo (1980).

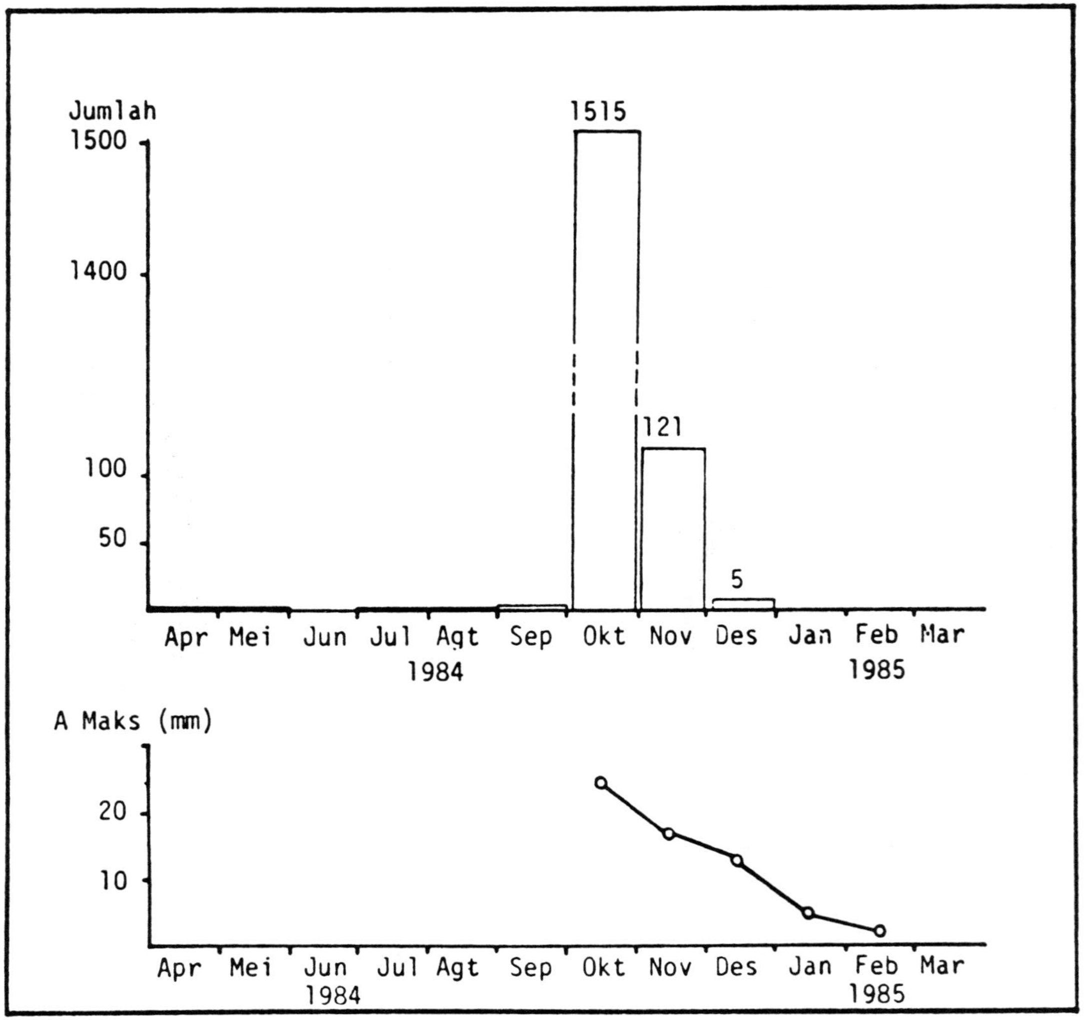

Figure 6.8.4. Monthly number of earthquakes and volcanic-tremor amplitude at Dieng from April 1984 to March 1985, from Volcanological Survey of Indonesia (1986).

PART 3: HISTORICAL UNREST AT LARGE QUATERNARY CALDERAS (continued)

See inside back cover for explanation and abbreviations

TENGGER/NGADISARI

CAVW number (active vent)	Latitude Longitude (degrees)	Diameter (km)	Local tectonic setting	Precaldera edifice	SiO_2 content (wt pct)	Age of collapse (yr B.P.)	Date of unrest	Type and duration of unrest ESTU STHF MGTF H Te	Eruption type
06-03-31	7.94S	16	Compr	Strato?/	R = 54-56		1804	---- ---- ---- - --	ex
(Bromo)	112.95E			LP-strat?	C =		1815	---- ---- ---- - --	ex
					intermediate?		1820	---- ---- ---- - --	ex
							1822-23	??-- ---- ---- - --	ex
							1825	---- ---- ---- - --	ex
							1829-30	---- ---- ---- - --	ex
							1835	---- ---- ---- - --	ex
							1837-38	---- ---- ---- x --	none
							1842-43	---- ---- ---- - --	ex
							1844	---- ---- ---- - --	ex
							1856-60	x--- ---- ---- - --	ex
							1865-68	?--- ---- ---- - --	ex
							1877	---- ---- ---- - --	ex
							1885-88	---- ---- ---Y - --	ex
							1890	---- ---- ---- - --	ex
							1893	---- ---- ---- - --	ex
							1896	---- ---- ---- - --	ex
							1898	x--- ---- ---- - --	none
							1906-10	---- ---- ---- - --	ex
							1915-16	---- ---- ---- - --	ex
							1921-22	---- ---- ---- - --	ex
							1928-30	---- ---- ---- - --	ex
							1935	---- ---- ---- - --	ex
							1939-40	---- ---- ---- - --	ex
							1948	---- ---- ---- - --	Ex
							1950	---- ---- ---- - --	ex
							1955-56	---- ---- ---- - --	ex
							1972	x--- ---- ---- - --	ex
							1980	---- ---- ---C - --	ex
							1983-84	---- ---- ---x - --	ex

TENGGER/NGADISARI (continued)

TECTONIC SETTING

The Tengger complex is at the northern end of the Semeru-Tengger group, formed by subduction of the Indo-Australian plate beneath east Java. Tengger is also immediately west of the anomalous Lamongan field of monogenetic basalt vents. The northeast margin of Tengger Caldera is notably linear, suggesting control by a NW-SE-trending fault. Neumann van Padang (1951) suggested two subtle alignments of vents within the caldera: NNE-SSW and E-W.

GEOLOGIC HISTORY

Tengger Caldera (fig. 6.9.1) was the subject of much early debate regarding the origin of calderas. Van Bemmelen (1937) concluded that the caldera formed just after paroxysmal eruptions of "very hot lavas, ... glass rich... with gas pores and flow structure," alternating with ash and lapilli layers from explosions. The "lavas" are actually welded tuffs. Van Bemmelen correctly inferred two episodes of caldera collapse: one in which the Ngadisari depression was formed, followed by later collapse to the southwest that formed the Tengger (Sandsea) Caldera. The Sapikerep Valley that extends northeast from the Ngadisari Caldera might be the relic southeast side of a third, oldest episode of caldera collapse, but it is more likely an erosional feature. The geology of the Tengger Mountains can be explained by south-southwestward migration of volcanism and caldera collapse without the gravitational-tectonic genesis inferred by van Bemmelen. Recent mapping is by Erfan and others (1988).

Bromo, the active vent, is a small tephra cone in the center of Tengger Caldera.

HISTORICAL ACTIVITY

Bromo has relatively frequent, small Strombolian eruptions; most reports discuss eruptions in progress rather than precursors.

1835-42: A small, blue lake formed in the crater of Bromo after its 1835 eruption, grew larger after its 1838 eruption, and became increasingly green from 1838 to 1841 (van Herwerden, 1844; Fennema, 1886). In early 1842(?) a part of the crater floor (a "half-fluid mass") swelled into a convex uplift that fissured and from which gases escaped; two months later, "the lava had sunken back and the floor ... had collapsed" (Fennema, 1886, citing van Herwerden, 1844). There is no report of concurrent changes on the caldera floor or elsewhere outside the small crater of Bromo.

1856-60: After long quiescence, Bromo became active again on 4 March 1858. A loud subterranean noise preceded that eruption. Another "terrible" eruption of Bromo on 18 October 1858 was also preceded by loud underground noise, and earthquakes were felt up to a distance of 1 paal (1.4 km) (Anonymous, 1858; Weitzel, 1859).

1865-68: At a time when tectonic earthquakes were commonly attributed to volcanic activity, an earthquake that "ravaged all of Java" on 17 May 1865 was attributed to Bromo, which was active during April and May of that year (Versteeg, 1867); few would now attribute such an earthquake to activity at Bromo, but this is not to rule out some relation between the two events.

TENGGER/NGADISARI (continued)

HISTORICAL ACTIVITY (continued)

1885-88: Strong sulfur gas emission with high steam clouds occurred at Bromo during 22-27 February 1888; further strong emission occurred in the last weeks of December, "but Bromo itself was quiet" (Figee and Onnen, 1890).

1972-present: Instrumental monitoring of seismicity at Bromo is discontinuous, but short seismic precursors have been noted for some eruptions (for example, that of 1972). Increased fumarolic activity was noted several days before eruptions resumed in 1980, after 8 years of quiescence. The first volcano observatory at Bromo was completed in August 1987.

REFERENCES

Anonymous, 1858, Berigten van verschillenden aard: Vulkanische verschijnselen in den Indischen Archipel: Nat. Tijdschr. v. Ned. Ind., pt. 17, p. 420-422.

Bemmelen, R.W. van, 1937, The volcano-tectonic structure in the Residency of Malang (Eastern Java): De Ing. in Ned. Ind., v. 4, no. 9, sect. IV, p. 159-172.

Bosch, C.A. van den, 1929, The history of the Tengger Mountains: Proc. 4th Pacific Sci. Congr., Bandung, v. IIA, p. 317-319.

Erfan, R.D., Hadisantono, R.D., Asmoro, P., and Mulyana, A.R., 1988, Geologic map of Bromo-Tengger volcano complex, East Java: Bandung, Volcanological Survey of Indonesia, 1 sheet, scale 1:66,667 [in press].

Fennema, R., 1886, De vulkanen Semeroe en Lemongan: Jaarb. Mijnwezen Ned. Ind., Wet. Ged., p. 5-130.

Figee, S., and Onnen, H., 1890, Vulkanische verschijnselen en aardbevingen in den O.I. Archipel waargenomen gedurende het jaar 1888: Nat. Tijdschr. v. Ned. Ind., pt. 49, p. 111-159.

Herwerden, J.D. van, 1844, Het Tenggersch Gebergte in deszelfs vulcanischen toestand beschouwd: Verh. Bataviaasch Genoot. Kunsten Wetensch., v. 20, p. 1-98.

Kemmerling, G.L.L., 1922, De G. Semeroe, de G. Bromo en de G. Lamongan in het begin van 1920: Vulkanol. Meded., no. 4, 40 p.

Kusumadinata, K., ed., 1979, Data Dasar Gunungapi Indonesia: Bandung, Volcanological Survey of Indonesia, 820 p.

Neumann van Padang, M., 1951, Catalogue of the active volcanoes of the world including solfatara fields, Pt. I, Indonesia: Rome, IAVCEI, 271 p.

Versteeg, W.F., 1867, Vervolg op de aanteekeningen omtrent aardbevingen en berguitbarstingen in den Indischen Archipel: Nat. Tijdschr. v. Ned. Ind., pt. 29, p. 210-236.

Volcanological Society of Japan, 1982, Bromo: Bull. Volcanic Eruptions, no. 20, p. 37 (contributions from A. Sudradjat and S. Siswowidjoyo).

Volcanological Survey of Indonesia, 1985, Laporan Tahunan Direktorat Vulkanologi, 1983-84: Bull. Volcanol. Surv. Indonesia, no. 112, 152 p.

----- 1986, Laporan Tahunan Direktorat Vulkanologi, 1984-85: Bull. Volcanol. Surv. Indonesia, no. 113, 259 p.

Weitzel, A.W.P., 1859, Berigten over aardbevingen en beruitbarstingen, vermeld in de Bataviasche Koloniale Courant, de Java-Governement Gazette, de Bataviasche Courant en de Javasche Courant, van 1810 tot 1830: Nat. Tijdschr. v. Ned. Ind., pt. 18, p. 152-244.

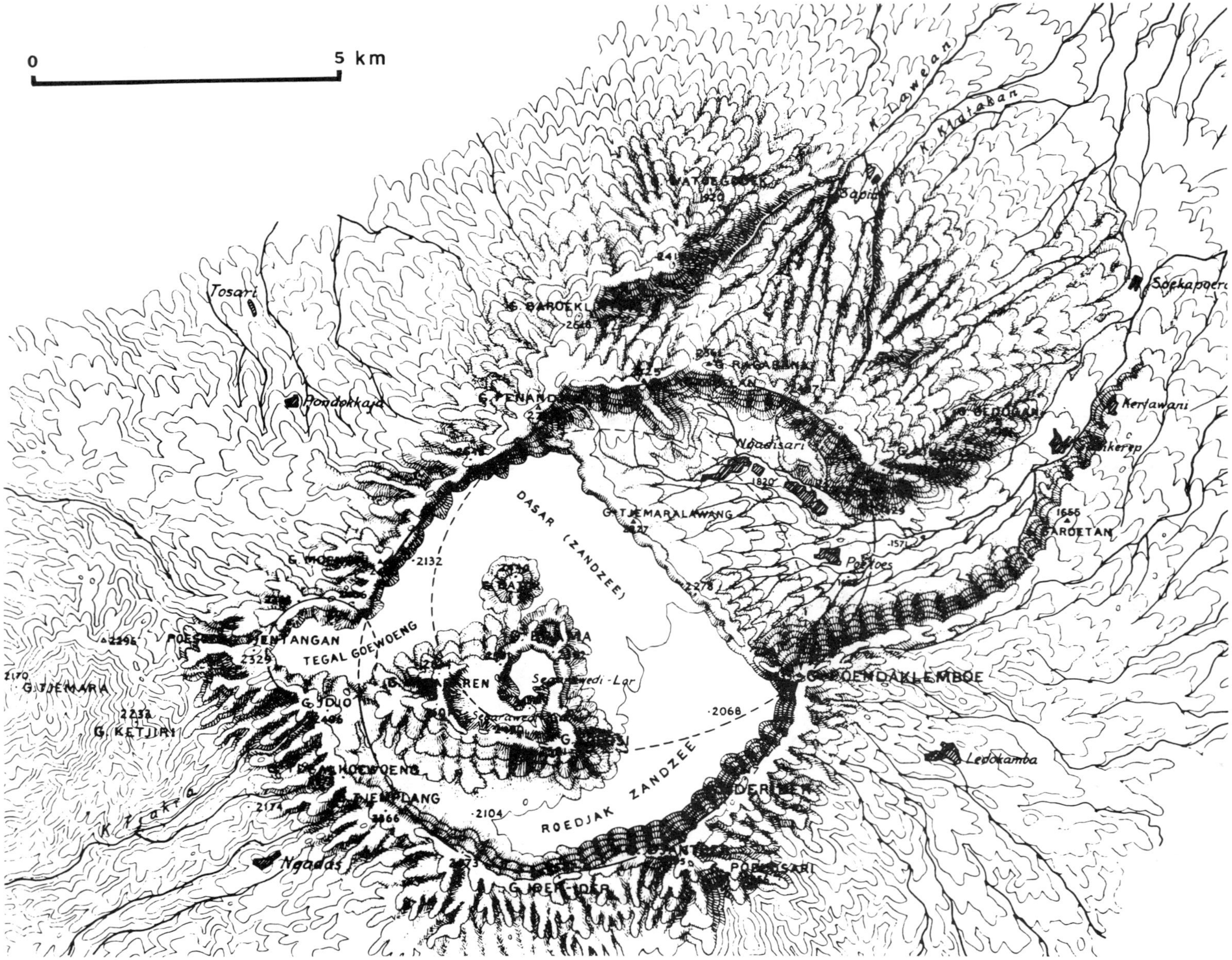

Figure 6.9.1. Topographic map of Tengger Caldera, showing at least two and probably three stages of caldera formation, from van den Bosch (1929).

PART 3: HISTORICAL UNREST AT LARGE QUATERNARY CALDERAS (continued)
See inside back cover for explanation and abbreviations

IJEN

CAVW number (active vent)	Latitude Longitude (degrees)	Diameter (km)	Local tectonic setting	Precaldera edifice	SiO_2 content (wt pct)	Age of collapse (yr B.P.)	Date of unrest	Type and duration of unrest ESTU STHF MGTF H Te	Eruption type
06-03-34	8.12S	20	Compr	Strato	R = 51-59		1586 (RA)	---- ---- ---- - --	Ex
(Raung= RA)	114.04E			complex	C = silicic		1593 (RA)	---- ---- ---- - --	Ex
				(3 or more			1597 (RA)	---- ---- ---- - --	Ex
06-03-35	8.06S			cones)			1638 (RA)	Y--- ---- ---- - --	Ex
(Merapi,	114.24E						1730 (RA)	---- ---- ---- - --	Ex
Kawah							between 1787-99 (RA)	---- ---- ---- - --	ex
Ijen= IJ)							1796 (IJ)	---- ---- ---- - --	pex
							between 1800-08 (RA)	---- ---- ---- - --	ex
							1812-15?(RA)	---- ---- ---- - --	Ex
							1817 (IJ)	---- ---- ---- - --	Ex or Pex
							1817 (RA)	---- ---- ---- - --	Ex
							1838 (RA)	---- ---- ---- - --	ex
							1859-60 (RA)	---- ---- ---- - --	ex
							1864 (RA)	---- ---- ---- - --	ex
							1881?(RA)	---- ---- ---- - --	ex
							1885 (RA)	---- ---- ---- - --	ex
							1890 (RA)	---- ---- ---- - --	ex
							1896-97 (RA)	---- ---- ---- - --	ex
							1902-04 (RA)	Y--- ---- ---Y - --	ex,lf
							1913-17 (RA)	---- ---- ---- - --	ex
							1917 (IJ)	---- ---- --DD - QU	pex
							1921 (RA)	---- ---- ---- - --	ex,lf
							1921 (IJ)	---- ---- --DD D --	none
							1924 (RA)	---- ---- ---- - --	ex,lf
							1927-29 (RA)	---- ---- ---- - --	ex
							1931 (RA)	A--- ---- ---x - --	none?
							1933 (RA)	---- ---- ---- - --	ex
							1936 (IJ)	---- ---- ---- - --	pex
							1936-45 (RA)	---- ---- ---- - --	ex,lf
							1938-39 (IJ)	---- ---- --xx - --	none
							1952 (IJ)	---- ---- ---- - --	ex

IJEN (continued)

CAVW number (active vent)	Latitude Longitude (degrees)	Diameter (km)	Local tectonic setting	Precaldera edifice	SiO_2 content (wt pct)	Age of collapse (yr B.P.)	Date of unrest	Type and duration of unrest ESTU STHF MGTF H Te	Eruption type
(continued from previous page)							1953 (RA)	---- ---- ---- - --	ex,lf
							1955-56 (RA)	---- ---- ---- - --	ex
							1958 (IJ)	x--- ---- --xx x --	none
							1971 (RA)	---- ---- ---- - --	pex
							1973-77 (RA)	---- ---- ---- - --	ex,pex
							1982 (RA)	B--- ---- ---- - --	ex
							1985 (RA)	x--- ---- ---- - --	ex

TECTONIC SETTING

Ijen Caldera is situated at the east end of Java, above the north-dipping subduction of the Indo-Australian plate beneath Java.

GEOLOGIC HISTORY

Raung (RA) and Merapi are stratovolcanoes that have grown on the ring fracture of the large Ijen Caldera (fig. 6.10.1). Raung has a large "avalanche caldera" that is open to the west (Neumann van Padang, 1939; Sutawidjaja and others, 1988). Kawah Ijen ("Ijen Crater") (IJ) is a lake-filled crater on the west (caldera-facing) flank of Merapi stratovolcano. Activity of Raung shows little if any relation to activity at Kawah Ijen.

HISTORICAL ACTIVITY

Most eruptions of Raung are small- to moderate-size ash ejections, with occasional lava extrusion; most eruptions of Kawah Ijen are phreatic explosions through the lake that fills that crater. Small earthquakes precede eruptions of Raung; seismicity, acoustic signals associated with gas bubbles, and the temperature and gas emissions of Kawah Ijen crater lake increase from time to time, sometimes leading to eruptions through the crater lake.

1638 (RA): "For seven days and seven nights, the populace was alarmed by a terrible sound which seemed to be coming from Raung; it was as if thousands of cannons were being let off simultaneously. The whole land of Matjanputeh was in continuous tremor, and the earthquakes were so strong that one could barely stand on one's feet. Now, all the water suddenly disappeared from out of the rivers and they also stayed dry for seven days and seven nights until the water all at once returned in an amount that made the rivers overflow and which nothing was able to resist...." (Bosch, 1858, citing J. Hageman).

IJEN (continued)

HISTORICAL ACTIVITY (continued)

1890 (RA): An ash eruption at Raung in 1890 was "not accompanied by earthquakes," but (continuous?) vibrations made houses tremble and rumbling sounds were heard. This eruption was said to be remarkable because residents at Pantjur, 17 km from the crater, could not recall having previously heard sounds from Raung (Figee and Onnen, 1891).

1902-04 (RA): Two vertical shocks were felt on the slope of Raung at 0330 hr on 16 February 1902, accompanied by rumbling and a violent crash like a cannon shot. At dawn, intense pulses of steam were seen at intervals of 5 to 6 minutes.

1913-17 (RA): Heightened activity was reported at Raung in 1913 and 1914 (Brouwer, 1913), including "heavy roaring." Two light quakes from Raung were felt near Besuki in May 1915.

1917 (IJ): As a result of a strong tectonic earthquake near Bali on 25 February 1917, acid water of Kawah Ijen's lake spilled over a sluice. By 28 February there was increased upwelling in the lake, and by 5 March there was spouting of water to 10 m, accompanied by much mud and noise. The temperature of the lake increased to 57 °C by (early?) March, then declined irregularly to 15 °C by the end of 1917 (Koninklijk Magnetisch en Meteorologisch Observatorium te Batavia, 1919).

1921 (IJ), 1936 (IJ), 1938 (IJ): About one week after a controlled sluicing (lowering of lake level) in February 1921, lake temperature at Kawah Ijen rose to 50 °C, bubbling increased, a strong sulfur smell was noted, and lake level rose well above normal for the amount of rainfall in that season (Kemmerling, 1921). Increased activity in November 1936 led to small eruptions through the lake, but similar increases in activity in March 1938 and February 1939 did not lead to eruptions.

1921 (RA): Increased steaming preceded an eruption at Raung in 1921.

1931 (RA): A dull bang and rumble at Raung on 19 April 1931 were followed by an earthquake that was felt at Sawodjadjar, Besuki, East Java, and by increased "smoke" emission from Raung's crater. This event is reported in Koninklijk Magnetisch en Meteorologisch Observatorium te Batavia (1932) but not in Netherlands Indies Volcanological Survey (1932) or subsequent compilations.

1952 (IJ): The most recent eruption of Ijen, in April 1952, began a "few moments" after several "heavy" local shocks. As soon as the eruption cloud cleared, the crater lake was seen to be "boiling." Several false alarms based on further temperature fluctuations occurred in January 1953, February 1957, February 1958, May 1959, and April 1962 (Hadikusumo, 1961; Suryo, 1961a,b, 1981).

1982 (RA): An eruption of Raung on 18 July 1982 was preceded by weakly felt tremor at 0300 hr of the same day.

1985 (RA): There were small ash explosions at Raung in November 1985 (no information about precursors).

See inside back cover for explanation and abbreviations

IJEN (continued)

COMMENTS

Increased upwelling in the lake at Kawah Ijen after a tectonic earthquake in 1917 might have been caused directly by shaking of the hydrothermal system beneath Kawah Ijen. However, in instances of controlled lake discharge, the reduction in confining pressure caused by lowering the lake level has repeatedly caused an increase in fumarolic activity beneath the lake, a rise in lake temperature, and in some instances phreatic or hydrothermal explosions. Therefore, the sudden seiche and uncontrolled discharge in 1917 were probably intermediate and necessary steps in the development of that unrest.

REFERENCES

Bemmelen, R.W. van, 1949, The geology of Indonesia, v. 1A: The Hague, Government Printing Office, 732 p.

Bosch, C.J., 1858, Uitbarstingen der vulkanen Idjen en Raun (Banjoe-wangi): Tijdschr. v. Ind. Taal-Land-en Volkenkunde, pt. 7, p. 265-286.

Brouwer, H.A., 1913, De Raoeng en zijn jongste eruptie: Nat. Tijdschr. v. Ned. Ind., pt. 73. no. 1, p. 84-97.

Figee, S., and Onnen, H., 1891, Vulkanische verschijnselen en aardbevingen in den O.I. Archipel waargenomen gedurende het jaar 1890: Nat. Tijdschr. v. Ned. Ind., pt. 51, p. 318-359.

Hadikusumo, D., 1961, Kawah Idjen: Bull. Volcanol. Surv. Indonesia, no. 100, p. 83-86.

Kemmerling, G.L.L., 1921, Vulkanologische Berichten I-XII: Nat. Tijdschr. v. Ned. Ind., pt. 81, no. 1, p. 33-66.

Koninklijk Magnetisch en Meteorologisch Observatorium te Batavia, 1917, Vulkanische verschijnselen en aardbevingen in den Oost-Indischen Archipel waargenomen gedurende het jaar 1915: Nat. Tijdschr. v. Ned. Ind., pt. 76, p. 65-113.

----- 1919, Vulkanische verschijnselen en aardbevingen in den Oost-Indischen Archipel waargenomen gedurende het jaar 1917: Nat. Tijdschr. v. Ned. Ind., pt. 78, p. 130-167.

----- 1932, Vulkanische verschijnselen en aardbevingen in den Oost-Indischen Archipel waargenomen gedurende het jaar 1931: Nat. Tijdschr. v. Ned. Ind., pt. 92, no. 2, p. 286-333.

Kusumadinata, K., ed., 1979, Data Dasar Gunungapi Indonesia: Bandung, Volcanological Survey of Indonesia, 820 p.

Netherlands Indies Volcanological Survey, 1932, Raoeng: Bull. Neth. Ind. Volcanol. Surv., no. 43-46.

----- 1936, Kawah Idjen: Bull. Neth. Ind. Volcanol. Surv., no. 78, p. 40-45.

----- 1938, Kawah Idjen: Bull. Neth. Ind. Volcanol. Surv., no. 83, p. 90-93.

Neumann van Padang, 1939, Ueber die vielen tausend Hugel im westlichen Vorlande des Raoeng-Vulkans (Ostjava): De Ing. in Ned. Ind., v. 6, no. 4, p. 35-41.

----- 1951, Catalogue of the active volcanoes of the world including solfatara fields, Pt. I, Indonesia: Rome, IAVCEI, 271 p.

Smithsonian Institution, Scientific Event Alert Network (SEAN), 1982, Raung Volcano: SEAN Bull., v. 7, no. 7.

Suryo, I., 1961a, Kawah Idjen: Bull. Volcanol. Surv. Indonesia, no. 101, p. 17-19.

----- 1961b, Kawah Idjen: Bull. Volcanol. Surv. Indonesia, no. 102, p. 15-16.

----- 1981, Kawah Ijen: Bull. Volcanol. Surv. Indonesia, no. 104, p. 50-53.

Sutawidjaja, I.S., Kastiman, S., and Suparman, 1988, Geologic map of Raung Volcano, East Java: Bandung, Volcanol. Surv. Indonesia, 1 sheet, scale 1:100,000 [in press].

Volcanological Society of Japan, 1980, Raung: Bull. Volcanic Eruptions, no. 18, p. 34.

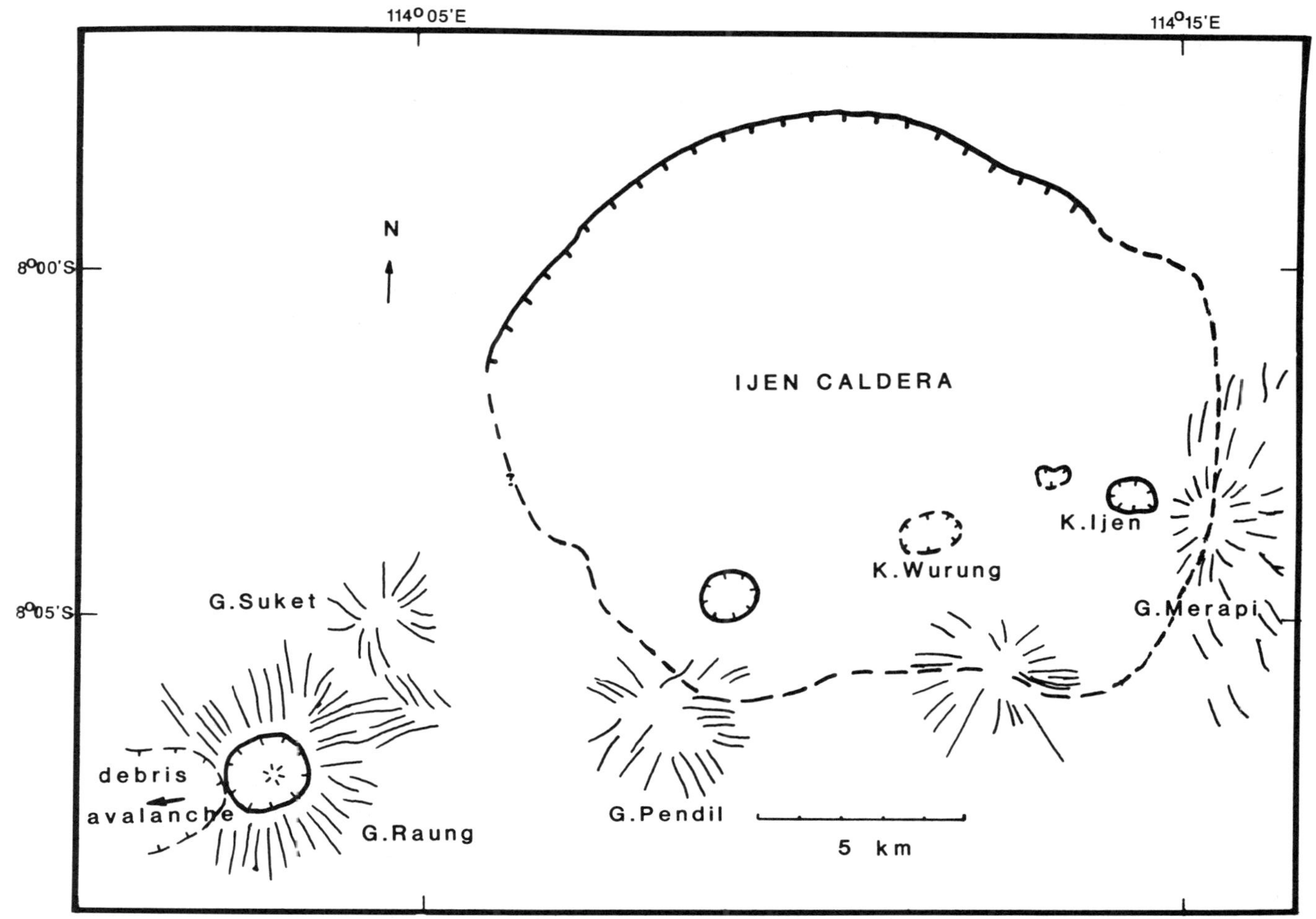

Figure 6.10.1. Approximate outline of Ijen Caldera.

PART 3: HISTORICAL UNREST AT LARGE QUATERNARY CALDERAS (continued)
See inside back cover for explanation and abbreviations

BATUR

CAVW number (active vent)	Latitude Longitude (degrees)	Diameter (km)	Local tectonic setting	Precaldera edifice	SiO_2 content (wt pct)	Age of collapse (yr B.P.)	Date of unrest	Type and duration of unrest ESTU STHF MGTF H Te	Eruption type
06-04-01 (Batur)	8.24S 115.38E	10 x 14 (outer) 7.5 (inner)	Compr	LL-strat	R = 50-55 C = 62-66	23,700	1804	---- ---- ---- - --	ex
							1821	---- ---- ---- - --	ex
							1849	---- ---- ---- - --	ex,lf
							1854	---- ---- ---- - --	ex
							1888	---- ---- ---- - --	ex,lf
							1897	---- ---- ---- - --	ex
							1902	?--- ---- ---- - --	none?
							1904-05	---- ---- ---- - --	ex,lf
							1917	xx-- ---- ---- - --	none
							1921-26	x--- ---Y --Y- Y QU?	ex,lf
							1959	---- ---- --?? ? --	none
							1963-64	B--- ---- ---- - QU	ex,lf
							1965-66	---- ---- ---- - --	ex
							1968	---- ---- ---- - --	Ex,lf
							1969	---- ---- --?? ? --	none
							1970-74	---- ---- ---- - --	ex,lf

TECTONIC SETTING

Batur is located on the island of Bali, Indonesia, near the middle of the Sunda-Banda volcanic island arc. The Indo-Australian plate is converging in a north-northeasterly direction toward the Eurasian plate at 5-6 cm/yr (Hamilton, 1979).

GEOLOGIC HISTORY

Stehn (1928), and later Marinelli and Tazieff (1968), described two Batur calderas: the inner, younger Batur II Caldera (7.5 km diameter), nested in the southeast end of the larger (10 x 14 km) Batur I Caldera. An active, small stratovolcano, also called Batur, has grown within the Batur II Caldera. Wheller and Varne (1986) studied the petrology of Batur products and correlated formation of the Batur Caldera (I and II undifferentiated) with the eruption about 23,700 yr ago of a thick dacitic ignimbrite sheet. They inferred that the caldera formed by collapse of an earlier Batur stratovolcano into a shallow magma chamber (figs. 6.11.1, 6.11.2).

BATUR (continued)

HISTORICAL ACTIVITY

1854: A small eruption occurred on 28 April: "Suddenly at about 1000 hours, the mountain began to expel black clouds from the uppermost crater; two lower openings remained steaming as usual" (de Seijff, 1855).

1888: Twenty-six shocks were felt at Buleleng, Bali (and "Lombok, Bali") between 2230 hr on 23 May and 0030 hr on 24 May 1888. Thirty more shocks were felt from 0030 hr on 24 May to 0030 hr on 26 May, the most severe at about 1400 hr on 25 May. The earthquakes were followed quickly by underground noises. Batur began to erupt on 30 May. More earthquakes were felt on 8 December (Figee and Onnen, 1890).

1902: Earthquakes and detonation sounds were noted during 9-11 October 1902, possibly from Batur (Koninklijk Magnetisch en Meteorologisch Observatorium te Batavia, 1904).

1904-05: A lava flow from the western-flank crater of Batur came to a standstill in the middle of the temple court of Batur village. The volcano then remained quiet until 1921.

1917: Strong tectonic earthquakes shook Bali in January and February 1917, but no changes occurred at Batur (Kemmerling, 1918).

1921-26: Eruptive activity resumed in the first days of April 1921, possibly but not certainly related to the 1917 earthquake. Incandescent, slaglike blocks were ejected, and violent detonations made the resthouse at Kintamani tremble. Lava began to flow on 9-10 April, then summit eruptions visibly diminished. But on 12 April, explosions were again numerous and as violent as before. Minor eruptions occurred during 1922-25. Then, preceded by earthquakes, a larger eruption began on the night of 2-3 August 1926. A fissure opened on the southwest foot of the active Batur cone, from which lava poured southwestward. On the following night, the fissure opened to the northeast, feeding lava flows to the north and southeast. The southeasterly flow engulfed most of the village of Batur. In the first days of the 1926 eruption, there was one continuous fissure at least 1300 m long, with about 20 vents. After 10 August, activity became focused at 3 centers along the fissure. The lavas were fluid and formed lava tubes (Stehn, 1928). Neither the central crater nor any of Batur's other parasitic craters showed any increase in activity during the 1926 eruption, but the temperature of hot springs at the northern shore of Lake Batur increased by 16 °C (Stehn, 1928).

1959: H_2S emissions from the northern one-third of the caldera lake increased for less than a month in June 1959. The lake water became white, and thousands of fish were "doped by this H_2S odour.... This phenomenon was probably due to solfataric action on the bottom of the lake" (Suryo, 1961).

1963-64: A volcano that lies about 10 km SE of the caldera, Agung, began erupting for the first time in more than a century on 19 February 1963. Approximately 3.4×10^8 m^3 of basaltic andesite lava flowed down the north slope of Agung (Suryo, 1981). Large explosive eruptions on 17 March and 16 May ejected an additional 2×10^8 m^3 of magma and killed about 1,100 persons, mostly by pyroclastic flows. On 18 May, several damaging tectonic(?) earthquakes occurred beneath the sea just north of Bali.

BATUR (continued)

HISTORICAL ACTIVITY (continued)

Weak earthquakes were felt at Kintamani and Penelokan (on the rim of Batur Caldera) at 0600 hr on 5 September 1963, and Batur began to erupt after 37 years of dormancy at 0945 hr on the same day. Approximately $4 \times 10^7 m^3$ of basalt and basaltic andesite was erupted from several vents on the west side of Batur cone, mostly as lava flows (Suryo, 1981) (see figs. 6.11.3 and 6.11.4 for the locations of the 1963 and other historical lava flows).

1965-66: Small explosions occurred weekly at Batur from August to December 1965. There was a brief increase in fumarolic activity on 28 April 1966 (Volcanological Society of Japan, 1966; Suryo, 1985).

1968: Frequent explosions occurred in January and February 1968, and a lava flow was extruded.

1969: In 1969, gas bubbles with a sulfur smell appeared on the surface of Lake Batur, the lake water turned white, and fish were again "doped," as in 1959 (Suryo, 1985).

1970-74: Minor explosions occurred in January 1970 (Suryo, 1985).

COMMENTS

Coinciding eruptions of Agung and Batur and earthquakes just north of Bali in 1963 can hardly be fortuitous. The exact relations between these events are unclear, but one possibility is that the eruption of Agung both reflected and triggered tectonic strain, and that Batur later responded to the same tectonic strain.

Fish kills like those at Batur in 1959 and 1969 and at Ranau in 1887, 1888, and 1903 are probably caused by convective overturns of caldera lakes, bringing oxygen-poor and hydrogen-sulfide-rich or carbon-dioxide-rich water to the surface. In tropical lakes, where there is little difference between surface and bottom temperatures, convective overturn might be caused by relatively minor fumarolic heating of bottom water.

BATUR (continued)

REFERENCES

Bemmelen, R.W. van, 1949, The geology of Indonesia, v. 1A: The Hague, Government Printing Office, 732 p.
Figee, S., and Onnen, H., 1890, Vulkanische verschijnselen en aardbevingen in den O.I. Archipel waargenomen gedurende het jaar 1888: Nat. Tijdschr. v. Ned. Ind., pt. 49, p. 111-159.
Hamilton, W., 1979, Tectonics of the Indonesian region, U. S. Geol. Surv. Prof. Paper 1078, 345 p.
Kemmerling, G.L.L., 1917, De vulkanen Batoer en Goenoeng Agoeng op Bali. Jaarb. Mijnw., v. 46, I, p. 50-73.
----- 1918, Beklimming van den G. Batoer, 18 Februari 1917: Nat. Tijdschr. v. Ned. Ind., pt. 77, p. 55-65.
Koninklijk Magnetisch en Meteorologisch Observatorium te Batavia, 1904, Vulkanische verschijnselen en aardbevingen in den Oost-Indischen Archipel waargenomen gedurende het jaar 1902: Nat. Tijdschr. v. Ned. Ind., pt. 63, p. 184-217.
Kusumadinata, K., ed., 1979, Data Dasar Gunungapi Indonesia: Bandung, Volcanological Survey of Indonesia, 820 p.
Marinelli, G., and Tazieff, H., 1968, L'ignimbrite et la caldera de Batur: Bull. Volcanol., v. 32, p. 89-120.
Neumann van Padang, M., 1951, Catalogue of the active volcanoes of the world including solfatara fields, Pt. I, Indonesia: Rome, IAVCEI, 271 p.
Purbo-Hadiwidjojo, M.M., 1971, Geologic map of Bali: Bandung, Geol. Surv. Indonesia, scale 1:250,000, 1 sheet.
Seijff, R.F. de, 1855, Togt naar den vulkaan Batoer, op het eiland Bali: Nat. Tijdschr. v. Ned. Ind., pt. 8, p. 119-126.
Stehn, Ch.E., 1928, De Batoer op Bali en zijn erupties in 1926: Vulkanol. Seismol. Meded., no. 9, p. 1-67.
Suryo, I., 1961, Gunung Batur: Bull. Volcanol. Surv. Indonesia, no. 102, p. 23.
----- 1981, Batur and Agung: Bull. Volcanol. Surv. Indonesia, no. 104, p. 53-93.
----- 1985, Batur: Bull. Volcanol. Surv. Indonesia, no. 106, p. 79-85.
Volcanological Society of Japan, 1963, Batur: Bull. Volcanic Eruptions, no. 3, p. 6.
----- 1966, Batur: Bull. Volcanic Eruptions, no. 6, p. 7.
Wheller, G.E., and Varne, R., 1986, Genesis of dacitic magmatism at Batur Volcano, Bali, Indonesia: Implications for the origins of stratovolcano calderas: Jour. Volcanol. Geotherm. Res., v. 28, p. 363-378.
Williams, H., and McBirney, A.R., 1979, Volcanology: San Francisco, Freeman Cooper, 397 p.
Yokoyama, I., and Siswowidjojo, S., 1970, Volcanological survey of Indonesian volcanoes, Part 5. A gravity survey on and around Batur Caldera, Bali: Bull. Earthq. Res. Inst., v. 48, p. 317-329.

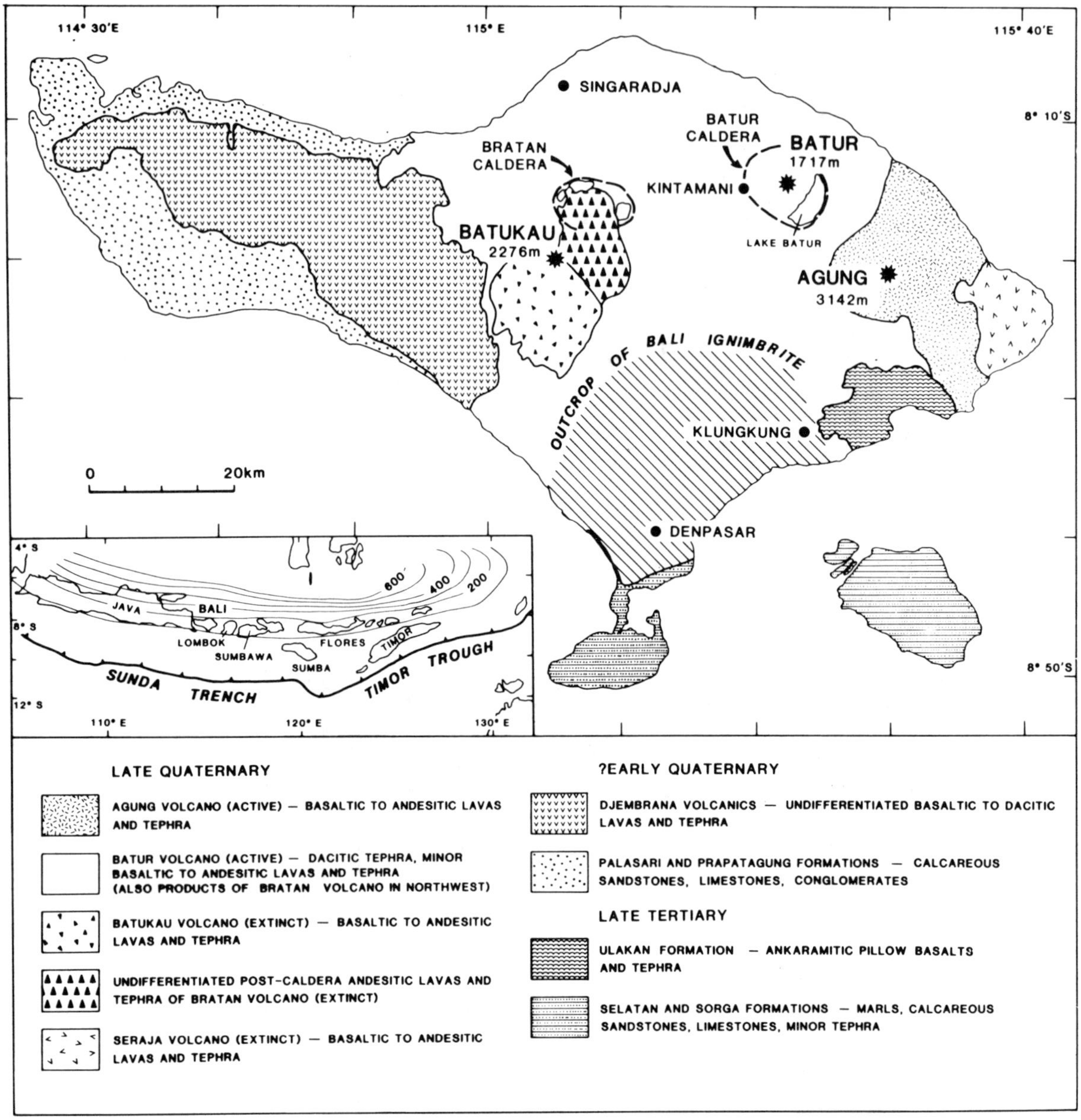

Figure 6.11.1. Geologic sketch map of Bali, showing location of Batur Caldera, from Wheller and Varne (1986), after Purbo-Hadiwidjojo (1971).

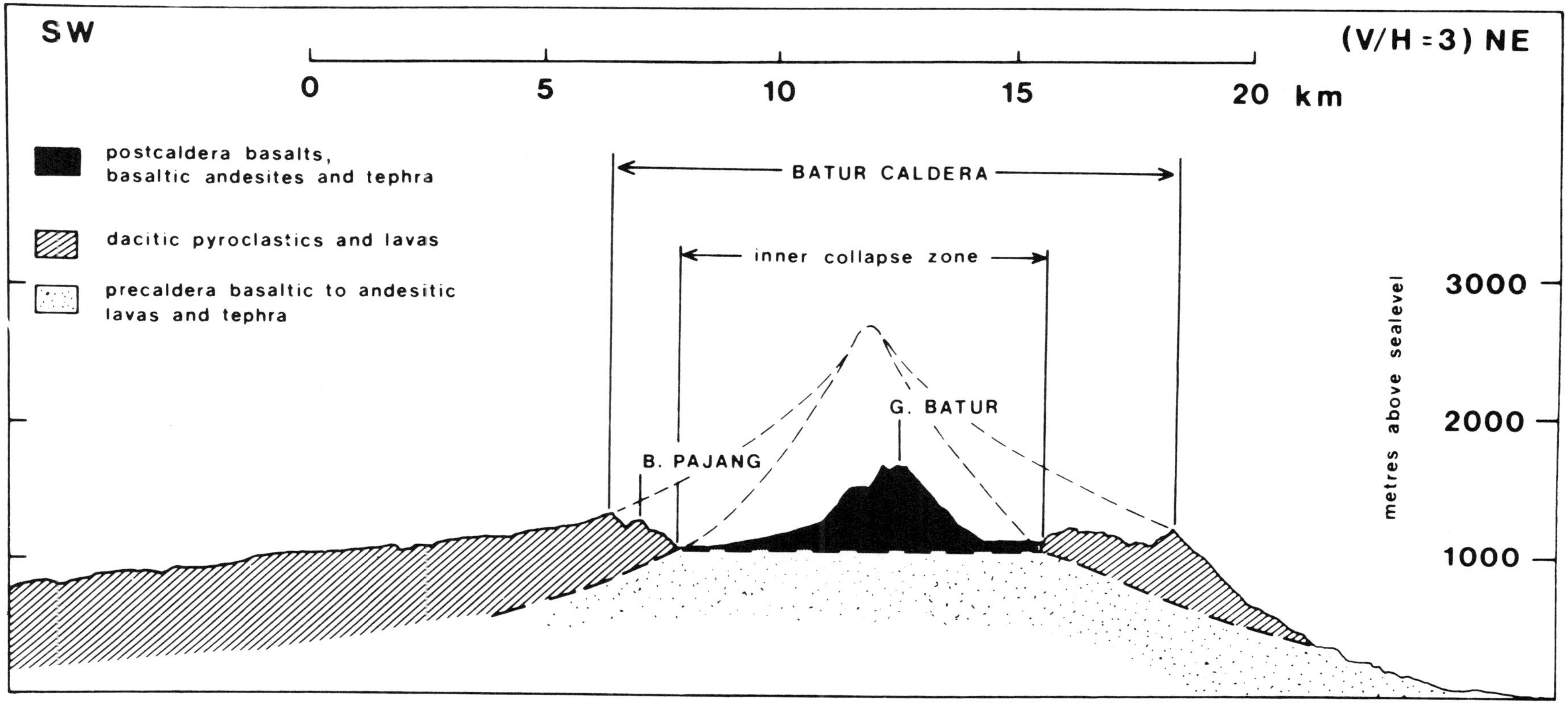

Figure 6.11.2. Southwest-northeast cross section and generalized stratigraphy of Batur Volcano, showing possible configuration before caldera formation, from Wheller and Varne (1986).

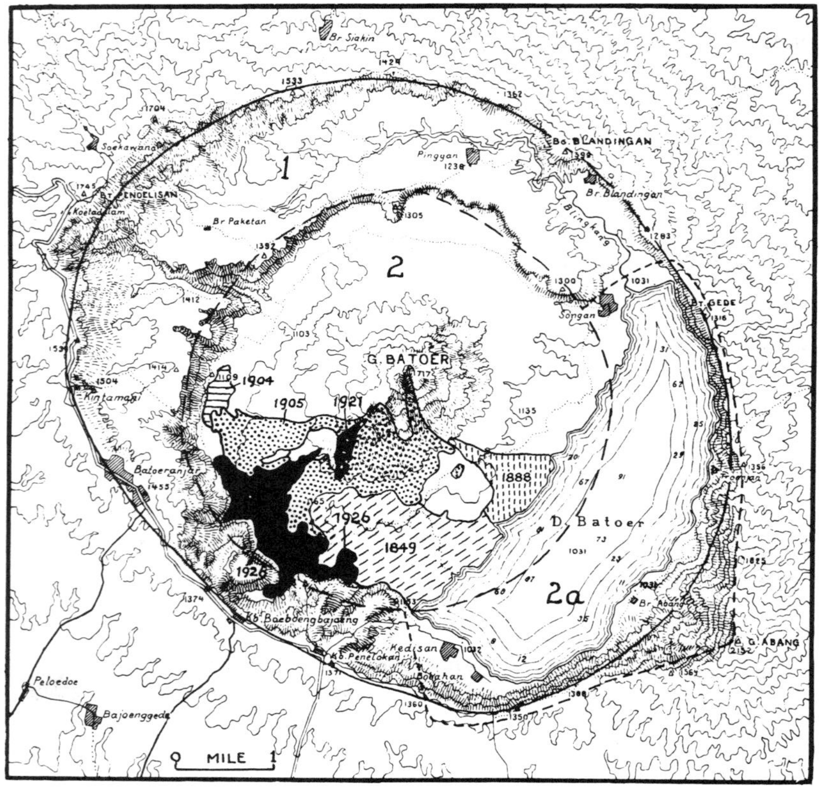

Figure 6.11.3. Historical lava flows at Batur Caldera, from Williams and McBirney (1979, p. 218). 1, original caldera; 2 and 2a, later collapse depressions. Lake depths are given in meters.

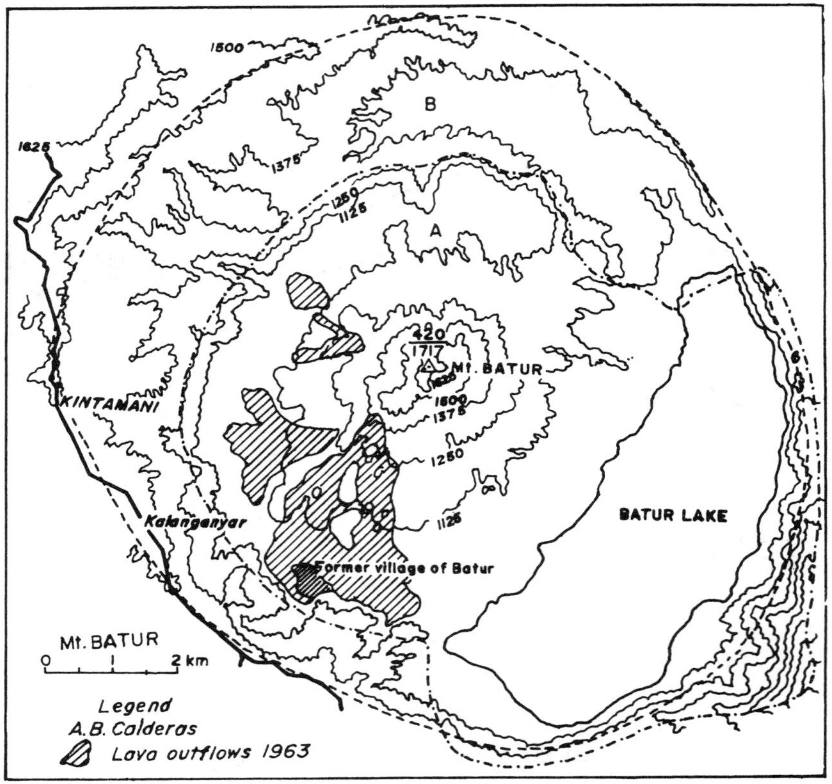

Figure 6.11.4. The 1963 lava flows at Batur, from Suryo (1981).

SEGARA ANAK

CAVW number (active vent)	Latitude Longitude (degrees)	Diameter (km)	Local tectonic setting	Precaldera edifice	SiO_2 content (wt pct)	Age of collapse (yr B.P.)	Date of unrest	Type and duration of unrest ESTU STHF MGTF H Te	Eruption type
06-04-03 (Rinjani, G. Baru)	8.42S 116.47E	5	Compr	Strato	R = 52-53 C = 65(?)		1847	---- ---- ---- - --	ex
							1884	Y--- ---- ---- - --	ex
							1900-01	---- ---- ---- - --	ex,lf
							1903	x--- ---- ---- - Q-	none
							1906	---- ---- ---- - --	ex
							1915	---- ---- ---- - --	ex
							1941?	---- ---- ---- - --	unknown
							1944	---- ---- ---- - --	ex,dm
							1949-50	---- ---- ---- - --	dm
							1965-66	---- ---- ---- - --	dm,ex

TECTONIC SETTING

Segara Anak is located on the island of Lombok, immediately east of Bali along the east-central Sunda island arc.

GEOLOGIC HISTORY

Gunung Rinjani is a large, composite stratovolcano on the east rim of Segara Anak Caldera; Gunung Baru ("New Mountain") is an intracaldera cone (fig. 6.12.1). The petrology of the calcalkaline lavas of Rinjani was studied recently by Foden (1983).

HISTORICAL ACTIVITY

1884: A relatively severe earthquake at 0640 hr on 8 August 1884 was felt across Lombok and Bali (10-sec duration). Thereafter, "Rinjani erupted smoke and flames for a few days." (De Commissie tot het Organiseeren en Verzamelen van Aardbevings-waarnemingen, 1885).

1898: An earthquake in 1898 of several seconds duration was possibly from Rinjani. It was felt in Bali and more strongly in Lombok (Koninklijk Magnetisch en Meteorologisch Observatorium te Batavia, 1899).

SEGARA ANAK (continued)

HISTORICAL ACTIVITY (continued)

1903: Two moderately severe earthquakes occurred beneath Rinjani on 14 November 1903 (Koninklijk Magnetisch en Meteorologisch Observatorium te Batavia, 1905).

1915: Smoke was seen at Rinjani on the morning of 4 November 1915 (Koninklijk Magnetisch en Meteorologisch Observatorium te Batavia, 1917).

1965-66: On 28 March 1966, people of Lombok felt an earthquake. An observer who climbed the volcano on 21 May reported lava extrusion from Gunung Baru (=Barujari) and a strong sulfur odor (Suryo, 1985).

REFERENCES

Bemmelen, W. van, 1919, Bezoek aan den Rindjani: Nat. Tijdschr. v. Ned. Ind., pt. 78, 45-54.

De Commissie tot het Organiseeren en Verzamelen van Aardbevings-waarnemingen, 1885, Uitbarstingen van vulkanen en aardbevingen in den O.I. Archipel waargenomen gedurende het jaar 1884: Nat. Tijdschr. v. Ned. Ind., pt. 45, no. 1, p. 458-471.

Foden, J.D., 1983, The petrology of the calcalkaline lavas of Rindjani Volcano, East Sunda arc: a model for island arc petrogenesis: Jour. Petrol., v. 24, no. 1, p. 98-130.

Koninklijk Magnetisch en Meteorologisch Observatorium te Batavia, 1899, Vulkanische verschijnselen en aardbevingen in den Oost-Indischen Archipel waargenomen gedurende het jaar 1898: Nat. Tijdschr. v. Ned. Ind., pt. 59, p. 79-139.

----- 1905, Vulkanische verschijnselen en aardbevingen in den Oost-Indischen Archipel waargenomen gedurende het jaar 1903: Nat. Tijdschr. v. Ned. Ind., pt. 64, p. 90-133.

----- 1917, Vulkanische verschijnselen en aardbevingen in den Oost-Indischen Archipel waargenomen gedurende het jaar 1915: Nat. Tijdschr. v. Ned. Ind., pt. 76, p. 65-113.

Kusumadinata, K., ed., 1979, Data Dasar Gunungapi Indonesia: Bandung, Volcanological Survey of Indonesia, 820 p.

Neumann van Padang, M., 1951, Catalogue of the active volcanoes of the world including solfatara fields, Pt. I, Indonesia: Rome, IAVCEI, 271 p.

Suryo, I., 1985, Rinjani: Bull. Volcanol. Surv. Indonesia, no. 106, p. 89-95.

Volcanological Society of Japan, 1966, Rinjani: Bull. Volcanic Eruptions, no. 6, p. 7.

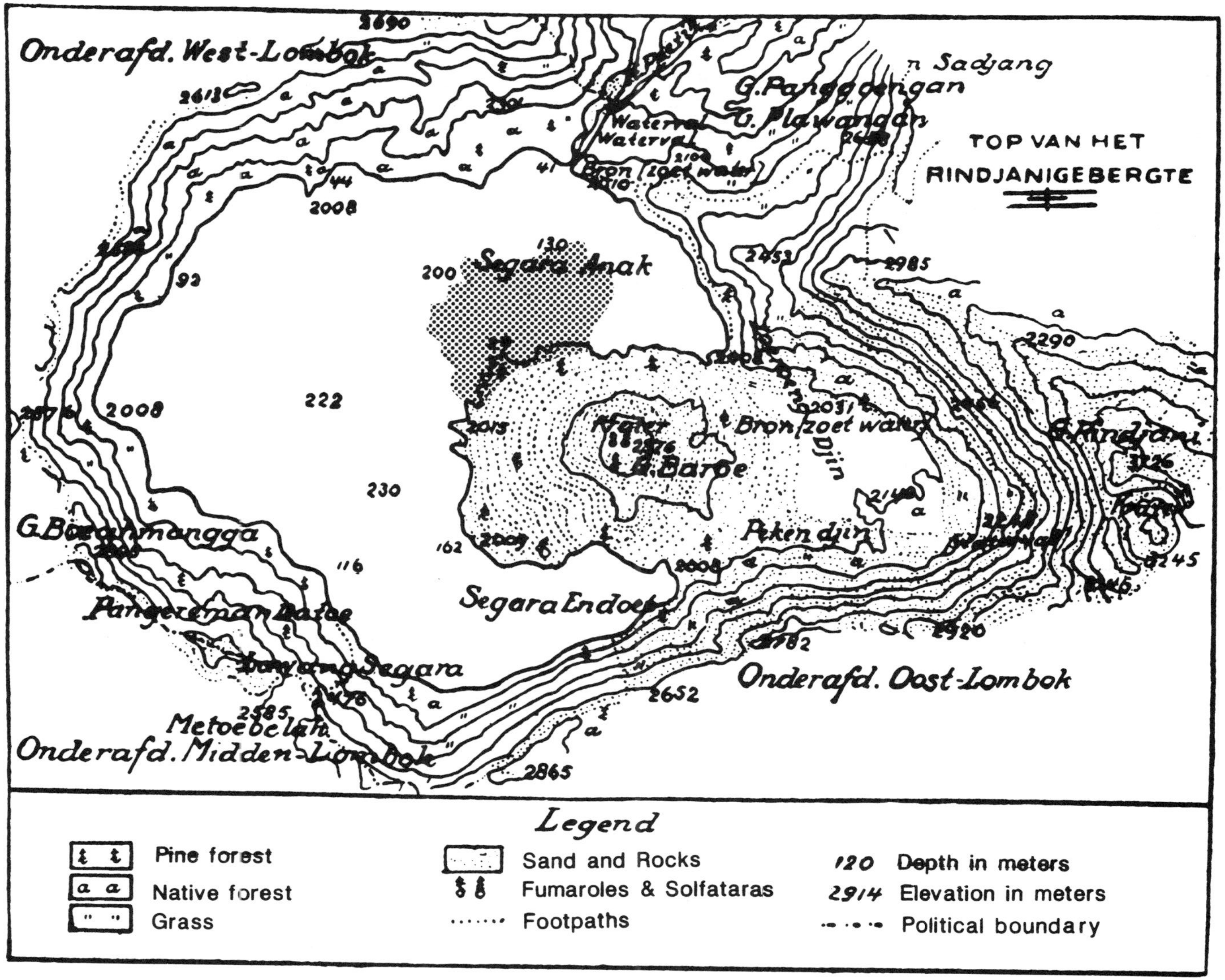

Figure 6.12.1. Segara Anak Caldera and postcaldera cones Gunung Baru (New Mountain) and Gunung Rinjani (eastern rim of caldera), after Neumann van Padang (1951) and Kusumadinata (1979). Original map was drawn in 1925; northward growth of Gunung Baru between 1925 and a survey in 1951 is shown by coarse stippling.

PART 3: HISTORICAL UNREST AT LARGE QUATERNARY CALDERAS (continued)
See inside back cover for explanation and abbreviations

TAMBORA

CAVW number (active vent)	Latitude Longitude (degrees)	Diameter (km)	Local tectonic setting	Precaldera edifice	SiO_2 content (wt pct)	Age of collapse (yr B.P.)	Date of unrest	Type and duration of unrest ESTU STHF MGTF H Te	Eruption type
06-04-04 (Tambora)	8.25S 118.00E	6	Back-arc exten?	Double strato on shield	R = 48-57 C = 56-57	A.D. 1815	1812-15	C--- ---- ---F - --	ex,EX,pex
							1819	---- ---- ---- - --	ex
						between	1847-1913	x--- ---- ---- - --	ex,lf

TECTONIC SETTING

Tambora is on Sumbawa Island along the east Sunda Arc (figs. 6.13.1, 6.13.2). It lies some 300 km behind the Sunda Trench, but the subduction zone in that area has a shallow dip and is less than 200 km deep beneath Tambora (Alzwar and others, 1981).

GEOLOGIC HISTORY

Tambora is a large stratovolcano composed dominantly of nepheline-normative, leucite-bearing trachybasalt and trachyandesite (Petroeschevsky, 1949; Foden and Varne, 1980; Alzwar and others, 1981; Barberi and others, 1983; Self and others, 1984; Foden, 1986). Before its large eruption in 1815 (discussed below), Tambora might have been in repose for as much as 5,000 years (Barberi and others, 1983).

HISTORICAL ACTIVITY

1812-15: At least 6 months and probably about 3 years of increased steaming and small phreatic eruptions preceded the 1815 Tambora eruption, the largest in historical time (Stewart, 1820; Zollinger, 1855; Crawfurd, 1856; Stothers, 1984; Sigurdsson and Carey, 1987). A moderately large explosive eruption occurred on 5 April 1815, from which ash fell in east Java and thunderlike sounds were heard up to 1400 km away. A still larger eruption occurred on 10-11 April, beginning as "three columns of fire rising to a great height" (Zollinger, 1855, p. 19) and ultimately ejecting about 50 km^3 of magma (dense rock equivalent) (Self and others, 1984; Sigurdsson and Carey, 1987). The eruption left a deep summit caldera where previously a much higher stratovolcano had stood. Earthquakes were felt as far away as Surabaya (500 km), possibly reflecting the caldera collapse.

1847-1913: A small, postcaldera cone and lava flow, Doro Afi Toi, originated sometime between 1847 and 1913 (Pannekoek van Rheden, 1918; Neumann van Padang, 1951). A strong earthquake on 13 January 1909, with an epicenter near Tambora (8.5°S, 117.4°E), was "presumably connected with Tambora" (Koninklijk Magnetisch en Meteorologisch Observatorium te Batavia, 1911). Might the earthquake have occurred during formation of Doro Afi Toi?

TAMBORA (continued)

COMMENTS

Will distinctive precursors give warning of the next caldera-forming eruption? Tambora's 1815 eruption probably had 3 years of precursory activity including small phreatic eruptions, but we find nothing in the sparse records that would be interpreted as evidence that a large, caldera-forming eruption was about to occur. On the contrary, there are many examples in this compilation of seemingly more intense caldera unrest that did not culminate in major eruptions.

REFERENCES

Alzwar, M., Barberi, F., Bizouard, H., Boriani, A., Cavallin, A., Eva, C., Gelmini, R., Giorgetti, F., Iaccarino, S., Innocenti, F., Marinelli, G., and Sudradjat, A., 1981, A structural discontinuity with associated potassic volcanism in the Indonesian island arc: First results of the CNR-CNRS-VSI mission to the island of Sumbawa: Rend. Soc. Geol. Ital., v. 4, p. 275-288.

Barberi, F., Bigioggero, B., Bizouard, H., Boriani, A., Cioni, R., Clocchiatti, R., Innocenti, F., and Marinelli, G., 1983, Tambora Volcano, Island of Sumbawa, Indonesia [abs.]: Programme and Abstracts, IUGG 18th Gen. Assembly, Hamburg, v. 1, p. 48.

Crawfurd, John, 1856, A descriptive dictionary of the Indian Islands & adjacent countries: London, Bradbury & Evans, 11, Bouverie Street, 459 p. (reprinted, 1971, Oxford in Asia Reprints: Kuala Lumpur, Oxford Univ. Press, 459 p.)

Foden, J., 1986, The petrology of Tambora Volcano, Indonesia: a model for the 1815 eruption: Jour. Volcanol. Geotherm. Res., v. 27, p. 1-41.

Foden, J.D., and Varne, R., 1980, The petrology and tectonic setting of Quaternary-Recent volcanic centers of Lombok and Sumbawa, Sunda arc: Chem. Geol., v. 30, p. 201-226.

Koninklijk Magnetisch en Meteorologisch Observatorium te Batavia, 1911, Vulkanische verschijnselen en aardbevingen in den Oost-Indischen Archipel waargenomen gedurende het jaar 1909: Nat. Tijdschr. v. Ned. Ind., pt. 70, p. 35-95.

Kusumadinata, K., ed., 1979, Data Dasar Gunungapi Indonesia: Bandung, Volcanological Survey of Indonesia, 820 p.

Neumann van Padang, M., 1951, Catalogue of the active volcanoes of the world including solfatara fields, Pt. I, Indonesia: Rome, IAVCEI, 271 p.

----- 1971, Two catastrophic eruptions in Indonesia, comparable with the Plinian outburst of the volcano of Thera (Santorini) in Minoan time: Acta 1st Int. Sci. Congr. on the Volcano of Thera, p. 51-63.

Pannekoek van Rheden, J.J., 1918, Geologische Notizen über die Halbinsel Sanggar, Insel Soembawa: Zeit. Vulkanol., v. 4, p. 85-192.

Petroeschevsky, W.A., 1949, A contribution to the knowledge of the Gunung Tambora (Sumbawa): Tijdschr. Kon. Ned. Aardr. Genoot., v. 66, p. 688-703.

Self, S., Rampino, M.R., Newton, M.S., and Wolff, J.A., 1984, Volcanological study of the great Tambora eruption of 1815: Geology, v. 12, p. 659-663.

Sigurdsson, H., and Carey, S., 1987, Dynamics of the 1815 Tambora eruption [abs.]: Eos, Trans. Am. Geophys. Union, v. 68, p. 1549.

Stewart, G.A., 1820, Description of a volcanic eruption in the island of Sumbawa: Trans. Literary Soc. of Bombay, v. 2, p. 109-114 (1874 reprinting).

Stothers, R.B., 1984, The great Tambora eruption in 1815 and its aftermath: Science, v. 224, p. 1191-1198.

TAMBORA (continued)

REFERENCES (continued)

Weitzel, A.W.P., 1859, Berigten over aardbevingen en beruitbarstingen, vermeld in de Bataviasche Koloniale Courant, de Java-Governement Gazette, de Bataviasche Courant en de Javasche Courant, van 1810 tot 1830: Nat. Tijdschr. v. Ned. Ind., pt. 18, p. 152-244.

Zollinger, H., 1855, Besteigung des Vulkans auf der Insel Sumbawa und Schilderung der Eruption desselben im Jahre 1815: Winterthur, Joh. Wurster & Co., 21 p.

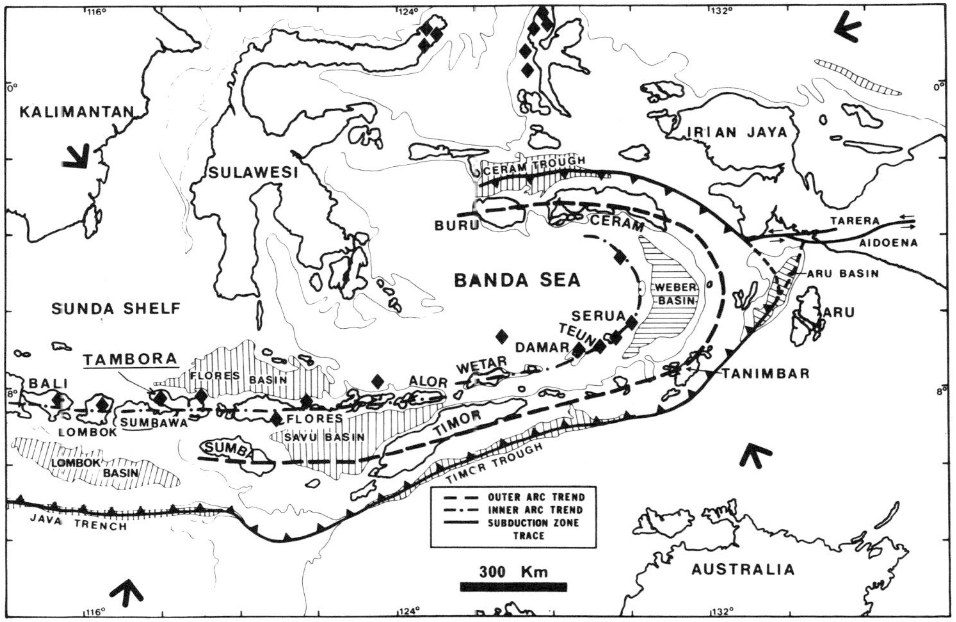

Figure 6.13.1. Locality map of east Sunda and Banda Arcs, showing location of Tambora Caldera on Sumbawa Island, from Foden (1986).

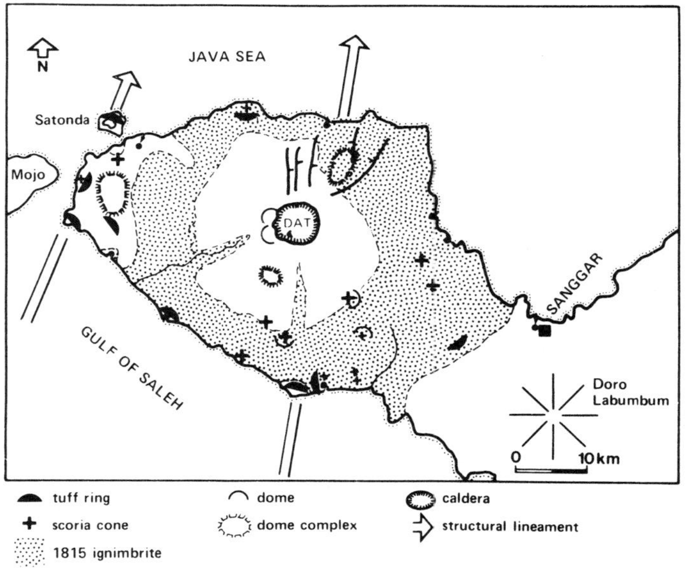

Figure 6.13.2. Sanggar Peninsula, Sumbawa Island, showing outline of Tambora Caldera and distribution of 1815 ignimbrite (from Self and others, 1984). DAT, post-1815 Doro Afi Toi cone and flow.

PART 3: HISTORICAL UNREST AT LARGE QUATERNARY CALDERAS (continued)
See inside back cover for explanation and abbreviations

LOBOBUTA/BADJAWA

CAVW number (active vent)	Latitude Longitude (degrees)	Diameter (km)	Local tectonic setting	Precaldera edifice	SiO_2 content (wt pct)	Age of collapse (yr B.P.)	Date of unrest	Type and duration of unrest ESTU STHF MGTF H Te	Eruption type
06-04-09 (Inielika)	8.73S 120.98E	7.5	Compr			Pleisto-cene(?)	1905	N--- ---- ---- - --	pex

TECTONIC SETTING/GEOLOGIC HISTORY

Lobobuta Caldera is located on the island of Flores along the Sunda island arc (figs. 6.14.1, 6.14.2). Inielika, a 1559-m-high stratocone, is in the northern part of Lobobuta Caldera. The caldera has an irregular shape, and no deposits have been correlated with caldera formation; thus the "caldera" might be an erosional feature.

HISTORICAL ACTIVITY

1905: According to Kemmerling (1929, p. 24), the 1905 eruption began suddenly, without the least warning of rumbling or shocks.

REFERENCES

Kemmerling, G.L.L., 1929, Vulkanen van Flores: Vulkanol. Seismol. Meded., no. 10, 138 p.

Koesoemadinata, S., Noya, Y., and Kadarsiman, D., 1978, Geologic map of the Ruteng Quadrangle, eastern Nusa Tenggara: Bandung, Geol. Res. Devel. Center, 1 sheet, scale 1:250,000.

Kusumadinata, K., ed., 1979, Data Dasar Gunungapi Indonesia: Bandung, Volcanological Survey of Indonesia, 820 p.

Neumann van Padang, M., 1951, Catalogue of the active volcanoes of the world including solfatara fields, Pt. I, Indonesia: Rome, IAVCEI, 271 p.

Pannekoek van Rheden, J.J., 1920, Einige Notizen über die Vulkane auf der Insel Flores (Niederlandisch Indien): Zeit. Vulkanol., v. 5, p. 109-163.

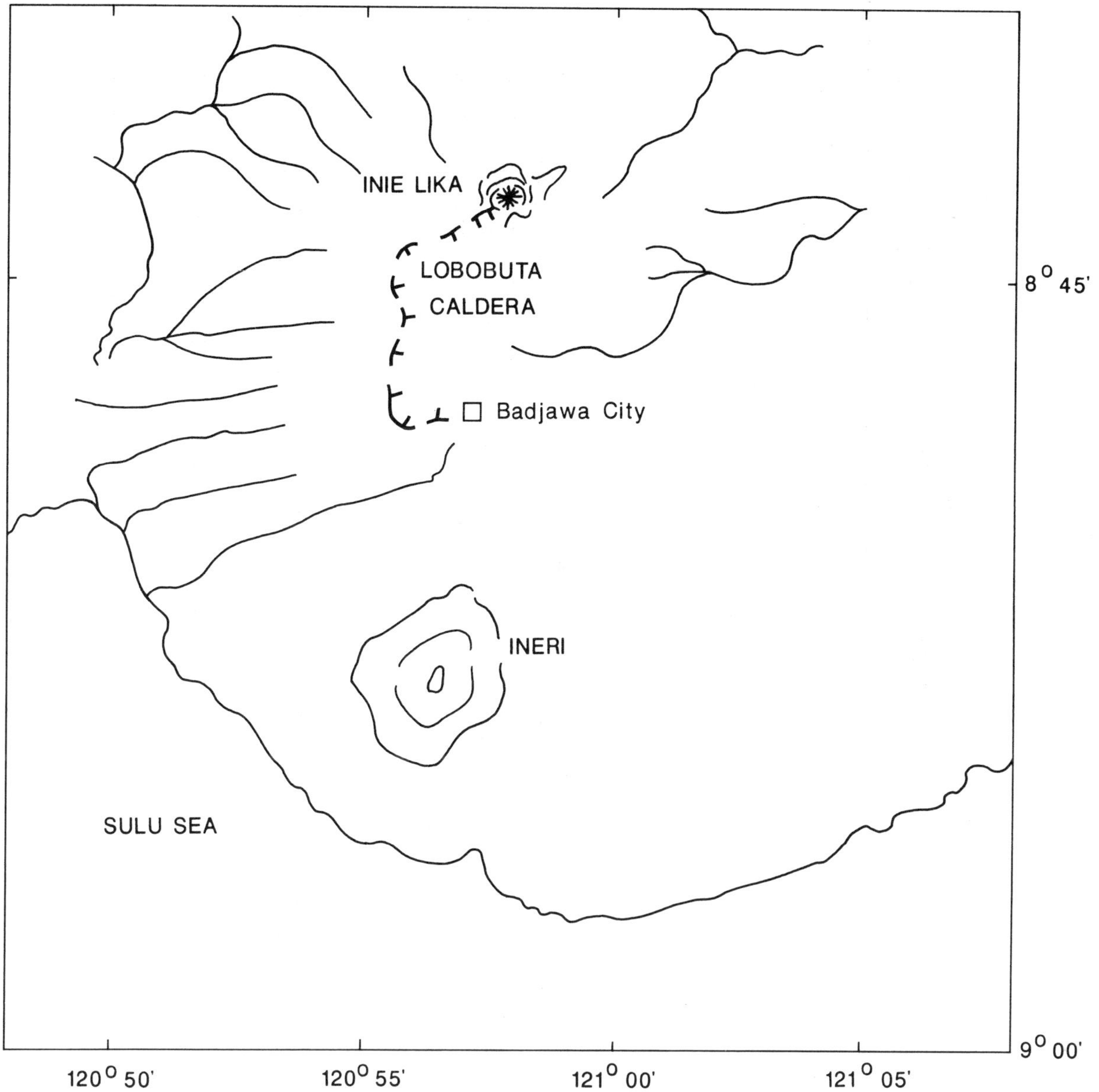

Figure 6.14.1. Sketch map of Lobobuta (or Badjawa) Caldera, Flores Island.

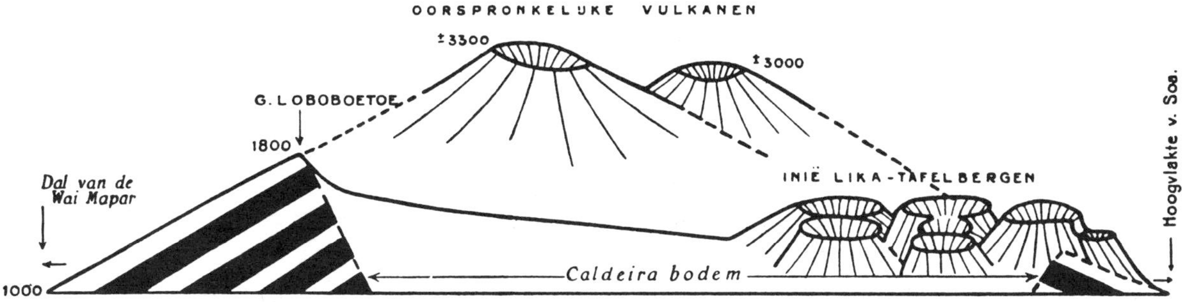

Figure 6.14.2. Schematic cross section of Lobobuta Caldera, from Kemmerling (1929).

PART 3: HISTORICAL UNREST AT LARGE QUATERNARY CALDERAS (continued)
See inside back cover for explanation and abbreviations

SUKARIA

CAVW number (active vent)	Latitude Longitude (degrees)	Diameter (km)	Local tectonic setting	Precaldera edifice	SiO_2 content (wt pct)	Age of collapse (yr B.P.)	Date of unrest	Type and duration of unrest ESTU STHF MGTF H Te	Eruption type
06-04-12 (Sukaria)	8.47S 121.46E	6 x 9 or 12 x 18?	Compr				1860's	---- ---- ---- - --	pex
06-04-13 (Ndete Napu)							1938	---- ---- ---E - --	pex
06-04-14 (Keli Mutu)							1968	---- ---- ---- - --	pex?
							1986	C--- ---- --?x - --	none

TECTONIC SETTING/GEOLOGIC HISTORY

Sukaria Caldera has been mentioned in geologic reports for several decades (Kemmerling, 1929) but its boundaries have not been satisfactorily reported. The sketch map in figure 6.15.1 is speculative, based largely on topography and an assumption that the Ndete Napu and Keli Mutu thermal areas are part of a single, large system. Nothing is known of a caldera-forming eruption. Keli Mutu has three varicolored lakes: These are, from west to east, Tiwu Ata Mbupu (bluish to "troubled" green; "lake of the old people"), Tiwu Nua Muri Kooh Poi ("troubled" green; "lake of young men and virgins"), and Tiwu Ata Polo (red; "lake of the bewitched people") (Ch.E. Stehn, in Netherlands East Indies Volcanological Survey, 1938; Kusumadinata, 1979).

HISTORICAL ACTIVITY

1938: During 1937-38, the color of one or more of the Keli Mutu lakes (Tiwu Ata Mbupu) changed from bluish green to "troubled" green, owing to addition of suspended sulfur and sulfuric acid (Netherlands East Indies Volcanological Survey, 1938; Smithsonian Institution, 1986).

1968: On 3 June 1968, water fountained 100 m high in the Tiwu Nua Muri (green, central) crater lake (Smithsonian Institution, 1986).

1986: Gas bubbling increased in the Tiwu Nua Muri crater lake on 27 April 1986, and an earthquake was felt on 28 April (Smithsonian Institution, 1986).

SUKARIA (continued)

REFERENCES

Kemmerling, G.L.L., 1929, Vulkanen van Flores: Vulkanol. Seismol. Meded., no. 10, 138 p.
Kusumadinata, K., ed., 1979, Data Dasar Gunungapi Indonesia: Bandung, Volcanological Survey of Indonesia, 820 p.
Netherlands East Indies Volcanological Survey, 1938 (January-March), Keli Mutu: Bull. Neth. Ind. Volcanol. Surv., no. 83, p. 95-96.
Neumann van Padang, M., 1951, Catalogue of the active volcanoes of the world including solfatara fields, Pt. I, Indonesia: Rome, IAVCEI, 271 p.
Pannekoek van Rheden, J.J., 1920, Einige Notizen über die Vulkane auf der Insel Flores (Niederlandisch Indien): Zeit. Vulkanol., v. 5, p. 109-163.
Smithsonian Institution, Scientific Event Alert Network (SEAN), 1986, Kelimutu Volcano: SEAN Bull., v. 11, no. 4, p. 21 (report from Olas, Suratman, Suparto, Kaswanda, and Adjat Sudradjat).

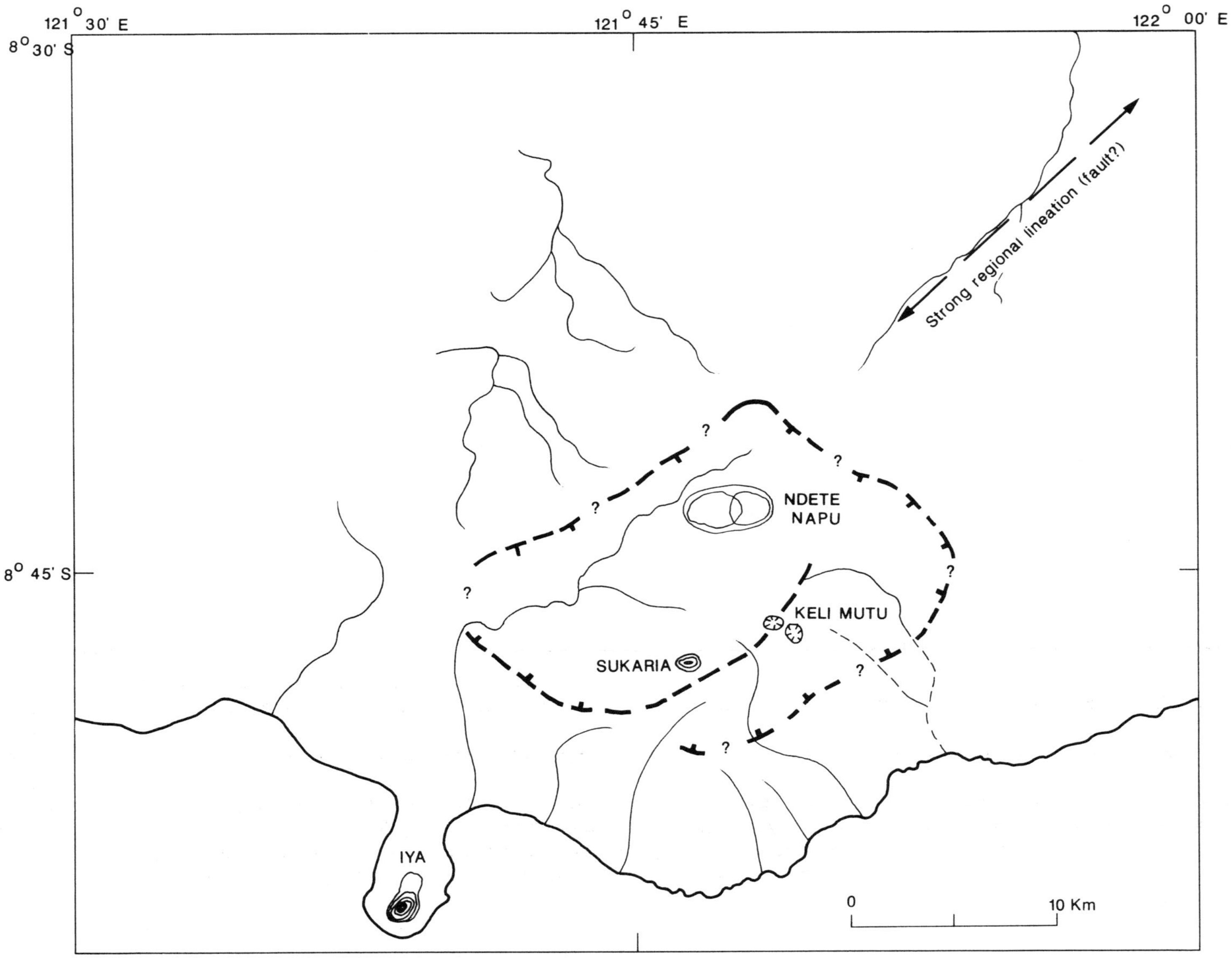

Figure 6.15.1. Sketch map showing possible boundaries of Sukaria Caldera. Smaller shape is from Kemmerling (1929); larger, rectangular shape is from this work, based on topography and assumption that Ndete Napu and Keli Mutu thermal areas are part of a single, large system.

BANDA API

CAVW number (active vent)	Latitude Longitude (degrees)	Diameter (km)	Local tectonic setting	Precaldera edifice	SiO_2 content (wt pct)	Age of collapse (yr B.P.)	Date of unrest	Type and duration of unrest ESTU STHF MGTF H Te	Eruption type
06-05-09 (Banda Api)	4.52S 129.87E	7 (outer) 3 (inner)	Compr	Submarine strato	R = 52-66 C = silicic?		1586	---- ---- ---- - --	ex
							1598-1602	---- ---- ---- - --	Ex
							1609	---- ---- ---- - --	ex
							1615	---- ---- ---- - --	Ex
							1632	x--- ---- ---- - --	ex
							1635?	---- ---- ---x - --	ex? or none?
							1683	x--- ---x ---- - --	Ex
							1690-96	---- ---- ---- - --	Ex
							1712	---- ---- ---- - --	ex
							1722(-23?)	---- ---- ---- - --	ex
							1749	---- ---- ---- - --	ex
							1762	---- ---- ---- - --	ex
							1765-66	---- ---- ---- - --	ex
							1773	---- ---- ---- - --	ex
							1775	---- ---- ---- - --	ex
							1778	---- ---- ---- - --	ex
							1816	x--- ---- ---- - --	ex
							1820	---- ---- ---- - --	Ex,lf
							1824	---- ---- ---- - --	ex
							1835	D--- ---- ---D - UQ?	none?
							1852-53	x--- ---- ---x - --	none
							1855-56	x--- ---- ---x - --	none
							1859-60	E--- ---- ---x - QU?	none?
							1877	E?-- ---- ---- - --	none
							1887	x--- ---- ---x - QU	none
							1890	E--- ---- ---E - --	ex
							1901-02	x--- ---- ---- - --	ex,lf
							1908	---- ---- ---x - --	none or ex?
							1987-88	C--- ---- ---x - QU	ex,lf

TECTONIC SETTING

The Banda Api area is seismically active with frequent tectonic earthquakes. Subduction beneath the Banda Arc is complex, but is probably dominated by northwest-dipping, halting subduction of the Indo-Australian plate beneath the Eurasian plate, complicated by a possible reversal of the sense of subduction (Hamilton, 1977; Cardwell and Isacks, 1978; Norvick, 1979) (fig. 6.16.1).

BANDA API (continued)

GEOLOGIC HISTORY

Ten islands comprise the Banda group: two mark the south and east rim of the Banda Caldera (P. Banda Besar, P. Pisang), and one is the top of an intracaldera cone (P. Gunungapi) (fig. 6.16.2). P. Neira marks the east rim of a smaller, nested caldera. Matahelumual (1988) likens the evolution of Banda Api to that of Krakatau: at least two episodes of caldera formation, each preceded by growth of an andesitic intracaldera stratovolcano.

HISTORICAL ACTIVITY

1820: "On the 11th of June...just before 12 o'clock, in an instant without the slightest warning, an eruption began which was so violent that all the people at once fled to the shore and crossed in boats to Banda Neira." (Bickmore, 1868).

1824: On 23 April 1824, a new crater suddenly opened, spewing out clouds of smoke and ash.

1835: A dense mist of sulfur vapor was observed in October 1835, followed by an unusually strong earthquake on 1 November that was felt on Amboina and the surrounding islands. That shock was followed by others until 4 November, and observers inferred an eruption of Banda Api (Reiche, 1859, p. 256).

1853: Light earthquakes occurred on 1, 14, and 19 October, and on 22 November, and the volcano steamed more than usual (Anonymous, 1854, p. 16).

1855-1856: Subterranean noises were heard on 29-30 December 1855, and a weak earthquake was felt on 18 January 1856. More subterranean rumbling was noted on 30 January 1856. During these subterranean noises, Banda Api would "smoke heavily and the air was very oppressive" (Anonymous, 1856, p. 245).

1859: Five earthquakes were reported during June-July 1859; one at 2000 hr on 20 July was accompanied by a cannonlike noise heard on the island of Great Banda and at Lonthoir, followed by a tsunami. Nothing else was observed on Banda Neira or on the rear wall of Great Banda. More earthquakes occurred on 11 August and 30 August, and an earthquake on 25 September was so severe that "it seemed as if an irresistible force wanted to change the locations of the islands of Great Banda and Neira. The sea dashed up against the southern beach of Banda Neira with terrible force and then gently relaxed to its usual level." Irresistible as the force might have been, no damage was reported (Anonymous, 1860a). A brief but severe earthquake followed on 18 October, and Banda Api ejected more dense "smoke" clouds than ordinary in September and October. Brief but severe earthquakes were reported again on 7-8 November and 29 December.

BANDA API (continued)

HISTORICAL ACTIVITY (continued)

1860: No earthquakes were reported in February 1860, but "smoking" was greater than normal (Anonymous, 1860b). A severe earthquake at 1206 hr on 19 July was accompanied by an underground roar that seemed to pass from southeast to northwest. Smaller earthquakes occurred on 30 July and on 12 August.

1877: Numerous earthquakes were felt in Neira from 8 January through 29 June. No eruption occurred. We suspect that these earthquakes were of local origin because no large tectonic earthquake was reported near Banda in 1977 (Soetardjo and others, 1985).

1887: A moderately strong earthquake on 15 May 1887 was apparently centered near Banda Api. After the earthquake, the volcano steamed more than usual. Additional, severe shocks occurred on 29-30 September, with no corresponding increase in activity of Gunung Api.

1890: On 12 August 1890, a violent earthquake was accompanied by strong underground rumbling. Shaking was of long duration, "as if Gunung Api was forcefully shaking all the wetness from itself and was tiring from the (heavy) rains (that had just occurred)." On 23 November, 3 days before the full moon, there was the strongest earthquake in the memory of all, exceeding that of 26 November 1852. Many aftershocks were felt for days thereafter. The earthquake left no house undamaged, and "was caused by Gunung Api..." An inspection climb of Gunung Api on 5 December revealed a new active crater, 5 m in diameter, a few meters below the west crater rim. Three fissures "shoot out from it like rays." To the south of the active crater, another new vent was lightly steaming. Also, a 100-m-long landslide scarp was noted on the inside of the cone, with a large opening in the middle about 5 m in diameter (another new crater?) (Figee and Onnen, 1891).

1901-02: Earthquakes were felt on 18 May 1901, with detonations at 1910 and 2000 hr. Then, at 2005 hr, violent earthquakes were felt and a bright glow was observed at the top of Gunung Api (Koninklijk Magnetisch en Meteorologisch Observatorium te Batavia, 1902). On 20 March 1902, underground rumbling was heard at intervals of 5-6 minutes, and one light earthquake was felt (Koninklijk Magnetisch en Meteorologisch Observatorium te Batavia, 1904).

1975: A <u>M</u>=5.9 tectonic earthquake destroyed or heavily damaged 81 houses in Banda Neira (Soetardjo and others, 1985).

1987-88: A <u>M</u>=6.7 earthquake occurred 150 km southeast of Banda Api on 17 June 1987. Shortly thereafter, a white steam plume rose 100 m above the crater rim. Eight small earthquakes were recorded in September 1987. On 2 May 1988, earthquakes increased sharply to a maximum of 110 events (11 of which were felt) on 5 May. An eruption began on 9 May (T. Casadevall, written commun., 1988).

BANDA API (continued)

REFERENCES

Anonymous, 1854, Berigten van verschillenden aard: Vulkanische verschijnselen in den Indischen Archipel: Nat. Tijdschr. v. Ned. Ind., pt. 6, p. 161-162.

----- 1856, Berigten van verschillenden aard: Aardbevingen en berguitbarstingen in den Indischen Archipel: Nat. Tijdschr. v. Ned. Ind., pt. 11, p. 245-250.

----- 1860a, Berigten van verschillenden aard: Vulkanische verschijnselen in den Indischen Archipel: Nat. Tijdschr. v. Ned. Ind., pt. 21, p. 396-406.

----- 1860b, Berigten van verschillenden aard: Vulkanische verschijnselen in den Indischen Archipel: Nat. Tijdschr. v. Ned. Ind., pt. 22, p. 136-137.

Bickmore, A.S., 1868, A description of the Banda Islands: Proc. Roy. Geogr. Soc., London, v. 12, p. 324-334.

Cardwell, R.K., and Isacks, B.L., 1978, Geometry of the subducted lithosphere beneath the Banda Sea in eastern Indonesia, from seismicity and fault plane solutions: Jour. Geophys. Res., v. 83, p. 2825-2836.

Figee, S., and Onnen, H., 1891, Vulkanische verschijnselen en aardbevingen in den O.I. Archipel waargenomen gedurende het jaar 1890: Nat. Tijdschr. v. Ned. Ind., pt. 51, p. 318-359.

Hamilton, W., 1977, Subduction in the Indonesian region, in Talwani, M., and Pitman, W.C., III, eds., Island arcs, deep trenches, and back-arc basins: Am. Geophys. Union, Maurice Ewing Ser., no. 1, p. 15-31.

----- 1978, Tectonic map of the Indonesian region: U.S. Geol. Surv. Map I-875-D, 1 sheet, scale 1:5:000,000.

Koninklijk Magnetisch en Meteorologisch Observatorium te Batavia, 1902, Vulkanische verschijnselen en aardbevingen in den Oost-Indischen Archipel waargenomen gedurende het jaar 1901: Nat. Tijdschr. v. Ned. Ind., pt. 62, p. 169-211 + errata.

----- 1904, Vulkanische verschijnselen en aardbevingen in den Oost-Indischen Archipel waargenomen gedurende het jaar 1902: Nat. Tijdschr. v. Ned. Ind., pt. 63, p. 184-217.

Kusumadinata, K., ed., 1979, Data Dasar Gunungapi Indonesia: Bandung, Volcanological Survey of Indonesia, 820 p.

Matahelumual, J., 1988, G. Banda-Api: Berita Berkala Vulkanologi, Bull. Volcanol. Surv. Indonesia, no. 115, 48 p.

Milsom, J., Audley-Charles, M.G., Barber, A.J., and Carter, D.J., Geological-geophysical paradoxes of the eastern Indonesia collision zone: Am. Geophys. Union, Geodynamics Ser., v. 11, p. 401-411.

Neumann van Padang, M., 1951, Catalogue of the active volcanoes of the world including solfatara fields, Pt. I, Indonesia: Rome, IAVCEI, 271 p.

Norvick, M.S., 1979, The tectonic history of the Banda arcs, eastern Indonesia: a review: Jour. Geol. Soc. London, v. 236, p. 519-526.

Reiche, M.Th., 1859, Berigten over aardbevingen en berguitbarstingen, vermeld in de Javasche Couranten van 1831 tot 1840: Nat. Tijdschr. v. Ned. Ind., pt. 18, p. 245-282.

Smithsonian Institution, Scientific Event Alert Network (SEAN), 1987, Banda Api Volcano: SEAN Bull., v. 12, nos. 6, 8 (contributions from Volcanological Survey of Indonesia).

Soetardjo, R., Untung, M., Arnold, E.P., Soetadi, R., Isail, S., and Kertapati, E., 1985, Indonesia: SEASEE Series on Seismology, v. 5, 199 p.

Verbeek, R.D.M., 1900-01, Geologische beschrijving van de Banda Eilanden: Jaarb. Mijnw. Ned. Ind., v. 29, p. 1-29.

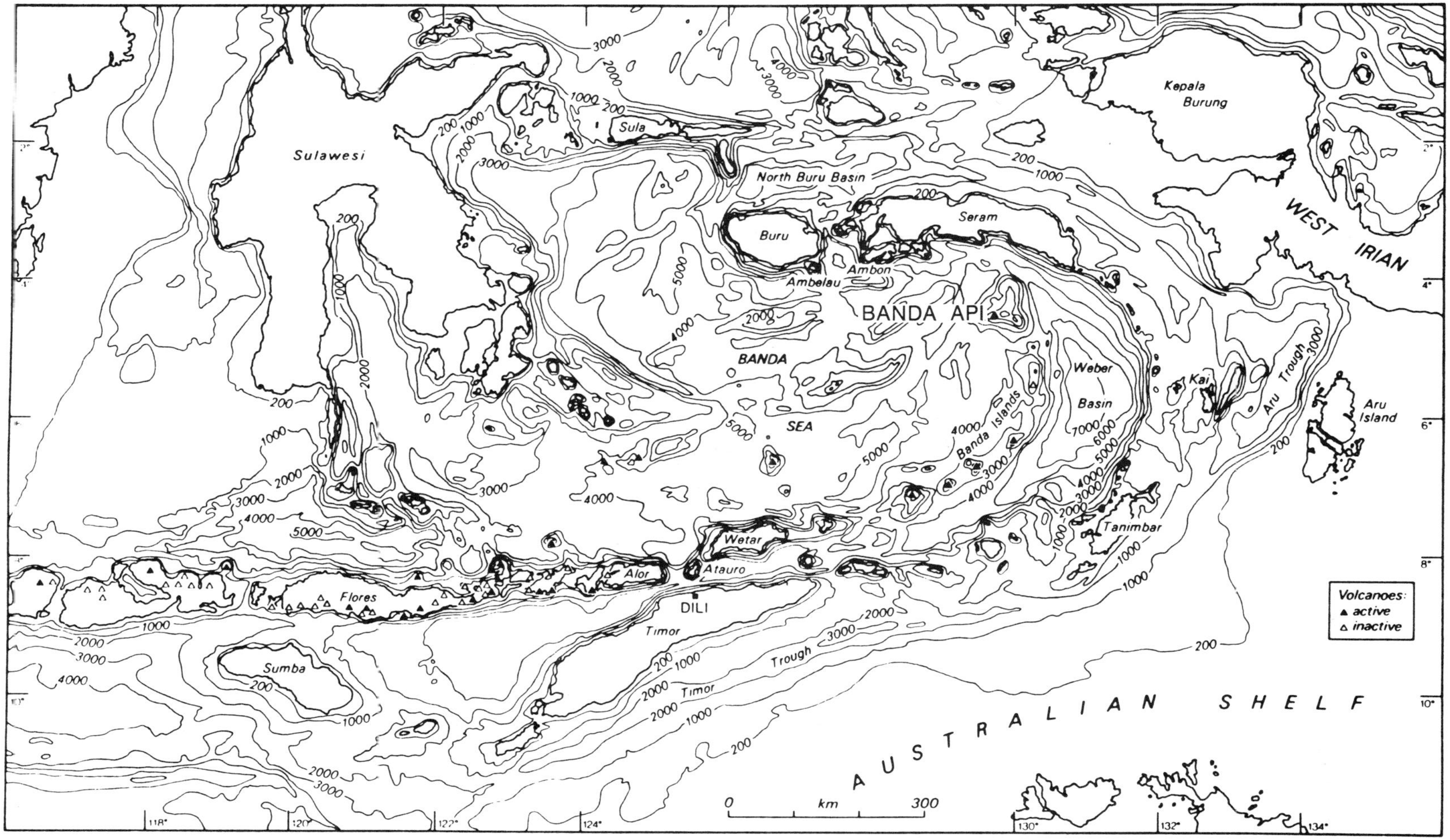

Figure 6.16.1. Bathymetric map of eastern Indonesia, from Milsom and others (1983) after Hamilton (1978), showing location of Banda Api Caldera. Isobaths in meters. Copyright by the American Geophysical Union.

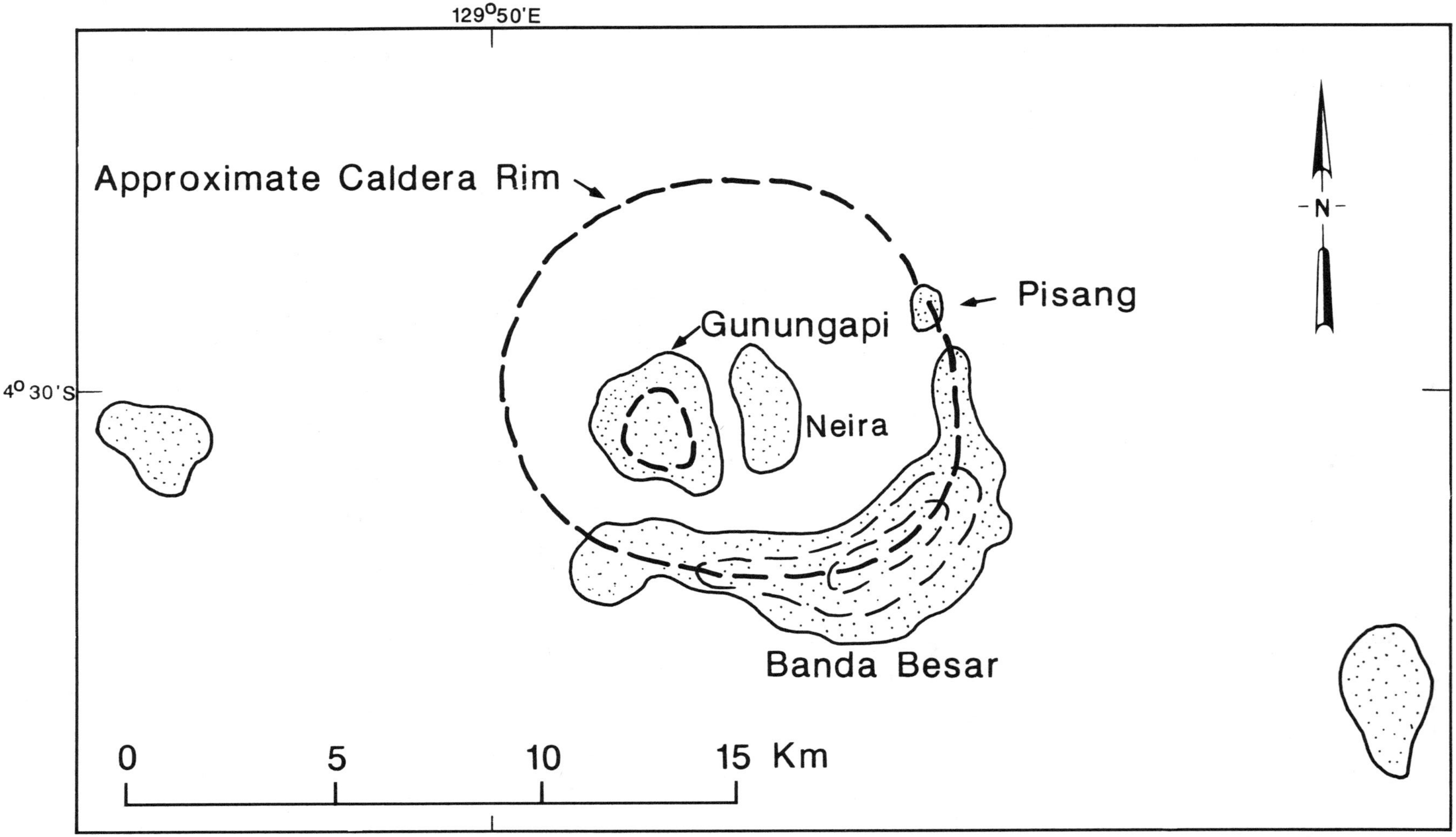

Figure 6.16.2. Approximate outline of largely submarine Banda Api Caldera. Gunungapi is an active intracaldera vent. Land masses stippled.

TONDANO

CAVW number (active vent)	Latitude Longitude (degrees)	Diameter (km)	Local tectonic setting	Precaldera edifice	SiO_2 content (wt pct)	Age of collapse (yr B.P.)	Date of unrest	Type and duration of unrest ESTU STHF MGTF H Te	Eruption type
06-06-03	1.25N	20 x 30?	Compr		R = b-a,		late 1300's(LO)	---- ---- ---- - --	Ex
(Soputan= SO)	124.85E				ol-d, r		late 1700's(LO)	---- ---- ---- - --	Ex
							1785 (SO)	---- ---- ---- - --	ex
06-06-04							before 1788 (MA)	---- ---- ---- - --	ex
(Sempu= SE)							1789 (MA)	---- ---- ---- - --	ex, pex?
							1819? (SE)	---- ---- ---- - --	ex?
06-06-10					(LO) 50-67		1819 (SO)	---- ---- ---- - --	ex
(Lokon-Empung= LO)							1829 (LO)	---- ---- ---- - --	ex
					C = silicic		1833? (SO)	---- ---- ---- - --	ex
06-06-11							1845 (SO)	---- ---- ---- - --	ex
(Mahawu= MA)							1846 (MA)	---- ---- ---- - --	pex
							1890 (SO)	---- ---- ---- - --	ex
							1893 (LO)	x--- ---- ---- - --	ex
							1894 (LO)	x--- ---- ---x - --	none
							1898? (SO)	---- ---x ---x - --	none
							1901 (SO)	xx-- ---x ---x x --	pex
							1904 (MA)	---- ---- ---- - --	ex
							1906-13 (SO)	DD-- ---D ---F - --	ex, lf
							1915 (SO)	---- ---- ---- - --	lf, dm
							1917 (SO)	---- ---- ---- - --	ex
							1923-24 (SO)	---- ---- ---- - --	ex, lf
							1930 (LO)	---- ---- ---- - --	ex
							1942 (LO)	---- ---- ---- - --	ex
							1947 (SO)	---- ---- ---- - --	ex
							1949 (LO)	---- ---- ---- - --	pex
							1951-53 (LO)	---- ---- --EE - --	ex
							1952 (MA)	---- ---- ---- - --	?
							1953 (SO)	---- ---- ---- - --	?
							1958-59 (MA)	---- ---- ---- - --	ex, pex
							1958-59 (LO)	xx-- ---- ---x - --	ex
							1961-62 (LO)	---- ---- ---x - --	ex, pex
							1963-66 (LO)	---- ---- ---- - --	ex
							1966-68 (SO)	YY-- ---- ---B - --	Ex,lf,pf,dm

TONDANO (continued)

CAVW number (active vent)	Latitude Longitude (degrees)	Diameter (km)	Local tectonic setting	Precaldera edifice	SiO_2 content (wt pct)	Age of collapse (yr B.P.)	Date of unrest	Type and duration of unrest ESTU STHF MGTF H Te	Eruption type
(continued from previous page)									
							1969-71 (LO)	YY-- ---- ---x - --	ex,pex,dm
							1970 (SO)	?--- ---- ---x - --	none
							1971 (SO)	---- ---- ---- - --	pex
							1973 (SO)	---- ---- ---- - --	ex
							1973-74 (LO)	---- ---- ---- - --	ex
							1975-79 (LO)	DD-- ---- ---- - --	ex,pf,dm
							1978 (MA)	---- ---- --x- - --	none
							1982 (SO)	A--- ---- ---- - --	Ex
							1984 (SO)	N--- ---- ---- - --	Ex
							1985 (SO)	EE-Y ---- ---- - --	Ex
							1982-87 (LO)	x--- ---- ---F - --	pex, ex
							1987 (MA)	E--- ---- --EE E --	none

TECTONIC SETTING/GEOLOGIC HISTORY

Tondano is a large caldera in north Sulawesi, near the southern end of the Sangihe Arc (Effendi, 1976; Morrice and others, 1983; Morrice and Gill, 1986). Several andesitic and basaltic stratovolcanoes lie on the caldera rim, and obsidian flows and thermal areas occur within the caldera (fig. 6.17.1). Its eastern margin is well defined, passing along the scarp east of Lake Tondano; its other margins are poorly defined. Tondano obsidian has 76 percent SiO_2 (Verbeek, 1908).

HISTORICAL ACTIVITY

1829 (LO): A new crater ("Tompaluan") formed in the saddle between Lokon and Empung in 1829; it is still an active vent.

1893 (LO): On 29 March 1893 at 1730 hr, "without any foresign," the Tompaluan vent began to smoke and continued to smoke without rumbling. Natives reported seeing fire twice. Earthquakes were felt near Tondano on 7 April, 8 August, and 18 September 1893, probably from Lokon. Lokon (Tompaluan) began to smoke again on 18 December (Figee and Onnen, 1895).

1898(?) (SO): An unconfirmed report (von Wolff, 1929) said that a 100-m-long "smoking" fissure developed at Soputan around 1898.

TONDANO (continued)

HISTORICAL ACTIVITY (continued)

1901 (SO): Earthquakes from 2 February until at least 14 February 1901 were felt most strongly in the district of Langowan and in Possan Rataha Ponosahen. They were felt almost continuously from the afternoon of 6 February through 9 February, with 20 separate shocks felt in Tomohon from 1330 to 2400 hr on 6 February ("almost continual motion"). The earthquakes were attributed to heightened activity of Soputan. No change was reported in the crater of Soputan, but two new mud pots appeared at its foot, and at least 22 fissures opened between Walelong and Kele Londej ridges (Koninklijk Magnetisch en Meteorologisch Observatorium te Batavia, 1902). Another observer in a different town saw ash ejection and steaming at Soputan as of 4 February (ibid). By early March, a new solfatara had formed 400 m upstream from the Rumerega warm spring and was jetting boiling mud to a height of 250 m; in contrast, there was no sign of increased activity in the main crater of G. Soputan (Kemmerling, 1923, p. 143-144).

1906-13 (SO): A new crater (Aeseput) formed at Soputan in 1906. "Activity in the ground" (=local earthquakes?), heavy fuming, and glow at night began on 5 June 1906 (Koninklijk Magnetisch en Meteorologisch Observatorium te Batavia, 1908). An earthquake on 17 June was felt in Tomohon (28 km north-northeast of Soputan) and was accompanied on the same day by an eruption of Soputan. Eruptions continued intermittently until 1913. The 1901 activity was a precursor to the 1906 eruption, because fumaroles formed in 1901 where activity broke out in 1906.

1915 (SO): Another new crater, Aeseput-weru, formed south of Aeseput at Soputan in 1915.

1932: A strong tectonic earthquake, felt with MMI=VIII in the Lake Tondano area, destroyed many houses but had no apparent effect on the nearby volcanoes (Soetardjo and others, 1985).

1958-59 (LO): Numerous small ash eruptions occurred at Lokon-Empung from February 1958 until December 1959 (Matahelumual, 1986).

1961-62 (LO): A small ash eruption occurred on 19 May; slightly stronger eruptions occurred on 24 October and 20 November 1961. Small phreatic explosions occurred from April to November 1962 (Matahelumual, 1986).

1963-66 (LO): There were repeated small explosions at Lokon-Empung from 17 December 1963 to March 1964, one in July 1965, and two in September 1966 (Suryo, 1985).

1966-68 (SO): At Soputan during 1966-68, there were weak explosions, lava outflow, and pyroclastic flows from the collapsing front of the lava flow. A seismic swarm occurred in August 1968, near the end of this activity (Suryo, 1985).

1969-71 (LO): More small explosions occurred from November 1969 to March 1970. All explosions during 1963-70 were from the Tompaluan (1829) crater, in the saddle between Lokon and Empung (Suryo, 1985).

1970 (SO): In February 1970, activity at Soputan increased again, with gray (ash-bearing?) gas from the main crater, continuous

TONDANO (continued)

HISTORICAL ACTIVITY (continued)

rumblings from Kawah Masem (5 km from the crater), and increased fumarolic activity through the 1966 lava flow. Activity returned to normal in May 1970 (Suryo, 1985).

1976 (LO): 373 (local?) earthquakes were recorded in January 1976, more than 10 times the normal monthly count, and an eruption began on 6 February (Volcanological Society of Japan, 1978).

1978 (MA): Crater lake temperature rose from a normal 20 °C to between 65 and 70 °C before January 1978. No eruption ensued (Casadevall, 1987; Volcanological Survey of Indonesia, 1987; I. Nairn, in Smithsonian Institution, 1987).

1982 (SO): A vigorous explosive eruption occurred on 26 August. Just 2 days before the eruption, geologists had seen no visible changes to indicate an impending eruption, and fumarole temperatures were low, from 76 °C to 84 °C. Further eruptions occurred in September, and a seismograph 6 km south of the volcano recorded "tremors" for 4 hours before an eruption on 9 November (Smithsonian Institution, 1982).

1984 (SO): No premonitory activity was recorded before the May 1984 eruption of Soputan. Seismicity increased from 6 August to 25 August 1984, then abruptly stopped for 6 days before a 5 hour explosive eruption on 31 August (Smithsonian Institution, 1984).

1985 (SO): From January to April the number of "tectonic" earthquakes increased, from 69 events/month to 130 events/month; the April count includes an $\underline{M}$=5.6 event on 4 April and its aftershocks. On 18 May the steam plume that normally rose 200 m above the crater rose to 300 m, and by 19 May the plume included ash and rose to 5 km above sea level. Interestingly, this eruption began at a time when "tectonic" seismicity was decreasing notably (Volcanological Survey of Indonesia, 1986) (fig. 6.17.2).

1982-87 (LO): Fumarolic activity at Lokon increased in November 1982 and continued through 1983 (Matahelumual, 1986). It further increased in April 1984 (a time of heavy rains and a rise in the level of the crater lake) (Volcanological Survey of Indonesia, 1986). In that same month, small "gas eruptions" through the lake sent "thick smoke" as high as 500 m above the crater rim. Vegetation as far away as 3.5 km downwind was killed, apparently by acid gas or acid rain. Occasional small earthquakes were recorded during 1984 and 1985. Then, on 21 March 1986, steam puffs rose 400 m above the Tompaluan crater rim, and on 22-24 March, phreatic eruptions ejected the crater lake and caused two small lahars (Matahelumual, 1986). In the latter half of 1986, several tens of explosions occurred each day; a few explosions per month continued through May 1987 (including 10 May and 13 May), but activity generally diminished during April-June 1987 (Smithsonian Institution, 1986-87; Casadevall, 1987; Volcanological Survey of Indonesia, 1987).

1987 (MA): From 17 April 1987 until early May 1987, a plume of whitish fume was observed about 100 m above Mahawu crater. By 30 June 1987, the temperature of the crater lake at Mahawu was 48 °C, 28 °C hotter than in October 1986 (Casadevall, 1987; Volcanological Survey of Indonesia, 1987). The volume of the lake increased during the same period. Both volcanic and tectonic earthquakes increased toward the end of April 1987 and continued in May and June 1987.

TONDANO (continued)

REFERENCES

Casadevall, T.J., 1987, Seventh report of the project Volcano Monitoring and Research in Indonesia, April 1, 1987 - June 30, 1987: Bandung, Volcanological Survey of Indonesia, 9 p.

Effendi, A.C., 1976, Geologic map of the Manado Quadrangle, North Sulawesi: Bandung, Geological Survey of Indonesia, 1:250,000.

Figee, S., and Onnen, H., 1895, Vulkanische verschijnselen en aardbevingen in den O.I. Archipel waargenomen gedurende het jaar 1893: Nat. Tijdschr. v. Ned. Ind., pt. 54, p. 205-235.

Kemmerling, G.L.L., 1923, De vulkanen van den Sangi-Archipel en van de Minahassa: Vulkanol. Meded., no. 5, 157 p.

Koninklijk Magnetisch en Meteorologisch Observatorium te Batavia, 1902, Vulkanische verschijnselen en aardbevingen in den Oost-Indischen Archipel waargenomen gedurende het jaar 1901: Nat. Tijdschr. v. Ned. Ind., pt. 62, p. 169-211 + errata.

----- 1908, Vulkanische verschijnselen en aardbevingen in den Oost-Indischen Archipel waargenomen gedurende het jaar 1906: Nat. Tijdschr. v. Ned. Ind., pt. 67, p. 54-86.

Kusumadinata, K., ed., 1979, Data Dasar Gunungapi Indonesia: Bandung, Volcanological Survey of Indonesia, 820 p.

Matahelumual, J., 1986, G. Lokon-Empung: Berita Berkala Vulkanologi: Bull. Volcanol. Surv. Indonesia, no. 114, 52 p.

Morrice, M.G., Jezek, P.A., Gill, J.B., Whitford, D.J., and Monoarfa, M., 1983, An introduction to the Sangihe arc: Volcanism accompanying arc-collision in the Molucca Sea, Indonesia: Jour. Volcanol. Geotherm. Res., v. 19, no. 1/2, p. 135-166.

Morrice, M.G., and Gill, J.B., 1986, Spatial patterns in the mineralogy of island arc magma series: Sangihe arc, Indonesia: Jour. Volcanol. Geotherm. Res., v. 29, p. 311-353.

Neumann van Padang, M., 1951, Catalogue of the active volcanoes of the world including solfatara fields, Pt. I, Indonesia: Rome, IAVCEI, 271 p.

Silver, E.A., McCaffrey, R., and Smith, R.B., 1983, Collision, rotation, and the initiation of subduction in the evolution of Sulawesi, Indonesia: Jour. Geophys. Res., v. 88, no. B11, p. 9407-9418.

Smithsonian Institution, Scientific Event Alert Network (SEAN), 1982, 1984-87, Soputan, Lokon-Empung: SEAN Bull., v. 7; v. 9, nos. 5, 8; v. 10, no. 5; v. 11, nos. 3-7, 12; v. 12, nos. 1-3, 6-8 (reports from A. Sudradjat, T. Casadevall, I. Nairn, and others).

Soetardjo, R., Untung, M., Arnold, E.P., Soetadi, R., Isail, S., and Kertapati, E., 1985, Indonesia: SEASEE Series on Seismology, v. 5, 199 p.

Suryo, I., 1985, Lokon and Soputan: Bull. Volcanol. Surv. Indonesia, no. 106, p. 125-133.

Verbeek, R.D.M., 1908, Molukkenverslag: Jaarb. Mijnw. in Ned. Ind. Wet. Ged., v. 37, p. 1-826.

Volcanological Society of Japan, 1978, Lokon: Bull. Volcanic Eruptions, no. 16, p. 33 (information from A. Sudradjat and J. Matahelumual).

Volcanological Survey of Indonesia, 1985, Laporan Tahunan Direktorat Vulkanologi, 1983-84: Bull. Volcanol. Surv. Indonesia, no. 112.

----- 1986, Laporan Tahunan Direktorat Vulkanologi, 1984-85: Bull. Volcanol. Surv. Indonesia, no. 113, 259 p.

----- 1987, Laporan Tahunan Direktorat Vulkanologi, 1985-86: Bull. Volcanol. Surv. Indonesia, preliminary edition, 434 p.

Wolff, F. von, 1929, Der Vulkanismus: Stuttgart, Ferdinand Enke, v. 2, p. 184.

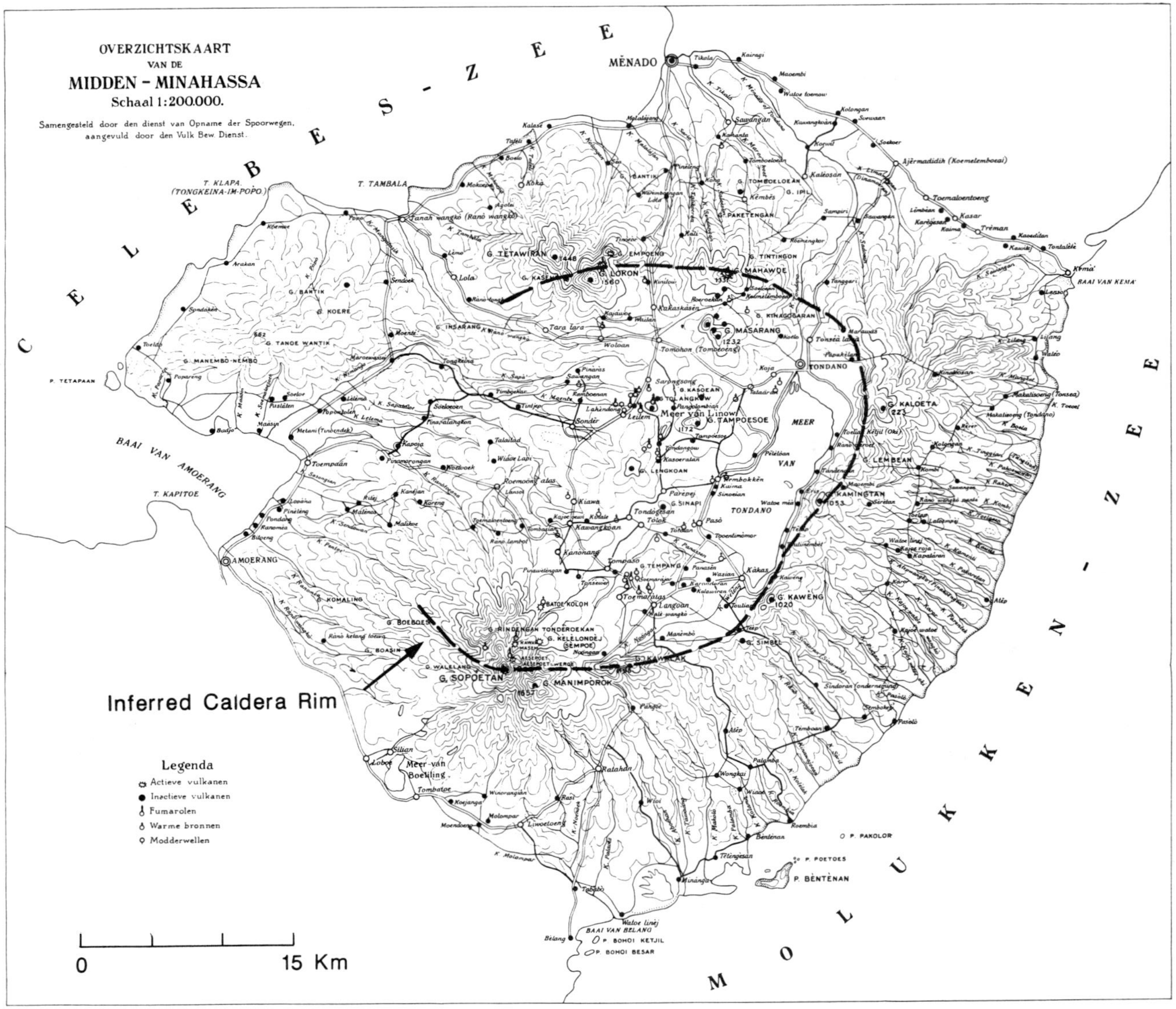

Figure 6.17.1. Topographic map of north Sulawesi, from Kemmerling (1923). We have added an approximate outline of Tondano Caldera, based on topography and distribution of vents and hydrothermal areas.

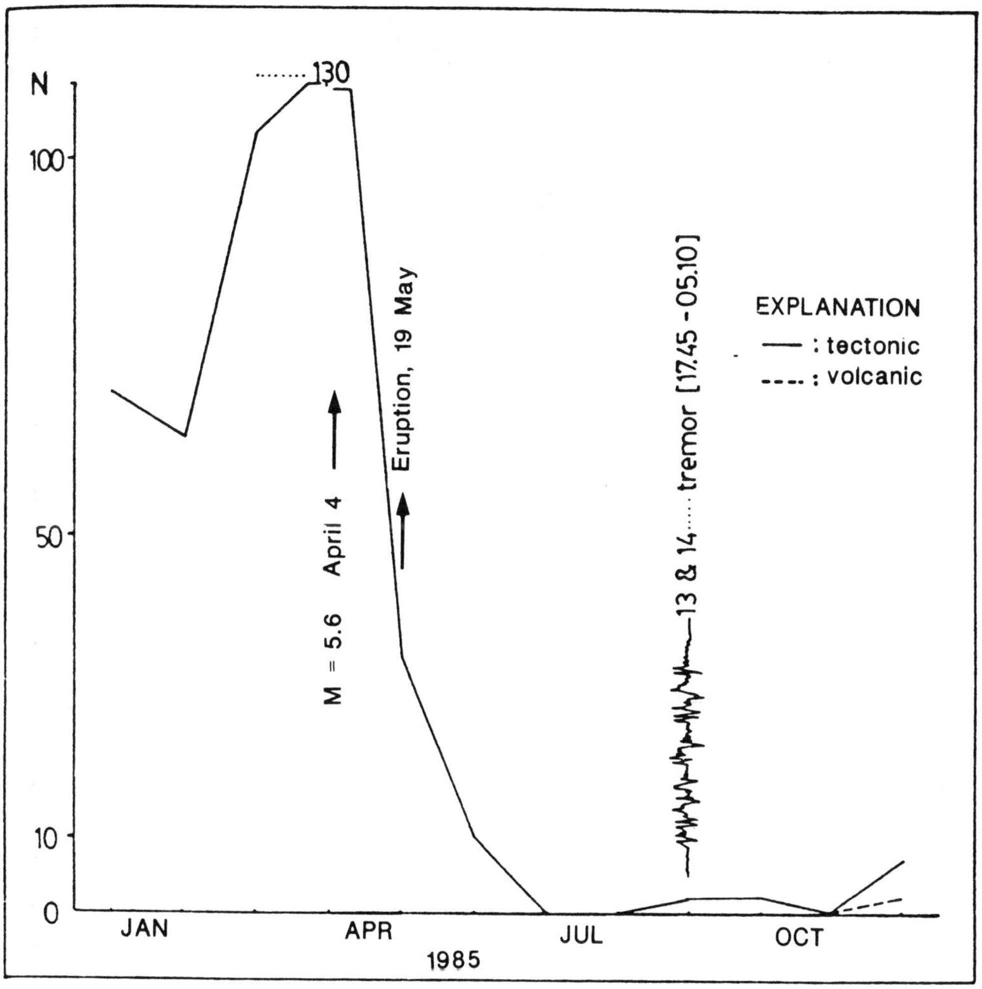

Figure 6.17.2. Earthquakes of Soputan Volcano before and during eruption of 19 May 1985 (Volcanological Survey of Indonesia, 1987). Note decrease in seismicity before eruption.

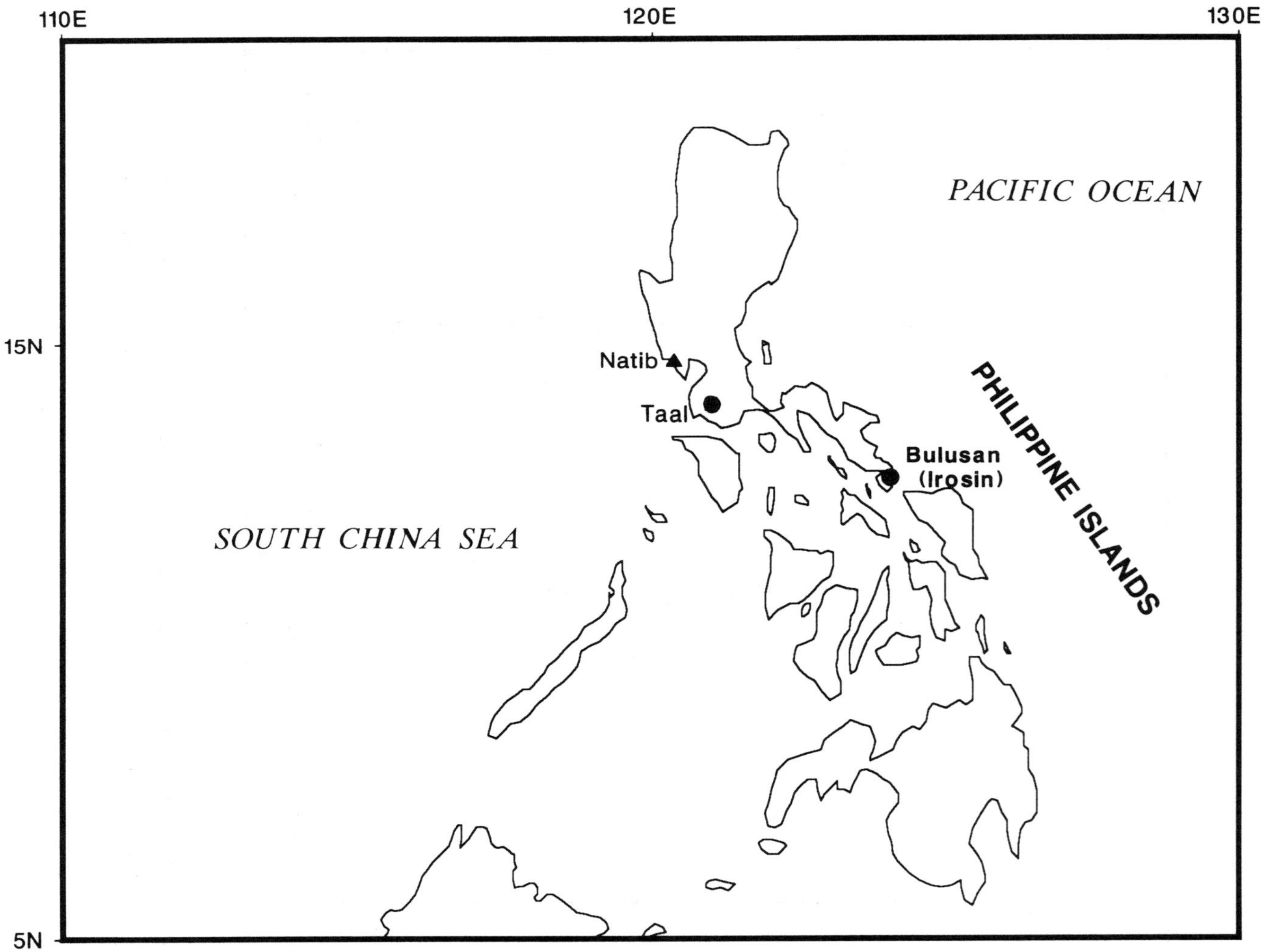

Figure 7. Locations of large, Quaternary restless calderas (solid circles) and nonrestless caldera (triangle) in Region 7.

BULUSAN (IROSIN)

CAVW number (active vent)	Latitude Longitude (degrees)	Diameter (km)	Local tectonic setting	Precaldera edifice	SiO_2 content (wt pct)	Age of collapse (yr B.P.)	Date of unrest	Type and duration of unrest ESTU STHF MGTF H Te	Eruption type
07-03-01 (Bulusan)	12.77N 124.05E	11	Compr	Strato	R = 59? or	>36,000	1852	---- ---- ---- - --	ex
					basaltic		1886	---- ---- ---- - --	ex
					andesite?		1889	---- ---- ---- - --	unknown
					(1918)		1892	---- ---- ---- - --	ex
							1894	---- ---- ---- - --	ex
					C = 69-71		1916	x--- ---- ---x - --	pex,ex
							1918-22	x--- ---- --xx - --	pex,ex,lf
							1928	---- ---- ---- - --	ex
							1933	---- ---- ---- - --	ex
							1966	---- ---- --x? ? --	none
							1978-83	Cx-- ---- --DD C --	pex,ex,Ex
							1986-88+	FF-- ---- ---F - --	pex

TECTONIC SETTING

Bulusan is an active stratovolcano and dome complex near the southern tip of Luzon Island in the Philippines. It is part of the Bicol Volcanic Chain, which is associated with the Philippine Trench.

GEOLOGIC HISTORY

A caldera-forming eruption more than 36,000 years ago (slightly older than could be measured by conventional ^{14}C methods; M. Rubin, written commun., 1985) produced extensive rhyolitic ash flows. Bulusan stratocone and domes have grown within the resulting Bulusan (or Irosin) Caldera (fig. 7.1.1; Philippine Institute of Volcanology and Seismology, 1984), and consist mainly of andesitic and dacitic lava flows, both from Bulusan and neighboring lava domes. A veneer of pyroclastic materials and secondary ash is exposed near the summit of Bulusan; pyroclastic flow and lahar deposits are found at its base (Philippine Institute of Volcanology and Seismology, 1984).

HISTORICAL ACTIVITY

Most historical eruptions have occurred from vents along an east-west-trending fracture in the summit area of Bulusan.

1978-83: Seismicity has increased for several days before recent eruptions; over the same time scale, temperatures of hot springs at the southwest base of Bulusan have decreased several degrees, recovered, then increased several degrees above previous

BULUSAN (IROSIN) (continued)

HISTORICAL ACTIVITY (continued)

levels (fig. 7.1.2). An eight-day seismic swarm preceded the 27 April 1981 eruption, but a swarm from 30 June to 4 July 1981 was not followed by an eruption. No seismic precursor to the June 1983 eruption was detected at a seismometer 4 km from the summit. Slightly elevated seismicity and small swarms continued through 1985. An intermittent increase in steaming from fumaroles in the summit crater and on the upper west flank of Bulusan was noted a month before an eruption in June 1983, and the temperature and discharge of the hot springs mentioned above increased by 2 °C and 4 L/sec (liters per second), respectively, one week before the same eruption.

1986-88+: A 10-hr-long seismic swarm on 19 April 1986 consisted of 229 instrumentally recorded and 3 felt earthquakes, tentatively located 7 km southeast of the summit crater, within the caldera. Earthquakes early in the swarm were of higher spectral frequency than those later in the swarm. Other swarms occurred at approximately 2-month intervals through at least February 1988 (fig. 7.1.3) (E.G. Ramos, written commun., 1987). Steaming and hot-spring temperatures did not change in 1986, but steaming from a summit fissure did increase slightly during the May 1987 earthquake swarm (Smithsonian Institution, 1986, 1987). A small ash eruption occurred on 20 February 1988 without immediate seismic or hot-spring precursor. A swarm of small earthquakes occurred within hours after the eruption and, probably coincidentally, a $\underline{M}$=6.9 regional earthquake occurred 80 km northeast of Bulusan on 24 February (R.S. Punongbayan, oral commun., 1988; Smithsonian Institution, 1988).

REFERENCES

Aguila, L.G., 1978, The 1978 eruptions of Bulusan Volcano: COMVOL Lett., v. 10, nos. 3-4, p. 4-8.

Commission on Volcanology, 1981, Catalogue of Philippine volcanoes and solfataric areas (3rd ed.): Quezon City, Commission on Volcanology, 132 p.

Commission on Volcanology, ca. 1981, Operation Bulusan: unpubl. rep., 23 p.

Goldsberry, J.P., 1916, Eruption of Bulusan Volcano: Philippine Jour. Sci., sect. A., v. 11, no. 5, p. 251-256.

Neumann van Padang, M., 1953, Catalogue of the active volcanoes of the world including solfatara fields, Pt. II, Philippine Islands and Cochin China: Naples, IAVCEI, 49 p.

Philippine Institute of Volcanology and Seismology, 1984, Bulusan Volcano: Quezon City, PHIVOLCS, 20 p.

Saderra Maso, M., Fr., 1919, Recent eruptions of the Bulusan Volcano, 1916 and 1918: Philippine Weather Bureau, Monthly Bull., Jan. 1919, 3 p.

----- 1923, Recent eruptions of the Bulusan Volcano, 1919 and 1922: Philippine Weather Bureau, Monthly Bull., undated, 3 p.

----- 1928, Steam as immediate cause of the last eruptions of Taal and Bulusan Volcanoes: Proc. 3rd Pacific Sci. Congr., v. 1, p. 761-767.

Smithsonian Institution, Scientific Event Alert Network (SEAN), 1979-88, Bulusan Volcano: SEAN Bull., v. 3, nos. 7-8; v. 5, nos. 1-4, 6-9, 11; v. 6, nos. 4-7; v. 8, no. 6; v. 11, no. 4; v. 12, nos. 5, 12; v. 13, no. 2 (information from Commission on Volcanology and Philippine Institute of Volcanology and Seismology).

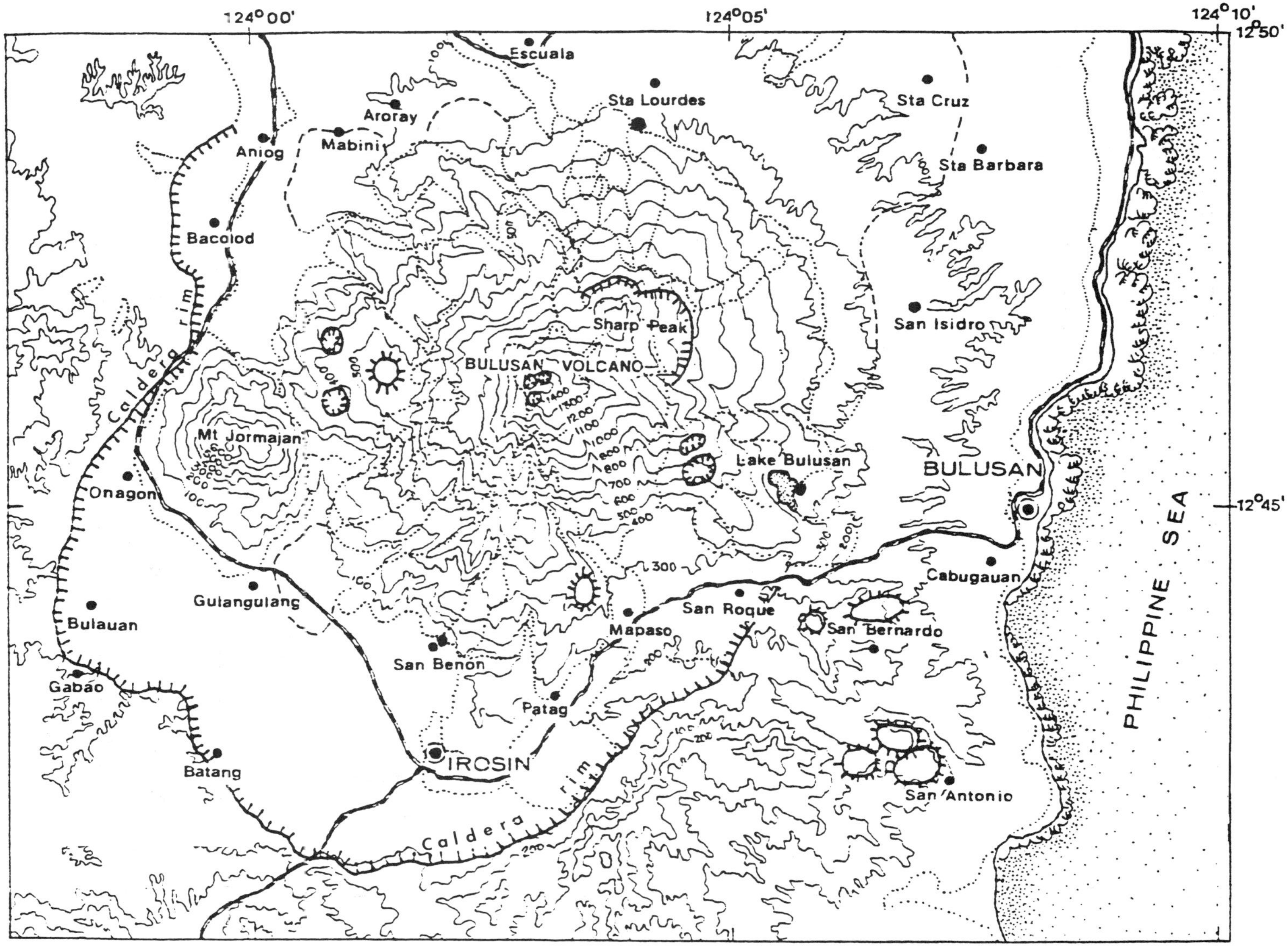

Figure 7.1.1. Bulusan (Irosin) Caldera, Bulusan Volcano, and surroundings, from Philippine Institute of Volcanology and Seismology (1984). The Philippine Institute of Volcanology and Seismology (PHIVOLCS) operates a permanent observatory at San Benon, and temporary observatories at Aniog, Lake Bulusan, and near San Isidro.

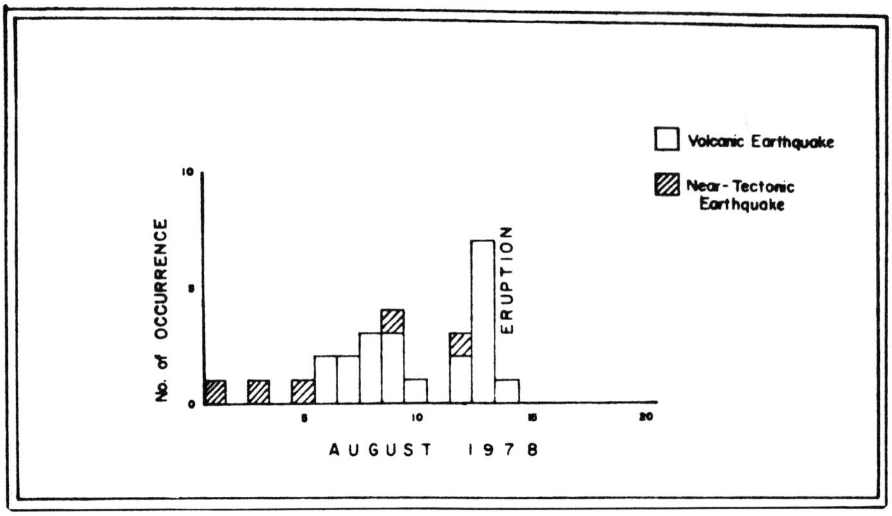

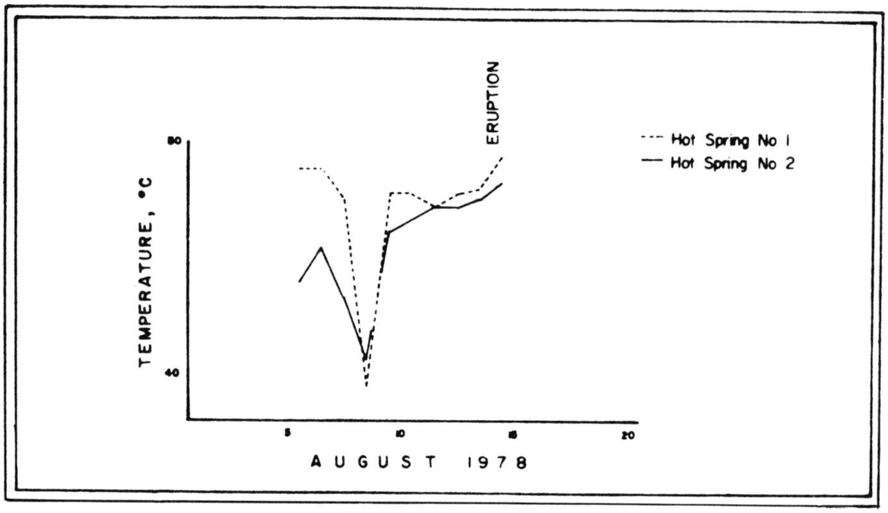

Figure 7.1.2. Daily number of earthquakes before an eruption of Bulusan volcano on 14 August 1978 (top). Variation in temperature of two hot springs at San Benon prior to the August 1978 eruption (bottom). Both figures from Aguila (1978).

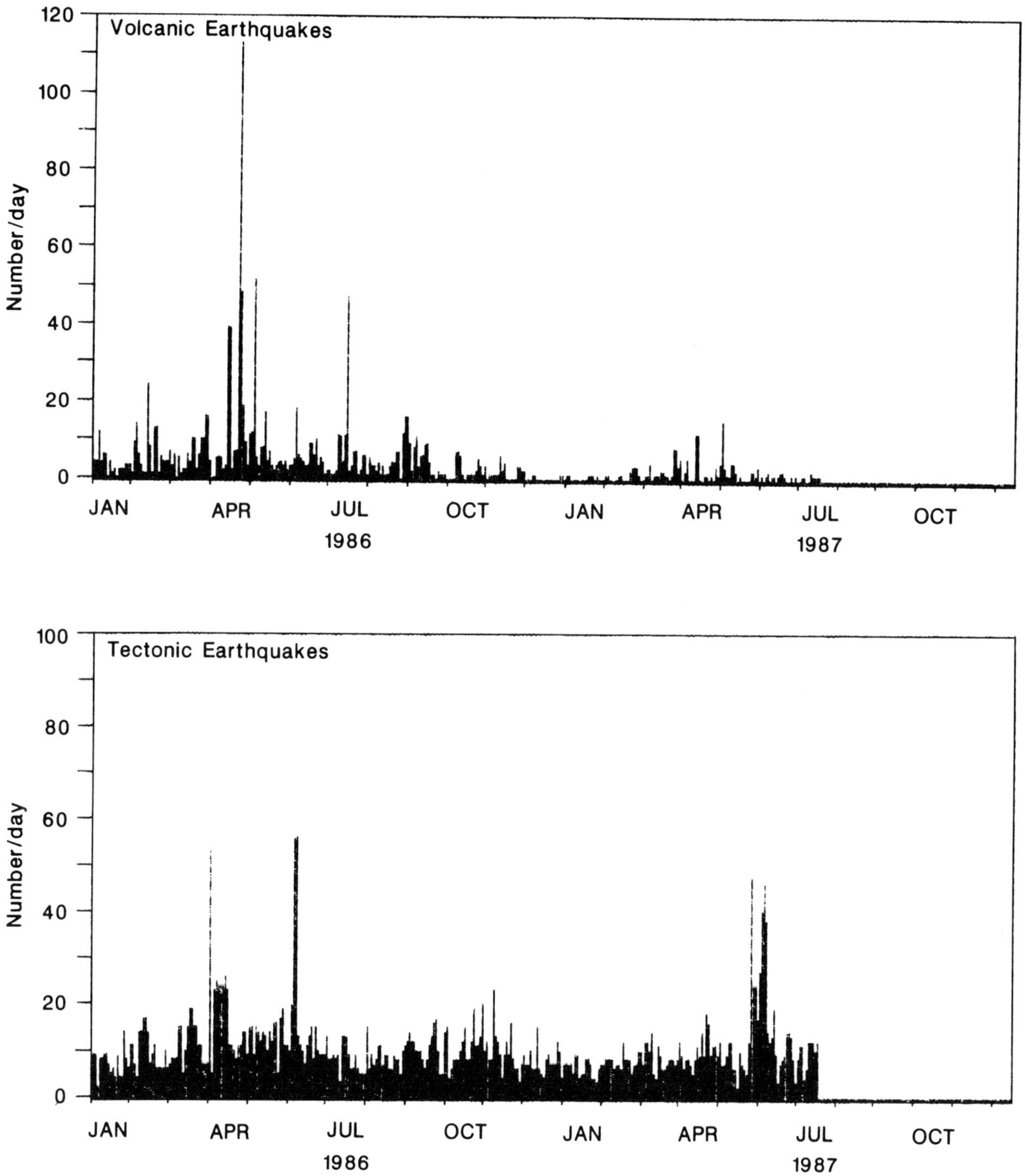

Figure 7.1.3. Daily numbers of volcanic (top) and tectonic (bottom) earthquakes near Bulusan Volcano from January 1986 to December 1987, from Smithsonian Institution (1987), courtesy of Philippine Institute of Volcanology and Seismology.

PART 3: HISTORICAL UNREST AT LARGE QUATERNARY CALDERAS (continued)

See inside back cover for explanation and abbreviations

TAAL

CAVW number (active vent)	Latitude Longitude (degrees)	Diameter (km)	Local tectonic setting	Precaldera edifice	SiO_2 content (wt pct)	Age of collapse (yr B.P.)	Date of unrest	Type and duration of unrest ESTU STHF MGTF H Te	Eruption type
07-03-07	14.00N	15 x 20	NE-SW	Over-	R=50-59		1572	---- ---- ---- - --	Ex
	121.00E		fault	lapping	C=58-64		1591	---- ---- ---- - --	ex
			normal	strato-			1605-11	---- ---- ---- - --	ex
			to	volcanoes			1634-35	---- ---- ---- - --	ex
			volcanic				1641	---- ---- ---- - --	Ex
			chain				1645	---- ---- ---- - --	ex
							1707	---- ---- ---- - --	ex
							1709	---- ---- ---- - --	ex
							1715-16	x--- ---- ---? - --	P-Ex
							1729	---- ---- ---- - --	ex
							1731	x--- ---- ---- - --	p-ex
							1749	xx-? Y-?Y --B- Y --	Ex
							1754	Y--? ---x ---Y - --	P-EX
							1790	---- ---- ---- - --	ex
							1808	---- ---- ---- - --	ex
							1825	---- ---- ---- - --	ex
							1842	x--- ---- ---- - --	ex
							1873-74	---- ---- ---- - QU?	ex
							1878	D--- ---- ---- - --	ex
							1880	BB-- ---- ---- - --	none
							1881	xx-- ---- ---- - --	none
							1883	A--- ---- ---- - --	none
							1887	xx-- ---- ---- - --	none
							1903-04	---- ---- ---E - --	ex,pex
							1911	CC-x x-?x ---C - --	ex,P-EX
							1953	---- ---- --EE E --	none
							1965-70	A--C -?-- x-EE E --	P-EX,lf
							1971-77	N--- ---C --FF - --	ex,pex
							1986-87	DD-- x-?D ---- - --	none

TAAL, Region 7, CAVW number 07-03-07

TAAL (continued)

TECTONIC SETTING

Taal Caldera lies directly astride a major NE-SW-trending zone of active faulting (the "Taal Line") (figs. 7.2.1, 7.2.2). The nature of that zone is not fully understood; it might be part of a transform boundary between east-dipping Manila Trench subduction and west-dipping subduction from the Philippine Trench, or else it might be a transverse break within that east-dipping subduction.

GEOLOGIC SETTING

Taal Caldera is either a single large caldera that is presently divided into two basins by a sublacustrine ridge extending southwest and northeast from Volcano Island, or else it is a composite of two or more calderas (fig. 7.2.3). Volcano Island is a postcaldera feature formed by coalescing low stratovolcanoes. Much of the caldera is occupied by Lake Taal; a smaller lake occupies the main crater within Volcano Island. According to Fray Gaspard de San Agustin (quoted by Le Gentil de la Galaisiere, 1779-81), Lake Taal in 1698 was a saltwater lake that supported "excellent tuna, though not as good as those of the Mediterranean Sea."

HISTORICAL ACTIVITY

1572, 1591, 1605-11: During 1572-1611, the volcano in Lake Bombon (Lake Taal) was "wont to spit forth many and very large rocks, which are glowing and destroy the crops of the natives." Accordingly, Fray Albuquerque, the first rector in the new town of Taal, "employed a kind of exorcism by celebrating mass on the very island of the volcano" (Saderra Maso, 1911). A similar mass was held by Fray Bartolome de Alcantara in 1591, "for the reason that volcano had begun to belch forth extraordinary masses of smoke." Fray Tomas de Abreu, the priest of Taal from 1605-11, responded to "frequent rumblings" of the volcano by having a big wooden cross carried up there and planted at the active crater. This cross was "so heavy, being made of very hard wood, that it took four hundred men to carry to the rim of the crater. Since that cross was placed on the top of the volcano, not only has it done no damage, but the island has recovered its fertility" (San Agustin, 1698, quoted in Le Gentil de la Galaisiere, 1779-81).

1634, 1645: Saderra Maso (1911), citing Semper, suggests that activity in these two years, and indeed from the colonial discovery of the volcano in 1572 until the first well-established eruption in 1707, was "only solfataric activity, or at most very unimportant outbreaks."

1707: Binintiang Malaki (northwest corner of Volcano Island) formed "in a tremendous display of thunder and lightning; but aside from fear and trembling, no damage was done ...on the shores of Lake Bombon" (Saderra Maso, 1911).

1709: Binintiang Munti cone (southwest corner of Volcano Island) formed during 1709-29(?) (Minoza, 1965).

TAAL (continued)

HISTORICAL ACTIVITY (continued)

1715-16: In 1716, there were magmatic or phreatomagmatic explosions at the Calauit vent in the SE part of Volcano Island. Earthquakes and venting shifted from there southeastward to beneath the lake (Saderra Maso, 1911, citing Fr. Manuel de Arce; Alcaraz, 1965). On 24 September, artillerylike sounds were followed shortly by fire fountaining. While the eruption maintained ash columns, strong earthquakes(?) convulsed the lake and produced big waves that eroded as much of 50 feet of soil on the shores of Taal town. The eruption continued strongly for 4 days, during which the "heat of eruption" (and lake overturn?) killed all the fish in the lake, and a strong sulfur smell permeated all towns near the lake (Zuñiga, 1803).

1731: In 1731, a sublacustrine eruption off Pira-piraso (northeast corner of Volcano Island) formed Babuing and Napayong Islands (Faustino, 1933; Alcaraz, 1965). "Vast and towering obelisks of earth and sand arose out of the water" to form the new islands. In Los Baños, a town 30 km NE of Volcano Island, Fr. Torrubia heard "a continuous fire of heavy artillery, as if two mighty armies were engaged in battle. This was followed by a terrible earthquake of long duration, after which we heard only isolated detonations, not with the former frequency, but very much sharper.... The terrible phenomenon lasted many days, accompanied by subterranean rumblings which caused the entire region to tremble..." (Saderra Maso, 1911).

1749: Unusually frequent earthquakes were felt in Manila, especially during the 6 weeks before a violent eruption from the main crater that began on night of 11 August (Saderra Maso, 1895). Four hours of "extensive glare" preceded the explosions. Then, at 0900 hr on 12 August, a strong earthquake damaged the convent, and a seiche destroyed several lakeshore villages. The strongest of these earthquakes, of probable tectonic origin, caused damage in towns as far way as Calamba, Laguna de Bay, and Antipolo. According to an eyewitness, Fr. Buencuchillo, O.S.A. (cited in Saderra Maso, 1911), "during these terrible convulsions of the earth fissures opened in the ground amid horrifying roars, said fissures extending from the northern and northeastern beach of the lake as far as the neighborhood of the town of Calamba (30 km northeast of Volcano Island, near Los Banos). Here as well as elsewhere, the whole shore of Lake Bombon has been disturbed. The entire territory of Sala and part of that of Tanauan have been rendered practically uninhabitable--the water courses have been altered, former springs have ceased to flow and new ones made their appearance, the whole country is traversed by fissures, and extensive subsidences have occurred in many places.... During my flight I saw a great many tall trees, such as coconut and betel-nut palms, either miserably fallen or so deeply buried that their tops were within reach of my hands. I likewise saw several houses which formerly, in accordance with Philippine custom, had their floors raised several yards above ground, but had sunk to such a degree that the same ladder which once served to ascend into them, was now used to descend to them."

1750?: Pye (1756) wrote: "In the year 1750 they had an earthquake here, which lasted for three months, with almost continual tremblings, which at last broke out in an eruption, in a small island in the middle of a large lake, all round which the bottom is unfathomable. The third day after the commencing of the eruption, there arose four more small islands in the lake, all burning; and about a mile distance from one there is a continual fire, which comes out of the water, where there is no ground, for upwards of an hundred fathoms deep. This I saw myself, and went very near it.... This happened but four years ago, and if you were to feel some of the shocks, you would think they were capable of producing any thing, for they are very terrible indeed." Confusion of dates is possible; an eruption in 1750 is not corroborated by other writers, and we suspect that Pye is referring to the eruption of 1749.

TAAL (continued)

HISTORICAL ACTIVITY (continued)

1754: In 1754, the volcano's main crater "quite unexpectedly commenced to roar and emit, sky-high, formidable flames intermixed with glowing rocks which, falling back upon the island and rolling down the slopes of the mountain, created the impression of a large river of fire" (Fr. Buencuchillo, in Saderra Maso, 1911). This was the largest eruption in the history of Taal; it lasted from just after 2100 hr on 15 May until 3 December. During those 7 months there were greater or lesser eruptions of "fire, rocks, and ash,... accompanied by submarine noises and earthquakes" (Saderra Maso, 1895). On the night of 28 November, during renewed eruptions, "the earth was in continuous, swaying motion," as observed about 2 km west of Taal town. "In the villages to the west of the lake, which were the greater and better part, all the houses have either collapsed under the load of material...or have disappeared completely, swept away by the waves which in these places were so violent that they dug three ditches or channels, too wide and deep to be forded, and thus rendered impassable the road which joins the town with Balayan. In other parts of the lake shore have likewise opened many cracks and occurred extensive slides. The worst of all is, that, the mouth of the river Pansipit having been blocked, the lake is rising and invading the towns of Lipa and Tanauan.... The bulk of the population left this neighborhood.... Thus out of 1,200 taxpayers whom Taal contained formerly, hardly 150 remain in the poorest and least respectable villages, which suffered little from the rain of ashes" (Fr. Buencuchillo, in Saderra Maso, 1911). Following this eruption, the towns of Taal, Lipa, and Tanauan were moved to their present sites.

1808: In 1808, explosions occurred in the main crater, which became larger and shallower during the eruption. Heavy ash fall (up to 84 cm) was reported "in the immediate vicinity of the volcano" (Saderra Maso, 1911).

1842: On 2 August, extraordinary subterranean noises were heard around Taal Lake, but no immediate volcanic activity or large earthquake occurred. Then, at 0930 hr on 24 October, there was a strong earthquake with E-W oscillation, lasting 2 minutes, followed by aftershocks on that day and the next. Perrey (1860, p. 97) says that one earthquake on 25 October was felt as far away as Macao, but Saderra Maso (1895) supposes that Perrey meant Macato, a town in Capiz province, Panay Island.

1852: On 24 December 1852, in the afternoon or evening, a strong earthquake occurred in the Balayan region, with an origin near Taal. The bell tower in the church of Taal town fell (Saderra Maso, 1895).

1872: At 1148 hr on 29 December, a very strong earthquake was felt throughout Luzon. It was preceded, at least in Manila and the epicentral region, by two smaller shocks at 0600 hr and 0900 hr. The epicenter, according to Saderra Maso (1895), was somewhere just west of Batangas, the entrance to Manila Bay, or Zambales.

1873-74: During 1873-74, eruptions occurred in the main crater. Activity may have been related to the strong 1872 earthquake.

1878: From the end of October until 12 November 1878, frequent noises from the volcano were heard, culminating during 12-15 November in ejections of ash that covered the entire island.

TAAL (continued)

HISTORICAL ACTIVITY (continued)

1880: On 1 September at 0100 hr, there were gentle earthquakes accompanied by subterranean noises (Saderra Maso, 1895, p. 70).

1881: The largest earthquake of a swarm in the Taal region occurred at 0148 hr on 24 October 1881. Many earthquakes were felt in Taal town in a 24-hr period, including some at 0055, 0148 (strongest), 0210, 0225,..., 0600, and 0808 hr. The stronger of these earthquakes were felt in Manila and Punta Santiago, 50 km to the north and 40 km to the SW of Taal Caldera, respectively (Saderra Maso, 1895).

1883: A strong earthquake at or just southwest of Taal occurred at 1751 hr on 27 July.

1887: On 6 April 1887, from 1258 hr through at least 1642 hr, there were earthquakes and subterranean noises that seemed to come from the volcano. However, it was a very stormy afternoon (Saderra Maso, 1895).

1903-04: Sometime between Centeno's visit of 1882 and Adams' visit during or shortly before 1910, the Green Lake (within the main crater) extended itself southward and one point in the lake began to boil vigorously. Elsewhere in the crater, fumaroles that were active in 1882 were quiet during most of Adams' visits. But in 1904 a new crater formed during small eruptions that produced "great columns of vapor" and intermittently ejected mud and steam to about 150-m height. Activity soon diminished, and in December 1905 the floor of the new vent was covered by a boiling lake. By March 1907 the lake had disappeared, leaving only some hot mud spots with weak activity (Adams, 1910).

1911: Moderately strong earthquakes and an unusually large "steam cloud" were noted on 27 January 1911, 3 days before the climactic eruption (fig. 7.2.4). Earthquakes on 27 January were felt as far away as Batangas City (30 km from Taal). By the following day the earthquakes had increased in frequency and intensity, and the steam cloud was "frequently blackened by the ejection of mud." Many of the earthquakes on 28 January were felt in Manila, 50 km to the north. On the afternoon of 29 January, ground cracks opened in a 9-km-wide zone extending from the towns of Lemery and Taal (20 km southwest of the 1911 vent) to Talisay (12 km northeast of the 1911 vent). Earthquakes as recorded in Manila decreased sharply in number in the afternoon of 29 January, but their intensity increased to maximum Rossi-Forel intensities of IV in Manila, VII near Taal town, and VIII or IX on Volcano Island (Saderra Maso, 1911). After a brief period of less frequent earthquakes, activity increased at about 1800 hr with violent outbursts and sharp earthquakes, and at about 2300 hr a very severe earthquake was followed closely by ejection of a massive black cloud above the volcano. At 0100 hr on 30 January, a still stronger outburst occurred, and at 0220 hr, without any severe earthquake but accompanied by a loud explosion noise, a climactic phreatomagmatic eruption occurred from the main crater. Descriptions of this eruption by Pratt (1911), Saderra Maso (1911), and Worcester (1912) are fascinating accounts of base surge, half a century before the term was introduced; these surges destroyed most of Volcano Island and, together with seiches, killed about 1,334 persons. Pratt (1911) reported that there was no juvenile material erupted; Saderra Maso (1911) and Worcester (1912) cite reports of incandescence and "breadcrusted" blocks, but they agree that most of the material erupted was older rock and mud.

TAAL (continued)

HISTORICAL ACTIVITY (continued)

1911 (continued): Gilchrist (1911) reported that on 29 January 1911 the water level of Taal Lake (the large caldera lake) fell by 6 or 8 inches and conjectured that "the entrance of the boiling water into the heated subterranean depths was the primary cause of the explosion." We have not found any other report that water levels decreased lake-wide and wonder if Gilchrist's report might not refer to localized uplift of the lake shore or of Volcano Island.

During the great eruption, earthquakes nearly ceased. Very few earthquakes were recorded at Manila from 0210 hr until 0357 hr on 30 January. After 0357 hr, earthquakes were "again as frequent (and intense) as they had been before the eruption" (Saderra Maso, 1911). "Damage due to the earthquakes was slight. The towns of Lemery and Taal were the most shaken. In Lemery the tower of Caysaysay Church was partly demolished. Masonry gate posts in the wall which surrounds the church were overturned to the west. This church is directly in the line of one of the fissures and some of the damage is evidently due to the vertical displacement along it as the west side dropped.... Shocks also cracked and displaced several old masonry walls in Talisay" (Pratt, 1911).

"Of the numerous N30°E trending fissures [fig. 7.2.5], none could be traced across Volcano Island, a fact scarcely explained by the greater fall of ash... Fissures are vertical, en echelon, and show only a vertical (no lateral) displacement.... The maximum displacement on a single fissure was 3 meters; more often displacement was a few tens of centimeters. Displacement did not take place completely at the time the fissures appeared; it increased perceptibly for several days, during which earthquakes were frequent <u>after</u> the opening of the fissures.... A rush of gas occurred when a large fissure along the west side of the Pancipit Valley opened, and near the seacoast a number of gushing springs developed in the general vicinity of the fissures. Both gas and water were at ordinary temperatures" (Pratt, 1911).

As a result of the 1911 eruption, the lower Pancipit Valley (about 5 km southwest of the southwest caldera rim) subsided by up to 3 m relative to sea level. Talisay subsided a few centimeters, and Volcano Island (in the center of the caldera, and including the active vents) subsided by about 3 m relative to lake level. In contrast, the south shore and the upper end of the Pancipit Valley (southwest shore of the caldera lake) were uplifted by as much as 1 m, and some points along the north shore of the caldera lake were uplifted by smaller amounts. The deformation was irregular and affected the entire caldera and points outside the caldera along the NE-SW fault zone.

1953: Solfataric activity beneath the Main Crater Lake and along a 100-m stretch of its southwest edge increased in May 1953, raising the temperature of the crater lake from 32 °C (February) to 36 °C (May) (Andal and Yambao, 1953; Macdonald, 1954). Later, the quantity of steam fluctuated and the fumarole field enlarged itself further to the northwest.

1965-70: There was only a very slight seismic precursor to the eruption of 28-30 September 1965: one earthquake on 22 June, one each on 16, 17, and 19 August, three on 22 August, one on 25 August, two on 30 August, one each on 1 and 17 September, and two on 27 September (Minoza, 1965). A short-term seismic precursor first appeared at 0200 hr, about 20 minutes before the start of the eruption (Volcanological Society of Japan, 1966; C. Andal, oral commun., 1982). Water-tube tilt and magnetic measurements showed

TAAL (continued)

HISTORICAL ACTIVITY (continued)

1965-70 (continued):

some erratic and probably insignificant changes before the eruption (Alcaraz, 1965). Leveling along the west shore of Volcano Island showed approximately 5 cm of uplift in the (few?) days preceding the 1965 eruption; unfortunately, a tide gauge on the north shore of the island was destroyed by a mudflow during the eruption (C. Andal, oral commun., 1982). The clearest precursor to the 1965 eruption was an increase in the temperature of the Main Crater Lake on Volcano Island (fig. 7.2.6), from 30-33 °C to 45 °C over the first three weeks of July 1965, 2 months before the eruption began (Alcaraz and Tan Cardoso, 1969) (fig. 7.2.7). The 45 °C temperature was maintained from late July until the start of the eruption. The water level of the Main Crater Lake rose 3 or 4 feet above normal between mid-May and late June 1965, after which it declined slowly; by the start of the eruption it was back down to the normal level for that time of year. A field inspection 3 days before the eruption found the Main Crater Lake quiet, without any reactivation of the solfataric activity at its southwest edge that had caused concern in 1953 but that had waned in subsequent years. The farmer on whose land the new vent formed stated that he had not noticed any change in his fields, except that his rice plants were growing better than in any previous year (Alcaraz, 1965). After the eruption, residents of Kaygabok and Saluyan villages reported that their cattle had become restless several hours before the eruption (C. Andal, written commun., 1984). The eruption began at approximately 0200 hr on 28 September with fountaining of incandescent lava; base surges and seiches began at about 0325 hr and killed approximately 190 persons.

The eruption of 5 July to 4 August 1966 was preceded by increases in bubbling and water temperatures in the 1965 lagoon and in a pond in the 1965 conelet. The temperature of the Main Crater Lake rose from 36 °C to 40 °C, and local seismicity also increased.

Approximately 2-3 months of tilt preceded the August 1967 eruption, including two weeks of quite irregular tilt directions 4-6 weeks before the eruption. Other precursors to the 1967 eruption included an increase in temperature of water ponded in the 1966 cone, from 40 °C to 57 °C, slight increases in the temperature of the Main Crater Lake, and harmonic tremor beginning a few hours before the eruption.

Thirty-five days of increased fumarolic activity on the 1966 cone, 38 days of more frequent volcanic earthquakes, and several hours of harmonic tremor preceded the January-April 1968 eruption. There were similar seismic, tilt, and fumarolic precursors to an eruption on 29 October 1969, and also changes in the temperature, color, and turbidity of the Main Crater Lake several months before that eruption (fig. 7.2.8). The peak of earthquake activity was in September (C. Andal, written commun., 1984), and harmonic tremor began 7.5 hours before the eruption. The 1965-68 eruption sites are shown in figure 7.2.6.

The SiO_2 content in the crater lake increased a few months before a small eruption in November 1970; steaming increased and harmonic tremor began a few hours before that eruption. Earthquakes continued after the eruption and peaked on 23 December, but they did not lead to a further eruption.

TAAL (continued)

HISTORICAL ACTIVITY (continued)

1971-77: In October 1971, tension cracks developed atop Mt. Tabaro, a few hundred meters from the active vent. At the same time a hot spring developed on the flank of Mt. Tabaro. Ground temperatures on Mt. Tabaro, monitored since the development of these cracks in 1971, increased slowly from October 1971 until January or February 1973, at which time they began to rise more rapidly. The increase in temperatures lasted only a few days or weeks, and no eruption followed immediately. Ground temperatures then increased by 73 °C in the 15 days before phreatic eruptions on 3-24 September 1976, and by 28 °C in the 46 days before a small phreatic eruption on 3-4 October 1977. New ground fractures were observed on Mt. Tabaro 4 days before the 1976 eruption; on the day before the eruption, fissures grew from 5 cm wide and 50 m long to 10 cm wide and 100 m long. There was a great increase in fumarolic activity and a slight increase in seismicity before the 1976 eruption, but only a slight increase in steaming before the 1977 eruption (C. Andal, written commun., 1984). In the active periods of 1976 and 1977, the Main Crater Lake showed an increase in pH from 2.3 to 2.6, an increase in dissolved SiO_2 from 300 to 450 ppm, an increase in the concentrations of SO_4 and Cl, and a decrease in the ratio of Na/K (Tabirara and others, 1983).

1986-87: A 1.5 km long, 30 cm wide, N65°E-trending fissure formed in Agoncillo town (southwest shore of Taal Lake) in August 1986 (R.S. Punongbayan, written commun., 1988) (fig. 7.2.5). Then, from 30 October through early November 1987, five earthquakes were felt by residents of Agoncillo with MMI_{max}=IV. Residents felt a strong vertical component, and heard earthquake sounds. The earliest earthquakes of this swarm were classed as tectonic; later earthquakes of the series were classed as local tectonic (LT) or volcanic (S) earthquakes (R.S. Punongbayan, written commun., 1988) (fig. 7.2.9). Instrumentally determined hypocenters for events on 6 November were at shallow depths beneath Manalao Point and along a line from Manalao Point to the 1965-77 eruption site on Volcano Island (figs. 7.2.5, 7.2.6). Lake-level monitoring stations around Taal Lake suggest approximately 10 cm of local subsidence near the earthquakes, along the west and southwest shores of the caldera lake (Smithsonian Institution, 1987; R.S. Punongbayan, written commun., 1988).

REFERENCES

Adams, G.I., 1910, Geological reconnaissance of southwestern Luzon: Philippine Jour. Sci., sect. A., v. 5, p. 57-116.
Alcaraz, A., 1965, Taal Volcano an enigma?: Proc. Natl. Sci. Technol. Week, Manila. Mineral Eng. Mag., v. 17, no. 4, p. 22-30.
----- 1969, Is Taal Volcano drawing closer to another eruptive activity?: COMVOL Lett., v. 3, no. 5, p. 1-10.
Alcaraz, A., and Tan Cardoso, M.C., 1969, Heat considerations in the 1965 Taal Volcano eruption: COMVOL Lett., v. 3, no. 4, p. 1-4.
Andal, G.A., and Aguila, L.G., 1977, Prognostic criteria of the 1976 eruption of Taal Volcano: COMVOL Lett., v. 9, nos. 1-2, p. 1-6.
Andal, G., and Yambao, B., 1953, The present activity of Taal volcano: Unpubl. rep., Commission on Volcanology, 5 p.
Centeno y Garcia, Jose, 1885a, Estudio geológico del volcán de Taal: Madrid, Tello, 53 p. (also in Bol. Comisión de la Mapa Geol. de España, v. 12, p. 169-208).
----- 1885b, Catalogo de las rocas del volcán de Taal y de los montes proximos a la laguna de Bombon: Bol. Comisión de la Mapa Geol. de España, v. 12, p. 219-221.

TAAL (continued)

REFERENCES (continued)

Commission on Volcanology, 1981, Catalogue of Philippine volcanoes and solfataric areas (3rd ed.): Quezon City, Commission on Volcanology, 132 p.

----- ca. 1982, Operation Taal, Unpubl. rep., 22 p.

Dannung, E., Crisologo, E., Garcia, D., and de Guzman, A., 1982, The new eruption site of Taal volcano: its genesis and development: Proc. First Seminar Workshop on Philippine Volcanoes and Volcanic Terranes, December 13-14, 1982, p. 18-36.

Faustino, L.A., 1933, Taal volcano and its eruptions: Proc. 5th Pacific Sci. Congr., v. 3, p. 2377-2378.

Gilchrist, C.A., 1911, The recent volcanic eruption in the Philippines: Bull. Geogr. Soc. Philadelphia, v. 9, p. 1-8.

Le Gentil de la Galaisière, Guillaume Joseph Hyacinthe Jean Baptiste, 1779-81, Voyage dans les mers de l'inde, fait par ordre du roi, à l'occasion de passage de Vénus sur le disque de Soleil, le 6 Juin, 1761, et le 3 du même mois 1769: Paris, Academie Royale des Sciences, Imprimerie Royale, v. 2, 366 p.

Lim, J.B.R., 1983, Correlating the orientation of tectonic stress with the flank eruptions of Taal Volcano: Philippine Jour. Volcanol., v. 1, no. 1, p. 41-65.

Macdonald, G.A., 1954, Philippine volcanoes during 1953 and early 1954: Volcano Lett., no. 523, p. 1-5.

Minoza, W.A., 1965, Seismic aspects of the Taal eruption of September 28, 1965: Proc. Natl. Sci. Technol. Week, Manila, Mineral Eng. Mag., v. 17, no. 4, p. 22-30.

Moore, J.G., Nakamura, K., and Alcaraz, A., 1966, The 1965 eruption of Taal Volcano: Science, v. 151, p. 955-960.

Neumann van Padang, M., 1953, Catalogue of the active volcanoes of the world including solfatara fields, Pt. II, Philippine Islands and Cochin China: Naples, IAVCEI, 49 p.

Perrey, Alexis, 1860, Documents sur les tremblements de terre et les phénomènes volcaniques dans l'archipel de Philippines: Memoires de l'Academie de Dijon, v. 7, 110 p.

Pratt, W.E., 1911, The eruption of Taal Volcano, January 30, 1911: Philippine Jour. Sci., v. 6A, p. 63-83.

Pye, G., 1756, Extract of a letter, dated Manila, October 1, 1754, upon an earthquake: Phil. Trans. Roy. Soc. London, v. 49, p. 458-459.

Ruelo, H.B., 1983, Morphological and crater development of Mt. Tabaro eruption site, Taal Volcano, Philippines: Philippine Jour. Volcanol., v. 1, no. 2, p. 19-68.

Saderra Maso, M., 1895, La seismología en Filipinas. Datos para el estudio de terremotos del archipiélago Filipino: Manila, Ramirez y Compania, 122 p. (Observatorio de Manila).

----- 1911, The eruption of Taal volcano Jan. 30, 1911: Weather Bureau Rep., Manila, 45 p.

San Agustin, Gaspar de, 1698, Conquistas de las islas Philipinas: la temporal, por las armas del Senor Don Phelipe segundo el prudente; y la espiritual, por los religiosos del orden de nuestro padre San Augustin: fundacion, y progressos de su provincia del santissimo nombre de Jesus, parte primera: Madrid: Impr. Manuel Ruiz de Murga, 544 p.

Santos, T.M., and Listanco, E., 1980, Petrographic variations in the lavas of the 1968 eruption of Taal volcano: Nat. Appl. Sci. Bull., Univ. Philippines, v. 32, p. 1-4.

TAAL (continued)

REFERENCES (continued)

Semper, K., 1869, Die Philippinen und ihre Bewohner; sechs Skizzen: Wurzburg, Stuber'sche Buchhandlung, 143 p.

Smithsonian Institution, Scientific Event Alert Network (SEAN), 1987, Taal Volcano: SEAN Bull., v. 12, no. 10-12.

Tabirara, E., Campita, N.R., Yabut, R.P., and Aquino, A.L., 1983, Geochemical parameter for predicting Taal volcano eruption: Unpubl. PHIVOLC rep., 34 p.

Tolentino, B.S., 1973, Taal Volcano, Ann. Rep. for 1972-73: Quezon City, Commission on Volcanology, p. 288-289.

Volcanological Society of Japan, 1966-70, Taal: Bull. Volcanic Eruptions, no. 6, p. 3-4; no. 7, p. 3; no. 8, p. 7-8; no. 10, p. 8-9.

Wolfe, J.A. and Self, S., 1982, Structural lineaments and Neogene volcanism in southwestern Luzon, in Hayes, D.E., ed., The tectonic and geologic evolution of Southeast Asian seas and islands, Pt. 2: Am. Geophys. Union Geophys. Monograph 27, p. 157-172.

Worcester, D.C., 1912, Taal Volcano and its recent destructive eruption: Natl. Geogr., v. 23, p. 313-367.

Zuñiga, Joaquín Martínez de, O.S.A., 1803, Historia de las Islas Filipinas, y descripción geografica de algunos provincias. Sampaloc: Fr. Pedro Arguelles de la Concepcion Religioso Francisco. 687 p. (English translation by J. Maver, 1814, Historical view of the Philippine Islands: London, J. Asperne, 2 v., 687 p.

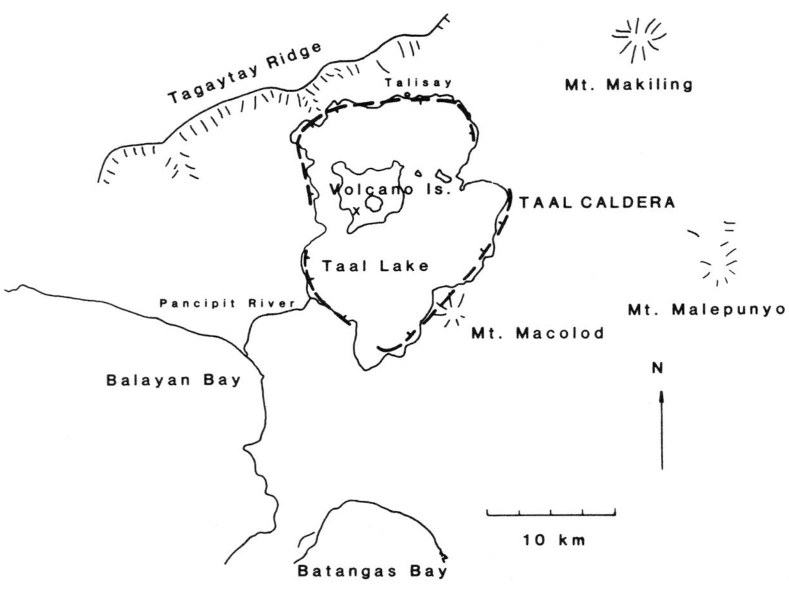

Figure 7.2.1. Sketch map showing outline of Taal Caldera.

LEGEND:

– 90 – LAKE BATHYMETRY IN FATHOMS

------- ZONE OF MONOGENETIC FLANK OF ERUPTIONS

ERUPTION CRATER/CONE

Figure 7.2.2. Partial bathymetry of Taal Lake, from Lim (1983). The two basins of Taal Lake may represent a single caldera with a horst or constructional ridge separating the basins, or they may represent two overlapping calderas with a constructional ridge on their common border.

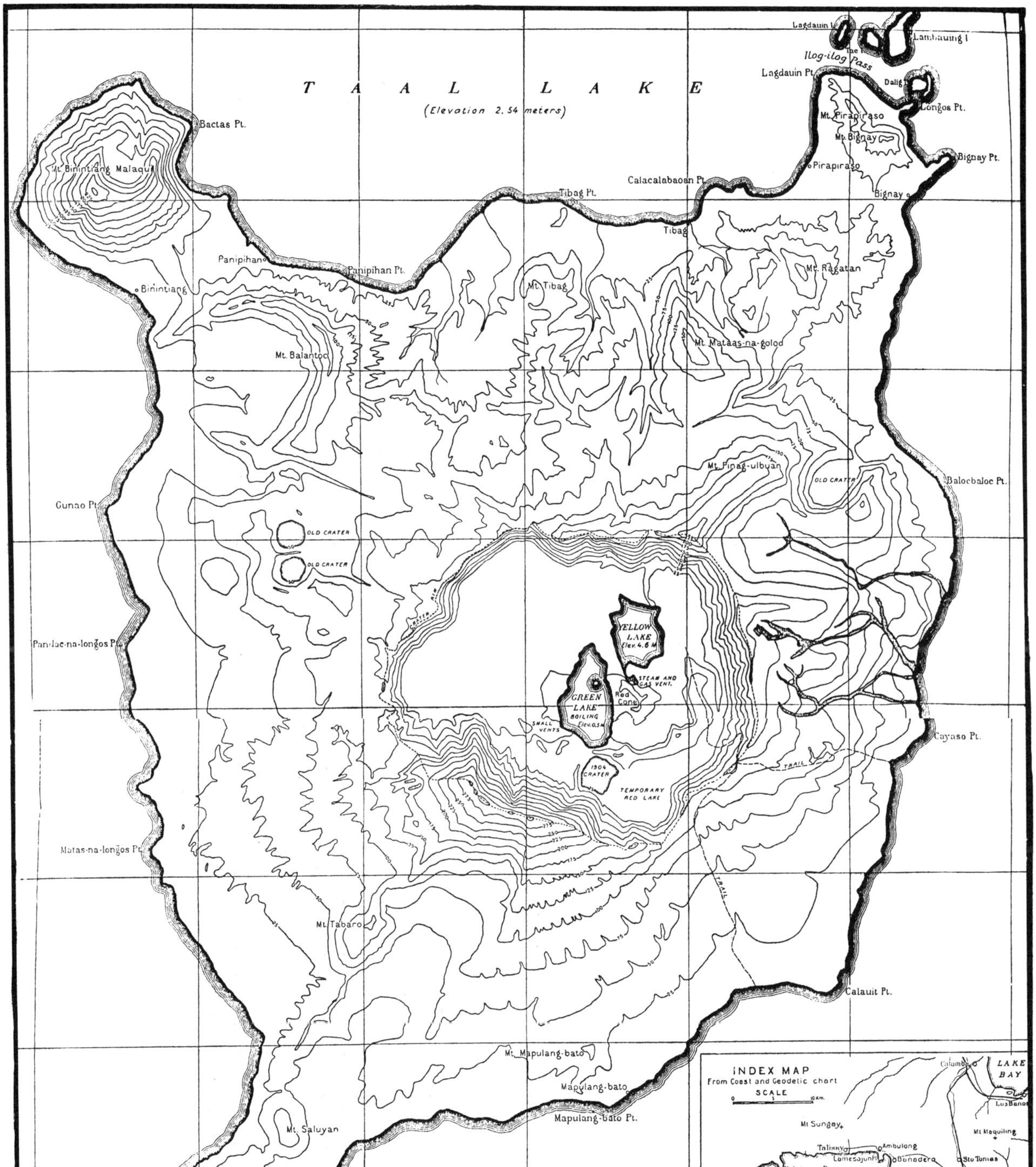

Figure 7.2.3. Taal Lake and Taal Island in 1909, before great eruption of 1911, from Adams (1910).

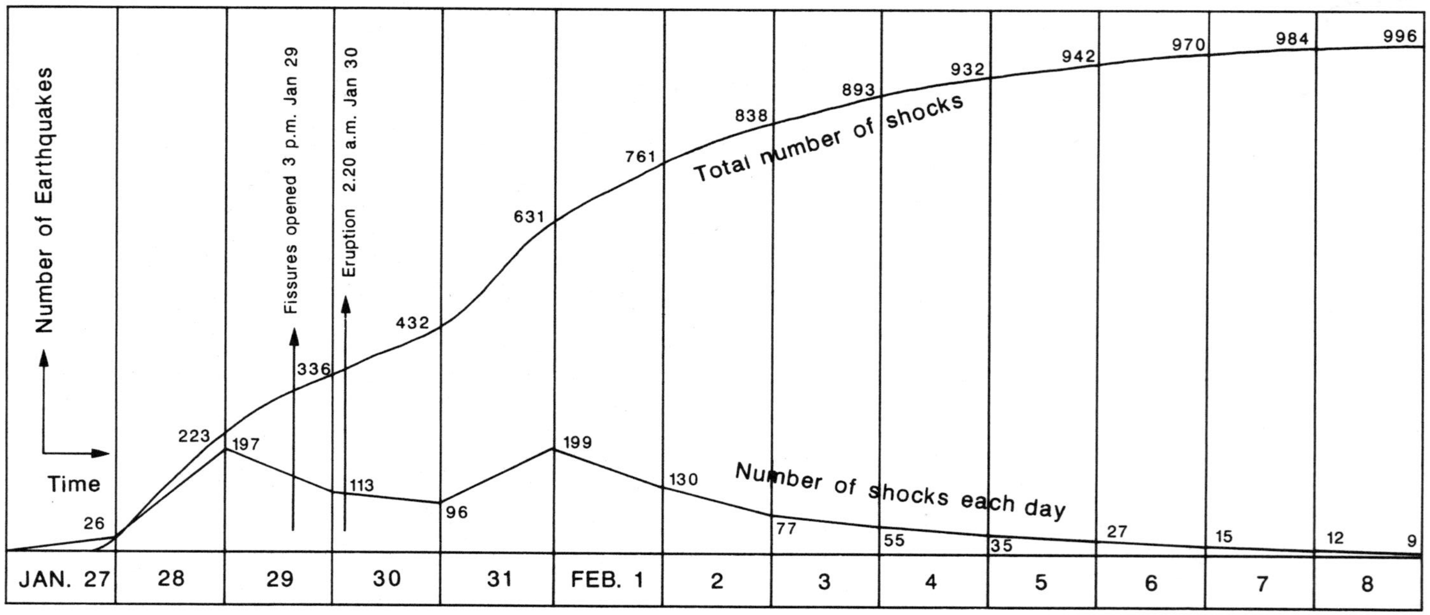

Figure 7.2.4. Cumulative and daily number of earthquakes preceding the great Taal eruption in January 1911, from Pratt (1911), after Saderra Maso (1911).

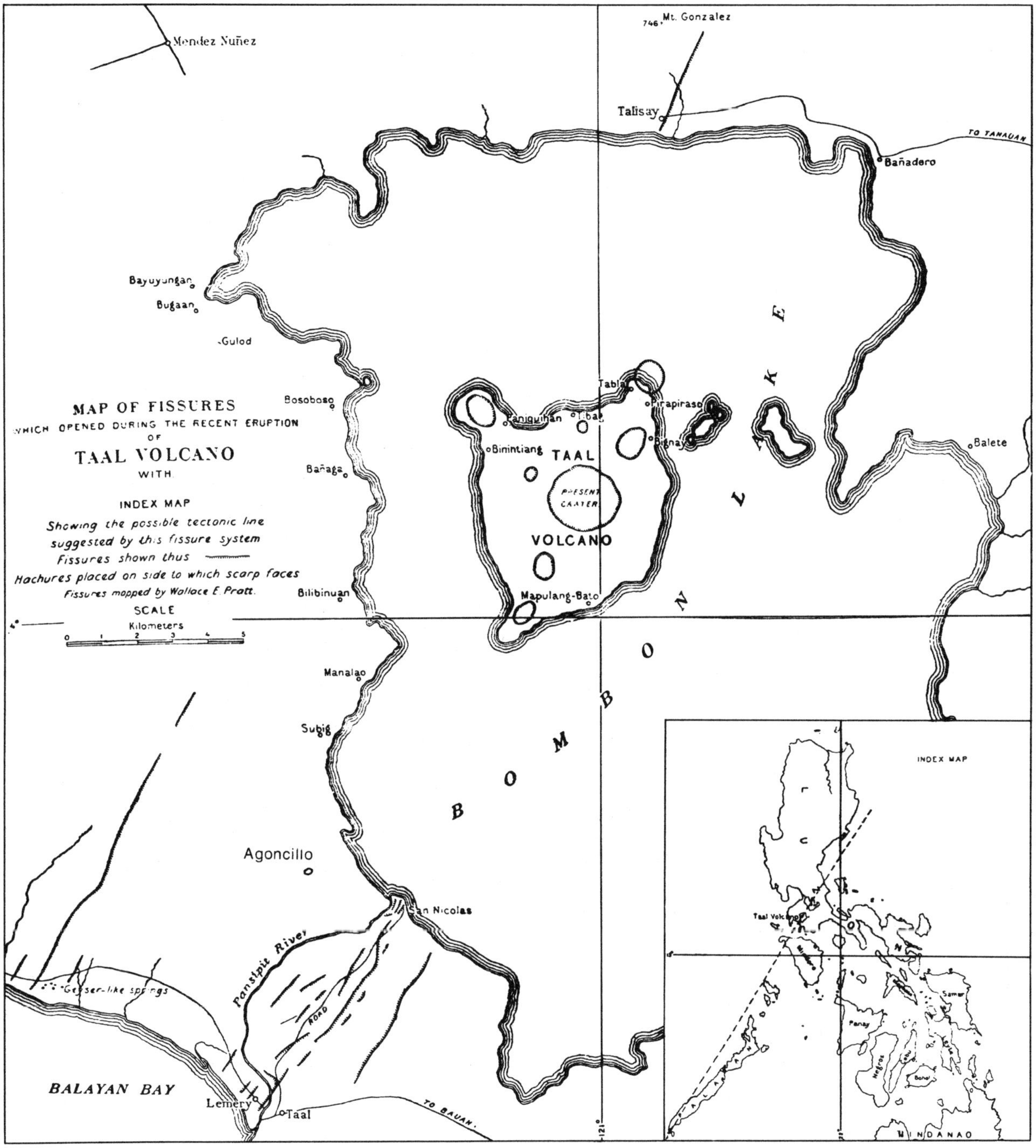

Figure 7.2.5. Locations of fissures that opened during the 1911 eruption of Taal Volcano, from Pratt (1911).

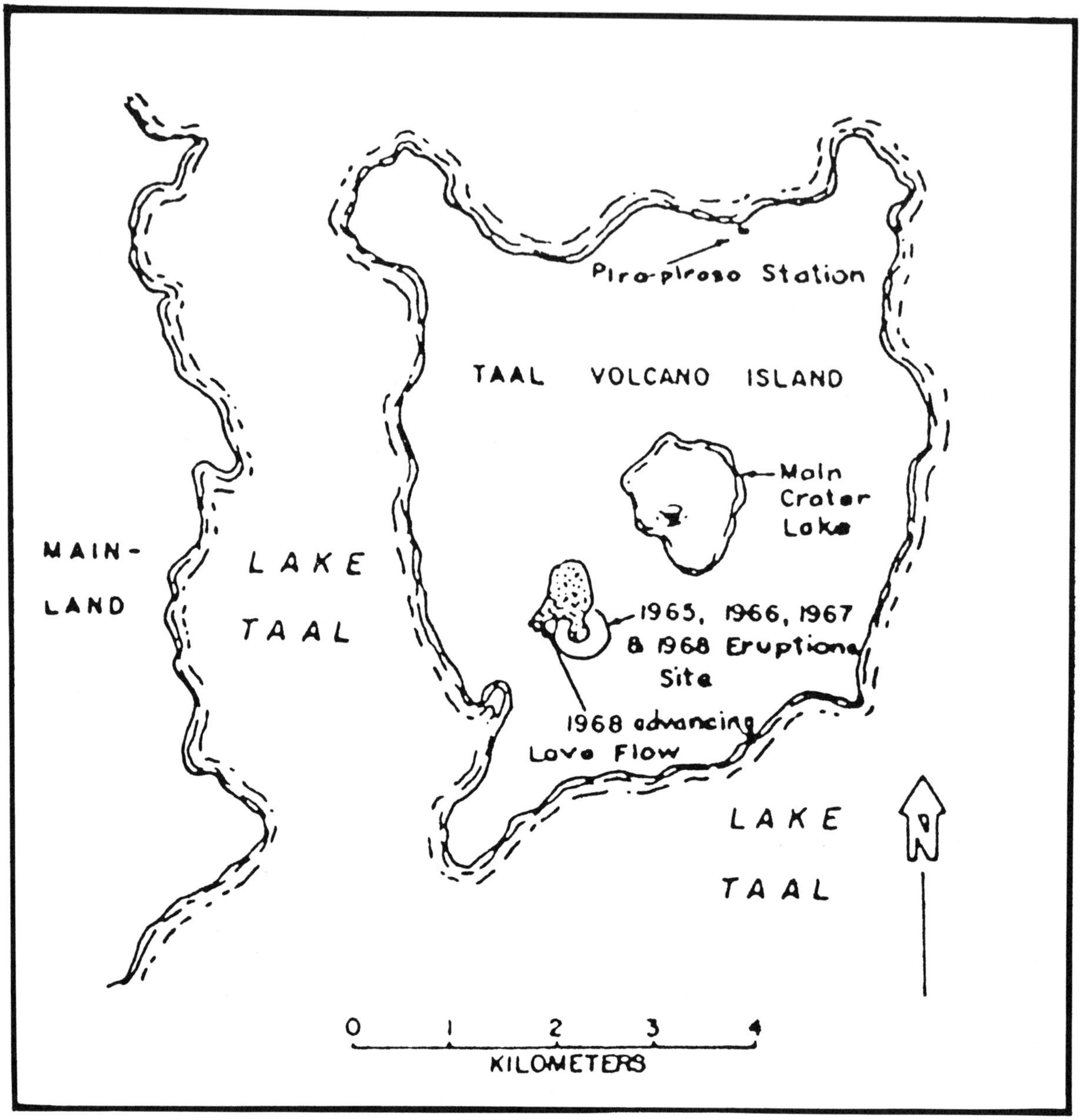

Figure 7.2.6. Volcano Island (Philippine Commission on Volcanology, written commun., 1975) showing Main Crater Lake in its center and the 1965, 1966, 1967, and 1968 eruption sites on its southwest flank. Stippled portion represents advancing lava flow as of 29 February 1968.

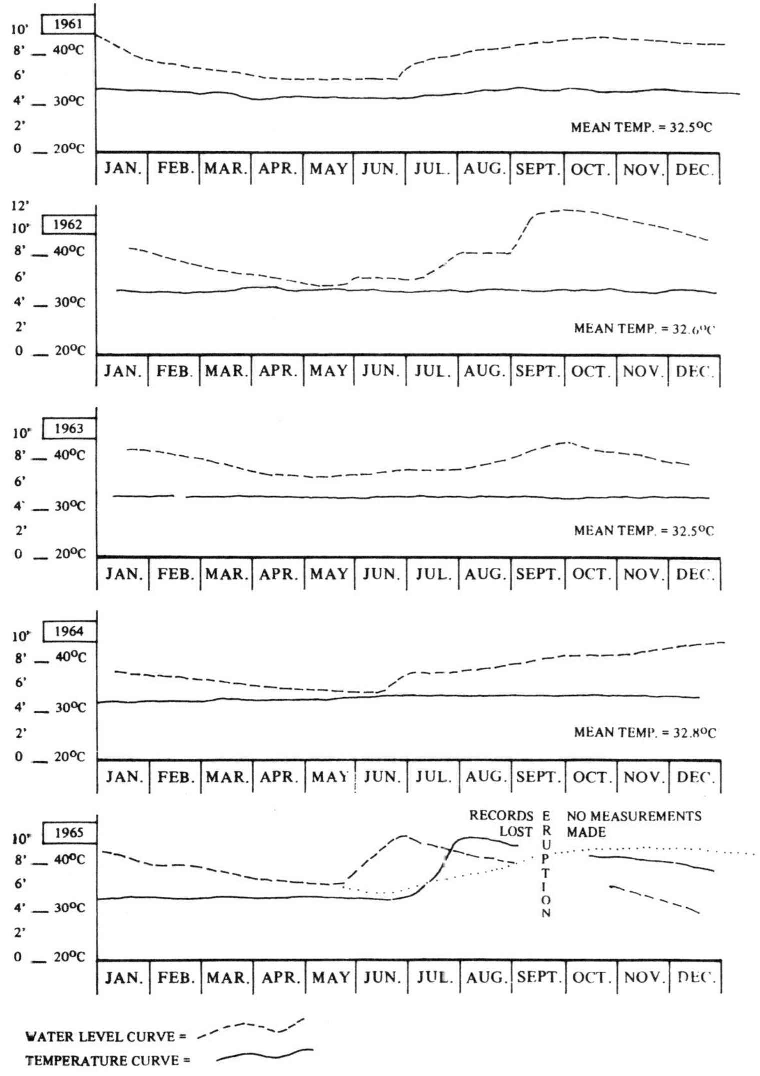

Figure 7.2.7. Temperature and water level at Taal Lake during 1961-65, from Dannung and others (1982).

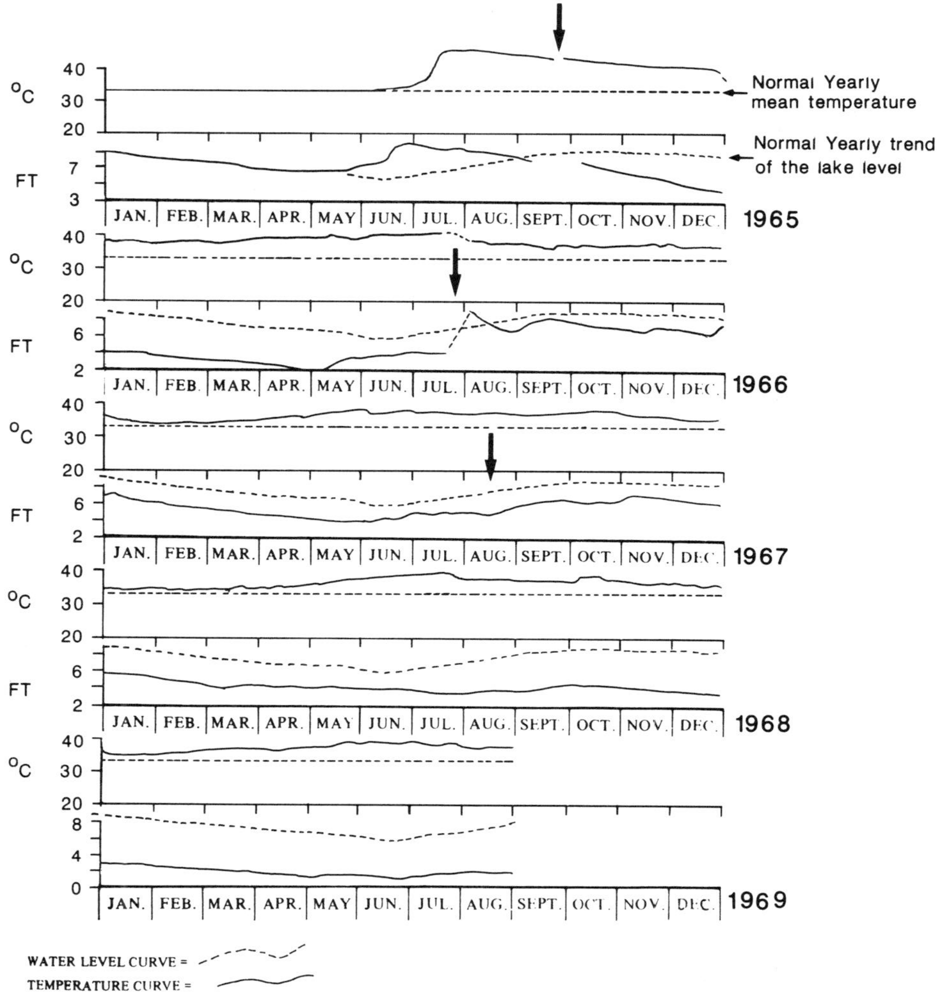

Figure 7.2.8. Temperature and water level at Taal Lake during 1965-69, after Alcaraz and Tan Cardoso (1969) and Alcaraz (1969). Arrows indicate eruptions in September 1965, July 1966, and August 1967 (mild).

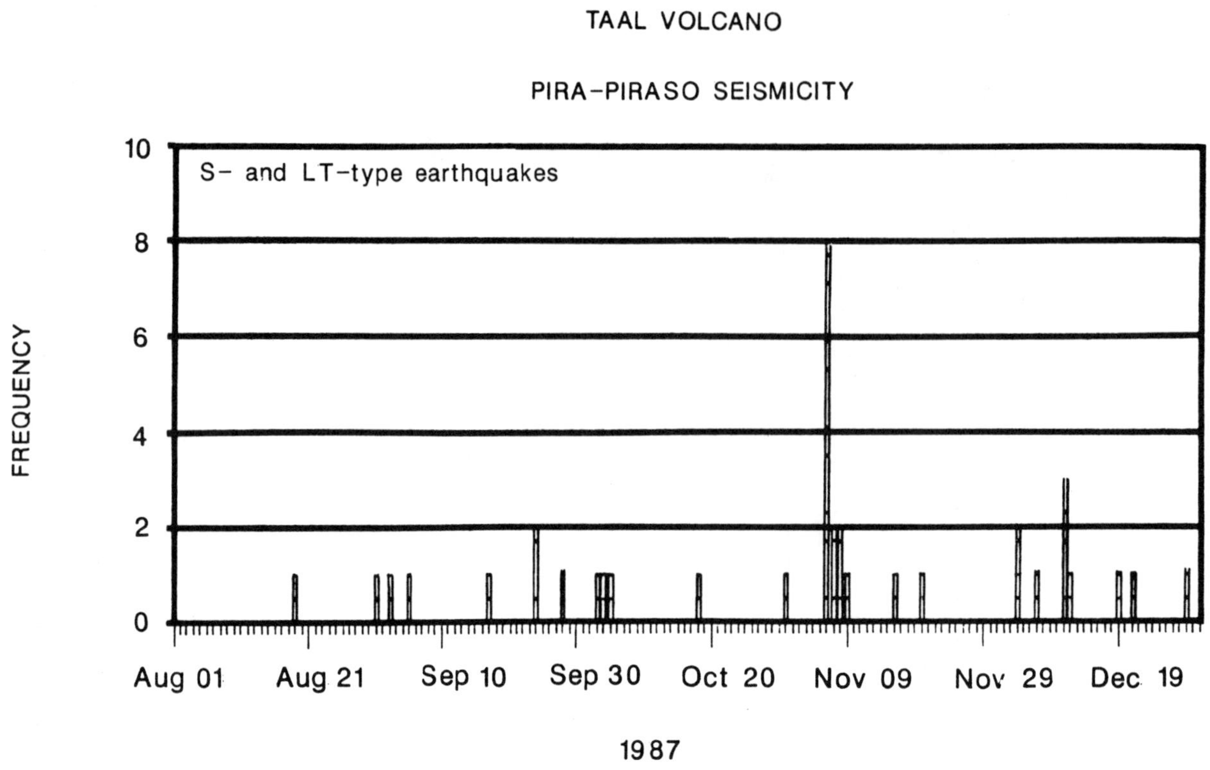

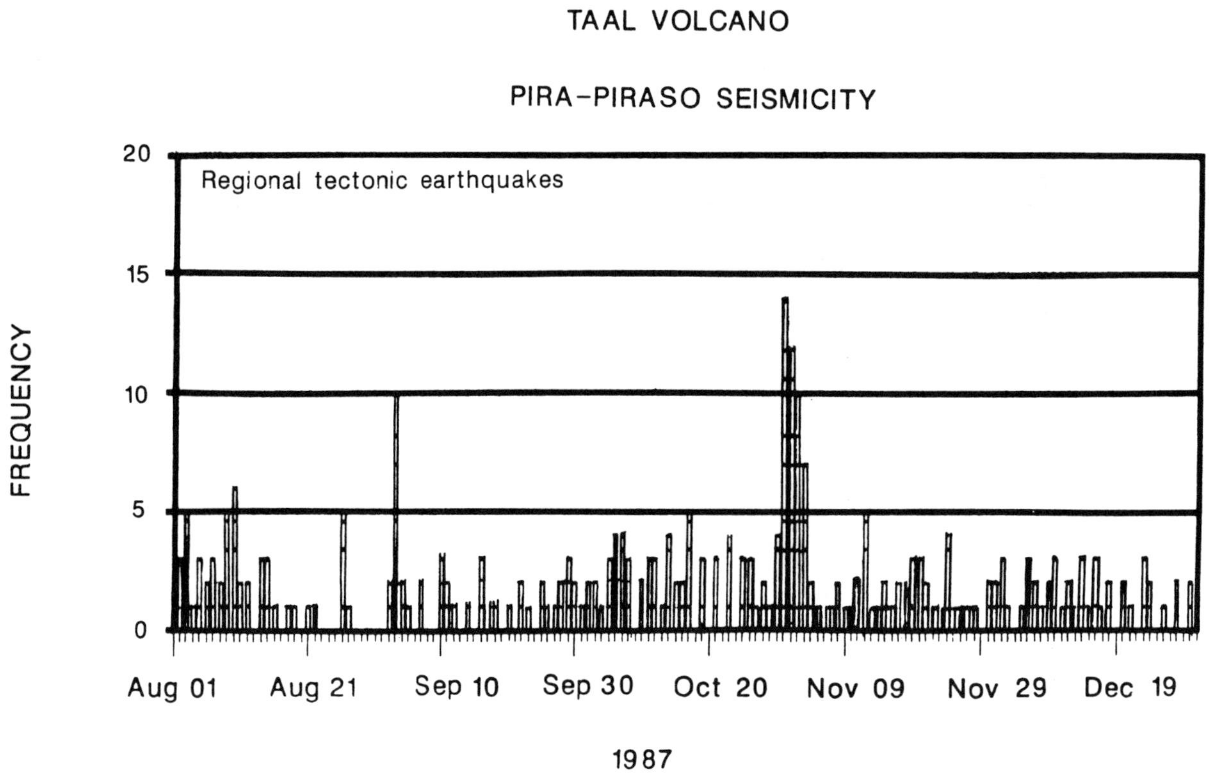

Figure 7.2.9. Daily number of earthquakes recorded at Taal Volcano (station Pira-Piraso) from August through December 1987 (R.S. Punongbayan, written commun., 1988). Top, impulsive volcanic earthquakes ("S" type, dominant frequency 3-4 Hz) and local tectonic earthquakes ("LT" type); bottom, regional tectonic earthquakes.

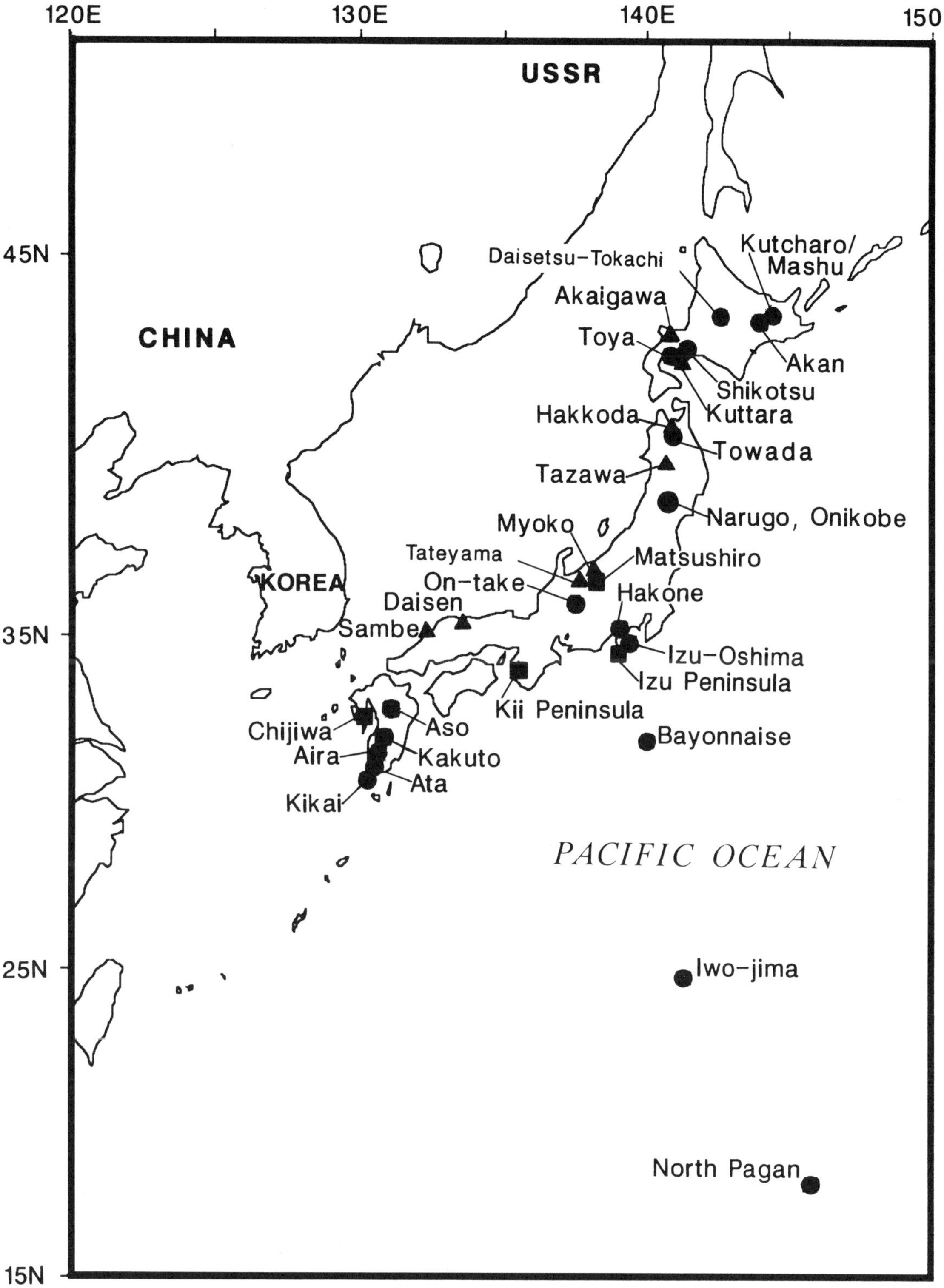

Figure 8. Locations of large, Quaternary restless calderas (solid circles), nonrestless calderas (triangles), and restless noncaldera areas (squares) in Region 8.

KIKAI

CAVW number (active vent)	Latitude Longitude (degrees)	Diameter (km)	Local tectonic setting	Precaldera edifice	SiO_2 content (wt pct)	Age of collapse (yr B.P.)	Date of unrest	Type and duration of unrest ESTU STHF MGTF H Te	Eruption type
08-02-06 (Tokara-Iwo Jima =Satsuma-Iwo-Jima; =Shin-Iwo-Jima)	30.78N 130.28E	19	Exten? (graben)	LP-strat?	R = 65-72 C = 71-72	6,300	1934-36 1988	CC-- Z--C --ZZ - -- ---- ---- ---Z - --	ex,lf ex

TECTONIC SETTING

Kikai Caldera lies at the southern end of a volcano-tectonic graben 20 to 30 km wide that trends NNE-SSW across the southernmost part of Kyushu, Japan. The graben contains several other calderas, including Ata, Aira, Kakuto, and Aso, which, like Kikai, formed by voluminous eruptions of dacitic to rhyolitic magmas (Aramaki, 1984).

GEOLOGIC HISTORY

The Kikai Caldera is mostly submarine, with three islands and a few reefs representing part of the caldera rim and central cones (figs. 8.1.1, 8.1.2). A double collapse structure for the caldera is inferred from submarine topography and a seismic profiling survey (K. Ono, written commun., 1983). Three cycles of voluminous silicic eruptions are known. The latest caldera-forming eruption, about 6,300 yr B.P., is the largest known Holocene eruption on earth (bulk volume of products=150 km^3). Pyroclastic flows from this eruption traveled at least 100 km across the sea to southern Kyushu (Ui, 1973); airfall ash from the same eruption blanketed most of Japan (Machida and Arai, 1983). Postcaldera products are bimodal in composition, including many silicic vents and one basaltic vent.

HISTORICAL ACTIVITY

1934-36: Two weak earthquakes were felt at 1500 hr on 12 September 1934, followed by a strong earthquake at 2320 hr and many more earthquakes in the ensuing days. Seismicity climaxed on 17 September (both in frequency and intensity), accompanied by rumbling sounds. Thereafter, shocks diminished in frequency. Also on 17 September, fishermen noted gas ebullitions above a site 2 km east of Satsuma-Iwo-Jima Island, where the sea was about 300 m deep. A huge column of "smoke" and abundant pumice blocks first appeared on

KIKAI (continued)

HISTORICAL ACTIVITY (continued)

20 September (Tanakadate, 1936a). After continuous eruption of ash and pumice through October and November, a new island emerged in early December. The early pyroclastic cone was eventually destroyed, and the present island, Shin-Iwo-Jima, was capped by a rhyolite lava flow in 1935.

After the eruption of Shin-Iwo-Jima, the sea bottom around the vent subsided. The subsided part, containing Iwo-Jima and Takezima, has a diameter of about 15 km if it is roughly circular. In March 1936, the subsidence was about 90 cm on the coast of Takezima and on Iwo-Jima part of the shore had subsided about 70 cm. Frequent earthquakes during 13-19 October 1936 preceded increased fumarolic activity or an unconfirmed eruption on 19 October 1936. A large earthquake occurred on 26 October 1936, followed or accompanied by an additional 33 cm of subsidence of Iwo-Jima. Strong rumbling occurred on 8 November 1936, followed by numerous shocks until the middle of November; steaming from the dome of Iwo-dake increased, and the temperature of its crater floor increased considerably. "After these convulsions, the shore near the hamlet of the island sank again by about 30 cm" (Tanakadate, 1939).

Vigorous solfataric activity, including fumaroles with temperatures up to 800 °C, occurs at present at Iwo-dake on the large island of Satsuma-Iwo-dake. Discharge of SO_2 and HF is reportedly so high that these gases cannot be derived solely from seawater; they must be resupplied by magma at least once every 1,000 years (Yoshida and Ozawa, 1981).

1988: Gray plumes rose 400-500 m above the Iwo-dake summit crater on 18 January, without known precursors. Temperature of a fumarole on 26-28 January was 788 °C, about the same as that measured in 1985 (Smithsonian Institution, 1988).

COMMENTS

Tanakadate (1936b) wrote that on 20 September, "as a cyclonic center was passing the scene a huge column of smoke rose on the sea surface, accompanied by an issuing of abundant pumiceous lava blocks." Typhoons are typically preceded or accompanied by storm waves of up to several meters height, and the temporary compression and decompression associated with a storm wave may have triggered this pulse of the eruption. The eruption and passage of a typhoon could be fortuitous, but we note that earth tides, which exert a slower, lesser disturbance, are thought to have triggered eruptions elsewhere. Records of precursory activity indicate that the volcano was already primed to erupt, if not already erupting beneath the sea, before the typhoon passed, so a small additional stimulus could reasonably be considered the trigger of this event.

REFERENCES

Aramaki, S., 1984, Formation of the Aira Caldera, southern Kyushu, about 22,000 years ago: Jour. Geophys. Res., v. 89, no. B10, p. 8485-8501.

Ehara, S., Yuhara, K., and Shigematsu, A., 1980, Heat flow measurements in the submarine calderas, southern Kyushu, Japan: Bull. Volcanol. Soc. Japan, v. 25, p. 51-61.

KIKAI (continued)

REFERENCES (continued)

Kamada, M., 1979, A birth of new island - Satsuma-Iwojima, a short comment on the newspaper articles: Seismicity and Volcanic Activity of Kagoshima Prefecture, v. 11, p. 102-117.

Kuno, H., 1962, Catalogue of the active volcanoes of the world including solfatara fields, Pt. XI, Japan, Taiwan and Marianas: Rome, IAVCEI, 332 p.

Machida, H., 1981, Tephrochronology and Quaternary studies in Japan, in Self, S., and Sparks, R.S.J., eds., Tephra studies: Dordrecht, D. Reidel, p. 161-191.

Machida, H., and Arai, F., 1983, Extensive ash falls in and around the Sea of Japan from large late Quaternary eruptions: Jour. Volcanol. Geotherm. Res., v. 18, p. 151-164.

Matsubaya, O., Ueda, A., Kusakabe, M., Matsuhisa, Y., Sakai, H., and Sasaki, A., 1975, An isotopic study of the volcanoes and the hot springs in Satsuma Iwojima and some areas in Kyushu: Bull. Geol. Surv. Japan, v. 26, p. 375-392.

Matsuo, S., Suzuoki, T., Kusakabe, M., Wada, H., and Suzuki, M., 1974, Isotopic and chemical compositions of volcanic gases from Satsuma-Iwojima, Japan: Geochem. Jour., v. 8, p. 165-173.

Matumoto, T., 1937, Submarine eruptions on the Tokara-Iwojima (off of Iwo-ga-shima) and the geology of the island and its surroundings: Bull. Volcanol. Soc. Japan, ser. 1, v. 3, no. 2, p. 144-162.

----- 1943, The four gigantic caldera volcanoes in Kyushu, Japan: Japanese Jour. Geol. Geogr., v. 19, p. 1-57.

Ono, K., Soya, T., and Hosono, T., 1982, Geology of the Satsuma-Io-jima district: Quadrangle series, scale 1:50,000, with text: Tsukuba, Geol. Surv. Japan, 80 p.

Sakai, H., Ueda, A., Casadevall, T.J., and Moore, J.G., 1981, The isotopic ratios and chemistry of sulfur at Kilauea Volcano, Hawaii, and Satsuma Iwojima, southern Kyushu: Abstracts, IAVCEI Tokyo Mtg., p. 312-313.

Smithsonian Institution, Scientific Event Alert Network (SEAN), 1988, Kikai (Satsuma Iwo-Jima) Volcano: SEAN Bull., v. 13, no. 1, p. 6 (information from Volcanological Center, Japan Meteorological Agency).

Tanakadate, H., 1935, A volcanological and petrographical note of the new volcanic islet near Io-zima: Proc. Imperial Acad., v. 11, p. 370-372.

----- 1935, Evolution of a new volcanic islet near Io-zima: Proc. Imperial Acad., v. 11, p. 152-154.

----- 1936a, Preliminary report on the eruption of Iwo-jima, Kagoshima Prefecture: Bull. Volcanol. Soc. Japan, ser. 1, v. 2, p. 188-209.

----- 1936b, Volcanic activity in Japan and vicinity during the period between July 1934 and October 1935: Japanese Jour. Astron. Geophys., v. 13, p. 121-139.

----- 1939, Volcanic activity in Japan and vicinity during the period between November 1935 and December 1938: Japanese Jour. Astron. Geophys., v. 16, p. 95-120.

Ui, T., 1973, Exceptionally far-reaching, thin pyroclastic flow in southern Kyushu: Bull. Volcanol. Soc. Japan, v. 18, p. 153-168.

Williams, H., 1941, Calderas and their origin: Univ. Calif. Publ. Geol. Sci., v. 25, p. 239-346.

Yoshida, M., and Ozawa, T., 1981, Amounts of some chemical components discharged from Satsuma-Iwo-jima volcano in relation to their feeding sources: Bull. Volcanol. Soc. Japan, v. 26, no. 1, p. 25-34.

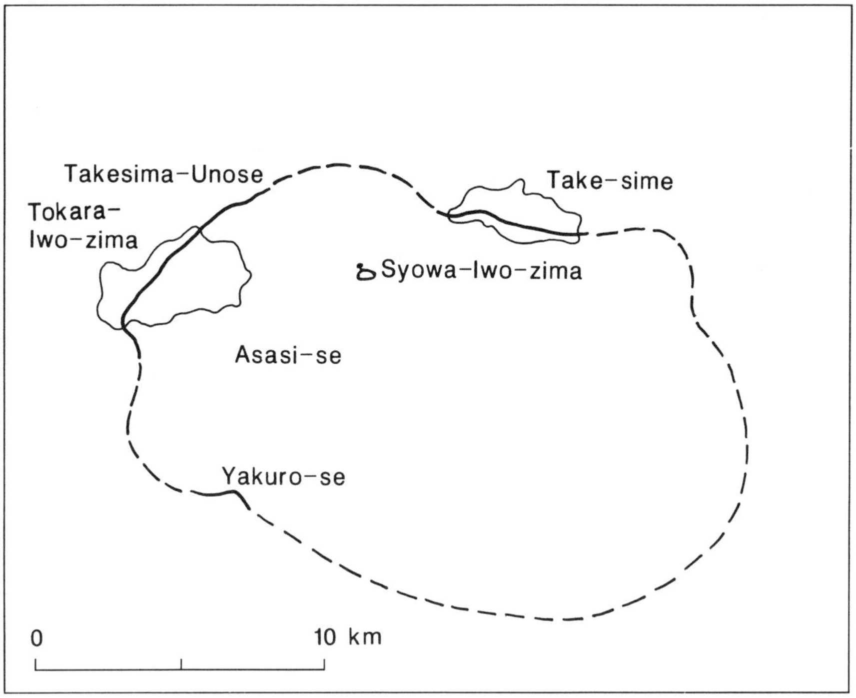

Figure 8.1.1. Kikai Caldera, including Tokara-Iwo-zima and Take-sima, from Kuno (1962), after Matumoto (1937, 1943). Outline of caldera is dashed where inferred. Syowa-Iwo-zima = Shin-Iwo-Jima.

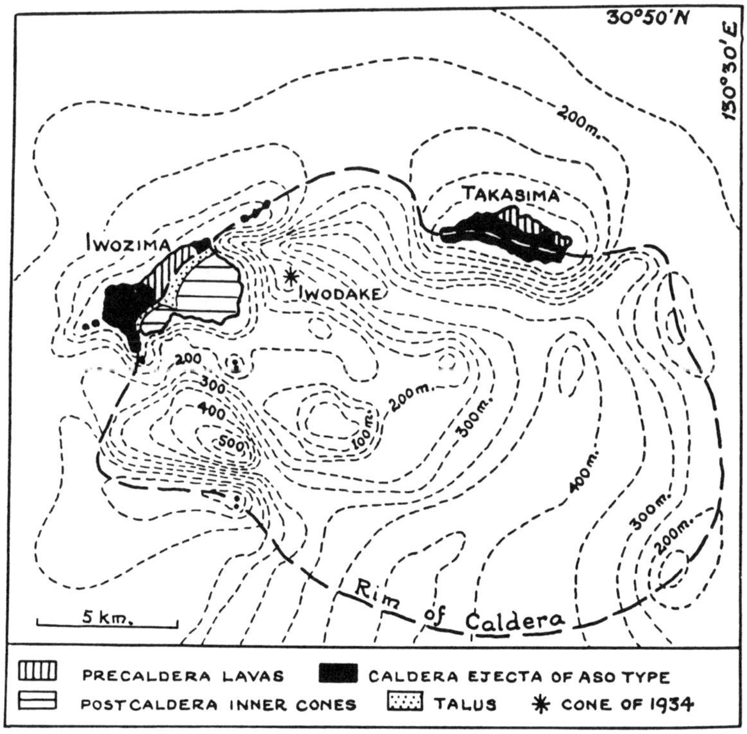

Figure 8.1.2. Geology and bathymetry of Kikai Caldera, from Williams (1941, p. 282), after Matumoto (1943).

ATA

CAVW number (active vent)	Latitude Longitude (degrees)	Diameter (km)	Local tectonic setting	Precaldera edifice	SiO_2 content (wt pct)	Age of collapse (yr B.P.)	Date of unrest	Type and duration of unrest ESTU STHF MGTF H Te	Eruption type
08-02-07	31.20N	15 x 25	Exten?	None?	R = 51-57	>40,000	A.D. 91	?--- ---- ---- - --	ex?
(Kaimon-dake;	130.55E				C = 68-73	(Ata);	860	---- ---- ---- - --	ex
Ibusuki						ca. 5,000	866	---- ---- ---- - --	ex
Volcanic Field)						(Ikeda)	874	---- ---- ---- - --	ex
08-02-071							882	---- ---- ---- - --	ex
(Ikeda Caldera)							885	---- ---- ---- - --	ex
							1615?	---- ---- ---- - --	ex?
							1967-68	CC-- ---- ---? ? --	none
							1970	EE-- ---- ---- - --	none
							1981?	BB-- ---- ---- - --	none

TECTONIC SETTING/GEOLOGIC HISTORY

The Ata Caldera as defined by Matumoto (1943, 1963) lies on the southern tip of the Satsuma Peninsula and beneath the entrance to Kagoshima Bay (figs. 8.2.1, 8.2.2). On the basis of a gravity anomaly and geological considerations, Aramaki and Ui (1966) relocated the Ata Caldera in Kagoshima Bay just north of Matumoto's location (fig. 8.2.3); the revised location has been confirmed by subsequent marine seismic profiling and corresponds to a known bathymetric depression in the bay. One result of the relocation is that Kaimon-dake, the only active volcano in the vicinity, no longer lies on the western rim of the new Ata Caldera, but we include Kaimon-dake and the rest of the Ibusuki Volcanic Field in this compilation because activity throughout the field appears to be closely related to the Ata Caldera. Lying as these do in a large volcano-tectonic graben that also includes the Aira and Kakuto Calderas, activity in one area can affect activity in adjoining parts of the graben. Matumoto (1943, 1963) identified two small calderas, Ikeda and Yamakawa, that are nested within his Ata Caldera (fig. 8.2.2); these two are now considered part of the Ibusuki Volcanic Field adjoining the Ata Caldera (Kobayashi and Naruo, 1986).

HISTORICAL ACTIVITY

91 A.D.: Ishikawa (1981) cites traditional accounts of a large earthquake, thunder, a storm, and a large wave in the winter of this year. The accounts claim, perhaps figuratively, that "during the confusion Kaimon-dake rose up at the site of Ikeda Lake."

9th century A.D.: Eruptions in the 9th century A.D. were explosive, and one or more covered the environs with ash up to 30 cm thick (Kuno, 1962; Nakamura, 1967, 1980; Ishikawa, 1981).

ATA (continued)

HISTORICAL ACTIVITY (continued)

1967-68, 1970: Earthquake swarms occurred on 5-8 August 1967 and in March 1968. The earthquakes were recorded at Sakurajima; their locations are poorly constrained but are probably around the east coast of Ibusuki City, on the boundary of or within Matumoto's Ata Caldera (Nishi, 1968). The maximum magnitude was 3.9, and the cumulative seismic energy release (of those earthquakes of $\underline{M}$ 2.2 or higher) was 2 x 10^{18} ergs. Another earthquake swarm occurred in April 1970 (fig. 8.2.4), with epicenters immediately east of Lake Ikeda (Yoshikawa and others, 1972).

1981: A more sensitive, temporary seismic network was deployed from 30 July to 5 August 1981, and it detected one small swarm on 4 August (1 day only, $\underline{M}_{max}$=2.0). Epicenters were at the northern foot of Mt. Tsuji-no-dake, one of the central volcanic cones of the Ibusuki district, along a tectonic line that had been inferred from a line of fumaroles and altered ground (fig. 8.2.5). Hypocenters were about 5 km deep, and earthquake spectra were indistinguishable from those of tectonic earthquakes (Kakuta, 1984).

REFERENCES

Aramaki, S., Fukuyama, H., Kamo, K., and Kamada, M., 1981, Sakurajima Volcano, in Kubotera, A., ed., Field excursion guide to Sakurajima, Kirishima, and Aso Volcanoes: Tokyo, Volcanological Society of Japan, p. 1-17.

Aramaki, S., and Ui, T., 1966, The Aira and Ata pyroclastic flows and related calderas and depressions in southern Kyushu, Japan: Bull. Volcanol., v. 29, p. 29-47.

Gutenberg, R., and Richter, C.F., 1954, Seismicity of the Earth and related phenomena: Princeton, N.J., Princeton Univ. Press, 310 p.

Ishikawa, H., 1981, History of volcanic activity, study of ejecta and hazard appraisals of Kaimondake volcano, in Shimozuru, D., ed., Volcanic disaster characteristics and hazard appraisals: Special Rep. Grant-in-Aid for Natural Disaster Science, Ministry of Education, p. 180-184.

Kakuta, T., 1984, Local earthquake swarm occurring in south Kyushu: Jour. Phys. Earth, v. 32, p. 143-160.

Katsui, Y., 1971, ed., List of the world active volcanoes: Tokyo, Volcanol. Soc. Japan/IAVCEI/IUGG, 160 p.

Kobayashi, T., and Naruo, H., 1986, Phreatomagmatic deposits erupted from Ibusuki Volcanic Center, Japan: Abstracts, Int. Volcanol. Congr., New Zealand, February, 1986, p. 108.

Kuno, H., 1962, Catalogue of the active volcanoes of the world including solfatara fields, Pt. XI, Japan, Taiwan and Marianas: Rome, IAVCEI, 332 p.

Matumoto, T., 1943, The four gigantic caldera volcanoes in Kyushu, Japan: Japanese Jour. Geol. Geogr., v. 19, p. 1-57.

----- 1963, Caldera volcanoes and pyroclastic flows of Kyushu: Bull. Volcanol., v. 26, p. 401-413.

Nakamura, M., 1967, On the volcanic products and history of Kaimon-dake Volcano: Bull. Volcanol. Soc. Japan, v. 12, no. 3, p. 119-131.

----- 1980, Possibility of new volcanic activity at Ibusuki volcanic field, Kyushu, Japan; Bull. Volcanol. Soc. Japan, v. 25, p. 195-205.

ATA (continued)

REFERENCES (continued)

Nishi, K., 1968, On the earthquake swarm in the Ata caldera district on August, 1967; and a note on the relation between the linear dimension of the aftershock region and the difference in magnitude between a mainshock and the largest aftershock: Ann. Rep. Disaster Prev. Res. Inst., Kyoto Univ., v. 11, p. 97-109.

Suwa, A., 1980, The surveillance and prediction of volcanic activities in Japan: GeoJournal, v. 4, no. 2, p. 153-159.

Yoshikawa, K., Eto, T., and Nishi, K., 1972, Recent subsurface activity in and around Kaimon-dake: Seismicity and Volcanic Activity in Kagoshima Prefecture, v. 4, p. 21-28.

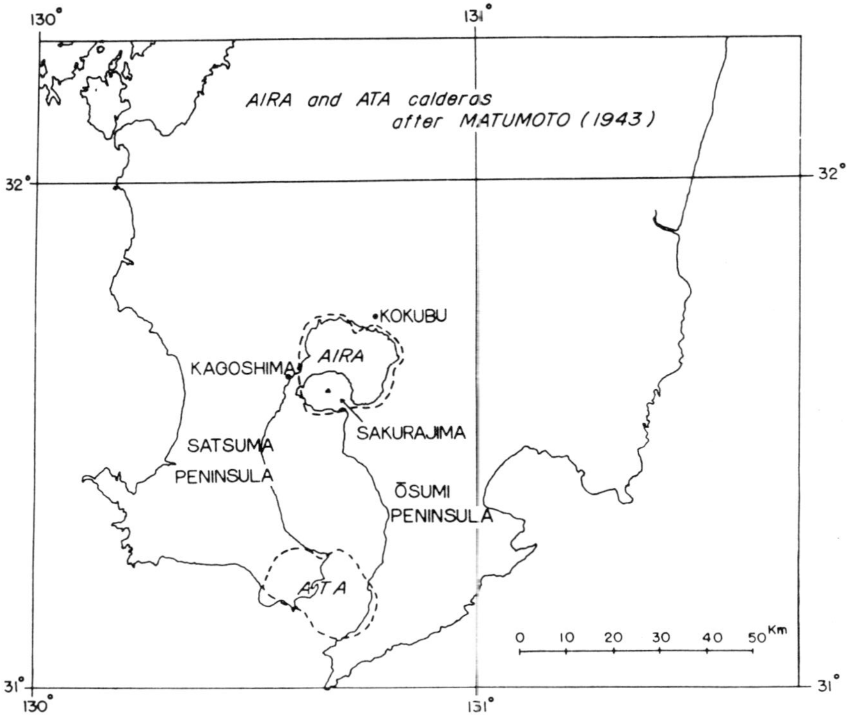

Figure 8.2.1. Locations of Ata and Aira Calderas (dashed lines) as proposed by Matumoto (1943, 1963), from Aramaki and Ui (1966). Revised location of Ata Caldera proposed by Aramaki and Ui (1966) is shown in figure 8.2.3.

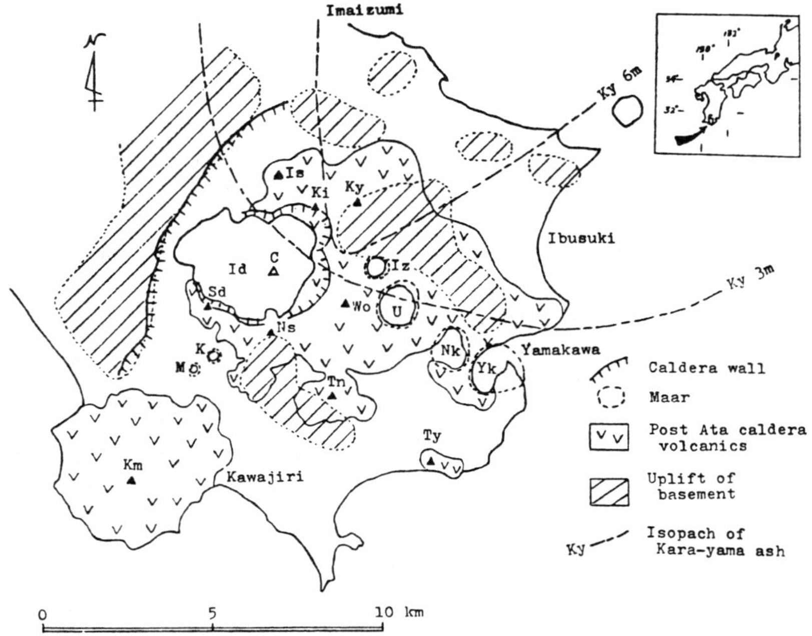

Figure 8.2.2. The Ibusuki Volcanic Field, from Nakamura (1980), showing small Ikeda Caldera nested within western rim of Ata Caldera (as proposed by Matumoto (1943, 1963)) and location of Kaimon-dake. C, crypto-volcanoes; Id, Lake Ikeda; Is, Ishi-mine; Iz, Ikezoko Maar; K, Pond Kagami; Ki, Kiyomi-dake; Km, Kaimon-dake; Ky, Kara-yama; M, Pond Mizunashi; Nk, Narikawa Maar; Ns, Nabeshima-dake; Sd, Senda-yama; Tn, Tsujino-dake; Ty, Take-yama; U, Pond Unagi; Wo, Washio-dake; Yk, Yamakawa Maar. The revised location of Ata Caldera proposed by Aramaki and Ui (1966) is shown in figure 8.2.3.

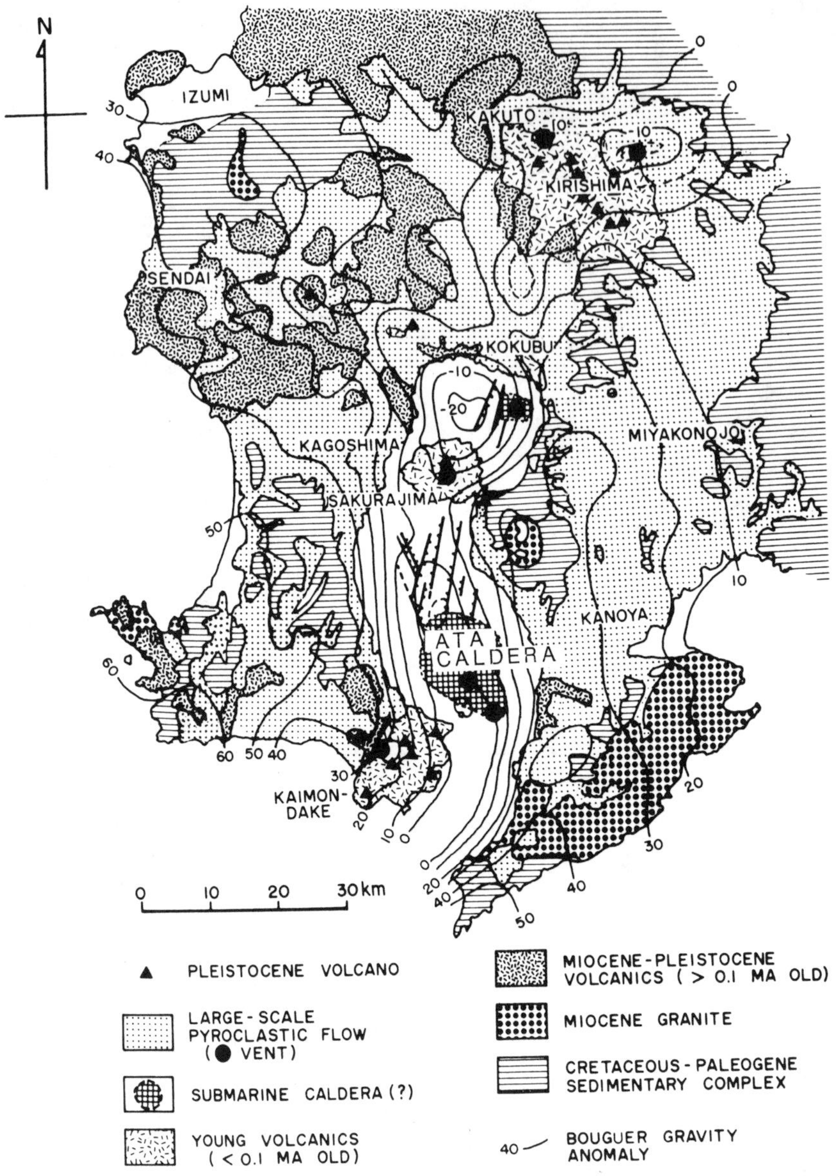

Figure 8.2.3. Geologic sketch map of Ata, Aira, and Kakuto Calderas, modified slightly from Aramaki and others (1981). All three calderas lie within a NNE-SSW-trending graben that is partly filled by Kagoshima Bay. Note revised location of Ata Caldera (after Aramaki and Ui, 1966) relative to location suggested by Matumoto (1943).

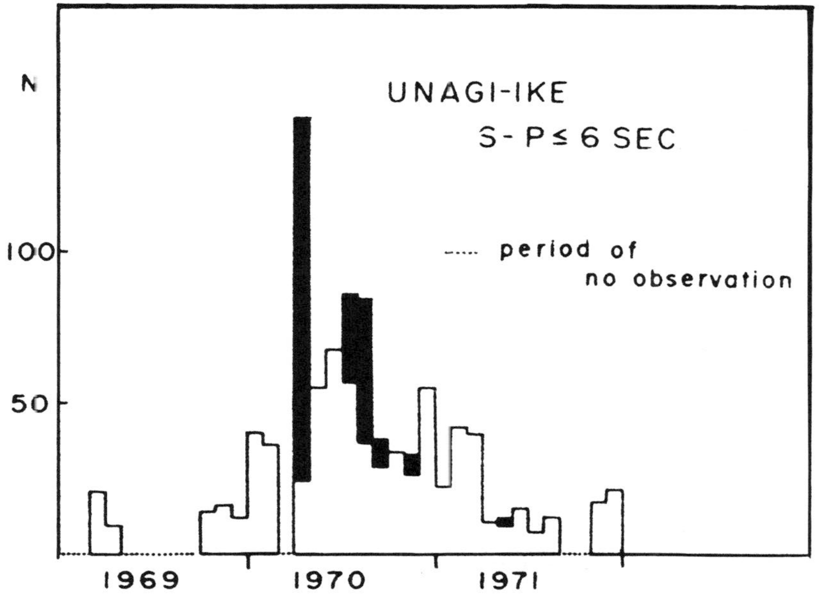

Figure 8.2.4. Monthly number of earthquakes recorded just west of Kagoshima Bay during 1969-71, from Yoshikawa and others (1972). Solid bars represent estimates for those periods for which instrumentral records were incomplete.

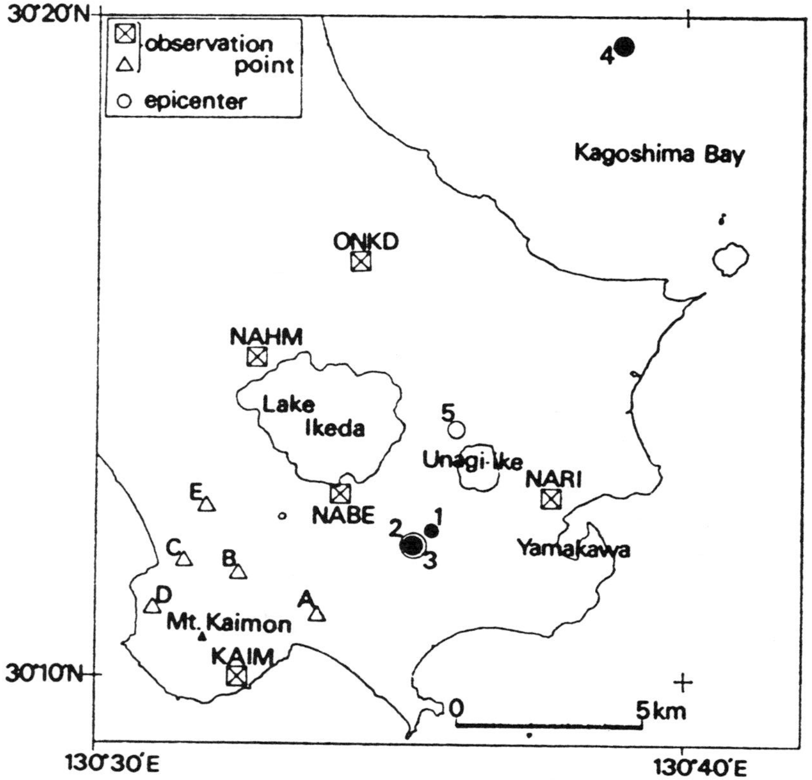

Figure 8.2.5. Epicenters of five earthquakes recorded by a temporary network of seismometers during 30 July - 5 August 1981, from Kakuta (1984). Sizes of circles indicate magnitude of earthquakes; largest event (3) was of magnitude 2.0.

PART 3: HISTORICAL UNREST AT LARGE QUATERNARY CALDERAS (continued)

See inside back cover for explanation and abbreviations

AIRA

CAVW number (active vent)	Latitude Longitude (degrees)	Diameter (km)	Local tectonic setting	Precaldera edifice	SiO_2 content (wt pct)	Age of collapse (yr B.P.)	Date of unrest	Type and duration of unrest ESTU STHF MGTF H Te	Eruption type
08-02-08	31.58N	20	Exten?	None?	R = 57-67	22,000	708	---- ---- ---- - --	Ex
(Sakurajima)	130.67E		(NNE-SSW		C = 74-75		712	---- ---- ---- - --	Ex?
			graben or				716-18	---- ---- ---- - --	Ex
			"volcano-				764	---- ---- ---- - --	Ex
			tectonic				766	x--- ---- ---- - --	Ex
			depression")				778	---- ---- ---- - --	unknown
							1468	---- ---- ---- - --	Ex
							1471	---- ---- ---- - --	Ex,lf
							1473	---- ---- ---- - --	ex
							1475-76	C--x ---- ---- - --	Ex,lf,cdm
							1478	---- ---- ---- - --	ex
							1642	---- ---- ---- - --	ex
							1670?	---- ---- ---- - --	unknown
							1678	---- ---- ---- - --	ex
							1706	---- ---- ---- - --	ex
							1742	---- ---- ---- - --	ex
							1749	---- ---- ---- - --	ex
							1756	---- ---- ---- - --	ex
							1766	---- ---- ---- - --	ex
							1779-81	C--C Z--- ---C C --	Ex,lf,cdm
							1785	---- ---- ---- - --	ex
							1790-92	---- ---- ---- - --	ex
							1794	---- ---- ---- - --	ex
							1797	---- ---- ---- - --	ex
							1799	---- ---- ---- - --	ex
							1860	---- ---- ---- - --	ex?
							1878-79	---- ---- --x- ? --	none
							1899	---- ---- ---- - --	ex
							1913-15	E--G Z--- --EB E UQ	EX,lf
							1931	---- x--- ---- - --	none
							1935	C--G ---- ---- - --	ex
							1938-42	---G Z--- ---- - --	pex,ex

AIRA (continued)

CAVW number (active vent)	Latitude Longitude (degrees)	Diameter (km)	Local tectonic setting	Precaldera edifice	SiO_2 content (wt pct)	Age of collapse (yr B.P.)	Date of unrest	Type and duration of unrest ESTU STHF MGTF H Te	Eruption type
(continued from previous page)							1946	---G Z--- ---- - --	ex,lf
							1948	---G ---- ---- - --	unknown
							1950	---G ---- ---- - --	pex,ex
							1954	---G ---- ---- - --	pex
							1955-88+	CCBG ZDx- ---D x --	pex,ex,lf

TECTONIC SETTING

The Aira Caldera is one of several that have developed within a NNE-SSW-trending graben in southern Kyushu; others include Kikai, Ata (+Ikeda), and Kakuto-Kobayashi (fig. 8.3.1). Each has formed after one or more large explosive eruptions of silicic magma.

GEOLOGIC HISTORY

About 22,000 years ago, a series of large-scale pyroclastic eruptions produced the Aira Caldera at the northern end of Kagoshima Bay in southern Kyushu (figs. 8.3.1, 8.3.2). The outline of the caldera was strongly controlled by faults bounding the volcano-tectonic graben forming Kagoshima Bay (Aramaki, 1984). Sakurajima ("cherry island") is a stratovolcano that has grown on the southern margin of the Aira Caldera since at least 13,000 yr B.P. (fig. 8.3.3). Of its three peaks, only Minami-dake (South Peak) is currently active.

HISTORICAL ACTIVITY

A.D. 708: Sakurajima Island is said to have "welled up and risen in a night" (Koto, 1916).

A.D. 764: In December 764: "The air was rendered dark by smoke and clouds... the sky cleared after seven days, and it was found that in the sea off Shinni-mura [of Kagoshima], sand and rock collected themselves into three new islets." The islets were shaken by earthquakes for several years thereafter, as in July 766, and the government sent relief to distressed people (Omori, 1916a, p. 55).

1475-76: Several small nonvolcanic(?) islands (Karasu-jima, Oko-shima) rose from the sea during 1475-76. Strong earthquakes were felt during the 5 days preceding an explosive eruption and lava extrusion in 1476 (Omori, 1916a, p. 59).

AIRA (continued)

HISTORICAL ACTIVITY (continued)

1779-81: Kagoshima and its surroundings within a 40-km radius were shaken by strong earthquakes on 7 and 8 November 1779. From 1000 hr to 1200 hr on 8 November, "the water in all the wells of Sakurajima boiled up, spouting at several points, and the color of the sea became purple" (Koto, 1916). Steam clouds rose from the summit at about 1100 hr on 8 November, and a major eruption began at 1400 hr on the same day (17 hours after the initial seismicity). On 11 or 12 November, several small islands began to grow off the northeast coast of Sakurajima in 125 m of water. The first rose above sea level on 21 November; others rose above sea level during December 1779 through May 1780. These islands formed principally by submarine eruptions, but some uplift, possibly localized as above a cryptodome, is required to explain sea shells in soil on the top of one of these new islands, 44 m above the sea and 165-170 m above the sea floor (Koto, 1916; Omori, 1916a; Kuno, 1962). A hot spring on the south side of the island ceased to flow (but revived in 1835-36); two new hot springs developed at the east and west ends of the island. The 1779-80 eruption was followed by a period of subsidence, made apparent as the sea flooded low ground along the coast, especially in the streets and vacant grounds of the city of Kagoshima (Imamura, 1930). Subsidence was first reported during June-August 1780, and again in April 1781; it continued at a slower rate for 40-50 years. The depth of flooding was 5-6 shaku (1.5-1.8 m) in Kagoshima and over 10 shaku (3 m) elsewhere. "The low streets... of Kagoshima, were at times of spring tides, covered by sea water such that it was impossible to wade through even with high geta [wooden clogs]....The efforts of the reigning lord of Satsuma to provide against the intrusion of sea water by the construction of embankments were ineffectual, and there were, in the neighborhood of Kajiki, some villages entirely destroyed by tides or obliged to remove to higher ground. The cause of the remarkable elevation of the sea surface may be in the depression of the ground..." (Omori, 1916a, p. 142).

1878-79: Increased steaming from the top of Minami-dake was reported from the end of 1878 until March 1879.

1913-15: A major eruption of nearly 2 km^3 of dacitic magma began on 12 June 1914. Precursory activity included an earthquake swarm in May and June 1913 in the Kakuto Caldera, 55 km north of Sakurajima, and eruptions of nearby Kirishima volcano on 8 November, 9 December 1913, and 8 January 1914. The Kakuto earthquakes were accompanied on 29-30 June 1913 by "sympathetic earthquakes" near Kagoshima and the town of Ijuin, 15 km west of Kagoshima, including two large earthquakes on 30 June of estimated magnitude $\underline{M}_s$=5.8 (Abe, 1979). Hot springs on the east coast of Sakurajima ceased to flow in the spring of 1913, "so that the inn-keepers were obliged to give up their business"; a little later, hot springs on the southern shore of Sakurajima became too hot for bathing during low tide (Koto, 1916). A week before the large eruption of Sakurajima on 12 January 1914, a small pond on the south flank of Sakurajima dried up, "the fishes in it all died, and some wells in the island dried up, or the water-table became considerably lower" (Koto, 1916). The earthquakes near Kagoshima and the drying of wells "were reported and the authorities replied that these were mere effects of the activity of Kirishima [Kakuto] already referred to" (Koto, 1916). At 0830 hr on 12 January 1914, cool water gushed out on the beach at Saido, on the north side of Sakurajima; hot springs on the southeast side of Sakurajima at Arimura spouted to a height of 1 m at about the same time, and very hot water began to pour from several places between Waki and Arimura, making it difficult for people to walk along the beach (Omori, 1914, p. 12). In one well in Yuno-saki (Saido), the water level rose nearly 3 m on the morning of 12 January, to within 1.6 m of the ground surface. A column of "white smoke" rose from the summit of Sakurajima at about 0800 hr on 12 January, and a major eruption began at about 1000 hr.

AIRA (continued)

HISTORICAL ACTIVITY (continued)

An immediate seismic precursor, earthquakes that were felt strongly in Kagoshima, began more than 30 hr before the start of the 1914 eruption (fig. 8.3.4). On Sakurajima proper, where 10 times as many earthquakes were felt as in Kagoshima, the first earthquakes were felt on the night of 10 January, and they became stronger from about 0300 hr on 11 January (Omori, 1915, p. 78). Earthquakes were most frequent from 2000 hr on 11 June to 0800 hr on 12 January, with a temporary quiescence between 0200 and 0300 hr on 12 January (fig. 8.3.4); the eruption began with small explosions at about 1000 hr on 12 January. The maximum earthquake magnitude before the eruption was $\underline{M}$=5.2 (Abe, 1979). A strong earthquake, one of the largest known "volcanic" earthquakes, occurred about 8 hours after the start of the eruption, on the caldera boundary just southwest of Sakurajima. Abe (1981) estimated its magnitude to be $\underline{M}$=7.0, and its depth to be 13 km. Significantly, this earthquake occurred 7 hours before explosive activity reached a climax and before lava extrusion began. After the earthquake, 1.56 km^3 of dacitic lava flowed down the eastern and western sides of the volcano. Omori (1914) called attention to the coincidence of the full moon on 12 January 1914.

Uplift that presaged the 1914 eruption began half a century or so earlier. At Yokoyama, a village on the west coast of Sakurajima Island, a road that in the 1850's and 1860's had been at the shoreline rose by the late 19th century to be 40 m inland (reflecting uplift of at least 1.5 m). Similarly, the lakeshore at rivers flowing into the northwest part of Kagoshima Bay rose by about 1 m (relative to previous sea level) during the same interval (Imamura, 1930). The rate of uplift increased in the years before the eruption, such that in the small harbor of Hamano-ichi, near Kokubu, "traffic... had recently to be carried on with wagons instead of boats" (Omori, 1916a, p. 146).

1935-present: Many small eruptions have occurred since 1914. In mid-1935, earthquakes of local origin were felt at Yumoto, on the south coast of Sakurajima. Ash eruptions began on 20 September 1935, and as these were the first since the disastrous eruption of 1914, 650 people fled to the opposite coast. Over 10 cm of ash was deposited on the south and east flanks of the volcano and crops were damaged (Tanakadate, 1936). Tanakadate (1939, p. 114) wrote of a "submarine eruption" (or, alternatively, a landslide?) on 14 March 1938, 300 m offshore. A column of water 3 m in diameter and 7 m high suddenly rose with a roaring sound, followed soon after by a second column at a point slightly distant from the first. Perhaps coincidentally, the first eruption at Sakurajima since 1935 began suddenly 14 days later. Small eruptions also occurred in 1939-42, explosions and lava extrusion occurred in 1946 (fig. 8.3.5), and more small explosions occurred in 1948(?), 1950, and 1954. Explosions reverted to the summit in 1955 (fig. 8.3.6) and continue there to the present. More explosions (474) occurred in 1985 than in any other year since 1955 (fig. 8.3.7); as of the time of this writing (February 1988), explosions seem to be increasing in frequency after a temporary lull.

Most of Aira's seismicity in recent decades has been beneath or slightly SSW of Sakurajima (fig. 8.3.8). Explosions bear an imperfect relation to A- and B-type earthquakes (figs. 8.3.9-8.3.12), although there are instances when B-type earthquakes clearly presage explosions. For example, in July 1983, the number of B-type earthquakes suddenly increased between 1800 hr on 29 July and 0300 hr on 30 July. Earthquake size then decreased but still remained higher than usual. Explosive activity resumed on 31 July, accompanied by a decrease in seismic activity. A study of the focal mechanisms of A-type earthquakes of Sakurajima (Nishi, 1978) showed that during calm periods the focal mechanisms are strike-slip, during early stages of revival of activity (relatively few

AIRA (continued)

HISTORICAL ACTIVITY (continued)

explosions/day) first motions indicate oblique-slip normal faulting, and during maximum activity (highest number of explosions/day) the earthquakes are along oblique-slip normal or reverse faults. In the waning stages of eruptive activity, focal mechanisms again reflect strike-slip or reverse faulting. Nishi (1984) found that the spectral content of B-type earthquakes that precede explosions was more nearly monochromatic than that of B-type earthquakes not followed by explosions (figs. 8.3.13, 8.3.14).

Immediately after the 1914 eruption, the floor and surroundings of Aira Caldera subsided rapidly--as much at 6.0 m on the floor of Kagoshima Bay (figs. 8.3.15-8.3.17). However, slow uplift began anew within a year, and that uplift continues to the present (figs. 8.3.18-8.3.25). Bench mark 2474, on the edge of the caldera 10 km west of Sakurajima, has shown the most uplift and subsidence of any point around the shore of Kagoshima Bay except on Sakurajima proper. B.M. 2474 rose about 30 cm between 1891 and the 1914 eruption, dropped about 85 cm immediately after that eruption, recovered about 55 cm between 1915 and 1946, dropped over 15 cm immediately after the 1946 eruption, and has been rising since that time, except for periods of relative stability at times of unusually strong eruptions of Sakurajima (1960 and 1974). The net change from 1946 to present is about 25-30 cm of uplift, and the net change since 1891 is about 10 cm of uplift. Bench mark 2469, just outside the west caldera rim, shows no remarkable change and is used for a reference; other points around the west and north caldera rims showed changes of 0.25-0.75 times those shown by B.M. 2474. Points that are 15-20 km northeast of Sakurajima, along the northeast caldera rim, and points that are closer and along the eastern rim show little uplift or subsidence. Following the 1914 eruption, the volume of subsidence was approximately 0.7 km^3, vs. 2 km^3 of magma that was erupted. Kamo (1978) estimated that the magma reservoir has expanded during the 20th(?) century at a rate of 10^6 m^3/yr. Since 1975, points on the west caldera rim have remained at a constant elevation, and a point on the northwest shore of Sakurajima proper has subsided at about 0.5 cm/yr, in response at least initially to frequent eruptions of Sakurajima. Horizontal strain has been consistent with this slow deflation (figs. 8.3.26, 8.3.27) (Eto and Ishihara, 1977; Ishihara and Eto, 1978).

Prior to an eruption on 24 February 1959, tiltmeters at Hakamagoshi showed outward tilt, as if Sakurajima was inflating, but then a reversal to inward tilt 1-2 days before the eruption began (figs. 8.3.28, 8.3.29) (Kubotera and Yoshikawa, 1963, p. 314-315). Similar reversals in tilt occurred in March 1961 (fig. 8.3.30) and in May 1976 (fig. 8.3.31). A high-sensitivity water-tube tiltmeter installed in 1985 has shown consistent, slight inflationary tilt (on the order of 0.1 microradians) 10 minutes to a few hours before explosions, followed by comparable deflation just after explosions. The tiltmeter system is being tested for use in public warnings (fig. 8.3.32) (Kamo and Ishihara, 1986a,b).

Mogi (1957) inferred that the overall pattern of uplift and subsidence at Aira Caldera could be related to a magma body situated 10 km beneath the center of the caldera; Yokoyama (1971, 1974) used slightly different assumptions to estimate a depth of 6 km (figs. 8.3.33, 8.3.34). Recently, Yokoyama (1986) reexamined deformation since the 1914 eruption and inferred two pressure sources directly beneath Sakurajima Volcano. One, at 8 km depth, is directed obliquely upward and toward the north, and the other, at 2 km depth, is directed vertically upward. Shallow, short-period deformation of Sakurajima is superimposed on the caldera-wide pattern.

Yoshikawa (1961) and Kubotera and Yoshikawa (1963) reported 60 cm of uplift of Hakamagoshi (west end of Sakurajima Peninsula) relative to the Kagoshima Harbor tide gauges (about 4 km west of Hakamagoshi) from July 1958 to August 1960 (figs. 8.3.23, 8.3.25).

AIRA (continued)

HISTORICAL ACTIVITY (continued)

The latter uplift is not easily reconciled with a single point source of uplift beneath Aira Caldera, and so these authors concluded that (a) some fault displacement was also occurring, or (b) there is a second point source of uplift 3 km beneath Sakurajima Volcano. Minakami (1962) cited a finding by Y. Yasui and K. Yoshikawa that the northern and southern parts of Sakurajima deform separately, as if an active fault passes from east to west through the summit of Sakurajima, coincident with the southern topographic margin of the Aira Caldera.

Localized uplift and subsidence have been noted from time to time since January 1914. Several small islets formed in February and March 1914, probably by inflation of submarine toes of lava flows. Pumice and broken pieces of straw, bamboo, and sticks rose with the surface of the lava, but bottom sediments were not uplifted in the same manner as they had been in 1779-80 (Omori, 1916b, p. 263 ff.). In 1931, 20 m off the north coast of Sakurajima near the village of Sirahama, an area only 36 m in diameter subsided 4 m (Tanakadate, 1934).

Shortly after the large eruption of 1914, as the caldera floor was subsiding, increased soil temperatures were noted in an orchard of Nishimoto, behind the village of Saido (600 m map distance and 70-80 m vertically from the northwest coast of Sakurajima). The first symptom was death of peach trees in March 1915, at the lowest, northern end of the area, and gradual spreading of the same effect upslope (southward). Subsequently *Quercus acuta*, *Quercus serrata*, *Eriobotrya japonica*, and other species also died. The soil substrate was pumice; maximum temperature was 64.6 °C. By 1 October 1915, the area of heat and tree kill was 150 m in diameter, and dilute chlorine and benzene fumes were noted. Similar fumes had been noted at the village of Take, on the western side of Sakurajima, where in the previous spring an ox had been killed and several people had taken ill (Omori, 1916b, p. 310 ff.). Soil temperature also increased at the coast at Ushine (on the mainland, east of the Seto Strait and more or less along strike of the line of vents through Sakurajima's summit and the southern wall of Aira Caldera). On 6 April 1915, dug-up soil was too hot to be handled. These effects diminished after 1915 and eventually disappeared.

Shimozuru and others (1957) observed that isostatic rebound in response to loading/unloading by oceanic tides and especially by typhoon storm waves occurs 4 1/2 to 20 hours after the maximum loading. The length of the lag is proportional to the period of loading and is in all cases longer than lags at nearby coastal areas, away from the caldera. Ono and others (1978) and Hirata and Uchiyama (1981) documented seismic-wave attenuation through a volume beneath the caldera.

During studies of the submarine Wakamiko "proto-caldera" that is nested within the Aira Caldera, CO_2 emission from the floor of that caldera during September 1977 (a time of frequent B-type earthquakes) may have been twice that in November 1977 (a time of seismic quiet) (Horibe and others, 1980). However, Sugiura and others (1981) used different assumptions to reestimate the CO_2 flux for the same periods, and they found only marginally greater flux in September than in November. Hirabayashi and others (1982, 1986) reported that, during a study period from 1976 to 1983, the ratio of HCl to SO_2 in fumaroles in the summit crater at Sakurajima and the concentration of H_2 in gas from a hot spring at the southwest foot of the volcano both increased during and up to several weeks before increases in the frequency of volcanic explosions (figs. 8.3.9, 8.3.35).

AIRA (continued)

COMMENTS

Aira and Sakurajima illustrate how eruptions and magma resupply at an andesitic stratocone can cause caldera-wide subsidence and uplift, even though the stratocone (Sakurajima) is near the margin of the caldera. Several interpretations of the depths and direction of pressure sources are possible, as noted above.

Numerous pioneering studies have been and continue to be done at Sakurajima and the Aira Caldera. Omori (1914-22) was the first to document coupling of ground deformation across an entire caldera with eruptions from a vent on its rim. Frequent explosive activity has enabled the Sakurajima Volcano Observatory to examine small precursory changes of many explosions and to discover promising new patterns in the spectral content of earthquakes and submicroradian tilt before explosions (Kamo, 1978; Nishi, 1984; Kamo and Ishihara, 1986a, and others).

REFERENCES

Abe, K., 1979, Magnitudes of major volcanic earthquakes of Japan 1901 to 1925: Jour. Fac. Sci., Hokkaido Univ., ser. VII, v. 6, p. 201-212.

----- 1981, Seismometrical re-evaluation of the Sakurajima earthquake of January 12, 1914: Geophys. Bull. Hokkaido Univ, v. 39, p. 57-62.

Aramaki, S., 1984, Formation of the Aira Caldera, southern Kyushu, about 22,000 years ago: Jour. Geophys. Res., v. 89, no. B10, p. 8485-8501.

Aramaki, S., Fukuyama, H., Kamo, K., and Kamada, M., 1981, Sakurajima Volcano, in Kubotera, A., ed., Field excursion guide to Sakurajima, Kirishima, and Aso Volcanoes: Tokyo, Volcanological Society of Japan, p. 1-17.

Aramaki, S., and Ui, T., 1966, The Aira and Ata pyroclastic flows and related calderas and depressions in southern Kyushu, Japan: Bull. Volcanol., v. 29, p. 29-47.

Decker, R.W., and Kinoshita, W.T., 1971, Geodetic measurements, in The surveillance and prediction of volcanic activity: Paris, UNESCO, p. 47-74.

Eto, T., 1967, Volcanic crustal deformations (III): Crustal deformation in the vicinity of Aira Caldera and the activity of V. Sakurajima: Bull. Volcanol. Soc. Japan, v. 12, p. 80-88.

----- 1974, Crustal deformation associated with volcanic activity around Sakurajima: Seismicity and Volcanic Activity of Kagoshima Prefecture, v. 6, p. 50-66.

----- 1985, Ground deformations associated with volcanic activity in the vicinity of Sakurajima volcano, in Okada, Hm., ed., Comparative study of physical background to volcanic activity and its relation to prediction of eruption disasters: Special Rep. Grant-in-Aid for Nat. Disaster Sci., Ministry of Education, publ. by Hokkaido Univ., Sapporo, in cooperation with Tohoku Univ., Univ. of Tokyo Earthq. Res. Inst., Nagoya Univ., Kyoto Univ., and Kyoto Univ. Disaster Prev. Res. Inst., p. 23-29.

AIRA (continued)

REFERENCES (continued)

Eto, T., and Ishihara, K., 1975a, Recent vertical crustal deformation around Sakurajima: Seismicity and Volcanic Activity of Kagoshima Prefecture, v. 7, p. 49-51.

----- 1975b, Crustal deformation of Aira caldera and volcanic activity of Sakurajima volcano, in Kamo, K., ed., Rep. Joint Geophys. Geochem. Obs. Sakurajima Volcano (I), p. 19-24.

----- 1977, Ground deformation of Sakurajima volcano and summit eruptive activity, in Kamo, K., and Kubotera, A., ed., Rep. Joint Geophys. Geochem. Obs. Sakurajima Volcano (V), November-December 1976, p. 21-33.

Eto, T., and Ohya, F., 1972, Observation of crustal deformation in the Sakurajima and Yoshimatsu areas by electro-optical means (1): Bull. Volcanol. Soc. Japan, v. 17, p. 135-143.

Fukuyama, H., and Ono, K., 1981, Geological map of Sakurajima Volcano: Geological Maps of Volcanoes, no. 1, scale 1:25,000, Geological Survey of Japan.

Hirabayashi, J., Ossaka, J, and Ozawa, T., 1982, Relationship between volcanic activity and chemical composition of volcanic gases -- A case study on the Sakurajima Volcano: Geochem. Jour., v. 16, p. 11-21.

----- 1986, Geochemical study on volcanic gases at Sakurajima Volcano, Japan: Jour. Geophys. Res., v. 91, no. B12, p. 12,167-12,176.

Hirata, T., and Uchiyama, T., 1981, Damping area in the Aira Caldera of South Kyushu: Zisin, v. 34, no. 3, p. 435-437.

Horibe, S., Gamo, T, and Baba, Y, 1980, Estimation of carbon dioxide flux from submarine volcano in Kagoshima Bay, Japan, in Goldberg, E.D., Horibe, Y., and Saruhashi, K., eds., Isotope marine chemistry: Tokyo, Uchida Rokakuho Publ. Co., p. 327-337.

Imai, I., Teraoka, Y., Ono, K., Matsui, K., and Okumura, K., 1980, Geological map of the Kagoshima Quadrangle: 1:500,000 series, Sheet 15, Geological Survey of Japan.

Imamura, A., 1930, Topographical changes accompanying earthquakes or volcanic eruptions: Publ. Earthq. Invest. Comm. in Foreign Languages, no. 25, p. 1-143.

Ishihara, K., and Eto, T., 1978, Surface deformation associated with the eruptive activity at the summit crater of Sakurajima volcano: Ann. Rep. Disaster Prev. Res. Inst., Kyoto Univ., no. 21B-1, p. 153-162.

Kamo, K., 1978, Some phenomena before the summit eruptions at Sakura-zima Volcano: Bull. Volcanol. Soc. Japan, v. 23, p. 53-64.

Kamo, K., and Ishihara, K., 1986a, A primary experiment on the automatic judgement of the stages of eruptive activity by a water-tube tiltmeter [abs.]: unpubl. abstract, Int. Volcanol. Congr., New Zealand, February 1986.

----- 1986b, Precursor of summit eruption observed by water-tube tiltmeters and extensometers: Ann. Rep. Disaster Prev. Res. Inst., Kyoto Univ., no. 29B-1, p. 1-12.

Koto, B., 1916, The great eruption of Sakurajima in 1914; Tokyo Imperial Univ., Jour. Coll. Sci. v. 38, art. 3, p. 1-237.

Kubotera, A., and Yoshikawa, I., 1963, Prediction of volcanic eruptions of Aso and Sakurajima and some related geophysical problems: Bull. Volcanol., v. 26, p. 297-317.

Kuno, H., 1962, Catalogue of the active volcanoes of the world including solfatara fields, Pt. XI, Japan, Taiwan and Marianas: Rome, IAVCEI, 332 p.

Machida, H., and Arai, F., 1983, Extensive ash falls in and around the Sea of Japan from large Late Quaternary eruptions: Jour. Volcanol. Geotherm. Res., v. 18, p. 151-164.

AIRA (continued)

REFERENCES (continued)

Matumoto, T., 1943, The four gigantic caldera volcanoes in Kyushu, Japan: Japanese Jour. Geol. Geogr., v. 19, p. 1-57

----- 1963, Caldera volcanoes and pyroclastic flows of Kyushu: Bull. Volcanol., v. 26, p. 401-413.

Minakami, T., 1960, Fundamental research for predicting volcanic eruptions, Pt. I: Bull. Earthq. Res. Inst., v. 38, p. 497-544.

----- 1962, Report on volcanic activity in Japan for the period from 1957 to 1959: Bull. Volcanol., v. 24, p. 7-22.

Minakami, T., Hiraga, S., Miyazaki, T. and Uchibori, S., 1969b, Fundamental research for predicting volcanic eruptions, Pt. II: Bull. Earthq. Res. Inst., v. 47, p. 893-949.

Mogi, K., 1957, On the relation between the eruptions of Sakurajima and the crustal movements in its neighbourhood: Bull. Volcanol. Soc. Japan, v. 1, no. 1, p. 9-18.

----- 1958, Relations between the eruptions of various volcanoes and the deformations of the ground surfaces around them: Bull. Earthq. Res. Inst., v. 36, p. 99-134.

Nishi, K., 1978, On the focal mechanism of volcanic earthquakes in Sakurajima volcano: Ann. Rep. Disaster Prev. Res. Inst., Kyoto Univ., v. 21, p. 145-152.

----- 1984, Volcanic B-type earthquake swarm preceding volcanic explosion: Ann. Rep. Disaster Prev. Res. Inst., Kyoto Univ., v. 27B-1, p. 29-34.

Nishi, K., and Iguchi, M., 1983, Preliminary report on the seismicity in south Kyushu: Ann. Rep. Disaster Prev. Res. Inst., Kyoto Univ., v. 26B-1, p. 23-29.

Omori, F., 1914-22, The Sakurajima eruptions and earthquakes: Imperial Earthquake Invest. Comm. Japan, Bull.; Pt. I., General Account, v. 8, no. 1 (1914), p. 1-34; Pt. II, On the sound and ash precipitation areas of, and on the level changes caused by, the eruption of 1914, with historical sketches of earlier Sakurajima outbursts, v. 8, no. 2 (1916a), p. 35-179; Pt. III, Course of activity in the Sakurajima eruption of 1914, v. 8, no. 3 (1916b), p. 181-321; Pt. IV, Results of the leveling surveys and the Kagoshima Bay soundings made after the Sakurajima eruption of 1914, v. 8, no. 4 (1920a), p. 322-351; Pt. V, Seismographical observations of the fore-shocks, after-shocks, and after-outbursts of the Great Sakura-jima Eruption of 1914, v. 8, no. 5 (1920b), p. 353-466; Pt. VI, Notes on the destructive earthquake of Jan. 12, 1914, and miscellaneous remarks on the eruptions of Sakura-jima, v. 8, no. 6 (1922), p. 467-525.

----- 1915, The eruption of Sakura-jima, 1914: Bull. Seismol. Soc. Am., v. 5, p. 71-95.

Ono, K., Ito, K., Hasegawa, I., Ichikawa, K., Iizuka, S., Kakuta, T., and Suzuki, H., 1978, Explosion seismic studies in south Kyushu, especially around Sakurajima Volcano: Jour. Phys. Earth, v. 26 supp., p. S309-S319.

Ossaka, J., Hirabayashi, J., and Ozawa, T., 1986, On the estimation of volcanic activity of Sakurajima from chemical composition of volcanic gases (5), <u>in</u> Kamo, K., ed., Rep. Joint Geophys. Geochem. Obs. Sakurajima Volcano (V), October-December 1982, p. 71-88.

Shimozuru, D., 1971, A seismological approach to the prediction of volcanic eruptions, <u>in</u> The surveillance and prediction of volcanic activity: Paris, UNESCO, p. 19-45.

Shimozuru, D., Goto, K., and Nozoe, S., 1957, Observations of the crustal deformations at Hakamagoshi, Sakurazima Volcano: Bull. Volcanol. Soc. Japan, v. 2, no. 1, p. 1-5.

Smithsonian Institution, Scientific Event Alert Network (SEAN), 1987-88, Sakurajima Volcano: SEAN Bull., v. 12, no. 6; v. 13, nos. 1, 2 (information from Japan Meteorological Agency).

AIRA (continued)

REFERENCES (continued)

Sugiura, Y., Kamada, M., and Nedachi, Y., 1981, Rate of carbon dioxide evolution from submarine volcano in Kagoshima Bay: Sci. Rep. Kagoshima Univ., no. 30, p. 93-102.

Tanakadate, H., 1934, Volcanic activity in Japan and vicinity during the period between June 1931 and June 1934: Japanese Jour. Astron. Geophys., v. 12, p. 89-108.

----- 1936, Volcanic activity in Japan and vicinity during the period between July 1934 and October 1935: Japanese Jour. Astron. Geophys., v. 13, p. 121-139.

----- 1939, Volcanic activity in Japan and vicinity during the period between November 1935 and December 1938: Japanese Jour. Astron. Geophys., v. 16, p. 95-120.

Taneda, S., 1961, Moving of the magna chamber of the Sakura-jima Volcano: Jour. Geol. Soc. Japan, v. 67, p. 593-605.

----- 1977, Significance of the change of magma composition and the upward moving of the magma chamber of Sakura-jima Volcano: Bull. Volcanol. Soc. Japan, v. 22, p. 61-64.

Wakita, H., 1982, Changes in groundwater level and chemical composition, in Asada, T., and Ohnuki, M., eds., Earthquake prediction techniques: Tokyo, Univ. Tokyo Press, p. 175-216.

Yasui, Y., 1963, Summary of forecasting the explosion of Volcano Sakurajima: Geophys. Mag., v. 31, p. 491-504.

Yokoyama, I., 1971, A model for the crustal deformations around volcanoes: Jour. Phys. Earth, v. 19, p. 199-207.

----- 1974, Crustal deformation associated with volcanic activities: Tectonophysics, v. 23, p. 349-360.

----- 1986, Crustal deformation caused by the 1914 eruption of Sakurajima Volcano, Japan, and its secular changes: Jour. Volcanol. Geotherm. Res., v. 30, p. 283-302.

Yokoyama, I., and Ohkawa, S., 1986, The subsurface structure of the Aira Caldera and its vicinity in southern Kyushu, Japan: Jour. Volcanol. Geotherm. Res., v. 30, p. 253-282.

Yoshikawa, K., 1961, On the crustal movement accompanying with the recent activity of the Volcano Sakurajima, Pt. 1: Bull. Disaster Prev. Res. Inst., no. 48, p. 1-15.

----- 1962, On the crustal movement accompanying with the recent activity of the Volcano Sakurajima, Pt. 2: Bull. Disaster Prev. Res. Inst., no. 50, p. 1-26.

Yoshikawa, K., and Nishi, K., 1966, Seismic observation at Volcano Sakurajima (5): Ann. Rep. Disaster Prev. Res. Inst., Kyoto Univ., v. 9, p. 47-54.

Yoshimura, J., 1957, On the volcanic activity of Sakura-zima and the soil temperature at Kagoshima: Quart. Jour. Seismol., v. 21, p. 181-191.

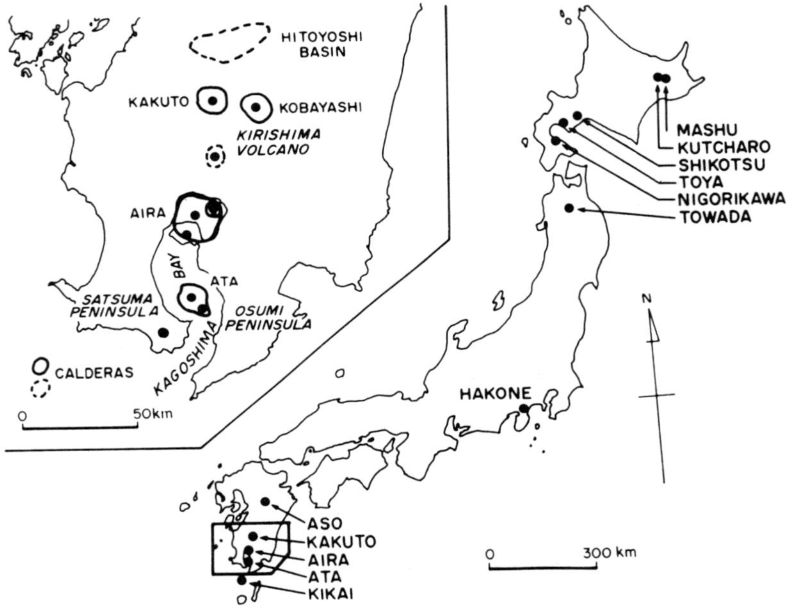

Figure 8.3.1. Index map showing location of Aira Caldera and other Japanese Quaternary calderas included in this compilation, from Aramaki (1984). Copyright by the American Geophysical Union.

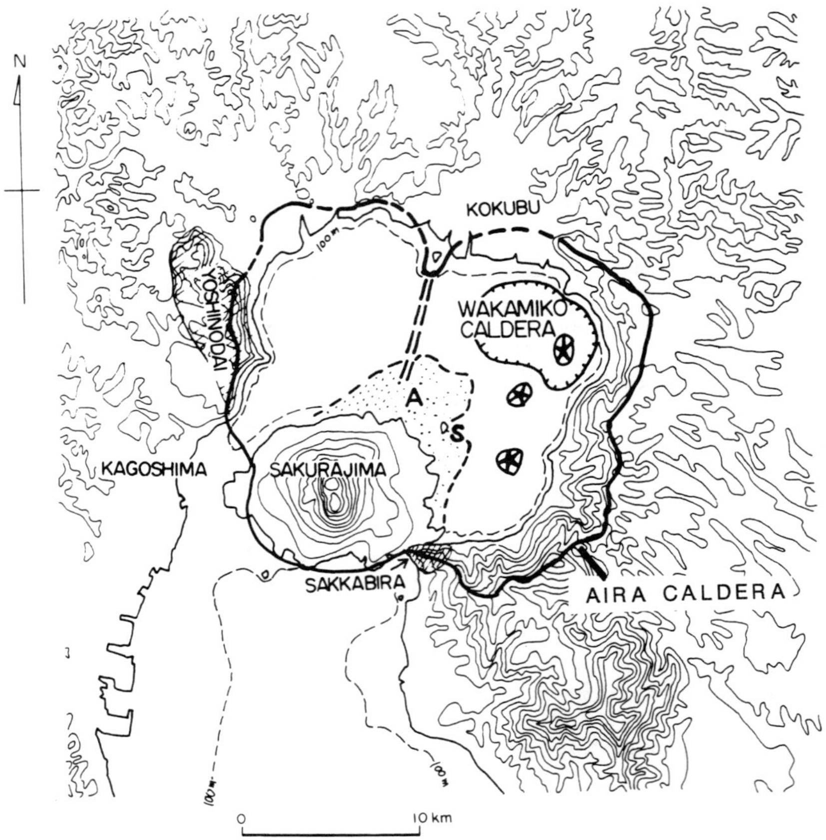

Figure 8.3.2. Generalized topographic map around Kagoshima Bay, showing location of Aira Caldera, from Aramaki (1984). Contour interval is 100 m. Thick line is rim of Aira Caldera originally proposed by Matumoto (1943). Stippled area marked A is a submarine extension of a 1779 lava flow from Sakurajima; S is a small island consisting of pumice breccia ejected from Wakamiko Caldera. Three postcaldera cones are shown with stars. Floor of Aira Caldera is divided by a north-south ridge (double broken lines). Copyright by the American Geophysical Union.

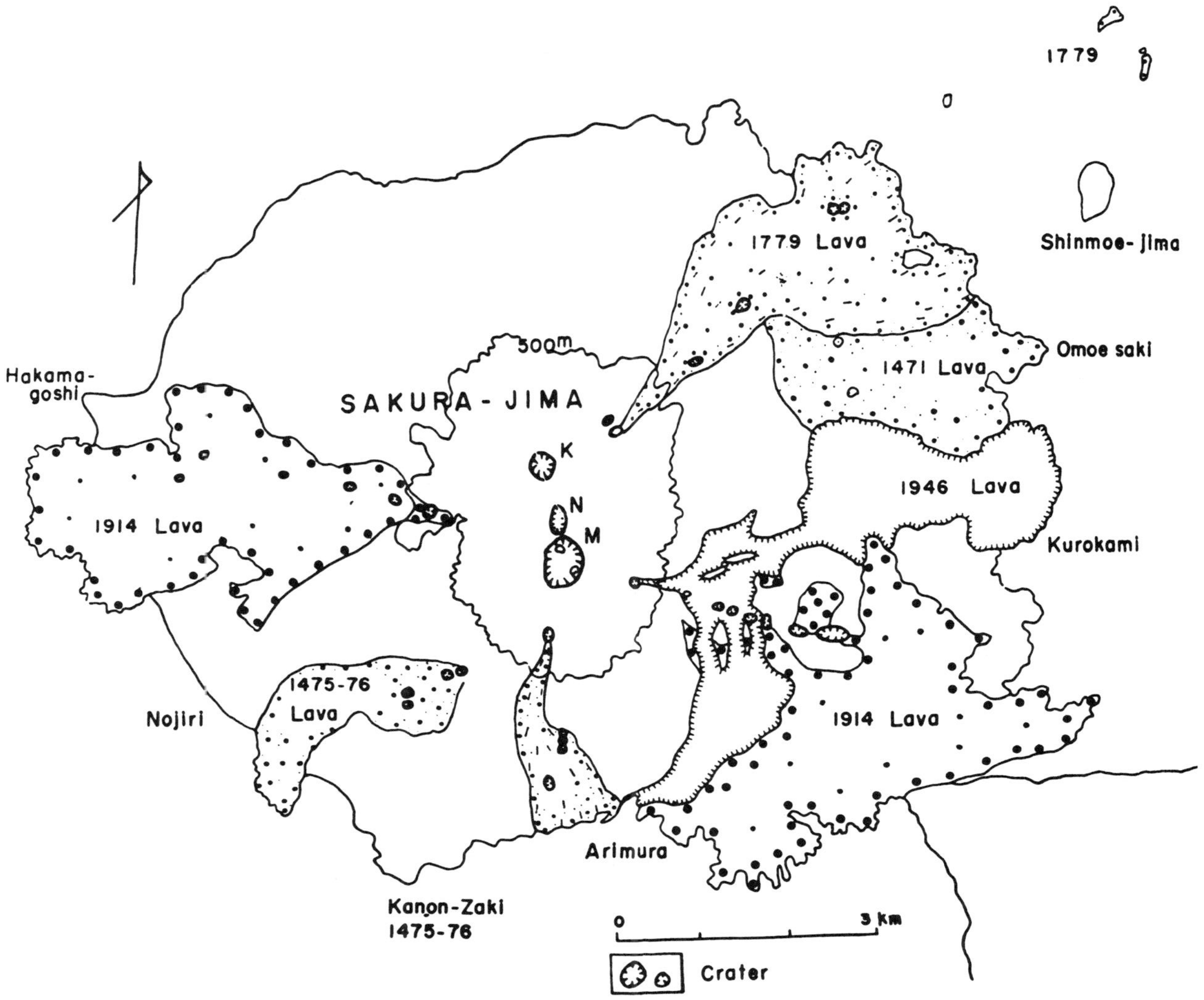

Figure 8.3.3. Historically active vents of Sakurajima volcano, from Taneda (1961). K, Kita-dake; N, Naka-dake; M, Minami-dake. Two sets of five vents each opened during the 1914 eruption, one set on west flank and one on east flank of Sakurajima. Within each set, activity migrated from higher to lower elevations (that is, from east to west in western set and from west to east in eastern set) (Yokoyama, 1986).

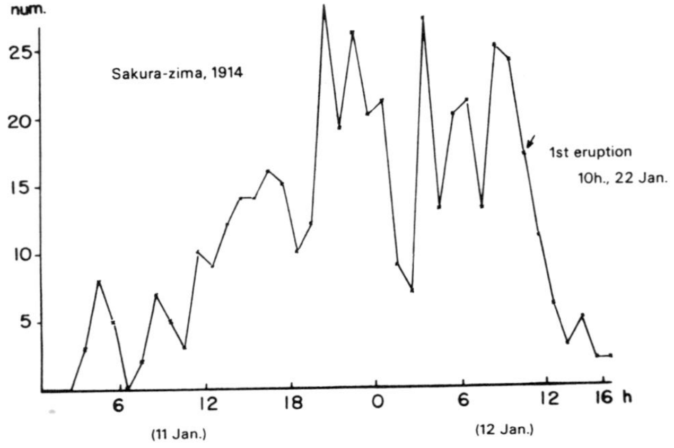

Figure 8.3.4. Hourly numbers of earthquakes recorded at Kagoshima Meteorological Observatory during the 1914 eruption of Sakurajima Volcano, from Shimozuru (1971).

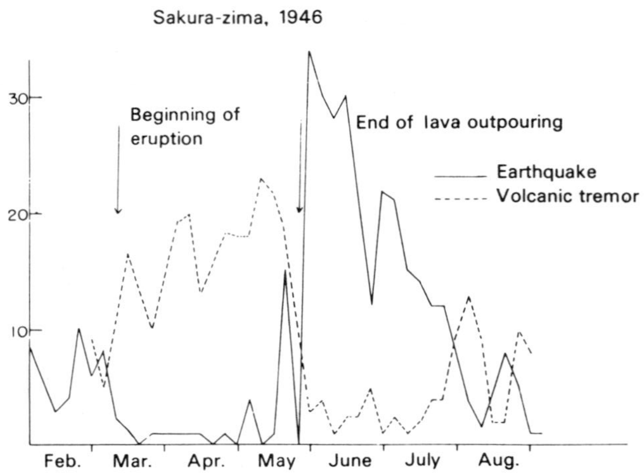

Figure 8.3.5. Numbers of earthquakes recorded in successive 5-day periods during the 1946 eruption of Sakurajima, from Shimozuru (1971).

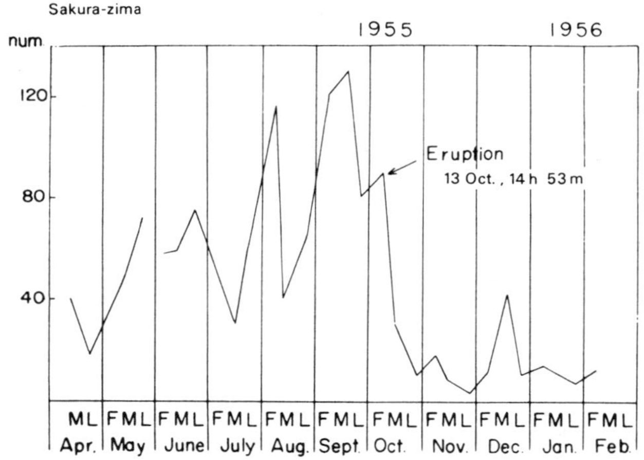

Figure 8.3.6. Numbers of earthquakes recorded at Kagoshima Meteorological Observatory in successive 10-day periods preceding the 1955 eruption of Sakurajima, from Shimozuru (1971).

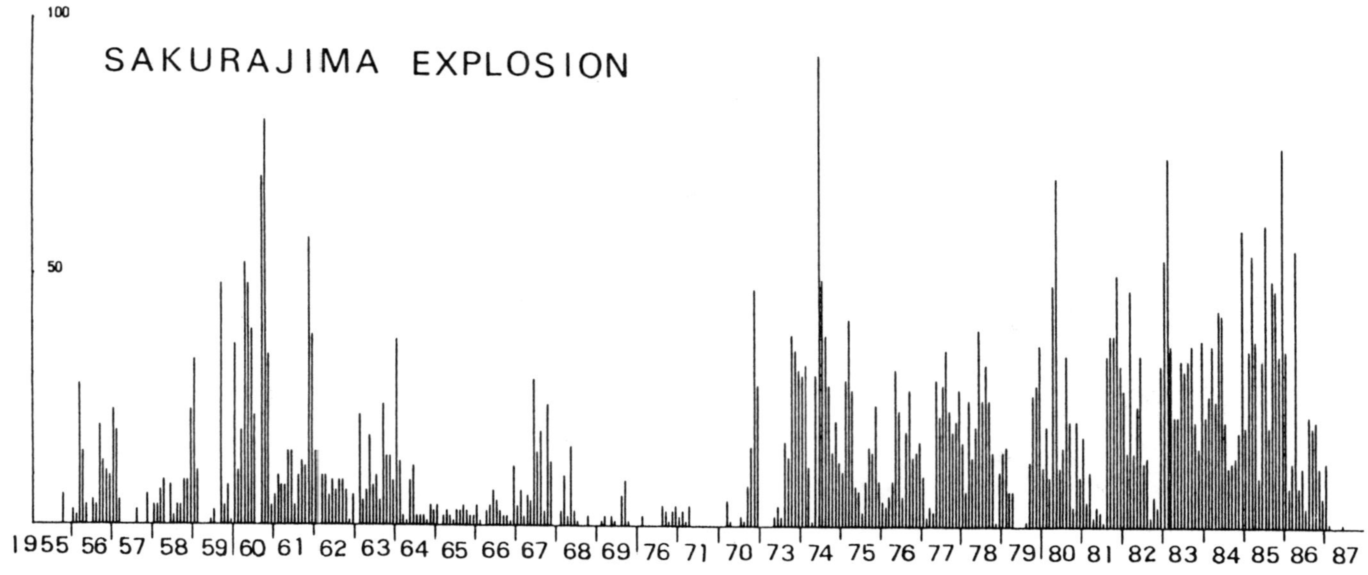

Figure 8.3.7. Monthly number of explosions at Sakurajima, 1955 to May 1987, from Smithsonian Institution (1987; courtesy of Japan Meteorological Agency).

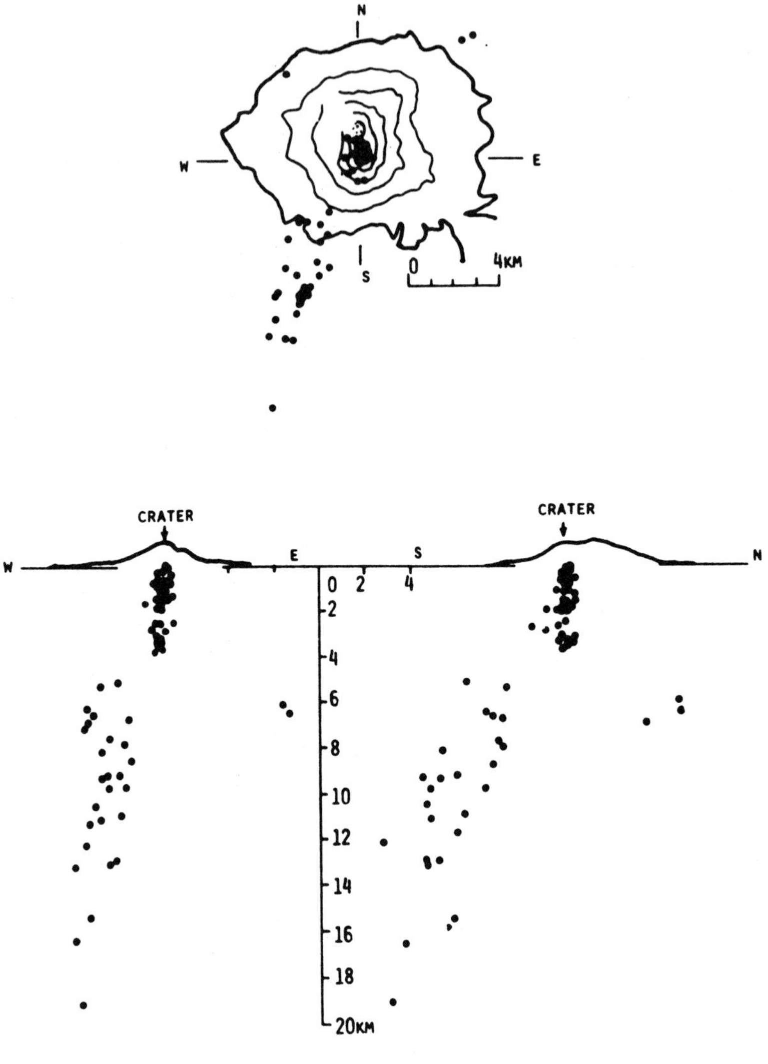

Figure 8.3.8. Epicenters and hypocenters of A-type volcanic earthquakes at Sakurajima Volcano, Aira Caldera, from August 1975 to December 1980 (from Aramaki and others, 1981), after Nishi (1978).

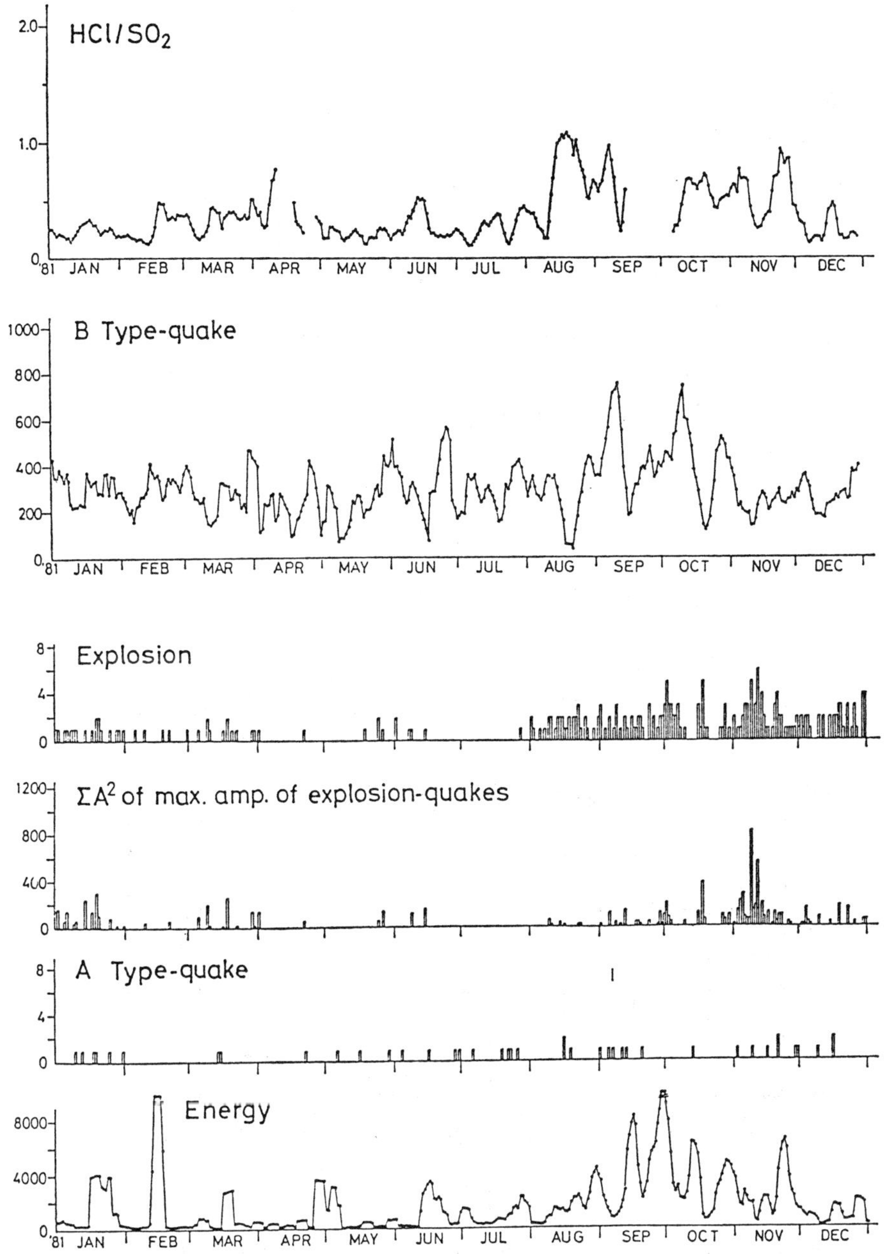

Figure 8.3.9. HCl/SO_2 molar ratio in the atmosphere at Harutayama versus several types of geophysical data on volcanic activity at Sakurajima Volcano during 1981, from Ossaka and others (1986). This plot is representative of similar data for 1979-83, also shown in Hirabayashi and others (1982, 1986).

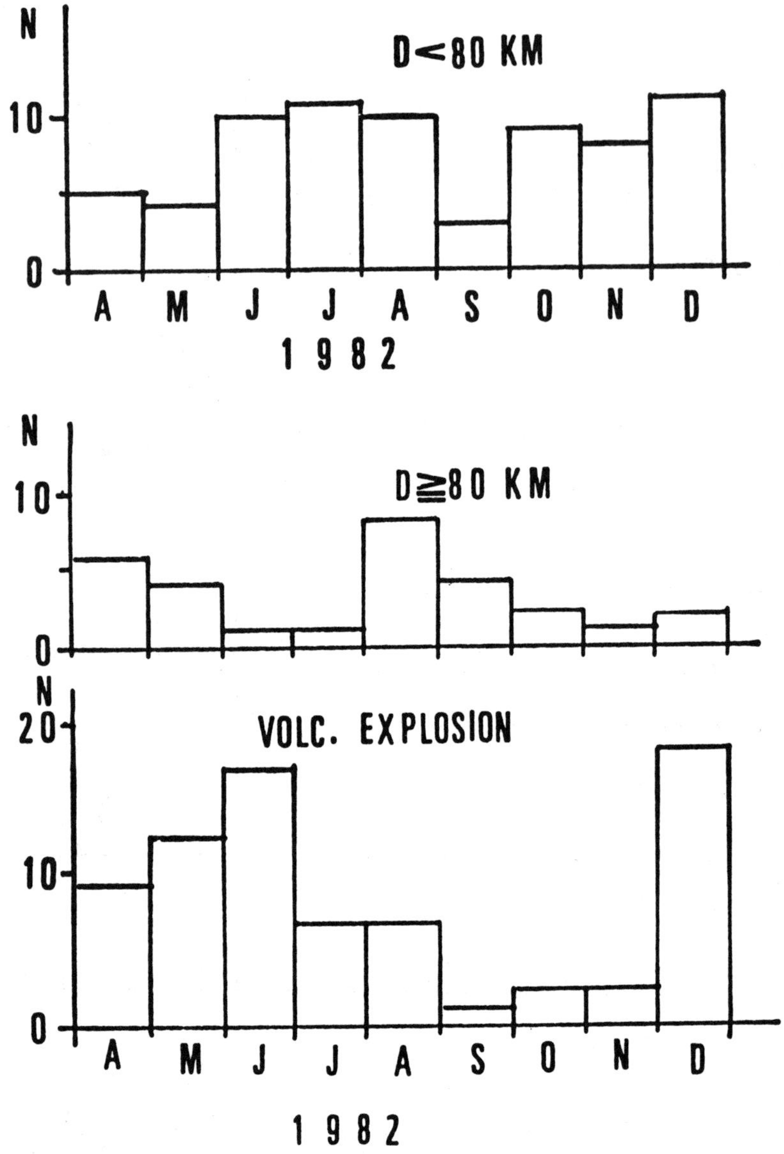

Figure 8.3.10. Relation between monthly seismic activity and volcanic activity at Sakurajima, from Nishi and Iguchi (1983). Top, monthly number of earthquakes less than 80 km deep; middle, monthly number of earthquakes more than 80 km deep; bottom, monthly number of volcanic explosions at Sakurajima.

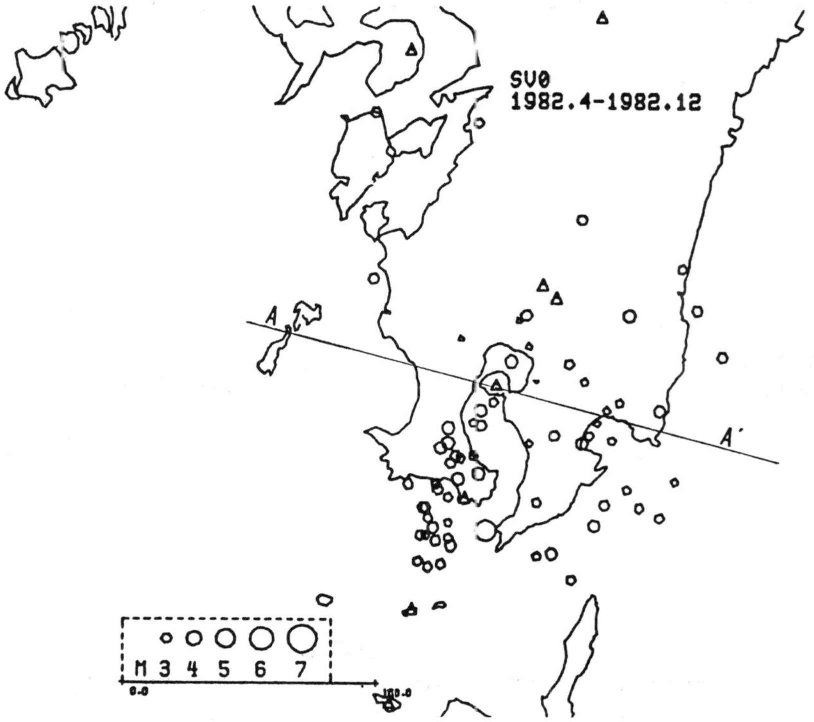

Figure 8.3.11. Epicentral distribution of earthquakes during April 1982 through December 1982, from Nishi and Iguchi (1983). Line A-A' indicates location of cross section perpendicular to volcanic front (figure 8.3.12). Nishi and Iguchi (1983) argued that deep earthquakes increased in mid-1982, before the number of explosions at Sakurajima increased in December 1982 (figure 8.3.10).

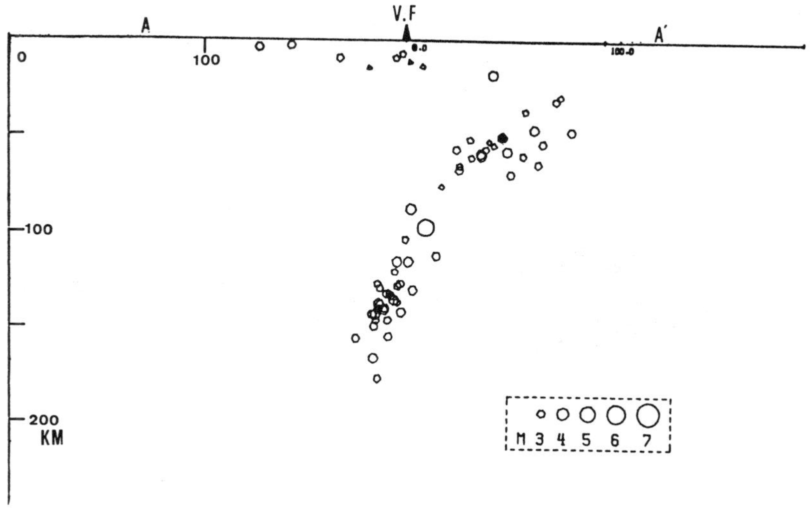

Figure 8.3.12. Vertical section perpendicular to volcanic front near Aira Caldera, from Nishi and Iguchi (1983). See figure 8.3.11 for location of line A-A'.

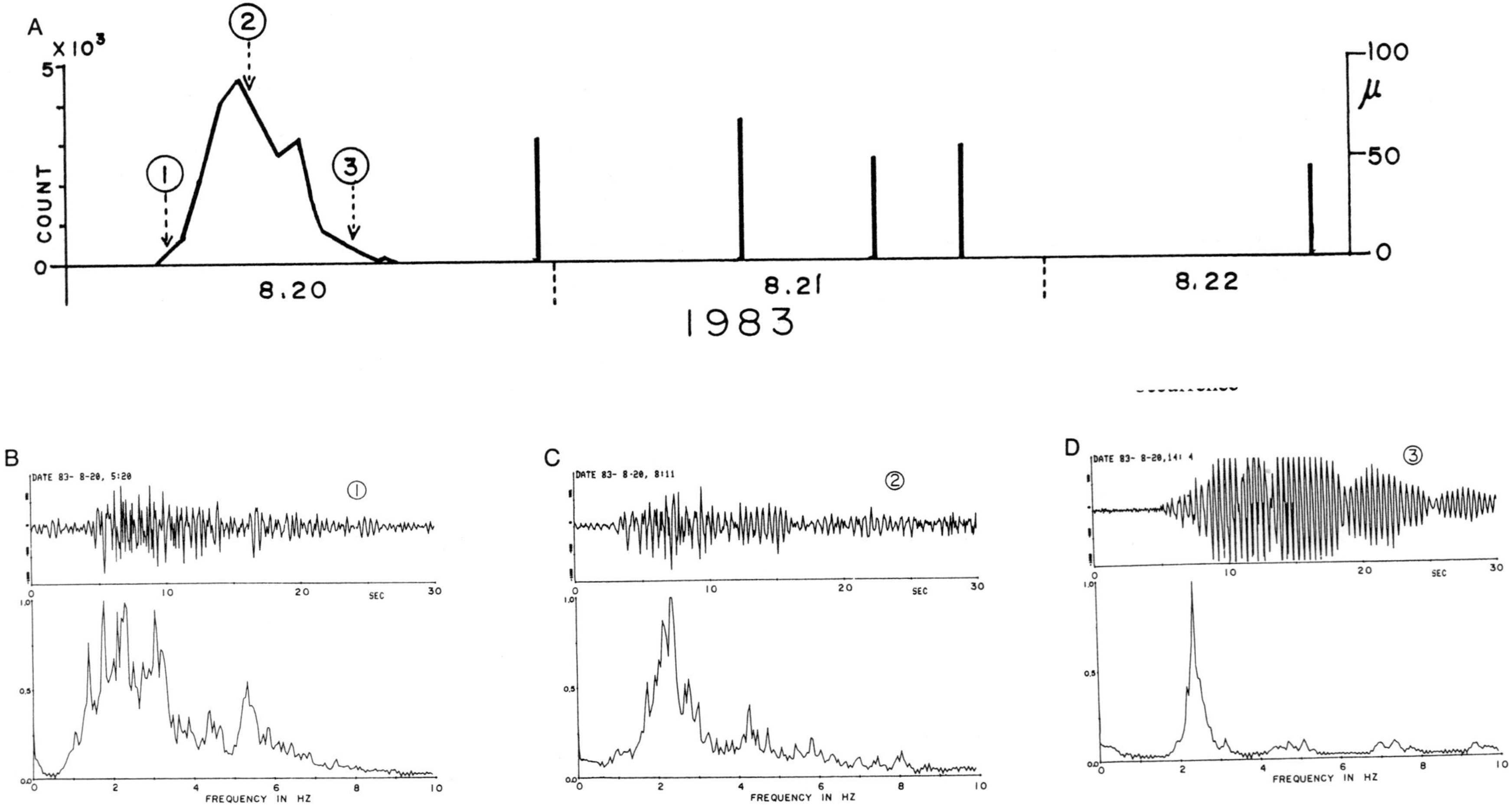

Figure 8.3.13. A, Time sequence of volcanic earthquake swarm and five volcanic explosions that followed at Sakurajima in August 1983. Numbers indicate time of occurrence of earthquakes whose waveforms are shown in B-D. B, Waveform and its Fourier spectrum of the earthquake whose time of occurrence is shown in A as event 1. C, Same as B except for event 2. D, Same as B and C except for event 3. All figures from Nishi (1984).

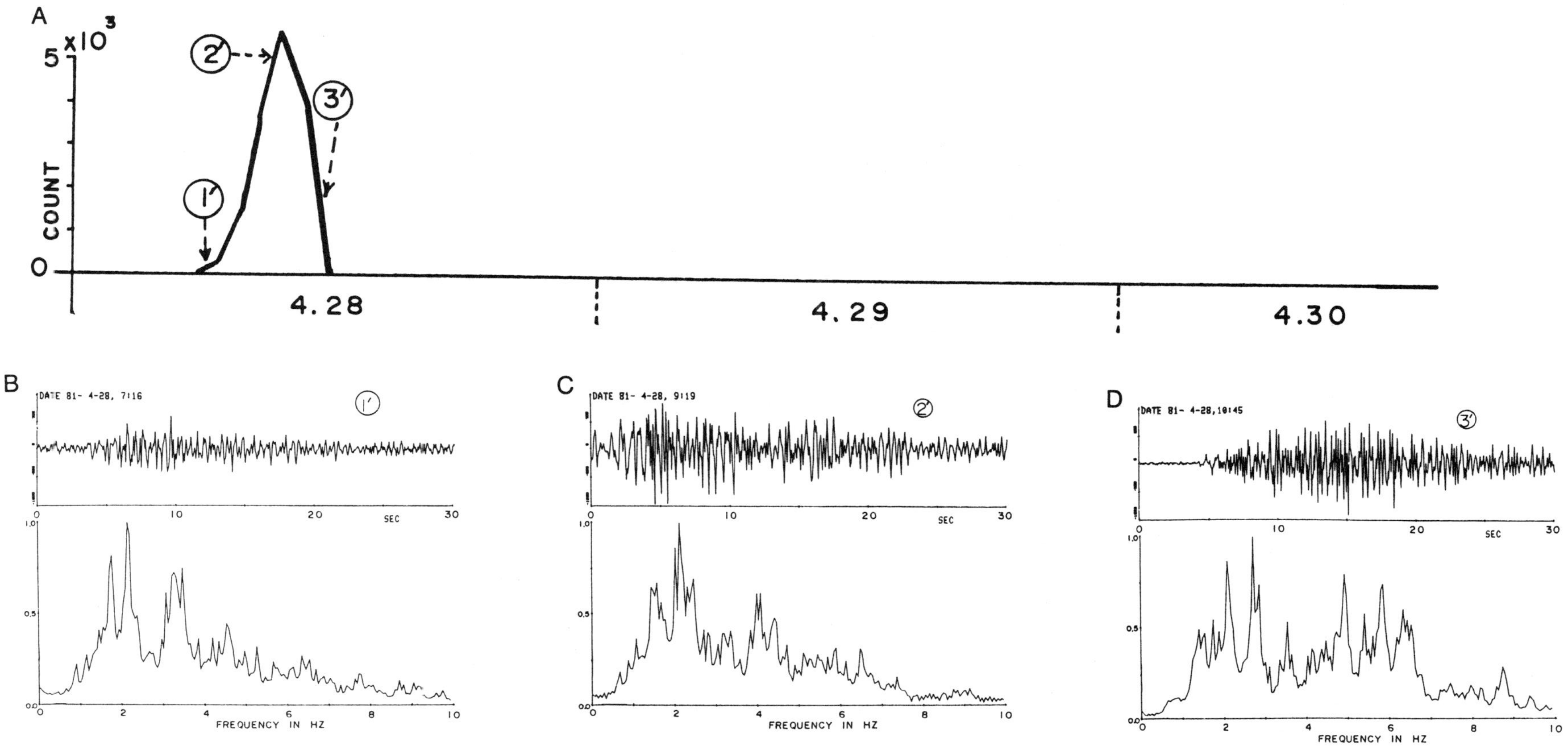

Figure 8.3.14. A, Time sequence of volcanic earthquake swarm at Sakurajima in April 1981. In this case, no volcanic explosion followed the swarm (compare with figure 8.3.13). Numbers indicate time of occurrence of earthquakes whose waveforms are shown in B-D. B, Waveform and its Fourier spectrum of the earthquake whose time of occurrence is shown in A as event 1. C, Same as B except for event 2. D, Same as B and C except for event 3. All figures from Nishi (1984).

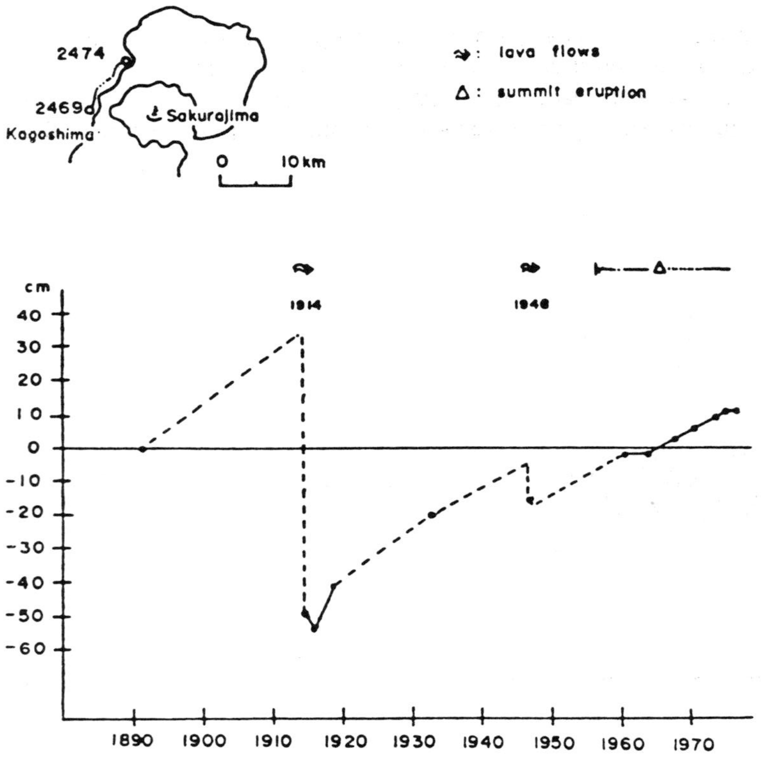

Figure 8.3.15. Elevation change at bench mark 2474 with respect to bench mark 2469 from 1891 to 1980, from Kamo (1978).

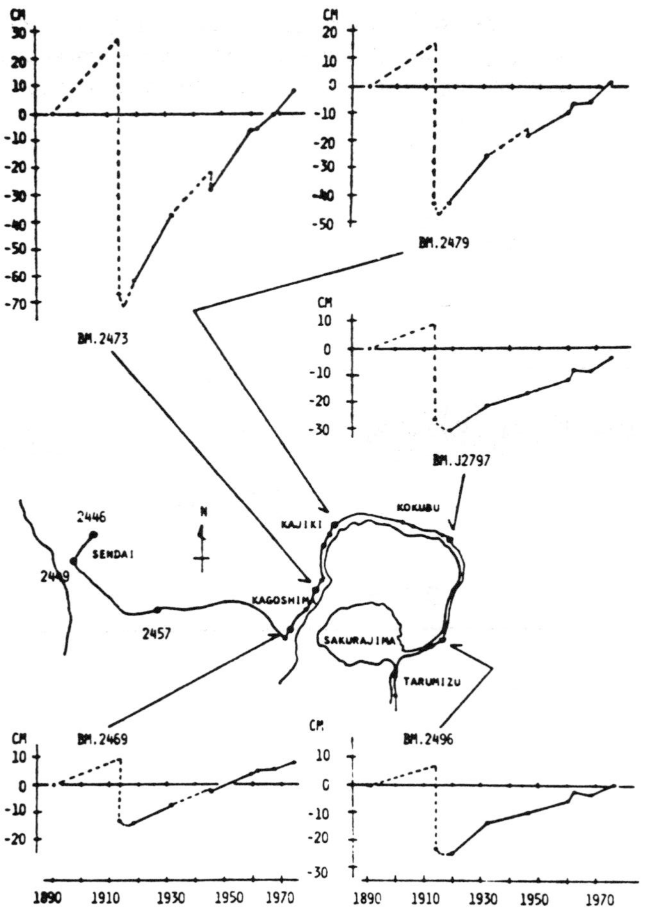

Figure 8.3.16. Vertical displacements at four bench marks along the rim of Aira Caldera from 1891 to 1975, from Kamo (1978).

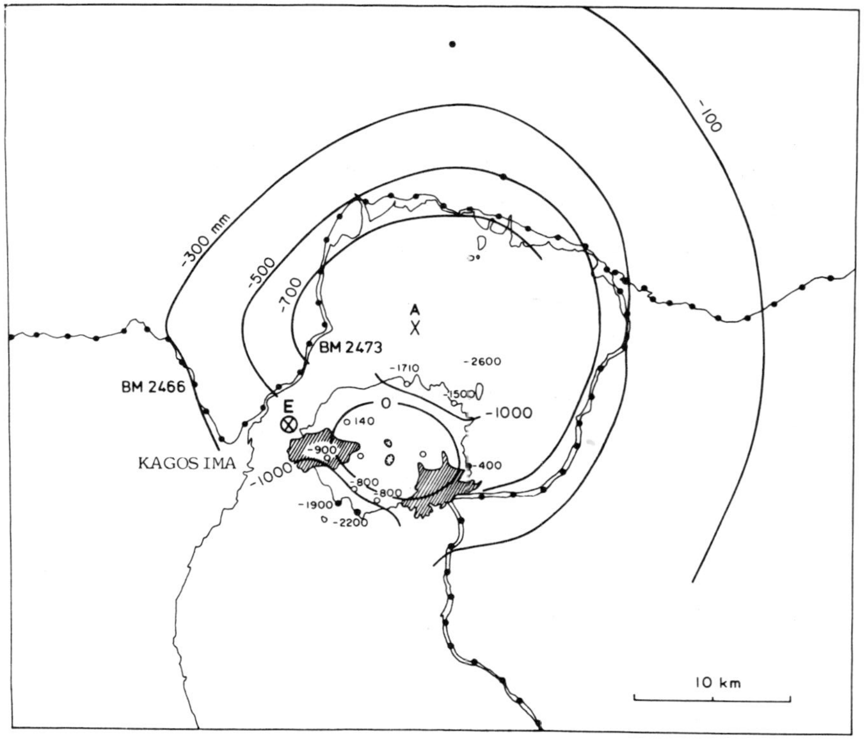

Figure 8.3.17. Vertical displacements in Aira Caldera between leveling surveys in 1894 and July 1914, 6 months after the 1914 eruption of Sakurajima, from Yokoyama (1974) after Omori (1916a). A, assumed center of subsidence; E, epicenter of large earthquake on first day of eruption.

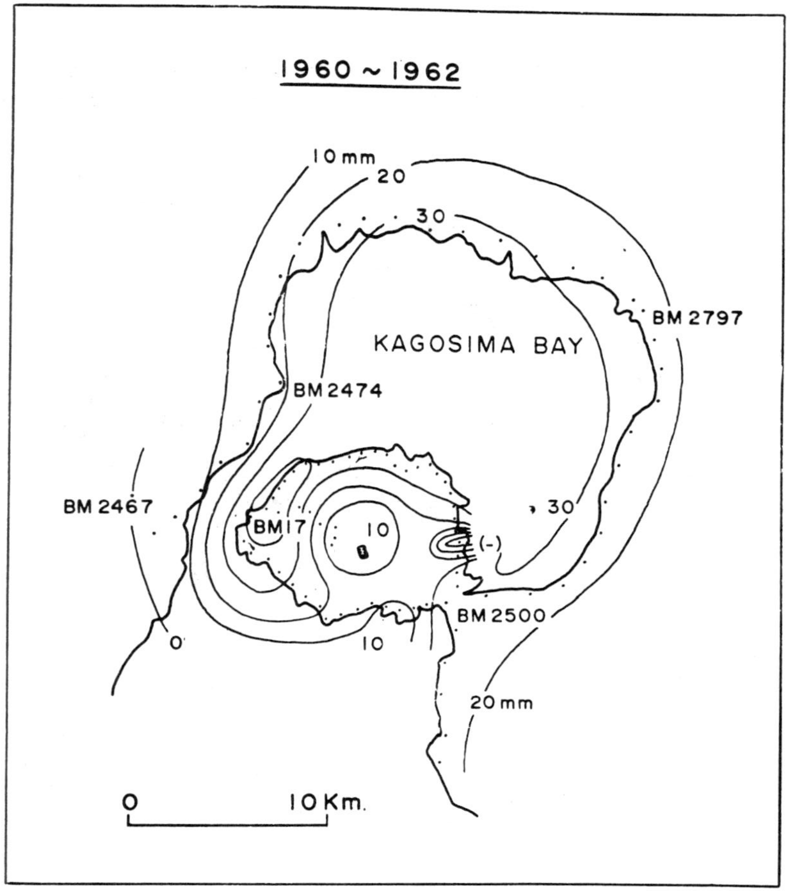

Figure 8.3.18. Uplift in Aira Caldera from 1960 to 1962, from Eto (1967). Contours in millimeters; bench mark 2467 assumed stable.

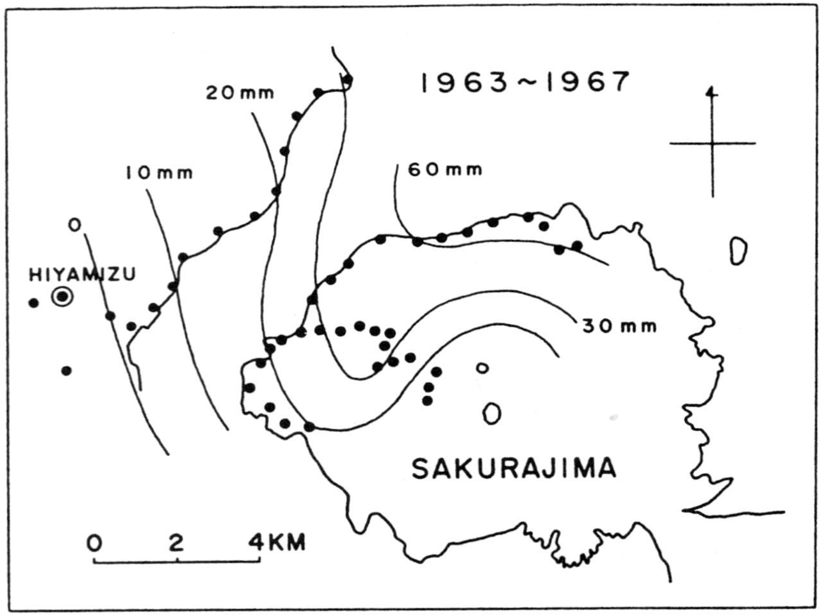

Figure 8.3.19. Elevation changes (uplift) at Aira Caldera during 1963-67, from Eto (1974). Note that scale of this figure is larger than scales of figures 8.3.18 and 8.3.20 -- this figure shows only Sakurajima and the southwest part of Aira Caldera.

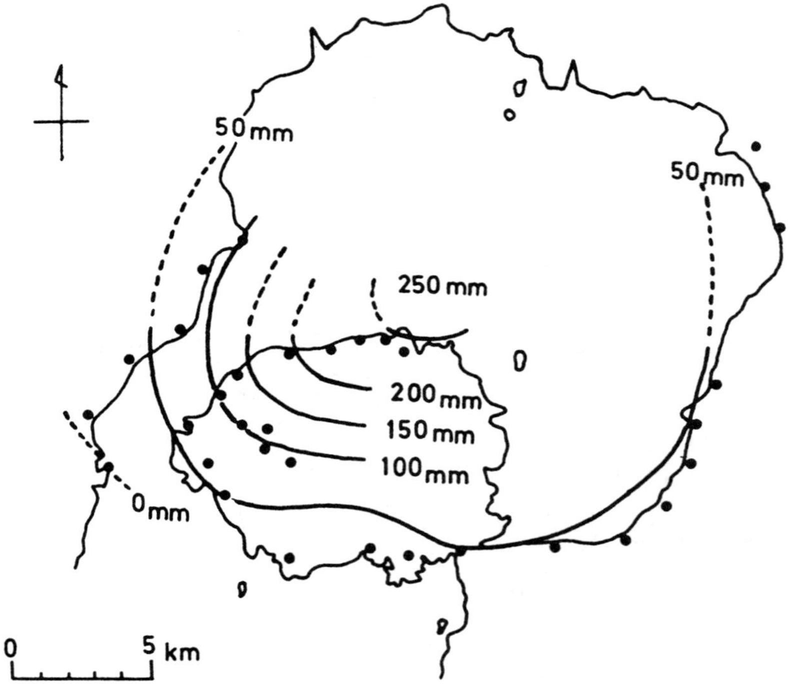

Figure 8.3.20. Elevation changes (uplift) at Aira Caldera between leveling surveys in 1965-68 and 1974-75, from Eto and Ishihara (1975a).

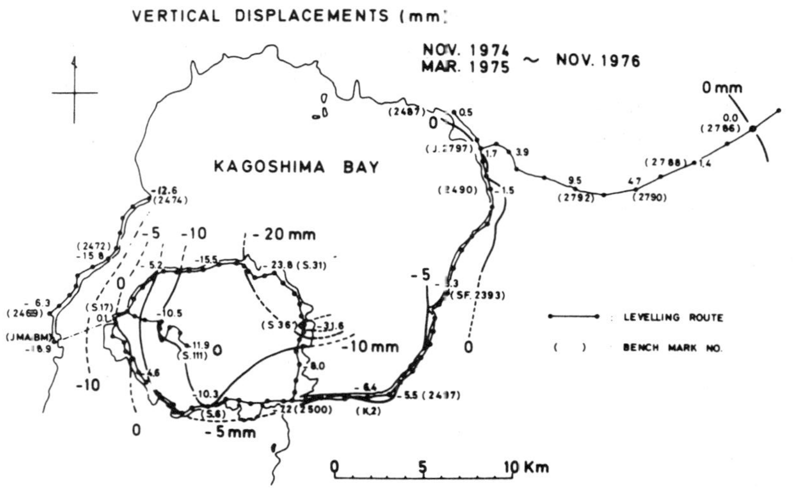

Figure 8.3.21. Vertical displacements at Aira Caldera from 1974 to 1976, from Eto and Ishihara (1977).

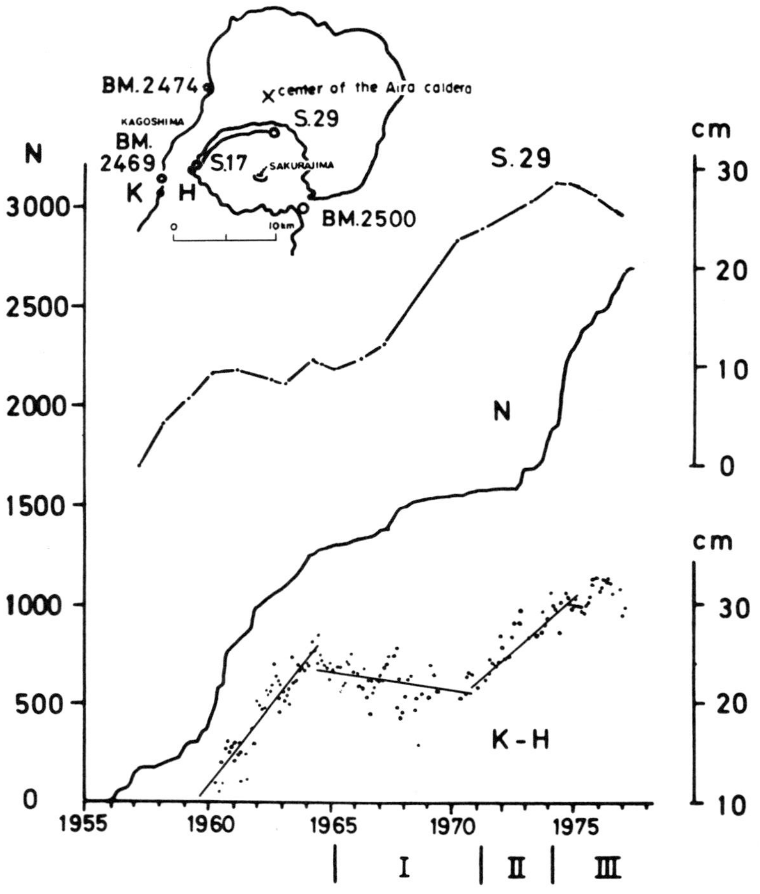

Figure 8.3.22. Elevation changes related to number of volcanic explosions at Sakurajima during 1955-75, from Eto and Ishihara (1977). Top, vertical displacement of bench mark 29 relative to bench mark 17 (H); middle, cumulative number of explosions; bottom, uplift at Sakurajima inferred from monthly mean-sea-level changes at Kagoshima tide gauge.

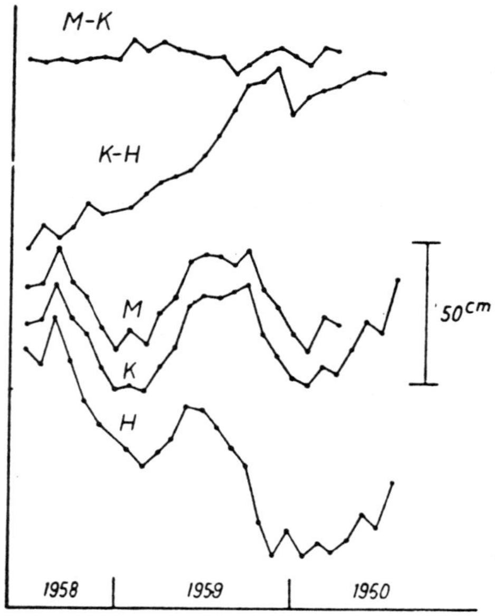

Figure 8.3.23. Changes in monthly mean-sea-level values at three stations, and changes in differences between stations, from Yoshikawa (1961). H, Hakamagoshi; K, Kagoshima; M, Makurazaki.

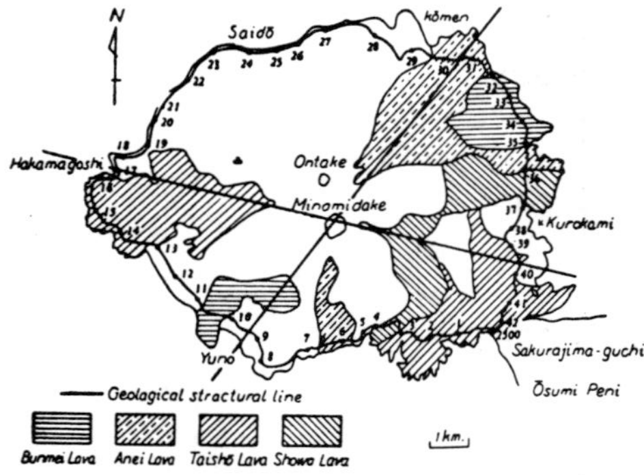

Figure 8.3.24. Locations of bench marks (numbers) and historical lava flows at Sakurajima, from Yoshikawa (1961).

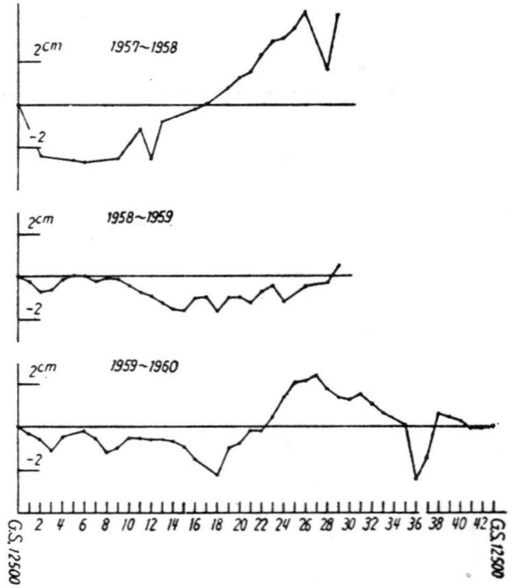

Figure 8.3.25. Vertical displacements of numbered bench marks during three periods, from Yoshikawa (1961). See figure 8.3.24 for locations of bench marks.

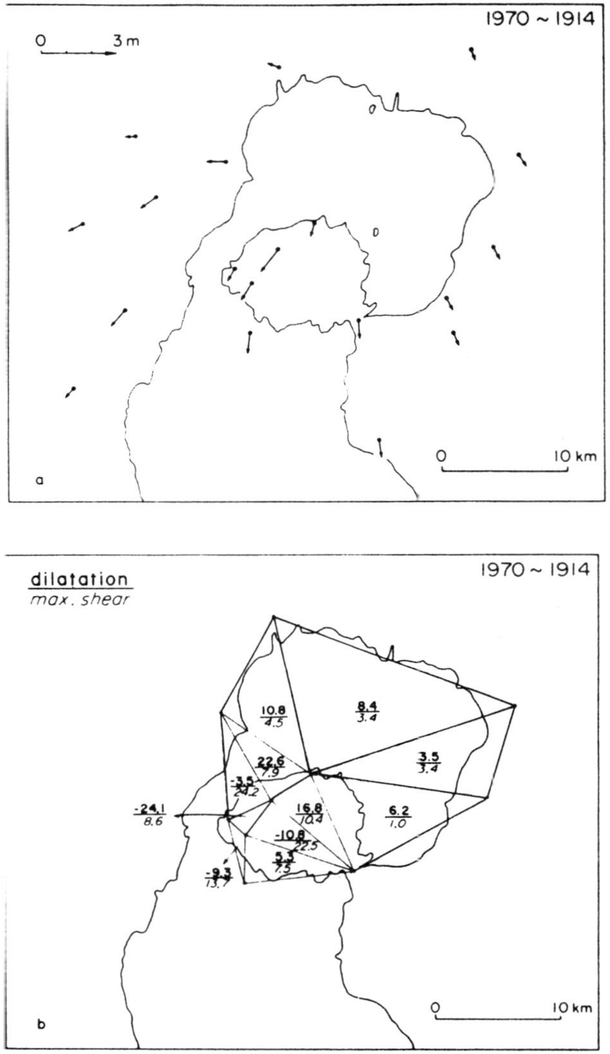

Figure 8.3.26. Results of triangulations around Aira Caldera from 1914 to 1970, from Yokoyama (1986). Top, horizontal displacements; bottom, dilatations and maximum shears, both in units of 10^{-5}.

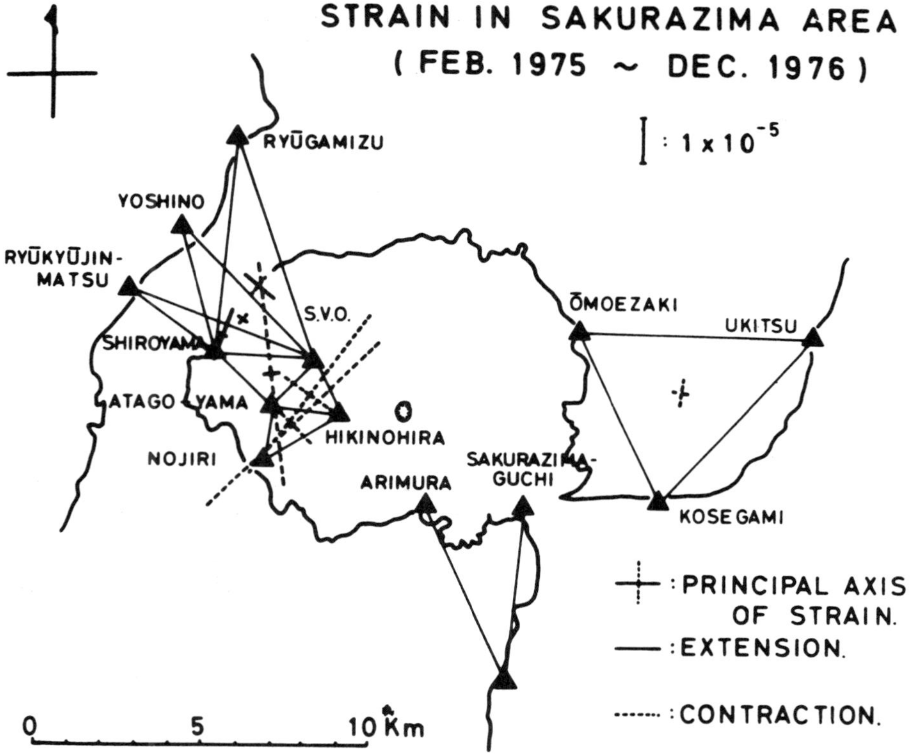

Figure 8.3.27. Horizontal strain in Sakurajima area from February 1975 to December 1976, from Eto and Ishihara (1977).

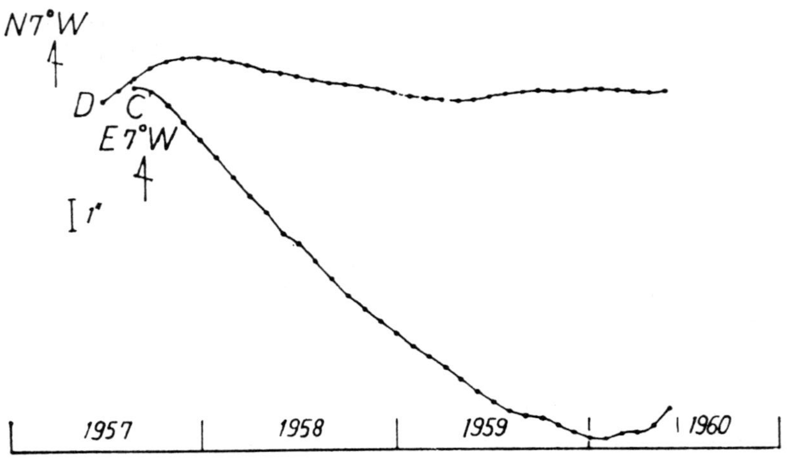

Figure 8.3.28. Secular variation of ground tilt at Hiyamizu (see figure 8.3.19 for location) during 1957-60, from Yoshikawa (1962).

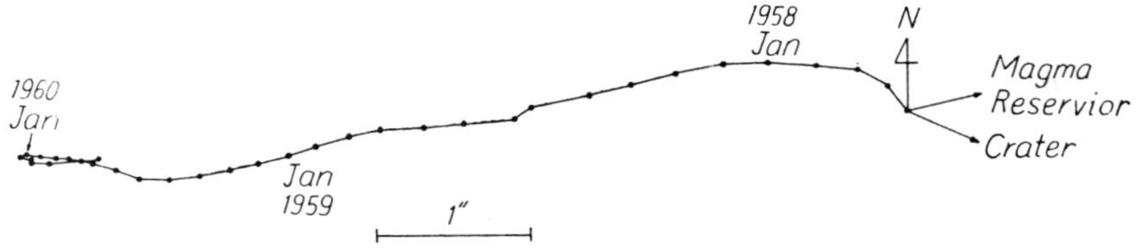

Figure 8.3.29. Vector diagram of secular tilt at Hiyamizu from 1957-60, from Yoshikawa (1962).

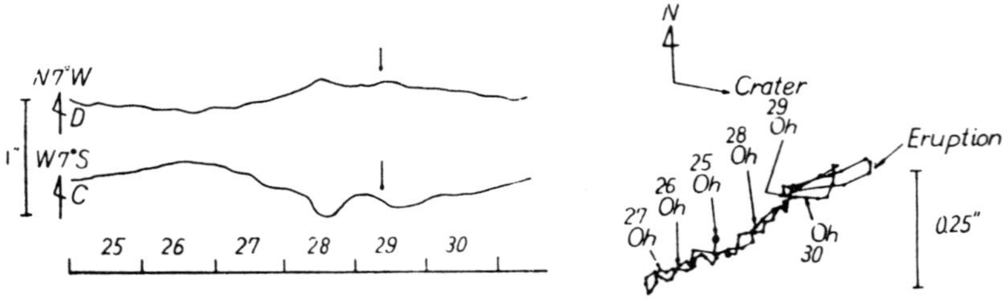

Figure 8.3.30. Anomalous tilt at Hiyamizu before the 29 March 1961 eruption of Sakurajima, from Yoshikawa (1962).

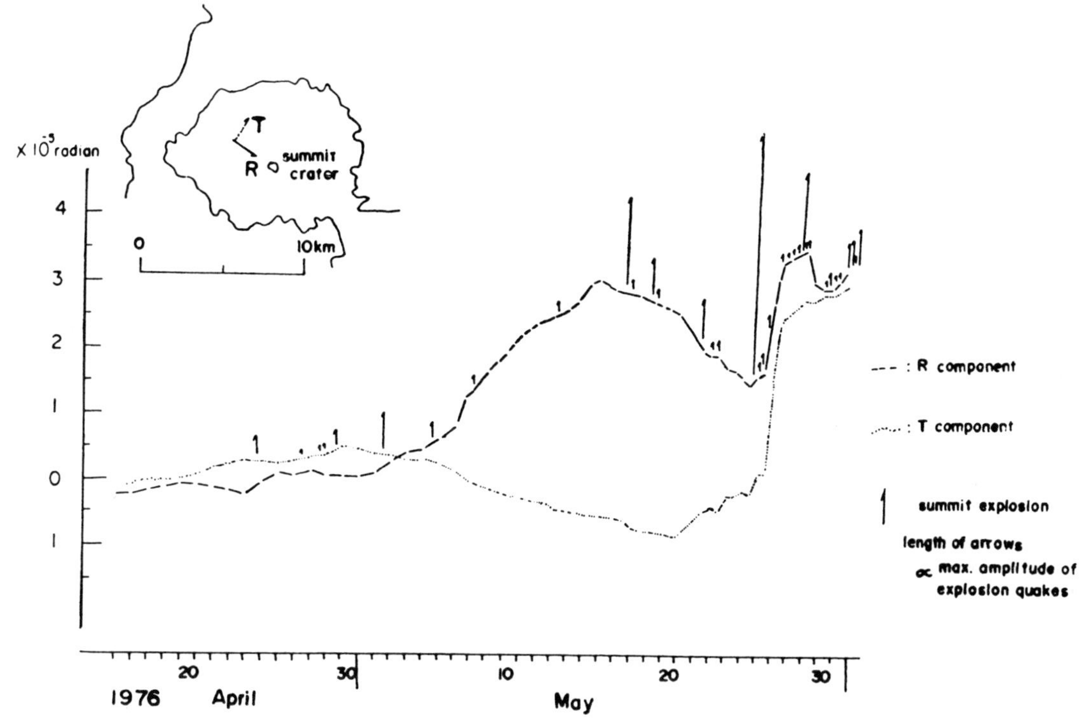

Figure 8.3.31. Radial and tangential components of tilt on northwest flank of Sakurajima, and eruptions during April-May 1976, from Kamo (1978).

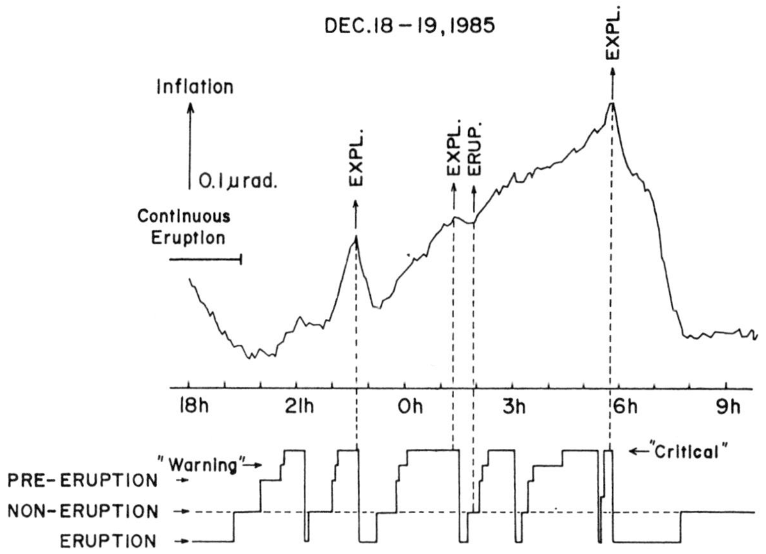

An example of the practice of the program. The stages judged by the computer in real time are compared with the actual eruptions and the tilt change in the radial direction.

Figure 8.3.32. An example of performance of an automatic warning system based on tilt changes at Sakurajima Volcano, from Kamo and Ishihara (1986a, b). During a test in December 1985, "critical" indications were as follows: 27 for 39 explosions, 4 for 5 eruptions, 7 for 22 continuous eruptions, and 21 for no remarkable event.

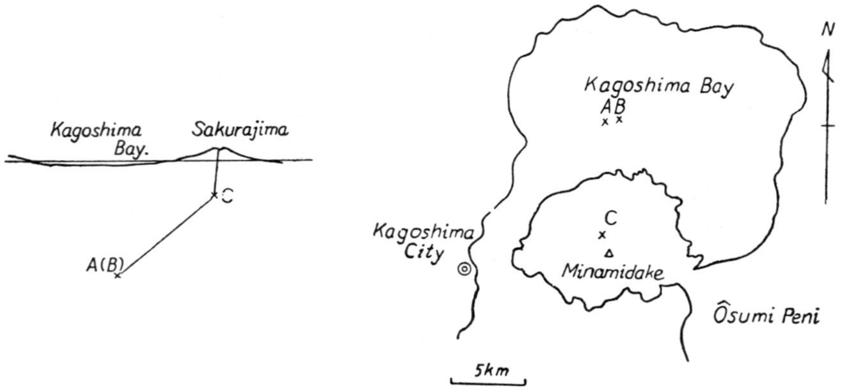

Figure 8.3.33. Deformation sources inferred from leveling at Aira Caldera, from Yoshikawa (1961). A, 1957-58; B, 1958-59 and 1959-60; C, 1959-60. Note apparent upward migration of pressure source.

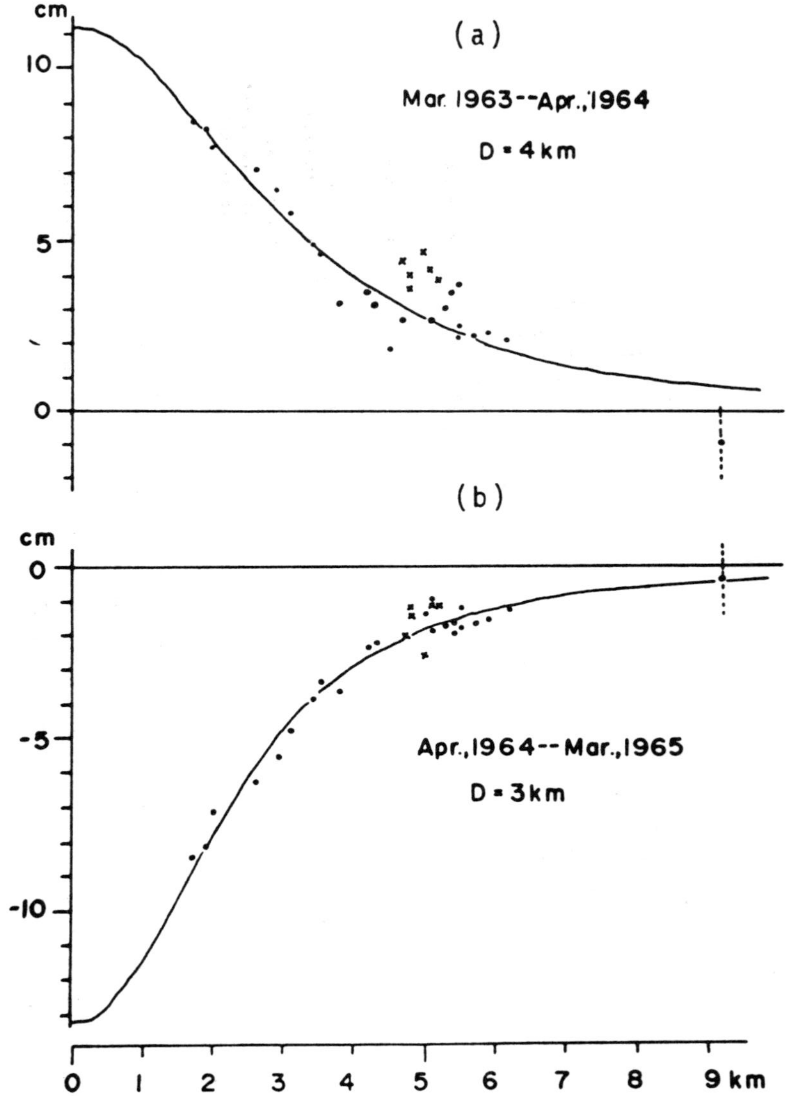

Figure 8.3.34. Radial distribution of vertical displacements around Sakurajima from (a) March 1963 to April 1964 and (b) April 1964 to March 1965 (Ishihara and Eto, 1978). Curves are calculated for a point source at indicated depths.

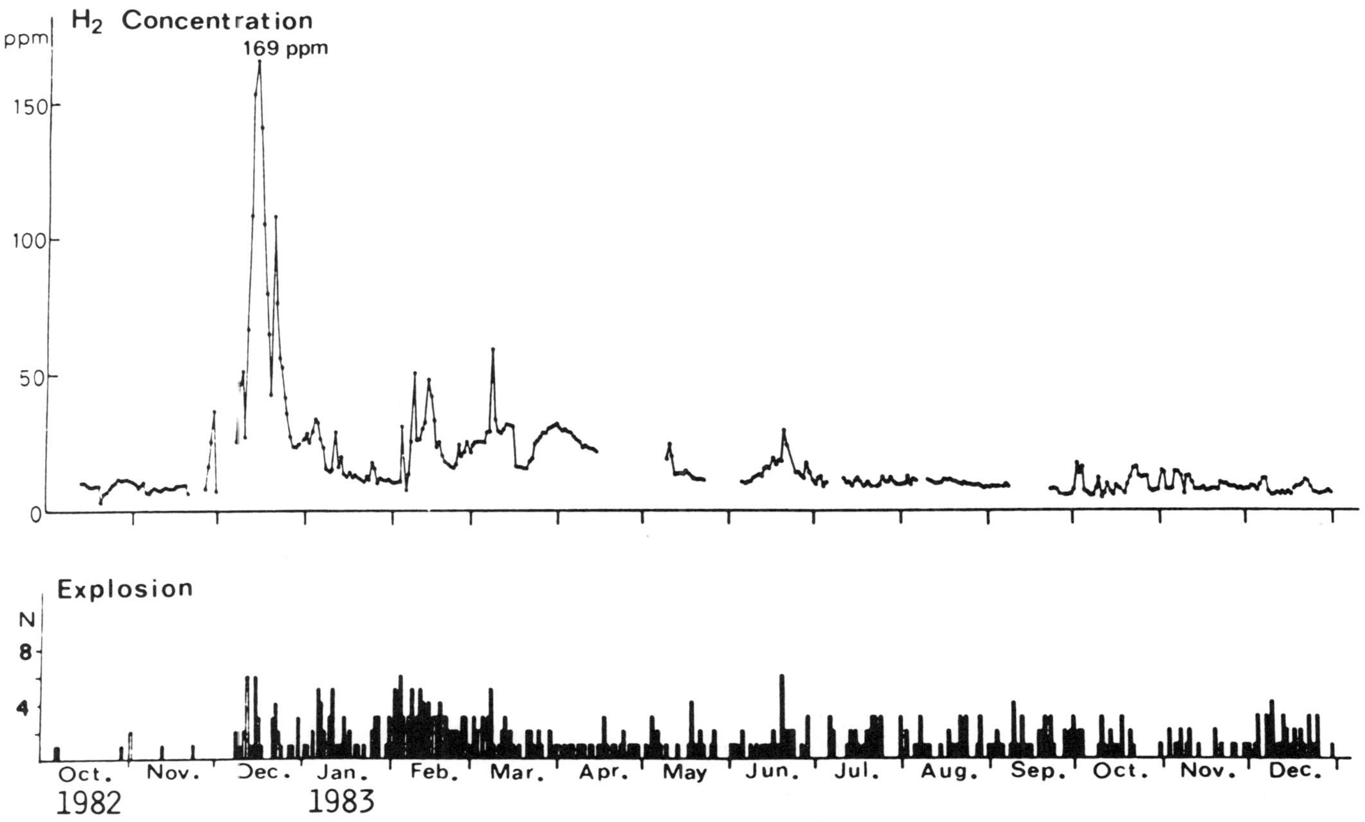

Figure 8.3.35. Top, variation of H_2 concentration in gas from a hot spring near southwest foot of Sakurajima; bottom, daily frequency of explosions in summit crater from October 1982 to December 1983 (Hirabayashi and others, 1986). Copyright by the Ame-ican Geophysical Union.

PART 3: HISTORICAL UNREST AT LARGE QUATERNARY CALDERAS (continued)
See inside back cover for explanation and abbreviations

KAKUTO

CAVW number (active vent)	Latitude Longitude (degrees)	Diameter (km)	Local tectonic setting	Precaldera edifice	SiO_2 content (wt pct)	Age of collapse (yr B.P.)	Date of unrest	Type and duration of unrest ESTU STHF MGTF H Te	Eruption type
08-02-09	32.03N	10 x 15	Exten?	Strato?	R = 50-67	110,000	742	---- ---- ---- - --	ex
(Kirishima)	130.82E	(Kakuto)			C = 73-75	65,000	788	---- ---- ---- - --	Ex
		12 x 16				12,000	837	---- ---- ---- - --	ex
		(Kobayashi)					843	---- ---- ---- - --	ex
		12 x 30					857-58	---- ---- ---- - --	ex
		(combined)					945	---- ---- ---- - --	ex
							1112	---- ---- ---- - --	ex
							1167	---- ---- ---- - --	ex
							1184	---- ---- ---- - --	ex
							1235	---- ---- ---- - --	ex
							1524	---- ---- ---- - --	ex
							1554	---- ---- ---- - --	ex
							1566	---- ---- ---- - --	ex
							1574	---- ---- ---- - --	ex
							1576-78	---- ---- ---- - --	ex
							1585	---- ---- ---- - --	ex
							1587-88	---- ---- ---- - --	ex
							1598-1600	---- ---- ---- - --	ex
							1617	---- ---- ---- - --	ex
							1628	---- ---- ---- - --	ex
							1637-38	---- ---- ---- - --	ex
							1659-64	---- ---- ---- - --	ex
							1677-78	---- ---- ---- - --	ex
							1706	---- ---- ---- - --	ex
							1716-17	---- ---- ---- - --	Ex,ex
							1719	---- ---- ---- - --	unknown
							1768	---- ---- ---- - --	lf,dm
							1771	---- ---- ---- - --	ex?
							1822	---- ---- ---- - --	ex
							1832	---- ---- ---- - --	ex
							1880	---- ---- ---- - --	ex

KAKUTO, Region 8, CAVW number 08-02-09

KAKUTO (continued)

CAVW number (active vent)	Latitude Longitude (degrees)	Diameter (km)	Local tectonic setting	Precaldera edifice	SiO_2 content (wt pct)	Age of collapse (yr B.P.)	Date of unrest	Type and duration of unrest ESTU STHF MGTF H Te	Eruption type
(continued from previous page)							1887-91	---- ---- ---- - --	ex
							1894-1900	---- ---- ---- - --	ex
							1903	---- ---- ---- - --	ex
							1913-14	EE-- ---- ---- - QU	ex
							1915	EE-- ---- ---- - --	none
							1923	---- ---- ---- - --	ex
							1934	---- ---- ---x x --	none
							1946	---- ---- ---- - --	ex
							1959	ZZ-- ---- ---- - --	pex
							1961	EE-- ---- ---- - QU	none
							1966-69	FFC- x-?- --F- F QU	none
							1971	NZ-- ---- --F- F --	pex,ex
							1974-76	EE-- xx-- ---- - --	none
							1977-79	FF-- xx-- --xx x --	pex
							1980-83	E--- ---- --FF - --	pex

TECTONIC SETTING/GEOLOGIC HISTORY

Kakuto Caldera, together with Kikai and Aira Calderas, lies within a NNE-SSW-trending graben that has been a remarkable locus of silicic volcanism during the Quaternary. Most workers have defined the Kakuto Caldera as an oval depression, 10 km x 15 km in diameter, northwest of Kirishima Volcano. However, Tajima and Aramaki (1980) discovered a gravity low just east of Kakuto Caldera, which they named the Kobayashi Caldera. Yokoyama and Ohkawa (1986) suggest that the Kakuto and Kobayashi Calderas are part of the same volcanic system and redefine the Kakuto Caldera as the figure-eight-shaped combination of the previously defined Kakuto and Kobayashi Calderas (fig. 8.4.1). Voluminous explosive eruptions are known to have occurred in the Kakuto area ca. 110,000 yr B.P., ca. 65,000 yr B.P., and ca. 12,000 yr B.P.. At least one of the first two eruptions corresponds to formation of the western caldera; the eastern (former Kobayashi) caldera could have formed in any of these three eruptions (see Machida, 1981, p. 175).

Kirishima Volcano is a chain of vents that extends southeastward from the southern rim of Kakuto Caldera. Historically active vents of Kirishima are aligned along a SE-NW axis (from southeast to northwest: Takachiho-mine, Ohachi, Shinmoe-dake, Karakuni-dake, and Tearai Hot Springs). Several geothermal fields are situated along the southwest side of Kirishima, and just north of Kirishima in the Kakuto Caldera.

KAKUTO (continued)

HISTORICAL ACTIVITY

Kirishima has erupted at least 60 times in the past 1,200 years, but there have been no historical eruptions within Kakuto Caldera. Activity at Kirishima seems to influence activity at Kakuto and vice versa.

1913-14: From May through December 1913, at least seven discrete seismic swarms occurred in the northwestern part of the caldera (Masaki area) (fig. 8.4.2), followed immediately by small eruptions at Takachiho-mine on 8 November and 9 December 1913, and on 8 and 13 January 1914. These small eruptions occurred, perhaps coincidentally, just before and during the large eruption of Sakurajima (60 km to the south-southwest) on 12 January 1914. Minakami and others (1970) also note a close temporal correlation with a large earthquake in the Hyuganada area, 100 km to the east (see also the descriptions of swarms in 1961 and 1968, below).

1915: The Kurino earthquake swarms, located in the southwestern part of the caldera about 2 km north of the Kurino-dake Spa (northwestern flank of Kirishima), began in July 1915 and lasted for one month. Three discrete swarms occurred, the second with more events than the first and the third with more than the second. The "intensity" of the Kurino swarms was greater than that of the Masaki swarms, but no eruption followed.

1923, 1934, 1946: Small explosive eruptions at Ohachi, Shinmoe-dake, and Karakuni-dake during 1923-46 were not accompanied by notable swarms in Kakuto Caldera. An event in July 1934, previously reported as a questionable eruption (Kuno, 1962), was an increase in gas emissions through the Shinmoe-dake crater lake. Beginning in the spring of 1934, the crater lake of Shinmoe became turbid from mud explosions on its floor, and fumarolic emissions from the crater floor reportedly killed water plants. Rumbling was heard on 25 July 1934, apparently from mud and gas explosions (Tanakadate, 1936).

1959: A small phreatic eruption of Shinmoe-dake occurred on 17 February 1959, with little or no precursory seismicity. This was the first eruption at Shinmoe-dake since 1832, as eruptions between 1832 and 1959 had been from the Ohachi vent. The 1959 eruption was followed shortly by swarms of earthquakes located 1-3 km beneath Shinmoe-dake and beneath another, older vent 5 km to the northwest. These swarms continued into March 1959, and then at a much reduced level for about one year.

1961: Following a $\underline{M}$=7.0 earthquake at Hyuganada (100 km east of Kirishima) on 27 February 1961, swarms began beneath Iimori-yama (northern foot of Kirishima, southern edge of Kakuto Caldera). The Iimori-yama swarms peaked about 17 days after the Hyuganada earthquake, continued until May 1961, and included in their later stages a few events beneath Shinmoe-dake and Takatiho-mine (fig. 8.4.6) (Minakami and others, 1970, p. 229).

1966-69: Small swarms occurred beneath Karakuni-dake in April 1966, beneath Kurino-dake in March and November 1967, and beneath Takatiho-mine in October-December 1967 (fig. 8.4.4). Thermal activity shifted westward from 1959 until at least 1968 (Minakami and others, 1968). Strong earthquake swarms began in early February 1968 near Ebino (figs. 8.4.2-8.4.4, 8.4.7), in the west-central part of the Kakuto Caldera. These peaked in late February and early March and declined nearly to background levels by early May. The

KAKUTO (continued)

HISTORICAL ACTIVITY (continued)

strongest earthquakes of these Ebino swarms occurred at 0851 and 1045 hr on 21 February 1968 (M 5.6 and M 6.1, respectively) and at 0059 hr on 25 February 1968 (M 5.6). Occasional small swarms were noted until at least the end of 1969. The epicentral area had a radius of about 4 km; hypocentral depths were between 2 and 9 km, mostly 5-6 km. Seismic b-values for various intervals during 1968-69 were between 1.0 and 1.5. Additional small swarms and volcanic tremor occurred beneath Shinmoe-dake on 6-8 March 1968; small swarms occurred beneath Nakadake (between Shinmoe-dake and Takachiho-mine) on 22-25 March and 15-18 April 1969, and beneath Ohachi and Takachiho-mine between 12 August and 31 December 1969. Fault-plane solutions for most of the larger 1968-69 earthquakes suggested either right-lateral shear on a NNW-SSE-trending fault or left-lateral shear on an ENE-WSW fault; solutions for a few earthquakes suggested point-source subsidence in the Kakuto Caldera (Minakami and others, 1969a). As in 1913 and again in 1961, there was a close coincidence between earthquakes at Hyuganada (100 km east of Kirishima), and earthquakes at Shinmoe-dake and the Kakuto Caldera; in the 1968 case, however, earthquakes began first in the Kakuto Caldera, were followed two weeks later by earthquakes at Shinmoe-dake, and six weeks later by a M 7.5 earthquake at Hyuganada (Minakami and others, 1969c, p. 729-730).

A short leveling line within the southern part of the caldera and another on the northwest flank of Kirishima Volcano showed small changes between March 1967 and November 1969. The greatest change along the caldera line was 1.6 cm of subsidence. However, this value may be deceptively small: tilts inferred from successive points along the leveling lines are as much as 100 microradians (down to the east), which, if extrapolated, implies as much as several tens of centimeters of uplift several kilometers to the west, or a similar amount of subsidence to the east. An EDM (electronic distance measurement) network was established in the area just after the 1968 earthquakes; early data did not yield a clear picture but suggested NE-SW extension (Eto and Ohya, 1972).

1971: A small phreatic explosion occurred at Tearai Hot Springs, at the northern foot of Kirishima, on 5 August 1971. No precursory seismicity was noted other than that which had been occurring across the entire Kirishima-Kakuto area since 1966. There was a small earthquake swarm beneath Kurino-dake on 21 November 1971. Also in the early 1970's, strong fumarolic activity developed at Ebino-kogen, on the west flank of Kirishima near the Kirishima Volcano Observatory and the Ebino-Kogen Visitor Center; by 1981, this activity had migrated to the southwest, and only a trace remained by the visitor center.

1974-76: In September 1974, seismicity in the caldera increased slightly to as many as 20 events/day, compared with a background level of 1-2 events/day. During October-December, activity shifted southeastward to the Kirishima complex (Miyazaki and Shimozuru, 1975). Much more notable seismic swarms occurred near the east end of the caldera, in the vicinity of the town of Iino, for several months in late 1975 and early 1976 (M_{max}=4.5, depths mostly 2-10 km) (fig. 8.4.5). A maximum of 0.77 cm of subsidence was recorded, corresponding to an inward tilt of 1.3 microradians.

1977-79: A prolonged swarm of small earthquakes occurred beneath Takachiho-mine from December 1977 to early July 1978, and a brief swarm followed at Shinmoe-dake in early July (Kirishima Volcano Observatory, 1979). This northwestward migration of seismicity (from Kirishima toward the caldera; fig. 8.4.8) culminated in four seismic swarms in Kakuto Caldera: (1) August-October 1978, with

KAKUTO (continued)

HISTORICAL ACTIVITY (continued)

epicenters in the eastern part of the caldera, near the 1975 epicenters; (2) November 1978 - February 1979, with epicenters extending from the west-central part of the caldera to the western rim; (3) March-May 1979, with epicenters concentrated in a small area on the western ring fracture; and (4) June-December 1979, with epicenters dispersed throughout the western half of the caldera (Ishihara, 1980). The most intense swarms were in August 1978 and March 1979 ($\underline{M}_{max}$=4.9). Tiltmeters at the then newly relocated Yoshimatsu Observatory (at the western end of the caldera) recorded more than 170 microradians of tilt, down toward the center of the caldera, between August 1978 and December 1979; tilt was continuous, and it accelerated before several of the larger earthquakes (figs. 8.4.10, 8.4.11) (Ishihara, 1980, p. 27). Nearly 50 microradians of tilt occurred in the two weeks preceding the March 1979 swarm, which included the $\underline{M}$=4.9 earthquake and a $\underline{M}$=4.0 earthquake, both centered 7 km from the observatory. The sense of tilt in the 1978-79 episode is consistent with that observed in the previous decade or more--subsidence in the eastern part of the caldera--suggesting a possible relation to a westward migration of thermal activity, noted above. A small sulfur flow (10 cm wide, 50 m long), surrounded by explosively ejected sulfur, was found at Shinmoe-dake in the spring of 1979 and must have been erupted during the winter of 1978-79.

1980-83: A pit inside Shinmoe-dake crater enlarged in 1980 and ejected "tar-like materials." Minor volcanic earthquake swarms occurred near Shinmoe-dake on 18-20 July 1980, and an isolated earthquake (JMA intensity=3) occurred near Karakuni-dake on 3 December 1980. Fumarolic activity and temperatures at Shinmoe-dake crater started to increase in December 1980; this activity culminated with a 6-m-long, 30-cm-wide, 2-cm-thick sulfur flow, erupted at some time before 8 January 1982. The flow issued from a fumarole that in December 1980 had a temperature of 184 °C, the highest observed since temperature measurements were begun in the area in 1979. Fumarolic activity also increased at the Io-dani area, 5 km west-southwest of Shinmoe-dake, from November 1980 through 1981.

Small earthquake swarms occurred at Shinmoe-dake in the latter half of 1981, in June 1982, and in January-December 1983. Another earthquake swarm occurred in August 1982 beneath the south flank of Mt. Kurino-dake (fig. 8.4.9), at the northwest end of the Kirishima chain and near the area of the 1915 earthquake swarm ($\underline{M}_{max}$=3-4, depths mostly 3-6 km) (Kirishima Volcano Observatory, 1983).

COMMENTS

A close temporal relation exists between swarms at Kakuto and both swarms and eruptions at Kirishima. In addition, unrest in 1913, 1961, and 1968 at Kakuto and Kirishima occurred just before or just after large earthquakes at Hyuganada (100 km east of Kirishima), and the unrest in 1913 occurred just before the large eruption and earthquake at nearby Sakurajima Volcano. The caldera and Kirishima seem to be responding to regional tectonic strain.

KAKUTO (continued)

REFERENCES

Aramaki, S., 1968, Geology of the Kakuto Basin, southern Kyushu, and the earthquake swarm from February, 1968: Bull. Earthq. Res. Inst., v. 46, p. 1325-1343.

Earthquake Research Institute, Univ. of Tokyo, 1983, Microearthquake swarm in the southern flank area of Mt. Kurino-dake in Kirishima Volcanoes: Rep. Coord. Comm. for Prediction of Volcanic Eruptions, no. 26, p. 49-50.

Eto, T., and Ohya, F., 1972, Observation of crustal deformation in the Sakurajima and Yoshimatsu areas by electro-optical means (1): Bull. Volcanol. Soc. Japan, v. 17, p. 135-143.

Hedervari, P., 1975, On the three dimensional distribution of aseismic domains beneath Volcano Kirisima, Kyushu, Japan: Pacific Geol., v. 9, p. 101-112.

Hiraga, S., 1972, Earthquake swarms of geothermal fields in Japan: Jour. Japanese Geotherm. Energy Assoc., v. 9, no. 1, ser. 32, p. 30-39.

Ishihara, K., 1980, A seismic observation at Kakuto caldera in southern Kyushu, Japan (II), Seismic activity at the western somma and ground-tilt in 1978 and 1979: Ann. Rep. Disaster Prev. Res. Inst., Kyoto Univ., v. 23:B-1, p. 21-28.

----- 1982, A seismic observation at Kakuto caldera in the southern Kyushu, Japan (III)--Abnormal tilt changes before small earthquakes: Ann. Rep. Disaster Prev. Res. Inst., Kyoto Univ., v. 25B-1, p. 11-22.

Ishihara, K., and Sudo, Y., 1979, A seismic observation at Kakuto Caldera in southern Kyushu, Japan (I)--A comparison of the locations of hypocenters obtained from tripartite observation at two sites: Ann. Rep. Disaster Prev. Res. Inst., Kyoto Univ., v. 22:B-1, p. 35-44.

Kagiyama, T., Uhira, K., Watanabe, T., Masutani, F., and Yamaguchi, M., 1979, Geothermal survey of the Volcanoes Kirishima: Bull. Earthq. Res. Inst., v. 54, p. 187-210.

Kaminuma, K., 1973a, Relation between volcanic activities of Sakurajima and Kirishima volcanoes: Seismicity and Volcanic Activity in Kagoshima Prefecture, v. 5, p. 29-36.

----- 1973b, Unusual shocks felt at Kirishima-machi, Kagoshima Prefecture: Seismicity and Volcanic Activity of Kagoshima Prefecture, v. 5, p. 37-39.

Kaminuma, K., Masutani, F., and Yamaguchi, M., 1974, The anomalous seismic activity in the Takachiho-no-mine, Kirishima Volcanoes, Kyushu, Japan: Bull. Volcanol. Soc. Japan, v. 19, p. 11-18.

Kaminuma, K., Yamaguchi, M., and Masutani, F., 1973, The seismic activity in the Ebino area, Kirisima Volcano: Bull. Volcanol. Soc. Japan, v. 18, p. 1-10.

----- 1974, The seismic activity in and near the Shinmoe-dake and the Takachiho-no-mine, Kirishima Volcanoes, Kyushu, Japan: Bull. Volcanol. Soc. Japan, v. 19, p. 1-10.

Kirishima Volcano Observatory, 1979, Seismic activities in and around the Kirishima Volcanoes: Rep. Coord. Comm. for Prediction of Volcanic Eruptions, no. 14, 1979, no. 3, p. 73-77.

----- 1983, Seismic activities in and around the Kirishima Volcanoes: Rep. Coord. Comm. for Prediction of Volcanic Eruptions, no. 25, 1983, no. 3, p. 49-50.

Kobayashi, T., Aramaki, S., Watanabe, T., and Kamada, M., 1981, Kirishima Volcano, in Kubotera, A., ed., Field excursion guide to Sakurajima, Kirishima and Aso Volcanoes: Tokyo, Volcanological Society of Japan, p. 17-32.

Kuno, H., 1962, Catalogue of the active volcanoes of the world including solfatara fields, Pt. XI, Japan, Taiwan and Marianas: Rome, IAVCEI, 332 p.

KAKUTO (continued)

REFERENCES (continued)

Machida, H., 1981, Tephrochronology and Quaternary studies in Japan, in Self, S., and Sparks, R.S.J., eds., Tephra studies: Dordrecht, D. Reidel, p. 161-191.

Minakami, T., 1974, Seismology of volcanoes in Japan, in Civetta, L., Gasparini, P., Luongo, G., and Rapolla, A., eds., Physical volcanology: Amsterdam, Elsevier, p. 1-27.

Minakami, T., Hagiwara, M., Yamaguchi, M., Koyama, E., and Hirai, K., 1970, The Ebino earthquake swarms and the seismic activity of the Kirishima volcanoes, in 1968-1969, Pt. 4, Shifts of seismic activity from the Kakuto Caldera to Simmoe-dake, Naka-dake and Takatiho-mine: Bull. Earthq. Res. Inst., v. 48, p. 205-233.

Minakami, T., Hiraga, S., Miyazaki, T., and Terao, H., 1969a, The Ebino earthquake swarms and the seismic activity of the Kirishima volcanoes, in 1968-1969, Pt. 2, Geographical distribution of initial motions and travel time curves along the Kirishima Volcanoes: Bull. Earthq. Res. Inst., v. 47, p. 745-767.

Minakami, T., Hiraga, S., Miyazaki, T., and Uchibori, S., 1969b, Fundamental research for predicting volcanic eruptions, Pt. II: Bull. Earthq. Res. Inst., v. 47, p. 893-949.

Minakami, T., Shimozuru, D., Miyazaki, T., Hiraga, S., and Yamaguchi, M., 1968, The 1959 eruption of Simmoedake and the 1951 Iimore-yama earthquake swarm: Bull Earthq. Res. Inst., v. 46, p. 965-992.

Minakami, T., Utibori, S., Yamaguchi, M., Gyoda, N., Utsunomiya, T., Hagiwara, M., and Hirai, K., 1969c, The Ebino earthquake swarms and the seismic activity of the Kirishima volcanoes, in 1968-1969, Pt. 1, Hypocentral distribution of the 1968 Ebino earthquakes inside the Kakuto Caldera: Bull. Earthq. Res. Inst., v. 47, p. 721-743.

Miyazaki, T., Hiraga, S., and Minakami, T., 1969, The Ebino earthquake swarms and the seismic activity of the Kirishima volcanoes, in 1968-1969, Pt 3., Crustal deformation inside the Kakuto Caldera relating to the 1968 Ebino earthquakes: Bull. Earthq. Res. Inst., v. 47, p. 769-781.

Miyazaki, T. and Shimozuru, D., 1975, Activity of Kirishima Volcano: Seismicity and Volcanic Activity of Kagoshima Prefecture, v. 7, p. 52-57.

Miyazaki, T., and Watanabe, T., 1979, Volcanic activity of Kirishima volcano--seismic activity near Ohachi crater, Takachicho: Seismicity and Volcanic Activity of Kagoshima Prefecture, v. 11, p. 65-70.

Miyazaki, T., Yamaguchi, M., Masutani, F., and Terao, H., 1976, The earthquake swarms in the northern area of the Kirishima Volcanoes, 1975-1976: Bull. Earthq. Res. Inst., v. 51, p. 115-149.

Smithsonian Institution, Scientific Event Alert Network (SEAN), 1979, 1982, Kirishima Volcano: SEAN Bull., v. 4, no. 6, p. 7; v. 7, no. 1, p. 8 (contributions from Manabu Komiya and Office of Volcanic Observations, Japan Meteorological Agency).

Tajima, H., and Aramaki, S., 1980, Bouguer gravity anomaly around Kirishima volcanoes, Kyushu: Bull. Earthq. Res. Inst., v. 55, p. 241-257.

Tanakadate, H., 1936, Volcanic activity in Japan and vicinity during the period between July 1934 and October 1935: Japanese Jour. Astron. Geophys., v. 13, p. 121-139.

Volcanological Society of Japan, 1982, Kirisima: Bull. Volcanic Eruptions, no. 20, p. 95.

Yokoyama, I., and Ohkawa, S., 1986, The subsurface structure of the Aira Caldera and its vicinity in southern Kyushu, Japan: Jour. Volcanol. Geotherm. Res., v. 30, p. 253-282.

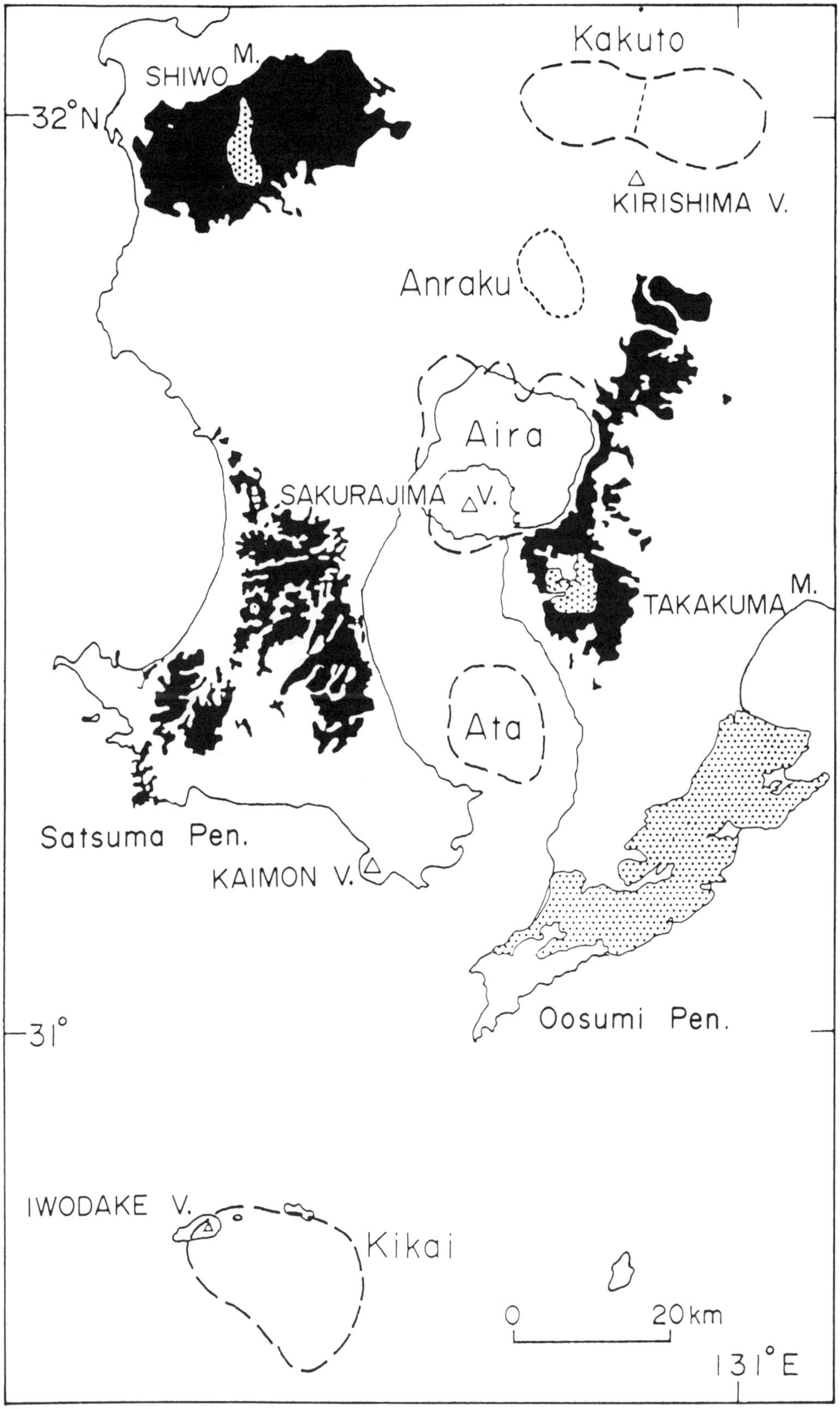

Figure 8.4.1. Geological sketch map of southern Kyushu, from Yokoyama and Ohkawa (1986). Kakuto and Kobayashi Calderas are shown as a single, twin caldera (Kakuto). Black, Mesozoic basement; dotted, Mesozoic granite; broken line, caldera boundary.

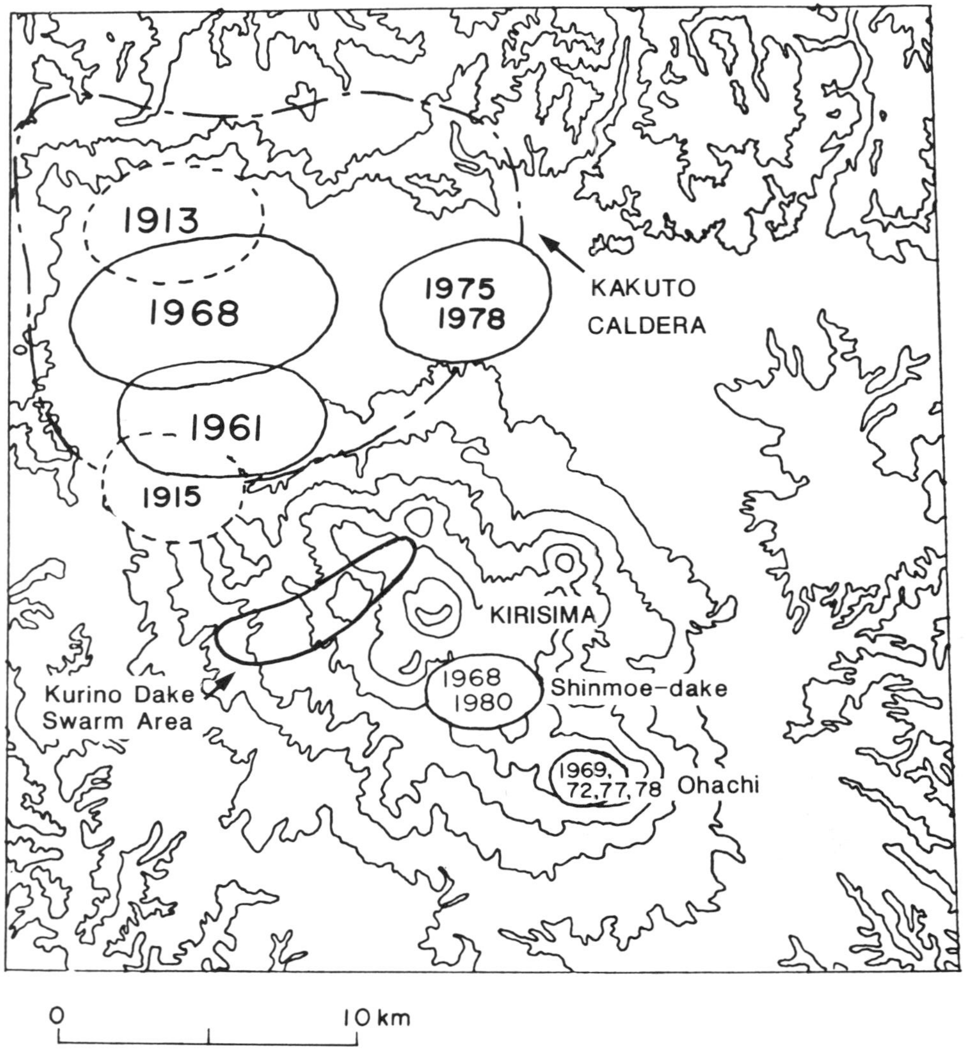

Figure 8.4.2. Epicentral areas of seismic swarms at Kakuto Caldera from 1913 to 1982, after Kobayashi and others (1981). Epicentral areas dashed where approximate.

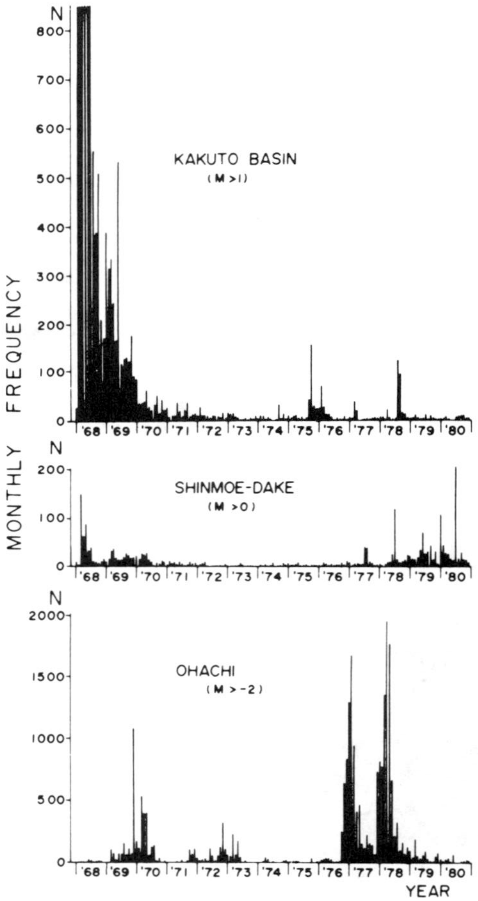

Figure 8.4.3. Monthly frequency of earthquakes in Kakuto basin, Shinmoe-dake, and Ohachi regions during 1968-80, from Kobayashi and others (1981).

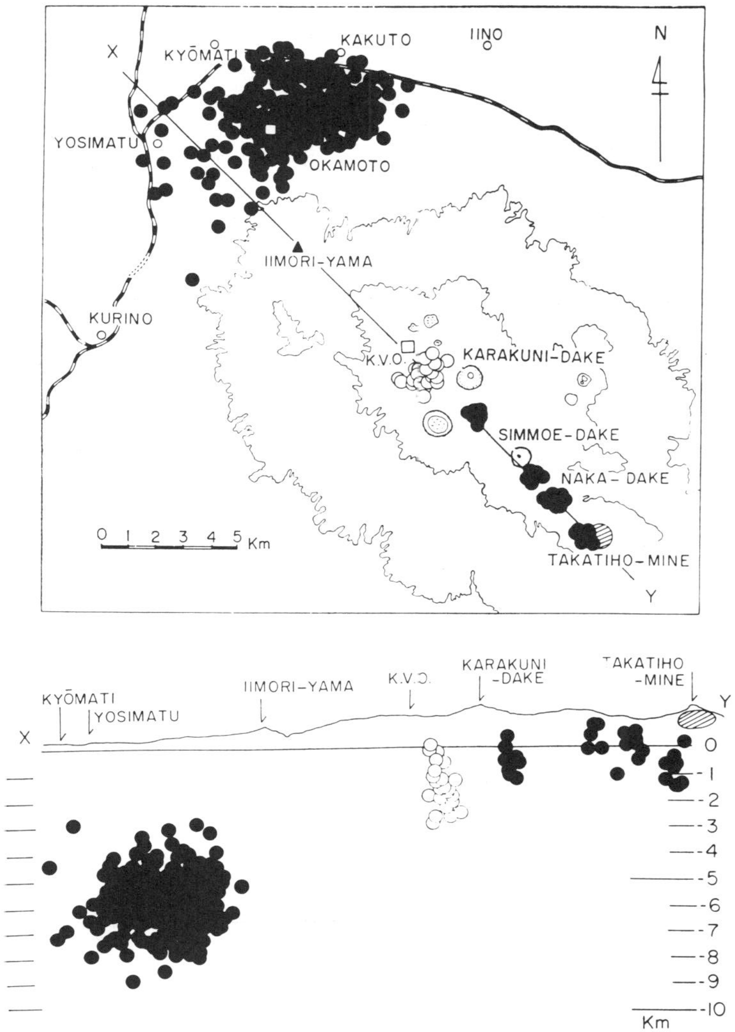

Figure 8.4.4. Hypocentral distributions of a series of earthquake swarms at the Kirishima volcanoes, from Minakami and others (1970). Solid circles: 1968-69 Ebino swarm, March 1968 Simmoe-dake swarm, March 1969 Naka-dake swarm, and October-December 1967 Takatiho-mine swarm. Hachured area: October-December 1969 Takatiho-mine swarm (B-type earthquakes). Open circles: April 1966 Karakuni-dake swarm.

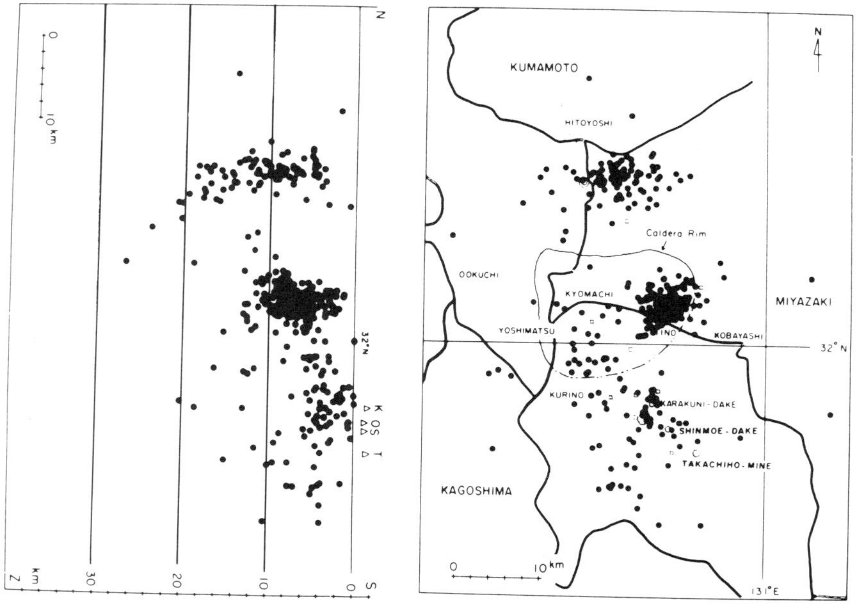

Figure 8.4.5. Distribution of earthquake hypocenters near Kakuto Caldera from 1 January 1975 to 31 March 1976, from Kobayashi and others (1981).

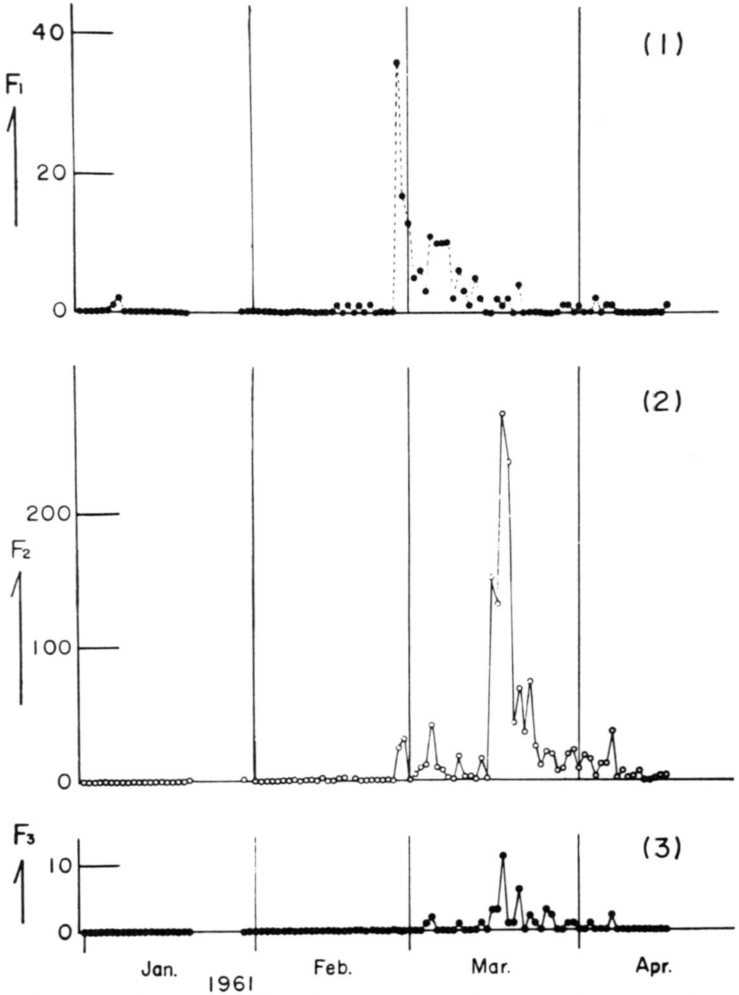

Figure 8.4.6. Seismicity near Kakuto Caldera during January-April 1961, from Minakami and others (1968). 1, daily number of earthquakes at Hyuganada (100 km east of Kirishima); 2, daily number of earthquakes at Kirishima Volcano; 3, number of earthquakes felt at Ebino. Note apparent correlation between tectonic earthquakes at Hyuganada and swarms at Kirishima Volcano and Kakuto Caldera (see also figure 8.4.7).

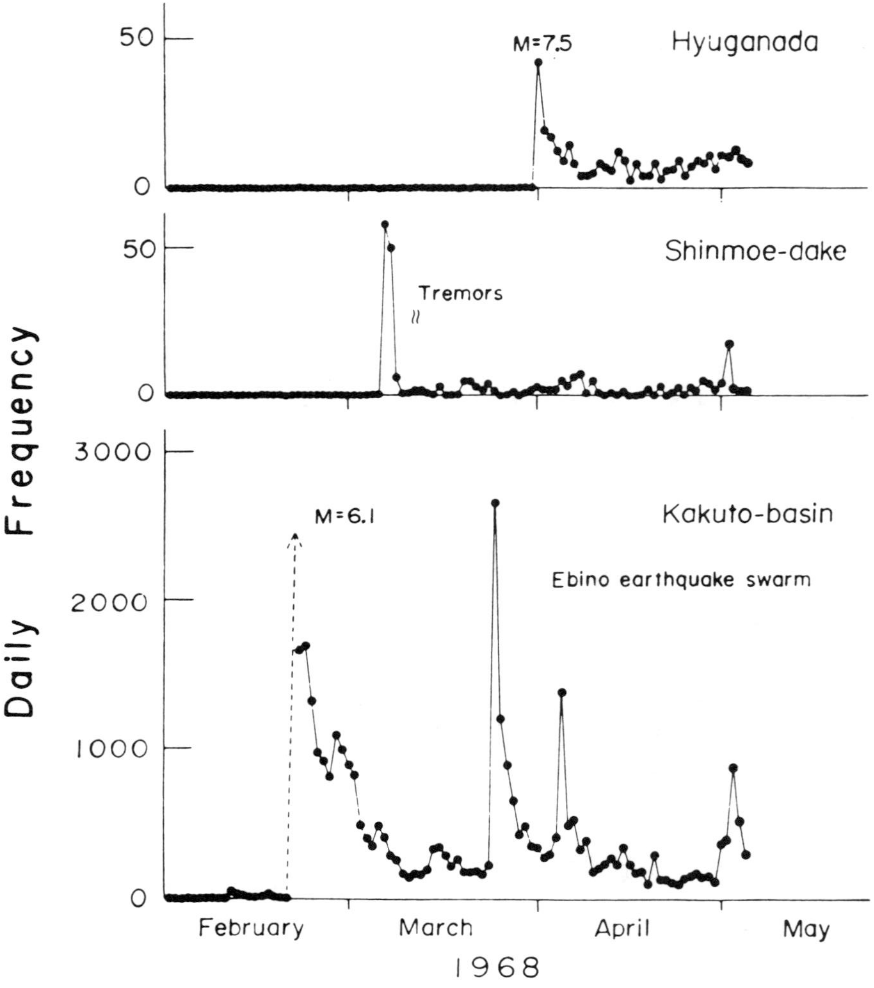

Figure 8.4.7. Seismicity at Hyuganada, Shinmoe-dake, and the Kakuto basin during February-May 1968, from Minakami and others (1969c). Earthquakes at Hyuganada preceded those at Kakuto and Kirishima in 1961 (figure 8.4.6), but followed those at Kakuto and Kirishima in 1968.

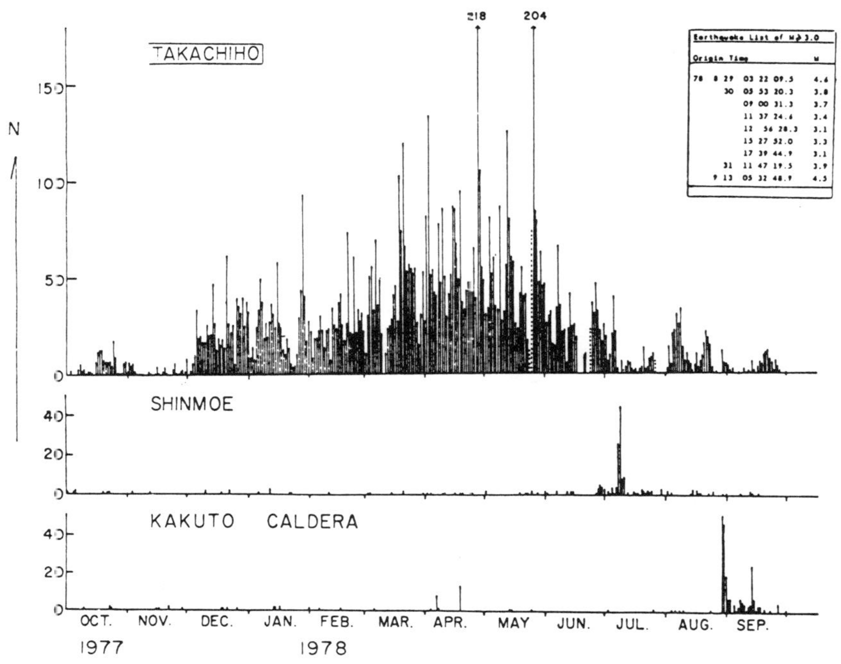

Figure 8.4.8. Numbers of earthquakes at Takachiho-mine, Shinmoe-dake, and Kakuto during October 1977-September 1978, from Kirishima Volcano Observatory (1979). Note migration toward Kakuto Caldera.

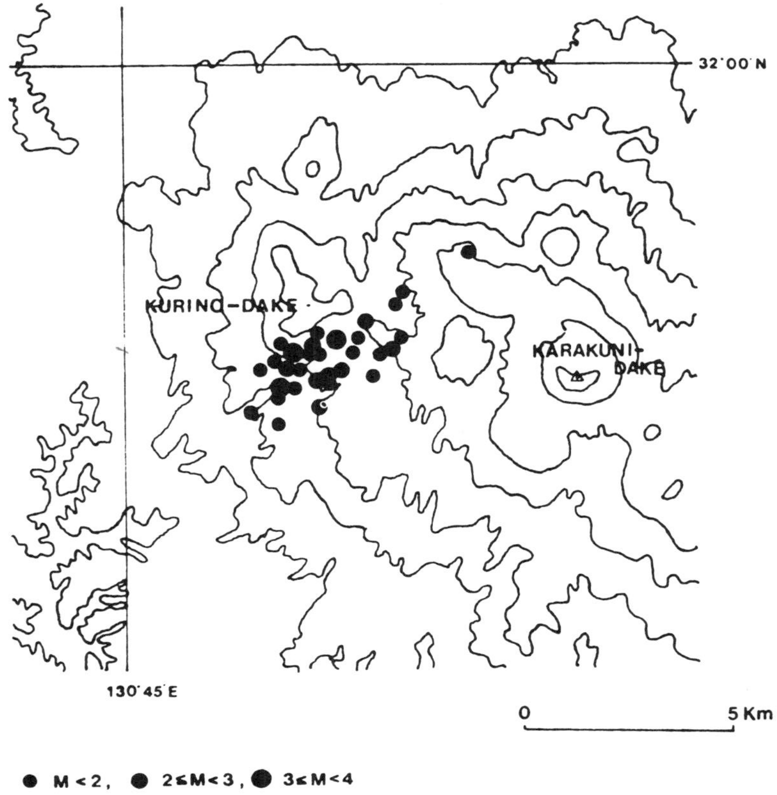

Figure 8.4.9. Microearthquake swarm in southern flank area of Mt. Kurino-dake (one of the Kirishima volcanoes) during 17-18 August 1982, from Earthquake Research Institute, University of Tokyo (1983). Depths ranged from 0 to 10 km.

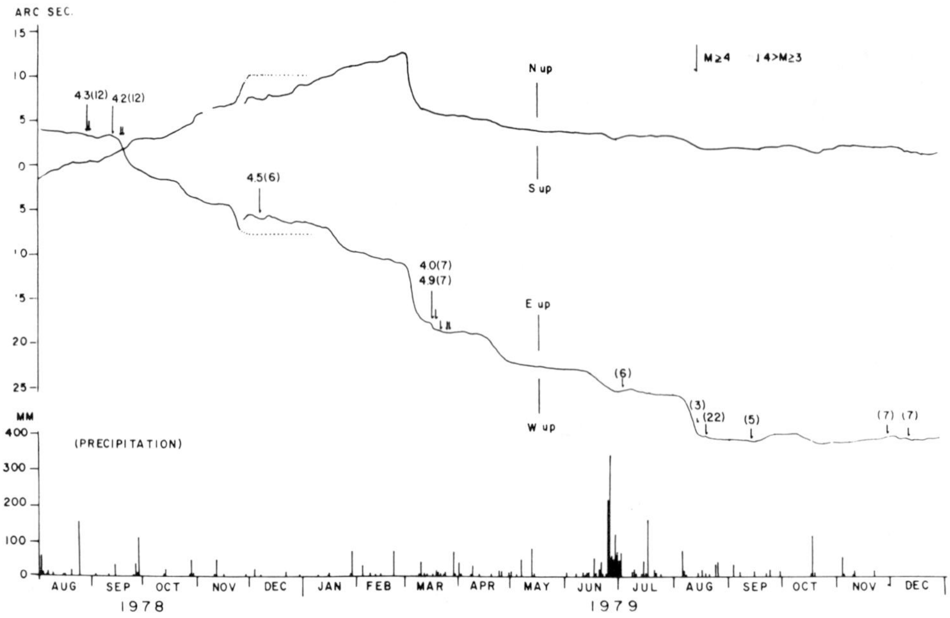

Figure 8.4.10. Ground tilts at Yoshimatsu Observatory (indicated by "+" in figure 8.4.11) and daily number of earthquakes larger than magnitude 3 during a swarm in March 1979, from Ishihara (1980). Numbers above arrows are earthquake magnitude and epicentral distance (parentheses) in kilometers. One arcsecond equals about 5 microradians. The swarm occurred along or just outside the west rim of the caldera. Yoshimatsu Observatory is located about 2 km inside the west caldera rim, 6 km east-southeast of the swarm area.

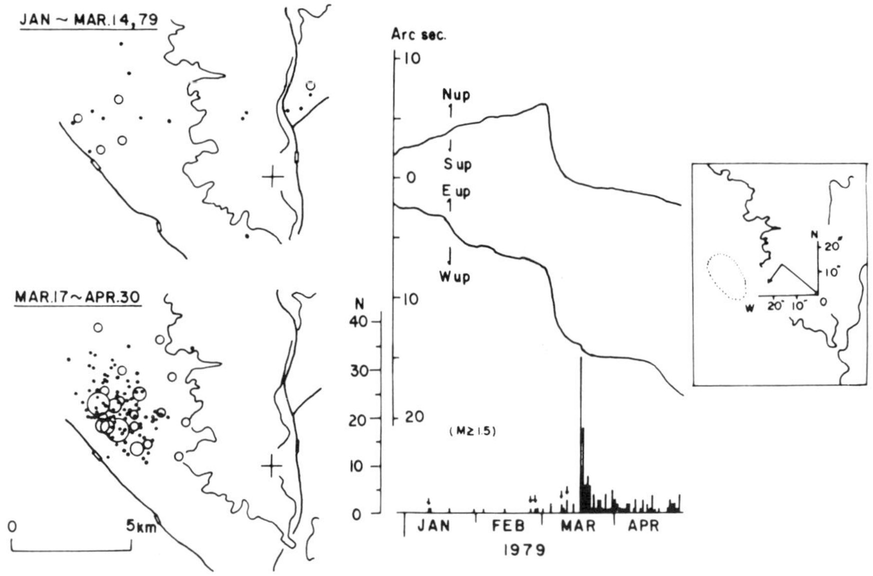

Figure 8.4.11. Right, relation between ground tilt and seismic activity in the western part of Kakuto Caldera; left, epicentral distributions of foreshocks and aftershocks. Figure from Ishihara (1980).

ASO

CAVW number (active vent)	Latitude Longitude (degrees)	Diameter (km)	Local tectonic setting	Precaldera edifice	SiO_2 content (wt pct)	Age of collapse (yr B.P.)	Date of unrest	Type and duration of unrest ESTU STHF MGTF H Te	Eruption type
08-02-11 (Naka-dake)	32.88N 131.10E	21	Right-lateral NE-SW fault with associated N-S exten	Strato (several cones)	R = 52-54 C = 70-50 (Aso-4 glass dominantly 69-70 pct SiO_2)	300,000 150,000 100,000 29,000(?)	pre-1884	---- ---- ---- - --	ex,Ex
							1884	---- ---- ---- - --	ex
							1894	---- ---- ---- - --	ex
							1897	---- ---- ---- - --	ex
							1906	---- ---- ---- - --	Ex
							1916	---- ---- ---- - --	ex
							1920	---- ---- ---- - --	ex
							1925-85+	EEEE FC-- --EE E --	ex,Ex,pex

TECTONIC SETTING

Aso Caldera (figs. 8.5.1-8.5.4) is situated along, and cut by, a major NE-SW-trending fault zone, the Oita-Kumamoto zone, along which the most recent activity has been right-lateral shear. An associated E-W graben also passes through Aso Caldera. The rate of vertical displacement along the Futagawa Fault, a main fault in this zone, has varied from 0.6 to 1.7 mm/yr for the past 350,000 yr, based on offset of the Aso-1 through Aso-4 ignimbrites. The highest rate (1.7 mm/yr or more) applies to the past 50,000-100,000 yr. The Oita-Kumamoto fault zone crosses the northwestern half of the Aso Caldera and is thought by Wada and Nishimura (1971, 1981), on the basis of anomalous P-wave ray paths, to split the basement of the caldera into two blocks, the north-northwest one dipping 15°E and the south-southeast one dipping 25°NNE.

GEOLOGIC HISTORY

Aso Caldera has been the site of four voluminous eruptions of silicic magma within the past 300,000 yr (Aso-1 through Aso-4). The latest of these (Aso-4) is now thought to have occurred ca. 29,000 yr B.P. and spread a marker ash throughout Japan (Machida and others, 1985). Postcaldera volcanism has built a complex of vents in the center of the caldera, including the historically active Naka-dake cone. The locus of activity within Naka-dake shifts by a few hundred meters from decade to decade; activity was in the north part of the crater at the beginning of the 20th century and then shifted to the south for about 20 years (until 1930); eruptions from 1932 to 1934 were in the middle of the crater, and activity has continued to shift since that time. Naka-dake erupts basaltic andesite (SiO_2=52-54 percent), and its existence together with other postcaldera mafic products suggests that if a large silicic magma body still remains beneath the Aso Caldera, that silicic magma must be effectively isolated from the basaltic magma as the latter is fed to the surface. Sassa (1936b) noted an anomalous deflection of ray paths across Aso Caldera but declined to

ASO (continued)

GEOLOGIC HISTORY (continued)

attribute this anomaly to the relatively small body of magma that was inferred beneath Naka-dake. Recently, Sudo (1987) found that S waves disappear and P waves are delayed by a large body of magma(?) beneath the central and eastern parts of Aso Caldera, at depths between 6 and 12 km.

A few small basaltic cinder cones occur along the Oita-Kumamoto fault zone west of the caldera, and the active central vent, Naka-dake, has grown above the E-W graben within the caldera.

HISTORICAL ACTIVITY

Caldera boundary unrest:

Aso Caldera has experienced numerous earthquake swarms during the past 50 years, centered on, just inside, and just outside its northern and western rims (figs. 8.5.5-8.5.7). These swarms occur along the Oita-Kumamoto fault zone and are not directly related to volcanic activity. First-motion studies have shown NW-SE compression for earthquakes occurring along an E-W-trending portion of the zone west of the caldera, and NE-SW compression for earthquakes occurring along the N-S-trending western edge of the caldera. Earthquake swarms occurred near Oguni in November 1928 - February 1929 (figs. 8.5.8, 8.5.9), near Higo-ozu in July 1930 and November 1933, and in the northeast part of Aso Volcano in June 1941. The largest earthquake during the 1928-29 swarm was of magnitude 5.4.

Four swarms of small earthquakes occurred beneath the western part of Aso Caldera from August 1970 - January 1972. Each of these swarms lasted less than 40 hours and was located along the west caldera boundary; they moved progressively from south to north (figs. 8.5.10, 8.5.11) (Wada and others, 1972). Seismic energy release in the four swarms was 10^{19}, 10^{17}, 10^{18}, and 10^{18} ergs, respectively. Depths were 4-8 km in the first swarm, southwest of the caldera rim; 4-6 km in the third and fourth swarms, on and just north of the northwestern crater rim; and 4-16 km in the second swarm, under the Tateno Valley, where a segment of the western caldera rim has been downdropped (Wada and others, 1972). Another swarm began on 20 January 1975 (figs. 8.5.12, 8.5.13); early epicenters ringed a small seismic gap that was later filled by a M=6.0 earthquake on 23 January 1975. The swarm ended on 30 January after a total of 50 felt quakes. The b-value before the M=6.0 earthquake was 0.73; afterward it was 1.01. Yamashina and Murai (1975) wrote that, contrary to a preliminary determination of strike-slip faulting, focal mechanisms for the January-April 1975 earthquakes indicated normal slip along an E-W fault. The most recent swarm (as of the time of this writing) occurred in June 1983 on the northwest rim of the caldera, at depths of 4-6 km with largest magnitude at least 3.0 (Aso Volcanological Laboratory, 1984) (figs. 8.5.14-8.5.16).

Intracaldera unrest:

Tremor: Aso is a classic locality for the study of volcanic tremor. Sassa (1935, 1936a) distinguished four types on the basis of period, mode of oscillation, and temporal relation to eruptions. These four types are I (or "first type"), surface (Love) waves, relatively monochromatic with period 0.8-2.0 sec, usually ca. 1.0 sec, generated several times per hour during noneruptive periods by

ASO (continued)

HISTORICAL ACTIVITY (continued)

internal explosions of volcanic gases; II (second type), relatively continuous, surface (Rayleigh) waves, period 3.5-8.0 sec, generated by explosions(?) and vibration of the magma reservoir; III (third type), body waves, period averaging 0.5 sec with a broad spectral content, generated by explosions within the magma during times when the volcano is erupting; and IV (fourth type), body waves, period 0.2 sec, associated with actual eruptions. The second type of tremor occurs infrequently and is usually preceded (0-20 sec earlier) by tremor of the first type. Both the first and the second types resemble low-frequency earthquakes--shallow, spindle-shaped, possibly like what Minakami later called "b-type" earthquakes (see fig. 10 of Sassa, 1935). Kubotera (1974) attributes the third type to microearthquakes just beneath the crater. Sassa (1936a, p. 16) describes how the first type decreased in frequency after the eruption of 7 November 1929 and finally appeared as intermittent wave groups rather than as continuous trains.

Kamo (1962) detected much longer period tremor that increased noticeably about 2 days before an unusually large eruption of Naka-dake on 24 July 1958 and then disappeared (together with Sassa's third type of tremor) about one hour before that eruption. The period of this long-period tremor increased from 40 sec to 55 sec over the 2 days before the eruption. Sassa also noted tremor caused by rainfall and stream flow over two waterfalls near Aso; such tremor is not included in types I-IV.

<u>Earthquakes</u>: Seismicity and ground deformation have been recorded before several eruptions of Naka-dake (figs. 8.5.17, 8.5.18). During a series of small explosions in November 1932 - May 1933, the amplitude of "tremor" more or less paralleled the intensity of explosions from early December 1932 through February 1933, but it increased sharply 2 days before an eruption on 26 March 1933 (Tanakadate, 1934; Sassa, 1935, 1936a). Occasional moderately strong (felt) earthquakes (for example, one on 21 December 1934) have been followed by renewed explosive activity. "Pulsating tremor" occurred continuously from 16 October - 25 December 1949, for 100 days prior to unusually strong Strombolian eruptions on 26 December; during the same period, no ordinary volcanic earthquakes were recorded (Minakami and Sakuma, 1953). Tremor decreased anomalously just before an eruption in 1950, and explosions on 29 April 1953 (part of a continuing series) were preceded by an increase in the amplitude of continuous tremor (of Sassa's first and second types; Shima, 1960). The amplitude increased on 6 April (Minakami, 1956) or 14 April (Shima, 1960) but then decreased several days before the eruption and became quiet for about 3 hours immediately before the eruption (Minakami, 1956; Shima, 1960). After two years' repose, on 17 November 1957, microtremors of the first and second types began to increase gradually, both in frequency of occurrence and in amplitude. The increase in the first kind of tremor was interrupted on 18 November, and that of the 2nd kind on 28 November, and an eruption occurred at 1900 hr on 1 December. Afterward, and also in 1953, the frequency and amplitude of the first type of tremor continued to increase for at least a month (Shima, 1960). The 1976 and 1979 eruptions were preceded by a sudden increase in tremor amplitude and seismic energy release 1 week and 3 months before the eruptions, respectively, and by a period of high seismicity and then a sharp decrease in seismicity 1-10 days before these eruptions. In late December 1984, during a series of small explosions, many persons felt earthquakes, although tremor amplitude generally was low throughout this period (Smithsonian Institution, 1984, 1985).

<u>Tilt</u>: Measurements at two stations near the active crater during 1931-32, 1937-41, and 1953-61 (and continuing today) have detected tilt changes that, although complicated, are useful for anticipating eruptions (figs. 8.5.17-8.5.21). About 5 microradians of tilt occurred at the Volcanological Laboratory about 8 km west of Naka-dake (location shown on figs. 8.5.3 and 8.5.4) between 7

ASO (continued)

HISTORICAL ACTIVITY (continued)

and 2 days before the eruption of 26 March 1933, but it then reversed by 10 microradians (in the sense of deflation) during the two days before the eruption (Sassa, 1935, 1936a). The magnitude of these changes is not remarkable, but the reversal might be useful in determining whether an eruption is imminent. Ito (1961) subtracted out seasonal variation and found that, in general, tilt measurements showed several hundred microradians of inflationary tilt at Hondo and Umanose for several months before large explosions in 1953 and 1958 (fig. 8.5.20). Deflation of a comparable amount occurred within several weeks after these eruptions. An active period from April 1965 to March 1966 included a slight inflationary tilt anomaly as measured at Sunasenri, 1 km south of the active crater; that inflation plus an increase in the mean amplitude of Sassa's first kind of volcanic microtremor and the daily number of volcanic earthquakes began approximately 4, 2, and 0.5 months before the largest eruptions of this active period (Wada and Sudo, 1967; Wada and others, 1968; fig. 8.5.21).

Uplift/subsidence: Yoshikawa (1954) showed a general correlation between uplift of the central cone and the amplitude of tremor (figs. 8.5.22, 8.5.23). Repeated precise leveling from 1937 until 1953 showed that the top of the central cone subsided by 1 cm from 1937 to 1938, rose 2.5 cm from 1938 to 1939, and subsided a total of 10 cm from 1938 until 1951; the amplitude of tremor was high in 1938-39 and declined irregularly until 1951. Tremor increased and minor uplift (2 cm) was noted from 1951 to 1953. Over the longer period of 1937-81, leveling from the northern floor of the caldera to the summit of Naka-dake has shown gradual subsidence following large explosions or periods of minor explosions, interrupted by rapid inflation during renewed intrusive episodes (Kikuchi and Ono, 1985) (fig. 8.5.24). Unfortunately, leveling information is only available for the central part of the caldera; a point on the caldera floor was assumed to be stable for the aforementioned studies, and we do not know whether this assumption is validated by leveling to points outside the caldera.

Horizontal strain: Slope distances between Naka-dake crater and points near the south and west base of the central cone contracted by 1-3 x 10^5 from March 1975 to March 1976; the same lines extended slightly from March 1976 to the end of 1981 (Tanaka and others, 1982). Measurements of slightly different lines in November 1977, April 1978, and June 1978 showed up to 2 cm (or 40 microstrain) of contraction (Ono and Sako, 1978). Changes along a line from the west caldera rim to a point on the west-southwest caldera floor showed about 3 cm of extension (ca. 5 microstrain) during the same November 1977 - June 1978 period, suggesting that inflation was localized at Naka-dake proper (Ono and Sako, 1978) (figs. 8.5.25, 8.5.26).

Gravity changes: Repeated gravity measurements argue against broad, caldera-wide uplift in the period 1965-73, which included an average number of small eruptions (Kubotera and others, 1969, 1974). Similarly, from 1973 to 1982, gravity changes were insignificant (Kubotera and Satomura, 1985).

Temperature changes and gas emission: Continuous temperature measurements in the crater of Naka-dake after the large 1958 eruption showed identifiable increases in temperature (by as much as 50 °C) 2 to 3 days before additional explosions (Taneda, 1963). The amount of CO_2 dissolved in nearby hot springs increased slightly about 8 months before the 1979 eruption (fig. 8.5.27); the temperatures of those springs also increased, though less convincingly and over a longer time period (less useful for eruption forecasting). After an eruption in June 1981, explosions temporarily ceased, but the temperature of water in a small pond in

ASO (continued)

HISTORICAL ACTIVITY (continued)

Naka-dake crater increased from late 1981 until late 1984, and water level in that pond decreased from April 1984 - January 1985. A warning of increased volcanic activity was issued on 11 October 1984; ash ejection began anew on 24 October and continued until at least January 1985. Initial explosions were phreatic, but by January 1985 the explosions were phreatomagmatic.

The SO_2 flux and Cl/S ratio in fumarolic gases bear a complicated relationship with seismicity and explosions (figs. 8.5.28, 8.5.29). Apparently, SO_2 and Cl/S sometimes increase before and especially during periods of explosions, and then decrease as the explosions decrease (Ota, 1984; Kubotera, 1984).

COMMENTS

Aso Caldera provides a good example of earthquakes focused at the intersection of regional shear faults and caldera ring faults. In this case, there seems to be relatively little relation between volcanic activity at the center of the caldera and swarms near its margin. If plots of earthquake epicenters at Aso and Long Valley, California, are superimposed with their regional stress directions parallel, strong similarities may be seen in the distribution of earthquakes at and just outside these calderas.

REFERENCES

Aso Volcanological Laboratory, Kyoto University, 1984, An earthquake swarm in the northern part of Aso Caldera, June 1983: Rep. Coord. Comm. for Prediction of Volcanic Eruptions, no. 29, p. 46-48.

Fuzimoto, M., 1960, Crustal deformation on and around Volcano Aso [abs.]: Abstracts, General Meeting of the Volcanological Society of Japan, May 10, 1960.

Ito, Y., 1961, Tilting motion of the ground as related to the volcanic activity of Mt. Aso, and micro-process of the tilting motion of ground and structure: Bull. Disaster Prev. Res. Inst., Kyoto Univ., v. 42, p. 2-15.

Kamo, K., 1962, Nature of the volcanic micro-tremors at the Volcano Aso: Pt. 2, Some natures of the volcanic micro-tremors of the 1st kind at the Volcano Aso: Disaster Prev. Res. Inst., Kyoto Univ., Bull. 55, p. 1-13.

Kikuchi, S., Hoka, T., and Yamada, T., 1978, Precise leveling at Volcano Aso: Rep. Joint Obs. Aso Volcano 1977, p. 16-18.

Kikuchi, S., and Ono, H., 1985, Observation of crustal deformation at volcano Aso, in Okada, Hm., ed., Comparative study of physical background of volcanic activity and its relation to prediction of eruption disasters: Special Rep. Grant-in-Aid for Nat. Disaster Sci., Ministry of Education, publ. by Hokkaido Univ., Sapporo, in cooperation with Tohoku Univ., Univ. of Tokyo Earthq. Res. Inst., Nagoya Univ., Kyoto Univ., and Kyoto Univ. Disaster Prev. Res. Inst., p. 55-63.

Kubotera, A., 1974, Volcanic tremors at Aso Volcano, in Civetta, L., Gasparini, P., Luongo, G., and Rapolla, A., eds., Physical volcanology: Amsterdam, Elsevier, p. 29-47.

----- ed., 1975, Report on earthquake swarm activity at Aso in January 1975, and the disasters: Special Rep., Ministry of Education, Aso, Kumamoto, Kyoto Univ., Aso Volcano Observatory, 91 p.

----- ed., 1978, Joint geophysical and geochemical observations of Aso Volcano, (I), August-December 1977: Special Rep., Ministry of Education, Aso, Kumamoto, Kyoto Univ., Aso Volcano Observatory, 75 p.

ASO (continued)

REFERENCES (continued)

----- ed., 1984, Joint geophysical and geochemical observations of Aso Volcano, (II), August-December 1981: Special Rep., Ministry of Education, Aso, Kumamoto, Kyoto Univ., Aso Volcano Observatory, 121 p.

Kubotera, A., and Mitsunami, T., 1980, An earthquake swarm in the northern part of the Aso Caldera and migration of their foci: Tectonophysics, v. 70, p. 223-236.

Kubotera, A., and Otsuka, M., 1970, Nature of non-volcanic microtremors on the Aso caldera: Jour. Phys. Earth, v. 18, no. 1, p. 115-124.

Kubotera, A., and Satomura, M., 1985, Repeated gravity measurements on Aso Volcano, in Okada, Hm., ed., Comparative study of physical background of volcanic activity and its relation to prediction of eruption disasters: Special Rep. Grant-in-Aid for Nat. Disaster Sci., Ministry of Education, publ. by Hokkaido Univ., Sapporo, in cooperation with Tohoku Univ., Univ. of Tokyo Earthq. Res. Inst., Nagoya Univ., Kyoto Univ., and Kyoto Univ. Disaster Prev. Res. Inst., p. 64-68.

Kubotera, A., Satomura, M., and Sumitomo, N., 1974, Repeated gravity measurements on Aso Volcano: Bull. Volcanol. Soc. Japan, v. 19, no. 2, p. 123-125.

Kubotera, A., Tajima, H., Sumitomo, N., Doi, H., and Izutuya, S., 1969, Gravity surveys on Aso and Kuju volcanic region, Kyushu District, Japan: Bull. Earthq. Res. Inst., v. 47, p. 215-255.

Kubotera, A., and Yoshikawa, K., 1963, Prediction of volcanic eruptions of Aso and Sakurajima and some related geophysical problems: Bull. Volcanol., v. 26, p. 297-317.

Machida, H., and Arai, F., 1983, Extensive ashfalls in and around the Sea of Japan from large late Quaternary eruptions: Jour. Volcanol. Geotherm. Res., v. 18, p. 151-164.

Machida, H., Arai, F., and Momose, M., 1985, Aso-4 ash: A widespread tephra and its implications to the events of late Pleistocene in and around Japan: Bull. Volcanol. Soc. Japan, v. 30, no. 2, p. 49-70.

Minakami, T., 1956, Report on volcanic activities and volcanological studies in Japan for the period from 1951 to 1954: Bull. Volcanol., v. 18, p. 38-76.

Minakami, T., and Sakuma, S., 1953, Report on volcanic activities and volcanological studies in Japan for the period from 1948 to 1951: Bull. Volcanol., v. 14, p. 79-132.

Mitsunami, T., and Kubotera, A., 1977, On the activity of the earthquake swarm in the northern part of the Aso Caldera, 1975--Interpretation of hypocenter migration: Zisin, v. 30, p. 73-90. [see also: Kubotera and Mitsunami, 1980].

Ono, H., 1985, Seismic activity near the crater of Mt. Nakadake, Volcano Aso, in Okada, Hm., ed., Comparative study of physical background of volcanic activity and its relation to prediction of eruption disasters: Special Rep. Grant-in-Aid for Nat. Disaster Sci., Ministry of Education, publ. by Hokkaido Univ., Sapporo, in cooperation with Tohoku Univ., Univ. of Tokyo Earthq. Res. Inst., Nagoya Univ., Kyoto Univ., and Kyoto Univ. Disaster Prev. Res. Inst., 37-42.

Ono, H. and Sako, M., 1978, Distance measurements at Volcano Aso, by means of electro-optical method: Rep. Joint Observ. Aso Volcano, 1977, p. 19-22.

Ono, K., Kubotera, A., and Ota, K., 1981, Aso Volcano, in Kubotera, A., ed., Field excursion guide to Sakurajima, Kirishima, and Aso Volcanoes: Tokyo, Volcanological Society of Japan, p. 33-52.

Ossaka, J., ed., 1984, Fundamental research for predicting volcanic eruptions through observations of volcanic gases: Special Rep., Grant-in-Aid Nat. Disaster Sci., no. A-59-4, 141 p.

ASO (continued)

REFERENCES (continued)

Ota, K., 1984, Geochemical observations of hot springs at Aso Volcano, in Ossaka, J., ed., Fundamental research for predicting volcanic eruptions through observations of volcanic gases: Special Rep., Grant-in Aid Nat. Disaster Sci., no. A-59-4, 141 p.

Sassa, K., 1935, Volcanic micro-tremors and eruption-earthquakes: Mem. Coll. Sci., Kyoto Imperial Univ., Ser. A, v. 18, no. 5, p. 255-294.

----- 1936a, Micro-seismometric study on eruptions of the Volcano Aso: Mem. Coll. Sci., Kyoto Imperial Univ., Ser. A, v. 19, no. 1, p. 11-56.

----- 1936b, Anomalous deflection of seismic rays in volcanic districts (Geophysical studies on the Volcano Aso, III): Mem. Coll. Sci., Kyoto Imperial Univ., Ser. A, v. 19, p. 65-78.

Shima, M., 1958, On the second volcanic micro-tremors at the volcano Aso: Disaster Prev. Res. Inst., Kyoto Univ., Bull. 22, p. 1-6.

----- 1960, Volcanic micro-tremors at the Volcano Aso: Disaster Prev. Res. Inst., Kyoto Univ., Bull. 34, p. 1-17.

Shimozuru, D., 1971, A seismological approach to the prediction of volcanic eruptions, in The surveillance and prediction of volcanic activity: Paris, UNESCO, p. 19-45.

Smithsonian Institution, Scientific Event Alert Network (SEAN), 1976-85, Aso Volcano: SEAN Bull., v. 1-10.

Sudo, Y., 1975, Seismometric study on earthquakes at the western region of the Aso Caldera--focal mechanisms and tectonic lines: Bull. Volcanol. Soc. Japan, v. 20, p. 1-12.

----- 1981, Seismic activities at the western region of the Aso Caldera: Bull. Volcanol. Soc. Japan, v. 26, no. 4, p. 263-279.

----- 1987, Attenuation of seismic waves in the Aso Caldera, Kyushu, Japan; evidence for an attenuating body in the crust: Abstracts, Hawaii Symposium on How Volcanoes Work, Hilo, January 1987, p. 243.

Sudo, Y., and Yamada, T., 1985, Characteristic seismic activity around the caldera of Aso Volcano, in Okada, Hm., ed., Comparative study of physical background of volcanic activity and its relation to prediction of eruption disasters: Special Rep. Grant-in-Aid for Nat. Disaster Sci., Ministry of Education, publ. by Hokkaido Univ., Sapporo, in cooperation with Tohoku Univ., Univ. of Tokyo Earthq. Res. Inst., Nagoya Univ., Kyoto Univ., and Kyoto Univ. Disaster Prev. Res. Inst., p. 49-54.

Tanaka, Y., Churei, M., Sawada, Y., and Shibata, T., 1982, Accuracy of distance measurement and detection of significant base lines in four active volcanoes in Japan: Pap. Meteorol. Geophys., v. 33, no. 3, p. 175-185.

Tanakadate, H., 1934, Volcanic activity in Japan and vicinity during the period between June 1931 and June 1934: Japanese Jour. Astron. Geophys., v. 12, p. 89-108.

Taneda, S., 1963, Volcanothermal investigation of active crater--temperature variation in the bottom ground of the Naka-dake Crater of Aso Volcano: Mem. Fac. Sci., Kyushu Univ., ser. D., v. 14, no. 1, p. 23-38.

Wada, T., 1981, Mechanism generating volcanic tremor (2): Bull. Volcanol. Soc. Japan, v. 26, no. 1, p. 1-7.

Wada, T., Kamo, K., Ono, H., and Sudo, Y., 1968, The nature of volcanic earthquake observed before and after an eruption: Ann. Rep. Disaster Prev. Res. Inst., Kyoto Univ., v. 11, p. 121-130.

Wada, T., Kamo, K., and Sudo, Y., 1972, A characteristic occurrence of earthquake swarms at the Volcano Aso caldera-rim: Contrib. Geophys. Inst., Kyoto Univ., v. 12, p. 145-155.

ASO (continued)

REFERENCES (continued)

Wada, T., and Nishimura, K., 1971, Apparent azimuths of P waves and a structure under the volcano Aso: Contrib. Geophys. Inst., Kyoto Univ., v. 11, p. 179-189.

----- 1981, On a structure under the Aso caldera: Bull. Volcanol. Soc. Japan, v. 26, no. 2, p. 83-92.

Wada, T., and Ono, H., 1963, Spectral study of micro-tremors (1): Propagation of the micro tremors of the 1st kind observed at volcano Aso: Bull. Volcanol. Soc. Japan, v. 8, no. 1, p. 1-10.

Wada, T., and Sudo, Y., 1967, Focal mechanism of volcanic earthquakes of Volcano Aso: Bull. Volcanol. Soc. Japan, v. 12, no. 2, p. 61-69.

Wada, T., Sudo, Y., and Kamo, K., 1973, Earthquake swarms at caldera-rim of the Volcano Aso: Bull. Volcanol. Soc. Japan, v. 18, p. 19-31.

Watanabe, K., Momikura, Y., and Tsuruta, K., 1979, Active faults and parasitic eruption centers on the west flank of Aso Caldera, Japan: Quaternary Res., v. 18, no. 2, p. 89-101.

Yamashina, K., and Murai, I., 1975, On the focal mechanism of the earthquakes in 1975 in the central part of the Oita prefecture and in the northern part of Aso, and their relation to the active fault system: Bull. Earthq. Res. Inst., v. 50, p. 295-302.

Yoshikawa, K., 1954, Crustal movement of Aso, Part 1: Jour. Seismol. Soc. Japan (ii) (Zisin), v. 7, p. 151-154.

Yoshikawa, S., Kamo, K., and Kitsunezaki, C., 1959, Seismic exploration in the vicinity of the Crater of Nakadake, Aso Volcano: Bull. Volcanol. Soc. Japan, v. 4, p. 20-32.

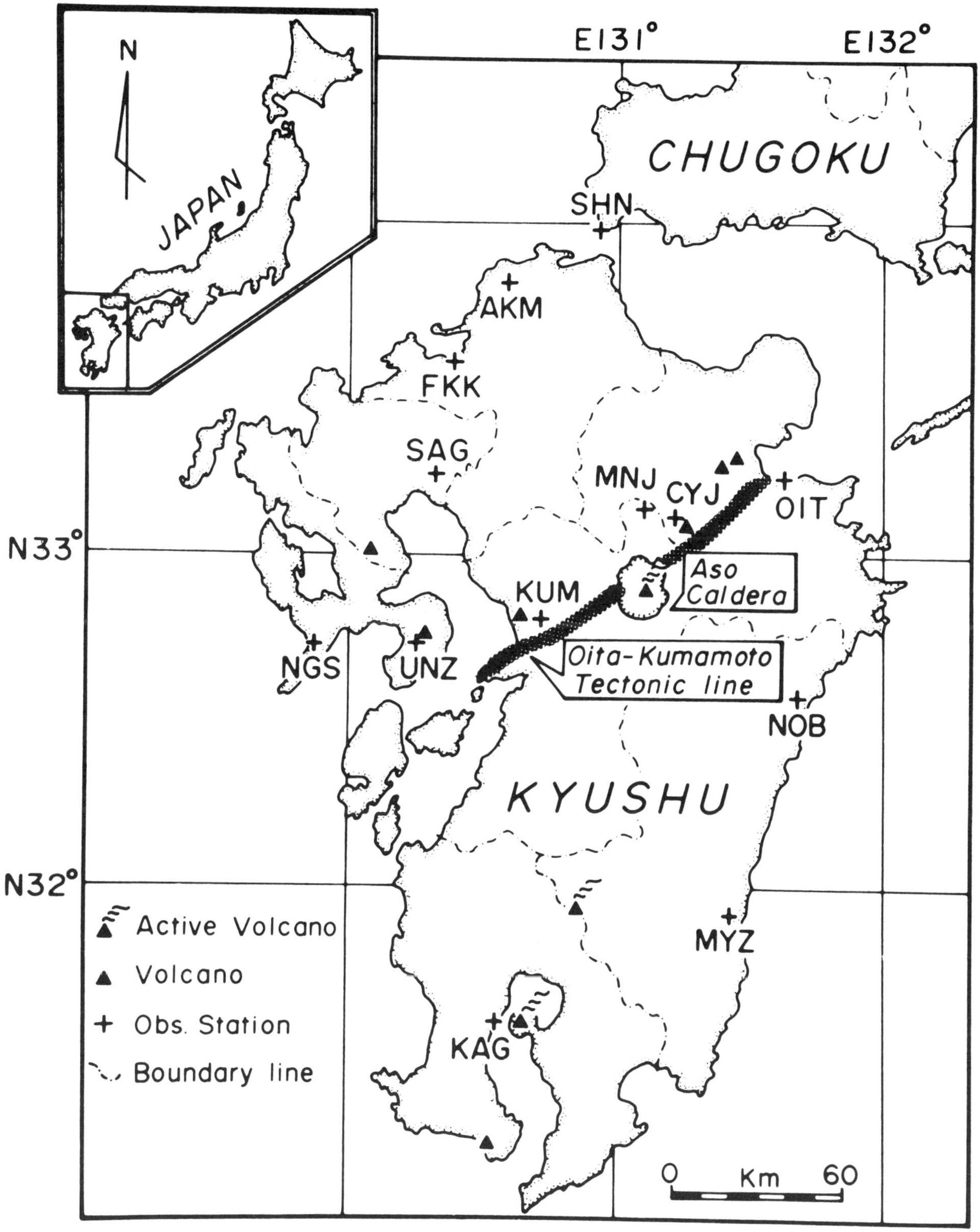

Figure 8.5.1. Relation between Aso Caldera and Oita-Kumamoto tectonic lineament, from Kubotera and Mitsunami (1980). Also shown is Shimabara Peninsula, including Unzen Volcano (solid triangle near UNZ). Crosses indicate seismometer stations.

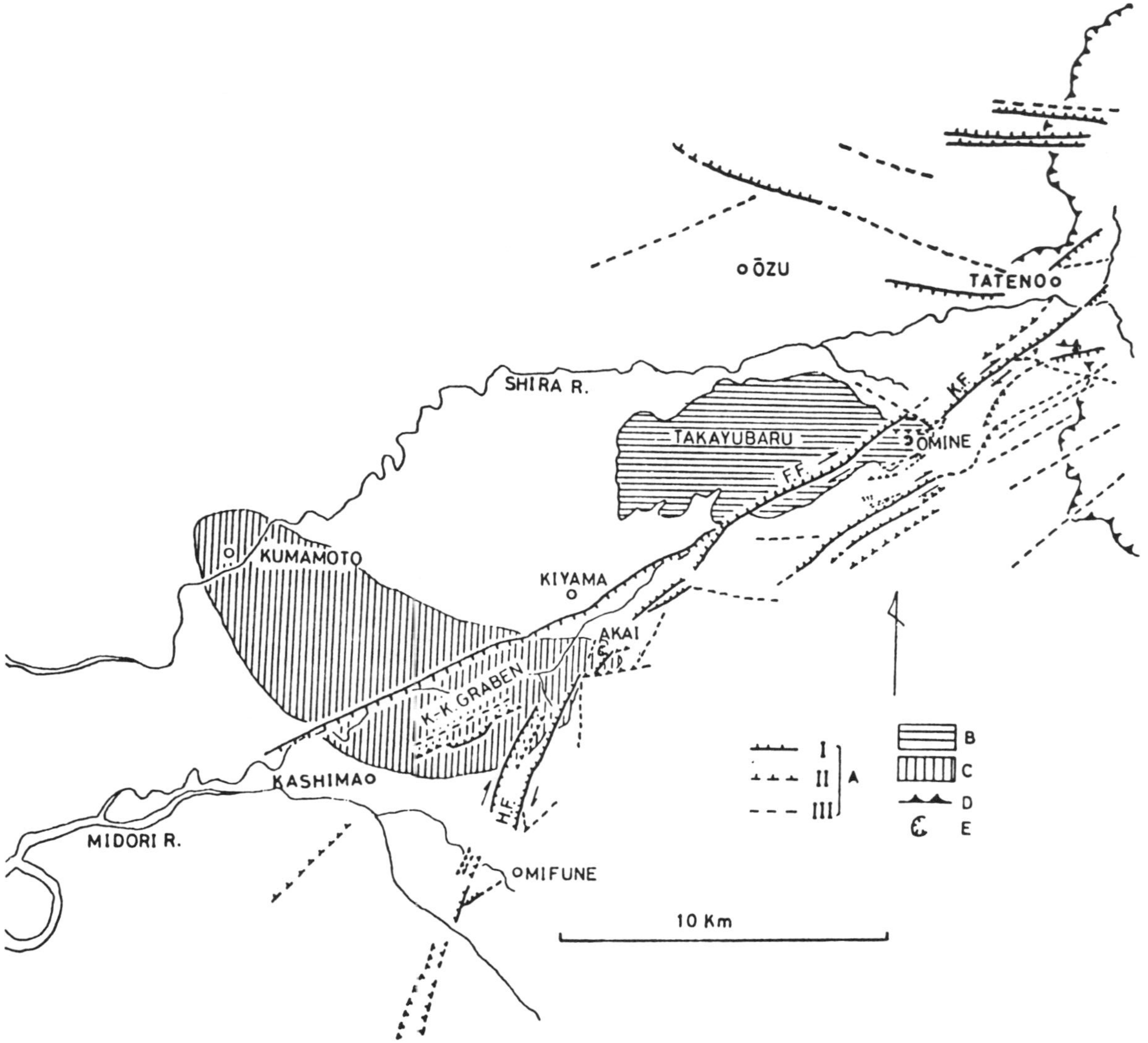

Figure 8.5.2. Active faults and parasitic volcanoes along Oita-Kumamoto fault zone immediately west of Aso Caldera, from Watanabe and others (1979). A, active faults (I, II, and III degrees of certainty); B, Takayubaru lava and Omine pyroclastic cone; C, Togawa lava and Akai pyroclastic cone (mostly buried); D, rim of Aso Caldera; E, craters of Omine and Akai pyroclastic cones; K.F., Kitamukiyama Fault; F.F., Futagawa Fault; H.F., Hinagu Fault.

Figure 8.5.3. Topographic map of Aso Caldera, from Ono and others (1981).

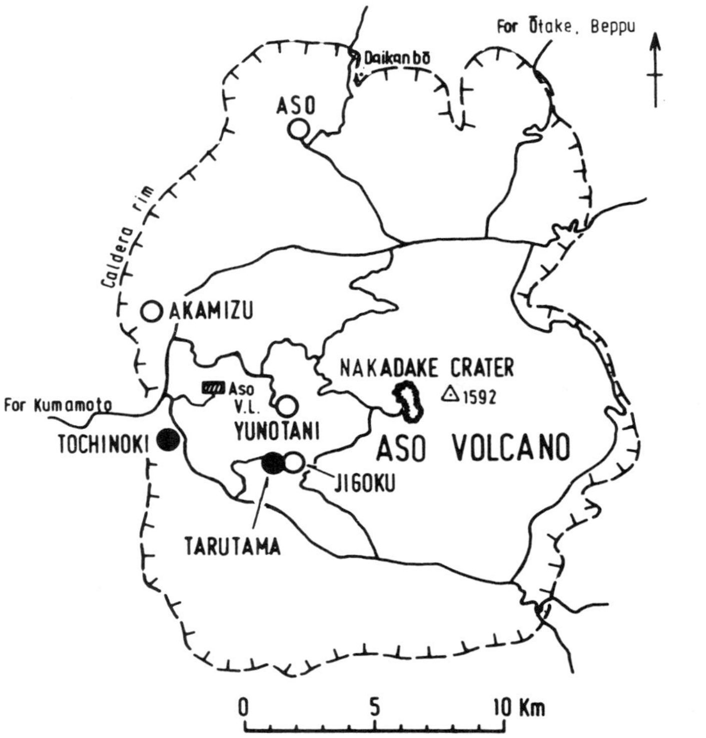

Figure 8.5.4. Aso Caldera showing locations of Naka-dake crater and hot springs (solid and open circles), from Ono and others (1981). Solfataras occur in Yunotani, Jigoku, and Tarutama hot-spring areas.

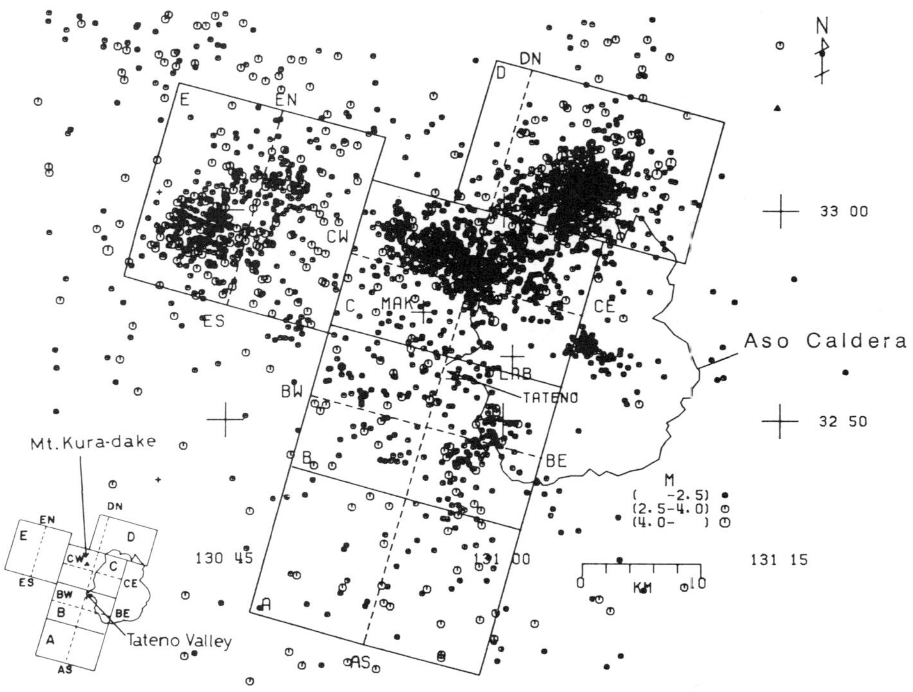

Figure 8.5.5. Epicenters for well-located earthquakes in and near Aso Caldera from 1962 to 1979, from Sudo (1981). The broken lines, AS-DN, BW-BE, and CW-CE, show locations of cross sections shown in figure 8.5.6.

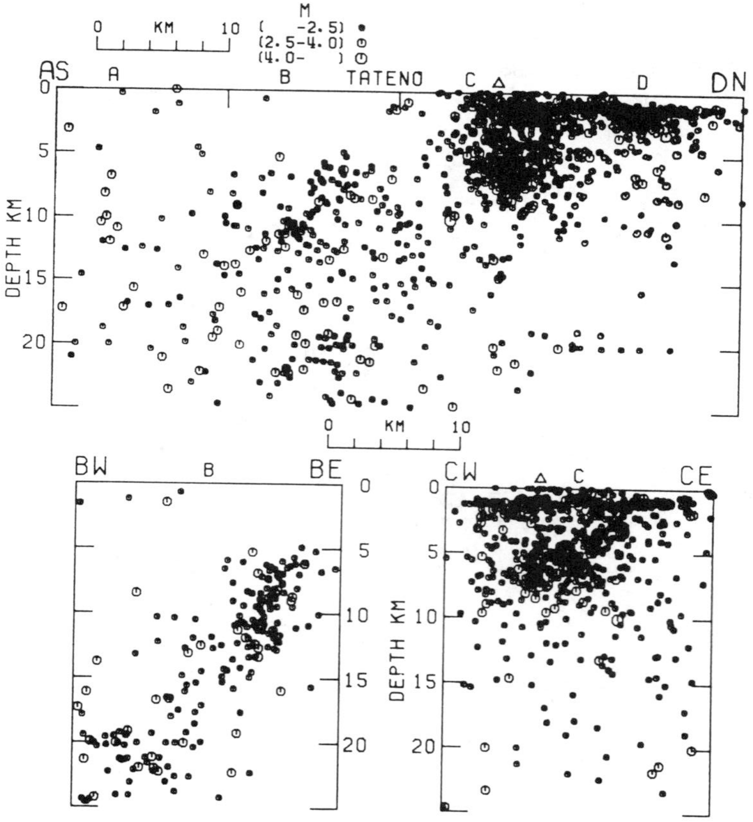

Figure 8.5.6. Vertical cross sections along lines shown in figure 8.5.5, from Sudo (1981). Triangle indicates Mt. Kura-dake.

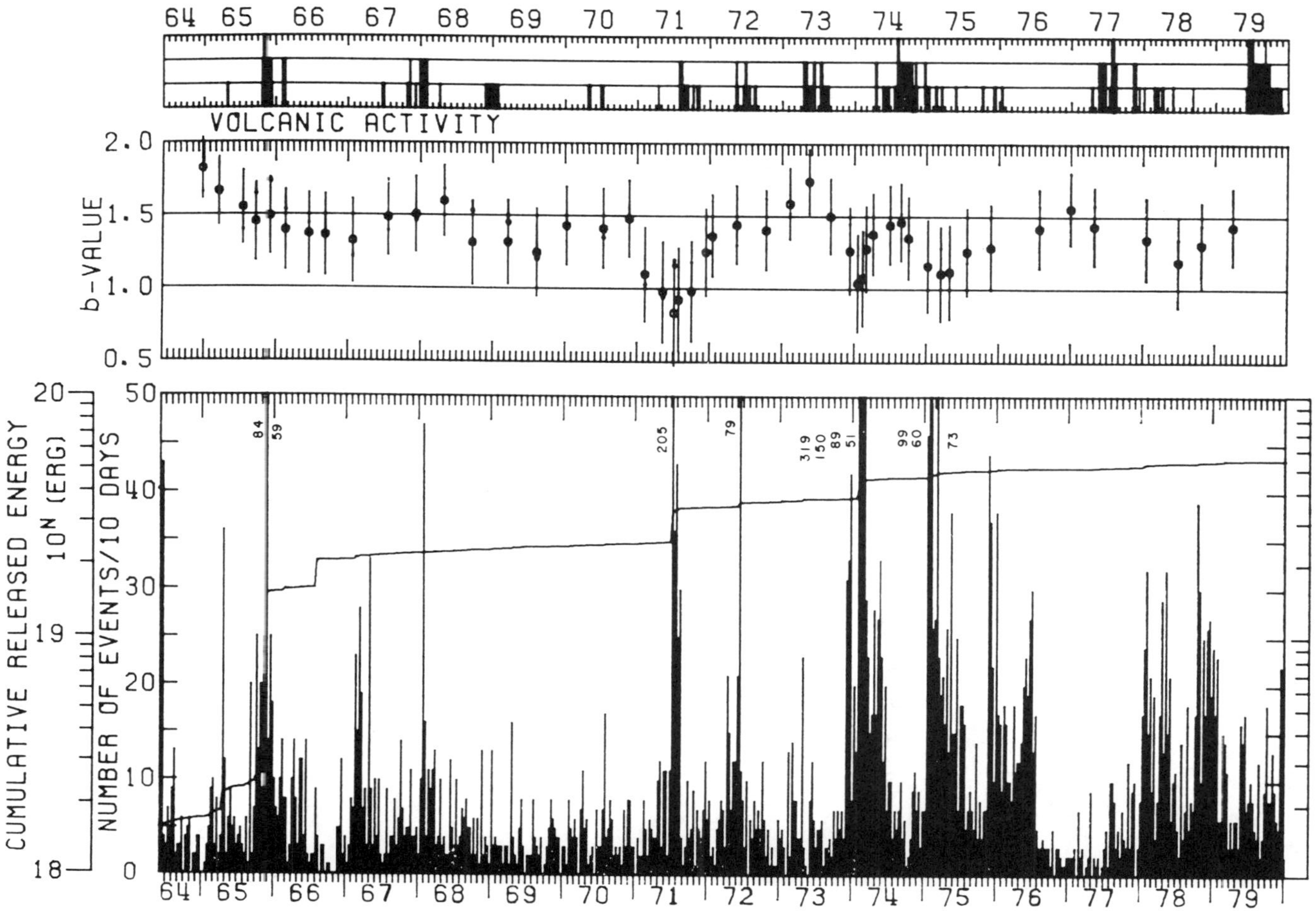

Figure 8.5.7. Temporal changes in seismicity at Aso Caldera from 1964 to 1979, from Sudo (1981). Top, volcanic activity at Mt. Naka-dake, represented by vertical bars, with length corresponding to intensity; middle, variation in b-value, estimated by the method of maximum likelihood for samples of 150 events (magnitude greater than 1.7), with error bars indicating 95-percent confidence limits; bottom, histograms showing the number of earthquakes per 10-day interval. Solid curve shows the cumulative seismic energy release; an abrupt increase in the number of events indicates swarms.

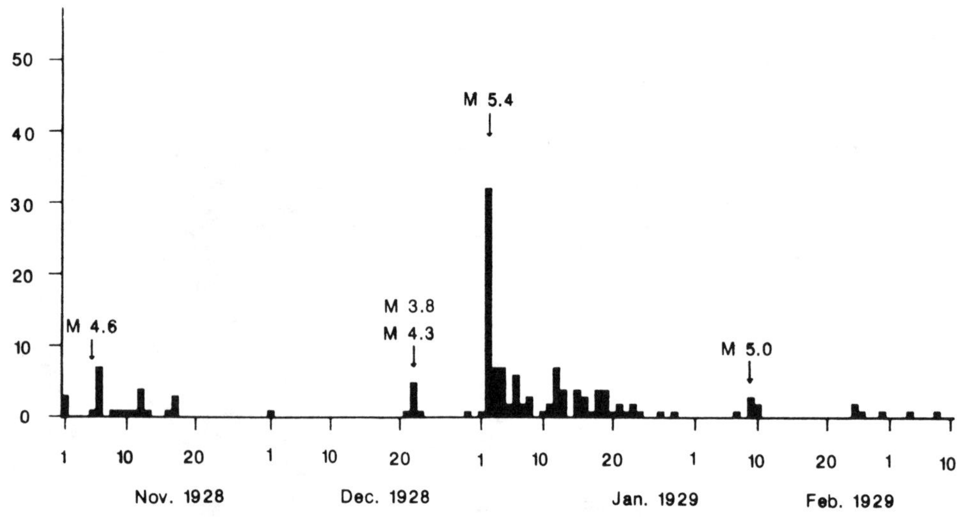

Figure 8.5.8. Daily number of felt earthquakes near Aso Caldera during a swarm from November 1928 to February 1929, from Kubotera (1975).

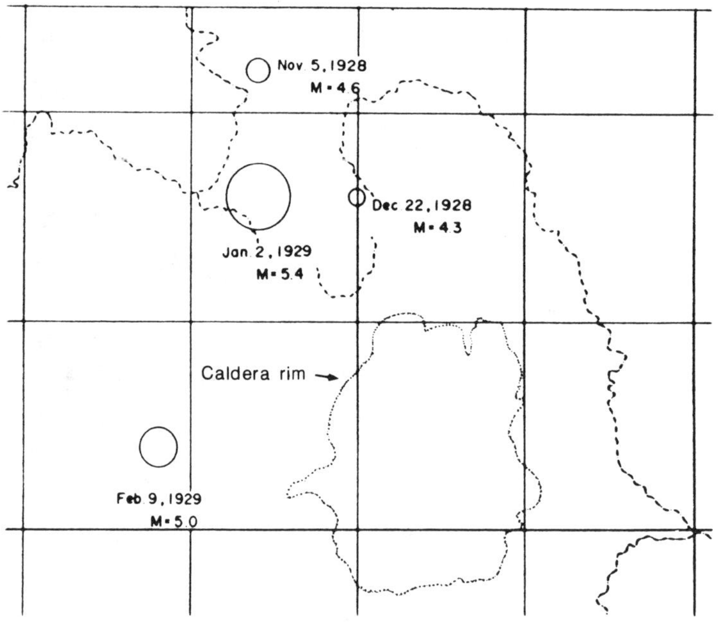

Figure 8.5.9. Epicenters of large earthquakes near Aso Caldera during 1928-29, from Kubotera (1975).

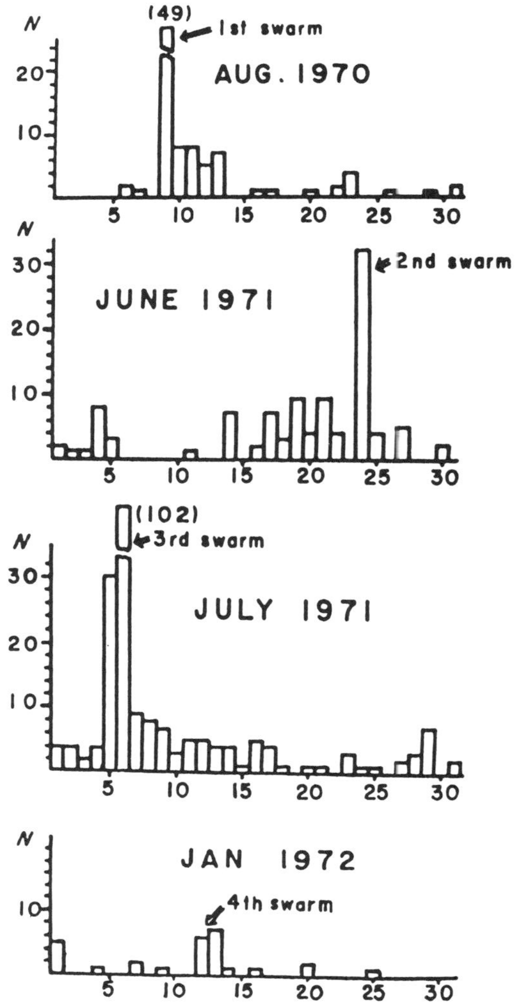

Figure 8.5.10. Daily number of earthquakes during four swarms along the northwest rim of Aso Caldera, 1970-72, from Wada and others (1972).

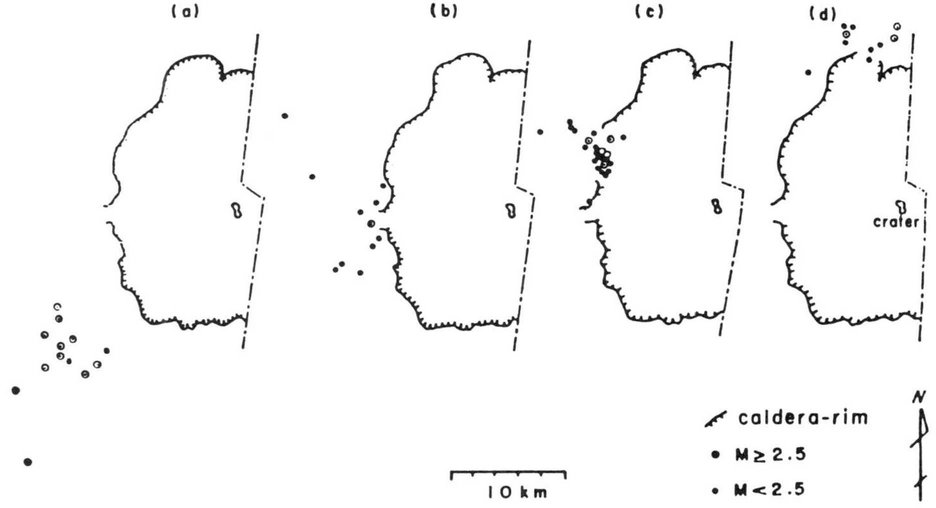

Figure 8.5.11. Epicenters of earthquakes shown in figure 8.5.10, from Wada and others (1972). Note northeastward migration.

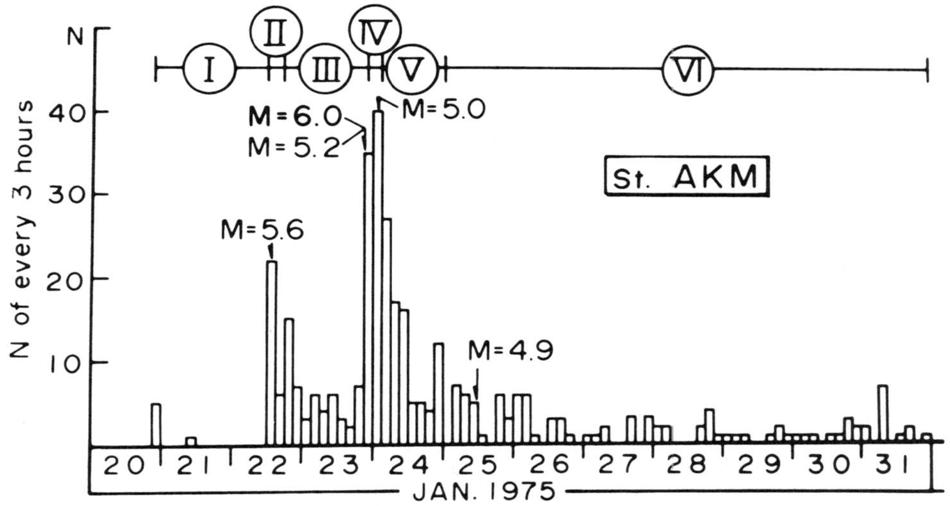

Figure 8.5.12. Number of earthquakes recorded at station AKM (see figure 8.5.1) at 3-hr intervals during a swarm in January 1975, from Kubotera and Mitsunami (1980). See figure 8.5.13 for epicentral locations.

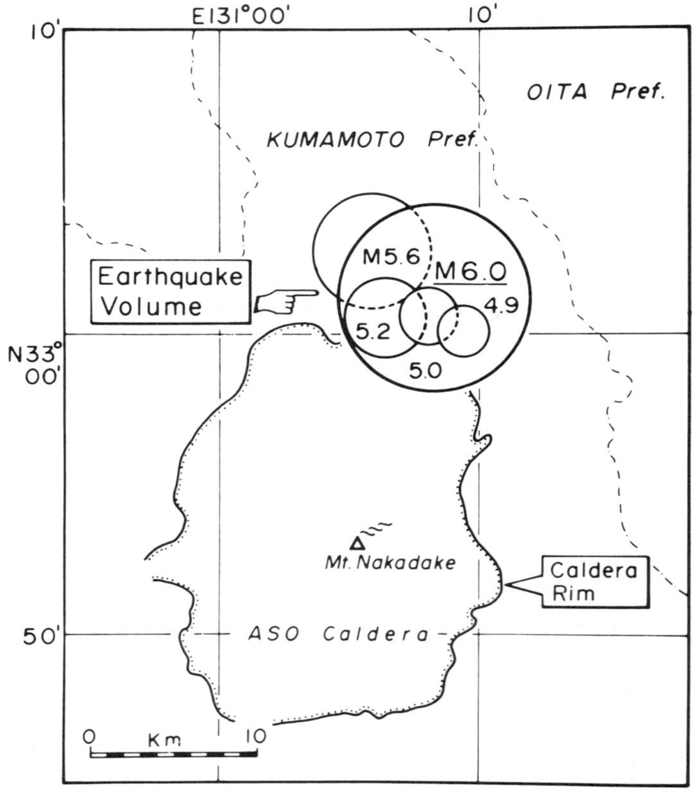

Figure 8.5.13. Distribution of epicenters of major shocks during an earthquake swarm near the rim of Aso Caldera in January 1975, from Kubotera and Mitsunami (1980). The size of each circle indicates the earthquake volume projected to the ground surface.

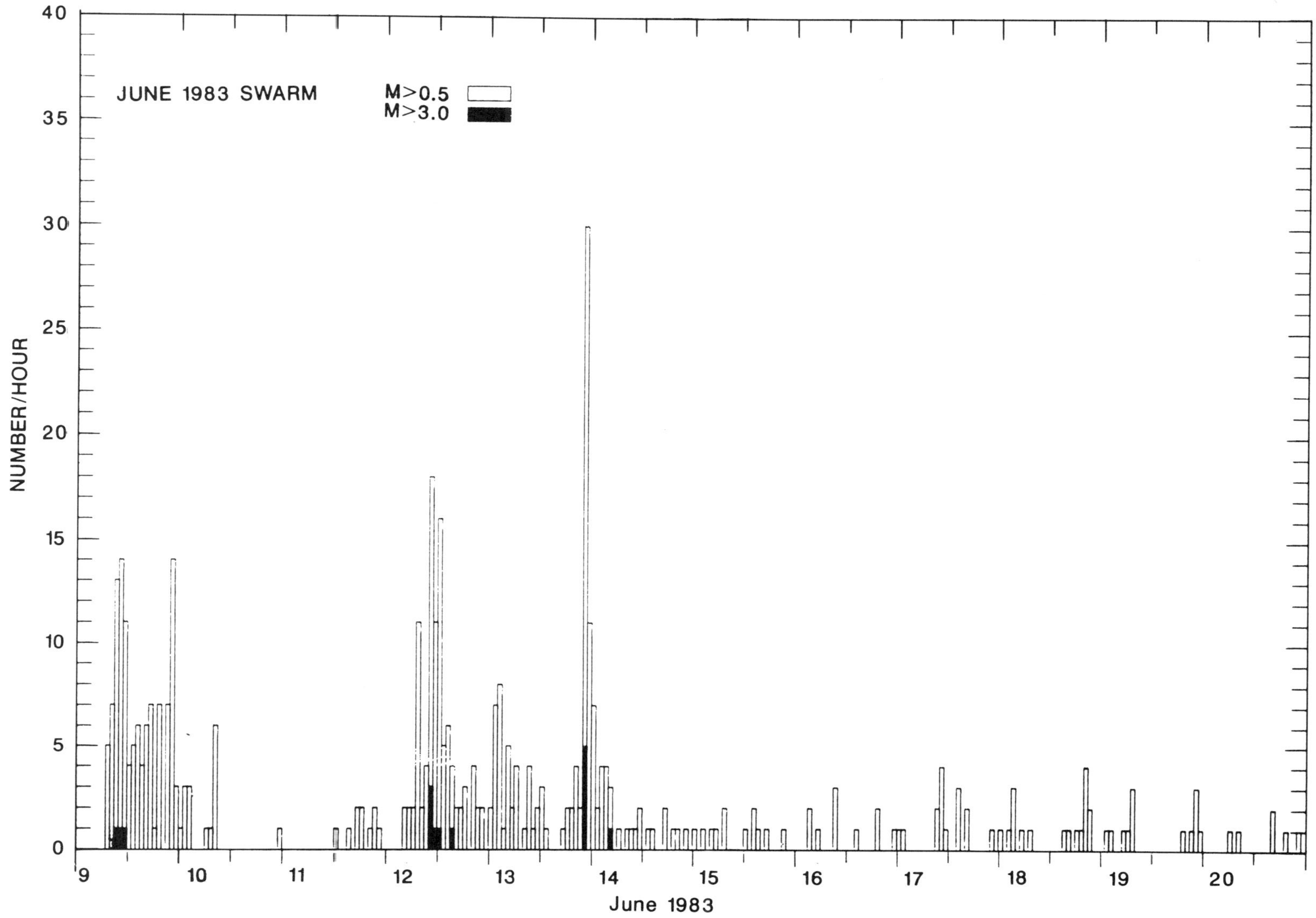

Figure 8.5.14. Hourly number of earthquakes during a swarm in the northern part of Aso Caldera during 9-14 June 1983, from Aso Volcanological Laboratory (1984). See also figures 8.5.15 and 8.5.16.

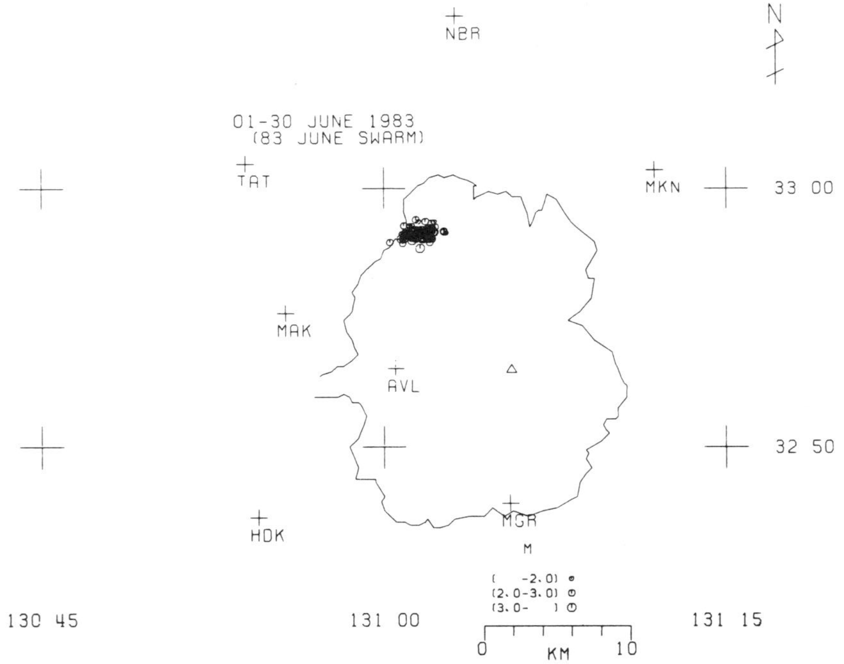

Figure 8.5.15. Epicenters of a swarm of earthquakes along the rim of Aso Caldera in June 1983, from Aso Volcanological Laboratory (1984). Crosses, seismic stations; hexagons, epicenters; triangle, Naka-dake.

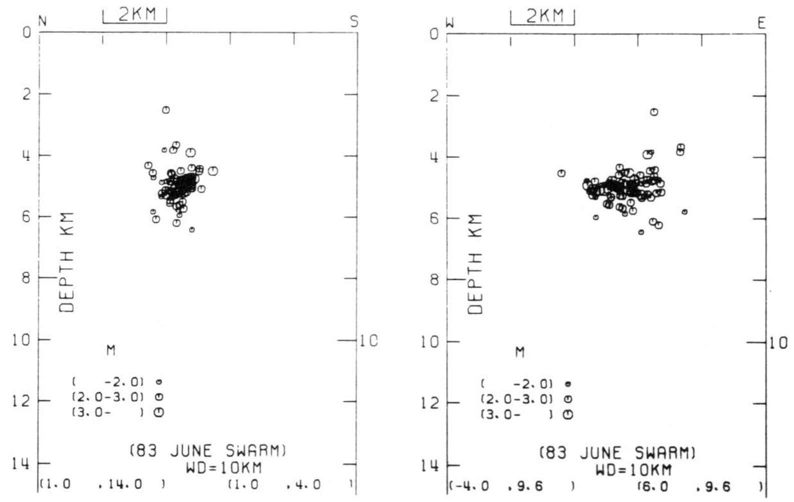

Figure 8.5.16. Vertical cross sections of earthquakes during a swarm at Aso Caldera in June 1983, from Aso Volcanological Laboratory (1984).

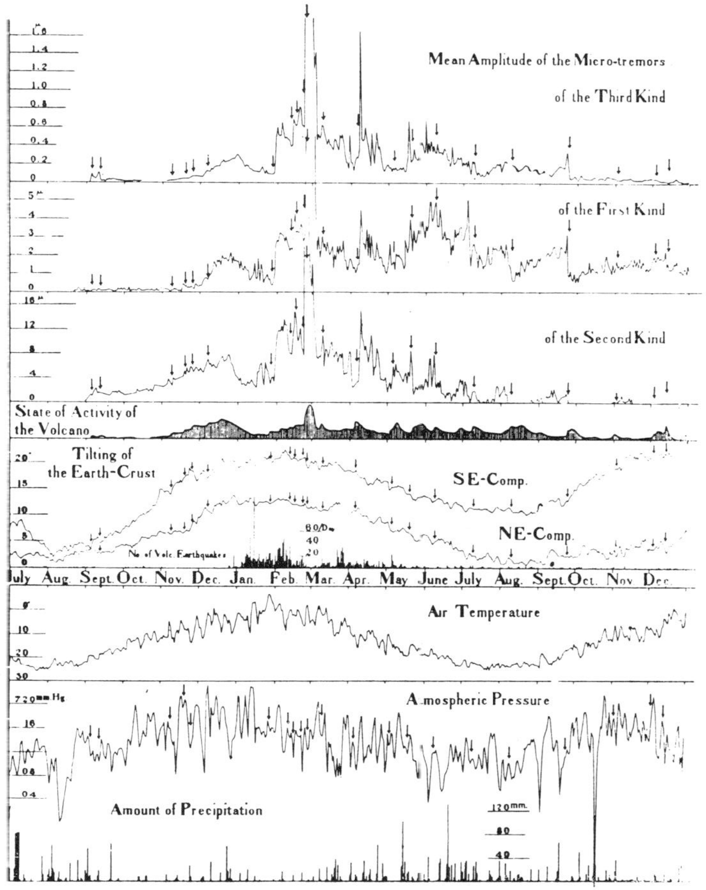

Figure 8.5.17. Variations in seismicity, eruptive activity, ground tilt, and environmental conditions at Nake-dake during 1932-33, from Sassa (1936a).

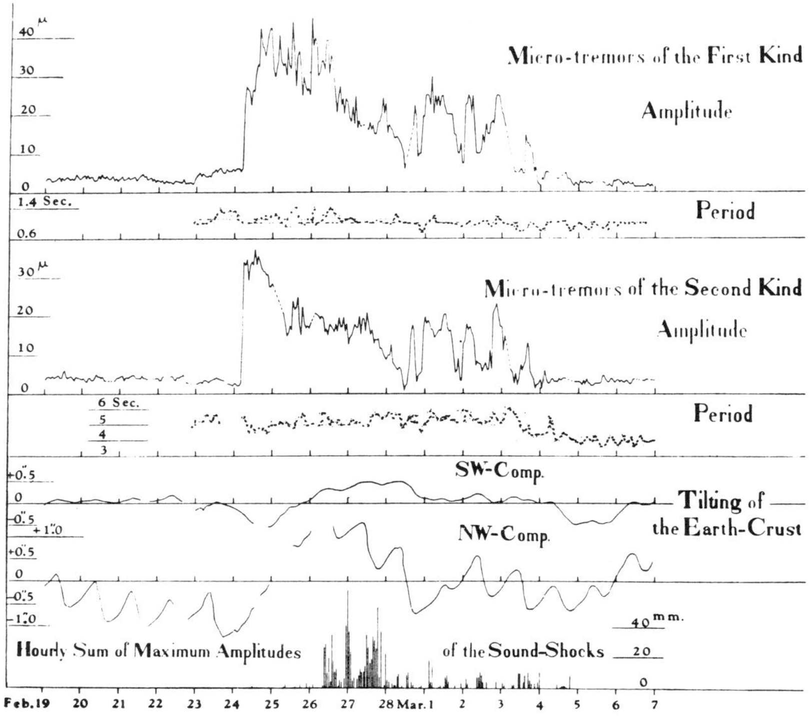

Figure 8.5.18. Variations in seismicity, ground tilt, and eruptive activity at Nake-dake during February-March 1933, from Sassa (1936a).

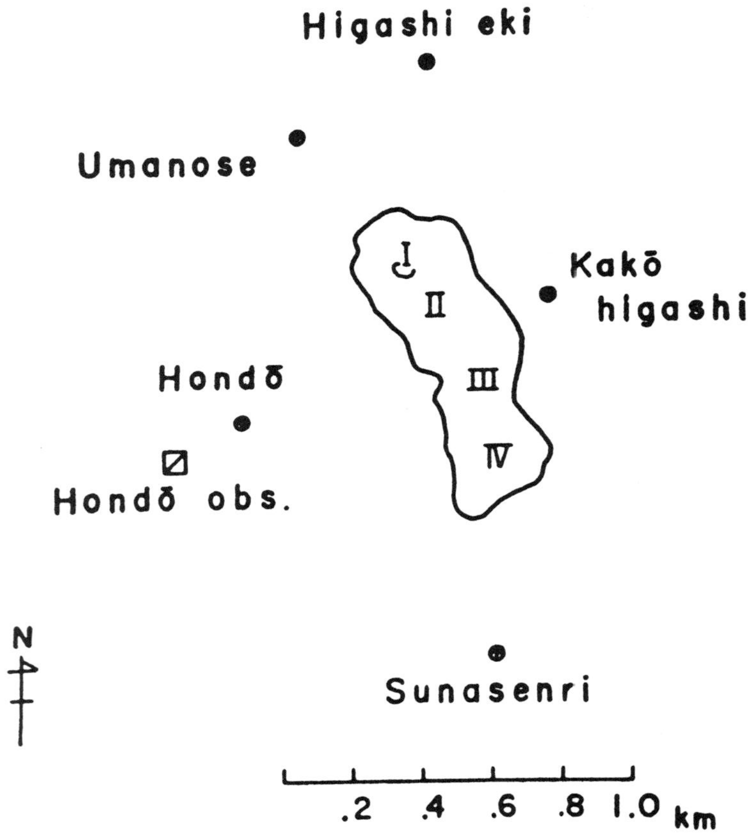

Figure 8.5.19. Locations of seismic and tilt stations near Naka-dake, from Wada and others (1968).

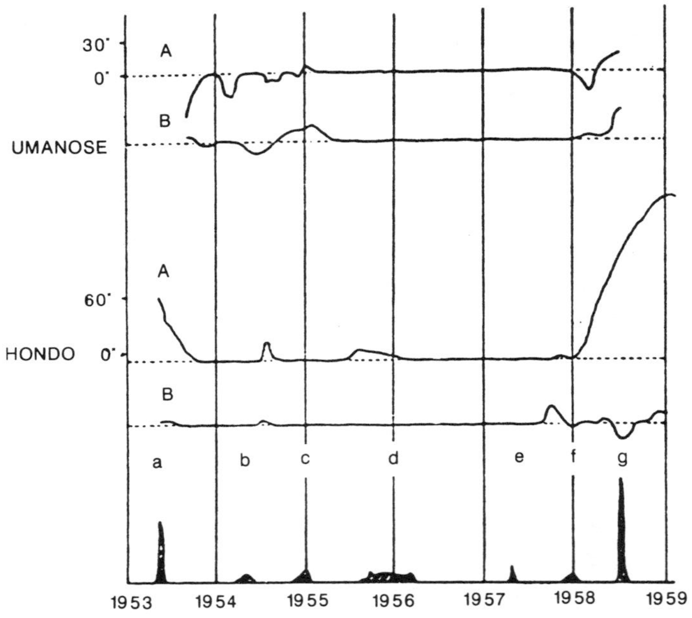

Figure 8.5.20. Residual ground tilt (in arcseconds) at Hondo and Umanose (see figure 8.5.19) during 1953-59, from Ito (1961). A and B are SW and NW components, respectively. Lower case letters: a, large eruption in April-May, 1953; b-f, small eruptions in 1954, 1955, 1956, April 1957, November-December 1957; g, violent eruption in June 1958. One arcsecond is about 5 microradians.

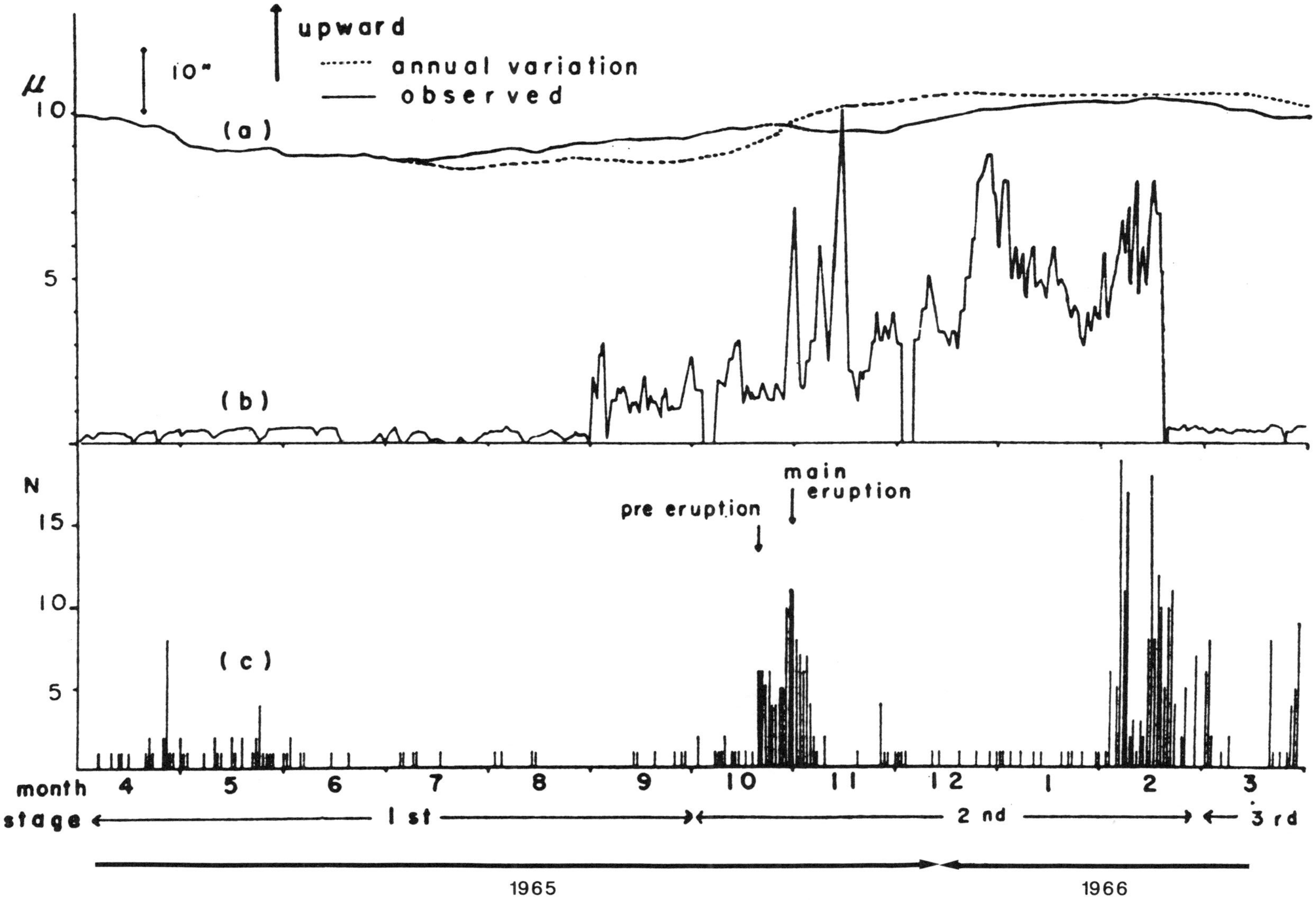

Figure 8.5.21. Variations in tilt and seismicity near Naka-dake during an active period from April 1965 to March 1966, from Wada and others (1968). Top, ground tilt at Sunasenri (see figure 8.5.19 for location); middle, mean amplitude of type I microtremors at Sunasenri; bottom, daily number of volcanic earthquakes. Stage 1, 6 April - 30 September; stage 2, 1 October - 28 February; stage 3, 1-31 March.

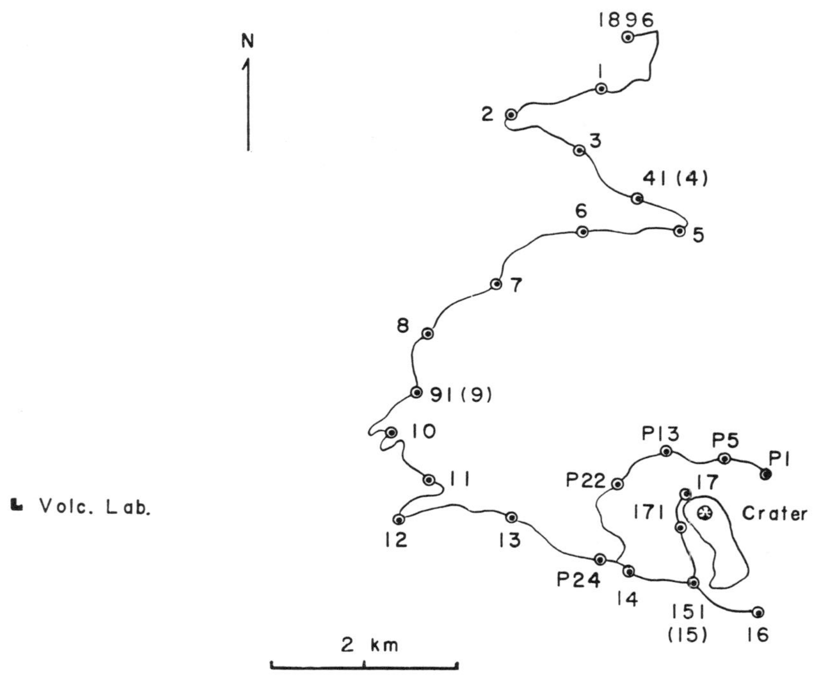

Figure 8.5.22. Leveling route and bench marks on floor of Aso Caldera and near Naka-dake ("crater"), from Kubotera (1984).

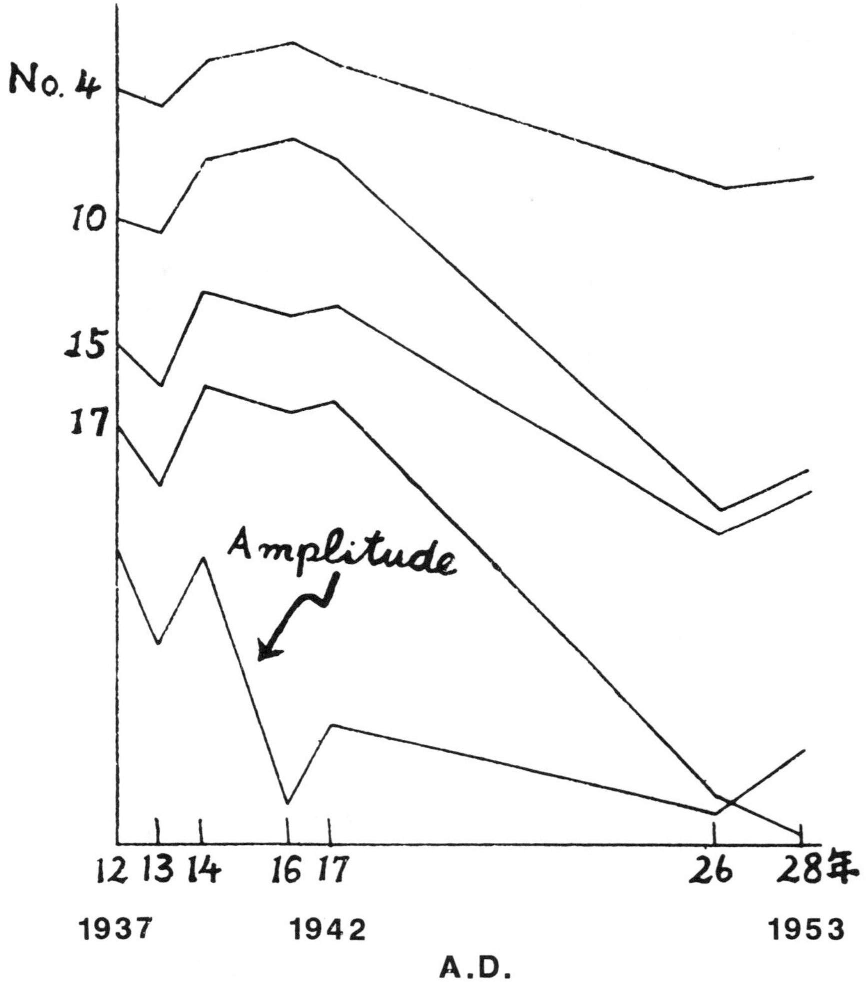

Figure 8.5.23. Relation between amplitude of type I volcanic microtremors (bottom curve) and elevations of four bench marks (numbers 4, 10, 15, and 17 in figure 8.5.22) at Aso Caldera, from Yoshikawa (1954). In general, decreases in mean tremor amplitude correspond to subsidence; the greatest subsidence occurred near summit of Naka-dake (bench mark 17).

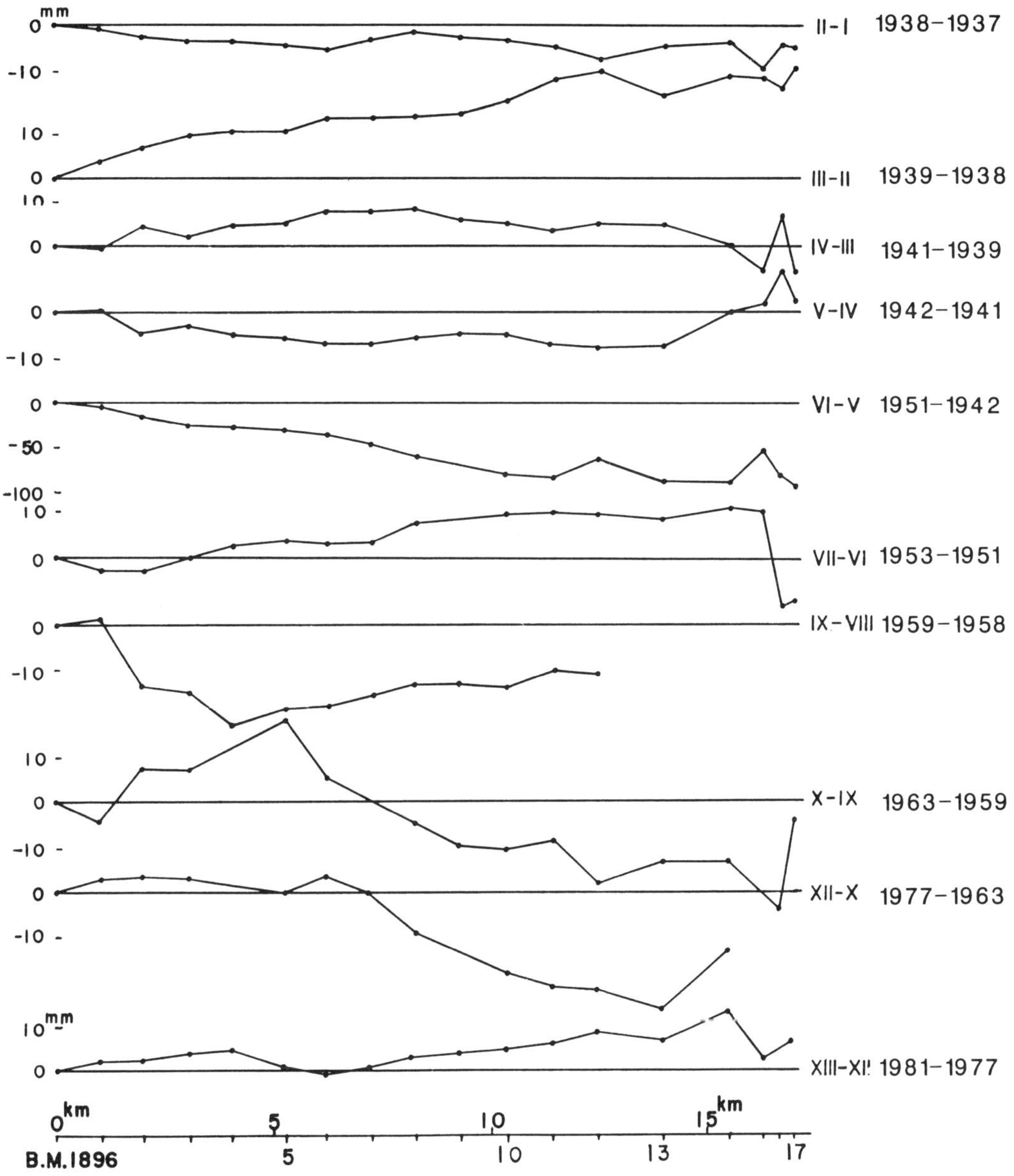

Figure 8.5.24. Elevation changes along leveling line at Aso Caldera from 1937 to 1981, from Kubotera (1984).

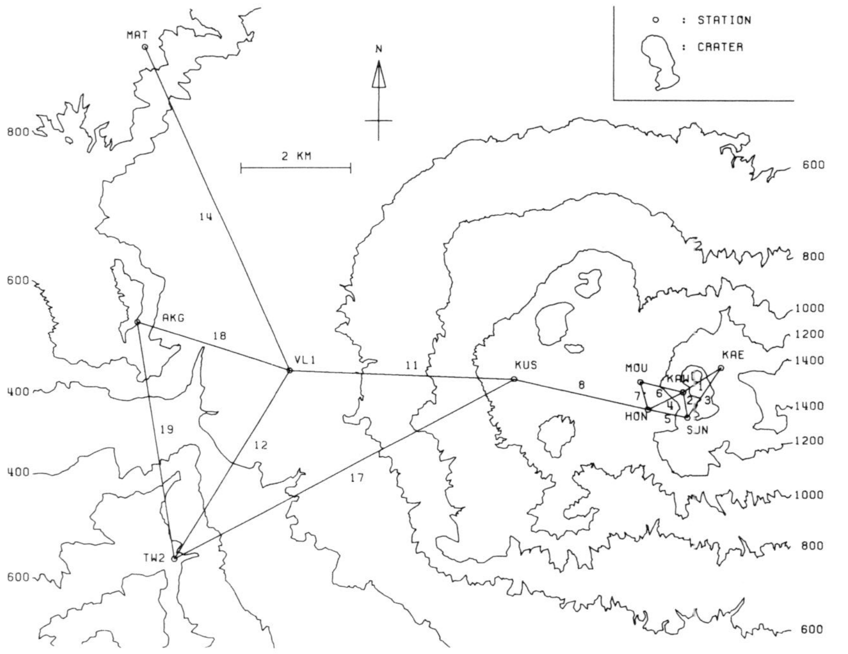

Figure 8.5.25. Network of lines for precise distance measurements at Aso Caldera, from Kubotera (1984).

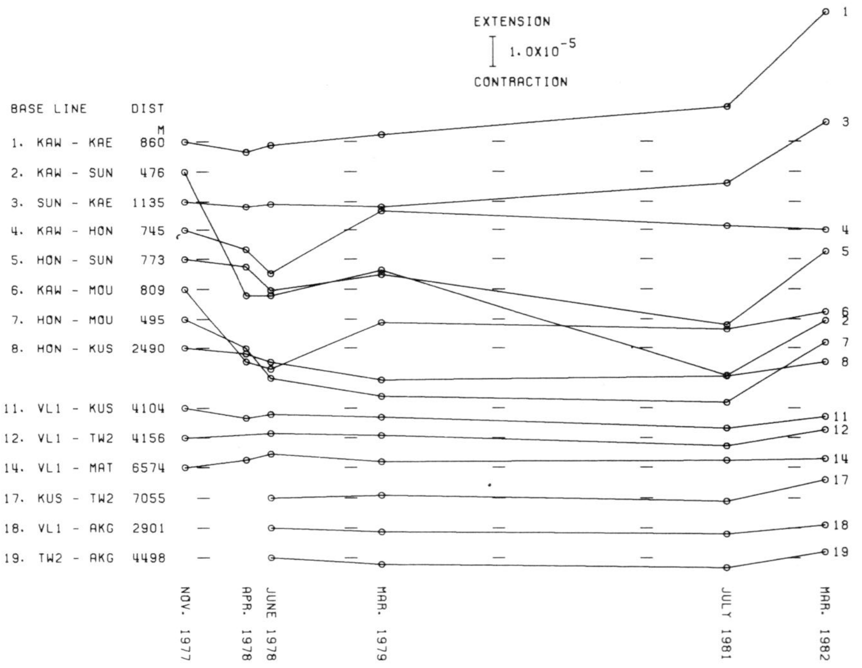

Figure 8.5.26. Line-length changes at Aso Caldera from 1977 to 1982, from Kubotera (1984). See figure 8.5.25 for location of lines.

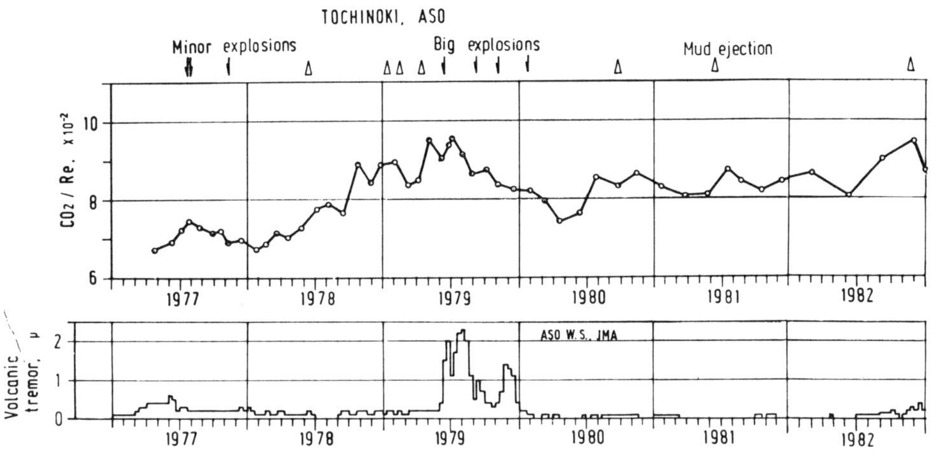

Figure 8.5.27. Variations in CO_2/Re ratio and amplitude of volcanic tremor at Naka-dake from 1977 to 1982, from Kubotera (1984). Re, evaporative residue.

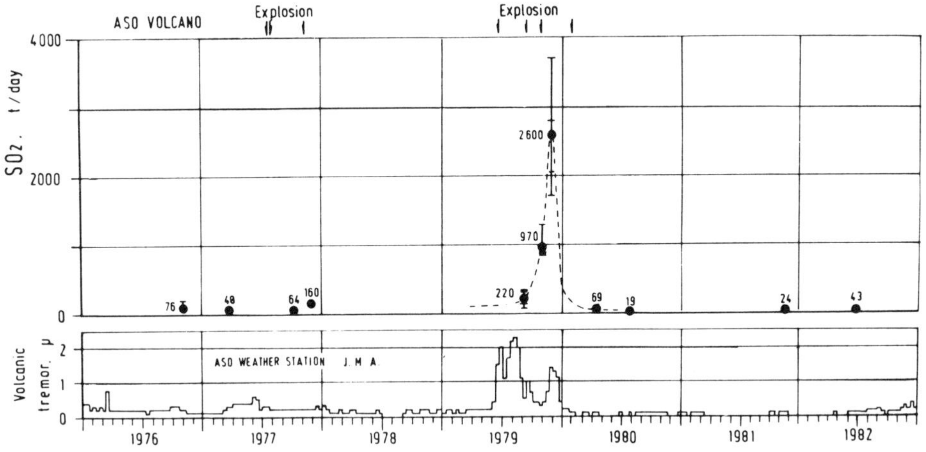

Figure 8.5.28. Variations in sulfur dioxide flux and amplitude of volcanic tremor at Naka-dake from 1976 to 1982, from Kubotera (1984). A decrease in SO_2 flux accompanied the end of explosions.

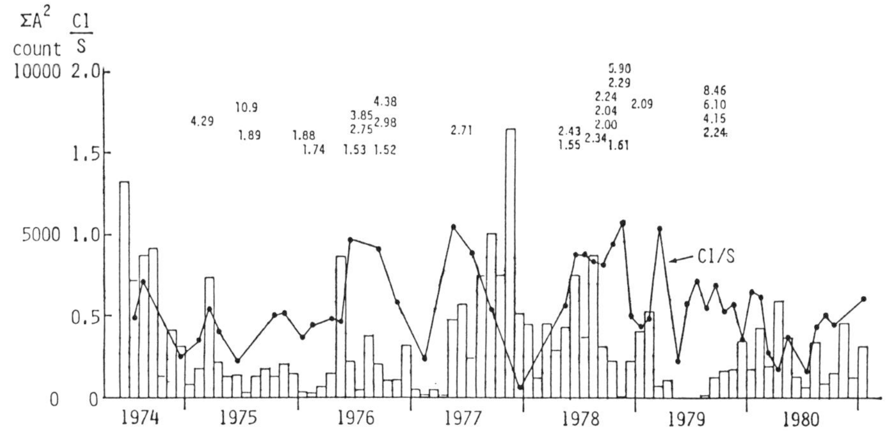

Figure 8.5.29. Variations in Cl/S ratio and seismicity at Aso Caldera during 1974-80, from Ota (1984).

PART 3: HISTORICAL UNREST AT LARGE QUATERNARY CALDERAS (continued)
See inside back cover for explanation and abbreviations

HAKONE

CAVW number (active vent)	Latitude Longitude (degrees)	Diameter (km)	Local tectonic setting	Precaldera edifice	SiO_2 content (wt pct)	Age of collapse (yr B.P.)	Date of unrest	Type and duration of unrest ESTU STHF MGTF H Te	Eruption type
08-03-02	35.22N	10	Compr	Strato	R = 54-59	180,000	1786	xx-- ---- ---- - --	none
(Kamiyama)	139.02E				C = 58-63	66,000	1917	xx-- ---- ---- - --	none
						49,000	1920	xx-- ---- ---- - --	none
							1923	---- ---- ---x - --	none
							1933-35	EE-- ---- --EE - --	none
							1940-46	FF-- ---- ---F F --	none
							1952-53	EE-- ---- --?? ? --	none
							1959-60	EE-- ---- ---- - --	none
							1966-67	EE-- ---- --FF F --	none
							1972-73	EE-- ---- --FF F --	none
							1981-83+	FF-- ---- ---- - --	none

TECTONIC SETTING

The northernmost tip of the Philippine Sea plate is squeezed between subduction of the Pacific plate beneath the Philippine Sea plate and subduction of the Philippine Sea plate beneath the Eurasian plate. Hakone, a product of subduction of the first-mentioned subduction, lies on the northernmost, squeezed tip of the Philippine Sea plate (Nakamura and others, 1984).

GEOLOGIC HISTORY

The geology of the Hakone area was described by Kuno (1950, 1950-51) and supplemental information may be found in Oki and others (1978) and Machida (1981). Hakone Volcano has experienced a long history of explosive eruptions and two major periods of caldera formation, 49,000-66,000 yr B.P. and ca. 180,000 yr B.P. (Machida, 1981). The modern caldera formed as a result of multiple explosive eruptions involving several tens of cubic kilometers of dacitic and andesitic magma (figs. 8.6.1, 8.6.2). Following the latest collapse, at least seven cones and domes have grown inside the caldera, including the most recently active vent, Kamiyama. About 3,000 yr B.P., a large phreatic (?) explosion and landslide (Oki and others, 1978) or pyroclastic flow from Kamiyama dammed part of the moat, allowing Lake Ashinoko to form; a lava plug and pyroclastic flow followed immediately. This was the last magmatic eruption that is known to have occurred at Hakone Volcano.

HAKONE (continued)

HISTORICAL ACTIVITY

Seismic swarms have occurred repeatedly during this century. Unrest in 1917 began with weak earthquakes during 17-24 January and on 30 January. Stronger earthquakes occurred on 31 January; 242 were noted in Ubako, northwest of the volcano, and some were recorded in Tokyo (85 km distant). "Rolling" and "vertical" movements were noted 18 times between 1300 hr and 1800 hr on 2 February, in the central part of Hakone Volcano. An intracaldera seismic network was established after swarms in 1959 and 1960 (figs. 8.6.3-8.6.5). For the periods 1950-60 and 1961-68, the b-value for Hakone earthquakes was 1.6. Focal depths in 1959-60 were shallower than 4 km, mostly between 1 and 2 km, but earthquakes were exclusively of the A-type; B-type earthquakes and volcanic tremor were not observed. Hypocenters in the May-July 1966 swarms were shallower than 1.5 km. Epicenters in 1959-60 were within a rather tight circle 1.5 km in diameter; epicenters in 1966 were also tightly clustered, but were shifted 0.9 km north of the 1959-60 epicenters. Seismicity in 1981-83 was elevated (figs. 8.6.5, 8.6.6); most events were shallow and beneath Kamiyama as in previous years, but one swarm (September 1982) was just northeast of the caldera at depths of 6-18 km (figs. 8.6.7-8.6.9) (National Research Center for Disaster Prevention, 1983; Hiraga and others, 1984).

Together, the 1959 and 1966 epicentral areas correspond to the 100 °C isotherm at sea level (figs. 8.6.10, 8.6.11). Strong correlations exist between the highest hydrothermal temperatures and earthquakes, and between increased seismic activity and increased fumarolic activity (fig. 8.6.12). All are greatest in the center of the caldera, beneath the Kamiyama postcaldera stratovolcano. Changes in the locus and intensity of solfataric activity occurred in 1923, in 1933 before the swarms of 1934-35, and in 1941 before the swarms of 1943-44.

On 10 May 1933, a solfatara in Oowakidani exploded with a loud detonation sound, and mud from a new pit gushed 2 m into the air. One worker was killed in this single explosion (Tanakadate, 1934). Between February 1934 and October 1935, or possibly longer, "trembling sounds" were heard from Komagatake (a central cone, near Kamiyama), and soil temperature in that area rose remarkably. Water of Lake Asinoko became noticeably warm at the foot of Komagatake, a number of mud cones formed on Komagatake, and 25 acres of bamboo and cryptomeria forest were killed by heat (Tanakadate, 1936). During 1941-45, the main site of the Owaku-dani solfatara migrated about 1 km southwestward, and since then the temperature of the Ubago hot spring, located still farther southwest, has risen (Kuno, 1962). A landslide and mudflow from a solfataric area occurred in July 1953, 6-8 months after several earthquake swarms (Kuno, 1962), and temperatures of hot springs rose soon after a swarm in 1966 (Oki and others, 1978). The temperature of Gora hot spring rose 20 °C in May 1967 (fig. 8.6.12), and earthquakes also increased (Hiraga, 1972). Temperatures continued to rise until at least 1971; an area of anomalous heating that includes this hot spring enlarged over the same period.

COMMENTS

Hiraga (1972) proposed that the correspondence between earthquakes and thermal areas is caused by steam or hot water rising from depth and expanding as it nears the surface, thereby causing local stresses and earthquakes. An alternate possibility, suggested to explain swarms in other areas, is that the high-temperature waters reduce the strength of the rock and thereby tend to focus regional

HAKONE (continued)

COMMENTS (continued)

strain release in thermal areas. Earthquakes occur in swarms rather than in main shock-aftershock sequences because stresses are concentrated in small volumes, and because high pore pressures cause stress to be released in small increments before it can build to cause a large earthquake.

REFERENCES

Hiraga, S., 1972, Earthquake swarms of geothermal fields in Japan: Jour. Japanese Geotherm. Energy Assoc., v. 9, no. 1, ser. 32, p. 30-39.

Hiraga, S., Ito, H., Odaka, S., and Hakamata, K., 1984, Seismometrical study in Hakone Volcano, 1981-82: Bull. Hot Spring Res. Inst., Kanagawa Prefecture, v. 15, no. 5, p. 149-182.

----- 1985, Seismometrical study in Hakone Volcano, 1983: Bull. Hot Spring Res. Inst., Kanagawa Prefecture, v. 16, no. 4, p. 101-128.

Kuno, H., 1950, Petrology of Hakone Volcano and the adjacent areas, Japan: Bull. Geol. Soc. Am., v. 61, p. 957-1020.

----- 1950-51, Geology of Hakone Volcano and adjacent areas: Jour. Fac. Sci., Tokyo Univ., sec. II, v. 7, p. 257-279, p. 351-402.

----- 1962, Catalogue of the active volcanoes of the world including solfatara fields, Pt. XI, Japan, Taiwan and Marianas: Rome, IAVCEI, 332 p.

----- 1964, Dike swarm in Hakone Volcano: Bull. Volcanol., v. 27, p. 53-59.

Machida, H., 1980, Tephra and its implications with regard to the Japanese Quaternary period, in The Association of Japanese Geographers, ed., Geography of Japan: Tokyo, Teikoku-Shoin, 440 p.

----- 1981, Tephrochronology and Quaternary studies in Japan, in Self, S., and Sparks, R.S.J., eds., Tephra studies: Dordrecht, D. Reidel, p. 161-191.

Minakami, T., 1960, Fundamental research for predicting volcanic eruptions, Pt. I: Bull. Earthq. Res. Inst., v. 38, p. 497-544.

----- 1974, Seismology of volcanoes in Japan, in Civetta, L., Gasparini, P., Luongo, G., and Rapolla, A.: Physical Volcanology: Amsterdam, Elsevier, p. 1-27.

Minakami, T., Hiraga, S., Miyazaki, T., and Uchibori, S., 1969, Fundamental research for predicting volcanic eruptions, Pt. II: Bull. Earthq. Res. Inst., v. 47, p. 893-949.

Mogi, K., 1963, Some discussion on aftershocks, foreshocks and earthquake swarms--the fracture of a semi-infinite body caused by an inner stress origin and its relation to the earthquake phenomena (third paper): Bull. Earthq. Res. Inst., v. 41, p. 615-658.

Nakamura, K., 1977, Volcanoes as possible indicators of tectonic stress orientation--principle and proposal: Jour. Volcanol. Geotherm. Res., v. 2, p. 1-16.

Nakamura, K., Shimazaki, K., and Yonekura, N., 1984, Subduction, bending and eduction. Present and Quaternary tectonics of the northern border of the Philippine Sea plate: Bull. Soc. Geol. France, v. 26, no. 2, p. 221-243.

Nasu, N., 1935, Recent seismic activities in the Izu Peninsula (Pt. II): Bull. Earthq. Res. Inst., v. 13, p. 400-416.

Nasu, N., Kishinouye, F., and Kodaira, T., 1931, Recent seismic activities in the Izu Peninsula, Pt. I: Bull. Earthq. Res. Inst., v. 9, p. 22-35.

HAKONE (continued)

REFERENCES (continued)

National Research Center for Disaster Prevention, 1983, The September 1982 earthquake swarm near Hakone-Gohra, Kanagawa Prefecture: Rep. Coord. Comm. for Prediction of Volcanic Eruptions, no. 26, p. 47-48.

Oki, Y., Aramaki, S., Nakamura, K., and Hakamata, K., 1978, Volcanoes of Hakone, Izu and Oshima: Hakone, Hakone Town Office, 88 p.

Oki, Y., Ogino, K., Hirano, T., Odaka, S., Awaya, T., Sugiyama, S., and Oyama, M., 1983, Thermal springs and mineral waters of Kanagawa Prefecture, Japan: Bull. Hot Springs Res. Inst., Kanagawa Prefecture, v. 14, no. 4, p. 99-216.

Tanakadate, H., 1934, Volcanic activity in Japan and vicinity during the period between June 1931 and June 1934: Japanese Jour. Astron. Geophys., v. 12, p. 89-108.

----- 1936, Volcanic activity in Japan and vicinity during the period between July 1934 and October 1935: Japanese Jour. Astron. Geophys., v. 13, p. 121-139.

Tsuboi, C., 1933, Vertical crustal displacement in the seismic region of Ito, on the East coast of the Idu Peninsula: Bull. Earthquake Res. Inst., v. 11, p. 488-499.

Yukutake, T., and Tanaoka, I., 1960, Magnetic survey on Hakone Volcano by use of a proton magnetometer: Bull. Earthq. Res. Inst., v. 38, p. 41-54.

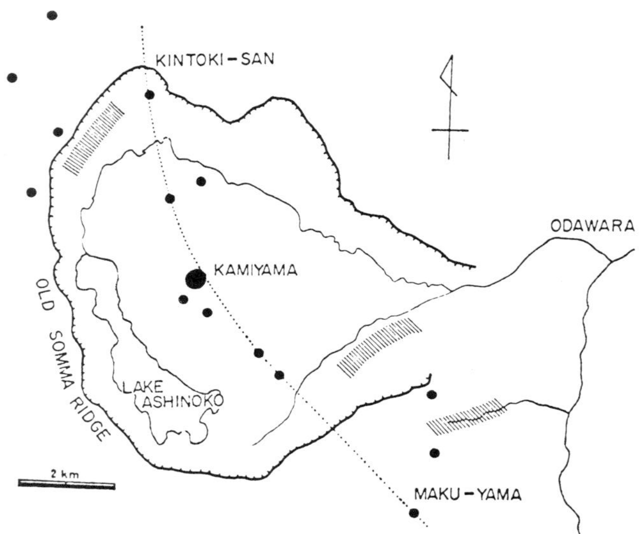

Figure 8.6.1. Sketch map of Hakone Caldera, from Oki and others (1978) after Kuno (1964) and Nakamura (1977). Small dots, parasitic cones; large dot, main postcaldera cone; hatched areas, dike swarms.

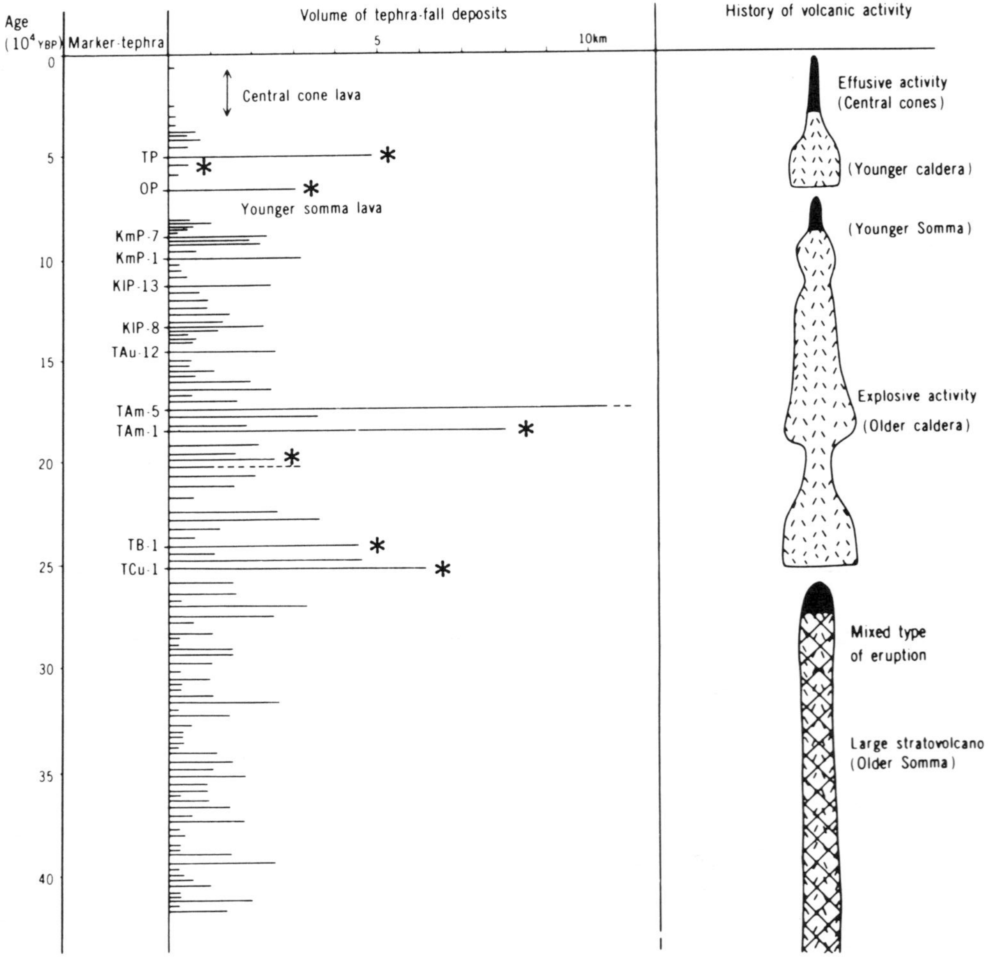

Figure 8.6.2. Eruptive history of Hakone Volcano, showing a long history of explosive eruptions and two major periods of caldera formation, 49,000-66,000 yr B.P. and ca. 180,000 yr B.P. (Machida, 1980). Asterisks indicate voluminous pyroclastic flows, whose volumes are not included in the bars for the volumes of tephra-fall units.

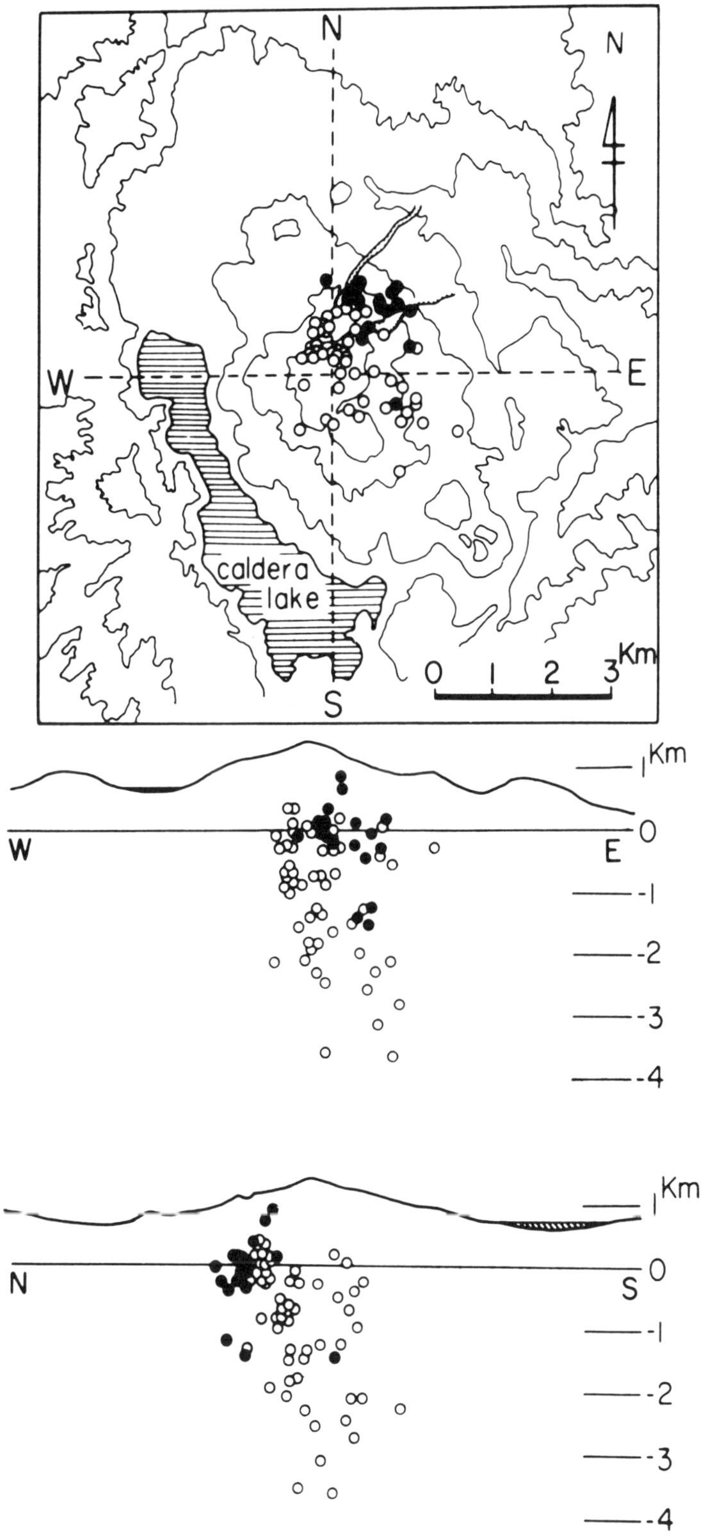

Figure 8.6.3. Hypocentral distribution of the 1959-60 (open circles) and 1966 (solid circles) earthquake swarms at Hakone Caldera, from Minakami and others (1969).

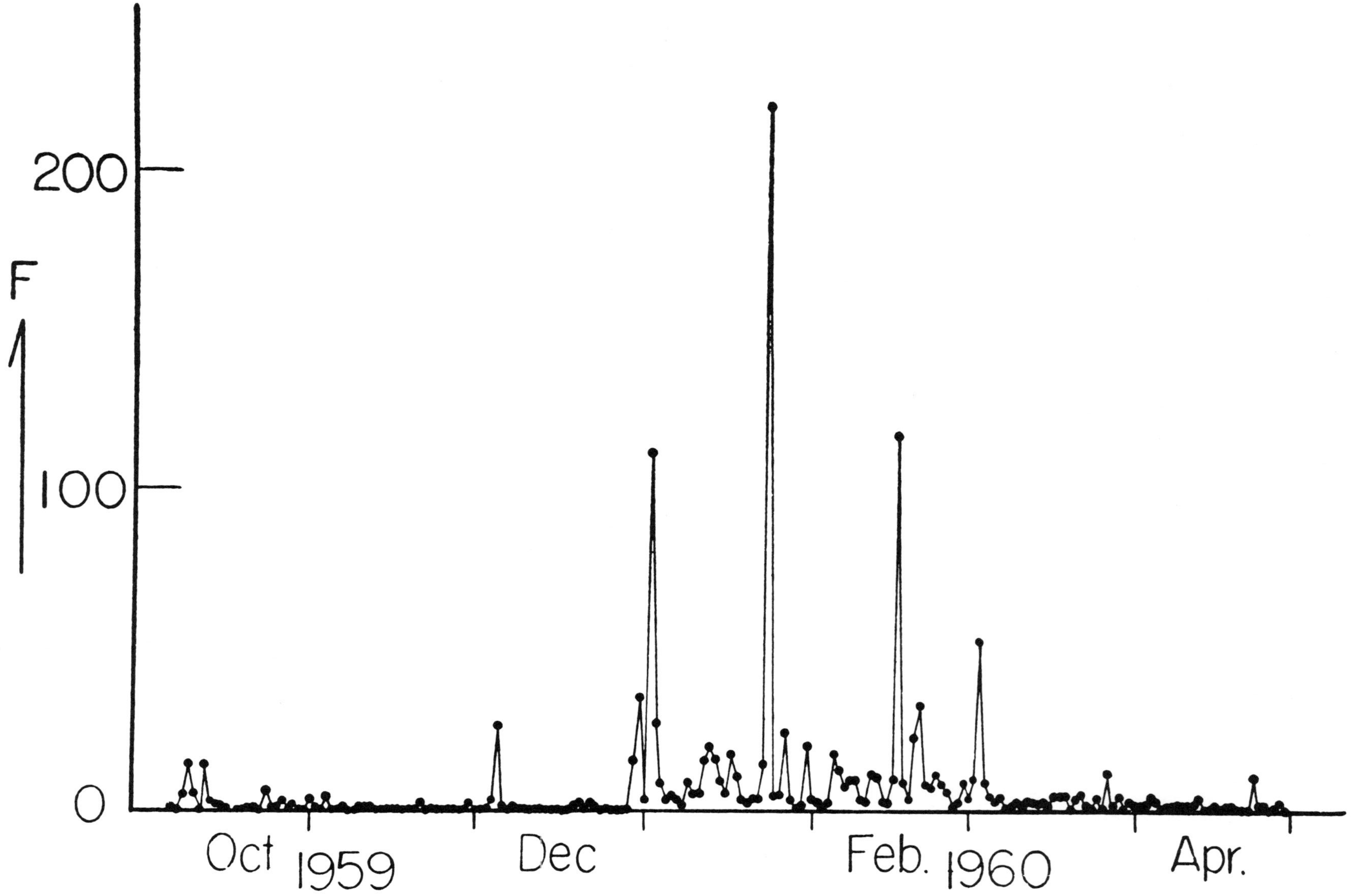

Figure 8.6.4. Daily frequency of earthquakes during the Hakone swarms of 1959-60, from Minakami (1960).

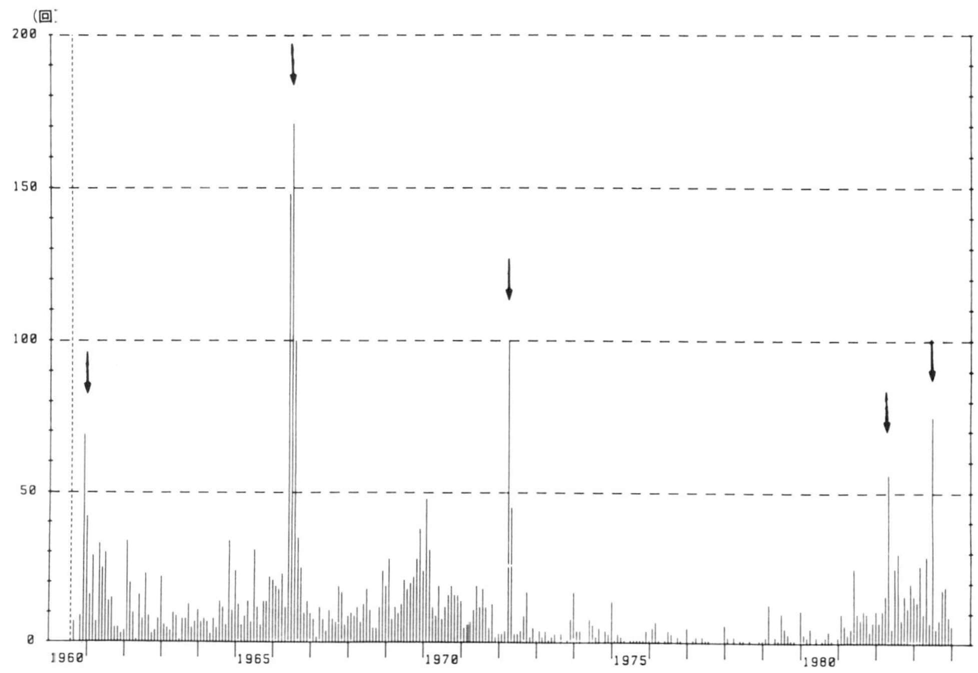

Figure 8.6.5. Monthly numbers of earthquakes at Hakone Volcano Observatory, 1960 to 1983, from Hiraga and others (1985). Arrows, earthquake swarms. Earthquakes just outside Hakone Caldera in 1982 (see figures 8.6.7 and 8.6.8) are excluded.

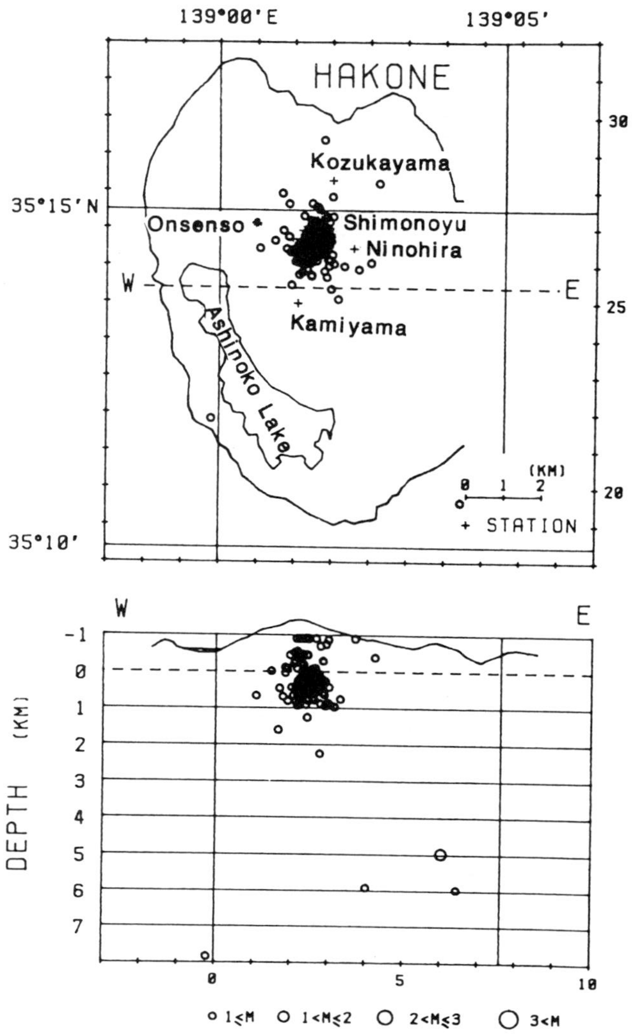

Figure 8.6.6. Hypocenters of earthquakes during 1983 at Hakone Caldera, from Hiraga and others (1985). Earthquakes that occurred just outside Hakone Caldera in 1982 (see figures 8.6.7 and 8.6.8) are excluded.

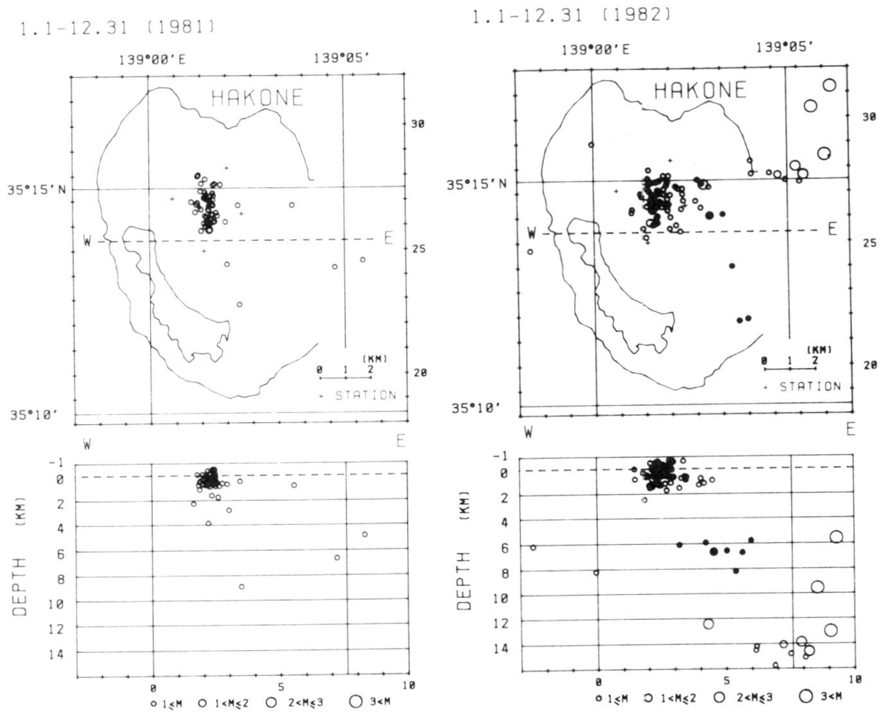

Figure 8.6.7. Earthquakes that occurred beneath or near Hakone Caldera, 1981-82, from Hiraga and others (1985). Relatively deep events northeast of the caldera occurred in September 1982.

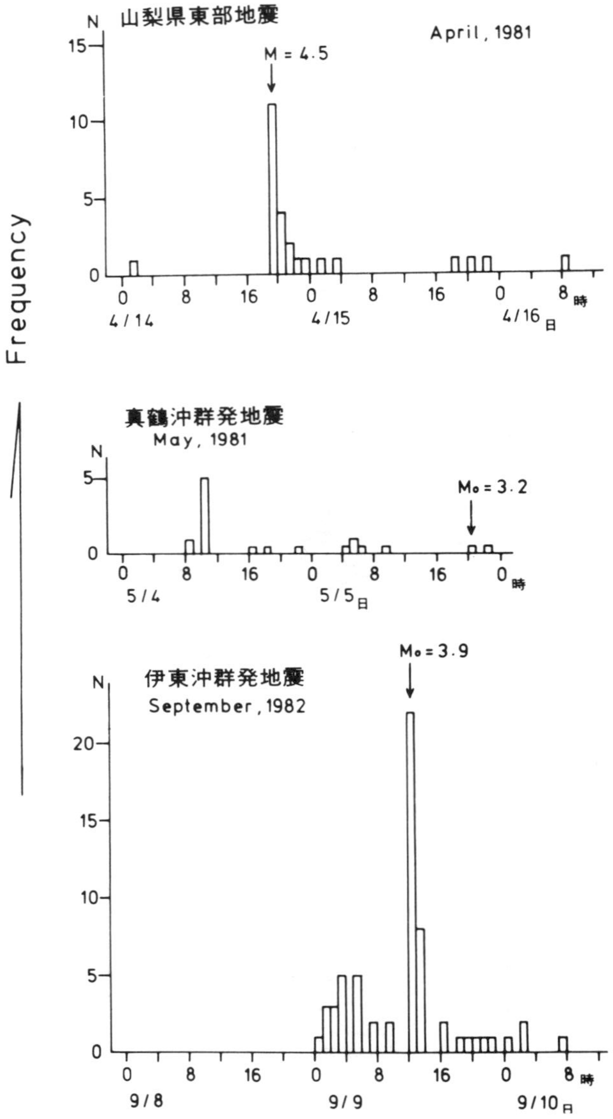

Figure 8.6.8. Time history of earthquake swarms beneath Hakone Caldera in April and May 1981, and immediately northeast of the caldera in September 1982, from Hiraga and others (1985).

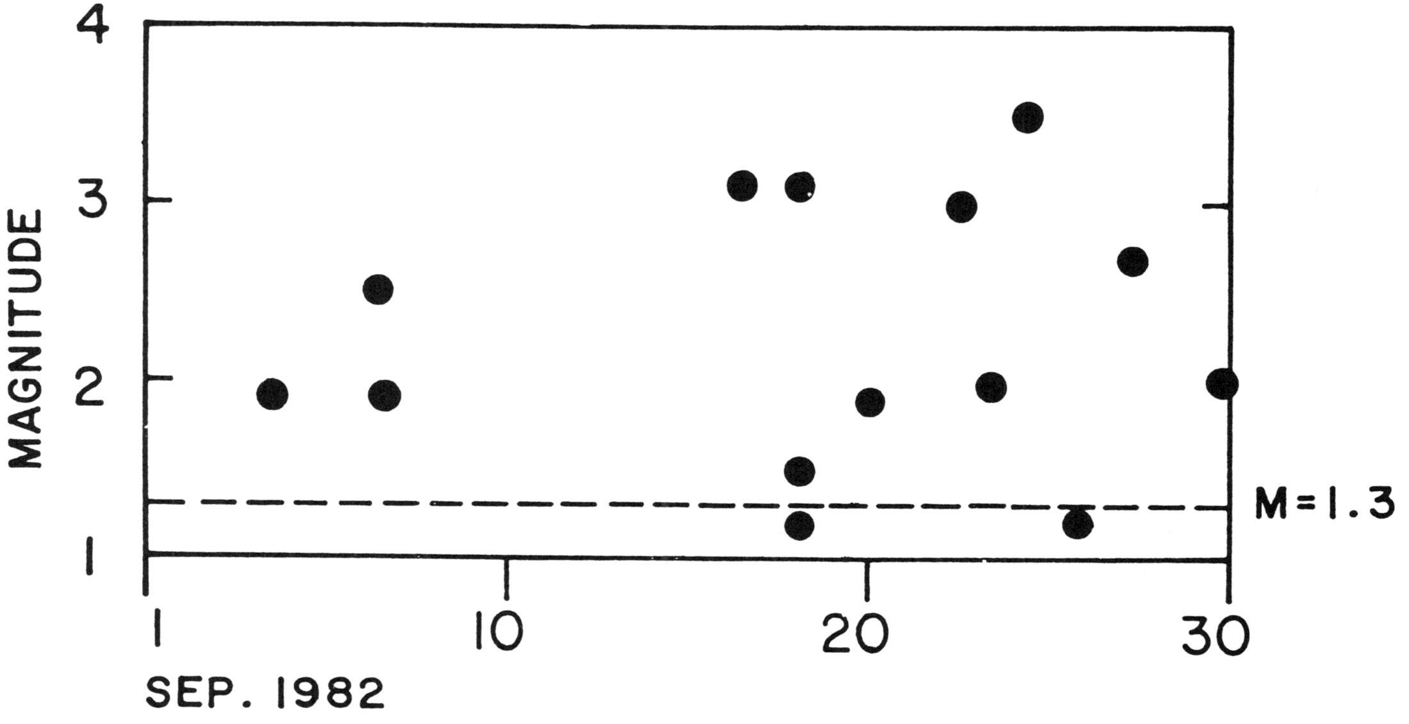

Figure 8.6.9. Magnitude versus time plot for a small earthquake swarm just northeast of Hakone Caldera in September 1982 (see figure 8.6.8), from National Research Center for Disaster Prevention (1983). Earthquakes were 6-18 km deep.

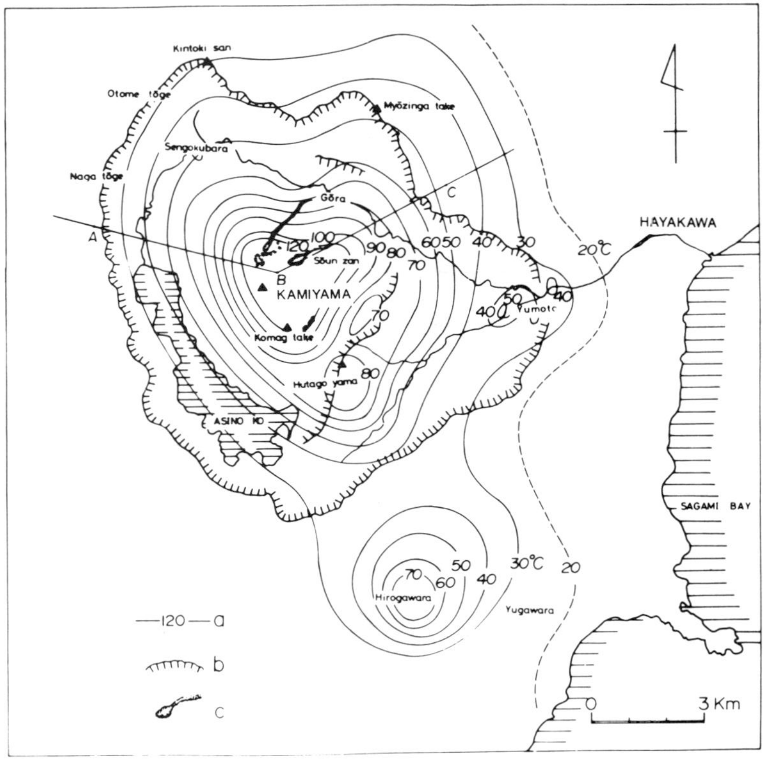

Figure 8.6.10. Thermal structure of Hakone Caldera at sea level, from Oki and others (1978). Contour interval, 10 $^{\circ}$C; a, isotherm; b, caldera rim; c, solfatara. Note locations of Kamiyama and Hayakawa for comparison with figure 8.6.11.

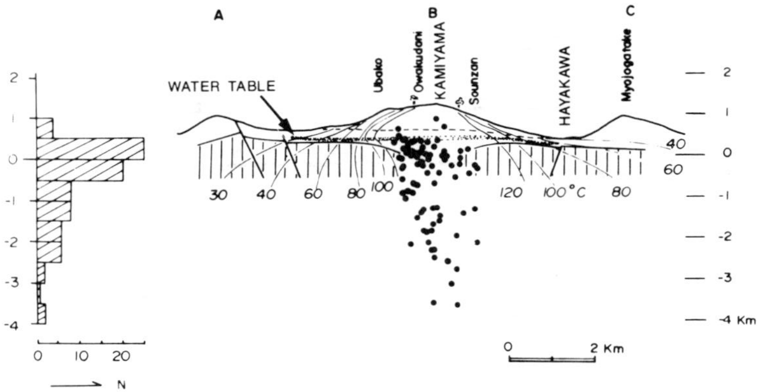

Figure 8.6.11. Distribution of microearthquakes beneath Hakone Caldera (right) and their depth-frequency relation (left), from Oki and others (1978). Note that all epicenters are within the 100 $^{\circ}$C isotherm.

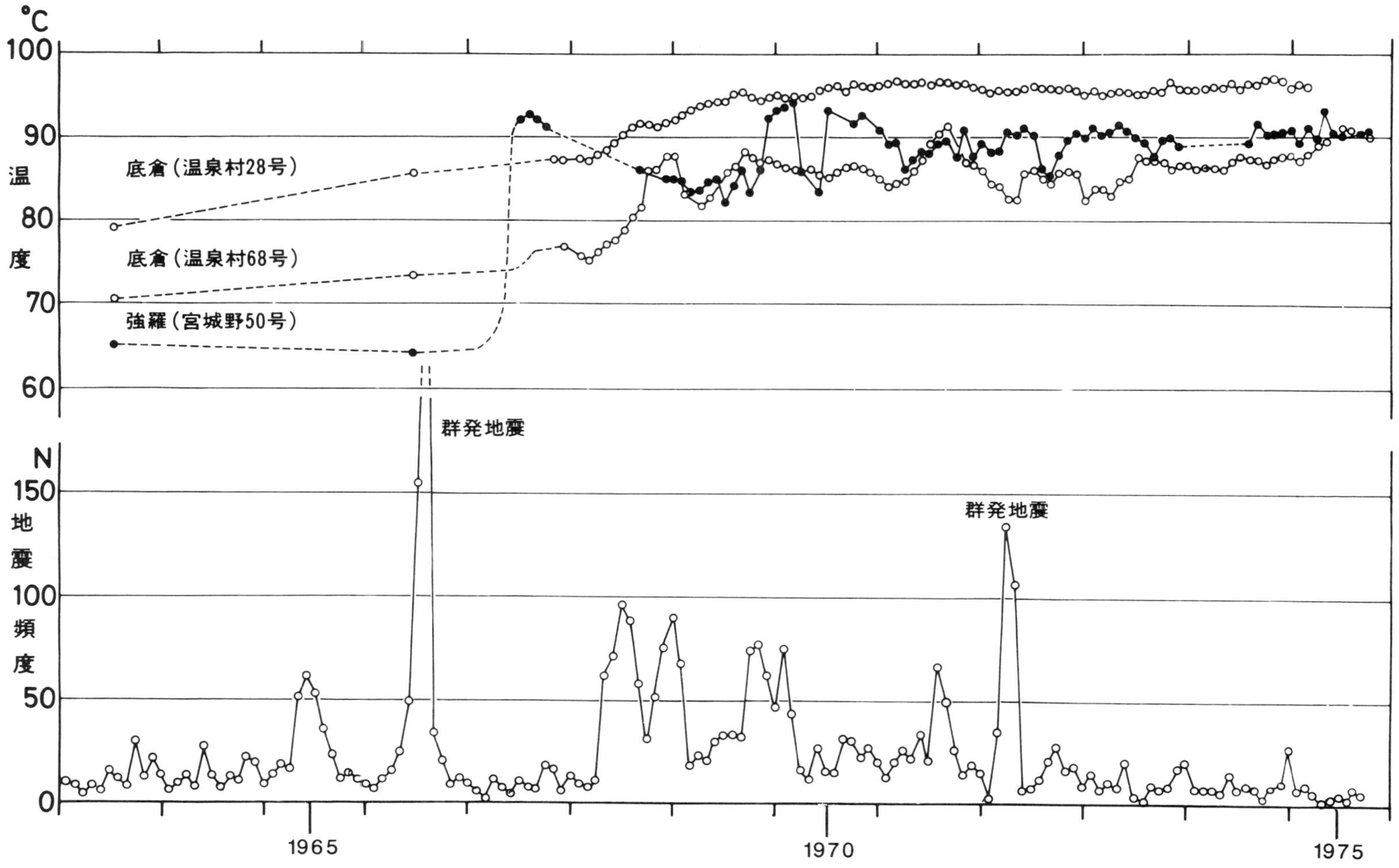

Figure 8.6.12. Relation between seismicity and geothermal water temperatures at three stations in Hakone Caldera, from Oki and others (1983). Note general increase in temperatures after (and perhaps before) 1966 earthquake swarm but little or no change associated with 1972 swarm.

PART 3: HISTORICAL UNREST AT LARGE QUATERNARY CALDERAS (continued)
See inside back cover for explanation and abbreviations

ON-TAKE

CAVW number (active vent)	Latitude Longitude (degrees)	Diameter (km)	Local tectonic setting	Precaldera edifice	SiO_2 content (wt pct)	Age of collapse (yr B.P.)	Date of unrest	Type and duration of unrest ESTU STHF MGTF H Te	Eruption type
08-03-04 (On-take)	35.90N 137.48E	4 x 5	Compr	Strato	R = 51-69 C = 69?	between 70,000 and 90,000	1976-84	FF-F ---- ---F F UQ	pex

TECTONIC SETTING/GEOLOGIC HISTORY

The On-take area includes several coalescing andesitic stratovolcanoes, from which roughly 13 km^3 of rhyolitic tephra was erupted between 70,000 and 90,000 yr B.P.. The modern On-take Volcano (figs. 8.7.1, 8.7.2) lies in the northern part of a caldera that probably formed during the rhyolitic eruption(s), but whose boundaries are largely buried (I. Moriya, written commun., 1987). A small summit caldera is approximately 2 km in diameter, and might be related to a moderately large eruption that occurred about 23,000 yr B.P. (Kobayashi and others, 1968). Before 1979, the only historical activity at On-take proper had been fumarolic. A striking, NE-SW (fault-controlled?) valley is the site of much fumarolic activity on the southwest flank of On-take, but recent unrest has been along a conjugate, NW-SE line crossing the summit.

HISTORICAL ACTIVITY

1976-84: Seismic swarms within a 5 km x 12 km area situated 7-15 km southeast of On-take began in August 1976, reached their peak in late 1978, and continued at least through 1984 (figs. 8.7.3-8.7.6, 8.7.9-8.7.11). Hypocenters were shallow, and $\underline{M}_{max}$=5.3. Focal-mechanism determinations made after an eruption in 1979 (see below) suggested strike-slip to oblique slip in the regional P direction (N70-80°W) and strike-slip along faults trending N40°W, oblique to regional structures and not obviously related to On-take Caldera. Low-amplitude S waves recorded west of Mt. On-take were interpreted to reflect an unusual focal mechanism rather than S-wave attenuation (Hori and others, 1982).

On-take erupted in October 1979 for the first time in recorded history (more than 1,000 yr), after only 6 hours of local seismicity that was recognized only in retrospect as a precursor to the eruption (figs. 8.7.5, 8.7.6). From October 1979 until April 1980 (or longer), a series of nine vents formed along a NW-SE-trending line, noted above. Because there is an earthquake-free zone between the above-mentioned swarms and the eruption site, and because the swarms neither increased just before the eruption nor decreased just after the eruption, the relation between these swarms and the eruption is unclear (Aoki and others, 1980;

ON-TAKE (continued)

HISTORICAL ACTIVITY (continued)

Volcanological Society of Japan, 1981-82). Smithsonian Institution (1979, v. 4, no. 12), citing a different source, states that the frequency of earthquakes in the swarm area decreased after the eruption. Very slight volcanic seismicity was detected in the summit (vent) area by seismometers installed after the initial eruption.

A 2-km-long electronic distance measurement (EDM) line on the east-southeast flank of On-take was measured in the month immediately following the 1979 eruption; it initially contracted by 1.5 cm in 2 weeks, but then recovered 0.5 -1 cm over next 2 weeks (Aoki, 1980, p. 98-100) (fig. 8.7.7). Leveling surveys along a NNE-SSW route about 25 km east of On-take Volcano showed as much as 4 cm of uplift from 1951 to 1967 and a comparable or slightly greater amount of fault-bounded subsidence from 1967 to 1979 (fig. 8.7.8) (Geographical Survey Institute, 1980). The maximum changes occurred about 40 km south of On-take and have no obvious relation to the volcano. Rather, they are probably related both to the earthquake swarms noted above and to a large earthquake (described next) that occurred about halfway between the largest level changes and On-take Volcano.

Based on the composition of water-soluble components of the ash erupted in 1979, Ossaka and others (1983) estimated the underground temperature in the vent area soon after the eruption to be higher than 250 °C, whereas that at the orifice was measured to be as low as 90 °C. They concluded that only high-temperature gases, not magma, rose to shallow depth during the eruption, triggering phreatic explosions. From 1980 to 1982, fumarolic temperatures in the vent area increased to 108-145 °C, but the ratio of SO_2 to H_2S decreased rapidly, indicating decreased temperature at depth. Therefore, the increase in surface temperatures was attributed to decreased cooling by ground water owing to evaporation during the eruption (Ossaka and others, 1983).

A $\underline{M}$=6.8 earthquake occurred on 13 September 1984, 10 km southeast of On-take Volcano at a depth of < 4 km (figs. 8.7.9-8.7.11). This is in the same area that had been affected by seismic swarms since 1976 (Fujii and Aoki, 1985). Fifty-one landslides occurred, including a large one (3.4×10^7 m^3) from the south flank of On-take Volcano (Inokuchi, 1985; Endo and others, 1986; Nagaoka, 1987). Fumaroles in the 1979 vent area were not visibly affected by the main shock or aftershocks (Smithsonian Institution, 1984), but Sugisaki and Sugiura (1986) showed measurable changes at a superheated fumarole on On-take's summit, from 1981 to a time just one week before the large 1984 earthquake, followed by a reversal of most of those changes after the earthquake. HCl and SO_2 concentrations increased severalfold from 1982 or 1983 to the time of the earthquake, as did the ratios of He, H_2, CH_4, and N_2 to Ar (fig. 8.7.12). These were interpreted as an increase in the ascent of deep gases of higher temperature (Sugisaki and Sugiura, 1986), probably as a result of crustal strain and the opening of cracks to greater depth beneath the volcano.

COMMENTS

On-take is of special interest for two reasons: (1) Its 1979 eruption is one of only a few eruptions in this compilation that followed a repose of several centuries or longer; and (2) although immediate precursors to the eruption were remarkably minor, the eruption was probably related in some way to 1976-84 seismic swarms that were located 7-15 km southeast of the eruption site, and to the 1984 $\underline{M}$=6.8 Nagano earthquake, centered 10 km southeast of On-take.

ON-TAKE (continued)

COMMENTS (continued)

We speculate that regional tectonic stresses built over the 1976-84 period, changed the ground water and/or magma plumbing beneath On-take Volcano enough to trigger the 1979 eruption, and eventually culminated in the 1984 shock. Sugisaki and Sugiura (1986) made a similar interpretation for the more limited time period 1981-84. An alternate speculation (Sano and others, 1986) is that the swarms, eruption, and large Nagano earthquake were all a result of diapiric magma intrusion beneath the area of the September 1984 shock.

REFERENCES

Aoki, H., ed., 1980, Report on the study of volcanic activity and disaster of the 1979 Ontake-san eruption: Special Rep., Ministry Education, Nagoya, Nagoya Univ., 168 p.

Aoki, H., Oida, T., Fujii, I., and Yamazaki, F., 1980, Seismological study of the 1979 eruption of Ontake-san, in Aoki, H., ed., Report on the study of volcanic activity and disaster of the 1979 eruption of Ontake-san eruption: Special Rep., Ministry Education, Nagoya, Nagoya Univ., p. 55-71.

Aramaki, S., and Ossaka, J., 1983, Eruption of Ontakesan, October 28, 1979: Report on volcano activities and studies in Japan for the period from 1979 to 1982, 18th IUGG Gen. Assembly, Hamburg and the Science Council of Japan, p. 1-7.

Earthquake Research Institute, University of Tokyo, 1980, Seismic activity of Ontake Volcano: Rep. Coord. Comm. for Prediction of Volcanic Eruptions, v. 18, p. 6-11.

Endo, K., Sumita, M., Machida, M., and Furuichi, M., 1986, Debris avalanche deposits by the 1984 collapse of Ontake Volcano, Central Japan: Proc. Symp. on Volcanic Products and Quaternary Geology, Nihon Univ., January 1986, p. 27-46.

Faculty of Science, Nagoya University, 1980, Seismic activity around Mt. Ontake: Rep. Coord. Comm. for Prediction of Volcanic Eruptions, Pt. 1, v. 18, p. 1-5; Pt. 2, v. 19, p. 1-5.

Fujii, I., and Aoki, H., 1985, Seismicity in the southern part of Hida Mountains, in Okada, Hm., ed., Comparative study of physical background of volcanic activity and its relation to prediction of eruption disasters: Special Rep. Grant-in-Aid for Nat. Disaster Sci., Ministry of Education, publ. by Hokkaido Univ., Sapporo, in cooperation with Tohoku Univ., Univ. of Tokyo Earthq. Res. Inst., Nagoya Univ., Kyoto Univ., and Kyoto Univ. Disaster Prev. Res. Inst., p. 88-91.

Geographical Survey Institute, 1980, Vertical movement around Mt. Ontake: Rep. Coord. Comm. for Prediction of Volcanic Eruptions, v. 18, p. 18-21.

Hori, S., Aoki, H., and Ooida, T., 1982, Focal mechanisms of the earthquake swarm southeast of Mt. Ontake, Central Honshu, Japan: Zisin, v. 35, p. 161-169.

Inokuchi, T., 1985, The Ontake rock slide and debris avalanche caused by the Naganoken-Seibu earthquake, 1984: Proc. IVth Int. Conf. and Field Workshop on Landslides, Tokyo, The Japanese Landslide Society, p. 329-338.

Kobayashi, K., Minagawa, K., Machida, M., Shimizu, H., and Kitazawa, K., 1968, The Ontake pumice-fall deposit Pm-1 as a late Pleistocene time-marker in central Japan: Jour. Fac. Sci., Shinshu Univ., v. 3, no. 2, p. 171-198.

Kobayashi, T., 1985, Slope failure caused by the Naganoken-seibu earthquake and geology of Ontake Volcano: Symposium for River Disaster, 1985, preprint, p. 1-15.

ON-TAKE (continued)

REFERENCES (continued)

Nagaoka, M., 1987, Geomorphological characteristics and causal factors of the 1984 Ontake landslide caused by the Naganoken-seibu earthquake: Bull. Geogr. Surv. Inst., v. 31, p. 72-89.

Ossaka, J., Ozawa, T., Sakai, H., and Hirabayashi, J., 1983, Geochemical study of the activity of Kiso-Ontake volcano, after the 1979 eruption: Bull. Volcanol. Soc. Japan, v. 28, p. 59-74.

Regional Center for Earthquake Prediction Observations, Nagoya University, 1980, Earthquake swarm occurring at south-eastern foot of Mt. Ontake: Rep. Coord. Comm. for Prediction of Volcanic Eruptions, no. 17, p. 1-5.

Sano, Y., Nakamura, Y, Wakita, H., Notsu, K., and Kobayashi, Y., 1986, $^3He/^4He$ ratio anomalies associated with the 1984 Western Nagano earthquake: possibly induced by a diapiric magma: Jour. Geophys. Res., v. 91, no. B12, p. 12,291-12,295.

Seismological Division, Japan Meteorological Agency, 1980, On the earthquake swarm occurring near Ohtaki, Nagano Pref.: Rep. Coord. Comm. for Prediction of Volcanic Eruptions, no. 17, p. 6-7.

Smithsonian Institution, Scientific Event Alert Network (SEAN), 1979-80, 1984, On-take: SEAN Bull., v. 4, nos. 10-12; v. 5, nos. 1, 4-6; v. 9, no. 9 (contributions from Y. Sawada and others).

Sugisaki, R., and Sugiura, T., 1986, Gas anomalies at three mineral springs and a fumarole before an inland earthquake, Central Japan: Jour. Geophys. Res., v. 91, no. B12, p. 12,296-12,304.

Volcanological Society of Japan, 1981-82, On-take: Bull. Volcanic Eruptions, no. 19, p. 49-51; no. 20, p. 55-56 (contributions from Y. Sawada).

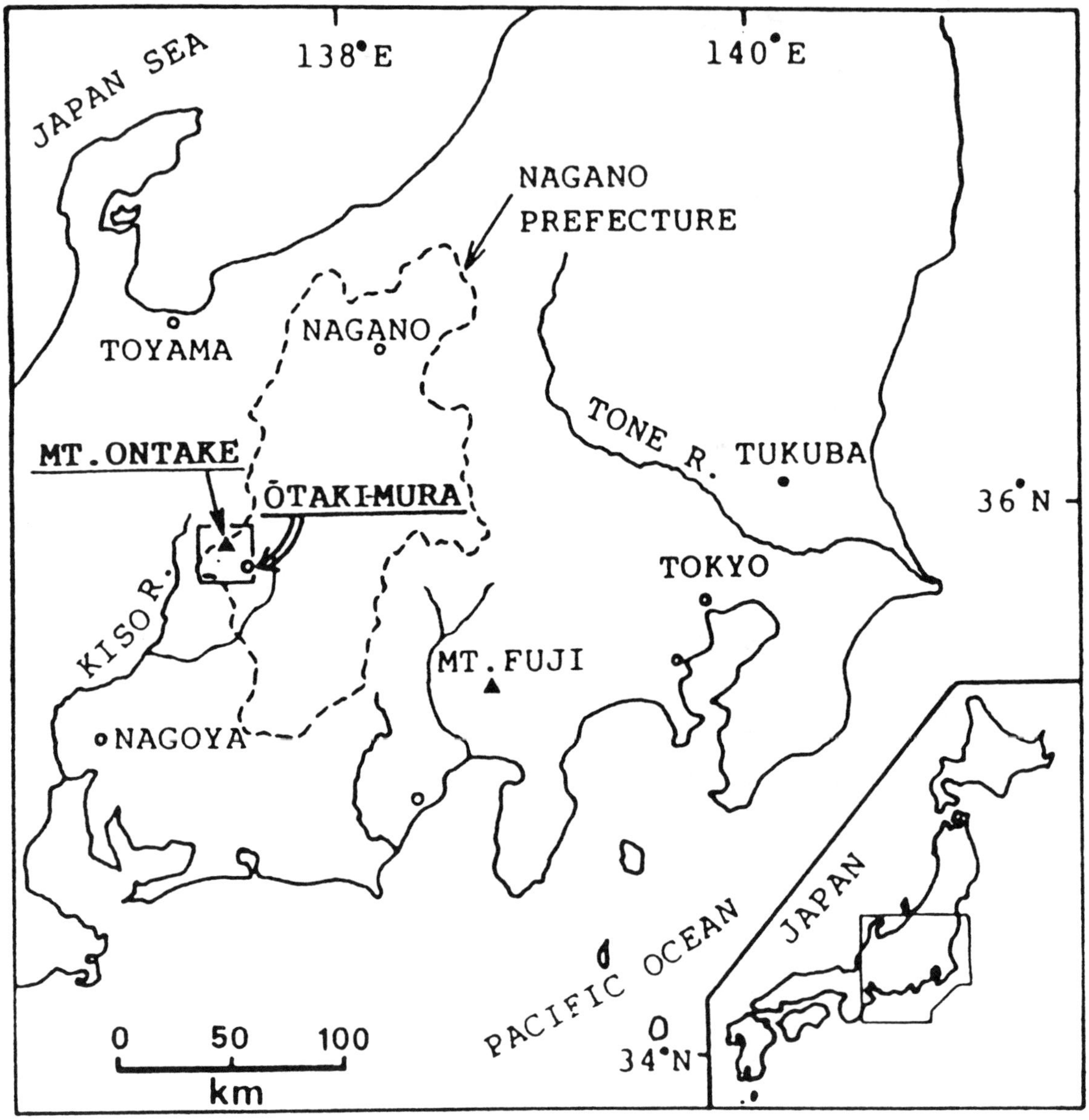

Figure 8.7.1. Sketch map showing location of Mt. On-take, from Inokuchi (1985).

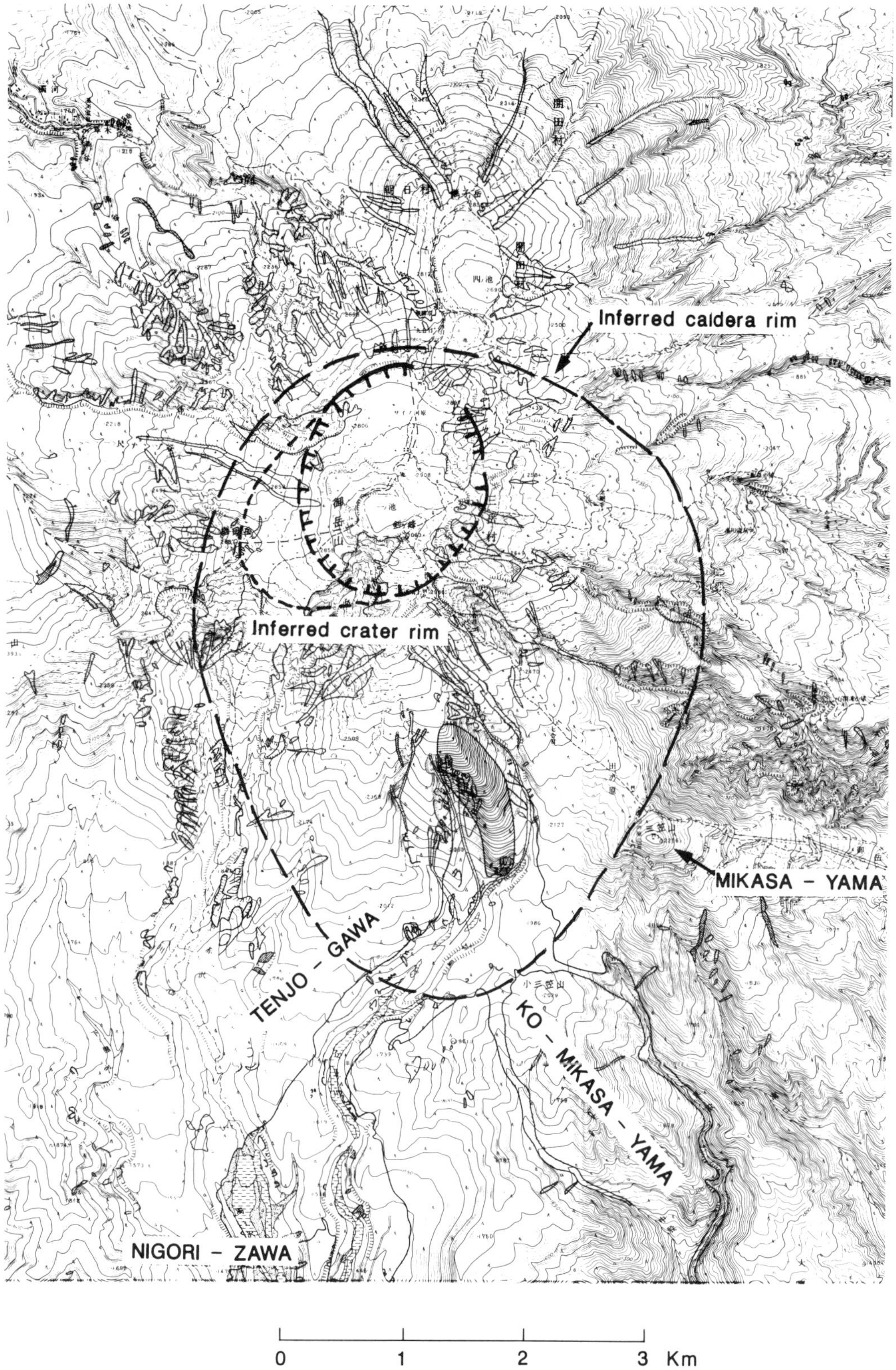

Figure 8.7.2. Topographic map of On-take Caldera and vicinity, with outline of caldera after I. Moriya (written commun., 1987).

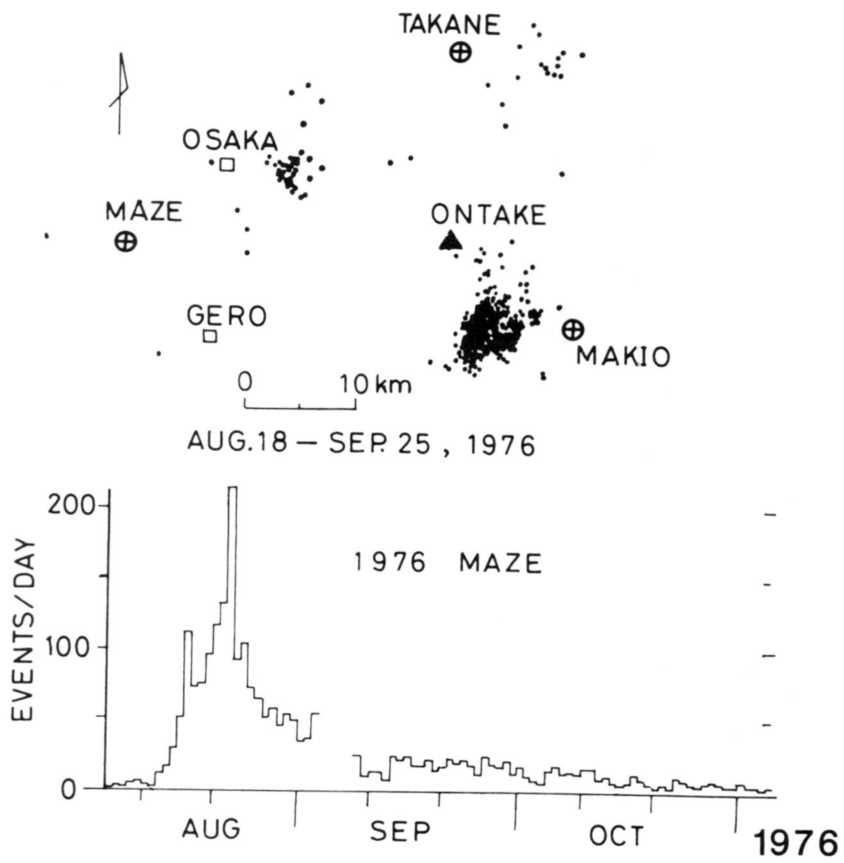

Figure 8.7.3. Epicenters and daily number of earthquakes near On-take during August-October 1976, from Regional Center for Earthquake Prediction Observations (1980).

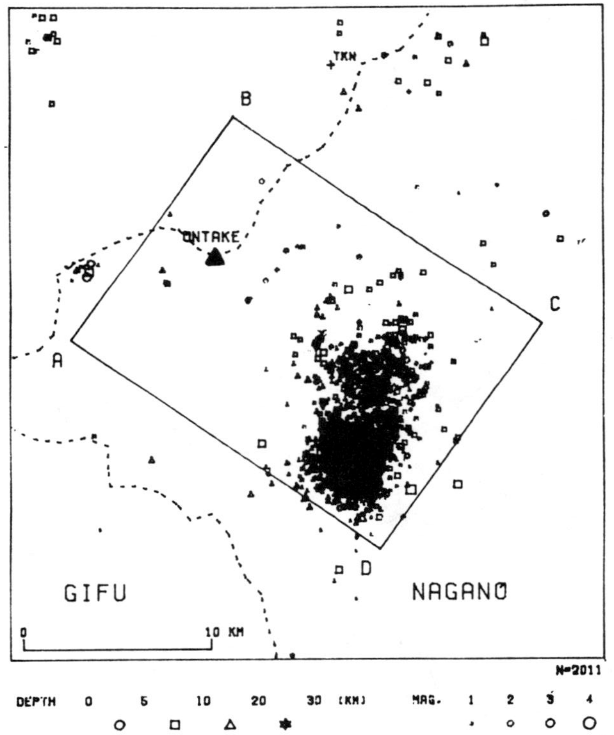

Figure 8.7.4. Epicenters for a swarm of earthquakes near On-take during 1978-79, from Regional Center for Earthquake Prediction Observations (1980).

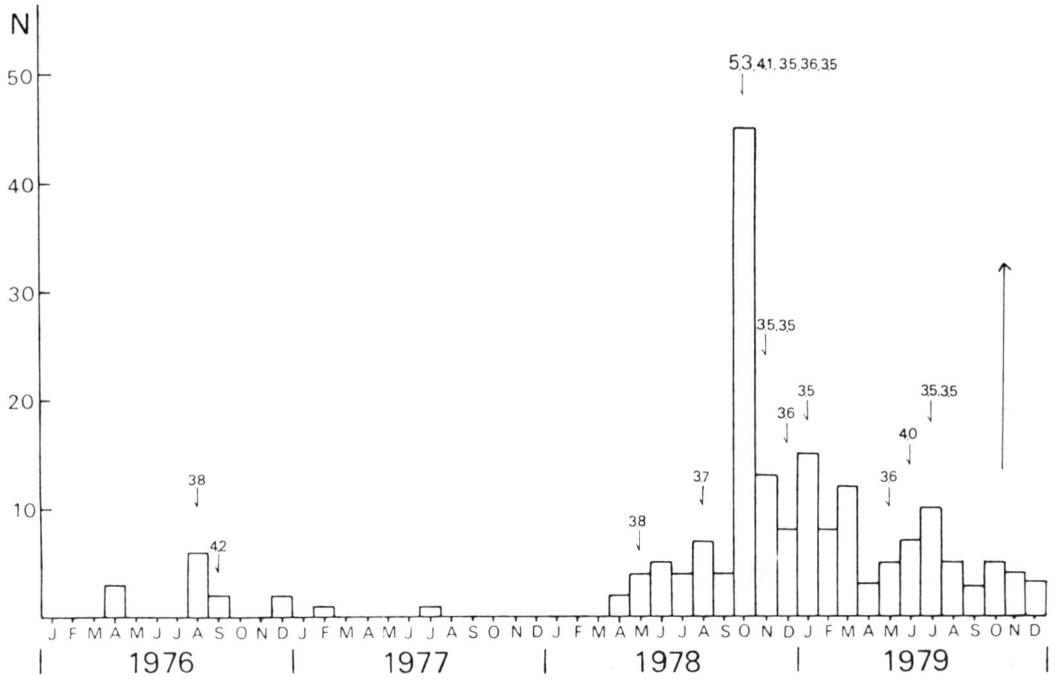

Figure 8.7.5. Monthly number of earthquakes near On-take Caldera during 1976-79, from Regional Center for Earthquake Prediction Observations (1980).

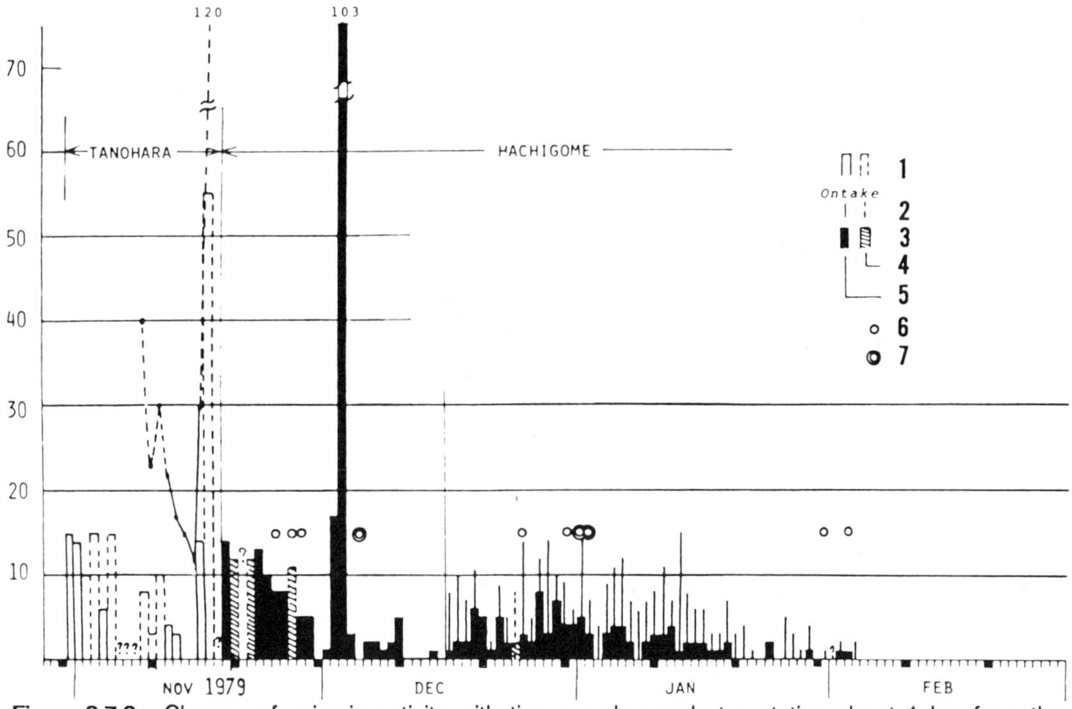

Figure 8.7.6. Change of seismic activity with time as observed at a station about 1 km from the 1979 vents at On-take, from Aramaki and Ossaka (1983). Bars indicate daily frequency of volcanic earthquakes and circles the occurrence of "tremors." Explanation: 1, estimated daily frequency of earthquakes larger than 600 microkines; 2, recorded frequency of earthquakes larger than 200 microkines; 3, recorded frequency of earthquakes larger than 600 microkines; 4, estimated; 5, observed; 6, tremors; 7, intense tremors.

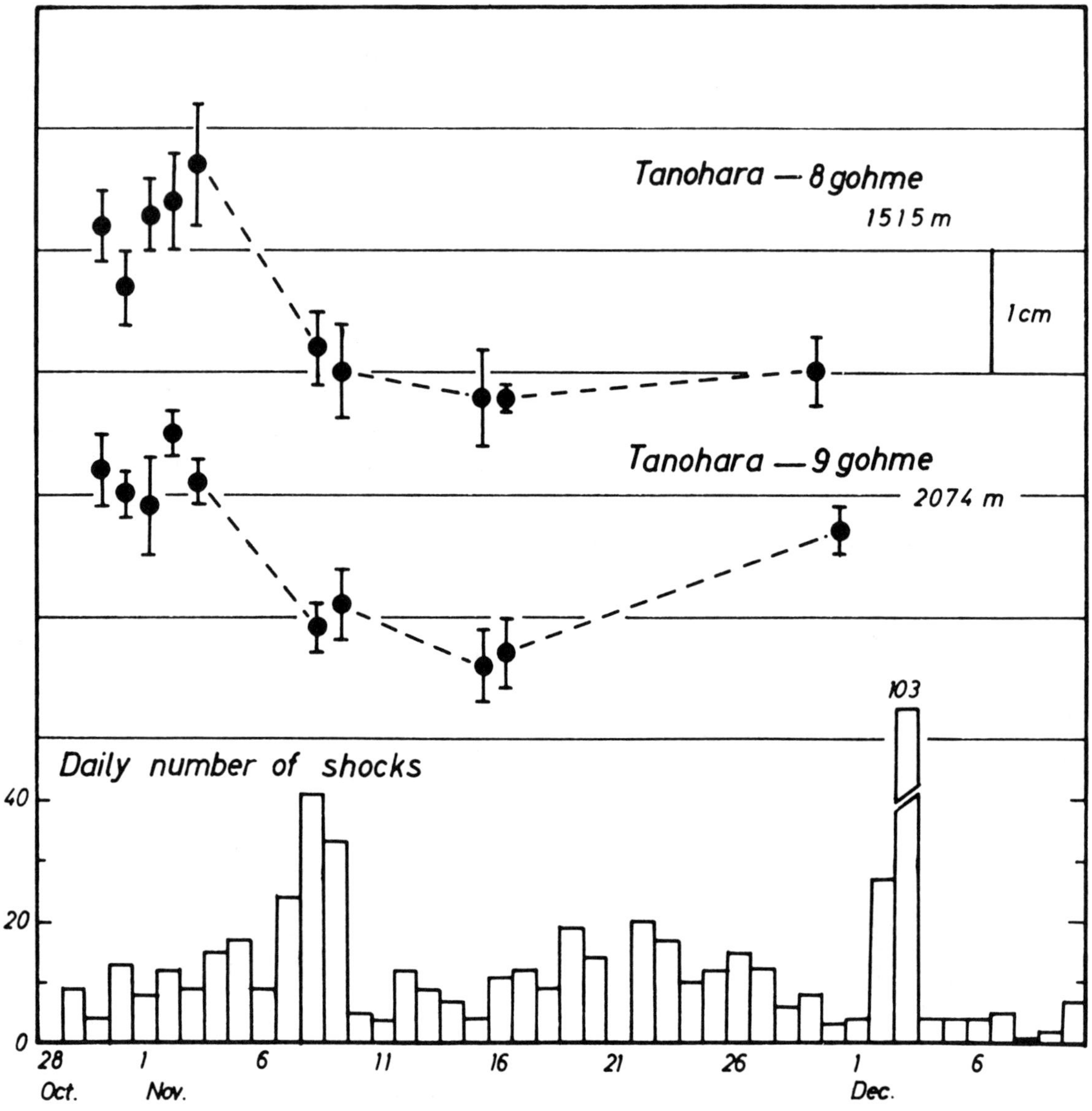

Figure 8.7.7. Changes in line length and daily number of earthquakes at Mt. On-take during October-December 1979, from Regional Center for Earthquake Prediction Observations (1980). Stations 9 gohme, 8 gohme, and Tanohara lie along a line on the WSW flank of Mt. On-take; distances between stations are shown in meters.

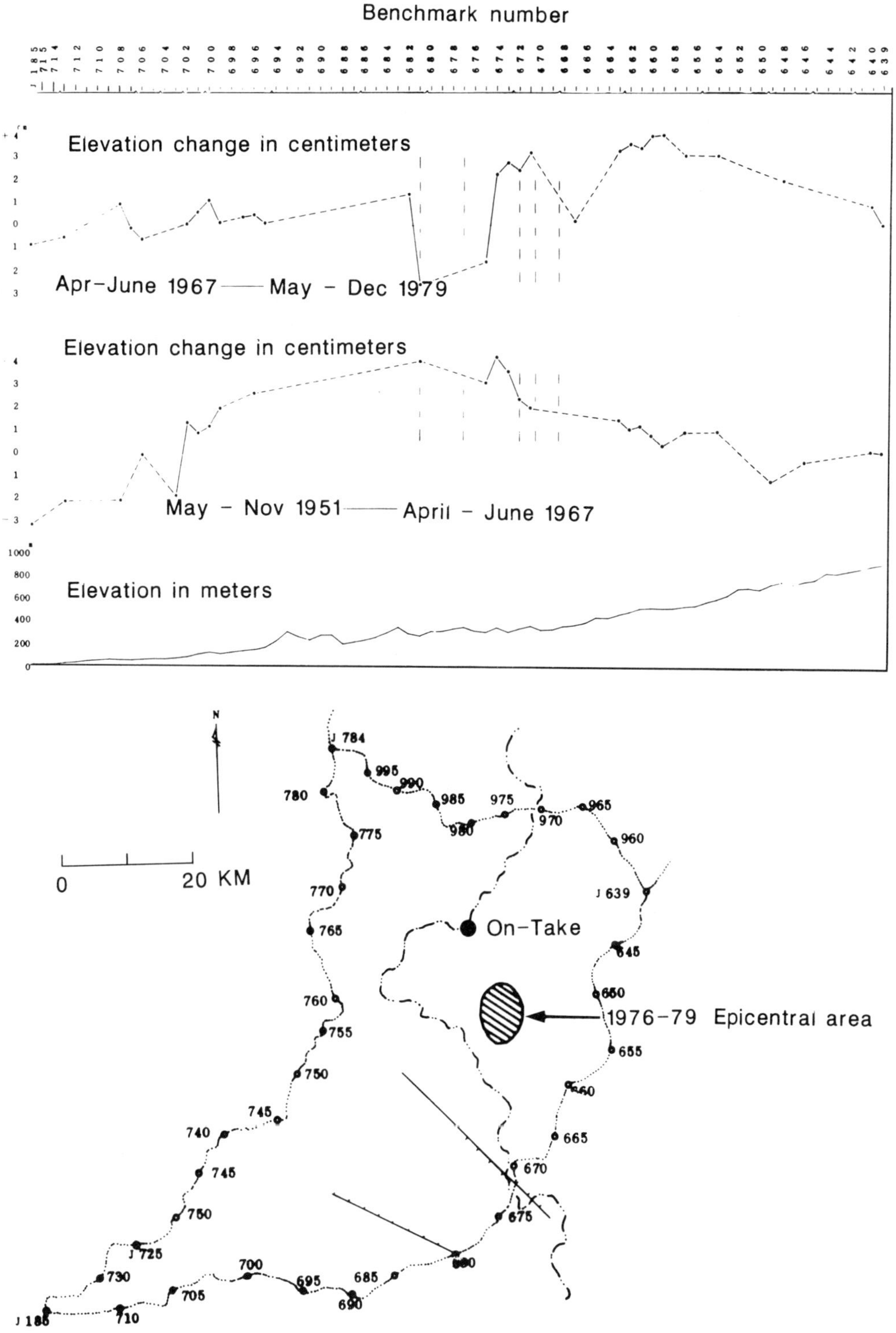

Figure 8.7.8. Leveling results (top) and route (bottom) in vicinity of On-take Volcano, May-November 1951 to April-June 1967 and April-June 1967 to May-December 1979 (Geographical Survey Institute, 1980). Note initial uplift, then downfaulting, of a northwest-trending block. Hachures represent epicentral area of 1976-79 earthquake swarms.

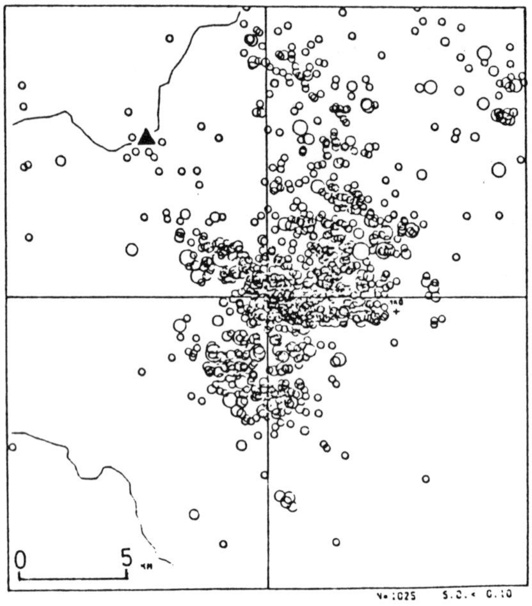

Figure 8.7.9. Epicenters of earthquakes near On-take Volcano, 1 January 1980 to 14 September 1984, from Fujii and Aoki (1985). Solid triangle at upper left marks On-take Volcano.

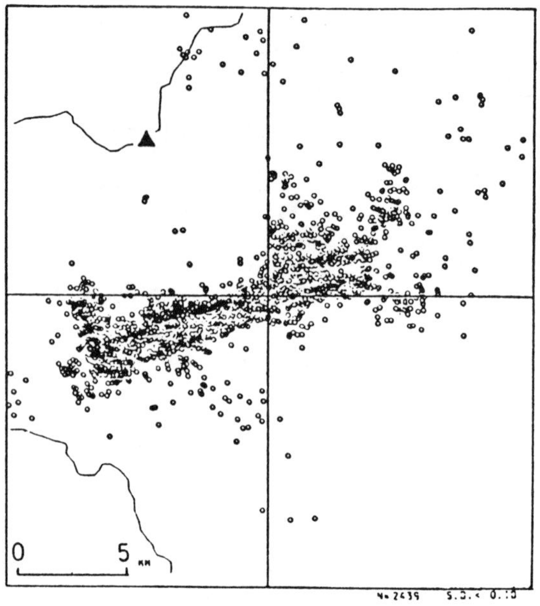

Figure 8.7.10. Aftershocks of Nagano-ken Seibu earthquake on 13 September 1984, from Fujii and Aoki (1985). Solid triangle at upper left marks On-take Volcano.

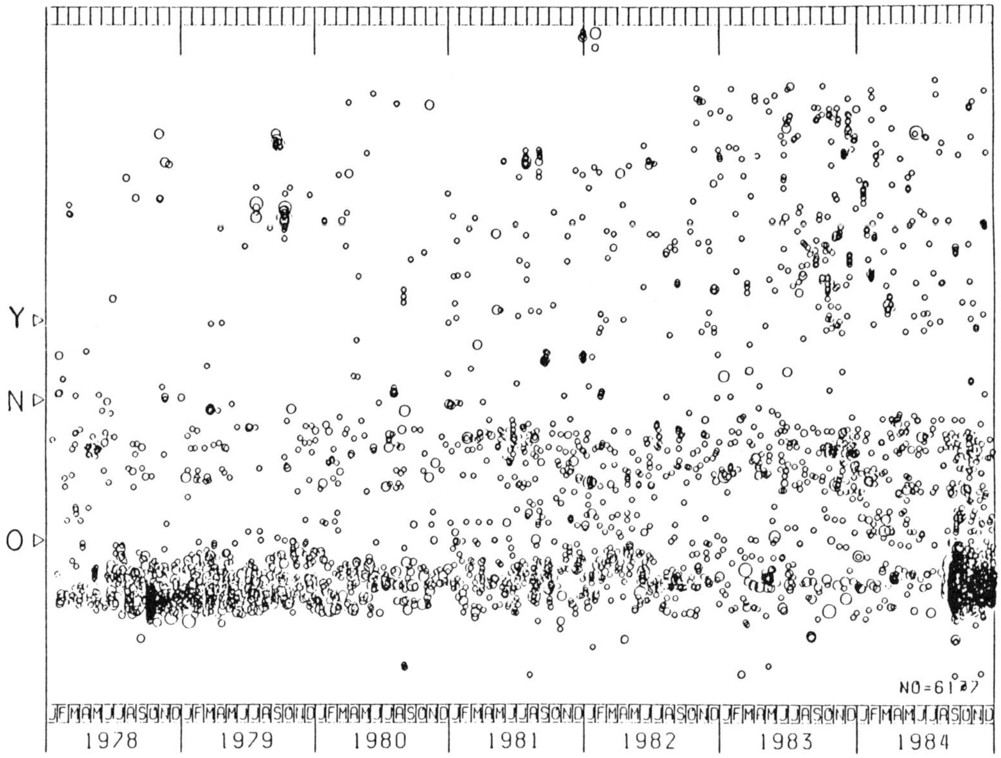

Figure 8.7.11. Space-time diagram of seismic activity along Hida Mountains near On-take during 1978-84, from Fujii and Aoki (1985). Ordinate, north-south distance; O, On-take; N, Norikura; Y, Yarigatake. A magnitude 6.2 earthquake on 13 September 1984 triggered a vigorous sequence of aftershocks (figure 8.7.10).

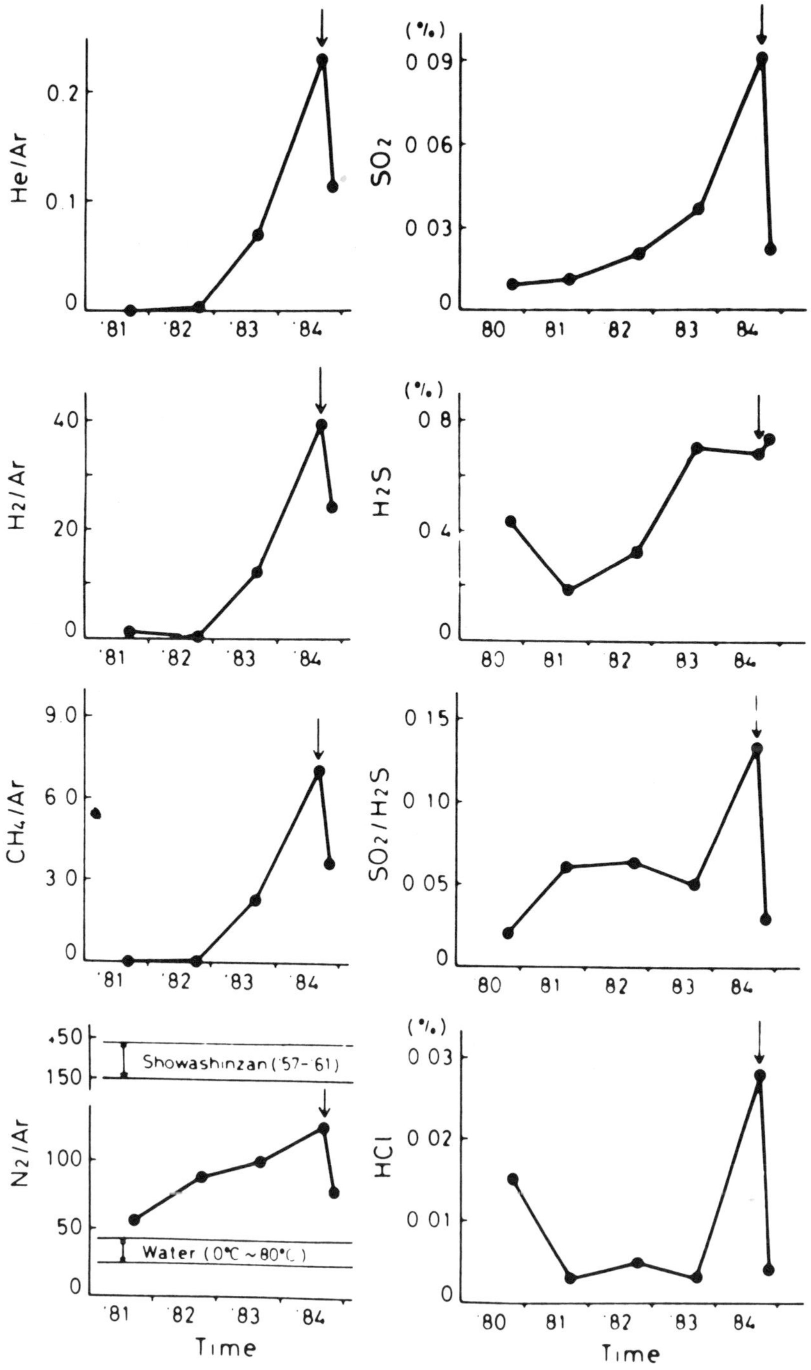

Figure 8.7.12. Temporal variation of gas chemistry at a fumarole on On-take Volcano during 1981-84, from Sugisaki and Sugiura (1986). Arrows indicate an earthquake that occurred 9 km southeast of On-take, one week after measurements in summer 1984. In the plot of N_2/Ar, "Showashinzan" shows the range observed in a fumarole (703 ^oC and 760 ^oC) on the Showashinzan Volcano, Japan, from 1957 to 1961, and "water" shows range observed in gases dissolved in water that is in equilibrium with atmospheric air at 0 ^oC and 80 ^oC.

PART 3: HISTORICAL UNREST AT LARGE QUATERNARY CALDERAS (continued)
See inside back cover for explanation and abbreviations

ONIKOBE and NARUGO

CAVW number (active vent)	Latitude Longitude (degrees)	Diameter (km)	Local tectonic setting	Precaldera edifice	SiO_2 content (wt pct)	Age of collapse (yr B.P.)	Date of unrest	Type and duration of unrest ESTU STHF MGTF H Te	Eruption type
08-03-20	38.73N	10			C = 56-65	2,300,000	1716-36	---- ---- ---- - --	pex
(Narugo)	140.73E	(Onikobe)			(Onikobe)		1744	---- ---- ---- - --	pex
08-03-20A	38.80N	5?			R = 69-71		1783	---- ---- ---- - --	pex
(Onikobe)	140.67E	(Narugo)			(Narugo)		1944	---- ---- ---- - --	pex
08-03-21	38.95N				R = 55-58		1946	---- ---- ---- - --	pex
(Kurikoma)	140.78E				(Kurikoma)		1950	---- ---- ---- - --	pex
							1957	x--- ---- ---- - --	none

TECTONIC SETTING

The Onikobe Caldera is in northern Honshu, in a region of E-W compression associated with subduction of the Pacific plate beneath the Eurasian plate.

GEOLOGIC HISTORY

Onikobe Caldera is one of only a few calderas in Japan at which ring fractures and a resurgent dome have been identified (Yamada, 1981, 1983). Postcollapse resurgent doming lifted a 2.5 km x 3 km block in the northwestern part of the caldera more than 1000 m and updomed surrounding rock. Kurikoma Volcano is an andesitic stratovolcano about 10 km northeast of the caldera rim; Narugo is a smaller caldera and cluster of silicic domes 5 km southeast of Onikobe's southeast rim (fig. 8.8.1). Onikobe, Narugo, and Kurikoma are treated here as a single volcanic center.

HISTORICAL ACTIVITY

We are not aware of any historical unrest at Onikobe Caldera itself, but we have included it in this compilation on the basis of historical eruptions and seismicity at Kurikoma.

1946, 1950: Eruptions produced only accidental material; that of November 1946 generated a mudflow.

1957: "Trembling" was reported, but apparently there was no eruption (Ueki and Suzuki, 1972).

ONIKOBE and NARUGO (continued)

HISTORICAL ACTIVITY (continued)

During a routine seismic surveillance in August-September 1971, a few earthquakes were observed within Onikobe Caldera -- presumably a normal occurrence.

REFERENCES

Harada, H., 1979, On the volcanic activity, Kurikomayama Volcano: Tohoku Dayori, no. 98, p. 16-32.

IAVCEI (International Association of Volcanology and Chemistry of the Earth's Interior, Working Group on the World Volcanological Map), 1973-84, Data sheets of the post-Miocene volcanoes of the world, with index maps: Rome, IAVCEI, unpaginated.

Katsui, Y., Oba, Y., and Soya, T., 1978, Records of volcanic eruptions in historic times and estimation of future eruptions: Bull. Volcanol. Soc. Japan, v. 23, no. 1, p. 41-52.

Kuno, H., 1962, Catalogue of the active volcanoes of the world including solfatara fields, Pt. XI, Japan, Taiwan and Marianas: Rome, IAVCEI, 332 p.

Research Group for the Geological Map of Kurikoma Geothermal Area, 1986, Geological map of Kurikoma Geothermal Area: Misc. Map Ser. (21-3), Geol. Surv. Japan, 1 sheet, scale 1:100,000 plus explanatory text, 27 p.

Sendai Meteorological Agency, 1982, Akita-Komagatake and Kurikoma-yama: Report of mobile volcanological observation team of Tohoku district, Japan Meteorological Observatory, v. 1, p. 1-37.

Sumi, K., and Takashima, I., 1976, Absolute ages of the hydrothermal alteration and associated volcanic rocks in some Japanese geothermal fields, in Proc. United Nations Development and Use of Geothermal Resources, San Francisco, v. 1, p. 625-634.

Ueki, S., and Suzuki, Z., 1972, Seismic observation at Kurikoma volcano, northern Honshu, Japan: Bull. Volcanol. Soc. Japan, v. 17, no. 3, p. 145-157.

Yamada, E., 1973, Subaqueous pumice flow deposits in the Onikobe Caldera, Miyagi Prefecture, Japan: Jour. Geol. Soc. Japan, v. 79, p. 585-597.

----- 1981, Volcanism and tectonism in the Onikobe Caldera, with special reference to the evolution of the magma reservoir: Abstracts, IAVCEI Symp. on Arc Volcanism, Tokyo, p. 421-422.

----- 1983, Geologic development of the Onikobe caldera with special reference to its hydrothermal system: Sendai, Inst. Geol. and Paleontol., Tohoku Univ., Ph.D. thesis, 128 p.

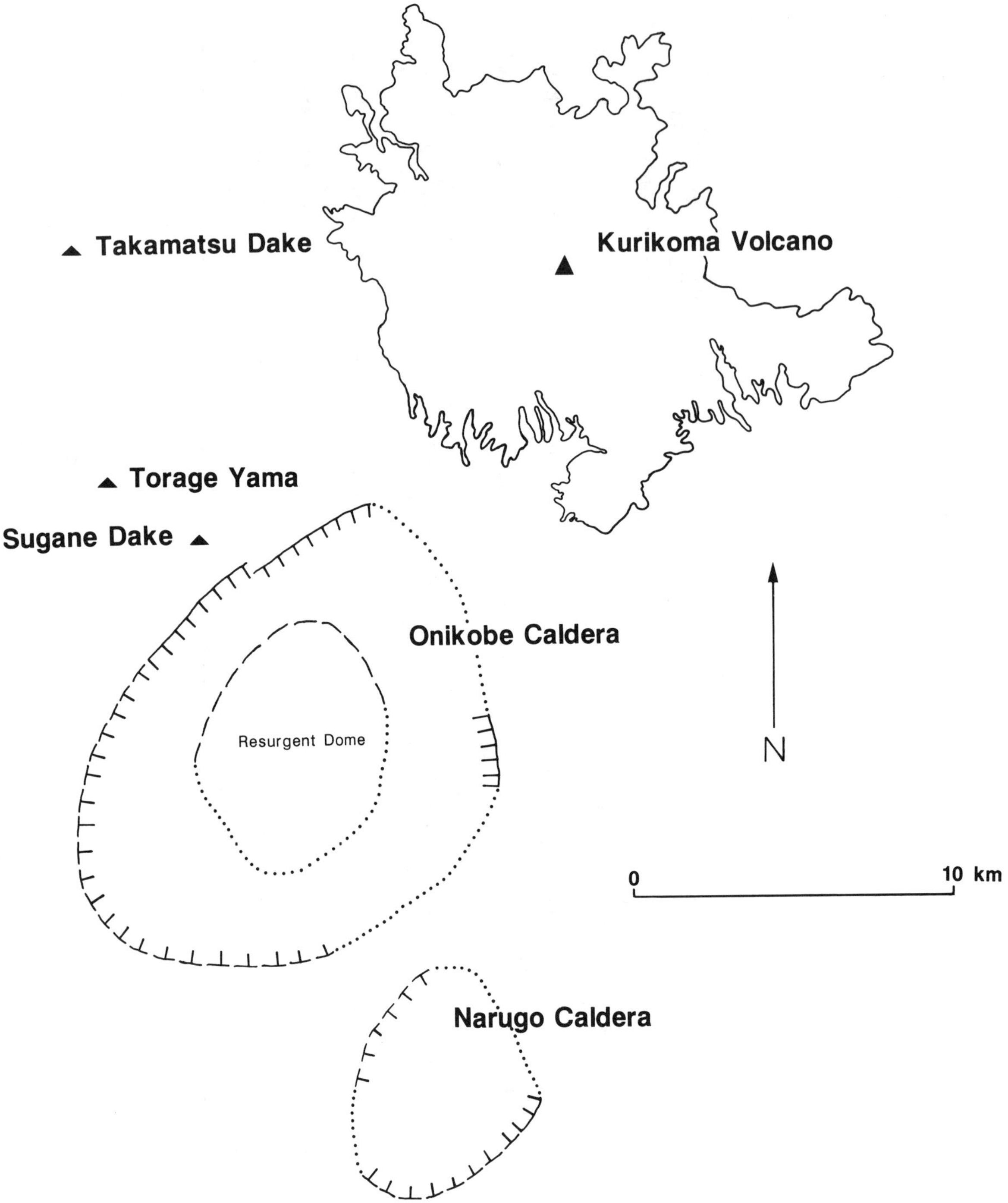

Figure 8.8.1. Sketch map of Kurikoma geothermal area, showing approximate locations of Onikobe Caldera, Narugo Caldera, and Kurikoma Volcano, based on a geologic map by Research Group for the Geologic Map of Kurikoma Geothermal Area (1986). Caldera boundaries inferred from configurations of mapped faults (solid lines), topographic rims (dashed), and eruptive products (dotted).

TOWADA

CAVW number (active vent)	Latitude Longitude (degrees)	Diameter (km)	Local tectonic setting	Precaldera edifice	SiO_2 content (wt pct)	Age of collapse (yr B.P.)	Date of unrest	Type and duration of unrest ESTU STHF MGTF H Te	Eruption type
08-03-271	40.50N 140.92E	11	Compr	Several small strato- volcanoes	C = 64-69 R = 68	11-13,000; 5,400; A.D. 915?	A.D. 915?	---- ---- ---- - --	EX,dm

TECTONIC SETTING/GEOLOGIC HISTORY

Towada Caldera, located in northern Honshu, is associated with subduction of the Pacific plate beneath the Eurasian plate along the Japan Trench. Activity has spanned at least 200,000 years and has included several voluminous explosive eruptions. An average of 1.1 km^3 of magma has been erupted per thousand years for the past 55,000 years (Hayakawa, 1985), about an order of magnitude less than at the more active calderas of this compilation. The main Towada Caldera formed between 11,000 and 13,000 yr B.P.; a small nested caldera ("Nakanoumi crater") is thought to have formed about 5,400 yr B.P (fig. 8.9.1).

HISTORICAL ACTIVITY

915 A.D.(?): The latest eruption of Towada blanketed most of the area in fig. 8.9.1 with pyroclastic flows and airfall tephra. Radiocarbon dates suggest an age of ca. 1,250 yr B.P.; an old historical document (Fuso Ryakki) notes an eruption on 17 August in A.D. 915 which Arai and others (1986) and Machida (1987) infer to be this latest eruption of Towada.

REFERENCES

Arai, F., Machida, H., Okumura, K., Miyauchi, T., Soda, T., and Yamagata, K., 1986, Catalog for late Quaternary marker tephras in Japan: II. Tephras occurring in northeast Honshu and Hokkaido: Geogr. Rep. Tokyo Metropol. Univ., v. 21, p. 223-250.

Chiba, M., 1966, Genesis of magmas producing pumice flow and fall deposits of Towada Caldera, Japan: Bull. Volcanol., v. 29, p. 545-558.

Hayakawa, Y., 1985, Pyroclastic geology of Towada Volcano: Bull. Earthq. Res. Inst., v. 60, p. 507-592.

Machida, H., 1987, Volcanoes, tephra, and large-scale destruction, chap. 3 in Japan Assoc. for Quaternary Res., eds., Quaternary maps of Japan: Tokyo, Univ. Tokyo Press, p. 11-16.

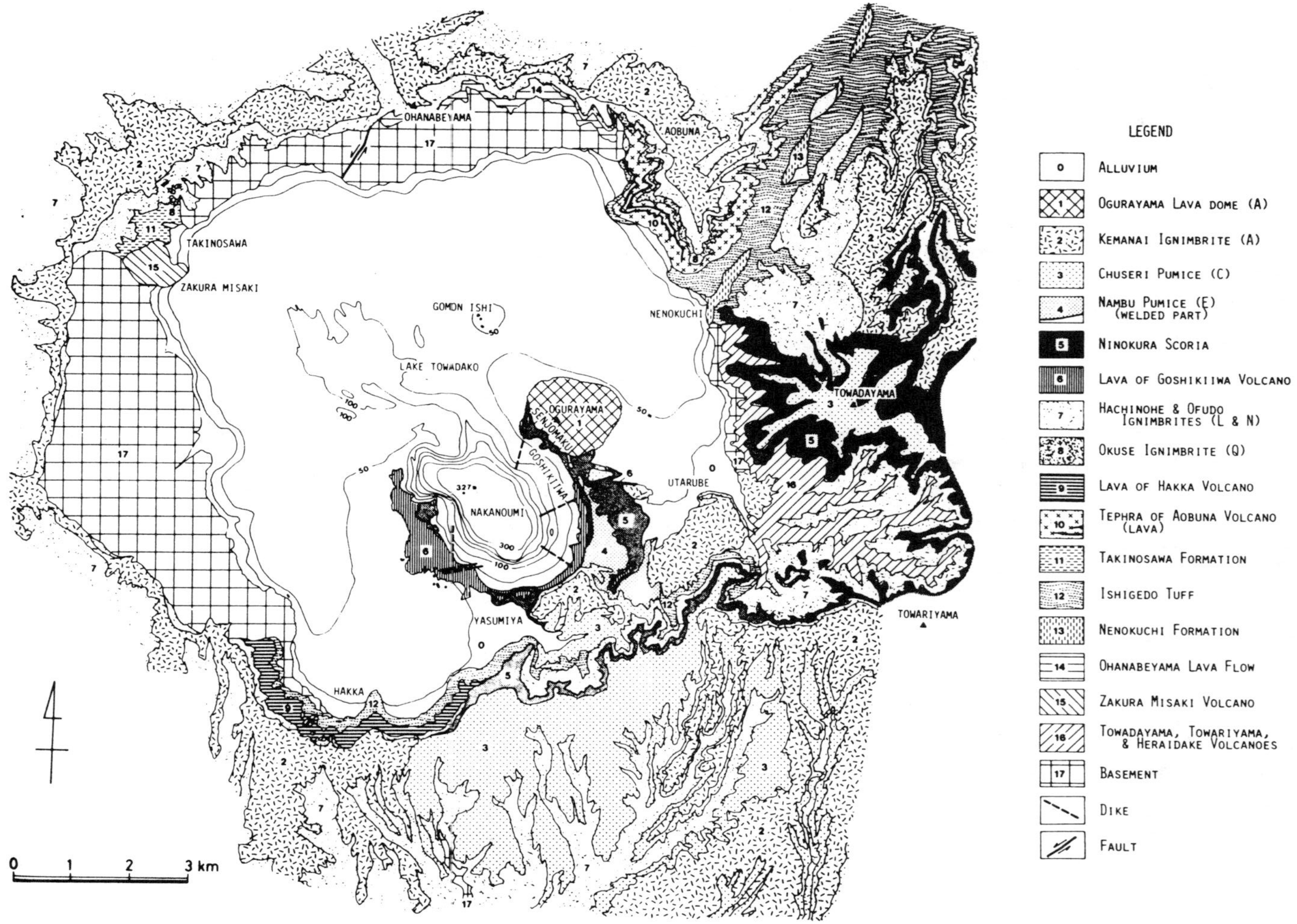

Figure 8.9.1. Geologic map of Towada Volcano, from Hayakawa (1985).

PART 3: HISTORICAL UNREST AT LARGE QUATERNARY CALDERAS (continued)
See inside back cover for explanation and abbreviations

IZU-OSHIMA

CAVW number (active vent)	Latitude Longitude (degrees)	Diameter (km)	Local tectonic setting	Precaldera edifice	SiO_2 content (wt pct)	Age of collapse (yr B.P.)	Date of unrest	Type and duration of unrest ESTU STHF MGTF H Te	Eruption type
08-04-01	34.73N	4	Compr	LL-strat	R = 52	550±250	684	---- ---- ---- - --	lf?
(Izu-Oshima)	139.38E				C = 52-54		1112	---- ---- ---- - --	Ex,lf
							1338	---- ---- ---- - --	EX,lf
							1415	---- ---- ---- - --	Ex
							1416	---- ---- ---- - --	Ex
							1421	---- ---- ---- - --	Ex,lf,pex
							1442	---- ---- ---- - --	unknown
							1552	Z--- ---- ---- - --	Ex,lf
							1600/01	---- ---- ---- - --	ex
							1612/13	---- ---- ---- - --	ex
							1636-38	---- ---- ---- - --	ex
							1684-90	Y--- ---- ---- - --	Ex,lf
							1695	---- ---- ---- - --	unknown
							1777-92	---- ---- ---- - --	EX,lf
							1803	---- ---- ---- - --	ex
							1822-24	---- ---- ---- - --	ex
							1837	---- ---- ---- - --	ex
							1846	---- ---- ---- - --	ex
							1870	---- ---- ---- - --	ex
							1876-77	---? ---- ---- - --	ex,lf
							1910	---- ?--- ---- - --	ex
							1912-14	---F ---- ---- - --	ex,Lf
							1915	---F F--- ---- - --	ex
							1919-20	---F F--- ---- - UQ?	lf
							1922-23	---F ---- ---- - UQ	ex,pex,lf
							1928	---- F--- ---- - --	ex
							1933-35	---F ---- ---- - --	lf
							1938-40	EE-G ---- ---- - --	ex,lf

IZU-OSHIMA (continued)

CAVW number (active vent)	Latitude Longitude (degrees)	Diameter (km)	Local tectonic setting	Precaldera edifice	SiO_2 content (wt pct)	Age of collapse (yr B.P.)	Date of unrest	Type and duration of unrest ESTU STHF MGTF H Te	Eruption type
(continued from previous page)							1950-51	??YG ---- Z--- - UQ	ex,lf,ll
							1953-54	---- ---- E--- - --	ex,lf
							1955-71	BB-F F--- ---- - UU	ex,lf
							1974	---F ---- ---- - UQ	ex
							1977-78	EE-- ---- FF-- - --	none
							1983-88+	FFEF FEEC ---N - U?	Ex,lf

TECTONIC SETTING

Izu-Oshima lies near the triple junction formed by subduction of the Pacific plate beneath the Eurasian and Philippine Sea plates, and by subduction of the Philippine Sea plate beneath the Eurasian plate (Nakamura and others, 1984). As a result of plate motions, compressional strain accumulates in the Oshima area before, and is relieved by, great subduction earthquakes. A regional NNW-SSE-trending fault zone cuts the island, and eruptions along the faults account for elongation of the island.

GEOLOGIC HISTORY

Izu-Oshima is a frequently active island volcano (fig. 8.10.1) located 30 km east of the Izu Peninsula (Tsuya and others, 1954; Isshiki, 1964). A summit caldera measuring 3 km x 4 km contains a young cone, Mihara-yama, which in turn has a small summit crater, 150 m in diameter (figs. 8.10.2, 8.10.3). Virtually all eruptions have been of basalt or basaltic andesite--even explosive eruptions represented in a well-documented tephra section (Nakamura, 1964). Unrest at Oshima Volcano is coupled at least indirectly with unrest on the Izu Peninsula, described in Part 4 of this compilation.

HISTORICAL ACTIVITY

The floor of Mihara-yama's summit crater rises and falls periodically by as much as 400 m in 20 years (Tsuya and others, 1956) (fig. 8.10.4). Periods in which the floor rises usually culminate with moderately large eruptions of Mihara-yama; such eruptions are sometimes followed a few years later by strong earthquakes ($\underline{M}$=8 or stronger). Examples during this century are the following (fig. 8.10.5):

Start of uplift	Eruption of Izu-Oshima	Eruption of Miyake-zima	Great earthquake
1908	1912-14, 1919	-	1923
1933	1950-51	1940	1953
ca. 1963	1974, 1986	1962, 1983	Expected

IZU-OSHIMA (continued)

HISTORICAL ACTIVITY (continued)

Kimura (1976) inferred that eruptions of Miyake-zima, 80 km south of Oshima, are followed and terminated, 2-17 years later, by eruptions of Mihara-yama (Oshima) as strain release propagates northward. Kimura also inferred that great earthquakes are likely within 3-11 years after those eruptions of Mihara-yama. Not all workers accept these far-reaching inferences.

Nakamura (1971, 1975), Kanamori (1972), Nakamura and Tazawa (1974), Kimura and Toyoda (1975), and Yamashina and Nakamura (1978) interpret the episodic uplift and subsidence of the crater floor as an indicator of volumetric strain, much like fluid in a straw, in which tectonic strain squeezes the reservoir and forces magma upward, causing the crater floor to rise and eruptions to occur. Then a large earthquake relieves the strain, and liquid retreats down the conduit. Alternatively, the magma might "retreat" into dikes in the edifice of Oshima, but seismicity normally accompanies dike intrusion and was not reported in 1953-54.

Yamashina and Nakamura (1978) presented a semiquantitative evaluation of this mechanism, comparing the amount of strain indicated by earthquakes to the amount of strain necessary to cause the observed uplift and subsidence of the crater floor (fig. 8.10.6). The cumulative volumetric strain along the southeastern coast of Honshu was $+9 \times 10^{-6}$ (dilatation) during the period 1922-31, and -7×10^{-7} (contraction) during the period 1933-50. The crater floor subsided 430 m and then rose 480 m during those two time intervals, respectively. By assuming a spherical magma reservoir with a radius of 5 km and a conduit with a cross-sectional area of 10^4 m^2 (56-m radius), Yamashina and Nakamura estimated that the observed tectonic strains would have caused 500 m of subsidence followed by 40 m of uplift--crudely consistent with the observed changes. Any weak portion of the crust, as might occur near a volcano, would tend to concentrate stresses and thereby increase local volumetric strain. Yamashina and Nakamura concluded that Oshima may indeed have acted as a volumetric strain gauge for the period 1923-50, but that a similar process could not be shown for the period 1951-75.

Shimozuru and others (1978) noted that a large earthquake that occurred between Oshima and the Izu Peninsula was preceded by earthquake swarms beneath Oshima in November-December 1977 (figs. 8.10.7, 8.10.8). Hypocenters migrated westward and downward, to the ultimate locus of the strong earthquake. A similar relation between earthquake swarms beneath Oshima and large tectonic earthquakes west of Oshima might also have occurred in June 1905 (Shimozuru and others, 1978). In the 10-year period before the 1978 earthquake, total magnetic field intensity at Nomashi (west coast of Oshima) decreased at a rate of about -1.6 nT/yr (Yukutake and others, 1978a). Gravity values on Oshima increased over the same 10-year period, and electrical resistivity changed in a way that was consistent with magma intruding into the edifice of Oshima (Yukutake and others, 1978a,b).

Most earthquake swarms on Oshima are neither strong nor clearly related to eruptions; however, swarms do precede some eruptions (figs. 8.10.9, 8.10.10). "A succession of minor shocks occurred in the neighborhood of the volcano during June 16-18 [1938], more than 2 months before the eruption [of 10 August 1938]" (Tanakadate, 1939). A similar earthquake swarm occurred in late 1948 and might be considered an early seismic precursor to the major eruption of 1950-51. However, no immediate seismic precursor to that eruption occurred (Minakami and Sakuma, 1953) (fig. 8.10.11). Two types of "tremors" occur--one at a depth of several hundred meters, yielding relatively high-frequency, short-period (0.2-0.4 sec) signals of relatively low amplitude; and a second originating

IZU-OSHIMA (continued)

HISTORICAL ACTIVITY (continued)

at the vent, with relatively longer period (0.3-0.9 sec) (fig. 8.10.11). Tremors with intermediate character were predominant for several days before and after each eruptive episode in the 1950-51 eruption (Minakami and Morimoto, 1956). In 1965-66, during a period of frequent small explosions, earthquakes occurred 3-10 hours before new explosions (Tanaka, 1968). Eight volcanic earthquakes occurred between 1113 hr and 1142 hr on 19 January 1969, and an eruption (accompanied by tremor) began at 1520 hr of the same day (Tanaka, 1969); tremor with a dominant period of 0.3-0.5 sec occurred between subsequent phases of this eruption.

Leveling during and after the 1950-51 eruption showed that Oshima Island did not subside relative to tide gauges, so there is no evidence for preeruption inflation of the island as a whole. However, Tazawa (1984) reports on shoreline studies that suggest that the island was low ("sea level was high") about 1,500 yr B.P., and that shorelines emerged remarkably about A.D. 550 (at or presumably just before the time of caldera formation). Since then, the island has been subsiding gradually. Electronic distance measurements across the caldera have shown little change except coseismic changes during the 1978 earthquake and partial recovery of these changes (Earthquake Research Institute, 1984) (figs. 8.10.12, 8.10.13).

Systematic changes in the inclination and declination of the magnetic field have also been noted (Rikitake and others, 1963; Yokoyama, 1971) (fig. 8.10.14). Changes could be explained by demagnetization of a spherical volume with a radius of 2 km, located 5 km beneath the crater. Several periods of decrease and recovery were noted; from figure 8.10.14 it appears that demagnetization began as much as a year before, and began to recover several months before, increased eruptive activity. Intermittent measurements of magnetic field strength show a corroborative pattern: from July to September 1950, the magnetic flux decreased by 6 x 10^{14} emu and recovered about 10 percent of that after the eruption. Total field-strength measurements from 1959-61 did not show any significant changes, but subsequent changes (Yukutake and others, 1978a) have been interpreted as a result of thermal demagnetization, similar to the demagnetization effect inferred earlier for changes in magnetic declination. Some (but not all) fumaroles in the crater of Mihara-yama show temperature changes that might correspond with the heating inferred from magnetic measurements (fig. 8.10.15) (Suwa and Tanaka, 1959).

Izu-Oshima has entered a period of renewed unrest as of the time of writing (December 1987). Small earthquake swarms occurred on 21 May - 1 June (figs. 8.10.16, 8.10.17), 22 October, and 22 and 30-31 December 1983. Additional swarms occurred during 16-24 August 1985. The first volcanic tremor in 12 years began on 7 July 1986 and continued intermittently (on some days, about every 2 hours) through at least early September 1986. A minor earthquake swarm occurred on 21 August, with epicenters on the north part of the island, 4-5 km north-northeast of Mihara-yama's summit crater (Smithsonian Institution, 1986).

Earthquake activity increased sharply in September 1986, with swarms on 11-13 September and 26-27 September, including a $\underline{M}$=4.0 shock on 11 September. Epicenters for the swarm of 11-13 September were on the west coast of the island, and those of 26-27 September were on the north coast. Seismicity then dropped to low levels in October and remained low through 15 November, except for tremor that occurred at irregular intervals through October and became continuous about the first of November. Higher amplitude

IZU-OSHIMA (continued)

HISTORICAL ACTIVITY (continued)

tremor was superimposed on the continuous tremor every 1-2 hours, beginning on the morning of 15 November, and a spectacular eruption began at 1718 hr on the same day--first as lava fountains and lava extrusion from the Mihara-yama crater (phase 1) (Aramaki and Nakamura, 1987; Nakamura, 1987). Earthquake activity began to increase after the start of the eruption, and several $\underline{M}$=4 earthquakes accompanied the start of phase 2, when eruption fissures opened and vigorous fire fountains rose from the caldera floor and northwest of the caldera on 21-24 November (figs. 8.10.18-8.10.20). A relatively small phase 3 lasted for 2 hours on 18 December, preceded by earthquakes on 1-9 December and tremor on 17 December.

Deformation measurements indicated slow deflation of the caldera and summit region until phase 2 of the eruption. After about 1 day of inflation, deflation resumed. Comparison of leveling results from December 1986 and October-November 1982 shows clearly a NNW-SSE-trending zone of net subsidence bisecting Oshima, with a maximum subsidence of 62 cm (Nakamura, 1987; Tada and Hashimoto, 1987).

Intermittent tremor resumed on 1 January 1987, occurring at 1- to 3-hour intervals. Several hundred earthquakes, with a maximum JMA intensity of 3, occurred 15 km northwest of Izu-Oshima at about 10-20 km depth during 6-17 May 1987, and other swarms occurred in the same month beneath the summit and southeast parts of Izu-Oshima (Smithsonian Institution, 1987) (fig. 8.10.21). Episodes of tremor lasting 1-5 hours occurred repeatedly in June-September 1987, as did small shallow summit quakes in July (Smithsonian Institution, 1987). An earthquake swarm with twice the number of earthquakes as the previous maximum for 1987 occurred in early October, and more than 1,000 earthquakes were recorded in the few days before renewed explosions on 16-19 November (Smithsonian Institution, 1987). Earthquake activity decreased to near zero after 24 November and remains low as of this time of writing (January 1988); tremor increased in mid-December 1987 and continues as of this time of writing (Smithsonian Institution, 1988). The crater floor of Mihara subsided about 175 m during or shortly after the November activity. Small ash eruptions occurred on 25 January and 27 January 1988.

COMMENTS

Because the 1986 eruption was the first to occur outside the caldera since the large "Y4" explosive eruption of ca. 1421 (Nakamura, 1964), and because fissures were observed at both ends of Oshima Island, some concern has been voiced about the possibility of another large explosive eruption. No such eruption has occurred as of this writing (January 1988).

Even though correlations between activity of Miyakezima, Mihara-yama, and great earthquakes are imperfect, they are nonetheless important to watch. The level of the crater floor of Mihara-yama is clearly a sensitive indicator of changes in Izu-Oshima's magma reservoir and may well be a sensitive indicator of regional compressive strain. As such, it may be a tool in forecasting great earthquakes in the Tokyo area. Earthquake swarms at Oshima might also forewarn of major earthquakes in the Izu Peninsula area, just west of Oshima.

IZU-OSHIMA (continued)

REFERENCES

Acharya, H., 1982, Volcanic activity and large earthquakes: Jour. Volcanol. Geotherm. Res., v. 13, p. 373-378.

Aramaki, S., and Nakamura, K., 1987, The 1986 eruption of Izu-Oshima Volcano [abs.]: Abstracts, 19th Gen. Assembly IUGG, Vancouver, British Columbia, v. 2, p. 422.

Earthquake Research Institute, University of Tokyo, 1984, Geodimeter survey of Mihara-yama, Oshima: Rep. Coord. Comm. for Prediction of Volcanic Eruptions, no. 31, p. 44-46.

Foster, H.L., and Mason, A.C., 1955, 1950 and 1951 eruptions of Mihara-yama, O shima Volcano, Japan: Bull. Geol. Soc. Am., v. 66, p. 731-762.

Iida, K., Hayakawa, M., and Katayose, K., 1951, Gravity change according to the eruption: Jour. Geogr. (Japan), v. 60, no. 3, p. 36-38.

Isshiki, N., 1964, Guidebook for excursion, 2., O-shima Volcano: Kawasaki, Geol. Surv. Japan, 24 p.

Kaminuma, K., Kayano, I., and Kubota, S., 1975, Earthquake swarms occurred near Oshima, Japan, from December 1964 to February, 1965: Bull. Volcanol. Soc. Japan, v. 20, no. 1, p. 37-47.

Kanamori, H., 1972, Relation between tectonic stress, great earthquakes and earthquake swarms: Tectonophysics, v. 14, p. 1-12.

Kimura, M., 1976, Major magmatic activity as a key to predicting large earthquakes along the Sagami trough, Japan: Nature, v. 260, p. 131-133.

Kimura, M., 1978, Significant eruptive activities related to large interplate earthquakes in the northwestern Pacific margin, in Uyeda, S., Murphy, R.W., and Kobayashi, K., eds., Geodynamics of the western Pacific, Jour. Phys. Earth, v. 26 supp., p. S557-S570.

Kimura, M., and Toyoda, J., 1975, Recent level change in the bottom-floor in the summit crater of Mihara-yama, Oshima Island, Japan: Bull. Volcanol. Soc. Japan, v. 20, p. 65-78.

Kizawa, T., and Tanaka, Y., 1972, Topographical survey of the Mihara crater, Volcano Izu-Oshima: Pap. Meteorol. Geophys., v. 23, no. 4, p. 411-428.

Minakami, T., Miyazaki, T., and Takahashi, T., 1951, The 1950-1951 eruption of Oo-sima and its seismometrical investigations (I): Bull. Earthq. Res. Inst., v. 29, p. 359-381.

Minakami, T., and Morimoto, R., 1956, Volcanic activities in Japan during the years 1949-1953: Eighth Pacific Sci. Congr., p. 241-252.

Minakami, T., and Sakuma, S., 1953, Report on volcanic activities and volcanological studies in Japan for the period from 1948 to 1951: Bull. Volcanol., v. 14, p. 79-132.

Nakamura, K., 1964, Volcanostratigraphic study of Oshima Volcano, Izu: Bull. Earthq. Res. Inst., v. 42, p. 649-728.

----- 1971, Volcano as a possible indicator of crustal strain: Bull. Volcanol. Soc. Japan, v. 16, p. 63-71.

----- 1975, Volcano structure and possible mechanical correlation between volcanic eruptions and earthquakes: Bull. Volcanol. Soc. Japan, v. 20, p. 229-240.

----- 1987, The 1986 eruption of Izu-Oshima Volcano: DELP News (Japanese Program of the International Lithosphere Project), no. 18, p. 27-31.

Nakamura, K., Shimazaki, K., and Yonekura, N., 1984, Subduction, bending and eduction. Present and Quaternary tectonics of the northern border of the Philippine Sea plate: Bull. Soc. Geol. France, v. 26, no. 2, p. 221-243.

IZU-OSHIMA (continued)

REFERENCES (continued)

Nakamura, K., and Tazawa, K., 1974, A physical interpretation for correlation between 1974 Izu-Hanto-oki earthquake and a minor eruption of Mihara-yama, Izu-Oshima: Bull. Volcanol. Soc. Japan, v. 19, p. 159-160.

Nishi, K., and Tazawa, K., 1984, Earthquake swarm on May 31, 1983 in Izu-Oshima: Bull. Volcanol. Soc. Japan, v. 29, no. 2, p. 109-111.

Oki, Y., Aramaki, S., Nakamura, K., and Hakamata, K., 1978, Volcanoes of Hakone, Izu and Oshima: Hakone, Hakone Town Office, 88 p.

Rikitake, T., Yokoyama, I., Uyeda, S., and Yukutake, T., 1963, Geomagnetic studies on Volcano Mihara: Bull. Volcanol., v. 26, p. 49-55.

Shimozuru, D., Takeda, T., Sawada, M., Osada, N., Koyama, E., and Kagiyama, T., 1978, Investigation of the activity of the Oshima Volcano associated with the Izu-Oshima-kinkai earthquake: Bull. Earthq. Res. Inst., v. 53, p. 1085-1099.

Smithsonian Institution, Scientific Event Alert Network (SEAN), 1986-88, Oshima: SEAN Bull., v. 11, nos. 8-11; v. 12, nos. 1, 5-7, 9-11; v. 13, no. 1 (information from Japan Meteorological Agency, Earthquake Research Institute, Yokohama Science Center, and others).

Suwa, A., and Tanaka, Y., 1959, The changes in the temperatures of the fumaroles in the crater of Miharayama, Oshima, in connection with the activities of the volcano: Bull. Volcanol. Soc. Japan, v. 3, p. 107-118.

Tada, T., and Hashimoto, M., 1987, The 1986 eruption of Izu-Oshima and crustal deformation: Earth Monthly, v. 9, p. 396-403.

Tanaka, Y., 1968, A seismometrical study of Izu-Oshima (I)--On the volcanic tremor and volcanic activity of Mihara-yama: Pap. Meteorol. Geophys., v. 19, no. 4, p. 627-650.

----- 1969, A seismometrical study of Izu-Oshima (II)- On the 4 kinds of volcanic tremors at the Volcano Mihara-yama: Pap. Meteorol. Geophys., v. 20, no. 4, p. 385-416.

Tanakadate, H., 1939, Volcanic activity in Japan and vicinity during the period between November 1935 and December 1938: Japanese Jour. Astron. Geophys., v. 16, p. 95-120.

Tazawa, K., 1984, Coastal deposits of Izu-Oshima Island: their implications in the volcanic activity and sea level change: Bull. Volcanol. Soc. Japan, v. 29, no. 1, p. 1-15.

Tsuya, H., Morimoto, R., and Ossaka, J., 1954, The 1950-1951 eruptions of Mt. Mihara, Oshima Volcano, Seven Izu Islands, Japan. Part I., The 1950 eruption: Bull. Earthq. Res. Inst., v. 32, p. 35-66; Part II, The 1951 eruption: Bull. Earthq. Res. Inst., v. 32, p. 289-312.

Tsuya, H., Okada, A., and Watanabe, T., 1956, Evolution of Mihara Crater, Volcano Oshima, Izu, in the course of its activities since 1874: Bull. Earthq. Res. Inst., v. 34, p. 33-60.

Volcanological Society of Japan, 1987, 1986 eruption of Izu-Oshima Volcano (set of postcard photographs, with accompanying figures).

Yamashina, K., and Nakamura, K., 1978, Correlations between tectonic earthquakes and volcanic activity of Izu-Oshima Volcano, Japan: Jour. Volcanol. Geotherm. Res., v. 4, p. 233-250.

Yokoyama, I., 1969, Anomalous changes in geomagnetic field on Ooshima volcano related with its activities in the decade of 1950: Jour. Phys. Earth, v. 17, p. 69-76.

----- 1971, Gravimetric, magnetic, and electrical methods, in The surveillance and prediction of volcanic activity: Paris, UNESCO, p. 75-101.

IZU-OSHIMA (continued)

REFERENCES (continued)

Yukutake, T., Yoshino, T., Otani, K., Kimoto, E., Shimomura, T., and Ishikawa, Y., 1978a, Anomalous secular variation in the geomagnetic total intensity on Oshima Volcano: Bull. Earthq. Res. Inst., v. 53, p. 881-892.

Yukutake, T., Yoshino, T., Utada, H., and Shimomura, T., 1978b, Time variations observed in the earth resistivity on the Oshima Volcano before the Izu-Oshima-Kinkai earthquake on January 14, 1978: Bull. Earthq. Res. Inst., v. 53, p. 961-972.

Yukutake, T., Yoshino, T., Utada, H., Shimomura, T., and Kimoto, E., 1981, A change in the apparent electrical resistivity of Mt. Mihara of Oshima Volcano observed in association with the Izu-Hanto-Toho-Oki earthquake, 1980: Bull. Earthq. Res. Inst., v. 56, p. 623-627.

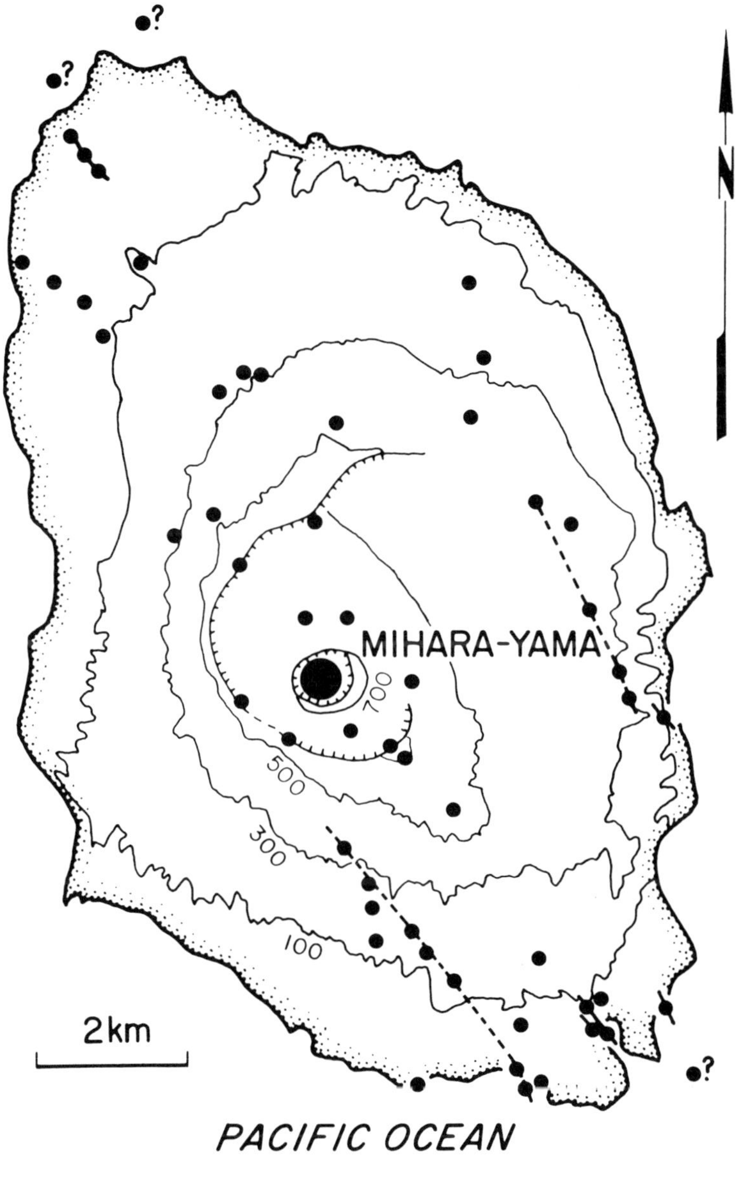

Figure 8.10.1. Oshima Island and Caldera, from Oki and others (1978), showing distribution of flank craters (circles), dikes (circles with bars), and fissures (dashed lines).

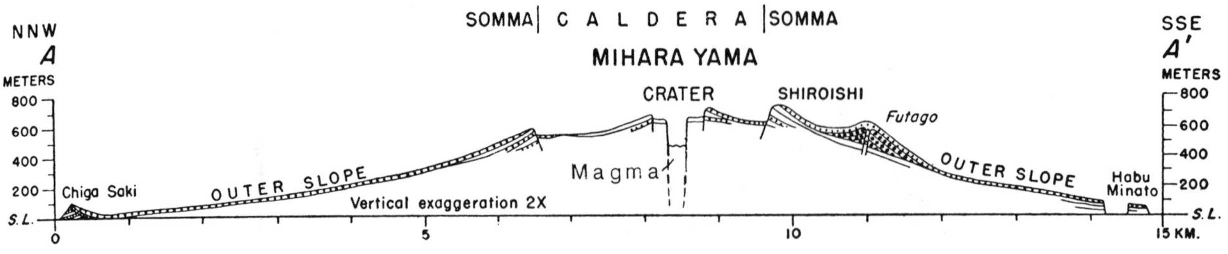

Figure 8.10.2. Cross section of Oshima Volcano, from Foster and Mason (1955), showing location of magma column in central pit.

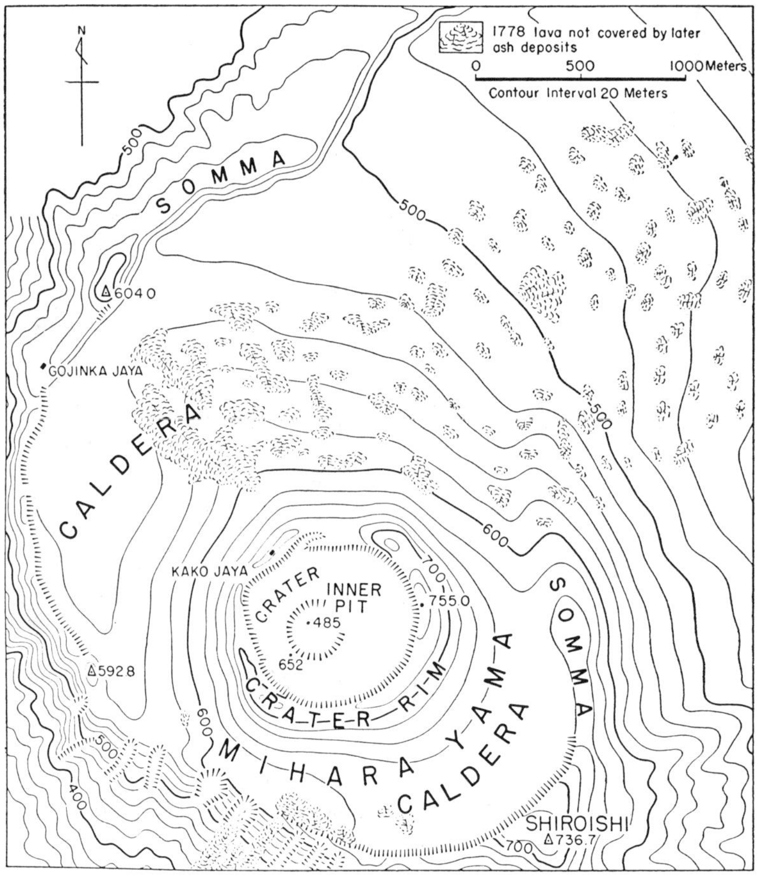

Figure 8.10.3. Topographic map of Mihara-yama prior to the 1950 eruption, from Foster and Mason (1955).

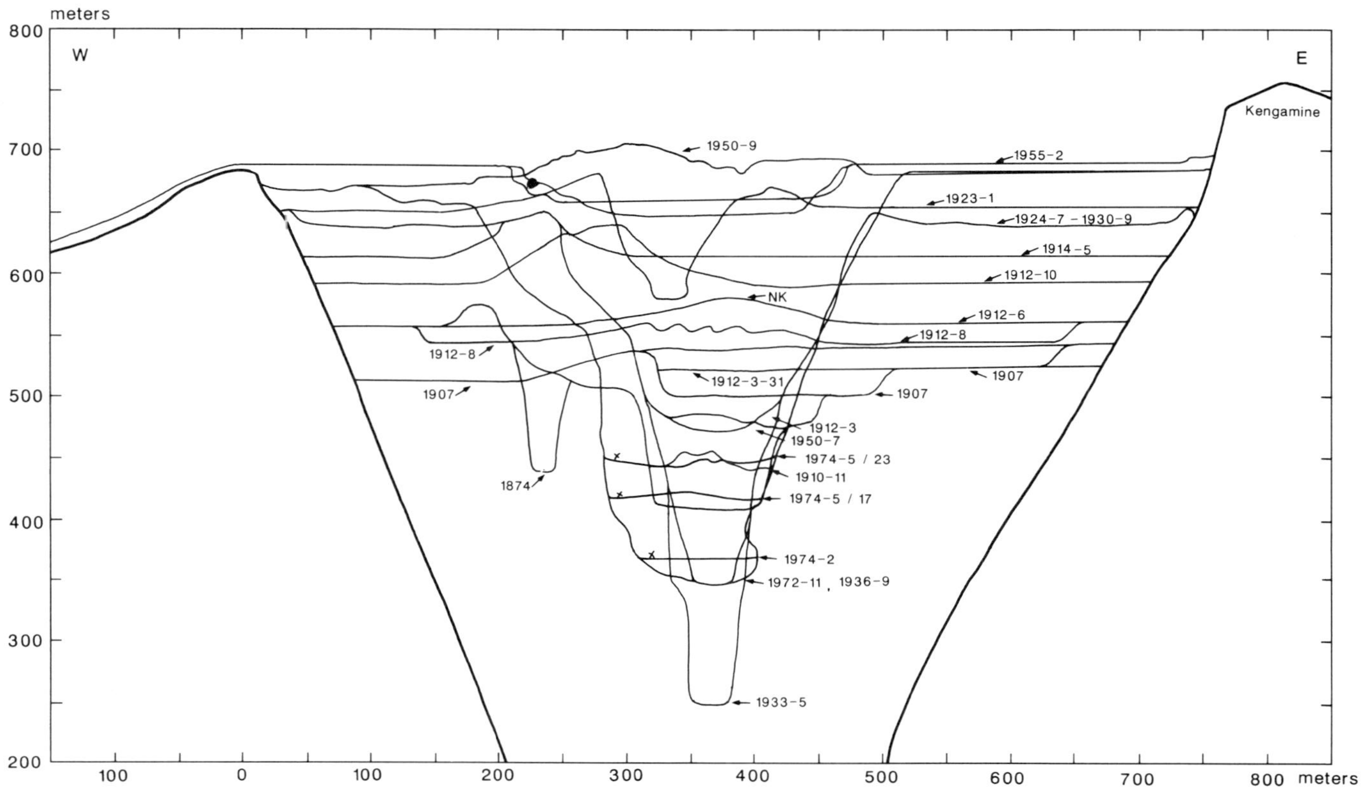

Figure 8.10.4. Changes in level of crater floor at Mihara-yama, after Kimura and Toyoda (1975) and Tsuya and others (1956). Dates (year-month) indicate configuration of crater floor at selected times from 1874 to 1974; other lines indicate other configurations during that interval.

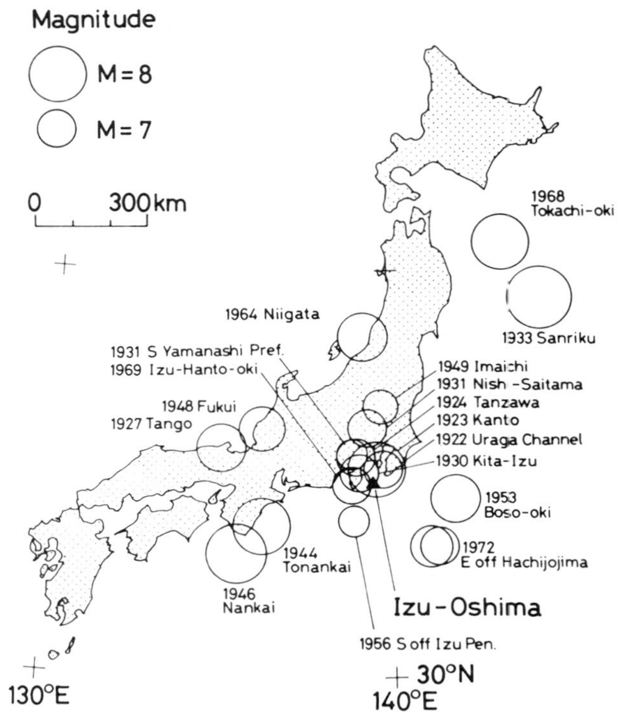

Figure 8.10.5. Distribution of earthquakes larger than magnitude 6.5 near Oshima from 1921 to 1975, from Yamashina and Nakamura (1978).

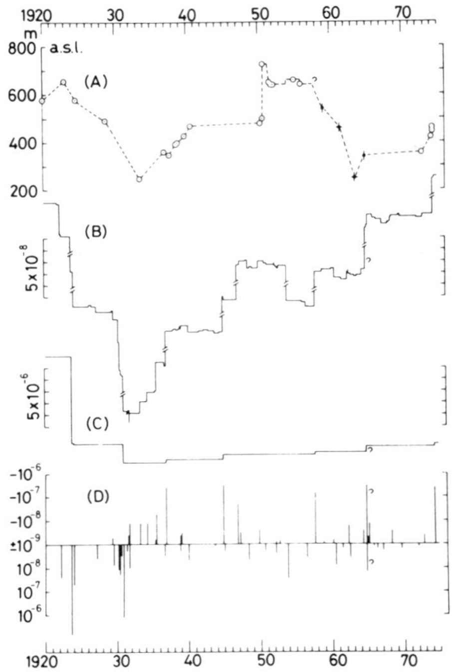

Figure 8.10.6. Volumetric strain and change in height of the crater floor (A) at Mt. Mihara-yama during 1920-75, from Yamashina and Nakamura (1978). Cumulative curves of volumetric strain (B,C). Slash marks represent out-of-scale. Contributions by smaller earthquakes (not shown in D) are included. Volumetric strain changes (D) caused by major earthquakes (larger than 10^{-9}). Positive, dilatation; negative, contraction.

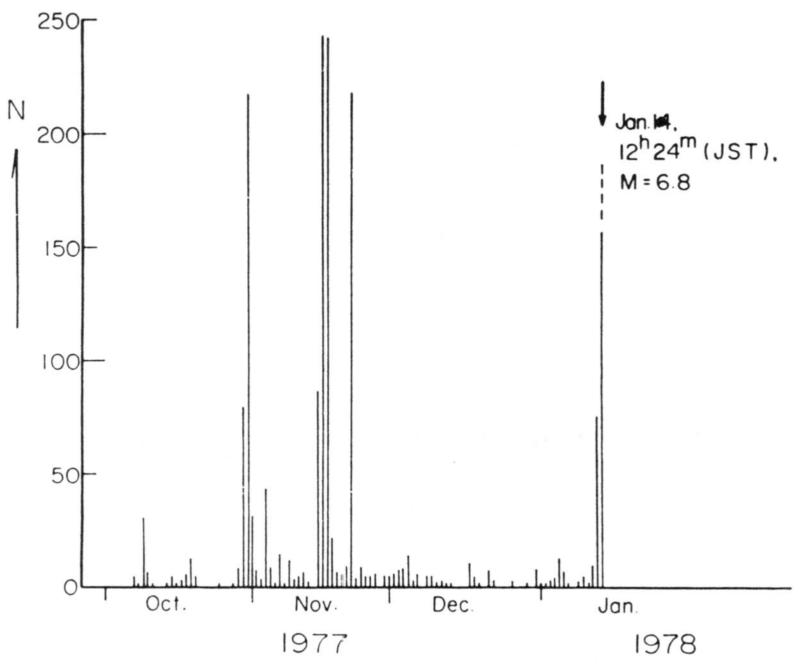

Figure 8.10.7. Daily frequency of earthquakes at Oshima from October 1977 to January 1978, from Shimozuru and others (1978).

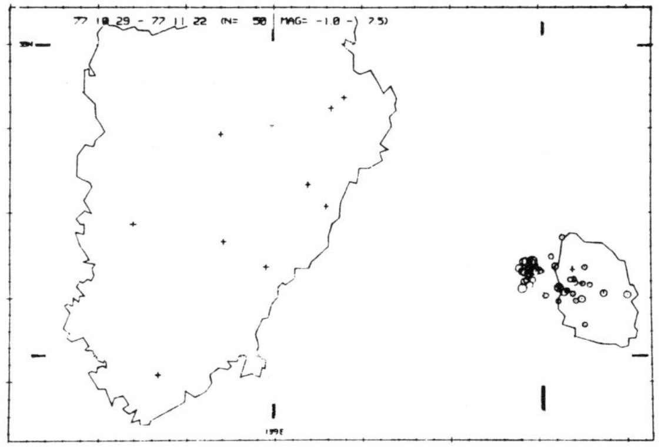

Figure 8.10.8. Plot of epicenters near Oshima from 29 October to 22 November 1977, from Shimozuru and others (1978).

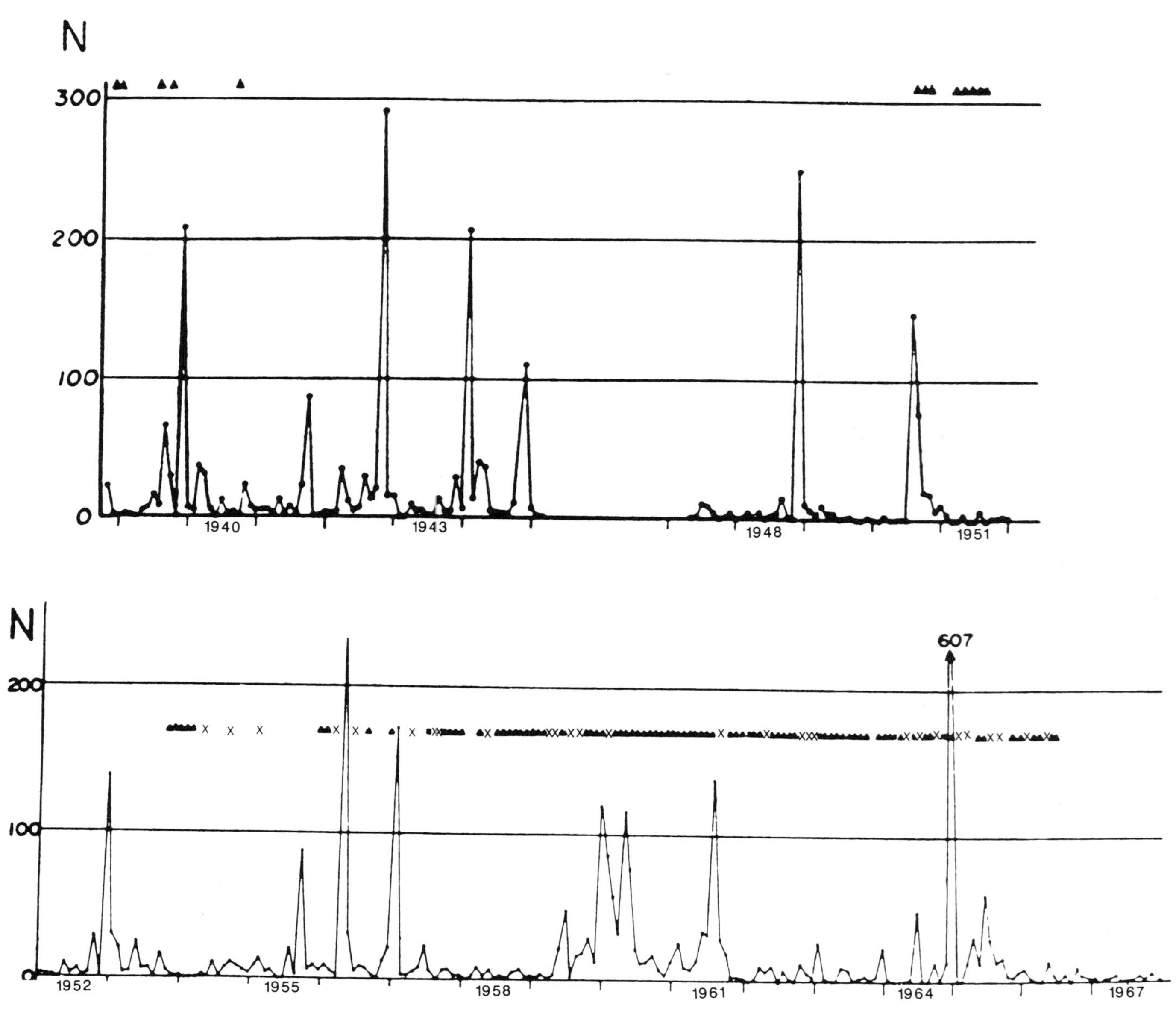

Figure 8.10.9. Monthly number of volcanic earthquakes at Oshima from 1940 to 1967, from Tanaka (1968). Triangle, eruptions; x, rumbling or explosive sound.

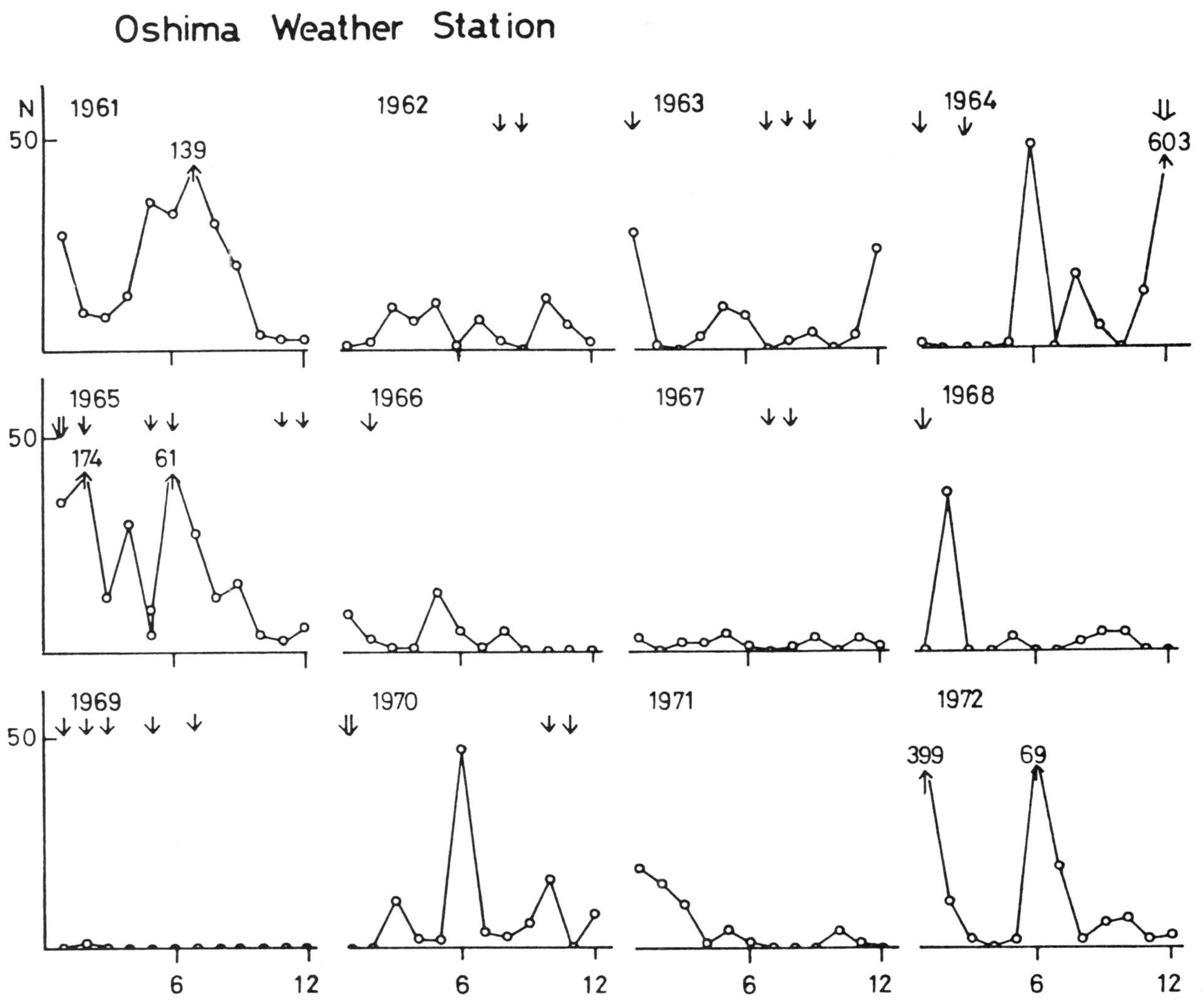

Figure 8.10.10. Monthly number of earthquakes at Oshima from 1961 to 1972, from Kaminuma and others (1975). Single arrcw, small eruption; double arrow, large eruption.

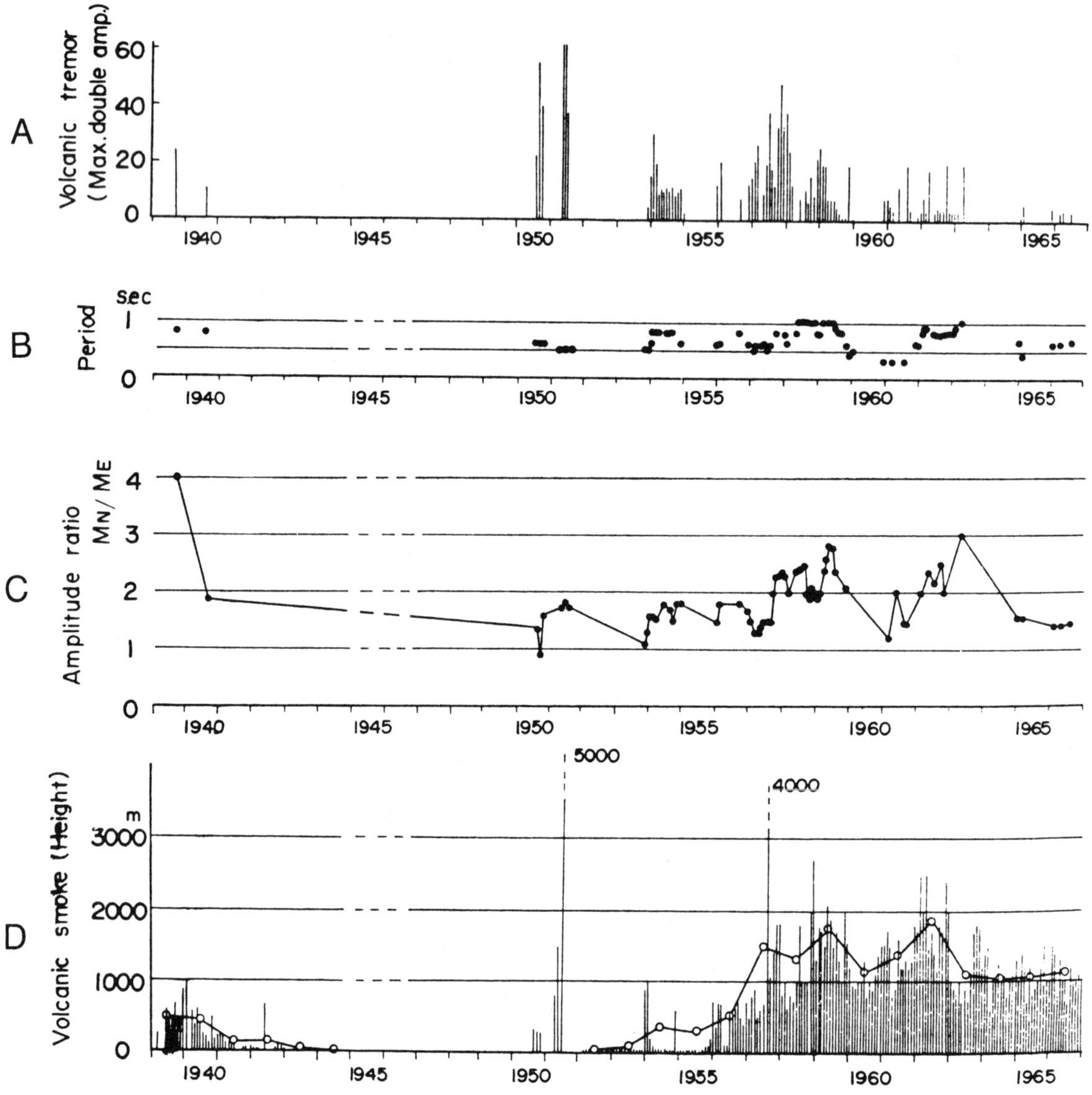

Figure 8.10.11. Variations in seismicity and eruptive activity at Oshima from 1940 to 1965, from Tanaka (1968). A, Monthly maximum amplitude of eruption tremor. B, Dominant period of tremor. C, Ratio of the amplitudes of N-S and E-W components of seismograms. D, Maximum height of "volcanic smoke" at Mihara-yama.

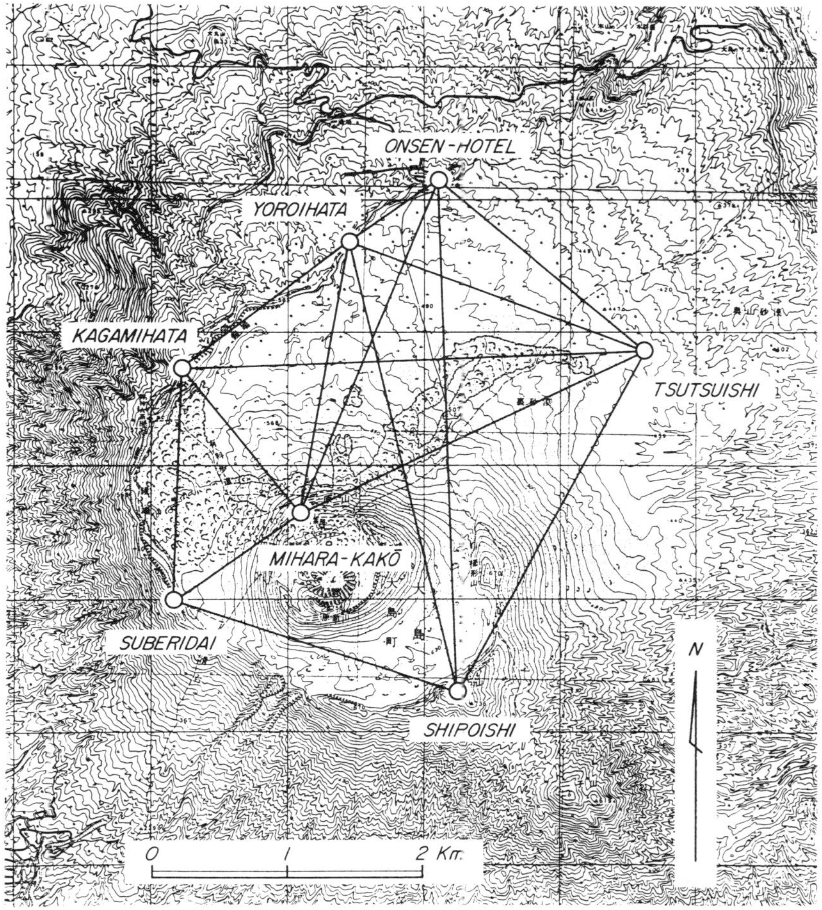

Figure 8.10.12. Trilateration network at Oshima, from Earthquake Research Institute (1984).

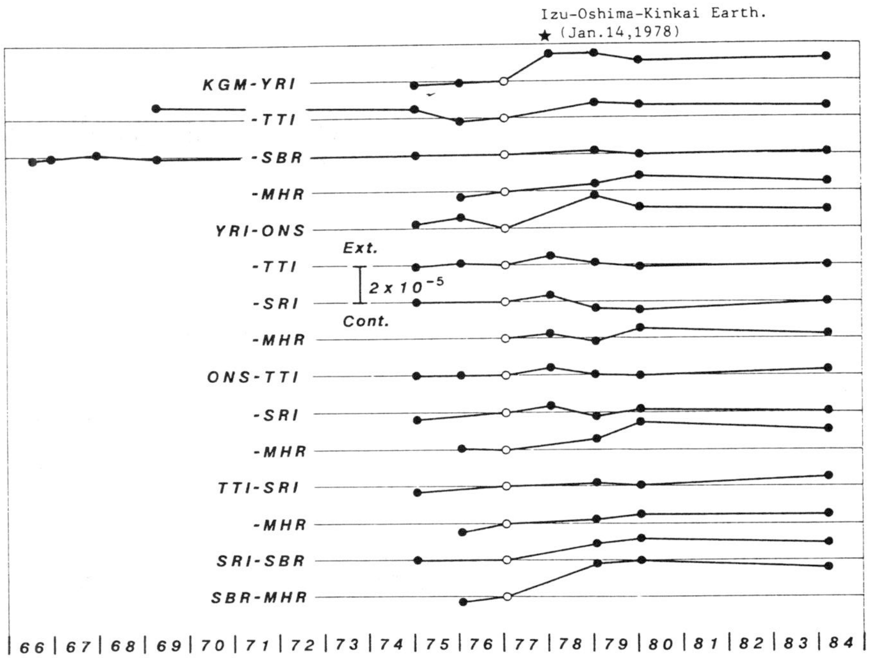

Figure 8.10.13. Cumulative strain during 1966-84 along lines shown in figure 8.10.12, from Earthquake Research Institute (1984).

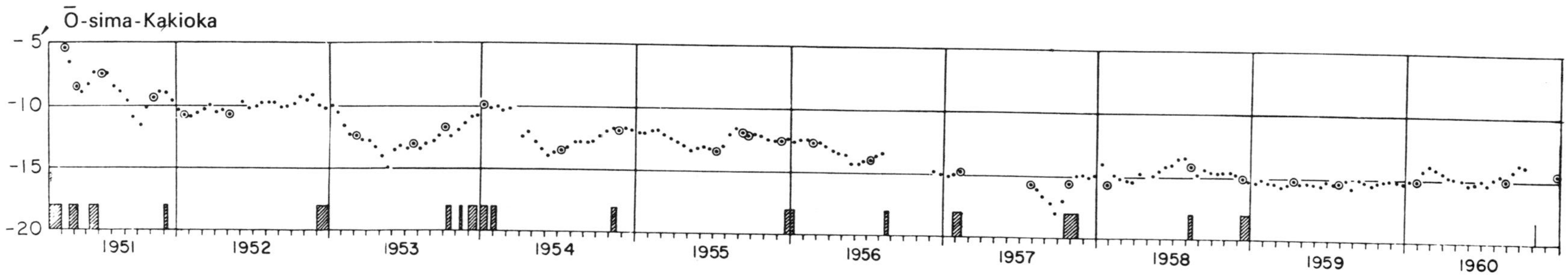

Figure 8.10.14. Variation in difference of semimonthly mean of westerly declination between Oshima and Kakioka, from Yokoyama (1971). Kakioka is 180 km northeast of Oshima. Circled dots denote values calibrated by absolute measurements. Hachured areas denote active periods at Oshima Volcano.

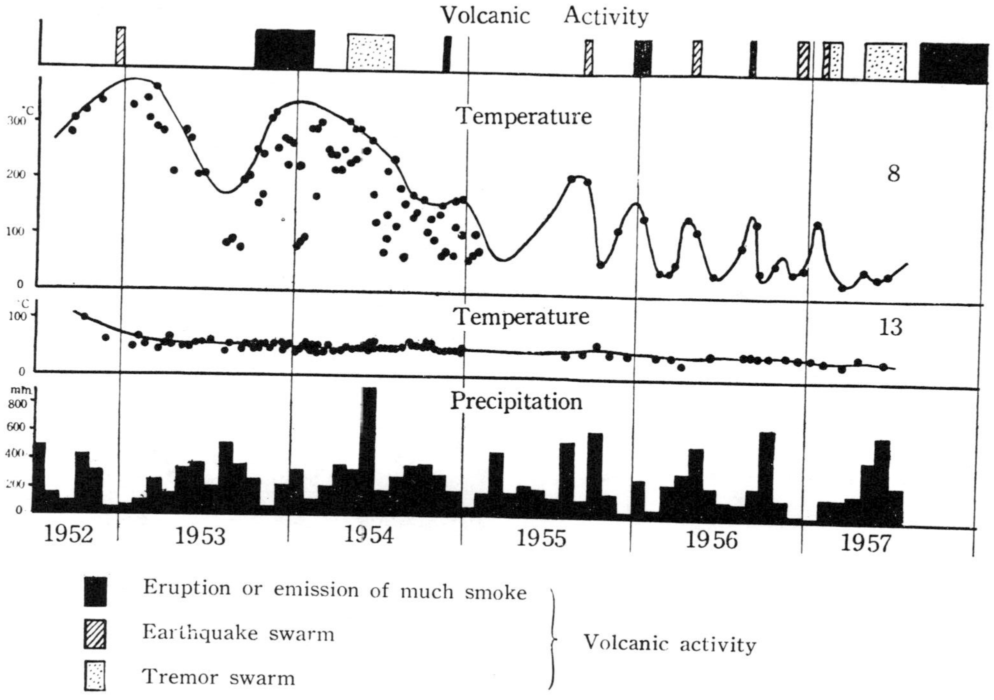

Figure 8.10.15. Activity at Mihara-yama from 1952 to 1957, from Suwa and Tanaka (1959). Temperature data are from a fumarole in summit crater.

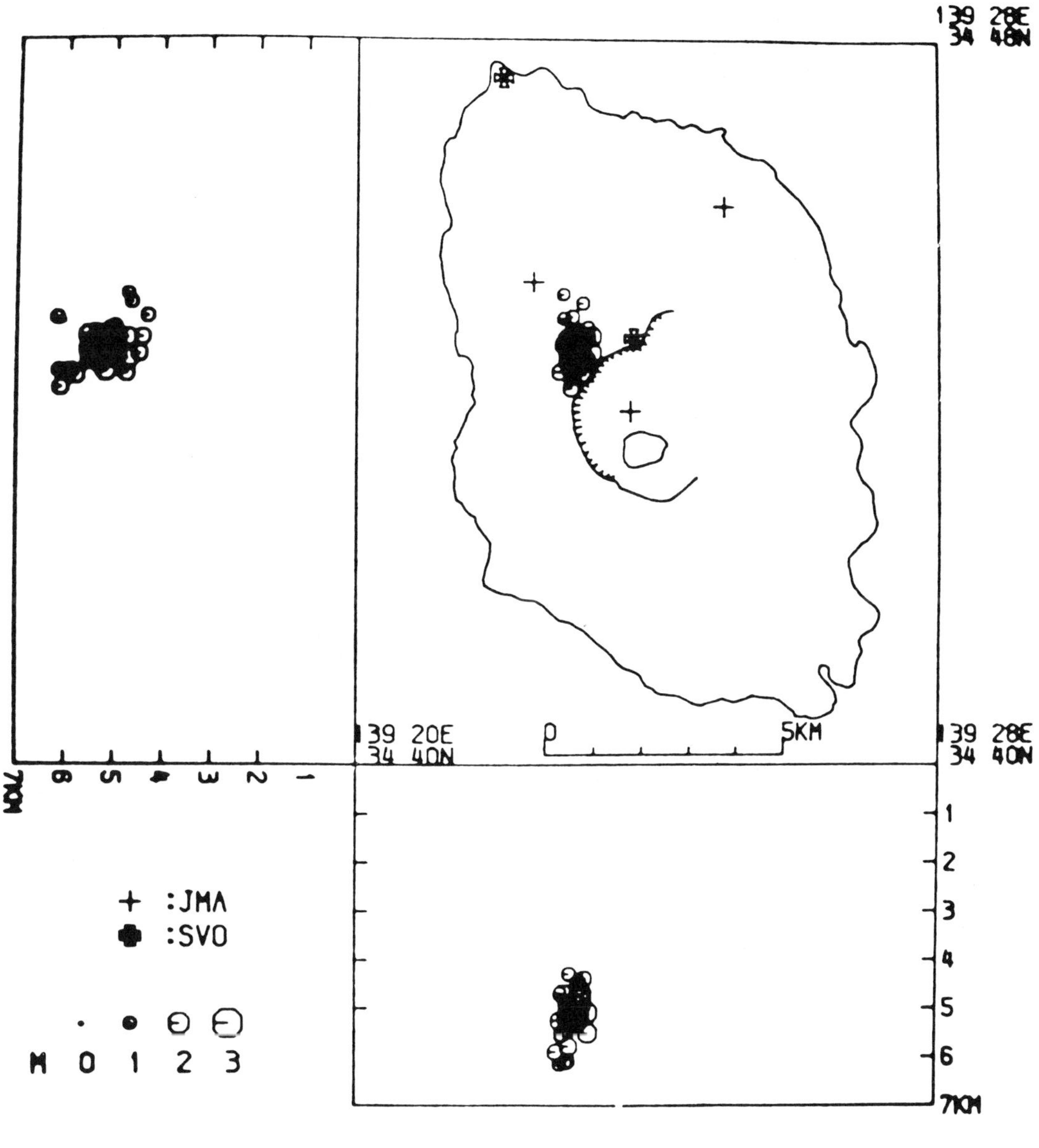

Figure 8.10.16. Distribution of hypocenters and seismic stations for an earthquake swarm northwest of Izu-Oshima in May 1983, from Nishi and Tazawa (1984).

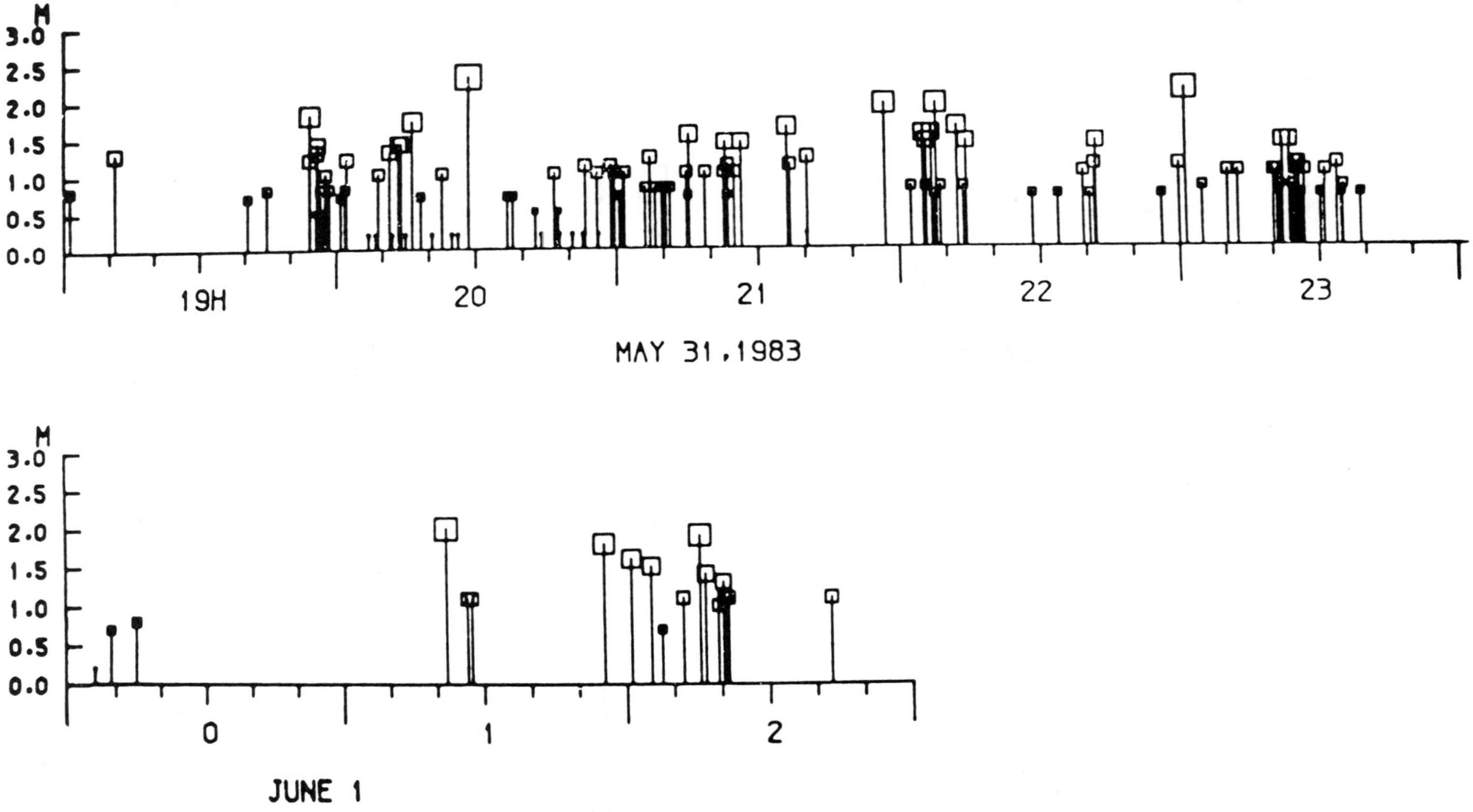

Figure 8.10.17. Magnitude-time diagram for an earthquake swarm about 4 km from Izu-Oshima Caldera in May 1983, from Nishi and Tazawa (1984).

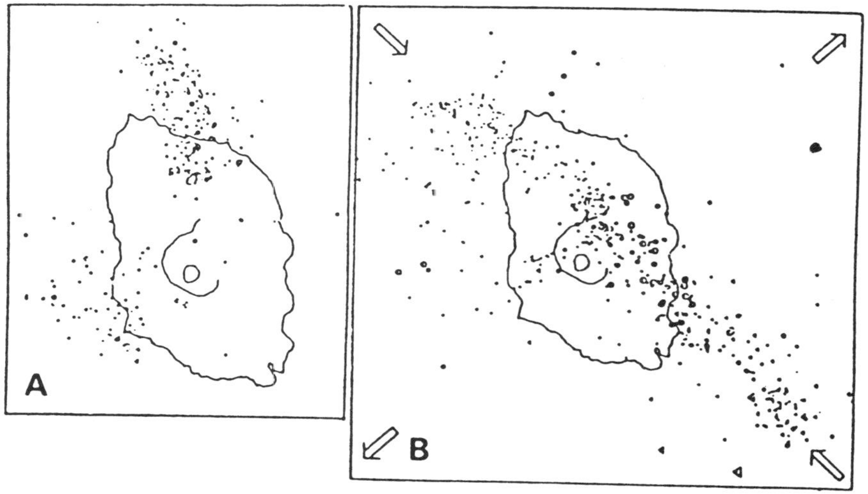

Figure 8.10.18. (Above) Epicenter distribution of 1986 earthquakes near Oshima Island, from Nakamura (1987). A, Earthquakes during summit eruption, 15-20 November (that is, before flank fissure eruption). B, Earthquakes during and after flank fissure eruption, 22-20 November. Arrows indicate P and T axes of ambient stress field.

Figure 8.10.19. (Right) Vertical crustal deformation (in centimeters) from October-November 1982 to December 1986, mostly associated with the 1986 flank fissure eruption of Izu-Oshima, from Nakamura (1987). Compiled by Tada and Hashimoto (1987) from data by the Geographical Survey Institute (squares) and the Earthquake Research Institute (open circles). The 0-cm and 30-cm contour lines are emphasized. Summit eruption site (solid circle) and B and C fissure vents (heavy line; see also figure 8.10.20) also are marked. Solid square marks a tide-gauge station that was assumed to be stable.

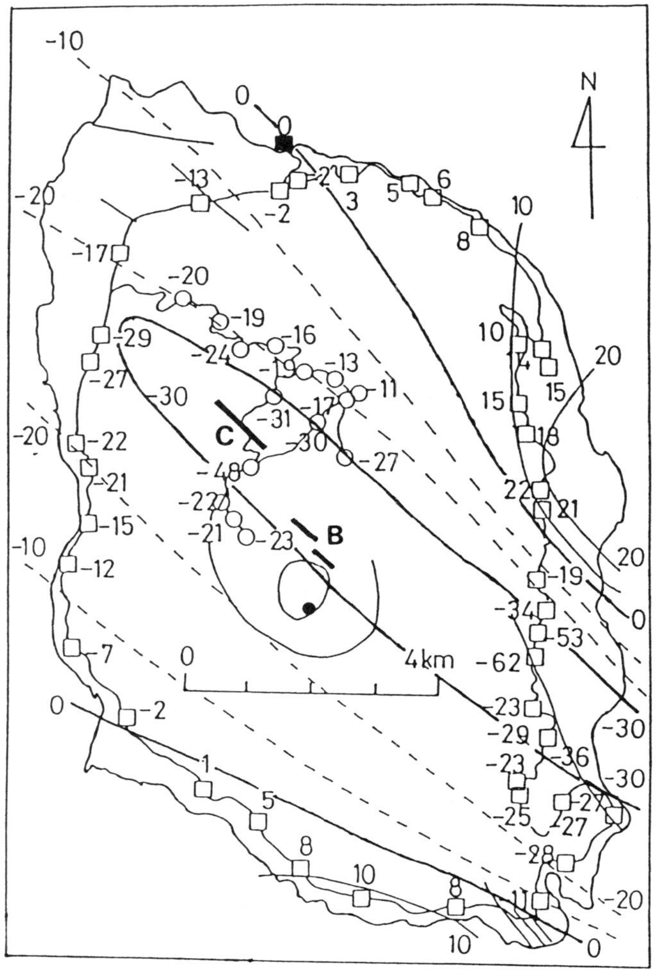

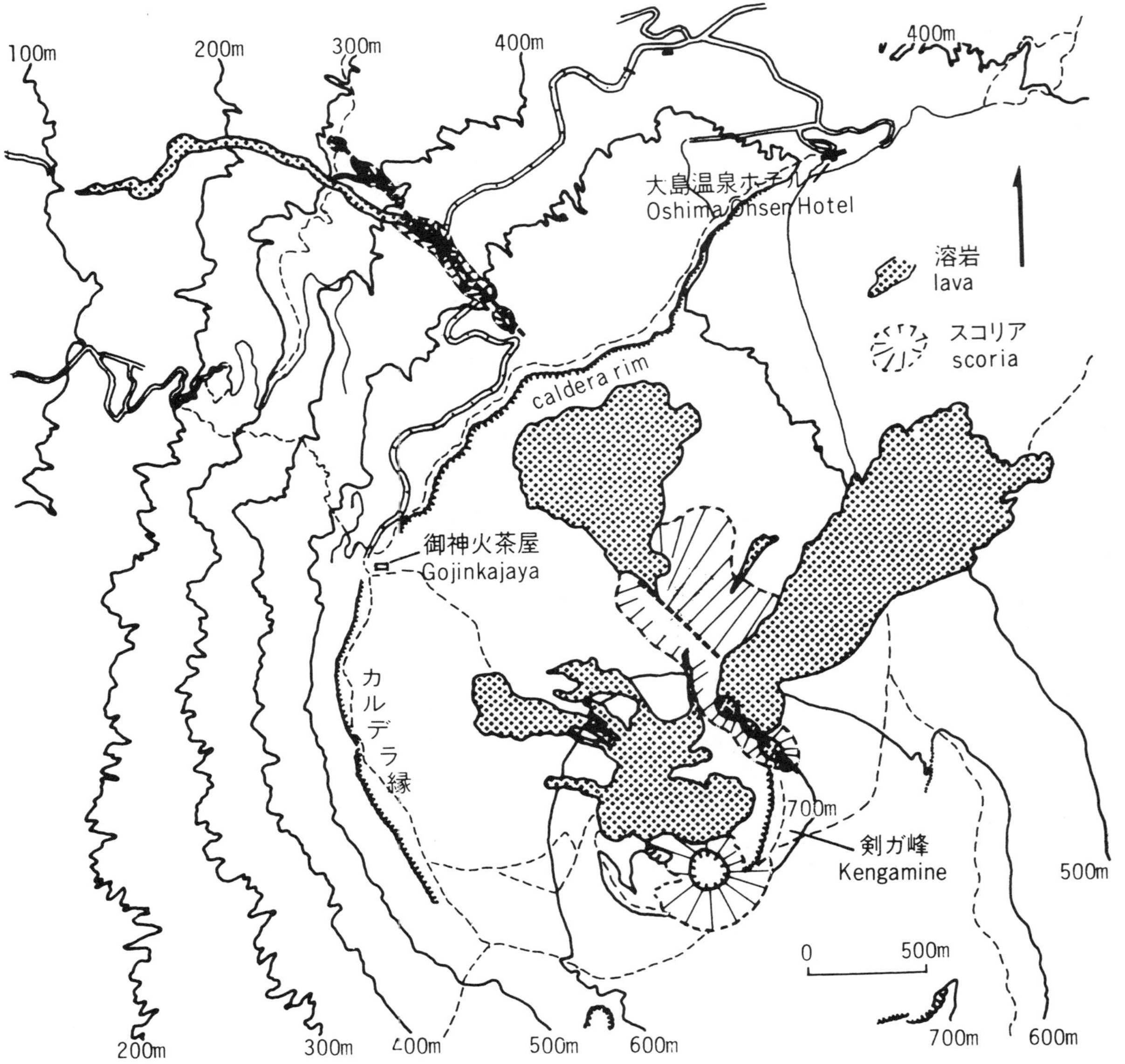

Figure 8.10.20. Fissure vents, lava flows, and tephra fall of the 1986 Izu-Oshima eruption, from Volcanological Society of Japan (1987).

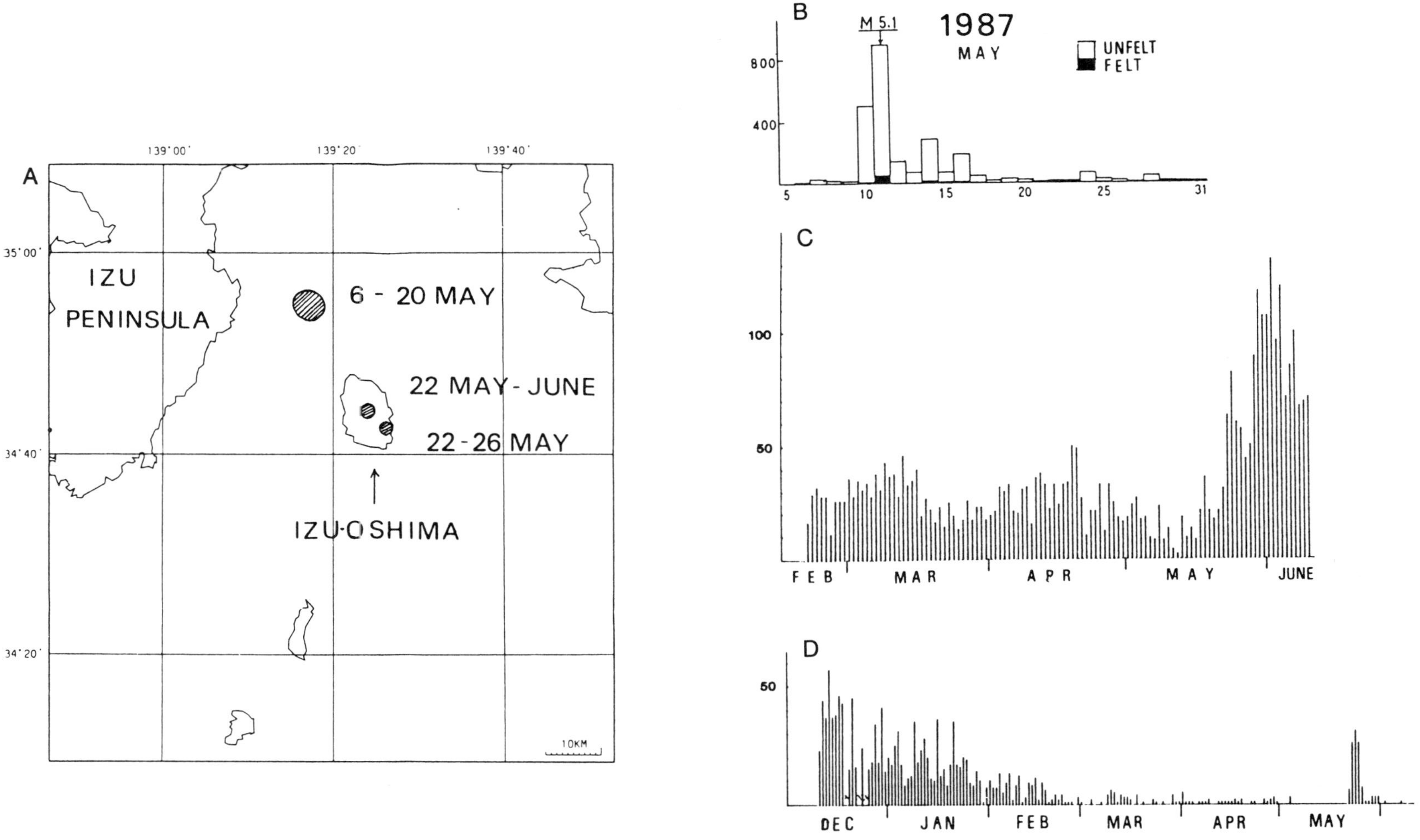

Figure 8.10.21. Seismicity at Oshima Island, December 1986 to June 1987 (Smithsonian Institution, 1987, contribution from Volcanological Center, Japan Meteorological Agency). A, Locations of earthquake swarms (shaded) in Oshima area, May 1987. B, Daily earthquake counts (felt and recorded) for northwest epicentral area in (A), 5-31 May 1987. C, Daily earthquake counts in summit area of Oshima, 20 February to 10 June 1987. D, Daily earthquake counts for southeast epicentral area, 10 December 1986 to 10 June 1987.

PART 3: HISTORICAL UNREST AT LARGE QUATERNARY CALDERAS (continued)
See inside back cover for explanation and abbreviations

BAYONNAISE

CAVW number (active vent)	Latitude Longitude (degrees)	Diameter (km)	Local tectonic setting	Precaldera edifice	SiO_2 content (wt pct)	Age of collapse (yr B.P.)	Date of unrest	Type and duration of unrest ESTU STHF MGTF H Te	Eruption type
08-04-07	31.95N	10	Compr	Submarine	R = 47-69		1896	---- ---- ---- - --	sub
(Myojin-sho)	140.01E			strato?	C = silicic		1906	---- ---- ---- - --	sub,ex
	(Myojin-sho)						1915	---- ---- ---- - --	sub
							1934	---- ---- ---- - --	sub
							1946	---- ---- ---- - --	sub,dm
							1952-54	---- ---- ---- - --	sub,ex,dm
							1955	---- ---- ---- - --	sub
							1957-59	---- ---- ---- - --	sub
							1960	---- ---- ---- - --	sub
							1970	---- ---- ---- - --	sub,ex
							1979?	---- ---- ---- - --	sub
							1980?	---- ---- ---- - --	sub

TECTONIC SETTING/GEOLOGIC HISTORY

Bayonnaise is a largely submarine caldera in the southern Izu Islands, formed above the subduction of the Pacific plate beneath the Philippine Sea plate. Myojin-sho, a lava dome, grew in the center of the caldera (see below and fig. 8.11.1) (Kuno, 1962). Bayonnaise Rocks, a high point on the western rim of the caldera, is basaltic in composition. Pumice that erupted in 1906 was white, but its phenocrysts (augite, hypersthene, bytownite) indicate a basaltic or andesitic composition (Wakimizu, 1908, Appendix); pumice that erupted in 1952 ranged from 63 to 68 percent SiO_2 (Kuno, 1962).

HISTORICAL ACTIVITY

Historical eruptions have occurred from several vents within, on the rim of, and outside the caldera. The eruption of 1896 occurred approximately 14 km north of Bayonnaise Rocks, while that of 1906 occurred in the vicinity of Bayonnaise Rocks. Eruptions in 1915 occurred from widely separated vents: Those in February and April occurred at a point 10-12 km east of Bayonnaise rocks, whereas those in June and July occurred 19 km northeast and 4 km southwest of the Rocks, respectively. The eruption of May 1934 occurred about 9 km east of Bayonnaise Rocks, that of February-December 1946 occurred 10 km east-northeast of Bayonnaise Rocks, and that of September 1952 (Myojin-sho) occurred nearby, about 9 km east-northeast of Bayonnaise Rocks. During the September 1952 eruption, a lava dome (Myojin Reef) rose above sea level on 17 September; within a week it had been destroyed by explosions. On 24 September, the research vessel Kaiyo-maru was destroyed by another explosion, killing all 31 persons aboard. Discolored water was noted in July 1979 and again in November 1980.

BAYONNAISE (continued)

REFERENCES

Dietz, R.S., and Sheehy, M.J., 1954, Transpacific detection of Myojin volcanic explosions by underwater sound: Bull. Geol. Soc. Am., v. 65, p. 941-956.

Kuno, H., 1962, Catalogue of the active volcanoes of the world including solfatara fields, Pt. XI, Japan, Taiwan and Marianas: Rome, IAVCEI, 332 p.

Morimoto, R., 1960, Submarine eruption of the Myojin reef: Bull. Volcanol., v. 23, p. 151-160.

Morimoto, R., and Ossaka, J., 1955, The 1952-53 submarine eruption of the Myojin reef near Bayonnaise Rocks, Japan (1): Bull. Earthq. Res. Inst., v. 33, p. 221-250.

Smithsonian Institution, Scientific Event Alert Network (SEAN), 1979-80, Bayonnaise Rocks: SEAN Bull., v. 5, no. 11.

Volcanological Society of Japan, 1981, Myojin-sho: Bull. Volcanic Eruptions, no. 19, p. 52.

Wakimizu, T., 1908, The ephemeral volcanic island group in the Iwo-Jima group: Imperial Earthq. Invest. Comm., no. 22C, p. 1-33.

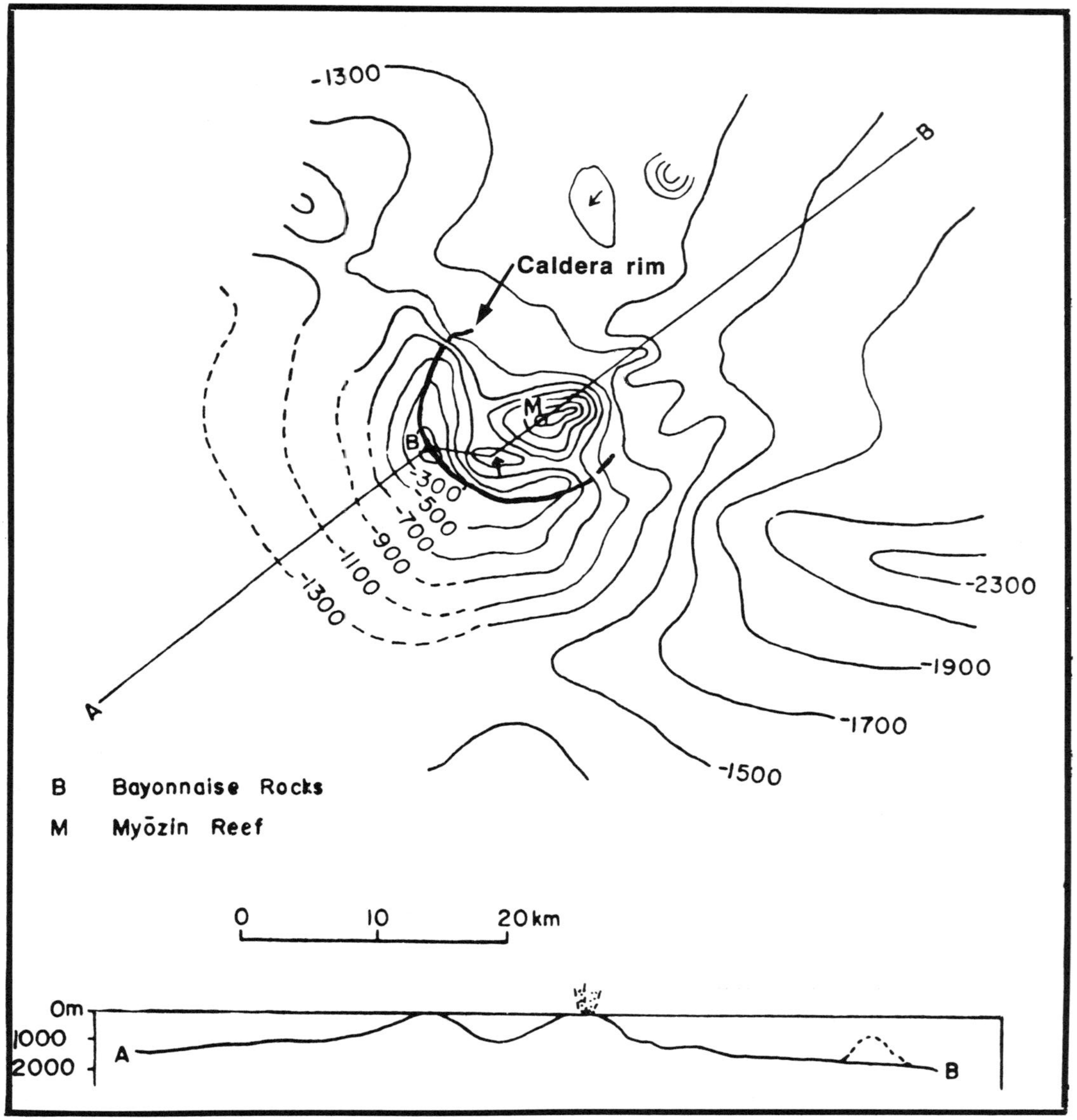

Figure 8.11.1. Submarine topography of Bayonnaise Rocks area, from Kuno (1962); a similar map appeared in Morimoto and Ossaka (1955). We have inferred approximate outline of caldera rim from bathymetry and distribution of postcaldera vents.

IWO-JIMA

CAVW number (active vent)	Latitude Longitude (degrees)	Diameter (km)	Local tectonic setting	Precaldera edifice	SiO_2 content (wt pct)	Age of collapse (yr B.P.)	Date of unrest	Type and duration of unrest ESTU STHF MGTF H Te	Eruption type
08-04-12	24.75N 141.33E	9	Compr	Strato	R = 58-61 C = ?	≥2,600	1889	---G ---- ---- - --	pex
							1922	---G ---- ---- - --	pex
							1935	---G ---- ---- - --	pex
							1944	---F ---- ---- - --	pex
							1955-57	---G ---G --F- F --	pex
							1967-69	---G ---- ---- - --	pex
							1975-76	---F ---- ---- - --	pex
							1978	---F ---- ---- - --	pex
							1980	xx-F -x-- ---- - --	pex
							1982	CC-F -x-- ---- - --	pex, sub

TECTONIC SETTING

Iwo-Jima is one of the Kazan Retto (Volcano) Islands in the Izu-Ogasawara arc, formed above the subduction of the Pacific plate beneath the Philippine Sea plate. Stern and others (1984) suggest that Iwo-Jima is situated at the current apex of back-arc spreading that is propagating northward. We accept this possibility but believe that the remarkable uplift of Iwo-Jima has a more immediate, caldera-related explanation (see below).

GEOLOGIC HISTORY

Iwo-Jima ("Sulphur Island") lies within a submarine caldera (Kaizuka and others, 1983, 1985) (figs. 8.12.1, 8.12.2). Pumiceous tuffs, syenitic intrusive rocks, and glassy trachyandesite lavas are the dominant rocks on Iwo-Jima (Oyagi and Inokuchi, 1985) (fig. 8.12.3). Wakimizu (1908) remarked that Motoyama, a low postcaldera cone that takes up most of the island, is formed almost exclusively of yellow tuff, in greater proportion to lavas than any other Japanese volcano with which he was familiar. Terraces and corals at the top of Motoyama suggest that it is not a constructional "tuff cone" at all, but rather a structural uplift (Tsuya, 1936), similar to a resurgent dome. The southern tip of the island is a small cone (called Suribachi or Mt. Pipe) consisting of lavas and tuff. As of World War II, fresh pumice lapilli deposits on the flanks of Suribachi were largely barren of vegetation, suggesting a young age. Numerous fumaroles occur on Iwo-Jima, and temperature gradients are high. "...one test hole drilled on the northwestern flank of Motoyama had to be abandoned at a depth of 242 feet, because the temperature of the rock took the temper out of the bits" (Swenson, 1948). High temperatures are also suggested by Oshima and others (1981, 1985) on the basis of a 20-km-diameter weakly magnetized core within what is otherwise a strongly magnetized (basaltic) shield.

IWO-JIMA (continued)

GEOLOGIC SETTING (continued)

Charcoal from low in the section of the Motoyama tuffs yielded an age of 2,600 yr B.P. (Stern and others, 1984); thus, the caldera is at least 2,600 yr old.

HISTORICAL ACTIVITY

Remarkable uplift of the island has occurred for at least the past 700 years and possibly much longer (figs. 8.12.4-8.12.9). Rates of uplift over the past 700 years, calculated from radiocarbon-dated terraces, have been 15-20 cm/yr. A 1,350±100-yr-old terrace is now 110 m above sea level; a 745±75-yr-old terrace is now 108 m above sea level. Eruptions from Suribachi buried a prominent shoreline terrace that was 110 m above sea level in 1945, but those eruptions predate another terrace that in 1945 was 37 m above sea level. A shoreline on which Captain Cook's surviving crew landed in 1779 is now 40 m above sea level.

Maximum uplift on the island between 1952 and 1968 was slightly over 9 m along the northeast coast (57 cm/yr). A domal uplift of 7 m occurred during the same period near the center of the island (44 cm/yr), and less than 1 m of uplift occurred at the southwest tip of the island (6 cm/yr) (Morimoto and others, 1968; Kaizuka and others, 1983, 1985). Leveling between 1968 and 1978 showed as much as 3 m of island-wide uplift (Ossaka and others, 1985); leveling between 1976 and 1980 showed as much as 1.9 m of uplift (National Research Center for Disaster Prevention, 1981; Kumagai and Takahashi, 1985). The pattern of uplift changes from year to year but is in broad form a domal uplift.

Seismicity associated with this active ground deformation has been recorded continuously since March 1976; several earthquakes are recorded each day, and small swarms (with up to 40 events/day) occur every few months (Kumagai, 1982, 1985; National Research Center for Disaster Prevention, 1985) (figs. 8.12.11-8.12.13). Hypocenters are dispersed beneath the entire island, at depths of 0-4 km. Tilt measurements since 1976 show tilts of as much as 250 microradians over periods of a few months (Kumagai, 1982).

Since the first recorded phreatic explosion in 1889, 15 additional phreatic eruptions have occurred. Some recent explosions have been preceded by seismic swarms; others have had no seismic precursor (Kumagai, 1985) (figs. 8.12.12, 8.12.13). Fault movements were observed for over a year before a January 1976 explosion, and the temperature of at least one fumarole increased by about 70 °C over the same period (figs. 8.12.10, 8.12.14). The floor and cold-water supply of a U.S. Air Force building became heated in 1955, two years before the 1957 explosion.

COMMENTS

Iwo-Jima is a superb example of active resurgent doming. Doming is episodic, not continuous, as evidenced by the uplifted, dated terraces and the historical geodetic measurements. Rates of uplift (several tens of centimeters per year) are among the highest in this compilation, and the uplift is among the most long lived. An important and unanswered question is whether the historical uplift is a rebound following the last caldera-forming eruption, or rather a precursory signal of another caldera-forming eruption to come.

IWO-JIMA, Region 8, CAVW number 08-04-12

IWO-JIMA (continued)

REFERENCES

Corwin, G., and Foster, H.L., 1959, The 1957 explosive eruption of Iwo-Jima, Volcano Islands: Am. Jour. Sci., v. 257, p. 161-171.

Hydrographic Department, Maritime Safety Agency, 1983, Submarine topographic and geological survey in the adjacent seas of Iwo-Jima: Rep. Coord. Comm. for Prediction of Volcanic Eruptions, no. 26, p. 64-67.

Kaizuka, S., Kato, S., Nagaoka, S, and Miyauchi, T., 1985, Geomorphology of Iwo-jima and surrounding sea floor: Jour. Geogr. (Tokyo), v. 94, no. 6, p. 424-436.

Kaizuka, S., Miyauchi, T., and Nagaoka, S., 1983, Marine terraces, active faults, and tectonic history of Iwo-Jima: Ogasawara Res., no. 9, p. 13-45.

Krauskopf, K.B., 1948, Notes on the geology of Iwo-Jima: Trans. Am. Geophys. Union, v. 29, p. 207-210.

Kumagai, T., 1985, Volcanism and seismicity in the Iwo-jima volcanoes: Jour. Geogr. (Tokyo), v. 94, no. 6, p. 455-463.

Kumagai, T., and Takahashi, H., 1985, Crustal deformations on Iwo-jima (II): Jour. Geogr. (Tokyo), v. 94, no. 6, p. 479-487.

Kumagai, T., ed., 1982, Observational data of volcanic activities, no. 1 (Iwo-jima and Kirishima Volcanoes): National Research Center for Disaster Prevention, Review of Research for Disaster Prevention, no. 74, p. 2-144.

Macdonald, G.A., 1948, Petrography of Iwo-Jima: Bull. Geol. Soc. Am., v. 59, p. 1009-1018.

Morimoto, R., Ossaka, J., Hatori, T., Izutsuya, S., Urabe, K., Takahashi, H., Okada, Y., Hirabayashi, J., Isa, K., and Isobe, H., 1968, The upheaval of an insular volcano Iwo-Jima, Ogasawara Islands and its recent state of activity: Jour. Earthq. Studies, v. 77, p. 255-283.

National Research Center for Disaster Prevention, 1981, Volcanic activities of Iwo-Jima (Sulphur Island), Kazan Retto (Volcano Islands): Rep. Coord. Comm. for Prediction of Volcanic Eruptions, no. 21, p. 1-4.

----- 1982, Observational data on volcanic activities, number 1, Iwo-Jima and Kirishima volcanoes: Review of Research for Disaster Prevention, no. 74.

----- 1985, Seismic activities in Iwo-Jima volcanoes (December 1982-June 1984): Rep. Coord. Comm. for Prediction of Volcanic Eruptions, no. 32, p. 56-60.

Oshima, S., Kaneko, T., Onondera, K., Nakagawa, H, and Tozaki, T., Mishina, M., and Ossaka, J., 1981, A three-component aeromagnetic survey of Iwo-Jima [abs.]: Abstracts, 1981 IAVCEI Symp. on Arc Volcanism: Tokyo and Hakone, p. 284.

Oshima, S., Kaneko, T., Onondera, K., Nakagawa, H, and Tozaki, T., Mishina, M., and Ossaka, J., 1985, Geomagnetic anomaly distribution of Iwo-Jima: Jour. Geogr. (Tokyo), v. 94, no. 6, p. 502-515.

Ossaka, J., Tsuji, S., and Ogura, H., 1985, Crustal deformations on Iwo-jima (I): Jour. Geogr. (Tokyo), v. 94, no. 6, p. 474-479.

Oyagi, N., and Inokuchi, T., 1985, Geology of Iwo-jima: Jour. Geogr. (Tokyo), v. 94, no. 6, p. 436-445.

Stern, R.J., Smoot, N.C., and Rubin, M., 1984, Unzipping of the Volcano Arc, Japan: Tectonophysics, v. 102, p. 153-174.

Swenson, F.A., 1948, Geology and groundwater resources of Iwo-Jima: Bull. Geol. Soc. Am., v. 59, p. 995-1008.

Takahashi, H., Kumagai, T., and Oyagi, N., 1976, Fault movement in Iwo-Jima (Sulphur Island), Volcano Islands: Bull. Volcanol. Soc. Japan, v. 22, p. 286-287.

Tsuya, H., 1936, Geology and petrography of Io-sima (Sulphur Island), Volcano Islands Group: Bull. Earthq. Res. Inst., v. 14, p. 453-480.

Wakimizu, T., 1908, The ephemeral volcanic island group in the Iwo-Jima group: Imperial Earthq. Invest. Comm., no. 22C, p. 1-33.

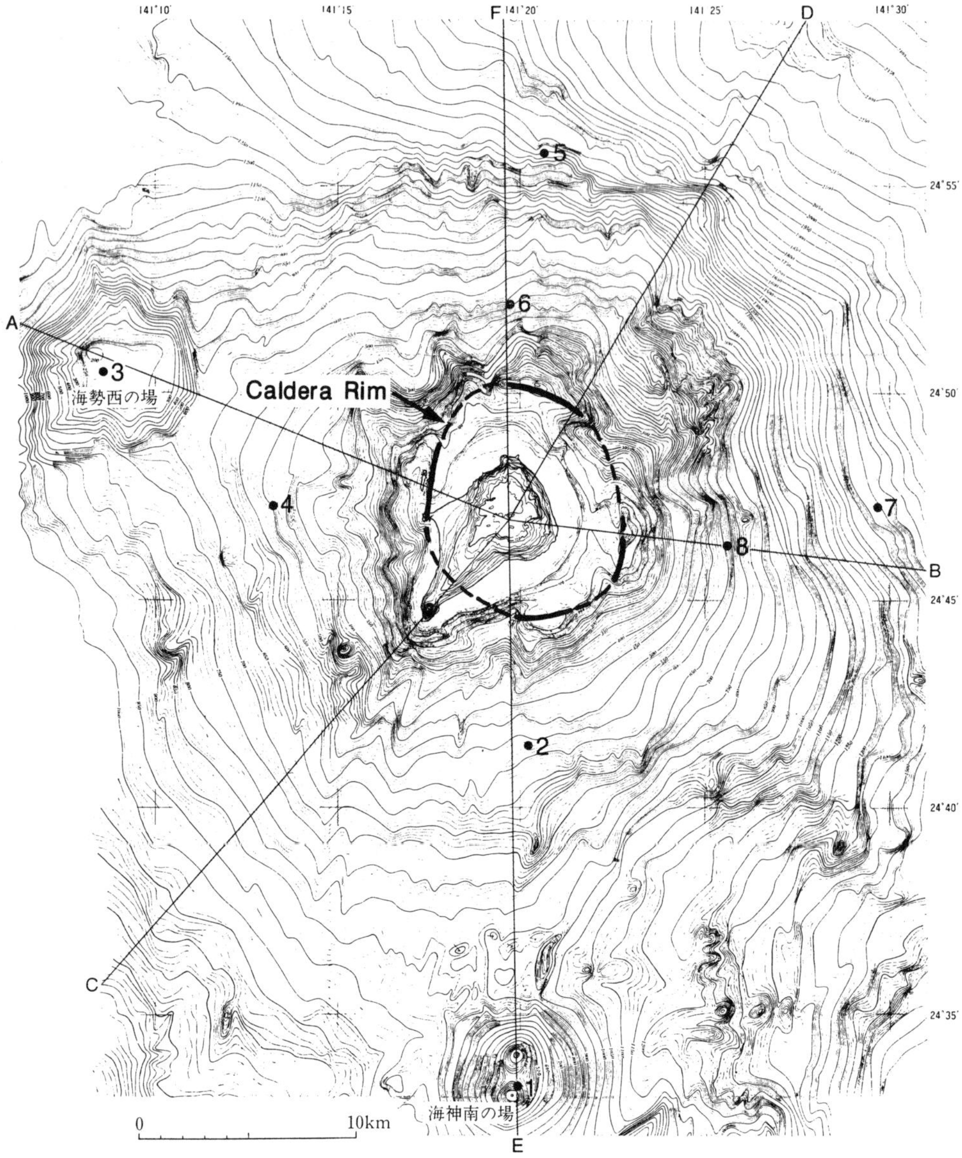

Figure 8.12.1. Submarine bathymetry around Iwo-Jima, from Kaizuka and others (1985). Contour interval is 10 m. Topographic profiles along lines A-B, C-D, and E-F are shown in figure 8.12.2.

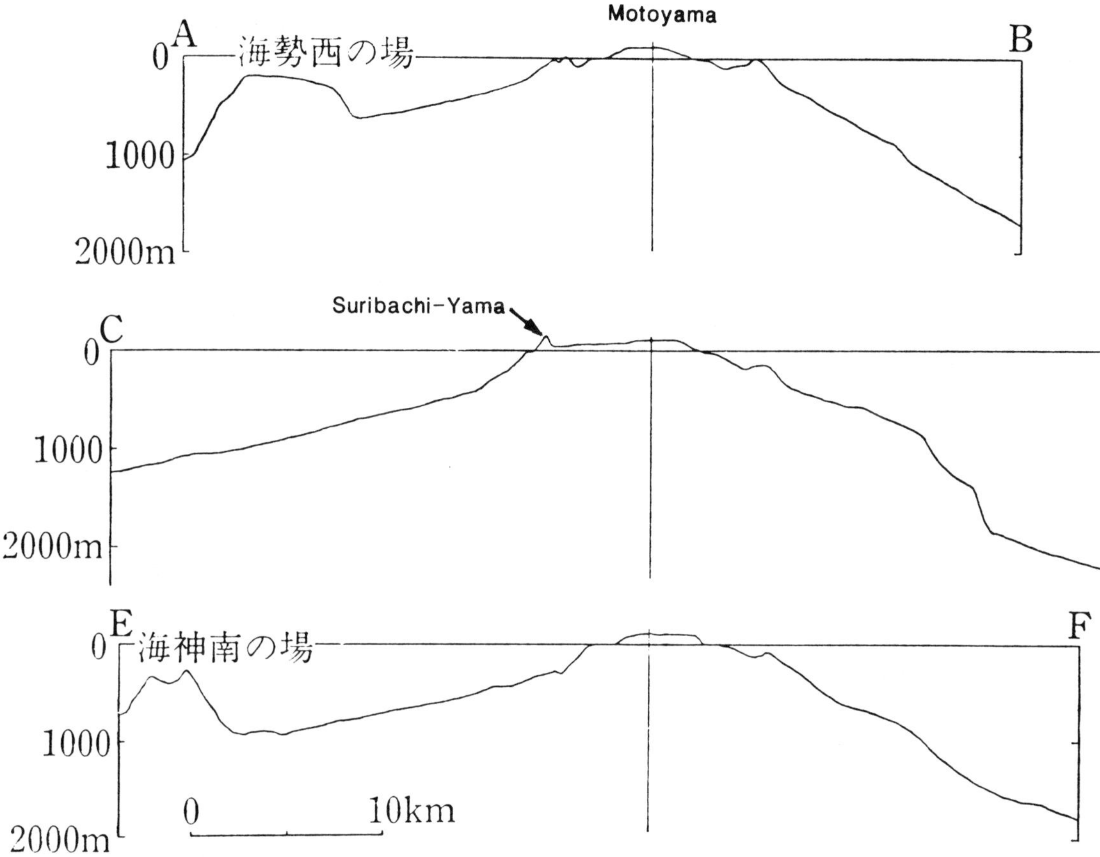

Figure 8.12.2. Three topographic profiles of Iwo-Jima Volcano along lines shown in figure 8.12.1, from Kaizuka and others (1985).

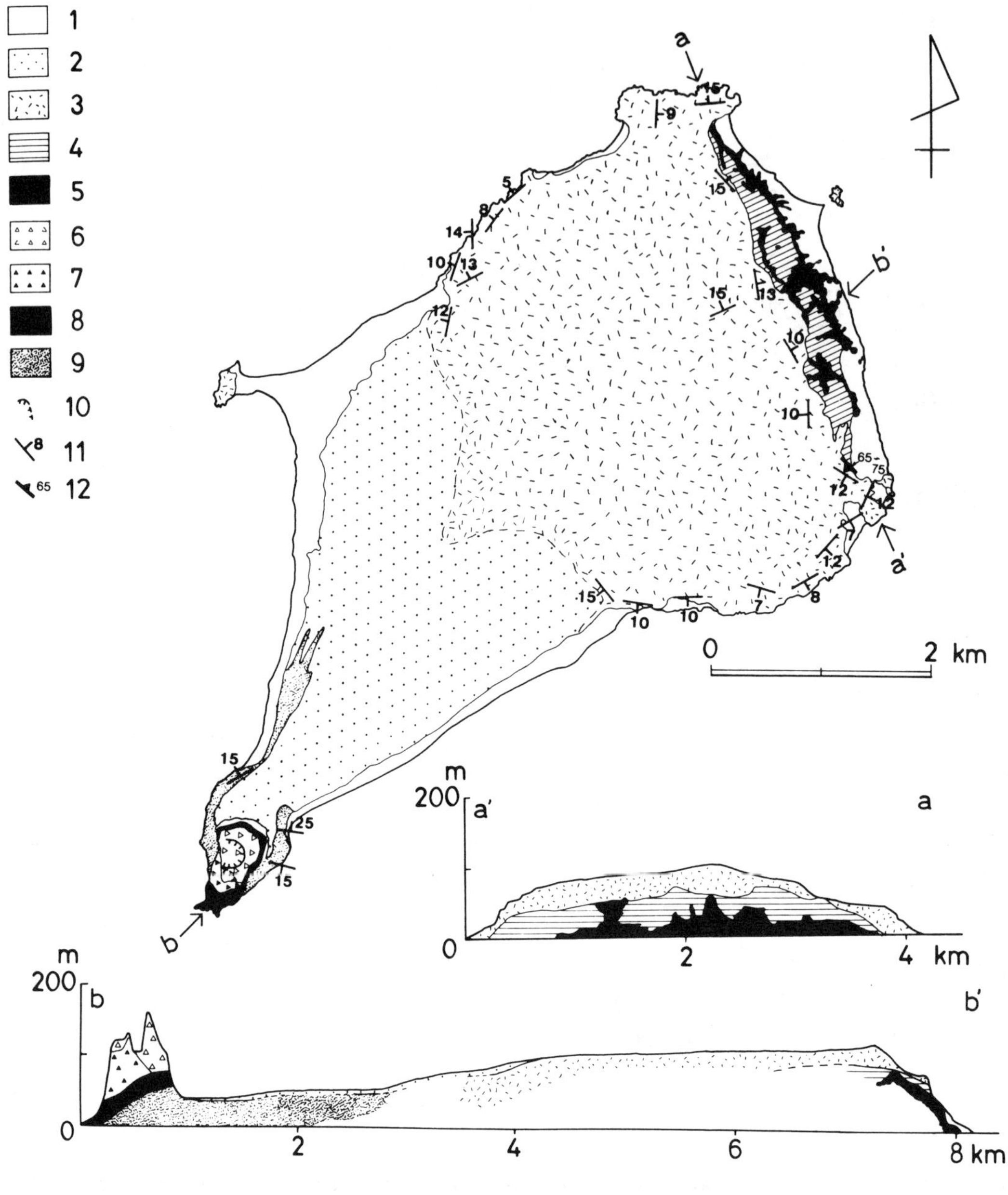

Figure 8.12.3. Geological map and sections of Iwo-Jima, from Kaizuka and others (1983). Marine-terrace deposits are omitted. Explanation: 1, recent beach sand (mostly after 1945); 2, Chidorigahara sand deposits; 3, upper part of Motoyama tuff; 4, lower part of Motoyama tuff; 5, lava in the lower part of Motoyama tuff; 6, younger cone pyroclastics of Suribachi-yama; 7, older cone pyroclastics of Suribachi-yama; 8, lava of Suribachi-yama; 9, lowest pyroclastics of Suribachi-yama; 10, crater; 11, strike and dip of tuff; 12, strike and dip of fault.

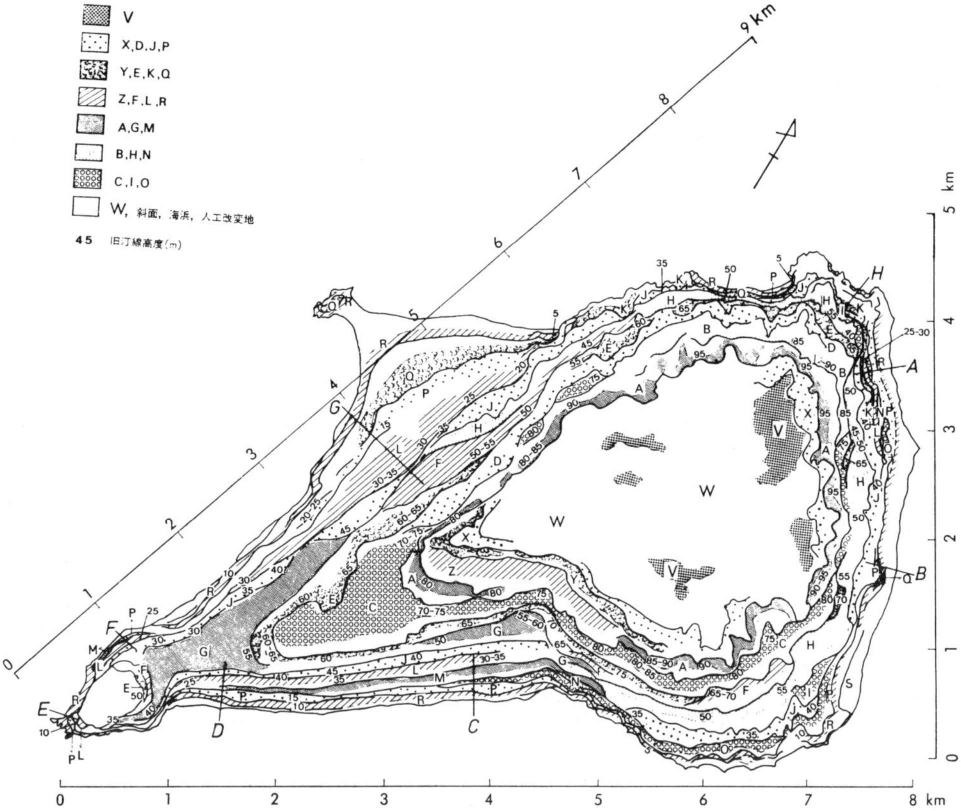

Figure 8.12.4. Uplifted paleoshorelines on Iwo-Jima (lettered), from Kaizuka and others (1985). The same shorelines are shown in profile in figure 8.12.5.

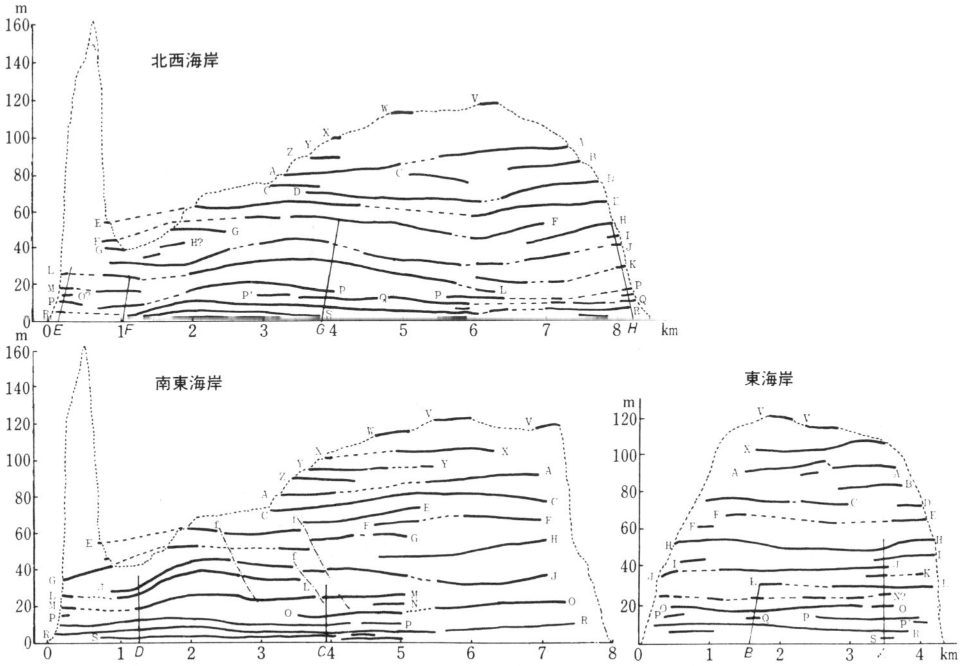

Figure 8.12.5. Projected profiles of uplifted shorelines and the outline of Iwo-Jima, from Kaizuka and others (1985). Terraces are lettered V-Z and A-S (oldest to youngest). Faults cut terraces at least as young as terrace M.

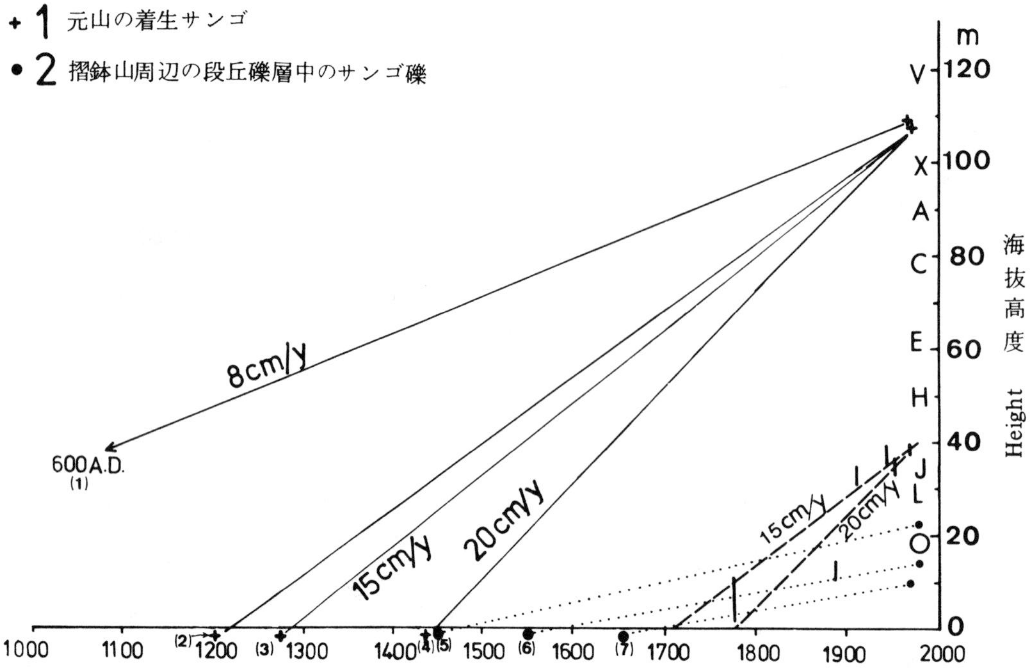

Figure 8.12.6. Long-term rates of uplift at Iwo-Jima, estimated from radiocarbon ages of uplifted terraces, from Kaizuka and others (1983). An age of 1,350±100 yr for a terrace at 100-m elevation implies an average uplift rate of 8 cm/yr. Ages of 745, 675, and 510 yr B.P. (all ±75 yr) for a terrace at 108-m elevation imply an average rate of 14-20 cm/yr, comparable to rate for terrace on which Captain Cook's crew landed in 1779. Solid lines are for terraces on Motoyama, dashed lines for terraces on isthmus, and dotted lines for terraces on Suribachi-yama.

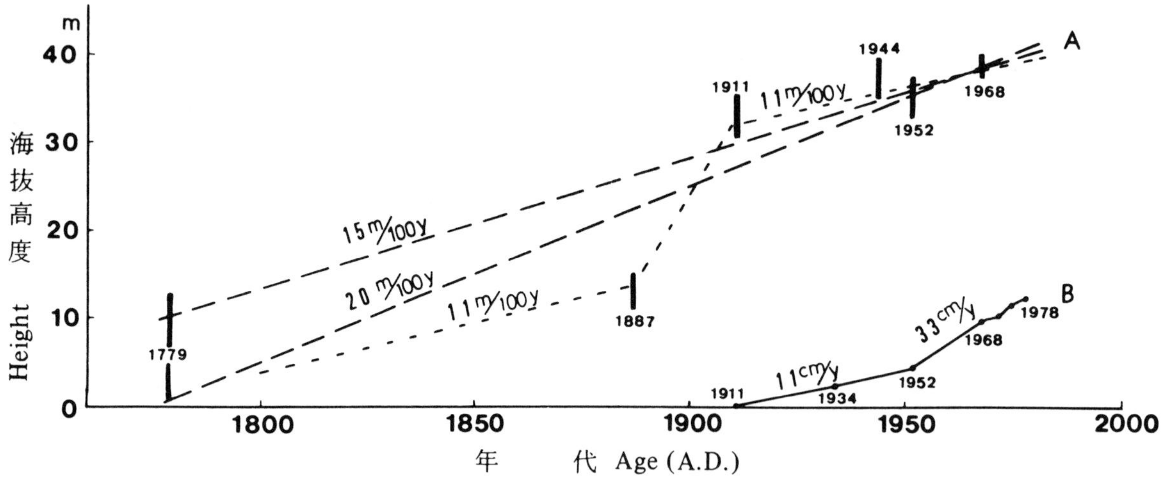

Figure 8.12.7. Historical uplift rates at Iwo-Jima, from Kaizuka and others (1983). A, Isthmus at eastern foot of Suribachi-yama since 1779. B, Same location(?) since 1911. Note apparent changes in rate of uplift since A.D. 1800.

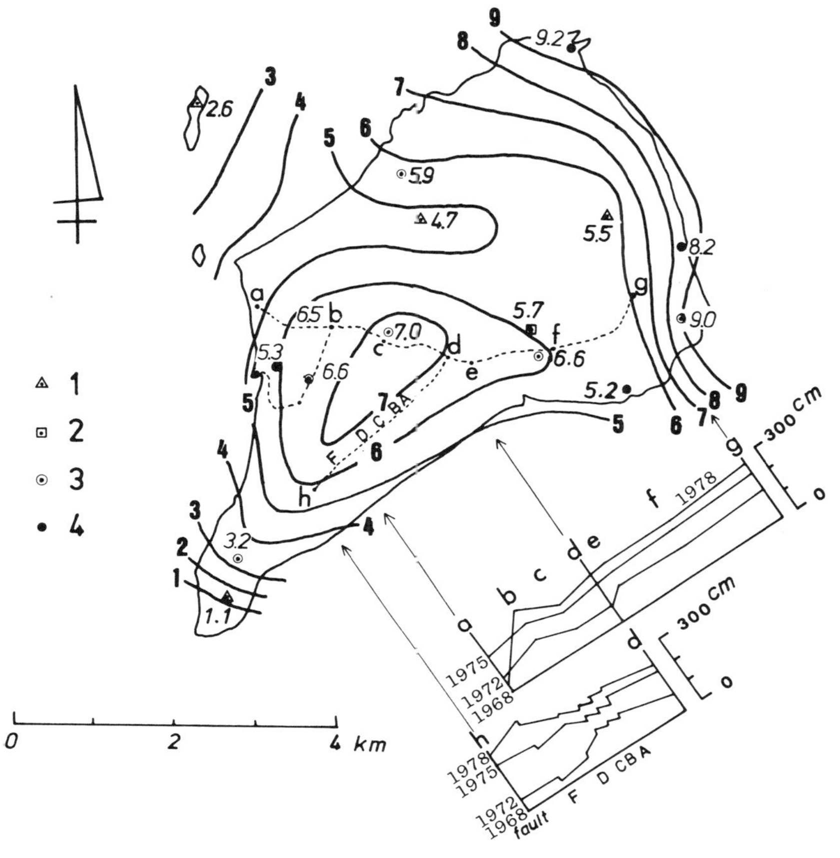

Figure 8.12.8. Historical uplift at Iwo-Jima measured geodetically, from Kaizuka and others (1983). Contours and individual values in meters, from 1952 to 1968. At lower right, graph showing vertical movements along leveling route (dotted line) from 1968 to 1978. Letters a-h, bench marks; A-F, active faults; 1, triangulation point; 2, bench mark; 3, elevation determined from aerial photograph; 4, surveyed point along the coast.

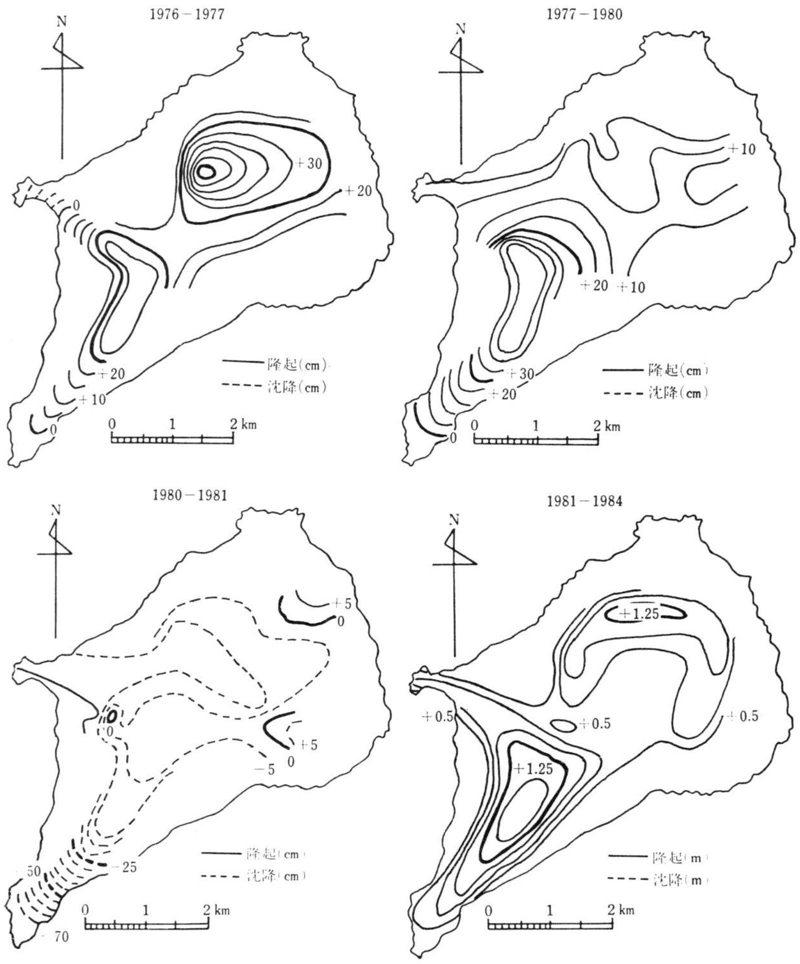

Figure 8.12.9. Contours of uplift at Iwo-Jima from March 1976 to February 1984, from Kumagai and Takahashi (1985). Note changes in pattern of uplift.

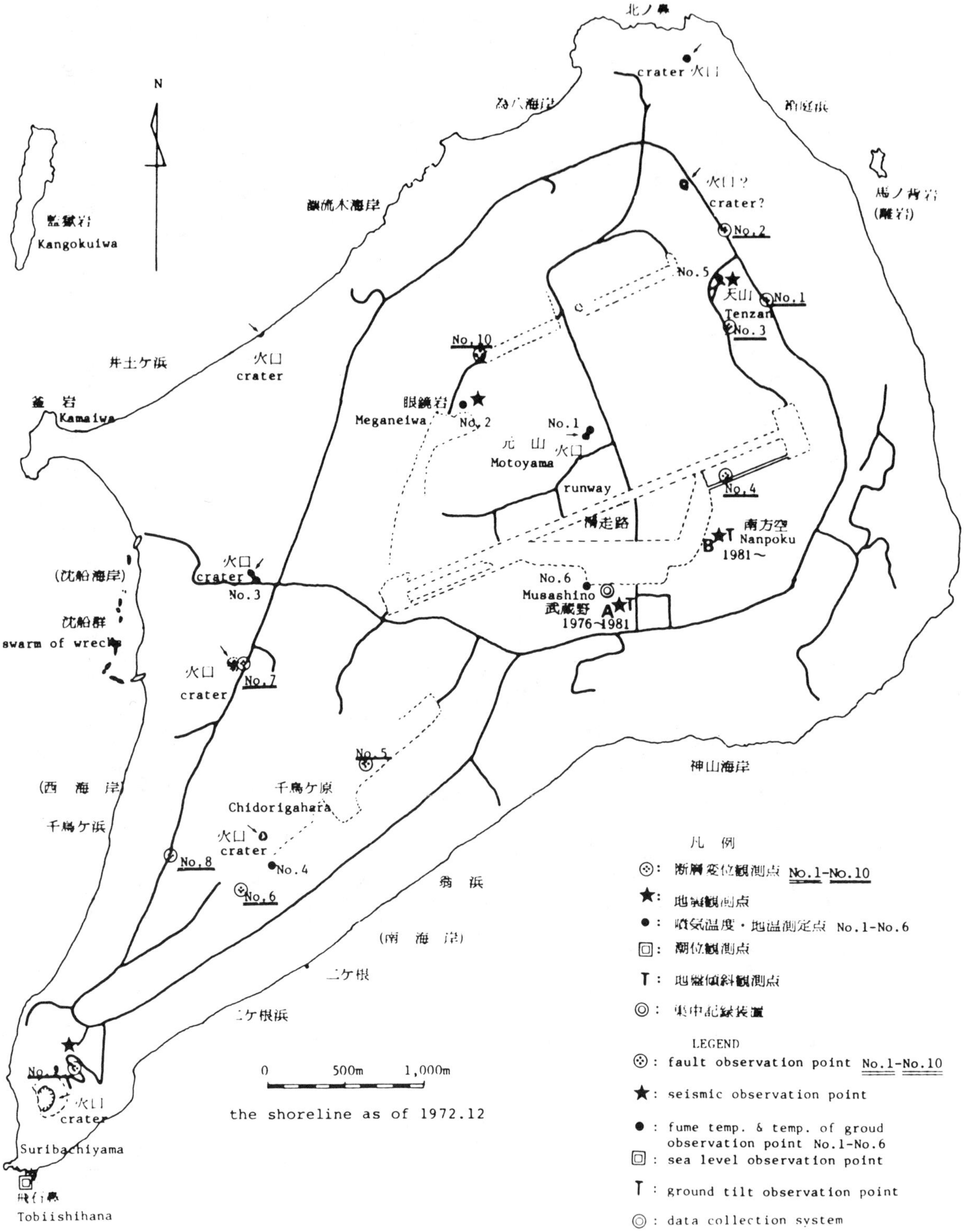

Figure 8.12.10. Locations of monitoring points on Iwo-Jima Island, from National Research Center for Disaster Prevention (1982).

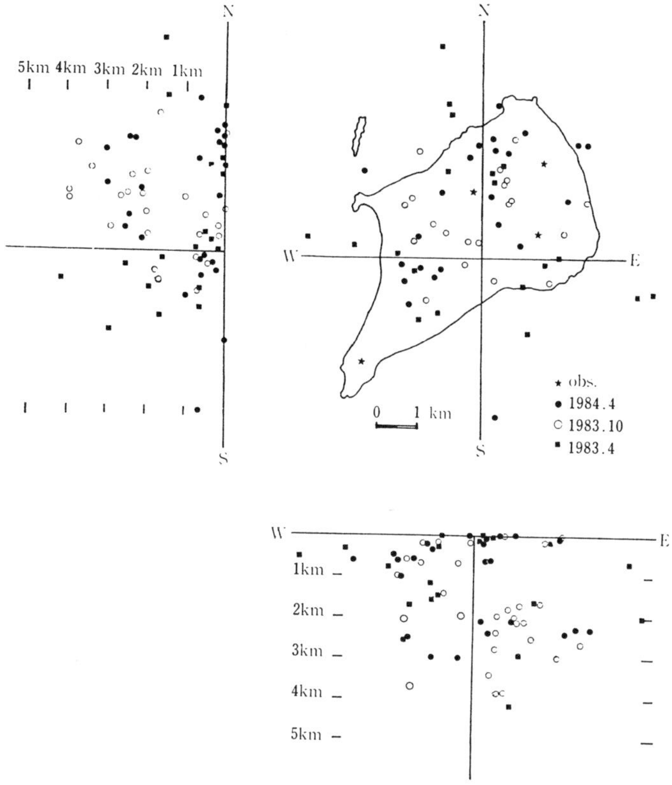

Figure 8.12.11. Hypocenters of earthquakes at Iwo-Jima, 1983-84, from Kumagai (1985).

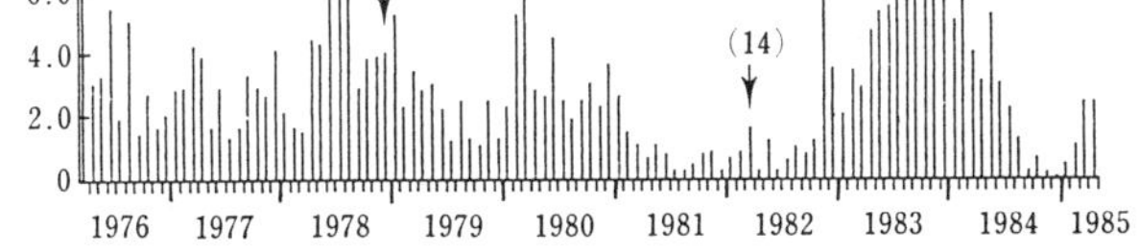

Figure 8.12.12. Mean daily number of A-type volcanic earthquakes at Iwo-Jima, by month, for the period 1976-85 (Kumagai, 1985).

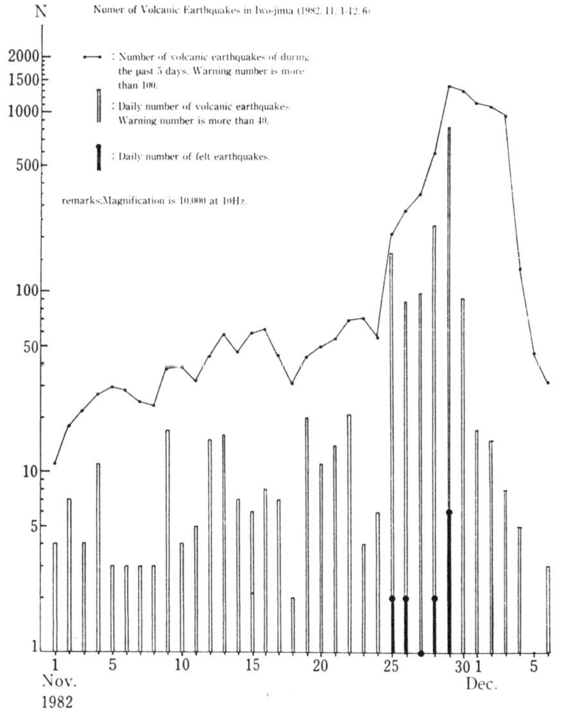

Figure 8.12.13. Daily earthquake count at Iwo-Jima, November 1982, from Kumagai (1985). Small phreatic explosions occurred on 28 and 29 November.

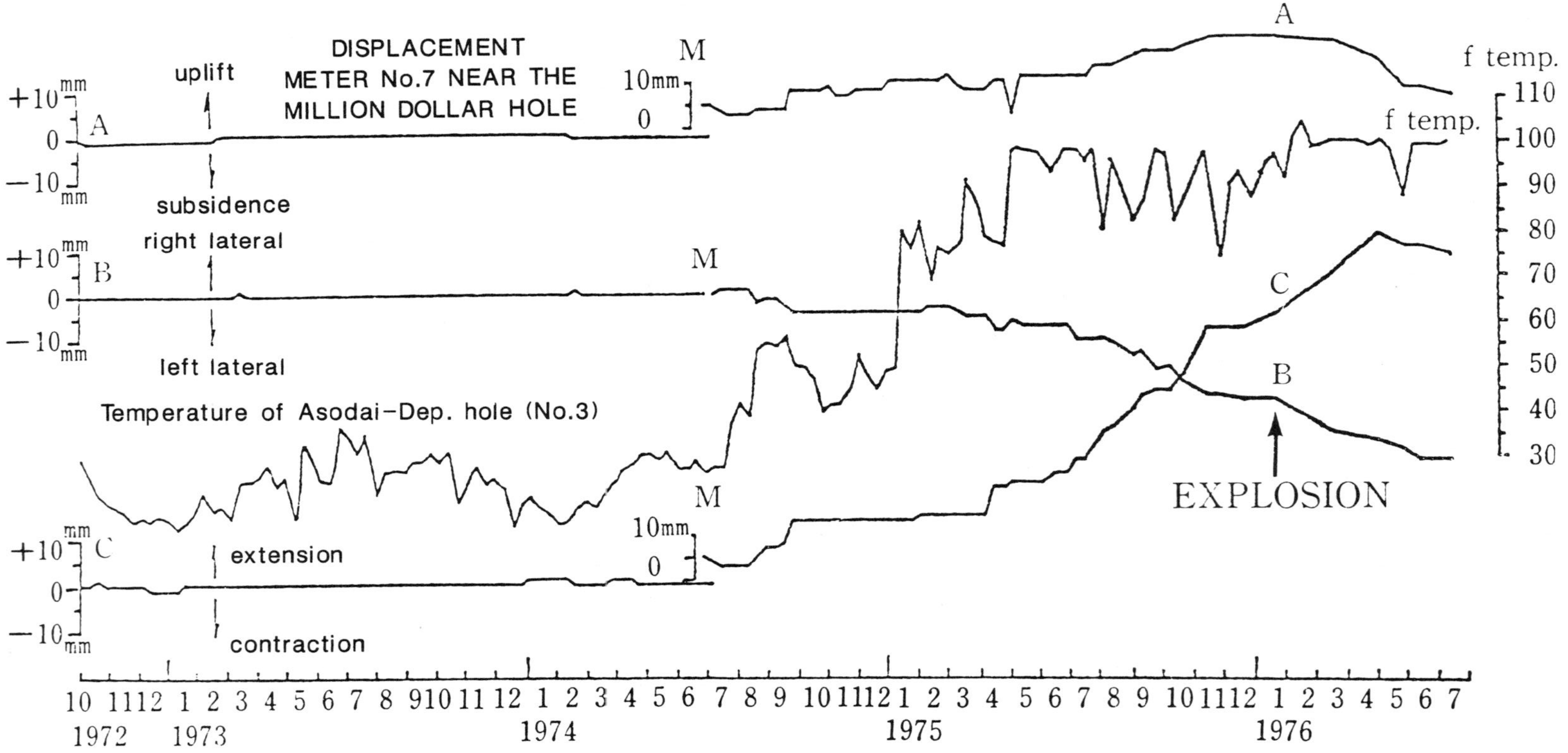

Figure 8.12.14. Components of fault displacement and changes in fumarolic temperature at Iwo-Jima, 1972-76, from Kumagai and Takahashi (1985). Displacements were measured at fault observation point 7 (see figure 8.12.10); temperature changes were measured at the Asodai-Dep. hole.

NORTH PAGAN

CAVW number (active vent)	Latitude Longitude (degrees)	Diameter (km)	Local tectonic setting	Precaldera edifice	SiO_2 content (wt pct)	Age of collapse (yr B.P.)	Date of unrest	Type and duration of unrest ESTU STHF MGTF H Te	Eruption type
08-04-17 (Mt. Pagan)	18.13N 145.80E	5 x 6		Strato	R = 48-52 C = mafic?		1669?	---- ---- ---- - --	Ex,lf
							1820's	---- ---- ---- - --	ex
							1872-73	---- ---- ---- - --	Ex,lf
							1909	---- ---- ---- - --	ex
							1917	x--- ---- ---- - --	ex
							1923-25	x--- ---- ---- - --	ex,lf
							1981-85	E--Y -Y-C ---C - --	Ex,ex,lf
							1987-88?	---- ---- ---- - --	ex?

TECTONIC SETTING

Pagan Island is in the northern Marianas arc, where the Pacific plate is subducted beneath the Philippine Sea plate at about 10.7 cm/yr (Circum-Pacific Map Project, Northwest Quadrant Panel, 1981).

GEOLOGIC HISTORY

Mt. Pagan is a small stratovolcano that has grown within an unnamed 5 km x 6 km somma caldera (Corwin and others, 1957). Another, slightly smaller stratovolcano--South Pagan--has grown within a 2.5-km-diameter somma caldera on the southern end of Pagan Island, 10 km southwest of Mt. Pagan (fig. 8.13.1). Several vents of South Pagan are aligned along a NE-SW axis, pointing toward Mt. Pagan, but the subsurface relation between Mt. Pagan and South Pagan is unknown. As of 1954, coral reefs around the north shore of Pagan Island were 0.3 to 1.3 m above mean sea level, indicating that this part of the island has been uplifted about 1.5-4.5 m relative to sea level; reefs at the south end of the island have been raised by a lesser amount (Corwin and others, 1957). Tayama (1936) also noted borings of mollusks 1-2 m above the high-tide line on the west shore of Pagan Island. Some of this apparent uplift is likely due to lowering of eustatic sea level, as reported at many islands in the Pacific. However, the difference between the northern and southern parts of the island suggests that (1) the north part of the island, around the North Pagan Caldera, has been uplifted relative to the south part of the island; (2) recent volcanic activity of South Pagan has obscured reefs that are correlative with those in the north; or (3) erosion of cliffs around South Pagan has removed any such reefs.

NORTH PAGAN (continued)

HISTORICAL ACTIVITY

North Pagan erupted (or is presumed to have erupted) in the years listed above. The eruption in 1872 was termed "major" (Marche, 1891). In 1917, "earth trembling and fire phenomena occurred from time to time"; in early 1923, similar "earth trembling and fire columns were frequent." The 1923 activity climaxed on 3 March 1923 (Tanakadate, 1940).

An airstrip built on the caldera floor during World War II was pitted with bomb craters and overgrown with vegetation but apparently not cut by any faults as of 1954 (Corwin and others, 1957).

In 1981, earthquakes were felt near North Pagan beginning in late March or early April, with magnitudes probably between 3 and 4. None was recorded at the nearest WWNSS station in Guam, 450 km south of Pagan. By 13 May, ground cracks and increased fumarolic activity were noted in the summit region of North Pagan. Earthquake activity increased markedly at 0745 hr on 15 May, and an explosive eruption began at 0915 hr on the same day. The eruption was from several new vents high on the north flank of North Pagan, aligned along a N-S fissure. From 21 May to 28 May 1981 (a period of observation), the level of a small lake 1 km west of the new vents dropped at a rate of about 24 mm/day; possible causes include tidal changes, slow draining of the lake, evaporation, or uplift of that part of the island. The lake measurements also hinted at slight tilt, consistent with inflation beneath the 1981 vents hours to a few days before an ash eruption on 24 May and lava extrusion on 25-26 May 1981 (Banks and others, 1984) (figs. 8.13.2, 8.13.3). Electronic distance measurements for the period 22-28 May 1981 showed contraction of only one line--to a point near the summit, indicating local inflation high in the cone (Banks and others, 1984) (fig. 8.13.4).

An earthquake on 15 May 1984 that was felt as far away as Agrigan Island (65 km north of Pagan) marked the start of a brief (1 day?) eruptive phase in 1984; intermittent activity continued through April 1985 and then paused until a small ash eruption in September 1987 (Smithsonian Institution, 1985, 1987). Further eruptions might have occurred in February 1988 (Smithsonian Institution, 1988).

REFERENCES

Banks, N.G., Koyanagi, R.Y., Sinton, J.M., and Honma, K.T., 1984, The eruption of Mount Pagan Volcano, Mariana Islands, 15 May 1981: Jour. Volcanol. Geotherm. Res., v. 22, p. 225-269.

Circum-Pacific Map Project, Northwest Quadrant Panel, 1981, Plate tectonic map of the Circum-Pacific region, northwest quadrant: Tulsa, Am. Assoc. Pet. Geol., 1 sheet, scale 1:10,000,000.

Corwin, G., 1961, Volcanic suite of Pagan, Marianas Islands: U.S. Geol. Surv. Prof. Paper 424-C, p. C204-206.

Corwin, G., Bonham, L.D., Terman, M.J., and Viele, G.W., 1957, Military geology of Pagan, Mariana Islands: U.S. Army Corps of Engineers, Intelligence Dossier, 252 p.

NORTH PAGAN (continued)

REFERENCES (continued)

Fosberg, F.R., and Corwin, G., 1958, A fossil flora from Pagan, Mariana Islands: Pacific Sci., v. 12, p. 2-20.

Kuno, H., 1962, Catalogue of the active volcanoes of the world including solfatara fields, Pt. XI, Japan, Taiwan and Marianas: Rome, IAVCEI, 332 p.

Larson, E.E., Reynolds, R.L., Merrill, R., Levi, S., Ozima, M., Aoki, Y., Kinoshita, H., Zasshu, S., Kawai, N., Nakajima, T., and Hirooka, K., 1974, Major element petrochemistry of some extrusive rocks from the volcanically active Mariana Islands: Bull. Volcanol., v. 38, no. 2, p. 361-377.

Marche, A., 1890, Mon voyage aux Iles Mariannes: Bull. Soc. Geogr. Marseilles, v. 14, p. 22-30.

----- 1891, Rapport general sur une mission aux Iles Mariannes: Nouvelles Arch. Miss. Sci. Lit., v. 1, p. 241-280.

Smithsonian Institution, Scientific Event Alert Network (SEAN), 1981-88, Pagan: SEAN Bull., v. 6, nos. 4-6, 9-11; v. 7, nos. 2, 11; v. 8, nos. 3, 8, 9, 10; v. 9, no. 4; v. 10, no. 4; v. 12, no. :8; v. 13, no. 2 (communications primarily from N. Banks, L.G. Eldredge, and R. Walter).

Tanakadate, H., 1940, Volcanoes in the Mariana Islands in the Japanese mandated South Seas: Bull. Volcanol., ser. II, v. 6, p. 199-223.

Tayama, R., 1936, Geomorphology, geology, and coral reefs of the North Marianas Group: Inst. Geol. Paleontol., Tohoku Imperial Univ., Contrib. in Japanese Language, no. 23, 88 p.

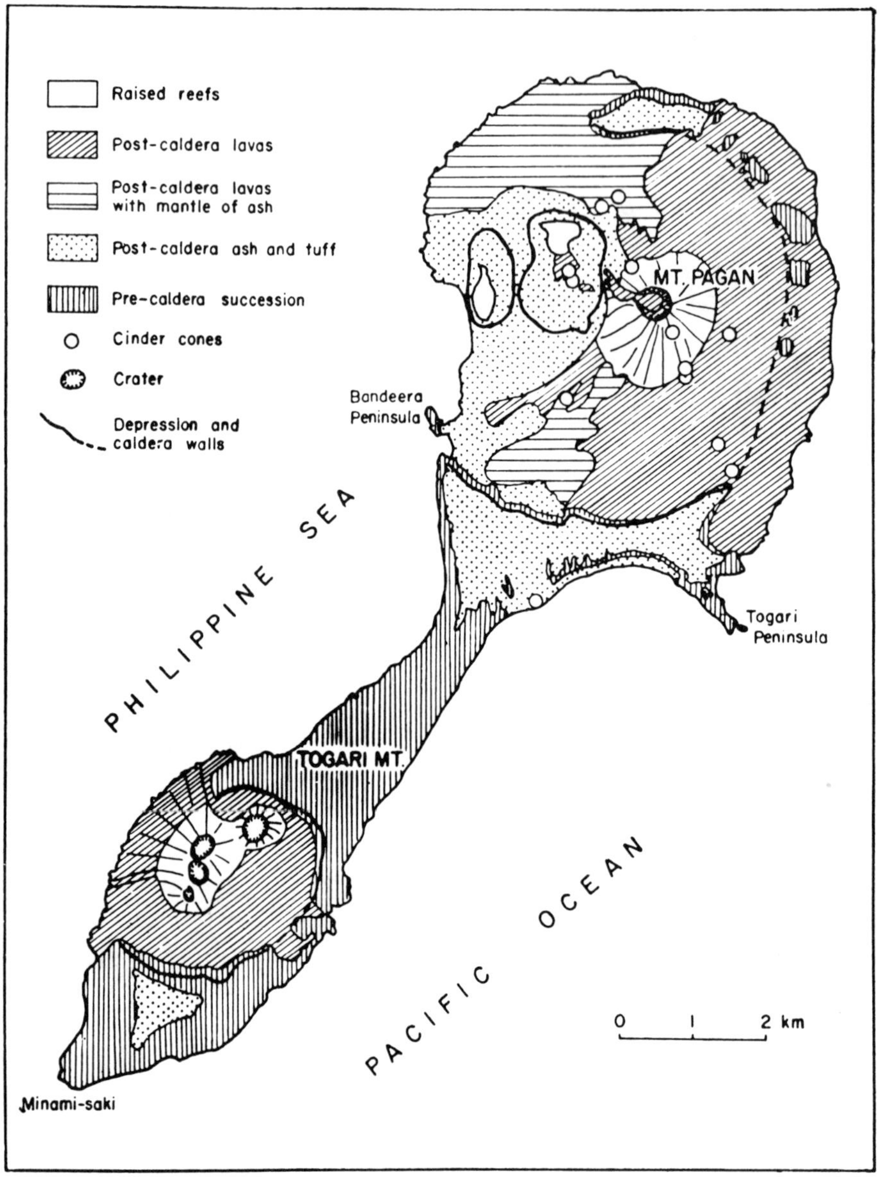

Figure 8.13.1. Geologic map of Pagan Island, from Kuno (1962).

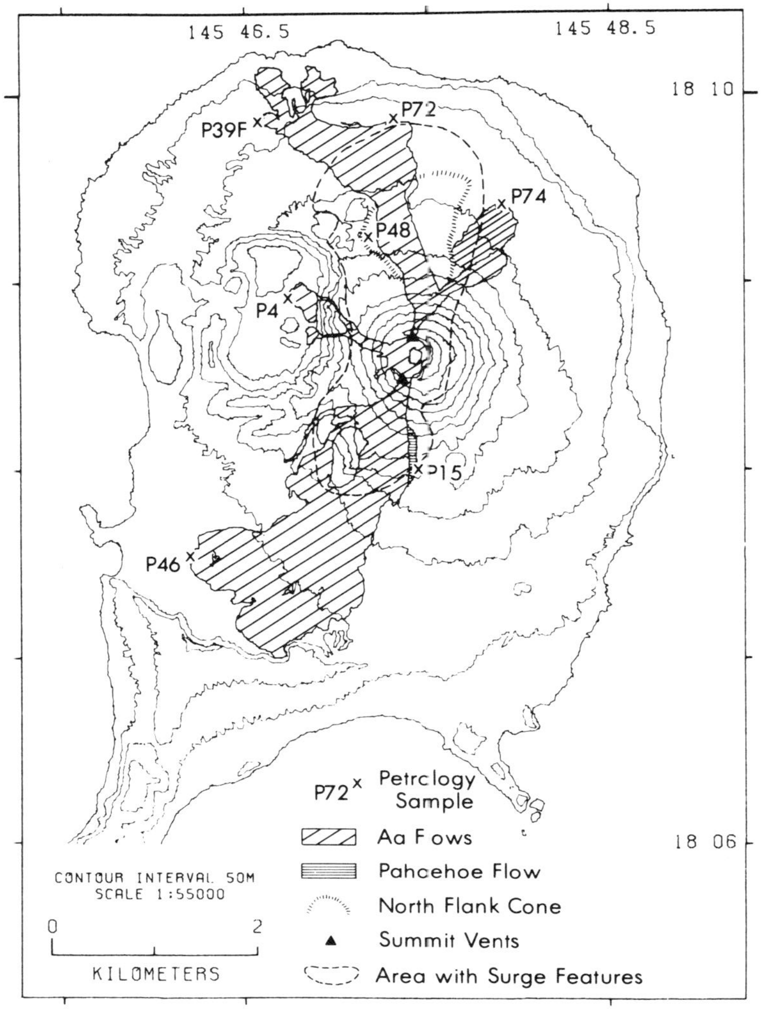

Figure 8.13.2. Locations of vents, flows, and surge features from the 15 May 1981 eruption of Mount Pagan, from Banks and others (1984).

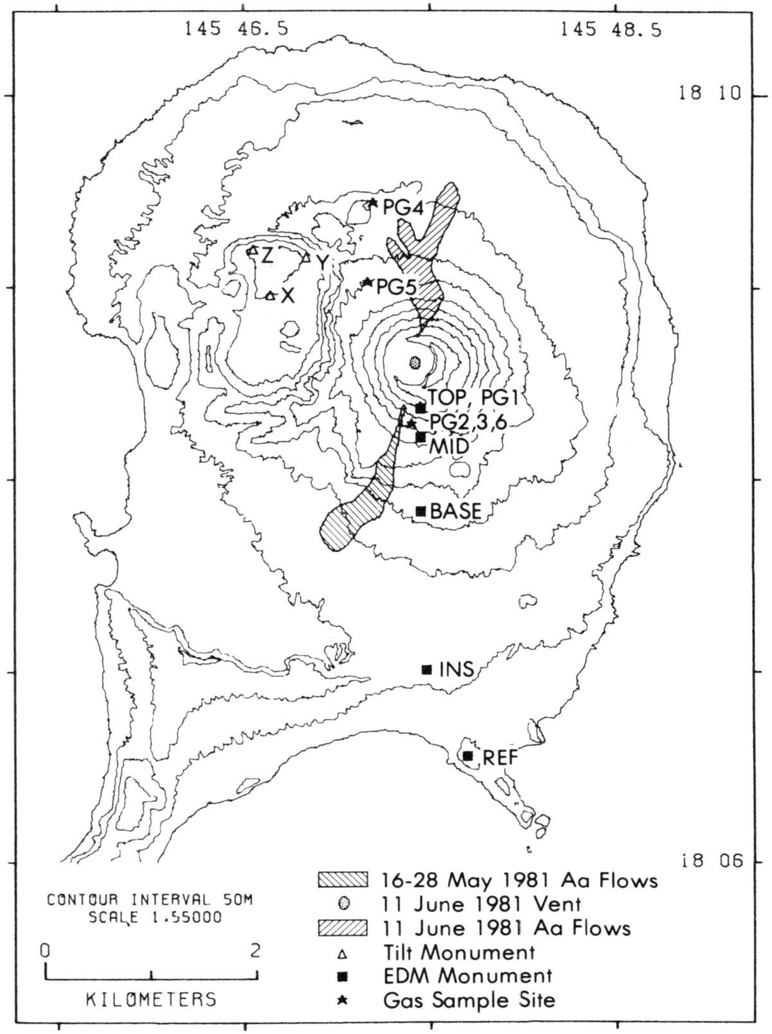

Figure 8.13.3. Lava flows of 16-28 May 1981 and 11 June 1981 at Mount Pagan, from Banks and others (1984). Squares mark locations of electronic distance measurement (EDM) stations (see figure 8.13.4).

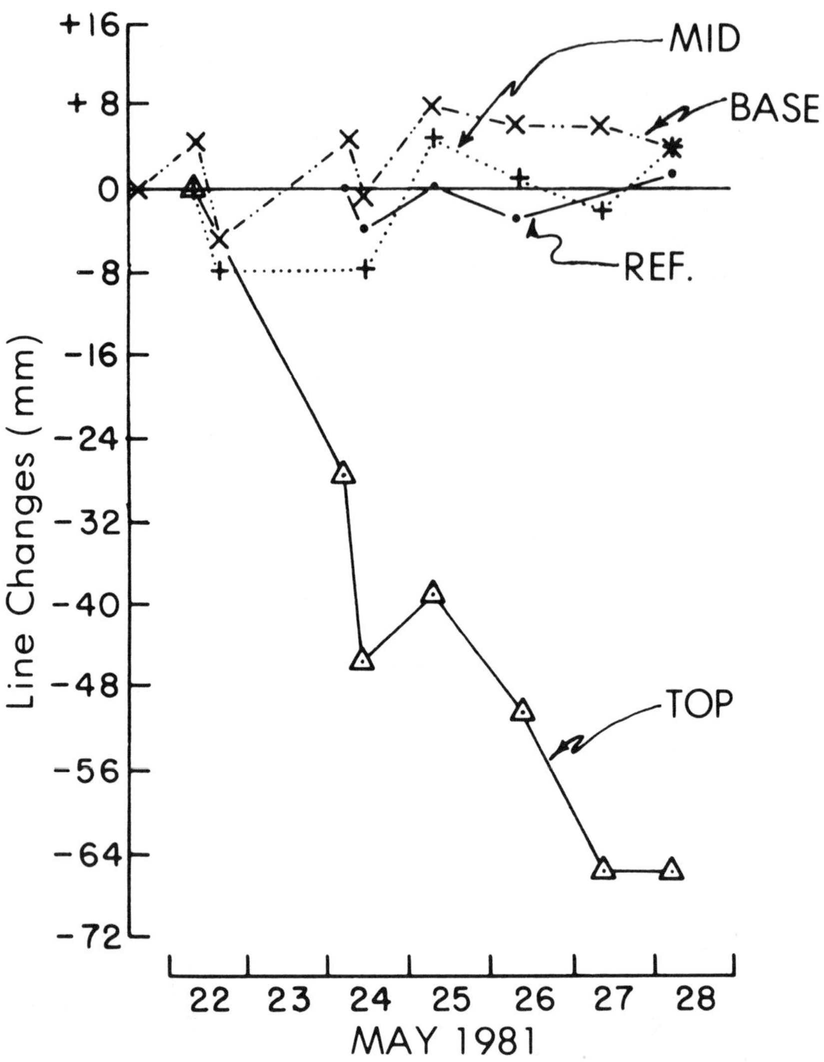

Figure 8.13.4. Changes in lengths of electronic distance measurement (EDM) lines versus time at Mount Pagan, from Banks and others (1984). See figure 8.13.3 for locations of stations.

TOYA

CAVW number (active vent)	Latitude Longitude (degrees)	Diameter (km)	Local tectonic setting	Precaldera edifice	SiO_2 content (wt pct)	Age of collapse (yr B.P.)	Date of unrest	Type and duration of unrest ESTU STHF MGTF H Te	Eruption type
08-05-03 (Usu)	42.53N 140.83E	10	Compr	?	R = 68-72	110,000	1663	C--- ---- ---- - --	EX,PEX,dm
					C = 73		1769	x--- ---- ---- - --	EX,dm
					precaldera		1822	C--- ---- ---- - --	EX,cdm
					= 49-53		1853	D--- ---- ---- - --	EX,dm
							1910	C--C ---Y ---Z Z QU?	pex,cdm
							1932	---- x--- ---- - --	none
							1943-45	EE-E ZE-E Z--- E --	pex,cdm,dm
							1977-82	CCYA F--A --FA F --	Ex,ex,cdm

TECTONIC SETTING

Toya Caldera lies in the V-shaped intersection of the Kurile and Japan arcs, along both of which the Pacific plate is subducted in a northwesterly direction beneath the Eurasian plate at a rate of approximately 10 cm/yr (Circum-Pacific Map Project, Northwest Quadrant Panel, 1981).

GEOLOGIC HISTORY

Toya Caldera formed by collapse about 110,000 yr B.P., based on the age of pyroclastic deposits that have been related to the caldera-forming event (Okumura and Sangawa, 1984; Machida and others, 1985; Ikeda and Katsui, 1986). Usu is a truncated stratovolcano with late-stage dacitic domes, located on the southern boundary of Toya Caldera (figs. 8.14.1-8.14.3). The main body of Usu was formed by eruptions of basaltic and andesitic magma (Usu somma lava). Then, 7,000 to 8,000 years ago, a violent explosion and a large debris avalanche formed a somma that is now largely filled by dacitic domes. Viscous dacitic magma has on occasion upheaved the land surface by more than 200 m directly above a cryptodome (for example, Showa-shinzan ("Roof Mountain") in 1943-45).

Most of Toya Caldera is occupied by Lake Toya; the Naka-jima (or Nakano-shima) group of andesitic domes formed an island in the center of Lake Toya before Usu Volcano began to grow. Two additional domes have been identified by bathymetric and magnetic studies of the caldera lake (Nishida, 1984). On the basis of the same magnetic survey, Nishida (1984) proposes that the diameter of the Toya Caldera structure is only half that of the present topographic depression; if so, Usu lies outside the main structure of the caldera, on its modern topographic rim.

Two or more terraces have been identified along the Sobetsu River, in addition to possible marine terraces (Katsui and others, 1978b, p. 386); these terraces might imply episodic uplift of the Usu area in recent geologic time.

TOYA (continued)

HISTORICAL ACTIVITY

1663-1853: The duration of seismicity before eruptions was 3 days in 1663, unknown in 1769, 3 days in 1822, and 10 days in 1853 (figs. 8.14.4, 8.14.8). That in 1822 included at least three felt shocks and rumbling on 9 March, 44 felt earthquakes on 10 March, 75 on 11 March, more than 100 on 12 March, and an uncounted(?) number on 13 March; the eruption began on 14 March (Omori, 1911; Imamura, 1930). Omori (1911) distinguished between earthquakes and "earth rumblings" ("jinaris"); premonitory rumbling sounds were more numerous than felt earthquakes, but at least some of the rumblings may have been small, unfelt earthquakes.

1910: Six days of precursory seismicity preceded explosions on 25 July 1910; the number of earthquakes/day increased as the outburst approached, with a few events on 21 July, more than 25 on 22 July, 110 on 23 July, and 351 on 24 July (Imamura, 1930; Abe, 1979) (figs. 8.14.4, 8.14.5, 8.14.7, 8.14.8). On the Sapporo seismograph record, earthquakes were most numerous between 1800 hr on 24 July and 0900 hr on 25 July, after which earthquake activity fell off sharply (about 13 hours before the eruption began at 2200 hr on 25 July) (see fig. 8.14.5) (Omori, 1911; Okada, 1982). The number of earthquakes felt at Nishi-Monbets, 8 km southeast of Usu, also decreased 16-20 hours before the eruption (Omori, 1911, p. 9). Strong shocks on the 24th and 25th caused local damage and were felt as far as 60 km to the northeast ("through" the caldera) and 140 km to the southwest; maximum M_s was estimated to be 5.1 (Abe, 1979) or 5.5 (Okada, 1982). The largest earthquakes occurred about 30 hours before the initial explosion. In addition to the magnitude 5.5 earthquake mentioned above, Okada (1982) estimated that 23 other events in July 1910 had magnitudes between 4.3 and 5.3. The first large precursory earthquake (M=4.7) was a high-frequency event; large events just before the eruption (up to M=5.5) were low-frequency events (Okada, 1982). Large lava blocks fell from O-usu dome on 23-24 July, probably owing to severe shaking by earthquakes.

Earthquakes in 1910 accompanied formation of cracks and faults, most trending E-W across the northwest foot of Usu. Explosions began on 25 July on the northwest side of Usu. As the locus of eruption migrated eastward across the north foot of Usu, 15 craterlets opened by 2 August, and 45 by the start of November (Ishikawa, 1964). Many small mud cones formed along the sea shore in the Abuta-Usu area, several kilometers southwest of the craterlets. This phenomenon is common at times of strong earthquake shaking in areas with thick, water-saturated soil layers. Most explosive activity stopped by the beginning of August and was followed by remarkable growth of a "new mountain," the Meiji-shinzan cryptodome. Maximum uplift of the cryptodome was 160 m, reached by the end of November (Imamura, 1930; Minakami, 1950); the area of major, cryptodome-driven uplift was roughly 2 km x 2 km, including a 2-km-long stretch of the southern shore of Lake Toya. The slope of land near that southern shore was formerly 5°, but it gradually rose to 30° near the top of the elevated area and to 22° on its flank. A home on the southern shore had been tilted 12° from vertical by 4 September. The uplifted land near Nishi-kohan began to subside on 18 September 1910 (Oinouye, 1917, citing Omori, 1911, 1913). The top of the "new mountain" (cryptodome) had subsided 36 m by April 1911 and 43 m by July 1911 (Oinouye, 1917).

A 30-cm rise in the level of Lake Toya along its northern shore was noted by 28 July 1910 (Oinouye, 1917); on 6 August, Omori discovered a lowering of lake level along its southern shore near Nishi-kohan (fig. 8.14.6). These two observations suggest that uplift of Usu caused at least 30 microradians of northward tilt of Toya Caldera. This is presumed to have begun about the time of the first explosion on 25 July (Omori, 1920). As measured by leveling, uplift was centered around Usu Volcano, extending west and

TOYA (continued)

HISTORICAL ACTIVITY (continued)

south to the coast of Volcano Bay as well as north to the shore of Lake Toya (Omori, 1913); Abuta-Tokotan (6 km west-southwest of Usu, on Volcano Bay) was uplifted 34 cm between leveling surveys in 1905 and 1911. Releveling in 1912 and 1919 along the western and northwestern sides of Lake Toya showed slight subsidence from previous elevations. In addition to the uplift localized around Usu, the floor of Lake Toya might have been uplifted by as much as 8 m in those parts north and northeast of Nakano-shima Island, and downdropped as much as 1 m northwest of the island (compare figs. 8.14.2 and 8.14.3). These estimates are based on a relatively small number of depth soundings in 1904 and 1917 and are therefore less certain than the land-based measurements. However, they are corroborated by temporary changes in the level of the lake itself (Omori, 1920). The level of the lake at all recording stations and in the Sobetsu River that drains the lake rose by about 60 cm from late July through mid-August 1910, and then, by draining through the narrow Sobetsu outlet, returned to its former level by November 1910 (Omori, 1920). The magnitude of this change could be explained if a 10 km^2 area of the lake bottom rose 45 cm, fully consistent with the observed soundings (Omori, 1920).

Omori (1911, p. 6) speculated that the 1910 eruption of Usu might in some way have been related to (triggered by?) a tectonic earthquake that occurred off the coast of Teshio (northwest Hokkaido) on 15 June 1910, 36 days before the start of the eruption. In the same discussion, Omori noted that eruptions of Tarumai in 1874 and 1908 and of Komagatake in 1856 had occurred shortly before or after similar large tectonic earthquakes.

No hot springs were known to have existed on the southwest side of Toya Caldera before 1910 (Ishikawa, 1964). In the latter part of 1910, when elevation of Nishi-kohan was at a maximum and the "new mountain" (cryptodome) was developing rapidly, the temperature of part of Lake Toya near the southern shore was 24 °C, several degrees above the normal temperature for that time of year. Apparently, new hot springs or fumaroles had formed on the lake floor. The first subaerial Toya hot spring developed just after the 1910 activity; its temperature gradually increased from 42 °C in 1910 to 44 °C in the winter of 1922. A hotel was built in 1919--the forerunner of today's popular thermal spa (note fig. 8.14.49). A similar posteruption temperature increase occurred on the east side of Usu, at Fukaba, where the temperature of a spring rose from cool in 1912 to more than 40 °C in 1914. Another hotel was built, but the temperature decreased gradually thereafter and the hotel was closed in 1917. Water levels in hot springs now mimic that of Lake Toya, rising and falling in accord with hydroelectric power generation, so it seems likely that the water of these springs is lake water (Ishikawa, 1964).

1932: No eruption occurred, but on 19 May an elliptical, 60-m-wide area off the north coast of Lake Toya subsided by as much as 11 m. Lake temperatures at the end of May were 11.3 °C within the area of subsidence, in contrast to 5.5-7.3 °C elsewhere in the lake and a 5.0 °C average lake temperature for May (Tanakadate, 1934).

1943-45: Earthquakes began on 28 December 1943, increased in severity and frequency to nearly 200 felt earthquakes/day on 29-31 December 1943, then declined sharply to only 10/day by 6 January 1944 (figs. 8.14.7, 8.14.8) (Minakami and others, 1951). The strongest of these earthquakes, on 5 January 1944, was felt more than 20 km from the volcano. Depths of these early earthquakes were approximately 10 km, and one might have been as deep as 20 km (Minakami and others, 1951; Kizawa, 1960). After 6 January, epicenters

TOYA (continued)

HISTORICAL ACTIVITY (continued)

shifted from the northwest side of Usu to its east side, a distance of 6 km. Depths of earthquakes east of Usu were initially 6-9 km and later rose to near the surface. By mid-1944, over 100 earthquakes/day were felt on the east side, increasing to over 250 earthquakes/day on 22 June (the day before explosions began).

Dramatic deformation occurred as the Showa-shinzan ("Roof Mountain") cryptodome began to grow in February or March 1944; by October, ground had been pushed laterally up to 60 m and vertically more than 100 m (Mimatsu, 1962) (figs. 8.14.11-8.14.15). The Sobetsu River was dammed by upheaval. Ultimately, over 200 m of uplift occurred, at rates of over 30 cm/day, with corresponding tilts of more than 480 microradians/day. Tilt was so rapid that it could be measured by wooden staffs along the edges of 10 m x 10 m water-filled rice paddies (Omote, 1950). Deformation was largely restricted to the Usu complex; leveling showed only 1.5 cm of uplift along the west caldera margin and none along the north and east sides of the caldera. From April to July 1944, measurements of staffs on the southwest, southeast, and north-northwest sides of Lake Toya and on the central island indicated 16.1 mm of uplift of Takainoue (southeast side) relative to Muko-toya (north-northwest side) (tilt of 20 microradians, down to N13°E) (Omote, 1950).

Explosions began on 23 June 1944, about 6 months after the start of rapid uplift and cryptodome emplacement, and continued through the end of October 1944. The cryptodome finally broke the surface in early November 1944 and continued to rise until September 1945. For comparison, explosions in 1910 and 1977 occurred only at the start of cryptodome growth. The initial explosions of 1944 were phreatic, in which ground water was being heated by magma of the cryptodome. The silica content of ash increased from 57 percent to 69 percent in late-stage phreatomagmatic explosions, as ash was derived partly from the new magma (Katsui, 1981). As the cryptodome was emplaced, numerous fissures formed at the surface and the water table was greatly disturbed; some wells dried up entirely, while others became artesian wells (Minakami and others, 1951).

The top of Showa-shinzan subsided about 25 m from 1945-68 and an additional 0.5 m from 1968-80, presumably owing to cooling (Katsui and others, 1981) (fig. 8.14.16).

1977-82: Seismicity increased slightly from 1972-77 and more in June 1977, but the level of seismicity was still within background limits and therefore was not regarded as an eruption precursor. On 6 August, seismicity increased sharply, followed 32 hours later by an explosive eruption of dacitic pumice (figs. 8.14.17-8.14.19, 8.14.21-8.14.27). In the last 9 hours before eruption, seismicity showed an exponential increase, and 4-7 hours before eruption, low-frequency earthquakes increased relative to high-frequency earthquakes. Initial hypocenters were 3.5 km below Ko-usu and O-usu; later they rose 2 km in 30 hours, most of that distance in the last 12 hours (Yokoyama, 1985). A few earthquakes occurred less than 1 km below Showa-shinzan during the same precursory period (Suzuki and others, 1980). At 0850 hr on 7 August, a farmer found a normal fault with a displacement of 40 cm across a road in the summit crater--a fault that had not existed at 1600 hr on the previous afternoon (Yokoyama, 1985) (fig. 8.14.32). A series of magmatic explosions began at 0912 hr on 7 August and continued intermittently until 14 August. The peak of seismicity occurred on the second day of the eruption, with more than 400 events of $\underline{M}$=2.5 or greater in one day. The seismic $\underline{b}$-value was 2.0 before the explosions, 1.5 immediately after the explosions, 2.0 three days later, and then gradually declined to 1.0 during

TOYA (continued)

HISTORICAL ACTIVITY (continued)

the next month. After the initial explosions, focal mechanisms were dip-slip, related to emplacement of the Usu-shinzan cryptodome. Maximum depth was 2 km, $\underline{M}_{max}$ was 4.3, and earthquakes were felt to distances of 8 km.

Phreatic explosions occurred on 16 November 1977, 13 January 1978, 25 February - 11 March 1978, and 24 April 1978; stronger phreatic explosions and some phreatomagmatic to magmatic ones occurred from 25 May to 27 October 1978 (Niida and others, 1980). Cryptodome growth (and associated earthquakes) continued until March 1982 (Yokoyama, 1985). A small, late earthquake swarm on 17 September 1982 included the only possible deep events of the entire 1977-82 unrest (Okada and others, 1984). Okada and others (1981), Takeo (1983), and Yokoyama (1986) distinguished several "families" of earthquakes, each with a distinct source mechanism. One family of the largest earthquakes ($\underline{M}$ >3.8) represented reverse faulting on the north-northwest side of a U-shaped fault block that was being pushed northward (family KB; fig. 8.14.35). Another family of large earthquakes (OU; fig. 8.14.35) represented normal faulting along the south-southeast side of the same fault block, and other families of smaller earthquakes occurred throughout the activity (1977-82).

As in 1910, deformation was fault bounded and localized on Usu proper (Katsui and others, 1985) (figs. 8.14.28-8.14.40); in fact, much of the damage to buildings in 1977-82 was along faults that had earlier been active in 1910. Uplift began at the same time as a sharp increase in seismicity; the maximum uplift rate of over 90 cm/day was observed shortly thereafter. Total uplift (relative to bench mark 1053 on the southeast shore of Lake Toya) was about 180 m in 5 years, bounded by a U-shaped, 2-km-long fault (fig. 8.14.35). A baseline between the north foot of the volcano and its north crater rim shortened by about 180 meters, and a ropeway/cableway down the east side of Usu shortened by approximately 18 m (Watanabe, 1983) (figs. 8.14.17, 8.14.28). Slackening of the cableway was an unusual but effective indicator of inflation, especially in the early stages of this eruption. After the rapid deformation in early August, deformation rates decayed exponentially for about a year, with only one small interruption (Yokoyama and Seino, 1979). In detail, contraction along a line HK-NR (north foot to north crater rim of Usu) occurred in coseismic steps superimposed on slower, continuous contraction (Okada, 1983) (figs. 8.14.34, 8.14.35). Both the 1910 and the 1977-82 episodes of cryptodome growth involved thrusting northward from Usu, a fact that Harada (1981) interprets as northward wedging of the Usu edifice as magma intrudes up the southern wall of the Toya Caldera (figs. 8.14.25, 8.14.26). No significant deformation was detected around the east, north, or west sides of the caldera, on Naka-jima (island, center of caldera), or on the south and southwest parts of the Usu edifice itself during 1977-82.

The temperature of a hot spring (S-4), located on the north flank of Usu near the 1910 explosion craters increased 10-20 °C during slightly more than a year before the 1977 eruption, and other hot springs increased in temperature just before the 1977 eruption (Abiko, 1984) (figs. 8.14.41, 8.14.42). Notsu and others (1980) described 0.2-1.0 °C transient increases in the temperature of a shallow well in Sohbetsu-Onsen immediately following large earthquakes caused by cryptodome emplacement; Watanabe (1983) described related, coseismic changes in well levels. Yamaguchi and Horitsu (1984) note a "rapid" temperature increase of several degrees in many Toya hot springs during the phreatic eruptions of 1978; by 1981, most had cooled gradually to pre-1977 temperatures.

TOYA (continued)

HISTORICAL ACTIVITY (continued)

Systematic changes in the composition of hot springs have been noted during activity at Usu (figs. 8.14.43-8.14.46). SO_2 emission rates in June and September 1978 and in November 1982 were progressively less each time, following the decline in seismic activity and ground deformation (Ohta and others, 1984) (fig. 8.14.47). The radon concentration in groundwater at a 7.4-m-deep well at the northeast foot of Usu showed a subtle decline through 1978 and early 1979 as seismicity and deformation also declined (Notsu and others, 1983) (fig. 8.14.48).

COMMENTS

Usu's moderately large explosive eruptions (>0.1 km^3 of dacitic magma) have begun after remarkably short precursory activity.

Historical deformation and seismicity have been localized around Usu, especially around its cryptodomes. Minor seismicity has been reported at Nakatoya along the northeast rim of Toya Caldera ($\underline{M}_{max}$=3.1; Okada and others, 1984); no other seismicity is recognized in the region except at Usu. Far-field deformation within Toya Caldera is roughly that which would be expected even if the caldera were not present.

Cryptodome growth at Usu creates new hot springs and with them, new opportunities for hot-springs resorts (fig. 8.14.49). Caldera unrest can foster tourism, but it can also bring large populations into close range of the volcano.

REFERENCES

Abe, K., 1979, Instrumental magnitude of Japanese earthquakes, 1901-1928: J. Seismol. Soc. Japan, v. 32, p. 341-352.

Abiko, T., 1984, Temporal variation in chemical compositions of hot springs at Toyako- and Sohbetsu-Onsen after the 1977-1978 eruption of Usu Volcano, in Ossaka, J., ed., Fundamental research for predicting volcanic eruptions through observations of volcanic gases: Special Rep., Grant-in Aid Nat. Disaster Sci., no. A-59-4, 141 p.

Circum-Pacific Map Project, Northwest Quadrant Panel, 1981, Plate tectonic map of the Circum-Pacific region, northwest quadrant: Tulsa, Am. Assoc. Pet. Geol., 1 sheet, scale 1:10,000,000.

Geographical Survey Institute, 1978, Crustal deformation around Usu Volcano: Rep. Coord. Comm. for Prediction of Volcanic Eruptions, no. 12, p. 19-22.

Harada, T., 1981, Stress field in Usu Volcano deduced from focal mechanism solutions: Bull. Volcanol. Soc. Japan, v. 26, p. 93-110.

Iizuka, S., 1963, Temperature measurements of Showa-shinzan fumaroles, (1954-1963): Geophys. Bull. Hokkaido Univ., v. 11, p. 53-58.

Ikeda, T., and Katsui, Y., 1986, Formation of Toya caldera and its pyroclastic deposits [abs.]: Bull. Volcanol. Soc. Japan, v. 31, p. 269.

Imamura, A., 1930, Topographical changes accompanying earthquakes or volcanic eruptions: Publ. Earthq. Invest. Comm. in Foreign Languages, no. 25, p. 1-143.

TOYA (continued)

REFERENCES (continued)

Inouye, W., 1948, On the earthquakes and topographic changes accompanied by the Usu eruption of 1944: Quart. Jour. Seismol., v. 14, p. 9-24.

Inouye, W., and Homma, N., 1948, On the Usu eruption of 1944: Quart. Jour. Seismol., v. 14, p. 25-32.

Ishikawa, T., 1964, Geological structure of Volcano Usu, Japan, considered from its 1910 and 1943-1945 activities and the genesis of Toya Hot Springs: Stromboli, Rivista di Vulcanologia, no. 9, new series, p. 13-35.

Kadomura, H., Okada, Hm, Imagawa, T., Moriya, I., and Yamamoto, H., 1983, Erosion and mass movements on Mt. Usu accelerated by crustal deformation that accompanied its 1977-1982 volcanism: Jour. Nat. Disaster Sci., v. 5, no. 2, p. 33-62.

Katsui, Y., Komuro, H., and Uda, T., 1985, Development of faults and growth of Usu-shinzan cryptodome in 1977-1982 at Usu Volcano, North Japan: Jour. Fac. Sci., Hokkaido Univ., ser. 4, v. 21, no. 3, p. 339-362.

Katsui, Y., Niida, K.,, Onuma, K., Oba, Y., Suzuki, T., and Kondo, Y., 1978a, Geological study of the 1977-1978 volcanic activity of Usu, Hokkaido, in Usu Eruption and Its Impact on Environment, Hokkaido Univ., December 1978, p. 3-12.

Katsui, Y., Oba, Y., Suzuki, T., Kondo, Y., Watanabe, T., Niida, K., Uda, T., Hagiwara, S., Nagao, T., Nishikawa, J., Yamamoto, M., Ikeda, Y., Katagawa, H., Tsuchiya, N., Shirahase, M., Nemoto, S., Yokoyama, S., Soya, T., Fujita, T., Inaba, K., and Koide, K., 1978b, Preliminary report of the 1977 eruption of Usu volcano: Jour. Fac. Sci. Hokkaido Univ., Ser. 4, v. 18, p. 385-408.

Katsui, Y., Yokoyama, I., Watanabe, H., and Murozumi, M., 1981, Usu Volcano, in Katsui, Y., ed., Field excursion guide to Usu and Tarumai Volcanoes and Noboribetsu Spa: Tokyo, Volcanological Society of Japan, 64 p.

Kizawa, T., 1957, A study of earthquakes in relation to volcanic activity (I): Earthquakes that occurred during the formation period of a new dome "Showa-Shinzan" and their forerunners near Volcano Usu: Pap. Meteorol. Geophys., v. 8, no. 2, p. 150-169.

----- 1959, A study of earthquakes in relation to volcanic activity (II): A study on the characteristics of the seismic waves in the earthquake swarm that occurred during the formation of Showa-Shinzan near Volcano Usu and the mechanism of volcanic activity: Pap. Meteorol. Geophys., v. 9, p. 204-239.

----- 1960, A study of earthquakes in relation to volcanic activity (III): Earthquakes during the period of activity of Volcano Usu (1943-1945), two new phases of the earthquakes, and earthquakes of the crypto-volcanic activity at Ito: Pap. Meteorol. Geophys., v. 11, no. 1, p. 30-96.

Kuno, H., 1962, Catalogue of the active volcanoes of the world including solfatara fields, Pt. XI, Japan, Taiwan and Marianas: Rome, IAVCEI, 332 p.

Machida, H., and Arai, F., 1981, Late Quaternary large eruptions recorded in distal areas around Japan: IAVCEI Symp. on Arc Volcanism, Tokyo, p. 214-215.

Machida, H., Okamura, K., Arai, F., and Miyauchi, T., 1985, Late Pleistocene widespread tephras in Hokkaido [abs.]: Bull. Volcanol. Soc. Japan, v. 30, p. 101-102.

Mimatsu, M., 1962, A diary on the growth of Showa-Shinzan: Sobetsu, 202 p. (sketch reproduced by many authors).

Minakami, T., 1950, Report on the volcanic activities in Japan during 1939-1947: Bull. Volcanol., v. 10, p. 45-58.

Minakami, T., Ishikawa, T., and Yagi, K., 1951, The 1944 eruption of Volcano Usu, Hokkaido, Japan: Bull. Volcanol., v. 11, p. 45-157.

Mogi, K., 1958, Relations between the eruptions of various volcanoes and the deformations of the ground surfaces around them: Bull. Earthq. Res. Inst., v. 36, p. 99-134.

TOYA (continued)

REFERENCES (continued)

Nakamura, K., 1975, Volcano structure and possible mechanical correlation between volcanic eruptions and earthquakes: Bull. Volcanol. Soc. Japan, v. 20, p. 229-240.

Niida, K., Katsui, Y., Suzuki, T., and Kondo, Y., 1980, The 1977-1978 eruption of Usu Volcano: Jour. Fac. Sci., Hokkaido Univ., ser. IV, v. 19, no. 3, p. 357-394.

Nishida, Y., 1984, Subsurface structure of Toya Caldera, Japan as revealed by detailed magnetic survey: Jour. Phys. Earth, v. 32, p. 423-434.

Nishida, Y., and Miyajima, E., 1984a, Magnetic survey on Usu Volcano and its parasitic cones by means of a proton precession magnetometer, in Yokoyama, I., ed., Report of Joint Geophysical and Geochemical Observations, Usu Volcano in 1982 and Tarumai Volcano in 1983: Usu, Hokkaido, Usu Volcano Observatory, p. 63-70.

----- 1984b, Subsurface structure of Usu volcano, Japan as revealed by detailed magnetic survey: Jour. Volcanol. Geotherm. Res., v. 22, p. 271-285.

Notsu, K., Abiko, T., and Wakita, H., 1980, Coseismic temperature changes of well water related to volcanic activities of Usu Volcano: Jour. Phys. Earth, v. 28, p. 617-624.

----- 1983, Radon concentration changes in groundwater related to the volcanic activities of Usu Volcano: Bull. Volcanol. Soc. Japan, v. 28, no. 3, p. 305-308.

Oba, Y., 1964, Petrological studies of the Usu volcano, with special reference to somma lavas: Jour. Japanese Assoc. Mineral. Petrol. Econ. Geol., v. 51, p. 53-66, 88-97.

----- 1966, Geology and petrology of Usu volcano, Hokkaido, Japan: Jour. Fac. Sci., Hokkaido Univ., ser. 4, v. 13, p. 185-236.

Ohta, K., Matsuwo, N., Hirokawa, M., and Sugoshi, T., 1984, Emission rates of sulfur dioxide from Usu Volcano: Joint Geophysical and Geochemical Observations of Usu Volcano in 1982 and Tarumai Volcano in 1983: Usu, Hokkaido, Usu Volcano Observatory, p. 127-131.

Oinouye, Y., 1917, A few interesting phenomena on the eruption of Usu: Jour. Geol., v. 25, p. 258-288.

Okada, Hm., 1982, Earthquake swarm activity of Usu volcano in 1910: Geophys. Bull. Hokkaido Univ., v. 41, p. 53-63.

----- 1983, Comparative study of earthquake swarms associated with major volcanic activities, in Shimozuru, D., and Yokoyama, I., eds., Arc volcanism: physics and tectonics: Tokyo, Terra Scientific Publ. Co., p. 43-61.

Okada, Hm., Watanabe, H., Yamashita, H., Maekawa, T., Suzuki, A., and Yokoyama, I., 1984, Seismometrical observation of Usu Volcano in 1977-1983, in Yokoyama, I., ed., Joint Geophysical and Geochemical Observations of Usu Volcano in 1982 and Tarumai Volcano in 1983: Usu, Hokkaido, Usu Volcano Observatory, p. 3-13.

Okada, Hm., Watanabe, H., Yamashita, H., and Yokoyama, I., 1981, Seismological significance of the 1977-1978 eruptions and the magma intrusion process of Usu Volcano, Hokkaido: Jour. Volcanol. Geotherm. Res., v. 9, p. 311-334.

Okumura, K., and Sangawa, A., 1984, Age and distribution of Toya pyroclastic flow [abs.]: Bull. Volcanol. Soc. Japan, v. 29, p. 338.

Omori, F., 1911, The Usu-san eruption and earthquake and elevation phenomena: Bull. Imperial Earthq. Invest. Comm., v. 5, no. 1, p. 1-38.

----- 1913, The Usu-san eruption and the earthquake and elevation phenomena (II): Comparison of the bench mark heights in the base district before and after the eruption: Bull. Imperial Earthq. Invest. Comm., v. 5, no. 3, p. 101-107.

TOYA (continued)

REFERENCES (continued)

Omori, F., 1920, The Usu-san eruption and the earthquake and elevation phenomena (III): Results of precise levelings in 1905-1919, with accounts of the observations relating to the level change of the Lake of Toya: Bull. Imperial Earthq. Invest. Comm., v. 9, no. 2, p. 41-76.

Omote, S., 1950, Precise leveling at the eastern foot of Volcano Usu: Bull. Earthq. Res. Inst., v. 28, p. 133-142.

Ossaka, J., Hirabayashi, J., and Ozawa, T., 1984, Chemical composition of volcanic gases after the 1977 eruption of Usu volcano, in Ossaka, J., ed., Fundamental research for predicting volcanic eruption through observations of volcanic gases: Special Rep., Grant-in-Aid for Nat. Disaster Sci., Ministry Education, Tokyo, Tokyo Inst. Technol., no. A-59-4, 141 p.

Seino, M., 1983, Seismic activity accompanying the 1977-1978 eruption of Usu volcano, Japan: Pap. Meteorol. Geophys., v. 34, no. 3, p. 105-141.

Soya, T., Katsui, Y., Niida, K., and Sakai, K., 1981, Geological map of Usu Volcano: Geol. Surv. Japan, Geological map of volcanoes no. 2, scale 1:25,000.

Suzuki, S., and Kasahara, M., 1979, Seismic activity immediately before and in the early stages of the 1977 eruption of Usu Volcano, Hokkaido, Japan: Jour. Fac. Sci., Hokkaido Univ., ser. 7, v. 6, no. 1, p. 239-254.

Suzuki, S., and Okayama, M., 1978, Seismic activity immediately before the 1977 eruption of Usu Volcano, in Sakai, Y., ed., Investigation of volcanic disasters caused by the 1977 Usu eruption: Special Rep., Ministry Education, Sapporo, Hokkaido Univ., p. 1-9.

Suzuki, S., Yamashita, H., Watanabe, H., Okada, Hm., and Nishida, Y., 1980, Hypocenters of the premonitory earthquakes to the eruption of Usu volcano: Bull. Volcanol. Soc. Japan, v. 25, no. 3, p. 181-193.

Takeo, M., 1983, Source mechanisms of Usu Volcano, Japan, earthquakes and their tectonic implications: Phys. Planet. Earth. Interior, v. 32, p. 241-264.

Tanakadate, H., 1934, Volcanic activity in Japan and vicinity during the period between June 1931 and June 1934: Japanese Jour. Astron. Geophys., v. 12, p. 89-108.

Wano, K., and Okada, Hm., 1980, Peculiar occurrence of Usu earthquake swarm associated with recent doming activity: Jour. Seismol. Soc. Japan, v. 33, p. 215-226.

Watanabe, H., 1983, Changes in water level and their implications to the 1977-1978 activity of Usu Volcano, in Shimozuru, D., and Yokoyama, I., eds., Arc volcanism: physics and tectonics: Tokyo, Terra Scientific Publ. Co., p. 81-93.

----- 1984, Gradual bubble growth in dacite magma as a possible cause of the 1977-1978 long-lived activity of Usu volcano: Jour. Volcanol. Geotherm. Res., v. 20, p. 133-144.

Watanabe, H., Maekawa, T., Suzuki, A., Yamashita, H., Eto, T., and Nakamura, S., 1984a, Results of precise levellings around Usu Volcano in 1982, in Yokoyama, I., ed., Joint Geophysical and Geochemical Observations of Usu Volcano (September-December, 1982) and Tarumai Volcano (September-October, 1983), p. 25-31.

Watanabe, H., Yamashita, H., and Maekawa, T., 1984b, Electrical study of the 1977-1982 activity of Usu Volcano, in Yokoyama, I., ed., Joint Geophysical and Geochemical Observations of Usu Volcano (September-December, 1982) and Tarumai Volcano (September-October, 1983): Usu, Hokkaido, Usu Volcano Observatory, p. 71-80.

TOYA (continued)

REFERENCES (continued)

Yamaguchi, H., and Horitsu, T., 1984, The influence of the activity of Usu Volcano in 1977-1978 on the Toya Hot Springs, in Yokoyama, I., ed., Joint Geophysical and Geochemical Observations of Usu Volcano (September-December, 1982) and Tarumai Volcano (September-October, 1983): Usu, Hokkaido, Usu Volcano Observatory, p. 105-116.

Yokoyama, I., 1978, The 1977 eruption of Usu volcano with a special reference to prediction of volcanic activities: Bull. Volcanol. Soc. Japan, v. 23, p. 65-82.

----- 1985, Volcanic processes revealed by geophysical observations of the 1977-1982 activity of Usu Volcano, Japan: Jour. Geodyn., v. 3, p. 351-367.

----- 1986, Explosions, earthquake swarms, and deformation in 1977-1982 activity of Usu Volcano, Japan: Roy. Soc. New Zealand Bull. 24, p. 141-153.

Yokoyama, I. Katsui, Y., and Abiko, T., 1983, The 1979-1982 activities of Usu Volcano, in Report on volcanic activities and volcanological studies in Japan for the period from 1979 to 1982: 18th Gen. Assembly, Hamburg, p. 15-19.

Yokoyama, I., Katsui, T., Oba, Y., and Ehara, Y., 1973, Usu-zan, its volcanic geology, history of eruption, present state of activity and prevention of disaster: Sapporo, Committee for Prevention of Disasters of Hokkaido, 254 p.

Yokoyama, I., and Seino, M., 1979, Prediction of development in the 1977-78 activities of Usu Volcano with consideration for energy discharge: Jour. Fac. Sci., Hokkaido Univ., ser. 7, v. 6, no. 1, p. 187-200.

Yokoyama, I., Yamashita, H., Watanabe, H., and Okada, Hm., 1981, Geophysical characteristics of dacite volcanism in the 1977-1978 eruption of Usu Volcano: Jour. Volcanol. Geotherm. Res., v. 9, p. 335-358.

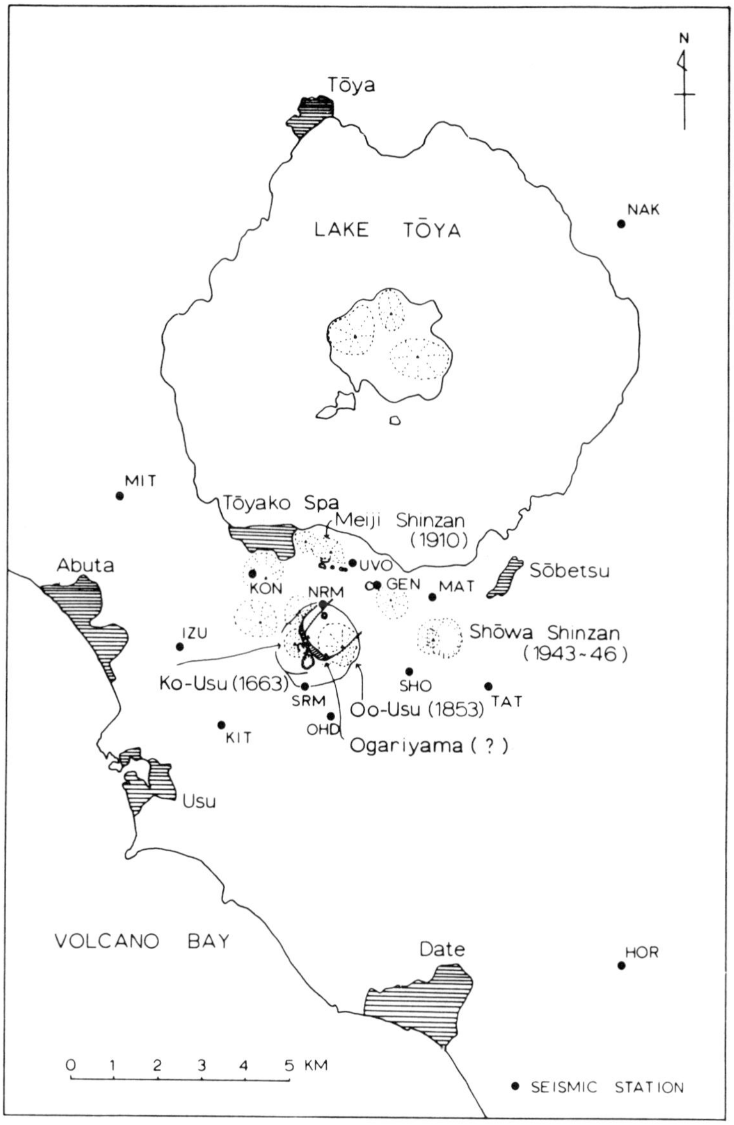

Figure 8.14.1. Location map of Toya Caldera and area around Lake Toya, from Okada and others (1981). Usu Volcano is located south of Lake Toya, in southwestern Hokkaido, Japan. Solid dots indicate seismometer network of Usu Volcano Observatory (UVO), to which all seismic signals are telemetered.

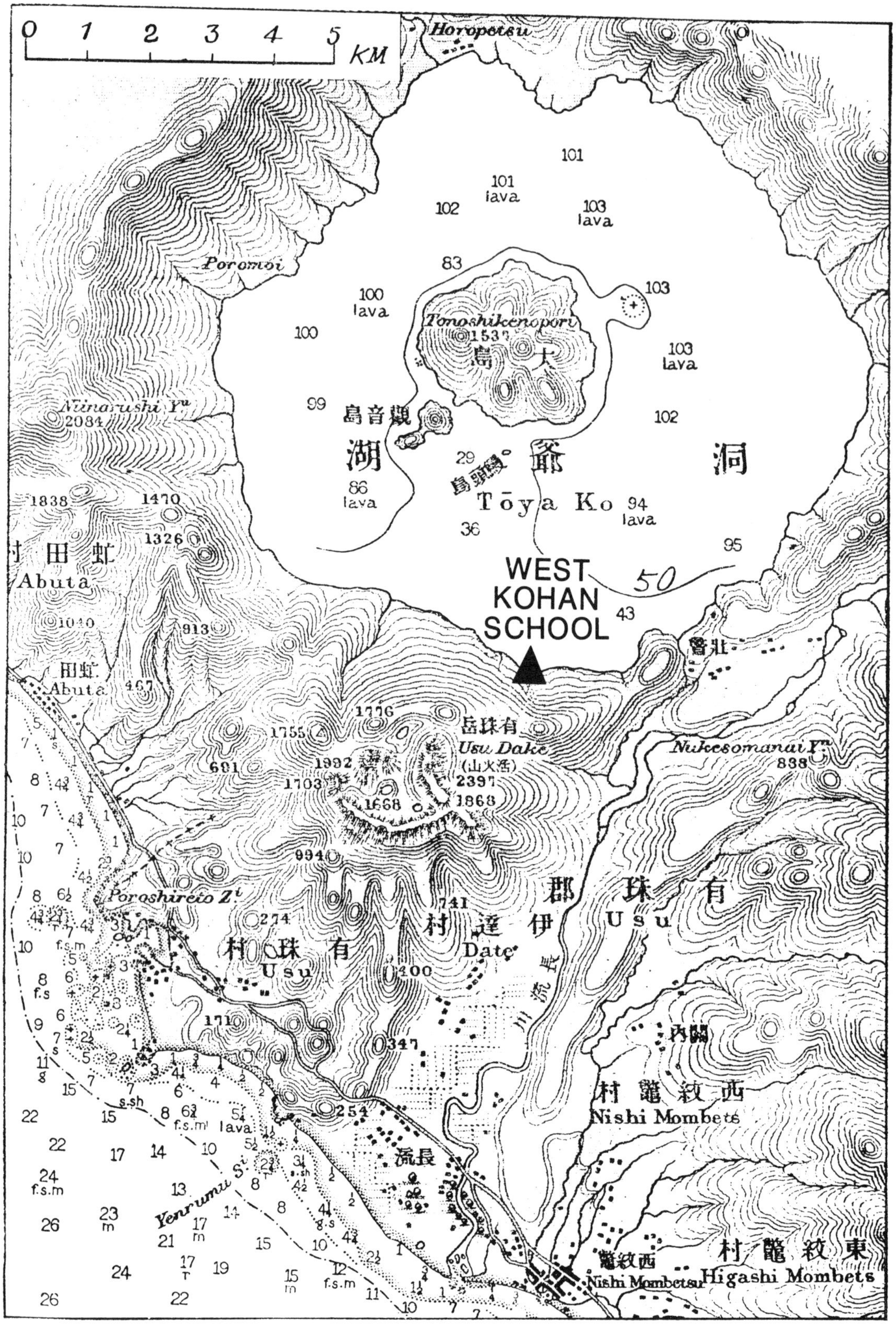

Figure 8.14.2. Topographic map of area around Lake Toya in 1904, from Omori (1920). Contours in feet above hot springs at Volcano Bay; soundings in fathoms. Triangle shows location of West Kohan School (see figure 8.14.6).

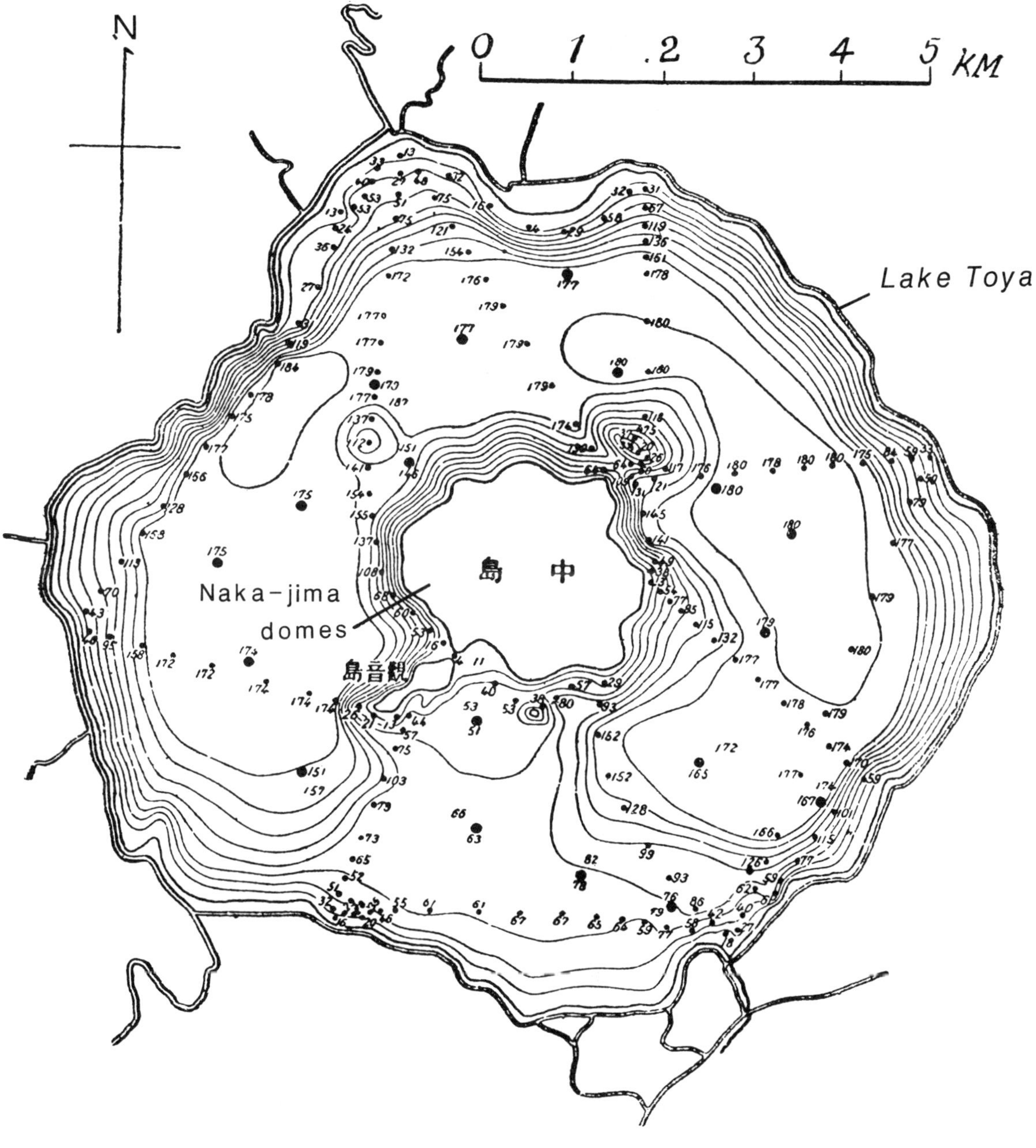

Figure 8.14.3. Bathymetry of Lake Toya in 1917, from Omori (1920). Soundings in meters; isobath interval 20 m.

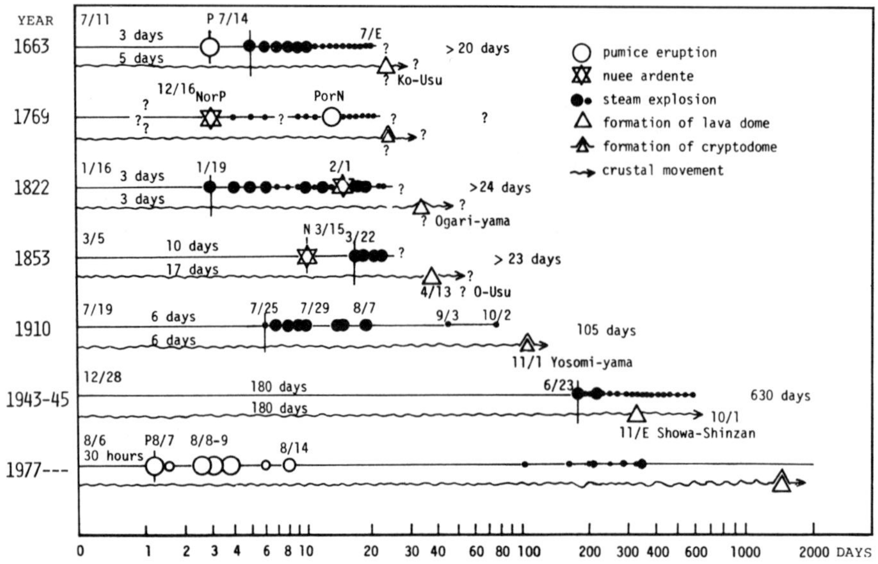

Figure 8.14.4. Schematic representation of activity at Usu Volcano since A.D. 1663, from Katsui and others (1978).

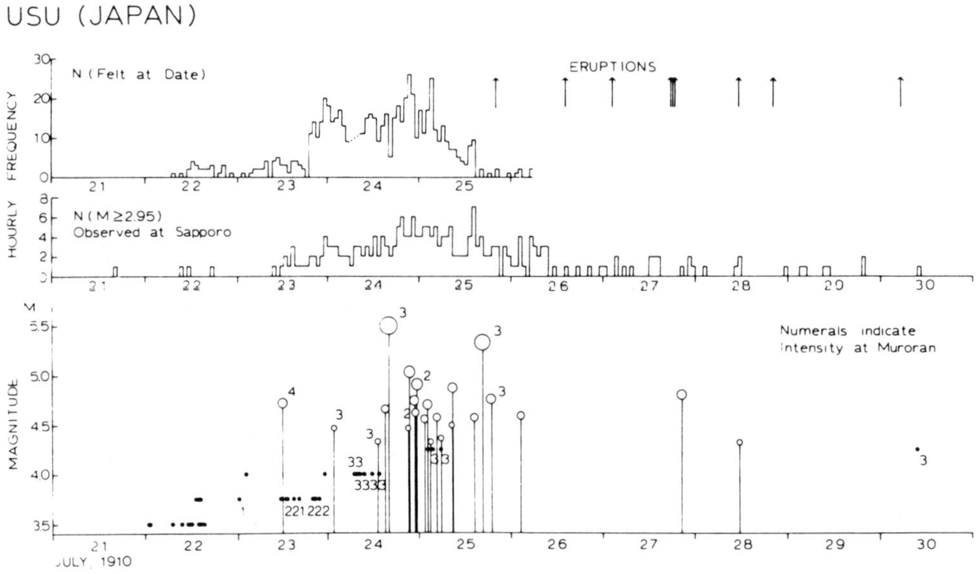

Figure 8.14.5. Time sequence of earthquake-swarm development at Usu Volcano in 1910, from Okada (1982). Note that three large earthquakes have particular significance: the first large (M 4.7) event initiated sudden increased seismicity, the largest (M 5.5) marked the start of most intense activity, and the second largest (M 5.3) occurred just before eruption. Number of felt earthquakes decreased drastically toward the M 5.3 earthquake and eruption, while large events (larger than M 2.95) continued at a moderate pace. Intensity data indicate that the first large event was a normal high-frequency earthquake (M 4.7, I=4), and that those in the later stages of buildup were low-frequency events (M 5, I=2-3).

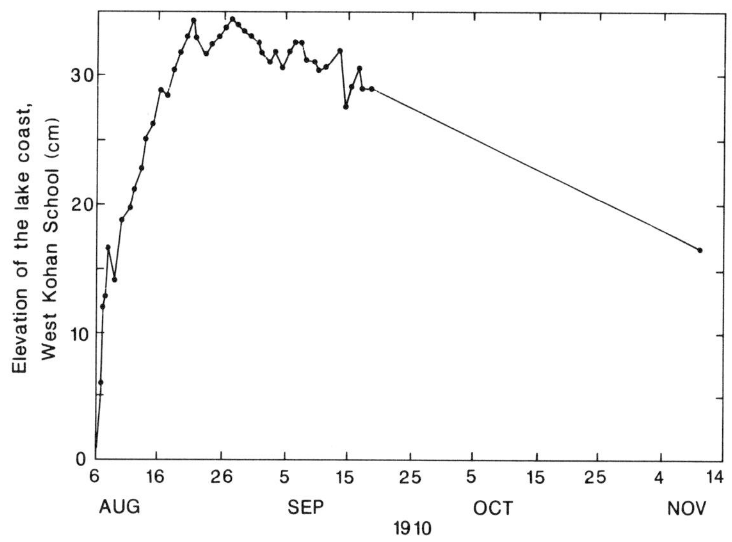

Figure 8.14.6. Elevation of shore of Lake Toya at West Kohan School, immediately northeast of Usu (see figure 8.14.2), during 1910 cryptodome growth, from Omori (1911).

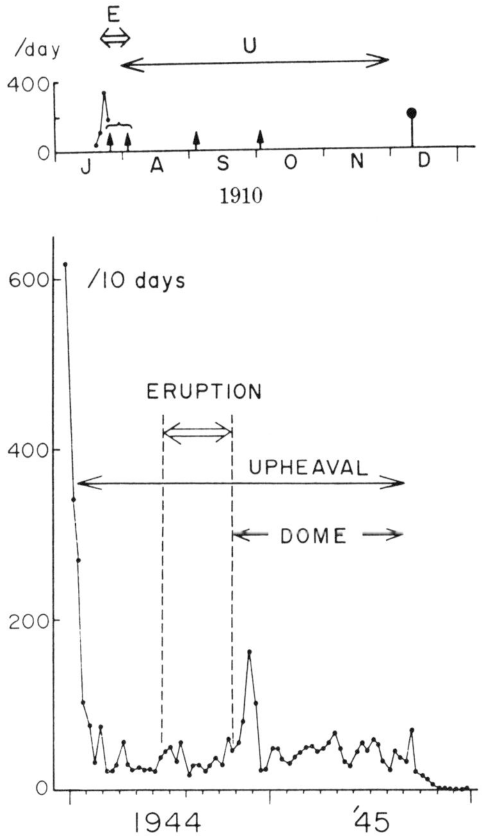

Figure 8.14.7. Earthquake counts and schematic diagrams of the 1910 and 1943-45 eruptions of Usu Volcano, from Yokoyama (1978).

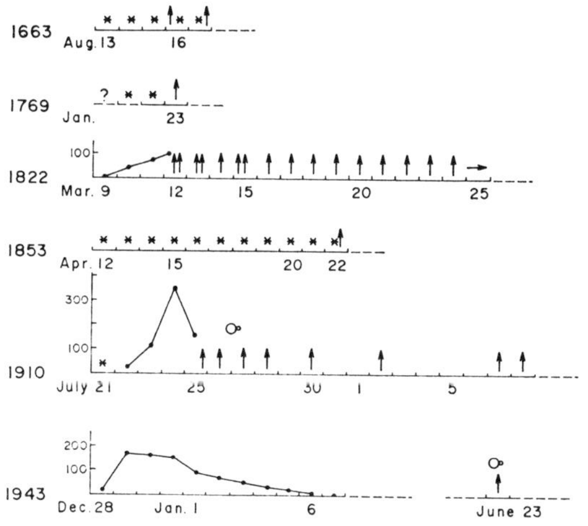

Figure 8.14.8. Schematic diagrams of historical activity at Usu Volcano, from Yokoyama (1978). Asterisks, days on which precursory earthquakes occurred. Vertical arrows, explosions. Double circles, flank eruptions.

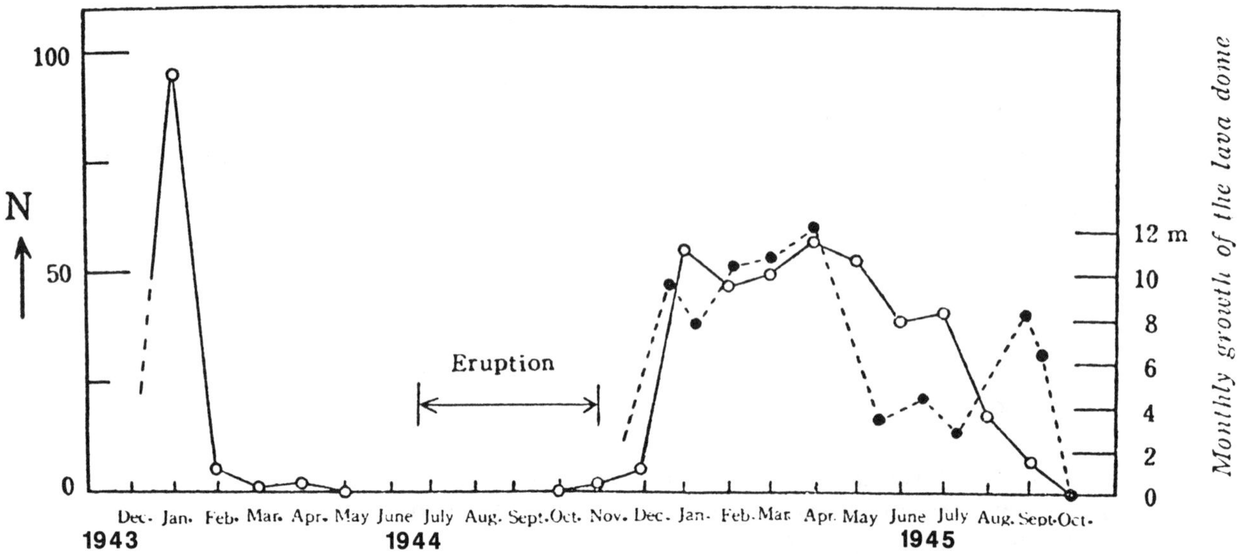

Figure 8.14.9. Relation between monthly frequency of earthquakes (open circles) recorded at Sapporo and monthly increment of growth of Showa-shinzan lava dome (filled circles), December 1943 to October 1945, from Kizawa (1957).

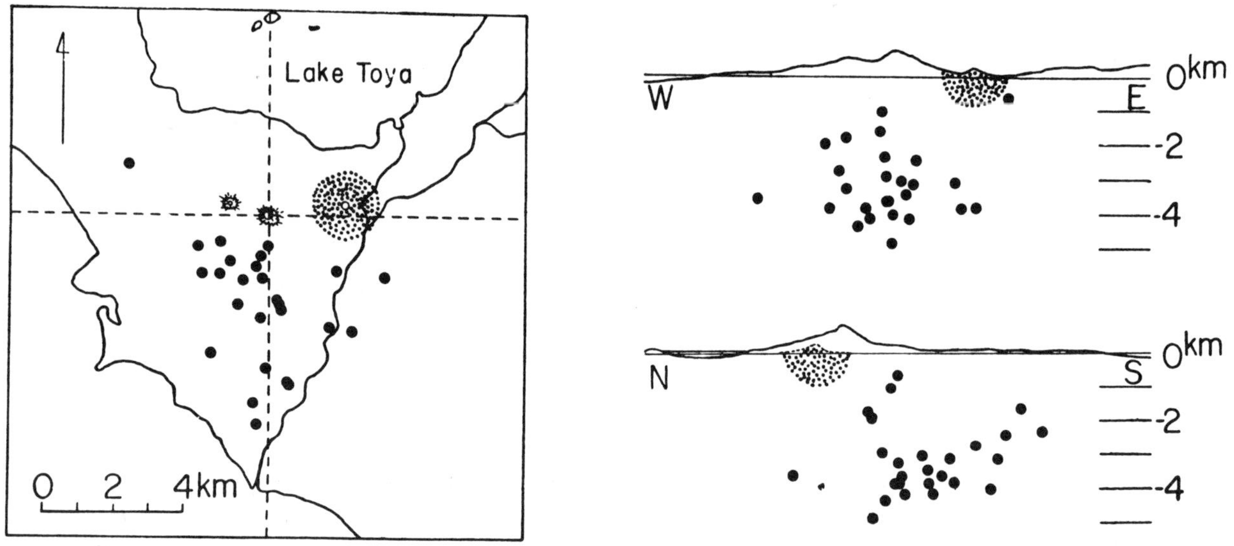

Figure 8.14.10. Hypocentral distribution of A-type and B-type earthquakes that took place during the 1944 activity of Usu Volcano, from Minakami and others (1951). Stippled pattern marks shallow seismicity associated with growth of Showa-shinzan cryptodome.

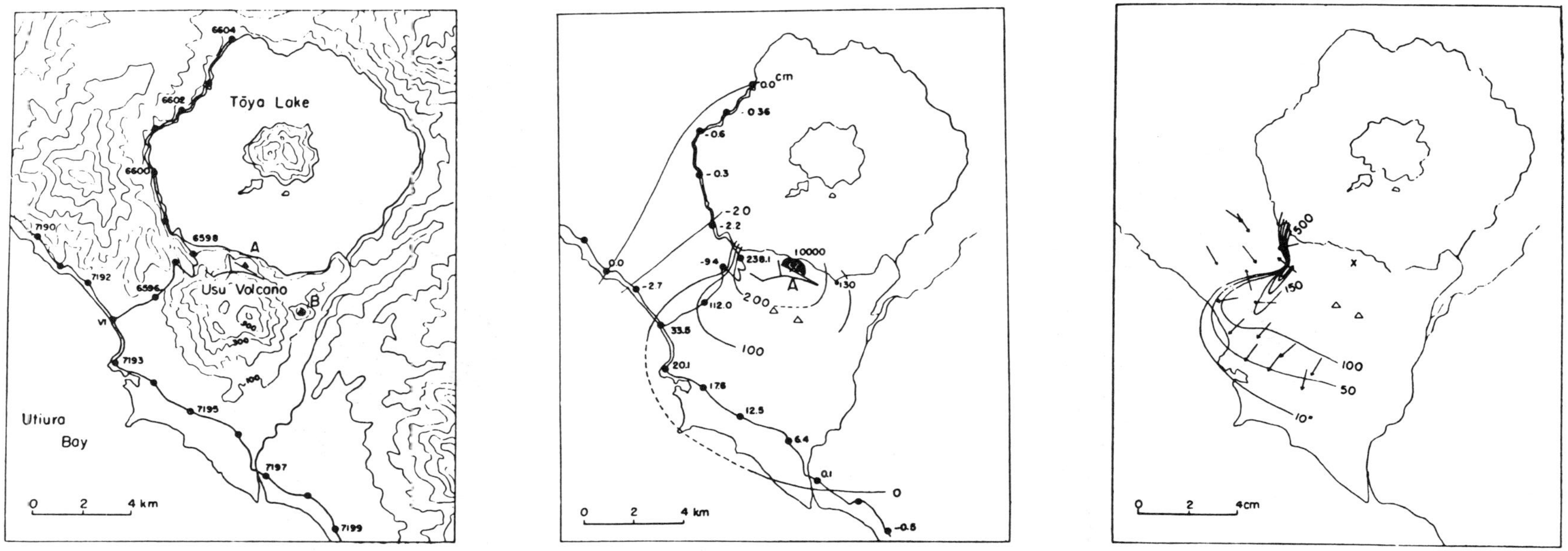

Figure 8.14.11. Topographic map (left) of Usu Volcano showing leveling route (marked with small numbers indicating bench marks). A, 1910 dome; B, 1943-45 dome. Vertical displacements (middle) of bench marks at Usu from 1905 to 1919 (including effects of 1910 eruption). A, 1910 dome. Tilting (right) of ground surface calculated from leveling results, 1905-19. Tilts in arcseconds (1 arcsecond is about 5 microradians). All from Mogi (1958).

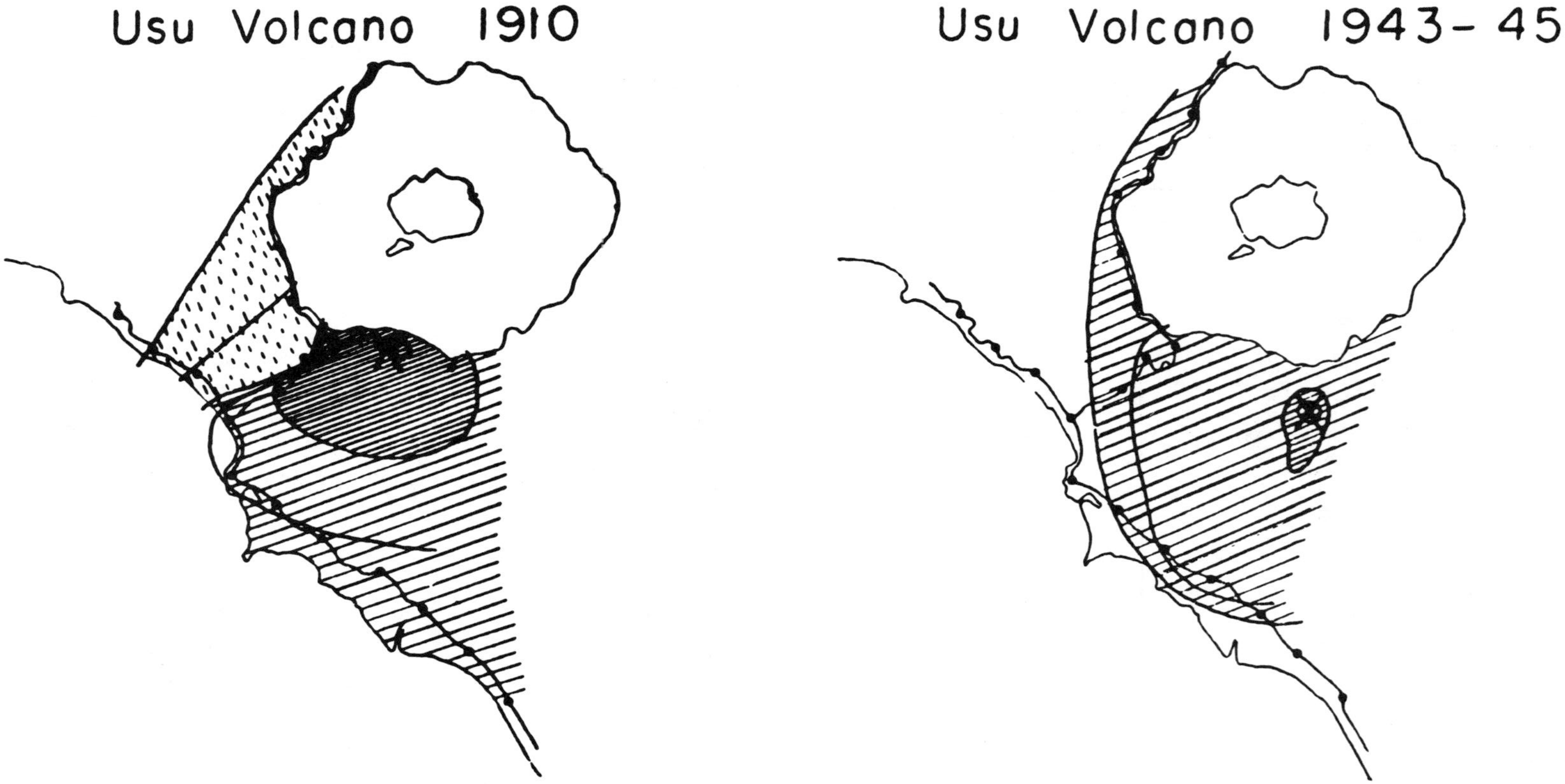

Figure 8.14.12. Areas uplifted during the 1910 (left) and 1943-45 (right) activity at Usu Volcano, from Mogi (1958). Denser patterns indicate more intense uplift.

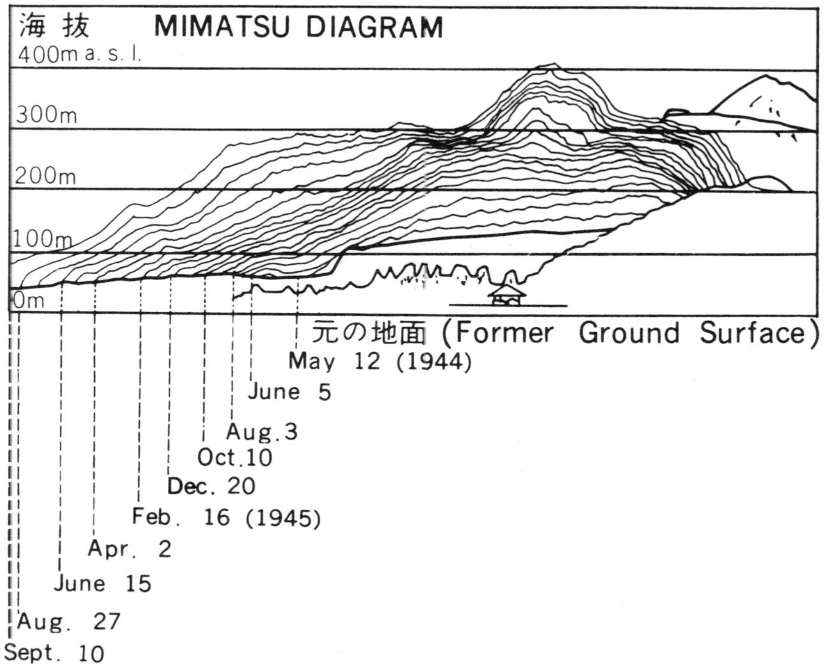

Figure 8.14.13. Mimatsu diagram showing growth of Showa-shinzan lava dome as viewed from 2.5 km east of the vent, after Mimatsu (1962).

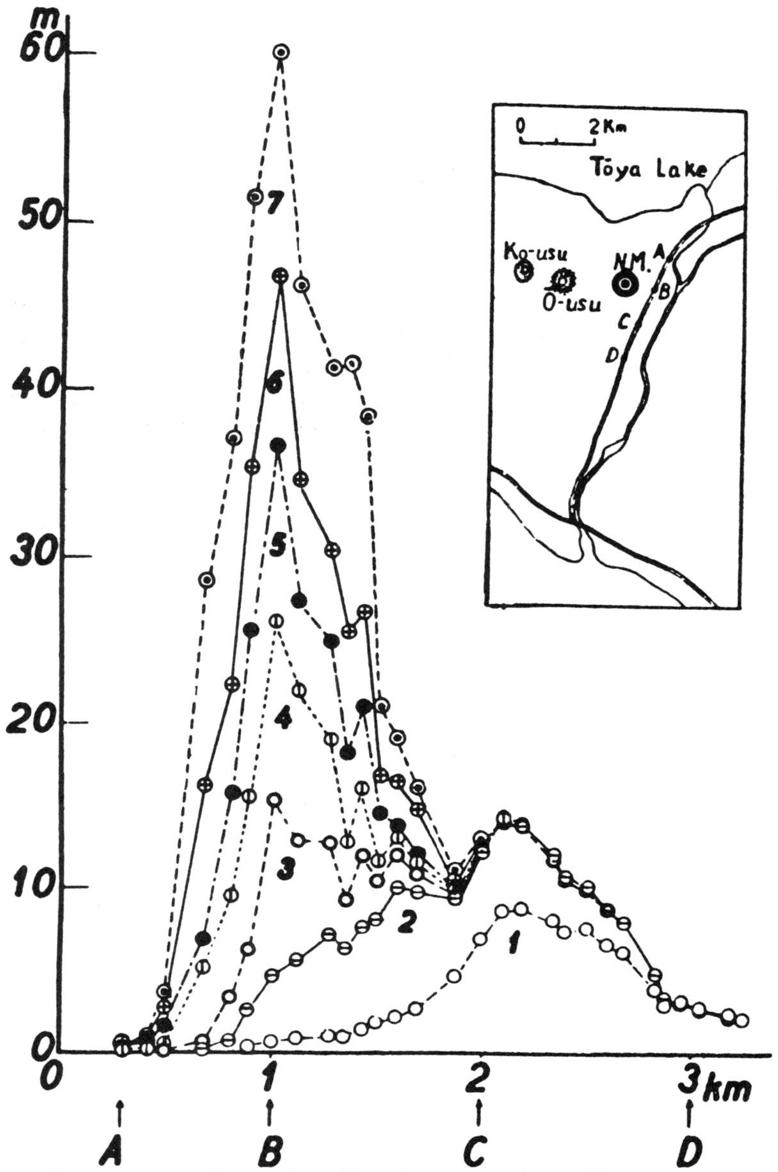

Figure 8.14.14. Dramatic uplift at the eastern foot of Usu Volcano during emplacement of Showa-shinzan cryptodome, from Minakami and others (1951).

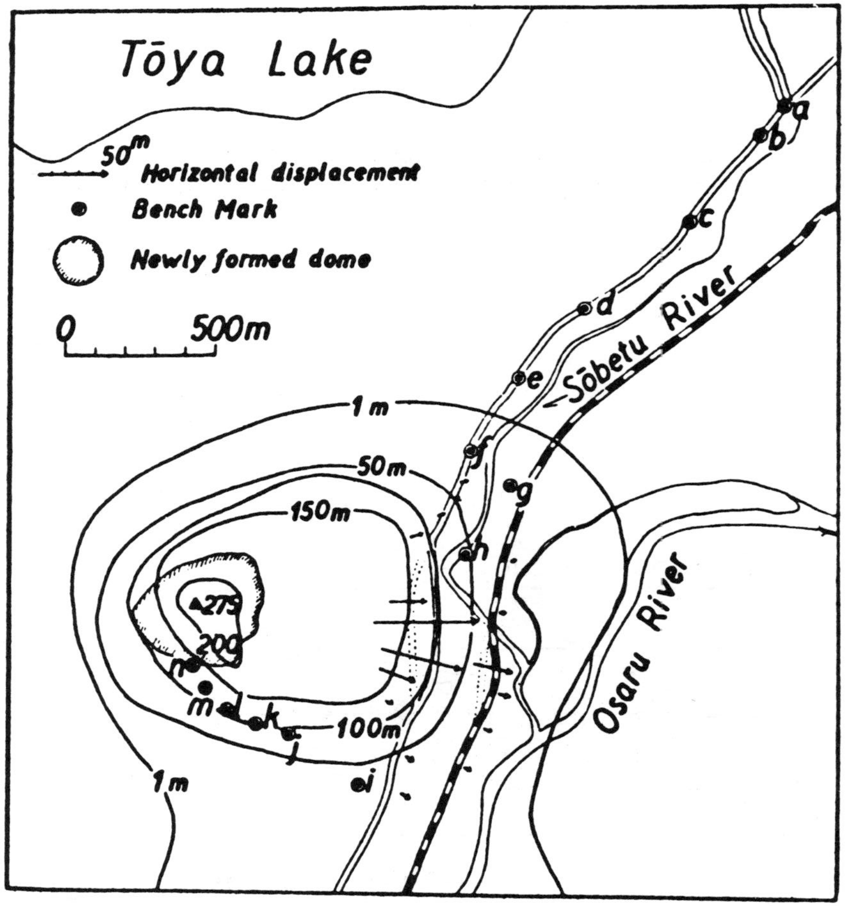

Figure 8.14.15. Uplift contours at Showa-shinzan lava dome, 1944-45, from Minakami and others (1951).

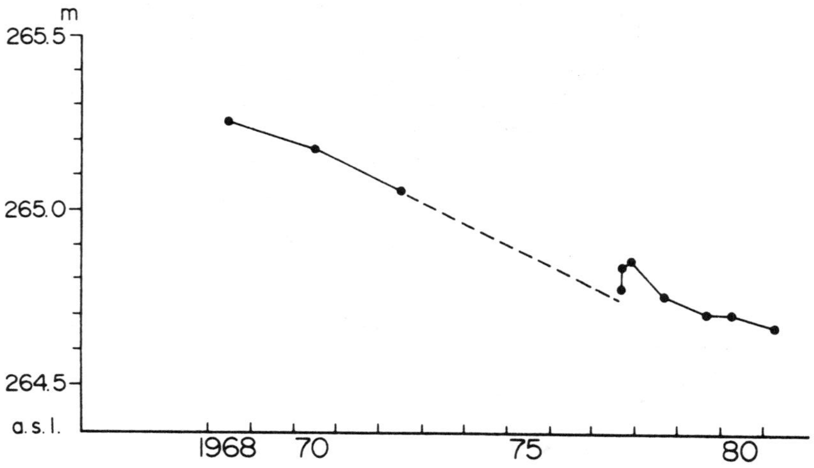

Figure 8.14.16. Subsidence at the top of Showa-shinzan dome, 1968-81 (and presumably continuing), from Katsui and others (1981). If subsidence began shortly after the dome stopped growing, the top subsided at an average rate of 43 cm/yr from 1945 to 1968, and at an average rate of 4 cm/yr from 1968 to 1981.

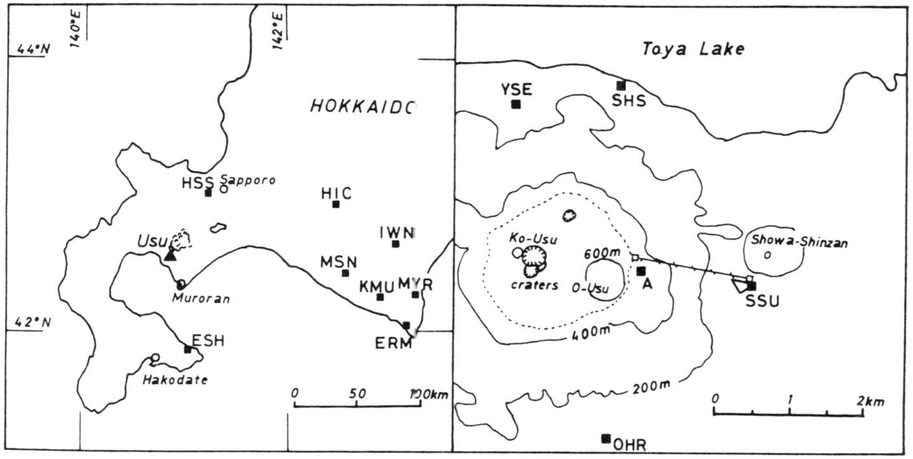

Figure 8.14.17. Locations of seismometer stations (HSS, HIC, and so forth) and a cableway (from near A to SSU) at Usu Volcano, from Suzuki and Okayama (1978).

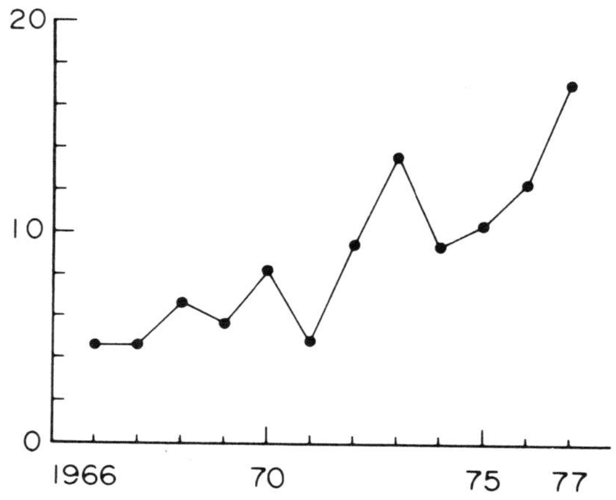

Figure 8.14.18. Annual means of monthly frequency of earthquakes originating at Usu Volcano before the 1977 eruption, from Yokoyama and others (1981).

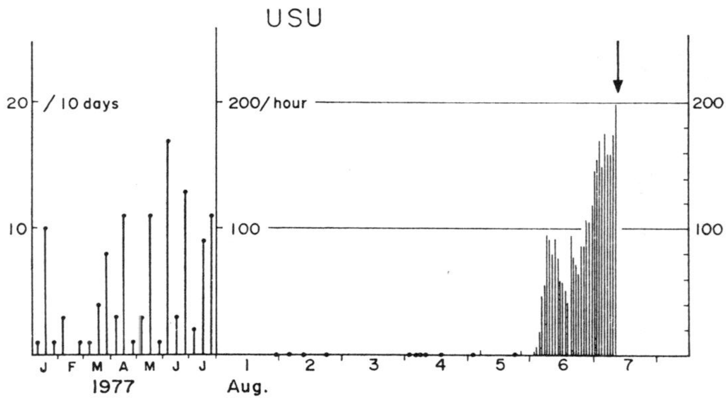

Figure 8.14.19. Seismic activity at Usu Volcano before the 1977 eruption (arrow), from Yokoyama (1978).

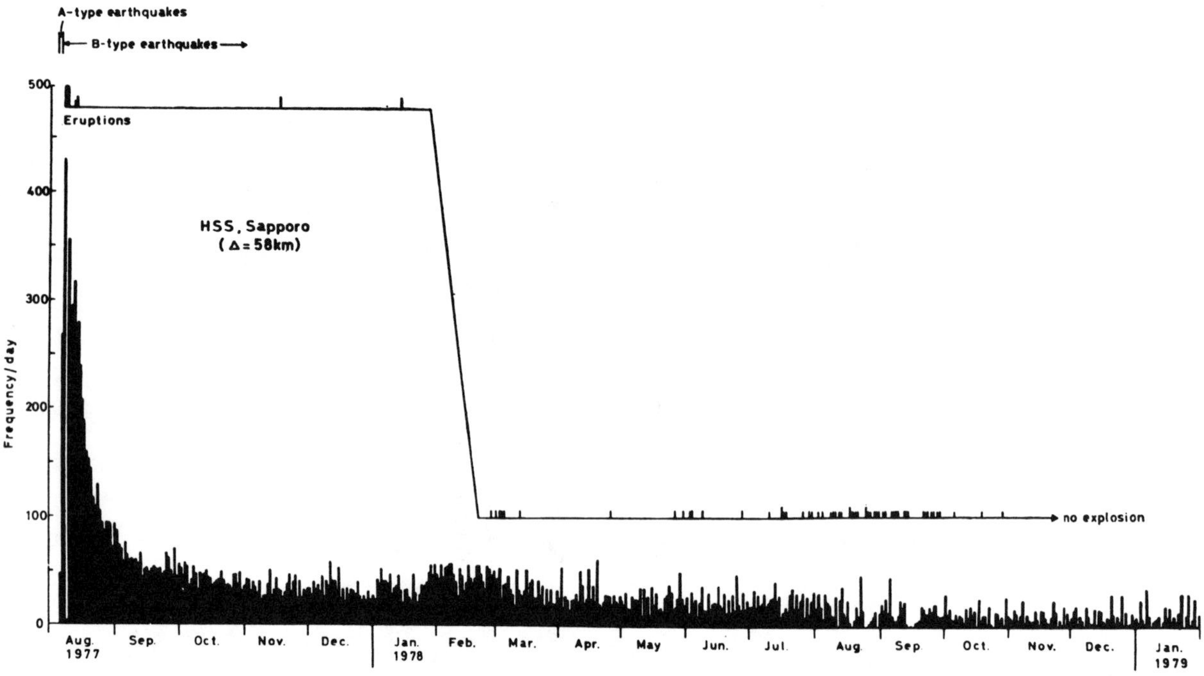

Figure 8.14.20. Daily frequency of earthquakes larger than M 2.5 at Usu Volcano, observed at station HSS (Sapporo Seismological Observatory; see figure 8.14.17), from Suzuki and Kasahara (1979). Seismicity increased sharply 32 hours before the first explosions in August 1977 (see figure 8.14.21).

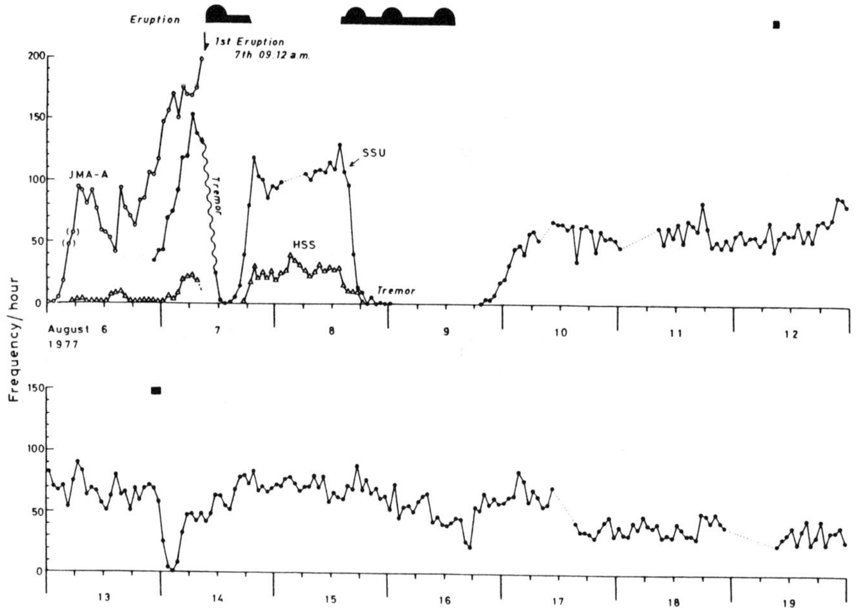

Figure 8.14.21. Hourly frequency of earthquakes at Usu Volcano (stations JMA-A, SSU, and HSS; see figures 8.14.17 and 8.14.32 for locations) during the August 1977 eruption, from Suzuki and Okayama (1978).

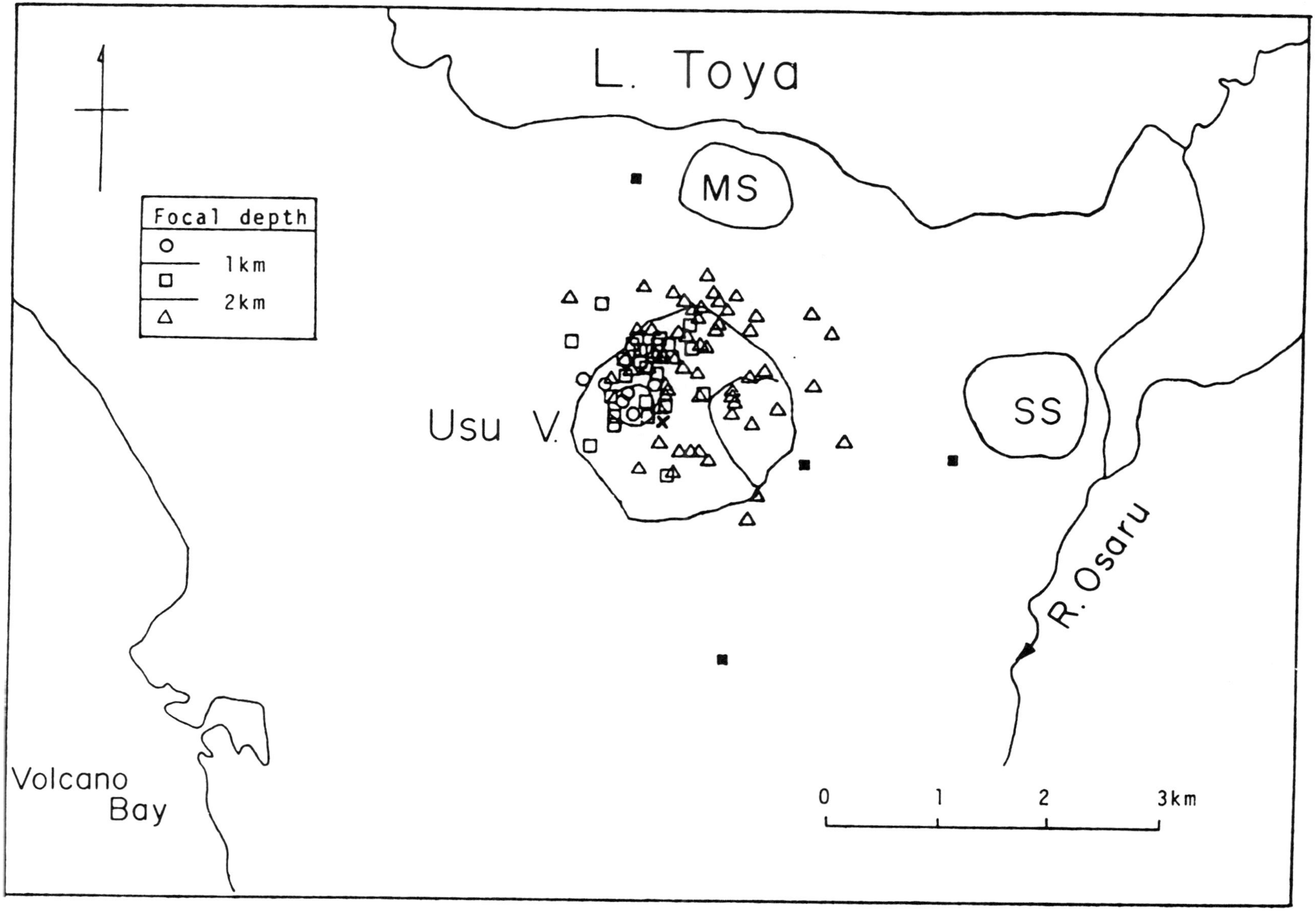

Figure 8.14.22. Epicenters of earthquakes preceding the 1977 eruption of Usu, from Yokoyama and others (1981). MS, Meiji-shinzan (1910 cryptodome); SS, Showa-shinzan (1943-45 lava dome); solid squares, seismometers; x, first eruption site.

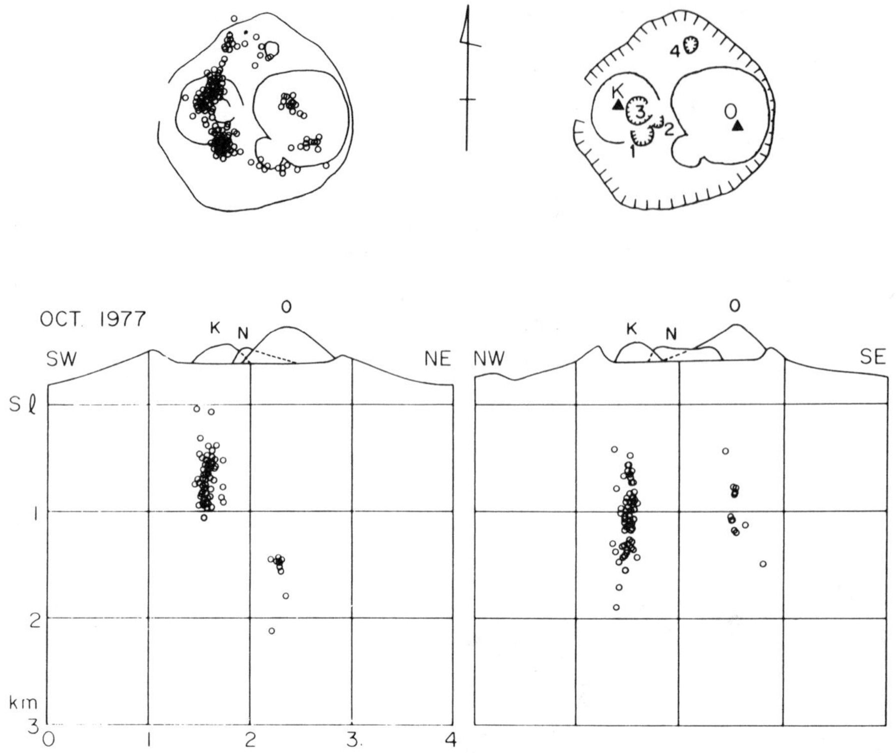

Figure 8.14.23. Distribution of hypocenters during October 1977 at Usu Volcano, from Yokoyama and others (1981). Note the relatively aseismic zone near center of crater that developed during active doming; in 1982-83, earthquakes occurred in this zone during subsidence of the cryptodome.

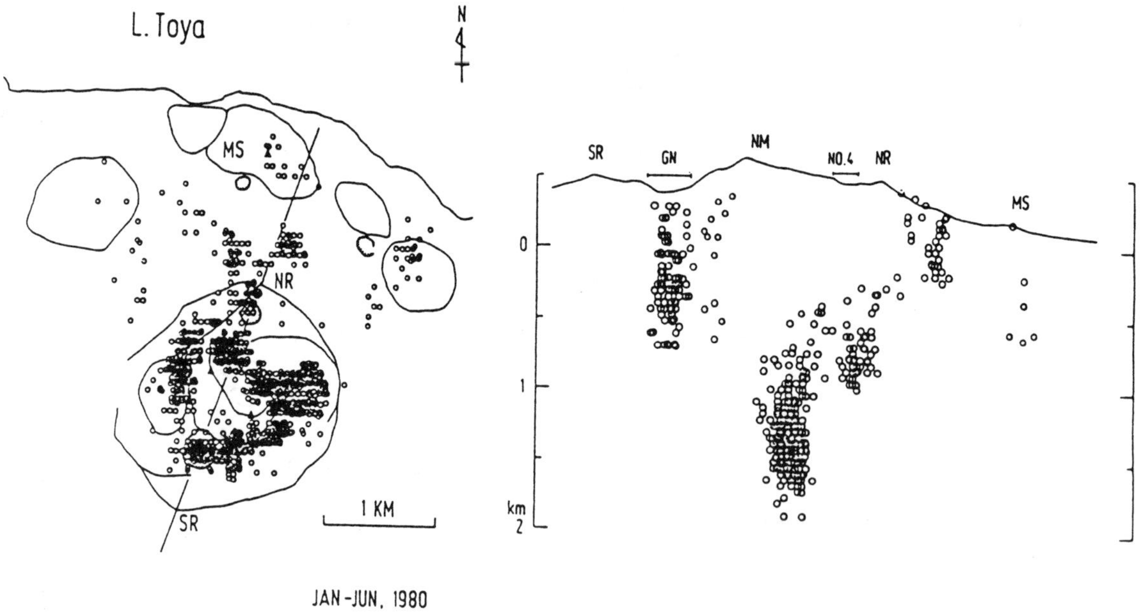

Figure 8.14.24. Distribution of hypocenters at Usu during January-June 1980, from Watanabe and others (1984b). Note earthquake-free zone beneath central part of summit crater, where dramatic upheaval was occurring.

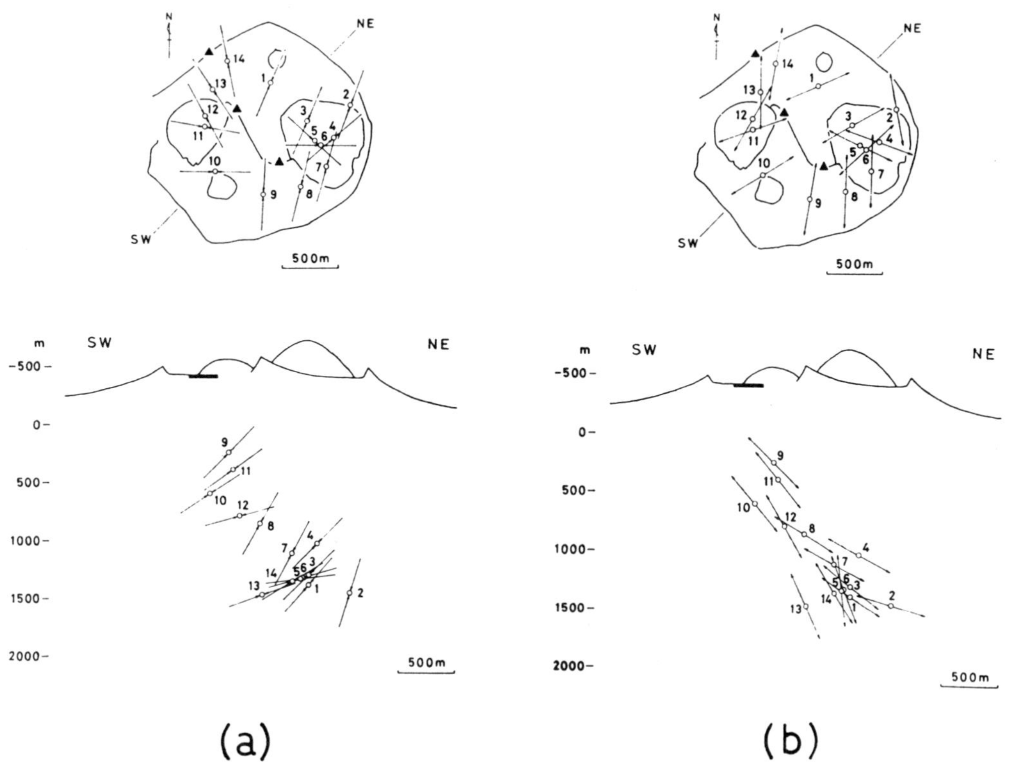

Figure 8.14.25. Spatial distributions of (a) compressional and (b) tensional stress axes during the 1977-78 activity at Usu Volcano, from Harada (1981).

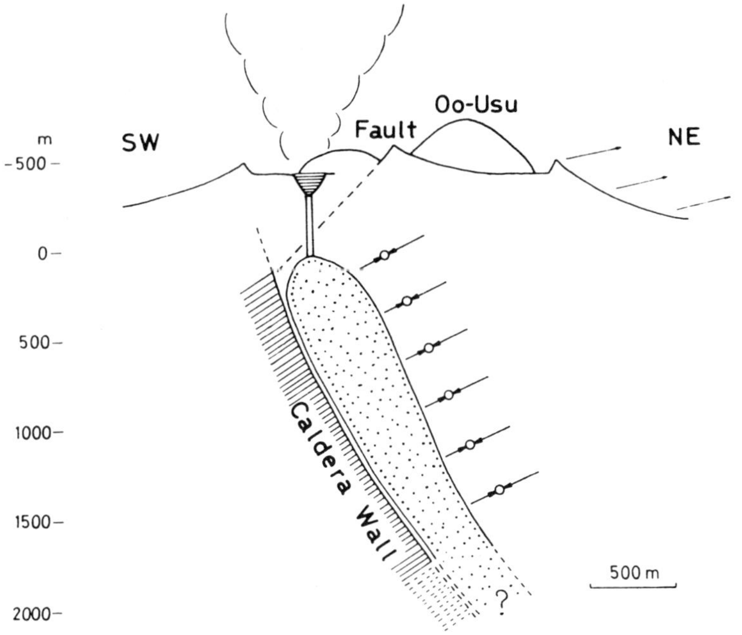

Figure 8.14.26. Schematic model of magma intrusion during the 1977 eruption of Usu Volcano, from Harada (1981).

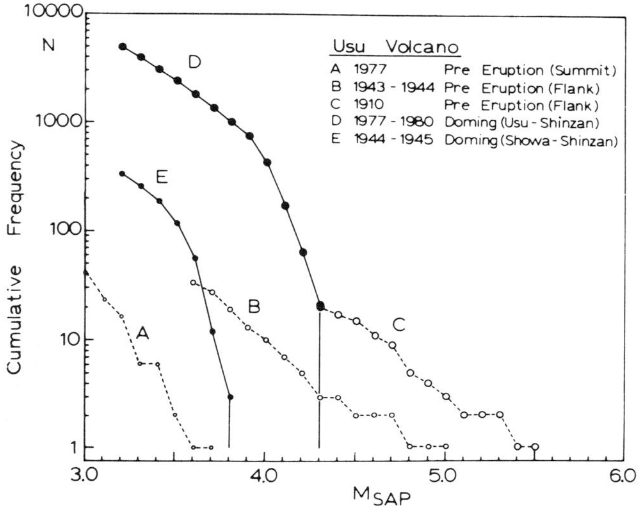

Figure 8.14.27. Top left, Cumulative-frequency versus earthquake magnitude for three recent eruptions of Usu Volcano, from Okada (1982). Data are all from Sapporo seismic station, 70 km from Usu, so they can be compared precisely. Preeruption activity follows roughly log-linear N(M) relation, but earthquakes during posteruption doming follow non-log-linear relation.

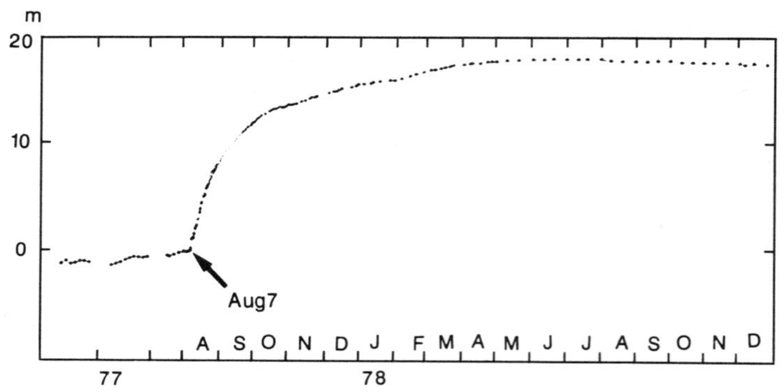

Figure 8.14.28. Top right, Extension of a slack cableway on the eastern side of Usu Volcano (see figure 8.14.17) during the 1977 activity, from Watanabe (1983). Extension is a measure of contraction between east rim and eastern foot of volcano. Anomalous extension began immediately before eruption began on 7 August 1977.

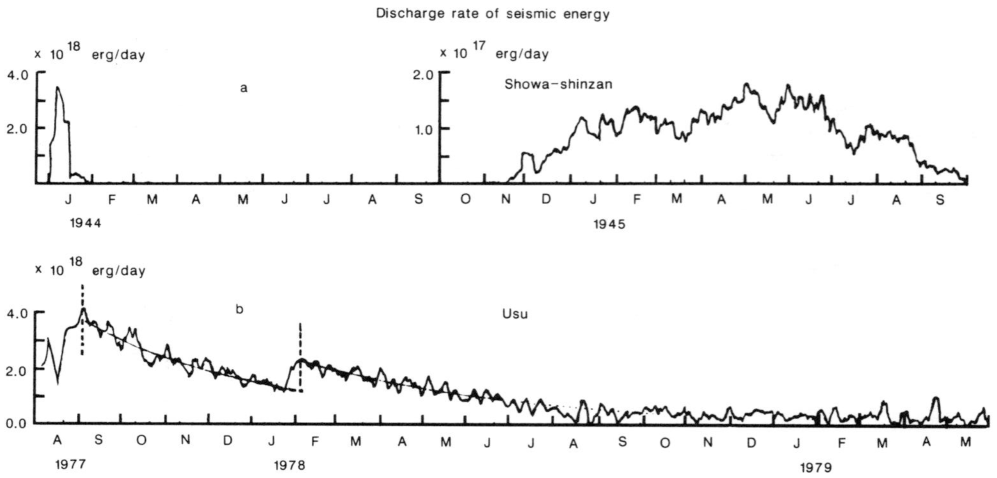

Figure 8.14.29. Bottom right, Rates of seismic energy release (ergs/day) during the 1943 (top) and 1977 (bottom) eruptions at Usu, from Seino (1983).

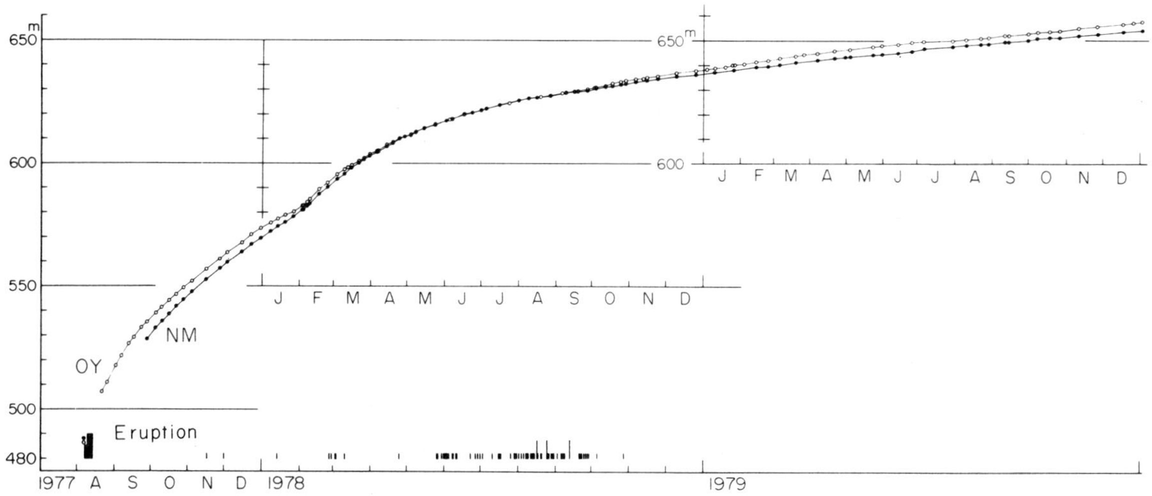

Figure 8.14.30. Doming of Ogariyama (OY) and "new mountain" (NM) during the 1977-79 activity at Usu Volcano, from Yokoyama and others (1981).

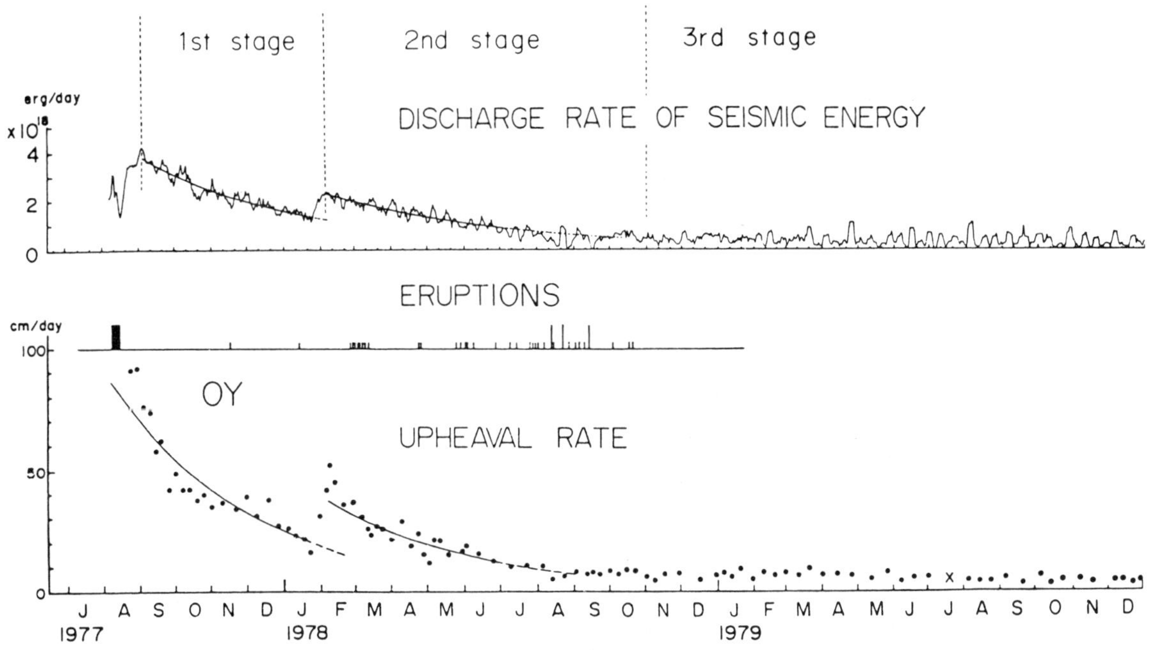

Figure 8.14.31. Daily seismic energy release averaged over 5-day intervals (top) and daily upheaval rate (bottom) of Ogariyama cryptodome during 1977-79, from Yokoyama and others (1981).

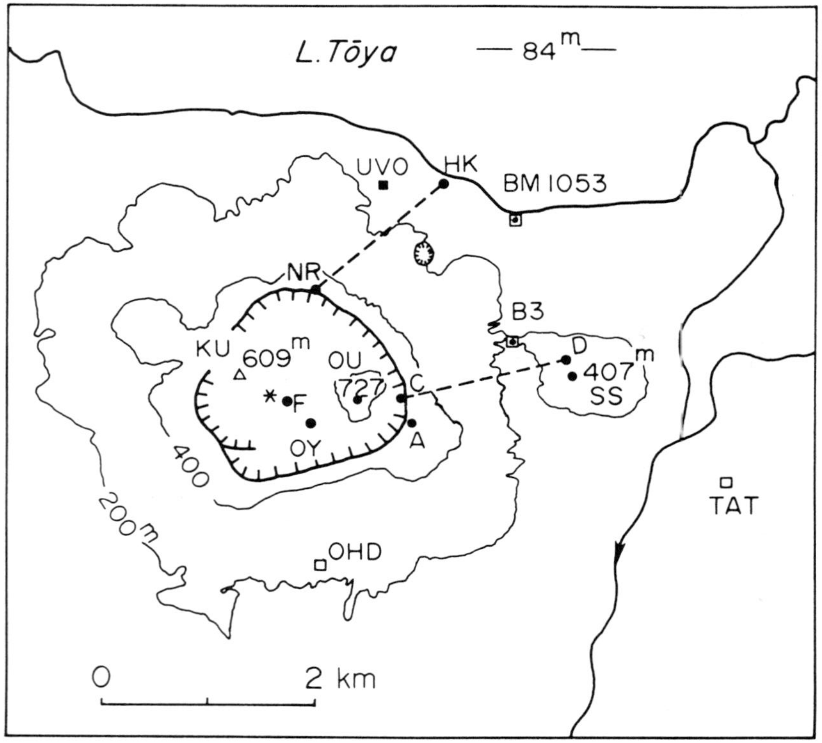

Figure 8.14.32. Usu Volcano at time of the 1977 eruption, from Yokoyama (1985). KU, Ko-Usu lava dome, formed in 1663; OU, O-Usu lava dome, formed in 1853; OY, Ogariyama cryptodome; SS, Showa-shinzan lava dome, formed in 1945; F, fault found on the morning of the 1977 outburst; *, first eruption site; A, JMA seismic station; UVO, Usu Volcano Observatory; NR and HK, endpoints of electronic distance measurement (EDM) line (see figure 8.14.33).

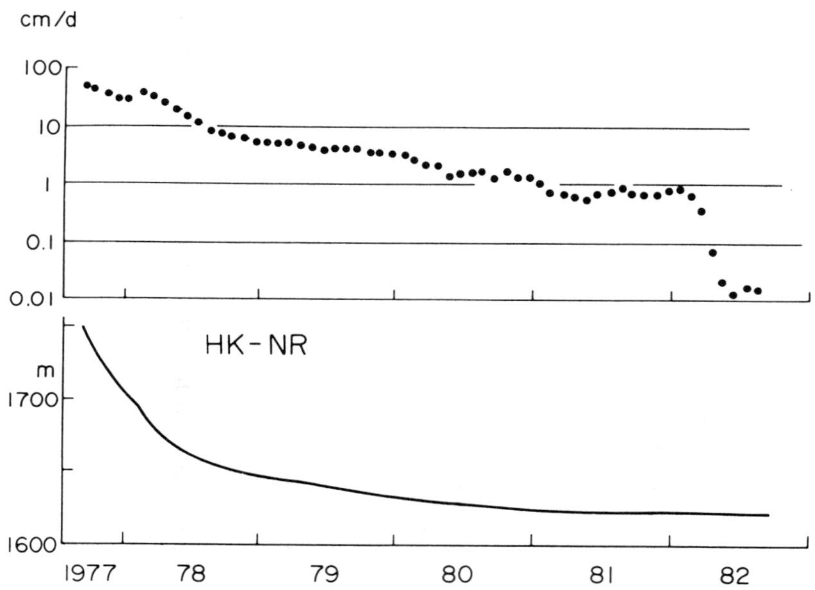

Figure 8.14.33. Contraction of base line HK-NR (see figure 8.14.32), from Yokoyama (1985). Top, rate in centimeters per day; bottom, cumulative contraction. Contraction of base line HK-NR over shorter periods is shown in figures 8.14.34 and 8.14.35.

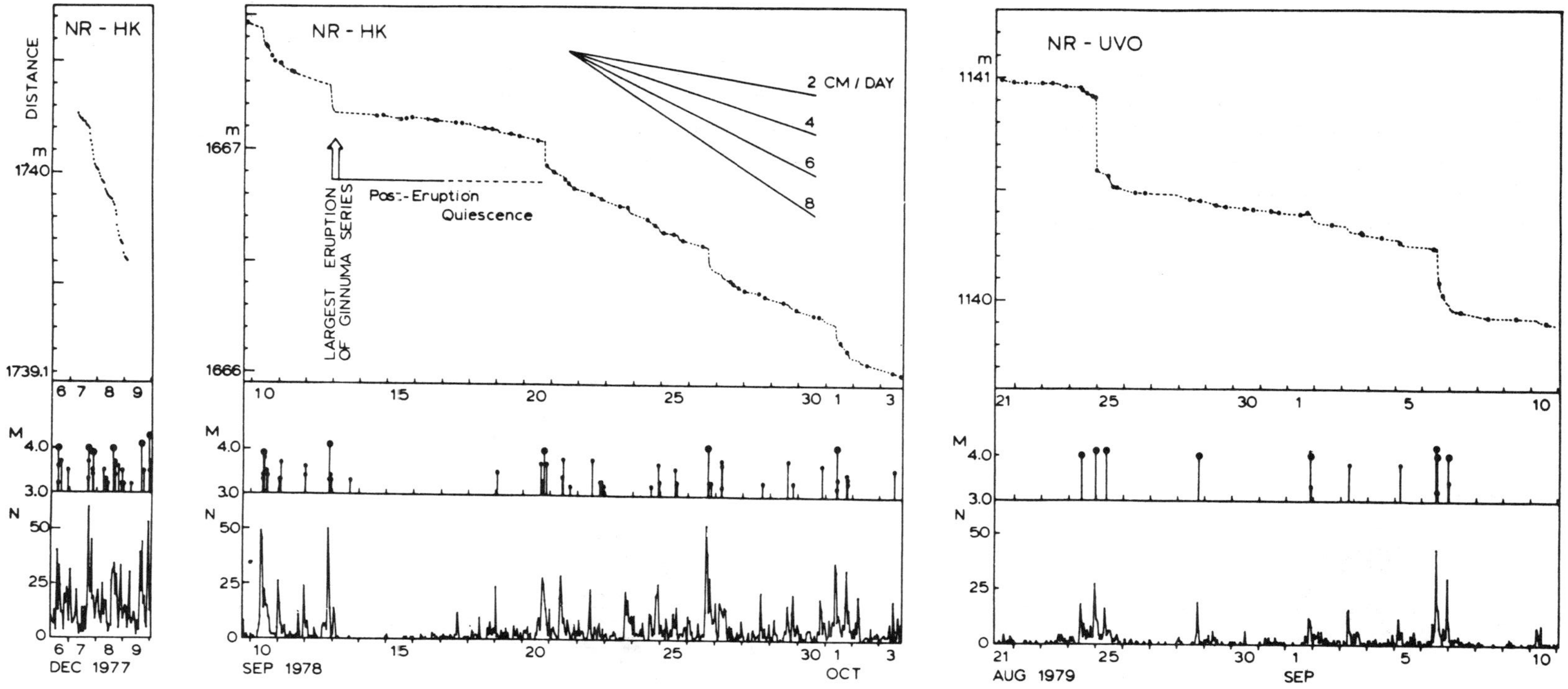

Figure 8.14.34. Relation between geodetic changes and seismicity at Usu, from Okada and others (1981). Left, 6-9 December 1977; middle, 9 September - 3 October 1978; right, 21 August - 10 September 1979. Distance measurement (top row) was repeated frequently between northern somma rim (NR; see figure 8.14.32) and southern shore of Lake Toya (HK) or between NR and UVO. Graph in middle row shows occurrence of larger earthquakes with time; graph in lower row shows hourly frequency of earthquakes at UVO. Note episodic nature of surface deformation and swarms. Episodic shortening became more significant as time went on. Largest eruption in second sequence occurred on 12 September 1978 from Ginnuma crater, immediately after large earthquakes. For about one week after that eruption, deformation and seismic activity were notably low.

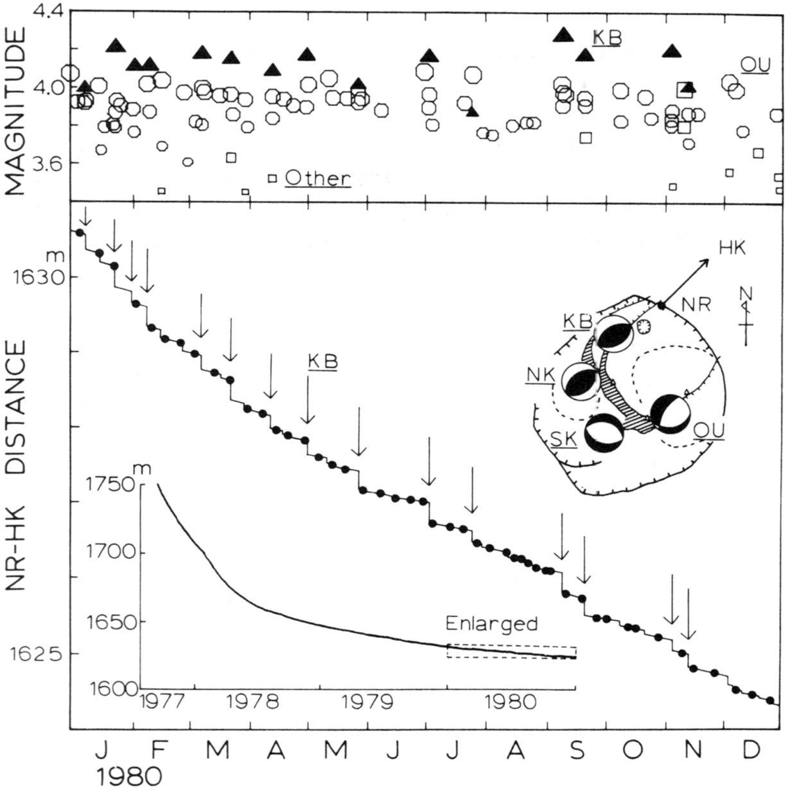

Figure 8.14.35. Correspondence between doming deformation data and earthquakes at Usu Volcano in 1980, from Okada (1983). Two major earthquake families are specified either by solid triangles (KB-family) or by open octagons (OU-family); open squares indicate other earthquakes. Although distance measurements are less frequent, they show large amounts of episodic deformation corresponding to KB-family earthquakes (arrows). Inset shows location and source mechanism (upper hemisphere, solid part compression) for major families of earthquakes.

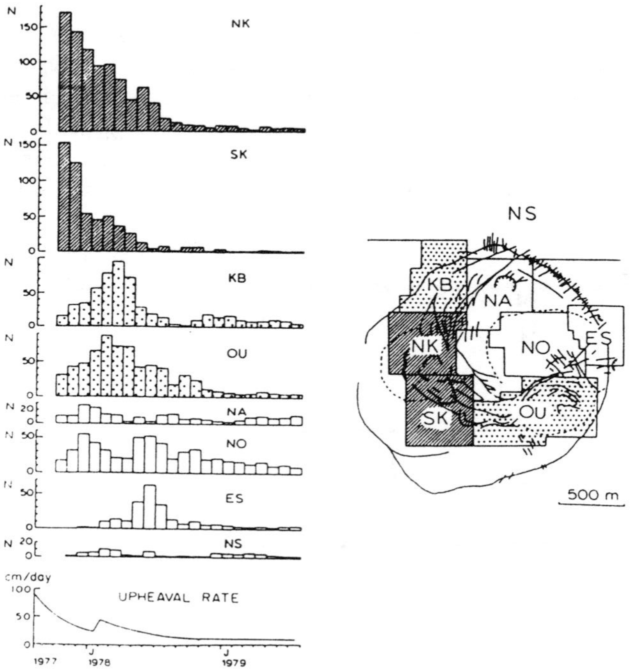

Figure 8.14.36. Division of earthquake families and monthly variation of seismic activity in each family, Usu Volcano, from Yokoyama (1986), after Okada and others (1981). The y-axes are normalized as mean daily frequency. Upheaval rate of summit is shown for comparison (bottom).

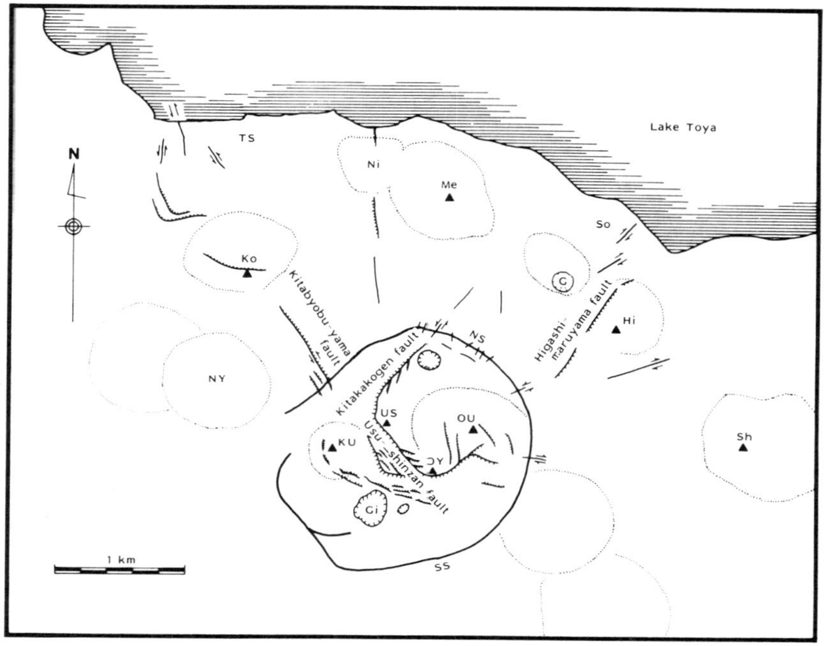

Figure 8.14.37. Major faults produced during the 1977-82 activity at Usu, from Katsui and others (1985). Ko, Konpira-yama; TS, Toyako Spa; Ni, Nishi-maruyama; Me, Meiji-shinzan; So, Sobetsu Spa; G, Genta-ana crater; Hi, Higashi-maruyama; Sh, Showa-shinzan; NS, north somma; SS, south somma; KU, Ko-Usu; US, Usu-shinzan; OY, Ogariyama; OU, O-Usu; Gi, Ginnuma crater; dotted lines, cryptodomes and lava domes.

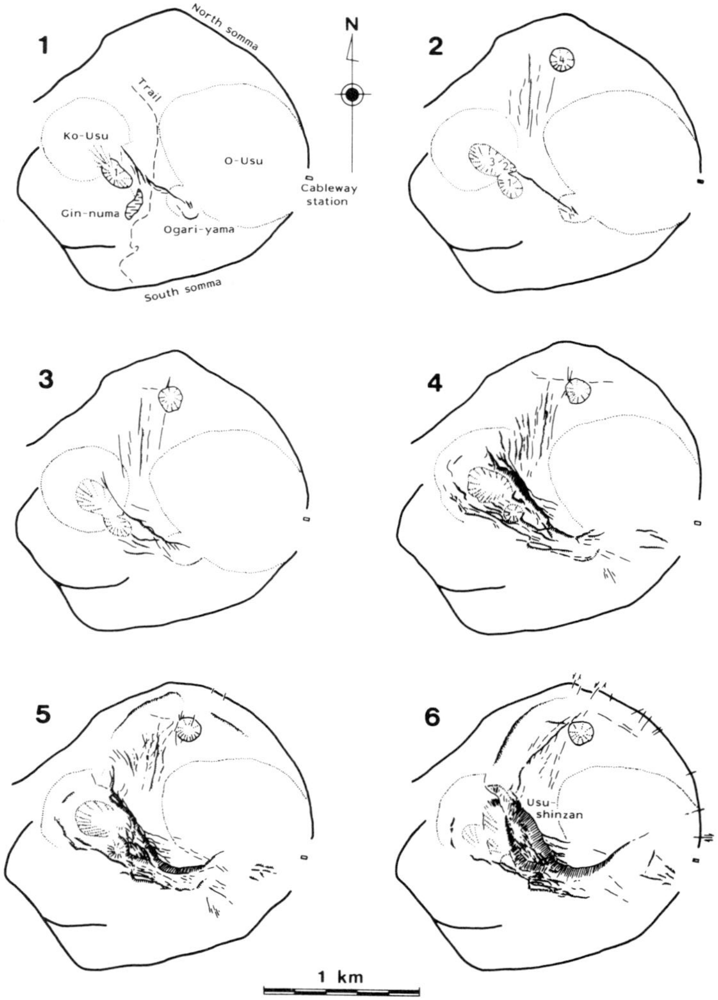

Figure 8.14.38. Sequence showing development of faults in summit area of Usu from August 1977 to November 1977, from Katsui and others (1985). (1) 8 August 1977; (2) 10 August; (3) 12 August; (4) 23 August; (5) 25 September; and (6) 29 November.

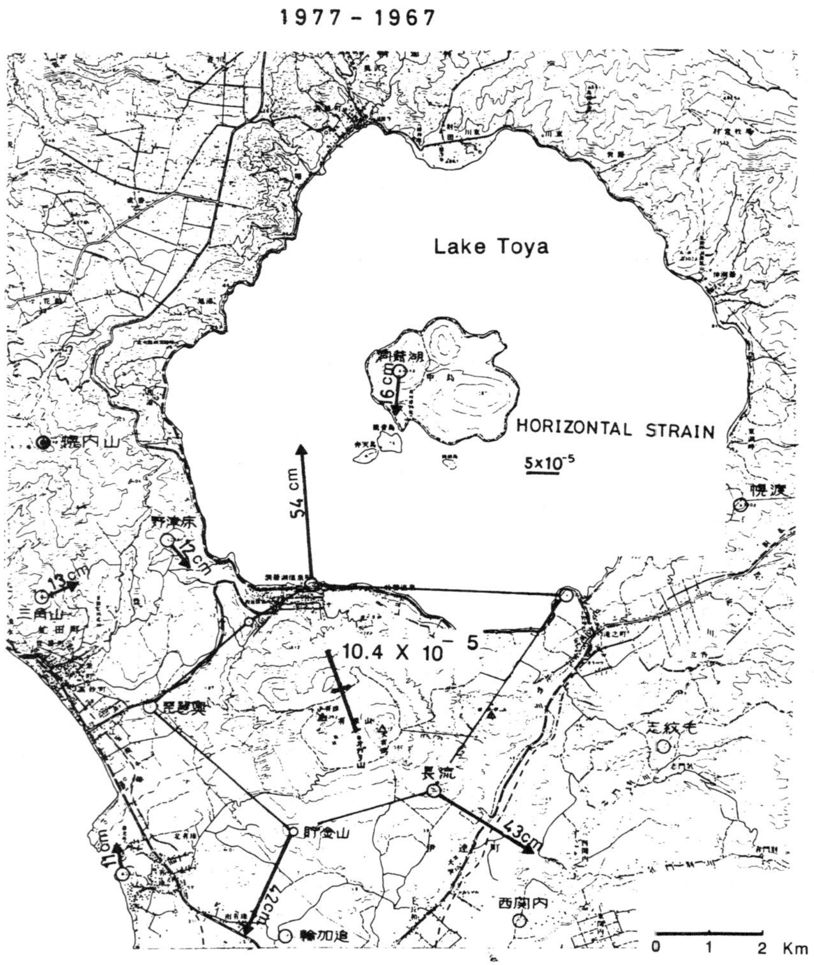

Figure 8.14.39. Horizontal strain changes at Usu from 1967 to 1977, from Geographical Survey Institute (1978).

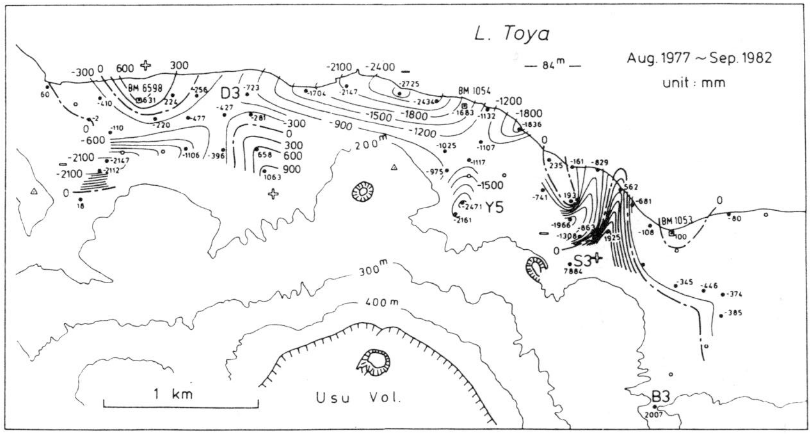

Figure 8.14.40. Vertical displacements at northern foot of Usu Volcano, relative to B.M. J6, from the start of eruption in August 1977 through September 1982, from Watanabe and others (1984a).

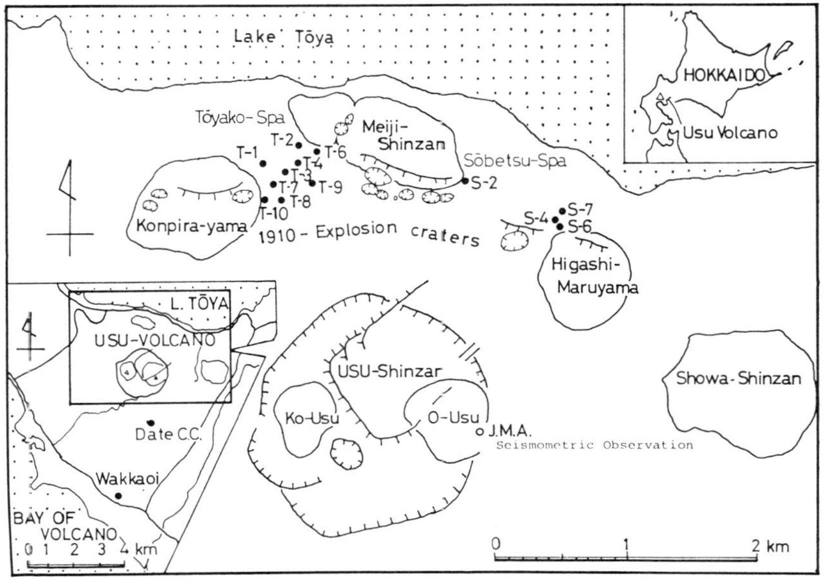

Figure 8.14.41. Locations of hot-spring temperature and chemical measurements at Usu Volcano, from Abiko (1984). Data are shown in next several figures.

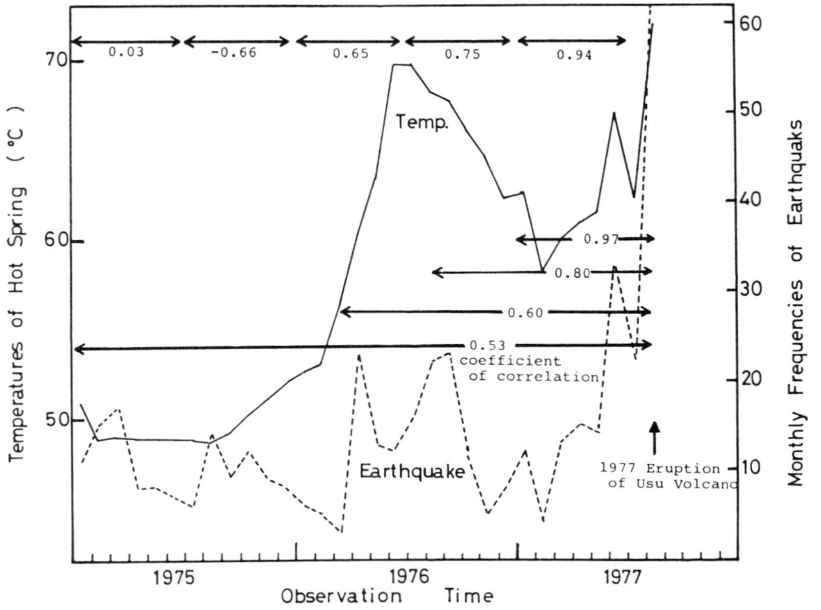

Figure 8.14.42. Hot-spring temperature at site S-4 (see figure 8.14.41) and seismicity at Usu during 1975-77, from Abiko (1984).

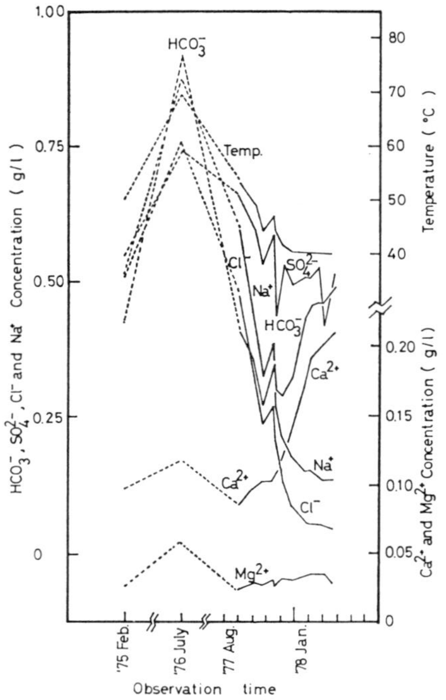

Figure 8.14.43. Variations in composition of spring S-4 from February 1975 to March 1978, from Abiko (1984).

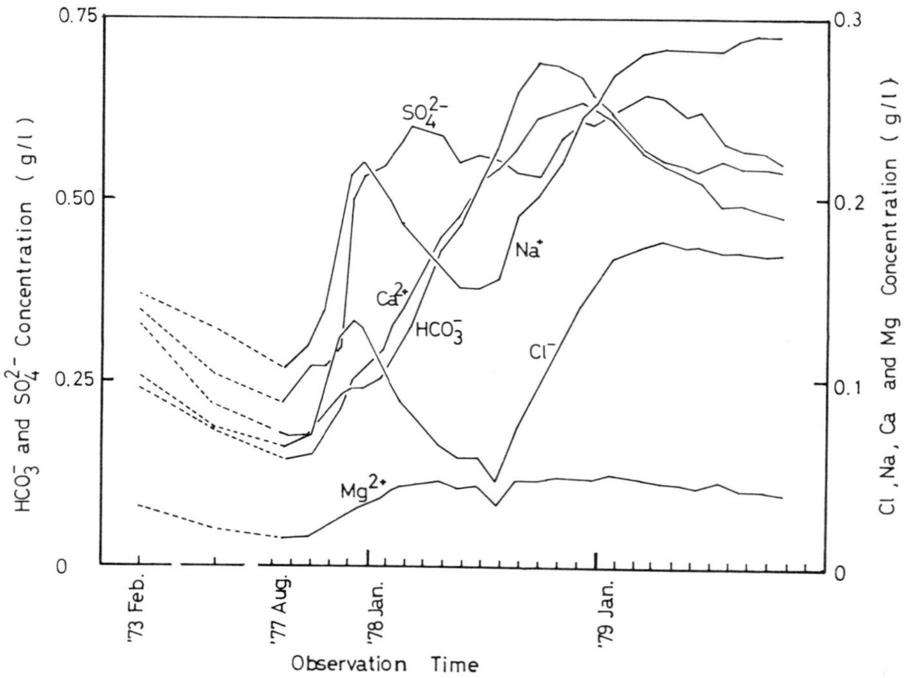

Figure 8.14.44. Variations in composition of spring S-7 from February 1973 to October 1979, from Abiko (1984).

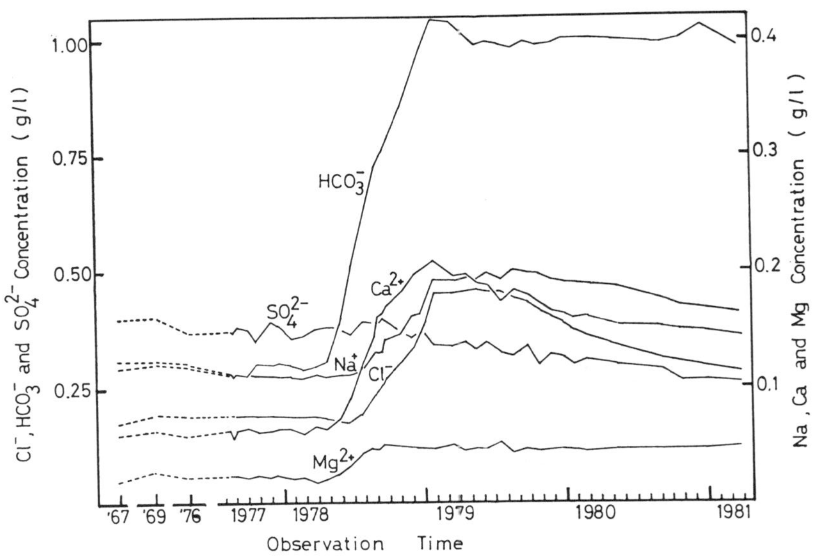

Figure 8.14.45. Variations in composition of spring T-7 from 1967 to 1981, from Abiko (1984).

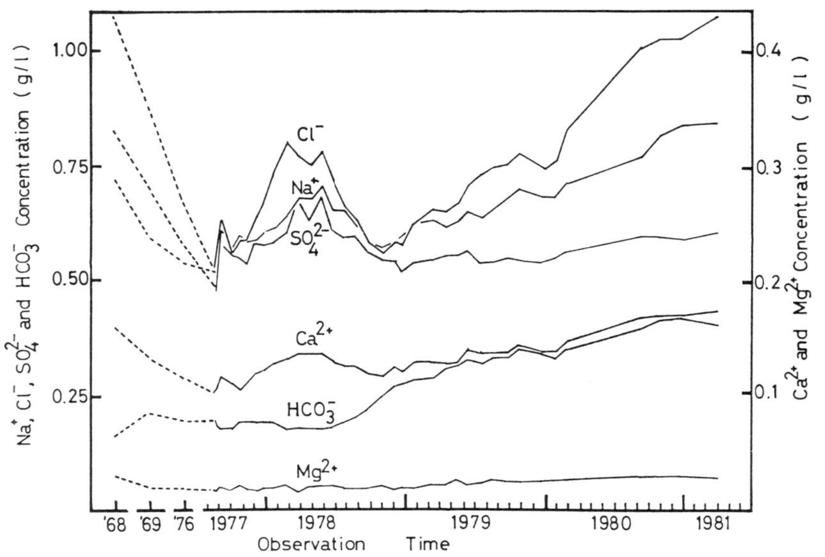

Figure 8.14.46. Variations in composition of spring T-9 from 1968 to 1981, from Abiko (1984).

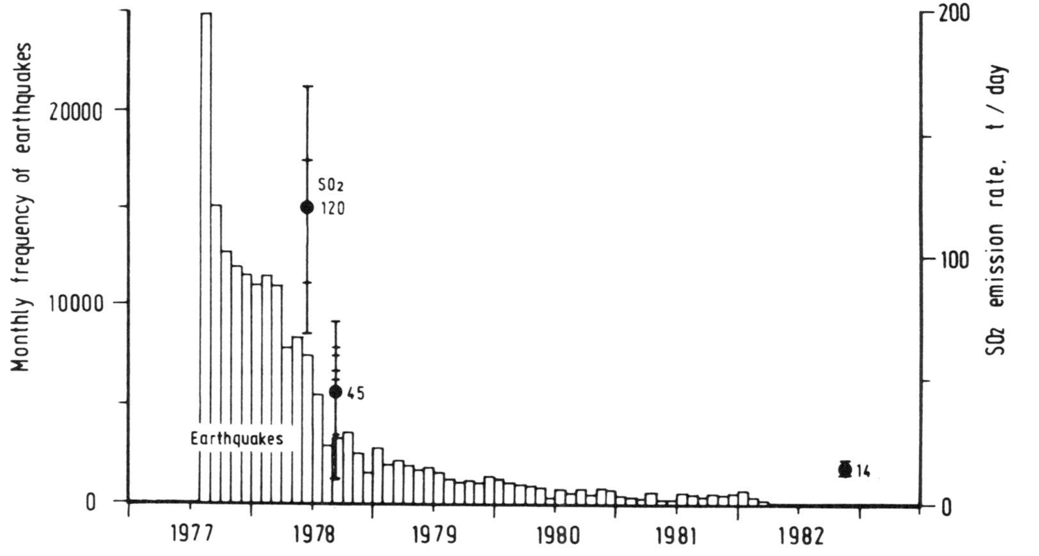

Figure 8.14.47. Variations in SO_2 emission rates and monthly frequency of volcanic earthquakes during 1977-82 at Usu Volcano, from Ohta and others (1984).

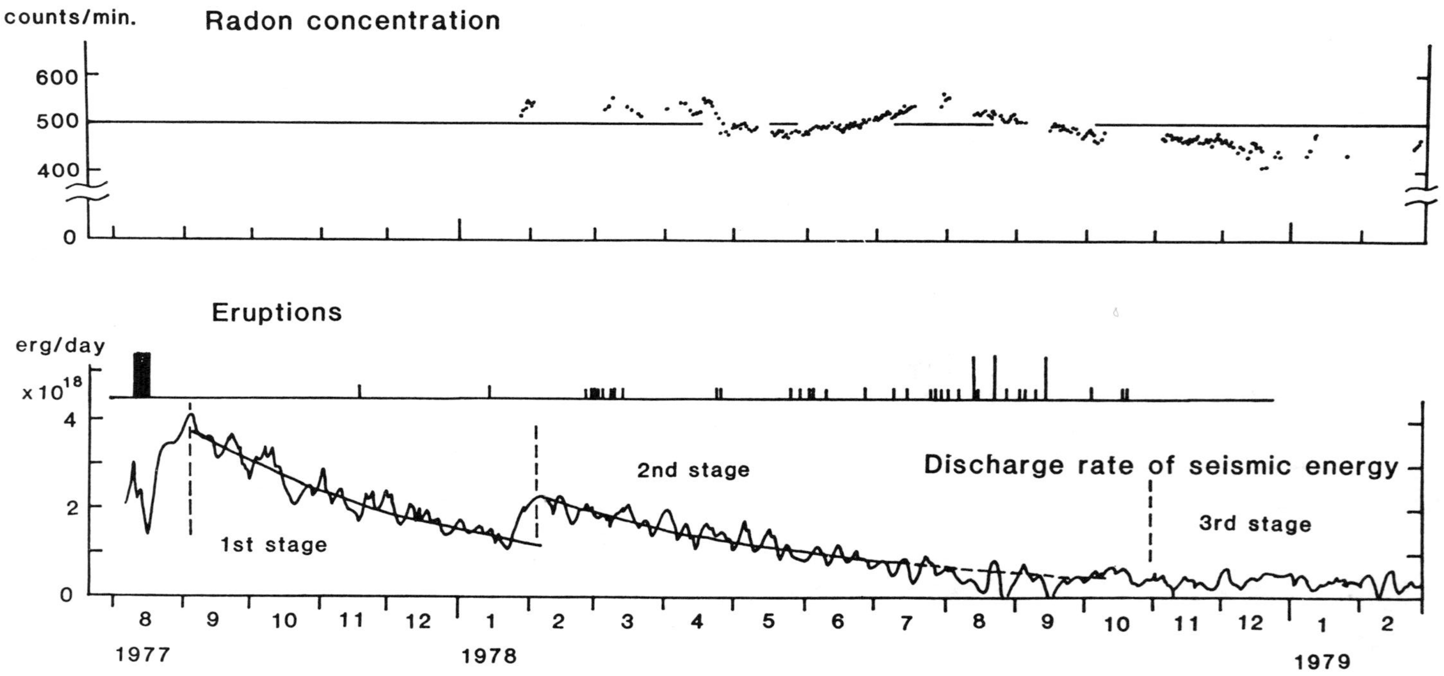

Figure 8.14.48. Radon-concentration changes in ground water, Usu Volcano, August 1977 to February 1979 (Notsu and others, 1983). Eruption information and rate of seismic energy release from Yokoyama and others (1981).

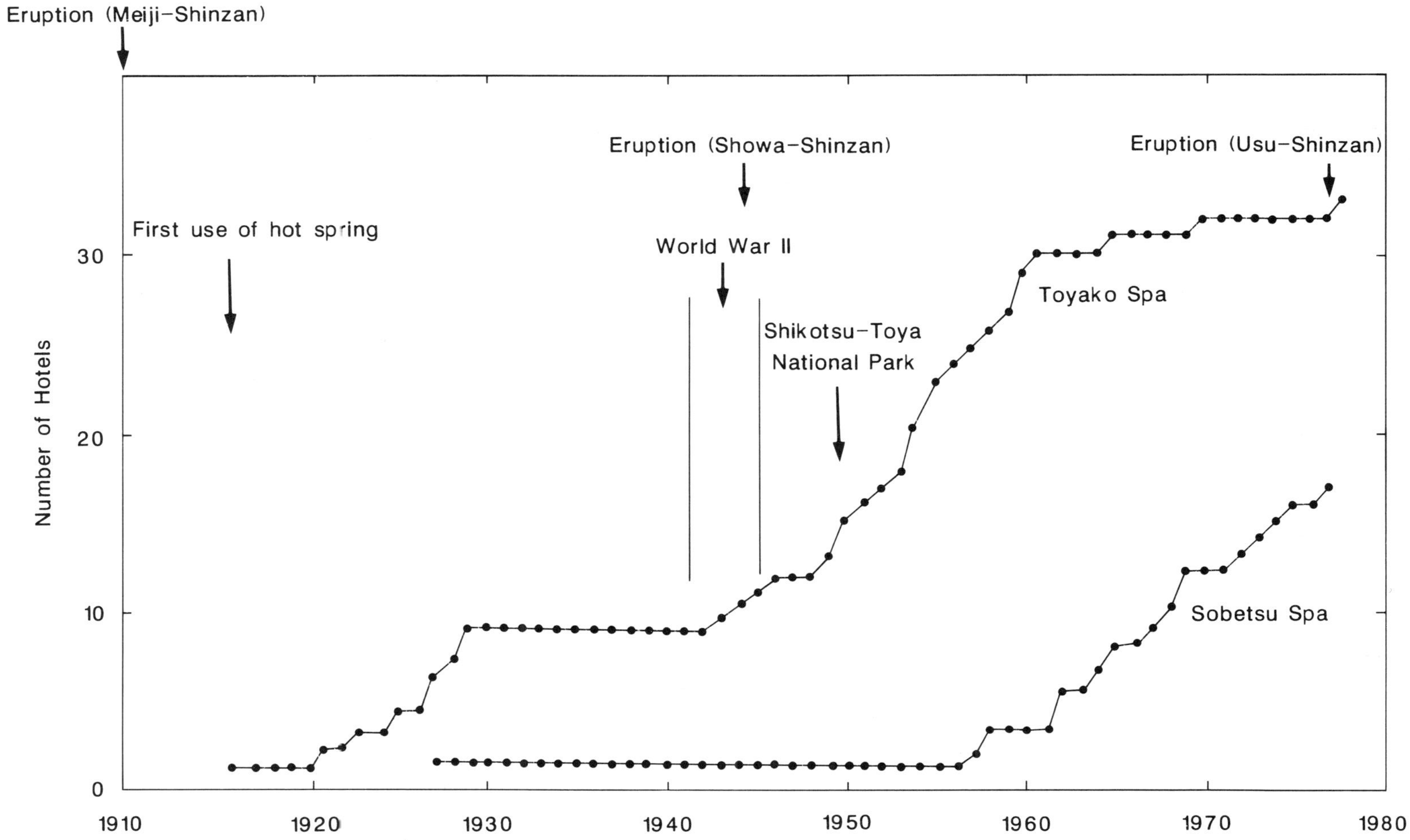

Figure 8.14.49. Number of resort areas at northern foot of Usu Volcano since the 1910 eruption. Cryptodome emplacement creates hot springs, which are loci for new hotels (unpublished supplement to field-trip guide, IAVCEI Symposium on Arc Volcanism, 1981).

PART 3: HISTORICAL UNREST AT LARGE QUATERNARY CALDERAS (continued)
See inside back cover for explanation and abbreviations

SHIKOTSU

CAVW number (active vent)	Latitude Longitude (degrees)	Diameter (km)	Local tectonic setting	Precaldera edifice	SiO_2 content (wt pct)	Age of collapse (yr B.P.)	Date of unrest	Type and duration of unrest ESTU STHF MGTF H Te	Eruption type
08-05-04 (Tarumai)	42.68N 141.38E	13 x 15	NW-SE line in caldera controls 3 vents	?	R = 55-62 C = 52-74	30,000-32,000	1667	---- ---- ---- - --	EX
							1739	x--- ---- ---- - --	EX
							1804-17	---- ---- ---- - --	Ex
							1867	x--- ---- ---- - --	ex,dm
							1874	---- ---- ---- - UQ	Ex
							1883-87	---- ---- ---- - --	ex,pex
							1894	---- ---- ---- - --	ex
							1909	x--- ---- ---- - QU?	Ex,dm
							1917-36	x--- --DD ---D - --	ex,pex
							1944-55	x--- ---- ---- - --	ex,pex?
							1967-81	FF-- --N- --?E - --	ex,pex
							1988	DD-- ---- ---? - --	none yet

TECTONIC SETTING

Shikotsu Caldera is situated near the southwestern end of the Kurile arc, along which the Pacific plate is subducted in a northwesterly direction beneath the Eurasian plate at a rate of approximately 10 cm/yr (Circum-Pacific Map Project, Northwest Quadrant Panel, 1981).

GEOLOGIC HISTORY

Shikotsu Caldera was the source of a large (125 km^3 bulk volume), compositionally zoned (rhyolite to andesite) tephra fall and ash flow about 30,000 years ago. Most of Shikotsu Caldera is occupied by Lake Shikotsu (fig. 8.15.1). Tarumai Volcano is situated on the southern rim of Shikotsu Caldera. The 1667 and 1739 eruptions of Tarumai were large and formed a 1.5-km diameter caldera at its summit (fig. 8.15.2); recent activity has included growth of a dome in that summit caldera, beginning in 1909.

HISTORICAL ACTIVITY

1874: A strong eruption of Tarumai on 8 February was followed on 28 February by a strong tectonic earthquake just west of Rumoe, at ca. 44°N latitude on the northwest coast of Hokkaido (Omori, 1911, p. 6).

SHIKOTSU, Region 8, CAVW number 08-05-04

SHIKOTSU (continued)

HISTORICAL ACTIVITY (continued)

1894: Minor ash ejections from the dome occurred on 8 February and 17 August (Y. Katsui, written commun., 1983).

1909: About one year after a large tectonic earthquake just west of the northern tip of Hokkaido (Omori, 1911, p. 2, 6), Tarumai had another moderately strong eruption. The eruption began on 30 April 1909, when a huge column of dark "smoke with fire" rose suddenly on the dome with terrible detonations and tremblings. A preexisting fissure at the south foot of the dome enlarged, and "smoke" (steam?) increased remarkably. On 12 May, a further outbreak took place with enormous "smoke and rolling and tremblings for about one-fourth hour" (Tanakadate, 1912, 1925).

1917-36: Small eruptions occurred each year from 1917 through 1923 and intermittently thereafter until 1936. A large explosion in July 1920 occurred on the eastern side of the dome at the fissure that was opened in 1917 (and later closed) (Tanakadate, 1925). The other eruptions were also along E-W fissures crossing the dome and inner crater wall (Tanakadate, 1934). One of these fissures extended laterally by 250 m in 1933, and the new portion opened to a maximum width of 7 m. Sulfur gases were emitted in several places along the fissure. An explosion occurred on 1 December 1933, followed by a strong earthquake on 5 December, by lesser volcanic earthquakes on 6 December, and by renewed eruption on 11 December (Tanakadate, 1934).

1967-81: An increase in seismicity, in the form of earthquake swarms, began in early 1967 (fig. 8.15.3). Seismicity increased about 3 months before JMA station A was installed. Small seismic swarms occurred in January-February 1975, May 1978, February 1979, and January-February 1981, with depths mostly less than 2 km and epicenters all within 1-2 km of the vent (no epicenters elsewhere in the caldera). The first eruptions since 1955 began in May 1978, and additional small eruptions occurred during December 1978 - May 1979 and in February 1981. Swarms that had been generally increasing since 1967 continued through the 1978 eruptions but stopped after the small eruption in February 1981. Some earthquakes occurred even after the eruption ended (figs. 8.15.4-8.15.6), including a $\underline{M}$=4.0 earthquake in an October 1931 swarm around Eniwa, on the north rim of Shikotsu Caldera (Okada and Watanabe, 1983).

During a period of routine observation in November-December 1975, Ehara and Yokoyama (1978) distinguished four types of seismic events at Shikotsu: one type beneath the summit dome of Tarumai, two types beneath the northeast flank of Tarumai, and a fourth type beneath the center of the northeast basin of Lake Shikotsu (fig. 8.15.7).

Line-length changes of about 10 cm across the summit crater occurred at the time of the February 1981 eruption (fig. 8.15.8) (Usu Volcano Observatory, 1979; Yamashita and others, 1984). Inflation of the summit dome of Tarumai was detected by detailed tilt-leveling starting in 1983 (fig. 8.15.9); slow inflation was continuing as of December 1987, in apparent association with new geothermal anomalies (Mori and Suzuki, 1988).

The temperatures of fumaroles in radial fissures within the summit dome declined from 300 °C to 100 °C or less from 1959 to 1971 (fig. 8.15.10) and might have declined further after 1971. These temperatures showed no change during the resumption of seismic

SHIKOTSU (continued)

HISTORICAL ACTIVITY (continued)

activity in 1967. However, at least one of these fumaroles did show a spike in CO_2 in 1968 (fig. 8.15.11) (Ishikawa and others, 1972). No thermal or fumarolic changes occurred before or during the 1978-81 eruptions, but temperatures and fuming increased 2 years after the 1981 eruption (Hm. Okada, written commun., 1988).

1988: Local earthquake swarms occurred during 24 January - 18 February, beneath the northwest somma (Smithsonian Institution, 1988). These swarms were small in comparison to those of February 1981 but are of special interest because the eruption of 1981 was very small despite notable seismic swarms in the preceding decade. Hence the question arises as to whether the 1988 swarms are a continuation of 1967-81 activity.

REFERENCES

Circum-Pacific Map Project, Northwest Quadrant Panel, 1981, Plate tectonic map of the Circum-Pacific region, northwest quadrant: Tulsa, Am. Assoc. Pet. Geol., 1 sheet, scale 1:10,000,000.

Ehara, S., and Yokoyama, I., 1978, Observation of volcanic earthquakes by radio-telemetric system: Geophys. Bull. Hokkaido Univ., v. 37, p. 31-41.

Ishikawa, T., 1952, The chemical characteristics of the lavas from Volcano Tarumai, Hokkaido, Japan: Jour. Fac. Sci., Hokkaido Univ., ser. 4, v. 7, p. 107-135.

Ishikawa, T., Yokoyama, I., and Katsui, Y., 1972, Tarumai-san, its volcanic geology, history of eruption, present state of activity and prevention of disaster: Sapporo, Committee for Prevention of Disasters of Hokkaido, 124 p.

Katsui, Y., 1981, ed., Field excursion guide to Usu and Tarumai Volcano and Noboribetsu Spa: Tokyo, Volcanological Society of Japan, 64 p.

Kuno, H., 1962, Catalogue of the active volcanoes of the world including solfatara fields, Pt. XI, Japan, Taiwan and Marianas: Rome, IAVCEI, 332 p.

Mori, H., and Suzuki, A., 1988, Tilt observation on the summit of Tarumai volcano: Hokkaido Geophys. Bull., Hokkaido Univ., v. 50, p. 47-53.

Oinouye, Y., 1909, Activity of Tarumai-dake in 1909: Rep. Imperial Earthq. Invest. Comm., v. 64, p. 1-23.

Okada, Hm., Mori, H., Watanabe, H., Suzuki, A., Maekawa, T., and Yokoyama, I., 1985, Activity of volcanic earthquakes and geothermal phenomena of Tarumai Volcano, in Okada, Hm., ed., Comparative study of physical background of volcanic activity and its relation to prediction of eruption disasters: Special Rep. Grant-in-Aid for Nat. Disaster Sci., Ministry of Education, publ. by Hokkaido Univ., Sapporo, in cooperation with Tohoku Univ., Univ. of Tokyo Earthq. Res. Inst., Nagoya Univ., Kyoto Univ., and Kyoto Univ. Disaster Prev. Res. Inst., p. 125-133.

Okada, Hm., and Watanabe, H., 1983, Development of the pre-eruption earthquakes of the 1977 eruption of Mt. Usu: Rep. Nat. Disaster Sci. Symp., no. 20, p. 354-357.

SHIKOTSU (continued)

REFERENCES (continued)

Omori, F., 1911, The Usu-san eruption and earthquake and elevation phenomena: Bull. Imperial Earthq. Invest. Comm., v. 5, no. 1, p. 1-38.

Smithsonian Institution, Scientific Event Alert Network (SEAN), 1978-81, 1988, Tarumai Volcano: SEAN Bull., v. 3, no. 6, v. 6, nos. 1-4; v. 13, no. 2 (information from Japan Meteorological Agency and others).

Tanakadate, H., 1912, Der Tarumai-Ausbrüch in Japan 1909: Zeit. Gesellschaft für Erdkunde zu Berlin, no. 9, p. 1-12.

----- 1925, Volcanic activity in Japan during 1914-1924: Bull. Volcanol., ser. 1, no. 3/4, p. 3-19.

----- 1934, Volcanic activity in Japan and vicinity during the period between June 1931 and June 1934: Japanese Jour. Astron. Geophys., v. 12, p. 89-108.

Umehara, H., Yamashita, H., and Maekawa, T., 1981, Volcanic earthquakes originating at Tarumai Volcano (March 1979-March 1980): Geophys. Bull. Hokkaido Univ., v. 39, p. 47-56.

Usu Volcano Observatory, Hokkaido University, 1979, Observations of Tarumai Volcano during the first half of 1979: Rep. Coord. Comm. for Prediction of Volcanic Eruptions, no. 16, p. 17-20.

Yamashita, H., Okada, Hm., Suzuki, A., Maekawa, T., Watanabe, H., and Yokoyama, I., 1984, Seismic activities of Tarumai Volcano from February 1980 through January 1984, in Yokoyama, I., ed., Joint Geophysical and Geochemical Observations of Usu Volcano in 1982 and Tarumai Volcano in 1983: Usu, Hokkaido, Usu Volcano Observatory, p. 147-156.

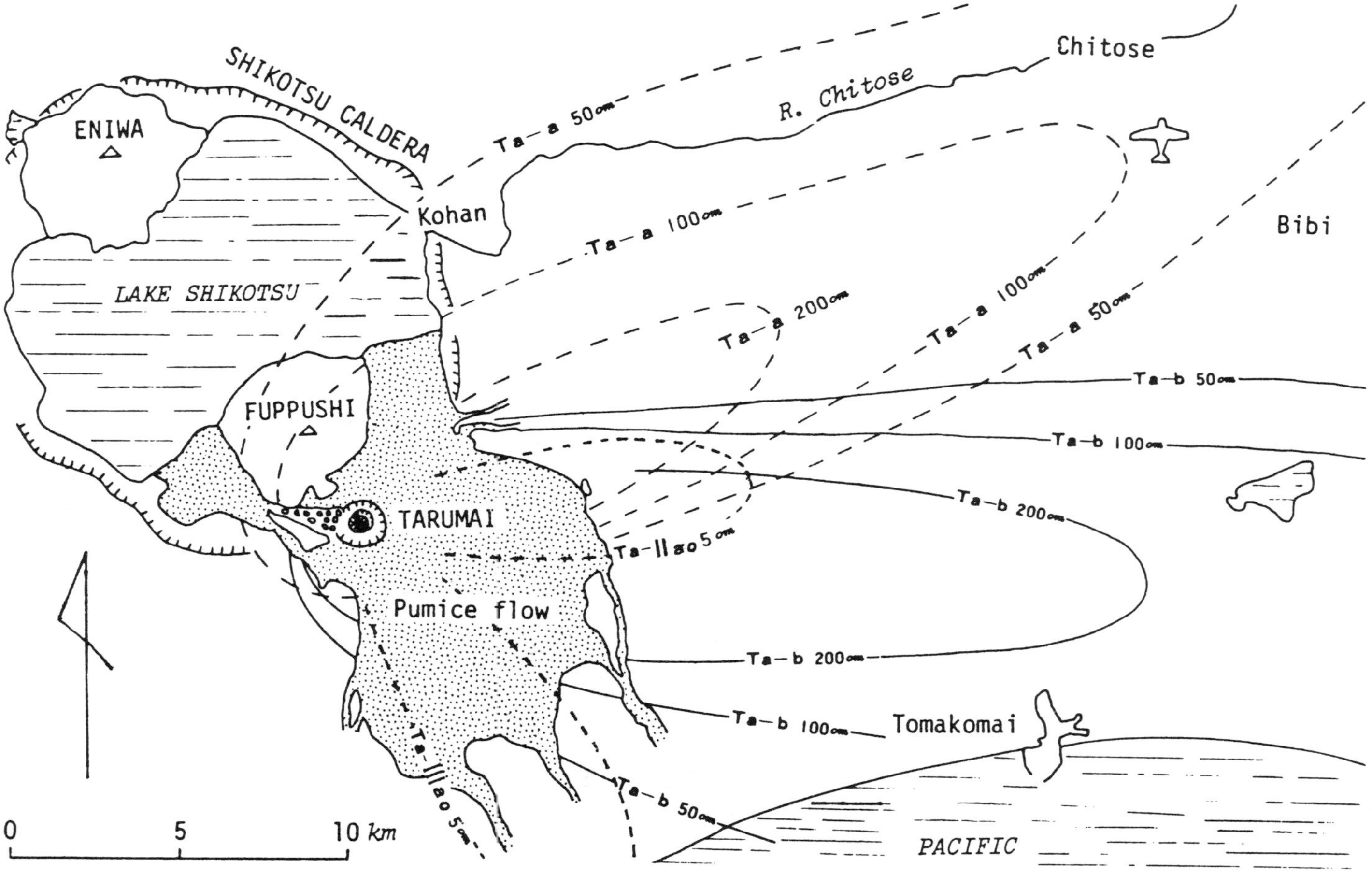

Figure 8.15.1. Sketch map showing location of Shikotsu Caldera and distribution of Ta-a (1739) and Ta-b (1667) pumice fall deposits and associated pumice flow deposits, from Katsui (1981). Distribution of later tephras, Ta-IIao (1804-17) and Ta-IIIao (1874), is also shown.

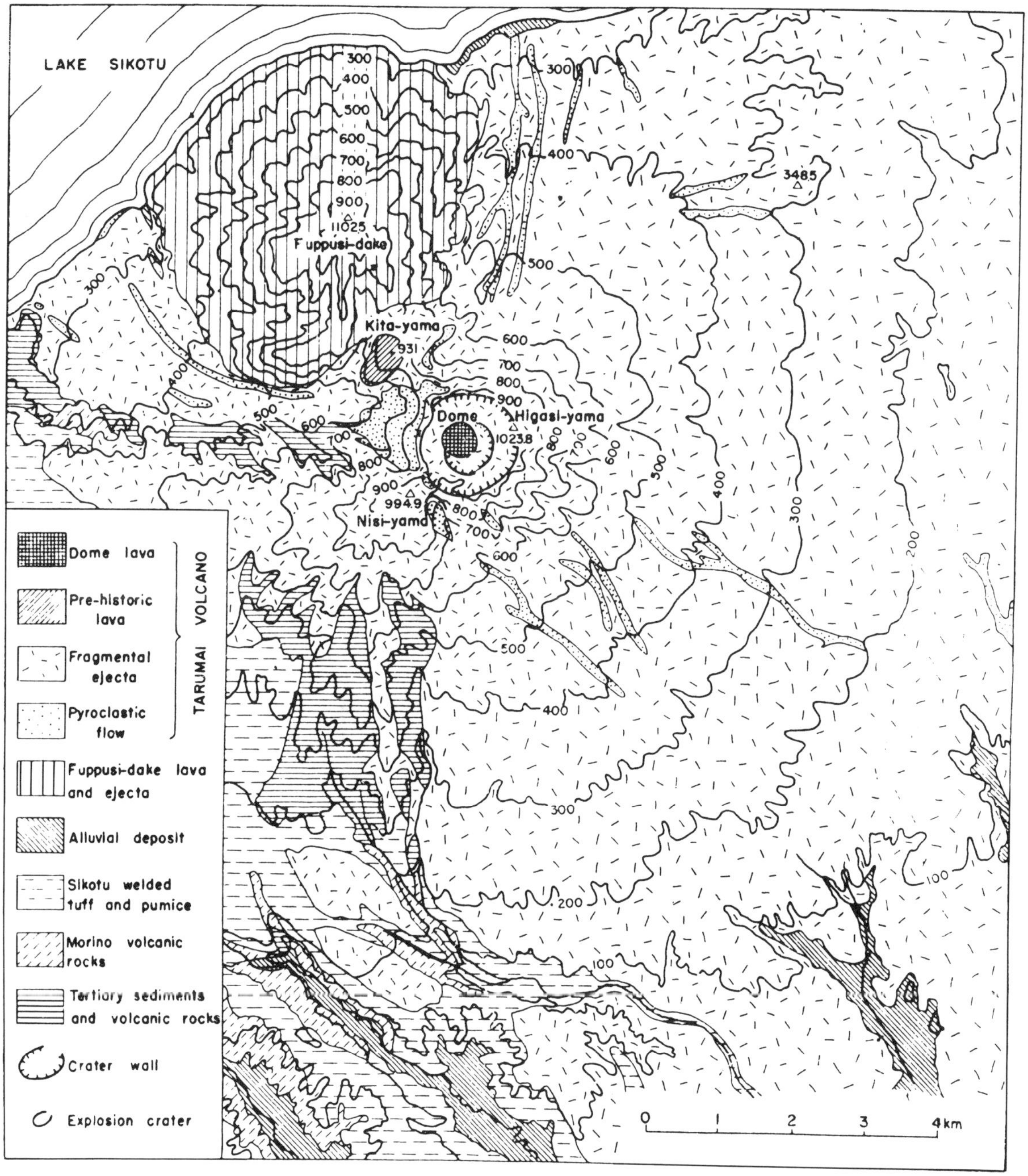

Figure 8.15.2. Geologic map of Tarumai Volcano, Shikotsu Caldera, from Katsui (1981).

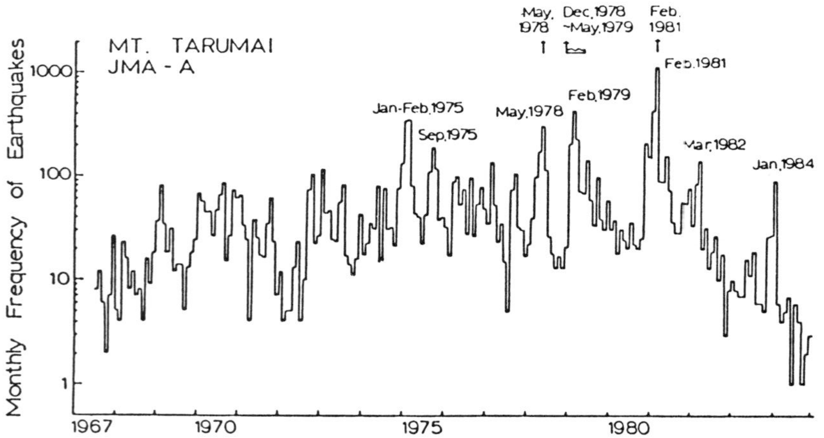

Figure 8.15.3. Monthly frequency of volcanic earthquakes at Tarumai during 1967-84, from Okada and others (1985).

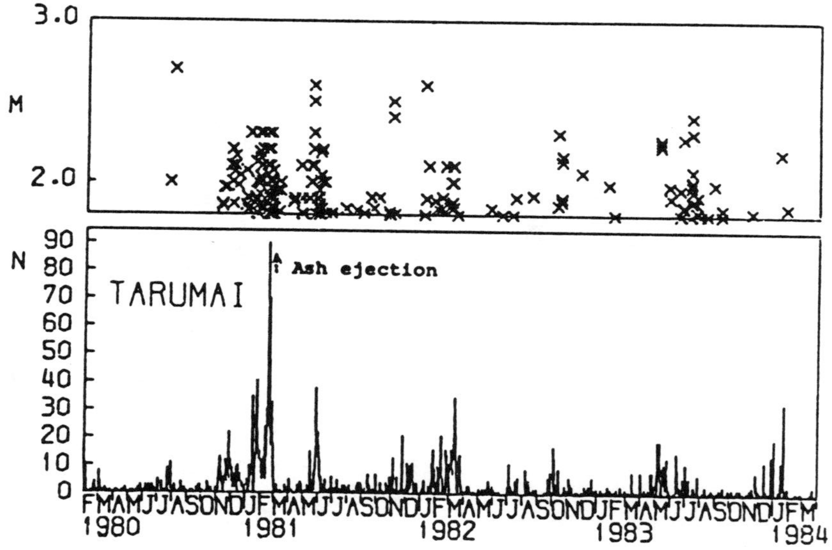

Figure 8.15.4. Daily frequency of volcanic earthquakes at Tarumai during 1980-84, from Yamashita and others (1984).

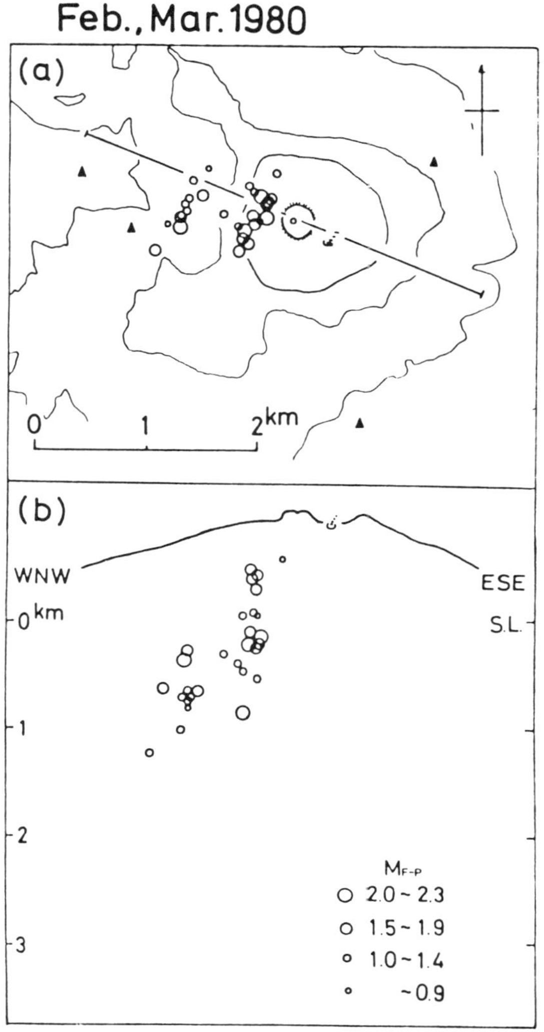

Figure 8.15.5. Epicenter distribution (a) and a WNW-ESE vertical cross section of hypocenters (b) in February and March 1980, from Umehara and others (1981). Diameters of circles are in proportion to magnitudes.

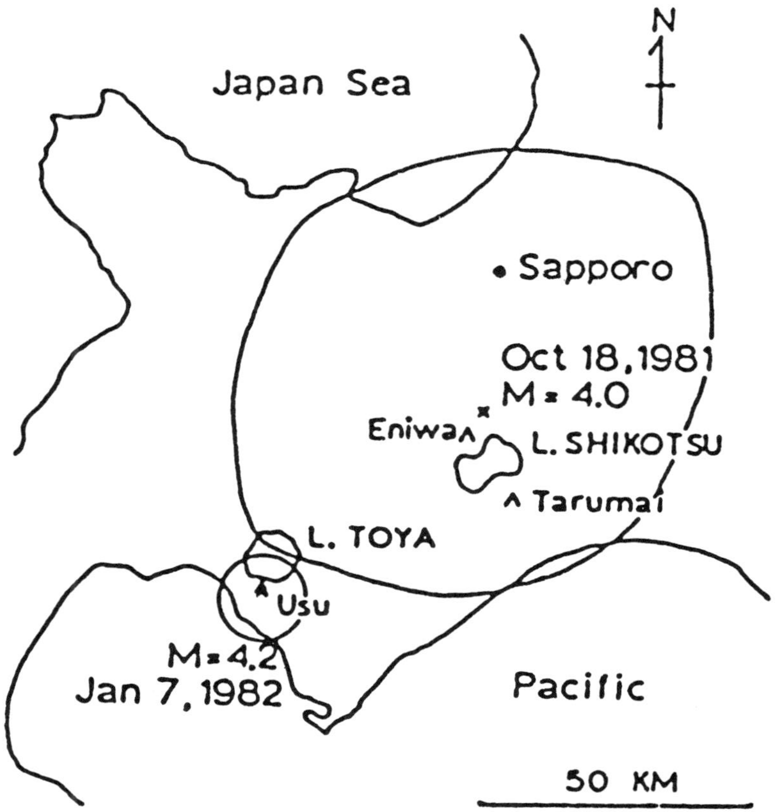

Figure 8.15.6. Epicenters of larger shocks near Shikotsu Caldera in 1981 and 1982, from Okada and Watanabe (1983).

Type	predominant period	order of arrival time	initial motion	epicentral areas preliminarily deduced
I	$0.1\sim0.2^{sec}$ at stalionn N	1 N — 2 D S	relatively sharp at station N. The senses are almost push.	near dome
II	$0.04\sim0.2^{sec}$ at stations D and S	1 D S — 2 N	sharp st stations D and S. The senses are not systematic.	the northeastern flank
III	$0.1\sim0.2^{sec}$ at station F	1 F — 2 S, 3 D N	relatively sharp at station F. The senses are all pull.	near the deepst part of Lake Shikotsu.
IV	$0.1\sim0.4^{sec}$ at stations N, D and S	1 S D — 2 N	relatively sharp at stations S and D. The senses are systematic.	the northeastern flank

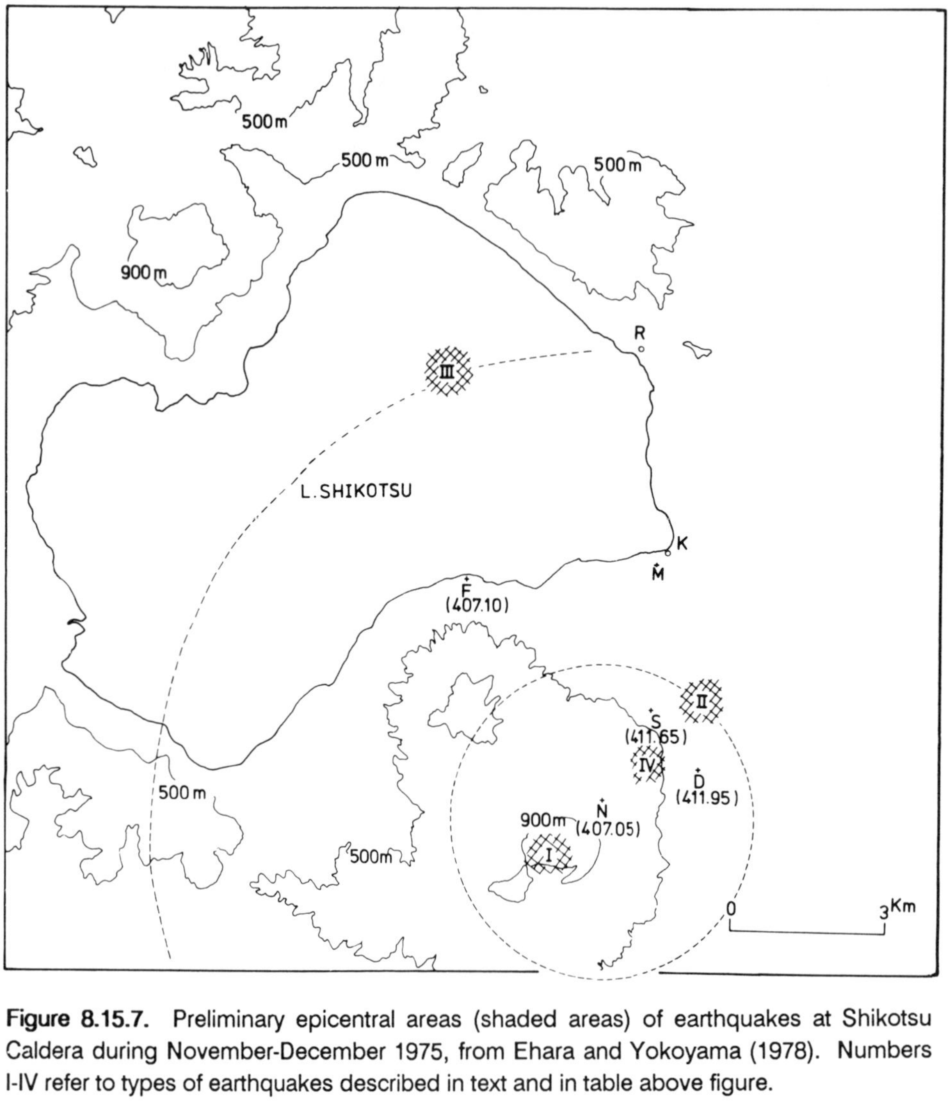

Figure 8.15.7. Preliminary epicentral areas (shaded areas) of earthquakes at Shikotsu Caldera during November-December 1975, from Ehara and Yokoyama (1978). Numbers I-IV refer to types of earthquakes described in text and in table above figure.

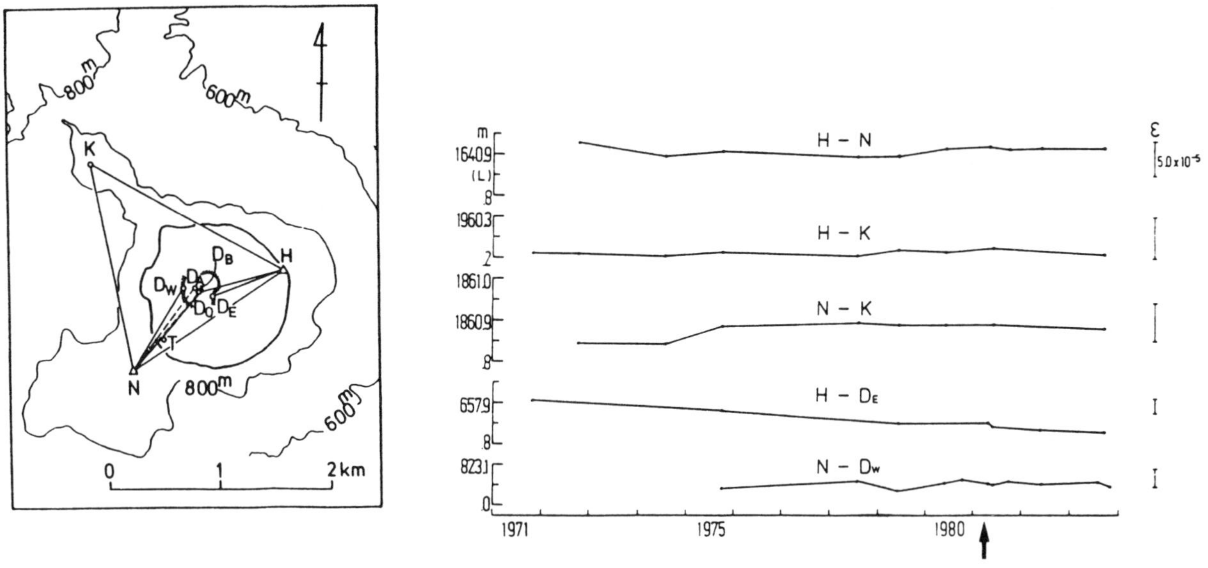

Figure 8.15.8. Base lines for distance and angle measurements at Tarumai Volcano (left). Solid lines, both angle and distance measurements; broken lines, angle measurements only. Secular changes in line length from 1971 to 1983 (right). Both from Yamashita and others (1984). Arrow marks February 1981 eruption.

Figure 8.15.9. Tilt changes from an automated tiltmeter in southwest part of Tarumai summit area (left), leveling network for tilt measurements (middle), and tilt changes from tilt levelings (Yamashita and others, 1984).

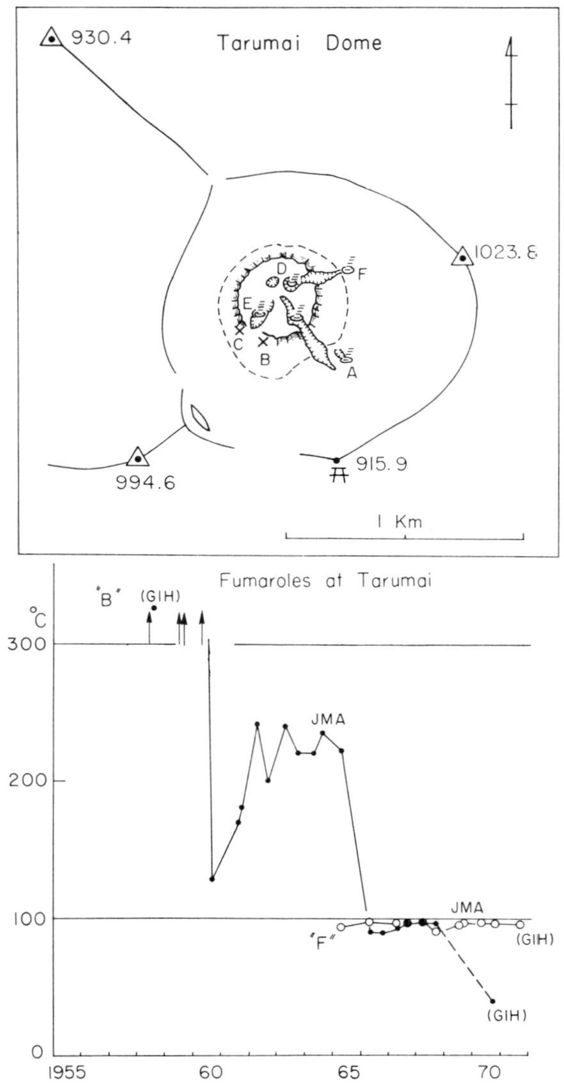

Figure 8.15.10. Location map and temperature changes in fumaroles B (solid circles) and F (open circles) at Tarumai Volcano from 1955 to 1971, from Ishikawa and others (1972).

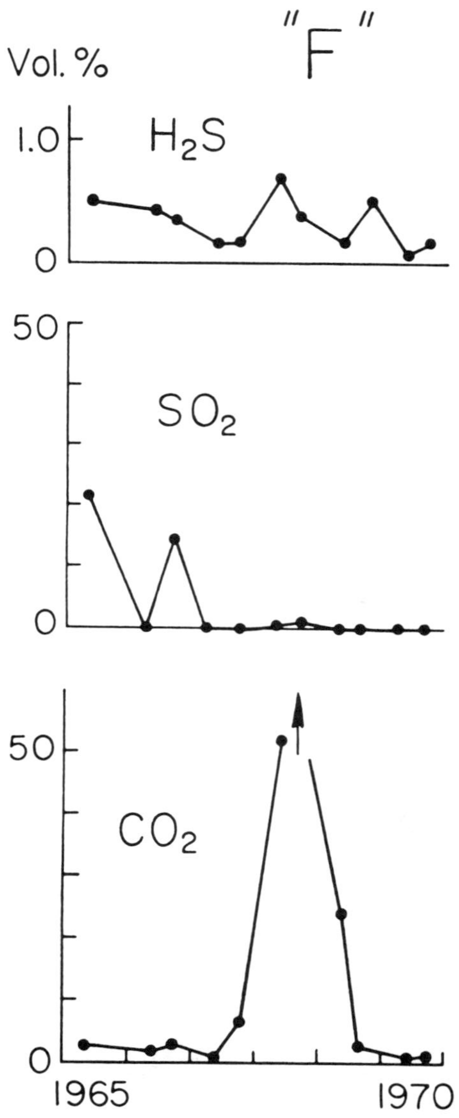

Figure 8.15.11. Changes in composition of fumarolic gases at Tarumai Volcano from 1965 to 1970, from Ishikawa and others (1972).

DAISETSU-TOKACHI GRABEN

CAVW number (active vent)	Latitude Longitude (degrees)	Diameter (km)	Local tectonic setting	Precaldera edifice	SiO_2 content (wt pct)	Age of collapse (yr B.P.)	Date of unrest	Type and duration of unrest ESTU STHF MGTF H Te	Eruption type
08-05-05	43.42N	20 x 30?	Exten	Pyroclastic	R = 47-61		1670	---- ---- ---- - --	lf
(Tokachi-dake)	142.68E	(Daisetsu)		plateau	C = 72		1854-57+	---- ---- ---? - --	ex
08-05-06	(Tokachi)						1887	---- ---- ---- - --	ex
(Daisetsu)	43.68N						1889	---- ---- ---- - --	ex
	142.88E						1923-28	YY-- ---- --YY Y QU	Ex,landsl,mf
	(Daisetsu)						1933	---- ---- ---x - --	none
							1936-37	---- ---- ---x - --	none
							1946-47	---- ---- ---x - --	none
							1952-62	EE-- ---F --EF B QU	pex, Ex
							1968-69	EE-- ---- ---- - QU	none
							1984-88	DDE- ---- --E- D --	pex

TECTONIC SETTING

The Daisetsu-Tokachi graben is a largely filled, NE-SW-trending graben in central Hokkaido, Japan. Tokachi-dake fills the southwestern end of the graben; the Daisetsu complex lies at the northeast end of the graben. A large negative gravity anomaly suggests a buried caldera at Daisetsu (Katsui and others, 1979, p. 29) (figs. 8.16.1, 8.16.2).

GEOLOGIC HISTORY

The Daisetsu-Tokachi graben is surrounded by an extensive plateau of rhyolite ash-flow tuffs that were erupted between 75,000 yr B.P. and 2.8 m.y. B.P. (Ono and others, 1981). Voluminous tephra layers can be traced to this plateau (Sasaki and others, 1974) and several filled calderas have been postulated, including Daisetsu Caldera (Katsui and others, 1979; Japan Association for Quaternary Research, 1987). Tokachi-dake and nearby cones are aligned along the southeast margin of the Daisetsu-Tokachi graben. North and South Daisetsu and some related cones filled the inferred Daisetsu Caldera, and the 2-km diameter Ohachidaira Caldera formed at the center of North Daisetsu Volcano about 30,000 years ago (Moriya, 1980). Asahi-dake is a late-stage cone of the Daisetsu complex, and vents along an E-W fissure through Asahi-dake may have been active as recently as 500-600 yr B.P. (Katsui and others, 1979).

DAISETSU-TOKACHI GRABEN (continued)

HISTORICAL ACTIVITY

1923-28: Emission of steam and other gases increased at Tokachi-dake in 1923; in the same year, Netunuma solfatara near the crater rim was replaced alternately by either a pond of molten sulfur or a pond of boiling water. The temperature of a hot spring on the northwest flank of Sulphur Hill, near the crater, increased remarkably. Fumarolic activity, puffing, and small explosions intensified during December 1925 - March 1926 and again in early April 1926, with a moderately strong explosion on 5 April. Fumarolic activity and puffing intensified still further on 4 May 1926, accompanied by rumblings and indistinctly felt earthquakes. "Fire phenomena [probably hot gases] were seen often " (Tanakadate, 1927). An explosion and small mudflow occurred around noon on 24 May, and a landslide, large explosion, and mudflow occurred at 1618 hr on 24 May 1926 (Murai, 1960). After the large explosion, activity gradually diminished in intensity until only weak puffing remained in December 1928. A large earthquake (M 6.8, 180 km from Tokachi) occurred on 5 September 1926 (Nishimura and Yamashita, 1982); three days later, Tokachi-dake erupted again.

Between 1928 and 1952, there were only occasional small increases in fumarolic activity in the Tokachi-dake crater.

1952-62: In August 1952, the usual background fumarolic activity began to intensify, and new pits (the Showa crater) formed. Frequent puffing was noted at Showa crater in September 1956, August 1957, October 1958, and August and November 1959. Distinct fissures opened a few years before 1962, on the east crater rim, and opened wider on 17 June 1962. Fumarole and ground temperatures in the Showa and nearby Ansei craters increased by 50 to about 150 °C from 1959 until early April 1962 (fig. 8.16.3), as they had also done in 1957-59 before minor eruptive activity (Ishikawa and others, 1971). A large earthquake off the southeast coast of Hokkaido on 23 April 1962 (M 7.0, 160 km from Tokachi) was followed immediately by a remarkable increase in fumarolic activity and by rocks falling from the crater walls. The pressure of sulfur vapor at fumaroles in the Showa crater increased, and more molten sulfur than usual was produced by the funnel condensers (sulfur mine). From early April to early June 1962, fumarolic temperatures rose from 150 °C to 300 °C; the ground temperature at Yunuma also increased markedly, from 45 °C in May 1960 to 76 °C in June 1962. Local earthquakes became more frequent (4, 9, 23, 27, and 28 May), and rockfalls from the east wall of the crater occurred on 29 April, 21 May, and 25 May. On 30 May, rumbling and steaming greatly intensified, and on 31 May many earthquakes occurred, including one of M 3.6 and one of M 3.3. Seismic energy release accelerated (fig. 8.16.4), and on 29 June hot muddy water gushed out of fumaroles in Kumanosawa, on the northwest slope of the central cone. A new fissure 4 m long formed on the ridge of Mae-tokati-dake. Then activity of fumaroles in Taisho crater decreased and felt earthquakes stopped, until a large explosion occurred from three vents around midnight of that same day. Earthquake counts increased slightly from early to mid-July 1962 and then returned to background level by the end of August (fig. 8.16.5); tremor stopped about 10 days after the 30 June explosion.

1968-69: A M 7.8 earthquake on 16 May 1968, 310 km from Tokachi, seems to have triggered unrest in the Tokachi area. A remarkable seismic swarm at Tokachi-dake started immediately after the large tectonic earthquake. Swarm activity declined temporarily, then increased again in December 1968 and peaked in January and March 1969 (figs. 8.16.6, 8.16.7). More than 10,000 volcanic earthquakes were recorded during the swarm activity. After March 1969, the wave form of the earthquakes changed to typical B-type and fuming activity increased slightly, but no eruption occurred (Hm. Okada, written commun., 1988). Yokoyama (1985) interpreted this episode as a stillborn eruption.

DAISETSU-TOKACHI GRABEN, Region 8, CAVW number 08-05-05 and 08-05-06

DAISETSU-TOKACHI GRABEN (continued)

HISTORICAL ACTIVITY (continued)

1984-88 and continuing: Small phreatic eruptions were observed on 19-20 June 1985, heralded by a minor seismic swarm (Katsui and others, 1987). Ground temperatures in the vicinity of the 1962 Tokachi-dake crater began to increase in 1984 and hot muddy water was ejected three times between 29 May and 31 July 1985. Temperatures continued to increase and were as high as 529 °C during measurements on 7 October 1986. Tremor was recorded on 1 September 1985, 20 December 1986, and several times in February 1987; tremor episodes continue as of March 1988. Swarms of small earthquakes occurred in November and December 1986, and as of March 1988 earthquake swarms continue and ground temperatures remain high around the 1962 crater (H. Okada, oral commun., 1988).

COMMENTS

The Daisetsu-Tokachi graben complex provides several good examples of hydrothermal unrest and eruptions apparently triggered by large tectonic earthquakes up to several hundred kilometers away. Calderas of this area are relatively old and it is not clear that any residual magma beneath the calderas has been involved in the unrest of this area.

REFERENCES

Ishikawa T., Yokoyama, I., Katsui, Y., and Kasahara, M., 1971, Tokachi-dake, its volcanic geology, history of eruption, present state of activity and prevention of disaster: Sapporo, Committee for Prevention of Disasters of Hokkaido, 136 p.

Japan Association for Quaternary Research, 1987, Quaternary maps of Japan: Tokyo, Univ. Tokyo Press, 110 p. and 4 map sheets.

Katsui, Y., and Takahashi, T., 1960, A note on the chemical composition of the lavas from the Daisetsu-Tokachi volcanic chain: Jour. Japan Assoc. Mineral. Petrol. Econ. Geol., v. 44, p. 142-151.

Katsui, Y., Yokoyama, I., and Ito, T., 1979, Asahi-dake, its volcanic geology, history of eruption, present state of activity, and prevention of disaster: Sapporo, Committee for Prevention of Disasters of Hokkaido, 42 p.

Katsui, Y., Yokoyama, I., Okada, Hm., and Oshima, H., 1987, Tokachidake (Supplement); Volcanic geology, history of eruption, present activity and prevention of disaster: Committee for Prevention of Volcanic Disasters Hokkaido, no. 11, 87 p.

Minakami, T., Hiraga, S., Miyazaki, T., and Uchibori, S., 1969, Fundamental research for predicting volcanic eruptions, Pt. II: Bull. Earthq. Res. Inst., v. 47, p. 893-949.

Moriya, I., 1980, Volcanoes in Hokkaido: Annex to Guidebook, Meetings and Field Studies of IGU Working Group on Perception of Environment, 24th Int. Geogr. Congr., 44 p.

Murai, I., 1960, On the mud-flows of the 1926 eruption of Volcano Tokachi-dake, Central Hokkaido, Japan: Bull. Earthq. Res. Inst., v. 38, p. 55-70.

----- 1963, A brief note on the eruption of the Tokachi-dake Volcano of June 29 and 30, 1962: Bull. Earthq. Res. Inst., v. 41, p. 185-208.

DAISETSU-TOKACHI GRABEN (continued)

REFERENCES (continued)

Nishimura, Y., and Yamashita, H., 1982, Earthquake swarms at Meakan volcano in March, 1982: Geophys. Bull. Hokkaido Univ., v. 41, p. 65-76.

Ono, K., Soya, T., and Mimura, K., 1981, Volcanoes of Japan (2nd ed.): Tokyo, Geological Survey of Japan, 1 map sheet, scale 1:2,000,000.

Sapporo District Meteorological Observatory, 1971, Report on the activities of volcano Tokachidake, Hokkaido: Tech. Rep. Japan Meteorological Agency, v. 74, p. 1-156.

Sasaki, T., Goto, K., Katayama, M., Katsui, Y., Kikuchi, K., Kimura, M., Kitagawa, Y., Oba, Y., Satoh, H., Tamura, S., and Tomioka, E., 1974, Distribution of the late Quaternary pyroclastic deposit in Hokkaido, Japan: Misc. Publ. Hokkaido Natl. Agric. Exp. Stn., no. 4, 1 map sheet with explanations, scale 1:600,000.

Smithsonian Institution, Scientific Event Alert Network (SEAN), 1985, 1987, Tokachidake Volcano: SEAN Bull., v. 10, nos. 6-7, 12; v. 12, no. 5 (information from the Volcanological Center, Japan Meteorological Agency).

Tada, F., and Tsuya, H., 1927, The eruption of the Tokachi-dake Volcano, Hokkaido, on May 24th, 1926: Bull. Earthq. Res. Inst., v. 2, p. 49-84.

Tanakadate, H., 1927, Explosive eruption of Tokachi-dake, Hokkaido, Japan: Bull. Volcanol. ser. 1, no. 13-14, p. 158-170.

----- 1928, The eruptive explosion of Tokati-dake, Hokkaido, in the year 1926: Proc. 3rd Pan-Pacific Sci. Congr., Tokyo, v. 1, p. 784-786.

Yokoyama, I., 1964, Seismometrical observation of the 1962 eruption of Volcano Tokachi, Hokkaido, Japan: Bull. Volcanol., v. 27, p. 217-223.

Yokoyama, I., 1985, Still-born eruption of Tokachi-dake Volcano in 1968-1969 [abs.]: Bull. Volcanol. Soc. Japan, v. 30, p. 98.

Yokoyama, I., Seino, M., Motoya, Y., Abe, T., Mishina, H., and Sasaki, Y., 1963, Seismometrical observation of the 1962 eruption of Volcano Tokachi, Hokkaido, preliminary report: Geophys. Bull. Hokkaido Univ., v. 10, p. 101-109.

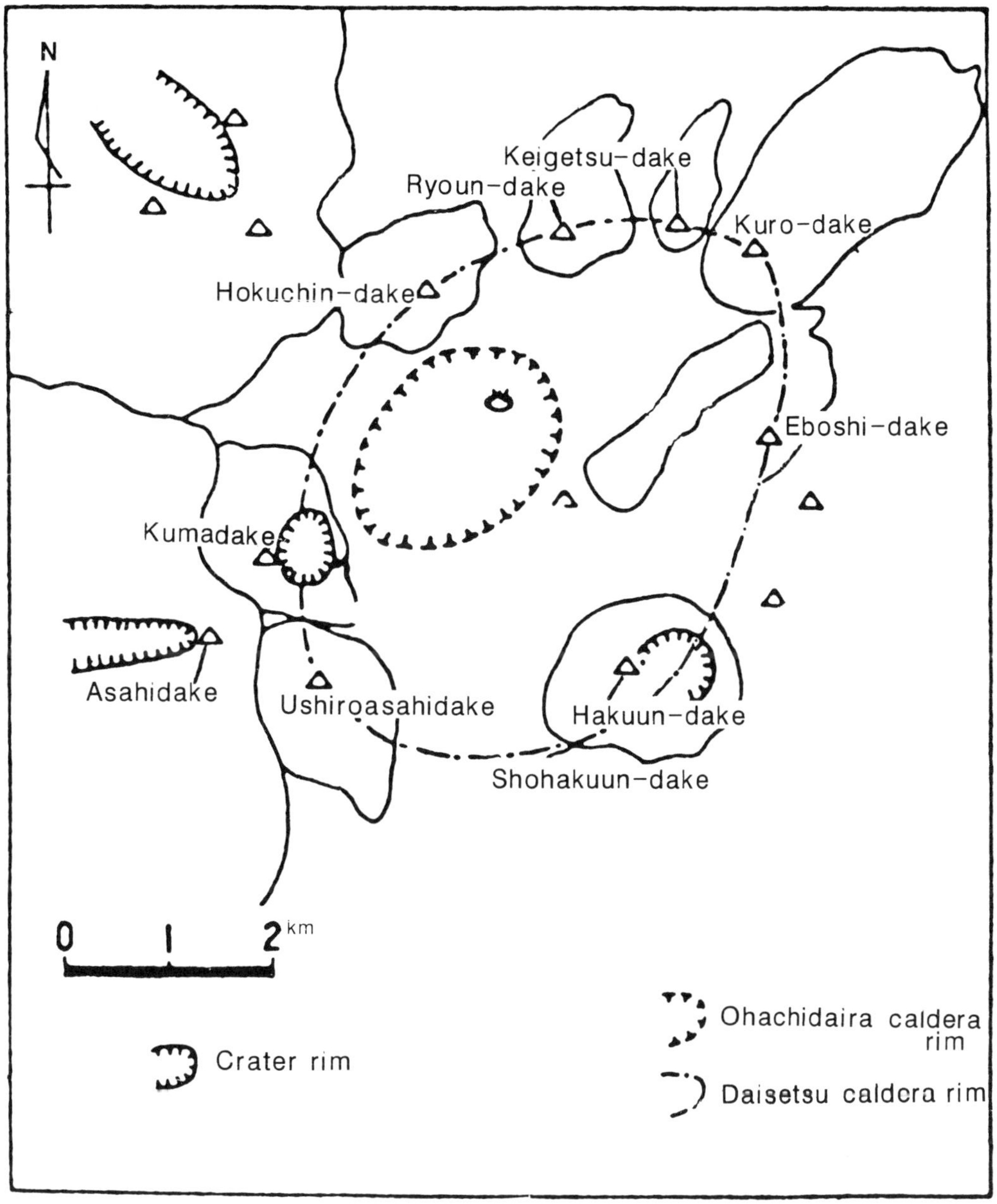

Figure 8.16.1. Simplified structure of composite Daisetsu Volcano and Daisetsu Caldera, from Katsui and others (1979). Older 6 km x 4 km caldera buried by later ejecta. Many domes extruded along caldera rim, including Hakuun-dake, Shohakuun-dake, Eboshi-dake, Kuro-dake, Keigetsu-dake, Ryoun-dake, and Hokuchin-dake. After those domes were built, another large eruption formed Ohachidaira Caldera inside Daisetsu Caldera. More recent activity occurred at Kumaga-dake and Ushiroasahidake.

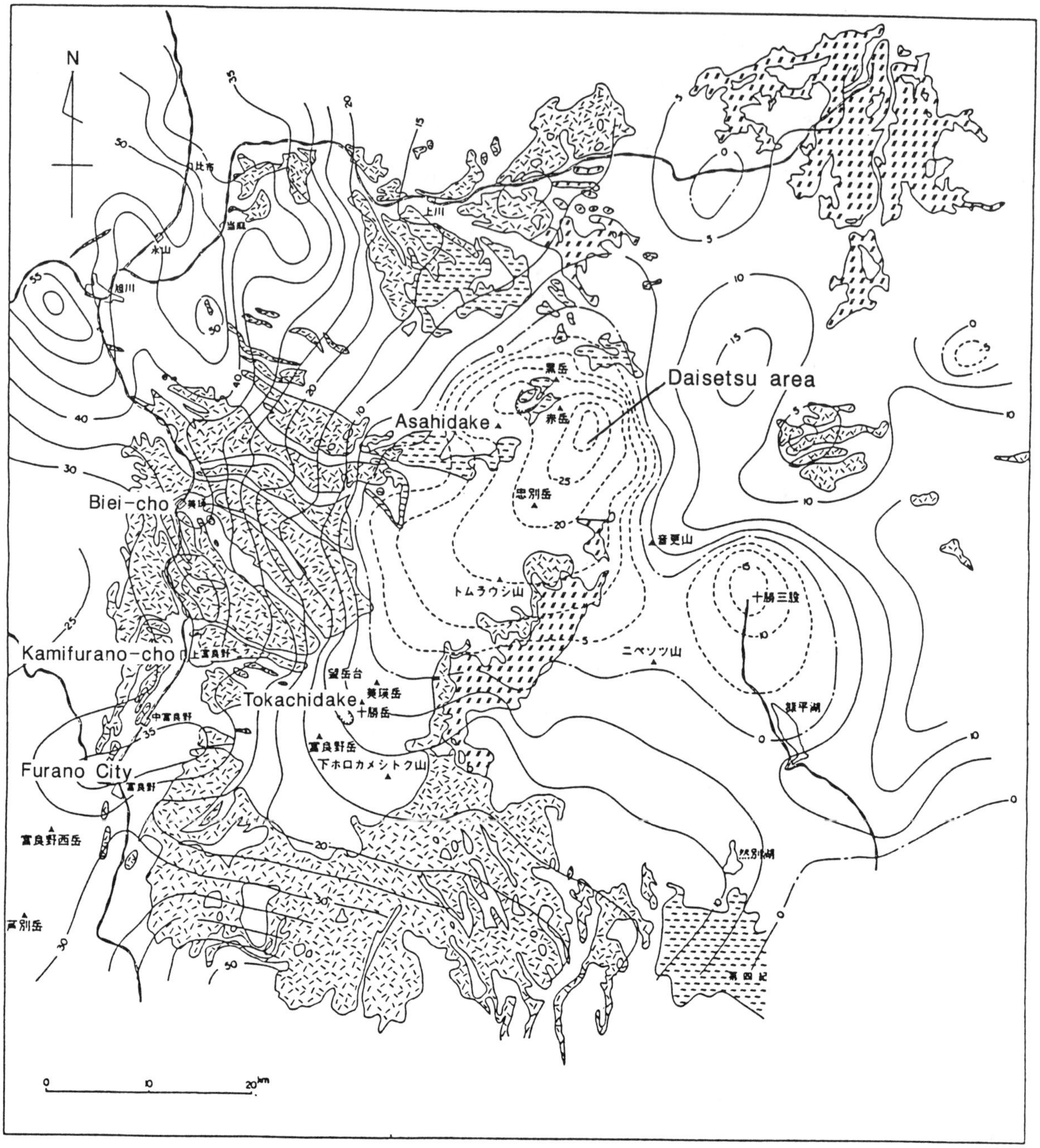

Figure 8.16.2. Bouguer gravity anomalies and distribution of 3 welded tuffs (stippling) in and around Daisetsu-Tokachi graben, after Katsui and others (1979). Volcanism occurred at both ends of graben, at Daisetsu in the northeast and at Tokachi in the southwest. Daisetsu Caldera marked by strong negative gravity anomaly; no similar anomaly exists around Tokachi. NE-SW elongation of gravity anomaly follows Daisetsu-Tokachi graben. Smaller proposed caldera about 35 km east of Daisetsu-Tokachi also marked by circular negative gravity anomaly.

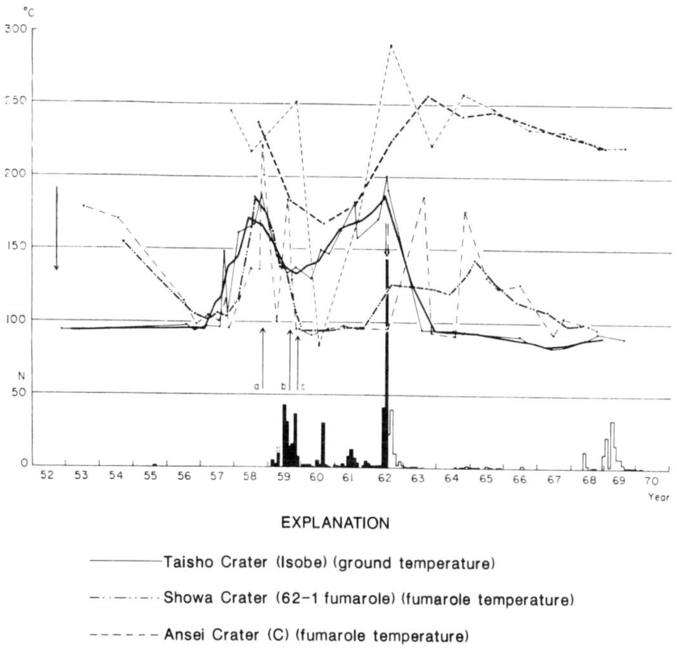

Figure 8.16.3. Seismic activity (bottom) and related thermal changes (top) at Tokachi-dake from 1952 to 1970, from Ishikawa and others (1971).

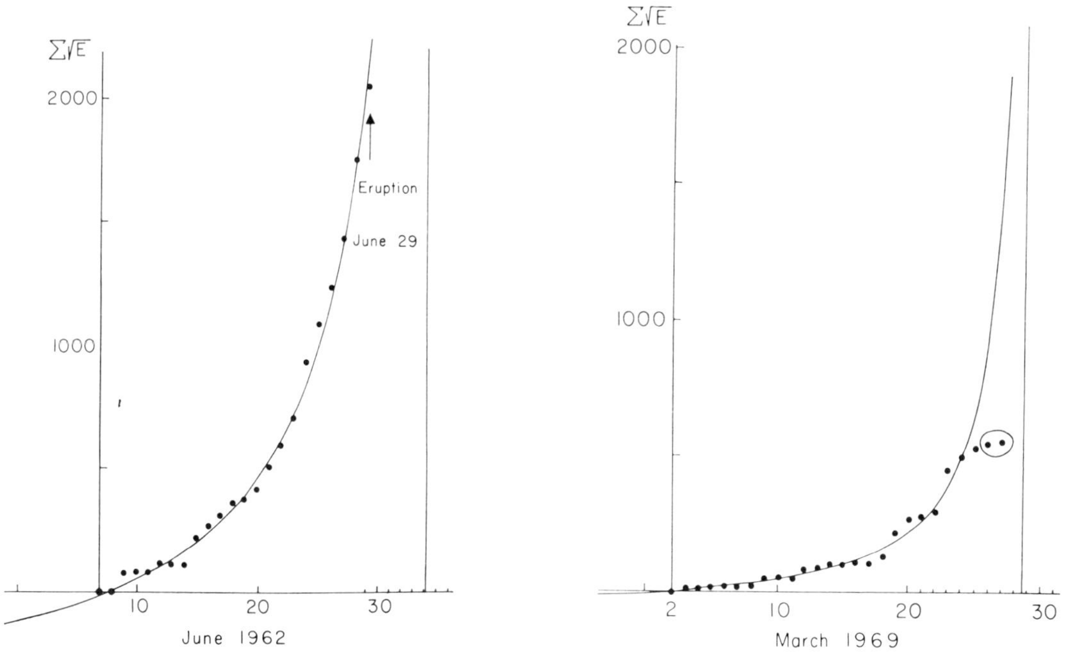

Figure 8.16.4. Cumulative seismic energy release associated with eruptions of Tokachi-dake in 1962 (left) and 1969 (right), from Ishikawa and others (1971). Circled points that fall below line at right might reflect saturation of seismic records immediately before eruption.

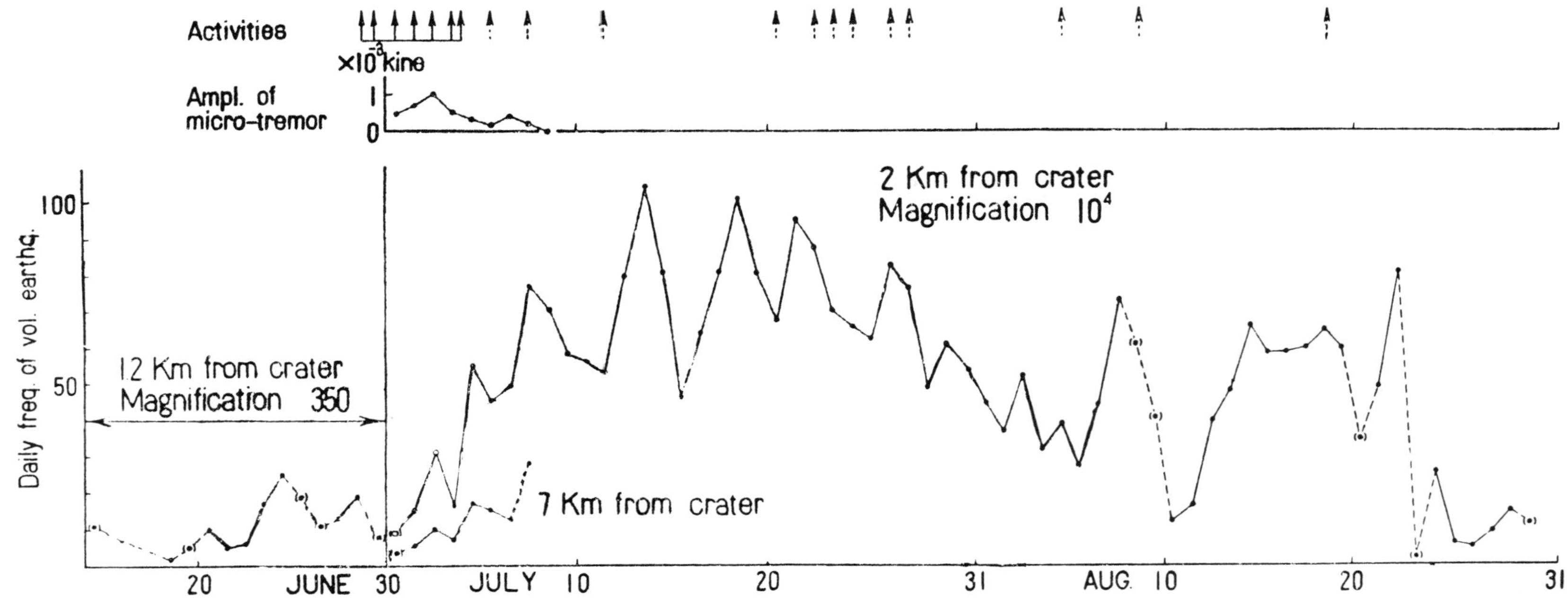

Figure 8.16.5. Daily frequency of volcanic earthquakes (bottom), amplitude of volcanic microtremors at Tokachi-dake (middle), and eruptive activity at Tokachi-dake (top) during 1962, from Yokoyama and others (1963).

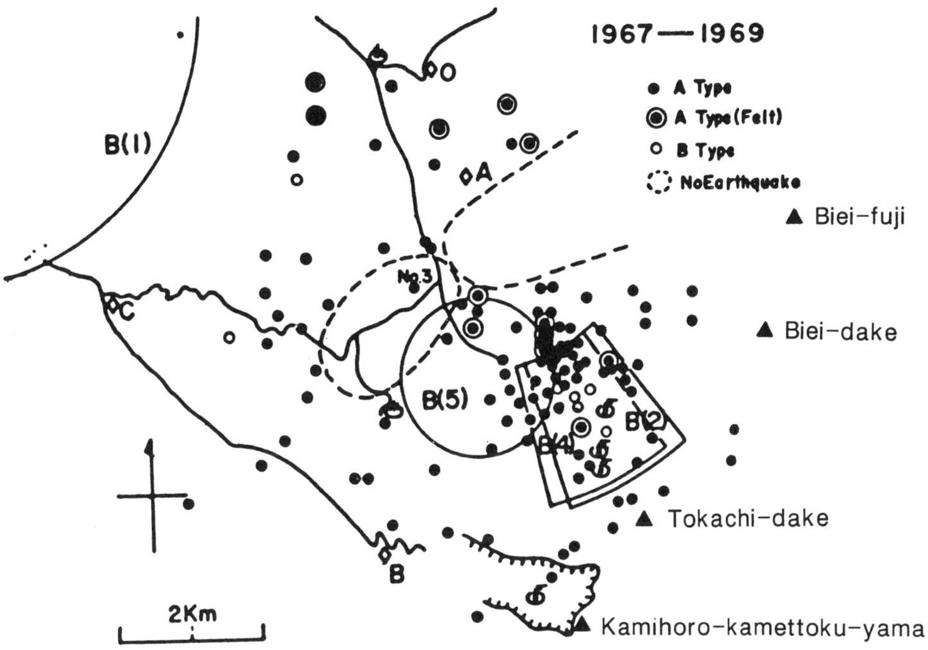

第3.2.20図　1967年～1969年の震央図

Figure 8.16.6. Epicenters of earthquakes near Tokachi during 1967-69, from Sapporo District Meteorological Observatory (1971).

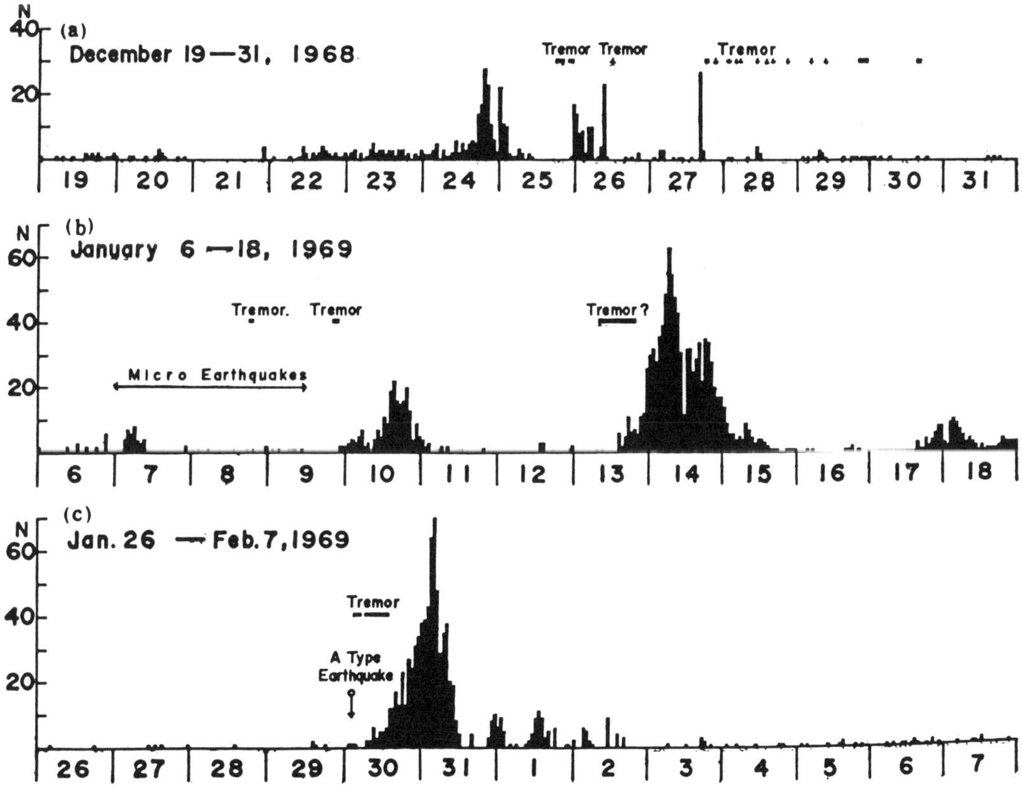

(a) 1968年12月19～31日　(b) 1969年1月6～18日　(c) 1969年1月26日～2月7日

Figure 8.16.7. Seismic activity near Tokachi-dake for three intervals during 1968-69, from Sapporo District Meteorological Observatory (1971).

PART 3: HISTORICAL UNREST AT LARGE QUATERNARY CALDERAS (continued)
See inside back cover for explanation and abbreviations

AKAN

CAVW number (active vent)	Latitude Longitude (degrees)	Diameter (km)	Local tectonic setting	Precaldera edifice	SiO_2 content (wt pct)	Age of collapse (yr B.P.)	Date of unrest	Type and duration of unrest ESTU STHF MGTF H Te	Eruption type
08-05-07 (Me-akan)	43.38N 144.02E	24 x 13	Exten?		R = 52-63 C = 67	>31,500	ca. 1800	---- ---- ---- - --	ex?
							1927	E--- ---- ---- - --	pex?
							1933	x--- ---- ---- - QU	none
							1937	---- ---- ---C - --	none
							1951-66	ECC- ---- --EE E UU	pex
							1973	CC-- ---- ---- - UU	none
							1982	DD-- ---- ---- - QU	none
							1987-88+	DDC- ---- --FF - --	pex

TECTONIC SETTING

Akan Caldera is near the southwestern end of the Kurile arc, along which the Pacific plate is subducted in a northwesterly direction beneath the Eurasian plate at a rate of approximately 10 cm/yr (see additional notes under Kutcharo Caldera, next section).

GEOLOGIC HISTORY

Akan Caldera (fig. 8.17.1) is an elongate depression that trends NE-SW and is partly filled by Lake Akan. Pyroclastic flows of the caldera-forming eruptions have a volume of approximately 100 km^3 (Ishikawa and others, 1969). Postcaldera cones including O-akan, Me-akan (fig. 8.17.2), Furebetsu, and Fuppushi have formed within the caldera; of these, only Me-akan has been historically active. Products of Me-akan have been pyroxene andesite and basaltic andesite (Moriya, 1980).

HISTORICAL ACTIVITY

1927: In April and May 1927, strong tremors and local shocks were felt in the main valley of the Siramizu-gawa, which is regarded as a fault line formed by the caldera depression (Tanakadate, 1939). Also, "rumblings" were reported at Me-akan-dake.

1933: "Rumblings" were reported "at the time of" the $\underline{M}$=8.3 Sanriku earthquake of 1933.

AKAN (continued)

HISTORICAL ACTIVITY (continued)

1937: On 2 December 1937, a smoke cloud rose on the east foot of Me-akan volcano; it was thought to be "an augmentation of the activity of the above-mentioned fumaroles, which were probably activated by the melted snow and the low atmospheric pressure, which was 757 mm in the city of Kusiro at that time" (Tanakadate, 1939).

1951-66: "Rumblings" (probably shallow earthquakes) were heard beginning on 31 July 1951. In August 1951, rumbling and activity of hot springs and solfataras increased. Thermal activity on the shore of Lake Akan declined during December 1951 - January 1952. Renewed seismic activity (400 local shocks) occurred for about 2 days after a M=8.2 earthquake beneath Tokachi-Oki (100 km southeast of Akan) on 4 March 1952. Earthquakes of these swarms were estimated to be 0-10 km deep (Minakami and Mogi, 1959). It is not clear whether changes at Akan before the earthquake were in some way precursors to the earthquake or unrelated to it. Small amounts of ash may have been emitted during activity in 1951 and 1952. The magnitude of a reported temperature increase in August 1951 is unknown; however, two fumaroles with temperatures of 166 and 157 °C in August 1951 had cooled to 142 and 101 °C, respectively, by July 1952 (Kizawa and others, 1953).

"Rumblings" were heard again in 1954, accompanied or followed by a small ash emission; a seismometer with 150x magnification was installed shortly thereafter. The first significant eruption in historical time began on 19 November 1955. The maximum temperatures of products (in and after 1955) was 300 °C, but there is no evidence that the ejecta were incandescent or juvenile (Sakuma and Murase, 1957). The low-magnification seismometer recorded only one earthquake during the period 23-30 November 1955. More sensitive seismometers (up to 15,000x magnification) were operated during June and August 1956 and showed increased swarm activity a few days before an explosion on 15 June 1956. The daily frequency of earthquakes, however, was low in comparison to that before magmatic eruptions of Asama and Sakurajima. Low-amplitude tremor (period=0.5 sec) occurred on 13-14 June 1956.

The water temperature at two Akan-kohan hot springs on the southwest shore of Lake Akan decreased by 3 °C from June to September 1958, an M=5.7 earthquake occurred on the rim of Akan Caldera on 22 January 1959, and a M=6.2 earthquake occurred about 20 km east of Akan and just south of the Kutcharo Caldera on 31 January 1959 (Motoya, 1981). An explosion of 3 October 1959 was preceded by about 20 days of gradually increasing seismicity and was followed by one day of sharply increased seismicity (fig. 8.17.3) (Murase and others, 1960). The precursory seismicity in 1959 was greater than that recorded at the same station immediately before the explosion in 1956. Hypocenters were not determined, but on the basis of S - P intervals, they were probably within 9 km of a station near the summit crater; after the explosion, earthquakes were shallower, within 2.5 km of the seismic station (Murase and others, 1960).

1973-79: In early and (especially) mid-June 1973, seismicity increased at Akan-kohan and at Me-akan hot springs (fig. 8.17.4). This increase was followed by a 2-day lull and then by a M=7.4 earthquake off Nemuro. Seismicity also increased on 24 June, the same day as a M=7.1 earthquake off Nemuro (exact temporal relation not clear from graph) (Yokoyama and others, 1976). A small swarm of earthquakes at Me-akan was associated with a nearby tectonic earthquake (M=3.9) in July 1976, and there were additional earthquake swarms in July 1979 (figs. 8.17.5, 8.17.6) (Nishimura and Yamashita, 1982).

AKAN (continued)

HISTORICAL ACTIVITY (continued)

1982: Seismicity increased slightly for 2 days near Akan (regional or local?) before a $\underline{M}$=7.1 earthquake off Urakawa (182 km southwest of Akan) on 21 March 1982. A sharp increase in local seismicity at Me-akan immediately followed that earthquake (figs. 8.17.5-8.17.7) (Nishimura and Yamashita, 1982). Local seismicity peaked on 25 March (84 events) and returned to background by the first week of April 1982. In March 1982, 411 seismic events were recorded, compared with totals of 97, 45, 491, 254, and 194 events in the years 1977-81, respectively (Smithsonian Institution, 1982). Some of the earthquakes are in a northwest-trending zone of regional seismicity that passes between the Akan and Kutcharo Calderas, close enough to Akan that stress continuity can be assumed (see fig. 8.18.2, Kutcharo Caldera, next section).

1987-88+: Geothermal and fuming activity gradually increased for several years; the highest temperature measured was 510 °C in October 1987. An earthquake swarm started on 9 December 1987; tremor started on 4 January 1988. Minor ash emission was observed occasionally starting on 6 January 1988. Tremor accompanied additional ash eruptions on 7-8 February and 18-19 February 1988 (Smithsonian Institution, 1987, 1988; Hm. Okada, written commun., 1988).

COMMENTS

Akan Caldera provides a good example of sensitivity of a magmatic system to regional tectonic earthquakes. Some earthquake swarms at Akan are apparently precursors to large tectonic earthquakes, whereas others are triggered by large earthquakes. In another case, temperature changes in hot springs at Akan immediately preceded a large tectonic earthquake.

REFERENCES

Ishikawa, T., Katsui, Y., Oba, Y., and Sato, H., 1969, Some problems of the calderas in Hokkaido: Bull. Volcanol. Soc. Japan, v. 14, no. 2, p. 97-108.

Katsui, Y., 1958, Akan-Kutcharo volcano group: Chikyu-Kagaku (Earth Sci.), no. 39, p. 19-29.

Kimura, M., 1978, Significant eruptive activities related to large interplate earthquakes in the northwestern Pacific margin: Jour. Phys. Earth, v. 26 supp., p. S557-S570.

Kizawa, T., Ono, Y., and Ide, S., 1953, Field investigation in the Akan Volcanic Region (in 1952): Quart. Jour. Seismol., v. 17, no. 4, p. 55-59.

Kuno, H., 1962, Catalogue of the active volcanoes of the world including solfatara fields, Pt. XI, Japan, Taiwan and Marianas: Rome, IAVCEI, 332 p.

Minakami, T., 1956, Report on volcanic activities and volcanological studies in Japan for the period from 1951 to 1954: Bull. Volcanol., v. 18, p. 39-76

----- 1962, Report on volcanic activities in Japan for the period from 1957 to 1959: Bull. Volcanol., v. 24, p. 7-22.

AKAN (continued)

REFERENCES (continued)

Minakami, T., and Mogi, K., 1959, Report on volcanic activities in Japan for the period from 1954 to 1957: Bull. Volcanol., v. 21, p. 127-152.

Minakami, T., and Sakuma, S., 1953, Report on volcanic activities and volcanological studies in Japan for the period from 1948 to 1951: Bull. Volcanol., v. 14, p. 79-132.

Moriya, I., 1980, Volcanoes in Hokkaido: Annex to Guidebook, Meetings and Field Studies of IGU Working Group on Perception of Environment, 24th Int. Geogr. Congr., 44 p.

Motoya, Y., 1981, Change in hot springs with relation to earthquakes in the Teshikaga region, eastern part of Hokkaido: Geophys Bull. Hokkaido Univ., v. 39, p. 71-80.

Murase, T., 1957, Geophysical studies of volcanoes in Hokkaido, Japan, Part 6: On the small earthquakes originated under the Volcano Me'akan-dake: Geophys. Bull. Hokkaido Univ., v. 5, p. 5-9.

Murase, T., Onda, I., Seino, M., and Nogoshi, M., 1960, Activity of Volcano Meakan-dake: Geophys. Bull. Hokkaido Univ., v. 7, p. 93-103.

Nishimura, Y., and Yamashita, H., 1982, Earthquake swarms at Meakan volcano in March, 1982: Geophys. Bull. Hokkaido Univ., v. 41, p. 65-76.

Ono, Y., and Minami, K, 1952, The investigation of the rumbling of Mt. Akan: Quart. Jour. Seismol., v. 16, no. 3-4, p. 99-108.

Ono, K., Soya, T., and Mimura, K., 1981, Volcanoes of Japan (2nd ed.): map with explanation: Geological Survey of Japan, Map Ser. no. 11, scale 1:2,000,000.

Sakuma, S., 1957, Volcanic tremor of Me'akan-dake: Jour. Fac. Sci., Hokkaido Univ., ser. 7, v. 1, p. 37-53.

Sakuma, S., and Murase, T., 1957, Recent activity of Volcano Me'akan-dake: Jour. Fac. Sci., Hokkaido Univ., ser. 7, v. 1, p. 21-36.

Smithsonian Institution, Scientific Event Alert Network (SEAN), 1982, 1987, 1988, Me-akan Volcano, Akan Caldera: SEAN Bull., v. 7, nos. 3, 8; v. 12, no. 12; v. 13, nos. 1, 2 (information from Japan Meteorological Agency).

Tanakadate, H., 1939, Volcanic activity in Japan and vicinity during the period between November 1935 and December 1938: Japan Jour. Astron. Geophys., v. 16, p. 95-120.

Yamaguti, T., and Ono, Y., 1953, Rumbling of Akan Volcano caused from the earthquake off the coast of Tokati: Quart. Jour. Seismol., v. 17, nos. 1-2, p. 69-76.

Yokcyama, I., 1958, Gravity survey of Kuttyaro caldera lake, Jour. Phys. Earth, v. 6, no. 2, p. 75-79.

Yokcyama, I., Katsui, Y., Ehara, S., and Koide, K., 1976, Me-akan-dake, its volcanic geology, history of eruption, present state of activity and prevention of disaster: Sapporo, Committee for Prevention of Disasters of Hokkaido, 138 p.

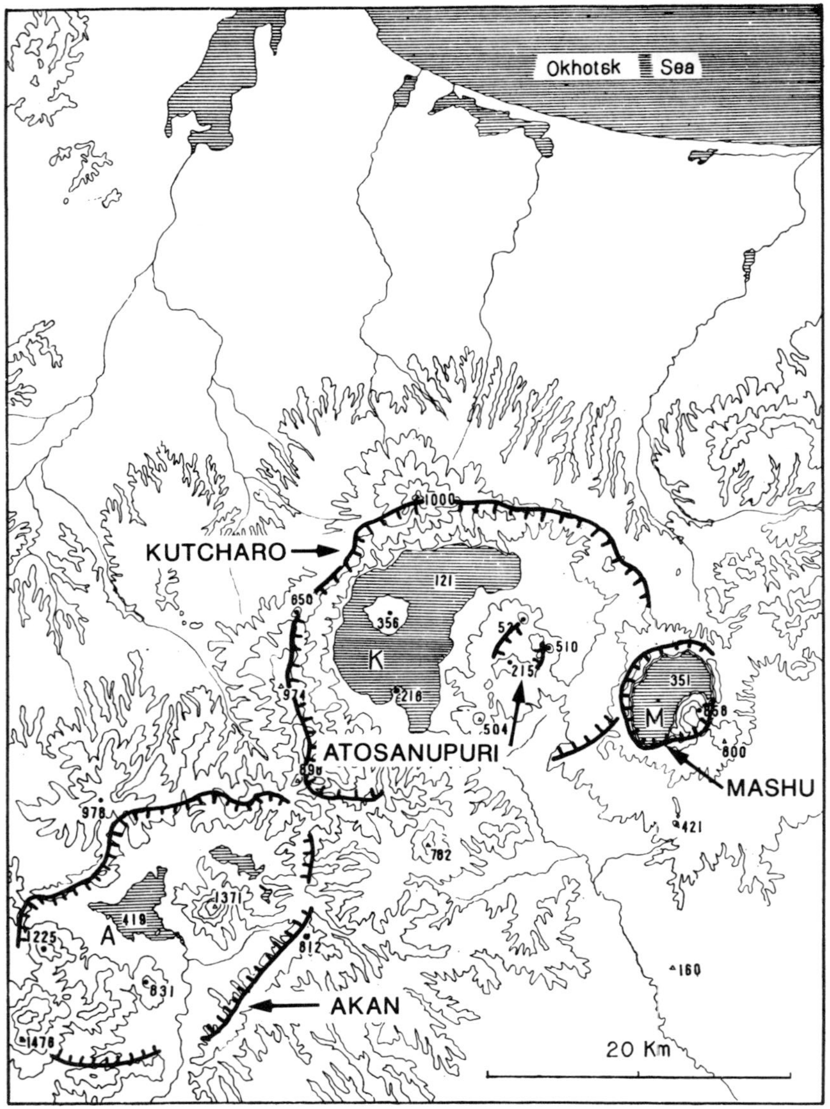

Figure 8.17.1. Topographic sketch map of Akan and Kutcharo Calderas, from Yokoyama (1958). Contour interval is 200 m.

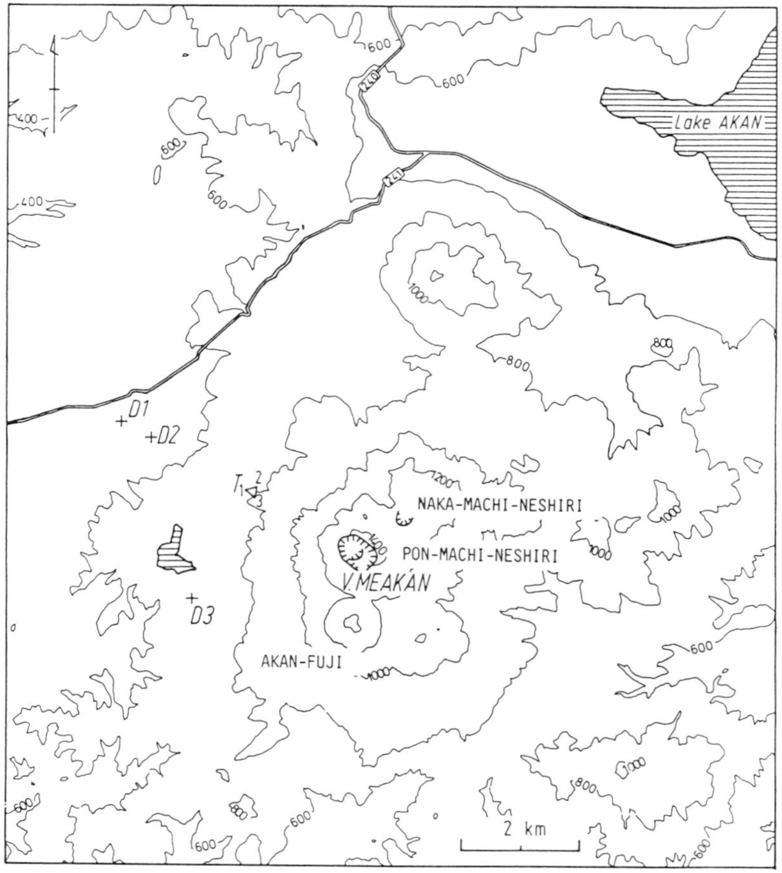

Figure 8.17.2. Topographic map of Me-akan Volcano, from Nishimura and Yamashita (1982). T, tripartite seismic array; D1-D3, seismometers with drum recorders.

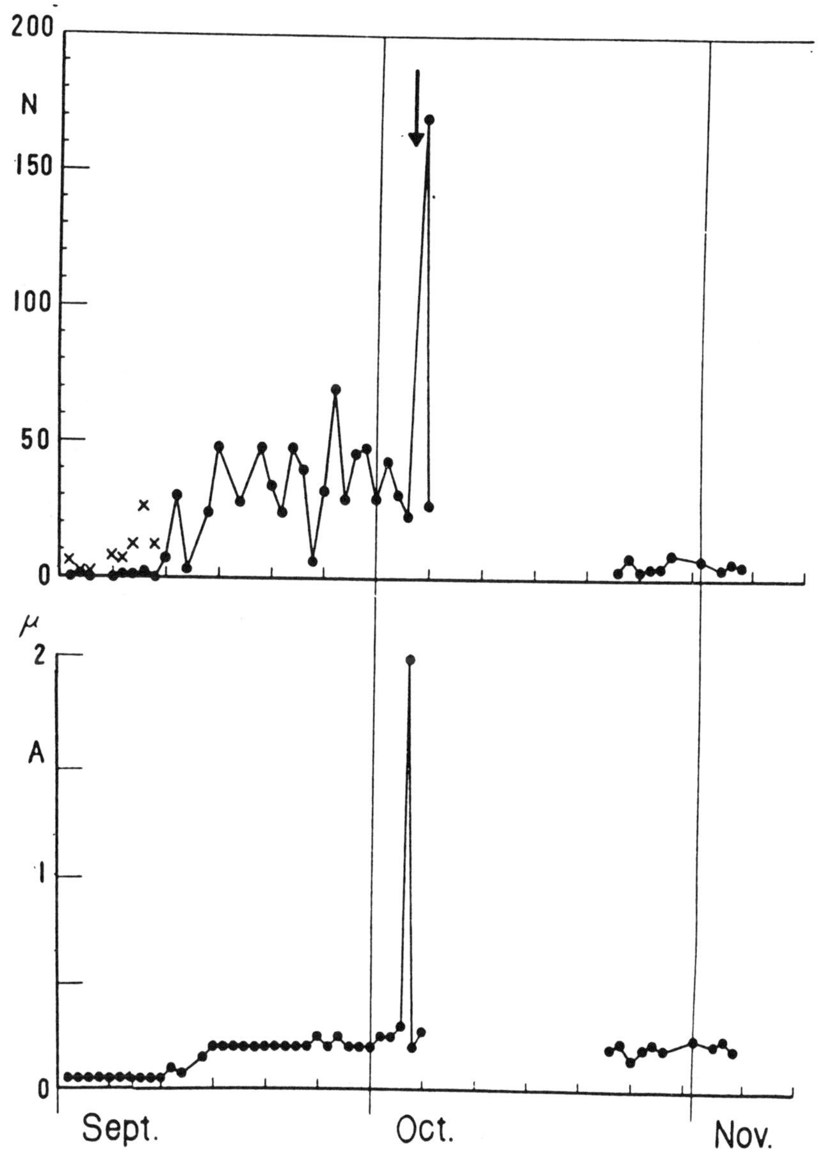

Figure 8.17.3. Diagram showing volcanic and seismic activity at Me-akan Volcano from 1 September to 4 November 1959 (Murase and others, 1960). Arrow indicates explosion. Top, frequency of earthquakes (x's, maximum amplitude greater than 0.2 microns; solid dots, maximum amplitude greater than 0.5 microns). Bottom, average maximum amplitude of volcanic tremor.

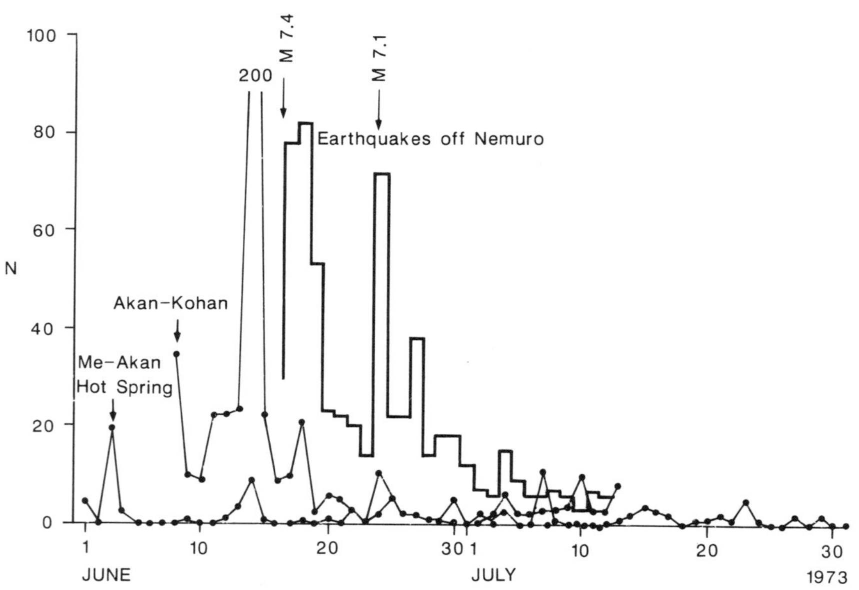

Figure 8.17.4. Relation between earthquake swarms at Akan Caldera and regional tectonic earthquakes, from Yokoyama and others (1976).

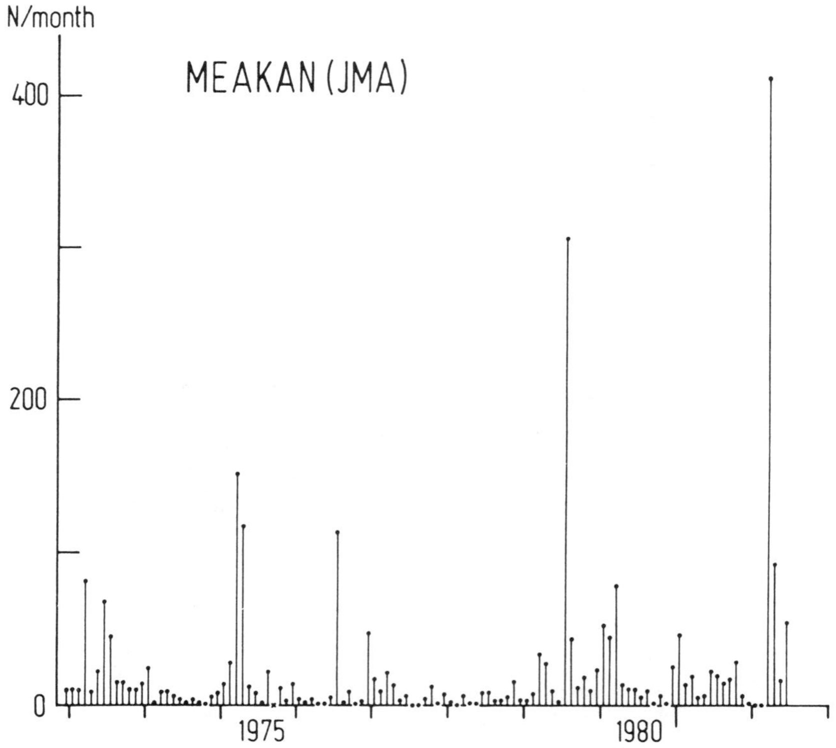

Figure 8.17.5. Monthly frequency of earthquakes at Me-akan Volcano during 1972-82, from Nishimura and Yamashita (1982).

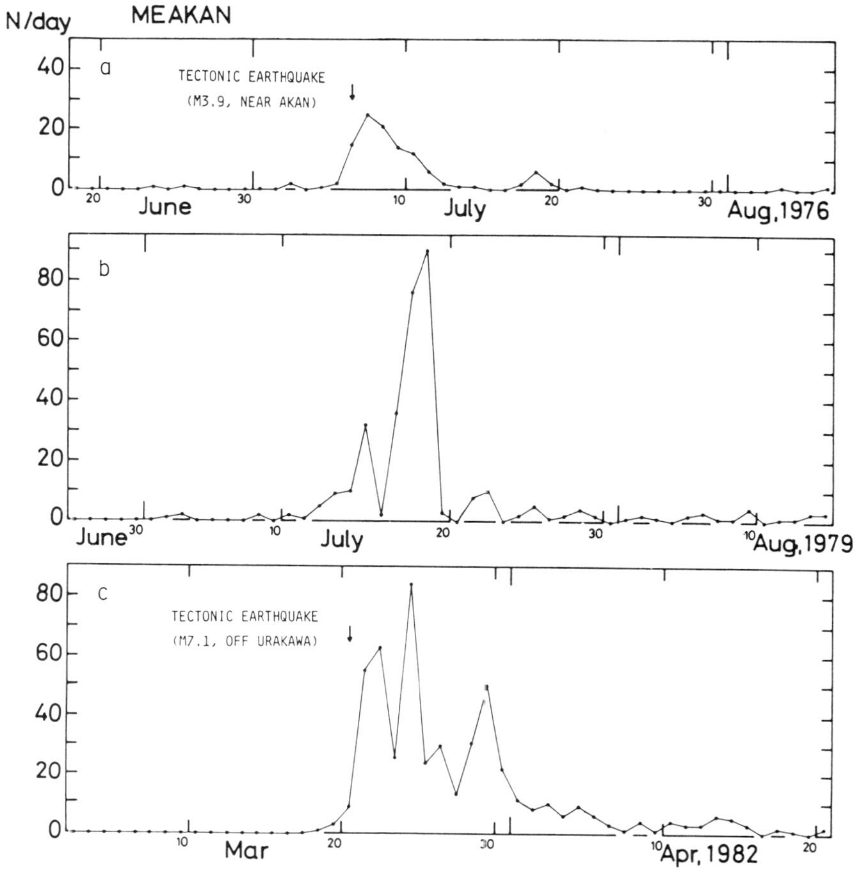

Figure 8.17.6. Three examples of earthquake swarms at Me-akan Volcano, two of which were associated with regional tectonic earthquakes, from Nishimura and Yamashita (1982).

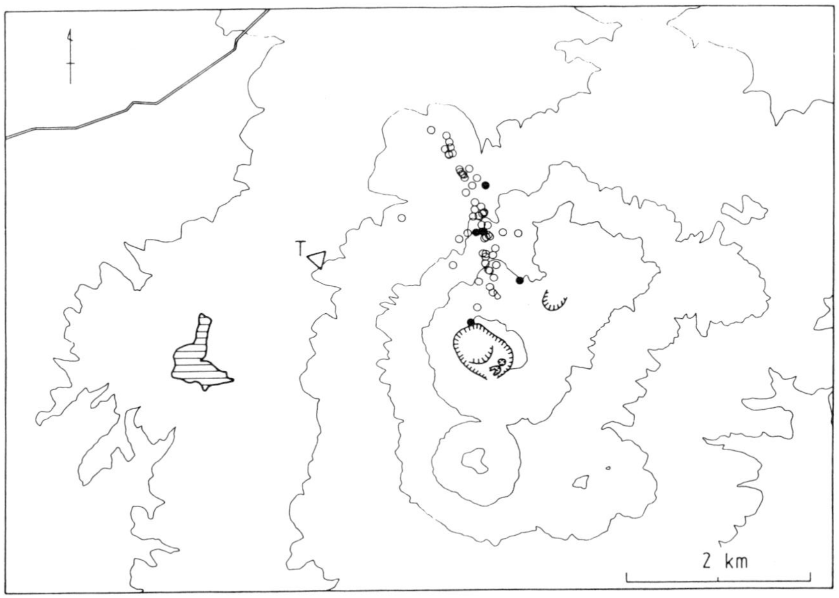

Figure 8.17.7. Distribution of epicenters near Me-akan Volcano in March 1982, from Nishimura and Yamashita (1982).

KUTCHARO (including MASHU)

CAVW number (active vent)	Latitude Longitude (degrees)	Diameter (km)	Local tectonic setting	Precaldera edifice	SiO_2 content (wt pct)	Age of collapse (yr B.P.)	Date of unrest	Type and duration of unrest ESTU STHF MGTF H Te	Eruption type
08-05-08	43.55N	20x26	NE-SW	?	R = 58-73	Approxi-	1938	xx-- ---x ---x x =?	none
(Atosanupuri)	144.43E	(Kutcharo)	alignment		C = 71-74	mately	1959	xx-- ---- ---- - =?	none
		6	of vents			30,000	1961	xx-- ---- ---- - --	none
		(Mashu)				(Kutcharo)	1963	xx-- ---- ---- - --	none
						7,000	1965	xx-- ---- ---- - --	none
						(Mashu)	1967	xx-- ---- ---- - --	none
							1968	xx-- ---- ---- - --	none
							1971	xx-- ---- ---- - --	none
							1975	xx-- ---- ---- - --	none

TECTONIC SETTING

Kutcharo Caldera (figs. 8.18.1, 8.18.2) is near the southwestern end of the Kurile arc, along which the Pacific plate is subducted in a northwesterly direction beneath the Eurasian plate at a rate of approximately 10 cm/yr. A historically active northwest-trending fault zone passes beneath the town of Teshikaga (fig. 8.18.2) along the southwestern edge of Kutcharo Caldera.

GEOLOGIC HISTORY

Kutcharo Caldera formed roughly 30,000 yr B.P. and is now mostly flooded by Kutcharo Lake (Katsui and others, 1975) (fig. 8.18.1). A postcaldera stratovolcano, Nakajima, forms an island in Kutcharo Lake near the west end of the caldera. A younger stratovolcano and cluster of domes, known collectively as Atosanupuri, lie near the center of the caldera. Mashu stratovolcano, with its own 6-km-diameter "Crater Lake"-type caldera, lies along and postdates the east-southeast rim of Kutcharo Caldera. The Mashu Caldera formed about 7,000 yr B.P. (Katsui and others, 1975). The latest activity of Mashu was a fairly large explosive eruption from a small intracaldera stratocone, Kamuinupuri, about 1,000 yr B.P. No present-day fumaroles or hot springs are known at Mashu.

KUTCHARO (including MASHU)

HISTORICAL ACTIVITY

A magnitude 6.0 earthquake occurred at a depth of 20 km beneath the southern end of Lake Kutcharo on 29 May 1938 (fig. 8.18.2) (Ishikawa, 1938; Tsuya, 1938; Suzuki and Hirota, 1973). Roughly 30 cm of normal-fault displacement, north-northeast (caldera) side down, was observed on a small peninsula near the southern end of the lake. About 4 km to the southeast, a roadway was offset by 2.4 m of left-lateral movement along a fault trending N25°E. During the earthquake, the top surfaces of two domes in the southern part of the caldera, Noboriondo and Maruyama, fractured in the direction of the chain (N60°W), and fumaroles on the north side of a third dome (Oakets, which had been dormant) became reactivated. Many hot springs spouted from new openings around the foot of Oakets dome. The lakeshore in the northern part of the chain of three domes rose about 1 m, while the southern part of that chain (including Oakets dome) appeared to have subsided by about the same amount. Maruyama dome also moved more than 1 m to the southeast, and many en echelon undulations of the land formed on its front (Tanakadate, 1939). Tanakadate (1939) believed that the earthquake was caused by magma intrusion beneath the domes, with some possible tectonic component; I. Yokoyama (written commun., 1983) finds no evidence for volcanic involvement with the 1938 events.

On 31 January 1959, a $\underline{M}$=6.2 earthquake occurred at a depth of 20 km, just southeast of the 1938 epicenter and just south of the caldera. A $\underline{M}$=5.7 foreshock occurred on the rim of neighboring Akan Caldera 9 days before the $\underline{M}$=6.2 earthquake near Kutcharo. Temperatures at Akan-kohan hot springs (in the Akan Caldera) increased by several degrees before the 31 January earthquake. "Zig-zag shaped" cracks formed at the east foot of Mt. Pekere and continued 2 km in a WNW-ESE direction, with 10 cm of vertical displacement and 25-30 cm of opening. Aftershocks were under the southern caldera rim (Seino and Okada, 1960), on the north side of the surface trace of the fault, and at depths of 5-8 km (Matumoto, 1959) or 10 km (Seino and Okada, 1960).

Additional earthquake swarms occurred beneath and near Kutcharo Caldera in 1961, 1963, 1965, 1967, 1968, 1971, and 1975 (I. Yokoyama, written commun., 1983). The largest events in 1965 were $\underline{M}$=5.1 and 5.0, at "0-km" depth (Nagamune, 1968). In November 1967, a $\underline{M}$=6.5 earthquake occurred at a depth of 20 km in the same general area as those in 1938 and 1959; it was followed by aftershocks in the vicinity of the main shock, beneath the Kutcharo Caldera, and along the rim of the Akan Caldera.

A few centimeters of regional subsidence in the Akan-Kutcharo area from 1955 to 1970 grades northeastward to as much as 20 cm of subsidence along the northeast tip of Hokkaido (Yokoyama and others, 1976). Suzuki and Hirota (1973) infer that the regional subsidence is a precursor to another great earthquake like that which last occurred offshore from the northeastern tip of Hokkaido in 1894 (fig. 8.18.4). They further infer that the earthquakes in the Akan-Kutcharo area from 1938 to the present reflect a slow, many-decades-long buildup of stress toward the next great earthquake, and that the stress is preferentially released in the Akan-Kutcharo area because the compressive strength of the crust there is "unexpectedly low, ... 13 or 14 bars."

Even though Mashu Caldera is relatively young and lies on the rim of the larger Kutcharo Caldera, it is not known to have been disturbed by recent, fairly strong earthquakes along the southern rim of Kutcharo Caldera (fig. 8.18.3).

KUTCHARO (including MASHU) (continued)

COMMENTS

Kutcharo and Akan Calderas seemingly respond to regional tectonic strain changes and therefore might be useful for predicting large tectonic earthquakes. The area may also release strain preferentially along a zone of weakness related to the caldera structures.

REFERENCES

Ishikawa, T., 1938, Report of the Kutcharo earthquakes (that) occurred on May 29, 1938: Kagaku (Science), Tokyo, v. 8, no. 10, p. 409-414.

Katsui, Y., Ando, S., and Inaba, K., 1975, Formation and magmatic evolution of Mashu Volcano, East Hokkaido, Japan: Jour. Fac. Sci., Hokkaido Univ., ser. 4, v. 16, no. 4, p. 533-552.

Katsui, Y., Yokoyama, I., Okada, Hm., Nishida, Y., Matsumoto, T., and Kawakami, N., 1986, Atosanupuri and Mashu (Kamuinupuri): its volcanic geology, history of eruption, present activity and prevention of disaster: Comm. for Prevention Volcanic Disasters Hokkaido, no. 10, p. 1-104.

Kuno, H., 1962, Catalogue of the active volcanoes of the world including solfatara fields, Pt. XI, Japan, Taiwan and Marianas: Rome, IAVCEI, 332 p.

Matumoto, T., 1959, Tesikaga earthquake of Jan. 31, 1959: Bull. Earthq. Res. Inst., v. 37, p. 531-544.

Motoya, Y., 1981, Change in hot springs with relation to earthquakes in the Teshikaga region, eastern part of Hokkaido: Geophys Bull. Hokkaido Univ., v. 39, p. 71-80.

Nagamune, T., 1968, The earthquake of November 4, 1967, near Lake Kuttyara, eastern Hokkaido: Travel time analyses for body waves: Zisin, v. 21, p. 237-240.

Ono, K., Soya, T., and Mimura, K., 1981, Volcanoes of Japan (2nd ed.): map with explanation: Geological Survey of Japan, Map Ser. no. 11, scale 1:2,000,000.

Seino, M., and Okada, Hm., 1960, Observation of aftershocks accompanied with Teshikaga-earthquake, by means of a seismograph of Ishimoto-type: Geophys. Bull. Hokkaido Univ., v. 7, p. 71-75.

Suzuki, S., and Hirota, T., 1973, Seismic activity in the eastern part of Hokkaido and its vicinity (Part 1)--Shallow earthquakes in the Teshikaga and Akan area: Zisin, v. 26, p. 77-92.

Tanakadate, H., 1939, Volcanic activity in Japan and vicinity during the period between November 1935 and December 1938: Japan Jour. Astron. Geophys., v. 16, p. 95-120.

Tsuya, H., 1938, Report of the study on the earthquakes in Kuttyaro district on May 29, 1938: Jour. Seismol. Soc. Japan, v. 10, p. 285-313.

Yokoyama, I., 1958, Gravity survey of Kuttyaro caldera lake: Jour. Phys. Earth, v. 6, no. 2, p. 75-79.

Yokoyama, I., Katsui, Y., Ehara, S., and Koide, K., 1976, Me-akan-dake, its volcanic geology, history of eruption, present state of activity and prevention of disaster: Sapporo, Committee for Prevention of Disasters of Hokkaido, 138 p.

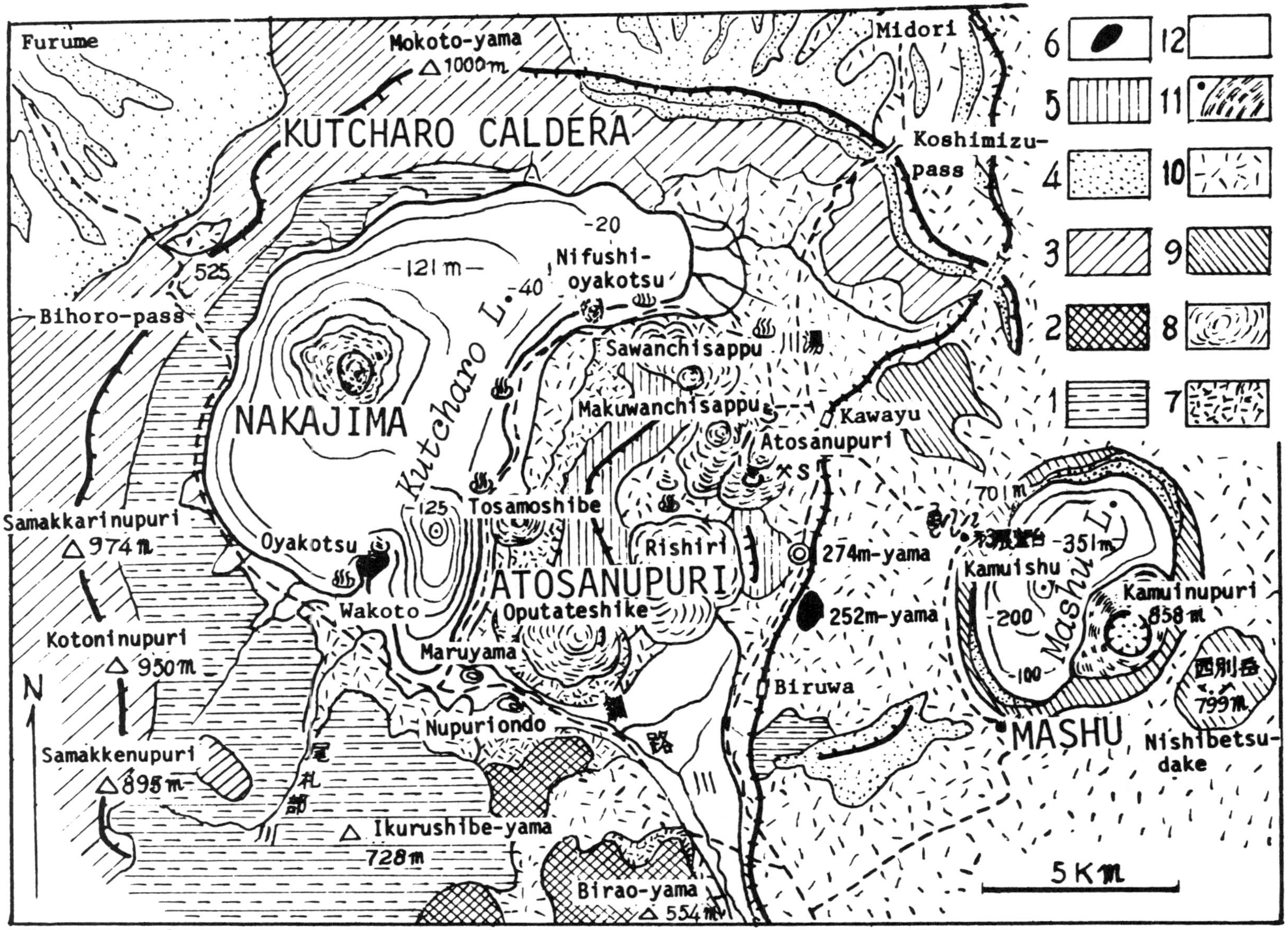

Figure 8.18.1. Geologic sketch map of Kutcharo Caldera area, from Katsui and others (1975). Explanation: 1, Neogene volcanic and sedimentary rocks; 2, Sattomonai volcanic rocks (Neogene?); 3, Kutcharo somma lava; 4, Kutcharo pyroclastic flow deposits; 5, Atosanupuri somma lava; 6, Oyakotsu and 252m-yama dome lavas; 7, Atosanupuri pyroclastic deposits; 8, Nakajima and Atosanupuri dome lavas; 9, Mashu somma lava; 10, Mashu pyroclastic deposits; 11, Kamuishu and Kamuinupuri lavas; 12, terrace deposits and alluvium.

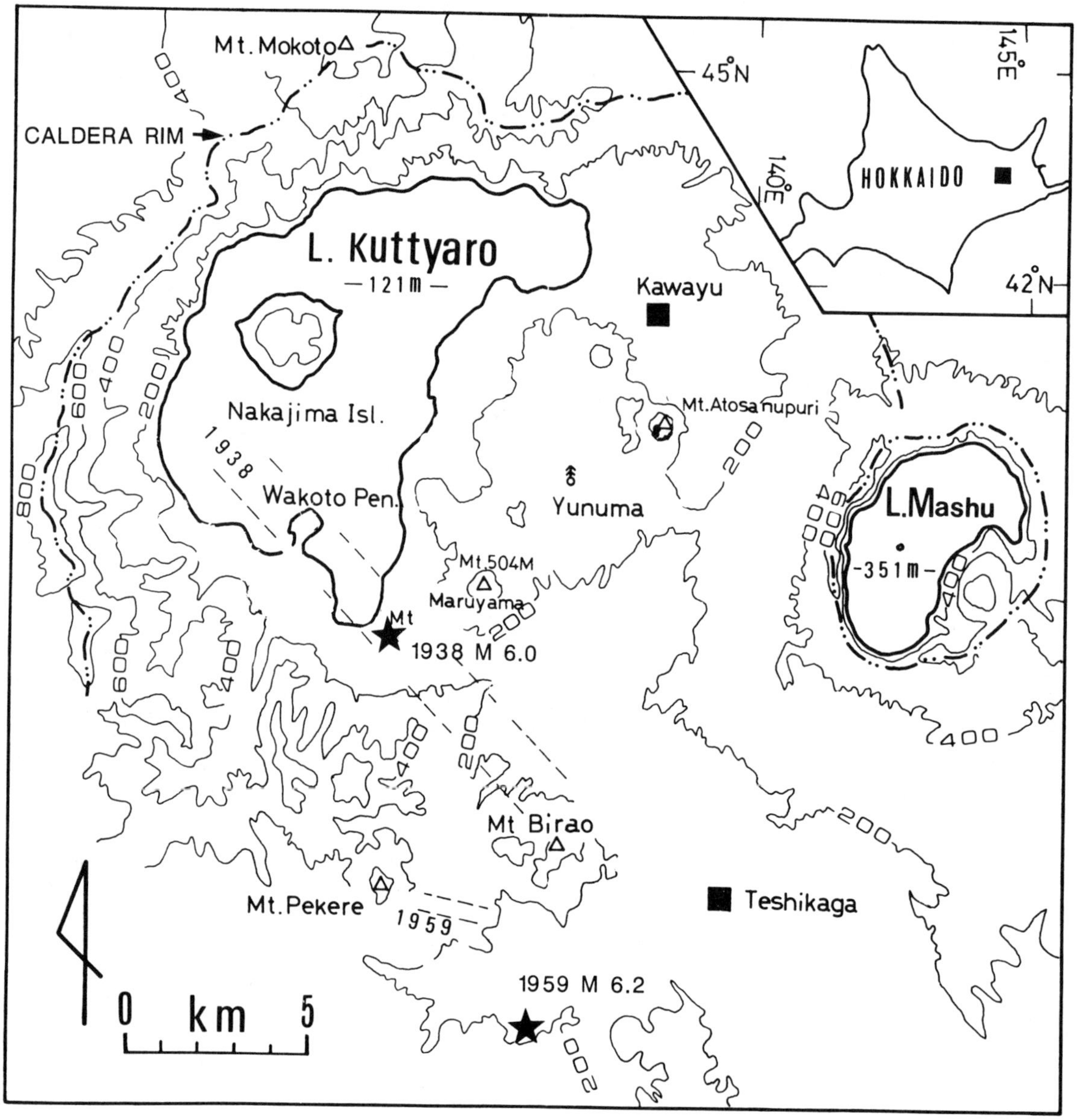

Figure 8.18.2. Topographic map showing Kutcharo and Mashu Calderas and epicenters of two magnitude 6.0 earthquakes in area. Dashed lines labeled 1938 and 1959 indicate surface breaks associated with earthquakes in those years (Y. Nishida, unpublished figures).

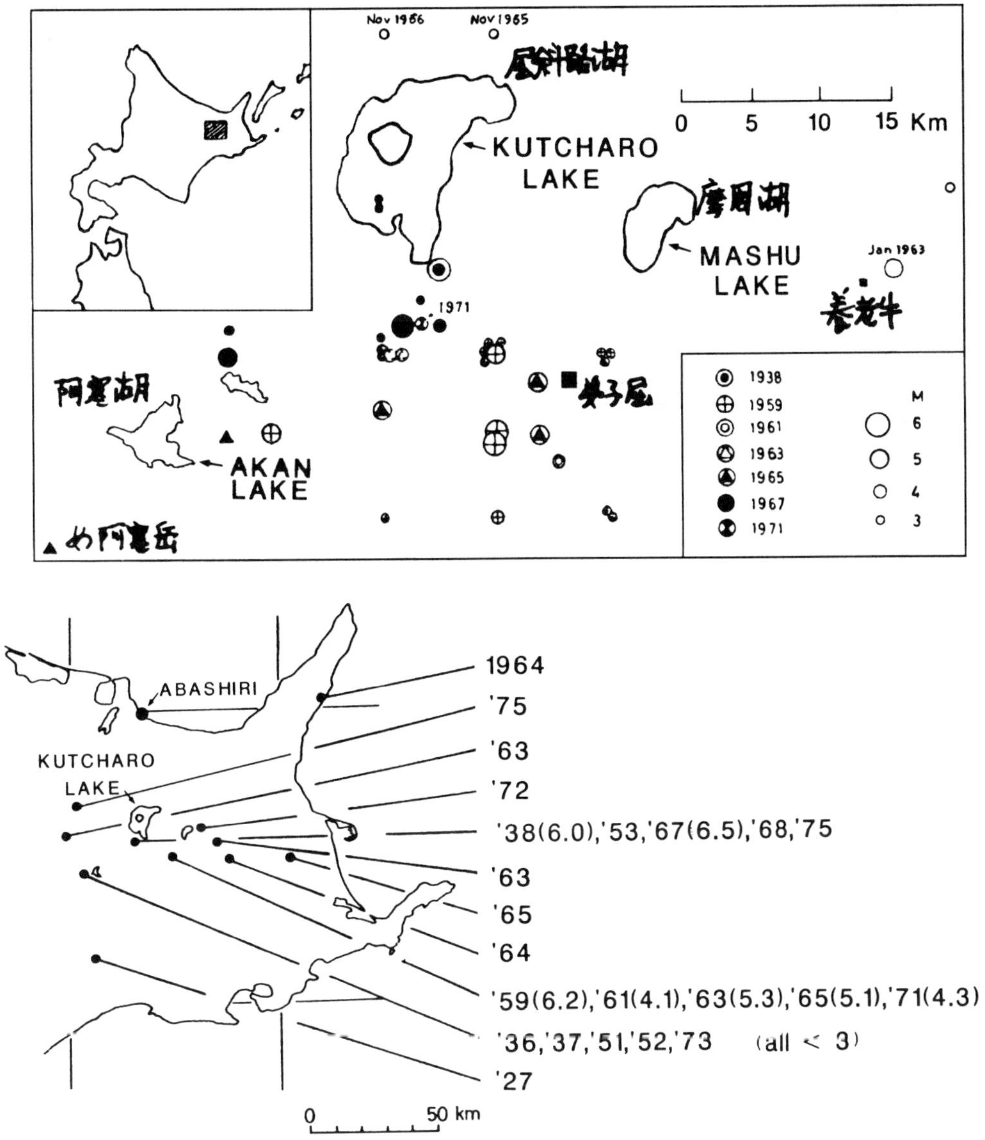

Figure 8.18.3. Epicenters of large historical earthquakes in Akan-Kutcharo-Mashu area of Hokkaido, after Yokoyama and others (1976) and Suzuki and Hirota (1973). Top, 1938 to 1971; bottom, 1927 to 1975.

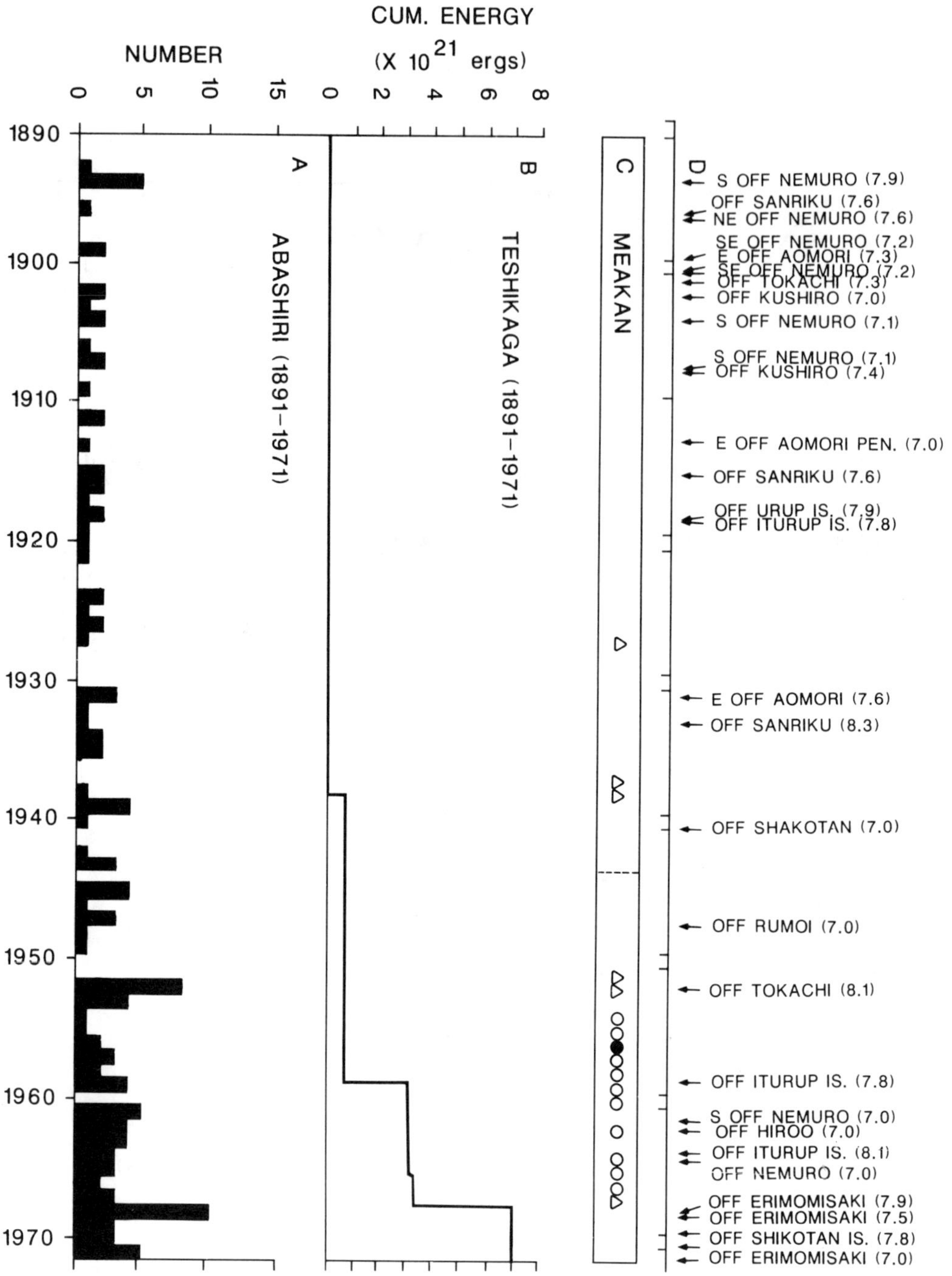

Figure 8.18.4. Seismic and volcanic activity near Kutcharo and Akan Calderas during 1891-1971, after Suzuki and Hirota (1973). A, Annual frequencies of felt earthquakes at Abashiri (45 km north of Kutcharo Caldera, on north coast of Hokkaido); B, cumulative seismic energy release in Teshikaga area since 1891; C, volcanic activity at Me-akan Volcano (solid circle, eruption with damage; open circle, eruption with slight damage; triangle, "anomalistic" volcanic phenomena (see preceding section on Akan)); D, earthquakes larger than M 7 near Hokkaido.

Historical Unrest at Large Calderas of the World

By CHRISTOPHER G. NEWHALL and DANIEL DZURISIN

Volume 2

U.S. GEOLOGICAL SURVEY BULLETIN 1855

DEPARTMENT OF THE INTERIOR
DONALD PAUL HODEL, Secretary

U.S. GEOLOGICAL SURVEY
Dallas L. Peck, Director

Library of Congress Catalog Number 88-600541

UNITED STATES GOVERNMENT PRINTING OFFICE, WASHINGTON : 1988

For sale by the
Books and Open-File Reports Section
U.S. Geological Survey
Federal Center, Box 25425
Denver, CO 80225

CONTENTS

Volume 1.

Volume 2.

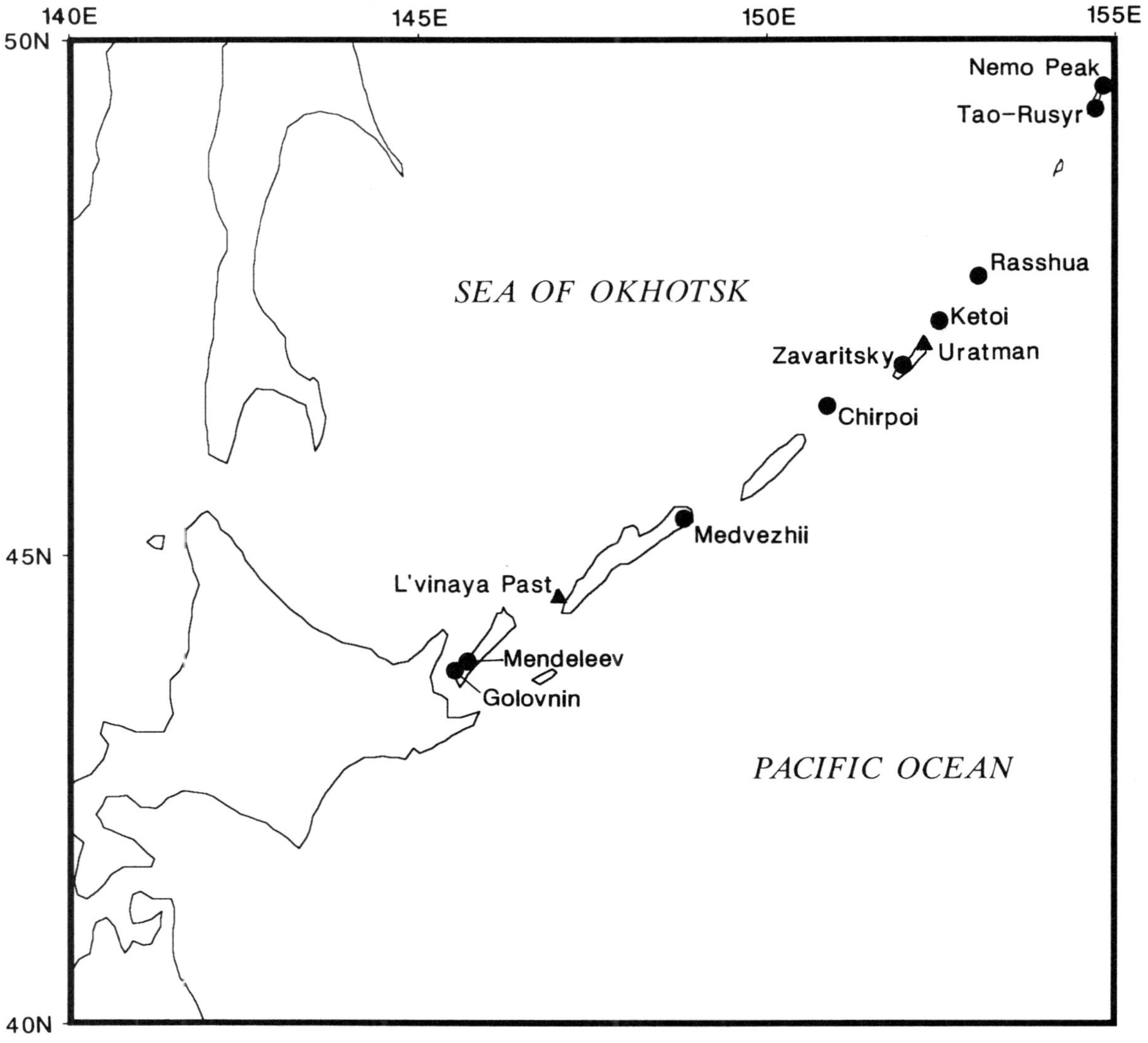

Figure 9. Locations of large, Quaternary restless calderas (solid circles) and nonrestless calderas (triangles) in Region 9.

PART 3: HISTORICAL UNREST AT LARGE QUATERNARY CALDERAS (continued)

See inside back cover for explanation and abbreviations

GOLOVNIN

CAVW number (active vent)	Latitude Longitude (degrees)	Diameter (km)	Local tectonic setting	Precaldera edifice	SiO_2 content (wt pct)	Age of collapse (yr B.P.)	Date of unrest	Type and duration of unrest ESTU STHF MGTF H Te	Eruption type
09-00-01 (Golovnin)	43.82N 145.53E	7 (inner, 4)	Compr	Strato	R = 58-65 C = 66-72		1848?	---- ---- ---- - --	ex

TECTONIC SETTING

Golovnin lies near the western end of the Kurile island arc, which stretches more than 1,200 km from Hokkaido in the west to Kamchatka in the east. The arc consists of two island chains, the Lesser and the Greater Kuriles (Gorshkov, 1970). The Kurile island arc is related to the Kurile-Kamchatka Trench, where the Pacific plate subducts northwestward beneath the Eurasian plate.

GEOLOGIC HISTORY

Golovnin (figs. 9.1.1, 9.1.2) has a 4-km-diameter summit crater within a larger ring structure that might be a caldera (Gorshkov, 1970) or, alternatively, a series of extrusive domes emplaced along a ring fracture (Erlich, 1986). The southwest part of the inner caldera is a bench that is raised 70-80 meters above the floor of the northern (lake-filled) part of the caldera. On the edge of the uplifted part, steeply dipping lacustrine sediments and radial and concentric fissures indicate resurgent doming (Erlich, 1986). Postcaldera andesite and dacite domes occur within the caldera.

HISTORICAL ACTIVITY

The only known historical eruption of Golovnin was in 1848 (Gushchenko, 1979, p. 239). No details of the eruption are known. The caldera contains numerous fumaroles, some as hot as 100 °C.

REFERENCES

Erlich, E., 1986, Geology of the calderas of Kamchatka and Kurile Islands with comparison to calderas of Japan and the Aleutians, Alaska: U.S. Geol. Surv. Open-File Rep. 86-291, 300 p.

GOLOVNIN (continued)

REFERENCES (continued)

Gorshkov, G.S., 1958, Catalogue of active volcanoes of the world including solfatara fields, Pt. VII, Kurile Islands: Naples, IAVCEI, 99 p.

----- 1970, Volcanism and the upper mantle: Investigations in the Kurile Volcanic Arc: New York, Plenum Press, 385 p.

Gushchenko, I.I., 1979, Eruptions of volcanoes of the world, a catalog: Acad. Sci. USSR Far Eastern Sci. Center, Moscow, Nauka Publ., 474 p.

Markhinin, E.K., 1959, Volcanoes of Kunashir Island, in Volcanism of Kamchatka and the Kurile Islands: Akad. Nauk USSR, Trudy Lab. Vulkanol., no. 17, p. 64-115.

Ostapenko, V.F., Fedorchenko, V.I., and Shilov, V.N., 1967, Pumices, ignimbrites, and rhyolite from the Great Kurile arc: Bull. Volcanol., v. 30, p. 81-92.

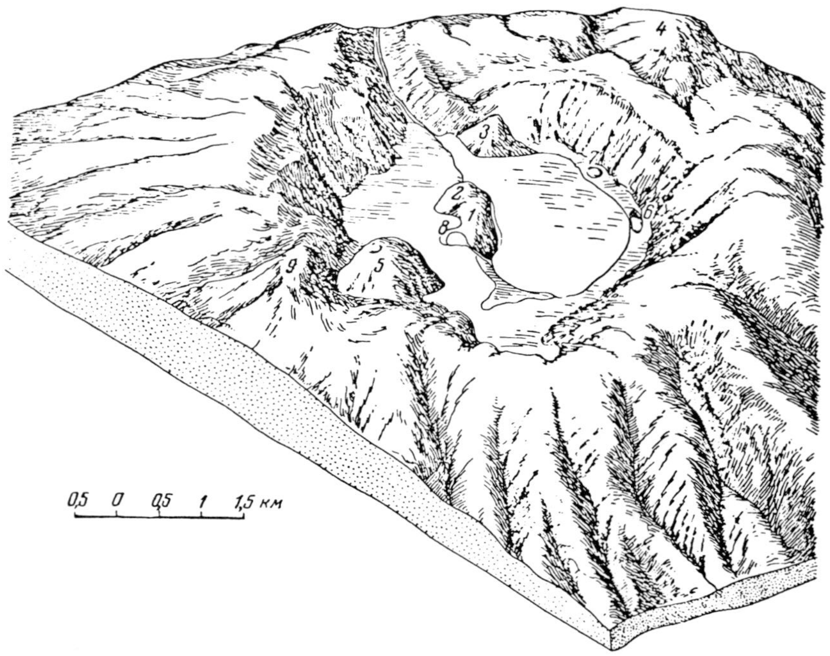

Figure 9.1.1. Perspective drawing looking northwestward at Golovnin Caldera and caldera lake, from Markhinin (1959). 1, east dome; 2, west dome; 3, lava dome; 4, outer dome; 5, steep dome; 6, "turtle" structure with solfatara area; 7, unnamed solfatara area; 8, warm part of the lake; 9, Mt. Golovnin.

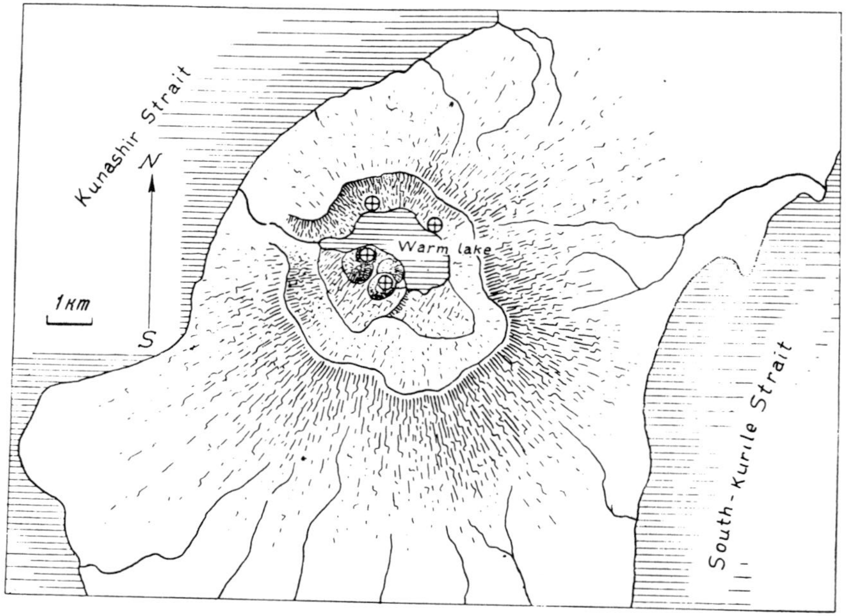

Figure 9.1.2. Sketch showing structure of Golovnin Caldera, from Gorshkov (1958). Cross in circle, solfatara.

MENDELEEV

CAVW number (active vent)	Latitude Longitude (degrees)	Diameter (km)	Local tectonic setting	Precaldera edifice	SiO_2 content (wt pct)	Age of collapse (yr B.P.)	Date of unrest	Type and duration of unrest ESTU STHF MGTF H Te	Eruption type
09-00-02 (Mendeleev)	43.90N 145.70E	6 (inner, 3)	"on 2 volcano-tectonic lines"	Strato (2 or more)	R = 53-66 C = 67?	pre-Würm inter-stadial	1880	---- ---- ---- - --	pex
							1900?	---- ---- ---- - --	pex
							1988	xx-- ---- ---x - --	none

TECTONIC SETTING

Mendeleev lies along the western part of the Kurile island arc (Gorshkov, 1970; see section on TECTONIC SETTING of Golovnin Caldera).

GEOLOGIC HISTORY

The precaldera edifice is a stratovolcano with dominantly andesitic and basaltic andesite products (figs. 9.2.1, 9.2.2). The caldera formed after eruption of dacitic magma, perhaps shortly before the Würm interstadial (that is, before about 39,000 yr B.P.) (Erlich, 1986). A new cone of basalt and basaltic andesite then formed during the interstadial. A "directed blast" (debris avalanche?) breached the west side of the new cone, and a dacite dome formed in its crater. Goriachy Pliazk geothermal field is located on the east flank of the Mendeleev complex, outside the caldera (Erlich, 1986).

HISTORICAL ACTIVITY

An eruption in 1880 was reportedly of gas and included weak explosions (Gorshkov, 1970). When Milne (1886) visited Mendeleev in 1880(?), he found a number of trees that had been killed by solfataric activity and thus inferred that activity to have been "not long in existence."

In hot springs at the base of Mendeleev Volcano, Cl, SO_4, and CO_2 increased by 10, 3, and 11 percent, respectively, from 1965 to 1966, during a period when seismic activity at Kunashir Island also increased (Markhinin and Bozhkova, 1975). NH_4 and HBO_2 increased 70 and 55 percent, respectively, in the same springs over the same time period. No measurements for subsequent years were reported.

MENDELEEV

HISTORICAL ACTIVITY (continued)

A summit earthquake swarm with focal depths of less than 10 km was recorded on 4-5 May 1987. The temperature at 74-m-depth in a geothermal bore hole 10 km from the summit of Mendeleev was found on 25-28 September 1987 to be 4 °C higher than in 1986. Temperatures increased 4-5 °C in hot springs on the east and south flanks, but fumaroles in the same area showed essentially no variation from 1986. Soviet geologists reported that the earthquakes seemed to be caused by exploitation of geothermal wells, used for heat production (Smithsonian Institution, 1988).

REFERENCES

Erlich, E., 1986, Geology of the calderas of Kamchatka and Kurile Islands with comparison to calderas of Japan and the Aleutians, Alaska: U.S. Geol. Surv. Open-File Rep. 86-291, 300 p.

Gorshkov, G.S., 1958, Catalogue of active volcanoes of the world including solfatara fields, Pt. VII, Kurile Islands: Naples, IAVCEI, 99 p.

----- 1970, Volcanism and the upper mantle: Investigations in the Kurile Volcanic Arc: New York, Plenum Press, 385 p.

Markhinin, E.K., and Bozhkova, L.I., 1975, Dependence of the chemical composition of thermal water upon seismic activity: Bull. Volcanol., v. 39, p. 104-111.

Milne, J., 1886, The volcanoes of Japan: Trans. Seismol. Soc. Japan, v. 9, pt. II, 184 p.

Smithsonian Institution, Scientific Event Alert Network (SEAN), 1988, Mendeleev volcano: SEAN Bull., v. 13, no. 1, p. 8 (information from G.S. Steinberg).

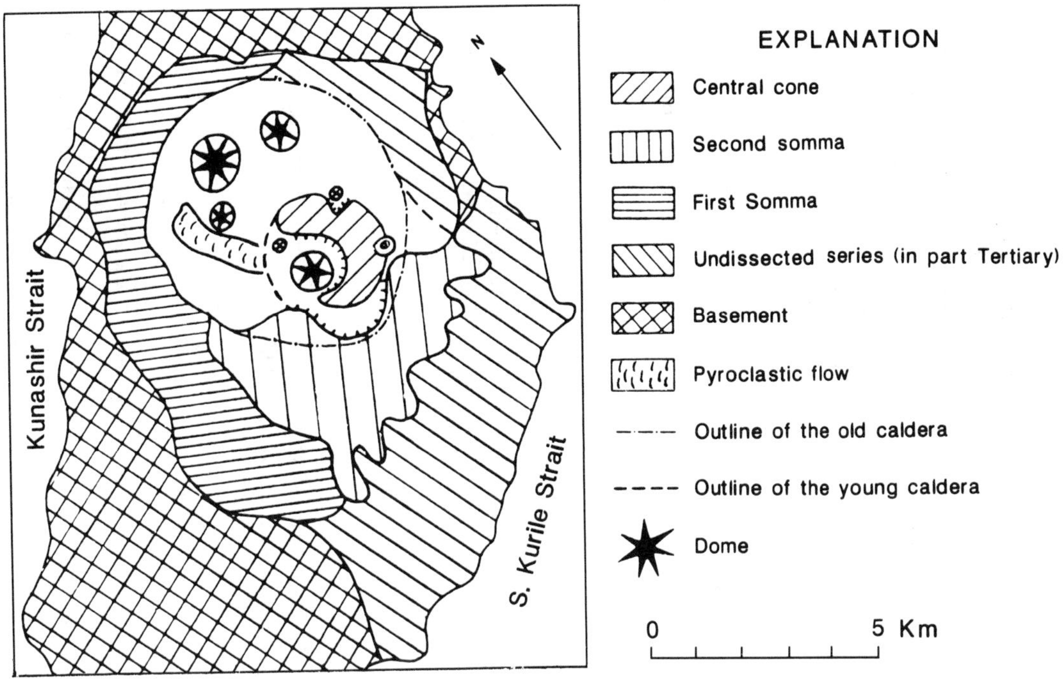

Figure 9.2.1. Sketch of structure of Mendeleev Volcano, from Gorshkov (1970).

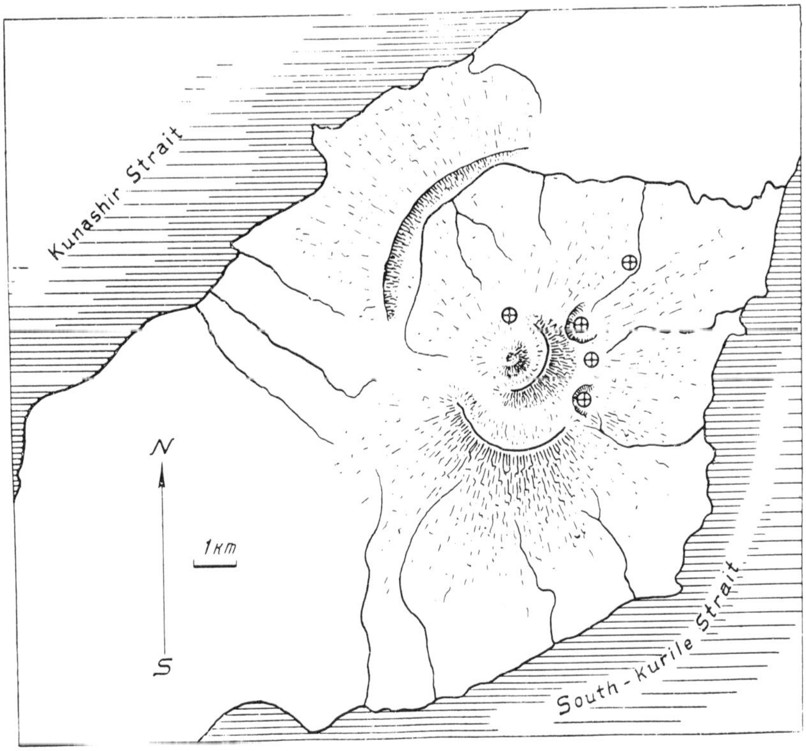

Figure 9.2.2. Sketch showing structure of Mendeleev Volcano, from Gorshkov (1958). Cross in circle, solfatara.

PART 3: HISTORICAL UNREST AT LARGE QUATERNARY CALDERAS (continued)

See inside back cover for explanation and abbreviations

MEDVEZHII

CAVW number (active vent)	Latitude Longitude (degrees)	Diameter (km)	Local tectonic setting	Precaldera edifice	SiO_2 content (wt pct)	Age of collapse (yr B.P.)	Date of unrest	Type and duration of unrest ESTU STHF MGTF H Te	Eruption type
09-00-10 (Kudriavy)	45.38N 148.80E	8	Compr	Strato	R = 49-55 C = 63-67	pre-glacial	1778 or 1779	---- ---- ---- - --	ex
							1883	---- ---- ---- - --	ex,lf
							1946	---- ---- ---- - --	pex
							1958	---- ---- ---? - --	unknown

TECTONIC SETTING

Medvezhii (figs. 9.3.1, 9.3.2) lies along the Kurile island arc (Gorshkov, 1970; see section on TECTONIC SETTING of Golovnin Caldera).

GEOLOGIC HISTORY

Most precaldera and postcaldera eruptions have been of basalt and andesite; the caldera-forming eruption was of dacite. Some postcaldera cones and domes occur along an E-W line through the caldera (Erlich, 1986). Ostapenko (1969) likened the precaldera edifice of Medvezhii to that of Hawaiian volcanoes.

HISTORICAL ACTIVITY

Explosions occurred in the summit crater of Kudriavy in 1778 or 1779, and in May-June 1883 (Gorshkov, 1958). Intense solfataric activity persists in the summit crater and at Kudriavy cone (Smithsonian Institution, 1981; Erlich, 1986).

REFERENCES

Erlich, E., 1986, Geology of the calderas of Kamchatka and Kurile Islands with comparison to calderas of Japan and the Aleutians, Alaska: U.S. Geol. Surv. Open-File Rep. 86-291, 300 p.

Gorshkov, G.S., 1958, Catalogue of active volcanoes of the world including solfatara fields, Pt. VII, Kurile Islands: Naples, IAVCEI, 99 p.

----- 1970, Volcanism and the upper mantle: Investigations in the Kurile Volcanic Arc: New York, Plenum Press, 385 p.

MEDVEZHII (continued)

REFERENCES (continued)

Ostapenko, V.F., 1967, Welded tuff of Medvezhya caldera (Iturup Island), Kurile Islands: Dokl. Akad. Sci. USSR, v. 174, no. 4, p. 941-944.

----- 1969, About the character of previous eruptions of Medvezhiy zone (Iturup Island), _in_ Volcanism and Izverzhenia: Moscow, Nauka Press, p. 122-126.

----- 1970, Morphology and structure of caldera-volcano Medvezhiy (Iturup Island, Kuriles): Izv. Sakhalin Div. of Geogr. Soc. USSR, iss. 1, p. 99-108.

Ostapenko, V.F., Fedorchenko, V.I., and Shilov, V.N., 1967, Pumices, ignimbrites, and rhyolite from the Great Kurile arc: Bull. Volcanol., v. 30, p. 81-92.

Smithsonian Institution, Scientific Event Alert Network (SEAN), 1981, Kudriavy Cone: SEAN Bull., v. 6, no. 12, p. 8 (information from G.S. Steinberg).

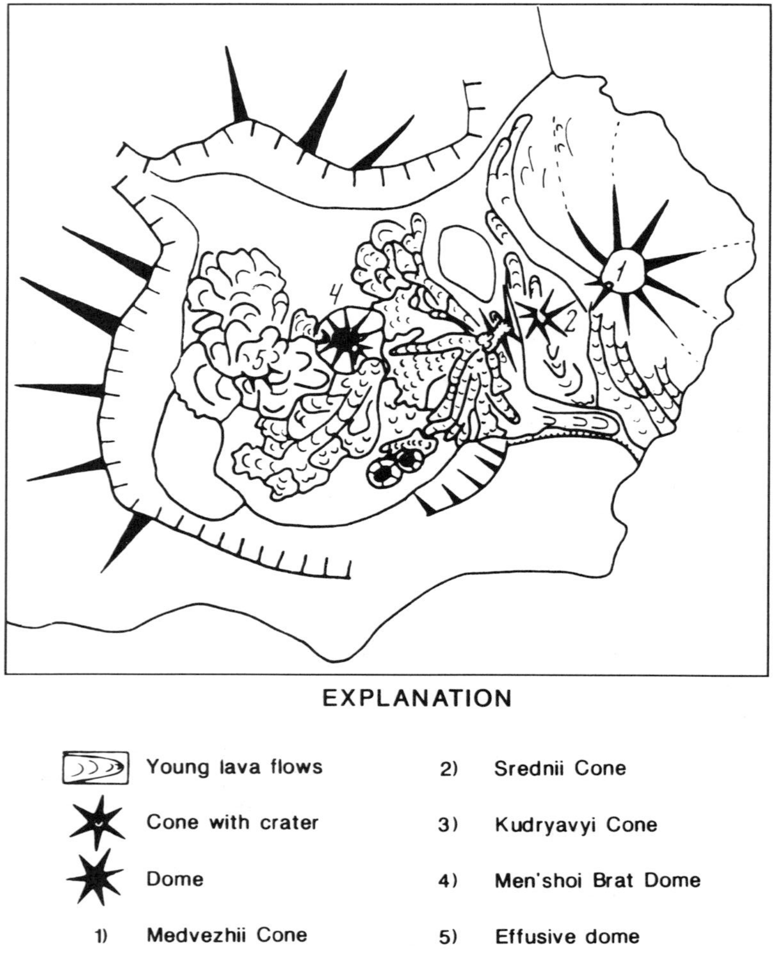

Figure 9.3.1. Sketch of structure of Medvezhii Caldera, from Gorshkov (1970).

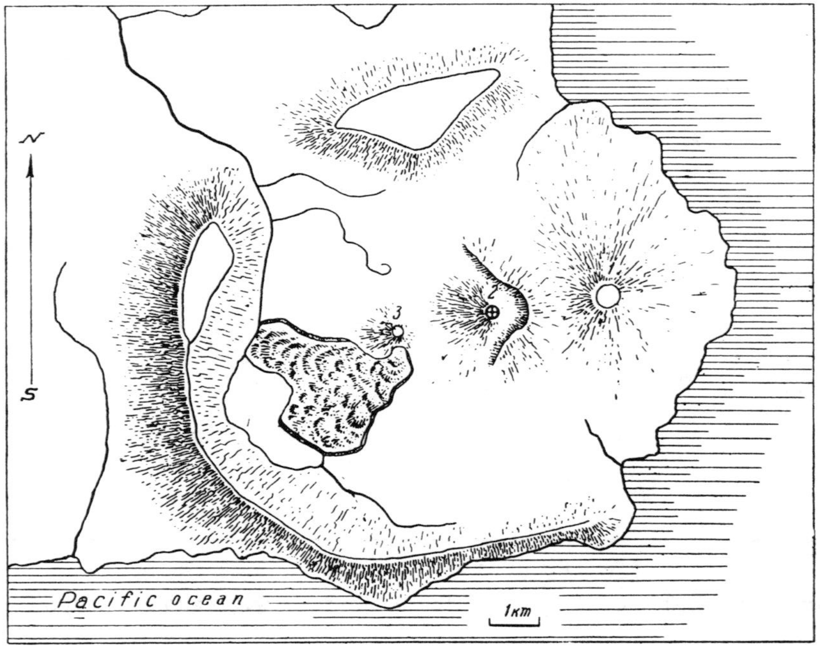

Figure 9.3.2. Sketch showing structure of Medvezhii Caldera with its active cone Kudriavy, from Gorshkov (1958). Explanation: 1, Medvezhii cone; 2, Kudriavy cone; 3, Menshoi Brat cone with lava flows; cross in circle, solfatara.

CHIRPOI

CAVW number (active vent)	Latitude Longitude (degrees)	Diameter (km)	Local tectonic setting	Precaldera edifice	SiO_2 content (wt pct)	Age of collapse (yr B.P.)	Date of unrest	Type and duration of unrest ESTU STHF MGTF H Te	Eruption type
09-00-15 (Snow= S)	46.52N 150.88E	8-9	Southern and western parts cut by faults; compr	Ol-px basalt cone, submarine	R = 59-60 C = 68?		ca. 1712 (C)	---- ---- ---- - --	EX?
09-00-16 (Cherny= C)							between 1770 and 1810 (S)	---- ---- ---- - QU?	ex,lf
							1811 (S)	---- ---- ---- - --	ex
							1854 (?)	---- ---- ---- - --	ex
							1857 (C)	---- ---- ---- - --	Ex
							1879 (S)	---- ---- ---- - --	ex,lf
							1960 (S)	---- ---- ---- - --	ex

TECTONIC SETTING

Chirpoi lies along the Kurile island arc (Gorshkov, 1970; see section on TECTONIC SETTING of Golovnin Caldera).

GEOLOGIC HISTORY

Subaerial and submarine topography suggests that Chirpoi Island has grown within two overlapping calderas that together are 8-9 km in diameter (fig. 9.4.1). Cherny Volcano lies at the center of Chirpoi Island, and Snow Volcano is a parasitic lava cone at Cherny's southern foot (Gorshkov, 1970). Brat Chirpoev Island, southwest of Chirpoi Island, is a fragment of the precaldera edifice, an olivine-pyroxene basalt cone (Erlich, 1986); however, it might also have had recent eruptions and showed solfataric activity in the mid-18th century (Gorshkov, 1970).

HISTORICAL ACTIVITY

Snow Volcano originated sometime between a visit by Cossack Captain Chernyi in 1770 and a visit by Captain Golovnin in 1811. A major tectonic earthquake on or about 2 January 1780 devastated the islands of Ketoi, Simushir, Chirpoi, and Urup and apparently triggered an eruption of Raikoke Volcano (Perrey, 1863, p. 70). It is possible but speculative that Snow began to grow in or shortly after 1780.

CHIRPOI (continued)

REFERENCES

Erlich, E., 1986, Geology of the calderas of Kamchatka and Kurile Islands with comparison to calderas of Japan and the Aleutians, Alaska: U.S. Geol. Surv. Open-File Rep. 86-291, 300 p.

Gorshkov, G.S., 1958, Catalogue of active volcanoes of the world including solfatara fields, Pt. VII, Kurile Islands: Naples, IAVCEI, 99 p.

----- 1970, Volcanism and the upper mantle: Investigations in the Kurile Volcanic Arc: New York, Plenum Press, 385 p.

Perrey, A., 1863, Documents sur les tremblements de terre et les phénomènes volcaniques dans l'archipel des Kouriles et au Kamtschatka: Lyon, Soc. Imp. Agric. Hist. Nat. Arts, 166 p.

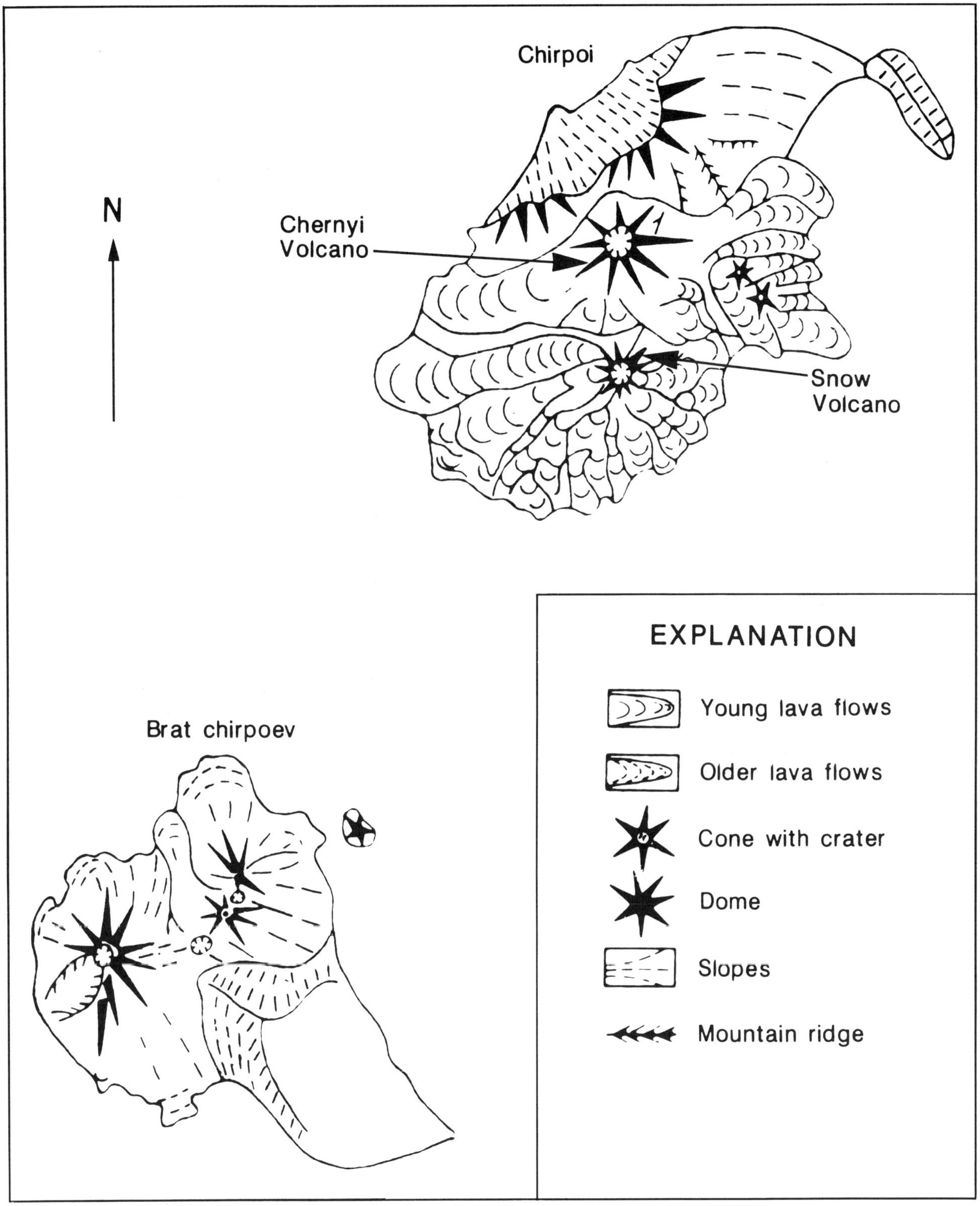

Figure 9.4.1. Sketch of structure of Chirpoi (top) and Brat Chirpoev (bottom), from Gorshkov (1970).

PART 3: HISTORICAL UNREST AT LARGE QUATERNARY CALDERAS (continued)
See inside back cover for explanation and abbreviations

ZAVARITSKY

CAVW number (active vent)	Latitude Longitude (degrees)	Diameter (km)	Local tectonic setting	Precaldera edifice	SiO_2 content (wt pct)	Age of collapse (yr B.P.)	Date of unrest	Type and duration of unrest ESTU STHF MGTF H Te	Eruption type
09-00-18 (Zavaritsky)	46.92N 151.95E	10 (outer) 8 (middle) 3 (inner) (concentric)	Fault, west side of outer caldera; compr	Strato (3 generations)	R = 54-59 C = 64-72	Second caldera, inter-glacial	1916/1930	---- ---- ---- - --	ex,dm
							1957	B--- ---- Z--E E --	Ex,dm,lf

TECTONIC SETTING

Zavaritsky lies along the central part of the Kurile island arc (Gorshkov, 1970; see section on TECTONIC SETTING of Golovnin Caldera).

GEOLOGIC HISTORY

The present Zavaritsky Caldera is the smallest and innermost of three successive calderas on Zavaritsky Volcano (figs. 9.5.1-9.5.3). Steep inward dips of lacustrine sediments from an earlier, higher level of the caldera lake in the innermost caldera suggest that sliding from the walls after later explosions deepened the innermost caldera (Gorshkov, 1970, p. 153).

HISTORICAL ACTIVITY

Strong fumaroles occur and thermal springs normally discharge on the southern edge of the caldera lake. Temperature of the lake is normally 30 °C. For several months before the start of the 1957 eruption, the activity of the fumaroles and thermal springs noticeably diminished. Earthquakes were felt in Kostodhka at 2200 hr on 11 November 1957; six more shocks were felt between 0200 and 0500 hr the next day. The two strongest earthquakes occurred at 0400 hr, separated by 15-20 minutes of thunderous roaring. A moderately large eruption began on 12 November 1957. During 10-12 August 1958, the relative difference in magnetic-field strength between points inside and outside the caldera increased by 1 gamma, then decreased by 2 gammas by 18 August (Bernstein, 1960).

In a September 1981 aerial inspection, no fumarolic activity was seen (Smithsonian Institution, 1981).

ZAVARITSKY (continued)

REFERENCES

Bernstein, V.A., 1960, On the magnetic field of Zavaritsky Volcano, Simushir Island, Kurile Islands: Bull. Vulkanol. Stantsii (Kamchatka), no. 30, p. 55-68.

Erlich, E., 1986, Geology of the calderas of Kamchatka and Kurile Islands with comparison to calderas of Japan and the Aleutians, Alaska: U.S. Geol. Surv. Open-File Rep. 86-291, 300 p.

Gorshkov, G.S., 1958, Catalogue of active volcanoes of the world including solfatara fields, Pt. VII, Kurile Islands: Naples, IAVCEI, 99 p.

----- 1970, Volcanism and the upper mantle: Investigations in the Kurile Volcanic Arc: New York, Plenum Press, 385 p.

Markhinin, E.K., 1960, The eruption of Zavaritsky on Simushir Is. in the autumn of 1957: Bull. Vulkanol. Stantsii (Kamchatka), no. 29, p. 7-14.

Ostapenko, V.F., Fedorchenko, V.I., and Shilov, V.N., 1967, Pumices, ignimbrites, and rhyolite from the Great Kurile arc: Bull. Volcanol., v. 30, p. 81-92.

Smithsonian Institution, Scientific Event Alert Network (SEAN), 1981, Zavaritsky Caldera: SEAN Bull., v. 6, no. 12, p. 8 (information from G.S. Steinberg).

Tovarova, I.I., 1960, Fumarolic activity in the Zavaritsky caldera: Bull. Vulkanol. Stantsii (Kamchatka), no. 30, p. 50-54.

Zelenov, K.K., and Kanakina, M.A., 1962, Biriuzovoye Lake (Zavaritsky Caldera) and changing of its water chemistry as a result of the eruption in 1957: Bull. Vulkanol. Stantsii (Kamchatka), no. 32, p. 24-33.

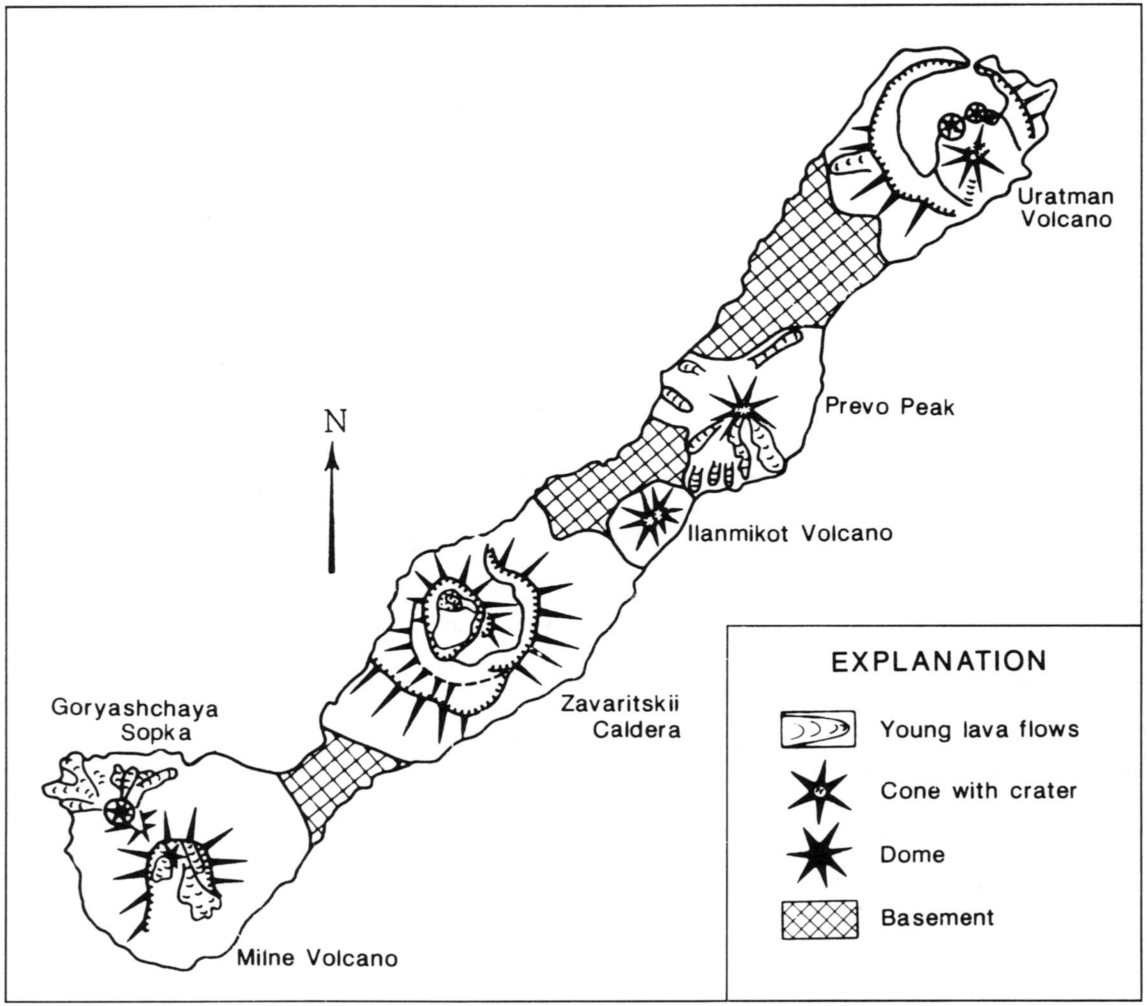

Figure 9.5.1. Sketch of distribution of volcanoes on Simushir Island, from Gorshkov (1970).

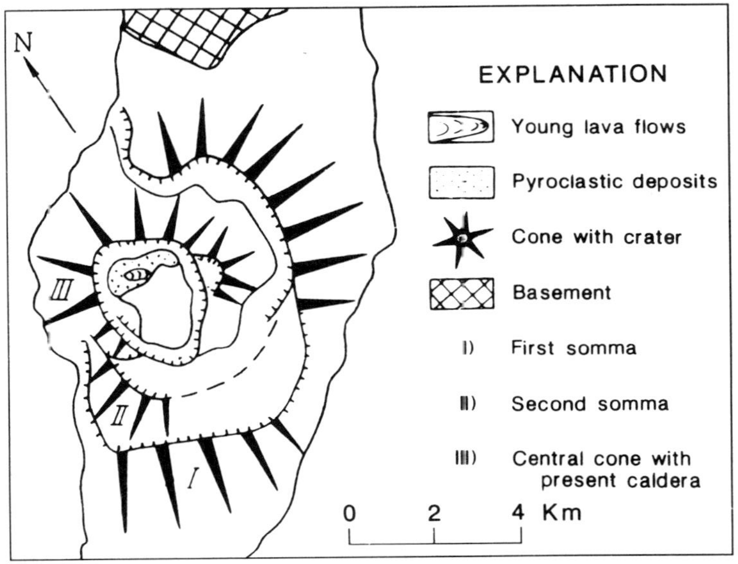

Figure 9.5.2. Sketch showing structure of Zavaritsky Caldera, from Gorshkov (1970).

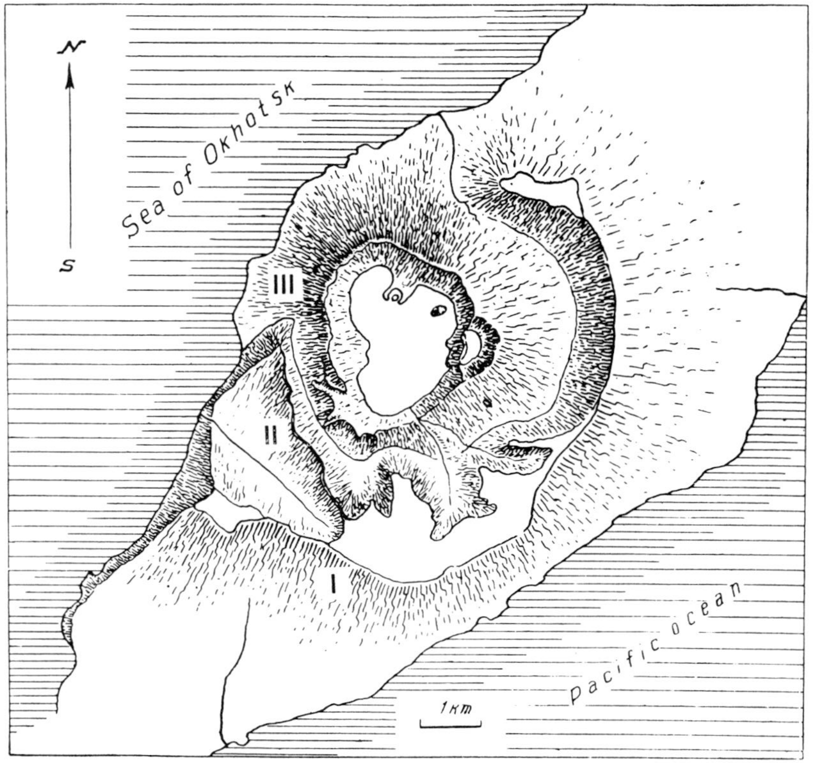

Figure 9.5.3. Sketch of structure of Zavaritsky Caldera, from Gorshkov (1958): I, first somma; II, second somma; III, central cone with present caldera.

PART 3: HISTORICAL UNREST AT LARGE QUATERNARY CALDERAS (continued)
See inside back cover for explanation and abbreviations

KETOI

CAVW number (active vent)	Latitude Longitude (degrees)	Diameter (km)	Local tectonic setting	Precaldera edifice	SiO_2 content (wt pct)	Age of collapse (yr B.P.)	Date of unrest	Type and duration of unrest ESTU STHF MGTF H Te	Eruption type
09-00-20 (Pallas Peak)	47.35N 152.48E	5 (outer) 1.5 (inner)	Compr	Shield	R= 58 C= silicic	pre-glacial?	1843-46	---- ---- ---- - QU?	ex,lf
							1924	---- ---- ---- - --	ex
							1960	---- ---- ---- - --	ex

TECTONIC SETTING

Ketoi lies along the central part of the Kurile island arc (Gorshkov, 1970; see section on TECTONIC SETTING of Golovnin Caldera).

GEOLOGIC HISTORY

Northeast-trending faults form a graben and horst that cross the northwest edge of the caldera. At least two generations of calderas are present, of which the older one is larger. Pallas Peak is a postcaldera cone on the southeastern margin of the older caldera (figs. 9.6.1, 9.6.2).

HISTORICAL ACTIVITY

Two "violent" earthquakes shook Urup Island, 1.5° of latitude (ca. 165 km) south of Ketoi, on 25 April 1843. A strong eruption of Ketoi began in July 1843, and "the whole island was on fire" (Perrey, 1863, p. 165).

In September 1981, an aerial inspection showed intense fumarolic activity on the outer north slope of the volcano (Smithsonian Institution, 1981).

REFERENCES

Erlich, E., 1986, Geology of the calderas of Kamchatka and Kurile Islands with comparison to calderas of Japan and the Aleutians, Alaska: U.S. Geol. Surv. Open-File Rep. 86-291, 300 p.

KETOI (continued)

REFERENCES (continued)

Gorshkov, G.S., 1958, Catalogue of active volcanoes of the world including solfatara fields, Pt. VII, Kurile Islands: Naples, IAVCEI, 99 p.

----- 1970, Volcanism and the upper mantle: Investigations in the Kurile Volcanic Arc: New York, Plenum Press, 385 p.

Perrey, A., 1863, Documents sur les tremblements de terre et les phénomènes volcaniques dans l'archipel des Kouriles et au Kamtschatka: Lyon, Soc. Imp. Agric. Hist. Nat. Arts, 166 p.

Smithsonian Institution, Scientific Event Alert Network (SEAN), 1981, Ketoi Caldera: SEAN Bull., v. 6, no. 12, p. 8 (information from G.S. Steinberg).

Vlasov, G.M., 1967, Kamchatka, Kuril, and Komandorskiye Islands: Geological description: Geology of the USSR, v. 31, Moscow, Izd.-vo Nedra, 827 p.

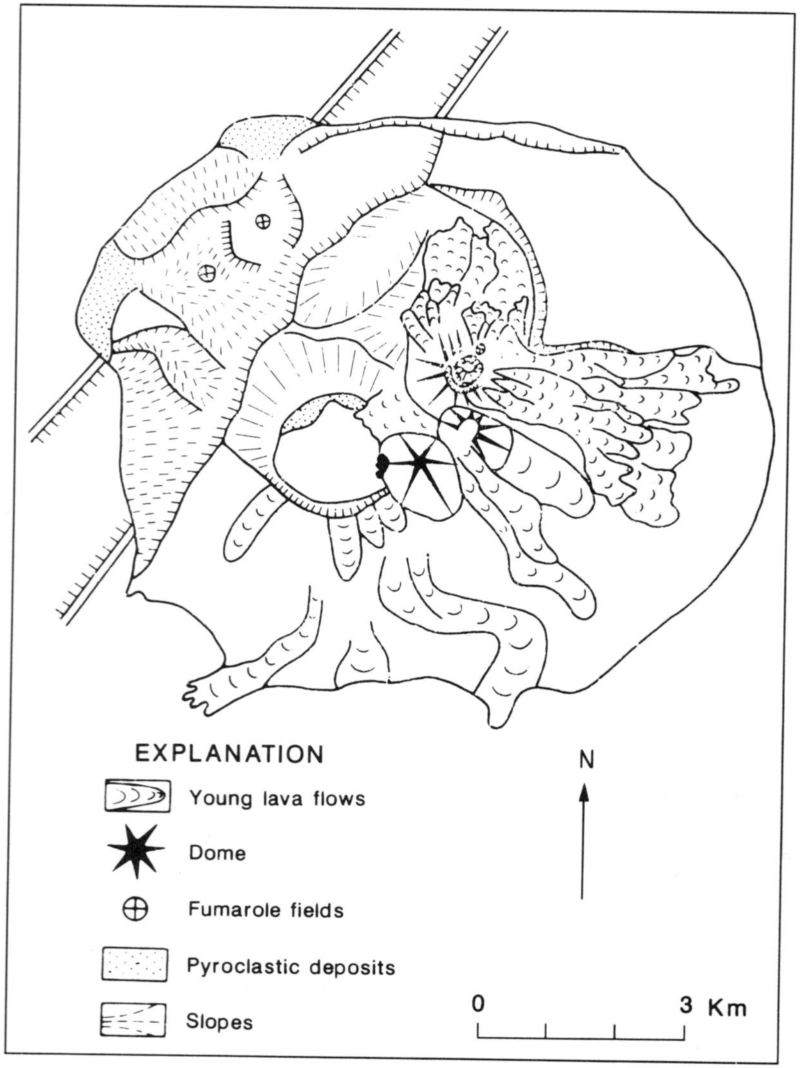

Figure 9.6.1. Sketch of structure of Ketoi Caldera, from Gorshkov (1970).

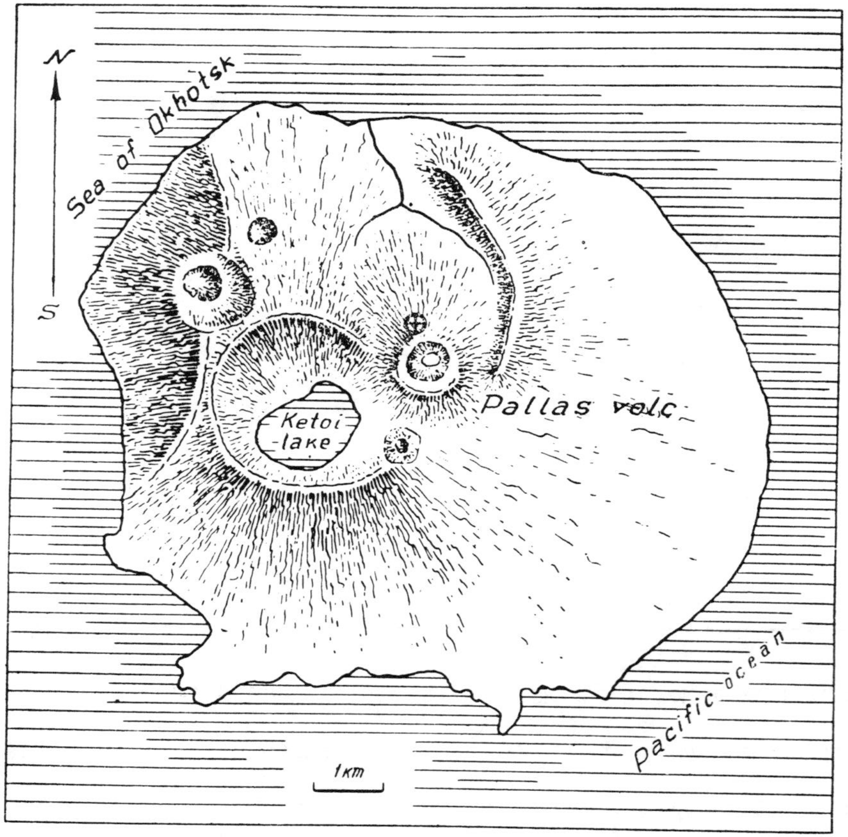

Figure 9.6.2. Sketch showing structure of Pallas Volcano and Ketoi Volcano, from Gorshkov (1958).

RASSHUA

CAVW number (active vent)	Latitude Longitude (degrees)	Diameter (km)	Local tectonic setting	Precaldera edifice	SiO_2 content (wt pct)	Age of collapse (yr B.P.)	Date of unrest	Type and duration of unrest ESTU STHF MGTF H Te	Eruption type
09-00-22	47.77N	6	Compr	Strato	R = 56±	pre-	1810?	---- ---- ---- - --	unknown
(Rasshua)	153.02E				C = ?	latest	1846	---- ---- ---- - --	Ex
						glaciation	1946	---- ---- --?C - --	none?
							1957	---- ---- --?x - --	pex?

TECTONIC SETTING

Rasshua lies along the Kurile island arc (Gorshkov, 1970; see section on TECTONIC SETTING of Golovnin Caldera).

GEOLOGIC HISTORY

Rasshua Volcano lies just north of the center of a 6-km-diameter somma (figs. 9.7.1, 9.7.2).

HISTORICAL ACTIVITY

1846: "A volcano on Rasshua smokes intensely and occasionally emits flames" (Doroshin, 1870).

1946: Steaming increased briefly on 4 November 1946, followed 3 days later by an increase in steaming at Sarichev, 28 km away, and 5 days later (on 9 November 1946) by an eruption of Sarichev.

1957: Fumarolic activity increased markedly and weak explosions occurred (Vlasov, 1967; Gorshkov, 1970).

REFERENCES

Doroshin, P., 1870, On some volcanoes, their eruptions and earthquakes in former Russian domains of America: Verhandlungen der Russisch-Kaiserlichen Mineralogischen Gesellschaft zu St. Petersburg, 2nd ser., p. 25-44.
Erlich, E., 1986, Geology of the calderas of Kamchatka and Kurile Islands with comparison to calderas of Japan and the Aleutians, Alaska: U.S. Geol. Surv. Open-File Rep. 86-291, 300 p.

RASSHUA (continued)

REFERENCES (continued)

Glavatsky, S.N., and Efremov, Yu.K., 1948, The eruption of the volcano Peak Sarichev in November 1946: Bull. Vulkanol. Stantsii (Kamchatka), no. 15, p. 8-12.

Gorshkov, G.S., 1958, Catalogue of active volcanoes of the world including solfatara fields, Pt. VII, Kurile Islands: Naples, IAVCEI, 99 p.

----- 1970, Volcanism and the upper mantle: Investigations in the Kurile Volcanic Arc: New York, Plenum Press, 385 p.

Gushchenko, I.I., 1979, Eruptions of volcanoes of the world, a catalog: Acad. Sci. USSR Far Eastern Sci. Center, Moscow, Nauka Publ., 474 p.

Vlasov, G.M., 1967, Kamchatka, Kuril, and Komandorskiye Islands: Geological description: Geology of the USSR, v. 31, Moscow, Izd.-vo Nedra, 827 p.

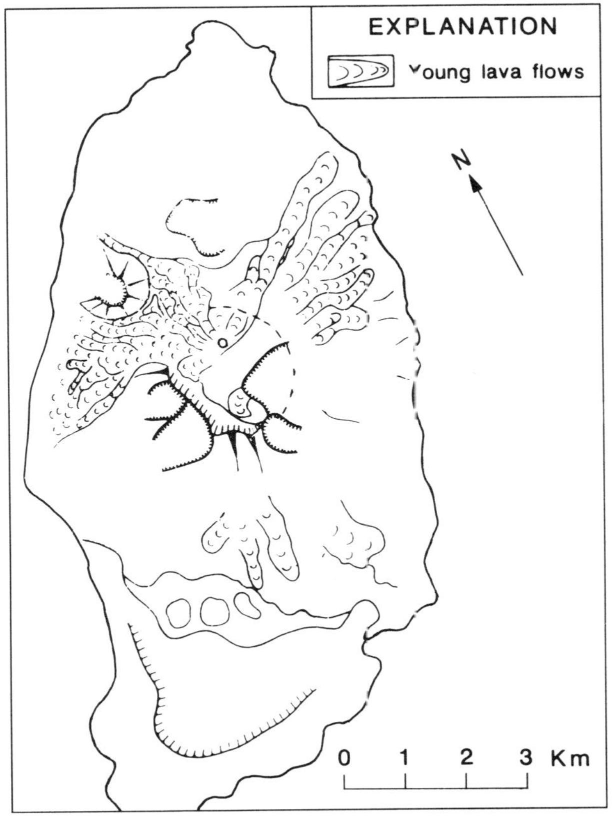

Figure 9.7.1. Sketch of structure of Rasshua Volcano, from Gorshkov (1970).

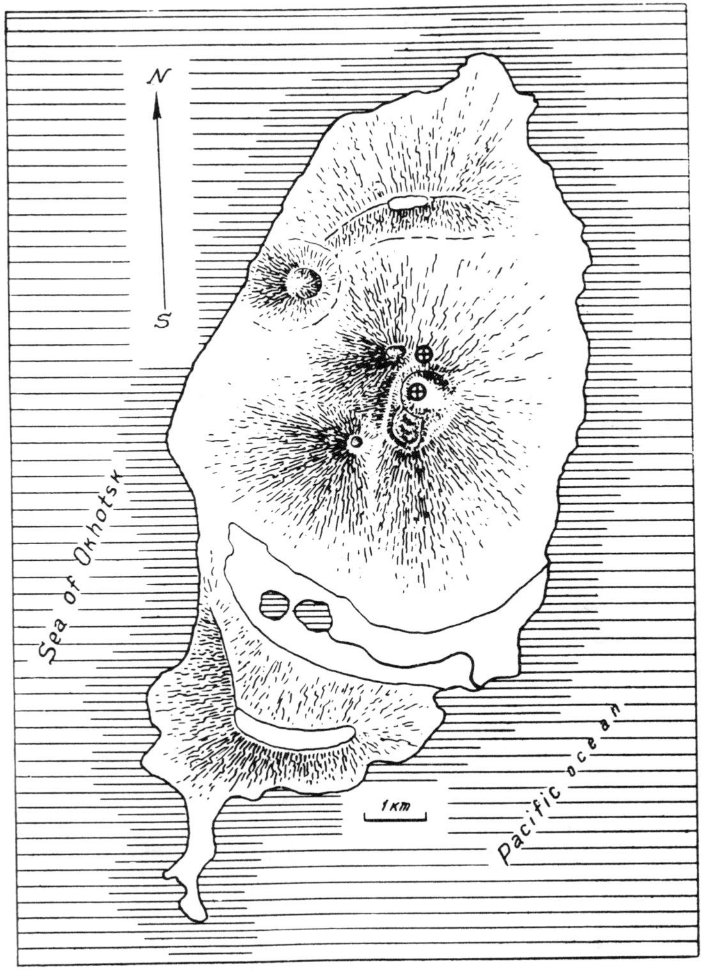

Figure 9.7.2. Sketch showing structure of Rasshua Volcano, from Gorshkov (1958).

PART 3: HISTORICAL UNREST AT LARGE QUATERNARY CALDERAS (continued)
See inside back cover for explanation and abbreviations

TAO-RUSYR

CAVW number (active vent)	Latitude Longitude (degrees)	Diameter (km)	Local tectonic setting	Precaldera edifice	SiO_2 content (wt pct)	Age of collapse (yr B.P.)	Date of unrest	Type and duration of unrest ESTU STHF MGTF H Te	Eruption type
09-00-31 (Krenitzyn Peak)	49.36N 154.71E	7.5	Inter-section of two volcano-tectonic lines	LL-strat	R = 49-64 C = 59-67	7040	1846	---- ---- ---? - --	none
							1879	---- ---- ---? - --	none
							1952	?--- ---- C--- - QU	Ex,dm

TECTONIC SETTING

Tao-Rusyr (figs. 9.8.1, 9.8.2) lies along the Kurile island arc (Gorshkov, 1970; see section on TECTONIC SETTING of Golovnin Caldera).

GEOLOGIC HISTORY

A precaldera edifice of basalt and basaltic andesite is capped by dacitic, pumiceous pyroclastic fall and flow deposits from the caldera-forming eruption. Krenitzyn Peak is an andesitic cone that grew in the caldera lake. The lake is heated from below and does not freeze over in winter.

HISTORICAL ACTIVITY

1846, 1879: Solfataric activity was noted, but it is not clear from the reports whether that was normal activity simply observed in these years or an abnormal increase in solfataric activity.

1952: An eruption occurred on 12 November 1952, from a new explosion crater on the east flank of Krenitzyn Peak. The eruption followed and may have been triggered by a strong tectonic earthquake (4 November 1952, $\underline{M}$ = 8.25, 50 km from the caldera) (Kimura, 1978). "Three days prior to the eruption, disturbances of the magnetic field were observed" (Gorshkov, 1958, p. 73). "A.A. Lettens, studying the horizontal structure of the magnetic field, detected strange behavior of a magnetic compass--a slow uneven deceleration in one direction, completely unlike normal measurements. These phenomena may have had some kind of relation and may have forewarned of the impending eruption" (Piip and Sviatlovsky, 1954). We have no details about where the disturbance was measured. Magma erupted in 1952 had an SiO_2 content of 62 percent (Gorshkov, 1954).

TAO-RUSYR (continued)

REFERENCES

Erlich, E., 1986, Geology of the calderas of Kamchatka and Kurile Islands with comparison to calderas of Japan and the Aleutians, Alaska: U.S. Geol. Surv. Open-File Rep. 86-291, 300 p.

Gorshkov, G.S., 1954, Chronology of the eruptions of the Kurile Range of volcanoes, 1713-1952: Akad. Nauk USSR, Trudy Lab. Vulkanol., no. 8, p. 58-99.

----- 1958, Catalogue of active volcanoes of the world including solfatara fields, Pt. VII, Kurile Islands: Naples, IAVCEI, 99 p.

----- 1970, Volcanism and the upper mantle: Investigations in the Kurile Volcanic Arc: New York, Plenum Press, 385 p.

Kimura, M., 1978, Significant eruptive activities related to large interplate earthquakes in the northwestern Pacific margin: Jour. Phys. Earth, v. 26 supp., p. S557-S570.

Piip, B.I., and Sviatlovsky, A.E., 1954, The eruption of Krenitzyn Peak in 1952: Bull. Vulkanol. Stantsii (Kamchatka), no. 20, p. 65-66.

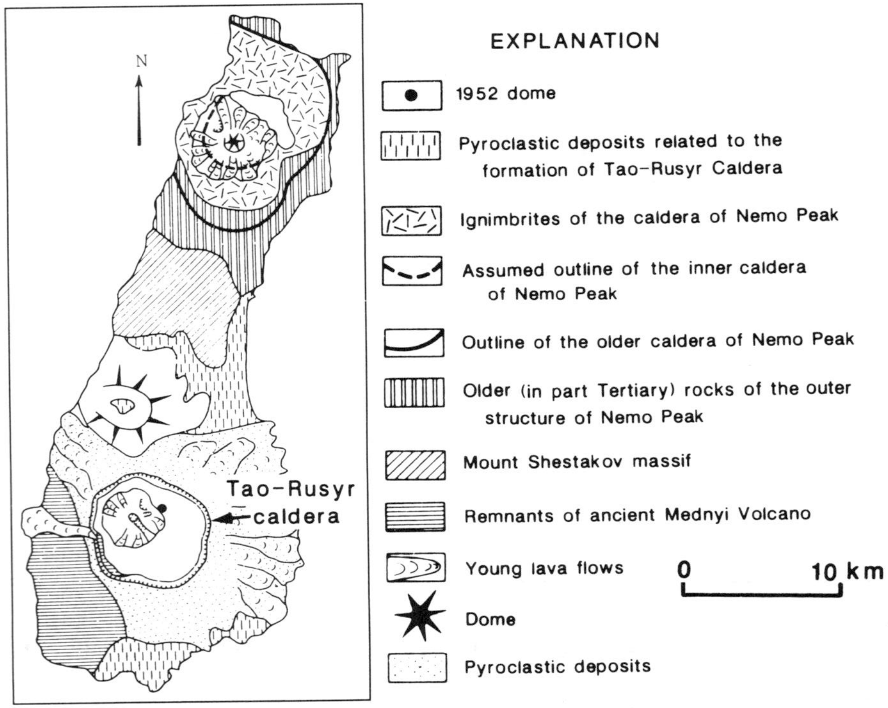

Figure 9.8.1. Sketch of structure of volcanoes of Onekotan, from Gorshkov (1970).

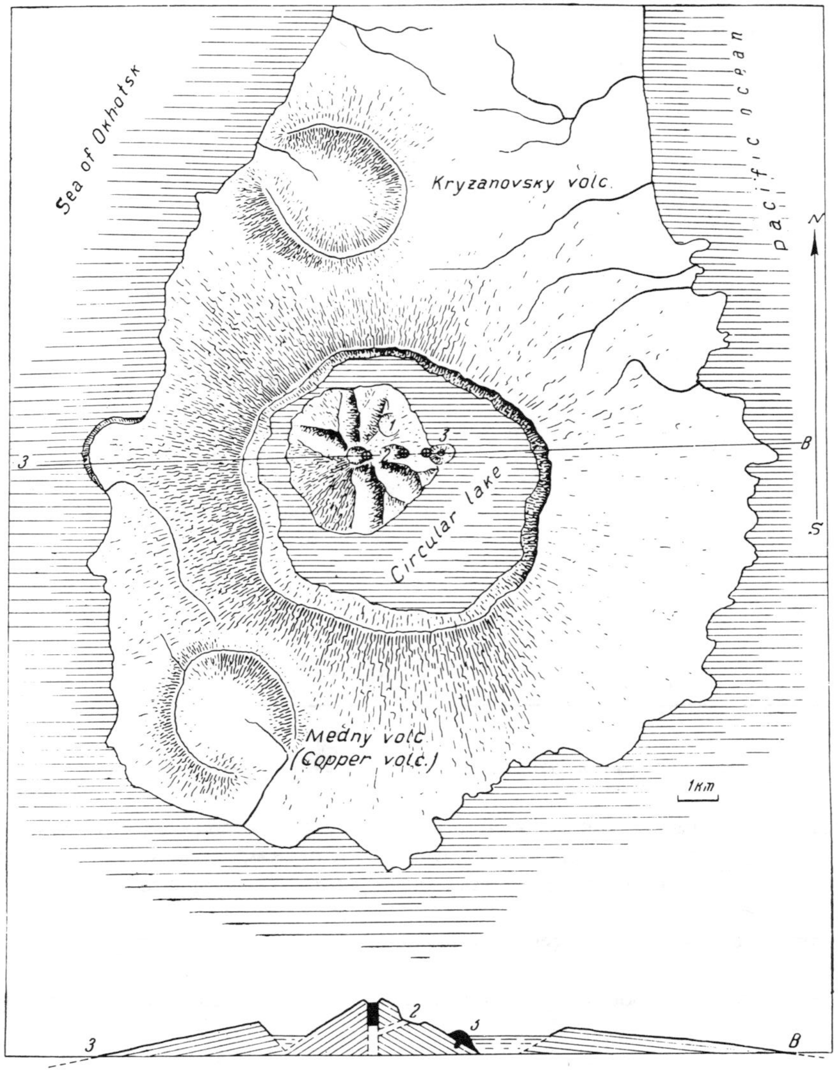

Figure 9.8.2. Sketch showing structure and a cross section of Krenitzyn Peak and Tao-Rusyr Caldera, from Gorshkov (1958).

NEMO PEAK

CAVW number (active vent)	Latitude Longitude (degrees)	Diameter (km)	Local tectonic setting	Precaldera edifice	SiO_2 content (wt pct)	Age of collapse (yr B.P.)	Date of unrest	Type and duration of unrest ESTU STHF MGTF H Te	Eruption type
09-00-32 (Nemo Peak)	49.57N 154.81E	11 each (older two) 5 (younger)	Two intersecting volcano-tectonic lines; compr	LL-strat	R = 54-59 C = 64	inter-glacial (younger); pre-glacial (older)	early 1700's	---- ---- ---- - --	ex
							1906	---- ---- ---- - --	ex,dm?
							1938	---- ---- ---- - --	ex

TECTONIC SETTING

Nemo Peak lies along the Kurile island arc (Gorshkov, 1970; see section on TECTONIC SETTING of Golovnin Caldera).

GEOLOGIC HISTORY

The Nemo Peak volcanic center has had repeated episodes of caldera formation. The system consists of two overlapping older calderas, each of diameter approximately 11 km, that formed in preglacial times. Then, a postcaldera cone grew on the overlapping boundaries of these two calderas until a smaller, 5-km-diameter caldera formed at its summit (figs. 9.9.1, 9.9.2). The presently active peak (Nemo Peak) grew inside the younger, smaller summit caldera.

HISTORICAL ACTIVITY

Few details are available. A young-looking dome might have formed during the eruption of 1906.

REFERENCES

Erlich, E., 1986, Geology of the calderas of Kamchatka and Kurile Islands with comparison to calderas of Japan and the Aleutians, Alaska: U.S. Geol. Surv. Open-File Rep. 86-291, 300 p.

Gorshkov, G.S., 1958, Catalogue of active volcanoes of the world including solfatara fields, Pt. VII, Kurile Islands: Naples, IAVCEI, 99 p.

----- 1970, Volcanism and the upper mantle: Investigations in the Kurile Volcanic Arc: New York, Plenum Press, 385 p.

Gushchenko, I.I., 1979, Eruptions of volcanoes of the world, a catalog: Acad. Sci. USSR Far Eastern Sci. Center, Moscow, Nauka Publ., 474 p.

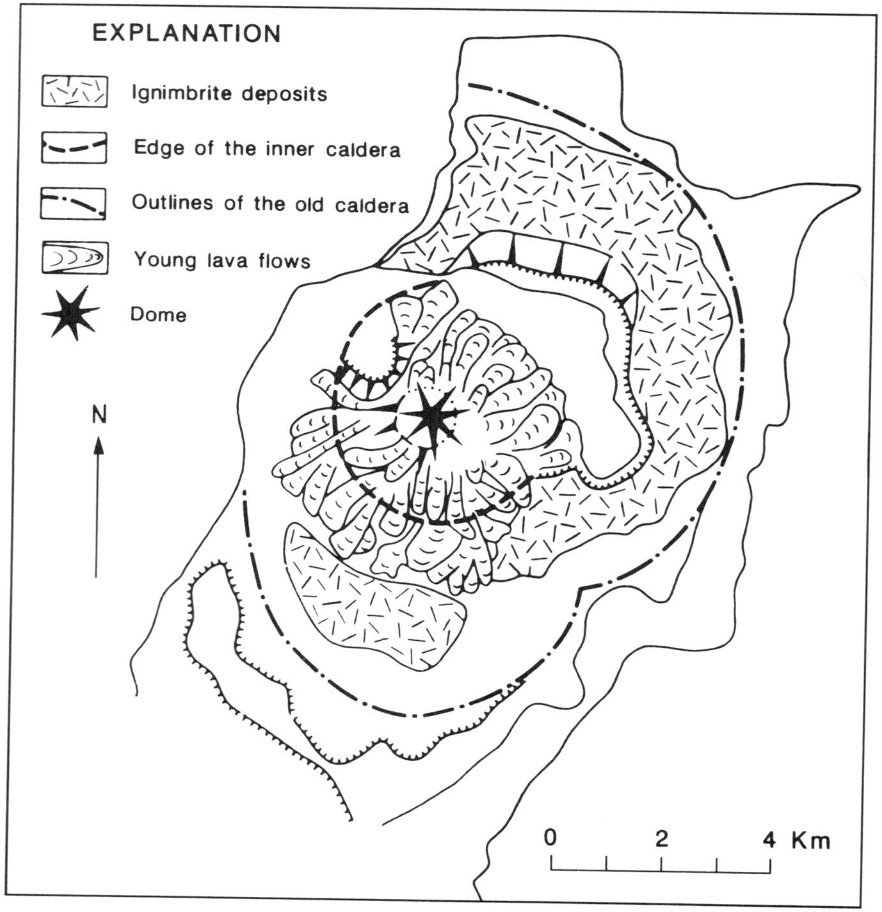

Figure 9.9.1. Sketch of structure of Nemo Peak, from Gorshkov (1970).

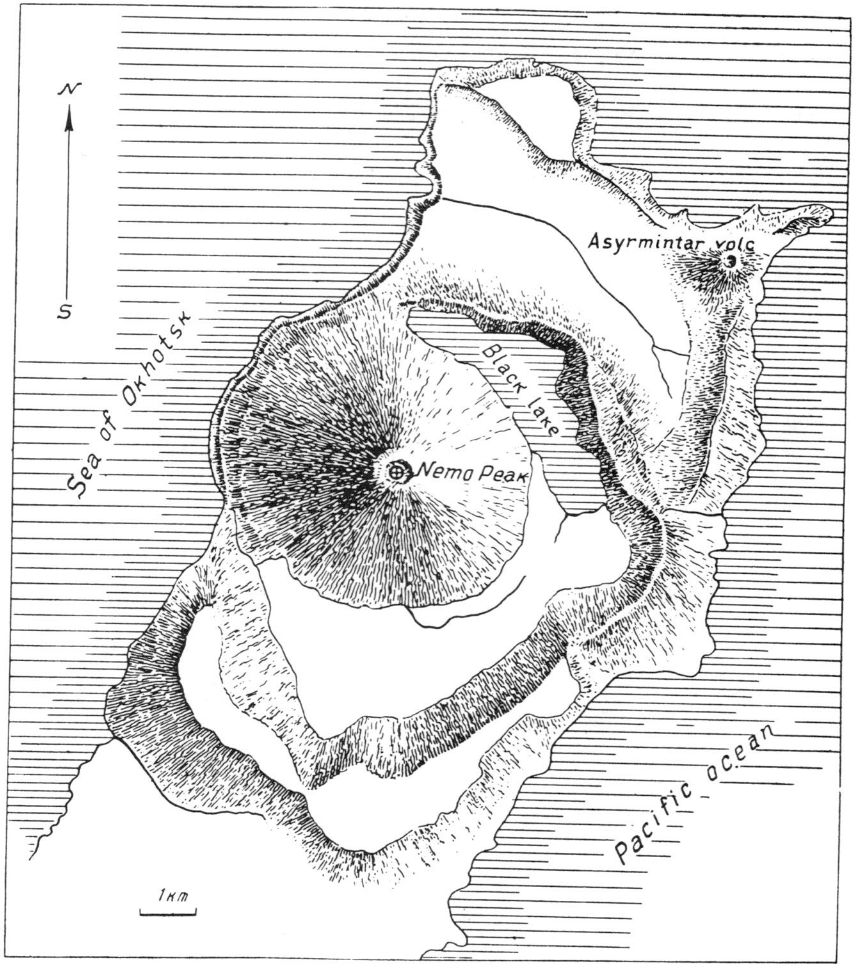

Figure 9.9.2. Sketch showing structure of Nemo Volcano, from Gorshkov (1958). Cross in circle, solfatara.

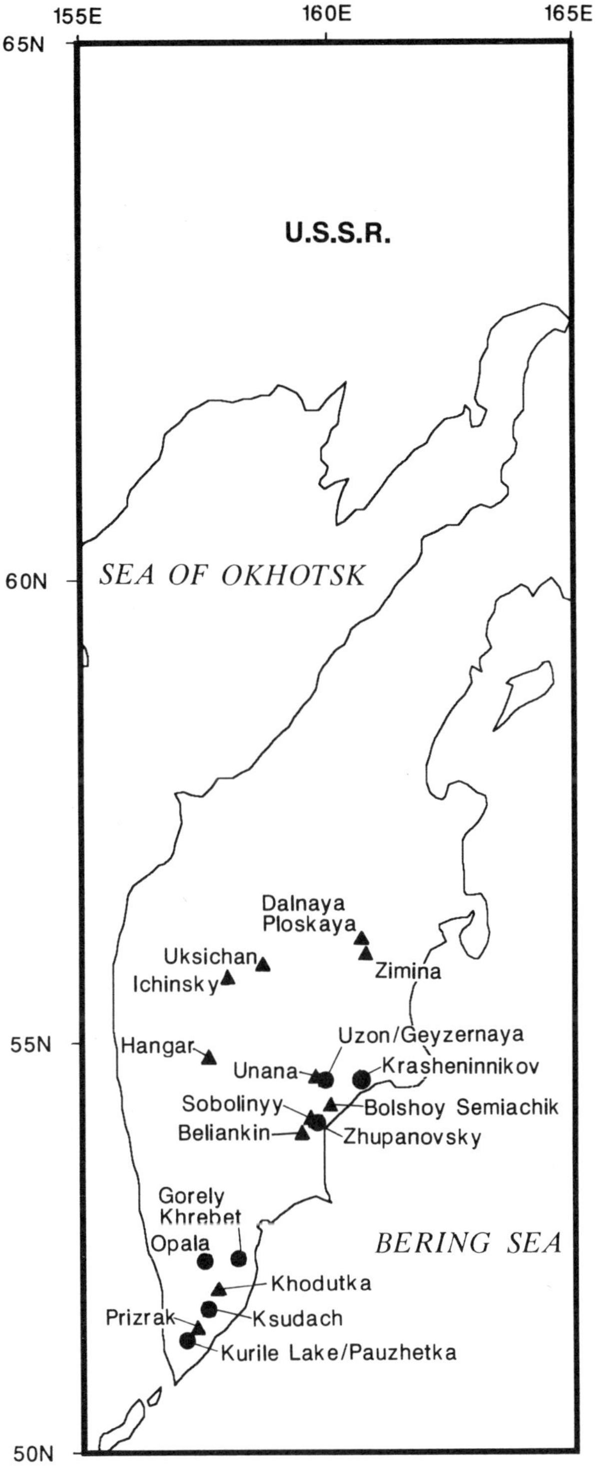

Figure 10. Locations of large, Quaternary restless calderas (solid circles) and nonrestless calderas (triangles) in part of Region 10, including Kamchatka and an adjacent part of Asian mainland (see also figure 10.8).

PART 3: HISTORICAL UNREST AT LARGE QUATERNARY CALDERAS (continued)
See inside back cover for explanation and abbreviations

KURILE LAKE/PAUZHETKA

CAVW number (active vent)	Latitude Longitude (degrees)	Diameter (km)	Local tectonic setting	Precaldera edifice	SiO_2 content (wt pct)	Age of collapse (yr B.P.)	Date of unrest	Type and duration of unrest ESTU STHF MGTF H Te	Eruption type
10-00-03 (Iliinsky)	51.45N 157.08E	20x25 (P) 10 (KL)	Inter-section of N-S graben and E-W strike-slip fault	None?	R = low 50's -63 C = d (KL); silicic (P)	8,000-8,340 (KL) 10,000-220,000 (P)	1901 1973-77+	---- ---- ---- - -- FF-- ---- ---- - --	pex, Pex none

TECTONIC SETTING

The Pauzhetka depression lies at the intersection of (a) N-S faults that form the east edge of the Kamchatka Median massif, the Golygin Mountains, with as much as 1000 m of displacement down to the east; and (b) the E-W trending Ozernovsky strike-slip fault.

GEOLOGIC HISTORY

The Pauzhetka depression is a large caldera ("volcano-tectonic depression") associated with the Golygin ignimbrite (10,000-220,000 yr B.P.) (figs. 10.1.1, 10.1.2). The presence of this ignimbrite in the Pauzhetka basin and on the Kambalny Ridge horst, as well as of lake terraces on the northeast side of Kurile Lake, suggests as much as 1000 m of resurgent doming within the recent geologic past (Beloussov, 1967). The Kurile Lake Caldera is nested within the Pauzhetka Caldera. Iliinsky is an active stratovolcano on the northeast rim of the Kurile Lake Caldera.

There has been a general migration of volcanism, including caldera-forming events, from southwest to northeast during the Quaternary. A major geothermal field has been developed in the Pauzhetka River graben - Kambalny Ridge horst area, part of the Pauzhetka depression.

Interestingly, $^{87}Sr/^{86}Sr$ ratios in ignimbrites of the Pauzhetka region (0.7023-0.7036) show little evidence of crustal contamination. Slightly higher values were found for rocks of Iliinsky Volcano (0.7054) and Cherniye Skaly Volcano (0.7054) (Erlich, 1986, Chap. 5, Table 11).

KURILE LAKE/PAUZHETKA (continued)

HISTORICAL ACTIVITY

Occasional seismic swarms, possibly seasonal in occurrence (March-April, July-August), have been reported by Levina and others (1980) (figs. 10.1.3, 10.1.4). Those in 1973 (10^{13}-10^{15} ergs of seismic energy release) were initially thought to be centered between Kambalny and Koshelev (Kirsanov and others, 1975); a more recent analysis places them along a northeast trend north of Kambalny Volcano (Levina and others, 1980). Focal mechanisms for two earthquakes suggest left-lateral strike-slip motion on a northeast-trending fault. Fumarolic activity had not changed at the conclusion of the December 1973 swarms (Kirsanov and others, 1975).

REFERENCES

Beloussov, V.I., 1967, Geological structure of the hydrothermal systems of Kamchatka: Bull. Volcanol., v. 30, p. 63-73.
Erlich, E., 1986, Geology of the calderas of Kamchatka and Kurile Islands with comparison to calderas of Japan and the Aleutians, Alaska: U.S. Geol. Surv. Open-File Rep. 86-291, 300 p.
Erlich, E.N., and Gorshkov, G.S., eds., 1979, Quaternary volcanism and tectonics in Kamchatka: Bull. Volcanol., v. 42, nos. 1-4 (spec. iss.), 298 p.
Erlich, E.N., and Melekestsev, I.V., 1972, Evolution of Quaternary volcanism and tectonics in the western part of the Pacific rim: Pacific Geol., v. 4, p. 1-22.
Erlich, E.N., Melekestsev, I.V., Tarakanovsky, A.S., and Zubin, M.I., 1972, Quaternary calderas of Kamchatka: Bull. Volcanol., v. 36, p. 222-237.
Kirsanov, I.T., Khrenov, A.P., and Vakin, Ye. A., 1975, Active volcanoes of Kamchatka and the Kurile Islands in 1973: Bull. Vulkanol. Stantsii (Kamchatka), no. 51, p. 3-18.
Kozhemyaka, N.N., 1979, Quaternary pumice, tuff-ignimbrite fields, and center of eruptions in Southern Kamchatka: Bull. Vulkanol. Stantsii (Kamchatka), no. 57, p. 26-38.
Krayevaya, T.S., Braitseva, O.A., Sheimovich, V.S., Egorova, I.A., and Lupikina, E.G., 1979, Deposits of Quaternary calderas of Kamchatka: Vulkanol. Seismol., 1979, no. 4, p. 3-11.
Krijanovsky, N., 1934, Volcanoes of Kamchatka: Bull. Geol. Soc. Am., v. 45, p. 529-550.
Leonov, V.L., 1981, Faults of the Pauzhetka volcano-tectonic structure: Vulkanol. Seismol., 1981, no. 1, p. 24-36.
Levina, V.I., Firstov, P.P., and Zobin, V.M., 1980, The seismicity of the Pauzhetka Geothermal Region in Kamchatka: Vulkanol. Seismol., 1980, no. 2, p. 81-97.
Luchitsky, I.V., ed., 1974, Development of rugged relief of Siberia: Far East: Kamchatka, Kurile, and Komandor Islands: Moscow, Nauka Press, 430 p.
Masurenkov, Yu.P., 1980a, The volcano-magma chamber-structure system (Geologic-petrologic aspect of the problem): Vulkanol. Seismol., 1980, no. 4, p. 34-49.
----- ed., 1980b, Volcanic center: structure, dynamics, and products: Karymsky structure: Moscow, Nauka Press, 299 p.
Naboko, S.I., 1954, Geysers of Pauzhetka: Bull. Vulkanol. Stantsii (Kamchatka), no. 22, p. 9-30.

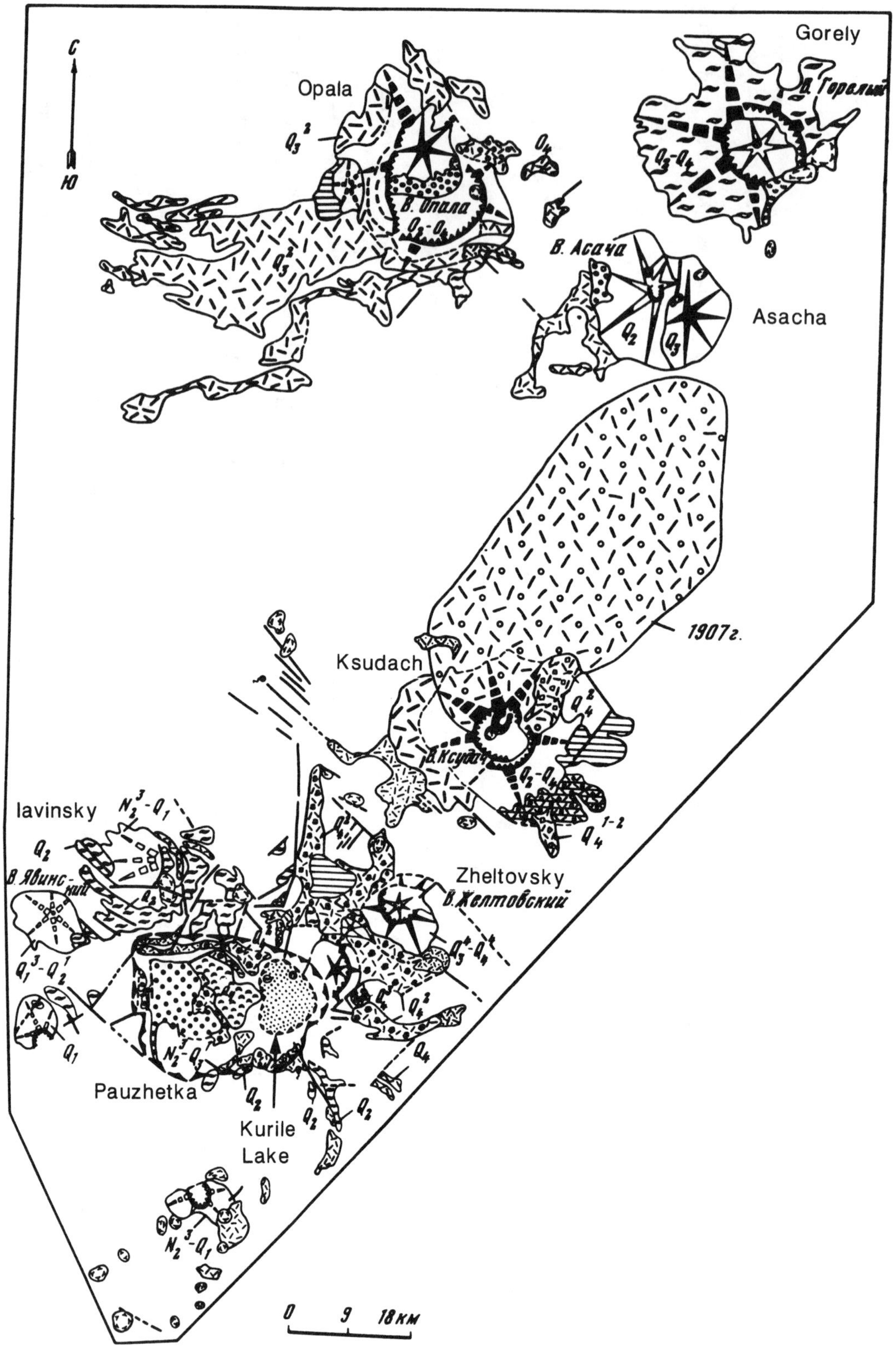

Figure 10.1.1. Volcanic centers of south Kamchatka, from Kozhemyaka (1979). Details of the geology of Kurile Lake/Pauzhetka center are given in figure 10.1.2.

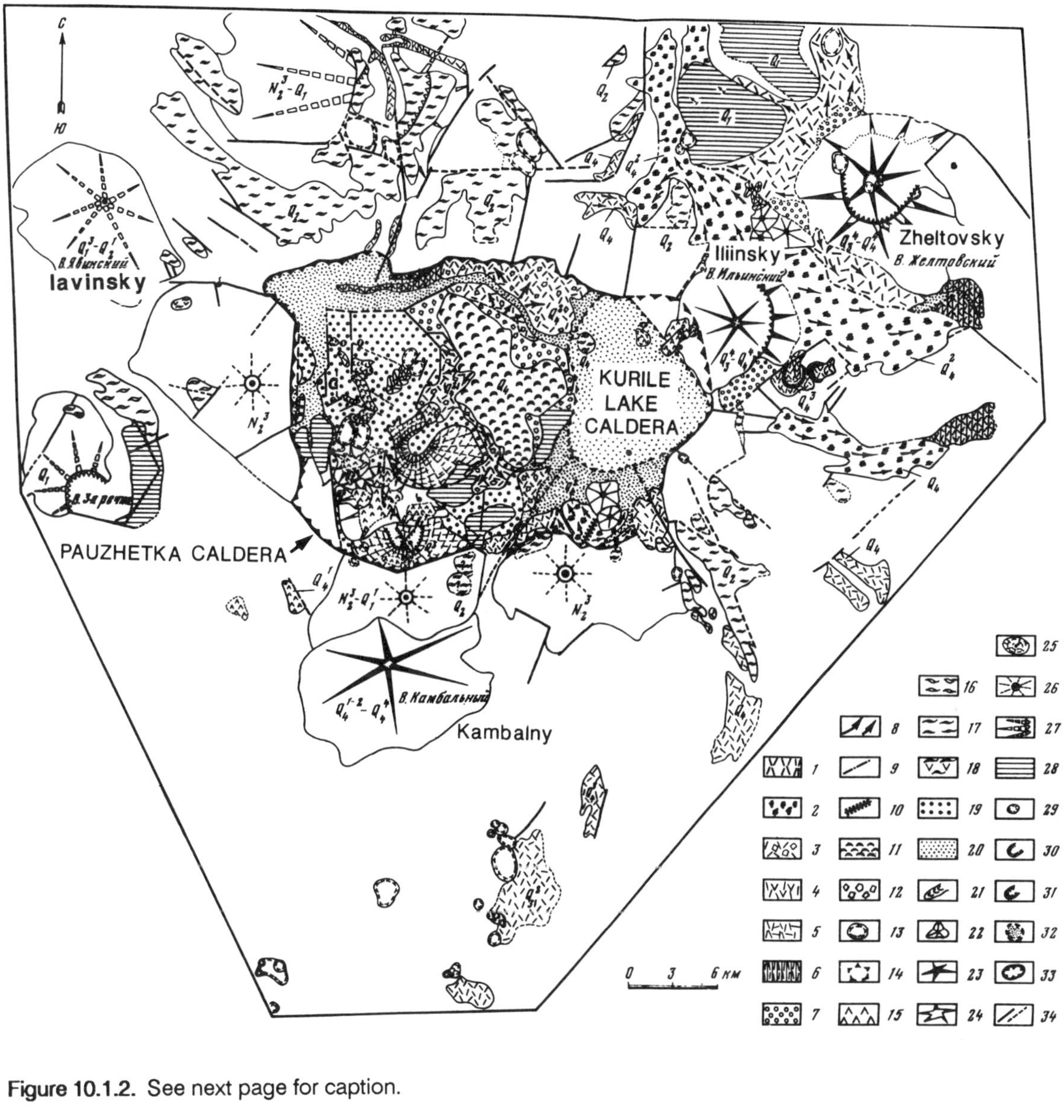

Figure 10.1.2. See next page for caption.

Figure 10.1.2. Geology of Pauzhetka volcano-tectonic depression, from Kozhemyaka (1979):

1. Pumiceous pyroclastic flow deposits from Zheltovsky Volcano -- Q_4.
2. Pumiceous pyroclastic flow deposits from Iliinsky Volcano -- Q_4.
3. Pumiceous pyroclastic flow deposits in the region of the Diky graben extrusive body - Q_4^2.
4. Modern pumice-cinder flows -- Q_4^3-Q_4^4.
5. Series of fine-grained pumiceous deposits of the fractured [unconsolidated?] type -- Q_4.
6. Reworked pumice.
7. Scoria blanket on stratovolcanoes -- Q_4.
8. Direction of pumiceous pyroclastic flows.
9. Inferred boundary between pyroclastic flows of Iliinsky and Zheltovsky volcanoes.
10. Fine-grained pumiceous deposit from a source (fissure-type) near Kirushutk -- Q_4^2.
11. Large polyphase extrusive complex of Diky graben -- Q_4.
12. Blast deposits (?) from the extrusive Diky graben complex -- Q_4.
13. Extrusive domes (cupolas) -- N_2^3.
14. Small extrusive domes -- specific extrusions of the ignimbrite tuff at Goliginsky mountain -- Q_2-Q_3.
15. Fine-grained dacitic lava.
16. Sheet of fine-grained ignimbrite tuff of dacitic composition -- Q_2-Q_3^1.
17. Platy ignimbrite tuff of liparitic dacite and liparite composition.
18. Fine-grained ignimbrite tuff of andesitic composition.
19. Thick pyroclastic tuff of the Pauzhetka sedimentary-volcanic depression -- N_2^3-Q_3^1.
20. Undifferentiated deposits of river valleys (alluvium).
21. Lava flows -- Q_3-Q_4.
22. Lava cones -- Q_3-Q_4.
23. Large conical stratovolcanoes -- Q_3^4-Q_4^4.
24. Important basaltic cones of lava volcanoes [shields?] -- Q_3.
25. Important pyroclastic strata of Kambalny Ridge -- Q_1-Q_2.
26. Important lava-shield volcanoes -- Q_1-Q_2.
27. Large, principal Pliocene volcanic centers -- N_2^3.
28. Plateau lava outliers -- Q_1.
29. Modern volcanic craters.
30. Newest explosion craters with pumice-scoria flows -- Q_4^3-Q_4^4.
31. Caldera rim volcanoes Iliinsky and Zheltovsky -- Q_4.
32. Kurile Lake volcano-tectonic depression.
33. Pauzhetka volcano-tectonic depression -- N_2^3-Q_1.
34. Tectonic shear zones -- a) known; b) hypothetical.

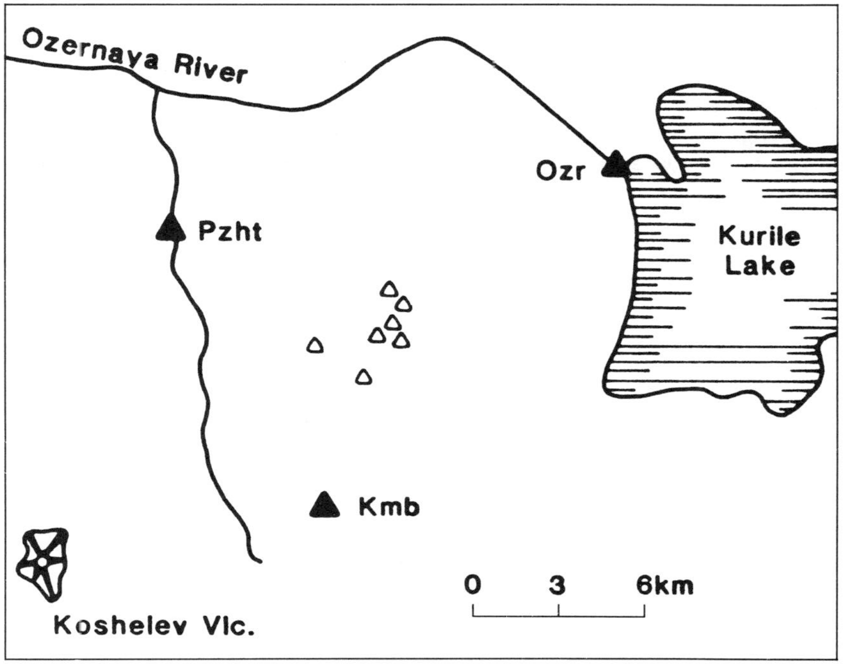

Figure 10.1.3. Epicenters of earthquakes within Pauzhetka geothermal field, from Erlich (1986) after Levina and others (1980). Solid triangles, seismic stations: Pauzhetka (Pzht), Kambalny (Kmb), and Ozernaya (Ozr). Open triangles, epicenters of earthquakes.

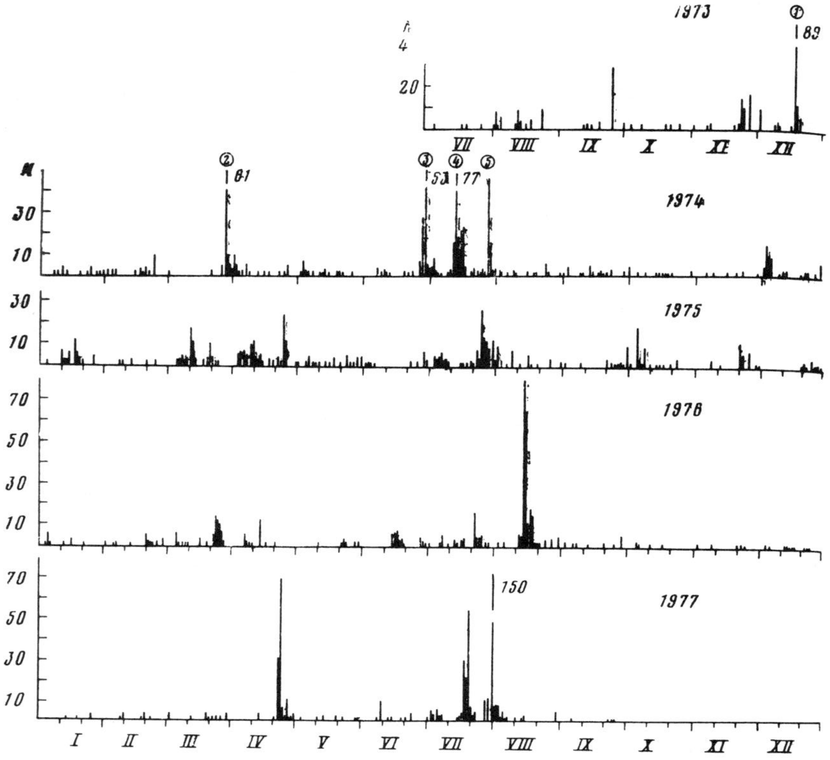

Figure 10.1.4. Daily counts of microearthquakes with energy class $\underline{K} \geq 5$, and S - P $\leq$ 2.5 sec, recorded on station Pzht in 1973-74 (detailed processing) and 1975-77 (preliminary processing), from Levina and others (1980).

PART 3: HISTORICAL UNREST AT LARGE QUATERNARY CALDERAS (continued)
See inside back cover for explanation and abbreviations

KSUDACH

CAVW number (active vent)	Latitude Longitude (degrees)	Diameter (km)	Local tectonic setting	Precaldera edifice	SiO_2 content (wt pct)	Age of collapse (yr B.P.)	Date of unrest	Type and duration of unrest ESTU STHF MGTF H Te	Eruption type
10-00-05 (Stübel cone)	51.83N 157.52E	7 x 13 (outer); 7 x 6, 4 x 5 4 x 4 4 x 4 (nested)	Inter-section of NE-trending graben and strike-slip fault	Shield	R = 60-69 C = d	>40,000 (outer); 30,000-40,000; 15,000; 8,800-3,600; 2,500-1,700 (nested)	1907	---- ---- ---- - --	EX

TECTONIC SETTING

Ksudach Caldera lies in the southern Kamchatka graben-syncline, at its intersection with a major, N45°E strike-slip fault that has offset the graben.

GEOLOGIC HISTORY

The Ksudach Caldera volcano is a shield of alternating basaltic lavas and dacitic pyroclastic deposits; this shield may have grown within a large but now poorly defined 7 km x 13 km diameter, 35-40 km^3 caldera (figs. 10.2.1-10.2.2) (Seliangin, 1987; Melekestsev and Sulerzhitsky, 1987). A 7 km x 6 km diameter main caldera and three 4-5 km diameter calderas that are nested in the north part of the 7 km x 6 km caldera all lie atop the shield. Postcaldera displacements (subsidence) and eruptions have occurred along north-northeast trending faults (Erlich, 1986). Dacitic pumice from the eruption that formed the youngest caldera is widely spread around Ksudach. A system of basaltic ring dikes that surround the rim of the youngest caldera was probably emplaced "late in the caldera-forming stage" (Erlich, 1986). At least three tephra layers are known to have originated from Stübel Cone, ca. 1650 yr B.P., ca. 700 yr B.P., and in A.D. 1907 (see below) (Melekestsev and Sulerzhitsky, 1987).

HISTORICAL ACTIVITY

More than 1 km^3 of dacitic magma was erupted in 1907 (Hulten, 1923, 1924; Piip, 1941); the silica content of compositionally banded pumice from this eruption varied from 60 to 69 percent.

KSUDACH (continued)

REFERENCES

Dubik, Yu.M., and Meniaylov, I.A., 1971, Gas-hydrothermal activity of Ksudach caldera: Bull. Vulkanol. Stantsii (Kamchatka), no. 47, p. 40-43.

Erlich, E., 1986, Geology of the calderas of Kamchatka and Kurile Islands with comparison to calderas of Japan and the Aleutians, Alaska: U.S. Geol. Surv. Open-File Rep. 86-291, 300 p.

Hulten, E., 1923, Some geographical notes on the map of South Kamchatka: Geogr. Annaler, v. 8, no. 5, 21 p.

----- 1924, Eruption of a Kamchatka volcano in 1907 and its atmospheric consequences: Geol. Foren. Forhandl., v. 46, no. 5, p. 407-417.

Melekestsev, I.V., and Sulerzhitsky, L.D., 1987, Ksudach Volcano (Kamchatka) in the last 10 thousand years: Vulkanol. Seismol., 1987, no. 4, p. 28-39.

Piip, B.I., 1941, On the intensity of the eruption of Ksudach volcano in March, 1907: Bull. Vulkanol. Stantsii (Kamchatka), no. 10, p. 12-23.

Seliargin, O.B., 1987, Geologic map of Volcano Ksudach: Vulkanol. Seismol., 1987, no. 5, p. 16-27.

Vlodavetz, V.I., and Piip, B.I., 1959, Catalogue of active volcanoes of the world including solfatara fields, Pt. VIII, Kamchatka and continental areas of Asia: Naples, IAVCEI, 110 p.

Volynets, O.N., 1979, Mixed lavas, (and) relationship of the melts forming these lavas: Bull. Volcanol., v. 42, nos. 1-4 (spec. iss.), p. 233-242.

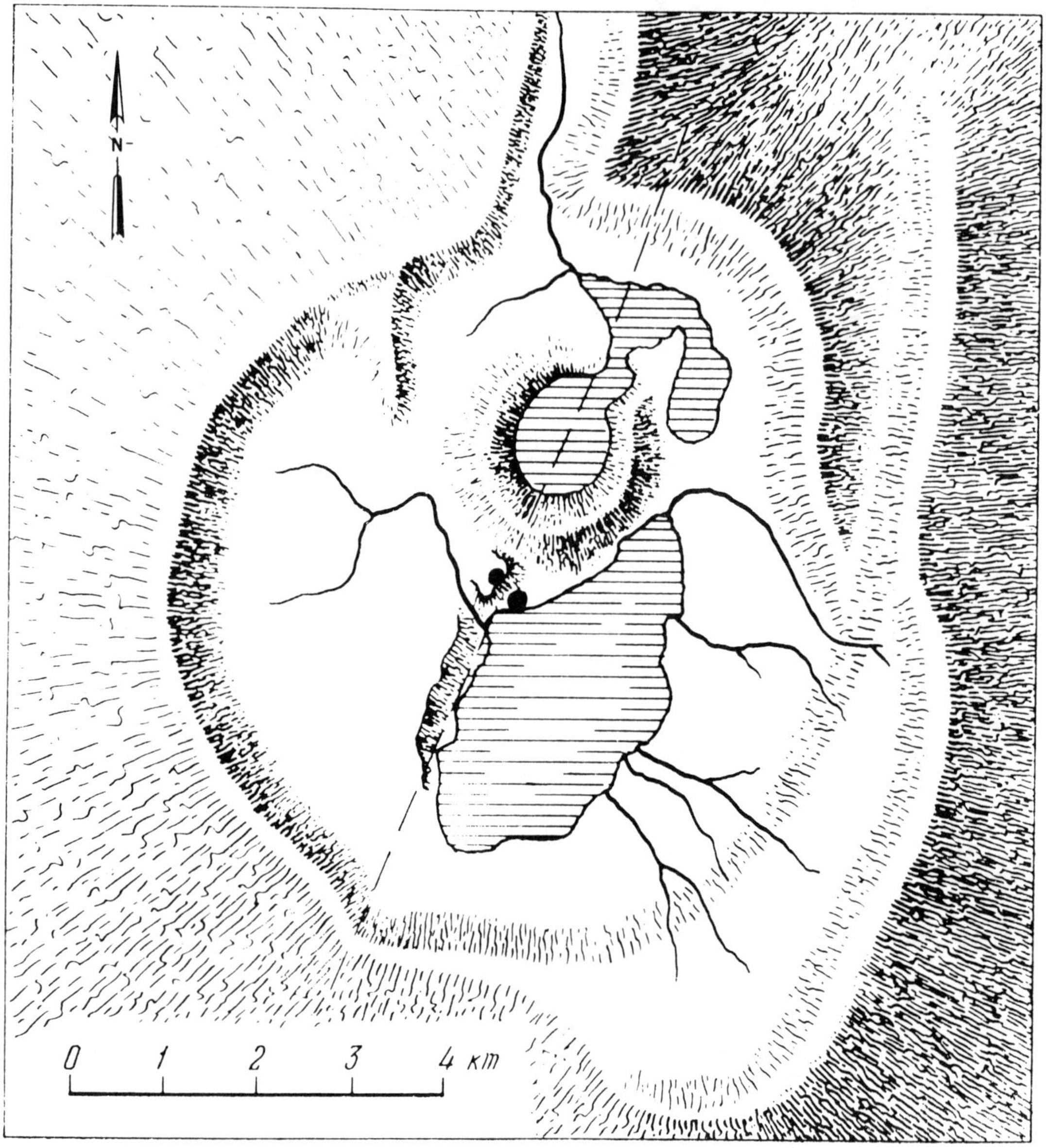

Figure 10.2.1. Sketch of top of Ksudach Volcano, from Vlovadetz and Piip (1959). Hot springs are indicated by dots, a fault by dashed line.

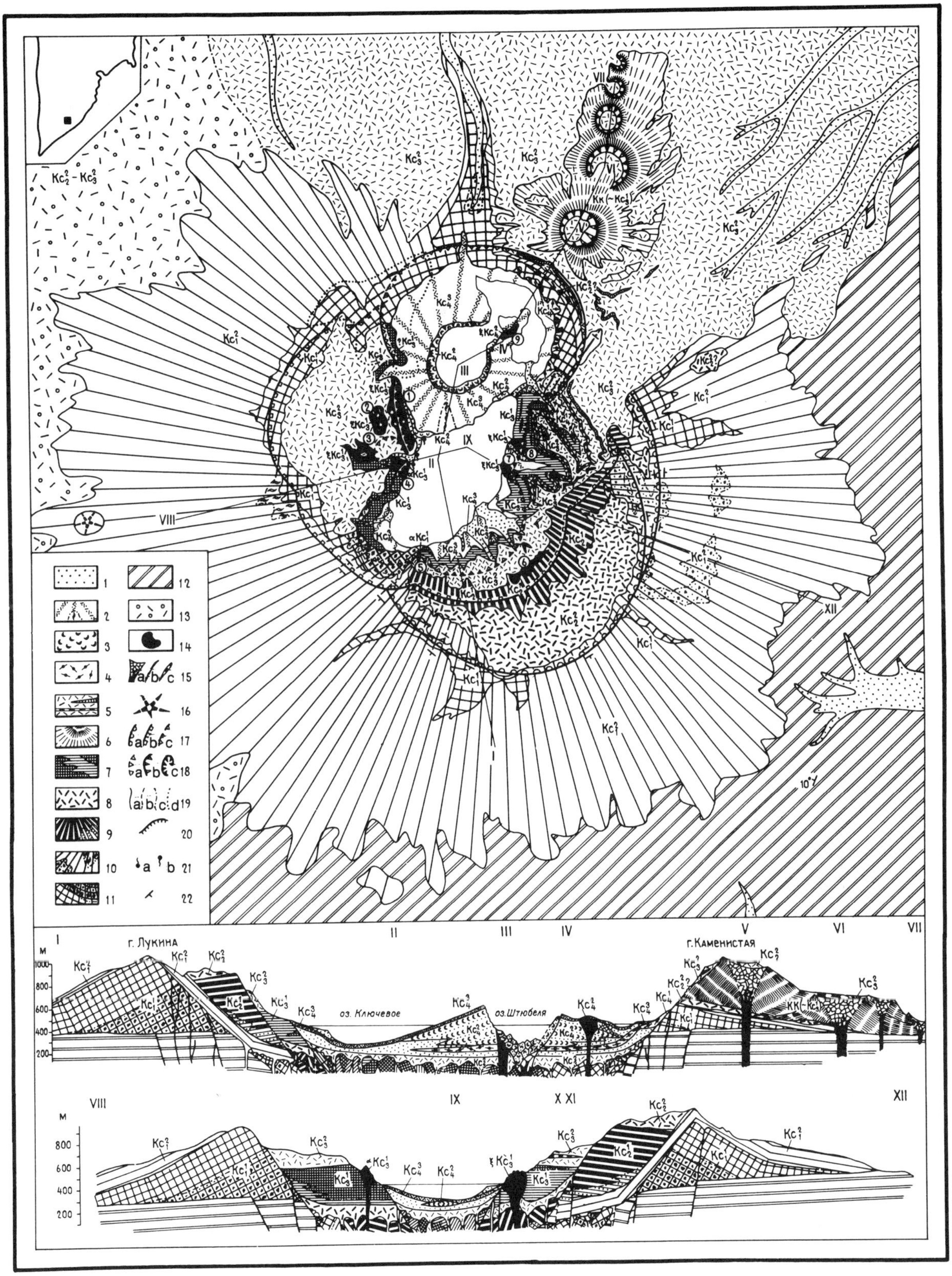

Figure 10.2.2. Geologic map of Ksudach Volcano, from Seliangin (1987). See next page for explanation.

Figure 10.2.2. Geologic map of Ksudach Volcano, from Seliangin (1987):

Kc_{1-4} -- Complexes of deposits corresponding to rhythms of volcanic activity. Most begin with phases (Kc_1^1, Kc_2^1) of basaltic-andesitic explosive-effusive volcanism, then have a break in volcanism, and conclude with phases (Kc_1^2, Kc_2^2, ...) that are explosive eruptions of mostly dacitic pumice (compositional range andesitic basalt to dacite), causing formation of great calderas and large craters.

1- alluvium.

Complex Kc_4:
2- andesitic scoria and dacitic pumices with local ignimbrite facies, blast deposits and nodules of allivalite and eucrite (olivine and augite gabbro) (Kc_4^3);
3- basaltic to andesitic lava and pyroclastic deposits of the Stübel cone, center of the most recent (1907) activity of Ksudach Volcano (Kc_4^2, complex Kc_3);
4- "early" dacitic pumice (Kc_4^1).

Complex Kc_3:
5- two horizons of pumiceous ignimbrites (the upper containing allivalite and eucrite nodules), dacite to andesite in composition (Kc_3^2);
6- complex of satellite volcanoes of the Kamenistaya Mountain (KK - Kc_3^1), basalt and basalt andesite in composition;
7- lavas of basaltic andesite to andesite composition, and sandy to gravelly volcanogenic lake deposits (closely spaced horizontal line pattern) (Kc_3^1).

Complex Kc_2:
8- andesitic and dacitic pumice (Kc_2^2);
9- lavas and pyroclastic deposits of basalt to basaltic andesite composition, and lake deposits (bold stripes in cross section) (Kc_2^1).

Complex Kc_1:
10- welded agglomerate tuffs of andesite and basaltic andesite composition (Kc_1^2);
11- tuffs of basaltic andesite, and lavas and pyroclastic rocks of basalt to andesite composition (Kc_1^1).

Basement complex:
12- sequence of basaltic lava sheets;
13- pumice deposits, undivided;
14- extrusive domes;
15- a) vents; b) dikes of basaltic andesite; c) ignimbritic dikes;
16- adventive cones;
17- boundaries of calderas of various ages: a) caldera Kc_1; b) caldera Kc_2; c) complex of calderas of the fourth generation Kc_3;
18- craters: a) destroyed and buried; b) with topographic expression; c) the most recent (Stübel crater);
19- geological contacts: a) observed; b) inferred; c) contacts between different facies of a single unit; d) boundary of recent eolian deposits;
20- faults;
21- modern geothermal manifestations: a) steam/gas; b) hot springs;
22- attitudes of layered rocks.

GORELY KHREBET

CAVW number (active vent)	Latitude Longitude (degrees)	Diameter (km)	Local tectonic setting	Precaldera edifice	SiO_2 content (wt pct)	Age of collapse (yr B.P.)	Date of unrest	Type and duration of unrest ESTU STHF MGTF H Te	Eruption type
10-00-07 (Gorely Khrebet)	52.45N 158.12E	10 x 13	WNW fracture normal to volcanic chain; compr	LL-strat	R = 51-71 (mostly b and a)	Upper Pleis- tocene	1828	---- ---- ---- - --	ex
							1832	---- ---- ---- - --	ex
							1855	---- ---- ---- - --	ex
							1869?	---- ---- ---- - --	unknown
					C = 60-66		1929-31	---- ---- ---- - --	Ex
							1947	---- ---- ---x - --	ex?
							1960-61	---- ---- --EE - --	ex
							1977-81	---- ---- --FF F --	Pex,ex
							1984-85	---- ---- ---- - --	pex

TECTONIC SETTING

Gorely Khrebet Caldera is located in the northeast corner of the southern Kamchatka graben-syncline, at the intersection of several E-W-trending "deep-seated faults" (Erlich, 1986).

GEOLOGIC HISTORY

Gorely Khrebet Caldera (figs. 10.3.1, 10.3.2) lies on the eastern margin of a large negative gravity anomaly (the Tolmachev anomaly, diameter 20-30 km, possibly representing a large buried caldera). The present volume of the upper Pleistocene ignimbrite sheet around the caldera is 120 km^3; its uneroded volume was probably greater (Luchitsky, 1974). The volume of caldera collapse was 200-250 km^3 (Erlich, 1986). Gorely Khrebet Volcano lies on an uplifted(?) area in the center of the caldera.

HISTORICAL ACTIVITY

1947: Strong solfatara activity was reported (Vlodavetz and Piip, 1959).

1960-61: Fumarolic temperatures increased, as during 1977-81 (see below). Then, after a series of minor ash eruptions (Kirsanov and others, 1964), fumarolic activity declined to a minimum during 1972-74 (Ivanov and others, 1980).

GORELY KHREBET (continued)

HISTORICAL ACTIVITY (continued)

1977-81: New fumaroles with temperatures up to 90 °C appeared in the summit crater of Gorely Khrebet in 1977 (fig. 10.3.4). During 1978-79, the number of fumaroles increased, maximum fumarole temperatures rose to 95 °C, and a warm lake formed on the crater floor. An area of hot ground appeared just west of the summit of Gorely Khrebet in 1978, and in 1979 fumarolic activity sharply increased in all craters of the massif. A green-colored lake formed in the central crater, and vigorous fumaroles became active at the base of the north wall. In August-September, gas plumes rose to heights of 500 m (Ivanov and others, 1980). Moderately large phreatic eruptions began in June 1980, and phreatomagmatic eruptions occurred in July and subsequent months. Juvenile blocks were alkali-rich basaltic andesite (Smithsonian Institution, 1980; Kirsanov and Ozerov, 1983). Earthquakes during 1980-81 occurred mainly in swarms, each lasting a few days to a few weeks (fig. 10.3.3) (Gabrilov and others, 1984).

1984-85: An eruption began on 27 December 1984. As in the 1980-81 eruption, initial phreatic activity was followed by phreatomagmatic explosions that ejected juvenile alkali-rich basaltic andesite (Kirsanov and Ozerov, 1983; Smithsonian Institution, 1985).

REFERENCES

Erlich, E., 1986, Geology of the calderas of Kamchatka and Kurile Islands with comparison to calderas of Japan and the Aleutians, Alaska: U.S. Geol. Surv. Open-File Rep. 86-291, 300 p.

Gabrilov, B.A., Gordeev, E.I., Ivanov, B.B., Ibshin, B.M., Stepanov, B.B., Farberov, A.I., Shirokov, B.A., and Yatsuk, B.B., 1984, Volcanic tremor and the Gorely Volcano earthquakes during the 1981 eruption: Vulkanol. Seismol., 1984, no. 6, p. 3-17.

Gushchenko, I.I., 1979, Eruptions of volcanoes of the world, a catalog: Acad. Sci. USSR Far Eastern Sci. Center, Moscow, Nauka Publ., 474 p.

Ivanov, B.V., Kirsanov, I.T., Chirkov, A.M., Khrenov, A.P., Andreyev, V.N., Vakin, Ye.A., and Trukhin, Yu.P., 1980, The status of the volcanoes of Kamchatka from May 1979 to January 1980: Vulkanol. Seismol., 1980, no. 3, p. 87-89.

Kirsanov, I.T., 1964, Volcanoes of southern and central Kamchatka in 1963: Bull. Vulkanol. Stantsii (Kamchatka), no. 37, p. 8-12.

----- 1967, Acid volcanism and thermal manifestations in the area of Mutnovsky and Gorely Volcanoes (southern Kamchatka): Bull. Volcanol., v. 30, p. 139-151.

----- 1981, Eruption of Gorely volcano in 1980: Vulkanol. Seismol., 1981, no. 1, p. 70-74.

Kirsanov, I.T., Ogorodov, N.V., and Chirkov, A.M., 1964, State of the Mutnovsky and Gorely volcanoes for the period from November 1960-June 1961: Bull. Vulkanol. Stantsii (Kamchatka), no. 36, p. 37-39.

Kirsanov, I.T., and Ozerov, A.Yu., 1983, Composition of products and energy effects of the eruption of Gorely Volcano, 1980-81: Vulkanol. Seismol., 1983, no. 1, p. 25-42 (English translation, Volcanol. Seismol., 1984, v. 5, p. 23-43).

Kozhemyaka, N.N., 1979, Quaternary pumice, tuff-ignimbrite fields, and center of eruptions in Southern Kamchatka: Bull. Vulkanol. Stantsii (Kamchatka), no. 57, p. 26-38.

GORELY KHREBET (continued)

REFERENCES (continued)

Luchitsky, I.V., ed., 1974, Development of rugged relief of Siberia: Far East: Kamchatka, Kurile, and Komandor Islands: Moscow, Nauka Press, 430 p.

Melekestsev, I.V., Braitseva, O.A., and Ponomareva, V.V., 1987, Dynamics of activity of Gorelyi and Mutnovskii volcanoes and volcanic hazard for adjacent areas (from tephrochronological studies): Vulkanol. Seismol., 1987, no. 3, p. 3-18.

Novograblenov, P.T., 1930, Eruption of Gorely volcano in 1929-30: Zapisky Russ. Geogr. Soc., v. 62, no. 4, p. 459-461.

Smithsonian Institution, Scientific Event Alert Network (SEAN), 1980, 1985, Gorely Khrebet: SEAN Bull., v. 5, no. 7; v. 10, no. 1 (information from Drs. Vakim and Kirsanov, and B.V. Ivanov).

Vlodavetz, V.I., and Piip, B.I., 1959, Catalogue of active volcanoes of the world including solfatara fields, Pt. VIII, Kamchatka and continental areas of Asia: Naples, IAVCEI, 110 p.

Volcanological Society of Japan, 1982, Gorely Khrebet: Bull. Volcanic Eruptions, no. 20, p. 62.

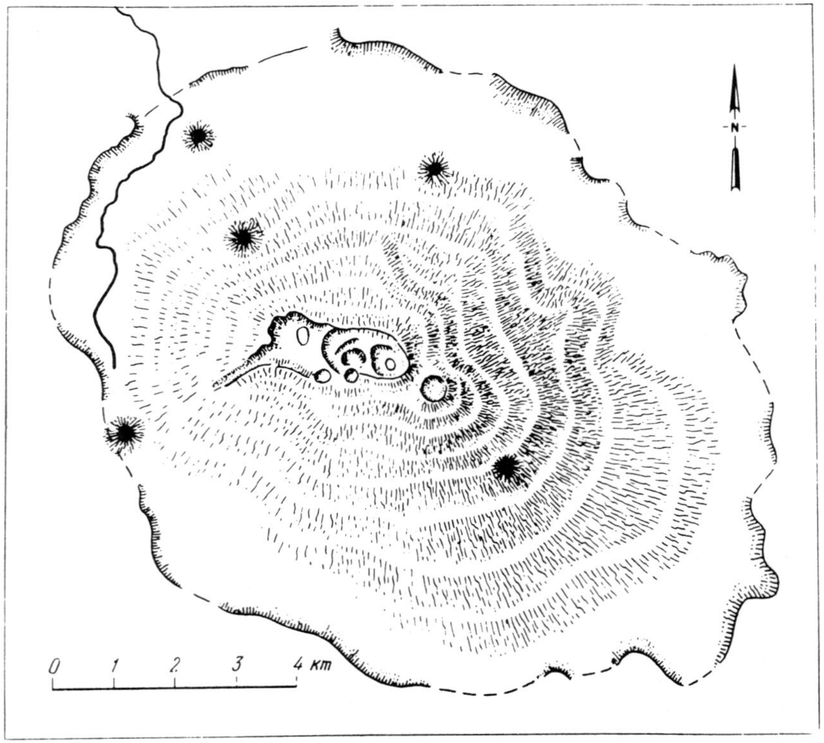

Figure 10.3.1. Sketch of caldera of Gorely Khrebet Volcano, from Vlovadetz and Piip (1959).

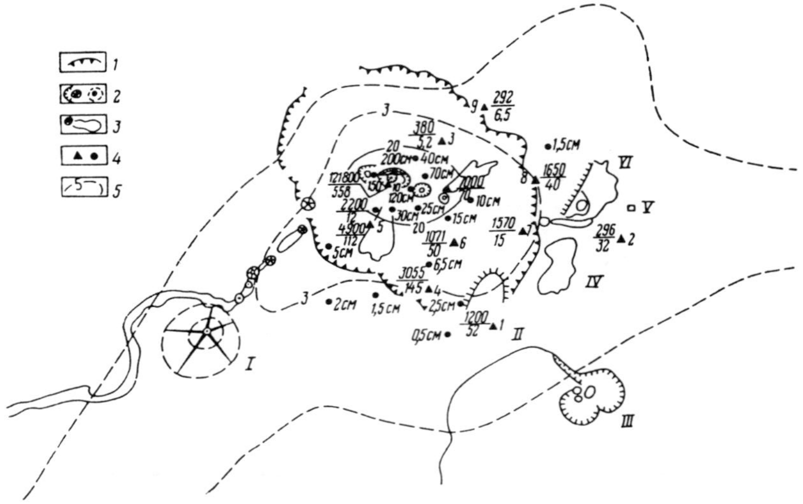

Figure 10.3.2. Sketch showing ash thicknesses and locations of sampling sites on flanks of Gorely Khrebet Volcano in September 1980, from Kirsanov and Ozerov (1983). Explanation: 1, caldera rim; 2, summit crater; 3, cinder cones and lava flows; 4, disposition of ash collectors and sites of cinder thickness measurements on flanks and in vicinity of volcano; 5, thickness contours at end of August - beginning of September 1980; fractions: above line, weight of ash in ash collector, below line, approximate duration of ash deposition; I, Mount Vysokaya; II, Mutnovskii station; III, Mutnovskii craters; IV, Mount Dvugorbaya; V, geologic camp Dachnye Istochniki; VI, Mount Skalistaya.

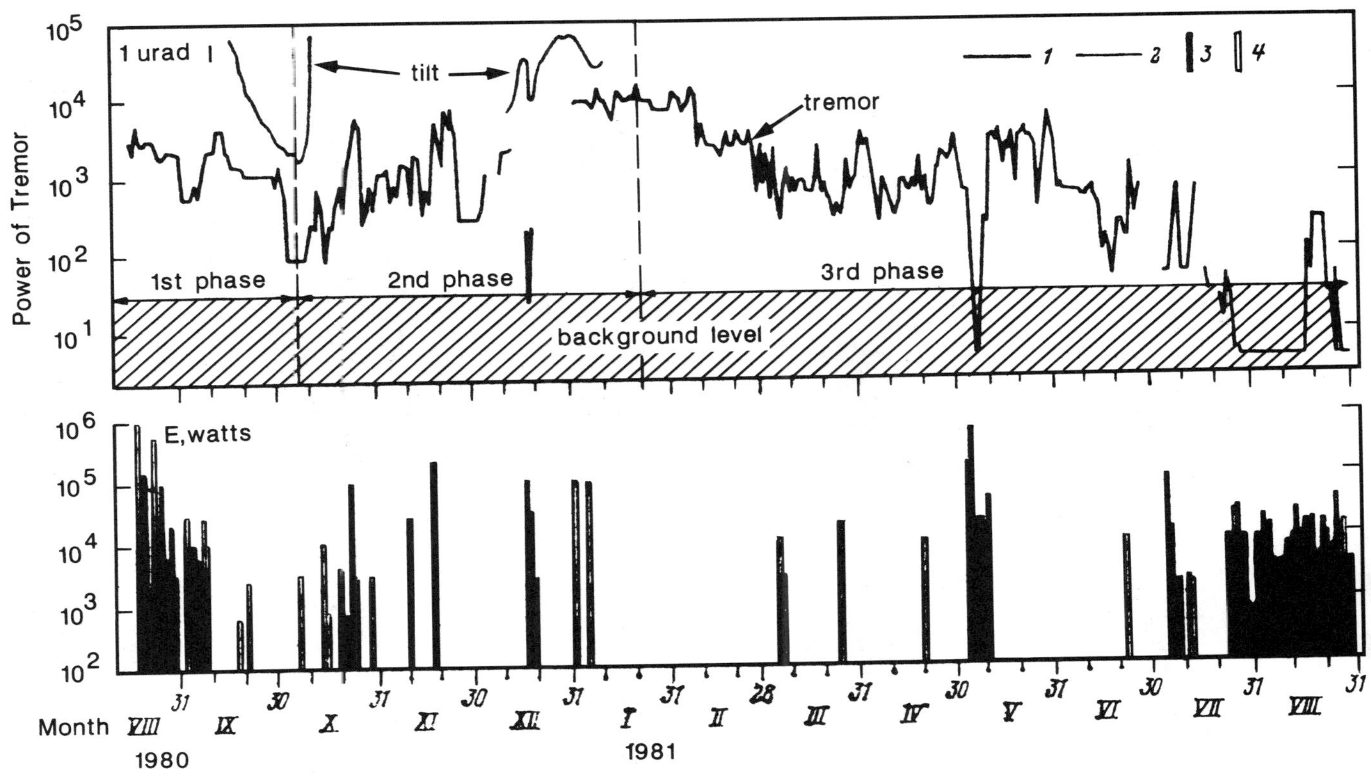

Figure 10.3.3. Graph of variability of daily averages of volcanic tremor power and earth tilt, from Gabrilov and others (1984). Top, (1) power of tremor [units=watts?]; (2) tilting motion, (small vertical bar = 1 microradian); (3) earthquakes and coseismic jumps in tilt; (4) explosion earthquakes. Bottom, daily magnitude of energy release of volcanic earthquakes, in watts.

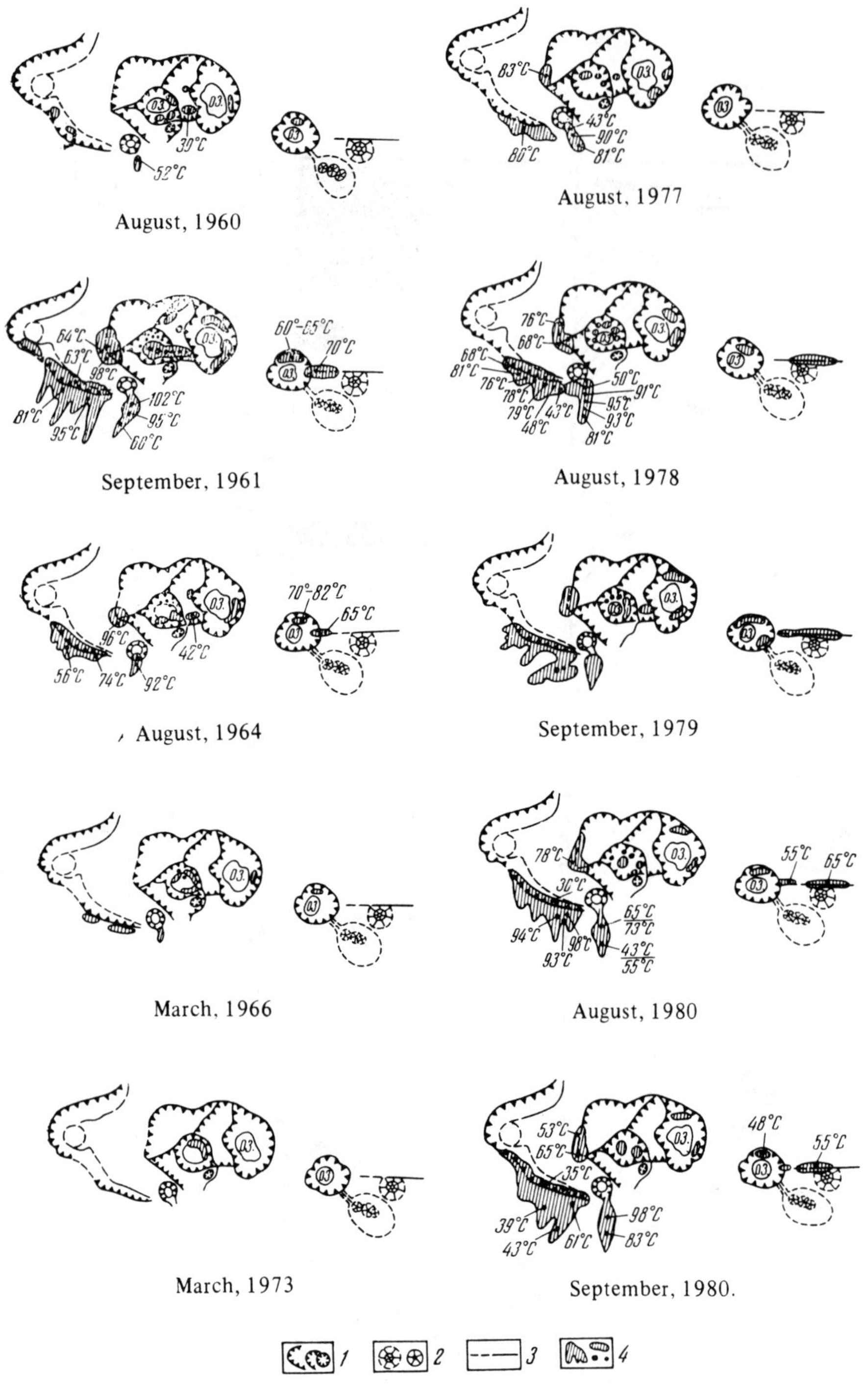

Figure 10.3.4. Changes in fumarole temperatures in summit crater of Gorely Khrebet Volcano during 1960-80, from Kirsanov and Ozerov (1983). Note that scarp at left (west) of each sketch is western rim of summit crater, not caldera. Explanation: 1, volcanic craters; 2, cinder cones; 3, fissures; 4, areas of active fumaroles, with temperatures indicated.

OPALA

CAVW number (active vent)	Latitude Longitude (degrees)	Diameter (km)	Local tectonic setting	Precaldera edifice	SiO_2 content (wt pct)	Age of collapse (yr B.P.)	Date of unrest	Type and duration of unrest ESTU STHF MGTF H Te	Eruption type
10-00-08 (Opala)	52.43N 157.45E	10 x 12	E-W faults at north end of N-S graben	Shield? LL-strat?	R = 52-66 C = silicic?	22,000	1776	---- ---- ---- - --	ex
							1827	---- ---- ---? - --	none
							1854	---- ---- ---? - --	none
							1894	---- ---- ---? - --	none

TECTONIC SETTING

Opala Caldera (figs. 10.4.1, 10.4.2) is situated in the northwest corner of the southern Kamchatka graben-syncline, where an E-W fault forms the north boundary of that graben. Opala Caldera is also located on the western margin of a strong negative regional gravity anomaly (the Tolmachev anomaly, diameter 20-30 km, possibly a buried caldera).

GEOLOGIC HISTORY

At least 90 km^3 of pumice (Erlich, 1986) was erupted about 22,000 yr B.P. (Krayevaya, 1967), forming the Opala Caldera. The postcaldera Opala stratovolcano lies on the north side of the caldera and shows a gradual evolution from basaltic to andesitic and dacitic magma (Erlich, 1986). Several postcaldera rhyolite domes lie on the east rim of the caldera. Explosions formed a 1.3 km x 2 km crater in the southeast part of the caldera about 1490 ± 90 yr B.P. (Melekestsev and others, 1967; Sheimovich and Patoka, 1979).

HISTORICAL ACTIVITY

The only known eruption in historical time (1776) was observed from Ust' Bolsheretsk, located on the western shore of Kamchatka about 100 km from Opala. Strong solfataric activity was noted in 1827, 1854, and 1894 (Vlodavetz and Piip, 1959).

REFERENCES

Erlich, E., 1986, Geology of the calderas of Kamchatka and Kurile Islands with comparison to calderas of Japan and the Aleutians, Alaska: U.S. Geol. Surv. Open-File Rep. 86-291, 300 p.

OPALA (continued)

REFERENCES (continued)

Erlich, E.N., Melekestsev, I.V., Tarakanovsky, A.S., and Zubin, M.I., 1972, Quaternary calderas of Kamchatka: Bull. Volcanol., v. 36, p. 222-237.

Kozhemyaka, N.N., 1979, Quaternary pumice, tuff-ignimbrite fields, and center of eruptions in Southern Kamchatka: Bull. Vulkanol. Stantsii (Kamchatka), no. 57, p. 26-38.

Krayevaya, T.S., 1967, New data on the age of young pumice deposits of the Kuril Lake: Probl. Kamchatka Geogr., no. 5, p. 43-56.

Luchitsky, I.V., ed., 1974, Development of rugged relief of Siberia: Far East: Kamchatka, Kurile, and Komandor Islands: Moscow, Nauka Press, 430 p.

Melekestsev, I.V., Braitseva, O.A., and Ponomareva, V.V., 1967, Dynamics of activity of Garelyi and Mutnovskii volcanoes and volcanic hazard for adjacent areas (from tephrochronological studies): Vulkanol. Seismol., 1987, no. 3, p. 3-18.

Popruzhenko, S.V., 1983, Genesis of the Opala Volcano Caldera: Vulkanol. Seismol., 1983, no. 6, p. 111-113.

Sheimovich, V.S., and Patoka, M.G., 1979, Opala Volcano and other Holocene centers of acid explosive eruptions in Southern Kamchatka: Vulkanol. Seismol., 1979, no. 5, p. 92-95.

Vlodavetz, V.I., and Piip, B.I., 1959, Catalogue of active volcanoes of the world including solfatara fields, Pt. VIII, Kamchatka and continental areas of Asia: Naples, IAVCEI, 110 p.

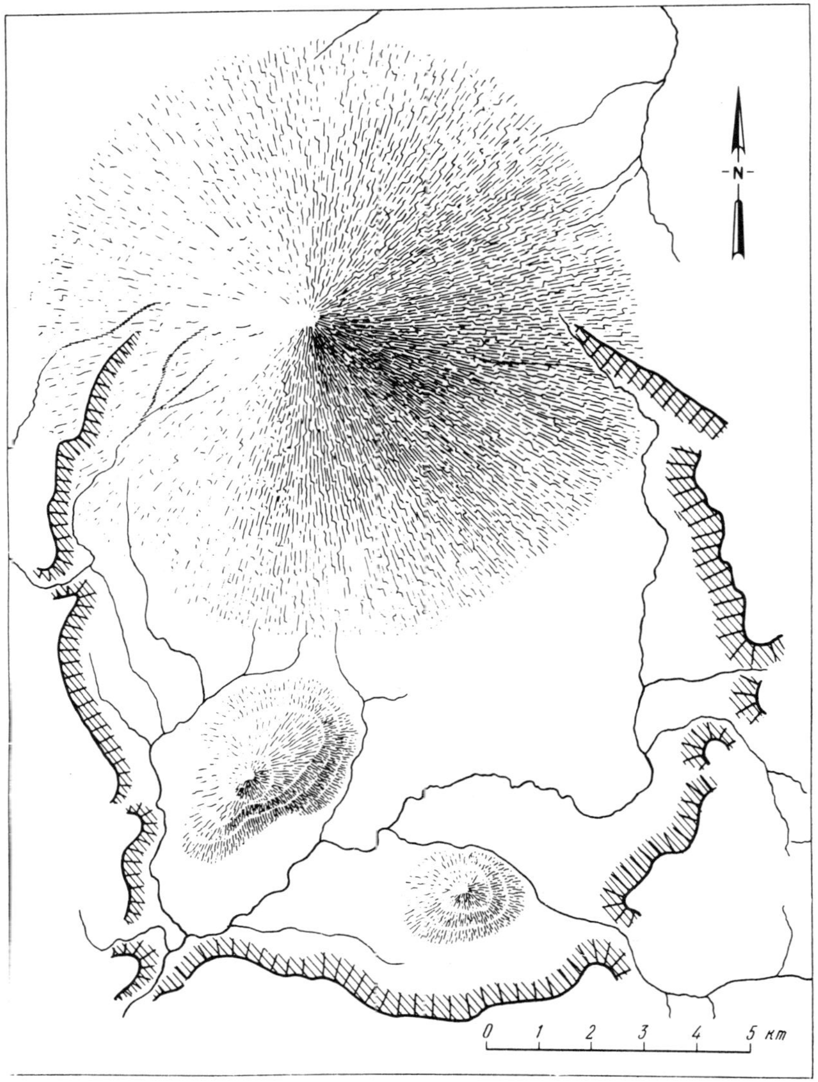

Figure 10.4.1. Sketch of caldera and cone of Opala Volcano, from Vlovadetz and Piip (1959).

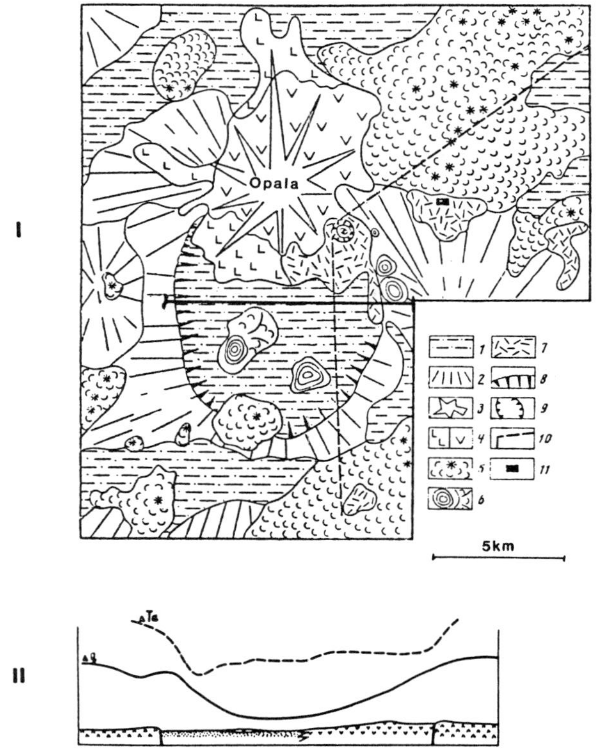

Figure 10.4.2. Geologic structure of Opala Caldera region, after Sheimovich and Patoka (1979). Explanation: I. 1, Quaternary fluvioglacial deposits; 2, lower-middle Pleistocene volcanics: basalts, andesites; 3, Holocene Opala Volcano; 4, andesite-dacite lavas of Opala Volcano; 5, Holocene basalts and scoria cones associated with areal volcanism; 6, Holocene andesitic and rhyolitic extrusive domes and associated lava flows; 7, directed blast deposits and rhyolite pyroclastic flows associated with Barany Amphitheatre crater; 8, scarps of caldera; 9, scarps of Barany Amphitheatre crater; 10, approximate boundaries of sector of spreading of pyroclastic formations of Barany Amphitheatre crater; 11, buried charcoal locality. II. Aeromagnetic (dashed) and Bouguer gravity (solid) profiles across Opala Caldera added from Erlich and others (1972).

PART 3: HISTORICAL UNREST AT LARGE QUATERNARY CALDERAS (continued)
See inside back cover for explanation and abbreviations

ZHUPANOVSKY (Z) COMPLEX
Including KARYMSKY (K), POLOVINKA (P), AKADEMII NAUK (AN), ODNOBOKY (O), STENA-SOBOLINY (SS), and MALY SEMIACHIK (MS) Calderas

CAVW number (active vent)	Latitude Longitude (degrees)	Diameter (km)	Local tectonic setting	Precaldera edifice	SiO_2 content (wt pct)	Age of collapse (yr B.P.)	Date of unrest	Type and duration of unrest ESTU STHF MGTF H Te	Eruption type
10-00-13	54.07N	Z = 40	Zone of	Shield?	K: R= a-d	8,400-	1771 (K)	---- ---- ---- - --	ex
(Karymsky)	159.60E		E-W right-		C= d	7,400	1804 (MS)	---- x--- ---- - --	Ex
	(K)	K = 5	lateral			(K)	1830 (K)	---- ---- ---- - --	ex
			strike-				1851-52(MS)	---- ---- ---- - --	ex
		P = 15	slip		P: C= a-d	mid-	1852 (K)	---- ---- ---- - --	ex
10-00-14			faulting;			Pleistocene	1854(K,MS?)	---- ---- ---- - --	ex
(Maly	54.12N	O = 4x5	compr			(P)	1908 (K)	---- ---- ---- - --	ex,LF
Semiachik)	159.88E						1911-12(K)	---- ---- ---- - --	ex
	(MS)	AN = 3x5			AN: C=	8,100	1915 (K)	---- ---- ---- - --	ex
					silicic	(AN)	1921 (K)	---- ---- ---- - --	ex
		SS = 15x21					1923 (K)	---- ---- ---- - --	ex
					MS: R= mafic	10,000-	1925 (K)	---- ---- ---- - --	ex
		MS = 6x7			C= 61-64	12,000	1929 (K)	---- ---- ---- - --	ex
						(MS)	1932-35(K)	---- ---- ---- - --	ex,lf
							1938 (K)	---- ---- ---- - --	pex
					SS: C= a-d	mid-	1940 (K)	---- ---- ---- - QU	ex
						Pleistocene	1943-47(K)	---- ---- ---- - --	ex,lf
						(SS)	1945-46(MS)	---- ---- ---- - QU	ex
							1952 (K)	---- ---- ---- - QU	ex
							1952 (MS)	---- ---- ---- - QU	ex
							1955-57(K)	?--- ---- ---- - --	ex
							1960-67(K)	E--- ---- ---- - QU	Ex,ex,lf
							1970-82(K)	YY-Y YY-- ---x E --	Ex,ex,lf,dm
							1985 (K)	---- ---- ---- - --	lf

TECTONIC SETTING

The Zhupanovsky complex lies in the eastern Kamchatka graben-syncline, in an area where that graben is crossed by a 6-8 km wide zone of E-W strike-slip faults. Earthquakes deeper than 50 km occur along these E-W faults. Parts of the Zhupanovsky complex that lie north of these faults are offset about 5 km to the east (Erlich, 1986).

ZHUPANOVSKY COMPLEX (continued)

GEOLOGIC HISTORY

A complex of several calderas lies within the large Zhupanovsky volcano-tectonic depression (figs. 10.5.1-10.5.3); terminology of calderas varies among authors. We use that of Masurenkov (1980) and Erlich (1986), who recognize the Polovinka Caldera (with Odnoboky and Akademii Nauk Calderas nested inside) and the Stena-Soboliny Caldera (with Maly Semiachik Volcano inside). The Stena-Soboliny and Polovinka Calderas may have formed about the same time--middle Pleistocene--after eruptions that together produced 280 km^3 of pyroclastic deposits. A "stable" block between the Polovinka and Stena-Soboliny Calderas contains the relatively small Karymsky and Dvor Calderas. Karymsky and Akademii Nauk are perhaps closely related, because they are separated by only 2 km and because most of the seismicity preceding eruptions in the Karymsky Caldera has been located beneath the Akademii Nauk Caldera. Karymsky Volcano is a young stratovolcano in the center of the Karymsky Caldera.

Precaldera eruptions of Maly Semiachik were of basalt and andesite; the caldera-forming eruption consisted of 6 km^3 of high-silica andesite and dacite. Recent eruptions have been of andesite and basalt. Hawaiian-like rift zones occur on the flanks of the Maly Semiachik volcanic edifice.

Major E-W strike-slip faults pass between the Polovinka and Stena-Soboliny Calderas, and much historical seismicity has occurred along these faults; other northeast- and northwest-trending faults are associated with right-lateral movement along the E-W faults.

HISTORICAL ACTIVITY

Karymsky Volcano:

Ivanov (1970) recognized two principal groups of historical eruptions of Karymsky Volcano: explosive and mixed effusive-explosive. Eruptive periods or "cycles" typically last several days to five years and are separated by reposes of 1-6 years. Most eruptions are explosive, but in 1908, 3×10^8 m^3 of dacitic lava was erupted onto the east slope of Karymsky. Magma of the 1962-65 eruption was also dacitic, with 61-63 percent SiO_2 (fig. 10.5.4) (Ivanov, 1970), and lava was again extruded in 1970 (6×10^6 m^3), 1979-82 (>5×10^6 m^3), and 1985 (small volume, as seen from U.S. Space Shuttle)(Ivanov and others, 1982; Smithsonian Institution, 1985). The total volume of tephra in 1962-65 was 0.21 km^3, and the total volume of lava for the same period was 0.06 km^3; half of the total volume was erupted in 1963 (Ivanov, 1970).

Eruptions of Karymsky are typically preceded by felt earthquakes (figs. 10.5.5-10.5.9). The November 1952 eruption followed and may have been triggered by a $\underline{M} = 8.25$ earthquake that occurred on 4 November 1952, 100-150 km from Karymsky. The April 1960 eruption followed and may have been triggered indirectly by a $\underline{M} = 8.2$ earthquake that occurred on 4 May 1959, 20 km from Karymsky.

"Not one quake was recorded from the region of Karymsky" before an eruption at 1900 hr on 10 May 1970; the first earthquake was recorded on 12 May at 0012 hr. Earthquakes increased sharply between 1200 hr on 12 May and 0400 hr on 13 May: "all quakes were

ZHUPANOVSKY COMPLEX (continued)

HISTORICAL ACTIVITY (continued)

accompanied by powerful explosions.... From 1600 hr to 0400 hr the earthquakes were so frequent that they appeared as spasmodic tremor" (Dubik and others, 1972). Frequent earthquakes occurred during July-August 1970, before dome growth in September 1970.

Earthquakes in the period 1970-73 and especially 1975, precursory to and accompanying eruptions at Karymsky Volcano, were concentrated beneath and just south of the Akademii Nauk Caldera (Firstov and others, 1977, 1978). In April and May 1975, seismicity increased 8-18 hr before eruptions and then decreased almost as sharply 2-6 hr before eruptions (fig. 10.5.9). No seismic precursor was detected before the 1976 eruption.

The area within 4 km of the summit of Karymsky Volcano subsided during 1972-81 relative to the stable block between the Karymsky and Akademii Nauk Calderas (Magus'kin and others, 1982). At a distance of 1.5 km from the crater, subsidence amounted to 3-4 cm; it was probably related to effusive activity. The volume of subsidence in a 9-year period ending in 1981 was 0.038 km^3, or 3 times the volume of lava discharged during the same period. General subsidence is interrupted occasionally by shallow-seated uplift before explosions at the volcano (figs. 10.5.11-10.5.12). Horizontal displacements of up to 5-10 cm have been noted along NW-SE faults after earthquake swarms (figs. 10.5.10, 10.5.13, 10.5.14).

A water-tube tiltmeter located 1.5 km from the crater of Karymsky Volcano was operated from 8 July to 12 September 1976, a period during which effusive activity changed to explosive activity and then back to effusive activity. Radially outward tilting began abruptly on 3 August, several days after an increase in explosion earthquakes (fig. 10.5.15). During 3-11 August, 82 arcseconds (about 410 microradians) of tilt was recorded, and an equal relaxation of that tilt occurred in the week after 11 August. During the period of inflation, explosions were frequent; then, after a series of "volcano-tectonic earthquakes," activity reverted to effusion and subsidence began (Zharinov and Firstov, 1985).

An increase in radon in thermal spring activity was noted 6-7 months before the May 1970 eruption and again before growth of the dome in September 1970 (fig. 10.5.16). The general pattern is an increase in radon 1-2 days before an increase in volcanic activity (Chirkov, 1975; Firstov and Chirkov, 1978).

Maly Semiachik Volcano:

An eruption of Maly Semiachik (fig. 10.5.17) in 1804 was the strongest historical eruption of that volcano and was accompanied by collapse of the summit (Erlich, 1986). Maly Semiachik was active again in 1851-52 and perhaps also in 1854. Additional eruptions occurred from autumn 1945 to spring 1946 and in early December 1952 (Vlodavetz and Piip, 1959). The 1952 eruption might have been a delayed response to the same M=8.25 earthquake (4 November 1952) that triggered the November eruption of nearby Karymsky (see above). Local earthquakes and early melting of snow on the southern part of Maly Semiachik have preceded one or more historical eruptions (Vlodavetz and Piip, 1959).

ZHUPANOVSKY COMPLEX (continued)

REFERENCES

Braitseva, D.A., Egorova, I.A., and Sulerzhitsky, L.D., 1979, Tephrochronological study of Karymsky Volcano: Vulkanol. Seismol., 1979, no. 1, p. 48-58.

Chirkov, A.M., 1975, Radon as a possible criterion for predicting eruptions as observed at Karymsky Volcano: Bull. Volcanol., v. 39, p. 126-131.

Dubik, Yu. M., Ogorodov, N.V., Abdurakhmanor, A.I., Khrenov, A.P., Tokarev, P.I., Firstov, P.P., Shirokov, V.A., and Lemzikov, V.K., 1972, The beginning of the new eruptive cycle of activity of Karymsky Volcano in May 1970: Bull. Vulkanol. Stantsii (Kamchatka), no. 48, p. 3-20.

Erlich, E., 1986, Geology of the calderas of Kamchatka and Kurile Islands with comparison to calderas of Japan and the Aleutians, Alaska: U.S. Geol. Surv. Open-File Rep. 86-291, 300 p.

Erlich, E.N., and Gorshkov, G.S., eds., 1979, Quaternary volcanism and tectonics in Kamchatka: Bull. Volcanol., v. 42, nos. 1-4 (spec. iss.), 298 p.

Farberov, A.I., Storcheus, A.V., and Pribylov, E.S., 1984, Low-intensity seismic activity of Karymskii Volcano in August, 1978: English transl., Volcanol. Seismol., v. 5, p. 323-334.

Firstov, P.P. and Chirkov, A.M., 1978, Radon in spontaneous gas of thermal spring of Karymsky Volcano (1966-1975): Bull. Vulkanol. Stantsii (Kamchatka), no. 54, p. 35-40.

Firstov, P.P., Lemzikov, V.K., and Feofilaktov, V.D., 1978, The seismic regime of Karymsky Volcano in 1975 and several questions on the forecasting of the time of thermal eruptions: Bull. Vulkanol. Stantsii (Kamchatka), no. 54, p. 27-34.

Firstov, P.P., Lemzikov, V.K., and Rulenko, O.P., 1977, The seismic regime of Karymsky Volcano, 1970-73, in Avdeiko, G.P., ed., Volcanism and geodynamics: Acad. Sci. USSR Far Eastern Sci. Center, Inst. Volcanol., Moscow, Nauka Publ., p. 161-178.

Firstov, P.P., Tokarev, P.I., Gusev, N.A., Rulenko, O.P., and Simbirev, I.B., 1978, Dynamics of eruption and seismic regime of Karymsky Volcano in 1976: Bull. Vulkanol. Stantsii (Kamchatka), no. 55, p. 27-34.

Gorelchik, V.I., 1964, Seismological study of activity of the Karymsky volcano: Bull. Vulkanol. Stantsii (Kamchatka), no. 36, p. 3-17.

Gushchenko, I.I., 1979, Eruptions of volcanoes of the world, a catalog: Acad. Sci. USSR Far Eastern Sci. Center, Moscow, Nauka Publ., 474 p.

Ivanov, B.V., 1970, The eruption of Karymsky Volcano in 1962-1965, and the volcanoes of the Karymsky group: Moscow, Nauka Publ., 135 p.

Ivanov, B.V., Andreev, V.N., Bogoyavlenskaya, G.E., Dubik, Y.M., Kirsanov, I.T., Rulenko, O.P., Firstov, P.P., and Chirkov, A.M., 1982, The activity of volcanoes of Kamchatka and Kurile Islands in 1981: Vulkanol. Seismol., 1982, no. 4, p. 103-108.

Ivanov, B.V., Chirkov, A.M., Ozerov, A.Yu., Khrenov, A.P., Dvigabo, V.N., Razina, A.A., Ovsyannikov, A.A., Gavrilenko, G.M., and Tokarev, P., 1984, Activity of the Kamchatka and Kurile Islands volcanoes in 1983: Vulkanol. Seismol., 1984, no. 6, p. 114-122.

Khrenov, A.P., Dubik, Yu.M., Ivanov, B.V., Ovsyannikova, A.A., Pilipenko, V.P., Taran, Yu.A., Firstov, P.P., and Chirkov, A.M., 1982, Eruptive activity of Karymsky volcano for 10 years (1970-1980): Vulkanol. Seismol., 1982, no. 4, p. 29-48.

Kimura, M., 1978, Significant eruptive activities related to large interplate earthquakes in the northwestern Pacific margin: Jour. Phys. Earth, v. 26 supp., p. S557-S570.

ZHUPANOVSKY COMPLEX (continued)

REFERENCES (continued)

Kirsanov, I.T., Ogorodov, N.V., Fedorov, M.V., and Chirkov, A.M., 1964, The state of Karymsky Volcano during 1960-61, and its eruption products: Bull. Vulkanol. Stantsii (Kamchatka), no. 35, p. 3-9.

Magus'kin, M.A., Enman, V.B., Syelyeznyev, B.V., and Shkryed, V.I., 1982, Particulars on shifts of the earth's surface at Karymsky Volcano, according to geodetic and photogrammetric data during the period 1970-1981: Vulkanol. Seismol., 1982, no. 4, p. 49-64.

Masurenkov, Yu.P., ed., 1980, Volcanic center: structure, dynamics, and products. Karymsky structure: Moscow, Nauka Publ., 299 p.

Polyak, B.G., Seliangin, O.B., and Braitseva, O.A., 1979, Dynamics of Maly Semiachik volcano activity: Dokl. Akad. Nauk USSR, v. 244, no. 1, p. 175-179.

Seliangin, O.B., 1977, The evolution of the caldera complex and the rhythms and direction of the volcanic process in the Karymsky volcano group in Kamchatka, in Avdeiko, G.P., ed., Volcanism and geodynamics, Moscow, Nauka Publ., p. 187-200.

Smithsonian Institution, Scientific Event Alert Network (SEAN), 1985, Karymsky Volcano: SEAN Bull., v. 10, no. 4, p. 4 (information from C.A. Wood).

Steinberg, G.S., 1966, On Maly Semiachik Caldera Volcano: Bull. Vulkanol. Stantsii (Kamchatka), no. 41, p. 29-31.

Tokarev, P.I., Firstov, P.P., and Lemzikov, V.K., 1969, Seismological investigations at Karymsky Volcano in 1966: Bull. Vulkanol. Stantsii (Kamchatka), no. 45, p. 12-21.

Vlodavetz, V.I., and Piip, B.I., 1959, Catalogue of active volcanoes of the world including solfatara fields, Pt. VIII, Kamchatka and continental areas of Asia: Naples, IAVCEI, 110 p.

Volcanological Society of Japan, 1973, 1978, Karymsky: Bull. Volcanic Eruptions, no. 11, p. 31-32; no. 16, p. 45 (information from Yu.M. Dubik).

Zharinov, I.A., and Firstov, P.P, 1985, Activity, seismic regime, and tilt of the earth's surface at Karymski Volcano during the summer of 1976: Vulkanol. Seismol., 1985, no. 2, p. 93-95.

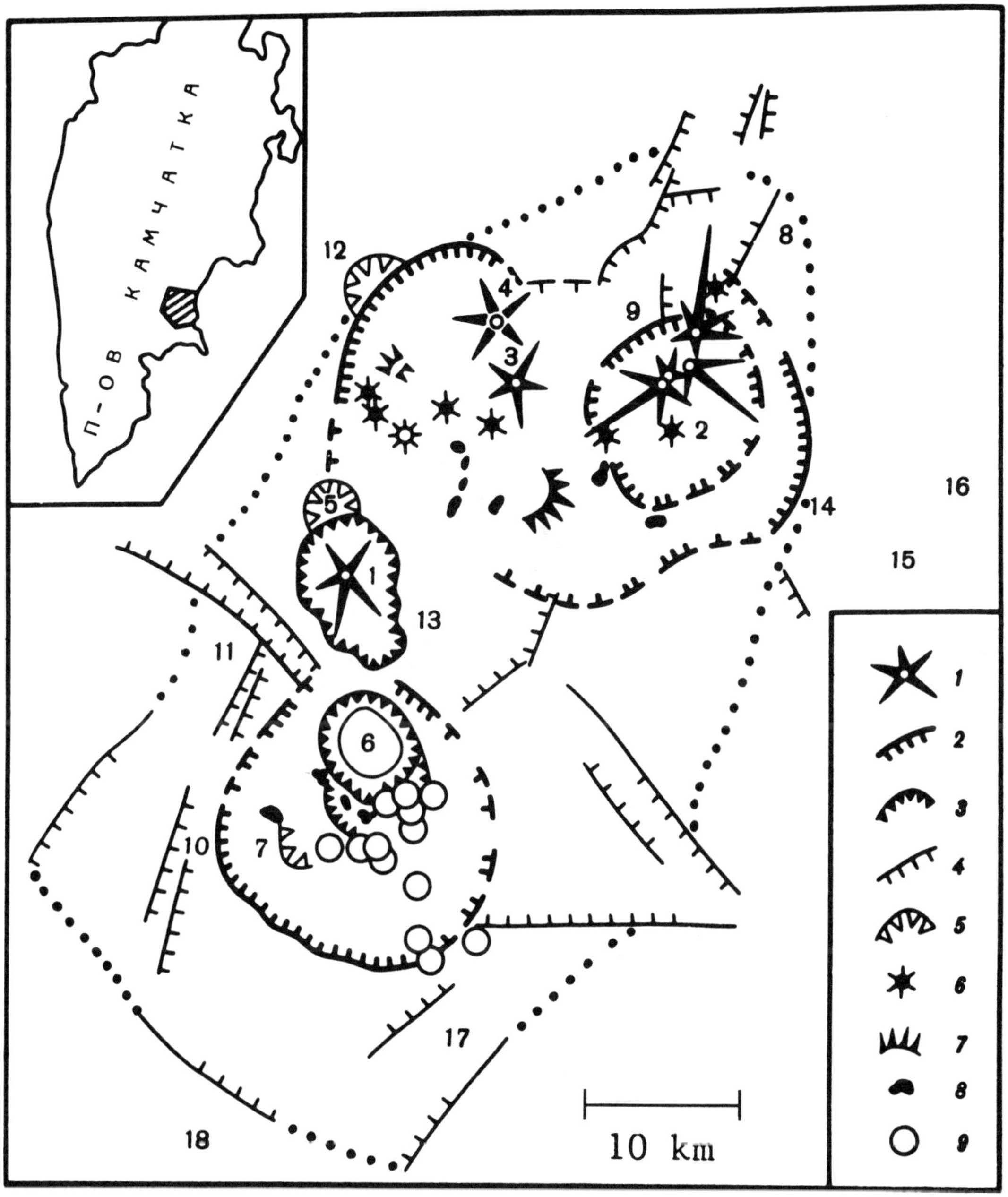

Figure 10.5.1. Volcanoes of the Karymsky complex, from Firstov and others (1978). Key to inset legend, with numbered symbols: 1, modern volcanoes; 2, caldera escarpments; 3, position of the ring fractures of ancient and young caldera groups; 4, fractures expressed in topography of displaced blocks; 5, explosion craters; 6, cinder cones and lava domes; 7, maar; 8, extrusions; 9, epicenters of volcano-tectonic earthquakes in 1975 with focal depths of 0-10 km. Key to numbers on map proper: 1, Karymsky Volcano; 2, Maly Semiachik Volcano; 3, Sukhoy; 4, Stupenchaty Bastion; 5, Dvor Caldera; 6, Akademii Nauk and Odnoboky Calderas; 7, Beliankina; 8, Drevny Karymsky; 9, Semiachik Caldera; 10, Krainy; 11, Razlaty; 12, Soboliny Caldera; 13, Berezovy; 14, Stena Caldera (Stena-Soboliny Caldera in figure 10.5.2 is a large structure encompassing both); 15, Pribezhny Iuzhy; 16, Pribezhny Severny; 17, Ditmara; 18, Zhupanovsky volcano-tectonic depression. Dotted line indicates approximate location of outer caldera rim.

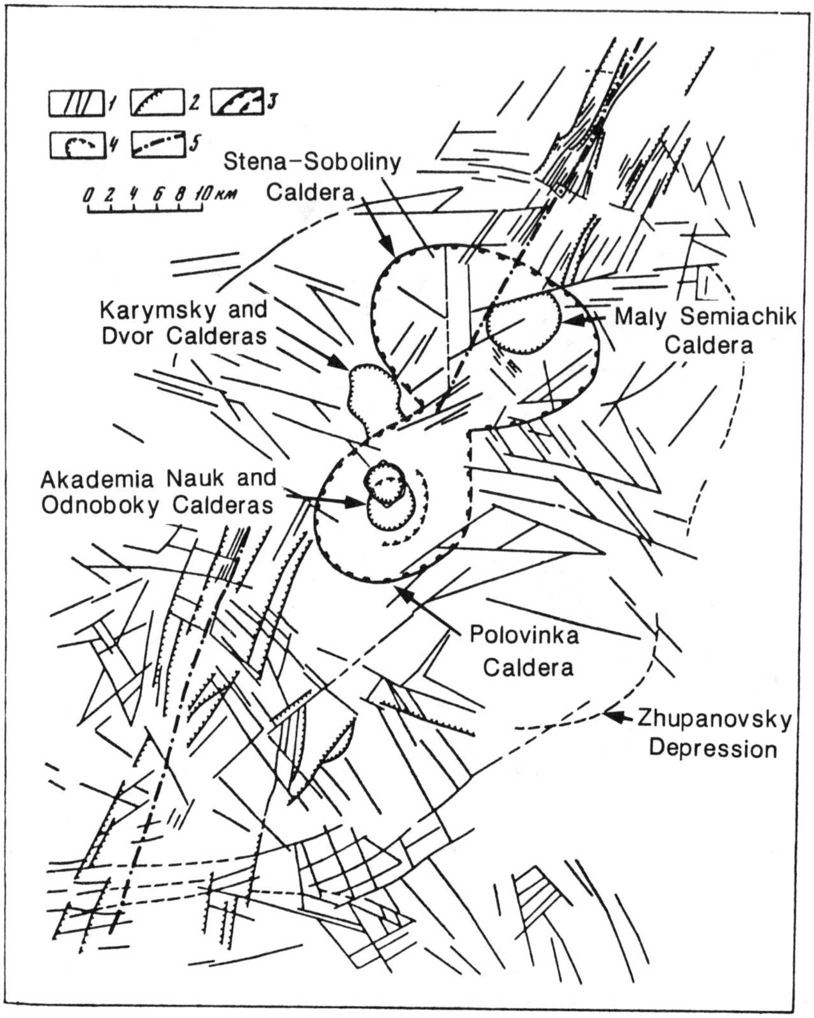

Figure 10.5.2. Structural map of Zhupanovsky volcano-tectonic depression, from Masurenkov (1980). 1, fracture without apparent displacement; 2, fracture with surface displacement; 3, mid-Pleistocene caldera fracture; 4, upper Pleistocene-Holocene caldera fracture; 5, rift axis.

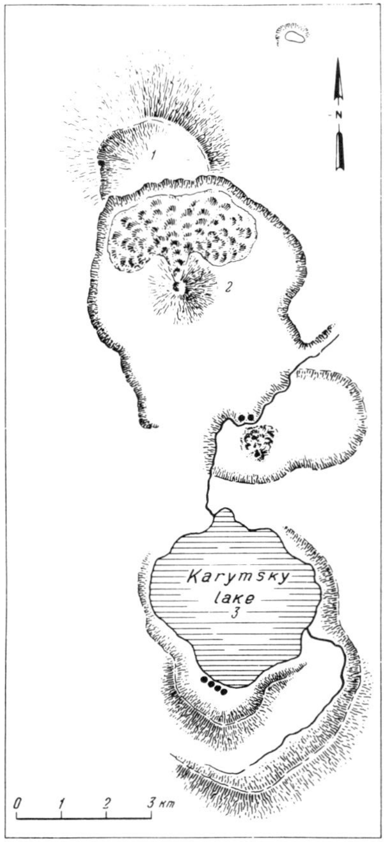

Figure 10.5.3. Karymsky and neighboring calderas, from Vlodavetz and Piip (1959). Explanation: 1, crater of extinct Dvor Volcano; 2, Karymsky Volcano and its caldera; 3, Karymsky Lake and Akademii Nauk Caldera; dots, hot springs.

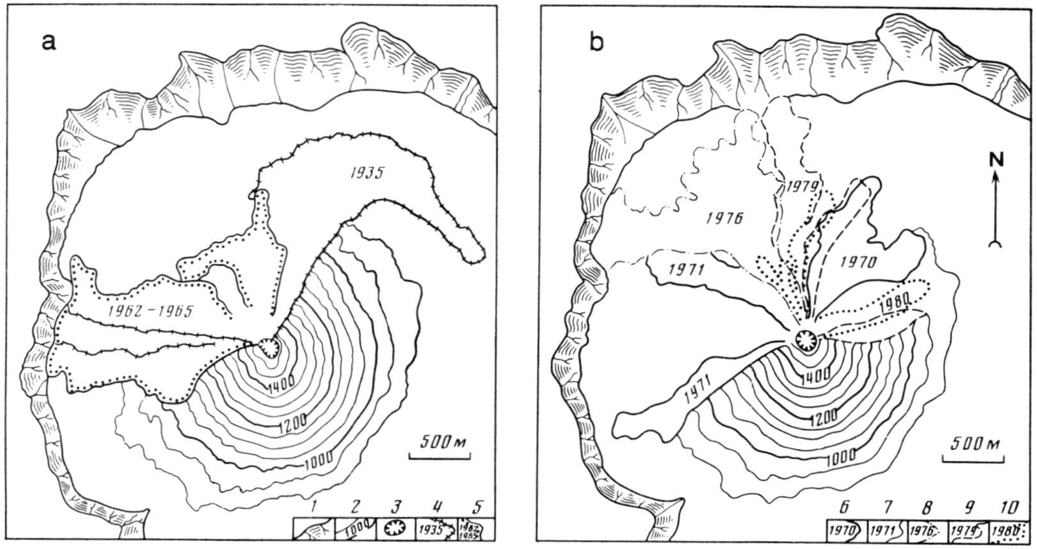

Figure 10.5.4. Lava flows at Karymsky Volcano during 1935-80, from Magus'kin and others (1982). 1, caldera boundary; 2, topographic contours, with elevations in meters; 3, volcanic crater; 4-10, margins of historical lava flows, with dates of extrusion (for example, 4 = 1935; 5 = 1962-65; and so forth).

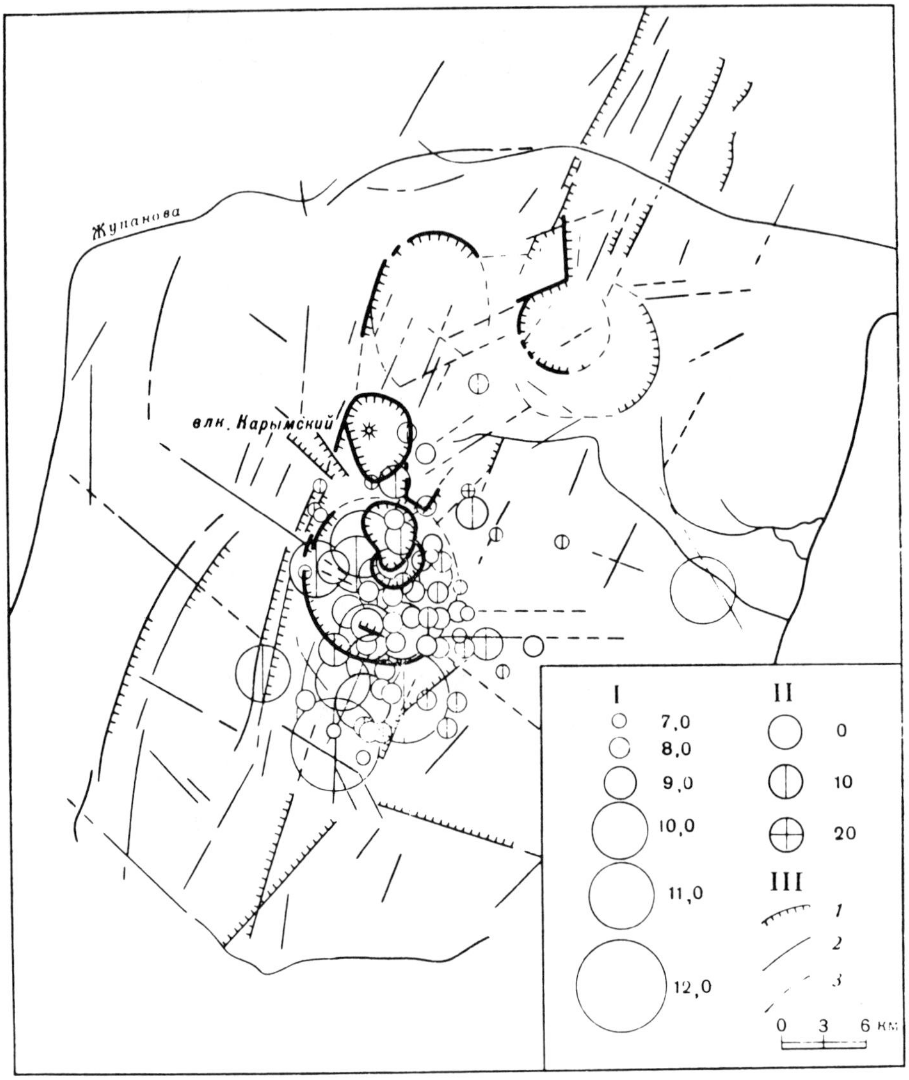

Figure 10.5.5. Epicenters near Karymsky Volcano during 1970-80, from Khrenov and others (1982). Explanation: I, energy class; II, depth (kilometers); III, faults: 1, topographic boundary; 2, minor; 3, buried and conjectured.

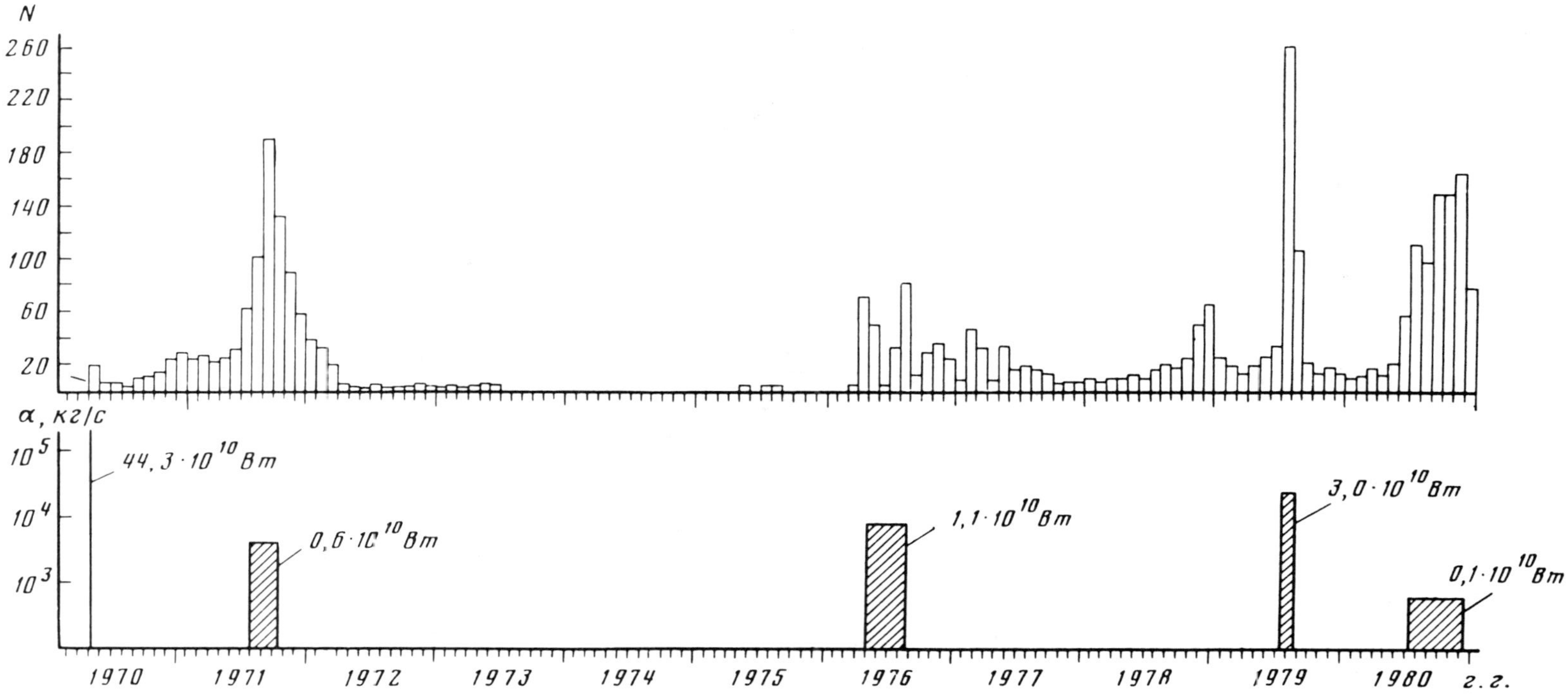

Figure 10.5.6. Monthly number of earthquakes (top) and rate of lava production (bottom) at Karymsky Volcano during 1970-80, from Khrenov and others (1982). Rate of lava production in units of kilograms per second; values beside each period of lava production are total number of kilograms(?).

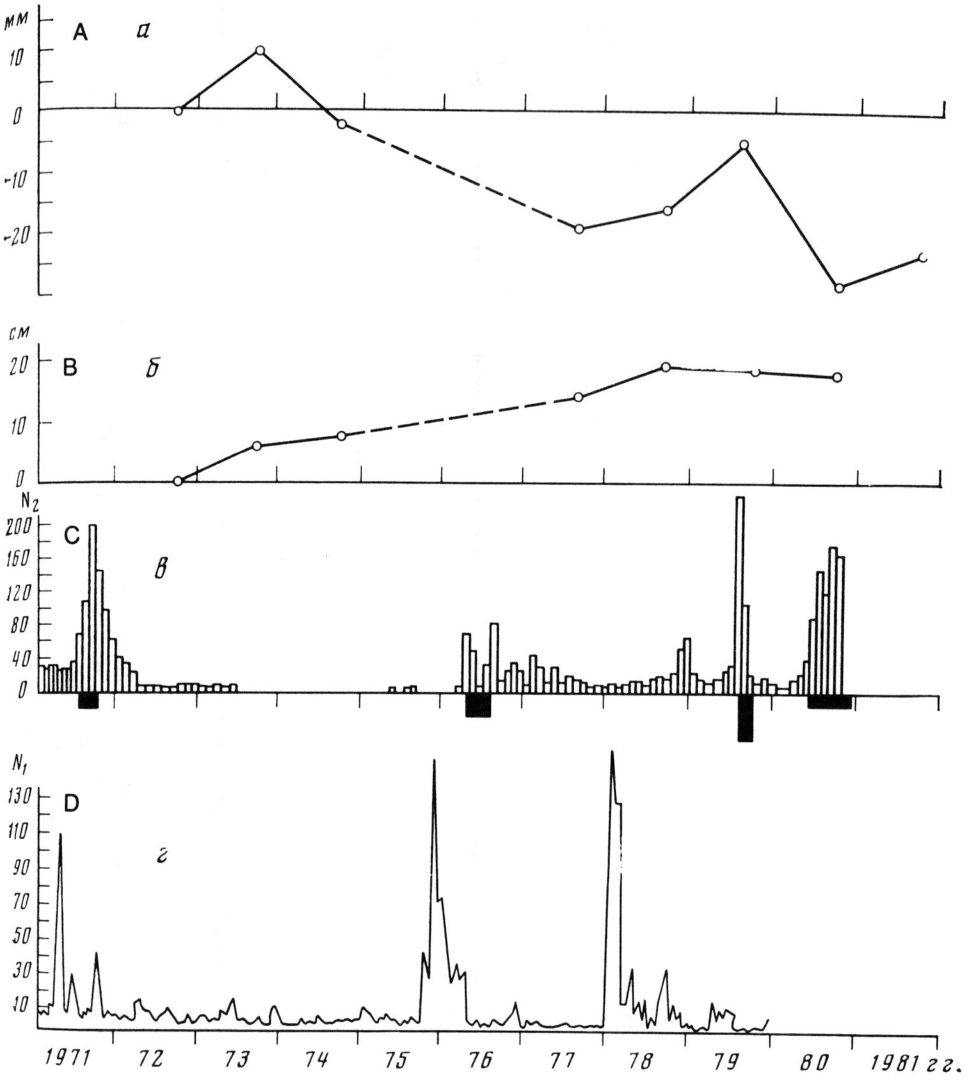

Figure 10.5.7. Activity at Karymsky Volcano from 1971 to 1981, from Magus'kin and others (1982). A, Average vertical displacement of several reference stations 1.5-1.7 km from the crater. B, Average horizontal displacement for whole area. C, Number of explosion earthquakes of energy class 5 or greater in 10-day intervals. D, Number of volcano-tectonic earthquakes of energy class K = 5 or greater for 15-day intervals. Black rectangles show times of lava outpouring; their area is proportional to volume extruded.

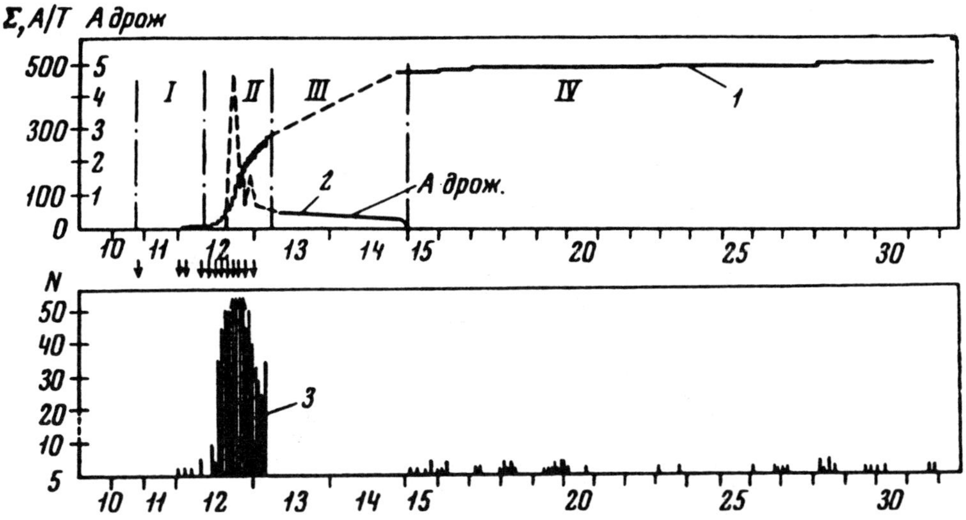

Figure 10.5.8. Seismic activity at Karymsky Volcano during May 1970, from Dubik and others (1972). Top, cumulative sum of A/T (micrometers per second), where A is maximum amplitude of surface wave and T is period: dashed line, spasmodic tremor; solid line, continuous tremor. Bottom, number of earthquakes with maximum amplitude greater than 0.5 micrometers. Arrows indicate explosions.

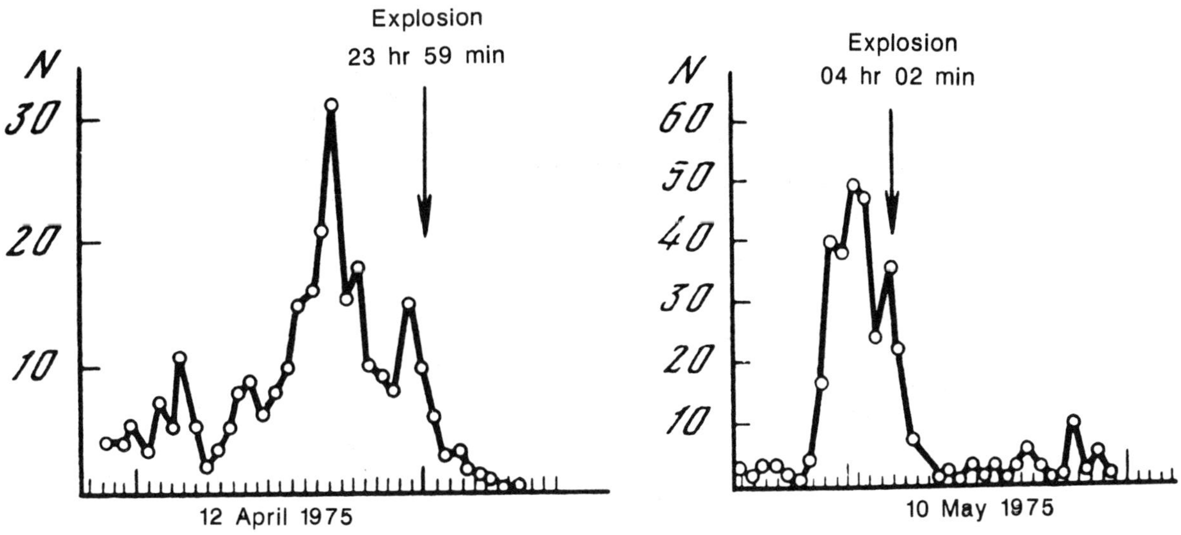

Figure 10.5.9. Hourly earthquake counts before eruptions (arrows) of Karymsky Volcano in April and May 1975, from Firstov and others (1978).

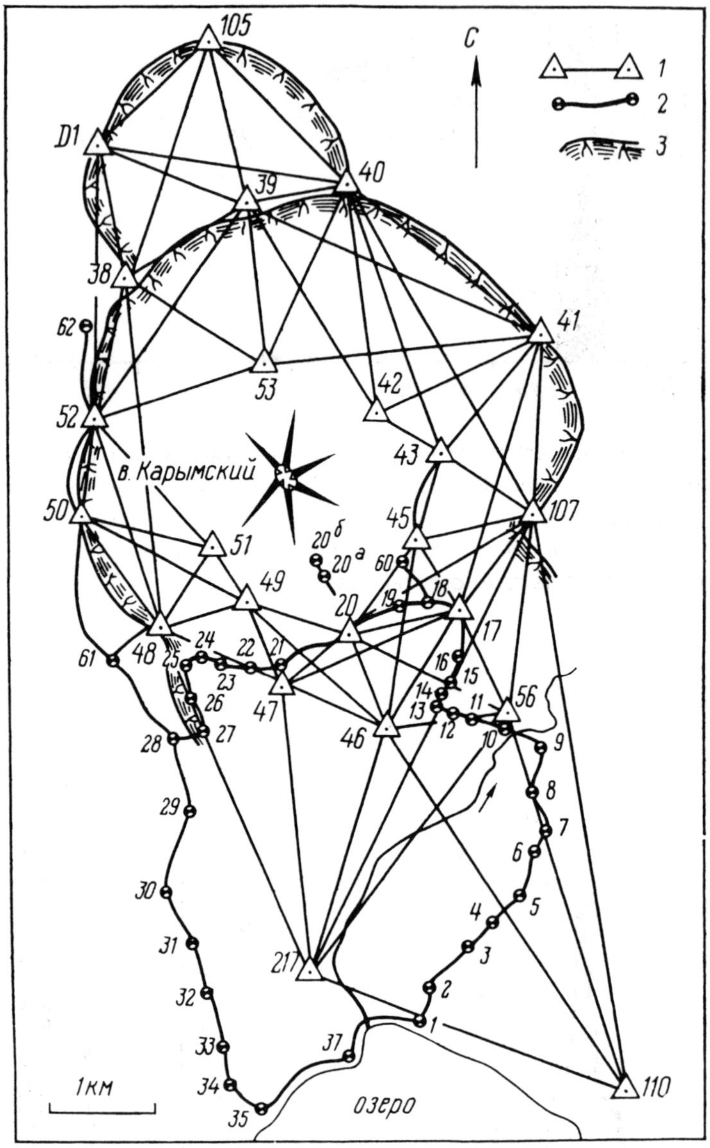

Figure 10.5.10. Geodetic networks at Karymsky Volcano, from Magus'kin and others (1982). 1, horizontal strain network; 2, leveling network; 3, crater rims. Sites 20a and 20b mark approximate location of tilt station (see figure 10.5.15).

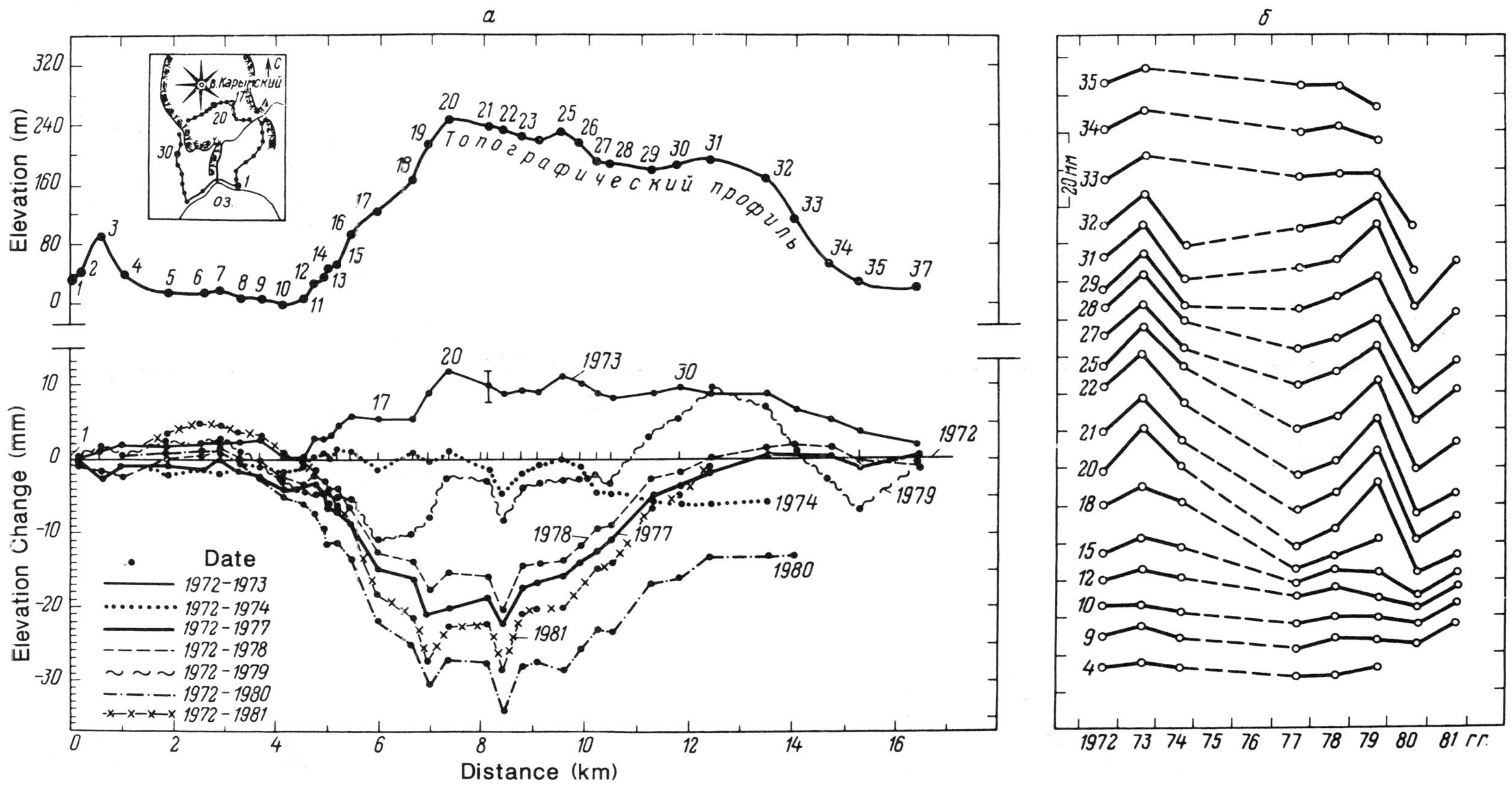

Figure 10.5.11. Elevation changes at Karymsky Volcano from 1972 to 1981, from Magus'kin and others (1982). Left, topographic profile and elevation changes measured along route shown in inset and in figure 10.5.10. Right, movements of numbered bench marks through time.

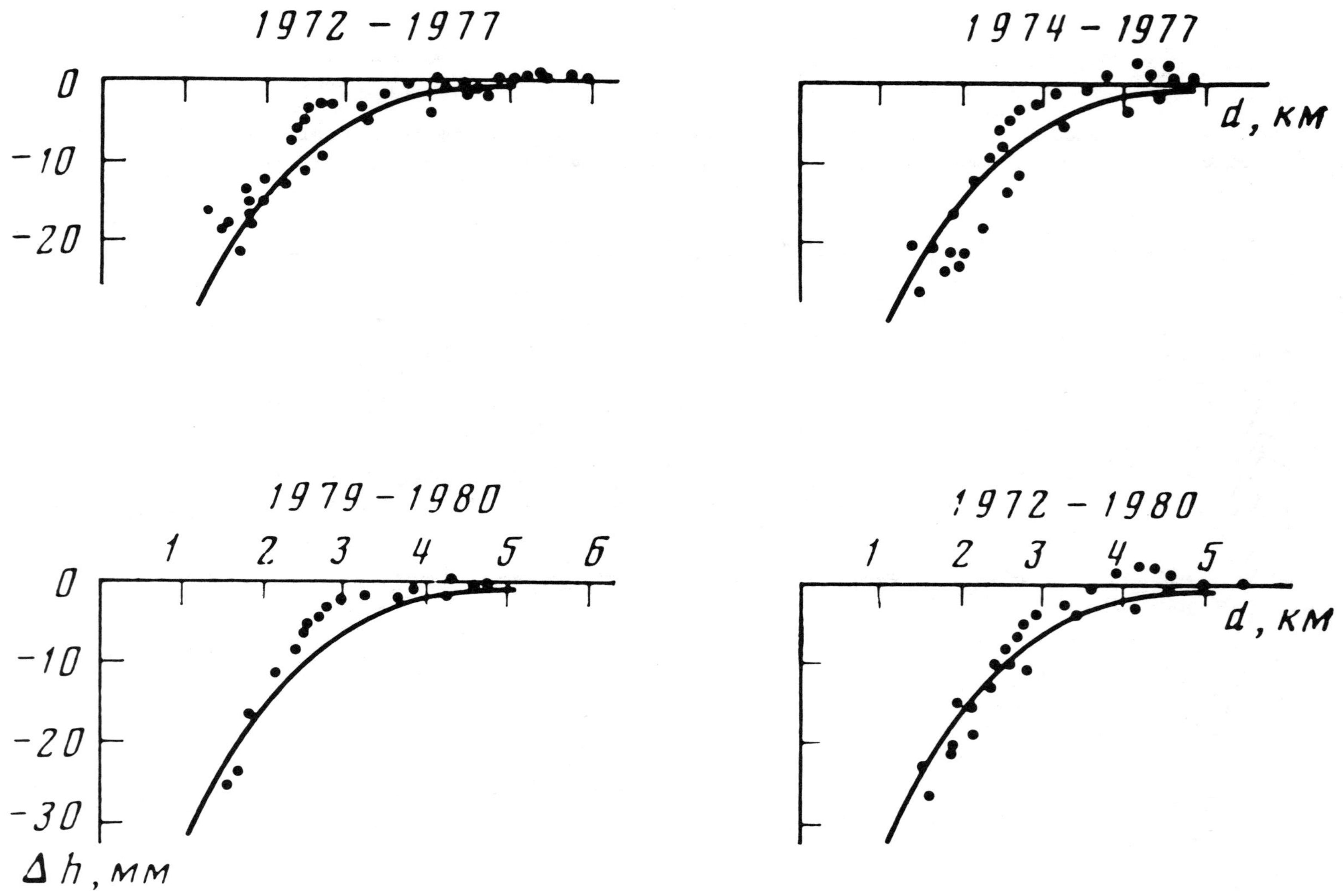

Figure 10.5.12. Subsidence profiles at Karymsky Volcano during indicated intervals, from Magus'kin and others (1982). Plot shows subsidence (ordinate) versus distance from crater of Karymsky Volcano (abscissa). Shape of profiles suggests a shallow source, possibly 1-1.5 km deep.

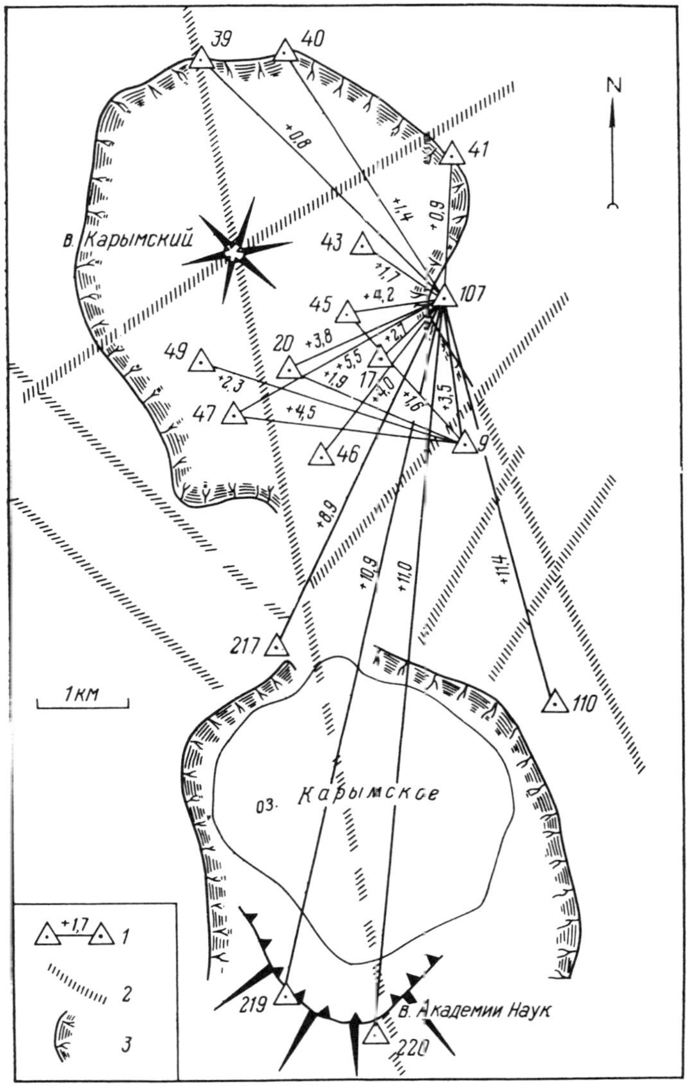

Figure 10.5.13. Horizontal distance changes at Karymsky Volcano from 1974 to 1981, from Magus'kin and others (1982). 1, endpoints of line (triangles), with change in centimeters; 2, faults; 3, crater rim.

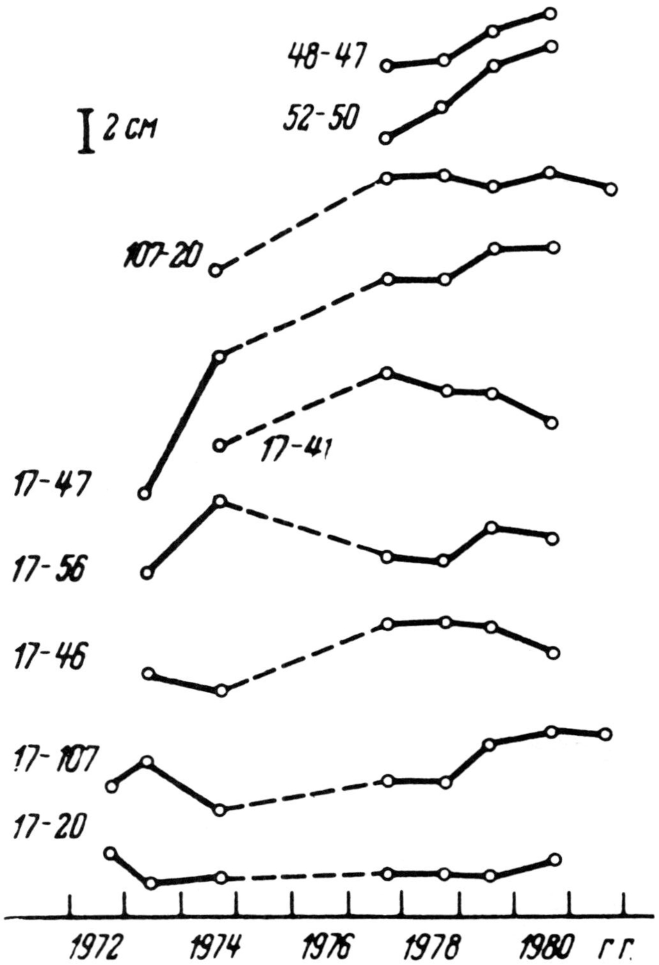

Figure 10.5.14. Horizontal distance changes as a function of time at Karymsky Volcano, from Magus'kin and others (1982). Bench-mark locations (numbers at left) are shown in figure 10.5.10.

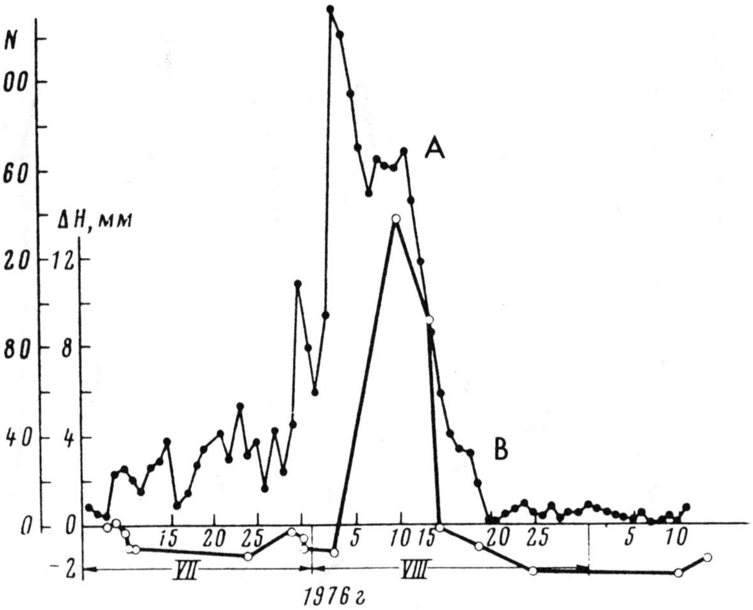

Figure 10.5.15. Activity at Karymsky Volcano during 1976, from Zharinov and Firstov (1985). A, Daily number of volcano-tectonic earthquakes at depths of 0-10 km. B, Level changes along 40-m base of radial tilt line.

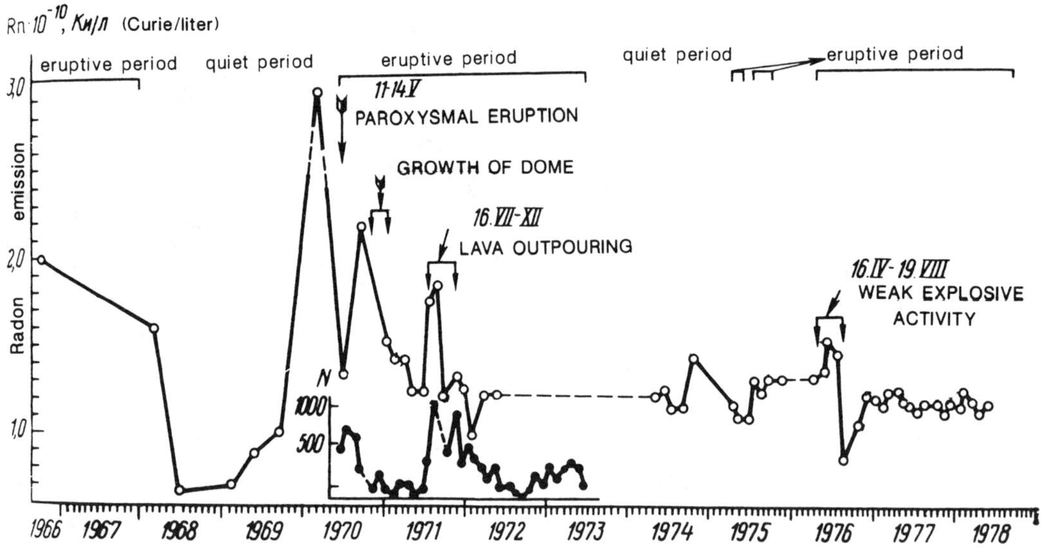

Figure 10.5.16. Radon emission and monthly number of earthquakes (inset, bottom) associated with eruptive at Karymsky Volcano during 1966-78, from Khrenov and others (1982).

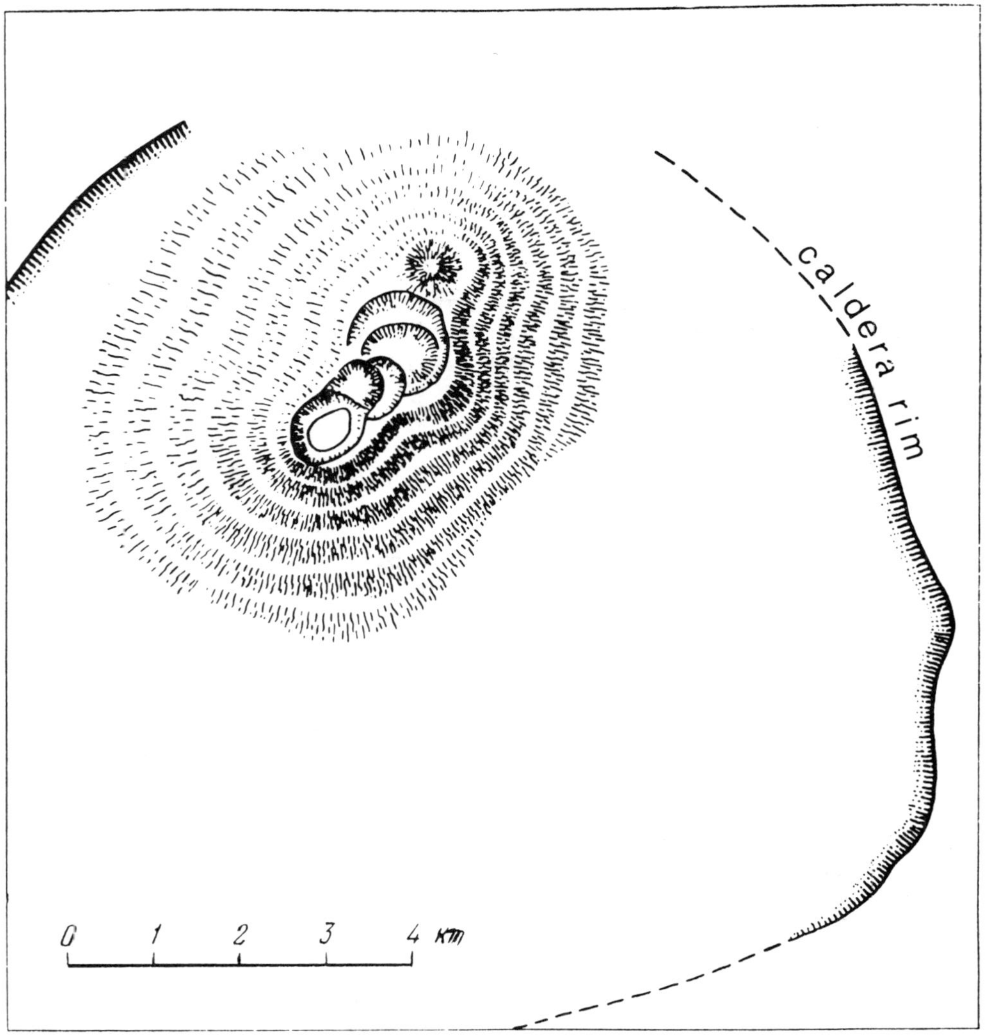

Figure 10.5.17. Sketch of Maly Semiachik Volcano and Caldera, from Vlodavetz and Piip (1959).

PART 3: HISTORICAL UNREST AT LARGE QUATERNARY CALDERAS (continued)

See inside back cover for explanation and abbreviations

UZON-GEYZERNAYA

CAVW number (active vent)	Latitude Longitude (degrees)	Diameter (km)	Local tectonic setting	Precaldera edifice	SiO_2 content (wt pct)	Age of collapse (yr B.P.)	Date of unrest	Type and duration of unrest ESTU STHF MGTF H Te	Eruption type
10-00-161 (Unana)	54.62N 159.70E	7 x 18 (Uzon and	Intersect. of E-W strike-	LL-strat? (Uzon)	R = 54-56+, some 65-71	mid-Pleistocene	19th cent.	---- ---- ---x - --	none
							20th cent.	xx-- ---- ---x - --	none
10-00-17 (Uzon)	54.53N 160.08E	Geyzernaya combined)	slip fault zone and NNE-SSW graben	Silicic edifice (Geyzernaya)	C = 68 (Uzon)				

TECTONIC SETTING

Uzon and Geyzernaya Calderas lie in the NNE-SSW-trending eastern Kamchatka graben-syncline, at its intersection with E-W faults (fig. 10.6.1). At least 3 to 5 km of left-lateral offset is apparent along the E-W faults.

GEOLOGIC HISTORY

Silicic volcanism has occurred in the Uzon area since at least the Pliocene; caldera-forming eruptions in the mid- to late Pleistocene produced a sequence of plateau-forming ignimbrites, some densely welded and others unwelded (Belousov and others, 1984; Erlich, 1986) (figs. 10.6.2-10.6.4). The Uzon and Geyzernaya basins are separate calderas, offset by 3-5 km of left-lateral movement along a major E-W fault zone that passes between the two calderas (Belousov and others, 1974; Erlich, 1986). A NNE-SSW-trending graben also passes through the Uzon-Geyzernaya area.

Together, the Uzon and Geyzernaya Calderas contain a major hydrothermal system (fig. 10.6.5). Numerous hot springs are located along northwest-trending fissures within the Uzon Caldera, just north of the main E-W fault. Other thermal features occur within and along the ring fractures of both calderas.

HISTORICAL ACTIVITY

In the middle of the 19th century, fumaroles and mud volcanoes were very active, making loud noises, and vegetation was stunted (Perrey, 1863). Similar conditions existed in 1986, so the condition described by Perrey may have been routine rather than

UZON-GEYZERNAYA (continued)

HISTORICAL ACTIVITY (continued)

anomalous. Seismic recording in this area has revealed a marked series of earthquakes along the E-W trending fault zone at a depth of 100-120 km (Naboko, 1974).

COMMENTS

The Uzon-Geyzernaya region in Kamchatka is similar in features, though not in size, to the Yellowstone region in the United States. Both occur in areas of regional crustal extension, have been sites for multiple caldera collapses, and currently exhibit vigorously active hydrothermal systems.

REFERENCES

Belousov, V.I., Grib, E.N., and Leonov, V.L., 1984, The geological setting of the hydrothermal systems in the Geysers Valley of Uzon Caldera: English translation, Volcanol. Seismol., 1984, v. 5, no. 1, p. 67-81.
Braitseva, O.A., Bogoyavlenskaya, G.E., Erlich, E.N., Pilipenko, G.F., and Karpov, G.A., 1974, Uzon-Semiachik volcanic region, in Volcanoes and hydrothermal areas of Kamchatka: Moscow, Nauka Press, p. 139-191.
Erlich, E., 1986, Geology of the calderas of Kamchatka and Kurile Islands with comparison to calderas of Japan and the Aleutians, Alaska: U.S. Geol. Surv. Open-File Rep. 86-291, 300 p.
Erlich, E.N., and Gorshkov, G.S., eds., 1979, Quaternary volcanism and tectonics in Kamchatka: Bull. Volcanol., v. 42, nos. 1-4 (spec. iss.), 298 p.
Florenskiy, I.V., and Trifonov, V.G., 1985, Recent tectonics and volcanism of the eastern volcanic zone of Kamchatka: Geotectonics, v. 19, no. 4, p. 314-320.
Naboko, S.I., 1954, Geysers of Kamchatka: Akad. Nauk USSR, Trudy Lab. Vulkanol., no. 8, p. 126-209.
----- ed., 1974, Volcanism, hydrothermal processes, and ore formation: Moscow, Nauka Press, 262 p.
Perrey, A., 1863, Documents sur les tremblements de terre et les phénomènes volcaniques dans l'archipel des Kouriles et au Kamtschatka: Lyon, Soc. Imp. Agric. Hist. Nat. Arts, 166 p.
Vlodavetz, V.I., and Piip, B.I., 1959, Catalogue of active volcanoes of the world including solfatara fields, Pt. VIII, Kamchatka and continental areas of Asia: Naples, IAVCEI, 110 p.

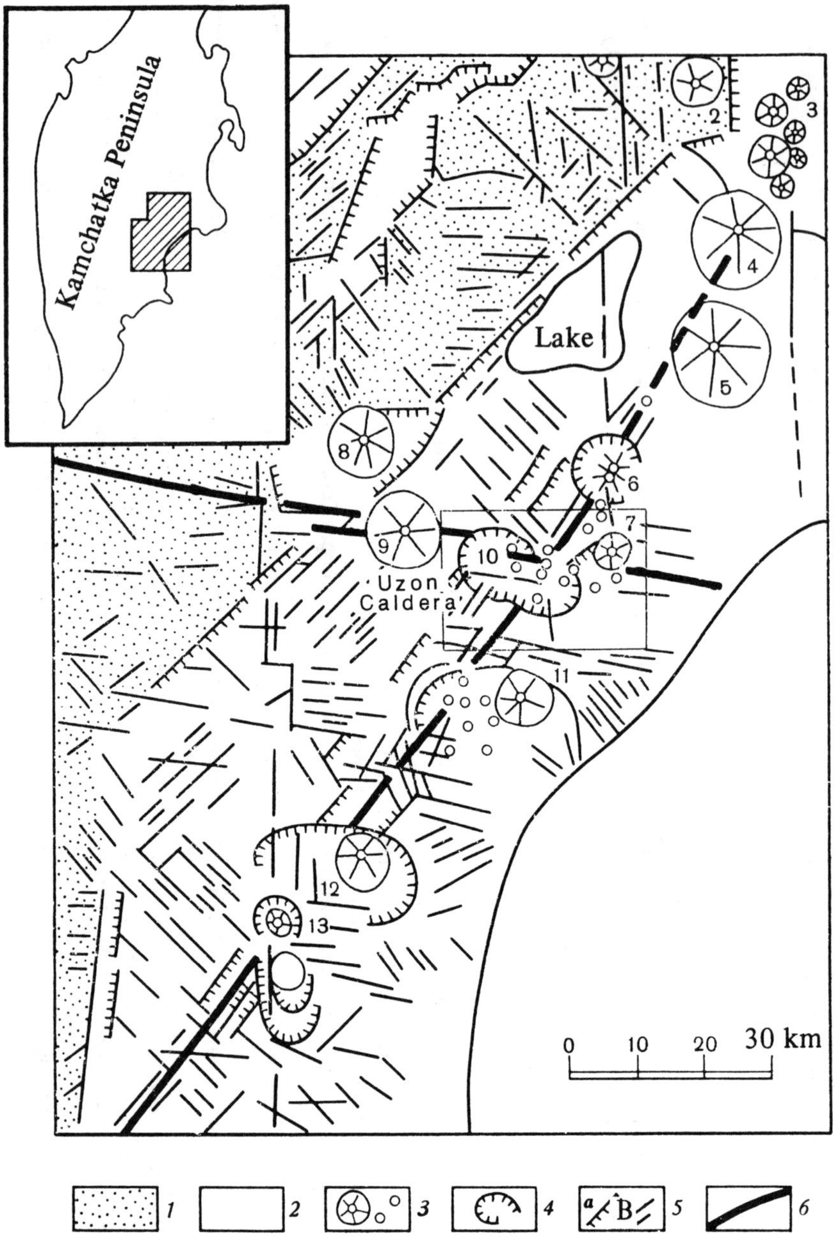

Figure 10.6.1. Structural map of eastern Kamchatka, from Belousov and others (1984). Explanation: 1, horst-anticline of Eastern Range in Kamchatka; 2, graben-syncline of eastern Kamchatka; 3, volcanoes (1, Kizimen; 2, Konradi; 3, Gamchensky ryad; 4, Shmidt's; 5, Kronotsky; 6, Krasheninnikov's; 7, Kikhpinych; 8, Unana; 9, Taunshits; 10, Uzon; 11, Greater Semyachik; 12, Lesser Semyachik; 13, Karymsky); 4, calderas and volcano-tectonic depressions; 5, faults (a, normal faults; b, cracks); 6, axial lines of northeast-trending and sublatitudinal regional fault zones intersecting Uzon-Geysernaya region. Inset, index map of the Kamchatka Peninsula showing the outline of figure 10.6.1.

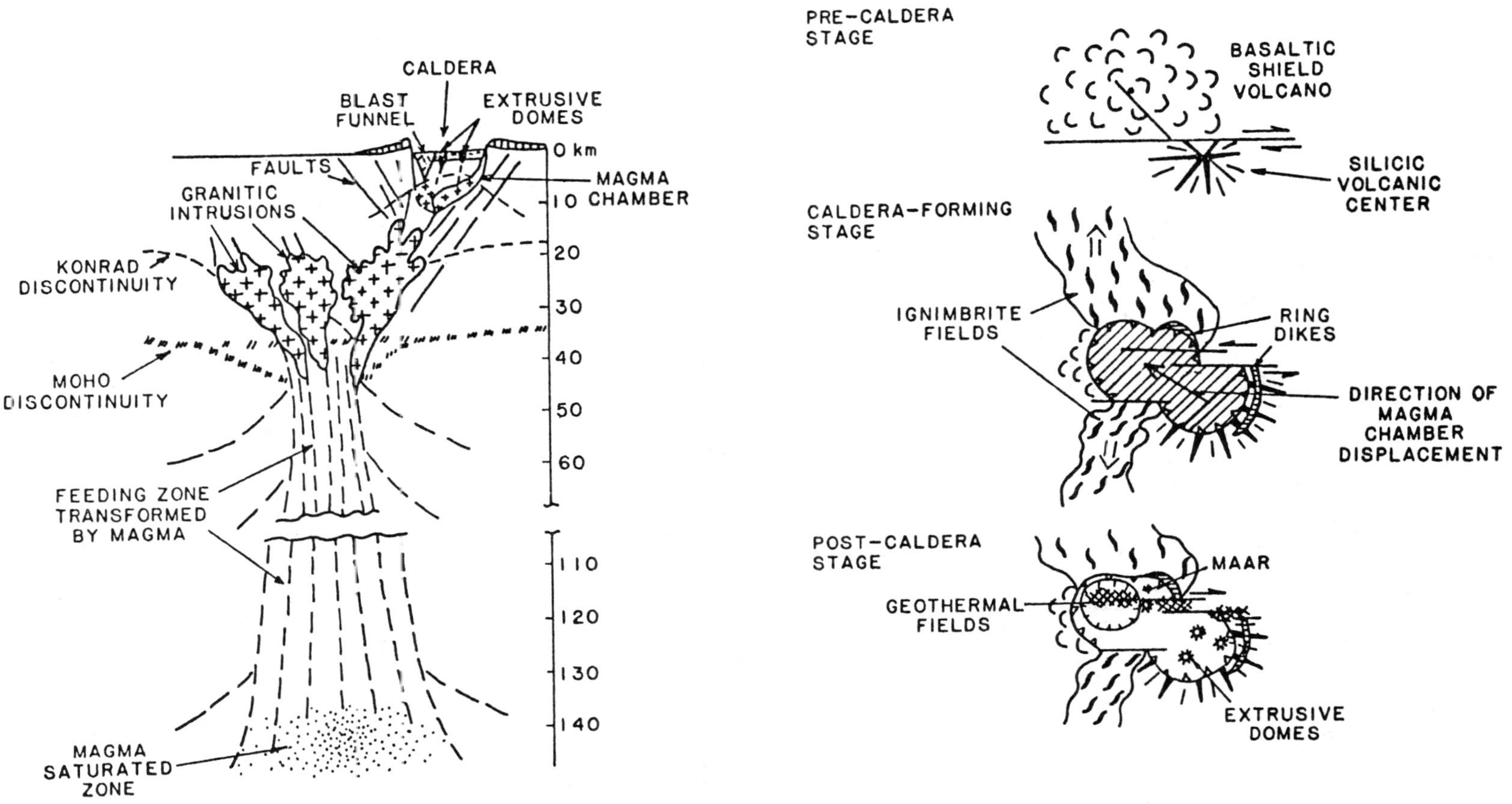

Figure 10.6.2. Model of formation of Uzon-Geyzernaya Caldera, Kamchatka, after Naboko (1974).

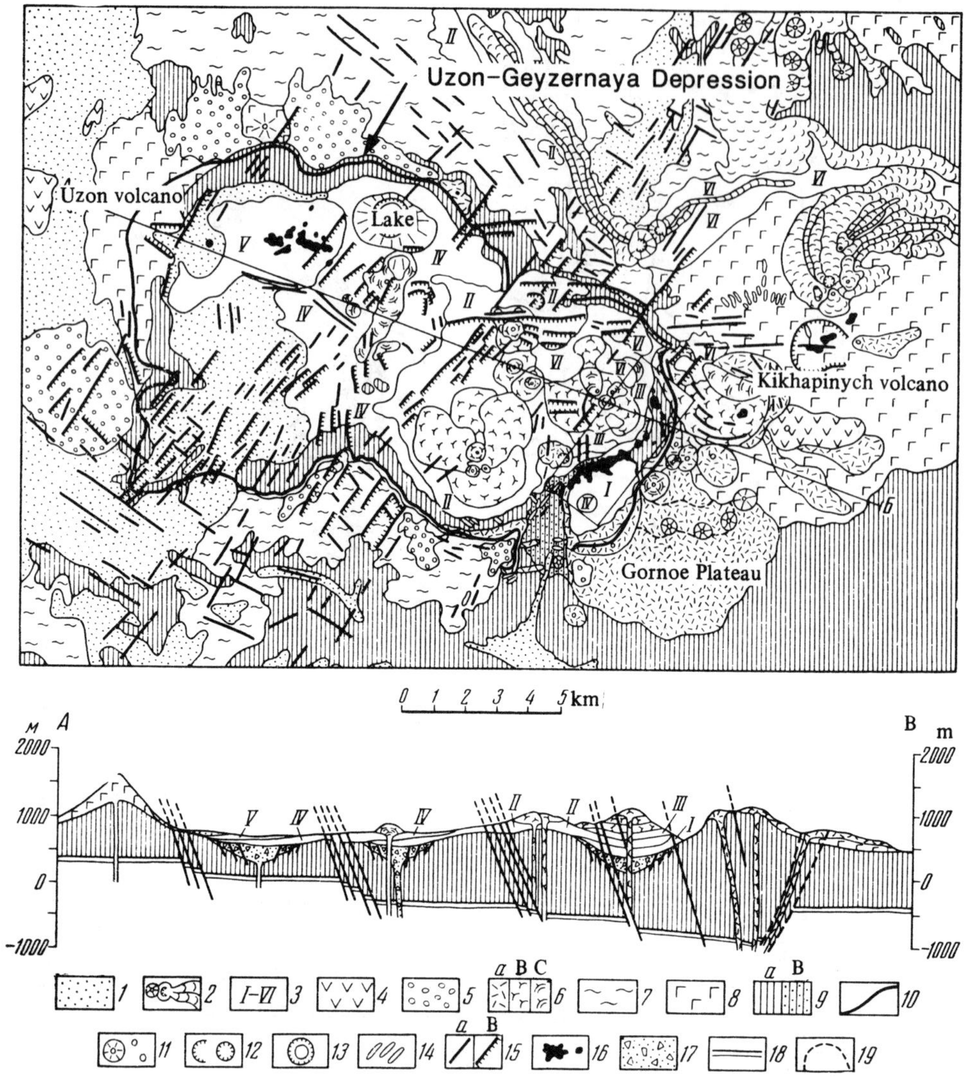

Figure 10.6.3. Schematic geologic map of Uzon-Geysernaya region, from Belousov and others (1984). Explanation: 1, alluvial, glacial, and talus-slope deposits and also cliff debris (Q4); 2, lavas and cinders of basaltic and andesite-basalt composition; 3, lacustrine sediments (members: I, Geyser; II, Pumice; III, Yellow Crags; IV, Second Lake and Southern Basin; V, Third Lake; VI, Colorado); 4, andesitic lavas; 5, explosion pumice of Uzon Caldera; 6, lavas of first (a), second (b), and third (c) cycles of Quaternary acid volcanism; 7, ignimbrite; 8, basaltic lavas; 9, Pliocene - lower Pleistocene deposits (members: a - old Lavas, Cape Tuff, Sand-Lava, Tuff-Ignimbrite, Flood Basalts; b - Estuary); 10, boundaries of erosion scarps bounding Uzon-Geysernaya depression; 11, cinder cones and other volcanic centers; 12, crater funnels; 13, maar; 14, dikes; 15, faults (a, fissures; b, normal faults); 16, areas of thermal anomalies (approximate outlines of 20 °C isotherm at 1-m depth); 17, sediments filling inferred calderas (in section); 18, tentative boundary of Pliocene - lower Pleistocene sediments (in section); 19, boundaries of separate beds and flows within identified members (markers).

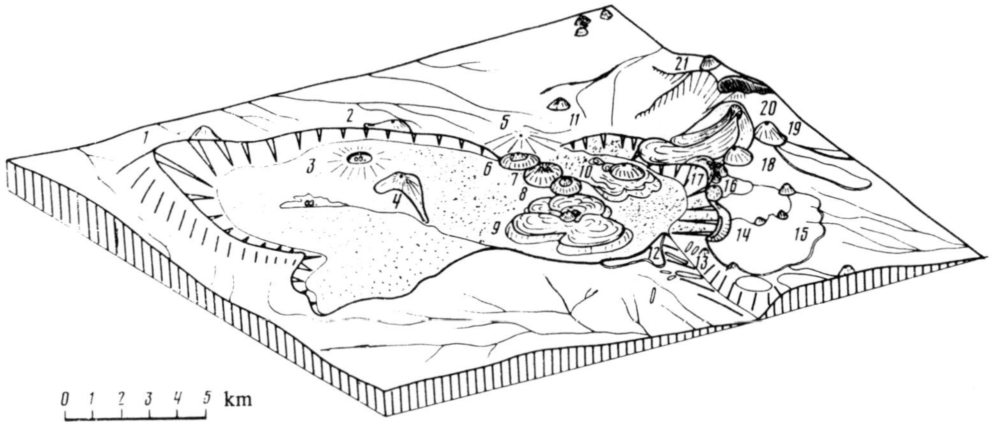

Figure 10.6.4. Block diagram of Uzon-Geysernaya region, from Belousov and others (1984). Dots indicate lacustrine sediments filling volcano-tectonic Uzon-Geysernaya depression. Explanation: 1, Mount Uzon; 2, Ozernaya extrusion; 3, maar of Lake Dal'nee; 4, Belaya extrusion; 5, pumice dome of Mount Otkrytaya, extrusions; 6, Tortik; 7, Ostanets; 8, Sestrenka; 9, Kruglaya; 10, Geyzernaya; 11, Mount Duga; 12, Shirokoe Plateau; 13, canyon of Shumnaya River, extrusions; 14, Greben; 15, Gornoe Plateau; 16, Rudich's; 17, Bortovaya; 18, First; 19, Mount Bezymyannaya; 20, Zheltaya extrusion; 21, Savich's Cone.

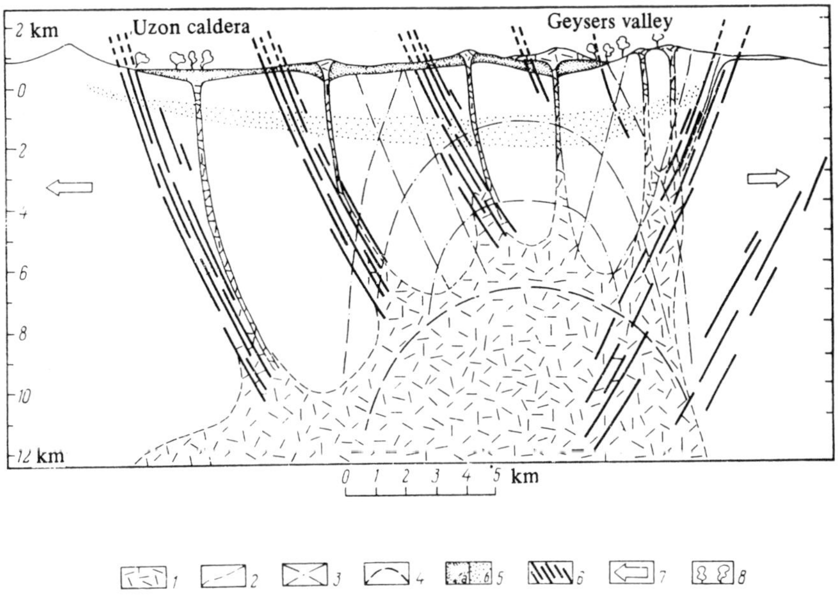

Figure 10.6.5. Schematic section showing structural control of hydrothermal system in Geysernaya Valley and Uzon Caldera, from Belousov and others (1984). Explanation: 1, area of crustal magma chamber and associated extrusions; 2, inferred boundaries of magma chamber; 3, system of ring-shaped and conical fractures above chamber, reconstructed from exposed dikes; 4, speculative upper boundary of chamber in the period of dike invasion; 5, water-bearing formations (a, upper; b, lower); 6, fault zones (arrows indicate direction of movement on faults); 7, direction of crustal extension in region; 8, manifestations of hydrothermal activity.

PART 3: HISTORICAL UNREST AT LARGE QUATERNARY CALDERAS (continued)
See inside back cover for explanation and abbreviations

KRASHENINNIKOV

CAVW number (active vent)	Latitude Longitude (degrees)	Diameter (km)	Local tectonic setting	Precaldera edifice	SiO_2 content (wt pct)	Age of collapse (yr B.P.)	Date of unrest	Type and duration of unrest ESTU STHF MGTF H Te	Eruption type
10-00-19 (Krasheninnikov)	54.62N 160.60E	9 x 11		Shield	C = d R = d,a,b	39,600	1963?	---- ---- ---x - --	none

TECTONIC SETTING

Krasheninnikov Caldera lies at the intersection of the eastern Kamchatka graben-syncline and a deep-seated, E-W fault (Erlich, 1986). In addition, a 21-km-long NNE-SSW set of fissures and small grabens cuts the caldera and localizes 15 parasitic lava vents and cinder cones (Erlich, 1986).

GEOLOGIC HISTORY

The precaldera edifice was a complex of one large stratovolcano and several smaller andesite and basalt volcanoes, extrusive domes, and scoria cones. The caldera formed ca. 39,000 yr B.P. after explosive eruption of >6 km^3 of dacitic pumice (Florensky, 1984; Erlich, 1986). Ignimbrite from this eruption overlies ignimbrite of the Uzon Caldera and is the youngest widespread pyroclastic unit in the Karymsky-Semiachinsky part of the eastern Kamchatka graben-syncline (Erlich, 1986). Krasheninnikov Volcano is a postcaldera stratocone in the center of the caldera (fig. 10.7.1). Voluminous lava flows issued from this cone 3,000 yr B.P. (Vodopadny flow), 1,500 yr B.P. (Ozerny flow), less than 600-700 yr B.P. (Yuzhny flow), and less than 400-500 yrs B.P. (Molodoy flow) (Ponomareva and Tsurupa, 1985). These flows would be described in historical accounts had they occurred in a populous area.

HISTORICAL ACTIVITY

Earthquakes occur at depths of >100 km beneath the northern margin of the caldera, possibly along a deep projection of the above-mentioned E-W fault (Erlich, 1986) or, alternatively, along the Benioff zone.

An active fumarole was discovered during aerial reconnaissance in 1963 (Steinberg, 1964). G.S. Steinberg (written commun., 1987) considers Krasheninnikov to be an active and possibly restless volcano.

KRASHENINNIKOV (continued)

REFERENCES

Erlich, E., 1986, Geology of the calderas of Kamchatka and Kurile Islands with comparison to calderas of Japan and the Aleutians, Alaska: U.S. Geol. Surv. Open-File Rep. 86-291, 300 p.

IAVCEI (International Association of Volcanology and Chemistry of the Earth's Interior), 1973, Data sheets on post-Miocene volcanoes of the world: Rome, IAVCEI.

Florensky, I.V., 1984, On the question of the age of Uzon and Krasheninnikov calderas: Vulkanol. Seismol., 1984, pt. 1, p. 102-106.

Ponomareva, V.V., and Tsurupa, A.A., 1985, Extended flows of liquid acid lava in Krasheninnikov Volcano: Vulkanol. Seismol., 1985, pt. 3, p. 85-91.

Skorokhodov, V.N., 1979, Dynamics of development of ring structures on the example of Kamchatka volcanoes Uksichan, Krasheninnikov and Maly Semiachik, <u>in</u> Problems of deep magmatism: Moscow, Nauka Press, p. 117-124.

Steinberg, G.S., 1964, On the activity and structure of the Krasheninnikov volcano: Bull. Vulkanol. Stantsii (Kamchatka), no. 37, p. 16-20.

Steinberg, G.S., and Rivosh, L.A., 1965, Geophysical study of the Kamchatka volcanoes: Jour. Geophys. Res., v. 70, p. 3341-3369.

Vlodavetz, V.I., and Piip, B.I., 1959, Catalogue of active volcanoes of the world including solfatara fields, Pt. VIII, Kamchatka and continental areas of Asia: Naples, IAVCEI, 110 p.

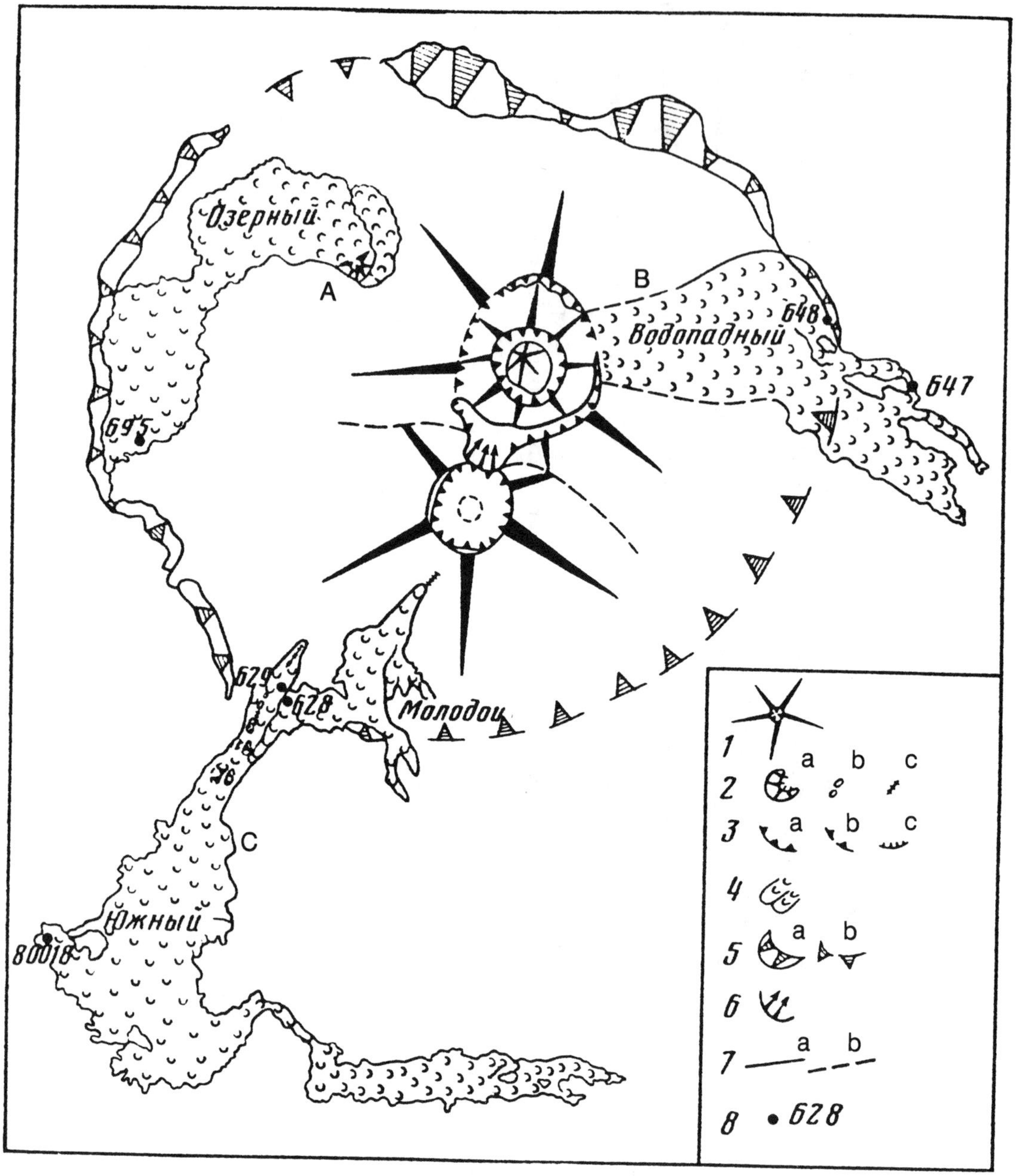

Figure 10.7.1. Geologic map of Krasheninnikov Volcano (Ponomareva and Tsurupa, 1985). A, Ozerny flow (ca. 1,500 yr B.P.); B, Vodopadny flow (ca. 3,000 yr B.P.); C, Yuzhny flow (600-700 yr B.P.); 1, stratovolcano cone; 2, Holocene eruptive vent (a, cinder cone; b, crater; c, chain of craters); 3, crater rim (a, well defined; b, inferred; c, cinder cone); 4, lava flow; 5, caldera rim (a, well defined by relief; b, inferred); 6, cirque wall; 7, fault (a, located; b, inferred); 8, lava sample location point.

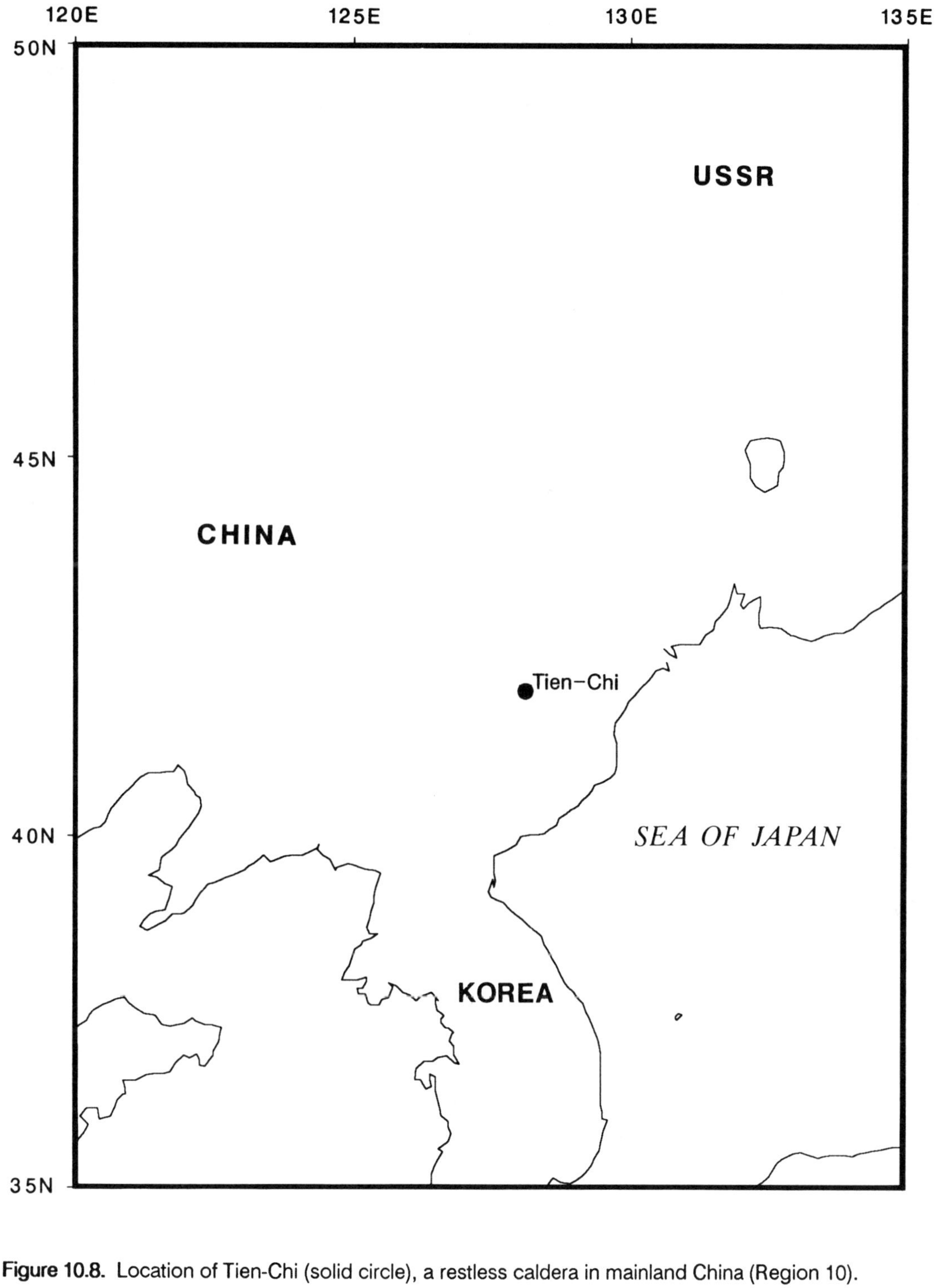

Figure 10.8. Location of Tien-Chi (solid circle), a restless caldera in mainland China (Region 10).

TIEN-CHI

CAVW number (active vent)	Latitude Longitude (degrees)	Diameter (km)	Local tectonic setting	Precaldera edifice	SiO_2 content (wt pct)	Age of collapse (yr B.P.)	Date of unrest	Type and duration of unrest ESTU STHF MGTF H Te	Eruption type
10-03-02 (Changpaishan including = Baegdusan = Pai-tou-shan = P'aektu-san = Hakuto-san)	41.98N 128.08E	5	Exten?		R = 49-?	800-900	1597?	---- ---- ---- - --	unknown
					C = 71		1702?	---- ---- ---- - --	unknown

TECTONIC SETTING/GEOLOGIC HISTORY

A large lava plateau and shield volcano on the Chinese-Korean border (Changpaishan) consists of tholeiitic to alkaline basalt flows with a cap of peralkaline silicic flows (mostly trachyte and pantellerite). A stratovolcano (or dome?)--P'aektu-san (also known as Hakuto-san)--has grown on top of the shield (fig. 10.8.1). A 5-km-wide, 850-m-deep caldera has developed at the summit of P'aektu-san (Ogura, 1969; Tomita, 1969; Whitford-Stark, 1983, 1987). A large-volume pumice deposit with a radius of 40 km is centered on the summit caldera and probably was erupted immediately before or during collapse of the caldera (Whitford-Stark, 1987). Tien-chi, a caldera lake, is 3 km x 4 km across and 375 m deep (Whitford-Stark, 1987).

This cluster of alkaline volcanoes is probably related to spreading behind the Japan arc.

HISTORICAL ACTIVITY

There are unsubstantiated reports of eruptions in 1597 and 1702 (Tomita, 1969; Ogura, 1969, p. 399).

REFERENCES

Machida, H., and Arai, F., 1983, Extensive ash falls in and around the Sea of Japan from large late Quaternary eruptions: Jour. Volcanol. Geotherm. Res., v. 18, p. 151-164.

Ogura, T., 1969, Volcanoes in Manchuria, in Ogura, T., ed., Geology and mineral resources of the Far East: Tokyo, Univ. Tokyo Press, v. 2, p. 373-413.

Tomita, T., 1969, Volcanic geology of the Cenozoic alkaline petrographic province of eastern Asia, in Ogura, T., ed., Geology and mineral resources of the Far East: Tokyo, Univ. Tokyo Press, Tokyo, v. 1, p. 139-202.

Whitford-Stark, J.L., 1983, Cenozoic volcanic and petrochemical provinces of mainland Asia: Jour. Volcanol. Geotherm. Res., v. 19, p. 193-222.

----- 1987, A survey of Cenozoic volcanism on mainland Asia: Geol. Soc. Am. Spec. Paper 213, 74 p.

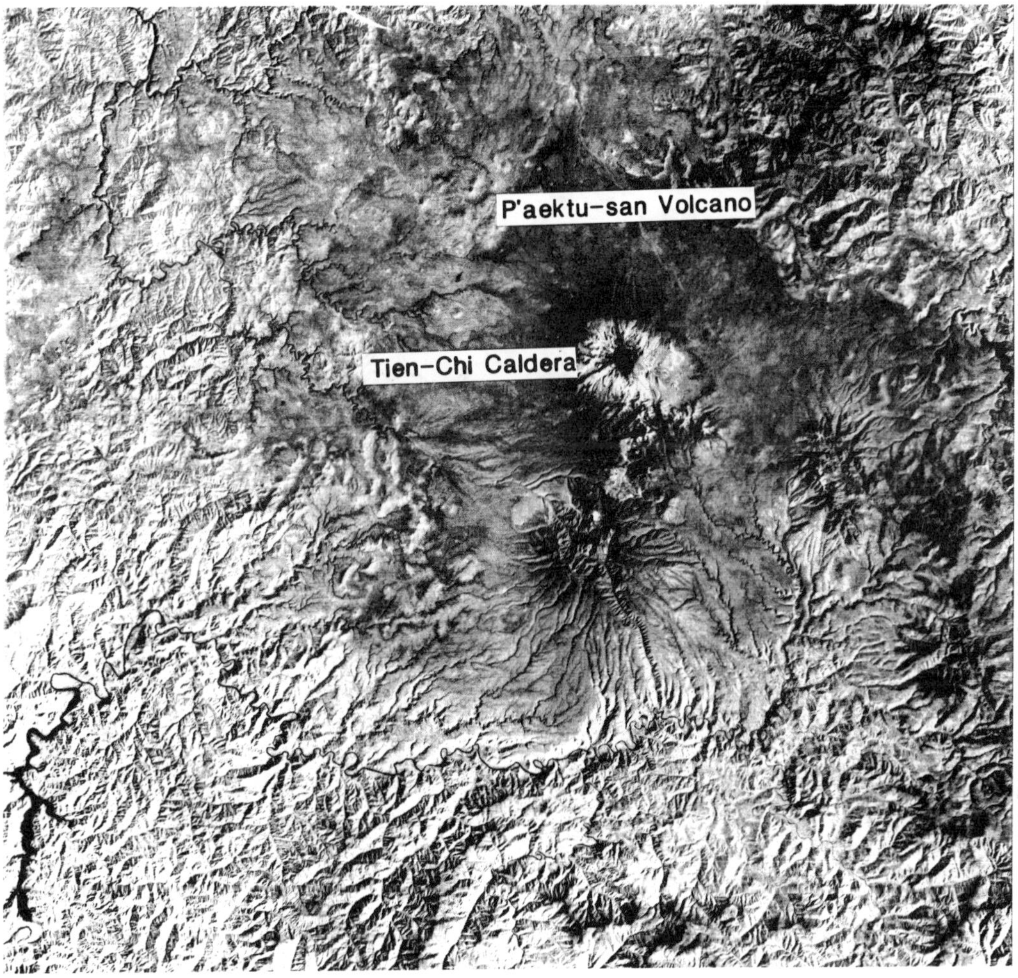

Figure 10.8.1. Landsat satellite image of P'aektu-san Volcano (northern cone) and its summit caldera, Tien-Chi (lake filled), as reproduced in Whitford-Stark (1987). Tien-Chi Caldera is approximately 5 km in diameter.

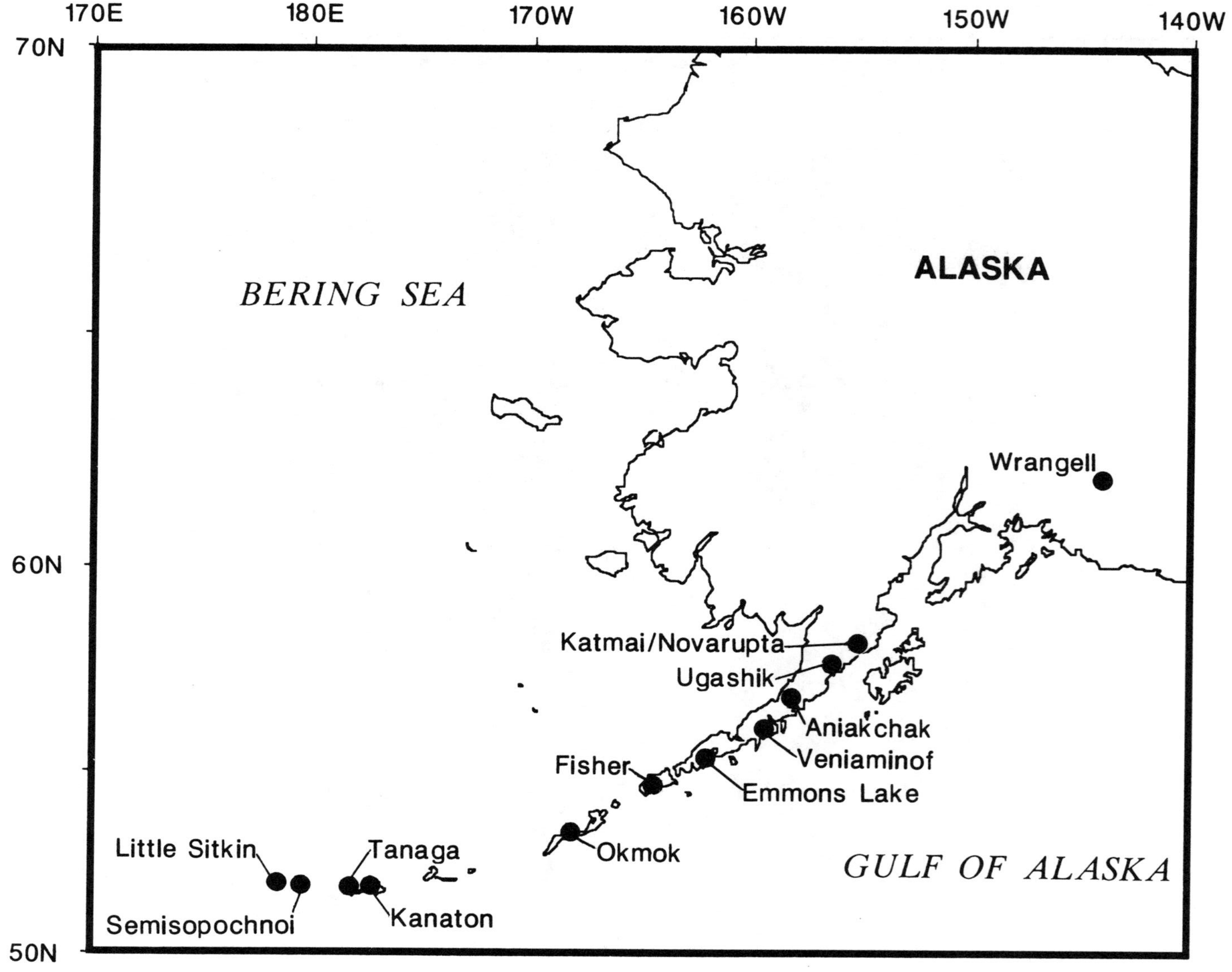

Figure 11. Locations of large, Quaternary restless calderas (solid circles) in Region 11.

LITTLE SITKIN, CALDERA ONE AND CALDERA TWO

CAVW number (active vent)	Latitude Longitude (degrees)	Diameter (km)	Local tectonic setting	Precaldera edifice	SiO_2 content (wt pct)	Age of collapse (yr B.P.)	Date of unrest	Type and duration of unrest ESTU STHF MGTF H Te	Eruption type
11-01-05	51.95N	5 (outer)	Compr	Strato	R = 55-69		1776	---- ---- ---- - --	unknown
(Little Sitkin)	178.53E	3 (inner)					1828-30	---- ---- ---- - --	ex
					C = 58 (One)				
					= 60-65 (Two)				

TECTONIC SETTING

The Aleutian volcanic arc is defined by over 65 major Quaternary volcanic centers and spans 2,500 km of the Alaskan mainland and the Aleutian Islands. At least 21 of these centers have calderas (Miller and Barnes, 1976), 19 of which may have formed in the Holocene.

Little Sitkin lies near the western end of the volcanically active part of the Aleutian arc. In this part of the arc, the Pacific plate is thrust beneath the North American plate at a rate of about 8 cm/yr. A submarine scarp, the Prokhoda scarp, connects the southeast part of Little Sitkin with Semisopochnoi Island, 56 km to the southeast.

GEOLOGIC HISTORY

Caldera One formed by collapse of the summit of a large basaltic-andesitic stratovolcano (figs. 11.1.1, 11.1.2) (Snyder, 1959). A low-silica dacite tuff overlies the edifice and might be the product of the caldera-forming eruption. Voluminous dacite lava flows built a cone within Caldera One. Miarolitic andesite inclusions are abundant in these lavas. Caldera Two, largely nested within Caldera One, formed after eruption of low-silica dacite pyroclastic flows (now welded tuff). Lavas ranging in composition from andesite to rhyodacite have been erupted since formation of Caldera Two, some along a zone (fissure) trending N67°W through the center of the caldera. One very young-looking flow (undated, but perhaps only 50-100 yr old) was erupted from the central vent of Caldera Two; another was erupted from a vent along the above-mentioned fissure, just outside the west rim of Caldera One (Snyder, 1959).

HISTORICAL ACTIVITY

Young-looking lava flows were probably erupted within historical time, but the exact date of their eruption is unknown. Coats (1950) attributed a 1776 eruption with "flames" to Little Sitkin, and Dall (1870) reported "smoke" above the island during 1828-30.

LITTLE SITKIN (continued)

REFERENCES

Coats, R.R., 1950, Volcanic activity in the Aleutian arc: U.S. Geol. Surv. Bull. 974-B, p. 35-49.
Dall, W.H., 1870, Alaska and its resources: London, Sampson Low, Son, and Marston, p. 467.
Miller, T.P., and Barnes, I., 1976, Potential for geothermal energy development in Alaska, in Halbouty, M.T., Maher, J.C., and Lian, H.M., eds., Circum-Pacific energy and mineral resources: Am. Assoc. Pet. Geol. Memoir 25, p. 149-153.
Snyder, G.L., 1959, Geology of Little Sitkin Island, Alaska: U.S. Geol. Surv. Bull. 1028-H, p. 169-210.
U.S. Geological Survey, 1951, limited revisions 1983, Rat Islands topographic map, Alaska topographic series, scale 1:250,000.

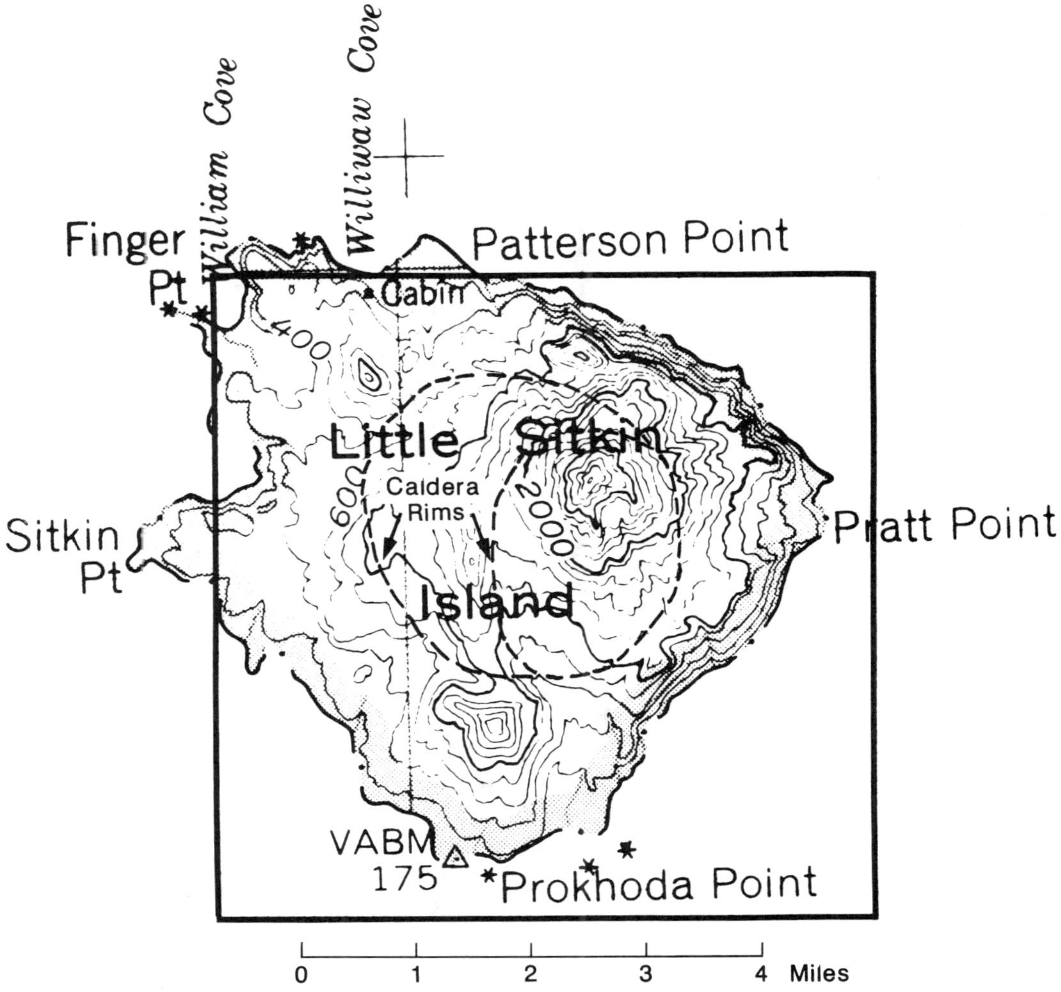

Figure 11.1.1. Topographic map of Little Sitkin Island, showing approximate outline of Little Sitkin Caldera (U.S. Geological Survey, 1951).

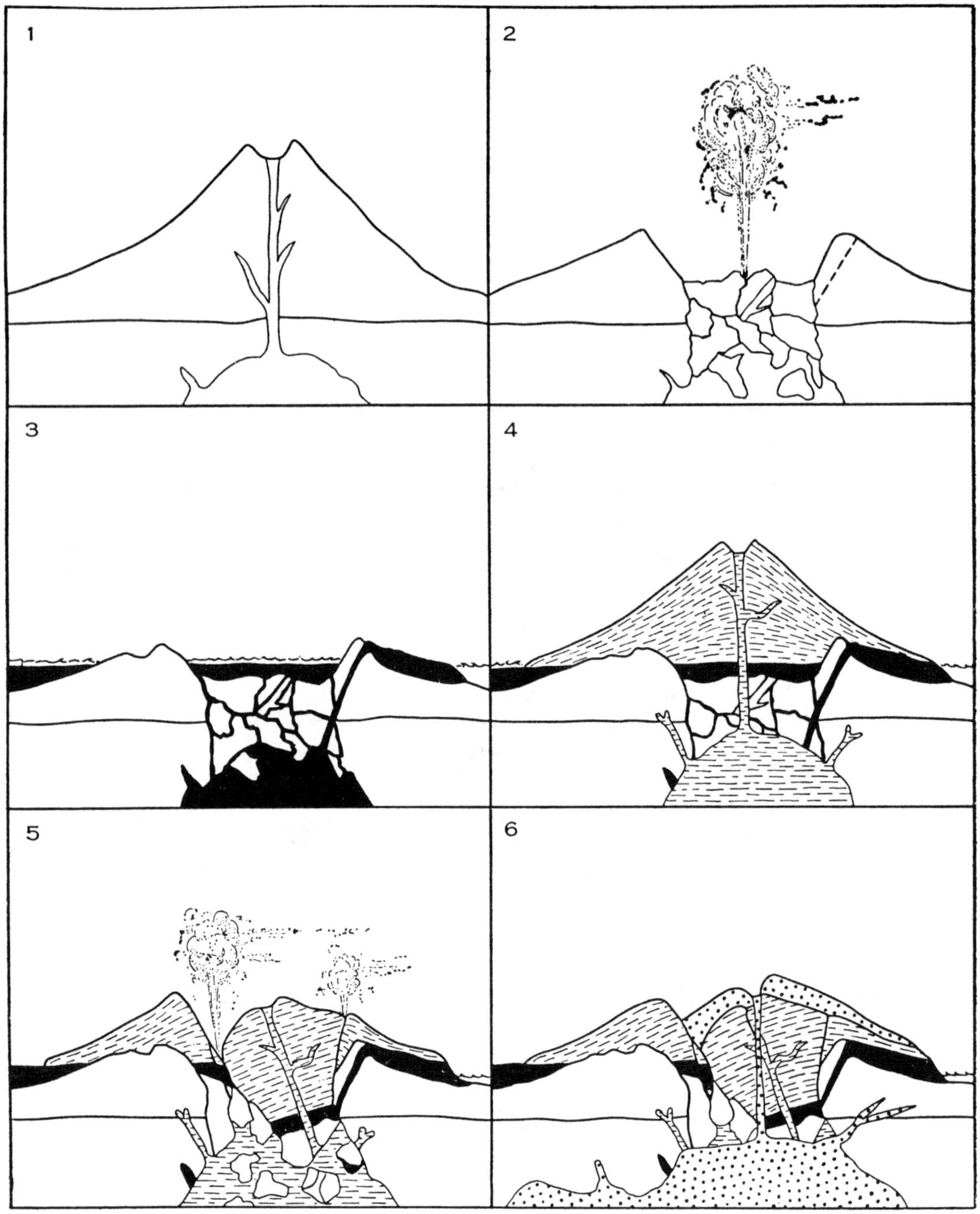

Figure 11.1.2. Schematic history of Little Sitkin Island, from Snyder (1959): 1, Erection of andesitic composite cone on platform 300 ft below sea level in late Tertiary or early Quaternary time. 2, formation of Caldera One after long period of erosion. 3, deposition of Sitkin Point formation in offshore basin and in caldera lake, and extrusion of flow sequence of East Point formation. 4, erection of high-silica dacite lava-flow cone of Double Point dacite in late Pleistocene or early Recent time. 5, formation of Caldera Two with attendant nuee ardente eruptions. 6, extrusion of low-silica dacite lava continuing to the present.

SEMISOPOCHNOI

CAVW number (active vent)	Latitude Longitude (degrees)	Diameter (km)	Local tectonic setting	Precaldera edifice	SiO_2 content (wt pct)	Age of collapse (yr B.P.)	Date of unrest	Type and duration of unrest ESTU STHF MGTF H Te	Eruption type
11-01-06 (Cerberus)	51.93N 179.60E	8	Compr	Strato	R = 56-63 C = 61-66	post-glacial	1772?	---- ---- ---- - --	"smoke"
							1790?	---- ---- ---- - --	"smoke"
							1792?	---- ---- ---- - --	"smoke"
							1830?	---- ---- ---- - --	"smoke"
							1873	---- ---- ---- - --	unknown
							1987	---- ---- ---- - --	ex

TECTONIC SETTING

Semisopochnoi Island lies near the southernmost part of the Aleutian arc. In this stretch of the arc, the Pacific plate is subducted beneath the North American plate at a rate of approximately 8 cm/yr.

GEOLOGIC HISTORY

Semisopochnoi Island (fig. 11.2.1) consists entirely of volcanic rocks and epiclastic volcanic sediments. A single, large andesitic-basaltic "shield" volcano ("Pochnoi") forms most of the present island. The central part of this edifice collapsed following eruption of a large volume of dacitic pumice. Mount Cerberus is a postcaldera stratocone, composed of basalt and basaltic andesite, that grew in the center of the caldera; Lakeshore Cone and Sugarloaf Peak are also postcaldera cones. Lava flows have nearly covered the caldera floor and have spilled onto the south flank of the precollapse edifice (Coats, 1959). One late flow from Cerberus may be less than 100 years old, because it is no more vegetated than a 40-year-old lava flow on nearby Kanaga Island (Coats, 1959).

HISTORICAL ACTIVITY

Coats (1959), citing Grewingk (1850), wrote that early explorers reported "smoke" in at least four years---1772, 1790, 1792, and 1830. Dall (1870) stated that the Semisopochnoi Islands lost their activity in 1772, but Becker (1898, p. 17) reported that a volcano on the island was again active in 1873. A plume was observed from the approximate location of Semisopochnoi on 13 April 1987, and ash-darkened snow was reported by a local pilot 11 days later (Smithsonian Institution, 1987).

SEMISOPOCHNOI (continued)

REFERENCES

Becker, G.F., 1898, Reconnaissance of the gold fields of southern Alaska, with some notes on general geology: U.S. Geol. Surv., 18th Ann. Rep., pt. 3, p. 13-24.

Coats, R.R., 1950, Volcanic activity in the Aleutian arc: U.S. Geol. Surv. Bull. 974-B, p. 35-49.

----- 1959, Geologic reconnaissance of Semisopochnoi Island, western Aleutian Islands, Alaska: U.S. Geol. Surv. Bull. 1028-O, p. 477-519.

Dall, W.H., 1870, Alaska and its resources: London, Sampson Low, Son, and Marston, p. 467.

Grewingk, C., 1850, Beitrag zur Kenntniss der orographischen und geognostischen Beschaffenheit der Nordwestküste Amerikas, mit den anliegenden Inseln: Verh. Russisch-Kaiserlichen Mineral. Gesellschaft zu St. Petersburg, 1848-49, p. 123-278.

Katsui, Y., ed., 1971, List of the world active volcanoes: Tokyo, Volcanol. Soc. Japan/IAVCEI/IUGG, 160 p.

Smith, R.L., Shaw, H.R., Luedke, R.G., and Russell, S.L., 1978, Comprehensive tables giving physical data and thermal energy estimates for young igneous systems of the United States: U.S. Geol. Surv. Open-File Rep. 78-925, 15 p. + tables.

Smithsonian Institution, Scientific Event Alert Network (SEAN), 1987, Semisopochnoi Volcano: SEAN Bull., v. 12, no. 4, p. 15 (information from J.W. Reeder, T. Miller, and W. Gould).

U.S. Geological Survey, 1951, limited revisions 1983, Rat Islands topographic map, Alaska topographic series, scale 1:250,000.

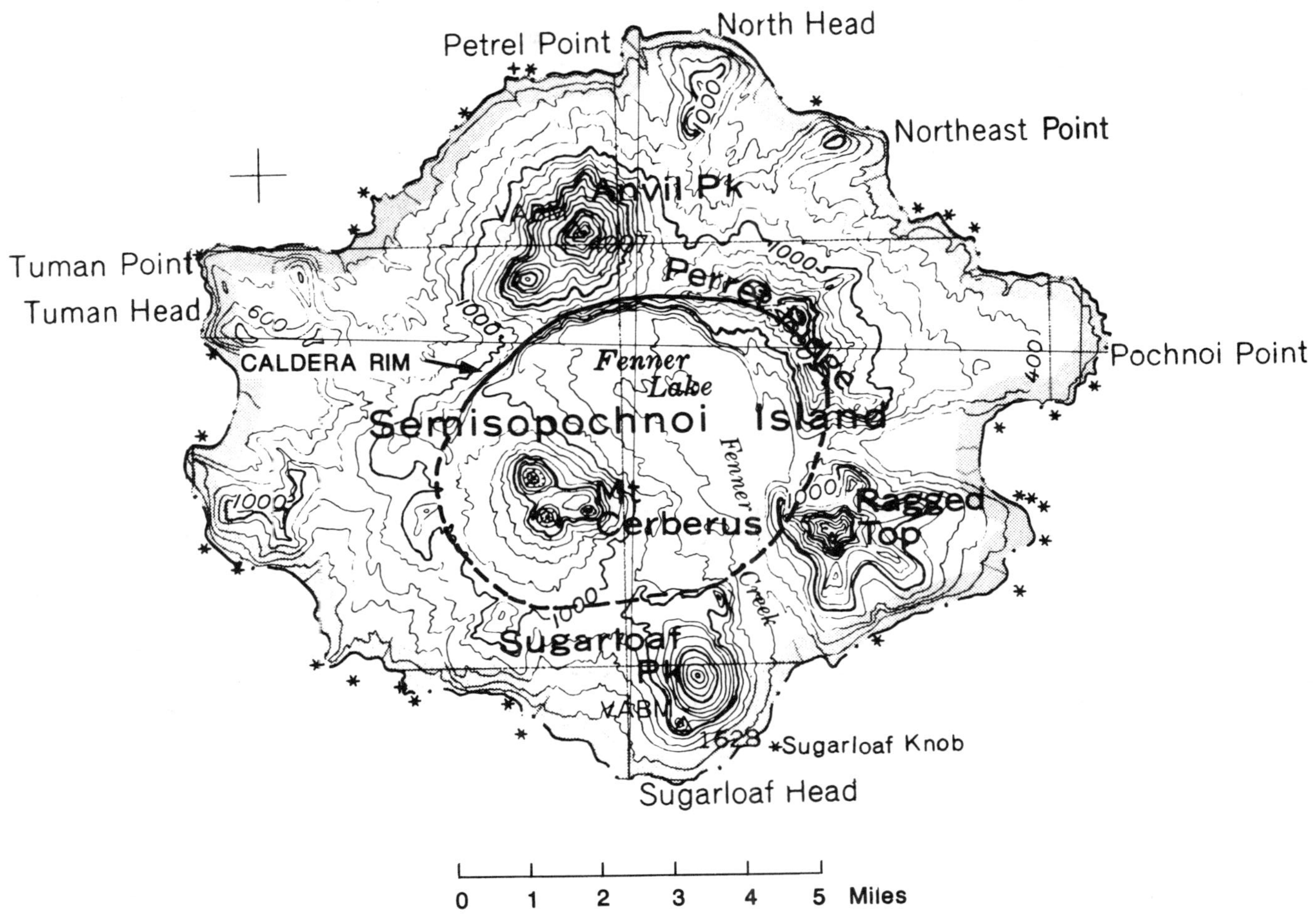

Figure 11.2.1. Topographic map of Semisopochnoi Island, showing approximate outline of Semisopochnoi Caldera (U.S. Geological Survey, 1951).

PART 3: HISTORICAL UNREST AT LARGE QUATERNARY CALDERAS (continued)
See inside back cover for explanation and abbreviations

TANAGA

CAVW number (active vent)	Latitude Longitude (degrees)	Diameter (km)	Local tectonic setting	Precaldera edifice	SiO_2 content (wt pct)	Age of collapse (yr B.P.)	Date of unrest	Type and duration of unrest ESTU STHF MGTF H Te	Eruption type
11-01-08 (Tanaga)	51.88N 178.13W	11	Compr	Strato	R = 52+		1763-70	---- ---- ---- - --	"active"
							1791?	---- ---- ---- - --	"smoke"
							1829	---- ---- ---- - --	ex?,"smoke"
							1914	---- ---- ---- - --	lf

TECTONIC SETTING

Tanaga Volcano and its caldera lie on the northwestern tip of Tanaga Island, near the southernmost part of the western Aleutian arc. Submarine ridges that are transverse to the arc suggest local segmentation into fault blocks. Convergence rates in this area are approximately 8 cm/yr.

GEOLOGIC SETTING

"Tanaga Volcano and its twin to the west are located within an arcuate structure, possibly a caldera. The volcano at the northwest tip of the island appears also to have been built up within a similar structure" (Fraser and Barnett, 1959). The postcaldera cone of Tanaga Volcano is unglaciated (fig. 11.3.1).

HISTORICAL ACTIVITY

No details available.

REFERENCES

Coats, R.R., 1950, Volcanic activity in the Aleutian arc: U.S. Geol. Surv. Bull. 974-B, p. 35-49.

----- 1956, Reconnaissance geology of some western Aleutian Islands, Alaska: U.S. Geol. Surv. Bull. 1028-E, p. 83-100.

Fraser, G.D., and Barnett, H.F., 1959, Geology of the Delarof and westernmost Andreanof Islands, Alaska: U.S. Geol. Surv. Bull. 1028-I, p. 211-248.

Katsui, Y., ed., 1971, List of the world active volcanoes: Tokyo, Volcanol. Soc. Japan/IAVCEI/IUGG, 160 p.

Smith, R.L., Shaw, H.R., Luedke, R.G., and Russell, S.L., 1978, Comprehensive tables giving physical data and thermal energy estimates for young igneous systems of the United States: U.S. Geol. Surv. Open-File Rep. 78-925, 15 p. + tables.

U.S. Geological Survey, 1954, limited revisions 1983, Gareloi Island topographic map, Alaska topographic series, scale 1:250,000.

----- 1957, limited revisions 1983, Adak topographic map, Alaska topographic series, scale 1:250,000.

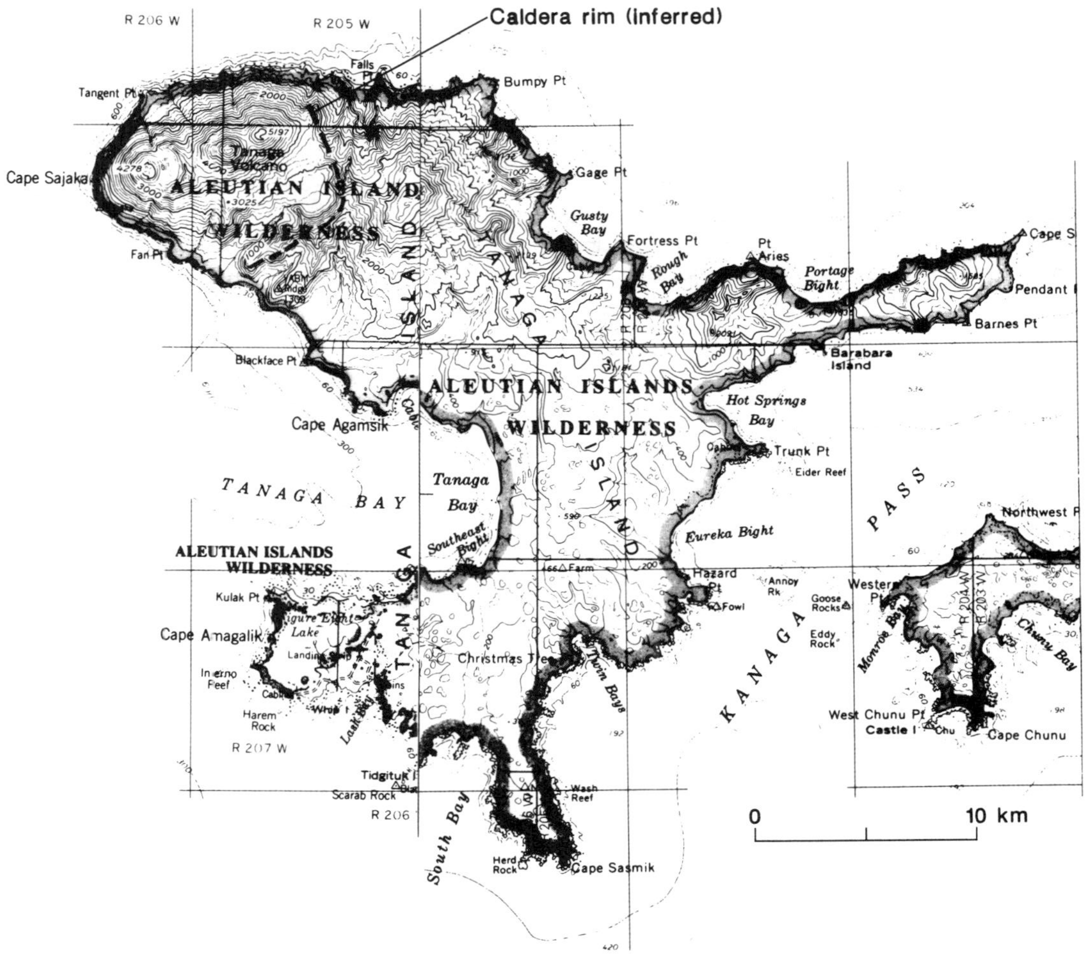

Figure 11.3.1. Topographic map of Tanaga Island and Tanaga Volcano, showing approximate outline of Tanaga Caldera (U.S. Geological Survey, 1954, 1957).

PART 3: HISTORICAL UNREST AT LARGE QUATERNARY CALDERAS (continued)
See inside back cover for explanation and abbreviations

KANATON

CAVW number (active vent)	Latitude Longitude (degrees)	Diameter (km)	Local tectonic setting	Precaldera edifice	SiO_2 content (wt pct)	Age of collapse (yr B.P.)	Date of unrest	Type and duration of unrest ESTU STHF MGTF H Te	Eruption type
11-01-11	51.93N	5.5	Compr	Strato	R = 55-60		1763?	---- ---- ---- - --	"active"
(Kanaga)	177.15W				C = 64		1768	---- ---- ---- - --	"active"
							1783-87	---- ---- ---- - --	"active"
							1790-91	---- ---- ---x - --	"smoke"
							1827-29?	---- ---- ---- - --	"smoke"
							1904	---- ---- ---- - --	lf
							1906	x--- ---- ---- - --	lf
							1933	---- ---- ---- - --	unknown
							1942	---- ---- ---- - --	unknown
							1951	---- ---- ---x - --	ex?

TECTONIC SETTING

Kanaga Volcano and the Kanaton Caldera lie on the north tip of Kanaga Island, near the southernmost part of the western Aleutian arc (figs. 11.4.1, 11.4.2). Fraser and Barnett (1959) show a major N-S fault, transverse to the arc, just west of Kanaton caldera. Convergence rates in this part of the arc are approximately 8 cm/yr.

GEOLOGIC HISTORY

An early stratovolcano (termed Mount Kanaton by Coats, 1956b) consists largely of andesitic and basaltic lava flows and tuff breccias. Collapse to form the Kanaton Caldera occurred after eruption of a dacitic tuff. The subaerial volume of that dacitic tuff is much smaller than would usually be associated with a 5.5-km-diameter caldera (Coats, 1956b), and because the northern wall of the caldera is not preserved, it is possible that this caldera formed by a very large debris avalanche rather than by collapse after a large explosive eruption. Bathymetric studies north-northwest of Kanaga would help to determine the origin of the caldera.

HISTORICAL ACTIVITY

According to a trapper who lived near Kanaga Volcano, earthquakes in May 1906 were followed by lava extrusion onto the east and west sides of the cone (Coats, 1956b). Some of the reported "smoke" from Kanaga has probably been fumarolic activity from hot

KANATON (continued)

HISTORICAL ACTIVITY (continued)

springs around the base of the volcano (Grewingk, 1850) and from conspicuous fumaroles near the summit (Coats, 1956b). Powers (1951) reported that "minor ash showers and fume emission" occurred at Kanaga and several other Aleutian volcanoes during 1951.

REFERENCES

Coats, R.R., 1950, Volcanic activity in the Aleutian arc: U.S. Geol. Surv. Bull. 974-B, p. 35-49.

----- 1952, Magmatic differentiation in Tertiary and Quaternary volcanic rocks from Adak and Kanaga Islands, Alaska: Geol. Soc. Am. Bull, v. 63, p. 485-514.

----- 1956a, Reconnaissance geology of some western Aleutian Islands, Alaska: U.S. Geol. Surv. Bull. 1028-E, p. 83-100.

----- 1956b, Geology of northern Kanaga Island, Alaska: U.S. Geol. Surv. Bull. 1028-D, p. 69-80.

Fraser, G.D., and Barnett, H.F., 1959, Geology of the Delarof and westernmost Andreanof Islands, Alaska: U.S. Geol. Surv. Bull. 1028-I, p. 211-248.

Grewingk, C., 1850, Beitrag zur Kenntniss der orographischen und geognostischen Beschaffenheit der Nordwestküste Amerikas, mit den anliegenden Inseln: Verh. Russisch-Kaiserlichen Mineral Gesellschaft zu St. Petersburg, 1848-49, p. 123-278.

Hantke, G., 1951, Uebersicht über die vulkanische tätigkeit, 1941-1947: Bull. Volcanol., v. 11, p. 161-208.

Katsui, Y., ed., 1971, List of the world active volcanoes: Tokyo, Volcanol. Soc. Japan/IAVCEI/IUGG, 160 p.

Powers, H.A., 1951, Aleutian volcanic activity: Volcano Lett., no. 514, p. 6.

Smith, R.L., Shaw, H.R., Luedke, R.G., and Russell, S.L., 1978, Comprehensive tables giving physical data and thermal energy estimates for young igneous systems of the United States: U.S. Geol. Surv. Open-File Rep. 78-925, 15 p. + tables.

U.S. Geological Survey, 1957, limited revisions 1983, Adak topographic map, Alaska topographic series, scale 1:250,000.

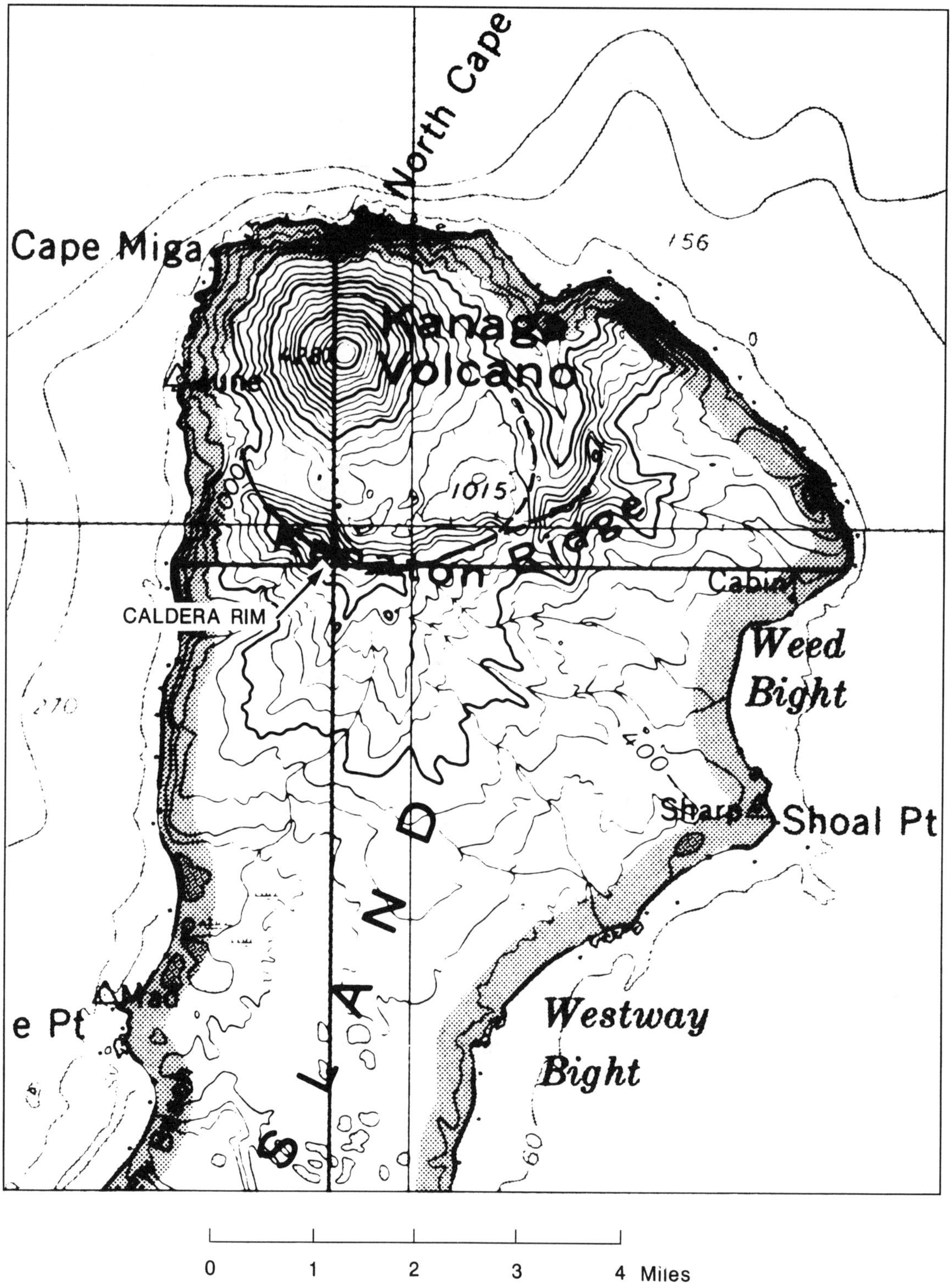

Figure 11.4.1. Topographic map of Kanaga Island, showing Kanaga Volcano and approximate outline of Kanaton Caldera (U.S. Geological Survey, 1957).

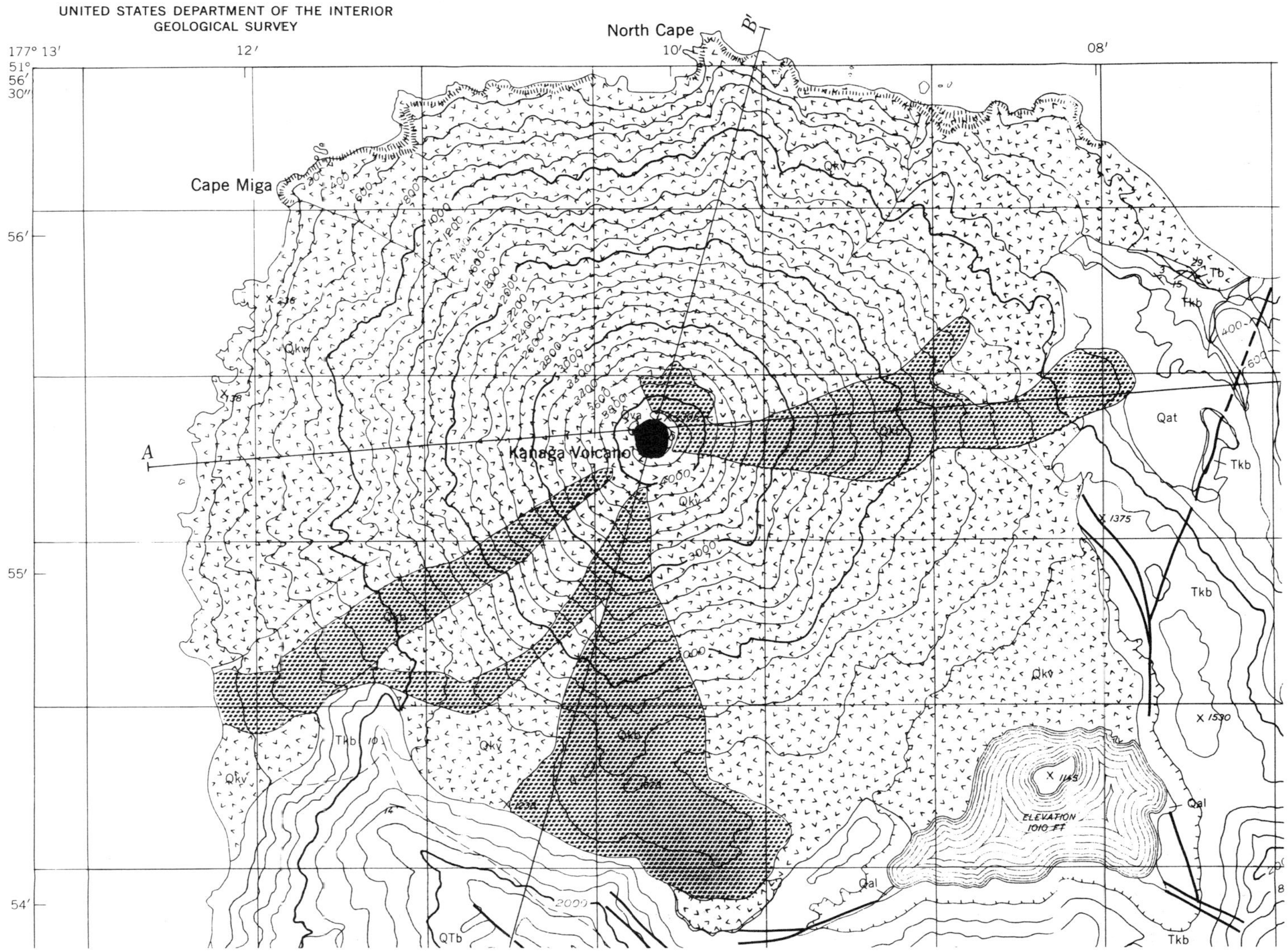

Figure 11.4.2. Recent products of Kanaga Volcano, from Coats (1956b). Stippled lava flows on volcano's flanks are historic and late prehistoric basalt flows; remainder of flanks consist of older basalt and andesite flows, scoria, and tuff breccia. See figure 11.4.1 for outline of Kanaton Caldera.

PART 3: HISTORICAL UNREST AT LARGE QUATERNARY CALDERAS (continued)
See inside back cover for explanation and abbreviations

OKMOK

CAVW number (active vent)	Latitude Longitude (degrees)	Diameter (km)	Local tectonic setting	Precaldera edifice	SiO_2 content (wt pct)	Age of collapse (yr B.P.)	Date of unrest	Type and duration of unrest ESTU STHF MGTF H Te	Eruption type
11-01-29 (Okmok)	53.42N 168.13W	10 (both calderas)	Compr	LL-strat	R = 51-54 C = 54-66	8,250, <2,400	1805	---- ---- ---- - --	unknown
							1817-20	---- ---- ---- - --	Ex
							1824-30	---- ---- ---- - --	unknown
							1878	Y--- ---- ---- - --	unknown, + tsunami?
							1899	---- ---- ---- - --	Ex
							1931	x--- ---- ---x - --	"smoke,"lf?
							1936?	---- ---- ---- - --	"smoke"
							1938?	---- ---- ---- - --	unknown
							1943	---- ---- ---- - --	ex
							1944-45	C--- ---- ---- - --	ex,lf
							1958	---- ---- ---- - --	lf
							1960-61	---- ---- ---- - --	Ex
							1983	---- ---- ---- - --	ex
							1986-88+	---- ---- ---x - QU?	ex

TECTONIC SETTING

Okmok Caldera is on Umnak Island, where the Pacific and North American plates converge at a rate of 7-8 cm/yr. The Aleutian arc is a single arc at Umnak Island, and Quaternary volcanoes have grown over Tertiary basement; to the east from Umnak Island, non-volcanic islands lie seaward of the chain of active volcanoes.

GEOLOGICAL HISTORY

Okmok is a basaltic composite volcano with a pair of overlapping summit calderas, each with an estimated diameter of about 10 km (fig. 11.5.1). The first formed about 8,250 yr B.P. (Black, 1975; Miller and Smith, 1975, 1987), the second less than 2,400 yr B.P. (Miller and Smith, 1983, 1987). The center of second collapse is about 1.5 km southwest of the center of the first (Byers, 1959; Miller and Smith, 1983). Each caldera formed after eruption of voluminous, dominantly andesitic ash flows (the Okmok Volcanics), locally separated by a lava flow and an erosional surface. Postcaldera activity has included mostly small, intracaldera basaltic extrusions and explosions (Byers, 1959).

OKMOK (continued)

HISTORICAL ACTIVITY

1817: A large explosive eruption covered some parts of Unalaska with 15 cm of ash. Some large stones were thrown out to a distance of 50 versts (53 km), and ash reached Unimak Island.

1878: A new crater, earthquakes, and a tsunami were reported (Hantke, 1951); their relation to Okmok Caldera is unclear.

1931: On 21 March, black smoke issued from Tulik on Umnak in place of the usual white steam. Two earthquakes on 29 March were recorded on a seismograph at Dutch Harbor, on Unalaska Island. An eruption occurred on Umnak on 13 May (Jaggar, 1931).

1944-45: Bradley (1948) reported that activity began anew in 1944. Wilcox (1959) reports that on 1 June 1945, "a sharp earthquake was felt at the Umnak Island army base on the east side of the mountain. Nothing otherwise unusual was noted, and the rim of the caldera was hidden by persistent low clouds." On 4 June, pilots reported a column of dark ash, and subsequent observations showed ash ejection and a lava flow from a cinder cone in the southern part of the caldera.

1958: A new lava flow was recognized from the air.

1960-61: Explosive eruptions occurred for several months, but no new lava flow was reported (Volcanological Society of Japan, 1986).

1983: A small ash eruption was noted by a local pilot at 0900 hr on 8 July, and on satellite imagery at 1616 hr of the same day (Volcanological Society of Japan, 1986).

1986-87: A small ash eruption was seen on 18 November 1986. Then on 5 January 1987, about 13 hours after a shallow, $\underline{M}_s$=6.6 earthquake 130 km southwest of Okmok, another small ash eruption occurred. Additional small ash eruptions occurred in January, June, July, and December 1987, and February 1988 (Smithsonian Institution, 1986, 1987, 1988).

REFERENCES

Black, R.F., 1975, Late-Quaternary geomorphic processes: effects on the ancient Aleuts of Umnak Island in the Aleutians: Arctic, v. 28, p. 159-169.
Bradley, C.C., 1948, Geologic notes on Adak Island in the Aleutian Chain: Am. Jour. Sci., v. 246, p. 214-240.
Byers, F.M., Jr., 1959, Geology of Umnak and Bogoslof Islands, Aleutian Islands, Alaska: U.S. Geol. Surv. Bull. 1028-L, p. 267-365.
----- 1961, Petrology of three volcanic suites, Umnak and Bogoslof Islands, Aleutian Islands, Alaska: Geol. Soc. Am. Bull., v. 72, p. 93-128.

OKMOK (continued)

REFERENCES (continued)

Byers, F.M., Jr., and Brannock, W.W., 1949, Volcanic activity on Umnak and Great Sitkin Islands, 1946-1948: Trans. Am. Geophys. Union, v. 30, p. 719-734.

Coats, R.R., 1950, Volcanic activity in the Aleutian arc: U.S. Geol. Surv. Bull. 974-B, p. 35-49.

Hantke, G., 1951, Uebersicht über die vulkanische tätigkeit, 1941-1947: Bull. Volcanol., v. 11, p. 161-208.

Jaggar, T.A., 1931, Difficulties of Alaskan earthquake study: Volcano Lett., no. 337, p. 1-3.

Katsui, Y., ed., 1971, List of the world active volcanoes: Tokyo, Volcanol. Soc. Japan/IAVCEI/IUGG, 160 p.

Keller, A., 1976, Okmok Caldera: Alaska Geogr., v. 4, no. 1, p. 75-83.

Miller, T.P., and Smith, R.L., 1975, Ash flows on the Alaska Peninsula: A preliminary report on their distribution, composition, and age [abs.]: Geol. Soc. Am. Abstracts with Programs, v. 7, no. 7, p. 1201.

----- 1983, Calderas of the eastern Aleutian arc [abs.]: Eos, Trans. Am. Geophys. Union, v. 64, p. 877.

----- 1987, Late Quaternary caldera-forming eruptions in the eastern Aleutian arc, Alaska: Geology, v. 15, p. 434-438.

Nye, C.J., 1983, Petrology and geochemistry of Okmok and Wrangell Volcanoes, Alaska: Univ. Calif. Santa Cruz, unpubl. Ph.D. diss.

Robinson, G.D., 1948, Exploring Aleutian volcanoes: Natl. Geogr., v. 94, no. 4, p. 509-522.

Smith, R.L., Shaw, H.R., Luedke, R.G., and Russell, S.L., 1978, Comprehensive tables giving physical data and thermal energy estimates for young igneous systems of the United States: U.S. Geol. Surv. Open-File Rep. 78-925, 15 p. + tables.

Smithsonian Institution, Scientific Event Alert Network (SEAN), 1983, 1986-88, Okmok: SEAN Bull., v. 8, no. 8; v. 11, no. 12; v. 12, nos. 1, 6, 7, 12; v. 13, no. 2 (information from J.W. Reeder).

U.S. Geological Survey, 1951a, limited revisions 1983, Umnak topographic map, Alaska topographic series, scale 1:250,000.

----- 1951b, limited revisions 1983, Unalaska topographic map, Alaska reconnaissance topographic series, scale 1:250,000.

Volcanological Society of Japan, 1986, Okmok: Bull. Volcanic Eruptions, no. 23, p. 35-36, 52-53 (information from J.W. Reeder, C.T. Madsen, and W. Gould).

Wilcox, R.E., 1959, Some effects of recent volcanic ash falls, with especial reference to Alaska: U.S. Geol. Surv. Bull. 1028-N, p. 409-476.

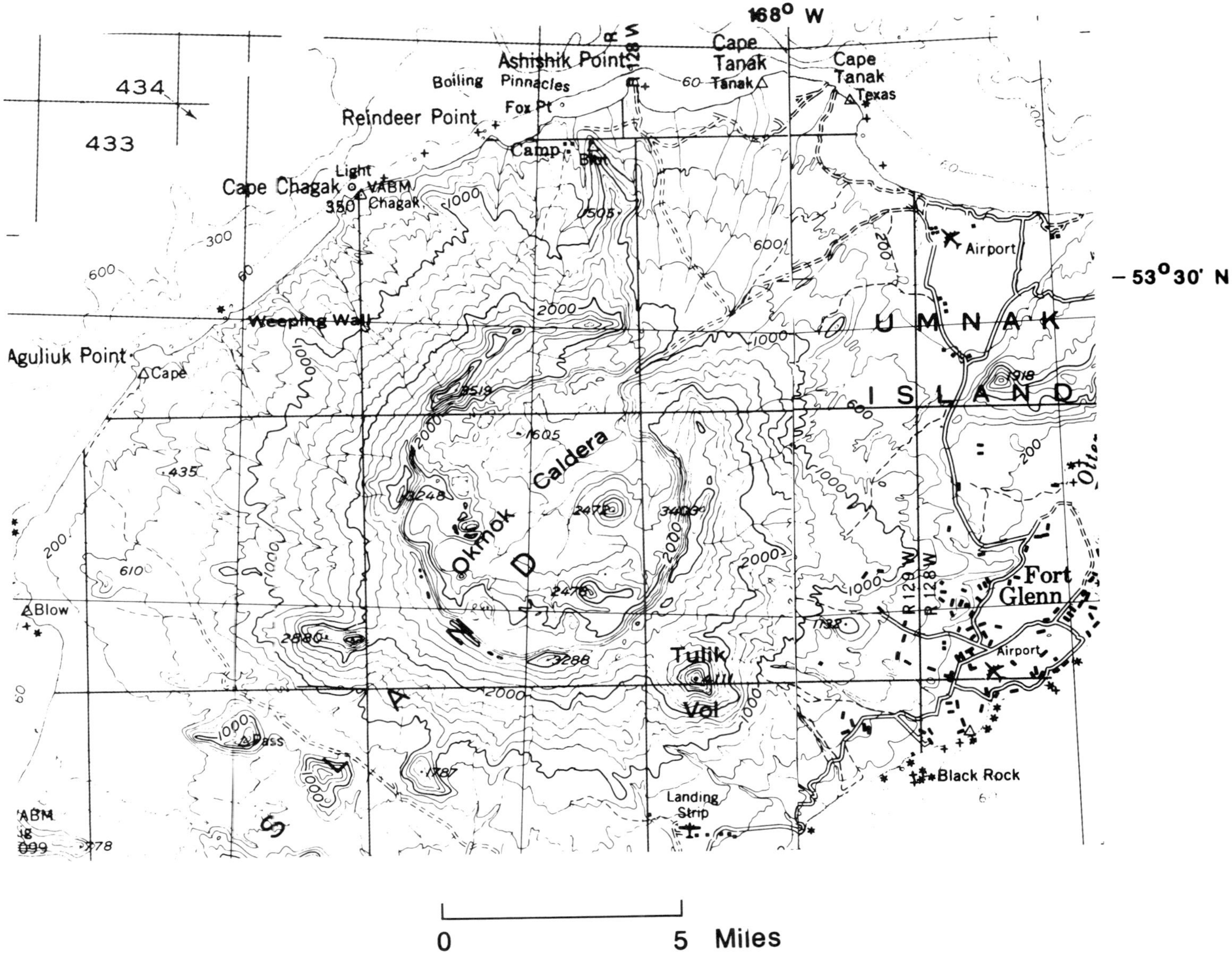

Figure 11.5.1. Topographic map of northeast Umnak Island and Okmok Caldera (U.S. Geological Survey, 1951a,b).

PART 3: HISTORICAL UNREST AT LARGE QUATERNARY CALDERAS (continued)
See inside back cover for explanation and abbreviations

FISHER

CAVW number (active vent)	Latitude Longitude (degrees)	Diameter (km)	Local tectonic setting	Precaldera edifice	SiO_2 content (wt pct)	Age of collapse (yr B.P.)	Date of unrest	Type and duration of unrest ESTU STHF MGTF H Te	Eruption type
11-01-35 (Fisher)	54.67N 164.35W	11 x 18	Compr	Strato	R = mafic? C = 67-68, minor 52-53	9,100	1795?	---- ---- ---- - --	ex
							1826-27	---- ---- ---- - --	Ex?

TECTONIC SETTING

Fisher Caldera is near the west end of Unimak Island, where the Pacific and North American plates converge at a rate of 7-8 cm/yr.

GEOLOGIC HISTORY

Fisher Caldera, one of the largest calderas in the Aleutian arc, formed about 9,100 yr B.P. by collapse of a large andesitic stratovolcano (fig. 11.6.1). Eruption of a large volume of dacitic pumice generated unusually mobile, high-velocity pyroclastic flows that climbed more than 400 m from lowlands to pass over distant ridges (Miller and Smith, 1977, 1987).

HISTORICAL ACTIVITY

Little is known of historical eruptions of Fisher Volcano. Grewingk (1850) wrote that a small volcano on the southwest side of Unimak Island exploded in 1795 and "fell in with a fearful noise." Coats (1950) attributed this to Pogromni and Isanotski, but T. Miller and J. Riehle (written commun., 1985) suggest that this eruption might have been from Fisher. Grewingk (1850) also mentioned eruptions and ashfall on the south end of Unimak Island in 1826. An active fumarole field occurs in the central part of the caldera (T.P. Miller, written commun., 1987).

REFERENCES

Coats, R.R., 1950, Volcanic activity in the Aleutian arc: U.S. Geol. Surv. Bull. 974-B, p. 35-49.

Grewingk, C., 1850, Beitrag zur Kenntniss der orographischen und geognostischen Beschaffenheit der Nordwestküste Amerikas, mit den anliegenden Inseln: Verh. Russisch-Kaiserlichen Mineral. Gesellschaft zu St. Petersburg, 1848-49, p. 123-278.

Katsui, Y., ed., 1971, List of the world active volcanoes: Tokyo, Volcanol. Soc. Japan/IAVCEI/IUGG, 160 p.

Miller, T.P., and Smith, R.L., 1977, Spectacular mobility of ash flows around Aniakchak and Fisher calderas, Alaska: Geology, v. 5, p. 173-176.

----- 1987, Late Quaternary caldera-forming eruptions in the eastern Aleutian arc, Alaska: Geology, v. 15, p. 434-438.

Smith, R.L., Shaw, H.R., Luedke, R.G., and Russell, S.L., 1978, Comprehensive tables giving physical data and thermal energy estimates for young igneous systems of the United States: U.S. Geol. Surv. Open-File Rep. 78-925, 15 p. + tables.

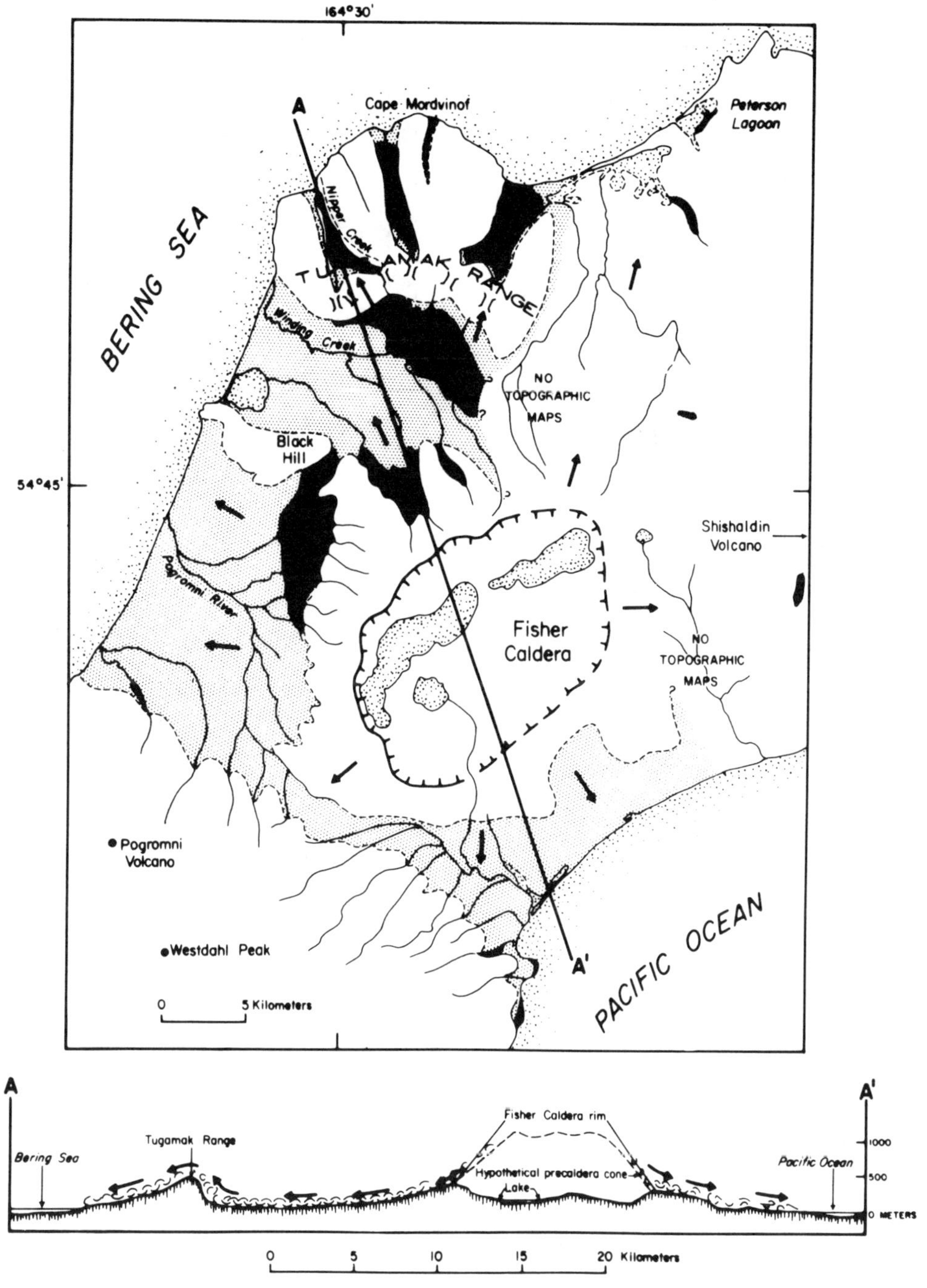

Figure 11.6.1. Fisher Caldera and distribution of ash flows from caldera-forming eruption, from Miller and Smith (1977). Outcrops, solid pattern; inferred original distribution of ash flows, shaded pattern; lakes, stippled pattern (for example, three areas on caldera floor). Arrows denote postulated flow direction of ash flows; brackets denote mountain passes. Cross section along line A-A' illustrates inferred movement of ash flows over Tugamak Range.

PART 3: HISTORICAL UNREST AT LARGE QUATERNARY CALDERAS (continued)
See inside back cover for explanation and abbreviations

EMMONS LAKE

CAVW number (active vent)	Latitude Longitude (degrees)	Diameter (km)	Local tectonic setting	Precaldera edifice	SiO_2 content (wt pct)	Age of collapse (yr B.P.)	Date of unrest	Type and duration of unrest ESTU STHF MGTF H Te	Eruption type
11-02-03	55.35N	19 x 10	Compr	Strato?	R = mafic	>10,000;	1762-86 PS?	---- ---- ---- - --	Ex,ex
(Pavlof)	162.02W				C = 72	possibly	1790	---- ---- ---- - --	unknown
						between	1817	---- ---- ---- - --	ex
11-02-04	55.45N					87,000	ca.1825?	---- ---- ---- - --	unknown
(Pavlof	161.87W					and	1838	---- ---- ---- - --	"smoke"
Sister=PS)						105,000	1844 or 1845	---- ---- ---- - ??	ex
							1846	---- ---- ---- - --	ex,lf
							1852?	---- ---- ---- - --	"smoke"
							1866?	---- ---- ---- - UQ?	ex
							1880?	---- ---- ---- - --	unknown
							1886	---- ---- ---- - --	unknown
							1892	---- ---- ---- - --	ex
							1894	---- ---- ---- - --	ex
							1901	---- ---- ---- - --	ex
							1906-11	YY-- ---x ---- - --	"smoke",Ex
							1914	---- ---- ---- - --	ex
							1917	x--- ---- ---- - --	ex
							1922-24	---- ---- ---- - --	Ex,ex
							1929-31	---- ---- ---- - --	ex
							1936-48	---- ---- ---- - --	ex,Ex,lf?
							1950-53	---- ---- ---- - --	ex
							1958	---- ---- ---- - --	ex,lf
							1960-63	---- ---- ---- - --	ex,lf
							1966	---- ---- ---- - --	ex,lf
							1973-77+	C-Y- ---- ---- - UQ	Ex,ex,lf
							1980-83	D-Y- ---- ---- - UQ	Ex,ex,lf
							1986-88+	C-Y- ---- ---x - --	Ex,ex

EMMONS LAKE (continued)

TECTONIC SETTING

Emmons Lake Caldera, Pavlof Volcano, and Pavlof Sister Volcano lie in a particularly active part of the Aleutian arc, where the Pacific and North American plates converge at a rate of 7.5 cm/yr, and where great earthquakes ($\underline{M}$ 7.8 and greater) occur on average every 60-80 years. The most recent great earthquake in the area was in 1903, and Kelleher (1970), Sykes (1971), and Davies and others (1981) have termed this area the Shumagin seismic gap. Beavan and others (1983) and McNutt (1987) suggested that tectonic compression in the area might increase eruptive activity of Pavlof. McNutt and Beavan (1987) also noted that the larger, magmatic eruptions of Pavlof occur in the fall, including four in the period 11-15 November of different years; they tentatively attributed this fact to a seasonal, nontidal ocean loading on the North Pacific continental shelf.

A strong northeast-southwest alignment of major vents suggests control by a regional fault of the same trend.

GEOLOGIC HISTORY

The outline of the Emmons Lake Caldera is poorly known; we have sketched an approximate outline based on topography and the distribution of young vents (fig. 11.7.1). A geologic map of the caldera is in preparation (T. Miller, written commun., 1988). Kennedy and Waldron (1955) suggested that the Emmons Lake basin might be a caldera, formed "when the lavas of the Dushkin basalt were being erupted." Alternatively, Miller and Smith (1987) suggest that Emmons Lake Caldera may have been the source of the Old Crow tephra, with an estimated age of 87,000-105,000 yr B.P. and an estimated bulk volume >50 km^3. McNutt and Jacob (1986) identified a low-velocity body beneath Emmons Lake Caldera, but their station spacing was not adequate to define its size and geometry precisely. Most of the postcaldera volcanism has been basaltic.

Pavlof Volcano lies approximately 2-3 km northeast of the rim of Emmons Lake Caldera, along a northeast-trending line of vents that also includes several in the caldera and Pavlof Sister (5 km northeast of Pavlof). Asymmetry of Pavlof suggests that its vent is migrating northeastward, away from the caldera. McNutt (1987) suggests that the magma column is continuous from the surface to a depth of 100-170 km, and that the occurrence of microearthquakes below 100 km depth 8 days before to 2 days after eruptions may indicate transmission of pressure through the magma column.

HISTORICAL ACTIVITY

Frequent, intermittent eruptions occur at Pavlof, including more than 30 in the historical record and more than 20 in this century (McNutt, 1987).

EMMONS LAKE (continued)

HISTORICAL ACTIVITY (continued)

1762-86: There were reports of "signs of activity" (Dall, 1870, p. 466) and that the volcano was "active" (Charisso, cited in Perrey, 1865, p. 21). These reports probably refer to activity of Pavlof Sister, which "became active, fell in, and after a violent earthquake [ceased] its activity..." (McNutt, 1987).

1844, 1846, 1854, 1866: An unusually strong earthquake in 1844, felt at Pavlof Harbor and on nearby Kodiak Island, occurred near the time of the 1844 eruption (exact relation not known). In 1846, a 1-day eruption occurred from a fissure on the flank of the volcano. Another, lesser earthquake was felt on 16 January 1854, and an "extraordinarily strong" earthquake occurred on 25 August 1866, five months after an eruption on 14 March 1866 (Doroshin, 1870). Beavan and others (1983) report great earthquakes in the Shumagin gap in 1788, 1847, and possibly 1903.

1911: The most powerful historical eruption of Pavlof occurred on 6-7 December 1911; announced by earthquakes, the north flank opened, large blocks were ejected, and lava flows issued from the fissure (McNutt, 1987).

1973-present: Seismic monitoring in the Pavlof-Shumagin area began in 1973. Eruptions have been preceded by a few hours to a few weeks of emergent "B-type" and impulsive explosion earthquakes (fig. 11.7.2) that, by analogy with signals of known explosions at the surface, are all attributed to near-surface degassing. In 1974, swarms of these very shallow earthquakes occurred several days before eruptions, sometimes decreasing in number just before an eruption. Again in 1981, these events became more frequent but then decreased in number and size in the 3 days immediately before an eruption (Smithsonian Institution, 1981).

A reversal in tilt direction in the Shumagin region occurred during 1977-80. During that period, regional microseismicity was high, Pavlof did not erupt and exhibited very low seismicity, and most teleseismically recorded deep earthquakes occurred in a cluster behind Pavlof (fig. 11.7.4; McNutt, 1987). Magmatic eruptions resumed at Pavlof in November 1980. McNutt and Beavan (1981) noted a tendency for swarms to occur at Pavlof near fortnightly tidal maxima; McNutt and Beavan (1987) showed that eruptions tend to occur preferentially at times of increased sea-level deviation in the fall and winter (fig. 11.7.5).

Low-frequency and explosion earthquakes increased again at Pavlof during 11-18 July 1983--apparently accompanied by small explosions. Then in October and early November (4, 9, and 11 November), up to 40 low-frequency earthquakes per day accompanied bursts of tremor lasting several minutes to half an hour. Small explosions were recorded seismically during 11-13 November. Tremor became continuous between 1700 and 1900 hr on 14 November, reaching its greatest amplitude between midnight and 1200 hr on 15 November (fig. 11.7.3). Glow was observed at about 2330 hr, and by 1220 hr on 14 November an airline pilot reported a plume rising through clouds. After 15 November, tremor declined but continued at a low level through approximately 16 December 1983, accompanied by low frequency (B-type) earthquakes (McNutt, 1986, 1987). Several bursts of tremor from 13 min to 4 hr in duration marked the end(?) of eruptions in December 1983. After an increase in tremor (and eruptions) on 17-21 December, seismicity decreased to a background level of several tens of events per day and remained at that level as of 2 April 1984. The spectral content of tremor at

EMMONS LAKE (continued)

HISTORICAL ACTIVITY (continued)

Pavlof remained within a narrow range (0.8-3.0 Hz) during more than a decade of recording; McNutt (1986, 1988) proposed that it results from resonance of magma and gas in the conduit.

A new period of eruptions began in mid-April 1986, and continues as of the time of this writing (February 1988). Detection of eruptions depends largely on seismic monitoring of tremor and sightings by commercial airline pilots and weather satellites when weather permits.

REFERENCES

Beavan, J., Hauksson, E., McNutt, S., Bilham, R., and Jacob, K., 1983, Tilt and seismicity changes in the Shumagin seismic gap: Science, v. 222, p. 322-325.

Coats, R.R., 1950, Volcanic activity in the Aleutian arc: U.S. Geol. Surv. Bull. 974-B, p. 35-49.

Dall, W.H., 1870, Alaska and its resources: London, Sampson Low, Son, and Marston, 627 p.

Davies, J., Sykes, L., House, L., and Jacob, K., 1981, Shumagin seismic gap, Alaska Peninsula: History of great earthquakes, tectonic setting, and evidence of high seismic potential: Jour. Geophys. Res., v. 86, p. 3821-3855.

Doroshin, P., 1870, Some volcanoes, their eruptions, and earthquakes in the former Russian possessions in America: Verh. Russisch-Kaiserlichen Mineral. Gesellschaft zu St. Petersburg, 2nd ser., 1870, p. 25-44 (English translation by J.B. Kisslinger).

Grewingk, C., 1850, Beitrag zur Kenntniss der orographischen und geognostischen Beschaffenheit der Nordwestküste Amerikas, mit den anliegenden Inseln: Verh. Russisch-Kaiserlichen Mineral. Gesellschaft zu St. Petersburg, 1848-49, p. 123-278.

Katsui, Y., ed., 1971, List of the world active volcanoes: Tokyo, Volcanol. Soc. Japan/IAVCEI/IUGG, 160 p.

Kelleher, H.A., 1970, Space-time seismicity of the Alaska-Aleutian seismic zone: Jour. Geophys. Res., v. 75, p. 5745-5756.

Kennedy, G.C., and Waldron, H.H., 1955, Geology of Pavlof Volcano and vicinity, Alaska: U.S. Geol. Surv. Bull. 1028-A, 18 p.

McNutt, S.R., 1986, Observations and analysis of B-type earthquakes, explosions, and volcanic tremor at Pavlof Volcano, Alaska: Bull. Seismol. Soc. Am., v. 76, p. 153-175.

----- 1987, Eruption characteristics and cycles at Pavlof Volcano, eastern Aleutians, and their relation to regional earthquake activity: Jour. Volcanol. Geotherm. Res., v. 31, p. 239-267.

----- 1988, Volcanic tremor at Pavlof Volcano, Alaska, 1973-1986: Pure Appl. Geophys. [in press].

McNutt, S.R., and Beavan, R.J., 1981, Volcanic earthquakes at Pavlof Volcano correlated with the solid earth tide: Nature, v. 294, p. 615-618.

----- 1987, Eruptions of Pavlof Volcano and their possible modulation by ocean load and tectonic stresses: Jour. Geophys. Res., v. 92, no. B11, p. 11,509-11,523.

McNutt, S.R., and Jacob, K.H., 1986, Determination of large-scale velocity structure of the crust and upper mantle in the vicinity of Pavlof Volcano, Alaska: Jour. Geophys. Res., v. 91, no. B5, p. 5013-5022.

EMMONS LAKE (continued)

REFERENCES (continued)

Miller, T.P., and Smith, R.L., 1983, Calderas of the eastern Aleutian arc [abs.]: Eos, Trans. Am. Geophys. Union, v. 64, p. 877.

----- 1987, Late Quaternary caldera-forming eruptions in the eastern Aleutian arc, Alaska: Geology, v. 15, p. 434-438.

Perrey, A., 1865, Documents sur les tremblements de terre et les phenomenes volcaniques des îles Aleutiennes, de la peninsule d'Aljaska et de la côte NO d'Amerique. Mem. Acad. Sci. Arts Belles-Lettres, Dijon, 131 p.

Simkin, T., Siebert, L., McClelland, L., Bridge, D., Newhall, C., and Latter, J.H., 1981, Volcanoes of the world: Stroudsburg, Pa., Hutchinson Ross Publ. Co., 232 p.

Smith, R.L., Shaw, H.R., Luedke, R.G., and Russell, S.L., 1978, Comprehensive tables giving physical data and thermal energy estimates for young igneous systems of the United States: U.S. Geol. Surv. Open-File Rep. 78-925, 15 p. + tables.

Smithsonian Institution, Scientific Event Alert Network (SEAN), 1981, 1983-87, Pavlof: SEAN Bull., v. 6, no. 9, v. 8, nos. 10-12; v. 9, nos. 1, 3; v. 11, nos. 3-8, 10, 12; v. 12, nos. 1, 3, 5-8 (principal correspondents: T. Miller, E. Yount, and S. McNutt).

Sykes, L.R., 1971, Aftershock zone of great earthquakes, seismicity gaps, and earthquake prediction for Alaska and the Aleutians: Jour. Geophys. Res., v. 76, p. 8021-8041.

Volcanological Society of Japan, 1977, Pavlof: Bull. Volcanic Eruptions, no. 15, p. 41.

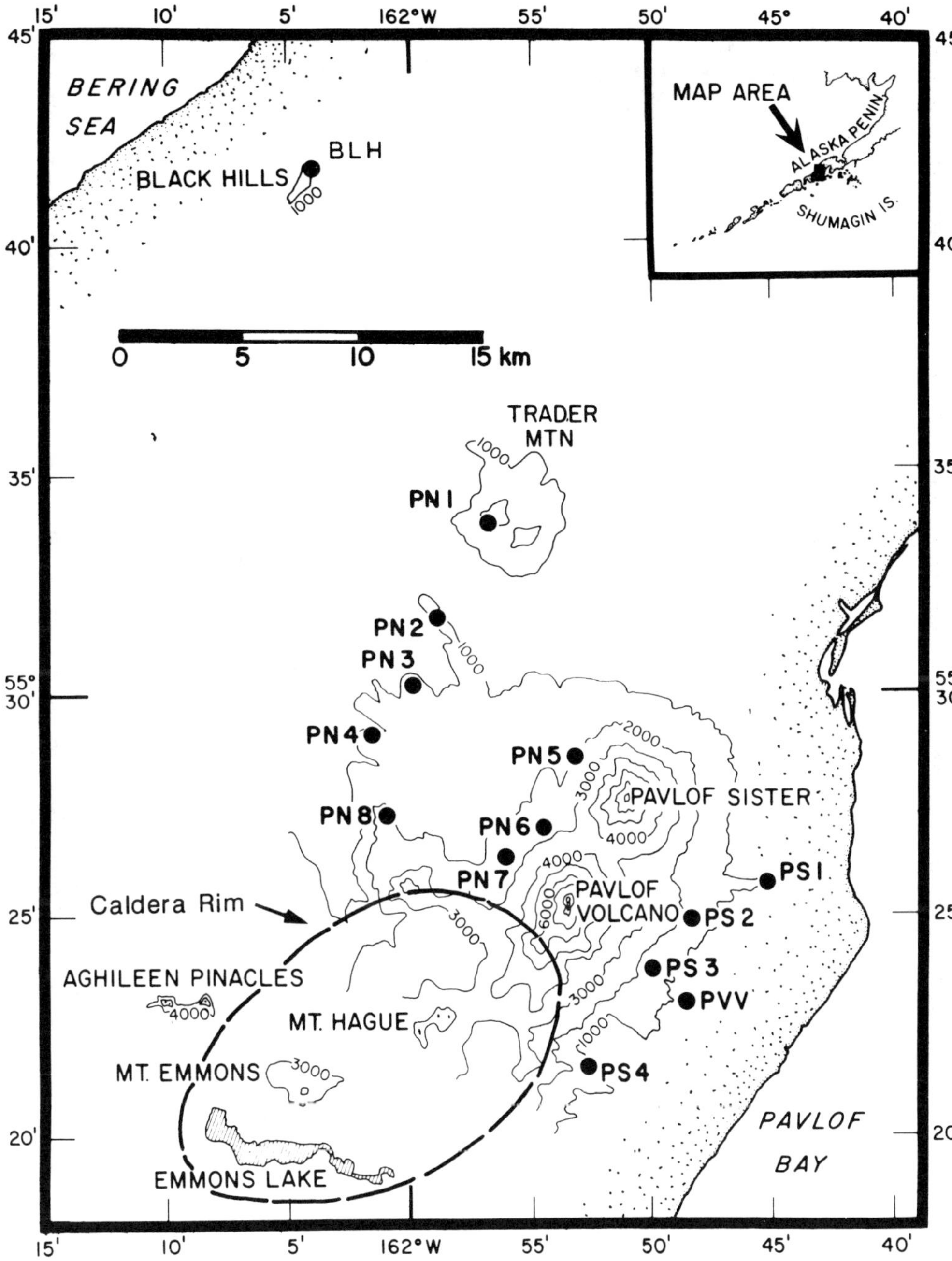

Figure 11.7.1. Sketch map showing locations of Emmons Caldera, Pavlof Volcano, and Pavlof Sister Volcano (topographic rim of Emmons Caldera added to base map of McNutt, 1987). Solid circles represent seismic stations.

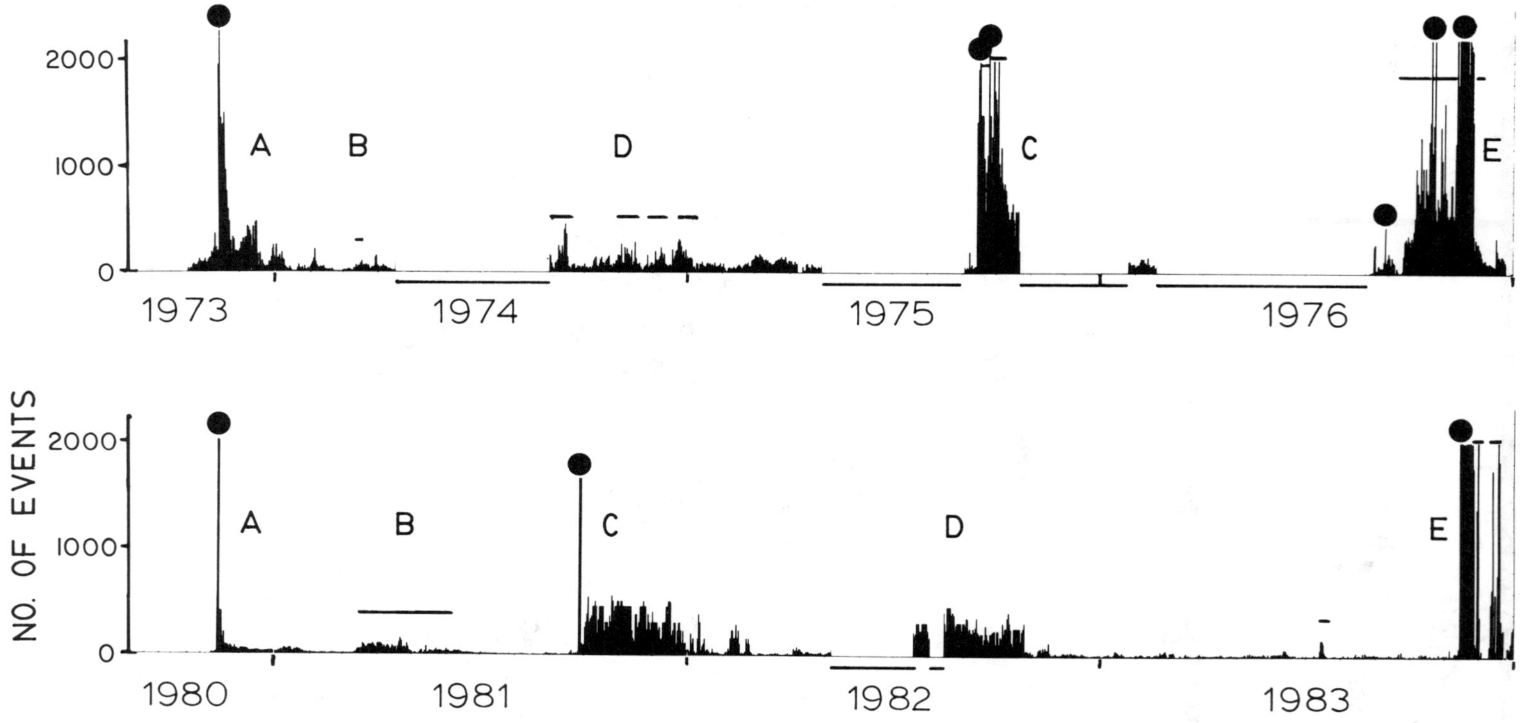

Figure 11.7.2. Number of B-type volcanic earthquakes and explosions per day at Pavlof Volcano during 1973-83, from McNutt (1987). Magmatic eruptions are indicated by solid circles, explosions by horizontal bars above data. Data gaps are indicated by horizontal bars beneath data.

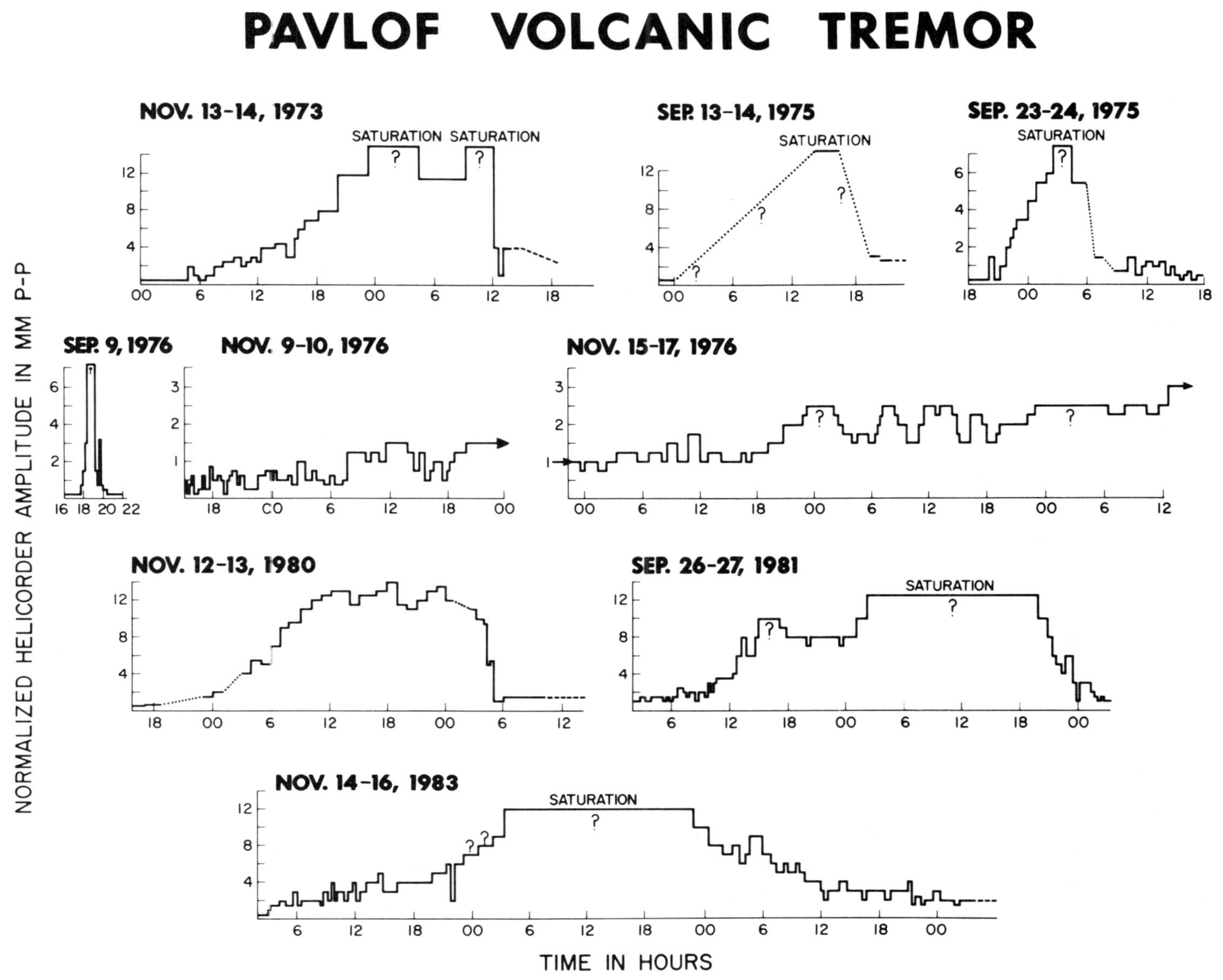

Figure 11.7.3. Tremor amplitude during magmatic eruptions at Pavlof Volcano, from McNutt (1987). Amplitudes normalized; numbers on horizontal axes, times in hours. Dotted where inferred (no seismograms available); dashed where intermittent.

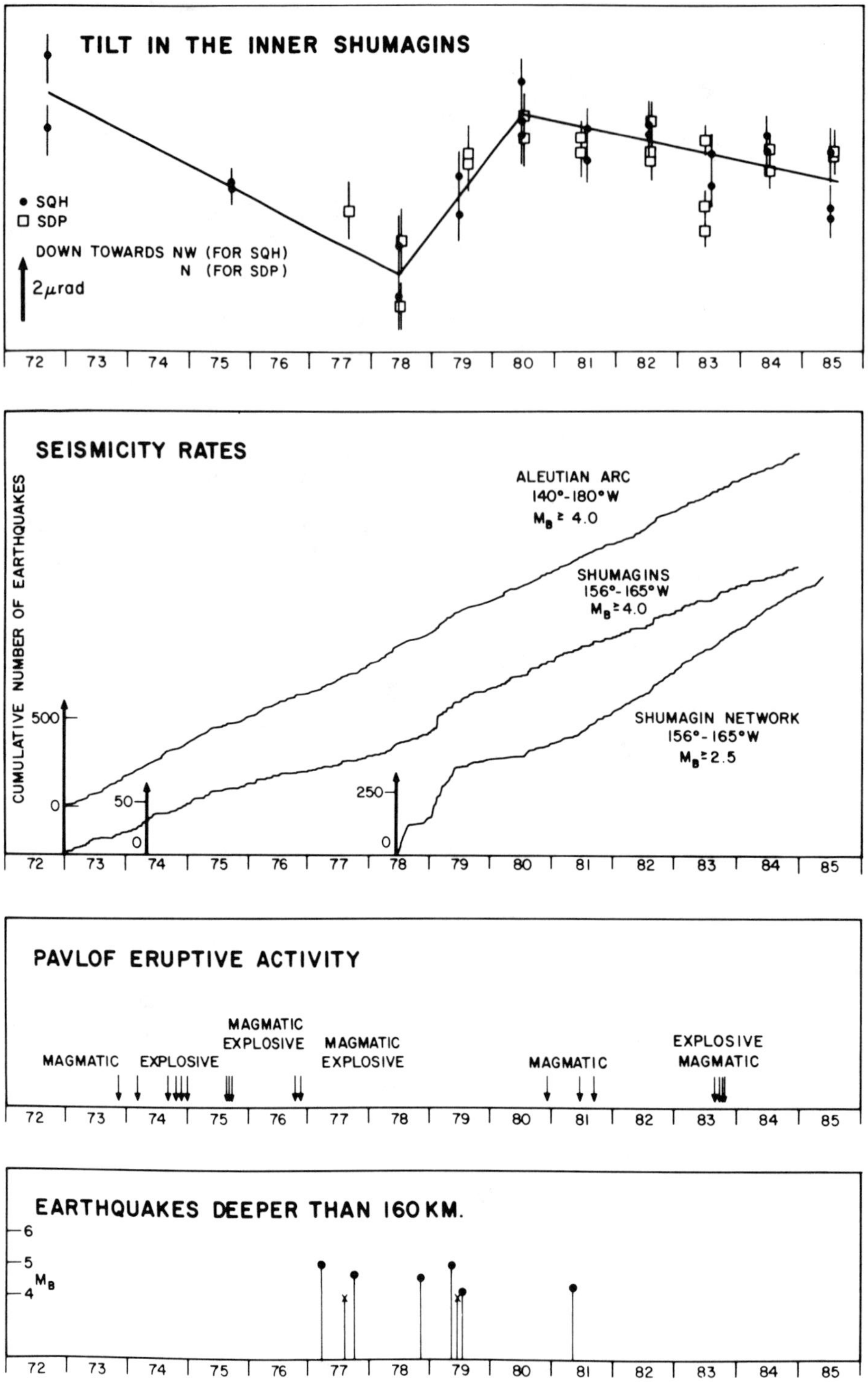

Figure 11.7.4. Tilt, seismicity rates, Pavlof eruptive activity, and deep earthquakes in the Shumagin region for time period 1972-85 (McNutt, 1987).

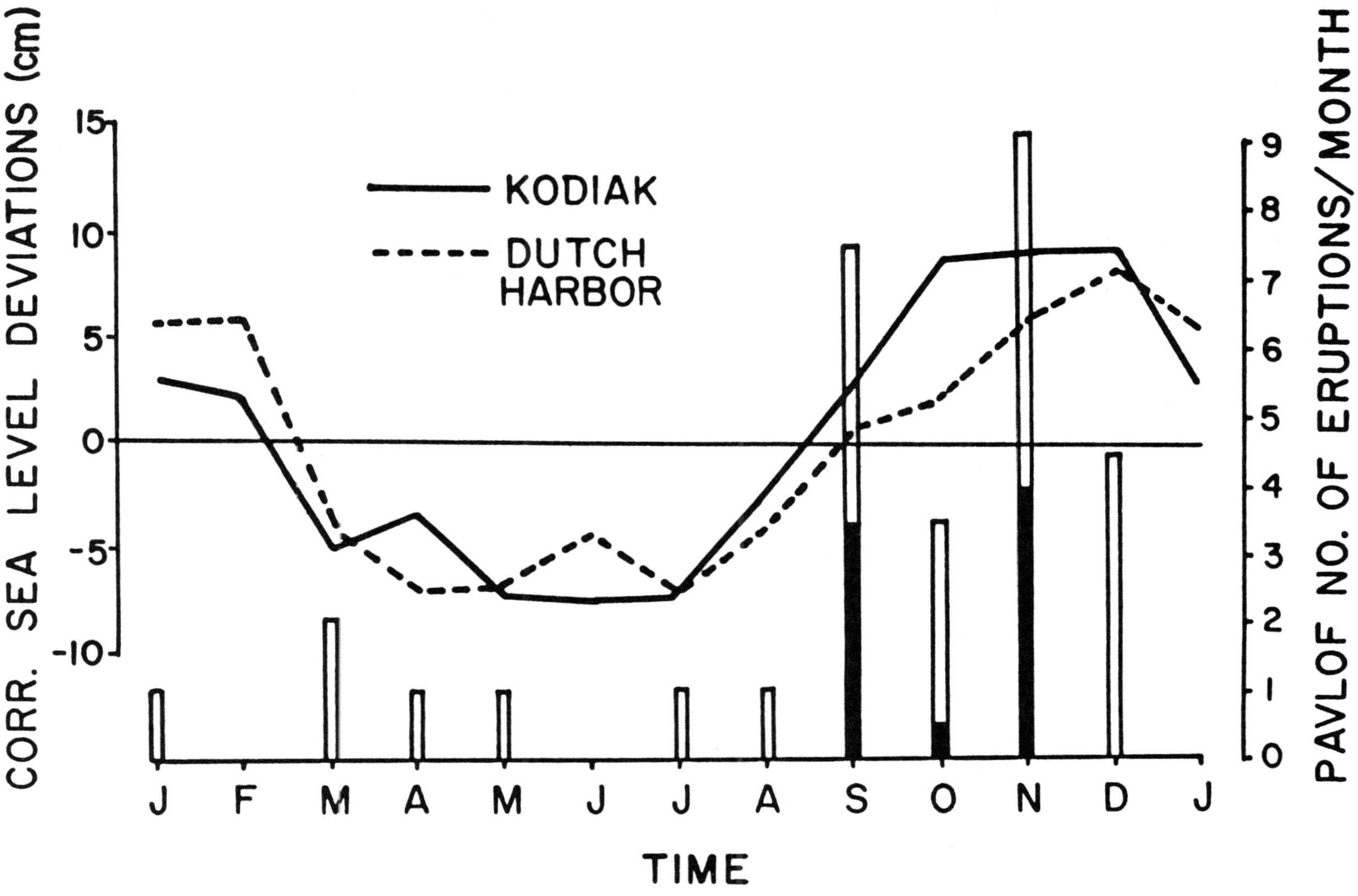

Figure 11.7.5. Deviations between monthly mean sea level and long-term mean sea level (1950-74), and number of eruptions per month at Pavlof Volcano (McNutt and Beavan, 1987). Monthly mean-sea-level data corrected for atmospheric pressure. Kodiak is about 500 km east-northeast of Pavlof; Dutch Harbor is about 350 km west-northwest of Pavlof. Histogram bars are solid for magmatic eruptions and open for explosions. Note correlation between eruptions and increased sea level in fall and winter. Copyright by the American Geophysical Union.

PART 3: HISTORICAL UNREST AT LARGE QUATERNARY CALDERAS (continued)
See inside back cover for explanation and abbreviations

VENIAMINOF

CAVW number (active vent)	Latitude Longitude (degrees)	Diameter (km)	Local tectonic setting	Precaldera edifice	SiO_2 content (wt pct)	Age of collapse (yr B.P.)	Date of unrest	Type and duration of unrest ESTU STHF MGTF H Te	Eruption type
11-02-07 (Veniaminof)	56.17N 159.38W	10	Compr	Strato	R = 53-54 C = 65	3,700	1830-39	---- ---- ---- - --	Ex?
							1852?	---- ---- ---? - --	"smoke"
							1874?	---- ---- ---? - --	"smoke"
							1892	---- ---- ---- - --	Ex
							1930?	---- ---- ---? - --	"smoke"
							1939	---- ---- ---- - --	ex
							1944	---- ---- ---- - --	unknown
							1956	---- ---- ---- - --	ex
							1983-84	Y-Y- ---Y ---Y - --	ex,lf
							1987	---- ---- ---- - --	ex

TECTONIC SETTING

Veniaminof Caldera is on the Alaska Peninsula, near the northeastern end of the Shumagin seismic gap (Kelleher, 1970; Sykes, 1971; Davies and others, 1981). McNutt (1987) suggests that activity of Veniaminof might be controlled in part by buildup of regional strain, as discussed below and under the Emmons Lake Caldera section.

GEOLOGIC HISTORY

The Veniaminof Caldera (figs. 11.8.1, 11.8.2) formed in the summit of a large (more than 400 km^3) basaltic-andesite stratovolcano approximately 3,700 yr B.P. (Miller and Smith, 1975, 1987). Ash flows of the caldera-forming eruption range from dacite (early) to andesite (late) (Miller and Smith, 1987). The active vent is one of several in a NW-SE-trending belt that bisects the caldera (Detterman and others, 1981; B. Yount, written commun., 1987).

HISTORICAL ACTIVITY

Early eruptions are known from reports by Grewingk (1850), Doroshin (1870), and Becker (1898). On 4 August 1838 an explosion occurred with a cracking sound and loud rumble; ash blew northeast as far as Katmai (Doroshin, 1870). The largest historical eruption was in 1892, when the skies of southwest Alaska were darkened (Becker, 1898). McNutt (1987) suggested that Veniaminof was active for a few years before and after the 1938 Shumagin earthquake, but not for 30 years before or after that period, until 1983;

VENIAMINOF (continued)

HISTORICAL ACTIVITY (continued)

therefore, the activity of Veniaminof since 1983 might foreshadow another large earthquake in this region, as might vigorous activity of Pavlof Volcano (McNutt, 1987). However, a 1944 eruption reported by Coats (1950) and an eruption in 1956 recently documented by Yount and others (1985) indicate that Veniaminof was not entirely quiet during this interval.

Before an eruption began in June 1983, the nearest seismometer was 30 km from the volcano; available reports describe seismicity only after the start of the eruption. Since June 1983, some earthquakes were felt at Perryville (25 km away) during and between eruptions. Tremor amplitude (and hence vigor of fountaining?) in early October 1983 was about half that during the June 1983 eruption. The eruption has included mild explosions, lava fountaining, lava flow(s), and growth of an intracaldera cone. Approximately 45 x 10^6 m^3 of lava was extruded from June 1983 to February 1984. Seismic activity declined in late March and early April 1984 but increased slightly in mid-April.

During field work in August 1984, occasional white vapor plumes and roaring noises lasting about 15 minutes alternated with 20-minute periods of quiet. No general increase in steaming was noted, but visibility was poor. Small earthquakes were felt during both the roaring and the quiet periods (B. Yount and T. Miller in Smithsonian Institution, 1984). A Lamont-Doherty Geological Observatory seismic station 30 km from the volcano recorded 3 events (low-frequency volcanic earthquakes and tremor) on 25 November; no events were large enough to trigger other stations, so magnitudes were less than 2.5 or 3.

Residents of Perryville were awakened by rumbling from the volcano at 0400 hr on 29 November 1984, and by 0800 hr a black ash cloud was rising from the vent. The column rose to a maximum height of 5.4 km; no incandescence was reported. Seismicity during this episode was minor compared with that at the onset of activity in June 1983.

A minor ash emission was noted on 19 March 1987 (Smithsonian Institution, 1987).

REFERENCES

Becker, G.F., 1898, Reconnaissance of the gold fields of southern Alaska, with some notes on general geology: U.S. Geol. Surv., 18th Ann. Rep., pt. 3, p. 13-24.

Coats, R.R., 1950, Volcanic activity in the Aleutian arc: U.S. Geol. Surv. Bull. 974-B, p. 35-49.

Davies, J., Sykes, L., House, L., and Jacob, K., 1981, Shumagin seismic gap, Alaska Peninsula: History of great earthquakes, tectonic setting, and evidence of high seismic potential: Jour. Geophys. Res., v. 86, p. 3821-3855.

Detterman, R.L., Miller, T.P., Yount, E., and Wilson, F., 1981, Quaternary geologic map of the Chignik and Sutwik Island quadrangles, Alaska: U.S. Geol. Surv. Misc. Invest. Ser. I-1292, 1 sheet, scale 1:250,000.

Doroshin, P., 1870, Some volcanoes, their eruptions, and earthquakes in the former Russian possessions in America: Verh. Russisch-Kaiserlichen Mineral. Gesellschaft zu St. Petersburg, 2nd ser., 1870, p. 25-44 (English translation by J.B. Kisslinger).

VENIAMINOF (continued)

REFERENCES (continued)

Grewingk, C., 1850, Beitrag zur Kenntniss der orographischen und geognostischen Beschaffenheit der Nordwestküste Amerikas, mit den anliegenden Inseln: Verh. Russisch-Kaiserlichen Mineral. Gesellschaft zu St. Petersburg, 1848-49, p. 123-278.

Kelleher, H.A., 1970, Space-time seismicity of the Alaska-Aleutian seismic zone: Jour. Geophys. Res., v. 75, p. 5745-5756.

Knappen, R.S., 1926, Geology and mineral resources of the Aniakchak district: U.S. Geol. Surv. Bull. 797, p. 161-223.

McNutt, S.R., 1987, Eruption characteristics and cycles at Pavlof Volcano, eastern Aleutians, and their relation to regional earthquake activity: Jour. Volcanol. Geotherm. Res., v. 31, p. 239-267.

Miller, T.P., and Smith, R.L., 1975, Ash flows on the Alaska Peninsula: A preliminary report on their distribution, composition, and age [abs.]: Geol. Soc. Am. Abstracts with Programs, v. 7, no. 7, p. 1201.

----- 1987, Late Quaternary caldera-forming eruptions in the eastern Aleutian arc, Alaska: Geology, v. 15, p. 434-438.

Smith, R.L., Shaw, H.R., Luedke, R.G., and Russell, S.L., 1978, Comprehensive tables giving physical data and thermal energy estimates for young igneous systems of the United States: U.S. Geol. Surv. Open-File Rep. 78-925, 15 p. + tables.

Smithsonian Institution, Scientific Event Alert Network (SEAN), 1983, 1984, 1987, Veniaminof: SEAN Bull., v. 8, nos. 5-12; v. 9, nos. 1-4, 8; v. 12, no. 3 (contributions from B. Yount, T. Miller, J. Taber, K. Hudnut, and J. Reeder).

Sykes, L.R., 1971, Aftershock zone of great earthquakes, seismicity gaps, and earthquake prediction for Alaska and the Aleutians: Jour. Geophys. Res., v. 76, p. 8021-8041.

U.S. Geological Survey, 1963, limited revisions 1983, Chignik topographic map, Alaska topographic series, scale 1:250,000.

Yount, E., and Miller, T.P., 1987, Hazard implications of the 1983-84 eruption of Mount Veniaminof, Alaska [abs.]: Alaska Geol. Soc. 1987 Geol. Hazards Symposium.

Yount, M.E., Miller, T.P., Emanuel, R.P., and Wilson, F.H., 1985, Eruption in an ice-filled caldera, Mount Veniaminof, Alaska Peninsula, in Bartsch-Winkler, S., and Reed, K.M., eds., The United States Geological Survey in Alaska: accomplishments during 1983: U.S. Geol. Surv. Circular 945, p. 59-60.

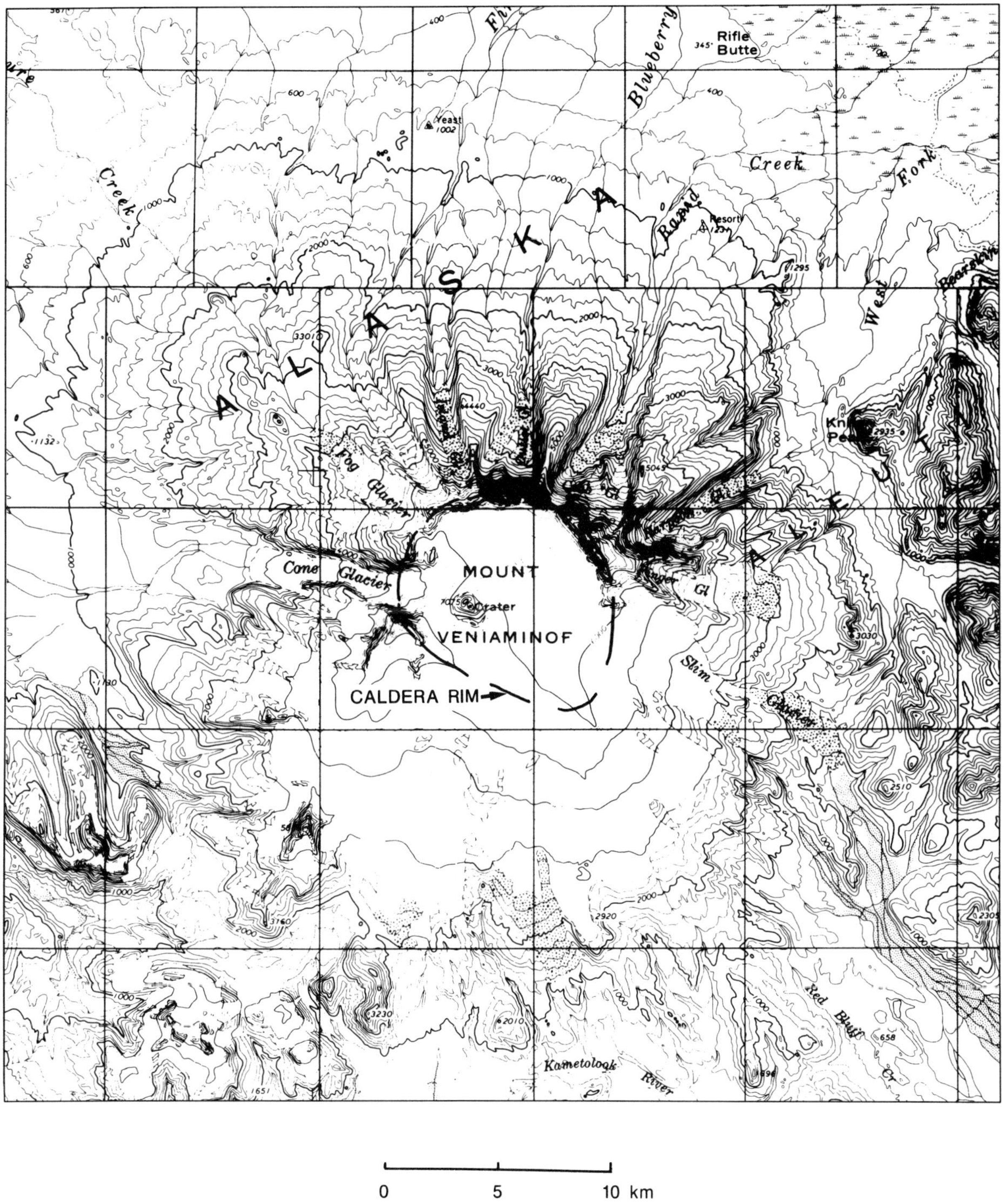

Figure 11.8.1. Topographic map showing Veniaminof Volcano and its summit caldera (U.S. Geological Survey, 1963).

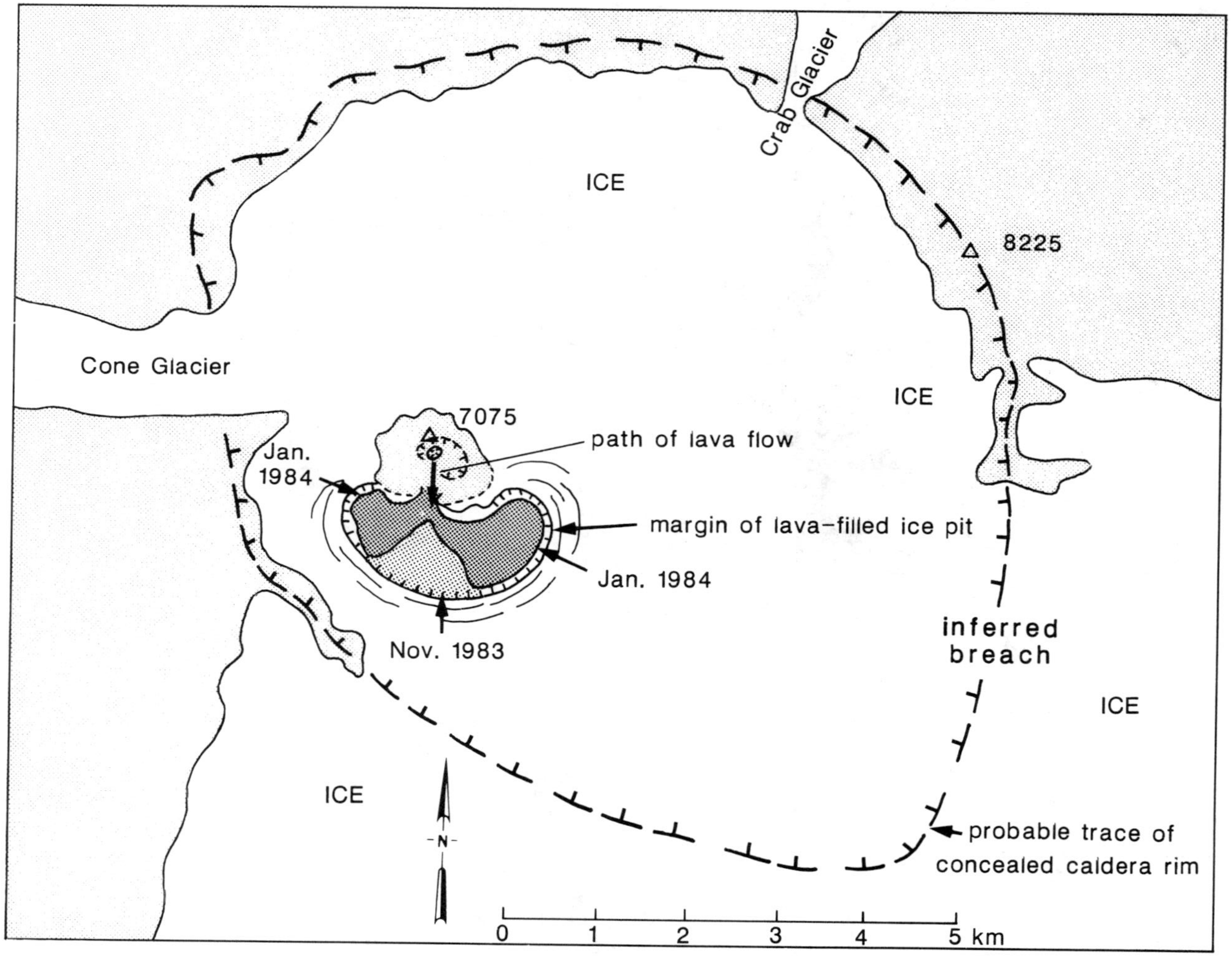

Figure 11.8.2. Sketch map of Veniaminof Caldera by Elizabeth Yount, showing eruptive activity on 15 June 1983 (Smithsonian Institution, 1983). Inferred breach in east caldera rim is after Yount and others (1985).

ANIAKCHAK

CAVW number (active vent)	Latitude Longitude (degrees)	Diameter (km)	Local tectonic setting	Precaldera edifice	SiO_2 content (wt pct)	Age of collapse (yr B.P.)	Date of unrest	Type and duration of unrest ESTU STHF MGTF H Te	Eruption type
11-02-09 (Aniakchak)	56.88N 158.15W	9.5	Compr	Strato	R = 64-66 C = 68, late andesite	3,400	1931	---- ---- ---- - --	Ex,dm

TECTONIC SETTING

Aniakchak Caldera lies along the eastern Aleutian arc, above a boundary between two subducted lithospheric segments (fig. 11.9.1) (Kienle and others, 1983). The caldera is ice free, about 10 km in diameter, and more than a kilometer deep.

GEOLOGIC HISTORY

Aniakchak Caldera (figs. 11.9.2, 11.9.3) formed in the summit of a glacially dissected andesitic stratovolcano that was built on the flank of a range of Jurassic-Tertiary basement. Ash-flow tuffs from the caldera-forming eruption extend 80 km from the caldera rim onto the Bering Sea lowlands. As at Fisher Caldera, these ash flows were extremely mobile; near Aniakchak they crossed over mountain passes that are 260 m higher than terrain closer to the volcano (Miller and Smith, 1977, 1987). The weighted mean average of three charcoal samples from the ash-flow tuff is 3,430±10 yr B.P. (Miller and Smith, 1987).

An older ash-flow tuff in isolated localities on the flanks of Aniakchak suggests the occurrence of an earlier violent eruption of unknown volume. The unit is postglacial; stratigraphic relationships constrain its age between 4,400 and 10,000 yr B.P. (Miller and Smith, 1987). No associated caldera has been identified.

Seven postcaldera cones and necks occur on the caldera floor; the largest of these (Vent Mountain) rises 670 m above the caldera floor.

HISTORICAL ACTIVITY

A series of violent explosions occurred during 1-11 May 1931 and again on 20 May; the source was apparently a cinder cone near the base of the northwest caldera wall (Jaggar, 1932). A dacitic dome was emplaced in the vent in the waning stages of the eruption.

ANIAKCHAK (continued)

REFERENCES

Coats, R.R., 1950, Volcanic activity in the Aleutian arc: U.S. Geol. Surv. Bull. 974-B, p. 35-49.

Hubbard, B.R., 1931, World inside a mountain: Aniakchak, the new volcanic wonderland of the Alaska Peninsula, is explored: Natl. Geogr., v. 60, p. 319-345.

Jaggar, T.A., 1932, Aleutian eruptions, 1930-1932: Volcano Lett., no. 375, p. 1-3.

Katsui, Y., ed., 1971, List of the world active volcanoes: Tokyo, Volcanological Soc. Japan/IAVCEI/IUGG, 160 p.

Kienle, J., Swanson, S.E., and Pulpan, H., 1983, Magmatism and subduction in the Eastern Aleutian arc, in Shimozuru, D., and Yokoyama, I., eds., Arc volcanism: physics and tectonics: Tokyo, Terra Sci. Publ., p. 191-224.

Knappen, R.S., 1926, Geology and mineral resources of the Aniakchak district: U.S. Geol. Surv. Bull. 797, p. 161-223.

Miller, T.P., and Smith, R.L., 1977, Spectacular mobility of ash flows around Aniakchak and Fisher calderas, Alaska: Geology, v. 5, p. 173-176.

----- 1987, Late Quaternary caldera-forming eruptions in the eastern Aleutian arc, Alaska: Geology, v. 15, p. 434-438.

Smith, R.L., Shaw, H.R., Luedke, R.G., and Russell, S.L., 1978, Comprehensive tables giving physical data and thermal energy estimates for young igneous systems of the United States: U.S. Geol. Surv. Open-File Rep. 78-925, 15 p. + tables.

Smith, W.R., 1925, Aniakchak Crater, Alaska Peninsula: U.S. Geol. Surv. Prof. Paper 132-J, p. 139-149.

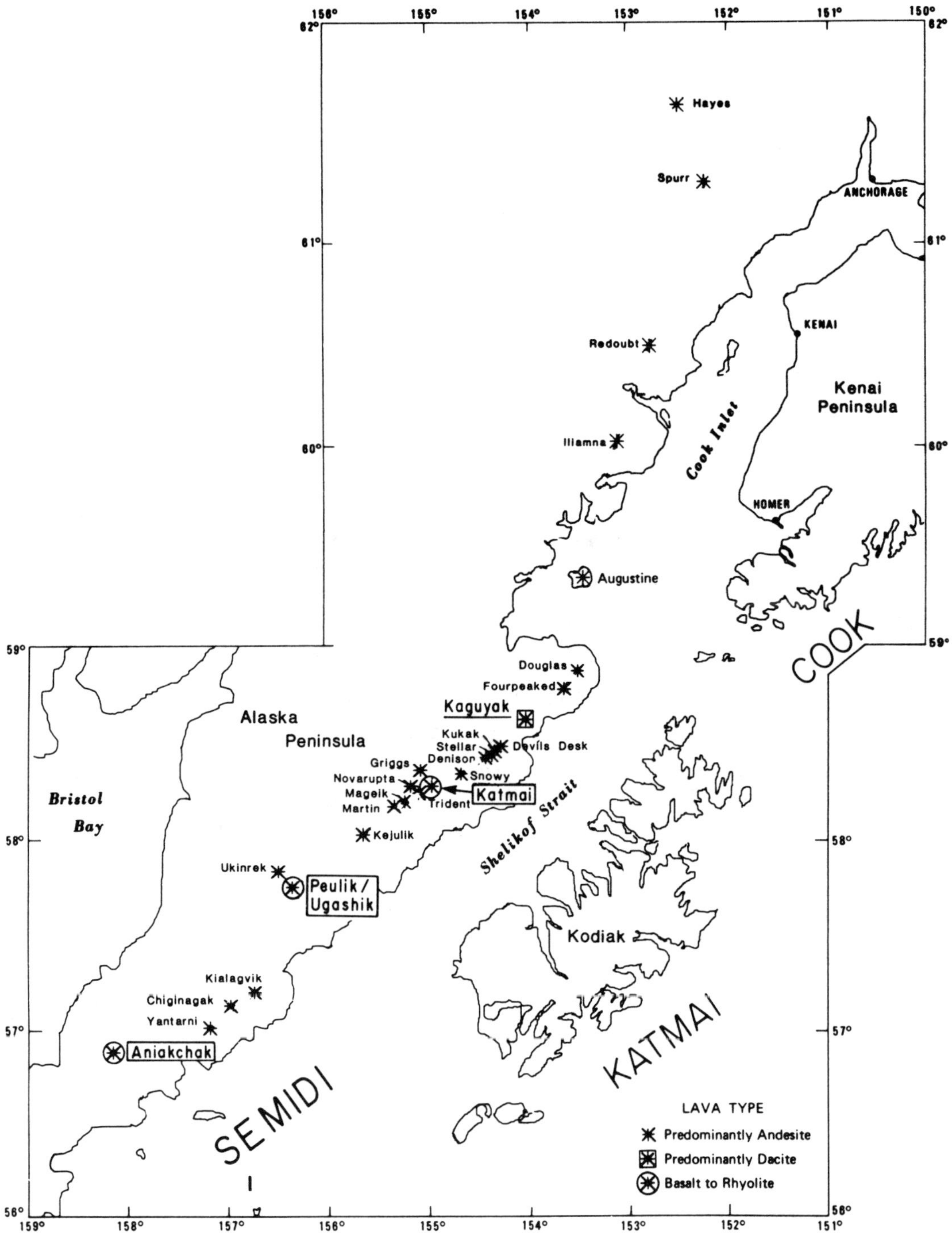

Figure 11.9.1. Volcanoes of eastern Aleutian arc, including Aniakchak, Ugashik, and Katmai Calderas (modified from Kienle and others, 1983).

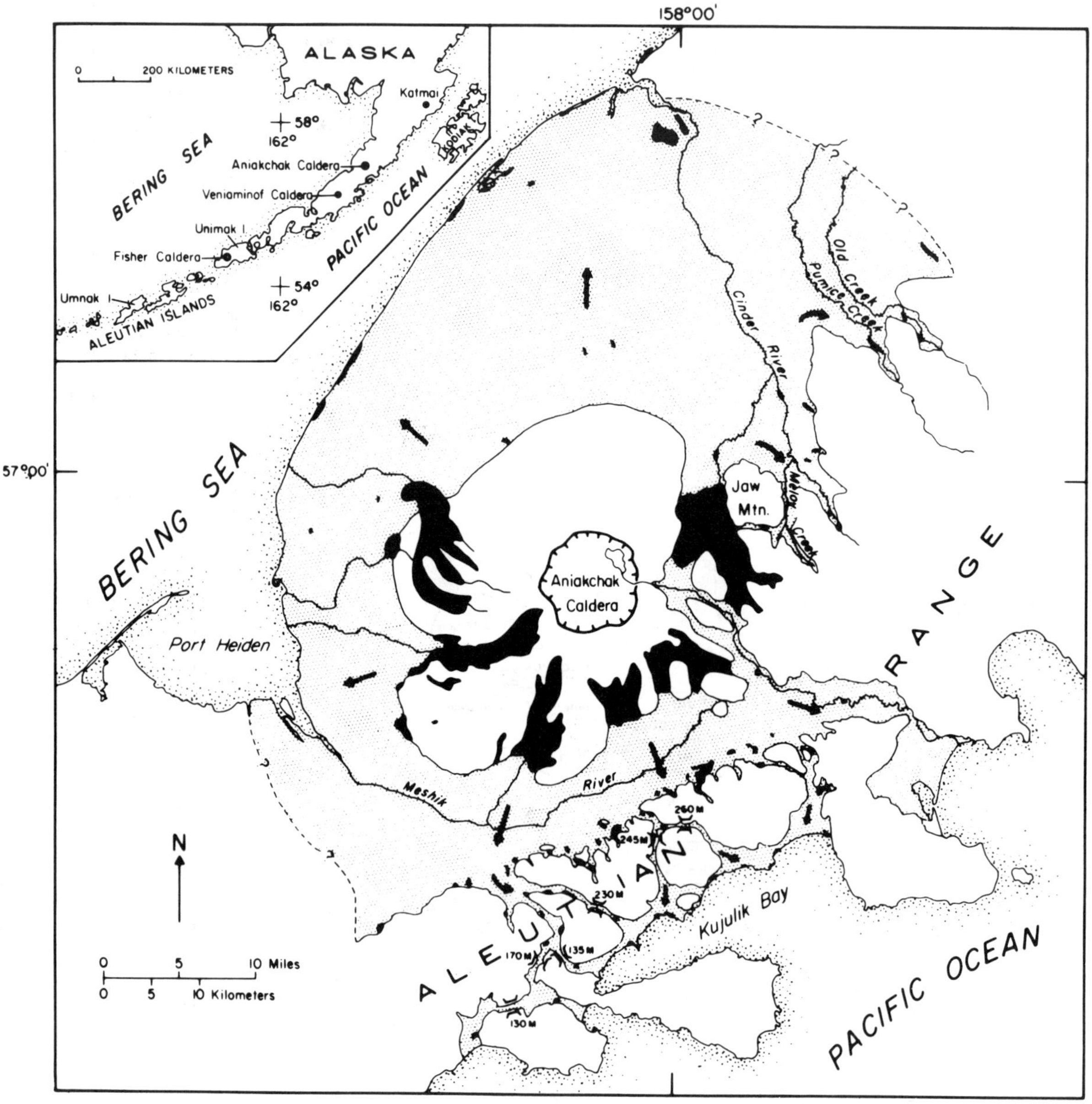

Figure 11.9.2. Aniakchak Caldera and distribution of surrounding ash flows, from Miller and Smith (1977). Outcrops shown in solid pattern; inferred original distribution of ash flows shown in shaded pattern. Arrows denote postulated flow direction of ash flows; brackets and numbers denote mountain passes and their altitudes in meters, respectively.

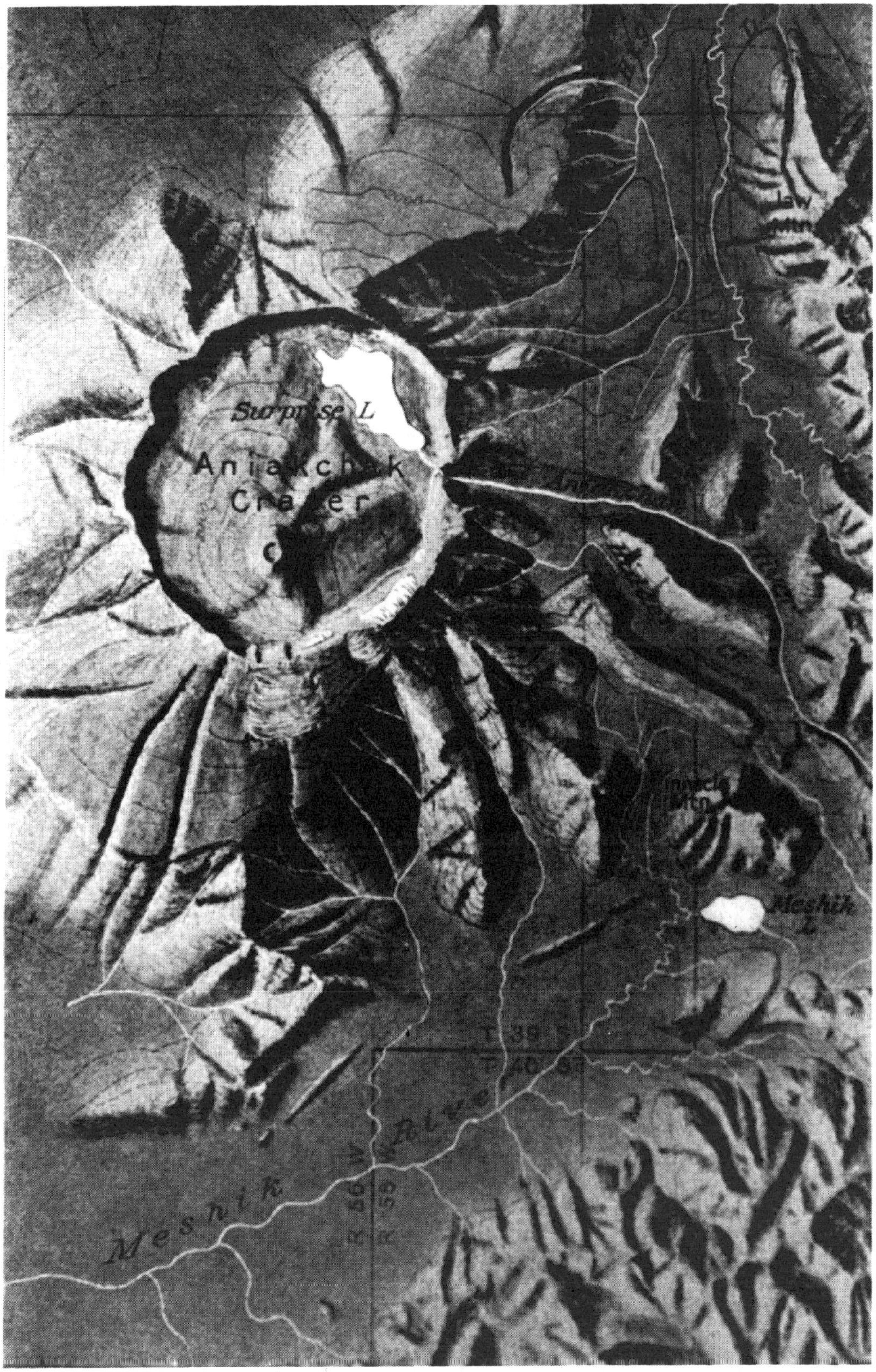

Figure 11.9.3. Relief map of Aniakchak Crater and vicinity, Alaska Peninsula, from Smith (1925).

PART 3: HISTORICAL UNREST AT LARGE QUATERNARY CALDERAS (continued)
See inside back cover for explanation and abbreviations

UGASHIK

CAVW number (active vent)	Latitude Longitude (degrees)	Diameter (km)	Local tectonic setting	Precaldera edifice	SiO_2 content (wt pct)	Age of collapse (yr B.P.)	Date of unrest	Type and duration of unrest ESTU STHF MGTF H Te	Eruption type
11-02-13A (Peulik=P)	57.75N 156.35W	5 x 6	Compr	Strato	R = b	40,000-171,000	1814 (P)	---- ---- ---- - --	Ex
							1852? (P)	---- ---- ---? - --	unknown
							1977 (U)	xx-- ---- ---- x --	Ex
11-02-13B (Ukinrek Maars =U)	57.83N 158.51W								

TECTONIC SETTING

Ugashik Caldera is located south of Lake Becharof on the Alaska Peninsula, on the boundary between the Semidi and Katmai segments of the Aleutian arc (fig. 11.10.1). This position astride a deep-seated cross-arc fault is thought to be favorable for formation of the caldera and back-arc maars (Kienle and others, 1983). The Ukinrek Maars lie at the intersection of this segment boundary with the NE-SW-trending Bruin Bay fault, a major regional reverse fault that dips steeply northwest with that side upthrown (Kienle and others, 1980).

GEOLOGIC HISTORY

A precaldera cone has been mostly removed by erosion, and the north rim of Ugashik Caldera has been breached by Peulik Volcano, a postglacial andesitic stratocone. Postcaldera nonglaciated silicic domes occupy the interior of the caldera (Miller and Smith, 1987). The caldera-forming eruption is tentatively assigned a late Pleistocene but pre-Holocene age, based on inferred relationships to deposits dated at about 40,000 yr B.P. and 171,000 yr B.P. (Miller and Smith, 1987).

The Ukinrek Maars (formed in 1977, 13 km north of Peulik) erupted alkali olivine basalt and are thought not to be connected to the Peulik magma plumbing system (Kienle and others, 1980).

HISTORICAL ACTIVITY

Little is known about historical eruptions of Peulik. Around 1814, "the summit of Peulik collapsed with a rumble, covering the base with enormous boulders. For about a week after this event, vapor rose from almost the entire surface of the mountain" (Doroshin, 1870).

UGASHIK (continued)

HISTORICAL ACTIVITY (continued)

In 1977, a pulse of alkali olivine basalt magma rose into water-saturated till and silicic volcanics 13 km north of Peulik. Ten days of phreatomagmatic eruptions ensued; these eruptions formed the Ukinrek Maars and spread ash north and east of the maars for at least 160 km (Kienle and others, 1980; Self and others, 1980).

COMMENTS

Ugashik Caldera is included in this compilation on the basis of Peulik's activity. The Ukinrek Maars are noted here with the Ugashik Caldera system because basaltic magma responsible for maar formation, even if not directly connected with the Ugashik magma feeder system, may have risen along a segment-boundary transverse fault connecting the two localities.

REFERENCES

Coats, R.R., 1950, Volcanic activity in the Aleutian arc: U.S. Geol. Surv. Bull. 974-B, p. 35-49.

Doroshin, P., 1870, Some volcanoes, their eruptions, and earthquakes in the former Russian possessions in America: Verh. Russisch-Kaiserlichen Mineral. Gesellschaft zu St. Petersburg, 2nd ser., 1870, p. 25-44 (English translation by J.B. Kisslinger).

Kienle, J., Kyle, P.R., Self, S., Motyka, R.J., and Lorenz, V., 1980, Ukinrek Maars, Alaska, I. April 1977 eruption sequence, petrology and tectonic setting: Jour. Volcanol. Geotherm. Res., v. 7, p. 11-37.

Kienle, J., Swanson, S.E., and Pulpan, H., 1983, Magmatism and subduction in the Eastern Aleutian arc, <u>in</u> Shimozuru, D., and Yokoyama, I., eds., Arc volcanism: physics and tectonics: Tokyo, Terra Sci. Publ., p. 191-224.

Miller, T.P., and Smith, R.L., 1987, Late Quaternary caldera-forming eruptions in the eastern Aleutian arc, Alaska: Geology, v. 15, p. 434-438.

Self, S., Kienle, J., and Huot, J.-P., 1980, Ukinrek Maars, Alaska, II. Deposits and formation of the 1977 craters: Jour. Volcanol. Geotherm. Res., v. 7, p. 39-65.

U.S. Geological Survey, 1963, limited revisions 1983, Ugashik topographic map, Alaska topographic series, scale 1:250,000.

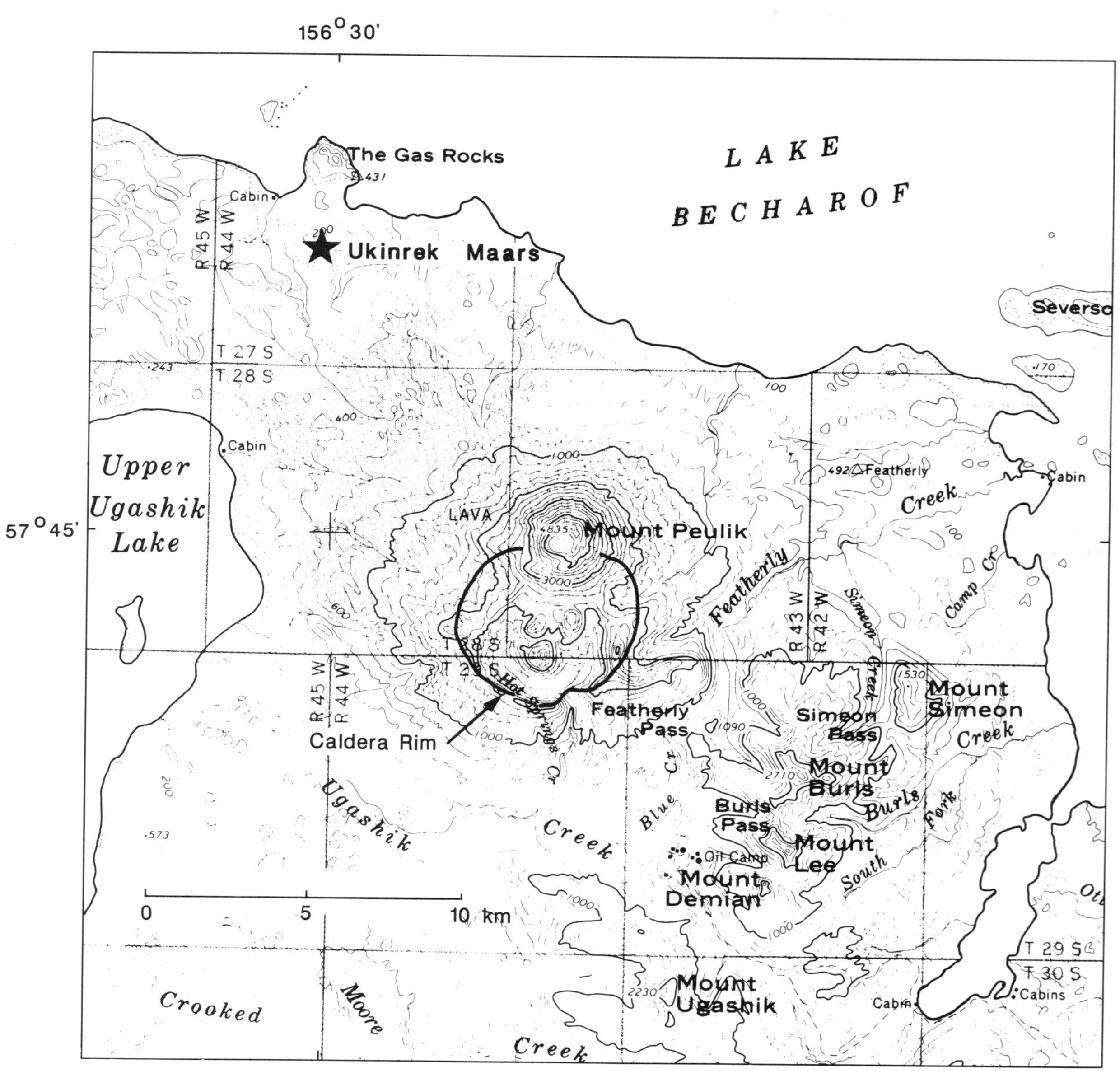

Figure 11.10.1. Topographic map showing Peulik Volcano, inferred outline of Ugashik Caldera, and approximate location of the Ukinrek Maars (star) (U.S. Geological Survey, 1963).

KATMAI/NOVARUPTA

CAVW number (active vent)	Latitude Longitude (degrees)	Diameter (km)	Local tectonic setting	Precaldera edifice	SiO_2 content (wt pct)	Age of collapse (yr B.P.)	Date of unrest	Type and duration of unrest ESTU STHF MGTF H Te	Eruption type
11-02-16 (Novarupta)	58.27N 155.16W	2 x 3	Compr	None?	R = C= r-a	A.D. 1912	1912	C--- Z--Z ---Z - --	EX,dm
11-02-19 (Katmai)	55.27N 154.98W	3 x 4		Strato	R = none C = d	A.D. 1912	1912 1929-31?	C--- Z--- ---Z - -- ---- ---- ---x - --	Pex none?
11-02-17 (Griggs)	58.35N 155.12W				R = ?		1965?	---- ---- ---x - --	none
11-02-18 (Trident)	58.23N 155.12W				R = a		1912-13? 1949-68 ca. 1974-75	---- ---- ---- - -- xx-- ---- ---F - -- ---- ---- ---- - --	ex? ex,Ex,lf Ex,lf
11-02-14 (Martin)	58.15N 155.38W				R = 63-64		1951? 1953?	---- ---- ---- - -- ---- ---- ---- - --	ex? ex?
11-02-15 (Mageik)	58.20N 155.25W						1912 1927? 1929? 1936? 1946?	---- ---- ---- - -- ---- ---- ---- - -- ---- ---- ---- - -- ---- ---- ---- - -- ---- ---- ---? - --	landsl ex? ex? Ex? none?

TECTONIC SETTING

Katmai and other volcanoes of the Aleutian arc are related to subduction of the Pacific plate beneath the North American plate, at a rate of approximately 6-7 cm/yr (Kienle and Swanson, 1983; Kienle and others, 1983). Both the Pacific plate and the overlying North American plate are segmented, and volcanoes above two of three main segment breaks (Ugashik and Aniakchak) are major silicic centers; Katmai and Kaguyak lie within a segment, possibly atop more subtle discontinuities in the arc (Kienle and others, 1983).

KATMAI/NOVARUPTA (continued)

GEOLOGIC HISTORY

The Katmai group of volcanoes includes five stratovolcanoes (Mt. Katmai, Trident, Mt. Griggs, Mageik and Mt. Martin), two dacite domes (Falling Mountain and Cerberus), and Novarupta, the principal vent for the Valley of Ten Thousand Smokes (VTTS) ash flow and now largely filled with ejecta and a rhyolite-dacite dome (figs. 11.11.1-11.11.3). The sizes, shapes, and interconnections of magma bodies beneath these vents are not well known. However, the eruption at Novarupta and collapse at Mt. Katmai in 1912 and petrologic and structural relations of those vents with Trident suggest that Novarupta and Mt. Katmai are hydraulically connected via Trident at a shallow depth (Hildreth, 1987). In addition, S-wave attenuation (Kubota and Berg, 1967; Matumoto, 1971) suggests one or more large magma bodies beneath the area. Matumoto (1971) inferred three 15-km-diameter lenses of S-wave attenuation--one less than 10 km beneath the Katmai-Novarupta-Trident-Griggs area, another beneath that at 20-30 km depth, and a third at 20-30 km depth beneath Mageik and Martin (fig. 11.11.4).

A -35 mgal Bouguer gravity anomaly centered near Katmai Pass (Decker, 1963; Berg and others, 1967; Kienle, 1969; Kienle and Swanson, 1983) may reflect the same magma bodies and/or low-density fill in the Valley of Ten Thousand Smokes (fig. 11.11.5).

Hildreth (1983, 1987) inferred that rhyolitic, dacitic, and andesitic magma coexisted, perhaps in related but physically discrete parts of a magma reservoir beneath the Novarupta-Trident-Katmai area, well before the 1912 eruption.

HISTORICAL ACTIVITY

1912: The eruption of Novarupta in 1912 was the largest eruption in this century. About 15 km^3 of magma was erupted; the adjacent Valley of Ten Thousand Smokes received 11-15 km^3 of ash-flow tuff, and about 20 km^3 of airfall tephra blew to the southeast for more than 1,000 km (Griggs, 1922; Hildreth, 1983). All or nearly all products were erupted from Novarupta (Curtis, 1968; Hildreth, 1983). The summit of Mt. Katmai collapsed to form a 3 km x 4 km caldera, apparently in hydraulic response to the eruption of Novarupta. On the basis of petrologic and structural arguments, Hildreth (1987) infers that magma from Trident moved to fill a posteruption void beneath Novarupta, in turn allowing magma from Katmai to move toward Trident and allowing the Katmai summit to collapse.

Few people lived in the Katmai area before the eruption, so reports of precursors are few. A guide told Doroshin (1870) that Mt. Katmai emits smoke, "but the inhabitants of the village of Katmai were not sure of this." J.E. Spurr made a geological reconnaissance of the area in 1898. He wrote: "Extensive hot springs emerge from the Katmai side of the mountains below [Katmai] pass, and there are very frequent earthquakes and other evidences of volcanic activity. Our party itself experienced a slight earthquake just after crossing [the pass].... One of these volcanoes is said by the natives to smoke occasionally" (Spurr, 1900).

Earthquakes were felt at Katmai village, 30 km from the volcano, at least 5 days before the 1912 eruption. The intensity of these quakes increased notably on 4 and 5 June. On 4 June, frightened by these earthquakes, the people of Katmai village moved 16 km

KATMAI/NOVARUPTA (continued)

HISTORICAL ACTIVITY (continued)

down the coast to camp at Cape Kubugakli. Earthquakes were also felt at Savonoski, about 30 km northwest of Katmai, although Fenner (1925) states that one resident of Savonoski (Mr. Bob Scott) did not feel earthquakes before the eruption. Some earthquakes were felt as far away as Kanatak (94 km to the southwest), Uyak (90 km to the southeast), and Nushagak (210 km west-northwest from Katmai) (Fenner, 1925). Strong gas emission (and phreatic eruptions?) began on 5 June, and more earthquakes and explosions occurred on the morning of 6 June (Martin, 1913; Griggs, 1922; Fenner, 1925). The climactic eruption began at about 1300 hr on 6 June, and continued for 2.5 days. A particularly strong earthquake occurred at 2200 hr on 6 June, and other strong earthquakes occurred through the night of 7 June. Lesser earthquakes occurred from 8 June to 15 August, with relatively strong events on 11 June, 21 June, and 30 July (Fenner, 1925).

Griggs (1920, 1922) found that a large slide or rockfall from Falling Mountain occurred during or shortly before eruption of the Valley of Ten Thousand Smokes ash flow, and inferred from other field relations that a large rockslide from Mageik occurred at about the same time.

Both Griggs (1922) and Fenner (1925) noted fissures and faults in the Valley of Ten Thousand Smokes, concluded that magma had been erupted from these fissures, and inferred that they had therefore formed above a magma intrusion. Griggs envisioned a batholith at shallow depth and thought that all of the faults except those near Novarupta were of shallow origin. Fenner concluded that a shallow sill had fed the eruption, through many faults and fissures. In the light of current knowledge of calderas, compaction of ash-flow tuffs, relatively rapid cooling of sills, and compositional zonation of the 1912 magma, Fenner's hypothesis is probably incorrect and Grigg's single batholith is probably an oversimplification and overestimation of the size of a near-surface body.

Following Curtis (1968), Hildreth (1983) concluded that virtually all of the 1912 eruption was from the Novarupta vent; further, all of the erupted magma is closely related to other Novarupta products and unlike those of Katmai. A corollary of this conclusion is that collapse to form the Katmai Caldera was caused neither by eruption from Katmai nor by eruption of Katmai magma from Novarupta. Rather, it must have been in hydraulic response to the eruption of Novarupta, from an unknown but complex magma reservoir between the two vents (Hildreth, 1983, 1987). Curtis (1968) cites stratigraphic evidence that the summit of Mount Katmai did not collapse instantaneously, but rather collapsed slowly during the entire period of the 1912 eruption. Many of the faults around the Novarupta vent are thought to be shallow-seated compaction features within a funnel-shaped vent, rather than faults formed by collapse into an evacuated magma reservoir (Hildreth, 1983), but scarps in mountains adjacent to Novarupta are difficult to explain by compaction and may indicate some collapse.

Other volcanoes of the Katmai group have been intermittently active throughout recorded history. An eruption on 14 March 1866, "possibly from the Katmai area," caused ash to fall at Katmai Village, Afognak Island, and Pavlof Harbor, but no earthquakes or noises were reported (Doroshin, 1870). Doroshin infers that the ash came from a newly formed volcano northwest of Katmai Village, in the direction of Mt. Katmai. Snyder (1954), Muller and others (1954), Eicher and Rounsefell (1957), Decker (1963), Ward and Matumoto (1967), and Wilson and Forbes (1969) report eruptions of a new vent on the southwest flank of Trident from 1951 (or 1949?) through

KATMAI/NOVARUPTA (continued)

HISTORICAL ACTIVITY (continued)

1968. This new vent on Trident was present as a linear group of ash-emitting fumaroles by 1951, and lava extrusion began from the same fissure in 1953 (Snyder, 1954; Decker, 1963). A lava dome that was growing in the summer of 1974 was destroyed before the summer of 1975 (Volcanological Society of Japan, 1976).

Years that are followed by a question mark in the table above, for example, "1929-31?" for Katmai, refer to sightings of plumes that might be routine fumarolic emission rather than eruptions. W. Hildreth (written commun., 1987) considers them to be vapor plumes because no young erupted products are found; we accept this as likely, but note that small ash eruptions even at well-observed volcanoes leave no enduring, identifiable deposit.

During 39 days of recording with a small tripartite array in 1965, Matumoto and Ward (1967) found the primary center of earthquake activity to be 10-20 km northeast of Mt. Katmai, near Snowy Mountain (fig. 11.11.6). However, results from a 10-station network for the period from September 1987 to early March 1988 showed that the primary center of activity was a few kilometers northeast of the rim of Katmai Caldera, at depths of 0.8 to 8.5 km. The earlier results could have been in error by 10-20 km, so the 1965 activity may also have been close to Mt. Katmai (Ward and Endo, 1988). From September 1987 to early March 1988, 22 events within 80 km of the Katmai network were large enough to locate; in addition, dozens of small events per day were recorded on a station near the lower southeast flank of Mt. Katmai. Profound attenuation of seismic waves was observed during the study, but it was not possible to distinguish between site, source, and path effects. A working hypothesis is that some of the attenuation was caused by a shallow magma body beneath Mt. Katmai.

Seismicity increased slightly in the Trident-Martin area on 3-4 August 1965, and Trident may have erupted during cloudy weather during this same period (Matumoto and Ward, 1967; Ward and Matumoto, 1967). During a recording period from July 1977 to June 1981, shallow earthquakes (depths less than 5 km) were recorded in the Martin-Mageik-Trident area (Kienle and others, 1983).

COMMENTS

Areas of collapse in 1912 (Mt. Katmai summit and around Novarupta vent) are less than 5 km in diameter. However, the unusual density of stratovolcanoes in the Katmai area, the leakage of about 15 km^3 of rhyolite and dacite magma in 1912, and the "hydraulic connection" between Novarupta, Trident, and Katmai prompt us to include the Katmai/Novarupta area in this compilation as a possible future site of a large, caldera-forming eruption (see Bacon, 1985).

The size and geometry of Katmai-area magma reservoir(s) remain poorly known. Small- to moderate-size magma reservoirs may exist at shallow depth beneath each of the vents in the Katmai area, and one or more large reservoirs may exist at greater depth beneath the Katmai-Novarupta-Trident-Griggs and Mageik-Martin areas, with hydraulic interconnections. Regional studies in support of a proposed research-drilling project at Katmai should help to constrain the existence and location of crustal magma reservoirs in the area.

KATMAI/NOVARUPTA, Region 11, CAVW number 11-02-14,15,16,17,18,19

KATMAI/NOVARUPTA (continued)

REFERENCES

Bacon, C.R., 1985, Implications of silicic vent patterns for the presence of large crustal magma chambers: Jour. Geophys. Res., v. 90, no. B13, p. 11,243-11,252.

Berg, E., Kubota, S., and Kienle, J., 1967, Preliminary determination of crustal structure in the Katmai National Monument, Alaska: Bull. Seismol. Soc. Am., v. 57, no. 6, p. 1367-1392.

Curtis, G.H., 1968, The stratigraphy of the ejecta from the 1912 eruption of Mount Katmai and Novarupta, Alaska: Geol. Soc. Am. Memoir 116, p. 153-210.

Decker, R.W., 1963, Proposed volcano observatory at Katmai National Monument: A preliminary study: unpubl. rep., Hanover, N.H., Dartmouth Coll., 54 p.

Doroshin, P., 1870, Some volcanoes, their eruptions, and earthquakes in the former Russian possessions in America: Verh. Russisch-Kaiserlichen Mineral. Gesellschaft zu St. Petersburg, 2nd ser., 1870, p. 25-44 (English translation by J.B. Kisslinger).

Eicher, G.J., Jr., and Rounsefell, G.A., 1957, Effects of lake fertilization by volcanic activity on abundance of salmon: Limnology and Oceanography, v. 2, no. 2, p. 70-76.

Fenner, C.N., 1920, The Katmai region, Alaska, and the great eruption of 1912: Jour. Geol., v. 28, p. 569-606.

----- 1923, The origin and mode of emplacement of the great tuff deposit of the Valley of Ten Thousand Smokes: Natl. Geogr., Contributed Technical Papers, Katmai Ser., v. 1, p. 1-74.

----- 1925, Earth movements accompanying the Katmai eruption: Pt. I. Regional earthquakes, Pt. II. Fracturing in the Valley of Ten Thousand Smokes: Jour. Geol., v. 33, p. 116-139, 193-223.

----- 1930, Mount Katmai and Mount Mageik: Zeit. Vulkanol., v. 13, p. 1-24.

----- 1938, The phenomenon of falling Mountain: Am. Jour. Sci., v. 248, p. 593-627.

Griggs, R.F., 1920, The great Mageik landslide: Ohio Jour. Sci., v. 20, p. 325-354.

----- 1922, The Valley of Ten Thousand Smokes: Washington, D.C., Natl. Geogr. Soc., 340 p.

Hildreth, W., 1983, The compositionally zoned eruption of 1912 in the Valley of Ten Thousand Smokes, Katmai National Park, Alaska: Jour. Volcanol. Geotherm. Res., v. 18, p. 1-56.

----- 1987, New perspectives on the eruption of 1912 in the Valley of Ten Thousand Smokes, Katmai National Park, Alaska: Bull. Volcanol., v. 49, p. 680-693.

Keller, A.S., and Reiser, H.N., 1959, Geology of the Mount Katmai area, Alaska: U.S. Geol. Surv. Bull 1058-G, p. 261-298.

Kienle, J., 1969, Gravity survey in the general area of the Katmai National Monument: Fairbanks, Univ. Alaska, Ph.D. thesis, 151 p.

----- 1970, Gravity traverses in the Valley of Ten Thousand Smokes, Katmai National Monument, Alaska: Jour. Geophys. Res., v. 75, p. 6641-6649.

Kienle, J., and Swanson, S.E., 1983, Volcanism in the eastern Aleutian arc: late Quaternary and Holocene centers, tectonic setting and petrology: Jour. Volcanol. Geotherm. Res., v. 17, p. 393-432.

Kienle, J., Swanson, S.E., and Pulpan, H., 1983, Magmatism and subduction in the Eastern Aleutian arc, <u>in</u> Shimozuru, D., and Yokoyama, I., eds., Arc volcanism: physics and tectonics: Tokyo, Terra Sci. Publ., p. 191-224.

Kubota, S., and Berg, E., 1967, Evidence for magma in the Katmai volcanic range: Bull. Volcanol., v. 31, p. 175-214.

Martin, G.C., 1913, The recent eruption of Katmai volcano in Alaska: Natl. Geogr., v. 24, p. 131-181.

KATMAI/NOVARUPTA (continued)

REFERENCES (continued)

Matumoto, T., 1971, Seismic body waves observed in the vicinity of Mount Katmai, Alaska, and evidence for the existence of molten chambers: Geol. Soc. Am. Bull., v. 82, p. 2905-2920.

Matumoto, T., and Ward, P.L., 1967, Microearthquake study of Mount Katmai and vicinity, Alaska: Jour. Geophys. Res., v. 72, no. 10, p. 2557-2568.

Muller, E.H., Juhle, W., and Coulter, H.W., 1954, Current volcanic activity in Katmai National Monument: Science, v. 119, p. 319-321.

Smith, R.L., Shaw, H.R., Luedke, R.G., and Russell, S.L., 1978, Comprehensive tables giving physical data and thermal energy estimates for young igneous systems of the United States: U.S. Geol. Surv. Open-File Rep. 78-925, 15 p. + tables.

Snyder, G.L., 1954, Eruption of Trident Volcano, Katmai National Monument, Alaska: February-June, 1953: U.S. Geol. Surv. Circular 318, 7 p.

Spurr, J.E., 1900, A reconnaissance in southwestern Alaska in 1898: U.S. Geol. Surv., 10th Ann. Rep., pt. 7, p. 43-263.

Volcanological Society of Japan, 1976, Trident, Bull. Volcanic Eruptions, no. 14, p. 52 (information from G.E. Blinn).

Ward, P.L., and Matumoto, T., 1967, A summary of volcanic and seismic activity in Katmai National Monument, Alaska: Bull. Volcanol., v. 31, p. 107-129.

Ward, P.L., and Endo, E., 1988, A preliminary report on seismic activity in Katmai National Park, Alaska: unpubl. rep., 16 p.

Wilson, C.R., and Forbes, R.B., 1969, Infrasonic waves from Alaskan volcanic eruptions: Jour. Geophys. Res., v. 74, no. 18, p. 4511-4522.

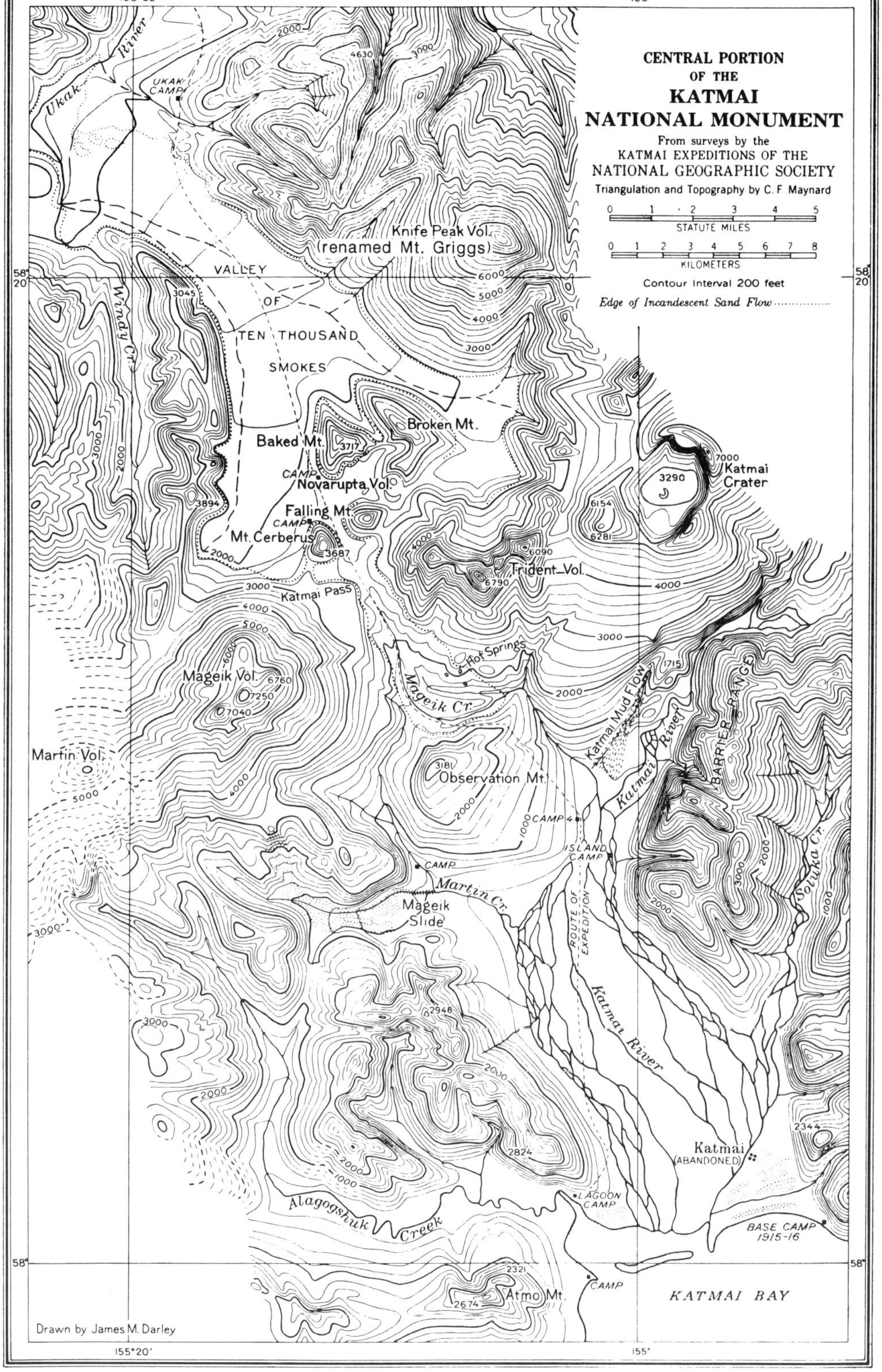

Figure 11.11.1. Topographic map of Katmai/Novarupta area, from Griggs (1922).

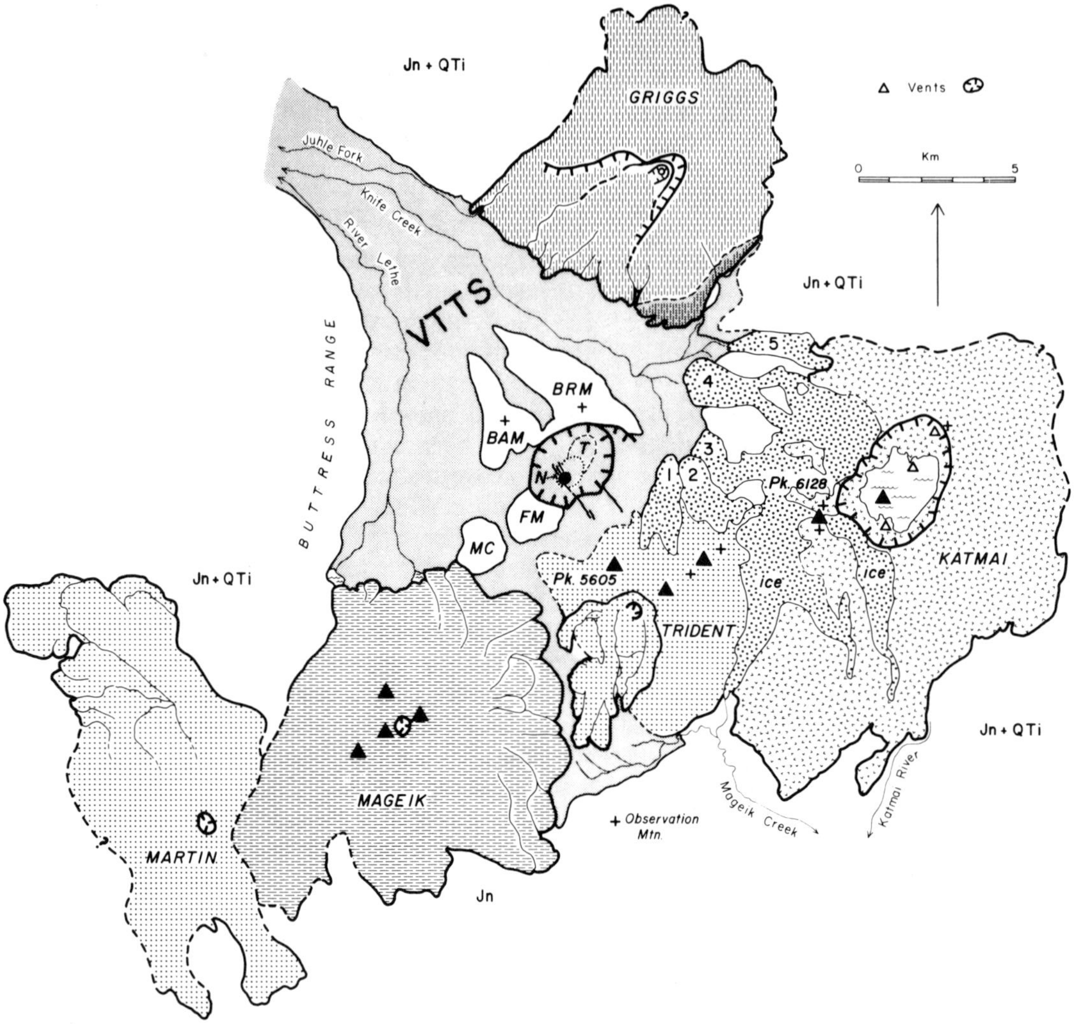

Figure 11.11.2. Outline map of Katmai area, from Hildreth (1987). Active andesite-dacite stratovolcanoes, Martin, Mageik, Trident, Katmai, and Griggs, surround the 1912 ash-flow sheet and a 2-km-wide vent depression within which The Turtle (T) ejecta mound and Novarupta rhyolite dome (N) were subsequently emplaced. Also near the 1912 vent are Holocene dacite domes Mount Cerberus (MC) and Falling Mountain (FM), and basement ridges Baked Mountain (BAM) and Broken Mountain (BRM). Presumed hydraulic transfer of magma from beneath Mount Katmai permitted summit collapse in 1912, forming a 600-m-deep caldera that now is occupied by a lake. VTTS, Valley of Ten Thousand Smokes. Discrete vent cones indicated by solid triangles, open craters by hachures; the 3 foundered summits of Mount Katmai (open triangles) are relocated from Griggs (1922; p. 270). The 5 Knife Creek glaciers are numbered for reference, but for clarity many other glaciers are omitted.

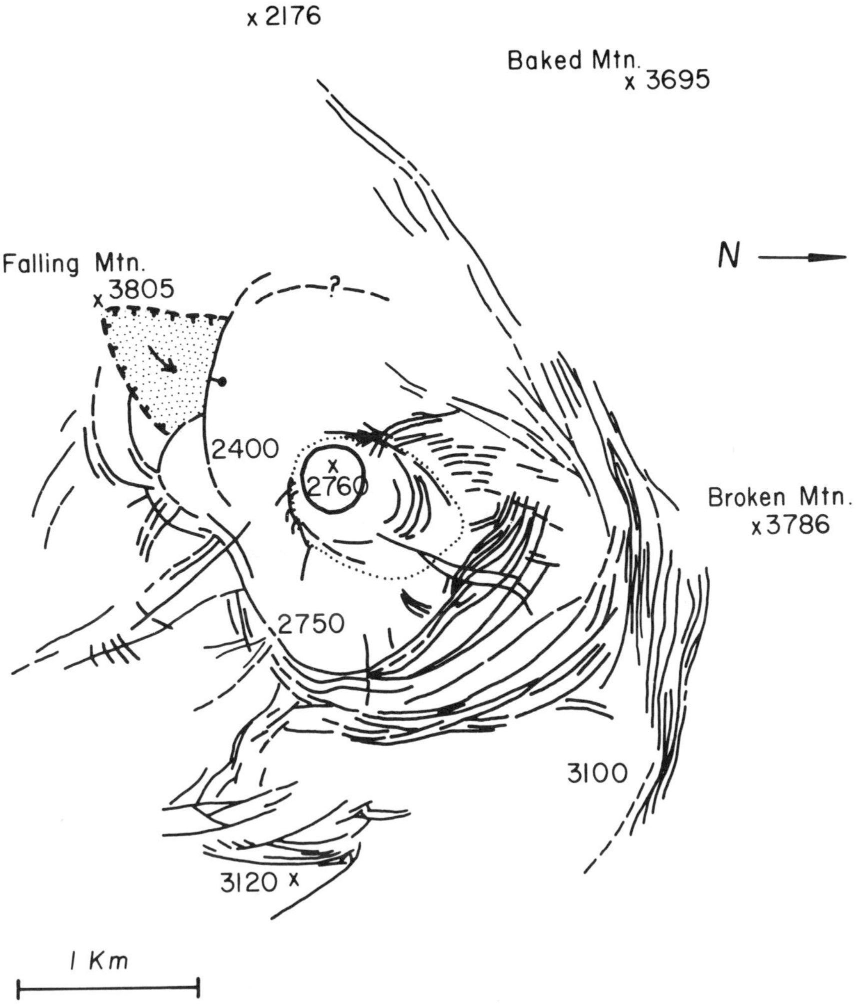

Elevations in feet (1m = 3.28 feet)

.... Crest of Novarupta tephra ring

⊥⊥⊥⊥ Margins of Falling Mtn. scarp

Heavy line outlines Novarupta

Figure 11.11.3. Fracture pattern outlining tephra-filled Novarupta Caldera, from Hildreth (1983). North is to the right.

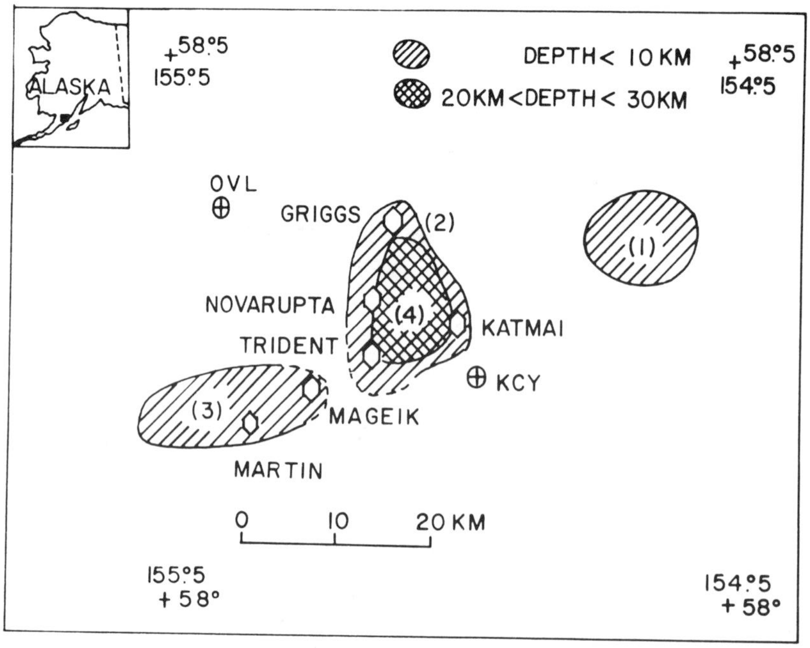

Figure 11.11.4. Magma chambers in Katmai region inferred from S-wave attenuation, from Matumoto (1971). Three chambers (1-3) may exist at shallow depths, probably less than 10 km; one chamber (4) is deeper, probably at 20-30 km depth.

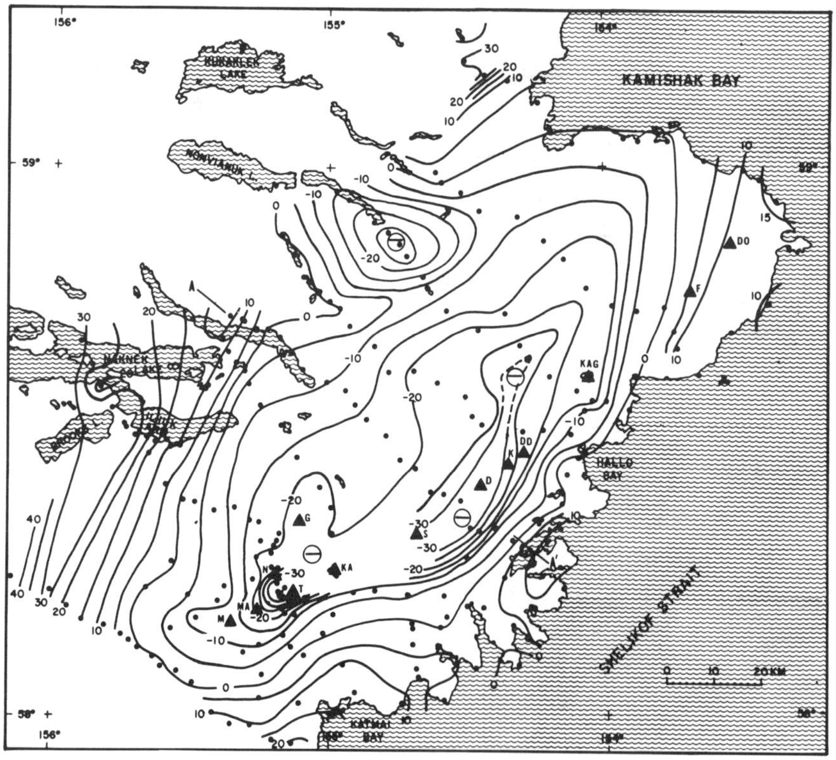

Figure 11.11.5. Bouguer gravity map of the Alaska Peninsula, from Kienle and Swanson (1983). Average crustal density of 2.67 g/cm^3 was assumed for reduction. Katmai area lies near center of gravity trough.

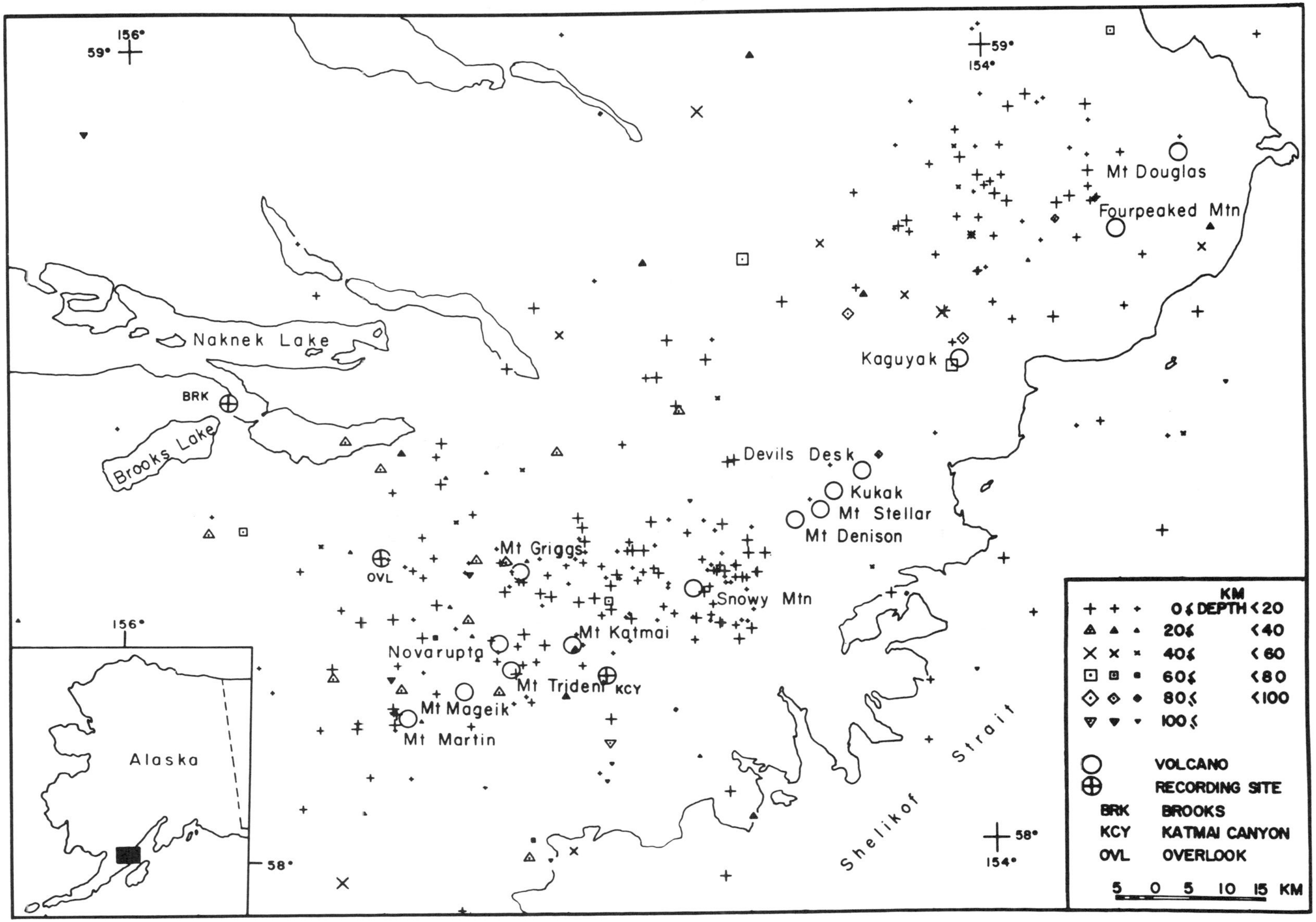

Figure 11.11.6. Seismic network and epicenters in Katmai region during summer of 1965, from Matumoto (1971). Three sizes of symbols indicate precision of hypocentral determination; larger symbol indicates more precise determination. On the basis of more accurate locations for earthquakes from September 1987 to March 1988, Ward and Endo (1988) suggest that many of the 1965 earthquakes shown near Snowy Mountain probably occurred closer to Mt. Katmai.

PART 3: HISTORICAL UNREST AT LARGE QUATERNARY CALDERAS (continued)
See inside back cover for explanation and abbreviations

WRANGELL

CAVW number (active vent)	Latitude Longitude (degrees)	Diameter (km)	Local tectonic setting	Precaldera edifice	SiO_2 content (wt pct)	Age of collapse (yr B.P.)	Date of unrest	Type and duration of unrest ESTU STHF MGTF H Te	Eruption type
11-05-02 Mt. Wrangell	62.00N 144.02W	4 x 6 (modern) 15 (ancestral)		Shield	R = 58-62 C = a?	late Quaternary	1819?	---- ---- ---- - --	ex?
							1884?	---- ---- ---- - --	unknown
							1899	---- ---- ---x - QU?	none
							1902	---- ---- ---- - --	ex
							1907	---- ---- --xx x --	none
							1965-present	xx-- ---- --xx - QU?	none yet

TECTONIC SETTING

The Wrangell Mountains are situated southeast of the main Aleutian arc, in a zone of major strike-slip faults along which narrow strips of the west coast of North America have moved northward. A triangular fragment of the continental margin (the Yakutat block) is moving with the Pacific plate and impinging on the continental block, so northeast-dipping subduction has stopped in that area. However, a weakly defined Benioff zone related to previous, northeast-dipping subduction still persists, parallel to the Wrangell chain and at an angle of about 80° to the direction of modern subduction in the Cook Inlet area (Circum-Pacific Map Project, Northeast Quadrant, 1981; Nye, 1983).

GEOLOGIC HISTORY

The Wrangell volcanoes (including Mt. Drum, Mt. Sanford, Mt. Wrangell, and Mt. Jarvis) are a group of large Pleistocene-Recent vents at the western end of the Wrangell Mountains. They are among the largest convergent margin volcanoes in the world and are thought to have grown at the unusually high rate of 30 km^3/m.y./km of arc (Nye, 1983). Mt. Wrangell itself has grown at a rate of at least several cubic kilometers per year (Nye, 1983).

On Mt. Wrangell, Benson and Motyka (1979) described an ice-filled, 1 km deep, 4 km x 6 km caldera, and an ancestral, 15-km-diameter caldera that extends northeast from Wrangell Caldera (Benson and Motyka, 1979). The age of Wrangell Caldera is unknown but is probably late Quaternary (T.P. Miller, written commun., 1987). Three postcaldera craters (North, East, and West Craters) have formed on the caldera rim (figs. 11.12.1, 11.12.2). Products of Mt. Wrangell are basaltic andesite to high-silica andesite. Some lava flows are as much as 35 km long, and the chemical compositions of one group of lava flows suggests a single zoned magma reservoir with a volume of at least a few tens of cubic kilometers (Nye, 1983).

WRANGELL (continued)

HISTORICAL ACTIVITY

Questionable eruptions occurred in 1819 and 1884 (Becker, 1898). On 3 September 1899, Mt. Wrangell steamed profusely just after a great earthquake and continued to "smoke with unusual animation for the rest of the season" (Benson and Motyka, 1979, quoting guide A.M. Powell). Crosby (1907, citing A.P. Porter) reported an increase in steaming and a warm, 2-day-long flood on 6-7 April 1907.

The best documented unrest in historical time has been from ca. 1965 to the present. Until 1965, glacial accumulation was in equilibrium with flow and basal melt, so the snow surface remained constant in altitude. Increased basal melting was first noted in 1965 and has continued, so that as of 1986 85 percent of the (pre-1965) ice in North Crater has melted (Benson and others, 1984) (fig. 11.12.3). Using the melting ice as a calorimeter, Benson and Motyka (1979) estimated that the minimum heat flow in North Crater was 1460 heat flow units (HFU), and heat flow for the caldera excluding North Crater was approximately 140 HFU.

All fumaroles were at or below the boiling temperature of water until the late 1970's, but by 1980 or 1981 some superheating was evident. By 1985, a temperature of 192 °C was measured at the edge of one fumarole. Fumarolic gases are mostly water, with CO_2 and SO_2 observed in the dry fraction. During April 1986, residents of the area reported occasional plumes rising 1 km above the summit (Benson and others, 1984; Smithsonian Institution, 1986).

COMMENTS

The cause of the current heating is not known, but Benson and Motyka (1979) and Benson and Follett (1986) speculate that it was triggered by the 1964 Prince William Sound ("Good Friday") earthquake, just as steaming of Wrangell was apparently triggered by another great earthquake in 1899 in Yakutat Bay.

REFERENCES

Becker, G.F., 1898, Reconnaissance of the gold fields of southern Alaska, with some notes on general geology: U.S. Geol. Surv., 18th Ann. Rep., pt. 3, p. 13-24.

Benson, C.S., and Follett, A.B., 1986, Applications of photogrammetry to the study of volcano-glacier interactions on Mt. Wrangell, Alaska: Photogrammetric Engineering and Remote Sensing, v. 52, no. 6, p. 813-827.

Benson, C.S., and Motyka, R.J., 1979, Glacier-volcano interactions on Mt. Wrangell, Alaska: Univ. Alaska Geophys. Inst., Ann. Rep., 1977-78, p. 1-25.

Benson, C.S., Sturm, M., Solie, D., Motyka, R., Mosely-Thompson, E., and Grootes, P., 1984, Glacier-volcano interactions, Mt. Wrangell, Alaska: Univ. Alaska Geophys. Inst., Biennial Rep., 1983-84, p. 102-104.

Circum-Pacific Map Project, Northeast Quadrant, 1981, Plate-tectonic map of the Circum-Pacific region, northeast quadrant: Tulsa, Circum-Pacific Council for Energy and Mineral Resources and the Am. Assoc. Pet. Geol., 1 sheet, scale 1,10,000,000.

WRANGELL (continued)

REFERENCES (continued)

Crosby, W.O., 1907, Volcanic activity in Alaska: Science, v. 26, p. 78.
Grewingk, C., 1850, Beitrag zur Kenntniss der orographischen und geognostischen Beschaffenheit der Nordwestküste Amerikas, mit den anliegenden Inseln: Verh. Russisch-Kaiserlichen Mineral Gesellschaft zu St. Petersburg, 1848-49, p. 123-278.
Miller, T.P., and Smith, R.L., 1975, Ash flows on the Alaska Peninsula: A preliminary report on their distribution, composition, and age [abs.]: Geol. Soc. Am. Abstracts with Programs, v. 7, no. 7, p. 1201.
Nye, C.J., 1983, Petrology and geochemistry of Okmok and Wrangell Volcanoes, Alaska: Santa Cruz, Univ. Calif., Ph.D. diss.
Smithsonian Institution, Scientific Event Alert Network (SEAN), 1986, Wrangell Volcano: SEAN Bull., v. 11, no. 4.

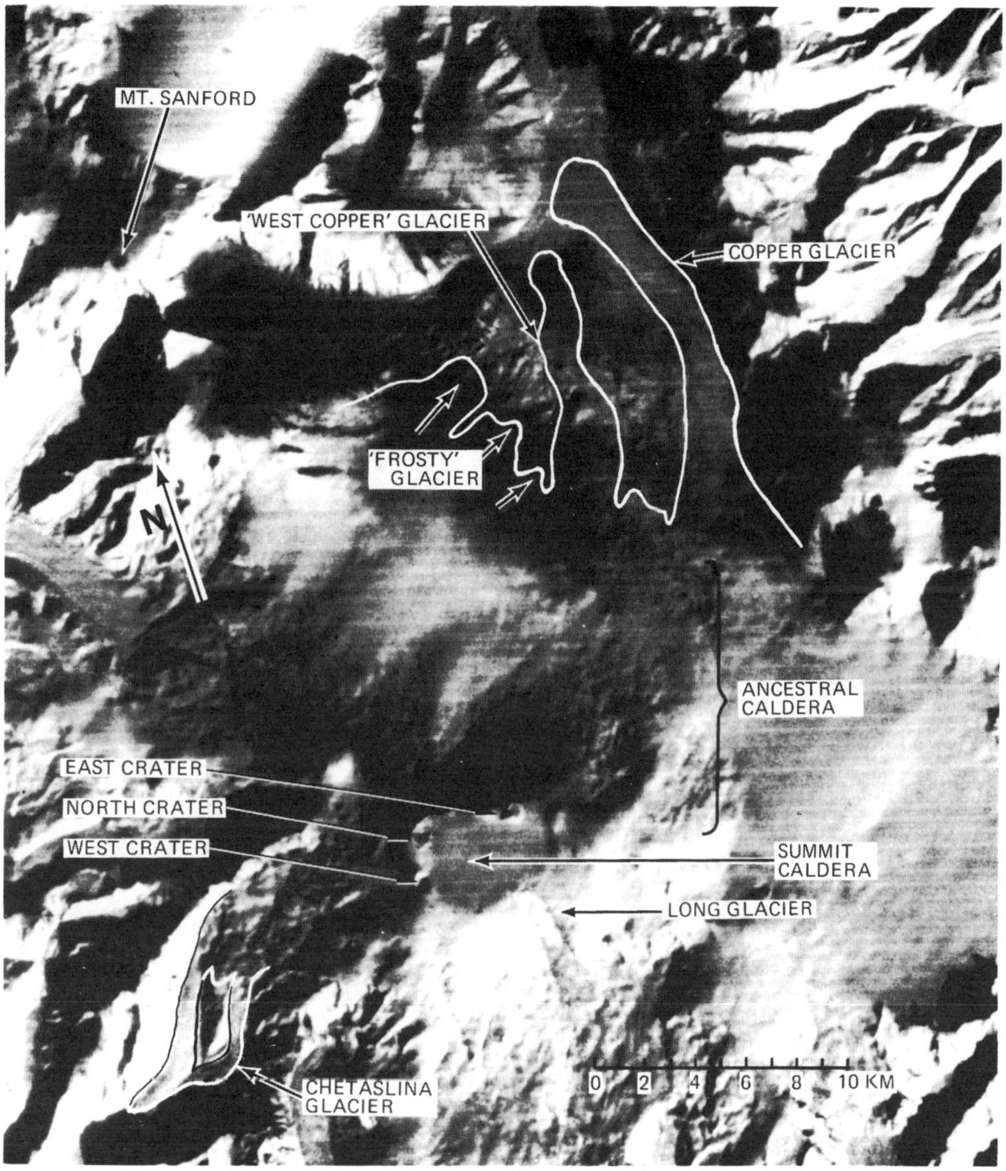

Figure 11.12.1. Mt. Wrangell area, showing summit caldera and approximate outline of ancestral caldera, after Benson and Motyka (1979).

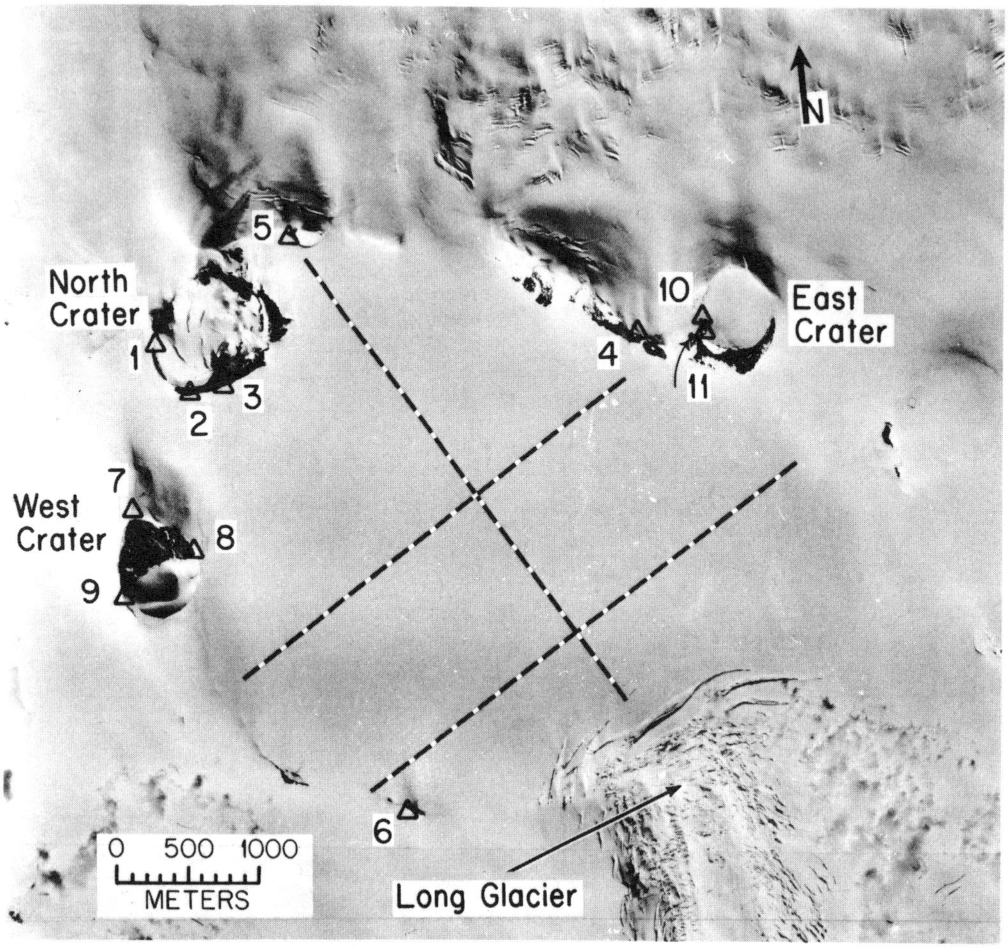

Figure 11.12.2. Summit caldera of Mt. Wrangell in 1975, after Benson and Motyka (1979). Numbers indicate control points for snow-and-ice survey; dashed lines indicate traverses for determination of ice thickness using a radio echo sounder.

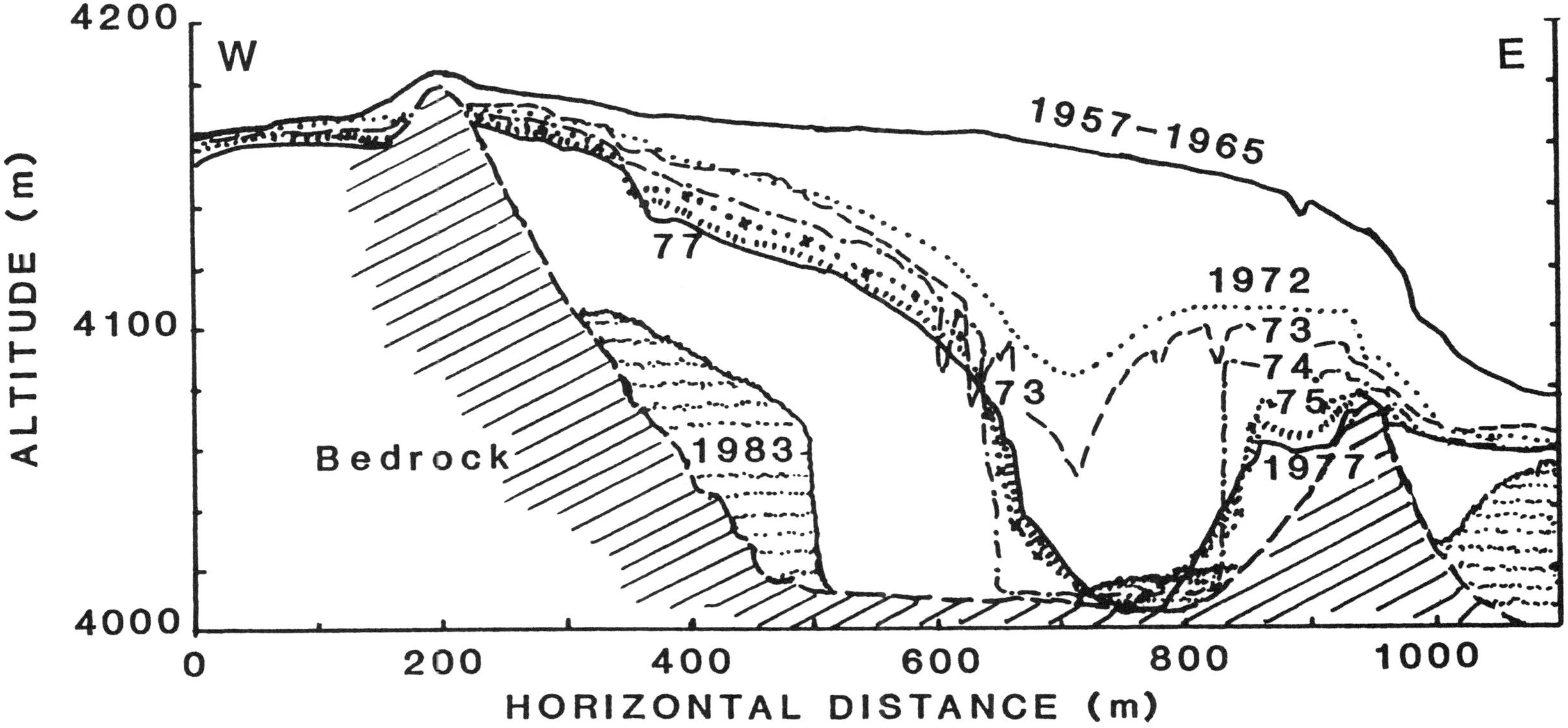

Figure 11.12.3. East-west cross section of Mt. Wrangell's North Crater, showing changes in ice volume between 1957 and 1983, after Benson and Motyka (1979).

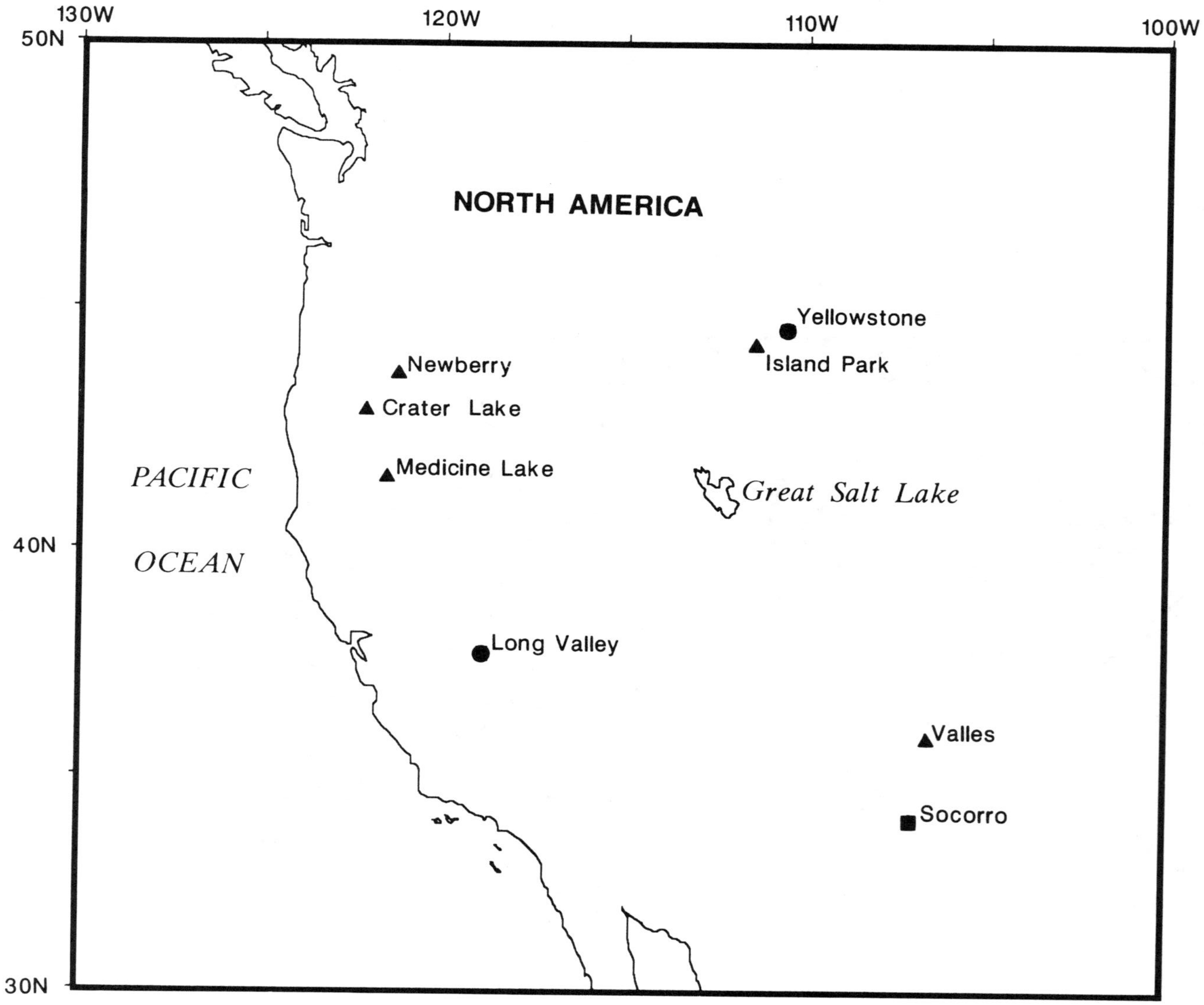

Figure 12. Locations of large, Quaternary restless calderas (solid circles), nonrestless calderas (triangles), and a restless noncaldera area (square) in Region 12.

LONG VALLEY

CAVW number (active vent)	Latitude Longitude (degrees)	Diameter (km)	Local tectonic setting	Precaldera edifice	SiO_2 content (wt pct)	Age of collapse (yr B.P.)	Date of unrest	Type and duration of unrest ESTU STHF MGTF H Te	Eruption type
12-03-08, (Mono Craters)	37.70N 118.87W	17 x 32	Exten	Domes	R = 73-74	700,000	1889	x--- ---- ---- - =?	none
					C = 75-77		1927	x--- ---- ---- - =?	none
							1941-42	x--- ---- ---- - =?	none
12-03-09 (Inyo Craters)							1978-present	FFxF -FFC FFFF F UU	none yet

TECTONIC SETTING

The Long Valley Caldera in eastern California is situated along the east edge of the Sierra Nevada tectonic block, and the western edge of the Basin and Range extensional province (figs. 12.1.1, 12.1.2). The entire region is one of high seismicity, ENE-WSW extension, and minor right-lateral strike-slip faulting. The Hilton Creek and Laurel Canyon Faults are major, recently active faults that cut the south-central part of the caldera; the Hartley Springs Fault and a set of related faults cut the northwestern part of the caldera (Pakiser, 1960; Pakiser and others, 1964; Bailey and others, 1976).

The Long Valley - Mono Craters region lies within a tectonically active zone that has produced four major earthquakes (M 6.8-8.0) since 1872, leaving three major gaps. The central gap, the White Mountains seismic gap, includes the Long Valley - Mono Craters region, and several authors have noted the potential for a major gap-filling earthquake in the coming years (Hill and others, 1985b).

GEOLOGIC HISTORY

Long Valley Caldera formed during and shortly after the Bishop tuff eruption of 700,000 yr B.P., and a resurgent dome formed shortly thereafter. Postcaldera eruptions produced moderately voluminous rhyolite domes and pyroclastic deposits, and less voluminous moat rhyolites, moat basalts, and rim quartz latites (fig. 12.1.2) (Bailey and others, 1976).

Recent eruptions in the area, from 40,000 yr B.P. to the present, have been at the Inyo domes (dominantly rhyodacitic) and the Mono domes and Mono Craters (dominantly rhyolitic) (Wood, 1977; Miller and others, 1982). The most recent eruptions in the Inyo-Mono chain occurred about 650-550 years ago, from multiple vents along nearly the full length of the Inyo chain and in the northern part of the Mono Craters chain, spanning a total distance of 40 km (Miller, 1985; Sieh and Bursick, 1986). Magmatic eruptions from the Inyo domes were followed shortly thereafter by phreatic explosions, and then by extrusion of rhyolite flows at three vents. Phreatic explosions occurred at other vents during this same period, including 12 within Long Valley Caldera (Miller, 1985). A likely

LONG VALLEY (continued)

GEOLOGIC HISTORY (continued)

explanation of this contemporaneous eruptive activity at multiple, linearly aligned vents is that a dike (or series of dikes) of silicic magma was emplaced, perhaps as part of an episode of regional crustal extension (Miller, 1985; Eichelberger and others, 1985; Fink, 1985). Recent drilling has confirmed the existence of a silicic feeder dike connecting several of the Inyo domes (Eichelberger and others, 1985). A suggestion that basaltic dike intrusion triggered these silicic eruptions 650-550 years ago (Eichelberger and others, 1987) is presently being reconsidered (J. Eichelberger, oral commun., 1988).

Seismic-refraction, heat-flow, and other studies suggest that a large body of magma exists beneath the western half of the caldera (fig. 12.1.3), overlain by an average of 1.4 km of Bishop tuff and 1 km of sedimentary and pyroclastic fill (Hill, 1976; Lachenbruch and others, 1976b; Ryall and Ryall, 1981). The fill is deepest in the northeastern and northwestern parts of the caldera. Anomalous ray paths and attenuation of S waves suggest one magma body beneath and slightly southeast of the resurgent dome, from 4.5 (mostly 7 or 8) km to at least 13 km below the surface. The same features suggest a separate cupola of magma just northwest of the resurgent dome, with its top at 5.5 km, probably connected to the central body below 8 km depth (Sanders, 1984).

A major hydrothermal system receives inflow from the west (Sierran front), and numerous hot springs occur in the center of the caldera. One of several possible models for subsurface hydrothermal reservoir temperatures (Sorey and others, 1978, model C, p. A50) has the area of highest temperatures in the southern caldera moat, just south of Casa Diablo--the area of pronounced seismicity in the current episode of unrest.

HISTORICAL ACTIVITY

1889: A M=5.7 earthquake was centered just south of Long Valley Caldera (fig. 1 of Cramer and Toppozada, 1980).

1927: Seismicity increased near Long Valley.

1941-42: Seismicity increased near the south boundary of the caldera (Cramer and Toppozada, 1980; D. Hill, written commun., 1987).

1978-present: Regional seismicity in the Long Valley/Sierran front area declined during fall 1977 - summer 1978 but began to increase in 1978 and remains slightly elevated at the time of this writing (late 1987) (figs 12.1.4-12.1.7). In the Long Valley area, this increase has been in the form of recurring seismic swarms, accompanied by broad uplift of the caldera floor, and local increases in fumarolic activity. Seismic swarms began in October 1978, and four of the five largest earthquakes thus far (four M=6 earthquakes) occurred during 25-27 May 1980. The first, third, and fourth of these large earthquakes occurred south of the caldera; the second occurred beneath the southern moat of the caldera, and the fifth (in 1986) occurred in Chalfant Valley, 40 km east-southeast of the caldera. Hypocentral depths of the four 1980 events were between 7 and 14 km (R. Cockerham, oral commun.; Julian

LONG VALLEY (continued)

HISTORICAL ACTIVITY (continued)

and Sipkin, 1985). Epicenters of these and other seismic events have been scattered along a roughly 30-km-long segment of a NNW-SSE-trending complex graben at the boundary between the Sierra Nevada and the Basin and Range.

The large earthquakes in 1980 were followed by at least 4 years of swarms and flurries, mostly just inside or just outside the south margin of the caldera (figs. 12.1.4-12.1.7). After May 1980, focal depths of swarms were between 2 and 11 km, with an apparent slight decrease in the mean focal depth of successive swarms, from approximately 8 km in mid-1980 to approximately 6 km in mid-1982 and between 5 and 6 km in 1983. The frequency of swarms, the numbers of earthquakes per swarm, and the energy release per swarm decreased irregularly from May 1980, increased briefly in January 1983, and have decreased since that time. Occasionally, small events in rapid succession produced "spasmodic tremor" signals. However, focal mechanisms are most consistent with ENE-WSW Basin and Range extension, with a right-lateral strike-slip component along a west-northwest-trending fault across the south margin of the caldera (Ryall and Ryall, 1983, Wallace, 1985; Hill and others, 1985a). An alternate but geologically less likely interpretation is that the focal mechanisms are compensated linear vector dipoles caused by dike intrusion along a northwest-trending fault just south of the caldera and perhaps along the southern ring fracture of the caldera (Julian, 1983; Savage and Cockerham, 1984; Julian and Sipkin, 1985; see also cautionary note by Aki, 1984). Ground fracturing during the May 1980 earthquakes was broadly consistent with E-W or NE-SW extension (Taylor and Bryant, 1980; Clark and others, 1982).

Since a swarm in July 1984, most seismicity has been outside the caldera. On 23 November 1984, a $\underline{M}_L$=5.7 earthquake occurred beneath Round Valley, 30 km southeast of Long Valley. A number of small earthquakes occurred in the Sierran block just south of the caldera in June-July 1986, and on 21 July 1986 a $\underline{M}_L$=6.4 earthquake occurred in the Chalfant Valley - White Mountains area, 40 km east-southeast of Long Valley. This foreshock-main shock sequence was followed by a typical set of aftershocks. These earthquakes, several tens of kilometers away from Long Valley, probably are not part of unrest at Long Valley, but the unrest at Long Valley might reasonably be considered part of a series of these regional earthquakes, beginning in 1978 (Ryall and Ryall, 1981, 1983).

The resurgent dome was uplifted by as much as 25 cm from 1975 (probably 1979) to October 1980, by another 8 cm by August 1982, and by yet another 7 cm by February 1983 (figs. 12.1.8-12.1.10) (Savage and Clark, 1982; Castle and others, 1984). However, by 1984 uplift had begun to slow (Hill, 1984; Hill and others, 1985a), totaling 4.5 cm from 1983 to 1984, and another 2.5 cm from 1984 to 1985 (Smithsonian Institution, 1985). The center of uplift apparently shifted several kilometers westward during 1983-85, from a site near the center of the resurgent dome to an area near the west caldera rim and the southern Inyo domes. Horizontal lines changed by as much as 45 microstrains (4.5 cm in 1 km; maximum change about 50 cm) from 1978 to mid-1982 (fig. 12.1.11), in patterns that may reflect both point-source uplift and regional extension (Savage and Clark, 1982; Denlinger and Riley, 1984). As of August 1984, extension across the south moat of the caldera and across the entire caldera continued at a rate of several microstrains per year (Estrem and others, 1985). By late 1986, rates had decreased to about 1 microstrain/yr.

Point-source models of the ground deformation suggest an inflating magma body at a depth of at least 7 km (Denlinger and Riley, 1984) or 10 km (Castle and others, 1984; Estrem and others, 1985). A model invoking two magma chambers postulates inflation of a

LONG VALLEY (continued)

HISTORICAL ACTIVITY (continued)

small reservoir at 5-km depth just east of Casa Diablo, and of a larger reservoir at 8-km depth beneath the north part of the resurgent dome (Rundle and Whitcomb, 1984). These and other models attribute inflation to injection of between 0.1 and 0.2 km^3 of magma into the Long Valley reservoir between 1975 and 1983 (Hill and others, 1985a). If one allows for uncertainties inherent in any model, all of the proposed deformation models are in reasonable agreement with the magma reservoirs postulated by Sanders (1984).

Horizontal deformation has been observed well beyond the margin of the caldera, and gravity data collected in early 1983 suggest that uplift might extend as far north as Lee Vining (Whitcomb and Rundle, 1983). However, leveling through the caldera from Tom's Place (southeast) to Lee Vining (northwest) shows no significant change in the elevation of Lee Vining relative to Tom's Place, suggesting that uplift has been confined to the caldera (Castle and others, 1984).

Magnetic field strength at sites on the southwest side of the caldera and along the Hilton Creek Fault increased systematically from 1977 to 1983, and as of May 1983 the cumulative increase was approximately 4 nT (Johnston and others, 1983). Johnston and others (1983) interpret this increase as a piezomagnetic response to increasing stresses from point-source uplift and (or) strain accumulation along the Hilton Creek Fault; they speculated that the magnetic changes will reverse if and when the stresses are relieved, as occurred at Matsushiro (see Part 4 of this compilation).

New fumaroles and vegetation kill were noted near Casa Diablo, on the south side of the resurgent dome (Casadevall and others, 1983). Hot spring pools at Hot Creek enlarged substantially in December 1982, possibly as a precursor to the January 1983 swarm (M.L. Sorey, written commun., 1983). Short-lived geysers and increased hot-spring discharge also occurred immediately after the earthquakes of May 1980 and January 1983. On at least four occasions in 1983, hydrogen gas emissions from the soil near Casa Diablo increased within a few hours or days before small earthquake swarms (K. McGee, written commun., 1983). Mercury concentrations in soils of Long Valley were significantly higher in 1982 than in a previous survey in 1975 (Varekamp and Buseck, 1984), and they were higher still in 1983 (Williams, 1985). In 1983, anomalously high radon was also found in soil gas from the area of uplift and the area of the January 1983 seismic swarm (Williams, 1985) (fig. 12.1.12).

COMMENTS

Unrest at Long Valley probably includes inflation of a residual silicic magma reservoir beneath the caldera, strain accumulation in the south moat of the caldera, and possibly heating of the shallow hydrothermal system, all in response to a regional tectonic strain event that began in 1978. Whether magma has intruded above the level of the residual magma reservoir at any time during the current unrest remains an open question. The unrest was initially considered tectonic, then magmatic or tectono-magmatic, and now certainly tectonic with shallow magma intrusion uncertain (or now, in our view, doubtful). Our doubts about shallow magma intrusion notwithstanding, caution about volcanic hazards is still required because crustal extension and a strong earthquake in the White Mountains seismic gap could trigger dike intrusion and phreatic or magmatic eruptions along the Mono Craters - Inyo domes chain (as

LONG VALLEY (continued)

COMMENTS (continued)

happened along the Mono-Inyo chain ca. 600 years ago and in the Cordón Caulle area of Chile immediately after the 1960 Chilean earthquake) (see section on Cordillera Nevada, CAVW number 15-07-14,15).

REFERENCES

Aki, K., 1984, Evidence for magma intrusion during the Mammoth Lakes earthquakes of May 1980 and implications of the absence of volcanic (harmonic) tremor: Jour. Geophys. Res., v. 89, no. B9, p. 7689-7696.

Archuleta, R.J., Cranswick, E., Mueller, C., and Spudich, P., 1982, Source parameters of the 1980 Mammoth Lakes, California, earthquake sequence: Jour. Geophys. Res., v. 87, no. B6, p. 4595-4607.

Bailey, R.A., 1982, Other potential eruption centers in California: Long-Valley-Mono Lake, Coso, and Clear Lake volcano fields, in Martin, R.C., and Davis, J.F., eds., Status of volcanic prediction and emergency response capabilities in volcanic hazard zones of California: Calif. Div. Mines Geol., Spec. Publ. 63, p. 17-28.

Bailey, R.A., Dalrymple, G.B., and Lanphere, M.A., 1976, Volcanism, structure and geochronology of Long Valley Caldera, Mono County, California: Jour. Geophys. Res., v. 81, p. 725-744.

Barker, J.S., and Langston, C.A., 1983, A teleseismic body-wave analysis of the May 1980 Mammoth Lakes, California, earthquakes: Bull. Seismol. Soc. Am., v. 73, p. 419-434.

Casadevall, T., Clark, M., McGee, K., Sutton, J., and Bezore, S., 1983, Monitoring hydrothermal activity at Long Valley, California [abs.]: Earthq. Notes, v. 54, no. 1, p. 76.

Castle, R.O., Estrem, J.E., and Savage, J.C., 1984, Uplift across Long Valley Caldera, California: Jour. Geophys. Res., v. 89, no. B13, p. 11,507-11,516.

Clark, M.M., Yount, J.C., Vaughan, P.R., and Zepeda, R.L., 1982, Map showing surface ruptures associated with the Mammoth Lakes, California, earthquakes of May 1980: U.S. Geol. Surv. Misc. Field Studies Map MF-1396, 1 sheet, scale 1:24,000.

Cramer, C.H., and Toppozada, T.R., 1980, A seismological study of the May 1980 and earlier earthquakes near Mammoth Lakes, California, in Sherburne, R.W., ed., Mammoth Lakes, California earthquakes of May 1980: Calif. Div. Mines Geol. Spec. Rep. 150, p. 91-130.

Denlinger, R.P., and Riley, F., 1984, Deformation of Long Valley Caldera, Mono County, California, from 1975 to 1982: Jour. Geophys. Res., v. 89, no. B10, p. 8303-8314.

Denlinger, R.P., Riley, F.S., Boling, J.K., and Carpenter, M.C., 1985, Deformation of Long Valley Caldera between August 1982 and August 1983: Jour. Geophys. Res., v. 90, no. B13, p. 11,199-11,209.

Dzurisin, D., Sylvester, A.G., and Cashman, K.V., 1983, Ground tilt since May 1982 at Long Valley Caldera, California [abs.]: Earthq. Notes, v. 54, no. 1, p. 75.

Eichelberger, J.C., Lysne, P.C., Miller, C.D., and Younker, L.W., 1985, Research drilling at Inyo Domes, California: 1984 results: Eos, Trans. Am. Geophys. Union, v. 66, p. 186-187.

LONG VALLEY (continued)

REFERENCES (continued)

Eichelberger, J.C., Younker, L.W., Vogel, T.A., and Miller, C.D., 1987, Coring beneath Inyo Craters, Long Valley, CA [abs.]: Eos, Trans. Am. Geophys. Union, v. 68, p. 1544.

Estrem, J.E., Lisowski, M., and Savage, J.C., 1985, Deformation in the Long Valley Caldera, California, 1983-1984: Jour. Geophys. Res., v. 90, no. B14, p. 12,683-12,690.

Fink, J.H., 1985, Geometry of silicic dikes beneath the Inyo Domes, California: Jour. Geophys. Res., v. 90, no. B13, p. 11,127-11,133.

Given, J.W., Wallace, T.C., and Kanamori, H., 1982, Teleseismic analysis of the 1980 Mammoth Lakes earthquake sequence: Bull. Seismol. Soc. Am., v. 72, p. 1093-1109.

Hermance, J.F., 1983, The Long Valley/Mono Basin volcanic complex in eastern California: Status of present knowledge and future research needs: Rev. Geophys. Space Phys., v. 21, no. 7, p. 1545-1565.

Hill, D.P., 1976, Structure of Long Valley Caldera, California, from a seismic refraction experiment: Jour. Geophys. Res., v. 81, no. 5, p. 745-753.

----- 1984, Monitoring unrest at a large silicic caldera, the Long Valley-Inyo Craters volcanic complex in east-central California: Bull. Volcanol., v. 47, no. 2, p. 371-395.

Hill, D.P., Bailey, R.A., and Ryall, A.S., 1985a, Active tectonic and magmatic processes beneath Long Valley Caldera, eastern California: An overview: Jour. Geophys. Res., v. 90, no. B13, p. 11,111-11,120.

Hill, D.P., Kissling, E., Luetgert, J.H., and Kradolfer, U., 1985b, Constraints on the upper crustal structure of the Long Valley-Mono Craters Volcanic Complex, eastern California, from seismic refraction measurements: Jour. Geophys. Res., v. 90, no. B13, p. 11,135-11,150.

Hill, D.P., Wallace, R.E., and Cockerham, R.S., 1985c, Review of evidence on the potential for major earthquakes and volcanism in the Long Valley-Mono Craters-White Mountains regions of eastern California: Earthq. Prediction Res., v. 3, p. 571-594.

Johnston, M.J.S., Mueller, R.J., Silverman, S.A., and Keller, V.G., 1983, Magnetic measurements across the Long Valley Caldera, 1972-1983: A comparison with observations and mechanisms during the Matsushiro earthquake swarm [abs.]: Earthq. Notes, v. 54, no. 1, p. 90.

Julian, B.R., 1983, Evidence for dyke intrusion earthquake mechanisms near Long Valley caldera, California: Nature, v. 303, p. 323-325.

Julian, B.R., and Sipkin, S.A., 1985, Earthquake processes in the Long Valley caldera area, California: Jour. Geophys. Res., v. 90, no. B13, p. 11,155-11,169.

Lachenbruch, A.H., Sass, J.H., Munroe, R.J., and Moses, T.H., Jr., 1976b, Geothermal setting and simple heat conduction models for the Long Valley Caldera: Jour. Geophys. Res., v. 81, no. 5, p. 769-784.

Lachenbruch, A.H., Sorey, M.L., Lewis, R.E., and Sass, J.H., 1976a, The near-surface hydrothermal regime of Long Valley Caldera: Jour Geophys. Res., v. 81, no. 5, p. 763-768.

Langbein, J., Linker, M., and Tupper, D., 1987, Analysis of two-color geodimeter measurements of deformation within the Long Valley Caldera: June 1983 to October 1985: Jour. Geophys. Res., v. 92, p. 9423-9442.

LONG VALLEY (continued)

REFERENCES (continued)

Luetgert, J.H., and Mooney, W.D., 1985, Crustal refraction profile of the Long Valley Caldera, California, from the January 1983 Mammoth Lakes earthquake swarm: Bull. Seismol. Soc. Am., v. 75, p. 211-221.

Miller, C.D., 1983, Chronology of Holocene eruptions at the Inyo volcanic chain, California [abs.]: Eos, Trans. Am. Geophys. Union, v. 64, p. 900.

----- 1985, Holocene eruptions at the Inyo volcanic chain, California: implications for possible eruptions in Long Valley caldera: Geology, v. 13, p. 14-17.

Miller, C.D., Mullineaux, D.R., Crandell, D.R., and Bailey, R.A., 1982, Potential hazards from future volcanic eruptions in the Long Valley-Mono Lake area, east-central California and southwest Nevada -- a preliminary assessment: U.S. Geol. Surv. Circular 877, 10 p.

Pakiser, L.C., 1960, Transcurrent faulting and volcanism in Owens Valley, California: Geol. Soc. Am. Bull., v. 71, p. 153-160.

Pakiser, L.C., Kane, M.F., and Jackson, W.H., 1964, Structural geology and volcanism of Owens Valley region, California--a geophysical study: U.S. Geol. Surv. Prof. Paper 438, 68 p.

Pitt, A.M., and Cockerham, R.S., 1983, Long Valley earthquake swarm: January 7, 1983 [abs.], Earthq. Notes, v. 54, no. 1, p. 73.

Rundle, J.B., Carrigan, C.R., Hardee, H.C, and Luth, W.C., 1986, Deep drilling to the magmatic environment in Long Valley Caldera: Eos, Trans. Am. Geophys. Union, v. 67, p. 490-491.

Rundle, J.B., and Hill, D.P., 1988, The geophysics of a restless caldera--Long Valley, California: Ann. Rev. Earth Planet. Sci., v. 16, p. 251-271.

Rundle, J.B., and Whitcomb, J.H., 1984, A model for deformation in Long Valley, California, 1980-1983: Jour. Geophys. Res., v. 89, no. B11, p. 9371-9380.

Ryall, A., and Ryall, F., 1981, Spatial-temporal variations in seismicity preceding the May 1980, Mammoth Lakes, California, earthquakes: Bull. Seismol. Soc. Am., v. 71, p. 747-760.

----- 1982, Propagation effects and seismicity associated with magma in Long Valley caldera, eastern California: Earthq. Notes, v. 53, no. 1, p. 46-47.

----- 1983, Spasmodic tremor and possible magma injection in Long Valley Caldera, Eastern California: Science, v. 219, p. 1432-1433.

Sanders, C.O., 1984, Location and configuration of magma bodies beneath Long Valley, California, determined from anomalous earthquake signals: Jour. Geophys. Res., v. 89, no. B10, p. 8287-8302.

Savage, J.C., and Clark, M.M., 1982, Magmatic resurgence in Long Valley Caldera, California--Possible cause of the 1980 Mammoth Lakes earthquakes: Science, v. 217, p. 531-533.

Savage, J.C., and Cockerham, R.S., 1984, Earthquake swarm in Long Valley, California, January 1983: Evidence for dike inflation: Jour. Geophys. Res., v. 89, no. B10, p. 8315-8324.

Savage, J.C., Cockerham, R.S., Estrem, J.E., and Moore, L.R., 1987, Deformation near the Long Valley Caldera, eastern California, 1982-1986: Jour. Geophys. Res., v. 92, no. B3, p. 2721-2746.

Sieh, K., and Bursick, M., 1986, Most recent eruptions of the Mono Craters, eastern central California: Jour. Geophys. Res., v. 91, no. B12, p. 12,539-12,571.

LONG VALLEY (continued)

REFERENCES (continued)

Smithsonian Institution, Scientific Event Alert Network (SEAN), 1982-85, Long Valley Caldera: SEAN Bull., v. 7, no. 12, v. 8, nos. 1, 11; v. 10, no. 10 (contributions from D. Hill, J. Savage, R. Cockerham, R.A. Bailey, F. Riley, and others).

Sorey, M.L., 1985, Evolution and present state of the hydrothermal system in Long Valley caldera: Jour. Geophys. Res., v. 90, no. B13, p. 11,219-11,228.

Sorey, M.L., Lewis, R.E., and Olmsted, F.H., 1978, The hydrothermal system of Long Valley Caldera, California: U.S. Geol. Surv. Prof. Paper 1044-A, p. A1-A60 (including preliminary geologic map of Long Valley).

Taylor, G.C., and Bryant, W.A., 1980, Surface rupture associated with the Mammoth Lakes earthquakes of 25 and 27 May, 1980, in Sherburne, R.W., ed., Mammoth Lakes, California earthquakes of May 1980: Calif. Div. Mines Geol. Spec. Rep. 150, p. 49-67.

Varekamp, J.C., and Buseck, P.R., 1984, Changing mercury anomalies in Long Valley, California: Indication for magma movement or seismic activity: Geology, v. 12, p. 283-286.

Wallace, T.E., 1985, A reexamination of the moment tensor solution of the 1980 Mammoth Lakes earthquakes: Jour. Geophys. Res., v. 90, no. B13, p. 11,171-11,176.

Whitcomb, J.H., and Rundle, J., 1983, Gravity variation in the Mammoth Lakes, Mono Lake and Owens Valley, California regions [abs.]: Earthquake Notes, v. 54, no. 1, p. 76.

Williams, S.N., 1985, Soil radon and elemental mercury distribution and relation to magmatic resurgence at Long Valley Caldera: Science, v. 229, p. 551-553.

Wood, S.H., 1977, Distribution, correlation, and radiocarbon dating of late Holocene tephra, Mono and Inyo Craters eruptions, eastern California: Geol. Soc. Am. Bull., v. 88, p. 89-95.

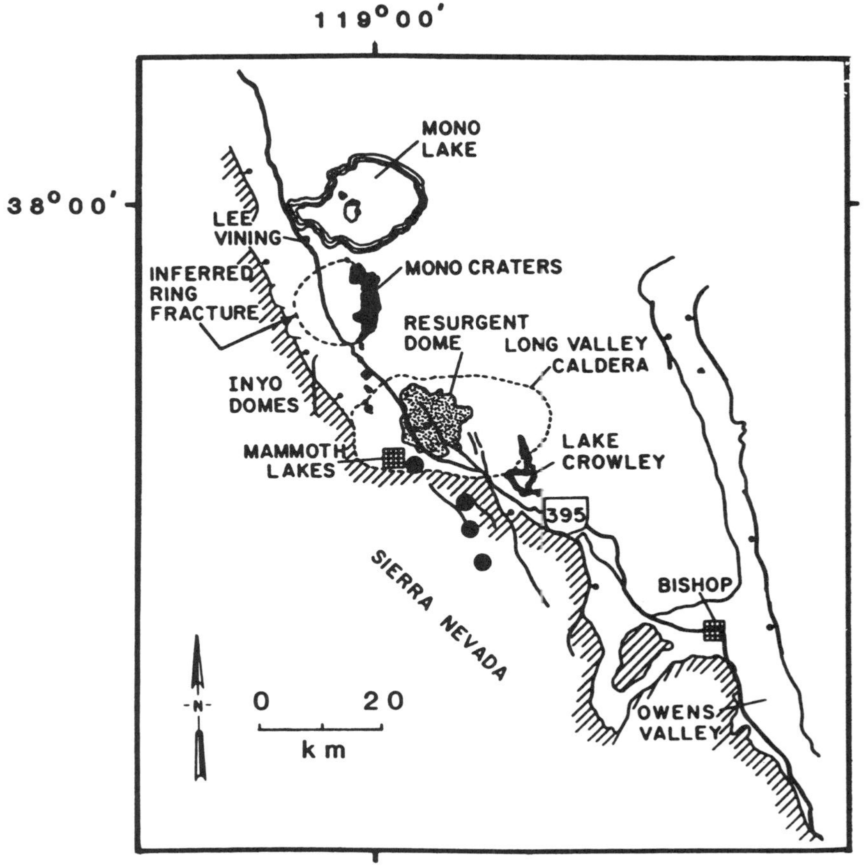

Figure 12.1.1. Generalized map of Long Valley/Mono Basin volcanic complex showing its position relative to Sierra Nevada to the west and Basin and Range province to the east (Hermance, 1983; after Bailey and others, 1976). Irregular black areas (Inyo domes, Mono Craters, and volcanic craters in Mono Lake) represent Holocene eruptive centers. Solid dots indicate epicenters of four magnitude 6 earthquakes during 25-27 May 1980. Copyright by the American Geophysical Union.

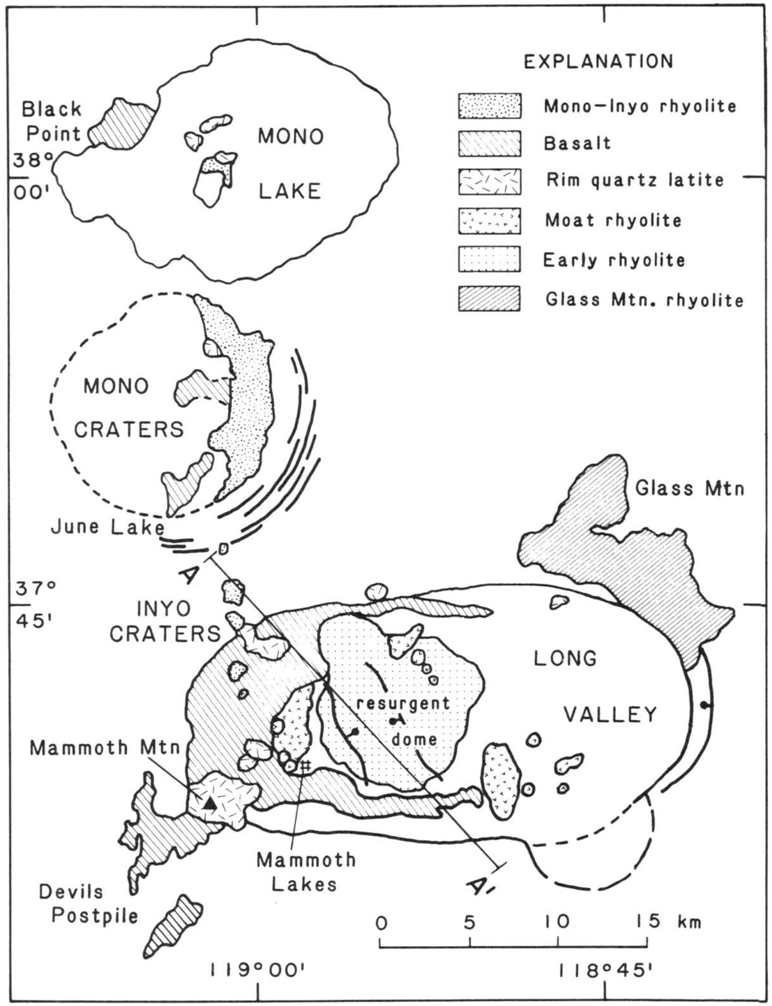

Figure 12.1.2. Generalized geologic map of Long Valley Caldera and Mono-Inyo Craters volcanic system, from Hill and others (1985c), after Bailey (1982). Line A-A' shows location of cross section in figure 12.1.3.

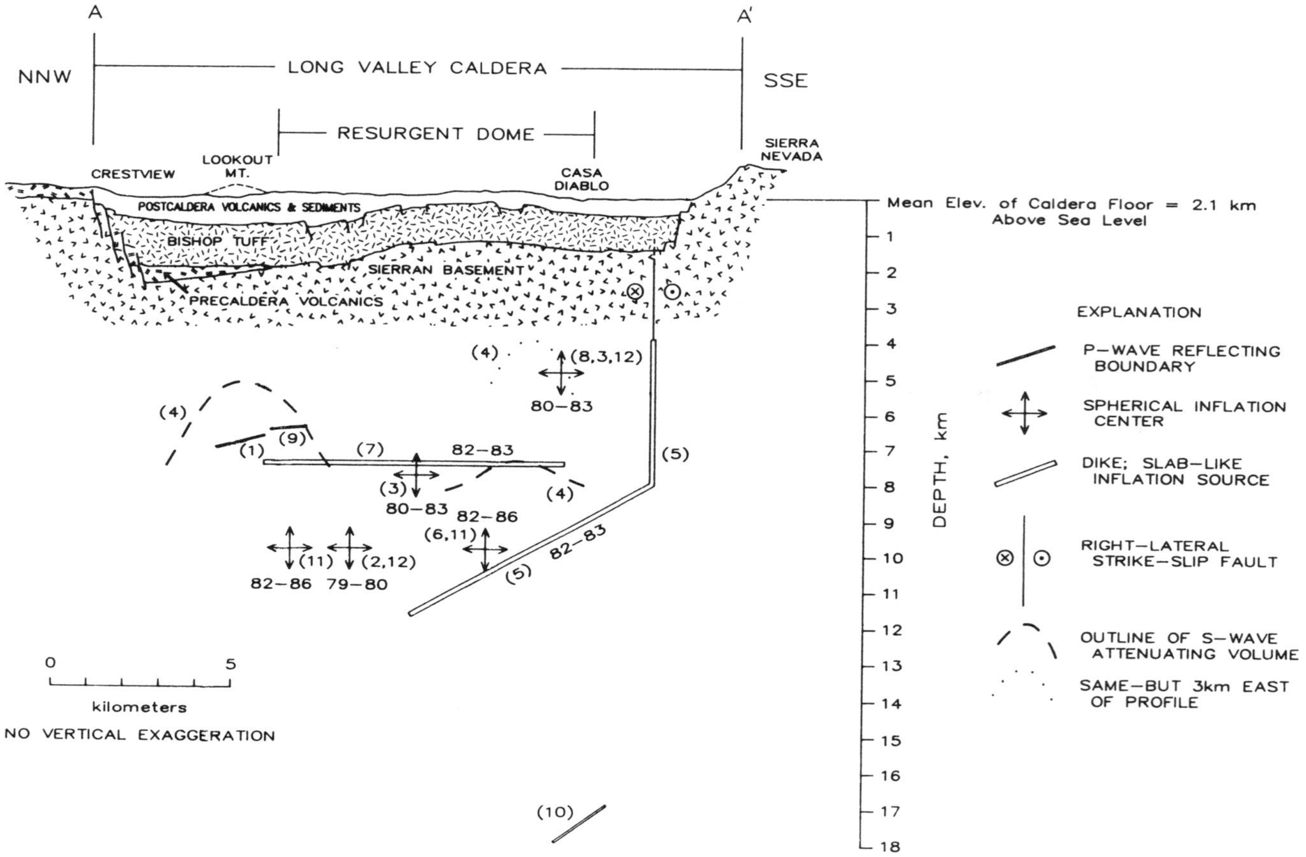

Figure 12.1.3. Vertical NW-SE section through Long Valley Caldera along line A-A' in figure 12.1.2, from Rundle and Hill (1988), after Hill and others (1985a) and Rundle and others (1986). Shallow caldera structure based on seismic-refraction results of Hill and others (1985b). Deeper structure beneath caldera shows location and geometry of inflation (dislocation) sources used to model deformation observed during years indicated (for example, 79-80), seismic P-wave reflecting boundaries, and volumes that produce anomalous S-wave attenuation. Parenthetic numbers give references: (1) Hill, 1976; (2) Savage and Clark, 1982; (3) Rundle and Whitcomb, 1984; (4) Sanders, 1984; (5) Savage and Cockerham, 1984; (7) Denlinger and Riley, 1984; (8) Denlinger and others, 1985; (9) Hill and others, 1985b; (10) Luetgert and Mooney, 1985; (11) Savage and others, 1987; (12) Langbein and others, 1987. Note that there is no seismic evidence for a large, interconnected magma reservoir at shallow depth beneath Long Valley Caldera (in contrast to the situation at Yellowstone Caldera, as shown in figures 12.2.6 and 12.2.7).

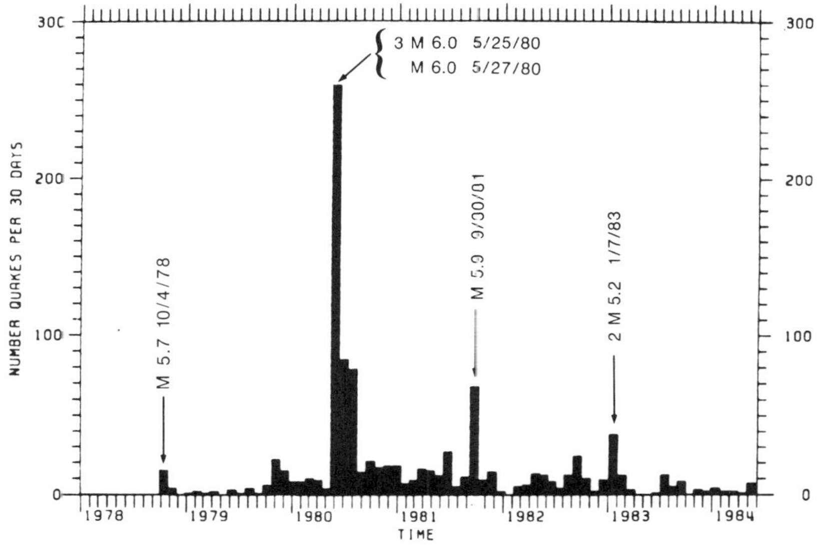

Figure 12.1.4. Temporal summary of seismic activity near Long Valley Caldera during 1978-84 in terms of occurrence rate of magnitude 3 and larger earthquakes per 30-day intervals (Hill and others, 1985c). Occurrence of principal earthquakes noted with arrows.

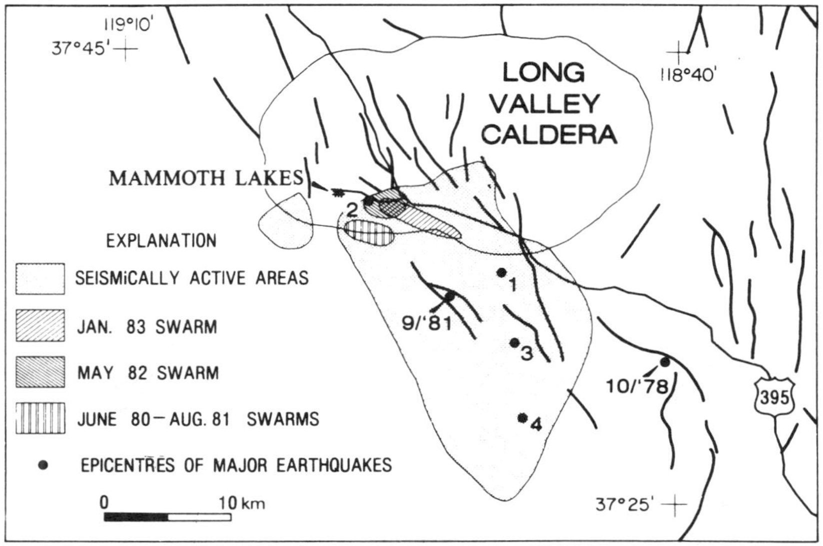

Figure 12.1.5. Map of Long Valley Caldera and vicinity, showing area of seismic activity and locations of earthquake swarms, from Julian (1983). Heavy lines, normal faults. Epicenters of four magnitude 6 earthquakes during 25-27 May 1980 are indicated with solid dots numbered in order of occurrence. Reprinted by permission from Nature; copyright 1983 by Macmillan Magazines, Ltd.

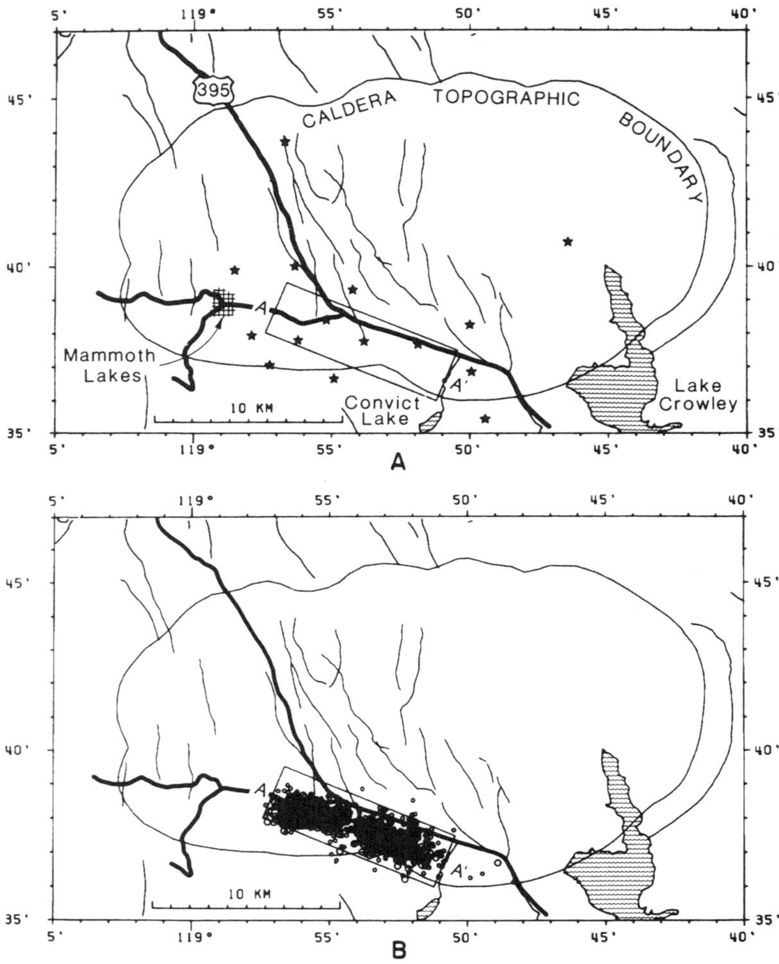

Figure 12.1.6. A, Map of Long Valley Caldera showing locations of 15 seismic stations (stars) used to locate epicenters shown below (Savage and Cockerham, 1984). B, Epicenters for 7-31 January 1983. Box encloses earthquakes plotted in cross-sectional views shown in figure 12.1.7. Outline of Long Valley Caldera and prominent faults are shown by light solid lines, principal highways by heavy lines. Copyright by the American Geophysical Union.

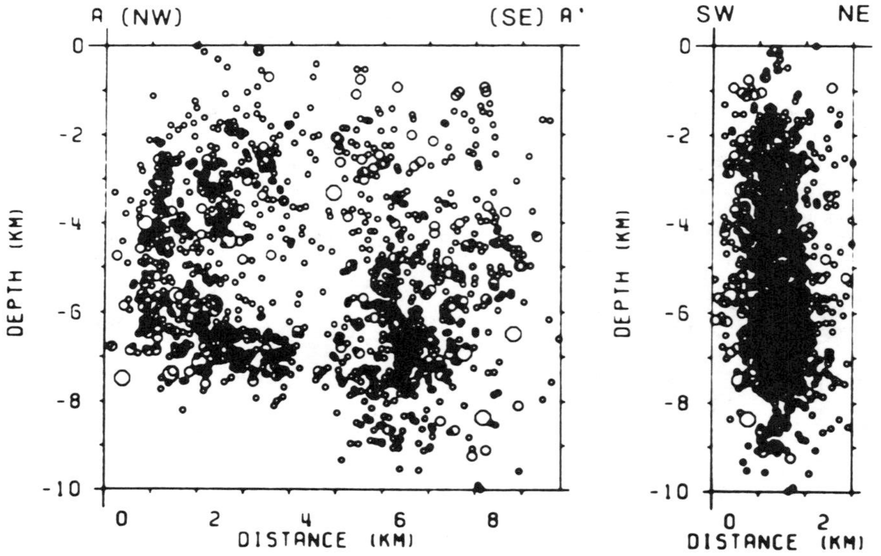

Figure 12.1.7. Projections of the hypocenters of earthquakes shown within box in figure 12.1.6 onto vertical planes striking west-northwest and south-southwest, from Savage and Cockerham (1984). Cross section A-A' is parallel to long dimension of box with A at west-northwest edge; other cross section is perpendicular to A-A'. Size of symbol is proportional to magnitude. Copyright by the American Geophysical Union.

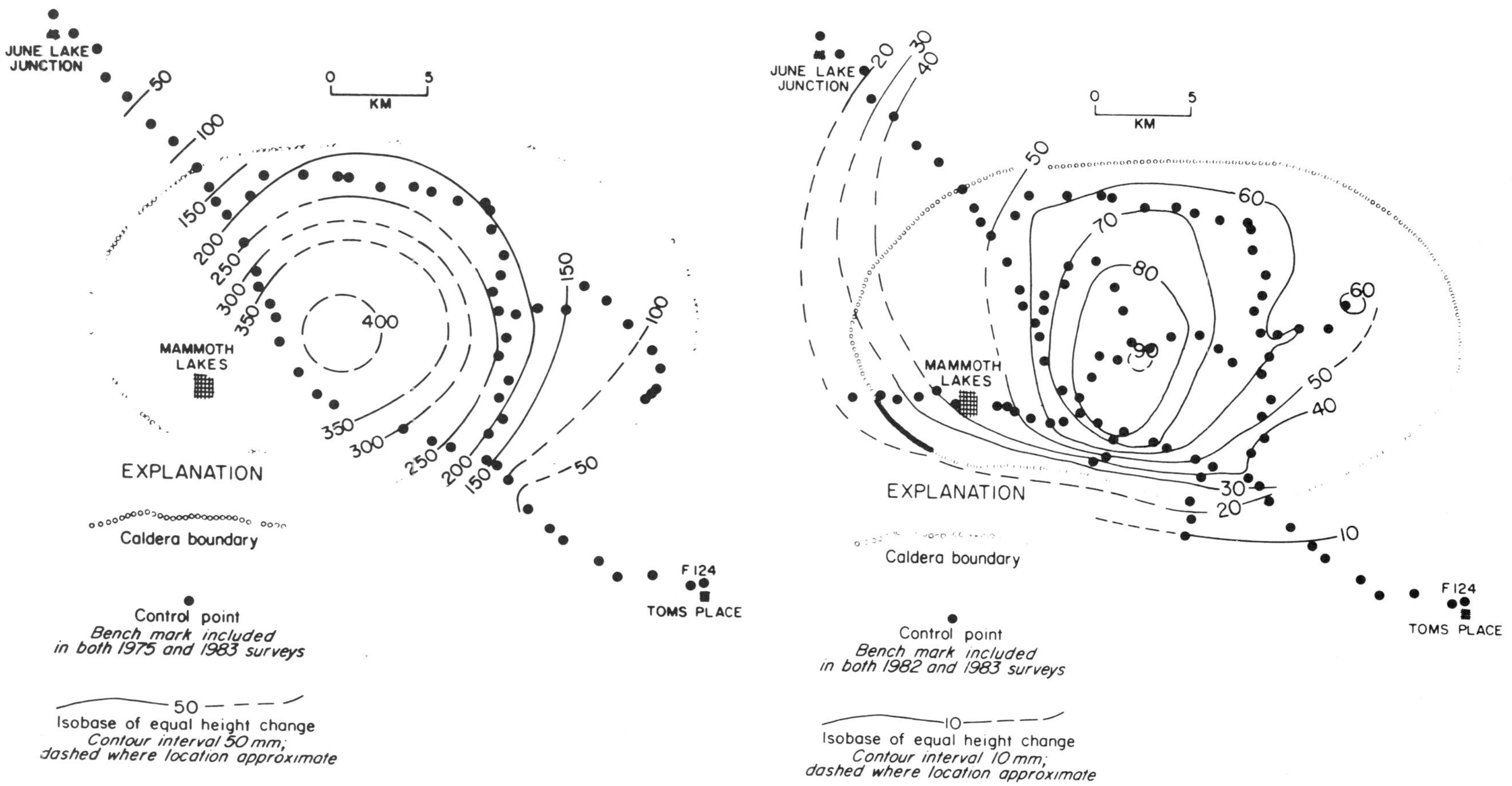

Figure 12.1.8. Uplift contours at Long Valley Caldera, 1975-83, from Castle and others (1984). Left, 1975-83; right, 1982-83. Both maps based on minimum constrained displacement adjustment with respect to bench mark F124. Copyright by the American Geophysical Union.

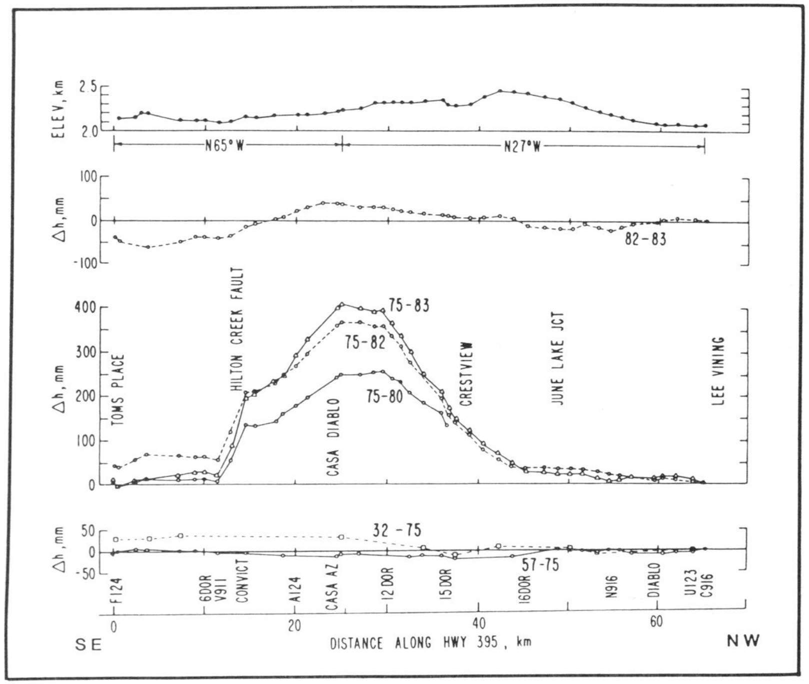

Figure 12.1.9. Elevation changes along U.S. Highway 395 across Long Valley Caldera from leveling surveys in 1932, 1957, 1975, 1980, 1982, and 1983. Note lack of appreciable deformation prior to 1975 and progressive uplift since 1980 (Hill and others, 1985c).

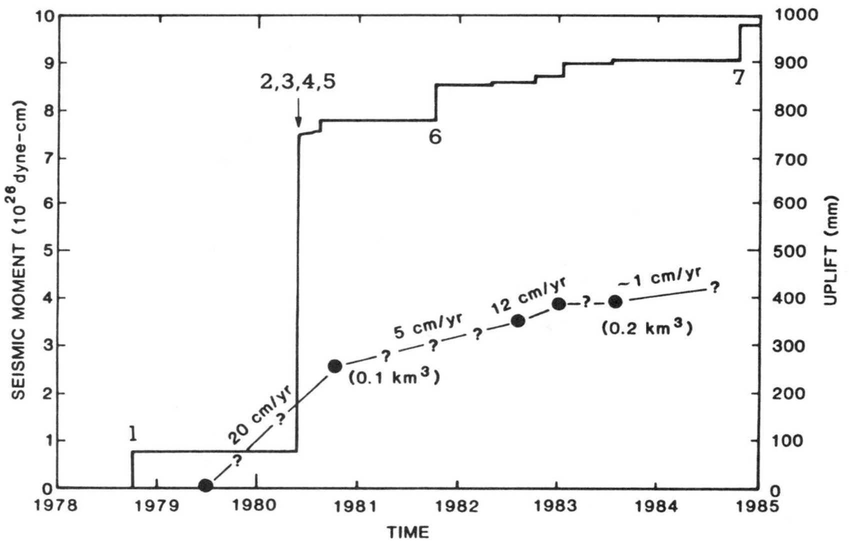

Figure 12.1.10. Summary of seismic activity and inflation of resurgent dome at Long Valley Caldera since 1978, from Hill and others (1985a). Solid line, cumulative seismic moment for earthquakes in Long Valley region from 1978 through mid-1984; solid dots, elevation history of bench mark CASA AZ near Casa Diablo Hot Springs (south part of resurgent dome). Copyright by the American Geophysical Union.

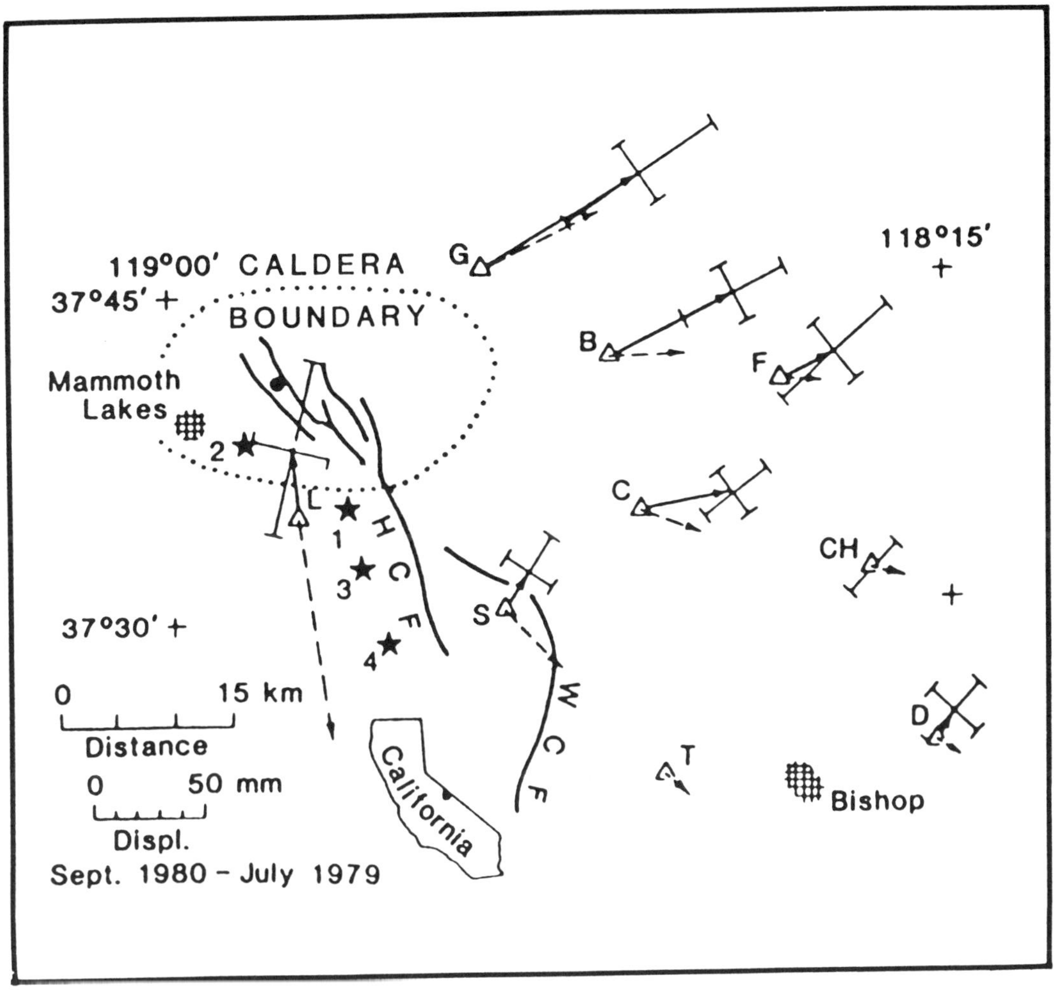

Figure 12.1.11. Horizontal displacements near Long Valley Caldera from precise trilateration surveys in July 1979 and September 1980, from Savage and Clark (1982). Solid arrows show observed displacements; dashed arrows show displacements expected from expansion of a spherical magma chamber at 11-km depth. Stars indicate epicenters of four magnitude 6 earthquakes in May 1980, in order of occurrence. Solid dot, center of resurgent dome; HCF, Hilton Creek Fault; WCF, Wheeler Crest Fault. Copyright by the American Association for the Advancement of Science.

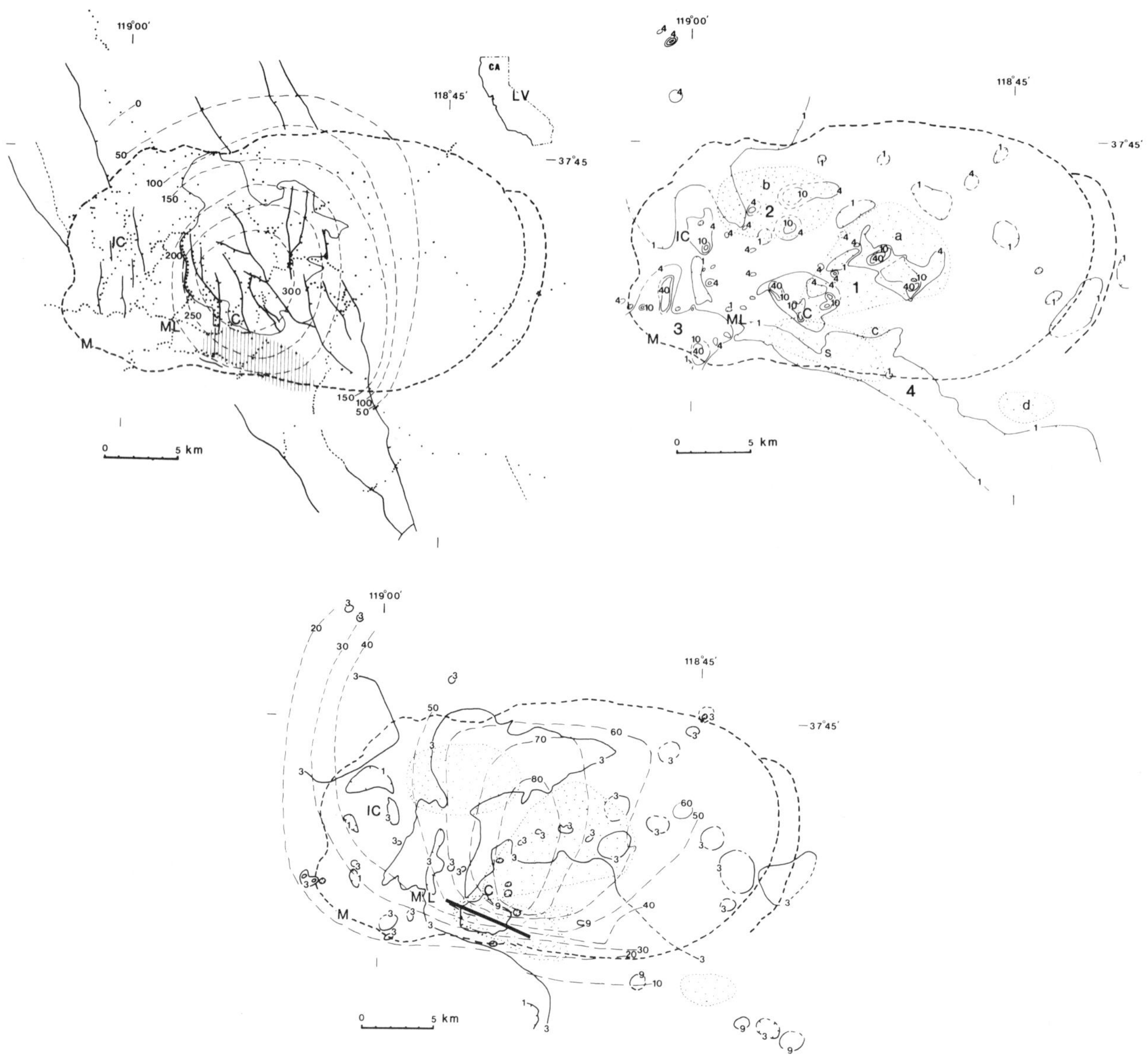

Figure 12.1.12. Top left, Long Valley Caldera (bold dashed lines) with principal faults (bold continuous lines), cumulative 1975-82 uplift in millimeters (light dashed lines), area of January 1983 seismic swarm (vertical ruling), outline of resurgent dome (light continuous line), and Hg^{O} and Rn sample sites (dots). Abbreviations: ML, city of Mammoth Lakes; IC, Inyo Craters; M, Mammoth Mountain; and C, Casa Diablo fumarole. Top right, contoured Hg^{O} anomalies (continuous lines, dashed where less certain; note irregular contour interval), magma bodies as inferred by Sanders (1984) (stippled areas a to d), and epicentral area of the January 1983 seismic swarm (hatched area labeled s). Hg^{O} contours are background (6.6 ppb), threshold (26.4 ppb), peak (66 ppb), and 264 ppb. Boldface numbers indicate broad Hg^{O} highs (areas 1 to 3) and an extensive Hg^{O} low (area 4). Bottom, contoured Rn anomalies (continuous lines, dashed where less certain), magma bodies (stippled areas), epicentral area of the January 1983 swarm (hatched), surface projection of an inferred dike (continuous heavy line), and 1982-83 uplift (light dashed lines), in millimeters. Radon contours are background (82 T/mm^2 per 30 days, threshold (246 T/mm^2 per 30 days), and peak (738 T/mm^2 per 30 days). Plots are from Williams (1985). Copyright by the American Association for the Advancement of Science.

YELLOWSTONE

CAVW number (active vent)	Latitude Longitude (degrees)	Diameter (km)	Local tectonic setting	Precaldera edifice	SiO_2 content (wt pct)	Age of collapse (yr B.P.)	Date of unrest	Type and duration of unrest ESTU STHF MGTF H Te	Eruption type
12-05-01	44.58N 110.53W	85 x 45	E-W exten with complications	None?	R = 70-77 C = 76-77	630,000	late 19th century and continuing	GG-G FG?x -GFx F QQ	none

TECTONIC SETTING

Yellowstone lies at the intersection of the Basin and Range tectonic province, dominated by E-W extension, and the eastern Snake River Plain, a linear downwarp or graben that has been a locus for basaltic volcanism since middle Miocene time (figs. 12.2.1, 12.2.2). According to one popular model, the rhyolitic Yellowstone Plateau marks the current location of a "hotspot" or melting anomaly in the upper mantle, and the basaltic Snake River Plain records the hotspot's northeastward track across the mobile North American plate (Smith and others, 1974, 1977).

Focal mechanisms (fig. 12.2.3) of a magnitude 7.5 earthquake and its aftershocks (Hebgen Lake, 1959, see below) suggested N-S extension near Hebgen Lake (about 70 km northwest of Yellowstone) and radial compression near the caldera (Ryall, 1962; Smith and others, 1974). Focal mechanisms of more recent earthquakes and geologic mapping in the caldera (Smith and others, 1977; Christiansen, 1984) suggest dominantly ENE-WSW or E-W extension.

GEOLOGIC HISTORY

Three times in the past 2 million years, large reservoirs of rhyolite magma have accumulated in the upper crust at Yellowstone, triggering cataclysmic eruptions and caldera collapses 2.0, 1.3, and 0.6 million years ago (fig. 12.2.1) (Christiansen, 1984). The first great eruption (2.0 m.y. B.P.) produced the Huckleberry Ridge Tuff (more than 2,450 km^3) and a composite caldera more than 75 km long, extending from Island Park on the west to central Yellowstone Park on the east. The second eruption (1.3 m.y. B.P.) produced the Mesa Falls Tuff (more than 280 km^3) and the Island Park Caldera west of Yellowstone Park; the third (0.6 m.y. B.P.) produced the Lava Creek Tuff (more than 1,000 km^3) and the present Yellowstone Caldera (fig. 12.2.2). Rhyolitic volcanism resumed within the Yellowstone Caldera after structural resurgence formed the Sour Creek and Mallard Lake resurgent domes. Renewed doming in the western caldera culminated with extrusion of 1,000 km^3 of intracaldera rhyolite flows between 150,000 and 75,000 yr B.P. (Christiansen, 1984).

YELLOWSTONE (continued)

GEOLOGIC HISTORY (continued)

There is abundant geophysical evidence for residual partial melt beneath Yellowstone Caldera (figs. 12.2.4-12.2.7) (Smith and others, 1974; Eaton and others, 1975; Lehman and others, 1982; Schilly and others, 1982), and the consensus among those who have studied the area is that the Yellowstone magmatic system will likely erupt again (Smith and Braile, 1983; Christiansen, 1984).

Deformed terraces along the shore of Yellowstone Lake and the Yellowstone River record a complex history of Holocene deformation that has continued to the present (fig. 12.2.8). Based on studies of the terraces, Hamilton (1985) proposed that intracaldera subsidence totaling more than 80 m occurred in stages during the early Holocene, and Meyer and Locke (1986) inferred net uplift of about 10 m in the northern Yellowstone Lake area during the late Holocene. The latter authors also cited geomorphic evidence that during the past several thousand years the lake fell to a level near the present, rose 6-7 m, fell to the present level or below, and now is rising again due to differential uplift of its outlet (see below).

Yellowstone Caldera currently contains one of the largest and most active hydrothermal systems in the world (fig. 12.2.9), and hydrothermal activity probably has been relatively constant for at least the past 10,000 years. The contemporary heat output of the Yellowstone magmatic system is 4×10^{16} cal/yr, or 5,500 MW (Fournier and Pitt, 1985; Fournier, 1986).

HISTORICAL ACTIVITY

Ferdinand V. Hayden, leader of the first scientific expedition to the Yellowstone area, reported that severe earthquakes were felt northeast of Yellowstone Lake in 1871. From that date until 1959, at least 76 earthquakes strong enough to be felt were reported in the Yellowstone Park - Hebgen Lake region. Smith and Braile (1983) report that at least seven of these exceeded magnitude 6 (fig. 12.2.10).

A magnitude 7.3 earthquake occurred in August 1959 near Hebgen Lake, 70 km northwest of Yellowstone Caldera. Most aftershocks and other earthquakes from 1959 to 1972 were just east of Hebgen Lake; a few shallow earthquakes occurred within the caldera, but none of these were in the northeast part of the caldera, where the shallowest geophysical anomalies occur. The Yellowstone Park earthquake ($\underline{M}$ 6.1) occurred on 30 June 1975 just inside the north caldera rim near Norris (Pitt and others, 1979; fig. 12.2.1).

Since 1972, many earthquakes have occurred near and within the caldera--more than can be attributed solely to an upgrading of the Yellowstone seismic network in late 1972. Events in the caldera typically occur in swarms less than 5 km deep, in contrast to events outside the caldera that are as much as 15 km deep. Many of the earthquakes occurring in the caldera are in areas of major hydrothermal activity. Swarms within the caldera include some just south of Norris Junction (1972, 1975), near the south end of Yellowstone Lake (1973), near Old Faithful (early 1978), Mud Volcano (late 1978), in West Thumb (several), and just outside the northwest caldera rim (late 1985-present). The 1985-present swarm includes 28 events larger than magnitude 3.5 from 16 October to 27 December 1985. Events of comparable size also occurred in 1986 and 1987, but the intensity of the swarm seemed to be waning as of late 1987.

YELLOWSTONE (continued)

HISTORICAL ACTIVITY (continued)

Leveling surveys in 1923 and 1975-77 detected a broad pattern of intracaldera uplift of up to 72 cm along a northeast-trending axis within Yellowstone Caldera (fig. 12.2.11) (Reilinger and others, 1977; Pelton and Smith, 1979, 1982). Annual surveys since 1983 show that the central caldera floor rose at an average rate of 15-17 mm/yr from 1976 to 1986, but that uplift stopped during 1984-85 and subsidence began during 1985-87 (Dzurisin and Yamashita, 1987 and unpubl. data) (figs. 12.2.12-12.2.14). The maximum uplift measured since the first survey in 1923 is about 86 cm near the Mud Volcano area (Dzurisin and Yamashita, 1987). No other geodetic measurements are available to determine whether the uplift occurred at a relatively constant rate between leveling surveys, or perhaps began around the time of the 1959 Hebgen Lake earthquake. However, gauging records from Yellowstone Lake suggest that tilting of the lakeshore has occurred at a relatively constant rate since 1923, with no discernible change associated with the 1959 earthquake (Hamilton, 1984). The 1923 and 1975-77 leveling surveys showed that deformation extended well beyond the caldera rim, indicating that the phenomenon is not wholly volcanic in origin. For example, doming of an 8,000 km^2 area surrounding the epicenter of the 1959 Hebgen Lake earthquake probably reflects postseismic viscoelastic relaxation of the crust in response to that earthquake (Reilinger and others, 1977; Reilinger, 1986).

Marler (1964) and Marler and White (1975) documented major changes in hydrothermal activity of the Yellowstone area as a result of the 1959 Hebgen Lake earthquake. Discharge at many hot springs increased within minutes to hours of the earthquake and in some cases was sustained for years after the earthquake. Increases in hot-spring and geyser activity were probably a result of new fractures formed in self-sealed caps that had kept water at elevated pressures, above hydrostatic pressure, and had moved boiling-point curves to abnormally high temperatures for given depths (White and others, 1975). The Yellowstone Park earthquake of 30 June 1975 also caused impressive hydrothermal changes throughout the caldera (R.A. Hutchinson, oral commun., 1987). The magnitude 7.3 Borah Peak, Idaho earthquake (250 km east-southeast of Yellowstone) on 28 October 1983 caused the average interval between eruptions at Old Faithful geyser to increase by more than 7 minutes from late October 1983 to early March 1984. The 1984 annual average interval (76.3 minutes) is the longest known in more than 115 years of observation (R.A. Hutchinson, written commun., 1985). Starting in April 1984, the average interval between eruptions started to decline toward its pre-earthquake value of 69.2 minutes. Then about 10 days after a small earthquake swarm beneath the Pitchstone Plateau (25 km south of Old Faithful) on 6 June 1987, the interval suddenly increased from 70.1 minutes to 75.7 minutes, apparently as a delayed response to the swarm (R.A. Hutchinson, written commun., 1988).

Hydrothermal changes in 1959, 1975, and 1983 followed large earthquakes, not vice versa. However, the relation between subsequent seismic swarms and hydrothermal changes has not been as clear, and hydrothermal activity may have influenced recent seismicity. Increased seismicity at Mud Volcano in 1978 preceded an increase in thermal activity there, but increased thermal activity in the Norris geyser basin in 1972 (and on several previous occasions) preceded seismic activity. At Norris, Trimble and Smith (1975) suggest that high fluid pressure and temperature reduce rock strength and induce earthquake swarms; Pitt and Hutchinson (1982) grant that this is plausible, but they note that thermal changes in the Norris basin are related to annual fluctuation of the water table and are only sometimes followed or accompanied by earthquakes.

YELLOWSTONE (continued)

COMMENTS

Yellowstone Caldera is the largest caldera in this compilation and is unique for several other reasons. The Yellowstone region has produced three caldera-forming eruptions in the past 2 million years, two of those among the largest eruptions known to have occurred on Earth (each more than 1,000 km^3). Yellowstone's hydrothermal system is among the largest and most active in the world, and its historical seismicity and uplift are comparable to those at the most active calderas in this compilation. In spite of this, the implications of historical unrest at Yellowstone for short-term tectonic and volcanic hazards are not well understood. For example, it is not known whether recent unrest is typical of quiescent periods or precursory to the next major earthquake or eruption. It is also not clear whether the Yellowstone magmatic system is in the waning stage of its third volcanic cycle that climaxed 630,000 years ago or in the early stage of a fourth cycle.

Possible causes of historical uplift at Yellowstone include (1) inflation owing to basaltic intrusions into the upper crust beneath the caldera (Pelton and Smith, 1979, 1982), (2) pressurization of a deep hydrothermal system by accumulation of magmatic fluids that are released during cooling and crystallization of a silicic magma body (Fournier, 1986), and (3) increased compressive strain that upwarps the hot and mechanically weak crust beneath the caldera (Meertens and Levine, 1985; Meertens, 1987). Over geological time scales, Yellowstone's vigorous hydrothermal system must derive its heat ultimately from basaltic intrusions, because the system's current heat flux would result in complete crystallization and cooling of about 10^4 km^3 of rhyolite in 10^5 years (Fournier, 1986). Stated differently, a 3-km-thick lens of rhyolite beneath the entire caldera would have completely cooled since the last magmatic eruption 75,000 yr B.P. unless there was additional heat input from basaltic intrusions sometime during that period. However, there is no compelling evidence that basalt has intruded beneath the caldera during historical time, because the current heat flux could be supplied entirely by heat derived from cooling of already crystalline rocks, or from crystallization of rhyolitic magma (Fournier, 1986). Evidence for the release of magmatic fluids from beneath the caldera includes high $^3He/^4He$ ratios and high chloride concentrations in hydrothermal waters (Fournier and Pitt, 1985). It is possible that both magmatic inflation and cooling/crystallization are occurring simultaneously, the first near the base of the silicic reservoir and the second at higher levels within the reservoir. With regard to the third mechanism listed above, Meertens and Levine (1985) and Meertens (1987) suggested that, although Yellowstone lies in zone of regional tectonic extension, compressive strain beneath the caldera could be a consequence either of magmatic inflation beneath the Hebgen Lake region or of subsurface movements during or after the 1959 $\underline{M} = 7.3$ Hebgen Lake earthquake.

Processes that might account for subsidence include (1) deflation owing to withdrawal of magma from beneath the caldera, most likely by lateral intrusion of basalt from beneath the silicic magma reservoir into preexisting zones of weakness along the caldera rim, (2) depressurization of the deep hydrothermal system caused by rupturing of an impermeable cap and upward migration of hydrothermal fluids, and (3) regional tectonic extension. The only suggestion that magma might have moved laterally from beneath the caldera since the onset of subsidence is a prolonged swarm of earthquakes just outside the northwest caldera rim that began in October 1985. However, that swarm occurred in the vicinity of the Hebgen Lake fault zone and probably was primarily tectonic in origin (R.B. Smith, oral commun., 1988). Measurements of total chloride flux from within the caldera show no definite increase during the period of subsidence, but an increase of 10 to 15 percent could have gone undetected (R. Fournier, oral commun., 1988).

YELLOWSTONE (continued)

COMMENTS (continued)

There is no evidence from precise trilateration measurements for unusual crustal extension in the Yellowstone region since the onset of subsidence: strain rates have been relatively linear during 1972-87 in the Hebgen Lake area, and only small movements that were consistent with the observed subsidence occurred during 1984-87 in the northeast part of Yellowstone Caldera (J.C. Savage, written commun., 1988).

REFERENCES

Christiansen, R.L., 1984, Yellowstone magmatic evolution: its bearing on understanding large-volume explosive volcanism, in Boyd, F.R., ed., Explosive volcanism: Inception, evolution, and hazards: National Research Council Studies in Geophysics, Washington, D.C., National Academy Press, p. 84-95.

----- 1988, The Quaternary and Pliocene Yellowstone Plateau volcanic field of Wyoming, Idaho, and Montana: U.S. Geol. Surv. Prof. Paper 729 [in press].

Dzurisin, D., and Yamashita, K.M., 1987, Vertical surface displacement at Yellowstone Caldera, Wyoming, 1976-1986: Jour. Geophys. Res., v. 92, no. B13, p. 13,753-13,766.

Eaton, G.P., Christiansen, R.L., Iyer, H.M., Pitt, A.M., Mabey, D.R., Blank, H.R., Jr., Zietz, I., and Gettings, M.E., 1975, Magma beneath Yellowstone National Park: Science, v. 188, p. 787-796.

Fournier, R.O., 1986, Hydrothermal activity in relation to late Holocene dynamics in the Yellowstone caldera: Proceedings, Science in the National Parks Conference, Colorado State University, Fort Collins, Colorado, July 1986.

Fournier, R.O., and Pitt, A.M., 1985, The Yellowstone magmatic-hydrothermal system, in Stone, Claudia, ed., Geothermal resources Council 1985 International Symposium on Geothermal Energy, Transactions: Int. Vol., p. 319-327.

Hamilton, W.L., 1984, Applying lake level gaging records to the investigation of uplift within the Yellowstone caldera, Yellowstone National Park [abs.]: Eos, Trans. Am. Geophys. Union, v. 65, p. 192-193.

----- 1985, Deformation models for interpreting warped Holocene terraces of Yellowstone Lake and the Yellowstone River at the Yellowstone caldera, Wyoming [abs.]: Eos, Trans. Am. Geophys. Union, v. 66, no. 46, p. 853.

Iyer, H.M., Evans, J.R., Zandt, G., Stewart, R.M., Cookley, J.M., and Roloff, J.N., 1981, A deep low velocity body under the Yellowstone caldera, Wyoming; Delineation using teleseismic P-wave residuals and tectonic interpretations: Summary: Geol. Soc. Am. Bull., v. 92, p. 792-798.

Lehman, J.A., Smith, R.B., Schilly, M.M., and Braile, L.W., 1982, Upper crustal structure of Yellowstone from seismic and gravity observations: Jour. Geophys. Res., v. 87, no. B4, p. 2713-2730.

Marler, G.D., 1964, Effects of the Hebgen Lake earthquake of August 17, 1959 on the hot springs of the Firehole Geyser basins, Yellowstone National Park: Geol. Soc. Am. Bull., v. 86, p. 749-759.

Marler, G.D., and White, D.E., 1975, Seismic geyser and its bearing on the evolution of geyser and hot springs of Yellowstone National Park: Geol. Soc. Am. Bull., v. 86, p. 749-759.

Meyer, G.A., and Locke, W.W., 1986, Origin and deformation of Holocene shoreline terraces, Yellowstone Lake, Wyoming: Geology, v. 14, p. 699-702.

YELLOWSTONE (continued)

REFERENCES (continued)

Meertens, C.M., 1987, Tilt tides and tectonics at Yellowstone National Park: Boulder, Univ. of Colorado, Ph.D. thesis, 236 p.

Meertens, C., and Levine, J., 1985, Compressive tectonic strain as a possible mechanism for long-term vertical deformation of the Yellowstone caldera: Eos, Trans. Am. Geophys. Union, v. 66, no. 46, p. 853.

Muffler, L.J.P., White, D.E., and Truesdell, A.H., 1971, Hydrothermal explosive craters in Yellowstone National Park: Geol. Soc. Am. Bull., v. 82, p. 723-740.

Nicholls, H.R., and Rinehart, J.S., 1967, Geophysical study on geyser action in Yellowstone National Park: Jour. Geophys. Res., v. 72, p. 4651-4663.

Pelton, J.R., and Smith, R.B., 1979, Recent crustal uplift in Yellowstone National Park: Science, v. 206, p. 1179-1182.

----- 1982, Contemporary vertical surface displacements in Yellowstone National Park: Jour. Geophys. Res., v. 87, p. 2745-2761.

Pitt, A.M., and Hutchinson, R.A., 1982, Hydrothermal changes related to earthquake activity at Mud Volcano, Yellowstone National Park, Wyoming: Jour. Geophys. Res., v. 87, p. 2762-2766.

Pitt, A.M., Weaver, C.S., and Spence, W., 1979, The Yellowstone Park earthquake of June 30, 1975: Bull. Seismol. Soc. Am., v.69, no. 1, p. 187-205.

Reilinger, R.E., 1986, Evidence for postseismic viscoelastic relaxation following the 1959 $\underline{M}$ = 7.5 Hebgen Lake, Montana earthquake, Jour. Geophys. Res., v. 91, no. B9, p. 9488-9494.

Reilinger, R.E., Citron, G.P., and Brown, L.D., 1977, Recent vertical crustal movements from precise leveling data in southwestern Montana, western Yellowstone National Park, and the Snake River Plain: Jour. Geophys. Res., v. 82, p. 5349-5359.

Ryall, A., 1962, The Hebgen Lake, Montana, earthquake of August 18, 1959: P-waves: Bull. Seismol. Soc. Am., v. 52, p. 235-271.

Schilly, M.M., Smith, R.B., Braile, L.W., and Ansorge, J., 1982, The 1978 Yellowstone-Eastern Snake River Plain seismic profiling experiment: data and upper crustal structure of the Yellowstone region, Jour. Geophys. Res., v. 87, no. B4, p. 2692-2704.

Smith, R.B., and Braile, L.W., 1983, Crustal structure and evolution of an explosive silicic volcanic system at Yellowstone National Park, in Boyd, F.R., ed., Explosive volcanism: Inception, evolution, and hazards: National Research Council Studies in Geophysics, Washington, D.C., National Academy Press, p. 96-109.

Smith, R.B., and Christiansen, R.L., 1980, Yellowstone Park as a window on the Earth's interior: Sci. Am., v. 242, p. 104-117.

Smith, R.B., Shuey, R.T., Freidline, R.O., Otis, R.M., and Alley, L.B., 1974, Yellowstone hot spot: new magnetic and seismic evidence: Geology, v. 2, p. 451-455.

Smith, R.B., Shuey, R.T., Pelton, J.R., and Bailey, J.P., 1977, Yellowstone hot spot: contemporary tectonics and crustal properties from earthquake and aeromagnetic data, Jour. Geophys. Res., v. 82, p. 3665-3676.

Trimble, A.B., and Smith, R.B., 1975, Seismicity and contemporary tectonics of the Hebgen Lake-Yellowstone Park region: Jour. Geophys. Res., v. 80, p. 733-741.

White, D.E., Fournier, R.O., Muffler, L.J.P., and Truesdell, A.H., 1975, Physical results of research drilling in thermal areas of Yellowstone National Park, Wyoming: U.S. Geol. Surv. Prof. Paper 892, 70 p.

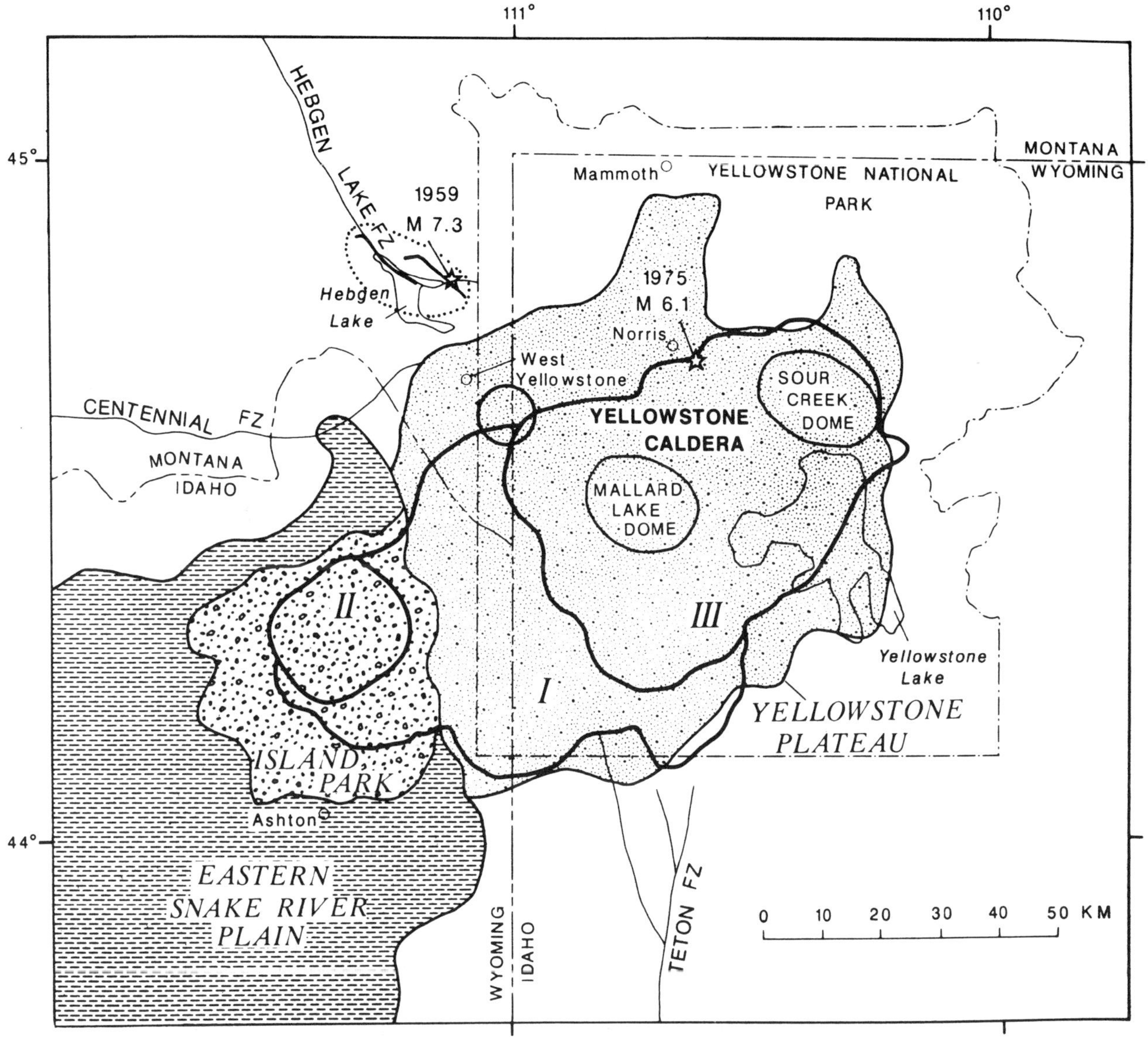

Figure 12.2.1. Yellowstone Plateau volcanic field, after Christiansen (1984). Rhyolitic Yellowstone Plateau lies along northeastward extension of basaltic Snake River Plain. Also shown are the Mallard Lake and Sour Creek resurgent domes, approximate outlines of three large calderas that formed in the past 2 million years (I, II, and III in order of formation), epicenters (stars) of the 1959 Hebgen Lake earthquake ($\underline{M}$ 7.3) and 1975 Yellowstone Park earthquake ($\underline{M}$ 6.1), and outline of Yellowstone National Park (irregular broken line). Reprinted with permission of National Academy Press, Washington, D.C.

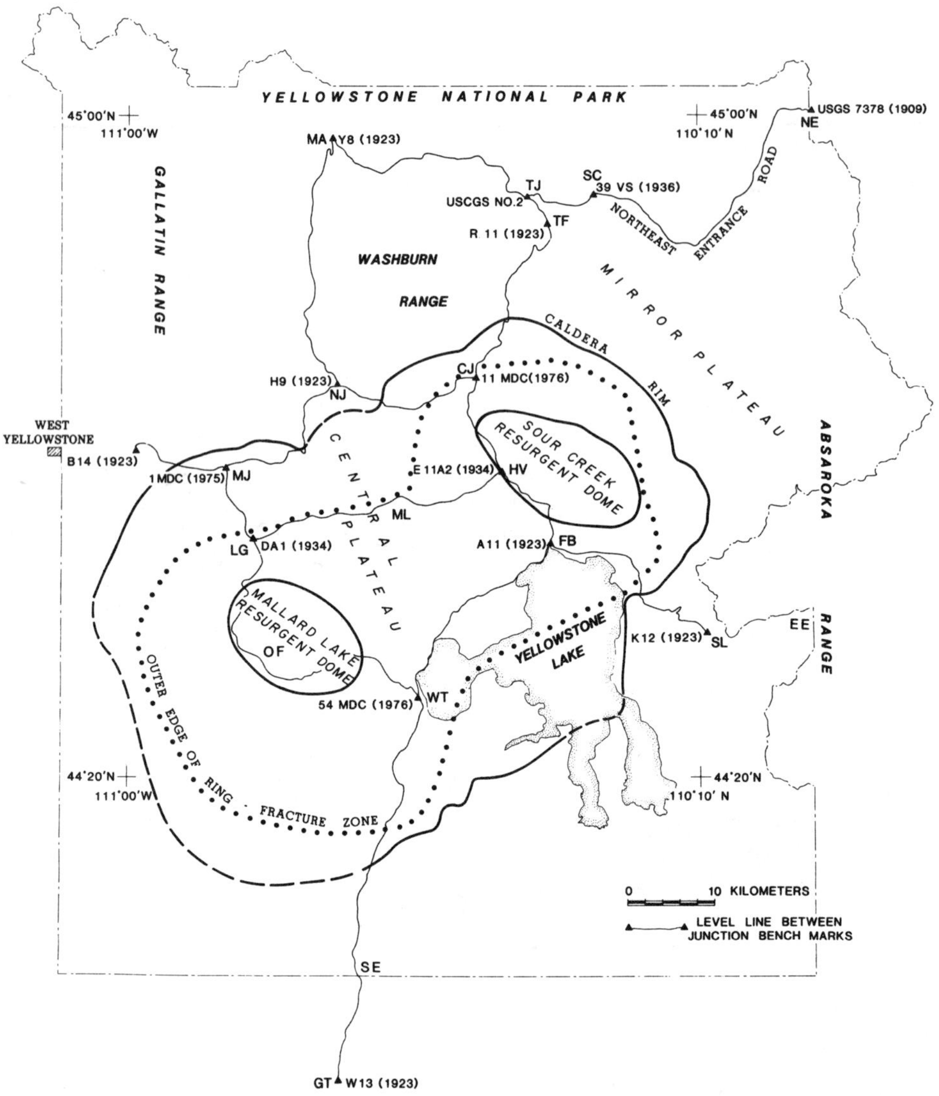

Figure 12.2.2. Yellowstone National Park (irregular broken line) and Yellowstone Caldera (topographic rim, heavy line; ring fracture, dotted line), from Pelton and Smith (1982). Also shown are Mallard Lake and Sour Creek resurgent domes, Yellowstone Lake, and leveling routes for surveys in 1923 and 1975-77 (fine lines). MA, Mammoth Hot Springs; TJ, Tower Junction; TF, Tower Falls; SC, Slough Creek turnoff; NE, Northeast Entrance; SL, Sylvan Lake; GT, Grand Teton National Park entrance; SE, South Entrance; WT, West Thumb; OF, Old Faithful; LG, Lower Geyser Basin; MJ, Madison Junction; NJ, Norris Junction; CJ, Canyon Junction; HV, Hayden Valley; ML, Mary Lake; FB, Fishing Bridge; EE, East Entrance. Lake Butte is near bench mark K12 (1923); Lewis Lake is near ring fracture south of WT (see figures 12.2.11 - 12.2.14 for leveling results). Copyright by the American Geophysical Union.

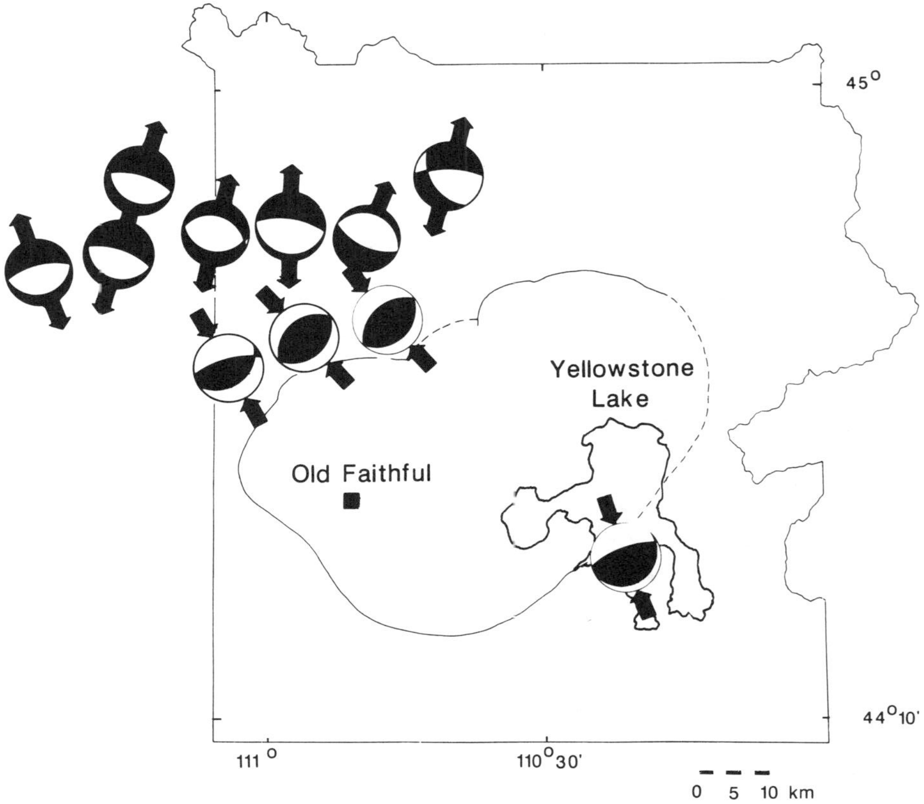

Figure 12.2.3. Fault-plane solutions for earthquakes in Hebgen Lake and Yellowstone regions, from Smith and others (1974). Upper hemisphere, equal-area stereographic nets. Dark quadrants, compression; light quadrants, dilatation. Inward arrows, P-axes; outward arrows, T-axes.

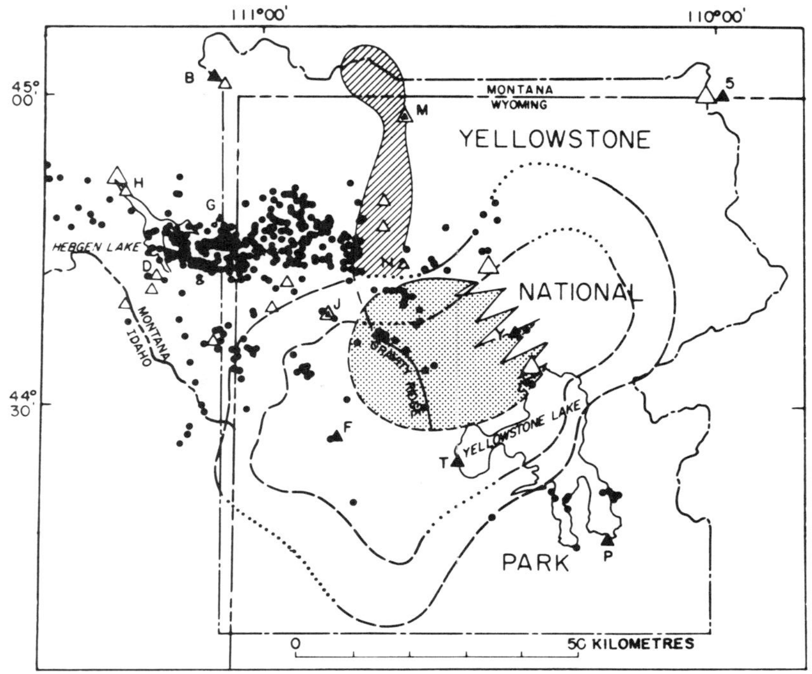

Figure 12.2.4. Seismicity and seismic attenuation in Yellowstone Plateau region, from Eaton and others (1975). Triangles identified by letters indicate seismograph stations operated from 1963 to 1975. Circles denote earthquake epicenters. Stippled area defines a zone of observed attenuation for compressional seismic waves from local earthquakes; shear waves are absent on some records for paths through this zone. Dashed curves denote top and bottom of local gravity gradient; dotted curves mark interrupted portions. Cross-ruling defines Norris-Mammoth corridor. Copyright by the American Association for the Advancement of Science.

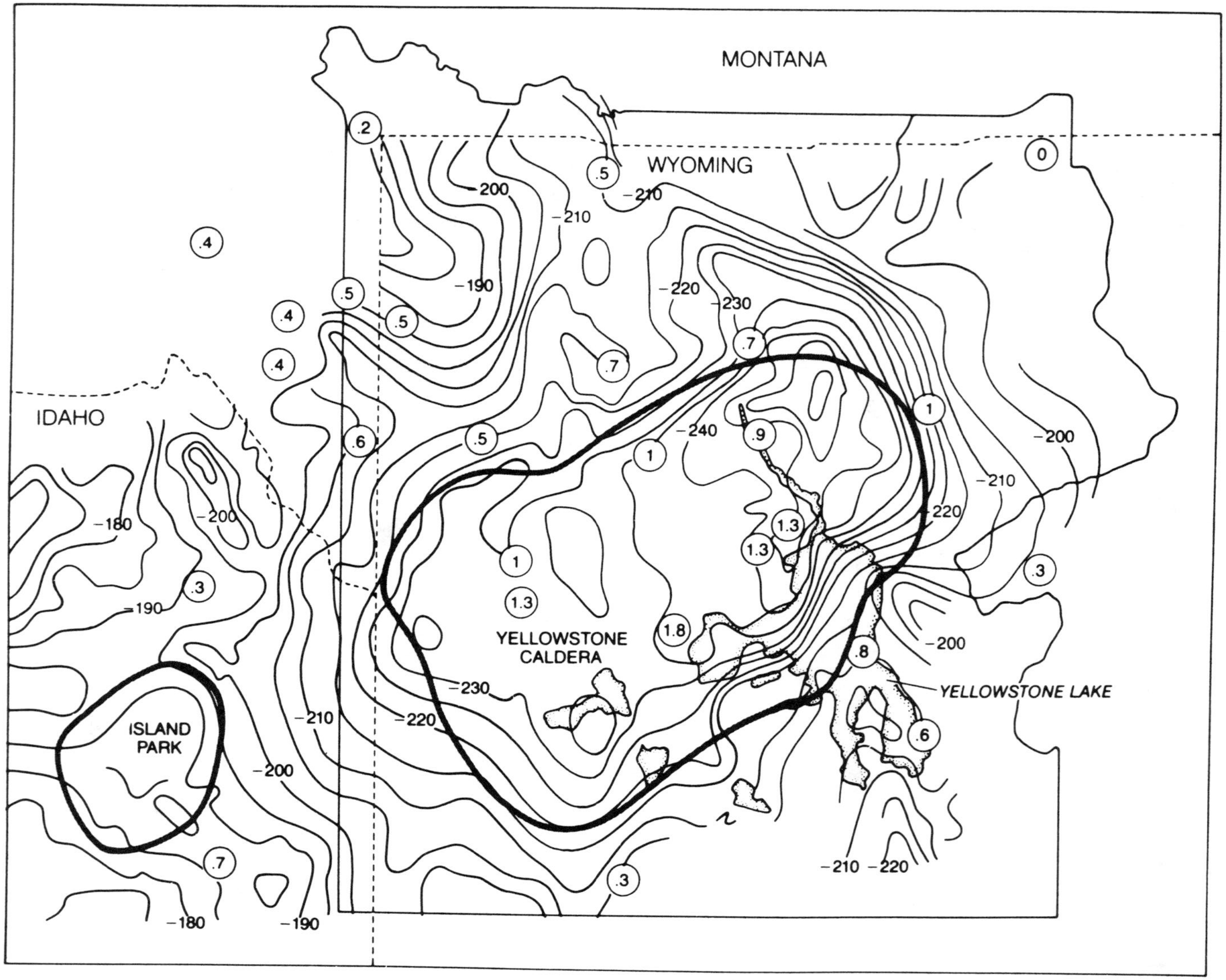

Figure 12.2.5. Bouguer gravity anomaly map and average P-wave delays in Yellowstone region, from Smith and Braile (1983). Contour interval, 5 milligals. Open circles, locations of seismograph stations that recorded teleseismic P-wave delays. Average delays are shown within circles in seconds. P-wave delays were normalized to seismograph station in northeast corner of Yellowstone National Park. Reprinted with permission of National Academy Press, Washington, D.C.

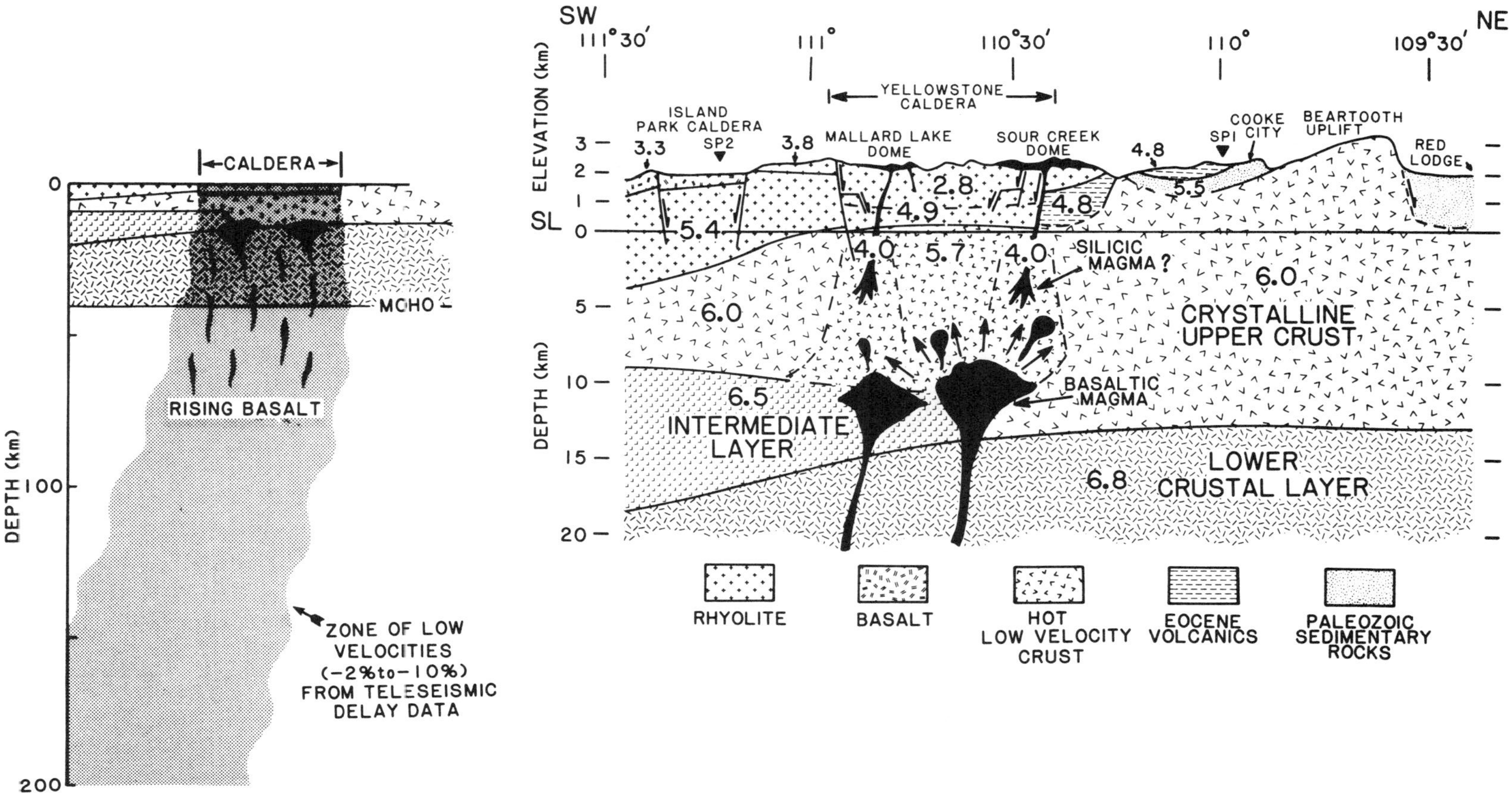

Figure 12.2.6. Idealized NE-SW geologic-seismic velocity model for crustal structure of Yellowstone - Island Park - Snake River Plain region, from Smith and Braile (1983). P-wave velocities are given in kilometers per second; low velocities in lower crust and upper mantle inferred from teleseismic delay studies (Iyer and others, 1981). Basaltic melt rising into crust from upper mantle provides thermal energy required to maintain silicic magma reservoir and active hydrothermal system in upper crust. Reprinted with permission of the National Academy Press, Washington, D.C.

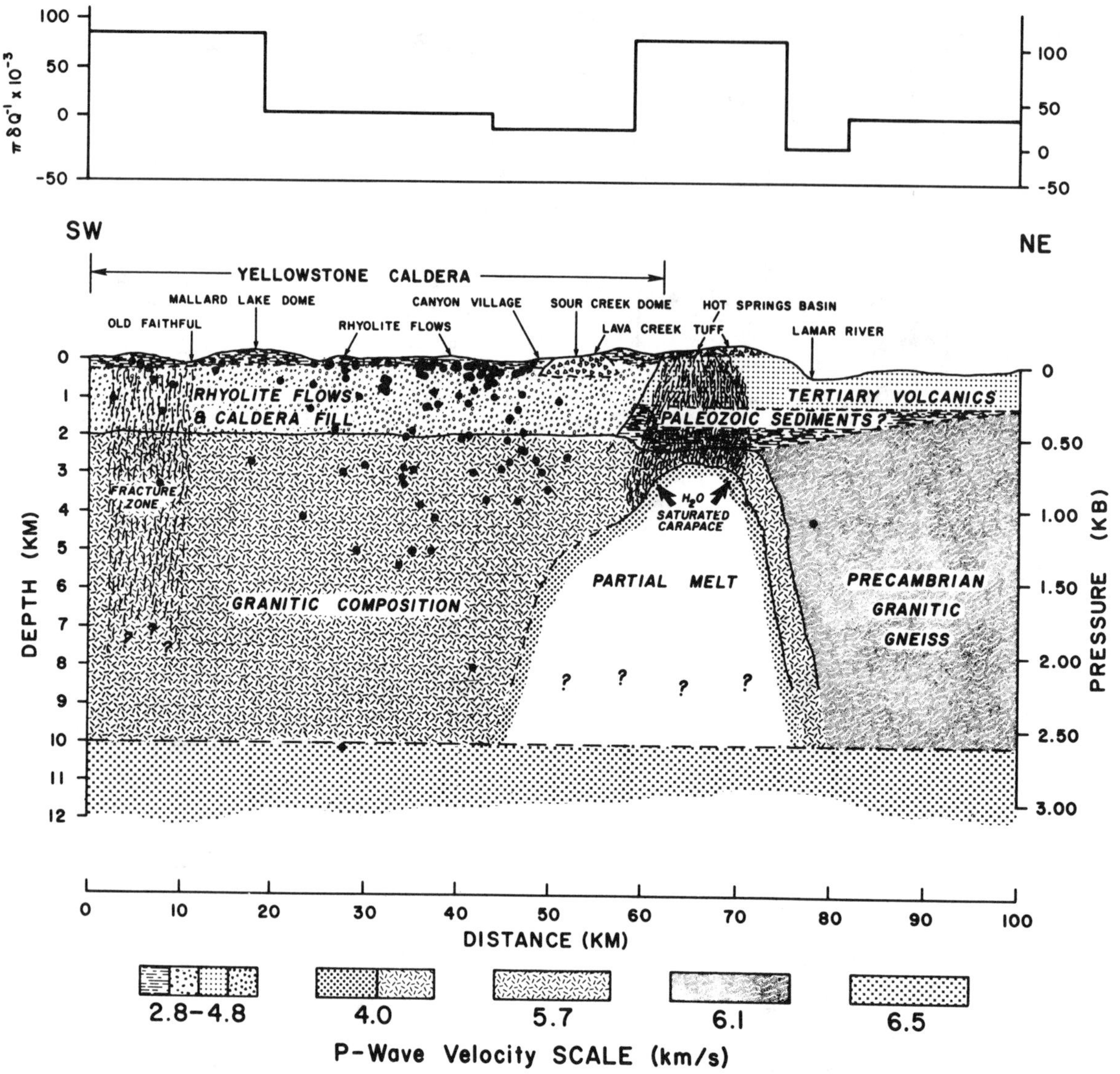

Figure 12.2.7. Relative P_g seismic attenuation (top) and generalized upper-crustal cross section (bottom) oriented SW-NE, parallel to axis of Yellowstone Caldera, from Smith and Braile (1983). Note in upper plot that apparent-Q is high over a low-velocity body beneath northeast caldera rim (inferred to be partial melt) and also over southwest caldera rim (inferred to be an area of fracturing and strong hydrothermal activity). Lowest P-wave velocities occur within caldera fill. Focal depths (black dots) were projected from outside low-velocity body. Reprinted with permission of the National Academy Press, Washington, D.C.

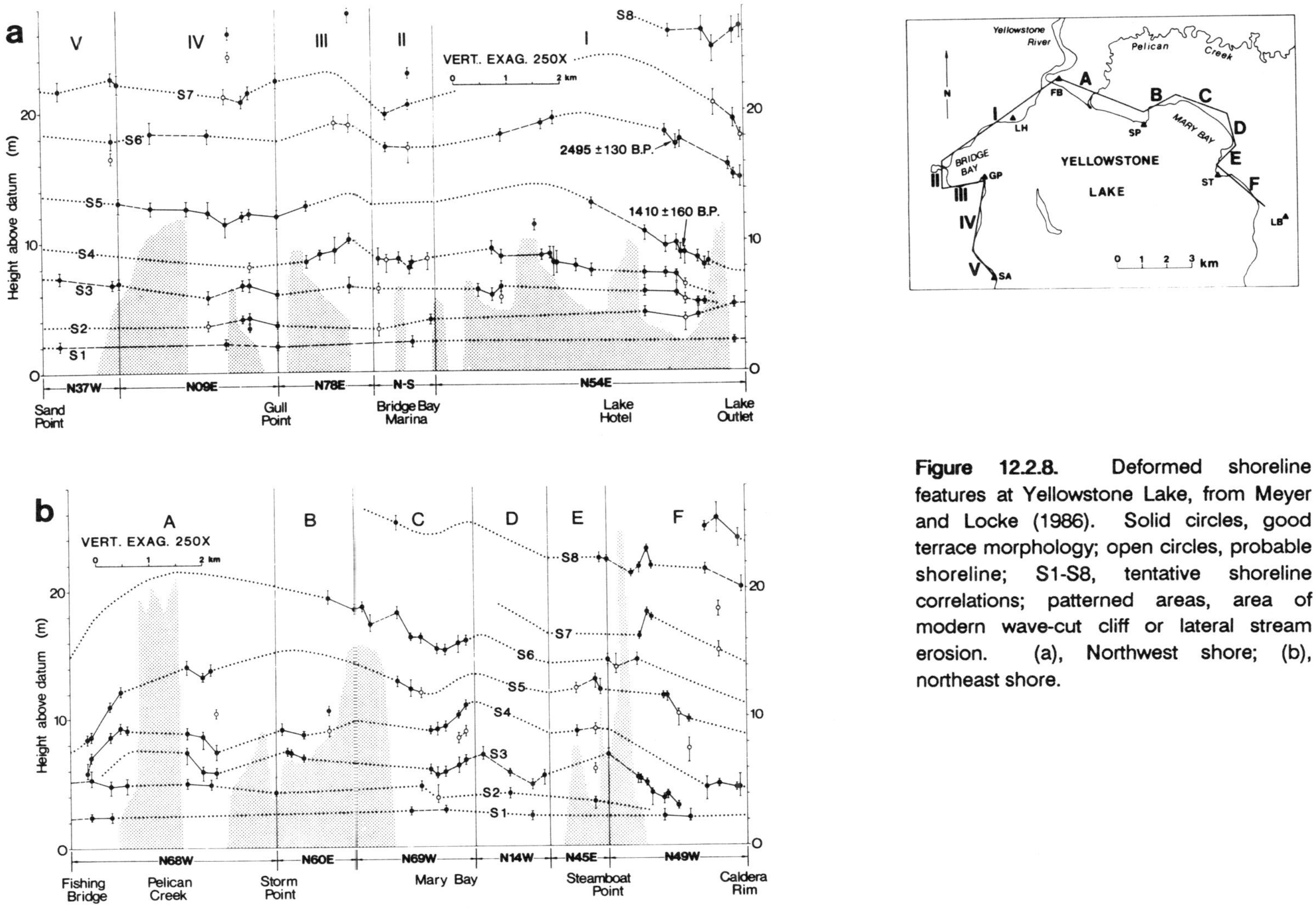

Figure 12.2.8. Deformed shoreline features at Yellowstone Lake, from Meyer and Locke (1986). Solid circles, good terrace morphology; open circles, probable shoreline; S1-S8, tentative shoreline correlations; patterned areas, area of modern wave-cut cliff or lateral stream erosion. (a), Northwest shore; (b), northeast shore.

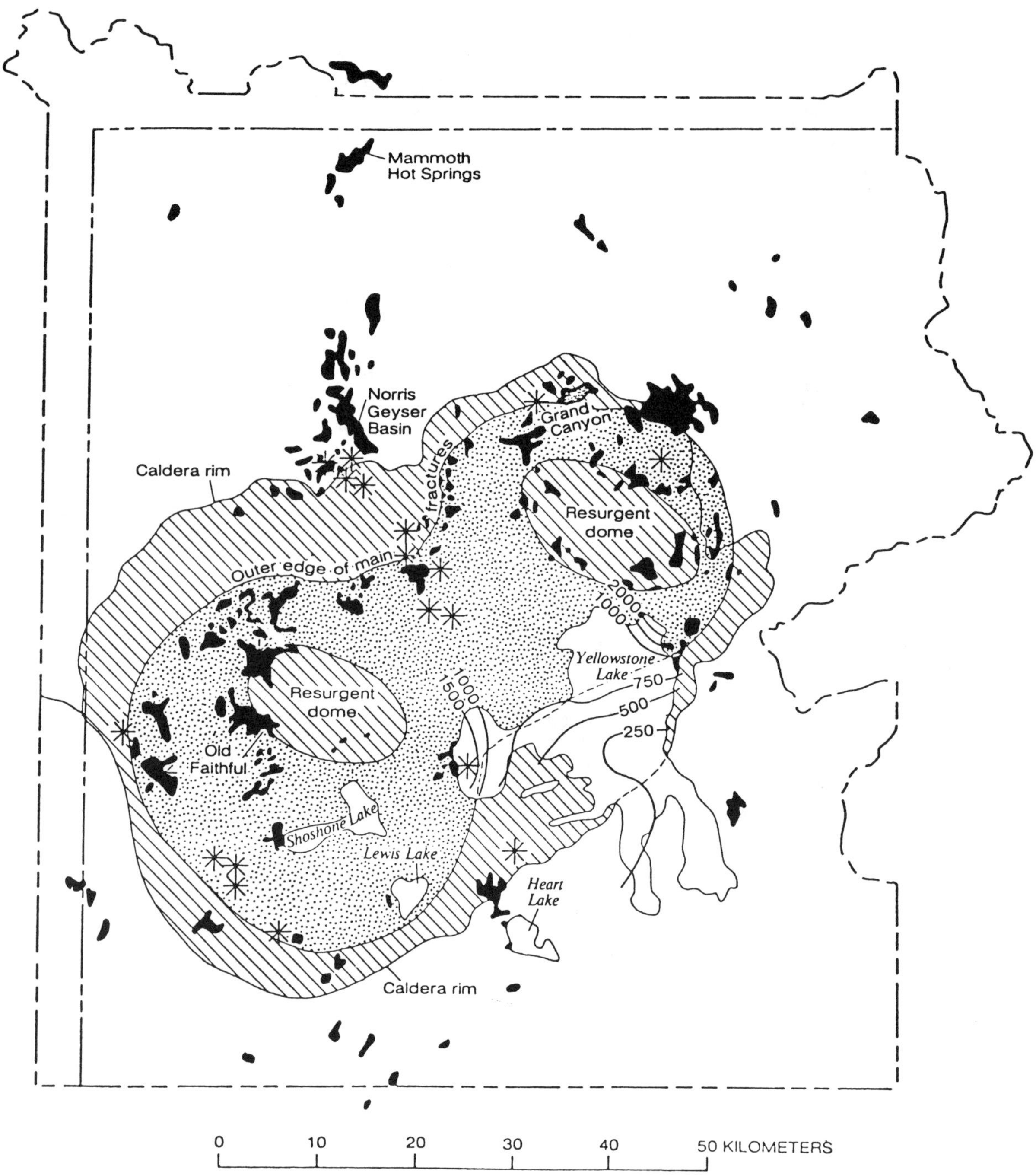

Figure 12.2.9. Sketch map of Yellowstone Caldera and its main thermal features, from Christiansen (1984). Stars, intracaldera rhyolite vents; irregular black areas, thermal areas. Contours of heat flow shown in Yellowstone Lake in units of milliwatts per square meter. Reprinted with permission of the National Academy Press, Washington, D.C.

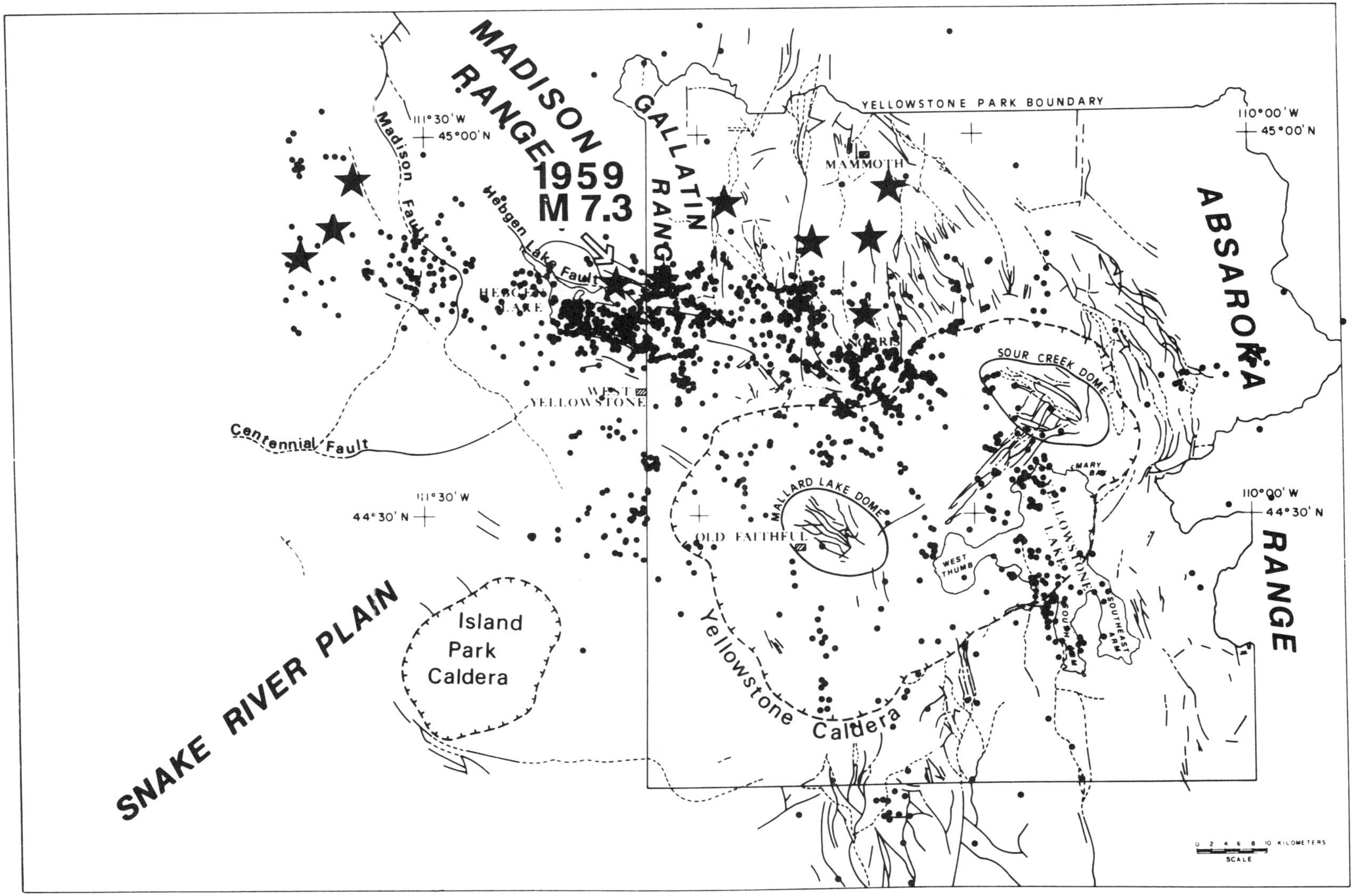

Figure 12.2.10. Earthquake distribution in Yellowstone - Hebgen Lake region from 1971 to 1979, from Smith and Braile (1983). Large stars correspond to epicenters of historical earthquakes larger than magnitude 6. The epicenter of the Yellowstone Park earthquake of 30 June 1975, which is shown just inside the north caldera rim near Norris in figure 12.2.1, is not shown here because estimates of its magnitude ranged from 5.7 to 6.1. Reprinted with permission of the National Academy Press, Washington, D.C.

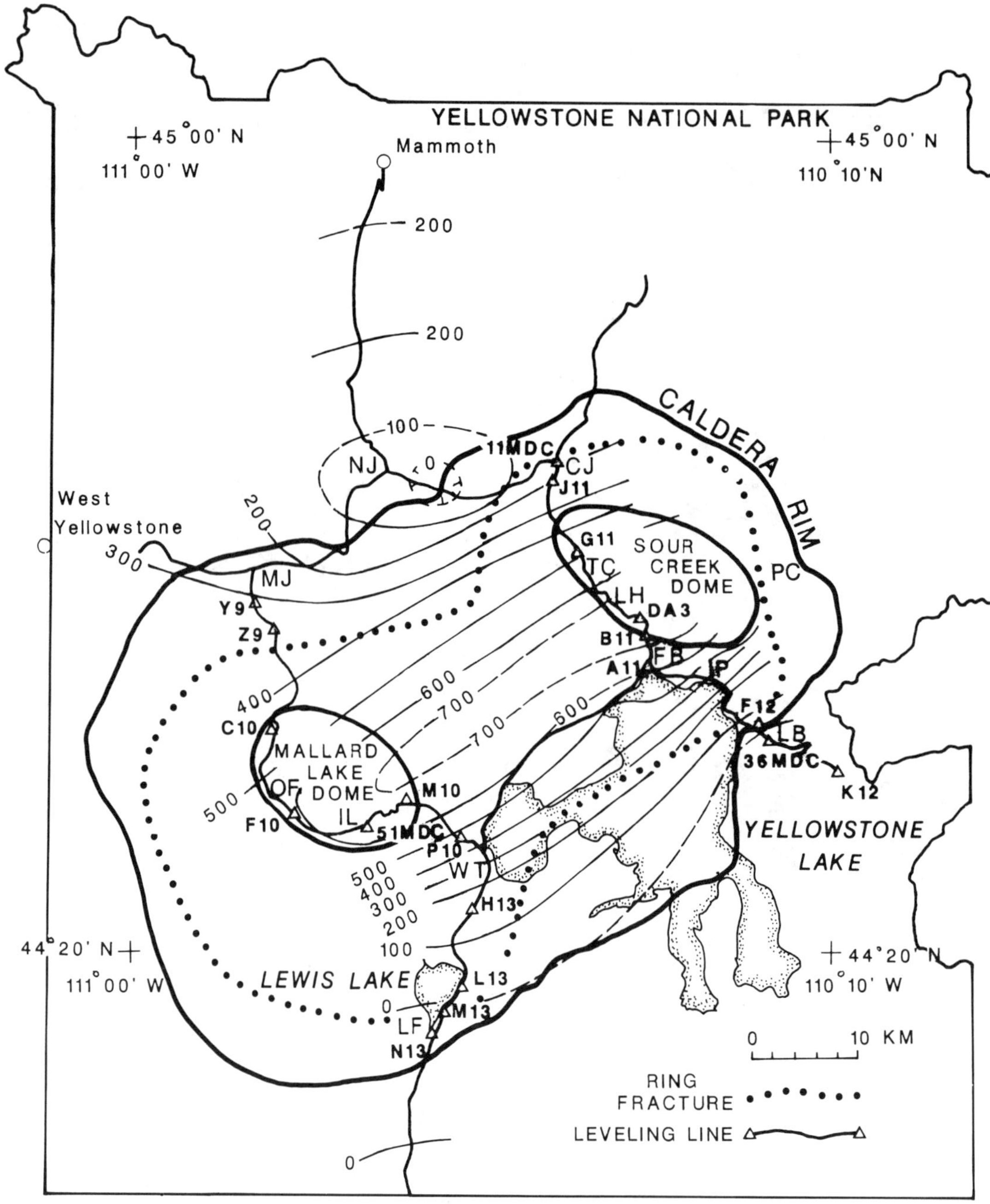

Figure 12.2.11. Contours of uplift at Yellowstone Caldera for the interval 1923-76, from Pelton and Smith (1982). Contour interval is 100 mm. Letter-number designations (for example, Z9, F10, DA3) indicate bench marks. MJ, Madison Junction; OF, Old Faithful; IL, Isa Lake; WT, West Thumb; LF, Lewis Falls; NJ, Norris Junction; CJ, Canyon Junction; TC, Trout Creek; LH, LeHardys Rapids; FB, Fishing Bridge; LB, Lake Butte. See figures 12.2.12 - 12.2.14 for 1983-86 leveling results between Lewis Falls and Madison Junction, and between Lake Butte and Canyon Junction. Copyright by the American Geophysical Union.

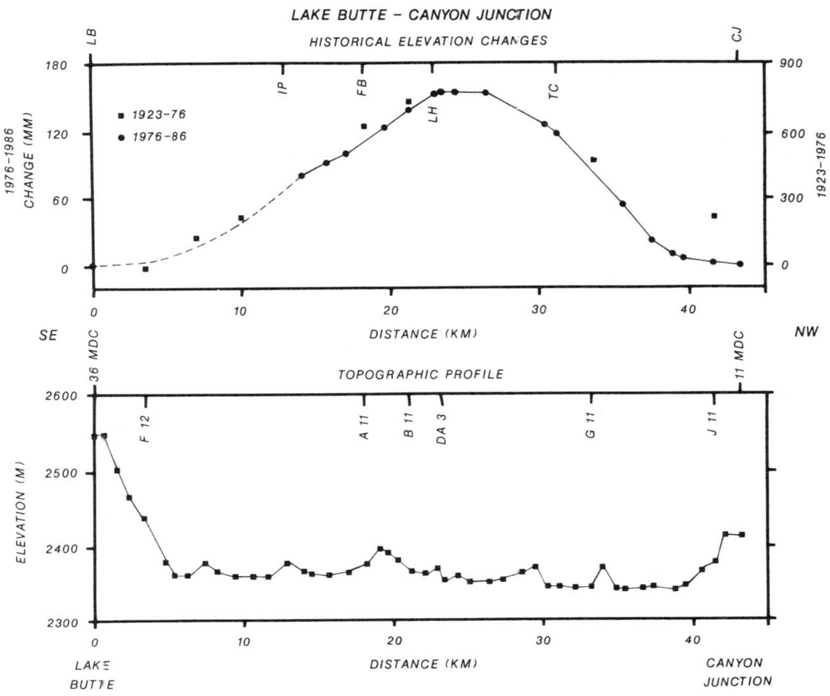

Figure 12.2.12. Historical elevation changes (top) and topography (bottom) along a line from Lake Butte to Canyon Junction (see figures 12.2.2, 12.2.11), from Dzurisin and Yamashita (1987). Copyright by the American Geophysical Union.

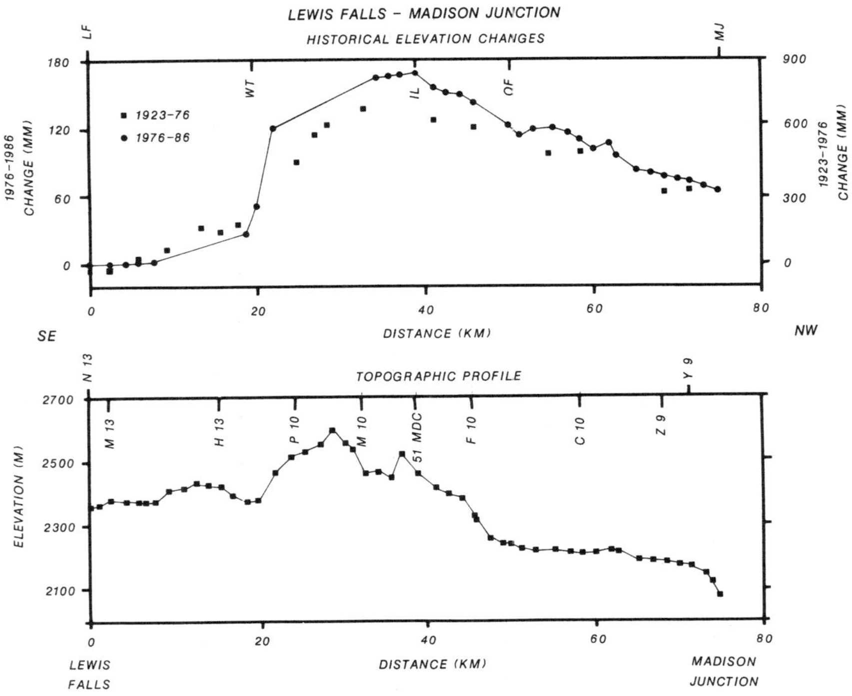

Figure 12.2.13. Historical elevation changes (top) and topography (bottom) along a line from Lewis Lake to Madison Junction (see figures 12.2.2, 12.2.11), from Dzurisin and Yamashita (1987). Copyright by the American Geophysical Union.

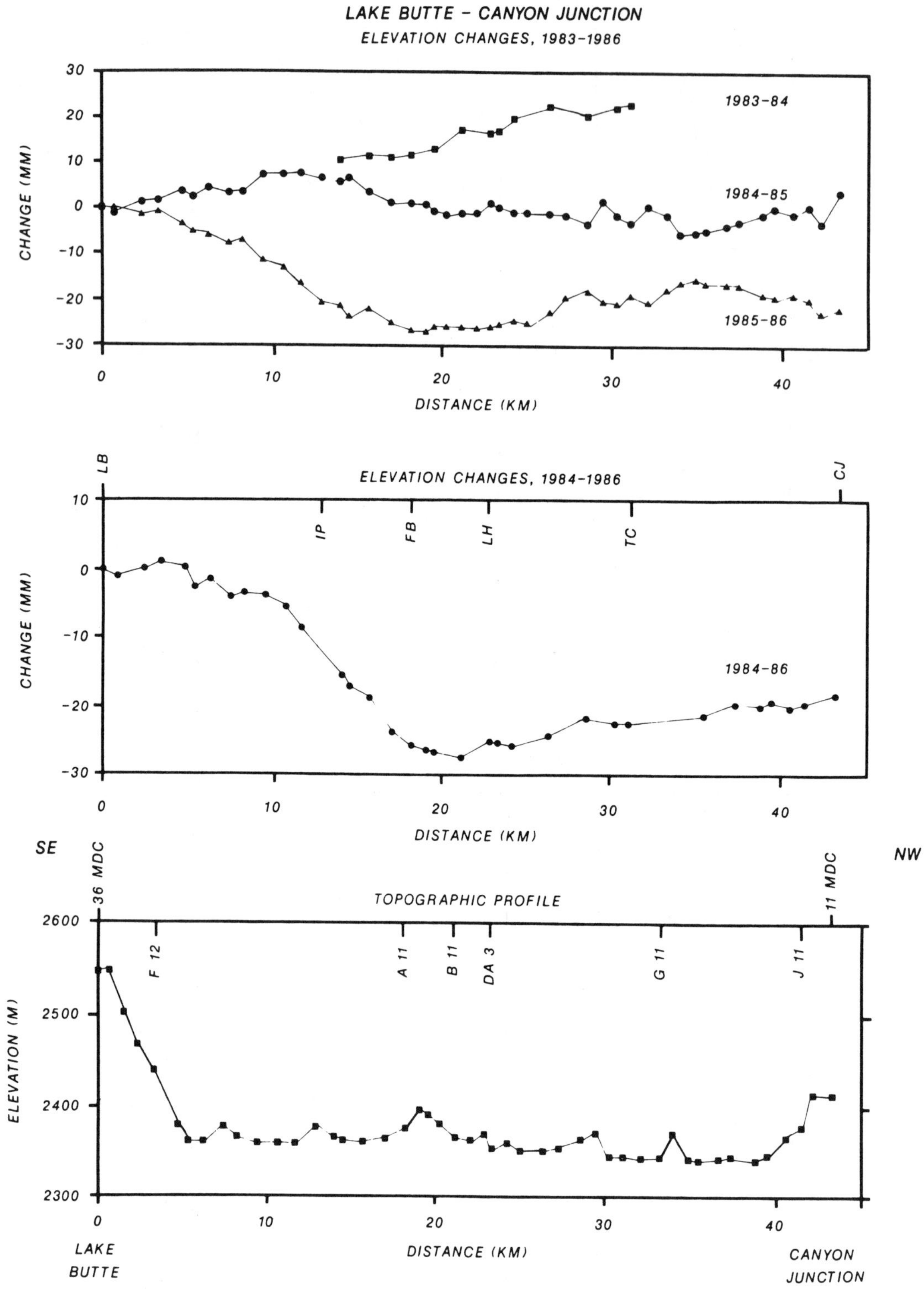

Figure 12.2.14. Annual elevation changes since 1983 along a line from Lake Butte to Canyon Junction (see figures 12.2.2, 12.2.11), from Dzurisin and Yamashita (1987). Copyright by the American Geophysical Union.

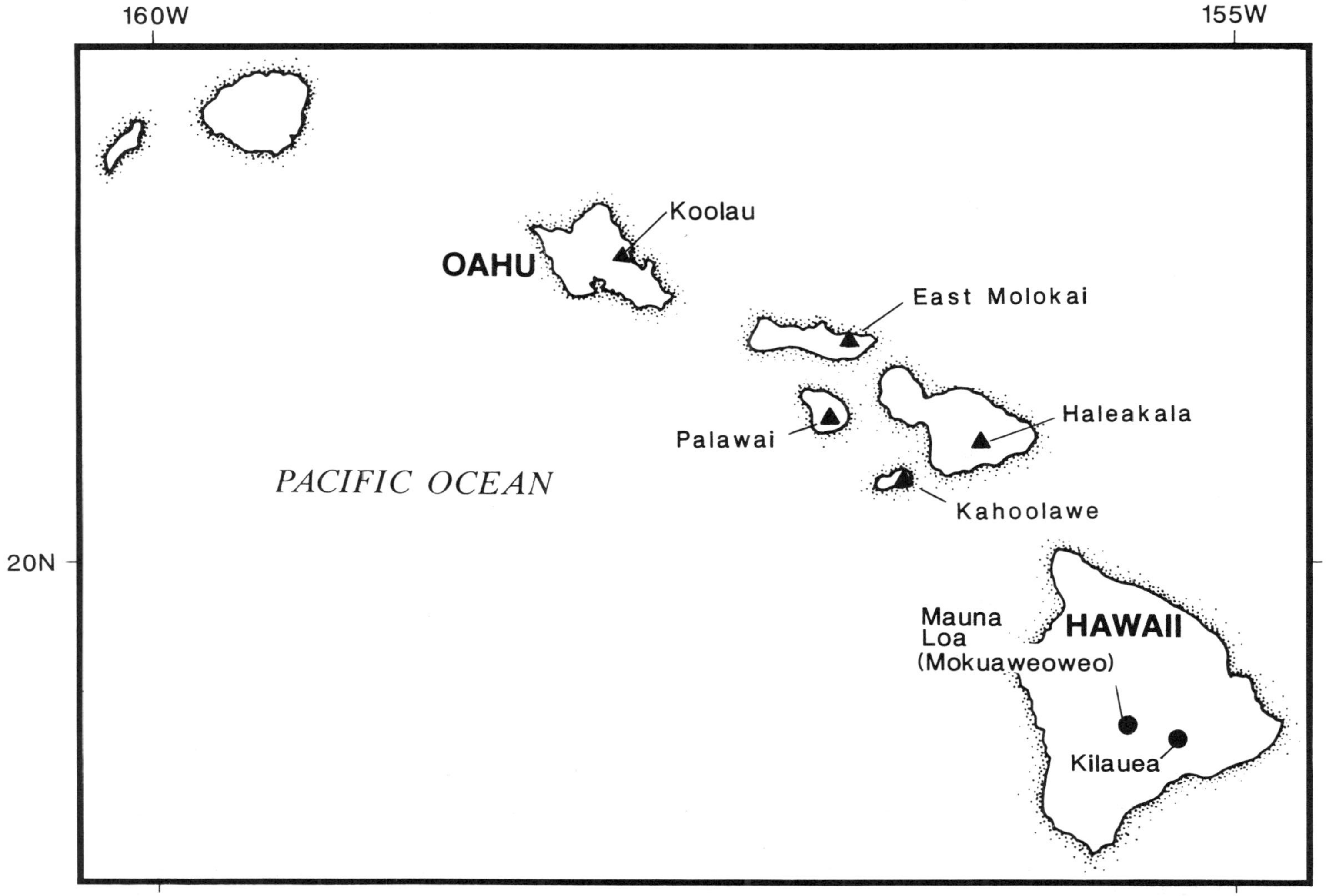

Figure 13. Locations of large, Quaternary restless calderas (solid circles) and nonrestless calderas (triangles) in Region 13.

PART 3: HISTORICAL UNREST AT LARGE QUATERNARY CALDERAS (continued)
See inside back cover for explanation and abbreviations

KILAUEA

CAVW number (active vent)	Latitude Longitude (degrees)	Diameter (km)	Local tectonic setting	Precaldera edifice	SiO_2 content (wt pct)	Age of collapse (yr B.P.)	Date of unrest	Type and duration of unrest ESTU STHF MGTF H Te	Eruption type
13-02-01 (Kilauea)	19.42N 155.29W	8 (outer) 4 (inner)	Hotspot	Shield	R = 47-51 C = 47-51	ca. 1500 200-250	1750(?)	---- ---- ---- - --	lf
							1790(?)	---- x--- ---- - --	lf,P-Ex
							1823	---- x--- ---- - --	lf
							1832	---- x--- ---- - --	lf
							1840	---- x--- ---- - --	lf
							1868	---- x--- ---- - UU	lf
							1877	---- ---- ---- - --	lf
							1884	---- ---- ---- - --	sub lf?
							1885	---- ---- ---- - --	lf,ll
							1886	---- x--- ---- - --	sub lf?
							1891	---- x--- ---- - --	none?
							Mar 1894	---- ---- ---- - --	lf
							Jul 1894	---- x--- ---- - --	lf
							1916	---- x--- ---- - --	none?
							1918	---- ---- ---- - --	lf
							Feb 1919	---- ---- ---- - --	lf,ll
							Dec 1919	---- x--- ---- - --	lf,ll
							1921	---- ---- ---- - --	lf
							1922	---- x--- ---- - --	lf
							1923	---- x--- ---- - --	lf
							May 1924	EED- ED-D ---- - --	pex
							Jul 1924	---- ---- ---- - --	lf
							1927	x-B- ---- ---- - --	lf,ll
							Feb 1929	B-x- -E-- ---- - --	lf,ll
							Jul 1929	x-B- -x-- ---- - --	lf
							1930	CBA- -C-- ---- - --	lf
							1931	D-Y- -E-- ---- - --	lf
							1934	B-B- -B-x ---- - --	lf
							Nov 1944	D--D xD-- ---- - --	none
							1950	x--E xE-x ---- - --	none
							1952	D--- -C-- ---- - --	lf,ll
							1954	EBB- -E-- ---- - --	lf

KILAUEA (continued)

CAVW number (active vent)	Latitude Longitude (degrees)	Diameter (km)	Local tectonic setting	Precaldera edifice	SiO_2 content (wt pct)	Age of collapse (yr B.P.)	Date of unrest	Type and duration of unrest ESTU STHF MGTF H Te	Eruption type
(continued from previous page)							1955	ECx- -C-x ---- - --	lf,cc
							1959	ECB- -E-- ---D - --	lf,cc
							1960	DCB- BD-B ---- - --	lf,cc
							Feb 1961	EBB- -E-- ---- - --	lf
							Mar 1961	D-Y- -D-- ---- - --	lf
							Jul 1961	D-Y- -E-- ---- - --	lf,ll
							Sep 1961	DBB- BE-Y ---- - --	lf
							1962	CBB- YE-- ---- - --	lf,ll
							May 1963	CxC- xD-x ---- - --	none
							Jul 1963	CCxx xC-x ---- - --	none
							Aug 1963	DDD- -E-x ---- - UU	lf,ll
							Oct 1963	B-B- BB-- ---- - --	lf
							Mar 1965	D-B- YEY- ---- - --	lf,ll
							Aug 1965	CC-- xD-- ---- - --	none
							Dec 1965	BBB- YE-Y ---- - --	lf
							1967-68	A-BF YFEY ---- - --	lf,ll
							Aug 1968	DCYx YDxx ---- - --	lf
							Oct 1968	BYBY YEx- ---x - --	lf
							1969-71	CYYE YEEY --YY - --	lf,ll
							1972-74	EYDY YEYY Y-YY - UU	lf,ll
							1975	DYY- YYxx -x-- - UU	lf
							Jun 1976	BB-- BBx- ---- - --	none
							Jul 1976	BB-- BB-- ---- - --	none
							Aug 1976	NN-x DDx- ---- - --	none
							Feb 1977	CBB- BBxx xx-- - --	none
							Sep 1977	BBB- BBxx -x-- - --	lf,cc
							May 1979	BB-- BB-- ---- - --	none
							Aug 1979	BBN- BB-- ---- - --	none
							Nov 1979	CCYY YDxx ---x - --	lf

KILAUEA (continued)

CAVW number (active vent)	Latitude Longitude (degrees)	Diameter (km)	Local tectonic setting	Precaldera edifice	SiO_2 content (wt pct)	Age of collapse (yr B.P.)	Date of unrest	Type and duration of unrest ESTU STHF MGTF H Te	Eruption type
(continued from previous page)							Mar 1980	CCY- YY-Y ---Y - --	lf (minor)
							Aug 1980	BB-x BB-- ---x - --	none
							Oct 1980	CBx- BB-- ---x - --	none
							Nov 1980	BB-- BB-- ---x - --	none
							Jan 1981	EE-- DDx- ---- - --	none
							Jun 1981	BBBx -B-- ---- - --	none
							Aug 1981	CCCx xCxx --xx - --	none
							Apr 1982	BBBY -BxY ---x - --	lf
							Jun 1982	CCC- CCxx ---C - --	none
							Sep 1982	BBBB -BxY ---Y - --	lf
							Dec 1982	BBN- BB-- --x- - --	none
							Jan 1983-88+	BBYY YYYY -xxx x --	lf,cc,ll

TECTONIC SETTING

Kilauea is one of five basaltic shield volcanoes that comprise the Island of Hawaii at the southeast end of the Hawaiian archipelago (figs. 13.1.1-13.1.3). The island marks the current location of the Hawaiian "hotspot" or melting anomaly in the upper mantle; northwestward migration of the Pacific plate relative to the hotspot has produced the Hawaiian-Emperor chain of volcanic islands and seamounts that stretches from Hawaii northwestward to the junction of the Kurile and Aleutian Trenches (Clague and Dalrymple, 1987).

Kohala and Mauna Kea volcanoes are the most mature on the Island of Hawaii; both have passed through a tholeiitic shield-building stage into an alkalic stage, and both have been dormant throughout historical time. Hualalai Volcano has erupted alkalic basalts repeatedly in the last 500 years, most recently in 1800-1801 (Moore and others, 1987). Mauna Loa and Kilauea Volcanoes have erupted tholeiitic basalts frequently during the 19th and 20th centuries and, with Loihi seamount off the southeast coast of Hawaii, form the current locus of hotspot volcanism along the Hawaiian-Emperor chain.

GEOLOGIC HISTORY

Kilauea has grown to its current summit elevation of 1,240 m above sea level through repeated extrusions of fluid basalt from its summit region and two prominent rift zones (figs. 13.1.3-13.1.6). Magma derived from a zone of partial melting in the upper

KILAUEA (continued)

GEOLOGIC HISTORY (continued)

mantle usually is stored in a shallow crustal reservoir 3-5 km beneath the summit region before being intruded into the volcanic edifice or extruded onto its surface. Explosive eruptions are rare at Kilauea, mainly because the basaltic magmas are very fluid. The few notable explosive eruptions have been phreatic or phreatomagmatic (for example, in 1790 and 1924) (Christiansen, 1979; Decker and Christiansen, 1984). The eruption of 1924 (relatively small, phreatic) occurred during a period of rapid subsidence of the summit region and notably of the Halemaumau lava lake; subsidence may have allowed ground water to enter the conduit area and "fuel" the phreatic explosions (Jaggar and Finch, 1924; Macdonald and Abbott, 1970). A similar mechanism may apply to large phreatomagmatic eruptions of Kilauea: catastrophic draining of the shallow magma reservoir during large flank eruptions or intrusions may trigger rapid summit subsidence, phreatomagmatic explosions, and caldera collapse (Peterson and Moore, 1987).

Two summit calderas have been identified at Kilauea--a prominent, 3 km x 5 km diameter caldera nested within an older, 8-km-diameter caldera that is largely filled by subsequent lava flows (figs. 13.1.4, 13.1.5) (Holcomb, 1980, 1987). Halemaumau is a small pit crater on the floor of the inner, modern caldera. Many historical eruptions have occurred both within the summit caldera and along rift zones that extend southwestward and eastward from the caldera. The main body of Kilauea is built on, and episodically slides away from, the much larger edifice of Mauna Loa. Such movements occur in response to gravitational and intrusive stresses (the latter induced by repeated forceful intrusions into the rift zones), causing major earthquakes (magnitude 7+) and ground displacements (many meters) (Fiske and Jackson, 1972; Swanson and others, 1976; Duffield and others, 1982).

During a magnitude 7.2 earthquake near Kalapana on the southeast coast of Hawaii in November 1975, parts of Kilauea's south flank moved as much as 3.5 m downward and 8 m seaward (figs. 13.1.8, 13.1.9); subsidence and horizontal movement were concentrated within the Hilina fault system (location shown in fig. 13.1.3). Historical geodetic measurements show that points between the summit area and the Hilina fault system moved progressively upward and southward in the years preceding the 1975 earthquake, suggesting that the earthquake was caused by gravitational slumping in response to accumulation of compressive strain during repeated intrusions into the east rift zone (figs. 13.1.7) (Lipman and others, 1985).

Similar ground movements probably occurred during large earthquakes at Kilauea in 1823 and 1868. Ellis (1963, p. 195; reprint of 1827 edition) reported that in 1823, "the ground, after being agitated some minutes with a violent tumultuous motion, suddenly burst open, for several miles in extent, in a direction from north by east, to south by west, and emitted, in various places at the same instant, a considerable quantity of smoke and luminous vapor, but none of the people were injured by it." After the earthquake, water in a well that was split by a crack became brackish, and the average water level in the well increased from about 0.5 m to nearly 1 m (suggesting at least 0.5 m of subsidence). Similarly, Brigham (1909, p. 481-482) reported that soon after the large earthquake at Kilauea in 1868, an area near the boat landing at Keauhou "appeared to have sunk, and the sea was flowing a fathom deep where houses formerly stood."

The modern Kilauea Caldera formed in the late 18th century, probably during the eruption(s) responsible for the Keanakakoi Ash, a widespread phreatomagmatic tephra in the Kilauea summit region that has been dated from oral accounts as having been deposited in

KILAUEA (continued)

GEOLOGIC HISTORY (continued)

about A.D. 1790 (Christiansen, 1979; Decker and Christiansen, 1984). The Uwekahuna Ash is an older, more widespread tephra that was deposited in at least two episodes of phreatomagmatic eruptions 1,000 to 2,200 years ago, possibly during earlier caldera-forming episodes (Dzurisin and Casadevall, 1986; Lockwood and Rubin, 1986).

Petrologic studies of Kilauea's eruptive products lead to the following conclusions concerning the petrogenesis of Kilauea lavas and the structure and dynamics of the magma plumbing system beneath the volcano (Wright and Fiske, 1971):

1. The chemical composition of magma erupted at the summit of Kilauea varies with the date of the eruption. Lavas erupted before 1750, during the eighteenth and nineteenth centuries, and in the twentieth century form groups that can be distinguished chemically.

2. During late prehistoric time, pockets of differentiated magma were formed within the rift zones by separation of the liquid remaining after partial crystallization of bodies of summit magma.

3. Lava from some eruptions has the composition of the liquid fraction as it is generated within the rift. Lava compositions of other eruptions are best explained by mixing of magma supplied from a central reservoir beneath the Kilauea summit with differentiated liquid in the rift.

4. The fact that components of "summit composition" appear in rift eruptions before they appear undiluted at the summit suggests that the central reservoir is vertically zoned. Rift eruptions are fed from lower levels, where younger magma is available; summit eruptions are fed from the relatively older magma above.

5. The unique and uniform composition of lava of each successive summit eruption suggests that summit eruptions end when all of the magma of one composition has been erupted. Magma erupted from the upper levels of the reservoir during one cycle is continually replaced from below by younger magma of different composition.

HISTORICAL ACTIVITY

Because unrest at Kilauea is essentially continuous, we have listed only the dates of eruptions and intrusions in the table above. Intrusions are usually characterized by rapid summit deflation, earthquake swarms along a rift zone, and inflation or ground cracking along that rift zone. Eruptions are defined to begin with the arrival of magma at the surface and to end when tremor at the vent ceases for at least several days.

KILAUEA (continued)

HISTORICAL ACTIVITY (continued)

Most historical eruptions of Kilauea until 1924 occurred in the inner summit caldera; from 1823 to 1894 and from 1907 to 1924, a lava lake was active almost continuously on the caldera floor or in Halemaumau, and eruptions along the rift zones were infrequent and shortlived. Flank eruptions in 1823, 1832, 1840, and 1868 were accompanied by pistonlike sinking of the caldera floor as magma drained from the summit area into the rift zones. The 1868 eruption was triggered by a magnitude 7.5 earthquake near the southern tip of Hawaii on 2 April, which also caused a tsunami and mudslide responsible for 79 deaths. The earthquake was preceded by a small eruption at the summit of Mauna Loa on 27 March, followed by a larger outbreak along Mauna Loa's southwest rift zone starting on 7 April. Phreatic explosions and collapse at Halemaumau in 1924 marked a change in the style of activity, and most eruptions since 1955 have been along the rift zones (Macdonald and Abbott, 1970; Peterson and others, 1976; Macdonald and others, 1983; Holcomb, 1987). Volumes of erupted lava are typically between 10^6 and 10^8 m^3 (Klein, 1982).

Seismicity and ground deformation at Kilauea are closely linked to the movement of magma into a shallow reservoir beneath the summit and into the rift zones (Klein and others, 1987). Infrequent deep earthquakes hint at supply of magma into the shallow magma storage system from depth (Eaton and Murata, 1960; Eaton and others, 1987; Klein and others, 1987) (figs. 13.1.10-13.1.12); more commonly, shallow earthquakes reflect inflation of the summit reservoir and the leading edge of a rift-zone intrusion (as in 1959 along the east rift zone; Eaton and Murata, 1960) (figs. 13.1.13-13.1.17). The hypocenters of historical earthquakes define in remarkable detail the magma conduit system beneath Kilauea, from a source of partial melting in the upper mantle through a near-vertical pipe to a shallow crustal reservoir and into the rift zones (figs. 13.1.12, 13.1.18) (Ryan and others, 1981). Earthquake swarms are common as magma intrudes the rift zones (fig. 13.1.19); swarms that are not clearly attributable to movement of magma (for example, ones in which regional tectonic stress release is modified to swarm activity in fractured and (or) high-pore-pressure environments) are not common at Kilauea. Inflation and deflation of the summit region are directly linked to filling and draining of the summit reservoir (Eaton and Murata, 1960; Fiske and Kinoshita, 1969; Ryan and others, 1983) (figs. 13.1.20-13.1.23). Ground deformation along the rift zones, including surface rupture, is directly attributable to intrusion along the rifts (Swanson and others, 1976; Pollard and others, 1983) (figs. 13.1.24, 13.1.25). In detail, the ground surface is uplifted along two ridges parallel to an intruding dike; these ridges move closer together as the dike nears the surface (Pollard and others, 1983). This pattern, combined with the distribution of earthquakes, can provide good indications of the depth of an intrusion and of any upward movement of the intrusion that might culminate in an eruption.

Tremor is often a short-term precursor to eruptions--beginning minutes to a few hours before the start of fountaining; a further, sharp increase in tremor accompanies the start of fountaining (figs. 13.1.34 and 13.1.35). Most eruptions are preceded by slow inflation of the summit area for a period of a few weeks to many months and are accompanied by rapid deflation of the summit (fig. 13.1.26). Typically, summit deflation begins a few hours (or occasionally as much as several days) before a flank eruption begins and is thus another, though less reliable, short-term precursor of eruptions. Many phases of the 1983-present eruption (see below) along the east rift zone have been preceded by several hours to several days of low-level activity at the Puu Oo vent, including a rise in the level of lava within that vent.

KILAUEA (continued)

HISTORICAL ACTIVITY (continued)

The magnitude of ground deformation before and after an eruption varies greatly in proportion to the size, depth, and shape of the source of the deformation. Typical deformation precursors to an eruption are several tens to several hundreds of microradians of inflationary tilt at the Hawaiian Volcano Observatory on the inner caldera rim (figs. 13.1.20, 13.1.21, 13.1.26), several centimeters to a few tens of centimeters of uplift and horizontal extension of the summit region (figs. 13.1.20, 13.1.21, 13.1.27, 13.1.28), and several tens of centimeters to a meter or more of outward displacement of points at and several kilometers outside the caldera rim. Contours of uplift are skewed toward one of the rifts if the magma migrates into that rift (figs. 13.1.24, 13.1.25).

The flux of SO_2 gas from Kilauea's summit area has been monitored regularly since June 1979. The average daily emission rate from 1979 to 1982 was 170 ± 70 t/d (metric tonnes/day); during 1983-84, the rate increased to 260 ± 90 t/d (Casadevall and others, 1987) (fig. 13.1.29). The increase coincided with the start of the long-lived Puu Oo eruption along Kilauea's east rift zone in January 1983. Casadevall and others (1987) found no evidence of precursory increases in SO_2 emission before eruptions during 1979-84, and indeed Gerlach (1986) has calculated that magmas at Kilauea will reach very close to the surface before "precursory" degassing of sulfur gases will be noticeable. Recent monitoring of the net flux of several magmatic gases including H_2, SO_2, and H_2S has yielded promising results (fig. 13.1.30); the flux of such gases may be an indirect reflection of strain and microfracturing before shallow magma movements within Kilauea (McGee and others, 1987). Occasionally, anomalous steaming or tree kill occurs during shallow intrusions or eruptions.

Although varied in detail, unrest at Kilauea follows a general pattern that is well illustrated by two recent periods of sustained activity along the east rift zone: Mauna Ulu during 1969-71 (Swanson and others, 1979) and 1972-74 (Tilling and others, 1987), and the Puu Oo area from 1983 to the present (December 1987) (Wolfe and others, 1987). In both cases, early fountaining episodes were preceded by slow inflation of the summit area for weeks or months and accompanied by rapid summit deflation and earthquakes at the summit and along the rift zone (figs. 13.1.31-13.1.39). Later episodes were characterized by progressively less seismicity and ground deformation until a quasi-steady state was established in which magma reached the extrusion sites continuously through well-developed conduit systems along the rift zone (fig. 13.1.40). Gravity, leveling, and trilateration data show that magma was stored elastically in the summit reservoir between fountaining episodes at Puu Oo during 1984-85 (fig. 13.1.41), but that at other times horizontal extension of the volcanic edifice owing to seaward movement of its south flank decreases lateral support of the fluid magma reservoir, thus lowering the pressure in the reservoir and allowing more magma to be injected without vertical uplift (Johnson, 1987).

Dzurisin (1980) and Klein (1982) noted that eruptions can be triggered by relatively small external disturbances such as fortnightly earth tides. Klein writes: "One useful concept considers the summit reservoir as a tank with drain valves for the rift zones and summit fissures. Forces small in comparison with the total magmatic pressures or tectonic strains can control flow by opening or partially opening the valves." Shimozuru (1987) reviewed earlier work on fortnightly tidal triggering of Hawaiian eruptions and showed that semidiurnal fluctuations in the level of the Halemaumau lava lake during 1919 were well correlated with solid-Earth tides.

KILAUEA (continued)

HISTORICAL ACTIVITY (continued)

Many intrusions at Kilauea do not culminate in eruptions (see listings of intrusions in Klein, 1982; Klein and others, 1987). Swanson (1972) concluded that about half of all magma supplied to Kilauea from depth is emplaced as intrusions and not erupted; Dzurisin and others (1984) concluded that from 1956 to 1983, about 35 percent of the total magma supply was extruded, 55 percent was intruded into the east rift zone, and 10 percent was intruded into the southwest rift zone.

COMMENTS

Kilauea is one of the most intensely studied volcanoes on Earth. The Hawaiian Volcano Observatory (HVO), perched on the northwest rim of Kilauea Caldera, was established in 1912 and recently commemorated its 75th anniversary. Numerous observations made by HVO staff through the years suggest the presence of a long-lived magma storage reservoir beneath the summit region. In fact, shallow magma reservoirs probably migrate upward within the volcanic edifice throughout much of the growth cycle of Hawaiian shields, giving rise to repeated episodes of caldera collapse and refilling (Decker, 1987). Most historical unrest at Kilauea can be linked directly to magma movements within the volcanic system, either upward from a zone of magma generation in the upper mantle or laterally within the volcanic edifice. Even large "tectonic" earthquakes such as those in 1868 and 1975 are probably triggered in large part by magmatic intrusions into Kilauea's rift zones.

REFERENCES

Brigham, W.T., 1909, The Volcanoes of Kilauea and Mauna Loa on the Island of Hawaii, their variously recorded history to the present time, Memoirs of the Bernice Pauahi Bishop Museum, v. 2, no. 4: Honolulu, Bishop Museum Press.

Casadevall, T.J., Stokes, J.B., Greenland, L.P., Malinconico, L.L., Casadevall, J.R., and Furukawa, B.T., 1987, SO_2 and CO_2 emission rates at Kilauea volcano, 1979-1984, <u>in</u> Decker, R.W., Wright, T.L., and Stauffer, P.H., eds., Volcanism in Hawaii, U.S. Geol. Surv. Prof. Paper 1350, v. 1, p. 771-780.

Christiansen, R.L., 1979, Explosive eruption of Kilauea Volcano in 1790 [abs.]: IAVCEI Hawaii Symp. on Intraplate Volcanism and Submarine Volcanism, Hilo, July 16-22, 1979.

Clague, D. A., and Dalrymple, G. B., 1987, The Hawaiian-Emperor volcanic chain, Part I, Geologic evolution, <u>in</u> Decker, R.W., Wright, T.L., and Stauffer, P.H., eds., Volcanism in Hawaii, U.S. Geol. Surv. Prof. Paper 1350, v. 1, p. 5-54.

Decker, R.W., 1987, Dynamics of Hawaiian volcanoes: An overview, <u>in</u> Decker, R.W., Wright, T.L., and Stauffer, P.H., eds., Volcanism in Hawaii, U.S. Geol. Surv. Prof. Paper 1350, v. 2, p. 997-1018.

Decker, R.W., and Christiansen, R.L., 1984, Explosive eruptions of Kilauea Volcano, Hawaii, <u>in</u> Boyd, F.R., ed., Explosive volcanism: Inception, evolution, and hazards: National Research Council in Geophysics, Washington, D.C., National Academy Press, p. 122-132.

KILAUEA (continued)

REFERENCES (continued)

Decker, R.W., Hill, D.P., and Wright, T.L., 1966, Deformation measurements on Kilauea Volcano, Hawaii: Bull. Volcanol., v. 29, p. 721-732.

Decker, R.W., and Kinoshita, W.T., 1971, Geodetic measurements, in The surveillance and prediction of volcanic activity: Paris, UNESCO, p. 47-74.

de St. Ours, P., 1982, Structural map of the summit area of Kilauea Volcano, Hawaii: U.S. Geol. Surv. Misc. Field Studies Map MF-1368, scale 1:24,000.

Duffield, W.A., Christiansen, R.L., Koyanagi, R.Y., and Peterson, D.W., 1982, Storage, migration and eruption of magma at Kilauea Volcano, Hawaii, 1971-1972: Jour. Volcanol. Geotherm. Res., v. 13, p. 273-307.

Duffield, W.A., Jackson, D.B., and Swanson, D.A., 1976, The shallow, forceful intrusion of magma and related ground deformation at Kilauea Volcano, May 15-16, 1970, in Gonzales Ferran, O., ed., Proc. Symp. on Andean and Antarctic Volcanology Problems: Rome, IAVCEI, p. 577-597.

Dvorak, J.J., and Okamura, A.T., 1987, A hydraulic model to explain variations in summit tilt rate at Kilauea and Mauna Loa volcanoes, in Decker, R.W., Wright, T.L., and Stauffer, P.H., eds., Volcanism in Hawaii, U.S. Geol. Surv. Prof. Paper 1350, v. 2, p. 1281-1296.

Dzurisin, D., 1980, Influence of fortnightly earth tides at Kilauea Volcano, Hawaii: Geophys. Res. Lett., v. 7, p. 925-928.

Dzurisin, D., Anderson, L.A., Eaton, G.P., Koyanagi, R.Y., Lipman, P.W., Lockwood, J.P., Okamura, R.T., Puniwai, G.S., Sako, M.K., and Yamashita, K.M., 1980, Geophysical observations of Kilauea Volcano, Hawaii, 2., Constraints on the magma supply during November 1975-September 1977: Jour. Volcanol. Geotherm. Res., v. 7, p. 241-269.

Dzurisin, D., and Casadevall, T.J., 1986, Stratigraphy and chemistry of the Uwekahuna Ash: product of prehistoric phreatomagmatic eruptions of Kilauea volcano, Hawaii [abs.]: Abstracts, Int. Volcanol. Congr., New Zealand, February 1986, p. 100.

Dzurisin, D., Koyanagi, R.Y., and English, T.T., 1984, Magma supply and storage at Kilauea volcano, Hawaii, 1956-1983: Jour. Volcanol. Geotherm. Res., v. 21, p. 177-206.

Eaton, J.P., and Murata, K.J., 1960, How volcanoes grow: Science, v. 132, p. 925-938.

Eaton, J.P., Richter, D.H., and Krivoy, H.L., 1987, Cycling of magma between the summit reservoir and Kilauea Iki lava lake during the 1959 eruption of Kilauea volcano, in Decker, R.W., Wright, T.L., and Stauffer, P.H., eds., Volcanism in Hawaii, U.S. Geol. Surv. Prof. Paper 1350, v. 2, p. 1307-1335.

Ellis, William, 1963 (reprint of 1827 edition), Journal of William Ellis, Narrative of a tour of Hawaii, with remarks on the history, traditions, manners, customs, and language of the inhabitants of the Sandwich Islands: Honolulu, Advertiser Publ. Co., Ltd., 342 p.

Fiske, R.S., and Jackson, E.D., 1972, Orientation and growth of Hawaiian volcanic rifts: the effect of regional structure and gravitational stresses: Proc. Roy. Soc. London, Ser. A, v. 329, p. 299-326.

Fiske, R.S., and Kinoshita, W.T., 1969, Inflation of Kilauea Volcano prior to its 1967-1968 eruption: Science, v. 165, p. 341-349.

Gerlach, T.M., 1986, Exsolution of H_2O, CO_2, and S during eruptive episodes at Kilauea Volcano, Hawaii: Jour. Geophys. Res., v. 91, no. B12, p. 12,177-12,185.

Greenland, L.P., Rose, W.I., and Stokes, J.B., 1985, An estimate of gas emissions and magmatic gas content from Kilauea Volcano: Geochimica et Cosmochimica Acta, v. 49, p. 125-129.

KILAUEA (continued)

REFERENCES (continued)

Holcomb, R.T., 1980, Kilauea Volcano: Chronology and morphology of the surficial lava flows: Stanford, Calif., Stanford Univ., Ph.D. diss., 321 p.

Holcomb, R.T., 1987, Eruptive history and long-term behavior of Kilauea volcano, in Decker, R.W., Wright, T.L., and Stauffer, P.H., eds., Volcanism in Hawaii, U.S. Geol. Surv. Prof. Paper 1350, v. 1, p. 261-350.

Jaggar, T.A., and Finch, R.H., 1924, The explosive eruption of Kilauea in Hawaii, 1924: Am. Jour. Sci., v. 8, p. 353-374.

Johnson, D.J., 1987, Elastic and inelastic magma storage at Kilauea volcano, in Decker, R.W., Wright, T.L., and Stauffer, P.H., eds., Volcanism in Hawaii, U.S. Geol. Surv. Prof. Paper 1350, v. 2, p. 1297-1306.

Kinoshita, W.T., Swanson, D.A., and Jackson, D.B., 1974, The measurement of crustal deformation related to volcanic activity at Kilauea Volcano, Hawaii, in Civetta, L., Gasparini, P., Luongo, G., and Rapolla, A., eds., Physical volcanology: Amsterdam, Elsevier, p. 87-115.

Klein, F.W., 1982, Patterns of historical eruptions at Hawaiian volcanoes: Jour. Volcanol. Geotherm. Res., v. 12, p. 1-35.

Klein, F.W., Koyanagi, R.Y., Nakata, J.S., and Tanigawa, W.R., 1987, The seismicity of Kilauea's magma system, in Decker, R.W., Wright, T.L., and Stauffer, P.H., eds., Volcanism in Hawaii, U.S. Geol. Surv. Prof. Paper 1350, v. 2, p. 1019-1185.

Koyanagi, R.Y., Chouet, B., and Aki, K., 1987, Origin of volcanic tremor in Hawaii, Part I, Data from the Hawaiian Volcano Observatory 1969-1985, in Decker, R.W., Wright, T.L., and Stauffer, P.H., eds., Volcanism in Hawaii, U.S. Geol. Surv. Prof. Paper 1350, v. 2, p. 1221-1257.

Lipman, P.W., Lockwood, J.P., Okamura, R.T., Swanson, D.A., and Yamashita, K.M., 1985, Ground deformation associated with the 1975 magnitude-7.2 earthquake and resulting changes in activity of Kilauea volcano, Hawaii: U.S. Geol. Surv. Prof. Paper 1276, 45 p.

Lockwood, J.P., and Rubin, M., 1986, Distribution and age of the Uwekahuna Ash, Kilauea volcano, Hawaii [abs.]: Abstracts, Int. Volcanol. Congr., New Zealand, February 1986, p. 112.

Macdonald, G.A., 1955, Catalogue of active volcanoes of the world including solfataric fields, pt. III, Hawaiian Islands: Naples, IAVCEI, 37 p.

----- 1972, Volcanoes: Englewood Cliffs, N.J., Prentice-Hall, Inc., 510 p.

Macdonald, G.A., and Abbott, A.T., 1970, Volcanoes in the sea: Honolulu, Univ. Hawaii Press, 441 p.

Macdonald, G.A., Abbott, A.T., and Peterson, F.L., 1983, Volcanoes in the sea (2nd ed.): Honolulu, Univ. Hawaii Press, 517 p.

McGee, K.A., Sutton, A.J., and Sato, M., 1987, Use of satellite telemetry for monitoring active volcanoes, with a case study of a gas-emission event at Kilauea volcano, December 1982, in Decker, R.W., Wright, T.L., and Stauffer, P.H., eds., Volcanism in Hawaii, U.S. Geol. Surv. Prof. Paper 1350, v. 1, p. 821-825.

Moore, R.B., Clague, D.A., Rubin, M., and Bohrson, W.A., 1987, Hualalai volcano: a preliminary summary of geologic, petrologic, and geophysical data, in Decker, R.W., Wright, T.L., and Stauffer, P.H., eds., Volcanism in Hawaii, U.S. Geol. Surv. Prof. Paper 1350, v. 1, p. 571-585.

Peterson, D.W., Christiansen, R.L., Duffield, W.A., Holcomb, R.T., and Tilling, R.I., 1976, Recent activity of Kilauea Volcano, Hawaii, in Gonzales Ferran, O., ed., Proc. Symp. on Andean and Antarctic Volcanology Problems: Rome, IAVCEI, p. 646-656.

Peterson, D.W., and Moore, R.B., 1987, Geologic history and evolution of geologic concepts, Island of Hawaii, in Decker, R.W., Wright, T.L., and Stauffer, P.H., eds., Volcanism in Hawaii, U.S. Geol. Surv. Prof. Paper 1350, v. 1, p. 149-189.

KILAUEA (continued)

REFERENCES (continued)

Pollard, D.D., Delaney, P.T., Duffield, W.A., Endo, E.T., and Okamura, A.T., 1983, Surface deformation in volcanic rift zones: Tectonophysics, v. 94, p. 541-584.

Ryan, M.P., Blevins, J.Y.K., Okamura, A.T., and Koyanagi, R.Y., 1983, Magma reservoir subsidence mechanics: theoretical summary and application to Kilauea Volcano, Hawaii: Jour. Geophys. Res., v. 88, p. 4147-4181.

Ryan, M.P., Koyanagi, R.Y., and Fiske, R.S., 1981, Modeling the three-dimensional structure of macroscopic magma transport systems, Application to Kilauea Volcano, Hawaii: Jour. Geophys. Res., v. 86, p. 7111-7129.

Shimozuru, D., 1987, Tidal effects on Hawaiian volcanism, in Decker, R.W., Wright, T.L., and Stauffer, P.H., eds., Volcanism in Hawaii, U.S. Geol. Surv. Prof. Paper 1350, v. 2, p. 1337-1343.

Swanson, D.A., 1972, Magma supply rates at Kilauea Volcano, 1952-1971: Science, v. 175, p. 169-170.

Swanson, D.A., and Christiansen, R.L., 1973, Tragic base surge at Kilauea volcano: Geology, v. 1, p. 83-86.

Swanson, D.A., Duffield, W.A., and Fiske, R.S., 1976, Displacement of the south flank of Kilauea Volcano: the result of forceful intrusion of magma into the rift zones: U.S. Geol. Surv. Prof. Paper 963, 39 p.

Swanson, D.A., Duffield, W.A., Jackson, D.B., and Peterson, D.W., 1979, Chronological narrative of the 1969-71 Mauna Ulu eruption of Kilauea Volcano, Hawaii: U.S. Geol. Surv. Prof. Paper 1056, 55 p.

Tilling, R.I., Christiansen R.L., Duffield, W.A., Endo, E.T., Holcomb, R.T., Koyanagi, R.Y., Peterson, D.W., and Unger, J.D., 1987, The 1972-1974 Mauna Ulu eruption, Kilauea Volcano: An example of quasi-steady-state magma transfer, in Decker, R.W., Wright, T.L., and Stauffer, P.H., eds., Volcanism in Hawaii, U.S. Geol. Surv. Prof. Paper 1350, v. 1, p. 405-469.

Wolfe, E.W., Garcia, M.O., Jackson, D.B., Koyanagi, R.Y., Neal, C.A., and Okamura, A.T., 1987, The Puu Oo eruption of Kilauea Volcano, episodes 1-20, January 3, 1983 to June 8, 1984, in Decker, R.W., Wright, T.L., and Stauffer, P.H., eds., Volcanism in Hawaii, U.S. Geol. Surv. Prof. Paper 1350, v. 1, p. 471-508.

Wright, T.L., 1971, Chemistry of Kilauea and Mauna Loa lava in space and time: U.S. Geol. Surv. Prof. Paper 735, 40 p.

Wright, T.L., and Fiske, R.S., 1971, Origin of the differentiated and hybrid lavas of Kilauea Volcano, Hawaii: Jour. Petrol., v. 12, no. 1, p. 1-65.

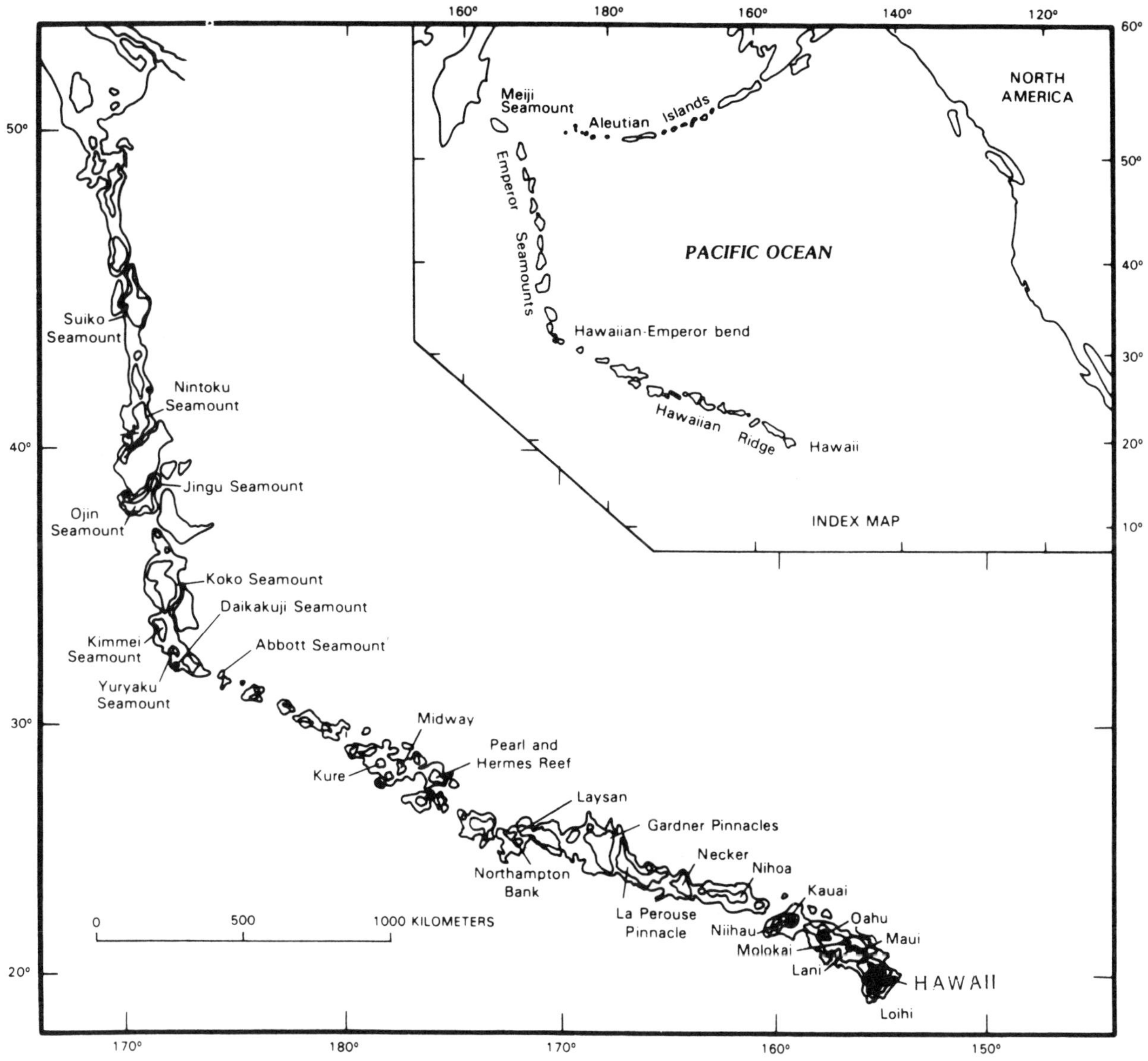

Figure 13.1.1. Bathymetry of the Hawaiian-Emperor volcanic chain, from Clague and Dalrymple (1987). Contours at 1-km and 2-km depths are shown in area of chain only. Inset shows location of chain (outlined by 2-km depth contour) in central North Pacific.

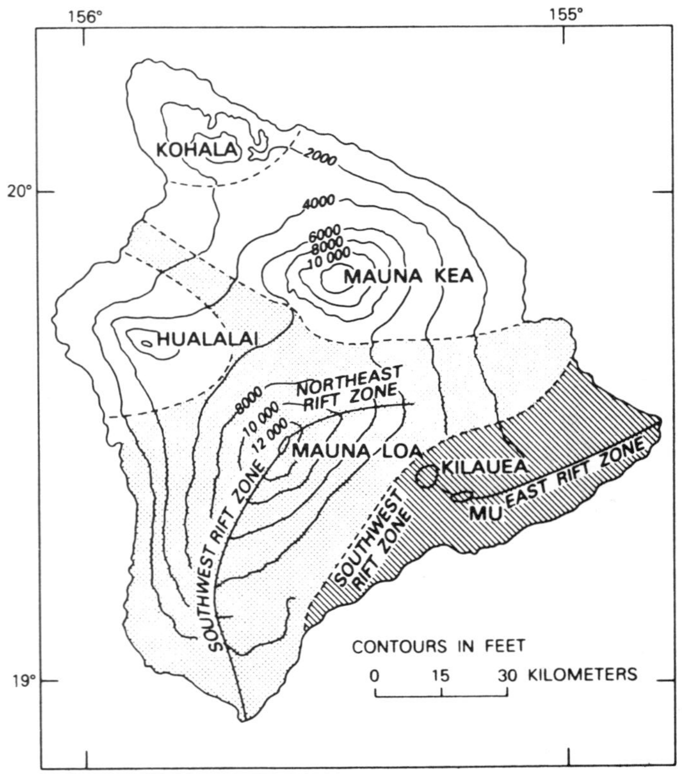

Figure 13.1.2. Index map showing the five volcanoes that make up Island of Hawaii and the rift zones of Kilauea and Mauna Loa Volcanoes, from Wright (1971). MU, site of the 1969-74 Mauna Ulu eruptions along Kilauea's upper east rift zone.

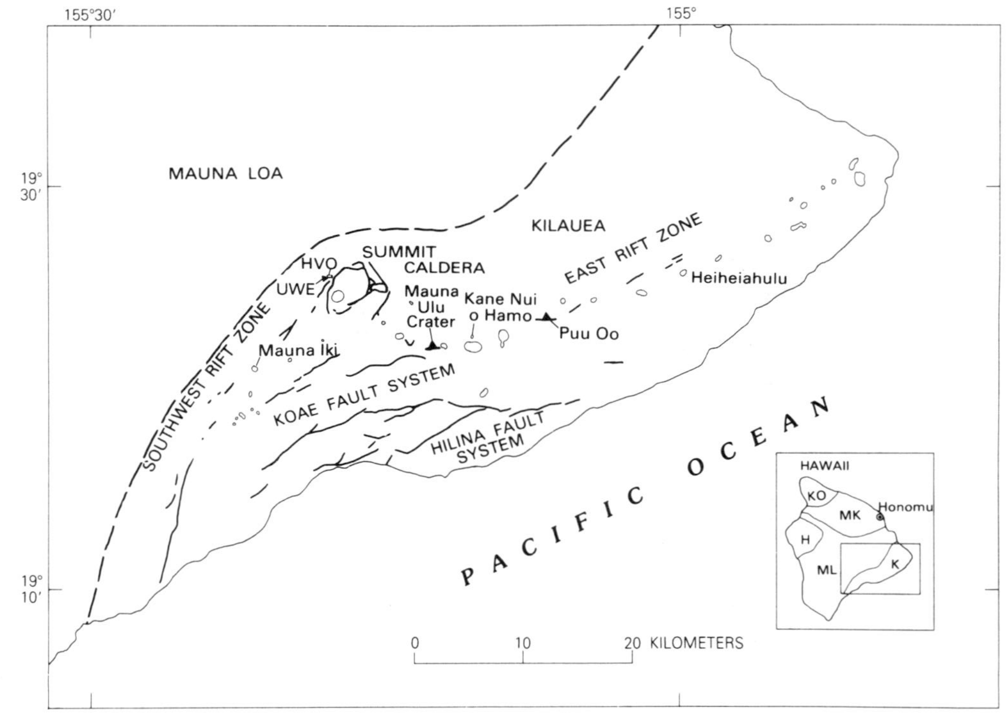

Figure 13.1.3. Kilauea Volcano, from Tilling and others (1987), showing summit caldera, rift zones, and principal fault systems; locations of the Hawaiian Volcano Observatory (HVO) and Uwekahuna vault (UWE) are also shown. Inset shows volcanoes making up the Island of Hawaii: KO, Kohala; MK, Mauna Kea; H, Hualalai; ML, Mauna Loa; and K, Kilauea. Mauna Ulu eruption site (1969-74) is along upper east rift zone; Puu Oo eruption site (1983-present) is along middle east rift zone (sites indicated by solid triangles). Mauna Iki (1919-20), Heiheiahulu (1750?), and Kane Nui o Hamo (prehistoric) are older volcanic shields along rift zones.

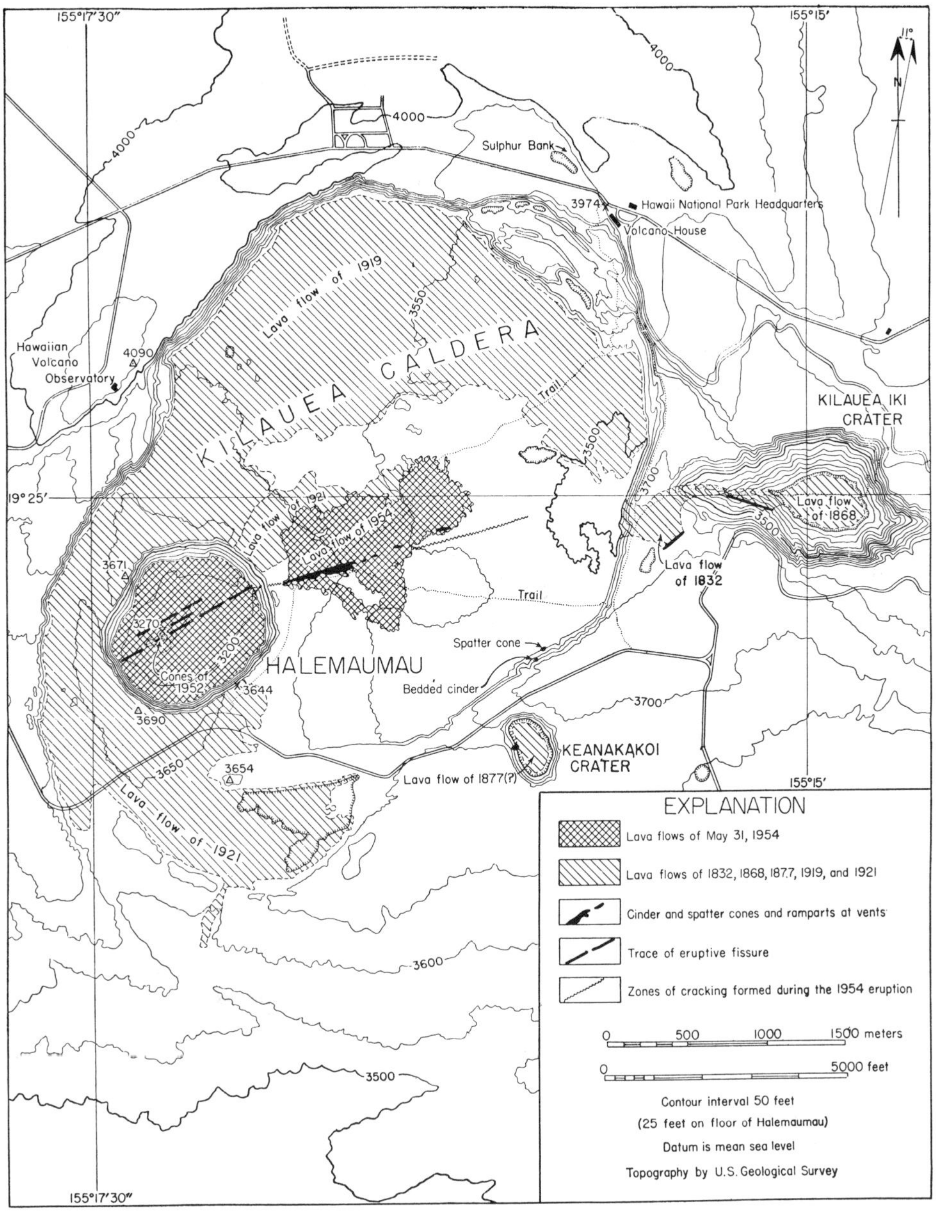

Figure 13.1.4. Topographic map of summit area of Kilauea Volcano, showing outline of Kilauea Caldera, Halemaumau (an inner pit crater), and recent lava flows (Macdonald, 1955).

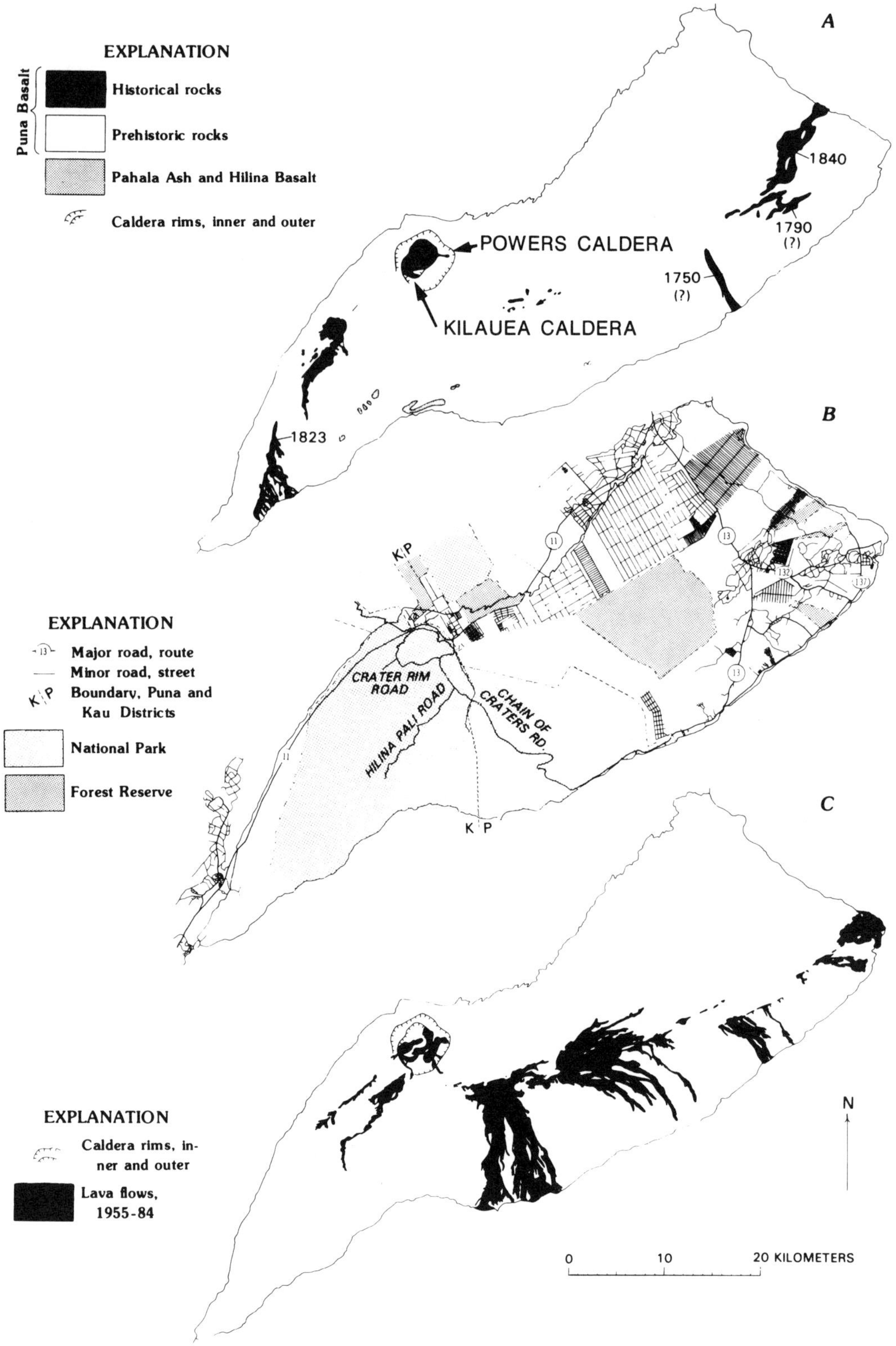

Figure 13.1.5. Historical eruptions and property development on Kilauea Volcano, from Holcomb (1987). A, stratigraphy of Kilauea, showing outlines of current caldera ("Kilauea Caldera") and an earlier, mostly buried caldera ("Powers Caldera"), plus historical lava flows erupted before 1954. B, Property subdivisions, developed mostly after 1950. C, Lava flows erupted during 1955-84.

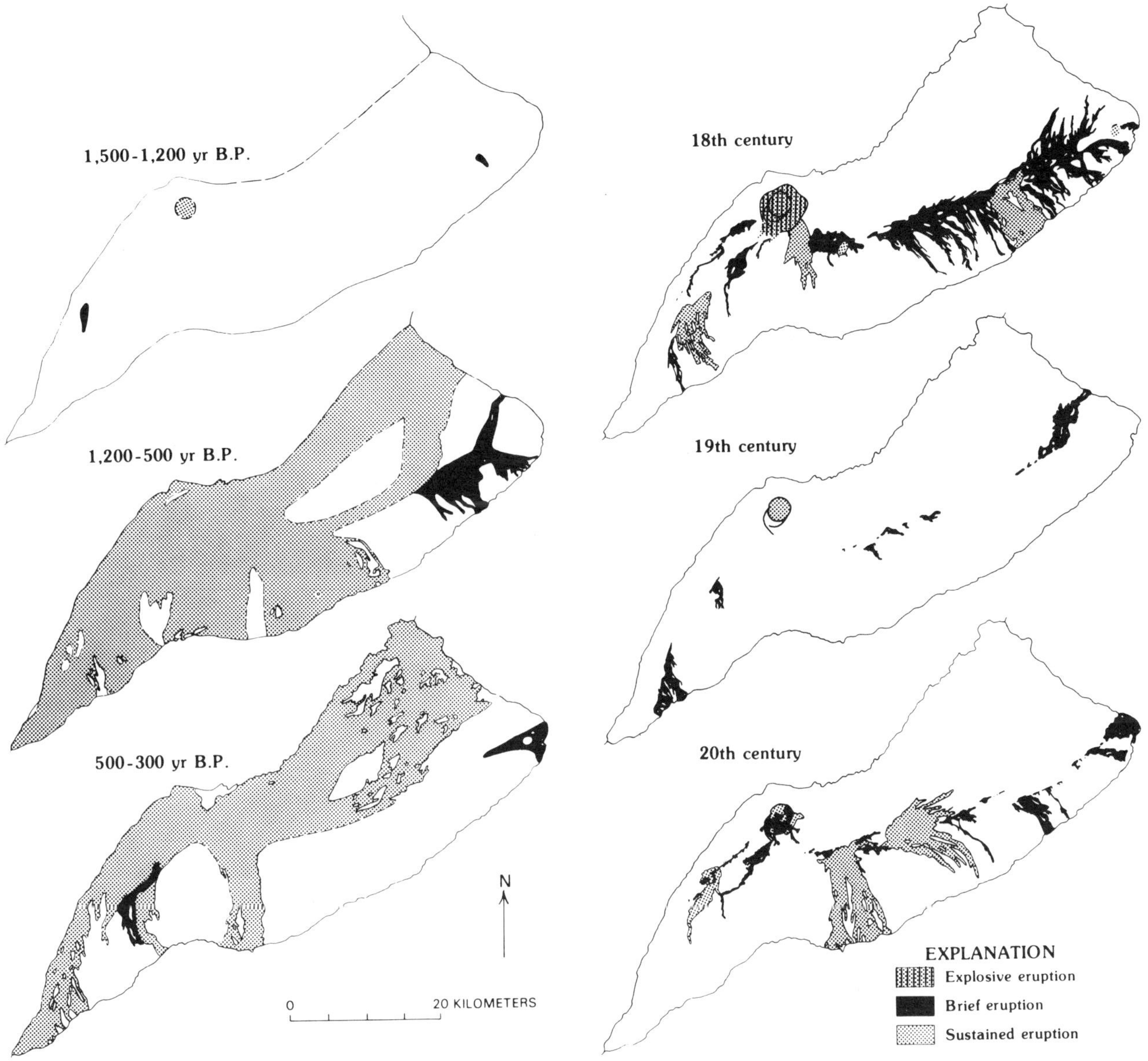

Figure 13.1.6. Series of maps summarizing Kilauea's eruption history during the last 1,500 years, from Holcomb (1987).

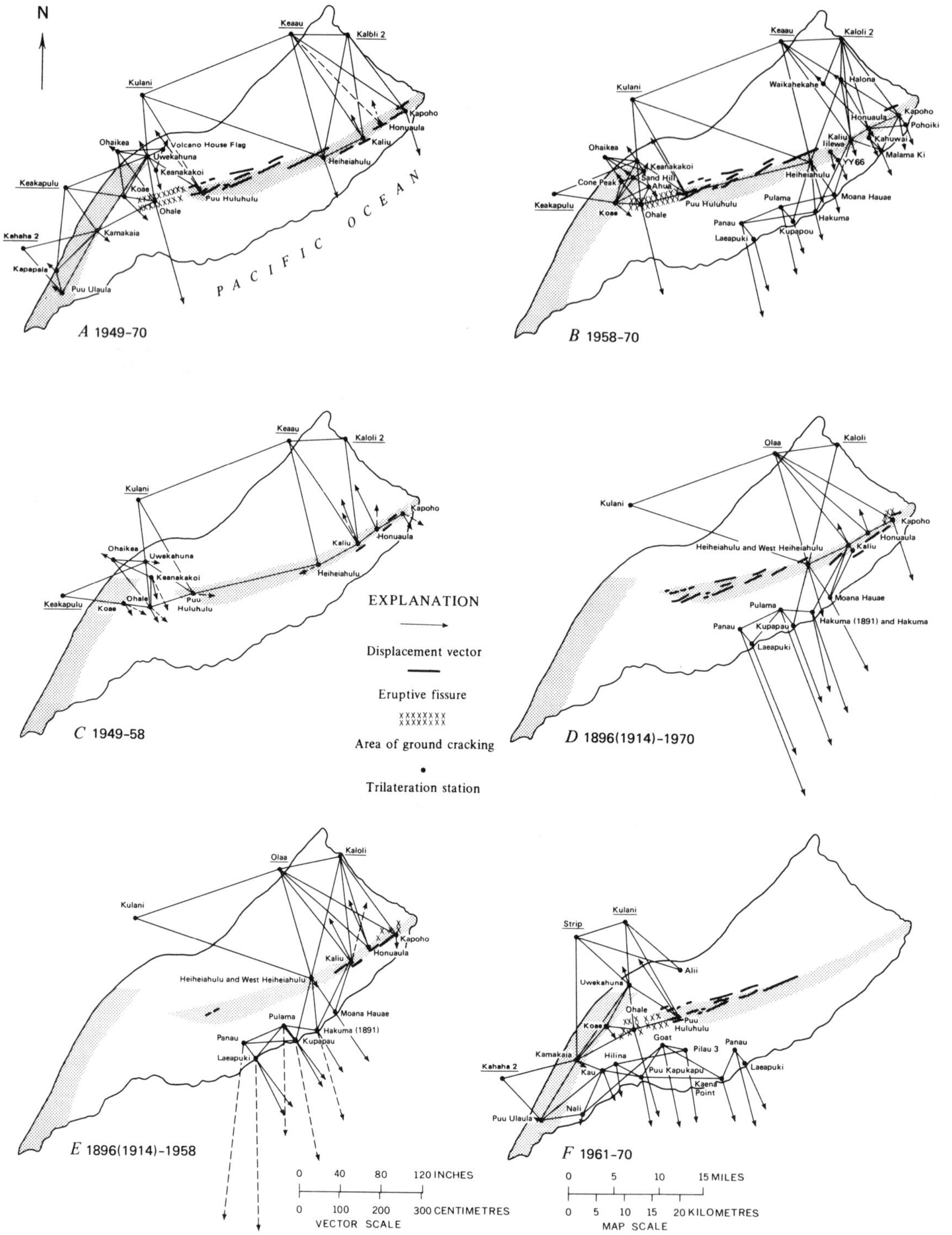

Figure 13.1.7. Horizontal ground displacements (arrows) at Kilauea Volcano derived from comparison of 1896 (1914), 1949, 1958, and 1961 triangulation surveys with 1970 trilateration survey (Swanson and others, 1976). Base stations considered stable are underlined; rift zones are shaded. Locations of rift eruptions and major episodes of ground cracking that occurred during a given survey are indicated.

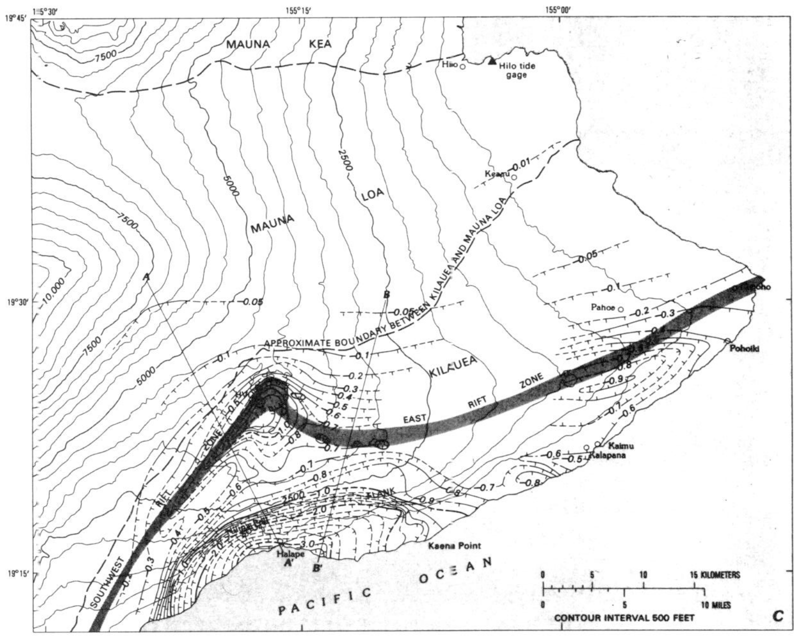

Figure 13.1.8. Interpreted contours of elevation changes associated with the November 1975 M 7.2 Kalapana earthquake, from Lipman and others (1985). Summit caldera and rift zones indicated by shaded pattern. Contour interval is 0.1 m, with supplemental 0.01- and 0.05-m contours; alternate contours omitted for values larger than -1.0 m. Hachures show direction of relative movement.

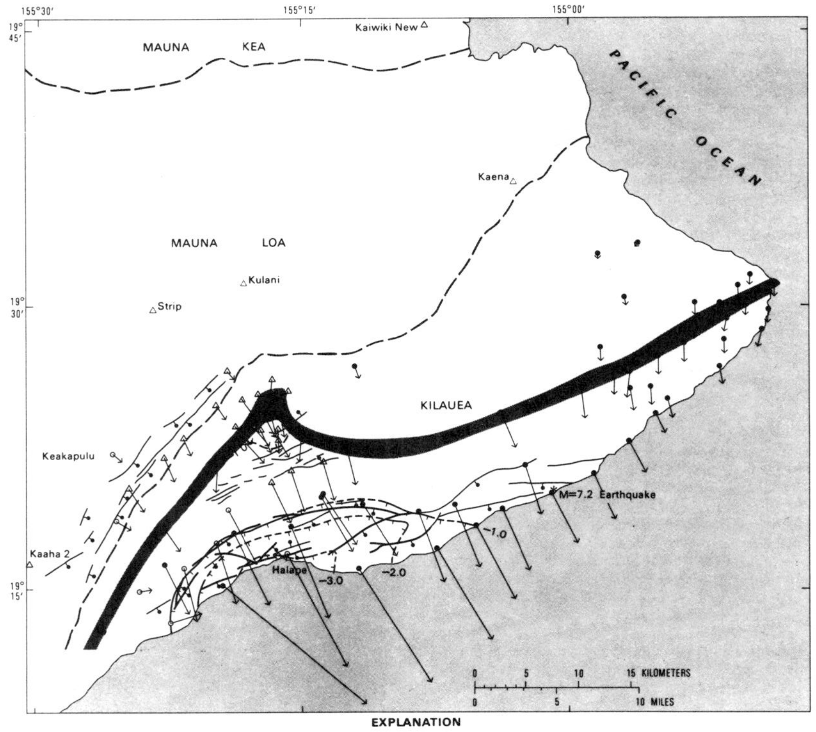

Figure 13.1.9. Interpretation of displacements associated with the November 1975 M 7.2 Kalapana earthquake, from Lipman and others (1985). Summit caldera and rift zones indicated by shaded pattern. Arrow tails at geodetic stations; lengths of arrows indicate amounts of displacement.

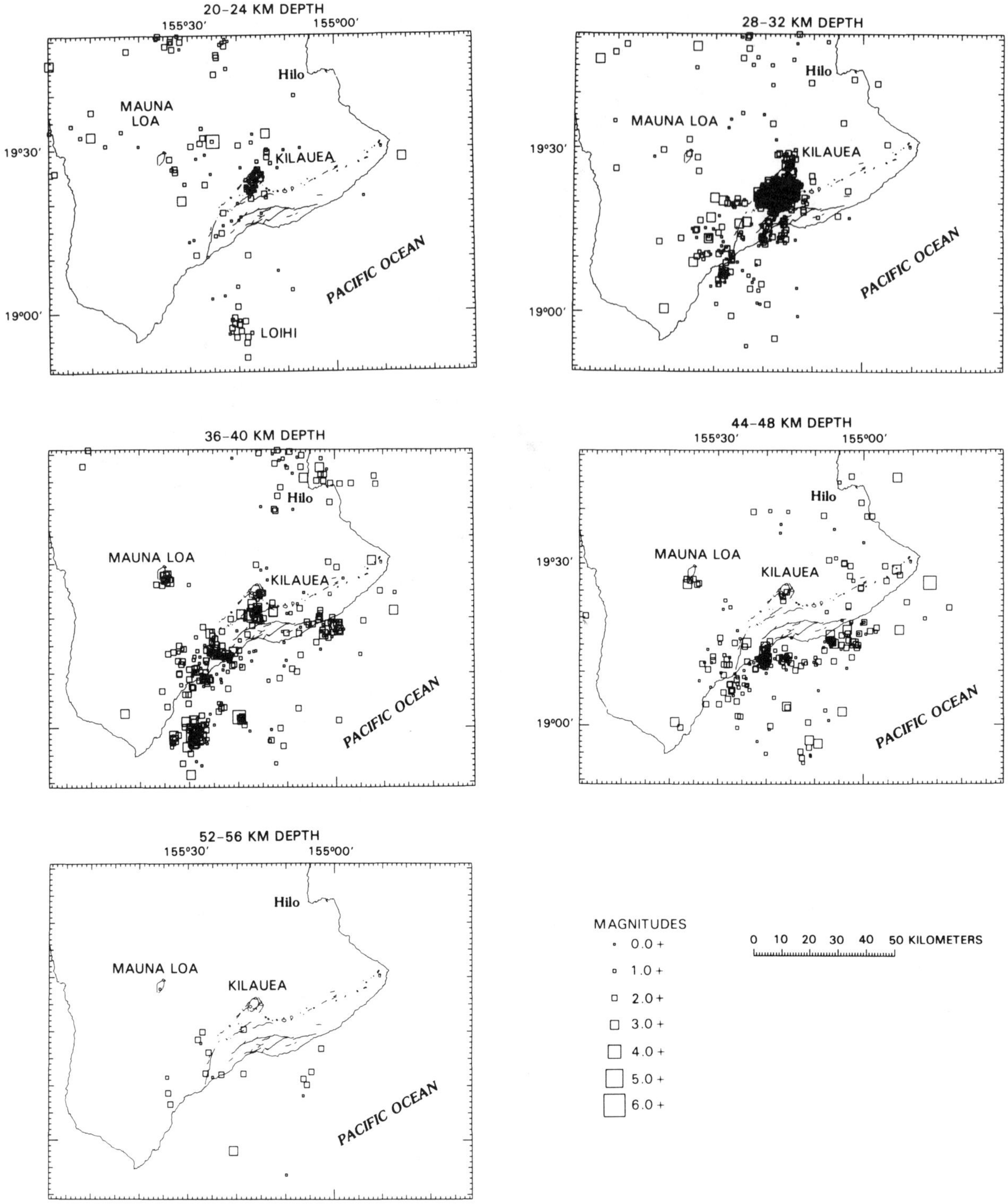

Figure 13.1.10. Selected depth slices from 20 to 60 km showing hypocenters of south Hawaii earthquakes, 1970-83, from Klein and others (1987). Note that magma conduit beneath summit caldera is nearly vertical, relatively narrow from 15- to 25-km depth, and more diffuse from 25- to 40-km depth (see also figure 13.1.11).

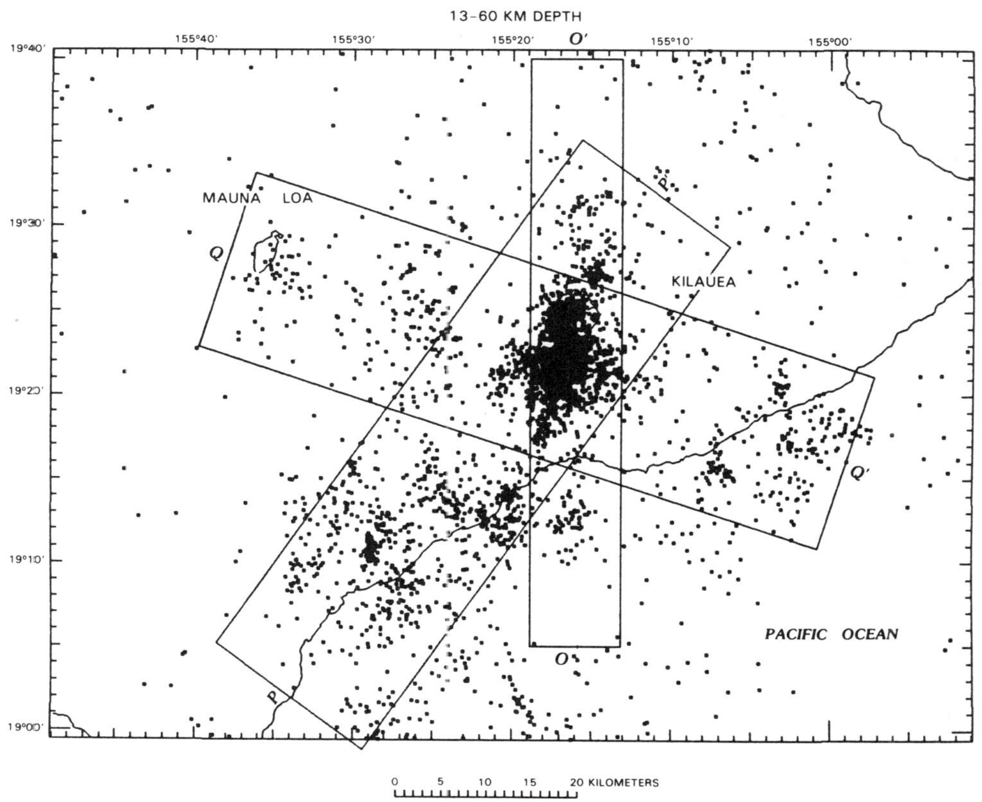

Figure 13.1.11. Above, Locations of 1970-83 earthquakes from 13- to 60-km depth beneath south Hawaii, from Klein and others (1987). Cross sections O-O' through Q-Q' are shown in figure 13.1.12. Only earthquakes within rectangular areas are plotted on sections. Although events above 13 km are plotted on sections, they are omitted from this map for clarity. All three sections include Kilauea's vertical magma conduit. Faults, fissures, and pit craters are shown in figure 13.1.3.

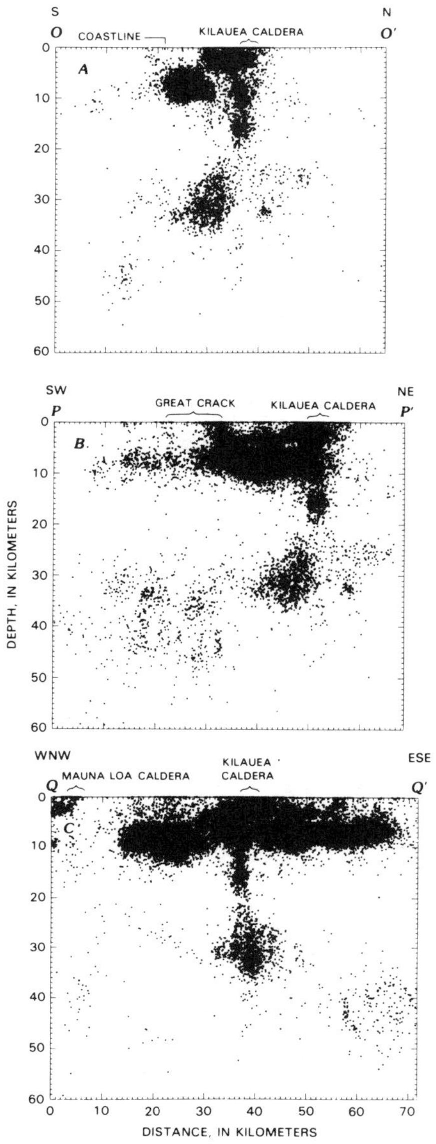

Figure 13.1.12. Right, Cross sections through Kilauea taken from surface to deepest seismicity at 60 km, from Klein and others (1987). Events during 1970-83 within areas of rectangles on previous figure are projected onto sections.

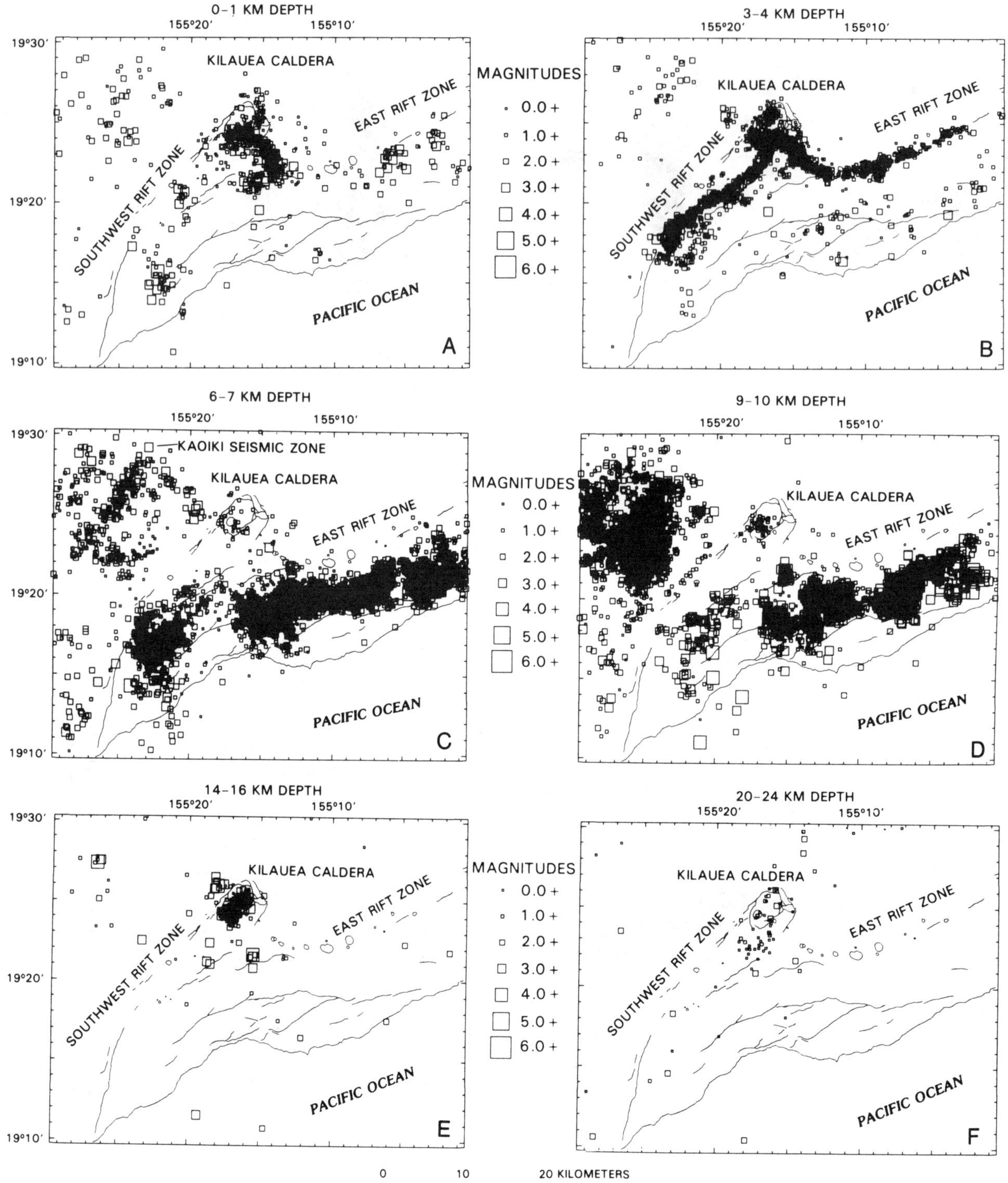

Figure 13.1.13. Selected depth slices from 0 to 24 km showing hypocenters of Kilauea earthquakes, 1970-83, from Klein and others (1987). Earthquake symbols are squares whose size depends on magnitude. Faults, fissures, and pit craters are shown in figure 13.1.3. From 0 to 2 km depth, most earthquakes occur in carapace rocks above summit magma chamber. From 2 to 4 km depth, earthquakes occur primarily in summit area and along two prominent rift zones. Kilauea's south flank and Kaoiki seismic zone between Kilauea and Mauna Loa are most active areas from 6 to 12 km depth; magma conduit beneath caldera is well defined by earthquakes from 14- to 40-km depth (see also figures 13.1.11, 13.1.12, and 13.1.14).

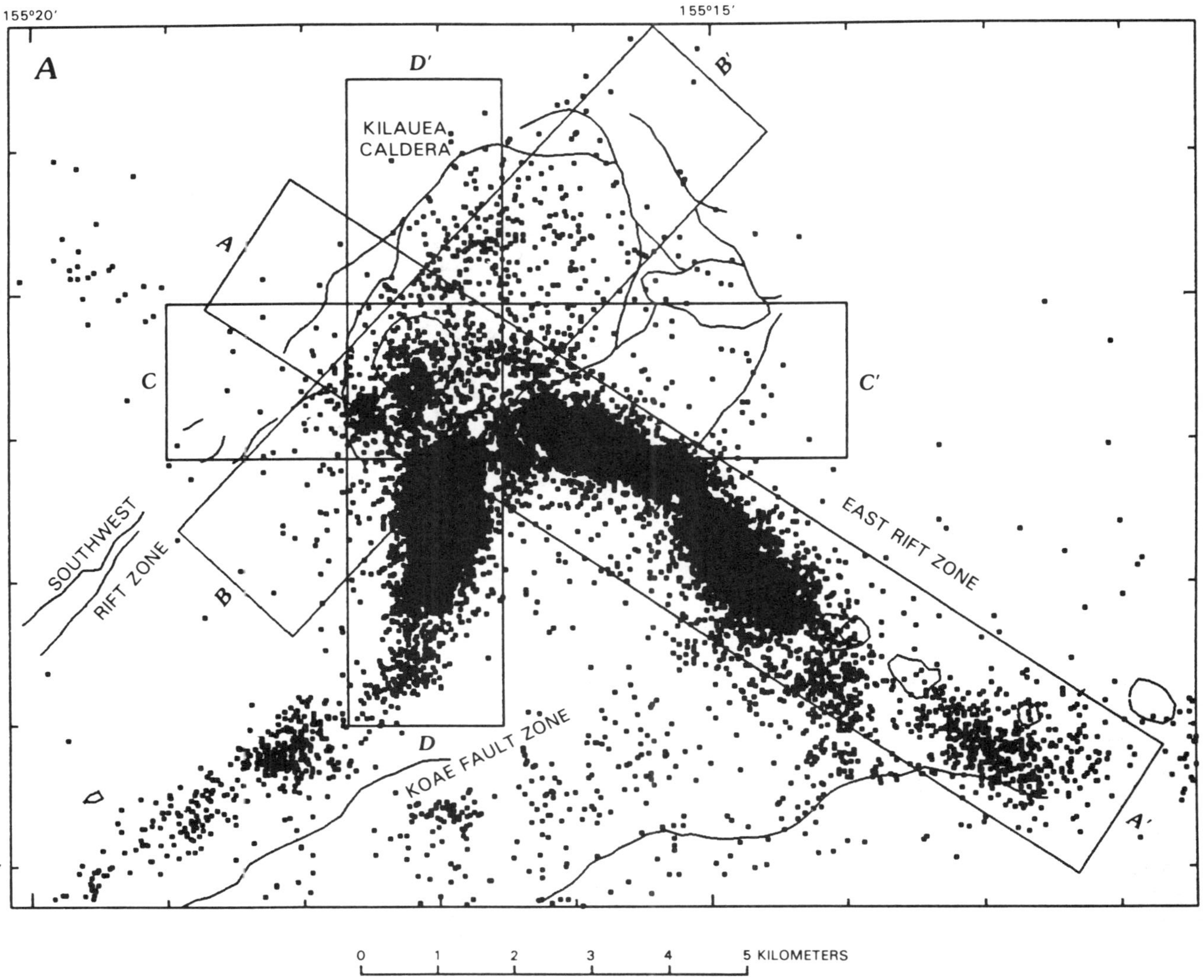

Figure 13.1.14. Epicenters of shallow 1970-83 earthquakes near Kilauea Caldera, from Klein and others (1987). Cross sections (rectangles) cross over buried magma reservoir. Note that sections in figure 13.1.15 extend to 12-km depth, but epicenters are shown here only for events in upper 5 km for clarity.

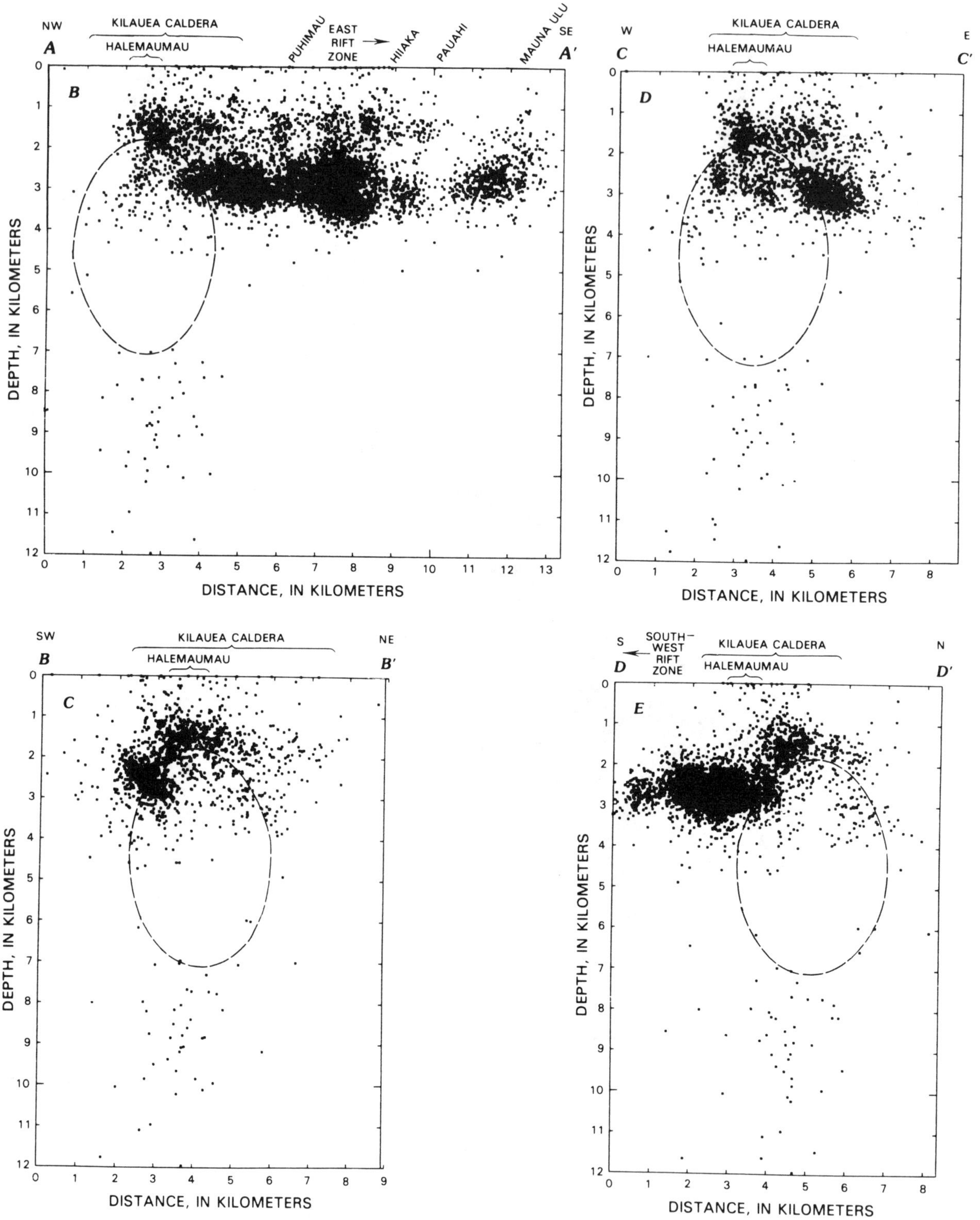

Figure 13.1.15. Cross sections of 1970-83 earthquakes shown in figure 13.1.14, from Klein and others (1987). Position of magma reservoir is approximated by dashed oval. Horizontal and vertical error of all hypocenters plotted is 1 km or less.

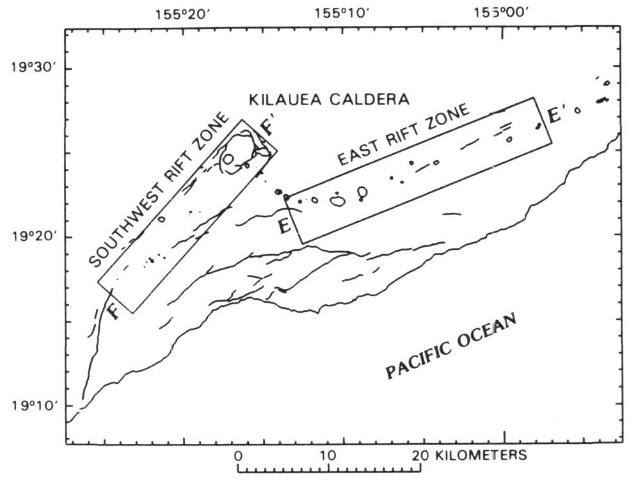

Figure 13.1.16. Areas included in longtudinal cross sections of rift zones (shown in figure 13.1.17), from Klein and others (1987). Only earthquakes within rectangles are plotted in the sections.

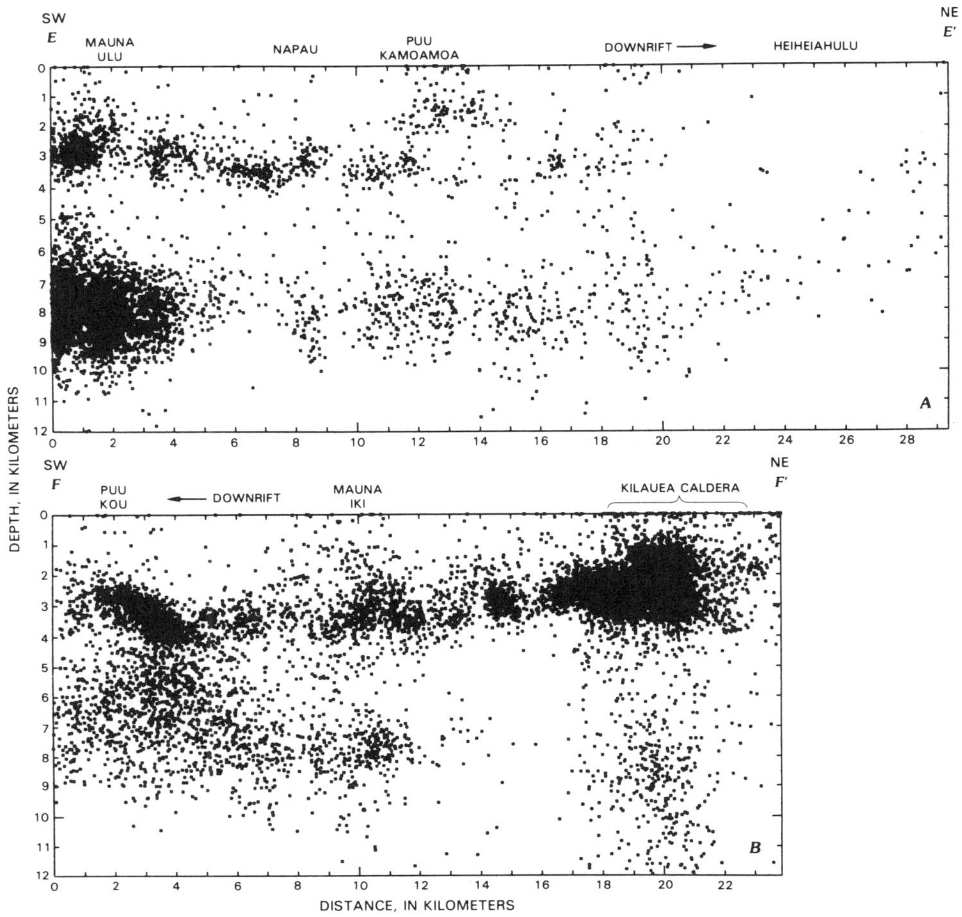

Figure 13.1.17. Longitudinal cross sections of earthquakes below east rift zone (A) and southwest rift zone (B), from Klein and others (1987). Positions and areas of sections are shown in figure 13.1.16. Main magma conduits (outlined by subhorizontal bands of seismicity) are at about 3-km depth.

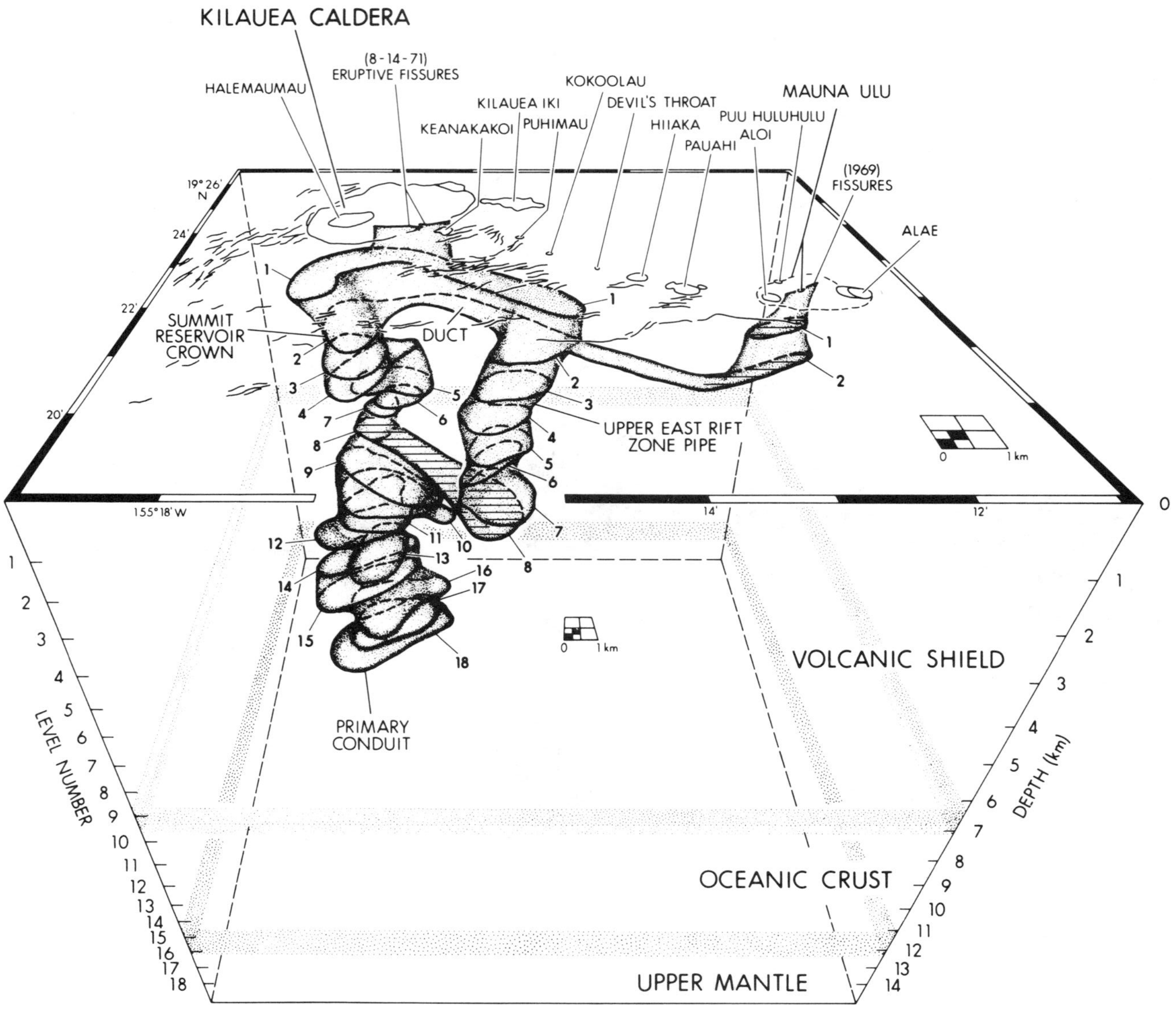

Figure 13.1.18. Perspective view northward and downward into interior of Kilauea Volcano, from Ryan and others (1981). Magma conduits are defined by zones of earthquake hypocenters that surround them over periods of several years. Numbers on magma conduits represent horizontal sections (slices) at depths corresponding to level-number scale on left side. Copyright by the American Geophysical Union.

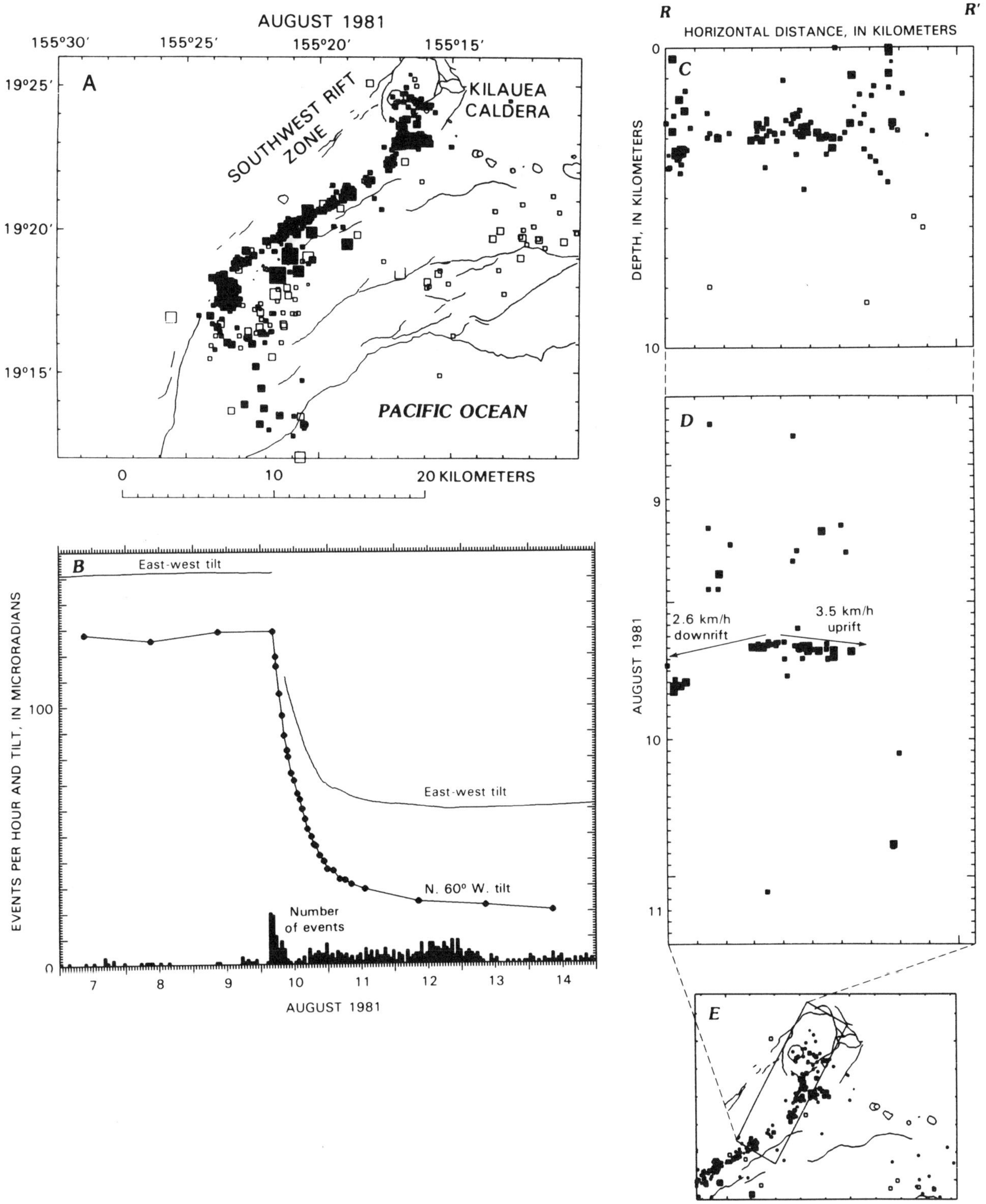

Figure 13.1.19. Seismicity and ground tilt during an intrusion into Kilauea's southwest rift zone on 10 August 1981, from Klein and others (1987). Map and cross section include earthquakes beginning on 9 August 1981. Solid lines in D indicate average lateral migration rates of earthquakes (uprift and downrift) during intrusion.

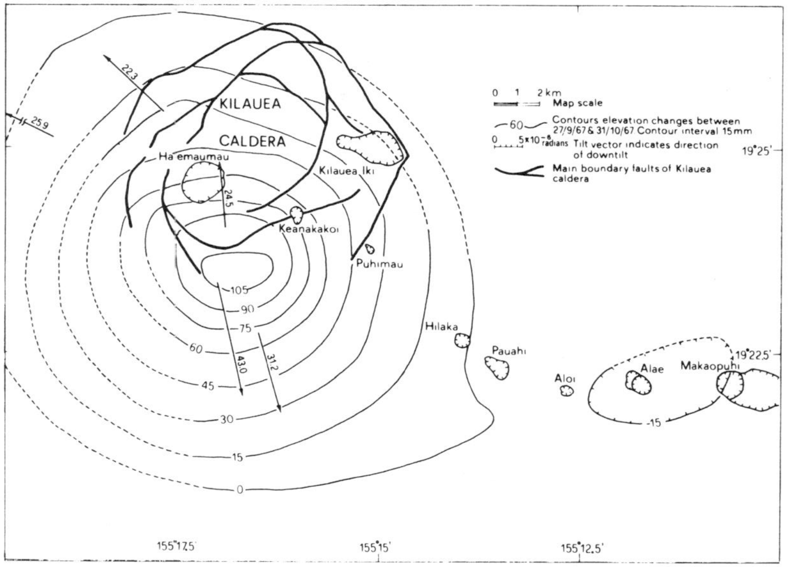

Figure 13.1.20. Contour map of vertical displacements from 27 September to 31 October 1967, from Kinoshita and others (1974). More than 100 mm of inflation during that interval was followed by a lava-lake eruption at Halemaumau from 5 November 1967 to 13 July 1968.

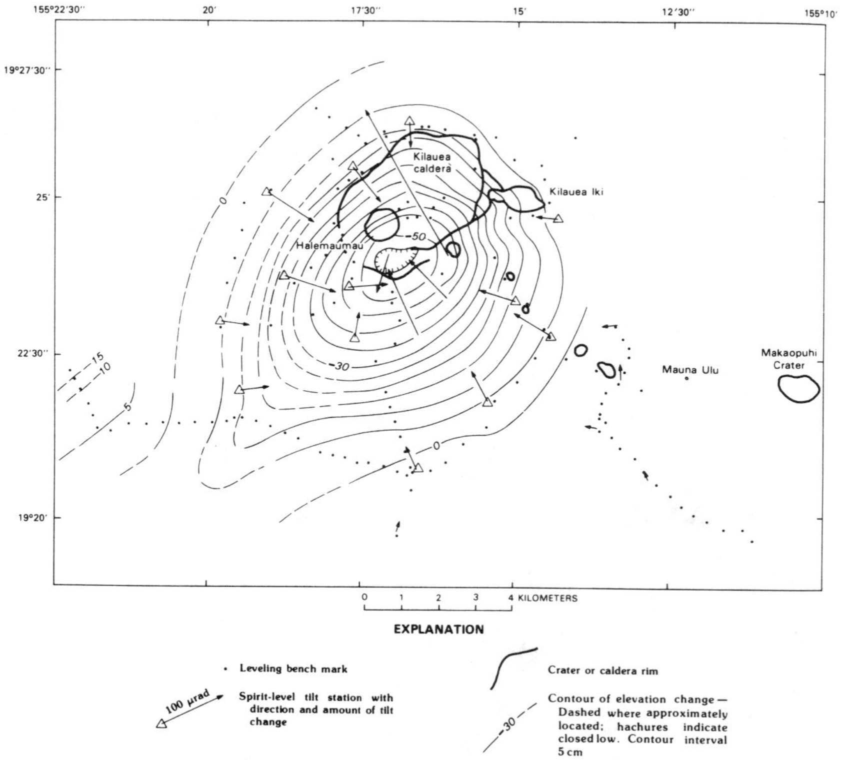

Figure 13.1.21. Subsidence and tilt in Kilauea summit region from 27 September - 5 October 1982 to 7-11 February 1983 (Wolfe and others, 1987). Contours show relative elevation change in centimeters.

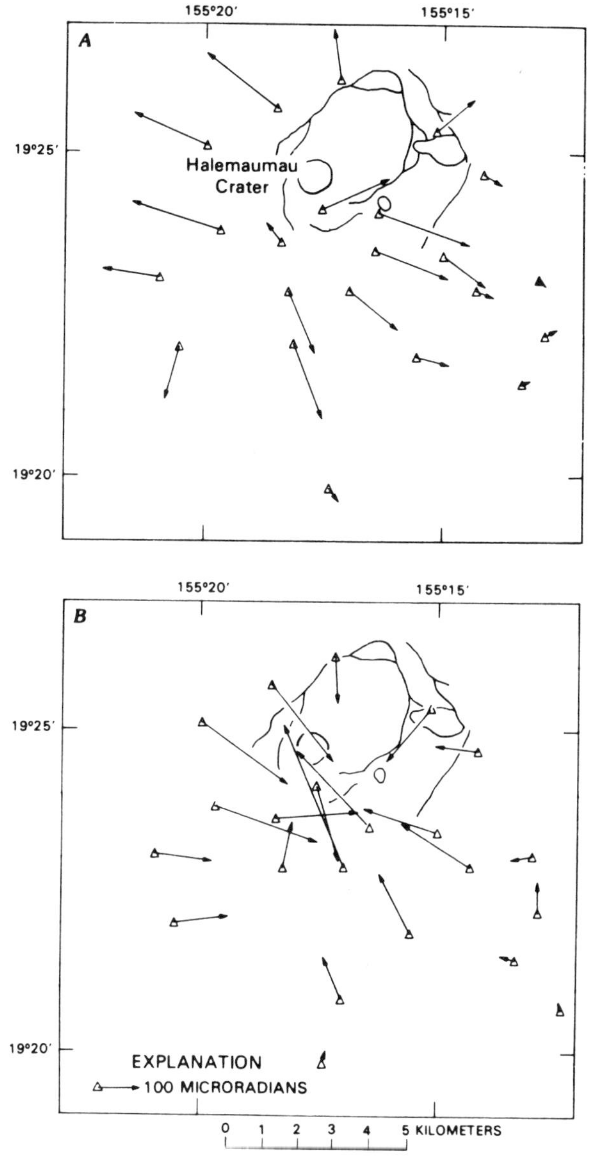

Figure 13.1.22. Tilt vectors measured in Kilauea summit region, from Dvorak and Okamura (1987). A, Uplift of 12-14 August 1981 to 3 March - 5 April 1982. B, Subsidence of 27 September - 14 October 1982 to 2-11 February 1983. Vectors were determined by measuring changes in relative elevation along 40-m-long triangles. Station locations shown by open triangles; tilt vectors point in direction of downward deflection of surface. General radial pattern of tilt vectors indicates that the most active part of summit magma reservoir lies beneath southern part of caldera (see also figure 13.1.23).

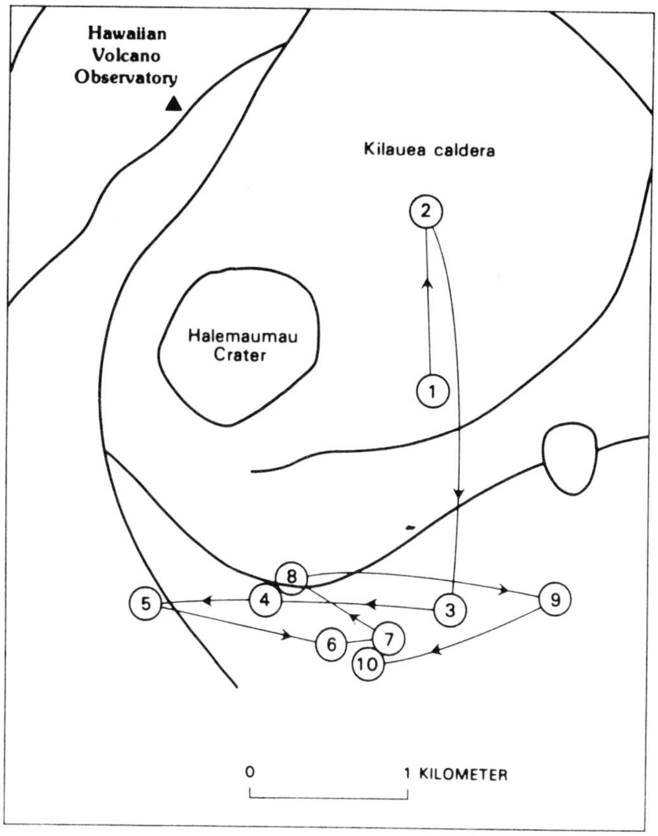

Figure 13.1.23. Lateral shift of apex of uplift of inflating bulge at Kilauea summit, modified from Fiske and Kinoshita (1969). Leveling surveys demonstrate shifting center of uplift: (1) January-July 1966; (2) July-October 1966; (3) October 1966 - January 1967; (4) January-February 1967; (5) during February 1967; (6) February-May 1967; (7) May-June 1967; (8) June-July 1967; (9) July-September 1967; (10) September-October 1967. Total uplift of the 10-km-wide bulge at its apex from January 1966 to October 1967 was approximately 70 cm; summit eruption of Kilauea began in November 1967. Copyright by the American Association for the Advancement of Science.

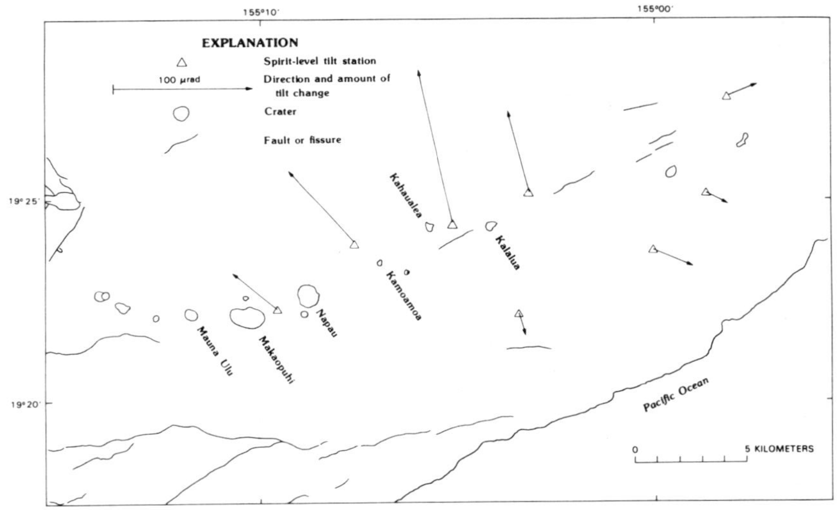

Figure 13.1.24. Tilt changes along east rift zone from spirit-level measurements, January 1978 to March 1981 (Wolfe and others, 1987). Inflation of rift zone occurred during a series of intrusions into rift following the September 1977 eruption. It culminated in a long-lived eruption on middle east rift zone that began in January 1983.

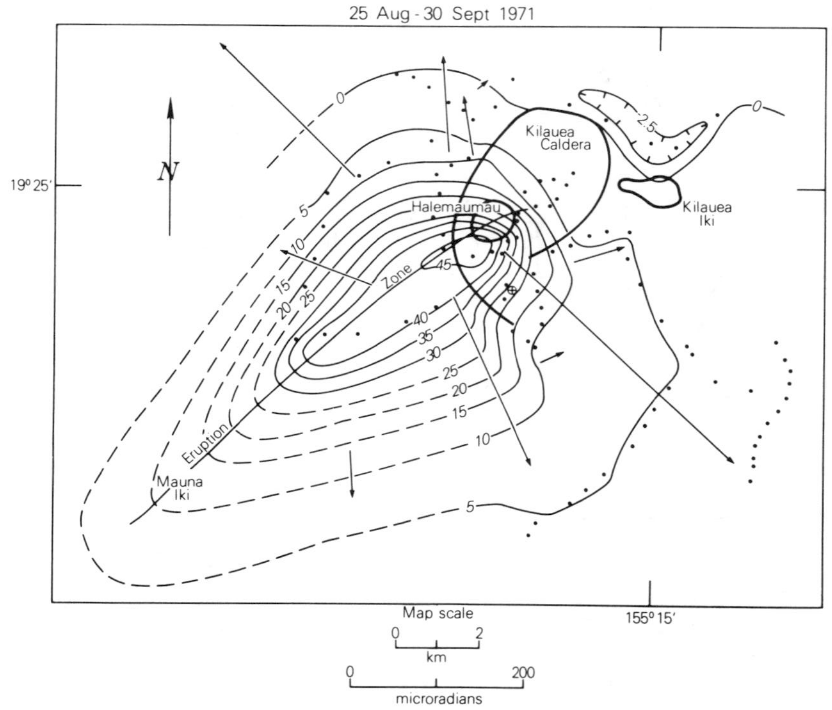

Figure 13.1.25. Vertical ground-displacement contours (in centimeters) and ground-tilt vectors for the period 25 August to 30 September 1971 (Duffield and others, 1982). Bench marks shown as dots; preeruption center of ground tilt shown as circle with cross. Zone of eruptive and noneruptive fissures generalized as solid line.

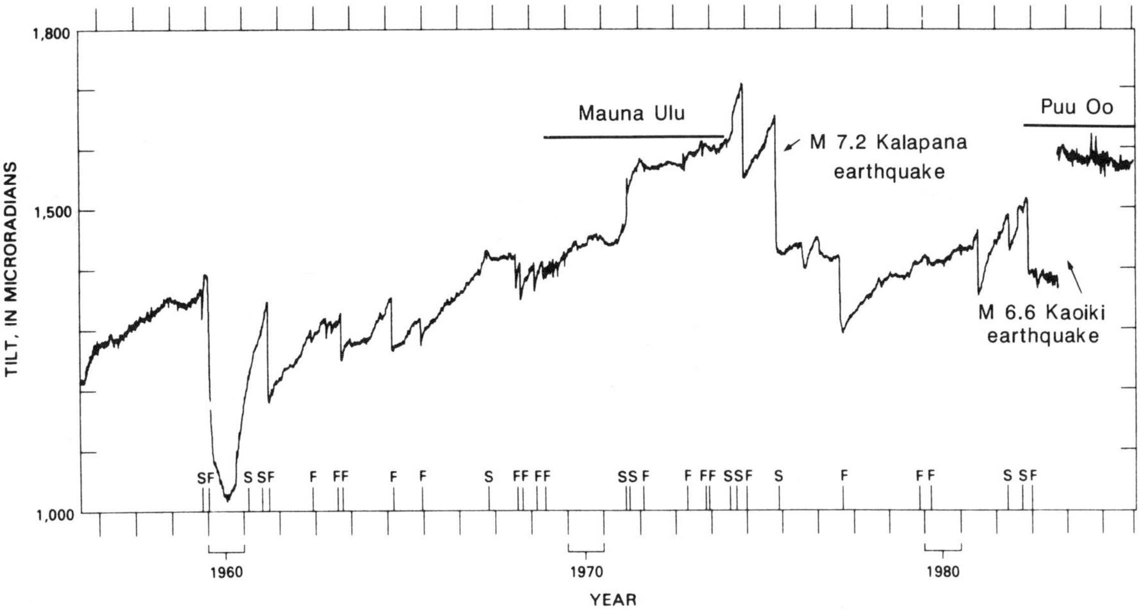

Figure 13.1.26. Radial tilt at Uwekahuna vault (see figure 13.1.3) on northwest rim of Kilauea Caldera for the period 1956-85, from Decker (1987). Increasing tilt (radially outward) indicates inflation of summit region; decreasing tilt indicates deflation. Magma added slowly to shallow reservoir system from depth between eruptions (slow inflation) is removed rapidly during flank eruptions and intrusions (sudden deflation). Vertical lines indicate eruptions; S, summit eruptions; F, flank eruptions. Offset in November 1983 caused by M 6.6 earthquake.

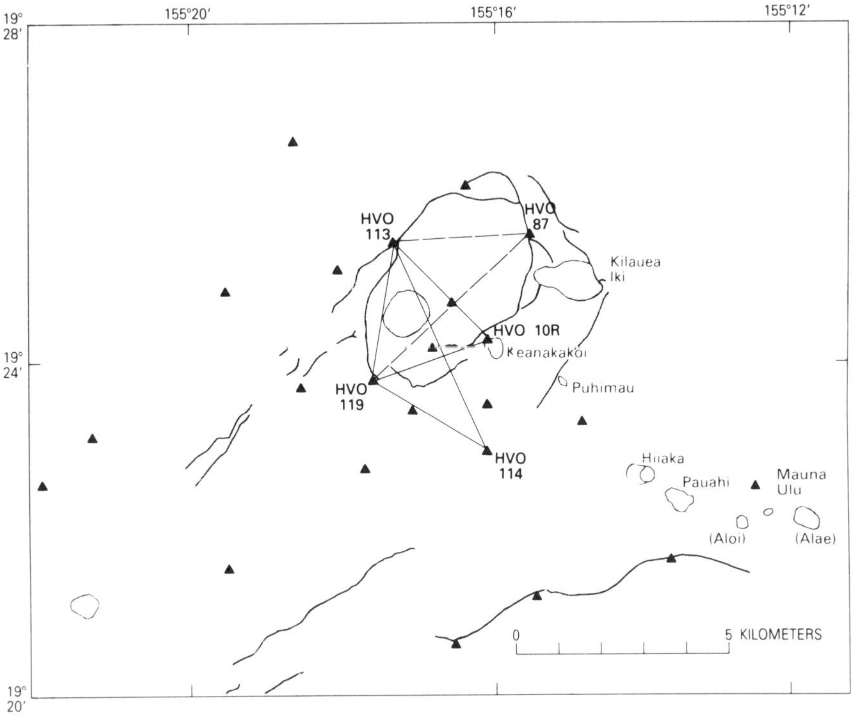

Figure 13.1.27. Location of electronic distance measurement (EDM) bench marks of Kilauea summit trilateration network, from Tilling and others (1987). Solid lines are measured more frequently than dashed lines (see figure 13.1.28).

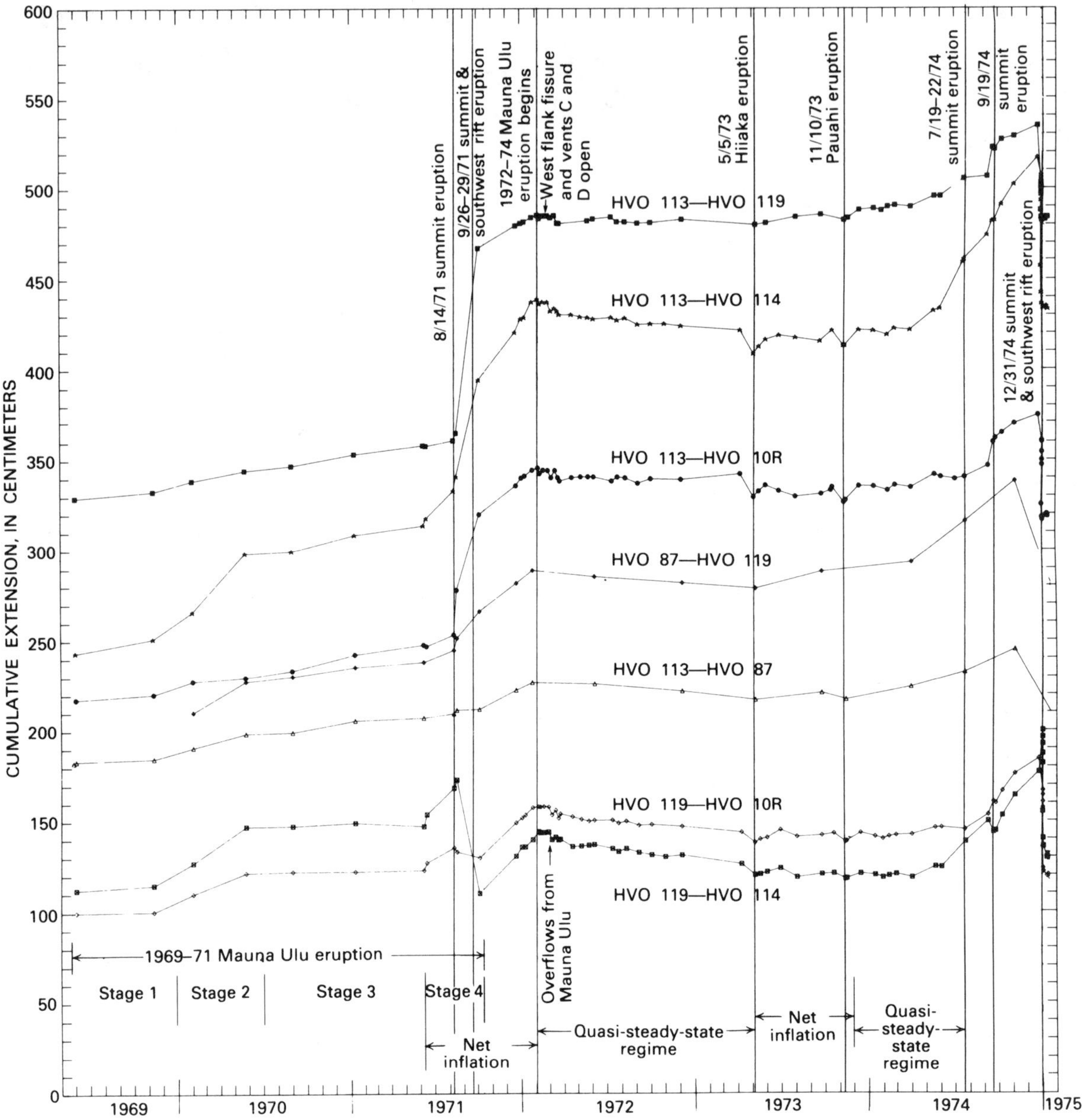

Figure 13.1.28. Variations of distances between electronic distance measurement (EDM) bench marks (figure 13.1.27) between May 1969 and January 1975, from Tilling and others (1987). Starting value for each cumulative extension curve is arbitrary. Frequency of measurements during 1969-71 was about the same as during 1972-75, but not all data are plotted here.

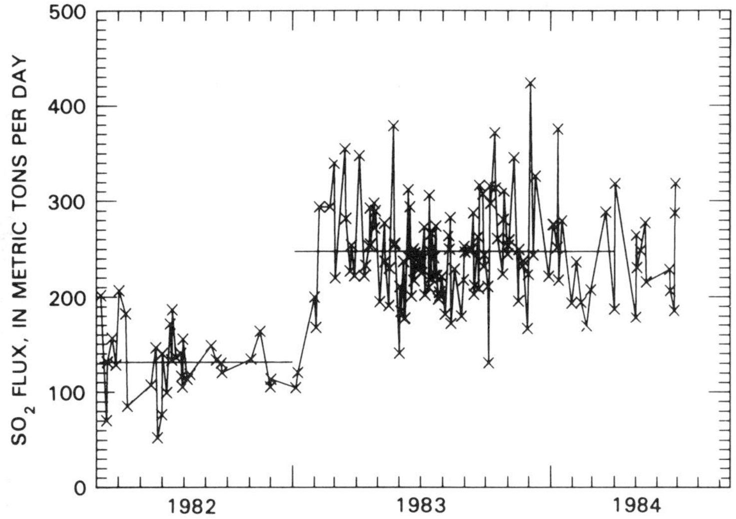

Figure 13.1.29. Total sulfur dioxide (SO_2) flux from Kilauea summit fumaroles, April 1982 through June 1984, based on ground measurements (Wolfe and others, 1987; modified from Greenland and others, 1985). Horizontal lines show change in average flux after start of Puu Oo eruption on middle east rift zone in January 1983.

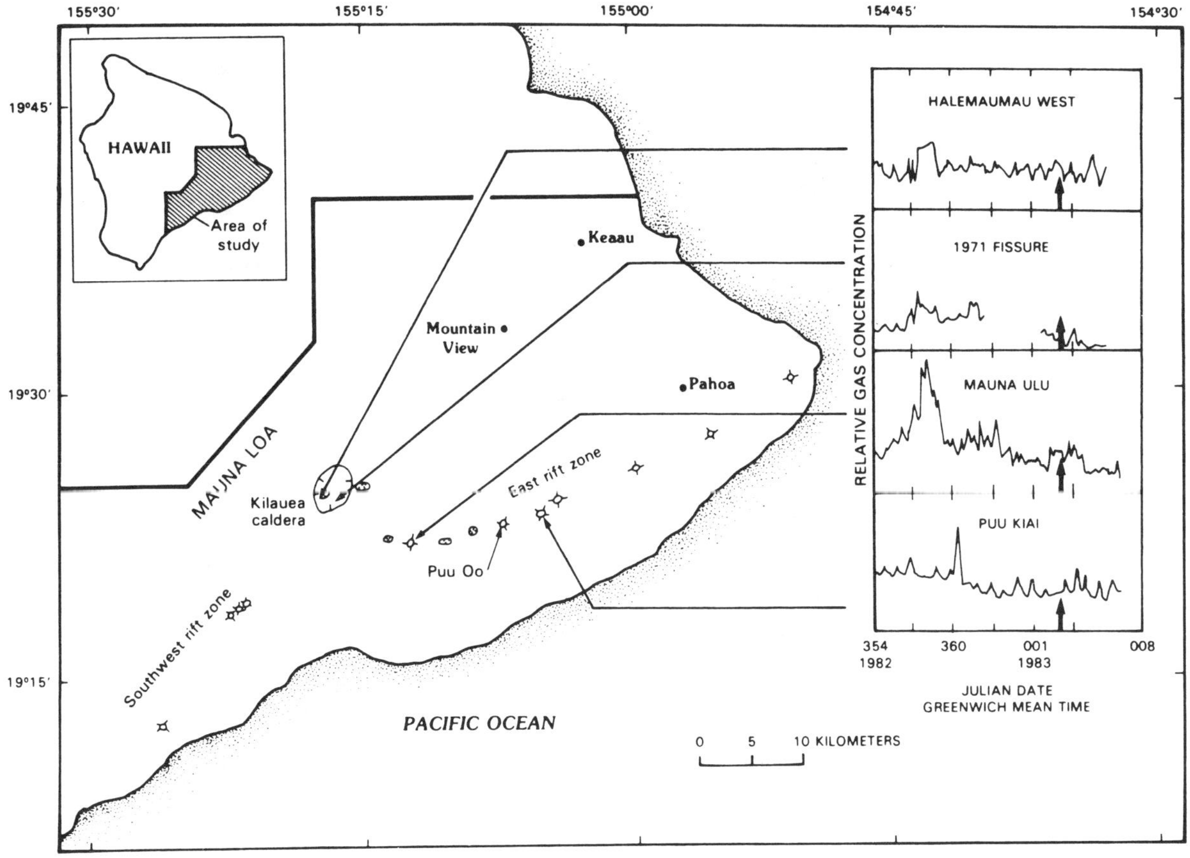

Figure 13.1.30. Results from satellite-telemetered gas-monitoring stations at Kilauea for late December 1982 through early January 1983, from McGee and others (1987). Increased gas emission in late December was detected simultaneously by two summit stations (Halemaumau West and 1971 Fissure) and upper east rift station (Mauna Ulu). Vertical arrow marks onset of eruptive activity at Puu Oo.

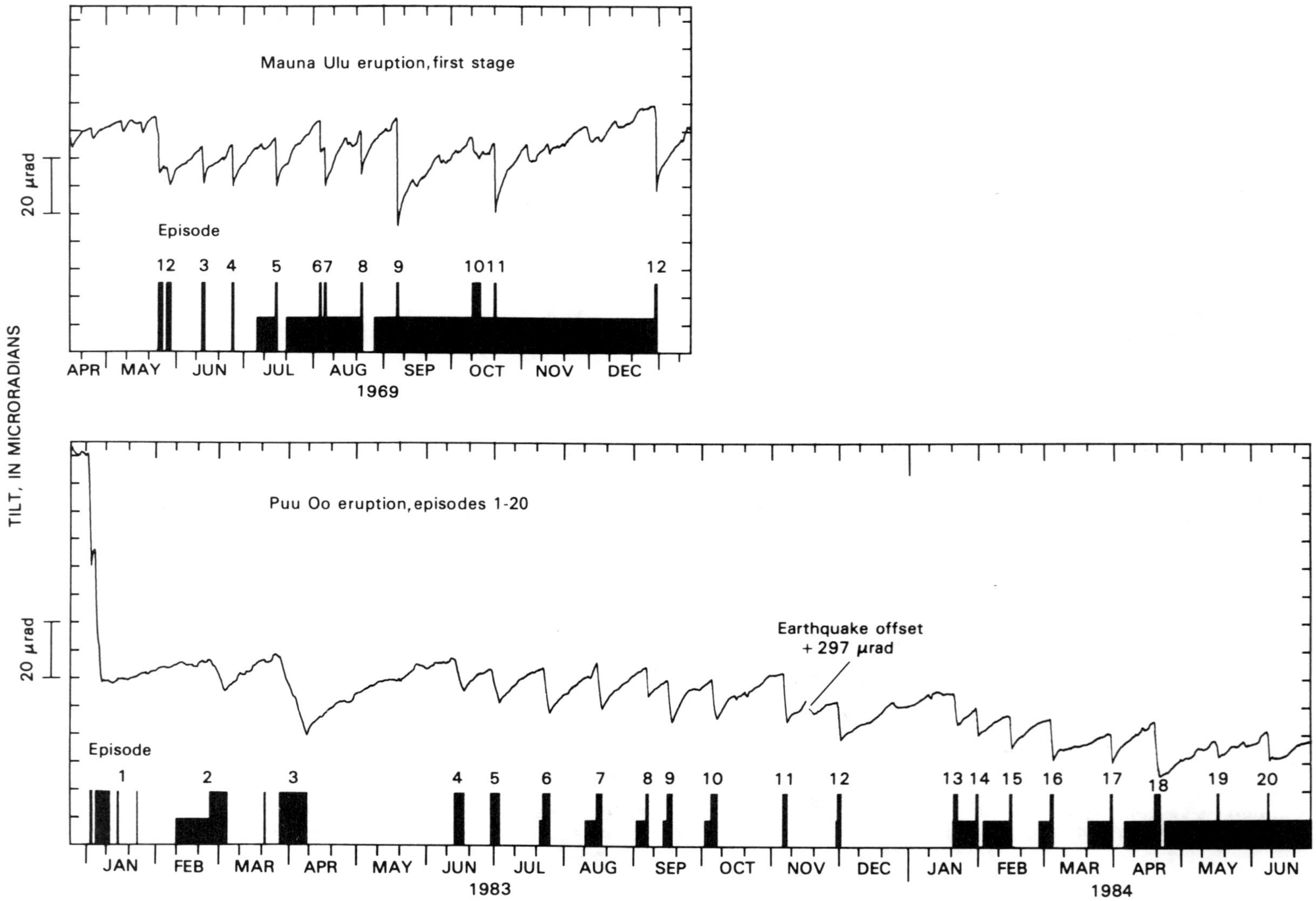

Figure 13.1.31. East-west tilt at Kilauea summit (measured at Uwekahuna vault; see figure 13.1.3), episodes of high lava fountaining (full-height bars), and periods of observed low-level eruptive activity (half-height bars) during first stage of the 1969-71 Mauna Ulu eruption (top) and during fountaining episodes 1-20 of the 1983-present Puu Oo eruption (bottom), from Wolfe and others (1987).

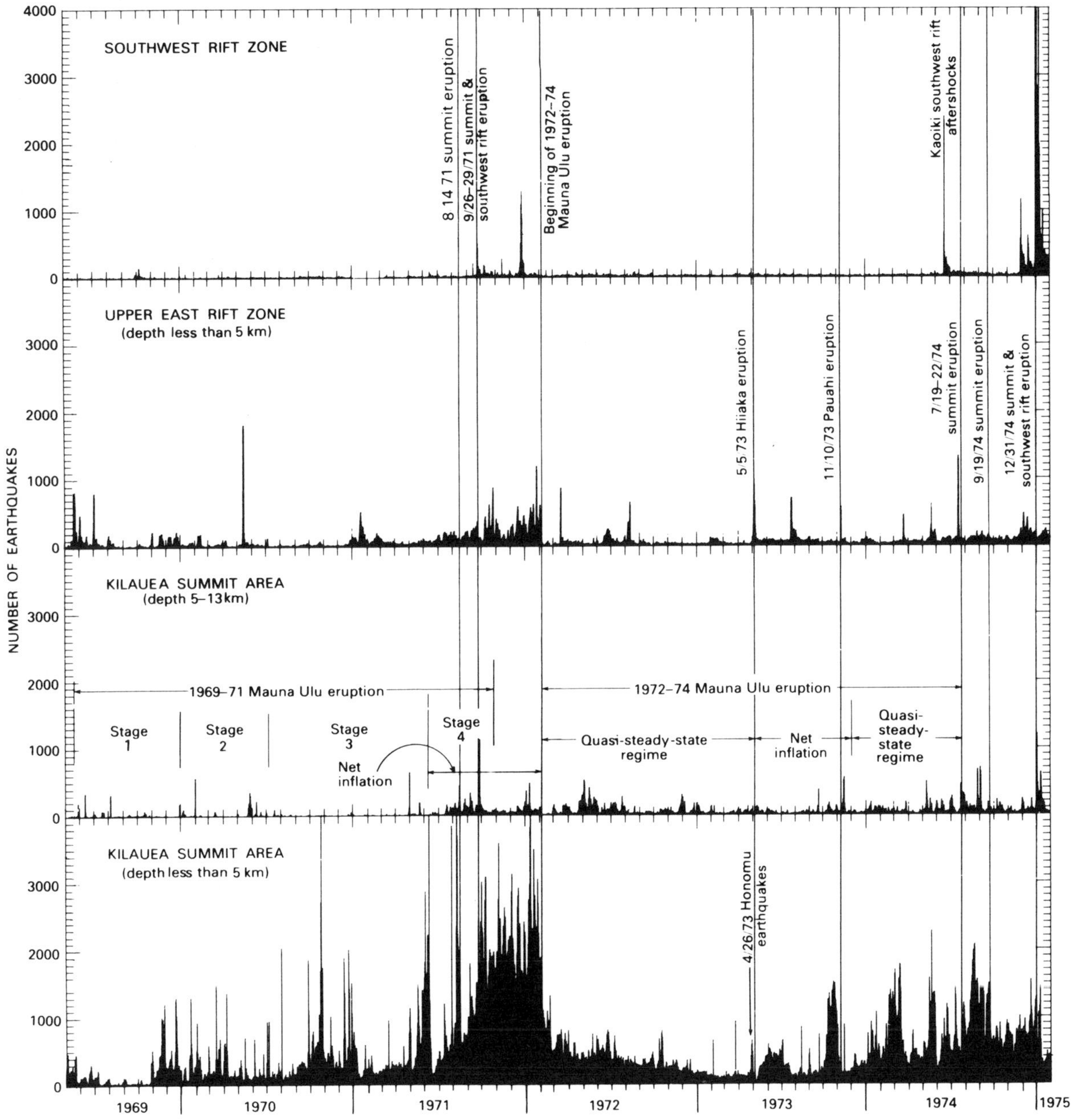

Figure 13.1.32. Numbers of earthquakes in summit area (shallow and crustal depth), in upper east rift zone, and in southwest rift zone of Kilauea from May 1969 to January 1975, from Tilling and others (1987). All events are counted, whether or not hypocenters are well determined; counts for southwest-rift earthquakes include those in Kaoiki fault system.

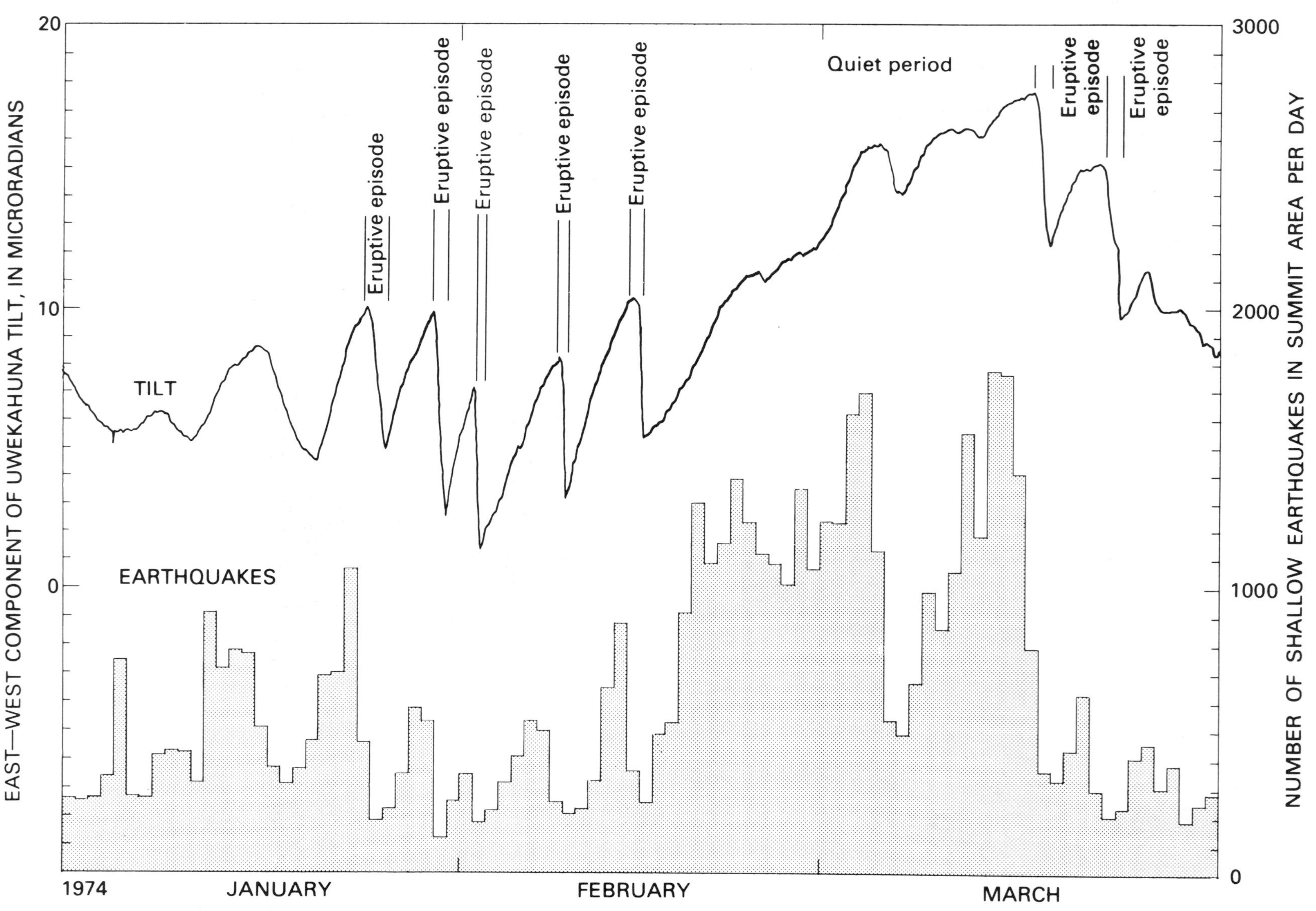

Figure 13.1.33. Correlation of eruptive episodes at Mauna Ulu during January-March 1974, characterized by vigorous fountaining and voluminous overflows, with variations in Uwekahuna tilt and daily number of shallow (depth less than 5 km) summit earthquakes (Tilling and others, 1987).

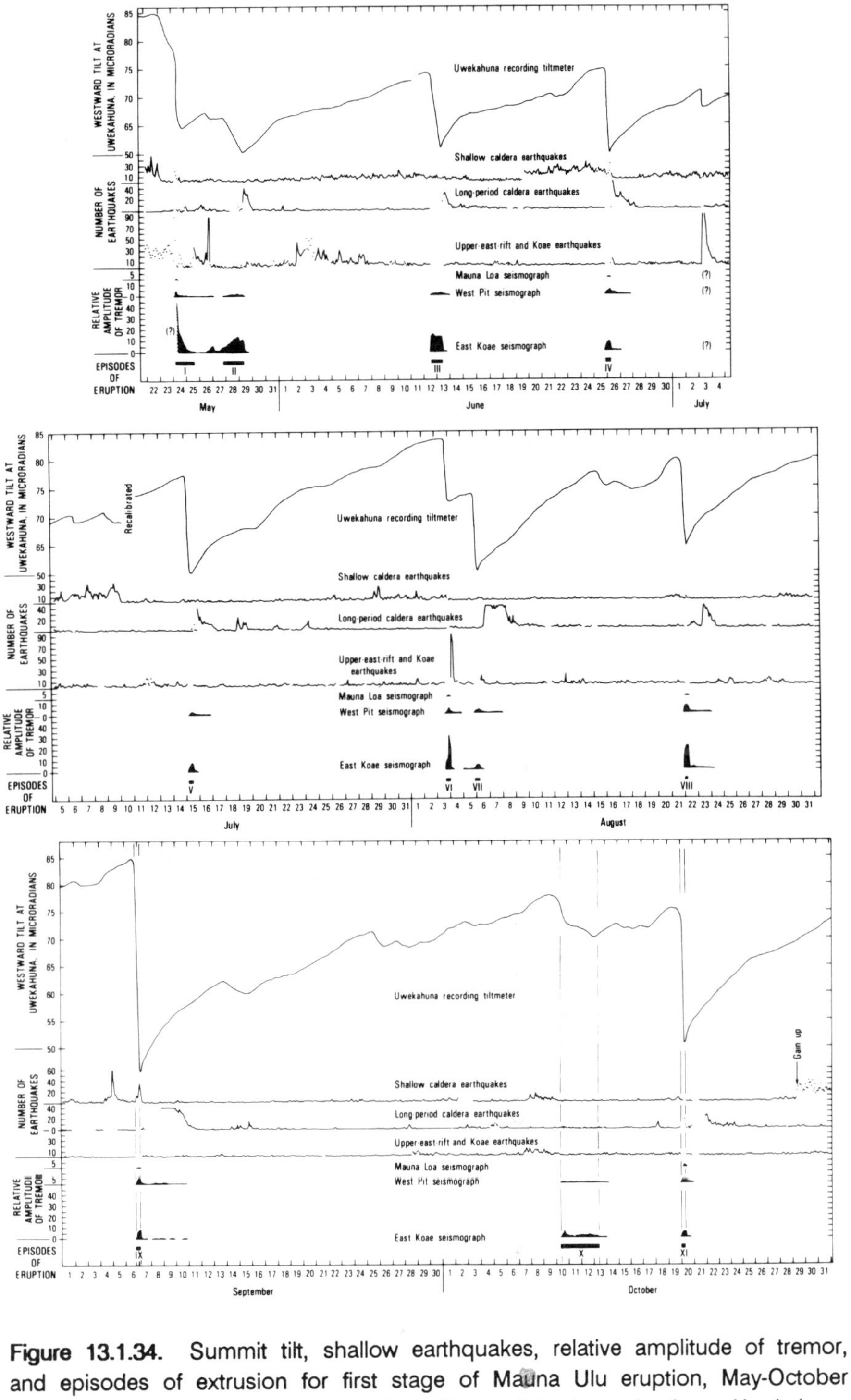

Figure 13.1.34. Summit tilt, shallow earthquakes, relative amplitude of tremor, and episodes of extrusion for first stage of Mauna Ulu eruption, May-October 1969 (Koyanagi and others, 1987). Tilt at 2-hr intervals from Uwekahuna tiltmeter (E-W) near northwest rim of Kilauea Caldera (see figure 13.1.3). Earthquake counts at 2-hr intervals from selected local stations (detection threshold about magnitude 0.1). Amplitude of tremor from adjusted hourly readings of smoked-paper records, plotted in relative units above background at three stations: Mauna Loa (MLO), 25 km from eruption site; West Pit (WPT), 11 km from eruption site; and East Koae (EKO), 5 km from eruption site. Times of major eruptive episodes shown by horizontal bars labeled with sequence numbers.

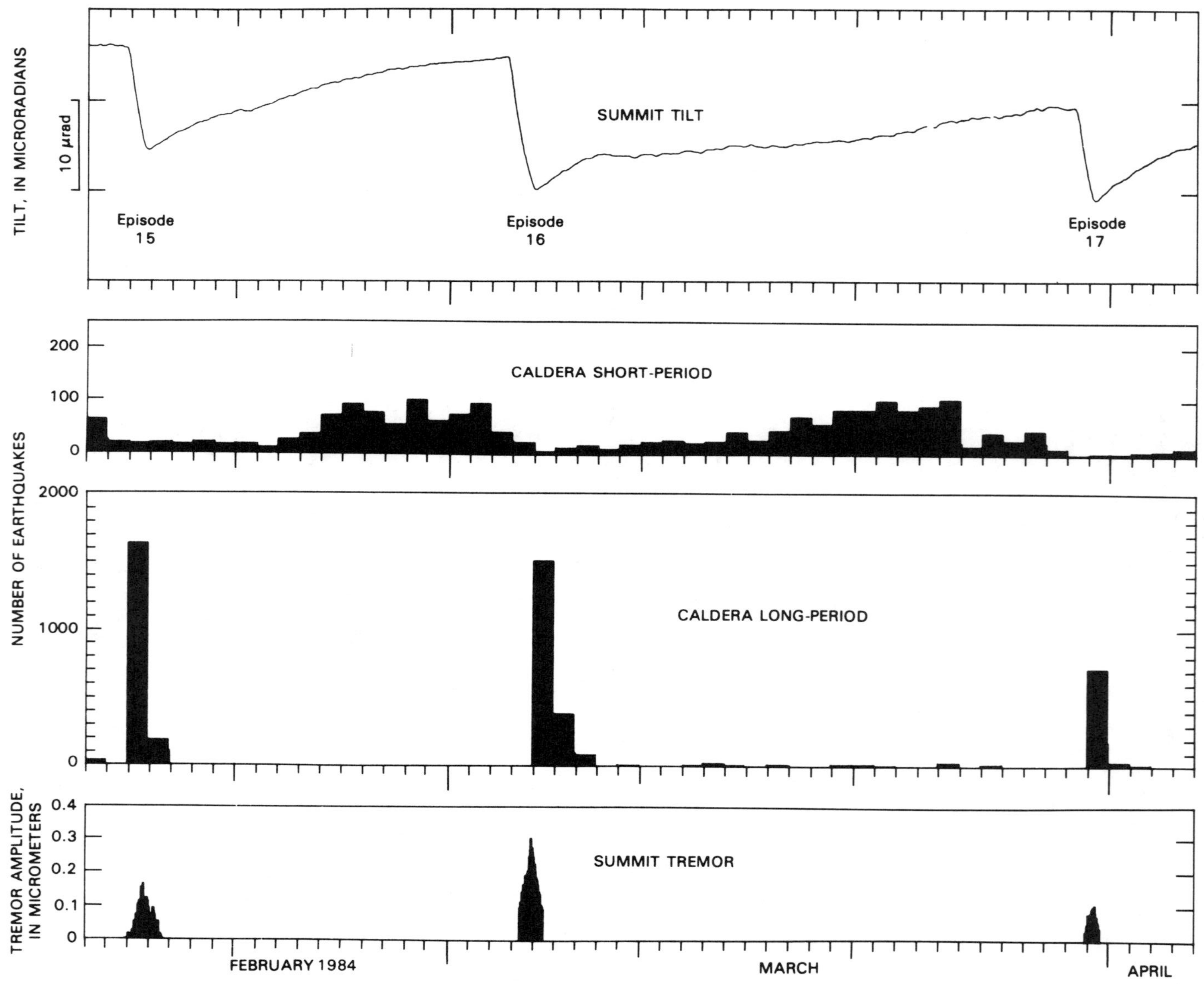

Figure 13.1.35. East-west summit tilt, daily frequency of short-period caldera earthquakes, and amplitude of shallow harmonic tremor beneath Kilauea Caldera for episodes 15-17 of Puu Oo eruption, from Wolfe and others (1987).

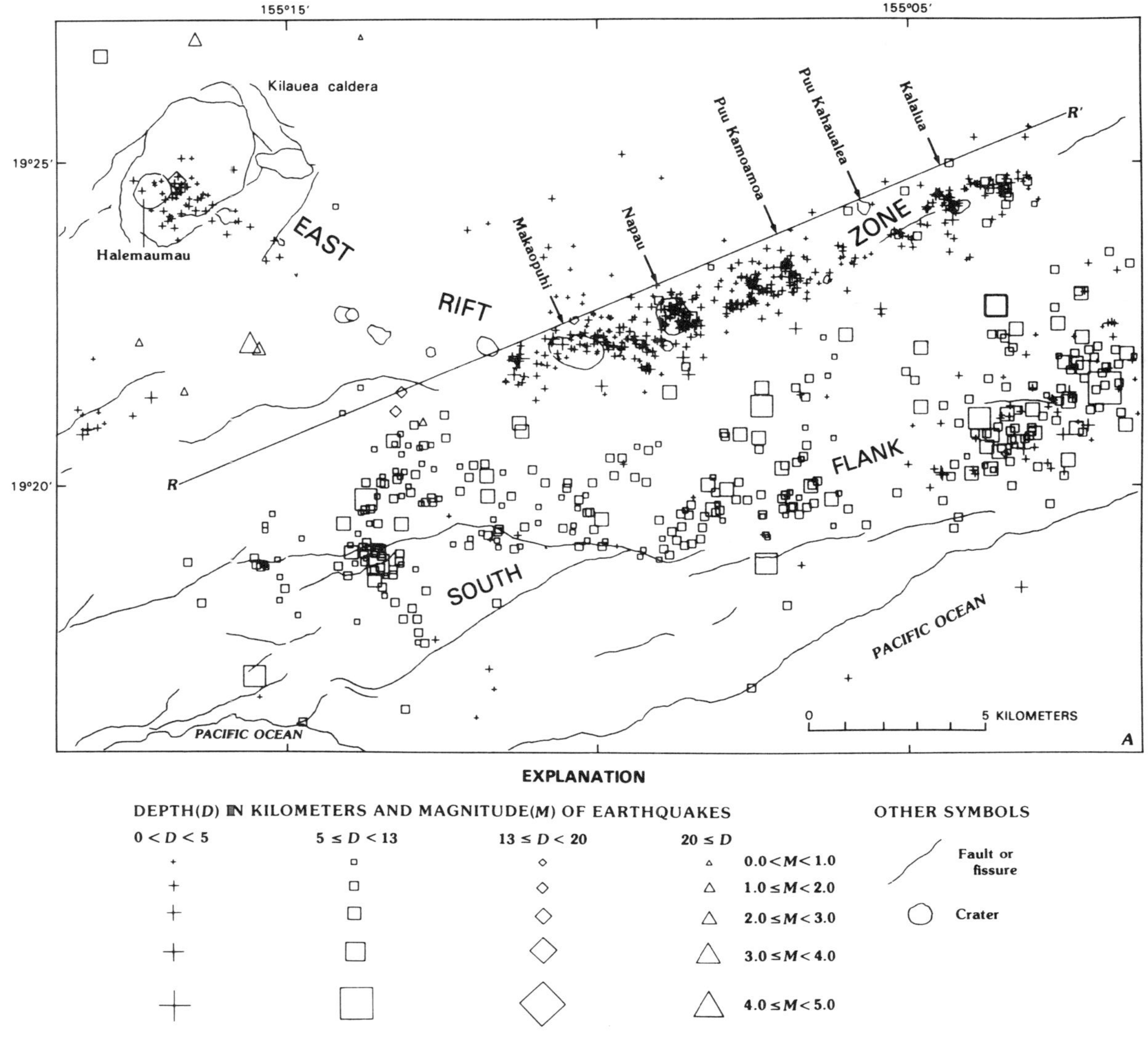

Figure 13.1.36. Located earthquakes at Kilauea during January 1983, from Wolfe and others (1987). A vertical cross section along line R-R' is shown in figure 13.1.37.

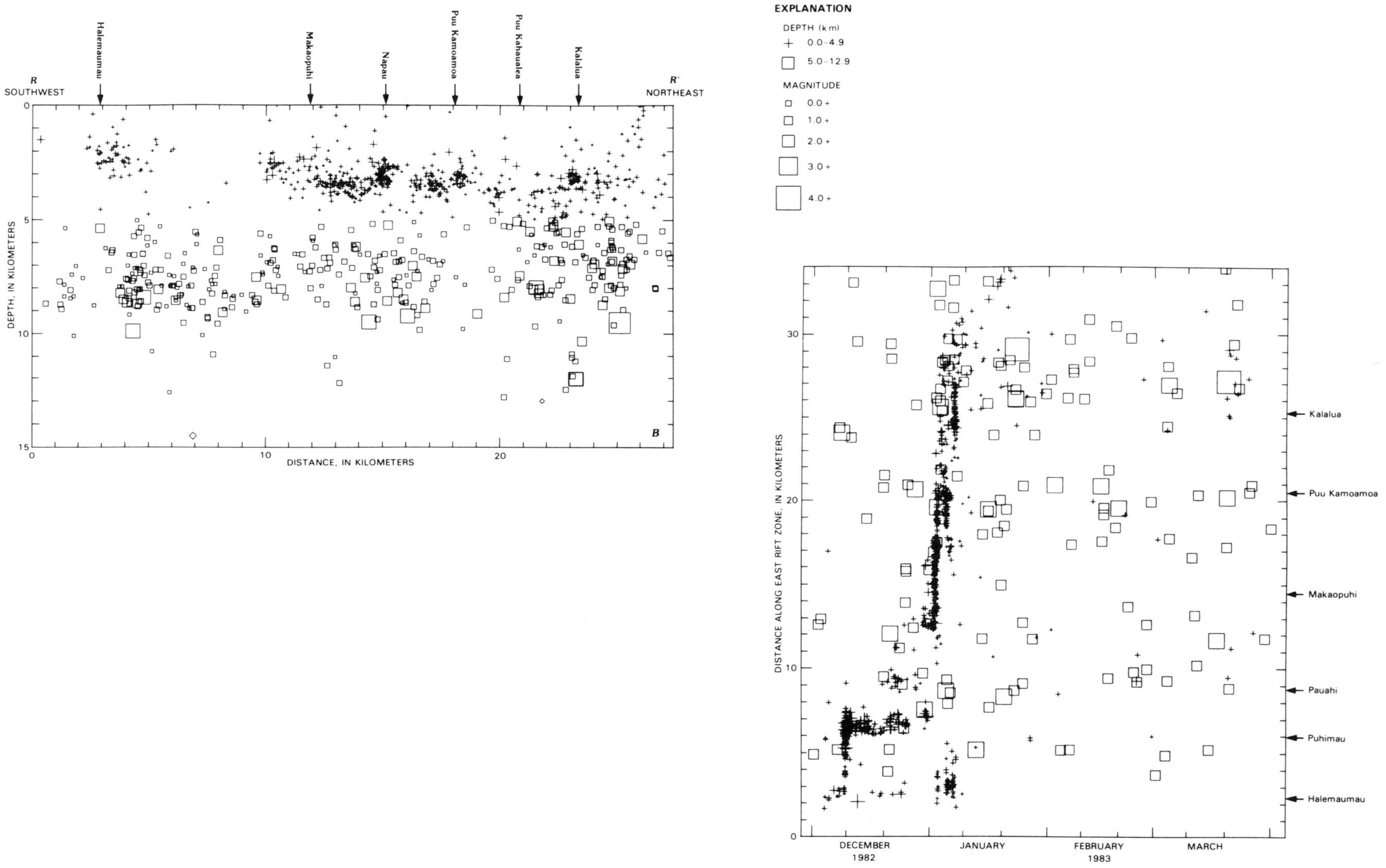

Figure 13.1.37. Depth-distance distribution (left) of earthquakes shown in figure 13.1.36, projected onto line R-R'. Space-time diagram (right) of Kilauea earthquakes, December 1982 to March 1983, with epicenters projected onto an east-west line. Both plots are from Wolfe and others (1987).

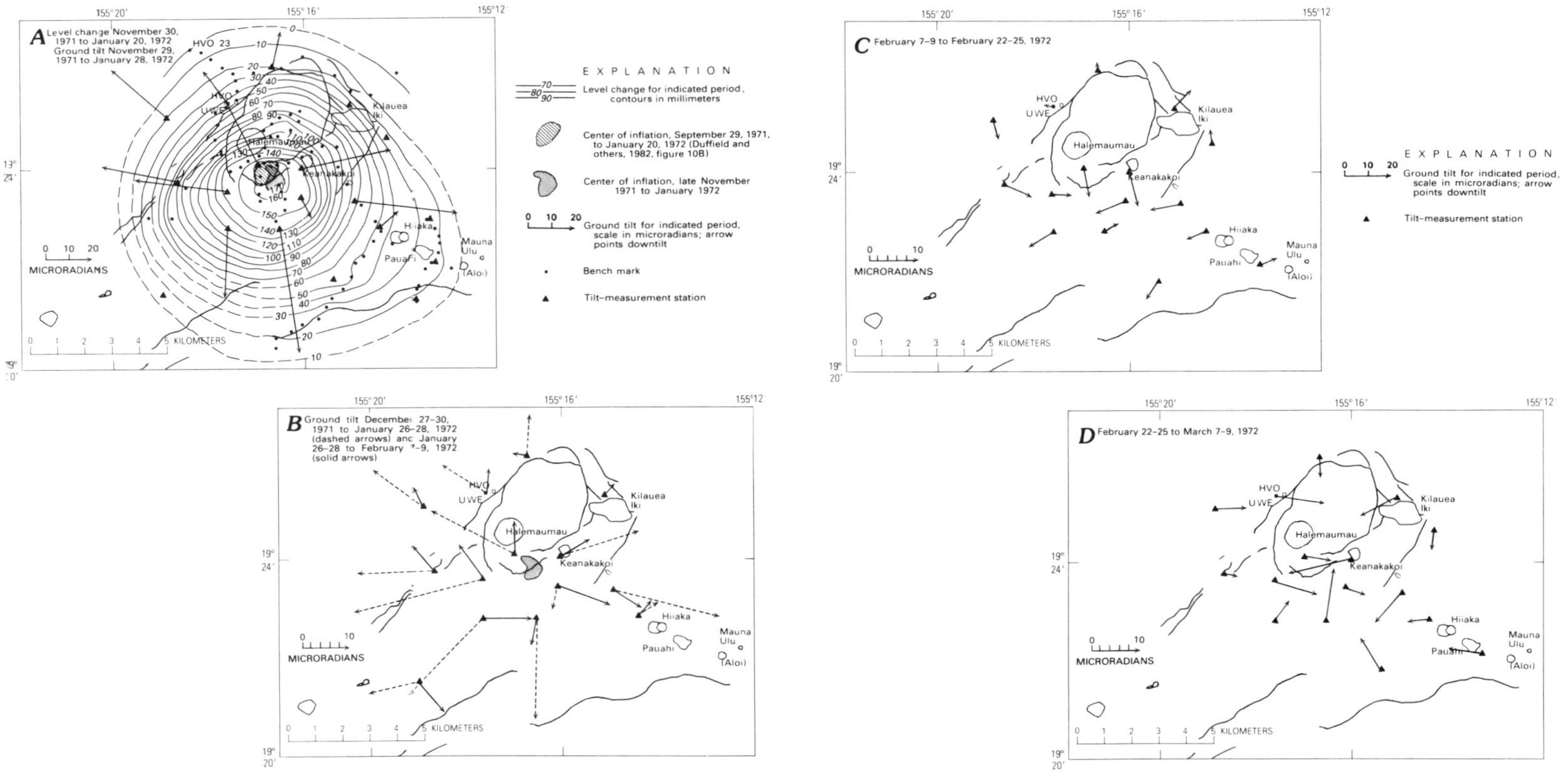

Figure 13.1.38. Results of leveling and tilt surveys on Kilauea before and during first two months of 1972-74 Mauna Ulu eruption, from Tilling and others (1987). Vector UWE calculated from E-W and N-S components of Uwekahuna tiltmeter. Contours of leveling data referenced to bench mark HVO 23, assumed to be stable. A, Level changes and ground tilt for the period November 1971 - January 1972. Inflation center defined for this 2-month period is virtually coincident with that obtained by Duffield and others (1982) for the 4-month period building toward the 1972-1974 Mauna Ulu eruption. B, Tilt changes for periods spanning the final month of preeruption inflation (dashed vectors) and first days of the eruption beginning on 3 February 1972 (solid vectors). C, Tilt changes from 7-9 February 1972 to 22-25 February 1972. D, Tilt changes from 22-25 February 1972 to 7-9 March 1972.

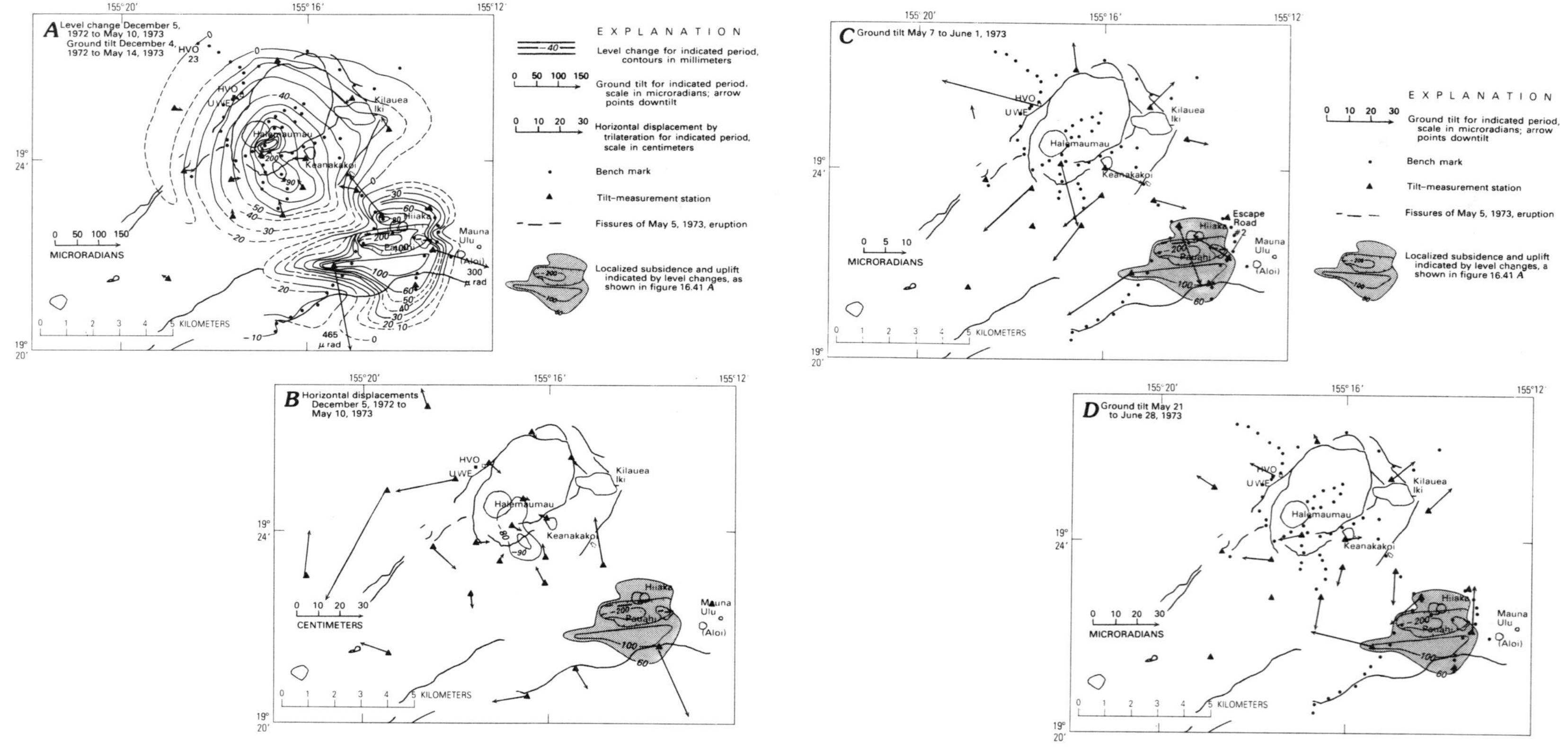

Figure 13.1.39. Results of leveling, tilt, and trilateration surveys from December 1972 to June 1973, from Tilling and others (1987). Vector UWE calculated from E-W and N-S components of Uwekahuna tiltmeter. Contours of leveling data referenced to bench mark HVO 23, assumed to be stable. A, Changes in elevation and tilt from December 1972 to May 1973, showing effect of the 5 May 1973 Pauahi-Hiiaka eruption and intrusion into eastern end of Koae fault system. B, Horizontal displacements from trilateration surveys between December 1972 and May 1973. Also shown are areas of localized subsidence and uplift determined from leveling data (A). C, Tilt changes between 7 May 1973 and 1 June 1973, showing relatively small changes (note change in vector scale from A) that generally reflect summit inflation following 5 May eruption. D, Tilt changes between 21 May 1973 and 28 June 1973; this period included the 9 June partial draining of lava lake at Mauna Ulu. Effect of draining and associated intrusion is apparently masked by net inflation.

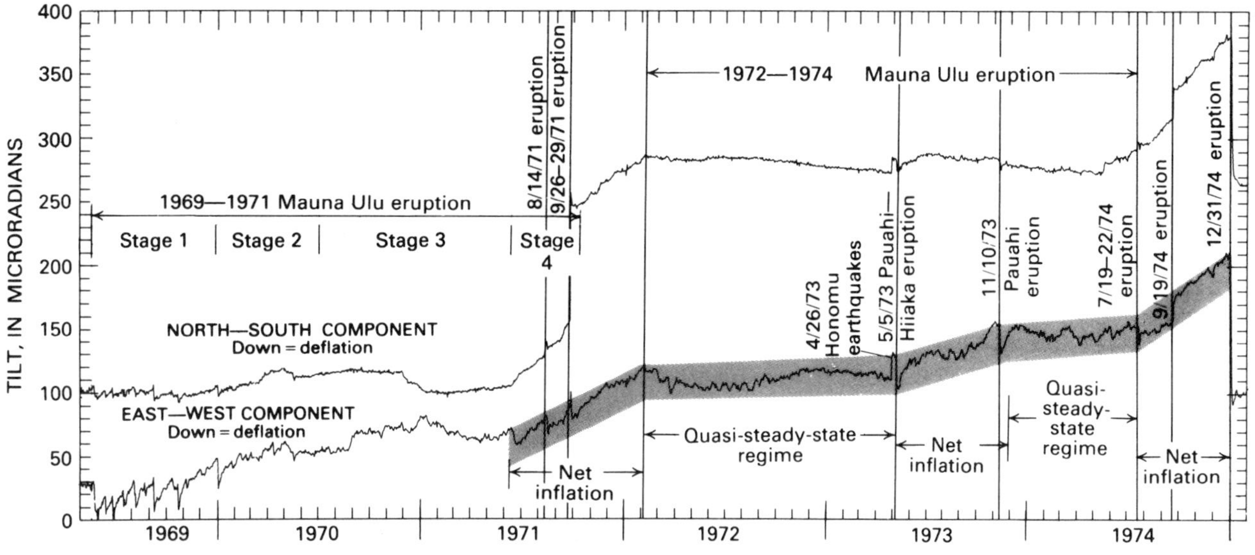

Figure 13.1.40. Variation in Kilauea summit tilt for the period May 1969 - January 1975 as measured by short-base water-tube tiltmeter at Uwekahuna vault (figure 13.1.3), from Tilling and others (1987). Microradian scale arbitrary, with value on 1 May 1969 taken as zero. Magnitude 6.2 Honomu earthquake (26 April 1973) caused an 18-microradian offset of E-W component of tilt. Tilt record for 1972-74 Mauna Ulu eruption is characterized by two distinctive plateaulike segments ("quasi-steady-state regime") separated by a period of net inflation.

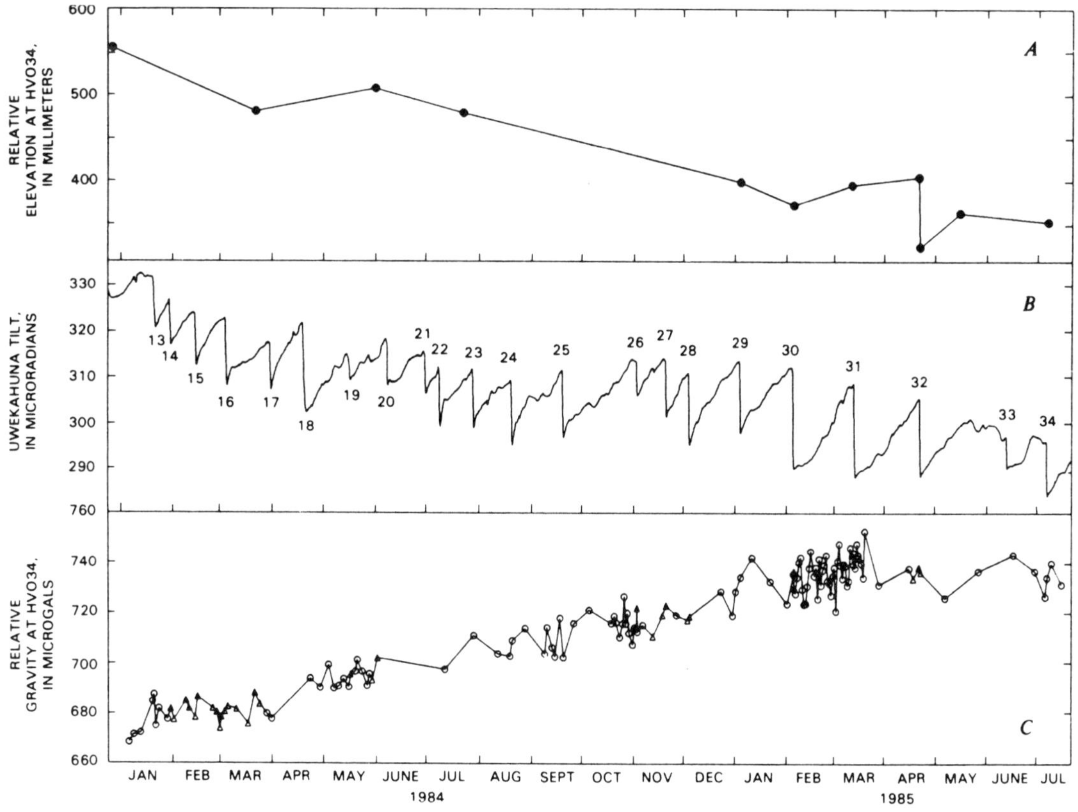

Figure 13.1.41. Relative elevation and gravity at bench mark HVO 34 (south part of Kilauea Caldera) and of E-W tilt at Uwekahuna, from Johnson (1987). A, Relative elevation at HVO 34. B, East-west tilt at Uwekahuna; eruptive episodes at Puu Oo indicated by numbers. C, Relative gravity at HVO 34, uncorrected for elevation changes. Triangles, slightly better precision than circles.

PART 3: HISTORICAL UNREST AT LARGE QUATERNARY CALDERAS (continued)
See inside back cover for explanation and abbreviations

MAUNA LOA/MOKUAWEOWEO

CAVW number (active vent)	Latitude Longitude (degrees)	Diameter (km)	Local tectonic setting	Precaldera edifice	SiO_2 content (wt pct)	Age of collapse (yr B.P.)	Date of unrest	Type and duration of unrest ESTU STHF MGTF H Te	Eruption type
13-02-02	19.48N	6 x 8	Hotspot	Shield	R = 48-53	< 590	1832	---- ---- ---- - --	lf
(Mauna Loa)	155.61W	(outer)			C = 48-53		1843	---- ---- ---- - --	Lf
		3 x 5					1849	---- ---- ---- - --	lf
		(inner)					1851	---- ---- ---- - --	lf
							1852	---- ---- ---- - --	lf
							1855	---- ---- ---- - --	lf
							1859	---- ---- ---- - --	LF
							1865	---- ---- ---- - --	lf
							1868	---- ---- ---- - UU	lf
							1870	---- ---- ---- - --	lf
							1871	---- ---- ---- - --	lf
							1872	---- ---- ---- - --	lf
							1873	---- ---- ---- - --	lf
							1875	---- ---- ---- - --	lf
							1876	---- ---- ---- - --	lf
							1877	---- ---- ---- - --	lf
							1880	---- ---- ---- - --	Lf
							1887	---- ---- ---- - --	Lf
							1892	---- ---- ---- - --	lf
							1896	---- ---- ---- - --	lf
							1899	---- ---- ---- - --	lf
							1903	---- ---- ---- - --	lf
							1907	---- ---- ---- - --	lf
							1914	---- ---- ---- - --	lf
							1916	---- ---- ---- - --	lf
							1919	---- ---- ---- - --	Lf
							1926	E-B- -E-- ---- - --	lf
							1933	--Y- ---- ---Y - --	lf
							1935	B-Y- ---Y ---Y - --	lf
							1940	E-Y- -E-- ---Y - --	lf
							1942	E-Y- ---Y ---Y - --	lf
							1943	D-Y- ---- ---D - --	lf

MAUNA LOA/MOKUAWEOWEO (continued)

CAVW number (active vent)	Latitude Longitude (degrees)	Diameter (km)	Local tectonic setting	Precaldera edifice	SiO_2 content (wt pct)	Age of collapse (yr B.P.)	Date of unrest	Type and duration of unrest ESTU STHF MGTF H Te	Eruption type
(continued from previous page)									
							1949	Y-Y- -E-Y ---E - --	lf,cc
							1950	D-Y- -E-Y ---E - --	LF
							1975	FEYE YEEY ---Y - --	lf
							1984	FFYF YFFY -xEE - UU	Lf

TECTONIC SETTING

Mauna Loa is one of five basaltic shield volcanoes that comprise the Island of Hawaii at the southeast end of the Hawaiian archipelago (fig. 13.2.1). Mauna Loa and Kilauea have erupted tholeiitic basalts frequently during historical time and, with Loihi seamount off the southeast coast of Hawaii, form the current locus of volcanism along the Hawaiian-Emperor chain. (See section on Kilauea Caldera, CAVW number 13-02-01).

GEOLOGIC HISTORY

Mauna Loa, which rises nearly 9 km above the isostatically depressed floor of the central Pacific (to 4,167 m above sea level), is the largest active volcano on Earth; its volume has been estimated at about 40,000 km^3 (Macdonald, 1972, p. 258), but that volume is probably too high because older, separate volcanoes are likely buried at depth (Lockwood and Lipman, 1987). Mauna Loa's lavas cover about 5,125 km^2 of the Island of Hawaii and approximately 4,000 km^2 of its submarine slopes.

Lavas of known historical age cover 13 percent of Mauna Loa's surface; they were erupted at an average rate of 45 x 10^6 m^3/yr from 1843 to 1876 but at less than half that rate since 1877 (fig. 13.2.3). This change coincides with an abrupt change in lava chemistry and may be related to effects of the magnitude 7.6 earthquake in April 1868 (see below). The average historical production rate is higher than average rates for the preceding 650 years, from shortly before the formation of Mokuaweoweo Caldera until the start of historical records in 1843. The preceding period of about 750 years was characterized by voluminous overflows from a summit lava lake. More than one-half of Mauna Loa has been buried by flows during the past 1,500 years, and almost 90 percent has been buried within the last 4,000 years (Lockwood and Lipman, 1987).

Mokuaweoweo Caldera is the younger and smaller of two calderas at the summit of Mauna Loa (Holcomb, 1980) (fig. 13.2.2). Major rift zones radiate southwestward and northeastward from Mokuaweoweo (fig. 13.2.1); most eruptions occur in Mokuaweoweo or along the rift zones (Lockwood and Lipman, 1987).

MAUNA LOA/MOKUAWEOWEO (continued)

HISTORICAL ACTIVITY

Except for a hiatus from 1950 to 1975, Mauna Loa has been continually restless, so we have listed only the dates of eruptions in the table above. Eruptions are defined to begin with the arrival of magma at the surface and to end when tremor at the vent ceases for at least several days.

Several historical eruptions have followed a pattern in which a summit eruption was followed by another summit eruption 2 to 3 years later, then by a flank eruption starting 1 to 4 days later (Lockwood and others, 1976; Lockwood and others, 1987, p. 544). The pattern seems clear for selected eruptions but is not statistically convincing if all historical eruptions are considered (Klein, 1982). Volumes of erupted lava typically are more than at Kilauea, often more than 10^8 m^3 (Lockwood and Lipman, 1987).

Relatively strong earthquakes occur every few years beneath the flanks of Mauna Loa, generally along arcuate fault zones that seem to bound large blocks of the shield that are shouldered aside by intrusions along the rift zones (figs. 13.2.4, 13.2.5). One particularly severe earthquake on 2 April 1868 (M 7.5) was preceded by a short summit eruption at Mauna Loa on 27 March and followed by a major flank eruption starting on 7 April (Wood, 1914; Macdonald and others, 1983; Endo, 1985). The most recent example included a large earthquake on 16 November 1983 (M 6.6, 12 km deep along the Kaoiki fault system; see fig. 13.2.16) (Koyanagi and others, 1984), followed by a major summit-flank eruption in March-April 1984 (Lockwood and others, 1985). Focal mechanisms for the earthquake and ground ruptures indicated right-lateral slip along the E-W-trending Kaoiki fault system, coupled with at least 15-20 cm of southward extension (Koyanagi and others, 1984).

In early 1974, intermediate-depth earthquakes and then shallow earthquakes increased beneath the summit area of Mauna Loa (Koyanagi and others, 1975) (figs. 13.2.6-13.2.8). Particularly intense swarms occurred in August and December 1974 and in February 1975. Most earthquakes were shallow and near the southwest rim of Mokuaweoweo Caldera; a few occurred northwest of the caldera at a depth of 5-7 km, and a few others occurred beneath the caldera at a depth of about 40 km. Geodetic measurements revealed that Mauna Loa also began to inflate during 1974 (figs. 13.2.12). Extension across the caldera amounted to more than 60 cm by mid-1975, an additional 20 cm by early 1976, and 2-5 cm/yr during 1976-83 (fig. 13.2.10). Summit tilt was on the order of several tens to almost 200 microradians during the same period. Uplift was centered near the southeast rim of Mokuaweoweo (Decker and others, 1983). Inversion of the geodetic data indicated a deformation source at 3-4 km depth, roughly coincident with an aseismic zone interpreted as a shallow magma reservoir beneath the southeast part of the summit area (fig. 13.2.14).

A small summit eruption on 5-6 July 1975 (Lockwood and others, 1987) ended the longest hiatus in eruptive activity at Mauna Loa in historical time (from 1950 to 1975). As the 1975 eruption waned, an intense swarm of earthquakes marked the progress of an intrusion into the upper northeast rift zone to a distance of 30 km from the summit (fig. 13.2.9).

Inflation continued and shallow and intermediate-depth earthquakes increased in frequency beneath Mauna Loa from 1980 to 1983, suggesting a slow but accelerating intrusion at 2-3 km depth (Decker and others, 1983). An intense swarm of intermediate-depth earthquakes (5-13 km) occurred beneath Mauna Loa's northwest flank in mid-September 1983 (figs. 13.2.6, 13.2.8). First-motion

MAUNA LOA/MOKUAWEOWEO (continued)

HISTORICAL ACTIVITY (continued)

analyses suggest that the earthquakes resulted from increasing lateral stresses generated in the summit and upper southwest rift zone, probably by intrusion of magma. On 16 November 1983, a magnitude 6.6 earthquake occurred beneath Mauna Loa's southeast flank. The earthquake generated an intense aftershock sequence that included events along Mauna Loa's upper northeast rift zone. Seismicity in that area remained high until it merged with eruptive seismicity starting on 25 March 1984. Shallow earthquakes in the summit area continued unchanged until a marked acceleration several weeks before the eruption (figs. 13.2.11, 13.2.19). The number of larger earthquakes ($\underline{M} > 1.5$) increased persistently as the eruption approached. The daily frequency of smaller earthquakes showed a more episodic increase and, in fact, decreased to below average on the day before the eruption (Lockwood and others, 1985, 1987). Continuous monitoring by electronic tiltmeters and field surveys showed that inflation continued at a steady, nonincreasing rate almost until the March 1984 outbreak (Lockwood and others, 1985).

The temperature of a fumarole along the 1975 eruptive fissure rose by 20 °C on 18 November 1983, concurrently with a brief increase in gas emission at the same site (figs. 13.2.15, 13.2.16B). These changes may have been a response to the $\underline{M}$ 6.6 Kaoiki earthquake two days earlier. Additional changes in temperature and gas emission along the 1975 fissure occurred on several occasions during December 1983 - March 1984. On 14 December 1983, an airborne observer saw an anomalous steam cloud rising to about 1 km above Mokuaweoweo, one day after an abrupt increase in temperature and gas emission along the 1975 eruptive fissure.

On 18 March 1984, a dull red glow was seen along a "crack" on the floor of Mokuaweoweo (1975 fissure?) by a hiker; on 23 March, several observers saw "steam clouds" rising above Mokuaweoweo. It is possible that the red color was a result of oxidation along the crack, and the level of steaming reflected atmospheric conditions and not changes related to the impending eruption. (Increased glow and steaming days in advance of an eruption seem unlikely, based on experience at Kilauea, but the observations cannot be refuted.) On 24 March, a day before the eruption, a professor with a geology class on the Mauna Loa summit saw "rocks and steam" being ejected from the 1975 fissure, but there was no way to alert authorities. Shallow seismicity beneath the caldera started to increase at 22:55 local time on 24 March; the onset of harmonic tremor accompanied a further increase at 23:30 (fig. 13.2.16). The earthquake swarm and tremor strengthened rapidly just before 01:00 on 25 March. Eleven earthquakes with $\underline{M}$=2.0-4.1 occurred between 00:51 and 02:10, and tiltmeters showed rapid summit inflation beginning at 01:00 (fig. 13.2.17). Magma apparently reached the surface at 01:25 that morning, based on a "strong signal" from an infrared sensor in a military satellite in Earth orbit. Eruptive fissures quickly migrated down the northeast rift zone and remained active until 15 April; a lava flow that temporarily threatened the city of Hilo stopped before doing significant damage. Phase 17 of the 1983-present Puu Oo eruption at Kilauea began on 30 March 1984, but the simultaneous eruptions did not seem to influence each other (Lockwood and others, 1987).

The amount of summit deflation accompanying the 1984 eruption was greater than the inflation since 1975, suggesting that some of the erupted magma might have been emplaced in a shallow reservoir before 1975, possibly before an electronic distance measurement (EDM) line across the caldera was first measured in 1964. Maximum subsidence during the eruption was 630 mm in the south part of the summit area. Typical electronic distance measurement lines across Mokuaweoweo contracted an average of 300 mm after initial dike-related extensions of about 600 mm, and tilt stations near the rim of Mokuaweoweo recorded 100-200 microradians of eruption-related

MAUNA LOA/MOKUAWEOWEO (continued)

HISTORICAL ACTIVITY (continued)

deflation (figs. 13.2.13, 13.2.17, 13.2.18). The summit area of Mauna Loa began to reinflate immediately after the 1984 eruption, suggesting recharge of the summit magma reservoir (Lockwood and others, 1985, 1987) (fig. 13.2.20).

COMMENTS

Mauna Loa was one of the most active volcanoes on Earth during the late 19th and early 20th centuries, but a hiatus in eruptive activity from 1950 to 1975 coincided with the development or improvement of many modern volcano-monitoring techniques. For this reason, relatively little is known about historical unrest at Mauna Loa compared with its diminutive neighbor, Kilauea (which has been intensely active since 1952). However, observations during the 1975 and 1984 Mauna Loa eruptions suggest many similarities with Kilauea, including the presence of a shallow crustal magma reservoir at 3-5 km depth. An important difference may be the relatively greater amount of magma storage along Kilauea's east rift zone, probably facilitated by dilatation across the rift zone during major south-flank subsidence events such as those in 1868 and 1975 (see section on Kilauea).

REFERENCES

Decker, R.W., Koyanagi, R.Y., Dvorak, J.J., Lockwood, J.P., Okamura, A.T., Yamashita, K.M., and Tanigawa, W.R., 1983, Seismicity and ground surface deformation of Mauna Loa Volcano, Hawaii: Eos, Trans. Am. Geophys. Union, v. 64, p. 545-547.

Endo, E.T., 1985, Seismotectonic framework for the southeast flank of Mauna Loa Volcano, Hawaii: Seattle, Univ. Washington, Ph.D. diss., 350 p.

Fiske, R.S., and Jackson, E.D., 1972, Orientation and growth of Hawaiian volcanic rifts: the effect of regional structure and gravitational stresses: Proc. Roy. Soc. London, Ser. A, v. 329, p. 299-326.

Holcomb, R.T., 1980, Kilauea Volcano: Chronology and morphology of the surficial lava flows: Stanford, Calif., Stanford Univ., Ph.D. diss., 321 p.

Klein, F.W., 1982, Patterns of historical eruptions at Hawaiian volcanoes: Jour. Volcanol. Geotherm. Res., v. 12, p. 1-35.

Koyanagi, R.Y., Endo, E.T., and Ebisu, J.S., 1975, Reawakening of Mauna Loa Volcano, Hawaii: A preliminary evaluation of seismic evidence: Geophys. Res. Lett., v. 2, p. 405-408.

Koyanagi, R.Y., Endo, E.T., Tanigawa, W.R., Nakata, J.S., Tomori, A.H., and Tamura, P.N., 1984, Kaoiki, Hawaii earthquake of November 16, 1983: A preliminary compilation of seismographic data at the Hawaiian Volcano Observatory: U.S. Geol. Surv. Open-File Rep. 84-798, 6 p.

Lockwood, J.P., Banks, N.G., English, T.T., Greenland, L.P., Jackson, D.B., Johnson, D.J., Koyanagi, R.Y., McGee, K.A., Okamura, A.T., and Rhodes, J.M., 1985, The 1984 eruption of Mauna Loa volcano, Hawaii: Eos, Trans. Am. Geophys. Union, v. 66, no. 16, p. 169-171.

Lockwood, J.P., Koyanagi, R.Y., Tilling, R.I., Holcomb, R.T., and Peterson, D.W., 1976, Mauna Loa threatening: Geotimes, v. 21, no. 6, p. 12-15.

MAUNA LOA/MOKUAWEOWEO (continued)

REFERENCES (continued)

Lockwood, J.P., and Lipman, P.W., 1987, Holocene eruptive history of Mauna Loa volcano, in R.W. Decker, T.L. Wright, and P.H. Stauffer, eds., Volcanism in Hawaii, U.S. Geol. Surv. Prof. Paper 1350, p. 509-535.

Lockwood, J.P., Dvorak, J.J., English, T.T., Koyanagi, R.Y., Okamura, A.T., Summers, M.L., and Tanigawa, W.R., 1987, Mauna Loa 1974-1984: A decade of intrusive and extrusive activity, in R.W. Decker, T.L. Wright, and P.H. Stauffer, eds., Volcanism in Hawaii, U.S. Geol. Surv. Prof. Paper 1350, p. 537-570.

Macdonald, G.A., 1972, Volcanoes: Englewood Cliffs, N.J., Prentice-Hall, Inc., 510 p.

Macdonald, G.A., and Abbott, A.T., 1970, Volcanoes in the sea: Honolulu, Univ. Hawaii Press, 441 p.

Macdonald, G.A., Abbott, A.T., and Peterson, F.L., 1983, Volcanoes in the sea (2nd ed.): Honolulu, Univ. Hawaii Press, 517 p.

Smithsonian Institution, Scientific Event Alert Network (SEAN), 1987, Mauna Loa: SEAN Bull., v. 12, no. 10.

Wood, H.O., 1914, On earthquakes of 1868 in Hawaii: Bull. Seismol. Soc. Am., v. 4, p. 169-203.

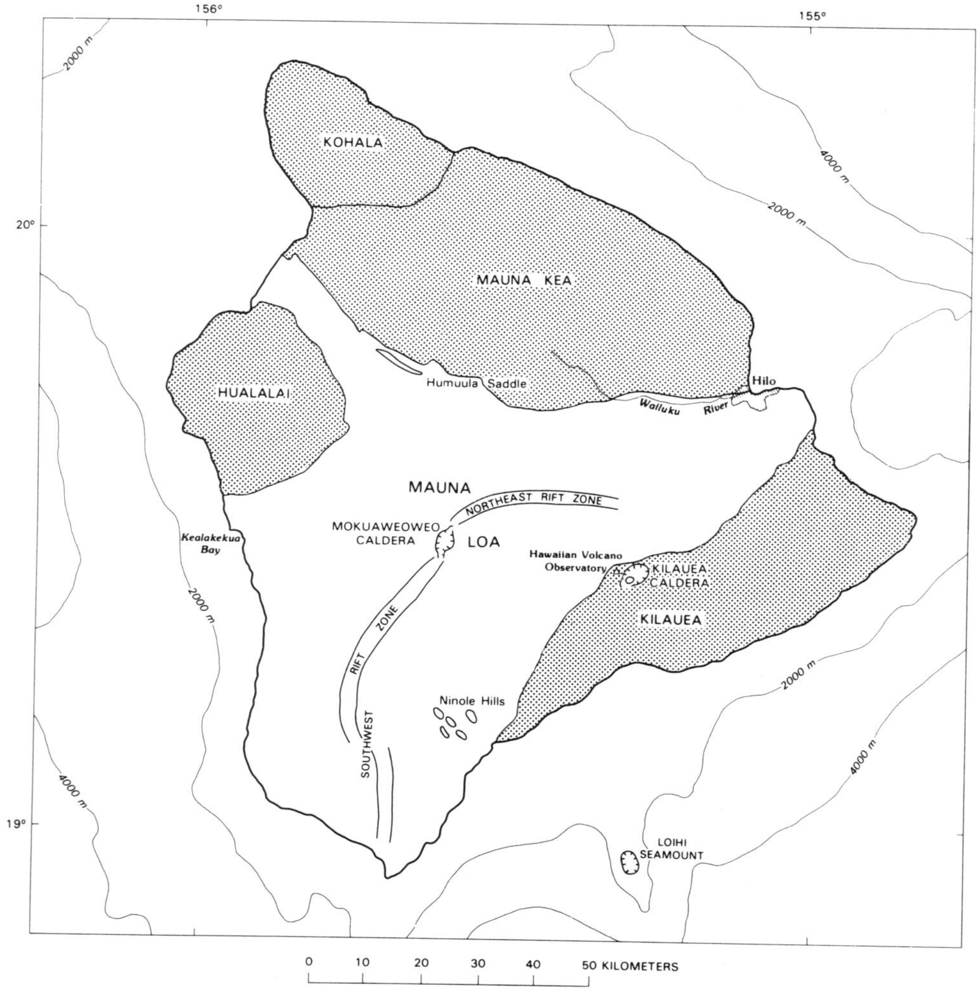

Figure 13.2.1. Island of Hawaii, showing five volcanoes (including Mauna Loa), major structural and geographic features, and bathymetry of adjacent sea (Lockwood and Lipman, 1987).

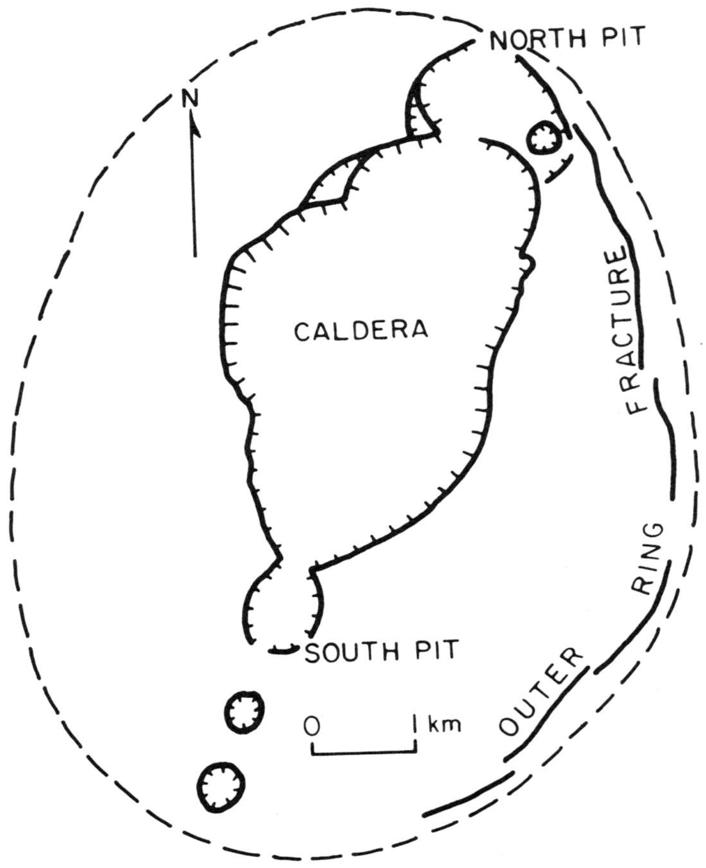

Figure 13.2.2. Outline of Mauna Loa's summit caldera and upper rift zones, from Decker and others (1983). Holcomb (1980) interprets outer ring fracture as a relic of a larger, buried caldera. Copyright by the American Geophysical Union.

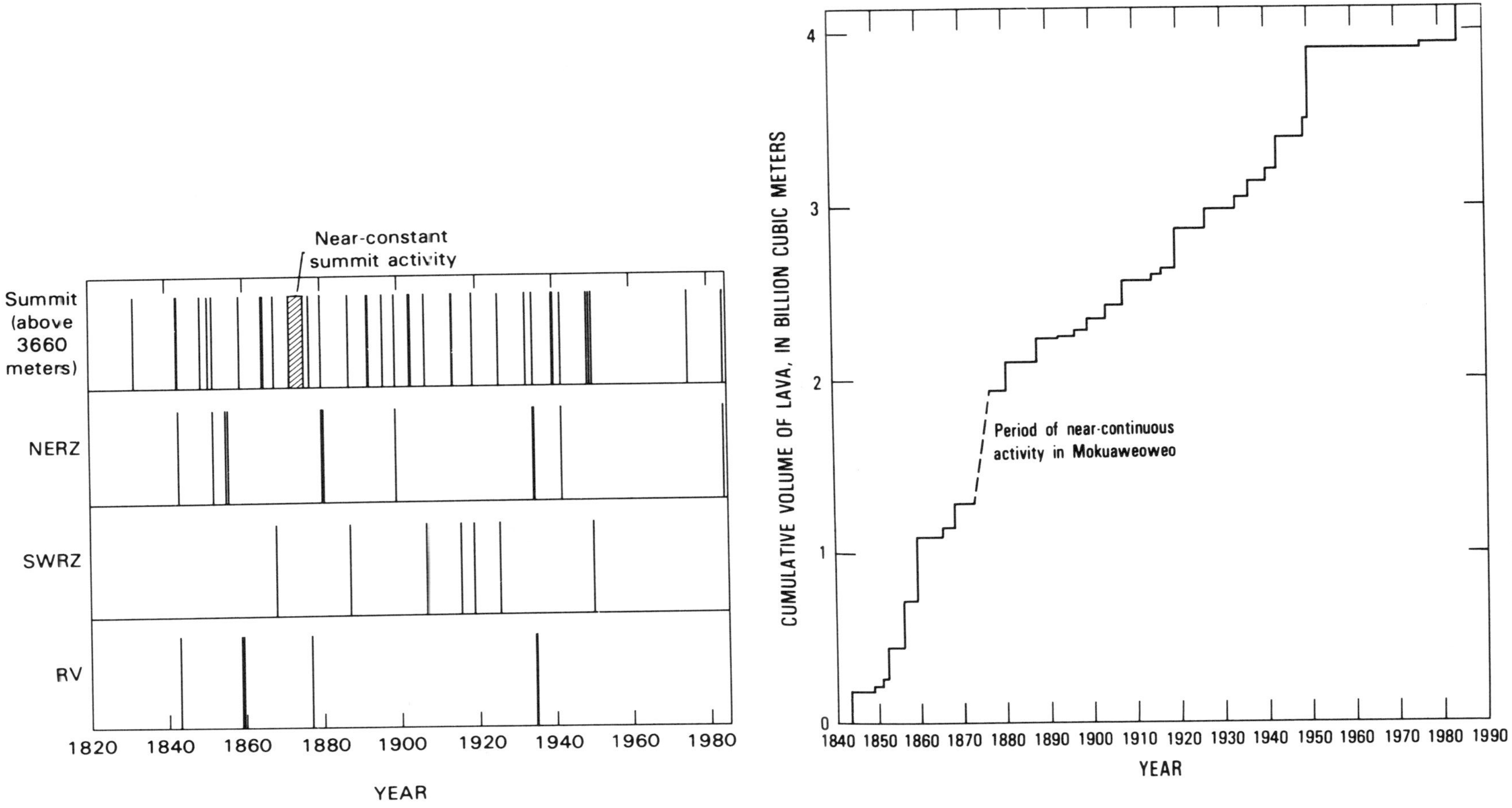

Figure 13.2.3. Chronology of historical eruptions of Mauna Loa (left), from Lockwood and Lipman (1987). NERZ, northeast rift zone; SWRZ, southwest rift zone; RV, northwest flank radial vent eruptions. Mauna Loa lava production, 1843-1984 (right), from Lockwood and Lipman (1987).

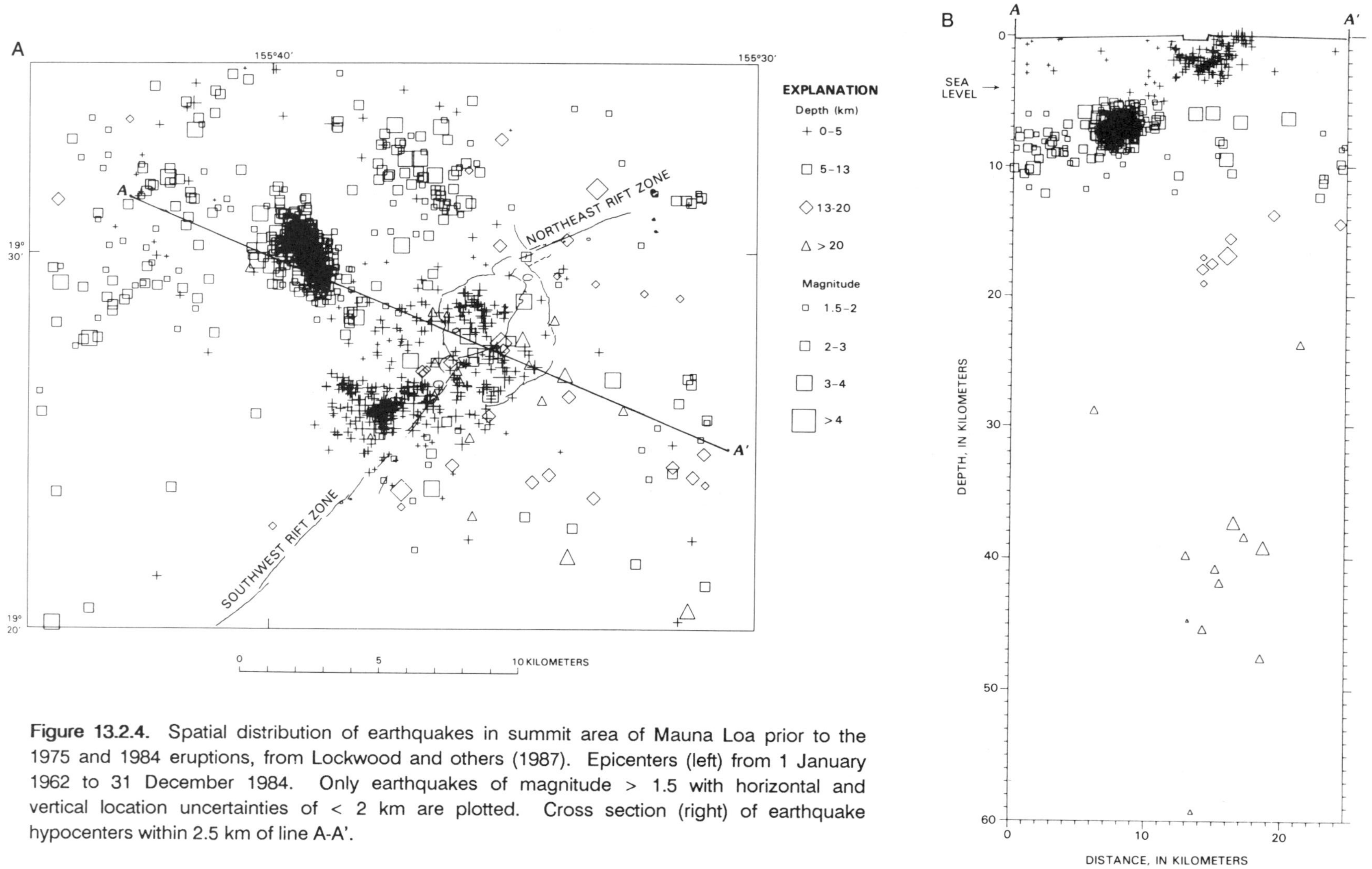

Figure 13.2.4. Spatial distribution of earthquakes in summit area of Mauna Loa prior to the 1975 and 1984 eruptions, from Lockwood and others (1987). Epicenters (left) from 1 January 1962 to 31 December 1984. Only earthquakes of magnitude > 1.5 with horizontal and vertical location uncertainties of < 2 km are plotted. Cross section (right) of earthquake hypocenters within 2.5 km of line A-A′.

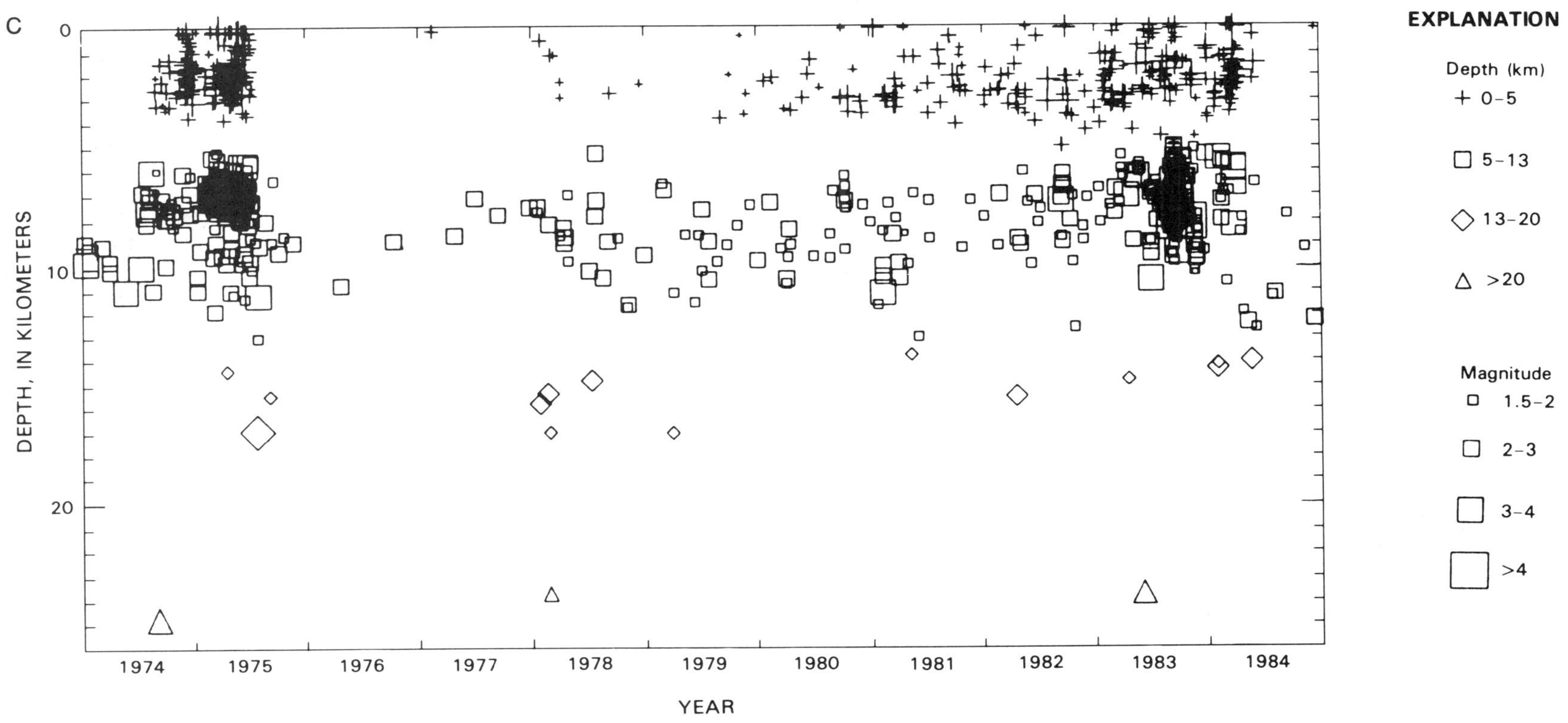

Figure 13.2.5. Depth-time plot of Mauna Loa summit earthquakes ($\underline{M}$ > 1.5) from 1 January 1974 to 31 December 1984 (Lockwood and others, 1987). Absence of earthquakes between 3- and 6-km depth suggests location of Mauna Loa's shallow magma storage reservoir.

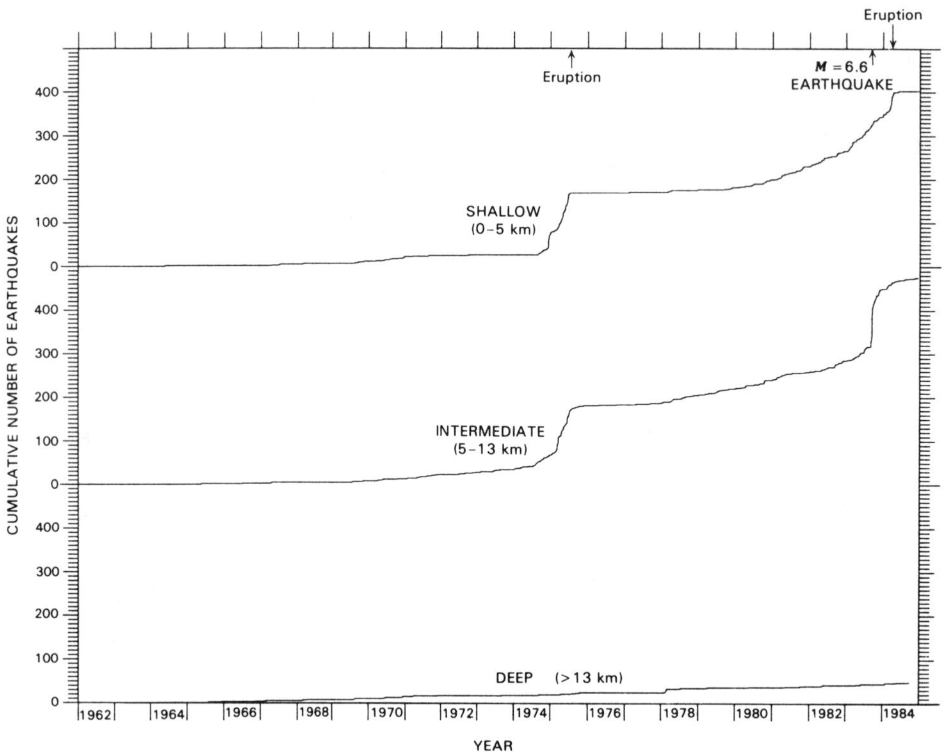

Figure 13.2.6. Earthquakes of M > 1.5 beneath Mauna Loa summit region from 1 January 1962 to 31 December 1984 (Lockwood and others, 1987). Note increase in intermediate-depth earthquakes associated with the M 6.6 earthquake of 16 November 1983.

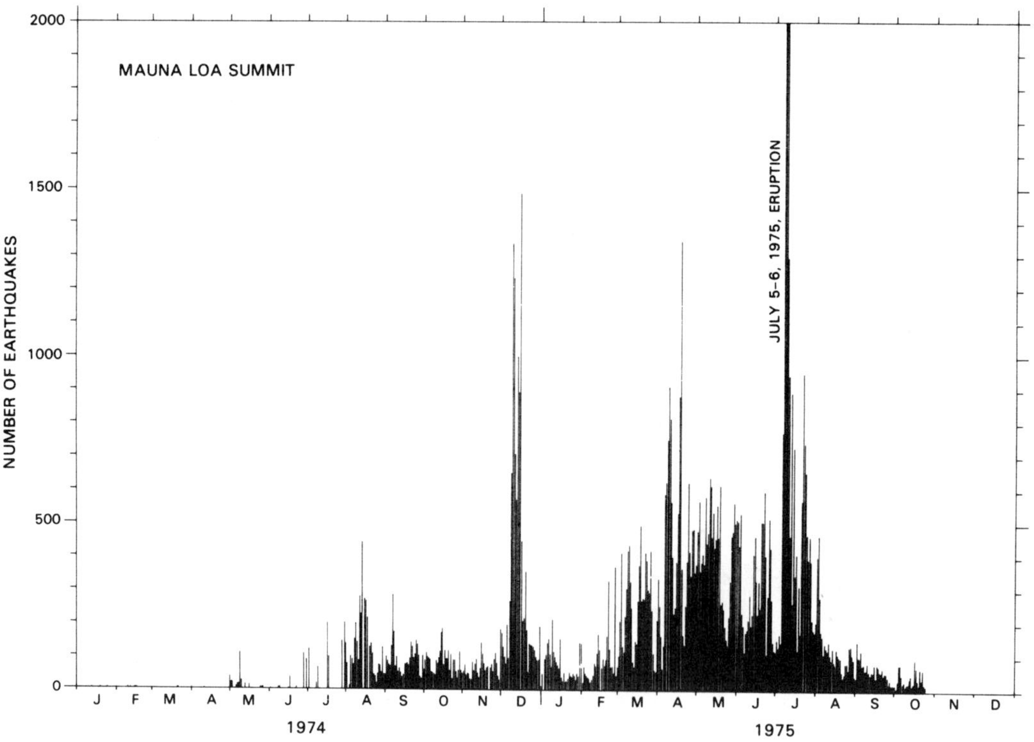

Figure 13.2.7. Daily frequency of Mauna Loa short-period earthquakes (M > 0.1), January 1974 to October 1975, from Lockwood and others (1976, 1987).

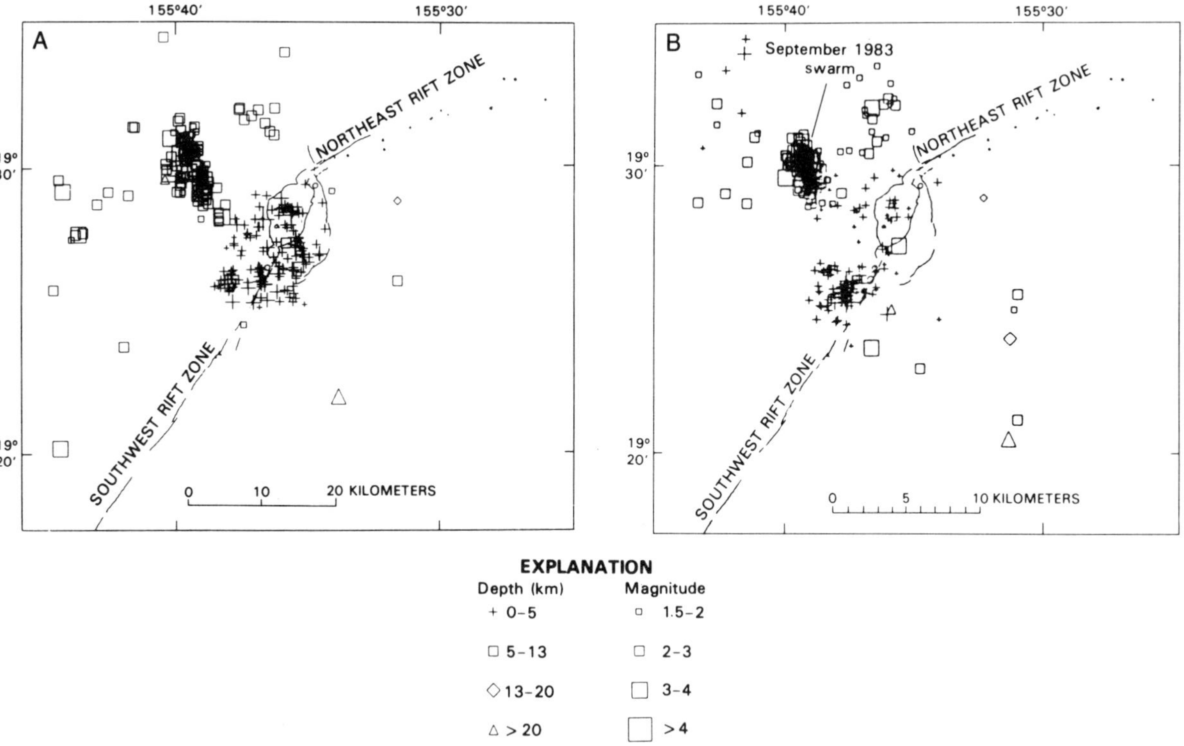

Figure 13.2.8. Earthquake foci beneath Mauna Loa summit region prior to the 1975 and 1984 eruptions, from Lockwood and others (1987). A, 1 January 1974 to 30 June 1975. B, 24 September 1982 to 24 March 1984.

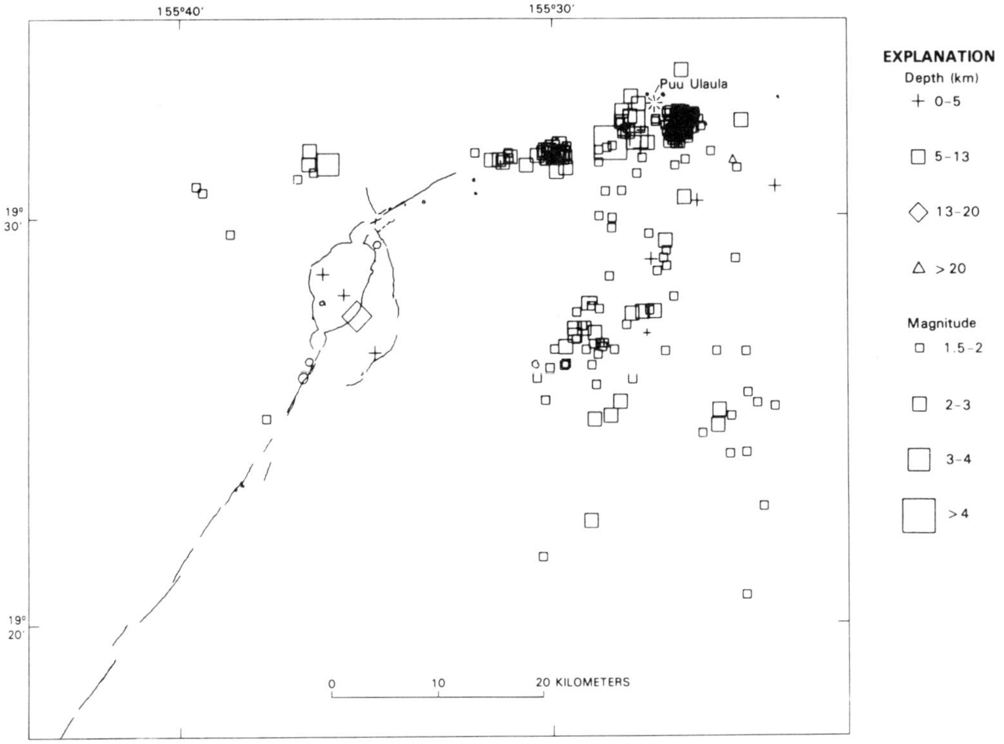

Figure 13.2.9. Earthquake foci ($\underline{M}$ > 1.5) in Mauna Loa summit region, 1-31 July 1975, from Lockwood and others (1987). Earthquakes west of 155°34′ W occurred before the 4-5 July eruption; those to east occurred mostly during 6-12 July as magma intruded from summit into upper northeast rift zone.

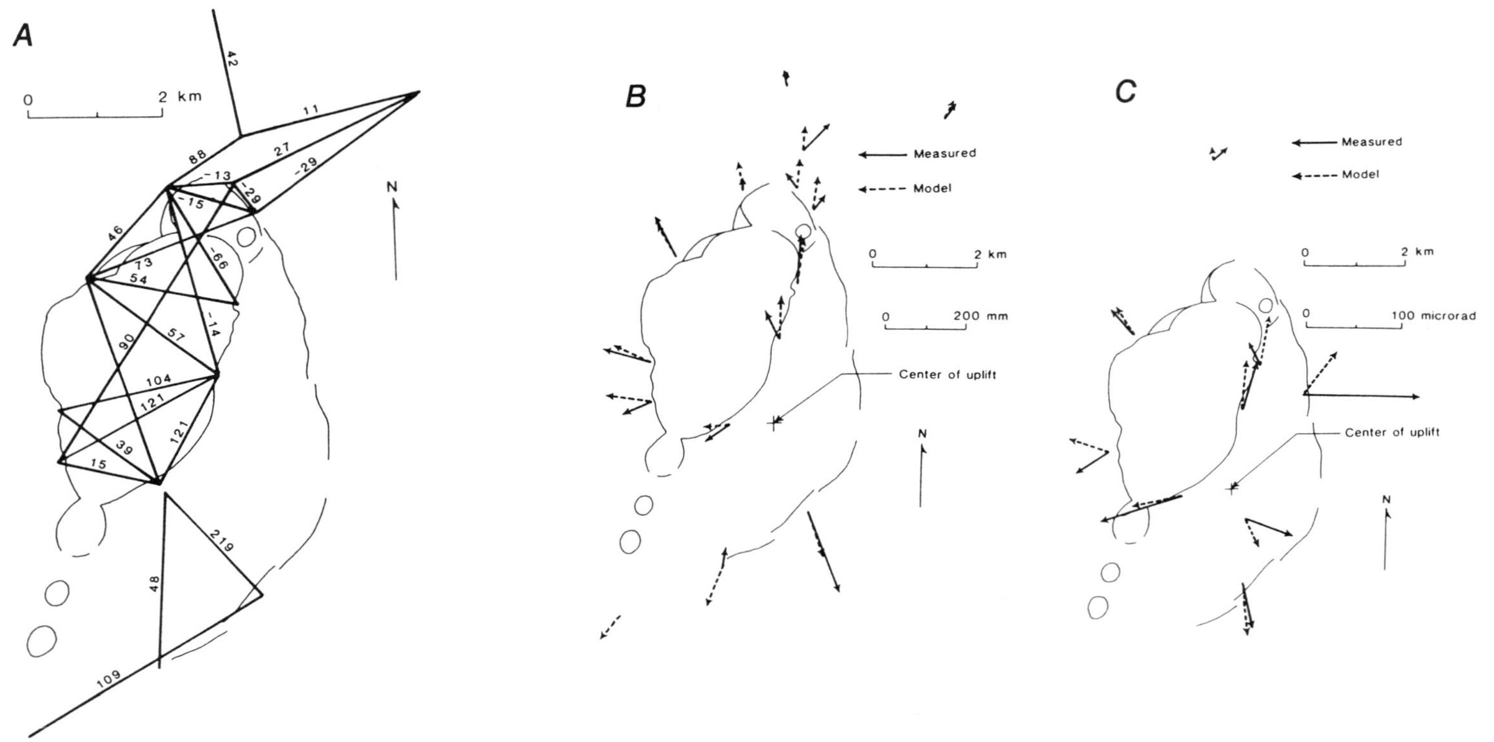

Figure 13.2.10. Horizontal displacements and tilt changes in Mauna Loa summit region from 1977 to 1981, from Decker and others (1983). A, Changes in line length (in millimeters). Positive values are extensions; negative values are contractions. Maximum measured strain is 100 microstrain on line southeast of caldera. B, Horizontal displacements (in millimeters) calculated from data in A (solid vectors) compared with those from a best-fit elastic deformation model (dashed vectors) using a point source at 3.2-km depth. C, Measured tilt vectors (solid) compared with best-fit elastic-deformation model vectors (dashed). Tilt scale shows vector amplitudes in microradians. Center of uplift is nearly identical to that determined from trilateration data in B, but source depth is 2.6 km from tilt data, compared with 3.2 km from trilateration data. Copyright by the American Geophysical Union.

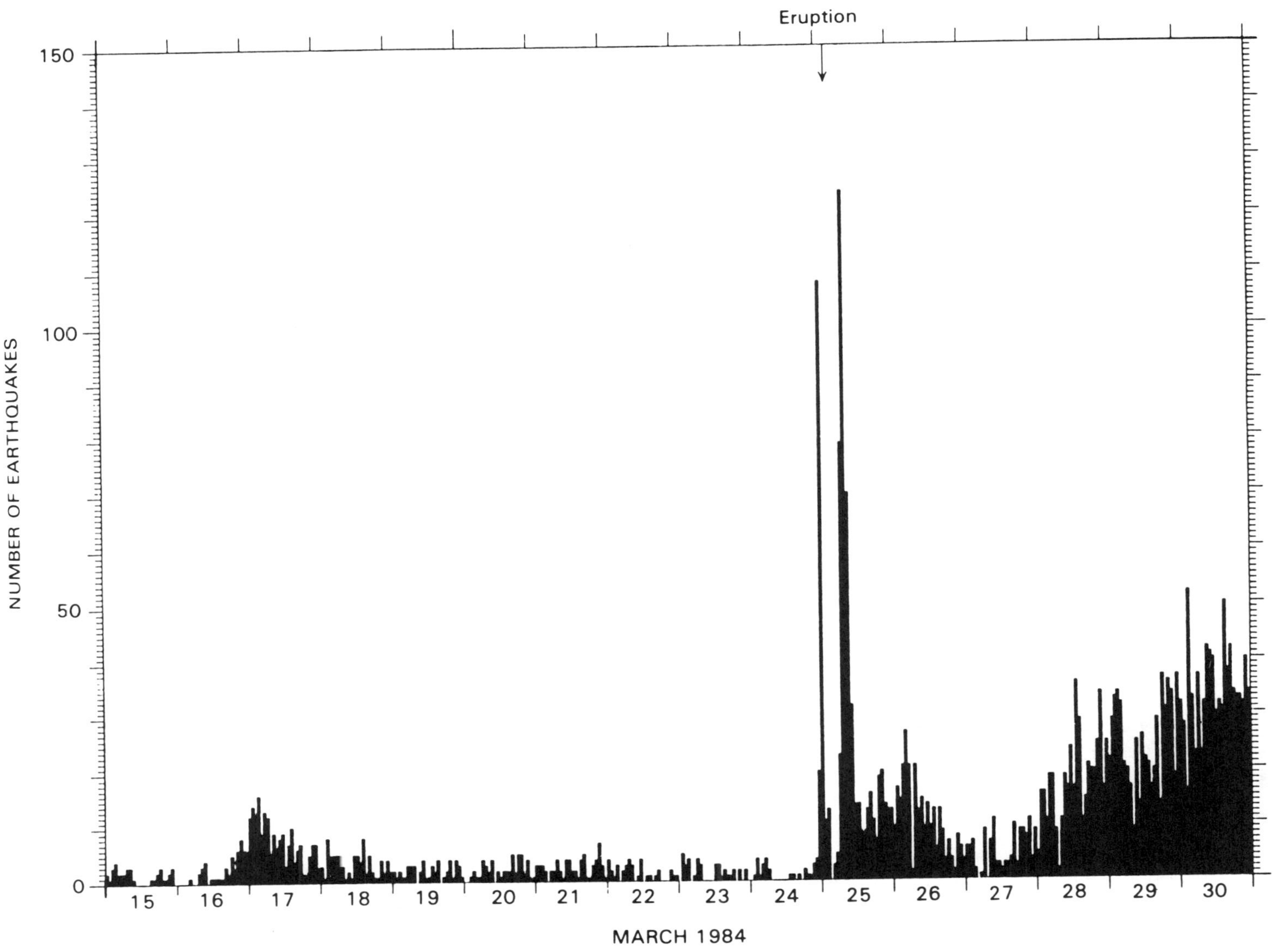

Figure 13.2.11. Hourly counts of Mauna Loa short-period summit earthquakes for the period 15-30 March 1984 (Lockwood and others, 1987). Increase in earthquake frequency after 27 March accompanied deflation of Mauna Loa summit during eruption. Low counts on morning of 25 March reflect masking by high-amplitude tremor.

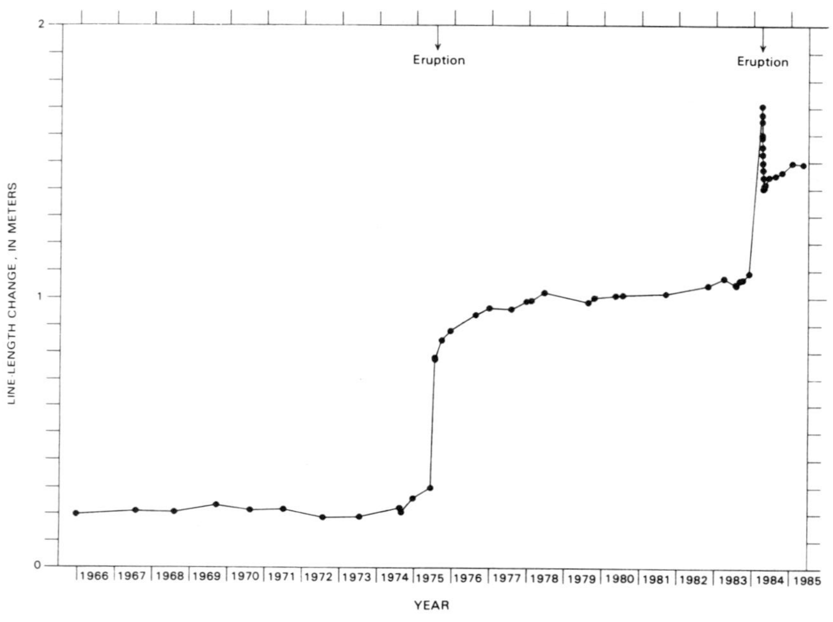

Figure 13.2.12. Horizontal extension across Mokuaweoweo Caldera, December 1965 to May 1985, as measured by trilateration surveys between stations HVO 92 and HVO 93 (see figure 13.2.16B). Abrupt extensions on 5 July 1975 and 25 March 1984 were associated with dike emplacement during eruptions. Reproduced from Lockwood and others (1987).

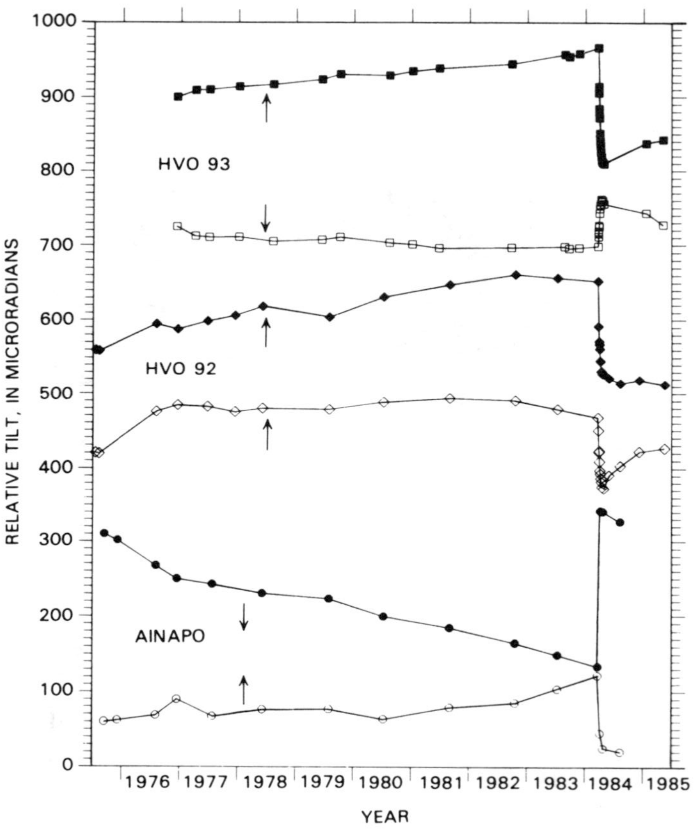

Figure 13.2.13. Spirit-level tilt changes at three Mauna Loa summit stations (see figure 13.2.16B), 1975 to 1985, from Lockwood and others (1987). Darkened symbols are N-S; open symbols are E-W component. Positive tilt changes indicate north or east down. Arrows show inflation direction for individual tilt components. Summit inflation is recorded at all stations prior to and after the eruption from 25 March to 15 April 1984.

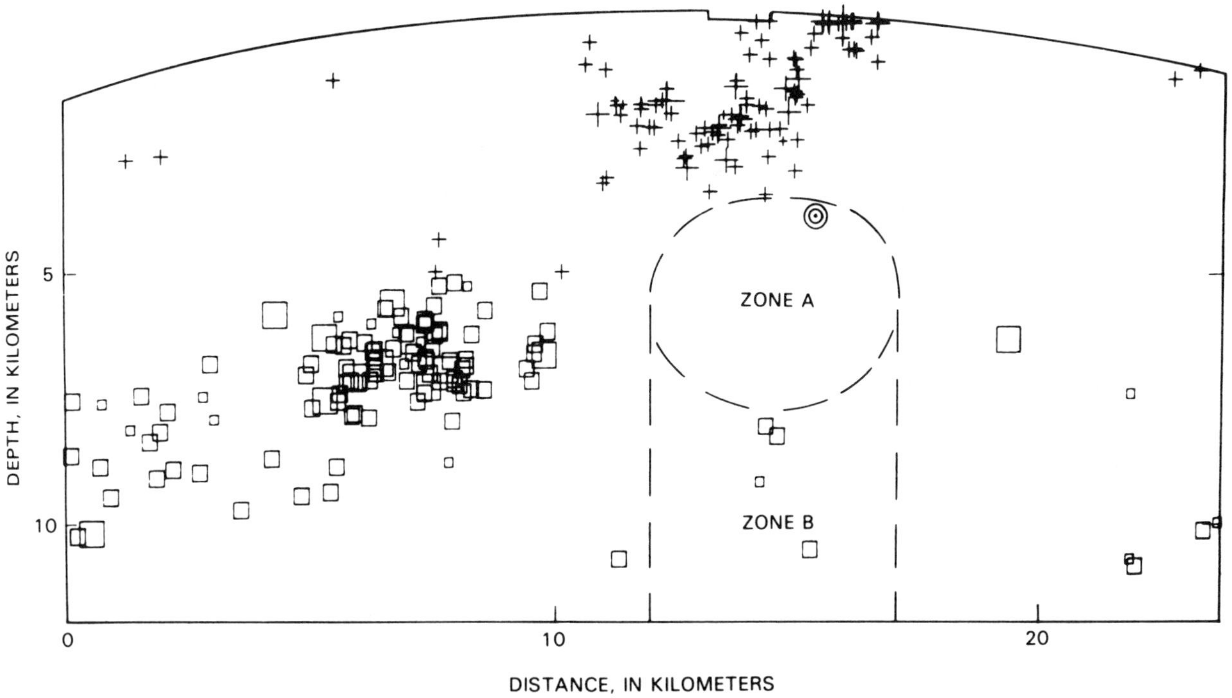

Figure 13.2.14. Approximately true-scale schematic cross section of proposed magma reservoir system beneath Mokuaweoweo, defined by aseismic zone and inversion of ground-deformation data (Lockwood and others, 1987, after Decker and others, 1983). Cross section and hypocenters same as figure 13.2.4 except earthquakes after May 1983 were omitted here. Double circle in zone A shows point-source inflation center calculated from 1977-81 deformation data.

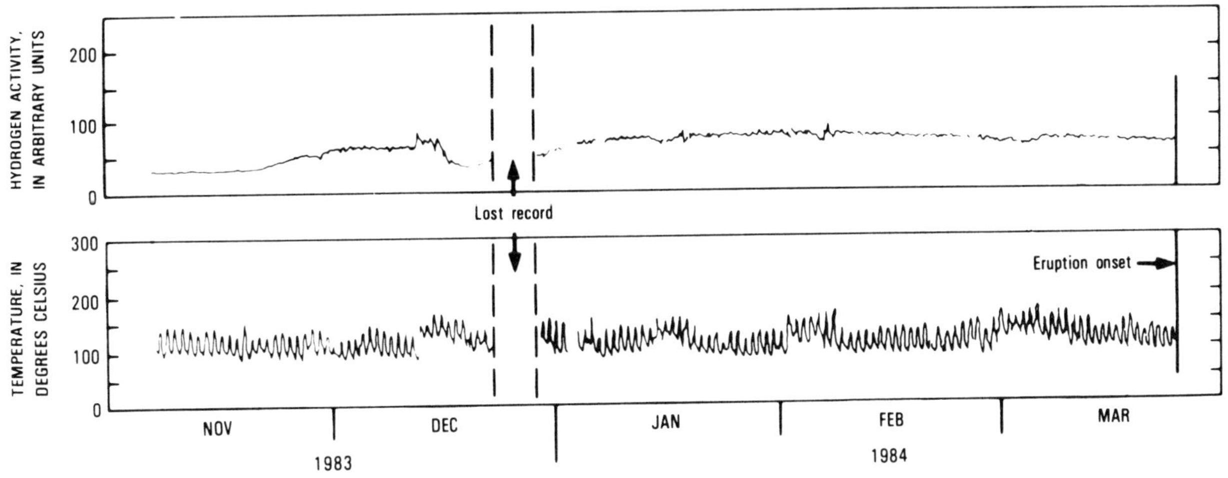

Figure 13.2.15. Variation of temperature and gas emission in a fumarole on the 1975 eruption fissure in Mokuaweoweo, between 5 November 1983 and 24 March 1984 (Lockwood and others, 1987). Large diurnal temperature variations may be caused by artificial instrumental amplification. Data from M. Sato and K. McGee, U.S. Geological Survey, in Lockwood and others (1985).

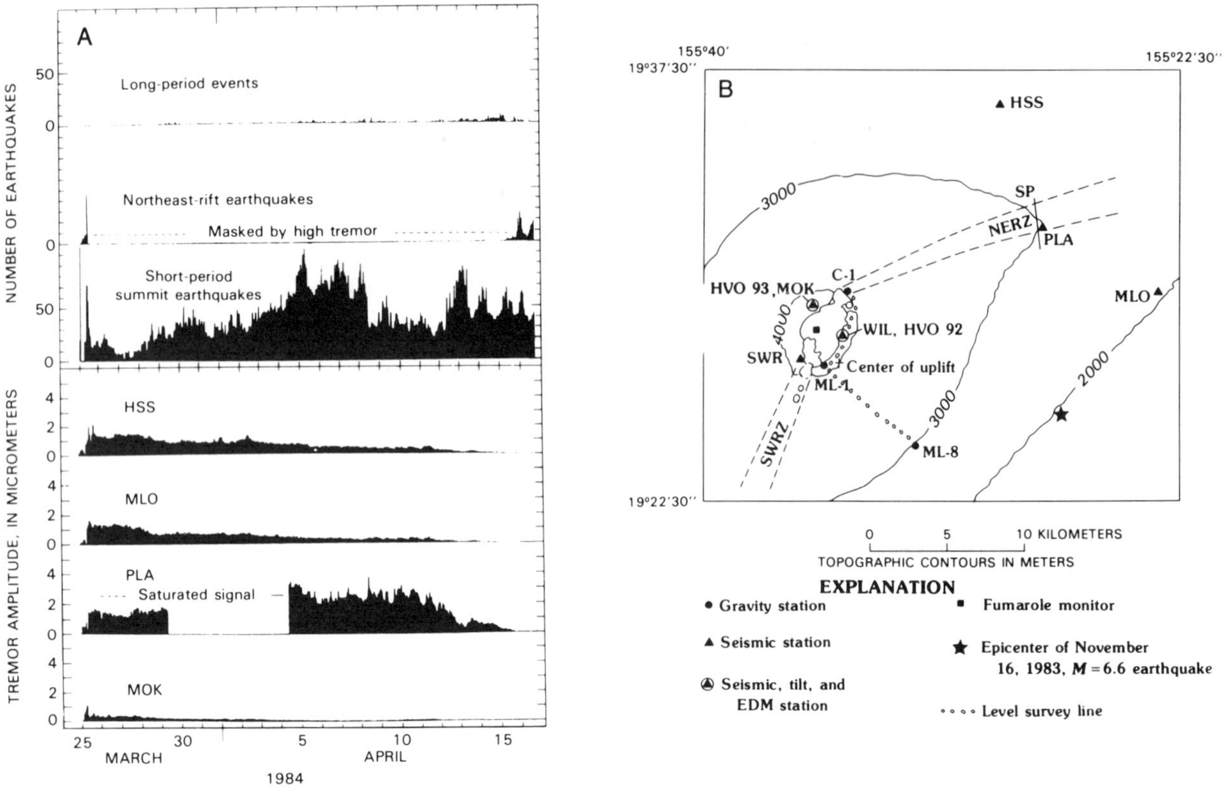

Figure 13.2.16. A (top left), Earthquake frequency (upper three plots) and tremor amplitude (lower four plots), 25 March - 16 April 1984 (Lockwood and others, 1985, 1987). Earthquakes as small as -1 magnitude were plotted when not masked by harmonic tremor. Tremor was read hourly at four seismic stations and plotted as micrometers of ground motion. Signals from station PLA were saturated until 4 April, when instrument sensitivity was lowered. B (top right), Locations of seismic stations and other monitoring sites in the Mauna Loa summit region. Note locations of fumarole monitor and epicenter of M 6.6 earthquake of 16 November 1983, both discussed in text. Copyright by the American Geophysical Union.

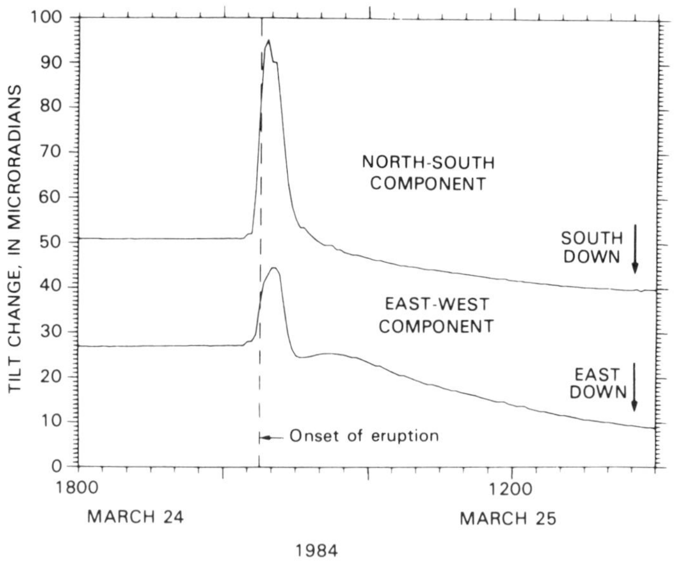

Figure 13.2.17. Bottom right, Continuously recording tiltmeter record from station MOK (see figure 13.2.16B) for a 24-hour period beginning at 1800 hr on 24 March 1984 (Lockwood and others, 1985, 1987). Positive tilt change (inflation) corresponds to downward deflection in north and west directions (that is, increasing values of both tilt components as plotted here; note different convention in figure 13.2.18). Copyright by the American Geophysical Union.

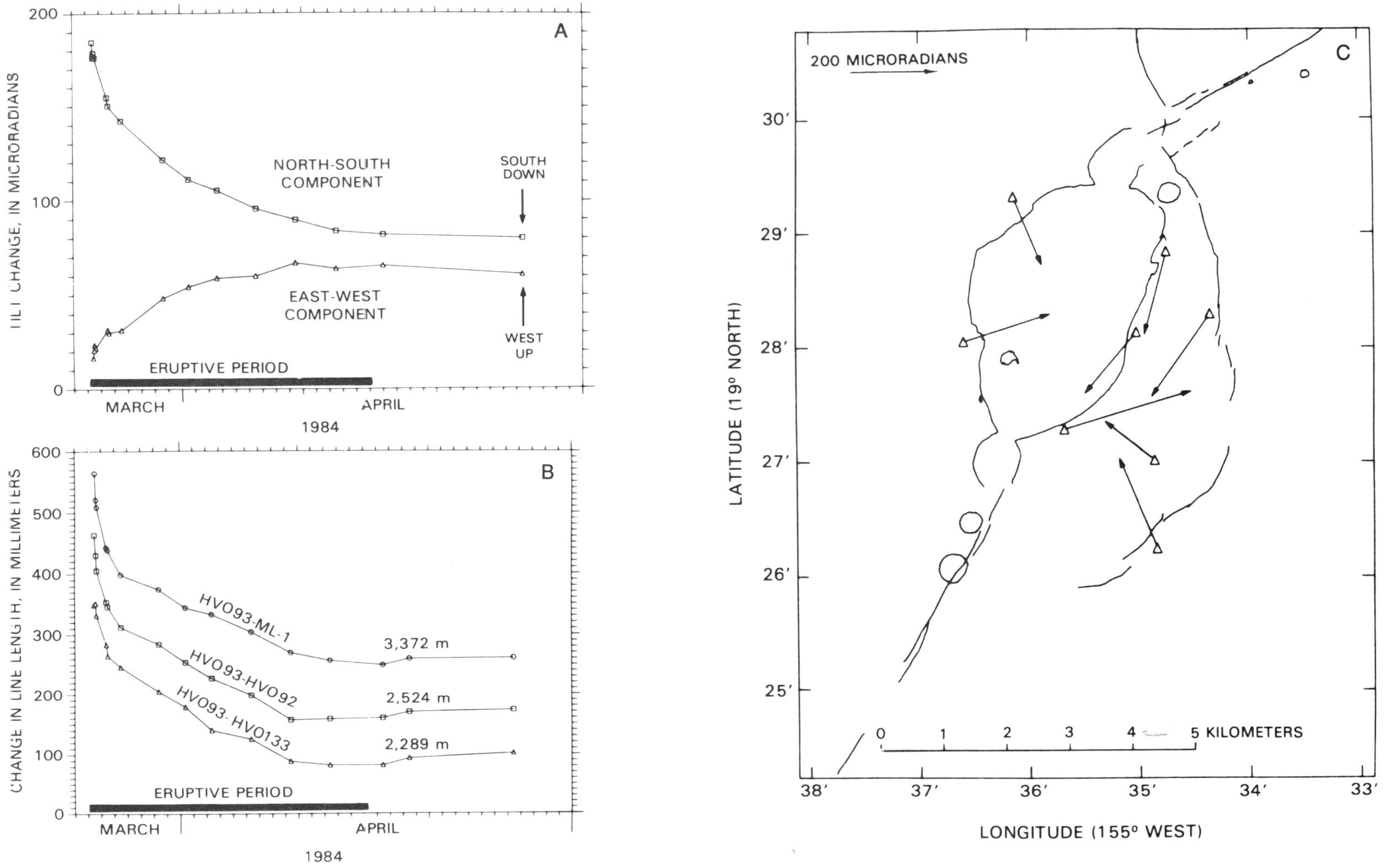

Figure 13.2.18. Tilt and horizontal-strain changes associated with the March-April 1984 eruption of Mauna Loa, from Lockwood and others (1985, 1987). A, Summit deflation curve from spirit-level tilt measurements made near station MOK (see figure 13.2.16B). A positive tilt change (inflation) corresponds to a downward deflection in north and east directions (that is, increasing value of N-S component and decreasing value of E-W component as plotted here; note different convention in figure 13.2.17). B, Summit deflation curve from measurements of horizontal distances across Mokuaweoweo Caldera. C, Net tilt changes in the Mauna Loa summit region between June 1983 and December 1984. Triangles indicate locations of stations; vectors drawn in direction of downward deflection. Copyright by the American Geophysical Union.

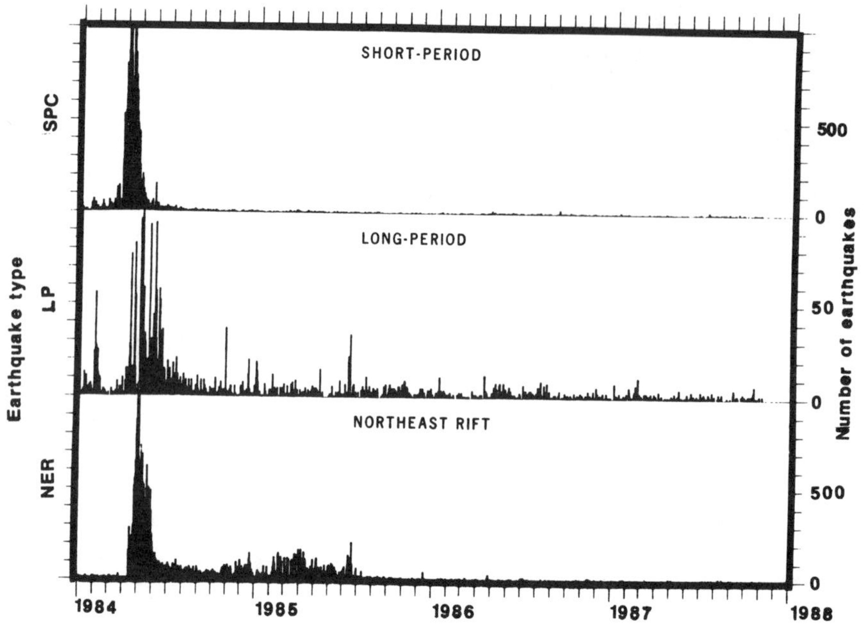

Figure 13.2.19. Daily number of recorded short-period (top) and long-period (middle) summit earthquakes, and northeast rift events (bottom) at Mauna Loa, January 1984 to 4 November 1987, from Smithsonian Institution (1987).

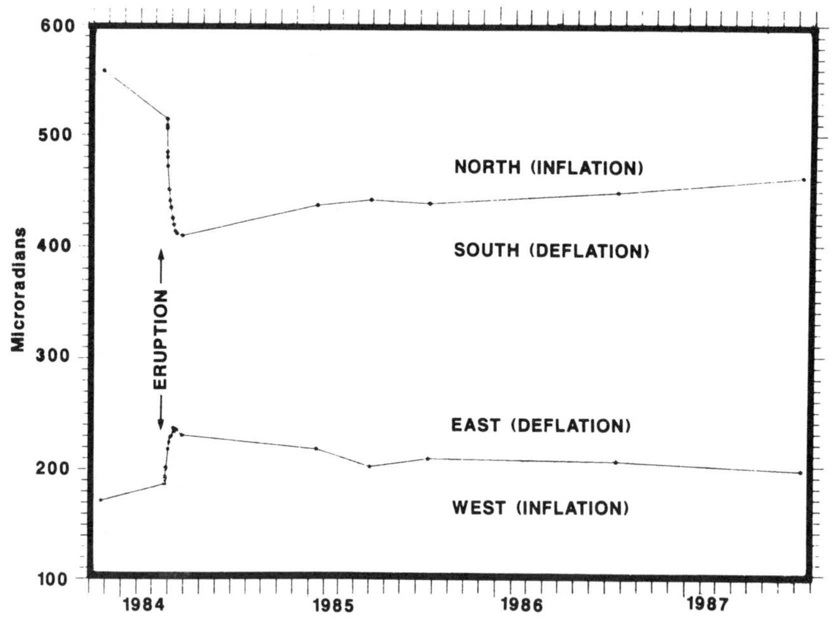

Figure 13.2.20. North-south (top) and east-west (bottom) components of summit tilt at Mauna Loa, November 1983 to August 1987, from Smithsonian Institution (1987). Note gradual reinflation since the March-April 1984 eruption.

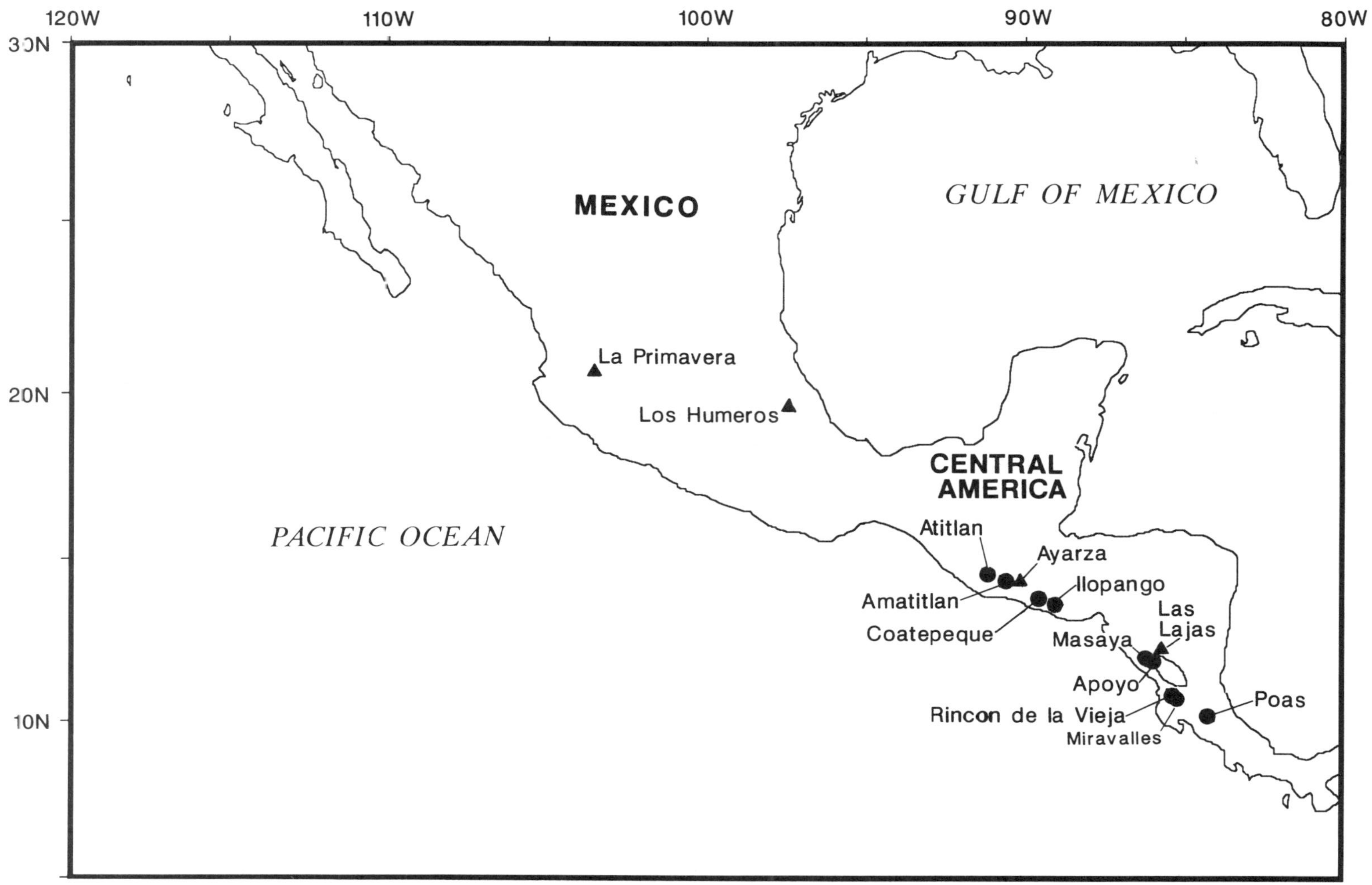

Figure 14. Locations of large, Quaternary restless calderas (solid circles) and nonrestless calderas (triangles) in Region 14.

PART 3: HISTORICAL UNREST AT LARGE QUATERNARY CALDERAS (continued)
See inside back cover for explanation and abbreviations

ATITLAN

CAVW number (active vent)	Latitude Longitude (degrees)	Diameter (km)	Local tectonic setting	Precaldera edifice	SiO_2 content (wt pct)	Age of collapse (yr B.P.)	Date of unrest	Type and duration of unrest ESTU STHF MGTF H Te	Eruption type
14-02-06	14.59N	17 x 20	NW-SE	Strato	R = 51-56	84,000	1469	---- ---- ---- - --	ex
(V. Atitlán)	91.18W		right-	(several)	C = 72-76		1717-21	---- ---- ---- - --	"active"
			lateral				1826-28	---? ---- ---- - --	Ex,lf
			shear with				1833?	---? ---- ---- - --	ex
			N-S tension				1837	---? ---- ---- - --	ex
			cracks				1843?	---? ---- ---- - --	ex
							1852?-53	---? ---- ---- - --	Ex,ex
							1856?	---? ---- ---- - --	ex

TECTONIC SETTING

Volcanism in Guatemala results from subduction of the Cocos plate beneath the Caribbean plate. The modern Atitlán Caldera is thought to have formed at the intersection of a NE-SW-trending fossil boundary between the western and central Guatemala segments of Cocos-Caribbean subduction and the modern volcanic front (Newhall, 1987). Today, the principal faults of the Atitlán area are right-lateral NW-SE strike-slip faults, conjugate left-lateral NE-SW strike-slip faults, and an inferred N-S tension fracture that is defined by a N-S alignment of the modern stratovolcanoes (see below). The modern caldera is about 35 km south of the Motagua (middle) part of the boundary between the Caribbean and North American plates.

GEOLOGIC HISTORY

Three large calderas (figs. 14.1.1, 14.1.2) have formed in the Atitlán region in the past 14 m.y. (Newhall, 1987). The modern Atitlán (III) Caldera formed approximately 84,000 yr ago, shortly after an eruption of about 250 km^3 of rhyolitic magma (Koch and McLean, 1975; Hahn and others, 1979; Drexler and others, 1980; Rose and others, 1987; Newhall, 1987). Three postcaldera stratovolcanoes--V. San Pedro, V. Toliman, and V. Atitlán--have grown on or just inside the southern boundary of the caldera (Rose and others, 1980).

ATITLAN (continued)

HISTORICAL ACTIVITY

The first recorded eruption of Atitlán was in 1469 (von Seebach, 1892). The volcano "awoke" in 1717 and remained "active" until 1721 (Sapper, 1925). Other eruptions occurred as noted in the table above; the only strong eruption among these was on 3 May 1853, when ash darkened the sky around Lake Atitlán from 1100 to 1400 hr (Sapper, 1925).

Aside from the eruptions of V. Atitlán, little is known about unrest at Atitlán Caldera. Seismicity has been low during the past decade or more (White and Harlow, 1979, 1980; Harlow and White, 1980). There is a hint of relatively long-period fluctuations in lake level (fig. 14.1.3); the lake level was reportedly low in the 1820's, 10-15 m higher in the 1870's, back down to its low level in the 1920's, and high again from the 1940's to the present. Most fluctuation is a result of variation in annual rainfall or the effects of regional earthquakes or the subsurface drainage from the lake; it is not known whether any uplift and subsidence of the caldera floor is involved. Short-term changes in lake level that are attributable to rainfall may be as great as the increase of 3.3 m during the exceptionally wet year of 1933; those attributable to earthquakes may be as great as the drop of 2 m that occurred within 1 month following the Guatemala earthquake ($\underline{M}$=7.5) of 4 February 1976 (Newhall and others, 1987). Sediments of the lake floor suggest episodic, decades-long increases in sublacustrine thermal activity: alternating decades-long periods with and without annual lake overturn are interpreted as periods of relatively high and low activity of fumaroles on the caldera floor (Newhall and others, 1987).

REFERENCES

Drexler, J.W., Rose, W.I., Jr., Sparks, R.S.J., and Ledbetter, M.T., 1980, The Los Chocoyos ash, Guatemala: A major stratigraphic marker in Middle America and in three ocean basins: Quaternary Res., v. 13, p. 327-345.

Hahn, G.A., Rose, W.I., Jr., and Meyers, T., 1979, Geochemical correlation of genetically-related rhyolitic ash-flow and air-fall ashes, central and western Guatemala and the equatorial Pacific, in Chapin, C.E., and Elston, W.E., eds., Ash-flow tuffs: Geol. Soc. Am. Spec. Paper 180, p. 101-112.

Harlow, D.H., and White, R.A., 1980, Preliminary catalog of seismicity prior to the Guatemala earthquake of February 4, 1976, from the area between Guatemala City and Lake Atitlán (March 1975-February 1976): U.S. Geol. Surv. Open-File Rep. 80-60, 27 p.

Koch, A.J., and McLean, H., 1975, Pleistocene tephra and ash-flow deposits in the volcanic highlands of Guatemala: Geol. Soc. Am. Bull., v. 86, p. 529-541.

Mooser, F., Meyer-Abich, H., and McBirney, A.R., 1958, Catalogue of active volcanoes of the world including solfatara fields, Part VI, Central America: Naples, IAVCEI, 146 p.

Newhall, C.G., 1980, Geology of the Atitlán Calderas, Guatemala: Hanover, N.H., Dartmouth Coll., Ph.D. dissertation, 364 p.

----- 1987, Geology of the Lake Atitlán region, western Guatemala: Jour. Volcanol. Geotherm. Res., v. 33, p. 23-55.

Newhall, C.G., Paull, C.K., Bradbury, J.P., Higuera-Gundy, A., Poppe, L.J., Self, S., Sharpless, N. Bonar, and Ziagos, J., 1987, Recent geologic history of Lake Atitlán, a caldera lake in western Guatemala: Jour. Volcanol. Geotherm. Res., v. 33, p. 81-107.

ATITLAN (continued)

REFERENCES (continued)

Rose, W.I, Newhall, C.G., Bornhorst, T.J., and Self, S., 1987, Quaternary silicic pyroclastic deposits of Atitlán Caldera, Guatemala: Jour. Volcanol. Geotherm. Res., v. 33, p. 57-80.

Rose, W.I., Jr., Penfield, G.T., Drexler, J.W., and Larson, P.B., 1980, Geochemistry of the andesitic flank lavas of three composite cones within the Atitlán Cauldron, Guatemala: Bull. Volcanol., v. 43, p. 131-153.

Sapper, K., 1925, Los volcanes de la América Central: Halle, Max Niemeyer, 116 p.

Seebach, K., von, 1892, Ueber Vulkane Zentralamerikas: Abhandlungen der Koniglichen Gesellschaft der Wissenschaften zu Göttingen, v. 38, 251 p.

White, R.A., and Harlow, D.H., 1979, Preliminary catalog of aftershocks of the Guatemala earthquake of February 4, 1976, from the area between Guatemala City and Lake Atitlán (February-June 1976): U.S. Geol. Surv. Open-File Rep. 79-864, 63 p.

----- 1980, Preliminary catalog of seismicity from south-central Guatemala (July-December 1976): U.S. Geol. Surv. Open-File Rep. 80-83, 39 p.

Williams, H., 1960, Volcanic history of the Guatemalan highlands: Univ. Calif. Publ. Geol. Sci., v. 38, p. 1-86.

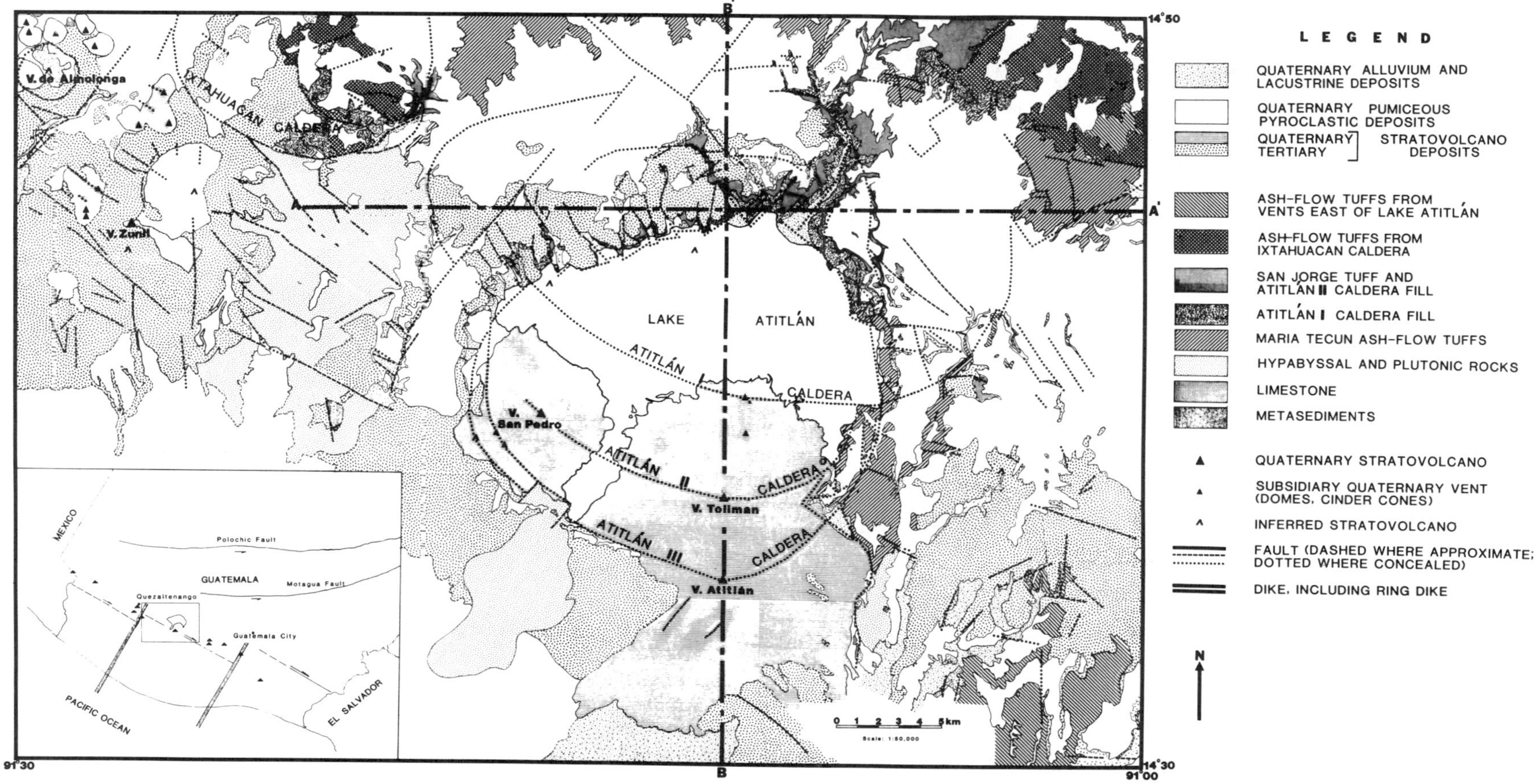

Figure 14.1.1. Geologic map of Atitlán area, showing trenchward (south-southwest) migration of successive calderas from 14 m.y. B.P. to the present, from Newhall (1987).

Figure 14.1.2. Topographic rim of modern, Atitlán III Caldera, and three postcaldera stratovolcanoes: V. San Pedro, V. Toliman, and V. Atitlán.

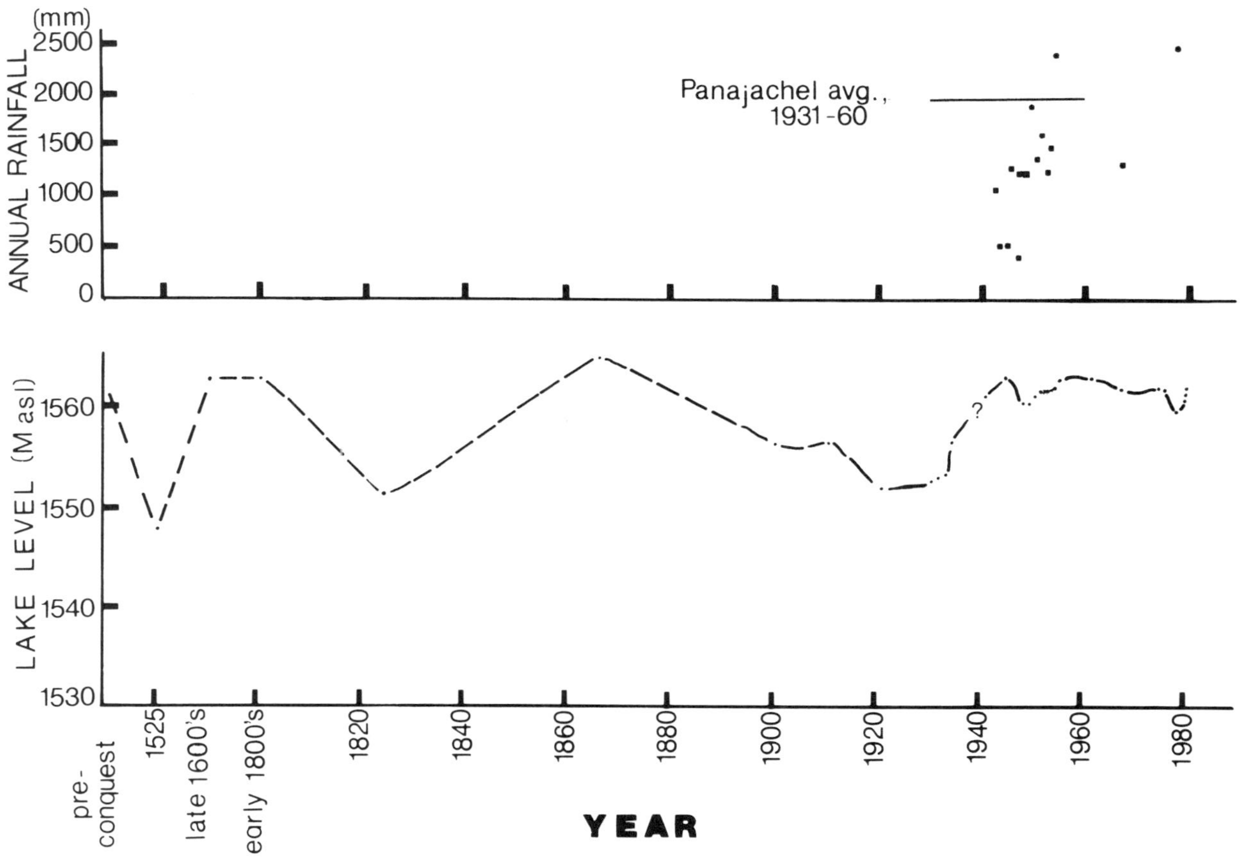

Figure 14.1.3. Annual precipitation and historical fluctuation of level of Lake Atitlán, from Newhall (1980). Annual precipitation at Panajachel (top) on northeast edge of Atitlán III Caldera. Estimate of historical fluctuations in lake level (bottom). Some of the fluctuation may be an artifact of imprecise altitude measurements. Fluctuation in lake level results from variable rainfall, earthquake-induced changes in subsurface drainage from lake, and (possibly) deformation of the caldera and its surroundings.

PART 3: HISTORICAL UNREST AT LARGE QUATERNARY CALDERAS (continued)
See inside back cover for explanation and abbreviations

AMATITLAN

CAVW number (active vent)	Latitude Longitude (degrees)	Diameter (km)	Local tectonic setting	Precaldera edifice	SiO_2 content (wt pct)	Age of collapse (yr B.P.)	Date of unrest	Type and duration of unrest ESTU STHF MGTF H Te	Eruption type
14-02-11	14.38N	14 x 16	NNE-SSW	?	R = 50-51	240,000	1565	x--- ---- ---- - --	Ex,1f?
(Pacaya)	90.60W		graben,		C = 68-75		1651	x--- ---- ---- - --	Ex
			transverse				1664-74	---- ---- ---- - --	Ex,ex
			break in				1690-99?	---- ---- ---x - --	ex
			volcanic				1775	N--- ---- ---- - --	ex,1f
			chain; inter-				1846	---- ---- ---- - --	ex
			section with				1891-1902	---- ---- ---x - --	none
			WNW-ESE trend				1961-88+	FFBx ---- -x-- - --	Ex,ex,1f

TECTONIC SETTING

Pacaya Volcano and the Amatitlán Caldera lie within a NNE-trending graben at its intersection with an ESE-WNW-trending left-lateral strike-slip fault (an extension of the Jalpatagua Fault). Pacaya and Amatitlán also lie astride the boundary between the central and eastern Guatemala segments of the Cocos-Caribbean subduction zone (Stoiber and Carr, 1973; Carr, 1976).

GEOLOGIC HISTORY

The Amatitlán area has been identified by Koch and McLean (1975) and Wunderman and Rose (1984) as the source of several voluminous silicic tephras in central Guatemala. At least nine major pyroclastic deposits, with a total volume of more than 70 km^3 (dense rock equivalent, DRE), were emplaced between about 300,000 yr and 23,000 yr B.P. (Wunderman and Rose, 1984). Wunderman and Rose (1984) inferred an Amatitlán Caldera with a resurgent dome (fig. 14.2.1). Postcaldera vents include a number of silicic domes in and around the caldera, and Pacaya Volcano, a basaltic and basaltic-andesite stratovolcano on the southern boundary of the caldera.

AMATITLAN (continued)

HISTORICAL ACTIVITY

Eruptions from Pacaya are frequent. Eruptions in 1565 and 1651, and perhaps since, were preceded or accompanied by strong earthquakes (Sapper, 1925). The most recent eruptions began in March 1983 and were continuing as of May 1987. Fumarolic activity from an area called Los Humitos, on Pacaya, began in 1891 and increased markedly in 1897 but eventually subsided without an eruption (Sapper, 1925, p. 34).

Regional seismic monitoring was upgraded in 1973, three years before the 1976 Guatemala earthquake. The upgraded monitoring showed a cluster of epicenters on what Wunderman and Rose (1984) have inferred to be a resurgent dome within the Amatitlán Caldera (figs. 14.2.2, 14.2.3). The same area remained active during and since the main shock and nearby aftershocks of February 1976. Most epicenters are along west-northwest-trending regional faults, near the center of inferred resurgence. Wunderman and Rose (1984) interpreted this seismicity to indicate that the dome is actively resurging. However, there are many other examples in this compilation of similar earthquake swarms, including some that have demonstrably shown no resurgence during the period of swarm activity, so confirmation of active resurgence would require the geodetic studies recommended by Wunderman and Rose (1984).

An area of the delta of the Rio Villalobos, on the north side of Lake Amatitlán, subsided irregularly in response to the February 1976 earthquake. Fractures up to 2 m wide and 1.5 m deep opened, and a number of houses were destroyed. This subsidence was apparently due to settling of water-saturated sediments and had no obvious relation to Amatitlán Caldera.

Recent eruptions of Pacaya have been preceded by several hours to a day or more of tremor--in some instances by banded tremor. A relatively strong explosion on 14 June 1987 was preceded by 15 hours of an unspecified increase in seismic activity (E. Sanchez, in Smithsonian Institution, 1987).

On 19-24 May 1988, a swarm of earthquakes ($\underline{M}_{max}$=4.5) occurred along the Mixco Fault Zone, slightly north of Amatitlán Caldera (E. Sanchez and D. Harlow, oral commun., 1988).

REFERENCES

Carr, M.J., 1976, Underthrusting and Quaternary faulting in northern Central America: Geol. Soc. Am. Bull, v. 87, p. 825-829.

Eggers, A.A., 1975, Mapa Geologico Amatitlán, Guatemala: map no. 2059-11G, scale 1:50,000, Instituto Geografico Nacional de Guatemala.

Eggers, A.A., and Chavez, D., 1979, Temporal gravity variations at Pacaya Volcano, Guatemala: Jour. Volcanol. Geotherm. Res., v. 6, p. 391-402.

AMATITLAN (continued)

REFERENCES (continued)

Eggers, A.A., Krausse, J., Rusk, H., and Ward, J., 1976, Gravity changes accompanying volcanic activity at Pacaya Volcano, Guatemala: Jour. Volcanol. Geotherm. Res., v. 1, p. 229-236.

Eichelberger, J.C., and McGetchin, T.R., 1976, Petrogenesis of the 1973 Pacaya lavas, Guatemala, in Gonzales-Ferran, O., ed., Proc. Symp. on Andean and Antarctic Volcanological Problems: Naples, IAVCEI, p. 435-449.

Harlow, D.H., 1976, Instrumentally recorded seismic activity prior to the main event, in Espinoza, A.F., ed., The Guatemalan earthquake of February 4, 1976, a preliminary report: U.S. Geol. Surv. Prof. Paper 1002, p. 12-16.

Harlow, D.H., and White, R.A., 1980, Preliminary catalog of seismicity prior to the Guatemala earthquake of February 4, 1976 from the area between Guatemala City and Lake Atitlán (March 1975-February 1976): U.S. Geol. Surv Open-File Rep. 80-60, 27 p.

Koch, A.J., and McLean, H., 1975, Pleistocene tephra and ash-flow deposits in the volcanic highlands of Guatemala: Geol. Soc. Am. Bull., v. 86, p. 529-541.

Mooser, F., Meyer-Abich, H., and McBirney, A.R., 1958, Catalogue of active volcanoes of the world including solfatara fields, Part VI, Central America: Naples, IAVCEI, 146 p.

Sapper, K., 1925, Los volcanes de la América Central: Halle, Max Niemeyer, 116 p.

----- 1927, Vulkankunde: Stuttgart, Engelhorns Nachf, 427 p.

Smithsonian Institution, Scientific Event Alert Network (SEAN), 1976-88, Pacaya Volcano: SEAN Bull., v. 1. nos. 5, 8; v. 2, nos. 8, 10; v. 3. no. 11; v. 5, nos. 2, 12; v. 6, nos. 2, 3; v. 7, nos. 2, 3; v. 8, nos. 1, 9, 11; v. 9, no. 8; v. 10, nos. 3-4; v. 11, nos. 3, 11, 12; v. 12, nos. 1, 5, 7, 8, 12; v. 13, nos. 1, 2 (contributions from A. MacKenney, E. Sanchez, E. Quevec, E. Molina, N. Banks, M. Doukas, and others).

Stoiber, R.E., and Carr, M.J., 1973, Quaternary volcanic and tectonic segmentation of Central America: Bull. Volcanol., v. 37, no. 3, p. 304-325.

White, R.A., and Harlow, D.H., 1979, Preliminary catalog of aftershocks of the Guatemala earthquake of February 4, 1976 from the area between Guatemala City and Lake Atitlán (February-June 1976): U.S. Geol. Surv. Open-File Rep. 79-864, 63 p.

----- 1980, Preliminary catalog of seismicity from south-central Guatemala (July-December 1976): U.S. Geol. Surv. Open-File Rep. 80-83, 39 p.

Wunderman, R.L., and Rose, W.I., 1984, Amatitlán, an actively resurging cauldron 10 km south of Guatemala City: Jour. Geophys. Res., v. 89, no. B10, p. 8525-8539.

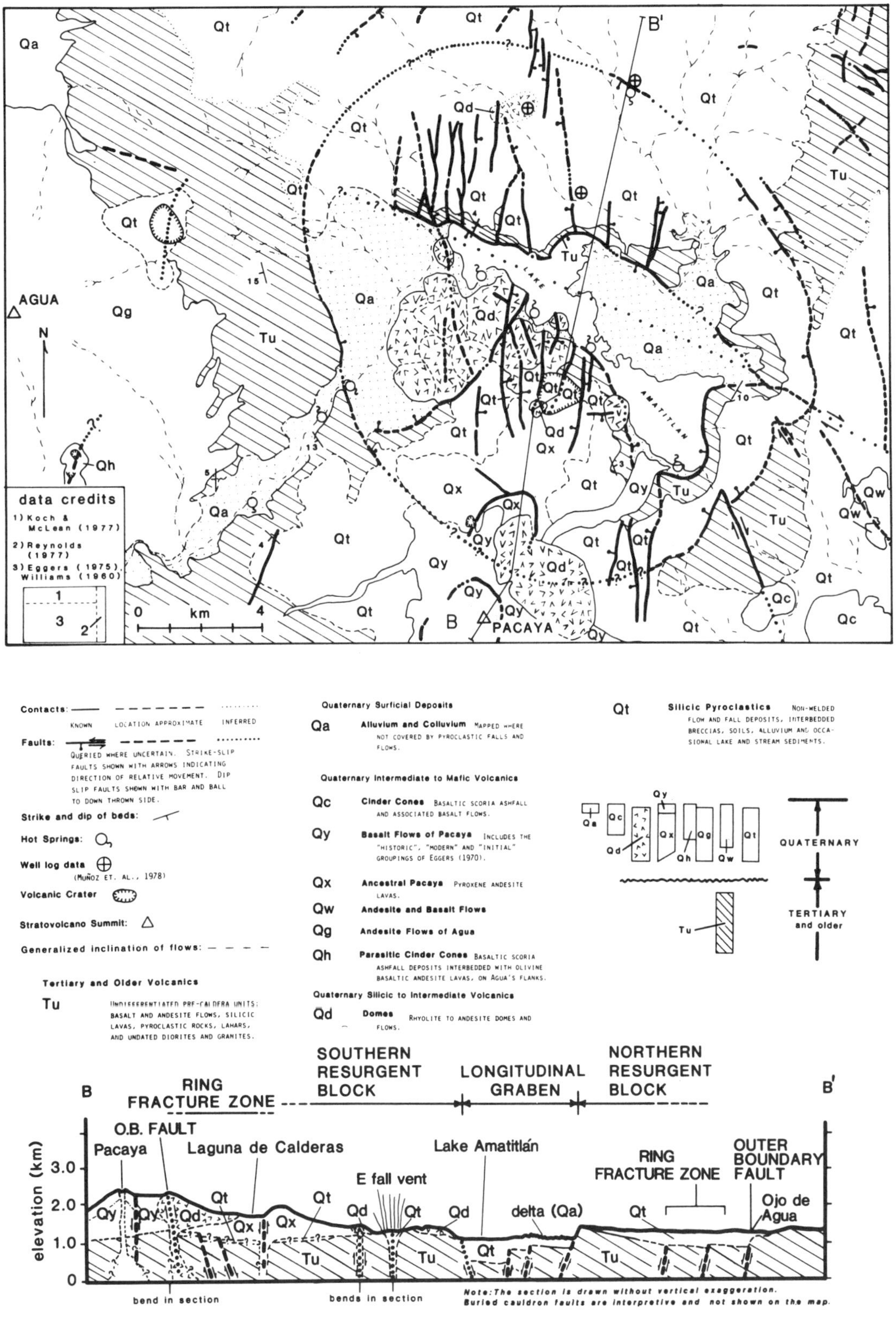

Figure 14.2.1. Generalized geologic map and cross section of Amatitlán Caldera, from Wunderman and Rose (1984). Copyright by the American Geophysical Union.

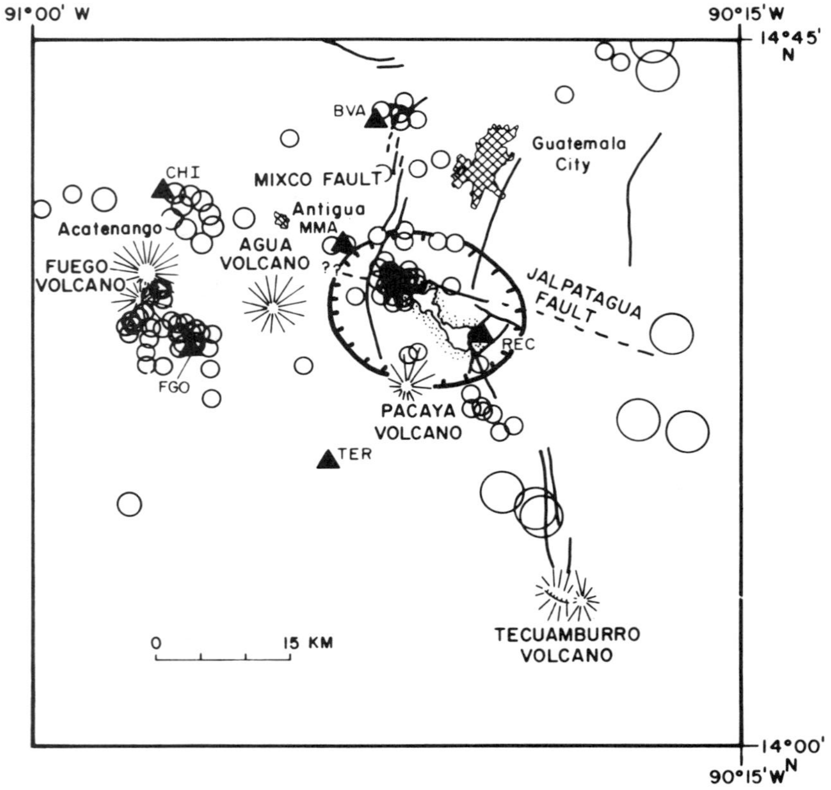

Figure 14.2.2. Epicenter locations (March-September 1975) and known faults in the Amatitlán region, after Harlow (1976). Seismic recording ended before the magnitude 7.5 Guatemalan earthquake of 4 February 1976. Solid triangles indicate seismometers; size of open circles reflects estimated error in epicentral location.

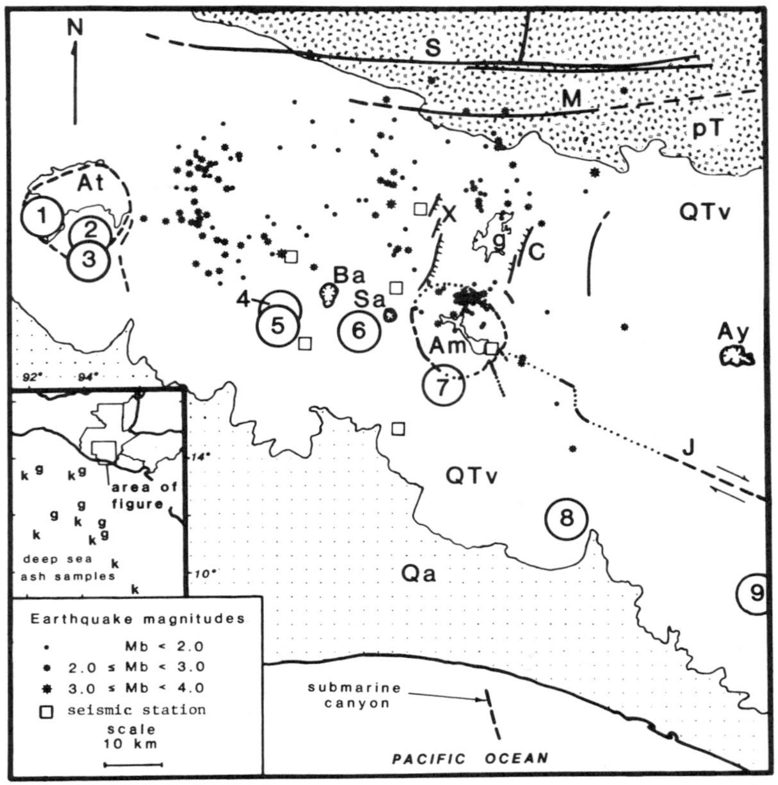

Figure 14.2.3. Regional setting, structure, and seismicity of Amatitlán Caldera, from Wunderman and Rose (1984) after Harlow and White (1980) and White and Harlow (1979, 1980). Qa, Quaternary alluvium; QTv, Quaternary and Tertiary volcanics; pT, pre-Tertiary rocks. Volcanic front composed of following strato-volcanoes: 1, San Pedro; 2, Atitlan; 3, Toliman; 4, Acatenango; 5, Fuego; 6, Agua; 7, Pacaya; 8, Tecuamburro; and 9, Moyuta. Slightly north of these stratovolcanoes are the following silicic centers: At, Atitlan; Ba, Barahona; Sa, Sabana Grande; Am, Amatitlan; and Ay, Ayarza. Major faults include: S, San Agustin; M, Motagua; J, Jalpatagua; X, Mixco; and C, Santa Catarina Pinula. Earthquake epicenters for May 1976 (after M 7.5 Guatemalan earthquake of 4 February 1976) demonstrate recent clustering of epicenters near Amatitlan Caldera. Guatemala City (letter g) is 10 km north-northeast of Amatitlán. Copyright by American Geophysical Union.

PART 3: HISTORICAL UNREST AT LARGE QUATERNARY CALDERAS (continued)
See inside back cover for explanation and abbreviations

COATEPEQUE

CAVW number (active vent)	Latitude Longitude (degrees)	Diameter (km)	Local tectonic setting	Precaldera edifice	SiO_2 content (wt pct)	Age of collapse (yr B.P.)	Date of unrest	Type and duration of unrest ESTU STHF MGTF H Te	Eruption type
14-03-02,	13.87N	7 x 10	Compr,	Strato?	R = 51-62	10,000	1520 (SA)	---- ---- ---- - --	Ex
(Santa Ana=SA)	89.55W		astride		C = 70	to	1524 (SA)	---- ---- ---- - --	Ex
14-03-03			segment			40,000	1570? (SA)	---- ---- ---- - --	Ex
(Izalco=I)			boundary				1576 (SA)	---- ---- ---x - --	Ex
14-03-04							1874 (SA)	---- ---- ---x - --	Ex?
(San Marcelino=SM)							1878 (SA)	---- ---- ---- - --	ex
						Santa	1879 (SA)	---- ---- ---- - --	ex
						Ana	1880 (SA)	---- ---- ---- - --	Ex
							1882? (SA)	---- ---- ---- - --	ex
							1884 (SA)	---- ---- ---- - --	ex
							1904 (SA)	D--- ---- ---- - --	ex
							1920 (SA)	---- ---- ---- - --	ex
							1770-72(I)	x--x ---- ---x - --	ex,lf
							1783 (I)	x--- ---- ---- - --	lf
							1793+ (I)	---- ---- ---- - --	ex,lf
							1798 (I)	---- ---- ---- - --	ex
							1802-03(I)	---- ---- ---- - --	ex,lf
							1805-07(I)	---- ---- ---- - --	ex
							1825? (I)	---- ---- ---- - --	ex
							1836 (I)	---- ---- ---- - --	ex
							1838 (I)	---- ---- ---- - --	ex
							1840 (I)	---- ---- ---- - --	ex
							1842 (I)	---- ---- ---- - --	ex
							1844 (I)	---- ---- ---- - --	ex,lf
						Izalco	1850 (I)	---- ---- ---- - --	ex
							1854 (I)	---- ---- ---- - --	ex,lf
							1856 (I)	---- ---- ---- - --	ex,lf
							1858-60(I)	---- ---- ---- - --	ex,lf
							1863-70(I)	---- ---- ---x - --	lf,ex
							1872-73(I)	---- ---- ---- - --	ex
							1879-85(I)	---- ---- ---- - --	ex,lf
							1887-1901(I)	---- ---- ---- - --	ex,lf

COATEPEQUE (continued)

CAVW number (active vent)	Latitude Longitude (degrees)	Diameter (km)	Local tectonic setting	Precaldera edifice	SiO_2 content (wt pct)	Age of collapse (yr B.P.)	Date of unrest	Type and duration of unrest ESTU STHF MGTF H Te	Eruption type
(continued from previous page)							1902 (I)	---- ---- ---- - QU?	ex,lf
							1903-05(I)	D--- ---- ---- - --	ex
							1912-16(I)	x--- ---- ---- - --	ex,lf
							1920-21(I)	N--A A--x ---x - --	ex,lf
							1925-28?(I)	---- ---- ---- - --	ex,lf
							1930-31(I)	---- ---- ---- - --	lf,ex
							1933-34(I)	---- ---- ---- - --	ex,lf
							1937-57(I)	---- ---- ---- - --	ex,Ex,lf
							1966 (I)	---- ---- -NZZ - --	lf
						San	1600's(SM)	---- ---- ---- - --	lf
						Marcelino	1722 (SM)	---- ---- ---- - --	lf

TECTONIC SETTING

Coatepeque Caldera is part of the chain of volcanoes resulting from Cocos-Caribbean subduction. Coatepeque lies astride the boundary between the eastern Guatemala and El Salvador segments of that subduction (Stoiber and Carr, 1973).

GEOLOGIC HISTORY

Coatepeque Caldera (figs. 14.3.1, 14.3.2) lies at the eastern foot of Volcán Santa Ana and probably formed after Santa Ana had grown to nearly its current size (Carr and Pontier, 1981). Several rhyolitic and dacitic domes grew within the caldera, and several basaltic and andesitic vents developed just south and west of Coatepeque. V. Santa Ana, V. Izalco, and V. San Marcelino are part of the same volcanic center as Coatepeque Caldera, although they lie just outside the caldera boundary.

HISTORICAL ACTIVITY

Santa Ana was particularly active during the late 1500's and again in the late 1800's. San Marcelino last erupted in 1772. Izalco, presently about 200 m high, began to grow in (or before?) 1770.

1770-1966: After precursory earthquakes and a rise in fumarolic activity, "the land swelled up and...the steam emissions increased amid a violent din" (Montessus de Ballore, 1888). Larde (1925) accepts the 1770 date given above for the birth of Izalco, but Ipiña (in Sapper, 1925, p. 43) places it at 1765. During eruptive periods of Izalco (listed above), explosions commonly occur at intervals of 1 minute to 1 hour. A relatively long period of explosions that began in 1887 stopped abruptly in January 1901, but then resumed on 10 May 1902, three weeks after a $\underline{M}$=7.9 earthquake off the south coast of western Guatemala (18 April) and only two

COATEPEQUE (continued)

HISTORICAL ACTIVITY (continued)

and three days after major eruptions of Mont Pelée and Soufriere St. Vincent, respectively (Sapper, 1925). In 1904, less than two weeks after an earthquake of "sufficient magnitude," both Izalco and Santa Ana again entered into eruption (Sapper, 1925). J. Larde, cited in Sapper (1925), watched the eruptive fissure from a distance of only a few tens of meters in 1920 and noted short-lived, local tumescence of up to 2 m immediately before explosions. The 21-year repose since Izalco's most recent eruption (1966) is slightly longer than any known previous repose of Izalco, but it is not sufficiently long to imply that Izalco's activity is decreasing.

Activity at these three stratovolcanoes is generally not accompanied by felt seismicity, ground deformation, or thermal activity in Coatepeque Caldera; an exception was agitation of Lake Coatepeque during activity of Izalco in mid-January 1925 (Larde, 1925). Records do not specify whether this agitation was of fumarolic or seismic origin.

REFERENCES

Carr, M.J., and Pontier, N.K., 1981, Evolution of a young parasitic cone towards a mature central vent; Izalco and Santa Ana Volcanoes in El Salvador, Central America: Jour. Volcanol. Geotherm. Res., v. 11, p. 277-292.

Larde, Jorge, 1925, El Volcán de Izalco: Bull. Volcanol., Ser. 1, nos. 5-6, p. 209-211.

Meyer, J., 1964, Stratigraphie des Bimskiese und -ashen des Coatepeque-Vulkans im westlichen El Salvador (Mittelamerika): Neues Jahrb. Geol. Palaontol., v. 119, p. 215-246.

Montessus de Ballore, F., 1888, Tremblements de terre et eruptions volcaniques au Centre-Amerique, depuis la conquete espagnole jusqu'a nos jours: Dijon, Soc. Sci. Natur. Saone-et-Loire, 293 p.

Mooser, F., Meyer-Abich, H., and McBirney, A.R., 1958, Catalogue of active volcanoes of the world including solfatara fields, Part VI, Central America: Naples, IAVCEI, 146 p.

Rose, W.I., Jr., Hahn, G.A., Drexler, J.W., Malinconico, M.L., Peterson, P.S., and Wunderman, R.L., 1981, Quaternary tephra of northern Central America, in Self, S., and Sparks, R.S.J., Tephra studies: Dordrecht, D. Reidel Publ. Co., p. 193-211.

Rose, W.I., Jr., and Stoiber, R.E., 1969, The 1966 eruption of Izalco Volcano, El Salvador: Jour. Geophys. Res., v. 74, p. 3119-3130.

Sapper, K., 1925, Los volcanes de la América Central: Halle, Max Niemeyer, 116 p.

Stoiber, R.E., and Carr, M.J., 1973, Quaternary volcanic and tectonic segmentation of Central America: Bull. Volcanol., v. 37, no. 3, p. 304-325.

Williams, H., and Meyer-Abich, H., 1955, Volcanism in the southern part of El Salvador, with particular reference to the collapse basins of Coatepeque and Ilopango: Univ. Calif. Publ. Geol. Sci., v. 32, p. 1-64.

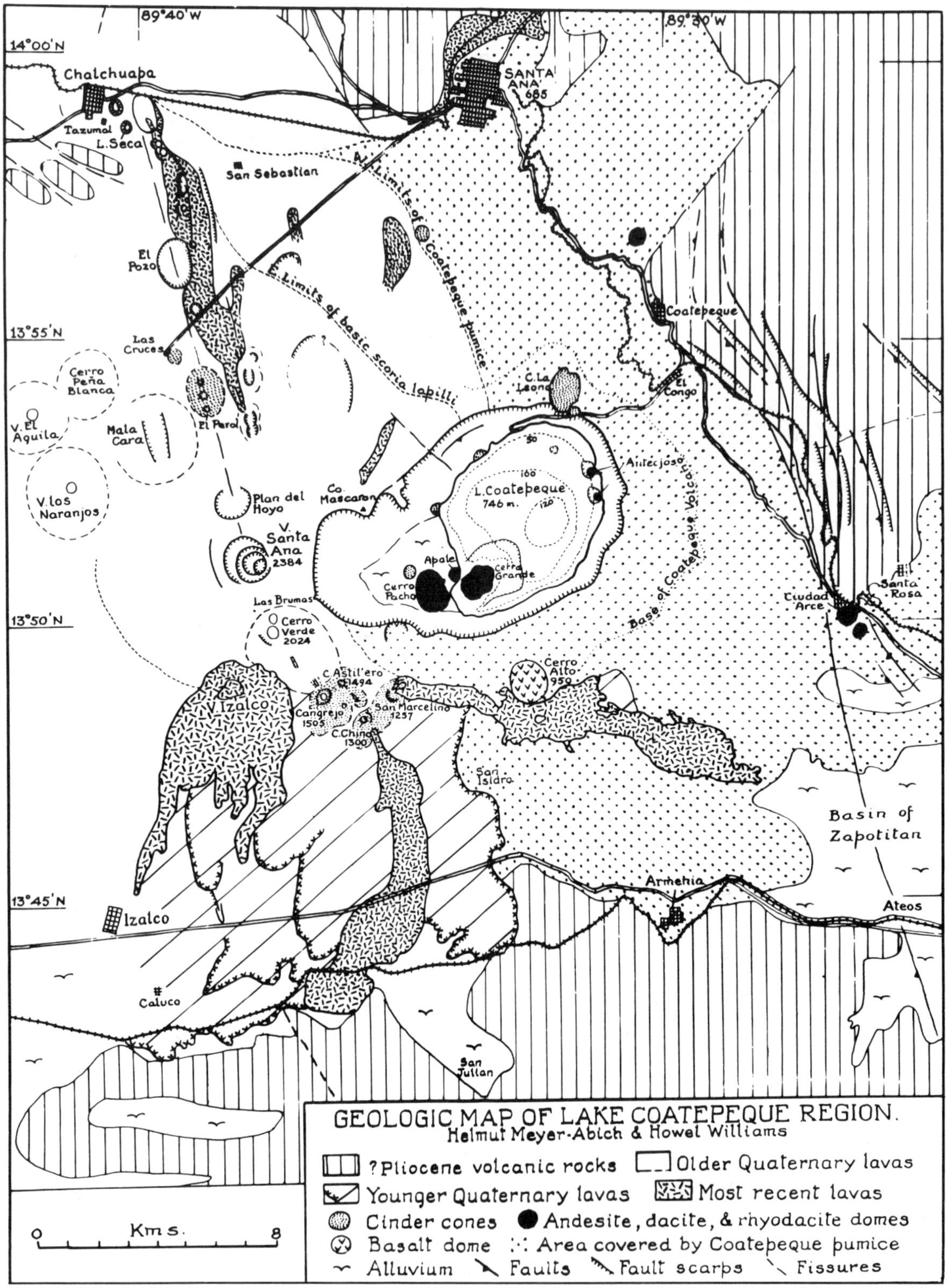

Figure 14.3.1. Geologic map of Lake Coatepeque region, from Williams and Meyer-Abich (1955).

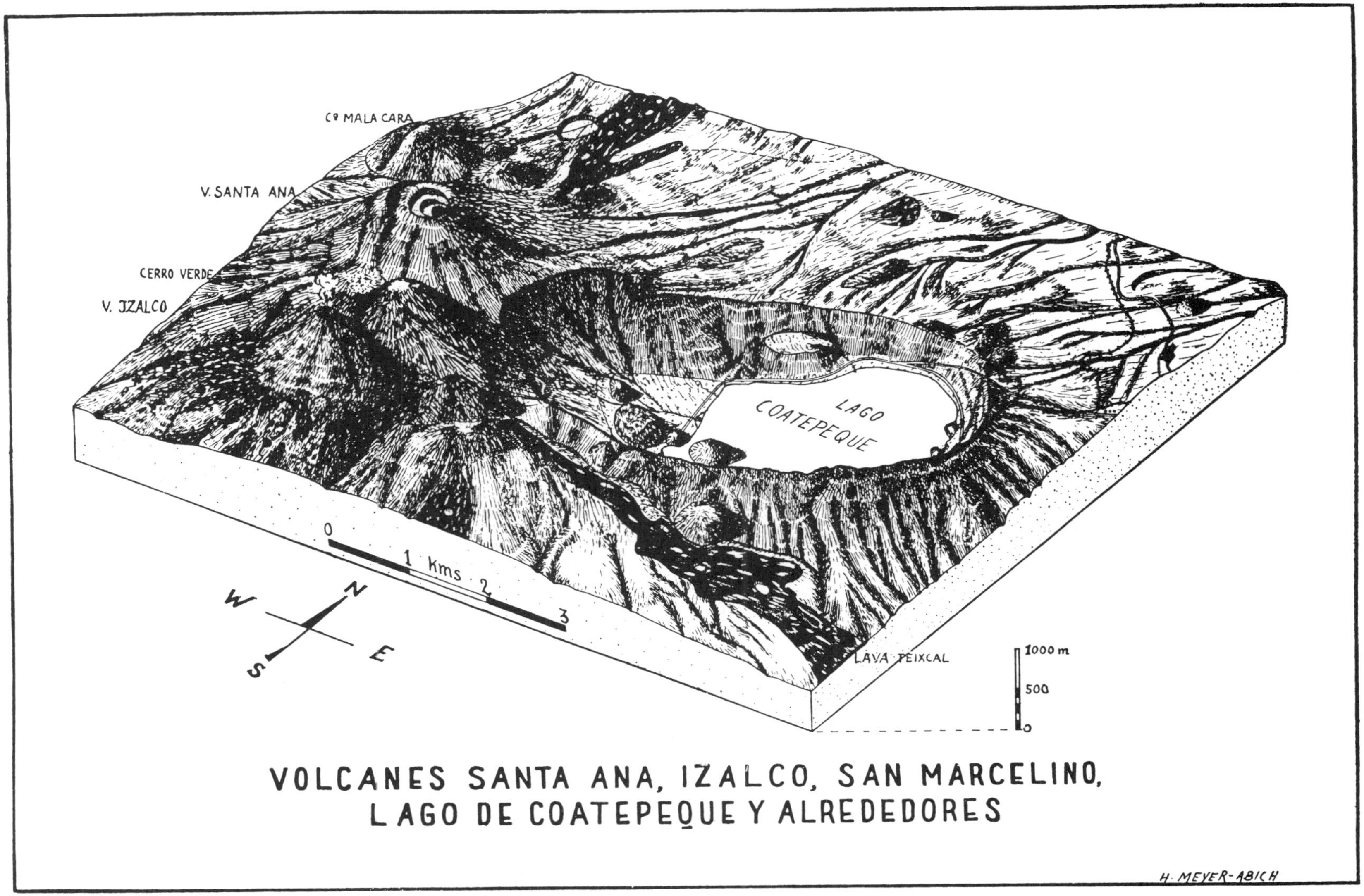

Figure 14.3.2. Perspective drawing of Lake Coatepeque region, from Mooser and others (1958). Note what appears to be a landslide scarp at the west end of caldera.

PART 3: HISTORICAL UNREST AT LARGE QUATERNARY CALDERAS (continued)
See inside back cover for explanation and abbreviations

ILOPANGO

CAVW number (active vent)	Latitude Longitude (degrees)	Diameter (km)	Local tectonic setting	Precaldera edifice	SiO_2 content (wt pct)	Age of collapse (yr B.P.)	Date of unrest	Type and duration of unrest ESTU STHF MGTF H Te	Eruption type
14-03-06 (Islas Quemadas)	13.67N 89.05W	8 x 11	Graben	?	R = 67 C = 63-68	A.D. 260	1765	xx-- ---- ---- - --	none
							1769-70	xx-- ---- ---- - --	none
							1873	---? ---- ---x - QU	none
							1879-80	D--? ?--- ---- - --	dm,ex
							1884	---- ---- ---? - --	unknown

TECTONIC SETTING

Ilopango is an elongate caldera bounded on the north and south by faults of a graben that parallels the volcanic chain throughout El Salvador and parts of Guatemala and Nicaragua.

GEOLOGIC HISTORY

The Ilopango Caldera (figs. 14.4.1, 14.4.2) formed as a result of the Tierra Blanca eruption of 260 A.D. (volume of pyroclastic deposits = 10-20 km^3 DRE) (Williams and Meyer-Abich, 1955; Sheets, 1979; Hart and Steen-McIntyre, 1983). Because Ilopango's shape is strongly controlled by regional faults, Williams and McBirney (1979) cite it as an example of a "volcano-tectonic depression." Islas Quemadas ("burnt islands") is a dacite dome that grew through Lake Ilopango in 1880 (see below).

HISTORICAL ACTIVITY

1765, 1769-70: Earthquakes were felt in the vicinity of Ilopango (Sapper, 1925, p. 54).

1873: At the time of an earthquake that struck nearby San Salvador, water of Lake Ilopango was lifted, bubbling, 2 to 3 ft, but it was not guessed that a volcano might be growing beneath the lake (Sapper, 1925).

1879-80: Severe earthquakes occurred from 20 December 1879 until 31 December 1879, and in January 1880 the Islas Quemadas dome grew up through Lake Ilopango. Sapper (1925) wrote that the lake began to rise on 6 January, and that by 9 January discharge down the Jiboa River was so strong that the entire valley was flooded, the town of Atuscatla was destroyed, and many cattle perished.

ILOPANGO (continued)

HISTORICAL ACTIVITY (continued)

Lake level began to drop on 11 January, and by 12 January emission of sulfur gas from the center of the lake was noted. On 20 January a huge column of ash and incandescent rocks rose from the center of the lake, and by 23 January a group of new islands, Islas Quemadas, had risen above the lake. Explosions and dome growth continued through mid-March (Sapper, 1925).

After the initial earthquakes in December 1879, earthquakes in January were few and rarely strong. However, on 23 February an earthquake (from Ilopango?) was felt throughout El Salvador, and immediately thereafter an unbearable sulfur smell permeated the area. The following day the column of steam increased, but by nightfall both steaming and several small parts of the islands had disappeared. The last of several phases of dome growth and explosions ended on 19 March 1880. Unspecified activity was observed again on 7-9 February and 13 March 1884 (Sapper, 1925), but no activity has been observed since 1884.

Goodyear (1880) calculated that the level of Lake Ilopango rose 1.2 m between 31 December and 11 January; that rise in lake level corresponded to a displacement of 0.13 km^3, or roughly the subaqueous volume of the Islas Quemadas dome, so there is no indication of major uplift of the caldera floor. Between 12 January and 20 January, the lake level fell 9.2 m (volume=0.96 km^3), at a maximum rate of nearly 10 cm/hr. By 8 February the lake level had fallen an additional 2.6 m, for an additional loss of 0.27 km^3, making a total loss of 1.2 km^3. The drop in lake level was due in part or in whole to rapid downcutting of the outlet channel, and towns downstream were flooded. Unfortunately, there is no documentation of the final depth of downcutting to indicate whether some of the decrease in lake level might also have been due to subsidence of the caldera floor.

Golombek and Carr (1978) noted a correlation between the earthquakes preceding the 1879-80 eruption and semidiurnal maxima and minima in earth tides. Five out of six phases of the eruption also showed some correlation with earth tides, with each of these phases beginning near fortnightly minima in tidal amplitude.

A $\underline{M}$=5.4 tectonic earthquake occurred at shallow depth beneath San Salvador in October 1986. Damage in San Salvador was severe, but the quake did not trigger any disturbance at the Ilopango Caldera.

REFERENCES

Anonymous, 1880, A lacustrine volcano: Nature, v. 22, p. 129.

Golombek, M.P., and Carr, M.J., 1978, Tidal triggering of seismic and volcanic phenomena during the 1879-1880 eruption of Islas Quemadas Volcano in El Salvador, Central America: Jour. Volcanol. Geotherm. Res., v. 3, p. 299-307.

Goodyear, W.A., 1880, Earthquake and volcanic phenomena, December 1979 and January 1880 in the Republic of Salvador: Panama, The Star and Herald, 56 p.

Hart, W.E.J., and Steen-McIntyre, V., 1983, Tierra Blanca Joven tephra from the A.D. 260 eruption of Ilopango Caldera, in Sheets, P., ed., Archaeology and volcanism in Central America: The Zapotitán Valley of El Salvador: Austin, Univ. Tex. Press, p. 14-34.

ILOPANGO (continued)

REFERENCES (continued)

Mooser, F., Meyer-Abich, H., and McBirney, A.R., 1958, Catalogue of active volcanoes of the world including solfatara fields, Part VI, Central America: Naples, IAVCEI, 146 p.

Sapper, K., 1925, Los volcanes de la América Central: Halle, Max Niemeyer, 116 p.

Sheets, P.D., 1979, Environmental and cultural effects of the Ilopango eruption in Central America, in Sheets, P.D., and Grayson, D.K., eds., Volcanic activity and human ecology: New York, Academic Press, p. 525-564.

Williams, H., and McBirney, A.R., 1979, Volcanology: San Francisco, Freeman Cooper, 397 p.

Williams, H., and Meyer-Abich, H., 1955, Volcanism in the southern part of El Salvador, with particular reference to the collapse basins of Coatepeque and Ilopango: Univ. Calif. Publ. Geol. Sci., v. 32, p. 1-64.

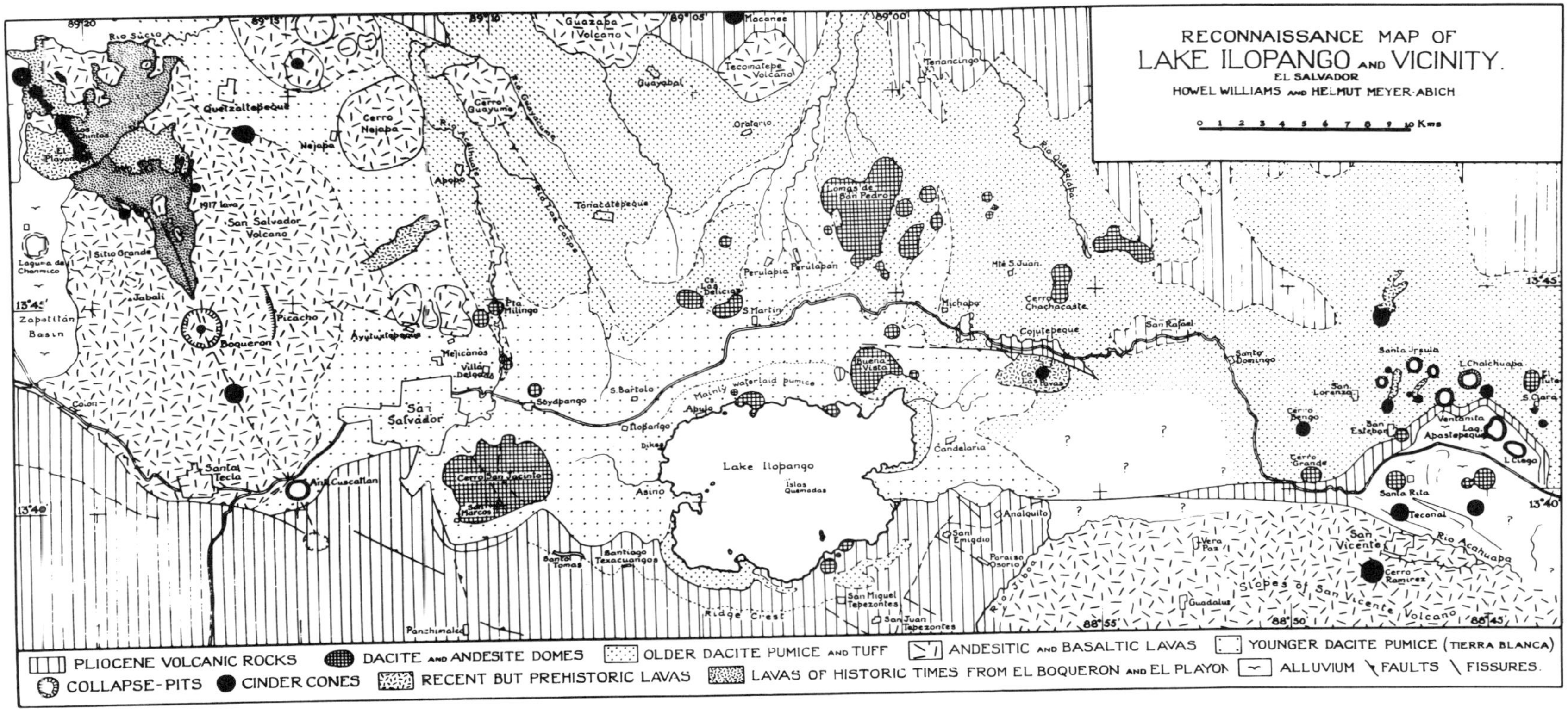

Figure 14.4.1. Geologic map of Ilopango Caldera, from Williams and Meyer-Abich (1955). Topographic rim of caldera is labeled "ridge crest."

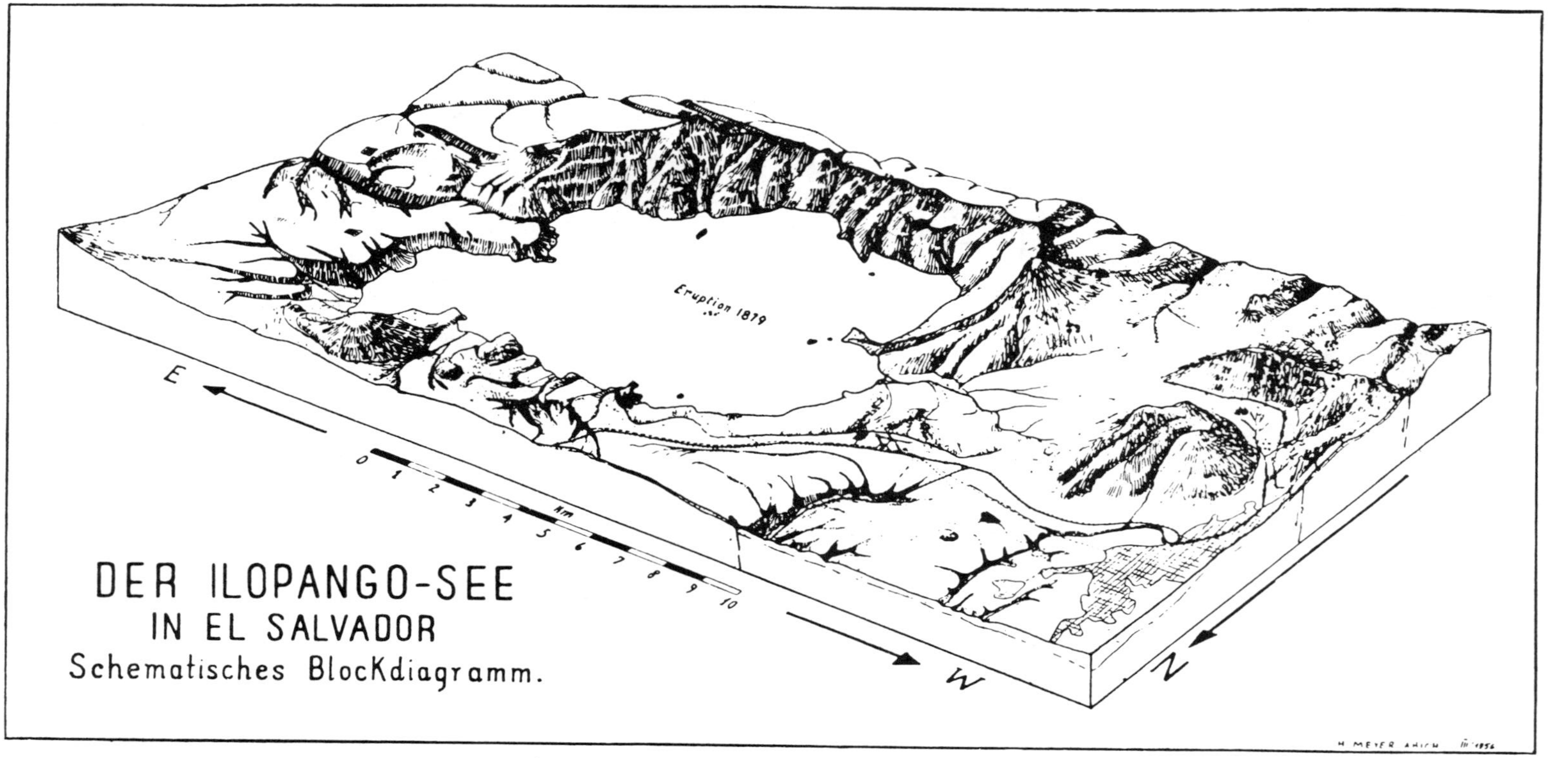

Figure 14.4.2. Perspective drawing of Ilopango Caldera, from Mooser and others (1958).

MASAYA

CAVW number (active vent)	Latitude Longitude (degrees)	Diameter (km)	Local tectonic setting	Precaldera edifice	SiO_2 content (wt pct)	Age of collapse (yr B.P.)	Date of unrest	Type and duration of unrest ESTU STHF MGTF H Te	Eruption type
14-04-10	11.98N	11 x 6	NW-SE-	Shield	R = 51	ca. 6000;	1524	---- ---- ---- - --	unknown
(Masaya)	86.15W		trending		C = 51-53	ca. 2200	1529-70	---- ---- ---- - --	ll
			fissure				1670	---- ---- ---- - --	Ex,lf
							1772-75	AA-- ---- ---- - --	lf,ex
							1852-59	EE-- ---- ---Y - --	ex,lf
							1902-06	D--- ---- ---Y - --	ex
							1913?	x--- ---- ---- - --	unknown
							1919-24	---- ---- ---Z - --	ex
							1946	---- ---- ---Z - --	lf
							1947-59	---- ---- ---F - --	none
							1965	---- ---- ---- - --	ex,lf
							1970-72+	---- ---- ---- - --	ex,ll
							1974-88+	x--- x--- ---Y - --	ll,ex

TECTONIC SETTING

Masaya is immediately southeast of a zone of transverse faulting that was active in 1972 ($\underline{M}_b$ 5.6 Managua earthquake on 23 December), and that is thought to be a boundary between western and eastern Nicaragua lithospheric segments (Stoiber and Carr, 1973). Masaya's magma is tholeiitic basalt, and the complex is in a section of Central America where the crust is relatively thin (Carr, 1984).

GEOLOGIC HISTORY

The Masaya Caldera (fig. 14.5.1) formed as a result of plinian eruptions of tholeiitic basalt (Bice, 1980, 1985; Williams, 1983a, b). Masaya is a basaltic shield volcano situated within the caldera and has several striking, steep-walled craters at its summit.

MASAYA (continued)

HISTORICAL ACTIVITY

Strong earthquakes were felt at 1030 hr on 16 March 1772, and at the same time observers saw fire and flying rocks (Sapper, 1925). After a series of earthquakes from 25 April to 11 May 1858, a small eruption occurred on 10 November 1858 (von Wolff (1929) says 10 September). After an earthquake on 25 June 1902, rumbling continued until 6 July, and on 15 July Santiago Crater began to emit steam and ash. Another eruption, on 3 or 5 January 1906, began after a series of earthquakes from 31 December 1905 through 5 January 1906 (Sapper, 1925; von Wolff, 1929). There is an unconfirmed report of an earthquake and activity at Santiago on 12 July 1913.

Episodes of greatly increased SO_2 emissions have occurred at Masaya about every 25 years since at least 1850. Strong outgassing is known to have occurred during 1852-59, 1902-06, 1919-27, 1947-59, and 1979 through the time of this writing (December 1987) (Williams and Stoiber, 1983; Stoiber and others, 1986; Smithsonian Institution, 1985-88). During each episode, crops (and human health?) in the area suffer considerable damage. Gases emerge from a NW-SE trending fissure system. These gas-emission events seem to occur after periods of eruptions and continue until the crater floor collapses and in most instances blocks off further gas emission. Relatively small collapses of the walls of Santiago's inner crater began in November 1986; by December 1987 and January 1988 a continuation of that activity together with collapse of the floor of adjacent Nindiri crater was merging the two craters into a single, larger crater. Gas emission remained vigorous and new magma may have been intruded beneath at least Santiago crater (Smithsonian Institution, 1987, 1988). Seismic swarms in and near the Masaya Caldera are relatively common (D. Harlow, oral commun., 1983).

COMMENTS

By comparing outgassing of SO_2 with the known concentration of sulfur in Masaya magma, Stoiber and others (1986) estimated that approximately 10 km^3 of basaltic magma had been degassed at Masaya since 1875. Comparison of this estimate with the volume of lava erupted since the Spanish Conquest (1524) (ca. 33 x 10^6 m^3) suggests that >99 percent of Masaya's magma is never erupted. Adding the erupted and intruded products, Stoiber and others (1986) estimated that basaltic magma supply to Masaya has been approximately 0.1 km^3/yr since 1875, a rate comparable to that of Kilauea and Etna.

REFERENCES

Bice, D.C., 1980, Tephra stratigraphy and physical aspects of Recent volcanism near Managua, Nicaragua: Berkeley, Univ. Calif., Ph.D. dissertation, 422 p.

----- 1985, Quaternary volcanic stratigraphy of Managua, Nicaragua: correlation and source assignment for multiple overlapping plinian deposits: Geol. Soc. Am. Bull., v. 96, p. 553-566.

Carr, M.J., 1984, Symmetrical and segmented variation of physical and geochemical characteristics of the Central American volcanic front: Jour. Volcanol. Geotherm. Res., v. 20, p. 231-252.

MASAYA (continued)

REFERENCES (continued)

McBirney, A.R., 1956, The Nicaraguan volcano Masaya and its caldera: Trans. Am. Geophys. Union, v. 37, p. 83-96.

Mooser, F., Meyer-Abich, H., and McBirney, A.R., 1958, Catalogue of active volcanoes of the world including solfatara fields, Part VI, Central America: Naples, IAVCEI, 146 p.

Sapper, K., 1925, Los volcanes de la América Central: Halle, Max Niemeyer, 116 p.

Smithsonian Institution, Scientific Event Alert Network (SEAN), 1985-88, Masaya Volcano: SEAN Bull., v. 10, no. 11, v. 11, no. 11; v. 12, no. 5; v. 13. no. 1 (contributions from D. Fajardo, P. Hradecky, and H. Taleno (INETER), B. van Wyk de Vries, H. Rymer, G. Brown, and others).

Stoiber, R.E., and Carr, M.J., 1973, Quaternary volcanic and tectonic segmentation of Central America: Bull. Volcanol., v. 37, no. 3, p. 304-325.

Stoiber, R.E., Williams, S.N., and Huebert, B.J., 1986, Sulfur and halogen gases at Masaya Caldera Complex, Nicaragua: Total flux and variations with time: Jour. Geophys. Res., v. 91, no. B12, p. 12,215-12,231.

Ui, T., 1972, Recent volcanism in Masaya-Granada area, Nicaragua: Bull. Volcanol., v. 36, p. 174-190.

Williams, S.N., 1983a, Geology and eruption mechanisms of Masaya Caldera complex: Hanover, N.H., Dartmouth Coll., Ph.D. dissertation, 159 p.

----- 1983b, Plinian airfall deposits of basaltic composition: Geology, v. 11, no. 4, p. 211-214.

Williams, S.N., and Stoiber, R.E., 1983, Volcanic hazards at Masaya Caldera complex, Nicaragua [abs.]: Abstracts, 18th Gen. Assembly, IUGG, Hamburg, v. 1, p. 359.

Wolff, F. von, 1929, Der Vulkanismus: Stuttgart, F. Enke, v. 2, p. 449.

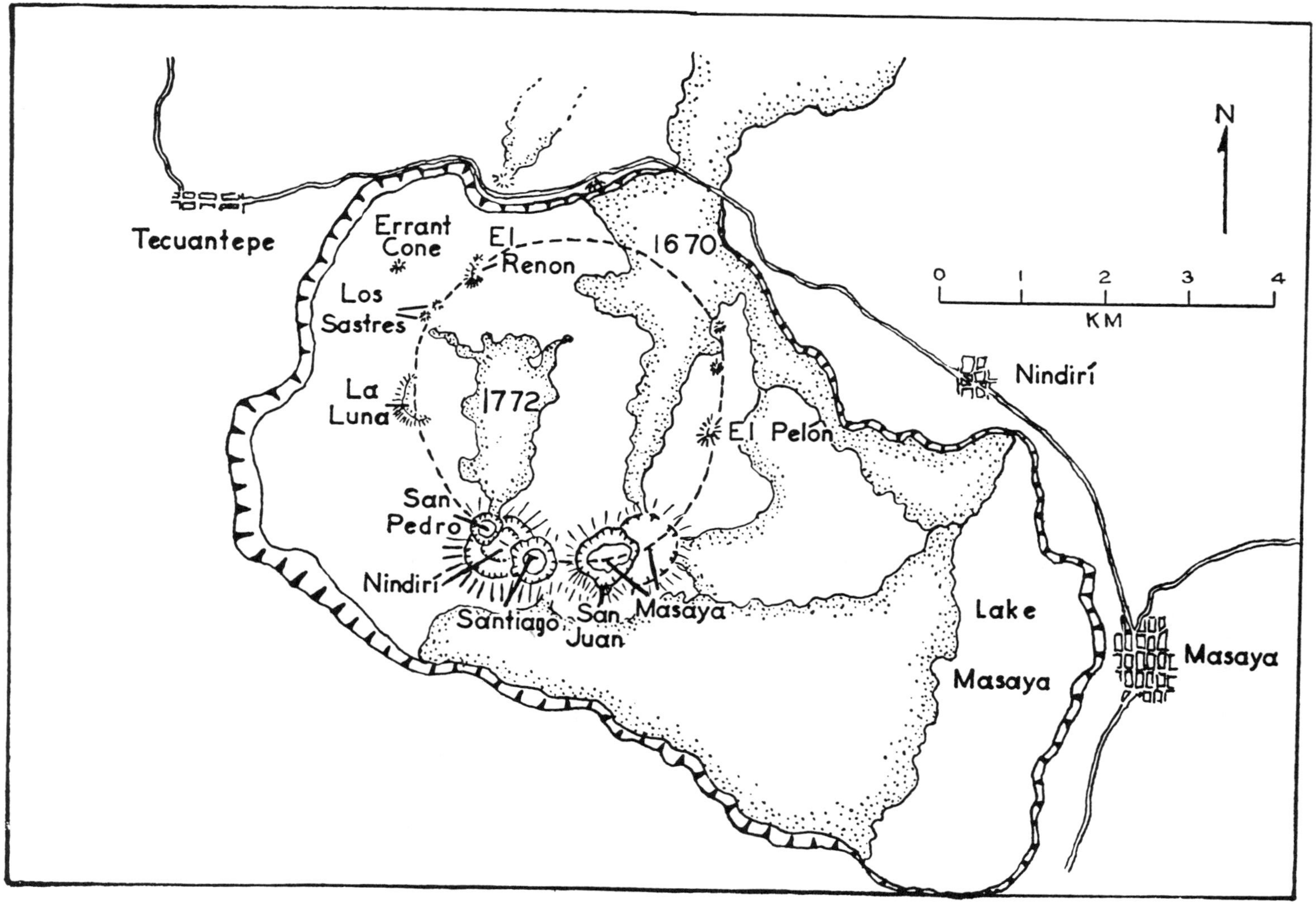

Figure 14.5.1. Sketch map of Masaya Caldera, from Mooser and others (1958). Dashed line shows circular pattern of young postcaldera vents.

APOYO

CAVW number (active vent)	Latitude Longitude (degrees)	Diameter (km)	Local tectonic setting	Precaldera edifice	SiO_2 content (wt pct)	Age of collapse (yr B.P.)	Date of unrest	Type and duration of unrest ESTU STHF MGTF H Te	Eruption type
14-04-101 (Apoyo)	11.92N 86.03W	6.5	Exten	Shield	R = a	23,000	1974	DD-- ---- ---- - QU?	none
					C = 66	21,000	1977	DD-- ---- ---- - --	none
							1978	DD-- ---- ---- - --	none
							1979	DD-- ---- ---- - --	none
							1981	DD-- ---- ---- - --	none
							1986	DD-- ---- ---- - QU?	none

TECTONIC SETTING

Apoyo Caldera (figs. 14.6.1, 14.6.2) lies within the NW-SE Nicaraguan Depression, at its intersection with a N-S graben (Sussman, 1985). Several of the N-S graben faults pass through and just east of Apoyo Caldera, between the caldera and the town of Granada.

GEOLOGIC HISTORY

Basaltic and andesitic lava flows built a broad shield during the Pleistocene, and dacitic domes intruded that pile, one approximately 90,000 yr B.P. Shortly before collapse, tholeiitic basalt lavas and scoria were erupted along N-S fractures cutting the shield and a pyroclastic plateau to the east of Apoyo. From about 23,000 to 21,000 yr B.P., about 11 km^3 of dacite magma was erupted in two Plinian eruptions, resulting in caldera collapse. A small volume of postcaldera andesitic lava was extruded along the ring faults, and the Granada cinder cones grew along a fissure about 2 km east of Apoyo (Walker, 1981, 1982; Sussman, 1985).

HISTORICAL ACTIVITY

Based on a catalog of Nicaraguan earthquakes (D.J. Leeds, cited in Martinez H., 1976), seismic activity occurred in the vicinity of Apoyo Caldera in 1529, 1536, 1570, 1772, 1856-59, 1870, and 1973. Some or even most of this seismicity was probably of regional tectonic origin, in the broader Masaya-Apoyo area. On 23 December 1972, three damaging tectonic earthquakes (largest, $\underline{M}_b$=5.6) struck Managua, 40 km northwest of Apoyo (Brown and others, 1973); no aftershocks occurred in the Apoyo area. However, seismic swarms at Apoyo proper began in or before September 1974 and continued intermittently at least until 1986. At least 100 earthquakes occurred in early September 1974, of which 27 were felt and had $\underline{M}$=3.7 to 4.0; 5 others had similar magnitude but were not felt (Martinez H., 1976).

APOYO (continued)

HISTORICAL ACTIVITY (continued)

Continuous instrumental monitoring began in 1977, and new earthquake swarms were recorded in 1977, 1978, 1979, and 1981 (Shepherd, 1986). Epicentral locations were clustered southeast of the caldera and were probably associated with movement along one of the N-S graben faults noted above (fig. 14.6.3).

A $\underline{M}$=6 earthquake occurred on 16 December 1985, several tens of kilometers south of Apoyo (Shepherd, 1986). Aftershocks were felt in the Apoyo area for a short time, followed by quiet. However, a new local earthquake swarm began in the first few days of January 1986, with many felt shocks and earthquake sounds. Activity may have peaked on 9-10 January, and the Apoyo area returned to relative quiet by mid-February. Felt reports supplemented by locations of some of the later earthquakes in the swarm suggested that the swarm was centered along the earlier-mentioned active fault about 3 km east of the caldera (near the point labeled E in fig. 14.6.2). A UNESCO/UNDP mission to Nicaragua concluded that the earthquakes were of tectonic origin, like those of swarms of the previous decade. A volcanic eruption was deemed not imminent (Shepherd, 1986). The relation of the 1972 Managua earthquake and the December 1985 earthquake to subsequent unrest at Apoyo is not known.

During or before earthquakes in September and October 1974, the temperature of Laguna de Apoyo rose from an unknown value (a warm spring at the lake's west-northwest edge had a temperature of 30.5 °C in 1925) to more than 35 °C. During October and early November 1974, a period between earthquake swarms, the lake temperature dropped by nearly 10 °C, before rising to a stable value of about 32-33 °C that was maintained for at least several months thereafter. The SO_4 content of lake water rose from 19 ppm (1969 baseline) to 180 ppm in September 1974, declined over the October - early November period of decreasing lake temperature, and recovered somewhat thereafter (fig. 14.6.4). The concurrence of temperature and chemical changes with seismicity led Martinez H. (1976) to conclude that the unrest at Apoyo was at least partly of volcanic origin.

As of January 1988, the lake's surface temperature ranged from 28.2-29.1 °C (Smithsonian Institution, 1988).

REFERENCES

Bice, D.C., 1985, Quaternary volcanic stratigraphy of Managua, Nicaragua: correlation and source assignment for multiple overlapping plinian deposits: Geol. Soc. Am. Bull., v. 96, p. 553-566.

Brown, R.D., Jr., Ward, P.L., and Plafker, G., 1973, Geologic and seismologic aspects of the Managua, Nicaragua, earthquakes of December 23, 1972: U.S. Geol. Surv. Prof. Paper 838, 34 p.

Martinez H., M.A., 1976, Riesgos volcanicos en Nicaragua y resultados preliminares sobre una metodologia experimental en la predicción de sismos en la Laguna de Apoyo: Rev. Geofis., no. 5, Mexico, p. 73-82.

Mooser, F., Meyer-Abich, H., and McBirney, A.R., 1958, Catalogue of active volcanoes of the world including solfatara fields, Part VI, Central America: Naples, IAVCEI, 146 p.

APOYO (continued)

REFERENCES (continued)

Sapper, K., 1925, Los volcanes de la América Central: Halle, Max Niemeyer, 116 p.

Shepherd, J.B., 1986, Report on a mission to Nicaragua, Feb 9 to Feb 22, 1986: unpubl. rep. to United Nations Development Programme (UNDP) and United Nations Educational, Scientific, and Cultural Organization (UNESCO), 13 p.

Smithsonian Institution, Scientific Event Alert Network (SEAN), 1988, Apoyo Volcano: SEAN Bull., v. 13, no. 1, p. 12 (information from B. van Wyk de Vries and others).

Sussman, D., 1985, Apoyo Caldera, Nicaragua: a major Quaternary silicic eruptive center: Jour. Volcanol. Geotherm. Res., v. 24, p. 249-282.

Ui, T., 1972, Recent volcanism in Masaya-Granada area, Nicaragua: Bull. Volcanol., v. 36, p. 174-190.

Walker, J.A., 1981, Petrogenesis of lavas from cinder cone fields behind the volcanic front of Central America: Jour. Geol., v. 87, p. 721-739.

----- 1982, Volcanic rocks from the Nejapa and Granada cinder cone alignments, Nicaragua, Central America: New Brunswick, N.J., Rutgers Univ., Ph.D. thesis, 143 p.

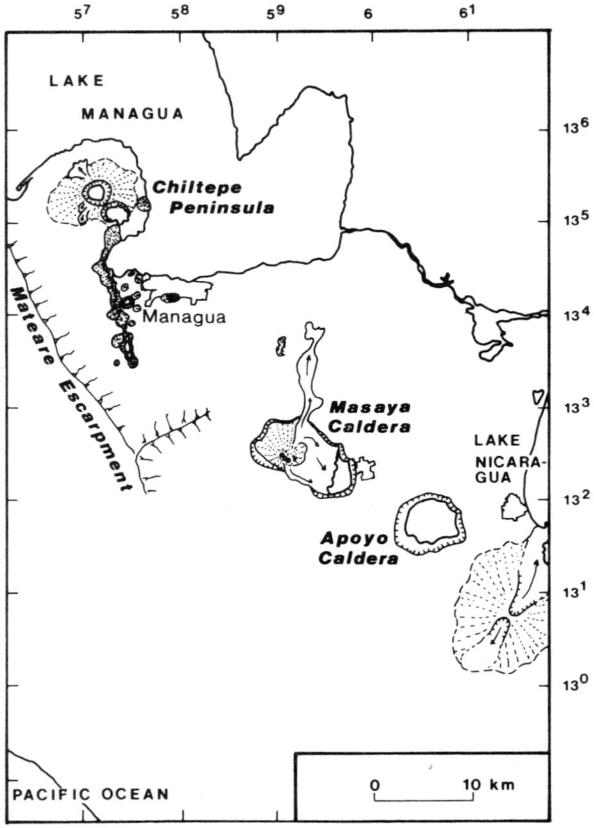

Figure 14.6.1. Principal geographic and volcanic features in central western Nicaragua, including Apoyo Caldera, from Bice (1985).

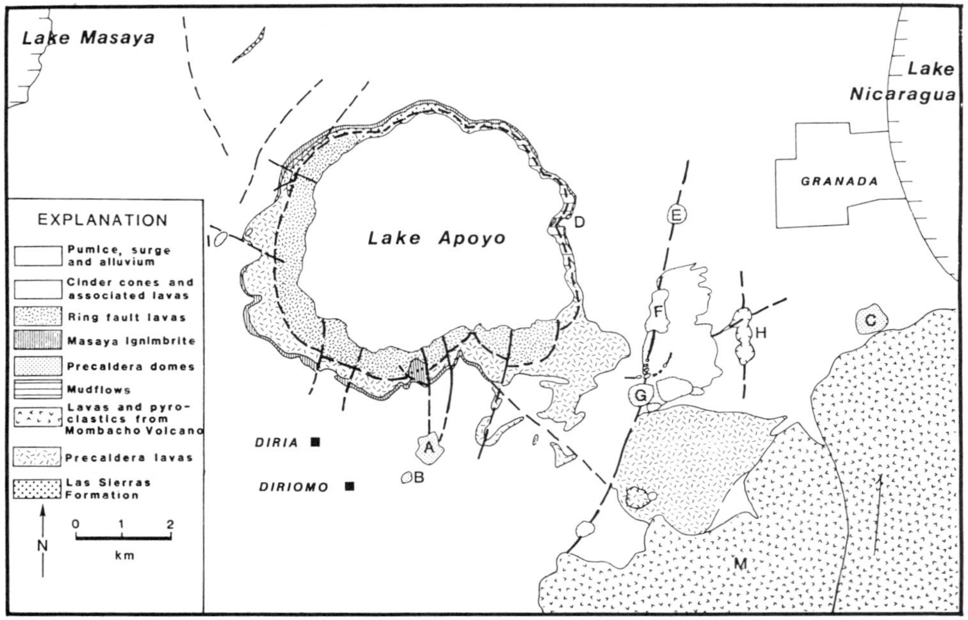

Figure 14.6.2. Simplified geologic map of Apoyo Caldera, from Sussman (1985). Heavy dashed lines denote faults. Patterns for units shown in inset; no pattern, undifferentiated Apoyo pyroclastic deposits (including airfall, ignimbrite, and pyroclastic surge), alluvium, and Masaya surge deposits. Letters refer to names of eruptive centers, as follows: A, El Cerrito dome; B, Cerritito dome; C, Lomo Poisentepe dome; D, Apoyoito dome; E, Granadito cinder cone; F, Cerro el Chanal cinder cone.

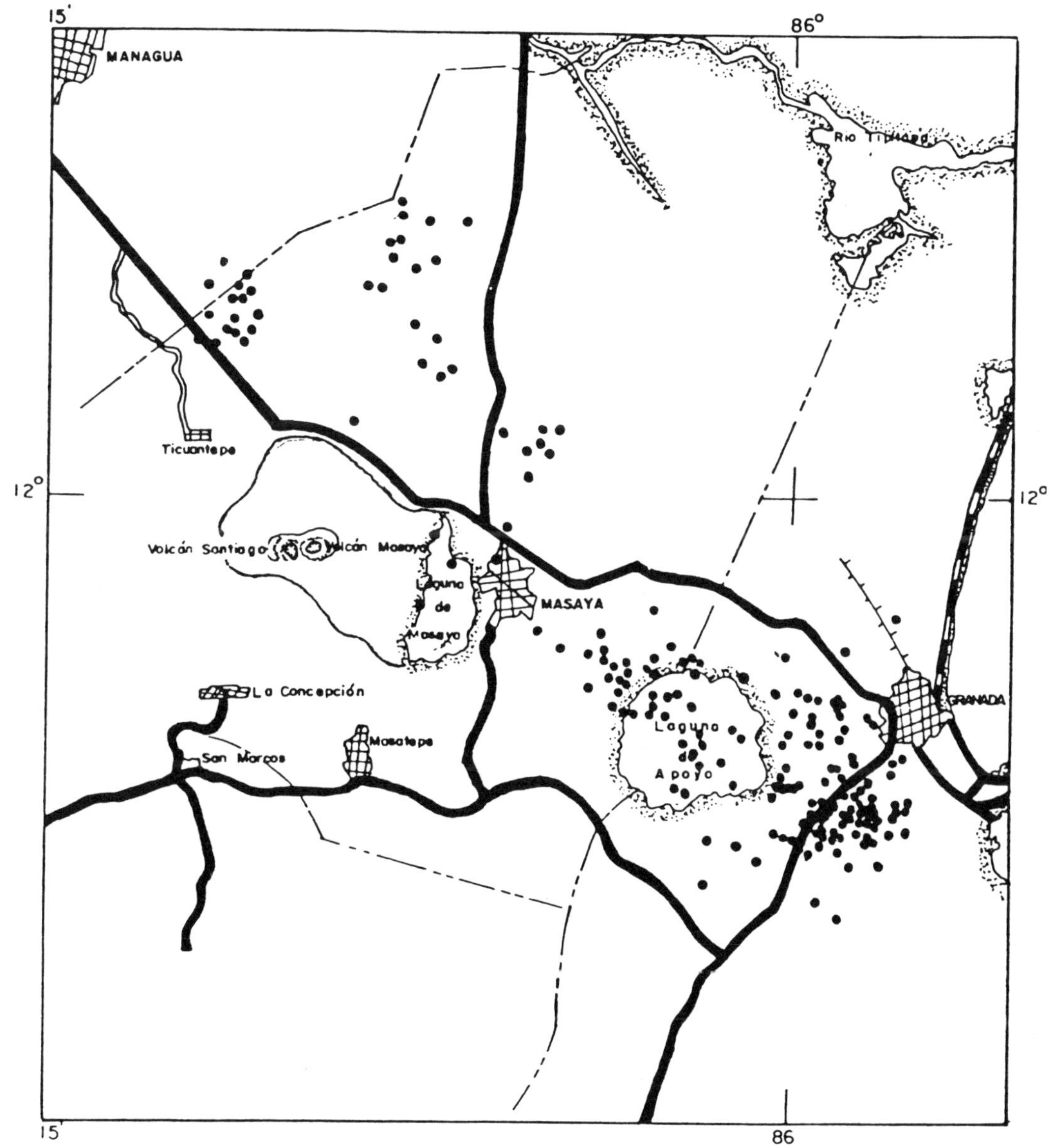

Figure 14.6.3. Location of earthquake epicenters near Apoyo Caldera, 1977 to 1981, after F. Seguro (unpubl. rep., 1982).

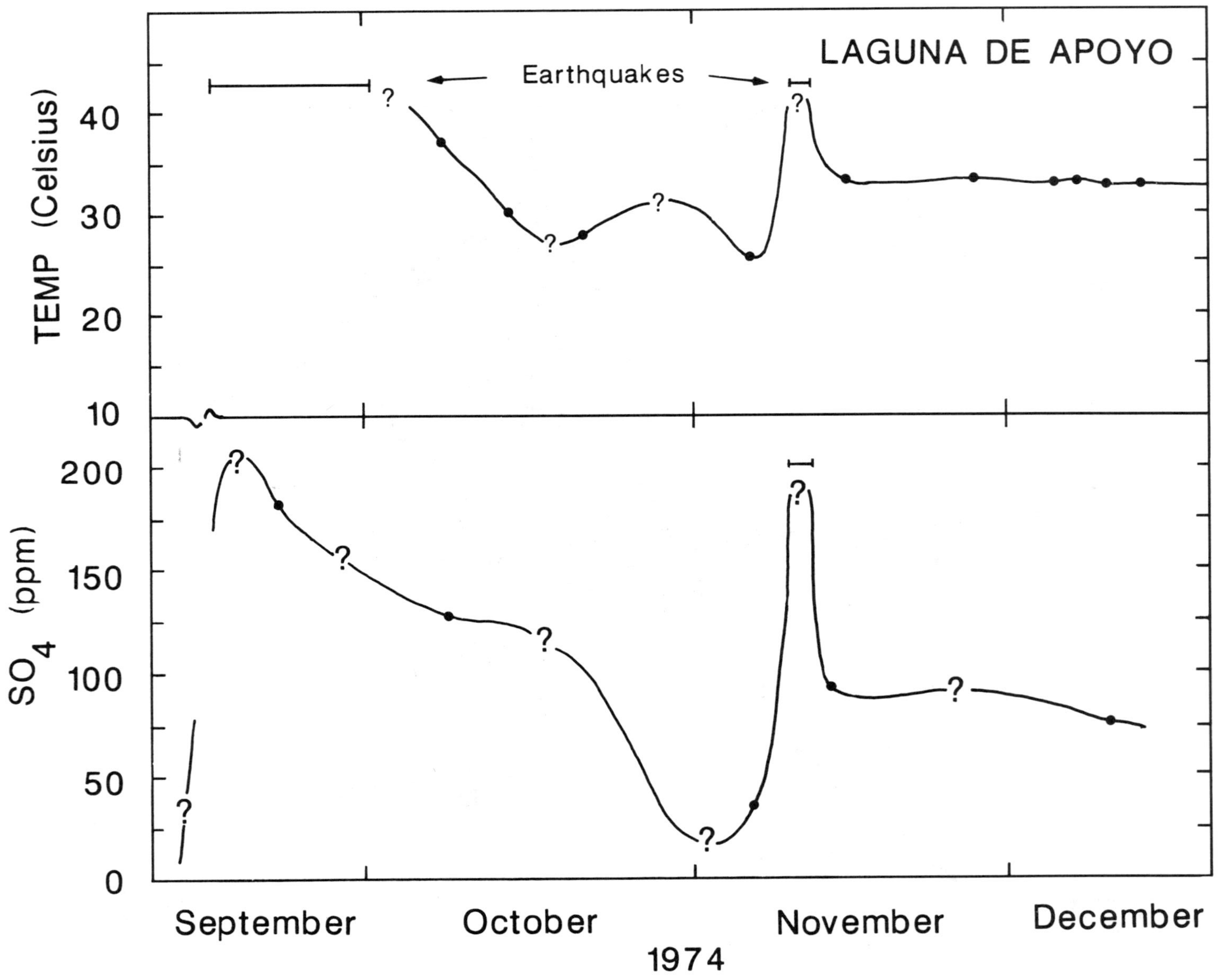

Figure 14.6.4. Temperature and SO_4 content at a sampling site at Lake Apoyo during September-December 1974, from Martinez H. (1976).

RINCON DE LA VIEJA

CAVW number (active vent)	Latitude Longitude (degrees)	Diameter (km)	Local tectonic setting	Precaldera edifice	SiO_2 content (wt pct)	Age of collapse (yr B.P.)	Date of unrest	Type and duration of unrest ESTU STHF MGTF H Te	Eruption type
14-05-02 (Rincón de la Vieja)	10.83N 85.33W	15-20 (outer)	Compr	Strato	R = 57-59		1860-63	---- ---- ---- - --	ex
					C = ?		1922	---- ---- ---C - --	ex
							1963	---- ---- ---x - --	none
		5 (inner)					1965-70	xx-- ---- ---E - --	pex, Ex
							1983-87+	---- ---- ---- - --	p-ex

TECTONIC SETTING

Rincón de la Vieja is situated along the volcanic front of Central America, at or just southeast of a postulated boundary between eastern Nicaragua and Costa Rican segments of the Cocos-Caribbean subduction zone (Stoiber and Carr, 1973).

GEOLOGIC HISTORY

Rincón de la Vieja, the largest volcano in northern Costa Rica, consists of six volcanic centers that form an elongate ridge, built on a shield of ignimbrites (fig. 14.7.1). Healy (1969) outlined a large, 15-20 km diameter caldera within which the present ridge-shaped volcano has grown. Carr and others (1986) concurred with this interpretation and outlined another nested, 5-km-diameter caldera on the Rincón de la Vieja edifice (fig. 14.7.1). A Rincón de la Vieja Caldera is not included in Barquero and Sáenz (1987).

HISTORICAL ACTIVITY

Most historical eruptions of Rincón de la Vieja have been vulcanian, Strombolian, or phreatic explosions from the central crater of the complex (Active Crater, see fig. 14.7.1).

1922: An observer (Don J. Fidel Tristán) found the volcano "active" on 11 April 1922 and with signs of a recent, grand eruption (Sapper, 1925). On 3 June of the same year, Tristán found strong SO_2 emission and a terrible noise from fumaroles; a large eruption occurred on 4 June.

1963: Strong steaming and sulfur gas emission were reported (Barquero, 1964).

RINCON DE LA VIEJA (continued)

HISTORICAL ACTIVITY (continued)

1965-70: Vapor "eruptions" in 1965 and September 1966 were followed by a moderately large ash eruption on 19 December 1966. On 23 February 1967, 28 eruptions occurred within 30 minutes, and nearby villages were evacuated (Thorpe and others, 1985). Bombs that were probably from this eruption have 57-59 percent silica (Carr and others, 1986). Small ash eruptions occurred until 1970.

1983-87+: Small phreatic explosions occurred in 1983 and 1984. Eruptions on 6 and 21 February 1983 produced one or more directed steam blasts that impacted areas up to 2 km south and east of the crater. An eruption on 31 March 1984 produced some blocks that might have been juvenile (Thorpe and others, 1985). Additional eruptions probably occurred between September and November 1985, judging from increased seismicity during that time. A small explosion occurred on 31 December 1986, and yet another on 1 April 1987 (Barquero and Segura, 1983; Thorpe and others, 1985; Smithsonian Institution, 1987).

REFERENCES

Barquero, J., 1964, El volcán Rincón de la Vieja: Inst. Geogr. Nac., Informe Semestral, San Jose, v. 10, p. 11-16.

Barquero H., J., and Sáenz R., R., 1987, Aparatos volcanicos de Costa Rica: Observatorio Vulcanologico y Sismologico de Costa Rica, Universidad Nacional, Heredia, 1 map sheet, scale 1:750,000.

Barquero, J., and Segura, J., 1983, La actividad del volcán Rincón de la Vieja: Bol. Vulcanol., no. 13, p. 5-10.

Carr, M.J., Chesner, C.A., and Gemmell, J.B., 1986, New analyses of lavas and bombs from Rincón de la Vieja volcano, Costa Rica: Bol. Vulcanol., no. 16, p. 23-30.

Healy, J., 1969, Notas sobre los volcanes de la sierra volcanica de Guanacaste, Costa Rica: Inst. Geogr. Nac., Informe Semestral, v. 15, p. 37-47.

Mooser, F., Meyer-Abich, H., and McBirney, A.R., 1958, Catalogue of active volcanoes of the world including solfatara fields, Part VI, Central America: Naples, IAVCEI, 146 p.

Sapper, K., 1925, Los volcanes de la América Central: Halle, Max Niemeyer, 116 p.

Smithsonian Institution, Scientific Event Alert Network (SEAN), 1987, Rincón de la Vieja Volcano: SEAN Bull., v. 12, nos. 2, 4 (information from Jorge Barquero H.).

Stoiber, R.E., and Carr, M.J., 1973, Quaternary volcanic and tectonic segmentation of Central America: Bull. Volcanol., v. 37, no. 3, p. 304-325.

Thorpe, R., Brown, G., Rymer, H., Barritt, S., and Randal, M., 1985, Recent volcano monitoring in Costa Rica: Earthq. Info. Bull., v. 17, no. 2, p. 44-49.

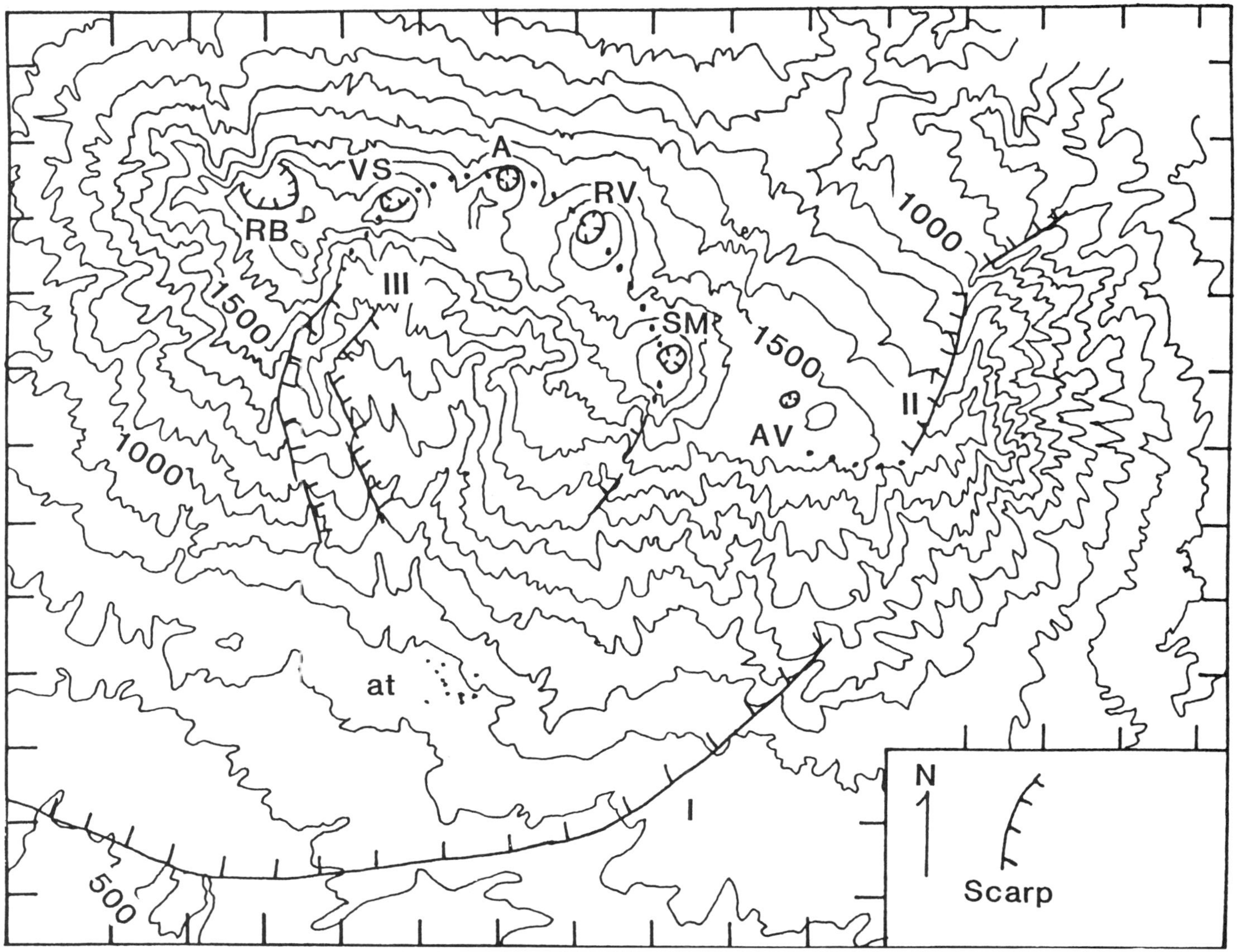

Figure 14.7.1. Topographic map showing outline of a large caldera (I and II) in which Rincón de la Vieja Volcano has grown and a nested caldera (III) on volcanic edifice (Carr and others, 1986). Vents: RB, deeply eroded cone drained by Rio Blanco; VS, Crater von Seebach; A, Active Crater; RV, Rincón de la Vieja; SM, Santa Maria; AV, double cone drained by Rio Aguas Verdes; at, prominent hot-spring area. Spacing between ticks at margins of map is 1 km.

PART 3: HISTORICAL UNREST AT LARGE QUATERNARY CALDERAS (continued)
See inside back cover for explanation and abbreviations

MIRAVALLES

CAVW number (active vent)	Latitude Longitude (degrees)	Diameter (km)	Local tectonic setting	Precaldera edifice	SiO_2 content (wt pct)	Age of collapse (yr B.P.)	Date of unrest	Type and duration of unrest ESTU STHF MGTF H Te	Eruption type
14-05-03	10.75N 85.15W	15 x 20	Compr; local extension, graben	Strato	R = 50-56 C = r	500,000 ±150,000	1946	xx-- ---- --?x - --	"steam eruption"

TECTONIC SETTING

Miravalles is situated along the volcanic front of Central America, just southeast of a postulated boundary between eastern Nicaragua and Costa Rican segments of the Cocos-Caribbean subduction zone (Stoiber and Carr, 1973).

GEOLOGIC HISTORY

Miravalles Caldera formed approximately 0.5 ± 0.15 m.y. ago (Tournon, 1984), after eruption of voluminous silicic ash flows. An andesitic stratovolcano of the same name grew within the caldera (Healy, 1969). At least five cones are recognizable in the summit area, distributed along a northeast-trending line (probable fault). The modern intracaldera cone may be the second of two; a previous postcaldera cone was destroyed (R. Fournier, oral commun., 1987). Melson and others (1986) note a preliminary age of 8,000(?) yr B.P. for a major plinian eruption from Miravalles, and an age of 7,000(?) yr B.P. for a large debris avalanche from Miravalles. A geothermal field is situated between the caldera scarp and Miravalles stratovolcano (fig. 14.8.1).

HISTORICAL ACTIVITY

No significant eruption has occurred at Miravalles in historical time. However, on 14 September 1946, a small steam eruption occurred high on the south flank of the volcano, near the summit crater. Weak earthquakes were felt at the time. An explosion crater 20 m in diameter formed, and steam, mud, and ash were thrown 100 m high (Hantke, 1951, citing a letter from Sigismund von Preussen of the town of La Barranca). The hydrothermal field is a hot-water-dominated system, with a vapor-dominated cap (few tens of meters thick) in some parts of the geothermal field. Changes in the vapor-dominated cap can give rise to surface hydrologic changes and small blowouts. Fumarolic activity has reportedly decreased in recent years (as of late 1987) (R. Fournier, oral commun., 1987).

MIRAVALLES (continued)

REFERENCES

Barquero H., J., and Sáenz R., R., 1987, Aparatos volcanicos de Costa Rica: Observatorio Vulcanologico y Sismologico de Costa Rica, Universidad Nacional, Heredia, 1 map sheet, scale 1:750,000.

Hantke, G., 1951, Uebersicht über die vulkanische tätigkeit, 1941-1947: Bull. Volcanol., v. 11, p. 161-208.

Healy, J., 1969, Notas sobre los volcanes de la sierra volcanica de Guanacaste, Costa Rica: Inst. Geogr. Nac., Informe Semestral, v. 15, p. 37-47.

Melson, W.G., Barquero H., J., Sáenz R., R., and Fernandez, E., 1986, Erupciónes explosivas de importancia en volcanes de Costa Rica: Bol. Vulcanol. (Costa Rica), no. 16, p. 15-19.

Stoiber, R.E., and Carr, M.J., 1973, Quaternary volcanic and tectonic segmentation of Central America: Bull. Volcanol., v. 37, no. 3, p. 304-325.

Tournon, J., 1984, Magmatismes du Mesozoique a l'actuale en Amerique Centrale: L'example de Costa Rica, des ophiolites aux andesites: Paris, Université Pierre et Marie Curie, D.Sc. thesis.

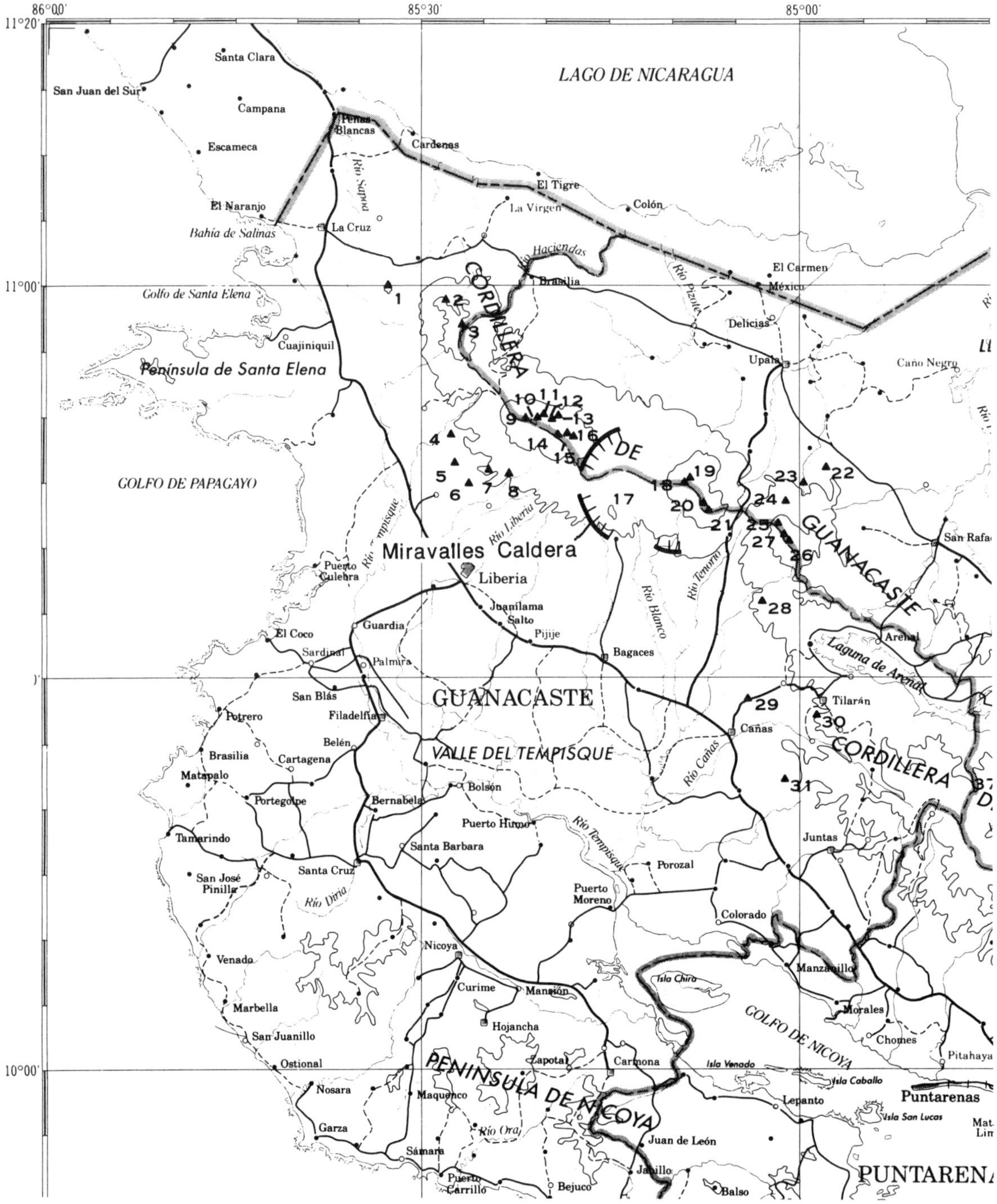

Figure 14.8.1. Miravalles Caldera and nearby volcanoes, from Barquero and Sáenz (1987). The postulated Rincón de la Vieja Caldera is located 5-10 km west-northwest of Miravalles Caldera. 1, El Hacha; 2, Orosi; 3, Cacao; 4, Fortuna; 5, Sn. Roque; 6, Cañas Dulces; 7, Cerro Gorgona; 8, Sn. Vicente; 9, Braun; 10, Von Seebach; 11, Rincón I; 12, Rincón II; 13, Rincón III; 14, Sta. María I; 15, Sta. María II; 16, Sta. María III; 17, Miravalles Caldera; 18, Miravalles I; 19, Miravalles II; 20, Miravalles III; 21, Miravalles IV; 22, Olla de Carne; 23, Bijagua I; 24, Bijagua II; 25, Montezuma; 26, Tenorio I; 27, Tenorio II; 28, C. Barrera; 29, C. Annunciación; 30, C. Tilarán; 31, C. Pelado y Delicias.

POAS

CAVW number (active vent)	Latitude Longitude (degrees)	Diameter (km)	Local tectonic setting	Precaldera edifice	SiO_2 content (wt pct)	Age of collapse (yr B.P.)	Date of unrest	Type and duration of unrest ESTU STHF MGTF H Te	Eruption type
14-05-04	10.20N	7 x 10	Compr	Strato	R = 50-63	<40,000;	1826-28	x--- ---- ---? - --	Pex,Ex?
(Poás)	84.22W	(outer)			C = 52-54	>7,540	1834	---- ---- ---- - --	Ex
		3					1838	---- ---- ---- - --	ex
		(inner)					1880	---- ---- ---- - --	Pex,Ex?
							1884	---- ---- ---- - --	pex
							1888-89	x--- ---- --xx x UQ/QU?	pex,geyser
							1899	---- ---- ---- x --	geyser
							1902-05	---- ---- ---- - --	geyser
							ca. 1907-08	---- ---- ---- - --	pex
							1910	x--- ---- ---- - UQ	Pex,Ex
							1912	?--- ---- ---- - QU	landsl
							1914-15	---- ---- ---- - --	pex,geyser
							1925	---- ---- ---- - --	pex
							1929?	---- ---- ---- - --	geyser?
							1946	---- ---- ---- - --	pex
							1952-57	B--- ---- ---- - --	geyser,pex,Ex
							1961	---- ---- ---- - --	pex
							1963-64	---- ---- ---- - --	pex, ex
							1967	---- ---- ---- - --	pex
							1969	---- ---- ---- - --	pex
							1972-73	---- ---- ---- - --	ex
							1974-75	---- ---- ---- - --	pex
							1976-80	---- ---- ---- - --	pex, ex
							1980-86	---- ---- --FF - --	none
							1987-88	x--- ---E --EE - --	pex

TECTONIC SETTING

Poás is situated near the southeastern end of the Central American arc, above the Cocos-Caribbean subduction zone. Cinder cones are strongly aligned in a N-S direction.

POAS (continued)

GEOLOGIC HISTORY

Two calderas, one nested within the other, developed in the summit area of Poás (fig. 14.9.1) (Thorpe and others, 1981). The younger, smaller collapse occurred less than 40,000 yr B.P. but more than 7,540 yr ago (a ^{14}C date on the postcaldera Botos cone) (Prosser, 1983, 1985; Prosser and Carr, 1987). The modern cone of Poás has grown within and nearly obscured the two calderas; only the flat-topped morphology of Poás belies its calderas. The summit of the modern cone has a main crater that may contain a sulfur-rich lake with a layer of molten sulfur at its base (Bennett and Raccichini, 1978; Francis and others, 1980).

The magma erupted to form the younger caldera has been tentatively identified as basaltic andesite (Prosser and Carr, 1987); magma associated with the earlier caldera-forming eruption is not known. A mafic composition is consistent with the lack of a major gravity low in the area of the caldera and the existence of a local gravity high at Poas's summit. Tournon (1984) presented analyses of basalt, andesite, and low-silica dacite--samples that may span the time(s) of caldera formation.

HISTORICAL ACTIVITY

Most historical eruptions of Poás have been small, phreatic or magmatic, sometimes including molten sulfur. In the 19th century, eruptions of 1828, 1834, and 1880 were relatively large, showering nearby towns with ash (Sapper, 1925). In the 20th century, relatively large eruptions occurred in 1910 and 1953. The eruption of 1910 began as a large geyser eruption on 25 January, when a plume rose to more than 8-km altitude (Sapper, 1925). The entire upper part of the volcano was covered with a thick layer of mud. A period of few observations followed, but by March at least 20 areas of small mud eruptions could be seen in the crater lake. On 13 April, strong earthquakes occurred (at least 23 from 0037 hr to 0815 hr), followed by quiet until additional earthquakes occurred at 1311 hr, 1500 hr, and 2200 hr. Twenty-four earthquakes were reported on 14 April, and 12 on 15 April. Then an especially strong earthquake struck on 4 May, causing severe damage in San Jose and Cartago. During the night of 13-14 May, some observers reported seeing flames from Irazu or Poás (in opposite directions from San Jose!). Two years later, on 6 June 1912, a strong earthquake occurred at shallow depth beneath an old crater, Laguna Vieja, 5 km west of Poás (Sapper, 1925).

During the 1953 eruption, phreatic explosions started at two sublacustrine vents, and the eruption gradually evolved from phreatomagmatic to Strombolian as the crater lake disappeared. The tallest eruption column reached 7 km above the crater and spread ash at least 50 km from the volcano.

An eruption in 1976 sent a column 3 km above the vent; activity in 1978-80 was limited to geyser eruptions. Fumarole temperatures then rose dramatically between September 1980 and March 1981--from about 90 °C to 960 °C (fig. 14.9.2) (Barquero, 1983). Glowing spots were first noted on the small intracrater cone in January 1981. Temperatures dropped only slightly from March 1981 to October 1983 (801 °C), but then cooled to 723 °C by December 1983, to 515 °C by December 1984, to 287 °C by December 1985, and to approximately 300 °C through early 1986 (Barquero and others, 1985; Fernandez and others, 1986; Barquero and Fernandez, 1986). The temperature of another fumarole, near the top of a cone inside the crater, was 530 °C in early 1986, rose to 680 °C by July 1986, declined gradually to 540 °C in January-May 1987, rose to 590 °C before the start of small phreatic eruptions in July 1987, and then

POAS (continued)

HISTORICAL ACTIVITY (continued)

declined to 470 °C or lower by February 1988 (the time of this writing). Small phreatic eruptions continued despite the decrease in temperature (Barquero and Fernandez, 1986; J. Barquero and R. Sáenz, written commun., 1987; Smithsonian Institution, 1988) (fig. 14.9.2).

The crater lake grew hotter in parallel with the subaerial fumarole temperatures, reaching 51 °C in June 1981 and 60 °C in early 1983 (some variability related to seasonal fluctuation in rainfall). Then the lake, too, cooled. Its temperature dropped to 45.5 °C by the end of 1985 and 38 °C in the first half of 1986, but it then rose again to 56 °C by December 1986 (Barquero and Fernandez, 1987). The maximum recorded lake temperature in January 1988 was 61 °C (Smithsonian Institution, 1988). The pH of the lake increased between May and September 1983 from 0.06 to 0.3; by early 1986 the pH of the lake was 0.5 and remained at that level through December 1986 (Barquero and Fernandez, 1986, 1987). SO_2 emission during the period of high fumarole temperatures was about 600 t/day (R.E. Stoiber, cited in Prosser, 1985).

Guendel (1981) showed an inverse correlation between the number of countable seismic events per day (20-140) and the duration of 3-4 Hz harmonic tremor (10^4 to 10^2 sec/day) during March-May 1981. He attributed that relation to rock fracturing (A- and B-type seismic events) under high gas pressure, followed by tremor as magma flows upward and degassing occurs. To better monitor these events, a permanent seismograph was installed in 1984.

Rymer and Brown (1984, 1987) noted fluctuation of gravity values on Poás, with a periodicity of about 30 days. Values changed by about 140 microgals at the crater rim and by 50 microgals on the south flank of Poás. Rymer and Brown interpret these variations as indicating movement of the magma column relative to the edifice or slight vesiculation of the magma--possibly tidally induced.

REFERENCES

Barquero, J., 1983, Termometria de la fumarola del Volcán Poas: Bol. Vulcanol., no. 13, p. 11-12.
Barquero H., J., and Fernandez, E., 1986, Estado de los volcanes de Costa Rica, Enero-Mayo 1986: Bol. Vulcanol., no. 17, p. 5-6.
----- 1987, Estado de los volcanes en Costa Rica, Junio-Diciembre 1986: Bol. Vulcanol., no. 18, p. 5-6.
Barquero H., J., Fernandez, E., and Van der Laat, R., 1985, Estado de los volcanes, Mayo 1983-Diciembre 1984: Bol. Volcanol., no. 15, p. 5-9.
Barquero, J., and Malavassi, E., 1984, Excursion al Volcán Poás, <u>in</u> Seminario vulcanologico conjunto USA-CR: Bol. Vulcanol., no. 14, p. 117-131.
Barquero H., J., Malavassi, E., and Araya, A., 1981, La actividad del volcán Poás en los años 1978-1980: Sumario de tres años del formulario de reporte diario de actividad: Bol. Vulcanol., no. 11, p. 10-12.
Bennett, F.D., and Raccichini, S.M., 1978, Subaqueous sulphur lake in Volcán Poás: Nature, v. 271, p. 342-344.
Boza, M.A., and Mendoza, R., 1981, The National Parks of Costa Rica: San Jose, Costa Rican Inst. of Tourism, 310 p.
Fernandez, E., Barquero, J., and Segura, J., 1986, Estado de los volcanes, Enero-Diciembre 1985: Bol. Vulcanol., no. 16, p. 5-14.

POAS (continued)

REFERENCES (continued)

Francis, P.W., Thorpe, R.S., Brown, G.C., and Glasscock, J., 1980, Pyroclastic sulfur eruption at Poás Volcano, Costa Rica: Nature, v. 283, p. 754-756.

Guendel, F., 1981, Estudio preliminar de las señales sismicas del Volcán Poás I.: Bol. Vulcanol., no. 11, p. 5-9.

Krushensky, R., and Escalante, G., 1967, Activity of Irazú and Poás Volcanoes, Costa Rica, November 1964-July 1965: Bull. Volcanol., v. 31, p. 75-94.

Locke, C.A., Brown, G.C., Thorpe, R.S., Cassidy, J., and Francis, P.W., 1985, Geophysical studies of Poas volcano: Bol. Vulcanol., no. 13, p. 89-94.

Mooser, F., Meyer-Abich, H., and McBirney, A.R., 1958, Catalogue of active volcanoes of the world including solfatara fields, Part VI, Central America: Naples, IAVCEI, 146 p.

Pittier, H., 1910, Costa Rica--Vulcan's Smith: Natl. Geogr., v. 21, p. 494-525.

Prosser, J.T., 1983, The geology of Poás Volcano, Costa Rica: Hanover, N.H., Dartmouth Coll., M.A. thesis, 165 p.

----- 1985, Geology and medium-term temporal magmatic variation found at the summit region of Poás Volcano, Costa Rica: Bol. Vulcanol., no. 15, p. 21-39.

Prosser, J.T., and Carr, M.J., 1987, Poás Volcano, Costa Rica: Geology of the summit region and spatial and temporal variations among the most recent lavas: Jour. Volcanol. Geotherm. Res., v. 33, p. 131-146.

Rymer, H., and Brown, G.C., 1984, Periodic gravity changes at Poás volcano, Costa Rica: Nature, v. 311, p. 243-245.

----- 1987, Causes of microgravity change at Poas volcano, Costa Rica: an active but non-erupting system: Bull. Volcanol., v. 49, p. 389-398.

Sáenz, R., 1982, Edades radiometricas de algunas rocas en Costa Rica: Bol. Vulcanol., no. 12, p. 8-10.

Sapper, K., 1925, Los volcanes de la América Central: Halle, Max Niemeyer, 116 p.

Smithsonian Institution, Scientific Event Alert Network (SEAN), 1987, 1988, Poas Volcano: SEAN Bull., v. 12, nos. 8, 12; v. 13, no. 1 (information from J. Barquero, R. Barquero, G. Alvarado, G. Soto, L.D. Morales, S. Mora, B. van Wyk de Vries, H. Rymer, and G. Brown).

Thorpe, R.S., Locke, C.A., Brown, G.C., Francis, P.W., and Randal, M., 1981, Magma chamber below Poás Volcano, Costa Rica: Jour. Geol. Soc. London, v. 138, pt. 3, p. 367-373.

Tournon, J., 1984, Magmatismes du Mesozoique a l'actuale en Amerique Centrale: L'example de Costa Rica, des ophiolites aux andesites: Paris, Université Pierre et Marie Curie, D. Sc. thesis.

Vargas, C.A., ed., 1979, Antologia: El Volcán Poás: v. 1, San Jose, Ed. Univ. Estatal a Distancia, 162 p.

Williams, H., 1952, Volcanic history of the Meseta Central Occidental, Costa Rica: Univ. Calif. Publ. Geol. Sci., v. 29, p. 145-180.

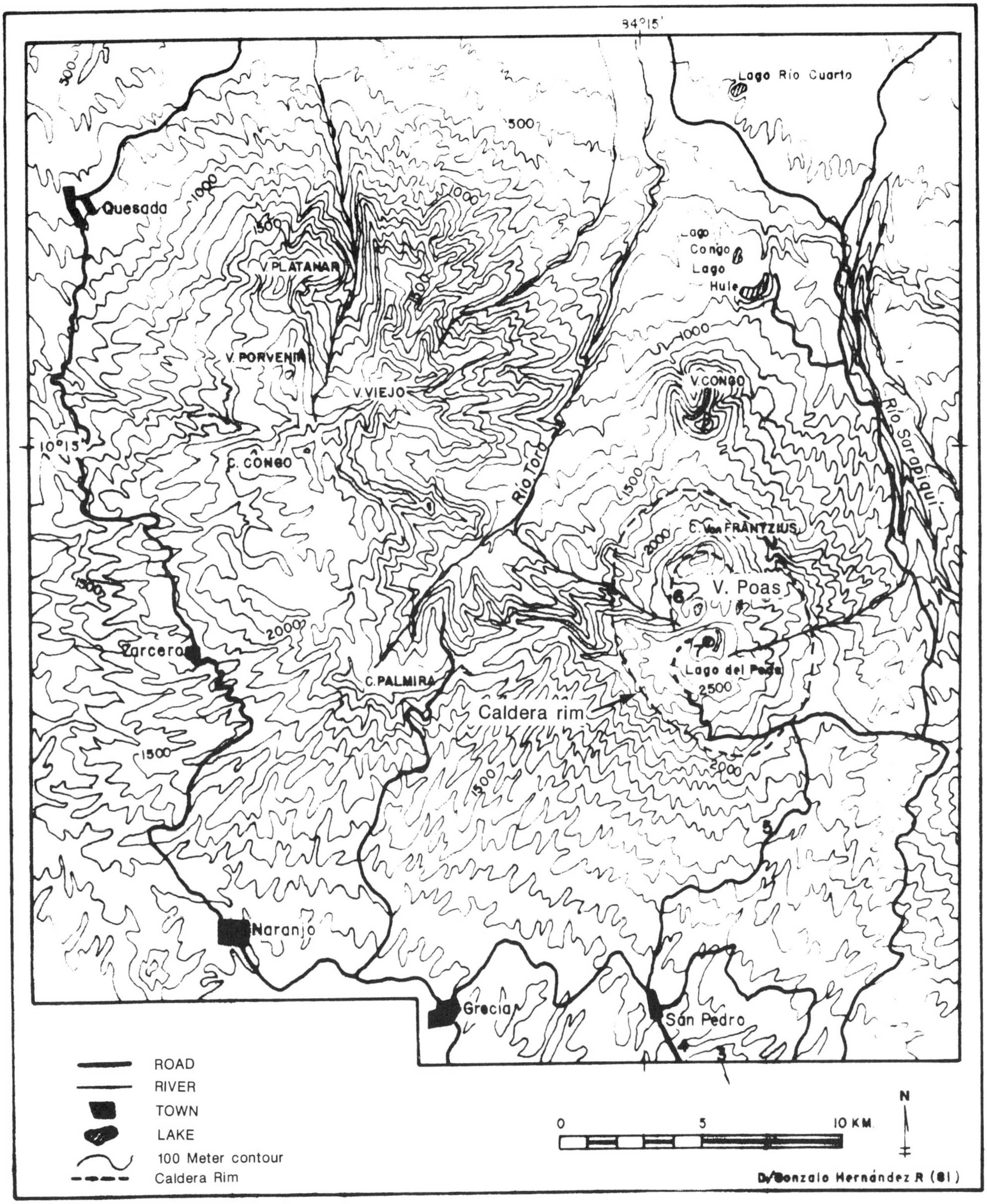

Figure 14.9.1. Location map showing Poás Caldera and V. Poás, after Barquero and Malavassi (1984).

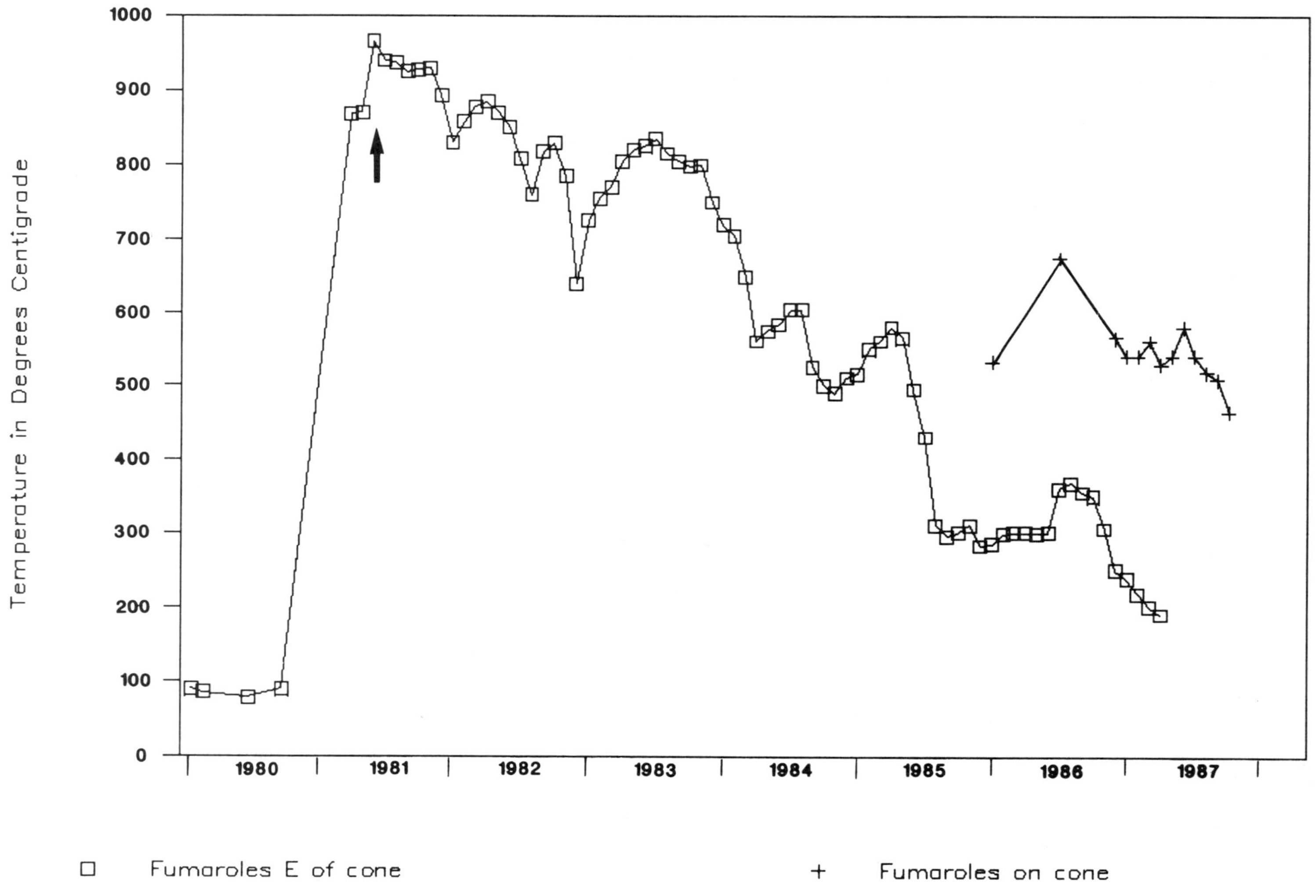

Figure 14.9.2. Changes in fumarole temperatures at V. Poás during 1980-87, from Barquero (1983) with more recent data provided by author. A small steam eruption occurred in May 1981 (arrow).

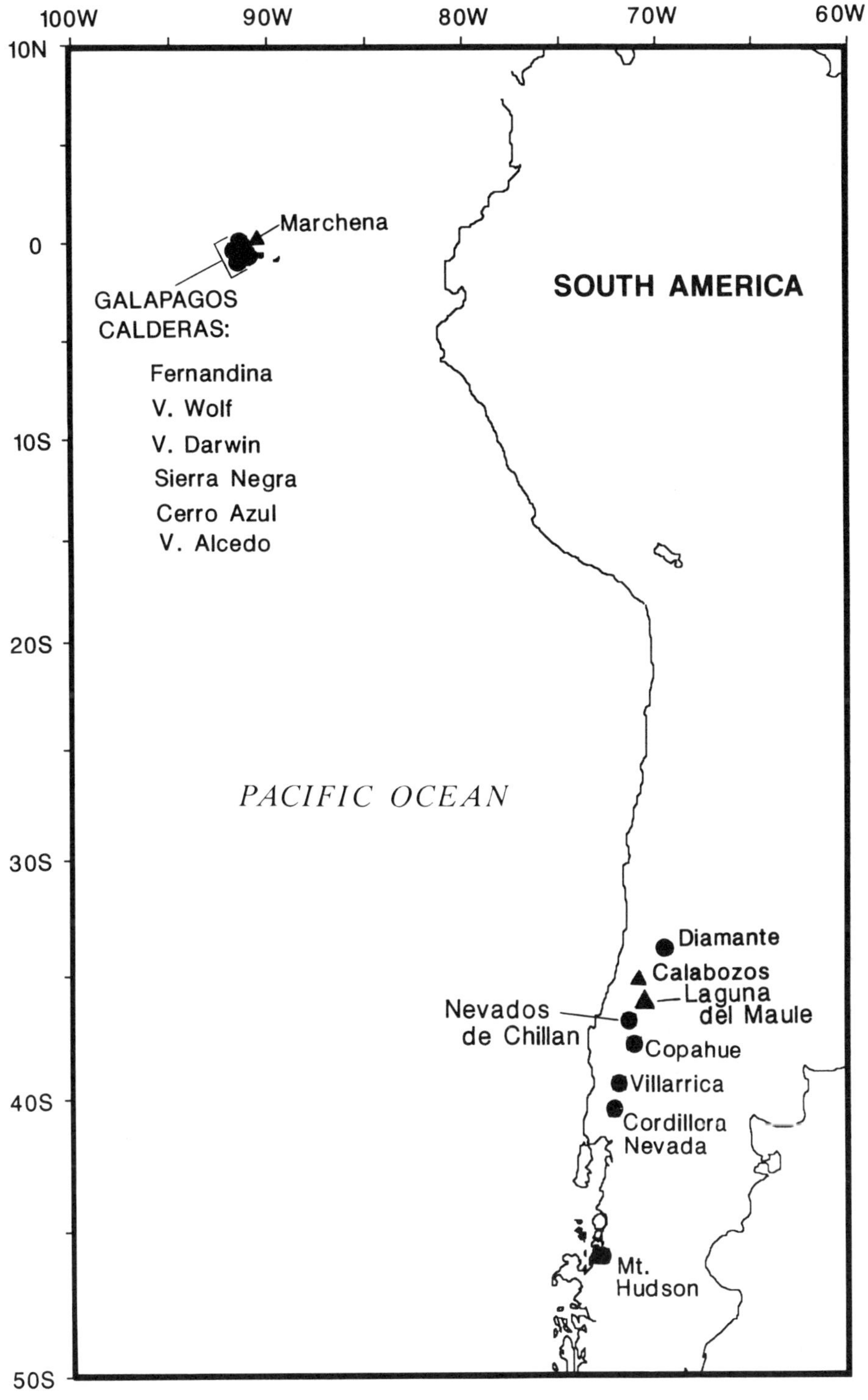

Figure 15. Locations of large, Quaternary restless calderas (solid circles) and nonrestless calderas (triangles) in Region 15.

PART 3: HISTORICAL UNREST AT LARGE QUATERNARY CALDERAS (continued)
See inside back cover for explanation and abbreviations

FERNANDINA

CAVW number (active vent)	Latitude Longitude (degrees)	Diameter (km)	Local tectonic setting	Precaldera edifice	SiO_2 content (wt pct)	Age of collapse (yr B.P.)	Date of unrest	Type and duration of unrest ESTU STHF MGTF H Te	Eruption type
15-03-01 (Fernandina)	0.37S 91.55W	4 x 6.5	Intersection of lineaments over "hotspot"	Shield	R = 49 C = mafic	Renewed collapse in A.D. 1968	1813-14	---- ---- ---- - --	lf,ex
							pre-1817	---- ---- ---- - --	ex,lf
							1819	---- ---- ---- - --	ex,lf
							1825	---- ---- ---- - --	Ex,lf
							1846	---- ---- ---- - --	lf
							1888	---- ---- ---- - --	unknown
							1926-27	---- ---- ---- - --	lf
							1937	---- ---- ---- - --	lf
							1958	---- ---- ---- - --	lf,ll
							1961	---- ---- ---- - --	ex,lf
							1968	C--- Z--Z ---Z Z QU	ex,lf,EX
							1972	---- ---- ---- - --	ex,lf
							1973	---- ---- ---- - --	lf
							1974-76	E--F ---- ---- - --	none?
							1977-78	B--? ---- ---- - QU	ex,lf
							1981	---- ---- ---- - --	lf
							1984	---- ---- ---Y - --	lf,ex

TECTONIC SETTING

The Galápagos Islands (fig. 15.1.1A) lie on the Nazca plate just south of the Galápagos Ridge spreading center. They are also thought to overlie a mantle hotspot, presently beneath the volcanically active western Galápagos Islands (Simkin, 1984, and references therein).

GEOLOGIC HISTORY

Fernandina is a single shield volcano with a well-developed summit caldera (figs. 15.1.1B, 15.1.1C). Benches at both ends of the caldera record prehistoric episodes of collapse. A swarm of arcuate, circumferential fissures occur on the benchlike caldera rim; radial fissures are rare (McBirney and Williams, 1969; Simkin and Howard, 1970).

FERNANDINA (continued)

HISTORICAL ACTIVITY

1968: A $\underline{M}$=4.5 earthquake occurred on 15 May 1968, 300 km north of Fernandina. A small flank eruption began 20 May and stopped within a few days. A slight increase in seismicity in the Galápagos region occurred in the last week of May and the first week of June 1968. The area was seismically quiet during 8-10 June; then, a few more earthquakes occurred on 11 June before a major explosive eruption from the caldera proper at 1040 hr on 11 June. Another sharp explosion occurred at 1708 hr on 11 June, and the eruption continued for another day or so. Beginning on 11 June, seismicity increased sharply as the caldera floor began to collapse (fig. 15.1.2). Collapse was asymmetric, like a trapdoor opening down to the southeast, where the maximum subsidence of 350 m was recorded. Seismicity reached its maximum on 19 June and ended on or about 23 June (Filson and others, 1973). The largest earthquake in the collapse event was of $\underline{M}$=5.4; several hundred earthquakes were of $\underline{M}$=4.0-5.4. Seismic energy release was remarkably constant over a 9-day period, indicating that the collapse took place regularly during this interval (Filson and others, 1973). Epicentral locations seemed to be skewed to the northeast of the caldera, between Fernandina and Isabela Island (fig. 15.1.3). However, Filson and others (1973) suggest that most were beneath the caldera and that their calculated epicenters are skewed to the northeast as an artifact of inaccurate estimates of seismic velocities between the caldera and distant recording stations. The volume of collapse was 1-2 km^3, much greater than the estimated 0.2 km^3 or less of erupted magma.

1974-76: Beginning in mid-1974, earthquakes and presumably concurrent uplift occurred at Punta Espinosa and across the straits at the foot of Darwin Volcano. Cumulative uplift at Punta Espinosa was 30 cm by the time it was first noticed in mid-September 1974, 6 weeks after a local earthquake. Uplift totaled 80-90 cm by February 1976, and no further uplift was noted between then and December 1980 (W.T. Kinoshita, oral commun., 1983; Simkin, 1984). This uplift is similar to but north of uplift that occurred in 1954 at Urvina Bay, at the west foot of V. Alcedo, 30 km south of V. Darwin and 30 km southeast of Punta Espinosa (see further discussion under V. Alcedo, CAVW number 15-03-04).

Another eruption of Fernandina occurred in August 1978, preceded and apparently triggered by a $\underline{M}$=4.5 earthquake 1 hour before the eruption. This eruption was from the northwest margin of the caldera, unlike other eruptions since the 1968 collapse that occurred in the southeast end of the caldera (the end that experienced the maximum subsidence in 1968). A small eruption occurred at the rim sometime between 4 December 1980 and 26 March 1982, unnoticed at the time (Simkin, 1984).

1984: Noise and glow from the summit were noted on 30 March 1984, and further activity occurred within the crater at least through 11 April 1984 (Smithsonian Institution, 1984). SO_2 flux as estimated from the TOMS instrument in the NIMBUS 7 satellite peaked on 1 April and dropped below its detection threshold by 3 April (Smithsonian Institution, 1984).

FERNANDINA (continued)

REFERENCES

Filson, J., Simkin, T., and Leu, L., 1973, Seismicity of a caldera collapse: Galápagos Islands 1968: Jour. Geophys. Res., v. 78, p. 8591-8622.

McBirney, A.R., and Williams, H., 1969, Geology and petrology of the Galápagos Islands: Geol. Soc. Am. Memoir 118, 197 p.

Richards, A.F., 1962, Catalogue of active volcanoes of the world including solfatara fields, Pt. XIV, Archipelago de Colon (Galápagos), Isla San Felix, and Islas Juan Fernandez: Rome, IAVCEI, 50 p.

Simkin, T., 1984, Geology of Galápagos, in Perry, R., ed., Galápagos (key environments): Oxford, Pergamon Press, p. 15-41.

Simkin, T., and Howard, K.A., 1970, Caldera collapse in the Galápagos Islands, 1968: Science, v. 169, p. 429-437.

Smithsonian Institution, Scientific Event Alert Network (SEAN), 1978, 1984, Fernandina: SEAN Bull., v. 3, no. 8, p. 4; v. 9, no. 3, p. 15-16.

Volcanological Society of Japan, 1975, Fernandina: Bull. Volcanic Eruptions, no. 13, p. 66-67.

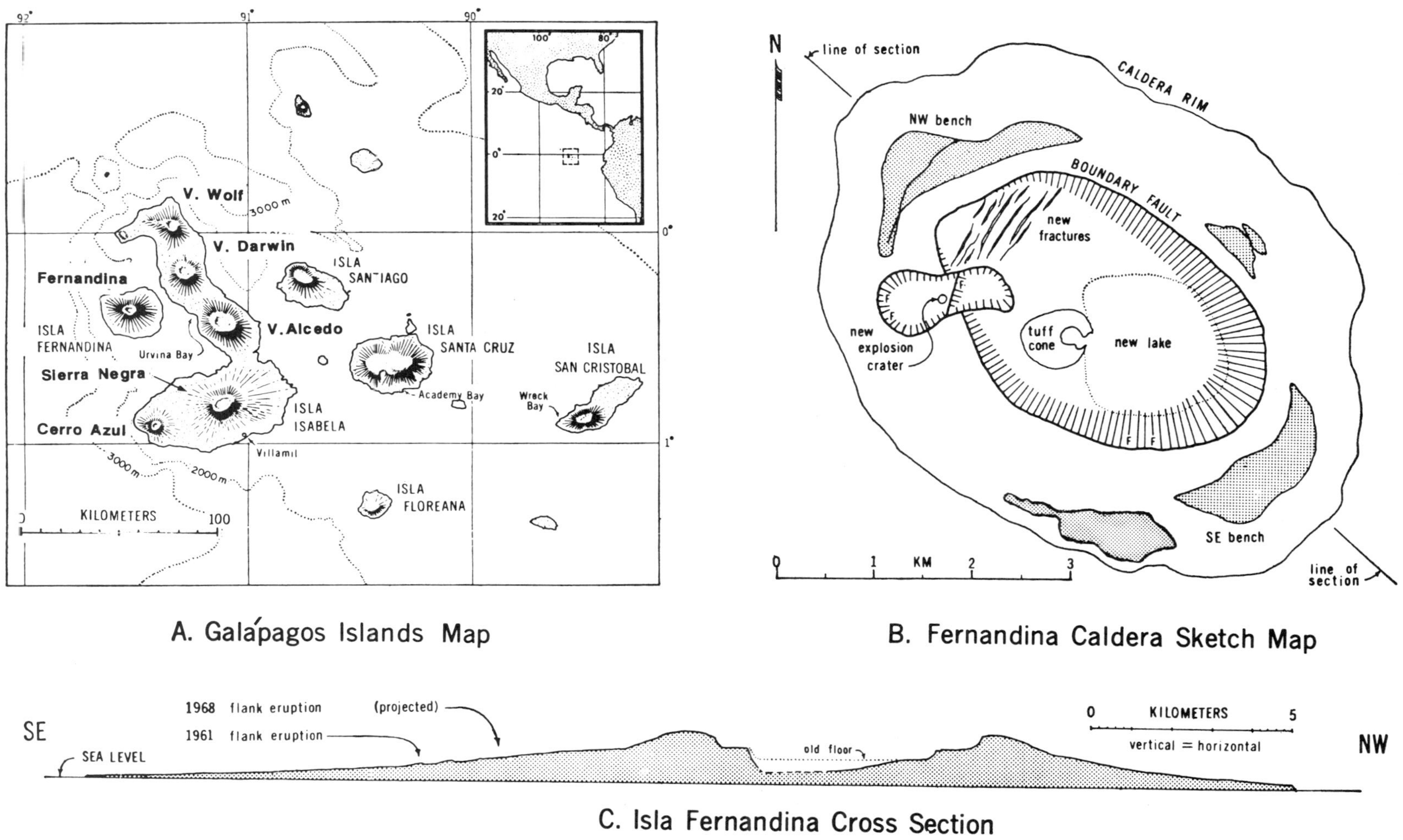

Figure 15.1.1. Location map (A), sketch map (B), and cross section (C) of Fernandina Caldera, from Simkin and Howard (1970). Additional information pertinent to sketch map (B) from Filson and others (1973). Subhorizontal benches are stippled. Prominent northwest and southeast benches, 350 m below the rim at 980-m elevation, mark the floor of an ancient collapse and were in turn isolated by a more recent (but prehistoric) collapse along the same boundary fault as that active in 1968. Copyright by the American Association for the Advancement of Science.

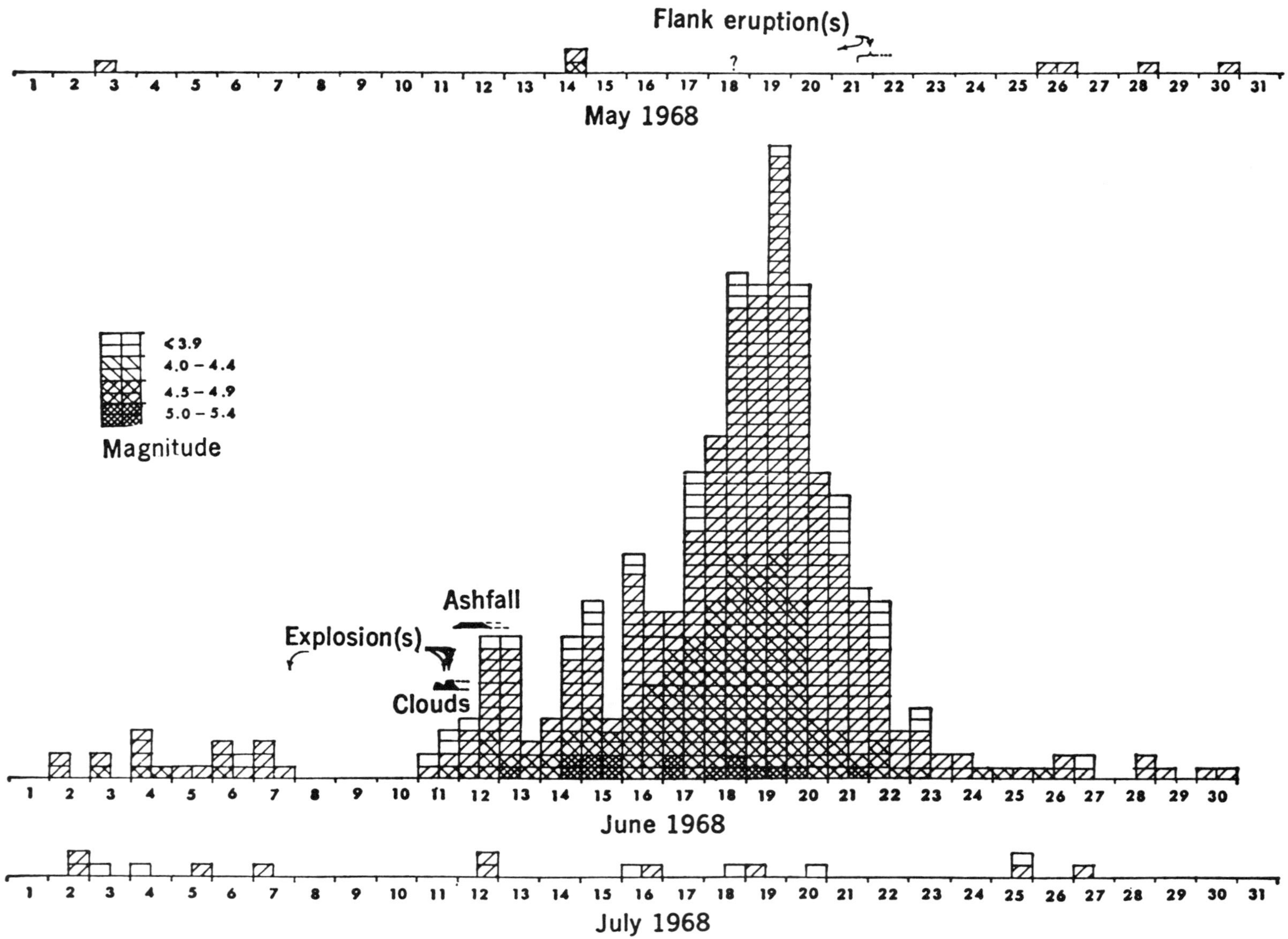

Figure 15.1.2. Earthquakes associated with eruption of Fernandina in June 1968, from Simkin and Howard (1970). Each block represents an earthquake believed to have taken place in Galápagos region during time interval shown. A similar histogram, plus energy release curves, are given in Filson and others (1973). Copyright by the American Association for the Advancement of Science.

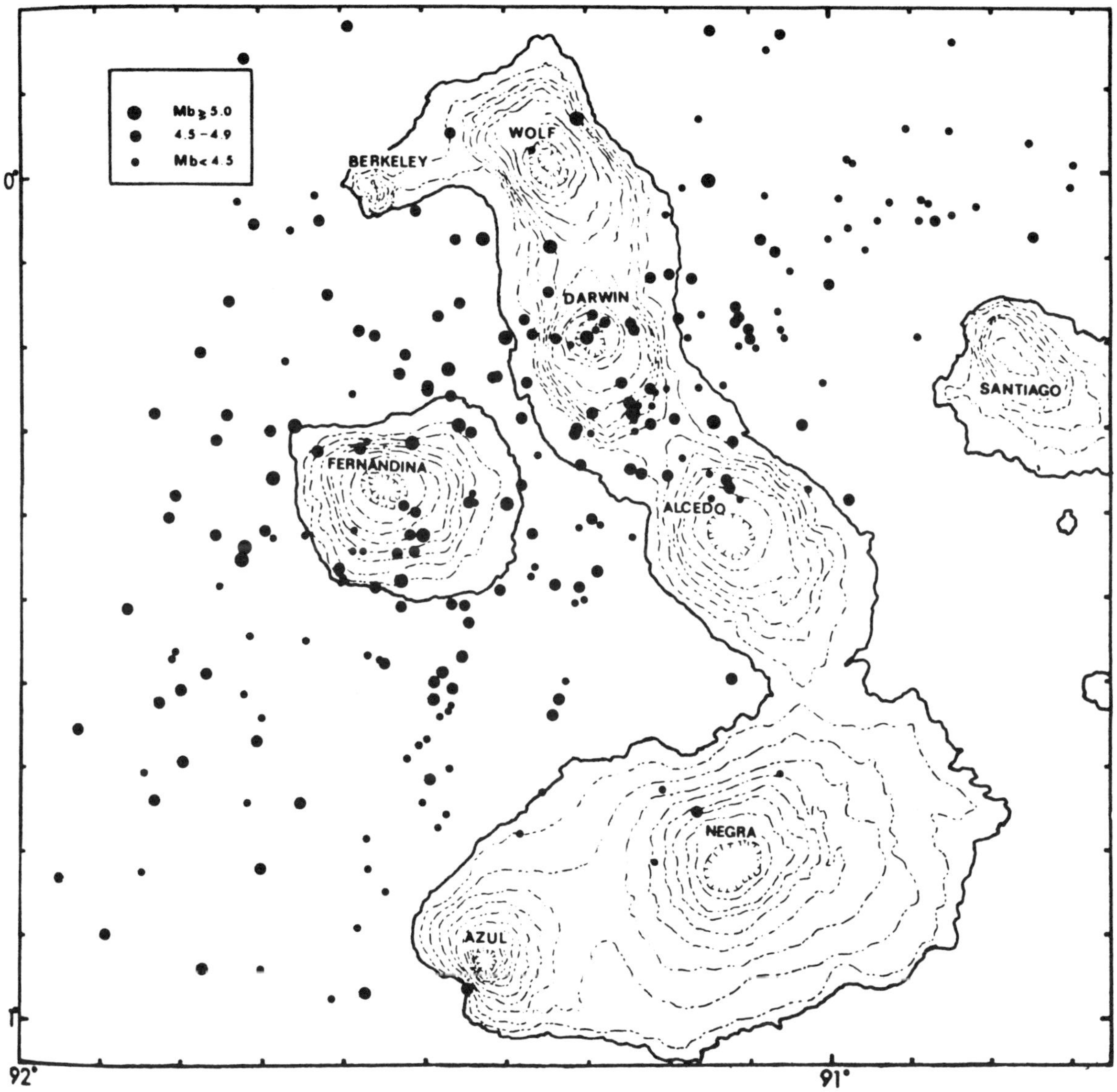

Figure 15.1.3. Epicenter locations of some of the 1968 Galápagos earthquakes, from Filson and others (1973). See text and Filson and others (1973) for discussion of probable errors in epicenters of earthquakes, most of which probably occurred at Fernandina during caldera collapse.

PART 3: HISTORICAL UNREST AT LARGE QUATERNARY CALDERAS (continued)
See inside back cover for explanation and abbreviations

V. WOLF

CAVW number (active vent)	Latitude Longitude (degrees)	Diameter (km)	Local tectonic setting	Precaldera edifice	SiO_2 content (wt pct)	Age of collapse (yr B.P.)	Date of unrest	Type and duration of unrest ESTU STHF MGTF H Te	Eruption type
15-03-02 (V. Wolf)	0.02N 91.35W	7 x 5.5	Intersection of lineaments over "hotspot"	Shield	R = mafic C = mafic		1797	---- ---- ---- - --	unknown
							1800	---- ---- ---- - --	ex
							1925-26	---- ---- ---- - --	lf
							1933	---- ---- ---- - --	lf
							1935	---- ---- ---- - --	unknown
							1938	---- ---- ---- - --	unknown
							1948	---- ---- ---- - --	ex,lf
							1963	---- ---- ---- - --	lf
							1973?	EC-- ---- ---- - QU	unknown
							1982	---- ---- ---Z - --	ex,lf

TECTONIC SETTING

The Galápagos Islands lie on the Nazca plate just south of the Galápagos Ridge spreading center. They may also overlie a mantle hotspot, presently beneath the volcanically active, western Galápagos Islands (Simkin, 1984, and references therein).

GEOLOGIC HISTORY

Wolf is a shield volcano at the northern end of Isabela (Albemarle) Island (see fig. 15.1.1A in section on Fernandina). Arcuate fissures occur on the walls of its summit caldera and on and just outside the caldera rim; some radial fissures exist on the west, north, and northeast flanks.

HISTORICAL ACTIVITY

1973: An earthquake swarm on 19 March 1973 included a $\underline{M}$=4.9 event located beneath the southeast flank of Wolf. A small eruption may have occurred in mid-November 1973, when a warden of Galápagos National Park heard strong and constant rumblings from the caldera. No eruption cloud was seen, and a ground party sent to investigate found no evidence of fresh lava or pyroclastic products (Smithsonian Institution, 1973).

1982: An eruption began on 28 August 1982. Lava fountains covered much of the summit caldera floor (in the Northern Hemisphere), and lava issued from the southeast flank (10 km away and in the Southern Hemisphere), making a two-hemisphere eruption (Smithsonian Institution, 1982; T. Simkin, written commun., 1987)!

V. WOLF (continued)

REFERENCES

McBirney, A.R., and Williams, H., 1969, Geology and petrology of the Galápagos Islands: Geol. Soc. Am. Memoir 118, 197 p.

Richards, A.F., 1962, Catalogue of active volcanoes of the world including solfatara fields, Pt. XIV, Archipelago de Colon (Galápagos), Isla San Felix, and Islas Juan Fernandez: Rome, IAVCEI, 50 p.

Simkin, T., 1984, Geology of Galápagos, in Perry, R., ed., Galápagos (key environments): Oxford, Pergamon Press, p. 15-41.

Simkin, T., Siebert, L., McClelland, L., Bridge, D., Newhall, C., and Latter, J.H., 1981, Volcanoes of the world: Stroudsburg, Pa., Hutchinson Ross Publ. Co., 232 p.

Smithsonian Institution, Center for Short-Lived Phenomena (SICSLP), 1973, Volcán Wolf: SICSLP Card 1749.

Smithsonian Institution, Scientific Event Alert Network (SEAN), 1982, Volcán Wolf: SEAN Bull., v. 7, no. 8, p. 4.

Volcanological Society of Japan, 1975, Wolf: Bull. Volcanic Eruptions, no. 13, p. 60.

PART 3: HISTORICAL UNREST AT LARGE QUATERNARY CALDERAS (continued)
See inside back cover for explanation and abbreviations

V. DARWIN

CAVW number (active vent)	Latitude Longitude (degrees)	Diameter (km)	Local tectonic setting	Precaldera edifice	SiO_2 content (wt pct)	Age of collapse (yr B.P.)	Date of unrest	Type and duration of unrest ESTU STHF MGTF H Te	Eruption type
15-03-03 (V. Darwin)	0.18S 91.28W	5	Intersection of lineaments over "hotspot"	Shield	R = 48 C = mafic		1813?	---- ---- ---- - --	ex

TECTONIC SETTING

The Galápagos Islands lie on the Nazca plate just south of the Galápagos Ridge spreading center. They may also overlie a mantle hotspot, presently beneath the volcanically active, western Galápagos Islands (Simkin, 1984, and references therein).

GEOLOGIC HISTORY

Darwin Volcano (see fig. 15.1.1A in section on Fernandina) is a typical Galápagos shield volcano with a caldera in the top of an "upside-down soup bowl" -shaped edifice. Arcuate fissures around the benchlike caldera rim have fed many lava flows, as have radial fissures on the flanks (McBirney and Williams, 1969). Two tuff cones, Tagus and Beagle, formed where one radial fissure extended to the sea.

HISTORICAL ACTIVITY

On 6 June 1813 a sea captain observed "a thick column of smoke rising rapidly", possibly from Darwin (Porter, 1822). "At night the whole atmosphere was illuminated by it, and yet we could perceive neither flames nor sparks thrown out by the centre," probably because activity was confined within the summit caldera.

REFERENCES

McBirney, A.R., and Williams, H., 1969, Geology and petrology of the Galápagos Islands: Geol. Soc. Am.Memoir 118, 197 p.
Porter, D., 1822, Journal of a cruise made to the Pacific Ocean (2nd ed.): New York, Wiley and Halsted, v. 1, 242 p.
Richards, A.F., 1962, Catalogue of active volcanoes of the world including solfatara fields, Pt. XIV, Archipelago de Colon (Galápagos), Isla San Felix, and Islas Juan Fernandez: Rome, IAVCEI, 50 p.
Simkin, T., 1984, Geology of Galápagos, in Perry, R., ed., Galápagos (key environments): Oxford, Pergamon Press, p. 15-41.

V. ALCEDO

CAVW number (active vent)	Latitude Longitude (degrees)	Diameter (km)	Local tectonic setting	Precaldera edifice	SiO_2 content (wt pct)	Age of collapse (yr B.P.)	Date of unrest	Type and duration of unrest ESTU STHF MGTF H Te	Eruption type
15-03-04 (V. Alcedo)	0.43S 91.12W	7	Intersection of lineaments over "hotspot"	Shield	R = mafic C = mafic		between 1945 and 1961	---- ---- ---- - --	lf
							1954	---x ---- ---- - --	none?

TECTONIC SETTING

The Galápagos Islands lie on the Nazca plate just south of the Galápagos Ridge spreading center. They may also overlie a mantle hotspot, presently beneath the volcanically active, western Galápagos Islands (Simkin, 1984, and references therein).

GEOLOGIC HISTORY

Alcedo is the lowest of five shield volcanoes on Isabela (Albemarle) Island (see fig. 15.1.1A in section on Fernandina). Many of the flows on the surface of Alcedo are from radial fissures, and the arcuate, circumferential fissures around the caldera rim are much less evident than at other Galápagos shields (McBirney and Williams, 1969). A bench on the caldera wall indicates at least two episodes of collapse.

HISTORICAL ACTIVITY

By comparing aerial photographs from 1945 and 1961, K. Howard (in Simkin, 1984) found one new lava flow low on the southeast flank of V. Alcedo.

A 6-km length of shoreline including Urvina Bay, at the western foot of V. Alcedo (see fig. 15.1.1A in section on Fernandina), was uplifted by as much as 4.6 m, probably in early 1954 (Couffer, 1956; Richards, 1957). More than 1 km^2 of coral reef was raised above sea level. Richards (1957) attributed the uplift to magma intrusion leading to an eruption of Alcedo in November 1954, but K. Howard (in Simkin, 1984) concluded that Alcedo could not have been the site of the November 1954 activity described by Richards (1957). Sierra Negra (40-50 km south-southeast of Urvina Bay) erupted in late 1953 and in 1954, but there are no reports that uplift of similar or greater magnitude occurred at the foot of that volcano. The cause of the 1954 Urvina Bay uplift is therefore unknown.

V. ALCEDO (continued)

REFERENCES

Couffer, J.C., 1956, The disappearance of Urvina Bay: Nat. Hist., v. 65, p. 378-383.

McBirney, A.R., and Williams, H., 1969, Geology and petrology of the Galápagos Islands: Geol. Soc. Am. Memoir 118, 197 p.

Richards, A.F., 1957, Volcanism in eastern Pacific Ocean basin, 1945-1955: Int. Geol. Congr., 20th, Vulcanologia del Cenozoico, Sect. 1, v. 1, p. 19-31.

----- 1962, Catalogue of active volcanoes of the world including solfatara fields, Pt. XIV, Archipelago de Colon (Galápagos), Isla San Felix, and Islas Juan Fernandez: Rome, IAVCEI, 50 p.

Simkin, T., 1984, Geology of Galápagos, in Perry, R., ed., Galápagos (key environments): Oxford, Pergamon Press, p. 15-41.

SIERRA NEGRA

CAVW number (active vent)	Latitude Longitude (degrees)	Diameter (km)	Local tectonic setting	Precaldera edifice	SiO_2 content (wt pct)	Age of collapse (yr B.P.)	Date of unrest	Type and duration of unrest ESTU STHF MGTF H Te	Eruption type
15-03-05 (Sierra Negra)	0.83S 91.17W	10.5 x 7	Intersection of lineaments over "hotspot"	Shield	R = 47-49 C = mafic		1813	---- ---- ---- - --	ex
							1817	---- ---- ---- - --	unknown
							1844	---- ---- ---- - --	lf
							1860	---- ---- ---- - --	ex
							1911 or 1912	---- ---- ---- - --	unknown
							1948	---- ---- ---- - --	ex,lf
							1953-54	---? ---- ---- - --	Ex,lf
							1957	---- ---- ---- - --	unknown
							1963	---- ---- ---- - --	ex,lf
							1979-80	A--- ---Y ---- - --	Ex,lf

TECTONIC SETTING

The Galápagos Islands lie on the Nazca plate just south of the Galápagos Ridge spreading center. They are thought to overlie a mantle hotspot, presently beneath the volcanically active western Galápagos Islands (Simkin, 1984, and references therein).

GEOLOGIC HISTORY

Sierra Negra is a large shield volcano on Isabela (Albemarle) Island, elongate in an east-northeast direction (see fig. 15.1.1A in section on Fernandina). Its summit caldera is shallower than those of other Albemarle shields. Recent eruptions have been from circumferential fissures just outside the northern rim of the caldera (McBirney and Williams, 1969; Delaney and others, 1973).

HISTORICAL ACTIVITY

Sierra Negra is known to have erupted from August 1953 until January 1954; another eruption in November 1954 has tentatively been attributed to Sierra Negra (Simkin, 1984). Uplift in early 1954 around Urvina Bay, 40-50 km north-northwest of Sierra Negra, bears an uncertain relation, if any, to the eruptions of Sierra Negra (see sections on V. Darwin and V. Alcedo).

A local earthquake was registered at Darwin Station at 0730 hr on 13 November 1979, and two larger, M=4.4 quakes occurred at 0750 hr and 0817 hr. An explosion was heard at 0845 hr, and during the succeeding days eruption clouds rose as high as 14 km, apparently introducing a globally detectable amount of SO_2 into the stratosphere. Seismicity increased on the second afternoon of the eruption, with M_{max} = 4.8. Lava flows issued from an 8-km-long fissure on the north flank (Smithsonian Institution, 1979-80).

SIERRA NEGRA (continued)

REFERENCES

Delaney, J.R., Colony, W.E., Gerlach, T.M., and Nordlie, B.E., 1973, Geology of the Volcán Chico area on Sierra Negra Volcano: Geol. Soc. Am. Bull., v. 84, p. 2455-2470.

McBirney, A.R., and Williams, H., 1969, Geology and petrology of the Galápagos Islands: Geol. Soc. Am. Memoir 118, 197 p.

Richards, A.F., 1962, Catalogue of active volcanoes of the world including solfatara fields, Pt. XIV, Archipelago de Colon (Galápagos), Isla San Felix, and Islas Juan Fernandez: Rome, IAVCEI, 50 p.

Simkin, T., 1984, Geology of Galápagos, in Perry, R., ed., Galápagos (key environments): Oxford, Pergamon Press, p. 15-41.

Smithsonian Institution, Scientific Event Alert Network (SEAN), 1979-80, Sierra Negra: SEAN Bull., v. 4, nos. 11-12; v. 5, no. 1 (with reports from H. Hoeck, A. Tupiza, and National Earthquake Information Service, U.S. Geol. Surv.).

CERRO AZUL

CAVW number (active vent)	Latitude Longitude (degrees)	Diameter (km)	Local tectonic setting	Precaldera edifice	SiO_2 content (wt pct)	Age of collapse (yr B.P.)	Date of unrest	Type and duration of unrest ESTU STHF MGTF H Te	Eruption type
15-03-06	00.90S	4 x 5	Hotspot	Shield	R = mafic		1932	---- ---- ---- - --	unknown
(Cerro Azul)	91.42W				C = mafic		1940	---- ---- ---- - --	lf
							1943	---- ---- ---- - --	Ex, lf
							1948	---- ---- ---- - --	lf
							1949?	---- ---- ---- - --	lf
							1951	---- ---- ---- - --	unknown
							1959	---- ---- ---- - --	lf
							1979	---- ---- ---- - --	ex,lf

TECTONIC SETTING

The Galápagos Islands lie on the Nazca plate just south of the Galápagos Ridge spreading center. They are thought to overlie a mantle hotspot, presently beneath the volcanically active, western Galápagos Islands (Simkin, 1984, and references therein).

GEOLOGIC HISTORY

Cerro Azul is the smallest of the Isabela (Albemarle) Island shield volcanoes (see fig. 15.1.1A in section on Fernandina) and has both circumferential and radial fissures. Benches on caldera walls indicate that several periods of subsidence have occurred.

HISTORICAL ACTIVITY

Historical accounts describe lava flows and eruption columns but nothing about precursors. The eruption in 1959 was from a new crater on the east slopes of Cerro Azul (T. Simkin, written commun., 1985).

REFERENCES

Banfield, A.F., Behre, C.H., Jr., and St. Clair, D., 1956, Geology of Isabela (Albemarle) Island, Archipelago de Colon (Galápagos): Geol. Soc. Am. Bull., v. 67, p. 215-234.

McBirney, A.R., and Williams, H., 1969, Geology and petrology of the Galápagos Islands: Geol. Soc. Am.Memoir 118, 197 p.

Richards, A.F., 1962, Catalogue of active volcanoes of the world including solfatara fields, Pt. XIV, Archipelago de Colon (Galápagos), Isla San Felix, and Islas Juan Fernandez: Rome, IAVCEI, 50 p.

Simkin, T., 1984, Geology of Galápagos, in Perry, R., ed., Galápagos (key environments): Oxford, Pergamon Press, p. 15-41.

Simkin, T., Siebert, L., McClelland, L., Bridge, D., Newhall, C., and Latter, J.H., 1981, Volcanoes of the world: Stroudsburg, Pa., Hutchinson Ross Publ. Co., 232 p.

DIAMANTE

CAVW number (active vent)	Latitude Longitude (degrees)	Diameter (km)	Local tectonic setting	Precaldera edifice	SiO_2 content (wt pct)	Age of collapse (yr B.P.)	Date of unrest	Type and duration of unrest ESTU STHF MGTF H Te	Eruption type
15-07-021 (Maipo)	34.2S 69.8W	15 x 20	Compr		R = 60 C = 74-76	450,000 ±60,000	1822?	---- ---- ---? - --	unknown
							1826?	---- ---- ---? - --	ex?
							1905?	---- ---- ---? - --	unknown
							1912?	---- ---- ---? - --	unknown

TECTONIC SETTING

Diamante Caldera is situated above the gently dipping (25°) subduction zone of south-central Chile, more or less between the loci of great earthquakes that destroyed Mendoza (Argentina) in 1861 and Valparaiso (Chile) in 1906.

GEOLOGIC HISTORY

Diamante Caldera formed approximately 0.45 m.y. ago after eruption of approximately 450 km^3 (bulk volume) of rhyolitic pyroclastic-flow deposits (Polanski, 1962; Stern and others, 1984; Harrington and Stern, 1987). The present surface area of Diamante deposits is 1,300 km^2, but the predissection, preburial area was approximately 15,000 km^2 (Stern and others, 1984) (fig. 15.7.1). Maipo Volcano is an andesitic stratovolcano that partially fills the western part of Diamante Caldera. Precaldera lavas are mostly andesite but include some rhyolite (Harrington and Stern, 1987).

HISTORICAL UNREST

We include Diamante Caldera among historically restless calderas because Maipo has reportedly erupted four times in the 19th and 20th centuries. However, some reported eruptions in this area have been occasional sightings of persistent steam plumes, and we cannot evaluate the accuracy of these particular reports. No deposits are known from any historical eruptions of Maipo.

DIAMANTE (continued)

REFERENCES

Harrington, R., Amini, H., Stern, C.R. and Charrier, R., 1984, The Maipo stratovolcano-caldera complex in the southern Andes of central Chile [abs.]: Eos, Trans. Am. Geophys. Union, v. 65, p. 1136.

Harrington, R., and Stern, C., 1987, The Diamante Caldera and Maipo stratovolcano complex: Acta, X Congr. Geol. Argentino, v. 4, p. 286-288 (Symp. on Andean Volcanism).

Polanski, J., 1962, Estratigrafía, nectectónica y geomorfología del Pleistoceno pedemontano entre los ríos Diamante y Mendoza: Asoc. Geol. Argentina, Rev., v. 27, nos. 3-4, p. 127-149.

Stern, C.R., Amini, H., Charrier, R., Godoy, E., Herve, F., and Varela, J., 1984, Petrochemistry and age of rhyolitic pyroclastic flows which occur along the drainage valleys of the Rio Maipo and Rio Cachapoal (Chile) and the Rio Yaucha and Rio Papagayos (Argentina): Rev. Geol. Chile, no. 23, p. 39-52.

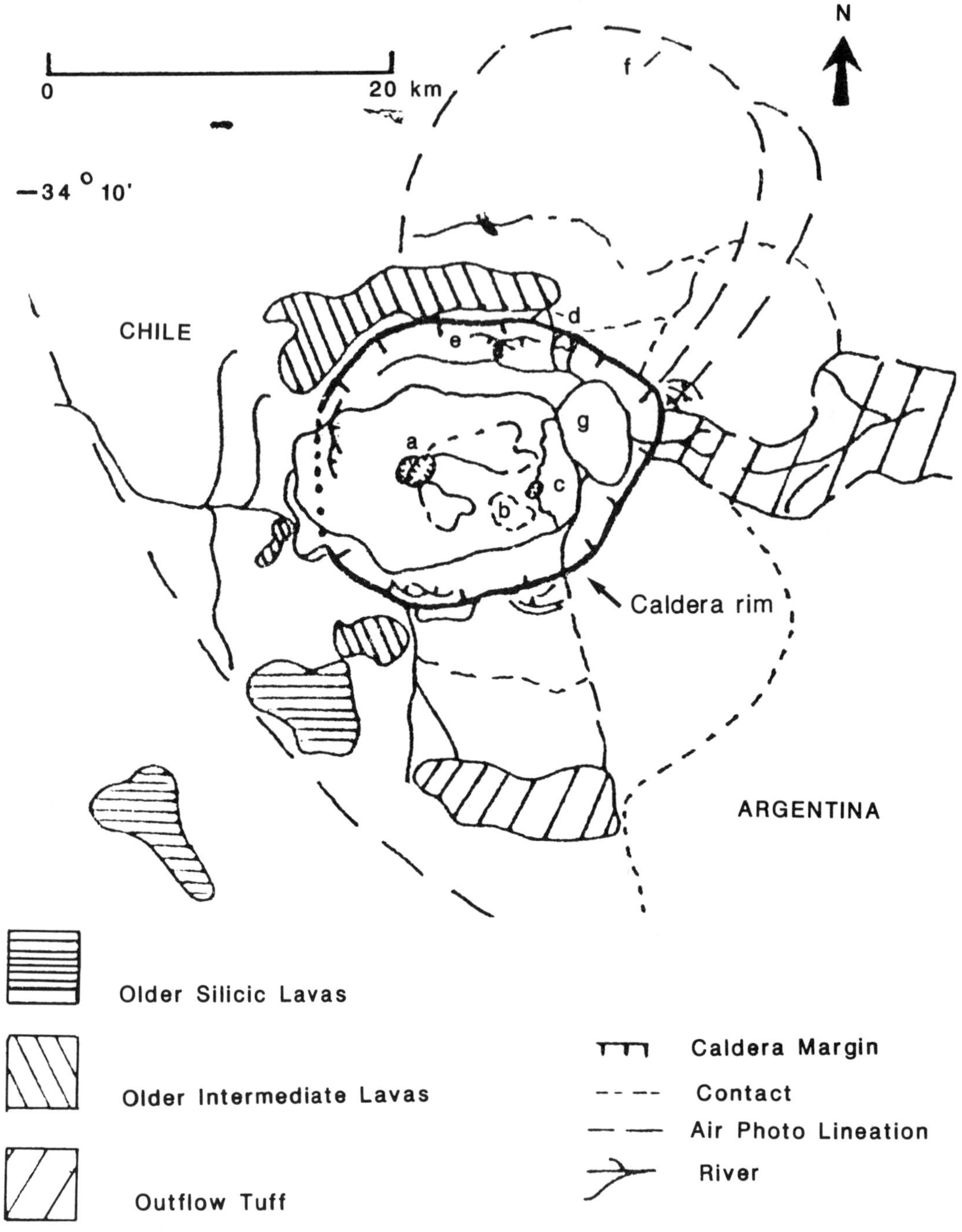

Figure 15.7.1. Geologic sketch map of Diamante Caldera, from Harrington and Stern (1987). Symbols: a, Maipo Volcano; b, mudflow; c, young flows; d, ring dome; e, ring fractures; f, fault; g, Laguna Diamante.

PART 3: HISTORICAL UNREST AT LARGE QUATERNARY CALDERAS (continued)
See inside back cover for explanation and abbreviations

NEVADOS DE CHILLAN

CAVW number (active vent)	Latitude Longitude (degrees)	Diameter (km)	Local tectonic setting	Precaldera edifice	SiO_2 content (wt pct)	Age of collapse (yr B.P.)	Date of unrest	Type and duration of unrest ESTU STHF MGTF H Te	Eruption type
15-07-07 (Nevados de Chillán; V. Nuevo=N)	36.87S 71.38W	7 x 4.5	Compr		R = 60-69 C = ?		1600's	---- ---- ---- - --	Ex
							ca. 1750	---- ---- ---- - --	Ex
							1835	---- ---? ---? ? QU	none
							1861-62	---- ---- ---x - QU	lf,Ex
							1864-65	---- ---- ---- - --	Ex
							1872	---- ---- ---x - --	none
							1877	---- ---- ---x - --	none
							1883	---- ---- ---- - --	ex
							1891	---- ---- ---- - --	ex
							1898	---- ---- ---- - --	ex
							1906 (N)	---- ---- ---x - UQ	ex
							1923 (N)	---- ---- ---x - --	ex
							1929 (N)	---- ---- ---x - --	ex,pex?
							1934 (N)	---- ---- ---x - --	ex
							1945-47(N)	---- ---- ---x - --	ex?
							1965-87+(N)	---- ---- ---F - --	pex,ex,lf,dm

TECTONIC SETTING

In south-central Chile, the Benioff zone dips 25° and has been the site of large, shallow earthquakes. Nevados de Chillán is about 100 km inland from the epicenter of a great earthquake in 1835, and several hundred kilometers south of the epicenters of several more great earthquakes of the 19th and 20th centuries.

GEOLOGIC HISTORY

Voluminous pyroclastic-flow deposits reached from the Nevados de Chillán area to the central valley of south-central Chile, and Deruelle and Deruelle (1975) suggest an 8 km x 10 km caldera at Nevados de Chillán. However, only a 7 km x 4.5 km diameter caldera is clear (fig. 15.8.1) (H. Moreno R., written commun., 1985, 1988). About 10 cones and domes are aligned along a northwest-trending axis through the Nevados de Chillán Caldera, of which Volcán Nuevo is the youngest.

NEVADOS DE CHILLAN (continued)

HISTORICAL ACTIVITY

Shortly after a great earthquake in 1835, centered near Concepción, immense cracks opened near Chillán, 100 km from the coast. Abundant mud and salt water were "erupted", and many new hot springs developed (Perrey, 1854, p. 145-146). Perrey does not indicate whether these changes occurred near the city of Chillán (about 70 km from the coast) or at Nevados de Chillán (about 140 km from the coast). Reiche (1896) described thermal areas of Chillán but did not mention any effect of the 1835 earthquake.

The 1861 eruption commenced a few months after a disastrous earthquake near Mendoza, Argentina, and the simultaneous extinguishing of V. Antuco (each a few hundred kilometers from Nevados de Chillán) (Stone and Ingerson, 1934). The vent for the 1861 eruption was Cerro Negro at the NW foot of Cerro Blanco (Philippi, 1861, 1863).

Stone (1935), Deruelle and Deruelle (1975), and H. Moreno R. (written commun., 1985) report that a small Strombolian eruption at Nevados de Chillán began on 6 August 1906, 10 days before a strong earthquake that destroyed Valparaiso. Volcán Nuevo formed during Strombolian activity of this 1906 eruption. A report by Casertano (1963a, b) of increased fumarolic activity in August and September 1906, after the strong, 16 August Valparaiso earthquake is possibly correct but fails to mention the eruption before the earthquake.

New(?) fumaroles have been observed since 1965, especially on the slopes of V. Nuevo. Remarkable gas emissions occurred in July 1973 along the axis of the caldera, and by December 1973 a new cone (dome) began to form on the slope of V. Nuevo (Deruelle, 1977; Volcanological Society of Japan, 1975, 1976, 1979; H. Moreno R., written commun., 1985). Small, rhythmic phreatomagmatic eruptions ended about June 1983 (H. Moreno R., written commun., 1985), and only a few explosions have been reported from June 1983 to the time of this writing (February 1988) (H. Moreno R., written commun., 1988).

REFERENCES

Casertano, L., 1963a, General characteristics of active Andean volcanoes and a summary of their activities during recent centuries: Bull. Seismol. Soc. Am., v. 53, p. 1415-1433.
----- 1963b, Catalogue of active volcanoes of the world including solfatara fields, Pt. XV, Chilean Continent: Rome, IAVCEI, 55 p.
Deruelle, B., 1977, Sur l'activité récente des Nevados de Chillán (Chili Central): Comptes Rendus Acad. Sci., Paris, Ser. D, v. 284, p. 1651-1654.
Deruelle, B., and Deruelle, J., 1975, Geologie des volcans Quaternaires des Nevados de Chillán (Chili): Bull. Volcanol., v. 38, no. 2, p. 425-444.
Moreno R., H., 1974, Airplane flight over active volcanoes of central-south Chile: Guidebook, Excursion D-3, Symp. on Andean and Antarctic Volcanology Problems: Naples/Santiago, IAVCEI, 56 p.
Perrey, A., 1854, Documents relatifs aux tremblements de terre au Chili: Ann. Soc. Imper. Agric. Hist. Nat. et Arts Utiles, Lyon, 206 p.

NEVADOS DE CHILLAN (continued)

REFERENCES (continued)

Philippi, R.A., 1861, Neuer Vulkan in Chile (Chillán): Petermanns Mitt., v. 8, p. 430.
----- 1862, Viaje a los baños i al nuevo volcán de Chillán: Anal. Univ. Chile, v. 20, p. 279-309.
----- 1863, Exkursion nach den Bädern und dem Neuen Vulkan von Chillán in Chile im Spätsommer 1862: Petermanns Mitt., v. 9, p. 241-257.
Reiche, K., 1896, Die thermen von Chillan in Chile: Globus, v. 69, no. 4, p. 60-62.
Simkin, T., Siebert, L., McClelland, L., Bridge, D., Newhall, C., and Latter, J.H., 1981, Volcanoes of the world: Stroudsburg, Pa., Hutchinson Ross Publ. Co., 232 p.
Smithsonian Institution, Scientific Event Alert Network (SEAN), 1977, Nevados de Chillán: SEAN Bull., v. 4, no. 7, p. 11-12 (information from O. Gonzalez).
Stone, J.B., 1935, The volcanoes of southern Chile: Zeit. Vulkanol., v. 16, p. 81-97.
Stone, J.B., and Ingerson, E., 1934, Some volcanoes of southern Chile: Am. J. Sci., v. 28, no. 166, p. 269-287.
Volcanological Society of Japan, 1975, 1976, 1979, Nevados de Chillán: Bull. Volcanic Eruptions, no. 13, p. 60-61; no. 14, p. 61; no. 15, p. 55; no. 17, p. 77 (information from B. Deruelle, J. Deruelle, and D.C. Shackelford).

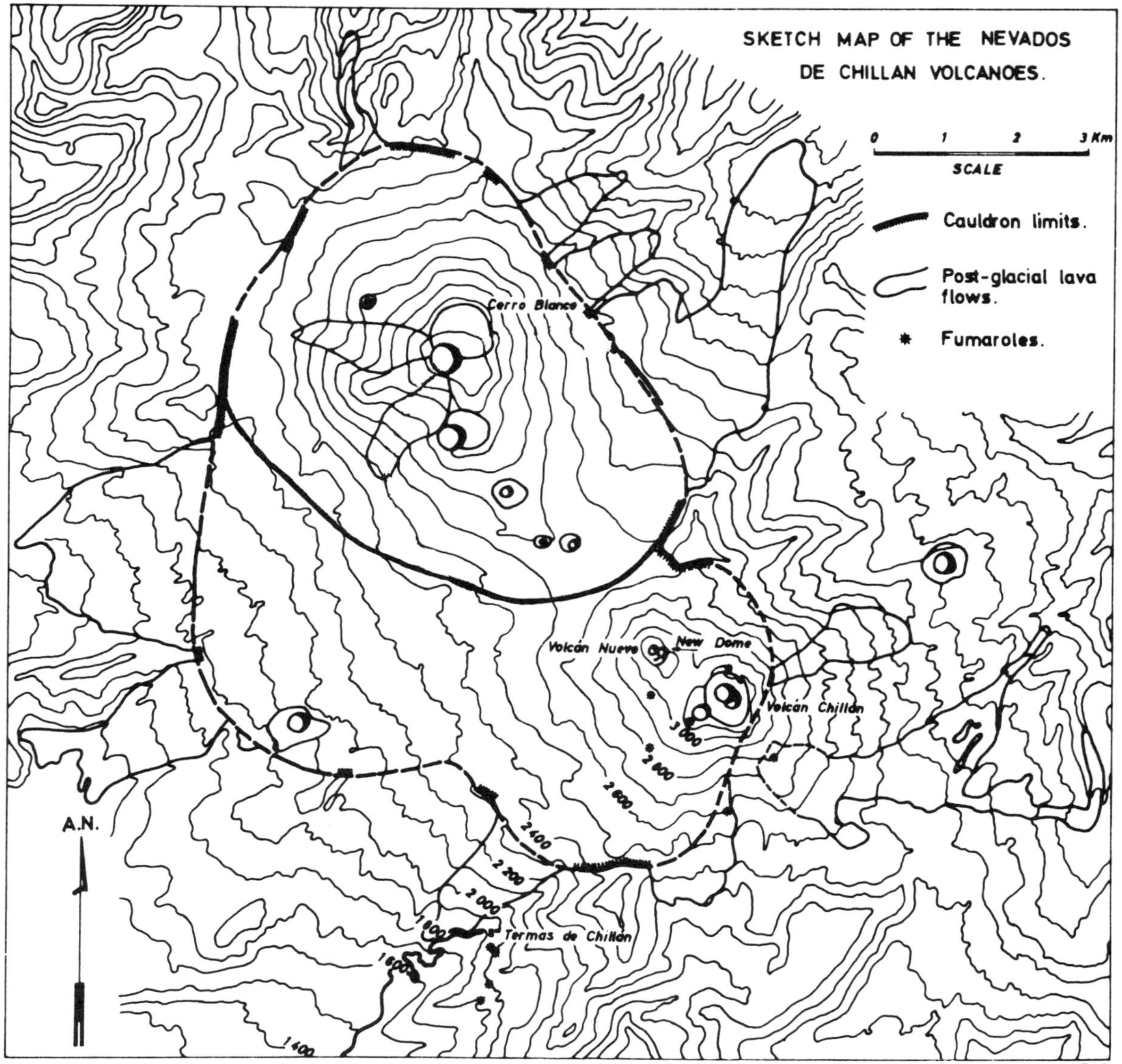

Figure 15.8.1. Approximate outline of Nevados de Chillán Caldera, from a contribution by B. Deruelle and J. Deruelle to Volcanological Society of Japan (1975). These authors propose the existence of an 8 km x 10 km diameter caldera, but Hugo Moreno R. (written commun., 1988) suggests that only a smaller 4.5 x 7 km diameter caldera (inner, solid line) is clear.

COPAHUE

CAVW number (active vent)	Latitude Longitude (degrees)	Diameter (km)	Local tectonic setting	Precaldera edifice	SiO_2 content (wt pct)	Age of collapse (yr B.P.)	Date of unrest	Type and duration of unrest ESTU STHF MGTF H Te	Eruption type
15-07-09 (Copahue)	37.85S 71.17W	ca. 10	Compr	Strato	silicic	Pleistocene	18th or 19th cent.	---- ---- ---- - --	unknown
							1867?	---- ---- ---- - --	unknown
							1937	---- ---- ---- - --	unknown
							1940-45	---- ---- --xx x --	none

TECTONIC SETTING

Copahue is on the Chilean/Argentinian border, at the northern end of the NNE-SSW-trending Liquiñe-Reloncaví fault zone and at or just south of the intersection of that fault zone with a transverse, NW-SE fault zone (Moreno R., 1976; Servicio Nacional de Geologia y Mineria, 1980).

GEOLOGIC HISTORY

Copahue consists of a central cone with several craters, built inside a large caldera (fig. 15.9.1). Intense solfataric activity has occurred in the eastern crater, which is filled with hot briny water.

HISTORICAL ACTIVITY

A notable increase in temperature and total acidity of the lake in Los Copahues crater occurred during 1940-45. A maximum temperature of 90 °C was measured in June 1944. The level of the crater lake dropped as much as 40 m relative to a fixed platform, in an inverse relation to temperature and acidity. Casertano (1964) concluded that drought in the early 1940's caused lake level to drop, and thus temperature and acidity to rise.

COPAHUE (continued)

REFERENCES

Casertano, L., 1964, Some reflections on the fumarolic manifestations of the Los Copahues crater: Bull. Volcanol., v. 27, p. 197-215.
Gonzales-Ferran, O., 1971, Geologia y evolucion del Volcán Copahue: unpubl. rep., Santiago, Chile, Dir. Fronteras y Limites.
Gonzales-Ferran, O., and Vergara, M.M., 1962, Reconocimiento geologico de la Cordillera de los Andes entre los paralelos 35° y 38° Lat. Sur.: Santiago, Dept. Geol., Univ. Chile, Publ. 24, p. 1-121.
Moreno R., H., 1974, Airplane flight over active volcanoes of central-south Chile: Guidebook, Excursion D-3, Symp. on Andean and Antarctic Volcanology Problems: Naples/Santiago, IAVCEI, 56 p.
----- 1976, The Upper Cenozoic volcanism in the Andes of Southern Chile (from 40°00′ to 41°30′S.L.), <u>in</u> Gonzales-Ferran, O., ed., Symp. on Andean and Antarctic volcanology problems: Naples, IAVCEI, p. 143-171.
Servicio Nacional de Geologia y Mineria, 1980, Mapa geologico de Chile: Santiago, Serv. Nac. Geol. Min., scale 1:1,000,000, sheet 4 of 6.

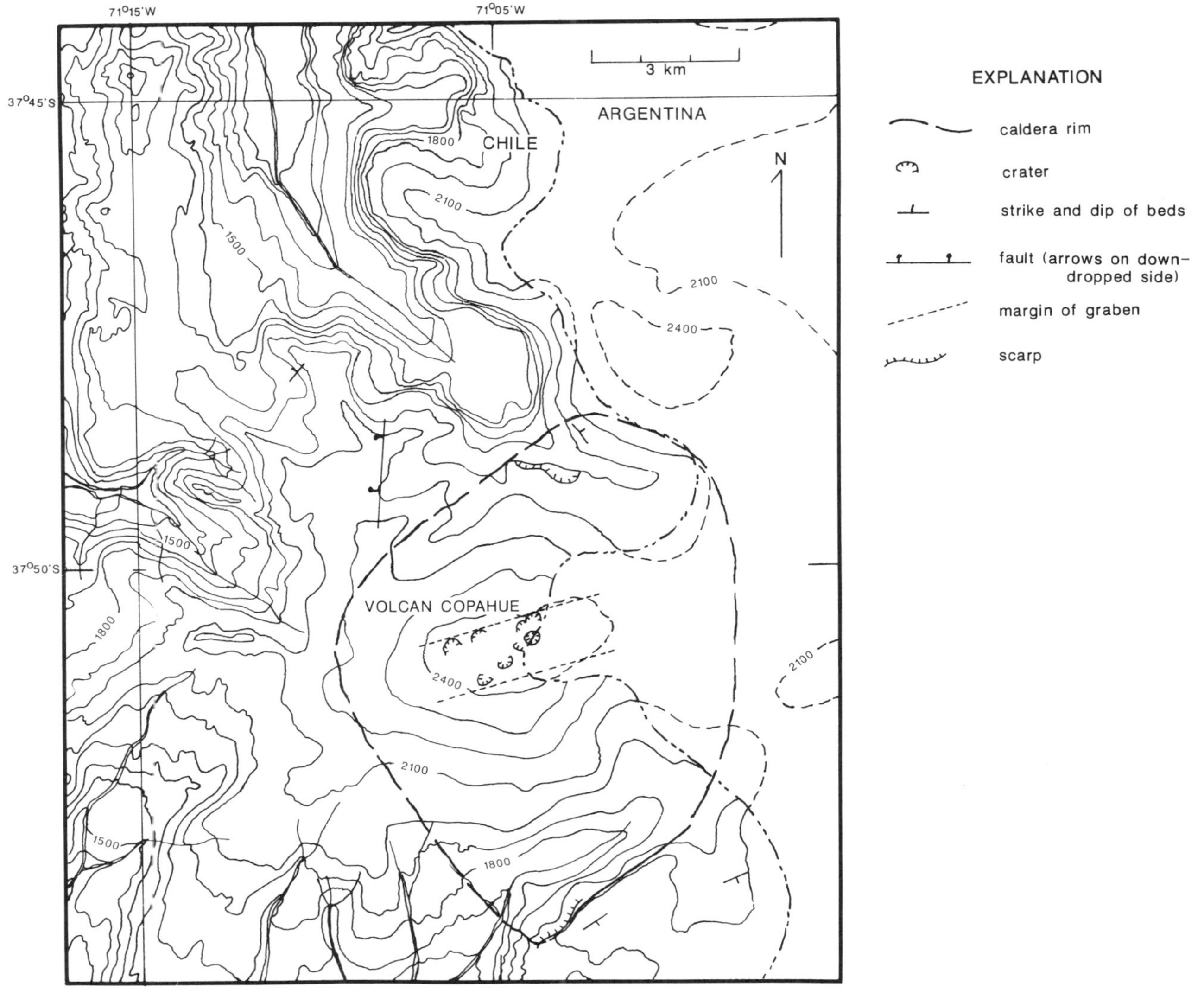

Figure 15.9.1. Volcán Copahue and approximate boundary of Copahue Caldera (Hugo Moreno R., written commun., 1988). Several craters are nested within a summit graben. Contour interval, 100 m within Chile and 300 m within Argentina (latter sketched from aeronautical chart). Topographic contours may not be accurate near border.

PART 3: HISTORICAL UNREST AT LARGE QUATERNARY CALDERAS (continued)
See inside back cover for explanation and abbreviations

VILLARRICA

CAVW number (active vent)	Latitude Longitude (degrees)	Diameter (km)	Local tectonic setting	Precaldera edifice	SiO_2 content (wt pct)	Age of collapse (yr B.P.)	Date of unrest	Type and duration of unrest ESTU STHF MGTF H Te	Eruption type
15-07-12 (Villarrica)	39.42S 71.95W	9 x 6 (outer) 2.2 (inner)	Compr	Strato	R = mafic C = 68-71		1558	---- ---- ---- - --	ex
							1575	---- ---- ---- - --	Ex?
							1640	---- ---- ---- - --	ex, lf
							1647	---- ---- ---- - --	lf, Ex
							1657	---- ---- ---- - --	unknown
							1730	---- ---- ---- - --	unknown
							1737	---- ---- ---- - --	unknown
							1777	---- ---- ---- - --	ex
							1787	---- ---- ---- - --	lf
							1790	x--- ---- ---- - --	ex
							1792	x--- ---- ---- - --	ex
							1796	x--- ---- ---- - --	ex
							1799	x--- ---- ---- - --	ex
							1801	x--- ---- ---- - --	ex
							1806	---- ---- ---- - --	ex
							1822*	---- ---- ---- - --	ex
							1832	---- ---- ---- - --	unknown
							1837	---- ---- ---- - --	unknown
							1852*	---- ---- ---- - --	ex
							1860*	---- ---- ---- - --	ex
							1867	---- ---- ---- - --	unknown
							1869*	---- ---- ---- - --	ex
							1874-76*	---- ---- ---- - --	ex
							1879	---- ---- ---- - --	unknown
							1883	---- ---- ---- - --	ex
							1893*	---- ---- ---- - --	ex
							1897-98	---- ---- ---- - --	ex
							1906-09	x--- ---- ---- - --	ex, lf
							1913	---- ---- ---- - --	Ex
							1915-18*	---- ---- ---- - --	ex
							1919	---- ---- ---- - --	unknown
							1920	B--- ---- ---- - --	ex

VILLARRICA, Region 15, CAVW number 15-07-12

VILLARRICA (continued)

CAVW number (active vent)	Latitude Longitude (degrees)	Diameter (km)	Local tectonic setting	Precaldera edifice	SiO_2 content (wt pct)	Age of collapse (yr B.P.)	Date of unrest	Type and duration of unrest ESTU STHF MGTF H Te	Eruption type
(continued from previous page)							1929*	---- ---- ---- - --	ex
							1933*	---- ---- ---- - --	ex
							1935*	---- ---- ---- - --	ex
							1938*	---- ---- ---- - --	ex
							1948-49	---- ---- ---- - --	Ex,lf
							1959-61	---- ---- ---- - --	ex
							1963-64	---- ---- ---- - --	Ex,lf
							1971-72	---- ---- ---- - --	Ex,lf
							1979-80	---- ---- ---E - --	pex,ex
							1983-85	DD-Y ---- ---- - --	lf,ex

* not shown in Riffo and others (1987)

TECTONIC SETTING

Villarrica is situated at the intersection of the major NNE-SSW Liquiñe-Reloncaví transcurrent fault zone and a northwest-trending alignment of volcanoes with Villarrica at its center (Moreno R., 1976; Servicio Nacional de Geologia y Mineria, 1980).

GEOLOGIC HISTORY

Villarrica stratovolcano, a symmetrical, mafic cone, has largely filled a 9 km x 6 km diameter caldera (fig. 15.10.1), which formed during or following eruption of silicic dacite (H. Moreno R., written commun., 1988). The rim of a smaller caldera appears at the 2400-m level on Villarrica stratovolcano as a slight irregularity in that volcano's symmetry (Moreno R. and others, 1980; R. Hickey-Vargas and others, written commun., 1988; H. Moreno R., written commun., 1988).

HISTORICAL ACTIVITY

Villarrica erupts frequently, with a monthly probability of 1.45×10^{-2}, and without regard to the length of time since the previous eruption (Muñoz, 1983, 1984). Typically, eruptions exhibit subterranean noises, loud and repeated bangs from the crater, columns of smoke and vapor, fountains of incandescent debris, ash columns that rise several kilometers above the summit, and lava flows (Flores, 1951).

1558, 1575: The town of Villarrica was destroyed twice; the second time, 350 persons were killed (Riffo and others, 1987).

VILLARRICA (continued)

HISTORICAL ACTIVITY (continued)

1640: Lava reached Lake Villarrica and the valley of the Rio Toltén. The area was visited by the Marquis of Bayde, who, aided by the eruption, negotiated peace with the indigenous Araucanian people (Riffo and others, 1987).

1906-08: In 1906, strong subterranean noises were heard, but little lava emission occurred (Riffo and others, 1987). Then, on 31 October 1908, "there was a great spout of fire that caused an avalanche or flow of rock, snow, and ice from the glacier on the east side of Villarrica...There was no earthquake at this time although there had been one in 1907" (Stone, 1935). A minor eruption occurred on 21 December 1908.

1920: "On December 9, 1920 a severe earthquake threw down the porch on the front of [a house 6 km north of the crater] and moved a large shed three inches off its foundations. Furniture in the house moved from east to west [obliquely down the mountain]. There were three short strong shocks, and slight shakes followed all the night of the ninth at intervals of about five minutes with a stronger shake every half hour. [Nearby] other houses and stables were knocked down. Early in the morning of the tenth began an explosive eruption whose cannonading rattled the windows in the house and lasted for about 36 hours. In the night of the 10th a great fountain of fire was seen..." (Stone, 1935). H. Moreno R. (written commun., 1988) described the 1920 eruption as "small tephra emissions, lava fountain(s), and several explosions."

1948-49: Eruptions occurred between 9 October 1948 and 3 February 1949. Moderately large explosions on 18 October, 1 January, and 31 January generated tephra and large avalanches (pyroclastic flows?) in five river drainages; 60 persons died (H. Moreno R., written commun., 1988). In January 1949 lava flows descended toward Coñaripe. Great forests were burned (Riffo and others, 1987).

1959-61: Weak explosions and minor tephra emission occurred.

1963-64: In 1963, avalanches of unspecified origin were directed to Lake Calafquén, cutting five bridges. Mudflows in March 1964 overwhelmed half of a small village, killing 22 persons (Riffo and others, 1987). H. Moreno R. (written commun., 1988) report tephra emission, lava flows, and lahars.

1971-72: Eruptions beginning on 29 October 1971 intensified in November and December, generating mudflows and tephra and killing 17 persons (H. Moreno R., written commun., 1988). Fluid lava flowed as much as 14 km from the vent (Riffo and others, 1987).

1979-80: Fumarolic activity increased in late September 1979 and small tephra eruptions occurred from 20 June-September 1980.

1983: Weak explosions and minor tephra emission occurred on 14 October (H. Moreno R., written commun., 1988). Reported pyroclastic flows onto ice-covered slopes are doubtful. A red glow was seen at the summit (Smithsonian Institution, 1983).

1984-85: A brief ash eruption, accompanied by an increase in seismicity, occurred on 11 August 1984. Lava extrusion and small explosions began on 30 October, and intense shallow seismicity (mostly B-type earthquakes) was recorded by a seismometer on the

VILLARRICA (continued)

HISTORICAL ACTIVITY (continued)

northwest flank of the volcano during 4-5 November 1984. Then tremor and underground rumbling were reported from Pucon, at a time when no further earthquakes were recorded by the seismometer. Strombolian explosions and lava extrusion occurred during 10-13 November, and weak eruptive activity and tremor continued until 17 November. On 18 November, the southwest slope between 2,200 and 2,800 m above sea level was seen to have bulged visibly, until more lava was extruded from the central crater and the northeast flank (fig. 15.10.2). Most eruptive activity ended in January 1985 (fig. 15.10.3), but minor ash emissions and red glow were observed until 18 April (Smithsonian Institution, 1984; Acevedo and Fuentealba, 1987, H. Moreno R., written commun., 1988). Seismicity in 1983 was dominated by A-type earthquakes, but B-type earthquakes and tremor became dominant in 1984 (Acevedo and Fuentealba, 1987).

Seismicity (B-type earthquakes and harmonic tremor) increased again from June to November 1985 (fig. 15.10.4) (Smithsonian Institution, 1985; H. Moreno R., written commun., 1985; Acevedo and Fuentealba, 1987).

REFERENCES

Acevedo A., P., and Fuentealba C., G., 1987, Antecedentes de la actividad microsismica del Volcán Villarrica relacionada con la erupción de Octubre de 1984: Bol. Vulcanol. (Costa Rica), no. 18, p. 13-17.

Flores S., E., 1951, La ultima erupción del Volcán Villarrica: Informaciones Geograficas, Santiago, v. 1, p. 59-62.

Moreno R., H., 1974, Airplane flight over active volcanoes of central-south Chile, in Gonzales-Ferran, O., ed., Guidebook, Excursion D-3, Symp. on Andean and Antarctic volcanology problems: Naples/Santiago, IAVCEI, 56 p.

----- 1976, The Upper Cenozoic volcanism in the Andes of Southern Chile (from 40°00′ to 41°30′S.L.), in Gonzales-Ferran, O., ed., Symp. on Andean and Antarctic volcanology problems: Naples, IAVCEI, p. 143-171.

Moreno R., H., Marangunic D., C., and Lahsen A., A., 1980, Recopilacion geologica de la zona de Pucon: Santiago, unpubl. rep., 18 p.

Muñoz, M., 1983, Eruption patterns of the Chilean volcanoes Villarrica, Llaima, and Tupungatito: Pure Appl. Geophys., v. 121, p. 835-852.

----- 1984, Probabilidad de erupción en el Volcán Villarrica en los proximos años: Tralka, Dept. Geofis., Univ. Chile, v. 2, no. 3, p. 323-325.

Riffo, P., Fuentealba, G., and Urra, L.H., 1987, Sintesis de las erupciónes del Volcán Villarrica (Chile): Bol. Vulcanol. (Costa Rica), no. 18, p. 8-11.

Servicio Nacional de Geologia y Mineria, 1980, Mapa geologico de Chile: Santiago, Serv. Nac. Geol. Min., scale 1:1,000,000, sheet 4 of 6.

Smithsonian Institution, Scientific Event Alert Network (SEAN), 1983, 1984, 1985, Villarrica: SEAN Bull., v. 8, no. 10, p. 12; v. 9. no. 11, p. 2-4; v. 10, no. 1, p. 3; v. 10, no. 11, p. 6 (information from O. Gonzalez-Ferran, G. Fuentealba C., P. Riffo A., P. Acevedo, and H. Moreno R.).

Stone, J.B., 1935, The volcanoes of Southern Chile: Zeit. Vulkanol., v. 16, p. 81-97.

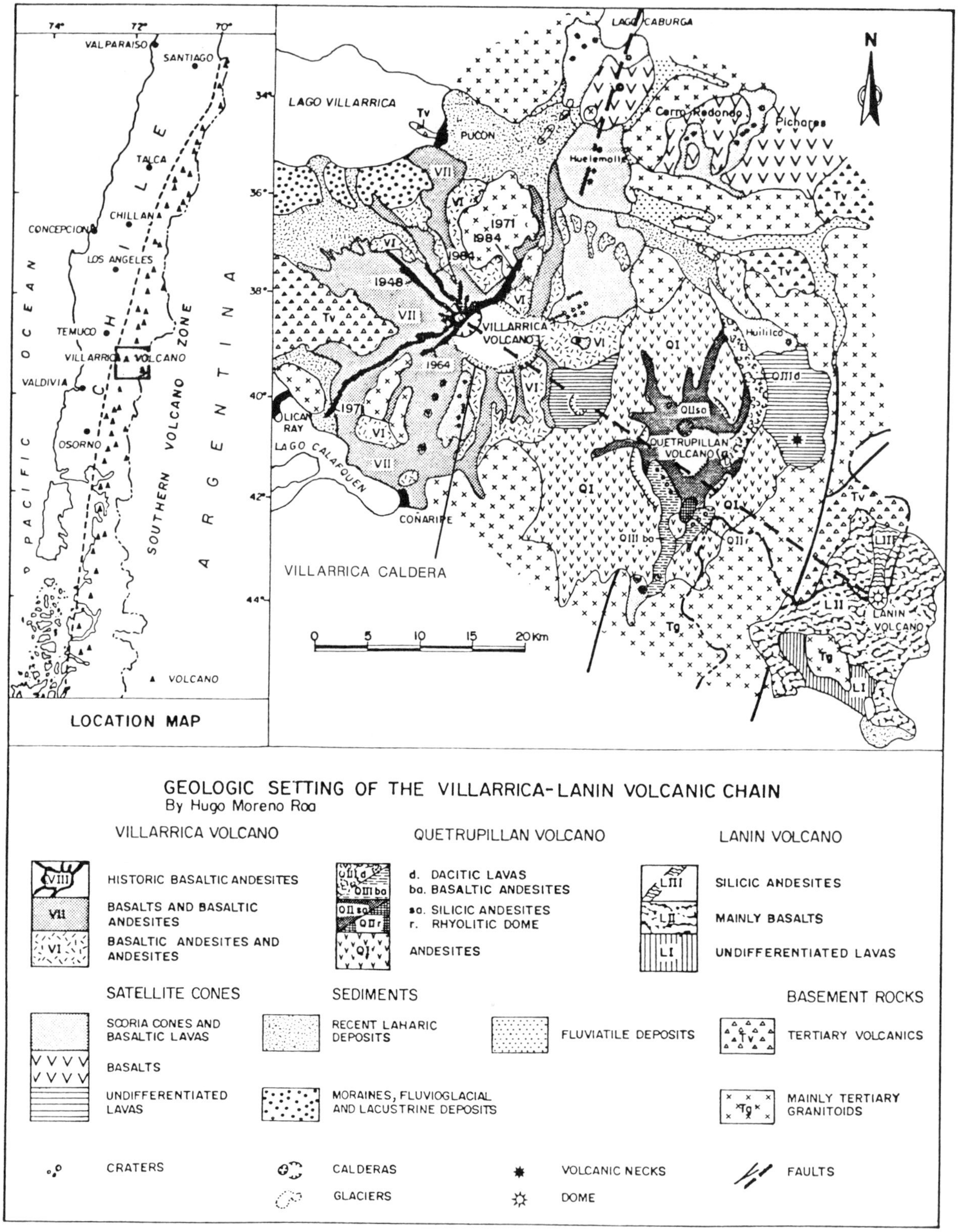

Figure 15.10.1. Geologic map of Villarrica-Lanin volcanic chain, showing locations of Villarrica Volcano and Caldera, from Moreno R. and others (1988).

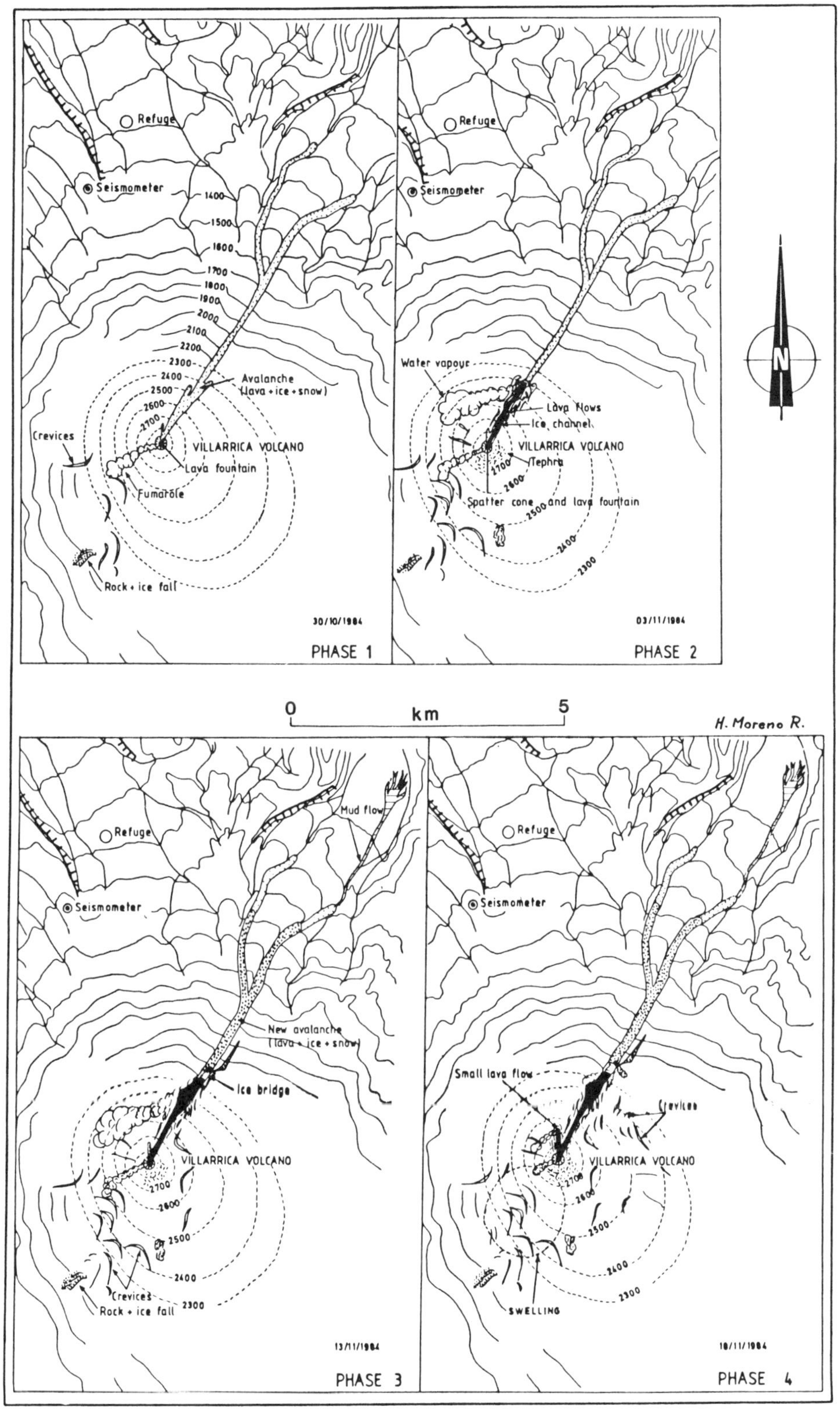

Figure 15.10.2. Sketch maps of Villarrica Volcano showing deposits of phases 1-4 of the 1984 eruption. Eruptive phases occurred on 30 October and 3, 13, and 18 November 1984, respectively (Smithsonian Institution, 1984; courtesy of Hugo Moreno R.). Note swelling of southwest flank during phase 4.

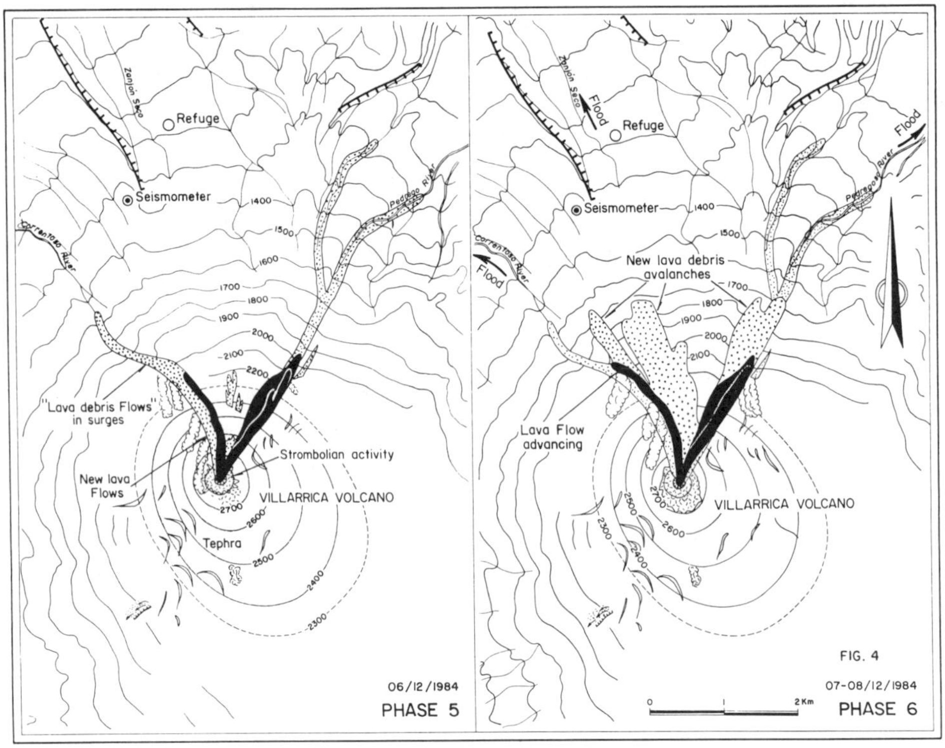

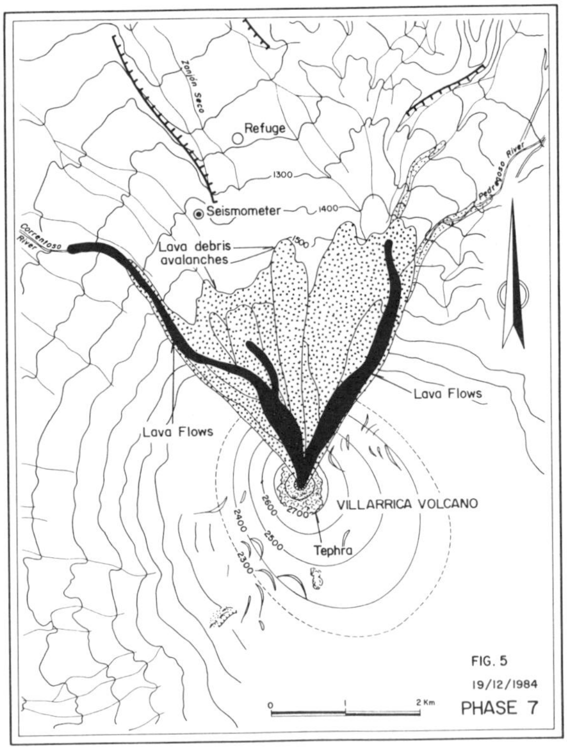

Figure 15.10.3. Sketch maps showing deposits of phases 5-7 of the 1984 eruption of Villarrica Volcano, from Moreno R. and others (1988).

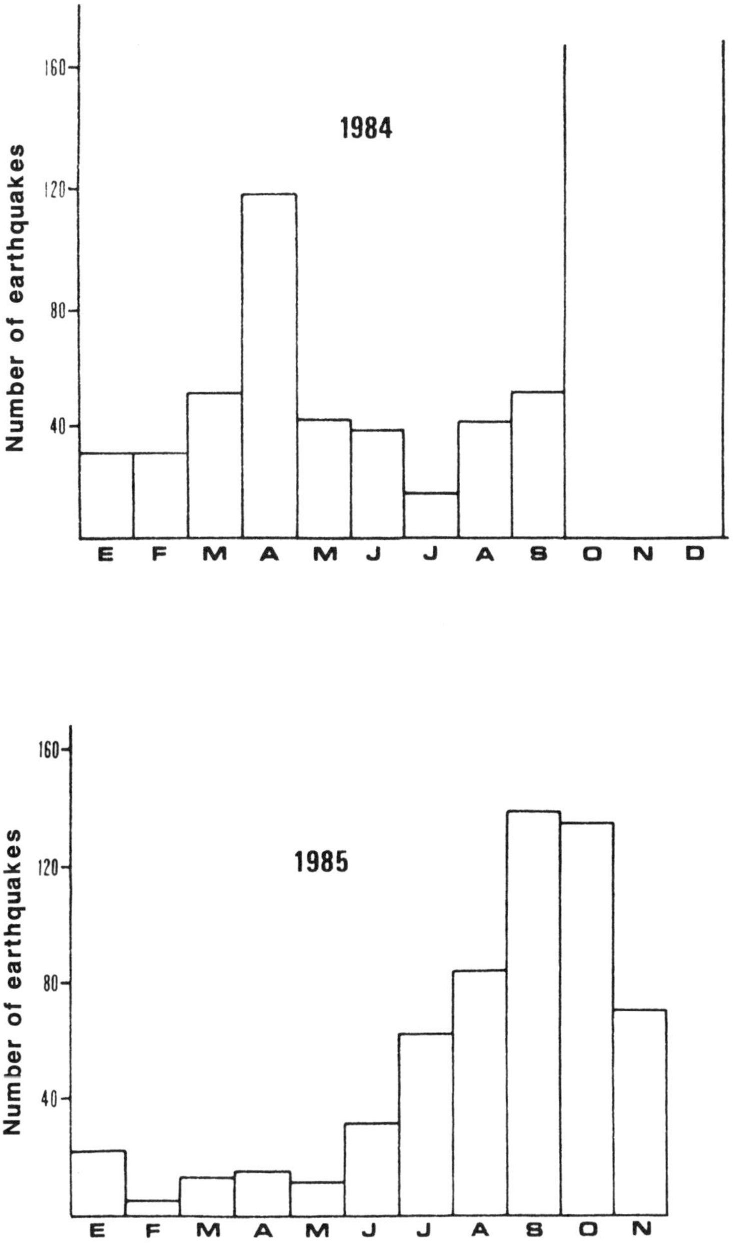

Figure 15.10.4. Monthly number of earthquakes at Villarrica during 1984 (top) and January-November 1985 (bottom), from Acevedo and Fuentealba (1987).

PART 3: HISTORICAL UNREST AT LARGE QUATERNARY CALDERAS (continued)
See inside back cover for explanation and abbreviations

CORDILLERA NEVADA

CAVW number (active vent)	Latitude Longitude (degrees)	Diameter (km)	Local tectonic setting	Precaldera edifice	SiO_2 content (wt pct)	Age of collapse (yr B.P.)	Date of unrest	Type and duration of unrest ESTU STHF MGTF H Te	Eruption type
15-07-141 (Cordón Caulle)	40.46S 72.25W	9	Compr		R = 64-71 C = 62-74	probably 100,000-200,000	1893?	---- ---- ---- - --	lf?
							1921-22	x--- ---- ---- - --	EX,lf
							1929	---- ---- ---- - --	ex
							1934	x--- ---- ---- - --	ex
							1960	C--- ---- ---- - QU	Ex,lf

TECTONIC SETTING

The Cordillera Nevada Caldera lies just west of the Liquiñe-Reloncaví line, a major N10°E strike-slip fault that spans much of the length of the south-central Chilean volcanic belt. It also lies at the apex of two segments of a conjugate set of N60°E and N50°W faults that are at least in part tension features associated with movement along the Liquiñe-Reloncaví Fault. Many of the volcanoes in this area, including the Cordillera Nevada center, the Cordón Caulle fissures, and Puyehue, have grown along this set of conjugate faults (Moreno R., 1976; Servicio Nacional de Geologia y Mineria, 1980).

GEOLOGIC HISTORY

The Cordillera Nevada Caldera (fig. 15.11.1), mapped by Moreno R. (1977), is the only known source of rhyolitic tuff in the southern Andes (Lopez-Escobar, 1984). The age of the Cordillera Nevada Caldera is unknown; extensive glaciation indicates that it is at least 16,000 years old, and probably 100,000-200,000 yr old (H. Moreno R., written commun., 1988). Cordón Caulle is a group of postcaldera silicic vents along WNW-ESE-trending fissures. The group has been classed with Puyehue Volcano (CAVW number 15-07-15) (Casertano, 1963), but Moreno R. (1977; oral commun., 1985) considers it to be independent of Puyehue. Some of the Cordón Caulle vents occur within the Cordillera Nevada Caldera, while others are outside the caldera--a relation not unlike that of the Inyo domes and the Long Valley Caldera in California, USA. Recent lavas of the Cordón Caulle group are dacitic, but subordinate basalt was erupted in the early development of this group of vents (Moreno R., 1977). It is not known if magma of the Cordón Caulle group is related to any beneath the Cordillera Nevada Caldera. An alternate name for Cordón Caulle, "Los Azufres," refers to solfataric activity that existed before the 1921-22 eruption.

Several maars, including Nilahue, occur in the nearby Carran group but bear no direct relation to the Cordillera Nevada Caldera or the Cordón Caulle group.

CORDILLERA NEVADA (continued)

HISTORICAL ACTIVITY

Simultaneous eruptions of dacite lava from several fissure vents occurred in 1921-22 and 1960, as shown in fig. 15.11.2. The 1921-22 eruption began with explosive activity on 13 December 1921. Ash and pumice were produced in great quantity, after earthquakes and a deafening roar. Further explosions on 19-21 December sent ash 62 km high. Ash fell in La Plata, 3000 km to the northeast, and was seen over Europe and southwest Africa. Lava extrusion began in February 1922 (Krumm, 1923; Hantke, 1940).

A small eruption occurred in 1934, after an earthquake (Hantke, 1940).

The 1960 eruption is one of the best-documented examples of an eruption triggered by a tectonic earthquake. A $\underline{M}$=8.3 earthquake occurred on 22 May 1960, some 300 km northwest of the caldera, and a strong explosive eruption began approximately 48 hours later, followed by extrusion of lava. Dacitic lava was extruded from multiple vents along a 14-km-long set of NW-SE-trending fissures. From the start of the eruption, and continuing for 6 days, ash obscured the sky (Katsui and Katz, 1967; Moreno R., 1977).

COMMENTS

Eruptions of Cordón Caulle are pertinent to unrest at Long Valley Caldera for two reasons. First, multiple, simultaneous dike-fed eruptions of dacitic magma might be a modern analog of the closely spaced eruptions along the Mono Craters - Inyo domes chain, ca. 550 yr B.P. Second, the eruption of Cordón Caulle in 1960 was triggered by a distant tectonic earthquake. Even though the likelihood of an eruption at Long Valley seems relatively low at the time of this writing (December 1987), a large earthquake is still expected in the White Mountains seismic gap. If part of a regional extension event, such an earthquake could conceivably trigger a repetition of the Mono Craters - Inyo domes dike intrusion and eruptions, just as the great 1960 Chilean earthquake apparently triggered the 1960 eruption of Cordón Caulle.

REFERENCES

Casertano, L., 1963, Catalogue of active volcanoes of the world including solfatara fields, Pt. XV, Chilean Continent: Rome, IAVCEI, 55 p.

Gerlach, D.C., Frey, F.A., Hickey, R., Moreno R., H., and Hildreth, W., 1983, Geochemistry of Puyehue Volcano and Cordón Caulle, Southern Andes (40.5°S): Eos, Trans. Am. Geophys. Union, v. 64, p. 326.

Hantke, G., 1940, Das Vulkangebiet zwischen den Seen Ranco und Puyéhue in Süd-Chile: Bull. Volcanol., ser. 2, v. 7, p. 75-83.

Katsui, Y., and Katz, H.R., 1967, Lateral fissure eruptions in the southern Andes of Chile: Jour. Fac. Sci., Hokkaido Univ., ser. IV, v. 13, no. 4, p. 433-448.

Krumm, F., 1923, Topographische und geologische Nachrichten über die Gegend östlich des Ranco-Sees in Süd-Chile, im besondern über den jüngsten Vulkan-Ausbruch "Los Azufres" (Dezember 1922)(sic): Geol. Rundschau, v. 14, p. 146-150.

CORDILLERA NEVADA (continued)

REFERENCES (continued)

Lopez-Escobar, L., 1984, Petrology and chemistry of volcanic rocks of the southern Andes, in Harmon, R.S., and Barreiro, B.A., eds., Andean magmatism--chemical and isotopic constraints: Nantwich, Cheshire, Shiva Publ. Ltd., p. 47-71.

Moreno R., H., 1976, The Upper Cenozoic volcanism in the Andes of Southern Chile (from 40°00' to 41°30'S.L.), in Gonzales-Ferran, O., ed., Symp. on Andean and Antarctic volcanology problems: Naples, IAVCEI, p. 143-171.

----- 1977, Geologia del area volcanica Puyehue-Carran: Santiago, Univ. Chile, Memoria de Titulo, 181 p.

Philippi, R.S., 1869, Die heissen Quellen am Puyehue--und Llauquihue-See in Chile: Petermanns Mitt., 1869, p. 459-461.

Servicio Nacional de Geologia y Mineria, 1980, Mapa geologico de Chile: Serv. Nac. Geol. Min., scale 1:1,000,000, sheet 4 of 6.

Stone, J.B., 1935, The volcanoes of Southern Chile: Zeit. Vulkanol., v. 16, p. 81-97.

Tazieff, H., 1962, Quelques observations sur la crise séismo-volcanique de mai 1960 au Chili central: Bull. Volcanol., v. 24, p. 83-86.

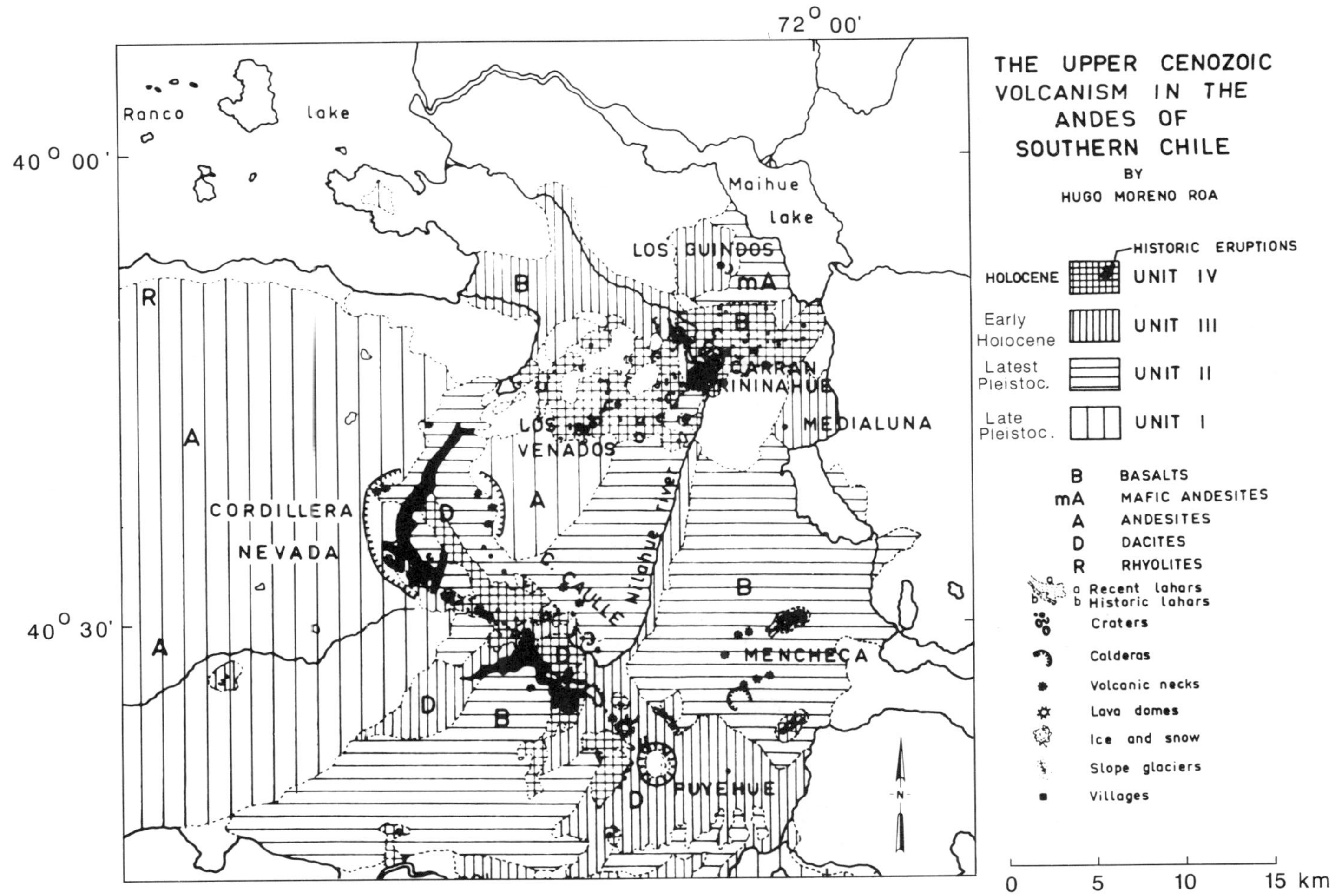

Figure 15.11.1. Geologic map of Cordillera Nevada Caldera, after Moreno R. (1976).

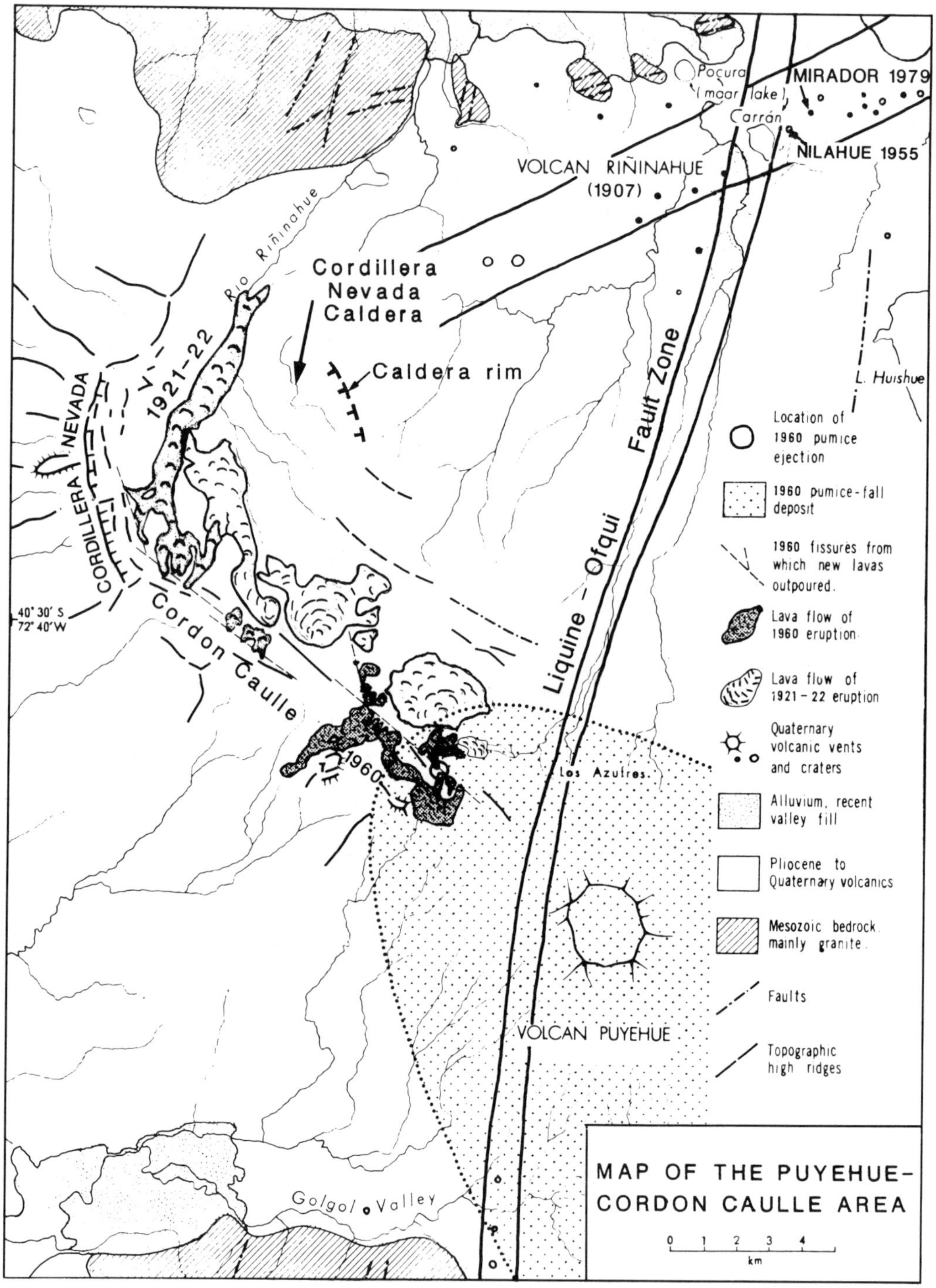

Figure 15.11.2. Vents and lava flows of the 1921-22 and 1960 eruptions of Cordón Caulle, which occurred along northwest-trending fractures that connect Puyehue with the Cordillera Nevada Caldera (see figure 15.11.1). 1955 Nilahue Maar is shown in upper right. Figure from Katsui and Katz (1967), modified by Hugo Moreno Roa (written commun., 1988).

MT. HUDSON

CAVW number (active vent)	Latitude Longitude (degrees)	Diameter (km)	Local tectonic setting	Precaldera edifice	SiO_2 content (wt pct)	Age of collapse (yr B.P.)	Date of unrest	Type and duration of unrest ESTU STHF MGTF H Te	Eruption type
15-08-057 (Mt. Hudson)	45.91S 72.92W	9	Several intersecting fault trends; youngest, E-W	Strato	R = 59-61 C = ?	Preglacial	1970-71	x--- ---- --EE E --	Ex
							1973	---- ---- --D- - --	none?, or subgl

TECTONIC SETTING

Mt. Hudson lies in a stretch of the southern Andes that has relatively few active volcanoes. A N-S fault (one of many extending much of the length of Chile) passes several kilometers west of Mt. Hudson (Servicio Nacional de Geologia y Mineria, 1980).

GEOLOGIC HISTORY

The caldera of Mt. Hudson (fig. 15.12.1) is ice filled and was not known until shortly before the unrest in 1970-71 (Fuenzalida Ponce and Espinosa, 1973).

HISTORICAL ACTIVITY

1970-71: Abnormally high level of a river was observed in April 1970, with marked fluctuations not correlated with rainfall. The caldera is ice filled, and melting of ice presumably was responsible for increased and fluctuating flow. An explosive eruption occurred on 12 August 1971. Shallow local seismicity, with MMI_{max} = 5, was reported, but there was no indication whether it preceded or accompanied the eruption.

1973: Ground temperatures rose suddenly in April, causing 70 percent of the ice cover of Mt. Hudson to melt. No eruption was observed, but it is possible that a small eruption occurred under ice.

REFERENCES

Fuenzalida Ponce, R., 1976, The Hudson Volcano, in Gonzales Ferran, O., ed., Symp. on Andean and Antarctic volcanology problems: Naples, IAVCEI, p. 78-87.
Fuenzalida Ponce, R., and Espinosa N., W., 1973, Hallazgo de una caldera volcanica en la Provincia de Aisen: Rev. Geol. Chile, no. 1, p. 64-66.
Godoy, E., Dobbs, M., and Stern, C., 1981, El volcán Hudson, primeros datos quimicos e isotopicos en coladas interglaciales: Communicaciones Dept. Geol. Univ. Chile, v. 32, p. 1-9.
Servicio Nacional de Geologia y Mineria, 1980, Mapa geologico de Chile: Serv. Nac. Geol. Min., scale 1:1,000,000, sheet 5 of 6.
Volcanological Society of Japan, 1973-75, Mt. Hudson: Bull. Volcanic Eruptions, no. 11, p. 43; no. 13, p. 62.

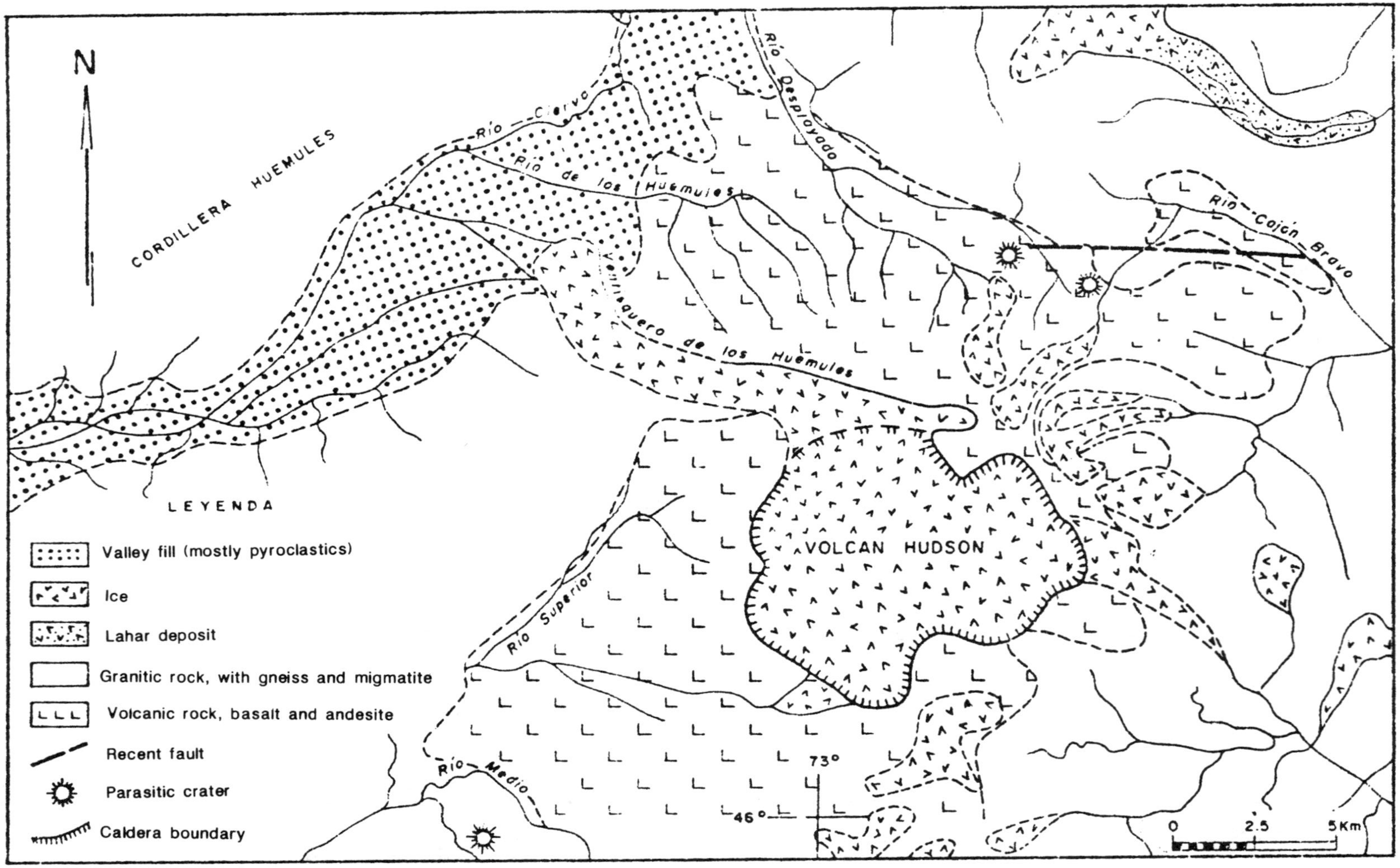

Figure 15.12.1. Geologic sketch map of Mt. Hudson and its caldera, after Fuenzalida Ponce and Espinosa (1973) and Fuenzalida Ponce (1976).

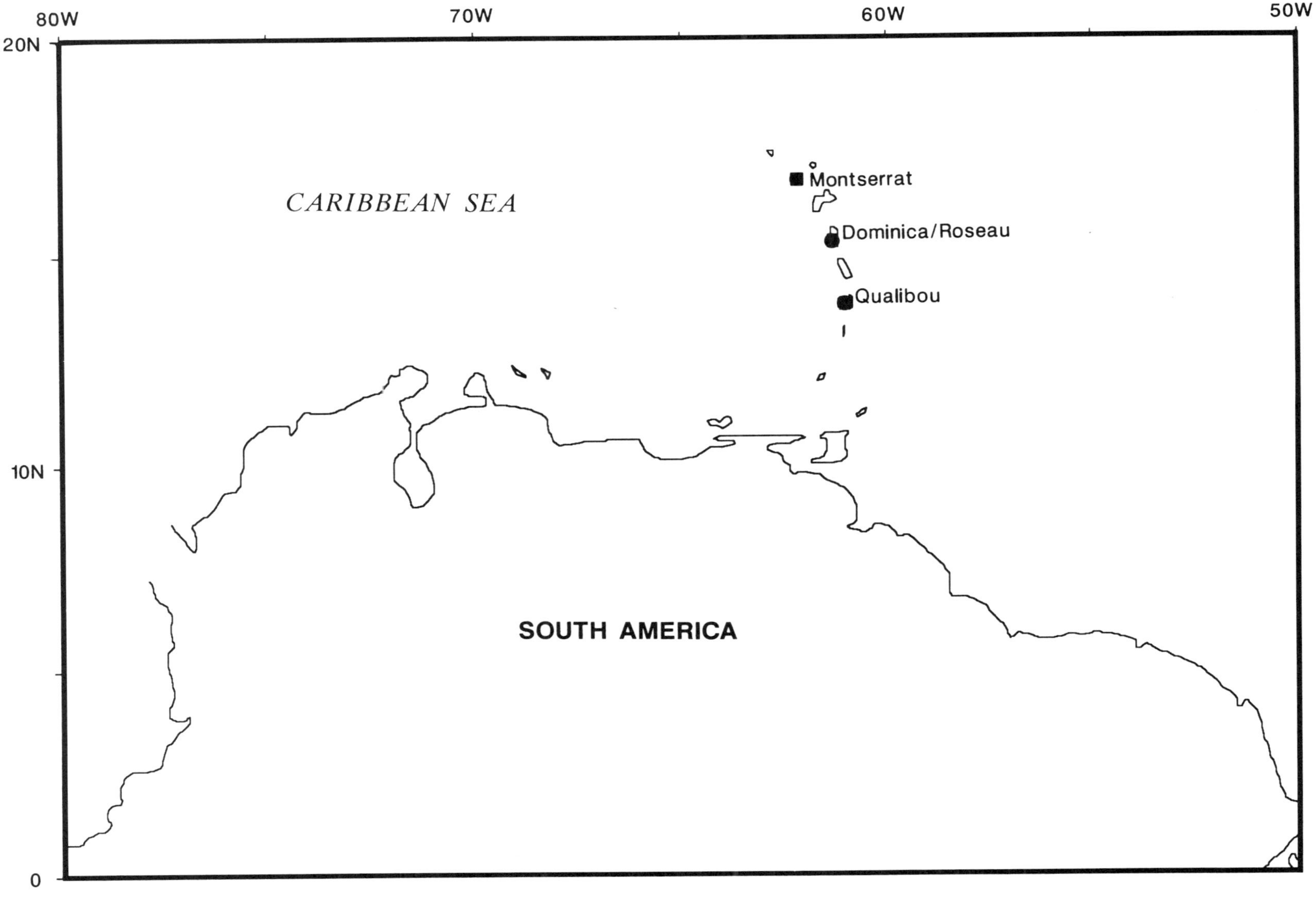

Figure 16. Locations of large, Quaternary restless calderas (solid circles) and a restless noncaldera area (square) in Region 16.

UNNAMED, near Roseau, Dominica

CAVW number (active vent)	Latitude Longitude (degrees)	Diameter (km)	Local tectonic setting	Precaldera edifice	SiO_2 content (wt pct)	Age of collapse (yr B.P.)	Date of unrest	Type and duration of unrest ESTU STHF MGTF H Te	Eruption type
16-00-10	15.30N	10	Compr	Strato?	R = 59-63	30,000	1673	?--- ---- ---- - --	none
(Valley of	61.30W				C = silicic		1765	xx-- ---- ---x - --	none
Desolation;							1843-49	xx-- ---- ---- - --	none
Micotrin)							1879-80	?--- ---- ---x - --	pex
							1967	CC-- ---- ---- - --	none
							1971	x--- ---- ---N N --	none
							1974	xx-- ---- ---N N --	none
							1976	xx-- ---- ---N N --	none

TECTONIC SETTING

Dominica is at the southern end of the northern Lesser Antilles lithospheric segment (Wadge and Shepherd, 1984). In this northern half of the arc, tectonic earthquakes occur along the Benioff zone and also above it, frequently at shallow depths, and swarms of tectonic earthquakes are relatively common (Shepherd, 1988).

GEOLOGIC HISTORY

A caldera roughly 10 km in diameter (fig. 16.1.1) has been postulated (Robson and Tomblin, 1966, referring to a depression bounded by an arcuate fault), rejected (Wadge, 1984; G. Wadge, written commun., 1984; H. Sigurdsson, written commun., 1984), reinterpreted as the relic of a gravitational collapse (Roobol and others, 1983), and repostulated as a normal collapse caldera (G. Heiken, written commun., 1986; J. Shepherd, 1987, written commun.). Robson and Tomblin (1966) considered Morne Trois Pitons, Morne Macaque (= Micotrin), Grande Soufriere Hills, Watt Mountain, and the Valley of Desolation (a solfatara field between the latter three centers) to be a single volcano, perhaps including two other nearby cones, Morne Anglais and Foundland; Wadge (1984) considers Morne Anglais and Foundland to be petrologically distinct from the earlier-mentioned vents, and Shepherd (1988) considers all of these vents to be closely spaced but independent.

HISTORICAL ACTIVITY

The only historical eruption in this area was a small steam explosion in 1880. Small earthquake swarms occur from time to time beneath the several vents of this center (Shepherd, 1988). Most of what is listed in the table and below is of regional tectonic origin and might not be related to the Roseau center; we include it because the precise epicenters are unknown.

UNNAMED, near Roseau, Dominica, Region 16, CAVW number 16-00-10

UNNAMED, near Roseau, Dominica (continued)

HISTORICAL ACTIVITY (continued)

1673: A strong shock (Perret, 1939), described by Robson (1964) as an earthquake of estimated intensity IV, was centered on or near Dominica.

1765: There was a report of "earthquakes and gas" (Perret, 1939); 150 shocks "more violent than any previously felt" (Robson, 1964), occurred in February and March, and a few more were felt until 30 June. Activity was centered somewhere on Dominica.

1816: Severe shocks on 15 August, with estimated intensity V, were centered on or near Dominica.

1838: During June, several "smart shocks", with maximum intensity V, were centered on or near Dominica.

1843-49: On 8 February, a strong tectonic earthquake occurred north(?) of Dominica. It was felt with intensity VIII in Roseau, and with intensity IX elsewhere in Dominica and on Guadeloupe, Antigua, Montserrat, Nevis, and St. Kitts. A relatively small earthquake in December 1843 was felt only in Roseau, and several more small earthquakes were felt during 1844-45, 1847, and 1849. Those in 1849 occurred in swarms and apparently were centered somewhere in the southern half of Dominica. Shepherd (1988) attributes the 1849 (and forerunning?) swarms to the Morne Patates area, which is south of, and probably unrelated to, the Roseau "caldera."

1879-80: An earthquake occurred beneath or near Dominica, with maximum intensity V, on 10 September 1879. A small phreatic explosion occurred from the Valley of Desolation, a solfataric area on the eastern rim of the caldera. The only reported precursor to this eruption was about 2 minutes of intermittent rumbling sounds before a large black cloud rose from the solfatara area. It is not known if the earthquake was precursory to the eruption.

1967-76: Since seismographs were first installed in Dominica in the 1950's, instrumentally recorded earthquake swarms have occurred in June 1967, January 1971, April-May 1974, September-November 1974, and March-April 1976 (Wadge, 1985). The first, between 19 and 21 June 1967, included 66 shallow earthquakes (Shepherd and others, 1971). The September-November 1974 swarm occurred about 1-2 km southwest of Micotrin, at depths mostly less than 3 km; other shallow earthquakes during the same period coincided with the area of the proposed caldera (J. Shepherd, written commun., 1987) (fig. 16.1.2). Events shown in figure 16.1.2 were shown by Wadge (1985) as being farther to the south but were recently relocated in the Roseau area (J. Shepherd, written commun., 1987).

From 10 February 1976 until May 1976, earthquakes 1-3 km deep occurred in a belt extending beneath the caldera and up to 2 km to the southwest. After the initial swarm on 10 February, fewer than 1 earthquake per day occurred until a M=5.9 earthquake occurred 170 km north of Dominica on 10 March. The largest event, on 14 March, was felt with a local MMI=VI but was not recorded on the worldwide seismic network. Solfataric activity at the nearby Watten Waven Soufriere did not change (Smithsonian Institution, 1976). The event has been interpreted as a magmatic intrusion (J. Tomblin, cited in Smithsonian Institution, 1976) or slumping in the Roseau delta (H. Sigurdsson, written commun., 1984).

UNNAMED, near Roseau, Dominica (continued)

COMMENTS

We include the Roseau basin here as a possible caldera and source of the voluminous (30 km^3 bulk volume) Roseau ash, because (1) we find it difficult to imagine that an eruption that large would not form a caldera, and (2) earthquake locations coincide with the area of the postulated caldera. The solfatara field and andesite domes noted above also occur within the proposed caldera.

REFERENCES

Carey, S.N., and Sigurdsson, H., 1980, The Roseau ash: deep sea tephra deposits from a major eruption on Dominica, Lesser Antilles: Jour. Volcanol. Geotherm. Res., v. 7, p. 67-86.

Perret, F.A., 1939, The volcano-seismic crisis at Montserrat, 1933-1937: Carnegie Inst. Washington Publ. 512, 76 p.

----- 1950, Volcanological observations: Carnegie Inst. Washington Publ. 549, 153 p.

Robson, G.R., 1964, An earthquake catalogue for the eastern Caribbean: Bull. Seismol. Soc. Am., v. 54, p. 785-832.

Robson, G.R., and Tomblin, J.F., 1966, Catalogue of the active volcanoes of the world including solfatara fields, Pt. XX, West Indies: Rome, IAVCEI, 56 p.

Roobol, M.J., Wright, J.V., and Smith, A.L., 1983, Calderas or gravity-slide structures in the Lesser Antilles Island arc?: Jour. Volcanol. Geotherm. Res., v. 19, p. 121-134.

Shepherd, J.B., 1988, Eruptions, eruption precursors and related phenomena in the Lesser Antilles: Bull. Volcanol. [in press].

Shepherd, J.B., Tomblin, J.F., and Woo, D.A., 1971, Volcano-seismic crisis in Montserrat, West Indies, 1966-1967: Bull. Volcanol., v. 35, p. 143-163.

Sigurdsson, H., 1972, Partly-welded pyroclast flow deposits in Dominica, Lesser Antilles: Bull. Volcanol., v. 36, p. 148-163.

Smithsonian Institution, Scientific Event Alert Network (SEAN), 1976, Micotrin/Dominica, West Indies: SEAN Bull., v. 1, nos. 7, 8, 10 (information from J. Tomblin, W. Person, and W. Aspinall).

Sparks, R.S.J., Sigurdsson, H., and Carey, S.N., 1980, The entrance of pyroclast flows into the sea, pt. 1. Oceanographic and geologic evidence from Dominica, Lesser Antilles: Jour. Volcanol. Geotherm. Res., v. 7, p. 87-96.

Wadge, G., 1984, A preliminary volcanic hazards study of Dominica, West Indies: Spec. Publ. 1984-5, Seismic Res. Unit, Univ. West Indies, Trinidad.

----- 1985, Morne Patates volcano, southern Dominica, Lesser Antilles: Geol. Mag., v. 122, no. 3, p. 253-260.

Wadge, G., and Shepherd, J.B., 1984, Segmentation of the Lesser Antilles subduction zone: Earth Planet. Sci. Lett., v. 71, p. 297-304.

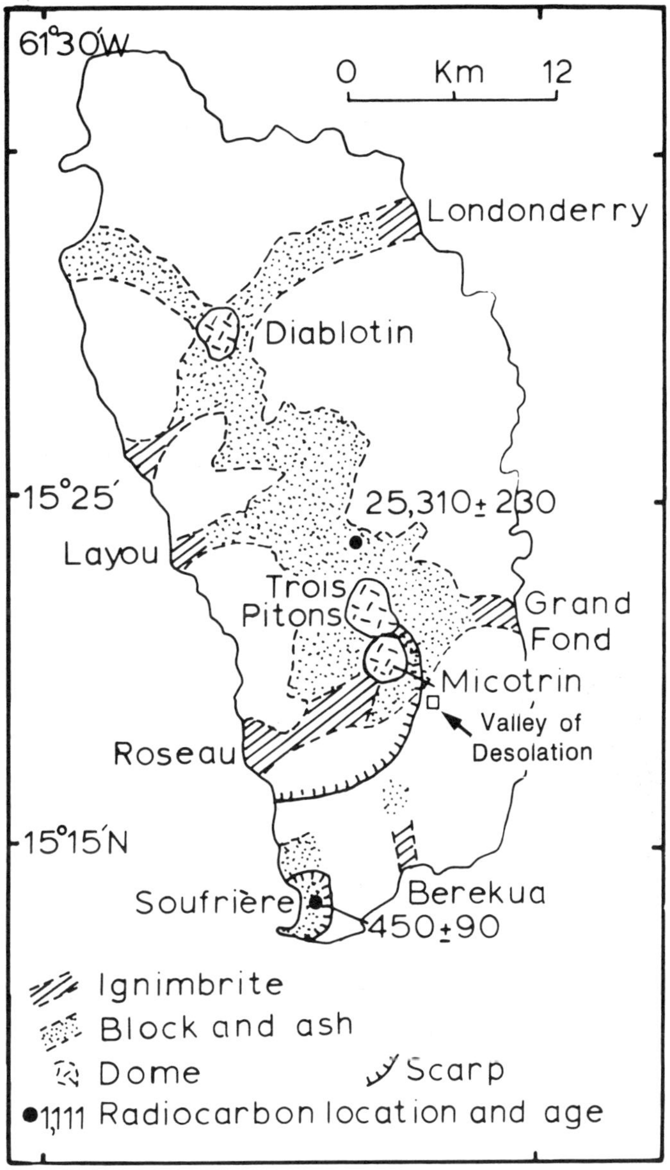

Figure 16.1.1. Unnamed caldera (suggested name, "Roseau") in Dominica, Lesser Antilles. Valley of Desolation is a solfatara field on caldera rim; Micotrin and Trois Pitons are postcaldera domes. Reproduced from Roobol and others (1983), who interpreted the feature as a gravity-slide structure rather than a collapse caldera.

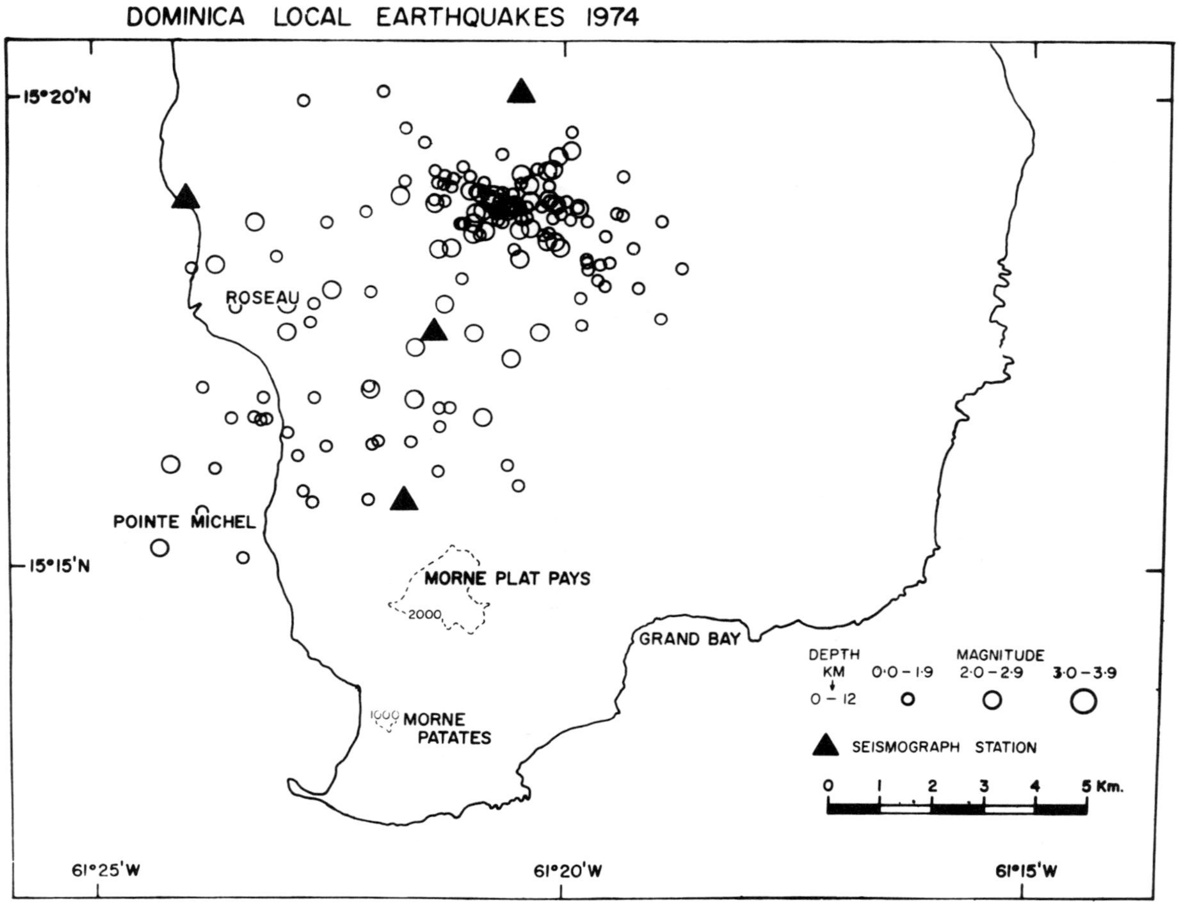

Figure 16.1.2. Locations of earthquakes that occurred during a swarm near Roseau "caldera" in 1974, from John B. Shepherd (written commun., 1987). See figure 16.1.1 for approximate outline of caldera.

QUALIBOU

CAVW number (active vent)	Latitude Longitude (degrees)	Diameter (km)	Local tectonic setting	Precaldera edifice	SiO_2 content (wt pct)	Age of collapse (yr B.P.)	Date of unrest	Type and duration of unrest ESTU STHF MGTF H Te	Eruption type
16-00-14 (Qualibou)	13.83N 61.05W	5 x 7	N60°E-trending faults; compr	Strato-cones	R = d C = silicic	ca. 39,000	1766	---- ---- ---- - --	pex
							1906-08	xx-- ---- ---- - --	none

TECTONIC SETTING

Qualibou is in the southern segment of the Lesser Antilles arc (Wadge and Shepherd, 1984). In this southern half of the arc, tectonic earthquakes are largely restricted to the Benioff zone, and swarms of shallow earthquakes are infrequent (Shepherd, 1988).

GEOLOGIC HISTORY

Qualibou is either a caldera on the west side of Saint Lucia (Tomblin, 1964; Wohletz and others, 1986) or a gravity-slide structure (Roobol and others, 1983) (figs. 16.2.1-16.2.3). The balance of recent work, including drilling and discovery of basalt on the west coast of Saint Lucia, suggests that Qualibou is a caldera that formed after eruption of approximately 6 km^3 (DRE) of silicic magma (Wohletz and others, 1986).

HISTORICAL ACTIVITY

A steam explosion occurred in or about 1766. A main shock - aftershock sequence or an earthquake swarm occurred from February 1906 to June 1908. A tectonic earthquake ($\underline{M}$=6.5, epicenter 13°N, 61.25°W, depth 160 km) shook Saint Lucia on 5 July 1940; numerous other tectonic earthquakes have also been felt on the island.

Occasional shallow earthquakes are detected by a seismometer at Qualibou (Shepherd, 1988). A survey of microseismicity by Aspinall and others (1976) showed repetitive small earthquakes about 2 km deep, which were inferred to reflect the zone of steam generation for this geothermal area (fig. 16.2.4).

QUALIBOU (continued)

REFERENCES

Aspinall, W.P., Michael, M.O., and Tomblin, J.F., 1976, Evidence for fluid bodies beneath the Sulphur Springs Geothermal Region, Saint Lucia, West Indies: Geophys. Res. Lett., v. 3, no. 2, p. 87-90.

Robson, G.R., 1964, An earthquake catalogue for the eastern Caribbean: Bull. Seismol. Soc. Am., v. 54, p. 785-832.

Robson, G.R., and Tomblin, J.F., 1966, Catalogue of the active volcanoes of the world including solfatara fields, Pt. XX, West Indies: Rome, IAVCEI, 56 p.

Roobol, M.J., Wright, J.V., and Smith, A.L., 1983, Calderas or gravity-slide structures in the Lesser Antilles Island arc?: Jour. Volcanol. Geotherm. Res., v. 19, p. 121-134.

Shepherd, J.B., 1988, Eruptions, eruption precursors and related phenomena in the Lesser Antilles: Bull. Volcanol. [in press].

Tomblin, J.F., 1964, The volcanic history and petrology of the Soufriere region, Saint Lucia: Oxford, England, Oxford Univ., Ph.D. thesis, 213 p.

----- 1965, The geology of the Soufriere volcanic centre, Saint Lucia: Trans. 10th Caribb. Geol. Congr., Trinidad, p. 367-376.

Vatin-Perignon, N., Chevallier, L., and Blanc, F., 1981, Les evenements dacitiques recents de la Soufriere de Sainte Lucie, Petite Antilles: PIRPSEV Bull. 90, Paris, 31 p.

Wadge, G., and Shepherd, J.B., 1984, Segmentation of the Lesser Antilles subduction zone: Earth Planet. Sci. Lett., v. 71, p. 297-304.

Westercamp, D., and Tomblin, J., 1980, Le volcanisme recent et les éruptions historiques dan la partie centrale de l'arc insulaire des Petites Antilles: Bull. Bur. Rech. Geol. Min. (BRGM), ser. 2, nos. 3-4, p. 293-319.

Wohletz, K., Heiken, G., Ander, M., Goff, F., Vuataz, F.-D., and Wadge, G., 1986, The Qualibou Caldera, Saint Lucia, West Indies: Jour. Volcanol. Geotherm. Res., v. 27, p. 77-115.

Wright, J.V., Roobol, M.J., Smith, A.L., Sparks, R.S.J., Brazier, S.A., Rose, W.I., Jr., and Sigurdsson, H., 1984, Late Quaternary explosive silicic volcanism on Saint Lucia, West Indies: Geol. Mag., v. 1212, p. 1-15.

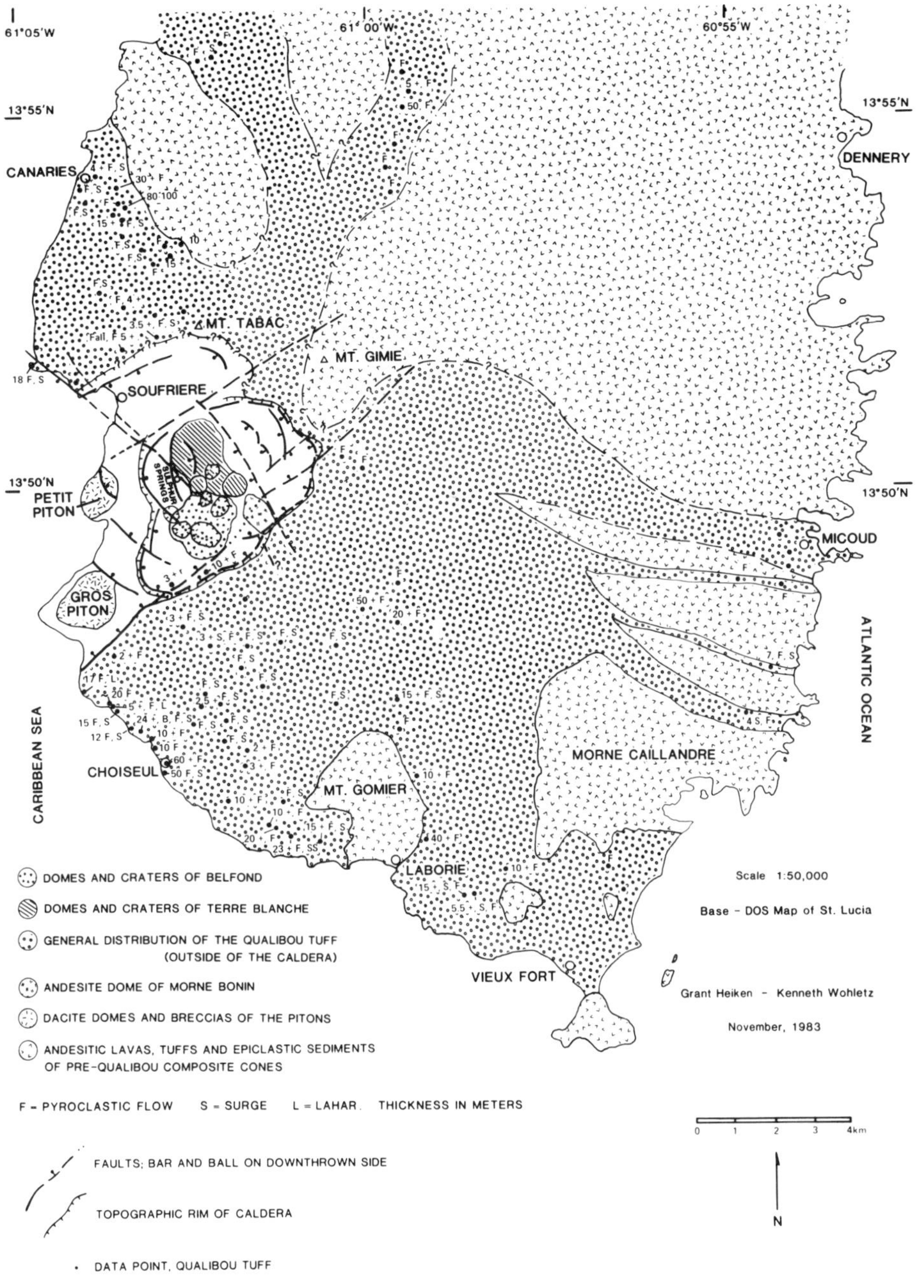

Figure 16.2.1. Reconnaissance geologic map, Qualibou Caldera and tuff, St. Lucia, from Wohletz and others (1986).

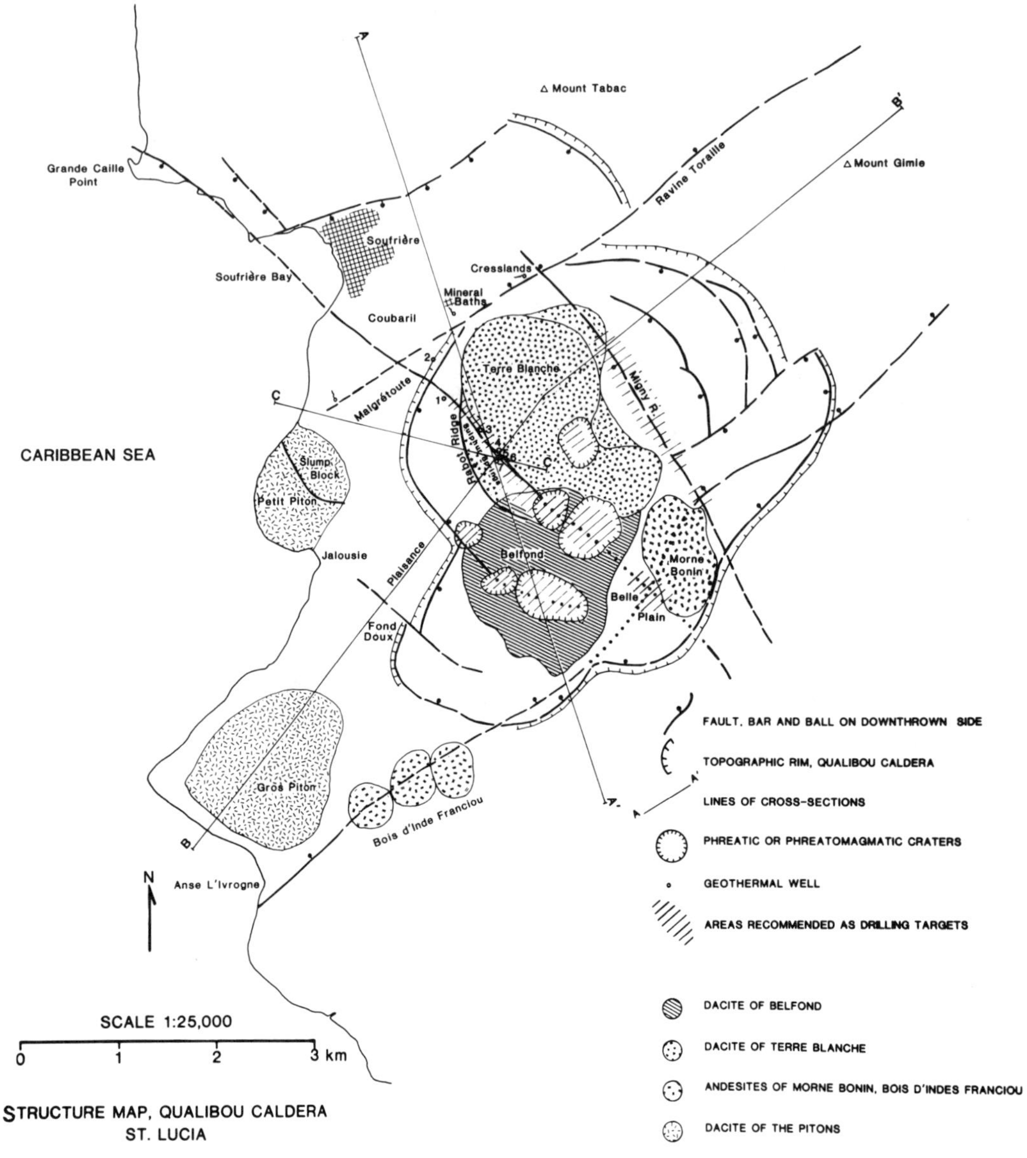

Figure 16.2.2. Structural geologic map of Qualibou Caldera, from Wohletz and others (1986).

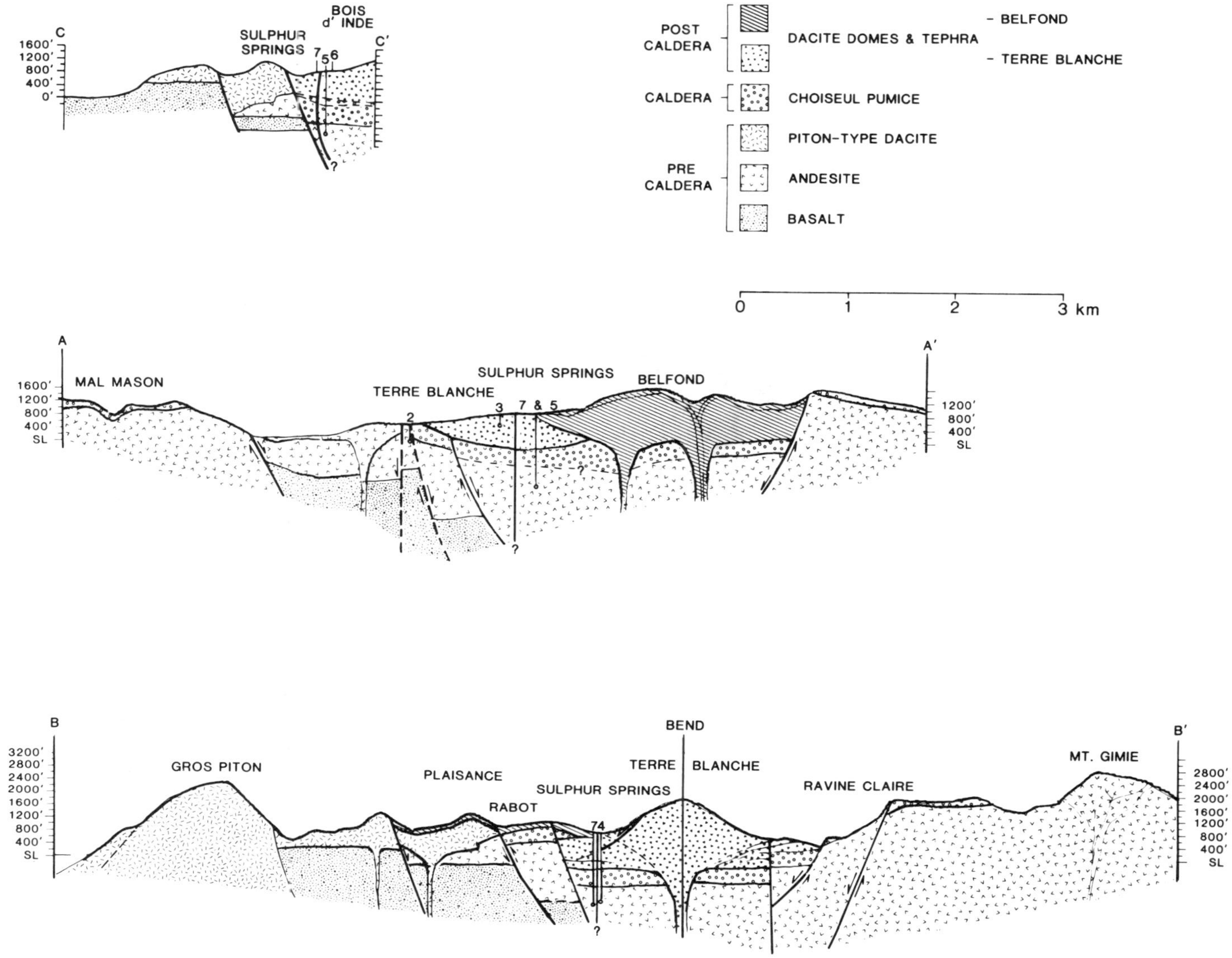

Figure 16.2.3. Cross sections through Qualibou Caldera, from Wohletz and others (1986). A-A': Mal Mason to Belle Vue (azimuth 342°); B-B': Anse L'Ivrogne to Mount Gimie (azimuths 40° and 50°); C-C': Malgretoute through Sulfur Springs (azimuth 282°). See figure 16.2.2 for locations of cross sections.

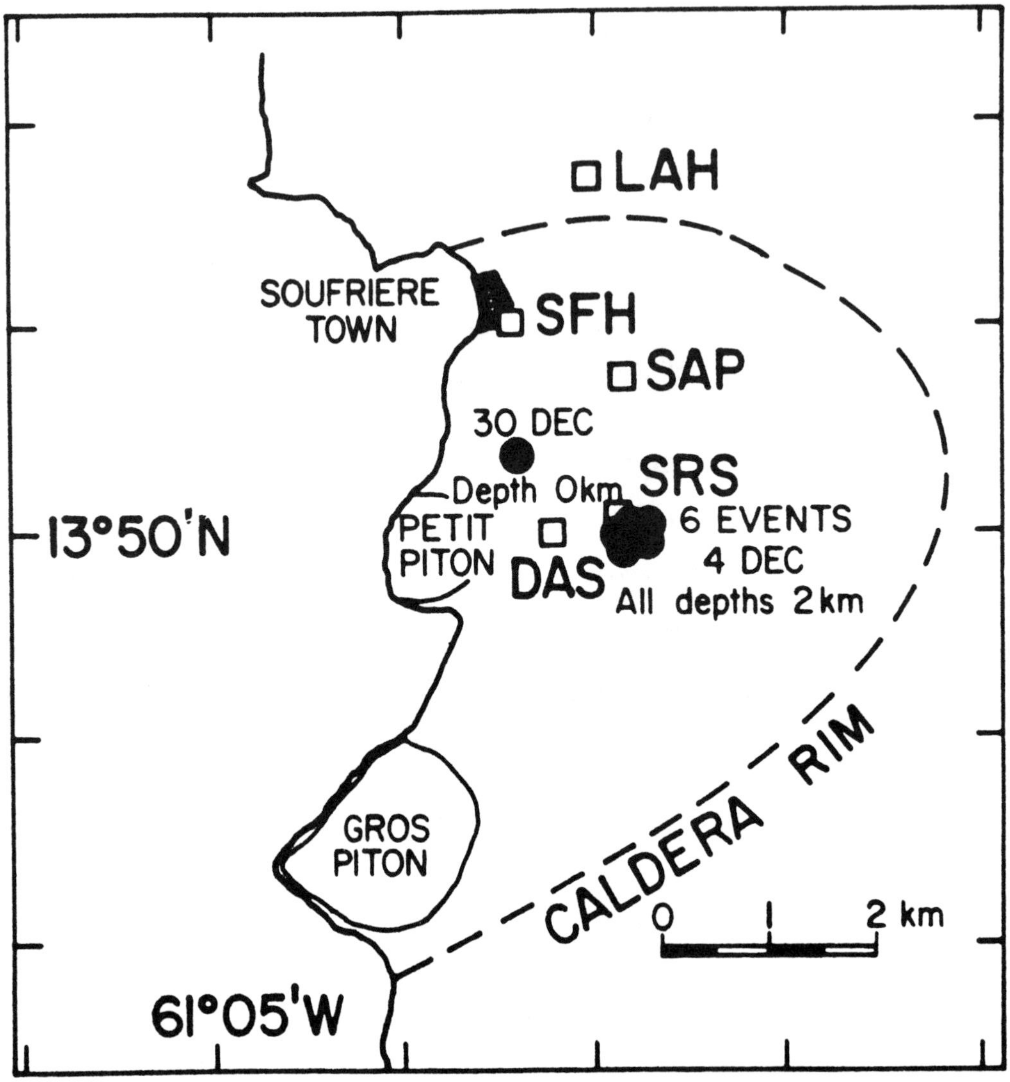

Figure 16.2.4. Location of seismic stations (open squares) and epicenters of local seismic events on 4 December 1974 and 30 December 1974 (solid circles) near Qualibou Caldera, St. Lucia, from Aspinall and others (1976). Copyright by the American Geophysical Union.

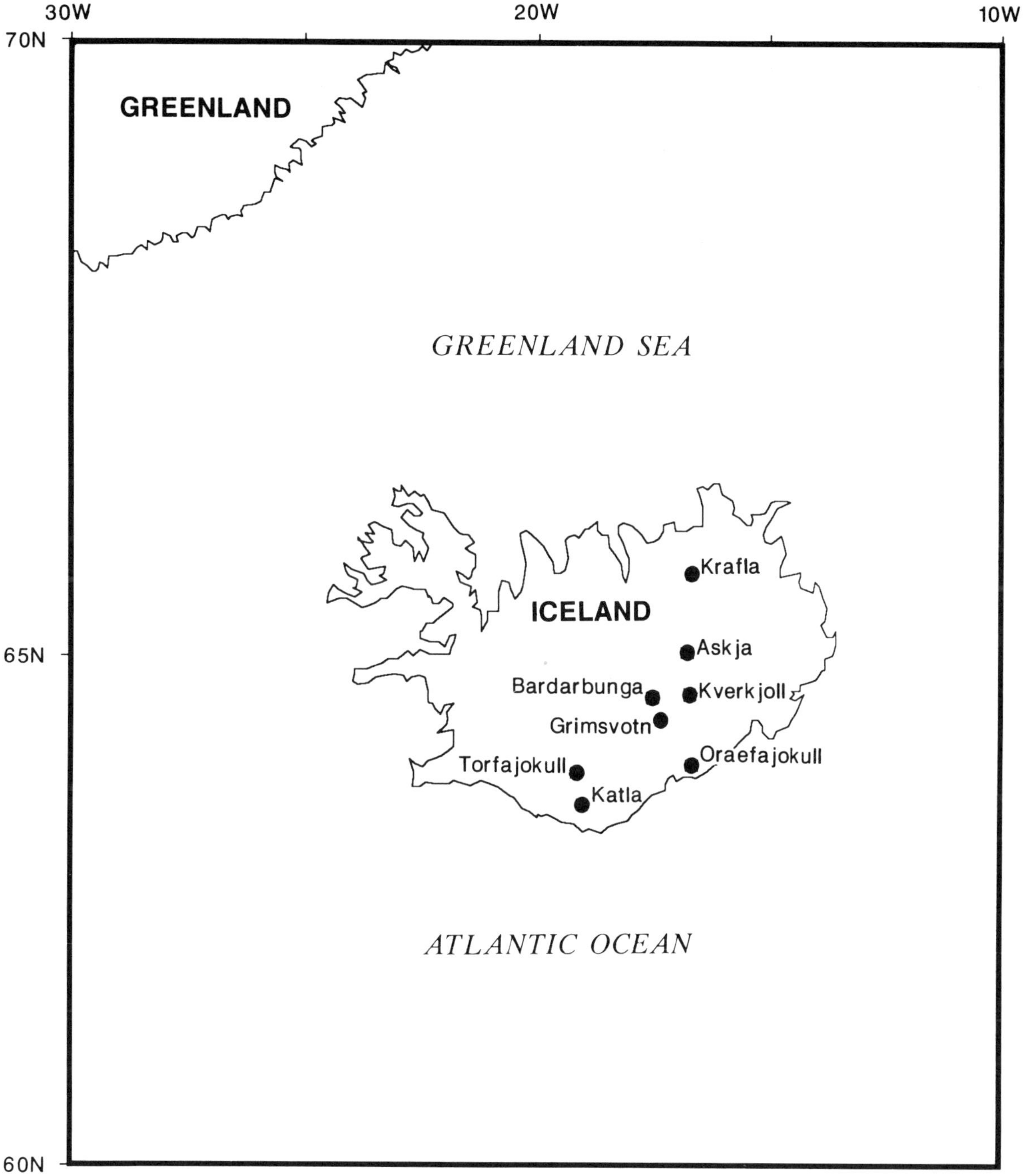

Figure 17. Locations of large, Quaternary restless calderas (solid circles) in Region 17.

PART 3: HISTORICAL UNREST AT LARGE QUATERNARY CALDERAS (continued)
See inside back cover for explanation and abbreviations

KATLA

CAVW number (active vent)	Latitude Longitude (degrees)	Diameter (km)	Local tectonic setting	Precaldera edifice	SiO_2 content (wt pct)	Age of collapse (yr B.P.)	Date of unrest	Type and duration of unrest ESTU STHF MGTF H Te	Eruption type
17-02-04 (Katla= K)	63.63N 19.03W	11 x 14	Intraplate	LL-strat (subglacial)	R = 47-49		A.D. 934 (E)	---- ---x ---- - --	Ex, LF
							ca. 950?	---- ---- ---- - --	lf, subgl
including 17-02-14 (Eldgja= E)							ca. 1179	---- ---- ---- - --	ex, subgl
							1245	---- ---- ---- - --	ex, subgl
							1262	---- ---- ---- - --	Ex, subgl
							1311	---- ---- ---- - --	Ex, subgl
							between 1354/1360	---- ---- ---- - --	Ex, subgl
							1416	---- ---- ---- - --	ex, subgl
							ca. 1490	---- ---- ---- - --	Ex, subgl
							1580	---- ---- ---- - --	ex, subgl
							1612	---- ---- ---- - --	ex, subgl
							1625	x--- ---- ---- - --	Ex, subgl
							1660-61	B--- ---- ---- - --	ex, subgl
							1721	x--- ---- ---- - --	EX, subgl
							1755-56	B--- ---- ---Y - --	EX, subgl
							1823	B--- ---- ---- - --	ex, subgl
							1860	B--- ---- ---- - --	ex, subgl
							1918	B--- ---- ---- - --	EX, subgl
							1951-88+	G--- ?--- ---- - --	subgl?

TECTONIC SETTING

The Mid-Atlantic plate boundary passes through Iceland and is reflected in two zones of crustal spreading and volcanism--an eastern zone, the site of most historical eruptions, and a western zone. Katla is situated near the south end of the eastern volcanic zone, south of its junction with the transform zone in South Iceland (fig. 17.1.1). It is thus in an intraplate setting where little or no extension is occurring. Eruptions of Katla are only indirectly linked with episodes of spreading if they are linked at all, in contrast to direct links between eruptions and spreading at Krafla (Gudmundsson and Sæmundsson, 1980).

GEOLOGIC HISTORY

Katla is hidden under the cap of the Myrdalsjökull ice sheet--on average about 400 m thick (fig. 17.1.2) (Thorarinsson, 1960; Rist, 1967; Sigbjarnarson, 1973). The only products of Katla that have been studied are distal tephra and glacial outburst flood (jökulhlaup) deposits (Thorarinsson, 1957; Einarsson and others, 1980; Jonsson, 1982). The Eldgja fissure swarm northeast of Katla, active in A.D. 934, is part of the Katla volcanic system (Thorarinsson and Sæmundsson, 1979).

KATLA (continued)

HISTORICAL ACTIVITY

Eruptions of Katla are known from as early as the 10th century, and since 1580 they have occurred on average every 42 years.

934: A fissure eruption from Eldgja was one of the largest in recorded history, comparable to that from Laki in 1783.

1625, 2 September and following: Small earthquakes were felt in Alftaver, ca. 35 km southeast of Katla, on the early morning of 2 September and were followed by an eruption and a jökulhlaup later the same morning. Earthquakes were reported to have occurred intermittently during the 12 days the eruption lasted.

1660, 3 November: Earthquakes were felt in Myrdalur ca. 20 km south of Katla for about 1 hour before an eruption cloud was observed.

1721: An earthquake was felt in Myrdalur at about 0900 hr on 11 May, followed at about 1300 hr by an eruption and a jökulhlaup. Earthquakes were reported to have occurred intermittently for several weeks.

1755: Earthquakes began to be felt at noon on 17 October in Myrdalur and continued for some hours before lightning (commonly reported in eruption clouds during Katla eruptions) and a jökulhlaup indicated that eruption had begun. Earthquakes were felt occasionally during the next few weeks.

1823: Earthquakes began to be felt at about 1800 hr on 26 June in Myrdalur, followed by an eruption about three hours later. Earthquakes were reported intermittently during the following two to three weeks.

1860: Earthquakes began to be felt in Myrdalur at 600-800 hr on 8 May. A jökulhlaup followed at 1730 hr that day, and an eruption cloud was observed early next morning.

1918: Earthquakes began to be felt in Myrdalur at about 1300 hr on 12 October. Shortly thereafter an eruption cloud was seen, and a jökulhlaup followed at about 1500 hr (Jonsson, 1982). Earthquakes continued to be felt at least until 28 October.

1951-88+: Earthquake activity at Katla apparently increased around 1951 (Tryggvason 1973b), and since then it has been moderately high. The seismicity has a pronounced annual cycle. The activity culminated in 1967 and 1976-77 with a few earthquakes approaching and exceeding magnitude 5. A small jökulhlaup, possibly caused by a small subglacial eruption, occurred on 25 June 1955. Because of the regular occurrence of Katla eruptions since 1580, another large eruption was expected about 1960, but none has yet occurred. Einarsson (1987) interprets the present seismicity as the result of deflation of the volcano, and speculates that the 1963-67 Surtsey eruptions and the 1973 Heimaey eruption might have relieved pressure in a deep zone of partial melt, thereby delaying the expected Katla eruption.

KATLA (continued)

HISTORICAL ACTIVITY (continued)

Ground tilt measurements made each year near the Myrdalsjökull ice cap (fig. 17.1.3) showed responses to variable snow load, and thus a relatively low (<10^{13} poises) viscosity of the underlying crust and mantle (Tryggvason, 1973a).

The only two historical eruptions of Eyjafjallajökull, a stratovolcano that lies about 30 km west of Katla, coincided approximately with eruptions of Katla in 1612 and 1821-23 (G. Larsen and P. Einarsson, written commun., 1988).

REFERENCES

Einarsson, E.H., Thorarinsson, S., and Larsen, G., 1980, The Solheimar tephra layer and the Katla eruption of 1357: Acta Nat. Islandica, v. 28, p. 3-20.

Einarsson, P., 1987, The anomalous mantle beneath Iceland and possible magma pressure connection between volcanoes [abs.]: Abstracts, Hawaii Symp. on How Volcanoes Work, Hilo, January 1987, p. 61.

Einarsson, P., and Sæmundsson, K., 1987, Earthquake epicenters 1982-1985 and volcanic systems in Iceland: map accompanying the festschrift "I Hlutarins Edli", Reykjavík, Menningarsjóour, 1 sheet, scale 1:750,000.

Gudmundsson, G., and Sæmundsson, K., 1980, Statistical analysis of damaging earthquakes and volcanic eruptions in Iceland from 1550-1978: Jour. Geophys., v. 47, p. 99-109.

Hid islenzka bokmentafelag, 1907-1915, Safn til sogu Islands, v. IV: Reykjavik, Kaupmannahofn, 1025 p.

Jakobsson, S.P., 1979, Petrology of Recent basalts of the Eastern Volcanic Zone, Iceland: Acta Nat. Islandica, v. 26, p. 1-103.

Johannesson, H., Jakobsson, S.P. and Sæmundsson, K., 1982, Geological map of Iceland, sheet 6, S-Iceland, second ed.: Reykjavik, Icelandic Museum of Nat. History/Iceland Geodetic Survey.

Jonsson, J., 1982, Notes on the Katla volcanoglacial debris flows: Jökull, v. 32, p. 61-68.

Kjartansson, G., 1962, Geological map of Iceland, sheet 6, South-Central Iceland: Reykjavik, Museum of Natural History, 1 sheet, scale 1:250,000.

Rist, S., 1967, The thickness of the ice cover of Mýrdalsjökull, southern Iceland: Jökull, v. 17, p. 237-242.

Sæmundsson, K., 1982, Oskjur a virkum eldfjallasvaedum a Islandi (Calderas in active volcanic areas in Iceland), in Thorarinsdóttir, H., Oskarsson, O.H., Steinthorsson, S., and Einarsson, Th., (eds.), Eldur er i nordri: Reykjavík, Sogufelag, p. 221-240.

Sigbjarnarson, G., 1973, Katla and Askja: Jökull, v. 23, p. 45-51.

Thorarinsson, S., 1957, The jökulhlaup from the Katla area 1955, compared with other jökulhlaups in Iceland: Jökull, v. 7, p. 20-25.

----- 1960, On the predicting of volcanic eruptions in Iceland: Bull. Volcanol., v. 23, p. 45-52.

----- 1973, Data sheets of the post-Miocene volcanoes of the world, Sheet II, Iceland: Rome, IAVCEI, 10 p.

Thorarinsson, S., and Sæmundsson, K., 1979, Volcanic activity in historical time: Jökull, v. 29, p. 29-32.

Thoroddsen, Th., 1925, Die Geschichte der Islandischen Vulkane: D. Kgl. Dansk Vidensk. Selsk. Skrifter, Naturvidensk. og Mathem. Afd. Copenhagen, 8th ser., v. 9, 458 p.

Tryggvason, E., 1973a, Surface deformation and crustal structure in the Myrdalsjökull area of south Iceland: Jour. Geophys. Res., v. 78, no. 14, p. 2488-2497.

Tryggvason, E., 1973b, Seismicity, earthquake swarms and plate boundaries in the Iceland region: Bull. Seismol. Soc. Am., v. 63, p. 1327-1348.

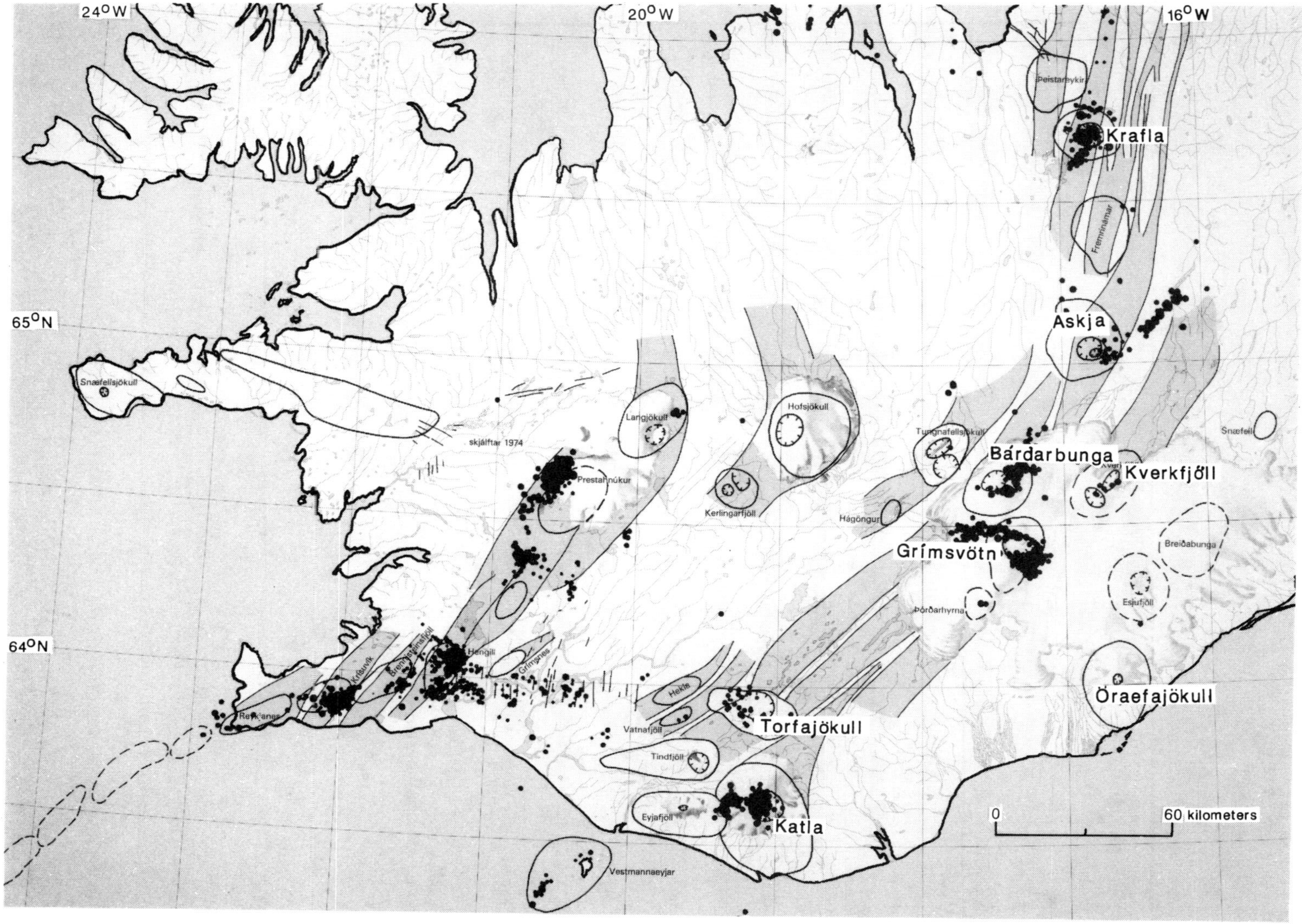

Figure 17.1.1. Volcanic centers and earthquake epicenters (1982-85) in Iceland, from Einarsson and Sæmundsson (1987). Restless calderas are labeled prominently.

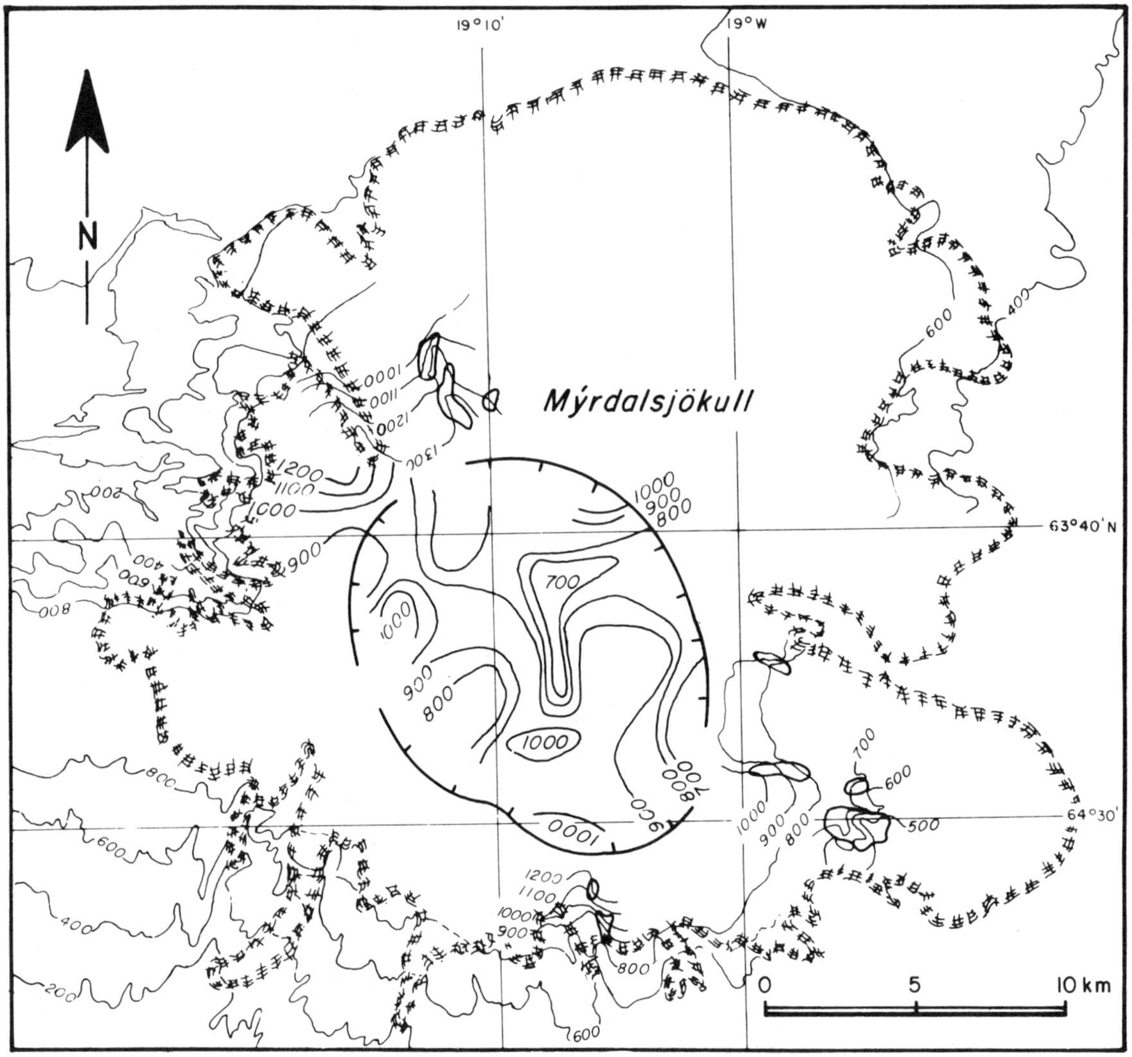

Figure 17.1.2. Katla Volcano and Caldera (both under the Myrdalsjökull ice sheet), from Sæmundsson (1982). Contour interval for subglacial topography, 100 m.

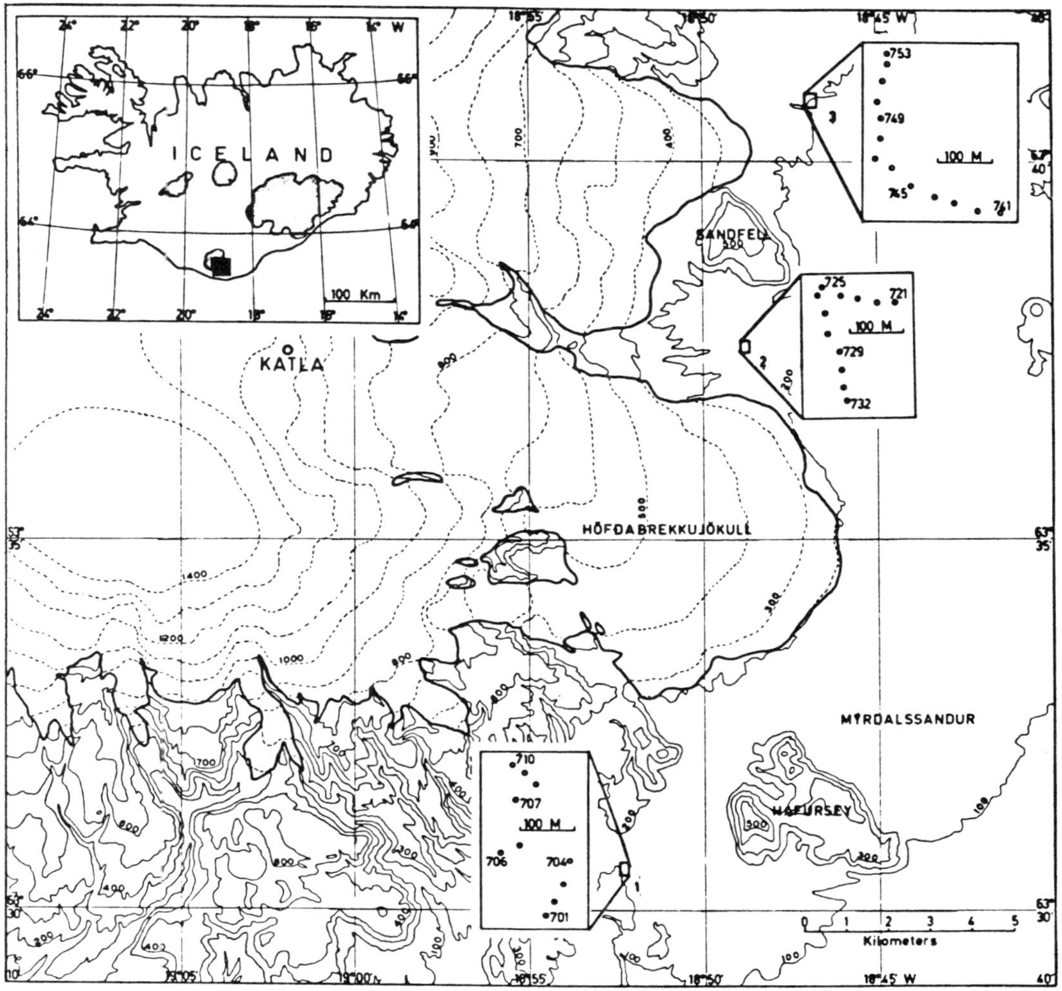

Figure 17.1.3. Southeast part of Myrdalsjökull ice field showing locations of Katla Volcano and bench-mark arrays for tilt measurements (Tryggvason, 1973a). Insets show location of map area and configurations of arrays.

ORÆFAJÖKULL

CAVW number (active vent)	Latitude Longitude (degrees)	Diameter (km)	Local tectonic setting	Precaldera edifice	SiO_2 content (wt pct)	Age of collapse (yr B.P.)	Date of unrest	Type and duration of unrest ESTU STHF MGTF H Te	Eruption type
17-02-21 (Oræfajökull)	64.00N 16.65W	5 x 4	Intraplate	Strato	R = 70(1362)	<700,000	1362	?--- ---- ---- - --	EX,subgl
					R = 56(1727)		1727-28	C--- ---- ---- - --	Ex,subgl

TECTONIC SETTING

The Mid-Atlantic plate boundary passes beneath Iceland and is reflected in two zones of spreading and volcanism--an eastern zone, the site of most historical eruptions, and a western zone. Oræfajökull is situated east of the eastern volcanic zone, in an intraplate setting (see fig. 17.1.1 in section on Katla). Eruptions of Oræfajökull are only indirectly linked with episodes of spreading, in contrast to direct links between eruptions and spreading at Krafla (Gudmundsson and Sæmundsson, 1980).

GEOLOGIC HISTORY

Oræfajökull is the highest central volcano in Iceland, 2119 m above sea level, and has a glacial cap (fig. 17.2.1). Most of its rocks are basaltic lava flows, palagonitic tuffs, and glacial deposits, but the high northwestern caldera rim (Hvannadalshnúkur) is a flow-banded rhyolite (liparite) (Thorarinsson, 1958). Most eruptions in the Holocene have been explosive summit eruptions; however, some Holocene lava flows from flank eruptions are also known (Thorarinsson, 1958).

HISTORICAL ACTIVITY

1362: At least 2 km^3 of rhyolitic magma (DRE) were erupted in 1362. Several settlements were destroyed by jökulhlaups and tephra fall. Because pumice fall penetrated farmhouses from nearly the beginning of the eruption, Thorarinsson (1958) inferred that strong earthquakes had occurred shortly before the eruption, damaging those houses.

1727: The second eruption during historical time, of intermediate composition magma, occurred in 1727. Earthquakes were felt at the Sandfell vicarage 9 km southwest of the center of the caldera approximately 1 day before the 1727 eruption began, but there is no historical record or archaeological evidence that the 1727 earthquakes caused damage (Thorarinsson, 1958).

ORÆFAJÖKULL (continued)

REFERENCES

Gudmundsson, G., and Sæmundsson, K., 1980, Statistical analysis of damaging earthquakes and volcanic eruptions in Iceland from 1550-1978: Jour. Geophys., v. 47, p. 99-109.

Perrey, A., 1845, Sur les tremblements de terre de la Peninsule Scandinave: Paris, Arthus Bertrand, 63 p.

Prestvik, T., 1980, Petrology of hybrid intermediate and silicic rocks from Oræfajökull, southeast Iceland: Geol. Foren. Stockh. Forh., v. 101, p. 299-307.

Thorarinsson, S., 1958, The Oræfajökull eruption of 1362: Acta Nat. Islandica, v. 2, no. 2, p. 1-99.

----- 1960, On the predicting of volcanic eruptions in Iceland: Bull. Volcanol., v. 23, p. 45-52.

----- 1973, Data sheets of the post-Miocene volcanoes of the world, Sheet II, Iceland: Rome, IAVCEI, 10 p.

Thoroddsen, Th., 1925, Die Geschichte der Islandischen Vulkane: D. Kgl. Dansk Vidensk. Selsk. Skrifter, Naturvidensk. og Mathem. Afd. Copenhagen, 8th ser., v. 9, 458 p.

Torfason, H., 1985, Geological map of Iceland, sheet 9, SA-Iceland: Reykjavik, Icelandic Museum of Natural History and Iceland Geodetic Survey, 1 sheet, scale 1:250,000.

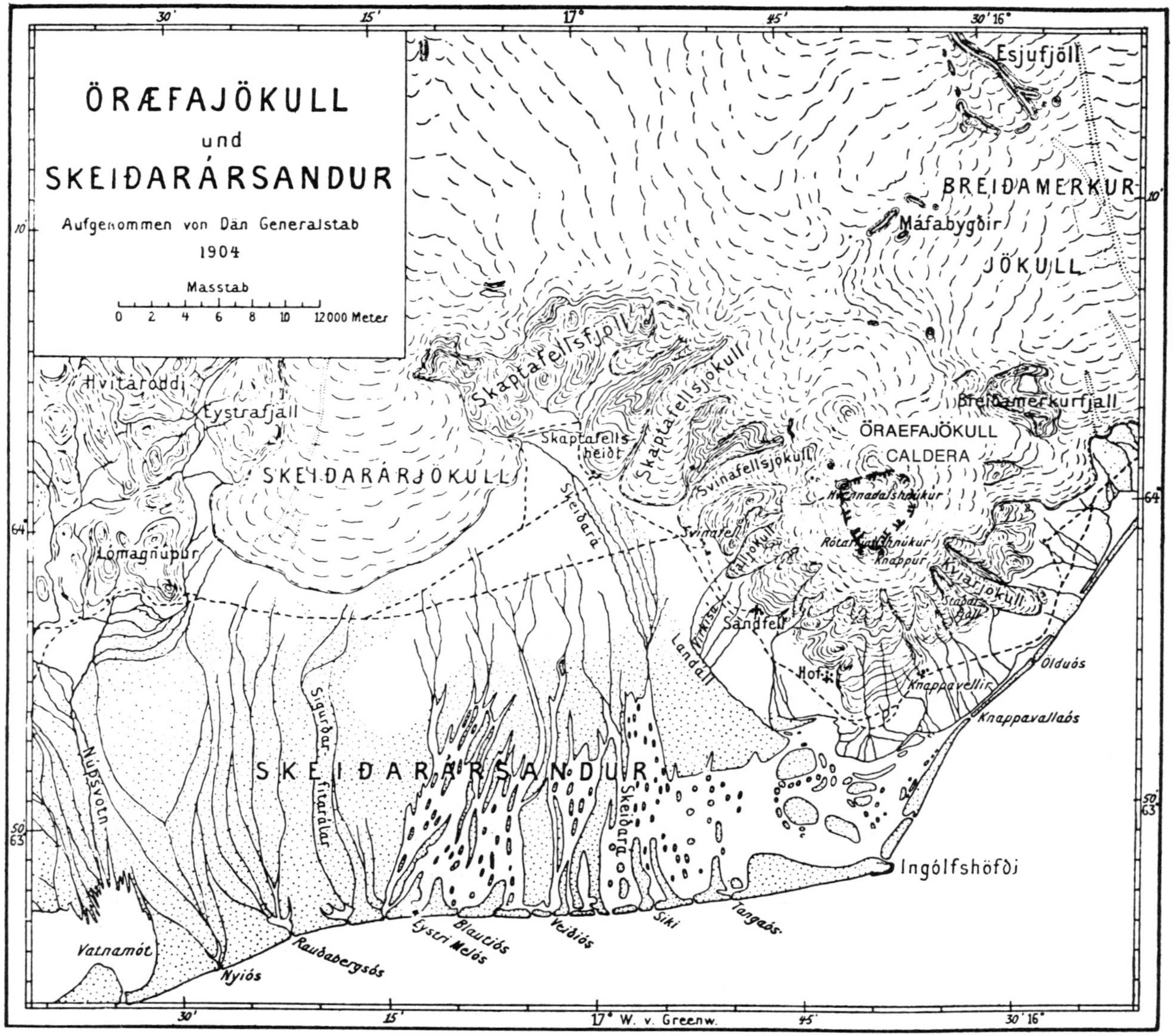

Figure 17.2.1. Location of Oræfajökull Volcano and Caldera on east border of eastern volcanic zone of Iceland, from Thoroddsen (1925). Caldera rim added from Torfason (1985).

BARDARBUNGA

CAVW number (active vent)	Latitude Longitude (degrees)	Diameter (km)	Local tectonic setting	Precaldera edifice	SiO_2 content (wt pct)	Age of collapse (yr B.P.)	Date of unrest	Type and duration of unrest ESTU STHF MGTF H Te	Eruption type
no CAVW number (Bárdarbunga)	64.62N 17.50W	7 x 11	Exten	LL-strat (subglacial)	R = b C = b		1477-ca.1480	---- --xx ---- - =?	EX
							1717	---- ---- ---- - --	EX
							1974-87	xx-- ---- ---- - --	none
17-02-18 (Trollagigar =Trollahraun)	64.43N 18.13W						1862-64	---- --xx ---- - --	lf,ex
within 17-02-11 (Vatnaöldur)	63.95N 19.28W						ca. 900	---- --xx ---- - --	EX,lf
17-02-11 (Veidivötn)	63.95N 19.28W						ca. 1480	---- --xx ---- - --	EX,lf

TECTONIC SETTING

The Mid-Atlantic plate boundary passes through Iceland and is reflected in two zones of spreading and volcanism--an eastern zone, the site of most historical eruptions, and a western zone. Bárdarbunga is situated in the central part of the eastern volcanic zone. A 150-km-long fissure swarm extends 40 km to the northeast and 110 km to the southwest of Bárdarbunga (figs. 17.1.1 in section on Katla, 17.3.1). As at Krafla, magma from a shallow magma chamber beneath Bárdarbunga has moved laterally into the fissure swarm during rifting episodes (Larsen, 1984). Alternatively, Jakobsson (1979) and Blake (1984) have suggested that a central feeder lies in the Klofnafell-Mani area, 70-80 km southwest of Bárdarbunga.

GEOLOGIC HISTORY

Bárdarbunga lies beneath the northwest corner of the Vatnajökull ice cap. It is the second highest central volcano in Iceland, with rocks exposed up to 1800 m above sea level. On the basis of historical accounts of eruptions and glacial outbursts, and distinctive chemical characteristics of tephra layers, at least ten tephra-producing eruptions are thought to have taken place beneath the northwest corner of the Vatnajökull ice cap during the last 1200 years (Thorarinsson, 1974; Larsen, 1982, 1984; Gudmundsson, 1986). Three eruptions took place on the fissure swarm southwest of Bárdarbunga during the same period. There is no direct evidence that eruptions have taken place within the Bárdarbunga caldera itself during historical time.

BARDARBUNGA (continued)

HISTORICAL ACTIVITY

1477 and 1717: Two large basaltic tephra layers have been correlated with historical accounts of tephra falls in 1477 and 1717. The 1477 tephra (tephra layer "a", formerly attributed to Kverkfjöll by Thorarinsson, 1958) was formed during a large event that included the Veidivötn eruption on the fissure swarm 40-110 km southwest of Bárdarbunga (Larsen, 1984). The 1717 tephra layer was formed by one of several eruptions reported during the first three decades of the eighteenth century, now attributed to the fissure swarm northeast of Bárdarbunga (Gudmundsson, 1986).

ca. 900 and ca. 1480: The two largest eruptions on the fissure swarm southwest of Bárdarbunga in historical time are the Vatnaöldur eruption, ca. 900 A.D., and the Veidivötn eruption, 1480 A.D.±11 yr (Larsen, 1984, 1987). The eruption fissures are up to 70 km long and each produced >1 km^3 of basalt (DRE) that was mostly erupted as tephra. In 1480, at least 1.6 km^3 of basalt was erupted from simultaneously active vents along the full 70-km length of the fissure (Larsen, 1987). Production was strongly controlled by elevation: Vents between 500- and 600-m elevation erupted 5 times as much magma per kilometer as did vents between 600- and 700-m elevation, and vents from 700- to 900-m elevation erupted very small volumes of magma (Larsen, 1987). The fissures continue into the Torfajökull caldera (see section on Torfajökull), where rhyolites were erupted simultaneously with the basalts elsewhere on the fissures. No historical accounts mention eruptions in this area, and they were identified through tephrochronological studies (Larsen, 1984, 1987).

1862-64: An eruption on the Trollahraun fissure, 30-45 km southwest of Bárdarbunga, produced 0.3 km^3 of lava and a minor amount of tephra (Thorarinsson and Sigvaldason, 1972). Contemporary documents do not mention earthquakes.

1974-88 and continuing: The seismic activity of Bárdarbunga increased markedly in 1974, and in the period 1974-86 nine earthquakes of magnitude 5 and larger occurred. No earthquakes of comparable magnitude occurred in the previous 50 years. Fault-plane solutions show reverse faulting that Einarsson (1986, 1987) suggested is the result of deflation of a shallow magma chamber beneath Bárdarbunga.

COMMENTS

Einarsson (1987) suggested that deflation at Bárdarbunga might be a pressure-drop response to concurrent inflation at Krafla, 110 km to the north, and speculated that the pressure changes were transmitted through a partially molten layer beneath the crust that has been inferred from seismic and magnetotelluric measurements.

BARDARBUNGA (continued)

REFERENCES

Blake, S., 1984, Magma mixing and hybridization processes at the alkalic, silicic, Torfajökull central volcano triggered by tholeiitic Veidivötn fissuring, South Iceland: Jour. Volcanol. Geotherm. Res., v. 22, p. 1-31.

Brandsdóttir, B., 1984, Seismic activity in Vatnajökull in 1900-1982, with special reference to Skeidararhlaups, Skaflarhlaups, and Vatnajökull eruptions: Jökull, v. 34, p. 141-150.

Einarsson, P., 1986, Seismicity along the eastern margin of the North American Plate, in Vogt, P.R. and Tucholke, B.E., eds., The geology of North America, v. 14, The western North Atlantic region: Boulder, Geol. Soc. Am., p. 99-116.

----- 1987, The anomalous mantle beneath Iceland and possible magma pressure connection between volcanoes [abs.]: Abstracts, Hawaii Symp. on How Volcanoes Work, Hilo, January 1987, p. 61.

Einarsson, P., and Bjornsson, S., 1979, Earthquakes in Iceland: Jökull, v. 29, p. 37-43.

Einarsson, P., and Brandsdóttir, B., 1984, Seismic activity preceding and during the 1983 volcanic eruption in Grímsvötn, Iceland: Jökull, v. 34, p. 13-23.

Gudmundsson, A., 1986, Eldgos a Dyngjuhalsi a 18. old (Eruptions in the Dyngjuhals area in the 18th century): Náttúrufrædingurinn, v. 56, p. 43-48.

Jakobsson, S.P., 1979, Petrology of Recent basalts of the Eastern Volcanic Zone, Iceland: Acta Nat. Islandica, v. 26, p. 1-103.

Larsen, G., 1982, Gjoskutimatal Jökuldals og nagrennis (Tephrochronology of the Jökuldalur area), in Thorarinsdóttir, H., Oskarsson, C.H., Steinthorsson, S. and Einarsson, Th., eds., Eldur er i nordri: Reykjavik, Sogufelag, p. 51-65.

----- 1984, Recent volcanic history of the Veidivötn fissure swarm, southern Iceland--an approach to volcanic risk assessment: Jour. Volcanol. Geotherm. Res., v. 22, p. 33-58.

----- 1987, The dike of the 1480 A.D. Veidivötn eruption, S-Iceland [abs.]: IUGG XIX Gen. Assembly, Vancouver, B.C., Abstracts, v. 2, p. 396.

Sæmundsson, K., 1972, Jardfraediglefsur um Torfajökulssvaedid: Náttúrufrædingurinn, v. 42, p. 81-99.

----- 1982, Oskjur a virkum eldfjallasvaedum a Islandi (Calderas in active volcanic areas in Iceland), in Thorarinsdóttir, H., Oskarsson, O.H., Steinthorsson, S. and Einarsson, Th., eds., Eldur er i nordri: Reykjavik, Sogufelag, p. 221-240.

Thorarinsson, S., 1958, The Oræfajökull eruption of 1362: Acta Nat. Islandica, v. II, 2, p. 1-99.

----- 1974, Votnin strid. Saga Skeidararhlaupa og Grimsvatnagosa (The swift flowing rivers. The history of Grímsvötn eruptions and jökulhlaups in Skeidara river): Reykjavik, Menningarsjodur, 254 p.

Thorarinsson, S., and Sigvaldason, G.E., 1972, The Trollagigar eruption 1862-64: Jökull, v. 22, p. 12-26.

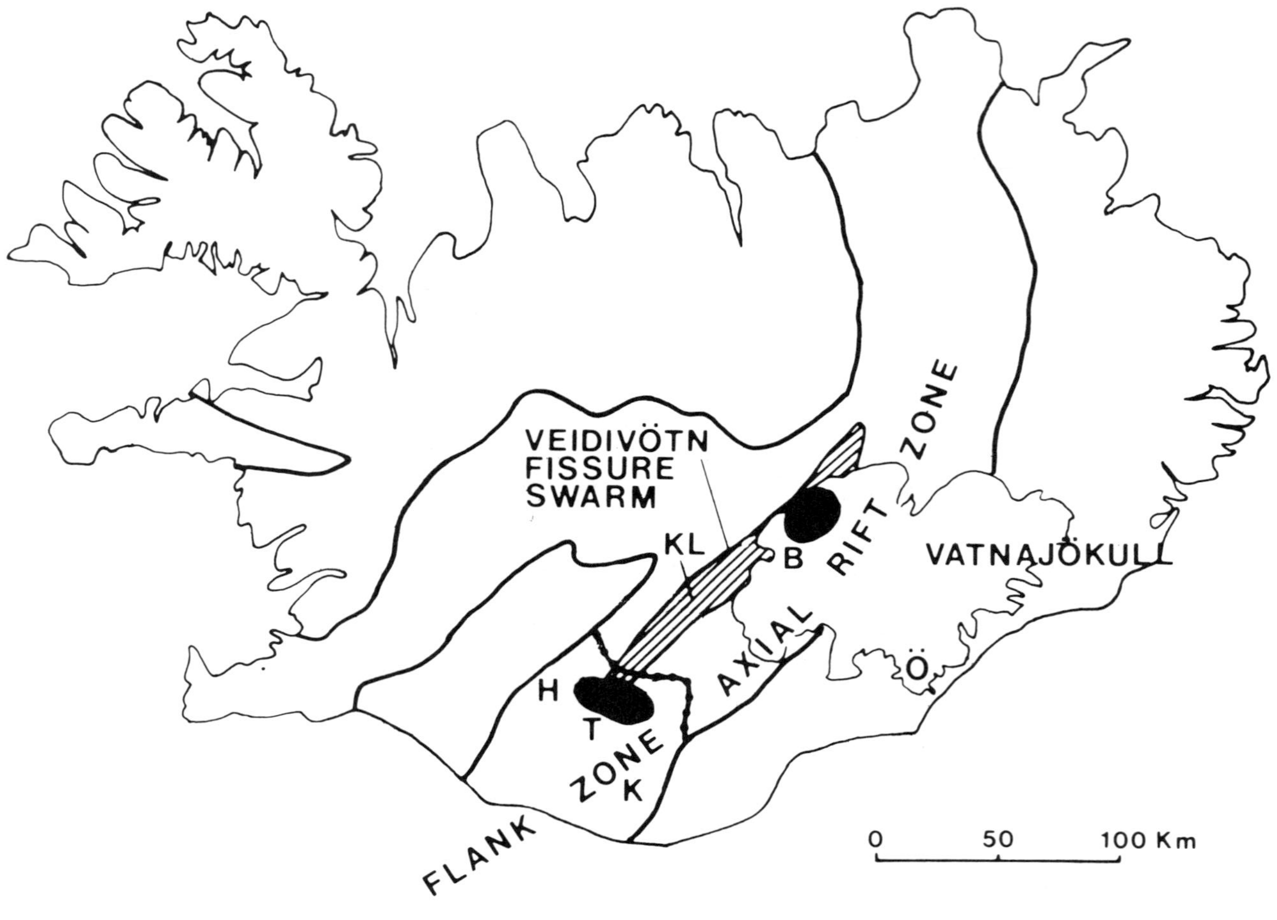

Figure 17.3.1. Veidivötn fissure swarm (ruled area) on eastern volcanic zone of southern Iceland, from Larsen (1984). B, Bárdarbunga central volcano; T, Torfajökull central volcano; KL, Klofnafell area; H, Hekla central volcano; K, Katla central volcano; Ö, Oræfajökull central volcano.

TORFAJÖKULL

CAVW number (active vent)	Latitude Longitude (degrees)	Diameter (km)	Local tectonic setting	Precaldera edifice	SiO_2 content (wt pct)	Age of collapse (yr B.P.)	Date of unrest	Type and duration of unrest ESTU STHF MGTF H Te	Eruption type
no CAVW number	63.93N	13 x 18	Intraplate		R = 70-72		ca. 900 A.D.	---- --xx ---- - --	lf,Ex
(Torfajökull)	19.15W				C = silicic		ca. 1480 A.D.	---- --xx ---- - --	lf,Ex
							1986-88+	CC-- ---- ---- - --	none

TECTONIC SETTING

The Mid-Atlantic plate boundary passes through Iceland and is reflected in two zones of spreading and volcanism -- an eastern zone, the site of most historical eruptions, and a western zone. The Torfajökull Central Volcano is located at the intersection where the eastern volcanic zone (a divergent plate boundary) joins with the south Iceland seismic zone (a transform plate boundary) (fig. 17.1.1 in section on Katla). Little or no extension is occurring south of this intersection.

GEOLOGIC HISTORY

Torfajökull Volcano is a feature of relatively low relief, and is an area of predominantly rhyolite volcanism and intense geothermal activity (fig. 17.4.1). Most of the rhyolite has been subglacially emplaced as domes or flows. A few postglacial flows of rhyolite exist in the western part of the caldera (Sæmundsson, 1972; Ivarsson, 1983; Blake, 1984; McGarvie, 1984).

HISTORICAL ACTIVITY

ca. 900 A.D.: Rhyolitic tephra and lava was erupted at Hrafntinnuhraun, inside the Torfajökull caldera, during a rifting episode on the Veidivötn fissure swarm. Basaltic tephra was erupted at the same time on the Vatnaöldur fissure (see section on Bárdarbunga).

ca. 1480 A.D.: Rhyolitic lava and tephra was erupted at Laugahraun and Namshraun, near the north rim of the Torfajökull caldera, during a rifting episode on the Veidivötn fissure swarm. These lavas lie on the southwesternmost 3-4 km of the ca. 70-km-long Veidivötn fissure (see section on Bárdarbunga), which produced basaltic tephra and lava (Larsen, 1984; McGarvie, 1984; Mörk, 1984).

The Torfajökull caldera is an area of moderate, persistent seismicity. A few earthquakes in the magnitude range 3-4 occur every year. The epicenters are mainly in the western part of the caldera, where the high temperature geothermal activity and postglacial volcanism is also concentrated. Since 1985, when a local seismograph was installed, low frequency earthquakes have been detected that appear to originate in the northern or eastern part of the caldera. Bursts of more than 40 low-frequency earthquakes per day occurred in August 1986, November 1986, March 1988, and April 1988 (Brandsdóttir and Einarsson, 1988).

TORFAJÖKULL (continued)

REFERENCES

Blake, S., 1984, Magma mixing and hybridization processes at the alkalic, silicic, Torfajökull central volcano triggered by tholeiitic Veidivötn fissuring, South Iceland: Jour. Volcanol. Geotherm. Res., v. 22, p. 1-31.

Brandsdóttir, B. and Einarsson, P., 1988, On low frequency earthquakes in the volcanic zones of Iceland [abs.]: Eos, Trans. Am. Geophys. Union, v. 69 [in press].

Einarsson, P., and Bjornsson, S., 1979, Earthquakes in Iceland: Jökull, v. 29, p. 37-43.

Ivarsson, G., 1983, The Torfajökull volcano, central south Iceland - chemical evolution and magma-mixing processes [abs.]: Eos, Trans. Am. Geophys. Union, v. 64, no. 45, p. 887.

Jakobsson, S.P., 1979, Petrology of Recent basalts of the Eastern Volcanic Zone, Iceland: Acta Nat. Islandica, v. 26, p. 1-103.

Larsen, G., 1984, Recent volcanic history of the Veidivötn fissure swarm, southern Iceland--an approach to volcanic risk assessment: Jour. Volcanol. Geotherm. Res., v. 22, p. 33-58.

McGarvie, D.W., 1984, Torfajökull - A volcano dominated by magma mixing: Geology, v. 12, p. 685-688.

Mörk, M.B.E., 1984, Magma mixing in the post-glacial Veidivötn fissure eruption, South Iceland: Lithos, v. 17, p. 55-75.

Sæmundsson, K., 1972, Jardfraediglefsur um Torfajökulssvaedid: Náttúrufrædingurinn, v. 42, p. 81-99.

----- 1982, Oskjur a virkum eldfjallasvaedum a Islandi (Calderas in active volcanic areas in Iceland), in Thorarinsdóttir, H., Oskarsson, O.H., Steinthorsson, S. and Einarsson, Th., eds., Eldur er i nordri: Reykjavik, Sogufelag, p. 221-240.

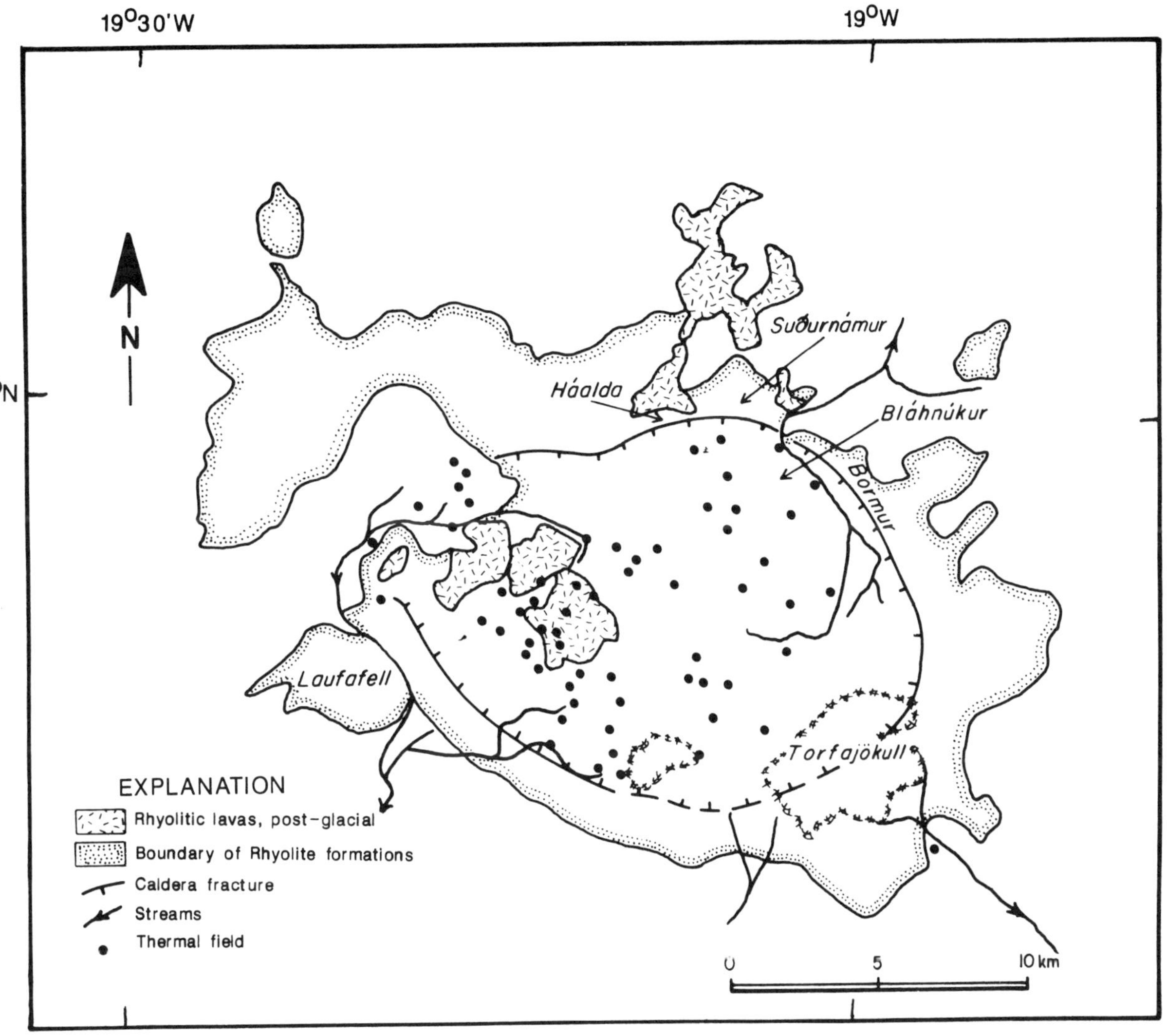

Figure 17.4.1. Outline map of Torfajökull central volcano and caldera, from Sæmundsson (1982). See figures 17.1.1 and 17.3.1 for location of Torfajökull at southwest end of Veidivötn fissure swarm.

PART 3: HISTORICAL UNREST AT LARGE QUATERNARY CALDERAS (continued)
See inside back cover for explanation and abbreviations

GRIMSVÖTN

CAVW number (active vent)	Latitude Longitude (degrees)	Diameter (km)	Local tectonic setting	Precaldera edifice	SiO_2 content (wt pct)	Age of collapse (yr B.P.)	Date of unrest	Type and duration of unrest ESTU STHF MGTF H Te	Eruption type
17-02-22	64.42N	6-8	Exten	LL-strat	R = mafic		1332	---- ---- ---- - --	ex, subgl
(Grimsvötn= G)	17.33W			(subglacial)			1341	---- ---- ---- - --	ex, subgl
17-02-15							1598	---- ---- ---- - --	Ex, subgl
(Lakagigar= L)							1603?	---- ---- ---- - --	ex, subgl
17-02-16							1619	---- ---- ---- - --	ex, subgl
(Sidhjökull= S)							1629	---- ---- ---- - --	ex, subgl
17-02-17							1638	---- ---- ---- - --	ex, subgl
(Hagongur= H)							1659	---- ---- ---- - --	ex, subgl
17-02-19							1681?	---- ---- ---- - --	subgl?
(Thordarhyrna= T)							1684-85	---- ---- ---- - --	ex, subgl
							1706	---- ---- ---- - --	ex, subgl
							1716	---- ---- ---- - --	ex, subgl
							1725-26	---- ---- ---- - --	subgl
							1753 (S)	---- ---- ---- - --	ex
							1766	---- ---- ---- - --	ex, subgl
							1774	---- ---- ---- - --	ex, subgl
							1783-84(L)	DD-- --xx ---- - --	EX, LF
							1783-85	---- ---- ---- - --	ex, subgl
							1797	---- ---- ---- - --	ex, subgl
							1816	---- ---- ---- - --	ex, subgl
							1823?(G,T)	---- ---- ---- - --	ex, subgl
							1838	---- ---- ---- - --	ex, subgl
							1854?	---- ---- ---- - --	ex, subgl
							1861?	---- ---- ---- - --	ex, subgl
							1867	---- ---- ---- - --	ex, subgl
							1873	Y--- ---- ---- - --	Ex, subgl
							1883	---- ---- ---- - --	ex, subgl
							1887? (T)	---- ---- ---- - --	ex, subgl
							1891-92?	---- ---- ---- - --	ex, subgl
							1902-4(G,T)	x--- ---- ---- - --	ex, subgl
							1910?	---- ---- ---- - --	ex, subgl

GRIMSVÖTN, Region 17, CAVW number 17-02-15,16,17,19,22

GRIMSVÖTN (continued)

CAVW number (active vent)	Latitude Longitude (degrees)	Diameter (km)	Local tectonic setting	Precaldera edifice	SiO_2 content (wt pct)	Age of collapse (yr B.P.)	Date of unrest	Type and duration of unrest ESTU STHF MGTF H Te	Eruption type
(continued from previous page)							1922	---- ---- ---- - --	ex, subgl
							1933-34	Y--- ---- ---- - --	ex, subgl
							1938	---- ---- ---- - --	subgl
							1939	---- ---- ---- - --	subgl
							1941	---- ---- ---- - --	subgl
							1945	---- ---- --x- - --	pex, subgl
							1948	---- ---- --x- - --	none?
							1954	---- ---- ---- - --	pex, subgl
							1982-83	EBA- ---- ---- - --	subgl, ex

TECTONIC SETTING

The Mid-Atlantic plate boundary passes through Iceland and is reflected in two zones of spreading and volcanism -- an eastern zone, the site of most historical eruptions, and a western zone. Grímsvötn is situated in the central part of the eastern volcanic zone (fig. 17.1.1 in section on Katla). A fissure swarm extends 70 km to the southwest. By analogy with Krafla it has been suggested that magma from a shallow magma chamber beneath Grímsvötn has moved laterally into the fissure swarm during rifting episodes (Sigurdsson and Sparks, 1978).

GEOLOGIC SETTING

Grímsvötn, together with Bárdarbunga and Kverkfjöll, lie beneath the vast Vatnajökull icecap in east-central Iceland (fig. 17.5.1). Due to its subglacial setting and remoteness, the geology of Grímsvötn is poorly known. The volcano has a composite central caldera, of which the southern rim is the only exposed part of the volcano. Only basaltic material is known to have erupted from Grímsvötn. Most known eruptions of Grímsvötn have occurred within the caldera, but the Laki eruption (12 km^3 of basaltic lava) from the fissure swarm 40-70 km southwest of the volcano may also have been fed from a shallow chamber beneath the caldera (Sigurdsson and Sparks, 1978; Sigurdsson, 1983). A large geothermal area in the caldera, with an estimated heat output of 5000 MW (Bjornsson, 1974, 1983), continually supplies meltwater to an ice-covered caldera lake. Approximately every 4-10 years the accumulated water is released in glacial outburst floods (jökulhlaups) when the water level is high enough to lift its ice dam (Thorarinsson, 1953; Björnsson, 1974, 1975; Björnsson and Kristmannsdóttir, 1984). Some jökulhlaups are associated with eruptions. In some instances it can be argued that the extra meltwater generated by a subglacial eruption triggered the draining of the caldera lake (for example, during the 1938 eruption). In other cases the sudden pressure release associated with draining may trigger an eruption (Thorarinsson, 1974).

GRIMSVÖTN (continued)

HISTORICAL ACTIVITY

Many of the eruptions listed in the table may be mislocated. They may have occurred in other volcanoes, known or unknown, beneath the northwest part of the Vatnajökull ice sheet.

1783-84: The largest basaltic fissure eruption in historical time began on 8 June 1783 on the Lakagigar fissure, 40-70 km southwest of Grímsvötn, after three weeks of earthquake activity. The eruption produced 11.9 km^3 of lava, 0.2 km^3 (DRE) of tephra, and a gas plume ("haze") that drifted across Europe. Beginning on about 8 June 1783, vigorous fountaining occurred from 135 craters along 10 en echelon segments of a 27-km-long rift (Thordarson and others, 1987). Eruptions began at the distal (southwest) end of the rift and migrated systematically uprift from 8 June until October 1783 (Thordarson and others, 1987). The eruption ended in February 1784. Eruptions also occurred at or near Grímsvötn on 26 September 1783 and possibly as early as on 18 July 1783, and continued into 1785 (Thorarinsson, 1969, 1984; Einarsson and Sveinsdóttir, 1984; Sigurdsson, 1982; Thordarson and others, 1987).

1873: During an eruption of Grímsvötn, an earthquake was felt in South Iceland, about 150 km away (Thorarinsson, 1974); it is uncertain whether the earthquake was associated with Grímsvötn.

1902-04: Jökulhlaups occurred north of Grímsvötn in December 1902 and May 1903. Another jökulhlaup began on 25 May 1903, southeast of Grímsvötn, during which earthquakes were felt in the Oræfi district on two (unspecified) days (Brandsdóttir, 1984). Earthquakes, rumbling, and crashing noises reached their maximum on 28-29 May 1903. Eruptions are thought to have occurred at Thordarhyrna (20 km south-southwest of Grímsvötn) and at Grímsvötn itself. The eruption(s?) continued until January 1904, but there are no further reports of felt earthquakes (Brandsdóttir, 1984).

1933-34: From 30 November to 10 December 1933, red glow and "fires" were reported at Grímsvötn, but no earthquakes were recorded at Reykjavik. Then, in 1934, a jökulhlaup occurred on 22 March. Five earthquakes of $\underline{M}$=3.5 or greater occurred on 30 March--one of $\underline{M}$=4.5 at 2004 hr and four of $\underline{M}$=3.5-3.75 within 2 hours after the start of an eruption at 2030 hr. Eruptions continued until 7 April, and a felt earthquake was reported on 30 April. Another eruption, possibly in Grímsvötn, occurred in December 1934 without any recorded earthquakes.

1938, 1941: Subglacial eruptions occurred, but there were no earthquakes large enough to be recorded (that is, none larger than $\underline{M}$=3.5).

1945: Increased geothermal activity was discovered in the south part of the caldera on 26-27 July. Jökulhlaups began on 16 September and lasted 11 days. A small eruption probably occurred, but no earthquakes were recorded or felt.

1948: The rate of ice melting at Grímsvötn was 50 percent higher than average in 1948, but no eruption or earthquakes were recorded.

GRIMSVÖTN, Region 17, CAVW number 17-02-15,16,17,19,22

GRIMSVÖTN (continued)

HISTORICAL ACTIVITY (continued)

1954 - mid-1982: Generally higher seismicity beneath the Vatnajökull ice sheet was associated with an increase in geothermal activity in the cauldron areas northwest of Grímsvötn. This and other short periods of heating contrast with a generally declining heat output during 1900-80 (Björnsson, 1983). Over the same period, a comparable increase in heat output has occurred in an area 10 km northwest of Grímsvötn, about halfway between Grímsvötn and Bárdarbunga (Björnsson, 1977, 1983).

1982-83: Earthquakes began in December 1982 and gradually increased during April-May 1983, centered under the southeast rim of Grímsvötn caldera (fig. 17.5.2). Intense seismicity began at 0400 hr on 28 May 1983, with $\underline{M}_{max}$ about 4. Earthquake activity declined at approximately 1000 hr, and bursts of tremor began at approximately 1200 hr. Tremor increased in intensity (amplitude) at about 1500 hr on 28 May and continued at a high level for about 1 day before decreasing. The eruption was first visible on the morning of 29 May, by which time it had broken and melted its way through the icecap (fig. 17.5.3) (Einarsson and Brandsdóttir, 1984; Grönvold and Jóhannesson, 1984).

REFERENCES

Björnsson, H., 1974, Explanation of Jökulhlaups from Grímsvötn, Vatnajökull, Iceland: Jökull, v. 24, p. 1-26.
----- 1975, Subglacial water reservoirs, Jökulhlaups and volcanic eruptions: Jökull, v. 25, p. 1-14.
----- 1977, The cause of Jökulhlaups in the Skaftá River, Vatnajökull: Jökull, v. 27, p. 71-78.
----- 1983, A natural calorimeter at Grímsvötn: an indicator of geothermal and volcanic activity: Jökull, v. 33, p. 13-18.
Björnsson, H., and Kristmannsdóttir, H., 1984, The Grímsvötn geothermal area, Vatnajökull, Iceland: Jökull, v. 34, p. 25-50.
Brandsdóttir, B., 1984, Seismic activity in Vatnajökull in 1900-1982, with special reference to Skeidararhlaups, Skaflarhlaups, and Vatnajökull eruptions: Jökull, v. 34, p. 141-150.
Einarsson, P., and Brandsdóttir, B., 1984, Seismic activity preceding and during the 1983 volcanic eruption in Grímsvötn, Iceland: Jökull, v. 34, p. 13-23.
Einarsson, Th., and Sveinsdóttir, E.L., 1984, in Gunnlaugsson, G.A., Gudbergsson, G.M., Thorarinsson, S., Rafnsson, S., and Einarsson, Th., eds., Skaftareldar 1783-1784: Reykjavik, Mal og Menning, p. 37-48.
Grönvold, K., and Johannesson, H., 1984, Eruption in Grímsvötn 1983, course of events and chemical studies of the tephra: Jökull, v. 34, p. 1-12.
Gudmundsson, G., and Sæmundsson, K., 1980, Statistical analysis of damaging earthquakes and volcanic eruptions in Iceland from 1550-1978: Jour. Geophys., v. 47, p. 99-109.
Johannesson, H., 1983, Gossaga Grimsvatna 1900-1983 i stuttu mali (A brief review of the volcanic activity of the Grímsvötn volcanic system 1900-1983): Jökull, v. 33, p. 146-147.
----- 1987, Grimsvatnagos 1816 (Evidence of an eruption in Grímsvötn in 1816): Náttúrufrædingurinn, v. 57, p. 157-159.

GRIMSVÖTN (continued)

REFERENCES (continued)

Kjartansson, G., 1962, Geological map of Iceland, sheet 6, South-Central Iceland: Reykjavik, Museum of Natural History, 1 sheet, scale 1:250,000.

Sigurdsson, H., 1982, Volcanic pollution and climate: the 1783 Laki eruption: Eos, Trans. Am. Geophys. Union, v. 63, p. 601-602.

----- 1983, The 1783 Lakagigar fissure eruption [abs.]: Eos, Trans. Am. Geophys. Union, v. 64, p. 887.

Sigurdsson, H., and Sparks, R.S.J., 1978, Lateral magma flow within rifted Icelandic crust: Nature, v. 274, p. 126-130.

Smithsonian Institution, Scientific Event Alert Network (SEAN), 1982-83, Grímsvötn: SEAN Bull., v. 7, no. 2; v. 8, no. 5.

Steinthorsson, S., 1977, Tephra layers in a drill core from the Vatnajökull ice cap: Jökull, v. 27, 27 p.

Thorarinsson, S., 1953, Some new aspects of the Grímsvötn problem: Jour. Glaciol., v. 2, no. 14, p. 267-274.

----- 1960, On the predicting of volcanic eruptions in Iceland: Bull. Volcanol., v. 23, p. 45-52.

----- 1969, The Lakagigar eruption of 1783: Bull. Volcanol., v. 33, no. 3, p. 910-929.

----- 1973, Data sheets of the post-Miocene volcanoes of the world, Sheet II, Iceland: Rome, IAVCEI, 10 p.

----- 1974, Votnin strid. Saga Skeidararhlaupa og Grimsvatnagosa (The swift flowing rivers. The history of Grímsvötn eruptions and jökulhlaups in Skeidara river): Reykjavik, Menningarsjodur, 254 p.

----- 1984, Annall Skaftarelda (The Laki Fires), in Gunnlaugsson, G.A., Gudbergsson, G.M., Thorarinsson, S., Rafnsson, S., and Einarsson, Th., eds., Skaftareldar 1783-1784: Reykjavik, Mal og Menning, p. 11-36.

Thordarson, Th., Self, S., Larsen, G., and Steinthorsson, S., 1987, Eruption sequence of the Skaftar Fires 1783-1785, Iceland [abs.]: Eos, Trans. Am. Geophys. Union, v. 68, p. 1550.

Thoroddsen, Th., 1925, Die Geschichte der Islandischen Vulkane: D. Kgl. Dansk Vidensk. Selsk. Skrifter, Naturvidensk. og Mathem. Afd. Copenhagen, 8th ser., v. 9, 458 p.

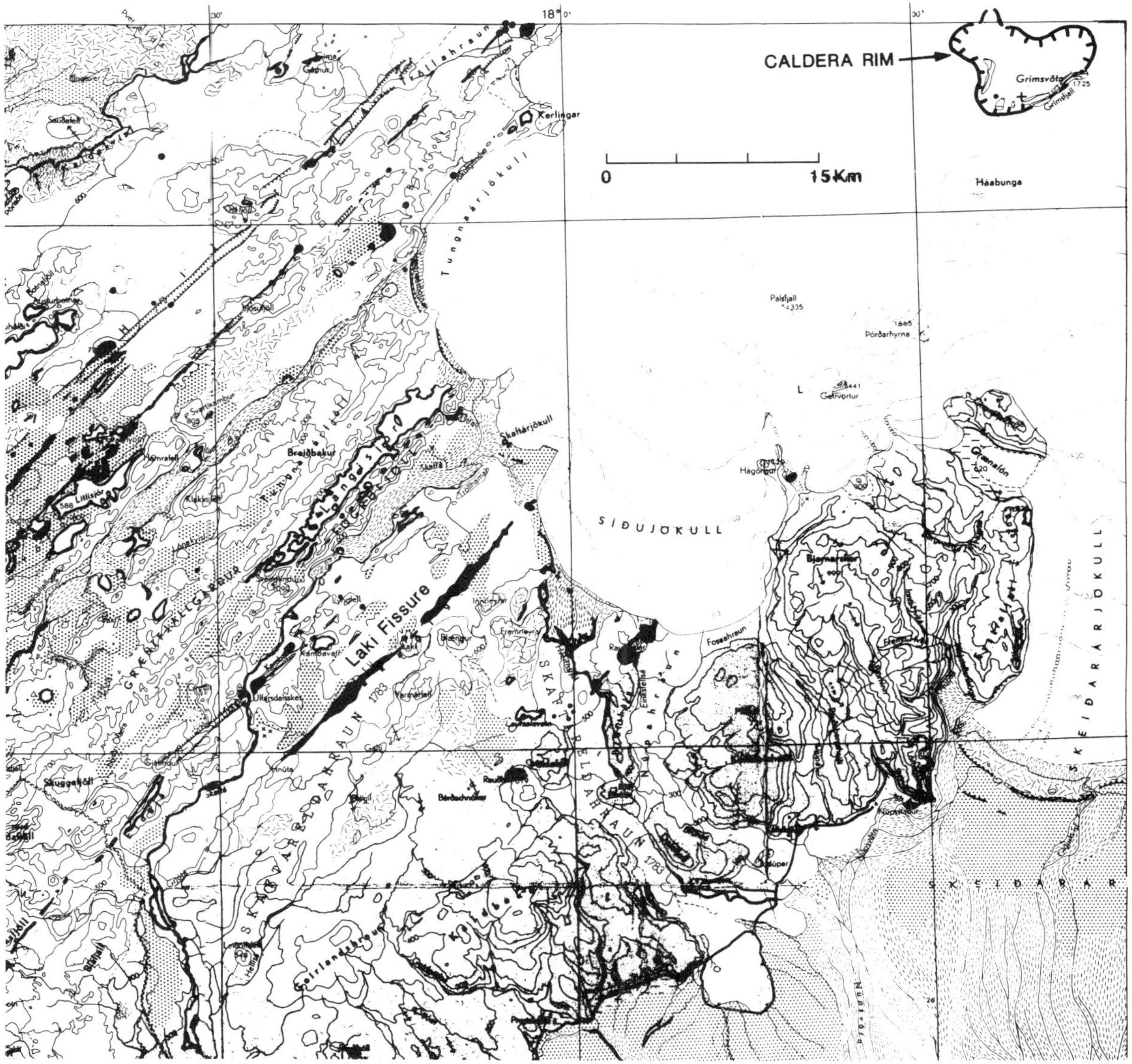

Figure 17.5.1. Grímsvötn Caldera, located under Vatnajökull icecap, from Kjartansson (1962). Voluminous basaltic lava flows were erupted from Lakagigar (Laki), southwest of Grímsvötn, in 1783.

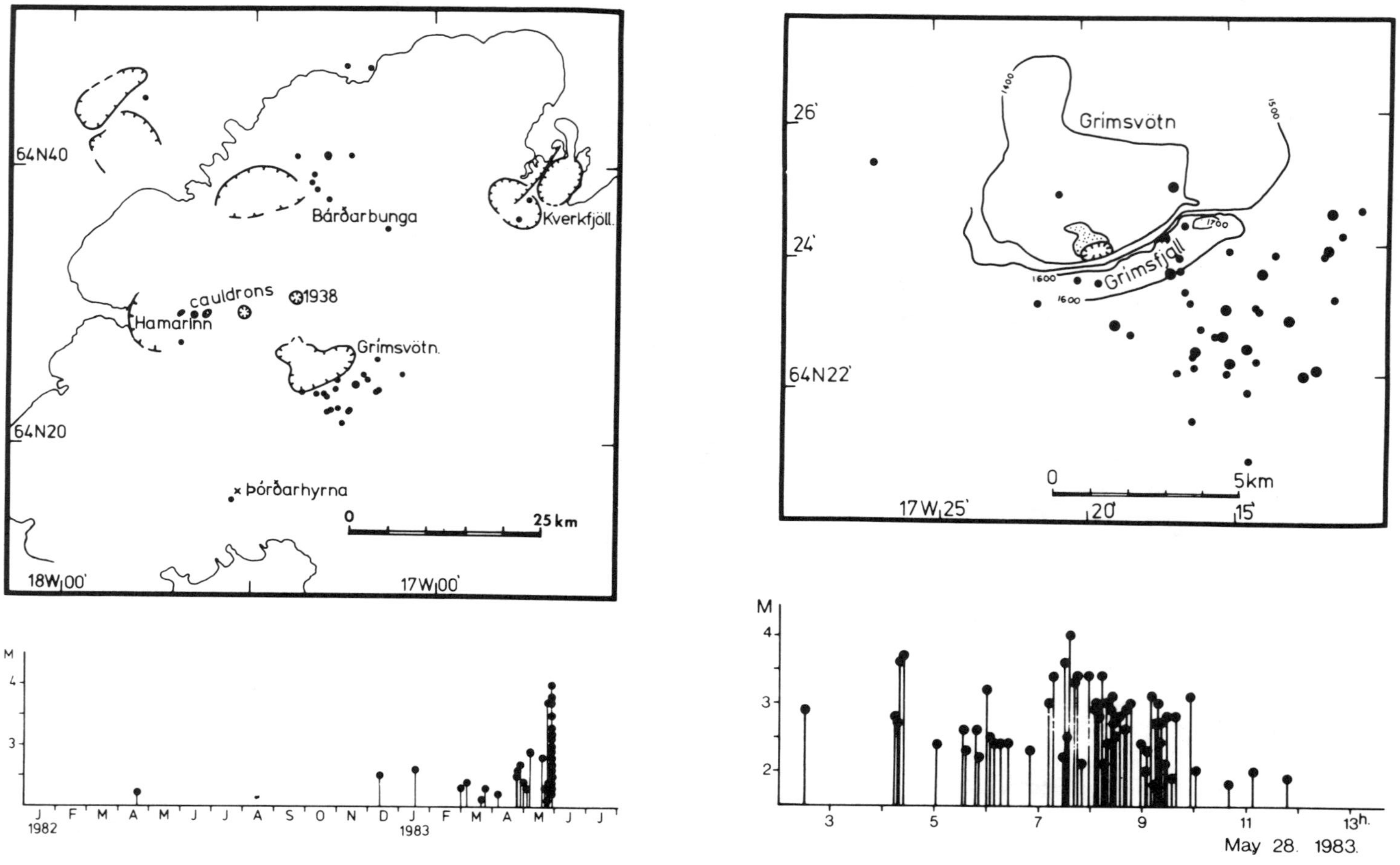

Figure 17.5.2. Epicentral map (left) of western Vatnajökull area for the period 1 January 1982 to 27 May 1983, from Einarsson and Brandsdóttir (1984). Small dots mark epicenters of earthquakes smaller than magnitude 3; larger dots denote events larger than magnitude 3. Calderas and inferred subglacial calderas are shown, as are ice cauldrons associated with jökulhlaups in Skafta River and presumed subglacial eruption of 1938. Almost all earthquakes southeast of Grímsvötn occurred in early 1983. Enlarged view (right) of Grímsvötn epicentral area and plot of earthquake magnitude versus time during May 1983.

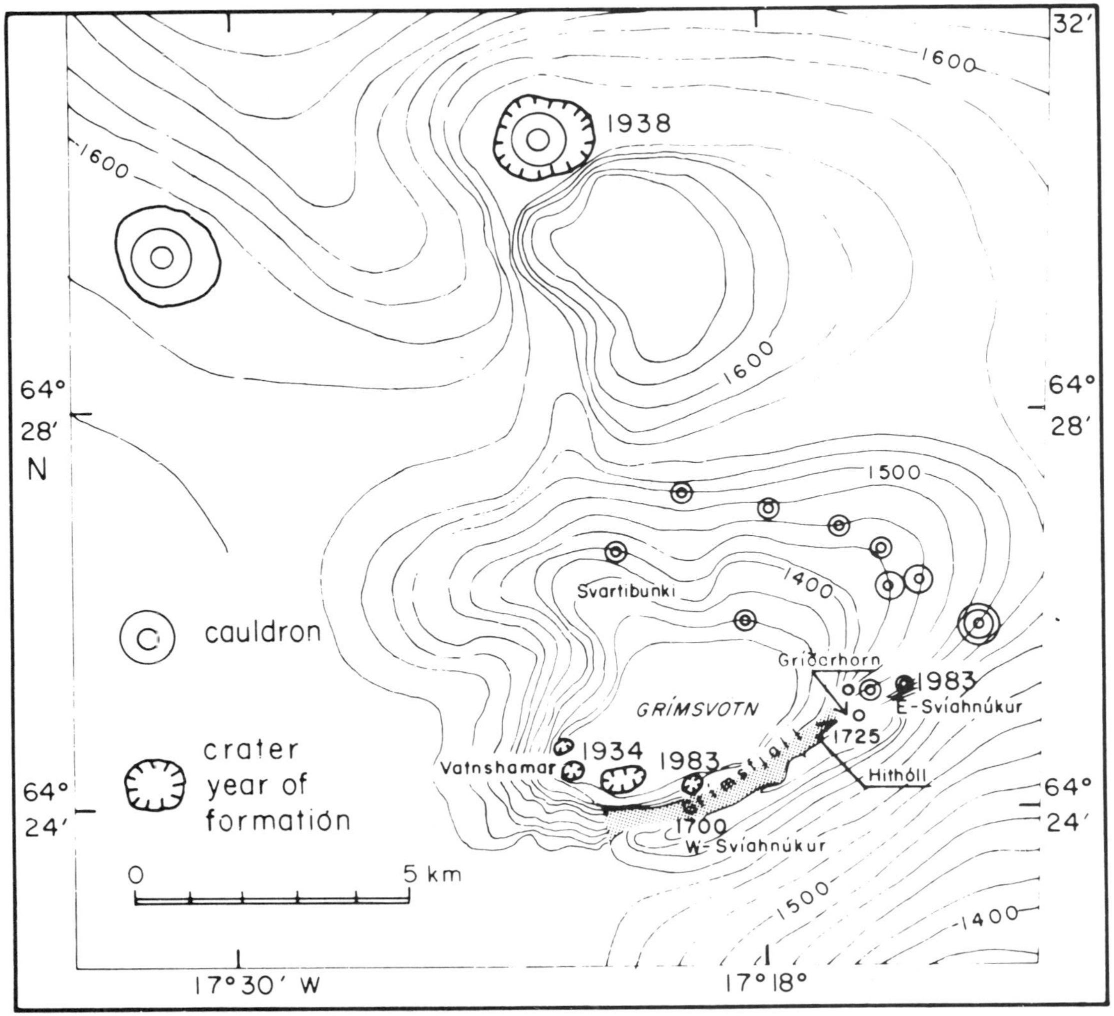

Figure 17.5.3. Grímsvötn area, from Björnsson and Kristmannsdóttir (1984). Shown are craters of 1934 and 1983, depression north of Svartibunki from 1938, and ice cauldrons. Cauldron that formed during jökulhlaup in 1983 is also marked. Cauldron northwest of Grímsvötn drains to Skafta River.

PART 3: HISTORICAL UNREST AT LARGE QUATERNARY CALDERAS (continued)
See inside back cover for explanation and abbreviations

KVERKFJÖLL

CAVW number (active vent)	Latitude Longitude (degrees)	Diameter (km)	Local tectonic setting	Precaldera edifice	SiO_2 content (wt pct)	Age of collapse (yr B.P.)	Date of unrest	Type and duration of unrest ESTU STHF MGTF H Te	Eruption type
17-02-23 (Kverkfjöll)	64.65N 16.70W	7.5 x 5.5 8 x 5	Exten	LL-strat, partly subglacial	R = mafic		1655	---- ---- ---- - --	subgl
							1684	---- ---- ---- - --	subgl, ex
							1711	---- ---- ---- - --	subgl
							1716?-17	---- ---- ---- - --	subgl, ex?
							1726	---- ---- ---- - --	subgl
							1726	---- ---- ---- - --	subgl
							1729	---- ---- ---- - --	subgl
							1874-75?	---- ---- ---- - --	subgl, ex

TECTONIC SETTING

The Mid-Atlantic plate boundary passes beneath Iceland and is reflected in two zones of spreading and volcanism--an eastern zone, the site of most historical eruptions, and a western zone. Kverkfjöll is situated in the central part of the eastern volcanic zone, just northeast of Grímsvötn (fig. 17.1.1 in section on Katla).

GEOLOGIC HISTORY

Kverkfjöll is a central stratovolcano on the northeast edge of the Vatnajökull ice sheet (fig. 17.6.1). Two ice-filled calderas have been inferred.

HISTORICAL ACTIVITY

No historical eruptions are known with certainty to have occurred at Kverkfjöll. Most possible eruptions of Kverkfjöll have been inferred from glacial outburst floods. The location of the 17th century eruptions is uncertain. The 18th century eruptions listed above have been attributed to Bárdarbunga as well as to Kverkfjöll (Gudmundsson, 1986; Larsen, 1982).

1874-75: An eruption column south of Askja may have originated at Kverkfjöll (Sigurdsson and Sparks, 1978; G. Larsen and P. Einarsson, written commun., 1988). This inferred eruption of Kverkfjöll occurred during major rifting in the Askja area (see section on Askja).

Small earthquakes have been recorded in Kverkfjöll in recent years.

KVERKFJÖLL (continued)

REFERENCES

Einarsson, P., and Brandsdóttir, B., 1984, Seismic activity preceding and during the 1983 volcanic eruption in Grímsvötn, Iceland: Jökull, v. 34, p. 13-23.

Gudmundsson, A., 1986, Eldgos a Dyngjuhalsi a 18. old (Eruptions in the Dyngjuhals area in the 18th century): Náttúrufrædingurinn, v. 56, p. 43-48.

Gudmundsson, G., and Sæmundsson, K., 1980, Statistical analysis of damaging earthquakes and volcanic eruptions in Iceland from 1550-1978: Jour. Geophys., v. 47, p. 99-109.

Larsen, G., 1982, Gjoskutimatal Jökuldals og nagrennis (Tephrochronology of the Jökuldalur area), in Thorarinsdóttir, H., Oskarsson, O.H., Steinthorsson, S. and Einarsson, Th., eds., Eldur er i nordri: Reykjavik, Sogufelag, p. 51-65.

Sæmundsson, K., 1982, Oskjur a virkum eldfjallasvaedum a Islandi (Calderas in active volcanic areas in Iceland), in Thorarinsdóttir, H., Oskarsson, O.H., Steinthorsson, S. and Einarsson, Th., eds., Eldur er i nordri: Reykjavik, Sogufelag, p. 221-240.

Sigurdsson, H., and Sparks, R.S.J., 1978, Rifting episode in North Iceland in 1874-1875 and the eruption of Askja and Sveinagja: Bull. Volcanol., v. 41, p. 149-167.

Thorarinsson, S., 1960, On the predicting of volcanic eruptions in Iceland: Bull. Volcanol., v. 23, p. 45-52.

----- 1973, Data sheets of the post-Miocene volcanoes of the world, Sheet II, Iceland: Rome, IAVCEI, 10 p.

Thorarinsson, S., and Sæmundsson, K., 1979, Volcanic activity in historical time: Jökull, v. 29, p. 29-32.

Thoroddsen, Th., 1925, Die Geschichte der Islandischen Vulkane: D. Kgl. Dansk Vidensk. Selsk. Skrifter, Naturvidensk. og Mathem. Afd. Copenhagen, 8th ser., v. 9, 458 p.

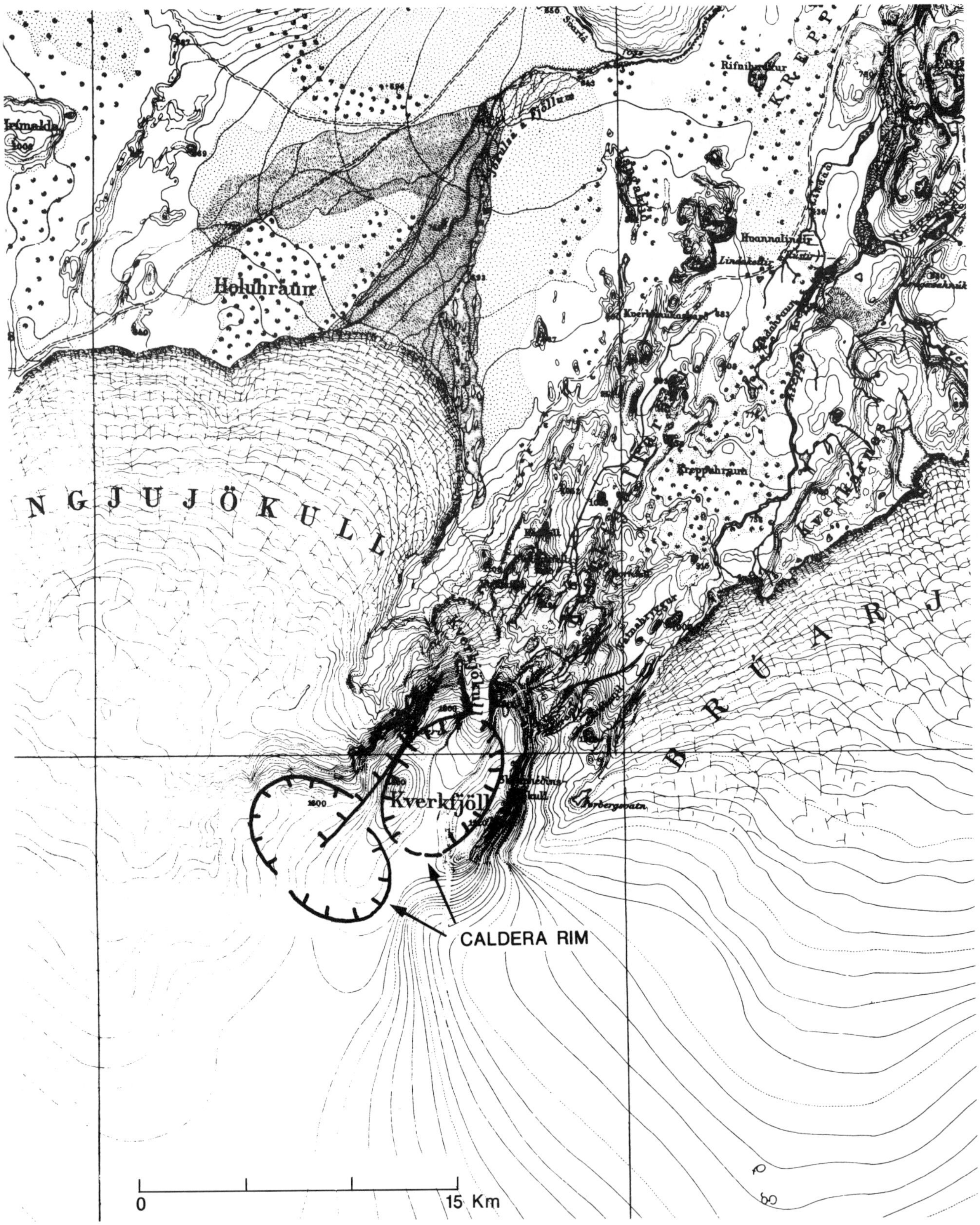

Figure 17.6.1. Inferred outlines of Kverkfjoll Calderas, located under Vatnajökull ice sheet. Outlines of calderas from Einarsson and Brandsdóttir (1984). See also figures 17.1.1 and 17.5.2.

ASKJA, OSKJUVATN

CAVW number (active vent)	Latitude Longitude (degrees)	Diameter (km)	Local tectonic setting	Precaldera edifice	SiO_2 content (wt pct)	Age of collapse (yr B.P.)	Date of unrest	Type and duration of unrest ESTU STHF MGTF H Te	Eruption type
17-03-C3 (Askja)	65.03N 16.75W	9 (Askja) 4 (Oskju-vatn)	Exten	LL-strat	R = 50-51 C = 51-75	A.D. 1875	1874-75	EE-- E--E --EE E QU	EX,1f
							1919?	---- ---- ---- - --	ex
							1921-26	---- ---- ---- - --	1f,ex
17-03-C9 (Sveinagja)							1929	---- ---- ---- - --	1f,ex
							1961	D--Z Z--D --DD D --	1f,ex
							late 1960's-present	---F FF-- ---- - --	none

TECTONIC SETTING

Askja is in the central part of the volcanic rift zone that marks the Mid-Atlantic plate boundary through Iceland. The 100-km-long Askja fissure swarm extends 25 km south-southwest and 75 km north-northeast of Askja (fig. 17.7.1). Two episodes of rifting are known historically (see below). Sveinagja, an area of venting during the 1874-75 unrest, is near the northern end of the Askja fissure swarm. The Tjornes Fracture Zone is an E-W-trending ridge-ridge transform zone, about 100 km north of Askja, connecting the eastern volcanic zone of Iceland and the Kolbeinsey Ridge.

GEOLOGIC HISTORY

Askja is a caldera in Dyngjufjoll, a central stratovolcano built mainly of palagonite tuff and pillow lavas. The Askja Caldera formed in prehistoric time; the Oskjuvatn Caldera formed in early 1875 (see below). The Askja Caldera apparently formed in response to migration of basaltic magma into the Askja fissure swarm; the Oskjuvatn Caldera formed in response to similar migration of basaltic magma and a late-stage explosive eruption of mixed basalt-rhyolite magma (Sigurdsson and Sparks, 1978; Sparks and others, 1981).

HISTORICAL ACTIVITY

1874-75: A rifting episode and large volcanic eruption occurred during this period (fig. 17.7.1). Severe earthquakes, including one with an estimated magnitude of 6 to 7, occurred on 17-18 April 1872 along the Tjornes Fracture Zone roughly 100 km north-northwest of Askja. A causal relationship with the Askja events is uncertain. Intense steam emission from Askja was reported in February 1874, and increased seismic activity and ground cracking occurred in the Sveinagja area, 50-70 km north of Askja in the fall of 1874. In mid-December 1874, seismic activity increased in the northern rift zone, and between Christmas and New Year's Day

ASKJA, OSKJUVATN (continued)

HISTORICAL ACTIVITY (continued)

earthquakes were felt continuously all over northeast Iceland (Sparks and others, 1981). Small explosive and effusive eruptions (of both basalt and rhyolite) began at Askja on 1 January 1875 and were soon joined by basaltic fissure eruptions at Sveinagja on 18 February. A paroxysmal, dominantly rhyolitic eruption of Askja occurred on 28-29 March. Eruptions continued until May 1875 at Askja and until October 1875 at Sveinagja.

The Oskjuvatn Caldera formed by collapse, principally as basaltic magma migrated into the fissures north of Askja. Collapse had begun by mid-February 1875, before the main explosion (28 March), and most collapse had occurred by July 1875. The volume of collapse was 2 km^3, compared with only 0.5 km^3 of magma erupted at Askja and Sveinagja combined; the balance was presumably intruded as dikes between Askja and Sveinagja (Sigurdsson and Sparks, 1978).

1919: A small eruption, for which the only evidence is a tephra layer in the Bárdarbunga ice core (Steinthorsson, 1977). Askja is a possible but uncertain source for this tephra.

1921-26: Lava flows issued from vents around Oskjuvatn's rim (fig. 17.7.2), but no regional rifting was observed.

1929: A basaltic fissure eruption occurred south of Askja about this time but the year is uncertain.

1961: An eruption occurred from vents in the northeast part of the Askja Caldera, but no regional rifting occurred. Small earthquakes began on 6 October 1961; large solfataras and geysers began to form on or about 9 October, together with many fissures on the caldera floor. The principal fissure was a N-S line in the eastern part of Askja Caldera, marked by a number of fumaroles in figure 17.7.2. This fissure opened from before 10 October until 15-17 October, accompanied by discharges of hot water. Violent geyser activity occurred from 17 October until 19 October or shortly thereafter, and this activity shifted to rhythmic fuming (without water discharge) until an eruption began on 26 October (Sigvaldason, 1964). About 0.1 km^3 of basaltic lava was erupted from an E-W trending fissure on the floor of Askja Caldera. After the eruption, the level of the caldera lake dropped 2-3 m and fissures formed along the caldera wall.

Late 1960's-present: Measurements of ground deformation in Askja were initiated in 1966. Precision leveling, lake-level measurements, and electronic distance measurements (EDM) show three periods of deflation (1966-67, 1968-70, and after 1972) and two periods of inflation (1967-68 and 1970-72) with a poorly defined center of uplift or subsidence in the central region of the Askja caldera. The rate of subsidence during deflation is about 8-9 cm/yr; the rate of inflation in 1970-72 was two or three times faster (20-30 cm/yr) in the central part of the caldera. Slow deflation occurred from 1972 until at least 1987 (Tryggvason, 1987).

ASKJA, OSKJUVATN (continued)

COMMENTS

Oskjuvatn is the only caldera in this compilation to have formed in conjunction with a major rifting episode in historical time, in 1874-75. Rifting and basaltic dike intrusion preceded and presumably triggered the eruption of silicic magma from the Askja central volcano.

REFERENCES

Björnsson, A., Sæmundsson, K, Einarsson, P., Tryggvason, E., and Grönvold, K., 1977, Current rifting episode in North Iceland: Nature, v. 266, p. 318-323.

Gudmundsson, G., and Sæmundsson, K., 1980, Statistical analysis of damaging earthquakes and volcanic eruptions in Iceland from 1550-1978: Jour. Geophys., v. 47, p. 99-109.

Sæmundsson, K., 1982, Oskjur a virkum eldfjallasvaedum a Islandi (Calderas in active volcanic areas in Iceland), in Thorarinsdóttir, H., Oskarsson, O.H., Steinthorsson, S., and Einarsson, Th., eds., Eldur er i nordri: Reykjavik, Sogufelag, p. 221-240.

Sigurdsson, H., and Sparks, R.S.J., 1978, Rifting episode in North Iceland in 1874-1875 and the eruption of Askja and Sveinagja: Bull. Volcanol., v. 41, p. 149-167.

Sigvaldason, G.E., 1964, Some geochemical and hydrothermal aspects of the 1961 Askja eruption: Beitrage Mineral. Petrog., v. 10, p. 263-274.

----- 1968, Structure and products of subaquatic volcanoes in Iceland: Contrib. Mineral. Petrol., v. 18, p. 1-16.

Sparks, R.S.J., Wilson, L., and Sigurdsson, H., 1981, The pyroclastic deposits of the 1875 eruption of Askja, Iceland: Phil. Trans. Roy. Soc. London, v. 299, Ser. A, p. 241-273.

Steinthorsson, S., 1977, Tephra layers in a drill core from the Vatnajökull ice cap: Jökull, v. 27, p. 2-27.

Thorarinsson, S., 1973, Data sheets of the post-Miocene volcanoes of the world, Sheet II, Iceland: Rome, IAVCEI, 10 p.

Thorarinsson, S., and Sigvaldason, G.E., 1962, The eruption of Askja in 1961, a preliminary report: Am. Jour. Sci., v. 260, p. 641-651.

Thoroddsen, Th., 1925, Die Geschichte der Islandischen Vulkane: D. Kgl. Dansk Vidensk. Selsk. Skrifter, Naturvidensk. og Mathem. Afd. Copenhagen, 8th ser., v. 9, 458 p.

Tryggvason, E., 1973, Seismicity, earthquake swarms and plate boundaries in the Iceland region: Bull. Seismol. Soc. Am., v. 63, p. 1327-1348.

Tryggvason, E., 1987, Surface deformation of the volcano Askja [abs.]: Abstracts, Hawaii Symposium on How Volcanoes Work, Hilo, January 1987, p. 257.

Volcanological Society of Japan, 1963, Askja: Bull. Volcanic Eruptions, no. 3, p. 7.

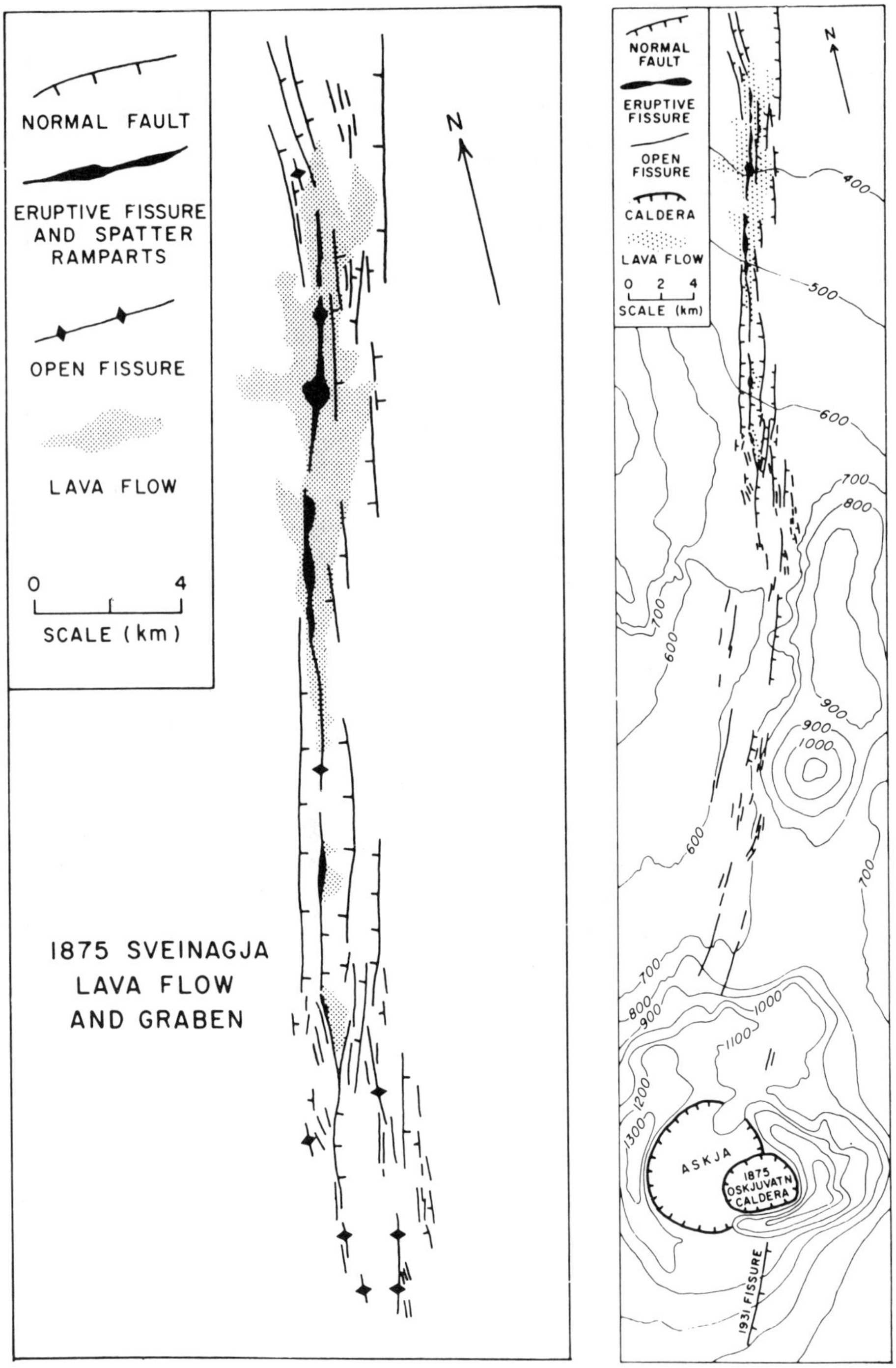

Figure 17.7.1. Askja-Sveinagja fissure swarm, active in the 1874-75 rifting episode, and location of Askja and Oskjuvatn Calderas (Sigurdsson and Sparks, 1978). Left, detailed faulting and distribution of 1875 lava in Sveinagja graben; right, position of graben and lava flow in relation to Askja and Oskjuvatn Calderas.

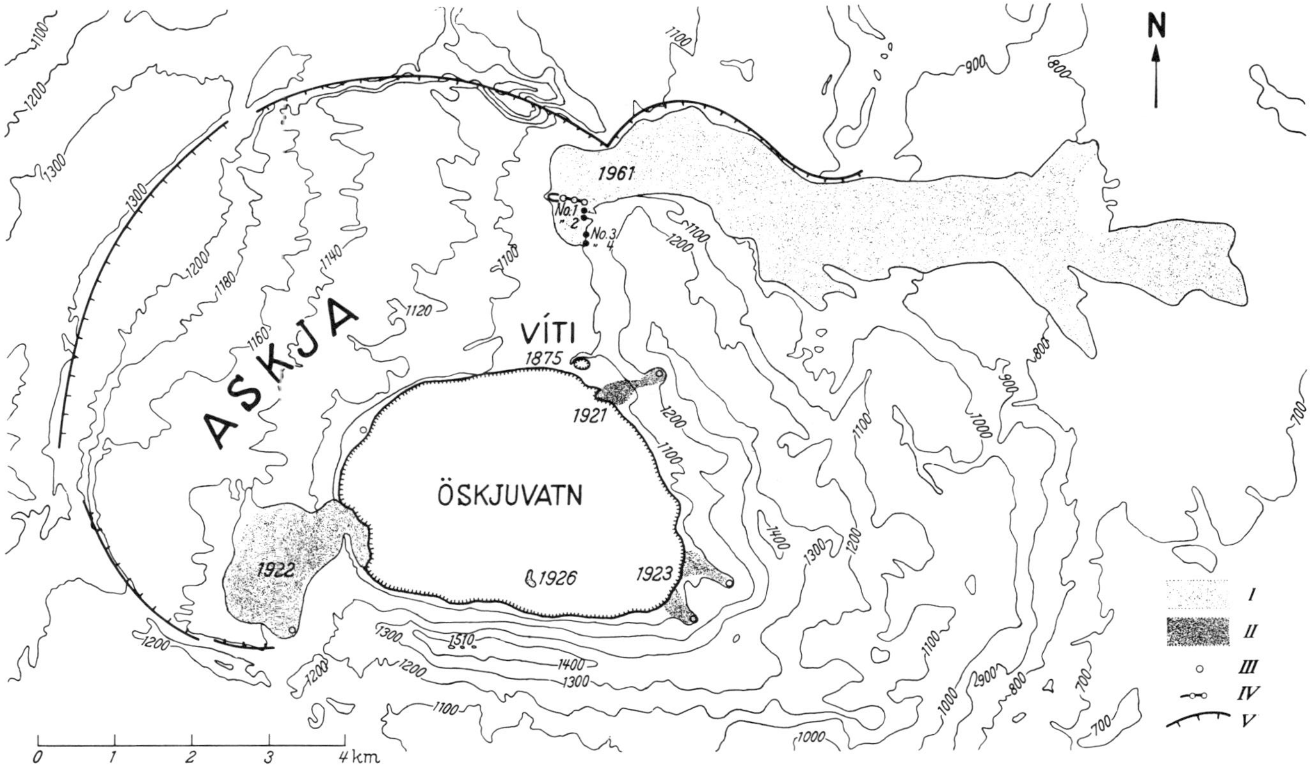

Figure 17.7.2. Askja Caldera, showing locations of recent lava flows, hot springs, craters, and major faults (Sigvaldason, 1964). I, 1961 lava flow; II, lava flows with indicated year of formation; III, preeruption hot springs; IV, lava craters; V, major tectonic fault escarpments.

KRAFLA

CAVW number (active vent)	Latitude Longitude (degrees)	Diameter (km)	Local tectonic setting	Precaldera edifice	SiO_2 content (wt pct)	Age of collapse (yr B.P.)	Date of unrest	Type and duration of unrest ESTU STHF MGTF H Te	Eruption type
17-03-11 (Krafla)	65.73N 16.68W	10, 8	Exten	Shield	R = b + minor r,d,a	<70,000	1724-29	BB-Y Y--Y --YY Y UQ	P-ex,ex,lf
							1746	Y--Y ---- ---- Y --	lf?
							1975-88+	EEBE BEEE -Exx Y UQ	lf,pex

TECTONIC SETTING

The Mid-Atlantic plate boundary passes through Iceland and is reflected in two zones of spreading and volcanism -- an eastern zone, the site of most historical eruptions, and a western zone. The axial rift zone of northeastern Iceland consists of five fissure swarms, each associated with and named after a central volcano (fig. 17.1.1 in section on Katla). These fissure swarms reflect spreading of the Eurasian and North American plates at a half-rate of about 1 cm/yr, and also updoming and spreading around an Icelandic hotspot. Rifting is episodic, occurring about once every hundred years; two of four historically recorded episodes have been along the Krafla swarm (figs. 17.8.1-17.8.3), in 1724-29 and 1975-present (Sæmundsson, 1974; Björnsson, 1985). Detailed observations of rifting in the Krafla area from 1975-1988+ have provided important insights into the processes of rifting and crustal accretion at ridges (see below).

GEOLOGIC HISTORY

Like most of the central volcanoes within the rift zones of Iceland, the Krafla Central Volcano is an indistinct structure. The weak, divergent crust does not support high volcanic edifices and subsidence plays a major role in the development of these volcanoes. Some of them have caldera structures and all have an associated fissure swarm running through them. The Krafla Central Volcano has produced mostly basaltic material in the form of interglacial or postglacial lavas, or subglacial formations of pillow lavas and palagonite tuffs and breccias. A few domes and ridges of rhyolite exist, and at least one catastrophic event has produced dacitic tephra. Krafla Caldera is marked by poorly defined arcuate structures where the dacite tephra is exposed. On the basis of S-wave shadows a crustal magma chamber is inferred to underlie the central part of the caldera. This chamber is thought to derive its magma from a subcrustal, partially molten layer indicated by seismic and magnetotelluric measurements.

The following is a conceptual model of Krafla activity from P. Einarsson and G. Larsen (written commun., 1988):

The activity of Krafla is primarily influenced by two processes, namely, inflow of magma into the crustal chamber and rifting of the plate boundary, as regulated by the magma pressure in the chamber and tectonic stress at the plate boundary. Tectonic stress

KRAFLA (continued)

GEOLOGIC HISTORY (continued)

presumably builds up slowly with time due to plate movements. When magma begins flowing into the magma chamber it can trigger the release of tectonic stress and thus initiate rifting. Fissures propagate away from the magma chamber near the center of the volcano, and magma flows laterally along the fissure swarm to the north and south of the volcano for tens of kilometers and forms a dike. Typical dike propagation velocities are 0.5-2.5 m/s. During emplacement of dikes the pressure in the magma chamber drops and the volcano deflates. This process can last hours to a few weeks depending on the length of the dike, and the deflation amounts to centimeters to a few meters in the center of the caldera. Extensive fissuring and normal faulting occurs above the dike, the central part of the swarm subsides, and the flanks are uplifted (figs. 17.8.16, 17.8.17). Horizontal and vertical displacements are of the order of tens of centimeters to meters. If magma inflow continues after a deflation event, the volcano reinflates at about 1-10 mm per day for weeks or months. Many such inflation - deflation cycles may occur during a rifting episode that may last several years to over a decade (figs. 17.8.10, 17.8.12, 17.8.15, 17.8.18). In the beginning of an episode the tectonic stress is high and most of the magma is emplaced in dikes in the crust. As the tectonic stress is released, the crust supports higher magma pressures, and an increasing proportion of the mobilized magma reaches the surface in eruptions.

Each phase of the activity is accompanied by its characteristic type of seismic activity:

Inflation earthquakes: During inflation periods earthquakes with magnitudes up to 4 occur in the magma chamber roof beneath the caldera. After stress in the roof exceeds a threshold level these earthquakes are well correlated with the rate of inflation. Inflation earthquakes stop immediately when inflation stops or deflation begins.

Deflation earthquakes: During large deflation events the deviatoric stress in the chamber roof is first relaxed, but then increases in the opposite sense. This can lead to earthquakes if deflation exceeds about 1 m. Deflation earthquakes occupy the same hypocentral volume as the inflation earthquakes, but have been as large as magnitude 5 at Krafla.

Rifting earthquakes: Intense swarms of earthquakes accompany rifting and dike intrusions in the fissure swarm of Krafla. The earthquakes begin near the caldera shortly after deflation sets in, and propagate along the fissure swarm away from the caldera, presumably marking the propagating edge of a dike (for example, figs. 17.8.4, 17.8.6). Both high and low frequency events occur--the latter apparently associated with surface faulting. Rifting earthquakes only seldom exceed magnitude 4, in spite of extensive surface faulting (fig. 17.8.5).

Intrusion tremor: Continuous tremor accompanies deflation and dike intrusion (fig. 17.8.13). The tremor begins as soon as deflation starts and its amplitude is roughly correlated with the rate of deflation. Amplitudes are uneven and spasmodic, and the frequency spectrum is broad. The tremor is mixed with rifting earthquakes.

Eruption tremor: When a dike reaches the surface and an eruption begins, a new type of tremor becomes the dominant seismicity. Eruption tremor has an even amplitude and a narrow frequency spectrum, with predominant frequency below 3 Hz. Its amplitude appears to be correlated with the height of lava fountains.

KRAFLA (continued)

GEOLOGIC HISTORY (continued)

During inflation periods almost all measured changes occur within the caldera or in its immediate vicinity. The inflation is centered beneath the caldera, as shown by elevation changes, tilt measurements, distance measurements and gravimetry (figs. 17.8.8, 17.8.9, 17.8.11, 17.8.17). These data fit well to a Mogi-type model of an inflating magma chamber at a depth of 2.6-3.0 km. Inflation is also reflected in a gradual widening of fissures within the caldera. Sometimes changes in geothermal activity and the composition of gases are detected during inflation, particularly if the previous deflation event was accompanied by intrusion into the magma chamber roof within the caldera.

During deflation events measurable changes occur both within the caldera and outside it. The onset of deflation is usually abrupt (figs. 17.8.12-17.8.14). Recording tiltmeters show a reversal in tilt within minutes or tens of minutes, and intrusion tremor appears simultaneously on the seismographs (figs 17.8.14, 17.8.15). The inflation bulge subsides and rifting earthquakes appear outside the caldera. The deflation rate reaches a maximum within hours and then slowly decreases as the dike lengthens. Eruptions in the current rifting episode have begun 1-7 hours after the onset of deflation, and have lasted 4-14 days. The lava is highly fluid basalt, erupted through 1-9-km-long fissures. Some very small phreatic eruptions have occurred, each lasting a few hours, and some small steam fields have developed in the fissure swarm days or weeks after an intrusion.

HISTORICAL ACTIVITY

The following is a summary of events of the current rifting episode, from P. Einarsson and G. Larsen (written commun., 1988) and references cited below (see also figs. 17.8.3-17.8.18):

1975: Seismic activity increased in the caldera region in the early part of the year. Each day 10-20 events were recorded, some reaching magnitude 4.

1975, 20 December: A major deflation event began, accompanied by a small fissure eruption in the caldera, and intrusion 10-15 km to the south and 60 km to the north (fig. 17.8.3). The northern intrusion extended all the way to the transform fault offshore and triggered a large earthquake swarm that culminated with a transform event of magnitude 6.5 on 13 January 1976. Both deflation of the caldera and rifting near the northern end of the fissure swarm continued until the middle of February with considerable seismicity in both areas. Total deflation in the caldera exceeded 2 m.

1976, 28 September: Small and slow deflation event lasted about 6 days. The intrusion was to the north, as shown by a swarm of very small earthquakes 10 km north of the caldera.

31 October: A rapid deflation was accompanied by intrusion and an intense earthquake swarm to the north of the volcano.

KRAFLA (continued)

HISTORICAL ACTIVITY (continued)

1977, 20 January: A deflation and intrusion event similar to, but shorter than, the last event.

27 April: A rapid deflation event was accompanied by a small eruption near the northern caldera rim and an intrusion to the south of the volcano (fig. 17.8.7).

8 September: A deflation event that was similar to that of April, except that the eruption near the caldera rim was larger and the dike to the south shorter. A small volume of basalt was erupted through a geothermal drill hole when the intrusion entered a producing geothermal area 10 km south of the caldera.

2 November: A very small deflation event occurred. The intrusion probably went a short distance to the north.

1978, 7 January: The second largest deflation event started off slowly but continued for 3 weeks. Earthquakes were initially small, as activity migrated to the north. After 20 hours the seismicity increased and culminated on 10-11 January with events of magnitude 4.6. Large scale rifting occurred 30-40 km north of the volcano, and the total deflation within the caldera exceeded 1 m.

10 July: A large deflation event was accompanied by a 30-km-long intrusion to the north (figs. 17.8.4-17.8.6).

10 November: A deflation event that was similar to the last one, but with a slightly shorter intrusion.

1979, 13 May: A deflation similar to the last two events was accompanied by a yet shorter intrusion to the north.

6 December: A small and slow deflation event was accompanied by small earthquakes slightly south of the caldera.

1980, 10 February: A relatively slow deflation was accompanied by a 5-7-km-deep intrusion into the southern fissure swarm. Very little surface rifting was observed.

16 March: A fast deflation was accompanied by a small lava eruption from a 5-km-long fissure extending northward from the center of the caldera. A larger volume of magma was intruded into the northern and southern fissure swarm. Significant rifting occurred along a 20-km-long segment of the swarm. This event marked the transition from the primarily intrusive phase to the eruption phase of the rifting episode.

10 July: A lava eruption followed the onset of deflation by 4 hours. The eruptive fissure zone was 4 km long, located about 9 km north of the caldera. A good part of the erupted lava flowed into open fissures within the fissure swarm, causing secondary rifting and intrusion from above. The eruption lasted 8 days and the lava volume is estimated to be 23 million m^3. Most of the mobilized magma is thought to have reached the surface in this event.

18 October: An eruption lasting 5 days followed the beginning of rapid deflation by about an hour. The eruptive fissure was 7 km long and extended northward from the center of the caldera. The volume of erupted lava is estimated to be 3.5 million m^3.

22 December: A small and slow deflation event was accompanied by small earthquakes about 13 km north of the center of the caldera, indicating intrusion.

KRAFLA (continued)

HISTORICAL ACTIVITY (continued)

1981, 30 January: A slow deflation began about 7 hours before the outbreak of a lava eruption. The eruptive fissure was 2 km long, located about 8 km north of the caldera center. The eruption lasted 6 days and the erupted volume is estimated to be 32 million m^3. A significant amount of lava is thought to have flowed into fissures during this eruption.

18 November: A lava eruption broke out a little over an hour after the onset of rapid deflation. The fissure was about 8 km long and extended from the caldera center to the north. The eruption lasted 5 days.

1982-84: The volcano reinflated after the November 1981 eruption but then stopped. A few periods of intermittent inflation occurred during the following two years, accompanied by moderate seismic activity within the caldera, demonstrating that the volcano was still alive.

1984: 4 September. An inflation period ended with deflation and lava eruption. The eruption began 3 hours after the onset of deflation. The fissure was 8.5 km long, extending from the center of the caldera to the north. After the initial outbreak the eruptive part of the fissure shortened and the flow rate decreased as in previous eruptions. After the second day, however, the eruption rate and the vigor of the eruption slowly increased, until the eruption suddenly came to an end on 18 September. The eruption was by volume significantly larger than the previous eruptions, and much lava flowed into fissures from above. Details of the surface deformation during this eruption led Tryggvason (1986) to suggest that magma was supplied from four reservoirs at depths between 2.6 and 20 km.

1984-88: The volcano reinflated after the 1984 eruption, and in early 1985 the inflation had nearly reached the highest previous level. Since then inflation has been intermittent, with long periods of no detectable movements. The most important periods of inflation were October 1986 - April 1987 and February - March 1988. A few small tilt events have suggested minor magma movements within the caldera. This unrest shows that the current episode of magmatism and rifting is not over yet, and the magma pressure in the shallow chamber is sufficient to bring magma to the surface.

The current unrest at Krafla has now lasted more than 12 years. Rifting during this episode has affected nearly 80 km of the Krafla fissure swarm. In the caldera region and immediately to the north the cumulative rifting exceeds 8 m, and in other areas the rifting amounts to at least 2-3 m. The total volume of erupted lava is about 0.25 km^3, and a minimum of 0.6 km^3 has been intruded into the fissure swarm.

The only significant historical activity before the current rifting episode occurred in the eighteenth century, mostly in the period 1724-29. Large scale crustal deformation was reported, and an estimated 0.45 km^3 of basaltic lava was erupted from fissures, mainly within the caldera. Two small eruptions occurred in the southern fissure swarm, and it appears that intrusive activity and rifting was mostly directed southward, to a distance of at least 30 km. As in the current episode, the first years were characterized by earthquakes and rifting, and most of the lava was erupted in the last two years of the episode.

KRAFLA, Region 17, CAVW number 17-03-11

KRAFLA (continued)

HISTORICAL ACTIVITY (continued)

The most prominent events in the 1724-29 unrest were:

1724, 17 May: A phreatomagmatic eruption in the caldera was preceded by strong earthquakes. The seismicity began a few hours before the eruption, which produced basaltic scoria mixed with silicic pumice. Changes in the water level of Lake Myvatn indicated that magma was intruded to the south.

1725, 11 January: Volcanic activity and changes in geothermal areas occurred within the caldera. Earthquakes were felt for several days prior to this event.
19 April: An eruption broke out on a short fissure 6 km south of the caldera, preceded by considerable seismic activity. Fissuring and vertical movements were observed in the fissure swarms as far as 30 km south of the volcano.
8 September: Strong earthquakes and changes in lake level indicate an intrusion to the south.

1727, 21 August: A lava eruption began in the caldera and lava flowed to the north and south.

1728, 18 April: A lava eruption began in the caldera after several hours of earthquake activity. A few hours later another eruption broke out 4 km south of the caldera. Changes in the level of Lake Myvatn indicated an intrusion to the south.
18 December: A lava eruption began in the caldera.

1729: 30 January: An intense lava eruption began in the caldera. Lava flows were reported several times during the following months, but it is unclear whether activity was continuous or intermittent. No activity was reported after September.

1746: 10 July: Volcanic activity of unspecified nature occurred in the caldera. It was accompanied by earthquakes and changes in the level of Lake Myvatn, indicating that an intrusion may have occurred.

COMMENTS

Recent unrest at Krafla is a superbly documented example of rifting and crustal accretion at a spreading ridge. Magma is supplied from depth to one or more central reservoirs (one atop another), and from those reservoirs the magma spreads laterally into rifts. Tectonic extension and magma intrusion are intimately linked; neither causes the other, but rather both are processes of spreading and crustal accretion.

KRAFLA (continued)

REFERENCES

Björnsson, A., 1985, Dynamics of crustal rifting in NE Iceland: Jour. Geophys. Res., v. 90, no. B12, p. 10,151-10,162.

Björnsson, A., Johnsen, G., Sigurdsson, S., Thorbergsson, G., and Tryggvason, E., 1979, Rifting of the plate boundary in North Iceland 1975-1978: Jour. Geophys. Res., v. 84, no. B6, p. 3029-3038.

Björnsson, A., Sæmundsson, K., Einarsson, P., Tryggvason, E., and Grönvold, K., 1977, Current rifting episode in North Iceland: Nature, v. 266, p. 318-323.

Brandsdóttir, B., and Einarsson, P., 1979, Seismic activity associated with the September 1977 deflation of the Krafla central volcano in NE Iceland: Jour. Volcanol. Geotherm. Res., v. 6, p. 197-212.

Einarsson, P., 1978, S-wave shadows in the Krafla caldera in NE-Iceland: evidence for a magma chamber in the crust: Bull. Volcanol., v. 41, p. 187-195.

Einarsson, P., and Björnsson, S., 1979, Earthquakes in Iceland: Jökull, v. 29, p. 37-43.

Einarsson, P., and Brandsdóttir, B., 1980, Seismological evidence for lateral magma intrusion during the July 1978 deflation of the Krafla volcano in NE-Iceland; Jour. Geophys., v. 47, p. 160-165.

Gudmundsson, G., and Sæmundsson, K., 1980, Statistical analysis of damaging earthquakes and volcanic eruptions in Iceland from 1550-1978: Jour. Geophys., v. 47, p. 99-109.

Hauksson, F., 1981, Episodic rifting and volcanism at Krafla in North Iceland; radon (222) emission from fumaroles near Leirhnjukur: Jour. Geophys. Res., v. 86(B), p. 11,806-11,814.

----- 1983, Episodic rifting and volcanism at Krafla in North Iceland: Growth of large ground fissures along the plate boundary: Jour. Geophys. Res., v. 88, no. B1, p. 625-636.

Johnsen, G.V., Björnsson, A., and Sigurdsson, S., 1980, Gravity and elevation changes caused by magma movement beneath the Krafla caldera: Jour. Geophys., v. 47, p. 132-140.

Larsen, G., Grönvold, K., and Thorarinsson, S., 1979, Volcanic eruption through a geothermal borehole at Námafjall, Iceland: Nature, v. 278, p. 707-710.

Marquart, G., and Jacoby, W., 1985, On the mechanism of magma injection and plate divergence during the Krafla rifting episode in NE Iceland: Jour. Geophys. Res., v. 90, no. B12, p. 10,178-10,192.

Oskarsson, N., 1984, Monitoring of fumarole discharge during the 1975-1982 rifting in Krafla Volcanic Center, North Iceland: Jour. Volcanol. Geotherm. Res., v. 22, p. 97-121.

Sæmundsson, K., 1974, Evolution of the axial rifting zone in northern Iceland and the Tjornes Fracture Zone: Geol. Soc. Am. Bull., v. 85, p. 495-504.

----- 1978, Fissure swarms and central volcanoes of the neovolcanic zones of Iceland: Geol. Jour., Spec. Iss. 10, p. 415-432.

Science Institute, University of Iceland, and the Icelandic Meteorologic Office, 1975-88, Skjalftabref (Seismological bulletin): v. 1-64, (various short articles on Krafla by Einarsson, P., Brandsdóttir, B., Grönvold, K., Tryggvason, E., Sæmundsson, K., and others).

Sigurdsson, H., 1980, Myvatn volcanism: Volcano News, no. 3, p. 1-8.

Sigurdsson, S., 1980, Surface deformation of the Krafla fissure swarm: two rifting events: Jour. Geophys., v. 47, p. 154-159.

KRAFLA (continued)

REFERENCES (continued)

Smithsonian Institution, Scientific Event Alert Network (SEAN), 1975-85, 1987, Krafla: SEAN Bull., v. 1-10, 12 (information from P. Einarsson, K. Grönvold, H. Sigtryggsson, G. Sigvaldason, E. Trygvasson, and others).

Thorarinsson, S., 1960, On the predicting of volcanic eruptions in Iceland: Bull. Volcanol., v. 23, p. 45-52.

----- 1973, Data sheets of the post-Miocene volcanoes of the world, Sheet II, Iceland: Rome, IAVCEI, 10 p.

Thoroddsen, Th., 1925, Die Geschichte der Islandischen Vulkane: D. Kgl. Dansk Vidensk. Selsk. Skrifter, Naturvidensk. og Mathem. Afd Copenhagen, 8th ser., v. 9, 458 p.

Torge, W., and Kanngieser, E., 1980, Gravity and height variations during the present rifting episode in northern Iceland: Jour. Geophys., v. 47, p. 125-131.

----- 1985, Regional and local vertical crustal movements in Northern Iceland, 1965-1980: Jour. Geophys. Res., v. 90, no. B12, p. 10,173-10,177.

Tryggvason, E., 1980, Subsidence events in the Krafla area, North Iceland, 1975-1979: Jour. Geophys., v. 47, p. 141-153.

----- 1984, Widening of the Krafla Fissure Swarm during the 1975-1981 volcano-tectonic episode: Bull. Volcanol., v. 47, no. 1, p. 47-69.

----- 1986, Multiple magma reservoirs in a rift zone volcano: ground deformation and magma transport during the September 1984 eruption of Krafla, Iceland: Jour. Volcanol. Geotherm. Res., v. 28, p. 1-44.

----- 1987, Myvatn lake level observations 1984-1986 and ground deformation during a Krafla eruption: Jour. Volcanol. Geotherm. Res., v. 31, p. 131-138.

Volcanological Society of Japan, 1977-86, Krafla: Bull. Volcanic Eruptions, nos. 15-23 (information from S. Thorarinsson, P. Einarsson, K. Grönvold, G. Sigvaldason, and others).

Wendt, K., Moller, D., and Ritter, B., 1985, Geodetic measurements of surface deformations during the present rifting episode in NE Iceland: Jour. Geophys. Res., v. 90, no. B12, p. 10,163-10,172.

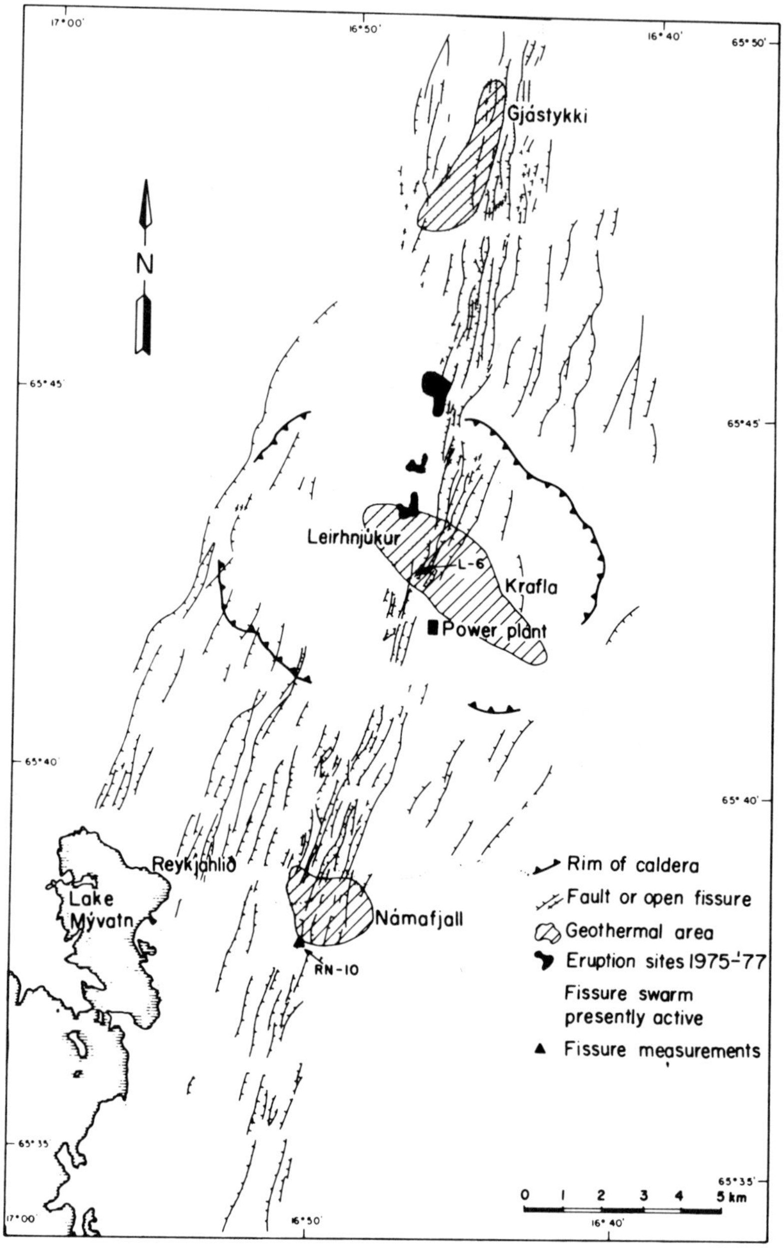

Figure 17.8.1. Outline geologic map of Krafla Caldera and associated fissure swarm. Mapped by K. Sæmundsson, from Björnsson and others (1979). Copyright by the American Geophysical Union.

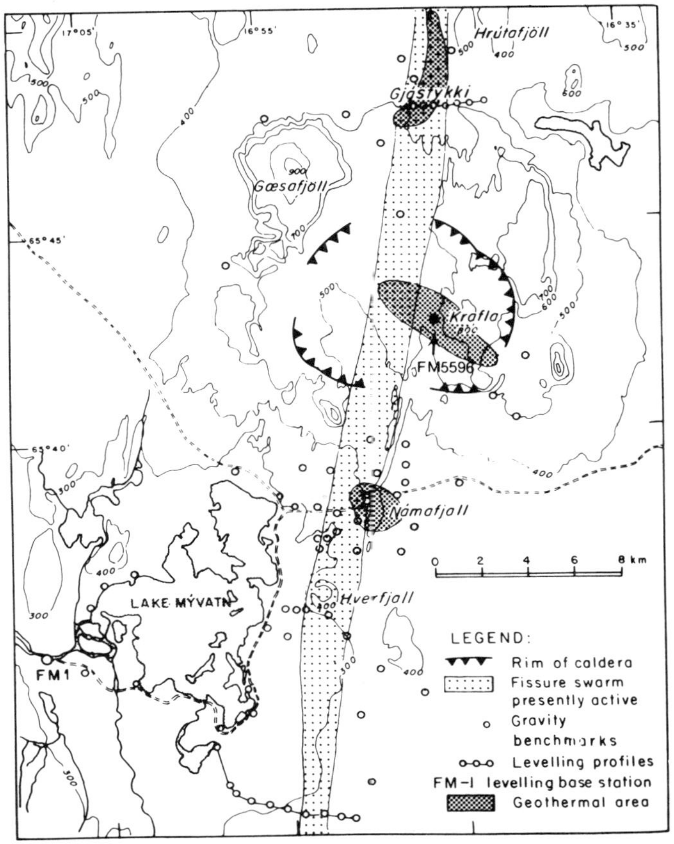

Figure 17.8.2. Krafla central volcano and its surroundings, from Johnsen and others (1980). Caldera and central part of active fissure swarm are shown. Open circles show gravity and leveling bench marks outside caldera.

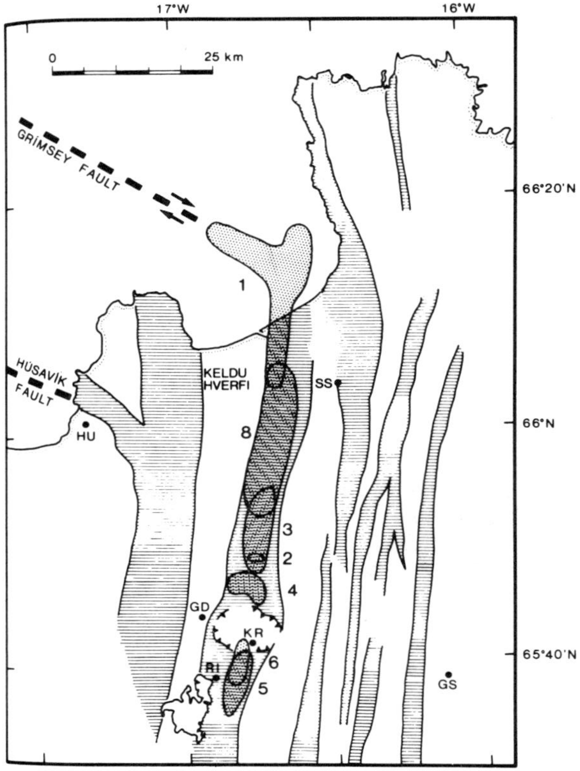

Figure 17.8.3. Index map of northern part of volcanic rift zone in northeast Iceland, from Einarsson and Brandsdóttir (1980). Dots, permanent seismograph stations. Hachured areas, fault swarms (from Björnsson and others (1977)). Krafla Caldera is located within Krafla fault swarm. Stippling, areas of maximum earthquake activity during different deflation events at Krafla. Area 1, epicentral area of first swarm of December 1975 - February 1976; 2, 1-2 October 1976; 3, 31 October - 1 November 1976; 4, January 1977; 5, April 1977; 6, September 1977; small swarm in November 1977 could not be located and is not shown; 8, January 1978.

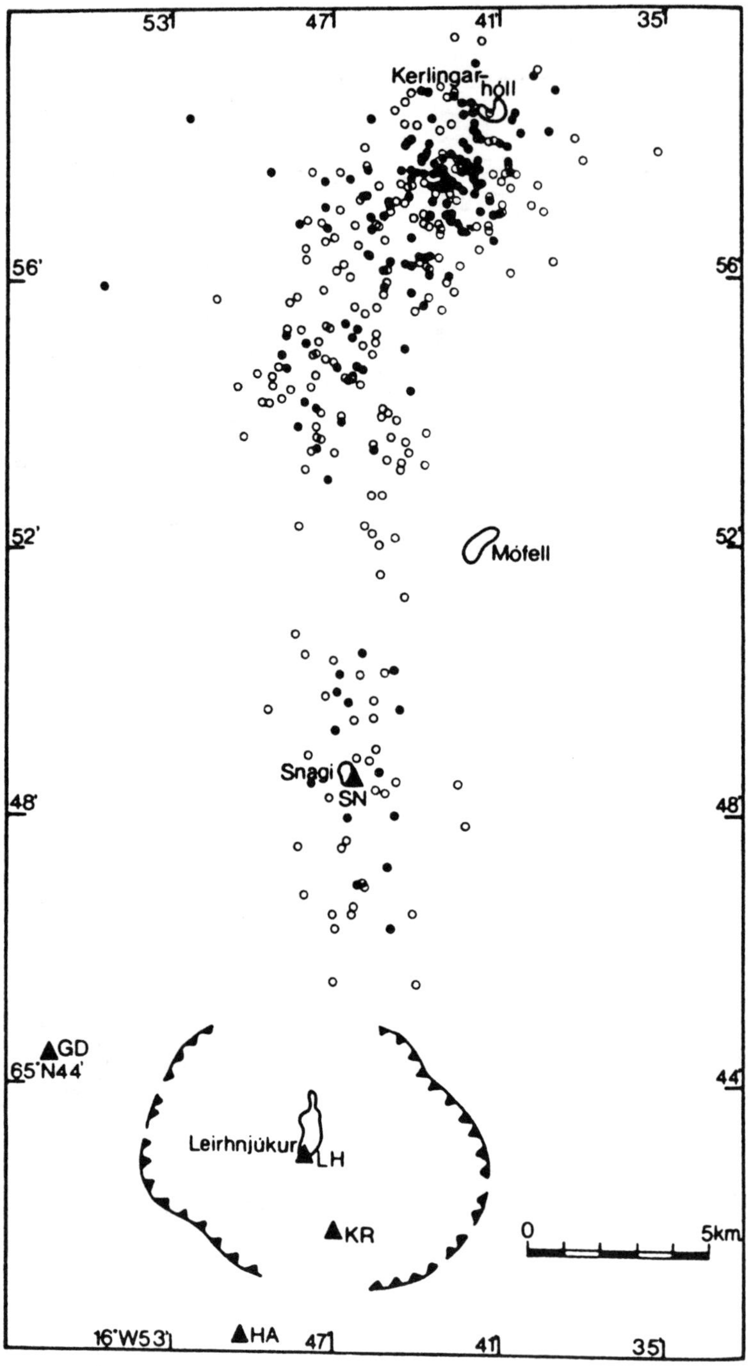

Figure 17.8.4. Epicentral map of July 1978 earthquake swarm, from Einarsson and Brandsdóttir (1980). Dots mark epicenters located with horizontal standard error of 1 km or less; circles denote epicenters with errors between 1 and 2 km. Seismograph stations are shown with triangles.

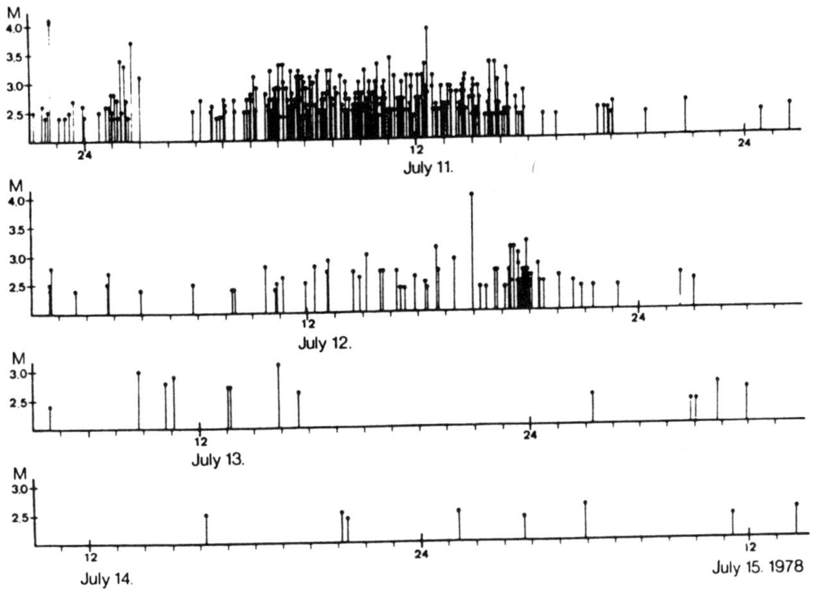

Figure 17.8.5. Time sequence of earthquakes during the July 1978 swarm at Krafla, from Einarsson and Brandsdóttir (1980). Magnitude is plotted as a function of time. Deflation began about 5 hours before start of plot.

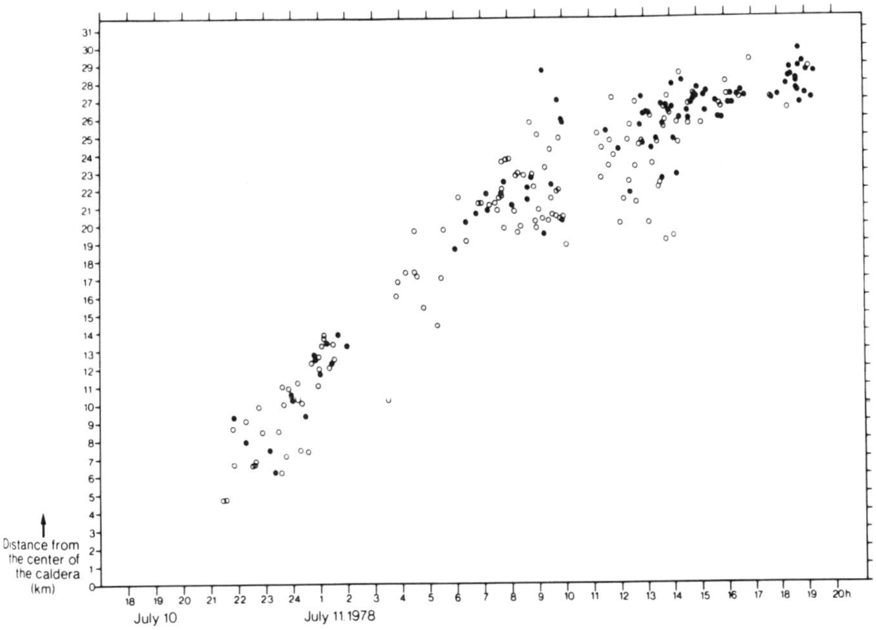

Figure 17.8.6. Migration of seismic activity during the July 1978 swarm, from Einarsson and Brandsdóttir (1980). Distance of epicenters from center of Krafla Caldera is plotted as a function of time. Dots and circles have the same meaning as in figure 17.8.4. Apparent gap in activity between 1000 hr and 1100 hr on 11 July was caused by a time signal failure. Tremor and deflation started at about 1700 hr.

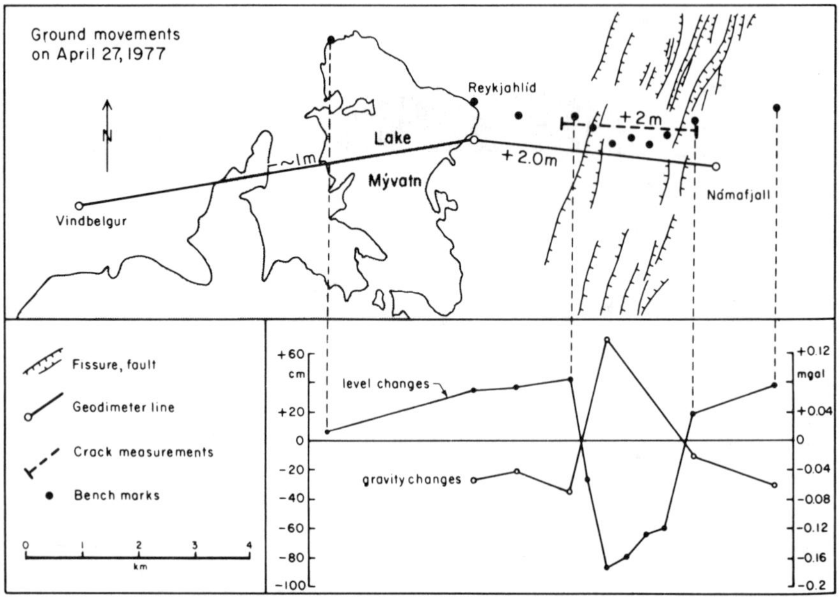

Figure 17.8.7. Horizontal and vertical ground movements across Krafla fissure swarm near Namafjall during subsidence event of 27 April 1977, from Björnsson and others (1979). Expansion was measured on individual cracks and also across swarms of cracks with an electronic distance meter (EDM). Elevation and gravity changes are shown in lower part of figure. Copyright by the American Geophysical Union.

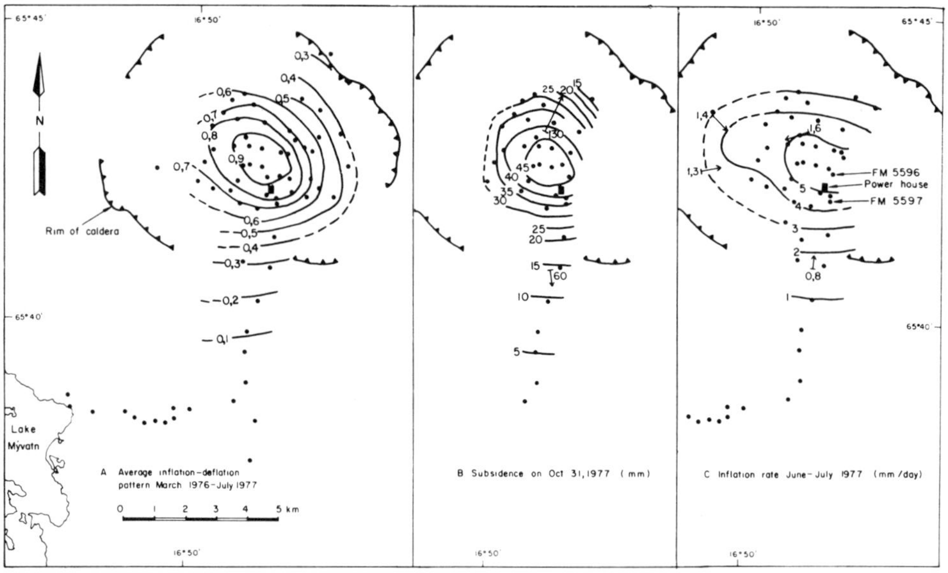

Figure 17.8.8. Patterns of uplift and subsidence in the Krafla area, from Björnsson and others (1979). Average vertical ground movement (left) throughout the period March 1976 to July 1977 in fractions of maximum movements. Total subsidence (middle) in centimeters during subsidence event of 31 October to 1 November 1976. Rate of uplift (right) in millimeters per day during a period of 1 month (June-July 1977). Arrows show tilt changes in microradians at four stations. Copyright by the American Geophysical Union.

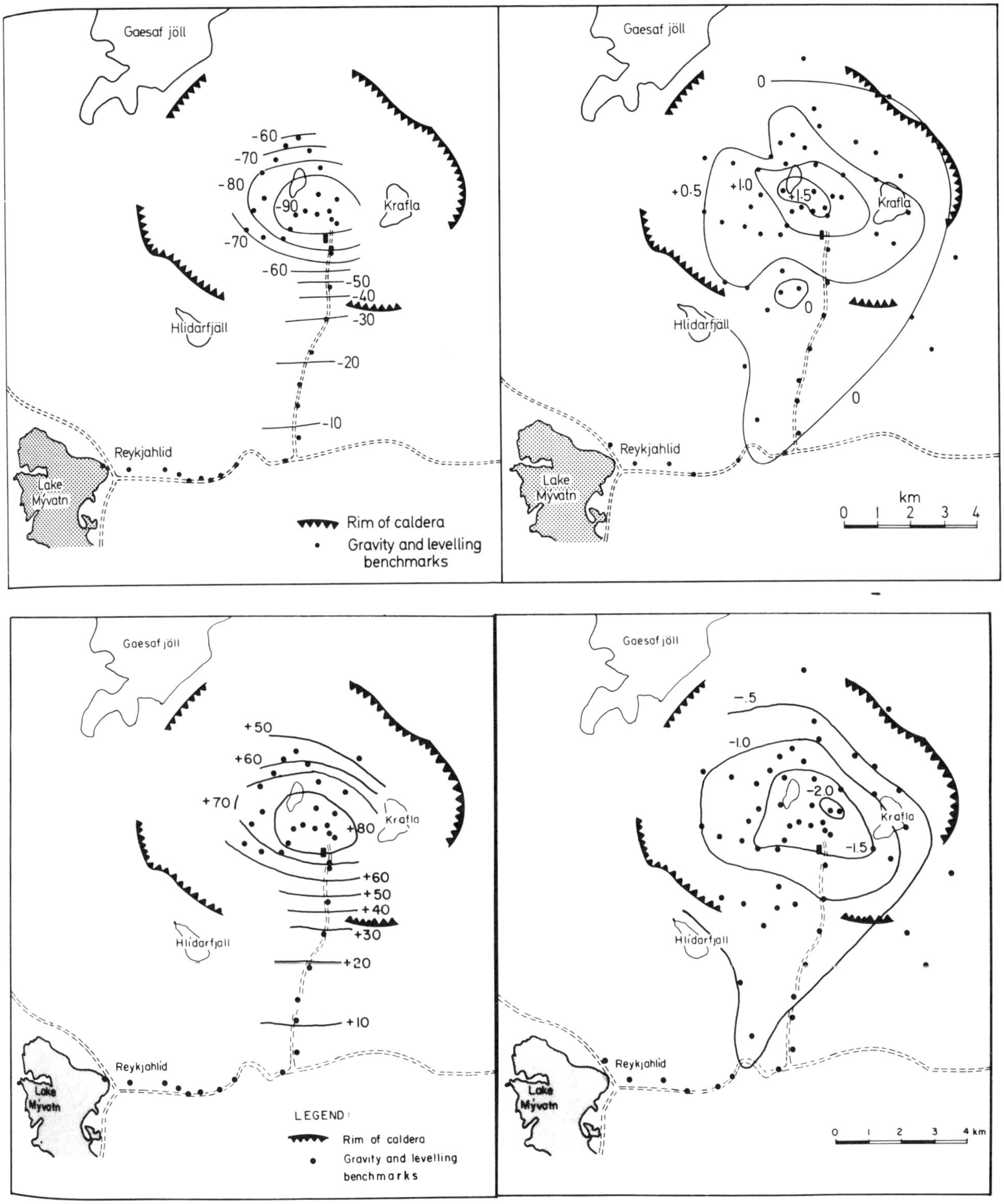

Figure 17.8.9. Measured elevation changes in centimeters (left) and gravity changes in milligals (right), from Johnsen and others (1980). Filled circles show bench marks occupied at the time. Subsidence event (top) in January 1978; inflation period (bottom) from January to June 1978.

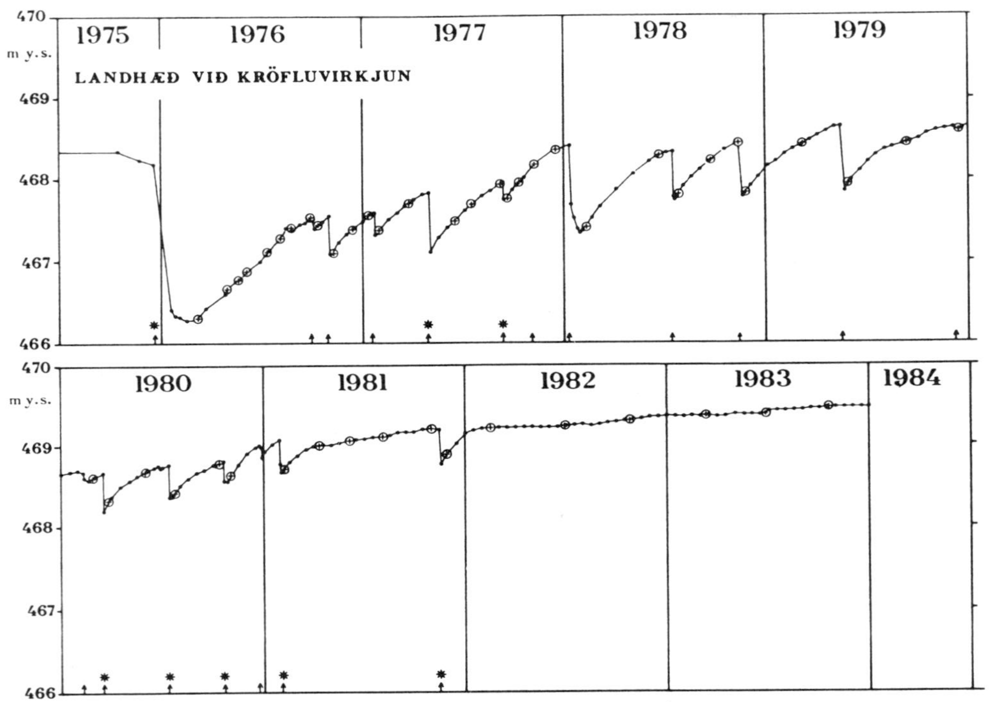

Figure 17.8.10. Land elevation changes at bench mark FM5596 about 1 km southeast of center of Krafla Caldera (see figure 17.8.2 for location), from Björnsson (1985). Reference bench mark (FM6414) is 20 km away, at southern end of Lake Myvatn. Crosses, leveling data; dots, interpolated values based on tilt measurements at Krafla power plant; arrows, rifting-deflation events; stars, volcanic eruptions. Copyright by the American Geophysical Union.

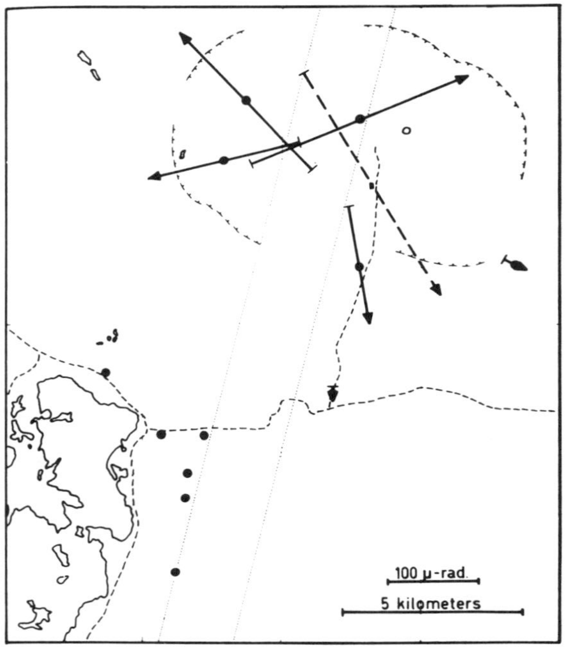

Figure 17.8.11. Observed ground tilt during inflation period 29 April to 8 September 1977, from Björnsson and others (1979). Arrows show direction and magnitude of observed tilt; if no arrow was drawn, tilt was less than error of the observations. Solid dots are dry-tilt stations. Direction of tilt at power station (dashed arrow) is based on an electronic tiltmeter that was operated for a few weeks during this period. Dotted lines outline most active part of Krafla fissure swarm. Copyright by the American Geophysical Union.

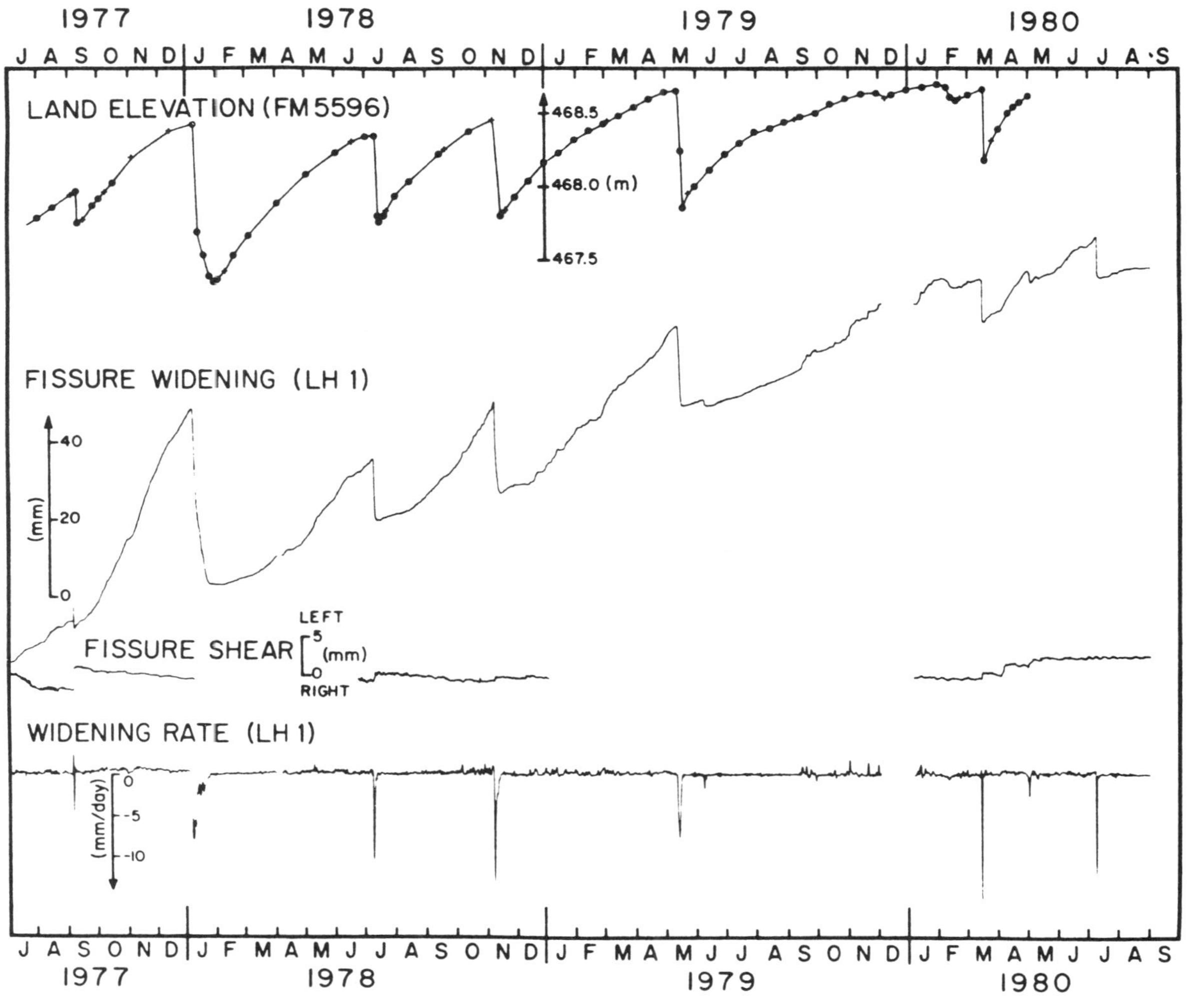

Figure 17.8.12. Time series of fissure width (LH1), fissure shear (LH2), and fissure widening rate at a site in Krafla Caldera, from Hauksson (1983). Land elevation at bench mark FM5596 (see figure 17.8.2 for location) is shown for comparison. Copyright by the American Geophysical Union.

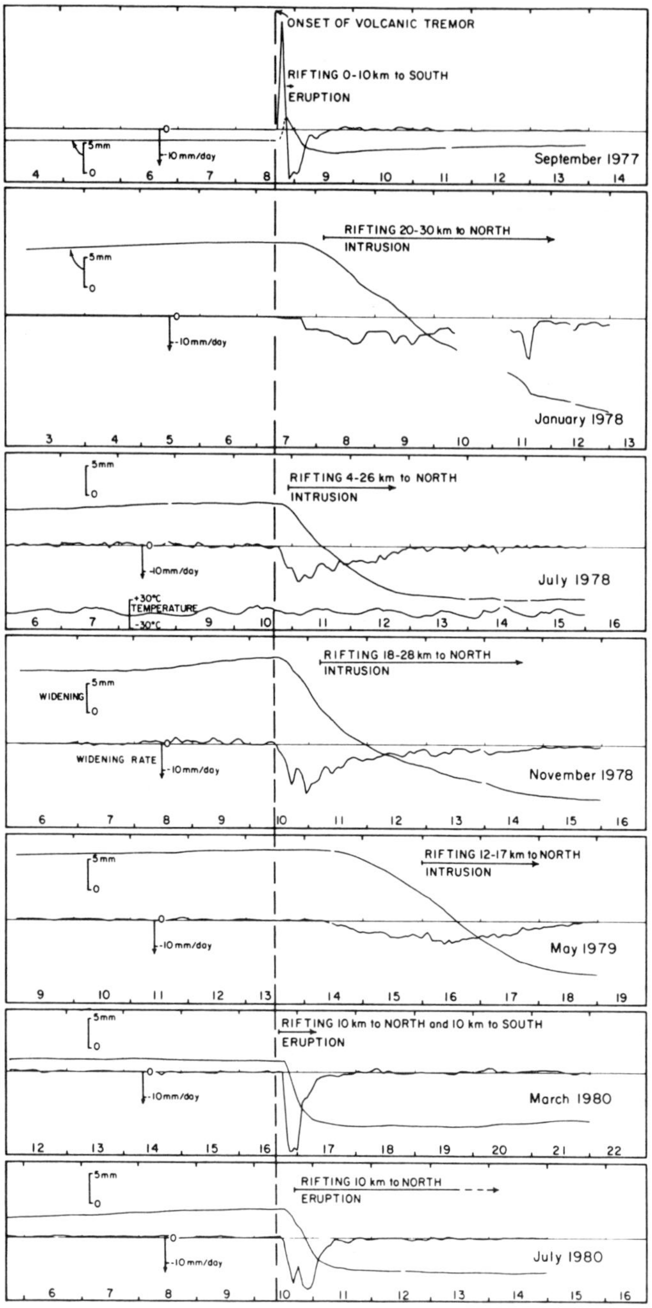

Figure 17.8.13. Rifting and tremor during subsidence episodes at Krafla, from Hauksson (1983). Rifting data from site LH1 (see figure 17.8.12). Different episodes are aligned so that onset of volcanic tremor coincides. Long horizontal arrow indicates direction and extent of rifting, and accompanying notes indicate whether subsidence was associated with an intrusion or an eruption. Rapid fluctuations in closure rate are real. Copyright by the American Geophysical Union.

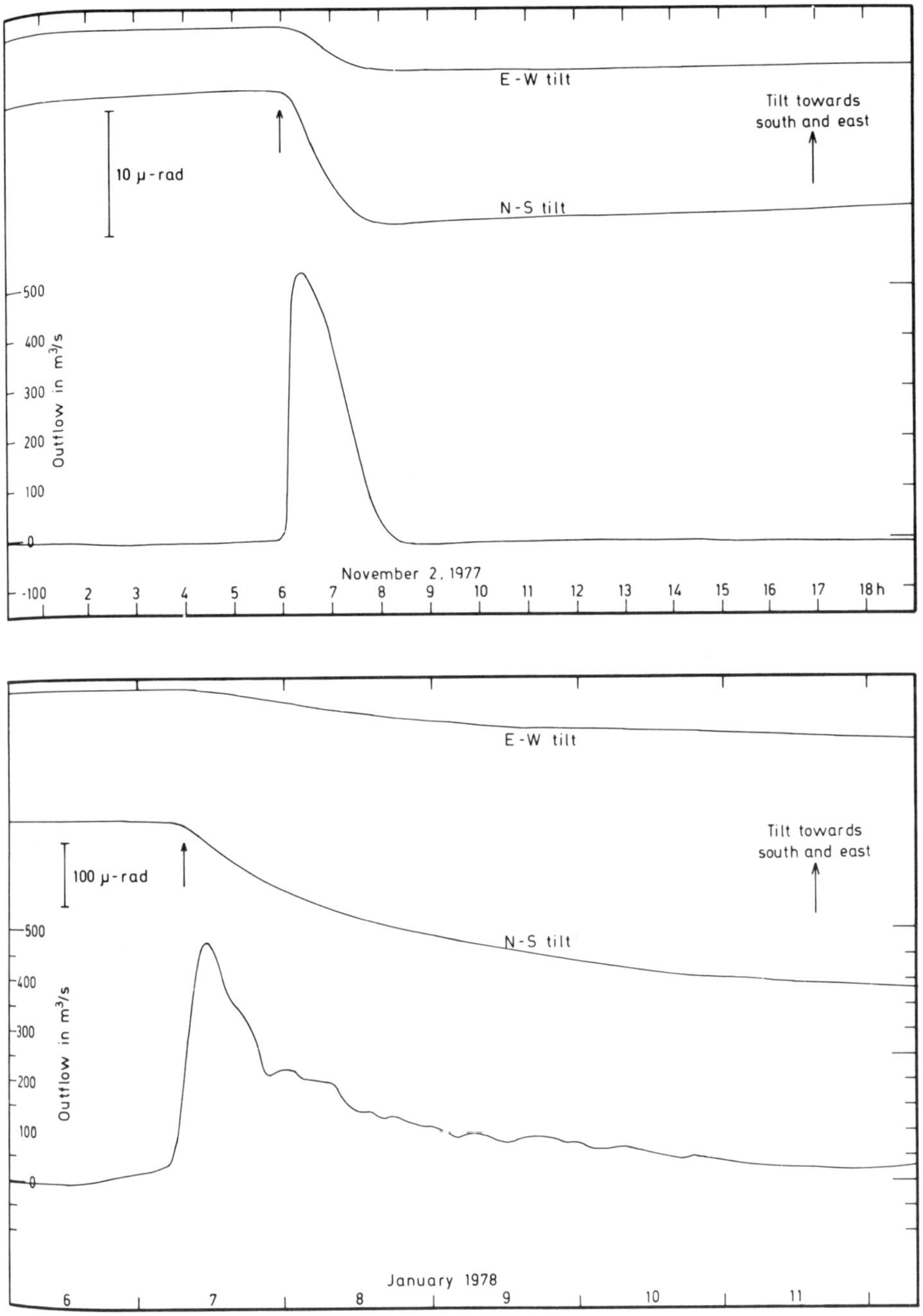

Figure 17.8.14. Tilt in Krafla power station as measured by an electronic tiltmeter, and rate of outflow from Krafla magma chamber during subsidence event of 2 November 1977 (top) and first days of subsidence event of January 1978 (bottom), from Tryggvason (1980).

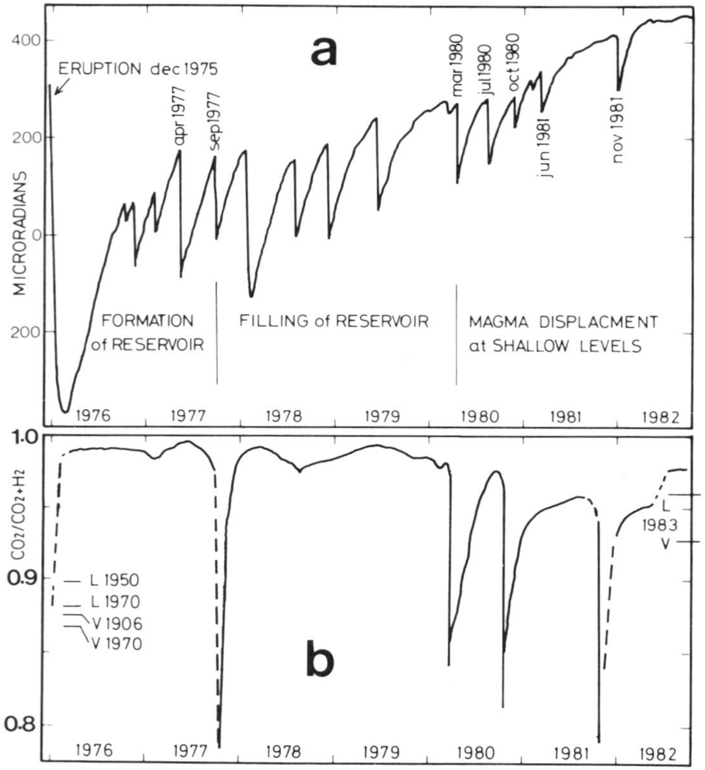

Figure 17.8.15. Ground tilt in Krafla power station (top), and carbon dioxide/hydrogen gas ratio at a monitoring station 1.2 km northwest of power station (bottom), both as a function of time (Oskarsson, 1984). Vertical scale at top shows relative tilt variation. Initial deflation, to left in figure, was estimated at about 2.5 m at center of caldera. Deflation events associated with volcanic eruptions are labeled with their dates. Bottom diagram shows $CO_2/(CO_2+H_2)$ ratio; dashed lines indicate missing data and are thus to be taken as inferred behavior of ratio. On left-vertical axis ratio for some older data is indicated by horizontal bars. Ratios for September 1983 are shown on right-vertical axis. Pulses of H_2 in conjunction with shallow intrusions and eruptions are evident in lower figure.

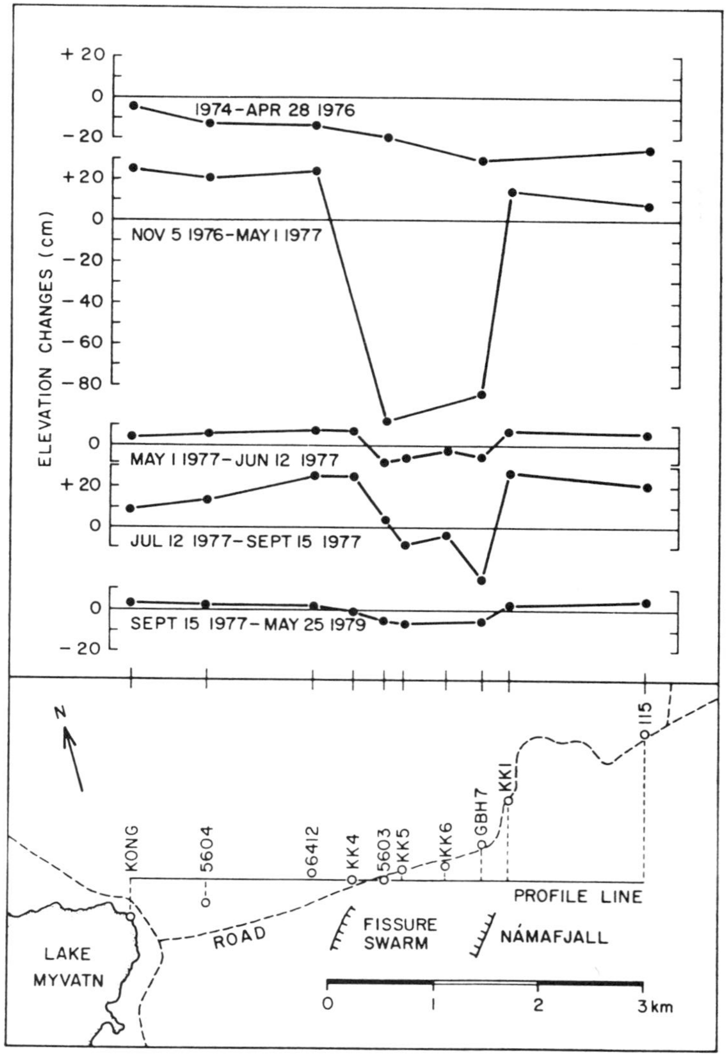

Figure 17.8.16. Elevation changes during 1974-79 on a profile across the Krafla fissure swarm at Namafjall, 10 km south of Krafla Caldera, from Björnsson (1985). Copyright by the American Geophysical Union.

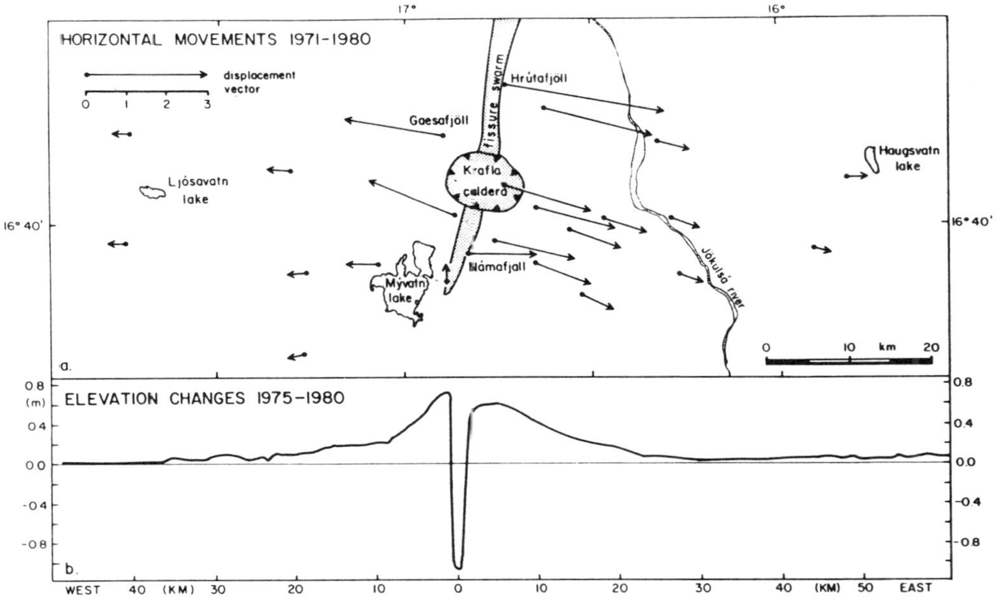

Figure 17.8.17. Horizontal movements (top), 1971-80, on a profile across axial rift zone in Krafla area. Elevation changes (bottom), 1975-80, on an E-W profile, projected onto latitude 16°40' (Björnsson, 1985). Copyright by the American Geophysical Union.

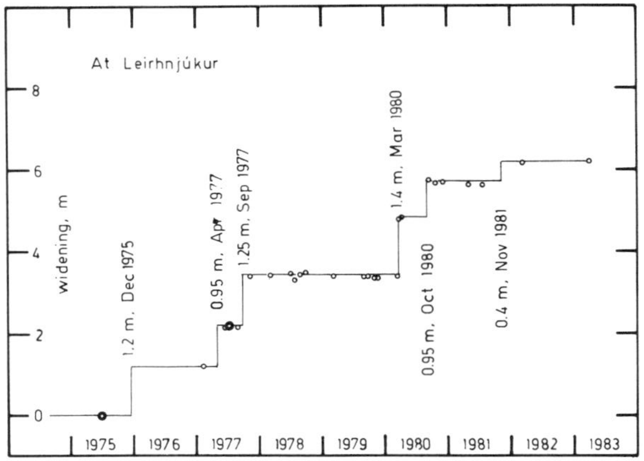

Figure 17.8.18. Accumulated widening of Krafla fissure swarm at Leirhnjukur (center of Krafla Caldera, see fig. 17.8.4) as a function of time, from Tryggvason (1984).

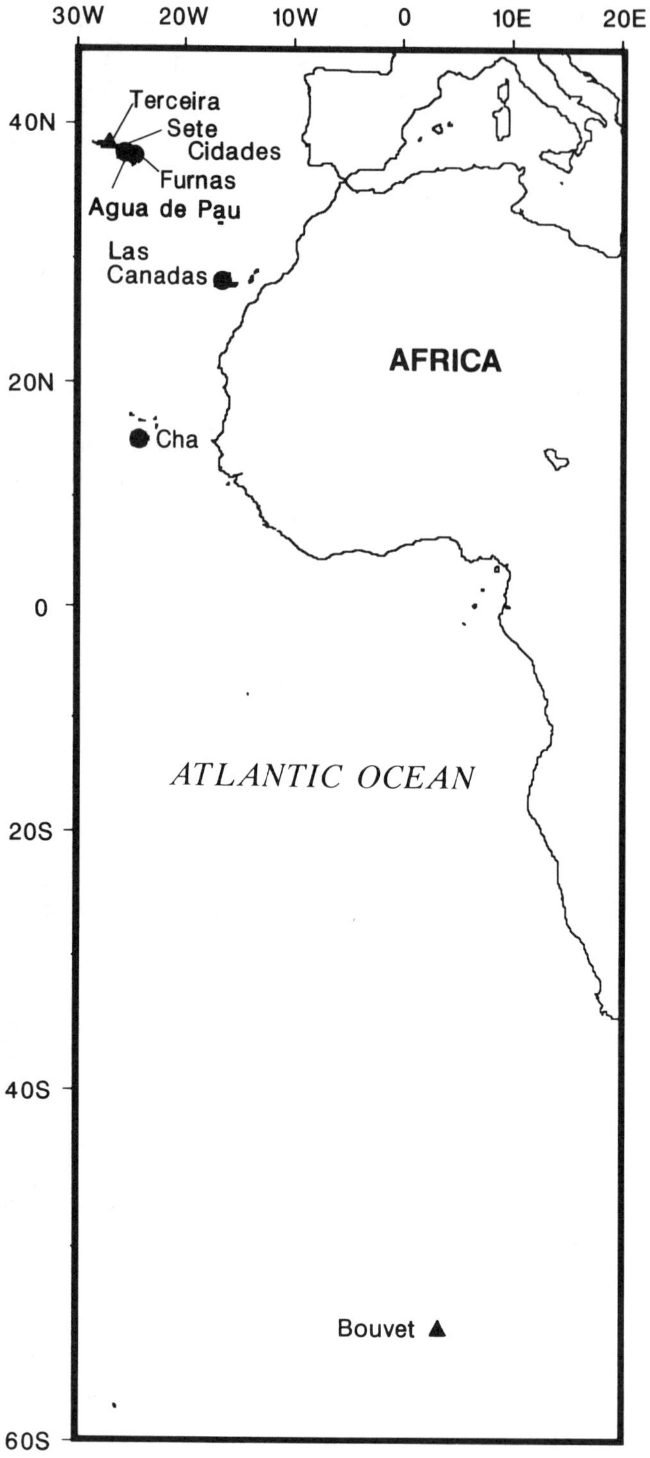

Figure 18. Locations of large, Quaternary restless calderas (solid circles) and nonrestless caldera (triangle) in Region 18.

SETE CIDADES

CAVW number (active vent)	Latitude Longitude (degrees)	Diameter (km)	Local tectonic setting	Precaldera edifice	SiO_2 content (wt pct)	Age of collapse (yr B.P.)	Date of unrest	Type and duration of unrest ESTU STHF MGTF H Te	Eruption type
18-02-08	37.87N 25.78W	5	Spreading center exten	LP-LL strat	R = 44-48 C = 65	ca. 22,000	ca. 1444	---- ---- ---- - --	P-ex
							1638	EE-- ---- ---- - --	ex, sub
							1682	xx-- ---- ---- - --	ex, sub
							1713	xx-- ---- ---- - --	sub?
							1810-11	EE-E ---E ---- - --	Ex, sub
							1861	---- ---- ---- - --	sub
							1880	---- ---- ---- - --	sub

TECTONIC SETTING

The Azores mark a possible hotspot along a spreading center, the WNW-ESE-trending Azores Fracture Zone. This fracture zone is responsible for the location and in part the caldera-bounding faults of Sete Cidades Caldera (Laughton and Whitmarsh, 1975; Moore, 1983, 1988).

GEOLOGIC HISTORY

The Sete Cidades Caldera (fig. 18.1.1) formed approximately 22,000 yr B.P., after eruption of about 6 km^3 of trachytic pumice (Moore, 1983, 1988). At least 22 postcaldera eruptions have occurred, and 12 postcaldera tephras have been identified by Booth and others (1978); these range in volume from 0.06 to 0.16 km^3 (DRE), and together they represent slightly over 1 km^3 of magma.

HISTORICAL ACTIVITY

ca. 1444: An eruption of trachyte occurred from a vent on the southwestern caldera floor.

1638: Strong earthquakes began on 26 May and continued for eight days. These were so strong that many persons abandoned their homes (Zbyszewski, 1963). A submarine basaltic eruption began on 3 July, from an offshore vent just west of Sete Cidades.

1682: Earthquakes on 16 December alarmed the populace, especially in the south part of the island, and a submarine eruption occurred 4 leagues (ca 15 km) from Ferraria (Zbyszewski, 1963).

SETE CIDADES (continued)

HISTORICAL ACTIVITY (continued)

1713: A seismic swarm and possible submarine eruption west of Sete Cidades were reported by Agostinho (1936).

1810-11: An earthquake on 17 June 1810 was followed by progressively more frequent and stronger earthquakes until 24 June. At 1400 hr on the latter date a very strong earthquake destroyed grand houses of the towns around Sete Cidades. In the town of Mosteiros, a crack opened that stretched from the caldera to the west. The crack was (on average) 1 large palm wide (22 cm), and at points was 20 to 30 palms (4.5-6.75 m) wide. Additional earthquakes occurred in August and October, and these were especially intense near Ponta Delgada. Residents noted that the land had been uplifted. During 29-30 January 1811, earthquakes intensified and began to occur at shorter and shorter intervals. On 31 January, there were 30 earthquakes or more within a few hours, and on 1 February a submarine eruption began several (2?) km offshore from the village of Ginetes, forming the ephemeral island of Sabrina. The volcano was quiet from 8 February to 13 June, when strong earthquakes were felt every 15-20 minutes on the western part of Ile São Miguel. Sulfur odor was unpleasant in Ponta Delgada, 35 km from Sabrina. On the night of the 14th, a new submarine eruption commenced from a point 1.8 km northwest of Pic de Camarinhas and 4-5 km west of the February eruption site (Zbyszewski, 1963).

1861, 1880: Submarine eruptions occurred off São Miguel (Mitchell-Thome, 1981). No details are available.

REFERENCES

Agostinho, J., 1936, The volcanoes of the Azores: Bull. Volcanol., v. 8, p. 123-138.

Booth, B., Croasdale, R., and Walker, G.P.L., 1978, A quantitative study of five thousand years of volcanism on São Miguel, Azores: Phil. Trans. Roy. Soc. London, Ser. A, v. 288, p. 271-319.

Laughton, A.S., and Whitmarsh, R.B., 1975, The Azores-Gibraltar plate boundary, in Kristjansson, L., ed., Geodynamics of Iceland and the North Atlantic area: Dordrecht, D. Reidel Publ. Co., p. 63-81.

Mitchell-Thome, R.C., 1981, Vulcanicity of historic times in the Middle Atlantic Islands: Bull. Volcanol., v. 44, no. 1, p. 57-69.

Moore, R.B., 1983, Preliminary geologic map of Sete Cidades Volcano, São Miguel, Azores: U.S. Geol. Surv. Open-File Rep. 83-742, 1 sheet, scale 1:18,000.

----- 1988, Geology of three late Quaternary stratovolcanoes on São Miguel, Azores: Bull. Volcanol. [in press].

Neumann van Padang, M., Richards, A.F., Machado, F., Bravo, T., Baker, P.E., and LeMaitre, R.W., 1967, Catalogue of active volcanoes of the world including solfatara fields, Pt. XXI, Atlantic Ocean: Rome, IAVCEI, 128 p.

Zbyszewski, G., 1963, Les phénomènes volcaniques modernes dan l'Archipel des Açores: Serv. Geol. Portugal, v. 47, p. 5-277.

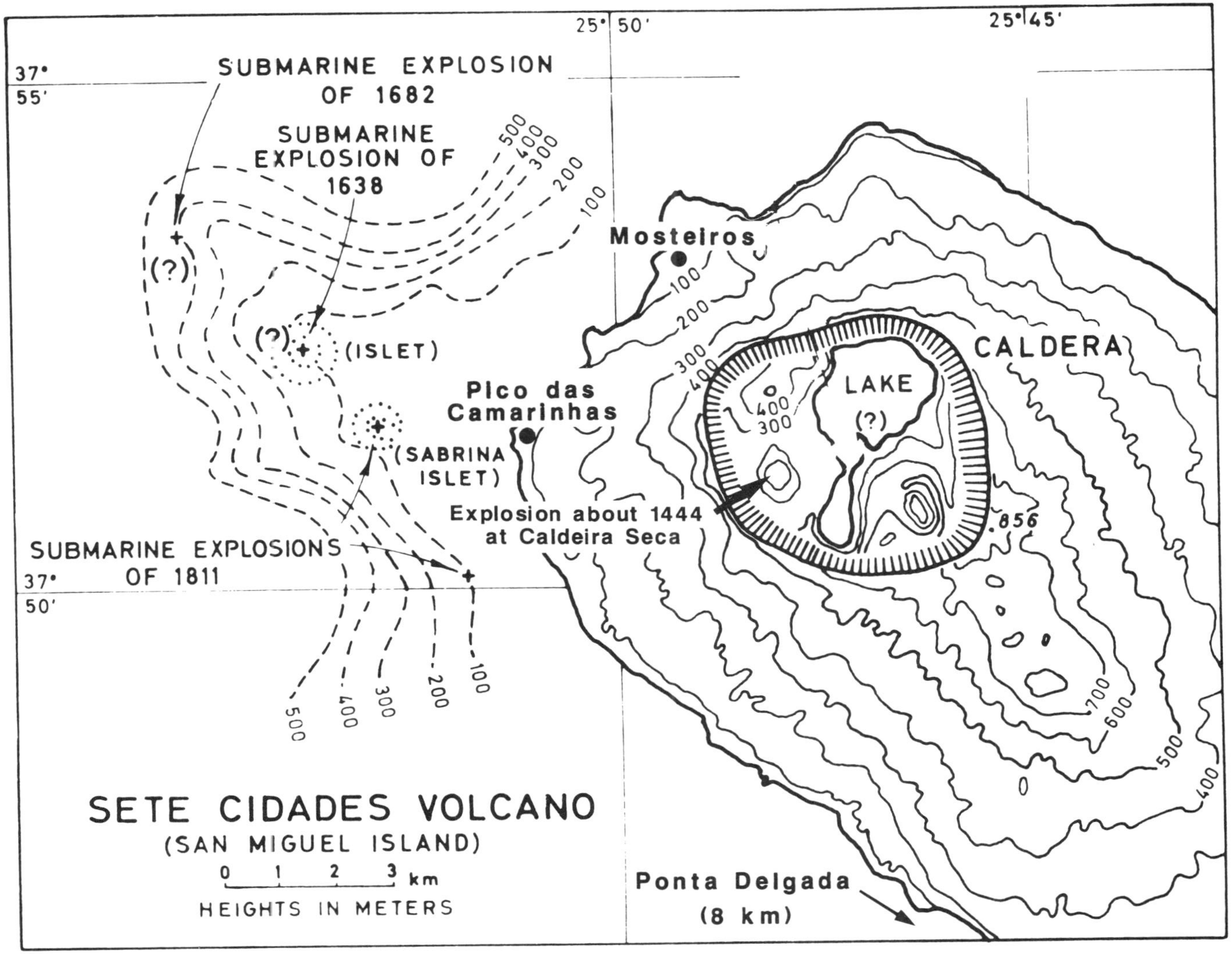

Figure 18.1.1. Topographic map of Sete Cidades Volcano, including submarine adventive vents, from Neumann van Padang and others (1967). Revised location of 1444 explosion from R.B. Moore (written commun., 1987)

PART 3: HISTORICAL UNREST AT LARGE QUATERNARY CALDERAS (continued)

See inside back cover for explanation and abbreviations

AGUA DE PAU

CAVW number (active vent)	Latitude Longitude (degrees)	Diameter (km)	Local tectonic setting	Precaldera edifice	SiO_2 content (wt pct)	Age of collapse (yr B.P.)	Date of unrest	Type and duration of unrest ESTU STHF MGTF H Te	Eruption type
18-02-09	37.77N 25.47W	4 x 7 (outer) 2.5 x 3 (inner)	Spreading center exten	LL-LP strat	R= 45-65 C= ca. 65	15,200 and between 30,000 and 45,000	1563-64	CC-- ---x ---- - --	EX, pex, lf
							1652	DD-- ---B ---Y - --	ex, lf
							1980's	xx-- ---- ---- - --	none

TECTONIC SETTING

The Azores mark a possible hotspot along a spreading center, the WNW-ESE-trending Azores Fracture Zone (Laughton and Whitmarsh, 1975).

GEOLOGIC HISTORY

Agua de Pau stratovolcano (fig. 18.2.1) consists of trachytic lava flows, domes, and pyroclastic deposits, built on a base of older volcanic rocks (Moore, 1986, 1988). Mafic eruptions have occurred on the flanks of the volcano but not within the caldera. An outer, 4 km x 7 km diameter caldera formed when at least 5 km^3 of trachyte was erupted sometime between 30,000 and 45,000 yr B.P. An inner, 2.5 km x 3 km caldera formed about 15,200 yr B.P., after a smaller eruption of trachyte (Moore, 1988). Several hot springs and fumaroles with temperatures near boiling occur on Agua de Pau, mainly on its northwest flank (Zbyszewski and others, 1958; Zbyszewski and Ferreira, 1959), and some of these are being explored for geothermal power potential (Moore, 1988).

HISTORICAL ACTIVITY

1563-64: A plinian eruption of trachyte pumice (on-land volume, 0.15 km^3) blanketed most of the island of São Miguel east of Agua de Pau (Walker and Croasdale, 1971). Earthquakes began on 24 June and became more frequent and more intense over the succeeding four days, until the eruption began on 28 June. On 29 June, strong earthquakes damaged houses of Ribeira Grande (8 km northwest of the caldera), and by 1 July most houses of that town were ruined. On 2 July, eruption from the main vent subsided, but extrusion of alkali olivine basalt began from the Cerro Queimado vent, 5.5 km northwest of the caldera vent and 5 km from Ribeira Grande (Zbyszewski, 1963; Moore, 1986, 1988). An additional, small phreatic(?) eruption occurred in the caldera on 10 February 1564 (Zbyszewski, 1963, p. 43). An intrusion of basalt into trachytic magma may have triggered the 1563 eruption (Moore, 1988).

AGUA DE PAU (continued)

HISTORICAL ACTIVITY (continued)

1652: Earthquakes beginning on 10 October 1652 continued until 16 October, causing some damage in the region of Lagoa. On 17-18 October the earthquakes became weaker, reintensifying for an hour and a half on 18 October. On the afternoon of 19 October, fissures opened at a vent at the west foot of Agua de Pau, about 10 km west of Lagoa de Fogo. A 1-week Strombolian eruption of alkali olivine basalt built a small cone and extruded a short lava flow (Zbyszewski, 1963; Moore, 1988).

1980's: Bursts of microearthquakes occur from time to time under the northeast flank of Agua de Pau (R.B. Moore, written commun., 1987)

REFERENCES

Booth, B., Croasdale, R., and Walker, G.P.L., 1978, A quantitative study of five thousand years of volcanism on São Miguel, Azores: Phil. Trans. Roy. Soc. London, Ser. A, v. 288, p. 271-319.

Laughton, A.S., and Whitmarsh, R.B., 1975, The Azores-Gibraltar plate boundary, in Kristjansson, L., ed., Geodynamics of Iceland and the North Atlantic area: Dordrecht, D. Reidel Publ. Co., p. 63-81.

Moore, R.B., 1986, Preliminary geologic map of Agua de Pau Volcano, São Miguel, Azores: U.S. Geol. Surv. Open-File Rep. 86-192, 1 sheet, scale 1:18,000.

----- 1988, Geology of three late Quaternary stratovolcanoes on São Miguel, Azores: Bull. Volcanol. [in press].

Neumann van Padang, M., Richards, A.F., Machado, F., Bravo, T., Baker, P.E., and LeMaitre, R.W., 1967, Catalogue of active volcanoes of the world including solfatara fields, Pt. XXI, Atlantic Ocean: Rome, IAVCEI, 128 p.

Walker, G.P.L., and Croasdale, R., 1971, Two plinian-type eruptions in the Azores: Jour. Geol. Soc. London, v. 127, p. 17-55.

Weston, F.S., 1964, List of recorded volcanic eruptions in the Azores with brief reports: Bol. Mus. Lab. Miner. Geol. Fac. Ciencas Lisboa, v. 10, p. 3-18.

Zbyszewski, G., 1963, Les phénomènes volcaniques modernes dan l'Archipel des Açores: Serv. Geol. Portugal, v. 47, p. 5-277.

Zbyszewski, G., Almeida, F.M. de, Ferreira, O. da V., and Assuncao, C.T. de, 1958, Carta geologica de Portugal na escala de 1/50,000; et Noticia explicativa da folia B, S. Miguel (Açores): Lisboa, Serv. Geol. Portugal.

Zbyszewski, G., and Ferreira, O. da V., 1959, Carta geologica de Portugal na escala de 1/50,000; et Noticia explicativa de folia A, S. Miguel (Açores): Lisboa, Serv. Geol. Portugal.

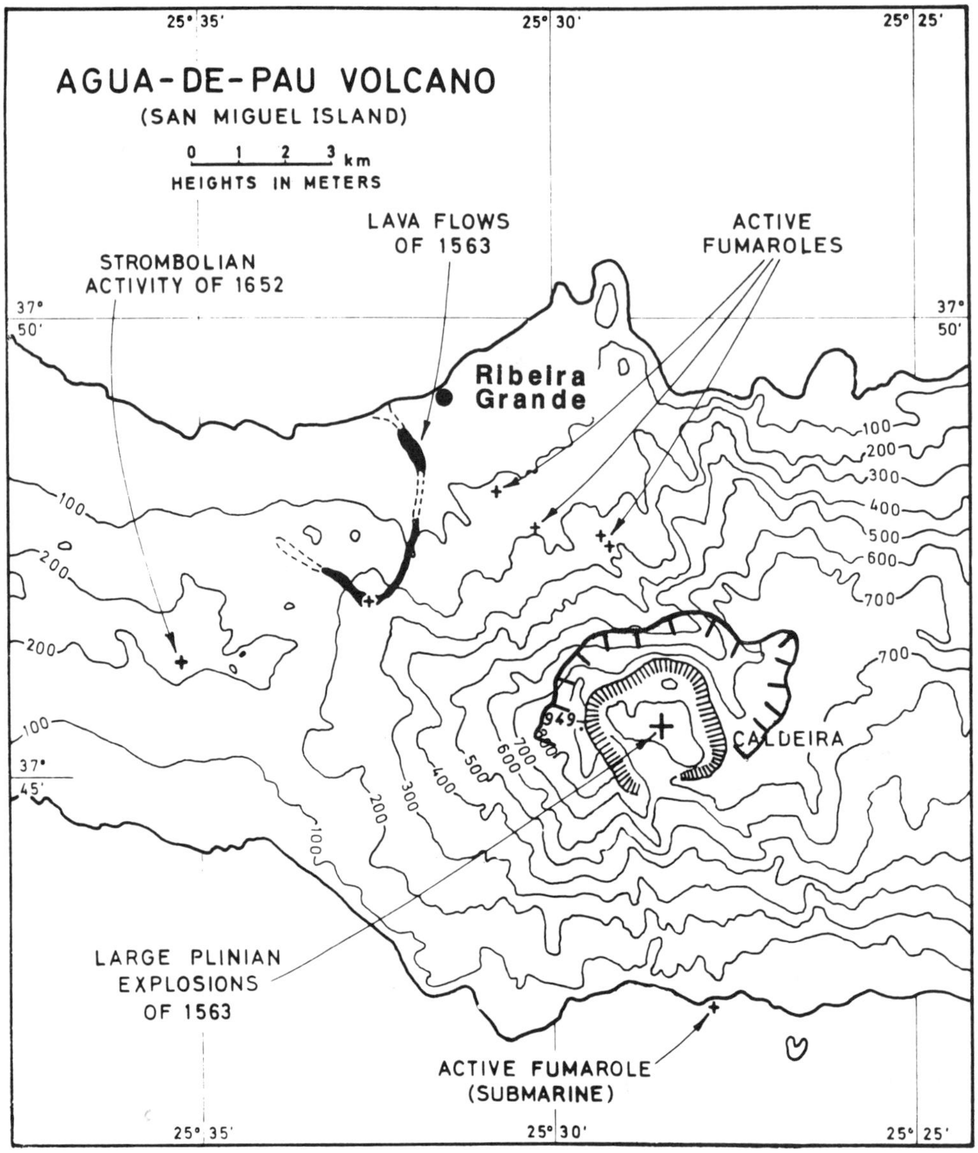

Figure 18.2.1. Activity of Agua de Pau during historical times, from Neumann van Padang and others (1967). Outline of outer caldera rim is after Moore (1986).

FURNAS

CAVW number (active vent)	Latitude Longitude (degrees)	Diameter (km)	Local tectonic setting	Precaldera edifice	SiO_2 content (wt pct)	Age of collapse (yr B.P.)	Date of unrest	Type and duration of unrest ESTU STHF MGTF H Te	Eruption type
18-02-10	37.77N 25.32W	6	Spreading center; exten	LP-strat	R = ca. 65 C = 65	ca. 12,000	1630	B--- ---B ---B B --	EX, dm

TECTONIC HISTORY

The Azores mark a possible hotspot along a spreading center, the WNW-ESE-trending Azores Fracture Zone (Laughton and Whitmarsh, 1975).

GEOLOGIC HISTORY

The modern caldera (fig. 18.3.1) formed about 12,000 yr B.P. after eruption of at least 7 km^3 of trachytic pumice (Moore, 1983, 1988). At least 11 layers of trachytic pumice, 4 with associated domes, postdate the caldera. All 11 eruptions occurred within the past 5,000 years (Booth and others, 1978), and 5 occurred within the past 1,000 years (Moore, 1988), most recently in A.D. 1630. Minor postcaldera eruptions have also produced lava flows and built cinder cones of hawaiite, west and south of the caldera (Moore, 1983, 1988). The older, 6-km diameter Povoaçao Caldera is contiguous with the east rim of Furnas Caldera.

HISTORICAL ACTIVITY

Earthquakes began on 2 September 1630, between 2100 hr and 2200 hr. The next morning, at 0800 hr, strong earthquakes were felt in the region of Furnas. Two small lakes dried out (or were emptied?) on account of the eruption. At 1100 hr, the earth continued to shake, and two fiery clouds arose from a vent on the southern caldera (Zbyszewski, 1963; Moore, 1988). Approximately 0.1 km^3 of magma (DRE) was erupted (Booth and others, 1978). At least 200 people were killed, most of them probably by mudflows in Ribeira Quente on the south coast (Zbyszewski, 1963; R.B. Moore, written commun., 1987).

REFERENCES

Booth, B., Croasdale, R., and Walker, G.P.L., 1978, A quantitative study of five thousand years of volcanism on São Miguel, Azores: Phil. Trans. Roy. Soc. London, Ser. A, v. 288, p. 271-319.

Laughton, A.S., and Whitmarsh, R.B., 1975, The Azores-Gibraltar plate boundary, in Kristjansson, L., ed., Geodynamics of Iceland and the North Atlantic area: Dordrecht, D. Reidel Publ. Co., p. 63-81.

FURNAS (continued)

REFERENCES (continued)

Moore, R.B., 1983, Preliminary geologic map of Furnas Volcano, São Miguel, Azores: U.S. Geol. Surv. Open-File Report 83-395, 1 sheet, scale 1:15,000.

----- 1988, Geology of three late Quaternary stratovolcanoes on São Miguel, Azores: Bull. Volcanol. [in press].

Neumann van Padang, M., Richards, A.F., Machado, F., Bravo, T., Baker, P.E., and LeMaitre, R.W., 1967, Catalogue of active volcanoes of the world including solfatara fields, Pt. XXI, Atlantic Ocean: Rome, IAVCEI, 128 p.

Zbyszewski, G., 1963, Les phénomènes volcaniques modernes dan l'Archipel des Açores: Serv. Geol. Portugal, v. 47, p. 5-277.

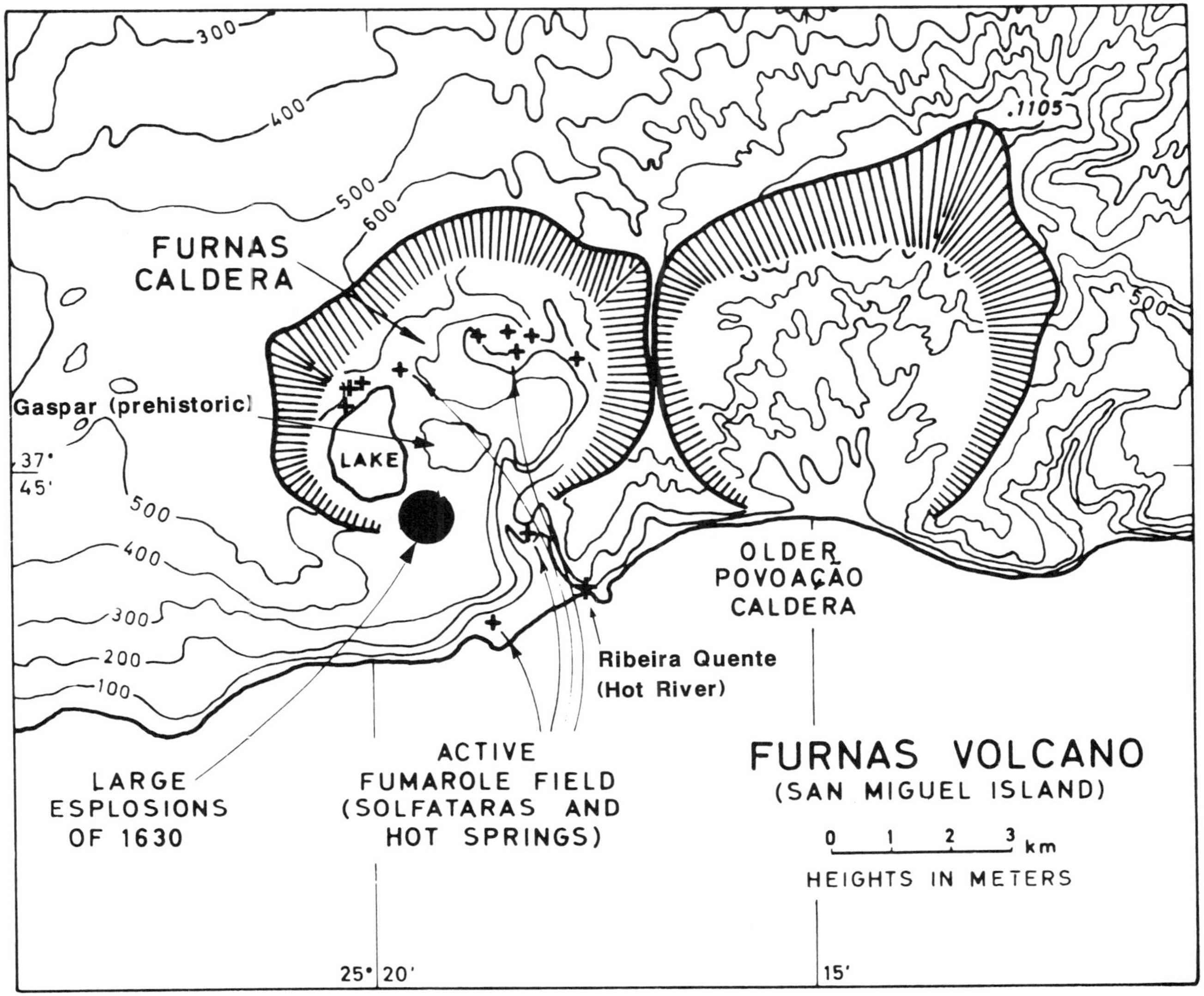

Figure 18.3.1. Furnas Volcano and extinct volcano of Povoacao, from Neumann van Padang and others (1967).

PART 3: HISTORICAL UNREST AT LARGE QUATERNARY CALDERAS (continued)

See inside back cover for explanation and abbreviations

LAS CAÑADAS

CAVW number (active vent)	Latitude Longitude (degrees)	Diameter (km)	Local tectonic setting	Precaldera edifice	SiO_2 content (wt pct)	Age of collapse (yr B.P.)	Date of unrest	Type and duration of unrest ESTU STHF MGTF H Te	Eruption type
18-03-03	28.30N	10 x 17	Hotspot?	LL-strat	R = 44-52	ca.	1341?	---- ---- ---- - --	unknown
(Pico de Teide,	16.63W				C = 60-61	150,000	1393/99?	---- ---- ---- - --	unknown
Chinyero,							1444?	---- ---- ---? - --	none?
and others)							1492	---- ---- ---- - --	lf
							1704-06	CC-- ---- ---- - --	ex,lf
							1798	---- ---- ---- - --	ex,lf
							1909	FF-- ---- ---- - --	ex,lf

TECTONIC SETTING

The Canary Islands lie in an intraplate setting, on the African continental shelf. Volcanism may result from a mantle hotspot beneath the islands. The alignment of vents and fissures in the Canary Islands suggests northeast-, northwest-, and north-south-trending faults; Las Cañadas Caldera is situated at the intersection of three such fissures or rift zones, called dorsals, of which the northwest- and northeast-trending ones have been dominant in recent geologic and historical time (Carracedo, 1985).

GEOLOGIC HISTORY

The Las Cañadas Caldera (figs. 18.4.1, 18.4.2) formed initially by collapse following a large plinian eruption that produced the Granadilla pumice (Booth, 1973), perhaps around 150,000 yr B.P. (Arana and others, 1985). Thick plinian and phreatoplinian pumice deposits occur on Tenerife (Fuster and others, 1968; Ridley, 1972; Booth, 1973; Arana and Carracedo, 1978). Subsequent gravitational failure of part of the north wall left the caldera open to the north (Booth, 1979). Pico Viejo and Teide Volcanoes have grown within the area of collapse, north of the escarpment.

HISTORICAL ACTIVITY

1492: The eruption of 1492 was probably from the Teide vent (Soler and others, 1984) (figs. 18.4.1, 18.4.2).

1704-06: An eruption of Siete Fuentes on 31 January 1704 (volume 0.4 x 10^6 m^3) was preceded by 7 days of premonitory seismicity; eruptions of Volcán Fasnia on 5 January 1705 (volume 2.5 x 10^6 m^3) and of Montana Arenas on 2 February 1705 (volume 24 x 10^6 m^3) were preceded by several days of seismicity, and an eruption of Montana Negra on 5 May 1706 (volume 66 x 10^6 m^3) was preceded by one evening of seismicity (data from F. Machado and A. Hernandez Pacheco, cited in Carracedo, 1985).

LAS CAÑADAS (continued)

HISTORICAL ACTIVITY (continued)

1909: A flank eruption along the northwest dorsal in 1909 was preceded by 18 months of seismicity, including earthquake swarms in July, August, and especially November 1908 (figs. 18.4.3-18.4.6). Maximum activity was 14 shocks in two hours, as felt in Orotava. "The top of a tall araucaria tree, seen reflected in a pool of water, was constantly agitated.... On 18 November 1909 detonations were heard, and these increased in violence until they were audible at Santa Cruz, 40 km from the scene of the outbreak, which was initiated on that date" (Perret, 1950, p. 149-151). Monge Montano (1983) also describes an increase in the intensity of earthquakes in November and December 1909, corresponding to an increase in the intensity of eruption (fig. 18.4.6). A short-term, very local precursor was noted by a farmer from Los Llanos who was "working 100 m away from the place of eruption when the earth began to tremble, and there was a sound like that of a flock of pigeons fluttering up... and then the earth burst forth amid a great noise" (von Wolff, 1929, v. 1, p. 520).

Geodetic and gravity networks have recently been established in the caldera (MacFarlane and Ridley, 1968; Sevilla and Martin, 1986; Sevilla and others, 1986; Viera and others, 1986), and a telemetered seismometer at the base of Teide has been operated by the Estacion Vulcanologica de Canarias since November 1987 (J. Carracedo, written commun., 1987).

REFERENCES

Arana, V., Barberi, F., and Ferrara, G., 1985, Petrology and geochemistry of Teide-Pico Viejo volcanic complex [abs.]: IAVCEI Symp., Catania, September 1985 (unpaginated).
Arana, V., and Carracedo, J.C., 1978, Los volcánes de las Islas Canarias: v. 1, Tenerife: Madrid, Ed. Rueda.
Booth, B., 1973, The Granadilla Pumice deposit of southern Tenerife, Canary Islands: Proc. Geol. Assoc., v. 84, no. 3, p. 353-370.
----- 1979, Assessing volcanic risk: Jour. Geol. Soc. London, v. 136, p. 331-340.
Carracedo, J.C., 1985, Geology of the Canary Islands, in Geografia de Canarias, Edit. Interinsular, Sta. Cruz de Tenerife, Canary Islands, 104 p.
Fernandez Navarro, L., 1924, Datos sobre el volcanismo canario: Bull. Volcanol., ser. I, no. 2, p. 129-155.
Fuster, J.M., Arana, V., Brandle, J.L., Navarro, M., Alonso, U., and Aparicio, A., 1968, Geologia y vulcanologia de las Islas Canarias--Tenerife: Madrid, Inst. "Lucas Mallada," 218 p.
MacFarlane, D.J., and Ridley, W.I., 1968, An interpretation of gravity data for Tenerife, Canary Islands: Earth Planet. Sci. Lett., v. 4, p. 481-486.
Monge Montano, F., 1983, Sismicidad en el Archipelago Canario, relacion con las erupciónes: Cuarta asamblea nacional de geodesia y geofisica, v. 1: Madrid, Inst. Geogr. Nac., p. 457-471.
Neumann van Padang, M., Richards, A.F., Machado, F., Bravo, T., Baker, P.E., and LeMaitre, R.W., 1967, Catalogue of active volcanoes of the world including solfatara fields, Pt. XXI, Atlantic Ocean: Rome, IAVCEI, 128 p.
Perret, F.A., 1950, Volcanological observations: Carnegie Inst. Washington Publ. 549, 153 p.
Ridley, W.I., 1972, The field relations of the Las Cañadas Volcanoes, Tenerife, Canary Islands: Bull. Volcanol., v. 35, p. 318-334.

LAS CAÑADAS (continued)

REFERENCES (continued)

Sevilla, M.J., and Martin, M.D., 1986, Geodetic network design for crustal deformation studies in the Caldera of Teide area: Tectonophysics, v. 130, p. 235-248.

Sevilla, M.J., Martin, M.D., and Camacho, A.G., 1986, Data analysis and adjustment of the first geodetic surveys in the Caldera of Teide, Tenerife, Canary Islands: Tectonophysics, v. 130, p. 213-234.

Soler, V., Carracedo, J.C., and Heller, F., 1984, Geomagnetic secular variation in historical lavas from the Canary Islands: Geophys. Jour. Roy. Astron. Soc., v. 78, p. 313-318.

Viera, R., Toro, C., and Arana, V., 1986, Microgravimetric survey in the Caldera of Teide, Tenerife, Canary Islands: Tectonophysics, v. 130, p. 249-257.

Wolff, F. von, 1929, Der Vulkanismus: Stuttgart: Ferdinand Enke, v. 1, no. 2, p. 520.

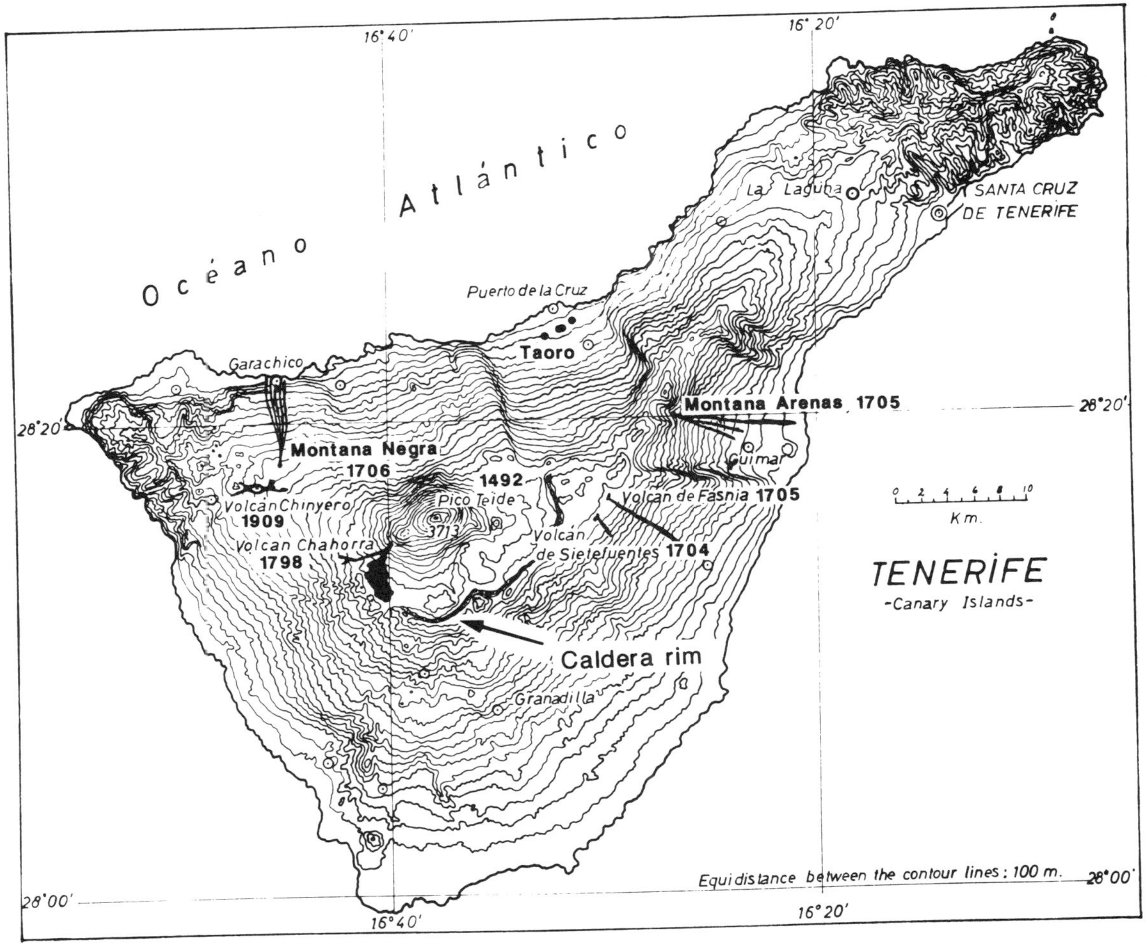

Figure 18.4.1. Island of Tenerife and Las Cañadas Caldera, with Pico de Teide at its center, from Neumann van Padang and others (1967), with eruption dates from Soler and others (1984).

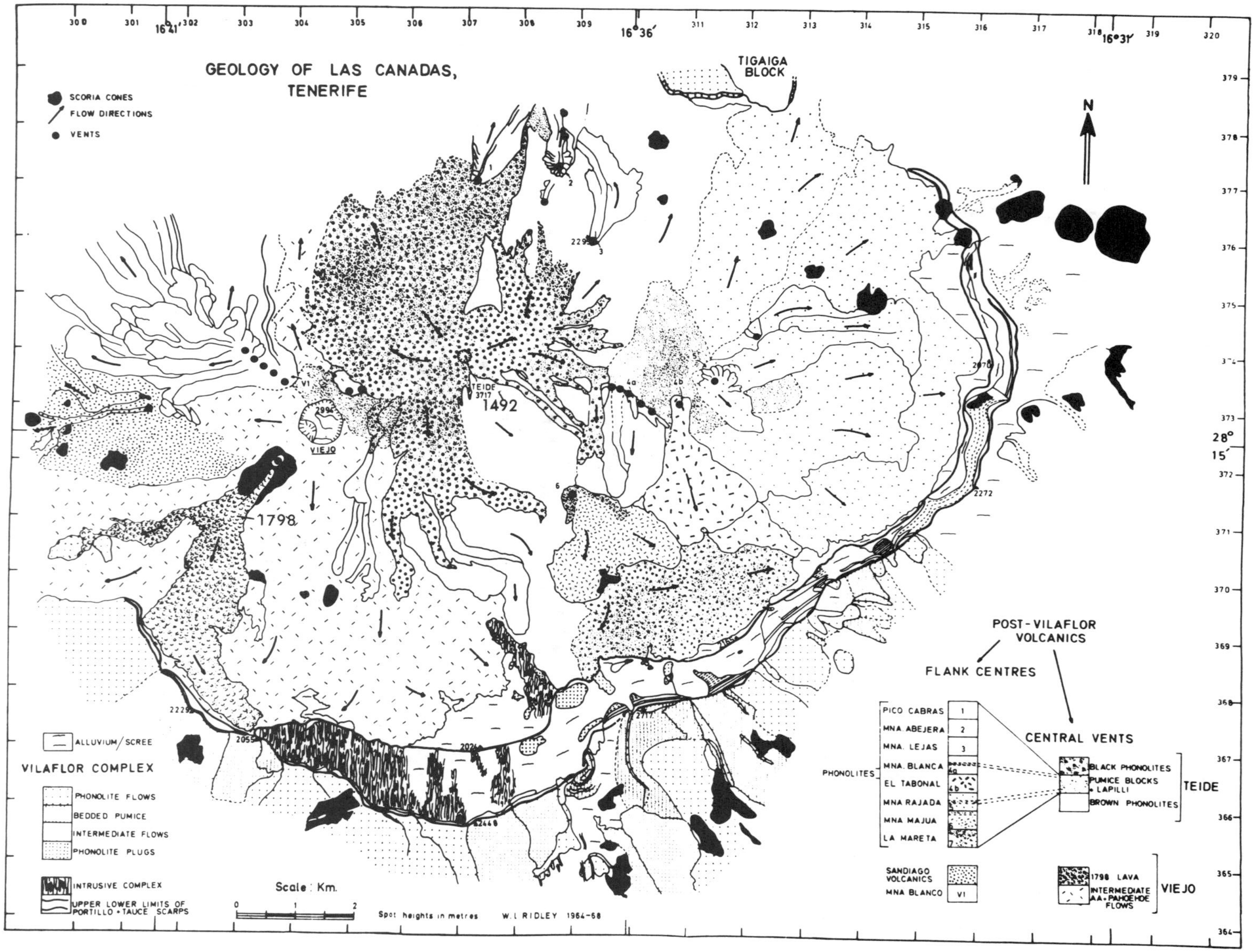

Figure 18.4.2. Geologic map of Las Cañadas volcanoes, including locations of the 1492 and 1798 vents, from Ridley (1972).

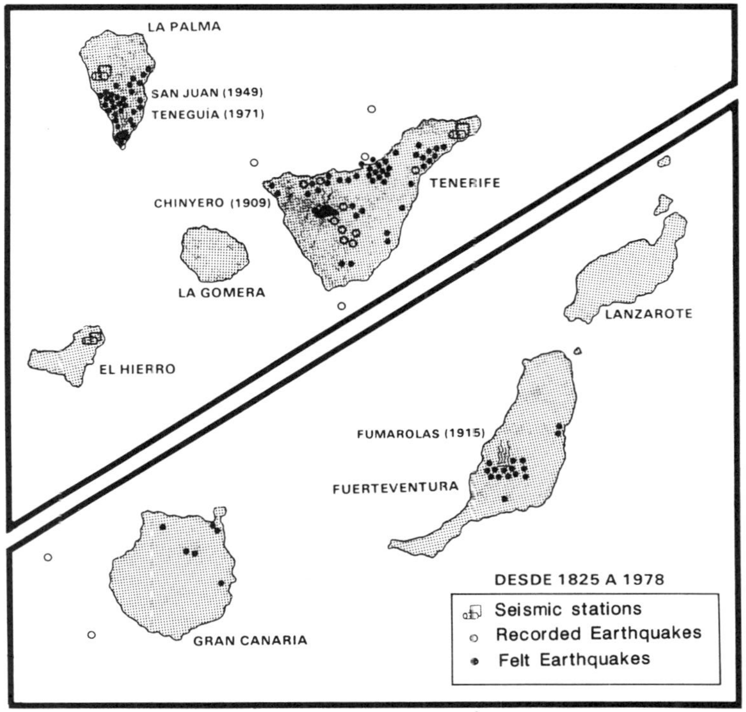

Figure 18.4.3. Seismic activity associated with volcanism in Canary Islands from 1825 to 1978, from Carracedo (1985). Note epicenters at Las Cañadas Caldera, Tenerife.

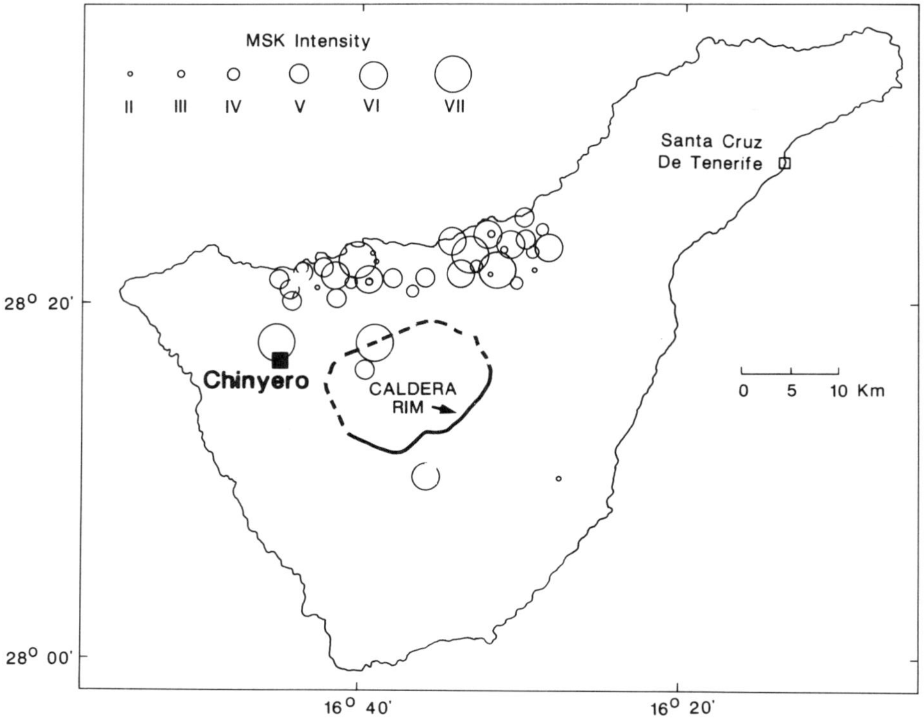

Figure 18.4.4. Distribution and intensity of earthquakes near Las Cañadas Caldera in 1909, from Monge Montano (1983). Caldera outline dashed where inferred.

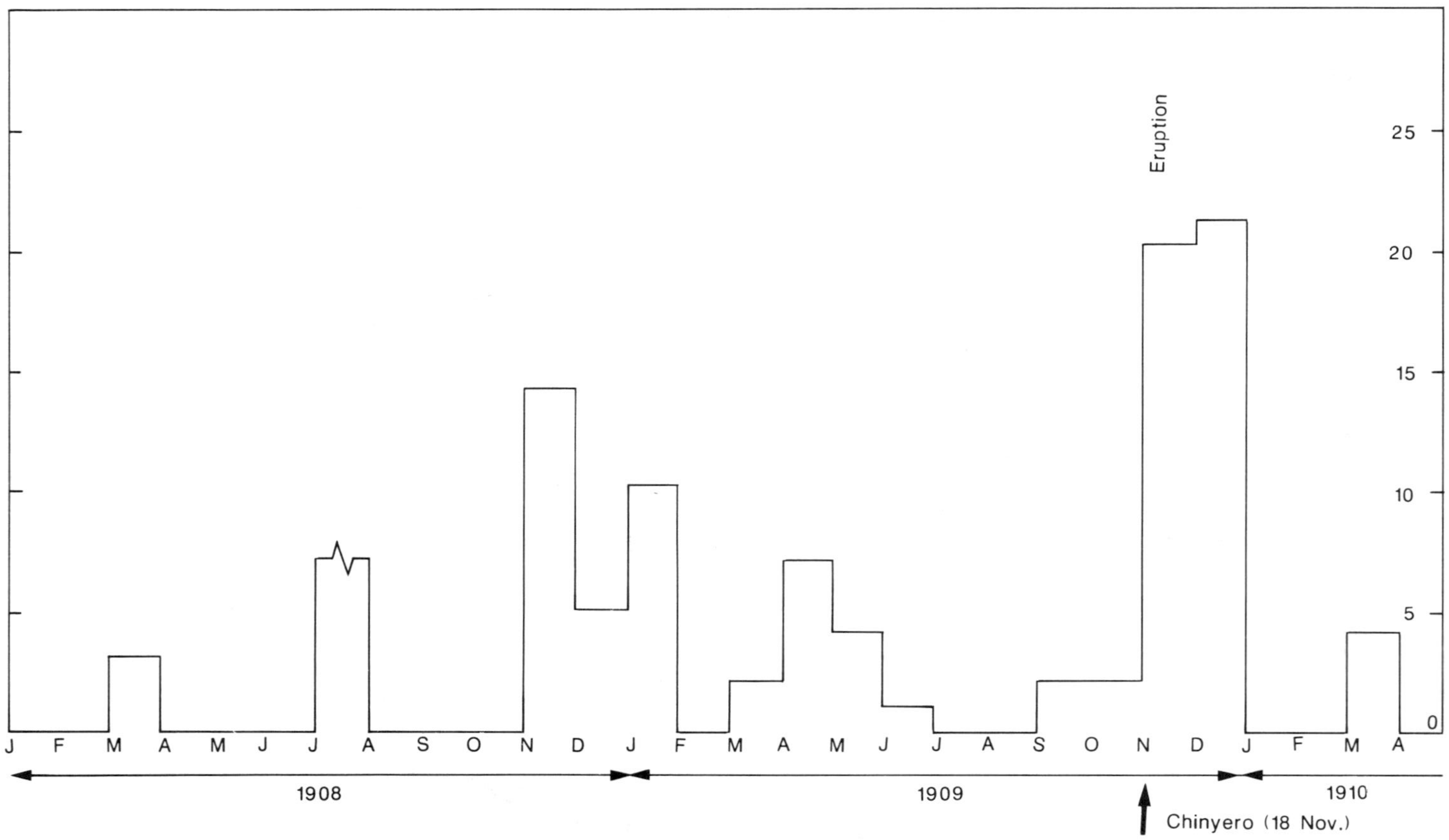

Figure 18.4.5. Temporal distribution of felt earthquakes associated with the 1909 Chinyero eruption (arrow) northwest of Las Cañadas Caldera, from Monge Montano (1983).

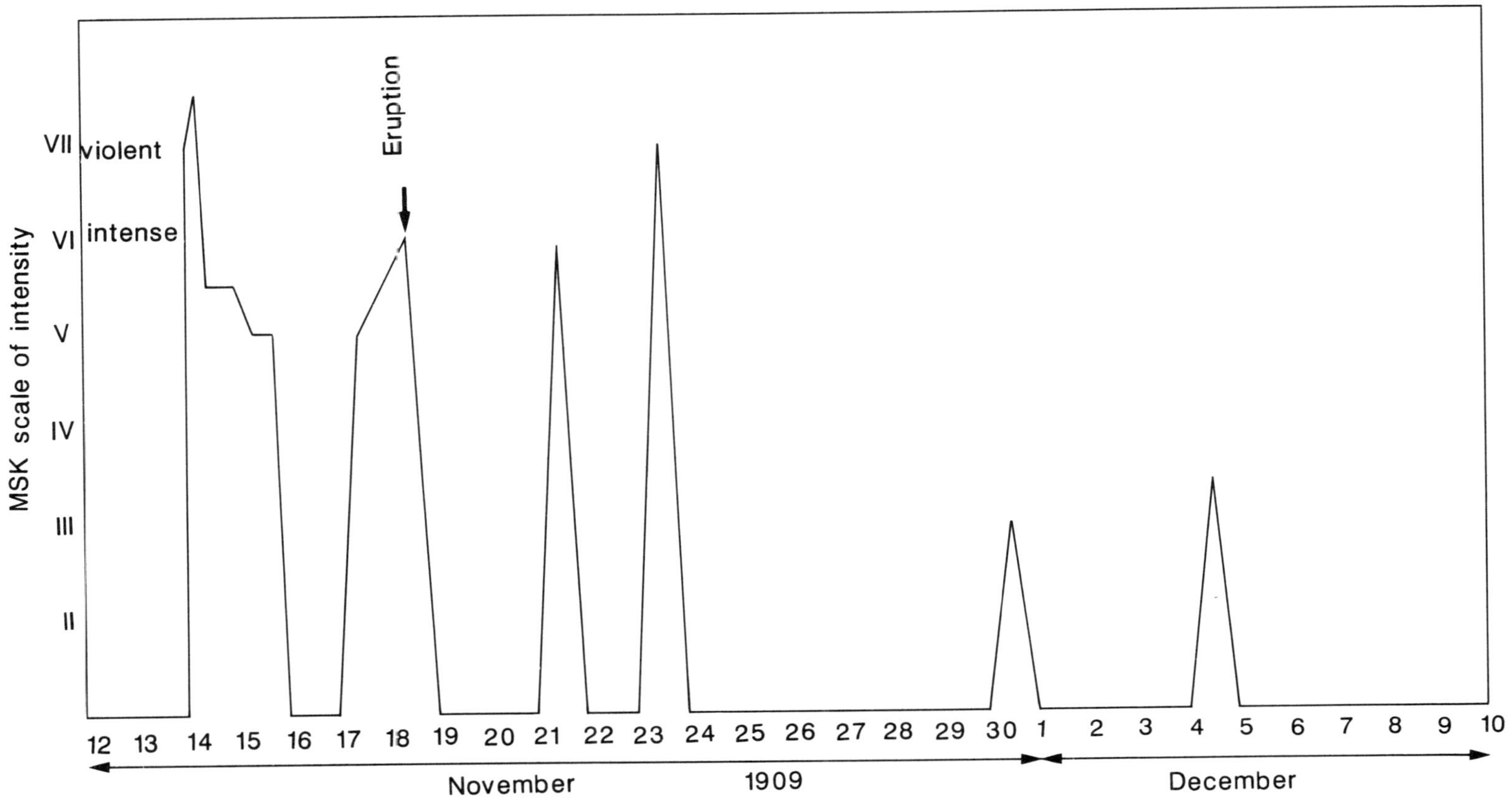

Figure 18.4.6. Temporal variation in intensity of earthquakes felt during an eruption of Volcano Chinyero, Las Cañadas Caldera, in 1909 (Monge Montano, 1983).

PART 3: HISTORICAL UNREST AT LARGE QUATERNARY CALDERAS (continued)
See inside back cover for explanation and abbreviations

CHA

CAVW number (active vent)	Latitude Longitude (degrees)	Diameter (km)	Local tectonic setting	Precaldera edifice	SiO_2 content (wt pct)	Age of collapse (yr B.P.)	Date of unrest	Type and duration of unrest ESTU STHF MGTF H Te	Eruption type
18-04-01 (Fogo)	14.95N 24.35W	8	Intra-plate	Shield	R = 40-50 C = mafic		<1500-1761	---- ---- ---- - --	ex,lf
							1769	---- ---- ---- - --	ex
							1785	---- ---- ---- - --	pex,lf
							1799	---- ---- ---- - --	ex,lf
							1817	---- ---- ---- - --	unknown
							1847	x--- ---- ---- - --	ex,lf
							1852	---- ---- ---- - --	ex,lf
							1857	---- ---- ---- - --	ex,lf
							1951	x--- ---- ---- - --	ex,lf

TECTONIC SETTING

Fogo Island (fig. 18.5.1) and the rest of the Cape Verde Islands are intraplate volcanic islands situated west of the margin of the African continent (Dash and others, 1971).

GEOLOGIC HISTORY

The Cape Verde Islands are known geologically for a strongly alkaline suite of volcanic and plutonic rocks (Mitchell-Thome, 1970; Gunn and Watkins, 1976).

HISTORICAL ACTIVITY

17th and 18th centuries: von Humboldt (1861-63) reports that Fogo was in continuous eruption from 1680 to 1713 and then in repose until 1798. Neumann van Padang and others (1967) and Mitchell-Thome (1981) cite different dates, which we adopt here, including continuous eruption from before 1500 until 1761, intermittent eruptions until 1857, and a subsequent period of repose until 1951.

1847: Fatalities occurred as a result of an earthquake.

1951: The eruption of 12 June was preceded by strong earthquakes and volcanic tremors (Ribeiro, 1954).

CHA (continued)

REFERENCES

Dash, B.P., Ball, M.M., Kings, G.A., Butler, L.W., and Rona, P.A., 1971, Geophysical measurements in the Cape Verde archipelago [abs.]: Eos, Trans. Am. Geophys. Union, v. 52, p. 235.

Gunn, B.M., and Watkins, N.D., 1976, Geochemistry of the Cape Verde Islands and Fernando de Noronha: Geol. Soc. Am. Bull., v. 87, p. 1089-1100.

Humboldt, A. von, 1861-63, Cosmos: (transl. from the German by E.C. Otté and W.S. Dallas): New York, Harper and Brothers, v. 5, p. 243.

Mitchell-Thome, R.C., 1970, Geology of the South Atlantic islands: Beitr. Regionalen Geol. Erde, v. 10, 367 p.

----- 1981, Vulcanicity of historic times in the middle Atlantic islands: Bull. Volcanol., v. 44, p. 57-70.

Neumann van Padang, M., Richards, A.F., Machado, F., Bravo, T., Baker, P.E., and LeMaitre, R.W., 1967, Catalogue of active volcanoes of the world including solfatara fields, Pt. XXI, Atlantic Ocean: Rome, IAVCEI, 128 p.

Ribeiro, O., 1954, A ilha do Fogo e as suas erupções: Lisbon, Ministério do Ultramar, Memórias, Série Geográfica, no. 1, 319 p.

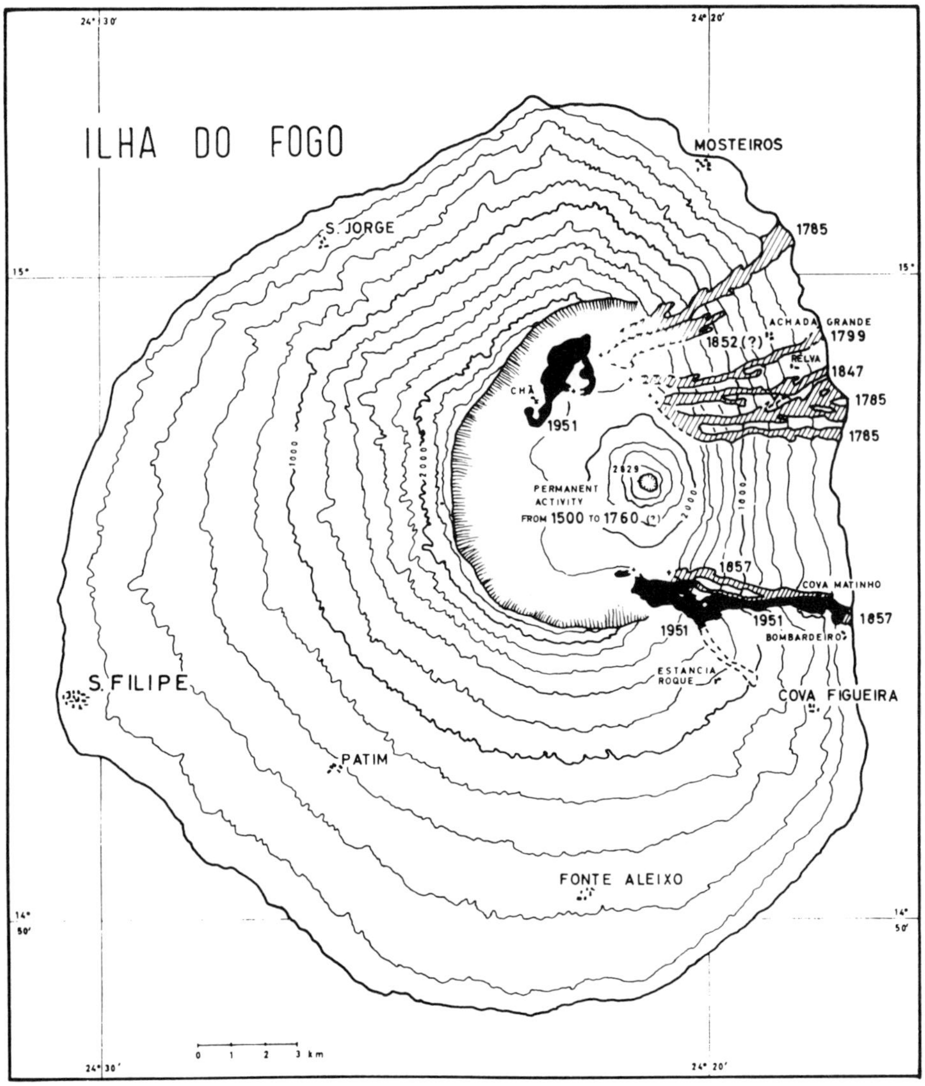

Figure 18.5.1. Island of Fogo, showing Cha Caldera and recent lava flows, from Neumann van Padang and others (1967).

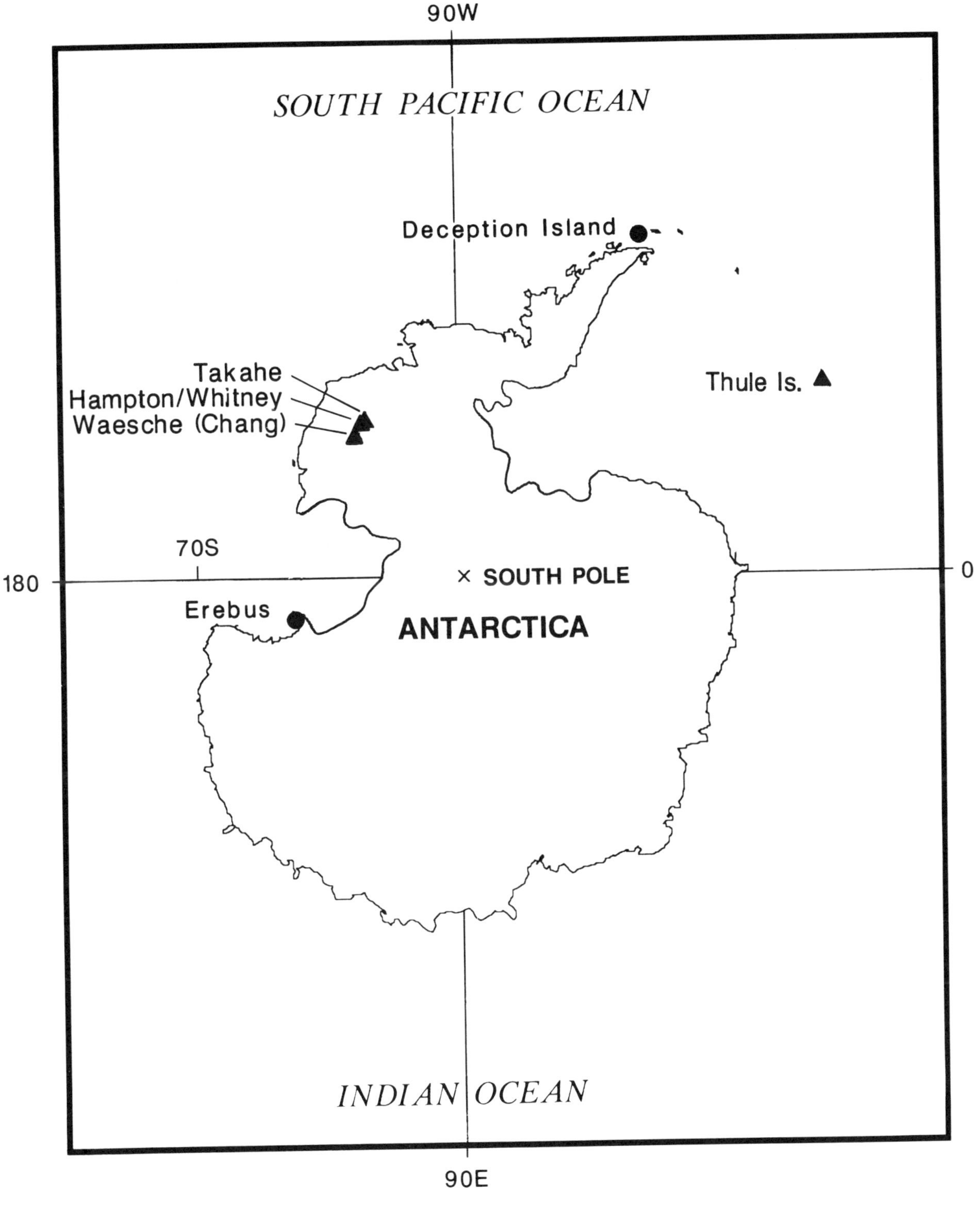

Figure 19. Locations of large, Quaternary restless calderas (solid circles) and nonrestless calderas (triangles) in Region 19.

PART 3: HISTORICAL UNREST AT LARGE QUATERNARY CALDERAS (continued)
See inside back cover for explanation and abbreviations

EREBUS

CAVW number (active vent)	Latitude Longitude (degrees)	Diameter (km)	Local tectonic setting	Precaldera edifice	SiO_2 content (wt pct)	Age of collapse (yr B.P.)	Date of unrest	Type and duration of unrest ESTU STHF MGTF H Te	Eruption type
19-00-02 (Erebus)	77.58S 167.17E	6 (older) 5 (younger)	Intra-plate	Strato	R = 56 (anorthoclase phonolite)		1841	---- ---- ---- - --	ex,lf
							1900?	---- ---- ---- - --	ex
							1908	---- ---- ---E - --	ll
							1911-12	---- ---- ---- - --	ex
							1915	---- ---- ---- - --	ex
							1947	---- ---- ---- - --	ex,lf(?)
							1955	---- ---- ---- - --	ex
							1963	---- ---- ---- - --	ll
							1972-87+	YYY- Y--- Y--- - --	ex,lf,ll

TECTONIC SETTING

Mt. Erebus is an intraplate volcano, situated at the southern end of the Terror Rift within the Victoria Land basin, a major sedimentary basin with over 12 km of fill and underlain by 21-km-thick crust. The basin is bounded to the west by the Transantarctic Mountains, with 40-km-thick crust (Cooper and Davey, 1985; Fitzgerald and others, 1986).

GEOLOGIC HISTORY

An older cone of Mt. Erebus (Fang Volcano), composed largely of benmoreite, was partly destroyed by an unknown event, leaving a caldera of ca. 6-km diameter (fig. 19.1.1). The modern cone of Mt. Erebus, composed largely of anorthoclase phonolite (Kyle, 1977), has largely filled the older caldera, leaving only Fang Ridge (north of Mt. Erebus) as a relic of the older cone. A roughly 5-km-diameter caldera developed at the top of the modern cone, and it too has been largely filled (Berninghausen and Neumann van Padang, 1960; P. Kyle, written commun., 1987).

HISTORICAL ACTIVITY

Most historical activity has been confined to the younger caldera. A reported exception was increased fumarolic activity in April-September 1908 and on 17 June 1908, near Abbott Peak, 10 km north-northeast of the summit of Mt. Erebus (cited in Berninghausen and Neumann van Padang, 1960). A seismic swarm in roughly the same area was recorded in October 1982 (Ueki and others, 1984; Kaminuma and others, 1985) (figs. 19.1.2-19.1.4).

EREBUS (continued)

HISTORICAL ACTIVITY (continued)

A lava lake discovered in 1972 grew slowly until 1976, when it was circular, about 60 m in diameter. It remained relatively constant until September 1984. Larger-than-usual Strombolian explosions occurred in September to December 1984. Earthquakes were felt, and glow and increased steaming were observed from McMurdo Sound (37 km southwest of the volcano) (P. Kyle and others, in Smithsonian Institution, 1984). The summit crater lava lake was buried by ejecta between 13 September 1984 and October 1984. When the lake was exhumed in December 1985 it was 15 m in diameter, and it grew to 20 m in diameter by December 1986 (P. Kyle, written commun., 1987).

SO_2 emission was 230 tonnes/day in December 1983, but less than 30 tonnes/day since then (P. Kyle, written commun., 1987); no remarkable deformation occurred during the same time period.

Dibble and others (1984) report that numerous small volcanic earthquakes occur in conjunction with volcanic activity (figs. 19.1.2, 19.1.3). Most events above 1 kJ in energy (ca. $\underline{M}$=0) are explosion earthquakes, with emergent rather than impulsive onsets. Numerous B-type earthquakes have energy greater than 10 kJ. Volcanic tremor is rare. Hypocenters shown in figures 19.1.2 and 19.1.3 are currently being reexamined and are shallower than shown (P. Kyle, written commun., 1987).

REFERENCES

Berninghausen, W.H., and Neumann van Padang, M., 1960, Catalogue of Active Volcanoes of the World including solfatara fields, Pt. X, Antarctica: Naples, IAVCEI, 32 p.

Cooper, A.K., and Davey, F.J., 1985, Episodic rifting of Phanerozoic rocks in the Victoria Land basin, western Ross Sea, Antarctica: Science, v. 229, p. 1085-1087.

Dibble, R.R., and Barrett, S.I.D., 1987, Time comparisons between video and seismic signals from explosions in the lava lake of Erebus Volcano, Antarctica [abs.]: Abstracts, 19th Gen. Assem. IUGG, Vancouver, Canada, v. 2, p. 397.

Dibble, R.R., Kienle, J., Kyle, P.R., and Shibuya, H., 1984, Geophysical studies of Erebus Volcano, Antarctica, for 1974 December to 1982 January: New Zealand Jour. Geol. Geophys., v. 27, no. 4, p. 425-455.

Fitzgerald, P.G., Sanford, M., Barnett, P.J., and Gleadow, A.J.W., 1986, Asymmetric extension associated with uplift and subsidence in the Transantarctic Mountains and Ross Embayments: Earth Planet. Sci. Lett., v. 81, p. 67-78.

Giggenbach, W.F., Kyle, P.R., and Lyon, G.L., 1973, Present volcanic activity on Mount Erebus, Ross Island, Antarctica: Geology, v. 1, no. 3, p. 135-136.

Kaminuma, K., Ueki, S., and Kienle, J., 1985, Volcanic earthquake swarms at Mt. Erebus, Antarctica: Tectonophysics, v. 114, p. 357-369.

Kienle, J., Rowe, C.A., Marshall, D.L., and Estes, S.A., 1984, Volcanic seismic studies of Mt. Erebus, Antarctica: Biennial Rep., Univ. Alaska Geophys. Inst., p. 147-149.

EREBUS (continued)

REFERENCES (continued)

Kyle, P.R., 1977, Mineralogy and glass chemistry of recent volcanic ejecta from Mt. Erebus, Ross Island, Antarctica: New Zealand Jour. Geol. Geophys., v. 20, no. 6, p. 1123-1146.

Kyle, P.R., and Cole, J.W., 1974, Structural control of volcanism in the McMurdo Volcanic group, Antarctica: Bull. Volcanol., v. 38, no. 1, p. 16-25.

Kyle, P.R., Dibble, R.R., Giggenbach, W.F., and Keys, J., 1982, Volcanic activity associated with the anorthoclase phonolite lava lake, Mount Erebus, Antarctica, in Craddock, C., ed., Antarctic geoscience, Madison, Univ. Wisconsin, p. 735-745.

Rose, W.I., Chuan, R.L., and Kyle, P.R., 1985, Rate of sulphur dioxide emission from Erebus volcano, Antarctica, December, 1983: Nature, v. 316, p. 710-712.

Rowe, C.A., and Kienle, J., 1986, Seismicity in the vicinity of Ross Island, Antarctica: Jour. Geodynam., v. 6, p. 375-385.

Shibuya, K., Baba, M., Kienle, J., Dibble, R.R., and Kyle, P.R., 1982, A study of the seismic and volcanic activity of Mount Erebus, Antarctica, 1981-1982, in Nagata, T., ed., Proc. 3rd Symp. on Antarctic geosciences, 1982, Tokyo, Nat. Inst. Polar Research, p. 54-66.

Smith, W.C., 1954, The volcanic rocks of the Ross archipelago: Brit. Antarct. Exped., 1910, Nat. Hist. Rep., Geology, v. 2, no. 1, p. 1-107.

Smithsonian Institution, Scientific Event Alert Network (SEAN), 1976-85, Erebus: SEAN Bull., v. 1-10 (contributions from P. Kyle and others).

Ueki, S., Kaminuma, K., Baba, M., Koyama, E., and Kienle, J., 1984, Seismic activity of Mount Erebus, Antarctica in 1982-1983: Memoirs Nat. Inst. Polar Research (Tokyo), p. 29-40.

U. S. Geological Survey, 1970, Ross Island topographic sheet, Antarctica 1:250,000 Reconnaissance Series, ST 57-60/6*(162°E-170°E), 1960, revised 1970.

Volcanological Society of Japan, 1974-83, Erebus: Bull. Volcanic Eruptions, nos. 12-20.

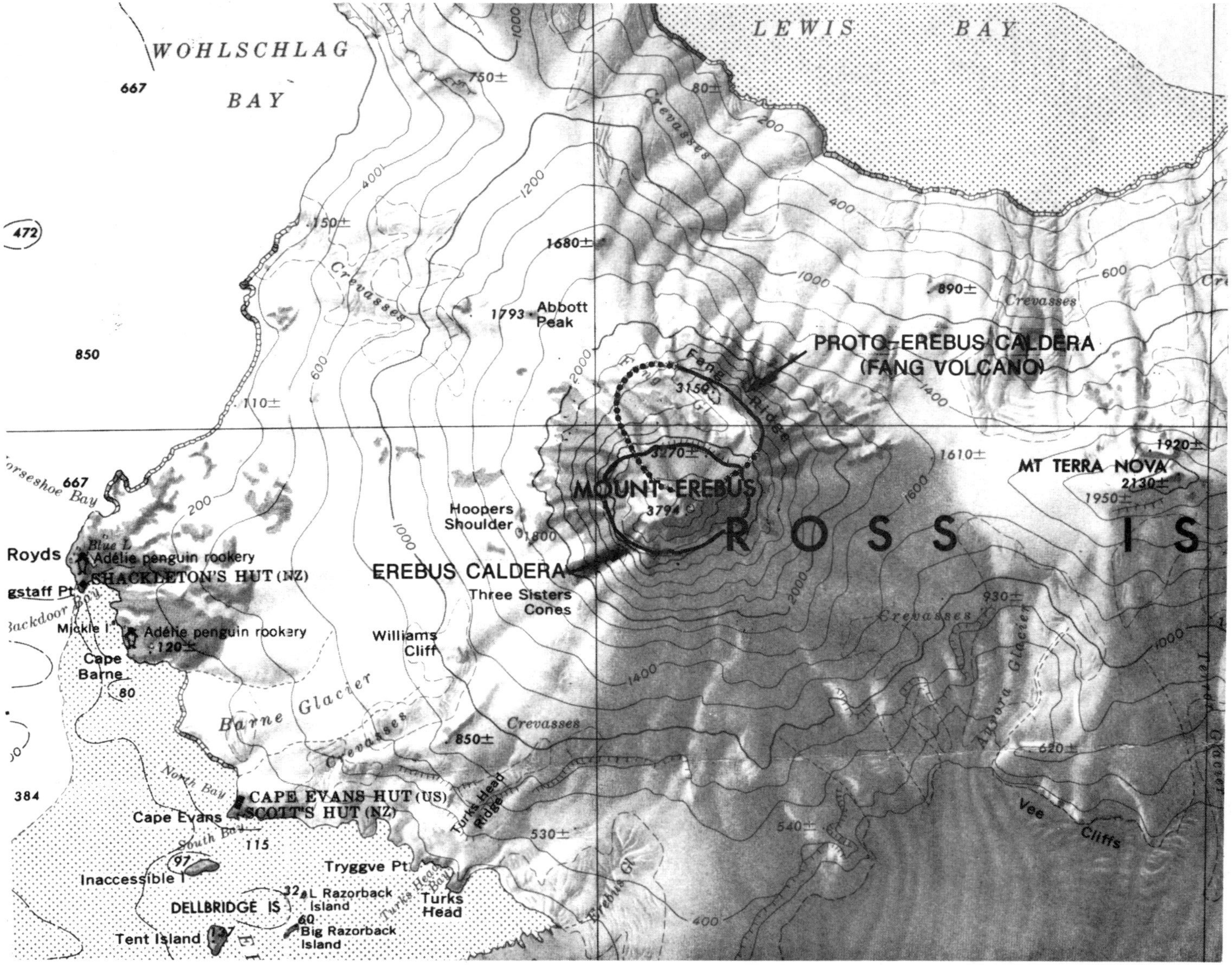

Figure 19.1.1. Topographic map showing outlines of Erebus Caldera and an older caldera (Proto-Erebus Caldera or Fang Volcano Caldera; dotted where inferred), which is now buried by Erebus except at Fang Ridge (U.S. Geological Survey, 1970; outlines courtesy of P.R. Kyle, 1987; dashed where inferred).

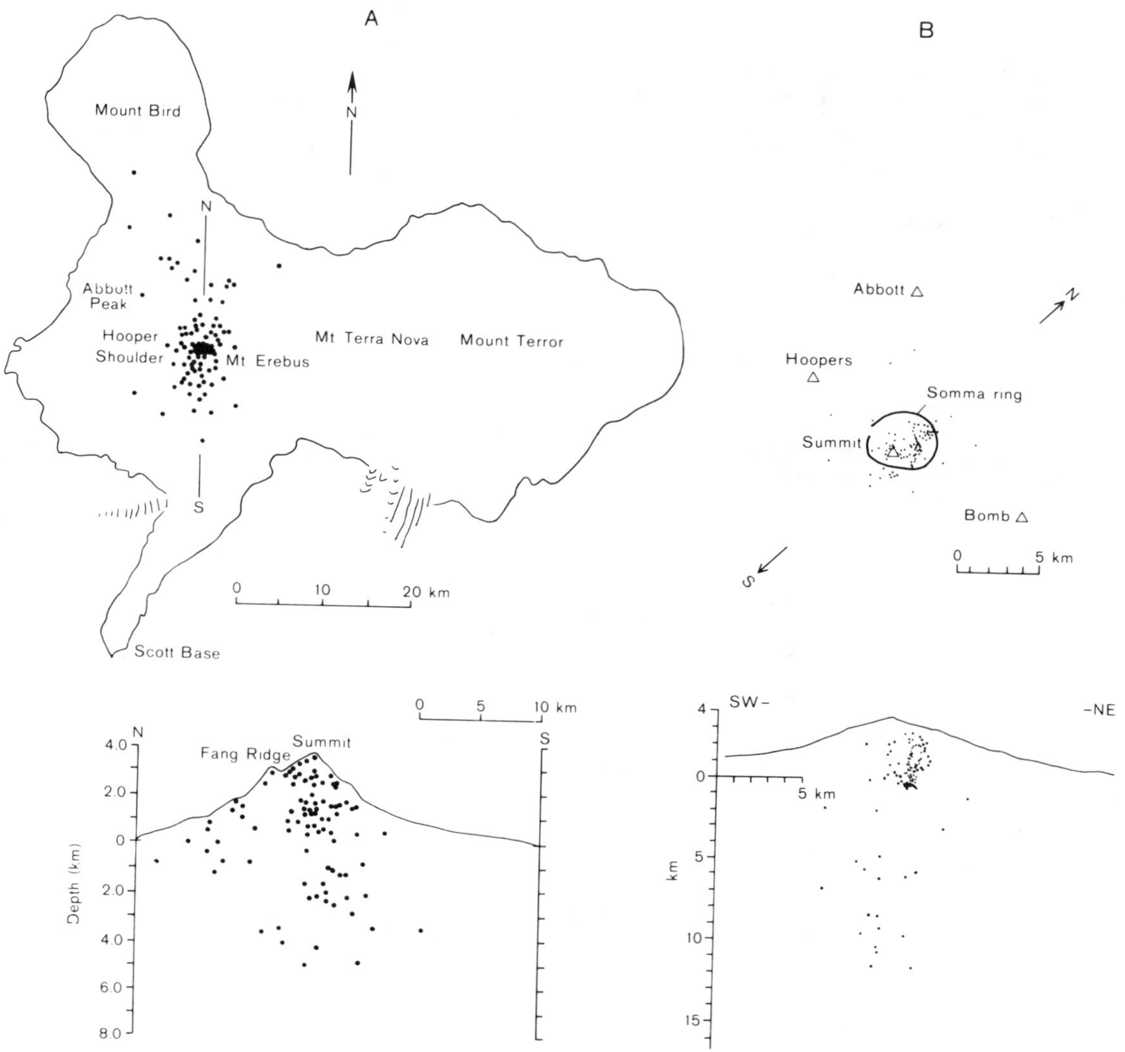

Figure 19.1.2. Epicenter map, 20 December 1980 to 6 January 1981, and projection of hypocenters onto a N-S section (left); epicenter map and projection of hypocenters during the 1981-82 field season onto a NE-SW section (right). Hypocenters below sea level are shown bold. Both diagrams from Dibble and others (1984).

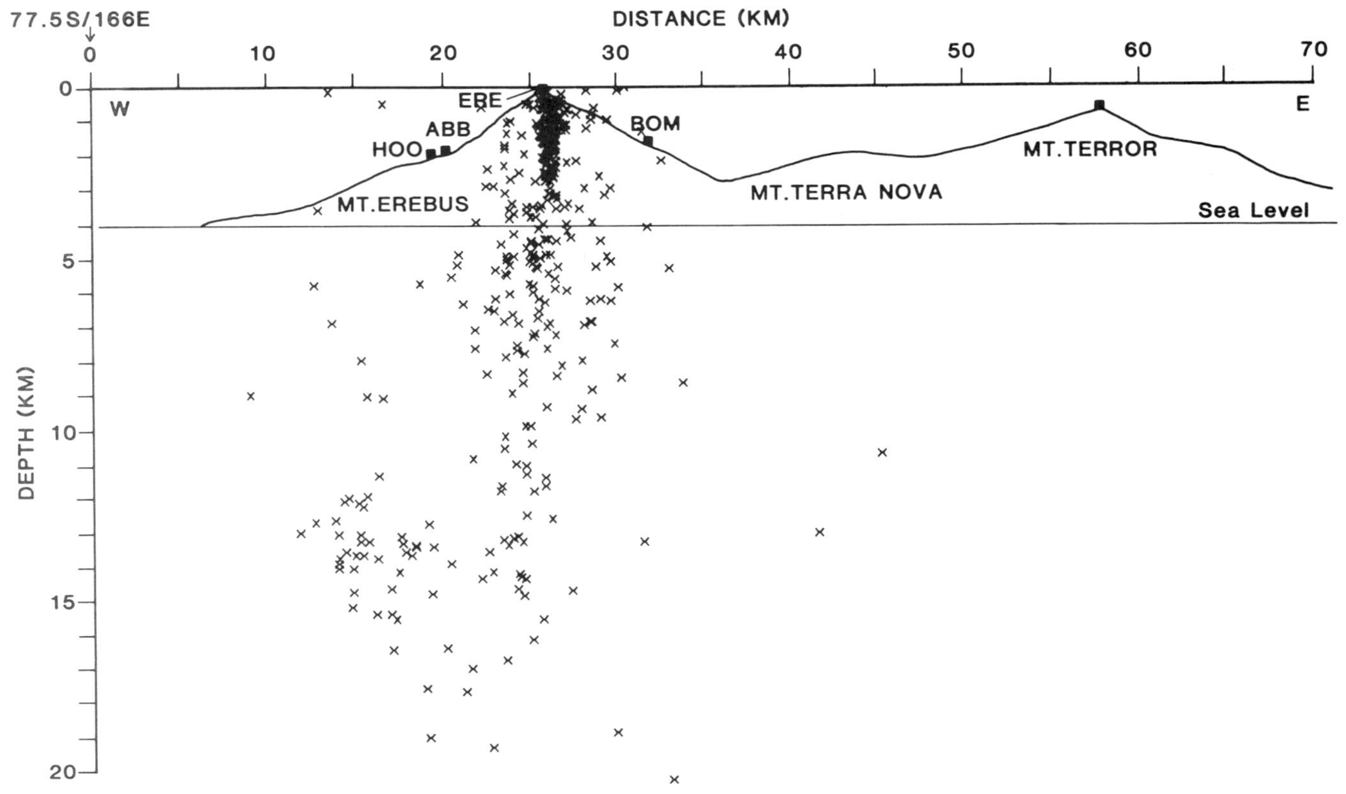

Figure 19.1.3. West-east hypocentral cross section through Mount Erebus and Mount Terror, Ross Island, Antarctica, showing 403 best-located events, December 1980 to February 1984 (Kienle and others, 1984). More recent data from video experiments suggest that most earthquakes plotted here at less that 5-km depth actually occurred at lava surface during Strombolian eruptions. Discrepancy probably reflects errors in seismic velocity models used to locate earthquakes in this area (Dibble, 1987; P. Kyle, written commun., 1987).

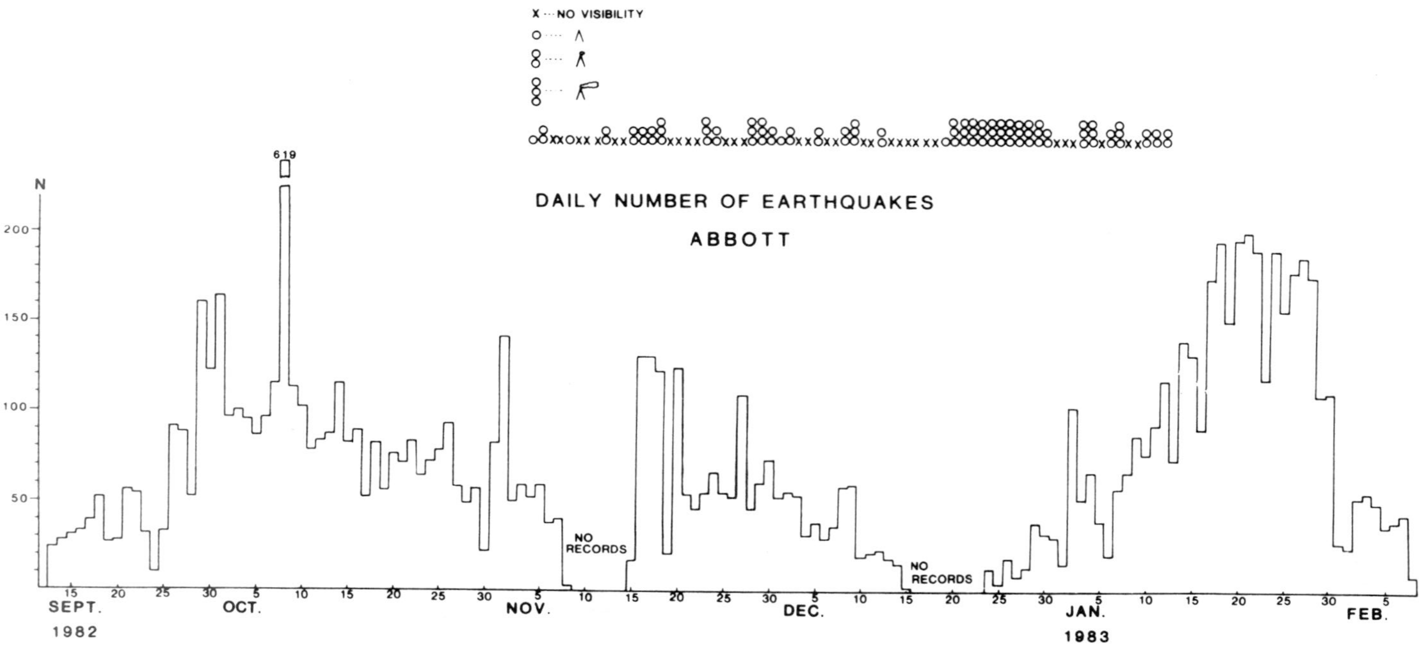

Figure 19.1.4. Plume observations of Erebus Volcano from McMurdo Station (top). Daily counts of earthquakes at Abbott Peak station (see fig. 19.1.2) from September 1982 to February 1983 (bottom; from Kaminuma and others, 1985).

DECEPTION ISLAND

CAVW number (active vent)	Latitude Longitude (degrees)	Diameter (km)	Local tectonic setting	Precaldera edifice	SiO_2 content (wt pct)	Age of collapse (yr B.P.)	Date of unrest	Type and duration of unrest ESTU STHF MGTF H Te	Eruption type
19-00-03 (Deception Island)	62.93S 60.57W	7	Exten	Strato	R = 50-70 C = 51-54		1820/29	---- ---- ---- - --	ex
							1829	x--- ---- ---x x --	none?
							1839-42	---- ---- ---F F --	ex, lf
							ca. 1912	---- ---- ---- - --	ex, Ex
							ca. 1917	---- ---- ---- - --	ex, Ex
							1923-27	x--- x--- --xx - --	none?
							1930	x--- x--- ---- - --	none
							before 1956	---- ---- ---- - --	ex
							1967	FF-x ---x ---- - --	pex, Ex
							1969	x--? ---Y ---E - --	Ex, ex, pex
							1970	x--x Z--x ---- - --	Ex, ex, pex
							1972	---- ---- ---- - --	sub
							1987	x--- ---- ---- - --	Ex?

TECTONIC SETTING

Deception Island is in the South Shetland Islands, on the Bransfield Rift where the rift is offset by a short ridge-ridge transform fault (Circum-Pacific Map Project, 1981).

GEOLOGIC HISTORY

A 7-km-diameter caldera in the center of Deception Island is largely flooded by the sea (figs. 19.2.1-19.2.3). The caldera apparently formed as a result of voluminous eruptions of sodic andesite or basaltic andesite (Baker and others, 1975). Numerous active vents are situated along the ring fractures, near the shoreline of the natural harbor. Slight chemical differences between magma erupted from various locations along the ring fracture in 1967, 1969, and 1970 were interpreted by Baker and others (1975) and Roobol (1979) as evidence of isolated intrusive pods of magma along the ring fracture.

HISTORICAL ACTIVITY

Deception Island has no permanent inhabitants, so apparent episodicity of eruption reports probably reflects episodic visits by sealing, whaling, and exploring expeditions.

DECEPTION ISLAND (continued)

HISTORICAL ACTIVITY (continued)

1829: The HMS Chanticleer visited Deception Island for 2 months to conduct pendulum experiments among other tasks. The captain wrote (reproduced in Roobol, 1979): "Subterranean sounds, as of mountain torrents, were frequently heard and on one occasion were so loud as to create some apprehension for the safety of the instruments, and accordingly the pendulums that had been experimented with were immediately taken (back) on board." Numerous fumaroles were observed throughout the caldera.

1842: The whole south side of Deception Island (13 vents) appeared "as if on fire." Hawkes (1961) correlated this report with young lavas in the Mount Kirkwood area; Roobol (1979) agrees with the location but interprets some of the 13 vents and "fire" to have been fumaroles and steam.

1912±5 yr, 1917±3 yr: Eruptions are inferred from pyroclastic deposits interbedded with ice. Vents were in or near "Lake Crater," south of Fumarole Bay (Klay and Orheim, 1969; Orheim, 1972).

1923: The shore of Deception Island subsided near a whaling station. Waters of the bay "boiled" and stripped paint from ship hulls (erroneously reported by some as 1921; Roobol, 1979).

1930: During or after an earthquake, the bottom of the harbor (caldera floor) dropped about 4.5 m.

1967: An eruption was preceded by 2-3 weeks of precursory seismicity (fig. 19.2.4); from the data of Lorca (1976), it appears that this seismicity was only slightly more than previous swarms in early, mid-, and late 1966 ($\underline{M}_{max}$ = 4.7; $\underline{d}_{max}$ ~ 20 km). The starting date for increased seismicity is not known; monitoring was begun on 16 January 1966. Eruption was from four vents along the northwest part of the ring fracture. A new island was built (Yelcho Island, or "New Island 1967" of fig. 19.2.3). In addition, Baker and others (1975, p. 25-26) report that a small area immediately south of the "Land Center" (fig. 19.2.3) was 55 m higher after the eruption than before and was intensely fractured and rifted, suggesting uplift. An alternate explanation--that 55 m of tephra was deposited at the site--was not supported by field observations.

1969: Almost 2 months of precursory seismicity greatly intensified 6 hours before eruptions on 21 February 1969. The temperature of Kroner Lake also increased from January to February. Eruptions were from 20-30 vents along a 5-km-long section of the caldera's east ring fracture (Baker, 1969; Baker and others, 1975).

1970: Shultz (1972) showed new land in the vicinity of the 1967 and 1970 craters (northwestern ring fracture; fig. 19.2.3). Some of this land was uplifted so much that the shoreline was displaced more than 500 m and Yelcho Island was rejoined to land, reducing the surface of Telefon Bay by nearly half (Gonzales-Ferran and others, 1971). Eleven or more 1970 vents developed along a 5-km-long arcuate graben, within and slightly outside the area of uplift.

DECEPTION ISLAND (continued)

HISTORICAL ACTIVITY (continued)

Roobol (1979) proposed five intrusive episodes culminating in eruptions since 1820, based on the historical record and particularly on information about areas of heat escape and eruption vents. These episodes were: 1. Mount Pond, 1829 (possible Pendulum Cove eruption); 2. Mount Kirkwood, 1839-42 (Mount Kirkwood fissure eruptions); 3. Pendulum Cove - Mount Pond - Whalers Bay sector, 1908-60 (1969 fissure eruption); 4. Telefon Bay, ca. 1956-70 (1967 and 1970 eruptions); and 5. Wensleydale Beacon, post-1962 to 1970 (1970 eruptions). Each of these intrusions emplaced a pod of magma at shallow depth, each in a slightly different state of chemical evolution.

1987: A plume was seen on satellite images on 23 July, and "recent, local" seismic activity was recorded at a station on King George Island, 100 km southeast of Deception Island (Smithsonian Institution, 1987).

COMMENTS

Deception Island is of special interest because on at least three occasions, eruptions have occurred simultaneously from multiple vents around a ring fracture. Because leaks of magma along ring structures are known to have preceded some large, prehistoric caldera-forming eruptions (see Bacon, 1985), we might expect that simultaneous eruptions from multiple vents like those at Deception Island could also lead to larger eruptions and caldera collapse, but so far that has not been the case.

REFERENCES

Bacon, C.R., 1985, Implications of silicic vent patterns for the presence of large crustal magma chambers: Jour. Geophys. Res., v. 90, no. B13, p. 11,243-11,252.
Baker, P.E., 1968, Comparative volcanology and petrology of the Atlantic island arcs: Bull. Volcanol., v. 32, p. 189-206.
----- 1969, A volcano erupts beneath the Antarctic ice: Geogr. Mag., v. 41, p. 115-126.
Baker, P.E., Davies, T.G., and Roobol, M.J., 1969, Volcanic activity at Deception Island in 1967 and 1969: Nature, v. 224, p. 553-560.
Baker, P.E., and McReath, I., 1971, Volcanic eruption of Deception Island: Nature, Phys. Sci., v. 231, p. 5-9.
Baker, P.E., McReath, I., Harvey, M.R., Roobol, M.J., and Davies, T.G., 1975, The geology of the South Shetland Islands: V. The volcanic evolution of Deception Island: Brit. Antarctic Surv. Sci. Reps., no. 78, 81 p.
Berninghausen, W.H., and Neumann van Padang, M., 1960, Catalogue of active volcanoes of the world including solfatara fields, Pt. X, Antarctica: Naples, IAVCEI, 32 p.
Circum-Pacific Map Project, Antarctic Panel, 1981, Plate-tectonic map of the Circum-Pacific region, Antarctica sheet, scale 1,10,000,000: Tulsa, Circum-Pacific Council for Energy and Mineral Resources/Am. Assoc. Pet. Geol., 1 sheet.
Clapperton, C., 1969, The volcanic eruption at Deception Island, December 1967: Brit. Ant. Surv. Bull. 22, p. 83-90.

DECEPTION ISLAND (continued)

REFERENCES (continued)

Gonzales-Ferran, O., Munizaga V., F., and Moreno R., H., 1971, Sintesis de la evolucion volcánica de isla Decepcion y la erupción de 1970: Univ. Chile, INACH Ser. Cient., v. 2, no. 1, p. 1-14.

Hawkes, D.D., 1961, The geology of the South Shetland Islands: II, The geology and petrology of Deception Island: Falkland Islands Dependencies Surv. Sci. Reps., no. 27, 43 p.

Klay, J.R., and Orheim, O., 1969, Glaciology and glacial geology on Deception Island: Antarctic Jour., v. 4, no. 4, p. 125-126.

Lorca, E., 1976, Deception Island: Seismic activity prior to the eruption of 1967, in Gonzales-Ferran, O., ed., Proc. Symp. Andean and Antarctic volcanology problems: Naples, IAVCEI, p. 632-645.

Orheim, O., 1972, Volcanic activity on Deception Island, South Shetland Islands, in Adie, R.J., ed., Antarctic geology and geophysics: Oslo, Universsitesforlaget, p. 117-120.

Roobol, M.J., 1973, Historic volcanic activity at Deception Island: Brit. Antarctic Surv. Bull., v. 32, p. 23-30.

----- 1979, A model for the eruptive mechanism of Deception Island from 1820-1970: Brit. Antarctic Surv. Bull., no. 49, p. 137-156.

Shultz, C.H., 1972, Eruption at Deception Island, Antarctica, August 1970: Geol. Soc. Am. Bull. v. 83, p. 2837-2842.

----- 1976, Petrology and petrochemistry of Deception Island, Antarctica, in Gonzales-Ferran, O., ed., Proc. Symp. Andean and Antarctic volcanology problems: Naples, IAVCEI, p. 498-517.

Smithsonian Institution, Scientific Event Alert Network (SEAN), 1987, Deception Island Volcano: SEAN Bull., v. 12, no. 8, p. 18 (information from W. Gould and G. Gutheridge).

Wilkes, C., 1845, Narrative on the United States Exploring Expedition, during the years 1838, 1839, 1840, 1841, and 1842: Philadelphia, Lea and Blanchard, 5 vols.

Williams, H., and McBirney, A.R., 1968, Geologic and geophysical features of calderas: Progress Rep., NASA Grant NGR-38-033-012, 87 p.

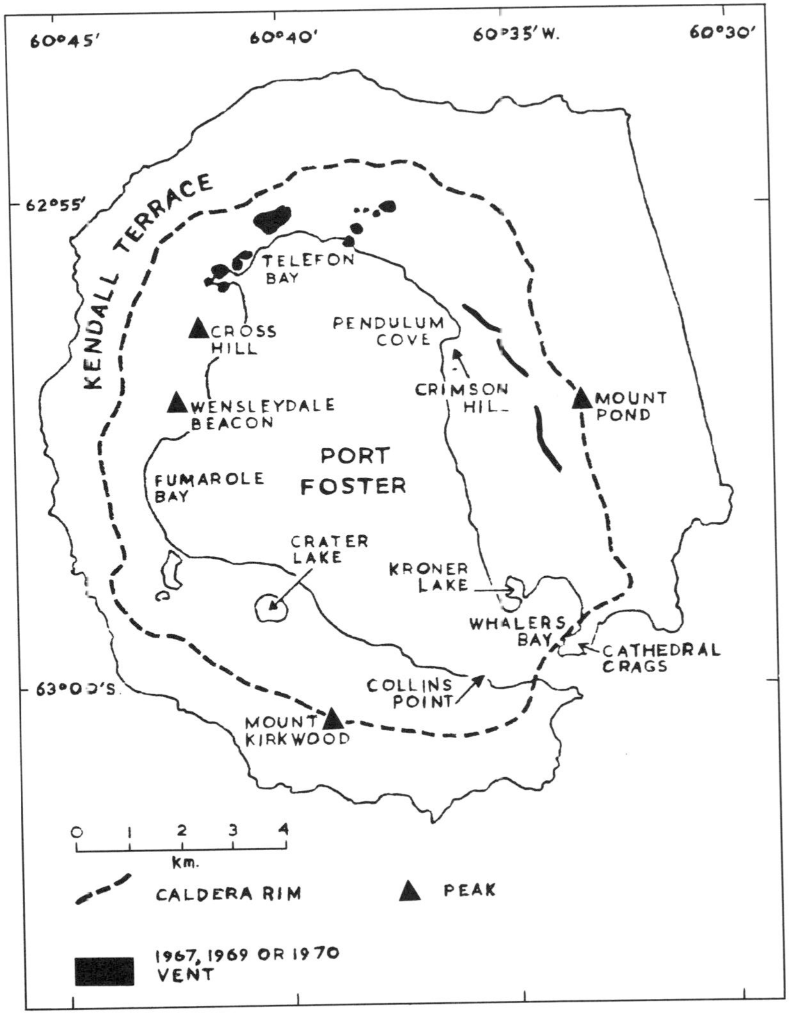

Figure 19.2.1. Location map of Deception Island, showing coastline after the 1970 eruption, from Roobol (1979).

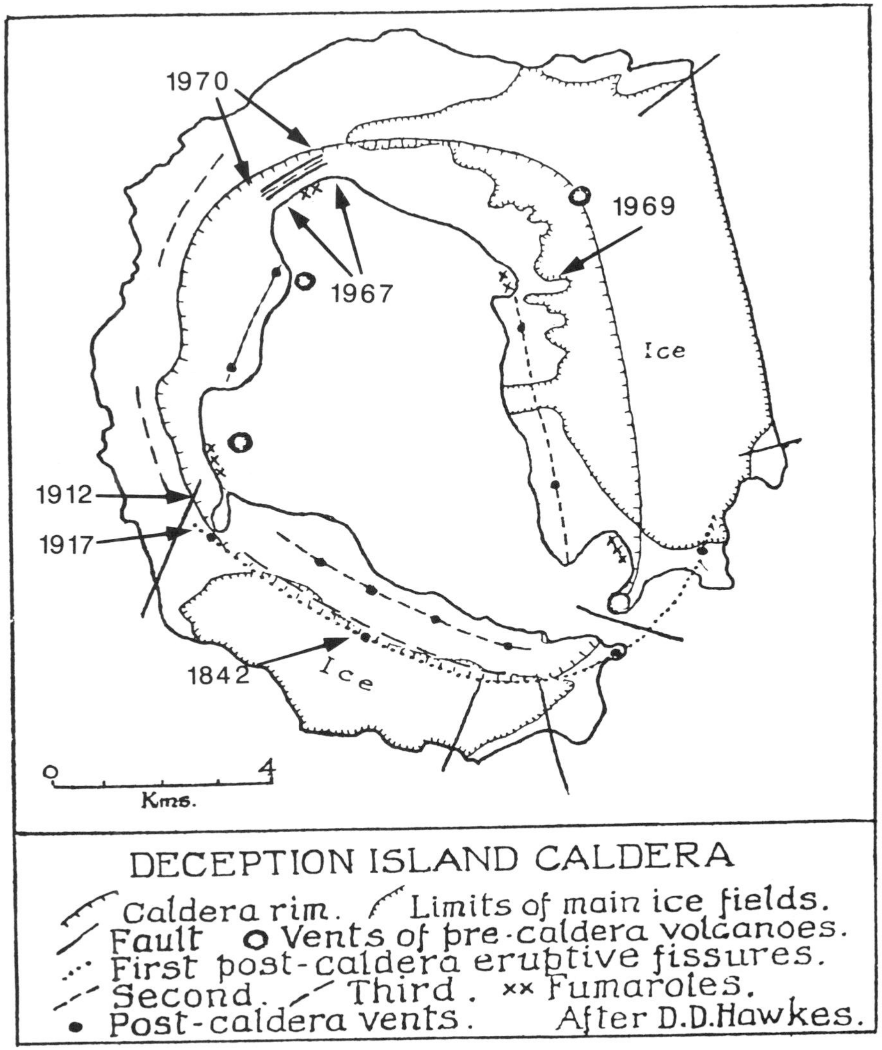

Figure 19.2.2. Sketch map of Deception Island Caldera, from Williams and McBirney (1968), with dates of historical eruptions from Gonzales-Ferran and others (1971).

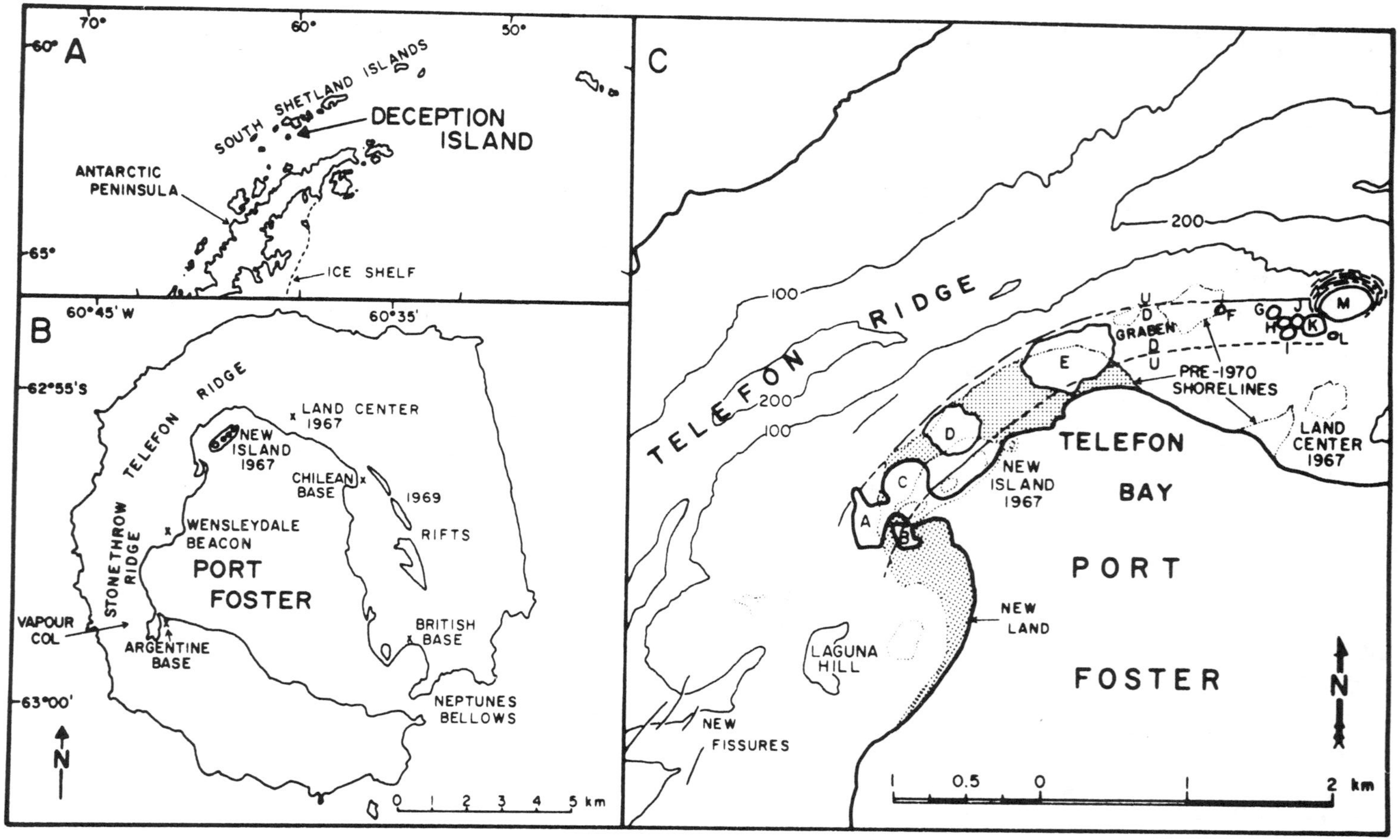

Figure 19.2.3. Deception Island showing recent eruption sites, from Schultz (1972). A-M are craters.

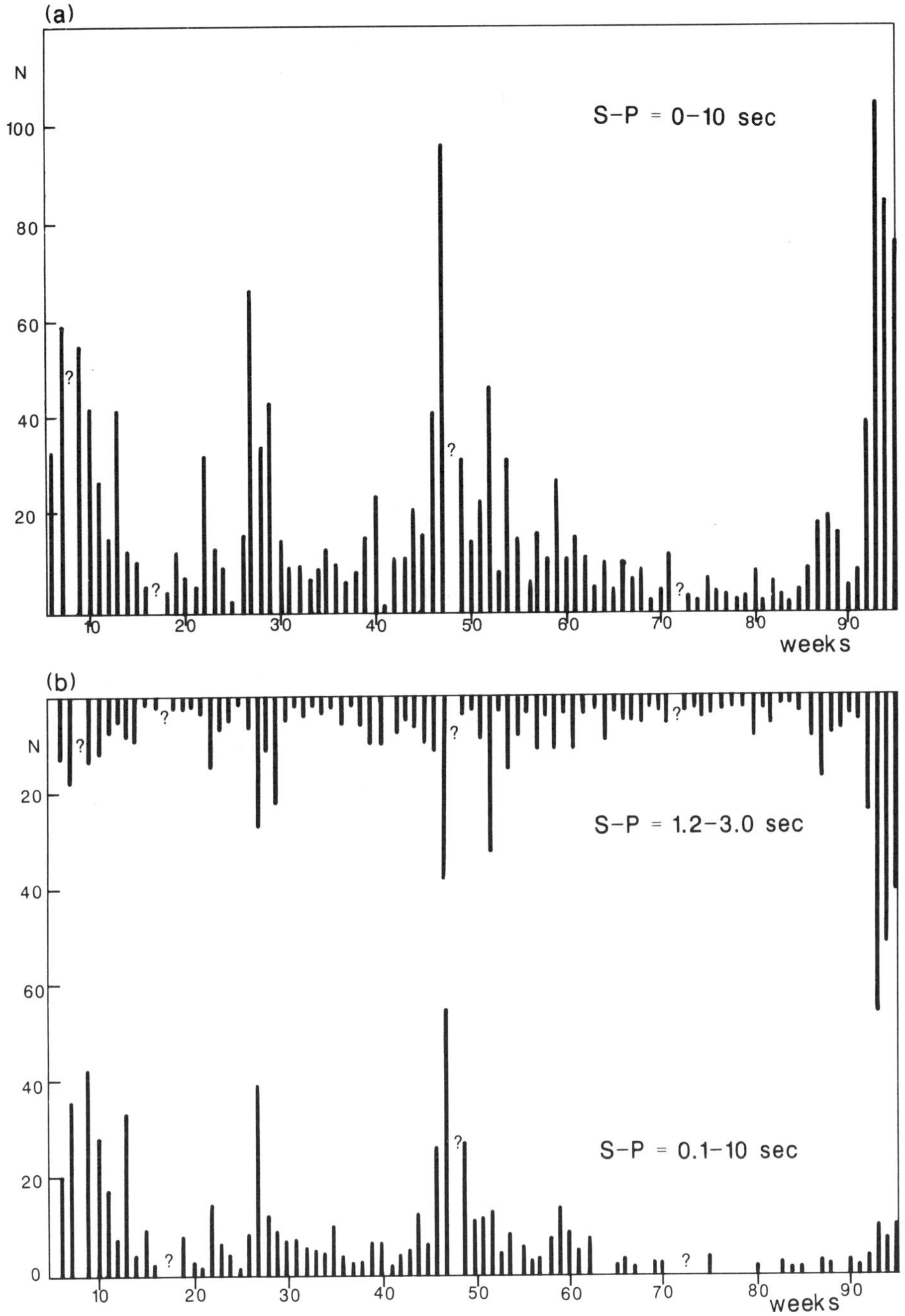

Figure 19.2.4. Seismic precursors to the 1967 eruption of Deception Island, from Lorca (1976). Number of events per week from 16 January 1966 to 30 November 1967 in each of three categories is shown: (a) events with S - P up to 10 s; (b) events with S - P of 0.1-1.0 s (bottom) and 1.2-3.0 s (top). Horizontal axes show number of weeks since the start of recording on 16 January 1966.

PART 4: Historical Unrest in Selected Noncaldera Settings

In Part 4 we demonstrate that unrest resembling that at calderas also occurs in noncaldera (and even nonvolcanic) settings. Several of these episodes have generated lively debate about the processes of unrest. We are persuaded by arguments that unrest at the first four of these locales (Chijiwa, Izu Peninsula, Kii Peninsula, and Matsushiro, all in Japan) has been fundamentally tectonic in origin, but we acknowledge and cite arguments for some involvement of magma in this unrest. Unrest at Socorro, New Mexico, is generally believed to be related to a large, deep magma reservoir in a tectonically active rift, and unrest at Montserrat is included as an example of unrest caused by episodic magma intrusion into a noncaldera volcano.

CHIJIWA

Latitude Longitude (degrees)	Tectonic setting	Age of latest volcanism	Youngest volcanic rock types	Restless years	Type and duration of unrest ESTU STHF MGTF H Te	Associated eruption?
32.7N 130.1E	East-west graben	A.D. 1791-92 1835?	a, d	860	---- ---- ---- - --	Ex?
				1657	---- ---- ---- - --	lf, ex?
				1663-64	---- ---- ---- - --	ex
				1791-92	EE-- ---Y ---Y Y --	ex,lf,LANDSL
				1835?	---- ---- ---- - --	ex?
				1922	CC-- ---- ---- - --	none
				1929-40	FF-- ---- ---- - --	none
				1951-66	FF-- ---- ---- - --	none
				1968-74	FF-- G-G- --F- F --	none
				1977-82	EE-- ---- ---- - --	none
				1984	EE-- ---- ---- - --	none

TECTONIC SETTING

An 8-km-wide E-W graben joins Chijiwa Bay and the central part of Unzen Volcano; recent volcanism in the area has occurred mainly within this graben. A largely submarine caldera has been postulated at the western foot of Unzen Volcano, about 90 km west of Aso Caldera (Ota, 1984) (figs. 20.1.1, 20.1.2). However, no products of a caldera-forming eruption are known, and most workers refer to the graben but not to a caldera. Sugiyama and others (1986) infer that the graben formed between 170,000 and 70,000 yr B.P.

GEOLOGIC HISTORY

Unzen Volcano (CAVW number 08-02-10) consists of a central, active cone (Fugen-dake) and eroded remnants of several older cones including Mayu-yama, a biotite-hornblende andesite dome.

CHIJIWA (continued)

HISTORICAL ACTIVITY

1657: In 1657 an olivine-bearing biotite-hornblende andesite lava flow (the Hato-ana or Koyake flow, formerly called the Furu-yake flow) was extruded from the Fugen-dake central cone of Unzen Volcano (Ogawa and Homma, 1926; Sendo and others, 1967). In the following year, thermal water and a lahar caused approximately 30 fatalities.

1663-64: Small explosions occurred in March 1663 and January 1664, and perhaps for as many as 10 to 12 years thereafter (Ogawa and Homma, 1926).

1791-92: Earthquakes that were precursory to the 1792 eruption of Unzen began on 2 November 1791 and were felt most strongly around the western coast of the Shimabara Peninsula (in the vicinity of Obama and Chijiwa). Shocks were felt 3-4 times daily, and increased in intensity about 6 December. Even though earthquakes were most intense on the west side of the Shimabara Peninsula, landslides occurred on the southeast flank of Mayu-yama. Detonation sounds, described as sounding like the firing of artillery, were heard in late January. Small explosions at the Fugen-dake vent of Unzen began on 9 or 11 February 1792, accompanied by strong earthquakes. Then, on 25 February, the ground at Biwa-no-bachi (1 km from the top of Fugen-dake and 4 km from Shimabara) began to tremble and "shook off earth and rocks" (Ogawa and Homma, 1926). The unrest at Biwa-no-bachi quickly led to extrusion of the Shinyake lava from a nearby flank vent (29 February or 1 March 1792; hypersthene biotite-hornblende dacite, 4-km-long flow, volume=0.11 km^3). On 21 March, another vent opened near the top of Fugen-dake, followed on 24 March by opening of seven more vents about 200 m northwest of the first. All were connected underground with the crater on Anasako-dani, and for this reason the top of Hando-iwa developed a rift 1-5 m wide (Ogawa and Homma, 1926, p. 13).

All the while that lava was flowing from vents on Fugen-dake, earthquakes continued. The strongest shock yet occurred on 21 April, with roaring and echoing noises from Mayu-yama "like the sound of guns on a Dutch frigate" (Ogawa and Homma, 1926). At least 300 earthquakes occurred on 21-22 April; $\underline{M}_{max}$ for 1792 (exact date unspecified) was 6.4. "Shaken by the shocks, masses of rocks and debris fell down from Mayu-yama and set the trees below on fire through the heat of their friction. The shocks were very severe and frequent for three days, more than 300 being counted during the night and more than 100 in each of the next two days. They were most intense from midnight to the morning of the second day... however, the earthquakes subsided gradually after the third day, although another severe shock occurred on the 29th of April" (Ogawa and Homma, 1926, p. 14). These earthquakes of late April caused a landslide near the village of Nakakoba, destroyed 61 houses and structures and 288 stables, and damaged the stone walls of Shimabara castle. Crevices formed in several places between Shimabara and Antoku, including two trending NW-SE and about 1 km long that crossed the castle ground. These crevices were at first only a few centimeters wide, but they later widened with increasing intensity of the shocks to more than 30 cm. The ground on the south side of these sank, and "springs gushed forth on the east of the castle close to the shore, while on the west up the hill beyond the castle, the fountain-heads dried up for a time" (Ogawa and Homma, 1926, p. 14). A larger crack, at least 1 m wide, opened from Iwomi-dake on the southern slope of Azuma-dake to the northern flank of Fugen-dake. This and parallel cracks trended E-W along the old Chijiwa fault scarp, south of Azuma-dake.

CHIJIWA (continued)

HISTORICAL ACTIVITY (continued)

On 29 April, a block from near the east foot of Mayu-yama reportedly slid about 200 m eastward (Katayama, 1974). Toward the middle of May, lava extrusion from Fugen-dake and earthquakes diminished, and people returned to their houses in Shimabara. But at about 2000 hr on 21 May, severe shocks resumed, accompanied or followed shortly by eastward collapse of a large portion of Mayu-yama (volume, 0.34 km^3). The resulting debris avalanche swept through two small villages and into the sea, causing a devastating tsunami. More than 14,000 people died, most from the tsunami. Kuno (1962) reports some difference of opinion as to whether the avalanche resulted from an explosion or was triggered by earthquakes; most workers have concluded that there were no major explosions, and that earthquakes were the immediate trigger of the avalanche. Katayama (1974), adapting a hydrologic model of Ota (1973), suggested that an influx of hydrothermal water from Chijiwa into Mayu-yama made the edifice unstable (related discussion below).

Two more eruptions followed from Fugen-dake--explosions on 19 July and 15 August--with lesser activity continuing until at least April 1793.

1922: A major earthquake swarm in December 1922 (fig. 20.1.3) caused 27 deaths, most on 8 December 1922. The largest event of this swarm, estimated to be $\underline{M}$=6.0, was preceded by a smaller event on 3 October and by several on 7 December in the 9.5 hours before the largest event (Ogawa and Homma, 1926). At least 1798 microseisms and 110 macroseisms were recorded at Nagasaki up to 13 January 1923. Omori (1928) considered these earthquakes to be volcanic, whereas Imamura (1928a, b) and others considered them to be tectonic. All workers agreed that they were centered beneath Chijiwa Bay.

1968-74: Several earthquake swarms occurred from 1968 to 1974, especially during 1970 and 1973-74 (figs. 20.1.4-20.1.8). New fumaroles appeared at the end of 1974(?). The $\underline{b}$-values for various swarms of this period ranged from 0.6 to 1.3. Hypocenters were approximately 10 km deep beneath the postulated caldera and progressively shallower to the east, reaching depths of 0-2 km beneath Unzen Volcano. Ota (1973) concluded that epicenters migrated upward and eastward, from 8 km to less than 5 km (possibly less than 1 km), during a swarm in June 1970, over a period of 3 weeks and a horizontal distance of 10-15 km. The sequence was not perfect, however, because the last earthquakes in the swarm were about 10 km deep at the western end of the supposed eastward migration (figs. 20.1.6, 20.1.7). A relocation of those hypocenters and a study of their first motions (Yamashina and Mitsunami, 1977) suggests movement along an E-W-trending, south-dipping normal fault (fig. 20.1.8).

The temperature and composition of hot springs change eastward toward Unzen and Shimabara Hot Springs (on the eastern flank of Unzen). Ota (1973) correlated these changes with the inferred eastward shallowing of earthquake hypocenters and suggested that the earthquakes were caused at least in part by migration of hydrothermal fluids from the postulated caldera eastward and upward to Unzen and Shimabara Hot Springs. Observation of this migration suggested to Katayama (1974) and Ota (1974) that the large 1792 debris avalanche from Mayu-yama occurred after a similar migration of hydrothermal fluids destabilized the east flank of Unzen.

1977-82: Earthquake swarms occurred in November 1977, January and December 1978, June and September 1979, and August and November 1980 (figs. 20.1.9-20.1.12). One shock in 1980 was felt nearby with JMA intensity 4.

CHIJIWA, Latitude 32.7N, Longitude 130.1E

CHIJIWA (continued)

HISTORICAL ACTIVITY (continued)

1984: Seismicity started increasing in May 1984 and climaxed in August with a swarm of 6,307 events, 409 of which were felt ($\underline{M}_{max}$=5.7, depth ca. 7 km, deepening from east to west; figs. 20.1.13-20.1.16) (Smithsonian Institution, 1984).

The east coast of the Shimabara Peninsula subsided 10-15 cm between leveling surveys in 1894 and 1958 (fig. 20.1.17). The net(?) horizontal strain across the Chijiwa area between 1890 and 1958 was approximately 10^{-5} (N-S extension--consistent with historical seismicity and mapped faults; fig. 20.1.18) (Yamashina and Mitsunami, 1977).

COMMENTS

Historical unrest at Chijiwa is a manifestation of active graben development superimposed on related volcanic activity at Unzen. The collapse of Mayu-yama in 1792 is probably an example of a large and devastating debris avalanche caused by graben faulting in conjunction with an eruption at a nearby vent.

REFERENCES

Homma, F., 1936, Unzen-dake: Bull. Volcanol. Soc. Japan, v. 3, p. 71-124.

Imamura, A., 1928a, Report on the changes in the land-level in connection with the Simabara earthquake of 1922: Bull. Imperial Earthq. Invest. Comm., v. 10, no. 2, p. 63-64.

----- 1928b, Seismographic study of the Simabara earthquake: Bull. Imperial Earthq. Invest. Comm., v. 10, no. 2, p. 66-70.

Katayama, N., 1974, Old records of natural phenomena concerning the "Shimabara catastrophe": Sci. Rep. Shimabara Volcano Obs., Kyushu Univ., no. 9, p. 1-45.

Komada, I., 1916, Report on geology of Onsendake (Unzen) Volcano: Rep. Imperial Earthq. Invest. Comm., v. 84, p. 1-102.

----- 1924, Geological considerations of the Shimabara earthquakes of December 8, 1922: Japanese Jour. Geol. Geogr., v. 3, p. 91-100.

Kubotera, A., 1975, Relation among earthquake swarm, gravity anomalies, and geothermal activities at Shimabara Peninsula: Jour. Japan Geotherm. Assoc., v. 12, no. 2, p. 9-13.

Kubotera, A., and Kikuchi, S., 1967, Seismic observation in Shimabara Peninsula--on the activity of Volcano Unzen: Spec. Contrib., Geophys. Inst., Kyoto Univ., no. 7, p. 161-168.

Kuno, H., 1962, Catalogue of active volcanoes of the world including solfatara fields, Pt. XI, Japan, Taiwan, and Marianas: Rome, IAVCEI, 332 p.

Matsuwo, N., 1979, Seismicity in and near the Shimabara Peninsula, April 1977-June 1978: Sci. Rep. Shimabara Volcano Obs., Kyushu Univ., no. 12, p. 17-28.

Mogi, K., 1963, Some discussion on aftershocks, foreshocks and earthquake swarms--the fracture of a semi-infinite body caused by an inner stress origin and its relation to the earthquake phenomena (Third paper): Bull. Earthq. Res. Inst., v. 41, p. 615-658.

CHIJIWA, Latitude 32.7N, Longitude 130.1E

CHIJIWA (continued)

REFERENCES (continued)

Ogawa, T., and Homma, F., 1926, Unzen Volcano: Guidebook for Excursion E-1,3,4, 3rd Pan-Pacific Sci. Congr., Tokyo, p. 1-30.

Omori, F., 1907, Note on the eruptions of the Unsen-dake in the 4th year of Kansei (1792): Bull. Tokyo Imperial Earthq. Invest. Comm., v. 1, p. 142-144.

----- 1928, On the Shimabara Earthquake of Dec. 8, 1922: Bull. Imperial Earthq. Invest. Comm., v. 10, no. 2, p. 43-62.

Ota, K., 1969, Study on the collapses in the Mayu-yama: 1. On the mechanism of collapse: Sci. Rep. Shimabara Volcano Obs., Kyushu Univ., p. 6-35.

----- 1970, On the ground water in the northern coast area of the Shimabara Peninsula: Sci. Rep. Shimabara Volcano Obs., Kyushu Univ., p. 14 ff.

----- 1971, Ground water in Chijiwa town and its adjacent area, the Shimabara Peninsula: Sci. Rep. Shimabara Volcano Obs., Kyushu Univ., p. 19 ff.

----- 1973, A study of hot springs on the Shimabara Peninsula: Sci. Rep. Shimabara Volcano Obs., Kyushu Univ., no. 8., p. 1-33.

----- 1974, Observation of volcanic hot-spring waters in the Unzen Volcano (I)-- An attempt for the prediction of volcanic eruption by monitoring the water qualities of volcanic hot-springs and the observation well of the Shimabara Volcano Observatory: Sci. Rep. Shimabara Volcano Obs., Kyushu Univ., no. 10, p. 21-33.

----- 1984, Unzen Volcano, geomorphology, geology and volcanic phenomena: Nagasaki Prefecture, 98 p.

Sawada, Yosihiro, 1978, Seismic activity at Unzen-dake Volcano and the unusual number of its occurrence frequency: Pap. Meteorol. Geophys., v. 29, no. 2, p. 83-96.

Sendo, T., Matsumoto, H., and Imamura, R., 1967, Geology and petrography of Unzen Volcano: Kumamoto Jour. Sci., ser. B, sec. 1, v. 7, p. 31-89.

Shimabara Earthquake and Volcano Observatory, Kyushu University, 1985, On the earthquakes (that) occurred on 6 August 1984 at Unzen Volcano, Shimabara Peninsula: Rep. Coord. Comm. for Prediction of Volcanic Eruptions, no. 32, p. 37-41.

Shimabara Volcano Observatory, Kyushu University, 1983, Seismic activity of Unzen Volcano in 1981 and 1982: Rep. Coord. Comm. for Prediction of Volcanic Eruptions, no. 28, p. 42-43.

Siebert, L., Glicken, H., and Ui, T., 1987, Volcanic hazards from Bezymianny- and Bandai-type eruptions: Bull. Volcanol., v. 49, p. 435-459.

Smithsonian Institution, Scientific Event Alert Network (SEAN), 1984, Unzen Volcano: SEAN Bull., v. 9, no. 9, p. 6-7.

Sugiyama, H., Hayashi, M., and Fujino, T., 1986, Fission track ages of the Unzen volcanic rocks in Western Kyushu, Japan: Bull. Volcanol. Soc. Japan, v. 31, no. 2, p. 85-94.

Volcanological Society of Japan, 1982, Unzen: Bull. Volcanic Eruptions, no. 20, p. 95.

Yamashina, K., and Mitsunami, T., 1977, Stress field in the Unzen volcanic area, Kyushu, Japan; Bull. Volcanol. Soc. Japan, v. 22, p. 13-25.

CHIJIWA, Latitude 32.7N, Longitude 130.1E

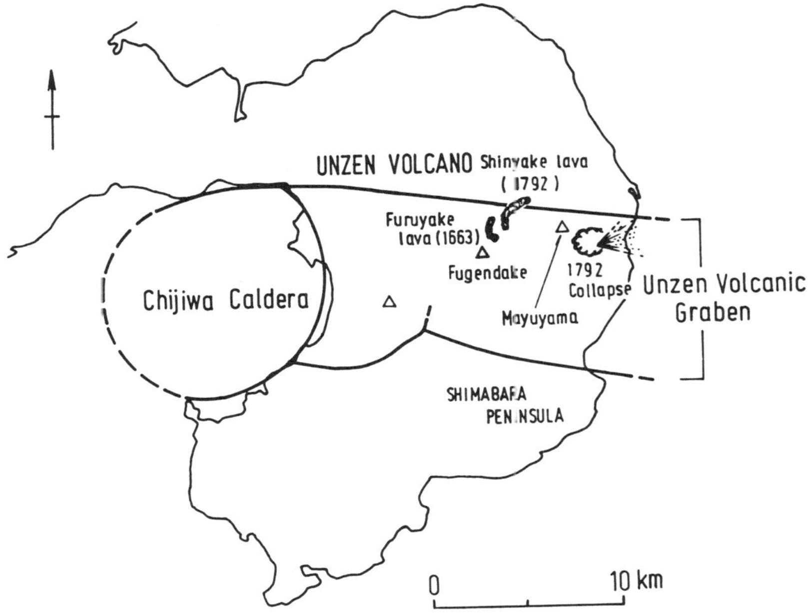

Figure 20.1.1. Locations of postulated Chijiwa Caldera and Unzen Volcano, courtesy of an anonymous reviewer. Most authors refer to Chijiwa graben but not to Chijiwa Caldera.

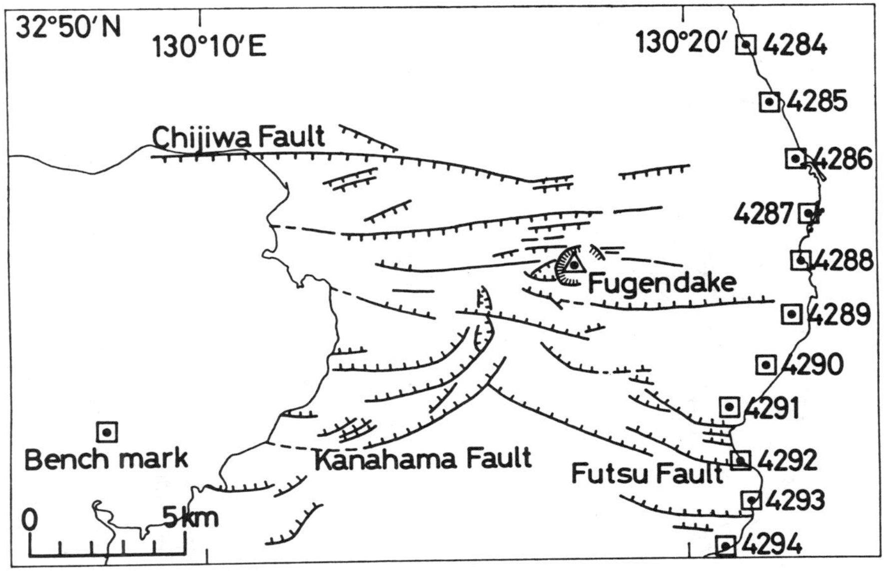

Figure 20.1.2. Active faults in Chijiwa area, from Yamashina and Mitsunami (1977). Numbers at right indicate bench marks (see figure 20.1.17).

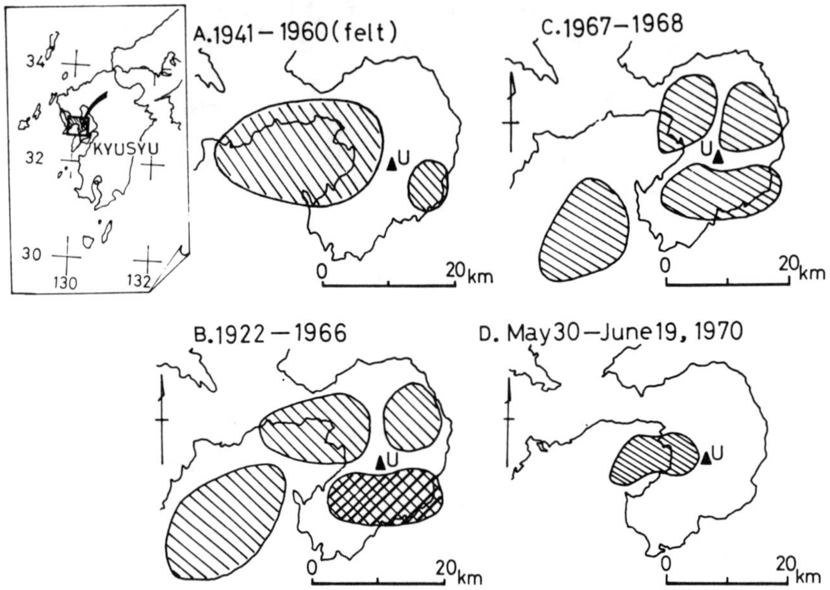

Figure 20.1.3. Epicentral areas near Unzen Volcano, from Sawada (1978). A, Felt earthquakes from 1941 through 1960. B, 1922 through 1966. C, 1967 through 1968. D, May-June 1970. U, Unzen Volcano. Cross hatching in B represents epicentral area of 1922 swarm, according to Omori (1928); Imamura (1928a,b) concluded that largest quakes of the 1922 swarm were centered northwest of area shown by Omori (1928), near head of Chijiwa Bay.

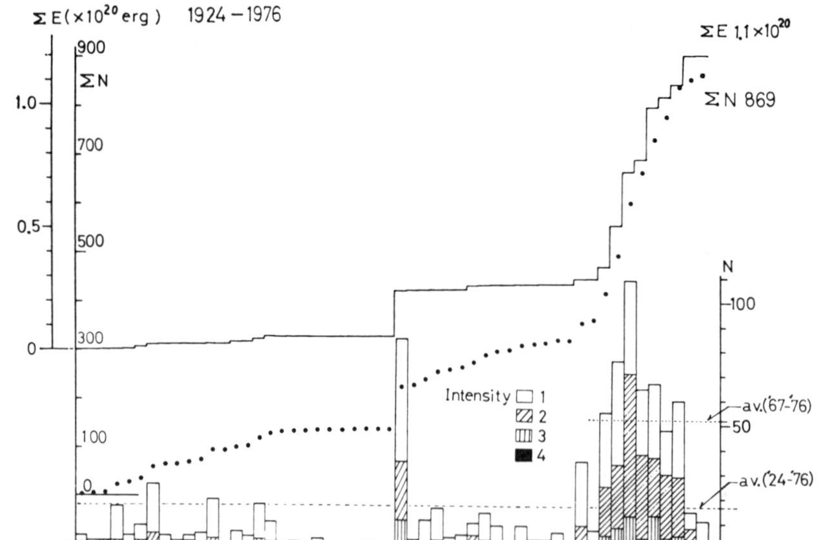

Figure 20.1.4. Seismicity in Chijiwa-Unzen area during 1924-76, from Sawada (1978). Bars, yearly number of felt earthquakes of indicated seismic intensities; dotted line, cumulative number of earthquakes; solid line, cumulative seismic energy release.

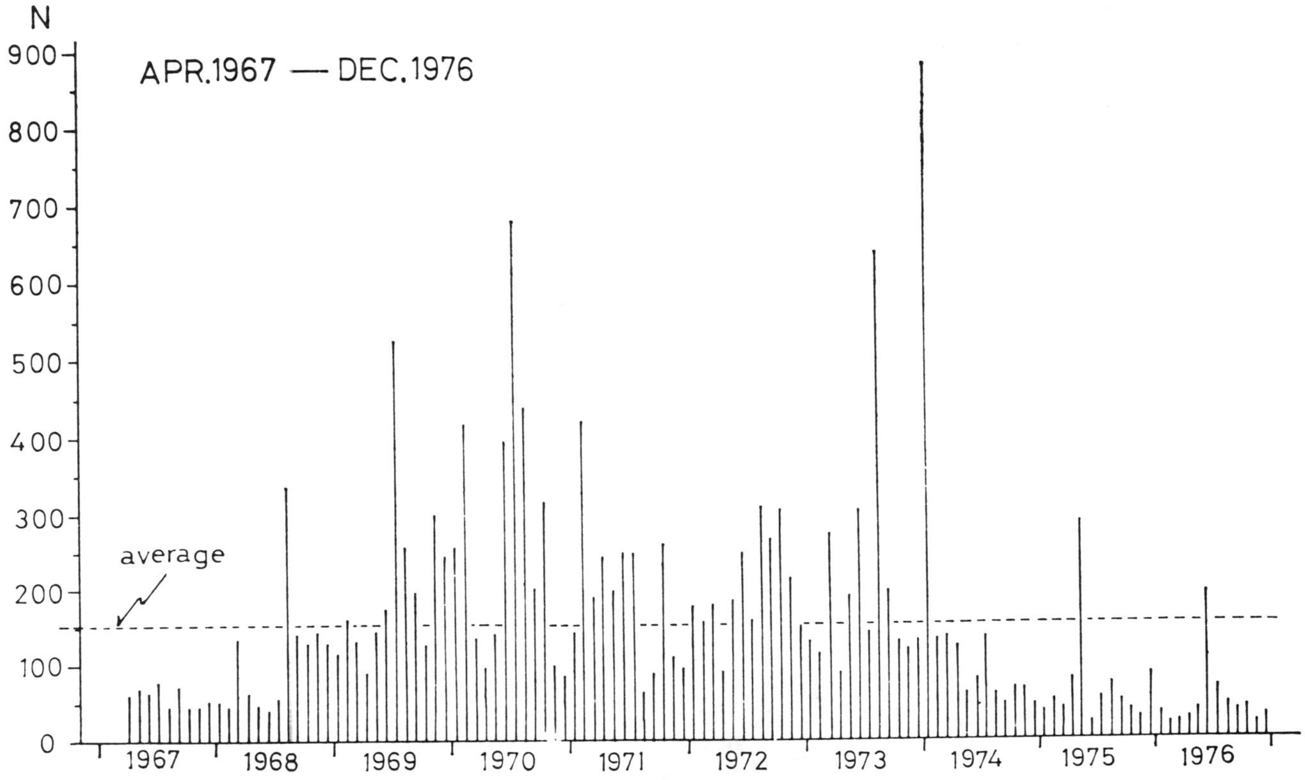

Figure 20.1.5. Monthly number of earthquakes near Unzen Volcano from April 1967 through December 1976, from Sawada (1978).

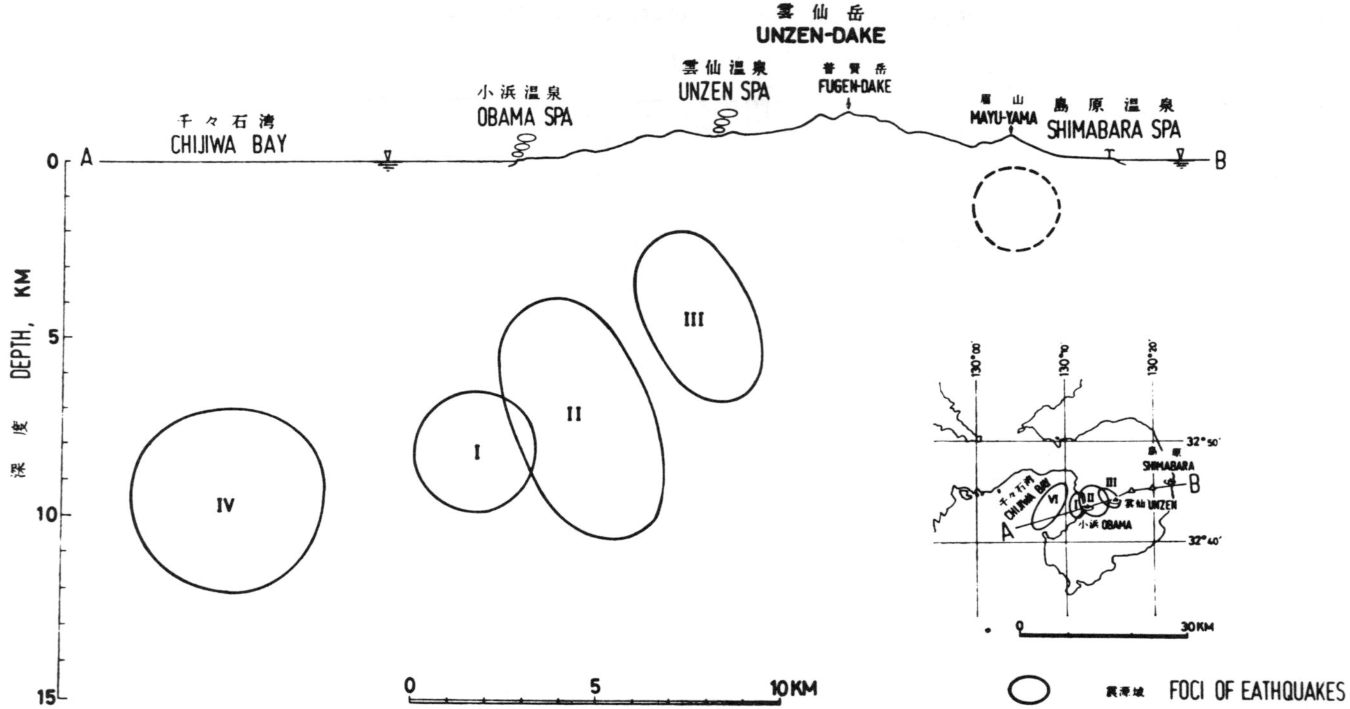

Figure 20.1.6. Distribution of hypocenters during an earthquake swarm in Chijiwa region during June 1970, from Ota (1973). Hypocenters initially migrated upward and eastward (from beneath Chijiwa Caldera toward Unzen Volcano), but last hypocenters were again beneath Chijiwa. Note revised locations shown in figure 20.1.8.

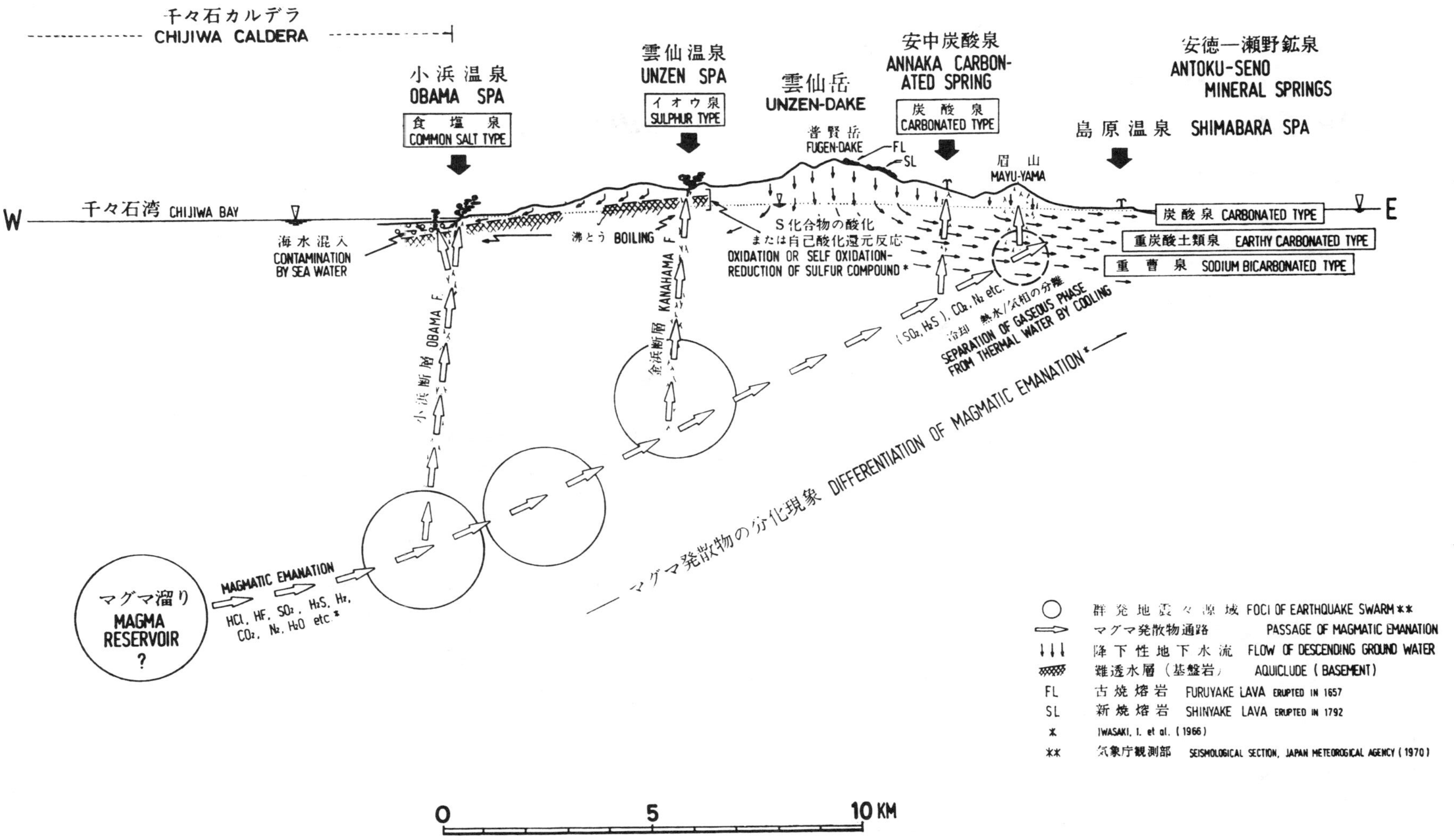

Figure 20.1.7. Schematic diagram showing a mechanism for earthquake migration resulting from fluid migration, from Ota (1973).

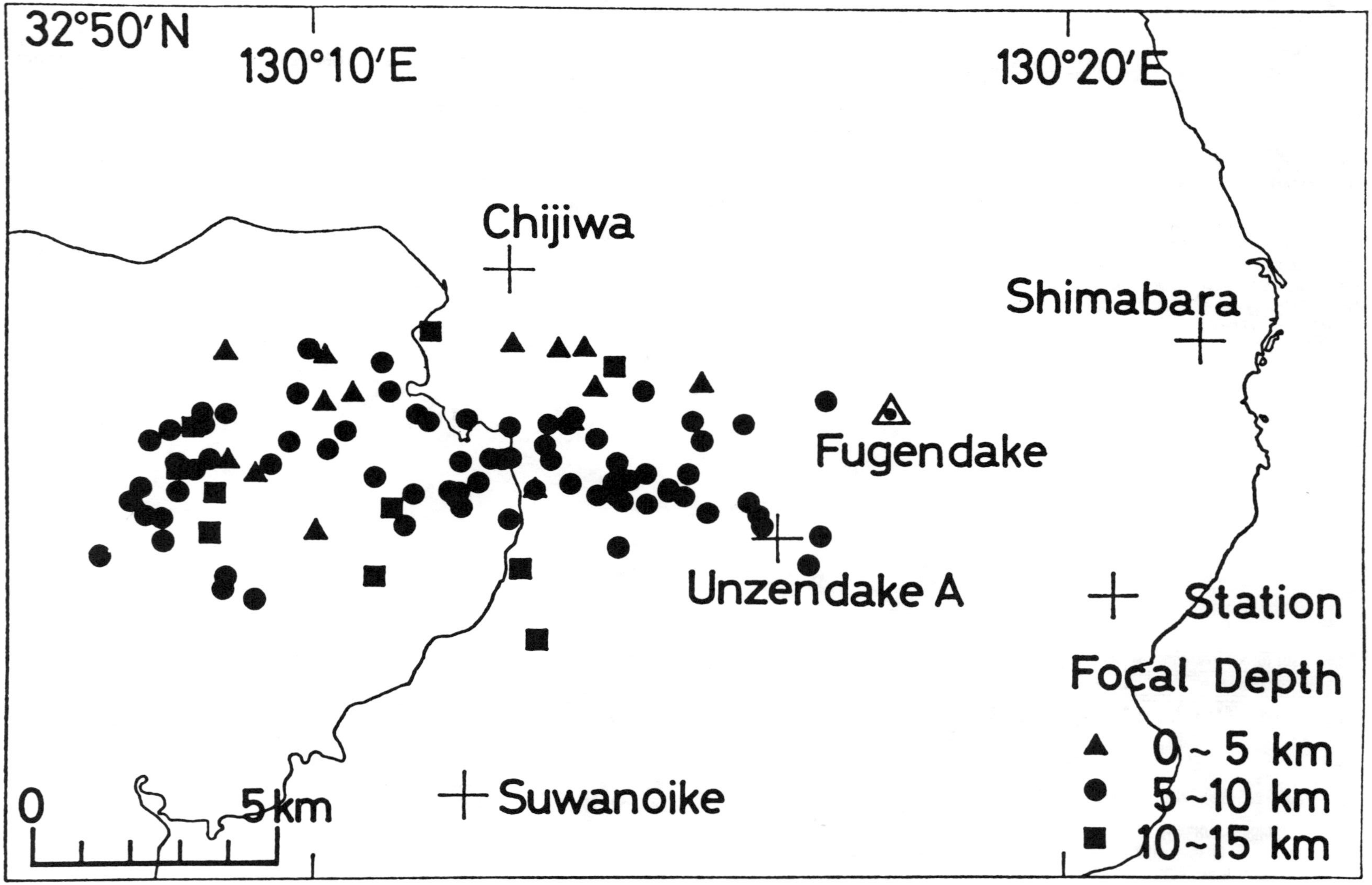

Figure 20.1.8. Relocated epicenters for same earthquakes as shown in figure 20.1.6, from Yamashina and Mitsunami (1977). No systematic migration pattern is obvious here.

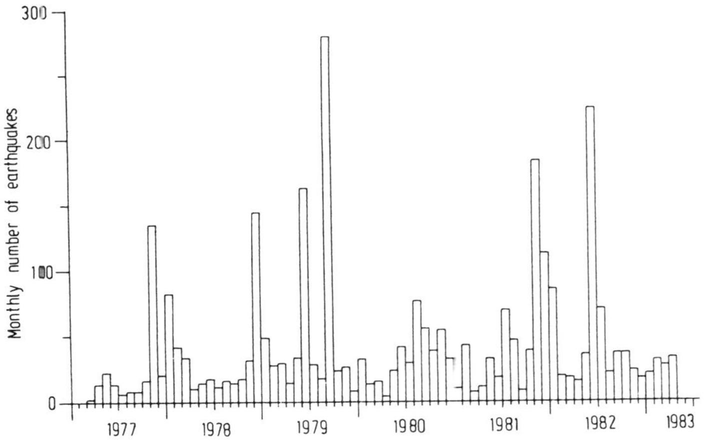

Figure 20.1.9. Monthly number of earthquakes at Unzen Volcano from March 1977 to April 1983, from Shimabara Earthquake and Volcano Observatory (1985).

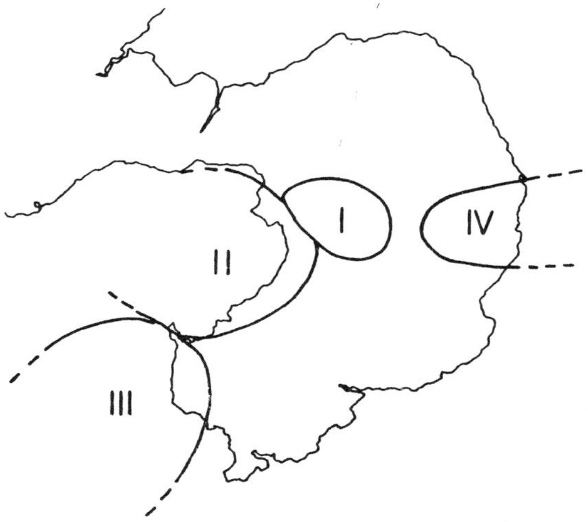

Figure 20.1.10. Epicentral zones for earthquakes in Chijiwa area from April 1977 to June 1978, from Matsuwo (1979).

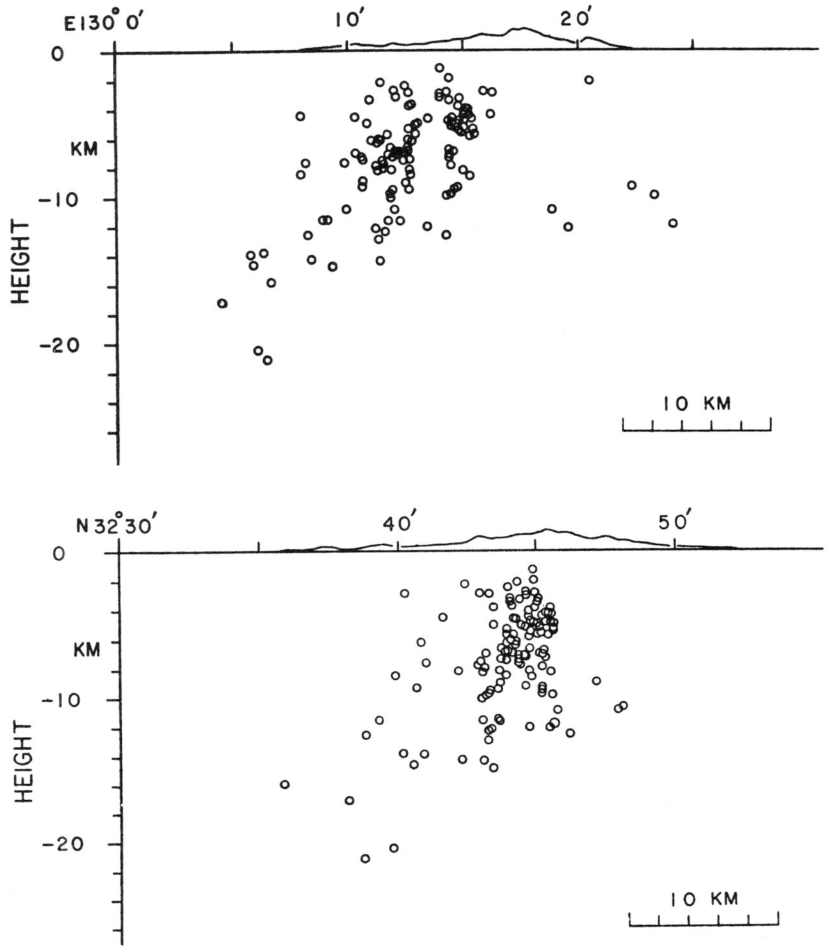

Figure 20.1.11. North-south (top) and east-west (bottom) vertical cross sections showing hypocenters of earthquakes in Chijiwa area from April 1977 to June 1978, from Matsuwo (1979).

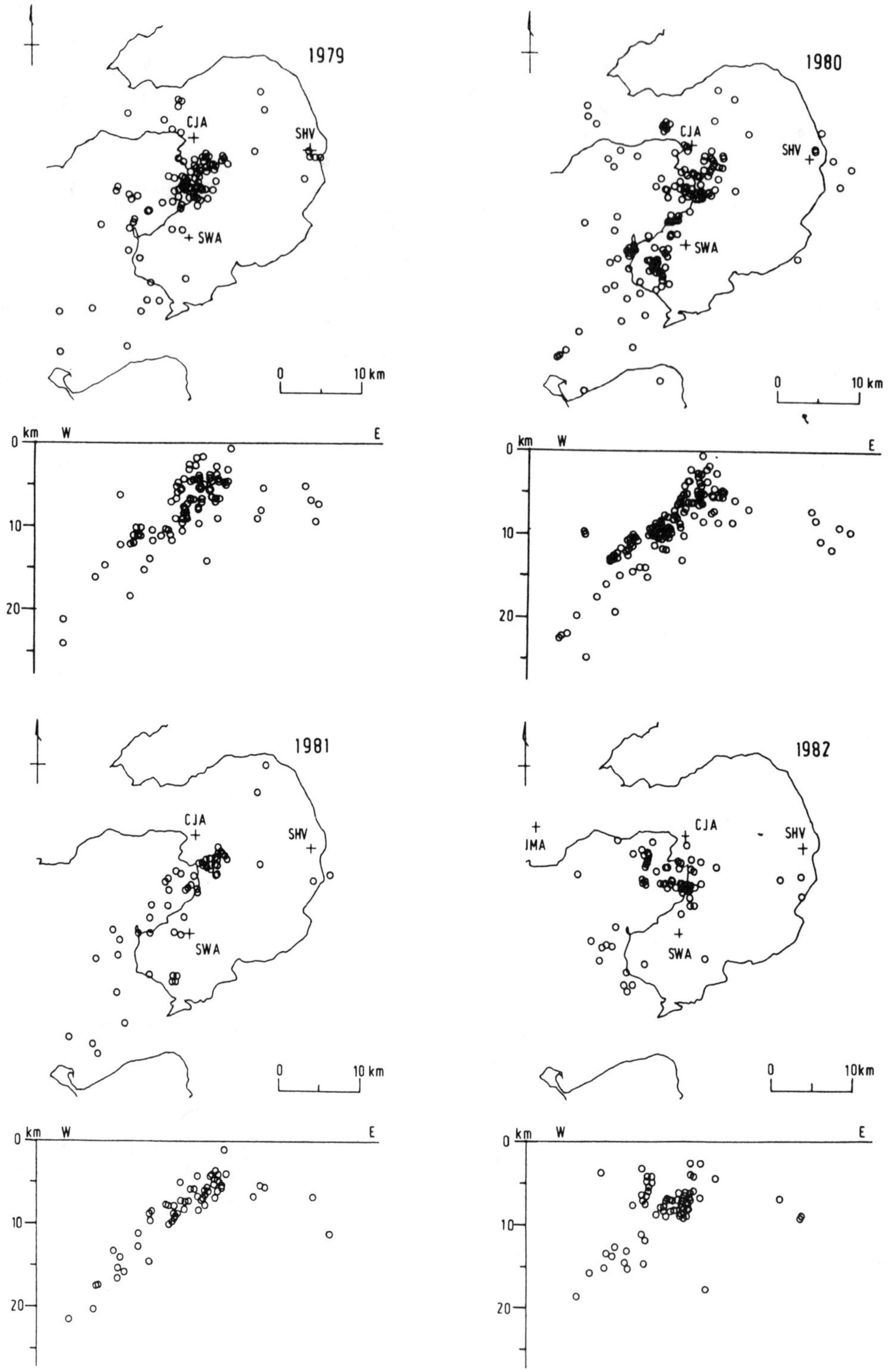

Figure 20.1.12. Epicentral distributions and vertical sections showing earthquakes in Chijiwa area from 1979 to 1982, from Ota (1984).

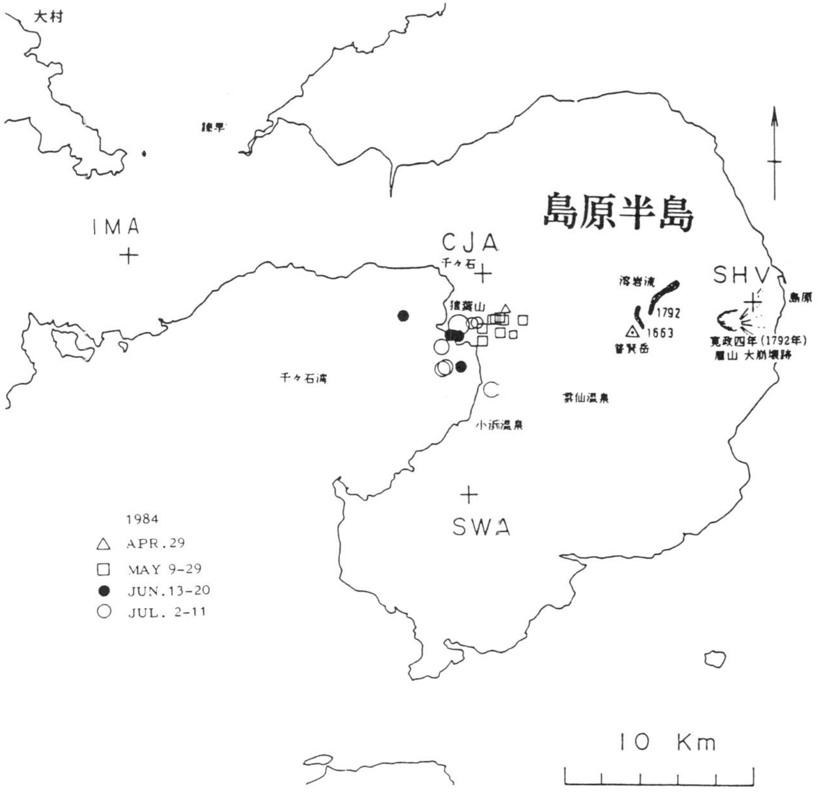

Figure 20.1.13. Epicenters of a series of foreshocks from 29 April to 11 July 1984 that preceded a swarm at Unzen Volcano in August 1984 (Shimabara Earthquake and Volcano Observatory, 1985). +, seismometer station; SHV, Shimbara; CJA, Chijiwa; SWA, Obama; IMA, Iimori.

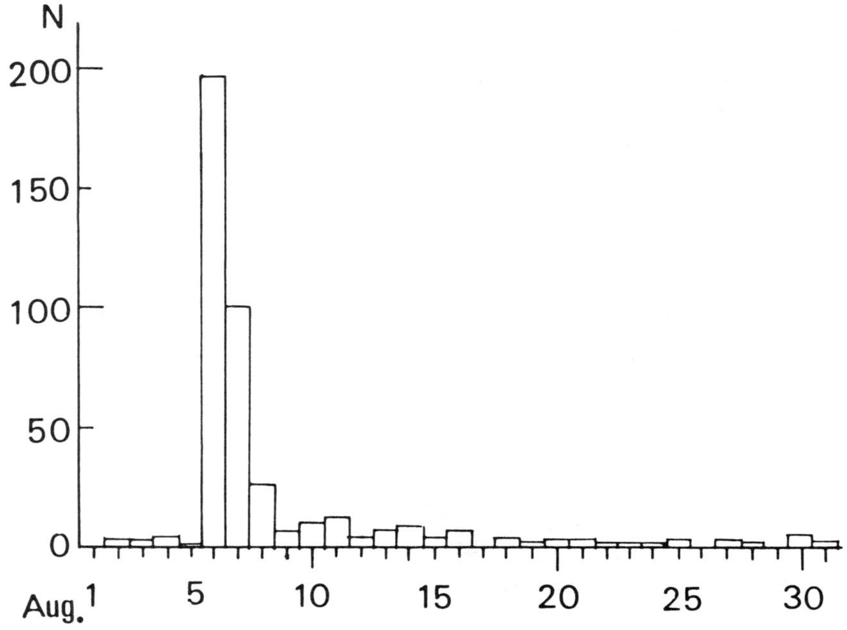

Figure 20.1.14. Daily number of felt earthquakes during a swarm at Unzen Volcano in August 1984, from Smithsonian Institution (1984).

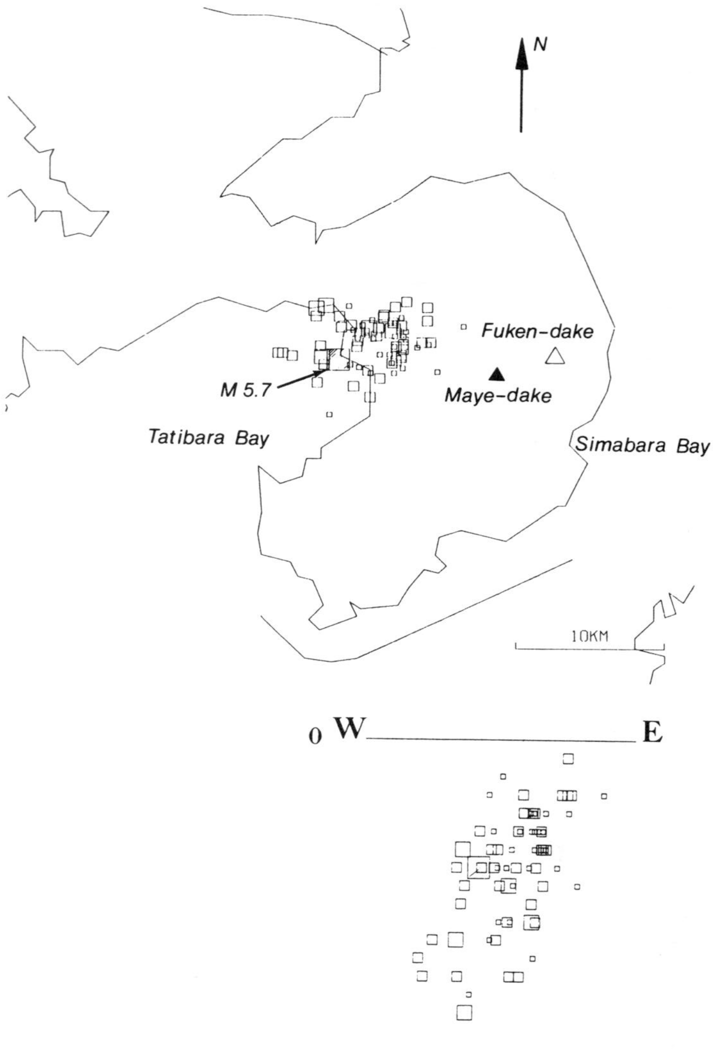

Figure 20.1.15. Epicenters and vertical section showing earthquakes at Unzen Volcano in August 1984, from Smithsonian Institution (1984).

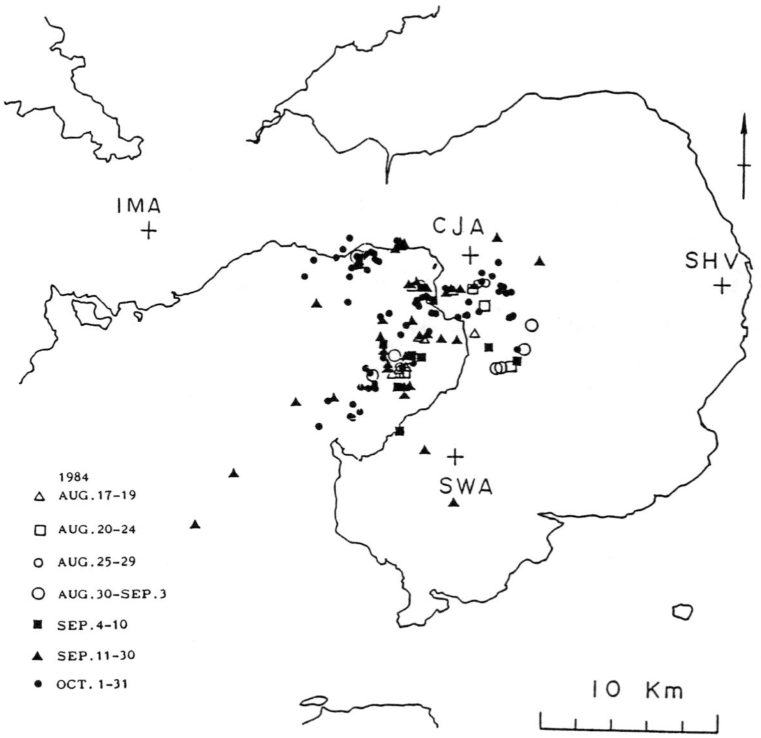

Figure 20.1.16. Epicenters of aftershocks near Unzen Volcano from 17 August 1984 to 31 October 1984, from Shimabara Earthquake and Volcano Observatory (1985).

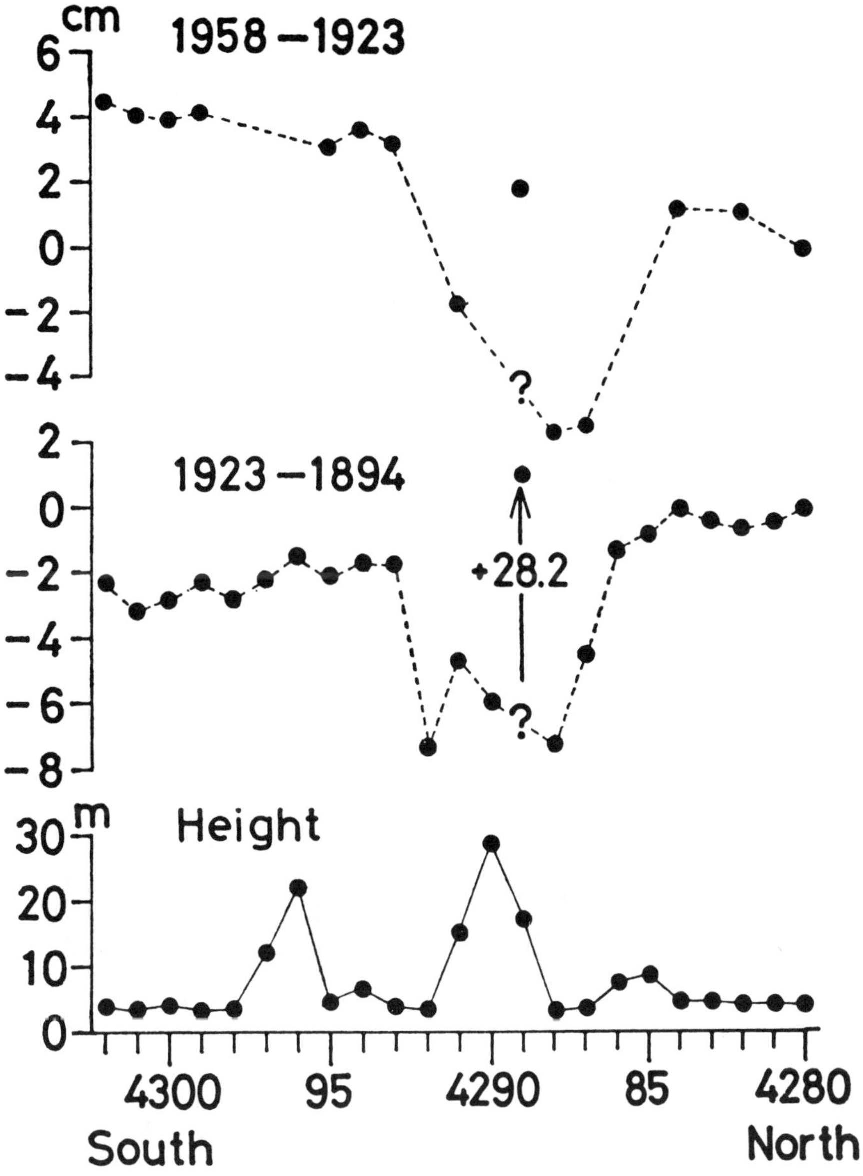

Figure 20.1.17. Leveling data showing subsidence in E-W-trending graben that crosses Shimabara Peninsula, from Yamashina and Mitsunami (1977). Bench-mark locations are shown in figure 20.1.2.

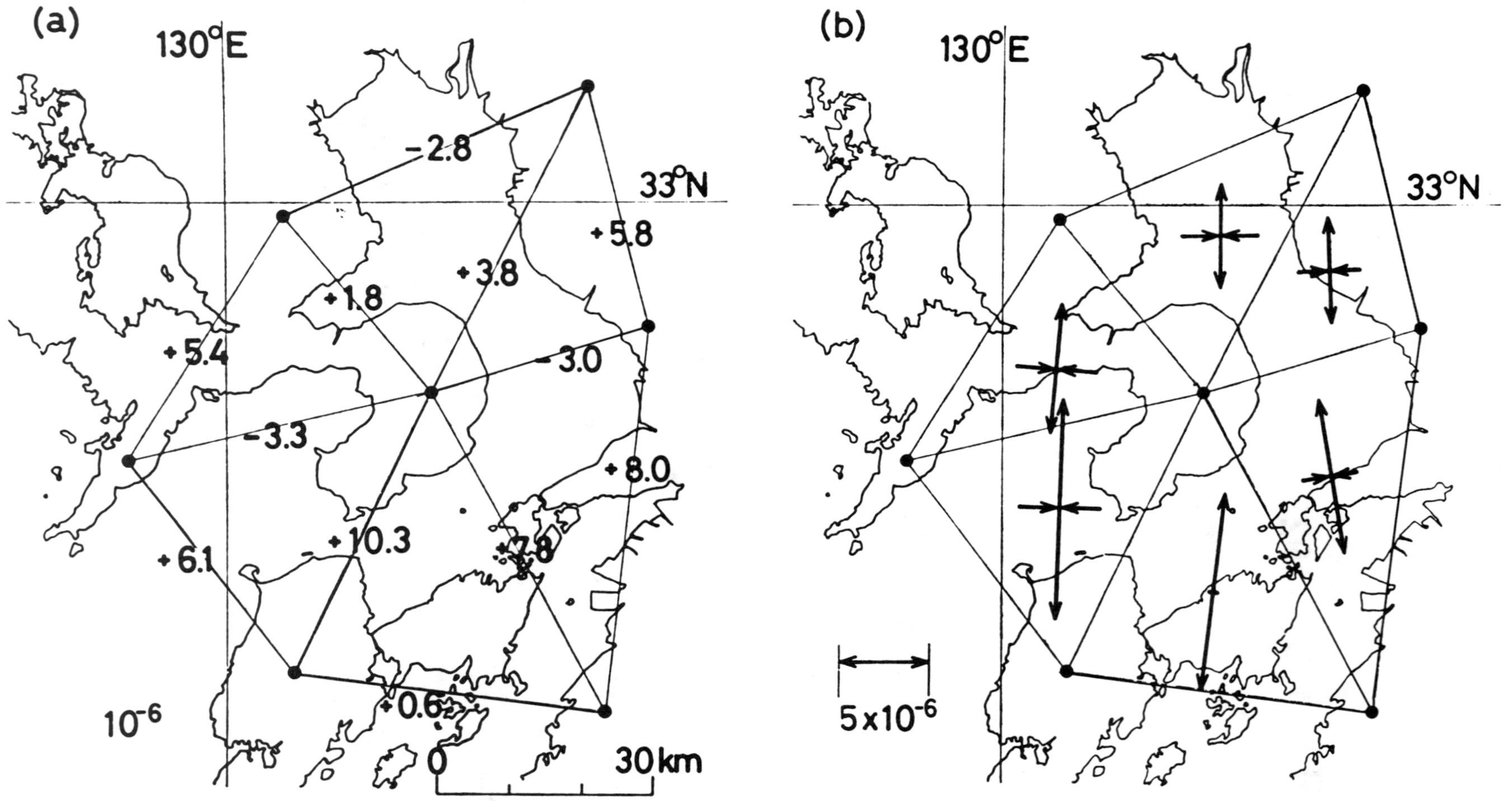

Figure 20.1.18. Horizontal strain (left, in units of 10^{-6}) and principal axes of horizontal strain (right) in Shimabara Peninsula from 1890 to 1958, from Yamashina and Mitsunami (1977). Positive values, extension; negative values, contraction. North-south extension shown here and subsidence in graben (from figure 20.1.17) suggest that dominant cause of repeated earthquake swarms in area is rifting (that is, graben formation).

IZU PENINSULA

Latitude Longitude (degrees)	Tectonic setting	Age of latest volcanism	Youngest volcanic rock types	Restless years	Type and duration of unrest ESTU STHF MGTF H Te	Associated eruption?
34-35N 139E	Compr	Late Pleistocene	b, a, d	1905	xx-- ---- ---- - =?	none
				1930	EE-G ---x ---- - =?	none
				1967-80+	FF-G --Gx x--- F =?	none

TECTONIC/GEOLOGIC SETTING

The Izu Peninsula lies at the northern tip of the Philippine Sea plate, where that plate is overriding the Pacific plate and being subducted beneath the Eurasian plate (Fujita and Tada, 1983) (see also section on Izu-Oshima Caldera, CAVW number 08-04-01). The Izu Peninsula has numerous volcanic vents, ranging from a small vent on the southern tip of the peninsula (Nanzaki) and Amagi Volcano in the south-central part of the peninsula to Hakone at its northern end (figs. 20.2.1, 20.2.2). Many small vents of the Higashi-Izu monogenetic volcano group, including 63 basaltic cinder cones, 4 dacite domes, and 9 phreatic craters (maars), lie in the general vicinity of a series of historical earthquake swarms (discussed below). Dacitic domes have grown as recently as 3,000 years ago, but most basaltic eruptions in the area occurred before about 40,000 years ago (Oki and others, 1978).

HISTORICAL ACTIVITY

Seismicity:

In 1905, an earthquake with estimated M=6.5 occurred slightly north of the 1978 epicenter (see below). It was preceded by swarm activity.

The great Kanto earthquake of 1923 occurred about 40 km northeast of the center of unrest on the Izu Peninsula (fig. 20.2.5).

Swarms occurred again in February-April 1930 and continued at a lower level until August 1930. From February to August 1930, 4,880 earthquakes were felt. New, intense swarms began on 7 November 1930, reached 690 earthquakes/day on 25 November, and culminated in the M=7.0 Kita-Izu earthquake on 26 November 1930, centered near the November swarm but 30-40 km from swarms of February-May 1930 (figs. 20.2.3, 20.2.5). The Kita-Izu earthquake had left-lateral slip on the N-S-trending Tanna Fault, indicating NW-SE compression. Earthquakes migrated upward immediately before the strong earthquake, from an average of about 6 km to an average of between 4 and 5 km. Most aftershocks were less than 5 km deep. Nasu (1935) interpreted this upward migration to indicate

IZU PENINSULA (continued)

HISTORICAL ACTIVITY (continued)

Seismicity (continued):

magmatic intrusion into a zone approximately 5 km deep, 2 hours before the strong earthquake. The final earthquake of this sequence was the M=5.5 Minami-Izu earthquake in March 1934, followed by 40 years of relative seismic quiet.

The Izu-Hanto-Oki earthquake of 1974 (M=6.9) at the southern tip of the Izu Peninsula involved right-lateral motion along a NW-SE-trending fault (figs. 20.2.3-20.2.5). Seismicity after the 1974 Izu-Hanto-Oki earthquake was high, with numerous swarms and a few shocks up to M=5, including the Amagi earthquake (M=4.9; 9 July 1974) and the Kawazu earthquake (M=5.4; 17 August 1976). The locus of swarms has shifted over a broad area, about 30-40 km in an E-W direction by about 20 km in a N-S direction, between Mt. Amagi on the west and Oshima Volcano on the east. All are generally 10-30 km south of the area of the 1930 swarms, and 5-15 km south of the center of the 1967-76 uplift (see below and figs. 20.2.3-20.2.7, 20.2.12).

A M=7.0 earthquake on 14 January 1978 was located between the Izu Peninsula and Oshima Island (figs. 20.2.3, 20.2.5, 20.2.8). The earthquake relieved strain on the east coast but not in the northern part of the area.

Another earthquake on 28 June 1980 (M=6.7, depth 16 km), was located near the east coast of the Izu Peninsula, between the 1978 epicenter and those of the 1975-76 swarms (fig. 20.2.7), and 10-20 km southeast of the center of recent uplift (figs. 20.2.3, 20.2.5, 20.2.6).

Fault-plane solutions for six earthquakes in this region (1930, 1934, 1974(2), 1976, and 1980) all show nearly vertical strike-slip faulting. The solution for the 1930 earthquake and observations of surface rupture indicate NW-SE compression; solutions for the subsequent earthquakes show N-S compression and movement along conjugate NW-SE- and NE-SW-trending faults (Abe, 1978).

Ground Deformation:

Releveling in April 1930 of a line previously leveled in 1923-24 showed a bulge of approximately 3 cm, with its center 2-4 km north of Ito-cho (Tsuboi, 1931a, 1933). Releveling in November-December 1930 showed that the bulge had grown to 10 cm just before and to 22 cm just after the 26 November earthquake, and that the growth from March 1930 until November 1930 was at a rate of 19 cm/yr (Tsuboi, 1931b). The maximum uplift was closer to the February-May swarms rather than to the November "main shock." Uplift continued after the November 1930 earthquake, and by 1933 the maximum uplift measured was 35 cm (Tsuboi, 1933; figs. 20.2.9-20.2.11).

From 1967 to 1980, maximum uplift was about 20 cm centered about 10 km west-southwest of Ito, in about the same place as the 1930 uplift (fig. 20.2.12). The half-width of the uplift was about 15 km, and its maximum extent was about 40 km. About 1 cm of subsidence occurred from 1967 to 1974, followed by 1 cm of uplift from 1974 to 1976, about 7 cm of additional uplift in the months before the 1978 earthquake, and about 8 cm of additional uplift before the 1980 earthquake (data reproduced in Thatcher and Savage,

IZU PENINSULA (continued)

HISTORICAL ACTIVITY (continued)

Ground deformation (continued):

1982). Viewed in a different way, about 8 cm of uplift followed the 1974 earthquake, and another 8 cm of uplift followed the 1978 earthquake. One interpretation of the uplift invokes inflation of a magma body 4 km in diameter at 10-km depth (Hagiwara, 1978) or, alternatively, by a volume increase of 0.15 km^3 in a magma body at 11-km depth (Thatcher and Savage, 1982).

In general, ground deformation on the Izu Peninsula is characterized by periods of subsidence lasting several decades alternating with intermittent uplift lasting about one decade. Strain energy in the northeast part of the Izu Peninsula has been released mainly in intermittent aseismic crustal uplift (Fujita and Tada, 1983). The gravitational potential associated with the uplift from 1974 to 1981 is equivalent to a M=7.5 earthquake, slightly larger than the actual seismic energy release (equivalent to a M=7.2 earthquake). Tada and Asano (1983) note that, as at Matsushiro, seismicity and uplift are centered where a "granitic layer" with a relatively high P-wave velocity (V_p=5.7 km/sec) lies at anomalously shallow depth (about 1 km).

Fujii (1977a, b) inferred a systematic migration in the locus of uplift, from east-northeast to west-southwest across the head of Sagami Bay, from 1923 to 1930 and from 1971 to 1974. By 1976, the locus of uplift had shifted back slightly to the north, between the loci of 1972 and 1974. Fujii interprets this migration as evidence for creep dislocation at depth, along a postulated "East-off-Izu subduction zone." The activity from 1923 to 1930 followed major right-lateral slip during the 1923 Kanto earthquake.

Hydrologic Changes:

Water-well levels rose in quadrants of contraction (compression) after the Izu-Hanto-Oki earthquake of 9 May 1974 and fell in quadrants of dilatation, thereby serving as a useful indicator of tectonic strain (Wakita, 1975). The water level in a well in Omaezaki, Shizuoka (95 km west-southwest of Ito) dropped by approximately 20 cm beginning on 28 December 1977, then returned to normal by 14 January 1978, at which time a large earthquake occurred just east of Ito (Wakita, 1982). The water temperature of one well and the Cl-content of water in another well, both near the center of uplift on the Izu Peninsula, decreased slightly several days before the same 1978 earthquake (Takahashi and Tsuneishi, 1978). A 0.2 °C decrease in the temperature of a thermal artesian well and a 1-2 cm increase in water levels in several wells were reported as much as 6 days before a M=6.7 earthquake on 29 June 1980 (Oki and Hiraga, 1980).

IZU PENINSULA (continued)

COMMENTS

Five general explanations have been advanced for recent unrest on the Izu Peninsula (figs. 20.2.13-20.2.15):

(1) Dilatancy before earthquakes (general concept of Scholz and others, 1973; Kaidzu and Tada, 1980);

(2) Aseismic creep dislocation (slow, aseismic thrust faulting at depth along the "East-off-Izu Line" (Ishibashi, 1977; Fujii, 1977a, b);

(3) Inflation or upward intrusion of a body of magma (Nasu and others, 1931; Hagiwara, 1978; Thatcher and Savage, 1982);

(4) NW-SE compression of a plastic volume, such as magma (Somerville, 1978); and

(5) Bending of the northern tip of the Philippine plate between the Sagami Trough and the Suruga Trough, compensatory movement of material at depth toward the axis of bending, and uplift caused by the resulting mass excess at depth (Nakamura and others, 1984; K. Nakamura, written commun., 1984).

REFERENCES

Abe, K., 1978, Dislocations, source dimensions and stress associated with earthquakes in the Izu Peninsula, Japan: Jour. Phys. Earth, v. 26, p. 253-274.

Fujii, Y., 1977a, Creep dislocation model of crustal upheaval in Izu Peninsula: Rep. Coord. Comm. Earthq. Prediction, v. 17, p. 68-70.

----- 1977b, Creep dislocation propagation along the East-off-Izu Tectonic Line, Japan, around the year of 1930 and 1976: Zisin, v. 30, p. 389-400.

Fujita, N., and Tada, T., 1983, Anomalous crustal uplift in the northeastern part of the Izu Peninsula, Central Honshu, Japan: Tectonophysics, v. 97, p. 147-152.

Hagiwara, Y., 1978, Gravity changes associated with seismic activities, in Kisslinger, C., and Suzuki, K., eds., Earthquake precursors: Tokyo, Japan Scientific Societies Press, p. 137-146.

Hiraga, S., Ito, H., Odaka, S., and Hakamata, K., 1982, Seismometrical observations in Hakone Volcano, 1980: Bull. Hot Springs Res. Inst., Kanazawa Prefecture, v. 13, no. 5, p. 111-125.

Ishibashi, K., 1977, Creep dislocation model of the Izu Peninsula uplift--significance of the East-off-Izu Tectonic Line: Rep. Coord. Comm. Earthq. Prediction, v. 17, p. 65-67.

Kaidzu, M., and Tada, T., 1980, Recent crustal uplift in the northeastern part of the Izu Peninsula: The Earth Monthly, v. 2, p. 132-136.

IZU PENINSULA

REFERENCES (continued)

Kimura, M., 1976, Major magmatic activity as a key to predicting large earthquakes along the Sagami through, Japan: Nature, v. 260, p. 131-133.

----- 1978, Significant eruptive activities related to large interplate earthquakes in the northwestern Pacific margin, in Uyeda, S., Murphy, R.W., and Kobayashi, K., eds., Geodynamics of the western Pacific: Jour. Phys. Earth, v. 26 supp., p. S557-S570.

Kimura, M., and Toyoda, J., 1975, Recent level change in the bottom-floor in the summit crater of Mihara-yama, Oshima Island, Japan: Bull. Volcanol. Soc. Japan, v. 20, p. 65-78.

Mogi, K., 1985, Earthquake prediction: Tokyo, Academic Press, 355 p.

Murai, I., and Kaneko, S., 1974, The Izu-Hanto-oki earthquake of 1974 and the earthquake faults, especially, the relationships between the earthquake faults, the active faults, and the fracture systems in the earthquake area: Spec. Bull. Earthq. Res. Inst., no. 14, p. 159-203.

Nakamura, K., 1971, Volcano as a possible indicator of crustal strain: Bull. Earthq. Res. Inst., v. 16, p. 63-71.

Nakamura, K., Shimazaki, K., and Yonekura, N., 1984, Subduction, bending and eduction. Present and Quaternary tectonics of the northern border of the Philippine Sea plate: Bull. Soc. Geol. France, v. 26, no. 2, p. 221-243.

Nakamura, K., and Tazawa, K., 1975, A physical interpretation for correlation between 1974 Izu-Hanto-oki earthquake and a minor eruption of Mihara-yama, Izu-Oshima: Bull. Volcanol. Soc. Japan, v. 19, p. 159-160.

Nasu, N., 1935, Recent seismic activities in the Izu Peninsula (Part II): Bull. Earthq. Res. Inst., v. 13, p. 400-416.

Nasu, N., Kishinouye, F. and Kodaira, T., 1931, Recent seismic activities in the Izu Peninsula, Part I: Bull. Earthq. Res. Inst., v. 9, p. 22-35.

Oki, Y., Aramaki, S., Nakamura, K., and Hakamata, K., 1978, Volcanoes of Hakone, Izu and Oshima: Hakone, Hakone Town Office, 88 p.

Oki, Y., and Hiraga, S., 1980, Water level anomalies prior to the magnitude 6.7 East-Off-Izu Peninsula earthquake, June 29, 1980: Bull. Hot. Springs Res. Inst., Kanazawa Prefecture, v. 12, no. 1, p. 9.

Scholz, C.H., Sykes, L.R., and Aggarwal, Y.P., 1973, Earthquake prediction: a physical basis: Science, v. 181, p. 803-809.

Shimazaki, K., and Somerville, P., 1979, Static and dynamic parameters of the Izu-Oshima, Japan earthquake of January 14, 1978: Bull. Seismol. Soc. Am., v. 69, p. 1343-1378.

Smithsonian Institution, Scientific Event Alert Network (SEAN), 1980, Earthquakes, Izu Peninsula: SEAN Bull., v. 5, no. 6.

Somerville, P., 1978, The accommodation of plate collision by deformation in the Izu block, Japan: Bull. Earthq. Res. Inst., v. 53, p. 629-648.

Tada, T., and Asano, S., 1983, Anomalous crustal uplift and crustal structure in the Matsushiro and the Izu Peninsula earthquake swarm regions: Tectonophysics, v. 97, p. 241-250.

Takahashi, H., and Tsuneishi, Y., 1978, Variations in chemical composition of thermal water before and after the 1978 Izu-Oshima-kinkai earthquake, at Tokunaga-minami and Tsukigase in the Izu Peninsula: Bull. Earthq. Res. Inst., v. 53, p. 987-994.

Thatcher, W., and Savage, J.C., 1982, Triggering of large earthquakes by magma chamber inflation, Izu Peninsula, Japan: Geology, v. 10, p. 637-640.

IZU PENINSULA

REFERENCES (continued)

Tsuboi, C., 1931a, A note on the results of the repeated precise levellings across the Ito earthquake area: Bull. Earthq. Res. Inst., v. 9, p. 151-158.

----- 1931b, On the results of repeated precise levellings around Idu Peninsula: Bull. Earthq. Res. Inst., v. 9, p. 271-289.

----- 1932, Investigation on the deformation of the earth's crust in Idu Peninsula connected with the Idu earthquake of November 26, 1930: Bull. Earthq. Res. Inst., v. 10, no. 2, p. 435-448.

----- 1933, Vertical crustal displacement in the seismic region of Ito, on the East coast of the Idu Peninsula: Bull. Earthq. Res. Inst., v. 11, p. 488-499.

Tsumura, K., 1977, Anomalous crustal activity in the Izu Peninsula, Central Honshu: Jour. Phys. Earth, v. 25 supp., p. S51-S68.

Tsumura, K., Karakama, I., Ogino, I., and Takahashi, M., 1978, Seismic activities before and after the Izu-Oshima-Kinkai earthquake of 1978: Bull. Earthq. Res. Inst., v. 53, p. 675-706.

Wakita, H., 1975, Water wells as possible indicators of tectonic strain: Science, v. 189, p. 553-555.

----- 1982, Changes in groundwater level and chemical composition, in Asada, T., and Ohnuki, M., eds., Earthquake prediction techniques: Tokyo, Univ. Tokyo Press, p. 175-216.

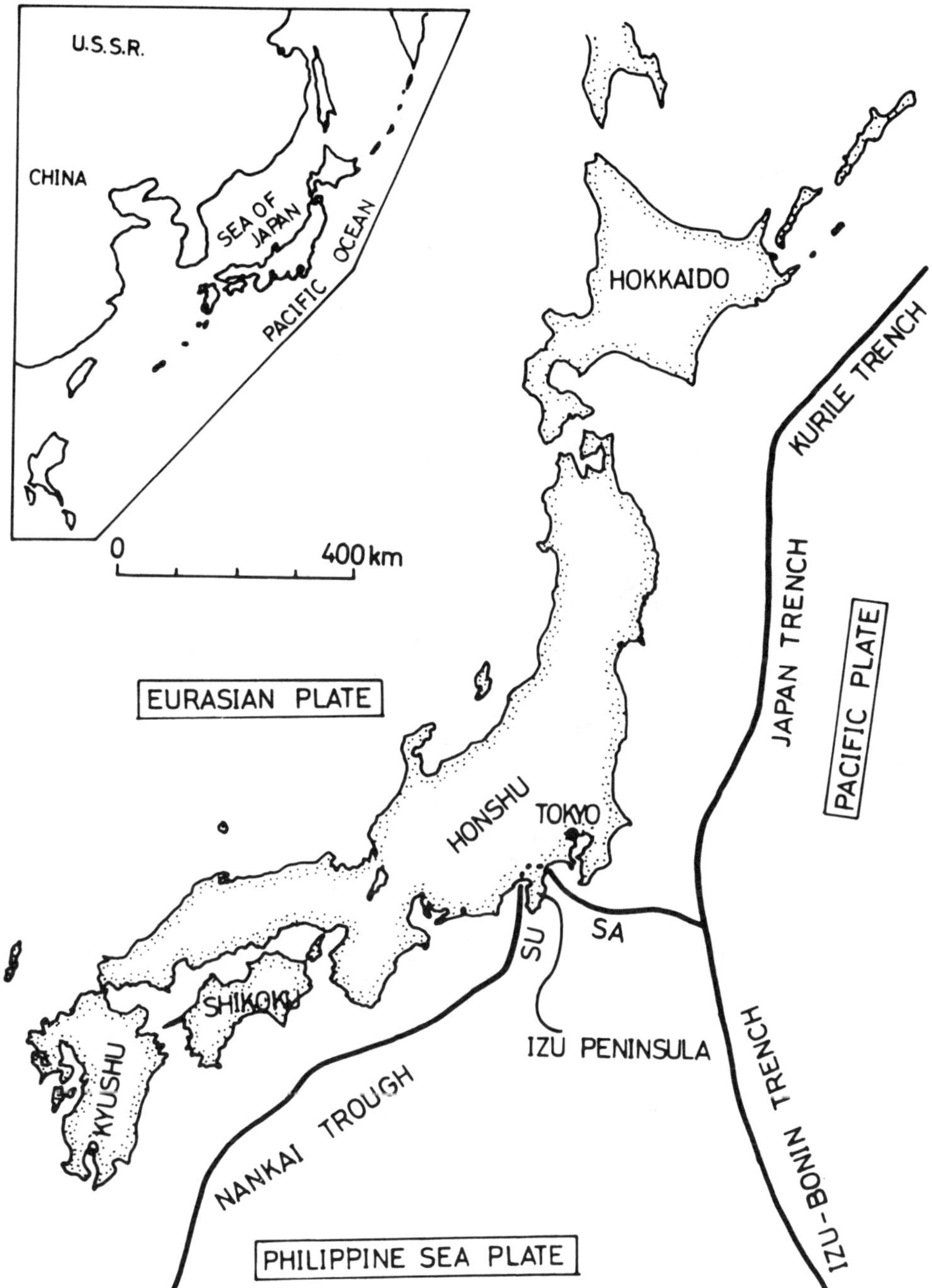

Figure 20.2.1. Location of Izu Peninsula and plate boundary around Japanese Islands, from Fujita and Tada (1983). SA and SU denote Sagami and Suruga Troughs, respectively.

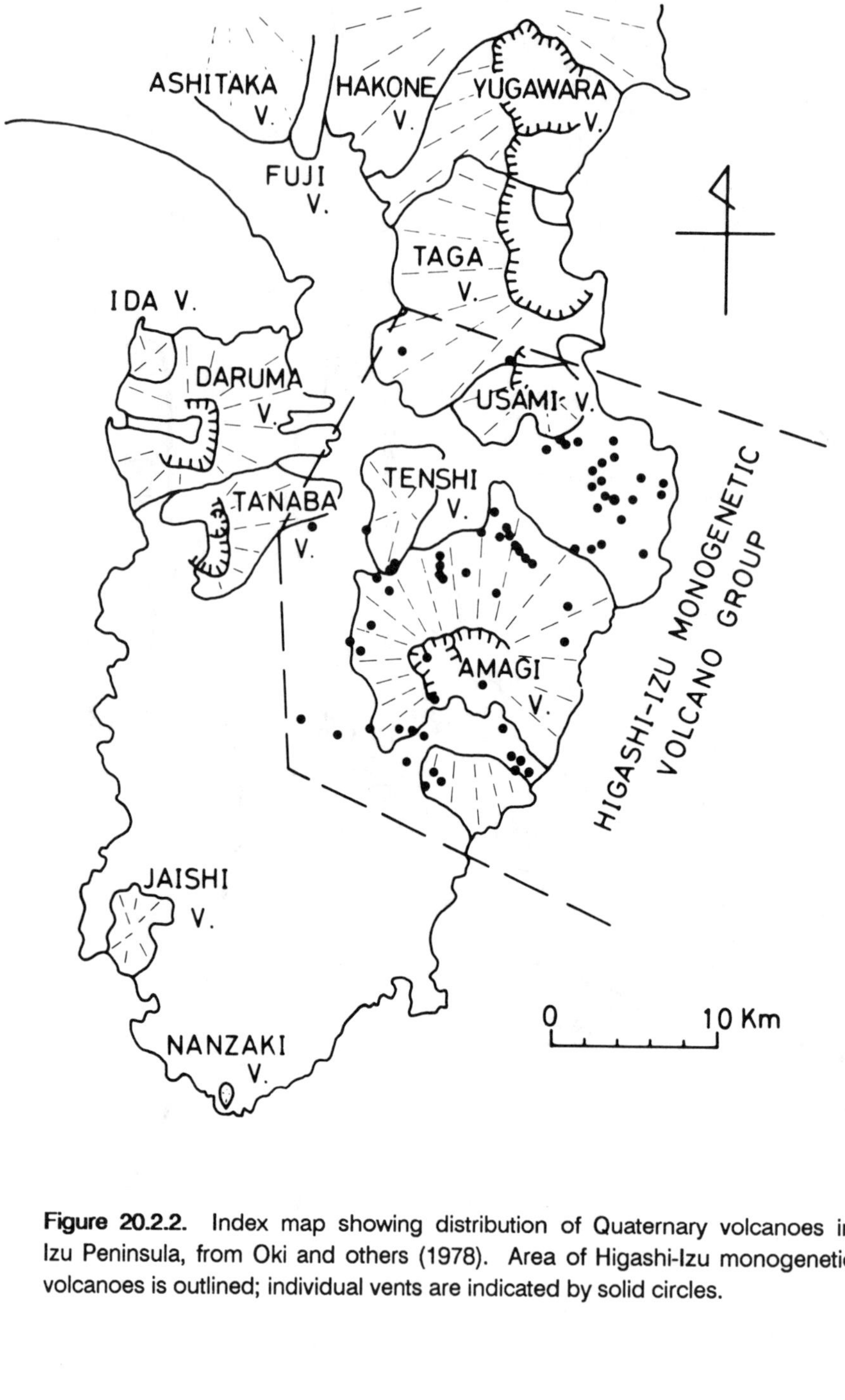

Figure 20.2.2. Index map showing distribution of Quaternary volcanoes in Izu Peninsula, from Oki and others (1978). Area of Higashi-Izu monogenetic volcanoes is outlined; individual vents are indicated by solid circles.

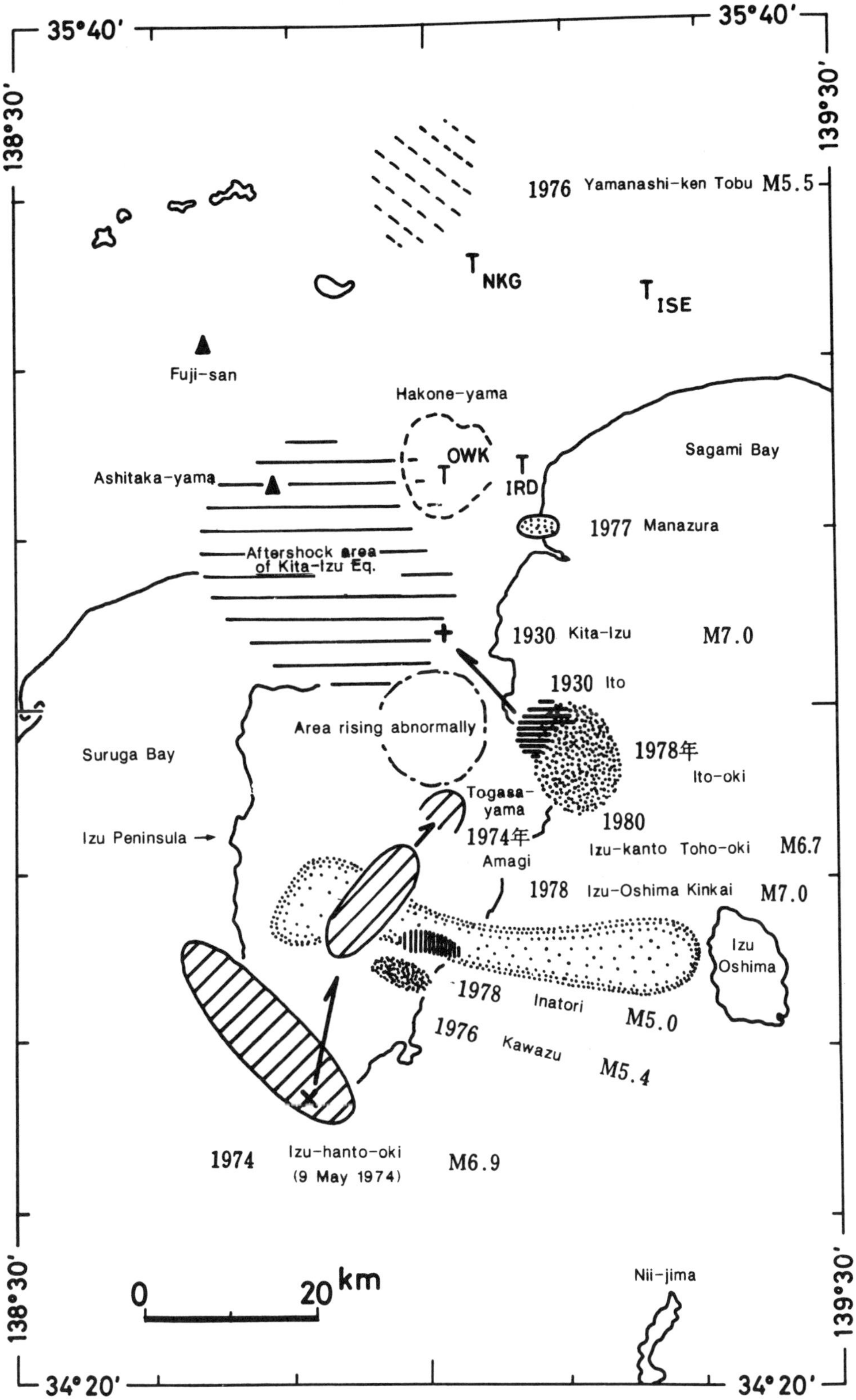

Figure 20.2.3. Locations of large earthquakes and aftershocks on Izu Peninsula, 1930-80, from Hiraga and others (1982). Note migration of activity during February-November 1930 (swarm culminating in main shock) and during May-June 1974 (main shock followed by swarms). Seismic stations are indicated by letter T.

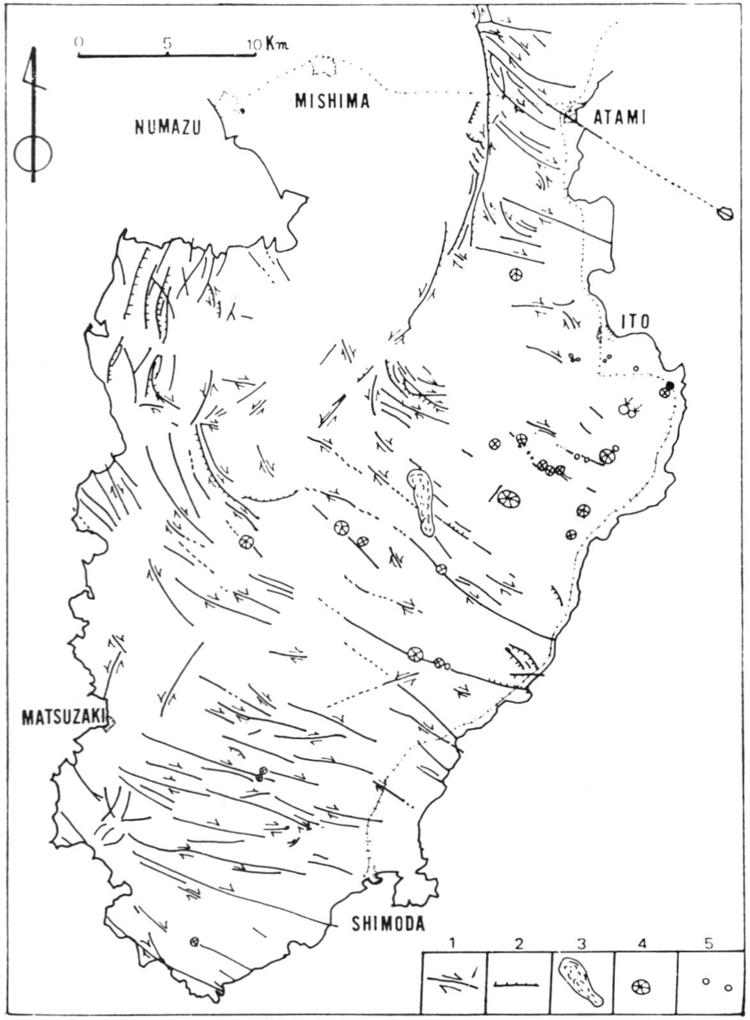

Figure 20.2.4. Active faults and lineaments in Izu Peninsula, from Tsumura (1977) after Murai and Kaneko (1974). 1, right-lateral fault; 2, scarplet; 3, lava flow; 4, volcanic cone; 5, explosion crater.

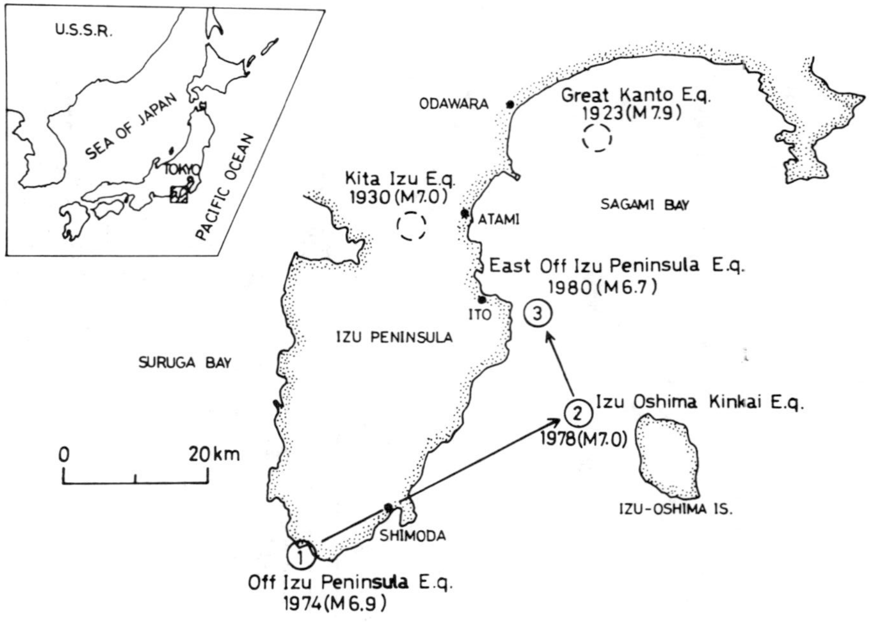

Figure 20.2.5. Epicenters of destructive earthquakes ($\underline{M} > 6.5$) in and around the Izu Peninsula since 1974, from Fujita and Tada (1983). Note that epicenters generally have migrated northward.

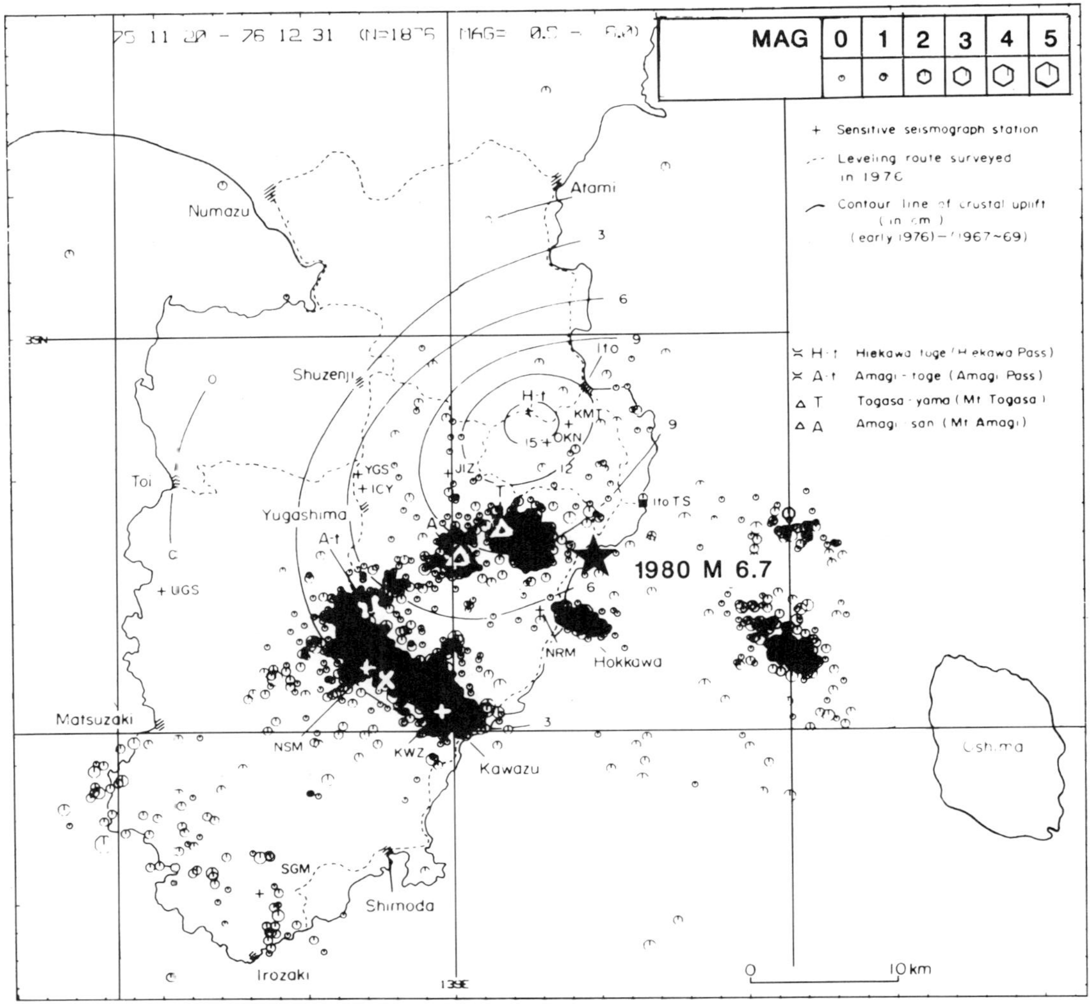

Figure 20.2.6. Distributions of crustal uplift (in centimeters) and epicenters of microearthquakes in Izu Peninsula, from Tsumura (1977). Contour lines are for interval from 1967-69 surveys to January-April 1976 surveys. Broken lines indicate leveling route along which surveys were carried out in 1976. Epicenters were determined by Earthquake Research Institute temporary seismographic network (stations OKN, ICY, NRM, and KWZ) from November 1975 to December 1976.

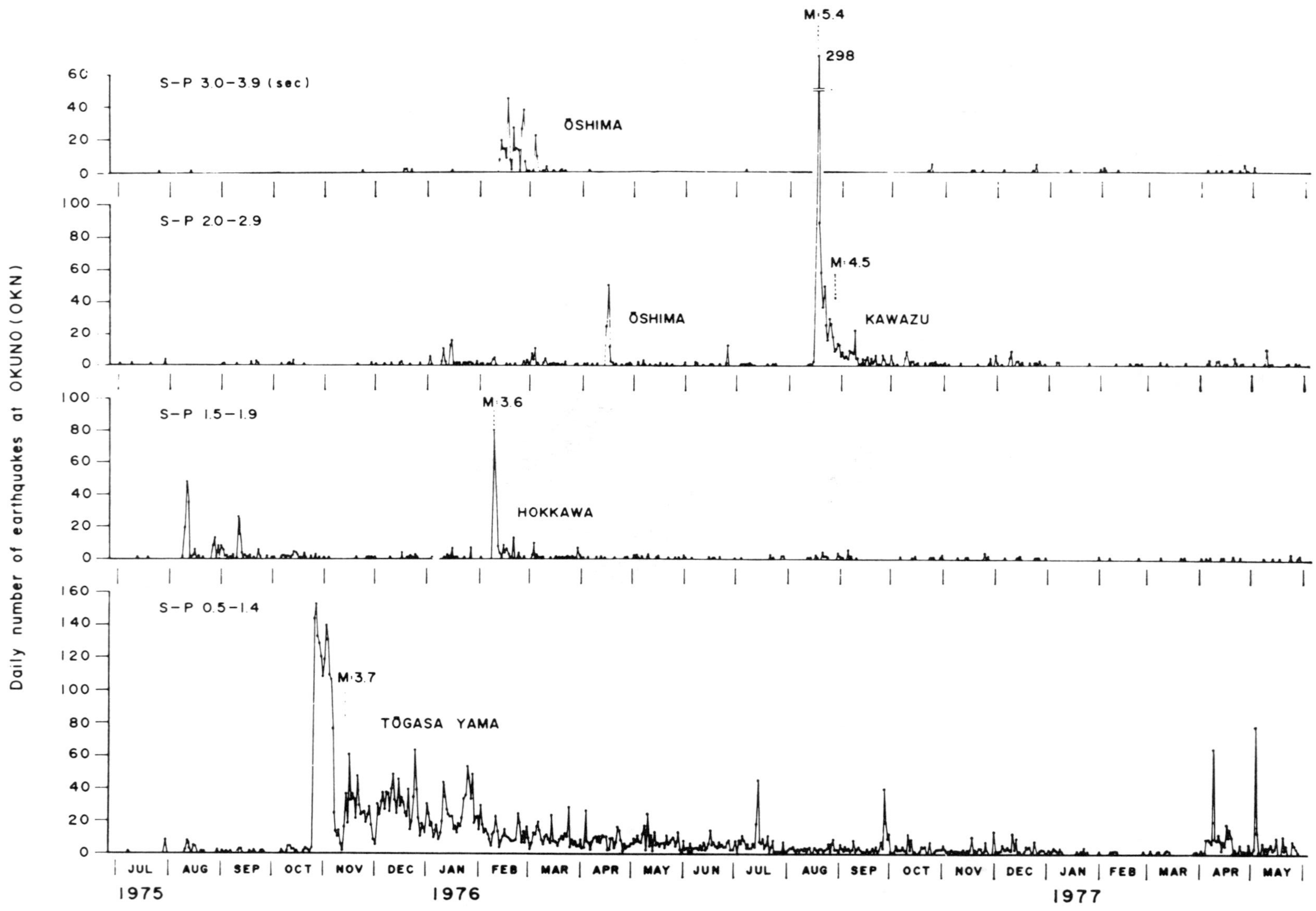

Figure 20.2.7. Number of microearthquakes recorded at Okuno station (OKN) during 1975-77 for different ranges of S - P time, from Tsumura (1977).

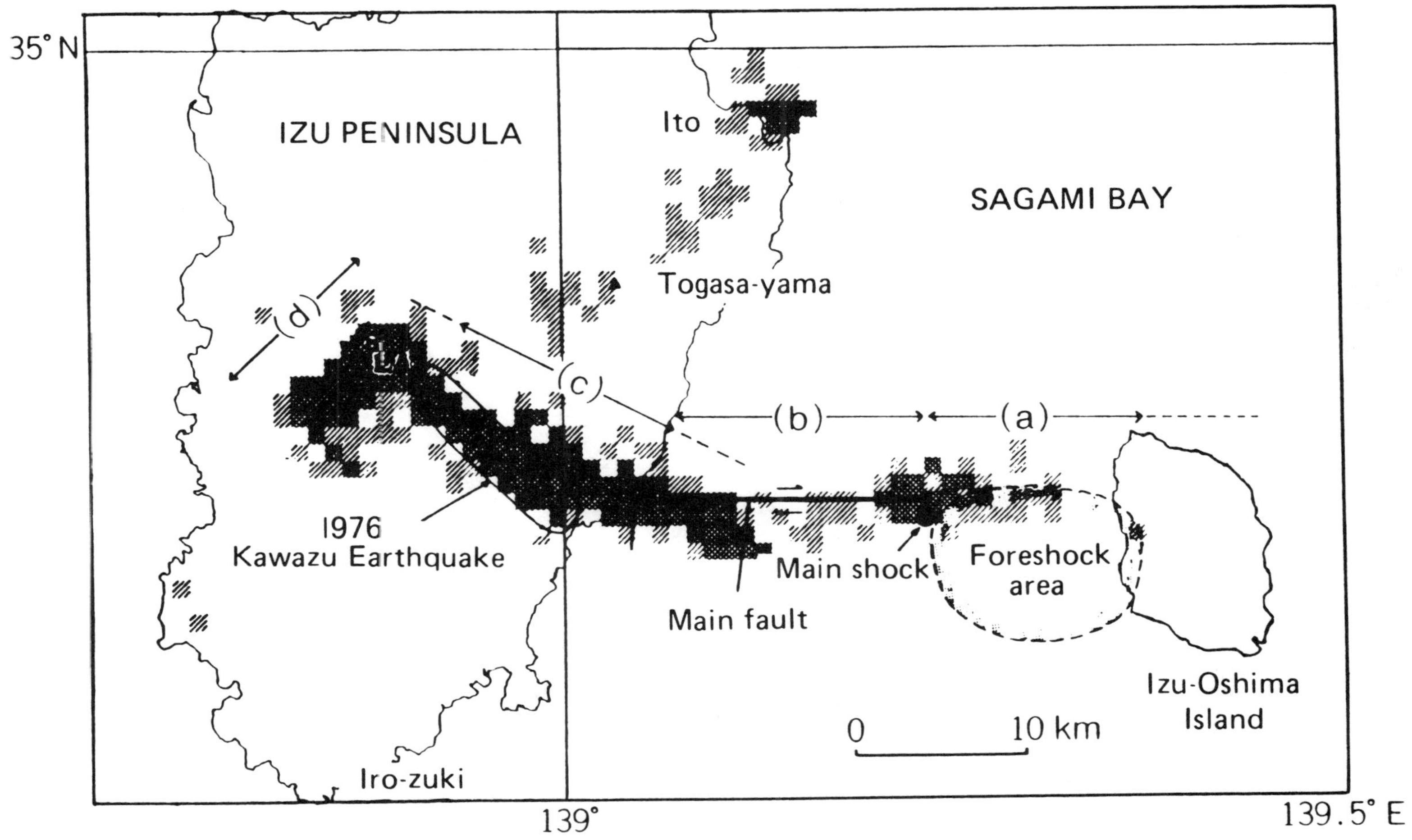

Figure 20.2.8. Main shock, principal fault, foreshock region, and aftershock region of the 1978 Izu-Oshima-kinkai earthquake (M 7.0), from Tsumura and others (1978).

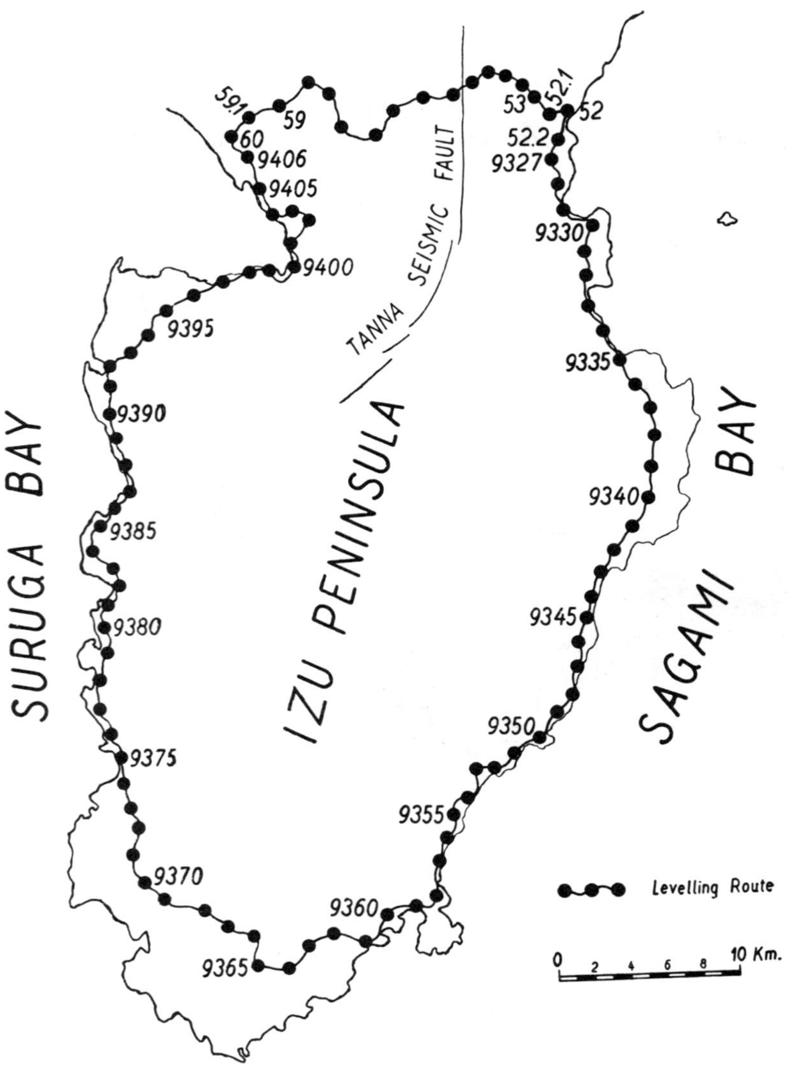

Figure 20.2.9. Leveling route around and across Izu Peninsula, from Tsuboi (1931b).

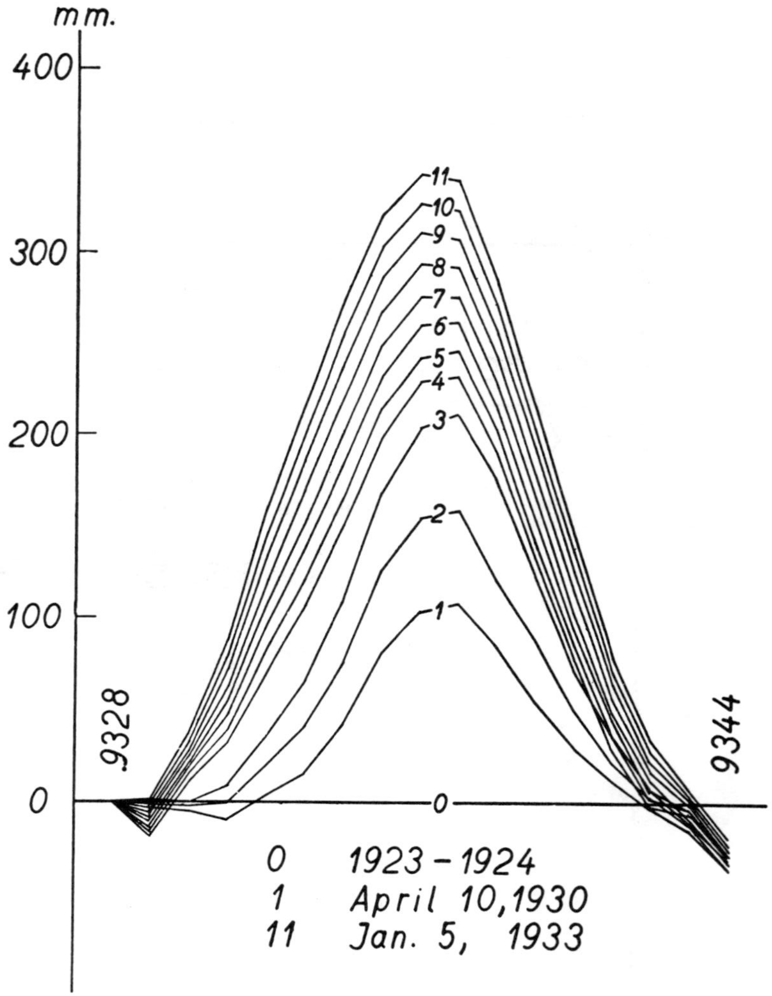

Figure 20.2.10. Successive heights of bench marks around Izu Peninsula at 100-day intervals starting on 10 April 1930 (Tsumura, 1977). Locations of bench marks 9328 and 9344 are indicated in figure 20.2.9.

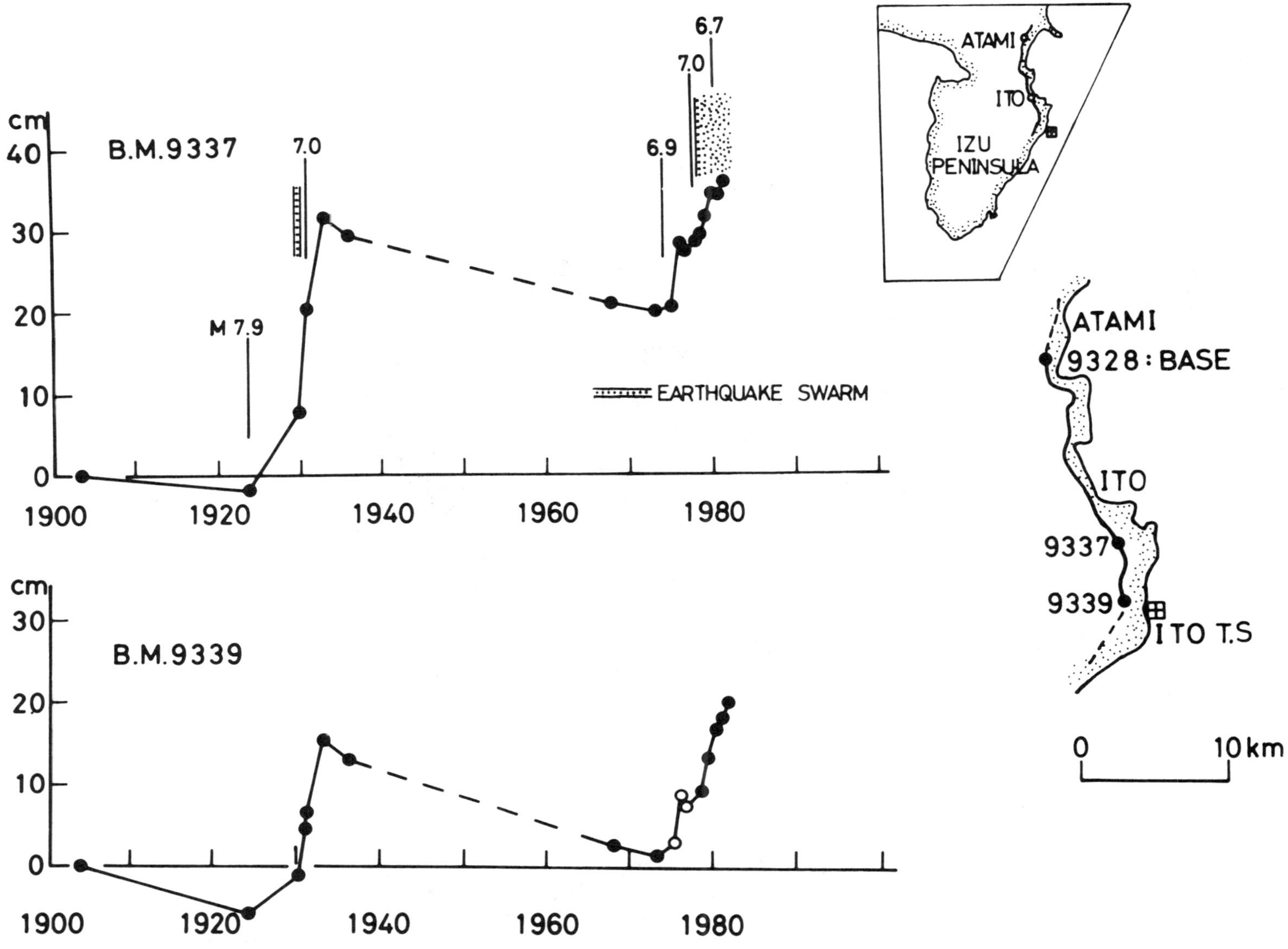

Figure 20.2.11. History of vertical crustal movement at bench mark 9339 near Ito (inset), from Fujita and Tada (1983). Note association of uplift and seismic activity during periods from 1924 to 1933 and from 1975 to 1983.

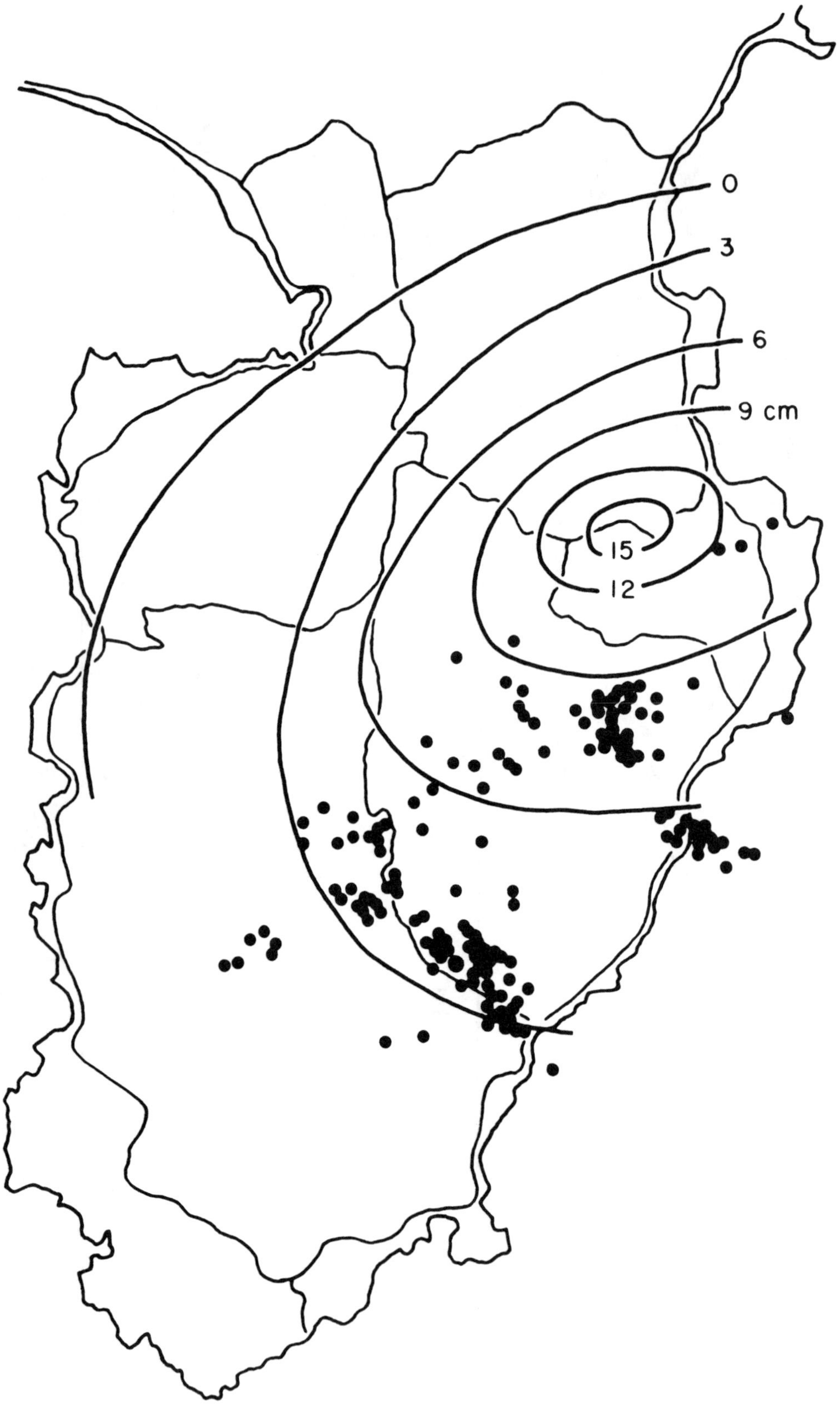

Figure 20.2.12. Ground uplift in eastern Izu Peninsula between 1967 and 1976, and earthquake swarms between 1975 and 1976 (Tsumura, 1977).

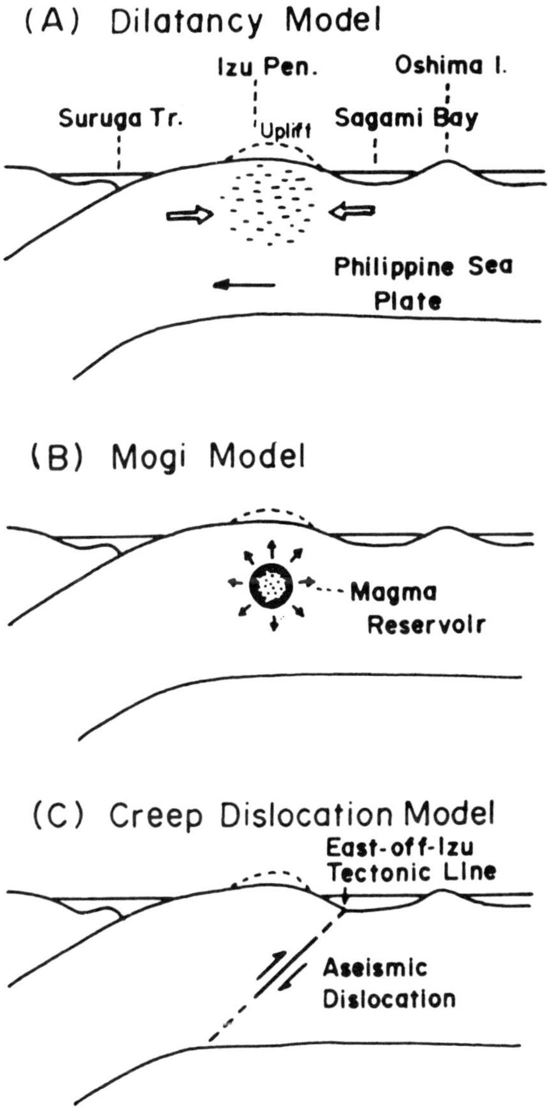

Figure 20.2.13. Three models for historical uplift of Izu Peninsula, from Tsumura (1977).

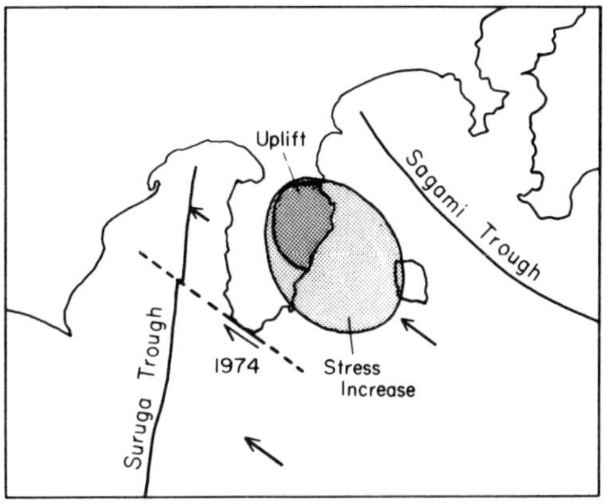

Figure 20.2.14. Illustration of hypothesis explaining increased crustal activity in eastern Izu Peninsula following the 1974 Izu-Hanto-oki earthquake (Mogi, 1985). Thick arrows, direction of movement of Philippine Sea plate. Compression of a plastic volume (magma?) is thought to focus uplift above that volume.

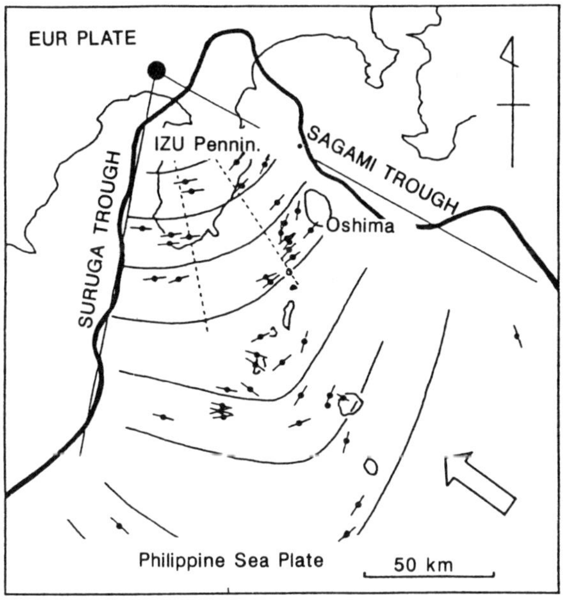

Figure 20.2.15. Concentric trajectories of T-axes of focal-mechanism solutions of shallow earthquakes in northern tip of Philippine Sea plate, from Nakamura and others (1984). Radial broken and solid lines show P-axis trajectories and general trend of troughs, respectively. Open arrow indicates direction of convergence between Philippine Sea and Eurasian plates.

PART 4: SELECTED EPISODES OF HISTORICAL UNREST IN NONCALDERA SETTINGS (continued)
See inside back cover for explanation and abbreviations

KII PENINSULA

Latitude Longitude (degrees)	Tectonic setting	Age of latest volcanism	Youngest volcanic rock types	Restless years	Type and duration of unrest ESTU STHF MGTF H Te	Associated eruption?
34N 135-136E	Compr	Cretaceous?	?	1899	x--- ---- ---- - --	none
				1920-30+	xx-x xx-- ---- - --	none
				1944	x--- ---- ---- - --	none
				1946	x--x ---- ---- - --	none
				1948	x--- ---- ---- - --	none
				1952-53	xx-x xx-- ---- - --	none

TECTONIC/GEOLOGIC SETTING

The Kii Peninsula, about 350 km west-southwest of the Izu Peninsula (fig. 20.3.1), has outcrops of serpentinite and other high-pressure, low-temperature metamorphic rocks in its central and southern parts, and Cretaceous(?) plutons and higher temperature metamorphic rocks in its northern part. No Quaternary volcanism is known, but hot springs are common. The tectonic setting is similar to that of the Izu Peninsula, with NW-SE compression at the interface between the Philippine Sea plate and the continental lithosphere. The Median Tectonic Line, a major structure with recent right-lateral strike-slip motion, passes ENE-WSW through the peninsula and separates the serpentinites and related rocks in the south from the plutons and related rocks in the north.

Kanamori and Tsumura (1971) distinguished two types of earthquakes on the Kii Peninsula: shallow earthquakes (less than 10 km deep) on the west coast near Wakayama, and deep mantle earthquakes (30-70 km deep) that define a NE-SW line across the central part of the peninsula (figs. 20.3.2, 20.3.3).

HISTORICAL ACTIVITY

At least three large earthquakes have occurred on the peninsula in the past 100 years: $\underline{M}$=7.6 in 1899, and $\underline{M}$=7.0 in 1948 and 1952. In addition, great earthquakes occurred just offshore from the Kii Peninsula in 1944 (the Tonankai earthquake, $\underline{M}$=8.3) and in 1946 (the Nankaido earthquake, $\underline{M}$=8.4) (figs. 20.3.1, 20.3.4, 20.3.5). Notable seismic swarms occurred in the vicinity of Wakayama, on the west coast of the peninsula and south of the Median Tectonic Line, in the early 1920's and 1952-53 (fig. 20.3.4). Interestingly, this swarm activity was greatest just before the Kanto earthquake (1923, $\underline{M}$=8.3, near Tokyo), and the Boso-Oki earthquake (1953, $\underline{M}$=8.3(?), 250 km southeast of Tokyo) (Kanamori, 1972).

KII PENINSULA (continued)

HISTORICAL ACTIVITY (continued)

Swarm activity on the peninsula increased from 1920 to 1923 and then declined (after the Kanto earthquake) for about 7 years. The shocks were shallow, less than 4 km deep, and scarcely felt beyond a radius of 20 km. "The worst do no more than cause cracks in plaster walls, overturn unstable objects, and only rarely throw down stone fencings, so that they could hardly arouse the interest of investigators in Japan, much less of those abroad" (Imamura and others, 1932). The increase in swarm earthquakes was accompanied by uplift of 9 cm at the midpoint of a 10-km-long N-S leveling line between Wakayama and Siotu/Kainan; the center of uplift was broadly coincident with the earthquake epicenters (figs. 20.3.6, 20.3.7). Although the line was established in 1898 and not remeasured until 1928, most of the uplift probably occurred after 1920 (Imamura and others, 1932).

After the peak swarm activity in 1923, the number of earthquakes felt at Wakayama did not return to the pre-1920 background level of about 10/yr but instead remained at approximately 100/year until the 1944 Tonankai earthquake. The number of felt earthquakes did decrease to about 20/yr in 1945 but rose again before the 1946 Nankaido earthquake, and then increased sharply to more than 300/yr in 1952-53 (Kanamori, 1972). Then, after the $\underline{M}$=8.3 Boso-Oki earthquake in 1953 the number declined again so that by 1958 (and until at least 1969), the number of earthquakes felt at Wakayama was again about 100/yr. Within each year, earthquakes were concentrated in discrete swarms separated by a month or more of inactivity.

While the area near Wakayama was being uplifted in the early 1920's, the southern tip of the Kii Peninsula was subsiding. A leveling line around the southern tip of the peninsula, roughly 60 km south-southeast of the swarm area, was measured in 1899 and 1931; at least 7 cm and probably 20-30 cm of subsidence of the southern tip of the peninsula occurred during that interval. Then the bench mark of maximum subsidence was uplifted suddenly by 72 cm at the time of the 1946 Nankaido earthquake. The time of uplift relative to the earthquake is not known from leveling, but tide gauges suggest that it occurred during or just after the earthquake. After the earthquake, slow subsidence of the southern tip of the peninsula resumed at a few millimeters per year, with a slight interruption in 1950-51 before the 1952-53 swarms (see below). Corresponding tilts were 12 microradians (down to the south) from 1900 until the 1946 earthquake, 24 microradians (up to the south) during the earthquake, and 4.8 microradians (back down to the south) from 1946 to 1958 (Okada, 1960) (fig. 20.3.8).

The N-S leveling line between Wakayama and Siotu/Kainan was releveled eight times between 1948 and 1967, following the Nankaido earthquake (figs. 20.3.6, 20.3.9, 20.3.10). The point of maximum uplift for the period 1898-1928 showed further uplift except during 1954-55, just after the 1952-53 swarms. Net uplift was 7 cm from 1948 to 1967; unfortunately, the shape of the uplift is not well known.

Wakita and others (1987) report anomalously high $^{3}He/^{4}He$ ratios in gases from hot springs on the Kii Peninsula (fig. 20.3.11). Because high ratios are associated with He from upper-mantle sources, Wakita and others interpret those high ratios to indicate a magmatic intrusion beneath the Kii Peninsula.

KII PENINSULA (continued)

COMMENTS

Most studies have concluded that the Kii swarms are of tectonic origin, in a region of NW-SE compression. Kanamori and Tsumura (1971) and Kanamori (1972) suggested that earthquakes occur in swarms where there is serpentinite and other rock that has been extensively fractured. The weak rocks cannot sustain enough stress to produce a large earthquake; rather, they fail in many small increments, as represented by the swarm earthquakes. As an alternative interpretation, Wakita and others (1987) suggested deep magmatic intrusion, as they did also for Matsushiro (see discussion in section on Matsushiro).

REFERENCES

Imamura, A., Kodaira, T., and Imamura, H., 1932, The earthquake swarms of Nagusa and vicinity: Bull. Earthq. Res. Inst., v. 10, p. 636-648.

Kanamori, H., 1972, Relation between tectonic stress, great earthquakes and earthquake swarms: Tectonophysics, v. 14, p. 1-12.

Kanamori, H., and Tsumura, K., 1971, Spatial distribution of earthquakes in the Kii Peninsula, Japan, south of the Median Tectonic Line: Tectonophysics, v. 12, p. 327-342.

Miyamura, S., 1960, Local earthquakes in Kii peninsula, central Japan: Bull. Earthq. Res. Inst., pts. 1-3 in v. 37, p. 347-358, 593-608, 609-635; pt. 4 in v. 38, p. 71-112.

Miyamura, S., Hori, M., and Matumoto, H., 1966, Local earthquakes in Kii Peninsula: Bull. Earthq. Res. Inst., pt. 5. in v. 44, p. 709-729.

Okada, A., 1960, Land deformation of the southern part of the Kii peninsula, southwestern Japan: Bull. Earthq. Res. Inst., v. 38, p. 125-143.

Okada, A., Izutuya, S., Ito, Y., and Kadono, K., 1968, Results of levelling resurvey between Wakayama and Kainan, Wakayama prefecture: Bull. Earthq. Res. Inst., v. 46, p. 413-416.

Sano, Y., and Wakita, H., 1985, Geographical distribution of $^{3}He/^{4}He$ ratios in Japan: Implications for arc tectonics and incipient magmatism: Jour. Geophys. Res., v. 90, no. B10, p. 8729-8742.

Wakita, H., Sano, Y., and Mizoue, M., 1987, High ^{3}He emanation and seismic swarms observed in a nonvolcanic, forearc region: Jour. Geophys. Res., v. 92, no. B12, p. 12,539-12,546.

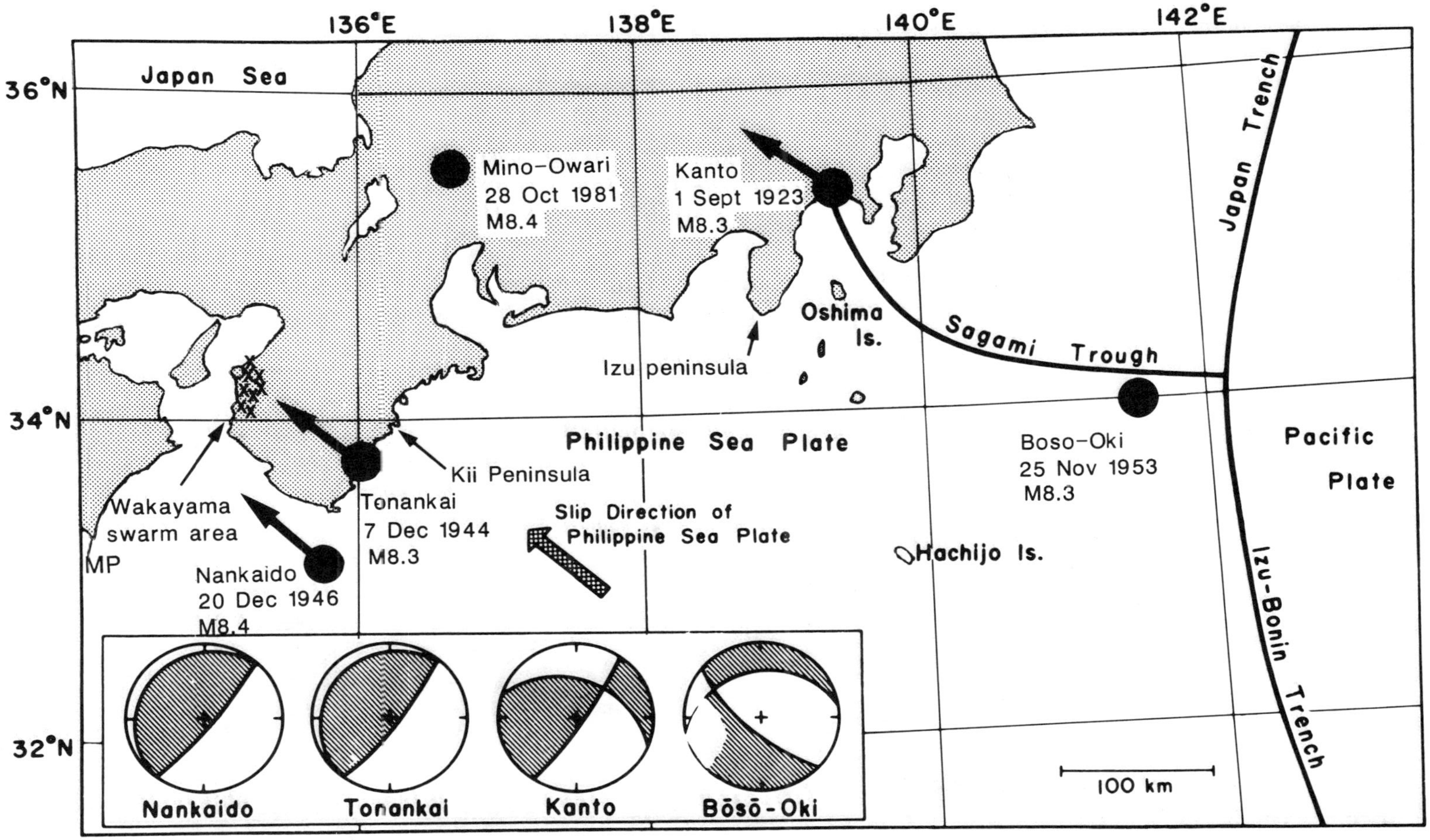

Figure 20.3.1. Locations of Kii Peninsula, Izu Peninsula, and great earthquakes ($\underline{M} > 8$) of southwestern and central Honshu since 1880, from Kanamori (1972). Arrows show direction of slip of footwall side for Kanto, Tonankai, and Nankaido earthquakes. Note location of Muroto Point (MP) on Shikoku Island for later reference in figure 20.3.8.

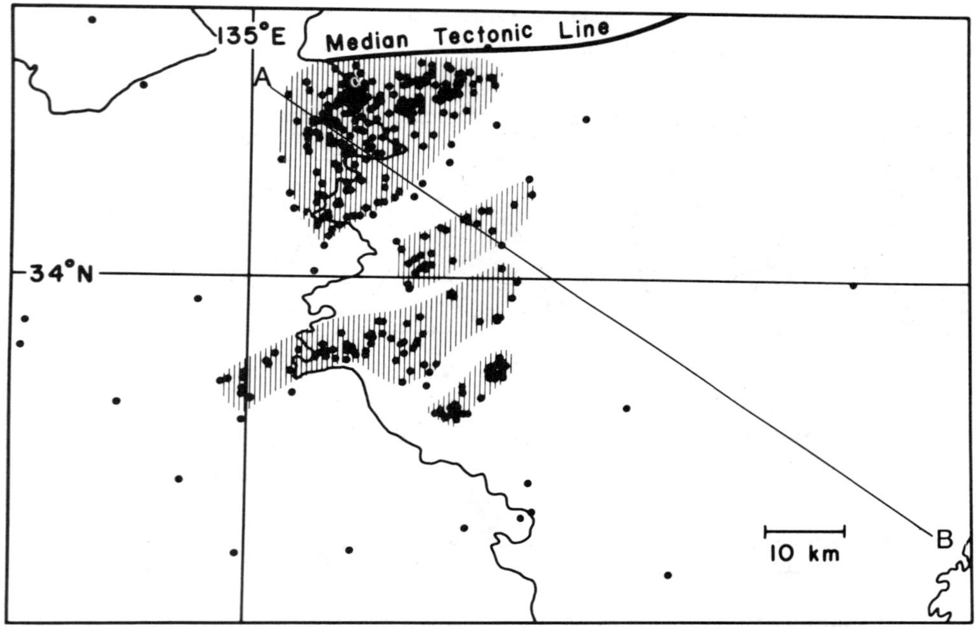

Figure 20.3.2. Epicenters of shallow earthquakes in vicinity of Wakayama, Kii Peninsula, from January to June 1975 (Kanamori and Tsumura, 1971). Note abrupt termination at Median Tectonic Line (see text).

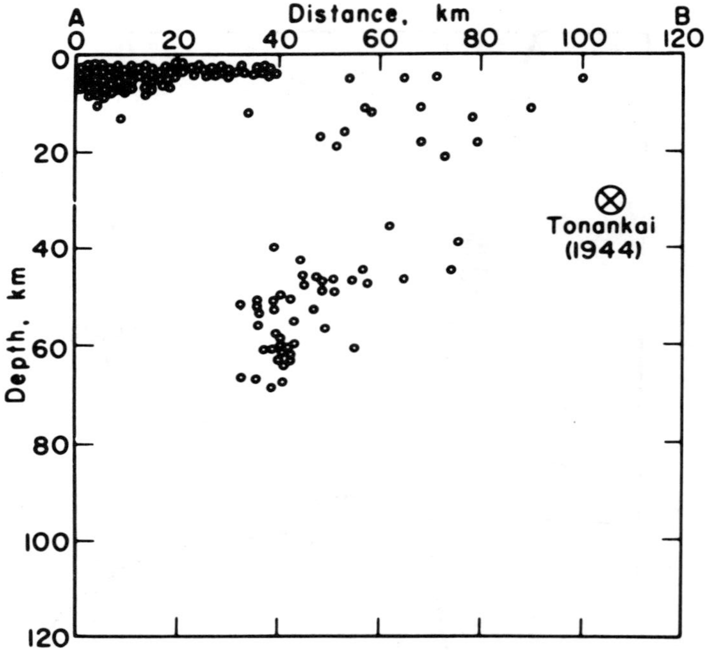

Figure 20.3.3. Hypocenters of earthquakes in vicinity of Wakayama, Kii Peninsula, from January to June 1975, projected onto a northwest-trending plane (Kanamori and Tsumura, 1971). Endpoint A is near Wakayama, and B is near epicenter of the 1944 Tonakai earthquake at southeast tip of Kii Peninsula. Note shallow swarm earthquakes near Wakayama and deep subduction earthquakes to southeast.

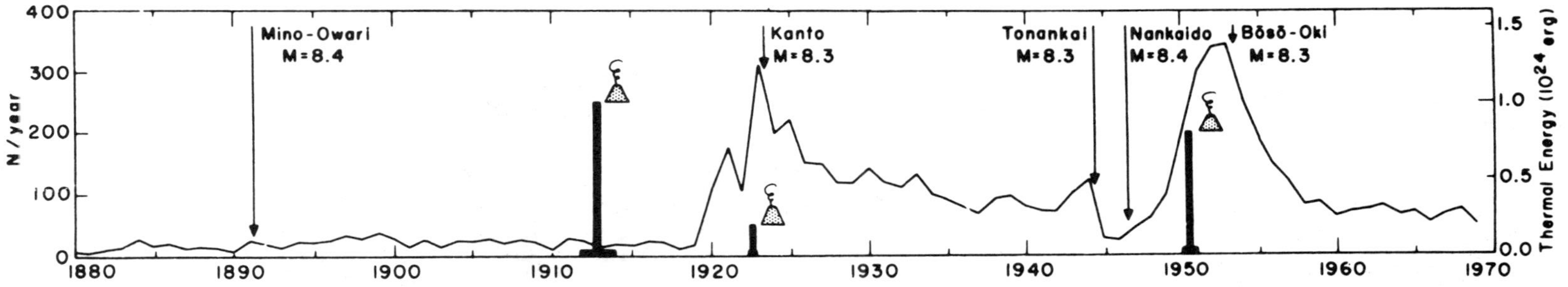

Figure 20.3.4. Annual number of felt earthquakes at Wakayama, 1880-1970 (Kanamori, 1972). Peaks represent swarms in Wakayama area (see figure 20.3.1). Arrows mark times of great earthquakes ($\underline{M} > 8$) in region. Vertical bars indicate thermal energy from eruptions of Oshima Island.

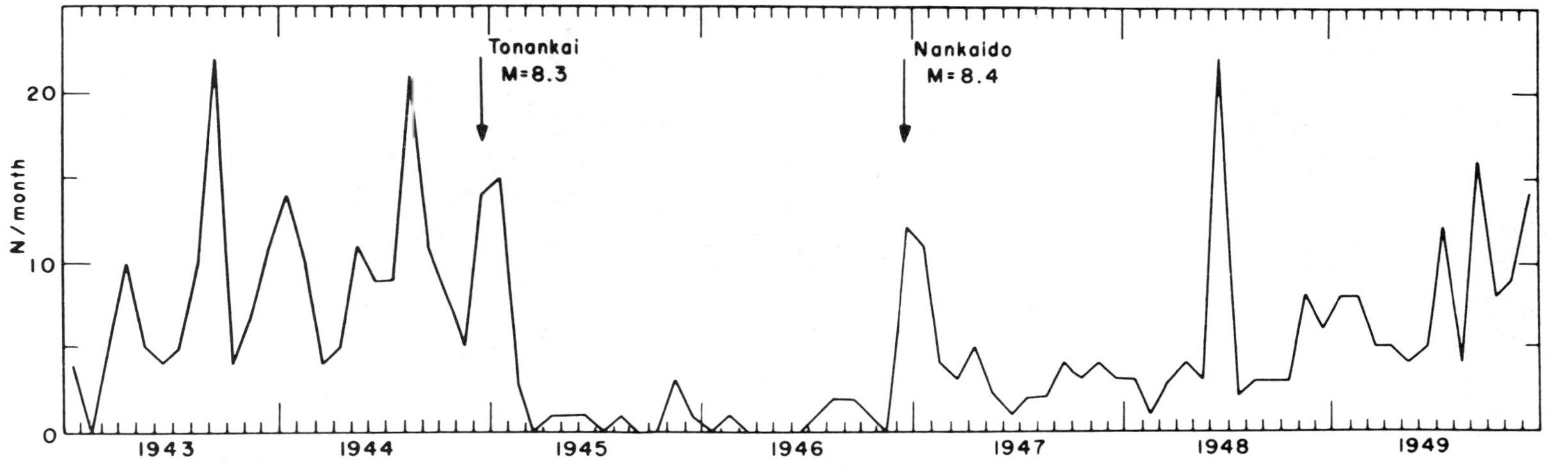

Figure 20.3.5. Monthly number of felt earthquakes at Wakayama, 1943-49 (Kanamori, 1972). Peaks represent swarms in Wakayama area.

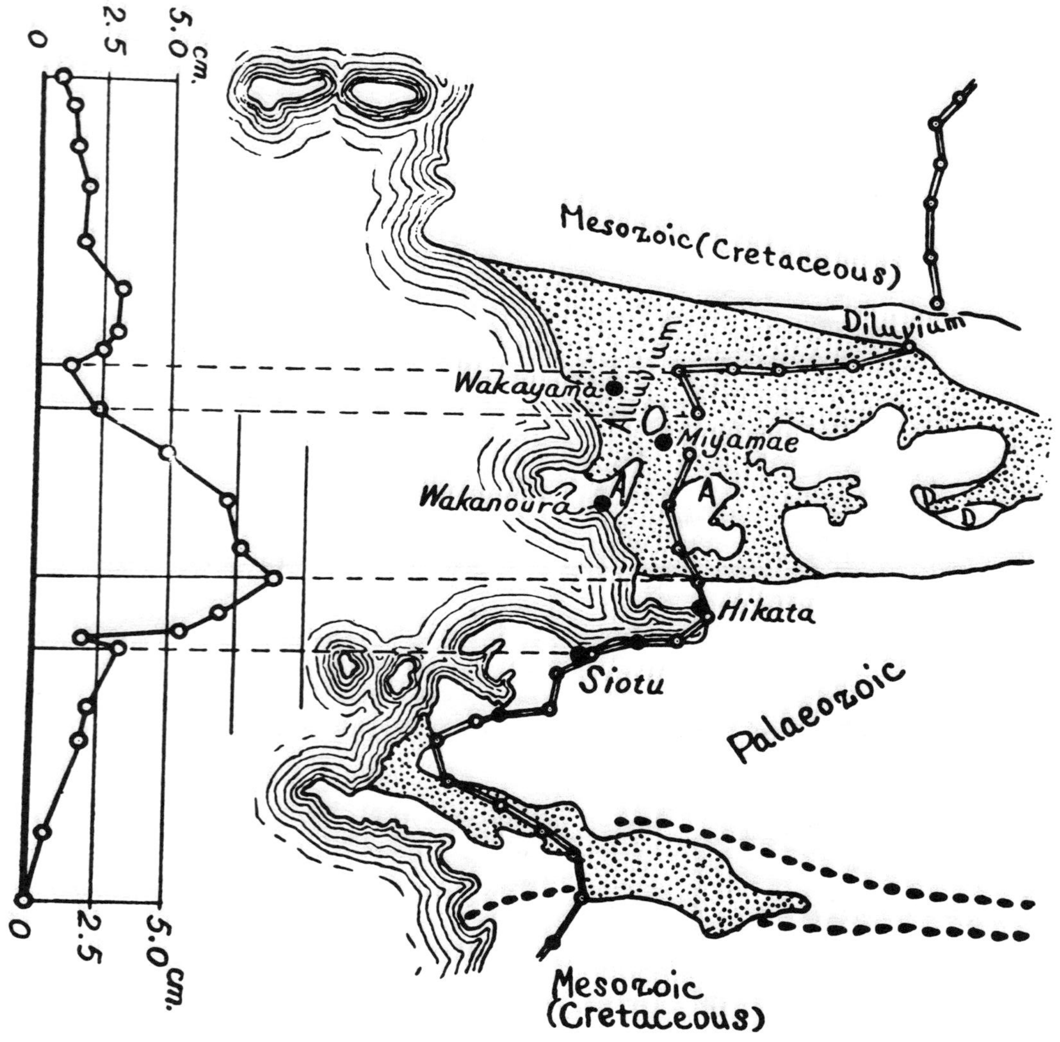

Figure 20.3.6. Location of leveling line along west coast of Kii Peninsula near Wakayama and uplift profile, 1895-1928 (Imamura and others, 1932).

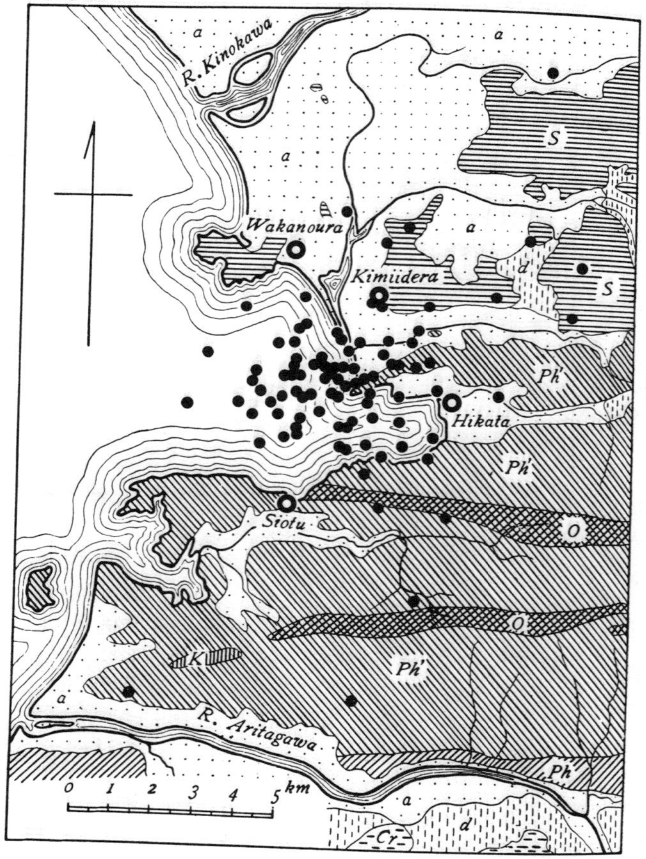

Figure 20.3.7. Earthquake epicenters along west coast of Kii Peninsula south of Wakayama, 1929-31 (Imamura and others, 1932). Bold open circles are towns; patterns show geologic units.

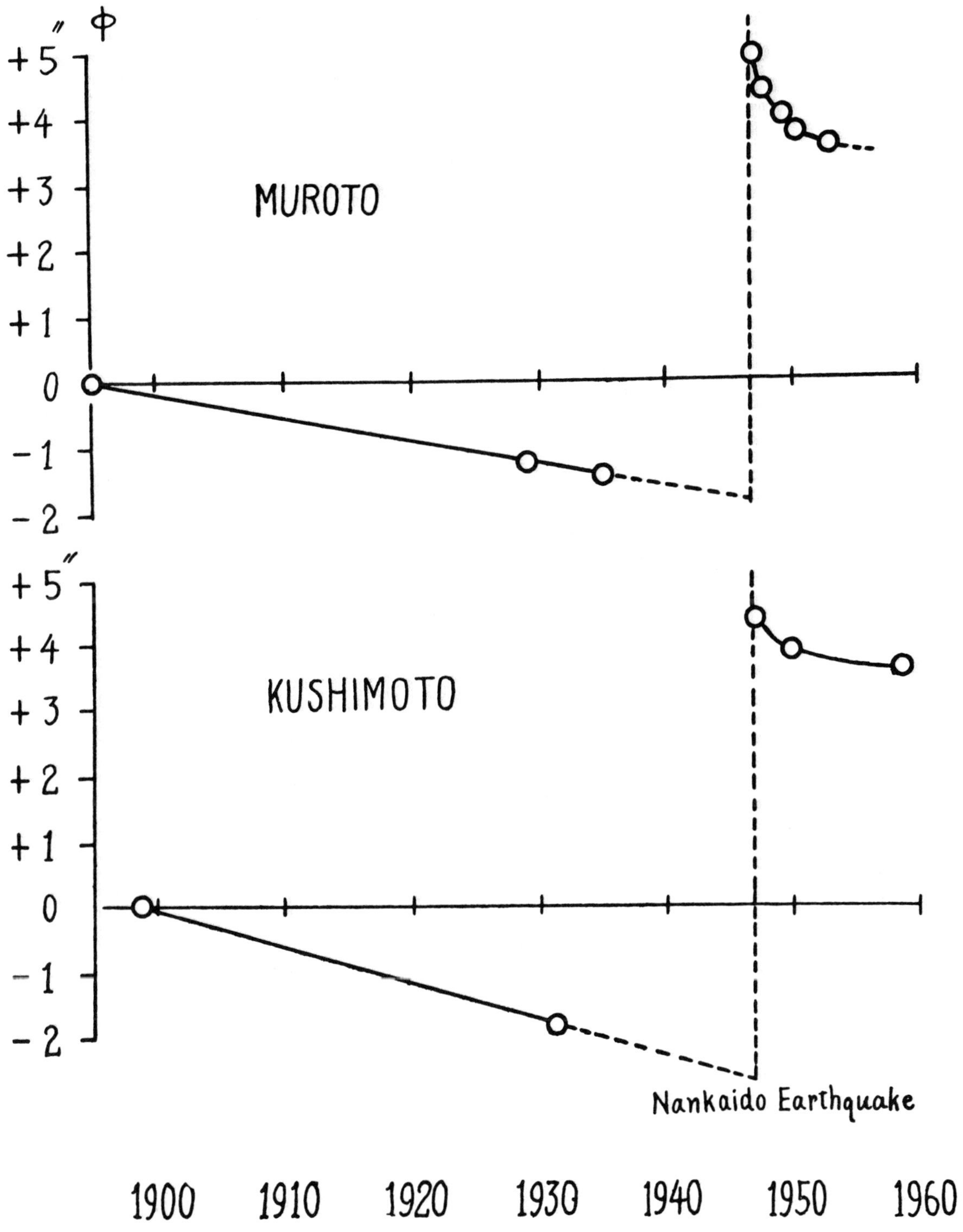

Figure 20.3.8. Southerly tilt in arcseconds (1 arcsecond is about 5 microradians) at Muroto Point, Shikoku Island (see figure 20.3.1) and Kushimoto (south tip of Kii Peninsula), 1928-50 (Okada, 1960). Tilting preceded the Tonakai (1944) and Nankaido (1946) earthquakes by at least two decades.

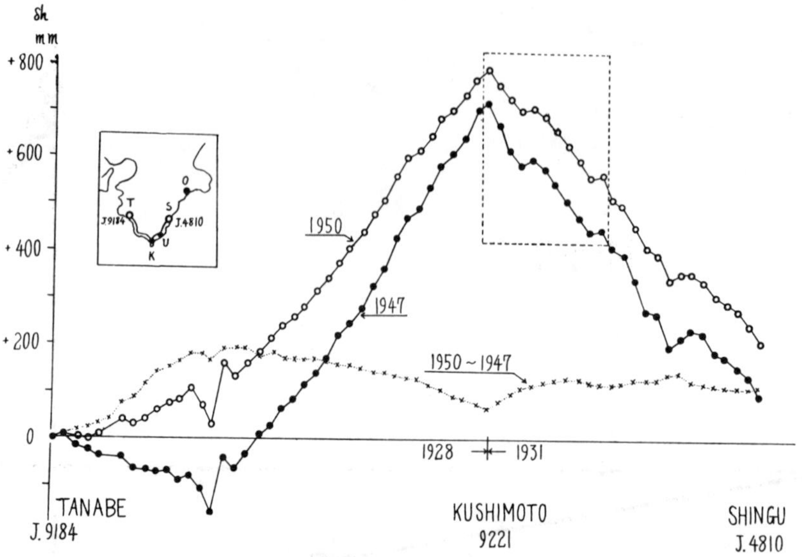

Figure 20.3.9. Results of leveling surveys around southern tip of Kii Peninsula, 1928-50 (Okada, 1960). Note uplift and northerly tilt shortly after great earthquakes of 1944 and 1946.

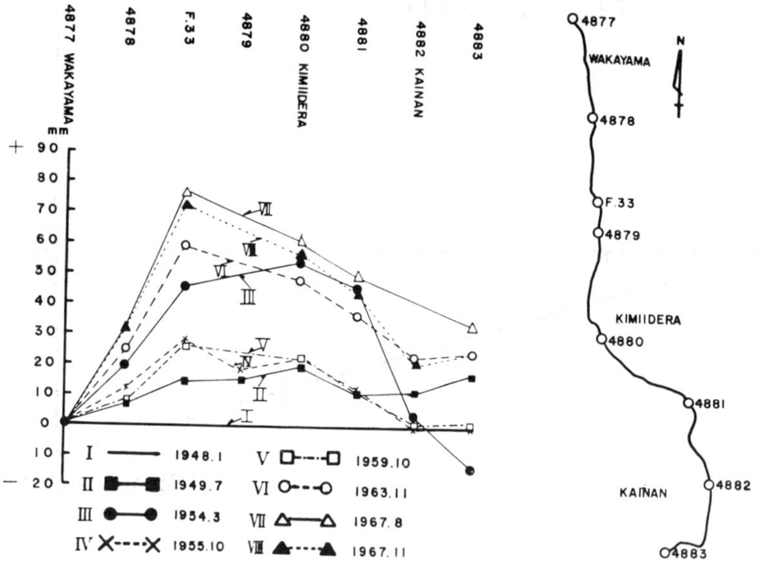

Figure 20.3.10. Profiles of elevation changes along west coast of Kii Peninsula, showing uplift and subsidence, 1948-67 (Okada and others, 1968). Bench mark 4877 at north end of leveling line was assumed stable.

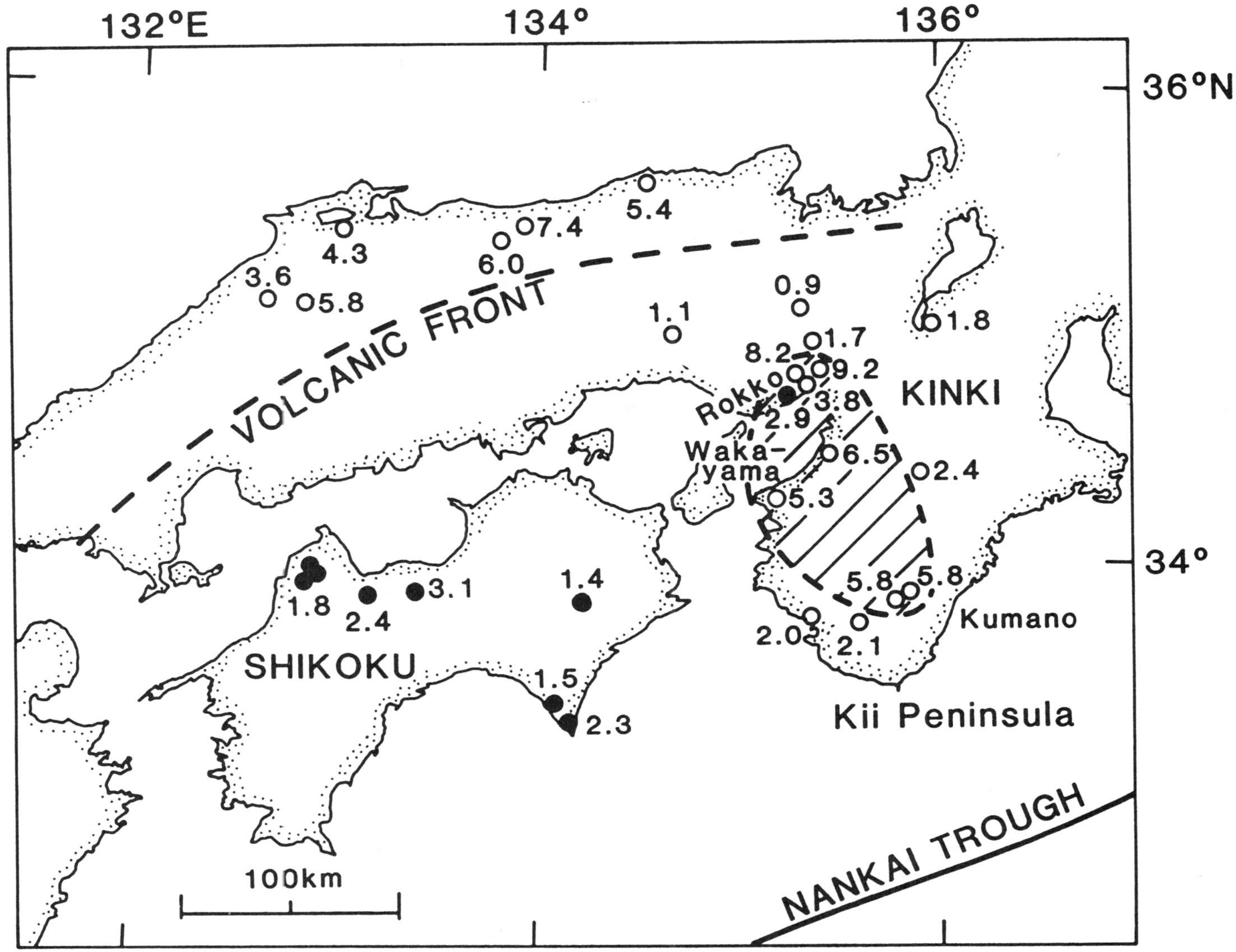

Figure 20.3.11. Locations of sampling sites and measured $^3He/^4He$ ratios (x 10^{-6}), Kii Peninsula (Wakita and others, 1987). Shown are volcanic front and outline of an area (hachured region) where $^3He/^4He$ ratios are higher than 2.8 x 10^{-6}, twice the atmospheric value. Closed and open circles represent data from studies by Wakita and others (1987) and Sano and Wakita (1985), respectively. Copyright by the American Geophysical Union.

MATSUSHIRO

Latitude Longitude (degrees)	Tectonic setting	Age of latest volcanism	Youngest volcanic rock types	Restless years	Type and duration of unrest ESTU STHF MGTF H Te	Associated eruption?
36.55N 138.20E	Compr	350,000 yr B.P.	d	887	x--- ---- ---- - --	none
				1718	x--- ---- ---- - --	none
				1847	x--- ---- ---- - --	none
				1853	x--- ---- ---- - --	none
				1858	x--- ---- ---- - --	none
				1897	EE-- ---- ---- - --	none
				1912	x--- ---- ---- - --	none
				1941-44	FF-- ---- ---- - --	none
				1952	x--- ---- ---- - --	none
				1958	x--- ---- ---- - --	none
				1965-68	FF-E EFFx FFF? F =?	none

TECTONIC/GEOLOGIC SETTING

An area around Matsushiro in northwest Honshu has experienced earthquakes, earthquake swarms, and other crustal unrest many times in recorded history (fig. 20.4.1). Earthquake swarms during 1965-68 are the best known and probably the strongest in the recent history of this area, but they are just the latest of at least three swarms in the past 100 years. The maximum compressional stress (regional) is WNW-ESE or E-W, and a number of NW-SE-trending left-lateral faults and NE-SW-trending right-lateral faults have been mapped. Nakamura (1983, 1984) inferred that Matsushiro, Niigata, and several other nearby areas prone to frequent seismic swarms lie near a nascent (newly forming) convergent boundary between the Eurasian and North American plates--a boundary that includes the Fossa Magna and passes along the northwest coast of Honshu.

No volcanic activity is known within historical time; one of the youngest volcanic vents in the area is Mt. Minakami, a dacite dome that last erupted about 350,000 yr B.P. (Matsuda, 1967; I. Yokoyama, written commun., 1983).

HISTORICAL ACTIVITY

An earthquake swarm occurred near Matsushiro in 1897 ($\underline{M}_{max}$=6.3) and another occurred in 1941-44 ($\underline{M}_{max}$=6.2) (fig. 20.4.2). In addition, earthquakes with magnitudes estimated to be greater than 5 occurred in the Matsushiro-Nagano area in 887, 1718, 1847, 1853,

MATSUSHIRO (continued)

HISTORICAL ACTIVITY (continued)

1858, 1912(?), 1952, 1958, and 1963 (Ichikawa, 1969). The largest earthquake in 1941 (July) had a $\underline{M}$=5.0 foreshock in March 1941 and many small foreshocks 1 hour before the main shock.

Unrest from 1965 to 1968, centered about 2 km north-northeast of Mt. Minakami, included increased seismicity, uplift, surface ruptures of local origin (landslides), and en echelon fractures with 50 cm of observable left-lateral strike-slip displacement and 30 cm of opening along a vertical fault trending N55°W (figs. 20.4.3-20.4.10). An inferred fault passes through the center of uplift, with about 20-30 cm of uplift on the southwest side of the fault relative to the northeast side (Nakamura and Tsuneishi, 1967). More than 700,000 earthquakes were recorded, with a total seismic energy release equivalent to a $\underline{M}$=6.3 earthquake (approximately 10^{23} ergs) (Tada and Asano, 1983). Hagiwara and Iwata (1968) described five periods of seismicity:

(1) August 1965 - February 1966, including a peak in November 1965 (relatively small epicentral area, less than 10-km diameter; mean depth about 4 km, 2-3 km during November 1965 - February 1966; first motion suggesting strike-slip(?), maximum compression N80-90°E) (fig. 20.4.3);

(2) March 1966 - July 1966, including a peak in March-April 1966 (still a relatively small epicentral area, just slightly expanded in a NE-SW direction; mean focal depth=4 km; included the strongest earthquakes of the entire swarm);

(3) August 1966 - December 1966, including a peak in August - September 1966 (noticeable extension of epicentral area NE-SW; perhaps a few more deeper earthquakes, mean depth about 5 km; earthquakes decreased on northeast side of Mt. Minakami, near the inferred fault; earthquakes were accompanied by landslides, ground fracturing, and new, gushing springs);

(4) January 1967 - May 1967 (earthquake epicenters became much more scattered along the same NE-SW zone, now measuring about 18 km x 34 km; a significant number of earthquakes occurred at depths of 8-14 km in the southwest part of the area) (fig. 20.4.4); and

(5) June 1967 - May 1968 (sparse number of earthquakes, with very few near Matsushiro; a few earthquakes at 2-8 km depths in the southwest end of the expanded epicentral area).

Fault-plane solutions for events until March 1966 indicated a consistent maximum horizontal stress direction, N80°W. Solutions for April 1966 to July 1966 gave more variable directions, averaging N87°W, and solutions from August 1966 onward indicated even more variable directions, averaging S88°W in the northern and central parts of the area and N74°W in the southern part of the epicentral area (Ichikawa, 1969). The source region, 15 km x 30 km and 0-10 km deep, is also an area of high seismic velocities at anomalously shallow depth ($\underline{V}_p$=6 km/sec, "granitic") surrounded by an area of anomalously low velocities and high attenuation ("weak country rock") (Okada and others, 1970; Tada and Asano, 1983; Nakamura, 1984). The "granitic layer" was more subject to brittle fracture

MATSUSHIRO (continued)

HISTORICAL ACTIVITY (continued)

than were the relatively weak rocks surrounding it. The area of uplift at the surface corresponded to the area in which this "granitic layer" lies at a shallow depth.

Anomalous rapid tilts of approximately 5 microradians magnitude, just south of Matsushiro town and just southwest of Mt. Minakami, occurred a few days prior to the relatively large earthquakes on 22 November 1965 and 10 March 1966 (Hagiwara and others, 1966). Maximum observed tilt between October 1965 and March 1966 was about 15 arcseconds (45-50 microradians), east and south up.

Uplift was especially rapid during periods 2 and 3. Maximum uplift (70-80 cm) was attained in October 1966. Subsidence from October 1966 onward reduced the net uplift to about 40 cm by the end of the swarms.

Maximum horizontal deformation (from electronic distance measurements) in the epicentral area was 116 cm of extension along a 3.1-km-long, N-S line (Sorobeku line), and 22 cm of compression on a 2.4-km-long, E-W line (Zozan line). Together, these data indicate between 1.5 and 2 m of left-lateral offset on the 10-km-long fault zone. Kasahara and others (1968) showed a change in the direction of contractional strain, from N70-80°W in late 1965 and early 1966 to S70-90°W in May-October 1966; this change correlates with the change in strain axes inferred from fault-plane solutions (Ichikawa, 1969).

The total magnetic intensity at Matsushiro relative to that at Kanozan (a control site, 180 km from Matsushiro) increased by 15 nT from November 1965 to September 1966. In October and November it declined by about 10 nT, and upon remeasurement 3 years later, it had declined by an additional 10 nT (fig. 20.4.13). Yamazaki and Rikitake (1970) infer that the increase in 1965-66 was a piezomagnetic effect due to stress of an unspecified origin, and that the subsequent decrease was due to a relaxation of that stress. Stuart and Johnston (1975) inferred that the initial increase was caused by "inflation of a shallow magma reservoir within a crust containing pre-existing horizontal shear stresses," and that the subsequent decrease was caused by "thermal demagnetization of host rocks." Johnston and others (1983) retained their interpretation of the increase at Matsushiro, but suggested that the decrease was due to stress relaxation.

Earthquakes triggered artesian activity at one or more hot-water wells. On two occasions, in March and August 1966, discharge at the Kagai thermal springs declined for several days, then increased on the day before principal earthquake swarms, with net increases of up to 10 percent of normal discharge (Kasuga, 1967) (figs. 20.4.11, 20.4.12). A remarkable peak in flow from both new and old springs--termed a "water eruption"--occurred during September-October 1966, just before the time of maximum uplift (October 1966) and just prior to expansion of the epicentral area. On two occasions, in July and mid-September 1966, strong flow from Well A of Kagai thermal springs carried sludge and rock fragments--in small hydrothermal eruptions. The temperature of discharged waters at Kagai rose 2 °C from November 1965 to April 1966, then rose another 1.5 °C by January 1967. These temperature changes bore an inverse relation to the discharge--higher rates of discharge corresponded to slightly lower temperatures, as would be expected if the ground water was of shallow, cool origin and was being heated in proportion to the time it convected through hotter rock at depth before being discharged, or if the normally hot water was being augmented by cool ground water from new sources.

MATSUSHIRO (continued)

HISTORICAL ACTIVITY (continued)

Centers of uplift, earthquakes, and unusual spring flow were all more or less coincident. The volume of subsidence following the peak uplift in October 1966 corresponds roughly to the total volume of anomalous spring discharge in the Matsushiro area (10^7 m^3, according to Iijima (1969) and Kisslinger (1975)).

COMMENTS

A lively debate on Matsushiro activity has included the following interpretations:

(a) magma intrusion into a region with preexisting horizontal shear stresses (modeled as a 1-km-diameter source at 3.5-km depth or as a 1-km-diameter cylinder at 2-km depth) (Stuart and Johnston, 1975; Wakita and others, 1978);

(b) dilatancy associated with formation of a new NW-SE-trending left-lateral fault, and conjugate right-lateral motion on one or more NE-SW-trending faults (Kasahara, 1970; Nur, 1974; Kisslinger, 1975);

(c) a combination of tectonic and hydrologic processes in which formation of the above-mentioned fault increased ground water pore pressures causing uplift until the "eruption" of ground water in September 1966 reduced pore pressures and permitted subsidence (Tsuneishi and Nakamura, 1970; Nakamura, 1971; Ohtake, 1974, 1976);

(d) a dominantly hydrothermal process, in which an influx of hot water caused increased ground water pore pressures or hydration of clay minerals and thereby caused uplift (Kasuga, 1967; Matsushima, 1967).

REFERENCES

Asano, S., Kubota, H., Okada, H., Nogoshi, M., Suzuki, H., Ichikawa, K., and Watanabe, H., 1969, Explosion seismic studies of the underground structure of the Matsushiro swarm area: Jour. Phys. Earth, v. 17, p. 77-90.

Hagiwara, T., and Iwata, T., 1968, Summary of the seismographic observation of Matsushiro swarm earthquakes: Bull. Earthq. Res. Inst., v. 46, p. 485-515.

Hagiwara, T., Yamada, J., and Hirai, M., 1966, Observation of tilting of the earth's surface due to Matsushiro earthquakes (I): Bull. Earthq. Res. Inst., v. 44, p. 351-361.

Healy, J.H., Rubey, W.W., Griggs, D.T., and Raleigh, C.B., 1968, The Denver earthquakes: Science, v. 161, p. 1301-1310.

Ichikawa, M., 1969, Matsushiro earthquake swarm: Geophys. Mag., v. 34, p. 307-331.

Iijima, H., 1969, Surface geology of Matsushiro area and disasters by the Matsushiro swarms, in Studies on the Matsushiro earthquake swarm: Tokyo, National Research Center for Disaster Prevention, p. 103-115.

Johnston, M.J.S., Mueller, R.J., Silverman, S.A., and Keller, V.G., 1983, Magnetic measurements across the Long Valley caldera, 1972-1983: A comparison with observations and mechanisms during the Matsushiro earthquake swarm [abs.]: Earthq. Notes, v. 54, no. 1, p. 90.

MATSUSHIRO (continued)

REFERENCES (continued)

Kasahara, K., 1970, The source region of the Matsushiro swarm earthquakes: Bull. Earthq. Res. Inst., v. 48, p. 581-602.

Kasahara, K., Okada, A., Shibano, M., Sasaki, K., Matsumoto, S., and Hirai, M., 1968, Electro-optical measurement of horizontal strains accumulating in the swarm earthquake area (4): Bull. Earthq. Res. Inst., v. 46, p. 651-661.

Kasuga, I., 1967, Aspect on the relation of thermal water and Matsushiro earthquakes in Kagai Hotspring area, Nagano Prefecture: Jour. Geogr. (Chigaku Zasshi), v. 76, p. 16-26.

Kisslinger, C., 1968, Energy density and the development of the source region of the Matsushiro earthquake swarm: Bull. Earthq. Res. Inst., v. 46, p. 1207-1223.

----- 1975, Processes during the Matsushiro, Japan, earthquakes as revealed by leveling, gravity, and spring-flow observations: Geology, v. 3, p. 57-62.

Matsuda, T., 1967, Geological aspect of the Matsushiro earthquake fault: Bull. Earthq. Res. Inst., v. 45, p. 537-550.

Matsushima, S., 1967, Water content in the crust and the hydrothermal reaction as a possible cause for the swarm type earthquake: Kyoto Univ. Geophys. Inst. Spec. Contrib., no. 7, p. 201-210.

Nakamura, K., 1971, A lesson from the Matsushiro earthquake swarm: Kagaku Asahi, no. 10, p. 127-133.

----- 1983, Possible nascent trench along the eastern Japan Sea as the convergent boundary between Eurasian and North American Plates: Bull. Earthq. Res. Inst., v. 58, p. 711-722.

----- 1984, On the hypothesis that the East Japan Sea - Fossa Magna represent a convergent boundary between the Eurasian and North American plates: Earth Monthly, v. 6, p. 25-28.

Nakamura, K., and Tsuneishi, Y., 1967, Ground cracks at Matsushiro probably of underlying strike-slip fault origin (II)--The Matsushiro earthquake fault: Bull. Earthq. Res. Inst., v. 45, p. 417-471.

Nur, A., 1974, Matsushiro, Japan, earthquake swarm: confirmation of the dilatancy-fluid diffusion model: Geology, v. 2, p. 217-221.

Ohtake, M., 1974, Seismic activity induced by water injection at Matsushiro, Japan: Jour. Phys. Earth, v. 22, p. 163-176.

----- 1976, The Matsushiro earthquakes, after 10 years: Kagaku (Science), v. 46, p. 306-313.

Okada, H., Suzuki, S., and Asano, S., 1970, Anomalous underground structure in the Matsushiro earthquake swarm area as derived from a fan-shooting technique: Bull. Earthq. Res. Inst., v. 48, p. 811-833.

Rikitake, T., Yamazaki, Y., Sawada, M., Sasai, Y., Yoshino, T., Uzawa, S., and Shimomura, T., 1967, Geomagnetic and geoelectric studies of the Matsushiro earthquake swarm (5): Bull. Earthq. Res. Inst., v. 45, p. 395-416.

Stuart, W.D., and Johnston, M.J.S., 1975, Intrusive origin of the Matsushiro earthquake swarm: Geology, v. 3, p. 63-67.

Tada, T., and Asano, S., 1983, Anomalous crustal uplift and crustal structure in the Matsushiro and the Izu Peninsula earthquake swarm regions: Tectonophysics, v. 97, p. 241-250.

Tsubokawa, I., Okada, A., Izutuya, S., Ito, Y., and Kadono, K., 1968, Levelling resurvey associated with the area of Matsushiro earthquake swarm (2): Bull. Earthq. Res. Inst., v. 46, p. 417-429.

Tsuneishi, Y., and Nakamura, K., 1970, Faulting associated with the Matsushiro swarm earthquakes: Bull. Earthq. Res. Inst., v. 48, p. 29-51.

Wakita, H., Fujii, N., Matsuo, S., Notsu, K., Nagao, K., and Takaoka, N., 1978, "Helium spots": Caused by diapiric magma from the upper mantle: Science, v. 200, p. 430-432.

Yamazaki, Y., and Rikitake, T., 1970, Local anomalous changes in the geomagnetic field at Matsushiro: Bull. Earthq. Res. Inst., v. 48, p. 637-643.

MATSUSHIRO, Latitude 36.55N, Longitude 138.20E

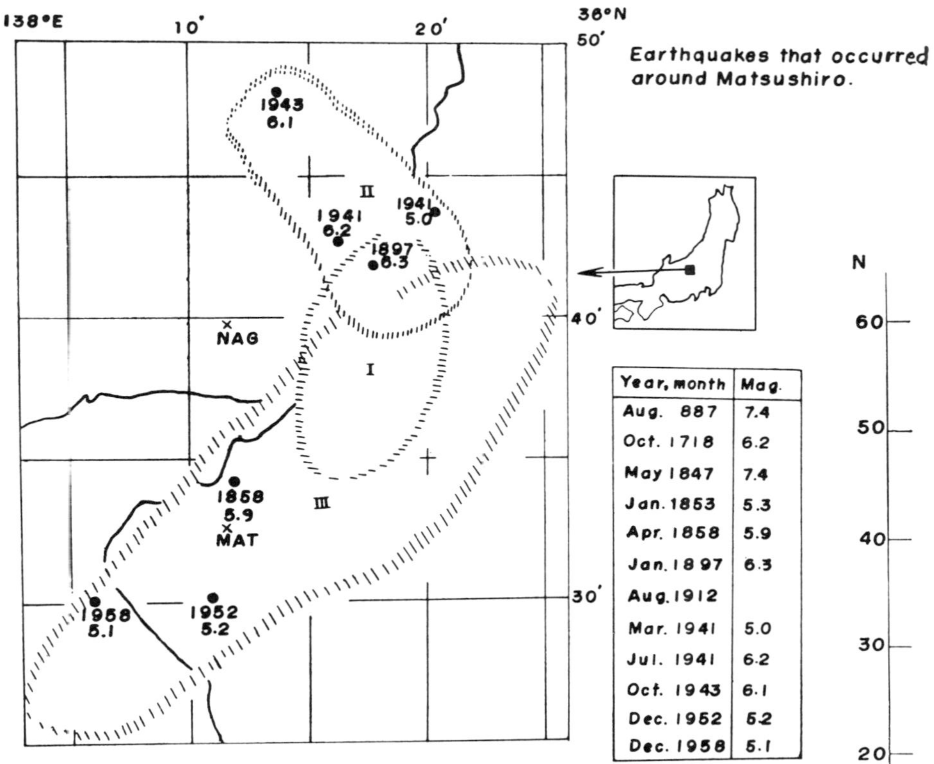

Year, month	Mag.
Aug. 887	7.4
Oct. 1718	6.2
May 1847	7.4
Jan. 1853	5.3
Apr. 1858	5.9
Jan. 1897	6.3
Aug. 1912	
Mar. 1941	5.0
Jul. 1941	6.2
Oct. 1943	6.1
Dec. 1952	5.2
Dec. 1958	5.1

Figure 20.4.1. Major earthquakes in Matsushiro area, A.D. 887-1958 (Ichikawa, 1969). NAG, Nagano Meteorological Observatory; MAT, Matsushiro Seismological Observatory; I, area of 1897 earthquake swarm; II, area of 1941-43 earthquake swarm; III, area of 1965-68 earthquake swarm.

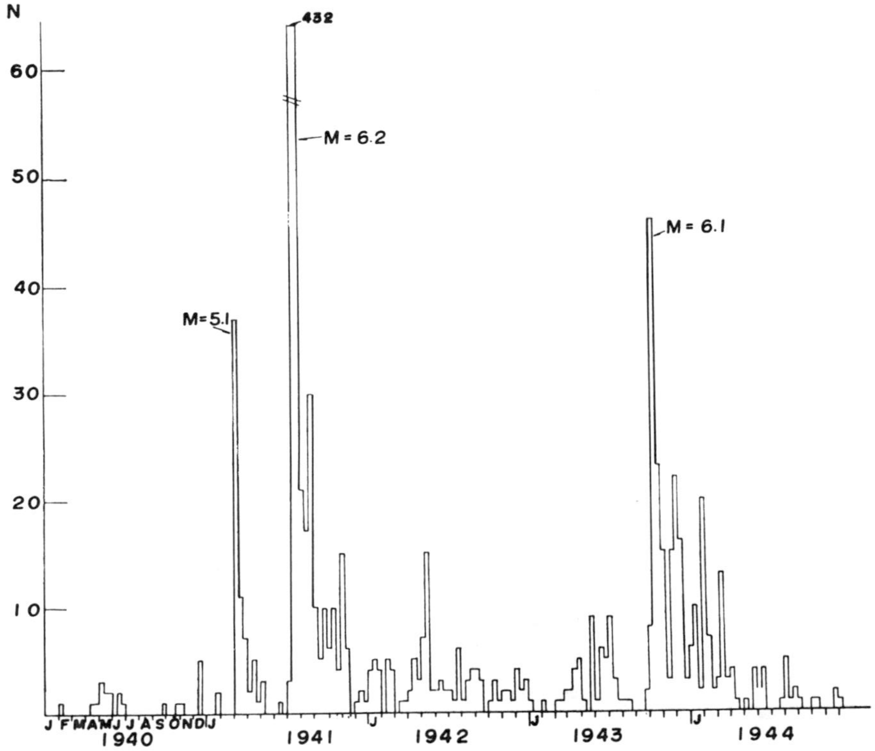

Figure 20.4.2. Number of earthquakes at Matsushiro per 10-day interval, 1940-44 (Ichikawa, 1969). Plotted is number of earthquakes with P - S intervals > 3 s on seismograms of Nagano Meteorological Observatory (see figure 20.4.1).

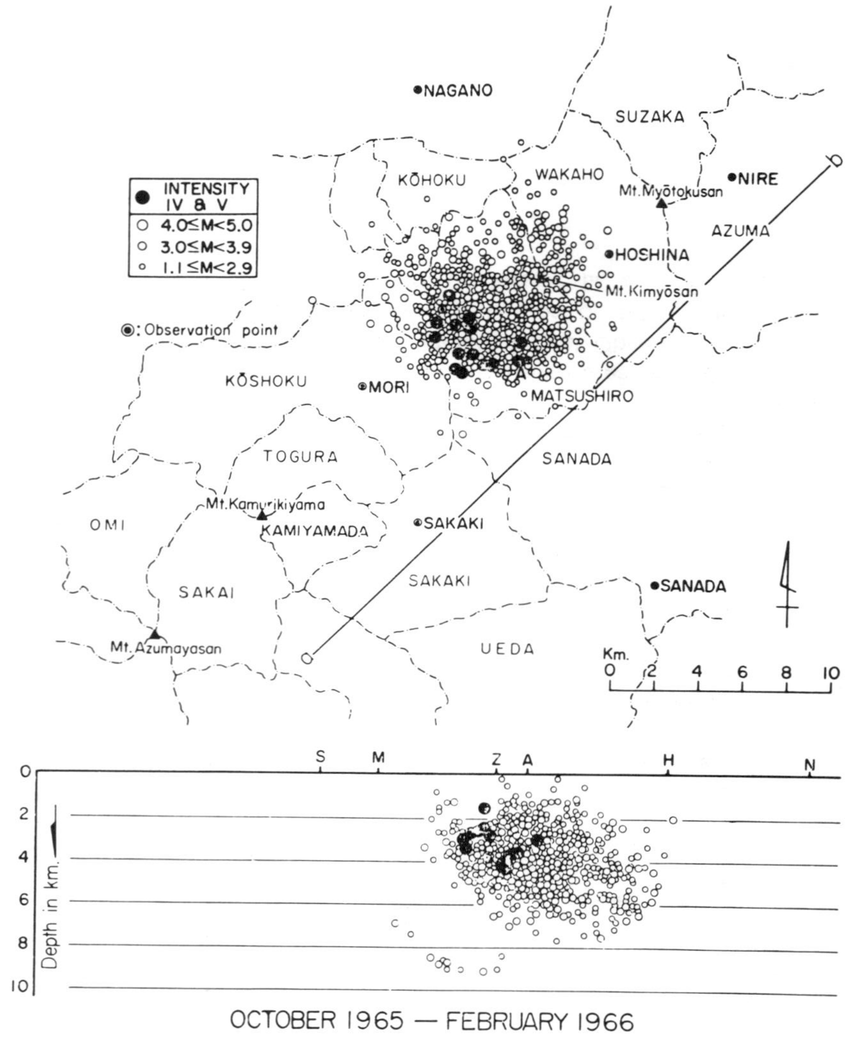

Figure 20.4.3. Epicenters and hypocenters during period 1 (see text), tightly clustered and relatively shallow, near Matsushiro and Mt. Minami-dake (Hagiwara and Iwata, 1968).

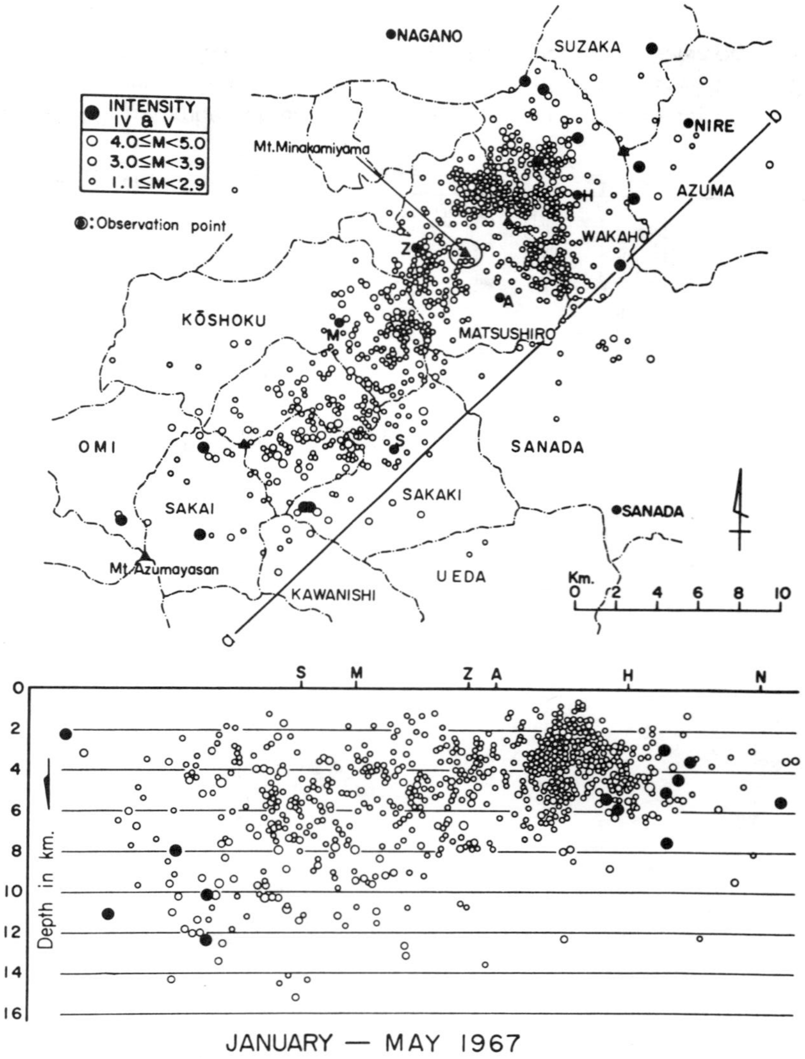

Figure 20.4.4. Epicenters and hypocenters during period 4 (see text), dispersed along a NE-SW trend and spanning a greater depth range in southwest part of swarm area (Hagiwara and Iwata, 1968).

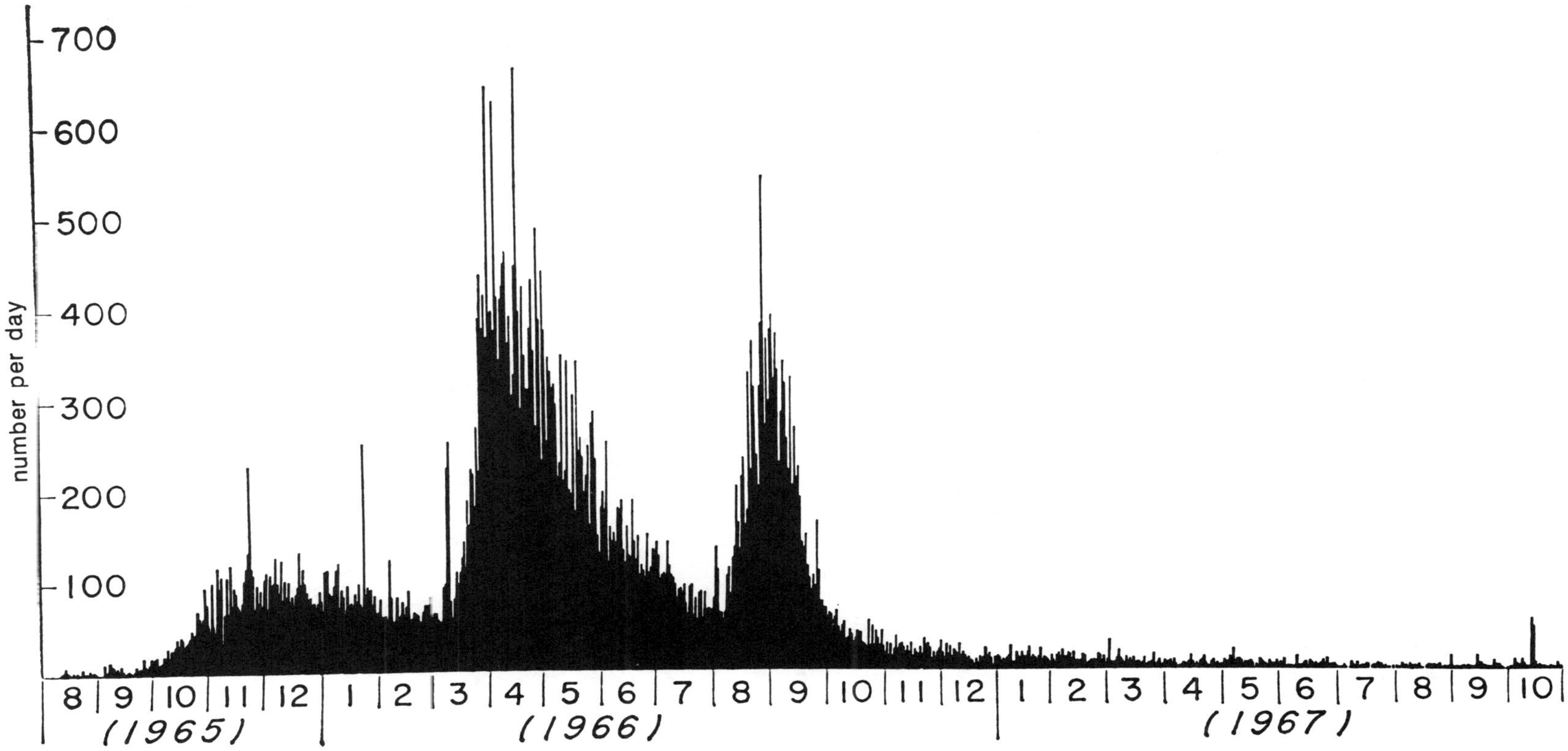

Figure 20.4.5. Daily frequency of felt earthquakes registered at Matsushiro Seismological Observatory, 1965-67, from Tsuneishi and Nakamura (1970).

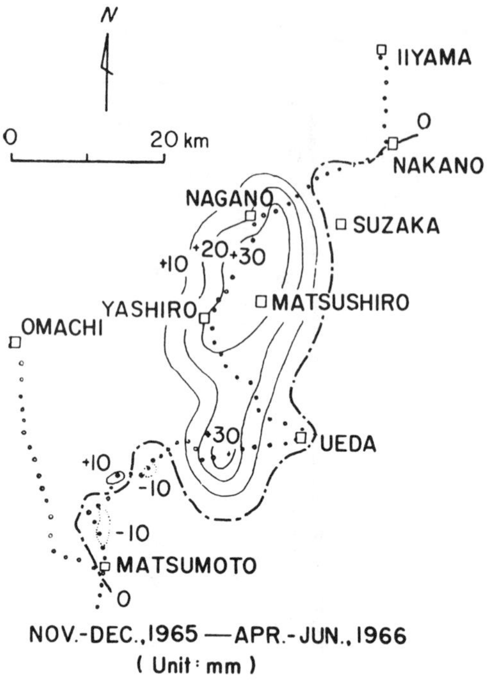

Figure 20.4.6. Contours of vertical land movements over a broad area around Matsushiro, late 1965 to mid-1966, from Kasahara (1970).

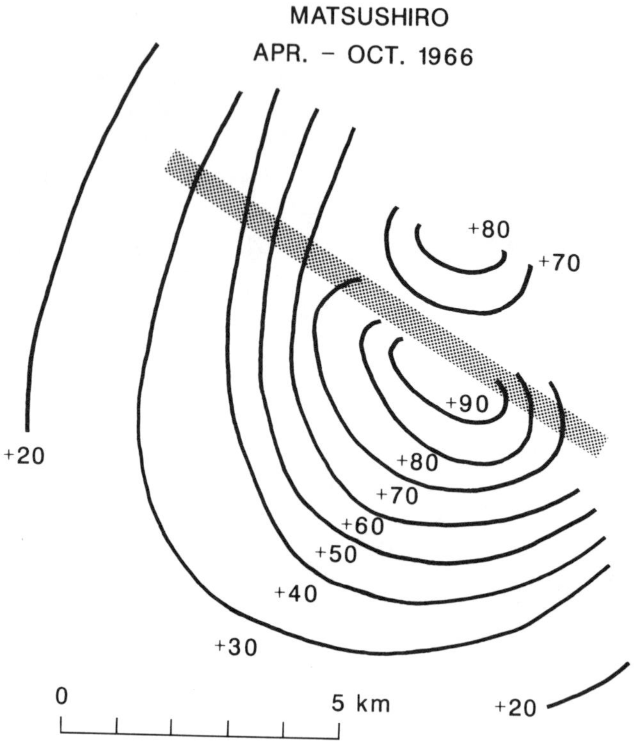

Figure 20.4.7. Contours of local uplift near Matsushiro town, April-October 1966, from Kasahara (1970).

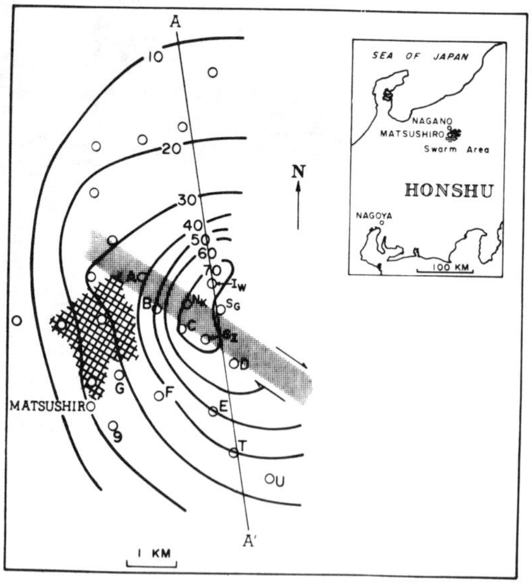

Figure 20.4.8. Contours of local uplift near Matsushiro, October 1965 - September 1966, from Stuart and Johnston (1975).

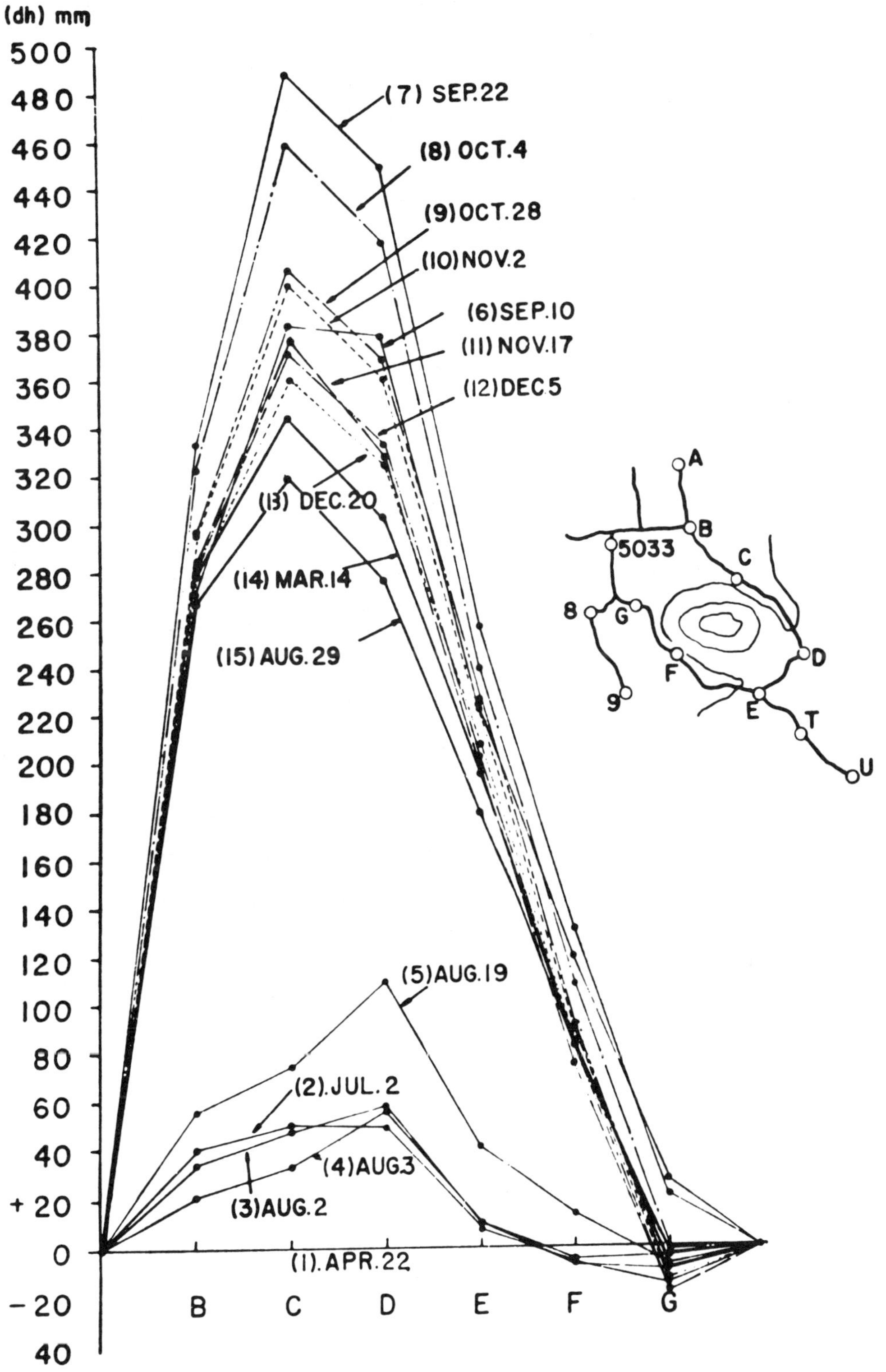

Figure 20.4.9. Sequential profiles of vertical ground displacements along a leveling route just east of Matsushiro, around Mt. Minakami, from 22 April 1966 to 29 August 1967. Numbers in parentheses indicate chronological order of leveling surveys. Reproduced from Tsubokawa and others (1968).

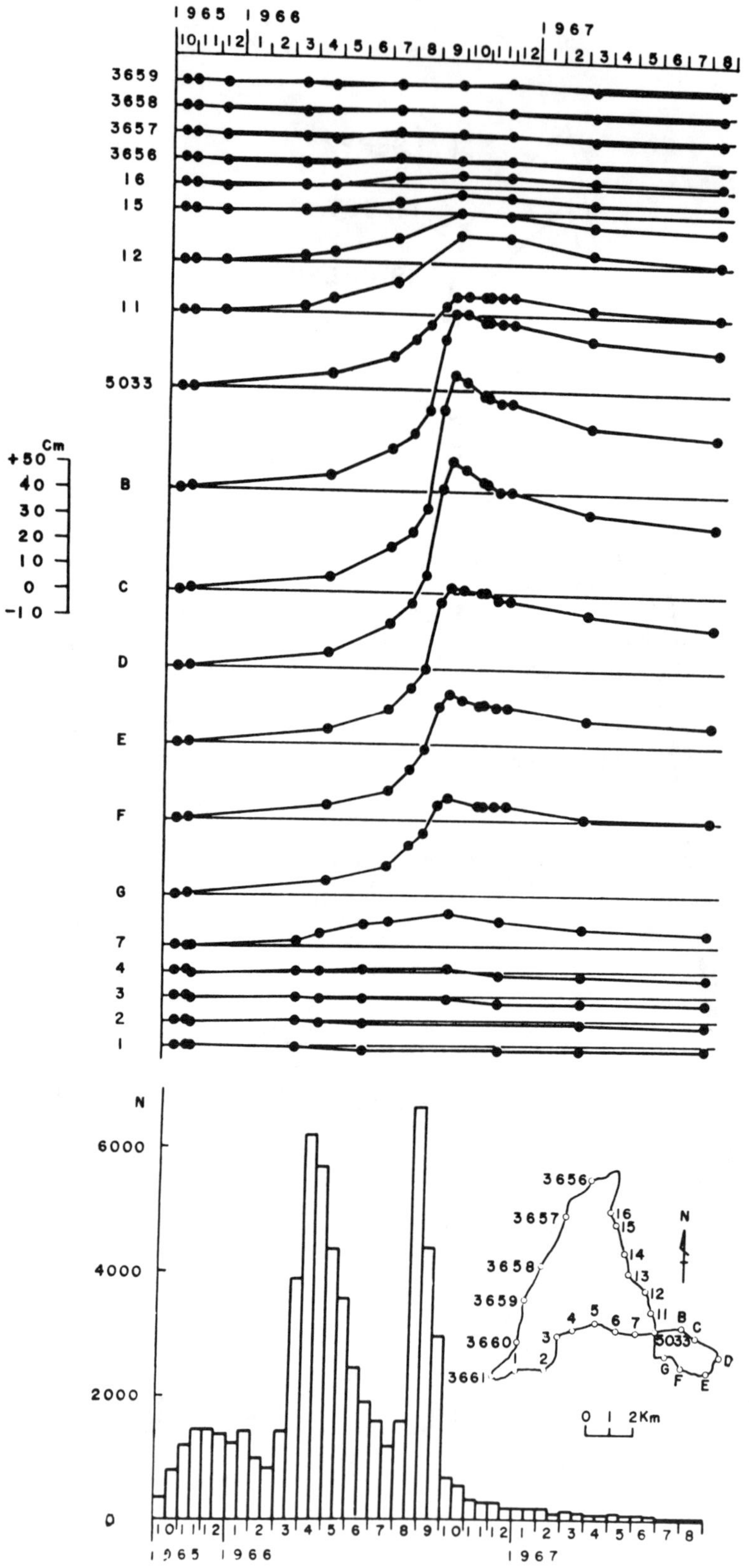

Figure 20.4.10. Top, history of vertical ground displacements at various numbered or lettered bench marks (inset). Bottom, biweekly numbers of earthquakes from October 1965 to August 1967. Both figures from Tsubokawa and others (1968). See figure 20.4.9 for profiles of vertical ground displacements through time.

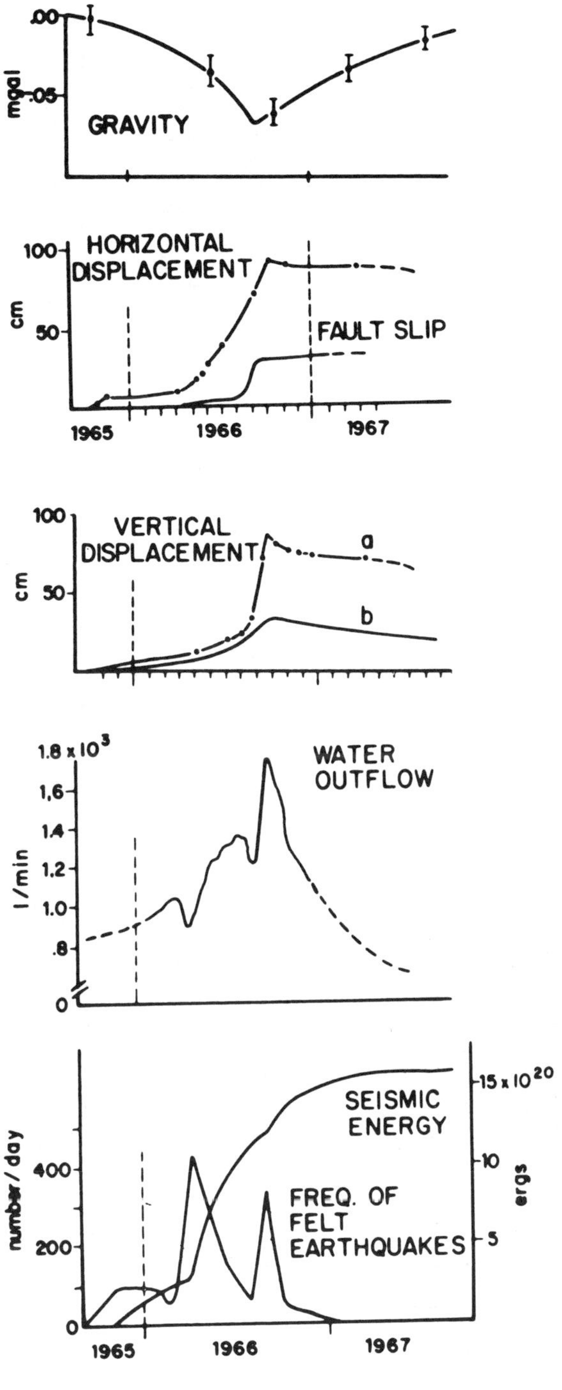

Figure 20.4.11. Temporal variations of deformation, water outflow, seismicity, and gravity at Matsushiro, 1965-68, from Nur (1974). Note that each seismic episode is preceded by declining outflow of ground water and followed by increased outflow. By the end of 1967, ground subsidence was 10-35 percent of peak uplift, whereas gravity had recovered almost completely, indicating net increase of mass.

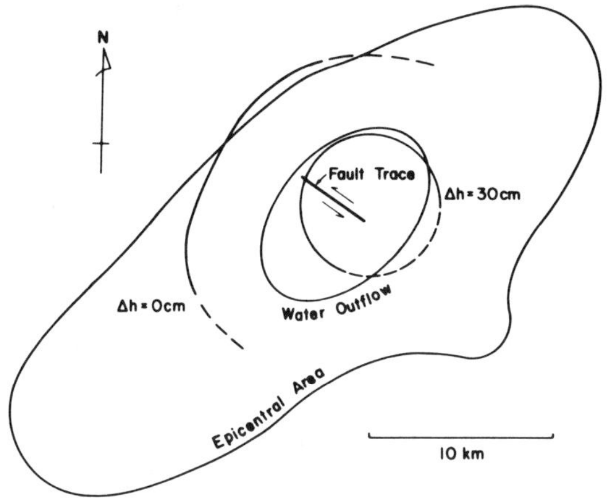

Figure 20.4.12. Spatial distribution of seismicity, water outflow, upheaval, and faulting around Matsushiro during the 1965-68 unrest (Nur, 1974).

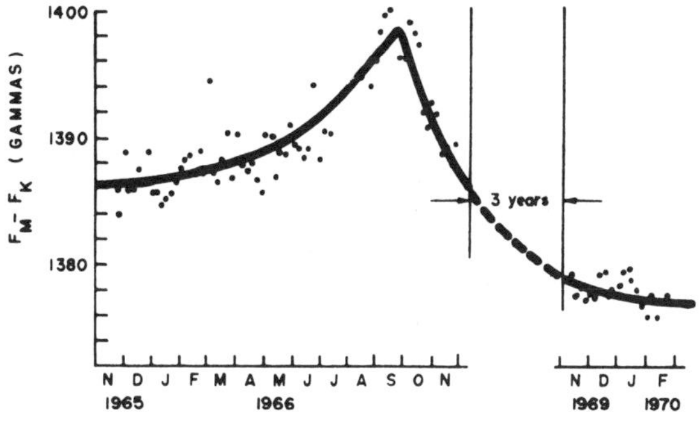

Figure 20.4.13. Five-day means of local magnetic-field differences between stations at Matsushiro and Kanozan, 1965-70 (Stuart and Johnston, 1975).

PART 4: SELECTED EPISODES OF HISTORICAL UNREST IN NONCALDERA SETTINGS (continued)
See inside back cover for explanation and abbreviations

SOCORRO

Latitude Longitude (degrees)	Tectonic setting	Age of latest volcanism	Youngest volcanic rock types	Restless years	Type and duration of unrest ESTU STHF MGTF H Te	Associated eruption?
34.00N 107.00W	Extension, with left-lateral shear on ENE-trending fault(?)	Pliocene	b	1849-50	xx-- ---- ---- - --	none
				1904	xx-- ---- ---- - --	none
				1906-07	xx-- ---- ---- - --	none
				1960's-87+	GG-G ---- --?- ? --	none

TECTONIC/GEOLOGIC SETTING

At least seven overlapping and nested cauldrons of Oligocene to Miocene age lie immediately west of Socorro, New Mexico (fig. 20.5.1) (Chapin and others, 1978). The easternmost and next to the youngest of these, the Socorro cauldron, formed approximately 27 m.y. B.P. It lies wholly within the Rio Grande rift and also lies atop the Morenci lineament, an inferred ENE-WSW-trending transverse shear zone (Chapin and others, 1978). The caldera also lies at the southern tip of an inferred, large, midcrustal magma body (Rinehart and others, 1979; Brocher, 1981). Neotectonic activity in the Rio Grande rift has been summarized by Riecker (1979).

According to Chapin and others (1978), "caldera collapse was followed by resurgent doming and accumulation of moat volcanics and sediments. After a long lull in magmatism...renewed rhyolitic magmatism occurred during the interval 12 to 7 m.y. ago. After another quiet spell, basaltic lava flows were emplaced at approximately 4 m.y. ago."

HISTORICAL ACTIVITY

Anomalous seismicity and uplift have occurred in an area approximately 85 km in diameter extending northward from Socorro (figs. 20.5.1-20.5.4). Swarms are known to have occurred in 1849-50, 1904, 1906-07, and from the early 1960's to the present. Microearthquake swarms have been persistent since a seismometer network was established in 1962. Since then, all earthquakes have been less than 13.5 km deep, mostly 5-10 km deep. Seismic activity at Socorro is significantly greater than in adjacent parts of the rift.

The center of uplift is approximately 25 km north of Socorro, within the Rio Grande rift but outside the Socorro cauldron (Reilinger and Oliver, 1976; Reilinger and others, 1980) (figs. 20.5.2-20.5.6). Maximum observed uplift is 20 cm relative to points outside the rift. The average rate of uplift was approximately 0.23 cm/yr from 1912 to 1951 and 0.18 cm/yr from 1951 to 1980 (Larsen

SOCORRO (continued)

HISTORICAL ACTIVITY (continued)

and others, 1986). Uplift in the nearby Belen-Albuquerque basin may be an extension of the Socorro uplift, or it may be a similar, related uplift in an adjacent segment of the rift.

No significant strain accumulation in the Socorro area was detected by precise trilateration measurements across the Rio Grande rift during 1972-84 (Savage and others, 1985) (figs. 20.5.7-20.5.9). The absence of detectable strain accumulation places an upper limit (2 standard deviations) of 3 mm/yr on east-west spreading across the 100-km-wide rift.

A roughly twofold increase in specific conductance and hardness of water from two wells near Socorro from 1951 to 1962 was tentatively interpreted by Bushman (1963) as related to pumping in the area.

COMMENTS

The seismicity and uplift occur directly above a crustal discontinuity inferred to be a sill-shaped magma body at a depth of 18-20 km, with a minimum areal extent of 1700 km^2 (Rinehart and others, 1979; Sanford and others, 1979; Brocher, 1981). Uplift and seismicity could be explained by an inflation of the inferred magma body at a rate of 0.01-0.02 km^3/yr. However, at this inflation rate, the crustal thickness would double in 10 m.y., which is not observed. Therefore, Reilinger and others (1980) infer that inflation of the magma body must be "oscillatory or episodic on the longer time scale."

Although the relation between historical activity and the Socorro cauldron is not entirely clear, recent activity is much more extensive than the old cauldron and seems to be more closely related to regional rifting and a larger magma reservoir than to any residual magma of the Socorro cauldron.

REFERENCES

Brocher, T.M., 1981, Geometry and physical properties of the Socorro, New Mexico, magma bodies: Jour. Geophys. Res., v. 86, no. B10, p. 9420-9432.

Bushman, F.X., 1963, Ground water in the Socorro Valley, in Kuellmer, F.J., ed., Guidebook of the Socorro region: New Mexico Geol. Soc., 14th Field Conf., p. 155-159.

Chapin, C.E., Chamberlin, R.M., Osburn, G.R., White, D.W., and Sanford, A.R., 1978, Exploration framework of the Socorro geothermal area, New Mexico: New Mexico Geol. Soc. Spec. Publ. 7, p. 115-129.

Hall, F.R., 1963, Springs in the vicinity of Socorro, in Kuellmer, F.J., ed., Guidebook of the Socorro region: New Mexico Geol. Soc. 14th Field Conf., p. 160-170.

SOCORRO (continued)

REFERENCES (continued)

Larsen, S., Reilinger, R., and Brown, L., 1986, Evidence of ongoing crustal deformation related to magmatic activity near Socorro, New Mexico: Jour. Geophys. Res., v. 91, no. B6, p. 6283-6292.

Reilinger, R., and Oliver, J., 1976, Modern uplift associated with a proposed magma body in the vicinity of Socorro, New Mexico: Geology, v. 4, p. 583-586.

Reilinger, R., Oliver, J., Brown, L., Sanford, A., and Balazs, E., 1980, New measurements of crustal doming over the Socorro magma body, New Mexico: Geology, v. 8, p. 291-295.

Riecker, R.E., ed., 1979, Rio Grande rift: Tectonics and magmatism: Washington, D.C., Am. Geophys. Union, 438 p.

Rinehart, E.J., Sanford, A.R., and Ward, R.M., 1979, Geographic extent and shape of an extensive magma body at mid-crustal depths in the Rio Grande rift near Socorro, New Mexico, in Riecker, R.E., ed., Rio Grande rift: Tectonics and magmatism: Washington, D.C., Am. Geophys. Union, p. 237-251.

Sanford, A.R., 1963, Seismic activity near Socorro, in Kuellmer, F.J., ed., Guidebook of the Socorro region: New Mexico Geol. Soc. 14th Field Conf., p. 146-151.

Sanford, A.R., Olsen, K.H., and Jaksha, L.H., 1979, Seismicity of the Rio Grande rift, in Riecker, R.E., ed., Rio Grande rift: Tectonics and magmatism: Washington, D.C., Am. Geophys. Union, p. 145-168.

Savage, J. C., Lisowski, M., and Prescott, W. H., 1985, Strain accumulation in the Rocky Mountain states: Jour. Geophys. Res., v. 90, no. B12, p. 10,310-10,320.

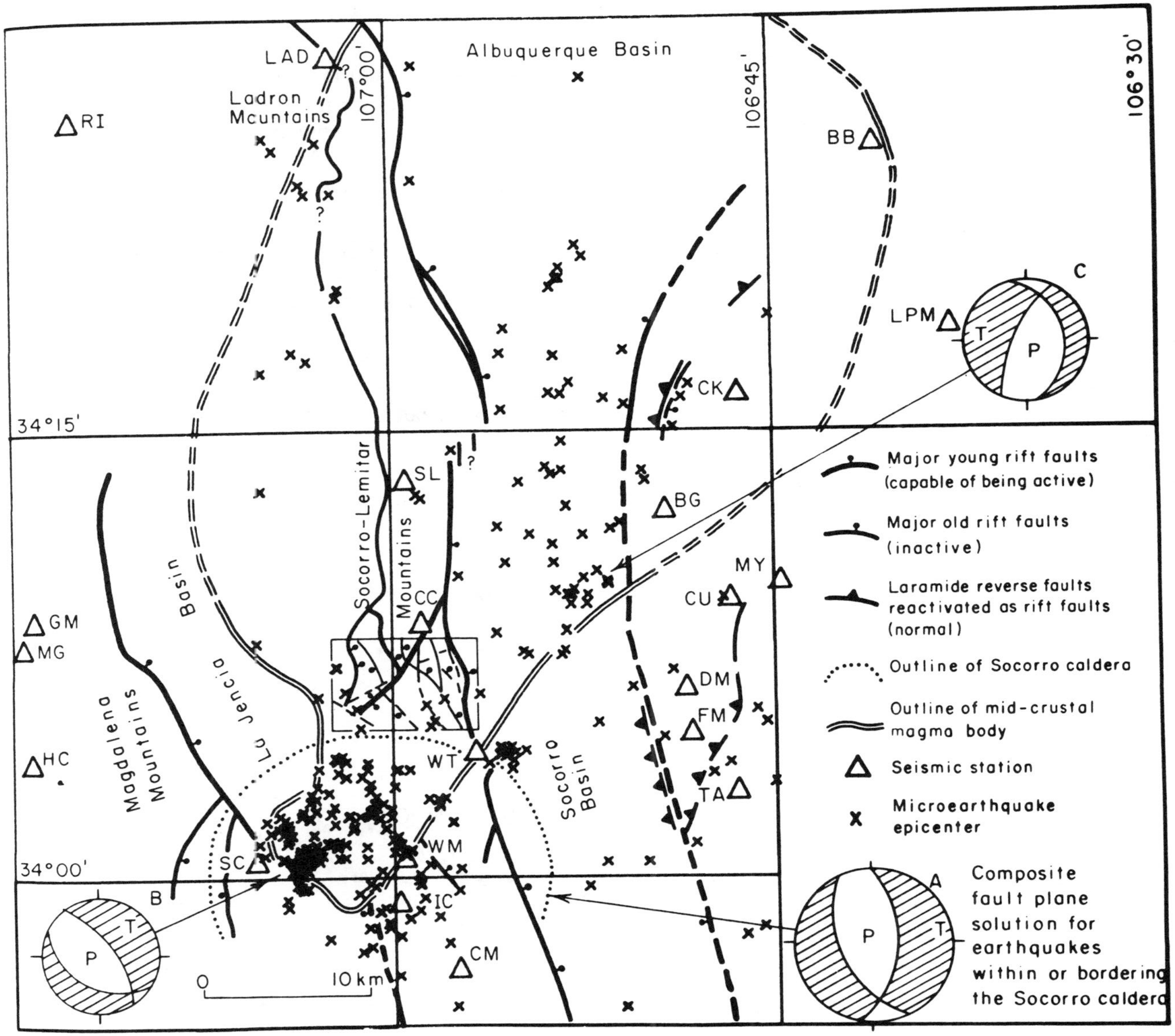

Figure 20.5.1. Rio Grande rift in vicinity of Socorro, New Mexico, showing epicenters of microearthquakes in relation to major faults, margins of a midcrustal layer of magma, and Tertiary Socorro Caldera (Sanford and others, 1979). Copyright by the American Geophysical Union.

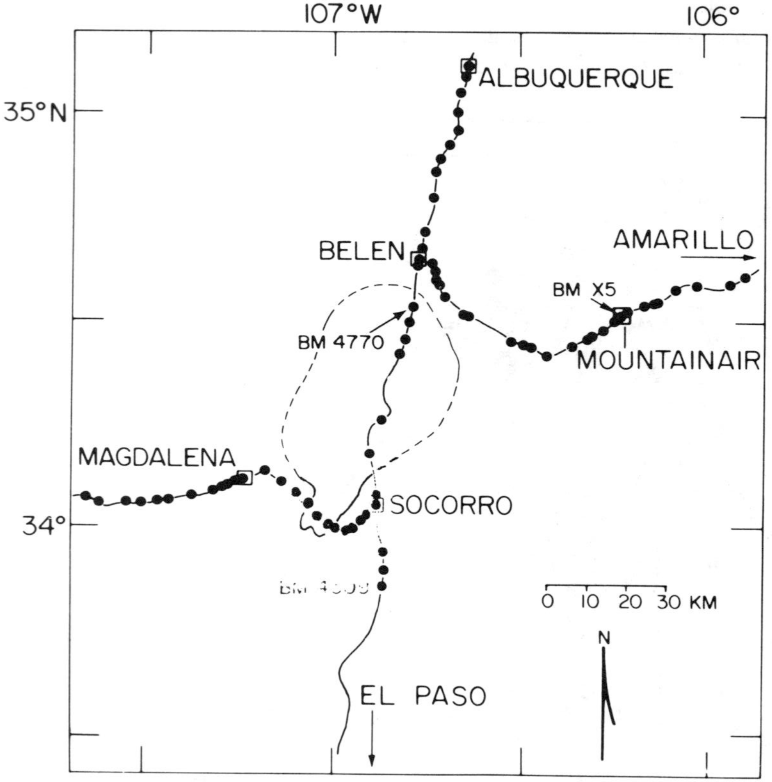

Figure 20.5.2. Top left, Locations of leveling routes and bench marks (dots) near Socorro, New Mexico, from Reilinger and others (1980). Outline of an inferred midcrustal magma body also is shown.

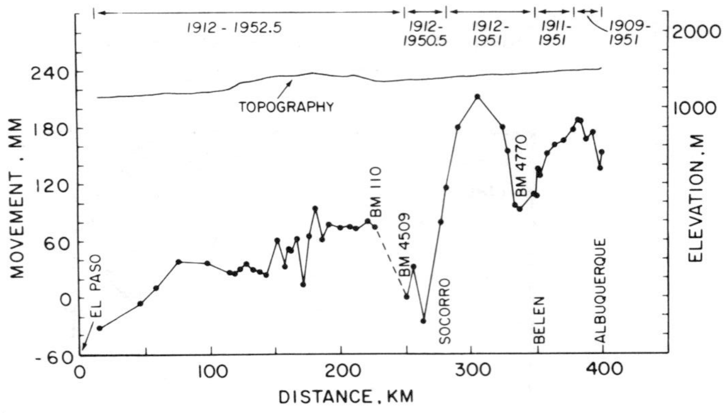

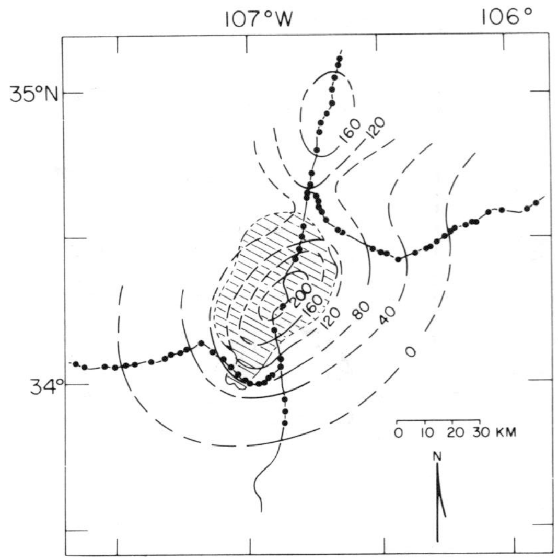

Figure 20.5.3. Top right, Elevation-change profiles and topography from El Paso, Texas, to Albuquerque, New Mexico, along route shown in figure 20.5.2 (Reilinger and others, 1980). Dates of leveling at top.

Figure 20.5.4. Bottom right, Contour map of vertical crustal movement in Socorro-Albuquerque area, from Reilinger and others (1980). Contours represent deformation that accumulated during a 40-year period or total movement between 1934 and 1958 (see text). Contour interval is 40 mm. Ruled area shows location of magma body; leveling routes and bench marks (dots) also shown.

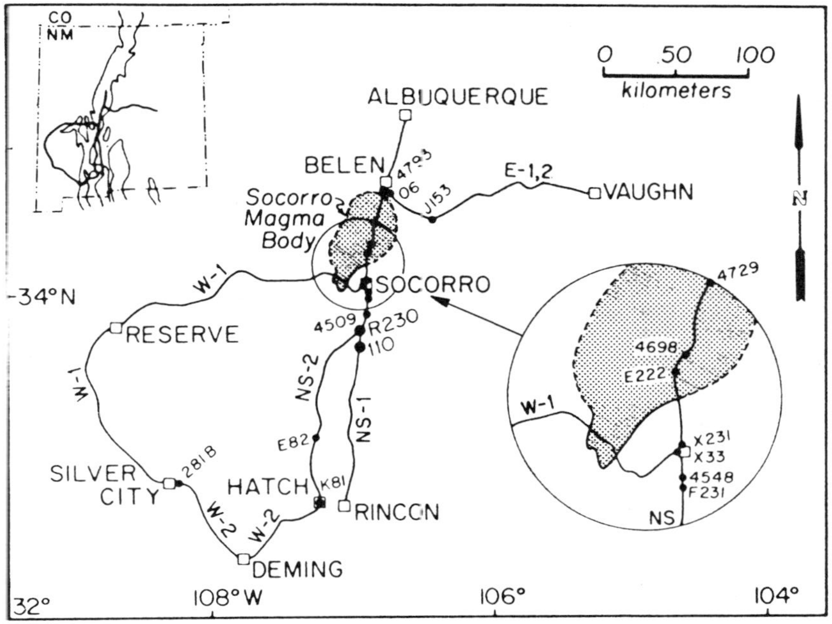

Figure 20.5.5. Leveling routes in Socorro area, from Larsen and others (1986). Inset at top left shows leveling lines relative to New Mexico and Rio Grande rift. Midcrustal Socorro magma body is shown by hachured pattern. Bench marks shown as solid dots. Copyright by the American Geophysical Union.

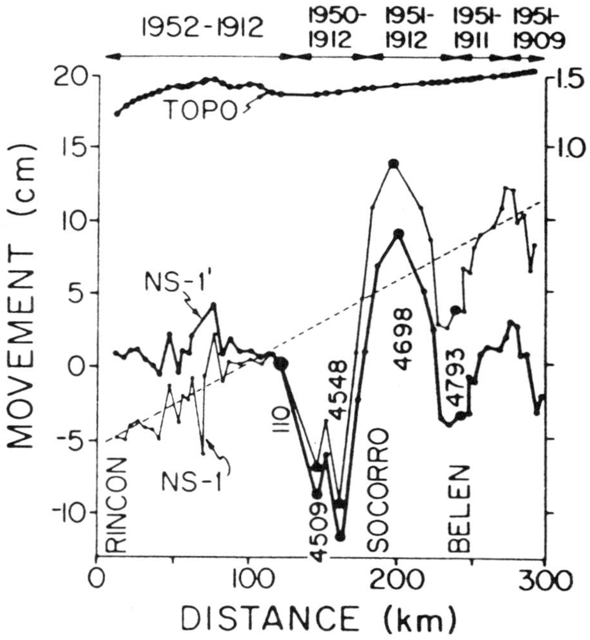

Figure 20.5.6. Elevation-change profiles and topography from Rincon to Albuquerque along the leveling route shown in figure 20.5.5 (Larsen and others, 1986). Dates of leveling are indicated at top of plot. Bench marks are shown as solid dots. Dashed line represents regional trend, possibly due to systematic error or long-wavelength tectonic tilt. Copyright by the American Geophysical Union.

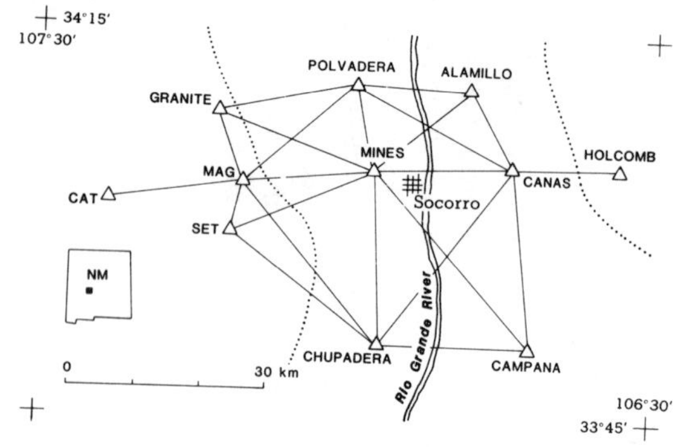

Figure 20.5.7. Socorro strain network, from Savage and others (1985). Extent of Rio Grande rift is indicated by dotted lines. Copyright by the American Geophysical Union.

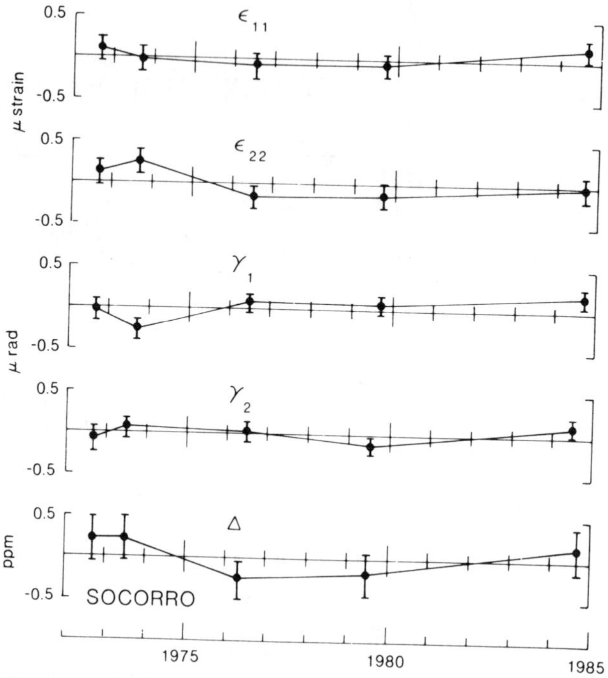

Figure 20.5.8. Line-length changes as a function of time for each of the lines in Socorro strain network (see figure 20.5.7), 1972-84 (Savage and others, 1985). Error bars, 1 standard deviation on either side of points. Copyright by American Geophysical Union.

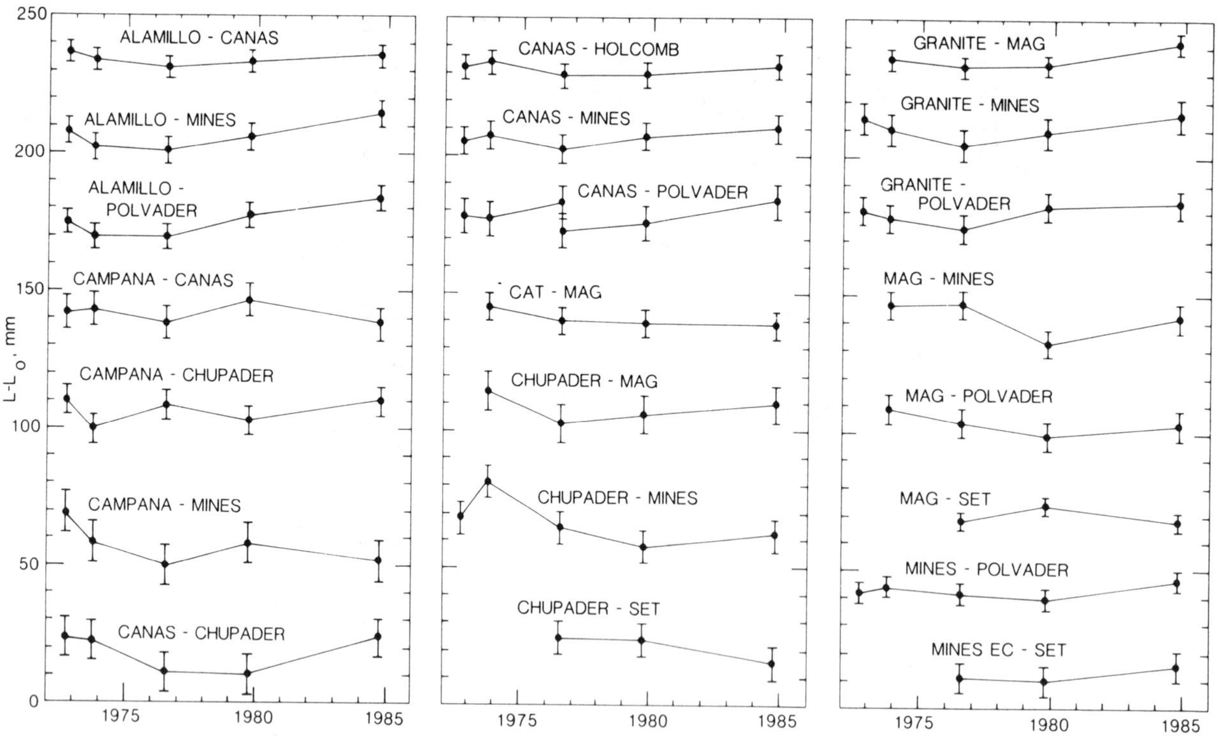

Figure 20.5.9. Strain accumulation in Socorro network, 1972-84 (Savage and others, 1985). Strains are referred to a geographic coordinate system (the 1-axis directed east and the 2-axis directed north). Error bars represent 1 standard deviation on either side of plotted points. Copyright by the American Geophysical Union.

MONTSERRAT
(Soufriere Hills)
CAVW # 16-00-05

Latitude Longitude (degrees)	Tectonic setting	Age of latest volcanism	Youngest volcanic rock types	Restless years	Type and duration of unrest ESTU STHF MGTF H Te	Associated eruption?
16.72N 62.18W	Compr	300 yr B.P.	58-63 pct SiO_2	1897-1900	FF-- ---- ---x x --	none
				1933-1937	FF-- ---- ---C x UQ	none
				1966-1967	FF-- -F-- ---F F --	none

TECTONIC/GEOLOGIC SETTING

Historical unrest on Montserrat has been in the vicinity of the Soufriere Hills, the youngest of many volcanic centers on this small island. The Soufriere Hills are a complex of andesitic domes; older andesitic domes and andesitic to basaltic lavas form nearby hills (fig. 20.6.1). In mapping to date (MacGregor, 1938; Rea, 1974), no evidence has been found that the Soufriere Hills lie within a caldera. Similarly, there is no suggestion from either mapping or geophysical studies that the Soufriere Hills are underlain by any large magma reservoir. A strong NW-SE lineament connects several vents, including the Soufriere Hills.

Lavas on Montserrat in general contain 50-63 percent SiO_2; three samples from the Soufriere Hills proper had 58-63 percent SiO_2. A few blocks of lava and pumice show compositional banding, 58 percent versus 62 percent SiO_2 (Rea, 1974).

HISTORICAL ACTIVITY

The southern half of Montserrat, in the vicinity of the Soufriere Hills Volcano, has experienced three volcano-seismic crises within the past century. A series of earthquakes occurred during 23-27 April 1897; one particularly strong event (MMI= 8) occurred on 25 April 1897. Even stronger earthquakes of this episode occurred in October 1900, near the end of the unrest. Houses, churches, and windmills were heavily damaged on and near the northern and western flanks of the volcano, and solfataric activity intensified about the time of the earthquakes. Another strong earthquake occurred near Guadeloupe on 29 April 1897, suggesting that the Montserrat activity was related to regional tectonic strain. Perhaps coincidentally, the 1897-1900 earthquakes followed the heaviest rains and flooding in recent years, in 1896.

MONTSERRAT (continued)

HISTORICAL ACTIVITY (continued)

Beginning in March 1933, solfataras on the lower slope of the Soufriere Hills began to show intensified activity. Earthquake swarms began in July 1933, and increased in frequency and intensity to a peak in November 1935 (figs. 20.6.2-20.6.4). Several thousand shocks were felt on the island during 1933-37. Those measured were shallow (1-2 km; Perret, 1939). At least 28 were of "the seventh or eighth magnitudes of the seismic scale" (Perret, 1939); the two strongest earthquakes of the series (12 December 1934 and 6 May 1935) had modified Mercalli intensities of 8 (Robson, 1964). Earthquake damage and solfataric activity were in essentially the same zone, north and northwest of the Soufriere Hills, as they had been in the 1897-1900 episode. The frequency of earthquakes varied seasonally, with maxima in May and November-December of each year. Increased H_2S emission often preceded the 1933-37 earthquake swarms by a few days, but it also intensified after the larger earthquakes. On 10 November 1935, a strong tectonic earthquake occurred 100 km north of Montserrat; activity at Montserrat decreased thereafter, although it did not fully return to background levels until late 1937.

Earthquakes were felt in January and February 1966 ($\underline{M}_{max}$=4.0), and a seismic array was installed in March and April 1966. The epicentral region was broad and crudely elongate in a NW-SE direction, coincident with the lineament connecting the Soufriere Hills and several other vents (figs. 20.6.5-20.6.8). It is not certain whether this represents a slight southward shift relative to the activity of 1933-37, because the latter activity was not precisely located. In the 1966-67 swarm, earthquake depths varied systematically, averaging 5.2 km in April and May 1966, rising to a minimum of 2.8 km (average) in July-September, deepening to an average of 13.1 km in April 1967, and ending at an average of 9.7 km in November 1967. The total seismic energy release was 3×10^{16} ergs, equivalent to a $\underline{M}$=4.5 earthquake. The $\underline{b}$-value of the earthquakes was 0.8 throughout the entire series. Seismic energy release reached seasonal maxima in May 1966, November-December 1966, and October-November 1967 (similar to the seasonality observed during 1933-37).

Three water-tube tiltmeters, located on the lower flanks of the Soufriere Hills about 3-4 km from the summit, indicated inflation centered just south of the summit area, coincident with the center of seismicity (figs. 20.6.9, 20.6.10). Inflation was greatest in July-October 1966 (maximum recorded rate of tilt=12 microradians/month). Inflation continued at a lower rate from October 1966 until January 1967, deflation occurred during January-March 1967, and inflation occurred again during March-September 1967. The net change was inflationary. Fumarolic activity increased sympathetically with inflation.

COMMENTS

The Royal Society mission (Powell, 1937, 1938), Perret (1939), and Shepherd and others (1971) have unanimously inferred that the unrest has been due to episodic, small magmatic intrusions beneath the Soufriere Hills. Perret thought a (phreatic) explosion from one of the solfataras would be the most likely outcome, whereas Shepherd and others considered an eruption from the summit of the Soufriere Hills to be more likely.

MONTSERRAT, Latitude 16.72N, Longitude 62.18W, CAVW # 16-00-05

MONTSERRAT (continued)

REFERENCES

MacGregor, A.G., 1938, The volcanic history and petrology of Montserrat, with observations on Mt. Pelee, in Martinique: Trans. Roy. Soc. (London), v. B229, p. 1-90.
Perret, F.A., 1939, The volcano-seismic crisis at Montserrat, 1933-1937: Carnegie Inst. Washington Publ. 512, 76 p.
Powell, C.F., 1937, Royal Society expedition to Montserrat, B.W.I.: Preliminary report on seismic observations: Proc. Roy. Soc. (London), v. A158, p. 479-494.
----- 1938, Royal Society expedition to Montserrat, B.W.I.: Final report: Trans. Roy. Soc. (London), v. A237, p. 1-34.
Rea, W.J., 1974, The volcanic geology and petrology of Montserrat, West Indies. Jour. Geol. Soc. London, v. 130, p. 341-366.
Robson, G.R., 1964, An earthquake catalogue for the eastern Caribbean, 1530-1960: Bull. Seismol. Soc. Am., v. 54, no. 2, p. 785-832.
Shepherd, J.B., Tomblin, J.F., and Woo, D.A., 1971, Volcano seismic crisis in Montserrat, West Indies, 1966-1967: Bull. Volcanol., v 35, p. 143-163.

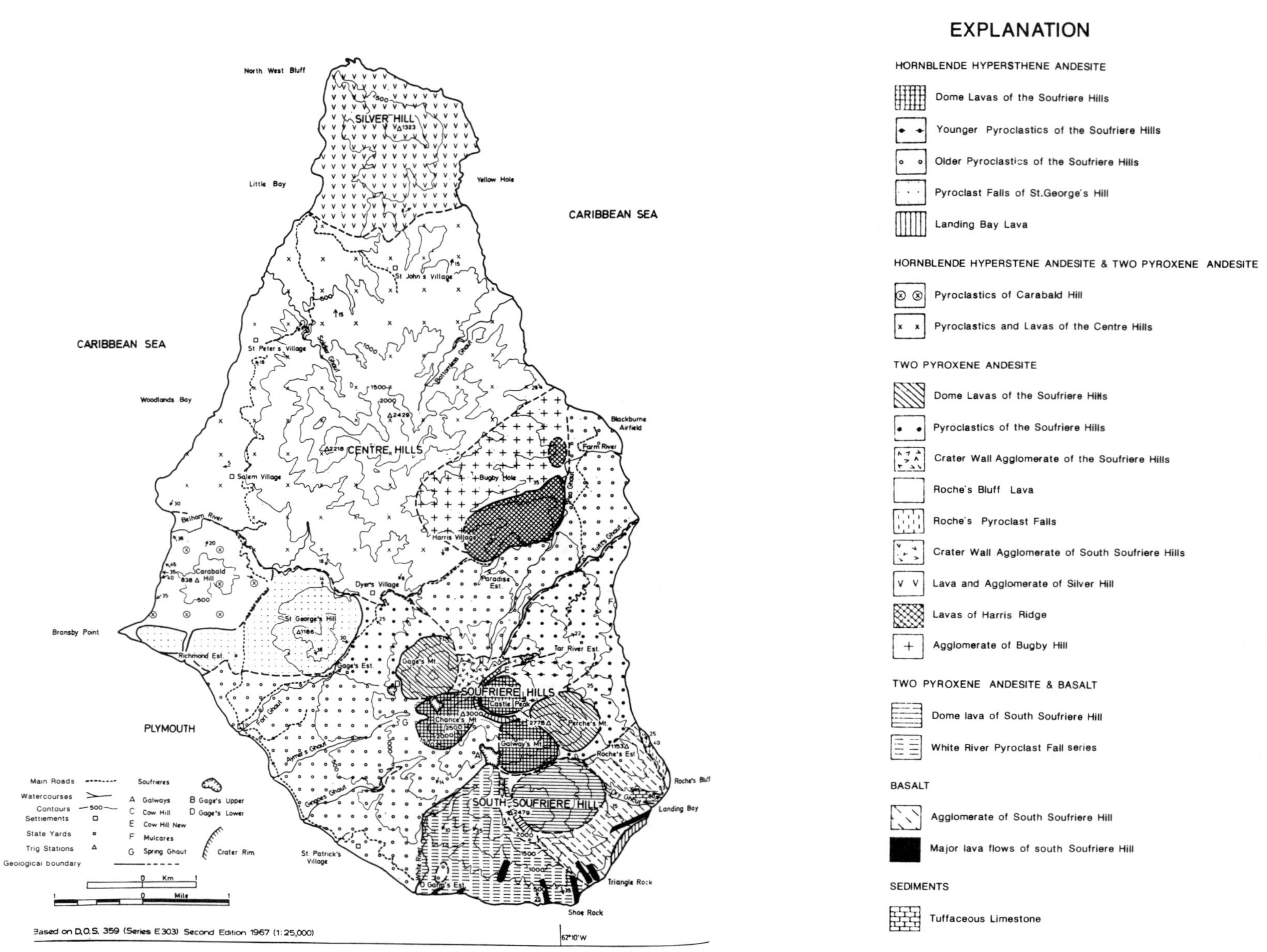

Figure 20.6.1. Geologic sketch map of Montserrat, Lesser Antilles, from Rea (1974). Most recent eruptions have been from Soufriere Hills. Reproduced by permission of the Geological Society of London.

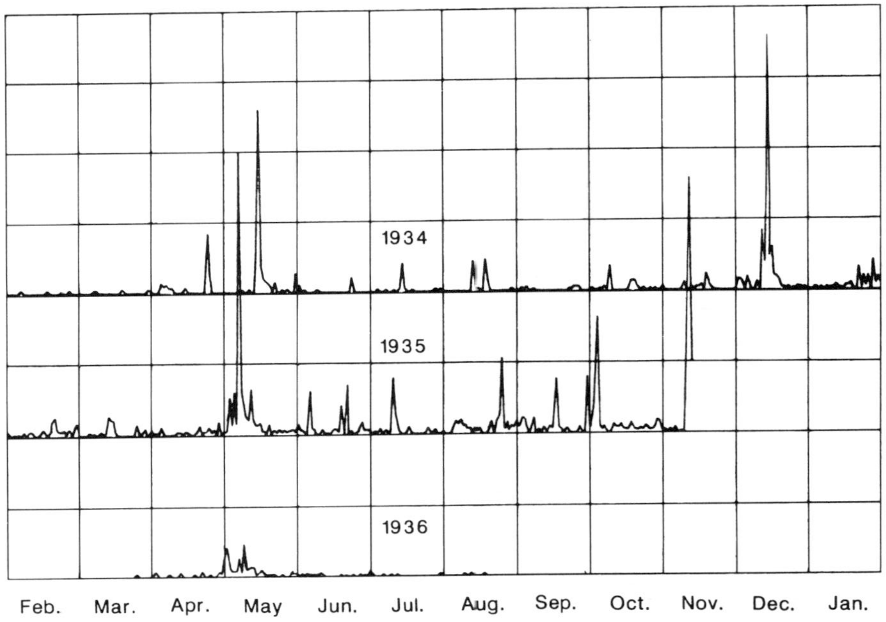

Figure 20.6.2. Daily number of earthquakes at Montserrat during 1934-36, from Powell (1937). Data for the period from 12 November 1935 to 24 March 1936 were unavailable, but seismic activity was minor.

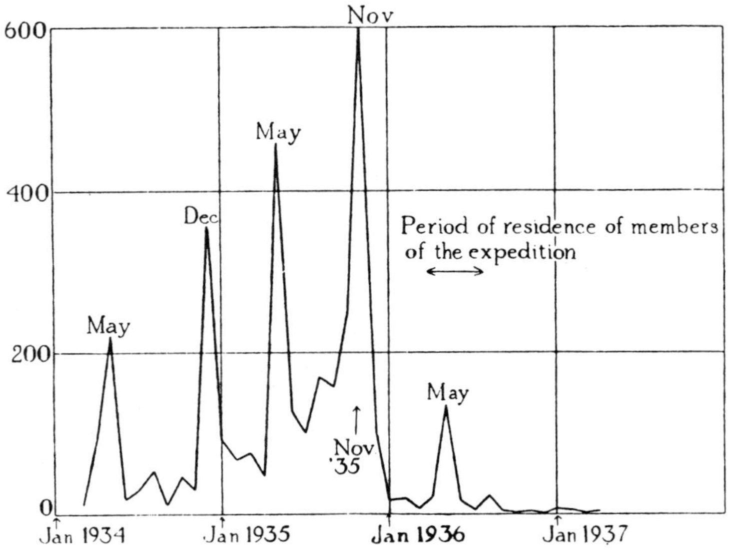

Figure 20.6.3. Monthly number of earthquakes at Montserrat during 1934-37, from Powell (1938).

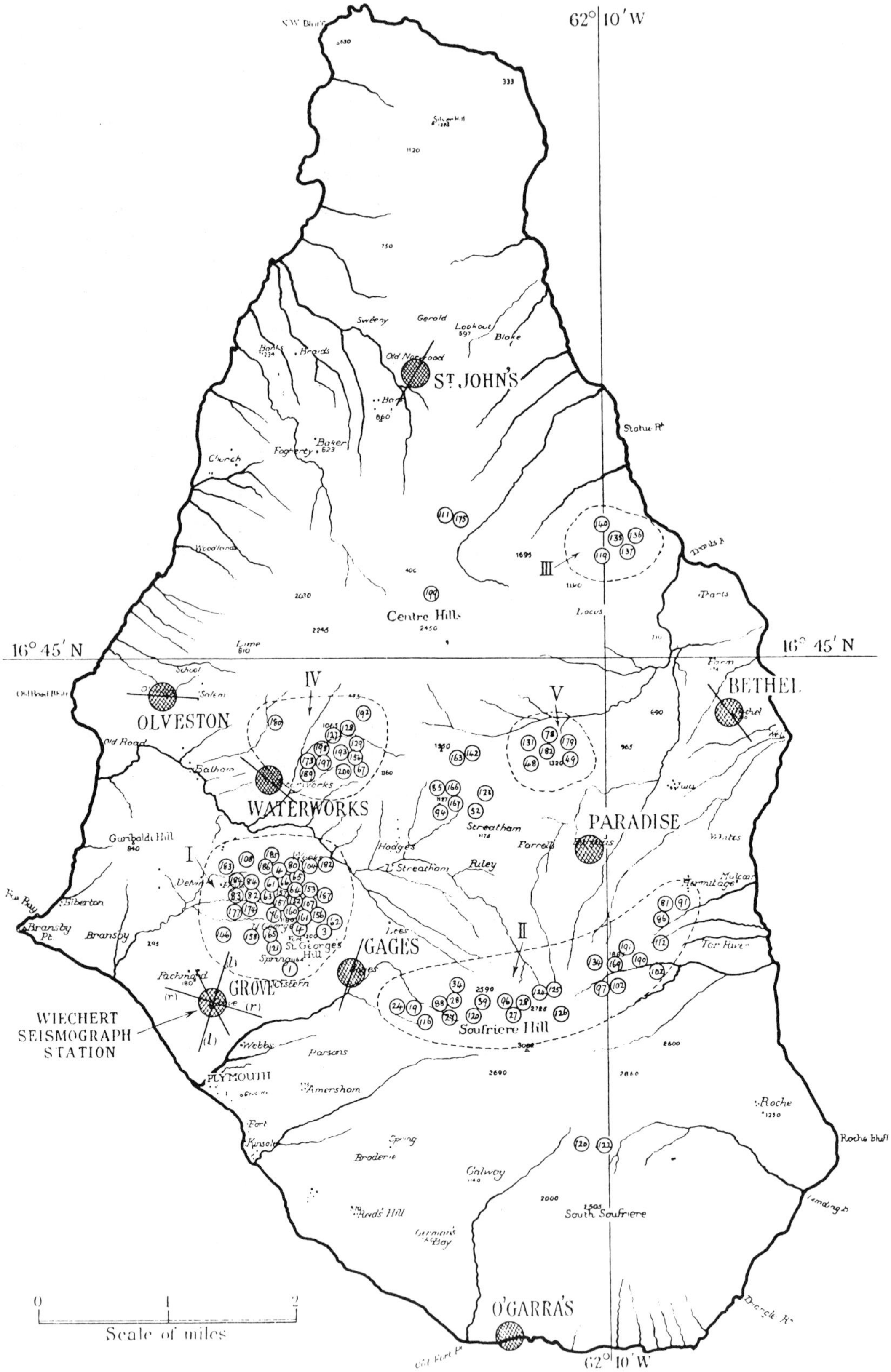

Figure 20.6.4. Epicenters for earthquakes at Montserrat during 1934-37, from Powell (1938).

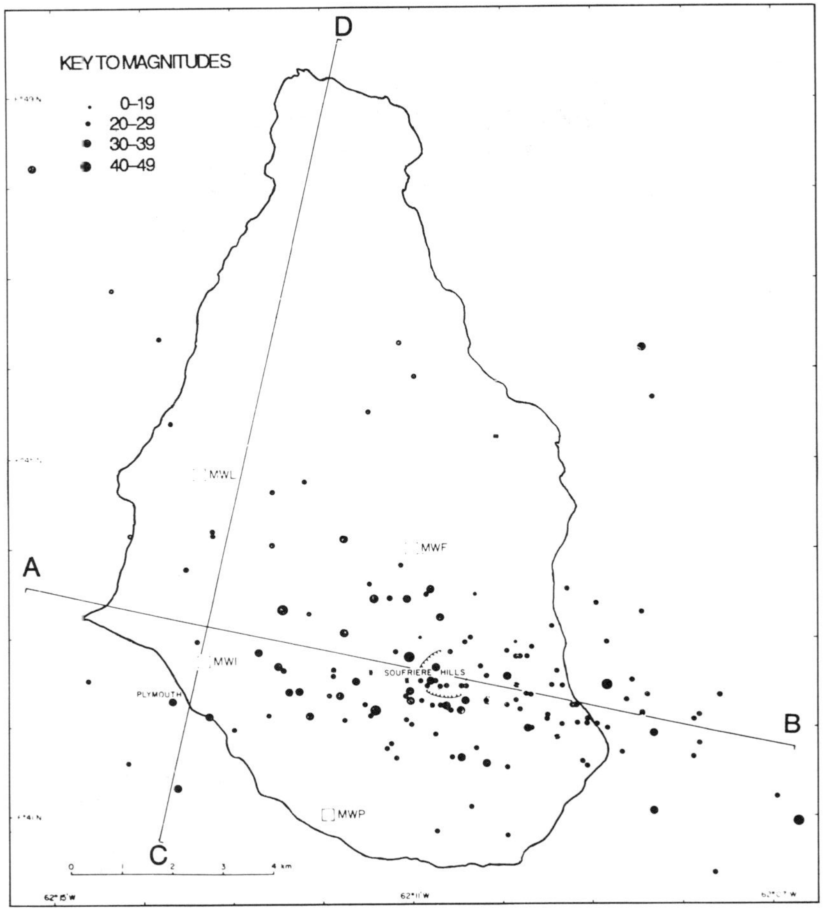

Figure 20.6.5. Distribution of epicenters during the 1966-67 seismic crisis at Montserrat, from Shepherd and others (1971). Solid circles, epicenters; squares, seismograph stations; MWI, Grove; MWL, Salem; MWF, Farrels; MWP, St. Patrick's.

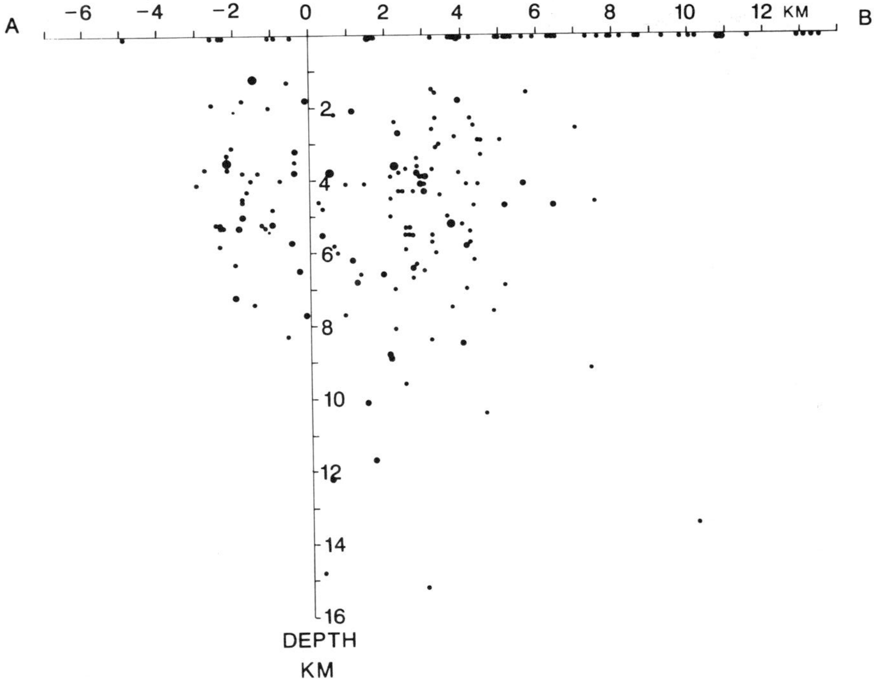

Figure 20.6.6. Cross section along line A-B in figure 20.6.5, showing all hypocenters projected onto a vertical plane (Shepherd and others, 1971). Earthquake magnitude is indicated by size of dot, as in figure 20.6.5.

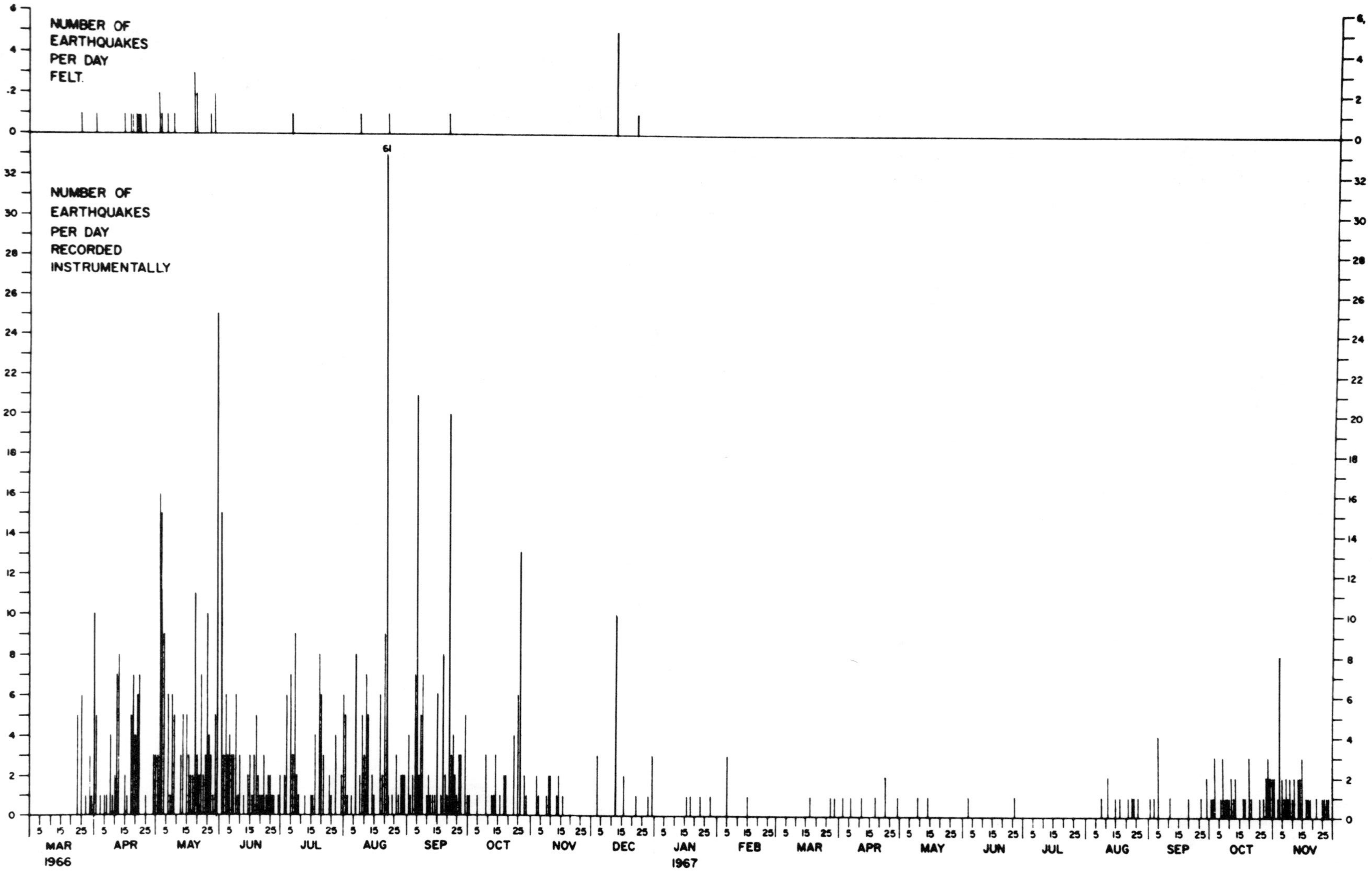

Figure 20.6.7. Daily number of local earthquakes during the 1966-67 seismic crisis at Montserrat, from Shepherd and others (1971).

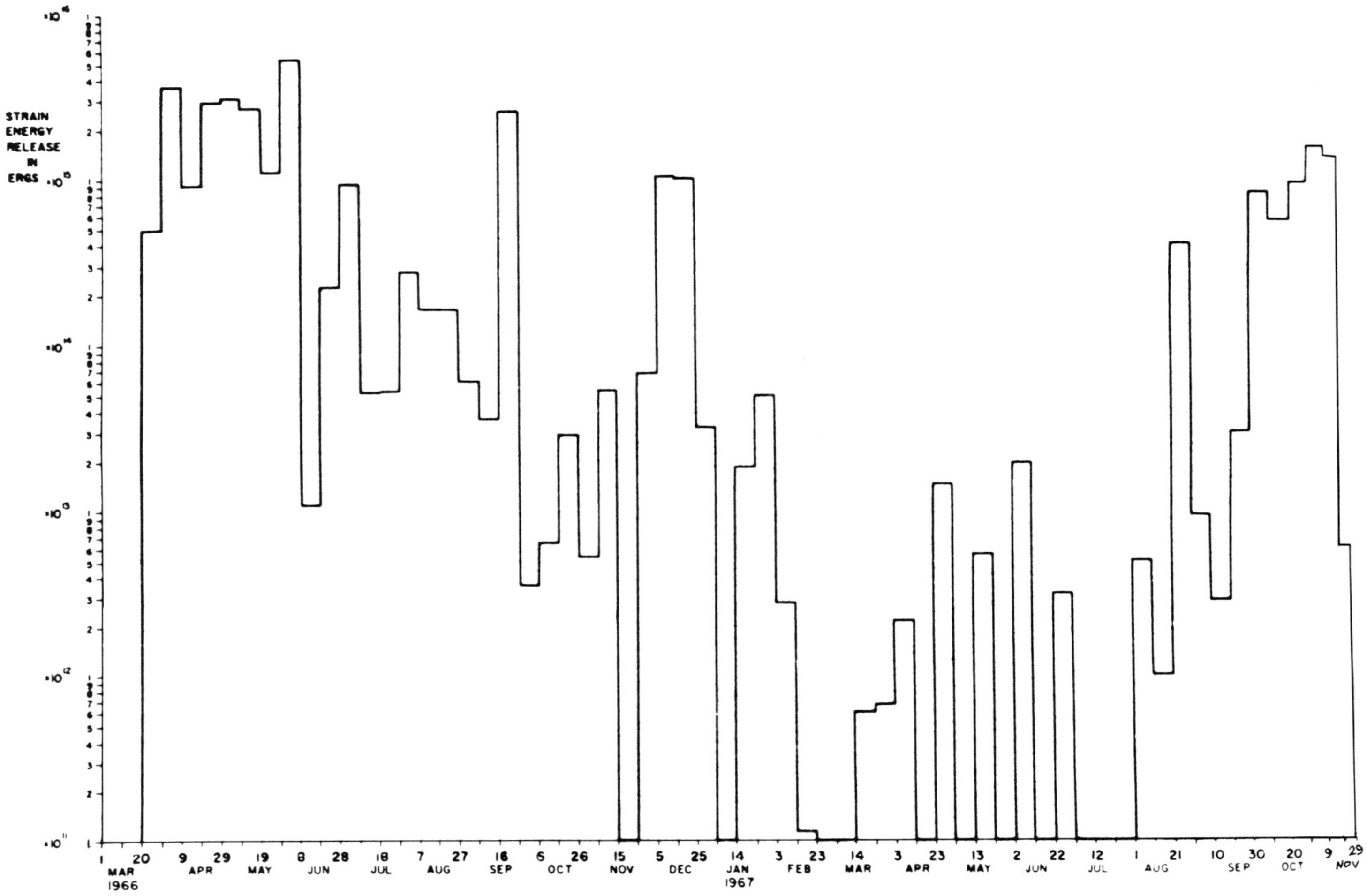

Figure 20.6.8. Earthquake strain energy release during the 1966-67 seismic crisis at Montserrat, from Shepherd and others (1971).

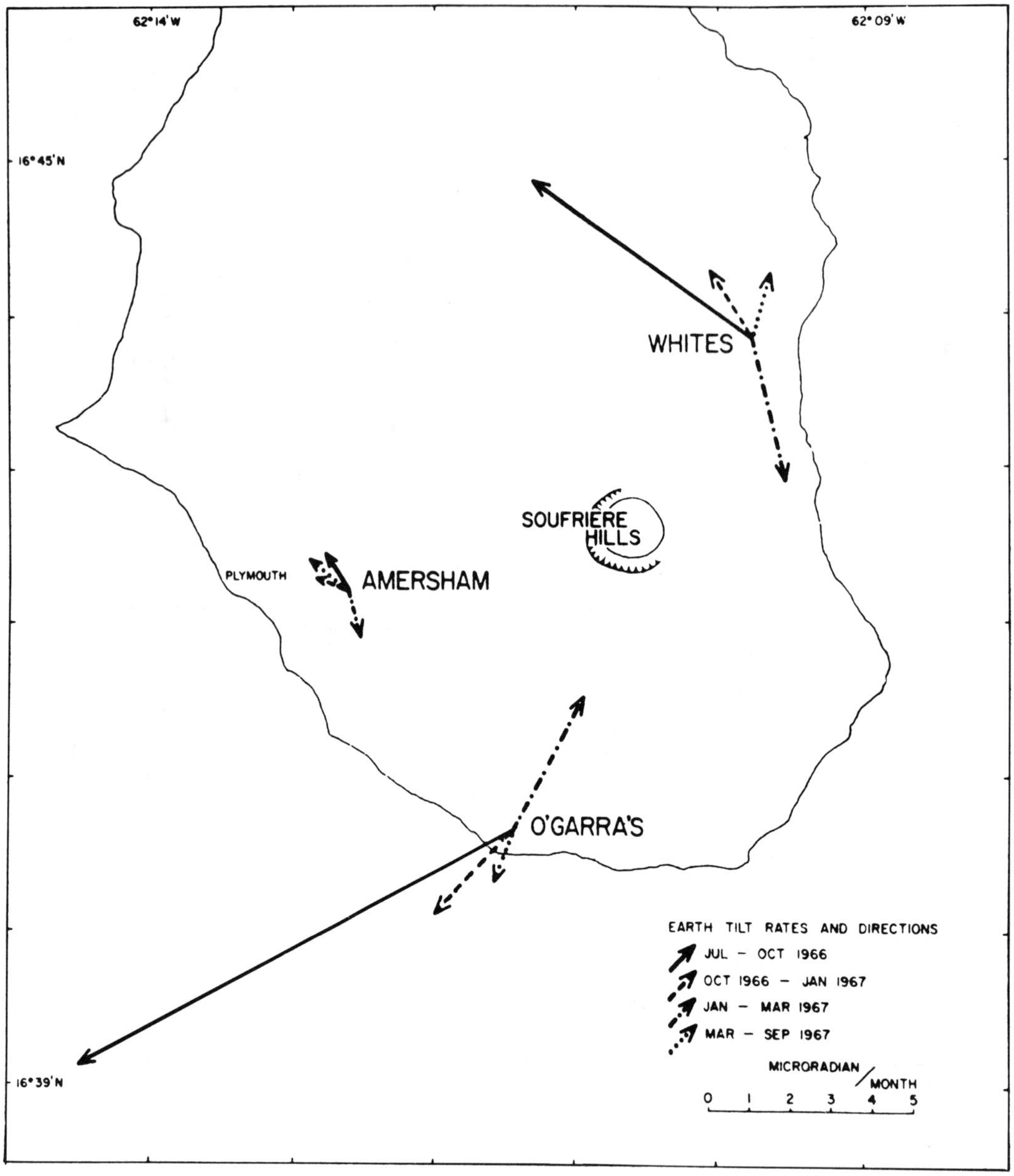

Figure 20.6.9. Earth-tilt rates and directions during the 1966-67 seismic crisis at Montserrat, from Shepherd and others (1971). Tilts were measured with a long-base water-tube tiltmeter.

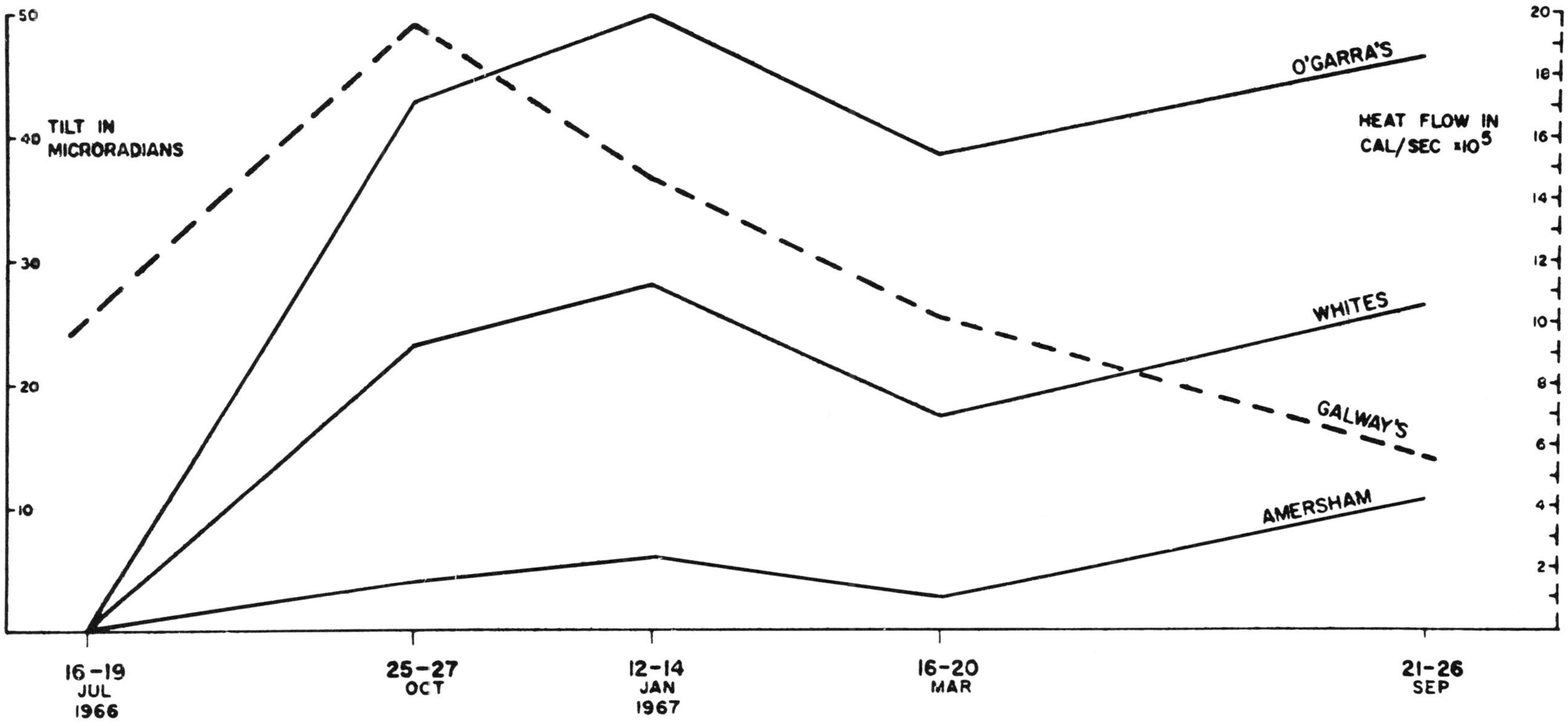

Figure 20.6.10. Comparison of rates of earth tilt (solid lines) with total heat flow at Galway's soufriere (dashed line) during the 1966-67 seismic crisis at Montserrat, from Shepherd and others (1971).

PART 5: Large Quaternary Calderas Not Known to Have Shown Historical Unrest

This section has two purposes:

(1) to provide a basis for comparing numbers of restless vs. quiet calderas of similar size and age;

(2) to elicit information from readers about unrest at some of the calderas listed, or about additional, quiet calderas. The list of calderas in this section is certainly incomplete, and your corrections and additions are welcome.

PART 5: LARGE QUATERNARY CALDERAS NOT KNOWN TO HAVE SHOWN HISTORICAL UNREST

Caldera	CAVW number	Latitude	Longitude	Diameter	Age of latest collapse	Dominant magma composition	References
Vico	None	42.3N	12.3E	7	ca. 0.15 Ma	tephrite, tephrite phonolite	Sollevanti (1983)
Sacrofano -Baccano	None	42.1 N	12.4E	25 x 13?	0.4-0.3 Ma	leucite tephrite, phonolite	de Rita and others (1983)
Nabro	02-01-101	13.35N	41.68E	8		rhyolite	IAVCEI (1973)
Mallahle	02-01-102	13.27N	41.65E	5		silicic	Mohr and Wood (1976)
Sorkale	02-01-103	13.27N	41.65E			rhyolite	IAVCEI (1973); CNR-CNRS Afar Team (1975)
Asavyo	02-01-104	13.07N	41.60E	8		silicic	Mohr and Wood (1976)
Oyma	02-01-104A	12.88N	41.62E	5		silicic	Mohr and Wood (1976)
Dabbahu (Moina)	02-01-113	12.63N	40.45E			basalt, trachyte, rhyolite	CNR-CNRS Afar Team (1973, 1975)
Groppo	02-01-116	11.73N	40.25E			silicic	IAVCEI (1973)
Afdem	02-01-17A	9.50N	40.85E			rhyolite	IAVCEI (1973)
Chana (Arussi)	02-01-19A	8.90N	40.22E			rhyolite	IAVCEI (1973)
Gadamsa	02-01-23	8.35N	39.18E	ca. 10			Di Paola (1972)
Bora-Bericcio (Bericha)	02-01-24	8.18N	38.98E			silicic	Di Paola (1972)
Gademota	None	7.9N	38.6E	ca. 20		silicic	G. Gabriel (written commun., 1987)
Aluto	02-01-27	7.77N	38.78E	ca. 8		silicic	Di Paola (1972)

PART 5: LARGE QUATERNARY CALDERAS NOT KNOWN TO HAVE SHOWN HISTORICAL UNREST (continued)

Caldera	CAVW number	Latitude	Longitude	Diameter	Age of latest collapse	Dominant magma composition	References
Shalla(=O'A)	02-01-28	7.50N	38.50E	17	0.24 Ma	rhyolite	IAVCEI (1973);Mohr and Wood (1976)
Duguna	None	7.1 N	37.9	ca. 10		silicic	G. Gabriel (written commun., 1987)
Emuruangogolak	02-02-051	1.50N	36.33E	outer, 7x9 inner, 3x4	ca. 0.5 Ma	trachyte	Weaver (1976-77); Williams and others (1984)
Silali	02-02-052	1.15N	36.23E	5 x 7.5		trachyte	McCall (1968); Williams and others (1984)
Menengai	02-02-06	0.23S	36.10E	8 x 12	29 ka	trachyte	Leat and others (1984)
Longonot	02-02-10	0.92S	36.45E	ca. 7.5	<0.5 Ma	trachyte, hawaiite	Williams and others (1984)
Suswa	02-02-11	1.18S	36.35E	outer, 8x12 inner, 5	ca. 0.24 Ma	phonolite, trachyte	Johnson (1969); Williams and others (1984)
Embagai	02-02-12A	2.92S	35.83E			alkali basalt, trachyte, nephelinite	IAVCEI (1973)
Olmoti	02-02-12D	3.00S	35.65E			alkali basalt, trachyte, nephelinite	IAVCEI (1973)
Ngorongoro	02-02-12H	3.17S	35.57E	16		alkali basalt, trachyte, nephelinite	Williams (1970); Searle (1972)
Ngurdoto	02-02-15A	3.30S	36.92E			nephelinite	IAVCEI (1973)
Ngozi	02-02-172	9.00S	33.55E			trachyte, phonolite, basalt	Harkin (1960); IAVCEI (1973)
Emi Koussi	02-05-021	19.70N	18.57E	14			Geze and others (1959)

PART 5: LARGE QUATERNARY CALDERAS NOT KNOWN TO HAVE SHOWN HISTORICAL UNREST (continued)

Caldera	CAVW number	Latitude	Longitude	Diameter	Age of latest collapse	Dominant magma composition	References
Tarso Yega	None			22			Geze and others (1959)
Tarso Tieroko	None	20.72N	17.86E	8		trachyte?	Geze and others (1959)
Tarso Voon	02-05-02	20.97N	17.17E	12 x 15		trachyte, andesite, rhyolite	Geze and others (1959)
Tarso Toon	None	21.05N	17.66E	12		trachyte?	Geze and others (1959)
Tousside	02-05-01	21.05N	16.42E	6			Geze and others (1959)
Maroa/ Whakamaru	04-01-24A	38.55S	176.05E	ca. 50	1.1-0.5 Ma	rhyolite	Wilson and others (1984, 1986)
Umboi (Talo)	05-01-06	5.59S	147.88E	13		silicic?	Johnson and others (1972)
Gallosuelo	05-02-10	5.33S	151.12E	13 x 10.5		andesite, dacite	Cooke and others (1976)
Bur ni Geureudong	06-01-04	4.82N	96.80E	30		andesite	Neumann van Padang (1951)
Singkarak	None	0.50S	100.50E	8 x 21		silicic	Verstappen (1973)
Hutapanjang	06-01-171	2.27S	101.60E	10 x 18		silicic	Posavec and others (1973) Rock and others (1982)
Gedongsurian	None	5.04S	104.23E	10 x 20		silicic	van Bemmelen (1949a)
Hulubelu	06-01-28	5.35S	104.60E	large		dacite, andesite	van Bemmelen (1931, 1949a, b) Neumann van Padang (1951)
Dano (Banten)	None	ca. 6.3S	106.0E	7 x 15		dacite	Katili (1983)
Buyan-Bratan	06-04-014	8.28S	115.13E	7 x 11		silicic	Purbo-Hadiwidjojo (1971)

PART 5: LARGE QUATERNARY CALDERAS NOT KNOWN TO HAVE SHOWN HISTORICAL UNREST (continued)

Caldera	CAVW number	Latitude	Longitude	Diameter	Age of latest collapse	Dominant magma composition	References
Poco Leok	06-04-07	8.41S	120.29E	6 x 6.75			Neumann van Padang (1972) Kusumadinata (1979)
Natib	07-03-082	14.72N	120.37E	7.5 x 4.5	ca. 70 ka	andesite, dacite	EBASCO (1978-79)
Sambe	08-03-001	35.15N	132.32E	6 x 8	16 ka	dacite	Ono and others (1981) Japan Assoc. Quat. Res. (1987)
Daisen	08-03-001B	35.37N	133.55E		46 ka	dacite	IAVCEI (1973)
Tateyama (Midagahara)	08-03-08	36.57N	137.60E	6 x 4	ca. 60 ka	andesite, dacite	Yamasaki and others (1966) Machida (1980) Kusakabe and others (1983)
Myoko	08-03-10	36.88N	138.12E		> 2 ka	andesite	Hayatsu (1976); Hayatsu and Arai (1980)
Tazawa	08-03-22B	39.75N	140.67E	ca. 6		dacite	IAVCEI (1973)
Hakkoda	08-03-28	40.65N	140.88E	8	2.1 Ma	andesite	Kuno (1962)
Akaigawa	08-05-04E	43.08N	140.82E	ca. 8	2.1 Ma	andesite, dacite	IAVCEI (1973); Ono and others (1981)
Kuttara	08-05-033	42.50N	141.18E	5?	11 ka	andesite-dacite	Katsui and others (1981)
L'vinaya Past	09-00-041	44.60N	147.00E	8	9.5 ka		Erlich and Melekestsev (1972) Erlich and others (1972)
Uratman	09-00-191	47.12N	152.23E	7.5		andesite	Gorshkov (1970)
Prizrak	10-00-041	51.6N	157.3E	5+			Luchitsky (1974)
Khodutka	10-00-051	52.07N	157.70E				Tomkeieff (1949) Vlasov (1967)
Beliankin	10-00-121	53.97N	159.42E				Sviatlovsky (1959)

PART 5: LARGE QUATERNARY CALDERAS NOT KNOWN TO HAVE SHOWN HISTORICAL UNREST (continued)

Caldera	CAVW number	Latitude	Longitude	Diameter	Age of latest collapse	Dominant magma composition	References
Sobolinyy	10-00-131	54.15N	159.60E				Vlasov (1967)
Bolshoy Semiachik	10-00-15A,B	54.3N	160.0E	25 x 30(?)			Erlich (1986)
Unana	10-00-161	54.62N	159.70E				Krijanovsky (1934); Tomkeieff (1949); Vlasov (1967)
Zimina	10-00-242	56.03N	160.72E				IAVCEI (1973); Vlasov (1967)
Dalnaya Ploskaya (Plosky)	10-00-261	56.20N	160.63E	5.6		basalt?	Erlich (1986)
Hangar	10-00-271	54.85N	157.53E	12 x 16 (?)	6.4 ka	basalt-dacite	Erlich (1986)
Ichinsky	10-00-28	55.77N	157.92E	inner, 6 outer,60		basalt, rhyolite	Erlich (1986)
Uksichan	10-00-30	55.92N	158.63E	ca. 10		basalt-rhyolite	Erlich (1986)
Newberry	12-02-11	43.68N	121.25W	6 x 8		basalt-dacite	MacLeod and others (1981)
Crater Lake	12-02-16	42.93N	122.12W	10	6.8 ka	andesite-dacite	Bacon (1983)
Medicine Lake	12-03-05	41.53N	121.53W	6 x 8	0.21 Ma(?)	andesite?	Donnelly-Nolan and others (1981) Mertzman (1981)
Island Park	12-04-05A	44.28N	111.40W	18 x 22	2, 1.3 Ma	rhyolite, basalt	Hamilton (1965); Christiansen (1984)
Valles	12-10-03	35.87N	106.57W	ca. 20	0.3 Ma	rhyolite	Smith and others (1961)

PART 5: LARGE QUATERNARY CALDERAS NOT KNOWN TO HAVE SHOWN HISTORICAL UNREST (continued)

Caldera	CAVW number	Latitude	Longitude	Diameter	Age of latest collapse	Dominant magma composition	References
Haleakala[1] (East Maui)	13-01-01	20.71N	156.25W	12 x 4		basalt	Stearns (1966); Macdonald (1972); Macdonald and others (1983); Clague and Dalrymple (1987)
Koolau (Oahu)	13-02-07	21.37N	157.30W	13 x 16	> 1.8 Ma	basalt	Clague and Dalrymple (1987)
Palawai (Lanai)	None	20.79N	156.91W	6 x 5	> 1 Ma	basalt	Stearns (1966); Clague and Dalrymple (1987)
E. Molokai (Molokai)	None	21.15N	156.85W	7 x 3	> 1 Ma	basalt	Stearns (1966); Clague and Dalrymple (1987)
Kahoolawe	None	20.55N	156.55W	5 x 4	> 1 Ma	basalt	Stearns (1966); Clague and Dalrymple (1987)
Los Humeros (including Los Potreros)	14-01-09B	19.68N	97.42W	15 x 21 (LH) 10 (LP)	0.46 Ma (LH) 0.1 Ma (LP)	rhyolite-rhyodacite	Ferriz and Mahood (1984)
Ayarza	14-02-113A	14.42N	90.12W	4 x 6	23 ka	rhyolite	Peterson and Rose (1985)
Las Lajas	14-04-17	12.30N	85.73W	7		basalt	McBirney and Williams (1965)
La Primavera	14-01-032C	20.7 N	103.53W	12	95 ka	rhyolite	Mahood (1980)
Marchena	15-03-08	0.33N	90.47W	6 x 4		tholeiitic + alkali basalt	McBirney and Williams (1969)
Calabozos	15-07-043	35.5 S	70.5 W	14 x 26	0.8 Ma 0.3 Ma 0.15 Ma	dacite, rhyolite	Hildreth and others (1984), Grunder and others (1987)

[1] Haleakala is thought by some to be primarily an erosional feature (Macdonald, 1972; Macdonald and others, 1983; Stearns, 1966, p. 43); others believe that the caldera formed initially by subsidence and was subsequently eroded. The most recent eruption of Haleakala occurred ca. A.D. 1790, from two vents along the southwest rift zone.

PART 5: LARGE QUATERNARY CALDERAS NOT KNOWN TO HAVE SHOWN HISTORICAL UNREST (continued)

Caldera	CAVW number	Latitude	Longitude	Diameter	Age of latest collapse	Dominant magma composition	References
Laguna del Maule	15-07-061	36.08S	70.52W	ca. 10?	postglacial	rhyolite	Lopez and Munizaga (1983), Frey and others (1984), W. Hildreth, written commun., 1987)
Terceira (Cinco Picos)	18-02-05=	38.70N	27.15W	7	>> 23 ka	alkali basalt, trachyte	Self and Gunn (1976)
Waesche (Chang)	19-00-026	77.17S	126.88W	6 x 8		silicic	Gonzales-Ferran and Gonzales-Bonorino (1972)
Hampton/ Whitney	19-00-027	76.48S	125.80W	Hamp: 6 Whit: 5?		silicic?	Gonzales-Ferran and Gonzales-Bonorino (1972)
Takahe	19-00-028	76.28S	122.08W	8		mafic?	Gonzales-Ferran and Gonzales-Bonorino (1972)
Thule Island	19-00-07	59.45S	27.32W	6-7		silicic?	Baker (1968)

REFERENCES

Bacon, C.R., 1983, Eruptive history of Mount Mazama and Crater Lake Caldera, Cascade Range, U.S.A.: Jour. Volcanol. Geotherm. Res., v. 18, nos. 1-4, p. 57-116.

Baker, P.E., 1968, Comparative volcanology and petrology of the Atlantic island arcs: Bull. Volcanol., v. 32, p. 189-206.

Bemmelen, R.W. van, 1931, Is de Oeloebeloe een vulkaan?: De Mijningenieur, v. 12, no. 3, p. 30-32.

----- 1949a, The geology of Indonesia, v. 1A+B, General geology of Indonesia and adjacent archipelagoes (w/ map folio): The Hague, Government Printing Office, 732 p. + maps.

----- 1949b, Report on the volcanic activity and volcanological research in Indonesia during the period 1936-1948: Bull. Volcanol., v. 9, p. 3-30.

Christiansen, R.L., 1984, Yellowstone magmatic evolution: its bearing on understanding large-volume explosive volcanism, <u>in</u> Explosive volcanism, Inception, evolution, and hazards: Washington, D.C., National Research Council, p. 84-95.

Clague, D.A., and Dalrymple, G.B., 1987, The Hawaiian-Emperor volcanic chain: Part I, Geologic evolution, <u>in</u> Decker, R.W., Wright, T.L., and Stauffer, P.H., eds., Volcanism in Hawaii: U.S. Geol. Surv. Prof. Paper 1350, p. 5-54.

CNR-CNRS Afar Team, 1973, Geology of northern Afar (Ethiopia): Rev. Geogr. Phys. Geol. Dynamics, v. 15, p. 443-490.

----- 1975, Geological maps of Afar: 1, Northern Afar (1971); 2, Central and Southern Afar (1975): La Celle St. Cloud, France, Geotechnip.

REFERENCES (continued)

Cooke, R.J.S., Baldwin, J.T., and Sprod, T.J., 1976, Recent volcanoes and mineralization in Papua New Guinea: Int. Geol. Congr., 25th, Excursion Guide no. 53, 30 p.

de Rita, D., Funiciello, R., Rossi, U., and Sposato, A., 1983, Structure and evolution of the Sacrofano-Baccano caldera, Sabatini volcanic complex, Rome: Jour. Volcanol. Geotherm. Res., v. 17, p. 219-236.

Di Paola, G.M., 1972, The Ethiopian Rift Valley (between 7 Deg 00 Min and 8 Deg 40 Min Lat. North): Bull. Volcanol., v. 36, p. 517-560.

Donnelly-Nolan, J., Ciamcanelli, E.V., Eichelberger, J., Fink, J.H., and Heiken, G., 1981, Roadlog for field trip to Medicine Lake Highland, in Johnston, D.A., and Donnelly-Nolan, J., eds, Guides to some volcanic terranes in Washington, Idaho, Oregon, and northern California: U.S. Geol. Surv. Circular 838, p. 141-150.

EBASCO, 1978-79, Volcanism and volcanic hazards [in association with the Philippine Nuclear Power Plant #1]: unpublished documents for the International Atomic Energy Agency, p. 56-86.

Erlich, E., 1986, Geology of the calderas of Kamchatka and Kurile Islands with comparison to calderas of Japan and the Aleutians, Alaska: U.S. Geol. Surv. Open-File Rep. 86-291, 300 p.

Erlich, E.N., and Melekestsev, I.V., 1972, Evolution of Quaternary volcanism and tectonics in the western part of the Pacific rim: Pacific Geol., v. 4, p. 1-22.

Erlich, E.N., Melekestsev, I.V., Tarakanovsky, A.S., and Zubin, M.I., 1972, Quaternary calderas of Kamchatka: Bull. Volcanol., v. 36, p. 222-237.

Ferriz, H., and Mahood, G.A., 1984, Eruption rates and compositional trends at Los Humeros Volcanic Center, Puebla, Mexico: Jour. Geophys. Res., v. 89, no. B10, p. 8511-8524.

Frey, F.A., Gerlach, D.C., Hickey, R.L., Lopez-Escobar, L., and Munizaga-Villavicencio, F., 1984, Petrogenesis of the Laguna del Maule volcanic complex, Chile (36°S): Contrib. Mineral. Petrol., v. 88, p. 133-149.

Geze, B., Hudeley, H., Vincent, P., and Wacrenier, Ph., 1959, Les volcans du Tibesti (Sahara du Tchad): Bull. Volcanol., v. 22, p. 135-172.

Gonzalez-Ferran, O., and Gonzalez-Bonorino, F., 1972, The volcanic ranges of Marie Byrd Land between long 100° and 140° W, in Adie, R.J., ed., Antarctic geology and geophysics: IUGS series-B no. 1, p. 261-275 (Andrus Caldera).

Gorshkov, G.S., 1970, Volcanism and the upper mantle: Investigations in the Kurile volcanic arc: New York, Plenum Press, 385 p.

Grunder, A., Thompson, J.M., and Hildreth, W., 1987, The hydrothermal system of the Calabozos Caldera, Central Chilean Andes: Jour. Volcanol. Geotherm. Res., v. 32, p. 287-298.

Hamilton, W., 1965, Geology and petrogenesis of the Island Park Caldera of rhyolite and basalt, eastern Idaho: U.S. Geol. Surv. Prof. Paper 504-C, p. C1-C37.

Harkin, D.A., 1960, The Rungwe Volcanics at the northern end of Lake Nyasa: Geol. Surv. Tanganyika, Memoir II, 172 p.

Hayatsu, K., 1976, Geologic study on the Myoko Volcanoes, Central Japan - Part 1. Stratigraphy: Kyoto Univ., ser. Geol. Min., v. 42, no. 2, p. 131-170.

Hayatsu, K., and Arai, F., 1980, Tephrochronological study on the Myoko Volcano tephra layers and their relation to activity of the volcano: Jour. Geol. Soc. Japan, v. 86, p. 243-263.

Hildreth, W., Grunder, A., and Drake, R.E., 1984, The Loma Seca Tuff and the Calabozos caldera: a major ash-flow and caldera complex in the southern Andes of central Chile: Geol. Soc. Am. Bull., v. 95, p. 45-54.

IAVCEI (International Association of Volcanology and Chemistry of the Earth's Interior), 1973-present, Data sheets on post-Miocene volcanoes of the world: Rome, IAVCEI.

REFERENCES (continued)

Japan Association of Quaternary Research, 1987, Quaternary maps of Japan: Tokyo, Univ. Tokyo Press, 119 p. + 4 map sheets.

Johnson, R.W., 1969, Volcanic geology of Mount Suswa, Kenya: Phil. Trans. Roy. Soc. London, v. 265, p. 383-412.

Johnson, R.W., Taylor, G.A.M., and Davies, R.A., 1972, Geology and petrology of Quaternary volcanic islands off the north coast of New Guinea: Aust. Bur. Min. Res. Record 1972/21, 127 p.

Katili, J.A., 1983, On the relation between tectonics and the formation of caldera with special reference to Krakatau: paper read at the 100th Centennial of Krakatau, Jakarta, August 1983.

Katsui, Y., Yokoyama, I., Murozumi, M., 1981, Noboribetsu Spa, in Katsui, Y., ed., Symposium on Arc Volcanism, Field Excursion Guide to Usu and Tarumai Volcanoes and Noboribetsu Spa, Pt. 3: Tokyo, Volcanological Society of Japan, p. 55-64.

Krijanovsky, N., 1934, Volcanoes of Kamchatka: Bull. Geol. Soc. Am., v. 45, p. 529-550.

Kuno, H., 1962, Catalogue of the active volcanoes of the world including solfatara fields, Pt. XI, Japan, Taiwan and Marianas: Rome, IAVCEI, 332 p.

Kusakabe, H., Hayashi, N., and Kobayashi, T., 1983, Genesis of banded sulfur sediments at the Jigokudani valley, Tateyama volcano, Japan: Bull. Volcanol. Soc. Japan, v. 28, no. 3, p. 245-261.

Kusumadinata, K., ed., 1979, Data Dasar Gunungapi Indonesia: Bandung, Volcanol. Surv. Indonesia, 820 p.

Leat, P.T., Macdonald, R., and Smith, R.L., 1984, Geochemical evolution of the Menengai caldera volcano, Kenya: Jour. Geophys. Res. v. 89, no. B10, p. 8571-8592.

Lopez E., L., and Munizaga V., F., 1983, Caracteristicas geoquimicas y petrogenesis del complejo volcanico Laguna del Maule, Andes del Sur, 36°00'S: Rev. Geol. Chile, no. 19-20, p. 3-24.

Luchitsky, I.V., ed., 1974, Development of Rugged Relief of Siberia: Far East: Kamchatka, Kurile, and Komandor Islands: Moscow, Nauka Press, 430 p.

Macdonald, G.A., 1972, Volcanoes: Englewood Cliffs, N.J., Prentice-Hall, Inc., 510 p.

Macdonald, G.A., Abbott, A.T., and Peterson, F.L., 1983, Volcanoes in the Sea (2nd ed.): Honolulu, Univ. Hawaii Press, 517 p.

Machida, H., 1980, Tephra and its implications with regard to the Japanese Quaternary period, in Association of Japanese Geographers, eds., Geography of Japan: Special Publ. 4, Assoc. Japanese Geographers, Tokyo, Teikoku-Shoin Co., p. 29-53.

MacLeod, N.M., Sherrod, D.R., Chitwood, L.A., and McKee, E.H., 1981, Newberry Volcano, Oregon: U.S. Geol. Surv. Circular 838, p. 85-103.

Mahood, G.A., 1980, Geological evolution of a Pleistocene rhyolite center--Sierra La Primavera, Jalisco, Mexico: Jour. Volcanol. Geotherm. Res., v. 8, p. 199-230.

McBirney, A.R., and Williams, H., 1965, Volcanic history of Nicaragua: Univ. Calif. Publ. Geol. Sci., v. 55, p. 1-65.

----- 1969, Geology and petrology of the Galapagos Islands: Geol. Soc. Am. Memoir 118, 197 p.

McCall, G.J.H., 1968, The five caldera volcanoes of the central rift valley in Kenya: Geol. Soc. London, Proc., no. 1647, p. 54-59.

Mertzman, S.A., 1981, Pre-Holocene silicic volcanism on the northern and western margins of the Medicine Lake Highland, California, in Johnston, D.A., and Donnelly-Nolan, J., eds., Guides to some volcanic terranes in Washington, Idaho, Oregon, and northern California: U.S. Geol. Surv. Circular 838, p. 163-169.

Mohr, P.A., and Wood, C.A., 1976, Volcano spacings and lithospheric attenuation in the eastern rift of Africa: Earth Planet. Sci. Lett., v. 33, p. 126-144.

REFERENCES (continued)

Neumann van Padang, M., 1951, Catalogue of the active volcanoes of the world including solfatara fields, Pt. I, Indonesia: Rome, IAVCEI, 271 p.

----- 1972, Catalogue of terrestrial crateriform structures: Part 2, Indonesia: Int. Astronom. Union, European Space Res. Organisation Publ. SP-93, 65 p.

Ono, K., Soya, T., and Mimura, K., 1981, Volcanoes of Japan (2nd ed.): Geol. Surv. Japan Map Series, no. 11, scale 1:2,000,000, 1 sheet.

Peterson, P., and Rose, W.I., Jr., 1985, Explosive eruptions of the Ayarza calderas, southeastern Guatemala: Jour. Volcanol. Geotherm. Res., v. 25, p. 289-307.

Posavec, M., Taylor, D., van Leeuwen, T., and Spector, A., 1973, Tectonic controls of volcanism and complex movements along the Sumatran Fault System: Geol. Soc. Malaysia, Bull. 6, p. 43-60.

Purbo-Hadiwidjojo, M.M., 1971, Geologic map of Bali: Bandung, Geol. Surv. of Indonesia, scale 1:250,000, 1 sheet.

Richards, A.F., 1962, Catalogue of active volcanoes of the world including solfatara fields, Pt. XIV, Archipelago de Colon (Galápagos), Isla San Felix, and Islas Juan Fernandez: Rome, IAVCEI, 50 p.

Rock, N.M.S., Syah, H., Davis, A.E., Hutchison, D., Styles, M.T., and Lena, R., 1982, Permian to Recent volcanism in Northern Sumatra, Indonesia: A preliminary study of its distribution, chemistry, and peculiarities: Bull. Volcanol., v. 45, p. 127-152.

Searle, R.C., 1972, A gravity survey of Ngorongoro Caldera, Tanzania: Bull. Volcanol., v. 35, p. 350-357.

Self, S., and Gunn, B.M., 1976, Petrology, volume and age relations of alkaline and saturated peralkaline volcanics from Terceira, Azores: Contrib. Mineral. Petrol., v. 54, p. 293-313.

Smith, R.L., and Bailey, R.A., 1968, Resurgent cauldrons: Geol. Soc. Am. Memoir 116 (Howel Williams Volume), p. 613-662.

Smith, R.L., Bailey, R.A., and Ross, C.S., 1961, Structural evolution of the Valles Caldera, New Mexico, and its bearing on the emplacement of ring dikes: U.S. Geol. Surv. Prof. Paper 424-D, p. D145-D149.

Sollevanti, F., 1983, Geologic, volcanologic, and tectonic setting of the Vico-Cimino area, Italy: Jour. Volcanol. Geotherm. Res., v. 17, p. 203-217.

Stearns, H.T., 1966, Geology of the State of Hawaii: Palo Alto, Calif., Pacific Books, 266 p.

Sviatlovsky, A.E., 1959, Atlas of the volcanoes of the Soviet Union: Moscow, Akad. Nauk Press, 172 p.

Tomkeieff, S.I., 1949, The volcanoes of Kamchatka: Bull. Volcanol., ser. 2, v. 8, p. 87-112.

Varekamp, J.C., 1980, The geology of the Vulsinian area, Lazio, Italy: Bull. Volcanol., v. 43, no. 3, p. 487-503.

Verstappen, H. Th., 1973, A geomorphological reconnaissance of Sumatra and adjacent islands (Indonesia): Verh. Roy. Dutch. Geogr. Soc., Gröningen, Wolters-Noordhoff Publ., 182 p.

Vlasov, G.M., 1967, Kamchatka, Kuril, and Komandorskiye Islands: Geological description: Geol. of the USSR, v. 31, Moscow, Izd.-vo Nedra, 827 p.

Weaver, S.D., 1976-77, The Quaternary caldera volcano Emuruangogolak, Kenya Rift, and the petrology of a bimodal ferrobasalt-pantelleritic trachyte association: Bull. Volcanol., v. 40, p. 209-230.

Williams, L.A.J., 1970, The volcanics of the Gregory Rift Valley, East Africa: Bull. Volcanol., v. 34, p. 439-465.

Williams, L.A.J., Macdonald, R., and Chapman, G.R., 1984, Late Quaternary caldera volcanoes of the Kenya Rift Valley: Jour. Geophys. Res., v. 89, no. B10, p. 8553-8570.

Wilson, C.J.N., Houghton, B.F., and Lloyd, E.F., 1986, Volcanic history and evolution of the Maroa-Taupo area, Central North Island, <u>in</u> Smith, I.E.M., ed., Late Cenozoic volcanism in New Zealand: Roy. Soc. New Zealand Bull. 23, p. 194-223.

REFERENCES (continued)

Wilson, C.J.N., Rogan, A.M, Smith, I.E.M., Northey, D.J., Nairn, I.A., and Houghton, B.F., 1984, Caldera volcanoes of the Taupo volcanic zone, New Zealand: Jour. Geophys. Res., v. 89, no. B10, p. 8463-8484.
Yamasaki, M., Nakanishi, N., and Miyata, K., 1966, History of Tateyama Volcano: Sci. Rep. Kanazawa Univ., v. 11, p. 73-92.